KU-489-677

Pathology of Domestic Animals

FOURTH EDITION Volume 1

Pathology of Domestic Animals

FOURTH EDITION Volume 1

EDITED BY

K. V. F. JUBB
School of Veterinary Science
University of Melbourne
Victoria, Australia

PETER C. KENNEDY
Department of Pathology
School of Veterinary Medicine
University of California, Davis
Davis, California, USA

NIGEL PALMER
Veterinary Laboratory Services
Ontario Ministry of Agriculture and Food
Guelph, Ontario, Canada

ACADEMIC PRESS, INC.
Harcourt Brace Jovanovich, Publishers
San Diego New York Boston
London Sydney Tokyo Toronto

This book is printed on acid-free paper. ♾

Academic Press, Inc.
1250 Sixth Avenue, San Diego, California 92101-4311

United Kingdom Edition published by
Academic Press Limited
24–28 Oval Road, London NW1 7DX

Library of Congress Cataloging-in-Publication Data

Jubb, K. V. F.
Pathology of domestic animals / K.V.F. Jubb, P.C. Kennedy, N.C. Palmer. – 4th ed.
p. cm.
Includes bibliographical references and index.
ISBN 0-12-391605-4 vol. 1
1. Veterinary pathology. I. Kennedy, Peter C. (Peter Carleton), date. II. Palmer, Nigel. III. Title.
SF769.J82 1992
636.089'607–dc20 92-12261
CIP

PRINTED IN THE UNITED STATES OF AMERICA
92 93 94 95 96 97 EB 9 8 7 6 5 4 3 2 1

Contents

CHAPTER 2

Muscle and Tendon

THOMAS J. HULLAND

CHAPTER 3

The Nervous System

K.V.F. JUBB AND C.R. HUXTABLE

CHAPTER 4
The Eye and Ear
BRIAN P. WILCOCK

CHAPTER 5
The Skin and Appendages
JULIE A. YAGER AND DANNY W. SCOTT WITH A CONTRIBUTION BY BRIAN P. WILCOCK

Contents of Other Volumes

Volume 2

Volume 3

Contributors

Volume 1

THOMAS J. HULLAND, Department of Pathology, Ontario Veterinary College, University of Guelph, Guelph, Ontario, Canada N1G 2W1.

C. R. HUXTABLE, School of Veterinary Studies, Murdoch University, Murdoch, Western Australia, Australia 6150.

K. V. F. JUBB, School of Veterinary Science, University of Melbourne, Werribee, Victoria, Australia 3030.

NIGEL PALMER, Veterinary Laboratory Services, Ontario Ministry of Agriculture and Food, Guelph, Ontario, Canada N1H 6R8.

DANNY W. SCOTT, Department of Clinical Sciences, New York State College of Veterinary Medicine, Cornell University, Ithaca, New York, USA 14853-6401.

BRIAN P. WILCOCK, Department of Pathology, Ontario Veterinary College, University of Guelph, Guelph, Ontario, Canada N1G 2W1.

JULIE A. YAGER, Department of Pathology, Ontario Veterinary College, University of Guelph, Guelph, Ontario, Canada N1G 2W1.

Volume 2

IAN K. BARKER, Department of Pathology, Ontario Veterinary College, University of Guelph, Guelph, Ontario, Canada N1G 2W1.

D. L. DUNGWORTH, Department of Pathology, School of Veterinary Medicine, University of California, Davis, California, USA 95616.

K. V. F. JUBB, School of Veterinary Science, University of Melbourne, Werribee, Victoria, Australia 3030.

W. ROGER KELLY, Department of Veterinary Pathology, University of Queensland, St. Lucia, Brisbane, Queensland, Australia 4072.

M. GRANT MAXIE, Veterinary Laboratory Services, Ontario Ministry of Agriculture and Food, Guelph, Ontario, Canada N1H 6R8.

NIGEL PALMER, Veterinary Laboratory Services, Ontario Ministry of Agriculture and Food, Guelph, Ontario, Canada N1H 6R8.

JOHN F. PRESCOTT, Department of Veterinary Microbiology and Immunology, Ontario Veterinary College, University of Guelph, Guelph, Ontario, Canada N1G 2W1.

A. A. VAN DREUMEL, Veterinary Laboratory Services, Ontario Ministry of Agriculture and Food, Guelph, Ontario, Canada N1H 6R8.

Volume 3

CHARLES C. CAPEN, Department of Veterinary Pathology, The Ohio State University, Columbus, Ohio, USA 43210-1093.

PETER C. KENNEDY, Department of Pathology, School of Veterinary Medicine, University of California, Davis, California, USA 95616.

P. W. LADDS, Graduate School of Tropical Veterinary Science and Agriculture, James Cook University, Townsville, Queensland, Australia 4811.

M. GRANT MAXIE, Veterinary Laboratory Services, Ontario Ministry of Agriculture and Food, Guelph, Ontario, Canada N1H 6R8.

RICHARD B. MILLER, Department of Pathology, Ontario Veterinary College, University of Guelph, Guelph, Ontario, Canada N1G 2W1.

B. W. PARRY, School of Veterinary Science, University of Melbourne, Werribee, Victoria, Australia 3030.

WAYNE F. ROBINSON, School of Veterinary Studies, Murdoch University, Murdoch, Western Australia, Australia 6150.

V. E. O. VALLI, College of Veterinary Medicine, University of Illinois, Urbana, Illinois, USA 61801.

Preface to the Fourth Edition

Thirty years will have elapsed since the publication of the first edition of "Pathology of Domestic Animals." In that time it has become like a living thing, changing and adapting as veterinary pathology has changed, yet retaining the generic character of the first edition. The fourth edition will be immediately recognizable to the many users of earlier editions, but it is changed in significant ways.

Yesterday's interesting new case becomes tomorrow's new syndrome. Accordingly, we have attempted to provide a more comprehensive textual inclusion of individual disease entities rather than relegate many diseases to the bibliographies as in earlier editions. This change respects and reflects current veterinary pathology practice. Our concern for the economically significant diseases remains and they are discussed in detail. However, recent rapid growth in knowledge has come from applying modern investigative techniques to comparative pathology and the diseases of companion animals.

Library resources are more readily available than when this work was first produced. The bibliographic listings have been reduced and placed closer to their subjects. These sources have been selected not only for intrinsic merit but for the comprehensive bibliographies they provide.

An essential part of the work has been the illustrations, the number and quality of which have been maintained. We are particularly grateful to our many colleagues who have provided photographs of high quality; they are acknowledged in the figure legends. We have to a large extent departed from the plate format for illustrations; this has allowed us to replace some illustrations and to locate others more appropriately in relation to the text.

The bringing together of this fourth edition has involved many people, far too many to name. Acknowledgments are made at the end of several chapters but we, as editors, are grateful to our contributing authors for their splendid and timely contributions, to Dr. Jennifer Anne Charles for meticulous editorial assistance, and to our word-processor operators Yvonne Pritchard and Kay Vincent who never allowed the way to seem too long or too weary. Edward W. Eaton of the University of Guelph prepared most of the new illustrations for this edition, and we thank him for his assistance. It remains a pleasure to work with Academic Press.

Melbourne, Australia
1993

K. V. F. Jubb
Peter C. Kennedy
Nigel Palmer

Preface to the Third Edition

Much has been happening in veterinary pathology between editions of this work. We had, from the time of the first edition, an expectation and a hope that as the number of scientists dedicated to this field of study grew, so also would the variety of publications to serve the diversity of interests. This anticipation has only partly been realized. We still have few books that address themselves to diseases of a single domestic species or to a single organ system of domestic animals. The need for a comprehensive treatment of diseases of domestic species, from the viewpoint of the pathologist, remains. The reception of earlier editions and the interest of our colleagues around the world have influenced us to try for the third time to produce a work of some universal usefulness.

The amount of information available on the pathology of animal disease has grown enormously, and the task of integrating so much new information into a coherent statement has grown on an equal or larger scale. Changes have become necessary in this book. This edition introduces the new generation of veterinary pathologists to a literary task that has grown much beyond what the original authors could handle. We take great satisfaction in this growth and in our colleagues' willingness to join us in the project. The contributors are identified with those chapters or parts of chapters for which they have been individually responsible, although this method does understate the contribution and the dedicated commitment of our coauthors to what has been very much a cooperative effort.

We have retained the original style and format. It was established as the medium for what were personal statements by the authors. We hope that we have been able to maintain some of that flavor. Some features of the style and the format have proven to be awkward in use by the busy working pathologist, and in recognition of this we have given attention to subdivisions in the text, to an expansion of tables of content, to details in indexation (including addition of a cumulative index, in Volume 3), and to an expanded selection of illustrations. The wish to preserve the original style has presented to us and to our contributing authors challenges on content and balance. We hope that these have reasonably been met. Inevitably, we have had to make choices in blending the contributions of our contributing authors into a whole. We have had to reduce excellent sections to keep these volumes within reasonable size, and we have expanded other sections for the sake of completeness. Inevitably too, some of our editorial judgments will be imperfect, and responsibility remains with us for deficiencies in the final compilation.

It is not possible adequately to acknowledge the many people who have contributed to this work; most of them will in this prefatory statement remain unnamed.

The contributors, all of whom volunteered effort without which this work could not have been completed, will find that the uses to which these volumes are put in the next few years will be a fuller tribute to their work than can be written here. The support of our many other colleagues in veterinary pathology is perhaps best indicated by their generosity in providing illustrative material. We have brought forward many of the plates or figures from the earlier editions and have added many new ones. Those brought forward or added are acknowledged in the legends, but many more excellent photographs were received than could be used. We are deeply grateful to those colleagues who offered them.

The institutions with which we are individually affiliated have made time and other resources available to us. The several chapters contain acknowledgments for assistance received, but we must here acknowledge Sandra Brown, Jean Middlemiss, and Edward W. Eaton of the University of Guelph for preparing most of the draft manuscript and many illustrations, Denise Heffernan, Lynette Magill, and Frank Oddi of the University of Melbourne for the preparation of final copy and illustrations, and Tammie Goates of that university for editing the bibliographies. We gratefully acknowledge a generous donation from Syntex Agribusiness toward the costs of preparation of the manuscript. We are grateful again to receive the courtesy and cooperation of Academic Press in this shared contribution to the study of animal disease.

Melbourne, Australia
1984

K. V. F. Jubb
Peter C. Kennedy
Nigel Palmer

Preface to the Second Edition

The first edition of "Pathology of Domestic Animals" went to press, not without some sense of satisfaction, with a philosophic acceptance of the many imperfections and a tentative hope that any future edition would provide an opportunity to refine our knowledge, understanding, and technique of communication. Alas, imperfections remain, different ones perhaps, inevitable products of the interaction of limited time, limited intellect, and unlimited supplies of scientific data.

We are impressed by the masses of data that weekly flood our libraries and by the short half-life of much of it, by the exponential increase in knowledge and the splintering of disciplines that proliferate therefrom, and by the inability of many disciplines relevant to medical science to be completely self-sustaining. More and more it is evident that the theme of pathology provides the central and connecting link in medical education and practice and the basis on which a multidisciplined structure can be supported. This is a difficult role for pathologists, but one which they will fill, not by virtue of superior intellects and capacious memories, desirable though these attributes may be, but rather by the proper application of the logic of the scientific method.

Therefore, in preparing this second edition we have attempted to incorporate new knowledge on the specific diseases of animals and, more earnestly, to find a theme of organ susceptibility and responsiveness. We do not doubt the validity of the approach even if we are unable as yet to apply it feasibly to all organs and systems. The format of this edition remains the same as for the first and for the same reasons; the logistics of suitable alternatives are too formidable.

Once again we must express our gratitude to the many people who have contributed in some way to the preparation of this edition. Especially, we are indebted to Professor T. J. Hulland of the University of Guelph for revising the chapter on muscle, to Dr. Anne Jabara, University of Melbourne for the section on mammary tumours, and to Dr. N. C. Palmer and Dr. J. S. Wilkinson of the University of Melbourne for material assistance and many fruitful conversations. As always, a heavy burden falls on those who convert our notes to manuscript and arrange the bibliography, a task shared and cheerfully and devotedly performed by Mrs. Sylvia Lewis and Miss Frances Douglas. We hope that we have done justice to those who have contributed illustrative material: Dr. A. Seawright, University of Queensland; Dr. D. Kradel, Pennsylvania State University; Dr. E. Karbe, University of

Zurich; Dr. B. C. Easterday, National Animal Disease Laboratory at Ames; Mr. J. D. J. Harding, Central Veterinary Laboratory, Weybridge; Dr. J. Morgan, University of California; Miss Virginia Osborne, University of Sydney.

Melbourne, Australia
October, 1969

K. V. F. Jubb
Peter C. Kennedy

Preface to the First Edition

The preface offers the opportunity to an author to present his excuses for having written the book and his justification of the content and mode of presentation. Our reasons for writing "Pathology of Domestic Animals" are as insubstantial but as compelling as those which committed Captain Ahab to the pursuit of Moby Dick, and we offer no excuses. Neither shall we attempt justification because a bad book cannot be justified and a good book is its own justification.

These volumes are based on our experience and on as much of the relevant literature of the world as we have been able to find and evaluate and we offer them to our colleagues and to all students of pathology in the hope that they will contribute to an understanding of animal disease. We anticipate some criticism in offering these as student texts but in doing so we indicate our confidence in teachers of pathology to guide students in the use of such volumes and in the ability of the student to profit from the exposure. Moreover, these volumes represent, it seems to us, a fair assessment of the needs of veterinary students in these times, since we realize as we should, that the knowledge of pathology possessed by most graduating students must serve them for the rest of their lives.

We should have preferred to write at greater length and in more detail of our chosen field, but practicality and economics have dictated that we can present here no more than a précis of the wealth of information that is the gift of our predecessors and contemporaries to the veterinary profession. In compensation, we have appended to each chapter an extensive but selected bibliography by the proper use of which the earnest seeker after further knowledge will be richly rewarded. Many valuable contributions from the old and foreign literature will not be listed in our bibliographies, perhaps because we have failed to appreciate their significance but largely because we have not obtained access to them.

We wish to emphasize to our younger colleagues that there exist vast gaps in our present knowledge, and we hope future work will do much to fill these gaps. For any errors of established fact that appear and for errors of interpretation of published information we tender, with our apologies, a request that they be drawn to our attention. We have not always attempted to distinguish between what we know and what we think we know, and in stating our position on many matters of controversy it is inevitable that we are sometimes in error; but we do prefer to state our positions while reserving our right to change our opinions when necessary.

The aim of the scientific method is to provide understanding, and the ultimate aim in all study of disease is to understand well enough to preserve the organism and prevent the disease. But disease and the temper of the community do not wait upon the languid spirit of most scientific enquiry; in the annals of veterinary science there are many endemic and epidemic diseases concerning which a broad search for understanding is necessarily postponed in the interest of quickly finding a way to avoid the disease or to face and exert some measure of control over it. Such hastily constructed controls are often satisactory but seldom enough, and usually they merely stem the tide while further enquiry can be made and understanding sought. It is from pathology, viewed broadly, that understanding comes and the need is great because there are old diseases still to be contended with, others now in existence but still to be recognized, and new ones to be anticipated. The pathologist is necessarily concerned with all matters pertaining to disease and we would enjoin him to remember this and meet his responsibilities in an age when urgency disturbs the spirit of the Groves of the Academy.

We have departed somewhat from tradition in the arrangement of these volumes. General pathology is well covered in many existing textbooks and we have not taken space for it, but have restricted our discussions to systemic or special pathology. Almost all we have to say on a particular subject or specific disease is said in one place under the organ system in which it appears most appropriate, although we have waived this general rule in an attempt to make the sections devoted to genitalia and special senses self-sufficient. A few diseases which resisted our systemic classification are relegated to an appendix in Volume 2. Detailed tables of contents are included for each volume to indicate the organization of the text and our classification of the diseases of the systems.

Guelph, Ontario
January, 1963

K. V. F. JUBB
PETER C. KENNEDY

CHAPTER 1 Bones and Joints

NIGEL PALMER
Ontario Ministry of Agriculture and Food, Canada

DISEASES OF BONES

I. General Considerations

Bones are dynamic organs growing and remodeling throughout life, and the study of skeletal disease is, perhaps more than in any other system, a study in tissue and organ dynamics. Bones are varied in form and function, and their construction involves a number of tissues whose existence and appearance are unfamiliar to many pathologists.

In order to provide a background for discussion of osseous diseases, several preliminary topics are considered; these are intended to give an approach to the study of bone and bones, some details of skeletal physiology and histology, and an overview of skeletal responses to disease.

Bibliography

Ackerman, L. V., Spjut, H. J., and Abell, M. R. (eds.). "Bones and Joints." Baltimore, Maryland, Williams & Wilkins, 1976.

Aegerter, E., and Kirkpatrick, J. A. "Orthopedic Diseases," 4th Ed. Philadelphia, Pennsylvania, W. B. Saunders, 1975.

Andrews, E. J., Ward, B. C., and Altman, N. H. (eds.). "Spontaneous Animal Models of Human Disease," Vol. 2. New York, Academic Press, 1979.

Barnett, C. H., Davies, D. V., and MacConaill, M. A. "Synovial Joints. Their Structure and Mechanics." New York, Longmans, Green, 1961.

Bourne, G. H. (ed.). "The Biochemistry and Physiology of Bone," 2nd Ed. New York, Academic Press, 1972.

Bronner, F., and Coburn, J. W. (eds.). "Disorders of Mineral Metabolism. Calcium Physiology." New York, Academic Press, 1982.

Duncan, C. P., and Shim, S.-S. The autonomic nerve supply of bone. *J Bone Joint Surg (Br)* **59B:** 323–330, 1977.

Enlow, D. H. "Principles of Bone Remodeling." Springfield, Illinois, Thomas, 1963.

Frost, H. M. (ed.). "Bone Biodynamics." Boston, Massachusetts, Little, Brown, 1964.

Frost, H. M. "Bone Remodeling and Its Relationship to Metabolic Bone Diseases." Springfield, Illinois, Charles C Thomas, 1973.

Grüneberg, H. "The Pathology of Development." Oxford, Blackwell, 1963.

Holick, M. F., Gray, T. K., and Anast, C. S. (eds.). "Perinatal Calcium and Phosphorus Metabolism." Amsterdam, Elsevier, 1983.

Jaffe, H. L. "Metabolic, Degenerative and Inflammatory Diseases of Bones and Joints." Philadelphia, Pennsylvania, Lea & Febiger, 1972.

Lacroix, P. "The Organization of Bones." London, Churchill, 1951.

Manolagas, S. C., and Olefsky, J. M. (eds.). "Metabolic Bone and Mineral Disorders." New York, Churchill Livingstone, 1988.

McKusick, V. A. "Heritable Disorders of Connective Tissue." St. Louis, Missouri, Mosby, 1960.

Nancollis, G. H. (ed.). "Biological Mineralization and Demineralization." Berlin, Springer-Verlag, 1982.

Nordin, B. E. C. (ed.). "Calcium in Human Biology." London, Springer-Verlag, 1988.

Resnick, D., and Niwayama, G. "Diagnosis of Bone and Joint Disorders." Philadelphia, Pennsylvania, W. B. Saunders, 1981.

Teitelbaum, S. L., and Bullough, P. G. The pathophysiology of bone and joint disease. *Am J Pathol* **96:** 283–354, 1979.

Weinmann, J. P., and Sicher, H. "Bone and Bones; Fundamentals of Bone Biology," 2nd Ed. St. Louis, Missouri, Mosby, 1955.

Woodard, J. C., and Montgomery, C. A. Musculoskeletal system. *In* "Pathology of Laboratory Animals," K. Benirschke, F. M. Garner, and T. C. Jones (eds.). New York, Springer-Verlag, 1978.

II. Examination of the Skeleton

The skeleton is neglected by many pathologists, often regarded as an inert structure, useful only for enclosing or supporting important organs, and, with the aid of neuromuscular devices, moving them from place to place. Several factors encourage this view: dissection of the skeleton is time-consuming and often unrewarding; bones, being hard, require special examination techniques; bones are anatomically diverse, and it is sometimes difficult to distinguish normal and abnormal form; and the histology of growing bones is complicated. Some of these are valid objections, but their importance can be minimized if the skeleton is examined in a systematic way.

A. Gross Examination

Complete dissection of the skeleton is rarely practical or necessary. An assessment of its status can be made during a routine autopsy. The jaws and teeth should be examined during removal of the tongue and larynx; the ribs, when thoracic viscera are exposed; and the cranium, on removal of the brain. Dissection of a femur and its articulations, followed by lengthwise bisection to expose marrow, growth plates, and trabecular bone, requires little effort. When the clinical history or gross appearance of the carcass suggests the need, a more detailed study should be made. Areas for special emphasis often can be determined by radiography. In animals with established or suspected congenital skeletal defects, a radiographic survey of the whole skeleton is important to characterize the lesions and establish a diagnosis.

A standard procedure should be used for gross examination of bones. Care is required to avoid slicing articular cartilage during dissection. Articular cartilages and synovial membranes must be examined before they dry. Long bones should always be sawn in the same way so that the normal appearance becomes familiar. Examination of the sawn bone must include appraisal of articular cartilages, epiphyses, physes (growth plates), metaphyses, and diaphysis (Fig. 1.1), with attention being given to thickness and regularity of cartilages, and density and amount of metaphyseal and diaphyseal bone. Doubts about the significance of an anatomical feature can often be resolved by comparison with the contralateral bone and a similar bone from another animal.

Radiographs of whole bones and of sawn slabs of whole bones are valuable for demonstrating and recording lesions. A machine such as the Faxitron (see Dunn *et al.*, 1975) can produce detailed x-rays of whole small bones and of sections through lesions in large ones.

Following gross examination, there are several techniques that may be usefully applied to bone and bones. Some of these are outlined and referenced here.

B. Laboratory Techniques

Measurement of bone mineral is sometimes indicated when metabolic or developmental disease is suspected. Determination of ash is done by heating bone at 600°C until there is no more weight loss. Results are usually expressed as a percentage of the weight of dry, fat-free bone, but this means that the proportion of one variable is expressed in terms of a dependent variable. It is better to express ash content in relation to the volume of the

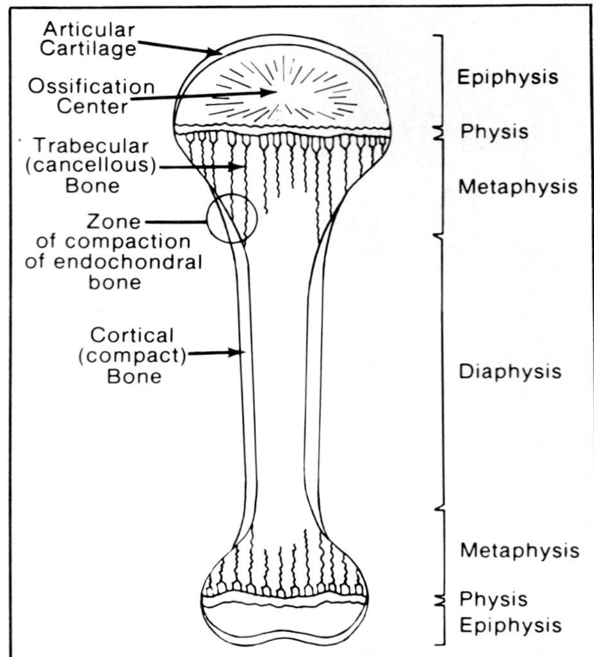

Fig. 1.1 Diagram of long bone showing main components and types of tissue.

whole bone or the volume of dry, fat-free tissue in the bone. Volumes are easily calculated using Archimedes' principle. The use of ash percentage is well established and is likely to persist, but its limitations should be recognized.

There are two major problems regarding measurement of bone ash. One pertains to the lack of normal values and the other to the time and labor required for analyses. Particularly in young animals there are very rapid changes in the degree of skeletal mineralization with aging. In normal lambs, the ash in the femur can increase by 8% from birth to 8 weeks. Also there may be differences of up to 4% in femoral ash between newborn lambs of different breeds raised under identical conditions. The labor requirement is partly due to the need to analyze the whole bone. There are such marked differences in mineral density between different areas of a bone, not only between metaphysis and diaphysis but also between different parts of the diaphysis, that analyses of segments of bones yield meaningless results unless anatomic landmarks are used scrupulously in selecting specimens. In general, ash determinations are useful in controlled experiments but will not be of much value in individual autopsies until an extensive range of normal values is assembled. Usually when values from an individual animal are low enough to appear significant, the abnormality is apparent without ashing.

Estimates of bone specific gravity are also useful for assessing metabolic bone diseases. Again, there is the problem of lack of normal values, but the determination is simple, requiring only weighing in air and in water and application of Archimedes' principle. Care must be taken to remove a standard amount of soft tissue and to weigh

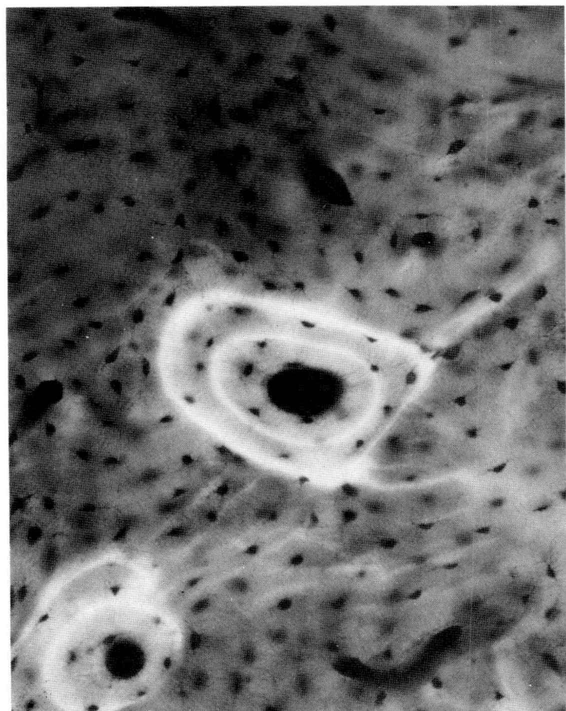

Fig. 1.2 Rings of tetracycline fluorescence in developing osteons. Dark vascular canals and numerous small osteocyte lacunae are visible. Fully mineralized ground section.

the bone in air in a tared container, immediately after dissection to prevent drying. This will provide an indication of the amount of bone tissue in the organ and the amount of mineral in the tissue.

A technique often employed in bone research, but with limited application in diagnostic pathology, is the use of fluorescent markers, such as tetracycline antibiotics, to label sites of mineralization. Periodic administration of such markers permits measurement of bone histomorphometry when undemineralized sections of bone are viewed in ultraviolet light (Fig. 1.2). A period of at least 10 days (longer in adults) is required between markers to get good separation of the bands of labeled bone. Animals should not be killed until 2 days after the last marker is given to allow time for clearance of nonspecific labeling. A recommendation for standardization of nomenclature, symbols, and units in histomorphometry has been published.

Microradiography of thick sections is useful for examining mineralization defects in bone tissue such as occur with osteofluorosis and some other metabolic diseases. Thick sections (60–100 μm) of mineralized bone are made by hand-grinding, or by sectioning bone embedded in hard plastic. The film is examined microscopically. Osteocyte lacunae, but not canaliculi, are visible, and variations in mineral density are apparent.

Fully mineralized sections of cortical bone are easily prepared by hand-grinding and can be stained to show osteoid, and distinguish degrees of mineralization.

Fully mineralized sections of growth plates are valuable for assessing endochondral ossification. Using special embedding media such as methacrylates, and heavy-duty sledge microtomes, 7- to 10-μm sections of dense cortical bone can be prepared. However, small pieces (1 × 2 cm) of mineralized physis and metaphysis from young animals can be processed using routine paraffin embedding if cortical bone is trimmed away. Excellent cytologic detail is attainable in well-fixed tissue.

The standard method for examination of bone uses formalin-fixed tissue, demineralized in acid solutions and embedded in paraffin. Rapid and complete demineralization, without excessive immersion in the acid solution, is necessary to get good sections. Slow demineralization in saturated ethylenediaminetetraacetic acid (EDTA) solutions also gives good results. With most young animals (younger than 4 to 5 months old), immersion of 5-mm slices of fixed bone in a large volume (40 times tissue volume) of acid solution for 24 to 48 hr, with constant agitation, is sufficient for demineralization. The solution can be re-used until the demineralization process starts to slow down. Testing for completion of demineralization can be done radiographically, but it is simpler to process a slightly larger block than is required and then trim the most densely mineralized part with a scalpel until it can be cut without a gritty sensation. The 5-mm block should then be trimmed to 3 to 4 mm, and the freshly cut surface used for sectioning. This eliminates some of the artifacts made by the saw blade. Rapid fixation, demineralization, and processing are necessary to ensure good sections. Tissue from mature animals may take much longer to demineralize.

Bone biopsies create special problems for pathologists. Often their interpretation is difficult, and may be impossible even with a precise history and details of the surgical site. The necessity for cooperation with the clinician and radiologist cannot be overemphasized. Most biopsies are taken to differentiate osteosarcoma and osteomyelitis in dogs. Surgeons should select the biopsy site after examining at least two radiographs of the skeletal lesion taken at right angles. Radiographs of medullary lesions of long bones should be used to determine the depth of penetration of the biopsy instrument necessary for sampling the center of the bone lesion. Unsuccessful biopsies usually result from sampling the borders of the lesion. These superficial biopsies contain little diagnostic tissue and mostly reactive endosteal and periosteal bone responses.

Bibliography

Anderson, C. "Manual for the Examination of Bone." Boca Raton, Florida, CRC Press, 1982.

Boivin, G., Anthoine-Terrier, C., and Obrant, K. J. Transmission electron microscopy of bone tissue. A review. *Acta Orthop Scand* **61:** 170–180, 1990.

Bonucci, E. *et al.* Technical variability of bone histomorphometric measurements. *Bone Miner* **11:** 177–186, 1990.

Bowes, D. N. and Dunn, E. J. A simple vacuum cassette for microradiography. *Stain Technol* **50:** 355–357, 1975.
Breur, G. J., Slocombe, R. F., and Braden, T. D. Percutaneous biopsy of the proximal humeral growth plate in dogs. *Am J Vet Res* **9:** 1529–1532, 1988.
Dunn, E. J. *et al.* Microradiography of bone, a new use for the versatile Faxitron. *Arch Pathol* **99:** 62, 1975.
Frost, H. M. Preparation of thin undecalcified bone sections by rapid manual method. *Stain Technol* **33:** 273–277, 1958.
Hahn, M., Vogel, M., and Delling, G. Undecalcified preparation of bone tissue: Report of technical experience and development of new methods. *Virchows Arch (A) Pathol Anat* **418:** 1–7, 1991.
Hwang, W. S., Ngo, K., and Saito, K. Silver staining of collagen fibrils in cartilage. *Histochem J* **22:** 487–490, 1990.
Jowsey, J. *et al.* Quantitative microradiographic studies of normal and osteoporotic bone. *J Bone Joint Surg (Am)* **47A:** 785–806 and 872, 1965.
Parfitt, A. M. *et al.* Bone histomorphometry: Standardization of nomenclature, symbols, and units. Report of the ASBMR histomorphometry nomenclature committee. *J Bone Min Res* **2:** 595–610, 1987.
Ralis, H. M., and Ralis, Z. A. Poorly mineralized osteoid bone and its staining in paraffin sections. *Med Lab Sci* **35:** 293–303, 1978.
Siemon, N. J., and Moodie, E. W. Reproducibility of specific gravity estimations on bone samples from different sites of cattle and sheep. *Calcif Tissue Res* **15:** 181–188, 1974.
Solheim, T. Pluricolor fluorescent labeling of mineralizing tissue. *Scand J Dent Res* **82:** 19–27, 1974.
Tothill, P. Methods of bone mineral measurement. *Phys Med Biol* **34:** 543–572, 1989.
Villaneuva, A. R., and Lundin, K. D. A versatile new mineralized bone stain for simultaneous assessment of tetracycline and osteoid seams. *Stain Technol* **64:** 129–137, 1989.

III. Bone Tissue

Two physically close cellular systems, the hematopoietic and the stromal fibroblastic, are involved in the production of the specific cells of bone. Osteoclasts are derived from the hematopoietic system, and osteoblasts, osteocytes, chondroblasts, and chondrocytes from the stromal fibroblastic system. Both systems contain true stem cells, which divide to yield one cell committed to the path of differentiation and another which retains stem cell functions.

A. Osteoprogenitor Cells

The stem cells of the stromal fibroblastic system are called osteoprogenitor cells (Fig. 1.3). A more precise term, determined osteogenic precursor cells, acknowledges that these are the only mesenchymal cells capable of independent osteogenesis. When transplanted, they are also capable of reconstituting the hematopoietic microenvironment characteristic of bone, and hematopoietic tissue derived from the host will develop in that bone. Mesenchymal cells in other tissues are called inducible osteogenic precursor cells in recognition of their osteogenetic ability when an appropriate inducer is present.

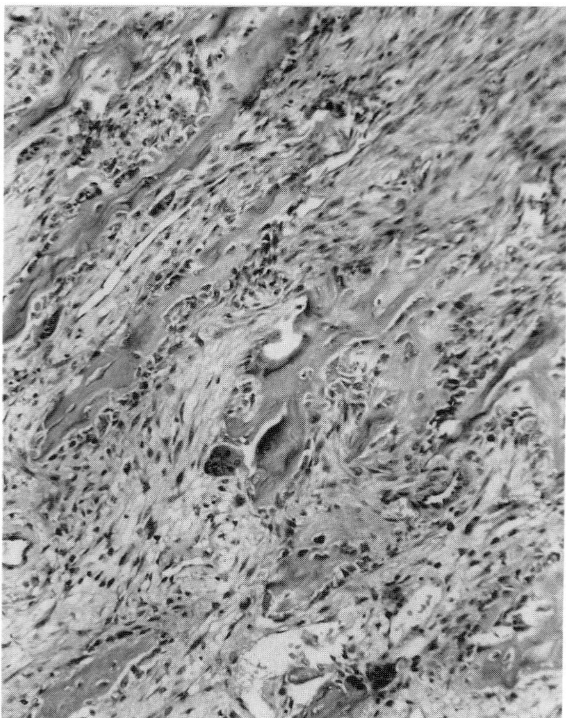

Fig. 1.3 Abundant osteoprogenitor cells, osteoblasts, and osteoclasts in hyperparathyroidism.

Osteoprogenitor cells are the source of chondroblasts and osteoblasts, and fibroblasts of bone. Genetic determinants presumably are responsible for the development of cartilage from condensed mesenchyme of embryonic limb buds, and certain physical factors can influence the direction of differentiation. For example, development of cartilage or bone at fracture sites is influenced by the adequacy of blood supply and by oxygen tension; ischemia and hypoxia favor cartilage production, whereas development of bone is stimulated by a good blood supply and high oxygen tension. The differentiation sequence from osteoprogenitor cell to active lamellar osteoblast takes about 60 hr. Development of fibroblasts from osteoprogenitor cells is stimulated by tension and by prolonged hyperparathyroidism. However, the precise mechanisms by which osteoprogenitor cells are stimulated to divide, and are directed to fibroblastic, chondroblastic, or osteoblastic streams, are not clear. There is compelling evidence, albeit much of it derived from *in vitro* experiments, and some of it conflicting, that **growth factors** are influential in the affairs of the skeleton. The precise significance and actions of the various growth factors are not known. It seems likely that there are complex interactions between factors, and that relative concentrations of activated factors are critical in controlling cell response.

Growth factors act primarily as local regulators of cell division and function, and are classified as autocrine or paracrine depending on whether they act on cells of the same, or of a different class. As well as their local actions, some of these substances may mediate the effects of systemic hormones. Several of the more important ones are mentioned below.

Transforming growth factor beta (TGF-β) is one of the growth factors synthesized by skeletal cells. It is the prototype of a family of factors with a range of biologic effects throughout the body. Evolutionally, it is a highly conserved molecule, having identical structure in many species, including humans, monkeys, cattle, pigs, and chickens. Almost all cell types have receptors for TGF-β, and almost all are dramatically affected by it. Its main sources are bone cells and platelets. Usually secreted in latent form, TGF-β is activated by pH extremes, either basic or acidic, by dissociating agents, and by enzymes such as plasmin and cathepsin D. Its major effects are regulation of mesenchymal cell replication and differentiation, and stimulation of the synthesis of bone matrix proteins including collagen, fibronectin, osteonectin, and proteoglycans. Cells of osteoblast lineage are very sensitive to its mitogenic activity, but epithelial cells are refractory. TGF-β inhibits bone resorption by preventing osteoclast formation and activation but, paradoxically, parathyroid hormone increases the release of TGF-β from bone, and also enhances its binding to osteoblast receptors. In addition, TGF-β can increase bone resorption through a prostaglandin-mediated mechanism. Variations of experimental protocols, and failure to distinguish latent and activated growth factors, may account for these apparent discrepancies. A role in the coupling of bone resorption and formation has been suggested for TGF-β.

Growth factors from bone marrow cells are classified as monokines and lymphokines according to their origin from monocytes or lymphocytes. Two monokines, interleukin-1 (IL-1, synonym catabolin) and tumor necrosis factor alpha (TNF-α, synonym cachectin) have quite well defined activity in skeletal tissue. IL-1 stimulates bone resorption, but, like parathyroid hormone and calcitriol, only in the presence of osteoblasts. Some of the activity of the substance known as osteoclast-activating factor is derived from IL-1, which also stimulates cell replication in bone cultures, and at low dosage stimulates collagen production; thus, like TGF-β, it may be instrumental in the linkage of resorption and formation. Synergism exists between IL-1 and both TNF-α and parathyroid hormone in stimulation of bone resorption.

Tumor necrosis factor α stimulates bone resorption and osteoblast production, but inhibits collagen synthesis by osteoblasts. It also stimulates production of IL-1 by macrophages; thus TNF-α and IL-1 have similar effects on osteoclasts and osteoblasts, with the exception of their mediation of collagen synthesis.

Prostaglandins, a group of unsaturated, oxygenated fatty acids synthesized from arachidonic acid, are members of the eicosanoid family of local factors or hormones that also includes prostacyclins, thromboxanes, and leukotrienes. Eicosanoids are unusual in that they are not stored but are produced rapidly as a result of mechanical or chemical injury to cell membranes in which arachidonic acid has a structural role. Prostaglandin production is stimulated by various growth factors and by parathyroid hormone, and prostaglandin E_2 (PGE_2) production in bone cell cultures is increased by mechanical forces. Prostaglandins influence the activity of bone cells in diverse, often paradoxic ways. Administration of PGE_2 to adult dogs induces formation of woven bone on periosteal surfaces and in the marrow cavity. However, high concentrations of prostaglandins inhibit osteoblastic collagen synthesis, and prostaglandins, including PGE_2, are potent mediators of bone resorption, and can enhance parathyroid hormone-stimulated resorption.

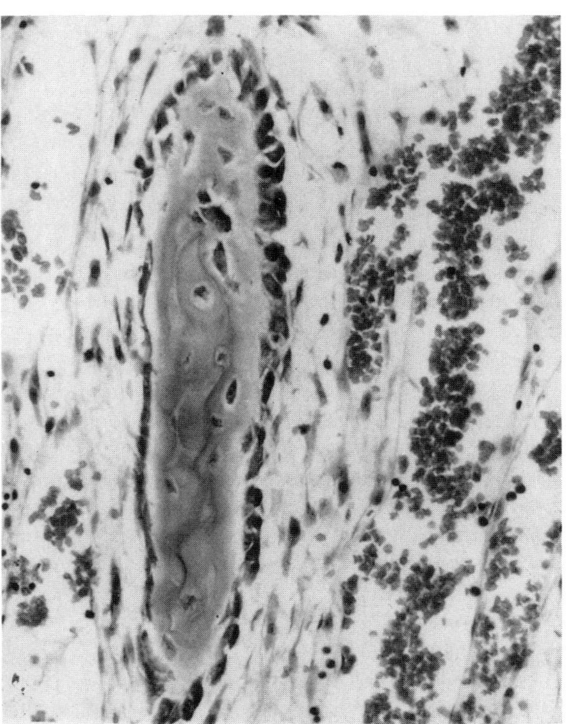

Fig. 1.4 Active osteoblasts depositing osteoid on medullary trabecula.

Until definitive roles and control mechanisms are better understood, prostaglandins and growth factors can be regarded as mediators that provide plausible explanations for the demonstrated ability of bone cells to react and interact locally, either to form or to resorb bone.

B. Osteoblasts

Osteoblasts produce and mineralize a matrix called osteoid. When making lamellar bone, they form a layer of plump, cuboidal to columnar cells with basophilic cytoplasm, on the bone-forming surface (Fig. 1.4). They are separated by narrow, matrix-free clefts. Their nuclei are located at the end of the cell which is away from bone and in contact with capillaries. Osteoblasts of woven bone are angular cells with a less-regular distribution. About 10% of osteoblasts are enclosed in matrix and become osteocytes; the others probably die, as mineral for the matrix passes through them. It is not clear when or how prospective osteocytes are chosen. Osteocytes have a surface antigen that osteoblasts lack, but the time of acquisition is not known. Osteoblasts on an active surface are replaced as required from the pool of osteoprogenitor cells. The productive life of a lamellar osteoblast in human bone is about 3 months.

Osteoblasts are responsive to parathyroid hormone, and may mediate the effects it has on osteoclasts. The immediate effect of parathyroid hormone is to inhibit the activity of preexisting osteoblasts, causing them to dedifferentiate and reduce collagen synthesis. Continued hormonal stimulation increases osteoblastic activity, and this probably represents the net response of newly differentiated cells. Osteoblasts, but not progenitor cells, have receptors for parathyroid hormone, and it is likely that the proliferation of progenitor cells stimulated by parathyroid hormone is mediated by osteoblasts. Osteoblasts, but not osteocytes, have receptors to calcitriol, but their response to it is variable, apparently depending, *inter alia,* on the hormone concentration and their degree of differentiation.

When bone surfaces are no longer the sites of active formation or resorption, they are covered by a thin, electron-dense, unmineralized layer of matrix called the **lamina limitans.** Covering the lamina limitans is a layer of flattened cells which are of osteoblast lineage and are called bone-lining cells. They are also known as **surface osteocytes,** resting or inactive osteoblasts, endosteal lining cells, and flattened mesenchymal cells. Bone-lining cells also cover the walls of the vascular channels which permeate cortical bone, but they are not present on periosteal surfaces. They are linked with each other and the deep osteocytes at gap junctions of their processes, and are involved in calcium homeostasis. When bone formation is restarted, the bone-lining cells possibly contribute to the deposition of bone, but for most of their life, they maintain a barrier between bone fluid and extracellular fluid.

Direct exchanges of calcium from bone crystals to blood, and replacement by ions such as sodium and magnesium, are important in calcium homeostasis and allow transfer of calcium, without phosphate, to blood. The bone-lining cells, under the control of parathyroid hormone, regulate this exchange by increasing the potential difference between bone fluid and extracellular fluid. Thus they determine the steady-state level of calcium in the blood, possibly by pumping hydrogen ion into the bone fluid and increasing solubilization of the bone mineral. Osteoclasts are not important in the fine regulation of calcium homeostasis.

C. *Osteocytes*

Osteocytes are formed almost entirely by burial of osteoblasts in osteoid, which is then mineralized. A few develop by metaplasia from fibrocytes, either directly or through an intermediate stage of chondrocytes, as in the chondroid bone where tendons and ligaments are inserted.

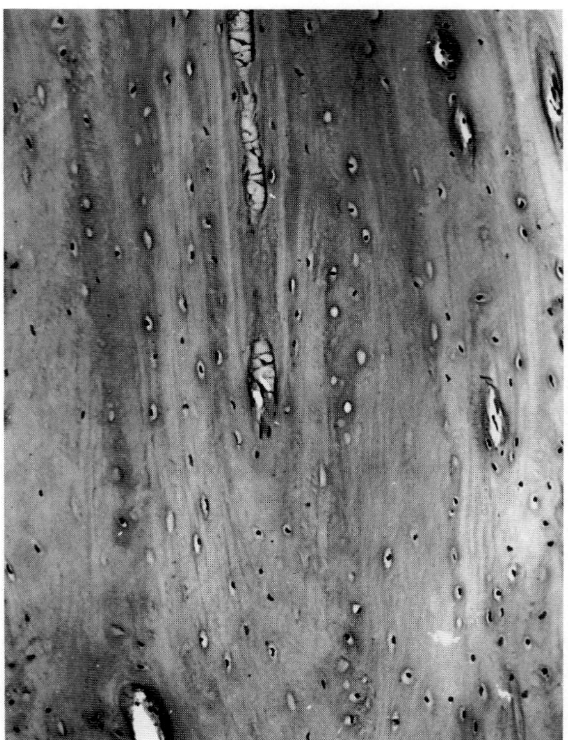

Fig. 1.5 Necrosis of bone in cortex of old dog. Empty, enlarged lacunae are visible in many areas.

Osteocytes of mature lamellar bone are flat or plump oval cells with many branching processes, which fill the canaliculi and anastomose with those of other osteocytes. They are evenly spaced and oriented with respect to lamellae. Osteocytes of woven bone differ little in appearance from the osteoblasts producing it. They have fewer processes than lamellar osteocytes and are less active metabolically. Osteocytes live in cavities called lacunae. There is a 1- to 2-μm layer of low-density bone around each lacuna (perilacunar bone), which contains fewer collagen fibers, and is less heavily mineralized than regular bone. This mineral is more labile than that deeper in the bone, and is important in calcium homeostasis.

The life span of an osteocyte in lamellar bone varies according to location, but in the cortex of an adult animal may be many years; in humans an osteocyte can survive for 35 years or more. Osteocyte death is recognizable by empty lacunae (Fig. 1.5). Prolonged demineralization of bone in acids for histologic processing may digest osteocytes, and the resultant empty lacunae can be mistaken for osteocyte death. Empty lacunae may also be due to displacement of osteocytes by the microtome knife.

The major activity of deep osteocytes is the demineralization and remineralization of perilacunar bone to help maintain calcium homeostasis. Parathyroid hormone, together with vitamin D, regulates this process, which involves removal of mineral from the perilacunar bone, leaving the matrix intact. Remineralization occurs under the influence of calcitonin, and the cycle may be repeated many times. Intermittently, the perilacunar bone is remodeled, that is, removed completely and replaced, by the osteocytes.

Osteocytes may also contribute in a small way to their own exhumation by resorbing bone when osteoclasts approach their lacunae. Disinterred osteocytes are able to migrate from their lacunae onto the surrounding bone.

Osteocytes are responsive to factors other than normal levels of parathyroid hormone, calcitonin, and vitamin D. Excess parathyroid hormone causes them to increase the volume of perilacunar bone, and extremely high levels stimulate enlargement of lacunae by removal of mineral and matrix. This activity is limited and does not progress with continued stimulation. Hypomineralization of bone around osteocytes is seen in fluorosis, in phosphorus deficiency, and in certain types of rickets and osteomalacia. As noted, calcitonin stimulates osteocytes to deposit mineral in perilacunar bone. Prolonged experimental calcitonin administration causes heavy mineralization of the perilacunar bone but, as with parathyroid hormone-stimulated activity, this does not progress beyond a certain level. Pituitary growth hormone, thyroxin, estrogens, and various growth factors also stimulate osteocytes under certain conditions.

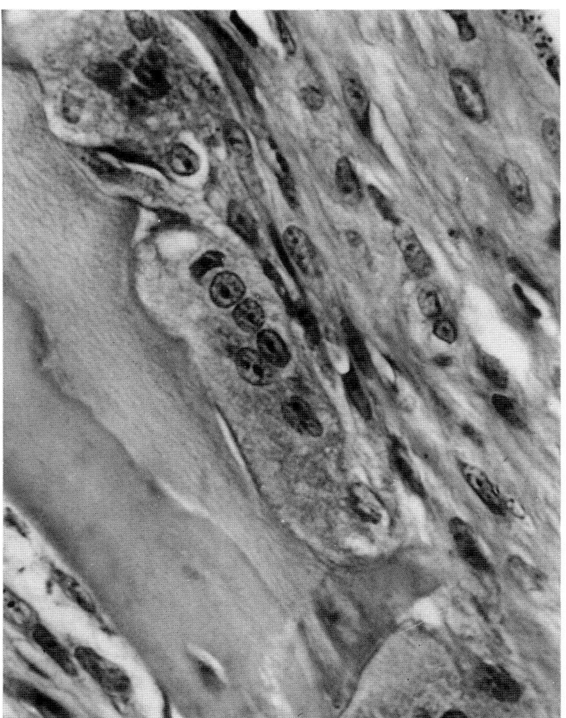

Fig. 1.6 Multinucleated osteoclasts in Howship's lacunae.

D. Osteoclasts

The products of hematopoietic stem cells are considered in detail in the Hematopoietic System (Volume 3, Chapter 2). Osteoclasts are probably derived from one branch of this system, but their precise origin is unclear. They may be related to monocytes and macrophages, but these cells cannot form osteoclasts, and contain tartrate-sensitive acid phosphatase, in contrast to osteoclasts, which possess a tartrate-resistant isozyme. Monoblasts and promonocytes may be the immediate osteoclast precursors, but it is also possible that they are derived from a distinct stem cell.

The typical osteoclast is a large cell with acidophilic cytoplasm and 2 or 3 to 100 nuclei, but usually several (Fig. 1.6). Osteoclasts are formed by fusion of precursor cells, fusion of precursor cells with an existing osteoclast, or by fusion of existing osteoclasts. In the resulting cells, only the nuclei remain distinct. Mononuclear phagocytes are sometimes associated with multinuclear osteoclasts, and may be either mononuclear osteoclasts, or phagocytes responsible for cleaning up organic debris left by osteoclasts. The longevity of osteoclasts is probably variable, and dependent on location and need; in basic multicellular units of human bone, they are active for about 1 month, and in beige osteopetrotic mice, less than 6 weeks. The life span of an individual osteoclast nucleus is about 10 to 14 days. The fate of osteoclasts is obscure; dead ones are found in certain diseases such as infectious canine hepatitis, but under normal circumstances, degenerate osteoclasts are rarely if ever seen. Disappearance through fission is probably their destiny.

Osteoclasts are found close to, or on the surface of, bone which is being resorbed, and they are responsible for bulk removal of bone including both mineral and matrix. They are not attracted to the matrices of cartilage or bone which have not previously been mineralized, but will resorb them when they come across them in the course of routine activities.

The concave cavities where osteoclasts work are Howship's lacunae (Fig. 1.7A,B). In these lacunae, osteoclasts appear as compact cells with acidophilic cytoplasm that is vacuolated near the bone-resorbing surface. When resorbing trabeculae, they often appear squamoid. Scanning electron microscopy shows that osteoclasts may branch extensively and may partially overlap adjacent osteoblasts and osteoclasts. The part of the osteoclast in contact with bone has a ruffled border, which moves continually during resorption. The ruffled border is surrounded by an organelle-free clear zone which seals the cell in contact with the mineralized tissue. A band of F-actin-containing podosomes probably contributes to the structure and adhesive properties of the clear zone. Carbonic anhydrase isozyme II, an enzyme present in the osteoclastic cytosol, generates protons by hydration of carbon dioxide to carbonic acid, and these are pumped across the membrane of the ruffled border into the sealed space between the cell and the bone. Primary lysosomes

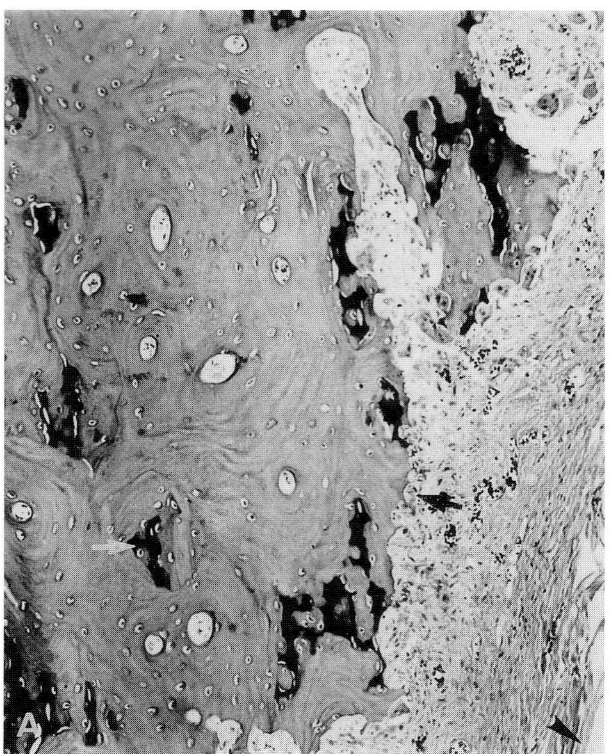

Fig. 1.7A Metaphysis in area of compaction of endochondral bone (see Fig. 1.1). Endochondral bone, which contains mineralized cartilage (white arrow), is incorporated into cortex and is modeled by osteoclasts. Each produces Howship's lacuna (arrow). Periosteal surface is at lower right (arrowhead).

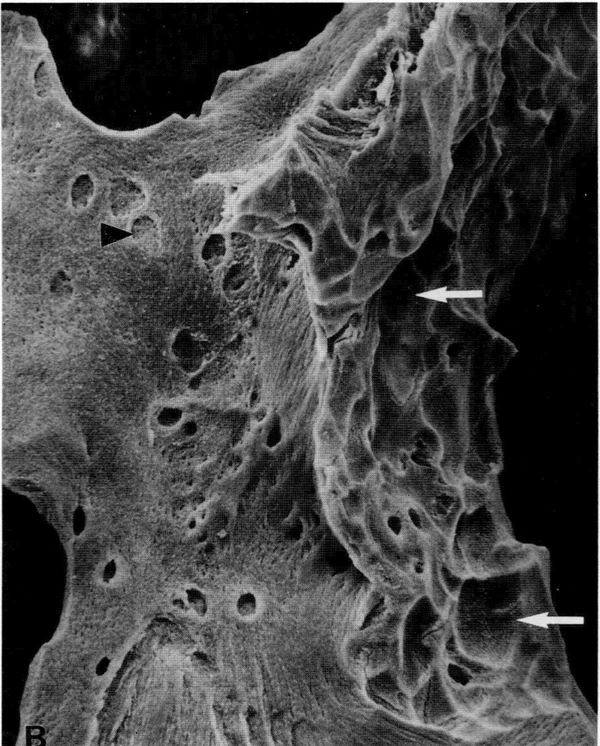

Fig. 1.7B Scanning electron micrograph of trabecula with soft tissue removed. Note fabric pattern of bone matrix, with lacunae of partly trapped osteoblasts (black arrowhead). There is a resorption cavity (right) composed of many Howship's lacunae (white arrows). Several osteocyte lacunae have been exposed by osteoclastic activity. (Courtesy of R. R. Pool.)

in the osteoclast fuse with the membrane of the ruffled border and release proteolytic enzymes, which include tartrate-resistant acid phosphatase and cathepsins, into the extracellular space. The mineral is solubilized first, and then the proteins are digested. Accordingly, the osteoclast is not a phagocytic cell, as the digestion of bone occurs in a sealed, extracellullar space that is analogous to a secondary lysosome. Osteoclasts can erode bone at the rate of $400\mu m^3$/hr and travel across bone surfaces at $100\mu m$/hr. In terms of bone processed per nucleus, an osteoclast is 10 to 50 times more efficient at removing bone than an osteoblast is at forming it, according to estimates in various species.

Osteoclasts do not have parathyroid hormone receptors, and their resorptive activity is probably controlled by a complex system that involves the response of cells of osteoblast lineage, which possess receptors to parathyroid hormone, to bone-resorbing agents. These osteoblastic cells may be the bone-lining cells, or possibly specialized parathyroid hormone target cells that lie in the intertrabecular spaces behind the active osteoblasts. They may interact with osteoclasts either through paracrine effects or by physical contact, and the effect of parathyroid hormone on this system probably exemplifies the interactions of other hormones and growth factors. Parathyroid hormone enhances the production by osteoblasts of latent collagenase and tissue plasminogen activator, and the latter converts plasminogen to plasmin. Plasmin activates the latent collagenase, which removes the thin layer of osteoid covering the bone surface, and exposes the mineralized bone to osteoclasts. Neighboring osteoclasts may be attracted to the site by exposure to collagen and noncollagenous proteins such as osteocalcin, which are chemotactic to osteoclasts. According to this theory, resorption stops, either when parathyroid hormone levels are reduced, or when calcitonin, a potent inhibitor of osteoclastic resorption, is produced.

Parathyroid hormone causes existing osteoclasts to become active and increase their nuclei, and also causes new osteoclasts to form. These changes can begin in 30 to 60 min, and continue for 48 to 96 hr, presumably as a result of fusion of preexisting precursors. Production of these precursors also may be stimulated by parathyroid hormone. In end-stage renal disease, the number of nuclei and cell size may increase dramatically. Not unexpectedly, large osteoclasts resorb more bone than small ones.

Mammalian, but not chicken, osteoclasts do possess

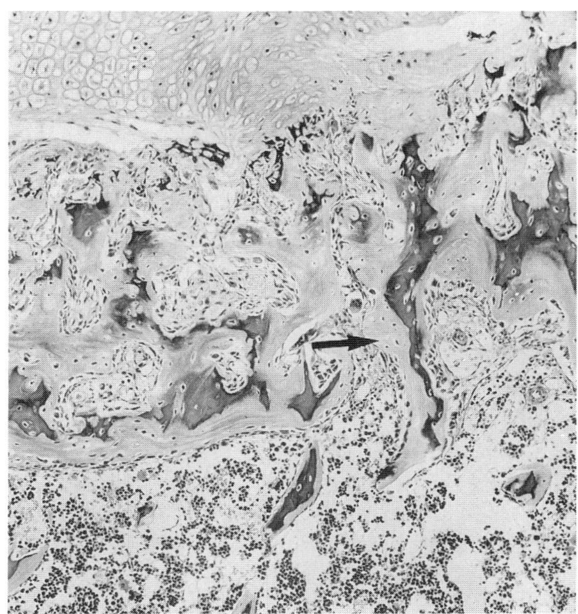

Fig. 1.8 Pale-staining unmineralized osteoid (arrow) in metaphysis of rachitic dog. (Courtesy of J. R. M. Innes.)

receptors for calcitonin, and respond within a few minutes by reduced activity of the ruffled border, and by moving away from the bone surface. Calcitonin also causes decreases in numbers of nuclei per osteoclast, numbers of osteoclasts, and numbers of osteoclast progenitors.

Osteoclasts are responsive to vitamin D, which causes their numbers and activity to increase. These responses are parathyroid hormone independent, and are probably mediated by other cells or factors, since osteoclasts do not have receptors for vitamin D. In excess, vitamin D may also stimulate generalized osteoclastic activity, and vitamin A has a similar effect, though its mechanism of osteoclast stimulation is different, but presently undetermined. Growth factors and prostaglandins are probably influential in stimulating the osteoclastic activity that occurs in some inflammatory and neoplastic diseases.

Some chemical poisons, such as phosphorus and lead, interfere with chondroclastic modeling of primary spongiosa, presumably by causing chondroblasts to produce an indigestible matrix. The heavy metal gallium has a direct inhibitory effect on osteoclasts, probably related to its adsorption onto bone surfaces.

E. Bone Matrix

The organic fraction of bone tissue is called **osteoid** or matrix, and it is composed of collagen and noncollagenous proteins, which are mainly glycoproteins and proteoglycans (Fig. 1.8). Few, if any, of these components are unique to bone. The collagen fibers of bone matrix are secreted in precursor form by osteoblasts. Lamellar osteoid contains about 90% type I collagen, chemically different forms of which also occur in skin and tendon. Chondrocytes produce the type II collagen found in cartilage. This type may also be present in notochord and in vitreous humor during ocular development. In contrast to the matrix of bone, cartilage matrix is collagen poor and proteoglycan rich.

The arrangement of collagen fibers is haphazard in woven (fibrous, primary) bone. This is primitive bone with coarse bundles of collagen best revealed by silver impregnation or birefringence. Fetal bone is of this type, as is the first bone formed in almost all pathologic ossifications, and in sites where there is a need for rapid structural reinforcement, whether pathologic or physiologic. Woven bone is later removed by osteoclasts and replaced by lamellar bone, which has thinner fibers arranged in concentric or parallel lamellae. The bone deposited in primary spongiosa is woven but fine-fibered, and it too is removed and replaced by lamellar bone. The expected trend, physiologically and in maturing pathologic ossifications, is from woven bone to lamellar bone. The trend is seldom reversed even in disease, although in Paget's disease of humans, in canine craniomandibular osteopathy, and in hypervitaminosis D, lamellar bone is removed and replaced by woven bone.

The second constituent of matrix, the noncollagenous proteins, make up the amorphous ground substance, the extracellular and interfibrillar material common to all connective tissues. In bone matrix there are several noncollagenous proteins, which can be classified as proteins containing γ-carboxyglutamic acid; phosphorylated and sulfated glycoproteins; proteoglycans, and sialoproteins. There are also protein constituents adsorbed from serum, and various growth factors (see preceding sections). In all probability, the propensity of bone matrix to mineralize is derived from its 10% content of noncollagenous protein. All of the noncollagenous proteins in mineralized matrix tend to be degraded with advancing age.

The γ-carboxyglutamic acid-containing proteins include osteocalcin (bone Gla protein, BGP) and matrix Gla protein. Osteocalcin is a vitamin K-dependent protein synthesized by osteoblasts, and its production is stimulated by calcitriol. It is present in serum, where it is accepted as a marker of osteoblast activity. Its concentration in bone increases as the bone matures. Osteocalcin is mineralized-tissue specific, being found only in bone and dentin. It is deposited in osteoid and predentin shortly before mineralization begins, and may either bind calcium or, less likely, retard mineralization until the appropriate time. The release of osteocalcin from bone when resorption is initiated may recruit osteoclasts to remodeling sites. At least in rats, vitamin K depletion has no discernible effect on the mechanical properties and osteocalcin content of bone. The other γ-carboxyglutamic acid-containing protein, matrix Gla protein, is not unique to mineralized tissues, also being found in cartilage, lung, heart, and kidney. Its function is obscure.

Osteonectin is a phosphorylated glycoprotein produced by many types of connective tissue cells, but mainly by osteoblasts. There is less osteonectin in woven than lamellar bone, and the concentration in bone increases with maturation. Osteonectin binds to type I collagen, and also to thrombospondin, and may be a nucleator in collagen-mediated mineralization. Alkaline phosphatase is a glycoprotein, and one of the three isozymes occurs in bone, as well as in liver and kidney. In bone it is produced mainly by osteoblasts, is found in matrix vesicles, and may be active in the preparation for matrix mineralization of woven bone and hypertrophic cartilage. Fibronectin is produced by all matrix-producing cells, including osteoblasts, and is involved in matrix organization as well as in cell migration and attachment. Thrombospondin is also present in many tissues. It is an osteoblast product that binds to connective tissues and serum proteins, and binds calcium to hydroxyapatite and to osteonectin. Tenascin is a matrix glycoprotein that is thought to have a role in chondrogenic, osteogenic, and odontogenic differentiation, and in tissue interactions.

Proteoglycans consist of a protein core with covalently attached chains of repeating disaccharide units called glycosaminoglycans. The glycosaminoglycans chondroitin sulfate, dermatan sulfate, heparan sulfate, and keratan sulfate are produced in the Golgi, whereas hyaluronic acid is produced at the plasma membrane. Mineralized bone contains two small chondroitin sulfate proteoglycans, biglycan (PG-1) and decorin (PG-2), that differ from the proteoglycans in cartilage in that they contain chondroitin-4-sulfate rather than chondroitin-6-sulfate. A large chondroitin sulfate proteoglycan is present in the osteoprogenitor cell zone of developing bone, but is removed before mineralization occurs. Biglycan and decorin are present in cells and matrix of newly formed bone, and biglycan is also found in lacunar walls and canalicular spaces. The synthesis of biglycan, but not decorin, is decreased by calcitriol. These proteoglycans are thought to play a key role in the organization, and possibly in the mineralization, of matrix.

Osteopontin (BSP-I, 44-kDa phosphoprotein, 2ar, pp66) is a sialoprotein that is found in a variety of cells in bone. Two other sialoproteins, bone sialoprotein II and bone acidic glycoprotein, also occur in bone but, as is the case for osteopontin, their function is unclear.

Serum proteins constitute up to one third of the organic matrix of bone. One of them, α-2HS-glycoprotein, is present in about 100 times higher concentration in bone than in serum. It is particularly concentrated in fetal and neonatal bone. Since it is an opsonic protein that induces phagocytosis by monocytes and macrophages, and also is a chemoattractant for monocytes, it may be active in recruitment of osteoclasts during bone resorption. It may also account for the replacement of woven bone by lamellar bone that is an integral part of skeletal maturation.

The multiple elements that make up a bone are joined by cement. This is a well-mineralized, collagen-free connective tissue with a high sulfur content and a higher Ca : P ratio than lamellar bone. In histologic sections, cement appears as basophilic lines of which there are two types—smooth and scalloped. The smooth lines are **resting lines,** which indicate that the previous cell activity was formative. The scalloped lines are **reversal lines,** which indicate that the previous cell activity was resorptive (Fig. 1.9). Cement lines are visible only in bone tissue covered by osteoblastic apposition, and they outline the mosaic of bony elements and indicate the scope of previous remodeling. Their origin is unclear, but their composition suggests they may be formed by mineralization of a layer of ground substance, and their location indicates they are formed at the end of a period of resorption or the initiation of a period of deposition. Cement lines may be important in resorption–formation coupling.

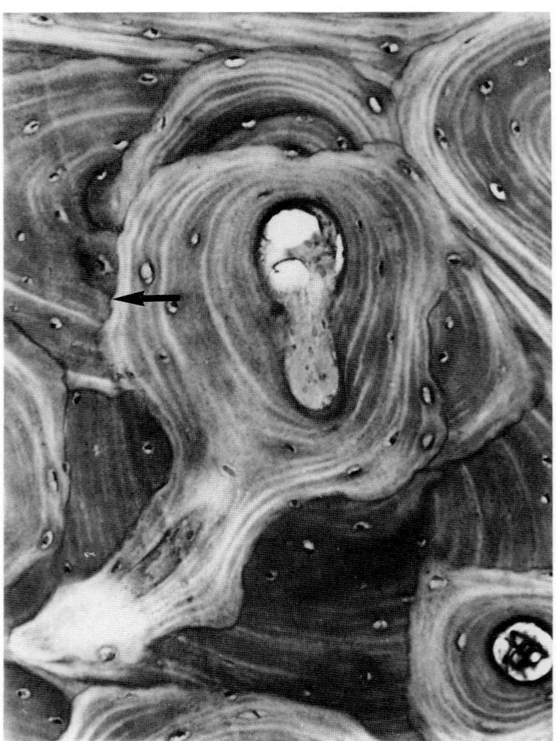

Fig. 1.9 Diaphysis. Mature horse. Secondary and tertiary osteons are outlined by basophilic reversal lines (arrow).

There are many factors that influence the quantity and quality of collagenous and noncollagenous proteins in the matrices of bone and growth cartilage. Examples of nutritional and genetic agents follow. For some of them the matrix component affected and the biochemical abnormality are known, but in many cases one or both are obscure or equivocal. Vitamin C is necessary to ensure an adequate supply of 4-hydroxyproline, which stabilizes the collagen molecule. When it is deficient, collagen either is not produced or is produced in defective form. Matrix deposition ceases, and differentiation of osteoblasts from osteoprogenitor cells is severely retarded. Deficiency of copper interferes with the cross-linking of collagen and elastin, probably due to a reduction in lysyl oxidase, a copper-dependent enzyme. Osteoblastic activity is depressed.

Aminonitriles, such as occur in sweet peas (*Lathyrus odoratus*), interfere with the formation of inter- and intramolecular cross-linkages during collagen synthesis and lead to skeletal deformities. The skeletal syndrome of odoratism is an experimental one and a useful model without a natural counterpart. There are inherited defects affecting type I collagen in animals, but these seem to affect the mechanical properties of dermal collagen rather than the osseous product, which is chemically different. Dermatosparaxis in Belgian calves is characterized by defective procollagen *N*-peptidase, an enzyme involved in the conversion of procollagen to collagen, but though bone lesions occur, fragility is not a clinical problem in affected calves.

Manganese deficiency has pathologic effects on both groups of matrix proteins. Manganese is required for the activation of xylosyltransferase, an enzyme involved in the synthesis of sulfated glycosaminoglycans, and "crooked calves" and premature synostoses in fetal calves may be related to deficient activity of this enzyme. Manganese is also active in the glycosylation of collagen, and deficiency may therefore be responsible for the decreased breaking strength of bones in "crooked calves." Papain, derived from *Carica papaya*, and some other proteolytic enzymes of plant origin, are active against the proteoglycans of cartilage matrix, but their effect on those in osseous matrix is not known. Papain causes loss of sulfated glycosaminoglycans in cartilage, as indicated by loss of metachromatic staining. Excess vitamin A also leads to degeneration of cartilage, probably due to the release from chondrocytes of lysosomal proteases which destroy the proteoglycans in matrix. In the mucopolysaccharidoses, lysosomal enzymes that metabolize various glycosaminoglycans are deficient, and the glycosaminoglycans accumulate in various tissues, including growth cartilage. Endochondral ossification is disrupted.

Vitamin D-deficiency rickets and osteomalacia exemplify diseases in which matrix defects occur, although even this is contested, but are ill defined. In the conditions named, osteoblasts produce a reduced amount of matrix, which does not mature at the normal rate, but does accumulate to excess on bone surfaces. The mineralization rate is retarded—in humans, the period between production and mineralization of matrix is increased from about 7 to 10 to over 100 days. In renal osteodystrophy, the limited affinity of matrix for mineral also is probably related to derangements of vitamin D metabolism and to other hormonal effects (see Section VI,F, Renal Osteodystrophy). Hypervitaminosis D, after an initial period of increased resorption, leads to the deposition of large amounts of osteoid matrix in the metaphyses, possibly in an overcompensated reversal effect, that has been called hypervitaminosis D rickets. The matrix is abnormal in that, initially, mineralization is delayed in onset although eventually complete and possibly accelerated. Fluoride in prolonged low but excessive dosage leads to disarray of osteoblasts and eventually to the production of a matrix in which mineralization is inhibited, the result qualifying as osteomalacia. Sulfated dextrans, experimentally, interfere with the production of osseous matrix but not of cartilage matrix.

Osteoblasts, especially lamellar osteoblasts, are target cells for endocrine action, and for some of the autocrine and paracrine influences discussed previously. They are profoundly affected by deficiency or excess of growth and thyroid hormones and are also influenced by them at physiological levels. Parathyroid hormone in excess depresses lamellar osteoblastic activity but enhances production of woven bone. Corticosteroids also depress lamellar osteoblasts but have much less effect on woven bone. In general, woven bone is relatively refractory to hormonal regulation. The fact that, in the postnatal animal, it tends to be produced in response to local requirements, suggests it may be relatively sensitive to growth factors.

F. Bone Mineral

Fully mineralized bone matrix is about 90% mineralized by volume; as the mineral accumulates, it replaces water in the matrix. Mineral in bone is a calcium-deficient apatite that contains carbonate and phosphate in the crystal lattice. Amorphous calcium phosphate, previously thought to be the initial form of bone mineral, is probably only a fleeting phase *in vivo*. Crystal maturation is associated with replacement of some of the phosphate by carbonate, and this accounts for the increased Ca : P ratio in aged bone. Maturation is also associated with increase in crystal size and a concomitant reduction of solubility. The molar Ca : P ratio in mature bone is about 10 : 6, and the ratio varies little in pathologic conditions, calcium accounting for about 27% and phosphorus for about 12% of the dried fat-free bone. The total amounts of calcium and phosphorus vary with the degree of mineralization of the matrix, being reduced in immature bone and in the examples of matrix abnormality referred to earlier.

There are several quantitatively important constituents of the inorganic part of bone in addition to calcium and phosphate. Carbonate is proportionately the next most significant component, substituting for phosphate in crystal structure and in surface adsorption; it has no obvious role in the economy of the tissue. Sodium is present to the extent of about one third the total body content, and approximately two thirds of bone sodium is nonexchangeable and therefore, by inference, part of the crystal structure. Bone is a large depot for magnesium. Its distribution in the mineral is uncertain, and there is debate on its availability for exchange; it is probably for the most part unavailable. Zinc is deposited in bone during mineralization of the matrix, and persists. It may be part of a metalloenzyme involved in the mineralization process. Bone normally contains a small amount of fluorapatite in which the fluoride ion is substituted for the hydroxyl ion of hydroxyapatite. Of the trace elements, manganese and copper are important in relation to matrix synthesis.

There is a time lag between the production of matrix and its mineralization. Thus a layer of unmineralized osteoid normally covers surfaces where bone is being formed. The layer may be relatively wide on lamellar bone because a time lag of several days is usual, but narrow and not obvious by light microscopy on woven bone, which mineralizes rapidly soon after it is formed. The rate of osteoid production and the lag between production and mineralization vary between species, and within species according to the age of the individual and the site in the bone. In general, osteoblasts deposit a 1- to 3-μm thick layer of osteoid per day. In lamellar bone, there is a clear demarcation between unmineralized osteoid and mineralized bone, which is called the **mineralization front** or the **phosphate ridge.** Tetracycline antibiotics, and other bone markers which fix to mineral, are bound at this location and are demonstrable in mineralized sections under ultraviolet light (Fig. 1.2). With continuous dosage in rapidly growing animals, enough tetracycline may be bound to produce yellow discoloration of bones and teeth in ordinary light. Gross discoloration of bones may also result from inherited and acquired porphyria. Red-brown discoloration of bones and teeth occurs in erythropoietic porphyria of cattle, pigs, and cats. Acquired porphyria causing pink bones of sheep in Australia was probably caused by exposure to lindane in a polluted stream. Several halogenated aromatic compounds possess porphyrinogenic properties.

Mineralization of lamellar osteoid is a two-stage and two-phase process. In the first stage, 70–75% of the total mineral capacity of osteoid is deposited by osteoblasts within a few days. The remaining mineral is deposited over several months and is dependent on osteocytic activity and chemical exchange between blood and bone crystals. The phases of mineralization of osteoid and cartilage matrix by hydroxyapatite crystals are those of nucleation and multiplicative proliferation. There are two mechanisms of nucleation, matrix vesicle and collagen mediated, but the final phase is always collagen mediated. The nucleation phase of crystal formation is not well understood. Extracellular fluids are supersaturated with respect to calcium phosphate and apatite, and it is not clear what prevents spontaneous, widespread mineralization. Magnesium is thought to play a role in inhibiting formation of hydroxyapatite. Alternatively, or additionally, other inhibitors of mineralization, such as pyrophosphate, and serum proteins that bind Ca^{2+}, may be destroyed at sites of mineralization by the actions of phosphatases and proteolytic enzymes.

Matrix vesicles are membrane-bound structures, rich in alkaline phosphatase and lipids, that are formed from microvilli on the surface of cells of hypertrophic cartilage and woven bone. Alkaline phosphatase enrichment in resting zone chondrocytes is stimulated by 24,25,dihydroxycholecalciferol, and in growth zone chondrocytes by calcitriol. The process of microvesicle formation involves detachment from F-actin in the cell membrane. Mineralization starts in widely distributed matrix vesicles, which ultimately are physically disrupted by the crystal formation, and progresses until the foci of mineral coalesce. At this time collagen-mediated precipitation of mineral ensues, and engulfs the matrix vesicles. The relationship between the two phases is not clear. When collagen-mediated mineralization occurs, the mineral is deposited in the hole zone regions of the collagen fibrils. Only later do small quantities of mineral form outside the collagen fibrils. It has been proposed that the initiating event in collagen-mediated mineralization is dependent on noncollagenous proteins, such as osteonectin, that adsorb onto collagen and act as nucleators. This is by no means certain.

Collagen-mediated mineralization is characteristic of lamellar bone and of the final stage of mineralization of woven bone. It may not begin in lamellar bone until 10 days after formation of osteoid. Mineralization of woven bone differs from that of lamellar bone in that it begins and is completed 1–3 days after osteoid formation. Woven bone contains less collagen and more proteoglycan than does lamellar bone, and its capacity for mineral is less. Since woven bone reaches maximum mineralization very rapidly, in microradiographs, for example of laminar bone, it may appear more densely mineralized than adjacent, newly formed lamellar bone. For this reason, the central core of woven tissue in laminar bone is sometimes called the hypermineralized primer.

Bibliography

Austin, L. A., and Heath, H. Calcitonin. Physiology and pathophysiology. *N Engl J Med* **304:** 269–278, 1981.

Band, C. A., and Boivin, G. Effects of hormones on osteocyte function and perilacunar wall structure. *Clin Orthop* **136:** 270–281, 1978.

Baron, R. Molecular mechanisms of bone resorption by the osteoclast. *Anat Rec* **224:** 317–324, 1989.

Beresford, J. N. Osteogenic stem cells and the stromal system of bone and marrow. *Clin Orthop* **240:** 270–280, 1989.

Bonewald, L. F., and Mundy, G. R. Role of transforming growth factor-beta in bone remodeling. *Clin Orthop* **250:** 261–276, 1990.

Canalis, E., McCarthy, T., and Centrella, M. Growth factors and the regulation of bone remodeling. *J Clin Invest* **81:** 277–281, 1988.

Eisenstein, R., and Kawanone, S. The lead line in bone—a lesion apparently due to chondroclastic indigestion. *Am J Pathol* **80:** 309–314, 1975.

Glimcher, M. J. Mechanism of calcification: Role of collagen–fibrils and collagen–phosphoprotein complexes *in vitro* and *in vivo*. *Anat Rec* **224:** 139–153, 1989.

Jackson, R. L., Busch, S. J., and Cardin, A. D. Glycosaminoglycans: Molecular properties, protein interactions, and role in physiological processes. *Physiol Rev* **71:** 481–539, 1991.

Legros, R., Balmain, N., and Bonel, G. Age-related changes in mineral of rat and bovine cortical bone. *Calcif Tissue Int* **41:** 137–144, 1987.

Marks, S. C., and Popoff, S. N. Bone cell biology: The regulation of development, structure, and function in the skeleton. *Am J Anat* **183:** 1–44, 1988.

McGrath, K. J., Heideger, W. J., and Beach, K. W. Calcium homeostasis III: The bone membrane potential and mineral dissolution. *Calcif Tissue Int* **39:** 279–283, 1986.

Mensink, J. A., and Strik, J. J. T. W. A. Porphyrinogenic action of tetrachloroazobenzene. *Bull Environ Contam Toxicol* **28:** 369–372, 1982.

Menton, D. N. *et al.* From bone lining cell to osteocyte—An SEM study. *Anat Rec* **209:** 29–39, 1984.

Nijweide, P. J., Burger, E. H., and Feyen J. H. M. Cells of bone: Proliferation, differentiation, and hormonal regulation. *Am Physiol Soc* **66:** 855–886, 1986.

Parfitt, A. M. The cellular basis of bone turnover and bone loss. A rebuttal of the osteocytic resorption–bone flow theory. *Clin Orthop* **127:** 236–247, 1977.

Robey, P. G. The biochemistry of bone. *Endocrinol Metabol Clin North Am* **18:** 859–902, 1989.

Rouleau, M. F., Mitchell, J., and Goltzman, D. Characterization of the major parathyroid hormone target cell in the endosteal metaphysis of rat long bones. *J Bone Miner Res* **5:** 1043–1053, 1990.

Sandhu, H. A., and Jande, S. S. Effects of beta-aminoproprionitrile on formation and mineralization of rat bone matrix. *Calcif Tissue Int* **34:** 80–85, 1982.

Schaffler, M. B., Burr, D. B., and Frederickson, R. G. Morphology of the osteonal cement line in human bone. *Anat Rec* **217:** 223–228, 1987.

Scherft, J. P. The lamina limitans of the organic matrix of calcified cartilage and bone. *J Ultrastruct Res* **38:** 318–331, 1972.

Tourtellotte, C. D., and Dziewiatkowski, D. D. A disorder of endochondral ossification induced by dextran sulphate. *J Bone Joint Surg (Am)* **47A:** 1185–1202, 1965.

Triffitt, J. T. Initiation and enhancement of bone formation. A review. *Acta Orthop Scand* **58:** 673–684, 1987.

Vaes, G. Cellular biology and biochemical mechanism of bone resorption. A review of recent developments on the formation, activation, and mode of action of osteoclasts. *Clin Orthop* **231:** 239–271, 1988.

van der Walt, J. G. Eicosanoids: A short review. *J S Afr Vet Assoc* **60:** 65–68, 1988.

von der Mark, K., and Conrad, G. Cartilage cell differentiation. *Clin Orthop* **139:** 185–205, 1979.

G. General Reactions of Bone Tissue to Injury

The responses of bones to injury can be considered at the level of the cell, tissue, or organ. Many examples of specific cell responses are given, and organ responses to mechanical stimuli are considered with adaptational deformities. References to general tissue reactions are made throughout the chapter, as for example, the preferential development of cartilage or bone according to oxygen tension and mechanical pressure, and the differentiation of woven bone in locations where rapid bone formation is required.

Bone cells and tissues respond to a variety of stimuli in a variety of ways. These may be manifested by variations in the type and number of cells that differentiate from the stem cell populations, as well as by the activities of these cells, once differentiated. Thus, osteoprogenitor cells may proliferate and produce osteoblasts, which may fabricate more or less normal or abnormal, matrix than usual, and the matrix may be adequately or inadequately mineralized, or they may produce woven bone. Many of the quantitative responses of bone cells cannot be measured on a routine basis, and the assessment of many qualitative features, such as the chemical composition of osteoid, requires techniques not available to most pathologists.

There are, however, certain principles, which apply to bone diseases in general, that are important analytic aids and do not require specialized techniques. The first of these, relating to endochondral ossification, is that disruption in the process at any point interferes with the normal production of metaphyseal bone. Intermittent insults, which are likely to result in modeling defects such as development of transverse trabeculae or growth-retardation lattices, will be discussed. When the cause is constitutional or metabolic, all fast-growing physes should be affected. In an inherited disease, extraskeletal hyaline cartilage may also be abnormal, as in the mucopolysaccharidoses. Careful examination, and comparison with the normal sequence, may identify the specific point in the process that is abnormal, but two considerations must be kept in mind: primary lesions are seldom all-or-none, and secondary lesions often complicate the initial change. Using rickets as an example, inhibition of mineralization and vascular invasion of cartilage is rarely complete, and the effects of normal activity on weakened rachitic metaphyseal structures almost always produce trabecular fractures and hemorrhages. These secondary effects can become the dominant histologic feature in the metaphyses in diseases characterized by disordered endochondral ossification.

The second principle involves the effect of normal weight-bearing, which influences not only the physeal–metaphyseal region that in the growing animal is most susceptible to injury, but also the epiphysis and diaphysis. Because of their major function as supporting structures, bones are at all times subjected to mechanical forces. This is so obvious, but is frequently overlooked, especially by microscopists. Mechanical forces acting on weakened metaphyses produce hemorrhages, trabecular fractures, and infractions. Hemorrhages are easily recognized in their early stages, but when they are extensive, organized by fibrous tissue, and being converted to woven bone, they must be distinguished from osteodystrophia fibrosa and other fibroplastic conditions. Fractured trabeculae are often most obvious when they contain a core of mineralized cartilage which is not aligned with the long axis of the bone. Infractions are compression fractures involving many trabeculae, but without gross discontinuity of the bone. Regardless of the number of trabeculae fractured, the sequel is the same: death of osteocytes and eventual resorption and replacement of the bone tissue.

A third principle involves the sequence of bone formation and resorption in normal and pathologic circumstances. In the modeling skeleton, formation and resorption are not coupled, whereas in the remodeling skeleton, resorption precedes formation. The latter has important implications because it means that, in the absence of iatrogenic effects, the mature skeleton may maintain its mass but cannot increase it, since new bone formation is restricted to those areas where old bone has been removed. In fact, with increasing age, many resorption cavities are less completely filled, and a gradual loss of skeletal mass occurs. A further consequence of coupling of resorption and formation in the remodeling skeleton is the effect it may have on short term metabolic studies. Since resorption precedes formation in the adult, examination of a given bone following the resorption phase will indicate osteoporosis, whereas examination after the formation phase may show normal skeletal mass. This is particularly relevant in pathologic situations when a "regional acceleratory phenomenon" develops in which many foci of remodeling are activated simultaneously (see Diseases of Bones, Section VI,B,2, Disuse Osteoporosis).

The resorption–formation sequence is reversed under some pathologic conditions, as when bone is infarcted. In Legg–Calvé–Perthes disease (see following sections), the healing phase following necrosis of the femoral capital epiphysis involves ingrowth of vascular and osteogenic tissue, which lays down woven bone on dead trabeculae. Logic suggests that when blood vessels grow in, dead bone would be exposed to osteoclast precursors and resorption would occur as readily as formation. Such is not the case. This is explicable in terms of the theory that bone resorption is initiated by viable cells of osteoblast lineage (see Diseases of Bones, Section III,C, Osteoclasts).

A fourth principle of bone behavior is that when bone is formed rapidly, woven bone is produced. This applies under physiologic conditions, as in the initial stages of laminar bone production, as well as in pathologic conditions, such as formation of callus. Woven bone, once formed, is almost invariably replaced by lamellar bone. Exceptions to this rule occur in most tumors, but in some benign neoplasms, woven bone is gradually replaced by lamellar bone.

A fifth principle of bone behavior involves the local or widespread production of new bone by injured or stimulated periosteum. This is usually woven bone, but the type depends on the rate of production. Formation of excess periosteal bone is a common lesion, which results from various stimuli. Perhaps the best known example is in the formation of Codman's triangle, a localized pyramidal region of periosteal bone which, though usually associated with osteosarcoma, is mimicked by osteomyelitis and some other lesions of bone. The reactive periosteal tissue is composed of trabeculae of woven bone formed at acute and right angles to the diaphysis. In its early stages it is not associated with the actual involvement of periosteum by tumor or inflammatory process but starts to develop some time after the endosteal surface is involved. This must be kept in mind when selecting and examining biopsy material. Release of diffusible growth factors, or interference with intracortical circulation, or both, may be responsible for activation of the periosteal cells.

Some conditions, such as hypertrophic osteopathy and porcine juvenile hyperostosis, are characterized by diffuse periosteal proliferation. Periosteal bone formation, involving both the axial and appendicular skeleton, is associated occasionally with disease in dogs caused by *Hepatozoon canis*, and *Leishmania donovani* sometimes causes periosteal proliferation (or osteolytic lesions) in dogs. The pathogenesis of such diffuse lesions is unknown, but hepatozoonosis is characterized by granulomatous myositis, and leishmaniasis by osteomyelitis, so interference with medullary or periosteal circulation may be a common factor. Evacuation of medullary contents of long bones stimulates diffuse periosteal new bone in experimental rabbits, as does obstruction of medullary circulation on some occasions.

H. Ectopic Mineralizations and Ossifications

Ectopic (heterotopic) mineralization (calcification) must be distinguished from ectopic ossification, and also from metaplastic ossification. **Ectopic mineralization** is the deposition of calcium salts of phosphate, silicate, etc., in abnormal locations. **Ectopic ossification** is the production of bone in an abnormal location. **Metaplastic bone** is bone that develops directly from another connective tissue by redifferentiation of mesenchymal cells. Metaplastic bone may be ectopic, but is not necessarily so.

Ectopic mineralization is divided into idiopathic, metastatic, and dystrophic forms, depending on the predisposing causes. Respectively, these are obscure, elevation of plasma calcium and/or phosphorus, and preexisting cellular injury. For example, idiopathic mineralization is seen in primary **tumoral calcinosis,** and metastatic mineralization develops in association with the hypercalcemia of uremia and hypervitaminosis D, and also in those uremic animals which have high blood phosphate; mineralization of soft tissue in dogs is expected when the serum Ca : P product, expressed in mg/100 ml, exceeds 60. Dystrophic mineralization does not depend on derangement of calcium metabolism, developing for example in the injured muscle fibers of nutritional myopathies and myocardial infarcts. Ectopic mineralization in diseases such as tuberculosis in cattle, pseudotuberculosis in sheep, and paratuberculosis in goats is probably related to production of calcitriol or a calcitriol-like substance by granulomatous tissue, such as occurs in tuberculosis and other granulomatous diseases of humans.

The mechanisms of ectopic mineralization may be analogous to the physiologic mineralization processes described previously. The concentration of Ca^{2+} is normally much higher in the extracellular fluid than it is in the cell, and the gradient is maintained by an adenosine triphosphate (ATP)-dependent pump in the cell membrane. Much of the Ca^{2+} that enters the cell normally is taken up by the mitochondria. When plasma Ca^{2+} is elevated, the concentration in the cell and in the mitochondria rises. Continued accumulation in mitochondria inhibits their ability to produce ATP, and the concentration of calcium in the cytosol approaches that in the extracellular fluid. Apatite crystals form in the mitochondria and act as nucleators for the more generalized mineralization that follows; thus, mitochondria function in a way similar to matrix vesicles. Dystrophic mineralization may proceed from apatite deposition in the mitochondria of injured cells, or it may utilize fragments of cell membranes from necrotic cells in lieu of matrix vesicles.

Calcinosis cutis is an example of ectopic mineralization, and is associated with hyperadrenocorticism, direct exposure to calcium chloride or wet concrete, and occurs in some normal poodle dogs (see Chapter 5, The Skin and Appendages). Another form of ectopic mineralization is that seen in **tumoral calcinosis** (calcinosis circumscripta). In dogs, lesions of tumoral calcinosis are usually solitary, are common in the skin of the extremities, and less common in the tongue and in the vertebral region. Single, midline lesions located between the dorsal tubercle of the atlas and the dorsal spine of the axis occur in young dogs and cause sudden or gradual onset of ataxia when they extend into the epidural space. Multiple lesions of tumoral calcinosis develop in some dogs, and are often attached to tendons, joint capsules, or periosteum. Lesions involving the processes of vertebrae C-4 and C-5 occurred in three littermate Great Danes aged 5–6 months and were thought to be congenital.

Tumoral calcinosis in horses is seen most often in animals about 2 to 4 years old, and 90% of lesions develop on the lateral aspect of the stifle. They are single or multiple, and not attached to skin, but some are fixed to the femorotibial joint capsule. These lesions are non-neoplastic and do not recur following removal.

There are several known and proposed causes of tumoral calcinosis. Breed and familial prevalences occur, German shepherds being prone to develop the solitary lesion, whereas in Great Danes and viszlas, lesions tend to be multiple. In humans, tumoral calcinosis can be inherited as an autosomal recessive trait, and is associated with hyperphosphatemia and multiple, periarticular deposits of amorphous calcium phosphate. In dogs (and horses), some lesions are thought to result from repetitive trauma with dystrophic mineralization of the injured tissues. Trauma causing cystic dilation, hyperplasia, and mineralization of apocrine glands may account for some of the canine dermal lesions. Occasionally, deeper lesions develop in association with buried polydioxanone suture material in dogs. The genesis of lingual lesions in dogs is obscure, but it is possible that heterotopic salivary glands degenerate, and provide a focus for mineral deposition.

Grossly, tumoral calcinosis consists of nodular masses usually one to a few centimeters across. Cutaneous lesions may ulcerate and discharge some of their content. When cut, they appear as white, multinodular soft-creamy to dry-gritty foci in a mass of dense connective tissue. Microscopically, the connective tissue surrounds amorphous and granular mineralized deposits, and there is often a granulomatous reaction with multinuclear giant cells around the mineralized foci. Metaplastic bone and cartilage occasionally develop in chronic lesions, which are then impossible to distinguish from pseudogout unless the mineral is analyzed. Pseudogout is a rare lesion of old dogs, characterized by deposition of calcium pyrophosphate dihydrate in periarticular tissues. Metaplastic bone and cartilage, and granulomatous inflammation, develop around the deposits.

A syndrome of idiopathic calcium phosphate deposition occurs in the Great Dane breed, and is distinct from the paravertebral lesions noted previously. It is characterized clinically by paraplegia and incoordination, starting in puppies aged 5–7 weeks, affecting up to two thirds of the litter with equal sex distribution. Deformity and dorsal displacement (pathologic spondylolisthesis) of the C-7 vertebral body, and mineral deposits within the diarthrodial vertebral joints are visible radiographically. In older dogs, fusion of articular processes near the cervicothoracic junction, mineralization of the costovertebral articulations, and arthropathy and periarticular mineralization of other joints may be present. Concentric compression of the cervical cord accounts for the clinical signs, and deposition of mineral in thickened, hypertrophic synovial membranes, articular and growth cartilages, and around joints, is responsible for the radiographic changes. Other lesions include shortened, bent long bones with thin cortices and excess medullary bone, and cartilaginous metaplasia in the thickened synovial tissue. Mineralization involving various organs including lung, heart, kidney, and spleen is also present in affected dogs, and similar but less severe changes may occur in clinically normal littermates. The condition is thought to be an autosomal dominant trait, and it is suggested that heterozygotes may show minimal lesions. Vitamin D metabolism appears to be normal, but there is a tendency to hypophosphatemia in affected dogs.

Mineralization of the supraspinatus tendon occurs in dogs, particularly hunting and working dogs, and may be associated with lameness. Lesions are often bilateral, originating as chondromucinous degeneration of collagen, and progressing to multifocal, then coalescent mineralization. The tendon must be incised to demonstrate the contained mineral. Lesions probably result from repetitive trauma, involving activities such as jumping and digging, to tissue whose blood supply is normally scant. Mineralization of the bicipital tendon also occurs in dogs in association with tenosynovitis.

Mineralization and ossification of the cranial part of the medial meniscus of the stifle joint are rare, incidental lesions in cats. Mineralization is probably secondary to trauma, whereas ossified menisci are thought to be anomalous.

Ectopic ossification occurs in many tissues and is usually of no pathologic significance. Small spicules of lamellar bone are often found in the pulmonary connective tissue of dogs and cattle, and occasionally there is multilobular involvement. Ossification of surgical scars is often seen in the abdominal wall of pigs. Ossification of the dura mater, ossifying pachymeningitis, develops in the lumbosacral and cervical regions of old dogs. Periarticular osteophytes are a common example of ectopic ossification. They, and synovial chondrometaplasia (synovial osteo/chondromatosis), are discussed with diseases of joints.

Widespread ectopic ossification of soft tissues occurs rarely in most species, but little is known of its cause. In humans, lesions occur in association with muscle trauma, or neuromuscular lesions without trauma, or without apparent predisposing cause. Progressive ossification of muscle, myositis ossificans, is sometimes familial. The lesions have a characteristic sequence of maturation in which fibroblastlike cells differentiate, and produce and mineralize osteoid, which ultimately is replaced by lamellar bone.

In animals, the syndromes of ectopic ossification have attracted a variety of names, depending on the distribution and identity of the tissues involved. Fibrodysplasia ossificans, with ossification, usually symmetrical, confined to the subcutis and epimysium of the neck, dorsum, groin, and extremities, is described in cats. The bone develops by osseous metaplasia of proliferative fibrous tissue, and by endochondral ossification of cartilaginous tissue. Diffuse idiopathic skeletal hyperostosis (DISH) is an idiopathic condition in humans and is described in dogs. It is characterized by extensive ossification around joints with eventual fusion of some axial and abaxial structures, but sparing of intra-articular structures. The bone develops largely by endochondral ossification of proliferative fibrocartilaginous tissue, with ultimate conversion to lamellar bone. The relationship of DISH to some of the canine disorders previously described is unclear. A familial disease in pigs associated with ossification of soft tissues, especially around the vertebrae, occurred in 34 of 115 offspring of an affected boar.

The prognosis for animals with widespread ectopic bone is generally poor. Interference with nerves and with joint function develop, and surgically removed bone is expected to redevelop. Localized, post-traumatic lesions sometimes regress, but if not, surgical removal is most successful when delayed until they are quiescent.

Some ectopic bone probably develops from inducible osteogenic precursor cells, which are present in many organs. The identity of the inducing substance(s) is not known, but growth factors must be suspected, as when muscle ossification follows injury to periosteum, or bone dust is scattered through the tissues during total hip arthroplasty. Ectopic bone often forms in lesions which have been mineralized for a long time, as in tumoral calcinosis, and in foci of pulmonary mineralization such as occur with poisoning by vitamin D and plants with vitamin D activity. In these situations, the cells responsible for mineralizing the connective tissues probably redifferentiate and produce metaplastic bone.

Metaplastic bone is not always ectopic. It normally is formed at some entheses (tendon insertions) and fibrous growth plates, probably in response to changes in local tissue environment.

Bibliography

Anderson, H. C. Calcific diseases. A concept. *Arch Pathol Lab Med* **107:** 341–348, 1983.

Bone, D. L., and McGavin, M. D. Myositis ossificans in the dog: A case report and review. *J Am Anim Hosp Assoc* **21:** 135–138, 1985.

Craig T. M. *et al. Hepatozoon canis* infection in dogs: Clinical, radiographic, and hematologic findings. *J Am Vet Med Assoc* **173:** 967–972, 1978.

Ellison, G. W., and Norrdin, R. W. Multicentric periarticular calcinosis in a pup. *J Am Vet Med Assoc* **177:** 542–546, 1980.

Flo, G. L., and Tvedten, H. Cervical calcinosis circumscripta in three related Great Dane dogs. *J Am Anim Hosp Assoc* **11:** 507–510, 1975.

Goulden, B. E., and O'Callaghan, M. W. Tumoral calcinosis in the horse. *N Z Vet J* **28:** 217–219, 1980.

Heimann, M., Carpenter, J. L., and Halverson, P. B. Calcium pyrophosphate deposition (chondrocalcinosis) in a dog. *Vet Pathol* **27:** 122–124, 1990.

Kirby, B. M. *et al.* Calcinosis circumscripta associated with polydioxanone suture in two young dogs. *Vet Surg* **18:** 216–220, 1989.

Lemann, J., and Gray, R.W. Calcitriol, calcium, and granulomatous disease. *N Engl J Med* **311:** 1115–1117, 1984.

Lewis, D. G., and Kelly, D. F., Calcinosis circumscripta in dogs as a cause of spinal ataxia. *J Small Anim Pract* **31:** 36–38, 1990.

Marks, S. L., Bellah, J. R., and Wells, M. Resolution of quadriparesis caused by cervical tumoral calcinosis in a dog. *J Am Anim Hosp Assoc* **27:** 72–76, 1991.

Morgan, J. P., and Stavenborn, M. Disseminated idiopathic skeletal hyperostosis (DISH) in a dog. *Vet Radiol* **32:** 65–70, 1991.

Puzas, J. E., Miller M. D., and Rosier, R. N. Pathologic bone formation. *Clin Orthop* **245:** 269–281, 1989.

Seibold, H. R., and Davis, C. L. Generalized myositis ossificans (familial) in pigs. *Pathologia Vet* **4:** 79–88, 1967.

Turrel, J. M., and Pool, R. R. Bone lesions in four dogs with visceral leishmaniasis. *Vet Radiol* **23:** 243–249, 1982.

Waldron, D. *et al.* Progressive ossifying myositis in a cat. *J Am Vet Med Assoc* **187:** 64–65, 1985.

Warren, H. B., and Carpenter, J. L. Fibrodysplasia ossificans in three cats. *Vet Pathol* **21:** 495–499, 1984.

Whiting, P. G., and Pool, R. R. Intrameniscal calcification and ossification in the stifle joints of three domestic cats. *J Am Anim Hosp Assoc* **21:** 579–584, 1984.

Woodard, J. C. *et al.* Calcium phosphate deposition disease in Great Danes. *Vet Pathol* **19:** 464–485, 1982.

Woodard, J. C. *et al.* Canine diffuse idiopathic skeletal hyperostosis. *Vet Pathol* **22**: 317-326, 1985.

VI. Development, Modeling, and Remodeling of Bones

In order to understand the pathology of bones, it is necessary to understand how they develop, mature, and maintain themselves—that is, their embryology, growth, modeling, and remodeling. These complex subjects are reviewed only briefly here. A list of references concludes this section.

A. Development

Long tubular bones such as the femur and tibia best demonstrate the common principles of growth and modeling, because changes occur in them at a faster rate, or over a longer period, than in other bones. In the embryo, there are two patterns of bone growth, each beginning with a condensation of mesenchymal cells. One of these is **intramembranous osteogenesis,** which involves deposition and mineralization of a plate of extracellular matrix by differentiated mesenchymal cells. (The flat bones of the skull are formed entirely by intramembranous osteogenesis.) The second pattern is **endochondral osteogenesis,** which involves differentiation of the mesenchymal condensations into chondroblasts and chondrocytes, and production of a mineralized scaffold upon which bone is then deposited. The actual process of bone deposition in endochondral osteogenesis is very similar to that in intramembranous osteogenesis—the main difference between the two patterns is the intermediation of the cartilage scaffold. Mesenchymal cells retain the ability to differentiate into osteoblasts or chondroblasts throughout life, and cartilage, formed for example in fracture callus or cartilage-forming tumors, is often replaced by bone via endochondral ossification.

Development and growth of a long bone involve both endochondral and intramembranous ossification. The mesenchymal condensation that gives rise to the bone first differentiates into cartilage. This anlage has a well-defined perichondrium which contributes to its transverse growth. Increase in size also depends on multiplication of the cartilage cells, production of intercellular matrix, and growth of the interzone tissue between bones, which later develops into joint structures.

At a certain stage of development the perichondrium is invaded by vessels, and a ring of intramembranous ossification forms around the shaft of the bone (in mammals, see Hall regarding avian long bones). At about the same time, cells including osteoclasts and vessels grow into the center of the cartilaginous anlage in a region where the cartilage cells start to enlarge and mineralize their intercellular matrix. Endochondral ossification begins in this, the primary ossification center of the bone. The center expands toward the ends of the bone in conjunction with the extension of the ends of the ring of perichondrial ossification. In bones destined to develop one or two or more growth plates, one or two or more secondary ossification centers develop at the ends of the bones and expand centrifugally. The secondary centers unite with the primary center when the bone stops growing. Until that time, endochondral ossification continues and contributes to growth of the bone in length.

Endochondral ossification entails the proliferation, maturation, hypertrophy, and modeling of cartilage. This process is best demonstrated and easiest to understand when studied in a long bone of a young animal which has developed an epiphysis with a terminal plate. The sequence is the same in slow growing bones but tends to be less orderly, and the various stages are less easily distinguished.

The **physis** (Fig. 1.1) is the principal seat of endochondral ossification. It is often called the growth plate or epiphyseal plate; the latter term should be avoided because it is easily confused with the growth cartilage of the epiphysis, which is subjacent to the articular cartilage and is also a site of endochondral ossification.

The terminal plate of bone separates the physis from the epiphysis. It abuts the reserve zone of the physis. There are perforations in the terminal plate through which arterioles from the epiphyseal artery pass in cartilage canals, through the reserve zone of the physis to the first part of the proliferative zone. The physis is nourished by diffusion from these vessels. Transverse growth of the physis is by perichondrial differentiation and apposition, and is separable from the interstitial growth which provides for increase in length. Chondroblasts produce cartilaginous matrix, and unlike their differentiated offspring, the chondrocytes, they retain the ability to divide. Both cell types produce a matrix that contains an anti-invasion factor, which prevents the synthesis of proteases by endothelial cells. Hypertrophic chondrocytes do not produce this factor. The perichondrium is supplied with blood from the perichondrial arteries.

The physis is divisible into three zones or layers. The zone next to the terminal plate of the epiphysis is the reserve, resting, or germinal zone. Distal to it are the zones of proliferation and hypertrophy. The cells of the reserve zone are spherical, occur singly or in pairs, and are separated by more extracellular matrix than are the cells in other zones. This zone contains abundant type II collagen, which is randomly distributed.

In the proliferating zone of the physis, the chondrocytes are flattened and aligned in columns. Mitoses are rarely seen in longitudinal sections. The true germinal cells of the physis are the first cells of the proliferating zone. Oxygen tension is higher in this zone than in any other part of the physis, and glycogen storage is a major function of these cells. They also produce proteoglycans. The collagen in the proliferating zone is condensed around the columns of cells. These columns are not always parallel with each other, or perpendicular to the physis. They line up more or less parallel with the long axis of the bone, whereas the physis is wavy, and arranged perpendicular to local lines of force, especially compression. The contribution of the physis to growth in length depends on cell division in the proliferative zone and cell enlargement in the hypertrophic zone. The effect of this growth pattern is to move the epiphysis and physis away from any fixed point in the diaphysis.

In the hypertrophic zone, the cells enlarge about 4 to 5 times in height and 10 times in volume by transporting water and electrolytes into the intracellular space and by increasing the absolute amount of matrix. Hypertrophic cells also produce type X collagen and reduced amounts of type II. They do not produce the anti-invasion factor, which prevents ingrowth of blood vessels, as do other cartilage cells. The calcium-binding proteins, chondrocalcin and calmodulin, are most concentrated in the lower part of the hypertrophic zone. In the first half of this zone, the cells accumulate calcium in their mitochondria and retain the glycogen they produced in the proliferative zone. Near the middle of the hypertrophic zone they lose the glycogen abruptly, and by light microscopy the cells appear vacuolated. Glycogen loss is associated with a change from aerobic metabolism as the cells get farther from their blood supply. When the glycogen is depleted, mitochondria release their calcium, which accumulates in matrix vesicles.

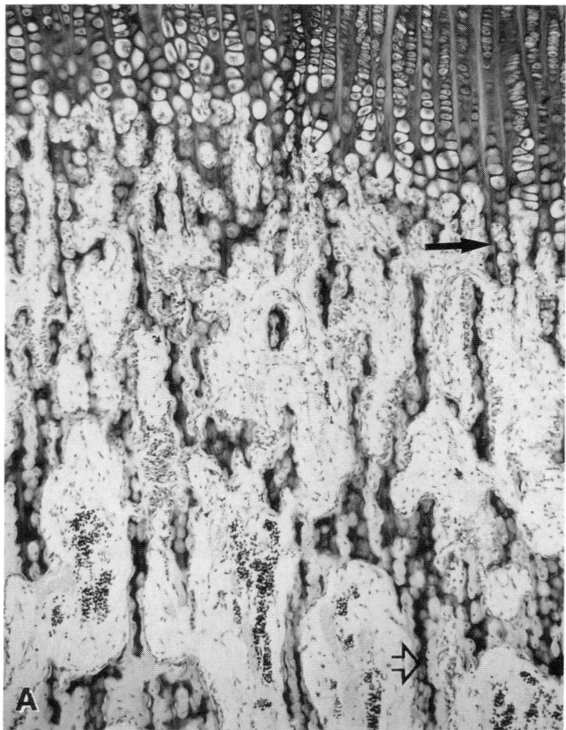

Fig. 1.10A Metaphysis. Normal endochondral ossification. Thin layers of bone are deposited on dark mineralized cartilage columns to form the trabeculae of primary spongiosa (arrow). These are resorbed or amalgamated to produce secondary spongiosa (open arrow).

The matrix proteoglycans in the hypertrophic zone differ from those in the proliferative zone, and this difference, which probably involves a change from neutral to acid glycosaminoglycans, permits the matrix to mineralize. The term "mineralize" is used instead of "calcify" because both phosphate and calcium are deposited in "calcified" tissues. An exception to this terminology is made in relation to the zone of provisional calcification (see the following sections.) Mineral is deposited mainly in the longitudinal columns of matrix surrounding the cell columns (the interterritorial matrix of chondrocytes); some is found in the transverse septa between cells in the territorial matrix.

The **metaphysis** (Fig. 1.10A) begins distal to the last intact transverse septum and extends to the point where metaphyseal narrowing ends at the metaphyseal–diaphyseal junction. The metaphysis comprises primary spongiosa, secondary spongiosa, and modeled trabeculae.

The longitudinal tubes of mineralized matrix which surround hypertrophic chondrocytes constitute the zone of provisional calcification. The fate of the hypertrophic chondrocytes depends on site and species. It is believed that they either die or transform into osteoprogenitor cells, but it is agreed that the penultimate cells are metabolically active up to the time of vascular invasion. Their fragmentation, as seen by light microscopy, can be avoided by fixation with ruthenium hexamine trichloride or osmium ferrocyanide. The transverse septa are eroded, and primitive capillary loops grow into the tubes of mineralizing cartilage, accompanied by perivascular cells and followed by osteoblasts. The capillaries come from two sources; those in the center of the metaphysis are from the nutrient artery, and those at the periphery are from metaphyseal vessels. Shortly after they arrive, osteoblasts start to form fine-fibered, woven bone on the mineralized cartilage of the primary spongiosa.

Primary spongiosa is the name given to the mineralized cartilage with its covering of bone. Modeling of the primary spongiosa is carried out by osteoclasts and osteoblasts as the growing physis leaves it behind in the metaphysis. This entails complete removal of some trabeculae of primary spongiosa and enlargement and fusion of the remainder by lamellar bone. The enlarged trabeculae compose the secondary spongiosa (Fig. 1.10A). Normally the trabeculae of the secondary spongiosa contain cores of mineralized cartilage. Depending on their location, the trabeculae are either eventually resorbed in the marrow and at the periosteal margins of the metaphysis, or are incorporated into the diaphysis (compaction of endochondral bone, Figs. 1.1, 1.7A). The combined primary and secondary spongiosae are sometimes called the chondro-osseous complex.

The tissues in the region of the primary spongiosa are fragile, and very susceptible to trauma, especially to shearing forces, which can produce slipped epiphysis (epiphyseal detachment). This is really epiphyseal–physeal detachment, which is fortunate because the separation occurs near the watersheds of the physeal and metaphyseal circulations. Thus repair is possible with minimal disruption of endochondral ossification. Abnormal compression often produces fractures of spicules, especially in those osteodystrophies which are characterized by osteoporosis.

Some structural support is provided for the zone of provisional calcification and the primary spongiosa by the **perichondrial (mineralized) ring** which surrounds some growth plates at the level of the reserve cartilage and extends a variable distance toward the metaphysis. At the physeal end of the perichondrial ring is the **ossification groove,** or groove of Ranvier, which supplies chondrocytes to the physis for diametric growth. The rest of the ring is fibrous and sometimes bony. Fibrous bone is present on some aspects of the inner face of the ring, around some metaphyses of some bones. The ring disappears when physeal growth ceases.

The growth of the **epiphysis** and its changing shape with age depend on interstitial growth of the epiphyseal cartilage and the articular cartilage. The rate is slower than that of the physis. The same principles which apply to an epiphysis also apply to the growth of bones without a physis, such as the cuboid bones, and the proximal ends of the metacarpals and metatarsals.

There are two "old" surfaces to a joint cartilage in an

immature animal: the outer articular cartilage, derived from the interzone tissues between adjacent embryonic bones, and the inner subarticular growth cartilage, derived from the cartilage of the epiphysis. The proliferative layer is located between these surfaces, and proliferation occurs simultaneously toward both surfaces, that is outward to the articular surface where the cells are continually shed into the joint space, and inward toward the epiphysis, where they participate in endochondral ossification, thereby contributing to the growth of the epiphysis.

The structure and properties of articular cartilage are discussed with Diseases of Joints.

B. Modeling

So long as the bone is growing, it must maintain its basic shape. The integration of continuous endochondral ossification with maintenance of shape involves a process of modeling which depends on resorption and formation of mineralized bone and cartilage. Modeling of secondary spongiosa was discussed with endochondral ossification. Physiologic activity by osteoclasts is the basis for the resorption phase of all modeling and remodeling of mineralized tissues (phagocytes of hard tissues are called chondroclasts, cementoclasts, etc., according to the substrate they are processing). Modeling requires intense, synchronized activity by osteoclasts and osteoblasts, and involves resorption and amalgamation of spongiosa, metaphyseal reconstruction, as well as the fabrication of cortical drifts throughout the skeleton. Modeling is driven by the A–R and A–F sequences of bone cell activity. These sequences involve either cell activation and resorption or cell activation and formation at a given site, but never both. The end result of modeling is the formation and shaping of growing bones. An incidental net effect of modeling is that only a few vestiges of the bone deposited originally on cartilage spicules remain; most of the final organ is membrane bone produced by the periosteum and the endosteum, and in primary osteons. Modeling does not necessarily end at maturity. If there is a need to reinforce parts of the skeleton as a response to mechanical stimuli (see adaptational deformities) new woven or lamellar bone may be produced. Since this is a bone-formative procedure, not preceded by resorption, the reinforcement is regarded as reactivation of modeling, rather than as remodeling.

Modeling depends on cellular activity in all parts of a bone, but the greatest activity occurs in the metaphysis, and it is here that disorders are most obvious. The metaphysis is wider than the diaphysis and must be reduced in diameter while retaining continuity with the diaphysis. The process is called funnelization or metaphyseal reduction, and is accomplished by subperiosteal resorption in the metaphysis, which reduces metaphyseal diameter until the metaphysis is incorporated into the diaphysis.

Modeling also involves complex diaphyseal changes. In a perfectly straight, cylindrical, tubular bone, enlargement of the diaphyseal diameter would require periosteal bone formation and endosteal bone resorption. However, bones are neither perfectly straight nor cylindrical, so modeling must also ensure that bones retain their characteristic shapes and that tuberosities retain their relative positions as bones grow. This involves coordinated resorption and formation in appropriate locations throughout the period of growth.

A special problem of bone modeling is the need to keep the metaphysis joined to the diaphysis during growth. This is achieved by incorporating metaphyseal trabeculae of the secondary spongiosa into the cortex. As the growth plate moves away from the diaphysis, it leaves behind a trail of trabeculae. These trabeculae are resorbed or enlarged according to the grand plan for the particular bone. As noted above, those at the periphery are resorbed during formation of the metaphyseal funnel. Those in the center of the growth plate are normally allowed to persist until they are resorbed at about the level of the metaphyseal–diaphyseal junction. The ones in between, which are in line with the diaphysis, are incorporated into the cortex. This process, **compaction of endochondral bone,** involves filling the spaces between the trabeculae with bone, and depositing endosteal and/or periosteal bone on the compacted bone as required. Remnants of trabecular bone often are seen in normal diaphyseal sections. They are easily recognized by their patterns and because they often have cores of mineralized cartilage (Fig. 1.7A).

Bones are not homogeneous organs, and bone is not an homogeneous tissue. The complexity of epiphyses, physes, and metaphyses is well recognized, but the structural diversity of diaphyses seldom is. Diaphyses are formed by compaction of endochondral bone and by deposition of endosteal and periosteal bone. Most of the bone formed during the growth period by the periosteum of farm animals and large dogs is **laminar bone** (Fig. 1.10B). This is a type not found in humans and therefore rarely described in histology texts. The production of laminar bone is necessary to accommodate the very rapid growth rates of the species mentioned. There is a limit to the rate at which osteoblasts can produce and mineralize osteoid, and this restricts the rate of diaphyseal expansion when bone is deposited on a single surface. The production of laminar bone overcomes this restriction.

Laminar bone formation is initiated by the production of a plate of woven bone within the periosteum some distance from the existing bone surface. This plate is linked at each end by osseous bars to the mineralized bone. Lamellar bone is then produced at equal rates on each side of the lamina until the space between the new lamina and the original bone is filled, save for a vascular channel. In this way the net rate of periosteal bone deposition is doubled. In fact, in the diaphyseal periosteum of a lamb, 8–10 laminae may form concomitantly, the development stage of each one being slightly behind that of its inner, and slightly ahead of its outer companion. Thus the accretion of periosteal bone could be 16–20 times more rapid than with conventional single-surface apposition. Laminar bone is also formed on the endosteal surface and on the periosteum of certain flat bones of the skull.

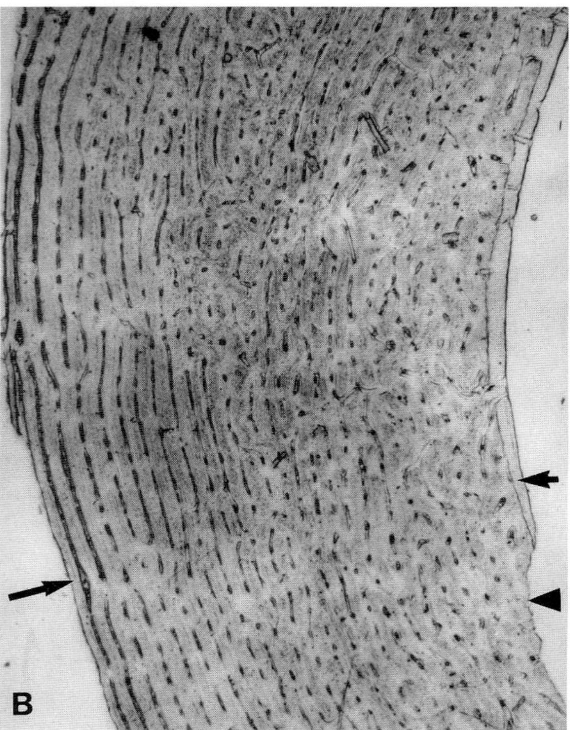

Fig. 1.10B Laminar bone. Femur. Lamb. Periosteal laminae (long arrow). Endosteal bone formed more slowly (short arrow). Endosteal resorption (arrowhead).

The classic structural unit of diaphyseal bone is the **osteon,** or **Haversian system.** This is a long, bony rod about 100 to 500 μm in diameter with a vascular canal (Haversian canal) in the center. Primary osteons are formed *de novo* in preexisting spaces between other bony structures, as for example between modeled secondary spongiosa during compaction of endochondral bone. Formation of secondary osteons is preceded by creation of resorption spaces (Fig. 1.9). Whereas primary osteons often have irregular shapes, secondary osteons tend to be ellipsoidal or circular in cross-section. There are very few osteons in the modeling skeleton of most domestic animals. With aging, more of them are formed as a result of remodeling.

Longitudinal growth of bones usually results from interstitial growth of cartilage. In certain sites, such as the proximal femur, it depends also on **fibrous growth plates.** Longitudinal growth at the proximal femur derives from cartilaginous growth of the head and the trochanter and the growth and metaplastic ossification of tendons that insert in the trochanteric fossa. Parallel bundles of woven bone derived from the tendon are as traceable in the metaphysis as the tubes of mineralized cartilage derived from the physes, and are modeled in the same way. Similar structures occur wherever tendons insert into the growing end of bones, in regions lacking cartilage.

In some places, such as the olecranon tubercle, tendons attach to cartilaginous structures and **chondroid bone** is formed. This tissue has histologic features of both cartilage and bone, and may develop in response to tension from frequently changing directions.

C. Remodeling

Bone remodeling is the process whereby the mature skeleton renews itself throughout life. Remodeling is driven by the "ARF" sequence of bone cell activity. This involves activation of bone-resorbing and bone-forming cells, followed by resorption and then formation at a given site by the activated cells. Abrogation of the sequence is called uncoupling, and is a rare event, but pathologic imbalances in the amounts of bone removed and added are common. The packet of cells that is involved at a remodeling site is called a basic multicellular unit, and the product of their labors is a bone structural unit. The reactivity of a skeleton depends on the number of multicellular units generated by the stem cell populations. Remodeling is a means of repairing the microdamage that is believed to occur throughout the skeleton as a result of materials fatigue from the stress and strain of constant use. As a result there is continuing removal of mature, often viable bone in the mature skeleton, and its replacement by young reactive bone. This is metabolically beneficial in that it provides sites for rapid transfer of electrolytes for homeostasis. Mature, fully mineralized bone is relatively dehydrated, the water replaced by mineral, and the crystals are large; thus, it has a limited capacity for electrolyte diffusion and transfer.

Remodeling activity occurs on three surfaces or envelopes, the **periosteal** surface, the **endosteal–trabecular** surface and the Haversian or **intracortical** surface. Normally, the periosteal surface is a growth surface throughout life, gradually expanding the diameter of the cortex. The endosteal–trabecular surface is predominantly resorptive, gradually expanding the diameter of the marrow cavity. Intracortical remodeling normally is balanced and acts primarily to maintain the structural integrity of the bone. In general, remodeling rates decline with increasing age, and a slow, inexorable loss of bone occurs with aging. Since bones stop growing at different times, the change from modeling to remodeling is gradual in terms of the skeleton as a whole.

For the pathologist, there are several important differences between a skeleton that is modeling and one that is remodeling. The modeling skeleton contains a high proportion of young bone, which is less densely mineralized and less rigid than old bone. This is advantageous when there is need for rapid ion exchange between bone mineral and blood, as in hypocalcemia, but not when resistance to deformation is reduced, as in rickets.

The modeling skeleton can produce new trabeculae by endochondral osteogenesis. The mature skeleton is severely limited in this respect; trabeculae can be made larger or smaller, but the production of completely new trabeculae which persist rarely if ever occurs in the adult.

Since trabeculae are more responsive to the needs of ionic equilibrium than is cortical bone, this is relevant to mineral homeostasis.

D. Abnormalities of Development and Remodeling

There are numerous adverse influences that interfere with the normal development, modeling, and remodeling of bones. In the absence of appropriate genetic determinants, hormonal environment, nutritional requirements, and mechanical stimuli, abnormalities are to be expected. Theoretically, every step in the complex series of events from mesenchymal condensation onward is susceptible to disruption. Developmental abnormalities may be genetic or teratogenic and may have localized or generalized effects (see Sections V,A and B, Generalized, and Localized, Developmental Disturbances).

Of the many inherited abnormalities, certain of the chondrodysplasias illustrate very well the effects of abnormal cartilage growth. Some of these disorders are characterized by decreased capacity for interstitial proliferation of chondrocytes, expressed grossly as reduced longitudinal growth. Palisading of chondrocytes is also irregular; the matrix, fragmented; mineralization, spotty; and vascular invasion, irregular. Appositional growth from the perichondrium may be unaffected, allowing development of characteristic short, wide epiphyses.

Normal growth of bone is controlled by the growth principles of the adenohypophysis acting directly or indirectly. Absence of growth hormone leads to pituitary dwarfism. Sequences in the plate are slowed, the plate is narrowed, and its metaphyseal surface may be sealed by a bony plate. Excessive growth of the physis is stimulated by excess of growth hormones and results in gigantism.

Reactive hyperemia may produce temporary stimulation of longitudinal growth, and local periosteal sectioning and injury also produce ipsilateral increased physeal growth. The periosteum is attached to the epiphyses at both ends of a long bone and creates pressure on the physes which, under normal conditions, restrains its growth rate. Release of this pressure permits an increase in growth rate and leads to metaphyseal deformity.

It is important to distinguish persistence of physeal cartilage, due to failed or retarded endochondral ossification, as in rickets or osteochondrosis, from physes showing rapid growth. Fast-growing physes are wider than those with slow growth, but the greater width is relatively constant across the plate.

Many diseases of bones are manifest by changes in endochondral ossification since growth cartilages are sensitive to general metabolic states. Endocrine disturbances, lack or imbalance of specific nutrients, malnutrition in both fetal and postnatal life, starvation, and cachexia all retard longitudinal growth to produce a narrow physis sealed by a layer of bone, as in hypopituitarism. When the causes are removed, normal activity returns in the growth cartilage and the sealing plate of bone is displaced into the metaphysis to remain for some time as a radiographically visible **line of arrested growth.** The line is represented by transverse trabeculae, which extend across the metaphysis, parallel to the growth plate (Fig. 1.11A,B). This is a nonspecific reaction to cessation of physeal growth, which also occurs with mineralization and removal of the physis (closure of the growth plate) at the time of maturity.

Another type of radiodense metaphyseal line, the **growth-retardation lattice,** also develops parallel to the physis (Fig. 1.12A,B,C) in some bovine fetuses infected with bovine virus diarrhea virus, and in animals poisoned by phosphorus or lead. There are other causes, presently undetermined. These lines are formed when there is a transient reduction in removal of mineralized cartilage, which could be the result of defective chondroclastic activity or indigestible matrix. Growth-retardation lattices and arrest lines may be multiple. Transverse trabeculae are usually more sharply defined in radiographs and more prominent histologically than are growth-retardation lattices.

Traumatic interference with epiphyseal vessels supplying the proliferative zone results in localized failure of growth. Vitamin A deficiency also is reputed to retard cartilage proliferation, but the effect is probably nutritionally nonspecific. Excess vitamin A, however, leads to hydrolysis and degeneration of cartilage matrix. Vitamin C is essential for collagen synthesis and maintenance of ground substance. In vitamin C deficiency, characteristic changes occur in the growth plate. Cartilage hypertrophy is slowed, and mineralization and thus ingrowth of vessels into the hypertrophic zone are irregular. The most prominent defect is deficient osteoblastic activity resulting in insufficient bone on the primary spongiosa. The spongiosa persists as the so-called scorbutic lattice, which fractures with minimal weight bearing.

Calcium deficiency, at least in domestic species, has little influence on the plate unless the deficiency is severe, in which case the plate may be narrowed and, especially in the center of the metaphysis, all the primary spongiosa removed by osteoclasis. In rickets, there is failure of mineralization of the cartilage matrix, and vascular invasion of the hypertrophic zone does not occur. The hypertrophic cartilage persists and, since cartilage proliferation continues, the growth plate becomes thicker than normal. The ultimate thickness of the plate is probably limited by the diffusibility of nutrients from the epiphyseal vessels. The matrix produced by osteoblasts also remains unmineralized in rickets and therefore is resistant to osteoclastic modeling, so that the metaphysis remains broad. Rickets is the result of vitamin D deficiency, or sometimes of phosphorus deficiency, and inherited forms of the disease occur in several species, including pigs and humans.

There are many causes of irregularly thickened physes, besides rickets. Theoretically, every step in the complex series of events leading up to mineralization of cartilage is subject to defects or disruptive influences. Since each step in the sequence must occur normally before the next one can begin, interference with normal endochondral growth is a feature of some chondrodysplasias, such as

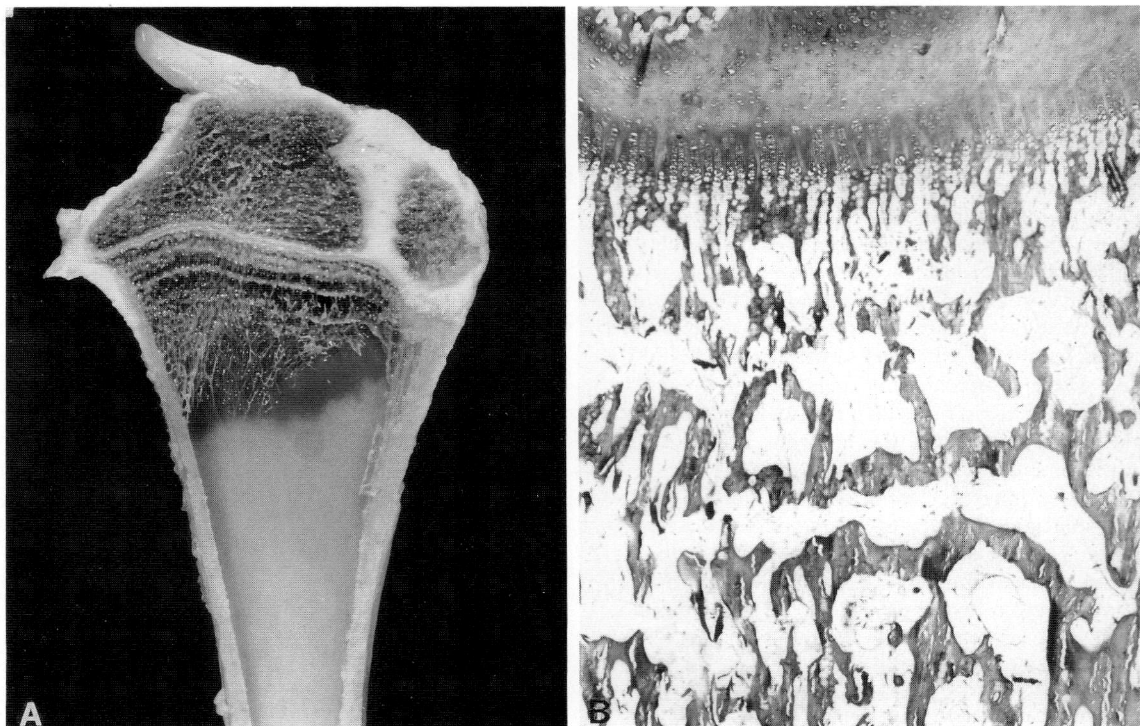

Fig. 1.11 Lamb. (A) Osteoporosis associated with recurrent episodes of malnutrition. There are several growth-arrest lines in proximal tibia. (B) Section of (A) showing transverse trabeculae formed during periods of growth arrest, and then left behind by growing physis.

that in Alaskan malamute dogs. Irregular thickening of the physes also occurs in the osteochondroses.

Whereas thickened physes may be manifestations of inherited or acquired defects of endochondral growth, they may also result from metaphyseal lesions. Interference with metaphyseal blood supply, commonly associated with fractures of the primary spongiosa, restricts mineralization and vascular invasion of the cartilage and allows temporary persistence of hypertrophic cells; complete hypertrophy and mineralization do not occur unless adjacent metaphyseal vasculature is normal, and vascular invasion does not occur unless the sequence of hypertrophy and mineralization is normal.

Traumatic or septic disruption of the physis is followed by the formation of a bridge of bone which unites epiphysis and metaphysis and acts as an anchor. Central anchors may eventually be removed after moderate inhibition of growth. Peripheral bridges are often more persistent and more effective inhibitors of growth, in which event continued physeal growth on the opposite aspect leads to angulation of the end of the bone. Such lesions occur in all species but particularly in young horses and dogs. Septic lesions of bone are quite common in large animals and occur in the epiphysis and metaphysis. They are almost always hematogenous. The relative frequency of metaphyseal localization may be related to the sinusoidal circulation in this site.

1. Angular Limb Deformities of Horses

Whereas angular limb deformity and physitis of horses are considered to be different clinical entities occurring in different clinical settings and presumably having different causes, the microscopic appearance of the physeal lesions is fundamentally similar, and it is not uncommon for young horses with physitis to develop angular limb deformity. The finding of similar histologic changes in the affected physes is not surprising since both disorders are centered on anatomic sites most vulnerable to injury in the cartilage model of the skeleton of the growing horse.

Angular limb deformities occur primarily in young foals and more commonly involve the carpal, tarsal, and fetlock joints of the limbs. Many foals are born with weakness of the supporting periarticular soft tissues of these joints and have minor structural abnormalities in the osteochondral junctions of developing joint surfaces and the physes of adjacent bones. These structural variations are often clinically manifest as angular limb deformities when the foals become weight bearing. Fortunately, most of these deformities spontaneously resolve within the first week or two. During this period the periarticular attachments strengthen, and the compressed portions of physes and developing joint surfaces are stimulated by intermittent compression from locomotion to undergo compensatory

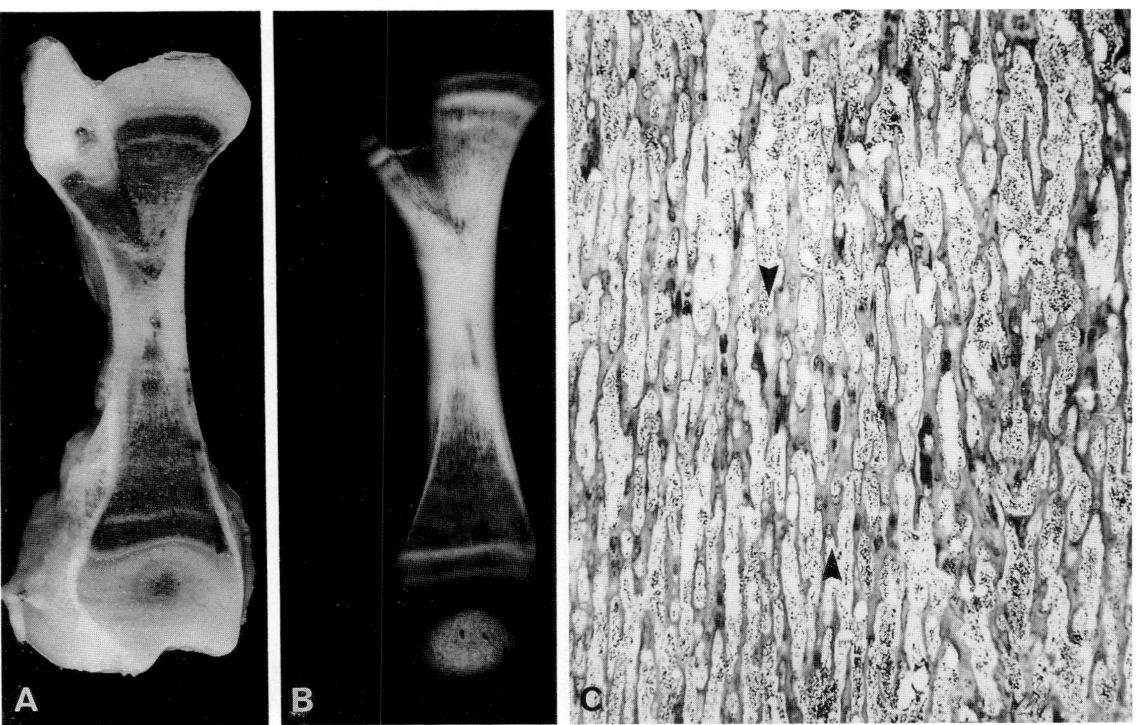

Fig. 1.12 Bovine fetus, bovine virus diarrhea. (A) Growth-retardation lattices in metaphyses, and persistence of medullary bone. (B) Radiograph of (A). (C) Section of metaphysis from (A) showing poorly defined bands composed of increased amounts of mineralized cartilage (between arrowheads). There is also persistence of primary spongiosa.

increases in the rate of endochondral ossification. This results in straightening the limbs.

When more significant abnormalities are present at birth, especially in certain carpal and tarsal bones due to immaturity, possible hypothyroidism or presumably, osteochondrosis, the angular limb deformity usually persists if not treated, because the capacity of the compensatory corrective mechanisms is exceeded. A few foals also acquire angular deformity as a result of trauma at birth or later. Other foals born with poor conformation, subjected to overnutrition or nutritional imbalance, or permitted to exercise excessively may also acquire angular limb deformities. Weakness of periarticular soft tissues and development of abnormally shaped bones affect normal anatomical joint surface alignment. Abnormal alignment secondarily alters normal biomechanical forces exerted on cartilaginous growth centers in the developing bones and joint surfaces of the limbs. Angular limb deformities, therefore, are initiated by a heterogeneous group of prenatal or postnatal factors acting directly or indirectly on the cartilage model of the skeleton of the foal and are promoted to clinical expression by the biomechanical forces of excessive compression resulting from muscle action, weight-bearing, and locomotion.

The recognized categories of angular limb deformities in foals include (1) joint laxity due to immaturity of periarticular supporting structures, (2) imbalance in growth in length of long bones of the limbs by uneven growth rates within portions of the distal physis of the radius, cannon bones, and tibia, (3) imbalance in growth in length of portions of the epiphyses of the aforementioned bones, (4) defects in endochondral ossification of the cuboidal bones of the carpus and tarsus, and (5) direct trauma to sites of endochondral ossification in any of the forming bones mentioned.

The usual stance of a foal with an angular limb deformity in the forelimbs resulting from joint laxity due to immaturity of periarticular supporting structures is one in which the carpi are close together and the distal ends of the limbs are turned out. In this stance, alignment of the limb from the elbow to the hoof describes part of the arc of a great circle with osteochondral junctions on the lateral side of the limb, i.e., the concave side of the arc, being set under compression and cartilaginous growth centers on the medial side of the limb, i.e., the convex side of the arc, being set under tension. If all of the osteochondral junctions, i.e., sites of active endochondral ossification of developing articular surfaces in the carpus and the physis of the distal radius, located on the lateral side of the carpus, are set under sustained, nonphysiologic forces of compression, endochondral ossification at these sites will be temporarily delayed or permanently stopped depending on the intensity and duration of the compressive forces. Corresponding gross and microscopic lesions develop at these sites.

The most vulnerable osteochondral junctions are those that have the highest growth rate, such as the distal radial physis. Physeal cartilage of the distal radial physis thickens on the lateral or compression side of the limb. Primary metaphyseal spongiosa on the lateral side of the limb is dense with many trabeculae being oriented parallel to the physis, whereas trabeculae of primary metaphyseal spongiosa on the medial side are long and thin and may sustain a tension fracture running parallel to the physis.

Skeletal sites having osteochondral junctions which give rise to valgus deformity of the carpus are located on the lateral side of the forelimb, and these include the metaphyseal physis of the distal radius, the articular cartilage of the distal radial epiphysis, and ulnar carpal bone, the lateral part of the third carpal bone and the fourth carpal bone. These carpal bones are also the ones most often affected by delays or disturbances in endochondral ossification associated with skeletal immaturity, hypothyroidism, and so-called osteochondrosis of the carpal bones.

2. *Physitis of Horses*

Physitis, or its synonyms, epiphysitis and physeal dysplasia, is a self-limiting disturbance of endochondral ossification affecting the metaphyseal physes of young horses mostly in one of two age groups. Most of the animals affected by this disorder are foals and weanlings ranging from 4 to 8 months of age, whereas the second group includes yearlings and horses in early training up to about 2 years of age. Sites most commonly affected clinically are the distal physes of the third metacarpal and metatarsal bones, radius, and tibia, and the proximal physis of the first phalanx. However, the physes of the other limb bones and vertebrae may also have lesions. Angular limb deformity can also be associated with this condition in young horses. Physitis of long bones forming the fetlock joints may cause flexural deformities and contracted tendons in some individuals.

The cause of physitis is undetermined, and the mechanisms involved are apparently complex. Several factors appear to be involved in producing lesions of physitis including (1) a genetic capacity for a rapid growth rate, (2) consumption of large quantities of a high-energy ration not corrected for calcium, phosphorus, and other nutrients, and (3) imbalanced mechanical loading of the physes resulting from abnormal conformation, being overweight, imposed excessive weight-bearing, or excessive activity. Less well understood in the pathogenesis of the lesions are the dietary roles of trace metals, especially copper and zinc. Other factors which should also be considered for possible roles in initiating physeal lesions are normal growth spurts, compensatory growth spurts following periods of skeletal stunting from malnutrition, stimulation of physes mediated by increased hormone release in response to the high intake of protein in the diet, or inherited abnormalities in endocrine mediation of cartilage development due to endocrine desynchrony following carbohydrate ingestion.

The histologic appearance of the physeal lesions in physitis is essentially similar to that found in foals and young horses with mild to moderate angular limb deformity. The morphologic changes located in the lateral one third of the distal radial physis of specimens from both disorders are those of disturbed endochondral ossification. The main lesion is a thickening of the metaphyseal aspect of the physeal cartilage, causing a delay or disruption in endochondral ossification. This lesion occurs in the lateral aspect of the radius where the physis is subjected to compression and retardation of vascular invasion of the zone of provisional calcification of the physis by metaphyseal blood vessels. The central one third of the physis shows little abnormality, whereas the medial one third of the physis has structural changes which reflect the abnormal tension that has developed on the medial side of the distal radius. These changes include transverse fractures and repair of the primary metaphyseal spongiosa. Except for the presence of angular limb deformity at the clinical level in foals and its absence in most older immature horses with physitis, the specimens from the two conditions have a similar appearance. Perhaps the major difference in the two disorders lies in the relative strengths of the periarticular attachments. The periarticular attachments are weak in young foals having a physeal disturbance. This laxity permits the deviation of the osseous structures from normal alignment and allows angular limb deformity to develop. Taut periarticular tissues would preclude development of limb deviation in most cases of physitis occurring in older foals and young horses.

Biomechanical forces of normal weight bearing and locomotion are probably not primary factors in initiating the cartilage lesions of physitis, but these forces apparently act to modify and cause progression of the initial lesions. However, sustained, unphysiologic biomechanical forces or trauma to physes can produce lesions indistinguishable from those of physitis.

E. Adaptional Deformities of the Skeleton

Embryonic bones are capable of a considerable degree of self-determination of their form as revealed in organ cultures. The final molding of shape and internal architecture, which begins in the fetus, is subservient to the demands of function, and is directed by pressure and tension. As a general rule, abnormal stresses on a young growing bone change both its shape and its structure, but abnormal stresses on a mature bone influence the internal reorganization of the structural elements rather than altering its shape. The underlying principle of adaptation to function is known as Wolff's Law of Transformation, which states that "every change in the form and function of bones, or of their function alone, is followed by certain definite changes in their internal architecture and equally definite secondary alteration in their external conformation, in accordance with mathematical laws" (Rasch and Burke, 1990).

Modifications to the law are now acknowledged. Whereas deviations in normal stresses, whether in strength or direction, can alter the shape and architecture of bones, certain rules apply. It is generally agreed that bone is strain responsive, and that after being subjected to certain types of strains, it responds in a way that will reduce strain when similar loads are reapplied. However, there is a minimal effective strain, below which there is no discernible effect on bone (except perhaps in terms of microdamage), and there are single, or rarely repeated, large strains that, short of fracture or tissue damage, have no effect on modeling. In between are the strains that are capable of invoking a modeling response, and the response is thought to be proportional to the average of the repeated peak strains that the bone sustains.

Adaptational deformities of the skeleton which develop in response to mechanical forces must be explained in terms of the influences such forces exert on osteoblasts and osteoclasts. There is considerable evidence suggesting that alterations in the electric potentials of osseous tissues are involved in these responses. The potentials which develop in the living animal may be classified as bioelectric, action, and strain related. Bioelectric potentials are dependent on a functioning population of bone cells and have a characteristic pattern along the bone surface, the metaphysis being negatively charged in comparison to the diaphysis. Action potentials are of short duration and are related to muscle contraction or nerve stimulation and therefore also depend on cell viability. Strain-related potentials differ from the others in that a viable cell population is not necessary for their generation. These potentials may be a product of the piezoelectric properties of bone, which reside in the ability of the organic components, mainly collagen and hyaluronate, to generate potentials when they are deformed. Alternatively, they may be streaming potentials that are created when fluid flows through deformed bone tissue. It is also possible that changes in bone other than those involving electrical potentials are responsible for controlling cellular responses to mechanical forces. When bone is intermittently deformed, the proteoglycans in the matrix are reoriented and can maintain their revised orientation for over 24 hr without further stimulus. Thus proteoglycan orientation may mediate the cellular response of bone tissue to stress.

When a whole bone is deformed, the convex surface, which is under tension, becomes electropositive, and the concave surface, which is compressed, is electronegative. Bone formation occurs on the concave (compressed) surface, and resorption, on the convex surface. Thus the response of mesenchymal cells appears to be related to the polarity of their environment, and the results of this response may be quite rapid, for example, being visible in radiographs within 2 weeks.

It may also be relevant that application of a direct current to bone activates osteoblasts, which contain large amounts of alkaline phosphatase, at the cathode where the environment is alkaline, whereas osteoclasts, which have abundant cytoplasmic acid phosphatase, are activated at the anode where acid conditions exist.

The mechanical effects of stress on the periosteum may also have significance for adaptational changes in bones. In the growing animal, the periosteum behaves as an elastic membrane on account of its fibroelastic component, which links the outer fibrous periosteum to the bone surface. When a bone is bent, the periosteum is moved away from the bone on the concave side of the bend, and against the bone on the convex side. Tension on the periosteum, short of causing rupture, promotes deposition of new bone, while pressure causes resorption. The movement of the outer layers of periosteum away from the osteoblast layer, which adheres to the bone surface, may be facilitated by collagenase production by the osteoblasts.

The apparently paradoxical effects of increased vascularity on the activity of mesenchymal cells are also explicable in terms of environmental potentials. Large dilated sinusoids characterized by slow blood flow tend to produce an environment which is relatively electropositive and favors bone resorption. Such conditions may exist adjacent to certain tumors, near the sinusoidal circulation of metaphyseal vessels or in the marrow cavity of calcium-deficient animals. Rapid blood flow favors increased electronegativity and promotes bone formation, such as may occur in the region of tumors containing vessels of small caliber, or in long bones of patients with hypertrophic osteopathy. Thus it seems probable that tissue vascularity, piezoelectricity, and other related factors exert significant control over the adaptational growth of bones, and that under certain circumstances the influence of one or the other of these factors is dominant.

When bone loses its function, it loses mass. The loss is the result of a local event called a regional acceleratory phenomenon. Such phenomena result from activation of a greater than normal number of basic multicellular units, and can also arise in response to any trauma, inflammation, or other stimulus. A common example is the atrophy of disuse seen in the jaws of old animals, sheep especially, as a localized osteoporosis and recession of bone where a tooth and its antagonist have been lost. Similar changes develop with the long-term ingestion of soft food. An additional example is provided by the stump of an amputated bone; there are acquired alterations in shape which result from the loss of some muscular attachments, but the severed end which is no longer functional becomes porotic and tapered, and the compact bone of the cortex is progressively replaced by light spongy bone.

Conversely, when a bone is exposed to frequent repetitive compression that exceeds the minimum effective strain, it increases in density. Such sclerosis is apparent in long bones which, during rickets, have become bent under the influence of ordinary mechanical forces. In the reparative phase of rickets, the cortex on the concave side of the bend becomes thickened and sclerotic by the apposition of new bone. Meanwhile, tension on the convex surface favors resorption, and this part of the bone becomes porotic and thinned. Adaptational changes also affect cancellous bone, causing selective removal of some trabeculae, particularly the horizontal plates, which emphasizes the vertical members grossly and in radiographs.

Bibliography

Acheson, R. M. Effects of starvation, septicaemia, and chronic illness on the growth cartilage plate and metaphysis of the immature rat. *J Anat* **93:** 123–130, 1959.

Balmain, N., Moscofian, A., and Cuisinier-Gleizes, P. The mineralized ring, a single structure peculiar to long bone growth. *Calcif Tissue Int* **35:** 232–236, 1983.

Bassett, C. A. L. Biophysical principles affecting bone structure. In "The Biochemistry and Physiology of Bone," G. H. Bourne (ed.), 2nd Ed. Vol. 3, pp. 1–76. New York, Academic Press, 1971.

Brighton, C. T. Structure and function of the growth plate. *Clin Orthop* **136:** 22–32, 1978.

Burr, D. B., and Martin, R. B. Errors in bone remodeling: Toward a unified theory of metabolic bone disease. *Am J Anat* **186:** 186–216, 1989.

Cowell, H. R., Hunziker, E. B., and Rosenberg, L. The role of hypertrophic chondrocytes in endochondral ossification and in the development of secondary centers of ossification. *J Bone Joint Surg* **69A:** 159–161, 1987.

Done, J. T. *et al.* Bovine virus diarrhoea–mucosal disease virus: Pathogenicity for the fetal calf following maternal infection. *Vet Rec* **106:** 473–479, 1980.

Editorial. Vital biomechanics: Proposed general concepts for skeletal adaptations of mechanical usage. *Calcif Tissue Int* **42:** 145–156, 1988.

Feik, S. A., Storey, E., and Ellender, G. Stress-induced periosteal changes. *Br J Exp Pathol* **68:** 803–813, 1987.

Friedenberg, Z. B. *et al.* The cellular origin of bioelectrical potentials in bone. *Calcif Tissue Res* **13:** 53–62, 1974.

Friedenberg, Z. B. *et al.* The response of nontraumatized bone to direct current. *J Bone Joint Surg (Am)* **56A:** 1023–1030, 1975.

Frost, H. M. Bone "mass" and the "mechanostat": A proposal. *Anat Rec* **219:** 1–9, 1987.

Gorham, L. W., and West, W. T. Circulatory changes in osteolytic and osteoblastic reactions. An experimental study utilizing two malignant mouse tumors. *Arch Pathol* **78:** 673–680, 1964.

Hall, B. K. Earliest evidence of cartilage and bone development in embryonic life. *Clin Orthop* **225:** 255–272, 1987.

Huffer, W. E. Morphology and biochemistry of bone remodeling: Possible control by vitamin D, parathyroid hormone, and other substances. *Lab Invest* **59:** 418–442, 1988.

Platt, B. S., and Stewart, R. J. C. Transverse trabeculae and osteoporosis in bones in experimental protein–calorie deficiency. *Br J Nutr* **16:** 483–495, 1962.

Rasch, P. J., and Burke, R. K. "Kinesiology and Applied Anatomy," Philadelphia, Pennsylvania Lea & Febiger, 1963. Cited by Frost, H. M. Skeletal structural adaptations to mechanical usage (SATMU): 1. Redefining Wolff's law: The bone-modeling problem. *Anat Rec* **226:** 403–413, 1990.

Shapiro, F., Holtrop, M. E., and Glimcher, M. J. Organization and cellular biology of the perichondrial ossification groove of Ranvier. *J Bone Joint Surg (Am)* **59A:** 703–723, 1977.

Singh, S., and Katz, J. L. Electromechanical properties of bone: A review. *J Bioelectr* **7:** 219–238, 1988–89.

Skerry, T. M. *et al.* Load-induced proteoglycan orientation in bone tissue *in vivo* and *in vitro*. *Calcif Tissue Int* **46:** 318–326, 1990.

Smith, J. W. The structure and stress relations of fibrous epiphysial plates. *J Anat* **96:** 209–225, 1962.

Spence, J. A., Mellor, D. J., and Aitchison, G. U. Morphology and radiopaque lines in bones of foetal lambs: The effects of maternal nutrition. *J Comp Pathol* **92:** 317–329, 1982.

Warrell, E., and Taylor, J. F. The role of periosteal tension in the growth of long bones. *J Anat* **128:** 179–184, 1979.

Wilsman, N. J., and Van Sickle, D. C. Cartilage canals, their morphology and distribution. *Anat Rec* **173:** 79–93, 1972.

V. Abnormalities of Development

Developmental errors, which ultimately are expressed in the skeleton, may be primary abnormalities of bone, cartilage, or primitive mesenchyme. They may be genetic or conditioned by the environment and, in either case, the anomalies may be local or systemic. Many prototypes in laboratory species are discussed by Gruneberg.

The complexity of the processes by which the skeleton is formed provides ample opportunity for error, and the classification of those errors provides ample opportunity for confusion. The International Nomenclature of Constitutional Diseases of Bone sought to bring some order and uniformity of terminology to the classification of human skeletal dysplasias, of which 80–100 types are recognized. Unfortunately, most anomalies of domestic animals are not described well enough to use this classification, which is based on clinical, genetic, and radiographic characteristics. Further definition depends on the demonstration of specific morphologic and biochemical defects. The ultimate aim is a classification based on recognition of the biochemical abnormalities and the defective, controlling gene.

Many dysplasias are named according to whether the epiphyses, metaphyses, or diaphyses are affected in radiographs. Involvement of the skull elicits the prefix *cranio,* and of the spine, *spondylo.* Some dysplasias are named for the segment of the limbs that is shortened, for example, rhizomelic (proximal), mesomelic (middle), and acromelic (distal) dysplasias. A few terms, such as achondroplasia, are based on concepts of pathogenesis.

It is apparent from the foregoing that in some of these disorders the entire skeleton may be involved; such is the case in some of the chondrodysplasias in which the defect is in the preparatory cartilage, and in some of the congenital osteodysplasias in which the defect is in the formation of definitive bone. The defect may be localized to certain regions or certain bones, producing local chondrodysplastic changes. The anlage of an individual bone or a group of bones may not develop. Abnormal division of cartilaginous precursors, failure of normal division of cartilaginous precursors, and the development of accessory centers of ossification provide a variety of abnormalities of size and shape. In addition, the plasticity of bone allows for a variety of adaptive modifications to primary abnormalities in associated bones or soft tissues. The development of the skull and axial skeleton is intimately related to the development of the neural groove, and anomalies of one often accompany anomalies of the other, those of the neural tissues apparently being primary in most cases.

It seems to be a general rule that, of the bewildering

variety of malformations, there are basic patterns to which are added relatively minor anomalous and adaptive variations. Few veterinary pathologists, however, have either the time or the intimate knowledge of embryology which are necessary for the analysis of anomalies, so usually they have been given cursory examinations. Fortunately, a burgeoning interest in animal models of human disease has led to a thorough investigation of some of these disorders. There is great need for further investigations of many of these conditions. Meticulous radiographic, anatomic, histologic, and where appropriate, clinical studies must be done, as a minimum, to delineate such disorders. Biochemical and genetic studies are required also to understand developmental abnormalities, since molecular heterogeneity among phenotypically similar chondrodysplasias is recognized in humans and other animals.

A. Generalized Developmental Disturbances

We are concerned here with developmental disorders that affect bones (the organs) as well as bone (the tissue). The conditions described are known to, or presumed to, have a genetic basis. Abnormalities similar to or identical with those of hereditary origin can be caused by environmental factors such as chemicals, viruses, and nutritional deficiencies.

Generalized developmental disturbances of bones are of considerable importance in animals, and many are reflections of a primary cartilaginous disorder. Formerly they tended to be lumped together as chondrodystrophies, although the term achondroplasia was often used synonymously. Strictly speaking, chondrodystrophy means defective or faulty nutrition of metabolism of cartilage, and achondroplasia refers to failure of growth of cartilage. We refer to this group of diseases as chondrodysplasias, and use more specific or less specific terminology when appropriate.

In humans, most of the chondrodysplasias are associated with histologic lesions, some of which are sufficiently characteristic to be useful diagnostic criteria. In other types the morphologic abnormalities are nonspecific and, in some, no lesions of chondro-osseous tissue are discernible. Thus the human chondrodysplasias are separable on a pathologic basis into those with normal endochondral ossification (this group includes achondroplasia and hypochondroplasia), those with a generalized defect in resting cartilage which secondarily affects the growth plate, and those with a primary defect in the growth plate cartilage. A similar range of morphologies can be expected in other species.

The disproportionate dwarfism of the chondrodysplasias must be differentiated from primordial dwarfism, which is characterized by proportionate body form. Primordial dwarfism is usually a recessive character fixed by selection, and is well known in cattle and dogs, the subjects often being classified as distinct breeds. Also classified as distinct breeds are the disproportionate canine dwarfs, such as the Pekingese and basset hound, which differ from the examples discussed later in being polygenic traits associated with abnormalities in receptors for hormones, matrix components, or other chondrocyte regulatory factors.

Morphologic analysis of chondrodysplasias is facilitated by an understanding of cartilage cells and their matrix. Cartilage grows by interstitial proliferation and by superficial apposition. There are some limitations on both processes. Interstitial growth is by mitotic division of chondrocytes, which then elaborate intercellular matrix. Appositional growth occurs on those surfaces of cartilage which are covered by perichondrium. The perichondrium consists of mesenchymal cells, and the chondrocytes which differentiate from them are capable of proliferating by mitosis. Thus increase in some dimensions can occur by interstitial and appositional growth. As a rule, longitudinal growth is interstitial only, and transverse growth is both interstitial and appositional.

In some chondrodysplasias, appositional and, therefore, transverse growth is normal, but interstitial growth of epiphyseal, articular, and basocranial cartilage is defective. Premature cessation of cartilaginous growth leads to premature closure of growth plates, and shortness of the affected bones results. The main sites of endochondral growth of the skull are the intersphenoid, spheno-occipital, and interoccipital synchondroses. Following their premature synostosis, the brain continues to grow and, to accommodate it, the cranium becomes enlarged and domed. In many instances, there is accompanying hydrocephalus, which adds to the cranial deformation. Unless hydrocephalus is extreme and develops rapidly, the cranial sutures and fontanels close normally, since in these the bone is formed in membrane. As a consequence of hypoplasia of the bones of the cranial base, the ethmoids, and the turbinates, the face is retruded with relative inferior prognathia, since the mandible develops in a cartilaginous skeleton that is formed by appositional differentiation of chondrocytes. Cleft palate may accompany the hydrocephalus, possibly because the normal-sized tongue physically interferes with development of the secondary palate. The spinal column is short because of diminished length of individual vertebrae, and the cartilages are more bulky than normal. The bony parts of the ribs are abnormally short, and the cartilaginous parts, abnormally long. The costochondral junctions may be enlarged since the cartilages continue appositional growth.

There is much variation in the severity of lesions at a given site in the different chondrodysplasias. For example, involvement of the basocranial synchondroses may be extreme in Dexter "bulldog" calves, moderate in "snorter" dwarf calves or minimal in pseudoachondroplastic miniature poodles. However, long-bone length relative to body size, which is dramatically reduced in Dexter bulldogs, and severely shortened in the miniature poodles, is only modestly affected in snorter dwarfs.

Bibliography

Bille, N., and Nielsen, N.C. Congenital malformations in pigs in a postmortem material. *Nord Vet Med* **29**: 128–136, 1977.

Byers, P. H. Invited editorial: Molecular heterogeneity in chondrodysplasias. *Am J Hum Genet* **45:** 1–4, 1989.
Greene, H. J., Leipold, H. W., and Huston, K. Bovine congenital skeletal defects. *Zentralbl Veterinaermed (A)* **21:** 789–796, 1974.
Huston, R., Saperstein, G., and Leipold, H. W. Congenital defects in foals. *J Equine Med Surg* **1:** 146–161, 1977.
Hwang, W. S. *et al*. The pathology of cartilage in chondrodysplasias. *J Pathol* **127:** 11–18, 1979.
Leipold, H. W., Dennis, S. M., and Huston, K. Congenital defects of cattle: Nature, cause, and effect. *Adv Vet Sci Comp Med* **16:** 103–150, 1972.
Minor, R. R., and Farnum, C. E. Animal models with chondrodysplasia/osteo-chondrodysplasia. *Pathol Immunopathol Res* **7:** 62–67, 1988.
Ponseti, I. V. Skeletal growth in achondroplasia. *J Bone Joint Surg (Am)* **52:** 701–716, 1970.
Poswillo, D. Mechanisms and pathogenesis of malformations. *Br Med Bull* **32:** 59–64, 1976.
Rieck, G. W., and Schade, W. Die Arachnomelie (Spinnengliedrigkei) ein neues erbliches letales Missbildungssyndrom des Rindes. *Dtsch Tieraerztl Wochenschr* **82:** 342–347, 1975.
Rimoin, D. L. International nomenclature of constitutional diseases of bone. *J Pediatr* **93:** 614–616, 1978.
Saperstein, G., Leipold, H. W., and Dennis, S. M. Congenital defects of sheep. *J Am Vet Med Assoc* **167:** 314–322, 1975.
Saperstein, G., Harris, S., and Leipold, H.W. Congenital defects in domestic cats. *Feline Practice* **6**:18-43, 1976.
Scott, C.I. Achondroplastic and hypochondroplastic dwarfism. *Clin Orthop* **114:** 18–30, 1976.
Sillence, D. O., Horton, W. A., and Rimoin, D. L. Morphologic studies in the skeletal dysplasias. A review. *Am J Pathol* **96:** 812–859, 1979.

1. Chondrodysplasias

a. CHONDRODYSPLASIAS OF CATTLE Several types of inherited disproportionate dwarfism occur in cattle. The number of gene loci involved is not clear; the possibility that several chondrodysplasias are controlled by a major achondroplastic-conditioning recessive gene with alleles at two other loci acting as modifiers has been suggested but not confirmed. Three bovine chondrodysplastic syndromes representative of the phenotypic range of disproportionate dwarfism are the Dexter bulldog, the Telemark lethal, and the snorter dwarf.

i. *Dexter Type* Dexter cattle are heterozygous for an incompletely dominant gene which, when homozygous, is lethal. The homozygote is the Dexter "bulldog." Dexter cattle probably originated from a mutation in the Kerry cattle of Ireland, and similar mutations may have occurred in other breeds, since comparable dwarfs occur in Charolais, Jersey, and Holstein cattle, as well as indigenous breeds in various countries.

Dexter bulldogs are usually aborted before the seventh month. They are characterized by extreme shortness of limbs, cranial base, and vertebral column. The limbs are abducted, variously rotated, and thick, since the soft tissues are of relatively normal volume and therefore, compared to the bones, excessive in volume. The tongue, of normal mass, protrudes, and the hard palate is absent.

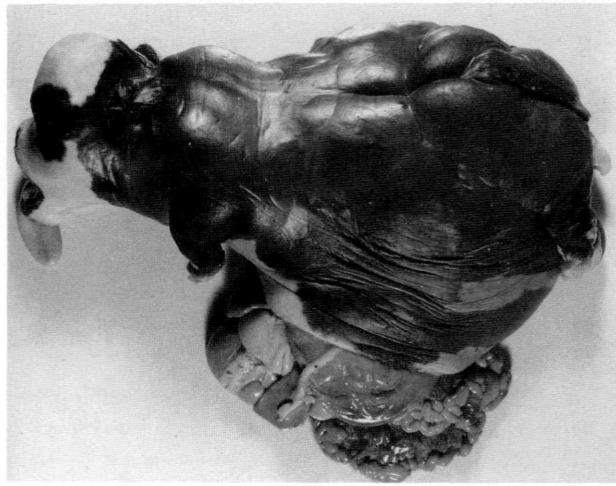

Fig. 1.13 Bovine chondrodysplasia. Bulldog calf. Cranium domed. Short head with retruded face, projecting mandible, and normal tongue. Short rotated limbs and large umbilical hernia.

There is a large umbilical hernia through which the abdominal viscera are eventrated (Fig. 1.13).

The epiphyses are normal but appear enlarged, and those of the appendages are mushroom shaped because of reduced interstitial growth and continued appositional growth. Histologically, the proliferative zone is irregularly separated from the hypertrophic zone by bands of fibrovascular tissue, which enclose small islands of cartilage. There is no orderly arrangement of chondrocyte columns, and the intercellular matrix is fibrillar. Provisional mineralization and subsequent ossification are irregular.

Holstein bulldogs have been examined in more detail than those of the Dexter breed. Stillborn calves are small, bones are misshapen, and some long bones and vertebrae lack growth plates. In the limbs the lesions are most severe in the distal bones, the femurs, scapulas, and humeri being relatively normal. Microscopically, there is an absence of growth plates in affected bones, and the resting cartilage shows irregular, atypical maturation and degeneration with a reduction of interterritorial matrix. Cartilage canals seem excessive, and increased type I collagen in cartilage analyses is likely due to the connective tissue around these canals. The condition has been likened to achondrogenesis type II (Langer–Saldino) of people.

ii. *Telemark Type* Telemark cattle are of Norwegian origin. The form of dwarfism which occurs in this breed is inherited as an autosomal recessive character; thus, the heterozygous parents are phenotypically normal. Chondrodysplastic calves are born alive at term, but they are unable to stand, and die of suffocation shortly after birth. The cranium is domed as a result of chondrodysplasia and hydrocephalus, the face is brachygnathic, the palate is cleft, the tongue protrudes, the neck is short, and the limbs are very short, bulky, and rotated to various degrees.

A similar autosomal recessive chondrodysplasia occurs in Jersey calves but, unlike chondrodysplastic Telemark calves, which are phenotypically uniform, there is great phenotypic variability in Jerseys, and many of the calves are viable. Defective Jersey calves show, in varying combinations and severity, a short, broad head, open fontanels, small ears, deformed mandible with disorder of the incisors, cleft palate, flexion of the interphalangeal joints, and short spiraled limbs. Those which are least affected have heads broader and somewhat shorter than normal and abnormally short limbs. Histologic characteristics are not known.

iii. *Brachycephalic (Snorter) Type* "Snorter" dwarfs were common in beef breeds, especially Herefords, in North America and New Zealand. The pattern of inheritance is not definitely established, but the trait appears to be an autosomal recessive factor, which produces a short-legged compact appearance in heterozygotes. Selection for this phenotype favored widespread dissemination of the chondrodysplasia gene in Hereford and in Aberdeen Angus cattle.

Snorter dwarfs have a short, broad head with a head-length : head-width ratio of approximately 1.75. The forehead bulges, the upper jaw is retruded, the mandible is slightly protruded with malocclusion, and the eyes are prominent and laterally displaced (Fig. 1.14). Lethargy, chronic ruminal tympany, and nasal dyspnea are characteristic. The vertebrae are longitudinally compressed, and the ventral borders of the vertebral bodies are more straight than concave. There is premature synostosis of the basocranial synchondroses, which is indicative of the basic chondroplastic defect. The heart is smaller and more globose than normal, but there are no other visceral abnormalities. The brain stem is shortened and the cerebellum, compressed. Distal long bones are proportionally shorter than proximal ones, but their diameter is equal to or greater than those in normal animals. The most reliable single indication of dwarfing is the ratio of total metacarpal length to diaphyseal diameter. Usually this is 4.0 or less in snorters and 4.5 or more in normal animals.

Histologically, growth plates are essentially normal, but the columns of palisading chondrocytes contain fewer cells and are more irregular, and the cells are less hypertrophied than in unaffected animals of the same age. Suggestions that snorter dwarfs have type I mucopolysaccharidosis are not substantiated. Snorter dwarfs may have a form of hypochondroplasia, since no qualitative defect of endochondral ossification is demonstrable.

The relationship of the dolichocephalic or long-headed dwarf to this type of chondrodysplasia is unclear. These animals are slightly larger than the short-headed variety, and the head, in proportion to the body, is exceptionally long with a length-to-width ratio of about two. The long head tapers to a rather fine muzzle, the limbs are slightly crooked, growth is slow, and the animals are unthrifty. Histologic examinations are not reported.

b. Chondrodysplasias of Sheep Hereditary disproportionate dwarfism is best known in the Ancon or Otter sheep, now extinct as a breed. Occasional recurrence of dwarfs of the Ancon type in other breeds of sheep is presumed due to new mutations. The major abnormality in Ancon sheep is a mesomelic, short-limbed dwarfism due to premature closure of certain growth plates. Their one-time popularity as a breed was based on the ease of confinement which this defect provided. The Ancon dwarf

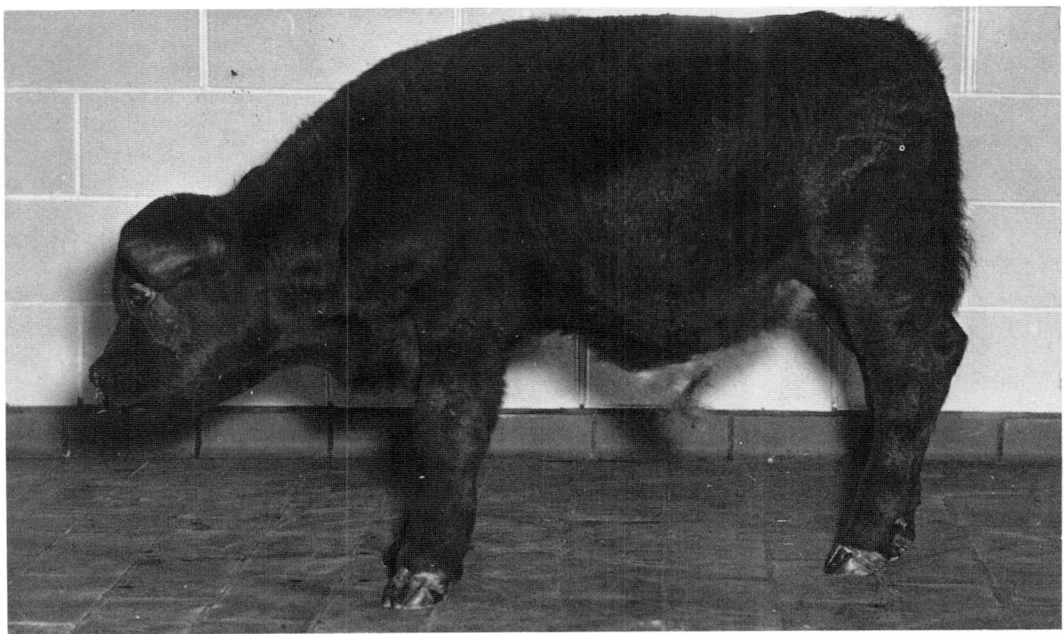

Fig. 1.14 Bovine chondrodysplasia. Brachycephalic snorter dwarf. Note bloated abdomen.

does not resemble the chondrodysplasia of Dexter bulldogs, although chondrodysplasia is present at birth.

A chondrodysplastic syndrome occurs in South Country Cheviot sheep in Scotland, but its incidence is minimized by selective breeding. The condition is present at birth and is characterized by achondroplasia of the head, protruding eyes, and short ears and tail. The forelimbs are short and the hooves, absent. Hindhooves are of reduced size and abnormal shape. An autosomal recessive trait is probably responsible.

An episode of disproportionate dwarfism occurred in a flock of mixed breeding in the United Kingdom. About 25% of 110 lambs born to 70 ewes were affected. There were no abortions, and most lambs were born alive but died in respiratory distress in a few minutes, possibly because of tracheal stenosis and pulmonary hypoplasia. The lambs resembled bulldog calves in many respects, but genetic causes were eliminated. A similar isolated episode occurred in a small flock in New Zealand. No teratogen was identified in either case.

Spider lamb chondrodysplasia (spider lamb syndrome, hereditary chondrodysplasia) occurs in North America as an inherited condition of Suffolk and Hampshire sheep and their crosses. It is so called because of the relatively long, thin, angular legs of affected lambs (Fig. 1.15A). The trait is an autosomal recessive that may have become disseminated by selection for long-legged animals. Spider lambs were first recognized in the late 1970s, and initial confusion regarding diagnostic criteria was associated with the concomitant occurrence of a nonheritable arthrogryposis–hydranencephaly syndrome. Deformed lambs may be aborted, stillborn, or born alive. Causes of death are not known. Other lambs are normal at birth but develop signs within a month. Both sexes are affected. Although spider lambs have a characteristic phenotype, there is a spectrum of gross lesions, and few are constant. Tall, finely boned, poorly muscled lambs with small heads, scoliosis, sternal deformity, and knock-knees are typical. Roman nose, deviation of the nose, deformed ears, kyphosis, and hind limb deformities producing a bowlegged stance are among the other gross lesions. Severe secondary degenerative arthropathy develops within a few months, and other adaptive deformities may result.

Fig. 1.15A Spider lamb. Long, spindly legs and lumbar kyphosis.

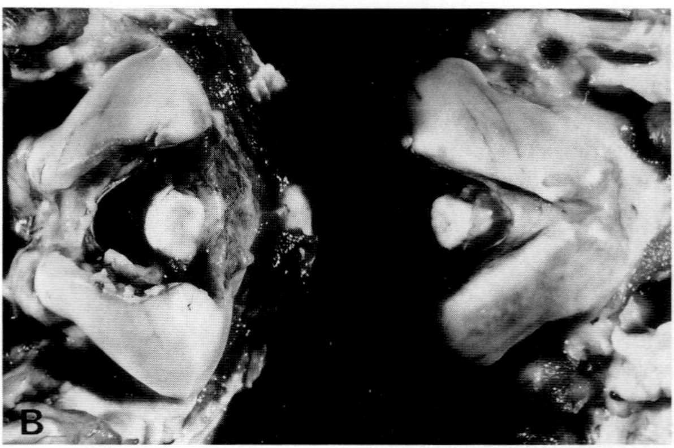

Fig. 1.15B Spider lamb (right). Elongation and increased angulation of occipital condyles. Decreased width of intercondyloid notch. Normal lamb (left).

Diagnostic criteria are radiographic and pathologic. Chromosomal and biochemical abnormalities have not been detected. In all lambs, there is elongation and increased angulation of the occipital condyles leading to diminution of the width of the intercondyloid notch (Fig. 1.15B). Almost all lambs have irregular ossification of the olecranon that is radiographically evident after 2 weeks of age as stippled densities and beneath the glenoid of the scapula (Fig. 1.16A,B). These changes will not be seen in fetuses prior to the age of onset of ossification. Scoliosis and pectus excavatum are associated with irregular ossification of the cervical and thoracic vertebral bodies and sternabrae respectively. Often there is malalignment or failure of fusion of the posterior part of the sternal bars. Pathologic fractures of the femoral head, distal tibial physis, and the talus occur occasionally. With the onset of degenerative joint disease, severe lesions develop in the atlanto-occipital, elbow, and carpal joints, among others.

Microscopically, there is abnormal development of centers of endochondral ossification in affected bones. Nodular hypertrophy of cartilage occurs, and multiple small ossification centers evolve, but their coalescence and expansion toward the articular surface are retarded, and the resultant lack of subchondral bone predisposes to arthropathy (Fig. 1.17). Nodular masses of hyaline cartilage remain in the metaphyses, and some of these may impinge on the cortex. Some foci of unresorbed hyaline cartilage fibrillate and undergo "cystic" degeneration (Fig. 1.18).

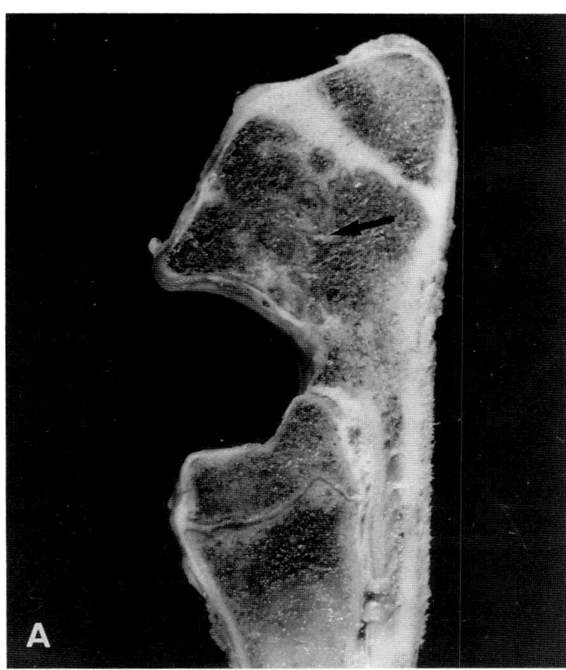

Fig. 1.16A Spider lamb. Irregular ossification of olecranon (arrow).

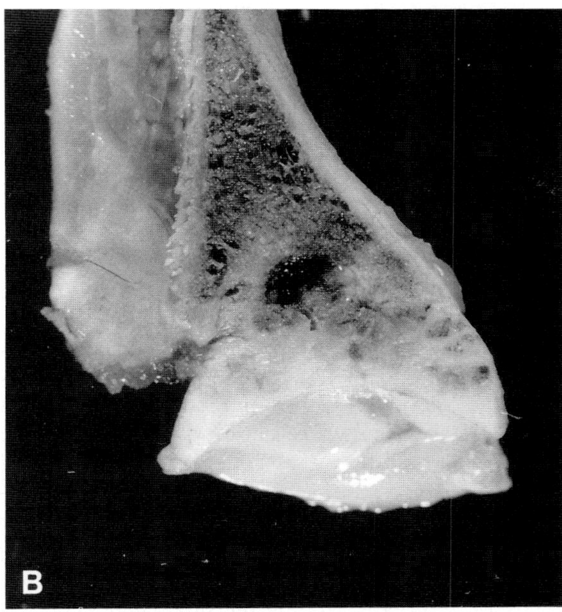

Fig. 1.16B Spider lamb. Irregular ossification of glenoid. Metaphyseal hemorrhage (above).

Mild irregularities of cartilage columns are present in some physes. That growth plates of long bones are largely unaffected is attested to by their extreme length, while their slender profile suggests reduced appositional activity of chondroblasts or osteoblasts.

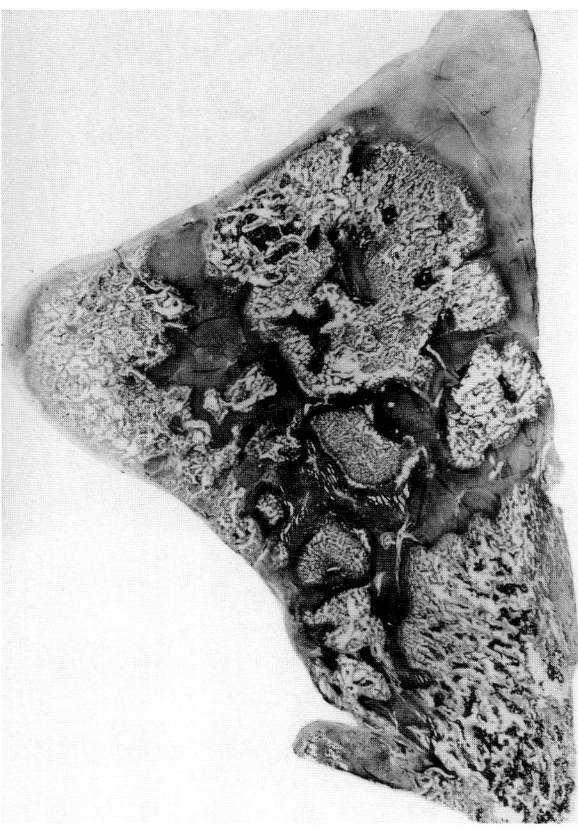

Fig. 1.17 Spider lamb. Olecranon. Multiple ossification centers.

The fundamental defect in spider lamb chondrodysplasia is not known. That the fault resides in the cartilaginous skeleton is accepted, and occasional failure of fusion of sternal bars indicates it is present early in embryogenesis, while the sporadic presence of auricular abnormalities suggests a generalized defect. The virtual absence of lesions in most long bones, which are usually most susceptible to the effects of cartilage dysplasias, is unusual. The demonstrable lesion in spider lambs is defective development of those ossification centers that develop around the time of birth; the clinical features of the syndrome are mostly related to effects of use on the defective bones. The centers that form *in utero* expand normally once established, and the long bones attain relatively normal length. At the time their primary centers are forming, there is little muscular activity to deform the skeleton so it is possible that all ossification centers develop abnormally, but many bones heal before birth without residual lesions. Such a sequence of recovery occurs commonly in certain chondrodystrophies as rickets.

c. CHONDRODYSPLASIAS OF PIGS A single occurrence of inherited chondrodysplasia is reported in the Danish

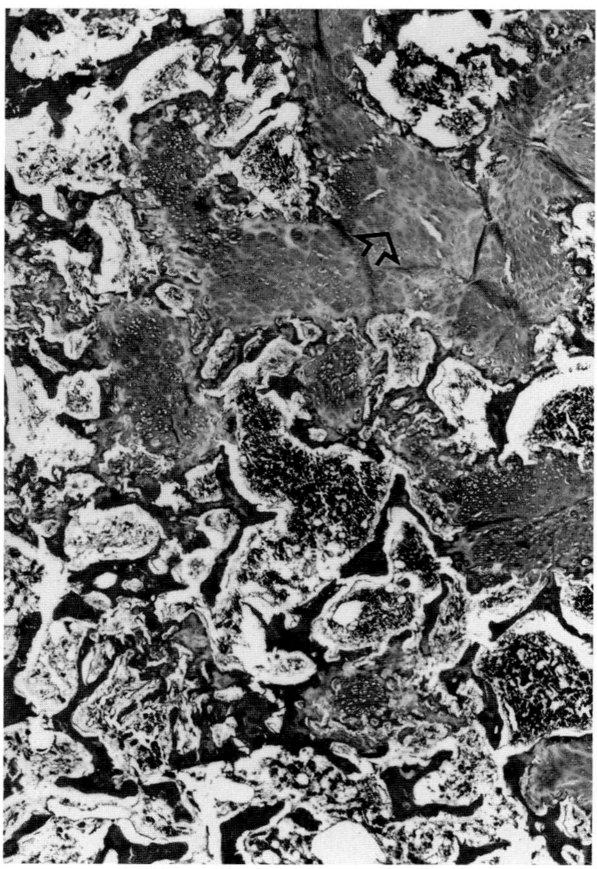

Fig. 1.18 Spider lamb. Nodules of residual cartilage. Some show degenerative changes (open arrow).

Landrace. Affected animals are short-limbed dwarfs with the forelegs shorter than the hind, the axial skeleton and skull being unaffected. First signs may be noted as early as 1 to 2 weeks, consisting of loose attachment of the limbs. Many pigs also have abnormally upright ears. Abnormal mobility of the joints and disturbances of gait are obvious at weaning. Later, arthrosis develops, and few animals reach breeding age. The physes of long bones are narrow due to a reduction in the depth of the proliferative zone and irregularity of subsequent zones. The cells in these zones appear normal. The condition is inherited as an autosomal recessive trait. No chromosomal abnormalities are detected.

Bibliography

Andresen, E. *et al.* Dwarfism in Red Danish cattle (Bovine achondroplasia: type RDM). *Nord Vet Med* **26:** 681–691, 1974.

Bowden, D. M. Achondroplasia in Holstein–Friesian cattle. *J Hered* **61:** 163–164, 1970.

Crew, F. A. E. The bull-dog calf: A contribution to the study of achondroplasia. *Proc R Soc Med* **17:** 39–54, 1924.

Duffell, S. J., Lansdown, A. B. G., and Richardson, C. Skeletal abnormality of sheep: Clinical, radiological, and pathological account of occurrence of dwarf lambs. *Vet Rec* **117:** 571–576, 1985.

Gregory, K. E., and Spahr, S. L. Miniature calves. *J Hered* **70:** 217–219, 1979.

Gregory, P. W., Tyler, W. S., and Julian, L. M. Bovine achondroplasia: The reconstitution of the Dexter components from non-Dexter stocks. *Growth* **30:** 343–369, 1966.

Gregory, P. W., Julian, L. M., and Tyler, W. S. Bovine achondroplasia: Possible reconstitution of the Telemark lethal. *J Hered* **58:** 220–224, 1967.

Hafez, E. S. E., O'Mary, C. C., and Ensminger, M. E. Albino-dwarfism in Hereford cattle. *J Hered* **49:** 111–115, 1958.

Horton, W. A. *et al.* Bovine achondrogenesis: Evidence for defective chondrocyte differentiation. *Bone* **8:** 191–197, 1987.

Jayo, M. J. *et al.* Bovine dwarfism: Clinical, biochemical, radiological, and pathological aspects. *J Vet Med A* **34:** 161–177, 1987.

Jensen, P. T. *et al.* Hereditary dwarfism in pigs. *Nord Vet Med* **36:** 32–37, 1984.

Jones, J. M., and Jolly, R. D. Dwarfism in Hereford cattle: A genetic, morphological, and biochemical study. *N Z Vet J* **30:** 185–189, 1982.

Mead, S. W., Gregory, P. W. and Regan, W. M. A recurrent mutation of dominant achondroplasia in cattle. *J Hered* **37:** 183–188, 1946.

Rook, J. S. *et al.* Diagnosis of hereditary chondrodysplasia (spider lamb syndrome) in sheep. *J Am Vet Med Assoc* **193:** 713–718, 1988.

Shelton, M. A recurrence of the Ancon dwarf in Merino sheep. *J Hered* **59:** 267–268, 1968.

Troyer, D. L., Thomas, D. L., and Stein, L. E. A morphologic and biochemical evaluation of the spider syndrome in Suffolk sheep. *Anat Histol Embryol* **17:** 289–300, 1988.

Vanek, J. A., Walter, P. A., and Alstad, A. D. Comparing spider syndrome in Hampshire and Suffolk sheep. *Vet Med* **82:** 430–437, 1987.

Wild, P., and Rowland, A. C. Systemic cartilage defect in a calf. *Vet Pathol* **15:** 332–338, 1978.

Wray, C., Mathieson, A. O., and Copland, A. N. An achondroplastic syndrome in South Country Cheviot sheep. *Vet Rec* **88:** 521–522, 1971.

d. Chondrodysplasias of Dogs

i. *Pseudoachondroplastic Dysplasia* This disease affects miniature poodles and has been called spondyloepiphyseal dysplasia, epiphyseal dysplasia, achondroplasia, and chondrodystrophia fetalis. The disease is probably inherited as an autosomal recessive trait and is usually recognized a few weeks after birth because of poor growth and gait. The skull is usually normal; mild inferior prognathism may be present. Endochondral bones of the spine and limbs are short with bulbous ends, and diaphyses are relatively wide (Fig. 1.19). The metaphyses of long bones are flared, but despite this extreme flaring, the cartilage of the epiphyses appears to droop over the edges of the metaphyses (Fig. 1.20A,B).

There is a defect in hyaline cartilage, including that of the trachea and nasal septum, which interferes with endochondral ossification in the epiphyses and growth plates. The defect appears to be decreased sulfation of the cartilage glycosaminoglycans, which could be primary in

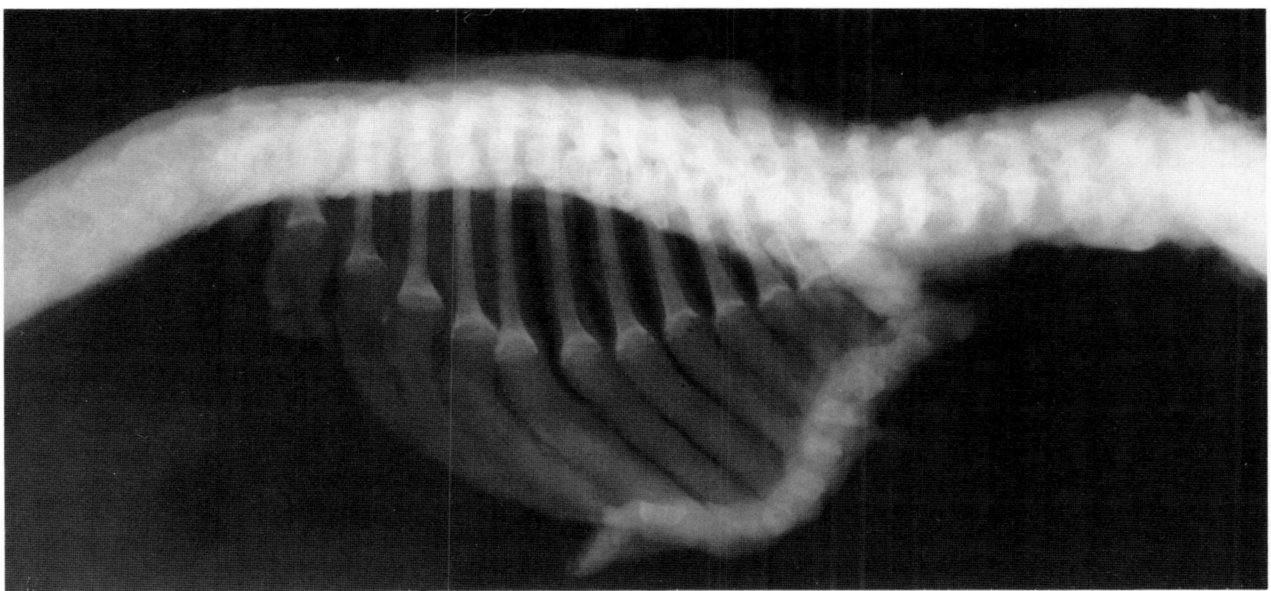

Fig. 1.19 Pseudoachondroplastic dysplasia in miniature poodle. Radiograph of thorax. Vertebrae are short, endochondral ossification is irregular, and costal cartilages are long. Costochondral junctions are large.

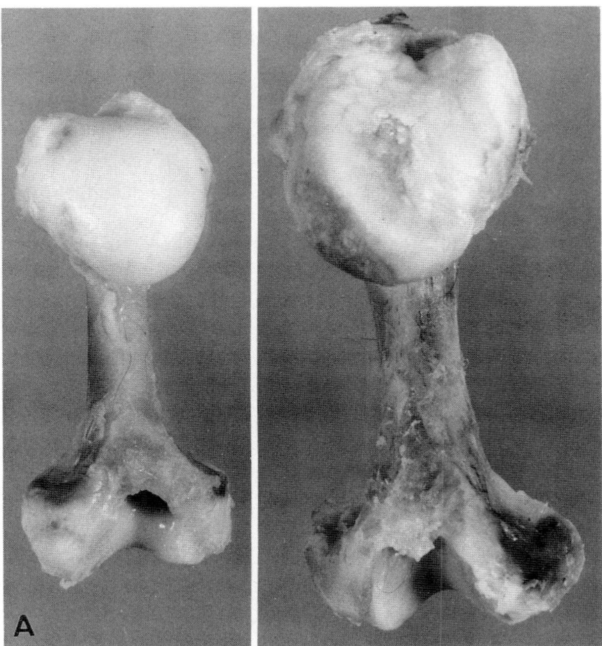

Fig. 1.20A Pseudoachondroplastic dysplasia in miniature poodle. Humerus of immature and mature dogs. Degeneration in articular cartilage of adult.

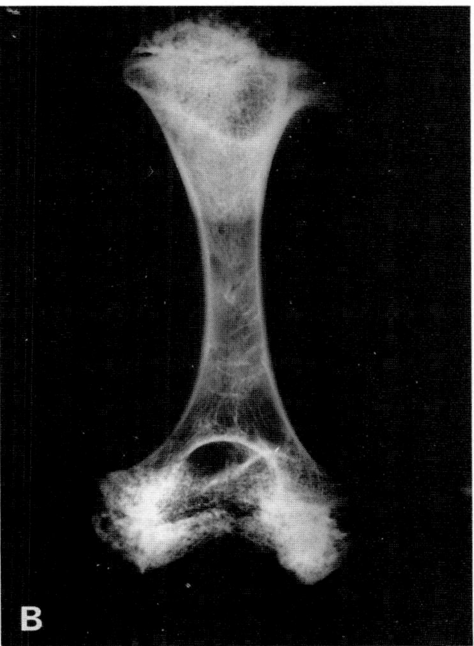

Fig. 1.20B Pseudoachondroplastic dysplasia. Radiograph of humerus showing irregular development of ossification centers.

the sulfation enzyme pathway or the result of increased sulfatase activity. The collagen is normal. Microscopically the intercellular substance is sparse and lacks basophilia. Chondrocytes are variable in size and are sometimes clumped together in enlarged lacunae. Many cells have vacuolated cytoplasm. The maturation sequence in growth zones is disordered and irregular (Fig. 1.21). Ossification centers develop multifocally and late and give a stippled appearance to radiographs of bones in affected puppies (Fig. 1.20B). Complete ossification occurs by about 2 years, but the limbs are only one half to two thirds normal length and are deformed. In contrast, the lumbar vertebrae are longer than in normal dogs of the same age. Degenerative arthropathy, especially of the hips, and spondylosis may develop at less than a year of age and are common in adult dogs (Fig. 1.20A).

ii. *Multiple Epiphyseal Dysplasia* Multiple epiphyseal dysplasia is a rare condition of beagle dogs, inherited probably as an autosomal recessive trait. Affected puppies walk with a swaying gait, but by maturity only slight lameness after exercise remains. Although microscopic lesions are present at birth, the earliest radiographic signs are not visible until 3 weeks of age. They consist of stippled mineralizations in the epiphyses, particularly of the femur and humerus and in the tarsal and carpal bones (Fig. 1.22A). Lumbar vertebrae may also be involved, but the tibia–fibula and radius–ulna are spared. The stippling gradually disappears until by 5 months it is incorporated into the normal ossification centers. There are no specific abnormalities in adult dogs, but evidence of osteoarthropathy may be present.

The abnormal foci of mineralization develop in a specific subarticular zone of the epiphyses and cuboid bones. The initial lesion is a floccular accumulation of chondroitin sulfate and glycoprotein in chondrocyte lacunae. This is followed by coalescence and liquefaction of the lacunae and then by mineralization of the contents of the resultant cysts (Fig. 1.22B). At no time is there evidence of extra ossification centers in the cartilage.

The stippled epiphyses in this condition give a superficial radiographic similarity to pseudoachondroplastic dysplasia of miniature poodles. Careful radiographic and microscopic examination readily differentiates them.

iii. *Chondrodysplasia in the Alaskan Malamute* Chondrodysplasia in the Alaskan malamute is a disproportionate short-limbed dwarfism with normal body length, which is inherited as an autosomal recessive trait with complete penetrance and variable expression. Macrocytic hemolytic anemia, characterized by a mean corpuscular hemoglobin concentration of less than 30%, may be a pleotropic effect of the chondrodysplasia gene. Anemia is present in all affected dogs, but heterozygotes which do not have

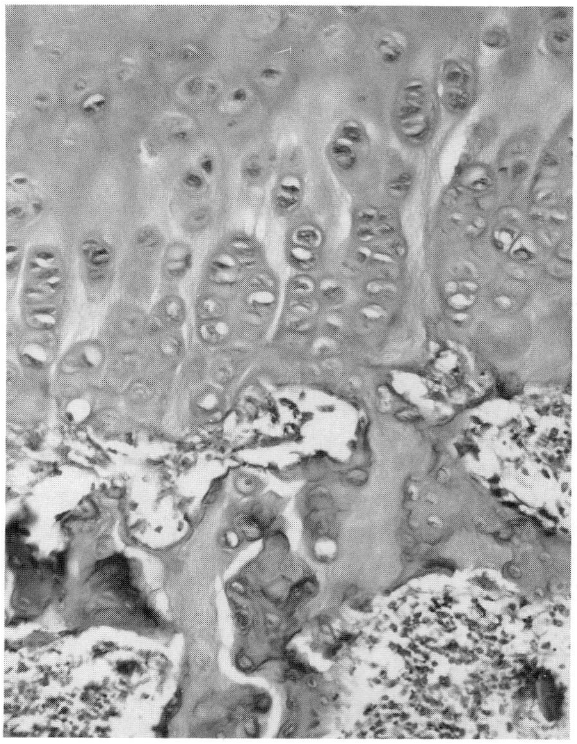

Fig. 1.21 Pseudoachondroplastic dysplasia. Miniature poodle. Microscopic appearance of growth plate. Note irregular columns, fibrillation of cartilage matrix, and abnormal germinal zone.

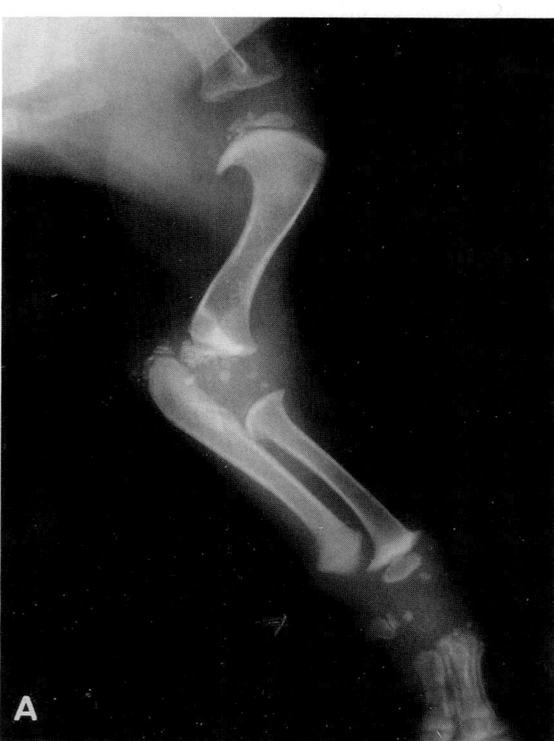

Fig. 1.22A Multiple epiphyseal dysplasia. Beagle dog. Radiograph showing stippled epiphyses in humerus, elbow, and carpals. (Courtesy of P. G. Rasmussen.)

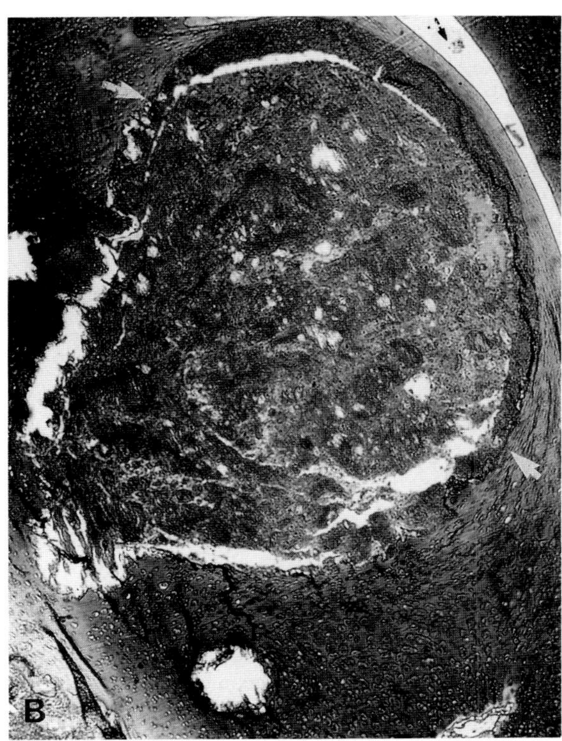

Fig. 1.22B Multiple epiphyseal dysplasia. Beagle dog. Area of degeneration in epiphyseal growth cartilage (arrows). Mineralization in these areas produces stippling. Joint space is marked by black arrow. (Courtesy of P. G. Rasmussen.)

skeletal lesions are usually intermediate between homozygotes and normal malamutes in hemoglobin concentration. Affected males exhibit delayed sexual maturity, and 45% of the spermatozoa they produce have acrosomal defects. The testicular condition is zinc responsive.

Abnormal endochondral ossification occurs throughout the body, but the most striking clinical and pathological effects are seen in the distal ulnar metaphyses where enlargement of the carpal joints and lateral deviation of the paws are evident. In radiographs, the normal V-shaped distal ulnar growth plate is replaced by an irregular, flattened plate with flared periphery (Fig. 1.23A,B). There is also delayed ossification of the cuboid bones. Clinical recognition of the condition is possible at 3 months of age; the best time for radiographic diagnosis is between 4 and 12 weeks.

Growth plates in affected puppies have normal germinal cartilage, but the proliferative zone is disorganized, and clusters of cells are separated by wide bars of intercellular matrix. Cells in some of the clusters are V-shaped, whorled, or rounded, and their rough endoplasmic reticulum is dilated with ruthenium red-positive material that is likely to be chondroitin sulfate. The dilation of the rough endoplasmic reticulum is consistent with inability to transport an abnormal product to the Golgi, and there is a deficiency of ruthenium red-positive material in the interterritorial matrix. The amount of proteoglycan monomer in the cartilage is increased, and the monomers have an increased hydrodynamic size, longer chondroitin sulfate chains, and an increased content of chondroitin-6-sulfate. There is less hydroxyproline in the cartilage, and the type II collagen has less stable cross-links than normal. The latter features may be responsible for the microfractures that are prominent in the metaphysis and result in an irregular increase in the length of the zones of maturation and degeneration. The resemblance to rickets is striking in demineralized sections, but mineralization of cartilage is normal (Fig. 1.24).

By 6 months of age, the radius and ulna show a distinct curvature with the concavity directed caudally. This deformity persists throughout life. Localization of signs to the distal ulna is probably related to the unique mode of growth of the radius–ulna combination. These bones must grow in perfect harmony in order to avoid conformational abnormalities and, since the forelimbs normally bear a greater proportion of the body weight than the hind limbs, they are likely to show more marked effects of growth plate lesions.

iv. *Chondrodysplasia in the Norwegian Elkhound* Disproportionate short-limb dwarfism in the Norwegian elkhound clinically resembles to some extent the disease in malamutes, but there are significant morphologic differences. In elkhounds the vertebral bodies are shortened, and there is a marked reduction in width of the zone of chondrocyte proliferation. The majority of chondrocytes in all zones contain multiple or multilocular intracytoplasmic inclusions which stain deep blue with Alcian Blue-periodic acid-Schiff (PAS) at pH 1.0 and 2.6 and are ruthenium red-negative (Fig. 1.25). Inclusions from degenerate chondrocytes persist and lie free in the lacunae. The composition of these inclusions, which are homogeneous in electron micrographs and are not located in endoplasmic reticulum, is not known. The cell columns in the zones of hypertrophy and degeneration are disorganized, and the trabeculae of primary and secondary spongiosa are coarse and also disorganized. Extensive horizontal bridging and excess osteoid are present in primary spongiosa.

v. *Osteochondrodysplasia in the Scottish Deerhound* A skeletal dysplasia, inherited probably as an autosomal recessive trait, occurs in Scottish deerhounds. It is notable in that it is characterized by periodic acid-Schiff-positive, diastase-resistant cytoplasmic inclusions in proliferative and hypertrophic chondrocytes, and by sporadic physeal lesions.

Signs of exercise intolerance and retarded growth first appear in 4- to 5-week-old puppies. Older dogs have a "bunny-hopping" gait, but neuromuscular lesions are absent. Later, kyphosis, marked bowing of the limbs, and joint laxity develop. Growth plates are irregular in width and the physeal–metaphyseal junctions are uneven. Mature dogs show severe deformities with shortened, porotic long bones, enlargement of sites of muscle attachment, and incongruencies of hip and elbow joints.

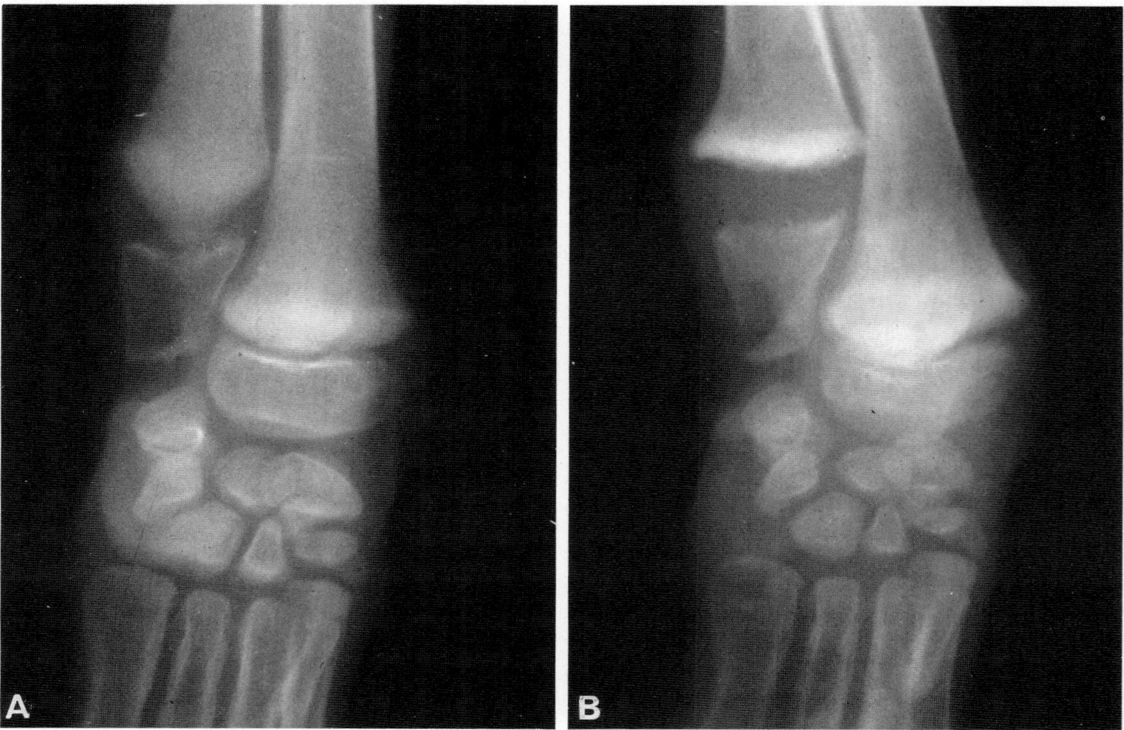

Fig. 1.23 Alaskan malamute chondrodysplasia. Normal (A) and affected (B) littermates. Increased depth of distal ulnar physis with loss of cone-shaped metaphysis in (B). (Courtesy of P. W. Pennock.)

Vertebrae are shortened, and carpal and tarsal bones are distorted.

Microscopically there are islands of basophilic, vacuolated chondrocytes in the epiphyses. Lesions of the physes are sporadic—some areas are relatively normal, while others are narrow and hypocellular or lack proliferative or hypertrophic zones. Metaphyseal lesions are also sporadic. In some areas the primary and secondary spongiosae are normal; in others, trabeculae are thin and sparse to the point of being almost replaced by fibrovascular connective tissue. The intermittent metaphyseal lesions may be secondary to the intermittent physeal lesions and to the effects of trauma.

vi. *Chondrodysplasia in the English Pointer* Chondrodysplasia in the English pointer is inherited probably as an autosomal recessive trait and is remarkable in that focal cavitation of some epiphyseal cartilages develops at an early age.

Affected animals are small and develop metaphyseal flaring and enlargement of the distal radial and ulnar growth plates at 5 to 7 weeks of age. Sometimes the tibial physes are also widened, and there may be inferior prognathism. Radiographs show that all growth plates are abnormal. Locomotory difficulties may be obvious at 5 weeks but usually develop later and typically are manifest by a bunny-hopping gait. Increased mineralization of laryngeal and tracheal cartilage occurs in some dogs. In 10-week-old dogs, articular cartilages appear grossly normal, but by 16 to 18 weeks, there are wrinkles and projections and some fibrillation of cartilage of all major limb joints. In the epiphyseal cartilage beneath these lesions, there are irregular cavities that contain strands of collagen and degenerate chondrocytes. In some joints the lesions appear to heal, leaving only mild fibrillation. In others a severe arthrosis develops that is characterized by thickening of the capsule, extensive loss of articular cartilage, and raised islands of surviving normal cartilage. Microscopically, degeneration of epiphyseal cartilage is apparent at 12 weeks, beginning as decreased basophilia and progressing to loss of interterritorial matrix and cavitation. The physeal lesion is variable in and between bones. The increased width of the physes is due to an increase in the hypertrophic zone. Development of primary spongiosa is apparently normal.

vii. *Chondrodysplasias in Other Dogs* A chondrodysplasia occurs in the **great Pyrenees** that produces a micromelic, short-limbed dwarf. Marked curvature of the forelimbs is present. Singular features of the condition are reduced diameter of the tracheal lumen, suggesting a generalized defect of hyaline cartilage, and vacuolation of many chondrocytes in the physes. The vacuoles represent irregularly dilated cisternae of rough endoplasmic reticulum. Short-limbed dwarfism in a **cocker spaniel** was associated with increased urinary excretion of dermatan sulfate and radiographic lesions similar to those of cats with mucopolysaccharidosis VI (see the following). Neutrophil lysosomal enzymes were normal, however. Physeal chon-

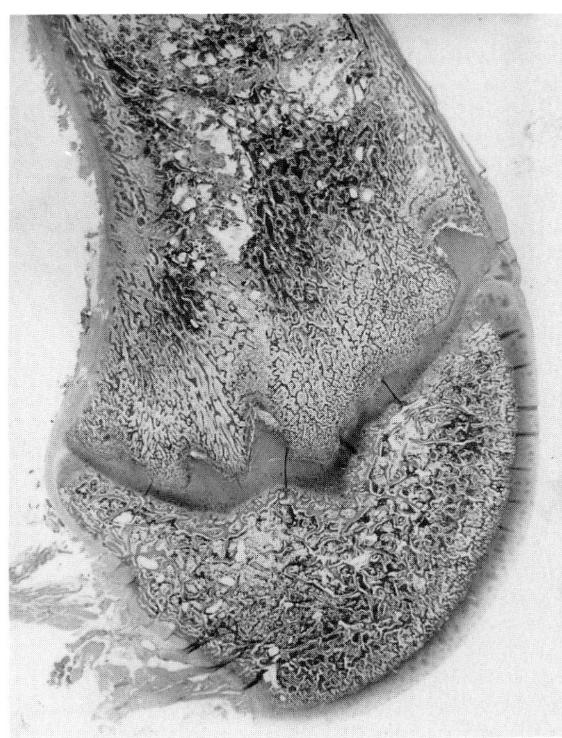

Fig. 1.24 Malamute. Irregular modeling of physeal cartilage. (Slide, courtesy of S. Fletch and W. Riser.)

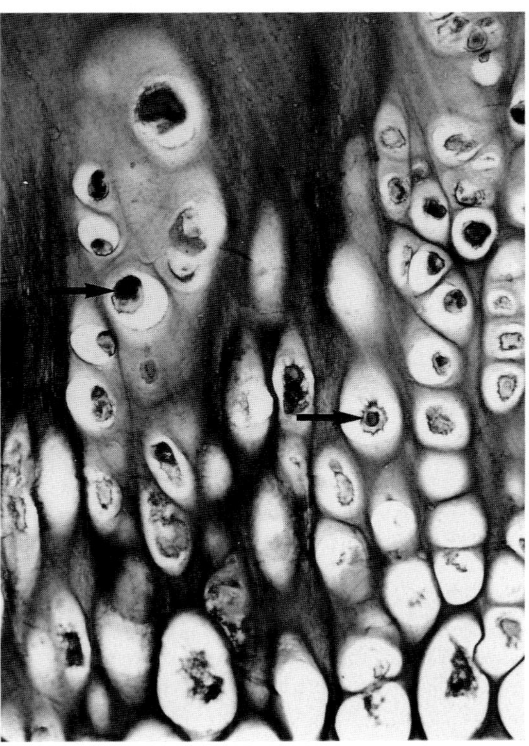

Fig. 1.25 Norwegian elkhound chondrodysplasia. Inclusion bodies in chondrocytes. (Slide, courtesy of S. A. Bingel.)

drocytes contained vacuoles, and the rough endoplasmic reticulum (RER) was dilated. Granular material was in the RER and in the territorial matrix.

viii. *Ocular and Skeletal Dysplasia* Syndromes combining ocular and skeletal dysplasias occur in the Labrador retriever and Samoyed breeds. In both breeds the ocular lesions include cataracts and retinal detachment. Retinal dysplasia also occurs in the retrievers (see Chapter 4, Eye and Ear). Bone lesions are evident at about 8 weeks and are confined to the axial skeleton. They include varus deformity of the elbow and valgus deformity of the carpus, and a "downhill" stance because of more severe involvement of the forelegs. Hip dysplasia and ununited anconeal process occur in both breeds. In the retrievers, the coronoid processes are also ununited, while both coronoid and anconeal processes are hypoplastic (see elbow dysplasia syndrome, this chapter). Detailed microscopy is not reported. Retarded growth of radius, ulna, and tibia occurs in the retrievers, and retained cartilage cores may be present in the distal ulna.

The mode of inheritance is probably autosomal recessive in Samoyeds. In Labrador retrievers both conditions also result from the effects of a single gene that has recessive effects on the skeleton and incompletely dominant effects on the eye. Thus heterozygotes have a clinically normal skeleton, but mild eye lesions are present.

e. Chondrodysplasias of Cats Pelger–Huet anomaly is an inherited condition, characterized by anomalous leukocytes, that occurs in humans, rabbits, and cats (see Volume 3, Chapter 2, The Hematopoeitic System). Homozygous and heterozygous forms are recognizable by the severity of leukocyte nuclear changes. Chondrodysplasia may occur concurrently in homozygous rabbits and is recorded in a stillborn homozygous cat and a live littermate. In both species, the skeletal anomaly may be a coincidental result of inbreeding.

Axial, abaxial, and cranial lesions occurred in the feline disease. A dome-shaped skull, and short nasal bones, vertebrae, ribs, and long bones were present. The metaphyses of ribs and long bones were enlarged. Microscopically, the chondrodysplastic cartilage stained irregularly and contained areas of cystoid degeneration. Endochondral ossification was retarded.

Bibliography

Bingel, S. A., and Sande, R. D. Chondrodysplasia in the Norwegian Elkhound. *Am J Pathol* **107:** 219–229, 1982.

Bingel, S. A., Sande, R. D., and Wight, T. D. Chondrodysplasia in the Alaskan malamute. Characterization of proteoglycans dissociatively extracted from dwarf growth plates. *Lab Invest* **53:** 479–485, 1985.

Bingel, S. A., Sande, R. D., and Wight, T. D. Undersulfated chondroitin sulfate in cartilage from a miniature poodle with spondylo-epiphyseal dysplasia. *Conn Tiss Res* **15:** 283–302, 1986.

Breur, G. J. *et al.* Clinical radiographic, pathologic, and genetic features of osteochondrodysplasia in Scottish deerhounds. *J Am Vet Med Assoc* **195:** 606–612, 1989.

Carrig, C. B., Schmidt, G. M., and Tvedten, H. W. Growth of the radius and ulna in Labrador retriever dogs with ocular and skeletal dysplasia. *Vet Radiol* **31:** 165–168, 1990.

Fletch, S. M. *et al.* Clinical and pathologic features of chondrodysplasia (dwarfism) in the Alaskan Malamute. *J Am Vet Med Assoc* **162:** 357–361, 1973.

Latimer, K. S., Rowland, G. N., Mahaffey, M. B. Homozygous Pelger–Huet anomaly and chondrodysplasia in a stillborn kitten. *Vet Pathol* **25:** 325–328, 1988.

Meyers, V. N. *et al.* Short-limbed dwarfism and ocular defects in the Samoyed dog. *J Am Vet Med Assoc* **183:** 975–979, 1983.

Riser, W. H. *et al.* Pseudoachondroplastic dysplasia in miniature poodles: Clinical, radiologic, and pathologic features. *J Am Vet Med Assoc* **176:** 335–341, 1980

Sande, R. D., and Bingel, S. A. Animal models of dwarfism. *Vet Clin North Am Small Anim Pract* **13:** 71–89, 1982.

Whitbread, T. J., Gill, J. J. B., and Lewis, D. G. An inherited enchondro- dystrophy in the English pointer dog. A new disease. *J Small Anim Pract* **24:** 399–411, 1983.

2. *Lysosomal Storage Diseases of Cats and Dogs*

The storage diseases are inherited and acquired conditions characterized by the accumulation of excess metabolites in lysosomes. They are classified according to the nature of the metabolites (see Chapter 3, The Nervous System). Skeletal lesions in animals are associated with the mucopolysaccharidoses (MPS), identified by the presence of excess glycosoaminoglycans (mucopolysaccharides), and the lipidoses, in which gangliosides accumulate.

The **mucopolysaccharidoses** of humans are genetic diseases associated with skeletal, cardiovascular, neurologic, and ocular abnormalities and usually with dwarfism. Most involve an abnormality in the lysosomal catabolism of dermatan sulfate and/or heparan sulfate. Similar inherited diseases marked by deficient activity of specific lysosomal enzymes are recognized in cats and dogs. The trypanocidal drug suramin inhibits a variety of lysosomal enzymes and produces MPS in rats and probably in other species.

Where the specific enzyme abnormality is known, the eponymous medical title is commonly used to describe the analogous veterinary condition. Thus MPS I and VI are known as Hurler's syndrome, involving α-L-iduronidase deficiency, and Maroteaux–Lamy syndrome, involving arylsulfatase-B deficiency, respectively. Both syndromes occur in cats, and MPS I also occurs in dogs.

Mucopolysaccharidosis VI occurs in Siamese and part-Siamese cats, and is inherited as an autosomal recessive trait; MPS I is reported in a domestic short-haired cat and probably has a similar mode of inheritance. The two feline syndromes differ morphologically only in that MPS I lacks epiphyseal dysplasia of long bones and metachromatic granules in circulating neutrophils. Meningiomas are recorded in 4 of 7 cats younger than 3 years that had MPS I.

Kittens with MPS VI can be recognized at 1 week of age by their excessive urinary dermatan sulfate. By 2 months of age the typical features of the syndrome, including broad flattened face, small ears, diffuse corneal clouding, large forepaws, and pectus excavatum are evident. Radiographic lesions are present in the axial and distal skeleton by 6 months of age and are progressive. Thoracic vertebral bodies are short, and there are fusions of cervical and lumbar vertebrae, absence or dysplasia of cervical and thoracic vertebral spinous processes, and increased width of intervertebral spaces (Fig. 1.26A). The ribs broaden at the costochondral junctions, and epiphyses are dysplastic. In older animals, the epiphyses and metaphyses of long bones are distorted by bone proliferation, and articular surfaces are irregular. Some animals develop posterior ataxia and paresis at younger than a year of age. The cause is compression of the thoracolumbar cord between T-12 and L-2 by osteophytes which project into the vertebral canal.

At autopsy, moderate dilation of lateral ventricles and thickening of the left atrioventricular valve are found. Membrane-bound cytoplasmic inclusions are present in hepatocytes, bone marrow granulocytes, vascular smooth muscle cells, and fibroblasts of skin, cornea, and heart valves. These inclusions are visible by light microscopy as clear vacuoles in smooth muscle of splenic trabeculae, and in fibroblasts of cornea and other ocular structures and of the heart valves. Chondrocytes of articular, aural, and growth cartilage are abnormal (Fig. 1.26B).

Mucopolysaccharidosis I also occurs in Plott hounds and is probably inherited as an autosomal recessive trait. The dogs, unlike cats with MPS I, are stunted in size, but otherwise the disease largely mimics the feline condition. Allogeneic bone marrow transplantation of irradiated dogs reduces the neurovisceral lesions of MPS I, but the effect is minimal in chondrocytes, except those near the articular surface.

GM_1 gangliosidosis occurs in cats, calves, sheep, and dogs and is due to a deficiency of β-galactosidase. In sheep, there is a concomitant deficiency of alpha-neuraminidase. In dogs, skeletal lesions are present in English springer spaniels, but in the beagle, only the neurovisceral changes are present. Also the stored unde-

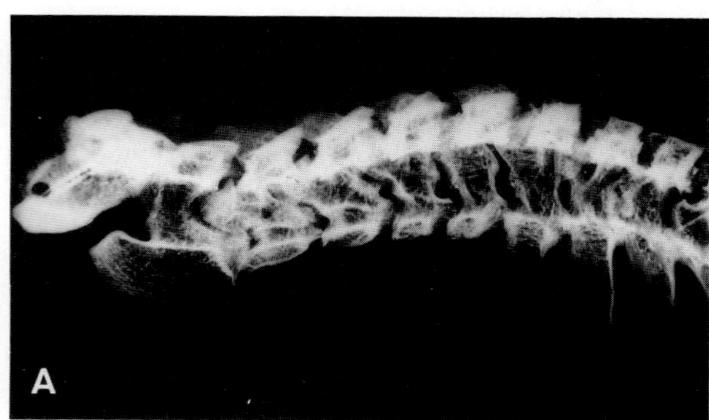

Fig. 1.26A Cat, mucopolysaccharidosis VI. Fusion and dysplasia of cervical vertebrae. Widening of vertebral joint spaces.

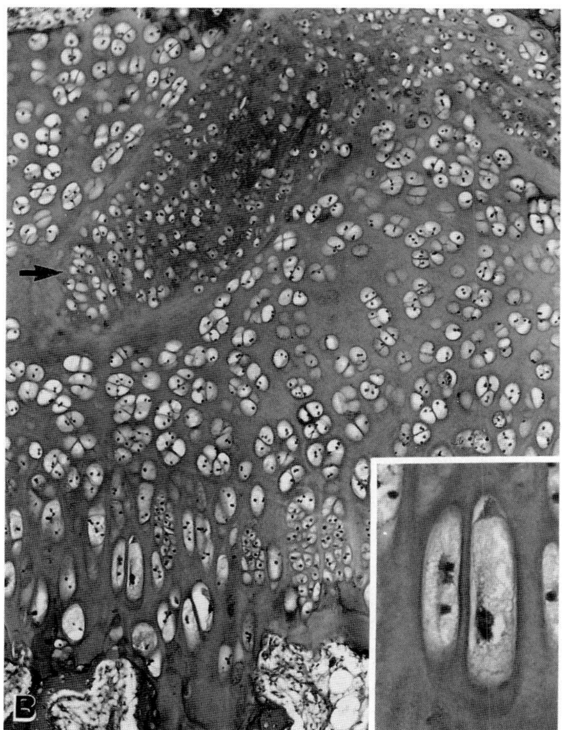

Fig. 1.26B Cat, mucopolysaccharidosis VI. Physis. Irregular cartilage columns and nodules of disorganized cells (arrow). Inset, fine vacuolation of chondrocytes and multicellular lacunae.

graded oligosaccharides differ between the two breeds. In both breeds the trait is autosomal recessive.

Affected springers develop neurologic signs at 4 to 6 months. Skeletal lesions may include proportional dwarfism with frontal bossing and hypertelorism, irregular intervertebral spaces, and degenerative changes in the femoral and tibial articular cartilages.

A storagelike disease with visceral and skeletal but not neural involvement occurred in a Rhodesian ridgeback dog. The dog was lame from an early age and showed multiple osteolytic lesions in radiographs of long bones and vertebrae. Osteoarthritic changes, especially in elbow, hip, and stifle, were accompanied by marked enlargement of acetabular and olecranon fossae. The liver, lymph nodes, and synovial villi were enlarged due to accumulation of cells with abundant vacuolated cytoplasm. Similar cells were seen in endocrine glands and vertebrae. The stored material, which appeared to be confined to the monocyte–macrophage system and synovial lining cells, was not identified but was variably oil red O-positive.

Bibliography

Alroy, J. *et al.* Neurovisceral and skeletal GM_1-gangliosidosis in dogs with β-galactosidase deficiency. *Science* **229:** 470–472, 1985.

Breider, M. A., Shull, R. M., and Constantopoulos, G. Long-term effects of bone marrow transplantation in dogs with mucopolysaccharidosis I. *Am J Pathol* **134:** 677–692, 1989.

Constantopoulos, G. *et al.* Suramin-induced storage disease. Mucopolysaccharidosis. *Am J Pathol* **113:** 266–268, 1983.

Doige, C. E., Farrow, C. S., and Gee, B. R. A storagelike disease with skeletal involvement in a dog. *Vet Radiol* **21:** 98–103, 1980.

Haskins, M. E., and McGrath, J. T. Meningiomas in young cats with mucopolysaccharidosis I. *J Neuropathol Exp Neurol* **42:** 664–670, 1983.

Haskins, M. E. *et al.* The pathology of the feline model of mucopolysaccharidosis VI. *Am J Pathol* **101:** 657–674, 1980.

Haskins, M. E. *et al.* Spinal cord compression and hindlimb paresis in cats with mucopolysaccharidosis VI. *J Am Vet Med Assoc* **182:** 983–985, 1983.

Neufeld, E. F. The biochemical basis for mucopolysaccharidoses and mucolipidoses. *Prog Med Genet* **10:** 81–101, 1974.

Shull, R. M. *et al.* Canine α-L-iduronidase deficiency. A model of mucopolysaccharidosis I. *Am J Pathol* **109:** 244–248, 1982.

3. Osteopetroses in Calves and Other Animals

The osteopetroses are diseases of bone characterized by accumulation of primary and/or secondary spongiosa in the marrow space and often by abnormalities of the bones of the head (Fig. 1.27A,B,C). Lesions of extraskeletal tissues often accompany the osseous disorder. Osteopetroses occur in humans, rat, rabbit, mouse, dog, cat, sheep, horse, pig, and ox. In humans, an autosomal recessive form that is manifest early in life, and an autosomal dominant form that may be asymptomatic, are recognized. Several other diseases characterized by excess skeletal mass occur in humans (see Beighton). Most of the osteopetroses of animals are inherited as autosomal recessive traits and resemble the analogous form in humans.

In those species in which the pathogenesis is known, defective osteoclastic resorption of mineralized tissue is responsible. This can be due either to reduced numbers or inefficiency of osteoclasts, or to abnormalities in the tissue being resorbed. Some forms of osteopetrosis in the mouse and rat are transmissible by mononuclear stem cells from spleen or thymus, and are curable with cells of similar origin from normal animals. In *op/op mice,* injections of recombinant human macrophage colony-stimulating factor cure the osteopetrosis, and normal levels of blood monocytes are established. Osteopetrosis in calves associated with failure of resorption of primary spongiosa and other bone tissue occurs in black and red Aberdeen-Angus, Hereford, Simmental, and Holstein breeds. Vertebral body osteosclerosis, osteopetrosis of mature bulls, is discussed in Volume 3, Chapter 3, The Endocrine System.

The condition in Angus calves is inherited as an autosomal recessive trait. Affected calves are small, premature (250–275 days gestation), and usually stillborn. Prominent clinical features are brachygnathia inferior, sloping forehead, impacted molars, and protruding tongue (Fig. 1.28 A, B).

The long bones are somewhat shorter than normal with narrow diaphyses. The metaphyses and epiphyses have a

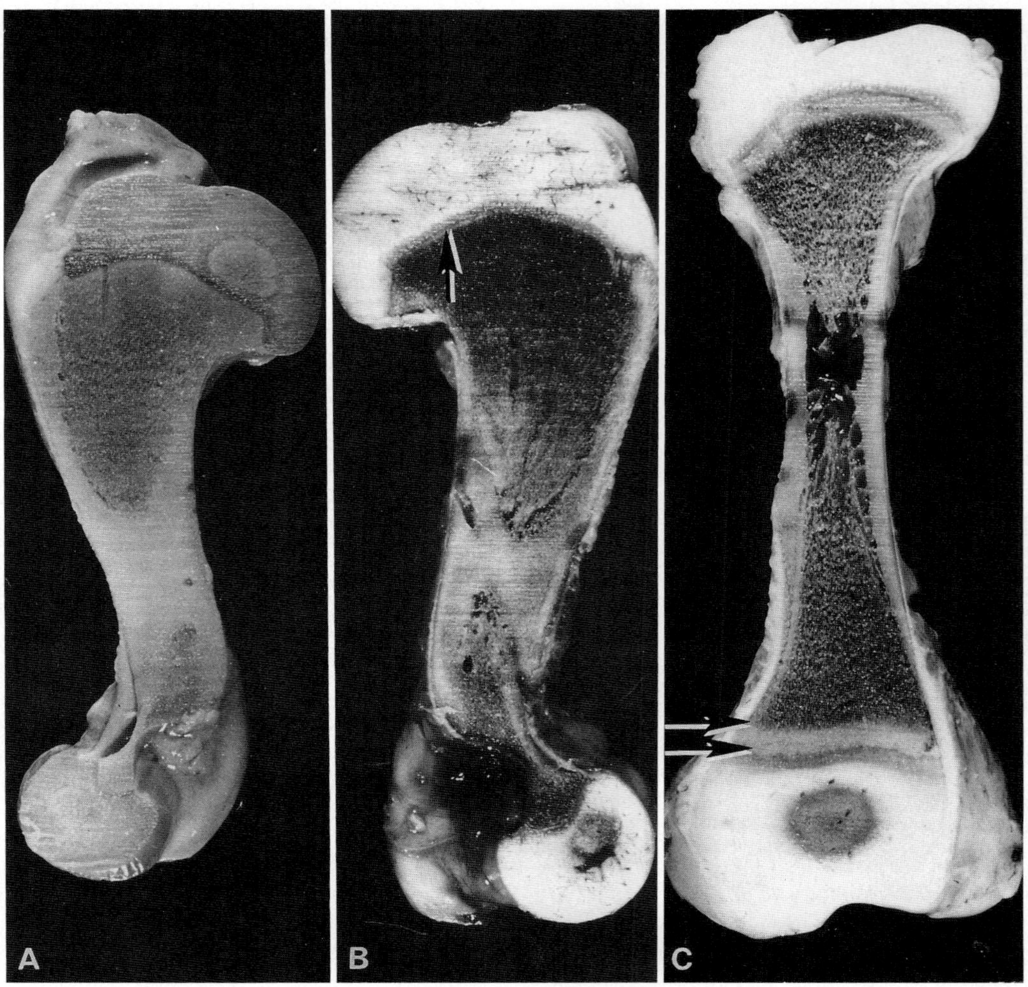

Fig. 1.27 Osteopetroses. Bovine fetuses. (A) Persistence of primary spongiosa. Holstein. Fine trabeculae fill the marrow cavity of the humerus. (B) Persistence of secondary spongiosa. Trabeculae and hematopoietic tissue fill marrow cavity of humerus. Note pale zone of primary spongiosa (arrow). (C) Persistence of primary and secondary spongiosa. A broad band of pale, primary spongiosa (between arrows) is replaced by secondary spongiosa with hematopoietic marrow.

normal shape but appear enlarged because of the diminutive diaphyses. The bones are very fragile. The vertebrae are compact, frontal and parietal bones are thick, and the cranial base, thick and dense. The marrow cavities of endochondral bones are filled with unresorbed primary spongiosa (Figs. 1.27A, 1.29). Although skeletal abnormalities are generalized, they may be variably expressed in individual bones. Intramembranous bones of the skull are dense and thick, and the diploë of the occipital, frontal, and parietal bones are absent. As a result, the cerebral hemispheres are flattened, and the cerebellum is coned by partial herniation through the foramen magnum. Many nutrient foraminae throughout the skeleton are abnormal, being hypoplastic, blind-ended, or absent. Optic nerves are hypoplastic.

Microscopic changes are prominent wherever endochondral bone is formed. Growth plates are relatively normal, although focal persistence of hypertrophic cartilage in the metaphysis may be seen. The metaphyses are relatively avascular, osteoclasts are sparse, and there is persistence of primary spongiosa covered by a thin layer of bone. Continued deposition of coarse woven bone contributes to the diaphyseal sclerosis. Cortical bone is apparently normal. Lesions are found in the brain and consist of mineralized vessel walls and/or neurons in thalamus, cerebellum, meninges, choroid plexuses, and around the aqueduct of Sylvius. Extramedullary hematopoiesis is present in liver and spleen.

A condition clinically and pathologically similar to that in Angus calves occurs in the Hereford and Simmental breeds and also has an autosomal recessive mode of inheritance. In Herefords, marked thickening of frontal bones

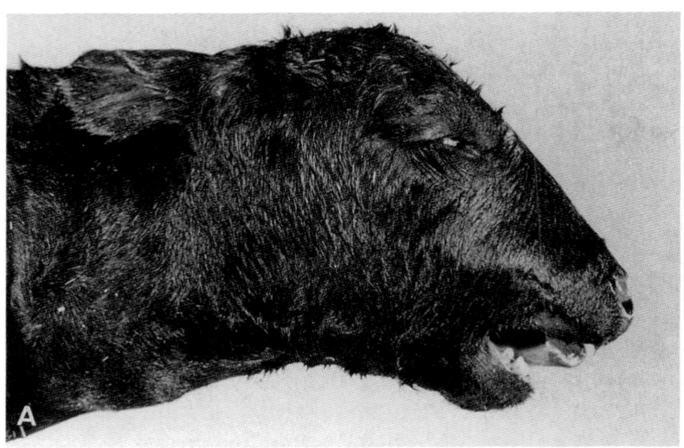

Fig. 1.28A Osteopetrosis. Aberdeen Angus. Note abnormal profile and inferior brachygnathia.

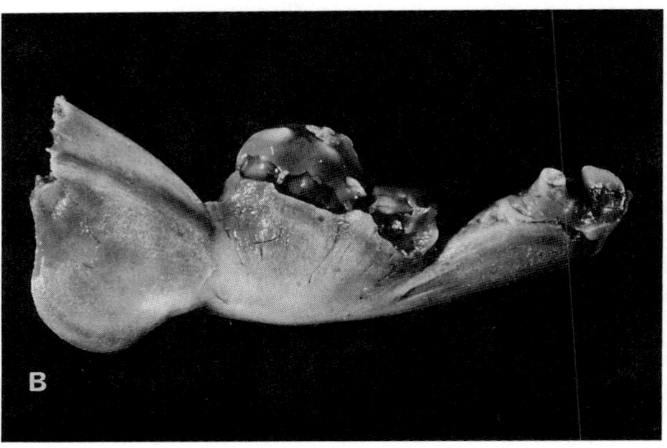

Fig. 1.28B Osteopetrosis. Aberdeen Angus. Impacted molars in deformed, fragile mandible.

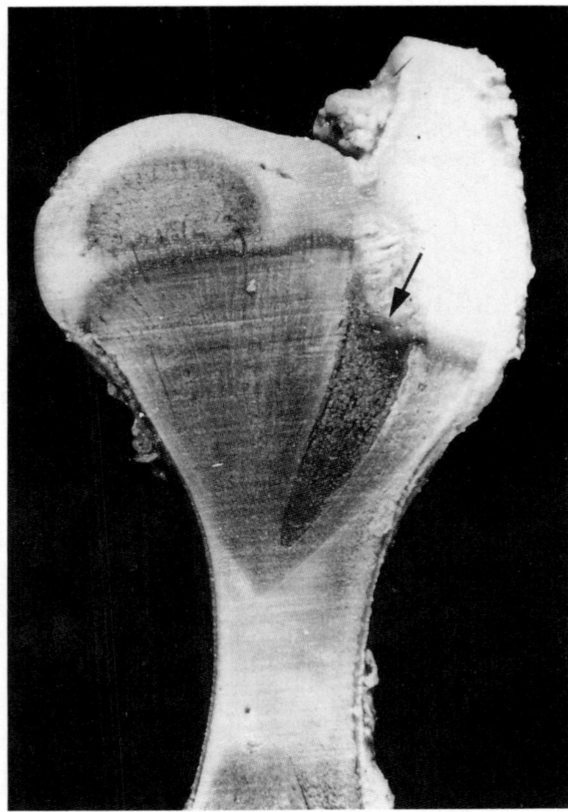

Fig. 1.29 Osteopetrosis. Aberdeen Angus. Persistence of primary spongiosa. Note contribution from fibrous growth plate (arrow).

results in doming of the forehead, and this gives a misleading resemblance to hydrocephalus. Cystic spaces, possibly from necrosis of bone, marrow, or fat, develop in frontal and long bones. In a herd outbreak of osteopetrosis involving 5 of 16 Angus and Angus–Charolais fetuses, similar cranial abnormalities occurred. Arthrogryposis of all limbs and kyphoscoliosis were also present. This condition, which was attributed to an autosomal recessive trait, appears distinct from the more common form of osteopetrosis seen in the Angus breed (see the previous sections).

The pathogenesis of osteopetrosis in calves apparently involves improper osteoclastic resorption and, to that extent, the disease resembles similar conditions in other species. Limited data suggest there may be a deficiency of osteonectin in bovine osteopetrotic bone, and this might interfere with the ability of osteoclasts to resorb it. The normal shape of affected endochondral bones emphasizes the differences between the mechanisms responsible for modeling the metaphyseal flare and those responsible for remodeling the spongiosa.

A second type of osteopetrosis, so-called metaphyseal dysplasia, occurs in Hereford, Holstein, and Japanese black calves and involves failure of resorption of secondary spongiosa (Fig. 1.27B).

Osteopetrosislike lesions occur in aborted bovine fetuses with growth-retardation lattices, cerebellar and cerebral abnormalities, and other stigmata of intrauterine infection with the virus of **bovine virus diarrhea** (Fig. 1.12A,B,C). **Feline leukemia virus** also produces varying degrees of medullary osteosclerosis in growing cats, along with nonregenerative anemia. It is possible that in both cases the lesions are caused by infection of osteoclast precursors. Idiopathic osteopetrosis in two adult cats, not associated with virus infection, was characterized by increased thickness of long-bone cortices and vertebral bodies. Bone resorption occurred but was insufficient to balance the increased deposition.

Osteopetrosis in **dogs** is reported in three littermate dachshund puppies, which had bones of normal size and shape but increased fragility. An osteopetrotic condition in a year-old Australian shepherd born of consanguineous parents is notable only in respect of the age of the dog and its survival for a further 15 months. The nature of the osseous lesion in this dog is unclear. Idiopathic osteopetrosis occurs in young pigs in association with anemia.

Osteopetrosis in three **foals,** including two of the Peruvian Paso breed, was characterized by weak bones, brachygnathia inferior, and irregular physeal cartilage resorption, and in this respect resembles the condition in Angus calves. In contrast, all foals were live born. The condition in Peruvian Paso foals might be inherited.

Bibliography

Beighton, P., Horan, F., and Hamersma, H. A review of the osteopetroses. *Postgrad Med J* **53:** 507–515, 1977.

Greene, H. J. *et al.* Congenital osteopetrosis in Angus calves. *J Am Vet Med Assoc* **164:** 389–395, 1974.

Kramers, P. *et al.* Osteopetrosis in cats. *J Small Anim Pract* **29:** 153–164, 1988.

Nation, P. N., and Klavano, G. G. Osteopetrosis in two foals. *Can Vet J* **27:** 74–77, 1986.

Ojo, S. A. *et al.* Osteopetrosis in two Hereford calves. *J Am Vet Med Assoc* **166:** 781–783, 1975.

Riser, W. H., and Fankhauser, R. Osteopetrosis in the dog: A report of three cases. *J Am Vet Radiol Soc* **11:** 29–34, 1970.

Umemura, T. *et al.* Persistence of secondary spongiosa in three calves. *Vet Pathol* **25:** 312–314, 1988.

Wolf, D. C., and Van Alstine, W. G. Osteopetrosis in five fetuses from a single herd of 16 cows. *J Vet Diagn Invest* **1:** 262–264, 1989.

4. Osteogenesis Imperfecta

There are many causes of fragile bones in domesticated animals, including osteopetrosis and arachnomelia; one of the rarest is osteogenesis imperfecta. The human disease, which is probably the most common inherited connective tissue disorder, occurs in several forms distinguishable by mode of inheritance, age of onset and, increasingly, the biochemical defect associated with the bone fragility. Frost has proposed a causative mechanism for osteogenesis imperfecta that is based on inappropriate cellular response to mechanical usage.

Several occurrences of bone disease, notably in kittens and puppies, have been reported as osteogenesis imperfecta, but most of these have a nutritional cause and are considered under metabolic diseases. However, other bone diseases that occur in fetal and neonatal calves and lambs are comparable to the congenital forms of human osteogenesis imperfecta. Also a kitten with fragile bones and teeth, and normal nutritional status, most probably had osteogenesis imperfecta.

a. Osteogenesis Imperfecta of Calves This condition is reported in Charolais in Denmark and in Holsteins in Australia and the United States of America. The disease is inherited, and the evidence in Holsteins suggests that it is an autosomal dominant trait which arose in separate spontaneous mutations but affects only a certain percentage of the spermatozoa. The diseases in Australian and American Holsteins are phenotypically similar but differ biochemically.

Affected Charolais calves are born alive and are bright and alert but unable to stand, apparently because of fractures. Callus formation indicates that some fractures occur *in utero*. Decreased radiodensity of diaphyseal spongiosa and cortex are described, and most bones have a thin cortex and wide marrow cavity. In a few long bones, the cortex is very thick, and the cavity is narrow.

The microscopic structure of the growth plate is normal, but the amount of bone deposited on the cartilage spicules is clearly deficient and the lace-like chondro-osseous complex persists well into the metaphysis. The deficiency of bone is even more apparent in the diaphyseal cortex where the laminae are narrow and irregular with a marked reduction in the amount of bone deposited on the hypermineralized primer.

The condition in Holsteins is very similar although a few calves are born dead, and some live calves can walk, albeit with difficulty because of hypermobility of the joints and tendon hypoplasia (Fig. 1.30). Joint hypermobility is present in all calves and probably contributes to the recumbency. Blue sclera, regarded as an hallmark of the human disease, and fragile teeth with translucent pink discoloration are also present in affected American Holsteins (Fig. 1.31A). The dental abnormality is due to defective formation of dentin, which is dysplastic and only about one quarter of normal thickness. Severely affected calves are stunted; fractures (Fig. 1.31B) and multiple growth arrest lines related to formation of transverse trabeculae occur in the metaphyses. At birth, growth plates are slightly irregular, and metaphyseal bone is dense and mainly fibrous. Trabeculae are not modeled, retaining their cartilage cores as in osteopetrotic calves, but are resorbed at the metaphyseal–diaphyseal junction. Older calves become osteopenic. Diaphyseal laminae are constructed normally, but compaction is incomplete, and the unfilled vascular spaces contain loose mesenchymal tissue.

Several chemical defects occur in the bone tissue of calves with osteogenesis imperfecta. In both American and Australian cattle, there is decreased apatite crystal size, probably due to arrested crystal growth rather than disordered structure. There is also a reduced amount of type I collagen in the tissues but apparently no qualitative defect. The levels of osteonectin, a noncollagenous protein which may be involved in mineralization of osteoid, are reduced to less than 10% of normal in the American Holsteins but are normal in the Australian cattle. Bone acidic plasma proteins are reduced in both types, as are bone proteoglycans. In addition, the American animals have reduced levels of bone sialoprotein. Phosphophoryn, a dentin-specific protein, and an osteonectinlike protein of dentin are also markedly deficient.

With respect to skin fragility, osteogenesis imperfecta contrasts with dermatosparaxis in Belgian calves which also have skeletal lesions. These animals exhibit defective activity of pro-collagen *N*-peptidase, one of the enzymes responsible for conversion of procollagen to collagen. The skin is poorly resistant to traction, but the bones, though abnormally brittle when processed, are grossly normal and not prone to fractures. Microscopically, the cortical width is somewhat increased, and there are remarkable

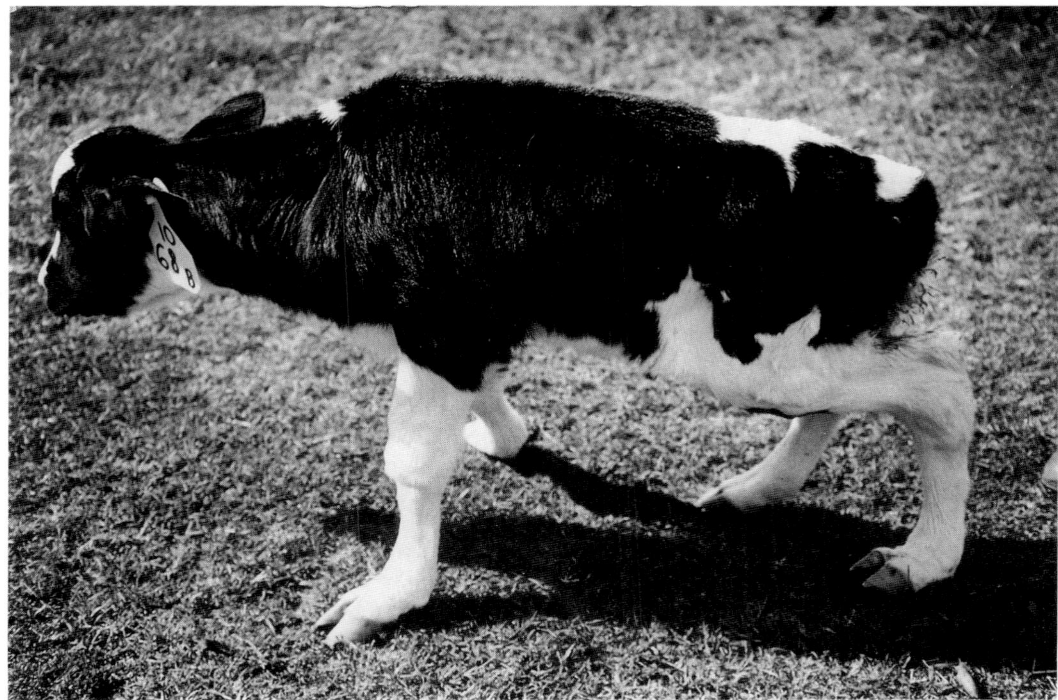

Fig. 1.30 Bovine osteogenesis imperfecta. Characteristic stance associated with joint laxity. (Courtesy of K. G. Thompson.)

abnormalities in the laminar bone. The hypercalcified primer is deficient or absent, but in microradiographs the laminae, which have surface irregularities and are disorganized, are hypermineralized in comparison with normal bone. Collagen fibers are sparse and disoriented. In 6-month-old animals, the cortical bone derived from periosteal laminae remains disorganized, whereas that originating in compaction of chondro–osseous complex is normal, indicating that this is an abnormality exclusively of periosteal (possibly laminar) bone. Detailed examination of bone is indicated in animals showing abnormal skin fragility, and vice versa, since similar mutations in different parts of the type I procollagen molecule can produce diseases that primarily affect either bone or skin.

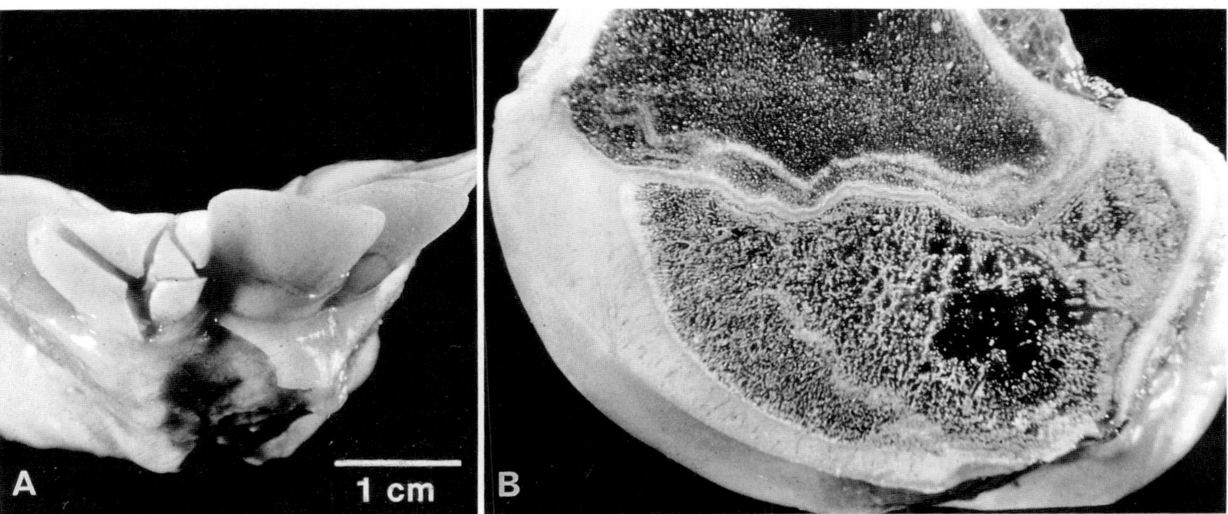

Fig. 1.31 Bovine osteogenesis imperfecta. (A) Fractured tooth due to dentin dysplasia. (Courtesy of K. G. Thompson.) (B) Distal femur. Microfractures in epiphysis and metaphysis. (Courtesy of K. G. Thompson.)

b. Osteogenesis Imperfecta of Lambs Osteogenesis imperfecta is inherited in Barbados blackbelly sheep, probably as an autosomal recessive trait. Lambs are affected at birth, and may have *in utero* fractures, but the disease is less severe than that in calves and in the lambs described later. Joint laxity, reduced tendon size, and increased fragility of long bones are present, but teeth are normal. Collagen and osteonectin levels are normal in affected lambs.

Osteogenesis imperfecta is reported in flocks from New Zealand and Britain. There were differences between the two conditions in that most of the New Zealand lambs were stillborn and had fragile teeth, whereas this was not so in the other flock. In both conditions, there was extreme fragility of bones, and histopathological changes similar to those described in Holstein calves occurred.

Bibliography

Brem, G. *et al.* Zum Auftreten des Arachnomelie-Syndroms in der Brown-Swiss Braunvieh Population Bayerns. *Berl Munch Tierarztl Wschr* **97:** 393–397, 1984.

Cohn, L. A., and Meuten, D. J. Bone fragility in a kitten: An osteogenesis imperfecta-like syndrome. *J Am Vet Med Assoc* **197:** 98–100, 1990.

Denholm, L. J., and Cole, W. G. Heritable bone fragility, joint laxity, and dysplastic dentin in Friesian calves: A bovine syndrome of osteogenesis imperfecta. *Aust Vet J* **60:** 9–17, 1983.

Dhem, A. *et al.* Bone in dermatosparaxis. I. Morphologic analysis. *Calcif Tissue Res* **21:** 29–36, 1976.

Fisher, L. W. *et al.* Two bovine models of osteogenesis imperfecta exhibit decreased apatite crystal size. *Calcif Tissue Int* **40:** 282–285, 1987.

Frost, H. M. Osteogenesis imperfecta. The set point proposal (a possible causative mechanism). *Clin Orthop* **216:** 280–297, 1987.

Holmes, J. R., Baker, J. R., and Davies, E. T. Osteogenesis imperfecta in lambs. *Vet Rec* **76:** 980–984, 1964.

Jensen, P. T., Rasmussen, P. G., and Basse, A. Congenital osteogenesis imperfecta in Charollais cattle. *Nord Vet Med* **28:** 304–308, 1976.

Nusgens, B., and Lapiere, C. M. Bone in dermatosparaxis. II. Chemical analysis. *Calcif Tissue Res* **21:** 37–45, 1976.

Paterson, C. R. Collagen chemistry and the brittle bone diseases. *Endeavour* **12:** 56–59, 1988.

5. *Congenital Hyperostosis of Pigs*

The deposition of radiating trabeculae by the periosteum occurs in many conditions. It is the distinguishing feature of hyperostosis of newborn pigs, which has been compared to Caffey's disease (diaphyseal dysplasia, infantile cortical hyperostosis) in children. The disease in pigs is rare, and is considered to be inherited as an autosomal recessive trait, but this mode of inheritance is not proven. Most affected piglets are born dead at term, and the rest usually die in 1 to 2 days. The cause of death is not known but has been attributed to starvation or cardiac insufficiency.

Skeletal changes are restricted to the limbs, which are thickened up to twice normal size, rigid, hard, and do not pit on pressure (Fig. 1.32). The forelimbs, one or both, are always involved, and the hind limbs may be normal or one or both affected. Affected piglets are well developed but the skin of the limbs is hyperemic, tense and shiny, and firmly fixed to the underlying tissue. The swelling is not sharply demarcated except when it extends to the coronet. Lesions are most severe in the major limb bones, especially the radius–ulna and the tibia (Figs. 1.33A,B; 1.34A). Involvement of metacarpals, scapula, and os calcis is less frequent. The soft tissues of the limbs, especially on exten-

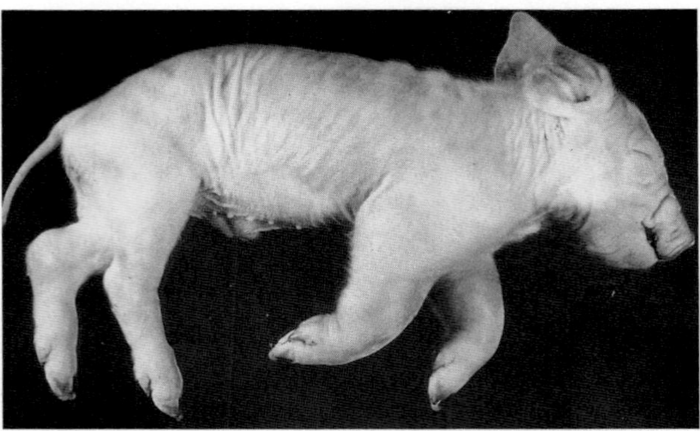

Fig. 1.32 Porcine cortical hyperostosis. Severe swelling of forelegs and mild involvement of left hindleg.

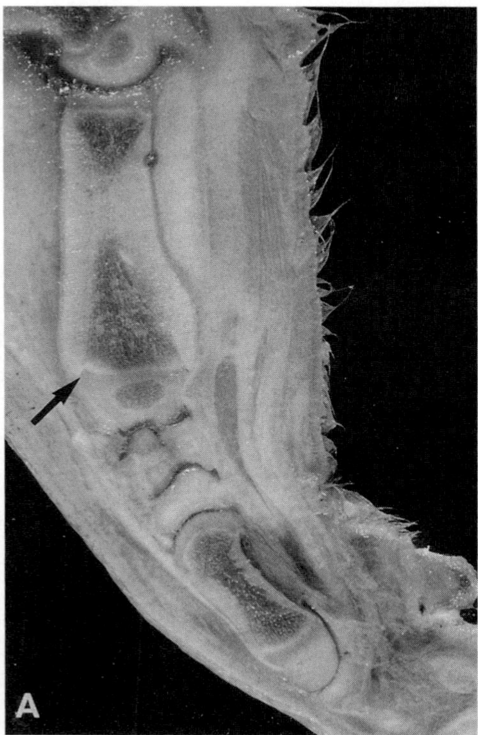

Fig. 1.33A Porcine cortical hyperostosis. Increased soft tissue and increased periosteal bone beginning at the ossification groove (arrow).

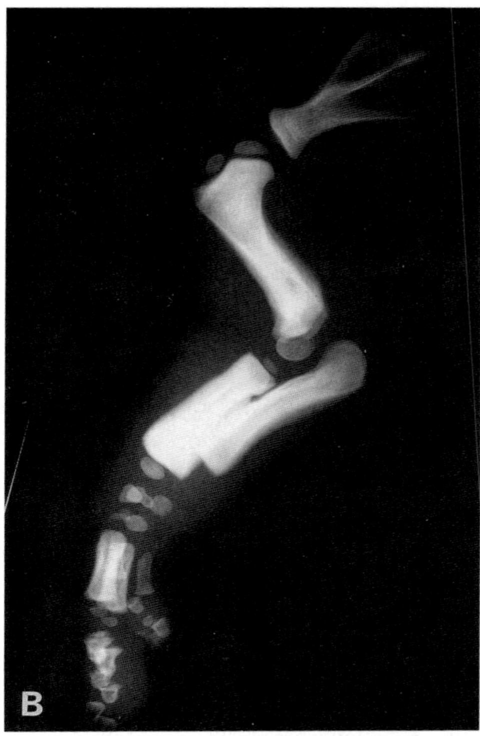

Fig. 1.33B Porcine cortical hyperostosis. Radiograph of foreleg showing increased density of bone and soft tissues.

sor aspects, are extensively infiltrated with edematous, relatively avascular connective tissue, which produces a fibrous fusion of fascia, tendons, sheaths, and joint capsules, and firm adhesion of the skin. The fibrous tissue also infiltrates the muscles, which are much reduced in mass, possibly from hypoplasia rather than from degeneration (Fig. 1.33A; 1.34B).

The joint surfaces and epiphyses are normal as are events in the metaphysis, and there is no excess of endosteal activity. The excess periosteal bone is deposited as eccentric sleeves around the diaphyseal cylinders, probably in areas in which osteoblastic activity in the periosteum is normally present, the abnormality being essentially one of excessive periosteal intramembranous bone formation. The new bone in delicate spicules is deposited on recognizable original cortex (Fig. 1.34A). There is no sign of excessive resorption of the original cortex or of remodeling of the new bone. The interstices of the new bone are occupied by a loose myxedematous tissue without marrow elements. The periosteum is thickened, and multiple layers of osteoblasts are present.

The pathogenesis of congenital hyperostosis of pigs is not known, but the extensive periosteal reaction is consistent with prolonged edema resulting from a local circulatory abnormality. Others have suggested the disease is a result of disorganization of the perichondrial ossification groove of Ranvier.

A disease somewhat resembling hyperostosis is reported in a West Highland white terrier.

Bibliography

Baker, J. R. Infantile cortical hyperostosis (Caffey's disease). *In* "Spontaneous Animal Models of Human Disease," E. J. Andrews, B. C. Ward, and N. H. Altman (eds.), Vol. 2., pp. 240–241. New York, Academic Press, 1979.

Doize, B., and Martineau, G.-P. Congenital hyperostosis in piglets: A consequence of a disorganization of the perichondrial ossification groove of Ranvier. *Can J Comp Med* **48:** 414–419, 1984.

Gibson, J. A., and Rogers, R. J. Congenital porcine hyperostosis. *Aust Vet J* **56:** 254, 1980.

B. Localized Developmental Disturbances

This section deals with localized anomalies of the limbs, sternum, ribs, and pelvis that occur more or less as isolated entities. Some of the anomalies of the neurocranium and vertebral column are described with abnormalities of the central nervous system and of joints, and certain facial anomalies with the oral cavity. Vertebral abnormalities including those causing wobbler syndrome are discussed separately below.

Limbs

The most frequent malformations of the limbs are the localized chondrodysplasias seen in chondrodystrophoid breeds of dogs which are inherited as polygenic traits. Variations in receptor activities for hormones and other factors that regulate physeal chondrocytes of mesodermal origin are responsible. The defect is usually confined to the distal bones, producing the sort of dwarfism seen in such breeds as the basset hound and dachshund. In some breeds, such as the Pekingese, chondrocytes of neural crest origin are also involved, and there are basocranial abnormalities. In other breeds, such as the Boston terrier and boxer, only the basocranium is affected. The cartilage abnormality may also involve the intervertebral disks, the central or perinuclear portions of which develop as immature hyaline cartilage with only a gradual transition at the periphery toward the normal laminated fibrocartilage of the annulus fibrosus. Such disks are predisposed to degeneration of the hyaline cartilage, followed by herniation of the disks (see disk degeneration and spondylosis, this chapter).

Hemimelia is a defect of the midsection of a limb and may be transverse or paraxial. For example, in transverse hemimelia the forearm may be absent, whereas in paraxial hemimelia there is aplasia of either the radius or ulna, or tibia or fibula. Bilateral tibial hemimelia and hypoplasia occur in calves, and unilateral radial hemimelia in dogs and cats. Bilateral hemimelia of Chihuahua dogs, and probably Galloway calves, is inherited as an autosomal recessive trait. Hemimelia also occurs in lambs and kids.

Syndactylia, or fusion of the toes, occurs in sheep, cattle, and swine. Any foot or all feet may be affected, but

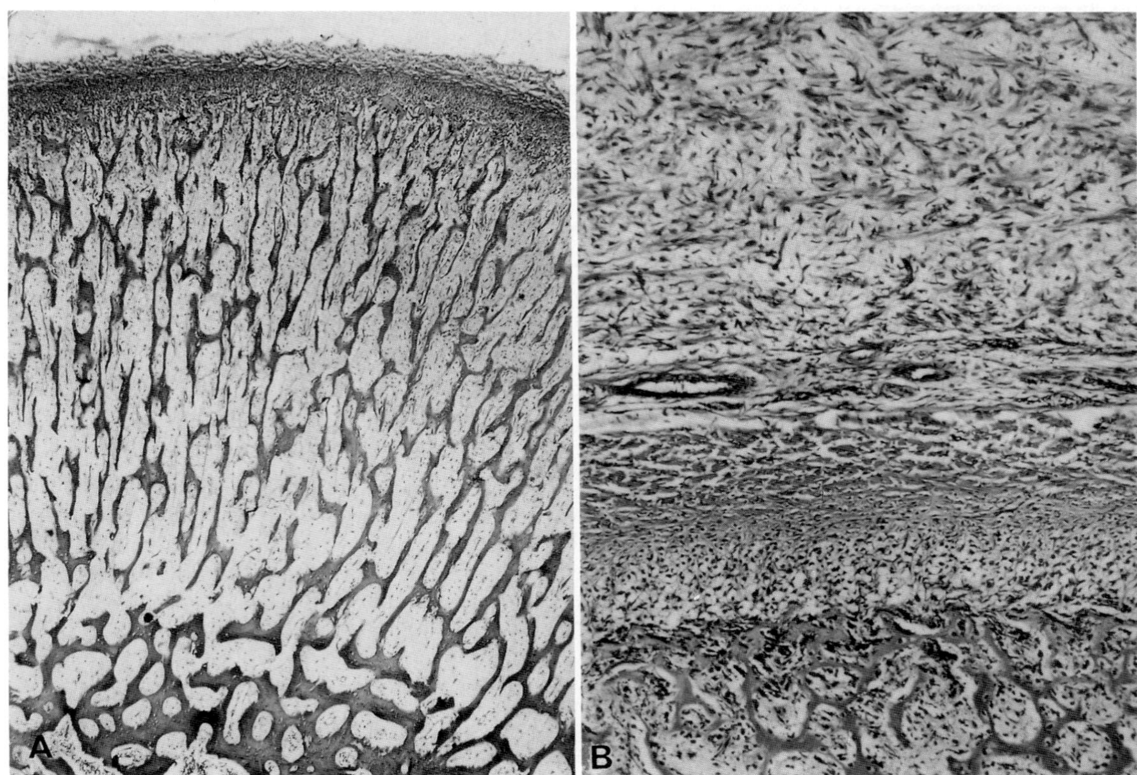

Fig. 1.34 Porcine cortical hyperostosis. (A) Cross section of bone showing radiating periosteal trabeculae. (B) Detail of periosteal reaction, and edema and atrophy of muscle.

the forefeet are most often involved. There is also considerable variation in the degree of fusion. The defect in swine is presumed to be due to a simple dominant gene, but in Holstein, Hariana, German Red Pied, Simmental, Chianina, and Angus cattle it is an autosomal recessive trait with incomplete penetrance. In some syndactylous Holsteins, there is an inability to survive, which is associated with development of hyperthermia.

Polydactylia is an increase in the number of digits. The anomaly is observed in all species but is perhaps best known in cats, dogs, horses, and cattle. An inherited syndrome of skeletal defects including polydactylia and syndactylia occurred in Australian shepherd dogs. An X-linked lethal gene was suspected. In horses, two forms of polydactylia occur. The common, atavistic type features an extra medial digit that articulates with a second metacarpal. The rare, teratogenic form is characterized by duplication of bones distal to the fetlock joint, producing a cloven hoof. Polydactylia is an inherited trait in various bovine breeds, but the inheritance pattern is not well understood. A polygenic mode requiring a dominant gene at one locus and two recessive genes at another locus is postulated in Simmental cattle.

Ectrodactylia, or split-hand deformity, is caused by incomplete fusion of the three rays that develop in the bud of the forelimb. It is reported in dogs and cats. In cats, a dominant gene with variable expressivity is responsible, and the defect is bilateral. In dogs, dominant inheritance is probable, but the defect is usually unilateral. The cleft in the paw mostly extends to the level of the metacarpals but occasionally reaches the carpus. Aplasia and hypoplasia of various carpal and metacarpal bones, duplication of digits, fusion of metacarpals, and elbow joint luxation may accompany the ectrodactylia. Teratogenic ectrodactylia can be produced in mice by cadmium and by carbonic anhydrase inhibitors such as acetazolamide. Conditions characterized by total or partial absence of phalanges have been reported in calves and lambs as ectrodactylia. The condition in lambs occurred in association with an achondroplastic syndrome. These anomalies are more appropriately classified with **adactylia**—absence of digits—which is also described as an isolated anomaly in Southdown lambs and may be inherited. Unilateral dysgenesis of the third phalanx accompanied by navicular bone agenesis is described in a Morgan filly. **Dactylomegaly,** a type of club foot (talipes), is reported in cattle. Abnormalities of the volar and plantar sesamoid bones, perhaps developmental, occur in the Rottweiler and, less often, in other dogs. Bilateral atrophy or absence of navicular bones of the forelegs is recorded in association with ankylosis of the caudal aspects of phalanges II and III in Simmental calves.

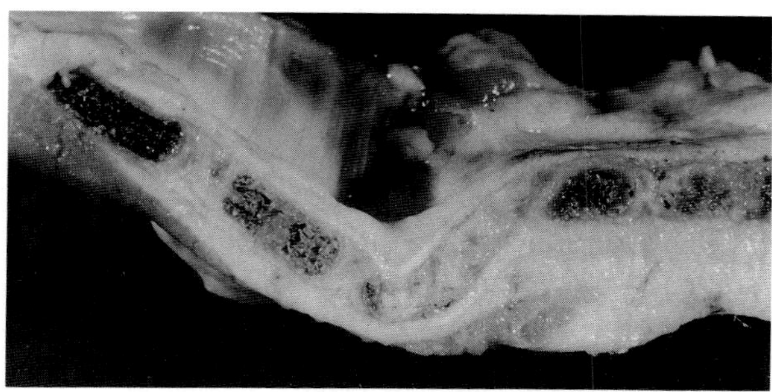

Fig. 1.35 Sternal deformity secondary to chondrodysplasia affecting sternebrae (center). Normal sternebra (left). Spider lamb.

2. *Sternum*

Lateral curvature of the sternum occurs in association with vertebral scoliosis, especially when the latter shows simultaneous torsion. Retraction of the caudal sternebrae and xiphoid is seen in lambs and calves, and is apparently due to shortness of the tendinous portions of the diaphragm. Clefts of the sternum may occur as isolated defects, but are usually accompanied by ectopia cordis or form part of the defect known as schistosomus reflexus, in which there is lordosis, dorsal reflection of the ribs with more or less total eventration, nonunion of the pelvic symphysis, and dorsal reflection of the pelvic bones. Sternal deformity occurs in spider lamb chondrodysplasia (Fig. 1.35) (see preceding sections).

3. *Ribs*

Costal abnormalities are usually secondary to malformation of the vertebral column or sternum. Absent or fused ribs correspond to absent or fused vertebrae and may accompany severe scoliosis.

4. *Pelvis*

The sacrum may be absent, or it may be hypoplastic or deviated in association with absence of the coccygeal vertebrae. Hypoplastic chondrodysplasia of the coccygeal vertebrae is a characteristic of French bulldogs and occurs occasionally, with kinking of the remnants, in cattle and cats. Malformations of the sacrum accompany other severe spinal defects. The pubic bones are separated and may be absent, in association with ectropion of the bladder.

Bibliography

Alonso, R. A. *et al.* An autosomal recessive form of hemimelia in dogs. *Vet Rec* **110:** 128–129, 1982.

Baum, K. H., Hull, B. L., and Weisbrode, S. E. Radial agenesis and ulnar hypoplasia in two caprine kids. *J Am Vet Med Assoc* **186:** 170–171, 1985.

Braund, K. G. *et al.* Morphological studies of the canine intervertebral disc. The assignment of the beagle to the chondroplastic classification. *Res Vet Sci* **19:** 167–172, 1975.

Carrig, C. B. *et al.* Ectrodactyly (split-hand deformity) in the dog. *Vet Radiol* **22:** 123–144, 1981.

Chapman, V. A., and Zeiner, F. N. The anatomy of polydactylism in cats with observations on genetic control. *Anat Rec* **141:** 205–217, 1961.

Dore, M. A. P. Teratogenic polydactyly in a halfbred foal. *Vet Rec* **125:** 375–376, 1989.

Feuston, M. H., and Scott, W. J. Cadmium-induced forelimb ectrodactyly: A proposed mechanism of teratogenesis. *Teratology* **32:** 407–419, 1985.

Gruneberg, H., and Huston, K. The development of bovine syndactylism. *J Embryol Exp Morphol* **19:** 251–259, 1965.

Hart-Elcock, L., Leipold, H. W., and Baker, R. Hereditary bovine syndactyly: Diagnosis in bovine fetuses. *Vet Pathol* **24:** 140–147, 1987.

Hawkins, C. D., and Grandage, J. Dactylomegaly, a type of club foot (talipes) in a herd of Shorthorn cattle. *Aust Vet J* **60:** 55–56, 1983.

Hawkins, C. D. *et al.* Hemimelia and low marking percentage in a flock of Merino ewes and lambs. *Aust Vet J* **60:** 22–24, 1983.

Hughes, E. H. Polydactyly in swine. *J Hered* **26:** 415–418, 1935.

Huston, K., and Wearden, S. Congenital taillessness in cattle. *J Dairy Sci* **41:** 1359–1370, 1958.

Johnson, J. L. *et al.* Hereditary polydactyly in Simmental cattle. *J Hered* **72:** 205–208, 1981.

Leipold, H. W. *et al.* Ectrodactyly in two beef calves. *Am J Vet Res* **30:** 1689–1692, 1969.

Leipold, H. W. *et al.* Anatomy of hereditary bovine syndactylism. I. Osteology. *J Dairy Sci* **52:** 1422–1431, 1969.

Leipold, H. W. *et al.* Adactylia in Southdown lambs. *J Am Vet Med Assoc.* **160:** 1002–1003, 1972.

Leipold, H. W. *et al.* Hereditary bovine syndactyly. II. Hyperthermia. *J Dairy Sci* **57:** 1401–1409, 1974.

Lewis, R. E., and Van Sickle, D. C. Congenital hemimelia (agenesis) of the radius in a dog and a cat. *J Am Vet Med Assoc* **156:** 1892–1897, 1970.

Martig, J., Riser, W. H., and Germann, F. Deforming ankylosis of the coffin joint in calves. *Vet Rec* **91:** 307–310, 1972.

Modransky, P. *et al.* Unilateral phalangeal dysgenesis and navicular bone agenesis in a foal. *Equine Vet J* **19:** 347–349, 1987.

Montgomery, M., and Tomlinson, J. Two cases of ectrodactyly and congenital elbow luxation in the dog. *J Am Anim Hosp Assoc* **21:** 781–785, 1985.

Nordby, J. E. *et al.* The etiology and inheritance of inequalities in the jaws of sheep. *Anat Rec* **92:** 235–254, 1945.

Ojo, S. A. *et al.* Tibial hemimelia in Galloway calves. *J Am Vet Med Assoc* **165:** 548–550, 1974.

Vaughan, L. C., and France, C. Abnormalities of the volar and plantar sesamoid bones in Rottweilers. *J Small Anim Pract* **27:** 551–558, 1986.

5. *Vertebrae*

The embryology of vertebrae and intervertebral disks is complex, involving interactions between ectoderm and mesoderm, and fusions and deletions of cell populations. The notochord, remnants of which persist in the disks, arises from mesoderm adjacent to the thickening in the ectoderm that becomes the neural plate and later the neural tube. Vertebrae develop from the somitic sclerotomes, which are of mesodermal origin. The sclerotomal cells of each somite form diffuse cranial and dense caudal populations, and a vertebra arises from the union of caudal cells of one sclerotome with the cranial cells of the adjacent sclerotome. These combined cell masses migrate to sur-

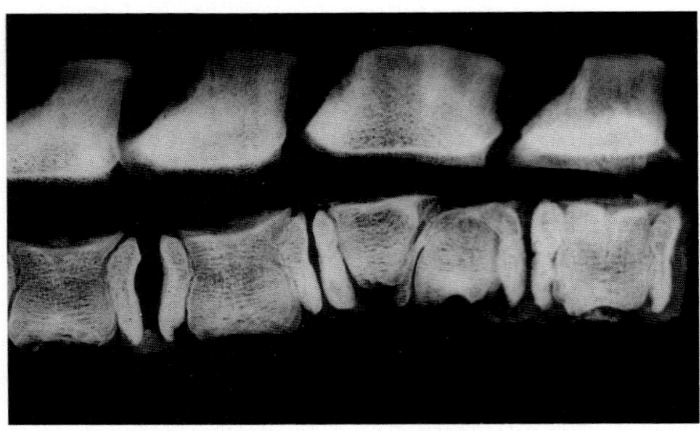

Fig. 1.36 Block vertebra. Two vertebral bodies are incompletely separated. Dorsal processes are fused.

round the neural tube and notochord, deleting the latter except for those cells that persist as the nucleus pulposus of the disk. In each vertebra the neural arch and the intervertebral disk (except nucleus) are formed by the caudal, dense aggregates of cells, while the cranial diffuse aggregates give rise to the vertebral body. An exception occurs in the development of the atlas and axis. The cranial cells of the fifth somite (somites 1–4 form occipital cartilages), which would be expected to form the body of the atlas, form the dens and the cranial articular surface of the axis. Many anomalies of the vertebral arches, such as spina bifida, arise under the aegis of the neural tube, while those of the bodies are caused by abnormal interactions between structures of mesodermal origin.

Block vertebra results from improper segmentation of the somites and may be complete or incomplete, in which case the arch, or all or part of the body, may be involved (Fig. 1.36). The anomalous structure may be equal to or shorter than the vertebrae it replaces. There are usually no clinical signs. **Butterfly vertebrae** are so called because of the appearance of the indented vertebral end plates on ventrodorsal radiographs. The cause is persistence of notochord, or a sagittal cleft of notochord, that leads to a dorsoventral, sagittal cleft in the vertebral body. The halves may spread laterally, but the condition is usually an incidental radiographic finding in screw-tailed dogs such as bulldogs, pugs, and Boston terriers. **Hemivertebrae** may arise as a result of displacement and inappropriate fusion of somites. In this case one member of a pair of somites fuses diagonally with a somite cranial or caudal to it, forming a vertebra but leaving the other members of the pairs to persist as hemivertebrae.

Subluxations, fusions, and other anomalies of cervical vertebrae are described with developmental disturbances of joints (this chapter).

6. *Cervical Vertebral Stenotic Myelopathy (Wobbler) in Horses*

There are several diseases which cause incoordination and locomotor disturbances in horses. Cervical vertebral stenotic myelopathy (cervical vertebral malformation, cervicospinal arthropathy) is one of these diseases, which are often gathered together into a clinical syndrome referred to as wobbles, equine incoordination, or equine sensory ataxia. Other diseases which cause the wobbler syndrome in horses include protozoal myeloencephalitis, equine herpesvirus-1 vasculitis, degenerative myeloencephalopathy, vertebral abscess, synovial cysts (especially in the lower cervical region), and certain congenital, traumatic, and idiopathic lesions (see Chapter 3, The Nervous System).

There are several lesions associated with cervical vertebral stenotic myelopathy, but they are not all present in all cases, and some of them are secondary, degenerative changes. The common, significant factor is morphologic or functional stenosis of the vertebral canal. The clinical syndrome is attributable to compression or stretching of the cervical spinal cord because of lesions in the vertebral column. Presently, estimates of the prevalence of wobbler horses with stenotic lesions vary from 12 to 80%.

Cervical vertebral stenotic myelopathy is divided into two pathologic syndromes, cervical vertebral instability and cervical static stenosis. In cervical vertebral instability there is narrowing of the spinal canal during flexion of the neck. Lesions are most common at vertebrae C-3 to C-5 in fast-growing, male thoroughbreds aged 8–18 months. Cervical static stenosis is characterized by compression of the cord, usually at C-5 to C-7, in large male horses of 1 to 4 years, and neck position is immaterial. Both syndromes are marked by ataxia in which the forelimbs are less severely and less consistently affected than the hind limbs. Mild cases may show only swaying of the hindquarters at slow gaits, while in severe cases, there are prominent errors of movement and disturbances of balance. Usually the onset is insidious, but vague signs are often exacerbated following trauma or strenuous activity. The disease is not progressive, but if the signs are severe, recovery does not occur. The lesions are confined to the spinal column and the spinal cord, the main changes being in the cervical region.

Cervical vertebral instability may be associated with any one of several lesions that produce narrowing of the vertebral canal and apparently damage the cord when the neck is ventroflexed (Fig. 1.37A,B). In a few cases the narrowing may be demonstrated by palpation of the anterior border of affected vertebra, where a lip is evident on the dorsal border of the vertebral body, apparently due to expansion of the anterior epiphysis. More often, a distinct lip is not palpable, but a midsagittal section of affected vertebrae shows that the floor of the spinal canal angles upward and the roof angles downward to produce a funnel-shaped canal with anterior stenosis. Occasionally, the cranial articular processes of the vertebral bodies project in a ventromedial direction and impinge on the spinal canal, compressing the dorsolateral parts of the cord. In a small proportion of cases, increased mobility of adjacent cervical vertebrae, which allows subluxation, is apparently responsible for the cord compression. Significant lesions in cervical vertebral instability occur most often at C-3 to C-

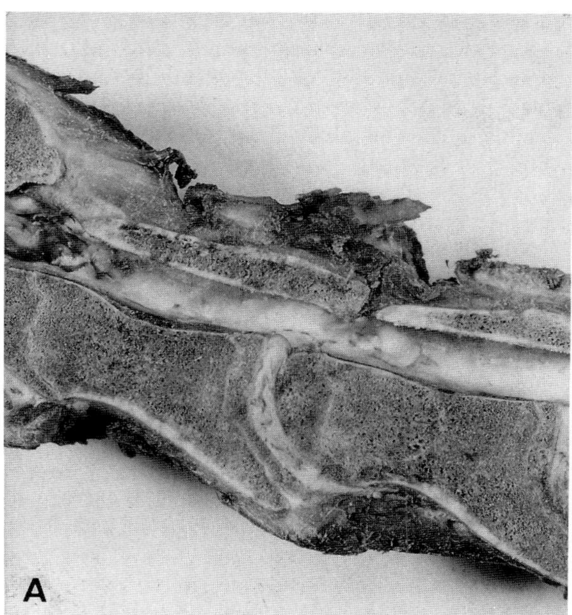

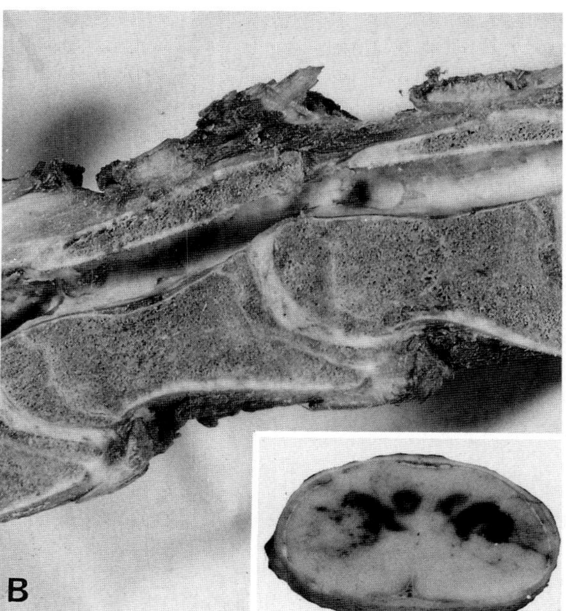

Fig. 1.37 Cervical vertebrae of "wobbler" horse. (A) Standing position. (B) With neck ventroflexed to show stenosis of C-3 to C-4. (Inset) malacia of cervical cord.

5 but other vertebrae are affected, sometimes more than one in a single animal.

Lesions are frequently present in the articular processes. They are often typical of degenerative arthropathy secondary to osteochondrosis. The cartilages are fibrillated and eroded with, in chronic cases, eburnation of the underlying bone. Sometimes there is notching and osteophytic "lipping" of the articular margins. Fractures through the articular facets, similar to those described in the medial coronoid process of the ulna in dogs (see osteochondrosis), are not uncommon. However, there is not a consistent relationship between this type of change and spinal cord lesions; thus, conservative interpretation of their clinical significance is necessary. Asymmetry of articular facets is also present in many affected horses, but this too is of variable significance, since many clinically normal horses have similar changes. Nonetheless, in some horses the spinal cord injury is due to vertebral malfunction created by asymmetry of articular facets (Fig. 1.38A,B).

The fundamental cause of cervical vertebral instability is not clear. There is growing evidence that the multiple genetic factors that permit rapid growth, and the dietary excesses that encourage it, are responsible. In particular the *ad libitum* feeding of high-protein, high-energy rations to weanlings to produce maximum-sized yearlings seems to be the major influence. Such feeding practices are commonly associated with osteochondrosis (see the following), and many affected horses have lesions of this disease, but the significance of this to the overall incidence of cervical vertebral instability is presently undetermined. The role of copper deficiency is also discussed with osteochondrosis.

Cervical static stenosis, the second, less common subdivision of cervical vertebral stenotic myelopathy, is caused primarily by thickening of the ligamenta flava and the dorsal laminae of the vertebral arches. The enlargement of these structures compresses the spinal cord, producing the typical syndrome. Histologically there is increased fibrocartilage at the attachment of the ligaments to the laminae, and deposition of fibrovascular tissue in the ligaments and the contiguous joint capsule. The thickening of the dorsal laminae is due to osteosclerosis, with the dense

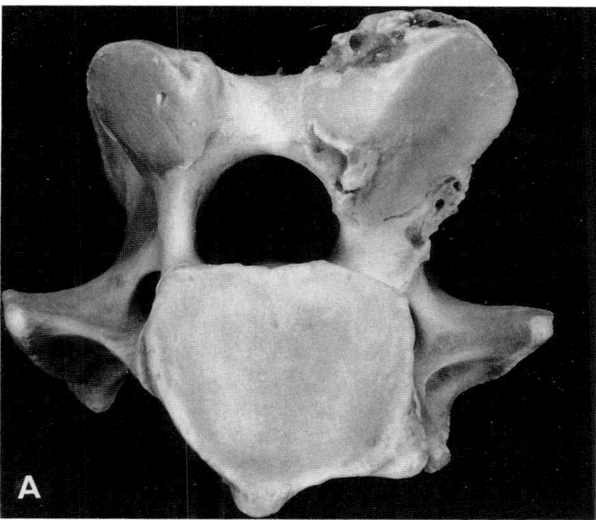

Fig. 1.38A Vertebral malformation. "Wobbler" horse. (Courtesy of J. D. Baird.)

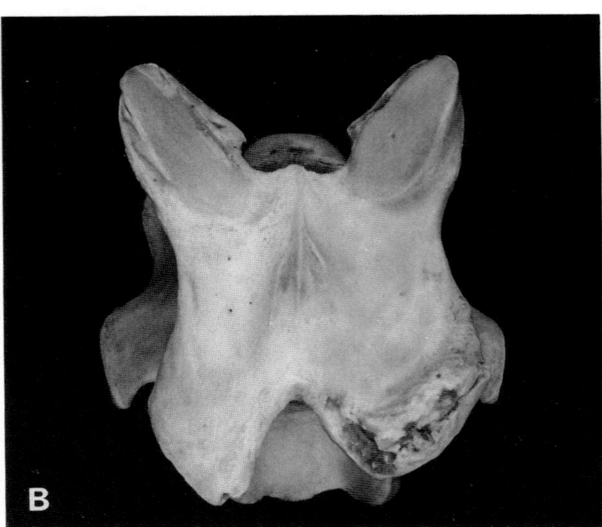

Fig. 1.38B Same vertebra as in (A). Dorsal view. (Courtesy of J. D. Baird.)

trabeculae being formed by compaction of the fibrocartilage of the ligamenta flava (see compaction of endochondral bone). Degenerative arthropathy of articular facets is usually present and, although its severity is not correlated with severity of signs, it may be the result of a primary vertebral instability that also subjects the ligaments and laminae to abnormal mechanical forces. The osteosclerosis and development of fibrocartilage are consistent with increased forces and joint movement.

In the final analysis, interpretation of osseous lesions depends on their correlation with spinal cord degeneration. There is variation in the suddenness and the severity of cord injury, and in the immediate pathogenesis of injury, which may occur directly by pressure on the nervous tissues, or indirectly if local blood vessels are compromised. Thus there may be one or more primary foci of softening in the cervical cord, with Wallerian degeneration of the ascending and descending tracts. Careful examination of the cord, with the dura mater reflected, sometimes reveals a slight depression in the contour of the nervous tissue at the affected site. Sometimes the primary foci of malacia are visible after fixation as brown areas in the gray matter, which usually overlies an articulation (Fig. 1.37B). The fixed cord should be sliced at intervals of no more than 2 mm to find them. Histologic examination is necessary to confirm and demonstrate lesions.

Microscopically there is more-or-less extensive demyelination in the cord. In severe cases with gross lesions, hemorrhagic and necrotic foci can be found in the gray matter in either the ventral or lateral horns and may be unilateral or bilateral. Even when lesions are absent from the gray matter, the primary focus or foci are visible in short segments of the white matter, which shows the moth-eaten appearance typical of demyelination. In the primary foci there are ragged microcavitations, which represent areas of liquefactive necrosis. Many adjacent demyelinated fibers are swollen and basophilic. About the primary focus, demyelination may be present in all tracts; cranial to the focus, Wallerian degeneration can be seen in ascending fibers, most of which are in the dorsal funiculi; caudal to the focus, the degenerating fibers are concentrated in the ventral and ventrolateral funiculi, symmetrical on either side of the ventromedian fissure, and extending as far as the lumbar region in some animals.

7. Wobbler Syndrome in Dogs

A condition similar to that described in horses also occurs in several breeds of large dogs and is particularly common in the Great Dane and Doberman pinscher. The syndrome is also known as cervical vertebral instability, cervical spondylopathy, and caudal cervical vertebral malformation–malarticulation. Male dogs appear to be most susceptible, and there is some evidence for an inherited predisposition. Age of onset of signs varies from a few weeks to several years, and, as in horses, signs are related to spinal cord compression or stretching. They may be mild or severe and may develop slowly or suddenly. Neck pain is sometimes manifest in dogs. In general, Great Danes show signs at an earlier age than Dobermans.

Any one of several anatomical or functional abnormalities of cervical vertebrae may account for spinal cord injury. Often more than one type of defect occurs in the same dog. Probably the most common defect in dogs, especially young dogs, is the funnel-shaped vertebral canal with anterior stenosis, but in general the lesions are similar to those described in horses, with a few differences. Cervical vertebral subluxations appear to be a relatively more common cause of cord injury in dogs than in horses, and posterior cervical vertebrae are quite often affected. Lesions of vertebral disks leading to dorsal protrusion of the annulus fibrosus are sometimes responsible for development of signs. In contrast to the usual forms of disk disease, these lesions are secondary to vertebral instability, do not involve primary disk degeneration, and usually are not associated with protrusion of nuclear material. In dogs, lesions similar to those seen in cervical static stenosis do occur, but cord injury is produced when the neck is extended, and it is due to elongation and thickening of the cranial tip of the vertebral arch. In extension, this projects into the vertebral canal, compressing the cord against the caudodorsal rim of the adjacent vertebra. Deformities of vertebral bodies are more common in dogs than horses, and are associated with early closure of the dorsal part of the caudal physis with resultant elongation of the ventral part of the body (Fig. 1.39). The misshapen vertebrae that result are prone to instability and secondary degenerative changes in the articular facets. They are also susceptible to deformation by the mechanical forces acting on them.

In dogs, localization of spinal abnormalities using contrast myelography is a valuable diagnostic aid, but, as in horses, correlation of skeletal and nervous tissue lesions is essential.

The causes of the lesions in dogs are probably similar to

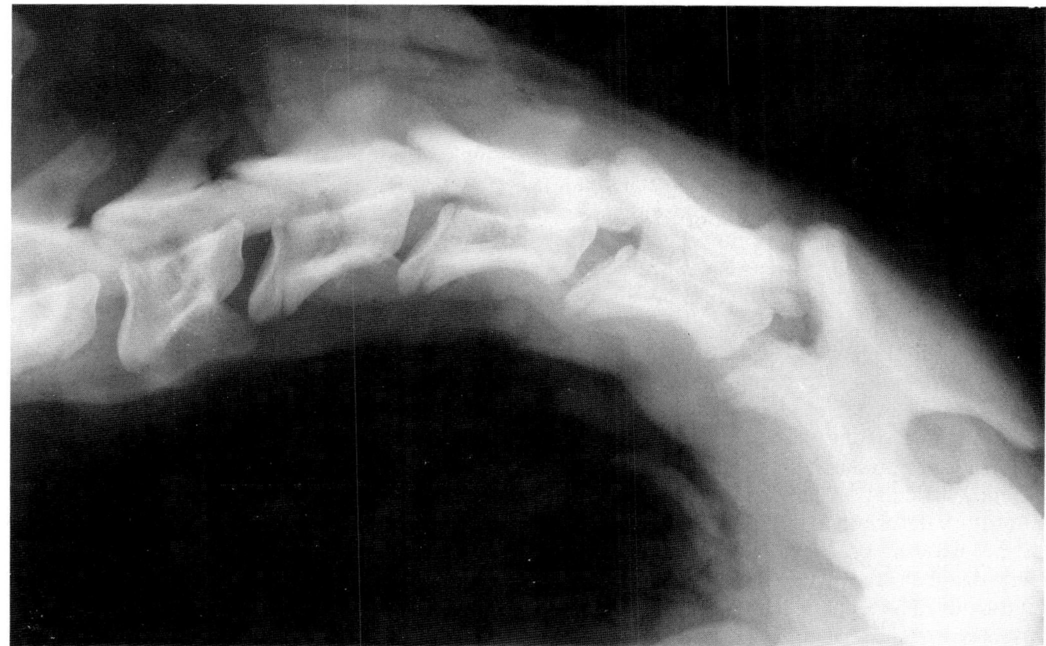

Fig. 1.39 Wobbler syndrome. Radiograph showing wedge-shaped joint spaces between vertebral bodies, and dorsal projection of anterior of bodies into vertebral canal. Doberman pinscher.

those proposed for horses. Production of vertebral deformities, including dorsoventral flattening of the vertebral canal and osteochondritis dissecans of articular processes, is possible by *ad libitum* feeding of rations high in protein, energy, and minerals to dogs of susceptible breeds. Thus a multifactorial etiology, in which environmental factors are combined with innate susceptibility, seems possible in this condition.

Stenosis of the vertebral canal, not associated with disk degeneration or other disease, occasionally involves the **lumbosacral** region of dogs. The lesions occur most often at L-6 to L-7 and L-7 to S-1. Stenosis is due to thickening of the articular facets and dorsal lamina of the vertebra and by hypertrophy of the ligamenta flava, which extend between the arches of adjacent vertebrae. Signs are related to compression of nerve roots and the cauda equina.

Stenosis of cervical vertebrae occurs in **sheep.** Similar in many respects to the condition in horses and dogs, the condition is associated with heavy feeding and probable trauma.

Bibliography

Bailey, C. S. An embryological approach to the clinical significance of congenital vertebral and spinal cord abnormalities. *J Am Anim Hosp Assoc* **11:** 426–434, 1975.

Fisher, L. F., Bowman, K. F., and MacHarg, M. A. Spinal ataxia in a horse caused by a synovial cyst. *Vet Pathol* **18:** 407–410, 1981.

Lewis, D. G. Cervical spondylomyelopathy ("wobbler" syndrome) in the dog: A study based on 224 cases. *J Small Anim Pract* **30:** 657–665, 1989.

Mayhew, I. G. *et al.* Spinal cord disease in the horse. *Cornell Vet* **68:** (Suppl.) 6, 1978.

Olsson, S.-E., Stavenborn, M., and Hoppe, F. Dynamic compression of the cervical spinal cord. A myelographic and pathologic investigation in Great Dane dogs. *Acta Vet Scand* **23:** 65–78, 1982.

Palmer, A. C., Kelly, W. R., and Ryde, P. S. Stenosis of the cervical vertebral canal in a yearling ram. *Vet Rec* **109:** 53–55, 1981.

Powers, B. E. *et al.* Pathology of the vertebral column of horses with cervical static stenosis. *Vet Pathol* **23:** 392–399, 1986.

Stewart, R. H., Reed, S. M., and Weisbrode, S. E. Frequency and severity of osteochondrosis in horses with cervical stenotic myelopathy. *Am J Vet Res* **52:** 873–879, 1991.

Tarvin, G., and Prata, R. G. Lumbosacral stenosis in dogs. *J Am Vet Med Assoc* **177:** 154–159, 1977.

C. Generalized Skeletal Abnormalities of Unclassified Type

In the diseases to be outlined, many malformations occur including various anomalies of the limbs, arthrogryposis with contractures, torticollis or scoliosis, and varying degrees of deformation of the skull. Most of the diseases are caused either by maternal ingestion of plant teratogens or by known or suspected deficiencies of essential nutrients. There are variations in the spectrum of lesions described within and between species for the same insult. These may be related to variations in period of exposure to the noxious influence, individual susceptibility, and rigor of pathologic investigation. The malformations produced may reflect primary aberrations in skeletal anlagen, but also may be caused by lesions which are primarily neural or muscular with secondary changes in the skeleton (see also Chapter 3, The Nervous System and Chapter 2, Muscles and Tendons).

1. Diseases of Foals

Contracted foals have congenital axial and appendicular contractures, but the nervous system, muscles, tendons, and ligaments are normal. The defects are present in fetuses aborted as early as the third month of gestation, indicating that the defect begins at least as early as the cartilage model. The spectrum of abnormalities consists of torticollis, scoliosis, asymmetry of the skull, and varying degrees of flexion from carpus and tarsus distally. The malformations occur in a variety of combinations, and, in cases of severe scoliosis, some are associated with failure or near failure of closure of the ventral abdominal wall. The axial deformities are due to asymmetrical hypoplasia of the intervertebral articulations; the scoliosis, accompanied by compensatory deformity of the thoracic wall. Malformations of the distal limbs are related to hypoplasia, principally of the distal or proximal epiphyses of the metacarpals or metatarsals. The epiphyseal hypoplasia may be evident as hypoplasia of the whole or part of the articular surface. The joint contractures are probably caused by muscular action attempting to stabilize joints that are grossly incongruent and unstable. Limb deformities (and abortions) in foals have been associated with maternal ingestion of *Astragalus mollisimus* (wooly loco) during gestation. Most of the deformities were of the flexion–extension type involving forelimbs or hind limbs.

Fetal ankylosis and abortion are recorded in foals whose dams graze hybrid sorghum and Sudan grass (*Sorghum sudanense*) during days 20–50 of gestation, and a similar association has been made in cattle. The lesions in foals involve many joints but are not well described. It is not clear whether the foals have ankylosis or arthrogryposis.

2. Diseases of Calves

Crooked calves result when pregnant cows ingest certain wild lupins, including *Lupinus caudatus, L. sericeus,* and *L. formosus,* especially between days 40 and 70 of gestation. Malformation of the limbs, notably the forelimbs, is the most common alteration, followed by involvement of the axial skeleton. The limb abnormalities consist of flexion contracture and arthrogryposis associated with disordered growth of joints, shortening, and variable rotation of the bones. Torticollis, and either scoliosis or kyphosis, are common, involvement of the thoracic spine being associated with costal deformities. There are various subtle and often asymmetric changes in the skull, where the most definable malformations are cleft palate, or brachygnathia superior. Depending on the nature and severity of the malformations, the calves, which are usually born alive, may survive. Growth is retarded and the malformations persist, often becoming more severe with age. Neuromuscular structure and function are regarded as normal, or at least not primarily responsible for the skeletal deformity.

Lupins contain several alkaloids, and some are teratogenic. The quinolizidine alkaloid anagyrine is present in *L. caudatus* and *L. sericeus* and is almost certainly a teratogen. In *L. formosus* the piperidine alkaloid ammodendrine may be responsible. Lack of fetal movement caused by a sedative or anesthetic effect of the alkaloids may be the basis for the skeletal deformities, cleft palates being associated with lack of tongue mobility. Alternatively, secondary effects of direct toxicity to the dam may be involved; sustained uterine contractions producing deformities in the fetus have been suggested.

Lupins cause two other distinct diseases in animals: a nervous disorder associated with the alkaloids in bitter lupins (see Chapter 3, The Nervous System) and a mycotoxicosis caused by *Phomopsis leptostromiformis* (see Volume 2, Chapter 2, Liver and Biliary System).

A **deficiency of manganese** has been incriminated in the production of abnormalities similar to those in crooked calves. In areas grazed by affected herds, the levels of manganese in soils, plants, and water are less than in areas where herds are unaffected. Pregnant cows fed 183 mg of manganese per day produced normal calves, and those fed 115–123 mg manganese per day produced calves with deformities of the limbs, including reduced length and breaking strength of humeri, enlarged joints, and twisted legs. Similar abnormalities occur in lambs and kids. There is no detailed microscopic study of the bone lesions in farm animals, and this hinders the assessment of suspected, naturally occurring diseases. In New Zealand, a naturally occurring chondrodystrophy in Charolais calves was thought to be caused by a maternal manganese deficiency precipitated by a diet of apple pulp and corn silage. The calves had short limbs with enlarged joints, and collapsed tracheas with thick cartilage rings. Microscopic lesions included irregular alignment and degeneration of physeal chondrocytes, absence of hypertrophic cells, and degeneration of matrix with unmasking of collagen fibers. Tracheal cartilage was also affected, and lesions were described in articular cartilages. However, since the calves improved clinically with age, and lesions of arthropathy were not seen, the "articular cartilage" lesions may have been in the growth cartilages of the epiphyses.

Generalized skeletal deformities involving most of the newborn calves in a herd of Aberdeen Angus cows occurred in Saskatchewan, Canada. Focal, premature synostosis of growth plates in vertebrae, basocranium, and long bones developed. Stenosis of affected vertebrae caused myelomalacia. Hypervitaminosis A, which can cause focal closure of growth plates, was thought to be the cause; however, in further occurrences of a similar disease in the same province, this diagnosis was rejected, and maternal manganese deficiency was suspected. Focal closure of growth plates is found in the offspring of manganese-deficient rats.

Manganese is an essential micronutrient, and syndromes associated with excess and deficiency are well recognized. In excess it is a neurotoxin, and experimentally causes rickets in rats, probably by precipitating insoluble manganese phosphate in the gut. Deficiency has been produced experimentally in various mammalian species, and the manifestations include impaired growth, skeletal anomalies of the type referred to, depressed reproductive function, and congenital ataxia in laboratory rodents and mink. The ataxic syndrome is related to reduction or absence of otoliths. A recessive pigment mutation, pallid *pa*, has a pleiotropic effect on otolith development in pallid mice. High concentrations of manganese at critical stages of gestation prevent otolith defects but do not alter pigmentation. It is suggested that melanocytes, normally present near the otoliths, are reservoirs for manganese. These cells are absent in pallid mice.

Manganese is necessary for the activation of xylosyltransferase, the first enzyme in the biosynthetic pathway of sulfated glycosaminoglycans and one that is unique to that process. A deficiency of manganese causes decreased production and increased degradation of cartilage glycosaminoglycans. Manganese is also required for the glycosylation of collagen, presumably the ultimate intracellular step in collagen synthesis, and it seems likely that its involvement in these processes is the basis of the lesions described above.

Veratrum californicum (false hellebore, wild corn, or skunk cabbage) is well recognized in the United States of America as a cause of cyclopian and other deformities in the progeny of ewes consuming the plant on day 14 of gestation. Similar anomalies occur in calves and can be produced in kids (see Chapter 3, The Nervous System and Chapter 4, Eye and Ear).

Ingestion of *V. californicum* by pregnant cows between days 12 and 30 of gestation may result in cleft palate, harelip, brachygnathia, hypermobility of hock joints, syndactylia, decreased length and diameter of all bones or shortening of metacarpals and metatarsals only, development of cornified skin dorsal to the coronary band, and reduced number of coccygeal vertebrae (brachyury). Between days 30 and 36 of gestation, the teratogens selectively inhibit growth in length of the metacarpal and metatarsal bones. Whereas the natural incidence of cyclopian deformities may be as high as 15%, that of the other anomalies is seldom more than 2%. *Veratrum californicum* contains the alkaloids cyclopamine, its glycoside cycloposine, and jervine. Cyclopamine is most abundant in the plant and produces the craniofacial defects, but the cause of the other anomalies is unknown.

Acorn calves born in the western United States were initially attributed to ingestion of acorns by pregnant cows. Later, nutritional deficiency in the dam, probably between months 3 and 6 of gestation, was incriminated. Up to 10% of calves may be affected in herds grazing native pastures of foothills. Similar conditions occur in Australia, where the incidence tends to be lower, and in Canada where over 40% of calves may be affected, and it is called congenital joint laxity and dwarfism. At birth generalized joint laxity, disproportionate dwarfism, and sometimes superior brachygnathia are present. Severely affected calves walk on the palmar and plantar surfaces of their phalanges, and many have varus and valgus deformities when standing. Diaphyseal curvature of long bones, spinal deformities, and cleft palate are not seen. Calves survive but do not prosper.

In Canada the disease is produced by feeding grass/legume silage during gestation and prevented by adding hay and grain to the silage ration. Timothy grass, fed as silage, produces the disease, while timothy from the same fields fed as hay does not. It is suspected that either deficiencies caused by leaching or fermentation of silage, or toxicosis produced by mycotic or bacterial activity, are responsible.

Other, isolated, examples of generalized abnormalities do occur sporadically, as for example, the forelimb abnormalities that affected up to 29% of calves in one herd in Australia. There was an axial malformation in one calf and congenital blindness in one, but otherwise the lesions were restricted to the forelimbs. The scapulae were present and usually wholly cartilaginous, the more distal portions of the limbs being absent or vestigial. A disproportionate dwarfism in two Dairy Shorthorn siblings also occurred in Australia and was characterized by marked shortening of the pelvic limbs, mild shortening of pectoral limbs, and a normal axial skeleton. Retardation of growth in the femurs, tibias, and metatarsals was prominent, and there was deformity of the femorotibial joints. The calves were affected at birth, which distinguishes the condition from **hyena disease,** a phenotypically similar affliction of cattle in western Europe and South America. These animals are affected at 3 to 6 months of age and have most severe lesions in the bones of the pelvic limbs and lumbosacral vertebrae, but bones of the pectoral limbs are also affected. Shortening and deformities of bones are associated with maturation failure in the physes. Toxicity due to excess vitamin D, and autoimmune disease associated with the bovine virus diarrhea agent, have been suggested as causes of hyena disease.

3. Diseases of Lambs

Pregnant ewes are at risk when consuming **locoweed,** *Astragalus* spp., and *Oxytropis* spp. Three syndromes related to selenium accumulation, acute intoxication, and locoweed poisoning are associated with ingestion of *Astragalus* plants. We are concerned here with certain features of locoweed poisoning, a disease characterized by neurological damage (see Chapter 3, The Nervous System), emaciation, habituation, abortions, and birth defects. The anomalies include brachygnathia, contractures or overextension of joints, limb rotations, osteoporosis, and bone fragility. The precise cause is not known. An early comparison with lathyrism was not confirmed, although there are similarities between the two conditions. There are lesions in nervous tissue of some lambs, but their relationship to the limb contractures is unclear.

Overdosing of ewes with the anthelmintic **parbendazole** between days 12 and 24 of gestation may also produce skeletal abnormalities in a high proportion of lambs. Neurologic lesions may also be present (see Chapter 3, The

Nervous System). The skeletal malformations include compression and/or fusion of vertebral bodies, fusion of proximal ribs, curvature of long bones, hypoplasia of articular surfaces, and absence of various bones such as ulna, humerus, and metacarpals. Lesions are reliably reproduced by a double dose of the drug and may even occur with less than twice the recommended dose.

Veratrum californicum causes numerous fetal anomalies in lambs. In addition to those described with diseases of calves, tracheal and laryngeal stenosis occurs in the offspring of ewes ingesting the plant on or about day 31 of gestation. Ingestion near day 29 produces marked shortening of the metacarpals, metatarsals, and tibiae, and sometimes the radii. There is great variation in degree of susceptibility, but lesions tend to be bilaterally symmetrical, and twins tend to be affected equally. Fusion of metacarpals may result in postnatal bowing of the legs, and fusion or deformities of joints may cause arthogryposis. Death of about 17% of embryos occurs with maternal ingestion of *V. californicum* on days 19 to 21 of gestation. Loss of 75% of embryos is recorded when the plant is ingested on day 14.

Syndromes described as **bent-leg** or bowie occur in various countries, and are attributed to various causes. In Australia ingestion of plants of the genus *Trachymene* (wild parsnip) by pregnant ewes is thought to be the cause of bent-leg in neonatal lambs. Gross deformities are usually most prominent in the forelimbs, and include flexion and lateral rotation of the knee joints and medial or lateral rotation of the fetlocks. In New Zealand, bowie is first seen after 3 to 4 weeks of age, and lateral displacement of the carpal joints (bowleg) is the prominent gross sign. Supplementation of the ewes' diet does not prevent the disease. In South Africa, a genetic susceptibility to a bent-leg syndrome is suspected.

4. Diseases of Pigs

Conium maculatum (poison hemlock, European hemlock, spotted hemlock) is toxic to various species and causes a syndrome characterized by excitation and subsequent depression. Cattle are most susceptible, and there is decreasing susceptibility through sheep, horses, and pigs. *Conium maculatum* is also teratogenic, especially for cattle, with pigs being less susceptible and sheep, relatively resistant. The plant causes arthrogryposis and spinal deformities in the offspring of sows and cows when fed between days 43 and 53 and 55 and 75 of gestation, respectively. Cleft palates occur in pigs exposed to the plant between days 30 and 45. Congenital limb abnormalities occur in lambs from exposed ewes, but they are undetectable 8 weeks after birth.

There are five piperidine alkaloids in *Conium maculatum,* the major ones being coniine and γ-coniceine. Both are thought to be toxic, and the latter, teratogenic. The composition and concentration of alkaloids can vary from day to day and from place to place.

Nicotiana tabacum (burley tobacco) causes arthrogryposis and sometimes brachygnathia and kyphosis in pigs exposed to it *in utero. Nicotiana glauca* (wild tree tobacco) causes similar lesions and also produces arthrogryposis and spinal deformities in calves exposed between days 45 and 75 of gestation. The teratogen is an alpha-substituted piperidine alkaloid, anabasine, which is present also in *N. tabacum.* It produces the lesions when given between days 43 and 53 of gestation, and also causes cleft palate when administered between days 30 and 37.

Jimsonweed (*Datura stramonium*) and wild black cherry (*Prunus serotina*) are suspected teratogens in sows, both being linked to arthrogryposis.

5. Diseases of Goats

Congenital skeletal malformations, including cleft palate, occur in kids whose dams are fed piperidine alkaloid-containing plants during gestation. *Conium maculatum* (poison hemlock), *Nicotiana glauca* (tree tobacco), and *Lupinus formosus* (lunara lupine) all produce congenital lesions similar to those of crooked-calf disease when does are exposed during days 30 to 60 of pregnancy. A quinolizidine-containing plant, *Lupinus caudatus* (tailcup lupine), that is also teratogenic for cattle, does not produce lesions in fetal kids, and is only mildly toxic for does. Possibly ruminal metabolism of the quinolizidine alkaloid anagyrine to a piperidine alkaloid does not occur in goats as is suggested may occur in cattle.

Bibliography

Abbot, L. C. *et al.* Crooked calf disease: A histological and histochemical examination of eight affected calves. *Vet Pathol* **23:** 734–740, 1986.

Barry, M. R., and Murphy, W. J. B. Acorn calves in the Albury district of New South Wales. *Aust Vet J* **40:** 195–201, 1964.

Carrig, C. B., Grandage, J., and Seawright, A. A. Disproportionate dwarfism in two bovine siblings. *Vet Radiol* **22:** 78–82, 1981.

Crowe, M. W., and Pike, H. T. Congenital arthrogryposis associated with ingestion of tobacco stalks by pregnant sows. *J Am Vet Med Assoc* **162:** 453–455, 1973.

Crowe, M. W., and Swerczek, T. W. Congenital arthrogryposis in offspring of sows fed tobacco (*Nicotiana tabacum*). *Am J Vet Res* **35:** 1071–1073, 1974.

Cunningham, I. J. Studies in the control of bowie in lambs. *N Z Vet J* **5:** 103–108, 1957.

Doige, C. E. *et al.* Congenital spinal stenosis in beef calves in western Canada. *Vet Pathol* **27:** 16–25, 1990.

Dyer, I. A., Cassatt, W. A., and Rao, R. R. Manganese deficiency in the etiology of deformed calves. *Bioscience* **14:** 31–32, 1964.

Dyson, D. A., and Wrathall, A. E. Congenital deformities in pigs possibly associated with exposure to hemlock (*Conium maculatum*). *Vet Rec* **100:** 241–242, 1977.

Espinasse, J. *et al.* Hyena disease in cattle: A review. *Vet Rec* **118:** 328–330, 1986

Harbutt, P. R., Woolcock, J. B., and Bishop, J. N. Congenital forelimb abnormalities in calves. *Aust Vet J* **41:** 173–177, 1965.

Hurley, L. S. Teratogenic aspects of manganese, zinc, and copper nutrition. *Physiol Rev* **61:** 249–295, 1981.

James, L. F. Syndromes of locoweed poisoning in livestock. *Clin Toxicol* **5:** 567–573, 1972.

Keeler, R. F. Toxins and teratogens of higher plants. *Lloydia* **38:** 56–86, 1975.

Keeler, R. F. Early embryonic death in lambs induced by *Veratrum californicum*. *Cornell Vet* **80:** 203–207, 1990.

Keeler, R. F., and Stuart, L. D. The nature of congenital limb defects induced in lambs by maternal ingestion of *Veratrum californicum*. *Clin Toxicol* **25:** 273–286, 1987.

Keeler, R. F., Young, S., and Smart, R. Congenital tracheal stenosis in lambs induced by maternal ingestion of *Veratrum californicum*. *Teratology* **31:** 83–88, 1985.

Leipold, H. W., Oehme, F. W., and Cook, J. E. Congenital arthrogryposis associated with ingestion of jimsonweed by pregnant sows. *J Am Vet Med Assoc* **162:** 1059–1060, 1973.

McIlwraith, C. W., and James, L. F. Limb deformities in foals associated with ingestion of locoweed by mares. *J Am Vet Med Assoc* **181:** 255–258, 1982.

Morgan, S. E., Brewer, B., and Walker, J. Sorghum cystitis ataxia syndrome in horses. *Vet Hum Toxicol* **32:** 582, 1990. (References to teratogenicity)

Niekerk, F. E., van *et al.* The bent-leg syndrome in sheep. I. The effect of pregnancy and age of the ewe on concentrations of plasma minerals. *J S Afr Vet Assoc* **61:** 119–123, 1990.

Panter, K. E. *et al.* Congenital skeletal malformations and cleft palate induced in goats by ingestion of *Lupinus, Conium,* and *Nicotiana* species. *Toxicon* **12:** 1377–1385, 1990. (Review with references to other species)

Philbey, A. W. *Trachymene glaucifolia* associated with bentleg in lambs. *Aust Vet J* **67:** 468, 1990.

Prozesky, L., Joubert, J. P. J., and Ekron, M. D. Paralysis in lambs caused by overdosing with parbendazole. *Onderstepoort J Vet Res* **48:** 159–167, 1981.

Rodriguez, J. I. *et al.* Effects of immobilization on fetal bone develop- ment. A morphometric study in newborns with congenital neuromuscular diseases with intrauterine onset. *Calcif Tissue Int* **43:** 335–339, 1988.

Rooney, J. R. Contracted foals. *Cornell Vet* **56:** 172–187, 1966.

Selby, L. A. *et al.* Outbreak of swine malformations associated with the wild black cherry, *Prunus serotina*. *Arch Environ Health* **22:** 496–501, 1971.

Shupe, J. L. *et al.* Lupine, a cause of crooked calf disease. *J Am Vet Med Assoc* **151:** 198–203, 1967.

Shupe, J. L., James, L. F., and Binns, W. Observations on crooked calf disease. *J Am Vet Med Assoc* **151:** 191–197, 1967.

Svensson, O. *et al.* Manganese rickets. A biochemical and stereologic study with special reference to the effect of phosphate. *Clin Orthop* **218:** 302–311, 1987.

Valero, G. *et al.* Chondrodystrophy in calves associated with manganese deficiency. *N Z Vet J* **38:** 161–167, 1990. (Reference list to Mn-associated chondrodystrophies)

VI. Metabolic Diseases of Bones

The physical rigidity of osseous tissues has long carried with it the implication of immutability. That these tissues are biologically plastic is now accepted and is the basis on which adaptional deformities are analyzed. Normal osseous tissues possess a high degree of metabolic activity directed principally toward maintenance of ionic equilibria, repair of damaged structures, and reaction to external stimuli. The changes that occur in diseased bone are largely a reflection of alterations in the processes by which bone is formed, so an understanding of the embryology and physiology of bone is a prerequisite for understanding the diseases.

In spite of numerous experiments and masses of data, there is still considerable confusion concerning the consequences for the skeleton of a lack or imbalance of specific nutrients, especially with relation to calcium and phosphorus in the diet. The difficulties are inherent in the interactions between calcium, phosphorus, vitamin D, and parathyroid hormone, and in the importance, not only of the absolute amounts of the various nutrients that are available, but in the balance between them. They are compounded by the frequent failure of researchers to specify precisely the morphologic changes in experimental osteodystrophies, and by the tendency to extrapolate *in vitro* results to living animals and to ignore differences between genera and even between classes.

The manifestations of nutritional osteodystrophy are osteoporosis, rickets, osteomalacia, and osteodystrophia fibrosa. These are morphologic diagnoses with relatively clear-cut distinguishing features but, since they are the result of disordered remodeling by tissues with a limited repertoire, they are not specific to a single nutritional osteodystrophy or even to nutritional disorders as a group. Osteoporosis, for example, is seen in the inherited disease osteogenesis imperfecta of calves, and changes virtually indistinguishable from rickets, at least by conventional methods, occur in the metaphyses in the chondrodysplasia of the Alaskan malamute. Uncomplicated osteodystrophia fibrosa is relatively characteristic in appearance, but may be due to chronic renal disease or parathyroid tumor, and may be a part of the disease caused by vitamin D deficiency. Although the morphology of a particular osteodystrophy is often regarded as a definitive entity whose only destiny is to get either better or worse, it might better be regarded as an anatomical milestone in the evolution of a disease; what begins as one disease manifestation may progress to another. This suggestion implies that changes in bone structure are influenced not only by the type of stimulus but also by its duration and severity and by the reactive capacity of the animal's tissues. Such a view may help explain why, for example, a deficiency of calcium sometimes produces osteoporosis and sometimes, osteodystrophia fibrosa or why phosphorus deficiency sometimes produces osteoporosis and, in other circumstances, osteomalacia. It is also important to realize that in metabolic processes, the proportions of nutrients are often much more important than their concentrations, and that optimal levels of a particular nutrient are optimal only when the levels of all other nutrients are also optimal. In the following descriptions, the effects of "uncomplicated" deficiencies and excesses are given, as far as this is possible. Under practical circumstances, most osteodystrophies result from more than one dietary imbalance, and this makes the etiologic diagnosis of such conditions extremely difficult. Even an educated guess frequently requires examination of several animals, with attention to their history and clinical chemistry, as well as a fairly comprehensive examination of skeletal tissues.

A. Hormonal Influences

The processes which determine the shape and internal structure of bones are subject to at least three types of regulation. Genetic determinants establish the limits of their size and shape; gravitational and mechanical forces influence the modeling and remodeling that determine their structural variation; and nutritional factors limit or permit the achievement of their ultimate potential. Superimposed on these varied influences are the effects of hormones. Although they are usually considered separately, they almost invariably interact, and hormonal influences are involved to some extent in most, if not all, metabolic bone diseases. For example, there is an interaction between androgen and mechanical stimuli in disuse osteoporosis in turkeys. In toms, disuse produces bone loss that is mainly endosteal, and the cortex is thinned, but in castrated males, bone loss following disuse is predominantly intracortical, and cortical thickness remains essentially normal. It is not known whether this pattern of loss occurs under similar circumstances in other species, but there are few hormones that do not influence some aspect of bone physiology. In some instances, excess or deficiency of a single hormone is the major influence in an osteodystrophy, and examples are noted.

1. Pituitary

Growth hormone is secreted by the somatotrophs of the anterior pituitary gland under the control of growth hormone-releasing hormone, which stimulates its synthesis and release, and somatostatin, which suppresses its secretion. In addition, estrogens, thyroid hormone, and insulinlike growth factor I (IGF-I, somatomedin C) help control secretion, the latter by a negative-feedback system. Growth hormone regulates the size of the skeleton and the soft tissues, both directly and by a growth-promoting effect mediated by IGF-I. In this regulation, growth hormone is thought to stimulate differentiation of prechondrocyte stem cells, while IGF-I stimulates proliferation of committed chondrocytes. Although IGF-I is produced in various tissues, including cartilage and pituitary, most of the activity in the circulation is from the liver. As well as its hormonal effects, IGF-I has autocrine or paracrine functions. Growth hormone is secreted in discrete episodes without periodicity throughout the day. Stimuli for its release may include exercise, stress, the postprandial fall in glucose, and other unknown factors.

There are two rare syndromes caused by abnormalities of growth hormone secretion that affect the development of the skeleton: dwarfism, produced by deficiency of hormone, and gigantism, by excess.

Pituitary dwarfism occurs rarely in domestic animals. It must be distinguished from constitutional dwarfism due to generalized genetic effects; in both conditions, growth, although greatly reduced, is proportional in osseous and soft tissues. Hypopituitary dwarfs have delayed bone development, and epiphyseal fusion is retarded. Two types of dwarfs occur in the German shepherd breed. In one type, the adenohypophysis and other endocrine organs are grossly and histologically normal, and in such dogs dwarfism may be due to end-organ resistance to hormonal action. In the other type, which is probably inherited as an autosomal recessive trait, the adenohypophysis is abnormal, there is a deficiency of growth hormone, and IGF-I levels are low. This type also occurs in the Carelian bear dog and some other breeds. Growth plates may remain open for up to 4 years, and there is disorganization of proliferative chondrocytes and degeneration of intercellular matrix in growth cartilage (see also Volume 3, Chapter 3, The Endocrine Glands).

Pituitary gigantism (acromegaly) also must be distinguished from constitutional gigantism seen in certain breeds of dogs. In humans, a prominent feature of acromegaly is overgrowth of the cranial bones, particularly those of the supraorbital region, but bony and soft tissue overgrowth of hands, feet, nose, and chin are also part of the syndrome. Thickening of the cranial bones has been reported in a dog with an acidophil adenoma of the pituitary gland. There is a similar thickening of the supraorbital portions of the frontal bones in St. Bernards and like breeds of dogs, which may be evidence of constitutional acromegaly. Normally blood levels of IGF-I are higher in large-breed dogs than in small, but growth hormone levels do not differ.

A syndrome described as acromegaly occurs in elderly bitches treated either with progestagens or spontaneously during the luteal phase of the estrus cycle. Diffuse increase in soft tissue in the throat causes inspiratory stridor, and the dogs have diabetes mellitus. Skeletal changes are not evident radiographically, but critical examinations have not been done. Enlargement of the interdental spaces may be present. Both growth hormone and IGF-I levels are increased. In cats, progestagens do not appear to stimulate growth hormone secretion, but an acromegalic syndrome occurs in association with growth hormone-secreting tumors of the pituitary in cats after middle age. Various clinical abnormalities occur in affected cats including insulin-resistant diabetes, consistently, and thickening of the nasal bones, the bones of the calvarium, and the rostral mandible in about half of them. Spondylosis and arthropathy also occur in about 40% of cats.

2. Thyroid

Thyroid hormones have a profound effect on the maturation of the skeleton through their effect on maturation of growth cartilage. A variety of congenital skeletal diseases, including bulldog calves and lambs, have associated thyroid abnormalities, but there are also syndromes in which the endocrine abnormality appears to be primary. Uncomplicated hypothyroidism in immature animals causes retardation of growth and development of endochondral bones. There is a marked decrease in rate of growth, and failure of proportional changes that normally take place as a result of maturation. The severity of the resultant infantile skeletal proportions depends on the age at which the thyroid deficiency began and the degree of deficiency. In addition to stunted growth, there is delay in the time of ossification of the epiphyseal centers (Fig. 1.40). Some animals may show failure of development, or dysgenesis, of secondary ossification centers and of centers in the carpal and tarsal bones, similar to the multifocal centers described in spider lambs.

Hypothyroidism in neonatal foals is associated with a variety of skeletal defects including angular limb deformities affecting the forelimbs or hind limbs, hypoplasia of some cuboid bones, and mandibular prognathism. The fourth and ulnar carpal bones are hypoplastic and/or incompletely ossified, and the third carpal may be similarly affected, or it may show hypoplasia of the lateral portion only. Osteochondritis dissecans of the third and fourth carpals, hypoplasia and/or collapse of the central and third tarsal bones, and rupture of the common digital extensor tendon may also occur. Foals with these lesions usually have hyperplasia and hypertrophy of thyroid epithelial cells although the gland is not grossly enlarged. Serum T_3 and/or T_4 levels are often reduced, and response to thyroid-stimulating hormone (TSH) may be poor. The basis of the hypothyroidism is not known—failure to respond to TSH may be due to a lack of receptors on the follicular epithelium or by primary hypoplasia or dysplasia of the gland.

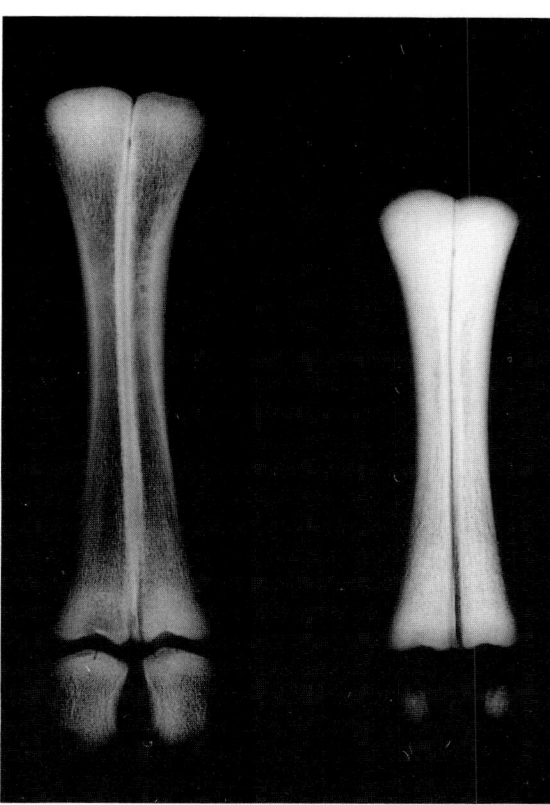

Fig. 1.40 Hypothyroid lamb (right). Metacarpal. Delayed development of ossification center and of longitudinal growth. Normal lamb (left).

Hypothyroidism with skeletal lesions also occurs in **Scottish deerhound** and **giant schnauzer** dogs. Puppies are small, weak, depressed, and somnolent, and have difficulty in walking when presented at a few weeks of age. There is absence or delayed development of secondary ossification centers in the axial and appendicular skeleton. Serum T_4 levels are reduced and hypercholesterolemia is present. Response to single injections of TSH is poor in both breeds, but three consecutive daily treatments corrected the thyroid hormone level in schnauzers. Follicles distended with colloid showing decreased resorption vacuoles are present in untreated schnauzers. In deerhounds, the follicles are small with little colloid, and many cells in the adenohypophysis have vacuolated cytoplasm consistent with "thyroidectomy cells." Genetic analysis in both breeds suggests an inherited condition, an autosomal recessive trait being suspected in schnauzers.

Hyperthyroidism in immature animals causes acceleration of normal processes of maturation. Severe degrees of hyperthyroidism are associated with osteoporosis, probably secondary to a negative metabolic balance. Hyperthyroidism, induced by feeding iodinated casein to dairy cows and swine in attempts to increase productivity, has resulted in osteoporosis. In humans, hyperthyroidism is associated with excess osteoid seams on bone surfaces. This is not true osteomalacia but a reflection of increased metabolic rate, the osteoid being normally receptive to mineralization. Hypercalcemia may be present, and this may be due to the increased metabolic rate or to primary hyperparathyroidism that often accompanies the thyroid disease. Hyperthyroid cats often have parathyroid hyperplasia, but neither concomitant hypercalcemia nor osteoporosis is recorded.

3. Gonads

Because of interest in postmenopausal osteoporosis in women, more is known of the skeletal effects of female than of male hormones, and of the female hormones, more of estrogen than of progesterone. Estrogens and androgens both affect the growing skeleton; osteoblasts possess receptors for both hormones; and both accelerate epiphyseal closure, producing appropriate skeletal dimorphism. The latter effects are related largely to the response of cartilage, and may be modulated via growth hormone and IGF-I, both of which are affected by estrogen. Hypogonadism, whether produced surgically or as a genetic or embryologic defect, produces delayed fusion of the epiphyses with disproportionate growth of tubular bones and failure of sexual differentiation of the pelvis. There is considerable variability between species in the expression of this change.

Estrogen has a direct effect on osteoblasts. There are nuclear receptors for the hormone which may be important in regulating the synthesis and maintenance of bone matrix by regulating the messenger RNA (mRNA) for type I procollagen and for TGF-β. A slight impairment of collagen synthesis due to estrogen deficiency would have significance for the mature skeleton in which hypoestrogenism is associated with reduced bone mass. Estrogen also influences osteoclasts, causing their disengagement from bone surfaces and ultimately their disintegration. Estrogen may be the principal regulator of bone mass in the mature skeleton and, according to this proposition, a major arbiter of postmenopausal osteoporosis in women. Estrogen has a sparing effect on calcium, which may account for its beneficial influence in this condition. Similarly, in hypogonadal males, calcium malabsorption and bone hypomineralization associated with low plasma calcitriol, are corrected by testosterone. Although not a clinical problem, changes in bone structure occur in spayed bitches, but these may be transient, and are apparently inconsistent. Mild osteoporosis, associated either with reduced trabecular bone volume in the vertebrae and a slight deficit in bone formation, or with increased numbers of resorption spaces in rib cortices, are among the findings in ovariohysterectomized bitches.

Hypergonadism in the growing skeleton produces premature epiphyseal closure and maturation of the skeleton. In cows, hyperestrogenism associated with functional follicular cysts produces relaxation of pelvic ligaments and causes the characteristic elevation to the tail head and predisposition to pelvic dislocation and fracture. There are no osseous changes in the adult associated with oversecretion of male hormone. Progesterone also has an effect on bone metabolism, but usually is ignored in discussions of hormones and postmenopausal osteoporosis. Progesterone modifies the skeletal effects of glucocorticoids, binding to glucocorticoid receptors on osteoblasts and displacing bound dexamethasone from them, suggesting that they are actually progesterone receptors. Limited data indicate that progesterone promotes bone formation, and in concert with estrogen, may be involved in the promotion of remodeling via effects on resorption–formation coupling. However, the significance of progesterone in the affairs of the skeleton is presently unclear.

4. Adrenal Cortex

Hyperadrenocorticism commonly causes osteoporosis in humans and dogs. Trabecular bone is more severely affected than cortical bone, and the ratio of diaphyseal to metaphyseal mass is increased. The pathogenesis and effects of excess corticosteroids appear to vary with species, with level and duration of exposure, and with the source of the compound, for example, iatrogenic versus endogenous, prednisone versus deflazacort. Glucocorticoids consistently reduce the rate at which bone is formed, and the amount produced, apparently by decreasing the time that osteoblasts spend making bone. Serum osteocalcin level, an indicator of bone formation, is decreased. Glucocorticoids also interfere with differentiation of osteoprogenitor cells to osteoblasts. In humans, there may be an increased bone-resorption rate that is associated with hyperparathyroidism secondary to steroid-induced reduction in calcium absorption from gut. Resistance to the intestinal action of calcitriol may be responsible. Parathyroid enlargement occurs in some dogs with hyperadrenocorticism, but increased bone resorption is not recorded. Dexamethasone, a synthetic analog of hydrocortisone, causes irregular growth plate vascularization in a variety of species, but the clinical significance of this is unclear.

Bibliography

Aaron, J. E. *et al.* Contrasting microanatomy of idiopathic and corticosteroid-induced osteoporosis. *Clin Orthop* **243:** 294–305, 1989.

Adams, P. H. Calcium-regulating hormones: General. *In* "Calcium in Human Biology," B. E. C. Nordin, (ed.), pp. 69–92. London, Springer-Verlag, 1988.

Bain, S. D., and Rubin, C. T. Metabolic modulation of disuse osteopenia: Endocrine-dependent site specificity of bone remodeling. *J Bone Min Res* **5:** 1069–1075, 1990.

Boyce, R. W. *et al.* Sequential histomorphometric changes in cancellous bone from ovariohysterectomized dogs. *J Bone Miner Res* **5:** 947–953, 1990.

Brown, R. A. *et al.* Microvascular invasion of rabbit growth plate cartilage and the influence of dexamethasone. *Bone Miner* **9:** 35–47, 1990.

Dempster, D. W. Bone histomorphometry in glucocorticoid-induced osteoporosis. *J Bone Miner Res* **4:** 137–141, 1989.

Eigenmann, J. E. *et al.* Growth hormone and insulin-like growth factor I in German shepherd dwarf dogs. *Acta Endocrin* **105:** 289–293, 1984.

Eigenmann, J. E. *et al.* Insulinlike growth factor I in the dog: A study in different dog breeds and in dogs with growth hormone elevation. *Acta Endocrin* **105:** 294–301, 1984.

Eriksen, E. F. *et al.* Evidence of estrogen receptors in normal human osteoblastlike cells. *Science* **241:** 84–86, 1988.

Greco, D. S. *et al.* Congenital hypothyroid dwarfism in a family of giant schnauzers. *J Vet Intern Med* **5:** 57–65, 1991.

Huntley, K. *et al.* The radiological features of canine Cushing's syndrome: A review of forty-eight cases. *J Small Anim Pract* **23:** 369–380, 1982.

Liu, C-C., and Howard, G. A. Bone-cell changes in estrogen-induced bone-mass increase in mice: Dissociation of osteoclasts from bone surfaces. *Anat Rec* **229:** 240–250, 1991.

Lund-Larsen, T. R., and Grondalen, J. Ateliotic dwarfism in the German shepherd dog. *Acta Vet Scand* **17:** 293–306, 1976.

McLaughlin, B. G., Doige, C. E., and McLaughlin, P. S. Thyroid hormone levels in foals with congenital musculoskeletal lesions. *Can Vet J* **27:** 264–267, 1986.

Melmed, S. Acromegaly. *N Engl J Med* **322:** 966–977,990.

Morris, H. A. *et al.* Malabsorption of calcium in corticosteroid-induced osteoporosis. *Calcif Tissue Int* **46:** 305–308, 1990.

Norrdin, R. W. *et al.* Trabecular bone morphometry in beagles with hyperadrenocorticism and adrenal adenomas. *Vet Pathol* **25:** 256–264, 1988.

Peterson, M. E. Acromegaly in 14 cats. *J Vet Int Med* **4:** 192–201, 1990.

Prior, J. C. Progesterone as a bone-trophic hormone. *Endocrine Rev* **11:** 386–398, 1990.

Raisz, L. G., and Bingham, P. J. Effect of hormones on bone development. *Annu Rev Pharmacol* **12:** 337–352, 1972.

B. Osteoporosis, Osteopenia, and Atrophy

Each of these names means a deficiency in the amount of bone tissue in the skeleton or part of the skeleton.

Osteoporosis and atrophy are interchangeable, and signify a pathologic loss of bone, with the remaining bone being chemically normal. Osteopenia also indicates that there is a reduction in bone mass, but makes no inference as to its quality. Thus it appropriately describes the radiographic appearance of conditions such as osteomalacia or osteogenesis imperfecta in which both quality and quantity are abnormal. Osteopenia is sometimes used to describe a reversible loss of bone, or physiologic changes such as occur with aging and lactation. Strictly speaking, all osteoporoses should be called osteopenias pending qualitative assessment.

Osteoporosis is a lesion, not a specific disease; osteoporotic describes the state of a bone or skeleton at a given time. It may be diagnosed subjectively by visual appraisal, or objectively by measurement of radiographs, sawn bones, or microscopic sections. Osteoporosis may be further defined according to its causes, but also on the basis of its pathogenetic mechanism, the anatomic location of bone loss, and the limiting factor in its development. The first definition is based on the particular imbalance in bone formation and resorption which causes the lesion to develop. The second defines, for example, the cortical or cancellous, axial or abaxial location of the deficiency, and the third considers the component of bone structure, either matrix or mineral with primary responsibility for the lack of bone.

Osteoporosis results from an imbalance between formation and resorption in favor of resorption. Theoretically this can occur when resorption and formation rates are reduced, increased, or within normal limits, so long as resorption predominates. The osteoporotic condition may become static if formation and resorption regain balance, and, within limits, if the cause is removed, the lesion may be reversed. The pathogenesis of osteoporosis, then, is comprehensible in terms of the rates and extent of formation and resorption. These rates have been calculated in some of the osteoporoses, and although they usually fluctuate over time and in different parts of the skeleton, repeated measurements using uniform techniques often allow a consensus to be reached.

An osteoporotic skeleton shows reduced radiologic density, but this usually represents advanced osteoporosis, since only gross alterations in the amount of mineral present, probably 30% or more of normal mass, are demonstrable in clinical radiographs. However, radiography of isolated bones can demonstrate much smaller changes, especially if standardized measuring techniques are used. Various forms of absorptiometry (single photon, dual photon, dual energy x-ray) and other noninvasive techniques, such as quantitative computed tomography and neutron activation analysis, permit much more precise estimates of bone mass but generally are available only in research settings.

In general, osteoporosis affects mainly those bones or portions of bone with a large component of cancellous tissue, such as the vertebral bodies, flat bones of the skull, scapula, and ilium, and metaphyseal trabeculae of long bones (Fig. 1.41). The cancellous resorption isolates those trabeculae which are most concerned with the transmission of weight-bearing stresses, those that are least affected by the resorption (Fig. 1.42). Usually cortical bone is resorbed most rapidly from the endosteal surface and along vascular channels so that the cortex becomes thinner and porous and the medullary cavity enlarges, but this is not always the case (see Section VI,B,2, Disuse osteoporosis).

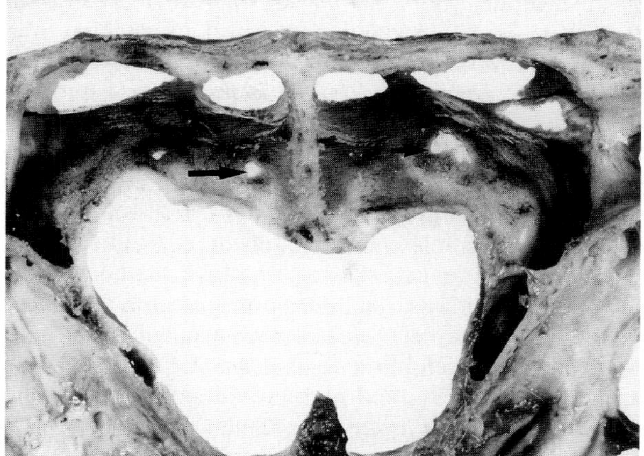

Fig. 1.41 Osteoporosis, calf. Enlarged paranasal sinuses and thin frontal bones with resorptive defects of inner plates (arrows).

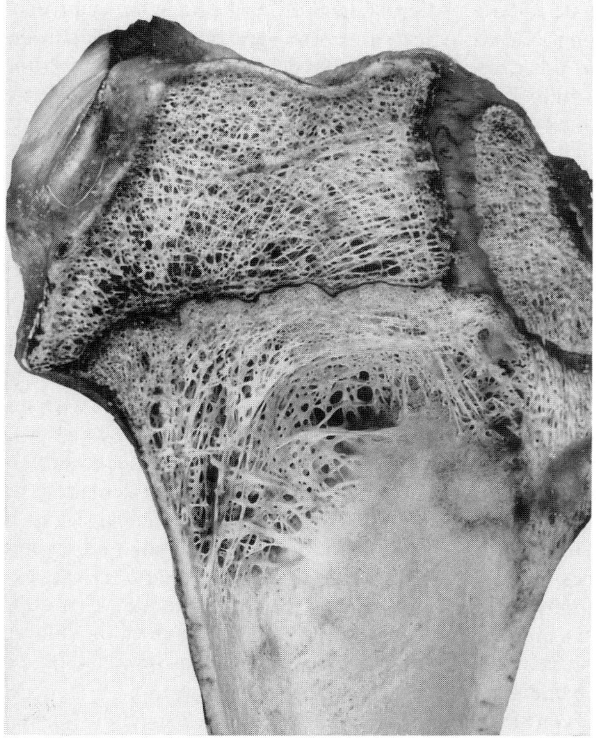

Fig. 1.42 Osteoporosis, calf. Proximal tibia. Thin trabeculae, enlarged marrow cavity, thin corticalis, distorted growth plate.

Osteoporotic bones are usually light and fragile, but often there is also a reduction in soft tissue mass and muscular power, which tends to preserve the skeleton against stress fractures. Sometimes there is a disproportionate relationship between soft tissue mass and skeletal mass, such as occurs in fast-growing pigs, that leads to trabecular and larger fractures, epiphyseal compression and articular collapse, distortion and fracture of growth plates, slipped epiphyses, and avulsion of ligaments.

Histologic examination is essential for the correlation and understanding of data obtained from other investigations as well as being necessary for quantitative morphometry. It is useful in assessing the site of resorption–formation imbalance, and may provide indications as to the pathogenesis of bone loss. There is evidence that osteoporosis evolving from increased resorption is characterized by a decline in trabecular numbers, while in osteoporosis caused by decreased formation, there are normal numbers of thin trabeculae. Also, as noted, all trabeculae are not treated equally when resorption predominates. The vertical, weight-bearing trabeculae tend to be preserved and, if necessary, repaired, while the horizontal cross-members are selectively removed. Ash analyses yield information as to the matrix–mineral relationship, but interpretation of results is potentially misleading when morphologic changes are ignored. In severely osteoporotic young animals, the hypertrophic zone of the growth cartilages may be narrowed or absent (Fig. 1.43), and when growth ceases, the physis is sealed by a plate of bone. Microscopically, trabeculae are reduced in number and/or size and often are fractured. Depending on the cause of the osteoporosis, evidence of abnormal activity of osteoblasts and/or osteoclasts is present. Intracortical resorption in long bones occurs along vascular channels and is produced by teams of osteoclasts working parallel to the long axes of the bones, the so-called cutting cones. Intracortical resorption is functionally important because a small loss of compact bone has a greater effect on bone stiffness than a comparable loss of cancellous bone.

There is no simple way of determining when a skeleton is significantly osteoporotic. Arbitrarily, a reduction in bone mass can be regarded as pathologic when the reduction is proportionately greater than that for soft tissue, and the skeleton cannot adequately serve the needs of muscular activity. In practice, a severely reduced skeletal mass is often adequate to meet mechanical demands, but conversely a skeleton of slightly reduced mass may be inadequate to support the rapidly increasing body weight as is seen, for example, in pigs and young beef cattle on highly nutritious diets. Thus the consequences of osteoporosis depend on the location of bone loss and the balance between hard and soft tissues, as well as on the degree of physical activity.

Osteoporosis is a common lesion, especially in farm animals, and is usually nutritional in origin. In grazing animals it is likely to be overlooked unless the prevalence of lameness or fractures is high. Calcium deficiency is often a major contributing factor in the development of osteoporosis, but few cases are due to uncomplicated deficiencies of a single nutrient, and in some areas of the world, phosphorus deficiency or protein–calorie malnutrition are more important than restricted calcium intake. Some specific causes of osteoporosis are discussed below. The roles of vitamin A and endocrine disturbances are referred to elsewhere.

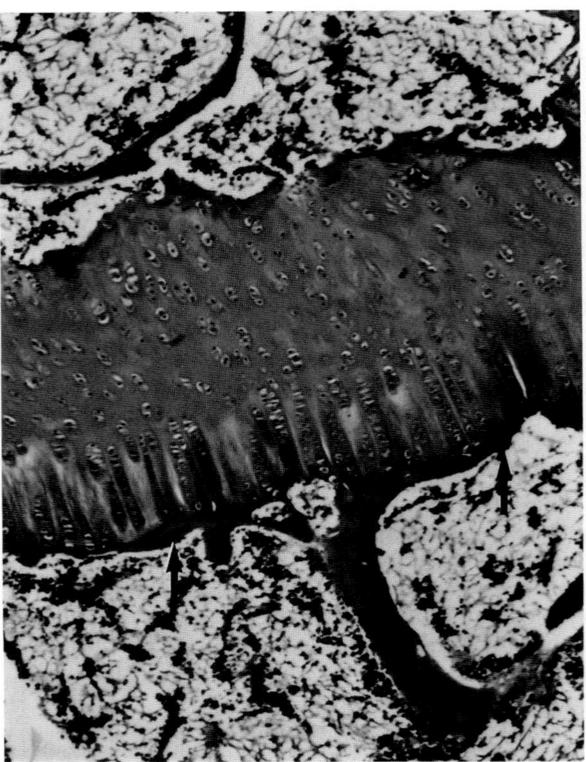

Fig. 1.43 Osteoporosis, calf. Osteochondral junction. Deficiency of metaphyseal bone (below), deficiency of epiphyseal bone (above). Growth plate is closed by plate of bone (arrows).

1. Osteoporosis Due to Starvation

Starvation, involving severely restricted intake of a balanced diet, includes components of protein and energy deficiency, since under such circumstances, available protein is utilized to provide energy. The effects on the skeleton are similar in either circumstance and may be exaggerated if the caloric value of the diet is increased by readily available carbohydrate. Starvation occurs commonly in farm animals, for example in drought-stricken areas, in calves consuming undigestible milk replacer, and in cattle fed low-quality hay *ad libitum* in cold climates when the requirements for energy are high. Protein deficiency and starvation have been well studied in experimental circumstances, especially in the pig. The effect of protein levels on the skeleton is complex; for example, in sheep the availability of dietary calcium is reduced by low protein intakes. On the other hand, high-protein diets produce hypercalciuria in rats, the effect being related to levels of sulfur-containing amino acids. Similar changes apparently occur in acute but not chronic experiments in young horses, indicating that adaptation to the high protein diet develops.

During starvation, growth of the skeleton is retarded and may cease. There is a matrix osteoporosis, with the quantity of bone ash being reduced, but the proportion of ash to matrix remaining about normal. The physes are narrowed, and the chondrocytes are small with a relative increase in the amount of matrix. The cortical and trabecular bone of the diaphyses is reduced in amount. The terminal plate on the epiphyseal aspect of the growth cartilage is deficient, while on the metaphyseal side, the trabeculae are deficient, and the physis is more or less sealed off by a layer of bone (Fig. 1.43). Characteristic transverse striations in the metaphysis may be demonstrated radiographically and consist of lateral fusions of the longitudinal trabeculae, few of which remain. Contrary to the usual situation in osteoporosis, the vertical trabeculae linking the transverse ones are smaller and sparser (Fig. 1.11A,B). The thicker transverse trabeculae are a result of intermittent cessation and reactivation of physeal growth, and represent a succession of sealing layers from the physis. Transverse radiodense striations also develop in limb bones and pelves of fetuses from ewes on a low plane of nutrition. In these lambs there may be considerable variation in the number of arrest lines between littermates. The number of lines is correlated positively with retardation of fetal size and skeletal maturity.

The pathogenesis of starvation osteoporosis in terms of remodeling imbalance is not completely defined. The histologic appearance of inactive osteoblasts and *a priori* reasoning both suggest that deficient bone formation is responsible. Support for this hypothesis is derived from findings in dogs with malabsorption syndrome, in which tetracycline-labeled sites of bone formation are sparse or absent. The level of bone resorption is not known.

The effect of severe undernutrition on the development of dentition is profound. Delayed formation and eruption of the whole dentition, with a relatively greater delay in the growth of the jaws, leads to overcrowding of the teeth, especially the permanent molars. This causes malocclusion and malalignment of teeth, and there is partial or complete elimination of the diastemata between incisors and canines and between canines and deciduous molars. Nutritional rehabilitation allows rapid growth of jaws and teeth, but malocclusion of permanent dentition persists, and often there is abnormal development of parts of some teeth.

2. *Disuse Osteoporosis*

Disuse osteoporosis is a loss of bone mass due to muscular inactivity and reduced weight bearing. It may be quite local, following paralysis or fracture of a limb, or more generalized, in association with prolonged recumbency or inactivity. Bones that normally carry the greatest loads suffer the greatest proportional loss from disuse. Disuse osteoporosis can occur in immobilized parathyroidectomized animals, and is seen in human fetuses with neuromuscular disease. Disuse osteoporosis is of considerable importance in humans, but less so in other animals. Measurable loss of mineral does occur in horses that have a limb cast for prolonged periods. This could cause clinical problems if maximum athletic activity were resumed immediately after removal of the cast.

In experimental long-term disuse osteoporosis of the forelimb in young adult beagle dogs, the lesion develops in several stages. The first is marked by rapid loss of tissue, which reduces the bone mass by about 16% in 6 weeks. This results from imbalanced resorption on endosteal, periosteal, and intracortical surfaces, and may be a nonspecific regional acceleratory phenomenon (see Diseases of Bones, Section I, General Considerations), associated with activation of numerous remodeling sites. In what is probably a continuation of the regional acceleratory phenomenon, there is marked increase in bone mass, which is presumably a result of net bone formation in the activated remodeling sites. This occurs between 8 and 12 weeks, and the bone mass approaches normal.

In the period between 8 and 12 and 24 and 32 weeks following immobilization, there is slower bone loss, 80–90% of which is subperiosteal bone. The reduction in bone mass may reach 30–50% of normal and be maintained at this low level. The distal bones of the limb lose more bone than the proximal, but in all cases, the subperiosteal bone is diminished.

The effects of remobilization on disuse osteoporosis are not clearly defined. Recovery from relatively short periods of immobilization do occur, but the age of the patient influences the response.

3. *Senile Osteoporosis*

Senile osteoporosis is common in humans and occurs in other animals. A loss of skeletal mass with age is regarded as physiologic. Various factors may mediate this loss including reduced formation of, and responsiveness to, active vitamin D metabolites, abnormal parathyroid hormone secretion, decrease in sex steroids, and reduced physical activity. When the loss of bone is much greater than normal, senile osteoporosis is diagnosed. Beagle dogs aged 7–8 years have about 8% less bone tissue in their forelimbs than young adults, but this is of no consequence to health. It is often difficult for veterinary pathologists to distinguish the effects of age and cachexia, especially in ancient herbivores in which peer competition and dental abnormalities may complicate the development of the osteoporosis.

Whether the inexorable loss of bone with aging is sufficient to cause senile osteoporosis probably depends on the amount of bone tissue in the skeleton at the time the decline begins, as well as the dietary, lactational, and gestational activities in the intervening years. There is little to distinguish senile osteoporosis from that of other causes except for the presence of increased numbers of dead osteocytes and empty lacunae, especially in interstitial lamellae of the cortex, which are characteristic of old bone tissue. Hypermineralization of lacunae and plugging of Haversian canals occurs in senile humans and to some extent in other animals.

4. Osteoporosis Associated with Intestinal Parasitism

Severely parasitized ruminants are often osteoporotic, and the lesions are usually considered to result from malabsorption. The effects of subclinical parasitism are often overlooked, but experimental low-level infections are capable of producing osteoporosis.

Continuous dosing of lambs with *Trichostrongylus colubriformis* over several months produces a mineral osteoporosis characterized by a reduction in cortical and trabecular bone. There is a reduction in length and volume of bones and in their degree of mineralization. Some animals show minor changes of rickets in growth plates and trabeculae. Hypophosphatemia develops in these lambs, and induced phosphorus deficiency, possibly complicated by calcium deficiency, is responsible for the osteoporosis. There is a possibility that reduced protein anabolism, mediated through increased corticosteroid production, is also involved. Concurrent coccidial infection may have an additive effect.

Subclinical infection of lambs with *Ostertagia circumcincta* also produces osteoporosis. This is a matrix osteoporosis, mediated apparently through restrictions in the availability of protein and energy for matrix deposition. Reduced bone volume is a consistent finding, but there is no effect on growth plates, length of bones being normal. In calves infected experimentally with *Ostertagia* spp. the utilization of phosphorus is adversely affected.

The effects of experimental parasite-induced osteoporoses are more severe in vertebrae than in tibiae. Subclinical infections probably contribute to the development of some of the obscure osteoporoses that afflict grazing sheep.

5. Osteoporosis Due to Calcium Deficiency

Uncomplicated calcium deficiency, that is, calcium deficiency associated with optimal dietary levels of phosphorus and vitamin D, rarely if ever occurs as a natural disease. Complicated calcium deficiency is common and is discussed below with osteodystrophia fibrosa, as is lactational osteoporosis in gilts.

In mature and immature animals of the species with which we are concerned, experimental calcium deficiency produces osteoporosis without evidence of osteomalacia or rickets. The osteoporosis is due to excess bone resorption (Fig. 1.44A,B) as a result of increased activity of parathyroid hormone, whose production and release are stimulated by decreased serum ionized calcium. In adult animals the bone loss probably results from an increase in activation of new remodeling sites. Cortical bone is lost as a result of increased endosteal resorption and, in severe calcium deficiency, intracortical resorption also. Cortical bone-formation rates may be increased or decreased slightly, depending on the stage of the process and the bone examined.

In both adult and growing animals there is severe loss of cancellous bone, and bones with a high trabecular component, such as vertebrae, are most severely affected. In immature metaphyses, primary spongiosa may be removed completely, so that there is no framework remaining for production of secondary spongiosa.

Unless the deficiency is extremely severe, physeal growth and mineralization of cartilage are normal. In young animals the remaining bone may have a remarkably low ash content; in lambs it may be reduced to 60% of normal after 8 weeks on a calcium-deficient ration. This does not justify a diagnosis of rickets or osteomalacia. Osteoid seams are of normal width and frequency, and mineralization occurs normally. Since there is less bone present than is normal, the unmineralized osteoid constitutes a relatively large proportion of bone tissue mass, tending to lower the average ash content. Also the older, more heavily mineralized, endosteal bone is resorbed, contributing to a reduction in the average ash content of the remaining bone.

In mature animals calcium deficiency is usually asymptomatic, but in the young, trabecular fractures and infractions and fractures of bones occur, and involve both the axial and appendicular skeleton (Fig. 1.45). Parathyroid glands are slightly enlarged due to hyperplasia. Osteodystrophia fibrosa sometimes develops but is less florid than in the osteodystrophy of calcium deficiency secondary to phosphorus excess. Only with severe and prolonged calcium deficiency does hypocalcemia develop, but rarely falls as low as with vitamin D deficiency. This is because, in the calcium-deficient animal, homeostatic mechanisms maximize absorption of dietary calcium, reabsorption of urinary calcium, and resorption of bone, in order to maintain serum ionized calcium. When vitamin D is deficient, homeostatic mechanisms are impaired, and hypocalcemia develops.

Prolonged calcium deficiency causes dental abnormalities in sheep, including retarded eruption and increased susceptibility to wear due to mild enamel hypoplasia. When adequate calcium becomes available, superior brachygnathia develops because of inadequate repair of the upper skull compared to the mandible.

Uncomplicated calcium deficiency is almost entirely an experimental condition, but it is important to bone pathology because it demonstrates the role of calcium in osseous disease, and also because it fails to produce rickets.

6. Osteoporosis Associated with Phosphorus Deficiency

Phosphorus deficiency occurs in areas of the world such as South Africa and northern Australia where there is a deficiency of the element in the soil and thus in the pasture. Hypophosphatemia may also be induced by *Trichostrongylus colubriformis* infections in sheep, and conceivably by excesses of dietary cations such as calcium, iron, and aluminum, which form insoluble complexes with phosphorus.

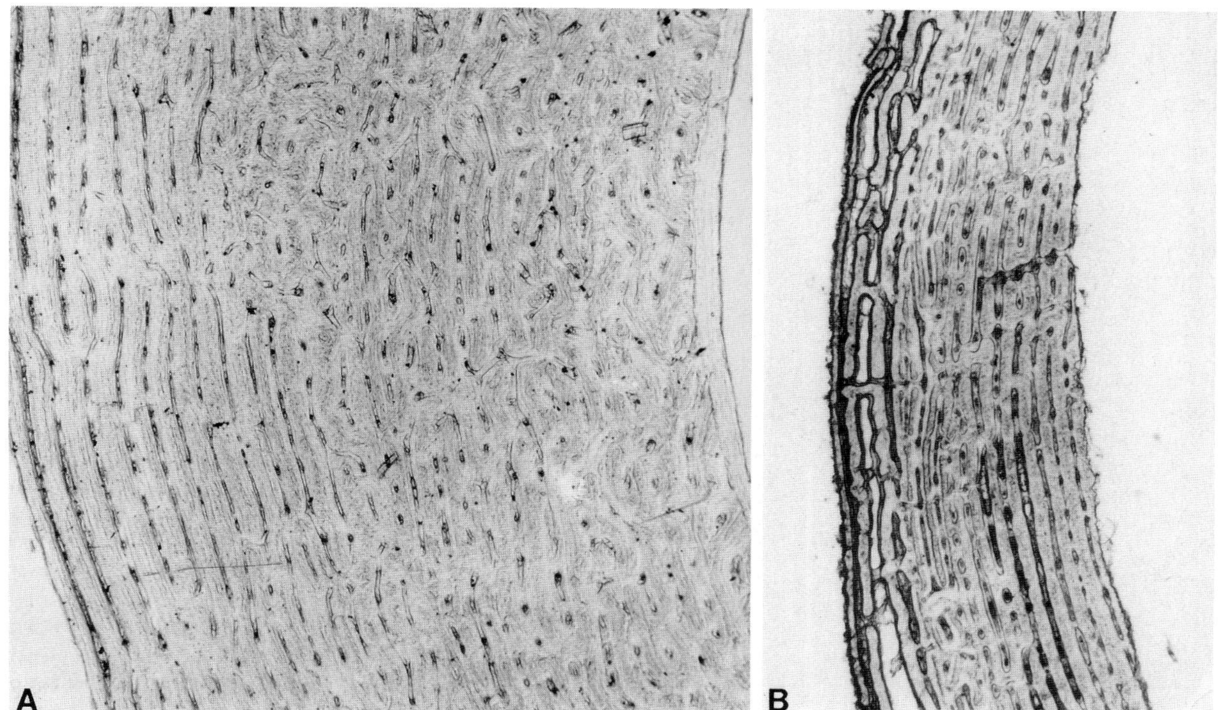

Fig. 1.44 Osteoporosis. Femoral cortex of lambs fed normal (A) and (B) calcium-deficient rations (B). Endosteal resorption has produced cortical thinning in (B). Both specimens show typical laminar bone, which makes up the entire cortex in (B).

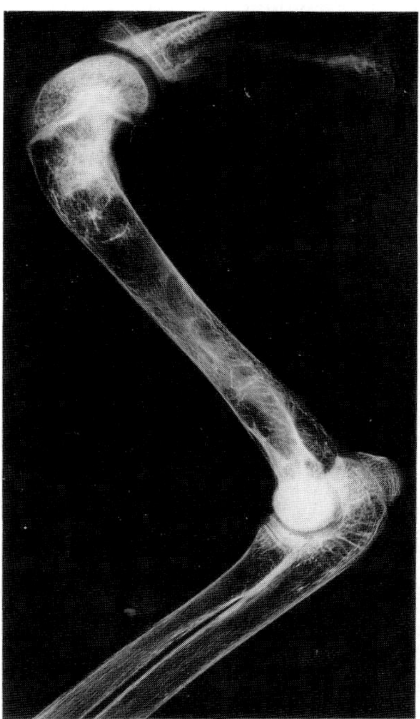

Fig. 1.45 Cat. Meat diet. Radiograph showing marked loss of cortical density with emphasis of trabecular patterns. Healing fracture in proximal humerus.

Classically, phosphorus deficiency produces osteomalacia in adults and rickets in growing animals, under both natural and experimental conditions. However, under some circumstances, it causes osteoporosis (Figs. 1.46, 1.47A,B). The reasons for this are not clear, but perhaps are related to the anorexia that often accompanies phosphorus deficiency, the age and growth rate of the animals, and the severity and duration of phosphorus deficiency. It is not clear whether osteoporosis and osteomalacia are interconvertible or whether osteoporosis is a stage in the development of osteomalacia. Phosphorus deficiency stimulates resorption of bone in experimental animals, by an unknown, parathyroid-independent mechanism.

7. Osteoporosis and Copper Deficiency

Bone fragility is often a feature of naturally occurring and experimental copper deficiency in lambs, calves, foals, pigs, and dogs. Copper catalyzes the activity of lysyl oxidase, an enzyme that is active in the cross-linking of collagen and elastin, and the bone lesions are probably related to a deficiency of this enzyme.

Simple and conditioned copper deficiency produce skeletal defects of a similar type. It is not clear whether the pathogenesis is the same in each case or to what extent the disease process is comparable in different species. In cattle, deficiency of copper combined with excess molybdenum results in negative phosphorus balance, and this may contribute to the brittleness of bones in this species.

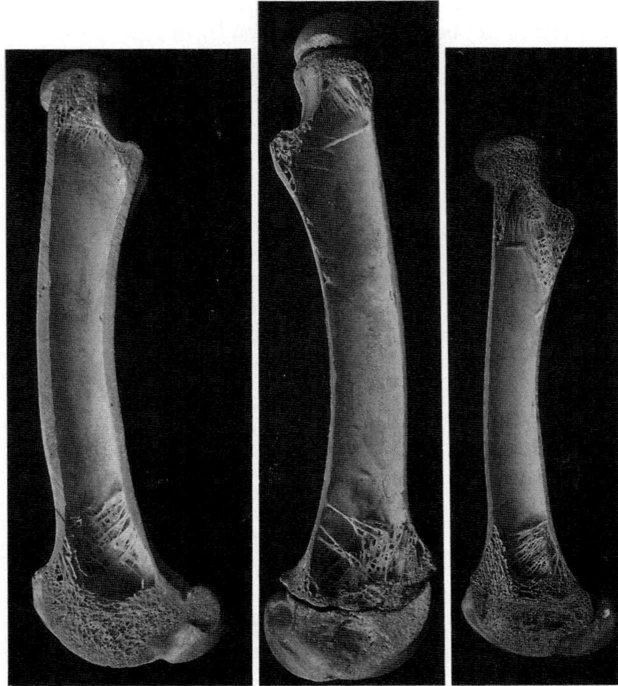

Fig. 1.46 Femurs. Sheep. Normal (left). Phosphorus-deficiency osteoporosis (center and right). (Courtesy of J. R. M. Innes.)

Clinically there may be bowing of long bones and enlargement of their ends, and sometimes a tendency to spontaneous fractures. In young horses the prevalence of osteochondrosis is increased in naturally occurring and experimental copper deficiency. Lesions can be reduced in number, but not eliminated, by copper-supplemented diets. There is considerable interest in the relationship of copper deficiency and osteochondrosis to that group of equine wobblers with cervical vertebral malformation (see previous sections).

Histologically, copper-deficient bones may show osteoporosis, cartilaginous abnormalities, or both. Osteoporosis, the lesion in naturally occurring deficiency in lambs (formerly attributed to excess dietary lead), and naturally occurring and experimental deficiencies in dogs, is apparently caused by reduced production of normally constituted osteoid. Thus it is a matrix osteoporosis. Endochondral ossification is qualitatively normal, but only a thin layer of osteoid is deposited on the mineralized cartilage in the primary spongiosa. The fragile trabeculae are covered by spindle-shaped, apparently hypoactive osteoblasts, and those trabeculae that are incorporated into the diaphysis are incompletely compacted. As a result the metaphyseal and diaphyseal bone is abnormally fragile.

The lesion in physeal cartilage is characterized by irregular alignment of cartilage columns and irregular persistence of cartilage in the metaphysis. It is not clear whether the latter is caused by a primary cartilage defect or is secondary to disruption of the metaphyseal circulation by trabecular fractures. It appears that the effect of weight bearing on these abnormal cartilaginous structures accounts for the clinically apparent enlargement of the metaphyses. The pathogenesis is somewhat analogous to that of rachitic deformities, and copper-responsive osteopathies in foals have been called rickets. As well as being malleable by compression, the persistent metaphyseal cartilage is also susceptible to shearing forces, and sheep with molybdenum-induced copper deficiency are prone to develop epiphysiolysis. They also form exostoses at sites of muscle attachment, probably

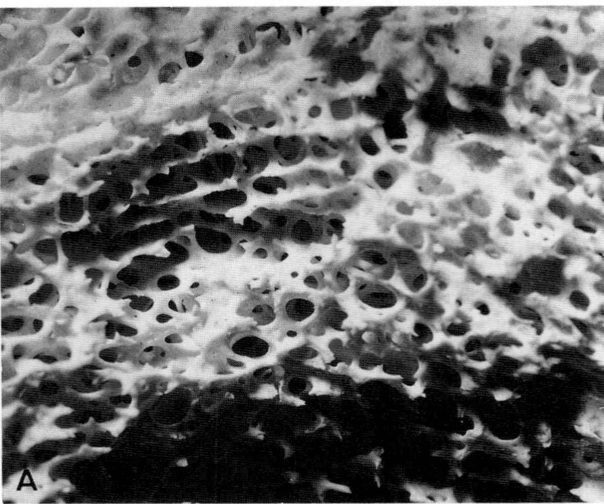

Fig. 1.47A Metaphyseal bone from femur shown in Fig. 1.46. Normal plates with few resorption cavities.

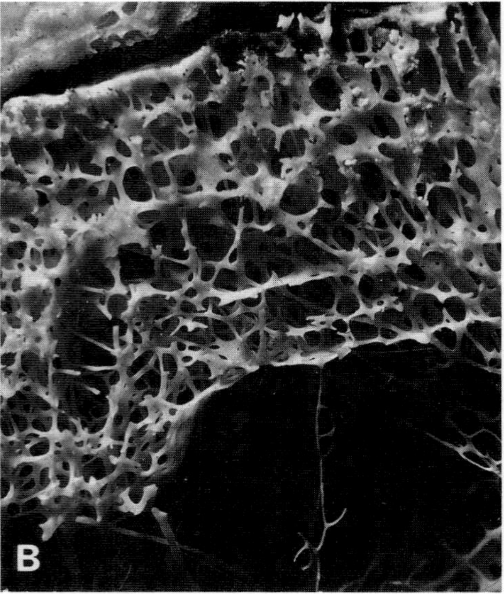

Fig. 1.47B Metaphyseal bone from femur shown in Fig. 1.46. Phosphorus-deficiency osteoporosis.

secondary to subperiosteal hemorrhage, as do rats fed excess molybdenum. It is not clear whether the effects of molybdenum are mediated entirely through induced copper deficiency or whether there is also a direct toxicity.

Bibliography

Coop, R. L. *et al. Ostertagia circumcincta* infection of lambs, the effect of different intakes of larvae on skeletal development. *J Comp Pathol* **91:** 521–530, 1981.

Cooper, C. Osteoporosis—an epidemiological perspective: A review. *J R Soc Med* **82:** 753–757, 1989.

Draper, H. H., and Scythes, C. A. Calcium, phosphorus, and osteoporosis. *Fed Proc* **40:** 2434–2438, 1981.

El-Maraghi, N. R. H., Platt, B. S., and Stewart, J. C. The effect of the interaction of dietary protein and calcium on the growth and maintenance of the bones of young, adult, and aged rats. *Br J Nutr* **19:** 491–509, 1965.

Frandsen, J. C. Effects of concurrent subclinical infections by coccidia (*Eimeria christenseni*) and intestinal nematodes (*Trichostrongylus colubriformis*) on apparent nutrient digestibilities and balances, serum copper and zinc, and bone mineralization in the pigmy goat. *Am J Vet Res* **43:** 1951–1953, 1982.

Hermann, H. J. Das histomorphologische Verhalten des Knockensystems des Jungrindes nach quantitativer Mangelernahrung. *Pathologia Vet* **2:** 468–492, 1965.

Jaworski, Z. F. G., Liskova-Kiar, M., and Uhthoff, H. K. Effect of long-term immobilisation on the pattern of bone loss in older dogs. *J Bone Joint Surg (Br)* **62:** 104–110, 1980.

Kunkle, B. N., *et al.* Osteopenia with decreased bone formation in beagles with malabsorption syndrome. *Calcif Tissue Int* **34:** 396–402, 1982.

Platt, B. S., and Stewart, R. J. C. Transverse trabeculae and osteoporosis in bones in experimental protein–calorie deficiency. *Br J Nutr* **16:** 483–495, 1962.

Schryver, H. F. *et al.* Growth and calcium metabolism in horses fed varying levels of protein. *Equine Vet J* **19:** 280–287, 1987.

Sykes, A. R., and Field, A. C. Effects of dietary deficiencies of energy, protein, and calcium on the pregnant ewe. I. Body composition and mineral content of the ewes. *J Agric Sci Camb* **78:** 109–117, 1972.

Sykes, A. R., Coop, R. L., and Angus, K. W. Experimental production of osteoporosis in growing lambs by continuous dosing with *Trichostrongylus colubriformis* larvae. *J Comp Pathol* **85:** 549–559, 1975.

Tonge, C. H., and McCance, R. A. Normal development of the jaws and teeth in pigs, and the delay and malocclusion produced by calorie deficiencies. *J Anat* **115:** 1–22, 1973.

Uhthoff, H. K., and Jaworski, Z. F. G. Bone loss in response to long-term immobilisation. *J Bone Joint Surg (Br)* **60:** 420–429, 1978.

Waymack L. B., and Torbert, B. J. Effect of *Ostertagia* species on phosphorus utilization and excretion in weanling calves. *Am J Vet Res* **30:** 1139–1144, 1969.

Wenham, G. A radiographic study of the changes in skeletal growth and development of the foetus caused by poor nutrition in the pregnant ewe. *Br Vet J* **137:** 176–187, 1981.

C. Osteomalacia

Osteomalacia, literally softening of bones, is a metabolic disorder of the adult skeleton that, in veterinary pathology, is usually defined in terms of its major causes, deficiency of phosphorus or vitamin D. In young, growing bone, deficiencies of phosphorus or vitamin D produce rickets, diseases in which lesions of growth cartilage are more prominent than those of bone. Osteomalacia is characterized by accumulation of excess unmineralized, and presumably unmineralizable, osteoid on trabecular surfaces. Fluorotic bones may also be osteomalacic (see osteofluorosis under Diseases of Bones, Section VI,H,3).

1. Phosphorus Deficiency

Osteomalacia due to deficiency of phosphorus is uncommon, except in areas of the world, such as South Africa, northern Australia, and the North Island of New Zealand, where the pasture is low in phosphorus, and supplemental feeding is rare. Cattle are more susceptible than sheep, and horses seem to be remarkably resistant to phosphorus deficiency.

Signs of phosphorus deficiency develop slowly, and the osteodystrophy may reach clinical magnitude only in cows in which deficiency is exaggerated by the requirements of pregnancies and lactations. Such animals lose condition, and develop transient and shifting lameness and a susceptibility to fractures. They may crave phosphorus-rich materials, and osteophagia and allotriophagia are characteristic signs of the deficiency. Fertility can be severely reduced. Estrum may be absent, irregular, or inapparent. Newborn calves are of normal size and development (at the expense of their dams), but while suckling, their growth is subnormal, and this is accentuated after weaning. Growth is also aberrant in that maturity is delayed, and deficient animals become long and lean. The hair is rough, slow to be replaced, and deficient in pigment to a variable degree.

Hypophosphatemia develops early in the disease, often in cattle showing few other signs of phosphorus deficiency. Blood calcium is normal or elevated. Anemia, and particularly the hemolytic crisis known as post-parturient hemoglobinuria (see Volume 3, Chapter 2, The Hematopoietic System), is sometimes associated with hypophosphatemia in cattle, and may occur independent of other signs of deficiency.

Osteomalacic bones have a diminished resistance to pressures and tensions, and increased susceptibility to the stresses and strains of ordinary activity. As a result there is excessive deposition of matrix where mechanical stimuli are strongest, such as at insertions of tendons and fascia, places of angulation and curvature, and on stress-oriented epiphyseal trabeculae. When the disease is advanced, the bones break easily, the marrow cavity is enlarged and may extend into the epiphysis, and the cortex is thin, spongy, and soft. Fractures are most common on the ribs, pelvis, and long bones of cattle, and in the vertebral column of swine. Deformities are often present, including kyphosis or lordosis, medial displacement of the acetabula with compression of the pubic bones, twisting of the ilia, and narrowing of the pelvis. The ribs and transverse processes of the lumbar vertebrae droop so that the thorax is narrowed and flattened and the sternum, prominent. Collapse of articular surfaces and degeneration of cartilages is sometimes observed, and tendons may separate from their attachments. Separation of the heads of the femurs within the articular capsules occurs occasionally and may be overlooked as a cause of posterior paralysis. In long-standing cases of osteomalacia, cachexia and anemia are often present.

The histological changes are characterized by active resorption of bone and by excess osteoid. Resorption first affects the trabeculae; later bone is resorbed from within the vascular canals and from beneath the endosteum. In advanced cases trabeculae are reduced in size and number, and cortices are thin and porous. The genesis of lesions depends to some degree on the vitiation of remodeling, involving the normal removal of aged well-mineralized bone, and its replacement by osteoid in which mineralization is absent or incomplete. This osteoid covers the trabeculae and lines the expanded Haversian canals, and localized accumulations occur at the aforementioned sites of mechanical stress.

The deposition of unmineralized matrix on remnants of mineralized bone probably protects them from osteoclastic resorption. It also results in localized resorptive activity with focally numerous osteoclasts that may form syncytia. There is less matrix deposited in adult than in growing animals, and the amount may be further reduced by intercurrent disease, and by deficiencies of protein and other nutrients, which often accompany phosphorus deficiency. Thus osteoporosis may be superimposed on osteomalacia, and in these cases, compensatory activity is likely to be minimal, osteoblasts few, and osteoid scant.

The pathogenesis of phosphorus-deficiency osteomalacia involves activities of various hormones including 1,25-dihydroxycholecalciferol (calcitriol) and parathyroid hormone (PTH). Hypophosphatemia stimulates renal production of calcitriol, which increases intestinal absorption of phosphorus by a PTH-independent mechanism, and provokes resorption of bone by osteoclasts. Both bone resorption and hypophosphatemia tend to increase plasma ionized calcium, thus suppressing PTH release and decreasing renal excretion of phosphorus. Hypophosphatemia also activates intrinsic renal mechanisms, independent of calcitriol and PTH, that reclaim phosphate from glomerular filtrate. The failure of osteoid to mineralize in phosphorus-deficiency osteomalacia is incompletely understood. Hypophosphatemia suppresses the production of 24,25-dihydroxycholecalciferol (24,25-DHCC), a vitamin D metabolite that, along with calcitriol, appears to be intimately involved in mineralization of cartilage and osteoid. One hypothesis holds that whereas calcitriol increases osteocalcin synthesis, 24,25-DHCC is required to direct it into the matrix where its calcium-binding quality can facilitate mineralization.

2. *Vitamin D Deficiency*

The active metabolites of vitamin D are now classified as hormones by merit of their central roles in calcium homeostasis and bone metabolism. In addition to these areas of activity, involvement in many other systems and in cell differentiation is postulated for the active metabolites. Both vitamin D_2 and D_3 are utilizable by domestic animals, but certain species, such as New World primates [Cebus monkeys, tamarins (*Saguinus* spp)], can utilize only D_3, whereas in chickens, D_3 has about 10 times greater potency than D_2. Vitamin D_3 (cholecalciferol) is formed in the epidermis beginning with the conversion of 7-dehydrocholesterol by ultraviolet light from sunshine to previtamin D_3. Light of about 300 nm is most efficient in driving the photolytic process, pigmented skin requiring more light than nonpigmented to achieve the same effect. The previtamin is in thermal equilibrium with the vitamin, and conversion occurs by thermal isomerization over a period of 3 days, but with prolonged exposure to sunlight, biologically inert sterols are formed. The vitamin is transported to the circulation by an α-1-globulin that binds poorly to the previtamin. Vitamin D_2 (ergocalciferol) is formed by irradiation of a plant sterol, ergosterol, and is absorbed in chylomicrons in the intestine. Both calciferols are converted in the liver by microsomal and mitochondrial enzymes to their 25-hydroxy forms. The regulation of this reaction is poorly understood. Plasma levels of the products tend to reflect dietary intake and to some extent, exposure to sunlight. In rats inactivation of cholecalciferol in the liver appears to be mediated by increased levels of calcitriol.

Further hydroxylation of the 25-hydroxy forms occurs in the mitochondria of the renal proximal tubules (and in the placenta) where the most active metabolite, calcitriol, is formed. This reaction is stimulated by low levels of plasma calcitriol, calcium, phosphorus, and parathyroid hormone. Increases in calcitriol, phosphorus, and parathyroid hormone have the opposite effect. Several other hormones also influence the 1-α-hydroxylation reaction. In addition, high levels of plasma calcitriol suppress synthesis and release of parathyroid hormone. Vitamin D is stored in the body in fat and muscle either as 25-hydroxycholecalciferol or as unhydroxylated cholecalciferol.

The biologic activity of calcitriol is not confined to intestine and bone, but it is there that its functions are best understood. Calcitriol stimulates the formation of several calcium-binding proteins, one of which, calbindin-D9k, is characteristic of mammalian intestine, whereas another, calbindin-D28k, is present in several sites, including avian intestine and avian and mammalian kidney. Intestinal calbindin is located in the enterocyte cytosol and is involved in transcellular transport of calcium by a sodium-dependent, energy-dependent process. Renal calbindin is found in the epithelial cells of the distal nephron and may be active in selective reabsorption of calcium. Calcitriol also stimulates intestinal absorption of phosphorus. Bone remodeling is influenced significantly by calcitriol and probably by other active metabolites of vitamin D. Calcitriol stimulates osteoclastic resorption, probably through the mediation of cells of osteoblast lineage since osteoclasts do not have receptors for calcitriol. It also increases osteoblastic numbers and activity. The activity of the calcium-binding protein osteocalcin is also moderated by vitamin D metabolites. Calcitriol stimulates synthesis of osteocalcin and raises plasma levels, whereas 24,25-dihydroxycholecalciferol, which is also synthesized in kidney, raises bone levels and may be required to move osteocalcin into bone.

Osteomalacia caused by vitamin D deficiency occurs in grazing animals where the combination of relatively high latitudes and relatively mild climates allow them to be pastured for much of the year. Such conditions occur in parts of the United Kingdom, New Zealand, and southern Australia. Critical factors in the development of osteomalacia are probably the unavailability of sun-cured hay, which contains ample vitamin D_2, the demands of pregnancy and lactation, and inadequate exposure to ultraviolet light. When the winter sun is at an angle of less than 30 degrees to the horizontal, the shorter ultraviolet rays are refracted into the atmosphere and are therefore unavailable for the activation of 7-dehydrocholesterol in the skin. In New Zealand and Tasmania, where green fodder is available for much of the year, the antivitamin D activity of carotene, combined with the low vitamin D content of the green fodder, also contribute to the development of osteomalacia. Sheep appear to be more susceptible to vitamin D-deficiency osteomalacia than cattle, possibly because much of their skin is covered by fleece. In grazing animals photobiosynthesis is more important than diet as a source of the vitamin.

Prolonged deficiencies of vitamin D (or phosphorus) are required to produce clinical and pathologic evidence of osteomalacia; thus, the drain of gestation and lactation may be needed to precipitate disease. The need for prolonged deprivation also militates against the development of florid vitamin D deficiency in housed adult animals fed hay and supplemental concentrates.

Osseous lesions in osteomalacia of vitamin D deficiency are probably very similar to those described for a deficiency of phosphorus. Experimental comparisons of the two seem not to have been made. Since hypocalcemia is an important manifestation of hypovitaminosis D, a degree of concomitant osteodystrophia fibrosa might be anticipated, particularly since low plasma calcitriol tends to increase parathyroid hormone levels; to what extent this occurs in osteomalacia is not known, but it does occur in vitamin D-deficiency rickets. The effect of age on proliferation of osteoprogenitor cells would no doubt reduce the likelihood in the adult animal.

D. Rickets

Whereas osteomalacia is a disease of mature bones, rickets is a disease of growing bones. The causes and pathogenesis are the same, but the vitality of youthful tissue and the transformations and vulnerability of growing bone introduce considerable complexities into the morphogenesis of rickets. Since bones mature at different times, the two diseases may coexist in a skeleton.

There is still controversy about the respective roles of vitamin D, calcium, and phosphorus in the etiology of rickets. The basic question is whether vitamin D prevents rickets solely by increasing blood calcium or whether it also changes the matrix (osteoid or cartilage) so it can accept mineral. The defect in matrix mineralization in both rickets and osteomalacia has been attributed solely to a local reduction in the ionic product of calcium and phosphorus. Application of the solubility product principle reveals inconsistencies, and it is based on the incorrect assumption that the matrix is freely bathed in tissue fluids. The fact that uncomplicated calcium deficiency does not produce rickets is inconclusive, because only in the terminal stages of this deficiency does severe hypocalcemia occur, at which time growth slows, obviating the development of rickets. There is also debate about the need for vitamin D, if any, when dietary calcium and phosphorus level and ratio are optimum. It is accepted that less vitamin D is required under such hypothetical conditions, but in practice, blanket statements that ignore species, age, growth rate, and other variables are best avoided. The failure of mineralization is probably due to defective matrix, which may be deficient in certain noncollagenous proteins such as osteocalcin. Osteocalcin is thought to participate in the mineralization of osteoid, and its production is calcitriol dependent. It seems probable that it is critically involved in the mineralization sequence.

1. Phosphorus or Vitamin D Deficiency

The basic lesions in mammalian rickets are failure of mineralization of osteoid and failure of mineralization of cartilaginous matrix. Both matrices accumulate, not because of excess production, which is in fact probably reduced, but rather as a result of the retarded mineralization. In the following discussion, emphasis is placed on the consequences of failure of mineralization of cartilage. The outcome as regards osteoid is similar to that in osteomalacia, with one significant difference—osteoid is deposited more rapidly in young bones than old and, since it is resistant to osteoclastic activity, resorption need not be a feature of rickets. Actually, resorption continues in areas in which it was active at the onset of the disease, but usually it is supplanted by deposition of matrix unless the deficiency is severe, in which case resorption continues apace.

The sequential changes in rickets consist of failure of mineralization and vascular invasion of cartilage, failure of replacement of growing cartilage by primary spongiosa, irregular persistence of cartilage, formation of osteoid on persistent cartilage with irregularity of osteochondral junctions, overgrowth of fibrous tissue in the metaphysis, and deformities of the shape and structure of the bones. We now consider these changes in more detail.

Failure of mineralization and vascular invasion of cartilage are the earliest morphologic manifestations of rickets. They seem to occur concomitantly, suggesting that they are influenced simultaneously by vitamin D metabolites. The effect of vitamin D on mineralization is discussed above. The normal metaphyseal vasculature is characterized by a constant supply of capillary sprouts, which

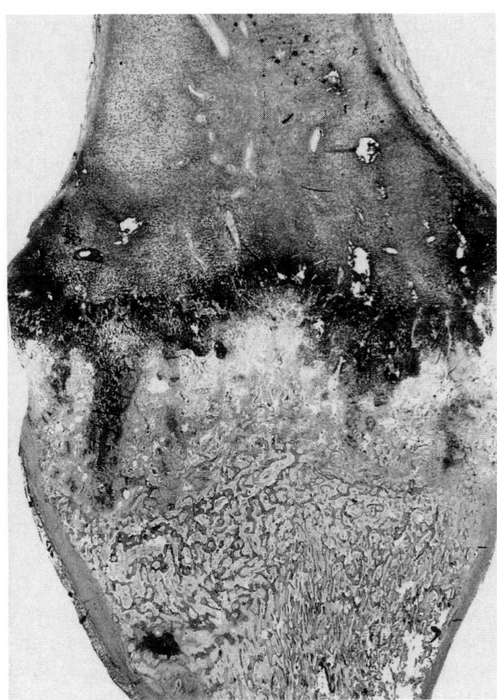

Fig. 1.48 Rickets, calf. Rib. Irregular masses of dark-staining unresorbed cartilage are left in the metaphysis. There is disorganization of metaphyseal bone.

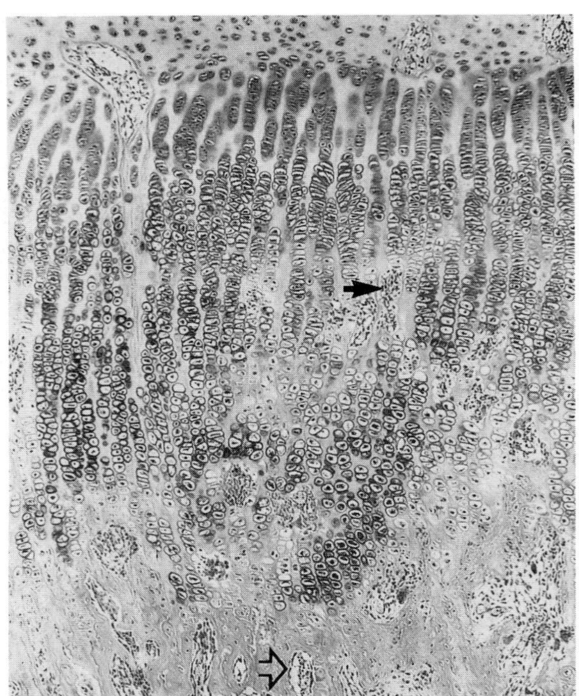

Fig. 1.49 Rickets, cat. Congenital atresia of bile duct. Humerus. Physis and metaphysis. There is irregular invasion of cartilage by blood vessels (arrow) into the irregularly lengthened hypertrophic cartilage. Pale-staining metaphyseal bone below (open arrow).

invade the mineralized cartilage accompanied by osteoblastic cells. In the vitamin D-deficient, rachitic rat, metaphyseal angiogenesis is halted or retarded, and mature arterioles, venules, and capillaries, which normally lie 350–500 μm from the physeal cartilage, are relocated in apposition to it. Vitamin D supplementation corrects this lesion within 96 hr. In the rachitic metaphysis in all species, hypertrophic chondrocytes accumulate and the nature of the vasculature is altered (Figs. 1.48, 1.49). Failure of mineralization is usually irregular; thus, persistence of cartilage is irregular, and the vascular tissue, which normally invades the columns of chondrocytes in regular loops, characteristically forms wide, irregular channels in the growth cartilage. As the cartilage continues to accumulate, its growth rate slows, possibly because of inadequate diffusible nutrient, and the hypertrophic cells form irregular clumps instead of the characteristic columns. Some of these clumps are bypassed by the irregular ingrowth of vascular and osteogenic tissue and ultimately come to lie in the metaphysis as unmineralized masses covered by osteoid. Occasionally persistent tongues of cartilage degenerate, probably as a result of abnormal stresses and inadequate diffusion of nutrients, and are replaced by connective tissue, in which metaplastic osteoid forms. Degenerate cartilage loses its basophilia and stains like osteoid, and is sometimes called pseudo-osteoid or cartilaginous osteoid. When the severity of the deficiency is reduced, plates of rachitic cartilage may be undermined by invading vessels, and left in the metaphysis, forming a grossly visible, translucent band parallel to the growth plate.

The consequences of failure of mineralization of cartilage which are important in the radiographic diagnosis of rickets are the increased depth of the physis (Fig. 1.50A,B), and the distortion of the unmineralized soft tissues of the metaphyses and epiphyses under the pressure of weight bearing. Important histologic criteria are the irregularity of the osteochondral junctions (Figs. 1.48, 1.49) and the irregular vascular invasion. Also, in vitamin D deficiency rickets, osteodystrophia fibrosa may develop as a result of the hypocalcemia which is part of the disease. In such cases the fibrous tissue is less dense and more diffuse than the fibrous callus that often forms in reaction to metaphyseal infractions and hemorrhages in rickets of any cause. The lesions in the diaphyses are similar to those in osteomalacia with accumulation of abundant unmineralized osteoid on bone surfaces.

The disturbances in osteochondral growth arising from the failure of mineralization and vascular invasion may result in gross skeletal deformities, depending on the degree of deficiency of vitamin D or phosphorus. In mild rickets the lesions may be so slight as to escape gross detection. Lesions are most prominent in those sites in which the cartilages contribute most significantly to skeletal growth. These are the physes of the proximal end of

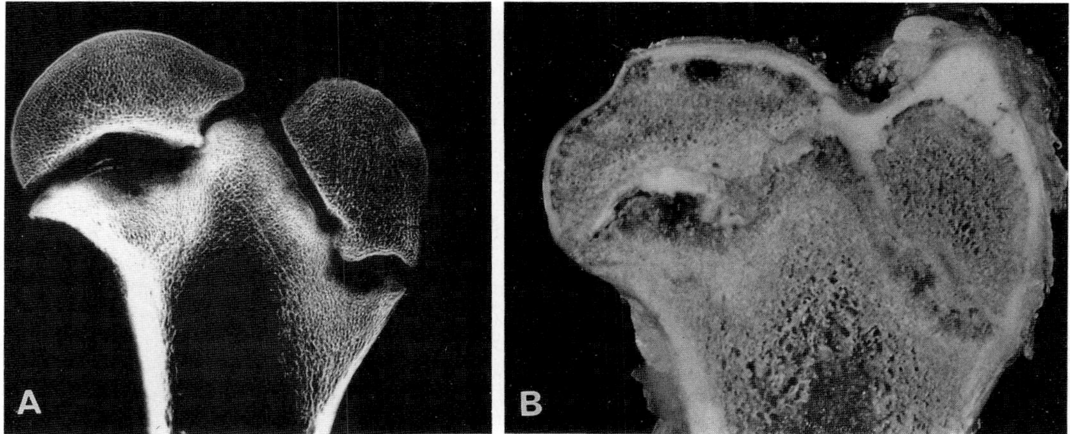

Fig. 1.50 Rickets. Pig. (A) Slab radiograph of proximal femur showing increased depth of physeal cartilage. (B) Proximal femur showing increased depth of physeal cartilage with hemorrhages in region of primary and secondary spongiosa. There is collapse of articular cartilage over the femoral head. (Courtesy of P. A. Taylor.)

the humerus, the distal ends of the radius, ribs, and femur, and both ends of the tibia and ulna, and, in ruminants especially, the distal growth plates of the metacarpal and metatarsal bones.

As a result of the primary changes in the growth plates, there is retardation or even temporary cessation of growth. This is most apparent in bones which normally grow at a comparatively rapid rate, and probably results from degeneration of physeal cartilage, which cannot be replaced. The diaphyses of these bones are shorter and broader than normal with a narrow marrow cavity. The alterations in diameter are in part compensatory and in part the outcome of normal appositional growth; compensatory thickening by the deposition of osteoid occurs in the concavities of the shaft which are most exposed to stress; normal appositional growth on the periosteal surface is not balanced by modeling resorption on the endosteal surface, since the bone is protected from osteoclasis by deposits of osteoid (Fig 1.51). The cortex, although increased in width, is soft and susceptible to weight bearing, so that curvature and fractures of long bones are common.

The spongy, highly vascular, subperiosteal osteoid may be visible grossly, particularly at the site of insertion of tendons and fascia, on the concavity of bends, and in fracture callus. The increased depth of the physeal plate is also visible grossly; the cartilage is semitranslucent and lacks the narrow chalky line of primary spongiosa. When persistence of cartilage is particularly irregular, projections from the physis protrude 1–2 mm into the metaphysis (Fig. 1.50A,B). The metaphysis is highly vascular and red, and a narrow zone of hemorrhage may be present as a result of multiple minute fractures of the trabeculae. Later the hemorrhagic zone is largely replaced by fibrous callus, and ultimately by bone. As a result of the weakening of the osteochondral junction, epiphyseal separation may occur; this most often affects the femoral head.

Shortening of the spinal column is relatively less than that of the long bones, possibly because of the greater number of growth centers in the column, each of which makes its small contribution to longitudinal growth. This means that less traumatic and ischemic damage to cartilage is likely, and reestablishment of growth patterns is possible when the nutritional deficiency abates. Spinal deformi-

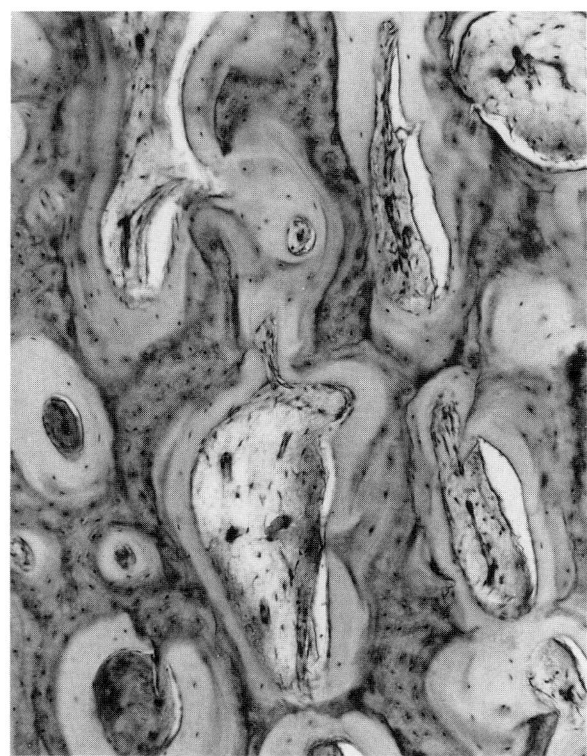

Fig. 1.51 Osteomalacia. Calf. Wide seams of pale-staining osteoid on bone surfaces.

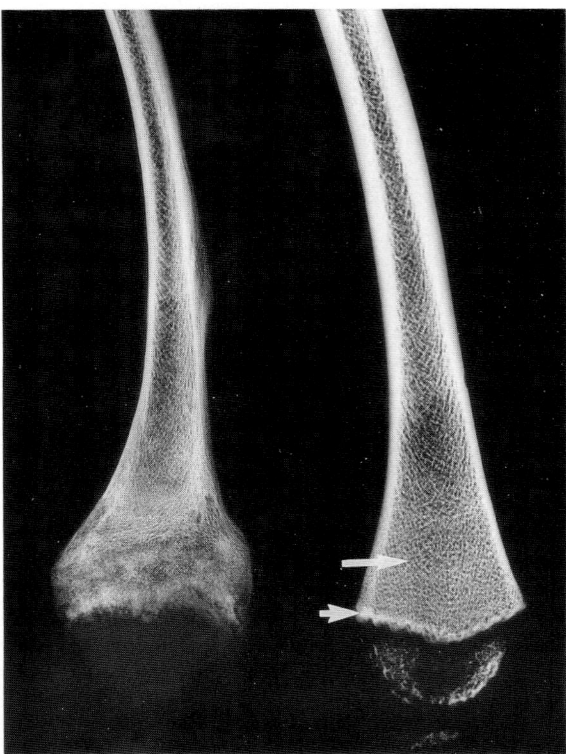

Fig. 1.52 Rickets, calf. Normal rib on right. There is metaphyseal flaring and trabecular disorganization in the rachitic bone. Note regular primary (short arrow) and secondary spongiosa (long arrow) in normal rib.

ties as kyphosis, lordosis, and scoliosis are common, with corresponding adaptational deformities of attached bones. The longer spinal processes may be awry, and the transverse processes may droop.

Enlargement of joints is a typical sign of rickets. It involves long bones and is usually accompanied by lateral or medial deviation. The enlargement is due partly to flaring of the metaphysis and partly to retarded longitudinal growth of the epiphysis and its flattening by weight bearing. Normally, the growth of bone at the physis is followed by subperiosteal resorption at the metaphysis. In rickets, osteoid (and unmineralized cartilage) persists in the metaphysis, and modeling of the shaft fails because the osteoid is resistant to resorption. This lack of resorption results in the clublike thickening in the metaphyseal region known as the rachitic metaphysis; it is most prominent at the costochondral junctions where the row of beaded metaphyses is called the rachitic rosary (Fig. 1.52). The enlargement of the ends of long bones is exaggerated somewhat by the collapse of the soft metaphysis under pressure. The flattening of the epiphysis gives it a mushroomlike shape and causes the margins to "lip" over the metaphysis.

The articular cartilages may be soft and unusually translucent over zones in which the epiphyseal trabeculae are compressed and fractured. In swine particularly, such subarticular collapse is common and causes grooving and folding of the articular surface and, ultimately, degenerative folding of the articular surface and degenerative arthropathy (Fig. 1.50B).

The articular capsules and ligaments relax, possibly as a result of hypocalcemia, and the volume of synovial fluid is much increased but can be distinguished from the product of synovitis by its lack of turbidity. When the hip joints are forcibly abducted at postmortem of a normal animal, the ligamentum teres usually ruptures; in a rachitic animal (and in one with osteochondrosis), the head of the femur or a large segment of its articular cartilage often detaches with the ligament. Distortion of the pelvis occurs as in osteomalacia.

The normal alignment of teeth depends on normal growth of the mandibular ramus. Growth of the mandibles is retarded in rickets and the spatial arrangement of the teeth is disorganized, sometimes to the extent that the jaws cannot be closed. However, the teeth are essentially normal, and eruption may be somewhat advanced.

The membranous bones of the skull show lesions similar to those under the periosteum of long bones. Osteoid is deposited and accumulates but is not mineralized. Growth at the base of the skull is subject to the same inhibitions as it is at any osteochondral junction; thus, the cranium appears more domed than normal. The fontanels may remain patent, and the sutures, wide.

The histological changes in rickets heal rapidly when the deficiency is corrected or growth ceases—in the latter case, even if the rachitogenic diet is continued. Indeed, diets grossly deficient in antirachitic factors do not produce rickets unless they also contain sufficient nutrients to permit growth. In healing rickets, remodeling may reduce the expression of mild gross deformities, but as a rule they are persistent and may cause later complications such as dwarfed stature. Pelvic deformities may cause constipation and dystocia, and epiphyseal lesions may progress to incapacitating arthropathy.

Rickets is almost always caused by a dietary deficiency of vitamin D or phosphorus. Sporadic cases are due to such lesions as bile duct aplasia, which results in failure to absorb fat-soluble vitamins. Rickets is also caused by acquired and constitutional defects in the metabolism of vitamin D (see renal osteodystrophy and vitamin D-refractory rickets).

In general, newborn animals do not have rickets, since they obtain adequate nutrients *in utero* at the expense of the dam. Cholecalciferol and 25-hydroxycholecalciferol cross the placenta, and fetal and maternal plasma levels are correlated. Calcitriol does not cross the placenta but is produced there from raw materials of maternal origin, and fetal levels are independent of, and lower than, maternal. The fetus is hypercalcemic relative to the dam, a parathyroid hormone-related protein secreted by the fetal parathyroid being necessary to maintain the gradient. Milk normally contains adequate supplies of calcitriol and its precursors, so rickets is rare in sucklings unless the dam dries up. The circumstances in which rickets occurs are discussed next in relation to the different species.

Rickets is rare in **foals** but authentic cases probably do occur. The low prevalence is probably due to the long period for which foals nurse and the relatively slow rate at which they grow.

Rickets in **calves** may be caused by phosphorus deficiency and by deficiency of vitamin D. Rickets of phosphorus deficiency is common in areas where the deficiency is enzootic, but unless the dam gives little milk, or the calf is orphaned, it is not severe. Mild rickets is prevalent in higher latitudes where animals are housed for long periods of the year, particularly when they are fed largely on grain and deprived of sunlight. Florid rickets is unusual, the more usual milder cases being characterized by enlargement of the joints, outward and forward bending at the knees and fetlocks, and overextension of the pasterns, with abnormal growth of the hooves. The forelimbs are affected more severely than the hindlimbs.

On phosphorus-deficient terrain, **sheep** develop osteophagia, but seldom clinical osteodystrophy. Rickets does occur in young sheep in the southernmost latitudes, where it is due to a deficiency of vitamin D. Two factors contribute to the deficiency: limited actinic radiation due to the low elevation of the sun and overcast skies in winter months, and the grazing of young sheep on green cereal crops during winter. Such crops are deficient in vitamin D and contain large amounts of carotene, which antagonize vitamin D. Sheep grazing under such conditions may develop rapidly fatal hypocalcemia, and many that are rachitic show neuromuscular hyperexcitability. The clinical syndrome and the osteodystrophy can be prevented and cured by vitamin D.

Piglets grow rapidly and, since they are weaned early and exposed to rachitogenic diets while the cartilages are still growing rapidly, the disease in this species tends to be florid. The usual rachitogenic factors are housing and inadequate sunlight, and lack of vitamin D in commercial feeds due to mixing errors or use of low-potency supplements. Phosphorus-deficiency rickets is not a recognized disease in pigs but is a potential complication of vitamin D deficiency. Phytate-phosphorus is the main source of the nutrient in pig feed and, unless supplemented, deficiency can develop. Absorption of phytate-phosphorus is increased by vitamin D, so when it is deficient, a concomitant reduction in absorption of phosphorus occurs.

With regard to **puppies** and **kittens** the same remarks apply as to piglets, except there is usually little phytate-phosphate in their postweaning diets.

The diagnosis of florid rickets is relatively easy, but many cases are marginal and can present problems. These problems arise because causes other than failure of mineralization permit hypertrophic cartilage to persist in the metaphyses of long bones. The most common of these are metaphyseal infractions, which tear blood vessels and interfere with the normal vascular invasion. These are likely to cause confusion when they are present in bones with other osteodystrophies. The costochondral junctions are particularly vulnerable. In pigs, lesions of osteochondrosis, especially when complicated by trauma, are sometimes difficult to distinguish from rachitic changes by routine histology. In general, visual appraisal of several bones, histologic examination of several sections including cortical bone, and attention to clinical history and serum chemistry will assist accurate diagnosis.

2. Vitamin D-Refractory Rickets

Several forms of rickets are related to idiosyncrasies in the metabolism of, or response to, vitamin D. The accurate diagnosis of such conditions often requires techniques not usually available in veterinary laboratories; nevertheless, these diseases should be considered when the usual causes of rickets are unlikely.

The vitamin D-refractory syndromes are characterized by clinical and morphologic lesions of rickets (or osteomalacia) in the presence of a normal diet and the absence of uremia, malabsorption, prolonged metabolic acidosis, or other antagonist to matrix mineralization. Several syndromes are recognized, and a few of these are mentioned next.

Vitamin D-resistant rickets (renal hypophosphatemic rickets) is characterized by hypophosphatemia, normocalcemia, decreased renal tubular reabsorption of phosphate, and skeletal deformities. Plasma calcitriol levels are inappropriately low. Hypophosphatemia results from impaired renal reabsorption and intestinal absorption of phosphorus. Combined treatment with calcitriol (to depress parathyroid hormone and thus phosphaturia) and phosphate has beneficial effects on bone lesions. The disease is inherited as an X-linked dominant trait in children and in mice.

Vitamin D-dependent rickets, types I and II, are syndromes refractory to physiologic doses of vitamin D, but are cured by pharmacologic doses. Type I is caused by deficient production of calcitriol and of 24,25-dihydroxycholecalciferol in the renal tubules. The disease is inherited as an autosomal recessive trait in humans and pigs, and sporadic cases have been suspected in dogs and cats. Heterozygotes are normal. The disease is characterized by high levels of serum parathyroid hormone and 25-hydroxycholecalciferol and low levels of active metabolites. Affected pigs are clinically normal until 4 to 6 weeks of age, but then plasma calcium and phosphate levels decrease, and alkaline phosphatase activity increases. Florid rickets develops over the next 3–4 weeks, and the pigs die unless treated.

Vitamin D-dependent rickets type II occurs in humans, and probably in the common marmoset, *Callithrix jacchus*. These animals, like other New World monkeys, cannot utilize vitamin D_2, but in addition, require unusually large amounts of D_3 and have very high levels of plasma calcitriol without hypercalcemia. Despite the elevated calcitriol, some of them develop a disease that is thought to be vitamin D-dependent rickets type II. The pathogenesis of this condition is unclear. End-organ resistance to calcitriol was suspected, but deficient production of 24,25-dihydroxycholecalciferol now seems more likely.

Bibliography

Anon. Vitamin D deficiency in the cat. *Nutr Rev* **16:** 140–141, 1958.

Audran, M., and Kumar, R. The physiology and pathophysiology of vitamin D. *Mayo Clin Proc* **60:** 851—866, 1985.

Barnes, J. E., and Jephcott, B. R. Phosphorus deficiency in cattle in the Northern Territory and its control. *Aust Vet J* **31:** 302–311, 1955.

Bonniwell, M. A. *et al.* Rickets associated with vitamin D deficiency in young sheep. *Vet Rec* **122:** 386–388, 1988.

Chan, J. C., Alon, U., and Hirschman, G. M. Renal hypophosphatemic rickets. *J Pediatr* **106:** 533–544, 1985.

Christakos, S. Gabrielides, C., and Rhoten, W. B. Vitamin D-dependent calcium-binding proteins: Chemistry, distribution, functional considerations, and molecular biology. *Endocrine Rev* **10:** 3–26, 1989.

Clements, M. R., Johnson, L., and Fraser, D. R. A new mechanism for indexed vitamin D deficiency in calcium deprivation. *Nature* **325:** 62–65, 1987.

Doige, C. E. Pathological findings associated with locomotory disturbances in lactating and recently weaned sows. *Can J Comp Med* **46:** 1–6, 1982.

Duckworth, J., Godden, W., and Thomson, W. The relation between rates of growth and rickets in sheep on diets deficient in vitamin D. *J Agric Sci* **33:** 190–196, 1943.

Duckworth, J. *et al.* Dental malocclusion and rickets in sheep. *Res Vet Sci* **2:** 375–380, 1961.

Freeman, S., and McLean, F. C. Experimental rickets. Blood and tissue changes in puppies receiving diets very low in phosphorus with and without vitamin D. *Arch Pathol* **32:** 387–408, 1941.

Gershoff, S. N. *et al.* The effect of vitamin D-deficient diets containing various Ca : P ratios on cats. *J Nutr* **63:** 79–92, 1957.

Grant, A. B. Carotene: A rachitogenic factor in green-feeds. *Nature (Lond)* **172:** 627, 1953.

Holick, M. F. The cutaneous photosynthesis of previtamin D_3: A unique photoendocrine system. *J Invest Derm* **76:** 51—58, 1981.

Hunter, W. L., Arsenault A. L., and Hodsman, A. B. Rearrangement of the metaphyseal vasculature of the rat growth plate in rickets and rachitic reversal: A model of vascular arrest and angiogenesis renewed. *Anat Rec* **229:** 453–461, 1991.

Johnson, K. A. *et al.* Vitamin D-dependent rickets in a Saint Bernard dog. *J Small Anim Pract* **29:** 657–666, 1988.

McRoberts, M. R., Hill, R., and Dalgarno, A. C. The effects of diets deficient in phosphorus, phosphorus and vitamin D, or calcium, on the skeleton and teeth of growing sheep. *J Agric Sci* **65:** 1–10, 1965.

Nisbet, D. I. *et al.* Osteodystrophic diseases of sheep. II. Rickets in young sheep. *J Comp Pathol* **76:** 159–169, 1966.

Pepper, T. A. *et al.* Rickets in growing pigs and response to treatment. *Vet Rec* **103:** 4–8, 1978.

Pointillart, A., Fontaine, N. and Thomasset, M. Effects of vitamin D on calcium regulation in vitamin D-deficient pigs given a phytate–phosphorus diet. *Br J Nutr* **56:** 661–669, 1986.

Shupe, J. L. *et al.* Clinical signs and bone changes associated with phosphorus deficiency in beef cattle. *Am J Vet Res* **49:** 1629–1636, 1988.

Smith, B. S. W., and Wright, H. Relative contributions of diet and sunshine to the overall vitamin D status of the grazing ewe. *Vet Rec* **115:** 537–538, 1984.

Stern, P. H. The D vitamins and bone. *Pharmacol Rev* **32:** 47–80, 1980.

Winkler, I., Schreiner, F., and Harmeyer, J. Absence of renal 25-hydroxy-cholecalciferol-1-hydroxylase activity in a pig strain with vitamin D-dependent rickets. *Calcif Tissue Int* **38:** 87–94, 1986.

Yamaguchi, A., *et al.* Bone in the marmoset: A resemblence to vitamin D-dependent rickets, Type II. *Calcif Tissue Int* **39:** 22–27, 1986.

E. Hyperparathyroid Disorders

1. Osteodystrophia Fibrosa

Osteodystrophia fibrosa (fibrous osteodystrophy) is a lesion that has acquired the status of a disease. It is characterized by extensive osteoclastic resorption of bone and formation of fibro-osseous tissue, and is caused by prolonged and excessive secretion of parathyroid hormone. Primary hyperparathyroidism due to idiopathic hyperplasia or functional tumor of the glands is extremely rare (see Volume 3, Chapter 3, The Endocrine Glands). More common is secondary hyperparathyroidism caused by various nutritional and metabolic derangements that lower plasma-ionized calcium, thereby stimulating synthesis and secretion of parathyroid hormone. (Parathyroid secretion is also sensitive to plasma magnesium levels, but the role of this control mechanism in osteodystrophic lesions is unclear, although minor involvement in the pathogenesis of renal osteodystrophy is suspected.)

2. Nutritional Hyperparathyroidism

The common causes of nutritional secondary hyperparathyroidism are deficiencies of dietary calcium and/or vitamin D, and excess dietary phosphorus. Each of these causes hypocalcemia, and usually two, sometimes all three, are involved in any animal with nutritional hyperparathyroidism. (Alone, vitamin D deficiency may also cause rickets and osteomalacia.) Secondary hyperparathyroidism also occurs in some animals with renal failure (see Diseases of Bones, Section VI,F, Renal Osteodystrophy).

In practice, nutritional hyperparathyroidism is caused by diets containing low calcium and a relatively high concentration of phosphorus and, with the exception of horses, affects young, rapidly growing animals. Horses seem to be remarkably sensitive to the effects of high-phosphorus diets, and remarkably resistant to the effects of rations low in phosphorus.

The role of phosphorus is not fully understood. High plasma phosphate depresses ionized calcium and thereby stimulates the release of parathyroid hormone. Plasma inorganic phosphate concentration does not directly influence parathyroid activity, but at high concentrations depresses the 1-α-hydroxylation of 25-hydroxycholecalciferol and stimulates the 24-hydroxylation. Parathyroid hormone stimulates the 1-α-hydroxylation reaction; thus, high plasma phosphate appears to stimulate the release of hormone while blocking some of its effects. Marked hyperplasia of the parathyroids sometimes occurs.

In all species, several factors influence the development and the severity of lesions in secondary hyperparathyroidism. These include the degree to which calcium is deficient and, perhaps more important, the degree to which phosphorus is in excess. This is no doubt related to the fact that, whereas the efficiency of calcium absorption decreases markedly at high intakes, that of phosphorus seems to be unchanged. Further, over a wide range of intakes, plasma calcium concentration is more sensitive to dietary phosphorus than to dietary calcium. Other factors include the age of the animal and the availability of an otherwise adequate diet, insofar as these permit or prevent optimal or maximal growth, and differences between species and individuals, insofar as these influence the sensitivity of parathyroid homeostasis. The rate of growth, the requirement for minerals, and the ability to become adapted to deficiencies and imbalances of the dietary minerals are also important.

The condition is sometimes called osteitis fibrosa and osteitis fibrosa cystica but it is not inflammatory, and cystic degeneration of the fibrous tissue is uncommon, except in humans and monkeys, and we prefer the descriptive term osteodystrophia fibrosa. In some species lesions are most severely expressed in the bones of the upper face and mandible, and for that reason, the disease is commonly known as bighead. Osteodystrophia fibrosa affects the horse and its relatives, goats, pigs, cattle, and sheep rarely, and dogs and cats, as well as a variety of more exotic species such as monkeys and iguanas.

Of the Equidae, the **horse** is more susceptible than the donkey or their hybrids. The disease occurs at any time after weaning, but the prevalence, and possibly the susceptibility, declines after about the seventh year. Horses require a calcium : phosphorus ratio of approximately 1 : 1. Diets in which the calcium : phosphorus ratio is 1 : 3 or wider can result in osteodystrophia fibrosa, depending to some extent on individual and familial susceptibility, and on alternative sources of calcium, such as the drinking water. The condition usually occurs after maintenance for some months on diets consisting largely of grain, corn, and grain by-products such as bran, hence the term bran-disease.

Osteodystrophia fibrosa also occurs in horses grazing tropical grasses high in oxalate, even though dietary calcium and phosphorus are normal. Prevalences of 10 to 15% occur in Sri Lanka and the Philippines, and the problem is also significant in northern Australia. Lush pastures are most hazardous, some containing as much as 7.8% oxalate on a dry-weight basis. Those with total oxalate over 0.5% and calcium : oxalate ratios below 0.5 are potentially dangerous because oxalate binds the calcium and makes it unavailable for absorption. Several grasses including *Setaria* spp., *Cenchrus ciliaris* (buffel grass), *Brachiaria mutica* (para grass), *Digitaria decumbens* (pangola grass), and *Panicum* spp., contain sufficient oxalate to produce clinical disease.

The early signs of osteodystrophia fibrosa consist of minor changes in gait, stiffness, transient and shifting lameness, and lassitude. Loss of appetite with progressive cachexia and anemia develop later. The anemia may be due to depression of erythropoiesis by parathyroid hormone or its metabolites. The diagnostic feature is swelling of the jaws that begins along the alveolar margins of the mandibles, producing cylindrical thickenings and reducing the intermandibular space. Almost simultaneously, the molar margins of the maxillae begin to swell, and the enlargement spreads to involve the palate, the maxillae, and the lacrimal and zygomatic bones; in severe cases, the nasal and frontal bones are swollen. Initially the swellings are soft, but later they harden. Involvement of the palate reduces the nasal passages and may cause dyspnea. Palatine and mandibular thickening causes reduction of the buccal cavity, and mastication is impaired. The face may be continually wet from occlusion of the lacrimal canals. The teeth loosen and are partially buried or exfoliated, and the softened bone yields to pressure, further impairing prehension and mastication; the food may be swallowed unchewed or allowed to drop out of the mouth. In advanced cases, the enlargement of the head may be extreme. Facial swelling is bilateral but not always symmetrical. At this stage some horses may be in good physical condition, but many are not.

Gross deformity in the remainder of the skeleton is evidence of advanced disease. The scapulae are thickened and curved so that the shoulder joint is displaced forward and the trunk droops in the pectoral girdle, giving undue prominence to the sternum. The vertebral column curves downward, sometimes upward, and the arch of the ribs is flattened.

There is a high susceptibility to fractures and to avulsion of ligaments. Fractures may result from slight trauma. Detachment of ligaments and tendons occurs chiefly in the lower limbs. Some animals escape the fractures and ligamentous detachments and pass into cachectic recumbency; attempts to force these animals to rise may result immediately in multiple fractures. Maceration of the bones from these animals reveals their porotic state; they are finely cancellous or pumicelike, and in advanced cases, have a foamy appearance and are brittle and crumbly. The weight of the macerated bones may be less than 50% of normal.

This description is of severe, untreated, nutritional hyperparathyroidism. Such cases are now rare, and gross distortion of the facial bones is not always present even in severe cases. Some animals show only loss of condition and obscure lamenesses. Many such lamenesses and many spontaneous fractures, such as of the sesamoids or phalanges, are probably expressions of mild, nonprogressive osteodystrophia fibrosa. Parathyroid hormone increases resorption of calcium, and decreases resorption of phosphate from the glomerular filtrate; thus, urinary levels of these substances are influenced by parathyroid activity. Analyses of urine from race horses in training, which are consuming rations high in grain, suggest that many of them have mild nutritional hyperparathyroidism.

Nutritional osteodystrophia fibrosa evolves from osteo-

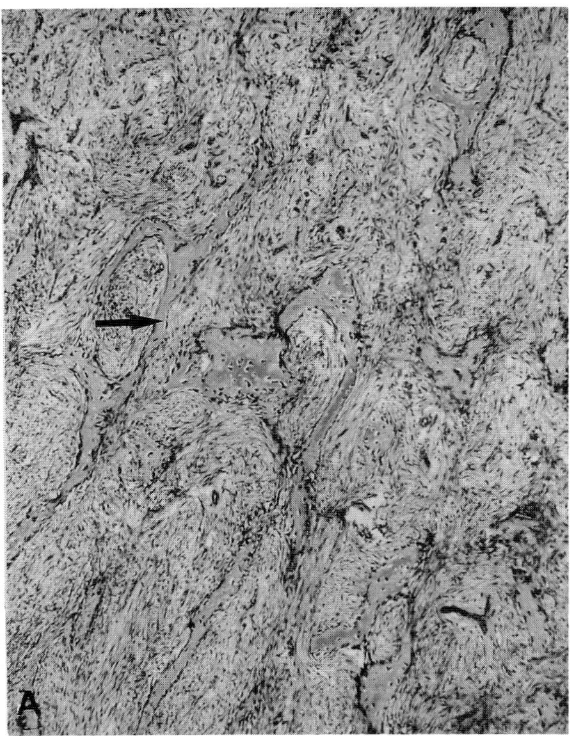

Fig. 1.53A Osteodystrophia fibrosa. Horse. Fine trabeculae of woven bone in dense osteoprogenitor tissue in facial bone.

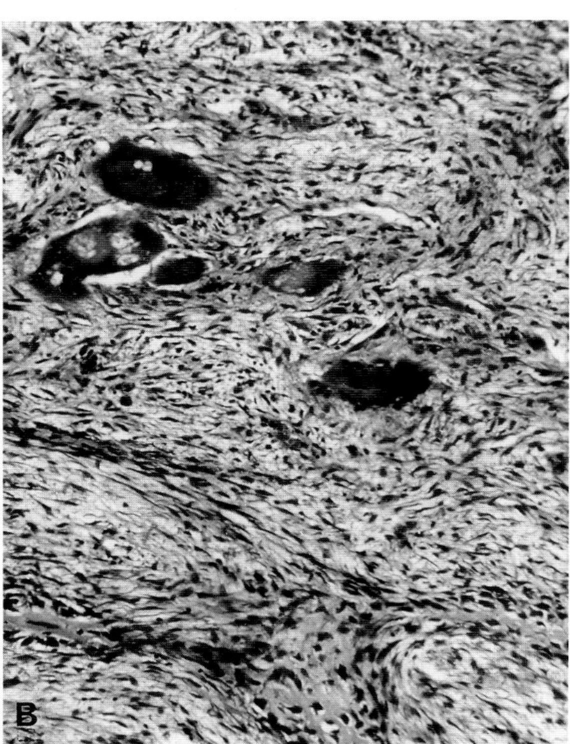

Fig. 1.53B Osteodystrophia fibrosa. Horse. Residual osteoclasts in osteoprogenitor tissue.

porosis, and evidence of this, which may be only microscopic, can be found in the long bones where there is extensive erosion of endosteal and periosteal cortical bone by osteoclasts. Parathyroid hormone causes an increase in the number of nuclei per osteoclast and then an increase in the number of cells. In severe nutritional hyperparathyroidism, osteoclasts may be extremely numerous and persistent. In the endosteum the resorbed bone is replaced by connective tissue, which initially is highly cellular and lightly fibrillar, and later more fibrous and less cellular. It originates from the osteoprogenitor cells that differentiate under the influence of parathyroid hormone and is in early stages visible as light cuffs around trabeculae. Later the medulla is replaced by fibrocellular tissue which contains irregular trabeculae. The original compact bone is resorbed from both surfaces and from within, and in places may be completely replaced by soft tissue. The abundance of fibrous tissue suggests fibroma, and the persistence within it of osteoclasts suggests giant-cell sarcoma (Fig. 1.53A,B). Hemorrhages discolor the connective tissues, which become brown as hemosiderin is formed; large hemorrhages occasionally result in cystlike cavities and, subsequently, fibrosis. Woven bone forms by metaplasia within this connective tissue.

In the periosteum, deposition of the laminar bone is begun in a normal manner, but the spaces between the laminae are filled with connective tissue (Fig. 1.54). The

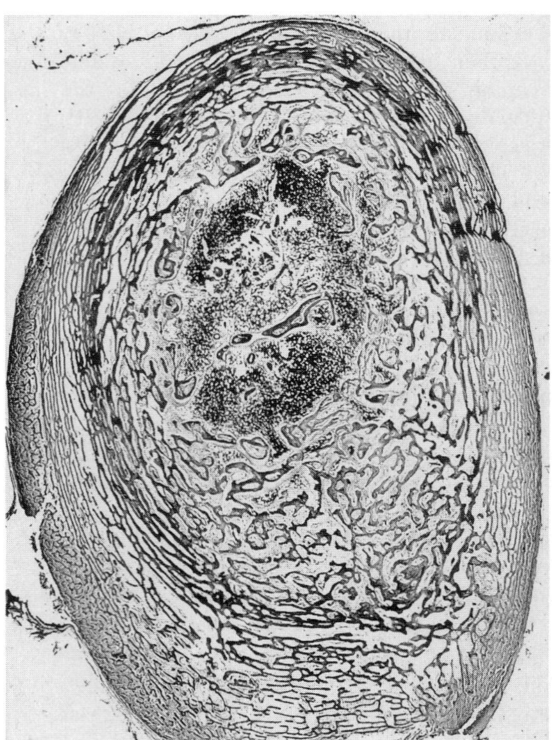

Fig. 1.54 Osteodystrophia fibrosa. Pig. Cross section of midshaft of humerus. Rarefaction of original cortex with eccentric deposition of periosteal new bone.

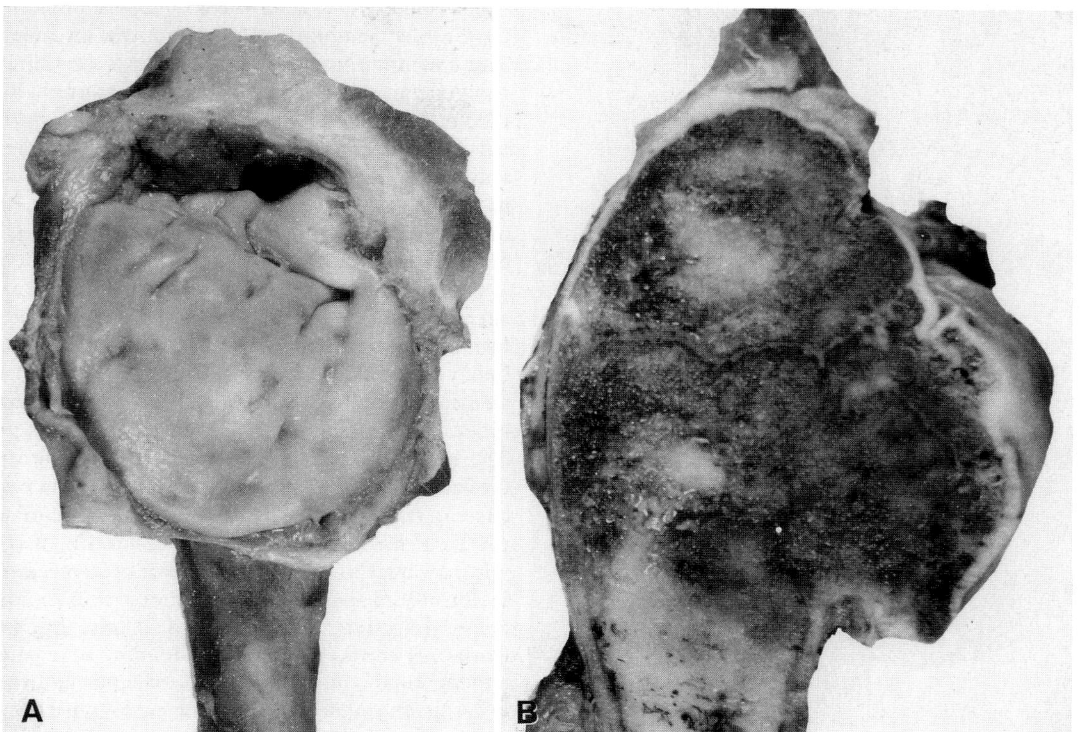

Fig. 1.55 Osteodystrophia fibrosa. Pig. (A and B) Collapse and downward deviation of articular head of humerus, wrinkling of cartilage, thickening of articular capsule.

new laminae may remain mineralized or may partially mineralize, only to be again resorbed and replaced, repeatedly and irregularly. Normally, woven bone is mineralized within a few hours of being deposited, but in secondary hyperparathyroidism the process is delayed, possibly because of reduced activity of individual osteoblasts.

The histologic changes in osteodystrophia fibrosa are most severe in the flat bones, especially those of the skull, but also in ribs, scapulae, and pelvic girdle; these too are increased in volume, and toward the costochondral junctions, the ribs frequently lose their flatness and become round or oval. The growth plates are normal unless other nutritional disease or traumatic complication occurs.

In severe cases, lesions are constantly present in the articular cartilages of the long bones (Figs. 1.55A,B, 1.56). These result from intense osteoclastic activity in the epiphyses by which not only the epiphyseal trabeculae, but also the calcified zone of the articular cartilages are extensively destroyed. The lack of support for the articular cartilage results in its herniation into the epiphysis. The depression of the cartilage is irregular without sharp margins, and the depressed surface degenerates and erodes.

Osteodystrophia fibrosa in **swine** usually occurs in young growing animals. Unsupplemented grain rations are the principal cause, and since pigs are often raised entirely indoors, vitamin D deficiency is often concomitant. Usually several animals in a group show signs, but only a few develop the pathognomonic enlargement of the skull. The mandibles may be affected first, and sometimes are the only site of gross lesions, but in other cases, the basocranium is also enlarged. Sometimes the mouth cannot be closed, and the tongue protrudes because of the swelling of the jaws and the shortness of the mandibular rami. The teeth are mobile and often deeply and obliquely embedded in the fibro-osseous connective tissue.

In gilts, **lactational osteoporosis** occurs when rations marginally deficient in calcium, and with normal or excess phosphorus, are fed over extended periods, prior to and during gestation, and during lactation. Generalized osteoporosis, and a tendency to fracture vertebrae, femurs, and phalanges, characterize the condition. The specific gravity of the bones is reduced, and that of a freshly dissected femur is usually less than 1.018 compared to a normal value of greater than 1.028 for a nonlactating young sow. Osteoporosis of this pathogenesis causes significant loss in some swine herds. The condition may be complicated by disuse osteoporosis, caused by confinement in farrowing crates, or by other dietary deficiencies. Histologic evidence of osteodystrophia fibrosa is minimal.

Severe osteodystrophia fibrosa occurs in **goats,** but lesions in **sheep** and **cattle** are usually mild. In goats, the fibroplastic enlargement chiefly involves the mandibles and maxillae and may cause respiratory distress. The relative severity of lesions in these bones in goats and in

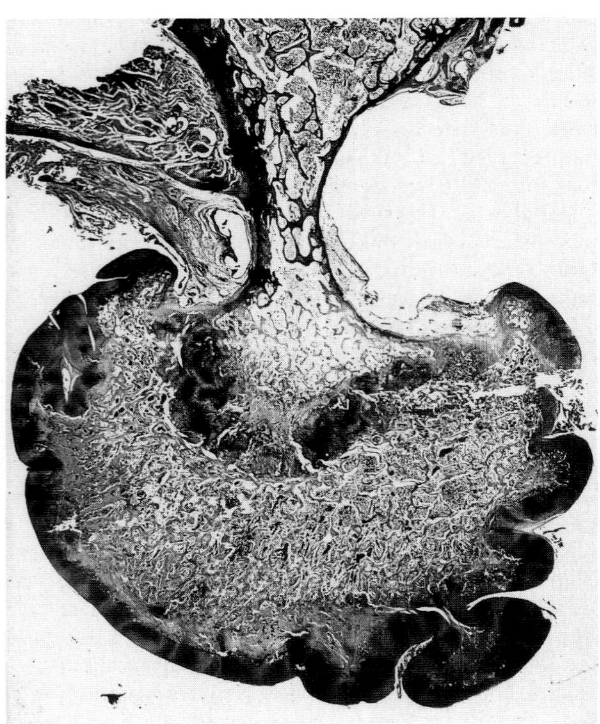

Fig. 1.56 Distal articular surface of humerus. Epiphyseal collapse with wrinkling of articular cartilage.

the other species previously described may be related to chewing, which exposes them to frequent and extreme mechanical stimuli.

In **dogs** and **cats,** which are more inclined to gulp than to chew, enlargement of the facial bones is less apparent, though facial swelling sometimes occurs. A more common lesion of the jaws in these species is loss of the lamina dura in radiographs due to resorption of the alveolar bone around the teeth (Fig. 1.57). Osteodystrophia fibrosa in dogs and cats is more fulminating than in herbivores, and signs usually begin a few weeks after weaning. It is caused by diets consisting largely or entirely of meat or offal. The calcium content of such feeds is very low, and the calcium : phosphorus ratio very wide—for example, in cardiac and skeletal muscle, the ratio is 1 : 20 and the calcium content is about 10 mg/100 g. The ideal ratio is about 1 : 1, and skeletal abnormalities develop when ratios are 1 : 2 or greater. In practice, in most animals on meat diets, the ratio is much wider. Provision of calcium alone corrects the balance.

Signs in kittens and puppies are similar to those in other species; obscure lamenesses, often precipitated by minor trauma, spontaneous fractures, lethargy, and recumbency. Radiologically and microscopically there is extreme porosity of the entire skeleton, and ventral displacement of the sternum is a characteristic finding in kittens. The bones are fragile, and fractures and deformities are common (Fig. 1.45). Attention is often drawn to the disease by the obstipation which complicates collapse of the pelvis. Deformity of the spinal column is also common, and curvature of the thorax tends to be severely angulated. Deformity of the ribs and sternum is in part adaptational to the vertebral deformity, and in part due to pressure from prolonged recumbency. Bending deformities of the appendicular skeleton, with the exception of the scapula, are unusual, at least in severe cases, probably because these animals walk very little. When they do walk, often the whole metatarsal is placed on the ground. Deformities of the long bones due to fracture may occur, and displacements, apparently due to laxity of ligaments and capsules, may be prominent. Fracture callus does form, but it is soft and never abundant, although osteogenesis in a callus is more productive than in physiological sites. Typically, in a callus, and to a lesser extent at points of tendinous insertion where some compensatory osteogenesis is attempted, cartilage is a prominent intermediary in the transition of soft tissues to bone. In severe cases, normal cortical bone is sparse, and bones are often discolored red-brown by the abundant periosteal fibrovascular tissue.

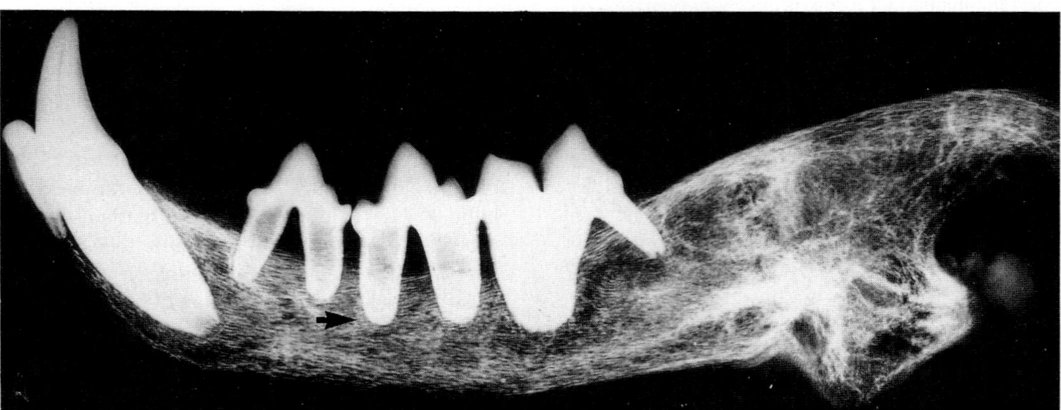

Fig. 1.57 Nutritional hyperparathyroidism. Kitten. Loss of most of the alveolar bone (arrow) and overall loss of density in mandible.

Unless complicated by rickets or by metaphyseal infractions, the growth plates are normal in depth, but resorption of primary spongiosa is excessive, and metaphyseal trabeculae are irregular in shape and do not contain mineralized cartilage cores since they are formed *de novo* in the medullary connective tissue. Numerous osteoblasts, 2–3 cells deep, often cover the spicules, and there are many osteoclasts, and moderate to abundant fibrous tissue between trabeculae. Hematopoietic marrow is displaced but not absent.

Because of the thin fragile bones seen in these young animals, this condition has been inappropriately called osteogenesis imperfecta. This is not an inherited disease, and though familial susceptibility may be a factor in some animals, these are probably a minority. With greater awareness of the importance of calcium and calcium : phosphorus ratios in pet nutrition, the incidence of osteodystrophia fibrosa is decreasing. Sometimes meat diets are deficient in vitamin D, and rickets accompanies osteodystrophia fibrosa. Correction of the vitamin deficiency repairs the rachitic changes, but apparently exacerbates the osteoporosis and osteodystrophia fibrosa, perhaps because vitamin D is essential for the action of parathyroid hormone. In the spontaneous disease, thyroid hyperplasia is also common, sometimes to the extent of exhaustion collapse of the gland. Meat is deficient in iodine, and dietary supplementation with iodine improves calcium balance and considerably slows the development of the osteodystrophy. Thyroid hormone facilitates the calcemic action of parathyroid hormone and this may account for the phenomenon.

Bibliography

Andrews, A. H., Ingram, P. L., and Longstaffe, J. A. Osteodystrophia fibrosa in young goats. *Vet Rec* **112:** 404–406, 1983.

Blaney, B. J., Gartner, R. J. W., and McKenzie, R. A. The inability of horses to absorb calcium from calcium oxalate. *J Agric Sci Camb* **97:** 639–641, 1981.

Caple, I. W., Bourke, J. M., and Ellis, P. G. An examination of the calcium and phosphorus nutrition of thoroughbred racehorses. *Aust Vet J* **58:** 132–135, 1982.

Parfitt, A. M. The actions of parathyroid hormone on bone: Relation to bone remodeling and turnover, calcium homeostasis, and metabolic bone disease. *Metabolism* **25:** 809–844, 909–955, 1033–1069, 1157–1188, 1976.

Scott, P. P., McKusick, V. A., and McKusick, A. B. The nature of osteogenesis imperfecta in cats. *J Bone Joint Surg (Am)* **45A:** 125–134, 1963.

Spencer, G. R. Porcine lactational osteoporosis. *Am J Pathol* **95:** 270–280, 1975.

Storts, R. W., and Koestner, A. Skeletal lesions associated with a dietary calcium and phosphorus imbalance in the pig. *Am J Vet Res* **26:** 280–294, 1965.

Walthall, J. C., and McKenzie, R. A. Osteodystrophia fibrosa in horses at pasture in Queensland: Field and laboratory observations. *Aust Vet J* **52:** 11–15, 1976.

F. Renal Osteodystrophy

Renal osteodystrophy is a complex syndrome, characterized by osteodystrophia fibrosa with or without osteomalacia, and sometimes occurs in animals with chronic renal failure. The descriptive synonyms, rubber jaw, renal rickets, renal osteitis fibrosa, and renal secondary hyperparathyroidism are sometimes used. We prefer the general term renal osteodystrophy because it encompasses all manifestations of this syndrome, which is well known in dogs but seldom recognized in other species.

Renal osteodystrophy results from abnormalities in the homeostasis of parathyroid hormone and, sometimes, vitamin D. Parathyroid hormone interacts principally with two organs, bone and kidney. In bone it stimulates resorption by osteoclasts and osteocytes thereby releasing calcium and phosphate to the blood. This activity is 1,25-dihydroxycholecalciferol (calcitriol)-dependent. In kidney it stimulates production of calcitriol via the 1-α-hydroxylase system, increases reabsorption of calcium and, most important, decreases reabsorption of phosphate from the tubules. Thus the net effect is elevation of plasma ionized calcium concentration.

The initial important change in renal failure, at least in relation to renal osteodystrophy, is retention of phosphate because of reduced glomerular filtration. This reduces plasma ionized calcium, thereby stimulating parathyroid hormone release, and diminishes renal production of calcitriol, which is necessary for parathyroid action on bone. In addition the underlying renal disease may render the kidney relatively ineffective in lowering blood phosphate, and prolonged hypersecretion of parathyroid hormone with enlargement of the glands (Fig. 1.58A,B) results. End-organ resistance to, and decreased breakdown of, parathyroid hormone, as well as an altered set-point (i.e., the level of ionized calcium at which PTH release is triggered) for PTH secretion, may also occur in uremic animals and be involved in the evolution of renal hyperparathyroidism.

The superimposition of osteomalacia on this lesion depends on further reduction of the glomerular filtration rate. Diminished renal production of calcitriol occurs due to high plasma phosphate, as noted above, and to loss of renal tubules. A further complication also involves the interaction of calcitriol and PTH. Acting directly on gene transcription, calcitriol at normal levels suppresses PTH synthesis and secretion. Since the parathyroid glands of uremic animals have a marked reduction in receptors for calcitriol, parathyroid response to calcitriol control tends to be impaired.

The actions and interactions of calcitriol and PTH are under constant study, as they are important in the treatment of renal failure and osteoporosis in humans, and the explanation given previously simplifies the complex series of events which leads to development of renal osteodystrophy. The mechanisms responsible for the development of the disease probably differ from case to case depending on chronicity and nature of the renal lesions, age, and individual responsiveness of the patient, and dietary levels of phosphate and vitamin D. For example, uremic dogs fed a normal phosphate diet have high plasma parathyroid hormone and florid renal osteodystrophy. Reduction of

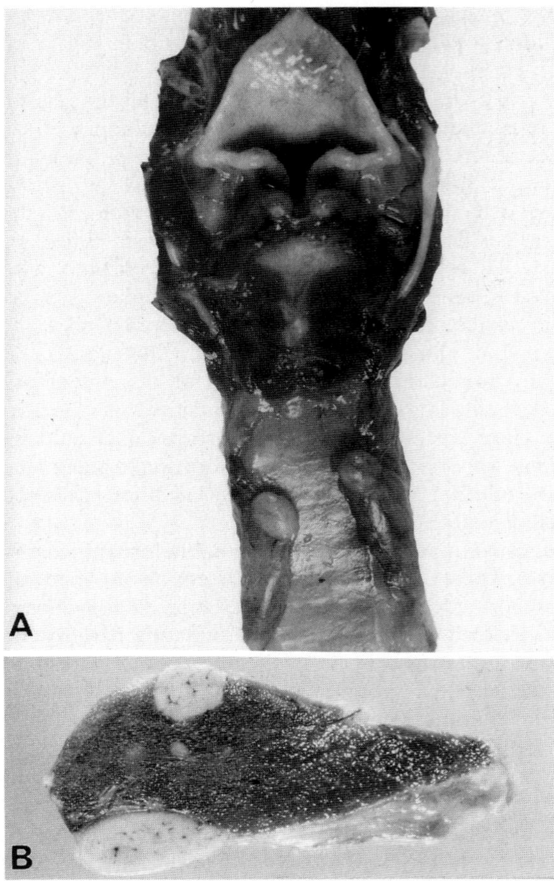

Fig. 1.58 Renal osteodystrophy. Dog. (A and B) Parathyroid hyperplasia.

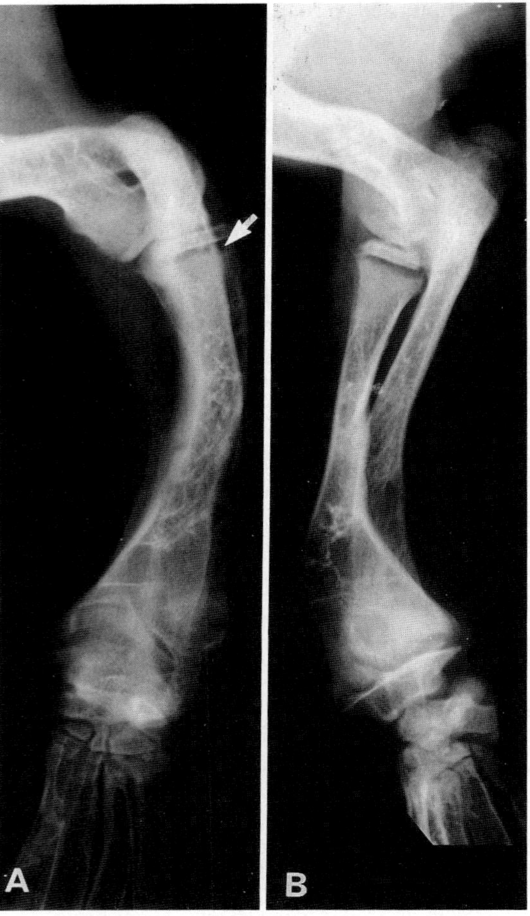

Fig. 1.59 Renal osteodystrophy. Dog. (A) Anterior radiogram of forearm showing rachitic and cystic changes. Note wide growth plates (arrow). (Courtesy T. J. Hage.) (B) Lateral radiogram of forearm showing rachitic and cystic changes. (Courtesy T. J. Hage.)

dietary phosphate reduces plasma PTH and severity of lesions. Addition of 25-hydroxycholecalciferol to the low-phosphate diet normalizes plasma PTH and prevents renal osteodystrophy.

Renal osteodystrophy is seldom diagnosed in animals unless clinical signs are advanced; the skeleton is rarely examined as a routine procedure in chronic renal disease, and signs of renal failure are usually predominant. Occasionally a stiff gait and arching of the back are seen and, in severe cases, facial swelling and marked demineralization and softening of the bones of the head may be apparent. Fractures may occur following minor trauma or manipulation.

Microscopic lesions are those of uremia (see Volume 2, Chapter 5, The Urinary System) plus skeletal disease (Fig. 1.59A,B). The osteodystrophy is generalized, but the process is more severe in the bones of the skull than elsewhere, and clinically only the skull may be involved. The severity of the changes in the skull varies considerably, but in the fully developed disease the bones are soft and resilient, and the canine teeth can be moved easily. In the hyperostotic form of renal osteodystrophy, the affected bones and face are swollen, but in the more common osteoporotic form, they are thin. Either the mandibles or the maxillae may be chiefly involved and, when severe, gross changes may be detectable in the nasal, frontal, and other bones of the face and in the zygomatic arches. The predominance of bone formation over resorption in the hyperostotic or osteosclerotic form of renal osteodystrophy is attributed to the preeminence of the anabolic effects of parathyroid hormone, but the basis for this is not clear.

Microscopic osseous lesions consist of osteodystrophia fibrosa, with or without osteomalacia. In humans with renal failure, relatively "pure" osteomalacia may occur, but this is not seen in dogs. Even in relatively young animals, histologic evidence of rickets is unlikely, since renal failure of long duration is necessary for development of renal osteodystrophy, and rickets develops only in association with fairly rapid growth; young animals with chronic renal failure typically are runts. Histologic evidence of bone resorption by numerous large osteoclasts is prominent, particularly around teeth where the alveolar bone may be removed completely (Fig. 1.60A,B). About

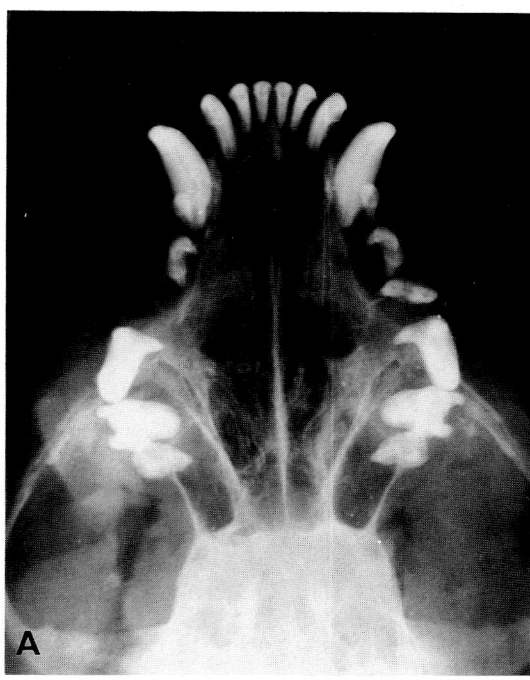

Fig. 1.60A Renal osteodystrophy. Dog. Loss of radiographic density and fracture of mandible. (Courtesy T. J. Hage.)

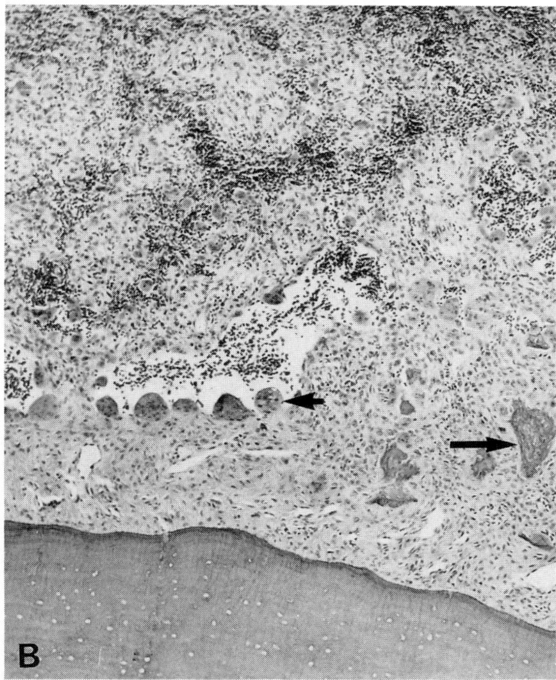

Fig. 1.60B Renal osteodystrophy. Dog. Hemorrhage and osteoclasts (short arrow) in connective tissue around cementum. Alveolar bone has been resorbed, and there is woven bone in connective tissue (long arrow).

25% of male beagle dogs that receive whole body irradiation at 2 days of age die with chronic renal failure between 8.5 and 24 months later. An increase in bone-remodeling units develops in these dogs, and there is increased intracortical porosity, more or less balanced by new bone formation. The cutting and closing cones are significantly longer in uremic dogs than in controls, and their bones contain complex remodeling sites in which the normal cone organization is disrupted. Increased resorption on endosteal and periosteal surfaces is not balanced by formation, and reduced cortical thickness results.

The renal lesion in these irradiated dogs is glomerulosclerosis, but no particular pattern of disease which leads to osteodystrophy is recognized.

Bibliography

Brodey, R. S., Medway, W., and Marshak, R. R. Renal osteodystrophy in the dog. *J Am Vet Med Assoc* **139:** 329—341, 1961.

Brown, A. J. *et al.* 1,25-(OH)2D receptors are decreased in parathyroid glands from chronically uremic dogs. *Kidney Int* **35:** 19–23, 1989.

Colussi, G. *et al.* Bone and joints alterations in uremic patients. Role of parathyroid hormone. *Contrib Nephrol* **77:** 157–167, 1990.

Cushner, H. M., and Adams, N. D. Review: Renal osteodystrophy—pathogenesis and treatment. *Am J Med Sci* **29:** 264–275, 1986.

Feinfeld, D. A., and Sherwood, L. M. Parathyroid hormone and 1,25(OH)2D3 in chronic renal failure. *Kidney Int* **33:**1049–1058, 1988.

Norrdin, R. W., and Shih, M. S. Profiles of cortical remodeling sites in longitudinal rib sections of beagles with renal failure and parathyroid hyperplasia. *Metab Bone Dis Rel Res* **5:** 353–359, 1983.

Slatopolsky, E. *et al.* How important is phosphate in the pathogenesis of renal osteodystrophy. *Arch Int Med* **138:** 848–852, 1978.

Villafane, F. *et al.* Bone remodeling in chronic renal failure in perinatally irradiated beagles. *Calcif Tiss Res* **23:** 171–178, 1977.

Weinstein, R. S., and Sappington, L. J. Qualitative bone defect in uremic osteosclerosis. *Metabolism* **31:** 805–811, 1982.

G. Scurvy

Ascorbic acid (vitamin C) is required for the hydroxylation of proline and lysine, which are essential to the formation of collagen. When ascorbic acid is deficient, there is reduction or failure in the secretion and deposition of collagen as well as an increase in the fragility and rate of degradation of that produced. Ascorbic acid also promotes cell differentiation, and in bone it facilitates the production of osteoblasts from progenitors and the hypertrophy of chondrocytes, both of which are critical to the production of endochondral bone. Most mammals synthesize ascorbic acid from glucose via glucuronic acid and gulonic acid. Some species, including humans, certain other primates, and guinea pigs, lack the hepatic microsomal enzyme L-gluonolactone oxidase and, in the absence of a dietary source of ascorbic acid, develop scurvy. Spontaneous mu-

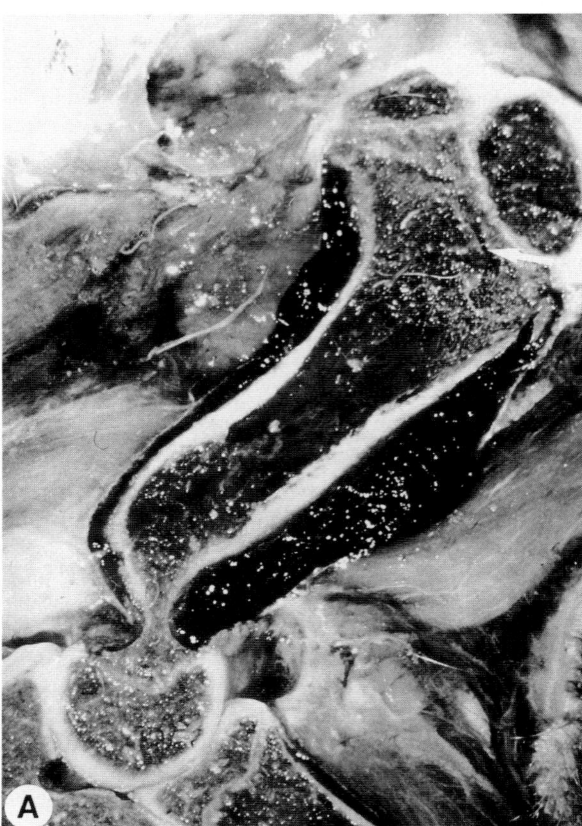

Fig. 1.61A Scurvy. Pig. Humerus. Massive subperiosteal hemorrhage and fractures across metaphysis (arrow). (Courtesy of J. Goltz.)

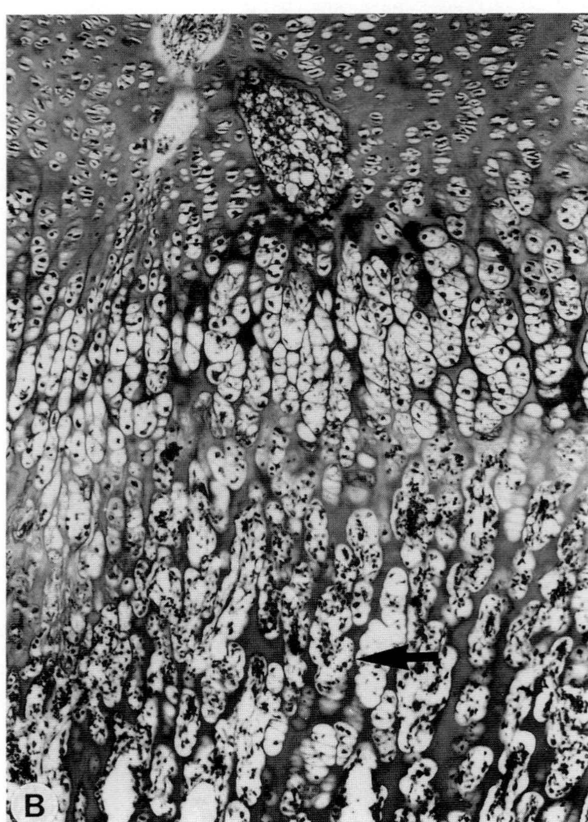

Fig. 1.61B Scurvy. Pig. Scorbutic lattice. Absence of bone deposition on cartilage frame (arrow). (Courtesy of J. Goltz.)

tations involving this enzyme are recorded in pigs and guinea pigs resulting respectively in a need for, or independence of, dietary ascorbic acid. In mutant pigs the trait is transmitted as an autosomal recessive characteristic, and the homozygotes develop classic scorbutic lesions shortly after maternal milk, a good source of ascorbic acid, becomes unavailable. Because ODS rats also lack L-gulonolactone oxidase, *od/od* homozygotes develop typical scurvy.

Pigs that lack the hepatic enzyme are normal at birth but lose condition, become reluctant to stand or move, and develop swellings around the joints 2–3 weeks following weaning. Gross lesions are dominated by massive subperiosteal accumulations of clotted blood around the shafts of the long bones, the scapulae, the bones of the head, especially the mandible, and on the ribs (Fig. 1.61A). The metaphyses are discolored by hemorrhage, fragile, and separate easily from the physes. Microscopic lesions reflect the interaction of vascular, osseous, and mechanical factors. The initial lesion is dilation of metaphyseal vessels, which tend to collapse late in the course of the disease and which are unduly fragile and prone to injury and hemorrhage. Physeal cartilage shows somewhat irregular columnization, hypertrophy, and mineralization, and deposition of bone on the mineralized template is reduced or absent because of defective collagen production (Fig. 1.61B). There is retarded differentiation of osteoblasts in the proximal metaphysis, and they are sparse and appear to have lost their polarity. Distally the metaphysis may be more or less populated by a loose network of mesenchymal cells, the Gerustmark. Osteoclasts, though individually competent, are also infrequent, and a reduced amount of cartilage matrix is modeled in the metaphysis. Thus the zone of provisional calcification accumulates, as the scorbutic lattice, until it succumbs to mechanical forces and collapses in a mass of hemorrhage, fibrin, and fragments of naked cartilage spicules, the Trummerfeldzone. Because of the high mineral content of the cartilage in the scorbutic lattice, it may be visible as a radiodense band distal to the radiolucent physis in clinical radiographs. A few millimeters distal to the scorbutic metaphyseal lesions lie the trabeculae formed before weaning (Fig. 1.62). They are generally unremarkable, save for the virtual absence of bone apposition on their surfaces. In severe scurvy, physeal growth slows or stops, probably as a result of pain, malaise, or inanition due to unwillingness to seek food.

Hemorrhages elevating the periosteum are conspicuous, but there is a marked deficiency in the expected

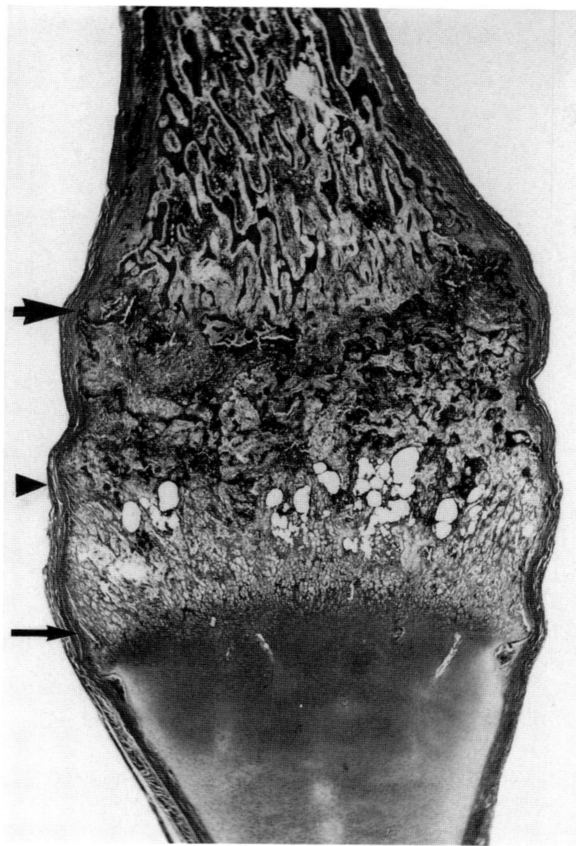

Fig. 1.62 Scurvy. Pig. Rib. Fragile spongiosa (between small arrow and arrowhead) collapses in the Trummerfeldzone. Note disoriented trabeculae. Bone distal to large arrow is normal. Cavities proximal to arrowhead are artifacts incurred during demineralization. (Courtesy of J. Goltz.)

fibroplastic and osteoplastic response. Poorly differentiated fibroblasts producing little collagen accumulate under the periosteum, and tendinous and ligamentous insertions are characterized by hypercellularity but a dearth of collagen. Thin or irregular spicules of basophilic woven bone populated by large numbers of atypical cells lie under the periosteum adjacent to the normal bone formed before weaning. The cortices are excessively narrow and porous because of the reduced bone apposition, and as a result, are excessively fragile.

Lesions are not prominent in other tissues. Hemorrhages are confined to subperiosteal regions, where they apparently result from trauma to fragile vessels. Blood coagulation is normal in scorbutic animals. Dental lesions, including arrested growth, resorption of dentin, and formation of osteoid "denticles" in the pulp, occur in the incisors of scorbutic guinea pigs, but apparently have not been sought in domesticated pigs.

Bibliography

Anon. Vitamin C regulation of cartilage maturation. *Nutr Rev* **48:** 260–262, 1990.

Bonucci, E. Fine structure of epiphyseal cartilage in experimental scurvy. *J Pathol* **102:** 219–227, 1970.

Jensen, P. T. *et al.* Congenital ascorbic acid deficiency in pigs. *Acta Vet Scand* **24:** 392–402, 1983.

H. Toxic Osteodystrophies

1. Vitamin D Poisoning

Most animals require low quantities of supplemental vitamin D, and the margin of safety, especially for cholecalciferol, is narrow. Although the median lethal dose for oral vitamin D_3 is 88 mg/kg body weight for dogs, some die from 10 to 20 mg/kg, and doses as low as 2 to 3 mg/kg can cause signs of toxicity. Cats may be even more sensitive. The increased potency of cholecalciferol over ergocalciferol is well established in chickens and New World primates. It appears that its toxicity is also augmented in some species including Old World monkeys, swine, and horses. Also, cattle and swine utilize cholecalciferol more readily than ergocalciferol, and this may partly explain the increased prevalence of accidental poisonings, since synthetic cholecalciferol is replacing ergocalciferol as the vitamin D source in feed. In contrast, rats discriminate against cholecalciferol in favor of ergocalciferol.

Fatal poisoning with vitamin D occasionally follows the over-enthusiastic or inadvertent dietary supplementation of calves, pigs, horses, dogs, and cats. When administered as one or two massive doses shortly before parturition, vitamin D appears to prevent postparturient hypocalcemia, milk fever, in cows, but mineralization of soft tissues may result, particularly in pregnant Jersey cows, and in this breed its use is contraindicated. Vitamin D is stored in muscle and fat, and meat from vitamin D-treated "downer cows" is a potential hazard to dogs and cats. In humans, individual susceptibility to toxicity exists, and depends on plasma concentration of free vitamin D metabolites, activity of 1-α-hydroxylase, and the efficiency of degradative metabolism. One or two large doses of vitamin D have also been used to prevent rickets in young sheep. Cervical vertebral ankylosis and scoliosis is reported in sheep following a single large dose of vitamin D, but the pathogenesis of these lesions is unknown. Cholecalciferol in rodenticides is a common source of acute poisoning in dogs and cats. Significant amounts of vitamin D may be secreted in milk, and soft tissue mineralization in suckling puppies has been attributed to high dietary levels in the bitch.

Several plants such as *Solanum malacoxylon* (syn. *glaucum, glaucophyllum*), *Cestrum diurnum,* and *Trisetum flavescens* cause diseases similar to hypervitaminosis D in grazing animals in various parts of the world. (See Enteque seco, Manchester wasting disease, etc. in Volume 3, Chapter 3, The Endocrine System). Traditionally vitamin D_2, ergocalciferol, is regarded as the plant form of vitamin D, but the toxic principle in these plants is 1,25-dihydroxycholecalciferol-glycoside. Lesions, including bone lesions, in poisoned animals are indistinguishable from those produced by excess vitamin D. Other calcinogenic plants in which the active principle is less well defined include *Solanum torvum, S. verbascifolium,* possibly *S. esuriale,* and *Dactylis glomerata*. Alfalfa (*Medicago sativa*) also contains vitamin D_3, as well as D_2, further contradicting the stereotype for vitamin D of plant origin.

In acute vitamin D poisoning such as occurs following fatal ingestion of a single dose, prominent gross lesions consist of gastric and small intestinal hemorrhage, and occasionally areas of myocardial discoloration. Microscopically, mucosal hemorrhage with necrosis of crypt cells, focal myocardial necrosis, and mineralization of intestinal mucosa, vessel walls, lung, and kidneys are seen. Mineralization tends to be more prominent with chronic or intermittent exposure.

The excess from a nontoxic dose of vitamin D is stored for a considerable time, mainly in fat, and is released as the fat is metabolized. The mechanisms and metabolism of toxic doses are not well understood, however. In general, skeletal effects and systemic toxicity are greatest when dietary calcium is increased, renal function or estrogen levels are diminished, or a vitamin D-hypersensitivity syndrome exists. Also, high levels of vitamin A may reduce the toxicity of vitamin D, and lesions consistent with vitamin A deficiency may develop with hypervitaminosis D. Single massive doses of vitamin D, especially with abundant available dietary calcium, may lead to persistent soft tissue injury. In renal tubules, the initial lesion is mitochondrial, and is followed by fatty and hydropic change, which leads to patchy epithelial necrosis. Administration of the vitamin promotes hypercalcemia and/or hyperphosphatemia, but the levels of these elements, or blood urea, as a reflection of renal injury, do not correlate well with severity of clinical signs of fever, anorexia, polyuria, malaise, and debility. The distribution of mineral deposits in soft tissues varies somewhat, but there is a predilection for the fibroelastic tissues of any organ, in particular the arteries, pulmonary alveolar septa, the mucosa and muscularis of the stomach, and the kidneys. The vascular deposits may extend from the internal elastic lamina through the media, and are followed by abundant fibrillary intimal proliferation, which eventually matures to a mixture of smooth muscle, collagen, and elastica. A similar sequence occurs in the endocardium. Death is usually attributable to renal failure. In rhesus monkeys (*Macaca mulatta*), three types of crystalline deposit occur with cholecalciferol toxicity—hydroxyapatite and calcium oxalate, and urates, which provoke a granulomatous reaction.

The changes in the skeleton in hypervitaminosis D are concomitant with changes in soft tissues and may be characterized by osteosclerosis or rarefaction of bone, depending on the level of dietary calcium and the pattern of exposure. The first response of bone to excess vitamin D is widespread, intense osteoclastic activity, which may remove most of the primary spongiosa and cause active resorption in other sites. With continued administration, the matrix produced by osteoblasts accumulates, sometimes in large amounts and in a distinctive pattern. It often has a tangled fibrillary arrangement and appears somewhat mucoid, floccular, and intensely basophilic (Fig. 1.63). Osteoblasts are abundant. Mineralization of this basophilic matrix is delayed, but it is gradually converted to a fairly homogeneous eosinophilic substance resembling osteoid. Initially, the maturation of the matrix is local and irregular in distribution, as is its earlier deposition, and it is unrelated to the normal patterns of osteogenesis. As toxicity is prolonged, the abnormal matrices continue to accumulate and virtually obliterate the marrow spaces. This produces a mosaic of basophilic matrix, acidophilic matrix, and newly formed mostly woven bone. The presence of abundant basophilic matrix is virtually pathognomonic for vitamin D toxicity and is diagnostically valuable when plasma levels of the vitamin are not known.

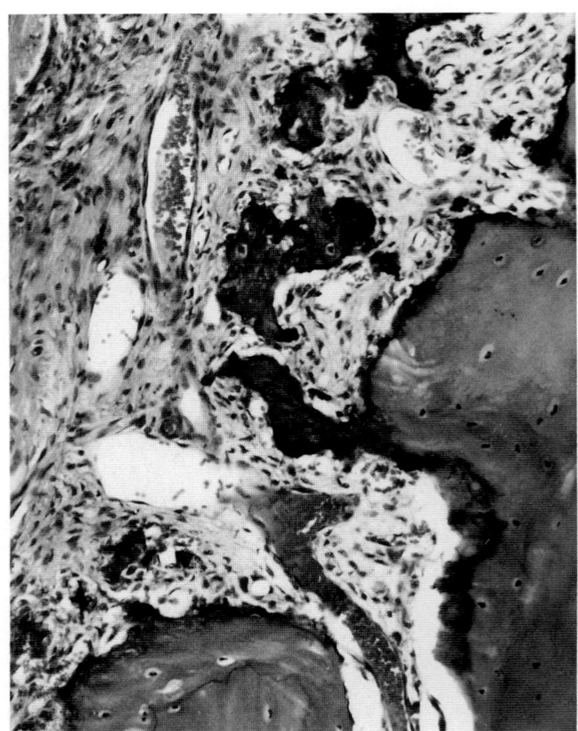

Fig. 1.63 Hypervitaminosis D. Abnormal basophilic matrix deposited on surface of trabeculae and in marrow spaces.

Intermittent administration of the vitamin, which is the more usual pattern, leads to surges of osteosynthetic activity characterized by rapid production of large amounts of abnormal matrix during periods of vitamin withdrawal. This matrix also can mature slowly and mineralize, but these processes are hastened by a further administration of the vitamin. With continued intermittent administration of vitamin D, there are further cycles of matrix deposition, maturation, and mineralization. The layers of bone produced in each cycle are separated by broad, basophilic, resting lines. Necrosis of osteocytes occurs with high doses of vitamin D, and groups of empty lacunae are often present in cortical bone and in the center of trabeculae.

Virtually all bones are affected to some extent, but the outstanding changes occur in long bones, especially in the ends where growth is most rapid. The epiphyses are usually normal and seem to escape even the resorptive changes. The growth plates are normal although metaphyseal erosion may be irregular and excessive. These abnormalities affect the metaphyseal spongiosa and extend well into the diaphysis to fill the medullary cavity. Active bone marrow is sparse, that remaining gradually being replaced by dilated veins and loose fibrous tissue, which is possibly osteoprogenitor, and which contains only a scattering of hematopoietic cells.

The sclerosis may alternate with transverse bands of resorption indicating intermittency, and current resorption of primary spongiosa may be accompanied by local intense neutrophilic reaction. Both the periosteum and endosteum are involved, and new periosteal bone contributes to the thickness of the metaphysis. Usually, the perichondrium is not affected, but sometimes it produces new cartilage and bone. In addition, there is rapid metaplastic bone formation in the thickened fibrous layer of the periosteum, which produces an almost complete new collar of bone around the metaphysis.

Experimentally in puppies, a single large dose of vitamin D produces malocclusion, pitting, irregular placement, and poor development of permanent teeth, but natural occurrence of these lesions is not recorded.

Bibliography

Boland, R. L. Plants as a source of vitamin D3 metabolites. *Nutr Rev* **44:** 1–8, 1986.

Capen, C. C., Cole, C. R., and Hibbs, J. W. The pathology of hypervitaminosis D in cattle. *Pathologia Vet* **3:** 350–378, 1966.

Clegg, F. G., and Hollands, J. G. Cervical scoliosis and kidney lesions in sheep following dosage with vitamin D. *Vet Rec* **98:** 144–146, 1976.

El Bahri, L. Poisoning in dogs by vitamin D_3-containing rodenticides. *Compend Cont Ed* **12:** 1414–1417, 1990.

Fooshee, S. K., and Forrester, S. D. Hypercalcemia secondary to cholecalciferol rodenticide toxicosis in two dogs. *J Am Vet Med Assoc* **8:** 1265–1268, 1990.

Harrington, D. D., and Page, E. H. Acute vitamin D_3 toxicosis in horses: Case reports and experimental studies of the comparative toxicity of vitamins D_2 and D_3. *J Am Vet Med Assoc* **182:** 1358–1369, 1983.

Howerth, E. W. Fatal soft tissue calcification in suckling puppies. *J S Afr Vet Assoc* **54:** 21–24, 1983.

Hunt, R. D., Garcia, F. G., and Walsh, R. J. A comparison of the toxicity of ergocalciferol and cholecalciferol in rhesus monkeys (*Macaca mulatta*) *J Nutr* **102:** 975–986, 1972.

Kasali, O. B. *et al. Cestrum diurnum* intoxication in normal and hyperparathyroid pigs. *Cornell Vet* **67:** 190–221, 1977.

Littledike, E. T., and Horst, R. L. Vitamin D_3 toxicity in dairy cows. *J Dairy Sci* **65:** 749–759, 1982.

Vieth, R. The mechanisms of vitamin D toxicity. *Bone Miner* **11:** 267–272, 1990.

2. *Vitamin A Excess and Deficiency*

Vitamin A in excess is mainly expressed as toxic injury to skeletal tissues. Vitamin A deficiency is also expressed mainly as a disturbance of hard tissues. It is convenient to consider the consequences of excess and deficiency here.

Dietary vitamin A originates from animal and plant sources. Retinyl esters of animal origin are hydrolyzed in the gut to retinol, which is bound to lipid and absorbed, then re-esterified and transported to the liver. Retinol is transported from the liver to target tissues bound to a specific transport protein, retinol-binding protein, an α-1-globulin. Provitamins such as carotenes and carotenoids of plant derivation are absorbed in the small intestine, where a molecule of β-carotene, for example, can be converted enzymatically in epithelial cells to two molecules of retinaldehyde (retinal). Most of this is reduced to the alcohol form, retinol, while a small proportion is oxidized to retinoic acid. Conversion of retinoic acid to retinol does not occur in the body. Retinoic acid is transported in plasma bound to albumin, and is not stored in the liver but is excreted in bile and urine.

Retinol is regarded as synonymous with vitamin A since it, or its derivatives, can supply all the requirements for reproduction, vision, and growth supplied by the vitamin. Retinoic acid does not support either the visual or some of the mammalian reproductive functions and it, with its natural and synthetic derivatives, are classified as retinoids.

The functions of vitamin A, other than that of retinal in relation to vision (see Chapter 4, The Eye and Ear), are incompletely understood. Retinol and retinoic acid stimulate osteoclast activity, causing them to increase acid phosphatase content and resorb bone. Development and differentiation of epithelial structures and bone are sustained by retinol, retinyl esters, retinal, and retinoic acid, but the mechanisms are unclear. It is postulated that retinol and retinoids may either act as carriers of sugar for glycoprotein biosynthesis or function through a specific receptor protein. The metabolically active form is also unknown—the model of 11-*cis*-retinal in the rhodopsin cycle does not apply to other functions. Retinyl phosphate, retinoic acid, or another metabolite, may be the active compound.

Just as the physiologic mediators of vitamin A activity are relatively obscure, so too are those of toxicity. The plasma level of retinol tends to remain constant even when toxicity (or deficiency) is imminent, and it is the predominant form of vitamin A in the blood. Retinyl esters constitute a very low proportion of total blood activity—about 5% in humans who have not recently consumed food high in vitamin A. Exposure to excess vitamin A markedly increases plasma levels of the retinyl esters once the storage capacity of the liver is exceeded, and there is evidence that they are responsible for the toxic effects. The role of retinoic acid in toxicity is less clear, but several synthetic retinoids can mimic the toxic effects of vitamin A in humans, as well as being teratogenic, although there is considerable variation in potency.

In discussing the toxicity of vitamin A and retinoids for postnatal bone, it is convenient to combine them under the heading of vitamin A, making distinctions where appropriate.

Poisoning by vitamin A is characterized by **injury to growth cartilage, osteoporosis,** and **development of exostoses.** Factors influencing the range of lesions in an animal include dose rate, length of exposure, species, age, and form of the vitamin preparation. The cartilage changes are manifest by decreased width and depth of the physes, which are due to reduced chondrocyte proliferation and reduced size of hypertrophic chondrocytes. Loss of proteoglycans and unmasking of collagen fibers occur, and sometimes with high doses, there is complete destruction of segments of growth plates. The fastest growing physes are most severely affected, and those parts that are subject to most compression seem to be most susceptible. As a result certain long bones are decreased in length. In organ culture vitamin A inhibits chondrocyte proliferation and reduces RNA and protein synthesis. Lysis of matrix may be the result of destabilization of lysosomal membranes with the release of acid proteases that attack matrix proteins.

The osteoporosis of vitamin A poisoning is associated with decreased numbers of osteoblasts and fewer, thinner osteoid seams than normal, and is most severe in cortical bone and some of the membranous bones of the skull. Periosteal formation is more severely affected than endosteal and intracortical formation, leading, in long bones, to thin cortices of reduced diameter with emphasized metaphyseal flare. Metaphyseal trabeculae tend to be fewer than normal but thicker than normal. Hypervitaminosis A interferes with cell differentiation in the embryonic skeleton and may have similar effects on osteoblasts in postnatal bone. The decrease in osteoid production also suggests a direct toxic effect on osteoblasts. The osteoporosis may be exacerbated by continued bone resorption, osteoclasts apparently being less sensitive than osteoblasts to the effects of increased vitamin A. Fractures are not a problem in farm animals and pets as they are in some laboratory species. Osteoporosis is reversible in animals removed from diets high in vitamin A, but the physeal damage is permanent, and the effects are emphasized with time.

The pathogenesis of osteophyte production is not clear but is thought to be associated with fragility of periosteal attachments, and provoked by tensions and minor trauma.

The experimental and naturally occurring syndromes produced by excess vitamin A are usefully separated according to mode of exposure, although there tends to be a carry-over of lesions between syndromes. Toxicity associated with single, large doses of vitamin A has been recognized in pigs and calves. In baby **pigs** given excess vitamin A orally as part of disease prevention programs, a characteristic syndrome develops. Grossly there is shortening of long bones and prolongation of the traction epiphyses of the tibial tuberosity, greater trochanter of the femur, and the humeral tuberosity (Fig. 1.64A,B). Rotation of epiphyses, particularly those of the distal

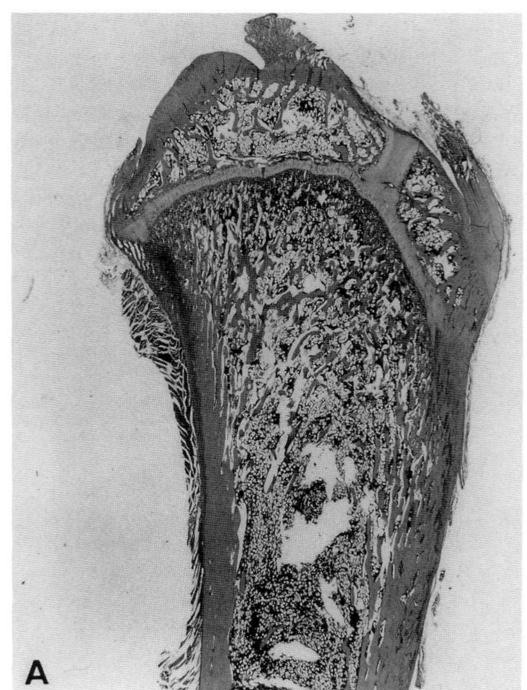

Fig. 1.64A Proximal tibia. Normal kitten. (Courtesy of L. Clark.)

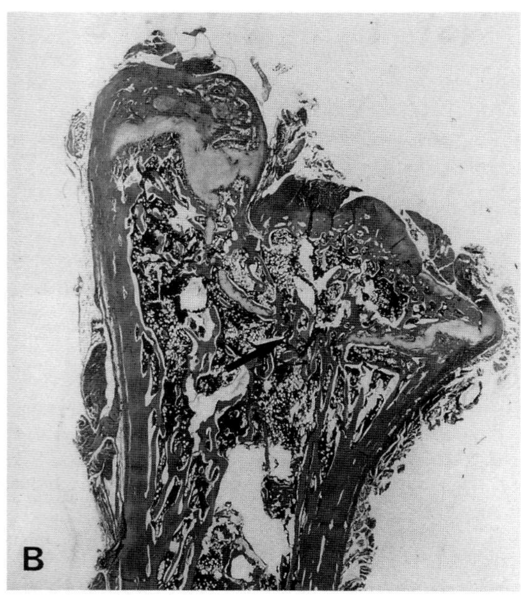

Fig. 1.64B Proximal tibia. Kitten. Vitamin A toxicity. Relative elongation of tibial tuberosity, and closure of center of growth plate (arrow). (Courtesy of L. Clark.)

femur and proximal and distal humerus, is also prominent, and tarsal and carpal bones may be distorted. Differences in length of the medial and lateral metatarsal bones, and to a lesser extent the metacarpal bones, also develop. Pigs treated shortly after birth show an abnormal gait and noticeably short legs by 6 to 8 weeks of age. The condition occurs with as little as twice the recommended dose of vitamin A. The basis for the gross lesions is destruction and focal closure of parts of the growth plates (Fig. 1.64B). Continued growth of the intact parts of the physis and superimposed adaptational deformities account for the characteristic appearance of the bones. Similar lesions in calves also result from a single exposure to high levels of the vitamin.

Comparable lesions occur in **kittens** given daily oral doses of 40 to 100 μg/g body weight of vitamin A for 4 to 5 weeks, followed by a 6 to 15 week recovery period. Lesions tend to be more severe in metacarpals, similar in tibia and distal femur, and less severe in other sites. In 7-week-old **pigs** given about 20 μg/g body weight for 5 weeks, then killed, osteoporosis is prominent. Lesions are especially severe in the squamous occipital bone, which may be so thin that the cerebellum can be crushed by manual pressure to the overlying skin. The bones are short and fragile with thin cortices, flared metaphyses, and typical physeal lesions, but the latter are less severe than at higher dose rates, and gross deformities do not occur. Exostoses may be present, however, especially near the insertion of the brachialis muscle on the proximal radius, and this lesion can develop at rates of 6.6 μg/g body weight.

Osteophyte formation is the hallmark of prolonged exposure to excess vitamin A in many species and is the outstanding feature of chronic poisoning in **cats** producing the syndrome of **deforming cervical spondylosis.** Poisoning results from prolonged feeding on bovine livers which, from grazing animals, ordinarily contain large amounts of vitamin A. The disease is common in Australia and Uruguay where beef livers are plentiful and relatively cheap but also occurs sporadically in other countries when domestic cats are given special diets.

The disease is seen in cats after 2 years of age and is characterized by postural changes, cervical ankylosis, forelimb lameness, and, in some animals, cutaneous hyperesthesia or anesthesia. The vertebral lesions can be confused with those of mucopolysaccharidosis VI. Extensive confluent exostoses develop, especially on the dorsal and lateral aspects of the cervical vertebrae and sometimes on the occipital bone, in general not involving the neural canal and only occasionally involving ventral aspects of the vertebrae (Fig. 1.65A,B). The intervertebral foraminae are considerably altered in shape and reduced in size, causing degeneration of nerves and denervation atrophy of muscles. Exostotic outgrowths occasionally occur on the anterior thoracic vertebrae, and these may be accompanied by lesions on the sternum and fixation of the ribs. Periarticular osteophytes also occur about the proximal joints of one or both forelimbs and, if extensive, may cause fixation of the shoulder and elbow joints, usually in the

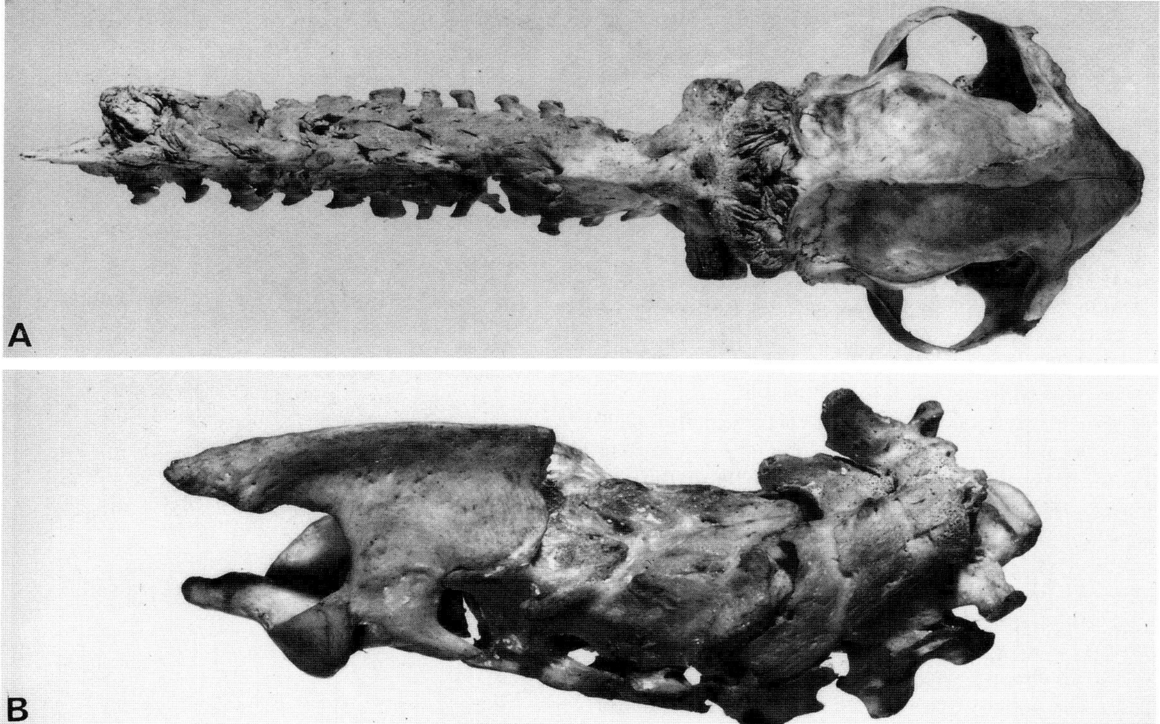

Fig. 1.65 Hypervitaminosis A. Cat. (A and B) Dorsoventral and lateral views of cervical vertebrae showing ankylosis. (Courtesy of A. A. Seawright.)

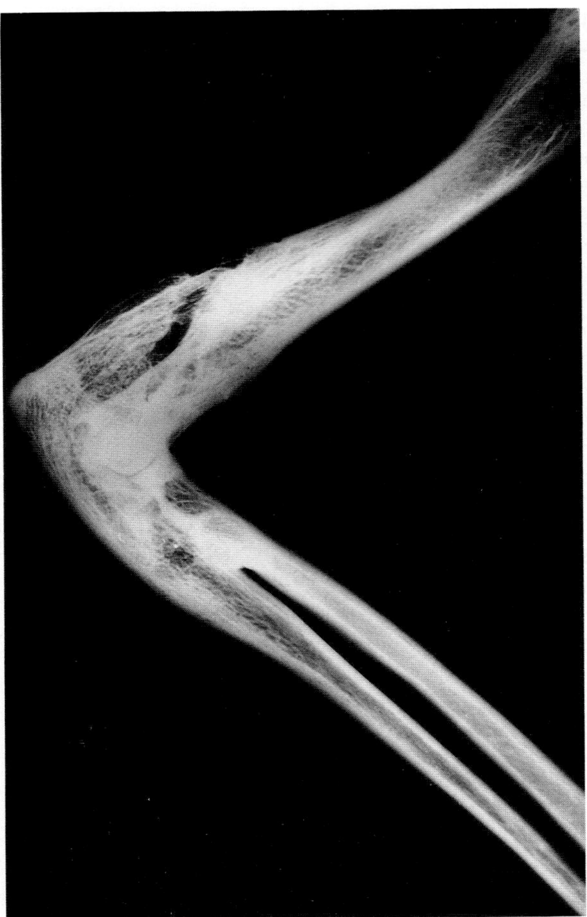

Fig. 1.66 Hypervitaminosis A. Cat. Radiographs of forelimb showing fusion of elbow joint. The lesion is chronic, and the contour of the ankylosis has been smoothed by osteoclasts.

flexed position (Fig. 1.66). The lumbar vertebrae and pelvic limbs are seldom affected. The osteophytes consist initially of osteocartilaginous tissue that overgrows joints and causes ankylosis, or of woven bone that develops at the site of muscle insertions and extends into the perimysium and sometimes encroaches on nerves. Later the woven bone and cartilage are replaced by lamellar bone. The cat's habit of grooming itself by licking its paws probably accounts for the concentration of osteophytes on the cervical spine and forelimbs. In animals with chronic lesions, patches of ungroomed fur may develop in inaccessible sites. In cats fed on bovine liver, the plasma levels of vitamin A tend to be higher and more persistent than in cats supplemented with the vitamin, even at higher equivalent dosage.

Many cats with cervical spondylosis have renal amyloidosis and dental disease. The pathogenesis of amyloidosis is not clear. Proliferative gingivitis develops in cats consuming beef liver or excess vitamin A for several months, and retardation of membranous bone formation in the jaws and alveolar processes also occurs. This leads to hypermobility of incisor teeth with subsequent exfoliation. Loss of molar teeth is not related to hypermobility and is probably secondary to gingival recession, plaque development, and the pocket formation of chronic periodontal disease. Older cats with cervical spondylosis often lose the molar crowns following carious destruction of the exposed roots, and the root remnants are left in the alveoli. Liver has a low calcium content, but excess vitamin A causes dental lesions even when dietary calcium is normal.

Several other lesions are associated with vitamin A toxicity. In some species, including humans, rats, rabbits, and dogs, it causes ectopic mineralization in internal organs. In cattle, pigs, and dogs there is decreased cerebrospinal fluid pressure and, in calves at least, this is associated with thinning of the fibrous cap of arachnoid granulations. Vitamin A is stored in the liver in Ito cells, and they increase in size with dietary excess. Peliosislike changes occur in the liver of people with hypervitaminosis A, and hemorrhages around joints and hair loss may also be seen. Alopecia, and also dermatitis, is a feature of toxicity in horses.

Excess vitamin A is **teratogenic,** the effects depending to some extent on dose, stage of gestation, and the compound administered. Teratogenicity is increased by protein or protein–energy malnutrition. Vitamin A is used extensively to treat skin disease in people and to a lesser extent in dogs. The half-life of some retinoids is many months, and anomalies in children exposed *in utero* are well recognized, but are not reported in puppies. Experimental hypervitaminosis A during pregnancy produces defects in many systems in a variety of animals. A continued high prevalence of cleft palate and lip, pulmonary hypoplasia, and abortions near term in a swine herd ceased following reduction of vitamin A supplementation to normal levels (Fig. 1.67A,B). In the embryo, excess vitamin

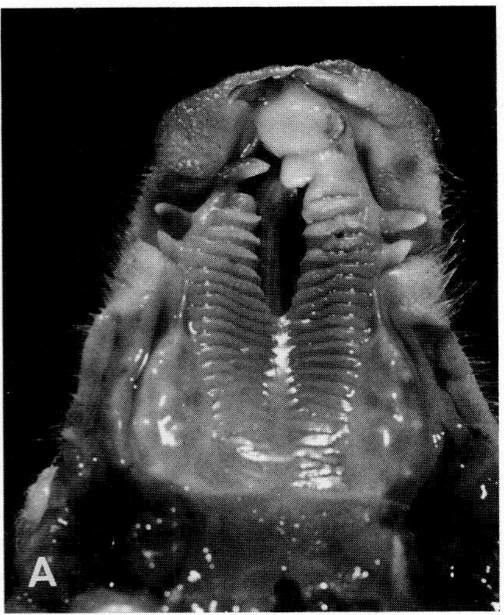

Fig. 1.67A Cleft palate and lip associated with hypervitaminosis A. Pig. Term fetus.

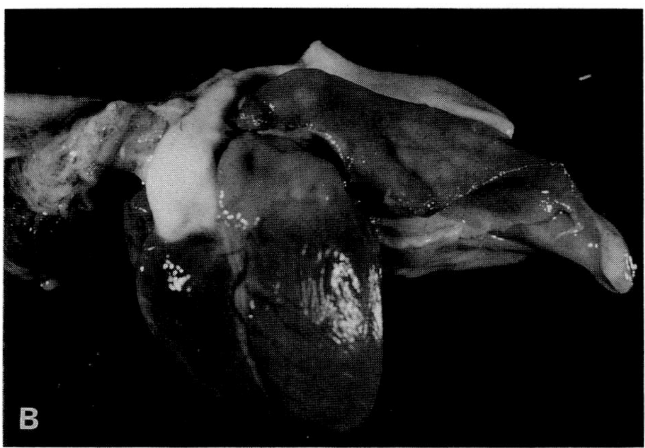

Fig. 1.67B Pulmonary hypoplasia associated with hypervitaminosis A. Pig. Term fetus.

A inhibits chondrogenesis and osteogenesis, possibly by modifying the differentiation of mesenchyme.

Vitamin A deficiency occurs in cattle and pigs fed unsupplemented rations of grain and/or old hay. Yellow corn, new hay, and fresh silage are adequate sources of carotene, but potency decreases with storage. Prominent signs of deficiency in growing cattle may include edema of the brisket and limbs, irreversible blindness, night blindness, and the neurologic effects of increased intracranial pressure. Clinical signs in growing pigs are incoordination and posterior paresis. Pigs younger than a week old may be affected if maternal supplies are low, but signs usually are seen in older animals. Gross neurologic lesions include osseous thickening of the tentorium cerebelli, cerebral edema, and coning of the cerebellum. Microscopically there is Wallerian degeneration in the spinal cord and brain (see Chapter 3, The Nervous System). The pathognomonic microscopic lesion in vitamin A-deficient cattle is squamous metaplasia of the parotid salivary duct, but this lesion resolves within a few weeks following vitamin repletion. In deficient pigs, focal squamous metaplasia in the urinary bladder may be grossly visible as 1- to 2-mm light yellow nodules on the mucosa. Sexual differences in susceptibility to vitamin A deficiency in cattle may be due to synthesis of beta-carotene in the corpus luteum and its conversion to retinol.

The nervous signs in growing animals with vitamin A deficiency may be related to cranial and spinal nerve degeneration and to hydrocephalus. The hydrocephalus probably is due mainly to decreased bulk absorption of cerebrospinal fluid into the blood, a process that occurs in the arachnoid granulations and villi. The granulations are located in the tentorium cerebelli, which is most severely affected by the thickening that occurs in the dura mater in vitamin A deficiency. This thickening results from accumulation of periodic acid-Schiff-positive glycosoaminoglycan in the connective tissue. In puppies, the tentorium cerebelli may be ossified.

The degeneration is caused by pressure on nervous tissue secondary to abnormalities of modeling of membranous bone. As a result of the modeling defects, the volumes of the cranial cavity and spinal canal are too small for the brain and spinal cord. Other sites of membrane bone are affected; the long bones, for example, may develop a coarse profile, but are less significant pathologically than the lesions in the skull. In the cranium the defect is particularly severe in the bones of the posterior fossa, and the cerebellum may herniate into the foramen magnum. In dogs the cranial nerves most severely affected are I, II, V (except the third branch) and VIII, while the remaining nerves largely or completely escape injury, some at least because of their site of exit from the cranium.

There are species differences in the nerves most severely affected. For example, in puppies deafness is a prominent sign because of changes in the internal auditory meatus, while in calves and pigs, blindness is outstanding, due to marked lesions of the optic foraminae (Fig. 1.68). The basis for these variations is not clear, but lesions are modified according to the severity of the deficiency and the stage of skeletal growth. Spinal nerve roots, especially in the cervical cord, may herniate into the intervertebral foraminae, although strangely, the lesions may not be bilaterally symmetrical. It appears that endochondral growth is not directly affected, but it can be difficult to distinguish the effects of vitamin A deficiency on bone growth from those of inanition, which may supervene in some species.

The pathogenesis of the disproportion of nervous system and skeleton is complex and is related to altered patterns of drift in bones that are growing during the period

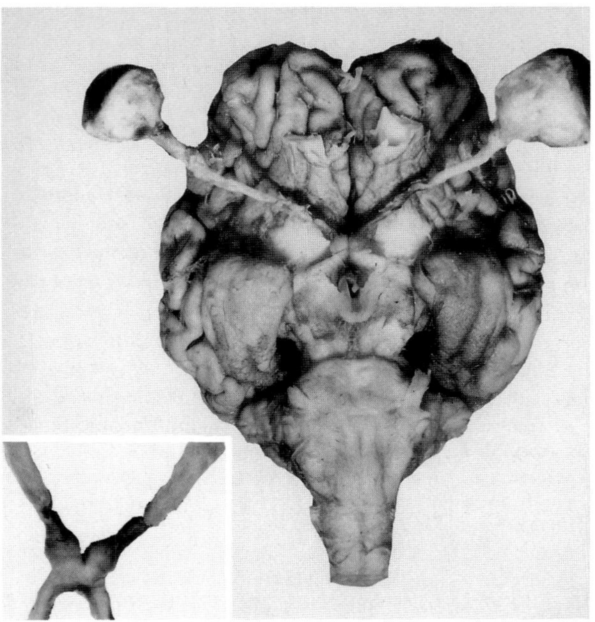

Fig. 1.68 Hypovitaminosis A. Calf. Compression of optic nerves following stenosis of the optic foramen. Inset, detail of nerves.

of deficiency. Normally osteoclasts are responsive to vitamin A, and in the cranium of deficient animals there is inadequate resorption of endosteal bone. Often bone is produced at sites where resorption should be occurring, thus exacerbating the retarded expansion of the cranial cavity and the various foraminae.

The effects of vitamin A deficiency on dental development is well studied in the continuously growing incisors of rats and guinea pigs but not in teeth of other species. Inadequate differentiation and spatial organization of odontoblasts lead to irregular formation and poor quality of dentin. Ameloblastic differentiation is also suboptimal and results in enamel hypoplasia. In addition, undifferentiated ameloblasts invade the dental pulp and stimulate the odontoblasts to form concretions there.

Neurologic signs also occur in adult cattle and in large carnivores with vitamin A deficiency and are thought to occur in adult swine. The papilledema (which is inconstant) and incoordination are attributed to increased pressure of cerebrospinal fluid. The pressure returns very rapidly to normal, and coordination is restored when vitamin A or carotene is administered, notwithstanding the Wallerian degeneration in the periphery of the spinal cord.

Vitamin A deficiency, like excess, is **teratogenic,** and swine and large Felidae appear to be very susceptible. Abortions and stillbirths and a variety of lesions including subcutaneous edema, microphthalmia, retinal dysplasia, hypotrichosis, supernumerary ears, polydactyly, arthrogryposis, cleft palate, pulmonary hypoplasia, diaphragmatic hernia, hepatic cysts, and cardiac, renal, and gonadal malformations may occur in the offspring of vitamin A deficient swine. Hydrocephalus and protrusions of spinal cord through intervertebral foraminae may also be present. The frontal and parietal bones are thin, while the basioccipital bone is thicker than normal. The latter bone is also thickened in neonatal calves from vitamin A-deficient dams, as are the squamous occipital, basisphenoid, and presphenoid bones. Stillbirths, and congenital blindness, incoordination, and thickened carpal joints are also seen in calves and, in addition, hydrocephalus with herniation of cerebellar vermis through foramen magnum, constriction and degeneration of optic nerves, and retinal dysplasia may be present. Vitamin A deficiency causes abortion in goats.

Bibliography

Biesalski, H. K. Comparative assessment of the toxicology of vitamin A and retinoids in man. *Toxicology* **57:** 117–161, 1989.

Carrigan, M. J., Glastonbury, J. R. W., and Evers, J. V. Hypovitaminosis A in pigs. *Aust Vet J* **65:** 159–160, 1988.

Clark, L. Hypervitaminosis A. A review. *Aust Vet J* **47:** 568–571, 1971.

Dobson, K. J. Osteodystrophy associated with hypervitaminosis A in growing pigs. *Aust Vet J* **45:** 570–573, 1969.

Doige, C. E., and Schoonderwoerd, M. Dwarfism in a swine herd: Suspected vitamin A toxicosis. *J Am Vet Med Assoc* **193:** 691–693, 1988.

English, P. B., and Seawright, A. A. Deforming cervical spondylosis of the cat. *Aust Vet J* **40:** 376–381, 1964.

Geelen, J. A. G. Hypervitaminosis A-induced teratogenesis. *CRC Crit Rev Toxicol* 351–375, 1979.

Hayes, K. C., and Cousins, R. J. Vitamin A deficiency and bone growth. I. Altered drift patterns. *Calc Tissue Res* **6:**120–132, 1970.

Hayes, K. C., McCombs, H. L., and Faherty, T. P. The fine structure of vitamin A deficiency. II. Arachnoid granulations and CSF pressure. *Brain* **94:** 213–224, 1971.

Mellanby, E. "A Story of Nutritional Research. The Effect of Some Dietary Factors on Bones and the Nervous System. The Abraham Flexner Lectures Series No 9." Baltimore, Maryland, Williams & Wilkins, 1950. (Details of vitamin A deficiency).

Palludan, B. The teratogenic effect of vitamin-A deficiency in pigs. *Acta Vet Scand* **2:** 32–59, 1961.

Paulsen, M. E. *et al.* Blindness and sexual dimorphism associated with vitamin A deficiency in feedlot cattle. *J Am Vet Med Assoc* **194:** 933–937, 1989.

Rosa, F. W., Wilk, A. L., and Kelsey, F. O. Teratogen update: Vitamin A congeners. *Teratology* **33:** 355–364, 1986.

Seawright, A. A., and Hrdlicka, J. Pathogenetic factors in tooth loss in young cats on a high daily oral intake of vitamin A. *Aust Vet J* **50,** 133–141, 174.

Seawright, A. A., English, P. B., and Gartner, R. J. W. Hypervitaminosis A and deforming cervical spondylosis of the cat. *J Comp Pathol* **77:** 29–39, 1967.

Van der Lugt, J. J., and Prozesky, L. The pathology of blindness in newborn calves caused by hypovitaminosis A. *Onderstepoort J Vet Res* **56:** 99–109, 1989.

3. Fluorine Poisoning

Fluorine rarely occurs free in nature but is found as fluorides in the atmosphere, soils, water, vegetation, and animal tissues. It is most common as a part of fluorite (fluorspar, calcium fluoride), fluorapatite (in rock phosphate), and in cryolite.

Acute poisoning by fluoride produces signs and lesions of gastroenteritis and may be associated with excitement, convulsions, and terminal depression. It was seen occasionally in pigs given sodium fluoride as a vermifuge, and may occur with accidental addition to feed of fluoride-containing compounds, or following ingestion of highly contaminated water. Superphosphate poisoning of sheep is due mainly to acute fluoride toxicity. Affected animals show gastrointestinal ulceration and nephrosis. Experimental acute poisoning produces renal tubular necrosis involving mainly proximal tubules. The renal toxicity of the inhalation anesthetic methoxyflurane is related to the release of inorganic fluoride following its biotransformation by hepatic cytochrome P-450 enzymes. Feed containing over 250 parts per million (ppm) or more of fluorine, on a dry matter basis, may cause acute toxicity.

Chronic ingestion of fluoride compounds causes dental abnormalities and osteodystrophy, called fluorosis. All species are susceptible, but because of the manner in which chronic poisoning occurs, fluorosis is virtually restricted to herbivorous animals, chiefly cattle and sheep. Dental fluorosis occurs in horses and wild ruminants as well as in guinea pigs and mink. The following discussion is based on fluorosis in cattle, which are more susceptible than sheep and horses.

Toxic levels may be obtained from subsurface waters, especially where rock phosphate is plentiful. Rock phosphates vary considerably in their fluoride content, and chronic poisoning has been observed in cattle and sheep given rock phosphates as "licks." Fluorine is volatile at high temperatures, and is part of the gaseous and particulate effluent from many industrial processes. Contamination of adjacent pastures, directly and by uptake of fluoride by plants, poisons grazing animals. Dust from volcanic eruptions also contains abundant fluorides. Because of the widespread distribution of fluorine in nature, many animals obtain small nontoxic doses; part may be obtained from drinking water, part from wind-blown dust that settles on pasture.

Fluoride is removed rapidly from the blood, by renal excretion and deposition in bones and teeth. A small amount is deposited in soft tissues. Some fluorine crosses the placental barrier, and although plasma levels are lower in the fetal than maternal circulation, under certain circumstances fetal fluorosis may develop. Evidence regarding fluorine levels in milk is conflicting. In ewes selective dilution or concentration of fluorine by the mammary gland are postulated. The deposition of fluoride in bone may be functionally comparable to that of other elements, such as lead and strontium, and may represent a detoxification mechanism. Unless the intake of fluoride is very low, there is a steady accumulation of the element since it is deposited as calcium fluoride or fluorapatite, which are of low solubility. The rapid deposition of fluoride in bone, such as may be observed with single high dosage, is probably due to ionic exchange on the surfaces of the crystal lattices, while the slower accretion of chronic poisoning is probably an incorporation of the element into the crystals. Many normal cattle have from 600 to 900 parts per million (ppm) in the bones, but fetal lesions apparently occur at much lower levels, for example about 100 to 200 ppm. Disease in older cattle does not develop until the level reaches 2500–3000 ppm. Forty to sixty ppm (dry matter) in feed will produce these levels in cattle after 2 to 3 years. In general, the toxicity of fluorine depends on the aqueous solubility of the compound fed; sodium compounds are more soluble and more toxic than calcium compounds.

Fluoride toxicity is enhanced by poor nutrition, and alleviated somewhat by high dietary intakes of calcium and aluminum. When cattle ingest toxic levels of fluorine, their urine usually contains 10 ppm or more; high concentrations may persist for some time after excess fluorine is removed from the diet. Therefore, diagnosis must not be based entirely on urine concentration. Plasma fluorine is also elevated during periods of exposure but declines a few days after removal of fluorine from the diet.

The pathognomonic changes of severe fluorosis occur in teeth and bones. The dental lesions result in increased abrasion of teeth, and the skeletal lesions cause shifting or severe lameness. They are accompanied by loss of production and a variety of nonspecific signs of debility.

Dental lesions develop only if intoxication occurs while teeth are in the developmental stages. Although the fetus does not accumulate high levels of fluorine, under some circumstances lesions may develop in the deciduous teeth of calves exposed during gestation. Microscopically the odontoblasts are disorganized and have vacuolated cytoplasm, and there is excessive predentin and formation of globular dentin. Fibrosis of the pulp cavity with ectopic bone formation is also seen. Lesions in permanent teeth, especially those that are last to erupt, are much more common, those that erupt first showing little or no damage. In cattle, the permanent teeth are sensitive to fluorosis between ~6 and 36 months of age. Dental fluorosis occurs with exposure during crown maturation (following formation), and when it precedes mineralization of the crown. The most severe effects occur when exposure coincides with initiation of crown formation. In areas of endemic fluorosis where ingestion of the element is usually more or less continuous, the lateral incisors of cattle show the most obvious lesions and, with the second premolars and second and third molars, the most severe lesions. The mildest macroscopic signs of dental fluorosis are small foci of mottling, which have a dry chalky appearance as compared to the normal glistening surface of enamel. The mottled areas are readily visible as opacities when the tooth is transilluminated. When the change is more severe, all the enamel in affected teeth may be chalky and opaque, predisposing to excessive attrition and chipping fractures (Fig. 1.69A,B). The teeth then wear irregularly, and in chronic cases may be worn to the gum line.

Yellow, dark brown, or black discoloration of the mottled enamel is very characteristic of fluorosis and may be a result of oxidation of the organic matrices of the teeth

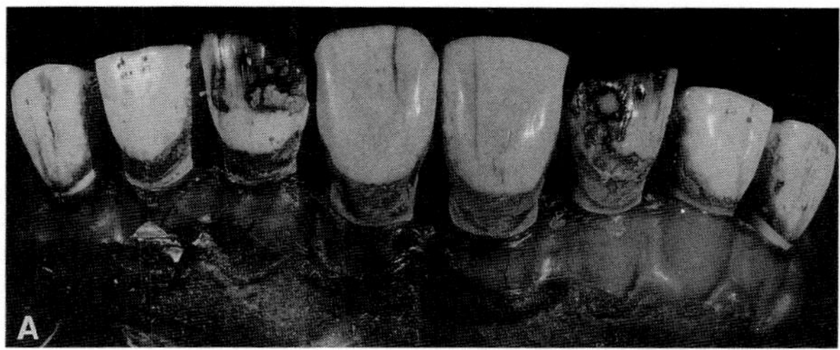

Fig. 1.69A Fluorosis. Incisors. Cow. Enamel discoloration and hypoplasia. (Courtesy of J. L. Shupe.)

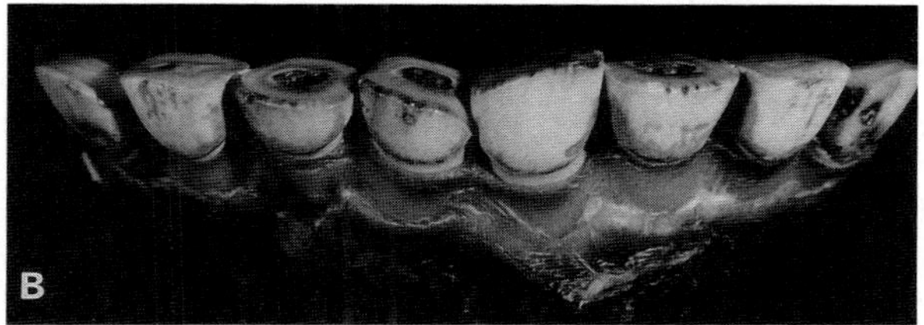

Fig. 1.69B Fluorosis. Incisors. Cow. As in (A). (Courtesy of J. L. Shupe.)

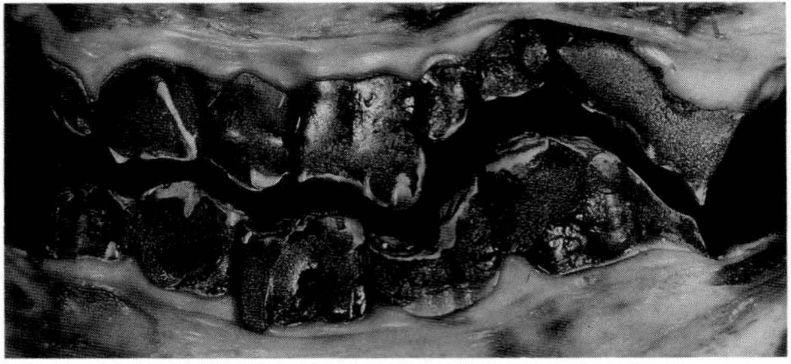

Fig. 1.70 Fluorosis. Molars. Cow. Excessive attrition. The discoloration is due to calculus and is unrelated to fluorosis. (Courtesy of J. L. Shupe.)

(Fig. 1.69A). The pigment is present in the enamel layer and possibly in the dentin; unlike the pigment of food stains and tartar, it is not limited to the surface and cannot be removed by scraping. Hypoplasia of enamel represents a more severe grade of fluorosis. Hypoplastic enamel is evident as punctate pits, or as horizontal grooves usually most prominent on the lateral aspect of the tooth. The horizontal disposition of the grooves and pits is attributable to periodic interference with mineralization of enamel during odontogenesis.

Because enamel accumulates fluorine in large quantities only during the relatively short developmental period, concentrations in fluorotic enamel are much lower than those in fluorotic bones. Normal enamel contains less than 200 ppm of fluorine, while chalky enamel has more than 400 ppm of fluorine; with very severe toxicity, concentrations greater than 2000 ppm of fluorine are reached. Fluorine concentrations in dentin approach those in bone.

Ameloblasts and odontoblasts are extremely sensitive to fluorine. It causes them to produce a matrix which mineralizes abnormally and is reduced in quality and quantity. The outer layer of enamel in particular is hypomineralized. The incremental lines in the enamel are disrupted, and the normal subsurface pigment band of bovine incisors is distorted. Dental fluorosis also causes cementum hyperplasia characterized by extension of cementum onto the coronal enamel.

In the examination of herbivores for evidence of dental fluorosis, it should be noted that the crowns of the teeth (except the incisors and three lower premolars of ruminants) are covered by cementum (see Volume 2, Chapter 1, The Alimentary System). Thus assessment of these teeth is based on the amount of attrition, and on correlations with lesions in incisors, which develop at the same time (Fig. 1.70). These associations are first incisor and

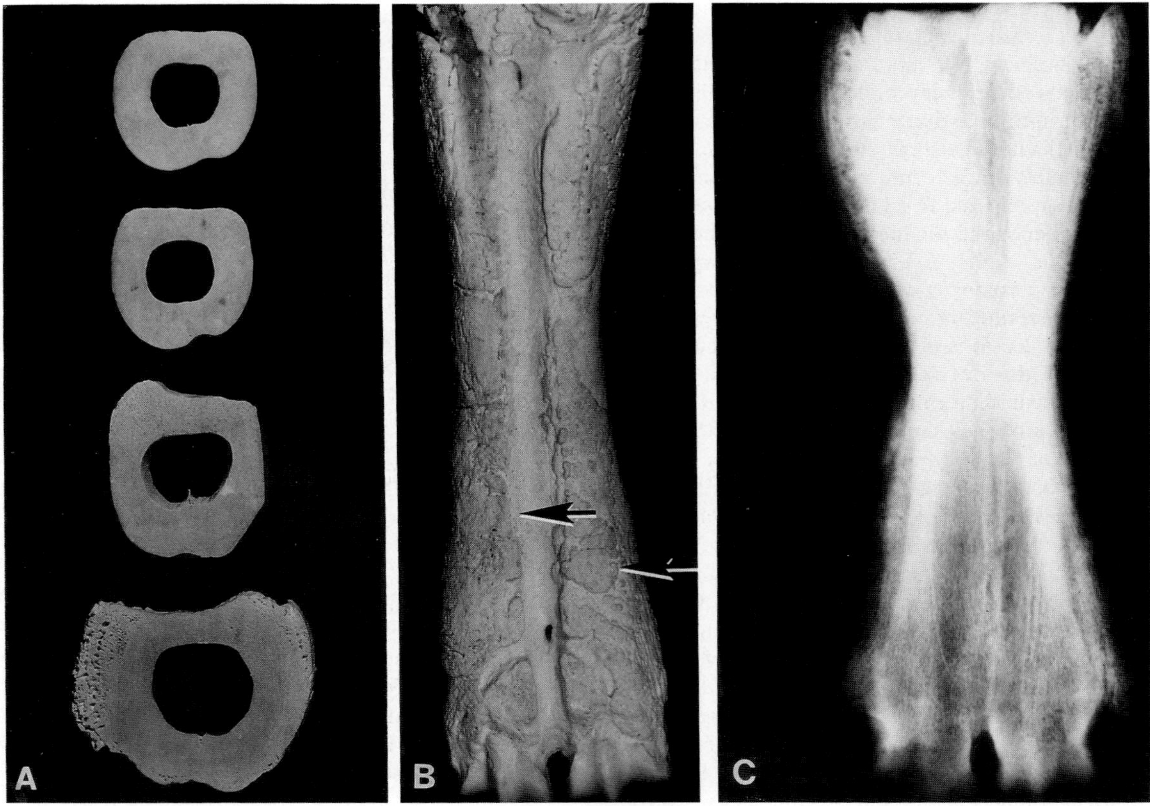

Fig. 1.71 Chronic osteofluorosis. Cow. (A) Cross sections of metatarsals. Normal (top) with increasing levels of osteofluorosis towards bottom (see text). (Courtesy of J. L. Shupe.) (B) Chronic osteofluorosis. Cow. Periosteal thickening of metatarsal (arrows), and (C) radiograph of similar bone. (Courtesy of J. L. Shupe.)

second molar (I1–M2), I2–M3 and I3–PM2 (premolar). Lesions in I2 must be severe before lesions are prominent in M3, and in general, incisor abrasion develops prior to molar abrasion. Lesions may develop in I4 in the absence of changes in other teeth.

Osteofluorosis is a generalized skeletal disturbance but not a uniform one. The obvious osseous change, development of exostoses, affects some bones, such as metatarsals and mandibles, more consistently than others. The incorporation of fluorine into skeletal mineral varies in different sites depending on relative metabolic activity, and this depends on a number of factors, including the age of the animal. Exposure of cattle after 3 years of age may produce osteofluorosis but not dental fluorosis. Young, growing bones may incorporate comparable amounts of fluorine in all parts, while adult bone may concentrate little in the relatively inert cortex, compared to the metabolically active trabeculae. Fluorine also tends to accumulate in the periosteum, and especially in the exostoses which subsequently develop. The preferential accumulation in certain regions of the bones is less evident as duration and degree of intoxication are increased.

In young, growing dogs and pigs, and presumably in other species, fluorine intoxication produces lesions which in many respects resemble rickets. The ends of long bones and the costochondral junctions are enlarged. The physes are usually increased in depth, softer than normal, and yield to pressure of weight bearing. The change at osteochondral junctions appears to be due to continued proliferation of chondrocytes, which fail to mature and align themselves. Associated with immaturity, there is a reduced amount of cartilaginous matrix and, although this appears to mineralize normally, albeit intermittently, the mineralized spicules are thin and fragile. The wide seams of osteoid which are deposited are poorly and irregularly mineralized.

Periosteal hyperostoses are widespread and bilateral; they give the macerated bones a chalky roughened appearance (Fig. 1.71). Lesions occur first on the medial surface of the proximal third of the metatarsal and later on the mandible, metacarpals, and ribs. The pelvis, vertebrae, and other bones of the distal limbs are also affected. A similar pattern of development occurs in horses. Although exostoses often develop at sites of tendinous or fascial insertions, this is not the rule, and the reasons for their localization are unknown.

Endostoses seldom occur in farm animals. Articular surfaces are normal, and lameness is due to involvement of the periosteum, encroachment of osteophytes on tendons and ligaments, and, in some cases, to mineralization and even ossification of the latter structures.

Three grades of osteofluorosis are recognized. In the first, elevated levels of fluorine are present in bone not showing structural or functional change (Fig. 1.71A). In general, for bovine bones, this represents concentrations up to approximately 2500 ppm.

The second grade is characterized by bone fluoride levels of 2500 to 5000 ppm. These bones are normal on gross

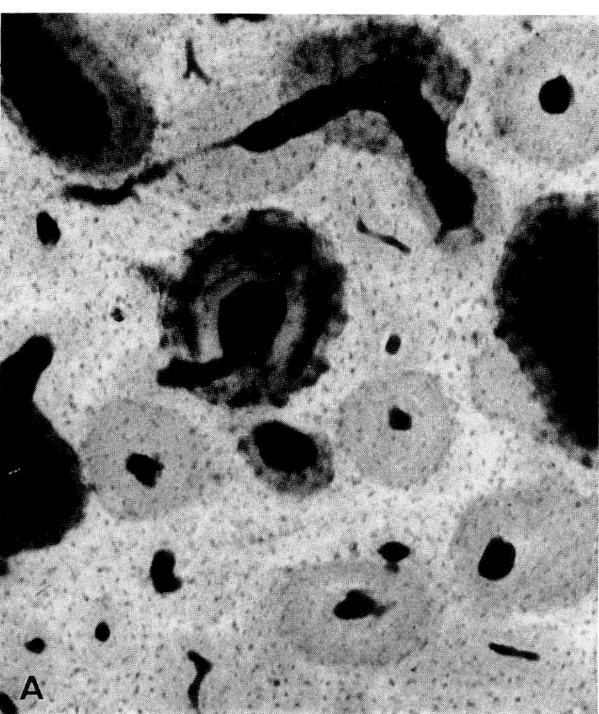

Fig. 1.72A Microradiograph showing mottled osteons of fluorosis. Sheep.

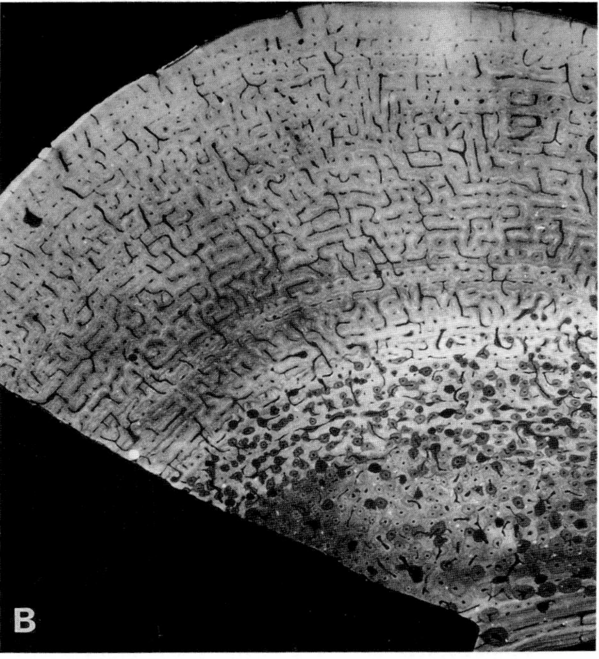

Fig. 1.72B Fluorosis. Microradiograph of bovine metatarsal. Cow. Mottled osteons in endosteal and periosteal bone. (Courtesy of J. L. Shupe.)

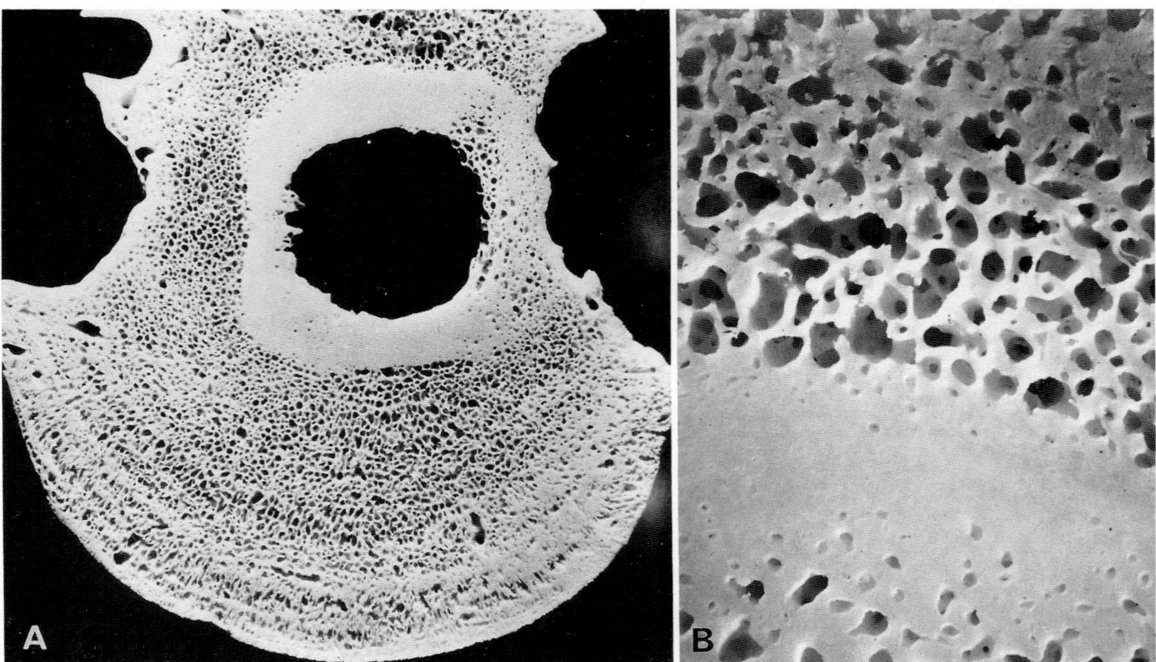

Fig. 1.73 Chronic osteofluorosis. Cow. (A) Cross section of metatarsus showing extent and nature of new periosteal bone. (Courtesy of J. L. Shupe.) (B) Close-up of (A) to show periosteal new bone above and intracortical resorption below. (Courtesy of J. L. Shupe.)

and routine radiographic examination, but microscopically there is mottling of osteons similar to that in enamel (Figs. 1.71A, 1.72A,B). The mottling is due to the effect of fluoride on osteoblasts, and its extent depends on the rates of growth or remodeling in the bone. Both endosteal and periosteal osteoblasts are affected and produce an abnormal matrix, which mineralizes abnormally. Mottling is distinguished by brown discoloration of some lamellae in some osteons or periosteal lamellae visible in unstained ground sections, and by large numbers of abnormal and tangled osteocytes in affected lamellae. In fetuses, brown mottling of laminar bone occurs at much lower concentrations of fluorine than in older cattle.

The third grade of osteofluorosis is associated with fluorine levels of 5000 to 6000 ppm or more, or with high dietary levels that may be expected to produce gross lesions rapidly (Fig. 1.71A,B,C). While new bone that is being formed is abnormal, preformed normal bone is altered in its mechanical properties and its life span reduced. The rate of remodeling is correspondingly increased both for fluoridated normal bone and for mottled bone formed under the influence of fluoride. The medullary cavity is enlarged, and resorption spreads progressively outward through the cortex, and may involve the laminar periosteal bone (Fig. 1.73A,B). Resorption cavities may be excavated much more rapidly than they can be refilled, and the cortex becomes porotic. The impaired mechanical properties induce periosteal reinforcement with laminar bone or, if the need is greater and the deposition of new bone more rapid, by cancellous coarse woven bone. When fluoride levels are very high, the new matrices produced are abnormal and remain unmineralized, as in osteomalacia.

The effects of fluorine on the skeleton vary with age, species, dose rate, and other factors as previously noted. Relatively low doses produce only an increased remodeling rate of trabecular bone in growing pigs, indicating stimulation of osteoclastic and osteoblastic activity. This may be the result of primary augmentation of resorption with coupled formation, but the effect is parathyroid independent. With relatively higher doses, fluorine is capable of interfering with mineralization of bone matrix, which accounts for the persistence and imperfect mineralization of the osteoid in periosteal osteophytes and in growing osteochondral junctions and the development of mottled osteons in cortical bone. The pathogenesis is unclear but may involve the ability of fluorine to alter the proportions of the glycosaminoglycans in bone and to interfere with the ability of proteoglycan monomers to form aggregates.

Bibliography

Editorial. Continuing controversy over dietary fluoride tolerance for dairy cattle. *Fluoride* **20:** 101–103,1987.

Johnson, L. C. Histogenesis and mechanisms in the development of osteofluorosis. *In* "Fluorine Chemistry," J. H. Simons (ed.), Vol. 4, pp. 424–441. New York, Academic Press, 1965.

Kragstrup, J., Richards, A., and Fejerskov, O. Experimental osteofluorosis in the domestic pig: A histomorphometric study of vertebral trabecular bone. *J Dental Res* **63:** 885–889, 1984.

Lucas, P. A., Ophaug, R. H., and Singer, L. The effect of vitamin A deficiency and fluoride on glycosaminoglycan metabolism in bone. *Conn Tissue Res* **13:** 17–26, 1984.

Maylin, G. A., Eckerlin, R. H., and Krook, L. Fluoride intoxication in dairy calves. *Cornell Vet* **77:** 84–98, 1987.

O'Hara, P. J., Fraser, A. J., and James, M. P. Superphosphate poisoning of sheep: The role of fluoride. *N Z Vet J* **30:** 199–201, 1982.

Shearer, T. R., Kostad, D. L., and Suttie, J. W. Bovine dental fluorosis: Histologic and physical characteristics. *Am J Vet Res* **39:** 597–602, 1978.

Shupe, J. L., Olson, A. E., and Sharma, R. P. Fluoride toxicity in domestic and wild animals. *Clin Toxicol* **5:** 195–213, 1972.

Shupe, J. L., *et al.* Relationship of cheek tooth abrasion to fluoride-induced permanent incisor lesions in livestock. *Am J Vet Res* **48:** 1498–1503, 1987.

Suttie, J. W., Clay, A. B., and Shearer, T. R. Dental fluorosis in bovine temporary teeth. *Am J Vet Res* **46:** 404–408, 1985. (No lesions found)

VII. Degenerative Diseases of Bone

A. Discontinuities of Bone and Healing of Fractures

Fractures of bone are usually the result of stress of short duration. If the bone is initially normal, the stress must be much greater than physiologic, or else be applied in a direction that is not physiologic. Fractures of bones which are initially normal are therefore termed traumatic. When the structure of bone is altered and its strength reduced by disease, fractures may occur in response to even mild physiologic stress; these fractures are termed pathologic or spontaneous.

Some types of fractures do not conform to this classification. Chip fractures are often a sequel to degenerative joint disease, and they are discussed with that topic. Fatigue (stress) fractures are the result of repetitive traumatic events that are individually innocuous but in sum cause bone failure. "Physiologic" loading of bone often leads to development of microfractures, which normally are repaired by basic multicellular units. When the capability of the repair processes is exceeded, because of failure of recruitment of enough new units or because too much bone is being remodeled, a fatigue fracture occurs. Although there is no displacement of bone, the lesion is painful and may be demonstrable in radiographs. This type of fracture is well recognized in horses with severe bucked shins. Lesions occur in the dorsal or dorsomedial cortex of metacarpal III and may be accompanied by mild periosteal proliferation. Young thoroughbred horses in training are most often affected, standardbred horses less so.

Fractures are classified further depending on their extent and nature. Thus, they may be complete or incomplete in terms of the degree of discontinuity; closed or compound when the overlying skin and soft tissues are intact or perforated, respectively; comminuted when multiple fragments occur at the site of fracture; compressed when the ends of the fragments are impacted into each other. Depressed fracture is a term usually restricted to the local displacement of fractures of flat bones. Many fractures are located in the cuboid bones of carpus and tarsus, and may be characterized by minimal or no displacement of the fragments. In contrast, avulsion fractures remove parts of bone attached to soft tissues from their natural site and usually involve apophyses or sesamoids.

Fractures may also be classified according to visibility as microfractures or gross fractures, with the former sometimes generating the latter. Microfractures involve both cortical and cancellous bone and are usually detectable only in histologic sections, although their presence may be inferred in areas of metaphyseal or subchondral hemorrhage. Multiple trabecular fractures without gross displacement of bone ends constitute an infraction.

Repair of fractures may be primary or secondary. Primary repair, either contact or gap in type, occurs when there is virtually no displacement of the bone fragments and is characteristic of microfractures in cortical bone, osteotomy sites, rigidly fixed incomplete fractures, etc. Primary repair by contact healing typically occurs at a microfracture where lamellar bone is deposited behind an osteoclastic cutting cone parallel to the bone cortex. Primary repair by gap healing, as in an osteotomy site, is characterized by deposition of lamellar bone by periosteal and endosteal osteoblasts perpendicular to the cortex. This bone is later replaced by lamellae that are parallel to the shaft. Secondary (classical) fracture healing is characterized by movement of the fragments, production of callus and deposition of fibrous bone. For descriptive purposes, the emphasis here is on the healing of a closed, complete, noncomminuted fracture in which good apposition of the ends of the fragments is obtained. Such fractures are usually incidental findings for most pathologists since healing commonly occurs without complication. Subsequently, the manner in which various complications impair healing will be discussed.

1. Repair of a Closed, Complete Fracture of a Long Bone

When the bone is fractured, numerous periosteal, intracortical, and medullary vessels are ruptured, and there is injury to adjacent soft tissues. The consequences of the vascular injury are, first, the formation of a blood clot, and, second, the cessation of local circulation with ischemia and necrosis. The severity of these consequences depends on the number and size of the vessels concerned, being maximal when the nutrient artery is involved. When the injury is solely to smaller vessels, circulation ceases back to the point of an anastomosis with an intact vessel. There is considerable regional and species difference in the richness of vascular anastomoses, but usually the extent of ischemic necrosis is small and its margins, very irregular; both osseous and medullary tissues are involved. The dead marrow is removed by liquefaction and phagocytosis; the dead bone, its extent indicated by pyknosis and eventually by dissolution of osteocytes, is removed by osteoclasis. Osteoclasts are attracted to areas of hemorrhage, but removal is a relatively slow process which continues long after the fragments are united by callus and for as long as the callus is being remodeled. However, the basal rate of cellular activity is increased 2–10 times in these areas, this being a manifestation of a regional acceleratory phenomenon activated by the injury. This enhancement of the normal remodeling processes results from an increase in the number of cells that differentiate.

Fracture repair, which includes the continuing removal of dead tissue, is arbitrarily divisible into five phases, which are of variable duration and tend to overlap. The initial phase of competent fracture repair is the direct result of the primary injury to surviving tissue and has two principal manifestations. It sensitizes the cells so that they are responsive to autocrine, paracrine, and systemic mitogens and growth factors, and it causes these substances to be released systemically, from injured tissues including bone, and from cells migrating into the damaged area. The latter include macrophages, which promote fibroplasia; platelets, which release various growth factors; leukocytes, which supply interleukin-1; and mast cells, which may be a source of angiogenic factors. The successful completion of the healing process depends largely on the timely and appropriate release of the various factors that promote production and differentiation of specialized cells at the fracture site.

The second phase is organization of the blood clot between the fragments by granulation tissue. As well as endothelial cells and fibroblasts, the invading cells include determined osteoprogenitor cells from periosteum and marrow, and pluripotent mesenchymal cells, the inducible osteoprogenitor cells derived from adjacent soft tissue and perivascular sites. Differentiated cells are induced by mediator proteins such as cartilage stimulatory activity and bone morphogenetic protein to produce cartilage or woven bone. Slight tissue alkalinity prevails at the fracture site, and this also favors bone formation.

Production of callus is the third stage of the healing process. The external callus, which is usually the most important one, is produced by the periosteum, and the internal callus, by the endosteum. The size of the callus is increased by increases in mobility of the bone fragments. The inner osteogenic layer of the periosteum proliferates, the gradient of activity decreasing away from the end of the fragment. Because of this gradient, the thickening of the periosteum, when viewed in longitudinal section, is roughly triangular, a finding which explains the spindle shape of the good callus. This periosteal growth and differentiation provide a collar of external callus around the end of each fragment. Periosteal proliferation may be under way within 24 hr of injury. Differentiation of osteoblasts with production and early mineralization of irregular trabeculae of fibrous bone is achieved within a week. Experimentally, periosteal activation agent derived from trabecular bone of growing animals can stimulate mineralization of bone in 3 days. The concentration of mineral in the matrix increases progressively but may not be visible in radiographs for 2 weeks. Even then, radiodensity is relatively low since the new fibrous bone has a low saturation point for mineral, trabeculae are relatively sparse, and mineralization tends to lag behind the formation of matrix in areas of rapid fibrous bone formation. As periosteal growth continues and the cells outstrip their blood supply, they differentiate into chondrocytes, resulting in production of hyaline cartilage in the callus, which often forms a roughly V-shaped structure more or less encircling the fracture site with the apex of the V pointing to the marrow. Cartilage formation is stimulated by movement of the bone fragments and minimized by fixation. The collars of callus grow toward each other and fuse, and the cartilage is invaded by vessels, undergoing endochondral ossification with production of woven bone in a manner analogous to that of a growth plate.

The internal callus is formed in the same manner as the external callus, but it rarely contains cartilage and usually it is smaller, although it may occlude the medullary canal temporarily.

Remodeling of the callus signals the onset of the fourth phase of fracture repair, which is dominated by osteoclastic resorption and lamellar bone production. Initially some lamellar bone is produced on the surface of fibrous spicules, but eventually these hybrid trabeculae are resorbed, and the entire callus is converted to lamellar bone with contributions from periosteal and endosteal apposition.

Modeling of the callus is the final phase of the repair. It progresses relatively rapidly and to completion in young animals, with restitution of original form and function. In older animals the process is less efficient, and residual lesions, such as a few trabeculae in the medullary cavity and a slight thickening on the periosteal surface, may persist. The stimuli for modeling activity include the stresses of weight bearing and muscle pull functioning under the aegis of cortical drift, and the effectiveness depends on the capability of the resident cells to generate basic metabolic units. The modeling process involves removal of the last necrotic bone and its replacement by lamellar bone, and the sculpting of the contours of the bone itself by removal of the last redundant trabeculae of the callus. Although osteoclastic and osteoblastic activity at the fracture site is greater than that in the contralateral bone, a continuation of the regional acceleratory phenomenon, this process takes months or years depending on the bone and the age of the animal.

2. *Complications of Fracture Repair*

Many complications may impair the repair process described. Such complications may be classified as technical and biologic, the former arising mainly from delayed or ineffective treatment, and the latter from delayed or ineffective tissue response. Technical complications account for most of the problems seen by veterinary pathologists, but some nonhealing fractures may be attributable to inexplicable failure of growth factors to be generated or to the refractoriness of cells exposed to them. When alignment of the fragments is not ideal, deformity, excessive callus, and prolonged remodeling of the callus in adaptation to functional stresses occur. The same complications attend comminution, with the additional feature that the dead fragments may be large and thus slowly resorbed. Compound fractures are infected, and if this is not adequately controlled, purulent periostitis and osteomyelitis supervene. Acute bacterial inflammation inhibits attempts at bony repair so that resorption of the necrotic bone and the formation of new bone are delayed. If the infectious process becomes chronic, osteogenesis may begin at the margins of healthy and diseased tissue and produce a large callus containing chronic inflammatory tissue and fistulae. If large pieces of bone become necrotic during the inflammatory process, or comminuted fragments are engulfed by the reaction, they may be sequestered as in osteomyelitis. When the duration of the infection is prolonged or much of the ends of the main fragment is destroyed, an osseous union will not occur, and instead, a pseudoarthrosis forms.

A pseudoarthrosis is a nonosseous union of the fragments, which permits continued mobility at the fracture site. There are many reasons for failure of osseous union including inadequate immobilization during repair with the continuance of twisting, shearing, and bending stresses; the interposition of soft tissues between the ends of the fragment to prevent the fusion of the collars of callus; osteomyelitis which inhibits reparative osteogenesis; excessive displacement of the ends of the fragments so that osseous union is not achieved before the reparative growth diminishes in activity and ceases; failure to reestablish circulation in one end of the bone; and a decrease in the regenerative capacity of the tissues, such as may occur in cachexia of any cause, senility, ischemia, and loss of neurotropism. Perhaps the most common of these causes are infection and avascularity of the fracture site and a consistent factor in many others is extensive tearing and injury to periosteum—the source of the external callus.

The simplest type of pseudoarthrosis following inadequate immobilization is a union by fibrous tissue. With time, the surfaces of the fragments become smooth and covered with compact bone. If mobility at the site of fracture continues, fibrous, or even hyaline, cartilage may form in the uniting fibrous tissue.

The highest degree of adaptation is a neoarthrosis or new joint. Clefts form in the fibrocartilaginous union and become lined by a membrane, with many of the features of the synovial membrane of a true joint, on the capsular connective tissue, and by hyaline cartilage on the apposed surfaces.

The healing of a pathologic fracture depends on the nature of the underlying disease. In a generalized osteodystrophy, the tissue of the callus is subjected to the same influences as the remainder of the skeleton. In osteodystrophia fibrosa and osteomalacia, callus forms quite rapidly and in adequate amount, but it is poorly mineralized and from the onset of mineralization is subjected to the same resorptive processes as the remainder of the skeleton. In rickets, callus forms, but the matrices remain poorly mineralized, and the replacement of the cartilage by bone is as irregular as is growth at the physes. Healing occurs if the basic osteodystrophy is corrected.

3. Repair of a Fracture In Which One Fragment is Dead

Probably the most common location for this type of fracture is the proximal femur. Slipped capital femoral epiphysis occurs in calves, foals, cats, dogs, and pigs. Most cases in calves are associated with forced extraction at birth. In dogs and cats, uncomplicated trauma usually is incriminated, whereas in pigs, sheep, and adult cattle, these fractures are presumably pathologic, being associated with osteochondrosis, molybdenum toxicity, and the osteomalacia of phosphorus deficiency respectively. In general, physes are resistant to most types of trauma except shearing and avulsion. Slipped epiphyses should be distinguished from Legg–Calvé–Perthes disease, in which epiphyseal separation, if it occurs, is a late event, and from fractures of the femoral neck.

The neck and head of the femur are vascularized chiefly by vessels of the articular capsule; in the young, the artery of the ligamentum teres makes some contribution, and in the adult, the metaphyseal vessels which reach the epiphyses also contribute, but these collaterals are of relatively minor importance. The long course of the capsular vessels from the point where the synovial membrane is reflected to the articular margin provides adequate opportunity for injury to the blood supply when the head or neck are displaced.

If there is complete separation of the head from the neck of the femur, the bone and marrow tissues of the head undergo avascular necrosis even if there is good apposition of the fracture by pinning. Gradually fibrovascular tissue, accompanied by osteogenic cells, invades across the fracture line and revascularizes the marrow spaces. Woven bone is deposited on the surface of the dead trabeculae, and some resorption of these trabeculae occurs. However, the net effect is an increase in the quantity of bone tissue in the femoral epiphysis, which is evident as increased radiodensity. The gradual invasion with bone formation across the fracture site by osteogenic tissue is often referred to as creeping substitution. This term is inappropriate to describe repair of dead cancellous bone, since initially bone tissue is added and only after a considerable time is live bone substituted for resorbed dead bone. Creeping substitution aptly describes primary repair of compact bone where tissue is removed by a cutting cone before space is available for its substitution.

As with the repair of a fracture in a long bone, various factors may interfere with the healing process, and various complications may occur. Peculiar to this type of fracture is the presence of an intact growth plate, which may provide a barrier to the movement of repair tissue. In such cases the neck, freed from many of the normal stresses of locomotion and weight bearing, may suffer disuse atrophy, whereas the avascular head, being isolated from resorptive tissue, retains its normal radiodensity for a long time. Osteoarthropathy commonly results from this type of injury as a result of resorption of cartilage and subchondral bone and subsequent collapse. Bone in which the cells are dead is structurally sound, and the microfractures are a sequel to weakening of bone by resorption.

Epiphyses often separate from metaphyses through the zone of hypertrophic cartilage where collagen fibers, which give hyaline cartilage its strength, are relatively least numerous. Growth plates suffer other types of injury, and the nature of the variations determines the pathologic and clinical outcome. Most of these injuries are not associated with complete disruption of blood supply, and in such cases, relatively complete healing is to be expected.

4. Premature Closure of a Growth Plate

Premature closure means earlier than normal cessation of activity by growth cartilage. Some lesions of this type are caused by hypervitaminosis A and radiation injury, and manganese deficiency may be responsible for physeal closure in the fetus. More often the lesions are believed to be caused by trauma, and many of them undoubtedly are due to simple trauma that injures the germinal layer of the physeal cartilage and/or the epiphyseal blood supply to the physis. Disruption of the metaphyseal circulation is not critical to physeal integrity but retards vascular invasion of the hypertrophic cartilage. In some cases there is probably an unusual susceptibility to injury because of very rapid growth, as in the case of large dogs, or a defect in the growth cartilage, as in swine and foals with osteochondrosis.

Premature closures in the distal ulnar growth plate of dogs, and the metacarpal growth plate of foals are common, and are well recognized as one cause of deformities of the distal forelimbs in these species. It is probably not coincidental that these are among the fastest growing sites in the skeleton at the time they are affected; thus, the effect of growth abnormalities is marked. In addition, the synchrony required of radius and ulna for normal development is disturbed by injury to the physis of either bone, and this emphasizes the effect of the lesion.

Closure of a growth plate usually results in the production of a bony bridge between epiphysis and metaphysis at the site where the growth cartilage has been destroyed. Surgical intervention to prevent the osseous union, by implanting a mass of fat in the defect, for example, may prevent a deformity. There is experimental evidence showing that the growth plate can proliferate latitudinally by interstitial growth to repair physeal defects, and this may be a rare natural occurrence. Often cessation of growth in combination with continued growth of the rest of the plate, or an adjacent or contralateral bone, results in skeletal deformity (Fig. 1.74).

In paired bones, such as the radius–ulna and tibia–fibula, patterns of growth deformities are more complex than those involving a single bone. The ultimate effect depends on the extent and location of the closure and the age at which it occurs. If the affected bone has almost attained its definitive length, then the effect is minimal, whereas a similar injury early in the growth period is potentially crippling. Lesions involving the periphery of the growth plate, whenever they occur, appear to be more detrimental to anatomical integrity than those that are centrally located. Premature, complete closure of the distal ulnar physis is described as an autosomal recessive trait in the Skye terrier. Details of pathologic changes are unreported. Nonhereditary premature closure of the ulnar physis, presumably caused by trauma, occurs commonly, but not exclusively, in large breed dogs. The combination of site and subject size is thought to be founded on the unusual susceptibility of the distal ulnar physis, whose conical shape predisposes it to post-traumatic closure, and the growth potential of the large breeds, which emphasizes anatomic imperfections.

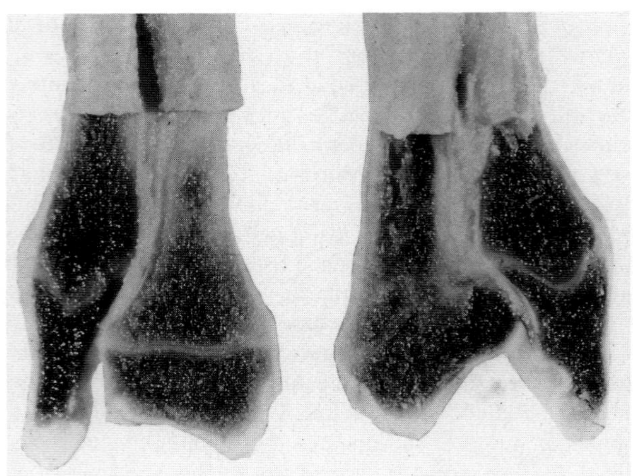

Fig. 1.74 Distal radius–ulna. Dog. Normal on left. Growth deformity following trauma to radial physis. (Courtesy of D. H. Read.)

In large dogs, especially Great Danes, closure of the distal ulnar growth plate can be confused with retarded endochondral ossification (retained endochondral cartilage core). Although the clinical changes are often similar, the radiologic, pathologic, and pathogenetic details are quite different, and prognosis and treatment are also different. Retarded endochondral ossification is bilateral, often clinically silent, and may heal uneventfully. The characteristic lesion is a cone-shaped mass of unmineralized hypertrophic cartilage with its base at the center of the ulnar growth plate and its apex projecting into the metaphysis (Fig. 1.75A,B). The periphery of the plate is normal. The cause of the abnormality is not known. It may be a manifestation of osteochondrosis or be caused by a temporary interruption or insufficiency of metaphyseal blood supply. Retained cartilage in the distal ulna occurred in 9 of 9 pups in two successive litters of captive Mackenzie Valley timber wolves. There were associated limb deformities in all animals.

5. Dorsal Metacarpal Disease of Racing Horses

Dorsal metacarpal disease encompasses a spectrum of lesions affecting the dorsal cortex of the third metacarpal bone of young racing horses and includes bucked shins

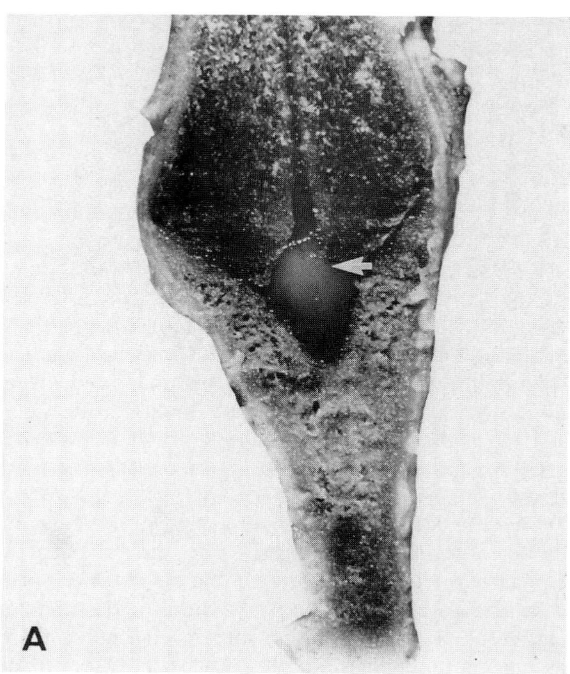

Fig. 1.75A Distal ulna. Dog. Retained cartilage core. A mass of unresorbed cartilage projects into the metaphysis (arrow).

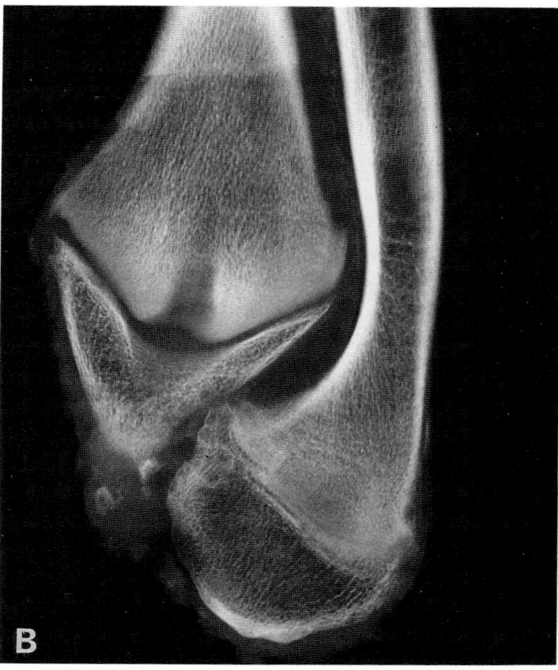

Fig. 1.75B Slab radiogram. Distal radius–ulna. Dog. Growth retardation and bone deformity with retained cartilage core.

and incomplete cortical fracture. While the two lesions occur at different anatomic sites in the dorsal cortex and primarily affect horses in slightly different age groups, both are thought to be caused by excessive compressive stress placed on the dorsal cortex on the third metacarpal bone of the growing horse and the resulting disturbances in the normal modeling and remodeling processes of that bone.

Bucked shins most commonly involves the middiaphysis of the dorsal cortex of both third metacarpal bones of 2-year-old and, sometimes, 3-year-old horses in their first year of race training. By contrast, incomplete cortical fracture is a stress fracture producing an incomplete fracture of either the middiaphyseal cortex or the distal diaphyseal cortex of the third metacarpal bone, and these fractures occur most commonly in 3- and 4-year-old racing horses. Incomplete cortical fractures occur about five times more commonly in the left than in the right third metacarpal bone. Differences in lesion sites and in ages of the affected horses may reflect differences in biomechanical stresses placed on the bones during training and racing with, perhaps, bucked shins resulting from the accumulative damage during conditioning at slower speeds and with incomplete cortical fractures possibly resulting from accumulative microdamage encountered at racing speeds. The dorsal cortex of the third metacarpal bone of the growing racehorse experiences marked cyclic loading on the dorsomedial cortex during training. Bone modeling accommodates for those stresses by causing the periosteum on the dorsal and, especially, the dorsomedial cortex to deposit large numbers of concentric layers of bone. However, remodeling of this laminar bone and its replacement by secondary osteonal bone is delayed by the stresses of training. It has been suggested that microdamage sustained during training and racing accumulates in the dorsal cortex because of the delay in the onset of bone remodeling, which should normally remove the damaged bone and replace it with new and biomechanically sound osteonal bone.

Bucked shins is characterized by one or more focal, smoothly contoured periosteal responses on the dorsal cortex of the third metacarpal bone. Active lesions are recognized by the accompanying hyperemia of the periosteum. The spectrum of morphologic changes has yet to be fully characterized, but the following summarizes observations on a number of lesions. Histologically, there is a periosteal bony response without unique features. Histologic and microradiographic sections of the affected subadjacent cortex may show abnormal patterns of bone remodeling in the compact cortical bone. Scanning electron microscopic studies have demonstrated what appear to be microfractures in cortical bone, and the abnormal patterns of bone remodeling may reflect attempts to belatedly repair accumulative microdamage.

An incomplete cortical fracture is recognized as a fracture line that courses obliquely from the periosteal surface of the dorsal cortex to a depth of about two thirds of the width of the dorsal cortex. Upon reaching this point, the fracture line may abruptly stop, it may extend vertically for a short distance and stop, or it may continue again at an oblique angle and return again to the dorsal cortex, producing the so-called "butterfly" fracture. Microradiographic studies have shown that the barrier for extension of the fracture line completely through the dorsal cortex is the original dorsal cortex of the term fetus, which has not undergone remodeling. Scanning electron microscopic studies have demonstrated microfractures within the path of the clinically apparent fracture line. The periosteal callus over the fracture line varies considerably in thickness.

Bibliography

Campbell, J. R. Bone growth in foals and epiphyseal compression. *Equine Vet J* **9:** 116–121, 1977.

Devas, M. B. Shin splints, or stress fractures of the metacarpal bone in horses, and shin soreness, or stress fractures of the tibia, in man. *J Bone Joint Surg* **49B:** 310–313, 1967.

Done, S. H., Meredith, M. J., and Ashdown, R. R. Detachment of the ischial tuberosity in sows. *Vet Rec* **15:** 520–523, 1979.

Ferguson, J. G., Dehghani, S., and Petrali, E. H. Fractures of the femur in newborn calves. *Can Vet J* **31:** 289–291, 1990.

Frost, H. M. The biology of fracture healing. An overview for clinicans. Part I and Part II. *Clin Orthop* **248:** 283–293, and 294–309,1989.

Glimcher, M. J., and Kenzora, J. E. The biology of osteonecrosis of the human femoral head and its implications: I. Tissue biology. II. The pathological changes in the femoral head as an organ and in the hip joint. III. Discussion of the etiology and genesis of the pathological sequelae; comments on treatment. *Clin Orthop* **138:** 284–309, 1979, **139:** 283–312, 1979, **140:** 273–312, 1980.

Henney, L. H. S., and Gambardella, P. C. Premature closure of the ulnar physis in the dog: A retrospective clinical study. *J Am Anim Hosp Assoc* **25:** 573–581, 1989.

Hulth, A. Current concepts of fracture healing. *Clin Orthop* **249:** 265–284, 1989.

Langenskiold, A. *et al.* Regeneration of the growth plate. *Acta Anat* **134:** 113–123, 1989.

Lau, R. E. Inherited premature closure of the distal ulnar physis. *J Am Anim Hosp Assoc* **13:** 609–612, 1977.

Lee, R. Proximal femoral epiphyseal separation in the dog. *J Small Anim Pract* **11:** 669–679, 1976.

Mann, F. A., and Payne, J. T. Bone Healing. *Semin Vet Med Surg* (*Small Anim*) **4:** 312–321, 1989.

Nunamaker, D. M. Stress fractures of the third metacarpal bone. *In* "Current Practice of Equine Surgery," N. A. White and J. N. Moore (eds.), pp. 622–626. Philadelphia, Pennsylvania, J. B. Lippincott, 1990.

Ossent, P., Mettler, F., and Isenbugel, E. Retained cartilage in the ulnar metaphysis with deformation of the forelegs in two litters of captive wolves. *Zbl Veterinaermed A* **30:** 241–250, 1984. (Good reference list)

Ramadan, R. O., and Vaughan, L. C. Disturbance in the growth of the tibia and femur in dogs. *Vet Rec* **104:** 433–435, 1979.

Richardson, D. W. Dorsal cortical fractures of the equine metacarpus. *Compend Cont Ed* **6:** S248–S254, 1984.

Stover, S. M. Bucked shins and stress fractures of the metacarpus in the horse. *In* "Large Animal Internal Medicine," B. P. Smith (ed.), pp. 1193–1196. St. Louis, Missouri, C. V. Mosby, 1990.

Vandewater, A., and Olmstead, M. L. Premature closure of the distal radial physis in the dog. A review of eleven cases. *Vet Surg* **12:** 7–12, 1983.

Watkins, J. P., and Auer, J. A. Physeal injuries. *Compend Cont Ed* **6:** S226–S234, 1984.

B. Osteosis

The most common cause of osteosis (degeneration and necrosis of osseous tissue) is ischemia. The pathogenesis of the ischemia most often involves trauma with fracture. Next in importance is the necrosis which occurs in areas of inflammation (osteitis); in fact, the extent of necrosis in osteitis largely determines the outcome of the inflammation. There are additional causes of necrosis, and these include excessive heat from horn-bud removal, especially in kids (Fig. 1.76), trauma with direct dislocation of the periosteum especially in horses, indirect dislocation by exudation, and peripheral ischemia such as occurs in ergotism, fescue foot, chronic anemia, and in dehydrated calves in cold climates. Necrosis and sequestration of bone, probably due to narrowing of medullary vessels, occur in association with osteosarcoma of dogs, particularly miniature schnauzers. Widespread focal necrosis of bone, most prominent in the metaphysis, is recorded in gray collies with cyclic neutropenia and is thought to be related to thrombosis of medullary vessels. Osteosis may

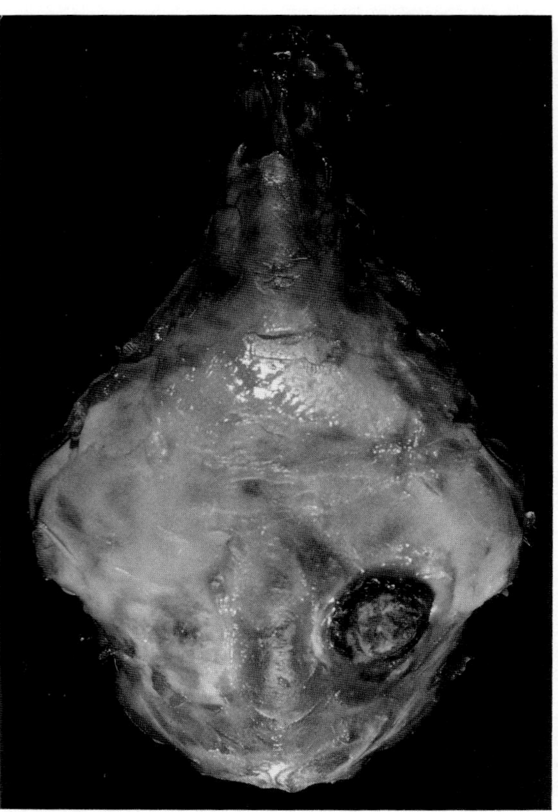

Fig. 1.76 Osteosis of bone following excess heat from dehorning. Kid. Underlying cerebrum was necrotic.

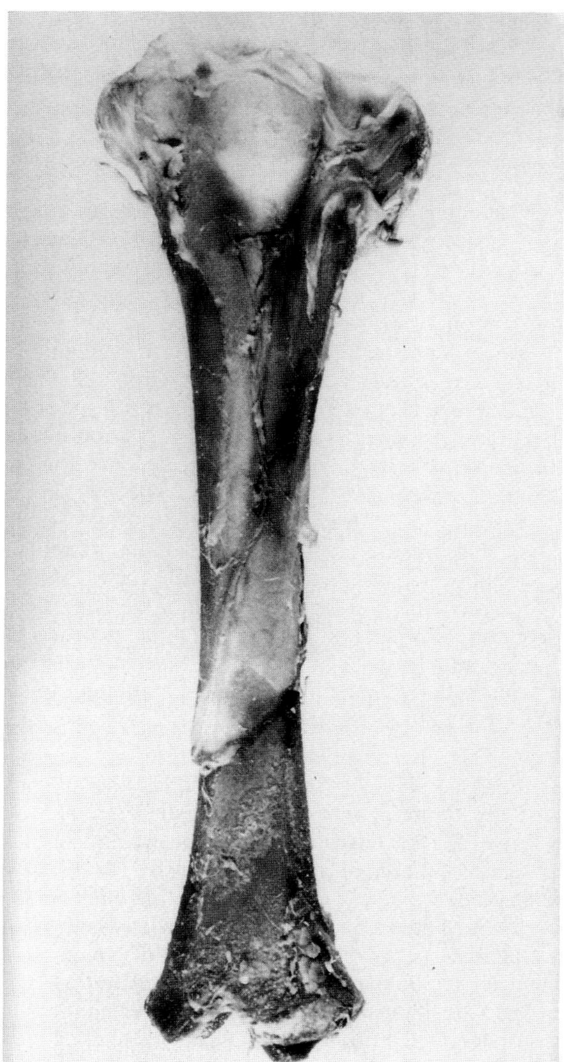

Fig. 1.77 Lamb. Ischemic necrosis of distal tibia following thrombosis of iliac artery. Viable bone has continued to grow.

also result from thromboembolic occlusion of individual vessels (Fig. 1.77). Infarction of marrow and sometimes of medullary bone occurs commonly in bovine lymphomatosis of the juvenile type.

1. Necrotic Bone

Necrotic bone is often impossible to recognize on gross examination. When the necrosis is very recent, its presence may be inferred only from discoloration or liquefaction of the marrow and adjacent soft tissue. The normal periosteum and cortical surface, except where muscles and fascia are inserted, are smooth, white, glistening, and firmly adherent. The earliest recognizable change in necrosis is often a change in the periosteum to a dull, dry, parchmentlike sheath which can be detached easily. Necrotic bone is slowly but progressively resorbed, and the irregularity of this process gives the cortex a chalky luster and produces a grating sensation when scratched with a metal rod. The necrotic bone may remain white or become light brown; later the limits of necrosis are indicated by marginal osteophytes or by progressive isolation by soft scar tissue.

Necrosis of bone is characterized histologically by the death and disappearance of osteocytes and, therefore, by the presence of empty lacunae (Figs. 1.5, 1.78A,B). Dead osteocytes may persist for several days up to 2 weeks; however, in such cases, their lack of viability is inferred from necrosis of adjacent marrow tissue and by comparison with living osteocytes elsewhere in the same section. Prolonged immersion in acidic demineralizing solutions can mimic osteocyte death but can be identified because the changes are nonselective. Necrosis of bone also leads to changes in the mineralized substance, as indicated by increased basophilia and friability of the necrotic tissue in histologic sections. The nature of the changes in the hard tissue is not known.

Necrotic bone may be removed completely, sequestered, or it may be covered by new bone and persist until it is replaced in the normal remodeling process. Its fate depends on its volume, its association with viable connective tissue, and on the normal healing processes. The sequestration of bone in pyogenic infections is discussed with osteitis.

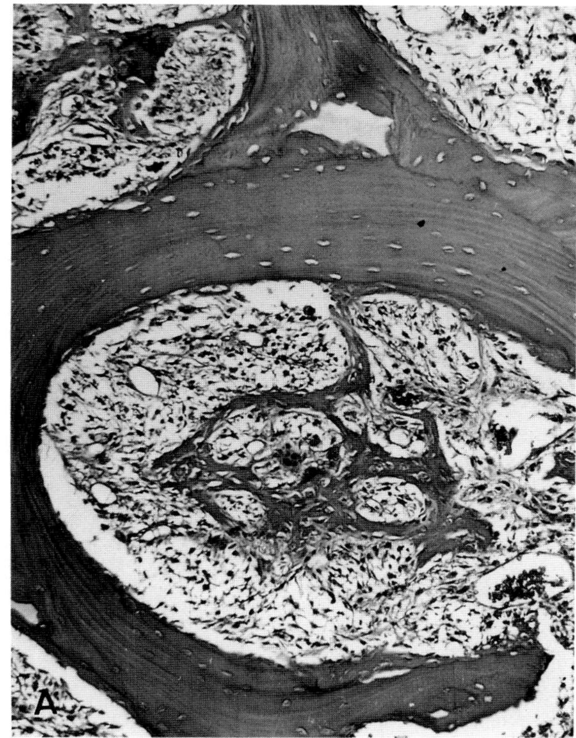

Fig. 1.78A Necrosis of bone following transient ischemia. Osteocyte lacunae in dead lamellar trabecula are empty. New woven bone is forming in marrow spaces.

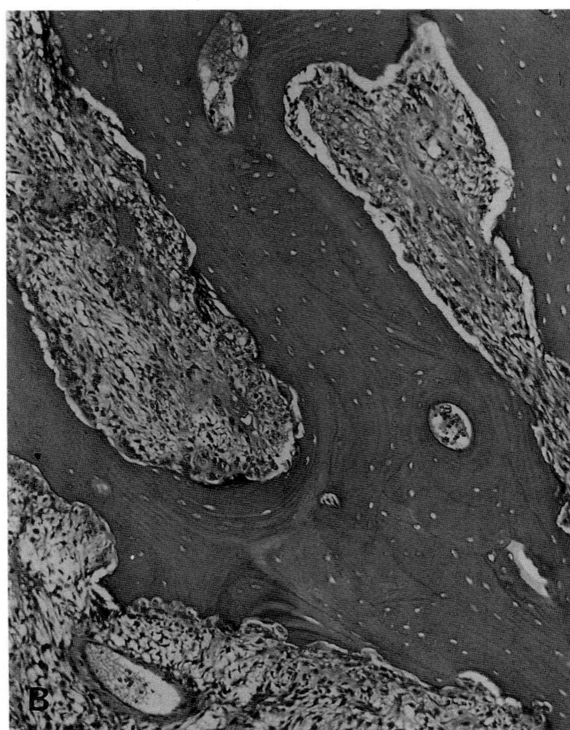

Fig. 1.78B Necrosis of bone. Dog. Femoral head. Tumor metastasis produced infarction. Dead bone being resorbed.

The simplest and most satisfactory process of healing consists of osteoclastic removal of dead bone and its ultimate replacement by lamellar bone. This is a common type of healing since it is the response to relatively minor trauma to cortical bone. When the injury is repetitive, regeneration may be excessive and result in the formation of local osteophytes. The healing process is less simple when part of the necrotic bone is disconnected from viable connective tissue, as occurs frequently with necrosis or detachment of the periosteum. In such lesions, resorption of necrotic bone may still be complete if its volume is small. If a relatively large volume of cortical bone is involved, sequestration may occur.

Resorption of necrotic bone in extensive areas of necrosis is often accompanied or preceded by deposition of fibrous or lamellar bone on some of the dead trabeculae. This produces a net increase in the amount of bone that persists until remodeling, months or years later, removes the repaired trabeculae. This type of healing process is typical of cancellous bone and is discussed with fractures in which one fragment is dead.

An important factor influencing the outcome of osteosis wherever it occurs is the efficiency of the collateral circulation. When the nutrient artery is occluded, large areas of the bone marrow become necrotic along with smaller areas of adjacent spongy and compact bone. The diaphyseal extremities are the most favorable sites for fractures, with respect to available blood supply for healing, because in these locations there is a dual blood supply of osseous origin—from the nutrient vessels on the diaphyseal side and from the vessels of the articular capsule and ligaments on the epiphyseal side—as well as that from adjacent soft tissues.

As a rule, the collateral circulation in cortical bone is inefficient because, in spite of abundant anastomoses, the anastomotic vessels are small and, being confined within narrow canals, are incapable of effective compensatory dilation. The resting and proliferative cartilage of the major growth plates depends on the epiphyseal circulation, while the sinusoidal circulation of the primary spongiosa is derived from metaphyseal vessels and is vulnerable to trauma. In contrast, the articular cartilages are relatively insensitive to regional ischemia since their nutrient is obtained by diffusion from the synovial fluid.

2. *Legg–Calvé–Perthes Disease*

The use of the name implies that this disease in animals is similar to the condition in children characterized by necrosis and collapse of the femoral capital epiphysis. Necrosis of the femoral head occurs under a variety of circumstances including dysbaric osteonecrosis of divers and compressed-air workers, and corticosteroid-induced avascular necrosis in patients receiving high doses of this drug for prolonged periods. In animals, necrosis of the femoral head is often a result of neck fractures or of slipped capital femoral epiphysis (see preceding sections). Fractures of the femoral head are rare except as part of Legg–Calvé–Perthes disease but have been seen in dogs, mainly female flat-coated retrievers, probably as a sequel to primary coxa plana. These conditions should not be confused with Legg–Calvé–Perthes disease, which is an ailment almost exclusively of small dogs, of which most are small terriers and poodles. Cats are affected rarely. Lameness of insidious onset, occurring between 4 and 8 months of age, is the usual clinical sign. Sex and leg incidence tend to be equal; bilateral involvement occurs in up to 15% of cases.

It is generally accepted that osteonecrosis in this disease is initiated by an episode, or repeated episodes, of ischemia. There is experimental and anatomical evidence supporting this theory. Intracapsular tamponade produces lesions similar to those in the naturally occurring disease, probably by occluding the veins which drain the femoral head. In highly susceptible poodles, many of the veins are subsynovial, whereas in resistant dogs they are mainly intraosseous. Delayed incorporation of vessels into bone prolongs the period of susceptibility to compression, and is a concomitant of suppressed skeletal growth in susceptible small dogs. It is postulated that under natural conditions trauma or transient synovitis causes effusion into the joint and leads to occlusion of the veins.

The initial lesion is capillary distension, followed by necrosis, then lysis of soft tissue, usually in the central part of the head of the femur. The articular cartilage and its subjacent growth cartilage as well as the proximal growth plate of the femur may be unaffected at this stage. Contin-

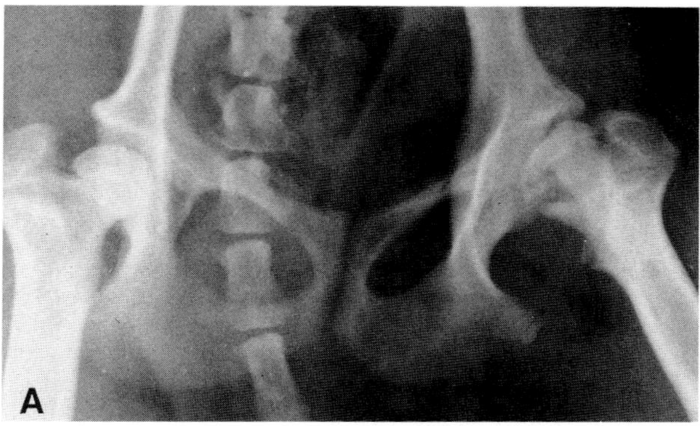

Fig. 1.79A Dog. Necrosis of femoral head (right). (Courtesy of T. J. Hage.)

ued proliferation of the subarticular growth cartilage without concomitant endochondral ossification is responsible for widening of the joint space in radiographs. Although the cells of bone and marrow are dead, the trabecular structure is unchanged, and thus radiodensity of the head is normal. This is the stage of potential Legg–Calvé–Perthes disease because the femoral head has not collapsed and, depending on the extent of the ischemic necrosis, may not do so. The repair of this lesion may involve two sequences: reestablishment of the circulation with proliferation of surviving mesenchymal cells in the area of necrosis, and invasion of vessels and mesenchymal cells from the margin of the dead tissue. The duration of hypoxia and the extent of necrosis probably determine whether one or both processes occur. In both cases, woven bone is deposited on dead lamellar trabeculae, and this is responsible for increased radiodensity of the head. Later the dead bone is removed by osteoclasts, and eventually lamellar bone replaces all the woven bone.

When repair tissue reaches the dead subchondral bone, it too is resorbed and to some extent replaced by living bone; however, there is net loss of subchondral bone. The amount of bone resorbed determines whether the head will collapse and thus whether the disease will develop.

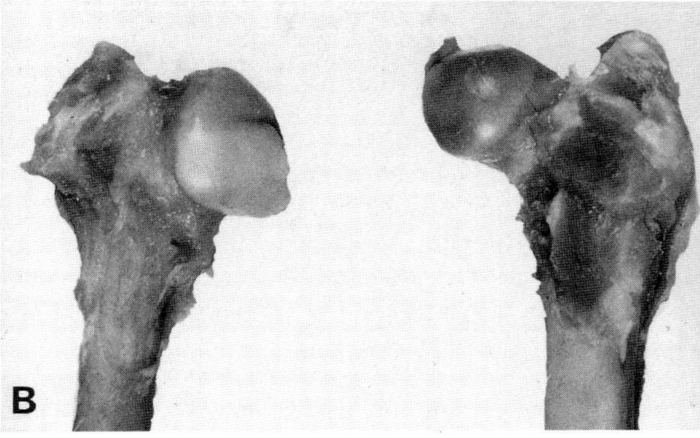

Fig. 1.79B Necrosis and collapse of femoral head (left.) Normal head (right).

In addition, the resorptive tissue may invade the cartilage of the head, and pannus from the periphery of the joint may destroy articular cartilage from above. These factors contribute to collapse of the head and development of coxa plana (Fig. 1.79A,B). Degeneration of the proximal growth plate of the femur occurs in some cases, probably because of hypoxic injury to germinal cells at the time of the initial insult. The expected outcome of Legg–Calvé–Perthes disease is degenerative arthropathy.

Bibliography

Anderson, J. F., and Werdin, R. E. Ergotism manifested as agalactia and gangrene in sows. *J Am Vet Med Assoc* **170:** 1089–1091, 1977.

Cheville, N. F., Cutlip, R. C., and Moon, H. W. Microscopic pathology of the gray collie syndrome. Cyclic neutropenia, amyloidosis, enteritis, and bone necrosis. *Pathol Vet* **7:** 225–245, 1970.

Dubielzig, R. R., Biery, D. N., and Brodey, R. S. Bone sarcomas associated with multifocal medullary bone infarction in dogs. *J Am Vet Med Assoc* **179:** 64–68, 1981.

Firth, E. C. Bone sequestration in horses and cattle. *Aust Vet J* **64:** 65–69, 1987.

Glimcher, M. J. Legg-Calvé–Perthes syndrome. Biological and mechanical considerations in the genesis of clinical abnormalities. *Orthopaedic Rev* **8:** 33–40, 1979.

Kawai, K. *et al.* Fat necrosis of osteocytes as a causative factor in idiopathic osteonecrosis in heritable hyperlipemic rabbits. *Clin Orthop* **153:** 273–282, 1980.

Kemp, H. B. S. Perthes disease in rabbits and puppies. *Clin Orthop* **209:** 139–159, 1986.

Moens, Y. *et al.* Bone sequestration as a consequence of limb wounds in the horse. *Vet Radiol* **21:** 40–44, 1980.

Olsson, S.-E., Poulos, P. W., and Ljunggren, G. Coxa plana-vara and femoral capital fractures in the dog. *J Am Anim Hosp Assoc* **21:** 563–571, 1985.

Salter, R. B., and Thompson, G. H. Legg-Calvé–Perthes disease. The prognostic significance of the subchondral fracture and a two-group classification of the femoral head involvement. *J Bone Joint Surg* **66A:** 479–489, 1984.

Slichter, S. J., *et al.* Dysbaric osteonecrosis: A consequence of intravascular bubble formation, endothelial damage, and platelet thrombosis. *J Lab Clin Med* **98:** 568–590, 1981.

Wang, G.-J., *et al.* Cortisone-induced intrafemoral head pressure change and its response to a drilling decompression method. *Clin Orthop* **159:** 274–278, 1981.

VIII. Inflammatory Diseases of Bones

Osteitis is inflammation of a bone or part of a bone. The process is necessarily confined to the vascular connective tissues of the periosteum, around intracortical vessels and in the medullary cavity. There are extensive vascular anastomoses between the periosteum, cortex, and the medulla; thus, progressive inflammation involves all three sites. Depending on whether the process begins in the periosteum or the medullary cavity, it may be classified as periostitis or osteomyelitis. The distinction is desirable because the pathogenesis is different, but unless the inflammation is early or non-progressive, it may not be possible. Epiphysitis, as often used, is an anatomically inaccurate term that describes lesions occurring in various physes and metaphyses of horses and cattle. It is thought to be caused by nutritional or mechanical factors and is discussed with equine osteochondrosis. In fact, epiphysitis is often part of hematogenous osteomyelitis. Expansion of the epiphysis depends on a roughly spherical (later hemispherical) growth plate that is served by metaphyseal vessels and thus is susceptible to blood-borne infections.

Osteitis, beginning as periostitis, is a common disease in animals. It may be slowly progressive from the outset, or develop chronicity from an initial acute lesion. Periostitis usually develops from a defect or focus of inflammation in the overlying soft tissues and, depending on its location, tends to be confined to one bone. Examples include infections of the feet of cattle with *Fusobacterium necrophorum* (footrot); necrotic stomatitis of cattle caused by the same organism; infections of the paranasal sinuses as an extension from perforating injuries; compound fractures; bite wounds that produce cellulitis or abscess; chronic paronychia; and pressure sores over bony prominences. Periostitis caused by *Hepatozoon canis* infection in dogs is secondary to myositis and does not progress to osteitis.

Osteitis may also follow the extension of inflammatory processes from synovial structures as in "poll evil" and "fistulous withers" of horses (see the following), or from joints as in bacterial polyarthritis. In the latter, it may be difficult or impossible to distinguish local spread and concomitant infection, since metaphyses share with synovial structures, serosae, choroid plexuses, and ciliary bodies a susceptibility to hematogenous localization of infection.

Other common examples of nonhematogenous osteitis are "bull nose" of swine caused by *Fusobacterium necrophorum,* which gains entry at the gumline when the canine teeth are cut, and mandibular osteomyelitis secondary to periodontitis, from which various anaerobic members of the oral flora may be isolated. In these examples, organisms probably spread via gingival lymphatics, which drain into alveolar bone. Infection of the pneumatic cavities of the skull, as in otitis media and following dehorning and rhinitis, may also lead to osteitis. Atrophic rhinitis of pigs is caused by *Bordetella bronchiseptica* and *Pasteurella multocida* and, like periodontal disease and rheumatoid arthritis, is characterized by an inflammatory reaction with loss of bone tissue (see Volume 2, Chapter 6, The Respiratory System).

It is convenient to include as examples of local primary periostitis the osteophytes which develop following trauma. **Exostoses** may develop from single insults if the periosteum is injured, especially if it is detached, or from repeated minor injuries. The usual types of trauma are knocks, cuts, and the tearing of ligamentous insertions. Small osteophytes may be resorbed completely, but large ones are converted from woven to lamellar bone and may persist indefinitely. Trauma without breakage of skin may also predispose to bacterial osteitis, presumably by injuring tissue that is then prone to infection.

Hematogenous osteitis is virtually synonymous with osteomyelitis and is seen much more often in young than old animals (Fig. 1.80A,B). Exceptions are the deep mycoses and *Leishmania donovani* infections, which tend to occur in mature dogs. Although hematogenous osteitis is often diagnosed, its real prevalence is underestimated because bones are seldom examined and cultured, and

Fig. 1.80A Hematogenous osteomyelitis. Pig. Bacteria (arrow) in cartilage canal of costal cartilage. There is acute inflammation in metaphysis.

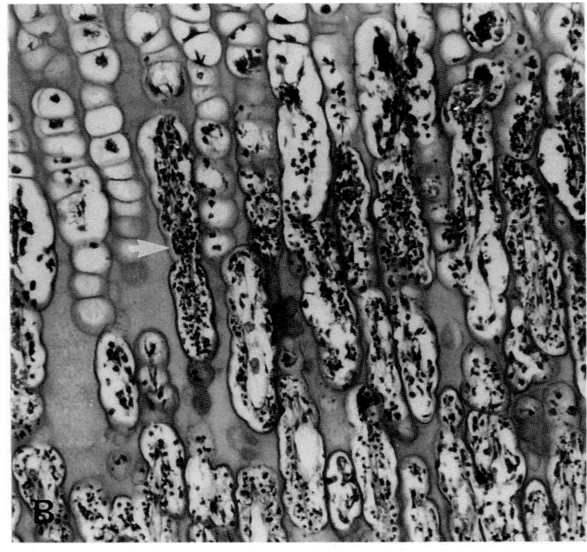

Fig. 1.80B Hematogenous osteomyelitis. Pig. Purulent inflammation in metaphysis (arrow).

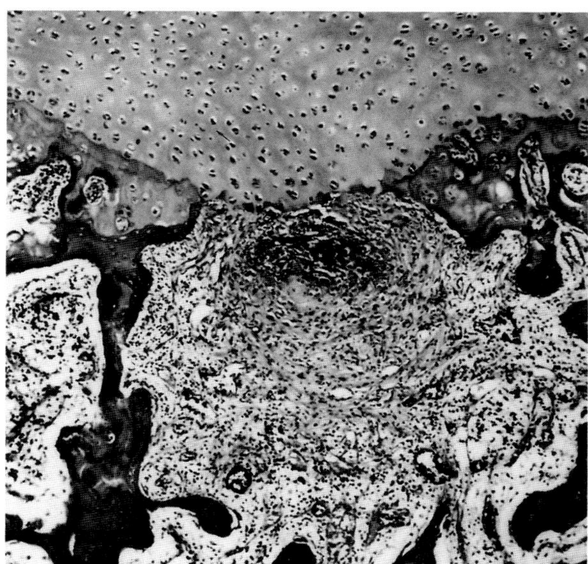

Fig. 1.81 Hematogenous osteomyelitis. Focus of acute inflammation in epiphysis. Articular cartilage is above.

many septicemic animals die before bone disease is prominent. For example, one survey found 70% of foals with joint-ill had osteomyelitis. Lesions are located in or near the area of the primary spongiosa and appear as white or yellow discolorations, often with central cavitations containing exudate or necrotic tissue. As well as metaphyses of long bones, the secondary ossification centers of the epiphyses are also common sites of primary hematogenous osteomyelitis (Fig. 1.81). In foals, lesions in the epiphyses tend to localize under the thickest part of the articular-growth cartilage and, in the metaphysis, they are often found in areas where the growth plate is irregular and not at right angles to the bone axis.

The metaphyseal localization of infections in young animals is influenced by characteristics of small vessels there, characteristics that are shared by vessels under the growth cartilage of the epiphyses. The capillaries which invade the mineralized cartilage make sharp loops before leading into wider sinusoidal vessels, which communicate with the medullary veins. These capillaries have endothelial gaps, permitting exit of bacteria, and the loops lack anastomoses; thus, their thrombosis tends to produce small areas of necrosis, which are necessary for infections to become established. Sluggish circulation in the sinusoidal system, and the relative inefficiency of the phagocytic cells lining it, also tend to favor the development and persistence of infection. Localization of bacteria in the vessels of cartilage canals is also common in hematogenous infections of young animals (Fig. 1.80A).

Bacterial factors also may be important in the pathogenesis of osteomyelitis. In *Staphylococcus aureus* infections, production of a fibrous exopolysaccharide glycocalyx by the organisms may protect against host defense mechanisms and prevent penetration of antibiotics. Also,

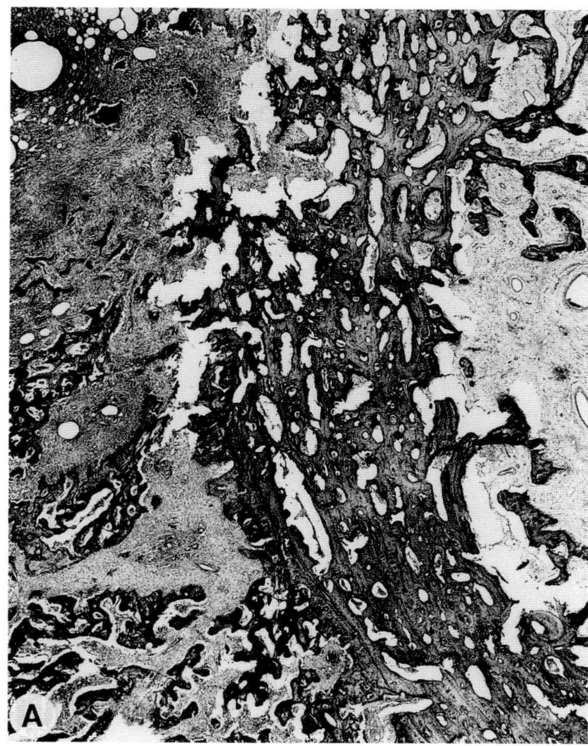

Fig. 1.82A Osteitis and periostitis. Suppurative reaction at upper left with resorption of cortex. Periosteal reaction at lower left.

S. aureus can adhere specifically to the cartilage matrix to which it can be exposed via the endothelial gaps in metaphyseal capillaries. This adhesive ability may also depend on the glycocalyx and could enhance establishment of infection during bacteremia. The same organism is able to oxidize arachidonic acid to prostaglandins, which may stimulate bone resorption and facilitate spread of an entrenched infection.

There are several possible sequelae to hematogenous infection of bone. The presence of solitary foci of chronic osteomyelitis, presumed hematogenous, in adolescent animals suggests that most infections are eliminated by host defences. Prompt and vigorous treatment with specific antibacterial agents may assist the animal in eliminating an early established infection. Alternatively, an infected area may be segregated by fibrous inflammatory tissue and woven bone with development of a metaphyseal abscess (Brodie's abscess), or the infection may spread from the metaphysis in any direction. The diaphysis provides the path of least resistance. From the diaphysis it may extend through vascular canals of the cortex to the periosteum, causing necrosis and sequestration (Fig. 1.82A). Infection may then spread under the periosteum and breach the joint capsule, but is much more likely to produce an abscess.

Arthritis secondary to osteomyelitis can occur in several ways (Fig. 1.82B). It usually develops by direct extension through the spongy trabecular bone at the phy-

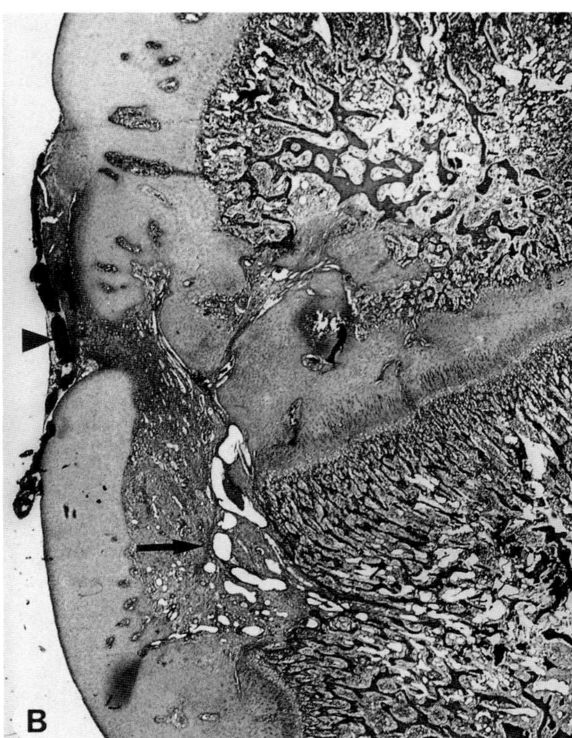

Fig. 1.82B Hematogenous osteomyelitis and arthritis. Pig. Chronic epiphyseal inflammation and fibrosis with breaching of physis (arrow) and extension to articular surface (arrowhead).

seal–metaphyseal junction in young animals or, in mature animals which lack the barrier of a growth plate, by direct extension through articular cartilage following necrosis and resorption of subarticular bone. Spread from growing epiphyses to the joint does occur as a sequel to epiphyseal abscess of hematogenous origin, but in young animals the growth plate normally provides an efficient barrier to extension of infection to the joint and from the metaphysis to epiphysis. Sometimes vessels cross the physis, offering a route for spread of infection between metaphysis and epiphysis. *Actinomyces pyogenes, Salmonella* spp., *Klebsiella* spp., *Escherichia coli,* and streptococci are the most common causes of hematogenous osteomyelitis in cattle. *Salmonella* lesions occur in the epiphyses, and are usually recognized in calves younger than 3 months old. *Actinomyces pyogenes* tends to localize in the metaphyses, causes signs in animals older than 6 months old, and usually originates from a primary focus in the lungs or elsewhere. Septic arthritis and secondary amyloidosis are complications. In foals, *E. coli* is a common cause of hematogenous osteomyelitis and usually is accompanied by polyarthritis. Osteomyelitis caused by *Salmonella* spp. often leads to sequestration of dead bone.

Bacterial osteomyelitis is rarely hematogenous in dogs. In surveys, 65–90% of cases are complications of fractures or fracture repair and involve staphylococci or streptococci. Many of the rest result from bite wounds, and

Fig. 1.83 Chronic sclerosing osteomyelitis. Cow. Hydatid disease.

anaerobes of the oral flora are incriminated. Mycotic osteomyelitis due to *Coccidioides immitis* and *Blastomyces dermatitidis* is important in dogs in some countries, and *Cryptococcus neoformans* and *Histoplasma capsulatum* occasionally spread to bone. These infections, and the osteitis of tuberculosis and brucellosis, are discussed elsewhere. Mandibular osteomyelitis in actinomycosis and virus infections of bone are discussed next. Other organisms such as *Pythium insidiosum,* various other fungi, and hydatid cysts (Fig. 1.83) localize sporadically in bone and cause osteomyelitis.

Vertebral osteomyelitis is perhaps recognized more often than hematogenous osteomyelitis of other parts of the skeleton because of its effect on the nervous system. Usually one or two adjacent vertebral bodies are involved. As in long bones, when localization is in the epiphyses, spread of infection is controlled by the resistance of the cartilages. Expansion occurs more readily from metaphyseal lesions and is usually in a transverse direction. The result is usually transverse pathological fracture and collapse of the vertebrae, followed by dorsal displacement of one or both fragments and compression of the spinal cord (Fig. 1.84). The suppurative process can permeate the cortex in any direction and may involve the pleura from the thoracic vertebrae, or the spinal muscles from the lumbar vertebrae. Pus, contained by the periosteum of the spinal canal, may protrude as an encapsulated abscess into the canal to compress the cord or it may penetrate the periosteum and then spread up and down the spinal canal. Seldom does it penetrate the dura mater and cause spinal myelitis (Fig. 1.84B).

As a rule the causative organism is *Actinomyces pyogenes,* and a primary focus of infection is not always found. The lesion may occur in any of the larger species but particularly in young cattle, pigs of any age after weaning, and lambs. In regions where brucellosis is endemic in swine, *Brucella suis* causes similar lesions. In dogs, vertebral abscesses are uncommon but are caused by *Nocardia* and other actinomycetes and may

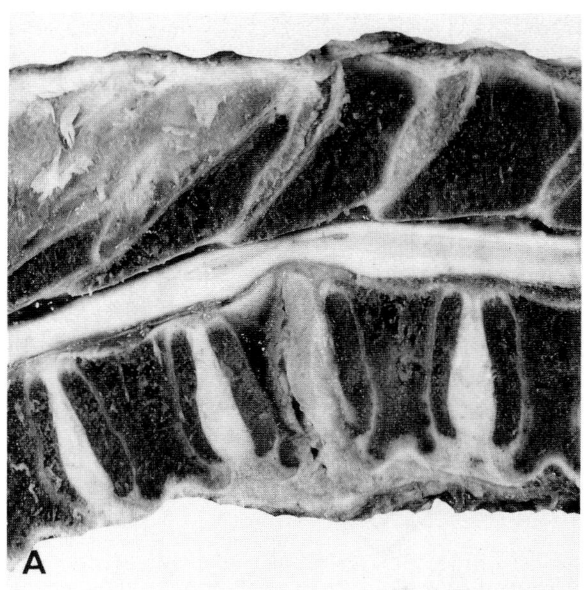

Fig. 1.84A Vertebral abscess. Calf. Collapse and angulation of vertebral column with compression of cord.

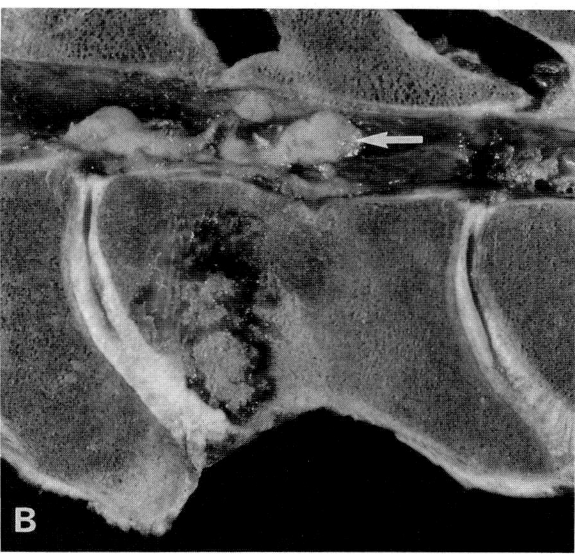

Fig. 1.84B Vertebral abscess. Cow. Fibrinopurulent exudate in spinal canal (arrow). Note sequestrum in vertebral body.

accompany pulmonary lesions or pleuritis caused by these bacteria. Vertebral abscess may be a sequel to diskospondylitis in dogs and other species (see the following). Such a progression is typical of *Brucella* spp. In horses, *Mycobacterium* spp. occasionally produce granulomatous and proliferative osteomyelitis, and the cervical vertebrae may be involved as part of a generalized disease. Streptococci, staphylococci, *E. coli*, *Salmonella* spp., and *Rhodococcus equi* are more common causes of vertebral osteomyelitis in horses.

The usual sequence of events in osteitis is necrosis, resorption, and production of bone, and these processes may occur and recur as the infection progresses. Necrosis of bone is due to the combined effects of diffusible bacterial toxins, enzymes from lysed neutrophils and of ischemia. The latter is probably due to vasculitis and collapse of the small vessels under the pressure of exudates within vascular canals. Usually necrotic bone is resorbed, but occasionally a sequestrum forms. Resorption and production are never perfectly balanced, and the predominance of one or the other is responsible for the qualification of osteitis as rarefying or sclerosing. Necrotic bone, that is bone in which all cellular elements are dead, is not structurally weakened, and resorption is required before pathologic fractures occur. Resorption is probably mediated through prostaglandins and local factors (see Diseases of Bones, Section I, General Considerations and Section III,D, Osteoclasts). The stimulus to bone production may be in part mechanical as an attempt to reinforce a weakened zone, or it also may be due to release of local factors secondary to the continuous mild irritation. In active, virulent infections, such as those caused by *Fusobacterium necrophorum,* there may be scant or no production. In less virulent but progressive infections, new bone is deposited by reactive periosteum or endosteum. This may be partly destroyed by progressive inflammation, and the cycle repeated. Ultimately, immature trabeculae are replaced by mature lamellar bone.

The fate of the necrotic bone depends largely on its architecture and its volume (see Diseases of Bones, Section VII,B, Osteosis), and on the nature of the inflammatory process. In chronic granulomatous osteitis, extensive necrosis is unusual, and the degenerate bone is removed. If, on the other hand, the inflammation is acute and suppurative, not only is necrosis of compacta more likely, but the necrotic portions also tend to be isolated from the surrounding viable tissue by a bath of exudate. Bone fragments which have lost contact with viable connective tissue are called sequestra, and the process of isolation is called sequestration (Fig. 1.85A,B).

A **sequestrum** may arise in one of three ways: (1) infection may be introduced directly or hematogenously into a comminuted fracture; (2) a pathologic fracture with fragmentation may occur in a focus of established osteitis; and (3) portions of necrotic bone may become separated from adjacent viable tissue following osteitis, particularly osteitis following penetrating and lacerating wounds. Even wounds that do not traumatize the periosteum may provoke sequestration if significant infection is established. The separation of necrotic bone is effected by osteoclasts, which undermine it, and the demarcation is completed by granulation tissue. The significance of sequestration lies in the fact that the necrotic fragment cannot be resorbed by osteoclasts and cannot be reached by antibodies or antibiotics. The sequestrum, if small, may dissolve in the exudate, or it may be discharged with pus on the surface. A large sequestrum requires surgical removal; otherwise, reactive osteogenesis may bury it, especially if the inflammation be-

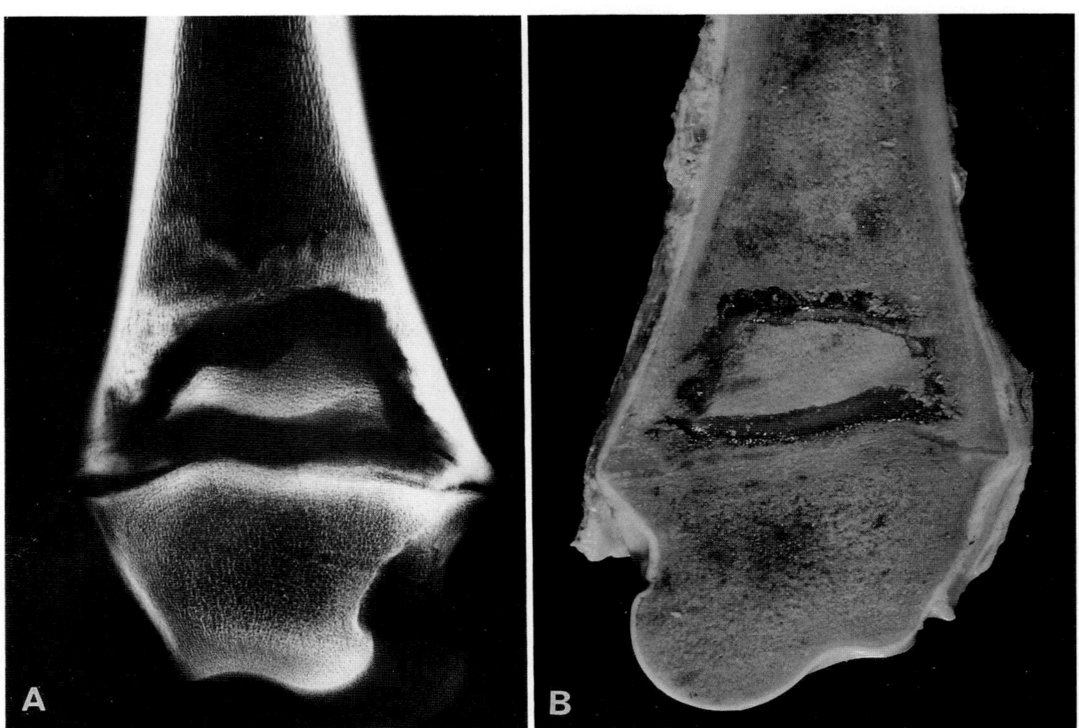

Fig. 1.85 (A and B) Slab radiograph and gross specimen. Distal radius. Salmonellosis. Horse. Sequestrum surrounded by radiolucent involucrum.

comes chronic. The new tissue which encapsulates a sequestrum is known as an **involucrum.** It may be permeated by fistulae from the central nidus of infection. The process of sequestration may be reversed if complete elimination of the infection allows granulation tissue to grow and enclose the fragment.

The histology and ease of diagnosis of osteitis vary according to the type and stage of infection. In acute bacterial osteomyelitis, radiographic changes are minimal, and histologic changes may be confined to necrosis of osteocytes, absence of osteoblasts, and acute inflammation in the marrow spaces. Acute osteomyelitis in young animals may be differentiated from myelopoiesis by the presence of fibrin, mature and fragmented neutrophils, and other necrotic cells in the primary spongiosa where hematopoietic tissue is usually absent.

When osteomyelitis is chronic, it may be granulomatous, is usually proliferative, and often is distinguished from neoplasia only with difficulty. Leukocytosis is rarely present (vertebral osteomyelitis seems to be an exception), radiographic findings may be inconclusive (although the presence of sequestration is strongly suggestive), and confirmation requires a biopsy specimen from the center of the lesion. The diagnosis is readily made, provided a representative section is obtained. If, as is often the case, biopsy yields mostly reactive woven bone and connective tissue, the presence of plasma cells and neutrophils suggests osteomyelitis rather than neoplasia. Discharging sinus tracts are often seen in chronic osteomyelitis, but culture of the exudate usually gives misleading results, and culture from the marrow is required. The lesions of osteomyelitis attain their zenith in chronic relapsing infections, presenting a kaleidoscope of microabscesses surrounded by connective tissue, basophilic bone criss-crossed by resting and reversal lines, and trabeculae of necrotic eosinophilic bone in the process of resorption and replacement. Such lesions are typical of secondary infection of comminuted fractures or primary infections complicated by fractures.

A. Actinomycosis

Actinomycosis is a specific infectious disease caused by *Actinomyces bovis*. This definition distinguishes it from chronic granulomatous infections caused by *Actinobacillus lignieresii, A. pyogenes, Staphylococcus aureus, Nocardia* spp., *Aspergillus fumigatus,* and other fungi that have similar pathologic features and are often labeled actinomycosis. Actinomycosis in species other than cattle is dealt with under mastitis (swine), bursitis (horses), and pleuritis (cats and dogs). Actinomycosis, classically, is a disease of cattle but, as mandibular osteomyelitis, it occurs occasionally in horses, pigs, sheep, and dogs, and as an oral or facial infection in man.

Actinomyces bovis is a gram-positive, branching filamentous organism which fragments readily in culture.

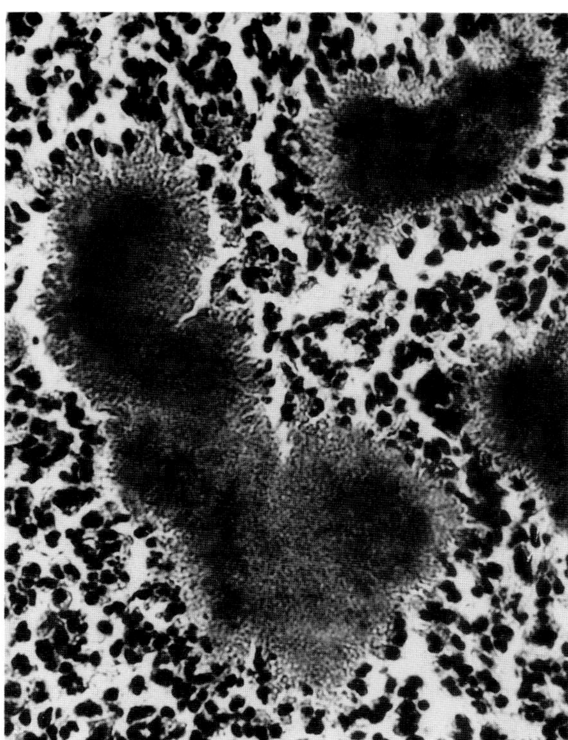

Fig. 1.86 Colonies of *Actinomyces bovis* with fuzzy clubs.

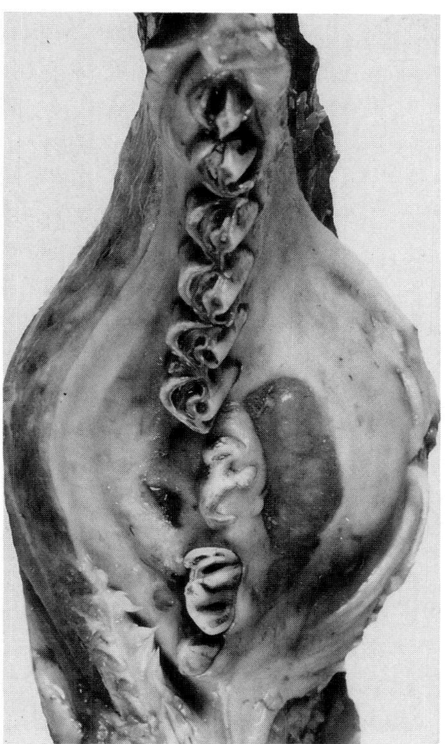

Fig. 1.87 Mandibular actinomycosis. Cow. Gingival ulceration, displacement of tooth, enlargement of mandible.

Fragmented *A. bovis* organisms resemble *A. pyogenes,* from which they must be differentiated, since the two sometimes occur together in pathologic material. *Actinomyces bovis* occurs in the pus of the characteristic lesions in colonies of up to 1 to 2 mm diameter, which are called sulfur granules and are light yellow and soft when viable but when old may be mineralized, opaque, and hard. The granules consist of an internal mass of tangled gram-positive filaments mixed with some bacillary and coccoid forms, and a periphery which consists of closely packed, club-shaped, gram-negative bodies (Fig. 1.86).

Actinomyces bovis is probably an obligatory parasite on the mucous membranes of the oropharynx of a number of animal species, and most infections involve the buccal tissues. The organism is not particularly virulent, and in most, and perhaps all, cases the surface tissues must be injured by some other agent for invasion to occur.

The classic lesion in cattle is "lumpy jaw," a mandibular osteomyelitis; the maxilla is rarely involved, and the organism rarely spreads even to regional lymph nodes which, though large and indurated, are not infected. The osteomyelitis follows direct extension of the infection from the gums and periodontium, presumably following injury by foreign bodies such as coarse vegetable material, or as a complication of periodontitis of other cause (Fig. 1.87).

Invasion via the gums produces typical lesions in the submucosal tissue. Extension to the periosteum causes actinomycotic periostitis, and the infection may not progress; similar lesions may be produced by *A. pyogenes. Actinomyces bovis* may invade bone directly though the periosteum, but osteomyelitis usually develops from periodontitis, presumably via lymphatics which drain into the mandibular bone.

The reaction of tissue to *A. bovis* is the same as to *Actinobacillus lignieresii* and is detailed under actinobacillosis. (Usually the colonies in actinomycosis are much larger, and the clubs are smaller and less discrete than those in actinobacillosis.) Basically the lesion consists of granulation tissue beset with small abscesses, sulfur granules, and suppurative tracts, which permeates the medullary spaces. The trabeculae are progressively destroyed or sequestered in pus as bone "sand." Large sequestra do not develop even when the cortex is invaded, probably because it is permeated slowly by numerous fistulae. These fistulae extend into overlying soft tissue and may discharge through skin or mucous membranes (Fig. 1.88A). Periosteal proliferation is excessive, and the bone may become enormously enlarged. The macerated specimen (Fig. 1.88B) reveals the porosity of the lesion and the local destruction of normal architecture. The teeth in the affected portion of the jaw become loosened, lost, or buried in granulation tissue.

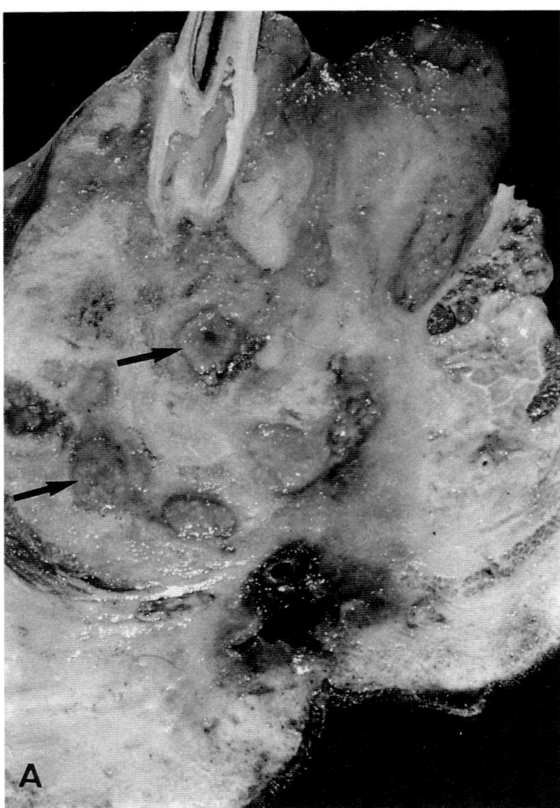

Fig. 1.88A Mandibular actinomycosis. Cow. Granulation tissue with pockets of granular pus (arrows). Dark, draining fistula, below.

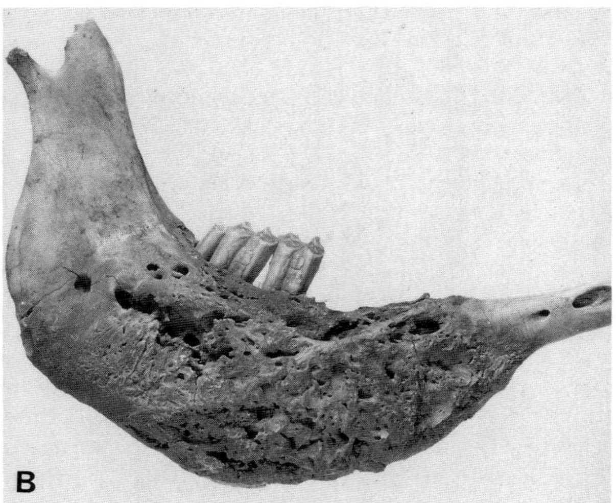

Fig. 1.88B Mandibular actinomycosis. Macerated specimen.

Fusobacterium necrophorum and a variety of nonspecific bacteria also cause osteomyelitis by spread from periodontitis, but the lesion is usually more destructive and less proliferative than with *A. bovis*.

B. Virus Infections of Bone

Several viruses produce distinctive lesions in bone. There are no specific diseases of bone that are known to be caused by virus infection.

Infectious canine hepatitis is characterized by metaphyseal hemorrhages, which are visible grossly in the intact bone at costochondral junctions (see Volume 2, Chapter 2, Liver and Biliary System). Acute hypertrophic osteodystrophy sometimes produces similar hemorrhages. The hemorrhages in infectious canine hepatitis are due to virus-induced injury to endothelial cells, the nuclei of which often contain typical inclusion bodies. Necrosis of all cellular elements in the metaphyses accompanies the hemorrhages; necrosis is probably due to hypoxia. Residual lesions are not described but may not have been sought. **Canine distemper** virus sometimes causes metaphyseal sclerosis in large, fast-growing dogs. In the proximal metaphysis there is retention of primary spongiosa, and in the midmetaphysis of some affected dogs, the trabeculae are thicker than normal. These are areas where modeling of trabeculae normally occurs, and virus-induced osteoclast necrosis is likely to be responsible for the lesions. Necrosis of other cells, including bone-lining cells, in the marrow also occurs. Distemper is discussed in Volume 2, Chapter 6, The Respiratory System.

Hog cholera also produces prominent lesions in bones. In peracute experimental infections, metaphyseal hemorrhages and necrosis of cells are prominent and are probably due to the action of the virus on vascular endothelium. All bones are affected, but the fast-growing ones are injured most, and gross lesions usually are visible only at costochondral junctions. In pigs surviving the acute disease there is irregular persistence of hypertrophic cartilage and a reduction in the amount of metaphyseal bone. These changes are attributable to hypoxic necrosis of osteoclasts and osteoblasts in the spongiosae, and likely predispose to metaphyseal infractions. Later, transverse striations develop in the metaphysis parallel to the growth plate. These striations appear to be formed by residual unmodeled growth cartilage, which is separated from the viable growth plate once resorptive mechanisms are reestablished, and by zones of irregular trabeculae of woven bone, which may represent callus that developed in the metaphyseal infractions. The viruses of hog cholera and bovine virus diarrhea (BVD) are closely related, and both produce metaphyseal striations (BVD in the fetus). The pathogenesis of both lesions is poorly understood. Hog cholera is discussed in Volume 3, Chapter 1, The Cardiovascular System.

Feline herpesvirus causes necrosis in the turbinate bones of germ-free cats following intranasal inoculation and also produces necrosis in the metaphyses and periosteum of growing bones when administered intravenously. Sites of active osteogenesis are most susceptible. The virus infects osteoprogenitor cells, osteoblasts, osteoclasts and endothelial cells, and typical intranuclear inclusions develop in them (see Volume 2, Chapter 6, The Respiratory System).

Medullary osteosclerosis may occur in cats infected with **feline leukemia virus,** and is associated with nonregenerative anemia (see Volume 3, Chapter 2, The Hematopoietic System). It is not clear whether the osteosclerosis is a direct result of the viral infection or is secondary to chronic anemia or hypoxia. Myelofibrosis may also be a feature of this infection. Viral particles are consistently present in the tumors of multiple ecchondromatosis in cats.

Bibliography

Alfaro, A. A., and Mendoza, L. Four cases of equine bone lesions caused by *Pythium insidiosum*. *Equine Vet J* **22:** 295–297, 1990.

Bennett, D. Pathological features of multiple bone infection in the foal. *Vet Rec* **103:** 482–485, 1978.

Booth, L. C., and Feeney, D. A. Superficial osteitis and sequestrum formation as a result of skin avulsion in the horse. *Vet Surg* **11:** 2–8, 1982.

Boyce, R. W. *et al.* Metaphyseal bone lesions associated with canine distemper virus infection. *Abstr Ann Meeting Amer Coll Vet Pathol,* 1983.

Caywood, D. D., Wallace, L. J., and Braden, T. D. Osteomyelitis in the dog: A review of 67 cases. *J Am Vet Med Assoc* **172:** 943–946, 1978.

Dunn, J. K., Farrow, C. S., and Doige, C. E. Disseminated osteomyelitis caused by *Clostridium novyi* in a cat. *Can Vet J* **24:** 312–315, 1983.

Finley, G. G. A survey of vertebral abscesses in domestic animals in Ontario. *Can Vet J* **16:** 114–117, 1975.

Firth, E. C. Bone sequestration in horses and cattle. *Aust Vet J* **64:** 65–69, 1987.

Firth, E. C., and Goedegebuure, S. A. The site of focal osteomyelitis lesions in foals. *Vet Q* **10:** 99–108, 1988.

Firth, E. C. *et al.* Haematogenous osteomyelitis in cattle. *Vet Rec* **120:** 148–152, 1987.

Hoover, E. A., and Griesemer, R. A. Bone lesions produced by feline herpesvirus. *Lab Invest* **25:** 457–464, 1971.

Hoover, E. A., and Kociba, G. J. Bone lesions in cats with anemia induced by feline leukemia virus. *J Natl Cancer Inst* **53:** 1277–1284, 1974.

Jang, S. S. *et al.* Paecilomycosis in a dog. *J Am Vet Med Assoc* **159:** 1775–1779, 1971.

Johnson, K. A., Lomas, G. R., and Wood, A. K. W. Osteomyelitis in dogs and cats caused by anaerobic bacteria. *Aust Vet J* **61:** 57–61, 1984.

Johnston, D. E., and Summers, B. A. Osteomyelitis of the lumbar vertebrae in dogs caused by grass-seed foreign bodies. *Aust Vet J* **47:** 289–294, 1971

Kahn, D. S., and Pritzker, K. P. H. The pathophysiology of bone infection. *Clin Orthop* **96:** 12–19, 1973.

Markel, M. D., *et al.* Vertebral body osteomyelitis in the horse. *J Am Vet Med Assoc* **188:** 632–634, 1986.

Norden, C. W. Lessons learned from animal models of osteomyelitis. *Rev Infect Dis* **10:** 103–110, 1988.

Stead, A. C. Osteomyelitis in the dog and cat. *J Small Anim Pract* **25:** 1–13, 1984.

Thurley, D. C. Disturbances in endochondral ossification associated with acute swine fever infection. *Br Vet J* **122:** 177–180, 1966.

Waldvogel, R. A., Medoff, G., and Swartz, M. N. Osteomyelitis: A review of clinical features, therapeutic considerations, and unusual aspects. *N Engl J Med* **282:** 198–206, 260–266, 316–322, 1970.

IX. Bone Diseases of Unknown Cause

A. Canine Hypertrophic Osteodystrophy

Hypertrophic osteodystrophy is a disease of unknown cause, affecting young, fast-growing dogs of the large and giant breeds. There are numerous synonyms including Moller–Barlow's disease, Barlow's disease, skeletal scurvy, hypo- or avitaminosis-C, metaphyseal osteopathy, and osteodystrophy II. Hypertrophic osteodystrophy is used most commonly, and although it makes implications about the cause of the disease and does not comfortably include those early cases that do not progress to the hypertrophic stage, we choose not to replace it at this time. As with any disease of unknown cause, there may be more than one etiologic agent.

Affected dogs are usually 3–6 months old but may be 2 months to nearly 2 years at the time of first presentation. In acute disease, fever, anorexia, and severe lameness with swelling in the metaphyseal regions of long bones are seen. The distal radius and ulna are usually most severely affected, but all fast-growing bones including ribs may be involved (Fig. 1.89A). Bones distal to the tarsus and carpus are almost always spared. On rare occasions the mandibles are involved, and there is one record of osteophytes around the optic foraminae that caused blindness. Radiographs show alternating radiodense and radiolucent zones in the metaphysis parallel to the growth plate. Usually there is also lipping of the metaphyseal margins. A few dogs die; Weimaraner dogs seem to be particularly susceptible, but most recover without progression of the lesion. In others the metaphyseal bands disappear, but periosteal and extraperiosteal ossification develops, and the ends of the long bones become swollen and hard. Occasionally

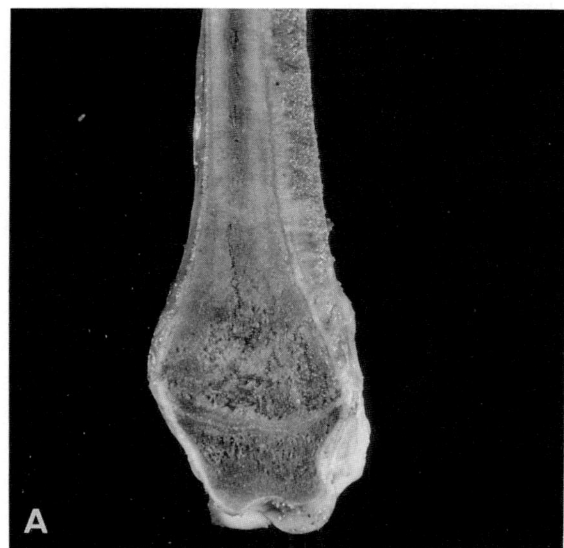

Fig. 1.89A Hypertrophic osteodystrophy. Dog. Metaphyseal necrosis in distal radius. Metaphysis shows diffuse inflammatory changes.

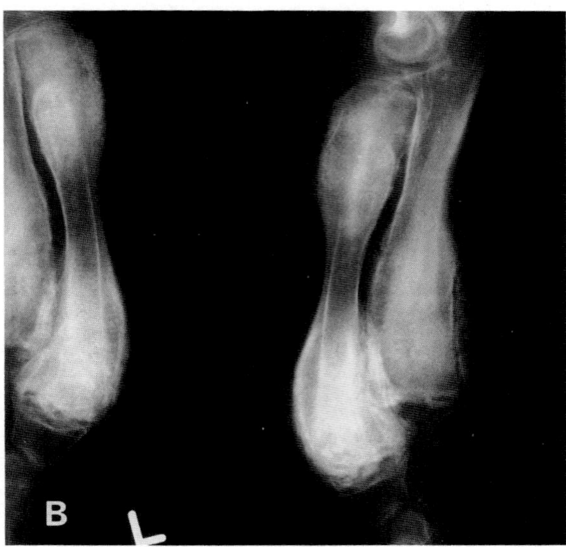

Fig. 1.89B Hypertrophic osteodystrophy. Dog. Extensive periosteal reaction in radius and ulna.

these swellings are extensive and involve almost two thirds of the length of the bone excluding only the middiaphysis and epiphyses (Fig. 1.89B). Remissions and exacerbations may occur over a period of weeks to months and, rarely, a chronically affected dog dies or, more commonly, is killed for humane reasons. Most animals recover completely without, or in spite of, treatment, and the excess metaphyseal bone is gradually removed.

The histologic lesions in acute cases are quite characteristic (Fig. 1.90A). The growth plates and primary spongiosa are structurally normal, but hemorrhages and necrosis of osteoblasts are prominent, and many neutrophils infiltrate the chondro–osseous complex extending up to the hypertrophic chondrocytes. Osteoclasts are active in this region. Distal to this zone, which is radiodense, is a band of hemorrhage and necrosis. In this region, which is relatively radiolucent, trabeculae are sparse, and those present are fractured. Masses of neutrophils infiltrate between the trabeculae. The relative widths of these two zones vary from case to case, but the zone of trabecular necrosis is usually located where primary spongiosa should be. In the diaphysis, away from the fracture zone, foci of necrosis and thrombosed veins sometimes occur. Numerous

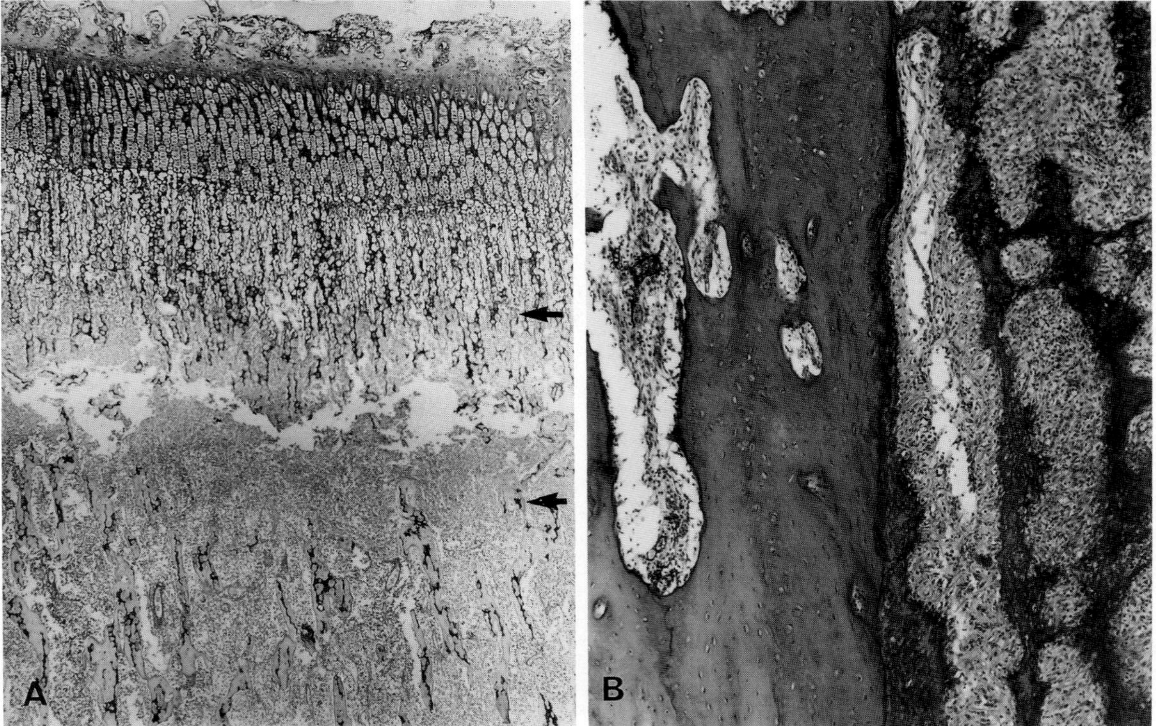

Fig. 1.90 Hypertrophic osteodystrophy. Dog. (A) Physis and metaphysis in acute disease. Physis (above) is normal. Resorption of cartilage and deposition of bone based on primary and secondary spongiosa has stopped because of zone of necrosis and inflammation (between arrows) in metaphysis. The trabecular bone formed some time ago (below bottom arrow) is normal, but there is inflammatory exudate in marrow. (B) Acute periostitis with periosteal and extraperiosteal new bone (same case as Fig. 1.89 (A). Note inflammatory exudate in vascular spaces of cortex.

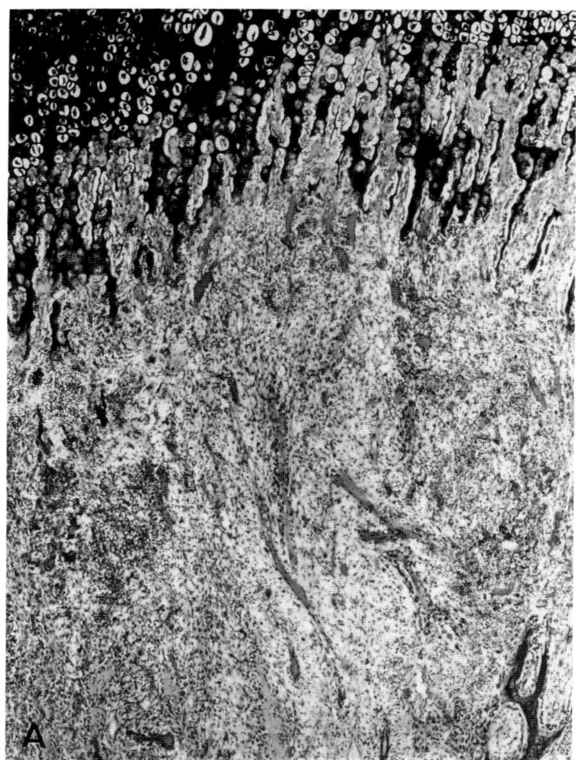

Fig. 1.91A Hypertrophic osteodystrophy. Dog. Exacerbation of disease superimposed on fibrous reaction [same case as Fig. 1.89(B)].

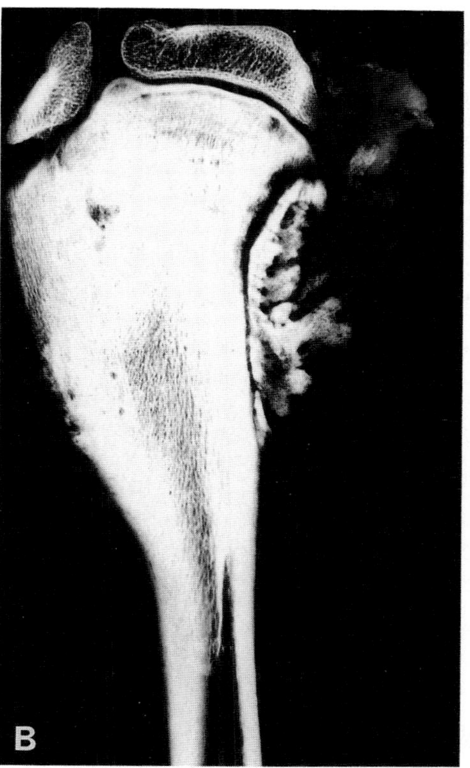

Fig. 1.91B Hypertrophic osteodystrophy. Dog. Proximal tibia. Slab radiograph showing metaphyseal sclerosis, periosteal and extraperiosteal new bone in chronic disease.

segmented neutrophils are often present in the periosteum, associated with hemorrhages, in cartilage canals of the epiphyses and in bones, such as vertebrae and turbinates, that are otherwise normal.

In some dogs subperiosteal and extraperiosteal woven bone is formed, probably in areas of previous hemorrhage and inflammation (Fig. 1.89B). At the same time there is proliferation of osteogenic fibrous tissue in the marrow spaces, which supplies cells for the reconstruction of the damaged tissues. In those animals that have recurrences of the disease, the histologic changes are superimposed on the osteogenic fibrous tissue in the marrow spaces, and radiologically, the metaphyseal lesions may be obscured by periosteal bone (Fig. 1.91A,B).

Extraskeletal lesions consisting of mineralization in various tissues and of arteries in various organs are present in acute and subacute cases. Some animals have myocarditis, pneumonia, interstitial nephritis, and necrosis or hyperplasia of splenic lymphoid follicles.

There have been numerous suggested causes for hypertrophic osteodystrophy, but none has been proven. Because of the unquestionable radiographic similarity between this disease and scurvy of children, vitamin C deficiency has been suggested as the cause. Scientific support for this hypothesis is lacking. Hypervitaminosis D and "overnutrition" have also been invoked as etiologic agents, but neither reproduces the lesions of metaphyseal osteopathy. The bulk of published support is for a nutritional cause, but where diets have been examined retrospectively, a variety of results, including excess phosphorus, low phosphorus, low calcium, excess cod-liver oil, excess mineral–vitamin supplements, or nothing unusual, have been found.

The possibility of an infectious cause seems not to have been examined, although the clinical signs and acute lesions suggest the need to do so (Figs. 1.89A, 1.90A). Many dogs have a history of diarrhea or other clinical abnormality 7–10 days prior to the onset of fever. In the acute disease, cell death at the chondro-osseous junctions caused by an infectious agent could be responsible for the deficiency of bone on the cartilage cores in the metaphyses. Normal physical activity would then be sufficient to fracture the fragile spicules and produce hemorrhages. Periostitis in the metaphyseal regions leads to the "hypertrophic" stage and along with the generalized osteomyelitis, is also consistent with an infectious agent.

Bibliography

Grondalen, J. Metaphyseal osteopathy (hypertrophic osteodystrophy) in growing dogs. A clinical study. *J Small Anim Pract* **17:** 721–735, 1976. (Twenty-six cases including a litter of Weimaraners, five necropsies)

Teare, J. A. *et al.* Ascorbic acid deficiency and hypertrophic osteodystrophy in the dog: A rebuttal. *Cornell Vet* **69:** 384–401, 1979.

Watson, A. D. J. Hypertrophic osteodystrophy: Vitamin C deficiency, overnutrition, or infection. *Aust Vet Pract* **8:** 107–108, 1978.

Woodard, J. C. Canine hypertrophic osteodystrophy, a study of the spontaneous disease in littermates. *Vet Pathol* **19:** 337–354, 1982. (Weimaraners)

B. Canine Craniomandibular Osteopathy

Craniomandibular osteopathy is a proliferative disease usually confined to the mandibles, occipital, and temporal bones, especially tympanic bullae, of young Scottish terriers and West Highland white terriers. Lesions, which are bilateral, may also involve the parietal, frontal, lacrimal, and maxillary bones and, rarely, dogs with craniomandibular osteopathy have subperiosteal and extraperiosteal new bone associated with long bones. Occasionally, lesions are confined to the mandibles. Various other breeds including Boston terrier, Labrador retriever, Great Dane, Doberman pinscher, and cairn terrier have been affected.

The disease is usually recognized at ~5 to 8 months of age because of either discomfort while eating or inability to open the mouth to eat. Proliferative changes may be palpable or grossly visible, and pain is elicited when the mouth is handled. Radiographic findings are characteristic, there being more or less symmetrical enlargement of the mandibles and tympanic bullae. Ossification of soft tissues, especially near the angular processes of the mandibles, may result in fusion of the processes to the bullae. The disease has an intermittent, progressive course lasting several weeks to a few months. It is often self-limiting and nonfatal; however, some dogs are killed because they cannot eat. During clinical exacerbations, which tend to recur every 2–3 weeks, fever is present, but laboratory data are normal.

At the height of the disease, atrophy of the muscles of mastication and increased fibrous and osseous tissue of the head are obvious. The full extent of the bone changes is best demonstrated in radiographs and macerated specimens (Fig. 1.92A,B,C). In the mandible, new bone deposition is usually most prominent in the region of the angle, involving medial, ventral, and lateral aspects but tending to avoid the alveolar bone (Fig. 1.92B). The tympanic bullae are often enlarged two to three times and filled with new bone (Fig. 1.92C).

Histologic changes in the mandible involve the endosteum, periosteum, and trabecular bone and present a complex pattern of intermittent and concurrent formation and resorption (Fig. 1.93). The mosaic of reversal and resting lines resulting from this cellular activity sometimes gives a resemblance to Paget's disease of humans. Further comparison is unjustified. The significance of this mosaic pattern is easily overemphasized, since the mandible of growing dogs is a site of intense bone modeling as teeth mature, erupt, and drift during normal development. Bone resorption in craniomandibular osteopathy appears to be random and disoriented, involving both preexisting lamellar trabeculae and newly formed woven bone. Similarly, bone formation occurs simultaneously in many areas, evolving from fibrous tissue, which frequently fills the marrow spaces and thickens the periosteum. Lymphocytes, plasma cells, and neutrophils are sometimes present in areas of active osteogenesis.

As the disease subsides, formation of woven bone may be superseded by lamellar bone production, and existing woven trabeculae are replaced by lamellar bone, which is then gradually resorbed. In some cases, especially where fusion of mandibles to tympanic bullae is present, resorption is insufficient to allow restitution of normal function.

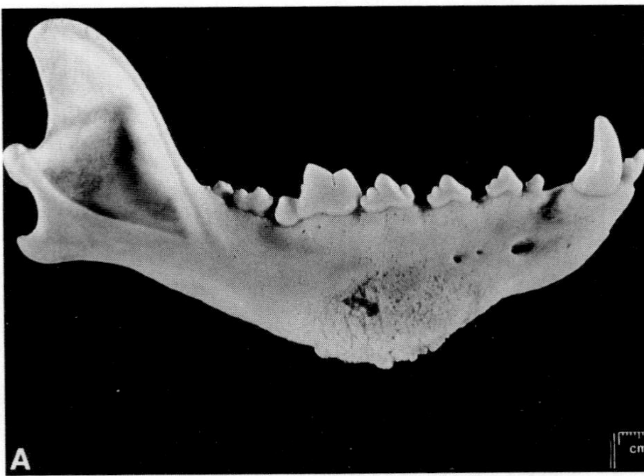

Fig. 1.92A Craniomandibular osteopathy. Dog. Mandible. Macerated specimen showing new bone on ventral margin. (Courtesy of K. G. Thompson.)

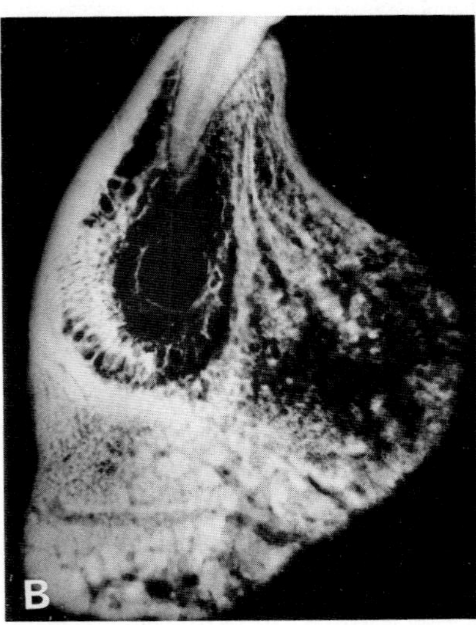

Fig. 1.92B Craniomandibular osteopathy. Dog. Slab radiograph of section through (A). Periosteal new bone on ventral and lateral sides, with endosteal proliferation medially. (Courtesy of K. G. Thompson.)

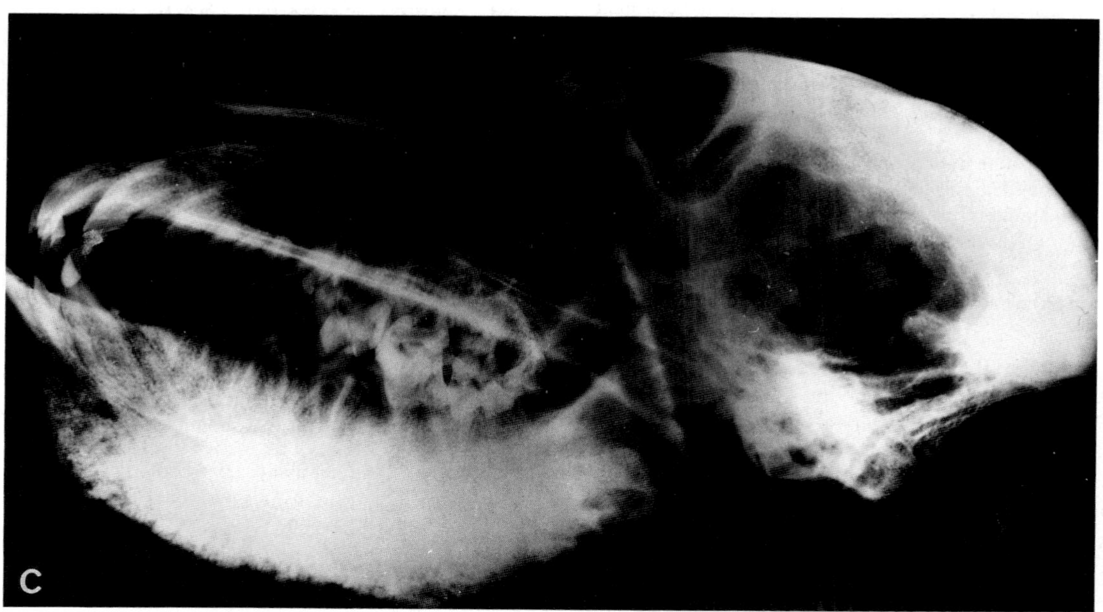

Fig. 1.92C Craniomandibular osteopathy. Dog. Lesions affect mandible, tympanic bulla, and occiput. (Courtesy of P. Olafson.)

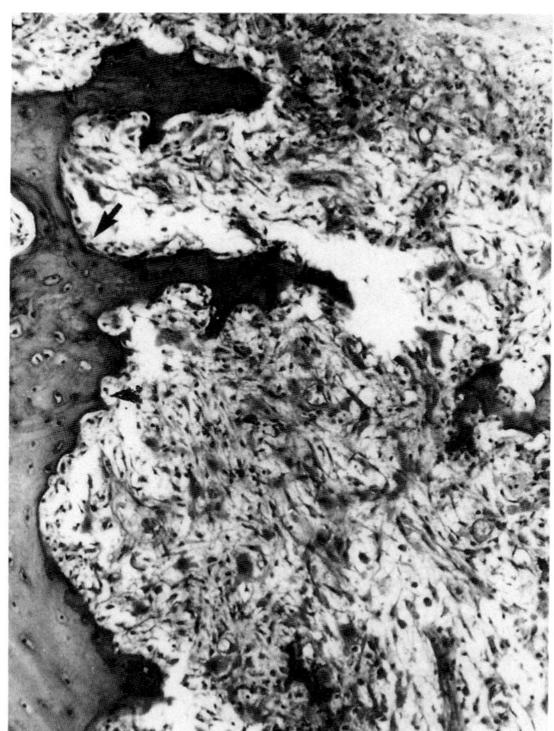

Fig. 1.93 Craniomandibular osteopathy. Dog. Resorption of trabecula (arrowhead). Bone deposition is occurring (arrow). Marrow space contains mesenchymal tissue, osteoclasts, and inflammatory cells. (Courtesy of K. G. Thompson.)

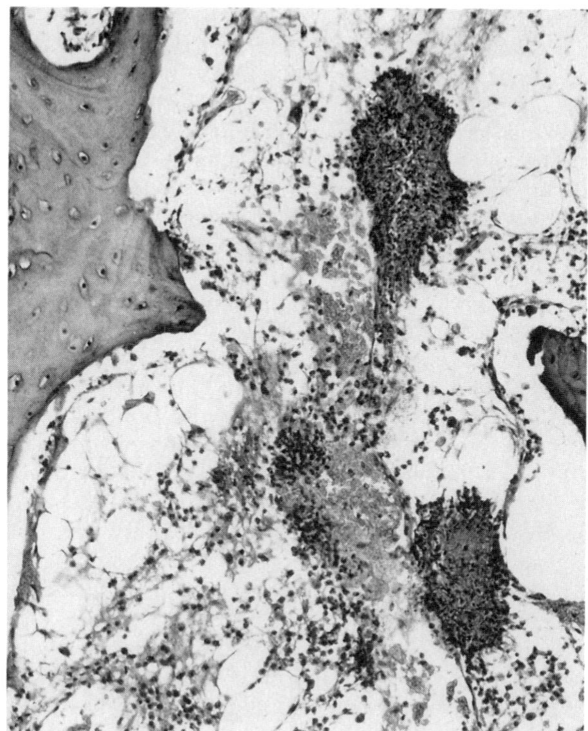

Fig. 1.94 Craniomandibular osteopathy. Dog. Diffuse and focal inflammation around thin-walled vessel (center). There are two dense foci of necrotic cells at right of the vessel. (Courtesy of K. G. Thompson.)

The cause of this disease is not known. The presence of inflammatory infiltrates (Fig. 1.94) suggests an infectious agent, but none has been demonstrated. The large number of cases in certain related breeds is consistent with a genetic predisposition, and an autosomal recessive trait has been suggested. There is convincing evidence for this mode of inheritance in the West Highland white breed. However, the occurrence of craniomandibular osteopathy in diverse breeds suggests that either there is enhanced susceptibility to a causative agent that is inherited or there is more than one cause of the condition.

A relationship has been described linking the bones affected, which are derived from cartilage precursors, and cessation of normal endochondral ossification with the clinical course of the disease. The nature of this relationship is unclear. Like several other proliferative diseases of bone in nonhuman mammals, this has been compared to infantile cortical hyperostosis of humans, which is itself an idiopathic condition.

Bibliography

Padgett, G. A., and Mostosky, U. V. The mode of inheritance of craniomandibular osteopathy in West Highland white terrier dogs. *Am J Med Genet* **25:** 9–13, 1986.

Riser, W. H., Parkes, L. J., and Shirer, J. F. Canine craniomandibular osteopathy. *J Am Vet Radiol Soc* **8:** 23–31, 1967.

Watson, A. D. J., Huxtable, C. R. R., and Farrow, B. R. H. Craniomandibular osteopathy in Doberman pinschers. *J Small Anim Pract* **16:** 11–19, 1975.

C. Canine Panosteitis

Canine panosteitis (panostosis, juvenile osteomyelitis, enostosis, eosinophilic panosteitis) is a disease of unknown cause that usually affects dogs of the large and giant breeds

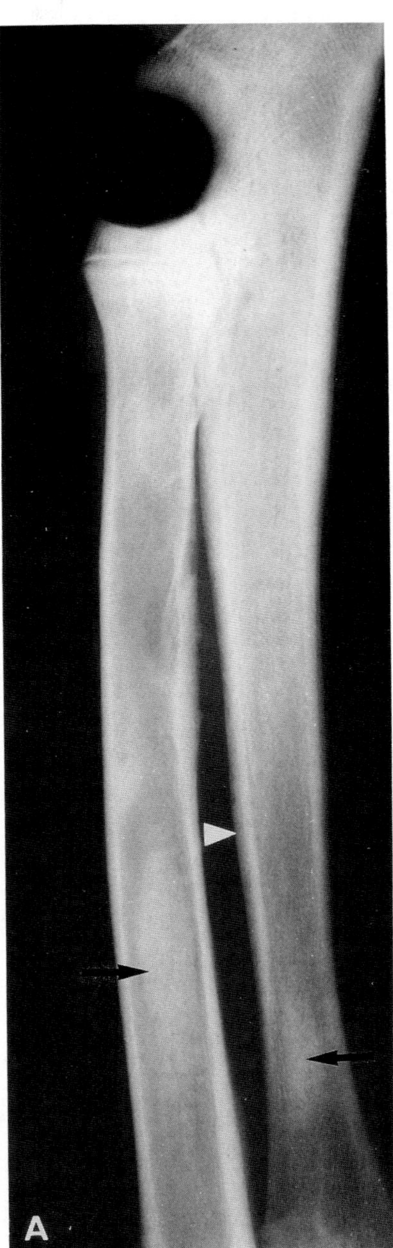

Fig. 1.95A Panosteitis. Dog. Radius–ulna showing focal and diffuse medullary densities (arrows) and periosteal reaction (arrowhead). (Courtesy of D. C. Van Sickle.)

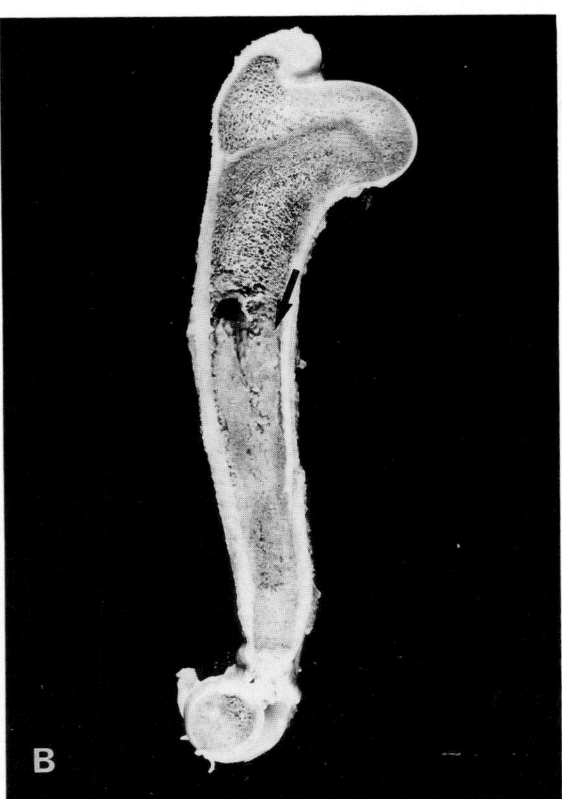

Fig. 1.95B Panosteitis. Dog. Humeral marrow distal to arrow is filled with woven bone.

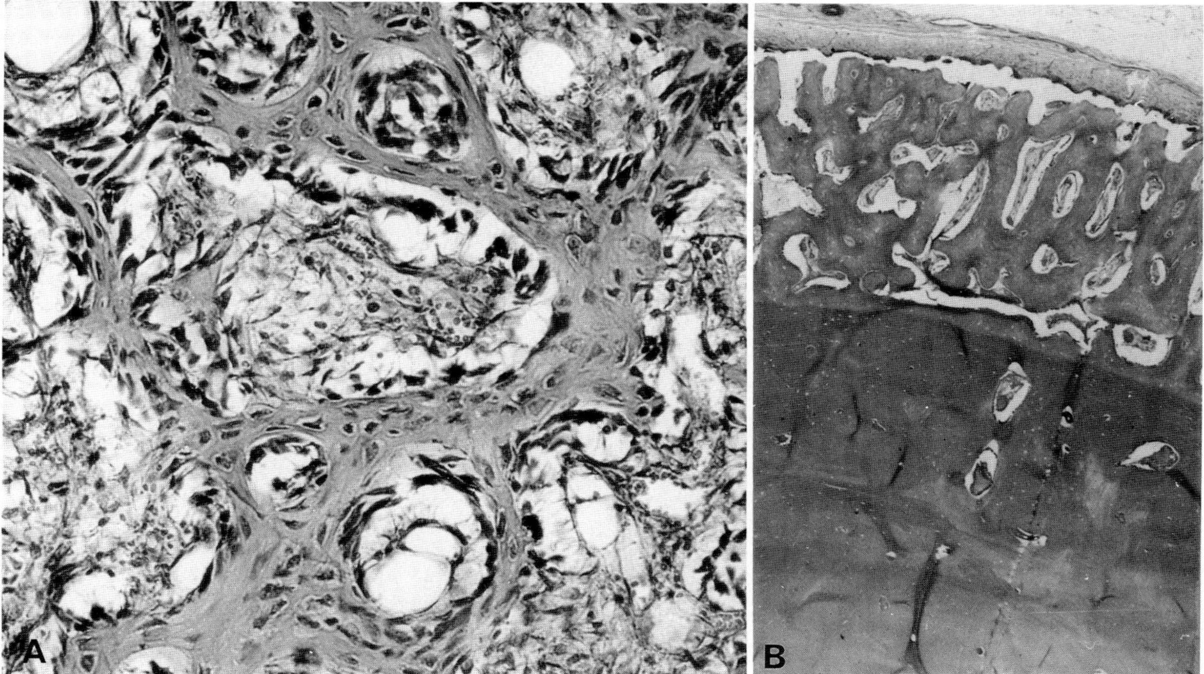

Fig. 1.96 Panosteitis. Dog. (A) Medullary bone formation. (Courtesy of D. C. Van Sickle.) (B) Periosteal bone formation. (Courtesy of K. A. Johnson.)

between 5 and 12 months of age. Occasionally small dogs such as miniature schnauzer and Scottish terrier are affected, and the age range extends from 2 months to 5 years. The disease was first described as eosinophilic panostitis of German shepherds, and 75% of reported cases involve this breed. Over two thirds of affected dogs are males.

Clinical signs are mild to extremely severe lameness, which tends to be shifting, prone to remission and exacerbation, and is usually self-limiting after 1 to several months. The lameness is associated with abnormalities in the diaphysis of a long bone, usually of a foreleg (Fig. 1.95). Multiple bone involvement occurs in about 50% of cases and rarely, other bones such as ilium and metatarsals are affected.

Because the disease is self-limiting and is well recognized by radiologists, pathologic studies are few, and the lesions are defined in terms of their radiograpic appearance. At the time when clinical signs begin, no lesions may be visible. About 10 days later, initial involvement of the long bone is seen as indistinct increased density or densities of the medulla in the region of the nutrient foramen. The density may increase in size and perhaps fill the whole medullary cavity, or may remain localized (Fig. 1.95A,B). If the reaction extends to the cortex, marked cortical thickening may result due to periosteal proliferation. When the animal recovers, the increased density disappears over a period of weeks to months.

The initial increased radiodensity is due to expanding areas of fibrovascular tissue in the bone marrow, which are rapidly replaced by woven bone (Fig. 1.96A). This bone is subject to consecutive episodes of formation and resorption; thus, reversal and resting lines are often present in the same trabecula, and osteoclasts and osteoblasts are active in the same microscopic field. Cartilage is sometimes formed in the fibrovascular tissue. Periosteal woven bone formation may be stimulated and this, like the medullary bone, is replaced by lamellar bone before being removed during the following months (Fig. 1.96B). The older, more mature bone in the center of the medullary lesions is removed first.

In most dogs, there is no evidence of inflammatory infiltrates in the lesions, but sometimes plasma cells and histiocytes are present. Eosinophilia is an inconstant feature of the hemogram, and serum chemistry is unremarkable. Abnormalities in clotting times are inconstant findings.

The cause of canine panosteitis is not known. Genetic and allergic factors, hyperestrogenism, filterable agents, and various bacteria have been proposed, but the evidence is inconclusive. In females, the onset of signs is sometimes associated with estrus, and it is thought that other physiological stress may precipitate the disease. The occurrence of panosteitis in dogs with hemophilia A suggests that sometimes the disease might be a sequel to intraosseous hemorrhage.

Bibliography

Bohning, R. H. *et al.* Clinical and radiologic survey of canine panosteitis. *J Am Vet Med Assoc* **156:** 870–883, 1970.

Grondalen, J., Sjaastad, O., and Teige, J. Enostosis (panosteitis) in three dogs suffering from hemophilia A. *Canine Pract* **16:** 10–14, 1991.

Stead, A. C., Stead, M. C. P., and Galloway, F. H. Panosteitis in dogs. *J Small Anim Pract* **24:** 623–635, 1983.

Van Sickle, D. C. Canine panosteitis. *In* "Selected Orthopedic Problems in the Growing Dog," R. B. Hohn (ed.), South Bend, Indiana, American Animal Hospital Association, 1975.

D. Hypertrophic Osteopathy

This syndrome, which occurs in humans and all species of domestic animals, is characterized by diffuse periosteal osteophytosis secondary to a chronic lesion, usually intrathoracic, which may be inflammatory or neoplastic. There are rare exceptions to the periosteum–thorax association, as in dogs with botryoid rhabdomyosarcoma of the urinary bladder, and mares with ovarian tumors, that sometimes develop typical osseous lesions. Occasionally human cases occur in association with chronic intestinal lesions and cirrhosis of the liver, but analogous cases in veterinary medicine are not recorded. These exceptions prove the rule, but hypertrophic osteopathy is often called pulmonary osteoarthropathy, reflecting the frequency with which lung is the site of the primary lesion. There are other synonyms including Marie's disease and acropachia.

The thoracic lesions associated with development of hypertrophic osteopathy are diverse. Formerly, it often accompanied pulmonary tuberculosis but now occurs most often with primary or secondary pulmonary neoplasms as well as with granulomatous pleuritis, granulomatous lymphadenitis of bronchial or mediastinal nodes, chronic bronchitis, *Dirofilaria immitis* infection, esophageal granulomas and tumors provoked in dogs by *Spirocerca sanguinolenta,* and with neoplastic disease of the thoracic wall. Hypertrophic osteopathy is present in only a minority of animals with thoracic lesions, but the incidence is much higher in some types than in others. For example, dogs with pulmonary metastases from osteosarcoma are much more likely to develop the syndrome than are dogs with metastatic carcinoma.

The initial clinical signs are often related to the secondary bone changes, and lesions in the thorax are discovered later. As noted, osteosarcoma may metastasize to lung and provoke hypertrophic osteopathy, but hypertrophic osteopathy is not commonly associated with other intraosseous neoplasms. Rarely, a primary extraskeletal neoplasm metastasizes to the new bone of hypertrophic osteopathy.

The periosteal osteophytes are usually confined to the limbs, and in the dog, the species most often affected, involve particularly the radius, ulna, tibia, metacarpals, and metatarsals, with relative sparing of the bones of the upper limbs and phalanges (Fig. 1.97A). Occasionally there are lesions of the vertebrae and skull, and minor endosteal lesions occur infrequently. Periosteal osteophytes involve the distal limb bones first and extend proximally. They are progressive but regress if the primary thoracic lesion is removed. The extent of the osteophytes

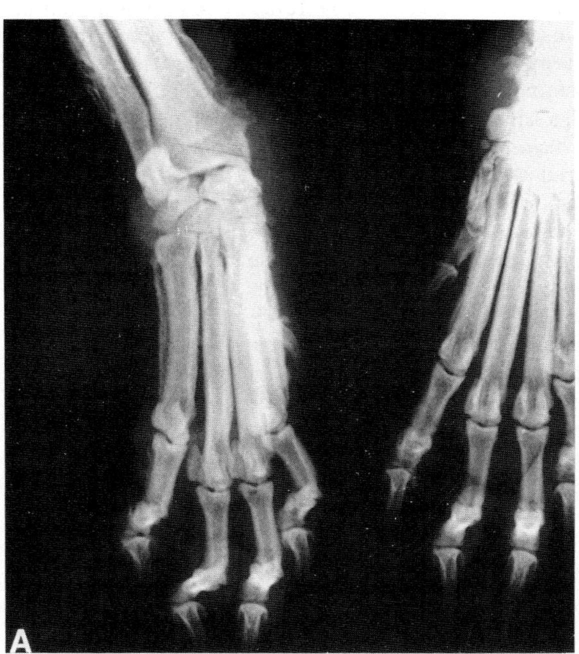

Fig. 1.97A Hypertrophic osteopathy. Dog. Periosteal reaction on metacarpals and radius–ulna. Relative sparing of phalanges. (Courtesy of T. J. Hage.)

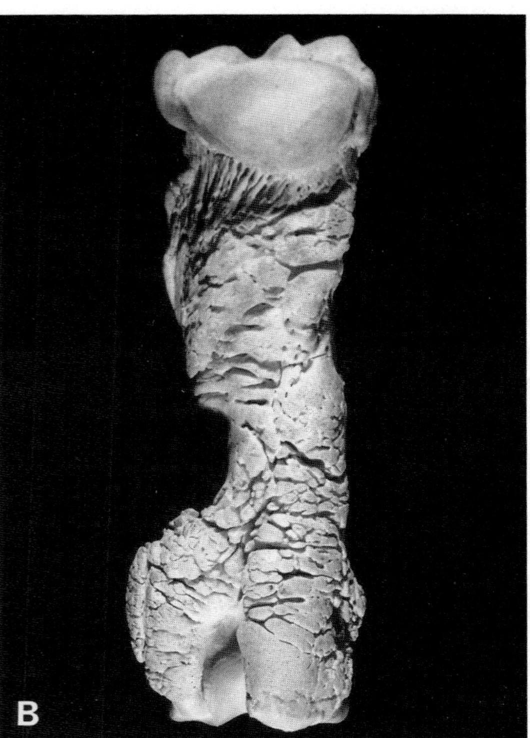

Fig. 1.97B Hypertrophic osteopathy. Horse. Humerus encrusted by periosteal new bone.

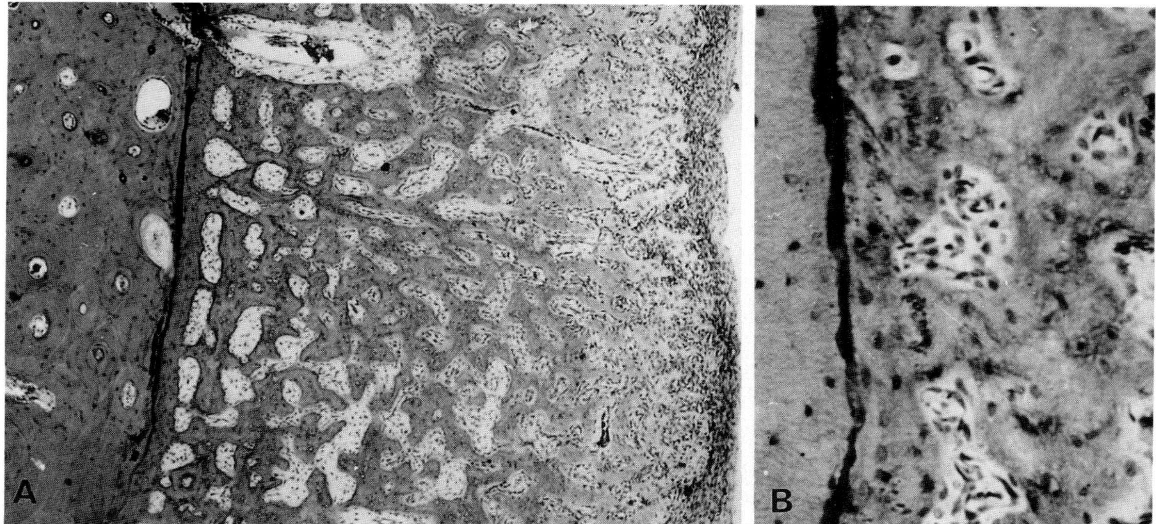

Fig. 1.98 Hypertrophic osteopathy. (A) Diffuse periosteal hyperostosis. (B) Detail of periosteal reaction on margin of original cortex shown by dark resting line.

is best demonstrated in radiographs and macerated specimens where, in florid cases, they appear as wart- or cauliflowerlike accretions extending along the whole length of the bone (Fig. 1.97B). Cross-sections of bone demonstrate the asymmetric development of these accretions, which tend to occur initially and most severely where the periosteum is free of tendinous insertions or adjacent bones and overlying tendons and can be elevated.

Histologically, the earliest changes are hyperemia and edema with proliferation of highly vascular connective tissue in the periosteum. These changes are associated with a rapid increase in peripheral blood flow in the distal half of the limbs. At this time there may be a light infiltrate of lymphocytes and plasma cells. Shortly thereafter osteoblasts deposit osteoid on the existing cortical bone and then, if the disturbance is severe, begin to lay down new trabeculae perpendicular to the original cortex. The new bone may be deposited with extraordinary rapidity, and the width of the cortex may be doubled in a few weeks. In the early and active stages of the disease, the new bone is clearly distinguishable from the original cortex because the former is trabecular, and the latter, compact (Fig. 1.98A,B). In the later stages, the distinction may not be possible because the original cortex is converted to spongy bone by resorption from within the vascular canals, with modest assistance from the endosteum.

In long-standing human cases, hyperemia, inflammation, and connective tissue proliferation in the synovial membranes occur, thereby justifying the name osteoarthropathy in humans. Similar lesions apparently have not been seen in other species and may represent secondary complications of long-standing disability.

The pathogenesis of hypertrophic osteopathy is obscure. Several hypotheses have been proposed, but none consistently fits the clinical and experimental observations. The neurogenic theory maintains that the skeletal changes result from reflex impulses that arise in the thoracic lesion and travel via the vagus nerve to cause vasomotor disturbances in the limbs. That the efferent path of the reflex is not neural is shown by the fact that denervation of an affected limb does not affect the increased blood flow. Limited support for this theory is provided by the regression of bone lesions in some, but not all, human and canine cases treated by vagotomy. The humoral theories propose either that a hormone or hormonelike substance is produced by the primary lesion or that arteriovenous anastomoses in the primary lesion prevent catabolism of such a substance in the lung. The hormone then produces the secondary skeletal lesions. Support for the lung-bypass aspect of the theories comes from the production of hypertrophic osteopathy following experimental venoarterial shunts in dogs, but general application of the hypothesis is ruled out by the absence of such shunts in most spontaneous cases. It is also suggested that reduced oxygenation of peripheral blood resulting from congenital or acquired arteriovenous shunts stimulates periosteal osteophyte production. Perhaps no single mechanism is responsible for the skeletal changes in this syndrome. The stereotypic lesions could simply reflect the limited range of periosteal reaction to diverse insults rather than its consistent response to a single stimulus.

Bibliography

Brodey, R. S. Hypertrophic osteoarthropathy in the dog: A clinico-pathologic survey of 60 cases. *J Am Vet Med Assoc* **159:** 1242–1256, 1971.

Hesselink, J. W., and van den Tweel, J. G. Hypertrophic osteopathy in a dog with a chronic lung abscess. *J Am Vet Med Assoc* **196:** 760–762, 1990. (Useful reference list)

Holling, H. E. Pulmonary hypertrophic osteoarthropathy. *Ann Intern Med* **66:** 232–233, 1967.

Sweeney, C. R. *et al.* Hypertrophic osteopathy in a pony with a pituitary adenoma. *J Am Vet Med Assoc* **195:** 103–105, 1989. (Lists other non-thoracic lesions)

Watson, A. D. J., and Porges, W. L. Regression of hypertrophic osteopathy in a dog following unilateral intrathoracic vagotomy. *Vet Rec* **93:** 240–243, 1973.

E. Osteochondrosis

Osteochondroses are diseases of pigs, dogs, horses, cattle, and sheep characterized by abnormalities of growth cartilage in both physeal and epiphyseal sites. Similar diseases occur in turkeys, chickens, and humans. The cause(s) of osteochondrosis is (are) not understood, and results of experimental investigations are often contradictory. There appear to be several contributing, perhaps related, factors including inheritance, rapid growth, physical activity, gender (males tend to be more susceptible), and nutrition (discussed in more detail in relation to species). If there is a common pattern in all species, it is in the tendency to develop fragile cartilage that is susceptible to injury by (usually) physiologic trauma. It is not certain, however, that there is a common etiology.

There are several manifestations and sequelae of osteochondrosis. Some of these have acquired the status of disease entities, so that the primary abnormality is sometimes overlooked, or a particular manifestation is confused with the disease itself. We discuss osteochondrosis of swine in subsequent sections. Similarities and differences in other species are then considered.

Osteochondrosis in **swine** is a generalized disease of growth cartilage, but significant changes tend to occur in specific locations, and are usually bilateral and symmetrical. There are three types of histologic lesions, which may be related. These lesions occur in growth cartilage of the physis and of the epiphysis and comprise eosinophilic streaks, foci of metaphyseal dysplasia, and zones of cartilage necrosis.

Eosinophilic streaks are narrow zones of acidophilic tissue in the cartilage matrix (Fig. 1.99). They are normally present in growth cartilage of young animals and increase in number and complexity with age. Some, probably most, of them are vestiges of cartilage canals, and in electron micrographs, they appear as zones of matrix with low fiber and high vesicle density. Others appear to be areas of matrix degeneration. The normal streaks are roughly parallel to the cartilage columns; in osteochondrosis they are often stellate, and often located near areas of cartilage necrosis or foci of metaphyseal dysplasia.

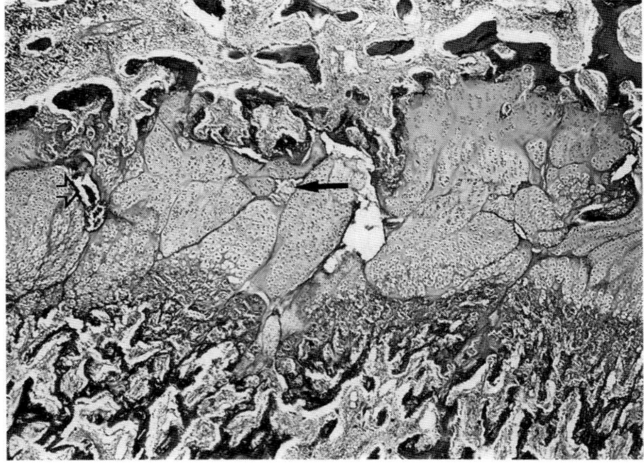

Fig. 1.99 Osteochondrosis. Pig. Eosinophilic streaks. The dark streaks form a reticulated pattern in physis. There are cavities in the physis (arrow).

Metaphyseal dysplasia is characterized by focal persistence of proliferative and/or hypertrophic cells, which produces a cone-shaped thickening in the growth cartilage (Fig. 1.100). Often there is an eosinophilic streak in the center or at one edge of the cone-shaped mass. The lesion is most common in the major growth plates and usually involves mid-hypertrophic cells. Often the chondrocytes fail either to form columns or to maintain orderly columns, but most appear viable and metabolically active. Ultrastructurally they develop nonspecific changes including accumulation of rough endoplasmic reticulum, lipid droplets, and mitochondria in their cytoplasm. More definitive changes occur in the pericellular matrix, which is highly condensed and contains large deposits of electron-dense material. These deposits are preserved only by using potassium ferrocyanide in the secondary fixative, and may consist of a protein with binding sites for the plasma membrane of the chondrocyte and several components of the

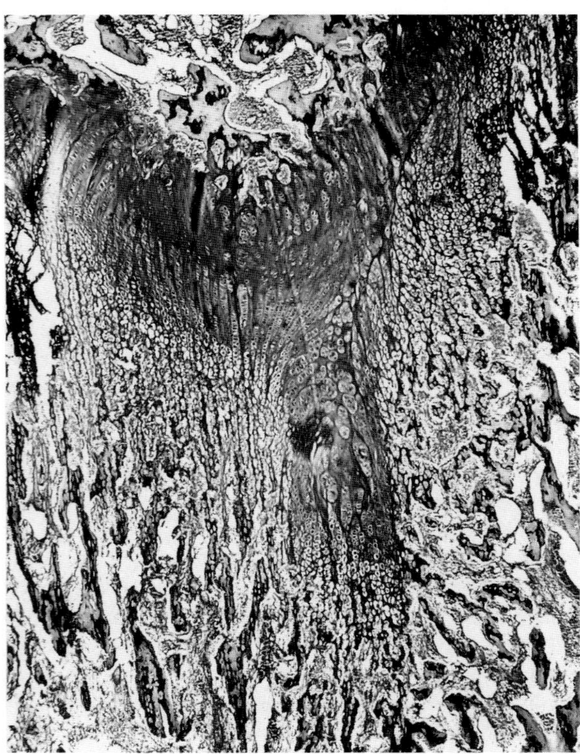

Fig. 1.100 Osteochondrosis. Pig. Focal metaphyseal dysplasia.

pericellular matrix. Farnum and Wilsman postulate that breakdown of the pericellular matrix components is an early lesion in the development of metaphyseal dysplasia, and is accompanied by failure of vascular penetration, despite continued mineralization of the affected cartilage.

The fate of the metaphyseal lesions is varied. Some undergo relatively normal ossification, others persist as nodules of cartilage in the metaphysis, and many degenerate and are replaced by fibrous tissue in which woven bone or fibrocartilage develops.

Zones of cartilage necrosis are most common and most significant in the articular–epiphyseal complex. They appear to arise following occlusion of cartilage canals and commonly involve the zone of proliferating cells. Normally, cartilage canals are eliminated during skeletal maturation by a process of chondrification, which apparently involves cartilaginous differentiation of their lining mesenchymal cells. In osteochondrosis of swine this process may be accelerated or enhanced, resulting in necrosis of chondrocytes due to hypoxia. This is followed by lysis and cavitation (Fig. 1.99). Alternatively the failure of chondrocytes to produce or maintain matrix proteoglycans may be responsible for cartilage necrosis. Necrosis of cartilage is apparently an early lesion and may not be visible at the time clinical signs develop. It is generally agreed that the areas of necrosis often lead to the development of osteochondritis dissecans. Less seriously, nodules of necrotic cartilage may be bypassed during endochondral ossification to remain in the epiphysis or metaphysis as a cartilage mass that often is covered by bone (Fig. 1.101).

Osteochondrosis does not primarily affect articular cartilage which, although nourished by diffusion from cartilage canals in the young animal, is not invaded by them. However, secondary involvement of articular cartilage, as in osteochondritis dissecans and arthropathy, is common. Also collapse of the articular–epiphyseal complex may produce indentations of intact articular cartilage.

These three abnormalities are responsible for the important pathologic changes in osteochondrosis of swine. Microscopic lesions occur in as many as 100% of pigs in affected herds but may progress to cause clinical disease in only a few animals. The prevalence of microscopic lesions in other species does not appear to have been studied. In most pigs, reestablishment of normal endochondral ossification and removal of defective cartilage allows the affected bones to develop in a reasonably normal manner. Lesions are most frequent and most severe in the joint cartilage of the medial femoral condyle, humeral condyles, anconeal process, lumbar vertebrae, mediodistal part of the talus, humeral head, glenoid of scapula, distal ulna, and dorsal acetabulum. In physeal growth cartilage, predilection sites are distal ulna and femur, costochondral junctions, femoral and humeral head, ischial tuberosity, and thoracolumbar vertebrae.

There are four important gross manifestations of osteochondrosis in swine, namely, osteochondritis dissecans, epiphysiolysis, deformities of bones, and arthropathy. The

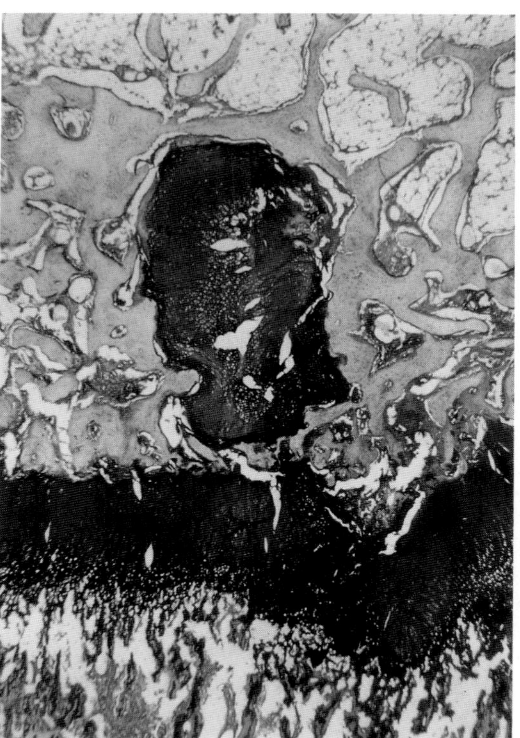

Fig. 1.101 Osteochondrosis. Pig. Nodule of cartilage partly encased in bone remains in epiphysis. Toluidine blue stain.

lesion of **osteochondritis dissecans** is a flap of cartilage that gives the appearance of having been dissected away from underlying bone (Figs. 1.102, 1.103A,B). It is usually attached to the adjacent cartilage at one end; if it separates and falls free into synovial fluid, it becomes a "joint mouse." Osteochondritis dissecans may develop when a fissure forms in or near an area of necrosis in the articular-epiphyseal cartilage (Fig. 1.104). These sites are susceptible to relatively mild trauma and, if the fissure extends through the articular cartilage, a flap is formed. Even if a dissecting lesion does not result, hemorrhage and fibrosis may occur, and the resulting localized radiolucent area may be the basis for one type of bone cyst in osteochondrosis (Figs. 1.103B, 1.105). The dissecting lesion does not involve bone, but sometimes flaps of cartilage contain osseous tissue. This probably develops by endochondral ossification following invasion of vessels and cells from the area of attachment. Osteochondritis dissecans in swine is almost always a manifestation of osteochondrosis and occurs most commonly in the medial condyles of the humerus and femur in animals of 5 to 7 months of age. Following loss of the piece of cartilage, the defect is filled by vascular fibrous tissue, which eventually is converted to fibrocartilage.

Epiphysiolysis is the separation of an epiphysis from metaphyseal bone. It is a traumatic lesion predisposed for by a defect in growth cartilage of the physis. In osteochondrosis, this defect probably develops either from an ex-

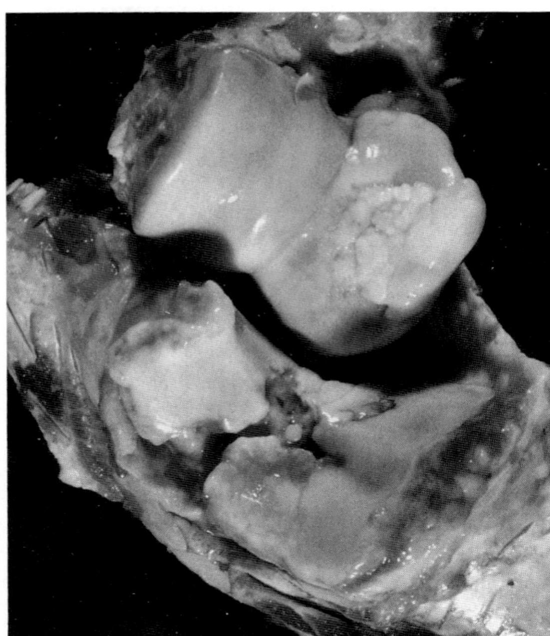

Fig. 1.102 Osteochondrosis. Pig. Osteochondritis dissecans and arthropathy of elbow.

Fig. 1.104 Horizontal split through cartilage in osteochondritis dissecans. Pig.

tended eosinophilic streak or from areas of necrosis in the growth cartilage, rather than from foci of metaphyseal dysplasia. The common sites for epiphysiolysis are femoral head, ischiatic tuberosity of females, and lumbar vertebrae. The distal epiphysis of the ulna and the anconeal process may also be involved, although strictly speaking, the lesion involving the latter should not be called epiphysiolysis. In pigs, the anconeal process does not develop from a separate ossification center, as in dogs, so **apophysiolysis** is a more appropriate term. Separation may be complete, as is often the case with the ischiatic tuberosity, or partial, as occurs with the head of the femur, which sometimes remains attached at its lateral margin (Fig. 1.106). Separation probably occurs when the process of endochondral ossification reaches or approaches the cartilage defect. The resulting fracture may extend in a jagged crack through primary and secondary spongiosa.

Bone deformities in osteochondrosis originate either

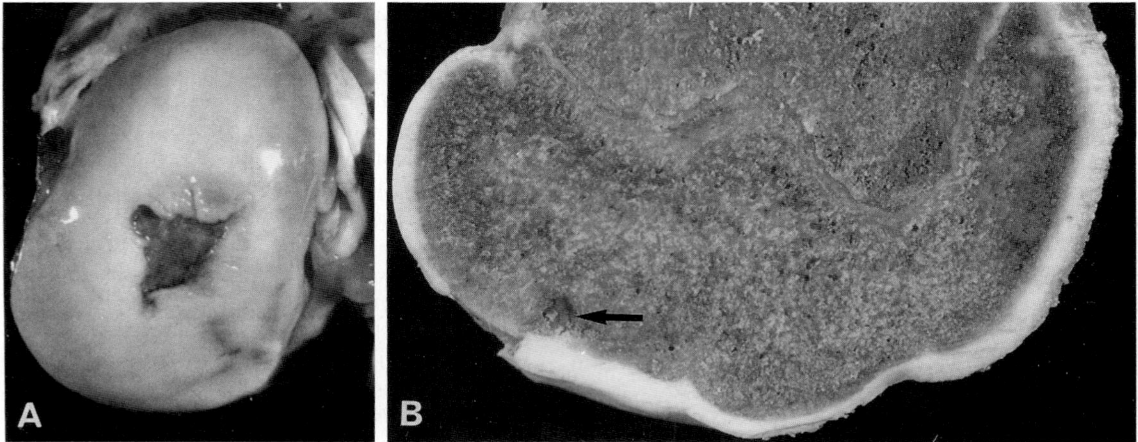

Fig. 1.103 Osteochondrosis. Pig. (A) Osteochondritis dissecans of distal femoral condyle. (B) Section through (A) showing thickening of cartilage, small subchondral "cyst" (arrow), and exposure of subchondral bone.

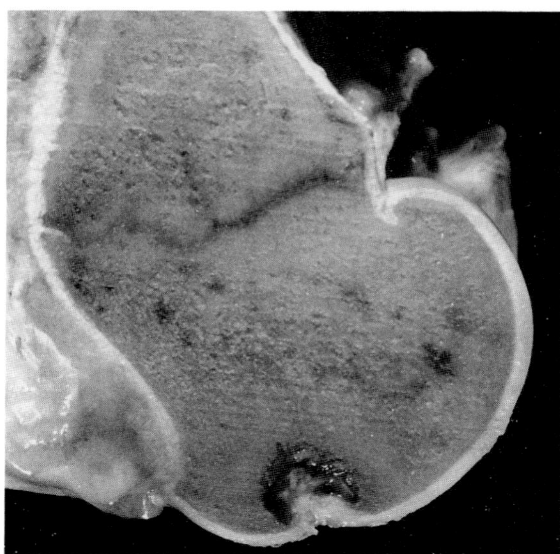

Fig. 1.105 Osteochondrosis. Horse. Bone "cyst" in epiphysis of distal femur.

from growth plate abnormalities of the types discussed or are secondary to epiphysiolysis or osteochondritis dissecans. Local growth plate defects originate from displacements of the plate along the plane of eosinophilic streaks (Fig. 1.107). Bony bridges develop between the ends of the plate, uniting epiphysis and metaphysis, in effect producing focal closures of the plate. Such lesions are relatively common in the femoral condyles and head.

Arthropathy develops at a relatively late stage of the disease and may be characterized by mild hypertrophy of

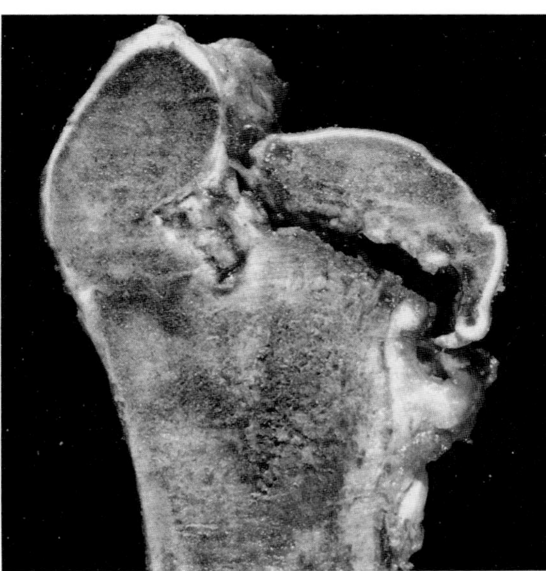

Fig. 1.106 Osteochondrosis. Pig. Epiphysiolysis of femoral head.

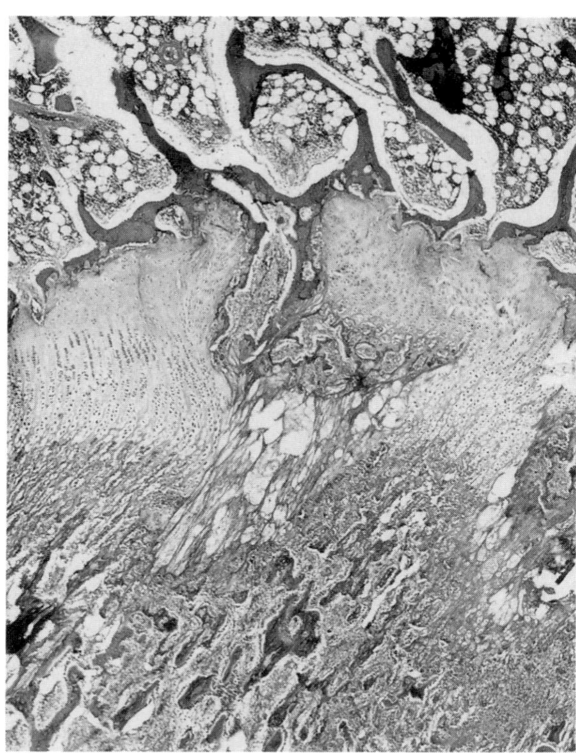

Fig. 1.107 Osteochondrosis. Pig. Degeneration of physeal cartilage and breaching of growth plate.

synovial villi and occasionally lymphoplasmacytic perivascular aggregates and scant fibrin exudate. The synovial reaction may be a response to release of degradation products from injured articular cartilage. In general, the synovial reaction following osteochondrosis is less severe than that associated with infectious arthritis. Grossly the arthropathy is nonspecific and may be attributable to osteochondrosis only by virtue of the location of lesions, residual microscopic changes in growth cartilage, and absence of other apparent cause. The lesions of arthropathy are detailed elsewhere; briefly, they consist of irregularities and fibrillation of joint cartilage, ulceration of articular surfaces with formation of fibrocartilaginous or osseous repair tissue, and marginal osteophytes with proliferation of synovial membrane and thickening of joint capsule. Sites commonly involved in porcine osteochondrosis are hock, elbow, lumbar spine, shoulder, hip, and stifle.

Invaginations of the articular–epiphyseal complex and nodules of cartilage in the epiphysis and metaphysis that are seen in osteochondrosis seem to be of minor pathologic importance, but they are of diagnostic significance in the differentiation of osteochondrosis and rickets. Cartilage invaginations are grossly visible, being up to several millimeters deep and a centimeter long. They are relatively common in the articular surface of the distal radius. Sometimes degenerative cartilage in the depth of the invagination is replaced by a mass of connective tissue or by a cystlike cavity, but there may be no apparent abnormality in the adjacent articular surface. Development of this type of osteochondrotic "cyst" apparently depends on access of synovial fluid to the degenerate tissue (Fig. 1.103B). Nodules of cartilage originating from necrosis in growth plates are most common in the epiphyses. They are composed of relatively acellular growth cartilage and are often encased by lamellar bone. Masses of cartilage separated from rachitic growth plates are most common in the metaphyses, are irregular in shape, and consist of hypertrophic cells.

Gross lesions in osteochondrosis are most spectacular when secondary manifestations have developed. Of the primary lesions, focal metaphyseal dysplasia is sometimes recognizable only by careful examination of growth plates. Undulations are normal in the physis, and the presence and extent of lesions often are difficult to assess grossly. Dysplasia in subarticular cartilage is more readily recognized, and areas of hemorrhage and fibrosis appear as dark red lines beneath the thickened cartilage. There is often little correlation between the extent of gross and microscopic lesions and clinical signs. This is true of both microscopic and gross lesions, including osteochondritis dissecans and epiphysiolysis.

In pigs, osteochondrosis appears to be one result of long-term selection for desirable carcass qualities combined with intensive management techniques. There is a positive correlation between these qualities and lesions of osteochondrosis, which is associated with high levels of growth hormones and somatomedins. Thus an inherent tendency to develop lesions exists in superior strains of breeding stock. All pigs with lesions do not develop the disease, although in some groups of breeding animals, disease prevalence approaches 100%. The reasons for such high morbidity often are not apparent, but in general, any factor that increases physical stress on the skeleton or which reduces the ability of the skeleton to resist that stress could be significant. In this regard, it is not uncommon for ischiatic epiphysiolysis to occur in young sows after breeding, and there is little doubt that rapid weight gain predisposes to development of the disease. There is no experimental evidence consistently incriminating mineral and vitamin deficiencies as causative factors in osteochondrosis, but it is conceivable that nutritional imbalances could precipitate disease in predisposed animals by increasing susceptibility to minor trauma. In the majority of field outbreaks, no nutritional deficiencies are demonstrable.

The sequelae to osteochondrosis are considered to be the major causes of the clinical syndrome of leg weakness of swine in Europe and North America. In Japan, a conformational defect of the proximal femur is thought to be responsible for more cases of leg weakness than is osteochondrosis. In pigs with this defect, the top of the greater trochanter is higher than the femoral head, instead of lower, and the angle between the two is wide.

Osteochondrosis in **dogs** here includes several orthopedic problems of young dogs, which are regarded as separate diseases. The inclusion of them as manifestations of osteochondrosis is for the present, and it is possible that reorganization of classifications will be necessary. As in swine, osteochondrosis in dogs is associated with rapid growth, and the important manifestations and sequelae are osteochondritis dissecans, epiphysiolysis, and arthropathy. In contrast to those for swine, large surveys and detailed histologic investigations of early lesions are sparse for dogs.

Osteochondritis dissecans is common in the humeral head of large and giant breeds, usually 4–8 months of age. Males are clinically affected three to four times as frequently as females, and lesions are bilateral in as many as 70% of affected dogs, even though lameness is often unilateral. Lesions develop from fissures in the posterior central aspect of the humeral head. It is widely accepted that these fissures result from trauma sustained during vigorous exercise and that osteochondritis dissecans is established when the defect extends to the articular surface. Fissures may exist for some time before this occurs, and reactions in the overlying cartilage and underlying bone consist of nodular proliferation of chondrocytes and diffuse proliferation of fibrovascular tissue respectively. Variations in the histologic appearance of the lesions are probably due to the degree of epiphyseal maturity when they develop and the amount of tissue reaction at the time they are examined. Common sequelae are formation of joint mice and degenerative arthropathy. Routine examination of the proximal humerus of giant breeds reveals a high prevalence of osteochondritis dissecans or its sequelae, which in most cases are mild and apparently asymptomatic. Degeneration of articular cartilage on the posterior central aspect of the humeral head sometimes is due to primary arthropathy (see Diseases of Joints, Section III,A,2, Arthropathy of the Canine Shoulder).

Osteochondritis dissecans also occurs in several other locations in dogs, including the medial part of the humeral condyle, the lateral, and less frequently the medial, condyle of the femur, and the medial trochlear ridge of the tibial tarsal bone.

Lesions in the humeral condyle form part of the **elbow dysplasia syndrome,** which includes ununited anconeal process and fragmented (ununited) medial coronoid process of the ulna. The latter conditions may also occur separately (see the following), or in other combinations. Together the three lesions, which are often bilateral, cause foreleg lameness in young, usually large, growing dogs and often lead to arthropathy associated with periarticular osteophytes. The elbow dysplasia syndrome is characterized by incongruity of the joint caused by a step between the cranial margin of the proximal ulna and the caudal margin of the proximal radius. An increase in length of the proximal ulna is responsible for the incongruity and results in disruption of the normal continuous curve made by the profile of the trochlear notch of the ulna and the articular cartilage of the radius. Trauma from normal use precipitates lesions. Radiographically there is increased width in

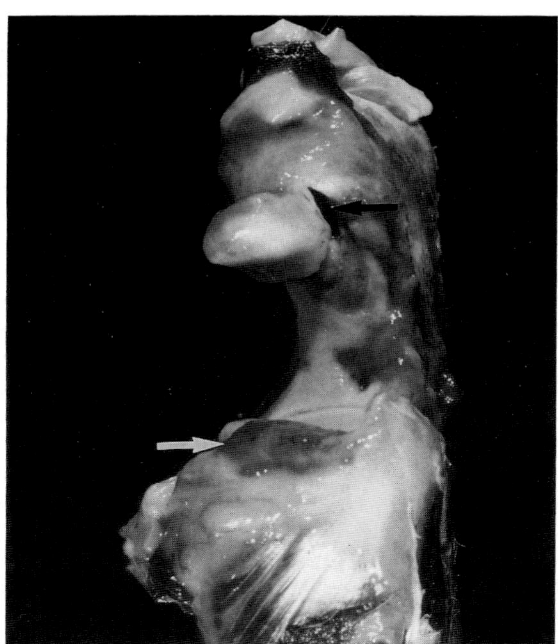

Fig. 1.108 Osteochondrosis. Dog. Ununited anconeal process (black arrow). The synovium is thick and dark (white arrow).

the humeroradial and humeroulnar joint spaces, and the distal humerus is displaced cranially on the radius.

Osteochondritis dissecans of the femoral condyles causes hind-limb lameness and usually progresses to arthropathy unless treatment is successful. Lesions are sometimes bilateral, and osteochondritis dissecans of the humeral head occasionally coexists. Lesions of the medial trochlear ridge of the tibial tarsal bone occur under similar circumstances and have a prognosis similar to those in the other locations.

Another manifestation of osteochondrosis in dogs is **ununited anconeal process** (Fig. 1.108). This lesion, which develops as an epiphysiolysis, is most common in the German shepherd, but also occurs in other large and giant breeds. The ossification center of the anconeal process develops at about 11 to 13 weeks and expands for about 6 weeks until osseous union with the ulna is achieved by 4 to 5 months. If extensive fissures develop in the bipolar growth cartilage between the process and the ulnar diaphysis, normal plate closure does not occur, and a fibrocartilaginous or fibrous union results.

Ununited (or fragmented) medial coronoid process of the ulna is less well known, but apparently more common, than ununited anconeal process. The coronoid process does not develop from a separate ossification center and is not easily seen in radiographs. Usually the first manifestations of disease are osteophytes around the anconeal process and the medial epicondyle of the humerus. Golden and Labrador retrievers and Rottweilers are particularly prone to this condition. In affected dogs, the medial coronoid process consists of one or several fragments, which may be in their normal position or completely separated from the articular surface of the proximal ulna as joint mice. Lesions begin as vertical fissures in the proximal face of the articular cartilage and extend into the subchondral bone. They may reach the vertical joint surface or the area of insertion of the annular ligament. When complete separation of the fragment or fragments does not occur, the walls of the fissure become lined by fibrocartilage.

Other conditions in dogs which are probably manifestations of osteochondrosis include retained cartilage core of the distal ulna (Fig. 1.75A,B) (which is contrasted with premature closure of growth plates in an earlier part of this chapter) and slipped capital femoral epiphysis (see epiphysiolysis in pigs and discussion of fractures). Retained cartilage cores also develop in the other long-bone metaphyses and may lead to growth deformities of the distal femur and proximal tibia. Canine hip dysplasia is classified as a manifestation of osteochondrosis by some workers, but we consider it separately (see Diseases of Joints).

The same can be said of causation of osteochondrosis in dogs as was said for pigs. Fast growth is almost invariably a characteristic of affected animals. Certain features of conformation probably account for the increased involvement of specific sites in some breeds.

Many young dogs have lesions of osteochondrosis, but few exhibit clinical signs. Also, many dogs with radiographic evidence of disease heal normally when growth slows and endochondral ossification is completed. The severity of lesions combined with the appropriate trauma or physiologic stress probably determine the onset of clinical disease. Experimental prolonged excess calcium intake increases the prevalence and severity of osteochondrosis lesions in susceptible dogs, but the relevance of this finding to naturally occurring disease is unclear. In pigs, increased dietary calcium does not affect the prevalence of osteochondrosis.

Osteochondrosis occurs in **horses** from a few weeks to about 2 years of age, and a genetic susceptibility is suspected. The disease is generally similar to that in pigs and dogs, although the terminology is somewhat different. In equine parlance, developmental orthopedic disease (DOD) encompasses the clinical/pathological manifestations and sequelae of osteochondrosis including osteochondritis dissecans and epiphysitis. The latter term indicates the lesion and traumatic sequel of metaphyseal dysplasia near major growth plates.

In horses, sites of significant lesions include the dorsal edge of the sagittal ridge of the distal tibia and the articular processes of cervical vertebrae where they are similar to those of the canine medial coronoid process. Also affected are the medial condyle and the trochlear ridges of the femur, the patella, and various sites in the tarsal and fetlock joints (Fig. 1.109). In these locations, the pathogenesis of lesions is basically similar to that previously described. Ossification of attached cartilage flaps seems to be more common in the horse than in other species. Foals with congenital and neonatal angular limb deformities

Fig. 1.109 Osteochondrosis. Horse. Fragment of cartilage from femoral lesion (white arrow) lies in synovial recess (dark arrow).

sometimes have concomitant hypoplasia of carpal bones and osteochondritis dissecans. These early lesions may both be manifestations of defective growth and maturation of cartilage. Many "wobbler" horses have severe osteoarthropathy of intervertebral joints which is often secondary to osteochondrosis (see Diseases of Bones, Section V,B,6, Vertebrae).

The relation between diet and osteochondrosis in horses is complex and unsettled. Experimental and naturally occurring copper deficiencies are associated with increased prevalence of cartilage lesions in foals, and copper supplements appear to reduce the prevalence of the disease, but not eliminate it. Copper catalyzes the activity of lysyl oxidase, which is required for the formation of the intermolecular cross-links of collagen, and cartilage from copper-deficient animals has an abnormally high proportion of soluble collagen, with reduced pyridinoline cross-linkages. Several protocols that produce osteochondrosis in foals and other animals can be related to copper deficiency: zinc toxicity may stimulate formation of metallothionein, which binds copper in intestinal epithelium and leads to its fecal excretion in sloughed cells. High-calcium diets, which may increase the prevalence of osteochondrosis in dogs, induce zinc deficiency by forming insoluble calcium–zinc–phytate complexes in the gut; dexamethasone, which in prolonged dosage is associated with osteochondrosis, also stimulates metallothionein production; molybdenum binds copper as thiomolybdates, reducing its absorption from the gastrointestinal tract and increasing its excretion from the tissues. [Molybdenosis in sheep causes subperiosteal hemorrhages of long bones, leading to formation of bony excrescences, and epiphysiolysis of the greater trochanter. In cattle, widening of the physes, osteoporosis, and pathologic fractures occur. Certain legumes, including trefoils (*Lotus* spp.) and sweetclover (*Melilotus officinalis* and *M. alba*), are molybdenum accumulators.]

Ingestion of feed high in soluble carbohydrate may predispose to development of osteochondrosis, possibly by regulating levels of circulating insulin and thyroxine. These hormones influence the release of growth hormone and somatomedins in several species including horses, and their inappropriate secretion could modify chondrocyte proliferation. There is a possibility that the usual postprandial synchrony of hormone release is disturbed, and the normal pattern of cartilage growth is altered, in horses given unrestricted access to high-energy feed. Experimental horses so treated have decreased levels of hexosamine and hydroxyproline, and increased levels of DNA in their growth plate cartilage, reflecting decreased matrix synthesis and relatively increased cellularity.

Lesions of the palmar surface of the distal end of the metacarpal in adult racing thoroughbreds are often discussed with the osteochondroses. The defects develop on the articular surface in contact with the proximal sesamoids. The initial lesion is fracture and necrosis of the subchondral bone, and the initiating factor is probably the trauma of hard running. With rest the fracture may heal, since articular cartilage above the fracture, nourished by synovial fluid, remains viable and probably holds the bone fragment in place. Collapse of subchondral tissues causes depression of the intact articular cartilage, but revascularization and healing may occur.

On occasion, the complete fragment breaks free, and this type of osteochondritis dissecans is analagous to some of the lesions described in humans. The fact that horses so affected are adults, and are athletes in hard training, suggests that these lesions of the palmar surfaces of metacarpals should be classified with stress fractures rather than with osteochondrosis.

Osteochondrosis in **cattle** is described infrequently. Significant lesions occur most commonly in the humeral head, distal radius, and distal femur, and changes are also found on the tibial tarsal and occipital condyles. "Epiphysitis" of metacarpal and metatarsal bones is a significant cause of lameness in fast-growing, grain-fed, housed calves and in many cases is due to osteochondrosis. The number and severity of lesions are increased by hard floors, indicating the significance of trauma to susceptible animals.

Osteochondrosis in **sheep** is rare. Microscopic lesions are common in growth plates of fast-growing lambs, but few of these progress to gross lesions. However, concerted efforts by producers to change conformation, reduce exercise, and maximize growth rates should eventually precipitate clinical disease. Valgus and varus limb deformities are common in ram lambs in feedlots and test stations in the United States of America and occur in other countries. Grossly, some affected rams have thickening of the distal radial physis on the medial side, but the deviation is usually lateral. The etiology and pathogenesis are unknown. Microscopic lesions in some lambs resemble those of osteochondrosis, but joint changes are absent.

Vascular and growth plate lesions mimicking osteochondrosis can be produced in lambs by procedures that increase weight bearing.

Bibliography

Bridges, C. H, and Harris, E. D. Experimentally induced cartilaginous fractures (osteochondritis dissecans) in foals fed low-copper diets. *J Am Vet Med Assoc* **193:** 215–221, 1988.

Carlson, C. S. *et al.* Effect of reduced growth rate on the prevalence and severity of osteochondrosis in gilts. *Am J Vet Res* **49:** 396–402, 1988.

Duff, S. R. I. Histopathology of growth plate changes in induced abnormal bone growth in lambs. *J Comp Pathol* **96:** 15–24, 1986.

Farnum, C. E., and Wilsman N. J. Ultrastructural histochemical evaluation of growth plate cartilage matrix from healthy and osteochondritic swine. *Am J Vet Res* **47:** 1105–1115, 1986.

Goedegebuure, S. A., and Hazewinkel, H. A. W. Morphological findings in young dogs chronically fed a diet containing excess calcium. *Vet Pathol* **23:** 594–605, 1986.

Grondalen, J., and Grondalen, T. Arthrosis in the elbow joint of young rapidly growing dogs. V. A pathoanatomical investigation. *Nord Vet Med* **33:** 1–16, 1981.

Hill, M. A. Causes of degenerative joint disease (osteoarthrosis) and dyschondroplasia (osteochondrosis) in pigs. *J Am Vet Med Assoc* **197:** 107–113, 1990.

Jensen, R. *et al.* Osteochondrosis in feedlot cattle. *Vet Pathol* **18:** 529–535, 1981.

Kincaid, S. A., Allhands, R. V., and Pijanowski, G. J. Chondrolysis associated with cartilage canals of the epiphyseal cartilage of the distal humerus of growing pigs. *Am J Vet Res* **46:** 726–732, 1985.

Kronfeld, D. S., Meacham, T. N., and Donoghue, S. Dietary aspects of developmental orthopedic disease in young horses. *Vet Clin North Am Eq Pract* **6**: 451–465,1990.

Nakano, T., Brennan, J. J., and Aherne, F. X. Leg weakness and osteochondrosis in swine: A review. *Can J Anim Sci* **67:** 883–901, 1987.

Olsson, S.-E. (ed.) Osteochondrosis in domestic animals. I. *Acta Radiol* (Suppl.) 358, 1978.

Pitt, M., Fraser, J., and Thurley, D. C. Molybdenum toxicity in sheep: Epiphysiolysis, exostoses, and biochemical changes. *J Comp Pathol* **90:** 567–576, 1980.

Visco, D. M. *et al.* Cartilage canals and lesions typical of osteochondrosis in growth cartilages from the distal part of the humerus of newborn pigs. *Vet Rec* **128:** 221–228, 1991.

Wind, A. P. Elbow incongruity and developmental elbow diseases in the dog: Part I. *J Am Anim Hosp Assoc* **22:** 711–724, 1986.

Wind, A. P., and Packard, M. E. Elbow incongruity and developmental elbow diseases in the dog: Part II. *J Am Anim Hosp Assoc* **22:** 725–730, 1986.

Woodard, J. C., Becker, H. N., and Poulos, P. W. Articular cartilage blood vessels in swine osteochondrosis. *Vet Pathol* **24:** 118–123, 1987.

Yamasaki, K., Ikeda Y., and Itakura, C. Bone lesions in pigs with locomotor dysfunction of hind legs. *J Comp Pathol* **100:** 313–322, 1989.

X. Neoplastic and Tumorous Conditions of Bones

In bone tumors of humans, many correlations between gross and histologic appearances, precise anatomic location, age of host, and other vital data, and the behavior of the tumors have been carefully documented. Most of the common and some of the rare tumorous conditions of humans have analogs, as far as microscopic appearance is concerned, in farm and pet animals, and there is a tendency to match pictures and name them accordingly. This is an acceptable procedure only if the diagnosis reflects the supposed origin of the tumor, if facts about behavior are not transferred uncritically along with the recognition of microscopic similarities, and if it is kept in mind that such diagnoses may well require later revision. Some of the following topics are discussed against this cautionary background. It should be noted that "tumor" includes but does not necessarily mean neoplasm.

In contrast to those in humans, metastatic neoplastic diseases of the skeleton are considered to be uncommon in animals, but this is partly due to a failure to look for them.

Primary skeletal tumors are common in dogs and somewhat less so in cats. In both these hosts, the tumors should be regarded as malignant until proved to be benign. Relatively few primary or metastatic neoplasms are recorded in other domestic species and generalizations are not possible.

There is disagreement about the classification of neoplasms of the skeleton because of their diverse origins and overlapping structural patterns. The varied origins include the ordinary connective tissues, the blood vessels, the cartilages, and the cells which really belong only in bone, the osteoclasts, osteoblasts, and their precursors. The overlapping structural patterns in tumors of osseous origin are reflections of embryogenetic transformations and include myxomatous, cartilaginous, and osseous tissues. Chondromatous tumors can ossify, and osteosarcomas can produce areas of cartilage or myxomatous tissue, and also can contain many osteoclastlike cells resembling those of giant-cell sarcomas.

A further difficulty in deciding the possible histogenesis of tumors which involve bone is imposed by the frequency with which new bone formation occurs. The presence of new bone both within and about a skeletal tumor does not mean that the tumor is an osteosarcoma. Osteosarcomas, by definition, produce osteoid or spicules of new bone, but bone also occurs in many other tumors in bone, whether primary or metastatic, as well as in a variety of non-neoplastic osseous lesions. It is important to attempt to distinguish the type or types of new bone associated with the tumor. The spectrum includes reactive periosteal, endosteal, or central bone, endochondral bone as in some cartilage tumors, metaplastic bone, and tumor bone or osteoid. In some cases the distinction is difficult or impossible to make in routine histologic sections.

When for any reason the periosteum is detached, new radiating spicules are deposited roughly perpendicular to the cortex. Thus the new periosteal bone has a triangular profile in clinical radiographs and is called Codman's triangle or spur. This sign is not pathognomonic for osteosarcoma but is characteristic of local subperiosteal irritation by primary or metastatic tumor, infection, or other stimulus. When the original periosteal membrane is breached, the expanding reactive bone produces the "sunburst" effect in clinical radiographs.

Most tumors growing within the bone marrow stimulate endosteal new bone formation until the neoplastic cells contact the endosteal surface, when resorption occurs. Among the tumors without this osteoplastic effect are myeloma and lymphosarcoma. Endosteal new bone is also stimulated by some tumors growing in the soft tissues around a bone in contact with the periosteum. The mechanisms of this osteoplastic effect is unclear, but the possibility that change in tissue potentials (see Diseases of Bones, Section IV,E, Adaptational Deformities of the Skeleton) or the activity of growth factors is involved, seems plausible.

Metastatic tumors also frequently stimulate bone resorption. Indeed formation and resorption usually occur concurrently with the same tumor and sometimes in different areas of the same trabecula. Thus designation of a tumor as osteolytic or osteoplastic depends to some extent on the tissue sample examined. The bulk of bone resorption in both primary and metastatic tumors is mediated by osteoclasts. Osteocytic osteolysis contributes in a minor way, and direct resorption by tumor tissue may occur occasionally, although it is difficult to rule out the possibility that mononuclear osteoclasts are involved in some cases where resorption by tumor tissue seems to be present. The conscription of osteoclasts is probably mediated by growth factors and prostaglandins (see Diseases of Bones, Section III,A, Osteoprogenitor Cells).

The reactivity of the host tissues and organs makes diagnosis of tumors in bone a very difficult and potentially humbling exercise for the pathologist, perhaps more so than for any other group of tumors. Furthermore, in no other type of lesion is the close cooperation of clinician, radiologist, and pathologist more necessary if an accurate diagnosis is to be made. This classification is histogenetic and includes tumors which may or may not be neoplasms. Some tumors of hematopoietic tissue and some tumors of tissues not specific to bone, such as hemangiosarcoma, are described briefly.

Bibliography

Jacobson, S. A. "The Comparative Pathology of the Tumors of Bone." Springfield, Illinois, Charles C Thomas, 1971.

Ling, G. V., Morgan, J. P., and Pool, R. R. Primary bone tumors in the dog: A combined clinical, radiographic, and histologic approach to early diagnosis. *J Am Vet Med Assoc* **165:** 55–67, 1974.

Liu, S.-K., Dorfman, H. D., and Patnaik, A. K. Primary and secondary bone tumours in the cat. *J Small Anim Pract* **15:** 141–156, 1974.

Misdorp, W., and van der Heul, R. O. Tumours of bones and joints. *Bull WHO* **53:** 265–282, 1976.

Patnaik, A. K. Canine extraskeletal osteosarcoma and chondrosarcoma: A clinicopathologic study of 14 cases. *Vet Pathol* **27:** 46–55, 1990.

Patnaik, A. K. *et al.* Canine sinonasal skeletal neoplasms: Chondrosarcomas and osteosarcomas. *Vet Pathol* **21:** 475–482, 1984.

Pool, R. R. Tumors of bone and cartilage. *In* "Tumors in Domestic Animals," J. E. Moulton (ed.), 3rd Ed., pp. 157–230. Berkeley, University of California Press, 1990.

A. *Fibrous Tumors of Bone*

1. Fibrosarcoma

Skeletal fibrosarcomas arise from connective tissue stroma of the medullary cavity and periosteum. Their histologic and gross appearances are essentially the same as that of fibrosarcomas elsewhere, although their innocuous microscopic appearance often belies their invasive tendencies. Central (medullary) fibrosarcomas are less common than those from periosteum. They do occur in mature male dogs of large and medium breeds but compose fewer than 5% of canine osseous neoplasms. This type is rare in other species.

In dogs, central fibrosarcomas arise most often in metaphyses of long bones and are destructive. They must be differentiated from osteosarcomas of the fibrosarcomatous type; the distinction may not be academic since it is suggested that, of the two, fibrosarcomas are slower or less likely to metastasize. It has been observed that the number of fibrosarcomas decreases and of fibrosarcomatous osteosarcomas increases as more areas of a medullary tumor are examined. If this observation is valid, then central fibrosarcomas are probably less common than is currently believed.

Periosteal fibrosarcomas occur in most species and most commonly involve flat bones, especially those of the head. In the dog, they often occur in the maxilla and mandible and are difficult to separate from fibrosarcomas arising in the subepithelial connective tissues of the oral cavity. Periosteal fibrosarcomas often exhibit low cellularity and appear innocuous. They may destroy the underlying cortex, however, and although they are more erosive than invasive, and can be dissected free from underlying bone, they often recur. In any event periosteal new bone is generally sparse. Some of these tumors metastasize to lung. The fibrosarcomas of nasal sinuses and dental arcades of horses may belong with this group of neoplasms.

2. Fibroma

Skeletal fibromas are extremely rare in pet and farm animals. Their microscopic appearance resembles that of nonskeletal fibromas. In humans, this is a common tumor, which is often given the prefix non-ossifying to distinguish it from the ossifying variety.

3. Ossifying Fibroma

Ossifying fibroma is a rare fibro-osseous tumor of the head, occurring most frequently in horses and cattle and primarily involving mandible or maxilla. Typically, ossifying fibroma is a sharply demarcated, expansile mass which distorts the normal contours of the affected bone. It is formed by a mixture of dense fibrous tissue and bone

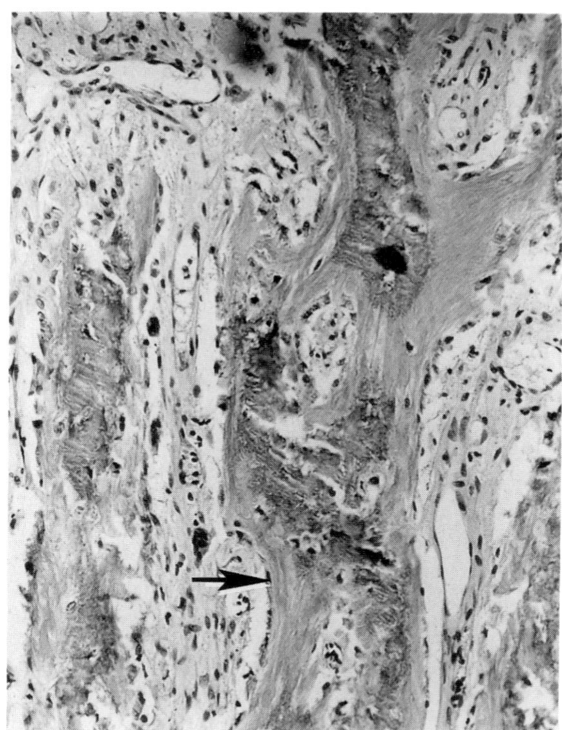

Fig. 1.110 Ossifying fibroma. Dog. Trabeculae of mineralizing woven bone rimmed by osteoblasts (arrow).

spicules and is tough and gritty or may be impossible to cut. The tumor consists of spindle-shaped fibroblasts which transform into osteoblasts and form irregular spicules of woven bone (Fig. 1.110). Lamellar bone is extremely rare in ossifying fibroma, and may be formed where woven trabeculae are resorbed and replaced. There is no cartilage.

There are several lesions from which this tumor must be distinguished, including fibrous dysplasia, osteoma, osteosarcoma, and, in horses, osteodystrophia fibrosa. Fibrous dysplasia may be polyostotic and is not confined to the head; however, the validity of the latter point of distinction could easily be negated as more cases of ossifying fibroma are recognized. The bone spicules of fibrous dysplasia are usually fairly uniform and are not rimmed by osteoblasts. In osteomas the spaces between trabeculae contain some marrow and less connective tissue, and the trabeculae are larger and denser with a far greater proportion of lamellar bone. The connective tissue cells of ossifying fibroma lack the pleomorphism and high mitotic index of osteosarcoma, and the bone is formed in a more regular and uniform manner. Confusion with equine osteodystrophia fibrosa is unlikely unless an inadequate history accompanies a small bone biopsy. In most areas of the dystrophic lesion, large osteoclasts are present along loose connective tissue, and these features readily distinguish it from ossifying fibroma.

The natural history of ossifying fibroma is obscure. Malignant transformation is not known to occur. Bone is not an inert tissue, and it seems reasonable that remodeling of woven trabeculae and replacement by lamellar bone could occur. If this is so, a maturing ossifying fibroma might come to resemble osteoma, with hematopoietic cells colonizing spaces between lamellar trabeculae.

4. Fibrous Dysplasia

Fibrous dysplasia is a rare, expansive, fibro-osseous lesion of bone, which is thought to be developmental rather than neoplastic. It occurs in horses, dogs, and cats and may be monostotic or polyostotic. Young or even newborn animals are usually affected. Monostotic lesions occur in the maxillary sinus of horses and are also reported in the jaw and infraorbital bones of dogs. In cats, lesions in various sites including mandible, distal ulna, and distal radius are described. Polyostotic fibrous dysplasia occurs in the long bones of dogs and may be an inherited trait in Doberman pinschers and Old English sheepdogs. The condition also occurs relatively commonly in German shepherds. Sometimes a single long bone is affected, and bilateral symmetry in long bones has been observed.

As in humans, there is considerable debate about the diagnostic criteria for fibrous dysplasia. It is an arbitrary decision to group the histologically diverse monostotic and polyostotic cystic lesions of canine bones under this classification. Some of the cases described belong here on proven morphologic grounds; the rest are included partly for convenience, but also because some of the microscopic variations could be due to degenerative and remodeling effects as the lesions evolve.

Fibrous dysplasia presents as a deformity of the affected bone, and in long bones of dogs, radiographs show smooth-contoured areas of radiolucency, which are often approximately circular or oval and may be divided by bony partitions (Fig. 1.111). Lesions in the equine maxilla are radiodense and distort adjacent structures.

Grossly, fibrous dysplasia is a gray, firm to hard, fibrous mass, which may be discolored by areas of hemorrhage and is gritty when cut. Some contain cavities filled with clear or bloody fluid, which are a few centimeters to less than a millimeter across; some are microscopic. The larger ones have bony ridges and plates in their walls and are lined by a fibrous membrane. They occur most often in the metaphysis, involving secondary spongiosa and periosteum, but are subject to relocation in the diaphysis as the bone grows. Fractures and infractions sometimes occur at the site of the large cavities which characterize long-bone lesions in dogs.

The unifying feature of the conditions described as fibrous dysplasia is histologic, and consists of areas of fibrous stroma containing trabeculae of woven bone, which replace normal osseous tissue. Typically the new, mineralized trabeculae develop in the fibrous tissue without the intercession of recognizable osteoblasts: a layer of osteoblasts on the surface of trabeculae is lacking in fibrous dysplasia (Fig. 1.112). The margins of the trabeculae merge with the surrounding stroma and are ill defined, and

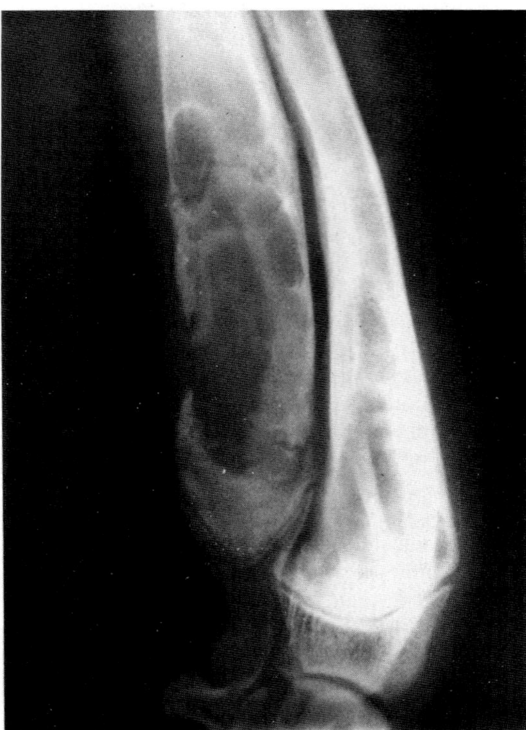

Fig. 1.111 Fibrous dysplasia. Doberman pinscher. Cystic lesions in distal foreleg. (Courtesy of R. Lavelle.)

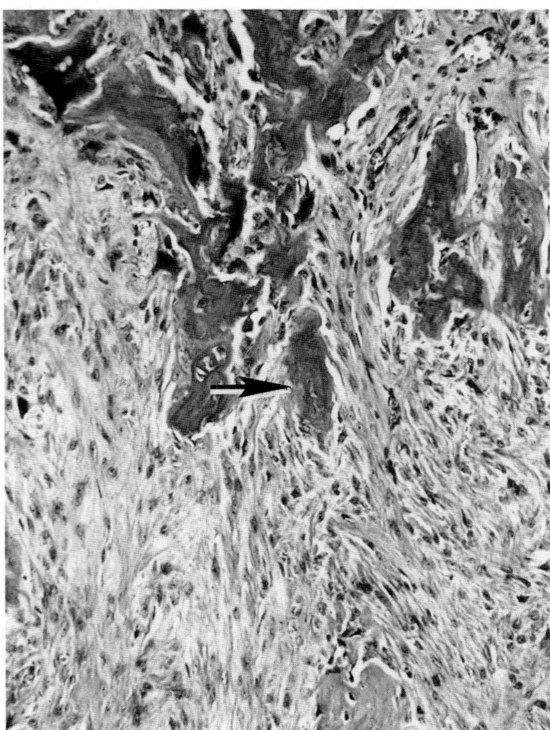

Fig. 1.112 Fibrous dysplasia. Doberman pinscher. Trabeculae of bone forming from fibrous tissue (arrow). There are no osteoblasts rimming trabeculae.

the shape and distribution of the trabeculae appear to be oriented about a meshwork of thin-walled vessels. The normal bone is removed by osteoclasis and as the fibrous mass expands through the cortex, a thin shell of periosteal bone forms around it. In cats, a moderate periosteal reaction sometimes develops; the presence or absence of a periosteal reaction may be a function of the rate of expansion of the primary lesion.

The pathogenesis of cavity formation is unknown. Degenerative or hemorrhagic changes in areas of the proliferative mass or in adjoining bone, with failure of normal reparative processes, might be responsible.

Malignant transformation of fibrous dysplasia is not reported in animals but occurs rarely in humans. In dogs the long-bone lesions usually resolve over several months with minimal residual deformity. Involvement of the bones of the head sometimes interferes with eating or causes pressure on the brain, necessitating surgical removal or euthanasia.

Bibliography

Carrig, C. B., and Seawright, A. A. A familial canine polyostotic fibrous dysplasia with subperiosteal cortical defects. *J Small Anim Pract* **10:** 397–405, 1969.

Hultgren, B. D. *et al.* Nasal-maxillary fibrosarcoma in young horses: A light and electron microscopic study. *Vet Pathol* **24:** 194–196, 1987.

Morse, C. C. *et al.* Equine juvenile mandibular ossifying fibroma. *Vet Pathol* **25:** 415–421, 1988.

Schrader, S. C., Burk, R. L., and Liu, S. Bone cysts in two dogs and a review of similar cystic bone lesions in the dog. *J Am Vet Med Assoc* **182:** 490–495, 1983.

Wilson, R. B. Monostotic fibrous dysplasia in a dog. *Vet Pathol* **26:** 449–450, 1989.

B. Cartilage-Forming Tumors

Tumors of the skeleton which consist solely or mainly of cartilage are rare in animals and in most species make up a small proportion of skeletal neoplasms. In dogs about 10% of tumors of bones are cartilaginous; in this species, cartilaginous tumors occur more often in the mammary gland. Sheep apparently are exceptions to the rule, a high proportion of their skeletal neoplasms being cartilaginous.

Many cartilaginous tumors contain areas of bone formed either by endochondral ossification of tumor cartilage or by metaplasia from mineralized tumor cartilage. If neoplastic cells form osteoid, which may or may not be mineralized, then the tumor is, by definition, an osteosarcoma, no matter how little osteoid or how much cartilage is produced.

Cartilage tumors are often very difficult to diagnose and classify, and the histologic distinction between benign and malignant varieties is seldom easy. Certain cytologic criteria of malignancy, established for human tumors, appear to have validity in other animals.

Synovial chondromatosis is a benign disease of humans, cats, and dogs in which nodules of hyaline cartilage form in the synovial lining, mainly in large joints. They are not associated with primary degenerative or inflammatory disease and are not pathogenically related to marginal osteophytes of degenerative arthropathy. They may, however, detach into the synovial fluid and irritate the synovium. The nodules, while still attached to the synovium, may undergo endochondral ossification; if these detach, the bony portion dies while the cartilagenous portion survives, sufficiently nourished from synovial fluid.

1. Chondroma

Chondroma is a benign tumor characterized by the formation of mature cartilage but lacking the histological characteristics of chondrosarcoma (high cellularity, pleomorphism, and presence of large cells with double nuclei or mitosis). Chondromas are separable into ecchondromas and enchondromas according to their location in the skeleton, enchondroma being a tumor arising from cartilage within the medullary cavity, whereas ecchondroma arises from cartilage elsewhere in the skeleton. Juxtacortical chondroma, which is thought to originate from mesenchymal cells, probably periosteal cells, adjacent to bone is not reported in species other than humans. These distinctions regarding anatomical location are necessary so that sufficient data can be accumulated to allow critical analysis of behavior. Currently there are not enough cases to justify separate discussion, other than to note that, rarely, multiple enchondromas occur in dogs, in which case the disease is called enchondromatosis. In humans, a similar condition, which often has a predominantly unilateral skeletal distribution, is regarded as a developmental anomaly without hereditary or familial influence.

Chondromas occur in dogs, cats, cattle, and sheep and are firm to hard, smooth or nodular, roughly spherical masses with a relatively thin fibrous capsule (Fig. 1.113). They more commonly involve flat bones and ribs than long bones. On cut surface, a lobular pattern is evident where the blue-white cartilage is divided by fibrous bands. Areas of mineralization or ossification appear as chalk-white stippling. Myxomatous tissue occurs in some tumors and has a gelatinous nature—it seems prone to hemorrhage and necrosis.

Histologically these benign tumors consist of irregular lobules of hyaline cartilage with cells that are, by definition, quite regular in size and appearance and typically chondrocytic. This is especially true of enchondromas. Foci of endochondral ossification and mineralization may be present, and lobules of myxomatous tissue sometimes develop.

Chondromas expand slowly, and a rapid change in size may indicate malignant transformation. Expanding tumors within the medullary cavity cause bone deformation and possibly fracture.

2. Osteochondroma

This is a cartilage-capped bony projection on the external surface of a bone. Tumors fitting this description occur

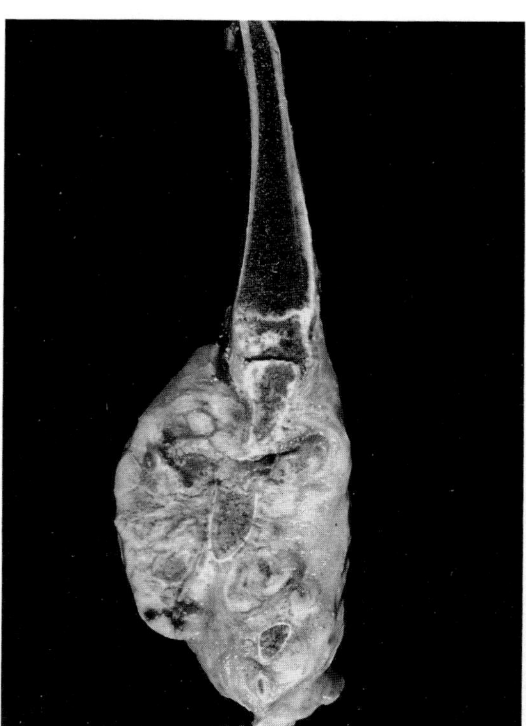

Fig. 1.113 Chondroma of costal cartilage. Cow. Tumor contains areas of endochondral bone.

singly or multicentrically; in multicentric cases the name multiple cartilaginous exostoses is used (Fig. 1.114A). This disease occurs in cats, humans, horses, and dogs. In the latter three species it is inherited as an autosomal dominant trait. There are many synonyms for the multicentric disease, including diaphyseal aclasis, metaphyseal aclasis, hereditary deforming chondrodysplasia, osteocartilaginous exostoses, and dyschondroplasia. Multiple cartilaginous exostoses are considered to be malformations, and although they occasionally give rise to neoplasms, usually they appear in young animals and stop growing at the same time as the parent bone. The disease in cats does not always follow this growth pattern for, in contrast to other species, it occurs in mature animals, long bones are seldom affected, and tumor sites include cranial bones formed by intramembranous ossification. It seems likely that some cases in cats are not cartilaginous malformations and not analogous to the disease in other species. A provocative role for C-type viruses is possible; some cats with multiple cartilaginous exostoses are serologically positive for feline leukemia virus, and viral particles are consistently present in the cells of the cartilage caps.

Osteochondromas are knobby protuberances usually involving endochondral bones and usually adjacent to growth cartilages, both physeal and subarticular. In the multicentric disease, lesions are sometimes present at birth but mostly become apparent at a few months of age and develop as cartilaginous masses, which are progres-

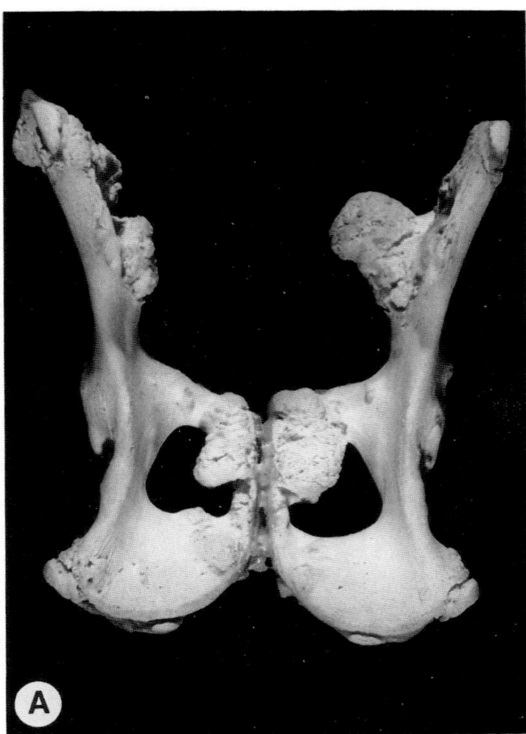

Fig. 1.114A Multiple cartilaginous exostoses. Dog. Macerated specimen. (Courtesy of P. W. Pennock.)

sively replaced from the base by spongy osseous tissue. The degree of osseous replacement varies, but characteristically the mass is capped by a thin layer of hyaline cartilage (Fig. 1.114B). The marrow spaces of the core communicate with those of the parent bone. The origin of the exostoses is not entirely clear, but from their location and natural history, it is deduced that they almost always arise from growth plates by disharmonious interstitial and appositional growth, in which the latter predominates in local situations. Implicit in this suggestion is a loss of normal polarity of the chondrocytes produced in the ossification groove, with the development of ectopic centers of cartilage growth. The rare involvement of an intramembranous skull bone in cats suggests that they can also develop from periosteal mesenchymal cells.

In most cases the exostoses are objectionable for esthetic reasons only, but occasionally a chondrosarcoma or osteosarcoma arises in an exostosis and, somewhat more often, expanding lesions interfere with normal gait or produce compression and degeneration of the spinal cord (Fig. 1.114B).

Multiple osteochondromas appear to be more common than solitary lesions, but this may reflect the incidence of reporting rather than of occurrence. The solitary lesions must be differentiated from the so-called osteochondromas, which are found in the periarticular region at the points of tendinous insertion and which are emphasized tuberosities and probably irritative, hyperplastic growths rather than neoplastic ones. Differentiation of the latter from ecchondroma with extensive endochondral ossification is based on the absence of a cartilage cap and absence of communication with the marrow cavity, and by the multicentric, unpolarized ossification process in the ecchondroma. Good clinical and radiographic support with proper orientation of the histologic specimen aid in making the distinction.

Nodular intratracheal osteochondromas sometimes occur in young dogs.

3. *Multilobular Tumor of Bone*

These tumors are locally aggressive, nodular masses which involve membranous bones particularly of the canine skull and rarely the vertebrae. The tumor is reported in cats and once in a horse. They are usually solitary lesions affecting mature dogs of medium and large breeds but may occur as early as 1 year of age. Though rarely

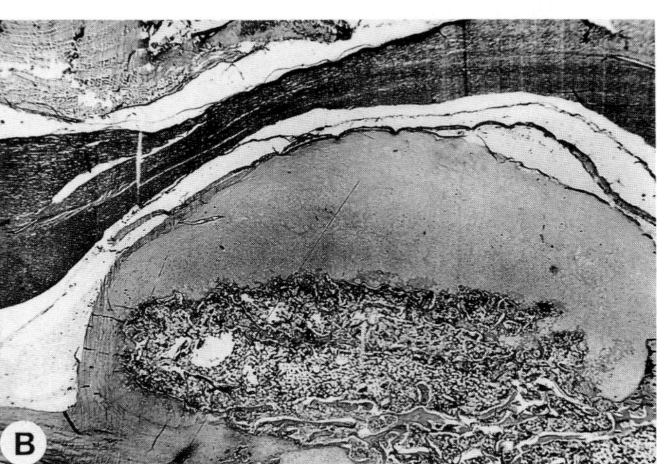

Fig. 1.114B Multiple cartilaginous exostoses. Vertebral tumor compressing spinal cord. Dog.

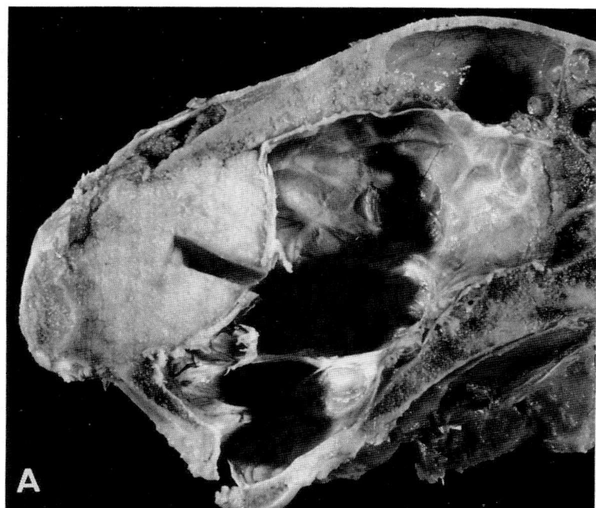

Fig. 1.115A Multilobular tumor of canine skull impinging on cranial vault.

reported, they are not uncommon in surgical material. Synonyms are chondroma rodens, calcifying aponeurotic fibroma, juvenile aponeurotic fibroma, and cartilage analog of fibromatosis. The last three aliases reflect their histologic resemblance to certain human neoplasms so named.

These tumors cause disfigurement or can produce clinical signs because of pressure on adjacent structures such as brain or lacrimal duct (Fig. 1.115A). The most common sites are the temporo-occipital areas and the zygomatic process. They are firm to very hard nodular masses, which on cut surface usually contain numerous gritty foci (Fig. 1.115B) set in a fibrocartilaginous matrix. Most tumors

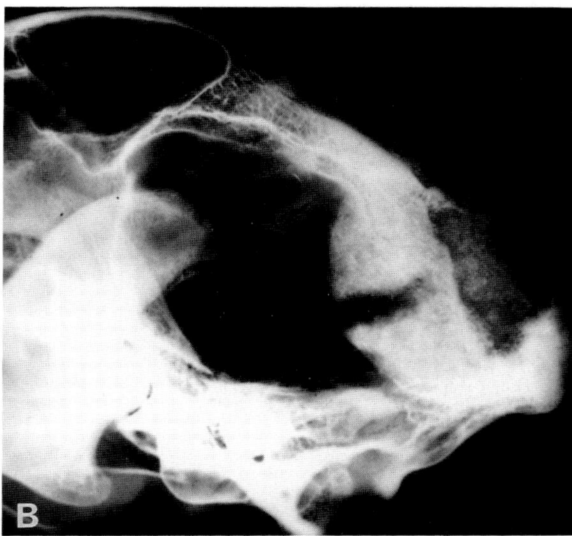

Fig. 1.115B Radiograph of (A). Note stippled mineralization in tumor.

are about 3 to 7 cm in diameter at the time of clinical examination, but smaller, strategically located masses may cause signs of disease.

The histologic appearance of the tumor is very characteristic, and consists of multiple circular, oval, or irregular-shaped nodules of cartilaginous, osseous, or osteocartilaginous tissue (Fig. 1.116A,B). These nodules are separated by thin, distinct fibrovascular septa or, in some areas of some tumors, by broader, ill-defined zones of mesenchymal tissue which merge with the nodules. The cartilaginous nodules usually are surrounded by spindle-shaped septal cells, which lie against the plump oval nuclei of the central mesenchymal tissue. The mesenchymal cells produce a pale eosinophilic, sometimes faintly basophilic, matrix, which mineralizes at the center of the lobule. Resorption of mineralized cartilage and endochondral bone formation sometimes occurs, and occasionally, osteoclastic remodeling of bone develops. In these nodules the matrix-producing cells are unequivocally chondrocytic in appearance. In bony nodules the mesenchymal cells produce a matrix that ossifies rather than mineralizes. In some tumors the bone is produced by angular osteoblasts with abundant basophilic cytoplasm, but in others, oval cells of indeterminate type elaborate a tissue reminiscent of chondroid bone. The mineralization process tends to progress centrifugally to involve the entire nodule, and the

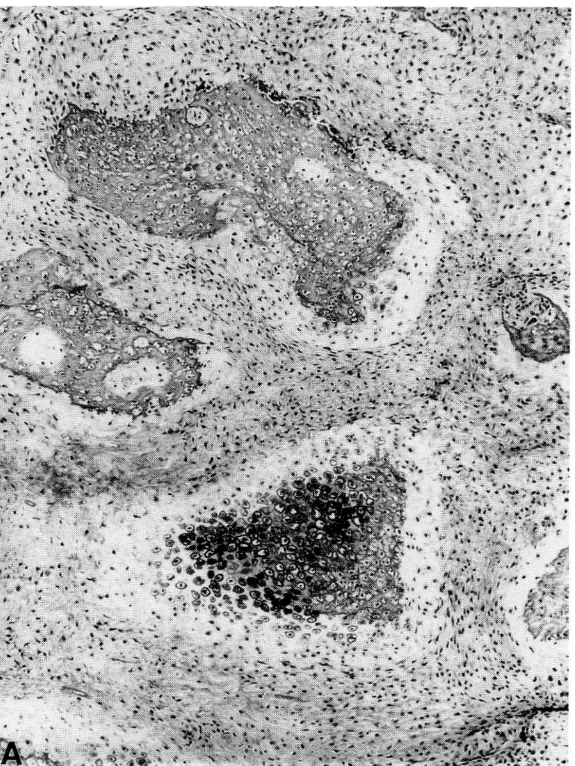

Fig. 1.116A Low power appearance of multilobular tumor of bone. Lobules of cartilage and chondroid bone separated by bands of stroma.

Fig. 1.116B High-power appearance of multilobular tumor of bone. Lobules of cartilage and chondroid bone separated by bands of stroma.

septal tissue then consists of dense collagen. In these areas the tumors are extremely hard. Some nodules contain osseous and cartilaginous tissue, and some tumors contain osseous and cartilaginous nodules. Although the nodular, formative pattern predominates, resorption of the hard tissues does occur, and nodules may be replaced by dense connective tissue in which cartilaginous metaplasia may occur.

Progressive malignant transformation is expected in long-standing or recurrent tumors. Locally aggressive growth and recurrence following surgery is the usual pattern of behavior. The presence of mitoses may presage malignancy as does necrosis, hemorrhage, and dominant overgrowth by one of the mesenchymal elements. The characteristic appearance is retained in metastases.

4. Chondrosarcoma

Chondrosarcoma is a malignant neoplasm in which cartilaginous matrix is produced by neoplastic cells. Neoplastic cells in chondrosarcoma never produce osteoid, but bone tissue may develop by metaplasia or by endochondral ossification.

Primary chondrosarcomas arise from existing normal cartilage, except articular cartilage, and from perichondrium. Secondary chondrosarcomas develop from abnormal cartilage such as exists in osteochondromas. Chondrosarcomas occur at sites such as pelvis, nasal cavity, sternum, and ribs more commonly than in long bones and are usually found in mature and aged animals.

Grossly, chondrosarcomas are firm or hard, lobulated tumors consisting largely of blue-white cartilage with perhaps some chalky areas of mineralization or ossification. They tend to grow expansively, and the surface then is smooth and capsulated, and the capsule merges with surrounding fibrous tissue. Those that are more malignant still resemble cartilage but contain areas of friability, degeneration, and hemorrhage, and these tend to infiltrate adjacent soft tissue and occasionally metastasize.

Histologically there may be no clear distinction between chondroma and chondrosarcoma, and some chondrosarcomas, and their metastases, appear quite orderly and benign. In these cases malignancy is a clinical criterion. Certain characteristics, namely, high cellularity, pleomorphism, and the presence of large cells with double or plump nuclei or mitoses are recognized as histologic criteria of malignancy. The tumor should be judged on the basis of its most malignant-looking area.

Chondrosarcomas occur in all species but are best known in dogs and sheep. Large but not giant breeds of dogs are most frequently affected. Compared to osteosarcoma, this tumor metastasizes late. Sometimes there is recurrence following surgical removal.

Bibliography

Brodey, R. S. *et al.* Canine skeletal chondrosarcoma: A clinicopathologic study of 35 cases. *J Am Vet Med Assoc* **165:** 68–78, 1974.

Brown, R. J., Trevethan, W. P., and Henry, V. L. Multiple osteochondroma in a Siamese cat. *J Am Vet Med Assoc* **160:** 433–435, 1972.

Diamond, S. S., Raflo, C. P., and Anderson, M. P. Multilobular osteosarcoma in the dog. *Vet Pathol* **17:** 759–763, 1980.

Doige, C. E. Multiple osteochondromas with evidence of malignant transformation in a cat. *Vet Pathol* **24:** 457–459, 1987.

Gambardella, P. C., Osborne, C. A., and Stevens, J. B. Multiple cartilaginous exostoses in the dog. *J Am Vet Med Assoc* **166:** 761–768, 1975.

McCalla, T. L. *et al.* Multilobular osteosarcoma of the mandible and orbit in a dog. *Vet Pathol* **26:** 92–94, 1989.

Pletcher, J. M., Koch, S. A., and Stedham, M. A. Orbital chondroma rodens in a dog. *J Am Vet Med Assoc* **175:** 187–190, 1979.

Pool, R. R., and Carrig, C. B. Multiple cartilaginous exostoses in a cat. *Vet Pathol* **9:** 350–359, 1972.

Richardson, D. W., and Acland, H. M. Multilobular osteoma (chondroma rodens) in a horse. *J Am Vet Med Assoc* **182:** 289–291, 1983.

Shupe, J. L. *et al.* Hereditary multiple exostoses. Hereditary multiple exostoses in horses. *Am J Pathol* **104:** 285–288, 1981.

Straw, R. C. *et al.* Multilobular osteochondrosarcoma of the canine skull: 16 cases (1978–1988). *J Am Vet Med Assoc* **195:** 1764–1769, 1989.

C. Bone-Forming Tumors

1. Osteoma

Osteomas are benign tumors consisting mostly of well-differentiated bone. These are uncommon lesions which usually affect the bones of the head and are best known on the jaws and in the nasal sinuses of horses and cattle. They also occur in other species including pig, dog, and cat. The distinction between osteoma, ossifying fibroma, fibrous dysplasia, heterotopic ossification, and exostosis is often difficult to make, and the decision is sometimes arbitrary.

Osteomas are hard masses which cause problems by virtue of expansive growth. They are covered by periosteal tissue, but the main masses are so hard that they must be sectioned with a saw.

Histologically, osteomas consist of trabeculae of bone formed by osteoblasts, remodeled by osteoclasts and ultimately converted almost entirely to lamellar bone. The trabeculae are gradually thickened by appositional growth so that ultimately the spaces between them are relatively inconspicuous, and the density of the tumor mass approaches that of mature cortex. Such tumors are virtually indistinguishable from exotoses; location and clinical history should be considered in differentiating the two. Evidence of endochondral ossification is inconsistent with osteoma but may be found in osteochondroma and fracture callus.

The type of intertrabecular tissue may be useful in establishing a diagnosis for benign bony masses. Typically, osteomas contain light connective tissue or fatty or hematopoietic marrow. Very dense fibrous tissue is more consistent with fibrous dysplasia and ossifying fibroma. Pool has discussed atypical osteomas and their range of microscopic variability.

2. *Osteoid Osteoma*

Osteoid osteoma is a benign tumor of human bone which has a characteristic histologic and radiographic appearance. It consists of an area of highly cellular, well-vascularized tissue containing osteoid and immature bone, which is surrounded by reactive bone. The central area, or nidus, is radiolucent; the periphery, radiodense. There are reports of vertebral lesions in cats with a somewhat similar appearance.

3. *Osteosarcoma*

Osteosarcomas are malignant tumors, arising from mesenchyme, in which the tumor cells produce osteoid or bone. These products are called tumor osteoid or tumor bone, but not malignant osteoid or malignant bone, since it is the cells which possess the characteristics of malignancy, not the matrices they produce. Osteosarcomas are the most common skeletal neoplasms in dogs and cats, composing about 80 and 50% of skeletal tumors, respectively. They are rare in other species.

Osteosarcomas are classified using various criteria including site of origin and histologic appearance. In some cases there is evidence that membership in a particular group has prognostic significance, but in general, this is not so. Most osteosarcomas metastasize to the lungs, often before the primary tumor is diagnosed. Nevertheless subgroupings of these tumors is potentially useful and should be employed wherever possible.

Most osteosarcomas arise centrally, especially in the metaphyseal region of long bones, but some arise in the periosteum and occasionally in extraskeletal tissues. Extraskeletal tumors are best known in dogs, especially in the mammary gland, but are also associated with mixed thyroid neoplasms and sometimes develop in organs such as liver, spleen, skin, and lungs; esophageal sarcomas associated with *Spirocerca lupi* in dogs are considered with tumors of the esophagus.

Two types of osteosarcoma arise in the periosteum—one behaves in the same way as do central osteosarcomas and is called periosteal osteosarcoma. The other type is the juxtacortical (parosteal) osteosarcoma, which is recognized in humans as having a high degree of structural differentiation, relatively slow growth, and a better prognosis than central osteosarcoma. A few of these tumors occur in domestic animals, and they behave similarly to their human counterparts.

Histological classification of osteosarcomas is based on the identity of the matrix produced by the tumor cells and on the appearance of the cells themselves. Simple osteosarcomas produce tumor osteoid and bone. A variant, the telangiectatic osteosarcoma, also produces tumor osteoid, but in some areas contains blood-filled spaces lined by malignant osteoblasts. This type, which produces a lytic lesion, has been confused with hemangiosarcoma or aneurysmal bone cyst (Fig. 1.117A). Compound osteo-

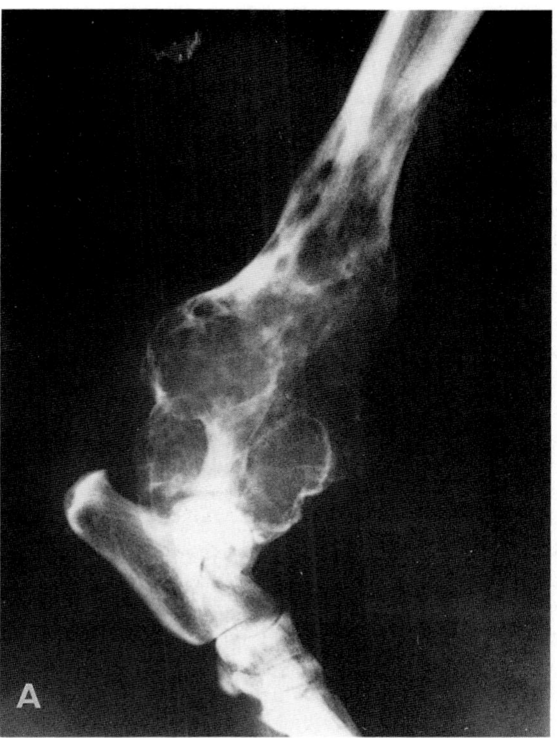

Fig. 1.117A Telangiectatic osteosarcoma. Dog.

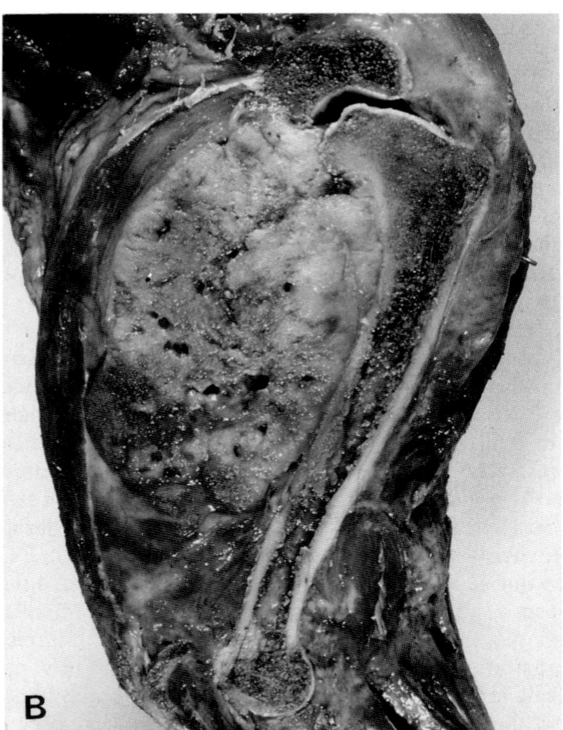

Fig. 1.117B Compound osteosarcoma. Humerus. Dog.

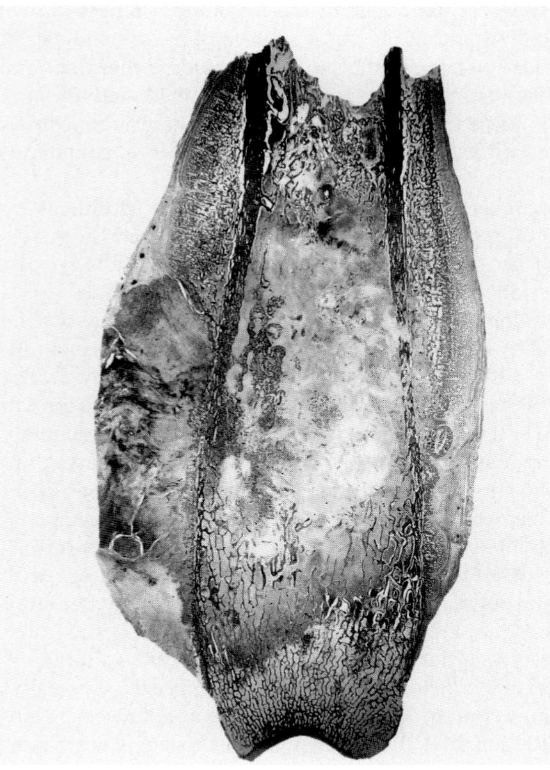

Fig. 1.118 Osteosarcoma, distal femur. Dog. (A.F.I.P. 572112. Contributor, L. Lieberman.)

sarcomas produce tumor cartilage in addition to tumor osteoid and bone (Fig. 1.117B). Pleomorphic osteosarcomas are highly cellular tumors which are distinguished from undifferentiated sarcoma only by merit of their sparse production of thin spicules of osteoidlike material. In dogs and cats ~50% of osteosarcomas are compound; the classification of many tumors probably depends on the number of sections examined.

Osteosarcomas classified according to cell type are divided into osteoblastic, chondroblastic, and fibroblastic varieties, depending on which type is preponderant. Fibroblastic osteosarcomas have a better prognosis than do the other types.

Osteosarcomas are often described as osteolytic or osteoblastic (sclerotic, productive) on the basis of their radiographic appearance. The designation may change as the tumor develops, but in general, those neoplasms which produce abundant mineralized matrix are radiodense, and those which are relatively nonproductive are radiolucent. Obviously the amount of necrosis and resorption of preexisting bone also affects the radiodensity.

Clinically, osteosarcomas present as painful swellings of bones. In dogs 60–80% involve the legs, the forelegs being affected nearly twice as often as the hindlegs. Almost all foreleg tumors are in either the proximal humerus (Fig. 1.117B) or distal radius; hind-leg tumors tend to be located in distal tibia and distal femur (Fig. 1.118). Affected animals are usually mature males of the large and especially the giant breeds. In cats, neither the preponderance of limb tumors over axial tumors nor the foreleg–hindleg disparity is apparent, but again, affected individuals are usually mature. In horses and cattle, most osteosarcomas involve the head. Osteosarcomas are rare in pigs but have been seen occasionally in extremely young animals.

Assessment of the gross features of osteosarcomas is enhanced by radiographs of dissected and bisected specimens. Radiolucent osteolytic tumors are usually quite hemorrhagic and soft, and they often contain light yellow areas of necrosis. Erosion of cortical bone, with production of pathologic fractures, is not uncommon, and osteolytic tumors tend to invade the adjacent soft tissues early in their course. Radiodense tumors are various shades of gray and contain areas of tumor bone and tumor cartilage, which give a stippled appearance to radiographs and a gritty texture when cut; some may have to be sawn. In compound osteosarcomas, tumor cartilage is sometimes recognizable by its texture and color, and it tends to be located at the periphery of the neoplasm. Productive tumors frequently provoke a more marked periosteal reaction than do the osteolytic tumors. Osteosarcomas of long bones commonly, perhaps invariably, erode cortical bone but rarely break through articular cartilage, even though, within the bone, they tend to extend into the epiphysis rather than the diaphysis.

The microscopic structure is variable both within and between tumors. Basically, they are sarcomas which pro-

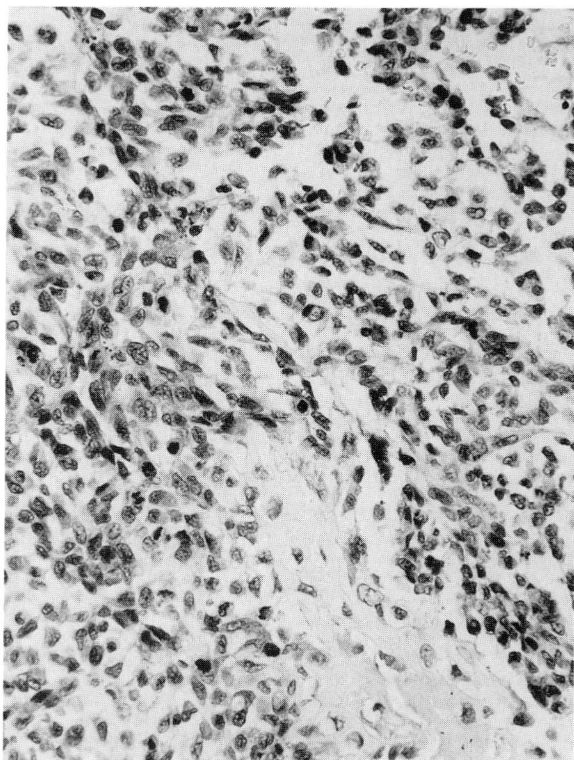

Fig. 1.119 Osteoid formation in osteosarcoma.

duce osteoid (Fig. 1.119). Electronmicroscopic examination of human osteosarcomas demonstrates that the majority contain several morphological cell types including anaplastic blast cells, osteoblastic, chondroblastic, and fibroblastic cells. A similarly wide spectrum exists in osteosarcomas of other animals. Depending on circumstances as yet poorly understood, one or more cell types prevail and produce a tumor which is either undifferentiated or mainly osteoblastic, chondroblastic, fibroblastic, or mixed (compound osteosarcoma). Some osteosarcomas also contain groups of multinuclear tumor giant cells, which may resemble those in giant-cell tumor of bone. Since osteoclasts are derived from stem cells in bone marrow and appear to circulate freely in blood as mononuclear cells, they can be attracted to all sites of bone formation whether neoplastic or otherwise. Many of the giant cells in bone tumors are normal osteoclasts. Confusion with giant-cell tumor is unlikely if several sections are examined, since tumor osteoid, the *sine qua non* of osteosarcoma, is present elsewhere in the tumor, but is not found or is very sparse in giant-cell tumor.

Osteoid formation in osteosarcoma is reminiscent of ordinary intramembranous osteogenesis, but in contrast to normal bone, mitoses may be present in osteoblastic cells. These neoplastic cells are plump and connect with each other by numerous cytoplasmic processes. The stroma between them condenses and is probably deposited in larger amounts than elsewhere in the tumor. In sections stained with hematoxylin and eosin, the osteoid appears homogeneous and is light pink. It may be irregularly mineralized.

Osteosarcomas of canine long bones metastasize to lung in as many as 75% of cases. Probably all of them would metastasize if the dogs were not killed. Pulmonary metastases tend to mimic the primary tumor. Osteosarcomas of the canine axial skeleton tend to metastasize less readily. In cats the opposite is true—feline axial osteosarcomas are more likely to metastasize than are those arising in long bones.

No specific cause of osteosarcoma is established, but several associations are recognized in dogs. It is known that the risk of developing osteosarcoma is 61 to 185 times greater for a dog weighing over 80 pounds compared to one weighing less than 20 pounds. Breeds most commonly affected are St. Bernard, Great Dane, boxer, German shepherd and Irish setter. Certain families of St. Bernards and also of Rottweilers are more susceptible than others. The site of development of osteosarcoma varies markedly between susceptible breeds. In general the ratio of forelimb to hindlimb tumors is about 2 : 1, but it varies from 1.2 : 1 in boxers to 5.7 : 1 in Great Danes. The long bone to flat bone ratio for the same breeds are 2.1 : 1 and 26 : 1 respectively (see Brodey for details). A sudden increased risk of developing osteosarcoma occurs after 5 years of age and decreases after 9 years of age. The possibility that an oncogenic virus is responsible for this epidemiologic pattern has been suggested.

Development of osteosarcoma following fixation of fractures using metallic devices is recorded. Osteosarcomas also occur in dogs in association with bone infarcts; there is evidence that infarction precedes neoplastic change. Miniature schnauzers seem to be unusually susceptible.

Bibliography

Bitetto, W. V. *et al.* Osteosarcoma in cats: 22 cases (1974–1984). *J Am Vet Med Assoc* **190:** 91–93, 1987.

Dubielzig, R. R., Biery, D. N., and Brodey, R. S. Bone sarcomas associated with multifocal medullary bone infarction in dogs. *J Am Vet Med Assoc* **179:** 64–68, 1981.

Foley, R. H. Osteoma in two young dogs. *J Am Anim Hosp Assoc* **14:** 253–255, 1978.

Gleiser, C. A. *et al.* Telangiectatic osteosarcoma in the dog. *Vet Pathol* **18:** 396–398, 1981.

Grundmann, E., Roessner, A., and Immenkamp, M. Tumor cell types in osteosarcoma as revealed by electron microscopy. Implications for histogenesis and subclassification. *Virchows Arch* **36:** 257–273, 1981.

Harcourt, R. A. Osteogenic sarcoma in the pig. *Vet Rec* **93:** 159–161, 1973.

Jeraj, J. *et al.* Primary hepatic osteosarcoma in a dog. *J Am Vet Med Assoc* **179:** 1000–1003, 1981.

Misdorp, W. Canine osteosarcoma. *Am J Pathol* **98:** 285–288, 1980.

Rosendal, S. Osteoma in the oral cavity in a pig (*Sus scrofa*). *Vet Pathol* **16:** 488–490, 1979.

Rumbaugh, G. E., Pool, R. R., and Wheat, J. D. Atypical osteoma of the nasal passage and paranasal sinus in a bull. *Cornell Vet* **68:** 544–554, 1978.
Smith, R. L., and Sutton, R. H. Osteosarcoma in dogs in the Brisbane area *Aust Vet Pract* **18:** 97–100, 1988.
Stevenson, S. *et al.* Fracture-associated sarcoma in the dog. *J Am Vet Med Assoc* **180:** 1189–1196, 1982.
Woog, J. *et al.* Osteosarcoma in a phthisical feline eye. *Vet Pathol* **20:** 209–214, 1983.

D. Giant-Cell Tumor of Bone

Giant-cell tumor is a very rare neoplasm recorded in dogs, cats, and cattle. The cases described tend to behave like the giant-cell tumors in humans, occurring most commonly as expansile osteolytic masses in the ends of long bones. Involvement of metacarpal bones and of the axial skeleton is also recorded in dogs and cats. As the mass expands, it destroys cortex but always tends to be, at least partly, circumscribed by a thin shell of bone. The osteolytic nature of the tumor combined with this bony shell gives the characteristic "soap-bubble" lesion in clinical radiographs. Giant-cell tumors are locally aggressive but usually do not metastasize. In humans, the tendency to metastasize cannot be predicted from histologic appearance; almost all of those that have metastasized did so following therapeutic irradiation.

Two types of cells are present in giant-cell tumors. The most numerous are mononuclear stromal cells, which may have a histiocytic or fibroblastic appearance. There are frequently many mitoses in these cells. The other type, which sometimes composes up to 35% of the cell population, is the multinuclear giant-cell from which the tumor derives its name. These cells resemble osteoclasts. Opinions differ as to whether the giant cells are formed by coalescence of stromal cells or by amitotic division or nuclear segmentation without cytoplasmic separation. There may be some collagen and osteoid in the tumor, but this is not a prominent feature. Hemorrhages and cavernous vascular spaces are common in giant-cell tumors, and degeneration with cavitation of the mass often occurs. The latter features may lead to confusion with certain bone cysts such as the aneurysmal bone cyst (see Diseases of Bones, Section X,G, Bone Cysts) and other tumors of bone, such as telangiectatic osteosarcoma. Giant cells occur in many lesions of bone, but in giant-cell tumor they are an integral part of the neoplasm. Their significance should be assessed in histologic fields free of hemorrhage, necrosis, or any other secondary or reactive change.

Giant-cell tumors probably arise from osteoclast stem cells of the bone marrow. The tumor giant cells have histochemical and ultrastructural features of osteoclasts, which supports an origin from bone-marrow cells. C-type particles budding from neoplastic cells in one feline giant-cell tumor may be coincidental, but in view of the proven oncogenic properties of morphologically similar particles in feline leukemia, it seems plausible that a similar oncogenic agent could produce giant-cell tumors under appropriate circumstances.

Extraskeletal giant-cell tumors resembling the skeletal neoplasms occur in humans and, rarely, other animals.

Bibliography

LeCouteur, R. A. *et al.* A case of giant-cell tumor of bone (osteoclastoma) in a dog. *J Am Anim Hosp Assoc* **14:** 356–362, 1978.
Popp, J. A., and Simpson, C. F. Feline malignant giant-cell tumor of bone associated with C-type virus particles. *Cornell Vet* **66:** 527–534, 1976.
Seiler, R. J., and Wilkinson, G. T. Malignant fibrous histiocytoma involving the ilium in a cat. *Vet Pathol* **17:** 513–517, 1980.

E. Miscellaneous Tumors of Bones

Tumors of the hematopoietic and lymphopoietic systems are considered with those systems. Bovine lymphomatosis of the juvenile type commonly involves long bones, causing infarction of marrow and necrosis of bone in more than 80% of affected calves. Myeloma, when it occurs in dogs, sometimes produces the multifocal "punched-out" radiolucent lesions characteristic of the disease in man. Lesions of this type are also produced occasionally in dogs by poorly differentiated histiocytic tumors.

Tumors of vascular tissue are described with the circulatory system. Intraosseous hemangioma is extremely rare in animals, but primary hemangiosarcoma of bone is not uncommon in dogs, particularly in flat bones.

Tumors of fat calls are described with tumors of skin. Primary tumors of fat cells are extremely rare in bones. Liposarcoma of bone occurs in dogs. Osteoliposarcoma, a form of malignant mesenchymoma, is a malignant neoplasm containing liposarcomatous and osteosarcomatous tissue and is recorded in the dog.

Bibliography

Bingel, S. A. *et al.* Haemangiosarcoma of bone in the dog. *J Small Anim Pract* **15:** 303–322, 1974.
Brodey, R. S., and Riser, W. H. Liposarcoma of bone in a dog: A case report. *J Am Vet Radiol Soc* **7:** 27–33, 1966.
Dueland, R., and Dahlin, D. C. Hemangioendothelioma of canine bone. *J Am Anim Hosp Assoc* **8:** 81–85, 1972.
Hugoson, J. Juvenile bovine leukosis. An epizootiological, clinical, pathoanatomical and experimental study. *Acta Vet Scand (Suppl.)* **22:** 5–108, 1967.
Linnabary, R. D. *et al.* Primitive multipotential primary sarcoma of bone in a cat. *Vet Pathol* **15:** 432–434, 1978.
Misdorp, W., and van der Heul, R. O. An osteo-(chondro-) liposarcoma ("malignant mesenchymoma") of the radius in a dog, with two types of metastases. *Zbl Veterinaermed (A)* **22:** 187–192, 1975.
Zachary, J. F., Mesfin, M. G., and Wolff, W. A. Multicentric osseous hemangiosarcoma in a Chianina–Angus steer. *Vet Pathol* **18:** 266–270, 1981.

F. Tumors Metastatic to Bone

Neoplasms which develop outside the periosteal envelope may spread to bone, either by direct extension (contact metastasis) or hematogenously. Hematogenous metastases are very rare in farm animals and, compared to those in humans, relatively rare in pet animals. Several explanations for this discrepancy have been proposed. The life span of most farm animals is abbreviated by slaughter, and that of many pets by euthanasia, thus limiting the time for tumors to develop, metastasize and, perhaps most important, produce signs of metastatic disease.

Widespread metastases occur quite often in dogs, yet bone metastases are thought to be uncommon. To some extent this reflects the lack of complete radiologic and pathologic survey of the skeleton in cases of malignant disease. One survey, in dogs, found that 24 of 141 tumors with visceral metastases also had skeletal metastases, and that the skeleton was surpassed only by lung, liver, and kidney as a preferential site of metastatic tumor. These figures do not include either contact metastases or tumors of hematopoietic and lymphoid tissues and are based on gross examination with histologic confirmation. Probably routine microscopic skeletal surveys in dogs with malignancies would increase the figure. In the series of 218 malignant tumors, only 26 originated in bone, indicating that skeletal metastases are about as common as primary tumors of bone in dogs, whereas in humans, metastatic bone lesions are far more common than primary bone tumors.

Most of the tumors that metastasize to the canine skeleton are carcinomas, and the most common sites of origin are mammary gland, liver, lung, and prostate. The humerus, femur, and vertebral column are the favored locations for metastases.

In humans, bone metastases are far more common than would be expected, based on the proportion of cardiac output going to the skeleton. This may be due to retrograde spread via the vertebral venous system and the sinusoidal vessels in hematopoietic marrow. Tumor metastases lodge initially in the spaces of cancellous bone and either extend along the surface of trabeculae, expand to produce osteolysis, or fill the intertrabecular spaces and cause ischemic necrosis.

In dogs and cats, squamous cell carcinomas, particularly those developing in the gingiva and digital skin, often invade bone by direct extension. Tumors of dental origin often involve bone. Intraosseous epidermoid cysts occur occasionally in dogs. These are not neoplastic but probably arise from traumatically displaced or heterotopic epidermal cells.

Bibliography

Goedegebuure, S. A. Secondary bone tumours in the dog. *Vet Pathol* **16:** 520–529, 1979.

Liu, S.-K., and Dorfman, H. D. Intraosseous epidermoid cysts in two dogs. *Vet Pathol* **11:** 230–234, 1974.

Liu, S.-K., and Hohn, H. B. Squamous cell carcinoma of the digit of the dog. *J Am Vet Med Assoc* **153:** 411–424, 1968.

G. Bone Cysts

A cyst has been defined as a collection of gas or fluid enclosed by a capsule. According to this definition, many of the "cysts" of bone are not true cysts. To radiologists, a bone cyst is a circumscribed, smooth-contoured area of radiolucency within a bone, which may be attributed to tumor, blood clot, granulation tissue or some other substance less dense than bone (Figs. 1.111, 1.117A, 1.120A,B). To pathologists, a bone cyst is a grossly visible, more or less discrete cavity often having a lining membrane and frequently containing fluid. The classification of bone cysts reflects their diversity of form and development and is based on location, anatomical complexity, and presumed pathogenesis. For example, cysts may be described as subchondral if they lie beneath articular cartilage, unicameral if single chambered, or aneurysmal if blood-filled and expansive.

Subchondral cysts are common in horses and pigs and are usually a manifestation of osteochondrosis (Figs. 1.103B, 1.105). They are visible in radiographs and in sectioned bones and are usually about 0.5 to 1.0 cm in diameter and roughly spherical. They originate either from degeneration of a nodule of residual epiphyseal cartilage, or from a focus of subarticular hemorrhage, which may

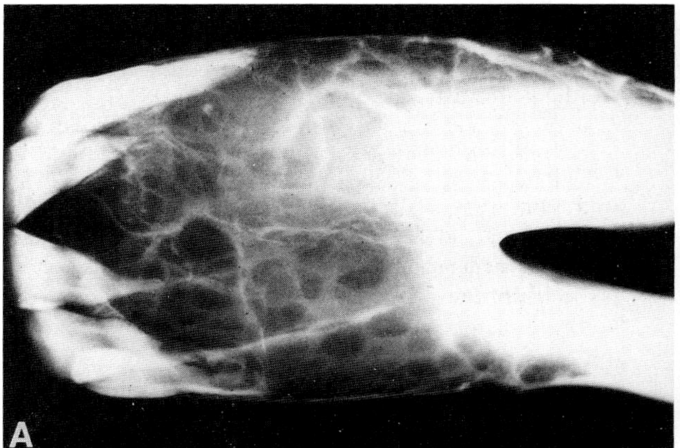

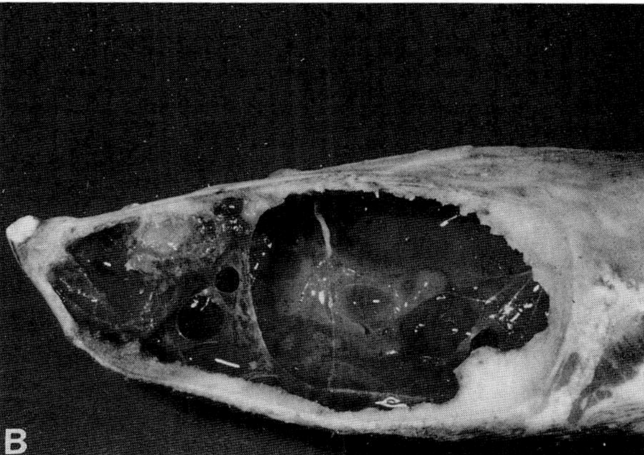

Fig. 1.120 (A) Radiograph of posthemorrhage "cyst" involving equine mandible. (B) Gross appearance of right half of (A). Lining of "cyst" contained abundant hemosiderin.

subsequently be organized and replaced by fibrous tissue or be filled with fluid of synovial origin. Cysts that develop from pinching off of invaginations of articular and growth cartilage tend to have a minute communicating channel to the articular surface. These are common in horses and occur most frequently in the phalanges and cuboid bones of the limbs, although virtually any limb bone may be involved. Focal subchondral hemorrhages often develop beneath articular cartilages of the femoral condyles, and less frequently in other sites of swine with osteochondrosis, and give rise to subchondral cysts. Subchondral cysts are often associated with lameness, but some disappear in the course of ossification and remodeling.

Subchondral cysts in humans are often associated with osteoarthritis. They do develop occasionally in chronic degenerative arthropathy of cattle and horses, but this type is far less common than the cysts of osteochondrosis, and tends to occur in older animals. The origin of cysts in osteoarthritis is obscure, but they appear to develop in areas containing hemorrhage and necrotic bone and are sometimes called detritus cysts.

Unicameral cysts, which are also known as solitary or simple bone cysts, are rare in animals. They are described with fibrous dysplasia. They may be polyostotic or monostotic and are most common in dogs.

Aneurysmal bone cysts (multilocular hematic bone cyst) are expansive lesions of bone which arise from disturbances in the vasculature of the bone marrow. They are not neoplastic lesions but may mimic certain tumors both radiographically and clinically and sometimes develop as complications of intraosseous neoplasms. They occur rarely in dogs, cats, and horses.

In radiographs, aneurysmal bone cysts appear as osteolytic expansive lesions which erode and deform the bone cortex to produce a fusiform or saccular defect, which is almost invariably surrounded by a shell of periosteal new bone and is shaped like a true aortic aneurysm. The radiographic appearance mimics giant-cell tumor of bone, but the usual initial location, in metaphysis or diaphysis rather than epiphysis, is different.

The gross appearance of an aneurysmal bone cyst is that of a blood-filled sponge. The blood is dark and may ooze continually from the lesion. The blood-filled spaces are separated by fibrous trabeculae of variable thickness. The broader trabeculae may contain osteoid, bone, and numerous osteoclastic giant cells and are sometimes lined by endothelium.

Multinuclear giant cells are a common feature of hemorrhagic lesions of bone, and their presence in aneurysmal bone cysts complicates their differentiation from giant-cell tumor of bone. Other hemorrhagic, osteolytic lesions of bone such as hemangioma, hemangiosarcoma, and telangiectatic osteosarcoma must also be considered in the differential diagnosis. To add further confusion, some aneurysmal bone cysts are secondary to intraosseous neoplasms, including giant-cell tumor; thus, numerous sections must be examined to eliminate this possibility.

There is probably no single cause of aneurysmal bone cysts. It appears that they may result from any insult that disrupts, but does not interrupt, the circulation of blood through the marrow. In humans arteriovenous fistulae and post-traumatic hemorrhage are postulated, with some justification, as causes of aneurysmal bone cysts. Lesions developing secondary to neoplasms probably result from massive intraosseous hemorrhage following erosion of vessels by the tumor. Presumably, the pathogenesis in other animals is similar.

The osteolytic nature of aneurysmal bone cyst is not adequately explained but seems to be related to the increased flow of blood through the spongy mass. (See Section IV,E, Adaptational Deformities of the Skeleton for discussion of blood flow and bone resorption.)

Bone cysts in hyperparathyroidism, which are well recognized in humans, (hence the term osteitis fibrosa cystica) are rare in other species, with the possible exception of nonhuman primates. They are seen in areas of exuberant bone resorption and fibrous repair and often contain hemorrhage or evidence of previous hemorrhages, and many osteoclasts.

Bibliography

Hanlon, G. F., Sautter, J. H., and Sherman, D. Multifocal osteolysis in a horse: A case report with special emphasis on the radiologic and pathologic findings. *J Am Vet Med Assoc* **178:** 238–241, 1981.

McIlwraith, C. W. Subchondral cystic lesions (osteochondrosis) in the horse. *Compend Cont Ed* **4:** S282–S291, 1982.

Renegar, W. R. *et al.* Aneurysmal bone cyst in the dog: A case report. *J Am Anim Hosp Assoc* **15:** 191–195, 1979.

Steiner, J. V., and Rendano, V. T. Aneurysmal bone cyst in the horse. *Cornell Vet* **72:** 57–63, 1982.

Walker, M. A. *et al.* Aneurysmal bone cyst in a cat. *J Am Vet Med Assoc* **167:** 933–934, 1975.

DISEASES OF JOINTS

I. General Considerations

Joints or articulations are structures in which two or more bones or cartilages are united. Depending on the tissue which unites them, they may be classified as either fibrous, cartilaginous, or synovial joints.

Fibrous joints (synarthroses) comprise sutures, syndesmoses, and gomphoses. Sutures occur in the skull, are formed by fibrous tissue, and permit continued growth of the bones they unite by virtue of the osteogenic cells at their borders. Syndesmoses unite skeletal structures by fibrous, elastic, or fibroelastic tissue. When the original tissue becomes ossified with age, a synostosis is formed. Such unions occur between the shafts of the tibia–fibula, and of the radius–ulna in some species. Gomphoses are really not joints, since the term refers to the fibrous unions of the teeth, which are not skeletal structures, to alveolar bone, by the periodontal ligament.

Cartilaginous joints (amphiarthroses) are united by hyaline or fibro-cartilage or a combination of the two. There are two types. A synchondrosis is a temporary joint in which hyaline cartilage is replaced by bone at maturity. Numerous examples exist including physeal cartilages, which unite two centers of ossification in long bones, and the joints where endochondral ossification occurs in the basocranium. The second type of cartilaginous joint is the symphysis, where the uniting tissue is fibrocartilage, at least during some stage of the joint's existence. Examples are the pelvic symphysis and the joints between vertebral bodies.

Synovial joints (diarthrodial joints) are characterized by a joint cavity, a synovial membrane lining a fibrous capsule, and mobility, which is facilitated by articular plates on the ends of the bones forming the joint. Simple synovial joints unite two articular surfaces; composite joints unite more than two. Some synovial joints also have ligaments, while others contain menisci (femorotibial joint) or marginal cartilage encircling the rim (acetabulum). Most, but not all, of the diseases with which we are concerned are diseases of synovial joints.

An understanding of the normal structure and function of synovial joints is essential for the interpretation of their reaction to disease. A brief summary of the synovial joint and, in general terms, its response to insult follows. As noted, the basic components of synovial joints are capsules and ligaments, synovial membrane, synovial fluid, and articular cartilages with some special ancillaries, including fat pads, synovial fossae, and menisci in some joints.

Articular capsules and ligaments are composed of fibrous tissues, interlaced collagen, and reticular fibers, with scant elastica and ground substance, and a few fat cells, histiocytes, and leukocytes. The tissue is poorly supplied with blood vessels, which, together with its unyielding density, limits its expression of inflammatory phenomena and its capacity for repair. Repair may, however, eventually be complete. The same characteristics also tend to limit the spread of infection or neoplasms through the capsule. Both ligaments and capsule are well provided with proprioceptive and pain end organs, the latter accounting for some of the pain of arthritic disease.

Fibrous capsules vary in their thickness and strength depending on the demands placed on them. They are not always complete, as, for example, at the anterior aspect of the stifle, where the capsule is partly replaced by the quadriceps tendon, and where it is deficient in the suprapatellar pouch. The fibrous capsules form sleeves around the joints, enclosing the articular structures. Their inelastic quality aids the proper location of bone ends and helps to restrict movement, especially near the extremes for the range of movement of a joint. In this the ligaments assist, their shape and disposition being in accord with the function of the joint. The limited elasticity of capsules and ligaments, although necessary for restriction of joint movement, means that excessive or prolonged tension may stretch the structures irreversibly.

Factors governing the tensile strength and elasticity of ligaments and capsule are largely unknown, except for the physiological softening of certain ligaments that occurs under the influence of hormones in pregnancy. Clinical impression has long been that unnatural laxity of capsules and ligaments occurs in a number of metabolic diseases, especially the osteodystrophies associated with calcium and phosphorus deficiencies, and that, provided degenerative changes are not advanced, tone is rapidly restored to these structures with correction of the deficiency. Abnormal mobility of joints may be prominent in osteogenesis imperfecta of calves and lambs and some of the syndromes of congenital limb deformity caused by poisonous plants and unidentified environmental factors.

Fibrous capsules and ligaments are attached to bone either directly or indirectly by attachment to each other or to other intra- and periarticular structures. Some of the fibers are attached to periosteum, and others become incorporated in bone as Sharpey's fibers. Whereas excessive tension on ligaments may rupture them, avulsion fracture or detachment of ligaments is more likely, especially in osteodystrophy. Much of the resistance to motion in stiff joints, including those of arthrogryposis, resides in the fibrous capsules rather than in synovial or articular structures. The significance of capsular thickening in lame animals is easily overlooked.

Synovial membranes, whether lining articular cavities, bursae, or tendon sheaths, are all biologically similar. They consist of two layers, an outer thin areolar, adipose, or fibrous layer merging with the articular capsule in the case of joints, and an inner lining intima, the critical component concerned with the production and turnover of synovial fluid. The intima is a smooth glistening layer, more cellular than fibrous, and extremely well vascularized. It is reflected on intra-articular bone, where it merges with the periosteum or the perichondrium, as the case may be. In the transition zone, it merges with the articular margins and spreads for a short distance over the nonbearing articular cartilage. The intima is also reflected over intra-articular ligaments and tendons. It is not a uniform structure, being well developed and slightly villous, especially in recesses of the joint, and attenuated over fixed positions, such as intra-articular ligaments.

The specialized lining synovial cells have a variable appearance and arrangement by both light and electron microscopy. Where the synovial membrane is subjected to pressure from adjoining structures, the subintimal layer is fibrous, and the lining cells generally resemble the underlying fibroblasts. The synovial cells overlying adipose and areolar tissue are similar to each other and may form one to three layers. Two main cell types are distinguished by electron microscopy. Type A cells are more numerous and tend to be located on the surface of the synovial membrane. They have irregular cytoplasmic and nuclear outlines, long cytoplasmic processes, and contain extensive Golgi apparatus, granules, and micropinocytotic vesicles. Type A cells are phagocytic and akin to macrophages. Type B cells tend to be located deeper in the synovial membrane and have smoother cytoplasmic and nuclear outlines. Their cytoplasm contains a large amount of endoplasmic reticulum and a relatively extensive Golgi apparatus. Many of them contain granules and vesicles similar to those in Type A cells. Type B cells produce hyaluronic acid. Other cells are intermediate between the A and B types, and it may be that the two types are different functional stages of the same cell. In proliferative synovitis, intermediate cells may outnumber type A cells. Some undifferentiated cells are also present in the synovial membrane, and it seems likely that these are the synovioblastic cells which differentiate to form the specialized lining cells and also give rise to synovial tumors.

The intimal cells do not form a lining membrane in the usual sense of an endothelium or epithelium. There is no basement membrane to the synovium, and there are no tight junctions or desmosomes between the synovial cells. Instead they are arranged in a granular intercellular matrix which contains fine fibrils, which are similar to fibrils in the cytoplasm of type A and B cells, and some collagen fibers. The discontinuous lining may be part of the reason that particulate matter and bacteria can so rapidly gain access to the subsynovial tissues and joint cavity. The abundance of capillaries and lymphatics probably also facilitates this movement.

The synovial membrane is freely permeable in either direction to molecules of small dimension, which may be removed by the capillaries and lymphatics. Particulate matter is also readily removed by phagocytes. Hyperemia of exercise and inflammation hastens the exchange. The removal of particulate matter from the joint and its deposition in the subintimal layer is a continuous process of removal of debris. When the volume of debris is large in diseased joints, its presence in the synovial membrane stimulates fibrosis of the capsule, which contributes to the swelling and fixation of diseased joints. Synovial membrane has great regenerative ability, following biopsy, for example, and great proliferative ability in certain disease states.

The transitional zone of the synovial membrane is of particular significance for pathological reactions. There is a gradual transition from the synovial membrane to the periosteum and cartilage margin, and the intima is well vascularized where it extends onto the cartilage surface. Its marked proliferative ability probably is correlated with its vascularity, and it is in this area that erosion, or alternatively lipping and osteophyte formation, occurs in arthritis and arthropathy. Hyperplasia and development of new villi occur with age and disease and give the synovial membrane a velvety appearance. Cartilaginous metaplasia sometimes occurs in the stroma of the synovial membranes, and ossification of the cartilage nodules may develop. If these nodules are broken off, they become "joint mice" in degenerate joints.

In chronic synovitis, lymphocytes, which may be arranged in follicles or as diffuse infiltrations, and plasma cells accumulate in the hypertrophic synovial villi. In septic arthritis, local antibodies may be secreted into the fluid. Transfer of serum antibodies to synovial fluid is probably insignificant in normal joints. Disruption of the blood–synovial barrier by inflammation permits the movement of serum constituents, including antibodies, to the synovial fluid. However, even in severe arthritis, such as that caused by *Mycoplasma bovis* in cattle, synovial fluid levels of immunoglobulin G (IgG) and IgM are only a fraction of serum levels, whereas synovial levels of IgA are usually greater than serum levels. Furthermore, the synovial level of IgA is greater than that of IgM, whereas in serum, it is lower. This suggests that there is preferential synovial synthesis of IgA in addition to a barrier to transport of IgA and IgM.

The transitional zone of the synovial membrane is also the initial site of synovial proliferation and of the vascular granulation tissue or pannus, which sometimes spreads across the articular plate in chronically diseased joints. **Pannus** may also develop from the marrow spaces of subchondral bone and extend onto the articular surface through defects in the cartilage. Pannus is important because it interferes with the normal nutrition of articular cartilage and is capable of eroding and destroying cartilage and bone by virtue of its collagenolytic activity. Collagenases may be produced by diseased synovial membrane, by granulation tissue, and by neutrophils in the synovial fluid. The collagens of the synovial membrane itself are generally resistant to these collagenases because of their cross-linking. Adhesion of pannus to opposing articular surfaces causes fibrous ankylosis and, if bone is formed in the connective tissue, bony ankylosis results. Few animals are permitted to survive to the stage of bony ankylosis, but fibrous ankylosis occurs occasionally in pigs with mycoplasma arthritis and in other chronic arthritides. Bony ankylosis sometimes involves the intercarpal joints of racing horses.

The **synovial fluid** is generally regarded as a protein-free dialysate of plasma to which hyaluronate, glycoprotein, immunoglobulins, lysosomal enzymes, and other unidentified macromolecules are added by synovial cells. There is, however, convincing evidence that some of the smaller plasma proteins are normally present in synovial fluid, and catabolic products of articular cartilage are also found there. The normal fluid does not contain fibrinogen or other macromolecules of plasma, but these proteins may accumulate if the joint is injured. There is very little information on selective transfer from blood to synovial fluid, but it is suggested that a blood–synovial barrier exists whose function depends on the presence of fenestrations in the synovial capillaries and probably on the intimal cells and the hyaluronate in their supporting stroma, which act as a filter for various molecules.

The volume, composition, and viscosity of synovial fluid vary markedly from joint to joint and between the same joints in different animals. Of the various indices for any normal joint, viscosity is perhaps the most constant. Other than the effects of exercise and inactivity, which tend to increase and decrease the amount of fluid, respectively, little is known of the factors which influence its volume in normal joints.

The number of cells in synovial fluid also varies between

normal joints, but the types are relatively constant. Most are mononuclear cells, and the presence of more than 6 to 10% neutrophils indicates infection or degeneration.

Normal synovial fluid is a viscous, clear, colorless, or slightly yellow fluid. It has two functions: the lubrication of the joint, and nourishment of articular plates. Synovial joints have two systems which must be lubricated, the familiar cartilage-on-cartilage system and a soft-tissue system in which synovial membrane slides on itself or on other tissues. Lubrication of the synovial system depends on hyaluronate molecules, which stick to the tissues and keep them apart—this is boundary lubrication. Depolymerization of the hyaluronate molecules of synovial fluid reduces its viscosity and thereby its effectiveness as a boundary lubricant for soft tissues. Cartilage-on-cartilage lubrication, on the other hand, is independent of hyaluronate. Synovial fluid contains a mucin, a mixture of hyaluronate and glycoprotein, and it is the glycoprotein which, under normal loading, is responsible for boundary-type lubrication of articular plates. This glycoprotein is specific to synovial fluid and is produced by the synovial cells. Under high loads, squeeze-film lubrication of articular plates occurs. In this type, the approaching surfaces generate pressure in the lubricant as they squeeze it out of the area of impending contact. The pressure thus generated keeps the surfaces apart. In joints, squeeze-film lubrication depends on joint fluid and on aqueous interstitial fluid, which weeps out of articular cartilage at the periphery of the loaded zone and merges with the joint fluid already present. It is aided by the presence of normal irregularities in the opposing articular surfaces, which tend to trap the pressurized lubricant, and possibly by the elastic properties of the cartilage as well. It is suggested that boundary lubrication is aided by bonding between lubricant and cartilage at specific receptor sites.

Normal synovial fluid is essential to the function of synovial joints. Changes in the fluid, such as reduced viscosity and hyaluronate content in joint effusion, and degradation of mucus by bacterial enzymes, occur, but little is known regarding the ability of intimal cells to restore the fluid to normal. There is, however, a significant reserve of lubricant in the normal joint.

The second equally significant function of synovial fluid is provision of nutrients for articular cartilage. There are two potential sources of nutrition for articular cartilage: diffusion from subchondral vessels and diffusion from synovial fluid. Both sources are probably important in immature animals, in which the cartilages are still growing, while in adults, the synovial source is more important. In immature animals, there is a rapid diffusion of radioactive material from the epiphyseal marrow to the joint cavity, which does not occur in the adult, probably because of the physical barrier offered by the layer of mineralized cartilage.

Certain abnormalities in synovial fluid in diseased joints have already been mentioned. Because of the wide and presently inexplicable variations in normal values, few significant general statements can be made regarding abnormal constituents. Increased amounts of fluid and changes in its physical properties from a clear viscous material to a cloudy or watery substance are consistent with inflammation or degeneration of synovial joints. Cytologic examination, and culture for bacteria, mycoplasmas, chlamydiae, and viruses, along with biopsy of synovial membrane may be helpful in allowing a diagnosis to be made. Specialized immunologic techniques and organ culture of synovial membrane may be indicated in some diseases.

The **articular cartilage** in all principal joints is hyaline. In young animals, it is smooth, bluish, and semitransparent, revealing the looped subchondral vessels which give it a stippled appearance. With advancing age it becomes yellowish, opaque, and less elastic. The cartilage tends to be thickest at points of maximum pressure, and the thickest parts tend to increase the curvature of the articular surface, being located at the center of convex articular plates, and the periphery of the concave ones.

There are important differences between the cartilage making up synovial joints in immature and mature animals. The articular plate is derived embryologically from mesenchymal tissue, which is present in the interzones between the cartilaginous anlagen of the definitive bones. The interzone develops into a three-layered structure; the two outer layers are destined to become the articular plates, while the middle layer cavitates and becomes the site of the joint space. The cartilages adjacent to the articular plates undergo endochondral ossification, thus contributing to growth of the epiphysis, which may continue throughout life or long after maturity. Normally endochondral ossification does not involve the cartilage of the articular plates, but ossification may occur in degenerative joint diseases. As the epiphyses grow, the articular plates also grow to adapt to the changing contours of the bone ends. The deep layer of the articular plate mineralizes (but does not ossify) and is separated from the more superficial radiate layer by the "blue line" or "tidemark," an undulating line, or, with aging and degeneration, series of lines, which stain blue with hematoxylin–eosin and red with Masson's trichrome. The mineralized layer of the articular plate is intimately attached to the subarticular bony plate.

This dual nature of the cartilage in immature animals is frequently overlooked. Routine histologic preparations do not distinguish the growth cartilage from the articular plate, although differences can be demonstrated histochemically. For purposes of discussion, **joint cartilage** means cartilage derived embryologically from both layers; **articular plate** means cartilage which develops from the interzone; and **subarticular growth cartilage** is that part of the epiphyseal cartilage model which underlies the articular plate.

The articular plate is divisible into four zones. The superficial, gliding zone contains oval or elongate cells and collagen fibers, which are tangential to the surface of the articular plate when viewed in a plane vertical to the articular surface. The deeper intermediate (transitional) zone contains oval or round cells, and the collagen fibers form an open, tangled meshwork when not under load, but tend to become oriented at right angles to the direction of prolonged loading. The radiate zone contains groups of large round cells, which tend to be arranged in short columns as in the physeal plate, and collagen fibers, arranged perpendicular to the articular surface. The mineralized zone, separated from the radiate zone by the tidemark, is the deepest stratum and is attached to the subarticular bone.

The subchondral bone, although embryologically and histologically distinct from the articular plate, is very important for normal function of synovial joints. Cartilage has a natural resilience, but it depends on the subchondral bony plate for support. In turn, the cartilage protects subchondral bone from fracture by deforming to produce a large contact area when it is loaded and by producing a more even distribution of stress at the level of the subchondral bone than at the articular surface. Thus defects in either the articular plate or the subchondral bone are potential causes of lesions in the other.

The cartilage matrix is produced and maintained by the chondrocytes, but, even in normal cartilage, the cells are diverse in structure and function, and proportionate composition of matrix varies with age, joint, and species. Water composes ~70% of the cartilage by weight. The remainder is composed of type II collagen and lesser amounts of type IX collagen and proteoglycans.

Covalent linkages between type II and Type IX collagen stabilize the meshwork, and other glycoproteins bind the type II collagen to the plasma membrane of the chondrocyte. An important function of the collagen is to constrain imbibition of water and swelling, which result from the powerful hydrophilic properties of the proteoglycans that fill the collagenous meshwork. Chondrocytes are capable of synthesizing collagen throughout the life of the animal, but the rate of synthesis does not compensate for wear in postmature joints and, with aging, traces of new collagens such as type I and variants of type II can be detected in the superficial zone. The proteoglycans specific to cartilage consist of a core protein with the glycosaminoglycan molecules keratan sulfate, chondroitin-4-sulfate, and chondroitin-6-sulfate attached to it. These are assembled in the Golgi apparatus, packaged in vesicles, and secreted into the extracellular matrix. The chondrocytes also synthesize, and extrude separately into the matrix, filaments of hyaluronic acid and link protein, a small glycoprotein. The hyaluronate binds to the amino-terminal end of the core protein, and the binding sites are joined and stabilized by the link protein. In this way, large proteoglycan aggregates are formed, comprising up to 40 monomers. The formation and stability of these aggregates are essential to the integrity of the cartilage.

A third proteoglycan of interest is dermatan sulfate. This is not specific to cartilage, being widely distributed in mesenchyme. The monomers are much smaller than the cartilage-specific proteoglycan and bind only to themselves to form small, unstable aggregates.

The distribution of proteoglycans in articular cartilage is not uniform. In immature cartilage, dermatan sulfate is displaced almost completely by chondroitin sulfate. In maturing cartilage the chondroitin sulfate is replaced by keratan sulfate, especially in the radial zone. In the postmature and aging cartilage, dermatan sulfates are increased in the superficial zone.

The metachromasia of cartilage matrix with stains such as toluidine blue is due to its glycosaminoglycan content. There is continual turnover of glycosaminoglycans in cartilage, and treatment with papain or hyaluronidase, which causes loss of proteoglycans, stimulates chondrocytes to synthesize new matrix constituents. When proteoglycans are lost, metachromasia is reduced, the intercellular substance becomes PAS positive, and collagen fibers are easier to demonstrate.

The cartilage of the articular plate contains neither nerves, lymphatics, nor blood vessels. Thus even severe injury to the cartilage does not cause pain unless there is concomitant synovial injury or injury to the subchondral bone. Evidence of this lack of innervation is commonly seen at autopsy, for severe lesions of cartilage are often asymptomatic, whereas minor soft-tissue lesions are often associated with severe lameness. The absence of blood vessels means that no vascular or exudative response to injury is possible, and not until subchondral tissue is exposed does a conventional inflammatory reaction occur.

Superficial lacerations of articular cartilage, which do not approach the tidemark, stimulate mitoses and synthesis of protein and glycosaminoglycan in chondrocytes near the lesions. However, this activity ceases after a week, and, at least in the short term, superficial lesions neither heal completely nor progress. Injuries that involve the subarticular bone cause hemorrhage, and the clot that results is organized by fibrovascular granulation tissue, which fills the defect in the cartilage in about 10 days. The proportion of collagen in the defect increases and, if the joint is exercised, the collagen is converted to fibrocartilage, which adheres to the edges of the defect as soon as 3 to 4 weeks after injury. The bone of the subarticular plate is repaired, and the overlying fibrocartilaginous scar persists and functions indefinitely. As with superficial defects, transient synthetic activity by chondrocytes occurs but is ineffective.

Traumatic lesions which involve the tidemark without directly involving the mineralized zone may stimulate vascular invasion of the defect from subarticular bone. Formation of granulation tissue, which is converted to fibrocartilage, then occurs.

Several chemical agents, such as papain, β-aminoproprionitrile, and excess vitamin A, have specific effects on the proteoglycans and glycosaminoglycans of cartilage matrix. Oxolinic acid, which belongs to the quinoline carboxylic class of antibacterial agents, is toxic to articular cartilage of juvenile dogs. All major diarthrodial joints are affected. Vesicles form in the intermediate zone of the cartilage as a result of chondromalacia. Coalescence of large vesicles can lead to shearing of the superficial layer. Intra-articular cortisol, administered because of its lympholytic and antiinflammatory action on synovial membranes, inhibits the synthesis of protein and glycosaminoglycans by chondrocytes and, if used over extended periods, can produce chondromalacia. Since articular cartilage lacks innervation, severe lesions may develop without clinical signs. It is often impossible to distinguish the relative effects of the original disease and the cortisol treatment, but the occurrence of unusually severe joint injury in racehorses in association with cortisol administration indicates that such treatment may exacerbate joint degeneration.

A single episode of intrasynovial hemorrhage usually has no effect on the articular plate. However, repeated episodes, such as occur in hemophilia, reduce proteoglycan concentrations and lead to erosion of cartilage. Phagocytosis of erythrocytes by synovial cells results in heavy deposits of hemosiderin in the synovial membrane and causes orange-tan discoloration. Proliferation of synovial cells with formation of villi, and thickening of the joint capsule, also develop with repeated hemarthrosis. The positive iron charges of hemoglobin bind to and neutralize the negative charges of hyaluronate in the synovium, altering its molecular sieve characteristics so that large plasma proteins including fibrinogen can enter the joint cavity.

Articular cartilage is metabolically active but at a very low rate, although, cell for cell, the rate is probably similar to that for other tissues. The metabolic requirements are probably met from the synovial fluid, which is capable of rapid exchange with blood for all but the largest molecules. The surface of the cartilage appears to be porous, at least in macromolecular dimensions, and the assumption is that nutrients diffuse into the cartilage aided by the massaging effect of normal movement. It appears to be true that the integrity of the cartilage is preserved only if joint movements are physiologic in their range and pressure. Sustained excesses of pressure and prolonged disuse accelerate degeneration.

The sequence of events that occurs when articular cartilage degenerates is similar for many diseases. The initial change is chondromalacia, a palpable softening associated with an increase in water content and a decrease in glycosaminoglycans. This paradox is explicable in terms of an increase in the domain of each of the hydrophilic glycosaminoglycan molecules as they become more widely separated in the collagen network. The earliest visible change is fibrillation of the cartilage, which loses its smooth, shiny nature and acquires a ground-glass appearance and a slightly velvety texture. Microscopically, there are vertical clefts in the superficial zone, which may extend to the mineralized layer; this implies that tearing of collagen fibers in the intermediate and radiate zones has occurred. Early fibrillation and loss of glycosaminoglycans is accompanied by death of some chondrocytes and reactive proliferation of others, which increase glycosaminoglycan production but change the proportions to resemble those of immature cartilage. Injured chondrocytes synthesize type I and abnormal type II collagens.

Proliferation of chondrocytes is recognizable histologically by the occurrence of cell clusters sometimes called brood capsules, clones, or multinuclear chondrones. Extensive fibrillation is accompanied by continued death of chondrocytes, reduced glycosaminoglycan production, and thinning of the articular plate. Horizontal splits of the junction of the radiate and mineralized zones, which approximately follow the tidemark, may result in complete loss of an area of the articular plate. Sometimes, extension of the mineralized zone into the radiate zone is associated with duplication of the tidemark.

Although loss of glycosaminoglycans is usually considered to be the initial abnormality in degeneration of cartilage, there is evidence that fatigue-induced changes in the collagen network may be central to the onset of degeneration. Whether these changes involve the structure of individual fibers or the linkages between them is unclear. Overhydration of the matrix occurs if it is not constrained by the collagen mesh, resulting in altered functional properties and loss of proteoglycans by diffusion into the joint space and disruption of linkages between matrix aggregates and cell membranes.

Fibrillation of cartilage is not always a progressive lesion. In aged animals, some degree of fibrillation of articular plates is common and apparently nonprogressive. For example, fibrillation is commonly found on the dorsomedial aspect of the head and the medial condyle of the femur in old dairy cows without any other evidence of joint disease. Often, however, fibrillation leads inexorably to ulceration and loss of cartilage with exposure of the subchondral bony plate, which becomes reinforced by new bone and develops a dense, lustrous surface. Such bone is often described as eburnated because of its ivorylike hardness and polish. Degenerative changes of such severity often are accompanied by grooves, oriented in the direction of motion, in the articulating bone surfaces, and by periarticular osteophytes and "osteochondromas," which develop in the transitional zone of the synovial membrane. Periarticular proliferations of this type are often regarded as late manifestations of joint injury. **Osteophyte development** begins, however, as early as 3 days following section of the anterior cruciate ligament in dogs. Osteophytes are grossly visible after 2 weeks and may be demonstrated in radiographs by 5 weeks. The cruciate lesion permits abnormal movement of the joint, and undoubtedly places unusual stresses on the joint capsule, which are probably responsible for the proliferative changes. In this condition, osteophytes appear to develop directly, and not by endochondral ossification of cartilage nodules, which is thought to be the more common pathogenesis. Usually, cartilage nodules develop by proliferation of mesenchymal cells beneath the reflection of the synovial membrane onto the perichondrium. Cartilage formed here is invaded by vessels from the underlying marrow, and endochondral ossification ensues. Similar structures may also be formed in the synovial membrane lining the joint capsule.

In chronic joint diseases, osteophytes occasionally become intra-articular loose bodies or "joint mice." **Joint**

mice are fragments of cartilage and/or bone that originate from synovial membranes or articular plates and are lying free in the joint space. There are two major sources: fragments displaced in osteochondritis dissecans (see Diseases of Bones, Section IX,E, Osteochondrosis), and reactive osteophytes or "osteochondromas" separated from the joint margin in chronic joint disease. Another potential source is the synovial membrane in osteochondromatosis (see the preceding sections).

Cartilage in joint mice survives and may grow, deriving its nourishment from synovial fluid. Ossification and bone resorption also occur occasionally in free joint mice without attachment to a blood supply. The osteoblasts probably arise from mesenchymal cells associated with the cartilage mass and the osteoclasts from mononuclear precursors in synovial fluids. Nutrition must also be supplied by synovial fluid. Rarely, reattachment of joint mice occurs.

Joint mice are important in that they may be confusing in radiographs, they usually indicate severe joint disease, and occasionally they interfere with joint movement.

A joint with ulceration of cartilage, eburnation of bone, periarticular osteophytes, proliferation of synovial villi, and thickening or distension of joint capsule may be regarded as an end-stage joint (Fig. 1.121A,B) and may result from any one of a multitude of insults. Speculation as to cause in an individual case should be cautious, although careful consideration of such factors as age of animal, species, group history, and joints affected often allows an informed guess to be made. Several nutritional, developmental, infectious, and idiopathic diseases which sometimes terminate in end-stage joints are discussed in the following pages.

Menisci, articular disks, and fat pads are specialized structures which occur in some joints. A **meniscus** is a semilunar or C-shaped structure with one flat and one concave surface. In domestic animals, menisci occur in the femorotibial joint. An **articular disk** is circular or oval,

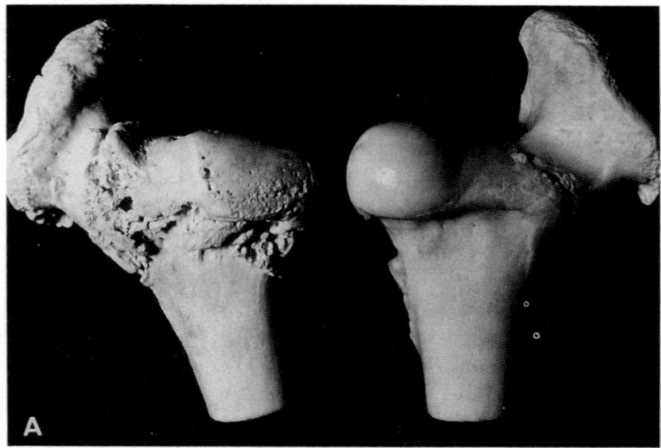

Fig. 1.121A Femurs. End-stage degeneration of joint (left). Normal (right). Note effects of remodeling, eburnation, and osteophyte production.

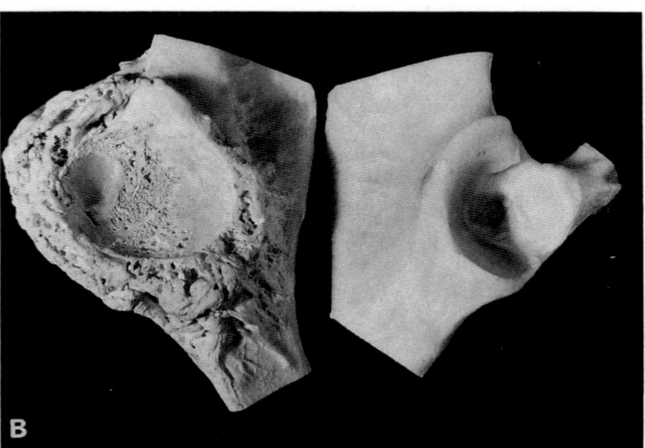

Fig. 1.121B Acetabula from (A).

flattened, and may have a central perforation. The temporomandibular joint contains an articular disk. The structure of articular disks and menisci varies somewhat between species, but generally they are composed of fibrocartilage. Unlike articular plates, they are innervated and have a blood supply, although the richness of the supply is variable between different areas. **Fat pads** are normally present in some joints, such as the coxofemoral joint of adult cattle.

Synovial fossae are normal depressions in the surfaces of articular cartilages of certain synovial joints. These bilaterally symmetrical depressions are located at or near the midline of the involved joints and are acquired during the first months of postnatal life as a result of joint modeling. During skeletal growth, acquired incongruities in central areas of certain joints cause progressive loss of contact between apposing articular surfaces. While the superficial or gliding layer of the articular cartilage in these areas persists, chondrocytes in the deeper zones of the subarticular growth cartilage are not maintained and gradually disappear. Gradually, synovial fossae appear as central depressions having distinct borders and smooth surfaces. The surfaces may have a bluish to pink color, reflecting the activity of the subchondral capillary bed. In most instances synovial fossae are important only in their recognition as normal joint structures; however, in horses they are structural points of weakness where infection can be passed between the joint cavity and the subchondral bone.

Bibliography

Cutlip, R. C., and Cheville, N. F. Structure of synovial membrane of sheep. *Am J Vet Res* **34:** 45–50, 1973.

Dieppe, P. Osteoarthritis. A review. *J R Coll Phys Lond* **24:** 262–267, 1990.

Fiala, O., and Bartos, F. Traumatic pannus. I. Macroscopical and microscopical changes after experimental reconstruction of the joint surface. *Acta Orthop Scand* **44:** 194–207, 1973.

Freeman, M. A. R. (ed.). "Adult Articular Cartilage," 2nd Ed., London, Pitman Medical, 1979.

Frost, G. E. Cartilage healing and regeneration. *J S Afr Vet Assoc* **50:** 181–187, 1979.

Gough, A. W. *et al.* Fine structural changes during reparative phase of canine drug-induced arthropathy. *Vet Pathol* **22:** 82–84, 1985.

Hamerman, D. The biology of osteoarthritis. *N Engl J Med* **320:** 1322–1330, 1989.

Havdrup, T., and Telhag, H. Mitoses of chondrocytes in normal adult joint cartilage. *Clin Orthop* **153:** 248–252, 1980.

Honner, R., and Thompson, R. C. The nutritional pathways of articular cartilage. An autoradiographic study in rabbits using 35S injected intravenously. *J Bone Joint Surg (Am)* **53A:** 742–748, 1971.

Meachim, G., and Roberts, C. Repair of the joint surface from subarticular tissue in the rabbit knee. *J Anat* **109:** 317–327, 1971.

Milgram, J. W. The development of loose bodies in human joints. *Clin Orthop* **124:** 292–303, 1977.

Mohr, W., and Wild, A. The proliferation of chondrocytes and pannus in adjuvant arthritis. *Virchows Arch (Cell Pathol)* **25:** 1–16, 1977.

Radin, E. L., and Paul, I. L. A consolidated concept of joint lubrication. *J Bone Joint Surg (Am)* **54A:** 607–616, 1972.

Redler, I. *et al.* The ultrastructure and biomechanical significance of the tidemark of articular cartilage. *Clin Orthop* **112:** 357–362, 1975.

Simon, S. R. *et al.* The response of joints to impact loading. II. *In vivo* behaviour of subchondral bone. *J Biomech* **5:** 267–272, 1972.

Sokoloff, L. (ed.). "The Joints and Synovial Fluid." New York, Academic Press, 1978.

Wilsman, N. J., and Van Sickle, D. C. Histochemical evidence of a functional heterogeneity in neonatal canine epiphyseal chondrocytes. *Histochem J* **3:** 311–318, 1971.

II. Developmental Disturbances of Joints

A. Luxations and Subluxations

Congenital luxations are rare in animals. **Atlantoaxial subluxations** occur in dogs, goats, cattle, and horses. Miniature and toy dog breeds are usually affected. The basic lesion appears to be failure of fusion of the odontoid process to the body of the axis. Signs vary from neck pain to tetraplegia, with age of onset being a few months to several years. In calves, absence or hypoplasia of the odontoid process occurs and atlanto-occipital fusion is often concomitant. Tetraplegia may be present at birth or develop at several months of age. In dogs and calves, fusion of the odontoid process with the axis normally occurs in the early months of life, so it is possible that, in some cases, postnatal influences are responsible for the condition.

Atlantoaxial subluxations occur in some Arabian foals with a familial, probably inherited, syndrome involving occipitalization of the atlas and atlantalization of the axis. Affected foals may be dead or tetraparetic at birth or develop progressive ataxia within a few months. They have atlanto-occipital fusion and atlantal condyles which articulate with a malformed axis. Thus the craniovertebral articulation is displaced caudally. Congenital atlanto-occipital fusion, with wedge-shaped vertebral piece and cervical scoliosis, also occurs in horses as a sporadic defect unrelated to the Arabian condition. Subluxations of cervical vertebrae in horses and dogs are discussed with wobbler syndrome.

Temporomandibular subluxation is seen in basset hounds and Irish setters. Clinically, open-mouth locking of jaws occurs when the zygomatic arch interferes with the normal movement of the coronoid process of the mandible. Developmental abnormalities in the condyloid process of the mandible and the mandibular fossa of the temporal bone are responsible for subluxation, but the cause of the dysplasia is unknown. The lesions are bilateral. A similar condition in American cocker spaniels may be bilateral or unilateral. In these animals, temporomandibular subluxation may be elicited by manipulation but is not associated with other clinical signs. Dysplasia of the mandibular fossa and hypoplasia or aplasia of the retroglenoid process are responsible for the excess joint mobility.

Subluxation of the carpus is reported as a sex-linked recessive trait in a colony of dogs in which hemophilia A also existed. Signs appear when the animals begin to walk. The two diseases do not develop in the same animals. Most carpal subluxations and luxations are secondary to trauma.

Patellar luxations and **subluxations** are common in dogs, less so in horses, and rare in other species. In dogs, most are associated with anatomical defects which are probably inherited on a polygenic basis. Females are affected 1.5 times as often as males. The condition may be uni- or bilateral, and luxations of variable severity may occur either medially or laterally and occasionally in both directions. Medial luxations account for about three quarters of the cases and are most common in small dogs, especially Pomeranians, Yorkshire terriers, Chihuahuas, miniature and toy poodles, and Boston terriers. Lateral luxations tend to occur in larger dogs, and some giant breeds, such as Great Dane, St. Bernard, and Irish wolfhound, may be affected. Signs commonly develop by a few months of age but sometimes are detected at birth. Various anatomical defects involving the femorotibial joint, and sometimes the whole limb, are present. Some of them are probably adaptational deformities. Patellar luxation in horses almost always occurs in a lateral direction and is associated with hypoplasia of the lateral ridge of the femoral trochlea. It is sometimes congenital and bilateral in foals, and also in calves. The condition probably is inherited in ponies, and may also be inherited in horses. The effects of luxations are discussed with traumatic injuries to joints.

Abnormal positions of joints in terms of overextension or overflexion occur in the muscular contractures and arthrogryposis; only in the most severe cases are the articular surfaces deformed. Congenital torticollis, whether or not it is associated with a branchiocephalic fold, is associated with articular abnormalities of the cervical vertebrae in sympathy with the concertinalike compression that occurs on the concave side of the deviation.

B. Hip Dysplasia

Dysplasia of the hip occurs in humans and is an important disease of large dogs. It is less a problem in cattle and is rare in other species, although the Norwegian Dole horse seems to be an exception to the rule. Two types of hip dysplasia are recognized in children: a congenital form, associated with laxity of the joint capsule and ligament, which is inherited as a dominant trait, and a form characterized by later onset and polygenic inheritance, which is related to acetabular dysplasia. The canine disease morphologically resembles the latter form, but the marked preponderance of females affected in the human disease is not observed in dogs, in which there is no gender difference. Although usually known as canine hip dysplasia, or congenital hip dysplasia, current theories of causation portray acetabular dysplasia as a more accurate term in dogs also.

The disease in dogs is inherited, and the mode of inheritance is probably polygenic, but estimates of heritability vary between only 0.2 and 0.5; thus hip dysplasia is the result of inheritance modified by environmental factors. The environmental influences are not completely understood, but rapid growth rate appears to be important.

Currently, there are two theories to explain the development of canine hip dysplasia. One states that low pelvic muscle mass is paramount in the development of the disease. The hip joint, probably more than any other joint, depends on muscle action to maintain the relations of articulating surfaces during weight bearing. In breeds of dogs predisposed to dysplasia, and in those affected by it, there is a relatively low pelvic muscle mass in relation to pelvic size. This discrepancy is exaggerated by feeding for maximal growth rate. The other theory states that the basic defect is intrinsic to the hip joint, and there is some suggestion that hip dysplasia is a manifestation of osteochondrosis.

Affected dogs are normal at birth, and definitive radiographic evidence of dysplasia may not appear until after a year of age. However, in severely affected dogs, changes occur much sooner and initial microscopic lesions may be seen as early as 30 days. These consist of edema, capillary hemorrhage, and tearing of a few fibers in the teres ligament, which seems to be largely responsible for congruency of the joint in the first month of life. The earliest radiographic abnormality, which is present by 7 weeks of age, is retarded development of the craniodorsal acetabular rim, and subluxation of the joint is palpable at this time. Marked changes occur between 60 and 90 days, consisting of increased subluxation and a lag in ossification of the acetabular rim. The cartilage on the dorsal surface of the femoral head degenerates where it contacts the acetabular rim during lateral subluxation. Progressive changes develop between 3 and 5 months, including medial deviation of the greater trochanter under the influence of the gluteal muscles. The acetabulum becomes shallow, distorted and wide, and there is drift of the capital epiphysis with remodeling of the femoral neck.

In the severe, early cases, degenerative changes may be present in the cartilages of the joint by 5 to 8 months of age, and thickening and ossification of the joint capsule reduce the range of motion of the joint. Microfractures in the dorsal acetabular rim occur at this time and are probably responsible for pain on rising and after exercise. Advanced degenerative joint disease, characterized by osteophytes at the attachments of the joint capsule, ulceration of cartilage, eburnation of bone on the articulating surfaces, and marked distortion of the acetabulum and the head and neck of the femur, may be present before a year of age (Fig. 1.122A,B).

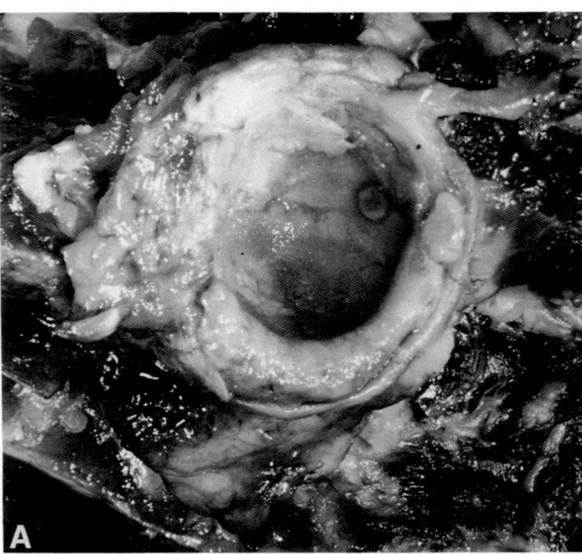

Fig. 1.122A Acetabulum. Congenital hip dysplasia. Dog. Degenerative arthropathy.

Lesions also develop in the soft tissues of the joint. Those involving the capsule are mentioned. Quite early in the course of the disease, proliferation of the synovial membrane, stretching, tearing, and edema of the teres ligament, and increased volume of synovial fluid are present. Later, mononuclear cell infiltration of synovial villi,

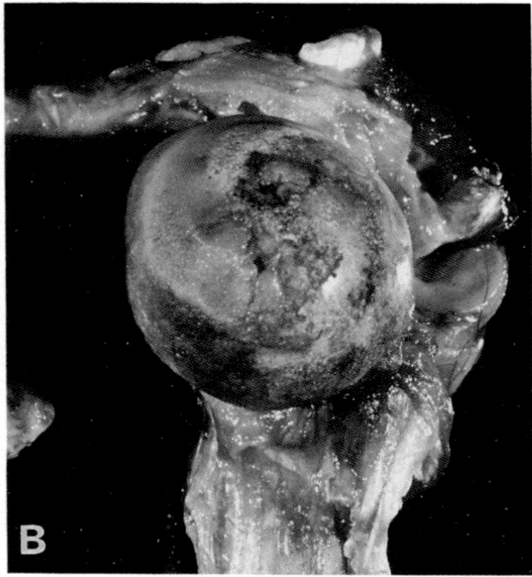

Fig. 1.122B Femoral head. Dog. From (A).

and continued edema and fibroplasia of the ligament are seen. Complete rupture of the teres ligament and luxation of the joint occur not uncommonly.

The essential changes in canine hip dysplasia appear to be those involving the dorsal rim of the acetabulum. Other lesions are secondary to subluxation of the femoral head and the changes which this engenders. Relatively few dogs develop the severe lesions described above by a year of age; indeed many show no radiographic evidence of hip dysplasia until after this time. Hip dysplasia is common in many large and giant breeds, and it is becoming simpler to list the breeds rarely if ever involved, as greyhound and Scotch collie, than those affected; however, the disease is not confined to large dogs, for such breeds as miniature poodle may also be affected.

Hip dysplasia in **cattle** is regarded as an inherited disease also. It is confined almost entirely to males, and the mode of inheritance is thought to be recessive and sex limited. The disease is best known in Herefords but also occurs in the Aberdeen Angus, Galloway, and Charolais breeds. In general, the pathologic changes mimic those in dogs, consisting of primary acetabular dysplasia with secondary degenerative joint disease. Some calves are affected at birth, and many are culled, because of lameness, before a year of age. Rapid growth and mineral deficiencies appear to exacerbate the condition in some cases.

Bibliography

Carnahan, D. L. *et al.* Hip dysplasia in Hereford cattle. *J Am Vet Med Assoc* **152:** 1150–1157, 1968.

Geary, J. C., Oliver, J. E., and Hoerlein, B. F. Atlanto axial subluxation in the canine. *J Small Anim Pract* **8:** 577–582, 1967.

Goddard, M. E., and Mason, T. A. The genetics and early prediction of hip dysplasia. *Aust Vet J* **58:** 1–4, 1982.

Grondalen, J. Malformation of the elbow joint in an Afghan hound litter. *J Small Anim Pract* **14:** 83–89, 1973.

Haakenstad, L. H. Chronic bone and joint diseases in relation to conformation in the horse. *Equine Vet J* **1:** 248–260, 1969.

Hedhammar, A. *et al.* Canine hip dysplasia: Study of heritability in 401 litters of German shepherd dogs. *J Am Vet Med Assoc* **174:** 1012–1016, 1979.

Hoppe, F., and Svalastoga, E. Temporomandibular dysplasia in American cocker spaniels. *J Small Anim Pract* **21:** 675–678, 1980.

Howlett, C. R. Pathology of coxofemoral arthropathy in young beef bulls. *Pathology* **5:** 135–144, 1973.

Hutt, F. B. Genetic defects of bones and joints in domestic animals. *Cornell Vet* **58:** 104–113, 1968.

Johnson, K. A. Temporomandibular joint dysplasia in an Irish setter. *J Small Anim Pract* **20:** 209–218, 1979.

Ladds, P. *et al.* Congenital odontoid process separation in two dogs. *J Small Anim Pract* **12:** 463–471, 1970.

Lust, G., and Farrell, P. W. Hip dysplasia in dogs: The interplay of genotype and environment. *Cornell Vet* **67:** 447–466, 1977.

Lust, G., and Summers, B. A. Early, asymptomatic stage of degenerative joint disease in canine hip joints. *Am J Vet Res* **42:** 1849–1855, 1981.

Martin, S. W., Kirby, K., and Pennock, P. W. Canine hip dysplasia: Breed effects. *Can Vet J* **21:** 293–296, 1980.

Mayhew, I. G., Watson, A. G., and Heissan, J. A. Congenital occipito-atlantoaxial malformations in the horse. *Equine Vet J* **10:** 103–113, 1978.

Meagher, D. M. Bilateral patellar luxation in calves. *Can Vet J* **15:** 201–202, 1974.

Pick, J. R. *et al.* Subluxation of the carpus in dogs. An X chromosomal defect closely linked with the locus for hemophilia A. *Lab Invest* **17:** 243–248, 1967.

Priester, W. A. Sex, size, and breed as risk factors in canine patellar dislocation. *J Am Vet Med Assoc* **160:** 740–742, 1972.

Riser, W. H. The dog as a model for the study of hip dysplasia. *Vet Pathol* **12:** 229–334, 1975.

Robinson, W. F., *et al.* Atlanto-axial malarticulation in Angora goats. *Aust Vet J* **58:** 105–107, 1982.

Rudy, R. L. Patellar luxation. *In* "Canine Surgery," 2nd Ed., J. Archibald (ed.), pp. 1117–1142. Santa Barbara, California, American Veterinary Publications, 1974.

Weaver, A. D. Hip dysplasia in beef cattle. *Vet Rec* **102:** 54–55, 1978.

White, M. E., Pennock, P. W., and Seiler, R. J. Atlanto-axial subluxation in five young cattle. *Can Vet J* **19:** 79–82, 1978.

III. Degenerative Diseases

A. Synovial Joints

Degenerative articular diseases are common disorders that affect mainly and most importantly the articulations of the limbs. They are known by such terms as degenerative joint disease, degenerative arthritis, osteoarthrosis, and osteoarthritis, but we prefer the term arthropathy that emphasizes the primary degenerative nature of the disorder and distinguishes the condition from arthritis with its primary inflammatory nature. In most arthropathies the initial disease process is not centered in the interstitium and capillary bed of the synovium and is not being driven from its outset by mediators of inflammation. Joint fluid of these joints is initially normal, and the joint disease is not related to a systemic or local soft-tissue disease process. However, chronicity of the particular form of arthropathy leads to progressive deterioration of joint surfaces and the accumulation of degenerative matrix elements and chondrocyte breakdown products within the joint cavity. Joint debris entrapped in the synovium has the capacity to provoke a low-grade synovitis so that synovial inflammation is the result rather than the initiator of the structural changes in the joint. Nevertheless, because of the limited range of reactions inherent in articular and synovial tissues, many of the gross morphological changes seen in chronic arthropathy may also occur in chronic arthritis, and in the end-stage joint, the differentiation between a primary inflammatory disease and degenerative joint disease may be impossible.

Degenerative joint diseases are classified as primary or secondary. Secondary arthropathies result from known diseases of nutritional, traumatic, developmental, or other cause. Virtually any insult that produces structural injury to joint cartilage or subchondral bone, or that is associated with prolonged or repetitive abnormal joint function, will ultimately produce degenerative joint disease.

Primary arthropathies are, by definition, of unknown cause, but the genesis of lesions is well documented, and generally occurs in a predictable sequence as outlined earlier. Briefly, articular plates soften, fibrillate, and ulcerate, and subchondral bone is eburnated. Osteophytes develop at joint margins, and remodeling of joint structures occurs (Fig. 1.123A,B).

Degenerative changes occur with age in articular cartilages and are so common as to be regarded as normal. Usually the aging of joints is paralleled by general processes of aging, and progresses slowly enough to allow some functional compensation. This is not degenerative joint disease. When the degeneration begins early in life, or progresses rapidly and reaches a degree which results in pain and locomotor disturbances, it is regarded as pathologic. Pain is related to involvement of synovial membrane and subarticular bone. The arbitrary division between aging changes and degenerative arthropathy is reaction by subarticular bone. Primary degenerative arthropathy is an exaggeration of the same degenerative process which occurs with aging, and it is probably, in large measure, due to the same causes—repeated trauma and nonphysiologic stresses.

Two basic factors influence the development of degenerative arthropathy. These are the inability of mature articular cartilage to repair itself effectively and the constant mechanical stresses to which it is exposed during normal use. Any increased susceptibility to injury, be it either a primary, perhaps inherent weakness, or a secondary change, increases the possibility of arthropathy.

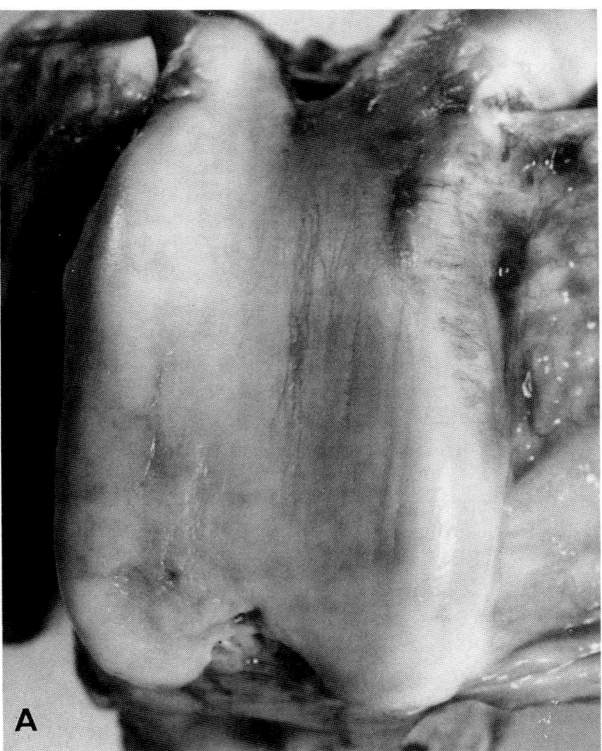

Fig. 1.123A Degenerative arthropathy. Pig. Linear grooving of articular cartilage.

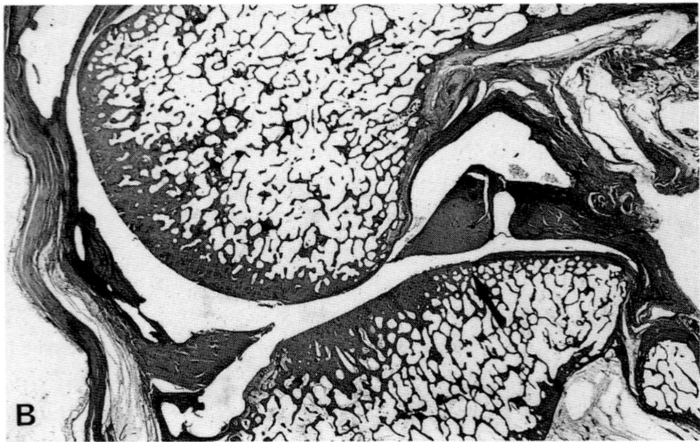

Fig. 1.123B Degenerative arthropathy in femorotibial joint. Dog. Note sclerosis of bone, degeneration of cartilage (arrow), and shredding of menisci. (A.F.I.P. 737597. Contributor, P. O'Connor.)

Many cases of arthropathy are attributable to antecedent diseases, but some are not. In these, the primary arthropathies, it seems possible that there is a metabolic defect in articular cartilage, and that some joints are more susceptible than others by virtue of their particular mechanical properties. Because the first change observed in degenerating cartilage involves the proteoglycans, much attention is directed toward the relationship of chondrocytes and matrix, and particularly to the proteolytic lysosomal enzymes, which are capable of splitting the proteoglycan molecules. These enzymes are present in chondrocytes and may be increased in degenerative cartilage. Whether the initial degenerative change is in collagen or proteoglycan, the early products of degeneration escape into the joint fluid and are taken up by the macrophages of the synovium. It is proposed, largely on the basis of *in vitro* studies, that cytokines such as tumor necrosis factor and interleukin-1, released from activated synovial macrophages, induce production of collagenases by chondrocytes together with altered proteoglycans, such as dextran sulfate, which have a limited capacity to aggregate.

In both primary and secondary arthropathies, the extent and distribution of the macroscopic changes are very variable, but the larger joints of the limbs, which are subjected to the greatest degree of weight bearing and movement, are usually the first to show degenerative changes and are most severely affected. In individual joints, those areas of cartilage which bear the most stress are the most vulnerable.

Just as bone which is exposed by fissures in the cartilage responds with the formation of callus, that which is exposed by large or small areas of ulceration proliferates. Owing to the continuous pressure and movement, the exposed bone remains flattened and is characteristically dense and polished, and in some instances, grooved in the direction of motion.

Certain variations occur, depending on the cause and location of lesions. When arthropathy is secondary to an osteodystrophy, there may be collapse rather than eburnation of subchondral bone. Also, in joints such as distal carpus and tarsus, which have restricted motion, ulceration of apposing articular plates may lead to fibrous or bony ankylosis without significant reaction at the joint margins. Usually, however, advanced degenerative changes in articular plates are accompanied by thickening of joint capsules and proliferation of synovial cells, with production of villi, cartilage nodules, and osteophytes. These proliferative changes and osteophytes probably develop because of abnormal movement of the joint with repeated trauma to the capsular attachment.

Degenerative changes in the menisci of the stifle joint are comparable to those in the cartilages. Fibrillation occurs, the collagenous fibers clump together, and the matrix diminishes in amount. Clefts develop near the insertions and on the concave border and run longitudinally, and small strips are progressively shredded off. In severe degeneration, the meniscus may be destroyed except for the outer portions, which are attached to the joint capsule. The larger medial meniscus usually shows more advanced changes than the lateral one.

Degenerative arthropathy, primary and secondary, is common in most domestic species, with the possible exceptions of sheep and cats. Certain types are either sufficiently well known or sufficiently important to merit individual consideration.

1. Ringbone, Spavin, and Navicular Disease

High ringbone, bone spavin, and navicular disease are three common disorders of horses that appear to arise because the musculoskeletal structures at the sites of these disorders share similar, anatomical, and biomechanical relationships which predispose these structures to undergo basically similar pathologic changes resulting in clinical disease. All three disorders develop at sites of high or sustained biomechanical loading and follow a reduction in the already limited motion characteristic of the location. High ringbone arises in the proximal interphalangeal joints, especially of forelimbs; bone spavin involves primarily the intertarsal joints of hind limbs; and navicular disease is centered in the navicular bones of the forelimbs.

Common pathologic findings in the three disorders include full-thickness necrosis of the cartilage covering bone surfaces at sites of sustained compression, remodeling of the subchondral bone underlying the site, and penetration of the necrotic cartilage by granulation tissue arising from the remodeling response in the subchondral bone. In advanced disease the granulation tissue produces fibrous or bony union with the apposing surface. In chronic cases of high ringbone and bone spavin, there is fibrous or bony ankylosis of the compressed regions of the apposing joint surfaces, while in navicular disease, there is fibrous ankylosis of the flexor cortex of the navicular bone with the dorsal surface of the deep digital flexor tendon.

If joint surfaces are fixed and set under sustained compression, full-thickness necrosis of the articular cartilage occurs within a few days. Cartilage necrosis is due to interference with its nutrition and metabolism; however, the subchondral bone beneath the site of sustained compression responds by activation of the bone-remodeling mechanism in an attempt to restructure the bony elements in order to dissipate the forces of compression. The sequence of tissue responses occurring in these three disorders is the same one that is responsible for primary bone healing in which a fracture line is immobilized and set under compression by use of a dynamic compression plate. Compression activates bone-remodeling activity on both sides of the fracture line, and healing occurs by the action of osteogenic granulation tissue, which arises from bone-remodeling sites and bridges the fracture line.

In the case of these three disorders, the necrotic cartilage is penetrated by the granulation tissue as it attempts to bridge and connect the apposing surfaces. In the cases of the two joint disorders, pain on movement of the joint, thickening of the periarticular tissue, and abnormal conformation ensure joint immobilization, while weight bearing furnishes the axial compression that is required. In navicular disease, faulty conformation is thought to provide the forces of compression and immobilization that initiate the cortical bone-remodeling mechanism which, in turn, gives rise to the tendon adhesions.

Ringbone is a degenerative arthropathy of the interphalangeal joints of ungulates, especially horses. The terms *high* and *low* refer to lesions centered respectively in the proximal or distal interphalangeal joints. This disorder typically has bilateral involvement and affects chiefly the forelimbs. Repeated episodes of minor trauma from athletic activity and mechanical stresses from faulty conformation placing strain on the insertion lines of the dorsal joint capsule are thought to be initiating factors. There is variation between affected individuals with respect to the relative severity of the articular and periarticular lesions. Degeneration and erosion of the articular surface may lead to fibrous or bony ankylosis, especially of the proximal interphalangeal joint. The periarticular response, which includes fibrous thickening of the dorsal joint capsule and bone formation beginning in the joint capsule insertion line, is a much more prominent feature than the cartilaginous changes. The periarticular bony response on the dorsal surface of the joint gives the lesion its name, ringbone.

High ringbone is found most often in adult horses used for western events or polo, where horses running at high speed make abrupt stops, turns, and twisting motions. Examination of necropsy specimens representing early examples of this disorder indicate that the joints have been immobilized by fibrous thickening of the dorsal joint capsule. There is only early cartilage degeneration of one or both condyles of the distal first phalanx and the apposing glenoid cavity of the proximal second phalanx. Joint fluid is normal, and the synovium is not congested. More advanced lesions show full-thickness necrosis of the cartilage followed by erosions in the subchondral bony plate and ankylosis. The lesion is modified if there is residual joint motion. In such specimens movement induces eburnation of the eroded joint surfaces, thickening of the subchondral bony plates, and inhibition of ankylosis.

High ringbone sometimes occurs in immature horses that have not performed any work. Prior to the development of the more characteristic radiographic changes of ringbone, the joint lesions have been referred to as juvenile arthrosis. While osteochondrosis and trauma have been suggested as the initiating causes, the examination of necropsy specimens representing early to late stages of this disorder indicates that the course of pathologic changes in young horses follows those in adults in which joint immobility due to fibrosis of the joint capsule precedes onset of lesions in the articular cartilage. Many of these young horses have faulty conformation due to a slightly backward broken axis of the proximal interphalangeal joint, which predisposes the insertion lines of the dorsal joint capsule to excessive strain.

Bone spavin is an arthropathy of the tarsus of the horse and occasionally the ox. Structural changes in this disorder are essentially the same as those occurring in high ringbone and presumably have a similar pathogenesis, which is initiated by sustained loading and joint immobilization. The major lesions develop on the medial side of the tarsus, primarily involving the distal intertarsal joint and less commonly the tarsometatarsal and proximal intertarsal joints. Examination of early cases of disease centered in the distal intertarsal joint indicates that periarticular fibrosis initially immobilizes the dorsomedial side of the joint and that "kissing lesions" subsequently develop in articular cartilage adjacent to the dorsomedial margins of the apposing articular surfaces of the central and third tarsal bones. Early lesions show full-thickness necrosis of the apposed cartilage surfaces and intense bone remodeling within the underlying thickened subchondral bony plate. Intermediate lesions show penetration of the necrotic cartilage by granulation tissue from the areas of intense bone remodeling in the subchondral plate. Late lesions are characterized by fibrous and bony union at sites of the initial cartilaginous lesions, loss of the normal width and density of the subchondral bony plate, bone formation in the periarticular fibrous tissue that often bridges the joint surfaces, and an extension of the degenerative processes across the central surfaces of the distal intertarsal joint as loss of joint space and fusion of the central and third tarsal bones progresses across the joint surface.

Navicular disease is a biomechanically induced, degenerative disease involving the distal half of the flexor surface of the navicular bones of the forelegs of mature horses. This disorder may be initiated when conformational malalignment of bones and attachments in the foot is sufficient to permit the deep digital flexor tendon to activate bone remodeling in the subchondral bony plate of the distal border of the flexor surface of the navicular bone. This site has the same biomechanical characteristics that are present in high ringbone and in bone spavin, i.e., sites of high load : low motion.

Horses having abnormal foot conformation, i.e., bone and joint alignments at either end of the normal range of conformation for the breed of horse, are predisposed to develop navicular disease. At one end of this spectrum are racing thoroughbred horses having very long toes and low heels. In such horses the deep digital flexor tendon places sustained pressure against the distal border of the flexor surface of the navicular bone. At the other end of the spectrum are those American quarter horses with an extremely upright conformation of the foot. The latter conformational abnormality causes the deep digital flexor tendon to exert repetitive concussive forces on the distal border of the flexor surface of the navicular bone during exercise. Either sustained pressure or repetitive concussive forces can activate the bone-remodeling processes in the subchondral bone of the navicular bone in an attempt to adapt the flexor cortex to these biomechanical stresses. If pressure or concussive forces are of low intensity and of limited duration, the subchondral bone and spongiosa thicken enough to absorb, diffuse, and redistribute the load. Such navicular bones are sclerotic, but the feet are asymptomatic. The fibrocartilage of the flexor surface of these horses often shows degenerative changes associated with increased loading.

When the sustained load or concussive forces acting on the distal border of the flexor surface of the navicular bone exceeds the physiologic limit, an exaggerated bone-remodeling response is initiated, which eventually produces lytic lesions in the subchondral bone of the flexor cortex of the navicular bone. As also found in high ringbone and bone spavin, granulation tissue arises within sites of intense bone remodeling in the sclerotic subchondral bone of the flexor cortex of navicular bone affected with navicular disease. The granulation tissue eventually penetrates the overlying layer of degenerative fibrocartilage covering the flexor surface. Unlike the lesions in high ringbone and bone spavin in which fibrous adhesions form between lesions of apposing joint surfaces, the granulation tissue in navicular disease forms adhesions to the apposing dorsal surface of the deep digital flexor tendon.

Horses with navicular disease have reduced venous drainage and venous hypertension in the medullary cavity of the navicular bone. Bone marrow fibrosis and distended veins can be demonstrated in the marrow spaces of the subchondral bone and spongiosa located in the distal halves of the navicular bones of horses with navicular disease. Humans having resting pain in joints affected with chronic degenerative arthropathy, i.e., osteoarthritis, also have dilated veins and venous hypertension in the subchondral spongiosa of those joints. Decompression of the cancellous bone in human patients having degenerative arthropathy results in immediate relief of pain, indicating that the pain is due to distended venous vessels. Presumably, this is also true concerning the pain in navicular disease.

In navicular disease it appears that vessels participating

in the intense bone-remodeling response leak edema fluid into the marrow spaces of the subchondral spongiosa supporting the flexor cortex. As in degenerative arthropathy, this fluid is organized by fibrous tissue which forms at the drainage angle lying in the medullary cavity of the distal border of the navicular bone. Here vessels are entrapped in the fibrous tissue, retarding venous drainage. This results in venous hypertension, dilated vessels, and pain.

The enlargement of the so-called vascular channels and "lollipop" formation seen in radiographs of horses with navicular disease result from the active hyperemia that supports the active bone-remodeling response. As the tiny nutrient arteries enter along the floor of the synovial fossae located on the distal border of the navicular bone, each main artery sends off small branches to the synovial lining of each of the respective synovial fossae. As an unfortunate consequence of active hyperemia associated with the remodeling process, the synovial lining of the fossae also undergoes active hyperemia. The hyperemic synovium attracts and activates osteoclasts. The osteoclasts enlarge the synovial fossae and eventually erode through the back walls of the fossae carrying the synovial lining with them. The resorptive process follows the pathway of the nutrient artery into the medullary cavity of the distal border of the navicular bone, creating deep synovial invaginations recognized on radiographs as enlarged vascular channels. Synovial-lined chambers sometimes occur at the ends of the synovial invaginations, creating the radiographic appearance of lollipops.

Ischemic necrosis of the navicular bone is reported from Europe and is associated with occlusion of arterioles entering the sesamoid bone from the distal suspensory ligament of the navicular bone. Anastomoses between intraosseous arterioles develop, and these too may become occluded. Clinical signs which are thought to be caused by bone necrosis, apparently do not occur before at least two arteries are involved, and in some horses, extensive occlusions do not cause navicular disease. Thus it appears that the rapidity of occlusion and development of anastomoses influence the occurrence of clinical disease.

The vascular changes in the navicular bone are part of a more widespread condition, since partial or complete occlusions of the digital arteries are often present in horses with navicular disease. Partial occlusions are usually caused by proliferation of subendothelial tissue, whereas organization of thrombi accounts for the complete occlusions. If occlusive lesions are sufficient to prevent collateral circulation between the digital arteries, or if they involve the entire lumen, then there is a decreased blood flow through the feet, and navicular disease may develop. Ischemia may be temporary, in which case it is followed by diffuse activation of osteoblasts on trabecular surfaces.

Hematogenous bacteria occasionally localize in the navicular structures of foals and cause an acute inflammation.

The relative significance of these lesions of navicular structures is not known. There are marked differences of fact and/or opinion between various parts of the world.

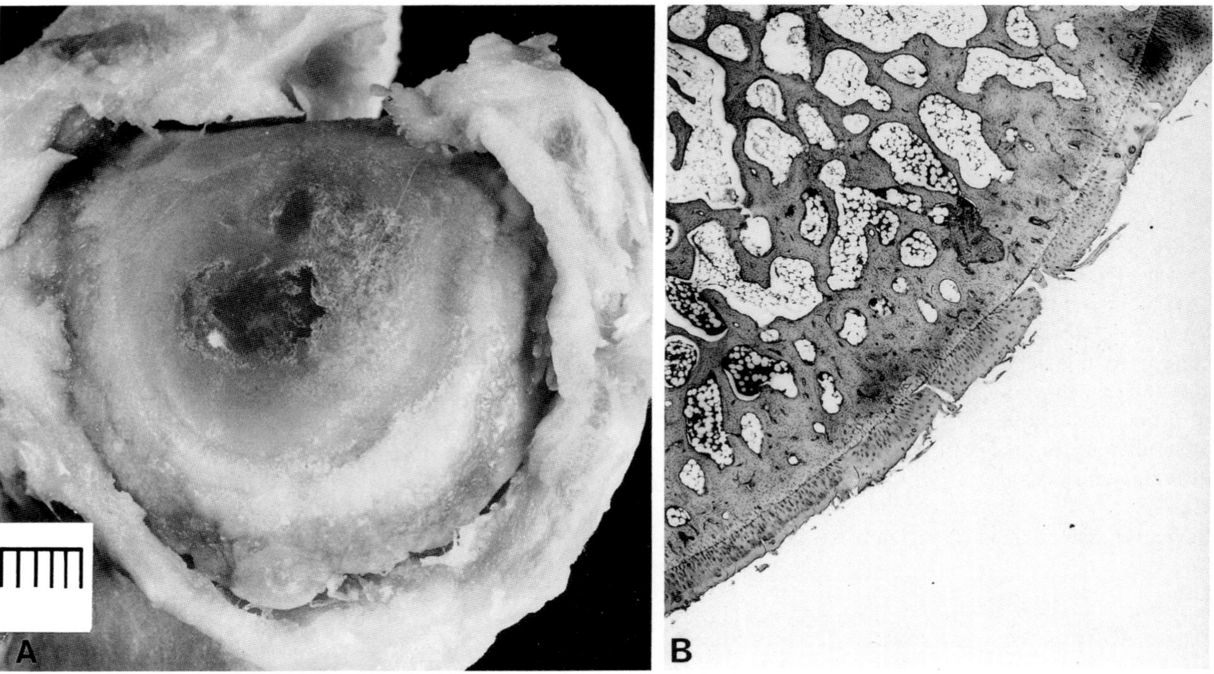

Fig. 1.124 Arthropathy of canine shoulder. (A) Thickened joint capsule, degeneration and fraying of cartilage and eburnation of subchondral bone. (Courtesy of D. C. Van Sickle.) (B) Degeneration, fraying, and loss of articular cartilage. (Courtesy of D. C. Van Sickle.)

The prevalence of vascular-based lesions in particular is unclear; some bones definitely show evidence of ischemia. The use of comprehensive vascular perfusion studies and critical microscopic examination in early cases of navicular disease is required.

2. Arthropathy of the Canine Shoulder

Arthropathy of the shoulder occurs quite commonly in middle-aged and old dogs. As many as 80% of animals older than 8 years of age are affected. Lesions are bilateral and develop slowly. They start at 5 to 6 years of age in the posterior central articular area of the humeral head as small blisters, and progress to severe cartilage ulceration, eburnation of subchondral bone, thickening of the joint capsule, and villous proliferation of synovium (Fig. 1.124A,B). Osteophytes develop along the posterior ventral rims of the articular surfaces and eventually encircle the margins of the joints. They also develop in the bicipital grooves. Changes in the glenoid of the scapula are minimal.

Involvement of the shoulder joint may be related to joint laxity. Like the hip joint, the shoulder is stabilized normally by muscular action so that atrophy of muscle with aging could permit excessive joint mobility and persistent mild trauma to the articular plate. The fact that the dog bears 65–70% of its body weight on its forelegs may account for the development of lesions in shoulders rather than hips.

Lesions in this disease develop in a similar location to those of osteochondritis dissecans. That they are not a sequel to this condition, which develops in immature dogs, is indicated by the virtual absence of early degenerative changes prior to 4 to 5 years of age, and by the histologic differences between the two diseases. In canine arthropathy the lesions are conventionally degenerative; those of osteochondritis dissecans are discussed with osteochondrosis.

3. Arthropathy of the Bovine Stifle

Arthropathy of the stifle occurs in dairy cows and is reported as an inherited trait in Holsteins and Jerseys. Signs of lameness and muscle atrophy occur in mature cows. Lesions are bilateral, and seem to develop in the conventional manner. Cartilage degeneration, bone eburnation, and osteophyte development occur on the distal femur and are most severe on the medial aspect (Fig. 1.125A,B). Complementary lesions are present in the proximal tibia, and shredding of the medial meniscus is common.

Stifle arthropathy also occurs in stud bulls. It may be secondary to poor conformation or ruptured meniscus during fighting.

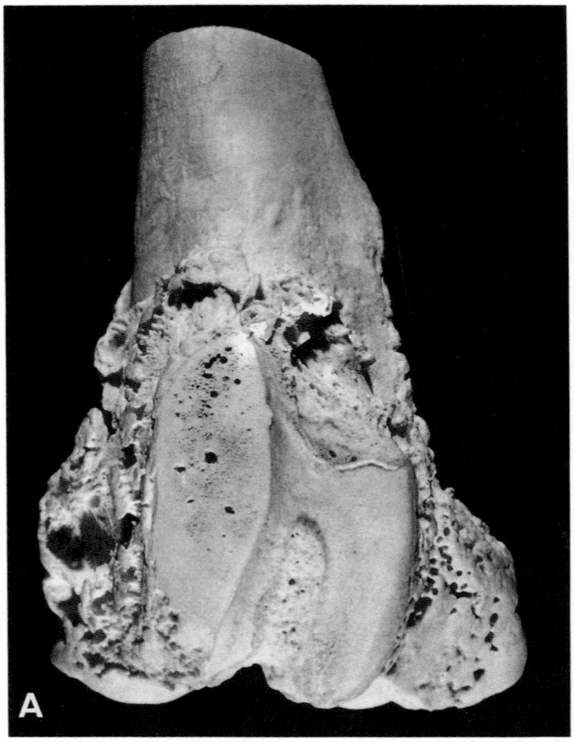

Fig. 1.125A Degenerative arthropathy. Cow. Macerated specimen. Degeneration, eburnation, and cyst formation in medial trochlear ridge. Remodeling and new bone formation on condyles.

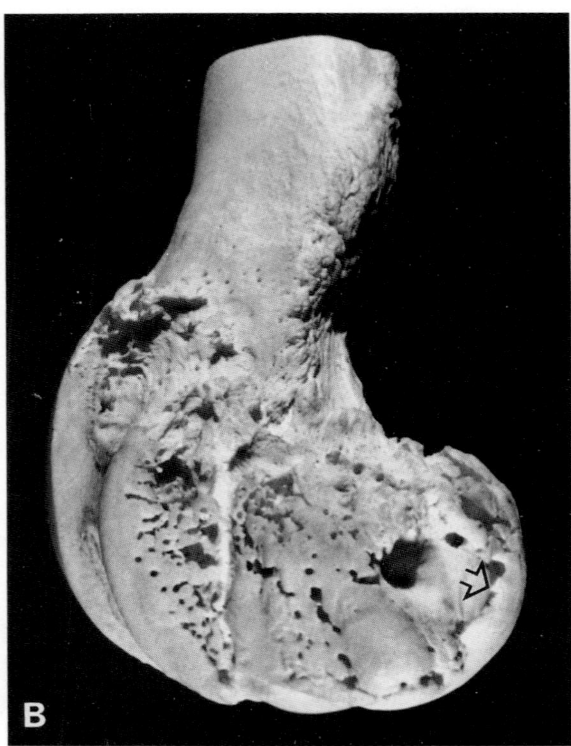

Fig. 1.125B Degenerative arthropathy. Cow. Lateral aspect of (A). Note subchondral resorption (arrow).

Bibliography

Alexander, J. W. Pathogenesis and biochemical aspects of degenerative joint disease. *Compend Cont Ed* **2:** 961–966, 1980.

Brookes, M., and Helal, B. Primary osteoarthritis, venous engorgement and osteogenesis. *J Bone Joint Surg (Br)* **50B:** 493–504, 1968.

Freeman, M. A. R. The fatigue of cartilage in the pathogenesis of osteoarthrosis. *Acta Orthop Scand* **46:** 323–328, 1975.

Hulth, A. Experimental osteoarthritis. A survey. *Acta Orthop Scand* **53:** 1–6, 1982.

Kendrick, J. W., and Sittmann, K. Inherited osteoarthritis of dairy cattle. *J Am Vet Med Assoc* **149:** 17–21, 1966.

Ljunggren, G., and Olsson, S.-E. Osteoarthrosis of the shoulder and elbow joints in dogs: A pathologic and radiographic study of a necropsy material. *J Am Vet Radiol Soc* **16:** 33–38, 1975.

Marshall, J. L. Periarticular osteophytes. Initiation and formation in the knee of the dog. *Clin Orthop* **62:** 37–47, 1969.

McDevitt, C., Gilbertson, E., and Muir, H. An experimental model of osteoarthritis; early morphological and biochemical changes. *J Bone Joint Surg (Br)* **59B:** 24–35, 1977.

Pool, R. R., Meagher, D. M., and Stover, S. M. Pathophysiology of navicular syndrome. *Vet Clin North Am (Equine Pract)* **5:** 109–129, 1989.

Shupe, J. L. Degenerative arthritis in the bovine. *Lab Invest* **8:** 1190–1196, 1959.

Stephens, R. W., Ghosh, P., and Taylor, T. K. F. The pathogenesis of osteo- arthrosis. *Med Hypotheses* **5:** 809–816, 1979.

Tirgari, M., and Vaughan, L. C. Clinicopathological aspects of osteo-arthritis of the shoulder in dogs. *J Small Anim Pract* **14:** 353–356, 1973.

Vaughan, L. C. Osteoarthritis in cattle. *Vet Rec* **72:** 534–538, 1960.

B. Cartilaginous Joints

1. Diseases Involving the Intervertebral Disks

The vertebral bodies, with the exception of the first two cervicals, are united by intervertebral disks. The disks, in association with the articular facets, permit some movement between the vertebrae, and the central portions, which are hydrophilic and normally turgid, are useful shock absorbers. A cross-section of a normal disk reveals that its center is composed of a tough jellylike mass, the nucleus pulposus, which is a remnant of the primitive notochord. The nucleus is surrounded in the transverse plane by the annulus fibrosus, an eccentrically lamellated fibrous ring, which is much broader ventrally than dorsally. The disks are attached to the ends of vertebrae in the same manner as articular cartilages are attached to epiphyses. The ventral longitudinal ligament fuses with the annulus fibrosus of each disk as it passes. The dorsal spinal ligament, lying in the vertebral canal, fuses with the dorsal part of the annulus fibrosus, except in the thoracic region between the second and tenth ribs; in this region the conjugal ligaments, which connect the heads of corresponding ribs, cross the floor of the canal between the annulus fibrosus and the dorsal longitudinal ligament. The conjugal ligaments fuse with and reinforce the dorsal portion of the annulus fibrosus. These anatomical relations have a considerable bearing on the consequences of degeneration of the disks.

Degeneration of the intervertebral disks advances with age so that there can be no precise distinction between a normal aging change and pathologic change. In cats, macroscopic signs of degeneration occur when the animal is almost senile. In dogs, there are breed differences in both the rate and the nature of the degeneration, and individual differences in the rate.

In the chondrodystrophoid breeds, exemplified by the dachshund and Pekingese, the mucinous nucleus pulposus degenerates first, and by about 1 year of age is beginning to be replaced by a cartilaginous type of tissue. This transformation begins at the periphery of the nucleus, is progressive, and is associated with a decline in glycosaminoglycan and water content and an increase in collagen. Beginning at the periphery, the chondroid tissue, which has replaced the normal nucleus, degenerates and mineralizes and becomes a crumbly mass. This change increases the likelihood that protrusion will occur; while the nucleus is a gel, it transmits pressures evenly to the annulus fibrosus; when degenerate, the nucleus no longer behaves physically as a fluid, and high pressures may then be transmitted to localized portions of the annulus. More or less simultaneous with the degeneration of the chondroid nucleus, there is degeneration of the inner lamellae of the annulus fibrosus, and this too is progressive until the outer lamellae are degenerate. The degenerative changes in the annulus fibrosus are the same as those in nonchondrodystrophoid breeds. In chondrodystrophoid breeds, prolapses are generally caused by complete rupture of the annulus fibrosus and are usually massive.

In nonchondrodystrophoid breeds, the initial degenerative changes occur later in life and consist of fissures in the annulus fibrosus, which stimulate vascularization. At this stage, the nucleus pulposus is normal. Degenerative changes in the annulus progress from within outward, independent of the nucleus, and the annulus is converted in foci to a structureless, granular mass with unmasked lamellae, which are thick and prominent and progressively frayed and disrupted. After middle age, the nucleus is gradually toughened by concentric deposition of collagen fibers. This fibrous transformation is regarded as a maturation change on which degeneration may be superimposed. The degeneration takes the form of necrosis in which dystrophic mineralization seldom occurs. Prolapses associated with this form of degeneration are associated with partial rupture of the annulus and bulging of the dorsal surface of the disk.

2. Dorsal Protrusion of Intervertebral Disks

The usual syndrome associated with the intervertebral disks involves dorsal protrusion of degenerate disk material in chondrodystrophoid dogs (Fig. 1.126). Prolapsed disks are the most common cause of spinal pain, paresis, and paraplegia in dogs. Affected animals are usually males aged 3–6 years. However, other syndromes do occur, including ventral herniation, embolism of disk material

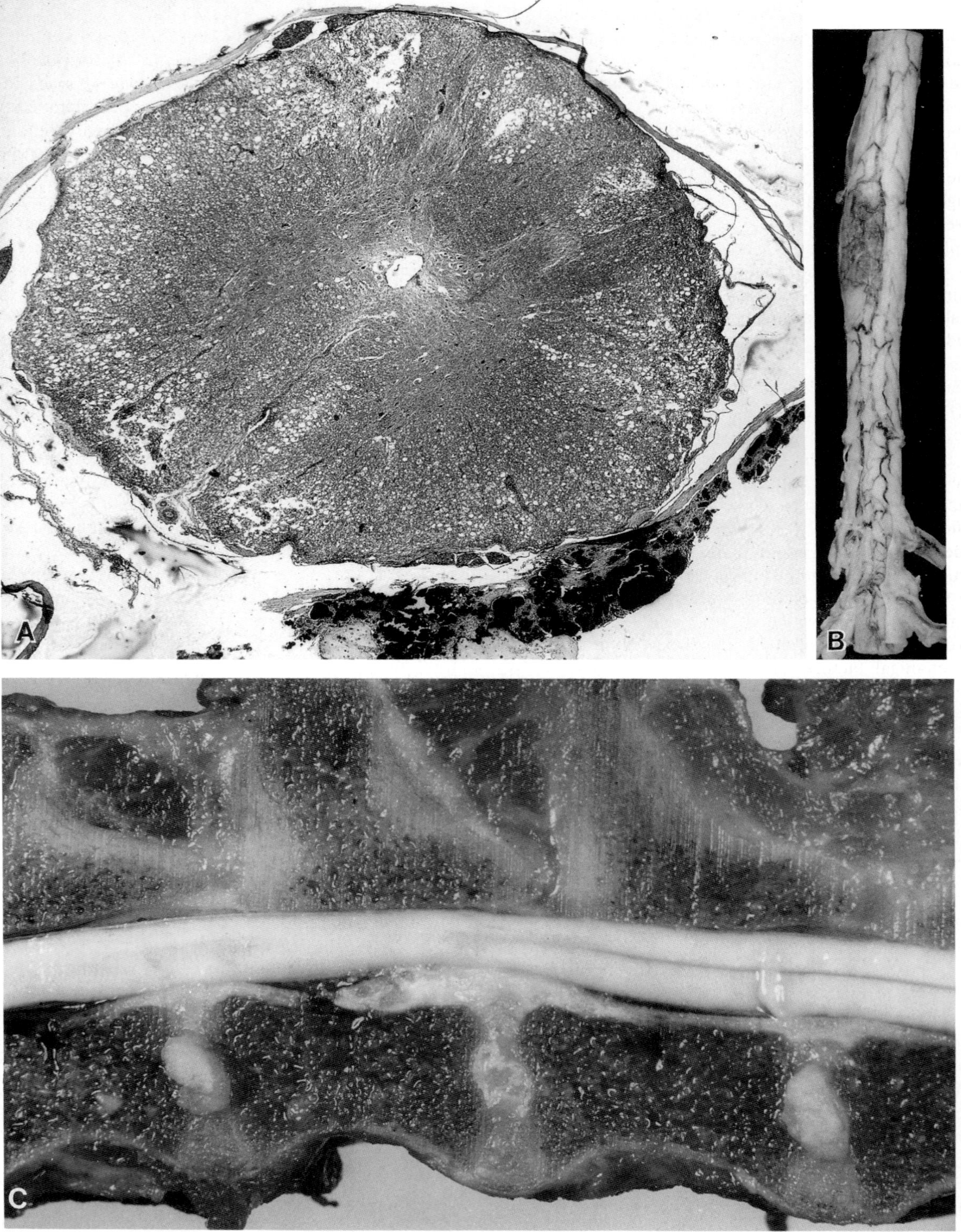

Fig. 1.126 Degeneration and herniation of intervertebral disks. Dog. (A) Extruded nucleus pulposus on dura mater with compression and degeneration of spinal cord. (B) Hemorrhage in cervical cord. (C) Degeneration of disks with dorsal extrusion of mineralized nucleus.

to the spinal cord, "disk explosions" associated with trauma, and, very rarely, cranial or caudal displacement of disk material into the end of a vertebra to produce a Schmorl's node.

The separation of chondrodystrophoid and nonchondrodystrophoid breeds according to susceptibility and type of disk degeneration is well recognized, but the separation according to disk-related disease is not so distinct. Beagles and cocker spaniels rank with dachshunds and Pekingese on the list of dogs with a very high incidence of disk-related disease. Some beagles have certain chondrodystrophoid characteristics, but there are genetic differences within the breed. Whether cocker spaniels have chondrodystrophoid characteristics, or not, is unclear. Dachshunds are most commonly affected with disk-related disease, being 10–12 times more prone than the canine population at large. Within the breed there is marked variation, and as many as 60% of some families are affected, compared to around 20% for the breed as a whole. The incidence of disk degeneration approaches 100% in these dogs. The pattern of disease in dachshunds is consistent with inheritance involving several genes, with no dominance or sex linkage, and subject to environmental modification.

A degenerate disk is predisposed to displacement of the nucleus pulposus into and through the annulus fibrosus. Since the dorsal part of the annulus is thinner than the ventral, the great majority of displacements occur dorsally toward or into the spinal canal. The displacement is caused by mechanical factors and may occur suddenly or by slower permeation through the annulus fibrosus. Because the thoracic spine is less mobile than other parts of the vertebral column, and because the dorsal part of the annulus fibrosus is reinforced in that region by the conjugal ligaments, herniation or protrusion of the nucleus pulposus is seldom observed in the thoracic area. Most disk disease involves the thoracolumbar and cervical regions and, in most breeds, is three to four times more common in the former than the latter. In beagles, cervical protrusions are more frequent, and in cocker spaniels the incidence is about equal in the two locations.

Dorsal herniations may, depending on direction and degree, cause injury to the roots of the peripheral nerves as they pass through the intervertebral foraminae, injury to the dura mater, and injury to the spinal cord with permanent or temporary functional deficit.

Dorsal protrusions may be complete or incomplete. If the protrusion is incomplete, the displaced nucleus is either covered by a narrow outer strip of the annulus fibrosus, or the annulus is completely ruptured, but the pulp still is contained by the dorsal ligament, which is displaced up (Fig. 1.127A). In either event, the displaced tissues bulge onto the floor of the vertebral canal. This type of protrusion occurs in nonchondrodystrophoid breeds.

A complete protrusion may pass directly up through the annulus fibrosus and dorsal longitudinal ligament, or it may pass obliquely and around the edge of the dorsal ligament. The dorsal protrusions are readily visible and palpable when the arches and spinal cord are removed. A straight pathway of protrusion may be evident when the disk is sliced transversely, but usually the pathway is slightly oblique or tortuous and detectable only by serially shaving the disks (Fig. 1.127B).

Most dogs with dorsal disk protrusions show evidence of localized spinal cord injury; some develop an extensive myelopathy (the ascending syndrome), and others are without signs. In the latter cases, the rate at which compression occurs is probably more important than the degree of compression in influencing the occurrence of neurologic signs.

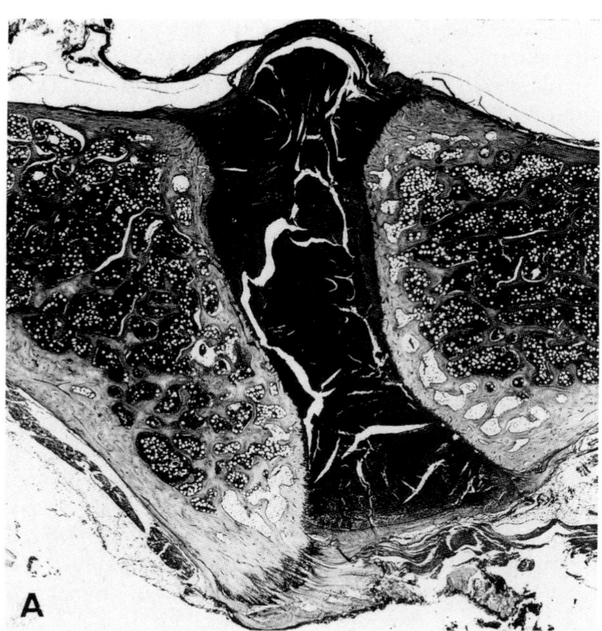

Fig. 1.127A Incomplete dorsal protrusion of nucleus pulposus. Dog.

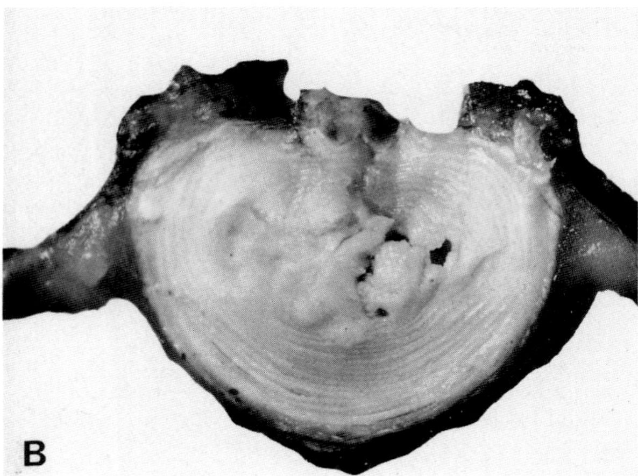

Fig. 1.127B Degenerate nucleus pulposus and ruptured annulus fibrosus. Dog.

The ascending syndrome is almost always associated with sudden complete protrusions of disk material, which extends widely in the epidural space to produce marked epidural and intradural hemorrhage. Extensive ischemic necrosis of the spinal cord occurs, probably secondary to vasospasm caused by subarachnoid hemorrhage, and to thrombosis of cord vessels.

The more usual syndrome is caused by a localized myelopathy involving two to four cord segments. Signs are related to myelomalacia due to direct compression of the cord by the disk material and to focal malacia or diffuse demyelination associated with vascular lesions. The malacic foci, which are often not related anatomically to the protruded disk, probably are ischemic, and a result of arterial compression by the protruded disk. Diffuse demyelination is likely to be secondary to venous obstruction.

In the overwhelming majority of dorsal protrusions, some part of the disk is degenerate. In so-called disk explosions, nondegenerate nucleus pulposus is extruded into the cervical vertebral canal as a result of trauma. The neurologic syndrome is characteristic, being dominated by hemiplegia, ipsilateral hyperthermia, and Horner's syndrome. Cervical pain is absent. Extensive unilateral malacia of the cord at the prolapse site is probably secondary to vascular disturbances.

A syndrome associated with necrotizing myelopathy, due to emboli from intervertebral disks, occurs in dogs and less often in cats, horses, pigs, and humans. Clinically, there is either acute quadriplegia or paraplegia, determined by the site in cervical or thoracolumbar cord of the myelopathy, often followed by gradual recovery with time and steroid therapy. Fibrocartilaginous emboli from the nucleus pulposus are present in arteries or veins, or both. Dogs with this condition are almost always mature members of the large or giant breeds.

The pathogenesis of embolization in this syndrome is intriguing, and involves extrusion of disk material into blood vessels in and around the annulus. The offending disk may be macroscopically normal. Venous emboli probably arise either by direct herniation of nucleus pulposus into the longitudinal venous sinuses, or into small vessels in the annulus, which drain into the sinuses. Arterial emboli probably develop by direct herniation into the small arteries associated with vascularization of a chronically degenerate annulus. Retrograde movement into the radicular artery and subsequent discharge to the spinal cord are then possible. Central to this proposed pathogenesis are the facts that nucleus pulposus material is sometimes present in vessels of the degenerating annulus of nonchondrodystrophoid dogs, that intradiskal pressures are considerably higher than systemic arterial pressure, and that normally hydrated nucleus pulposus is semifluid, and is therefore amenable to this type of movement.

Arterial embolization associated with this pathogenesis requires prior degeneration and vascularization of the annulus fibrosus. However, if trauma is involved, preceding degenerative changes theoretically are not necessary for direct herniation into the longitudinal venous sinuses. Such an occurrence seems feasible with a disk explosion.

Changes in the dura mater are often associated with disk herniation. Degenerative nuclear substance is irritative (whereas normal nucleus pulposus is not), and it provokes an inflammatory reaction in the periosteum, dorsal longitudinal ligament, and dura mater. Initially the cellular response is neutrophilic, but this soon subsides and gives place to fibrosis. The dura mater is often adherent tightly to the disk. The protruded nuclear substance mineralizes and may even, in long-standing cases, ossify. The grit is palpable on the dura mater and is sometimes called ossifying pachymeningitis. Many cases of ossifying pachymeningitis apparently are not related to disk protrusions.

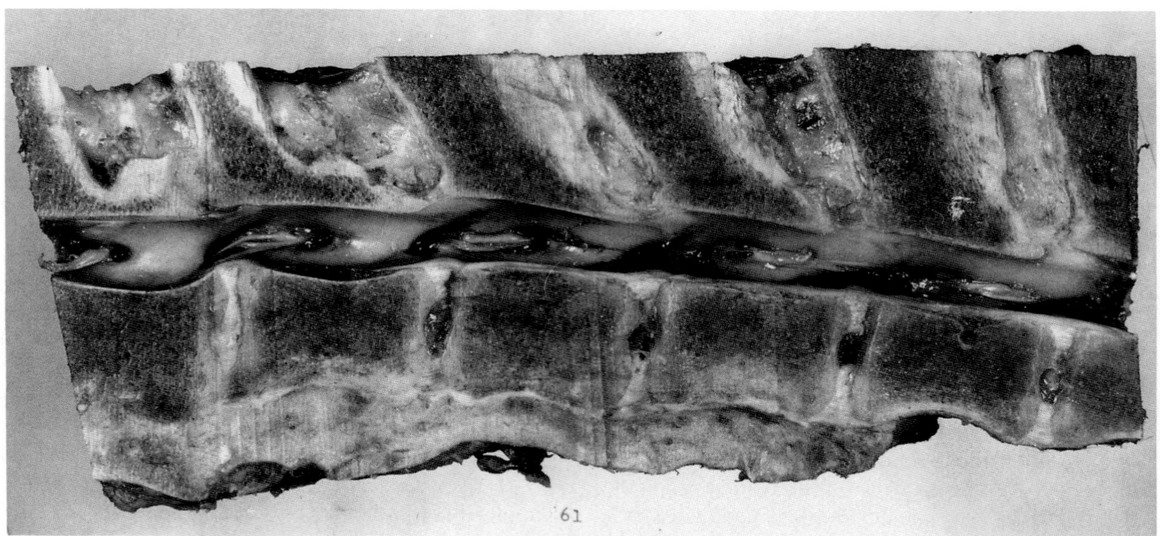

Fig. 1.128 Ankylosing spondylosis. Vertebrae are fused by new bone on ventral aspect of the bodies.

3. *Spondylosis*

Spondylosis (spondylosis deformans, ankylosing spondylosis) is a common condition of the vertebral column in which osteophytes develop at the intervertebral space as spurs or as complete bony bridges uniting the bodies of vertebrae (Fig. 1.128). They develop on the ventral and lateral aspects of the vertebral bodies. Osteophytes may be found dorsolaterally projecting into the vertebral canal, but these are small and uncommon. The development of osteophytes is induced by degenerative changes of the type found in nonchondrodystrophoid dogs in the annulus fibrosus. There may also be a degenerative arthropathy of the articular facets, and the reactionary osteophytes may produce ankylosis of these articulations also. Spondylosis implies that the lesion is primarily degenerative; similar gross changes may be produced by an inflammatory reaction, and these require microscopic differentiation (see diskospondylitis). Spondylosis is a common condition of bulls, pigs, and dogs, and is seen less often in other species. The disease is important in bulls (Fig. 1.129) kept in artificial-breeding studs and is to be expected in any animal past middle age. The cause is no doubt related to their duties, but the immediate pathogenesis is not clear; it may be related to the force of thrust at collection and the amount of pressure exerted on the intervertebral disk. Osteophytes develop mainly on the posterior end of thoracic vertebrae and the anterior end of lumbar vertebrae, and the incidence and size of osteophytes tend to decrease in either direction from the thoracolumbar junction. Thus the greatest number and size of osteophytes is in the area of greatest spinal curvature, where maximal pressure on the disks would be expected.

The sequence of osteophyte development begins with severe degeneration of the annulus fibrosus, which no doubt interferes with the function of the disk and probably allows abnormal movement between vertebrae. As a result there is formation of bone in the annulus fibrosus at the vertebral corners and stimulation of periosteal new bone on the ventral surface of the vertebral body. The initial degenerative changes in the annulus fibrosus may be present at 2 years of age. Irregular, basophilic, degenerate foci develop in the collagen. The osteophytes in the annulus originate by metaplasia of collagen with intermediation of fibrocartilage. They develop first in the outer annulus fibrosus and at its insertion to the rim of the vertebra. Growth is also in part by periosteal apposition and osseous metaplasia of ligaments. The trabecular bone of the osteophytes becomes continuous with that of the vertebral body and eventually densely sclerotic. In late stages, the heads of many ribs and the articular processes of the vertebrae bear large irregular osteophytes, which frequently cause ankylosis of the corresponding joint.

As much as 3 inches of new bone may be deposited on the ventral and ventrolateral aspects of the vertebral bodies, and new bone is also deposited within the bodies. The new bone is extremely hard and dense, and rather brittle.

Affected bulls show posterior weakness and ataxia, or paralysis, after dismounting from service. They may continue to be mildly ataxic or recover clinically to be affected again later. The spondylosis develops gradually, and although osteophytes occasionally form in the spinal canal, they seem to be without much effect. The onset of signs is usually associated with fracture of the vertebral bodies and of the ankylosing new bone. The line of fracture tends to follow a large penetrating vessel to the intervertebral disc, which is frequently separated, and then to diverge across the dorsal corner of one or other vertebra. There is little displacement of the fractured ends in most cases, which accounts for the incomplete spinal syndrome. Trauma to the spinal cord is usually mild, and paralysis is usually an accompaniment of hemorrhage or the result of repeated trauma.

In dogs, the incidence of osteophytes increases with age after about the fifth year, and they are a common incidental finding at autopsy. The development of spondylosis in both dog and cat is similar to that in the bull, with the primary morphologic change in the annulus fibrosus (Fig. 1.130). In dogs, most lesions occur at L-2 to L-3, and the lumbosacral articulation is often involved. The end result in horses is comparable to that in other species, but the evolution of the osteophytes has not been studied in detail.

Ankylosing spondylosis is not uncommon in adult sows. There is no indication of the cause; it may affect the anterior thoracic vertebrae or the usual caudal thoracic–anterior lumbar region. The ankylosis may be confined to the ventral aspect of the bodies, but in some cases there is extensive new-bone formation in the arches, which fuses the articular processes and enroaches on the spinal canal, giving it an irregular or triangular shape.

If the pathogenesis suggested for bulls is accurate, and applicable to other species, then any factor which permits abnormal mobility of intervertebral articulations has the potential to stimulate osteophyte formation and spon-

Fig. 1.129 Ankylosing spondylosis. Bull. Fusion of vertebrae by new bone on lateral aspect of the bodies.

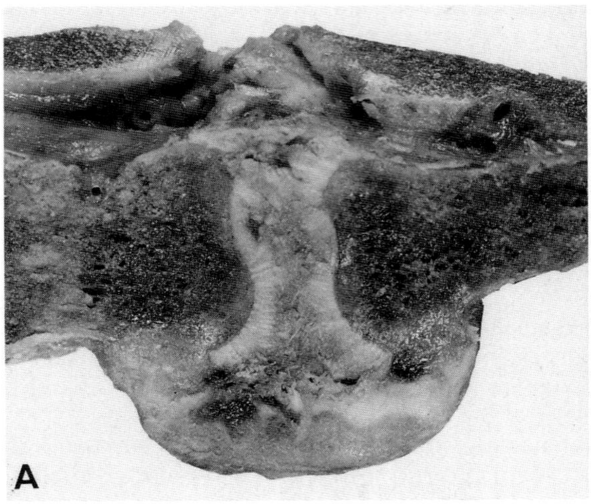

Fig. 1.130A Degeneration of intervertebral disk with dorsal and ventral herniation and ankylosing spondylosis. Dog.

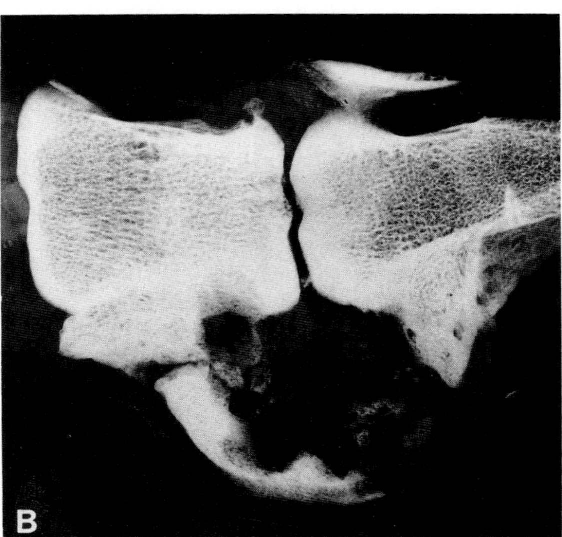

Fig. 1.130B Radiograph showing ossification in spondylosis. Dog.

dylosis. The occurrence of vertebral osteophytes in some dogs with vertebral malformations is consistent with this proposal.

Vertebral ankylosis in cats with chronic vitamin A toxicity is considered elsewhere.

Bibliography

Ball, M. U. *et al.* Patterns of occurrence of disk disease among registered dachshunds. *J Am Vet Med Assoc* **180:** 519–522, 1982.

Doige, C. E. Pathological changes in the lumbar spine of pigs: Gross findings. *Can J Comp Med* **43:** 142–150, 1979.

Goggin, J. E., Li, A., and Franti, C. E. Canine intervertebral disk disease: Characterization by age, sex, breed, and anatomic site of involvement. *Am J Vet Res* **31:** 1687–1692, 1970.

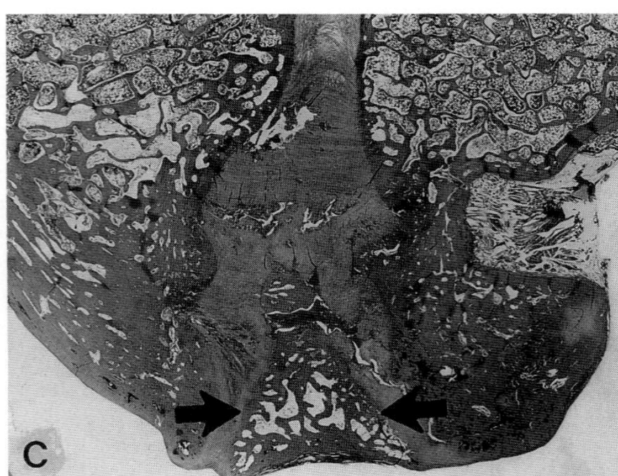

Fig. 1.130C Degeneration and ossification (arrows) in annulus fibrosus. Dog. (Courtesy of J. P. Morgan.)

Griffiths, I. R. A syndrome produced by dorsolateral "explosions" of the cervical intervertebral discs. *Vet Rec* **87:** 737–741, 1970.

Griffiths, I. R. Some aspects of the pathology and pathogenesis of the myelopathy caused by disc protrusions in the dog. *J Neurol Neurosurg Psychiatry* **35:** 403–413, 1972.

Griffiths, I. R. The extensive myelopathy of intervertebral disc protrusions in dogs ("the ascending syndrome"). *J Small Anim Pract* **13:** 425–437, 1972.

Griffiths, I. R. Spinal cord infarction due to emboli arising from the intervertebral discs in the dog. *J Comp Pathol* **82:** 225–232, 1973.

Hansen, H. J. A pathologic–anatomical study of disc degeneration in dog. *Acta Orthop Scand* (*Suppl.*) **11:** 1–130, 1962.

King, A. S., and Smith, R. N. Diseases of the intervertebral disc of the cat. *Nature* (*Lond*) **181:** 568–569, 1958.

Morgan, J. P., Ljunggren, G., and Read, R. Spondylosis deformans (vertebral osteophytosis) in the dog. *J Small Anim Pract* **8:** 57–66, 1967.

Olsson, S.-E., and Hansen, H. J. Cervical disc protrusions in the dog. *J Am Vet Med Assoc* **121:** 361–370, 1952.

Pass, D. A. Posterior paralysis in a sow due to cartilaginous emboli in the spinal cord. *Aust Vet J* **54:** 100–101, 1978.

Priester, W. A. Canine intervertebral disc disease—occurrence by age, breed, and sex among 8117 cases. *Theriogenology* **6:** 293–303, 1976.

Thomson, R. G. Vertebral body osteophytes in bulls. *Pathol Vet* (*Suppl.*) **6:** 1–46, 1969.

IV. Traumatic Injuries of Joints

Traumatic injuries to joints can be discussed as two main types; those which are single, sudden, and severe, and those which are multiple and minor. The latter types, which are chronic injuries with chronic effects, are considered under Diseases of Joints, Section III,A. (Acute injuries also frequently lead to degenerative arthropathy because of the structural and functional changes they induce.) Trauma to intervertebral disks is discussed with disk herniations.

Acute injuries to joints including sprains, subluxations,

and luxations—these are but varying degrees of a single type of injury incurred when the joint is forced to move beyond the limits of elasticity of its ligaments. The synovial tissues and capsule may be stretched, lacerated, or torn; the ligaments incur the same injuries and may, in addition, be detached from their bony anchorage. The menisci may be dislocated or split, and the articular cartilages may be compressed, split, or even detached from the underlying bone. The menisci and the articular cartilages, if injured, do not heal. If the articular cartilage is split or the capsule torn, hemorrhage occurs into the joint; the clot may be removed entirely or be organized to produce a fibrous ankylosis.

Minor sprains usually heal rapidly and leave no impairment. More severe injuries with laceration of the capsule and ligaments seldom repair satisfactorily but instead tend to be associated with persistent laxity of the articulation—an invitation to degenerative arthropaτny.

The most severe injuries result in luxation of the joint. (Luxation and subluxation can also occur spontaneously when the capsule and ligaments are loosened in chronic articular diseases.) The consequences vary with species, surgical accessibility, and the joint. Patellar luxations result in little injury to soft tissues, and they recur readily after reduction. Dislocations of other joints are usually associated with injury to the adjacent soft tissues as well as tearing of the capsule and ligaments; if reduced, reluxation or subluxation often occurs by permission of the stretched periarticular tissues.

In persistent luxation, if the articular processes lie in soft tissue, they atrophy, and the cartilage degenerates and is replaced by fibrous tissue and bone. If the dislocated end lies against bone, it incites a periosteal reaction. If movement is slight or absent, a fibrous or bony fixation may develop, but if movement continues, a new joint may form, which is sometimes quite efficient. Fibrous tissue about the dislocated end is organized into an articular capsule, which then acquires a synovial lining. The articulating surfaces are covered by fibrous tissue or polished bone, or in favorable cases, by cartilage.

The preceding discussion assumes that the traumatized structures are normal prior to the traumatic event. In many diseases, notably the osteodystrophies and osteochondrosis, where the articular structures are already abnormal, trauma precipitates the development of joint disease. A well-known example occurs in osteochondritis dissecans of the humeral head (see Osteochondrosis). This lesion probably develops when the thickened juvenile articular cartilage of the posterior central region of the humeral head is subject to compression by the posterior rim of the glenoid when the joint is in full extension.

V. Circulatory Disturbances in Joints

Intra-articular hemorrhage occurs frequently in the hemorrhagic diatheses (see Volume 3, Chapter 2, The Hematopoietic System). These hemorrhages occur in response to normal articular stresses, are polyarticular, and cause stiffness or lameness.

VI. Articular Foreign Bodies

True foreign bodies which penetrate the articular capsule frequently lead to septic arthritis, the nature and consequences of which are discussed later. Even if aseptic, spontaneous closure is inhibited by the continuous leakage of synovial fluid. The so-called joint mice are endogenous hard or soft foreign bodies. The hard type is discussed in Section VII,A, General Considerations; soft ones consist of a coagulum of blood, fibrin, or detached villi. Detachment of villi occurs after they hypertrophy in degenerative arthropathy or arthritis. Joint mice are usually found in pouchings of the articular capsule. They occasionally become wedged between articular surfaces and cause sudden stiffness.

Bibliography

Nelson, D. R., and Smith, R. M. Intermittent lameness in two cows from detachment of the anterior horn of the medial meniscus. *Vet Rec* **127:** 333, 1990.

Nelson, D. R., Huhn, J. C., and Kneller, S. K. Peripheral detachment of the medial meniscus with injury to the medial collateral ligament in 50 cattle. *Vet Rec* **127:** 59–60, 1990.

VII. Inflammatory Diseases of Joints

A. General Considerations

Arthritis is inflammatory joint disease. The primary and initiating disease process in the joint is centered in the walls of capillaries located in the interstitium of the synovium or other joint structures and is driven from its outset by mediators of inflammation located in those tissues. By comparison the much milder inflammatory response that occurs in traumatic or degenerative arthropathy and which may include synovial hyperemia, serous effusion, hemorrhage, and an elevated white cell count, is a secondary response to mediators released following traumatic damage to joint structures or to products released slowly from deteriorating joint surfaces. Arthritis may be subclassified as infectious or noninfectious. Noninfectious arthritis may be further subdivided into immunologically mediated arthritis and crystal-induced arthritis. The latter form of arthritis will not be discussed further and is represented by gout (synovial deposition or urate crystals) in avian and reptilian species and by a single case of pseudogout (synovial deposition of calcium phosphate crystals) in a dog.

Infections tend to cause simultaneous arthritis and tendovaginitis; each is primarily a synovitis (Fig. 1.131). Microorganisms may enter the joint hematogenously, by direct extension from a focus of infection in adjacent soft tissue or bone, or through a penetrating wound. Pressure sores are comparable to penetrating wounds.

Hematogenous infections often localize in other sites,

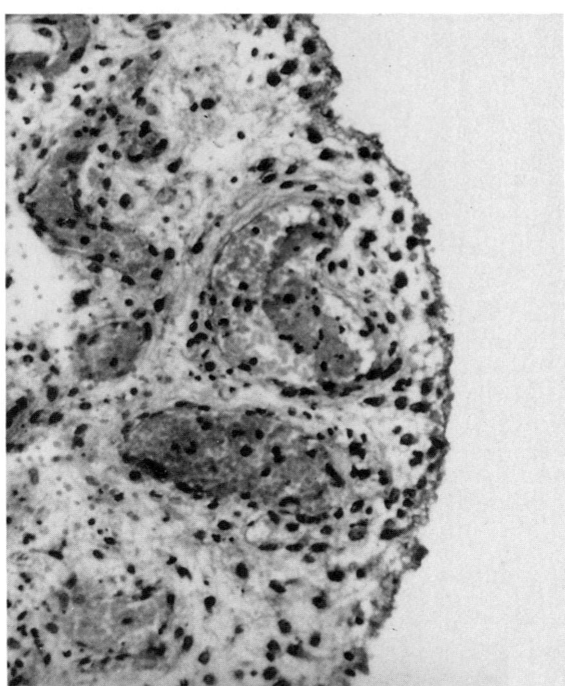

Fig. 1.131 Acute articular synovitis. Pig. Synovial membrane shows hyperemia and edema of acute inflammation.

including the bone marrow, but lesions in this site are not immediately evident. There may be direct spread of inflammation from a focus of osteomyelitis in bone to a joint. The line of least resistance is through the cortex at the margin of the physis, where in growing bones the cortex is being remodeled and is incomplete, and then through the attachment of the articular capsule. Direct spread from bone to joint is much less common than simultaneous hematogenous localization, but it is the pathway taken by suppurative infections such as those caused by *Actinomyces pyogenes* and staphylococci (Figs. 1.82B, 1.132A,B). Direct spread of infection from adjacent soft tissues to joints is also uncommon; it occurs in necrobacillosis as it does in footrot of cattle, but to most organisms, the fibrous layer of the articular capsule provides an efficient barrier. Penetrating wounds usually introduce a variety of bacteria, and the inflammation is suppurative and destructive (Fig. 1.33A,B).

In dogs and cats, most infectious arthritis is monoarticular, caused by penetrating wounds, and involves adolescent or older animals. In farm animals, most infectious arthritis is polyarticular, hematogenous, and affects neonates. The synovial layer of the articular capsule is a highly favorable site for the hematogenous localization of bacteria, along with other papillary structures such as the choroid plexus of the brain and the ciliary processes of the eye. It is not clear why bacteria establish themselves so readily in these structures, but the fluid medium in which they find themselves is nutritious and normally con-

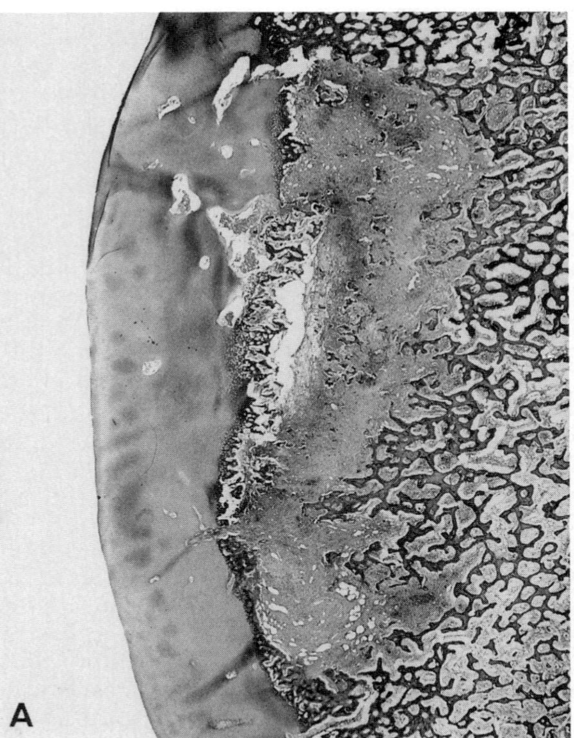

Fig. 1.132A Chronic hematogenous osteomyelitis interfering with endochondral ossification in epiphysis, and underrunning cartilage. Incipient arthritis.

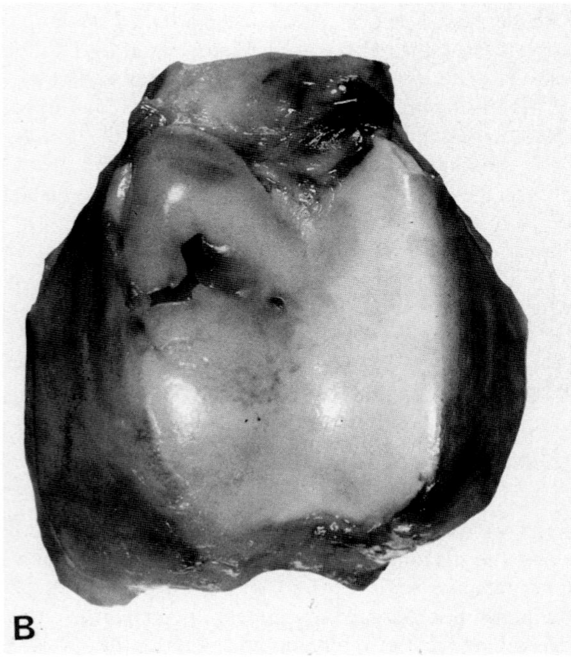

Fig. 1.132B Extension of epiphyseal osteomyelitis to cause arthritis. Distal femur. Horse.

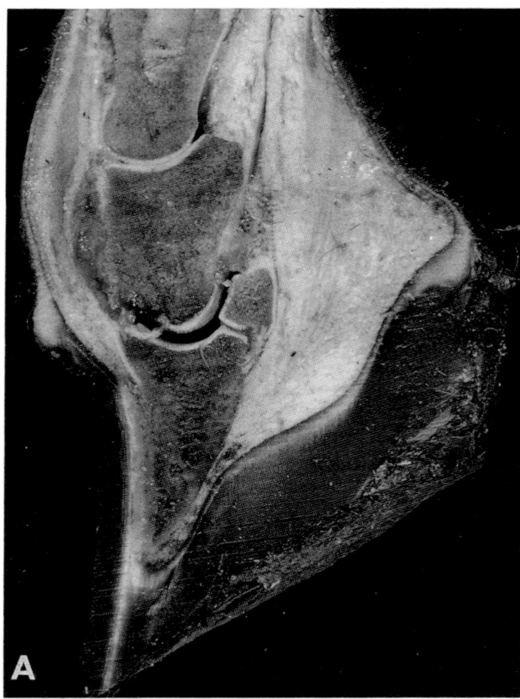

Fig. 1.133A Foal. Periostitis and osteomyelitis, and arthritis of distal interphalangeal joint, caused by penetrating wound.

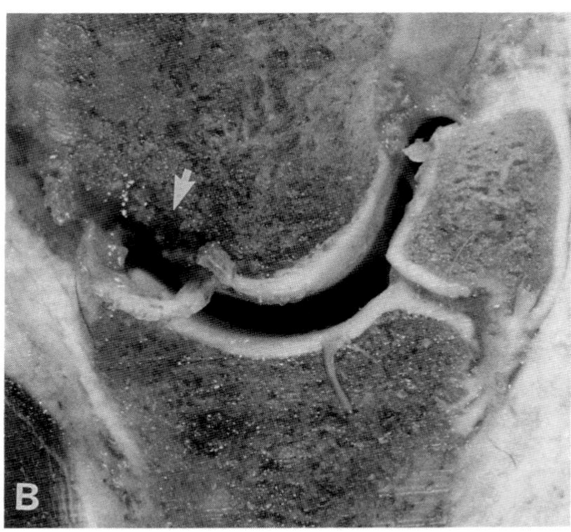

Fig. 1.133B Detail of (A). Destruction of articular cartilage and subchondral bone (arrow).

tains few leukocytes. The richness of the synovial vasculature and the discontinuous nature of the synovial membrane probably permit rapid entry to the joint. In septicemias, microorganisms appear in the joint before there is either gross or microscopic change in the synovial tissue.

Hematogenous arthritis is initially polyarticular. After a brief period, inflammation subsides in many of the joints,

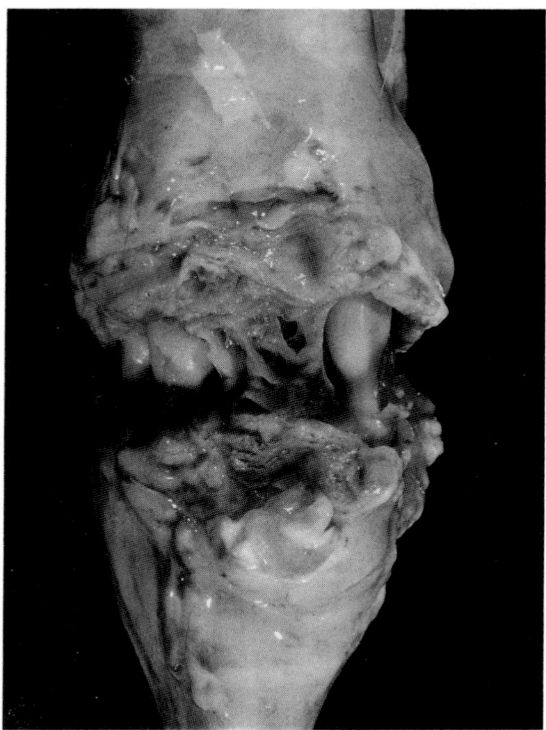

Fig. 1.134 Fibrinous arthritis. Calf. Periarticular edema with fibrin wads filling hock joints. (Courtesy of R. B. Miller.)

while in others it progresses. The large joints of the limbs are particularly susceptible to progressive development of the inflammation (Fig. 1.134), and their involvement tends to be bilateral. In most diseases, the inflammation is serofibrinous. However, the pyogenic staphylococci and *Actinomyces pyogenes* produce a purulent reaction. The pyogenic streptococci usually produce a fibrinous exudate because usually the animal dies before the reaction has time to become purulent. The course and consequences of fibrinous inflammation differ somewhat from those of purulent inflammation, and they are discussed separately. They do merge, however, depending on the nature and persistence of the causative organisms.

Fibrinous arthritis begins with the emergence of bacteria from blood vessels into the loose stroma of the synovial layer of the articular capsule, and their entry into the synovial fluid. The first macroscopic changes occur in the villi, which become prominent due to hyperemia and edema. (In a normal joint the villi are not visible unless the membrane is immersed in water.) The enlarged villi are most obvious where they are most numerous, which is at places of ligamentous insertion and in the transition zone where the articular capsule meets the perichondrium, but the synovitis is diffuse. The synovial fluid is increased in volume and is slightly turbid and mucinous. The membrane may be studded with petechiae. In this early stage of synovitis, the histological changes are edema and vascular engorgement (Fig. 1.131) with a sparse infiltration of lym-

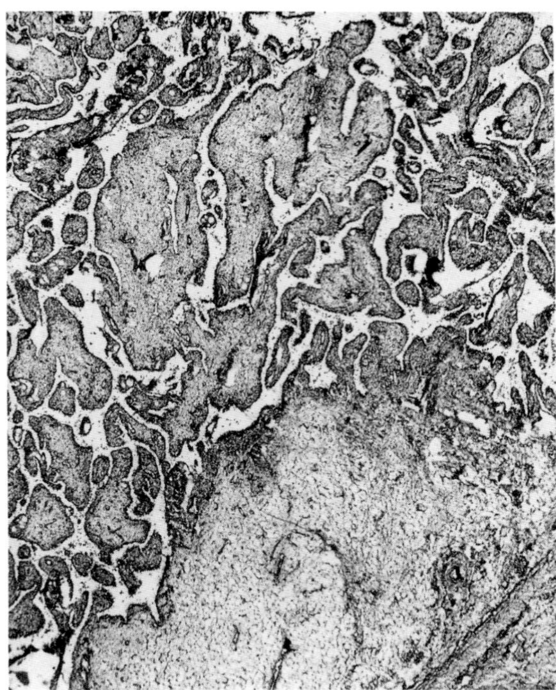

Fig. 1.135 Chronic hypertrophic articular synovitis. Horse.

phocytes and plasma cells. This serous or serofibrinous inflammatory exudate often infiltrates the fibrous layer of the articular capsule and the adjacent periarticular tissue. When the initial inflammatory reaction is severe, there may be gross edema of the periarticular tissue and a predominantly fibrinous effusion in the joint (Fig. 1.134), visible as a slimy, gray or yellow deposit in the synovial membrane, the distal articular cartilage, and in the recesses of the joint.

In arthritis of longer duration, there is less effusion in the synovial tissues. The stroma proliferates as does the synovial membrane. The villi continue to enlarge as a result of the cellular proliferation, and become complexly branched (Fig. 1.135). They resemble young, polypous, granulation tissue with an increasing population of mononuclear leukocytes, but there is no or minimal ulceration and very few neutrophils; extravasated neutrophils pass quickly into the synovial fluid but seldom in sufficient numbers to give the fluid a purulent character. The hypertrophy of the villi is greatest in the transition zone, and the proliferating fibrous stroma is joined by proliferating perichondrium to produce a fringe of granulation tissue which, as it grows, spreads across the articular cartilage. Such granulation tissue is known as pannus; it is important because it tends to adhere to the articular surface (Fig. 1.136A,B). The origin of the increased cells in the synovial membrane is, at least partly, bone marrow, as some of them contain leukocyte common antigen and carry markers characteristic of monocyte—macrophage lineage. Proliferation of deep stem cells also contributes to the increase.

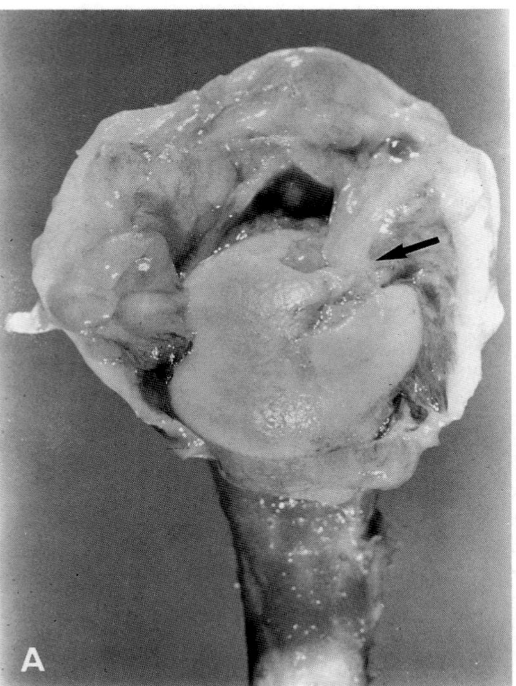

Fig. 1.136A Chronic synovitis. Humerus. Lamb. Synovial proliferation with thickened capsule. Cartilage is degenerate, and pannus from synovial membrane has invaded bone (arrow).

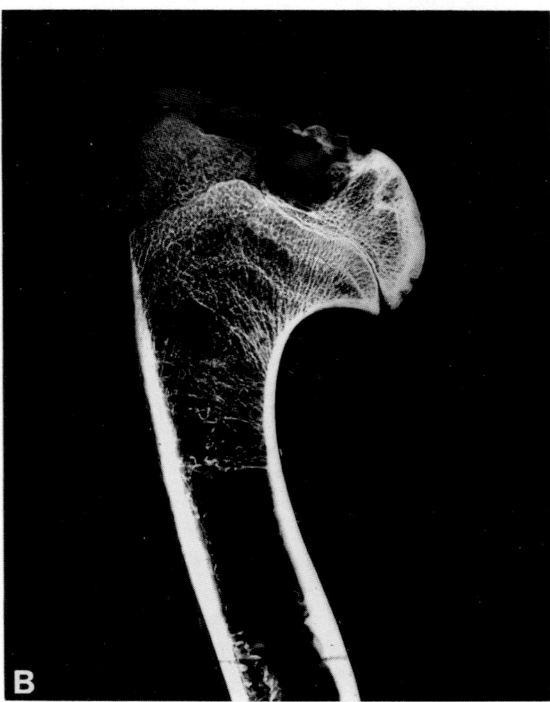

Fig. 1.136B Radiograph of (A) showing destruction of proximal humeral epiphysis.

It is at about this stage that the infection can be expected to be overcome and the joint sterilized, unless it is intermittently reseeded from the blood. When a joint becomes sterile, the changes in the tissue may take one of three courses depending on the severity of the acute reaction and on the development of structural and functional alterations in the supporting articular and periarticular tissues. First, the lesions may resolve completely. As stated earlier, complete and early resolution occurs in many joints, especially the smaller ones. Second, healing may occur by fibrosis. The process of organization is described in following sections and is to be expected when the injury is severe. Third, even in the absence of continued infection, which may be difficult to confirm, continued but low-grade activity with intense lymphocytosis may persist in the synovial membrane in conjunction with healing of other articular lesions. The maintenance of synovial activity may be due to inapparent infection, but alternatively ineffective disposal of the peptidoglycan components of microbial cell wall by macrophages may be responsible. For example, the cell wall of group A streptococci is relatively resistant to degradation by mammalian lysosomal enzymes, and is capable of provoking persistent inflammation in synovial tissues. Cell wall peptidoglycans from various other organisms, including *Erysipelothrix rhusiopathiae* and *Corynebacterium rubrum,* also have this ability. All bacterial cell walls contain peptidoglycans, but there is considerable structural heterogeneity between species, and possibly the types of side chain on the molecule determine its arthritogenic potential.

In the healing phase of fibrinous arthritis, the fibrin which is deposited on the synovial membrane and within the layers of the articular capsule and periarticular tissue is progressively organized. The synovial lining is repaired by proliferation of synovioblasts. The organization of exudates and the irritative hyperplasia of stromal connective tissues contribute much to the clinical enlargement of these joints and to permanent stiffness. Organization of the pannus is often associated with its fusion to the articular cartilage. The organized tissue may remain fibrous or it may undergo chondrification and ossification. Adhesion may occur between apposed articular surfaces via the organized pannus, and depending on the extent and location of the adhesions, there may be ankylosis.

Purulent arthritis may arise by local extension but is usually hematogenous. It is often monoarticular, but it may be bilateral and symmetrical and also tends to involve the large joints of the limbs. The progressive inflammatory changes are more severe and destructive than those of fibrinous arthritis. The exudate contains large numbers of neutrophils, and in the early stages, the synovial fluid is thin and cloudy. As the reaction progresses, the synovial lining ulcerates, and the exudate is converted to thick pus. In fibrinous arthritis, the articular cartilages survive except where pannus is adherent, but often in purulent arthritis, they are destroyed (Fig. 1.137). The mechanisms of destruction are complex and probably involve the combined actions of neutrophilic enzymes, bacterial lipopolysaccharides, and cytokines of macrophage origin, such as interleukin-1 and tumor necrosis factor. These agents are stimulators of chondrocyte-mediated degradation of cartilage, which involves release and activation of proteolytic enzymes.

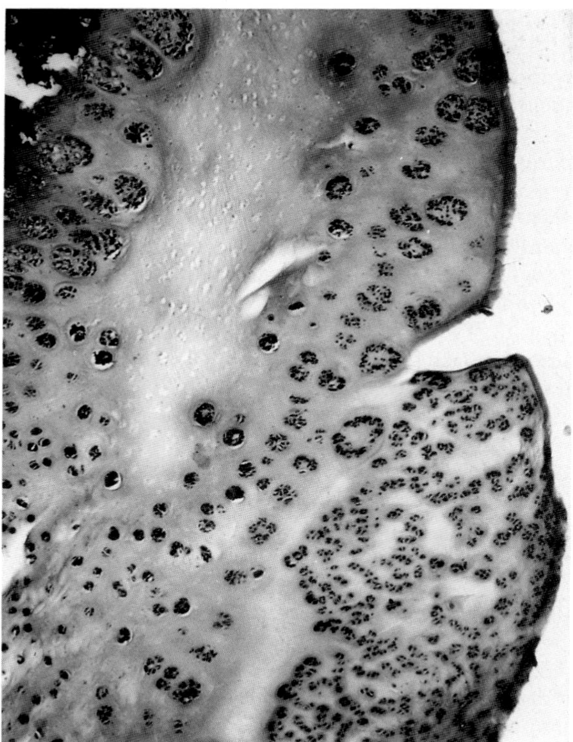

Fig. 1.137 Degeneration and reaction of articular cartilage in suppurative arthritis. Note multinuclear chondrones in injured cartilage.

The degenerative changes occur chiefly in the sites of weight bearing. Clefts in fibrillar cartilage and erosion of cartilage allow infection to gain entry to the deep mineralized layer or the subchondral bone. When this happens, the suppurative process dissects the articular cartilage from the bone and may detach it entirely. Dissection also begins irregularly about the periphery of the articular cartilage at places in the transitional zone where the synovial membrane is ulcerated. Purulent osteomyelitis then supervenes with more or less extensive destruction of the epiphysis.

The suppurative process may extend outward as well as inward from the synovial layer of the articular capsule to produce phlegmon in the periarticular tissues and to extend sometimes to adjacent tendon sheaths. The articular region is then greatly enlarged. Localization of the phlegmon as an abscess may be followed by fistulation on the skin. Empyema of the joint may also rupture and fistulate to the skin.

If the suppurative process is aborted, either spontaneously or therapeutically, before the articular cartilages degenerate, healing may be complete, or an active aseptic synovitis may persist. Periarticular phlegmon heals by dense scarification and results in permanent stiffness. Degeneration of the cartilage is followed by the ingrowth of granulation tissue from the subchondral bone and eventually by intra-articular ankylosis. Death may occur from suppurative processes in other organs, which arise as part of the primary dissemination of the infection or as metastases from the joints. Chronic suppurative arthritis is most frequently seen in cattle as an infection with *Actinomyces pyogenes*; some of these animals die in uremia of secondary amyloidosis.

The rate of lesion development in infectious arthritis is important to diagnostic pathology. It obviously varies from case to case depending on host–agent interactions, but useful information is provided by experimental infections. *Staphylococcus aureus* injected into rabbit knee joints produces purulent arthritis characterized by reduction of glycosaminoglycans in superficial cartilage by 2 days and total loss in 2 weeks. Erosion at the joint margins is visible in 5 days, and complete destruction of the joint can occur in 5 weeks. Irreversible damage to the joint occurs in 5 to 7 days. Intradermal *Neisseria gonorrhoeae* peptidoglycan causes severe hind-paw arthropathy in rats, in which there is pannus formation and erosion of cartilage and bone by 24 days, tendon sheath adhesions and periarticular fibrosis by 32 days, and fibrous ankylosis by 40 days.

The progressive pathology of tendovaginitis parallels that of articular synovitis. In the healing stages, adhesions frequently occur between the tendon and sheath.

B. Infectious Arthritis

1. Causes

Arthritis is a possible complication of many systemic infections involving bacteria, fungi, and even parasites such as *Micronema deletrix*. Listed below are microorganisms which consistently produce arthritis in sheep, goats, swine, cattle, horses, and dogs. This consistency does not necessarily imply predilection, but rather opportunistic localization in an infection that becomes systemic and is not too rapidly fatal. Infectious arthritis is rare in cats.

Sheep
- *Erysipelothrix rhusiopathiae*
- *Streptococcus* spp.
- *Haemophilus agni*
- *Staphylococcus* spp. (in tick pyemia)
- *Escherichia coli*
- *Mycoplasma* spp.
- *Chlamydia psittaci*

Goats
- Retrovirus
- *Mycoplasma* spp.

Swine
- *Erysipelothrix rhusiopathiae*
- *Streptococcus* spp.
- *Haemophilus* spp.
- *Actinomyces pyogenes*
- *Salmonella* spp.
- *Mycoplasma* spp.
- *Brucella suis*
- *Escherichia coli*
- *Actinobacillus suis*

Cattle
- *Streptococcus* spp.
- *Escherichia coli*
- *Actinomyces pyogenes*
- *Salmonella* spp.
- *Mycoplasma bovis*
- *Haemophilus somnus*

Horses
- *Actinobacillus equuli*
- *Streptococcus* (Group C)
- *Escherichia coli*
- *Klebsiella* spp.
- *Rhodococcus equi*
- *Salmonella* spp.

Dogs
- *Staphylococcus* spp.
- *Streptococcus* spp.
- *Escherichia coli*
- *Borrelia burgdorferi*
- *Blastomyces dermatitidis*

Some generalizations may be made about the prevalence of these infections relative to age. The ovine infections are, with the exception of infection by *Mycoplasma* spp., chiefly diseases of lambs. In cattle, streptococcal and coliform polyarthritis are neonatal; infections due to *A. pyogenes* and *Salmonella* spp. occur at any age. Streptococcal polyarthritis in swine is often a neonatal disease; the other infections listed for this host are seldom observed in sucklings. In horses, the organisms listed, other than *Salmonella* spp., cause intrauterine or neonatal infections. *Salmonella abortus-equi,* a rare infection, also occurs in the same age group. Neonatal polyarthritis of foals is traditionally known as joint ill.

2. Types

a. Erysipelas *Erysipelothrix rhusiopathiae,* the cause of porcine erysipelas and erysipeloid in man, is a gram-positive bacillus which has a wide geographic distribution and a wide host range. It causes outbreaks of disease in pigs, lambs, and birds, and sporadic disease in the other domestic species. **Porcine erysipelas** can be produced by ingestion of the organisms, by contamination of cutaneous wounds, and by bites of infected flies; these, presumably, are the routes of natural infection. The acute disease is a highly fatal septicemia, and the mild to chronic disease is characterized by necrosis of the skin, endocarditis, and polyarthritis. In epidemics, the septicemic form predominates, whereas in endemic areas the disease tends to be sporadic with cases of septicemia, polyarthritis, or endocarditis occurring in varying proportions.

Erysipelothrix rhusiopathiae is widespread in nature and is capable of survival and perhaps growth in decaying material of animal origin. It is present in the soil of pig pens and in pit slurry, and survives for 2 to 3 weeks on pasture spread with slurry. It is relatively resistant to many disinfectants, and is capable of infecting many species, some of them in epidemic proportions. In spite of these epidemiological features, pigs are probably the principal source of infection for pigs. The organism can persist for many months in the lesions of diseased pigs, and it is often carried in the tonsils, intestine, bone marrow, and gall bladder of healthy swine.

Erysipelas occurs in pigs of all ages, but the most susceptible are those from 2 months to 1 year of age and pregnant sows. The latter may abort or give birth to stillborn young, from which the organism can be cultured.

The **acute disease** is a febrile septicemia which usually develops within 24 hr of exposure to virulent organisms and produces disseminated intravascular coagulation. Endothelial cells of capillaries and venules throughout the body swell, monocytes adhere to them, and by 2 to 3 days development of hyaline thrombi is widespread. By 4 days, accumulation of fibrin within and around vessels, bacterial invasion of endothelium, and diapedesis of erythrocytes are prominent. Perivascular fibrin incites connective tissue proliferation in sites such as synovial membranes. Grossly, there is purple discoloration of the skin due to congestion and sometimes thrombosis of dermal capillaries and venules. Death, when it occurs in this stage, may not be accompanied by specific lesions. Petechiae or ecchymoses may be present on serous membranes and in other tissues, the spleen is almost always swollen and red from hyperemia, and there may be congestion and infarction in the gastric mucosa, as in many acute infections in pigs. Subsidence of the acute disease, or a milder initial course, often leads to swelling of joints and lameness, and to characteristic erythematous lesions in the skin.

The cutaneous lesions are rhomboidal in outline and slightly raised like welts (Fig. 1.138). They are readily visible in light-skinned pigs and palpable when not visible in dark-skinned pigs. The skin within the rhomboid may be uniformly bright red or purple or only the margins and the center, which appears as an insect bite, may be discolored. The latter lesions may progress to complete discoloration, or may return to normal within a few days. The uniform, bright red lesions may also resolve with a little scurfiness as a residue. The dark or purple lesions undergo dry necrosis; they may eventually peel off or, if forcibly detached, expose a raw base. Occasionally the rhomboid lesions coalesce over large areas of skin and lead to extensive cutaneous necrosis. The ends of the tail and ears may also die from ischemia and become dark, shrunken, and leathery. The discrete rhomboid lesions are associated with an arteritis; the smaller arterioles show acute cellular infiltration in their walls, and contain cellular thrombi. The inflammatory cells are chiefly neutrophils, but suppuration does not occur. The causative organisms are present in the cutaneous lesions and can be recovered if the inoculum is obtained from the subcutaneous side. Regularly, there is suppurative inflammation of the sweat glands, but otherwise the reaction is most severe at the junction of dermis and subcutis.

The discrete cutaneous lesions take 3 to 4 days to develop, by which time lesions may be found in the viscera. There may still be signs of septicemia with additional gross or microscopic signs of localization in many organs. The spleen is enlarged, and the pulp, red. Numerous petechiae may be visible in the renal cortices, and there is sometimes intense interstitial hemorrhage in the medulla. The cortical petechiae arise in part from small venules but mainly from the glomeruli. Glomerular changes are almost always present but vary in type. There may be diffuse glomerulitis characterized by swelling or proliferation of the glomerular endothelium. Many of the glomeruli show a sludging of erythrocytes in a part of the tuft, and this presages the development of the characteristic renal lesion of erysipelas—focal fibrinoid necrosis of the tufts with intracapsular hemorrhage. Occasionally, bacterial colonies are found in the necrotic tufts or in intertubular capillaries, where they provoke a tiny, intense focus of neutrophils (Fig. 1.139).

Synovitis is evident in many of the large joints as an increase in volume and viscosity of the fluid, and hyperemia of the synovial membrane. The articular lesions are those of fibrinous arthritis (see the preceding). In some

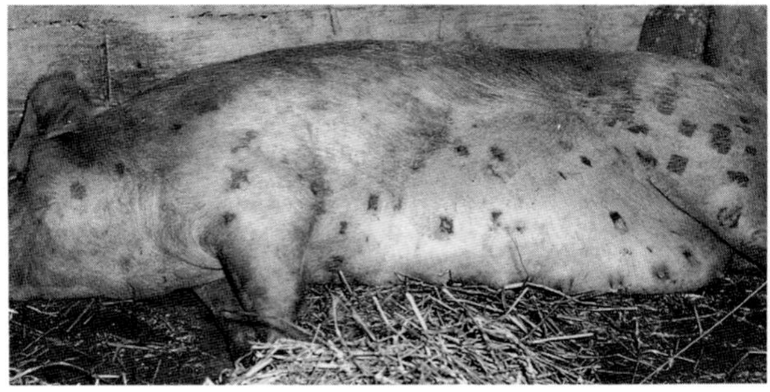

Fig. 1.138 Cutaneous diamonds in porcine erysipelas.

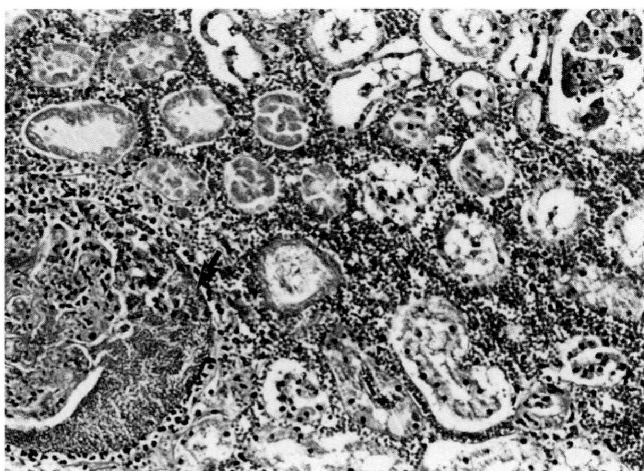

Fig. 1.139 Bacteria and acute inflammation (arrow) in Bowman's space, and intertubular hemorrhage, in porcine erysipelas.

pigs with erysipelas, the synovial arterioles show necrotizing inflammation and extensive plugging by cellular thrombi.

Lesions in other tissues in the acute disease are not specific. There is leukocytosis with many mononuclear cells in the hepatic sinusoids. The lungs show typical lesions of bacteremia in which organisms are sequestered in the pulmonary capillaries, but without the focal localization of embolic pneumonia. The alveolar vessels are intensely hyperemic and contain numerous sequestered leukocytes. Later, the alveolar walls tend to thicken as a result of cellular infiltration, mononuclear cells predominating, and this is accompanied by alveolar flooding and increased alveolar macrophages. The skeletal musculature often shows extensive degeneration, although it may be too recent to be visible grossly. Whether the muscular degeneration is an example of the classic change described as Zenker's degeneration or is mainly due to peripheral arteriolitis is difficult to decide. Of the remaining tissues, the brain and eyes appear to harbor lesions most consistently (Fig. 1.140). The organisms localize in the choroid and the ciliary process and provoke an intense infiltration of neutrophils. In the brain, there is cerebral leukocytosis and degeneration of the vessels of the cerebral white matter. Embolic colonies may be found in small vessels of brain and cord with a tiny surrounding zone of neutrophils and a few eosinophils; these focal infiltrations are quite common and are not always associated with visible organisms. The neural lesions do not produce neurological signs.

Chronic erysipelas is manifested as either endocarditis or arthritis or both. Localization of the organism in heart valves and joints occurs either as a sequel to acute septicemia, which subsides, or as a sequel to a mild or inapparent systemic course. The mitral valves are most often involved, and pigs that die of endocarditis have congestive heart failure and embolism; large infarcts in the spleen and kidneys are particularly common. The valvular vegetations may be very large and, while active, masses of bacteria can be found near the surfaces of the thrombi. In older lesions in which bacteria are dead, or at least not cultivable, the colonies are buried more deeply in the thrombi; in either case, the bacteria usually lie side by side in a radial alignment as a palisade.

Lameness due to arthritis is a common expression of erysipelas in pigs, and it may be unassociated with earlier acute or subacute signs of infection. This pattern of infection can be produced by injections of the organism, and also occurs in vaccinated swine. Although vaccination seems useful in preventing the acute syndrome and mortality, it appears to enhance susceptibility to polyarthritis but, since there are 22 serotypes, this is difficult to assess. The number of joints affected following injection of susceptible pigs depends on the number of organisms used and the arthritogenicity of the bacterial strain. In pigs with acute arthritis, organisms may be isolated from grossly normal as well as inflamed joints, but isolation from affected joints may be difficult in the chronic disease, even following systemic infection with virulent strains. Possibly the prolonged inflammation in such cases is due to persistence of bacterial antigens with continuing immunogenic reactions within synovial tissues. Antigens persist for up to 18 months in arthritic joints, and specific antibodies found in synovial fluid may be produced by plasma cells in the synovial membrane. Culture of pellets from centrifuged synovial fluid, however, usually yields a few organisms even in very chronic disease.

The lesions in chronic erysipelas arthritis, developing as a primary disease, are often not particularly severe. There is excess fluid, but the articular capsule may appear normal. The synovial membrane is hyperplastic and villous, with hyperemia and an infiltration of mononuclear cells including plasma cells. In severe arthritis with extensive pannus, deformation of the articular surfaces and erosion of the cartilage are common (Fig. 1.141A,B). Diskospondylitis often occurs in chronic erysipelas.

Erysipelas in sheep is almost always a percutaneous infection, entry being gained through the umbilicus, dock-

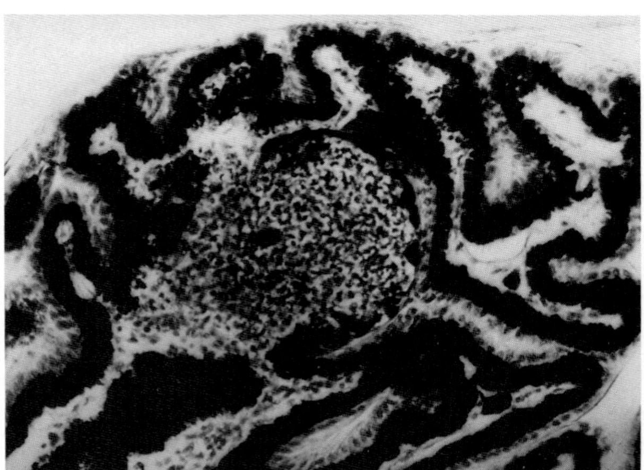

Fig. 1.140 Microabscess with bacterial colony in ciliary body in porcine erysipelas.

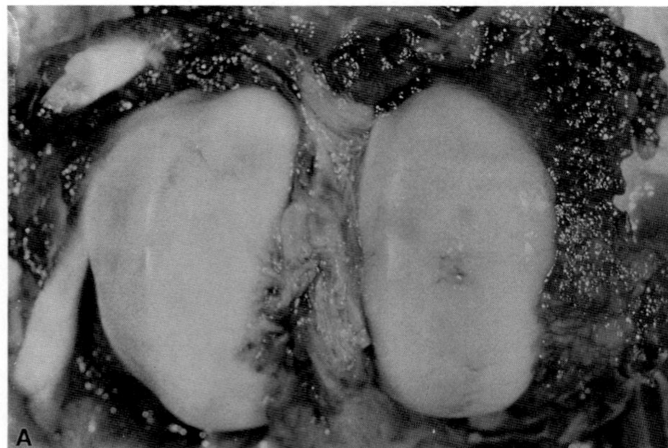

Fig. 1.141A Erysipelas. Distal femur. Pig. Severe proliferative synovitis and early degeneration of cartilage.

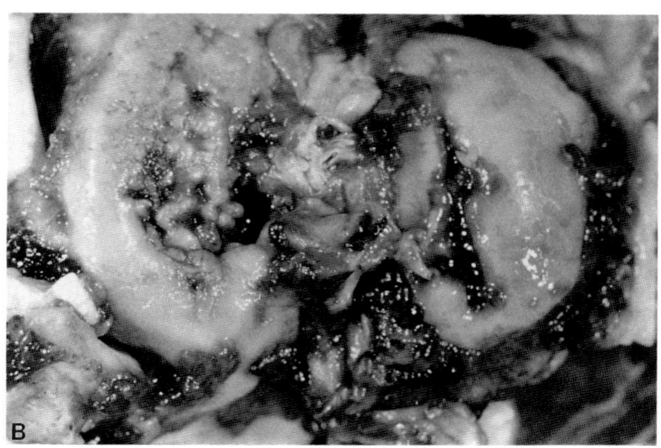

Fig. 1.41B Erysipelas. Distal Femur. Pig. Severe proliferative synovitis, and destruction of cartilage. [(A) and (B) courtesy of K. M. Johnston.]

ing and castration wounds, shear wounds, and cuts or abrasions acquired during dipping. Seldom is there any history of contact with pigs. Occasionally polyarthritis occurs in lambs following inoculation of contaminated serum of equine origin. Horses have been found to be naturally infected with *E. rhusiopathiae,* and natural infections in donor horses may be responsible for some of the contamination.

The lesion in sheep may be confined to the skin and subcutis at the point of entry, or there may be bacteremia with localization in joints. Rarely, death may occur from septicemia.

Fibrinopurulent polyarthritis, and osteomyelitis, is the usual form of erysipelas seen in young lambs after docking or castration, and sometimes following umbilical infections. The arthritis is subacute or chronic, and the morbidity may be as high as 50%. The mortality is low and is a consequence of severe lameness rather than an immediate result of the infection.

The main limb joints are involved. In the early stages, there is synovitis with an increased volume of turbid synovial fluid. Later, the fibrinous exudate may coagulate into firm pads. Initially, there is no change in the articular cartilage, but as a result of focal epiphyseal osteomyelitis, the cartilage recedes into pits in the epiphysis to produce what appear superficially to be ulcerations lying on medullary granulation tissue. In the chronic stages, the joints are stiffened and deformed by periarticular fibrosis and by periosteal and perichondrial osteophytes.

Cutaneous infections following dipping are associated with contamination of nonbactericidal dips with *E. rhusiopathiae*. Lesions occur most often about the fetlocks, but invasion may occur anywhere the skin is injured and contaminated. Postdipping infections, which mimic cutaneous erysipelas, are occasionally caused by both *Actinomyces pyogenes* and *Corynebacterium pseudotuberculosis,* and sometimes cutaneous erysipelas occurs when sheep are confined in wet, contaminated pens.

Lameness is severe and out of proportion to the gross changes in affected feet. The disease is febrile in some animals and associated with rapid wasting, but recovery occurs in 2 to 3 weeks. The affected pasterns are hot and painful, and there is regional lymphadenitis. The coronary band may be swollen, and the swelling, which is always slight, may extend to the metacarpal or metatarsal regions. The affected areas are progressively depilated. Incision reveals moderate erythema and slight edema in the subcutis.

Histologically, there is acute inflammation very similar to the rhomboidal lesions of swine erysipelas. There are superficial pustules in the epidermis. In the outer layers of the dermis, there are cellular thrombi within vessels and perivascular edema and accumulations of neutrophils. The reaction is more severe in the deeper layers of the dermis, in which the sweat glands suppurate, and larger vessels show acute necrotizing inflammation of their walls and cellular thrombosis. The same changes occur in the sensory laminae of the foot and are responsible for the severe lameness. In the course of 2 to 3 weeks, the inflammatory reaction resolves completely except for residual lesions in the larger blood vessels.

b. Streptococcal Polyarthritis and Meningitis in Swine Streptococci cause a variety of sporadic infections in **swine,** but there are also well-defined syndromes of streptococcal septicemia, with localization in synovial structures, meninges, and elsewhere, caused by *Streptococcus suis*. This organism shares cell-wall antigens with streptococci of Lancefield's group D but is not genetically related to them. Twenty-three serotypes, 1–22 and 1/2, are recognized, but many *S. suis* isolates, identified by morphology and biochemical reactions, are untypable. Associations between serotypes and site of isolation vary but, in general, serotype 1 bacteria tend to affect suckling pigs and are not pathogenic for other species, whereas serotype 2 bacteria tend to cause disease in weaner and feeder pigs and also cause meningitis and septicemia in humans. Serotype 7 may also have zoonotic significance. However, some serotype 2 strains cause only nonfatal pneumonia in gnotobiotic pigs, even following dexamethasone treatment. Increased virulence may depend on certain high-molecular-weight, cell wall proteins that are released by muramidase.

Streptococcus suis type 1 causes disease in piglets of 2 to 10 weeks, but in most cases affected piglets are 2–3 weeks of age. The disease may occur in one or all pigs in a litter, and a number of litters may succumb over a period of weeks.

The infection is initially bacteremic or septicemic with fever, and death may occur at this stage. There may be no sign of localization, and the lesions are those of a septicemia; the spleen is slightly enlarged and blue, the renal cortex bears few or many petechiae, and the lungs are hyperemic and show small pleural hemorrhages. There may be a slight excess of fluid in the serous cavities.

When the course is slightly longer, there are signs of synovial localization, including lameness and a slight puffiness of many joints, and signs of meningeal localization including ataxia, stupor, or coma, cutaneous hypersensitivity, and nystagmus. Every joint may show synovitis, but this is most evident in limbs. The synovial fluid is increased, slightly mucinous, and turbid; smears reveal a few neutrophils and a few bacteria. The meningitis is fibrinopurulent and affects mainly the basal meninges. The choroid plexuses are consistently involved, and yellow fibrinopurulent exudate covers them and floats in clumps in the ventricles. The cerebral aqueduct or foramina of Lushka may be occluded by exudate or by ependymal reaction, and internal hydrocephalus develops acutely.

Streptococcus suis type 1 probably is carried in the nasopharynx of sows and transferred to their litters shortly after birth. The portal of entry is the palatine tonsil. Piglets that are infected experimentally may have bacteremia for 2 to 3 weeks, after which developing immunity eliminates infection from blood and internal organs. The organism is relatively harmless unless it becomes established in joints or meninges.

Disease caused by *S. suis* type 2 may be clinically and pathologically similar, but often the range of lesions is greater. The organism is carried in the palatine tonsils of pigs, and infection is probably by the respiratory route. In infected herds, it is isolated from as many as 80% of normal pigs and is commonly found in nasal turbinates and pneumonic lungs, where it is probably a secondary invader. Limited outbreaks do occur, but sporadic, isolated disease is more common. The incubation period varies from 1 to 14 days, and the clinical course, from 4 to 48 hr. Affected pigs are usually about 10 to 14 weeks of age, and the most significant lesion is purulent meningitis, which, along with polyserositis, is visible grossly in about 50% of pigs. The bacteria probably enter the cerebrospinal fluid within monocytes via the choroid plexuses. Occasionally, *S. suis* type 2 causes septicemia in newborns. Pigs dying acutely may show only marked congestion of lungs and meninges with perhaps a few strands of fibrin adherent to the pleura, but purple discoloration of the skin and other signs of septicemia are usually present.

Some animals develop a more chronic disease, usually as a result of unsuccessful treatment, and they show subacute or chronic meningitis in which mononuclear cells may dominate the cellular exudate. Often there is encephalitis as inflammation extends into the brain via Virchow–Robin spaces or directly from the leptomeninges or the ventricles. Necrosis of submeningeal parenchyma is common, particularly in the cerebellar folia. Inflammation of the trigeminal ganglion and optic perineuritis also occur in a minority of pigs.

A few animals, usually those at the lower end of the age range, have purulent arthritis, and occasionally renal infarcts, hemorrhagic pancreatic necrosis, and hydronephrosis are seen. Fibrinopurulent pericarditis, endocarditis, or hemorrhagic, necrotizing myocarditis occur in some pigs. Endocarditis most often affects the mitral valve, but various combinations of involvement are seen. The myocarditis grossly may resemble mulberry heart disease, but the histologic lesions of necrotizing vasculitis and diffuse inflammation associated with bacteria are distinctive. *Streptococcus suis* is occasionally isolated from suppurative lesions in ruminants.

In experimental infections, other streptococci such as *S. pyogenes* (Lancefield's group A) and group C streptococci produce arthritis, and osteomyelitis, in swine. The latter organisms are occasionally isolated from field cases of arthritis in pigs.

Streptococci also cause polyarthritis and meningitis in **calves.** Metastatic iridocyclitis occurs in many of those infections in which polyarthritis occurs, but in none is it so consistent or prominent as in the streptococcal disease of neonatal calves. This is probably an umbilical infection in most cases but, because ocular lesions may be visible grossly by the twenty-fourth hour of life, there is a possibility that some of them are intrauterine. The clinical signs are of hypopyon, corneal opacity, and meningitis; the arthritis is acute, but there is no obvious joint swelling. Because of the very great tendency of the organisms to localize in the meninges and eyes, the disease is without a chronic phase. The morbid picture is the same as for the comparable disease in swine.

Streptococcus pneumoniae, the pneumococcus, of one or other serotype, occasionally produces in calves the pattern of infection previously described. This is the usual pattern in the experimental disease, but the natural disease is usually a fulminating septicemia in which petechiae and marked splenomegaly and turgidity are the only lesions. A similar disease occurs occasionally in kids.

Suppurative polyarthritis in **lambs** is caused by *Actinomyces pyogenes* as essentially a sporadic infection, *Staphylococcus aureus* as a sporadic infection or, as a complication of tick-borne fever, by *C. pseudotuberculosis,* and by *Streptococcus* spp. Next to *E. rhusiopathiae,* streptococci infections are perhaps the most frequent cause of polyarthritis in lambs. The navel is accepted as the route of entry, which is consistent with the maximal prevalence of the disease in sucklings. There is some variety of the combination of lesions seen at postmortem, and some of this may be due to different species of the organism; specific typing of *Streptococcus* spp. is seldom a routine procedure.

Localization in the joints and other organs is a sequel to bacteremia. Some lambs succumb in the early bacteremic phase, and if these show lesions, they are the changes of septicemia. Localization occurs in the course of 1 to 2 days and may involve any one or a combination of sites, which include the uvea, the cerebrospinal meninges, the valvular endocardium, myocardium, kidneys, and the joints. Meningeal localization seldom occurs in the absence of polyarthritis, but the latter may not be clinically evident in these cases because the clinical course is short and fatal. Polyarthritis may occur alone; initially, many joints are affected but, as the disease subsides, the inflammation may persist and lead to chronic changes only in the larger joints of the limbs. In approximately 20% of cases of subacute or chronic polyarthritis, there is coincident valvular endocarditis.

c. COLIFORM POLYARTHRITIS IN CALVES *Escherichia coli* often localizes in the joints or meninges, or both, in calves. In septicemic colibacillosis of neonatal calves, polyarthritis and tendovaginitis are particularly common but are early and easily overlooked lesions. *Escherichia coli* may enter the blood from either the gut, oropharynx, or umbilicus. Localization in the meninges and joints requires that the course of the disease be less fulminating than usual. The lesions are similar to those of the described streptococcal infection, with the exception that iridocyclitis and, especially, grossly visible intraocular lesions are quite unusual. Sometimes polyarthritis is chronic, with lesions restricted to one or two of the larger limb joints and tending to be symmetrical. Chronic coliform arthritis is often coincident with interstitial nephritis (white spotted kidney), which may develop into descending pyelonephritis and be fatal on this account.

d. HAEMOPHILUS SEPTICEMIA OF LAMBS Fastidious gram-negative organisms that have been isolated from several diseases of ruminants include *Haemophilus somnus* (cattle) and *H. agni* and *Histophilus ovis* (sheep). These bacteria are regarded as host-specific strains of a single species, which, according to priority, should be named *Histophilus ovis*. The strain known as *Haemophilus agni* produces an acute, fulminating, septicemic disease in lambs and more chronic disease in older sheep. There is an apparent regular association of high levels of feeding and the development of the disease, but the mechanism of this association is unknown. Little is known of the biology of the organism or the pathogenesis of the infection. *Haemophilus agni* has been isolated from preputial washings of normal rams in Canada, and *Histophilus ovis*, which causes *inter alia* mastitis in ewes and epididymitis in rams (see Volume 3, Chapter 5, The Male Genital System), and may be transmitted by preputial contamination at birth, can survive for several days in manure. Immunity to *H. ovis* epididymitis can be acquired via skin infection.

Animals with *Haemophilus agni* septicemia are usually not seen sick and are found dead, the course being less than 12 hr. Clinically, there is fever, depression, and extreme reluctance to move. The disease tends to remain enzootic in a flock, and losses may continue for several months.

The most constant postmortem findings are multiple hemorrhages throughout the carcass. They vary in severity, but are usually most obvious in lambs dying most acutely. The intermuscular connective tissue is wet and slightly stained with blood. The large blotchy hemorrhages of the serosa lack specificity, but tiny streaks of hemorrhage that are quite diagnostic can usually be seen in the muscles. They are best visualized by viewing the intact muscle through the perimysium, and are usually most common near the tendinous attachments.

In all cases except the most acute, small areas of necrosis are present in the liver. These appear as minute pale foci surrounded by a red halo. The spleen is enlarged; the pulp, soft and juicy. The absence of pulmonary lesions usually allows differentiation of this infection from *Pasteurella haemolytica* septicemia in which, in addition to pulmonary foci, there are focal hepatic lesions similar to those produced by *Haemophilus*. Animals that survive 24 hr or more, which is unusual unless the lambs are treated, develop fibrinopurulent arthritis, choroiditis, and basilar meningitis. In mature animals, a more extended course is usual, and lesions resemble those of *H. somnus* in cattle, and include myocarditis, embolic nephritis, and meningoencephalitis.

Histologically, there is evidence of overwhelming intravascular bacterial multiplication. Many vessels in many tissues are plugged with bacteria, and some of the emboli give rise to acute vasculitis and secondary infarctions. All tissues share to some extent in this reaction, but there is preferential involvement of liver and muscle. In these organs, there is often severe inflammation in the parenchyma adjacent to the damaged vessels.

In spite of the fact that the tissues often teem with the organisms, bacteriological confirmation may not be obtained unless samples are taken very soon after death and cultured in media enriched with blood or tissue fragments, and preferably in an atmosphere of 10% carbon dioxide. Even so, previous antibiotic treatment significantly impairs isolation of this organism.

e. GLÄSSER'S DISEASE OF SWINE Glässer's disease is a fibrinous meningitis, polyserositis, and/or polyarthritis caused by *Haemophilus parasuis*. The organism is part of the normal flora of the upper respiratory tract of conventional pigs and is often isolated from pneumonic lungs and regarded as a secondary invader. It is acquired from the dam shortly after birth, but colostral antibodies usually protect against disease, and active immunity develops by 7 to 8 weeks. It occasionally produces severe outbreaks of disease in animals aged around 5 to 12 weeks, often following transportation, or other management stress. Caesarian-derived pigs are susceptible at any age and often develop Glässer's disease after introduction to a boar test-station or a conventional herd.

Glässer's disease is defined on an etiological basis, as there are other organisms, such as *Streptococcus suis*, which produce similar lesions and similar diseases. *Mycoplasma* spp. also produce serositis and arthritis in swine, but the diseases are more chronic, and meningitis is either absent or lymphocytic depending on the species of mycoplasma involved.

Glässer's disease is peracute with high fever, lameness, and neurological disturbances which consist of paresis, stupor, and hyperesthesia. As in other septicemic diseases of pigs, the skin may show purple discoloration. The course is 1–2 days, and without treatment, the mortality rate is very high.

Usually the morbid picture is characterized by serofibrinous inflammation of the cranial and spinal meninges, the pericardium, the pleura, the peritoneum, and the synovial membrane of many joints. In individual cases, all these tissues or any combination of them may be inflamed;

occasionally the lesions occur in only one site, and in some pigs and some outbreaks, gross lesions are absent. Usually they are found in meninges, joints, peritoneum, pleura, and pericardium, in that descending order of frequency. Meningitis, which is more severe in the cranial than the spinal meninges, occurs in more than 80% of pigs. Bacterial meningitis is usually most obvious in the basal meninges, but, if the course is prolonged, the exudation may spread in the sulci over the hemispheres; in Glässer's disease, in spite of the short course, prominent changes often occur not only in the basal meninges, but also over the cerebellum and the posterior lobes of the hemisphere. Polyarthritis is most severe in the atlanto-occipital and large limb joints. The synovial fluid is increased in volume and contains enough leukocytes to make it cloudy. The synovitis is characterized by gray-yellow fibrin, which covers the membrane or accumulates as a meniscuslike pad between the articular surfaces.

The serosal exudate is usually serofibrinous and often more fibrinous than serous. The fibrin which mats the membranes is gray or yellow and rather dry. A small excess of straw-colored fluid is present in the cavities, and may contain yellow wads of fibrin. There is little or no hyperemia of the subserosa. Intense congestion of the gastric fundic mucosa is present, and sometimes there are petechiae in the renal cortex. Histologic changes are those of septicemia with fibrinous inflammation. Thrombosis of vessels in skin, meninges, and renal glomeruli is often prominent.

Haemophilus parasuis can be isolated from visceral pleura, provided its cultural needs are met, and the post-mortem interval is not too long. It is seldom isolated from other sites.

f. *Borrelia burgdorferi* Infection *Borrelia burgdorferi* is a spirochete that causes multisystemic disease in humans (Lyme disease, Lyme borreliosis) and is associated with disease in several species of animals. The disease in humans is long known with a wide distribution (North America, Europe, East Asia, possibly Australia, Africa), but the causative agent and the occurrence of infections in animals were unrecognized for many years.

The organism is transmitted by ticks, mainly the three-host tick *Ixodes dammini* in North America. Eggs are deposited by engorged females in summer and hatch in about 1 month. Larvae engorge for a few days in the period from May to September and about 2 months later molt to the nymphal stage. In the following spring–summer period they feed, drop off, and molt to the adult stage. This 2-year cycle may be extended to 4 years, since unfed larvae and nymphs may overwinter. Larvae and nymphs feed mainly on birds and small mammals—in North America, *Peromyscus* spp. (white-footed mouse and deer mouse) are the main mammalian hosts. The adults attach mainly to large mammals, particularly *Odocoileus virginianus*, the white-tailed deer, but raccoons, dogs, horses, cattle, and humans are among the potential victims. *Borrelia* infection in the principal wild-life hosts is essentially asymptomatic. Similar epidemiological patterns involving ticks of the *Ixodes ricinus* group, and various hosts exist in other parts of the world. In North America, *B. burgdorferi* has been isolated from deer flies and horse flies, which are potentially effective mechanical vectors, and from mosquitoes, which are likely to be less important. Such vectors are significant only in areas where efficient tick vectors maintain infection among wildlife populations.

Borrelia burgdorferi has been isolated from various other ticks in North America, including *Rhipicephalus sanguineus*, and the possibility of concomitant infection with *Ehrlichia canis* and the spirochete might account for the polyarthritis, which has been reported as a rare manifestation of canine ehrlichiosis.

Both transovarial and transstadial transmission of *Borrelia burgdorferi* occur, but only the latter is epizootiologically important. Intrastadial and interstadial transmission, which occur when several larvae and nymphs feed on an infected host, is of greatest significance to spread of infection. Transmission of infection to large mammals is usually by the bite of an adult tick and requires attachment for about 24 hr. The organism may be present in the urine of *Peromyscus* spp. and cattle, but the prevalence of infection by direct contact is not known.

Disease caused by *Borrelia burgdorferi* occurs in humans, dogs, horses, cattle, and possibly cats. In all species, infection is far more common than disease, but the host–bacterium interactions responsible are not well defined. Arthritis is a common feature of disease in animals with, in addition, myocarditis and nephritis in dogs, ocular disease and probably encephalitis in horses, and abortion in cattle. Since this is a septicemic disease, other manifestations of infection might be expected.

In dogs, the prevalence of antibody to *B. burgdorferi* and signs of arthritis are both high in areas where Lyme disease is common in humans, and the reverse is also true. Lameness, anorexia, fever, and fatigue are prominent signs, whereas lymphadenopathy and nephritis occur in a small proportion of dogs. The arthritis is typically migratory, intermittent, and oligoarticular. Detailed pathologic descriptions are unavailable since antibiotic treatment appears to be successful, and experimental reproduction of disease in dogs is not. In humans, the arthritis may be recurrent, leading eventually to ulceration of cartilage. In Lewis rats, acute exudative arthritis, tendonitis, and bursitis occur by 30 days postinfection, then regress, but exacerbations develop in some animals. Synovial, but not cartilaginous, ulceration occurs in rats, and there is villus hyperplasia with lymphoplasmacytic and macrophage infiltrates in the synovial membranes. Lymphoplasmacytic myocarditis also occurs in rats, and nonsuppurative myocarditis develops in a small proportion of dogs. Clinical evidence of renal disease is also seen in a few dogs, and organisms can be demonstrated in the kidney, but the lesions are poorly defined.

In horses, signs of lethargy, low-grade fever, painful and swollen joints, and reluctance to move are seen. Laminitis and panuveitis are less common. The organism has been demonstrated in anterior chamber fluid. A probable case of encephalitis is recorded. Few equine necropsies are reported, but it appears that, grossly and microscopically, arthritis is present and resembles that in dogs, with the addition of marked thickening and hyalinization of arterioles in synovial tissues.

Limited numbers of reports suggest the disease in cattle is associated with abortion, and may have a chronic course with relapses following treatment and the eventual development of emaciation. Otherwise, signs resemble those in dogs and horses. Arthritis is a common feature of the disease, and there is limited evidence of myocarditis and nephritis. In humans, a characteristic skin lesion, *erythema chronicum migrans,* may develop at the site of the tick bite, and represents the first stage of Lyme disease. Similar areas of discoloration have been described on bovine mammary skin.

Borrelia burgdorferi has been isolated from blood, urine, colostrum, and synovial fluid of cows, and the blood of a newborn calf. Antibodies in the blood of an aborted fetus suggest the organism may cross the placenta.

Diagnosis of borreliosis is made to some extent by exclusion. *Borrelia burgdorferi* may be demonstrable by darkfield or phase-contrast microscopy in urine or synovial fluid and by silver, immunoperoxidase, or immunofluorescent stains in synovial membranes or other tissues. Isolation of the organism may be possible by using special media. Often the diagnosis is based on typical signs in an animal with a high titer and potential or known exposure to the organism. There is serological cross-reactivity between *B. burgdorferi* and some other spirochetes, but apparently not with leptospires.

g. Chlamydia Polyarthritis in Calves and Lambs

The genus *Chlamydia* contains two, or possibly three species of gram-negative bacteria. *Chlamydia trachomatis* mainly infects humans, and *C. psittaci* mainly infects other animals, while a third species, *C. pneumoniae* that causes acute respiratory infections in humans, is proposed. Chlamydia belong to a specific eubacterial group of obligate intracellular parasites characterized by the absence of cell wall peptidoglycans. There are several biotypes of *C. psittaci,* and biotypes 1 and 2, corresponding to serotypes 1 and 2, are important in cattle and sheep. Serotype 1 is isolated from abortions and pneumonia (the abortion agent also causes epididymitis, orchitis, and seminal vesiculitis) and serotype 2, from polyarthritis, encephalitis, and conjunctivitis. Within each serotype, isolates from cattle and sheep are antigenically alike. Chlamydia may also cause enteritis in calves and lambs, and this may precede development of arthritis in calves. Sporadic or epidemic chlamydial polyarthritis occurs in calves and lambs, and occasionally in other animals. The natural habitat of chlamydiae is the intestinal tract, where they multiply in the mucosal epithelium. Organisms enter the lamina propria, and nonvirulent strains rarely progress past the mesenteric nodes. Following oral exposure of calves to virulent strains, chlamydiae spread to the liver and mesenteric nodes via blood vessels and lymphatics of the lower small intestine. Multiplication in these sites is followed by a primary, low-level chlamydemia with localization and multiplication in spleen, liver, lungs, and kidneys. The joints are infected during a secondary chlamydemia, and arthritogenic strains produce their most severe effects in synovial membranes, with clinical arthritis developing about 10 days after oral inoculation. They also produce severe watery or bloody diarrhea a few days prior to the development of arthritis.

The disease in calves is severe, both naturally and experimentally causing a high mortality rate. Affected calves may be weak at birth, suggesting intrauterine infection, and in these animals, fever, anorexia, reluctance to stand or move, and swelling of joints develop in 2 to 3 days, and death occurs 2 days to 2 weeks after onset of signs. All or many joints are affected, those of the limbs most severely. The subcutaneous and adjacent periarticular tissues are edematous with clear fluid, and this extends also around tendon sheaths. Surrounding muscles are hyperemic and edematous with petechiae in the fascia. Joint cavities are distended with turbid yellow-gray fluid, and strands or wads of fibrin adhere to the synovium. Viscera may show changes attributable to systemic infection.

In contrast to that in calves, the ovine syndrome is characterized by high morbidity and negligible mortality even though the infection is systemic. Conjunctivitis is part of the clinical picture in lambs, in which there is also depression, reluctance to move, joint stiffness which disappears with exercise, anorexia, and loss of weight. Most lambs recover, but a few are permanently lame. The articular lesions are similar to those in calves although modified by the milder reaction, the longer course allowing fibrotic thickening of tendon sheaths and articular capsules and hyperplasia of synovial villi. In soft tissues, including the central nervous system, histologic traces of inflammation are present. The organism is sometimes demonstrable in cytoplasm of synovial cells with Giemsa, Gimenez, or modified Stamp stains.

Bibliography

Appel, M. J. G. Lyme disease in dogs and cats. *Compend Cont Educ* **12:** 617–625, 1990.

Bellah, J. R., Shull, R. M., and Shull Selcer, E. V. *Ehrlichia canis*-related polyarthritis in a dog. *J Am Vet Med Assoc* **189:** 922–923, 1986.

Bennett, D., and Taylor, D. J. Bacterial infective arthritis in the dog. *J Small Anim Pract* **29:** 207–230, 1988.

Burgess, E. C., Gendron-Fitzpatrick, A., and Wright, W. O. Arthritis and systemic disease caused by *Borrelia burgdorferi* infection in a cow. *J Am Vet Med Assoc* **191:** 1468–1470, 1987.

Cohen, N. D., and Cohen, D. Borreliosis in horses: A comparative review. *Compend Cont Educ* **12:** 1449–1458, 1990.

Curtiss, P. H. The pathophysiology of joint infections. *Clin Orthop* **96:** 129–135, 1973.

Cutlip, R. C., and Ramsey, F. K. Ovine chlamydial polyarthritis: Sequential development of articular lesions in lambs after intraarticular exposure. *Am J Vet Res* **34:** 71–75, 1973.

Eamens, G. J., and Nicholls, P. J. Comparison of inoculation regimens for the experimental production of swine erysipelas arthritis. I. Clinical, pathological and bacteriological findings. *Aust Vet J* **66:** 212–216, 1989.

Eugster, A. K., and Storz, J. Pathogenetic events in intestinal chlamydial infections leading to polyarthritis in calves. *J Infect Dis* **123:** 41–50, 1971.

Gogolewski, R. P., Cook, R. W., and O'Connell, C. J. *Streptococcus suis* serotypes associated with disease in weaned pigs. *Aust Vet J* **67:** 202–204, 1990.

Higgins, R. *et al.* Isolation of *Streptococcus suis* from cattle. *Can Vet J* **31:** 529, 1990.

Kennedy, P. C. *et al.* A septicemic disease of lambs caused by *Haemophilus agni* (new species). *Am J Vet Res* **19:** 645–654, 1958.

Koning, J. de, Bosma, R. B., and Hoogkamp-Korstanje, J. A. A. Demonstration of spirochaetes in patients with Lyme disease with a modified silver stain. *J Med Microbiol* **23:** 261–267, 1987.

Liven, E., Larsen, H. J., and Lium, B. Infection with *Actinobacillus suis* in pigs. *Acta Vet Scand* **19:** 313–315, 1978.

Marsh, H. *Corynebacterium ovis* associated with an arthritis in lambs. *Am J Vet Res* **8:** 294–298, 1947.

Moulton, J. E., Rhode, E. R., and Wheat, J. D. Erysipelatous arthritis in calves. *J Am Vet Med Assoc* **123:** 335–340, 1953.

Pedersen, K. B., Henrichsen, J., and Perch, B. The bacteriology of endocarditis in slaughter pigs. *Acta Path Microbiol Immunol Scand* **92:** 237–238, 1984.

Roberts, E. D. *et al.* Pathologic changes of porcine suppurative arthritis produced by *Streptococcus equisimilis*. *Am J Vet Res* **29:** 253–262, 1968.

Sanford, S. E. Gross and histopathological findings in unusual lesions caused by *Streptococcus suis* in pigs. I. Cardiac lesions. *Can J Vet Res* **51:** 481–485, 1987.

Sanford, S. E. Gross and histopathological findings in unusual lesions caused by *Streptococcus suis* in pigs. II. Central nervous system lesions. *Can J Vet Res* **51:** 486–489, 1987.

Shewen, P. E. Chlamydial infection in animals: A review. *Can Vet J* **21:** 2–11, 1980.

Shupe, J. F., and Storz, J. Pathologic study of psittacosis-lymphogranuloma polyarthritis of lambs. *Am J Vet Res* **25:** 943–951, 1964.

Sikes, D., Neher, G. M., and Doyle, L. P. The pathology of chronic arthritis following natural and experimental *Erysipelothrix* infection of swine. *Am J Pathol* **32:** 1241–1251, 1956.

Simpson, R. M., Hodgin, E. C., and Cho, D.-Y. *Micronema deletrix*-induced granulomatous osteoarthritis in a lame horse. *J Comp Pathol* **99:** 347–351, 1988.

Smart, N. L. *et al.* Glässer's disease and prevalence of subclinical infection with *Haemophilus parasuis* in swine in southern Ontario. *Can Vet J* **30:** 339–343, 1989.

Steinberg, J. J., and Sledge, C. B. Chondrocyte mediated cartilage degradation: Regulation by prostaglandin E_2, cyclic AMP, and interferon alpha. *J Rheumatol* (Suppl. 27) **18:** 63–65, 1991.

Szczepanski, A., and Benach, J. L. Lyme borreliosis: Host responses to *Borrelia burgdorferi*. *Microbiol Rev* **55:** 21–34, 1991.

Thomson, R. G., and Ruhnke, H. L. *Haemophilus* septicemia in piglets. *Can Vet J* **4:** 271—275, 1963.

Waxler, G. L., and Britt, A. L. Polyserositis and arthritis due to *Escherichia coli* in gnotobiotic pigs. *Can J Comp Med* **36:** 226–233, 1972.

Williams, A. E., and Blakemore, W. E. Pathogenesis of meningitis caused by *Streptococcus suis* type 2. *J Inf Dis* **162:** 474–481, 1990.

Wood, R. L. Swine erysipelas—a review of prevalence and research. *J Am Vet Med Assoc* **184:** 944–949, 1984.

Wood, R. L., Cutlip, R. C., and Shuman, R. D. Osteomyelitis and arthritis induced in swine by Lancefield's Group A streptococci (*Streptococcus pyogenes*). *Cornell Vet* **61:** 457–470, 1971.

h. Mycoplasma Infections Mycoplasmas are ubiquitous organisms, difficult to isolate, propagate, and identify, and often difficult to associate with specific diseases. Polyserositis and polyarthritis produced by these organisms are problems of considerable magnitude, which often complicate mycoplasma (enzootic) pneumonia, or atrophic rhinitis. The diseases, of themselves, are seldom fatal, but produce a lingering debility from which animals do not completely recover.

Three *Mycoplasma* species are pathogenic for swine; *Mycoplasma hyopneumoniae* (*suipneumoniae*) causes enzootic pneumonia; *Mycoplasma hyorhinis* causes polyserositis and polyarthritis, usually in animals about 3 to 10 weeks of age; *Mycoplasma hyosynoviae* also causes polyarthritis, but not polyserositis, and usually in pigs older than 10 weeks of age. Other mycoplasmas are sometimes associated with various diseases of swine, but they are not primary pathogens.

Mycoplasma hyorhinis is a commensal or parasite of the nasal passages in most pigs, in which location it is essentially harmless. It commonly colonizes pneumonic lesions, especially those of *Mycoplasma* pneumonia, without obviously altering the lesion, but the lungs might be the main route for entry of this agent to the blood. The pathogenicity of *M. hyorhinis* is increased in intercurrent disease and in a variety of stressful circumstances.

After an incubation period of 3 to 4 days, the infection produces an erratic febrile reaction, anorexia, depression, and loss of body weight. Growth rate remains retarded, although after a month or so there may be signs of recovery. The infection is more likely to be fatal in suckling piglets or those with intercurrent pneumonia.

The organism has a predilection for collagen-containing structures including serous and synovial membranes and cartilage, but evidence of meningeal injury is either absent or mild and lymphocytic. The systemic infection persists after experimental inoculation for up to 2 months and probably longer. The serosal lesions may vary, and any combination of epicarditis, pleuritis, or peritonitis is possible. The serosal reaction is serofibrinous, and the cellular component is mononuclear with focal accumulations of neutrophils. The epicardium is uniformly affected, whereas pleuritis may be confined to the ventral margin of the lung and the pleura between lobes, and may extend to the outer surface of the pericardial sac. The peritoneal exudate is patchy and may be concentrated in the anterior abdomen, especially between liver and diaphragm, but often involves loops of intestine and the coiled colon. Progressive organization of exudates produces tough adhesions, but even after many weeks, there is still histologic evidence of active inflammation, which is consistent with the persistence of the organism.

Early in the disease, the effects of polyserositis predominate; later, signs of polyarthritis become evident as the initial synovial edema and hyperemia are replaced by villous hypertrophy, synovial thickening, and organization of fibrinous exudate. In chronic disease, thickening and brown discoloration of joint capsules, erosion and discoloration of articular cartilages, and pannus are observed. Discoloration of joint structures is probably a sequel to synovial hemorrhage.

Early microscopic changes consist of hyperemia, diffuse lymphocyte and macrophage infiltration, and mild villus hypertrophy followed by ulceration of synovial membranes. Fibrinopurulent exudate is consistently present in virtually all joints, and is most abundant in the recesses where the synovial membrane is reflected from the articular cartilage. Later, perivascular lymphoplasmacytic accumulations and more pronounced villus hypertrophy become evident, and fibrinous exudate is less prominent. Early in experimental infections, mycoplasmas invade the lacunae of superficial chondrocytes, and degeneration of chondrocytes combined with the enzymic activity of adherent neutrophils produces erosions of articular cartilage (Fig. 1.142A,B).

In chronic mycoplasma synovitis, there is marked villus hypertrophy and proliferation of synovial cells. Perivascular cuffs are less prominent, but nodules of lymphocytes, often surrounded by lymphatics, occur frequently in the villi. Dense fibrosis of some villi is evident. Erosions, developing first at the periphery of the articular cartilages, later occur centrally, and erosion of cartilage and subarticular bone by pannus is not unusual in prolonged survivors. Joint contractures are associated with the chronic arthritis and thickened joint capsules. Chronic lesions re-

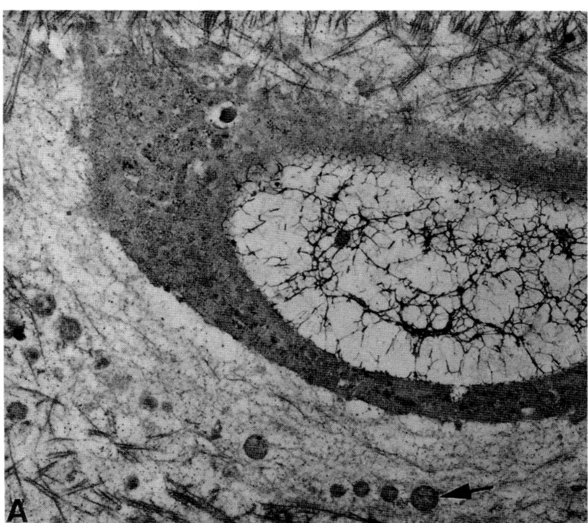

Fig. 1.142A *Mycoplasma hyorhinis* arthritis. Pig. Organisms in lacuna (arrow). Chondrocyte is degenerating.

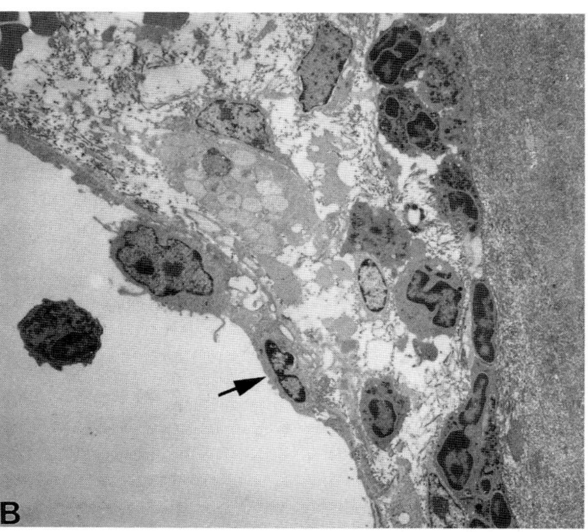

Fig. 1.142B *Mycoplasma hyorhinis* arthritis. Pig. Leukocytes apposed to (arrow) and invading degenerate cartilage. [(A) and (B) courtesy of K. M. Johnston.]

main active for many months, and *M. hyorhinis* can often be cultured from surface exudate in these joints.

Mycoplasma hyosynoviae is carried indefinitely in the pharynx and tonsils of many sows, who transmit it to their offspring. For some reason, transmission to pigs younger than 6 weeks of age is uncommon, and separation of pigs from their mothers by 5 weeks of age usually prevents infection.

The disease often develops following management stress and is more frequent and more severe in pigs with poor conformation and gait. Joint infection occurs in many pigs, but arthritis occurs most often in predisposed pigs.

Following experimental infection, there is an incubation period of 5 to 10 days. Lameness, lasting 3 to 10 days, occurs in many pigs and may involve one joint or several. Many pigs recover completely but, in 5 to 15% of diseased swine, lameness persists indefinitely.

The early pathologic changes resemble those caused by *M. hyorhinis*; chronic lesions are not completely described but are probably qualitatively similar to those produced by *M. hyorhinis*.

Failure to isolate organisms from many pigs with chronic *Mycoplasma* arthritis requires comment. It may be that the cultural techniques utilized are inadequate to demonstrate low numbers of organisms. Alternatively, the continued inflammatory reaction may be due to an immune response to persisting *Mycoplasma* antigen. Support for this view is derived from the observation that rabbits sensitized to *M. arthritidis*, either by infection or hyperimmunization, develop chronic arthritis when challenged with nonviable *M. arthritidis* antigen. Antibody in the synovial fluid of sensitized animals could retain antigen in immune complexes. These complexes may localize in synovial tissues and articular cartilage and stimulate a cell-mediated immune response, which causes the tissue destruction. The nature of the antigens is undetermined; since mycoplasmas lack a cell wall, and lack the ability to synthesize peptidoglycans, the antigen must differ from those responsible for chronicity of some bacterial arthritides. Whether this hypothesis is correct, or relevant to the natural disease, is undetermined.

Several syndromes in **goats** are caused by mycoplasmas. There is some overlap in their clinical and pathological features, which does not allow them to be defined pathologically. Because these organisms are difficult to study, an etiologic classification is also imprecise.

At present, it appears that three syndromes are separable. **Contagious caprine pleuropneumonia** (see Volume 2, Chapter 6, The Respiratory System) is typically a pulmonary infection and, atypically, a fatal septicemia or cellulitis, the latter manifestation being known as edema disease and occurring in Greece. The cause is *Mycoplasma* F38, which is related antigenically to *Mycoplasma mycoides* subsp. *capri* (formerly thought to be the cause), and to *M. capricolum*.

Contagious agalactia is caused by *M. agalactiae;* the disease is characterized by localization in the eyes, joints, and mammary glands. The name is too restrictive, but it is entrenched, and this disease is described with Diseases of the Mammary Glands in Volume 3, Chapter 5, The Female Genital System. Limb lesions affect primarily the periarticular connective tissues, and synovitis develops by extension. The third syndrome, that of septicemia–arthritis, is described here.

The **septicemia–arthritis** syndrome occurs in goats in North America, Sweden, and Australia. Two Mycoplasma species, *M. mycoides* subsp. *mycoides* and *M. capricolum,* are responsible, and a third, *M. putrefaciens,* occasionally causes polyarthritis and mastitis. The disease produced by *M. mycoides* subsp. *mycoides* and *M. capricolum* is similar, although pulmonary lesions may be more common and severe, and other lesions less severe, with the former. Sheep and pigs can be infected experimentally, producing a peracute septicemia or an acute disease with synovitis, serositis, and meningitis. *Mycoplasma mycoides* subsp. *mycoides* is separable into two groups on the basis of colony type—one of these groups is pathogenic for cattle but has very low virulence.

The septicemia–arthritis syndrome in goats is acutely febrile and associated with a drop in milk yield due to mastitis. In the course of ~2 days, affected animals develop acute polyarthritis. In suckling kids, the disease is peracute, and localization in joints is less obvious. During the systemic phase of the infection, there is a generalized acute lymphadenitis, splenitis, histiocytic meningitis, glomerulitis and renal tubular degeneration, and focal coagulative necrosis in the liver. Fibrinous peritonitis and pleuritis are sometimes present. Capillaries in many organs may contain thrombi, probably a manifestation of disseminated intravascular coagulation. Diffuse interstitial pneumonia, characterized by neutrophils or monocytes in alveolar walls and leakage of protein-rich fluid into alveoli, and sometimes accumulation of neutrophils in bronchioles, develops in kids. Pneumonia in adults tends to be less diffuse and more chronic.

As the infection subsides with articular localization, the draining lymph glands remain large and chronically hyperplastic. The articular lesion is a severe fibrinopurulent inflammation; any, and usually many, synovial structures may be involved.

The epidemiology of the septicemia–arthritis syndrome is incompletely understood. Some does probably acquire *M. capricolum* via the teat canal, and infection can spread to the opposite gland. Some kids probably are infected via milk, and horizontal transmission of both *M. capricolum* and *M. mycoides* subsp. *mycoides* occurs among young kids housed together.

Septicemia and arthritis in **lambs** that had exposure to goats' milk occurred in California, *M. capricolum,* or a closely related organism, being responsible, and causing a disease similar to that which occurs in kids.

Mycoplasma mycoides, subsp. *mycoides,* the agent of contagious bovine pleuropneumonia, causes polyarthritis and endocarditis in a proportion of **calves** in which it is used as a vaccine. Sporadic outbreaks of polyarthritis in cattle, not associated with *M. mycoides* vaccination, also occur in North America and Australia and are probably all caused by *M. bovis* (formerly *M. agalactiae* var. *bovis* and *M. bovimastitides*). This organism also causes mastitis, with or without arthritis, and is often isolated from bovine fibrinous pneumonia, where its significance is not well understood.

Mycoplasma bovis probably enters the bloodstream from the lungs and localizes in synovial membranes. Oral infection from the milk of a mastitic dam may also occur. Outbreaks of disease usually occur in feedlots and are often preceded by management stress. Usually 15% of animals show signs of lameness, followed within 1 to 2 days by severe swellings around joints. The stifle, shoulder, hock, and elbow joints are most often affected, and usually, severe lesions develop in one limb only.

In acute disease the joints are distended by opaque cream-colored fluid which contains flakes of fibrin. Synovia are red and swollen. Microscopic changes include light to diffuse infiltrations of lymphocytes, macrophages and plasma cells, and hyperplasia of synovial cells and of synovial villi. Focal perivascular necrosis and ulceration of synovial membrane may be present and attract numerous neutrophils, which accumulate in the joint spaces.

Most animals recover in 1 to 2 months, but a few develop chronic arthritis and are culled. These animals have proliferative and erosive arthritis with well-vascularized pannus extending from the thickened synovial membrane across the articular cartilage and eroding and undermining it from the joint margins. Spread of inflammation to the subchondral bone with development of osteomyelitis occurs in some animals. The predominant cells in the chronic synovial lesions are lymphocytes and plasmacytes with fewer neutrophils and macrophages. Isolation of mycoplasmas is relatively easy in the acute disease but is often more difficult in chronic stages.

Acute serofibrinous arthritis involving one or two joints, usually including the atlanto-occipital joint, occurs regularly in cattle with meningoencephalitis caused by *Haemophilus somnus*. These animals die suddenly, usually have hemorrhagic necrosis in the brain, and yield *Haemophilus somnus* on culture (see Chapter 3, The Nervous System).

Occasionally mycoplasma arthritis occurs in **cats** and **dogs**; *M. gateae* caused polyarthritis and tenosynovitis in a cat, *M. felis* was recovered from arthritis in a cat with suspected immune deficiency, and *M. spumans* was isolated in Australia from a greyhound with polyarthritis, and is a possible cause of the infectious arthritis seen in that breed in that country.

Bibliography

Barton, M. D. *et al.* Isolation of *Mycoplasma spumans* from polyarthritis in a greyhound. *Aust Vet J* **62:** 206, 1985.

DaMassa, A. J., and Porter, T. L. Acute disease in lambs caused by a *Mycoplasma* species. *Vet Rec.* **121:** 166–167, 1987.

DaMassa, A. J., Holmberg, C. A., and Brooks, D. L. Comparison of caprine mycoplasmosis caused by *Mycoplasma capricolum, Mycoplasma mycoides* subsp. *mycoides,* and *Mycoplasma putrefaciens. Israel J Med Sci.* **23:** 636–640, 1987.

Hjerpe. C. A., and Knight, H. D. Polyarthritis and synovitis associated with *Mycoplasma bovimastitidis* in feedlot cattle. *J Am Vet Med Assoc* **160:** 1414–1418, 1972.

Hooper, P. T., Ireland, L. A., and Carter, A. Mycoplasma polyarthritis in a cat with probable severe immune deficiency. *Aust Vet J* **62:** 352, 1985.

Moise, N. S. *et al. Mycoplasma gateae* arthritis and tenosynovitis in cats: Case report and experimental reproduction of the disease. *Am J Vet Res* **44:** 16–21, 1983.

Roberts, E. D., Switzer, W. P., and Ramsey, F. K. The pathology of *Mycoplasma hyorhinis* arthritis produced experimentally in swine. *Am J Vet Res* **24:** 19–31, 1963.

Rosendal, S., Erno, H., and Wyand, D. S. *Mycoplasma mycoides* subspecies *mycoides* as a cause of polyarthritis in goats. *J Am Vet Med Assoc* **175:** 378–380, 1979.

Ross, R. F., and Duncan, J. R. *Mycoplasma hyosynoviae* arthritis of swine. *J Am Vet Med Assoc* **157:** 1515–1518, 1970.

Ruhnke, H. L. *et al.* Isolation of *Mycoplasma mycoides* subspecies *mycoides* from polyarthritis and mastitis of goats in Canada. *Can Vet J* **24:** 54–56, 1983.

Ryan, M. J. *et al.* Morphologic changes following intraarticular inoculation of *Mycoplasma bovis* in calves. *Vet Pathol* **20:** 472–487, 1983.

Swanepoel, R., Efstratiou, S., and Blackburn, N. K. *Mycoplasma capricolum* associated with arthritis in sheep. *Vet Rec* **101:** 446–447, 1977.

Washburn, L. R., Cole, B. C., and Ward, J. R. Chronic arthritis of rabbits induced by mycoplasmas. III. Induction with nonviable *Mycoplasma arthritidis* antigens. *Arthritis Rheum* **25:** 937–946, 1982.

i. Caprine Arthritis-Encephalitis This syndrome is caused by a persistent infection with a nononcogenic retrovirus (lentivirus), family Retroviridae, subfamily Lentivirinae. The neurologic, mammary, respiratory and systemic features are described in Chapter 3, The Nervous System. In many herds, arthritis is the major or only clinical manifestation of infection, with signs of lameness often associated with carpal hygromas, weight loss, and reduced milk production. The prevalence of infection in a herd may reach 100%, but expression of disease is variable and rarely exceeds 25–30%.

A high prevalence of unilateral or bilateral carpal hygroma is characteristic, but virtually all joints have microscopic evidence of infection. The hygromas are chronic lesions that appear as flattened, cystic, subcutaneous distensions over the anterior carpus, and are filled with yellow or bloody fluid which contains fibrinous or gelatinous masses. Usually there is no communication with the carpal joint or tendon sheaths. The tendon sheaths and joint capsule are thickened by fibrous tissue in which collagen necrosis and mineralization may occur. Often the major joints, especially carpus and stifle, are distended with clear yellow fluid. Villous hypertrophy of synovia, fibrillation and erosion of cartilage, and destruction of joint structures by pannus are also most often seen in the carpus and stifle, but arthritis is demonstrable microscopically in many joints.

The arthritis–encephalitis virus is readily transmitted to kids in colostrum and milk, and horizontal transmission probably occurs. Colostrum may also contain antibody, but this does not influence viral transmission. In blood, the virus is present in monocytes and is activated when they mature into macrophages. Following infection, there is vascular injury to synovia-lined structures with exudation of plasma proteins into synovial fluid. Synovial villi hypertrophy, develop edema, and accumulate focal and diffuse infiltrations of plasma cells and lymphocytes and scattered macrophages, sometimes multinucleated. Other villi may be fibrosed, and there is hyperplasia of synoviocytes on their surfaces. Hyalinized masses of fibrin also form villuslike structures in the inflamed joints, and fibrin may be layered over the synovial membranes.

A related lentivirus of sheep, the progressive pneumonia virus, causes arthritis in its natural host and also produces arthritis when inoculated into kids. Similarly, the caprine arthritis–encephalitis virus causes lesions in lambs. Natural cross infections are not known to occur. In lentivirus arthritis, macrophages and lymphocytes are the predominant cells in synovial fluid. A small number of the macrophages are infected with the virus, but there is a reduction in virus gene expression that may be due to interferon in the synovial fluid. Plasma cells in the inflamed synovia produce immunoglobulin G1, presumably in response to persistent infection there, and the concentration in synovia is higher than that in serum. The relevance of this to lesion development is unclear, since the predominant lymphocytes in the synovia are T cells, and there is some suggestion that interaction between these cells and macrophages may be responsible for the chronic inflammatory response.

Lameness occasionally occurs in cats in association with **feline calicivirus** infection or 5–7 days following live-virus vaccination. Viral antigens are demonstrable in synovial macrophages 7 but not 4 days after experimental infection, but lesions are minimal and the disease process is poorly understood.

Bibliography

Bennett, D. *et al.* Detection of feline calicivirus antigens in the joints of infected cats. *Vet Rec* **124:** 329–332, 1989.

Bulgin, M. S. Ovine progressive pneumonia, caprine arthritis–encephalitis, and related lentiviral diseases of sheep and goats. *Vet Clin North Am Food Anim Pract* **6:** 691–704, 1990.

Robinson, W. F., and Ellis, T. M. Caprine arthritis–encephalitis virus infection: From recognition to eradication. *Aust Vet J* **63:** 237–241, 1986.

Woodard, J. C. *et al.* Caprine arthritis–encephalitis: Clinicopathologic study. *Am J Vet Res* **43:** 2085–2096, 1982.

C. Miscellaneous Inflammatory Lesions of Joint Structures

1. Bursitis

Bursitis often occurs as part of the synovial localization in hematogenous infections. There are also some isolated forms of bursitis, especially in horses, that are probably initiated by trauma and are characterized by accumulation of excess serous fluid and by varying degrees of synovial hypertrophy. They represent loci of lowered resistance and are predisposed to the localization of organisms which may alter the nature of the inflammatory reaction while contributing to its progression. The bursae which cover bony prominences are especially liable to traumatic injury. Capped elbow and capped hock are examples of serous bursitis. Carpal bursitis is also known as hygroma and is not uncommon in cattle; large hygromas should be suspected of being secondarily infected with *Brucella abortus*. Carpal hygromas are common in goats infected with caprine arthritis–encephalitis virus.

Fistulous withers and poll evil are the most serious forms of bursitis in horses and mules, affecting the supraspinous and atlantal bursae, respectively. Inflammation of the navicular bursa is discussed with navicular disease. The affected bursae are those between the nuchal ligament and the second, or less often, the third, fourth, or fifth thoracic spine in the case of fistulous withers, and between the nuchal ligament and the dorsal arch of the atlas in the case of poll evil. In each case there is a suppurative, granulomatous bursitis with a distinct tendency to produce fistulae onto the skin, and a lesser tendency to advance acutely as a phlegmon or to cause destructive osteitis of adjacent bones.

The primary lesions are probably either traumatic or, more often, inflammatory. Trauma may be penetrating, with introduction of infection, or blunt, with production of contusions in which organisms localize hematogenously. Inflammatory lesions are associated with a variety of organisms, including the nematode *Onchocerca cervicalis*, which lives in the nuchal ligament and environs, and provokes mineralized granulomas there when it dies. In areas where the prevalence of *O. cervicalis* infection is high, the prevalence of disease is not, and its etiologic role is questionable. In brucellosis-endemic areas, *Brucella abortus* is often isolated from closed lesions of bursitis, and vaccination is apparently a useful treatment. *Brucella abortus* lesions are unusual in other sites in affected animals. *Actinomyces bovis*, a common member of the bovine oral flora, sometimes can be cultured from closed lesions, and exposure to cattle is considered a predisposing factor. In North America and other locations where the incidence of brucellosis and the exposure of horses to cattle are declining, isolation of *B. abortus* and *A. bovis* from equine bursitis has also waned. *Streptococcus zooepidemicus* is now the most common isolate from closed lesions. The infections are probably hematogenous, but it is not clear whether a predisposing bursal lesion is required. *Actinomyces bovis*, in particular, is an unlikely primary pathogen, since in its own niche it is an uncommon opportunistic invader of traumatized oral mucosa, and it does not establish itself in regional lymph nodes draining those lesions of osteomyelitis in which it flourishes.

2. Diskospondylitis

Diskospondylitis is intervertebral disk inflammation with osteomyelitis of contiguous vertebrae. It occurs occasionally in dogs and pigs, and less often in cats, horses, cattle, and sheep. Usually it is caused by bacteria, mostly as a result of primary bacteremia but also secondary to a chronic infection. Sometimes, fungi or migrating foreign bodies are responsible. In dogs, *Staphylococcus aureus* and *Brucella canis* are the most common causes, while, in pigs, *Erysipelothrix rhusiopathiae* or *Brucella suis* is usually incriminated.

In dogs, diskospondylitis occurs most often in large males, and usually involves the lumbosacral spine, causing hyperesthesia and stilted gait and/or pelvic limb lameness. Often there is concomitant disease such as a urogenital infection. In horses, the cervical vertebrae are most often affected, and signs of neck pain predominate. Sometimes there is a history of a penetrating wound to the neck. Medical or surgical therapy is often successful in dogs but is rarely so in horses.

Organisms probably localize initially in the outer part of the annulus where they stimulate an inflammatory reaction. Alternatively, localization in the vertebral body with extension to the disk may occur. Early radiographic changes are reduction of the intervertebral disk space and loss of density in the vertebral endplates. Gross lesions appear as soft, gray areas of discoloration and disruption in the disk, which often extend into the vertebrae and therefore must be distinguished from Schmorl's nodes (Fig. 1.143). Complete destruction of disks, with fibrous replacement and vertebral osteophytes, occurs late in the disease. Such lesions are easily confused with spondylosis due to primary degeneration of the annulus fibrosus. The inflammatory reaction may extend dorsally into the spinal canal causing meningomyelitis, or laterally causing paravertebral abscess.

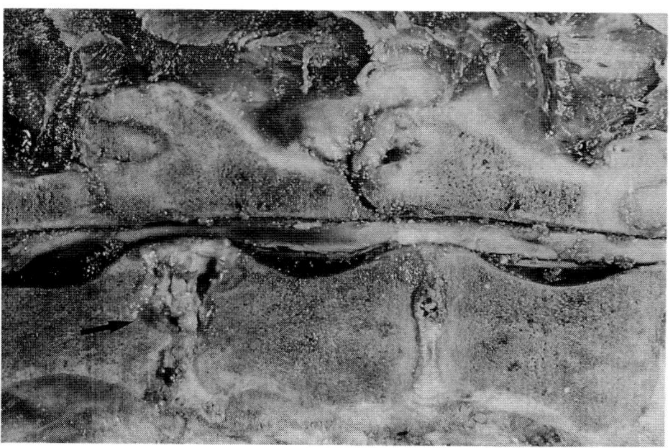

Fig. 1.143 Diskospondylitis. Note involvement of intervertebral disk and vertebral bone (arrow).

Bibliography

Adams, S. B., Steckel, R., and Blevins, W. Diskospondylitis in five horses. *J Am Vet Med Assoc* **186:** 270–272, 1985.

Brennan, K. E., and Ihrke, P. J. Grass awn migration in dogs and cats: A retrospective study of 182 cases. *J Am Vet Med Assoc* **182:** 1201–1204, 1983.

Doige, C. E. Diskospondylitis in swine. *Can J Comp Med* **44:** 121–128, 1980.

Gilmore, D. R. Lumbosacral diskospondylitis in 21 dogs. *J Am Anim Hosp Assoc* **23:** 57–61, 1987.

Malik, R., Latter, M., and Love, D. N. Bacterial discospondylitis in a cat. *J Small Anim Pract* **31:** 404–406, 1990.

Rashmir-Raven, A. *et al.* Fistulous withers. *Compend Cont Ed* **12:** 1633–1641, 1990.

Turnwald, G. H. *et al.* Diskospondylitis in a kennel of dogs: Clinico-pathologic findings. *J Am Vet Med Assoc* **188:** 178–183, 1986.

3. *Pigmented Villonodular Synovitis*

Pigmented villonodular synovitis is a disease of humans characterized by villous and nodular synovial proliferations. These lesions have a fibrous stroma, depositions of blood pigments and lipid, and histiocyte and multinuclear giant-cell infiltrations. Erosions of bone, and invasion of bone and soft tissues by the proliferative tissue is reminiscent of neoplastic behavior. The cause is unknown, but chronic hemarthrosis is said to produce similar lesions in experimental animals.

Conditions bearing some resemblance to pigmented villonodular synovitis occur in the horse and dog. In horses, there is involvement of the metacarpophalangeal or hock joint. The lesions are smooth-surfaced nodules attached to the margin of the synovial membrane. Bilateral lesions of the canine coxofemoral joint are described. Proliferation of synovial membranes and erosions of bone structures of the joint are the outstanding pathologic manifestations. Microscopic appearances are variable; in general, the greater the amount of hemorrhage, the more abundant the giant cells. In horses, chronic trauma to the synovial membrane may be responsible for the lesion.

Many chronic arthritides feature proliferations of pigmented, nodular, synovial villi. The term pigmented villonodular synovitis should be reserved for solitary masses that are large enough to interfere physically with the normal range of joint movement.

D. Noninfectious Arthritis: Immunologically Mediated

Immunologically mediated, noninfectious arthritis is a form of inflammatory joint disease in which the stimulus for inflammation is either a local and persistent immunological disturbance centered in the synovium of the affected joint or joints or an immune-mediated disease centered elsewhere in the body. In the latter instance, synovial involvement of a joint or joints is secondary to products of the immune response that are transported through the blood to the joints, where they become lodged in the walls of the synovial capillaries and initiate a local inflammatory response.

Because similar levels and types of inflammatory mediators can be released in both infectious and immunologically mediated, noninfectious forms of inflammatory joint disease, both disease processes can provoke similar morphologic patterns of joint destruction. Both disease processes can stimulate pannus formation, which may arise from granulation tissue in the inflamed capillary bed of the synovium and subchondral bone. Pannus may destroy joint margins and articular cartilage and may produce fusion of joint surfaces. One difference that is sometimes helpful in distinguishing the two major types of inflammatory joint diseases is that joint fluid removed from joints affected with infectious arthritis often contains toxic neutrophils, presumably an effect caused by the microbial agent. By comparison, the presence of toxic neutrophils is not a feature of joint fluid in cases of immune-mediated joint disease.

Most syndromes of immunologically mediated, noninfectious arthritis fit one of two morphologic patterns, i.e., erosive arthritis or nonerosive arthritis. In the erosive form the immunologic disease is centered in the joint. This results in chronic stimulation of pannus activity, resulting in loss of joint margins, joint instability, luxation, and fusion of low motion joints of the carpus and tarsus. In the nonerosive form, the primary immunologic disease is located elsewhere in the body, and the products of immunological disease are transported to the capillary bed of the synovium of affected joints. Since the primary disease may be transient, cyclic, or responsive to treatment, and it is not centered in the joint, products of the immune response that initiate the synovitis can be cleared periodically or permanently from the joint. Consequently, there is no chronic stimulation of pannus and destruction of joint surfaces.

The erosive pattern includes rheumatoidlike arthritis of dogs, polyarthritis of greyhounds, and feline chronic progressive polyarthritis. The nonerosive pattern includes polyarthritis of dogs and cats with systemic lupus erythematosus; idiopathic polyarthritis, which is identical to systemic lupus erythematosus except for an absence of serologic abnormalities found in that disease; chronic enteropathies; arthritis secondary to chronic infections; and possibly arthritis associated with sulfonamide treatment in dogs.

1. Erosive Arthritis

Rheumatoid arthritis is a potentially deforming and crippling disorder of humans that is defined by a combination of clinical, pathological, and laboratory criteria. The American Rheumatism Association lists 11 features of the disease of which at least 7 must be present to establish the diagnosis. Also, certain diseases, such as bacterial arthritis, bacterial endocarditis, and systemic lupus erythematosus, must be excluded in order to permit the diagnosis. The cause of rheumatoid arthritis is not known, though many aspects of its pathogenesis are understood. A relatively rare disease occurs in **dogs,** which resembles **rheumatoid arthritis** of humans in many ways. Many of the dogs are members of the small and toy breeds, and there is a significant preponderance in Shetland sheepdogs, but no gender or age predilection. Affected animals have episodes of anorexia, depression, and fever with generalized or shifting lameness associated with swelling around joints. The clinical course is progressive.

The joints most commonly affected are carpal, tarsal, elbow, and stifle. Involvement is usually bilateral, and the axial skeleton is usually spared. The earliest radiographic changes are soft-tissue swellings, loss of bone density around the joint and, often, focal radiolucent areas in the subchondral bone. Erosion of cartilage and subchondral bone, beginning in the transitional area of the synovial membrane, leads to disappearance of the joint space. Fibrous ankylosis occurs in some joints, especially in the tarsus and carpus. Deformities and luxations of severely affected joints occur in the chronic stages of the disease.

The prominent gross features of this disease in the early stages are thickening and brown discoloration of the joint capsule, and proliferation and enlargement of synovial villi with some fibrin layered on synovial surfaces. Later, erosion of articular cartilage is visible at the margins of the joint where fibrovascular tissue invades the epiphysis and spreads over the joint surface, destroying both cartilage and bone (Fig. 1.144). Articular structures in the center of the joints may be destroyed simultaneously by pannus developing in the underlying marrow.

Histologic changes in the synovium are characterized by villous hyperplasia with dense lymphoid and plasma cell infiltrations and patches of fibrin on the surface. The synovial lining cells are hyperplastic. Extravasated erythrocytes are often seen, and there are hemosiderin-laden macrophages in the connective tissue; these account for the brown discoloration seen grossly. In some areas there is focal loss of synovial lining cells, scattered neutrophils, and foci of connective tissue necrosis with more intense neutrophil infiltrations. Rheumatoid nodules, which occur in humans as areas of fibrinoid necrosis surrounded by palisading fibroblasts and histiocytes cells, are not seen in dogs. Necrotizing and obliterating vasculitis is apparently not a feature of the canine disease.

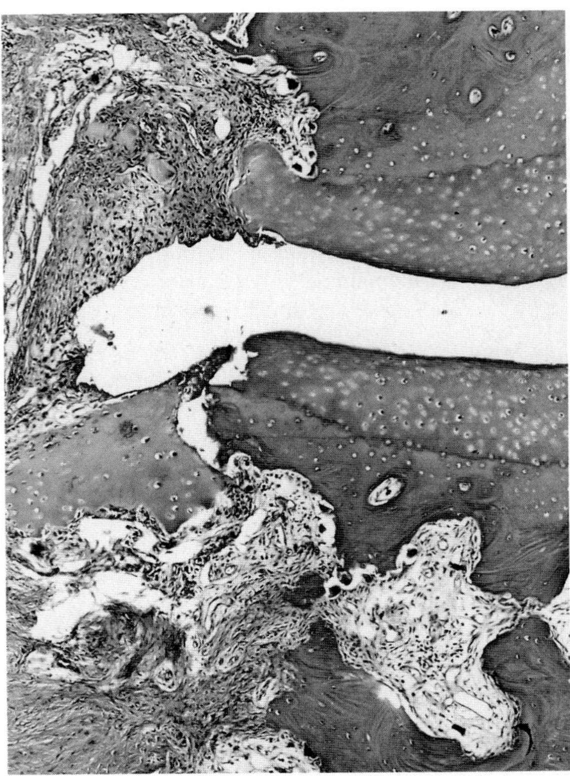

Fig. 1.144 Rheumatoidlike arthritis. Dog. Intercarpal joint. There is destruction of the joint margin by granulation tissue arising from inflamed synovium. (Courtesy of R. R. Pool.)

The dominant characters of canine rheumatoidlike arthritis are proliferative synovitis and tissue destruction by fibrovascular pannus. Occasionally periosteal new bone develops at the attachment of thickened joint capsules, but productive changes are always minimal.

Besides rheumatoid nodules and vasculitis, there are other typical, though not pathognomonic, features of the human disease. Two of these are ragocytes and rheumatoid factors. Ragocytes are polymorphonuclear cells which have phagocytized immune complex. They occur in synovial fluid of many human cases but are rare in dogs. Rheumatoid factors are IgG and IgM antibodies. They react with an antigen, which is an altered endogenous IgG protein, to form immune complexes in the joints. The development of these complexes is central to the pathogenesis of rheumatoid arthritis, for they are responsible for activating complement and thus stimulating leukotaxis. Leukocytes enter the joint and phagocytize the complexes but also release lysosomal enzymes, which are ultimately responsible for the joint lesions. It is not known why the IgG protein is antigenically altered. Rheumatoid factor of the IgM type is found in the serum of about 25% of dogs with rheumatoidlike arthritis. The IgG rheumatoid factor may be the more important of the two, but it is not known how often it develops in canine rheumatoidlike arthritis. Rheumatoid factors are found in a small proportion of dogs and humans that do not have joint disease.

The search for an infectious agent responsible for altered endogenous IgG protein in human rheumatoid arthritis is unremitting. Several diseases, including erysipelas and mycoplasma arthritis of swine, have been regarded as models for rheumatoid arthritis. These diseases demonstrate that the mechanisms and manifestations of disease are similar in various animals, but it is unclear whether either one is involved in the human arthritis.

a. Polyarthritis of Greyhounds An erosive polyarthritis of young greyhounds affects animals of both sexes and usually begins insidiously at 1 to 2 years of age. Mild to severe lameness, and joint swellings involving limb joints distal to and including elbows and stifles, are observed clinically. Superficial nodes are enlarged.

At autopsy there are excess turbid joint fluid, sometimes containing fibrin; yellow-brown discoloration and thickening of synovial membranes, sometimes with adherent fibrin plaques, sometimes with focal hemorrhages; and erosions of articular cartilage. Lesions are common in the first and second phalangeal joints, and in carpal, tarsal, elbow, and stifle joints. Occasionally, shoulder, hip, and atlanto-occipital joints are involved.

Microscopic lesions of proliferative synovitis are regularly observed. The usual changes are thickening of the synovial membrane by vascular fibrous tissue, proliferation of synovial cells, and inflammation confined to the superficial stroma. The inflammation is characterized by lymphoid aggregations and diffuse infiltrations of plasma cells and neutrophils. There is no evidence of vasculitis. Pannus formation, though often present, is never extensive, and changes in subchondral bone are usually minimal.

The cause of the disease is not known. There is no evidence incriminating an immune-mediated pathogenesis, and the disease does not seem to be related to athletic activities.

b. Feline Chronic Progressive Polyarthritis Two forms of this disease are recognized in cats. The erosive form is the feline counterpart of rheumatoidlike arthritis of dogs. This form occurs in older cats and can affect either sex. Like the canine disease the incidence is very low as compared to the other immunologically mediated, inflammatory joint diseases affecting the two species. The disease has an insidious onset, chronic progression, and essentially no clinical signs except for symmetric deformities of the small joints of carpus, tarsus, and digits caused by loss of joint margins, subluxation, collapse of joint space, and ankylosis. Joint fluid contains low to moderate numbers of neutrophils and lymphocytes. Synovial membranes are heavily infiltrated with lymphocytes and plasma cells. Large synovial villi often contain prominent lymphoid follicles with germinal centers. Rheumatoid factor is absent. Infection with feline leukemia virus is present in about half of the affected cats.

The periosteal proliferative form occurs only in relatively young intact or neutered male cats usually between 18 months and 5 years of age. The disease is etiologically linked to infection with feline syncytia-forming virus and feline leukemia virus. Evidence of feline syncytia-forming virus infection is present in all cats, and 60% also have evidence of feline leukemia virus infection. The disease is not reproducible by experimental infection with either of the two agents or with cell-free synovial tissue from affected cats. Because many cats are infected with both viruses without ever becoming ill, it is thought that the periosteal proliferative form of feline chronic progressive polyarthritis is an unusual manifestation of feline syncytia-forming virus infection in cats that are somehow predisposed. The role of feline leukemia virus is unclear, but it may potentiate the activity of the other viral agent.

Disease onset is recognized by sudden onset of high fever, severe pain in joints and tendon sheaths, and a lymphadenopathy associated with affected joints. Synovitis is initially characterized by fibrin and marked exudation of neutrophils (Fig. 1.145); later, plasma cells and lymphocytes predominate. Rheumatoid factor is absent. Synovial lining cells are hyperplastic throughout the course of the disease, but villus hypertrophy and pannus formation are

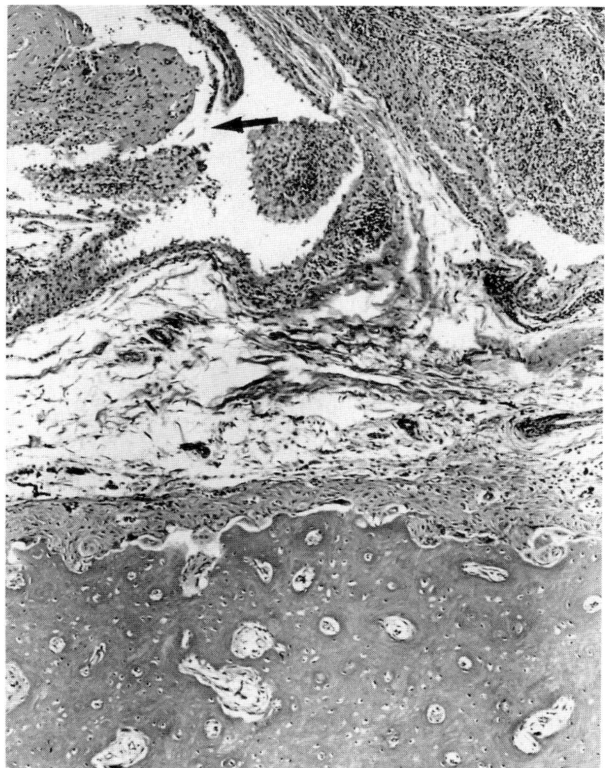

Fig. 1.145 Feline progressive polyarthritis. Periosteal proliferative form. There is fibrinopurulent synovitis (arrow) and periarticular new bone below. (Courtesy of R. R. Pool.)

uncommon until late in the disease. Swollen lymph nodes biopsied early in the disease have exuberant lymphoid hyperplasia, including marked extracapsular extension, and create a pattern easily mistaken for lymphosarcoma.

Skeletal lesions are bilateral and most frequently affect the tarsometatarsal and carpometacarpal joints; however, affected joints may include the elbow and knee joints and the articular facets of the thoracic and lumbar spine. Osteopenia of spongiosa and periosteal new bone production are prominent early manifestations of affected joints. The periosteal new bone is formed of trabeculae of woven bone tissue, and the intertrabecular spaces often are infiltrated with lymphocytes and plasma cells. Pannus formation is mostly a late feature and leads to periarticular erosions of joint margins, collapse of joint spaces, and fibrous ankylosis. However, instability and subluxations do not occur. This pattern of disease shares similarities to Reiter's arthritis of humans.

2. Nonerosive Arthritis

This category includes several distinctive syndromes of noninfectious, nonerosive arthritis affecting dogs and cats. While these forms of arthritis apparently have different causes, they are presumably mediated by similar immunopathologic mechanisms. These conditions occur under one of three circumstances; as a primary or idiopathic form, a form in which arthritis is also a feature of systemic lupus erythematosus, and a form in which the arthritis may occur in animals affected with a variety of diseases including infections by certain parasites, bacterial, and viral agents; chronic infections such as pyometra, endocarditis, and inflammatory bowel disease; drug hypersensitivity and neoplasia. The presenting clinical signs of the joint disease in this group are similar. Joint disease is typically cyclic in nature, and clinical signs may wax and wane. Periods of remission may occur. One or several joints of the distal parts of the limbs may be involved, with the smaller joints of the manus and pes being most often affected.

There are essentially no radiographic changes except for joint distension and soft tissue swelling, although signs may have been present for months. Occasionally, slight narrowing of the joint space is evident, and distension of the joint cavity is demonstrable by contrast arthrography.

Histologically, there is edematous thickening and inflammation of the synovial membrane. Neutrophils are concentrated at the intimal layer of the synovium and may be quite numerous, depending on the stage of the disease. Sparse accumulations of plasma cells and lymphocytes are present in the deeper layers of the synovium and fibrous joint capsule. Villus hyperplasia and pannus are not seen.

While the possibility that infection may be involved is virtually impossible to disprove, the weight of evidence suggests that joint disease is mediated by the deposition of circulating immune complexes in the synovial capillary bed; however, the inflammatory process apparently does not persist long enough at any given time to activate pannus formation and to maintain its destructive activity. In the most common primary or idiopathic form, the origin and nature of the antigen responsible for initiating the disease is unknown. The antigen in systemic lupus erythematosus is primarily nucleic acid. The antigens in the other forms of nonerosive arthritis are thought to be derived from the agent of disease or from damaged host tissues. Similar conditions probably occur in other domestic animal species.

Noninfectious polyarthritis, thought to be immune mediated, develops in a small proportion of dogs treated with sulfonamides. Combination drugs containing sulfadiazine are usually responsible, but sulfamethoxazole is also incriminated. Doberman pinschers seem to be unusually prone to the condition.

Signs of polyarthritis and fever, sometimes with lymphadenopathy, anemia, leukopenia, and thrombocytopenia, develop after 8 to 20 days of treatment. There is almost always a history of previous treatment with the drug. Signs regress within 2 to 5 days of drug withdrawal and, in one report, recurred within 2 hr to 5 days in 6 of 6 dogs that were reexposed.

Joint fluid is cloudy or bloodstained and shows leukocytosis with many neutrophils. There are no reports of morphologic investigations.

Bibliography

Bennett, D., Gilbertson, E. M. M., and Grennan, D. Bacterial endocarditis with polyarthritis in two dogs associated with circulating autoantibodies. *J Small Anim Pract* **19:** 185–196, 1978.

Cribb, A. E. Idiosyncratic reactions to sulfonamides in dogs. *J Am Vet Med Assoc* **195:** 1612–1614, 1989.

Ghadially, F. N., Lalonde, J.-M. A., and Dick, C. E. Ultrastructure of pigmented villonodular synovitis. *J Pathol* **127:** 19–26, 1979.

Glynn, L. E. Pathology, pathogenesis, and aetiology of rheumatoid arthritis. *Ann Rheum Dis* **31:** 412–420, 1972.

Huxtable, C. R., and Davis, P. E. The pathology of polyarthritis in young greyhounds. *J Comp Pathol* **86:** 11–21, 1976.

Krum, S. H. *et al.* Polymyositis and polyarthritis associated with systemic lupus erythematosus in a dog. *J Am Vet Med Assoc* **170:** 61–64, 1977.

Kusba, J. K. *et al.* Suspected villonodular synovitis in a dog. *J Am Vet Med Assoc* **182:** 390–392, 1983.

Monier, J. C. *et al.* Antibody to soluble nuclear antigens in dogs (German shepherd) with a lupuslike syndrome. *Dev Comp Immunol* **2:** 161–174, 1978.

Newton, C. D. *et al.* Rheumatoid arthritis in dogs. *J Am Vet Med Assoc* **168:** 113–121, 1976.

Nickels, F. A., Grant, B. D., and Lincoln, S. D. Villonodular synovitis of the equine metacarpophalangeal joint. *J Am Vet Med Assoc* **168:** 1043–1046, 1976.

Pedersen, N. C., and Pool, R. Canine joint disease. *Vet Clin North Am* **8:** 465–493, 1978.

Pedersen, N. C. *et al.* Noninfectious canine arthritis: Rheumatoid arthritis. *J Am Vet Med Assoc* **169:** 295–303, 1976.

Pedersen, N. C. *et al.* Noninfectious canine arthritis: The inflammatory nonerosive arthritides. *J Am Vet Med Assoc* **169:** 304–310, 1976.

Pedersen, N. C., Pool, R. R., and O'Brien, T. Feline chronic progressive polyarthritis. *Am J Vet Res* **41:** 522–535, 1980.
Schumacher, H. R., Newton, C., and Halliwell, R. E. W. Synovial pathologic changes in spontaneous canine rheumatoid-like arthritis. *Arthritis Rheum* **23:** 412–423, 1980.
Van Pelt, R. W., and Langham, R. F. Nonspecific polyarthritis secondary to primary systemic infection in calves. *J Am Vet Med Assoc* **149:** 505–511, 1966.
Wright-George, J. *et al.* A rheumatoid-like arthritis in calves. *Cornell Vet* **66:** 110–117, 1976.

VIII. Tumors of Joints

Neoplasms of articular cartilage do not occur. The so-called osteochondromas of joints are nodules of cartilage and sometimes bone, which develop in synovial structures of chronically irritated joints. The cartilage develops by metaplasia from connective tissue and the bony component of the nodules by metaplasia or by endochondral ossification of metaplastic cartilage.

Neoplasms of synovial structures, synoviomas, and synovial sarcomas, are relatively uncommon, but do occur in dogs, cats, horses, and cattle, usually in adult animals and usually involving the limbs.

Synovial tumors usually present as soft-tissue masses in the region of joints, which cause lameness. In dogs, the stifle and elbow are most often affected. It appears that most synovial tumors are malignant, but behavior is not readily predicted on histologic grounds. Synovial sarcomas destroy and cross joint structures, permeating bones and provoking moderate periosteal reaction; they are difficult to remove and often grow back following surgery. Metastases to regional lymph nodes and lung usually occur as late events, therefore differentiation from osteosarcomas is of practical importance.

Synovial tumors in humans have certain histologic characteristics which allow their differentiation from other tumors of soft tissues. They are often distinctly biphasic, having a sarcomatous component, which often resembles fibrosarcoma, and an epithelioid component which may form pseudoacini and resemble metastatic carcinoma. The biphasic characteristic usually is not well developed in dogs. Synovial sarcomas in dogs often contain clefts lined by the epithelioid cells and sometimes containing PAS-positive amorphous material, which is thought to be analogous to components of synovial fluid. Pseudoacini may be represented by clearly demarcated cavities lined by orderly cuboidal cells or, more commonly, by irregular spaces surrounded by unoriented stellate cells. Usually the stromal, fibrosarcomatous cells predominate, and it may be necessary to examine many sections to establish the identity of the tumor. Some synovial tumors, especially in cats, contain many multinuclear giant cells. Observations on limited numbers of synovial tumors in cats suggest that although they tend to recur, they are unlikely to metastasize. Some synovial tumors of dogs form papillary structures which resemble synovial villi. Intra-articular masses with a predominantly villous structure are more likely to be manifestations of reactive hyperplasia than neoplasia.

Synovial tumors arise from synovioblasts beneath the synovial membrane rather than the differentiated synovial cells. In dogs they are also associated with tendons and synovial sheaths. Electron microscopic examination of several histologically diverse soft-tissue tumors of humans such as monophasic synovial sarcoma, epithelioid sarcoma, and chondroid sarcoma, supports their classification as forms of synovial sarcoma. Awareness of the heterogeneous nature of synovial tumors under light microscopy should alert the pathologist faced with an obscure soft-tissue tumor intimately associated with a joint. Hemangiopericytoma, malignant fibrous histiocytoma, and giant-cell tumor of soft tissues are soft-tissue tumors which may resemble synovial sarcoma.

Bibliography

Cooney, T. P. *et al.* Monophasic synovial sarcoma, epithelioid sarcoma and chordoid sarcoma: Ultrastructural evidence for a common histogenesis, despite light microscopic diversity. *Histopathology* **6:** 163–190, 1982.
Davies, J. D., and Little, N. R. F. Synovioma in a cat. *J Small Anim Pract* **13:** 127–133, 1972.
Dungworth, D. L. *et al.* Malignant synovioma in a cow. *J Pathol Bacteriol* **88:** 83–87, 1964.
Griffith, J. W., Frey, R. A., and Sharkey, F. F. Synovial sarcoma of the jaw in a dog. *J Comp Pathol* **97:** 361–364, 1987.
Lipowitz, A. J., Fetter, A. W., and Walker, M. A. Synovial sarcoma of the dog. *J Am Vet Med Assoc* **174:** 76–81, 1979.
Mitchell, M., and Hurov, L.I. Synovial sarcoma in a dog. *J Am Vet Med Assoc* **175:** 53–55, 1979.

ACKNOWLEDGMENTS

Kathy Johnston and Louise Ruhnke reviewed the section on diseases caused by *Mycoplasma* spp. Ian K. Barker reviewed *Borrelia burgdorferi* infections. J. A. Lynch reviewed bacterial arthritides. R. R. Pool, Davis, California, contributed the sections on angular limb deformities, dorsal metacarpal disease, synovial fossae and ringbone, spavin and navicular disease. He also made valuable revisions to several other topics, principally noninfectious arthritides. The use of case material from the files of the Ontario Veterinary Laboratory Services, the Ontario Veterinary College, and the University of Melbourne is acknowledged.

CHAPTER 2 Muscle and Tendon

THOMAS J. HULLAND
University of Guelph, Canada

MUSCLES

I. General Considerations

Development of striated muscle in the embryo is from mesodermal somites, which give rise to myotomes. Within each myotome, which corresponds roughly to a vertebral body segment, with its spinal nerve, the individual muscles develop by a process of aggregation and migration of presumptive myoblastic cells. It is very likely that the undifferentiated mesodermal cells will be committed to a muscle destiny some time before significant structural change is visible, and the earliest detectible modification to presumptive myoblasts is a cross-sectional rounding of the spindle shaped cells. By this time as well, there may be subtle distinctions between those cells which will become myofibers and those which will become satellite cells.

The first clear sign of differentiation is the migration of presumptive myoblasts, destined to become myofibers, into the regions where future muscles will appear; this occurs before any nerve influence is exerted. The direct connection of the nerve to the myotome determines the subsequent route of innervation, but because migration has occurred, the muscle may receive nerve sprouts (as it has received myoblast group components) from more than one myotome.

The second phase of muscle development is incompletely separated from the first and subsequent phases. It begins with the early development of sarcoplasmic components, such as myofibrils, which identify the cells as muscle. The commitment to myogenesis is mutually exclusive with cell replication and, with possible rare exceptions, committed myoblasts are postmitotic. Myoblasts begin to fuse into elongated multinucleated cells about the time myogenesis begins.

Subsequent development of the myotube allows it to become a well-developed muscle fiber and consists of stepwise construction of actin and myosin filaments, the formation of the Z band into which the thin actin filaments insert, and the evolution of the tubular systems. The last of these steps, the invagination of the T tubular system from caveolae or other small, regular recesses on the outer sarcoplasmic membrane, provides an elaborate system of tubules, which run parallel to the Z bands and make contact with all developing myofibrillar units.

The fourth phase of development is one in which the evolving fiber grows, increases the number of myofibrils and nuclei, and moves the latter to the subsarcolemmal position. During the final phase of development, the development of a basal lamina, and an additional sheath of collagen, fibroblasts, and capillaries invest each developing myotube as orientation of the fiber into its final position of tension takes place. Development of fibers up to this point, just after the end of the first trimester, is independent of any neural connection, but subsequent fiber enlargement and the considerable increase in the number of fibers which occurs during the immediate prenatal and postnatal periods is dependent on a functional neural connection. A great deal is now known about the elements which induce development and promote connections between the approaching nerve and the specializing sarcoplasmic surface.

The muscle fiber increase in late gestation in domestic animals is probably comparable in all species, but it has been particularly well studied in the pig and is more orderly in that species than in others. Within what will become a primary bundle of muscle fibers, one to six primary or template myofibers become the focal point of the proliferation which, by early postnatal life, has produced one or

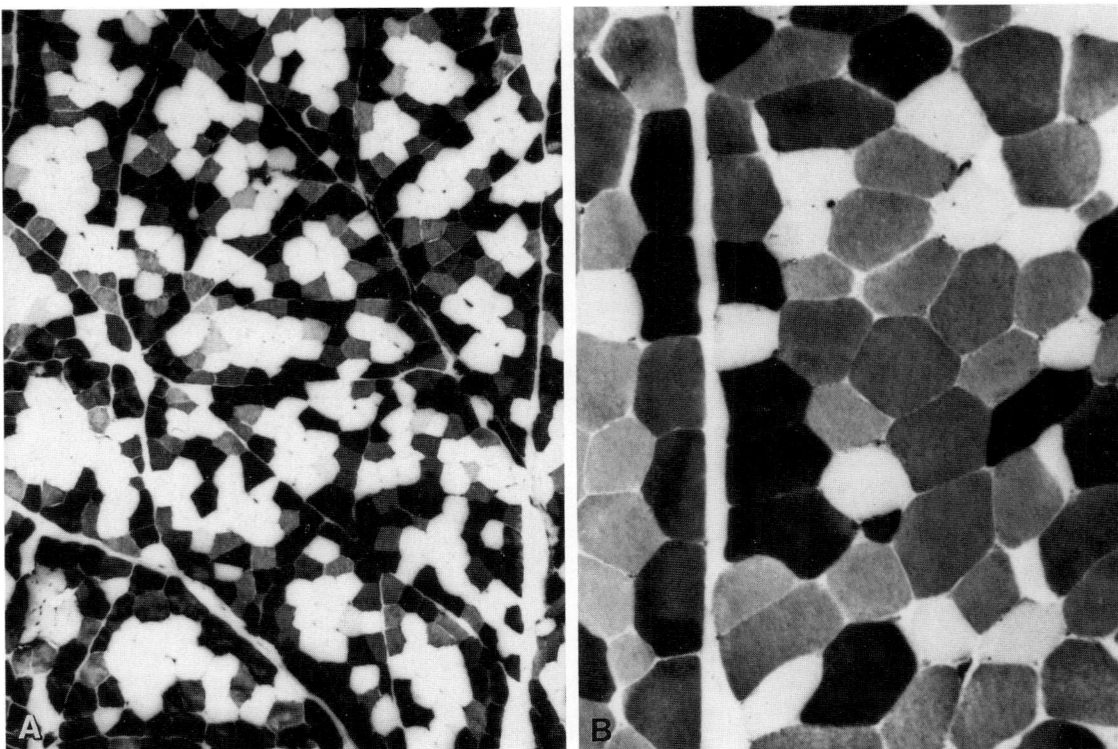

Fig. 2.1 (A) ATPase (alkaline) histochemical preparation of normal calf muscle to show light type 1 sublobular "template clusters." (B) Dark type 2 fibers clustered along edges of primary bundles.

more sublobular clusters of fibers within each primary bundle. It is not yet clear whether new fibers are produced next to the template fiber by a process of template fiber budding, or by incomplete or complete longitudinal division. Alternatively they may be derived from local myoblastic or satellite cells, which begin development within the basal lamina of the primary fiber and then separate and develop their own completely investing basal lamina. What is clear is that they are produced in waves, with subsequent fibers pushing those formed earlier to the periphery of the sublobule. Template fibers are always histochemically type 1, but the newly formed fibers are type 2 until near the time of birth, when a proportion of them (few or several depending on the muscle and the species) which lie closest to the template fiber will begin to modify to become type 1 fibers (Fig. 2.1A,B). The latter process may continue for 6 months or more after birth, and it may be that the process reverses again in early postnatal life in some muscles of the body.

During the period when templating is occurring, new fibers rapidly become innervated and, it seems, do not later change neural connection. In view of the described changes of fiber type, thought to be associated with exercise, it is difficult to reconcile other evidence which suggests that a single motor nerve axon can serve only one histochemical type of muscle fiber. The paradox may be resolved if it is accepted that some fibers have, or develop, different kinds of energy metabolism, which they use selectively depending on work demands. Alternatively, a change in the activity of the nerve may induce biochemical changes in the muscle fiber clusters, and hence a change in fiber type identity. This explanation is consistent with the fact that the functional capacity of a muscle fiber is limited by its nerve connection and its cytoplasmic components and of these, probably only the latter can be modified.

Muscle structure is arranged around muscle fibers. Muscle fibers are variable in size depending on age, exercise, nutritional status, position and function of the muscle in question, and on species, although the fibers of a mouse are only a little smaller than those of a horse. Muscle fibers in the extrinsic muscles of the eye are consistenly small (10–30 μm in diameter) and round; those of the major limb muscles vary from an average least diameter of 40–65 μm and, although rounded in the fresh state, often appear angular when fixed. Size of fibers increases with age until puberty, at which time males have slightly larger fibers than females. In old age, with certainty, but perhaps beginning in early maturity, the fiber diameter slowly decreases. In those domestic animals which have been studied, the size distribution of muscle fibers conforms more or less to a biological distribution curve, although in most postural muscles, the curve is skewed by a small population of large fibers extending upward from the expected upper

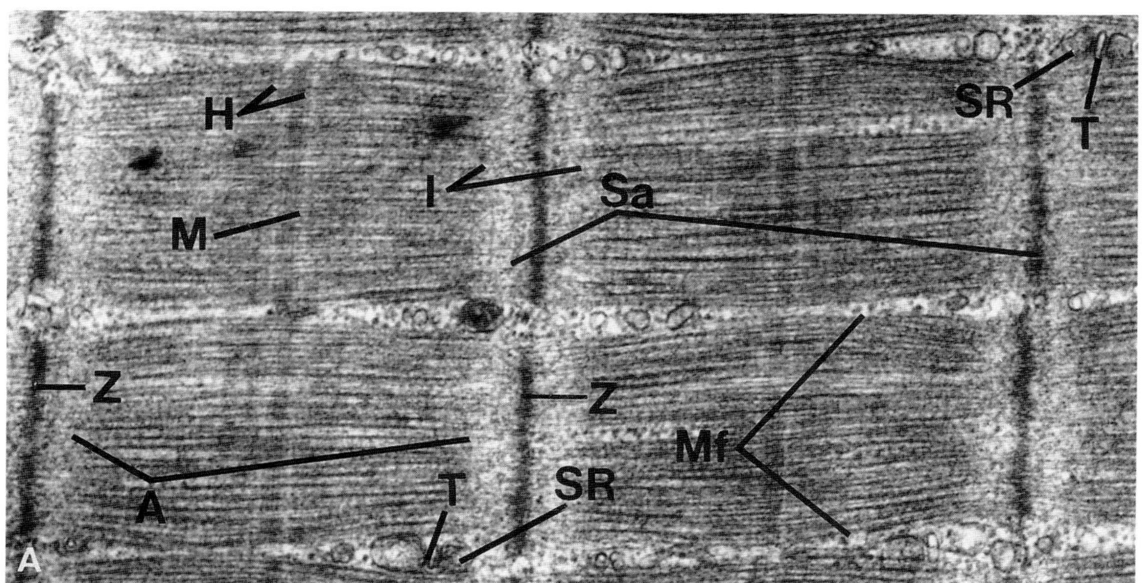

Fig. 2.2A Normal horse muscle. Anisotropic band (A), H band (H), isotropic band (I), M line (M), myofibril (Mf), one sarcomere (SA), sarcoplasmic reticulum (SR), T-tubule (T), Z band (Z).

size limit. In most instances, these large fibers are from the type 1 fiber population, but in some muscles in some species, they may be from the type 2 group. A distinctly bimodal curve readily develops in a muscle when disease, pregnancy, or nutritional status prompts withdrawal of muscle protein for general maintenance. This also increases the likelihood that large fibers will be round rather than polygonal in routine preparations.

The **structure of the muscle cell** or muscle fiber is quite well defined (Fig. 2.2A). The outer component is a thin, amorphous, but apparently quite tough, basal lamina consisting of three layers which, on most muscle cell surfaces, seems to be thrown into gentle folds (Fig. 2.2B). Atrophy of the fiber leads to much more obvious accordianlike folds of the basal lamina. Within this basal lamina are two separate cell populations with very similar nuclei; the multinucleate myofiber and the small, more numerous satellite cells, which play an important role in fiber repair and regeneration (Fig. 2.3A,B). The nuclei of both cell types are oriented to the long axis of the muscle fiber and are distributed regularly in a spiral manner. In normal muscle, less than 3% of the nuclei of the multinucleate myofiber cell are displaced internally, but the number tends to be higher adjacent to points where muscle and tendon interdigitate, or where muscle and bone meet through a short ligament or tendon. Nuclei are slender, oval, have evenly distributed chromatin and single, small nucleoli. The satellite cells consist of a simple cell membrane thrown around a nucleus, a minimum of cytoplasm with mitochondria, and a scant tubular system, all of which

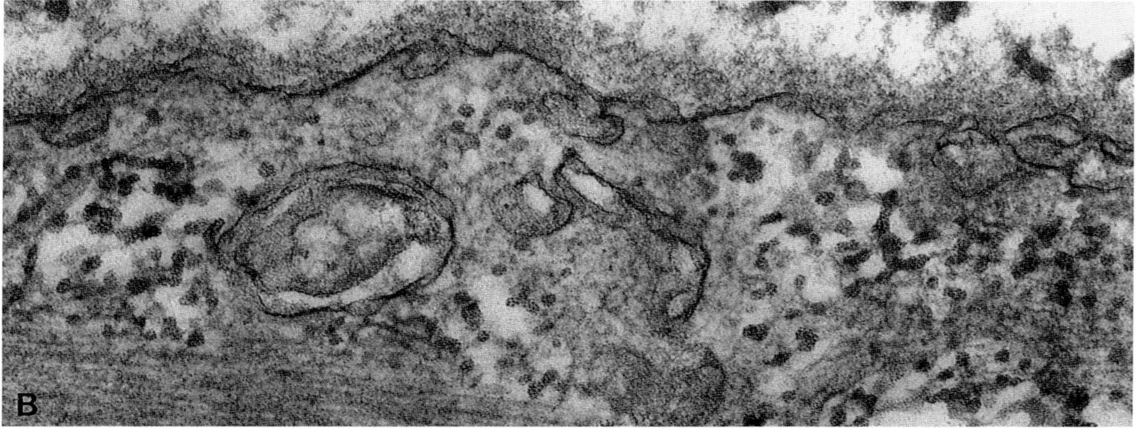

Fig. 2.2B Sarcoplasmic membrane invaginated to form caveolae; basement membrane on the upper aspect.

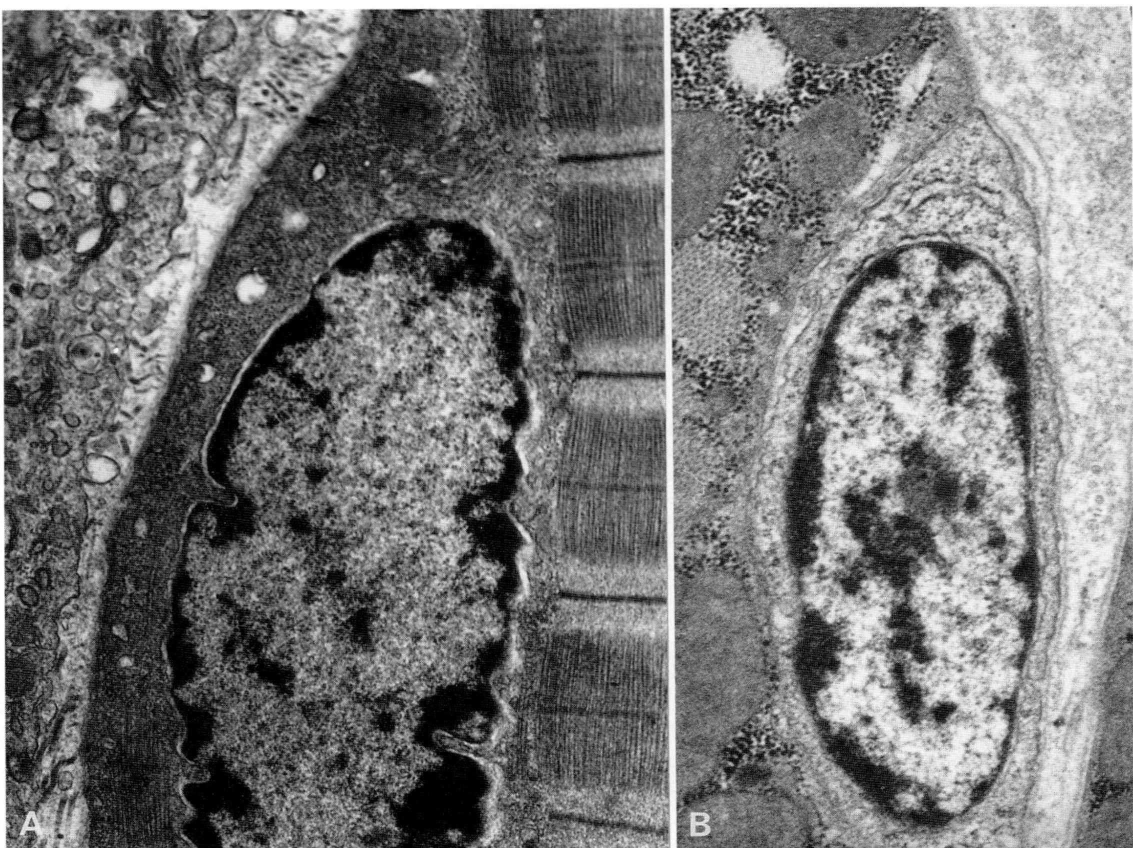

Fig. 2.3 (A) Myofiber nucleus and adjacent myofibril. (B) Satellite cell nucleus.

lie in a shallow indentation on the myofiber surface within the basal lamina. Satellite cells constitute 1–30% of the visible nuclei associated with a muscle fiber; the higher figure describes the neonatal state; the lower, old age. In mature muscle, 3–5% of the nuclei are satellite cells; a modest concentration of them occurs adjacent to motor end-plates.

Satellite nuclei are the only ones in muscle capable of mitotic division in postnatal life, and each time these cells divide in a growing muscle, they contribute one myocytic nucleus and a minute amount of cytoplasm to the growing pool of nuclei in the enlarging multinucleate, myofiber cell. The other daughter cell remains as part of the satellite cell pool outside the myocyte and retains mitotic capability. It is clear that this mechanism will make the satellite cells a progressively smaller proportion of the total nuclear population within the basal lamina, since their number remains constant from birth, whereas the nonreplicating postmitotic myofiber nuclei increase in number cumulatively.

The most distinctive characteristic of skeletal muscle cells is the presence of striated myofibrils, approximately 0.5–1.0 μm in diameter, consisting, in cross section, of thousands of regular size and regularly oriented myofilaments (Fig. 2.4). In longitudinal section, contractile units or sarcomeres about 20 μm long extend from one Z band to the next, and consist of bands and lines created by the stacking of thick and thin filaments with protein aggregates at certain places. The thin filaments (60 Å wide; actin, tropomyosin, and troponin) are fixed into the electron-dense Z band of noncontractile protein and incompletely overlap the thick filaments (160 Å wide; myosin). The zone of no overlap adjacent to the Z band is referred to as the I band. The central zone between the ends of the thin filaments in which only thick filaments are visible is the H band, and this is divided by a thin dense line of M substance on each filament; collectively these create the M line. The A band is the wide central zone which extends from one end of the thick filaments to the other and alternates with the I band. In sarcomere contraction, Z bands move closer together, reducing I and H band widths. In cross section at a central point where thick and thin filaments are present, one myosin filament is surrounded at regular intervals by six thin filaments, or twelve when the sarcomere is in strong contraction.

Myofibrils are surrounded by sarcoplasm, which makes up 30–40% of the fiber volume, and in it are elements of the T-tubular system, the sarcoplasmic reticulum, mito-

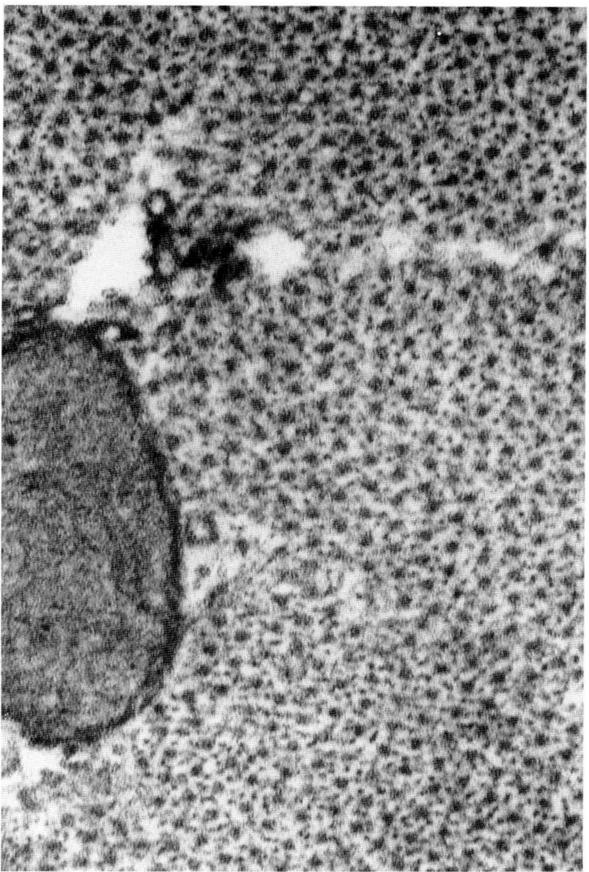

Fig. 2.4 Myofilaments in cross section.

chondria, lysosomes, glycogen granules, and often, fat droplets. The endoplasmic tubular systems in the muscle fibers of all mammals regularly come together to surround myfibrils and meet in triads with the T tubule between two tubular segments of sarcoplasmic reticulum at frequent regular intervals. In many animals these triads can be found at each end of each sarcomere, an arrangement which ensures very extensive direct surface membrane contact between the cell surface, the sarcomeres, and the endoplasmic reticulum. Sarcoplasm between the myofibrils and the outer cell membrane is often rich in mitochondria and glycogen granules.

Quite apart from their ability to shorten in active contraction, myofibrils are capable of easy and rapid length adjustment, accomplished by shedding or adding one or more sarcomeres. Under ideal experimental circumstances, up to 25% of the sarcomeres may be lost in a 24-hr period, a statistic which emphasizes that even in the healthy animal, there is potential for a considerable flux of muscle protein, which allows for rapid dismantling and reconstruction of myofibrils. The breadth of myofibrils increases or decreases equally rapidly by building or discarding myofilaments. When new myofibrils are required, an existing large myofibril splits longitudinally, but when atrophy reduces the need for some myofibrils, they are simply dismantled sarcomere by sarcomere or reduced in size by destruction of the most peripheral myofilaments.

Outside the muscle fibers and their satellite cells are three magnitudes of connective tissue framework. Intimately applied to the basement membrane of each muscle fiber is the netlike endomysium, which carries the capillary network with its longitudinal orientation. Around each primary bundle of 40 to 150 fibers is the thicker perimysium in which run larger vessels, nerve trunks, and sensory neuromuscular spindles. Around the outside of the muscle, or a major head of a muscle, lies the epimysium, which carries tendon organ sensory endings and sometimes prominent tendinous bands.

Types of muscle fibers may be distinguished by many histochemical and immunohistochemical techniques now available for studying the enzyme components; each new technique demonstrates an ever-expanding heterogeneity of muscle fibers. Many of these techniques subdivide the muscle fiber population in the same ways, and in many instances the subdivisions seem to bear little relationship to the biological response created by disease. The two techniques which seem best to identify different biological activities are the alkaline- and acid-resistant adenosine triphosphatase (ATPase) procedures and the nicotinic acid dehydrogenase tetrazolium reductase (NADH-TR) test for oxidative enzyme activity. These allow the division of the muscle population into two major groups and several subgroups (Figs. 2.1A,B, 2.5): type 1 fibers rich in oxidative enzymes and showing slow-twitch, red color characteristics (myoglobin), and the type 2 fibers rich in glycogen showing a fast-twitch response and being paler in color. It is generally assumed that these identify, respectively, fibers with predominantly oxidative energy source appropriate for sustained activity and those fibers with a predominantly glycolytic energy source appropriate for short-term, strong activity; but both types of fiber contain a broad spectrum of enzymes, which makes them capable of a wide range of biochemical activity and, in certain circumstances, an interconversion of working types.

Some species, particularly the pig and the horse, have fibers apparently rich in both oxidative and glycolytic ingredients, and many domestic animals have fibers which lack the clear-cut divisions that seem evident in the muscles of humans and laboratory animals. The type 2 fibers can be histochemically subdivided by ATPase response into subtypes but, for purposes of predicting biological activity and identifying diseases, the two main types seem to have the greatest relevance. A special subgroup of the glycolytic, type 2 group found in the masticatory muscles of the dog represent an exception to this, because they display antigenic differences from other muscles, and this makes them vulnerable to a unique disease process. Examination of a muscle in which about equal numbers of type 1 and type 2 fibers are present might suggest a random checkerboard distribution of the types; however, type 2 fibers are disproportionately found on the periphery of primary bundles, and type 1 fibers are disproportionately

Fig. 2.5 Type 1 fiber above with large mitochondria (arrow) and little glycogen. Type 2 fiber below. The mitochondria are smaller, and glycogen is abundant (larger arrow).

at the center of nests of fibers within the primary bundle (Fig. 2.1A,B). These observations are consistent with knowledge of the late fetal fiber evolution.

Muscles used repetitively but slowly and persistently should have high levels of type 1 fibers, while those used for short bursts of activity should have high levels of type 2 fibers. This is often the case; for example, the diaphragm of cattle often has 80–95% of type 1 fibers, while most other skeletal muscles have 10–60% type 1 fibers. On the other end of the scale, the longissimus and semitendinosus muscles, the two muscles which are most enlarged in racing greyhounds and thoroughbred horses, have ~80–95% type 2 fiber content. Obligatory postural muscles, such as the supraspinatus, tend to have a 40–50% type 1 component, but so do muscles which are used relatively sparingly. Individual variation in fiber proportions is quite wide, and this is magnified by breed, muscle region, and species differences. The anatomic position and the function of a muscle in an animal can provide only a general indication of its fiber composition.

Sensory muscle spindles are found in all skeletal muscles in and anchored to the perimysial connective tissue and associated with a small nerve radicle which contains 6–20 large sensory fibers. They are more numerous in some muscles than in others and are generally difficult to find in the larger muscles of mature animals. Spindles are about 0.5 to 3.0 mm long and 200 to 500 μm wide (Fig. 2.6). The intrafusal (central) space enclosed within the fibrous, multilayered outer sheath is continuous with lymphatic

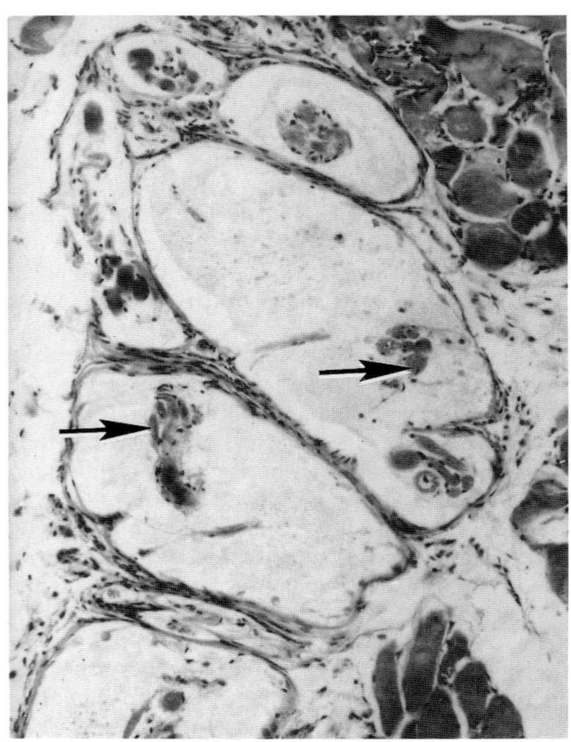

Fig. 2.6 Compound sensory spindle in normal muscle with specialized intrafusal muscle fibers (arrows).

Fig. 2.7 Motor end-plates (cholinesterase stain) showing terminal axons (arrow) and synaptic gutters (arrowhead).

space and contains 2–20 specialized small intrafusal muscle fibers. The intrafusal fibers are histochemically different from the extrafusal fibers but share the same motor nerve. The intrafusal fibers have a high content of nuclei often contained in a central "nuclear bag" region, or linearly distributed up the middle of the small round fibers. Abundant sensory axons entering the spindle have "flower-spray," "grape-cluster," or spiral endings on the intrafusal fibers, and these structures collectively give rise to impulses which record muscle stretch and contribute to proprioception. Complex spindles consisting of 2 to 10 spindles, incompletely overlapping end on end, are sometimes found in skeletal muscles and may be an expression of anomalous development. Intrafusal fibers generally do not participate in the pathologic processes which involve extrafusal fibers, with the possible exception of nutritional myopathy. Intrafusal fibers are also resistant to atrophy.

Motor end-plates are the sites of synaptic transmission of acetylcholine from nerve ends to muscle receptors, at which sites a surface-membrane-conducted, polarizing impulse is initiated. The interface takes the form of a complex, pretzel-shaped neural termination, which is pressed into a shallow but convoluted, cholinesterase-laden, synaptic gutter of matching shape on the myofiber surface. The end-plate is contained within the endomysium, and extensions of the endomysial connective tissue ensheath the bare terminal axon for a short distance until its myelin sheath is reached (Fig. 2.7). A little more than one end-plate for each muscle fiber exists in skeletal muscle, and although supernumerary endings (subterminal sprouts) can develop, such structures are rarely found in normal muscle. One motor neuron gives rise to an extremely variable number of terminal axons with one end-plate each, the number being inversely proportional to the fineness of motor movement required of the muscle, and directly proportional to the relative diameter of the muscle fibers. Thus, in the extrinsic ocular muscle, the axon : end-plate ratio may be as low as 1 : 10 on small, round muscle fibers, while in the gastrocnemius or gluteus muscles, the ratio may be 1 : 2000 on muscle fibers near the upper size limits. Terminal axons are given off at nodes of Ranvier in clusters of 5 to 30 per node to make muscle end-plate contact locally via a short, fine terminal axon. Although a low neural impulse level to skeletal muscle is translated into contraction of only a proportion of the muscle fibers, this is clearly not a form of muscle regulation which can lead to finely controlled movement, and consequently fine movement must be controlled by a low neural ratio.

Bibliography

Anon. The myofibroblast. *Lancet* **8103:** 1290–1291, 1978.

Allbrook, D. Skeletal muscle regeneration. *Muscle Nerve* **4:** 234–245, 1981.

Armstrong, R. B. Distribution of fiber types in locomotory muscles of dogs. *Am J Anat* **163:** 87–98, 1982.

Baker, J. H., and Hall-Craggs, E. C. B. Changes in length of sarcomeres following tenotomy of the rat soleus muscle. *Anat Rec* **192:** 55–58, 1978.

Bradley, R., and Duffell, S. J. The pathology of the skeletal and cardiac muscles of cattle with xanthosis. *J Comp Pathol* **92:** 85–97, 1982.

Bradley, R., and Fell, B. F. Myopathies in animals. *In* "Disorders of Voluntary Muscle," 4th Ed. J. Walton (ed.), pp. 824–872. Edinburgh and London, Churchill-Livingstone, 1981.

Cardinet, G. H., and Holliday, T. A. Neuromuscular diseases of the domestic animals: A summary of muscle biopsies from 159 cases. *Ann N Y Acad Sci* **317:** 29–313, 1979.

Carlson, B. M., and Faulkner, J. A. The regeneration of skeletal muscle fibers following injury: A review. *Med Sci Sport Exer* **15:** 187–198, 1983.

DeGirolami, U., and Smith, T. W. Pathology of skeletal muscle diseases. *Am J Pathol* **107:** 235–276, 1982.

Duncan, C. J. Role of calcium in triggering rapid ultrastructural damage in muscle: A study with chemically skinned fibers. *J Cell Sci* **87:** 581–594, 1987.

Duncan, C. J., and Jackson, M. J. Different mechanisms mediate structural changes and intracellular enzyme efflux following damage to skeletal muscle. *J Cell Sci* **87:** 183–188, 1987.

Edström, L., and Grimby, L. Effect of exercise on the motor unit. *Muscle Nerve* **9:** 104–126, 1986.

Essén-Gustavsson, B. *et al.* Muscular adaptation of horses during intensive training and detraining. *Equine Vet J* **21:** 27–33, 1989.

Fischman, D. A. Part I. Morphogenesis of skeletal muscle. *In* "Myology," A. G. Engel and B. Q. Banker (eds.), pp. 5–37. New York, McGraw-Hill, 1986.

Hadlow, W. J. Myopathies of animals. *In* "The Striated Muscle," Int. Acad. Pathology, Monograph No. 12, pp. 364–409. Baltimore, Maryland Williams & Wilkins, 1973.

Horwitz, A. F., and Schotland, D. L. Part I. The plasma membrane of the muscle fiber. *In* "Myology," A. G. Engel and B. Q. Banker (eds.), pp. 177–207. New York, McGraw-Hill, 1986.

Hulland, T. J. Histochemical and morphometric evaluation of skeletal muscle of cachectic sheep. *Vet Pathol* **18:** 279–298, 1981.

Ihemelandu, E. C. Genesis of fiber-type predominance in canine pectineus muscle hypotrophy. *Br Vet J* **136:** 357–363, 1980.

Kelly, A. M., and Rubinstein, N. A. Development of neuromuscular specialization. *Med Sci Sport Exerc* **18:** 292–298, 1986.

Lexell, J., Taylor, C. C., and Sjöström, M. What is the cause of the ageing atrophy? *J Neurol Sci* **84:** 275–294, 1988.

Riley, D. A., and Allin, E. F. The effects of inactivity, programmed stimulation, and denervation on the histochemistry of skeletal muscle fiber types. *Exp Neurol* **40:** 391–413, 1973.

Sivalchelvan, M. N., and Davies, A. S. Induction of relative growth changes in the musculoskeletal system of the sheep by limb immobilisation. *Res Vet Sci* **40:** 173–182, 1986.

Tabary, J-C. *et al.* Experimental rapid sarcomere loss with concomitant hypoextensibility. *Muscle Nerve* **4:** 198–203, 1981.

Wilson, C. J., and Dahners, L. E. An examination of the mechanism of ligament contracture. *Clin Orthop* **227:** 286–291, 1988.

II. General Reactions of Muscle

Rigor mortis is a result of contracture of the skeletal muscle that develops after death. It is preceded by a period of postmortem relaxation. Rigor mortis is characterized by a stiffening of the muscles and immobilization of the joints. It proceeds in orderly fashion from the muscles of the jaw to those of the trunk and then those of the extremities, and it passes off in the same order. It does not recur in the natural course of events or after the muscles are forcibly stretched. The time of onset, in average circumstances, is 2–4 hr after death; maximal rigor is achieved in 24 to 48 hr, after which it disappears. The intensity of rigor varies considerably, as does the time of onset. It is slight or absent in cachectic or chronically debilitated animals and occurs with extraordinary rapidity in animals which die during or shortly after intense muscular activity or rapid exsanguination, but it is long delayed in well-rested, well-fed animals. It is hastened in onset and disappearance in a warm environment and retarded in a cold environment. Rigor also affects the myocardium and the smooth muscles, but rigor of smooth muscle begins and ends more rapidly than that of striated muscle. Probably all smooth muscle, including that of the muscular arteries, is involved, and the effect of this rigor of smooth muscle should be considered in the interpretation of postmortem appearances.

The factors influencing the time of onset and degree of rigor are the glycogen reserves, the pH of the muscles at time of death, and the environmental temperature. When glycogen stores are low, the onset is rapid, as after intense exertion; if the pH is low, as also occurs after intense activity, the onset is rapid. Low pH, however, in the presence of high glycogen reserves, does not lead to accelerated rigor mortis.

The chemical events in rigor must be a modification of those occurring in normal contraction, and they continue for some time after "death" of the body. Immediately after death, glycogen is converted to lactic acid by anaerobic glycolysis, and creatine phosphate is broken down to produce creatine. These are both mechanisms for the resynthesis of adenosine triphosphate from adenosine diphosphate. The capacity for resynthesis of adenosine triphosphate determines the period of delay in the onset of rigor, and rigor will occur only when the rate of adenosine triphosphate degradation exceeds its rate of synthesis, and tissue adenosine triphosphate levels are low and dropping rapidly. The eventual disappearance of rigor, or its failure to develop in cachectic animals, is probably due to complete exhaustion of the chemical systems which produce energy.

Muscles may undergo alteration of color during rigor and putrefaction. In a well-fed animal which dies suddenly, the muscles become pale, somewhat like the flesh of fish, a change which is probably attributable to a leaching of myoglobin by lactic acid produced after death; this too is probably the explanation, excepting dystrophic calcification, for the pallor of recent necrosis. In other animals, particularly those whose glycogen stores have been depleted by chronic disease or malnutrition, the muscles become unusually dark after death. It should be remembered that not all normal muscles have the same depth of color. Pallor of muscles also is present in animals which are chronically anemic from iron deficiency or blood loss, an observation which indicates that in iron deficiency there is depletion of the myoglobin of muscles as well as the hemoglobin of blood.

The histologic appearance of muscle is not greatly altered through rigor, dissolution of rigor, or even through early putrefaction. During the hyperexcitable period immediately after death, asynchronous contraction of segments may lead to some artifacts if the sample is fixed at that stage. Sarcomeres in some fiber segments are shorter than in others, a characteristic which, under other circumstances, might suggest acute antemortem degeneration of a few hours standing. Samples taken during or after rigor and fixed at that stage generally show regularity of sarcomere length with all sarcomeres in a state of moderate to strong contraction. If it is desirable to minimize artifact, sampling of muscle is best done at the mid- to late-rigor phase. If it is desired to examine muscle in an uncontracted state, a fresh sample taken prior to or shortly after death must be clamped prior to fixation, then fixed while still clamped to a fixed extended length. Disintegration of myofilaments and sarcoplasmic reticulum begins in 4 to 6 h but is not detectable in the light microscope until many hours later when Z bands begin to fade and the central A bands become homogeneous, leaving simply alternate pale and dark I and A bands. Interpretation of early degenerative change and its distinction from autolytic change sometimes requires all available information about time and circumstances of death as well as clinical signs and biochemical changes shown prior to death. The changes characteristic of muscle reaction, atrophy, and hypertrophy are outlined both as examples of the capacity of muscle to change and adapt, and then as parts of specific diseases.

Bibliography

Bartos, L. *et al.* Prevention of dark-cutting (DFD) beef in penned bulls at the abattoir. *Meat Sci* **22:** 213–220, 1988.

Elliott, R. J. Postmortem pH values, and microscopic appearance of pig muscle. *Nature* **206:** 315–317, 1965.

Fabiansson, S., and Reuterswärd, A. L. Ultrastructural and biochemical changes in electrically stimulated dark cutting beef. *Meat Sci* **12:** 177–188, 1985.

Fjelkner-Modig, S., and Ruderus, H. The influence of exhaustion and electrical stimulation on the meat quality of young bulls: Part 1. Postmortem pH and temperature. *Meat Sci* **8:** 185–201, 1983.

Fjelkner-Modig, S., and Ruderus, H. The influence of exhaustion and electrical stimulation on the meat quality of young bulls: Part 2. Physical and sensory properties. *Meat Sci* **8:** 203–220, 1983.

Honikel, K. O., Roncales, P., and Hamm, R. The influence of temperature on shortening and rigor onset in beef muscle. *Meat Sci* **8:** 221–241, 1983.

A. Atrophy

Atrophy of muscle is broadly defined as reduction in size, but in the context of muscle disease processes, it means a reduction in muscle fiber diameter or cross-sectional area. It does not normally include the reduction in muscle length, weight, or bulk, which is related to adaptational shortening of muscle fibers by reducing the number of sarcomeres. It has already been indicated that such shortening can occur rapidly when a muscle is detached from its insertion, thereby reducing tension, while at the same time, it is stimulated repeatedly through its nerve connection. Similarly, lengthening by the addition of sarcomeres can be achieved by increasing the stretch length of an otherwise normal muscle.

Atrophy of muscle occurs in a variety of circumstances and, although progressive loss of contractile components and fiber size are common to all, each type has a slightly different impact on the fibers, and leads to different long-term responses. Denervation creates a situation in which not only is the muscle not contracting against resistance, but the fibers are also deprived of their normal low level of tonic stimulus. In respect to the latter, denervation may create a different muscle environment from that created by a neural lesion proximal to the lower motor neuron. Disuse atrophy induced by limb fixation or joint ankylosis deprives the muscle of much movement but does allow tonic stimuli through an intact nerve, whereas disuse atrophy caused by pain might result in much less muscle stimulation in the form of voluntary motor impulses. Cachectic atrophy on the other hand, like the atrophy of old age, occurs in a muscle with an intact nerve and a full range of movement, and creates a fiber environment quite unlike some of the other causes of atrophy. Atrophy of scattered single muscle fibers or small groups of fibers accompanies many muscle diseases which cause local pressure, impairment of blood supply, or extensive myolysis.

The early stages of generalized atrophy in a muscle may not be easy to detect, and even later stages of atrophy, which affect fibers more or less uniformly, may be difficult unless morphometry is used to determine mean fiber sizes. Histochemical techniques may highlight differences of response in different types of fiber. All fibers undergoing a reduction in size are losing myofilaments and myofibrils by a process of more or less simultaneous disaggregation of actin and myosin filaments and disintegration of the protein of Z bands. Briefly, in the sequence, a slight sarcoplasmic space appears between the shrinking myofibrillar mass and the cell membrane, but the membrane (and the basal lamina) soon begin to wrinkle and condense to accommodate the new slimmer lines. After a brief lag, mitochondria, glycogen granules, and lysosomes are reduced in number and size, closely to parallel the reduction in cell mass. After a slightly longer lag phase, the tubular systems are also reduced in volume appropriately. All forms of muscle fiber atrophy follow this sequence, although the rate at which it proceeds and the type and location of fiber involved may vary considerably. Provided that the interval is relatively short, the influences inducing atrophy are removed, and the muscle environment is returned to normal, the sequence is reversed, and fibers are restored to normal size.

Denervation atrophy, usually referred to in contradictory terms as neurogenic atrophy, is moderately common. It is part of many congenital dysplasias which involve skeletal muscle, with or without contracture or arthrogryposis. Denervation atrophy is always accompanied by muscle paralysis, but the clinical evidence may be obscured if the nerve is small.

Some of the best-known examples of denervation atrophy in animals are laryngeal hemiplegia in horses caused by injury to the left recurrent laryngeal nerve, injury to the supraspinatus nerve by the pressure of a poorly fitting collar in a work horse, symmetrical or asymmetrical gluteal atrophy in the horse, and radial or brachial paralysis in dogs due to trauma incurred in automobile accidents. Lesions of the cord or the major nerve radicles emerging from the spinal canal are also common antecedents of denervation atrophy. Occasionally a brain lesion may cause denervation muscle atrophy, but this requires a long interval of survival. The more remote, functionally, the nervous lesion is from the muscle, the longer the interval required for evolution of muscle atrophy because of the greater opportunity for some tonic impulses to be fed into the local spinal–muscle loop. Local spinal osteomyelitis, disk protrusion, chronic meningitis, traumatic lesions, metastatic tumors, or localized spinal malacia are the types of central lesions most likely to cause denervation atrophy of muscle. The lesion will be most marked in the muscles served by the cord segment in which the nerve damage occurs; however, muscle fibers served by cord segments distal to the lesion will ultimately show atrophic changes.

Following peripheral denervation, the reduction in volume of the muscle is likely to be very rapid. Within 2 to 3 weeks, two thirds of the muscle mass will be lost, although fiber measurement may be necessary to demonstrate this change histologically. Subsequent gross changes in the atrophic muscle will consist only of a further moderate reduction in muscle mass with increasing prominence of adjacent bones. In well-fed animals, even this muscle deficit may be sometimes obscured by superficial and intramuscular fat deposits. Clinical attention is claimed more by the signs and consequences of dysfunction than by changes in muscle mass.

The sequence of events which takes place at the muscle level in denervation atrophy are well defined for several species and, in some subtle respects, differ from events which attend other forms of atrophy. The morphologic sequence was described previously. The features unique to denervation atrophy are the rapidity with which the changes take place, the nuclear changes that occur, and the masking of the differences between the two fiber types with time.

During the first few days of denervation atrophy, type 1 fibers diminish in size more rapidly than the type 2 fibers because of the dependence of the former on a low level of continuous trophic stimulation. The type 1 fibers begin with a slightly larger mean diameter, but this distinction is soon lost, and the entire fiber population curve remains monophasic and moves uniformly toward progressively smaller size, except in junctional areas where denervated and innervated fibers mix. Oxidative enzymes increase, or perhaps fail to decrease proportionately to the reduction in other fiber components, and soon both type 1 and type 2 fibers have a heavy content of these enzymes. Distributional characteristics of ATPase remain normal until very late in the atrophic process, so it appears that the type 2 glycolytic fibers acquire a greater capability for oxidative phosphorylation as denervation atrophy proceeds. After about 2 weeks, and as the fibers increasingly round up, there is an apparent increase in the frequency of muscle nuclei and certainly an increased prominence of nuclei. The changes seem to be those associated with a relatively slow rate of nuclear disappearance compared with that for other structures, and diminished competition for space within the myofiber or within the basal lamina. Nuclei round up and seem to become larger, although no mitotic activity takes place in satellite cells, and myonuclei are incapable of mitosis. Satellite cells sometimes migrate at this stage and, in so doing, may alter their longitudinal orientation to one which is circumferential or spiral. This conceivably could explain the association between denervation with subsequent reinnervation and aberrant ring fibers (see Ringbinden), since many satellite cells have long polar cytoplasmic "tails," which may act as a weak tether in such migration (see Muscles, Section IV,B, Regeneration and Repair of Muscle).

Characteristic of denervation atrophy is the grouping of muscle fibers which have retained a neural connection between those which have not. This occurs at the margin of the area of denervation; it is not found in muscles in which the entire motor fiber contribution has been severed, and is most likely to be found in muscles which receive motor axons from two or more myotomes. The general pattern is of groups of 3 to 15 adjacent large fibers separated by 3 to 15 or more atrophic fibers (Fig. 2.8). The clusters of large and atrophic fibers relate to clusters of terminal axons given off at the nodal points, and constitute parts of a single trophic unit. It is generally believed that one axon innervates one fiber type, although with training, the superficial identifiers of fiber type may change. In muscles like the longissimus or semitendinosus, in which type 1 fibers are often solitary, the denervated or innervated "groups" may consist of one fiber. This denervation pattern involving single type 1 fibers (groups) is sometimes encountered in large animals. Type 2 fibers occur in clusters in skeletal muscle, with the possible exception of the bovine diaphragm; therefore, they rarely if ever show this pattern of denervation atrophy.

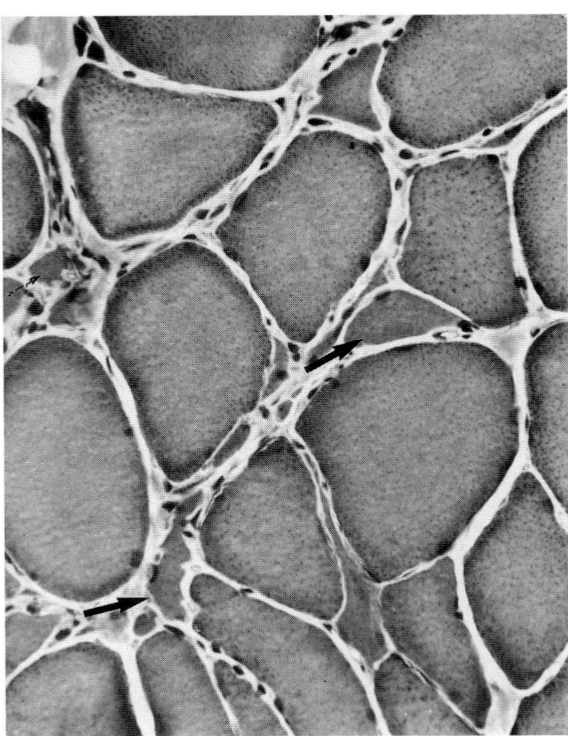

Fig. 2.8. Angular atrophic muscle fibers of denervation (arrows). Dog.

Very small fibers are characteristic of the advanced stages of denervation. As more space becomes available within the endomysium, all fibers round up, and because clusters of large fibers prevent architectural collapse, the atrophic fibers have an abundance of space, which in time becomes partly occupied by strands of endomysial collagen. The small fibers get smaller, and the large fibers get larger, not only because space is available, but presumably

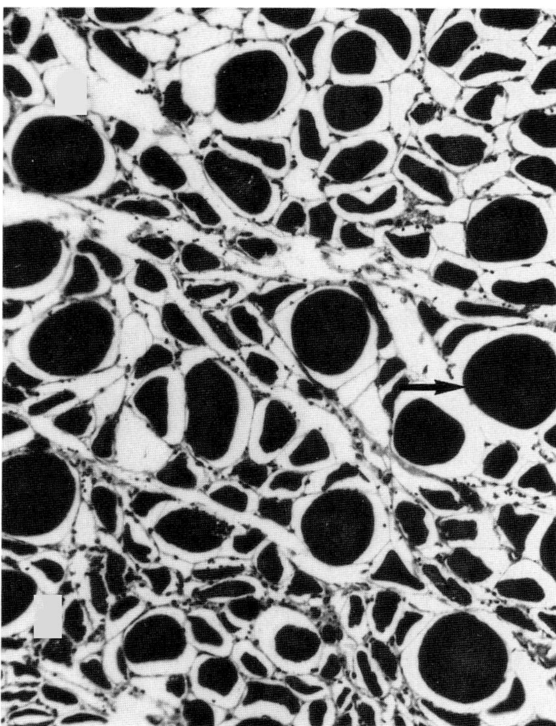

Fig. 2.9 Denervation pattern of atrophy in a muscle of a horse with cervical disk degeneration. There is a low level of type 1 fibers (arrow).

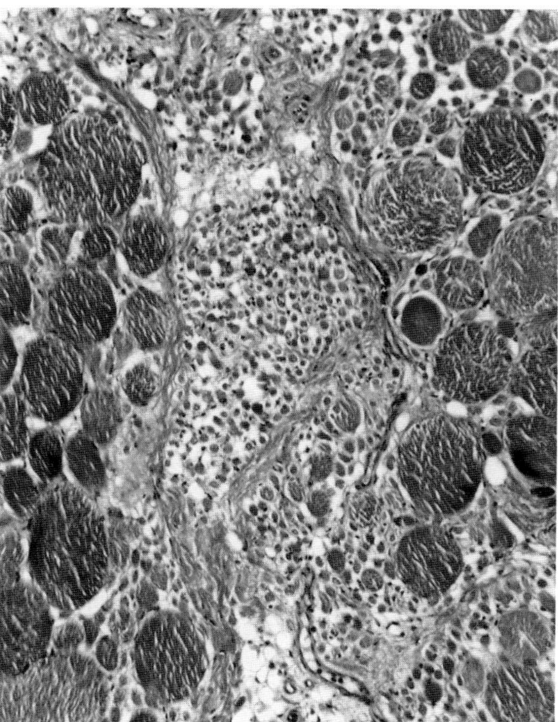

Fig. 2.10 Large and small fiber clusters of denervation atrophy of several months duration. Birth trauma. Calf.

also because there is a stimulus for compensatory hypertrophy (Fig. 2.9).

The late picture of fiber-group clusters is usually much clearer than the early one, particularly where only a few motor nerve fibers are damaged or where the denervation has taken place within the muscle at the level of the terminal axons. The groups of denervated fibers may then be very small and appear to be somewhat scattered in muscle in which the general framework is almost unaltered (Fig. 2.10). Fibers undergoing atrophy and losing myofibrillar mass may not become rounded because of pressures from adjacent normal fibers. They may be compressed by adjacent fibers and acquire a rounded angularity, which is sometimes the first indication of denervation atrophy.

During the more advanced stages of denervation atrophy, all denervated muscle fibers approach the level, somewhere around 20% of the original fiber mass, where further diminution occurs only very slowly. At this point the mean diameter is between 10 and 20 μm, and the largest fibers are less than 30 μm. A few of the thinnest fibers begin to dedifferentiate into elongated cells without myofibrils or myofilaments, or show evidence of degeneration in which the cell contents coagulate and are removed by macrophages. Both of these states probably are irreversible, even with reinnervation. End-stage muscle of this type results from 1 to 2 or more years of denervation (Fig. 2.11A).

Denervated muscles may be reinnervated readily if the original nerve damage has not disrupted the nerve sheath (Fig. 2.11B). Nerves grow back toward muscle at the rate of about 1.5 mm per day. If motor axons are able to regenerate and make contact with muscle fibers, and if the motor end-plate region on the muscle has not been destroyed, restoration of muscle end-plate connection can take one of two forms: an efficient and rapid return to the original end-plate in a manner indistinguishable from that of the original; or, more slowly, connection by subterminal sprouting. In the latter event, a terminal axon which has already made contact with one muscle fiber sends out a delicate branch to another fiber. The branch, which arises from the point between myelin sheath and end-plate, induces an end-plate formation at the point of contact or at a nearby receptive surface zone. A single terminal axon may, by sprouts, make effective subterminal contact with up to five muscle fibers, and each fiber innervated by the same nerve fiber will develop the type identity appropriate for the nerve, regardless of the original muscle fiber type. Thus, a denervated mosaic of type 1 and type 2 fibers might be completely reinnervated by an axon originally serving type 1 fibers, and the muscle fibers would all become biochemically and structurally type 1 fibers (see Development of Striated Muscle). An apparent example of this phenomenon is the type 1 fiber dominance found in the pectineus muscle adjacent to the acetabulum of German shepherd dogs with hip dysplasia. This change

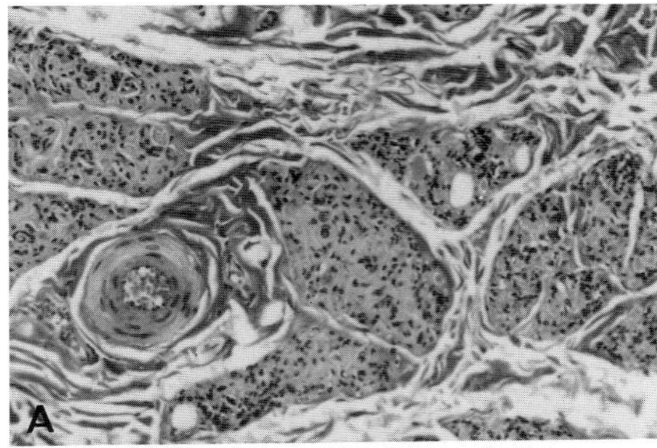

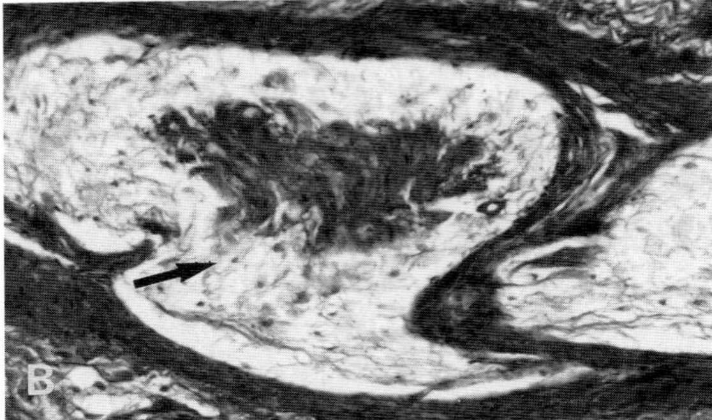

Fig. 2.11. "End-stage" muscle. Dog. (A) Fiber sheaths no longer contain myofibrils. (B) Muscular nerve trunk with long-standing denervation in which lost axonal space has been replaced by myxoid connective tissue (arrow).

from a checkerboard distribution to a monotonous dominance of one fiber type to the exclusion of the other(s) is known as fiber-type clustering. It is often a characteristic of denervation followed by reinnervation but is not an invariable part of lost nerve connections, nor is it only a result of denervation, as will be discussed in myotonic syndromes. It is important to recognize that fiber-type cluster-producing events, such as denervation or innervation failure, may occur in fetal life without visible effect grossly or functionally.

Disuse atrophy is the result of either reduced stimulation or reduced movement of a normally innervated muscle, or of an absence of normal muscle tension. It is probably inappropriate to place these in one category since their influence on muscle might be quite variable. A limb which is painful may be consciously protected against stimulation and use, whereas fixation without pain might permit vigorous isometric contraction. Thus the rate of early disuse atrophy is likely to be determined by the similarity of the muscle environment to that of denervation, and the sequence of events and state of the muscle will reflect this. Because the cause of disuse atrophy is likely to be transient, such as fracture or casting, often the muscle will not reach the stages of advanced atrophy. With paralysis and permanent joint ankylosis, however, end-stage muscle may be the fate of at least some muscles.

Disuse atrophy due to tenotomy or similar states may induce a very different set of reactions in muscle. As already pointed out, release of muscle tension is more likely to shorten muscle than to cause marked atrophy, but prolonged, reduced tension may cause slow atrophy as well. The sequence of events has not been explored well in domestic animals, but it appears that tenotomy leads to disproportionate atrophy of type 1 fibers with maintenance, or even hypertrophy, of type 2 fibers, the opposite effect to that seen in cachectic atrophy.

Atrophy of cachexia and malnutrition occurs when an animal is unable to supply enough dietary nutrients to maintain muscle; muscle becomes the source of nutrients for the rest of the body. This is in part because there can be a large protein flux into and out of muscle. Of the contractile muscle substance, 1–5% is dismantled each day, and in normal animals, an equal or greater amount is reconstructed. In view of the large bulk of the body muscle, this represents a very large amount of protein which can be borrowed on a daily basis. Net loss of muscle protein probably starts hours after negative nitrogen balance has been reached. A monograstric animal like the dog probably begins a net withdrawal from muscles after 24 hr of starvation. Ruminants reach the same point 2 days later, but calves and lambs reach it shortly after the monogastric animal would. The net withdrawal of muscle protein in undernutrition or poor alimentation parallels the effects of increased demands for protein and energy by a large conceptus late in pregnancy or those imposed by rapidly growing tumor masses in the body. Depot fat can provide body fuel to some extent, and one of the most efficient ways of using it is to metabolize it through muscle, but the withdrawal of muscle can only be delayed by this alternative source of energy and not prevented. In these circumstances, trophic neural contact is present, and although body movement may be reduced or slowed, use is retained. The atrophy of cachexia is gradual, and the result of circumstances not found in other atrophic conditions. The end result in some respects resembles that in other forms of atrophy, but the process may take years if the net withdrawal of muscle protein is subject to fluctuations, or is irregular or slight. Whether it occurs very slowly or not, the sequence seems to be the same. Essential muscles are preserved, while less essential muscles contribute to the maintenance systems, and their fibers atrophy.

The earliest changes are not difficult to monitor in the living animal because the back and thigh muscles are the first to undergo gross atrophy. This is not easily appreciated microscopically without measurement. Histochemically, it soon becomes clear that in general the type 2 fibers are depleted preferentially (Figs. 2.12A,B and 2.13A). In some "athletic adaptational" muscles like the longissimus, which are predominantly type 2, both fiber types are reduced at about the same rate (Fig. 2.13B). Generally after a month or two of borrowing from muscle, the gross evidence of emaciation becomes obvious as one third to one half of the muscle mass and most of the body fat will have been lost. By this time, in many muscles, the type 2

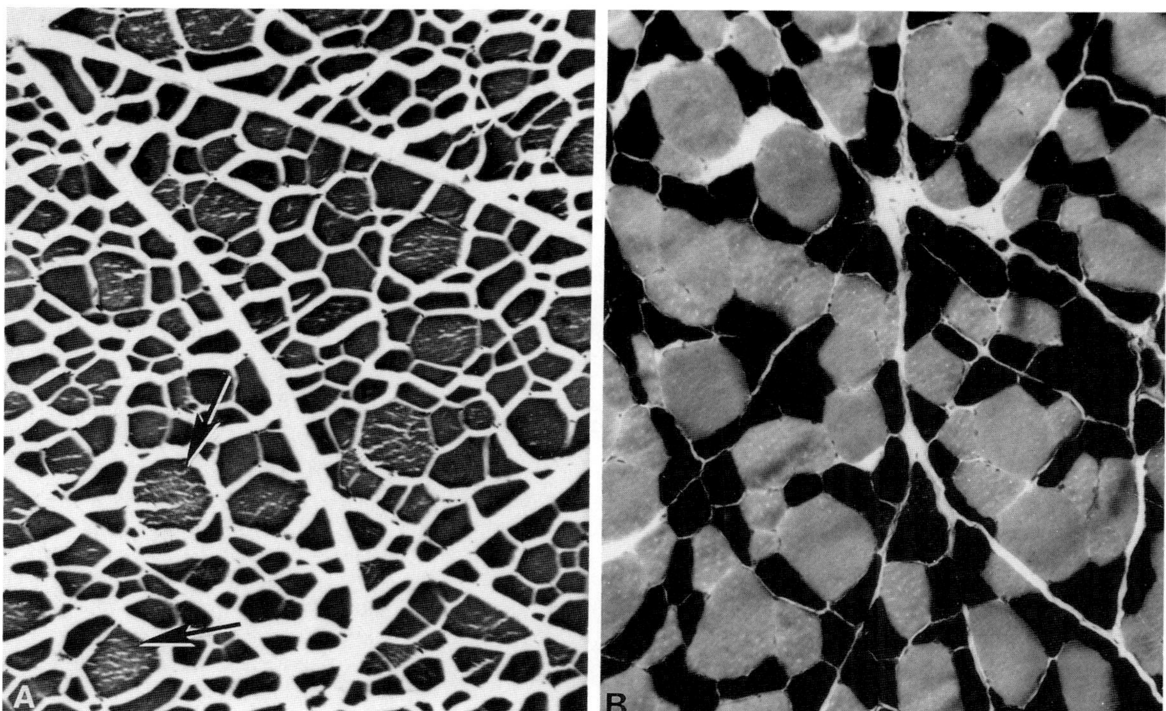

Fig. 2.12 Cachectic atrophy. Calf. (A) Atrophy-resistant type 1 fibers are prominent (arrows). (B) Cachectic atrophy. Emaciated sheep. Dark type 2 fibers. ATPase (alkaline) preparation.

fibers are steadily diminishing in size, and their angular shape contrasts with the much larger, plump type 1 fibers which, although reduced in size, show clear signs of resistance to atrophy (Fig. 2.13A). A histogram of muscle fiber size reveals an emerging biphasic curve with the mean diameters of the two types of fibers moving in the same direction but separating from each other.

Not all muscles show the same degree of fiber atrophy. The essential postural muscles like the supraspinatus show some differences in fiber size but relatively little absolute atrophy of type 1 fibers. The nonpostural muscles like the sternocephalicus show the greatest reduction, and because both types of fiber are affected, the tendency is for all fibers to be round and small though increasingly variable in size.

Throughout the early stages of atrophy, ATPase activity is unaltered, and oxidative enzyme content is reduced gradually. Later, the oxidative enzyme level of all fibers diminishes as does the glycogen content. Fat droplets can be seen in increasing numbers in type 1 fibers. No distinctive modifications of nuclei are apparent, and as muscle fiber size drops, so does the number of nuclei associated with each fiber. A condensation of endomysial connective tissue can be seen in the most atrophic muscles.

In the next, and usually the final stage of cachexia, muscles are slowly reduced over months to about one third or less of their original bulk. The muscle texture is soft, dark, and sticky. No fat remains. In nonpostural muscles, both fiber types are reduced further in size, but a biphasic curve still exists. Postural muscles have retained their mean fiber diameter somewhere just below normal, and a great variation in type 2 fibers is obvious. In all muscles, but particularly in postural muscles and those with a low type 1 content, a small part of the type 1 fiber group undergoes an absolute increase in fiber size and even longitudinal splitting. This has the effect of distorting the distribution curve by adding a long tail toward the upper end of size. Although these large fibers are not numerous, they are so large by comparison to the average, that they contribute a lot to the cross-sectional area of the muscle mass. The whole effect is greatly to reduce the glycolytic dependence in favor of an increase in the oxidative dependence and, generally, to reduce the number of fibers doing most of the work. A postural muscle might, in terms of cross-sectional areas, be switched from a 70% glycolytic muscle to one which is 80% oxidative. This phenomenon might explain why cachectic animals move so slowly, appear dull in their movements, and tend to have lost fine control of movements.

The type 1 atrophy-resistant (or hypertrophic) fibers are the largest of the type 1 group but are not otherwise a distinct population. Their ability to enlarge when most fibers are undergoing atrophy is unexplained. No indication of the signals that these fibers receive has yet emerged, nor of the characteristics on which they are selected, but the cachectic process seems to indicate a participation by trophic factors separate from neural or nutritional influences. The sequences sketched are the usual ones, but in some

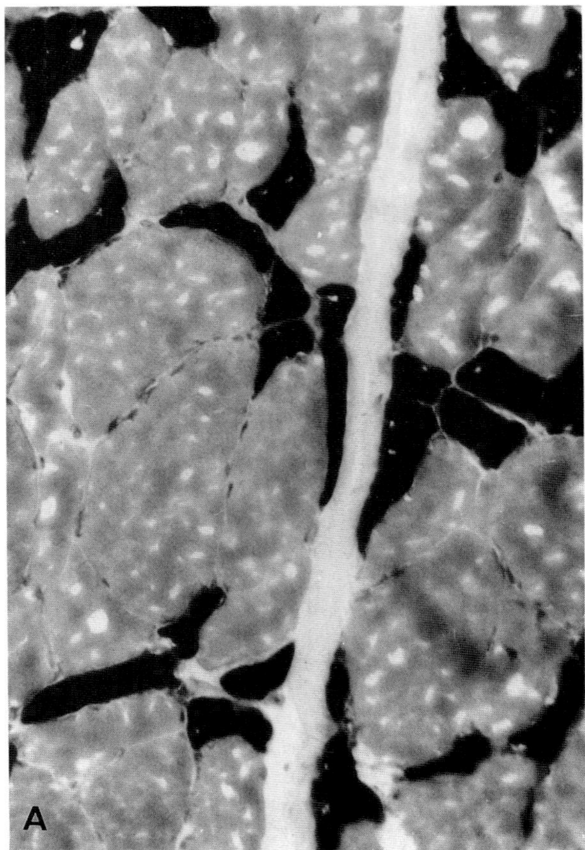

Fig. 2.13A Advanced type 2 cachectic atrophy. Sheep. Scrapie. ATPase (alkaline) preparation.

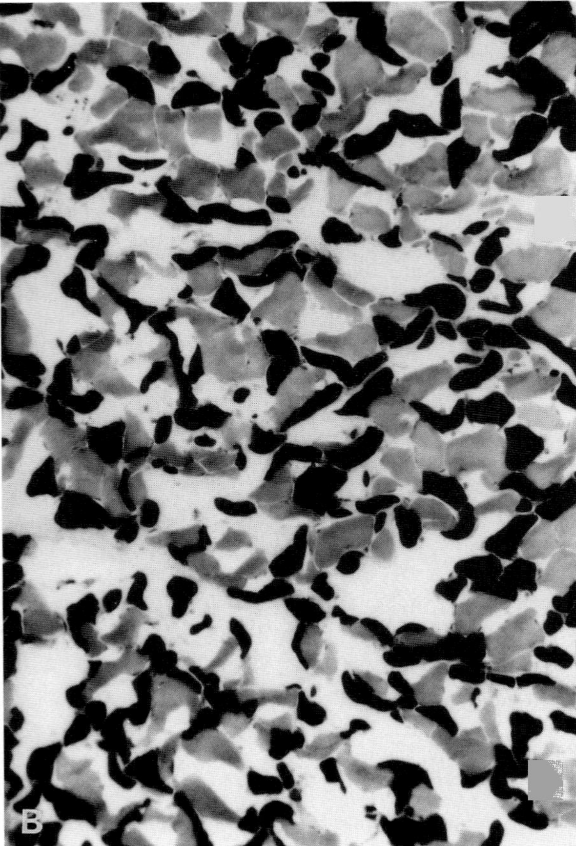

Fig. 2.13B Advanced cachectic atrophy of both fiber types. ATPase (alkaline) preparation.

muscles in some circumstances, the type 2 fibers are the ones which resist atrophy (Fig. 2.14).

B. Hypertrophy

Hypertrophy of muscle fibers is often physiologic and desirable, as is the muscle hypertrophy which often accompanies it. Racehorses and greyhounds have many more fibers in strategic places than do nonracing animals of the same species, but they also have slightly larger fibers when they are highly trained. This process of work hypertrophy of fibers is accomplished by adding sarcomeres, by adding myofilaments to the periphery of myofibrils (Fig. 2.15A), and by adding new myofibrils to the existing ones by a process of longitudinal splitting (Figs. 2.15B, 2.16A), and perhaps by *de novo* construction. There are sharp upper limits for this type of hypertrophy, and for animals it lies somewhere between 80 and 100 μm in lesser diameter. Absolute limits to the process are implied by the fact that in racing animals longitudinal fiber splitting is rare. Those present may represent new fiber building by budding of existing fibers or their satellite cells, or may be the product of regenerative repair. The hypertrophic products of repair can often be distinguished by a somewhat bizarre arrangement of the clefts in a semicircular, rose-petal type of conformation, which is the result of late fusion of myoblasts/myotubes (Fig. 2.16B). As one myotube fuses with one or more partially investing myotubes, capillaries and nuclei may be left in an interior position at the ends of partly sealed clefts. Such bizarre forms can be seen in reinnervation of normal muscle, myotonic syndromes, and sometimes following ischemia and/or muscle transplantation.

Muscle fiber hypertrophy occurs also when there is a deficiency of fibers in an area. This may be a compensatory phenomenon. It occurs in congenital neuromuscular diseases, in chronic postnatal denervation atrophies, in the abdominal muscles of multiparous animals repeatedly producing large conceptuses, in nutritional or primary dystrophies and in myotonic syndromes in which many muscle fibers have been irretrievably lost and, as indicated above, in advanced cachectic atrophy. These compensatory hypertrophies sometimes produce remarkably large muscle fibers: fibers of 150 to 200 μm least diameter often develop in animal muscles. In many cases these large fibers show clear evidence of incomplete longitudinal division, which allows one or more capillaries to be located near the center of the muscle fiber. When division is more or less com-

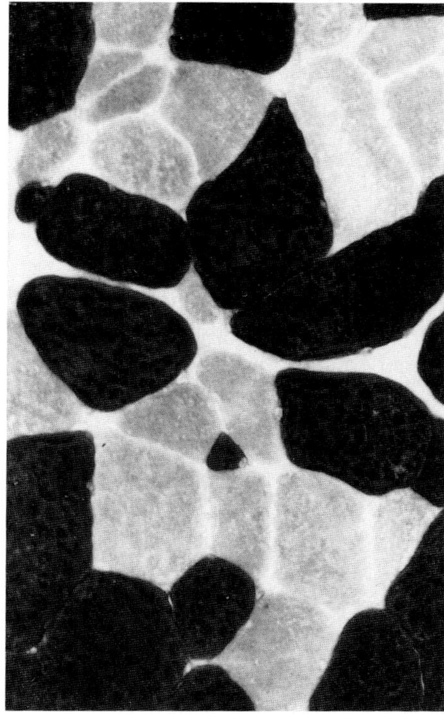

Fig. 2.14. Atypical selective atrophy of type 1 fibers in cachexia. A reversal of usual pattern. ATPase (alkaline) preparation.

plete, two or often more fibers become arranged in an "orange section" array and show a fiber-type grouping cluster. Sometimes very large fibers have hollow cores, and this suggests that one stimulus for the cleaving of fibers is central fiber hypoxia or malnutrition as the mid segment gets farther and farther from the vascular pipeline. No adequate explanation of the stimuli driving the process of hypertrophy has been proposed.

III. Congenital and Inherited Defects of Muscle

The terms congenital and inherited present problems of definition at any one time, and diseases or syndromes may change their status as a result of investigation and new information. There are few descriptions of developmental abnormalities of skeletal muscles in animals and even fewer descriptions which are yet adequate in respect of descriptive pathology. This should not imply that such abnormalities are rare; they are quite common and in some instances even quite important. They may be afflictions of muscle alone or part of a variety of more generalized malformations, especially those involving the spinal axis. Some of the defects are inherited, but probably fewer of them are inherited than are claimed to be.

Reference to defects of the spinal axis recognizes the primacy of neuroectoderm in the orderly development of the axial skeleton and the segmental development of myotomes. Neuroectodermal defects, some of which are described in this chapter with arthrogryposis and some in Chapter 3 with diseases of the nervous system, are

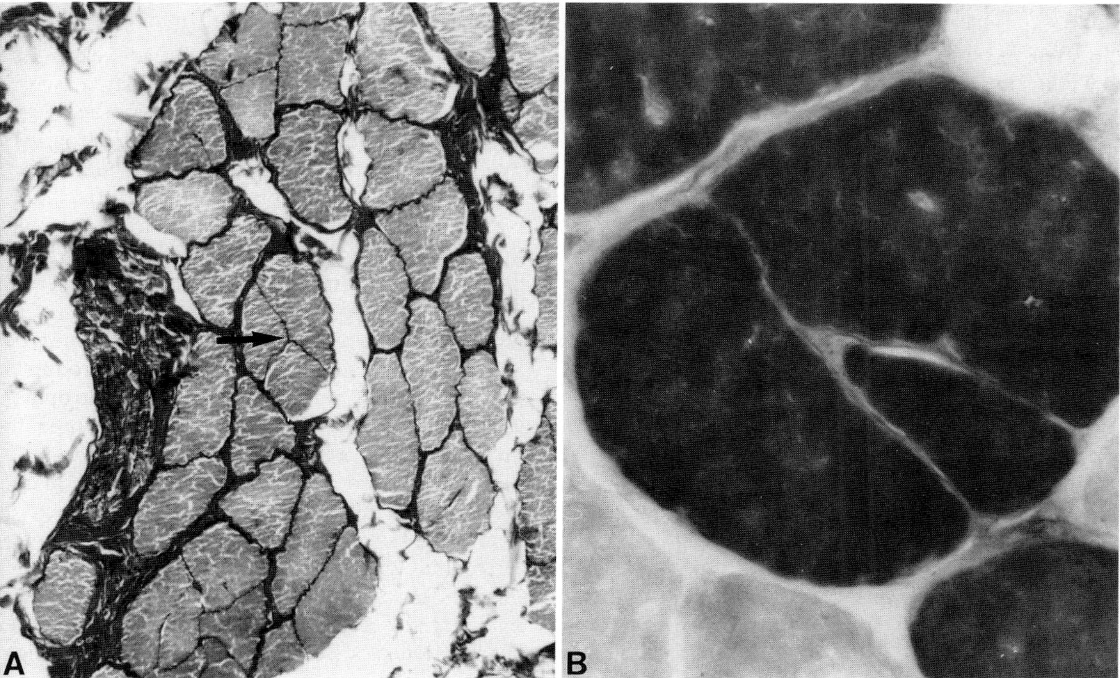

Fig. 2.15 (A) Fiber hypertrophy with fiber subdivisions (arrow). (B) Hypertrophy with early subdivision.

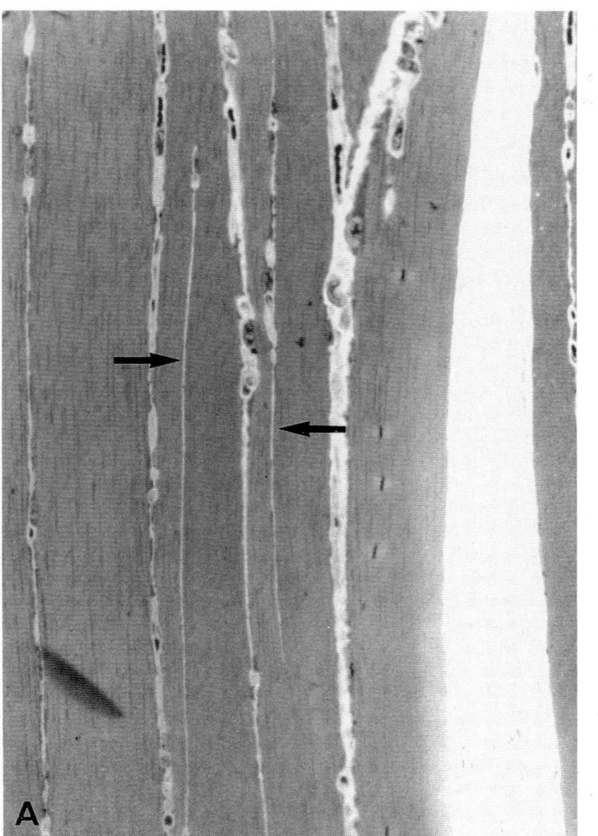

Fig. 2.16A Longitudinal clefts in two adjacent fibers (arrows).

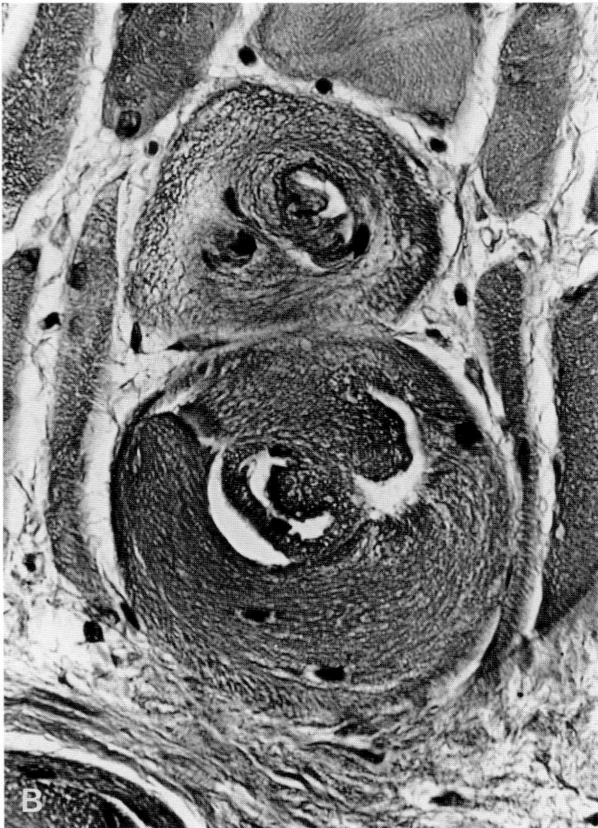

Fig. 2.16B Bizarre but transient regenerative fibers in a denervated muscle.

important. Inherited defects may be congenital or of early or delayed postnatal onset.

A. Neuroectodermal Defects

Selected for description here are syndromes which are, or which are reasonably presumed to be, consequent on aberrations of development of the neuroaxial skeleton and on the segmental patterns of innervation of the myotomes. The discussions in an earlier section on the plasticity of neuromuscular innervation, and of reinnervation following nerve injury, suggest that in the localization of spinal defects, some allowance may be made for variety in patterns of nerve growth. It is also likely that in the turbulent period of organogenesis, segmentation of myotomes and their derivatives will not be strictly preserved and that spinal myotomes will not be strictly separated from muscles which develop from cranial mesoderm. In fact, committed cells in the myotomes do have extensive migratory ability in some circumstances.

B. Arthrogryposis and Dysraphism

Arthrogryposis literally means crooked joint. The terms congenital articular rigidity and arthrogryposis multiplex congenita are sometimes also used synonymously to describe this syndrome, but the disease is, by definition, present at birth and therefore congenital, joints are not necessarily rigid, and frequently only one or two joints or only one limb may be involved (Fig. 2.17). Crooked calf

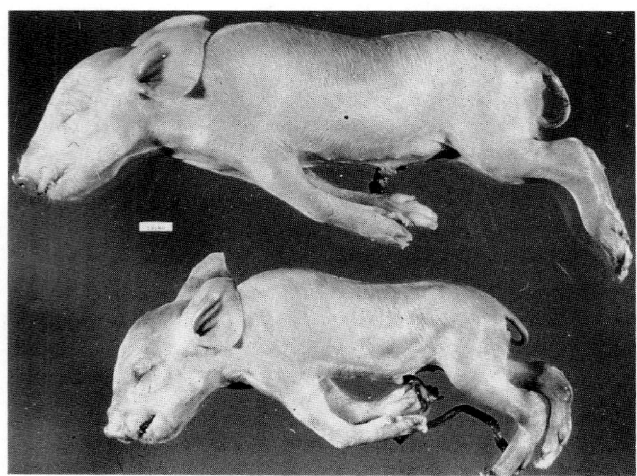

Fig. 2.17 Arthrogryposis. Pigs.

disease and acorn calf (or lamb) are environmental syndromes characterized by limb deformities (see Chapter 1, Bones and Joints.)

The causes of arthrogryposis often are not clear, although the syndrome can result from prenatal joint fixation by myofibroblasts around joints and coincident contracture of the tendon–muscle linkages around those joints, both of which can be induced by limb or limb part immobility. The cause of the limb immobility is usually denervation; in 90% or more of cases in both humans and animals, this seems to be the fact, although clearly such diseases as congenital dystrophy or myotonia, congenital muscular atrophy, or rare diseases which lead to muscle agenesis could also account for a lack of limb movement in the intrauterine environment. Such diseases have not been reported as causes of arthrogryposis in domestic animals.

Unequivocal links exist between arthrogryposis and the recognizable lesions caused by arrest or delay of neural tube closure (dysraphism). These include, in the extreme, spina bifida and cord agenesis, but also anomalies of the dorsal septum, anomalies of the central canal such as hydromyelia and syringomyelia, anomalies of the dorsal, central, or ventral gray matter (Fig. 2.18) and anomalies of the ventral median fissure (see Chapter 3, The Nervous System). Dysraphism does not always lead to arthrogryposis; segmental lesions are found sometimes in clinically normal animals. The lack of muscle lesions is explicable in terms of regional compensatory migrations of developing myoblastic or neural cells in the course of their development.

Such obvious spinal cord lesions as those described are present in a proportion, perhaps a majority, of arthrogrypotic domestic animals. In the rest, although an absence of primary myogenic or osteogenic lesions seems to point to a neurologic cause, the identity of the failed neural component is not obvious on routine investigation. Several careful studies of unilateral arthrogryposis in animals and children have indicated that the neural changes may be subtle and varied. The number of motor neurons in an apparently normal cord may be segmentally reduced, particularly in the caudal portion of the thoracic and lumbar eminences. Or the number of motor neurons may be increased, or the neurons may be disoriented in a segment subject to dysgenesis at an early stage of differentiation. Where comparisons are possible, peripheral nerve trunks in affected limbs are demonstrably smaller and often lacking some fascicles, although individual axons are normal. Beyond this, conjecturally, there could be failure of neural direction, or connection, or of end-plate development.

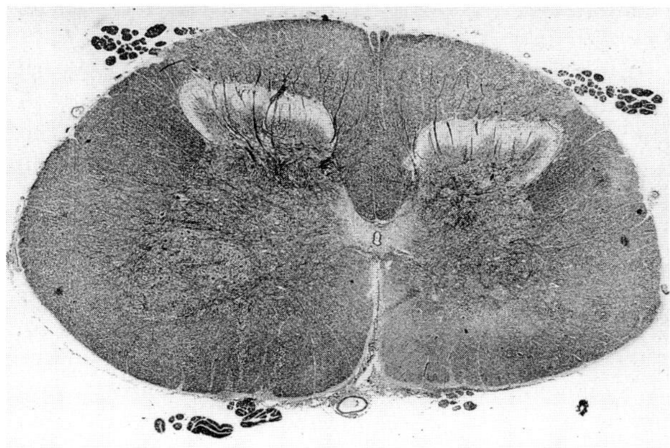

Fig. 2.18 Myelodysplasia. Calf. Failure of development of right ventral horn.

For proper *in utero* development, muscles require normal innervation. Without it, they atrophy or fail to develop and are more or less replaced by fibrous tissue and fat. Associated with each embryonic spinal cord segment is a myotome, a collection of cells with myoblastic potential, which migrate as the fetus develops and which ultimately form muscle. Committed myoblastic cells from one myotome may contribute to several muscles. Later, nerves from the same spinal segment follow the route of the myoblastic cells and innervate the same muscles. Most definitive muscles receive cells with myoblastic potential, and subsequently nerves, from two to six cord segments, but a few small muscles of the distal limbs receive contributions from only one. This explains why some muscles subtended from only one segment are vulnerable to denervation or lack of innervation resulting from a cord lesion involving a single segment, while a cord lesion of similar extent serving muscles which receive cells and nerves from more than one myotome can be asymptomatic. In the latter case collateral innervation from nerves arising in normal cord segments can develop. Evidence supporting this explanation is available from several sources. In children with club foot, those muscles innervated from the shortest length of cord (fewest cord segments) are the ones which fail to innervate in late fetal or early postnatal life, while other muscles suffering initial failure are restored to normal by a process of collateral innervation. The morphologically and functionally normal muscles can be detected as having had an abnormal history only by the presence of a characteristic histochemical fiber grouping (Fig. 2.19) (see denervation atrophy).

Arthrogrypotic fixation of the joints of lambs, calves, piglets, and foals, and less frequently of kittens and puppies, occurs with some regularity, and sometimes in epidemics. The extent of the lesions varies considerably; thus, one to four limbs and the axial skeleton may be involved. In mild cases two limbs, usually the hind, are fixed in flexion. Among those cases which have demonstrable spinal cord dysraphism defects, it is likely that hind-limb involvement will be more prevalent than forelimb lesions if only one set is involved. Where such cord lesions are not readily evident, it appears that selective forelimb involvement is more likely. This difference contributes to the growing conclusion that dysraphism is not the only predisposing event, even where a neurologic origin of the arthrogryposis is beyond doubt. Severely affected animals may have scoliosis, kyphosis, and torticollis, and the limbs or parts of limbs may be rotated,

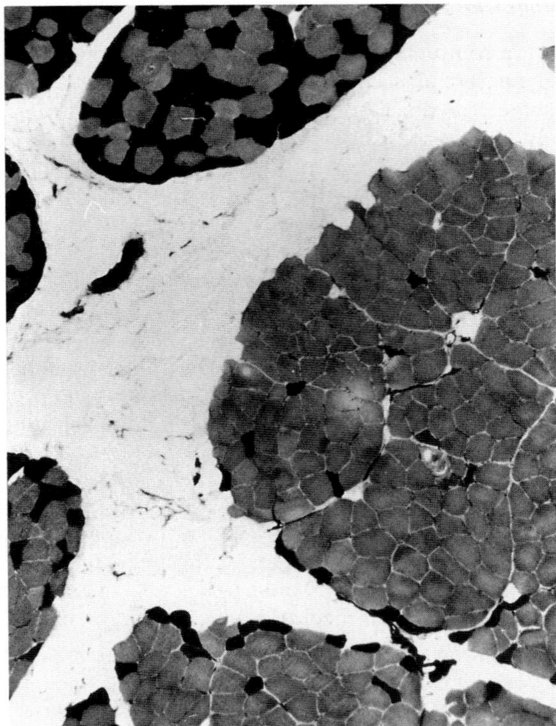

Fig. 2.19 Collateral innervation producing fiber type grouping. Sheep. ATPase (alkaline) preparation. Muscle bundle on the right shows type 1 fiber dominance.

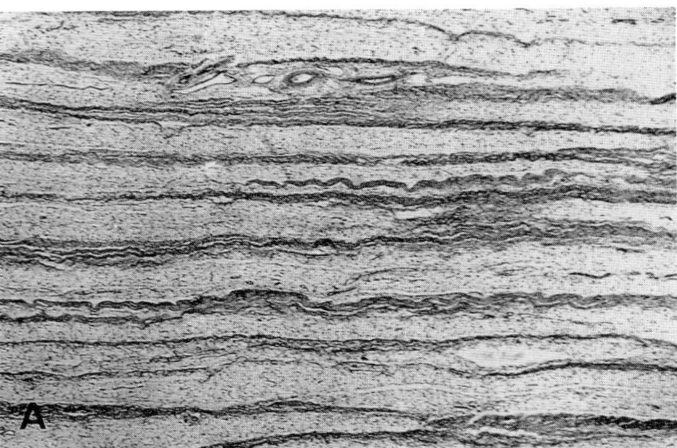

Fig. 2.20A Muscle hypodevelopment and fat replacement in arthrogryposis due to myelodysplasia. Calf.

Fig. 2.20B Higher power of 2.20A.

abducted, or curled backward or forward in grotesque positions.

Newborn animals afflicted with arthrogryposis are sometimes stillborn or may show evidence of death 2–4 days earlier, as indicated by autolytic changes. Hydroamnios is disproportionately associated with these, and so is dystocia. Many, perhaps most, are of normal skeletal size, although some are considerably smaller than normal or smaller than littermates. Where single limbs are affected, bones in the fixed limb may be smaller. On the other hand, litters of affected pigs may show only a splayleg tendency clinically. Limb muscle deficit is usually apparent, but in some cases the afflicted limb or limbs may reveal reduced muscle bulk only as a result of careful topographical examination. In others the knobbly joints, which are normal in size, but thrown into relief by the lack of adjacent muscle, signal the nature of the problem and may be the ones in greatest fixation. This wide variation, even between cases with apparent similarity of involvement, is in part due to a sometimes abundant replacement of the lost muscle mass by local masses of neonatal fat (Fig. 2.20A,B), a tissue lacking in thinner or less mature fetuses. In the rare cases where the limb involvement is unilateral, the animal is likely to be born alive. Fibrous tissue from atrophic muscle contributes to joint fixation. The increased volume of collagen has been regarded either as a secondary development, or as an incidental change. The demonstration of actin fibrils in fibroblasts (myofibroblasts) around fixed joints demonstrates that some of the fixation and perhaps most of it is related to tightening due to scarification of ligaments. It appears that this fibrosis occurs regularly only around a joint which is rendered incapable of movement by denervation or other immobilizing effect, and that it can occur very rapidly.

Dissection of major nerve trunks and muscle often is very difficult. Deficits seldom consist of the absence of one whole muscle, and it is often impossible to establish normal muscle landmarks because deficit and partial restoration of muscle tissue blend irregularly (Fig. 2.21). Nerve trunks often appear to be normal.

Histologically, it is quite easy to recognize irregularity of muscle development, but much of the mass of recognizable muscle will be superficially normal either as a result of limited original unimpeded development or due to collateral innervation. At the edges of the normal area, small groups of normal or small round muscle fibers, each of which is separated from like fibers by lipocytes or empty endomysial sheaths, may occur. Rarely, a typical denervation pattern of clusters of large and clusters of small fibers

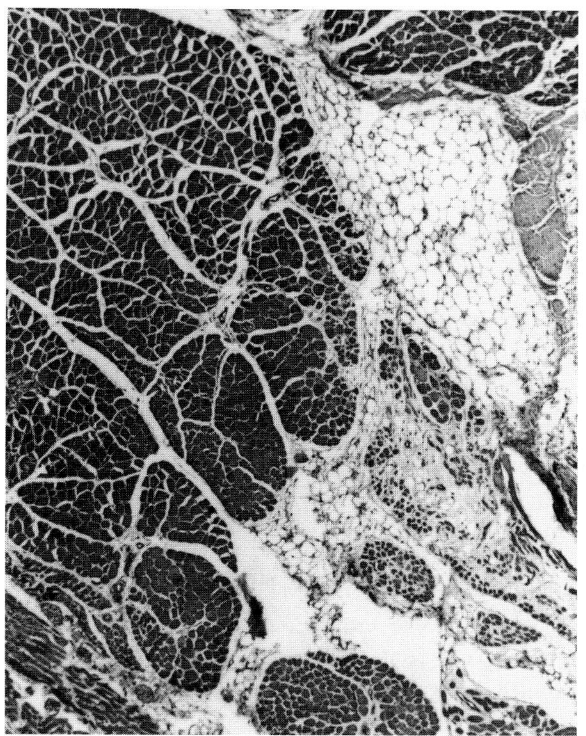

Fig. 2.21 Pattern of muscle fiber hypodevelopment and out-of-phase development in arthrogryposis.

can be seen. Variable portions of the thigh or shoulder mass are usually made up of soft depot fat tissue. This tissue, which is derived from improperly developed muscle, contains an unusually well developed vascular system.

The fixed joints are usually surrounded by a moderate increase in fibrous tissue, and articular surfaces are often somewhat distorted in adaptation. Occasionally, the skin forms a sling or web filling in the angle of the carpal or tarsal joints, and this is assumed to be a secondary lesion.

As noted above, it is often impossible to relate lesions to a specific cause. The critical event must occur early in pregnancy. Genetic causes are postulated in calves, sheep, and pigs, but the establishment of such a relationship does not require a different pathogenetic mechanism since the gene effect would apparently be directed at the neural component. Syndromes in the Charolais, Friesian, Swedish, and Red Danish breeds of cattle, sometimes associated with cleft palate, are consistent with that of a simple recessive or modified recessive characteristic. Environmental toxins or viruses probably could create a similar disease pattern. The viruses of Akabane disease, Cache Valley fever, bluetongue, and border disease cause outbreaks of arthrogryposis in cattle and/or sheep. *Lupinus laxiflorus* (wild lupine), and other plant toxins, also cause deformities in fetuses when they are ingested by the dam during early pregnancy (see Chapter 1, Bones and Joints).

C. Congenital Flexures

Pastern contracture and immobility by itself may sometimes be part of arthrogryposis. If the flexure is combined with some degree of distal limb rotation, the pathogenesis is likely to be associated with neural and muscle changes as a minor expression of arthrogryposis. Many lambs, foals, and calves are born with an apparent inability to straighten the fetlock and sometimes other distal joints, yet deformities are minimal, and muscle loss is not apparent. The problem in these cases seems to relate to the holding of the affected joint or joints in flexion without relief during a period in late gestation. The affected muscles appear to have reacted as they would following tenotomy by losing sarcomeres from the ends of myofibrils. This makes the joint initially incapable of extension, but under more or less constant tension, the lost sarcomeres are quickly returned in hours or days. Such an explanation seems applicable because many foals and lambs are restored to normal by the application of a thick fabric bandage for a few days. The original joint-fixing event can only be guessed at, but it would be easy to understand how it might happen in species in which the fetal size and limb length are high in relation to dam size.

D. Spinal Atrophies

In dogs there is a growing collection of diseases characterized by progressive motor paralysis in early postnatal life. Most appear to be congenital. Some are unique to one breed, but a few have been reported in crossbred dogs with little evidence of a gene-related causation. The progressive nature and fatal outcome of these syndromes indicate that the pathogenetic mechanisms are associated with progressive neural lesions. Similar diseases occur in human infants and are associated with abiotrophies, storage diseases, failed chemical pathways, and hereditary defects of cell structure. Similar causes may be anticipated in dogs. In all of the canine diseases, the effect on muscle is that of a progressive denervation atrophy, which has already been described in general terms. There seems little point in more detailed description of subtle variations of lesions except those of the nerves, which are included in Chapter 3, The Nervous System. Syndromes in which there appears to be a primary muscle fiber component will be described under dystrophies and myopathies.

E. Muscular Defects

Defects in the form and disposition of muscles associated with neuraxial deformities are quite common, but there are three syndromes of importance that are not associated with lesions in nervous tissue.

1. Congenital Clefts of the Diaphragm

Congenital diaphragmatic clefts affect all species either alone or as part of a more generalized malformation. They are formed as a result of incomplete fusion of the opposing

Fig. 2.22 Splayleg. Piglet.

pleuroperitoneal folds or, less frequently, by failure of fusion of one of these folds with the septum transversum. Most clefts are therefore in the dorsal, tendinous portion of the diaphragm and appear as enlargements of the normal esophageal hiatus. The clefts may permit the herniation of tissues ranging in size from a small button of liver to extensive aggregations of viscera. In very large congenital clefts, the diaphragm may be represented by only a narrow rim of muscle, which does little more than mark the diaphragmatic origin. Congenital clefts must be differentiated from acquired but healed lacerations of the diaphragm.

2. *Myofibrillar Hypoplasia in Piglets*

Myofibrillar hypoplasia, often called splayleg or spraddle-leg, is principally a disease of neonatal piglets. It occurs in all countries with well-developed pig-rearing industries. An estimated 60,000 pigs a year in Britain suffer from the disease, and about half of these die. The incidence of the disease fluctuates inexplicably but frequently appears as a farm or regional outbreak. The disease incidence within litters is variable, and a sow may produce consecutive litters with the disease. Genetic predisposition and infectious or nutritional causes are postulated but not proven. Intoxications by mycotoxins, and abnormalities of potassium–sodium metabolism have been suggested also.

Piglets with splayleg assume a posture with the hind legs or all four legs laterally extended (Fig. 2.22). They are apparently unable to adduct them or retrieve them from a forward or backward position. Mobility is reduced, and affected pigs are susceptible to accidents or may starve due to an inability to compete to suckle. The locomotor defect is transient and within a week, or at most two, survivors appear normal.

Not all piglets suffering from splayleg have myofibrillar hypoplasia; other conditions cause similar signs, but much less frequently. Developmental abnormalities of spinal closure may be responsible (see Muscles, Section III,B, Arthrogryposis and Dysraphism), and myopathy induced by the injection of saccharated iron may produce a similar syndrome in one or more litters (see Muscles, Section VII, Nutritional Myopathy). The splayleg syndrome, which is unrelated to either of the above causes, and as described here, is associated with an abnormally small mass of myofibrils within individual muscle fibers. The severity of the lesion varies, and all the pigs in a litter may be affected to some degree, but there is a threshold above which they show clinical effects of the disease.

In the late intrauterine development of skeletal muscles in pigs, the muscle mass is increased by the stem line-template method of new muscle fiber production. A large type 1 muscle fiber centrally located within a sublobule (a segment of a primary bundle) of developing muscle apparently divides longitudinally to create a smaller daughter fiber, or acts as a template beside which a new fiber evolves from myoblasts. The new fiber has characteristics of a type 2 fiber. The central template fiber continues to spawn new fibers, gradually pushing those formed earlier to the periphery and thereby creating a sublobule of 10 to 20 fibers. Beginning at about the time of birth, a few of the most recently formed type 2 fibers adjacent to the primary or template fiber take on the characteristics of the central type 1 fiber. This early postnatal period also seems to mark the end of vigorous new fiber production, although some increase may continue in some circumstances. Fibers formed late in gestation seem to have a high water content, and it is quite characteristic of young neonatal pig muscle that fixation creates a marked shrinking of the myofibrillar mass within the sarcolemmal sheath and endomysium. About 25 to 30% of the cross-sectional area within the fiber therefore appears to be empty. This characteristic is the basis for the diagnosis of myofibrillar hypoplasia, but the difference which exists in myofibrillar shrinkage or fiber size between affected and nonaffected piglets from the same herd often seems not to be significant. Muscles selected from splayleg pigs at postmortem examination on the basis of softness and paleness do show focal areas of greater than normal myofibrillar shrinkage and also an abnormally large glycogen content. By 7 days most of the space within the fibers will have been filled by myofibrils, and fibers will have increased in size by 60 to 80% over birth size in both splayleg and normal piglets. At this time also, splayleg pigs cease to show locomotor difficulties clinically, and glycogen granule distribution of both splayleg and normal pigs is of the adult type. In the critical period just before and just after birth, splayleg pigs show an irregularity or retardation in the transition of type 2 fibers to type 1 fibers in the central sublobular region. This retardation seems to be corrected at 7 to 8 days in synchrony with the disappearance of clinical signs and the rapid postnatal increase in fiber size in surviving splayleg pigs.

The characteristic lesions of myofibrillar hypoplasia are only partly revealed by light microscopy. Affected muscles, chiefly the longissimus, semitendinosus, and triceps, are incompletely or irregularly involved, and the most obvious change is lightness of staining with eosin due to the low volume of myofibrils within outer cell membranes of normal volume. The remainder of the fiber sheath appears to be empty but sometimes contains a nucleus or thin pink proteinaceous fluid. In longitudinal section, myofibrils appear to be diluted by watery fluid or clear strips of space. Occasional fibers may show segmental myolysis; these may be due to abnormal stretching of muscles attached to uncontrolled limbs or may indicate an unusually vigorous shifting of muscle proteins. As previously indicated, the changes seen in splayleg pigs may also occur less commonly and less obviously in normal littermates.

Electron microscopic changes in the muscles of splayleg pigs may provide an explanation for the clinical signs. A series of degenerative changes occur segmentally in muscle fibers. They consist of Z-line distortion with dispersal of the electron-dense material, disarrangement and loss of filaments, loss of register and alignment of sarcomeres, and thinning of myofibrils. These lesions develop in more than 70% of fibers in splayleg pigs but in only 30% of fibers in normal littermates and normal herdmates, a highly significant and consistent difference. In some fibers, phagolysosomes are present in increased numbers in affected pigs, vesicles form, and in a few fibers, segmental degenerative destruction of myofibrils occurs. By 7 days very few degenerative changes are present, and there are no significant differences between normal and splayleg pigs.

The characteristic changes in piglets with myofibrillar hypoplasia appear to be part delayed development and part degenerative or myolytic. One feature of the disease which is not understood is the oversized muscle fiber sheath, which is not consistent with hypoplasia. The underfilled sheath, in both splayleg and normal pigs, suggests that something, probably myofibrillar mass, has been removed in late gestation. Protein (myofibrils) can be withdrawn rapidly from muscle, and in some species which produce a relatively large conceptus, the dam lives off her own muscles, and perhaps those of her fetuses, in late pregnancy even when she is well fed. It is not inconceivable that myofibrillar hypoplasia results from myolytic atrophy triggered by the high demand for fetal and placental structural protein. It is possible that premature removal of feed from the sow, in the interest of emptying the intestinal tract in preparation for farrowing, would markedly increase protein withdrawal from fetal as well as maternal muscles.

Myofibrillar hypoplasia has been reported as an isolated case in a 6-week-old calf. Lesions very similar to those in splayleg pigs were seen.

3. Hyperplasia of Muscle Fibers in Calves and Lambs

This congenital and hereditary anomaly of muscle affects several of the European breeds of cattle and is reported in sheep in one flock. It is characterized by an increase in the number of fibers in affected muscles. The individual fibers are of normal size and structure. In homozygotes showing the full effect of the hyperplasia, the increase in gross muscle size is substantial, although not twice normal as the common name "double muscle" might suggest. In fact, the contours of the large muscles are regular and normal except for rather prominent topographical clefts between muscles, associated with large individual muscle size, reduced body fat proportion, and thin skin (Fig. 2.23). Bones may be lighter than normal and are often marginally shorter. Probably all striated muscles of the body are affected to some degree, but the ones showing the most obvious change are those of the thighs, rump, loin, and shoulder, giving the calf a very athletic appearance and predisposing to dystocia.

Attempts to exploit the disease in producing a meatier carcass have prompted an enormous amount of investigation into the heritability and stability of the disease charac-

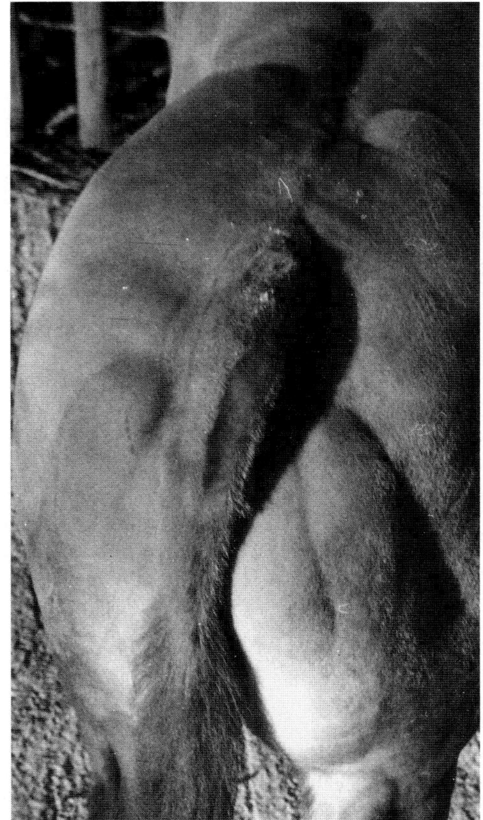

Fig. 2.23 Congenital muscular hyperplasia. Calf. (Courtesy of J. C. MacKellar.)

teristic. Some fiber increase is manifest by heterozygotes, but the level of heritability seems to be low and subject to capricious variability of penetrance or expression. Apparently the hyperplasia-inducing genes interact in an undetermined manner with several others.

The only qualitative changes observed with any regularity are a higher than normal level of small fibers with tapered ends, and a variable increase in the proportion of type 1 fibers, changes which would not be readily observed in routine muscle examinations. Attempts are being made to produce a more standardized animal, but the dystocia-producing potential of the calves has acted as a deterrent for most beef producers.

Bibliography

Abbott, L. C. *et al.* Crooked calf disease: A histological and histochemical examination of eight affected calves. *Vet Pathol* **23:** 734–740, 1986.

Andersson, B., and Andersson, M. On the etiology of "Scotty cramp" and "splay"—two motoring disorders common in the Scottish terrier breed. *Acta Vet Scand* **23:** 550–558, 1982.

Banker, B. Q. Arthrogryposis multiplex congenita: Spectrum of pathologic changes. *Hum Pathol* **17:** 656–672, 1986.

Bradley, R. A primary bovine skeletal myopathy with absence of Z discs, sarcoplasmic inclusions, myofibrillar hypoplasia, and nuclear abormalities. *J Comp Pathol* **89:** 381–388, 1979.

Bradley, R., and Wells, G. A. H. Developmental muscle disorders in the pig. *In* "The Veterinary Annual, Eighteenth Issue," C. S. G. Grunsell, and F. W. G. Hill (eds.), pp. 144–157. Bristol, England, Scientechnia. 1978.

Bradley, R., and Wells, G. A. H. The Pietran "creeper" pig: A primary myopathy. *In* "Animal Models of Neurological Disease," R. C. Rose, and P. O. Behan (eds.), pp. 34–64. Tunbridge Wells, England Pitman Medical, 1980.

Bradley, R., Ward, P. S., and Bailey, J. The ultrastructural morphology of the skeletal muscles of normal pigs and pigs with splayleg from birth to one week of age. *J Comp Pathol* **90:** 433–446, 1980.

Browne, L. M., Robson, M. J., and Sharrard, W. J. W. The pathophysiology of arthrogryposis multiplex congenita neurologica. *J Bone Joint Surg* **62:** 291–296, 1980.

Butterfield, R. M. Muscular hypertrophy of cattle. *Aust Vet J* **42:** 37–39, 1966.

Cho, D. Y., and Leipold, H. W. Congenital defects of the bovine central nervous system. *Vet Bull* **47:** 489–504, 1977.

Cork, L. C. *et al.* Pathology of motor neurons in accelerated hereditary canine spinal muscular atrophy. *Lab Invest* **46:** 89–99, 1982.

Cummings, J. F. *et al.* Focal spinal muscular atrophy in two German shepherd pups. *Acta Neuropathol* **79:** 113–116, 1989.

Ducatelle, R. *et al.* Spontaneous and experimental myofibrillar hypoplasia and its relation to splayleg in newborn pigs. *J Comp Pathol* **96:** 433–445, 1986.

Dyer, K. R. *et al.* Peripheral neuropathy in two dogs: Correlation between clinical, electrophysiological, and pathological findings. *J Small Anim Pract* **27:** 133–146, 1986.

Edwards, J. F. *et al.* Ovine arthrogryposis and central nervous system malformations associated with *in utero* Cache Valley virus infection: Spontaneous disease. *Vet Pathol* **26:** 33–39, 1989.

Engel, A. G. Part II. Basic reactions of muscle. *In* "Myology," A. G. Engel and B. Q. Banker (eds.), pp. 845–907, and 909–1043. New York, McGraw-Hill, 1986.

Handelsman, J. E., and Badalamente, M. A. Club foot: A neuromuscular disease. *Dev Med Child Neurol* **24:** 3–12, 1982.

Hartley, W.J. *et al.* Serological evidence for the association of Akabane virus with epizootic bovine congenital arthrogryposis and hydranencephaly syndromes in New South Wales. *Aust Vet J* **51:** 103, 1975.

Haughey, K. G. *et al.* Akabane disease in sheep. *Aust Vet J* **65:** 136–140, 1988.

Inada, S. *et al.* Canine storage disease characterized by hereditary progressive neurogenic muscular atrophy: Breeding experiments and clinical manifestation. *Am J Vet Res* **47:** 2294–2299, 1986.

MacKellar, J. C. The occurrence of muscular hypertrophy in South Devon cattle. *Vet Rec* **72:** 507–510, 1960.

Manktelow, B. W., and Hartley, W. J. Generalized glycogen storage disease in sheep. *J Comp Pathol* **85:** 139–145, 1975.

Mayhew, I. G. Neuromuscular arthrogryposis multiplex congenita in a thoroughbred foal. *Vet Pathol* **21:** 187–192, 1984.

Nawrot, P. S., Howell, W. E., and Leipold, H. W. Arthrogryposis: An inherited defect in newborn calves. *Aust Vet J* **56:** 359–364, 1980.

Nes, N, Lømo, O.M., and Bjerkås, I. Hereditary lethal arthrogryposis ("muscle contracture") in horses. *Nord Vet-Med* **34:** 425–430, 1982.

Rooney, J. R. Contracted foals. *Cornell Vet* **56:** 172–187, 1966.

Russell, R. G. *et al.* Variability in limb malformations and possible significance in the pathogenesis of an inherited congenital neuromuscular disease of Charolais cattle (syndrome of arthrogryposis and palatoschisis). *Vet Pathol* **22:** 2–12, 1985.

Seibold, H. R., and Davis, C.L. Generalized myositis ossificans (familial) in pigs. *Path Vet* **4:** 79–88, 1967.

Swatland, H. J. Developmental disorders of skeletal muscle in cattle, pigs, and sheep. *Vet Bull* **44:** 179–202, 1974.

Ward, P. S., and Bradley, R. The light microscopical morphology of the skeletal muscles of normal pigs and pigs with splayleg from birth to one week of age. *J Comp Pathol* **90:** 421–431, 1980.

F. Muscular Dystrophy

Muscular dystrophy is a term which describes a group of myopathies in both animals and humans, which have in common a collection of characteristics, none of which is distinctive. They are primary skeletal muscle diseases which are genetically transmitted and are characterized by progressive, degenerative changes leading to a loss of muscle fibers.

The dystrophies can be distinguished from the denervation atrophies because, at least in the early stages, the neural connections are normal, and from the nutritional, exertional, or toxic myopathies because there is evidence of progressive or repetitive disease. This may result from compromised attempts at regeneration or from repetitive initiation of the repair sequence. In later stages of the disease, fat and fibrous replacement become obvious where muscle maintenance has failed and where secondary complications such as denervation, and vascular and inflammatory events have occurred. Because muscle responds to disease in a limited number of ways, the late stages of various diseases, particularly the chronic myopathies, are often difficult to distinguish.

Although dystrophies are not common, the fact that they are inherited often leads to the appearance of clusters of cases in the same region. Both sexes are more or less equally involved except in the X-chromosome-linked dystrophies, in which males are predominantly (but not exclusively) involved. Some dystrophies are the result of double recessive traits; those which are sex-linked are the result of a single, recessive female transmitted trait, and some are apparently the result of a dominant gene. In some cases, dystrophies that are clinically identical may be inherited as either recessive or dominant traits. The mode of inheritance and family patterns are seldom distinctive in animals because there are too few data available for proper interpretation of single cases.

Muscular dystrophy in **cattle** is described in Meuse–Rhine–Yssel breed in the Netherlands with signs referable primarily to the diaphragm and intercostal muscles. The disease is regional and familial and affects cows 2–10 years of age. The most severe changes are found in the diaphragm, which is swollen, pale, and inflexible. Similar changes are visible grossly in intercostal muscles. Microscopic changes of dystrophy occur in many muscles, becoming more obvious as the disease progresses. They are generally quite typical of dystrophies with initial enlargement of both types of fibers, followed by atrophy and rounding up of fibers. There is evidence of resistance to atrophy or even of hypertrophy among some of the type 1 fibers. Some large fibers show clefting or subdivision, and an increasing number of fibers contain centrally located nuclei. Examples of segmental hypercontraction, mineralization, and phagocytosis of degenerate segments are common, and this is followed by visible evidence of regenerative repair by small basophilic fibers. More chronic lesions are increasingly fibrotic. A feature of this bovine dystrophy, which is common to dystrophic muscle in some other species, is the presence of large elongated segments of sarcoplasm which are free of myofibrils (central cores). Muscles in advanced disease are also markedly vacuolated.

A congenital dystrophylike myopathy has been reported in a single Friesian calf, which showed progressive weakness from birth. Muscle fibers were poorly developed, round in profile, and lacking adequate numbers of myofibrils. Many nuclei were present in and around these hypoplastic fibers. The most striking abnormality was the absence of Z discs with consequent disorientation of sarcomeres and an absence of reticular membrane triads.

The **Brown Swiss weaver syndrome** is a disease characterized by unsteadiness of the hind limbs. Muscle lesions are characterized by fiber atrophy and extensive fat replacement. It is not clear whether this is an example of dystrophy or arthrogryposis.

Muscular dystrophy in **sheep** of the Merino breed occurs in widely separated flocks in Australia and affects about 1 to 2% of the progeny each year after they reach 1–4 months of age. The limitation of the disease to one reed and to particular flocks with relatively closed breeding programs suggests that the myopathy is inherited, but specific data have not been gathered to confirm genetic transmission. The sexes are equally affected, and initial signs are lack of normal growth and reduced flexion of the hind limbs which leads to aberrations of gait or, in mild cases, just stiffness. Locomotor problems are increased with exercise and age, and under field conditions, the animals die of inanition at 6 to 18 months of age. With care, some animals can be maintained for up to 5 years, during which time the clinical disease is slowly progressive.

On postmortem examination, apart from emaciation, the prominent changes are in the vastus intermedius and medialis heads of the triceps of the anterior thighs, where normal muscle may be replaced by mature adipose tissue. The muscles in advanced cases appear gray, and are hard and atonic. Involvement of other muscles is limited to microscopic changes only, although the degree of pallor is an indication of the extent of the histologic changes.

The muscles most consistently involved, in addition to the vastus intermedius, are the soleus in the hind limb and the triceps brachii and anconeus of the forelimbs. These muscles have a high proportion of type 1 fibers, and it appears that this accounts for the lesion distribution since type 1 fibers are first, but not exclusively involved. Some other muscles, which also have a high proportion of type 1 fibers, are uninvolved or slightly involved.

The initial stages of the dystrophic lesion are heralded by a general increase in the diameter of type 1 fibers, and this is followed by an irregular atrophy of both types of fibers and then by characteristic degenerative changes. The fibers at this stage have rounded up, and although fiber proportions remain normal, type 1 fibers develop sarcoplasmic masses at the periphery of the cells or centrally. The central regions in such fibers contain nemaline rodlike structures or granules, or disoriented myofibrillar components, along with mitochondrial membrane complexes and glycogen granules. Nuclei may line up in the central space and may subsequently degenerate. Hypercontractive fiber segments are observed sporadically, but the progressive fiber fallout, which may reach 80% to 100% in 2-year-old lesions, seems related to a failure to sustain myofibrils rather than to repeated episodes of segmental destruction with compromised or even normal regenerative repair (Fig. 2.24A,B). Consistent with this view is the observation that in advanced lesions, severely atrophic fibers contain adequate numbers of satellite cells, all of which are in a quiescent state. Basophilic regenerative fibers are rare.

Type 2 fibers undergo atrophic changes accompanied by a progressive fibrosis of the endomysium and later by depot fat cells (Fig. 2.25A,B,C). Type 1 fibers are similarly affected but, in addition to the typical sarcoplasmic gaps in the myofibrils (Fig. 2.26), some become larger again and show signs of incomplete fiber clefting and splitting. These groups of incompletely separated fibers appear as nests or "orange section" clusters and are distinctive. They are

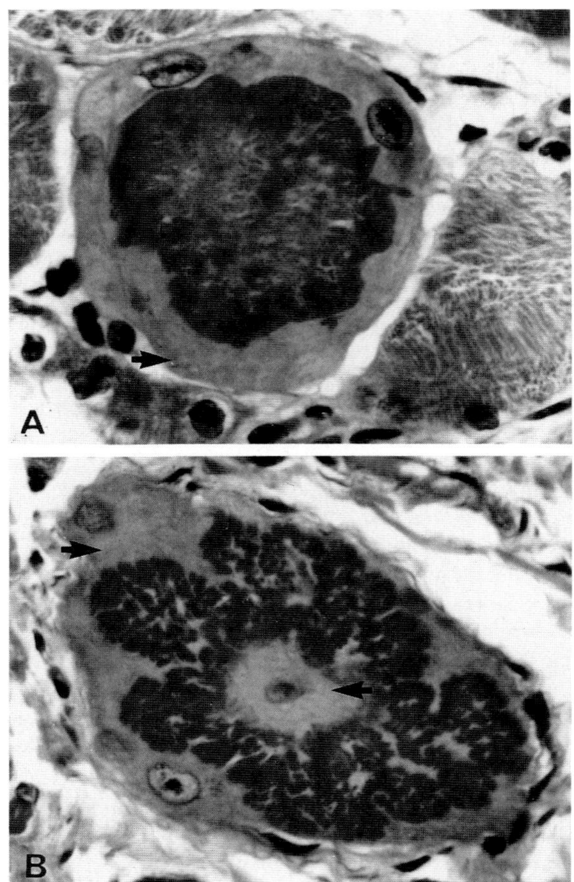

Fig. 2.24 Muscular dystrophy. Sheep. (A) Vastus intermedius muscle. Fiber with peripheral sarcoplasmic mass (arrow) and large vesicular nuclei. (Courtesy of M. D. McGavin.) (B) Vastus intermedius muscle. Fibers with central and peripheral sarcoplasmic masses (arrows) and large vesicular nuclei. (Courtesy of M. D. McGavin.)

thought to represent examples of compensatory hypertrophy. At the same time a few fibers develop annular fibrils (Ringbinden). In advanced muscle lesions, aggregates of lymphoid and histiocytic cells occur, but these seem to be unimportant in lesion genesis. Some fibers acquire centrally located nuclei reminiscent of myotonic dystrophy, but the characteristic electromyographic features associated with that disease are absent in affected sheep. At all stages of the disease, nerves, capillaries, end-plates, and neuromuscular spindles are unaffected except, occasionally, in a secondary way.

Early reports of muscular dystrophy in the **dog** described a progressive disease, characterized by weakness and abnormalities of gait which could not be related to changes in peripheral nerves, central nervous system, or end-plates and which provided clear evidence of cumulative fiber destruction. The uncertainty of disease definition has been removed by the confirmation of X-linked Duchenne-like dystrophy in the golden retriever in North

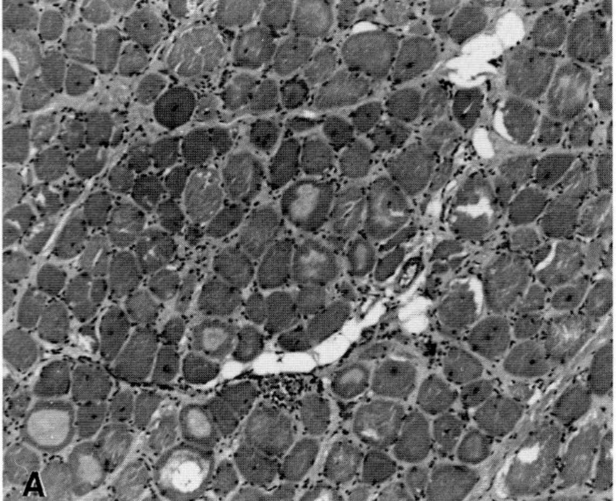

Fig. 2.25A Muscular dystrophy. Sheep. Vastus intermedius muscle. Dystrophy progressive to fat replacement. (Courtesy of M. D. McGavin.)

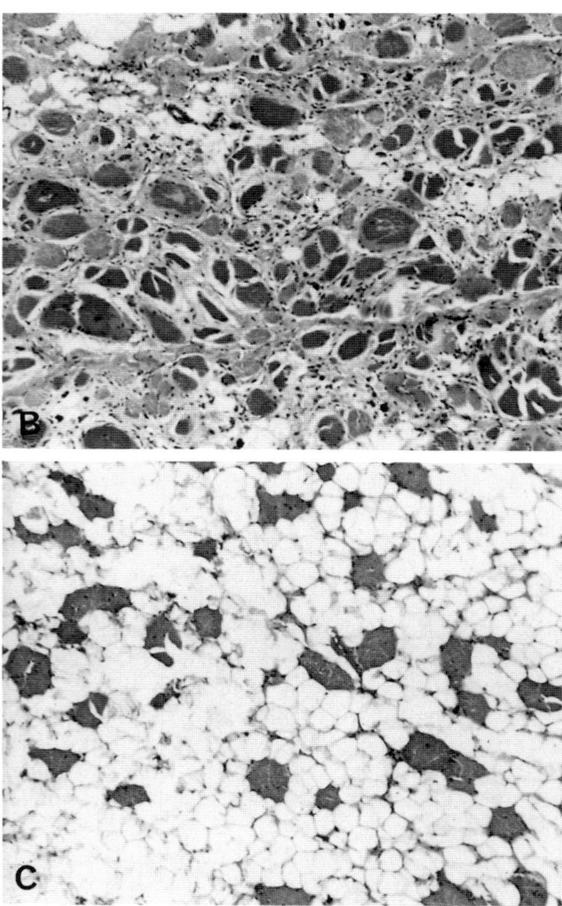

Fig. 2.25(B and C) Muscular dystrophy. Sheep. As in (A).

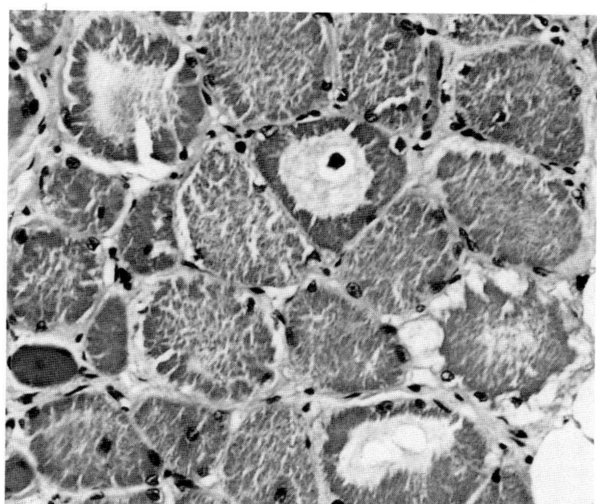

Fig. 2.26 Muscular dystrophy. Sheep. Vastus intermedius muscle. Rounded fibers with regions of fibers devoid of myofibrils. (Courtesy of M. D. McGavin.)

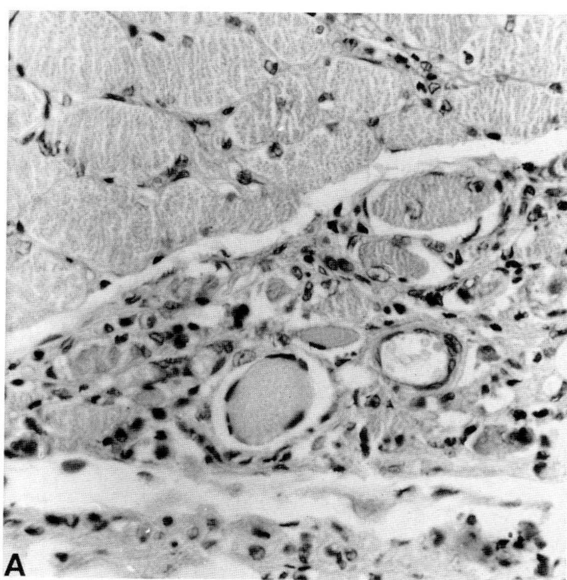

Fig. 2.27A Dystrophy with variation in fiber size and irregularity of fiber involvement. Dog.

America and in the Irish terrier in Europe. Further examples of this type of dystrophy can be anticipated, because the affected gene locus has a clear instability across several species. When the gene for the manufacture of the large muscle protein, dystrophin, is missing or defective, the gene product is progressively lost from its normal locale just under the sarcolemmal sheath in a mosaic pattern across the muscle fiber surface. The disease is primarily one of male dogs, but female puppies, presumably obligate carriers with both X-chromosomes defective, have equally damaging muscle degeneration at birth. Early deaths are probably related to severe diaphragmatic involvement, a feature which is uncommon in the human form of this dystrophy. Clinical signs are progressive limb weakness, progressive atrophy of muscles, and rarely, hypertrophy of thigh muscles. Eventually contractures and very limited limb movements lead to cachexia. Blood creatine kinase levels vary from 10 to 20 times normal at birth to 500 to 1000 times normal a month or two later.

At postmortem or biopsy of the affected muscles, paleness or yellow streakiness may be seen in the soft, somewhat atrophic muscle, which may also be irregularly discolored by hemorrhage. The muscles most frequently and most extensively affected are the tongue, temporalis, deltoideus, supraspinatus, trapezius, sartorius, and the diaphragm. Other muscles may contain microscopic lesions. The esophageal muscles are involved with some regularity and, in young dogs, may be hypertrophied.

Electromyography often reveals spontaneous high-frequency activity in some muscles, but not the prolonged activity seen in myotonia.

Histologically, the most outstanding feature of affected muscles is disproportion of fiber size and irregular, but often marked, atrophy of regional groups of fibers of both types (Fig. 2.27A). In spite of the general reduction in size of most fibers, individual type 1 fibers show a pronounced hypertrophy and may measure 250 μm in least diameter. Some of these fibers are in a state of hypercontraction. As the disease progresses, some large fibers split longitudinally or develop swirled or disorderly myofibril contents. The intrafusal fibers of neuromuscular spindles remain unaffected.

Consistently present, but rarely obvious, are segments of individual muscle fibers in varying stages of degeneration, sometimes hyaline, sometimes floccular. Histiocytes move into degenerating fibers, but the reaction often seems to be subdued or even disorderly. Evidence of regenerative repair is present but reduced in older lesions. The pattern seems to be one of repeatedly initiated degeneration followed by competent repair. It is uncertain whether the growing number of fibers with centrally located nuclei are immature regenerative fibers or fibers in which the nuclei migrate inward. Fat replacement of muscle is present, but sometimes not conspicuous (Fig. 2.27B).

Fibrous tissue in muscles may be absolutely increased in some focal areas, but most of the apparent increase is due to condensation of endomysium. Blood vessels, peripheral nerves, and end-plates are normal except for those changes of shape of end-plates and axons which are an inherent feature of fiber atrophy. In areas of fat replacement, motor nerve terminations appear as small buttons on the surface of depot fat cells. Electron microscopy of dystrophic fibers reveals a variety of subtle changes in common with Duchenne's dystrophy in children. Mitochondria and myelin body fibers collect in subsarcolemmal blebs of sarcoplasm along with rare dilated or honeycomb tubular structures (see Fig. 2.31B). Z-line streaming is present in older dogs, and assorted myofilament structures in disoriented array may be seen in central swirls or cores.

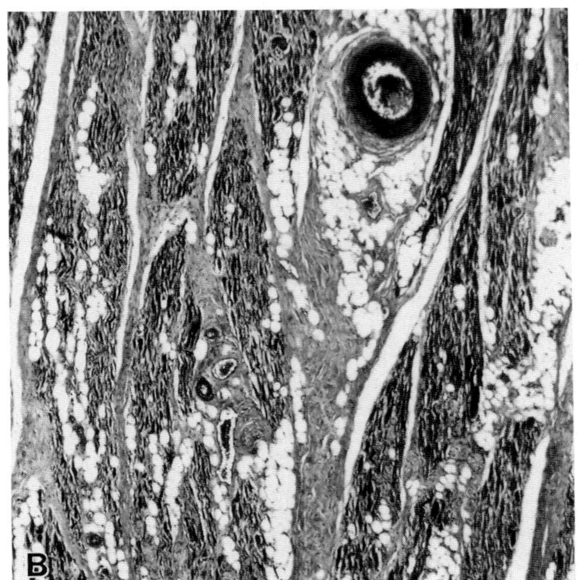

Fig. 2.27B Dystrophy with advanced fiber atrophy and fat replacement. Dog.

The relationship between loss of dystrophin and progressive disease of muscle fibers is not completely understood. Present consensus links dystrophin to active maintenance of, or structural participation in, the cell membrane. In the absence of dystrophin, gaps develop in the membrane. This allows the influx of unmanageable amounts of calcium, which precipitates hypercontraction and is followed by cleanup and repair of the degenerate segment of muscle fiber. Calcium studies confirm increased levels of calcium in dystrophic fibers and in hypercontracted segments. Gaps in the cell membrane adjacent to degenerative episodes can be demonstrated, but sometimes similar gaps are present in normal dog muscle. It is suggested that exercise, free radicals, and potassium or sodium influx/efflux might influence the progress of this disease.

Nemaline myopathy has been described in the cat and the dog where it is associated with weakness, wobbly gait, fatigue, and tremor. Histologic lesions consist of marked variation in fiber size, fiber splitting, and some central migration of nuclei. Elongated electron-dense rods with marked periodicity give the lesion its name, and these structures clearly are similar to, and an extension of, the Z-band material. The distribution of rods within muscle fibers is uneven.

Hereditary muscular dystrophy of Labrador retrievers has been reported from Europe, North America, and Australia, and the syndrome has been characterized to a limited extent. Pups of both sexes show a reduced exercise tolerance and a stiff gait beginning about 8 weeks of age. Muscle atrophy becomes apparent, but locomotor limitation, although progressive, may be slow for several years. Electromyography may reveal spontaneous, high-frequency activity, but both electrical and histological study of nerves indicate that they are normal.

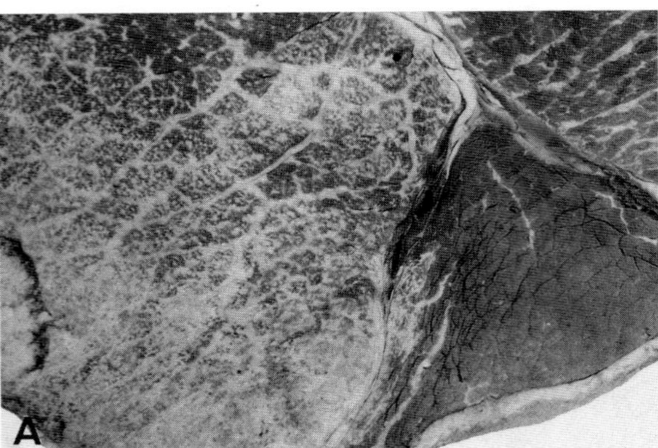

Fig. 2.28A Steatosis. Ox. Gross appearance of affected muscles.

Histologically, muscles reveal marked fiber size disparity, which increases with age. Both muscle fiber types participate in progressive atrophy/hypertrophy, but in time, type 2 fibers constitute a smaller than normal proportion of fibers in some muscles and are more atrophic. Hypertrophic fibers progressively display splitting or a whorled appearance of the myofibrils.

In all muscles, at all stages, muscle fibers reveal segmental degenerative changes with evidence of competent repair. Increased collagen deposition in the endomysium is an expected late change.

Although comparisons have been made between this dystrophy in Labrador retrievers and limb-girdle dystrophy in humans, the pathogenesis has not been explored thoroughly. The atrophy, particularly of type 2 fibers, suggests there may be an endocrine influence.

X-linked dystrophy in **cats,** although not well documented, has many similarities to the comparable disease in dogs, including a gene-caused deficiency of dystrophin.

Muscular steatosis is a disease characterized by too much fat deposited within muscles. It carries an inference that the fat is where muscle once was or ought to have been—effectively, fat replacement of muscle fibers. Steatosis is not associated with any clinical disability. It appears in clinically healthy animals, and it is usually a problem only in meat inspection.

Sometimes the steatosis affects several muscles of one limb, but more often it affects only one or several muscles in one region. It may be bilaterally symmetrical or asymmetrical. Rarely does it affect all of one muscle, and the dividing line between normal and fatty muscle is not sharp (Fig. 2.28A). Surviving muscle fibers in the marginal areas may be normal or smaller than normal but are often angular in fixed tissue as a result of adjacent pressure from turgid lipocytes.

The fat cells are now thought not to be dedifferentiated muscle fiber cells but rather special lipocytic cells or endomysial cells which retain the ability to differentiate into lipocytes (Fig. 2.28B). Studies on steatosis in normal market pig carcasses indicate that about 1 to 5% of pigs have

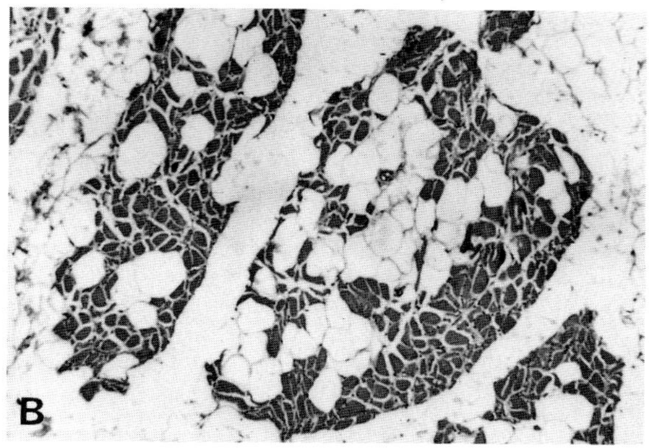

Fig. 2.28B Microscopic appearance of muscle in (A).

small steatotic lesions and that a smaller proportion have extensive lesions of the anterior thigh or loin muscles (Fig. 2.29). There are several causes for these lesions. About one third of the small lesions show evidence of minimal muscle development failure (see Section III,B, Arthrogryposis). About one quarter show a selective, progressive fat replacement of only type 2 fibers and are probably examples of failed regeneration following exertional myopathy. A few seem to be examples of failed regeneration following nutritional myopathy. Large loin lesions are more likely to have an ischemic cause.

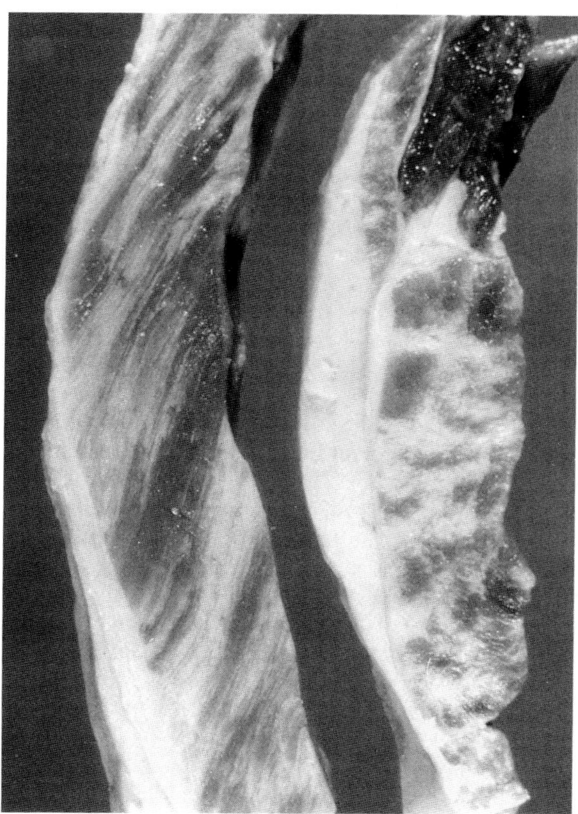

Fig. 2.29 Steatosis of the loin muscle of a clinically normal pig.

Steatotic lesions are occasionally found in the thigh or loin muscles of old dogs. They are usually bilateral and may extensively involve parts of several adjacent muscles. These lesions sometimes occur at sites of previous surgery but no cause has been found for their apparently progressive character.

Bibliography

Baker, M. S., and Austin, L. The pathological damage in Duchenne muscular dystrophy may be due to increased intracellular OXY-radical generation caused by the absence of dystrophin and subsequent alterations in Ca^{2+} metabolism. *Med Hypotheses* **29:** 187–193, 1989.

Bartlett, R. J. *et al.* The canine and human dys genes are highly conserved. *In* "Proceedings (Suppl.) of the Journal of Neurological Sciences," Vol. 101, p. 165, 1973.

Cooper, B. J. *et al.* Implications of mosaicism for dystrophin for understanding its function. *In* "Proceedings (Suppl.) of the Journal of Neurological Sciences," vol. 101, p. 228, 1973.

Cooper, B. J. *et al.* Canine muscular dystrophy: Confirmation of X-linked inheritance. *J Hered* **79:** 405–408, 1988.

Cooper, B. J. *et al.* The homologue of the Duchenne locus is defective in X-linked muscular dystrophy of dogs. *Nature* **334:** 154–156, 1988.

Dent, A. C., Richards, R. B., and Nairn, M. E. Congenital progressive ovine muscular dystrophy in western Australia. *Aust Vet J* **55:** 297, 1979.

Engel, A. G. Part III. Duchenne dystrophy. *In* "Myology," A. G. Engel and B. Q. Banker (eds.). New York, McGraw-Hill, 1986.

Funkquist, B., Haraldsson, I., and Stahre, L. Primary progressive muscular dystrophy in the dog. *Vet Rec* **106:** 341–343, 1980.

Goedegebuure, S. A., Hartman, W., and Hoebe, H. P. Dystrophy of the diaphragmatic muscles in adult Meuse–Rhine–Yssel cattle: Electromyographical and histological findings. *Vet Pathol* **20:** 32–48, 1983.

Hoffman, E. P. Dystrophin deficiency: Vignettes of cats and dogs, mice and men. *In* "Proceedings (Suppl.) of the Journal of Neurological Sciences," Vol. 101, p. 12, 1973.

Hoffman, E. P. Molecular diagnostics of Duchenne/Becker dystrophy: New additions to a rapidly expanding literature. *J Neurol Sci* **101:** 129–132, 1991.

Kornegay, J. N. *et al.* Muscular dystrophy in a litter of golden retriever dogs. *Muscle Nerve* **11:** 1056–1064, 1988.

Kramer, J. W. *et al.* A muscle disorder of Laborador retrievers characterized by deficiency of type II muscle fibers. *J Am Vet Med Assoc* **169:** 817–820, 1976.

Kramer, J. W., Hegreberg, G. A., and Hamilton, M. J. Inheritance of a neuromuscular disorder of Laborador retriever dogs. *J Am Vet Med Assoc* **179:** 380–381, 1981.

Leipold, H. W. *et al.* Weaver syndrome in Brown Swiss cattle: Clinical signs and pathology. *Vet Med Small Anim Clin* **68:** 645–647, 1973.

McGavin, M. D., and Baynes, I. D. A congenital progressive ovine muscular dystrophy. *Pathol Vet* **6:** 513–524, 1969.

McKerrell, R. E., and Braund, K. G. Hereditary myopathy in Labrador retrievers: A morphologic study. *Vet Pathol* **23:** 411–417, 1986.

Mehta, J. R. *et al.* Intracellular electrolytes and water analysis in dystrophic canine muscles. *Res Vet Sci* **47:** 17–22, 1989.
Richards, R. B., and Passmore, I. K. Ultrastructural changes in skeletal muscle in ovine muscular dystrophy. *Acta Neuropathol* **77:** 168–175, 1988.
Richards, R. B., Passmore, I. K., and Dempsey, E. F. Skeletal muscle pathology in ovine congenital progressive muscular dystrophy. *Acta Neuropathol* **77:** 95–99 and 161–167, 1988.
Valentine, B. A., Cummings, J. F., and Cooper, B. J. Development of Duchenne-type cardiomyopathy. *Am J Pathol* **135:** 671–678, 1989.
Valentine, B. A. *et al.* Canine X-linked muscular dystrophy: Morphologic lesions. *J Neurol Sci* **97:** 1–23, 1990.
Vos, J. H., van der Linde-Sipman, J. S., and Goedegebuure, S. A. Dystrophylike myopathy in the cat. *J Comp Pathol* **96:** 335–341, 1986.
Watson, A. D. J. *et al.* Myopathy in a Labrador retriever. *Aust Vet J* **65:** 226–227, 1988.
Weller, B., Karpati, G., and Carpenter, S. Dystrophin-deficient mdx muscle fibers are preferentially vulnerable to necrosis induced by experimental lengthening contractions. *J Neurol Sci* **100:** 9–13, 1990.
Wentink, G. H. *et al.* Myopathy with a possible recessive X-linked inheritance in a litter of Irish terriers. *Vet Pathol* **9:** 328–349, 1972.
Wentink, G. H. *et al.* Myopathy in an Irish terrier with a metabolic defect of the isolated mitochondria. *Zentralbl Veterinaermed* [*A*] **21:** 62–74, 1974.

G. Myotonic Syndromes

The myotonic syndromes have a variety of names, including myotonic myopathy and myotonic dystrophy, but they share a single unusual characteristic: the presence of a myotonic response in muscle to stimulation or percussion, or simply to the insertion of an electromyographic needle. The response is characterized by a delayed relaxation time, which allows the contraction to become cumulative in a typical crescendo–decrescendo, which lasts several seconds as the muscle returns to relaxation. The dimpling of muscle on percussion is the visual equivalent and the "dive bomber" response is the acoustic version of the electromyographic response. This gathering together of a somewhat varied group of diseases on the basis of a single characteristic inevitably leads to classification problems, but it does identify a group of phenomena, all of which are primarily muscle fiber centered, and all of which can be distinguished quite readily in the living animal.

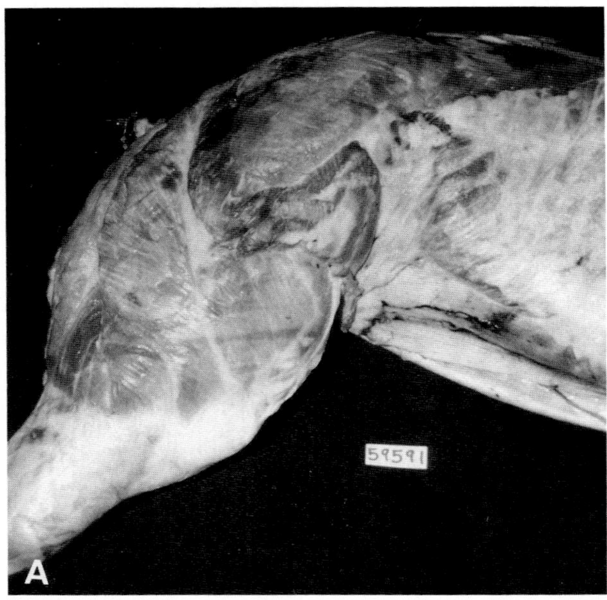

Fig. 2.30A Myotonic dystrophy. Hindquarters of a young foal showing muscle hypertrophy.

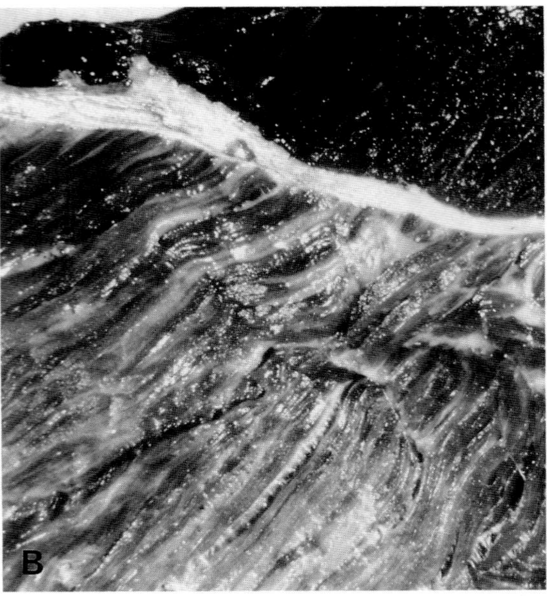

Fig. 2.30B Myotonic dystrophy. Hindquarters of a young foal showing coarseness of the cut muscle caused by fibrosis and fat replacement of fibers.

In domestic animals the myotonic syndromes can be divided, somewhat arbitrarily, into four subgroups which have features in common. **Heritable myotonia** fulfils all criteria for the definition of a dystrophy because it is inherited, primarily myogenic, progressive, and destructive. The first animal myotonia of this type was described in angora goats (fainting goats) over 100 years ago. When startled, a seriously affected animal involuntarily becomes muscle-rigid and may fall over like a tipped sawhorse, only to recover a few seconds later. The only other myotonic syndrome which seems to be a true dystrophy occurs in certain blood lines and families of chow chow dogs, particularly those in New Zealand, Australia, and Europe. Puppies of both sexes first show hind limb ataxia or stiffness at 1 to 3 months of age. At this time and later, most begin to show an enlargement of thigh and shoulder muscles, which gives them a very athletic appearance. The locomotor difficulties progress slowly with increasing time

Fig. 2.30C ATPase. Myotonic dystrophy.

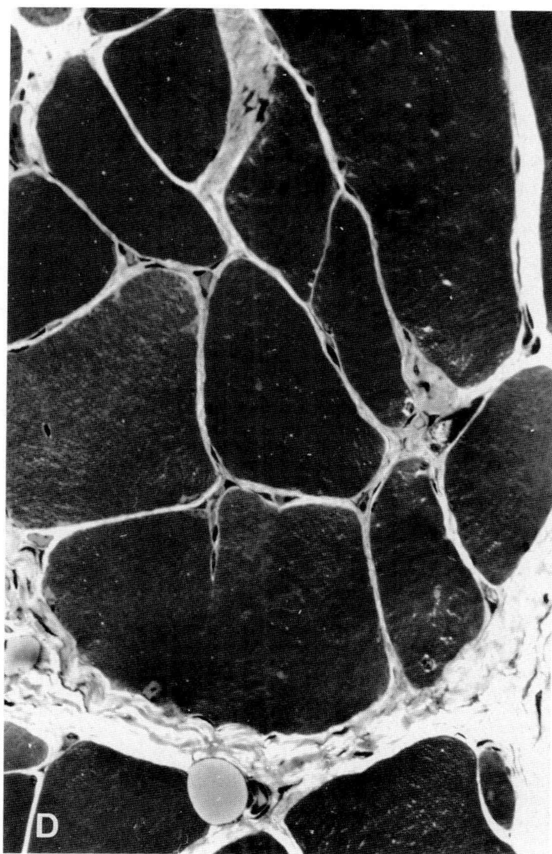

Fig. 2.30D Myotonic dystrophy. Horse. Clefting of a very large fiber.

required to rise after rest or after enforced or lateral recumbency. With activity, many of the signs disappear in minutes. Creatine phosphokinase blood levels are slightly elevated.

Gross changes are usually not observed in hereditary myotonias and histologic lesions are not distinctive in the early stages. In both goats and chow chow puppies, fibers may be slightly enlarged, and occasional fibers with segmental degeneration may be seen throughout the course of the disease. Fine structural lesions consist of enlarged abnormal mitochondria and a moderate increase in lysosomal particles. Some fibers are diastase resistant, PAS positive. In the chronic stages muscle fibers become increasingly disparate in size, and fibrous tissue increases in the endomysium. It appears that the functional changes in muscle in heritable myotonia are related somehow to the mitochondrial changes and to a decrease in resting chloride conductance.

Myotonic dystrophy fulfils the definition of dystrophy, except that it is not an inherited disease. This form of myotonia affects all types of horses and dogs other than chow chows.

In the horse, clinical signs of stiffness, and abnormalities of gait occur in both sexes and are first seen from birth to 2 or more years of age, but most often at 2 to 4 months. Animals affected at birth are unable to rise. The disease is progressive with most rapid progression in young foals. Stiffness increases and movements are slow, stilted, and halting. Often the body weight is abnormally shifted over the forelegs with the hind limbs partially dragging. Percussion dimpling occurs over the major thigh or shoulder muscles and in these regions muscles often become enlarged, firm, and more clearly defined (Fig. 2.30A). Creatine phosphokinase levels are moderately elevated (2-fold to 15-fold).

On postmortem, or when muscles are removed for biopsy, the muscle appears coarser than normal (Fig. 2.30B), as groups of primary bundles are separated by fibrous tissue and linear streaks of depot fat. Severe lesions are pale and yellow-white. Microscopically, early lesions consist of a growing fiber size disproportion as some type 2 fibers become larger, and some fibers of both types become smaller. With time these changes are exaggerated, and it is clear that regionally, one or the other fiber type dominates, and the opposite fiber type undergoes atrophy (Fig. 2.30C). The largest fibers assume peculiar shapes, and some show clefts or incomplete subdivision (Fig. 2.30D). Scattered fibers of all sizes undergo segmen-

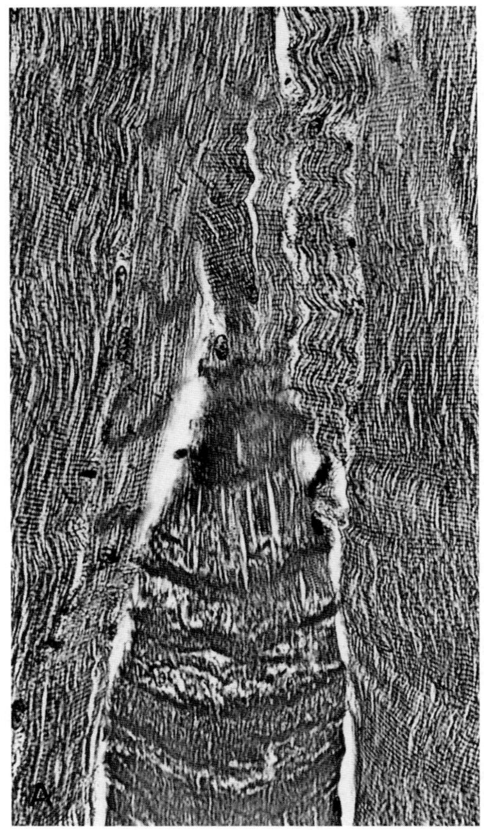

Fig. 2.31A Myotonic dystrophy. Horse. A complex fiber with three tails is undergoing hypercontraction and segmental degeneration.

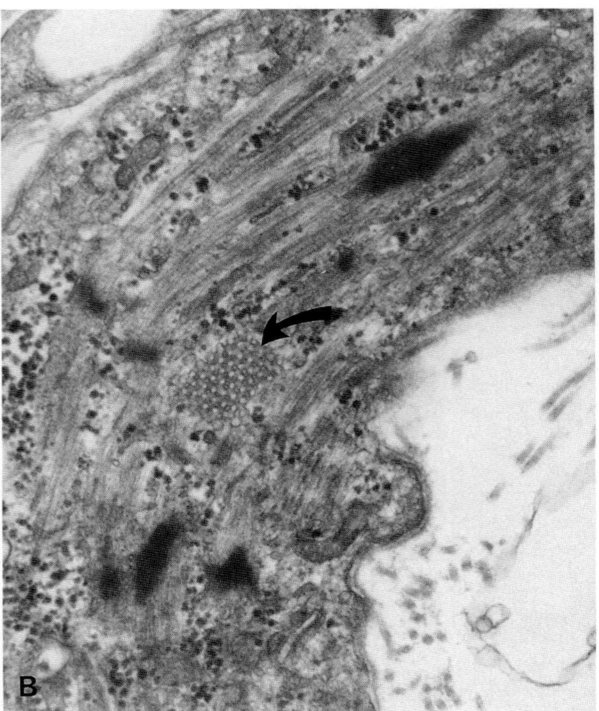

Fig. 2.31B Myotonic dystrophy. Horse. T-tubule aggregate (arrow) in a degenerating muscle fiber.

tal degeneration with repair following slowly (Fig. 2.31A,B,C). Myoblasts seem to suffer from a delayed fusion in repair, and the end result is often a cluster of myotubes or incompletely fused myotubes replacing what was formerly a single fiber (Fig. 2.31C). Some regenerative repair fails, and the gap is filled with fibrous tissue (Fig. 2.32A).

One of the most distinctive features of all forms of myotonic dystrophy is a progressive increase of centrally located muscle nuclei as the lesions mature, and the occurrence of nuclei in chains is often considered pathognomonic. Large numbers of nuclei, either scattered, clustered, or in rows can be seen both in fibers which have recently been repaired and in those that have not (Fig. 2.32B). Another distinctive feature, which may not be obvious in some cases, is the presence of annular rings (Ringbinden). These may become numerous and complex in some muscles (Fig. 2.32C).

Uneven collagen deposition is apparent in maturing lesions. It builds up particularly around areas of failed repair and advanced atrophy. The development of lipocytes follows shortly (Fig. 2.31C). The amalgamation of incompletely fused multiple myotubes in replacement of a single fiber, segmental amputations in fibers, and ringed fibers with subsequent hypertrophy create bizarre zigzag muscle units. It is possible that such physical alterations alone explain some or most of the clinical myotonia.

Nerves to muscles and end-plates are mostly normal, but terminal axons may be very long and tortuous as they thread between enlarged distorted fibers. Electromyography is distinctive, but a typical response may not be elicited in the center of advanced lesions. Vessels are also normal.

Myotonic dystrophy in the dog, apart from that in chowchows, is similar to that described in the horse. Early lesions consist of fiber size disparity with marked rounded enlargement of some fibers. More advanced lesions provide multiple examples of internal nuclear increase and hyalinization of fibers followed slowly by fragmentation and phagocytosis of debris. This leads in turn to regeneration or alternatively, fibrosis. As in the horse, myotube fusion in repair may be delayed and prone to immediate new episodes of degeneration. The very large fibers in dogs may contain central cores, distorted myofibrillar whorls, cytoplasmic masses, or focal myolysis (Fig. 2.32D).

Acquired myotonia occurs in the dog and cat as well as humans. In this rare disease, myotonic electromyographic signals and typical myotonic clinical manifestations follow prolonged drug administration. Prolonged, high levels of corticosteroid therapy is most frequently implicated, but several experimental drugs and 2,4-D are also known to

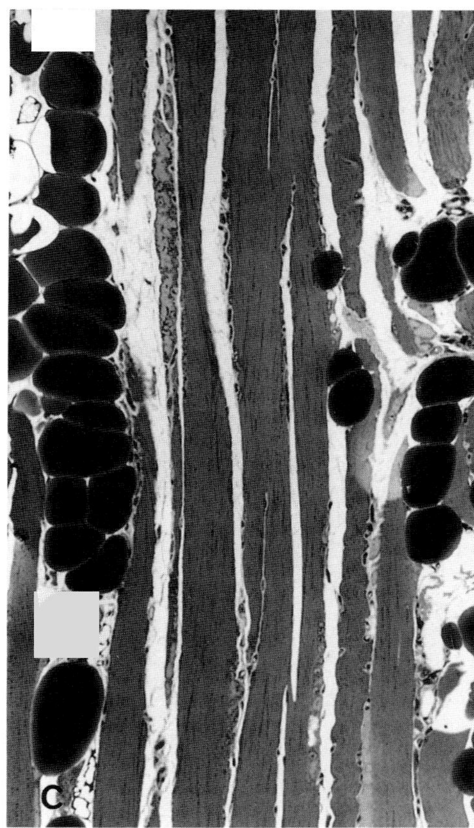

Fig. 2.31C Myotonic dystrophy. Horse. A single complex fiber in the middle shows evidence of delayed fusion and possible clefting. Plastic-embedded, semi-thin section. Toluidine blue stain.

cause the syndrome in some experimental animals. Histologically, some of the features of myotonic dystrophy may be seen, such as a central nuclear increase and segmental degeneration with incompetent repair.

Paramyotonic syndromes are diseases of muscle characterized by a myotonic electromyographic response and dimpling, but lacking the progressive and destructive characteristics. This group of abnormalities is likely to be mediated by electrolyte gating abnormalities at the muscle cell surface. Congenital paramyotonia occurs in the cavalier King Charles spaniel. Horses, dogs, and cats suffer from hyperkalemic periodic paralysis, and cats and dogs may develop paramyotonia as a result of hypokalemia. The clinical syndromes of stiffness and reluctance to walk are more readily apparent in a cold environment and, in contrast to other myotonic syndromes, the signs increase in severity with exercise. Histological changes in muscle are minimal.

Bibliography

Amann, J. F., Tomlinson, J., and Hankison, J. K. Myotonia in a chow chow. *J Am Vet Med Assoc* **187:** 415–417, 1985.

Anderson, J. R. Myotonia. *In* "Atlas of Skeletal Muscle Pathology," G. A. Gresham (ed.), Vol. 9, Chapter 11, pp. 89–96. Lancaster, Boston, The Hague, Dordrecht, MTP Limited, 1985.

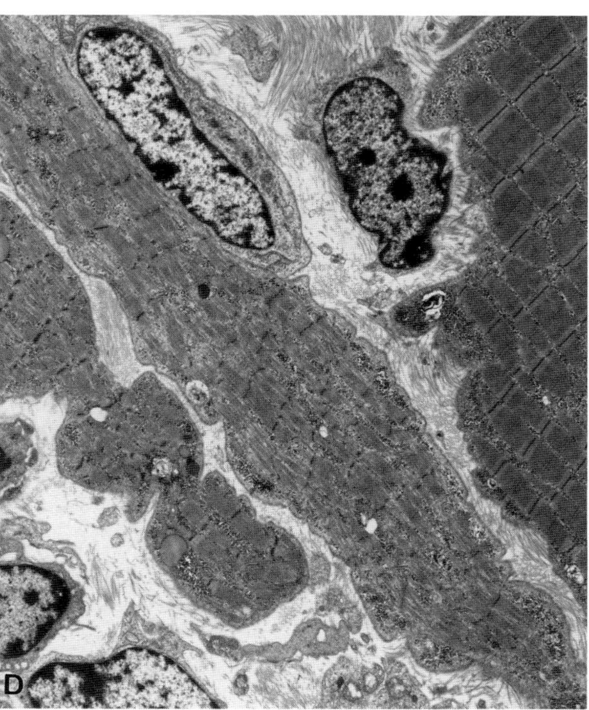

Fig. 2.31D Myotonic dystrophy. Horse. Two adjacent, recently formed myofibers, one of which has a satellite cell, undergoing myolysis, and atrophy. Increased collagen is visible above.

Andrews, F. M., Spurgeon, T. L., and Reed, S. M. Histochemical changes in skeletal muscles of four male horses with neuromuscular disease. *Am J Vet Res* **47:** 2078–2083, 1986.

Atkinson, J. B., Swift, L. L., and Lequire, V. S. Myotonia congenita. A histochemical and ultrastructural study in the goat: Comparison with abnormalities found in human myotonia dystrophica. *Am J Pathol* **102:** 324–335, 1981.

Barchi, R. L. A mechanistic approach to the myotonic syndromes. *Muscle Nerve* **5:** S60–S63, 1982.

Crews, J., Kaiser, K. K., and Brooke, M. H. Muscle pathology of myotonia congenita. *J Neurol Sci* **28:** 449–457, 1976.

Cros, D. *et al.* Peripheral neuropathy in myotonic dystrophy: A nerve biopsy study. *Ann Neurol* **23:** 470–476, 1988.

Degen, M. Pseudohyperkalemia in Akitas. *J Am Vet Med Assoc* **190:** 541–543, 1987.

Duncan, I. D., and Griffiths, I. R. A myopathy associated with myotonia in the dog. *Acta Neuropathol* **31:** 297–303, 1975.

Farrow, B. R. H., and Malik, R. Hereditary myotonia in the chow chow. *J Small Anim Pract* **22:** 451–465, 1981.

Gott, P. S., and Karnaze, D. S. Short-latency somatosensory evoked potentials in myotonic dystrophy: Evidence for a condition disturbance. *Electroencephalogr Clin Neurophysiol* **62:** 455–458, 1985.

Grafe, P. *et al.* Enhancement of K^+ conductance improves *in vitro* the contraction force of skeletal muscle in jypokalemic periodic paralysis. *Muscle Nerve* **13:** 451–457, 1990.

Griffiths, I. R., and Duncan, I. D. Myotonia in the dog: A report of four cases. *Vet Rec* **93:** 184–188, 1973.

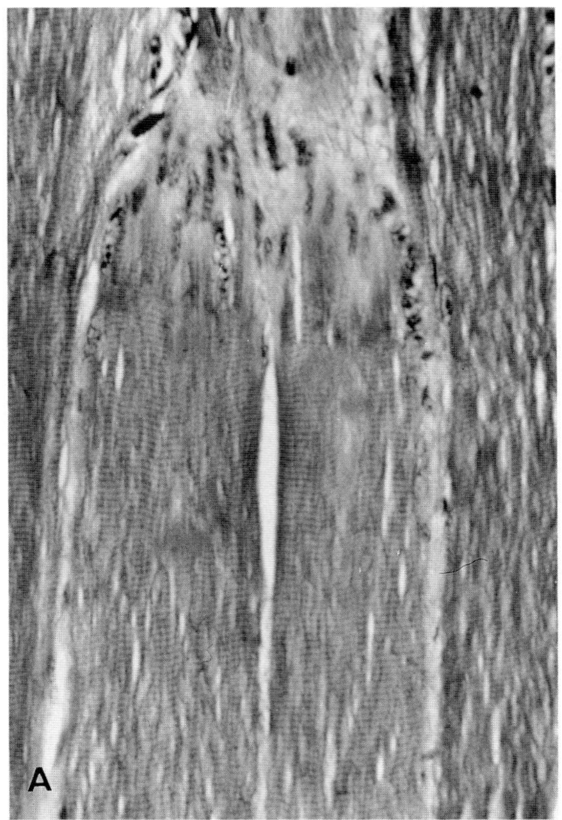

Fig. 2.32A Myotonic dystrophy. Horse. Fibrous connection replacing degenerate segments of three adjacent muscle fibers.

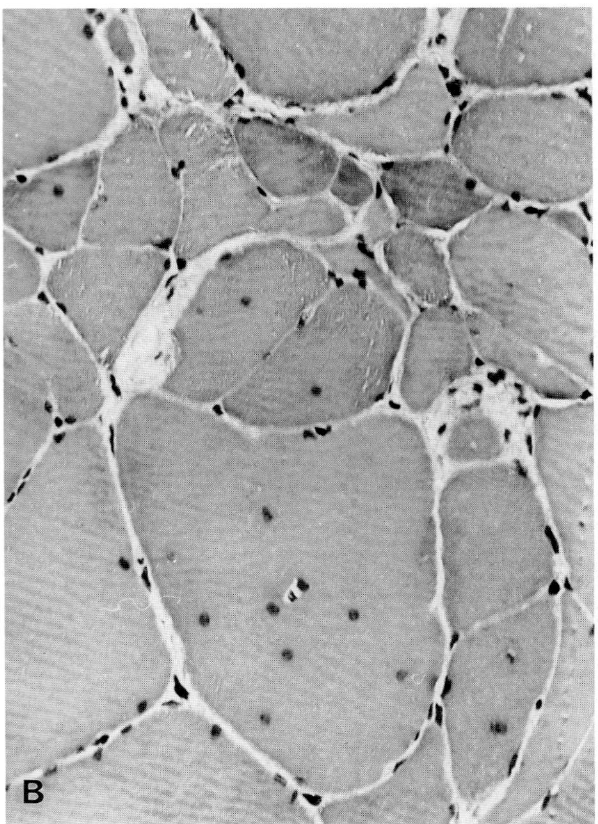

Fig. 2.32B Myotonic dystrophy. Horse. Many centrally located nuclei in large and small fibers.

Hegreberg, G. A., and Reed, S. M. Skeletal muscle changes associated with equine myotonic dystrophy. *Acta Neuropathol* **80:** 426–431, 1990.

Herrtage, M. E., and Palmer, A. C. Episodic falling in the cavalier King Charles spaniel. *Vet Rec* **112:** 458–459, 1983.

Iannaccone, S. T. *et al.* Muscle maturation delay in infantile myotonic dystrophy. *Arch Pathol Lab Med* **110:** 405–411, 1986.

Jamison, J. M. *et al.* A congenital form of myotonia with dystrophic changes in a quarter horse. *Equine Vet J* **19:** 353–358, 1987.

Kuhn, E. *et al.* The autosomal recessive (Becker) form of myotonia congenita. *Muscle Nerve* **2:** 109–117, 1979.

Jones, B. R., and Johnstone, A. C. An unusual myopathy in a dog. *N Z Vet J* **30:** 119–121, 1982.

Jones, B. R. *et al.* Myotonia in related chow chow dogs. *N Z Vet J* **25:** 217–220, 1977.

Martin, A. F., Bryant, S. H., and Mandel, F. Isomyosin distribution in skeletal muscles of normal and myotonic goats. *Muscle Nerve* **7:** 152–160, 1984.

McKerrell, R. E. Myotonia in man and animals: Confusing comparisons. *Equine Vet J* **19:** 266–267, 1987.

Reed, S. M. Progressive myotonia in foals resembling human dystrophia myotonica. *Muscle Nerve* **11:** 291–296, 1988.

Robinson, J. A., Naylor, J. M., and Crichlow, E. C. Use of electromyography for the diagnosis of equine hyperkalemic periodic paresis. *Can J Vet Res* **54:** 495–500, 1990.

Ronéus, B., Lindholm, A., and Jönsson, L. Myotoni hos häst. *Svensk Veterinärt* **35:** 217–220, 1983.

Shirakawa, T. *et al.* Muscular dystrophy-like disease in a thoroughbred foal. *J Comp Pathol* **100:** 287–294, 1989.

Shires, P. K., Nafe, L. A., and Hulse, D. A. Myotonia in a Staffordshire terrier. *J Am Vet Med Assoc* **183:** 229–232, 1983.

Simpson, S. T., and Braund, K. G. Myotonic dystrophy-like disease in a dog. *J Am Vet Med Assoc* **186:** 495–498, 1985.

Wentink, G. H., Hartman, W., and Koeman, J. P. Three cases of myotonia in a family of chows. *Tijdschr Diergeneesk* **99:** 729–731, 1974.

Wright, J. A., *et al.* A myopathy associated with muscle hypertonicity in the cavalier King Charles spaniel. *J Comp Pathol* **97:** 559–565, 1987.

H. Spastic Syndromes

Spastic paresis of cattle affecting one or both hind legs is seen in calves 3–7 months of age, although older cattle may occasionally show signs. The disease appears first as an exaggerated straightness of the hock, which increases to the point of inability to flex the hock because of more or less continuous tension of the gastrocnemius muscle and the Achilles tendon. In the later stages, in an animal with unilateral involvement, the affected limb may swing stiffly like a pendulum or be held out behind or ahead. The

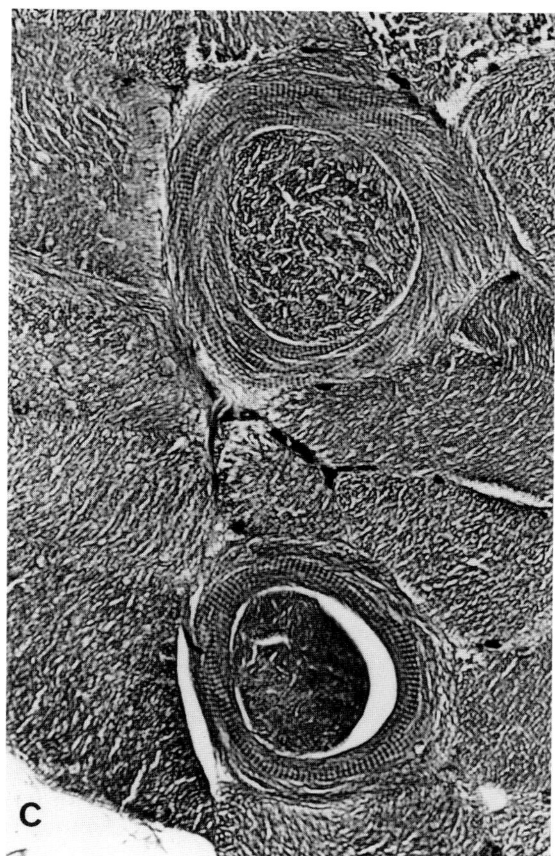

Fig. 2.32C Myotonic dystrophy. Horse. Typical Ringbinden with myofibrils escaping the endomysium.

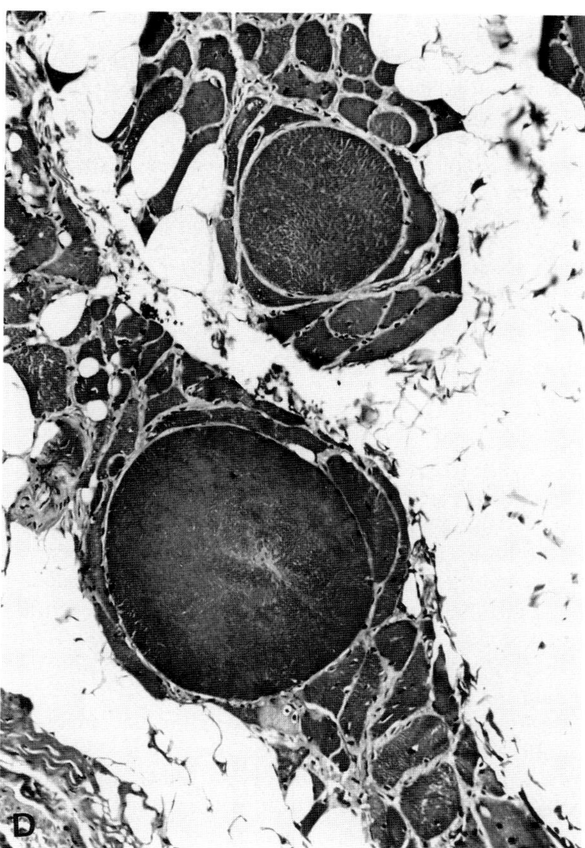

Fig. 2.32D Acquired myotonic dystrophy in a mature dog with fiber disparity, central nuclei, and fat replacement.

disease, also called contraction of the Achilles tendon, straight hock, or Elso heel (the latter derived from the name of the Friesian bull some of whose progeny had the disease), is worldwide in distribution. It affects Holstein–Friesian cattle primarily but has been found in several other dairy and beef breeds, and has a distinct familial pattern. Deafferentation (cutting of the sensory nerve root) of the gastrocnemius muscle relieves the effect, so it appears that the defect is an unmodulated or uncontrollable sensory–motor reflex loop.

A related syndrome, referred to as spastic syndrome or crampy spasticity or stretches affects mature animals, usually bulls, in a way which makes them periodically and suddenly extend and stiffen both hind legs, arch the back down, and extend the neck for a period of seconds to minutes. The disease is progressive, and it is reportedly associated with straight hocks and a variety of other abnormalities of bone, joints, or spinal column.

Scotch or **scotty cramp** is an episodic motor disease, primarily of Scottish terrier dogs, which is usually first seen after 2 to 6 months of age, but sometimes much later. It is characterized by hind limb and pelvic muscle hypertonicity along with upward arching of the back for periods of 1 to 30 min, and is precipitated by exercise or excitement. No consistent abnormalities are to be found in the central nervous system, peripheral nerves, or skeletal muscles.

Muscle hypertonicity precipitated by exercise or excitement has been described in several cavalier King Charles spaniels. Vacuolation of some muscle fibers and several minor ultrastructural changes occur, but no clear indication of a disease mechanism has been uncovered.

Dancing Doberman disease has been described as a spastic disease of one or both hocks in Doberman pinschers of any age. The animals are reluctant to bear weight continuously, and they shift hind legs frequently. This is a nonspecific sign, since some of these dogs apparently have a localized myotonic dystrophy and others have neurogenic muscle diseases.

Bibliography

Baird, J. D., Johnston, K. G., and Hartley, W. J. Spastic paresis in Friesian calves. *Aust Vet J* **50:** 239–245, 1974.

Bijlevald, K., and Hartman, W. Electromyographic studies in calves with spastic paresis. *Tijdschr Diergeneesk* **101:** 805–808, 1976.

Bradley, R., and Wijeratne, W. V. S. A locomotor disorder clinically similar to spastic paresis in an adult Friesian bull. *Vet Pathol* **17:** 305–315, 1980.

Chrisman, C. L. Dancing Doberman disease: Clinical findings and prognosis. *Prog Vet Neurol* **1:** 83–90, 1990.

De Ley, G., and De Moor, A. Bovine spastic paralysis. *Zentralbl Veterinaermed* [*A*] **23:** 89–96, 1976.

De Ley, G., and De Moor, A. Bovine spastic paralysis: Results of surgical deafferentation of the gastrocnemius muscle by means of spinal dorsal root resection. *Am J Vet Res* **38:** 1899–1900, 1977.

Meyers, K. M. *et al.* Muscular hypertonicity: Episodes in Scottish terrier dogs. *Arch Neurol* **25:** 61–68, 1971.

I. Metabolic Myopathies

Muscular dystrophy characterized by glycogen storage occurs as part of systemic glycogenosis in cattle and dogs. These syndromes are considered with the lysosomal storage diseases (see Chapter 3, The Nervous System) and have been shown to be due to deficient lysosomal glucosidase activity. There are similar syndromes in sheep and horses, but these are not characterized biochemically. Histologically, increased levels of glycogen can be demonstrated in muscle, but the changes in muscle do not reflect the greatly reduced exercise tolerance, tremors, and incoordination.

A polysaccharide storage disease of muscle, sometimes without storage lesions in other tissues, has been described in the horse (both heavy and light), cow, and in English springer spaniel. As in humans, relatively massive deposits of a PAS-positive glycoprotein substance, which is a processed product of glycogen but which is no longer biologically accessible, can be seen in basophilic rows of angular packets between myofibrils (Fig. 2.32E,F). The accumulated material may be either finely dispersed or in relatively large, membrane-bound masses. Fibers vary in size, and some may be vacuolated. In dogs, there is an association with phosphofructokinase deficiency, but not all dogs with the deficiency have muscle deposits. This is also true of phosphofructokinase deficiency in humans. In some humans, and presumably in horses, the muscle deposits are not dependent on the presence of a generalized carbohydrate-processing deficiency. It is not clear whether disorders of muscle function can be attributed to the polysaccharide stored in muscle.

Lipid storage diseases, with one exception, have not been defined for animals. The exception is carnitine deficiency in cats in which the lack of a fat-processing enzyme leads to abnormal storage of neutral fat in small globules between myofibrils. The lesion is particularly apparent in type 1 fibers. Groups of muscle fibers heavily laden with fat occur in the horse, cow, and sheep, but specific disease associations are elusive. Emaciated animals, and very young animals which have not digested food for a day or two, may have impressive amounts of lipid in some fibers.

Endocrine myopathies are uncommon in domestic animals, although selective atrophy of type 2 fibers occurs in hypothyroidism and perhaps in other endocrinopathies. It may be related to a selective neuropathy. This lesion is difficult to distinguish from cachectic atrophy. In humans, abnormal clusters of subcellular components are thought to be characteristic of endocrine myopathy.

Bibliography

Carpenter, S., and Karpati, G. Lysosomal storage in human skeletal muscle. *Hum Pathol* **17:** 683–703, 1986.

Edwards, J. R., and Richards, R. B. Bovine generalized glycogenosis type II: A clinico-pathological study. *Br Vet J* **135:** 338–348, 1979.

Giger, U. *et al.* Metabolic myopathy in canine muscle-type phosphofructokinase deficiency. *Muscle Nerve* **11:** 1260–1265, 1988.

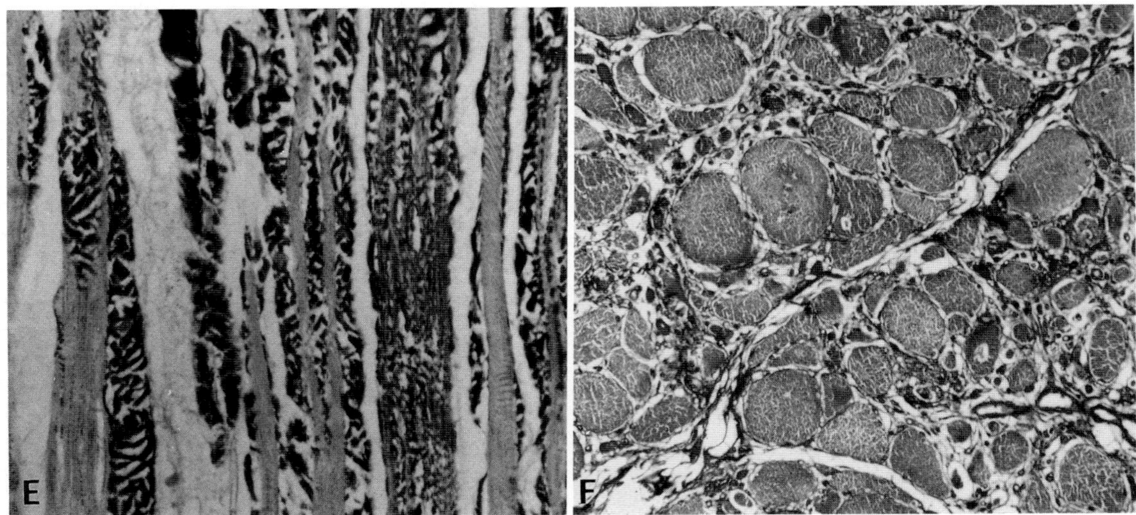

Fig. 2.32 Dystrophy. Horse. (E) Glycogen storage disease. PAS stain demonstrates dark, comma-shaped "packets" of glycogen. (F) Cross section of muscle in (E), showing variation in fiber size and aggregation of bodies in fibers.

Gullotta, F. Metabolic myopathies. *Path Res Pract* **180:** 10–18, 1985.

Harvey, J. W. *et al.* Polysaccharide storage myopathy in canine phosphofructokinase deficiency (type VII glycogen storage disease). *Vet Pathol* **27:** 1–8, 1990.

Ho, K.-L. Crystalloid bodies in skeletal muscle of hypothyroid myopathy. *Acta Neuropathol (Berl)* **74:** 22–32, 1987.

Howell, J. McC. *et al.* Infantile and late-onset form of generalised glycogenosis type II in cattle. *J Pathol* **134:** 266–277, 1981.

Howell, J. McC., Dorling, P. R., and Cook, R. D. Generalised glycogenosis type II. *Comp Pathol Bull* **15:** 2–4, 1983.

Indrieri, R. J. *et al.* Neuromuscular abnormalities associated with hypothyroidism and lymphocytic thyroiditis in three dogs. *J Am Vet Med Assoc* **190:** 544–548, 1987.

Jolly, R. D. *et al.* Generalized glycogenosis in beef Shorthorn cattle—heterozygote detection. *Aust J Exp Biol Med Sci* **55:** 141, 1977.

O'Sullivan, B. M. *et al.* Generalized glycogenosis in Brahman cattle. *Aust Vet J* **57:** 227–229, 1981.

Rafiquzzaman, M. *et al.* Glycogenosis in the dog. *Acta Vet Scand* **17:** 196, 1976.

Thompson, A. J. *et al.* Polysaccharide storage myopathy. *Muscle Nerve* **11:** 349–355, 1988.

J. Myasthenia

Two clearly defined types of myasthenia gravis occur in animals; a congenital disease in which acetylcholine end-plate receptors are few in number, and the acquired disease in which a normal number of end-plate receptors are blocked by circulating anticholinesterase receptor antibodies. In the **congenital** group, which accounts for 10% of cases, are puppies of four breeds which show clinical signs in early postnatal life: Jack Russell terriers, springer spaniels, smooth fox terriers, and the Gammel Dansk hønsehund. The trait is autosomal recessive. It leads to a failure of formation of acetylcholine receptors and development of shallow secondary clefts in the end-plate synaptic gutters.

Among the cases of **acquired myasthenia** are a disproportionate number of dogs and cats with thymoma, and this is also true in humans. The thymus normally contains cells with internal muscle components, and they apparently present the muscle antigen when the immune system becomes aberrant or when thymoma develops. Animals become very weak and may develop a voice change. The disease will be fatal naturally, so only in treated or well-nursed cases will there be muscle atrophy. Muscle fiber degeneration is not normally a feature of myasthenia.

Bibliography

Bors, M., Valentine, B. A., and de Lahunta, A. Neuromuscular disease in a dog. *Cornell Vet* **78:** 339–345, 1988.

Cuddon, P. A. Acquired immune-mediated myasthenia gravis in a cat. *J Small Anim Pract* **30:** 511–516, 1989.

Dardenne, M., Savino, W., and Rach, J-F. Thymomatous epithelial cells and skeletal muscle share a common epitope defined by a monoclonal antibody. *Am J Pathol* **126:** 194–198, 1987.

Darke, P. G. C., McCullagh, K. G., and Geldart, P. H. Myasthenia gravis, thymoma, and myositis in a dog. *Vet Rec* **97:** 392–394, 1975.

Flagstad, A., Trojaborg, W., and Gammeltoft, S. Congenital myasthenic syndrome in the dog breed Gammel Dansk Hønsehund: Clinical, electrophysiological, pharmacological, and immunological comparison with acquired myasthenia gravis. *Acta Vet Scand* **30:** 89–102, 1989.

Indrieri, R. J. *et al.* Myasthenia gravis in two cats. *J Am Vet Med Assoc* **182:** 57–60, 1983.

Johnson, R. P. *et al.* Myasthenia in springer spaniel littermates. *J Small Anim Pract* **16:** 641–647, 1975.

Lennon, V. A., Lambert, E., and Palmer, A. C. Acquired and congenital myasthenia gravis in dogs—a study of 20 cases. *In* "Myasthenia Gravis—Pathogenesis and Treatment," E. Satoyoshi (ed.), pp. 41–54. Tokyo, University of Tokyo Press, 1981.

Miller, L. M. *et al.* Congenital myasthenia gravis in 13 smooth fox terriers. *J Am Vet Med Assoc* **182:** 694–697, 1983.

Palmer, A. C. *et al.* Autoimmune form of myasthenia gravis in a juvenile Yorkshire terrier × Jack Russell terrier hybrid contrasted with congenital (nonautoimmune) myasthenia gravis of the Jack Russell. *J Small Anim Pract* **21:** 359–364, 1980.

Pflugfelder, C. M. *et al.* Acquired canine myasthenia gravis: Immunocytochemical localization of immune complexes at neuromuscular junctions. *Muscle Nerve* **4:** 289–295, 1981.

Wilkes, M. K. *et al.* Ultrastructure of motor endplates in canine congenital myasthenia gravis. *J Comp Pathol* **97:** 247–256, 1987.

IV. Degeneration of Muscle

A distinction must be made between the egress and ingress of muscle protein as muscle fibrils are daily broken down and reconstructed (physiologic myolysis), and the reactive events of longer duration which constitute muscle fiber degeneration (pathological myolysis) (Fig. 2.33A, B,C). In veterinary medicine, the word **myolysis** often is used imprecisely and therefore, its use here in a pathologic context will be avoided.

Degeneration of skeletal muscle fibers can occur at different levels of destructiveness and at different levels of completeness for any single myofiber. The myofibrils alone, or myofibrils and sarcoplasm, may undergo degenerative change leaving the sarcolemmal lamina, myonuclei, and satellite cells viable and intact. The next level of segmental degeneration leaves only the satellite nuclei and the basal lamina in place, and the third level destroys the satellite cells. The fourth level of segmental destruction destroys endomysial connective tissue cells and capillaries as well. Regenerative responses differ at each of these levels. Because the myofiber giant cells are so long, it is quite possible for a segment or several segments of a fiber to be destroyed without all of the cell being adversely affected, although subsequent reparative events sometimes fail to reconstruct the original fiber completely. It is distinctly possible, on the other hand, for a fiber to suffer repeated segmental destruction throughout its life span, and to be restored each time to completeness. Thus, complete muscle cell death is an exceptional event even in extensive lesions, but destructive events usually have

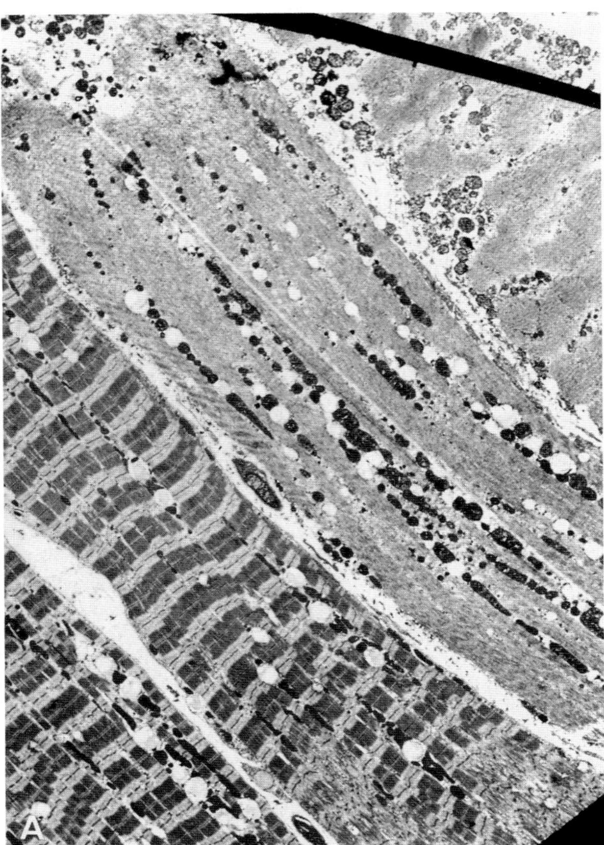

Fig. 2.33A Lipid droplets (incidentally) in the muscles of an anorectic foal also showing coagulative necrosis and focal myolysis (arrow).

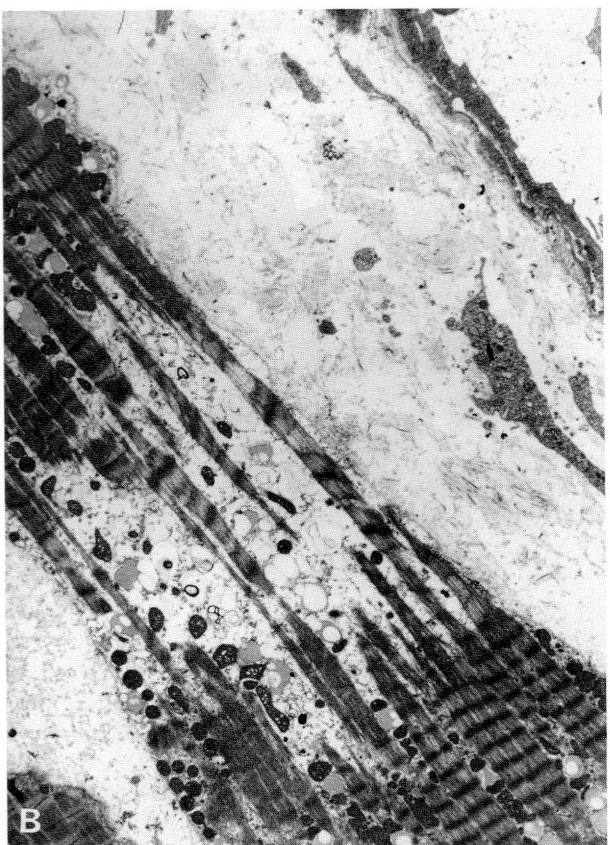

Fig. 2.33B Lipid droplets (incidentally) in the muscles of an anorectic foal also showing hypercontraction with tearing of myofibrils.

some lasting effect on the shape or number of fibers (see Regeneration).

Muscle fiber degeneration is frequently a disease of cellular membranes, and membrane failure takes the form of segmental myofiber degeneration. The events which initiate contraction are completed by nerve-triggered gating of Na^+ or K^+ or by passive release of calcium ions adjacent to sarcomeres from the endoplasmic reticulum or the extracellular space, but energy is required to recapture calcium ions to allow muscle fiber relaxation. Exhaustion of phosphate-bonded energy reserves often leaves the muscle fiber in a state of calcium-abundant contraction or hypercontraction, which soon reduces the complex myofilaments to a coagulum of contractile proteins (Fig. 2.34A). There appears to be a final common degenerative pathway of mitochondrial calcium overload, which begins with many different causes of membrane or energy failure and ends with hypercontraction and coagulation of the contractile proteins (Fig. 2.34B,C). Most animal myopathies seem to fit well into this sequence of events, particularly the nutritional myopathies. Gross visible mineralization occurs several hours after protein coagulation has been confirmed and striations have been lost. Calcium is released belatedly to degenerate fibrils from calcium-saturated mitochondria as they disintegrate. Although the almost chalky mineralization seen grossly in many cases of nutritional myopathy is usually referred to as dystrophic, it is not at all certain that such accumulations can occur as a result of a purely passive collection of mineral from the blood or dead tissue protein. At the very least it seems to require that mitochondria become overloaded with calcium early in the degenerative process.

Some nutritional myopathies and some toxic myopathies which show very similar changes are not characterized by mineralization to any appreciable extent, even though blood calcium levels are normal and lesions are as extensive as those in "white muscle disease." Degenerate muscle may be very difficult to detect grossly when it is uncalcified, and consequently is likely to escape detection. The event which determines whether mineralization occurs, in addition to the variable of the amount of blood calcium available, may well be the interval after hypercontraction over which cellular membranes allow calcium into the cell, and mitochondria

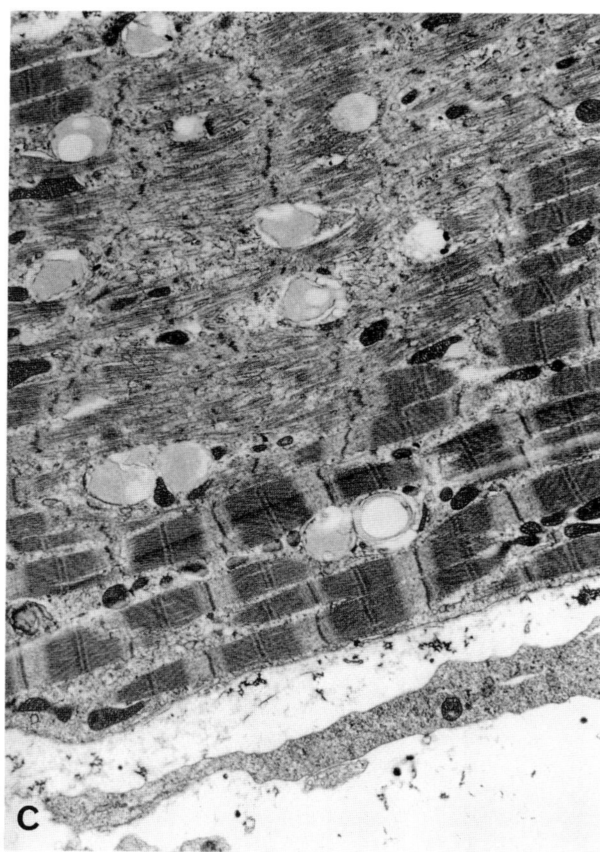

Fig. 2.33C Lipid droplets (incidentally) in the muscles of an anorectic foal also showing focal myolysis (reversible).

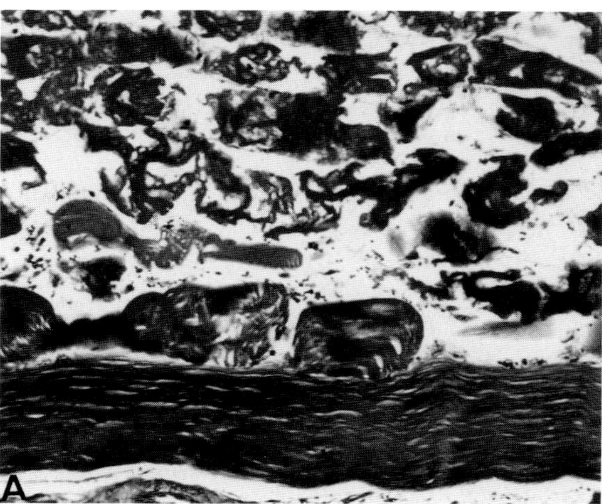

Fig. 2.34A Floccular or granular form of fiber degeneration. Cow. Nutritional myopathy.

continue to collect calcium from the cytosol. Eventually such activity will be ended by membrane disintegration and a release of calcium to the cytosol again, but calcium

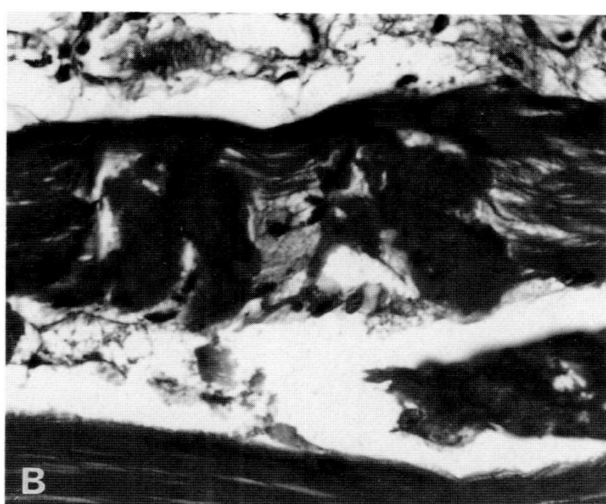

Fig. 2.34B Fiber degeneration. Segmental hypercontraction. Pig. Porcine stress syndrome.

Fig. 2.34C Fiber degeneration. Retraction cap or cup. Calf. Nutritional myopathy.

gathering could continue for a period of several hours. In fact, it is not clear that a segment of the myofiber does "die" in any real sense after membranous organelles cease to function, as long as myofiber nuclei are intact and perhaps even if they are not. The rapidity with which repair takes place suggests that muscle cell viability is not ended at any stage short of necrosis of all components.

The hypercontracted eosinophilic coagulum is readily removed by macrophages both in its earlier, premineralized state and in the later mineralized stage of granular degeneration (Fig. 2.34D). The early pale, hyaline stage (Fig. 2.34E) sometimes undergoes Zenker's degeneration. It is then a highly eosinophilic, glassy mass, which looks like a swollen muscle fiber cast. It sometimes retains faintly visible, tightly contracted striations, although these are eventually lost. Muscle fibers which show Zenker's

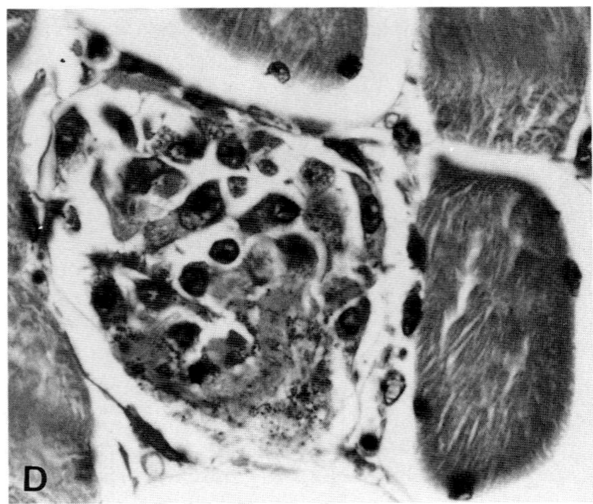

Fig. 2.34D Regenerative repair. Early clean-up of granular degeneration by macrophages.

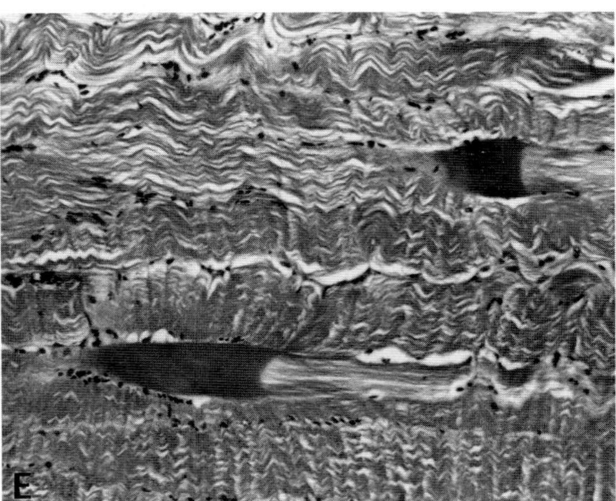

Fig. 2.34E Fiber degeneration. Hyaline. Horse. Tying-up syndrome.

degeneration seem to be very resistant to removal, or perhaps have no chemotactic attraction, and consequently act as a temporary physical block to repair. The persistent hyaline masses seem to have no staining or structural characteristics different from those which are removed readily. Mineralization of degenerating muscle fiber seems to enhance removal by macrophages.

Discoid degeneration develops when hypoxia plays a major part in the production of muscle lesions. The muscle striations, but not all of their fine features, are preserved, and the fiber begins to separate at the Z bands (Fig. 2.34F). The Z bands of several adjacent sarcomeres may break, but more often 3–10 sarcomeres remain intact before the next break occurs, perhaps on the opposite side of the fiber. The appearances suggest that the fiber is mummified and brittle.

Vacuolar degeneration of muscle fibers (Fig. 2.35) is

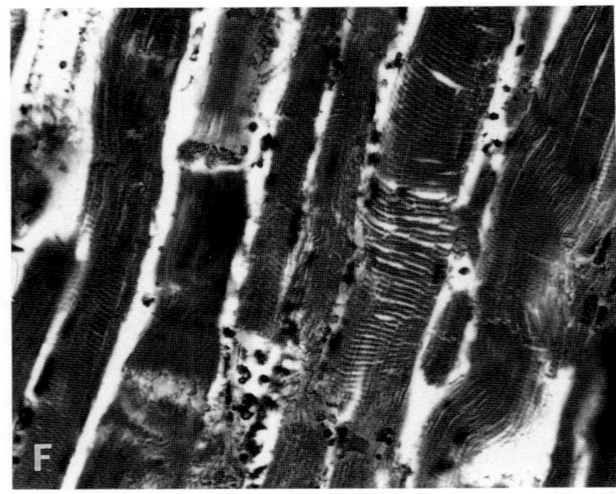

Fig. 2.34F Discoid fiber degeneration. Downer cow.

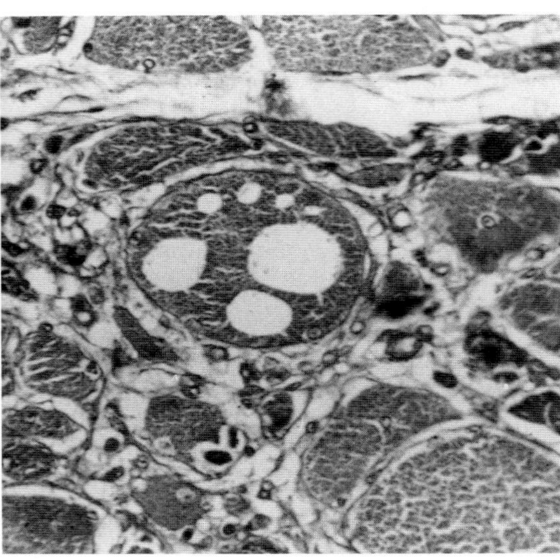

Fig. 2.35 Vacuolar fiber degeneration. Horse. Glycogen storage syndrome.

often associated with excess endogenous or exogenous corticosteroids. The horse may be particularly susceptible to this lesion. Occasionally, vacuolation is unexpectedly encountered in a few fibers in other circumstances. Small fat vacuoles are a very common finding in emaciated or anorectic animals because muscle is able to use triglycerides directly as a source of energy. Rows of minute droplets are held in considerable quantity by type 1 (oxidative) fibers, but type 2 fibers often contain neutral fat also (Fig. 2.33A). The presence of these fat droplets may be more evident as myofibrils disappear in cachectic atrophy, but it apparently reflects no pathologic change in muscle, and even well-nourished animals carry some droplets in their muscles. Excess fat in muscle can be an indication of carnitine deficiency. Quite separate from these aggregations of fat droplets, larger fat droplets suggestive of a fatty degeneration may be seen in muscle fibers in the dog and horse. Their cause is not known.

A variety of inclusionlike structures or focally abundant or altered structural components occur in the muscles of animals. Similar structures in the muscles of humans sometimes are associated with specific diseases, but patterns are not established in animal diseases except for the dystrophies (Fig. 2.31B). **Nemaline bodies** have been found in muscles of a dog with intestinal carcinoma and also associated with tenotomy and muscle dystrophy. It has been well described in a cat (see Muscles, Section III,F, Muscular Dystrophy). **Targetoid fibers** and ragged red fibers have been seen in myotonic muscle disease in dogs, and sarcoplasmic masses and central vacuolations have been observed in the muscles of sheep and cattle in dystrophy. Abnormal **glycogen accumulations** may develop in the muscles of dogs, sheep, and horses (see also Metabolic Myopathies). In all three species, these may be membrane-bound collections or may be partly membrane bound. The deposits are characterized by an abundance of PAS-positive material usually just under the sarcolemma. A case of exercise intolerance in a horse has been associated with abnormal accumulations of alkaline phosphatase in clusters of muscle fibers.

Ectopic ossification of skeletal muscle occurs rarely in all species of domestic animals and at times shows increased local incidence in meat-producing animals, particularly pigs (see Chapter 1, Bones and Joints).

A. Pigmentation of Muscle

Congenital **melanosis** of muscles occurs in calves with a low incidence similar to that of melanosis of liver and lungs. The melanin pigment is in connective tissue, not myofibers.

Xanthomatosis is the term used to describe the accumulation of abnormal amounts of yellow-brown to bronze pigment in some skeletal muscles and myocardium, and often also in adrenal cortex and distal convoluted tubules of the kidney. Black hair coats may become brown, and white skin may become yellow, but the presence of the pigment appears not to be detrimental to the health of the animal. The buildup of wear and tear pigments in old animals, particularly high-producing dairy cows, is indistinguishable morphologically from xanthomatosis, but it is sometimes associated with myocardial failure. There appears to be a disproportionately high incidence of xanthomatosis in Ayrshire cows and among mature cows of all breeds slaughtered for human consumption; in one study the incidence was about 0.1 to 0.5% overall compared to 10 to 25% for Ayrshires.

Skeletal muscles most frequently or most extensively involved are the masseter muscles and the diaphragm. Muscle fibers are of normal size or smaller, but normal in most other respects. Pigment granules, some of which stain positively with lipid and trichrome stains, tend to be located perinuclearly under the sarcolemma or centrally in the fiber. Particles are irregular in size and shape and on electron microscopic examination show a wide variation in shape and organization. All are electron dense, and some are membrane bound; it is presumed that they represent degraded lipid substances which have been released from phagolysosomes.

B. Regeneration and Repair of Muscle

The ability of muscle to repair itself is remarkable considering its high specialization and the great length and great vulnerability of individual fibers. The ability to rapidly repair the damaged segment of a fiber, without apparent complication to the rest of the fiber, is without parallel in other cells of the body. A muscle cell consists of a single large multinucleate myofiber and a large population of uninuclear satellite cells, coexisting within the tough basal lamina. The major participants in the regenerative sequence are macrophages of the blood; the satellite cells because only they, within the fiber, have retained the capability for mitotic division; and the basal lamina, which acts as a very efficient gate keeper. The integrity of the basal lamina determines from the very beginning whether the outcome will be regenerative repair, fibrous replacement, or a mixture of the two. An intact basal lamina effectively keeps myonuclei, satellite nuclei, and myoblastic cells inside. It is equally efficient in keeping fibroblastic cells out, but allows phagocytic cells easy entry and exit.

If damage to the fiber is segmental and such that only the myofibrils and the sarcoplasm along with its subcellular components have been injured, the first visible events consist of progressive mineralization (see Degeneration of Muscle), and an early rounding up of satellite cells as they prepare to undergo mitotic division. Daughter cells become apparent in hours as they leave their former sublaminal site. At about 12 hr, macrophages appear, and they may be accompanied by neutrophils in a short-lived wave. Both of these cell types receive free access to the mineralized contents of the damaged fiber. Macrophages dissolve and remove the debris (Fig. 2.34D); neutrophils disappear unless infection complicates the process. Once removal of sarcoplasmic debris is under way, and some space has been created within the collapsing sarcolemma, some of the proliferating satellite cells enter the myofiber (Fig. 2.36A,B). These are now myoblasts, which mix freely with the remaining (but not dividing) myonuclei.

Myoblasts increase in number until a critical myoblastic mass is reached. Only then do myoblasts begin to fuse, and this triggers the next sequence of events. Five or six days after degeneration occurred, fusion of myoblasts has produced unpolarized myotube giant cells, which begin to send out cytoplasmic processes as their nuclei divide (Fig. 2.36C,D). Some processes make contact with remaining viable segments of the original fiber within the basal lamina, and perhaps even within the original myofiber cell membrane. They make effective union either as a side-to-side, cell-to-cell splice or, more likely, they dissolve a cell membrane and unite the cytosols into a single myofiber.

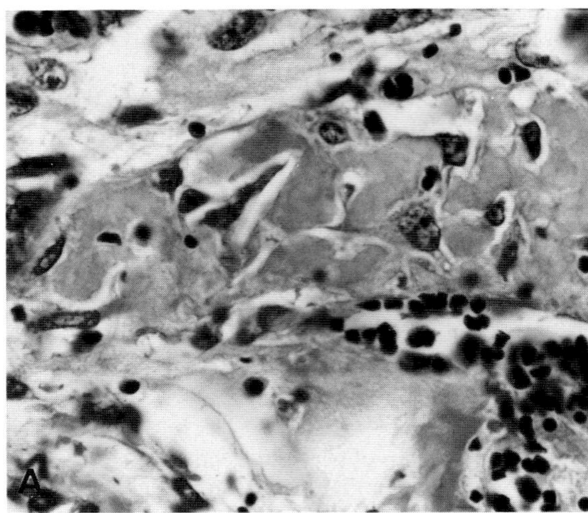

Fig. 2.36A Regenerative repair. Invasion of a hyaline fiber by macrophages and satellite cells.

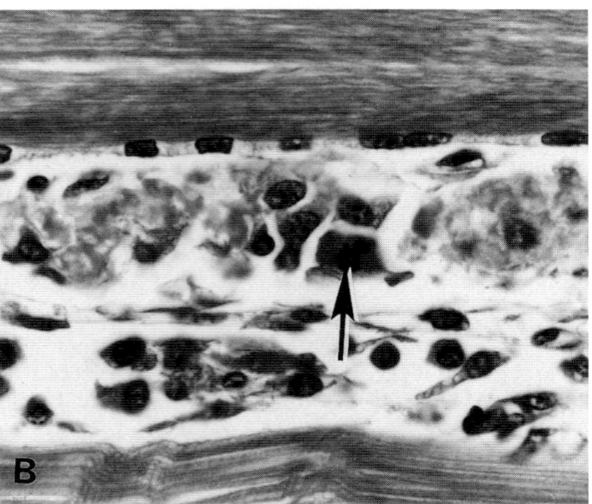

Fig. 2.36B Regenerative repair. Satellite cells in mitosis (arrow).

The free pole of the myotube giant cell then becomes a polarized regenerative probe, which grows within the cleared and collapsed sarcolemmal sheath (Fig. 2.36D). Other myotube cells may not make contact for some time, and remain as unpolarized giant cells until an adjacent myotube process contacts them. A growing myotube cell with its basophilic cytoplasm and central row of nuclei can grow for relatively long distances as long as the outer sheath is intact and no fibrous obstruction intrudes. Outside the sheath, a regenerating fiber can grow competitively with fibrous tissue for only 2–5 mm.

When contact has been made with the next viable segment of the original fiber at about 10 to 14 days, changes in the fiber consist mostly of production of new sarcomeres in the enlarging fiber. The new fiber has to find growing space between existing fibers but seems able to do this efficiently. As myofibrils, sarcoplasm, and organelles are added, the fiber becomes less basophilic, and at about 15 to 17 days, the nuclei migrate from their core position to lie just under the new cell membrane. Only in the most ideal of circumstances are all the parts restored to the predegeneration state by the efficient regeneration described. At the very least, two basal laminae, one inside the other, persist; later they may fuse (Fig. 2.16B). Any proliferating satellite cell/myoblasts which do not enter the original cell, or any myotube processes which escape from it, may produce a second fiber or second branch of a fiber within the original basal lamina.

Considering the possibility of breaks in membranes, the potential for myoblast formation, which may be inside or outside the original myofiber (which in turn is inside or outside the original basal lamina), and the potential for distortion of development by fibrous tissue invasion, it is easy to understand that the final result may not be perfect (see Figs. 2.16A, 2.37). Branched and split fibers, two or three small fibers replacing one large one over a segment, and bridges between two or more parts of one original fiber, often occur in repaired muscle. The longer and more numerous the gaps are, the more potential there is for aberrations in repair. The most important factors in preventing such aberrations seem to be the viability of nuclei and the wholeness of the basal lamina. Nutritional, exertional, and most toxic myopathies, although sometimes extensively destructive, are likely to repair reasonably well. Primary dystrophies repair well but degenerate again, and ischemic damage repairs badly or not at all, because satellite nuclei and endomysial cells may have been killed. Ischemia with early reflow permits some regenerative repair, particularly around the outside of primary bundles where blood flow returns first.

The reduced regenerative capability that develops with maturity may be related to low numbers of satellite cell myoblasts as the proportion of satellite cells to the total nuclear population in the fiber drops with age.

Ringbinden is one of the unusual structures which may be left behind when denervation or tenotomy is combined with atrophy and regenerative repair. It is a formation of aberrant myofibrils, which wrap themselves around a muscle fiber in a tight spiral fashion (Fig. 2.38A,B). Satellite nuclei can migrate, and they sometimes orient their long axes transversely rather than longitudinally. They can also proliferate, form myoblasts and myotubes, and develop into new fibers. It is possible that the annular fibrils represent a second small fiber, which is competing for space within one original basal lamina.

Ring fibers most often occur naturally in animals showing evidence of denervation atrophy, with or without evidence of ischemic muscle damage. Horses cast and anesthetized for several hours may be particularly prone to develop these lesions.

V. Circulatory Disturbances of Muscle

Skeletal muscle is a highly vascular tissue with an abundant capillary bed, which forms an extensive system of

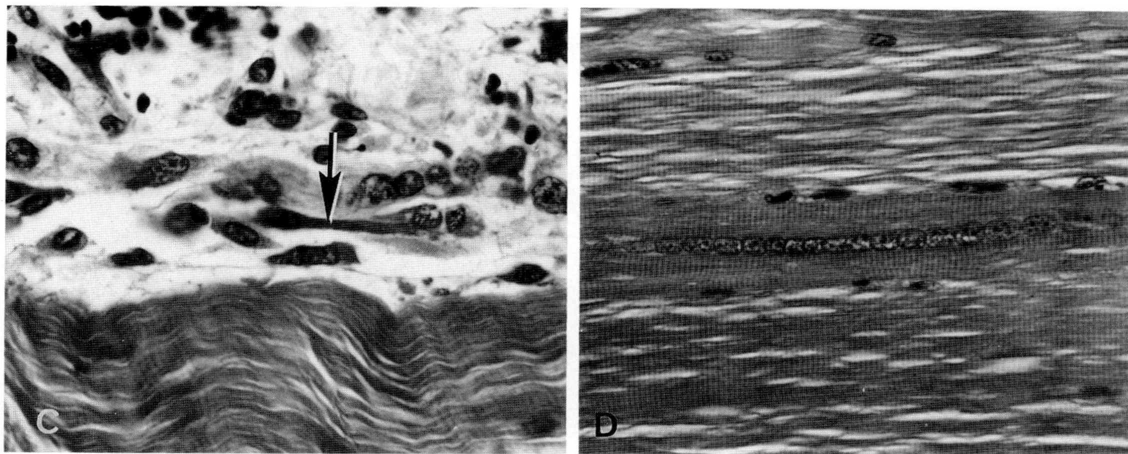

Fig. 2.36 Regenerative repair. (C) Elongating myoblasts (arrow). (D) Regenerative fiber with a row of central nuclei.

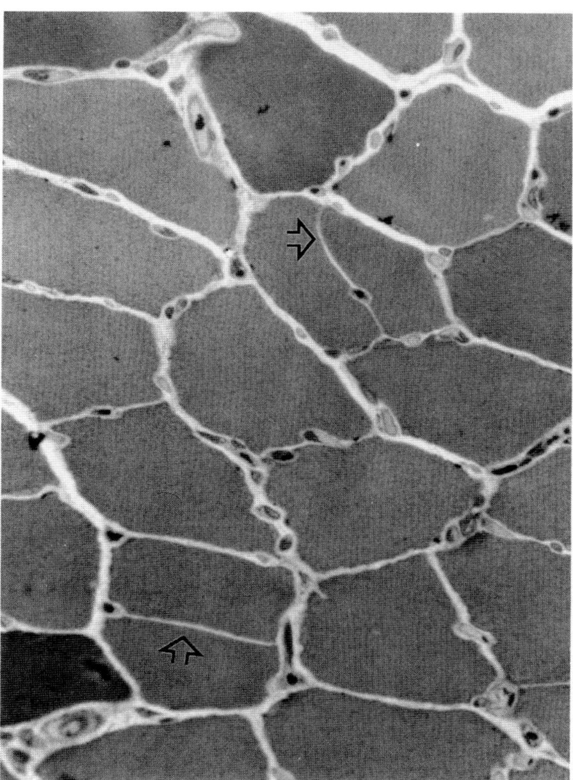

Fig. 2.37 Cleft fibers (arrows) in actively regenerating muscle. Horse subject to repeated episodes of tying-up.

anastomoses. Attempts to induce muscle fiber necrosis by ligating intermuscular arteries frequently fail, ostensibly because most muscles receive small collateral branches from tendons, fascial sheaths, and major nerve trunks, which carry their own limited arterial blood supply. Naturally occurring examples of ischemic muscle necrosis are not uncommon, although few of them are related to thromboses.

Each muscle fiber is served, at any given level, by 3 to 12 capillaries which run mainly longitudinally in the endomysium. Type 1 fibers are served by slightly more capillaries than are type 2 fibers of comparable size. Muscles like those of mastication are particularly well supplied; nearly twice as many capillaries serve each of these fibers than is the case for the major thigh or shoulder muscles.

One of the factors which seem to exert some control on muscle fiber size is the distance from capillary to the center of the fiber(s) served. When the distance becomes abnormally great, the fiber is likely to form a longitudinal cleft down one side into which a capillary slips, effectively serving the fiber interior (Figs. 2.15B, 2.16A). Complete or incomplete longitudinal fiber division appears to be a mechanism primarily initiated to improve the capillary to fiber ratio.

The effect of ischemia on muscle fibers, and their capacity to respond, are influenced by the completeness and duration of oxygen and nutrient deprivation. Least damaging is transient hypoxia of a few hours' duration, which causes coagulation of contractile proteins but does little harm to the other cellular components or to the satellite cells. Muscles are ordinarily restored to normal function by myoblast and myotube formation in about 16 to 20 days. The next level of injury is induced by episodes of ischemia lasting 6–24 hr, which cause death of both myofiber and satellite cell nuclei, and coagulation of long segments of muscle fiber content. The capacity for regenerative muscle fiber repair is lost or greatly reduced, and reaction is limited to local proliferation of fibroblasts of the endomysium or blood vessel walls. The repair process is quite rapid but leads to a randomly distributed mixture of regenerated or original fibers interspersed with segmental scars replacing muscle in adjacent fiber sheaths. The third level of ischemic injury is that lasting for more than

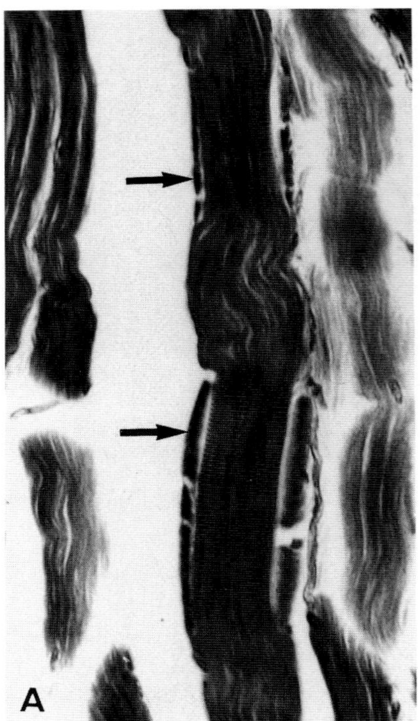

Fig. 2.38A Two ring bands in one fiber (arrows).

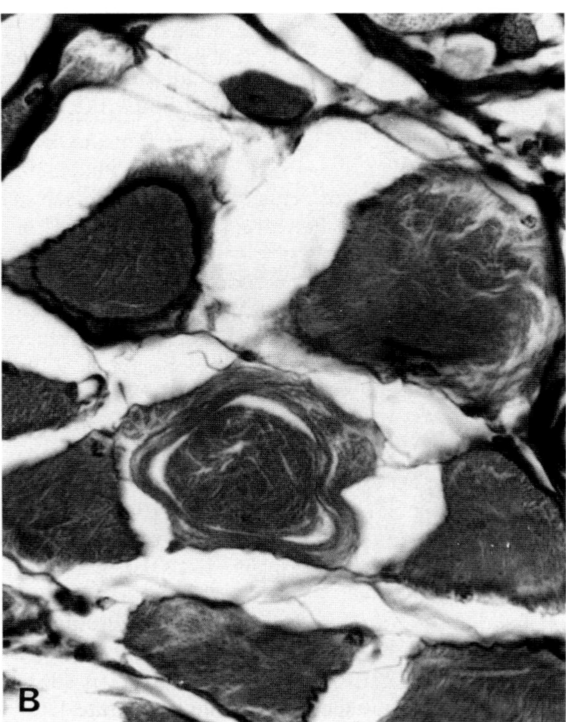

Fig. 2.38B Cross section of ringed fiber.

18–24 hr, which leads to death of all cells within an area of muscle. The response in this case is confined to a slow peripheral sequestration followed by phagocytic removal of dead muscle and fibroblastic proliferation, which may take months to complete. The end result is a mass of scar tissue relieved only at the periphery by a narrow zone of distorted muscle fiber proliferation.

During the acute phase of ischemic muscle destruction, a relatively orderly sequence of functional and morphologic changes can be expected. After as little as 1 hr of ischemia, the ability of the muscle fiber to contract in response to a neural stimulus is lost, and anesthesia is complete. Ultrastructurally, changes in subcellular components begin after about 6 to 7 hr of ischemia when the tubular systems become dilated and begin to disintegrate. Some mitochondria may show evidence of swelling and disarray at 8 hr, but even after 24 hr, many mitochondria may be well preserved. Z bands become less dense than normal at about 12 hr, and this is reflected in light microscopy by an increased distinctness of striation, which persists for several days. Apart from the changes in the I band, other components of the sarcomere, including myofilaments, retain their separate identity for at least 48 hr, and it is only after this time that the amorphous hyaline mass forms, free of cross or longitudinal striations. Discoid fracture of the fiber along Z bands occurs after 48 hr and may be an extension of the earlier Z-band disintegrative changes.

Regardless of the duration and extent of the ischemic change, the muscle fiber membrane becomes permeable to enzymes and myoglobin. Creatine phosphokinase levels reach a peak in 1 to 2 days and return to near normal in 4 to 5 days, even when destruction of muscle is extensive. Myoglobin may be released from muscle at 12 to 24 hr even in cases of minimal destruction, but the amount released will vary directly with severity and extent of damage. Even in those situations in which scar repair is greatly delayed, myoglobin loss occurs almost entirely in the first few days.

When a large mass of muscle is physically injured or when ischemic degenerative changes are extensive, large amounts of myoglobin are released into the blood stream. Much of this is excreted by the kidneys, but in the process, many myoglobin casts may be formed. A direct toxic effect of myoglobin on convoluted tubules may occur, but renal ischemia due to shock is probably responsible for the oliguria and anuria which sometimes result. Acidosis initiated by the release of lactate from damaged muscle may be exacerbated by oliguria and anuria. Hyperkalemia presents a risk in the early stages of massive muscle fiber breakdown.

The natural situations in which ischemic muscle necrosis occurs in domestic animals vary but little from one species to another. Four syndromes initiated by different processes are recognized, although the distinctions between them in both humans and animals are not always clear. In the compartment syndrome, the downer syndrome, and the muscle crush syndrome, ischemia is caused by increasing intramuscular pressure, while the vascular occlusive syndrome results from physical obstruction of the blood supply to muscle.

A. Compartment Syndrome

Muscles which are surrounded by either a heavy aponeurotic sheath or by bone and sheath are vulnerable to ischemia when muscle fibers well filled by myofibrils are subject to moderately vigorous but not exhaustive contraction. The syndrome occurs in well-conditioned athletes, but nowhere is the syndrome more clearly primary and specific than in the infarction which occurs in the supracoracoid muscles of some breeds of broiler chickens and in some breeds of turkeys. In these birds, a brief, vigorous flapping of the wings increases intramuscular pressure of the supracorticoid muscle within the inelastic breastbone and the outer muscle sheath. Muscle in full contraction increases in volume up to 20%, and this causes partial or transient collapse of the venous outflow. At the same time, muscle activity increases arterial blood flow to the muscle. Subsequent muscle contractions tend to build internal pressure until the intramuscular pressure exceeds first venous, then later, arterial blood pressure. Metabolites of the muscle fibers exerting an increased osmotic tension, coupled with increased arteriolar blood pressure, produce an accumulation of interstitial water early in the process, and this further increases intramuscular pressure. Once blood flow has stopped, ischemic changes of both muscle and vessels begin, and further water escapes from damaged endothelial cells. Pressure builds for 1 to 4 hr after muscle exercise, and the extent and severity of damage to muscle increases with time. In both humans and birds, early fasciotomy releases intramuscular pressure and restores the potential for complete regenerative repair.

The so-called spontaneous rupture of the gastrocnemius muscle of Channel Island breeds of cattle may represent an example of compartment syndrome leading to ischemia and subsequent rupture (Fig. 2.39A). In humans, the use of the term compartment syndrome is not limited to the nontraumatic syndrome outlined above, but it seems worthwhile to make such a distinction between syndromes initiated by external pressure and those initiated only by normal muscle activity.

B. Downer Syndrome

Humans and most of the domestic species share a muscle ischemia syndrome, which is initiated by external pressure of objects or by pressure created by the weight of body, torso, or head on a limb tucked under the body for prolonged periods. The condition in humans is usually related to drug overdose, while in animals it is induced by prolonged anesthesia, muscle, joint, or bone damage causing prostration, or metabolic or neurologic disease causing paresis. Animals in good condition are particularly susceptible and conversely, thin animals seldom suffer from ischemic muscle necrosis. Absolute size has some

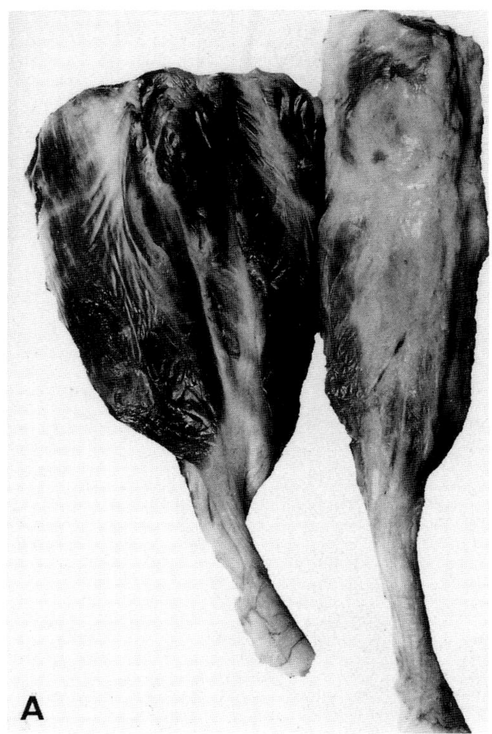

Fig. 2.39A Muscle ischemia. Rupture of gastrocnemius in a Channel Island cow. (Affected muscle on the right, normal on left). Probably due to compartment syndrome.

influence on the incidence of the disease, but rams and heavy ewes, boars, and sows and even large dogs are occasionally susceptible. Cows are most frequently affected as downer animals, partly because of their weight and their muscle bulk and partly because they are subject to diseases in which paresis is common (Fig. 2.39B). Horses usually suffer from this type of muscle ischemia as a result of having been anesthetized for several hours while lying on a hard surface.

The pathogenesis of the downer syndrome depends on the fact that pressure within muscles caused by the weight of the body on them may rise to levels considerably higher than both venous and arterial pressure. Muscles of limbs in a flexed or tuck position are particularly susceptible. Some regions of muscle which are in closest contact with the floor or with bones may be pressure blanched, but the intramuscular pressure created within skin and fascial sheaths of the limbs soon serves to collapse veins, causing congestion, and then collapses arteries. This blood-exclusive ischemia creates changes in tissue, which soon resemble the changes described for the compartment syndrome, and leads to further edema and more intramuscular pressure. In cows and horses, extensive ischemic lesions are sometimes created by a period of inertia as short as 6 hr, while some cows seem to be able to tolerate 12 hr or even more immobility with minimal residual lesions. Even 6 hr of anesthesia or comparable inactivity in horses and

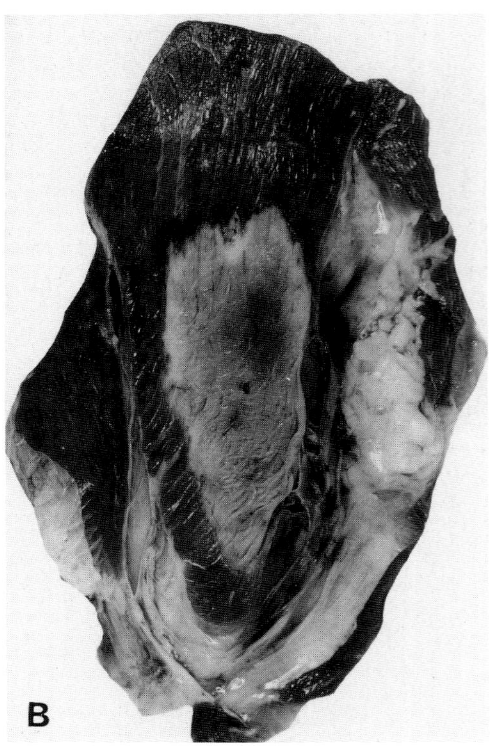

Fig. 2.39B Muscle ischemia. Infarcted muscle of paralyzed cow. Thigh muscle.

cows may cause sciatic or other nerve damage leading to peroneal nerve paralysis and a flexed rear fetlock or a dropped shoulder. As time passes and as the pressure is removed, the affected limb continues to swell as edema fluid increases under returned arterial flow. The limb usually extends involuntarily perhaps because the flexor muscles are rendered uncontractible by ischemic anesthesia. Swelling is reduced slowly and may be accompanied by fixation of skin to the underlying necrotic muscle. The extent of lesions within a muscle mass is quite variable but seldom involves more than half of the mass. This reflects the extent, duration, and severity of the blanching episode as well as the extent and rapidity of reflow (Fig. 2.40A,B,C). If the return of arterial flow is more or less complete, degeneration is minimized, but several factors may impede effective reflow. The formation of venous or arterial thrombi may slow or obstruct return. Persistent pressure within muscle sheaths may compromise venous flow. The formation of hemorrhages and hematomas as arterial blood pressures return to damaged vessels may further slow effective revascularization. Hemorrhage which seeps between muscle fibers may exert a satellite cell-sparing effect as least for a short period. Finally, although muscle repair is not retarded by denervation, effective recovery and mobilization may be delayed until nerves to muscles are regenerated. Exertional or spontaneous muscle infarction with muscle rupture in the cow

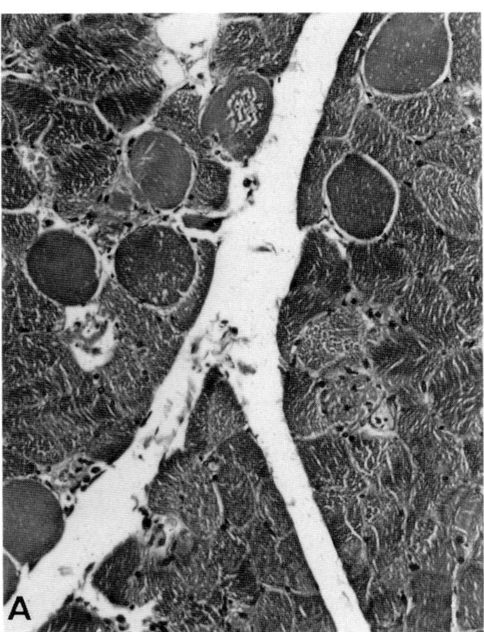

Fig. 2.40A Ischemia with early reflow with degenerate fibers after 4 days. Aortic thrombosis. Dog.

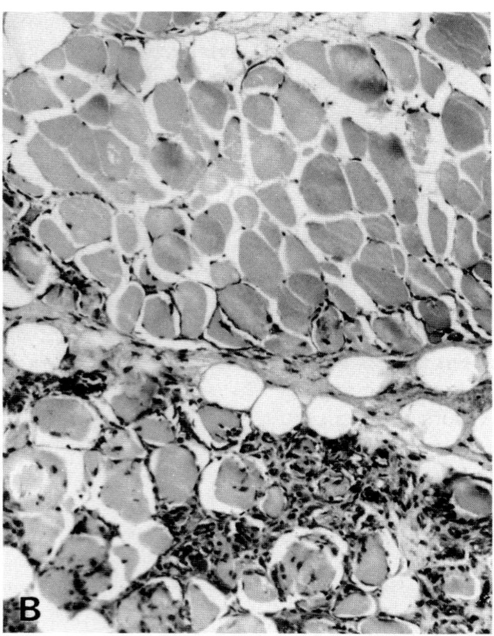

Fig. 2.40B Ischemia with delayed reflow. Only fibrous components are preserved. Downer cow.

leads to a massive outpouring of eosinophils, which impart a green tinge to the gross lesion.

A special set of circumstances is required for the postanesthetic downer syndrome (postanesthetic myopathy or myositis, postanesthetic compartment syndrome, postan-

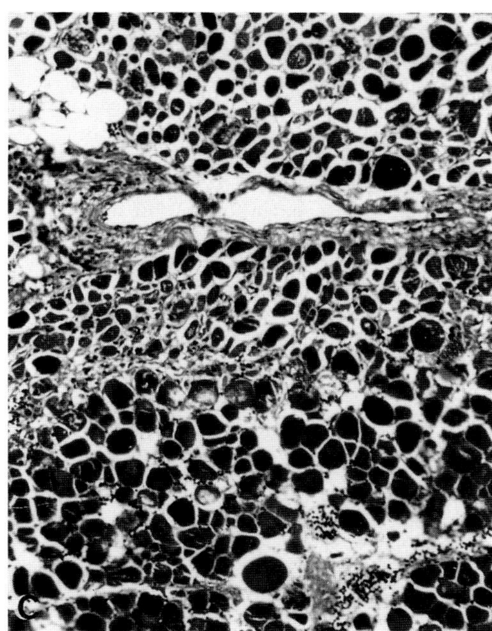

Fig. 2.40C Mixed pattern of damage at margin of an ischemic lesion. Only areas of total ischemia (below) preserve original fiber size.

esthetic myonecrosis) in the horse. Although it usually follows halothane anesthesia, it is not analogous to malignant hyperthermia in pigs, changes in blood and muscle pressure being of primary importance. Halothane may contribute to the pathogenesis, but the lesions seem to be explicable in terms of ischemia.

The disease affects well-muscled horses of all ages, particularly those over 600 kg, which undergo halothane anesthesia and recumbency in lateral, ventral, or dorsal positions. There is a marked rise in intramuscular pressure and blood pressure in the down-side muscles that is only partly alleviated by underpadding on that side. There is also a transient drop in blood pressure, which is common following induction of anesthesia with halothane, that in susceptible horses precipitates a sequence of pressure changes in the muscle. It is not clear whether the duration of anesthesia influences this sequence. The changes are similar to those described for compartment syndrome except that irreversible muscle damage, that usually requires 4–5 hr of ischemia, can occur with much shorter periods of halothane anesthesia with transient or persistent hypotension. Dramatic increases in creatine phosphokinase levels indicate that the major muscle fiber destruction occurs in the early recovery period and becomes apparent clinically as the halothane anesthetic and the ischemic anesthesia of muscle wear off. Muscles are then swollen, hard, and painful. The extent and level of ischemia determine whether muscle damage is reversible and reparable.

On postmortem, muscles are dark and often hemorrhagic. Several days after the initiation of lesions, the boundaries of necrotic muscles show competent regenerative repair. Deeper regions of the muscles are most severely damaged, but no muscle is entirely affected. Advanced lesions consist of irregular patches of scar tissue side by side with repaired muscle, which may contain evidence of denervation fiber atrophy and Ringbinden.

C. Muscle Crush Syndrome

This form of muscle ischemia has characteristics in common with downer syndrome. It is usually initiated by acute accidental trauma, often including bone fracture. It occurs less frequently in animals than in humans, but has been seen in the dog and perhaps the cow. Initial events center around the traumatic laceration of muscle, which leads to a combination of high osmotic tension and hyperemia, and brings abundant edema fluid to the area. Inflammatory fluids may complicate the local situation. If the damaged muscle and bone are still confined within a relatively firm sheath, conditions resembling those of compartment syndrome are set up. The limb swells and extends, and becomes turgid. Ischemia of variable extent ensues, but because of the great amount of myoglobin released, renal dysfunction may dominate the syndrome.

D. Vascular Occlusive Syndrome

When a major vessel to a limb is occluded the limb becomes cool, the arterial pulse is lost, skin over the limb loses its ability to sweat, and some limitation of movement may be apparent. When this occurs as a result of iliac artery thrombosis in the horse, the effects are usually transient because an effective collateral circulation will have begun to develop early. Some muscle degeneration probably occurs but is repaired rapidly and completely. In the cat with aortic–iliac thrombosis, more of the aorta is likely to be occluded than is the case in the horse, and collateral circulation is delayed. Muscle lesions are almost inevitable, but they are likely to be minor, and repair is possible if the cat survives. The anatomic pattern of degeneration varies from one case to another, and more distal muscles are not necessarily more vulnerable. Occlusion of major arteries is discussed in Volume 3, Chapter 1, The Cardiovascular System.

An ischemic lesion of muscle, seen in sheep in advanced pregnancy, appears also to be caused by arterial occlusion. Ewes carrying twins or triplets sometimes suffer from ischemic necrosis of the internal abdominal oblique muscle without evidence of congestion or hemorrhage. This muscle does not have a confining sheath; thus, the necrosis is not part of a compartment syndrome. The arterial supply to the abdominal oblique is via a tortuous branch of the internal iliac artery, which turns back on itself inside the iliac tuberosity, and it may be vulnerable to stretch and/or trauma. Ischemia of the muscle is followed by rupture, and subsequently the other abdominal muscles also rupture. In spite of all of this, the ewes sometimes lamb at term without difficulty.

VI. Physical Injuries of Muscle

Traumatic injuries of muscle are common and may be the result of external trauma, or they may be a result of a muscle rupture or tear of the fascia, as occurs occasionally in violent contraction, rarely in overextension. The effects of external trauma are very variable and depend on the qualities of the applied force, the presence or absence of concomitant fracture of the bones, and especially on the degree of hemorrhage, injury to blood vessels, and injury to motor nerves. The principles governing the outcome of these types of lesions have been discussed.

Violent contraction of muscle may result in either hernia or rupture with hemorrhage. A **hernia** occurs when the belly of the muscle protrudes through a rent in the overlying fascia and epimysium. The hernia can be reduced by pressure when the muscle is relaxed, and it hardens and bulges further when the muscle contracts, providing thereby a point of useful distinction between hernia and tumor of soft tissue.

An actual **rupture** of muscle tissue may also occur during violent exercise and is probably more common than rupture of the tendon. The muscle bulges at the end opposite to that which is torn. Such ruptures are not necessarily complete from the outset, but may become so later from additional strain or degeneration caused by infiltrating hemorrhage with pressure and ischemia. Regeneration in large defects is ineffective, and the gap is filled in by scar tissue. The muscle most frequently ruptured in animals is the diaphragm. Trauma, with acute abdominal compression, is the usual excitant in the dog. In cattle, it tends to follow diaphragmatic myositis secondary to traumatic reticulitis. In horses it is a consequence (in foals) of abdominal compression at parturition, and diaphragmatic rupture occurs occasionally with acute gastric dilation. Acquired ruptures must be differentiated from congenital and postmortem ruptures by attention to the edges of the cleft and the disposition of abdominal viscera. Muscles subjected to continuous pressure undergo degeneration, necrosis, and fibrous replacement. It is possible that pressure can have a direct effect on the fibers by disorganizing their internal structure, but the principal effect is to interfere with circulation.

The histologic appearance of traumatic injury includes sarcolemmal rupture with adjacent intact fibers showing hyaline and granular degeneration. The initial cellular reaction is neutrophilic accompanied by edema and hemorrhage. Reparative changes are conventional, but if the endomysial tubes are extensively disrupted, fibrosis will interfere with regeneration. Bleeding into muscle, whether caused by trauma or spontaneous hemorrhage, as in some hemorrhagic diatheses, is often sufficient in volume to result in hematoma, the fate of which will depend largely on its volume. Metaplastic bone may form in the capsule of the hematoma.

Bibliography

Bindseil, E. Eosinophilic leucocyte infiltration in ruptured necrotic thigh muscle of downer cows. *Vet Rec* **120:** 183–184, 1987.

Cox, V. S., McGrath, C. J., and Jorgensen, S. E. The role of pressure damage in pathogenesis of the downer cow syndrome. *Am J Vet Res* **43:** 26–31, 1982.

Friend, S. C. E. Postanesthetic myonecrosis in horses. *Can Vet J* **22:** 367–371, 1981.

Ghins, E., Colson-Van Schoor, M., and Marechal, G. Implantation of autologous cells in minced and devitalized rat skeletal muscles. *J Muscle Res Cell Motil* **7:** 151–159, 1986.

Grandy, J. L. *et al.* Arterial hypotension and the development of postanesthetic myopathy in halothane-anesthetized horses. *Am J Vet Res* **48:** 192–197, 1987.

Lindsay, W. A., McDonell, W., and Bignell, W. Equine postanesthetic forelimb lameness: Intracompartmental muscle pressure changes and biochemical patterns. *Am J Vet Res* **41:** 1919–1924, 1980.

Lindsay, W. A. *et al.* Induction of equine postanesthetic myositis after halothane-induced hypotension. *Am J Vet Res* **50:** 404–410, 1989.

Martindale, L., Siller, W. G., and Wight, P. A. L. Effects of subfascial pressure in experimental deep pectoral myopathy of the fowl: An angiographic study. *Avian Pathol* **8:** 425–436, 1979.

Mubarak, S., and Owen, C. A. Compartmental syndrome and its relation to the crush syndrome: A spectrum of disease. *Clin Orthop* **113:** 81–89, 1975.

Norman, W. M. *et al.* Postanesthetic compartmental syndrome in a horse. *J Am Vet Med Assoc* **195:** 502–504, 1989.

Rorabeck, C. H., and McGee, H. M. J. Acute compartment syndromes. *Vet Comp Orthop Traumatol* **3:** 117–122, 1990.

Steffey, E. P., Woliner, M. J., and Dunlop, C. Effects of five hours of constant 1.2 MAC halothane in sternally recumbent, spontaneously breathing horses. *Equine Vet J* **22:** 433–436, 1990.

Tirgari, M. Ventral hernia in the sheep. *Vet Rec* **106:** 7–9, 1979.

VII. Nutritional Myopathy

Nutritional myopathies (nutritional myodegeneration, nutritional muscular dystrophy, white muscle disease) are principally diseases of pigs and herbivores, and only in exceptional circumstances do they affect humans or other primates. They infrequently affect carnivores and irregularly affect herbivorous and some omnivorous zoo animals. The nutritional deficiencies involved in this form of muscle disease are principally those of selenium and vitamin E. Toxicants and other environmental factors and perhaps some other deficiencies may, at times, contribute to the muscle lesions historically associated with selenium/vitamin E deficiency. Muscle fiber degeneration is discussed in Section IV, Degeneration of Muscle; it is a selective, segmental degeneration of contractile components of the muscle cell, which leaves intact the ensheathing structures and satellite cells, and therefore enables a rapid and efficient regenerative repair to take place.

Nutritional myopathy is a problem around the world, but it occurs most often in those countries with well-developed livestock agriculture. Earlier attempts to link the disease primarily with certain soil types have lost validity: the disease occurs on soils as varied as volcanic, deep sedimentary, calcareous, and basaltic, particularly in North America, Europe, and Australasia. Those soils which produce pasture grasses and legumes low in selenium content are capable of predisposing grazing animals to deficiency. Yet the disease is not confined to regions with selenium-deficient soils, and diseased animals with low selenium levels in blood and tissues are found on farms where plant selenium levels appear to be adequate. Furthermore, animals with low tissue levels of both selenium and vitamin E may not develop lesions; conversely, addition of selenium and vitamin E to the ration may not protect all animals from disease. Most cases or enzootics of nutritional myopathy are circumstantially linked with either selenium deficiency or vitamin E deficiency, or both, and although the patterns of deficiency and lesions are extremely variable, there can be little question about the primary cause of the muscle disease being a deficiency.

Of the domestic mammals, pigs, cattle, and sheep are most susceptible. Horses and goats are moderately susceptible, and occasional cases have been reported in dogs and cats. Most zoo ungulates should be regarded as susceptible to the disease. Historically, nutritional myopathy has been thought of as a disease of young animals with emphasis on the very young. Rapid postnatal growth seems to predispose to disease, a problem perhaps of outgrowing a scarce resource or of biochemical transition as fiber types adjust to adult patterns. Although nutritional muscle degeneration does occur in mature or even old animals, it is predominantly a disease of the young.

In cattle, spontaneous myopathy occurs *in utero* in 7-month-old fetuses, and muscle lesions are seen in lambs and calves at birth. Lesions may not occur in calves or lambs born of cows or ewes which themselves have extensive lesions of nutritional myopathy prior to, or at the time of, parturition. It is equally true that dams of calves or lambs with extensive lesions seldom show clinical disease or even clinicopathological evidence of muscle fiber breakdown.

Nutritional myopathy occurs in all of the susceptible domestic species on widely variable planes of nutrition. Neonatal disease usually affects the thrifty, well-grown suckling animal, and the disease in yearlings and adults usually occurs in animals in good physical condition. The adult disease affects animals on marginal-quality rations, such as turnips or poor-quality hay, and can appear as clinical or subclinical disease in animals in very poor condition because of neglect or chronic disease. Animals with nutritional myopathy often lose condition rapidly and appear to be very unthrifty, even when the muscle disease is not clinically severe.

One of the most perplexing aspects of these myopathies is the irregularity and unpredictability of their occurrence. Natural disease is seldom a serious problem in consecutive years, yet some cases will occur in most years in any given region. A good deal of correlative and circumstantial evidence indicates that climate-related conditions, such as the length and the amount of sunshine of the growing season, and the length of the housing season, may be very important. Since the disease often occurs while animals are consuming stored feeds or shortly thereafter, the condition and duration of storage of the fodder material can be relevant. Detailed investigation sometimes reveals comparable levels of vitamin E and selenium in forage from one year to the next, yet the incidence of nutritional myopathy in animals consuming it may be quite different. Dry pastures or fields may be associated with an increase in the level of disease and may also have an influence later through the stored hay or grain harvested from them. On the other hand, lush pasture may also cause problems. Some pasture plants have moderately high levels in their leaves of polyunsaturated fatty acids, which are absorbed largely intact in herbivores. They are almost certainly antagonistic to selenium and vitamin E; thus, the requirements for these nutrients may be raised considerably when animals first graze new pasture. In most parts of the world, nutritional myopathy occurs in late winter or early spring, but in sheep it may occur more often in the fall in both pastured and feedlot animals that are immature.

The patterns of nutritional myopathy seem, in a general way, to obey the rules of straightforward deficiency of one or two essential nutrients. The metabolism of vitamin E and selenium is incompletely understood. Work on membrane integrity and membrane alterations in disease has improved understanding of the subcellular changes which seem to be a basic result of deficiency of these substances.

Vitamin E and selenium-containing enzymes are required in many cells as physiologic antagonists to a group of chemically varied substances known as free radicals. Free radicals are molecules with an odd number of electrons; they can be either organic or inorganic. Some free radicals are products of normal cell function, and several participate in, or are products of, oxidative metabolism. They may also be produced outside the cell as products of tissue radiation, drug reactions, and inflammation. One of the major sources of free radicals is the cell detoxification process, which renders materials less harmful by converting them to epoxides. Many intracellular and extracellular free radicals contain oxygen, and are involved in electron transfer reactions. They are highly reactive, and this is responsible for their rapid alteration (instability), which occurs in oxidation–reduction reactions within a wide range of cellular structures and enzyme systems.

Free radicals may initiate cellular injury by causing peroxidation of membrane lipids and by causing physicochemical damage to protein molecules, including those of mitochondria, endoplasmic reticulum, and cytosol. Protection against the effects of free radicals is provided partly by the constant presence of small scavenger molecules like tocopherols, ascorbate, and beta-carotene, which quench free radicals should they accumulate; both radical and scavenger are consumed in the process. Protection is also provided in part by selenium-containing enzymes of the glutathione peroxidase/glutathione reductase system. This system is capable, under normal circumstances, of more or less constant renewal by a complex sequence which makes use of several enzymes, although some consumption of the selenium-containing component does occur.

From the preceding, certain conclusions seem to emerge. Although vitamin E and the selenium-containing glutathione system perform many similar functions at the cellular level in the quenching of destructive metabolites and by-products, they may function independently. The circumstantial clinical evidence that more of one can relieve the need for the other in the prevention of muscle disease is also reasonably explained. The need for some of both at all times within the cell, and vitamin E outside the cell, perhaps is explained by the fact that the two mechanisms quench a different array of free radicals and that tocopherol operates both outside and inside the cell, whereas the glutathione system operates only inside the cell. The practical interpretation of deficiency of vitamin E or selenium should relate to the consumption of these elements during a steady intracellular production of free radicals, rather than an interpretation of these nutrients as structural cellular components, which may be deficient.

In the absence of sufficient protection, cellular membranes are modified by free radicals, and the ability of those membranes to maintain essential differential gradients for one or more ions is diminished or lost. The inward flow of calcium ions from the extracellular compartment to the cytosol, where calcium levels are one quarter to one half as high, causes a greatly increased demand for energy to move calcium away from the calcium-sensitive myofilaments and into mitochondria, which act as sumps. As a result, mitochondria may accumulate up to 50 times the normal amount of calcium, and there is depletion of the energy system. This initiates the sequence of events identified earlier (see Section IV, Degeneration of Muscle) as mitochondrial calcium overload, which goes on to calcium-induced hypercontraction of myofibrils and degeneration of myofibers.

If these explanations of intracellular dynamics and fiber degeneration are sustained, they would indicate that segmental muscle fiber degeneration with mineralization is not a specific lesion. Any event which can trigger the cascade of degenerative events in the muscle fiber could produce a similar lesion. Starvation, or poor-quality rations, or rations in which vitamin E and/or selenium have been destroyed, could produce muscle lesions simply by not providing enough protective scavenger molecules. Metallic toxicants or other chemicals, including therapeutic agents, might produce disease by binding selenium and inducing a higher than normal demand for antioxidant molecules such as tocopherols or glutathione. The relationship between unaccustomed exercise or cold weather and nutritional myopathy, which is frequently observed in all ages and all susceptible species, may be explained by an energy depletion at the intracellular level rather than by membrane alteration, especially in those instances of nutritional myopathy in which levels of tissue vitamin E and selenium are near normal. Excepting those cases associated with driving, transport, or other forced exercise, serum and tissue levels of vitamin E and selenium generally correlate well with levels in the feed of the animals concerned. This by itself seems to indicate that the naturally occurring disease is a true deficiency.

Once cell membranes have been altered, the stage has been set for leakage of intracellular enzymes. The level of the muscle enzymes in blood is a rough indication of the extent of muscle fiber destruction. Suggestions have been made that elevation of enzymes in plasma may accompany athletic use of muscles and indicate physiologic myolysis of muscle fibers. When monitored by muscle biopsies in animals and humans, however, such enzyme elevation is associated with degeneration of fibers or at least of some fiber segments.

Creatine phosphokinase in plasma is the enzyme which seems to indicate muscle damage most specifically although, in animals, glutamic oxalic transaminase is quite specific for muscle as well. It tends to rise later and disappear much later than creatine phosphokinase and therefore, its evaluation is most useful in establishing the time and duration of the muscular injury. Serum glutathione peroxidase determinations are reliable indicators of selenium availability. Tissue levels of selenium may not reflect availability if selenium is stored in complexed form. Mineral supplements often contain a variety of metals as contaminants. Those that have a myopathy-inducing effect experimentally include iron, silver, cobalt, copper, zinc, and cadmium. The complicity of such metals in the production of natural disease is undetermined but worthy of examination.

For each of the domestic species, there are common patterns of nutritional myopathy that include disease in mature animals. Recognition of myopathy in the older group may be inadequate because mineralization of degenerate muscle in older animals may be scant, and degeneration may therefore not be appreciated grossly. Also, the presenting signs of illness may not appear to have a muscular origin, and subtle signs of deficiency may be evident only as unthriftiness or by altered levels of serum enzymes. In some herds the incidence of elevated serum enzyme levels corresponds to a low level of muscle fiber degeneration and repair. Only a very small proportion of such animals develop extensive myopathic lesions.

A. Cattle

Nutritional myopathy occurs, sometimes in endemic proportions, in calves, mostly of beef type and 4–6 weeks old. It is also common in animals up to 6 months of age and occurs regularly in older cattle. In calves there is often a typical history indicating that the dam has been housed at least 3–4 months and had been fed poor-quality hay or not enough hay. Similar problems occur in calves in the United States of America, Australia, and New Zealand when legume hay alone or irrigated legume pasture is fed. Sulfur fertilizers for pasture, and copper deficiency in the dam, may be contributing factors. The precipitating event in many cases is physical activity, which converts subclinical to clinical disease. The feeding of cod liver oil which has become rancid has been blamed for producing the disease. The presenting sign in calves is often stiffness or dyspnea, but a shuffling gait, a dropping of the chest between the shoulders, and outward rotation of the forelimbs have also been noted. Some calves become recumbent and die rapidly with signs of respiratory failure. Calves older than 3 to 4 months may show myoglobinuria.

Postmortem lesions in calves are usually dominated by marked mineralization of skeletal and/or cardiac muscle. When the heart is extensively affected, intercostal muscles and the diaphragm are usually also affected, but other skeletal muscle lesions may not be widespread. Heart lesions in calves usually involve the left ventricle more than the right. Small lesions just under the epicardium or endocardium may be scattered as white brush strokes. The mineralized lesions are creamy white and opaque. Small streaks of hemorrhage may also be seen. Lungs are often filled with pink frothy fluid, and an excess of fluid may be present in the thorax. Acute pneumonic changes sometimes develop as a complication of pulmonary edema.

In those cases in which skeletal lesions predominate, the most extensive lesions can be found in the large muscles of the thigh and shoulder, but many others are affected, and the lesions are bilaterally symmetrical. Young animals often have extensive lesions in the tongue and in neck muscles, and occasionally in the voluntary muscles of rectum, urethra, and esophagus. Affected muscles are pale, irregularly opaque, and yellow to creamy white. The longitudinal divisions of muscle (primary bundles of 40 to 200 fibers) often have an indistinct banded appearance in most severely affected areas, and sometimes streaks of hemorrhage and a moderate local edema are present. Minimal myocardial lesions are sometimes present, but not accompanied by evidence of cardiac failure.

In older cattle, the patterns of disease vary considerably and are unrelated to age. Dairy and beef calves 6–12 months of age show stiffness and lethargy just after winter housing, or sometimes while they are housed. The disease is often related to poor-quality feed. In similar circumstances extensive myopathy occurs in pregnant heifers, and they may suffer a high incidence of abortion, stillbirth, and parturient recumbency. Feedlot steers fed high-moisture corn stored for 6 to 8 months show initial signs of diarrhea and unthriftiness and become recumbent for 2 or 3 weeks. Many show lesions in muscles at slaughter. Deficiency of vitamin E or selenium or both may be unintentionally induced by substituting urea for dietary protein, by the treatment of grain with propionic acid, and by the feeding of a high-carbohydrate diet or sugar cane waste with molasses. Pregnant cows may retain the placenta, and most mature animals with extensive lesions show myoglobinuria.

Postmortem lesions in yearling or mature animals may be difficult to detect because mineralization is slight. Muscles are usually pale, but this is not distinctive. Parts of muscles may be abnormally flabby and pale or wet, and rigor mortis of affected muscles is usually incomplete. Cardiac lesions are rarely present either grossly or microscopically, but on occasion, the Purkinje fibers may be selectively degenerate and mineralized.

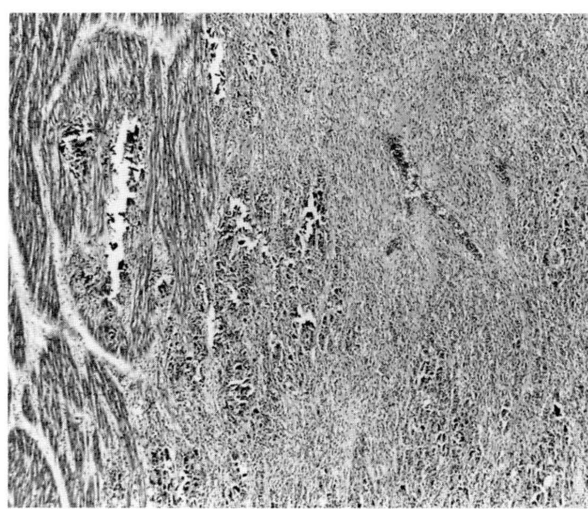

Fig. 2.41 Chronic nutritional myopathy. Fibrosis and mineralization. Yearling calf.

Histologic lesions of nutritional myopathy are varied only slightly by age. The earliest lesion visible by light microscopy is hypercontraction, recognizable in fiber segments with tightly contracted sarcomeres or no striations. In cross section, hypercontracted fibers are large, round, and stain strongly with eosin. This change is usually described as hyaline, implying an amorphous structure, but striations may be visible for 3 to 4 days when examined by electron microscopy. Eventually these hypercontracted fiber segments will be removed but not as quickly as other fibers which undergo fragmentation with or without mineralization. In both types of degeneration, the processes of degeneration and repair are essentially those described earlier, but allowance needs to be made in histologic examination for the incremental changes of chronic deficiency (Fig. 2.41).

By electron microscopy, the earliest detectable change is degeneration of mitochondria, and this is followed by loss of some parts of the sarcomere and then disintegration of the tubular systems. By histochemical examination it can be determined that type 1 fibers in the interior of primary bundles degenerate preferentially but not exclusively. Apart from this preference the pattern of degeneration appears to be random. This random pattern is helpful in distinguishing this from other muscle diseases such as azoturia or ischemic degeneration.

Changes in myocardium are very similar to those seen in skeletal muscle. Mineralization of fibers is often pronounced, and the coagulated, mineralized myofibrils appear to be rapidly removed by macrophages. Hyaline change in degenerate segments of myocardium rarely occurs, and this may be related to the much shorter length of the individual myocardial fibers. Degenerate myocardial fibers are not repaired; they are replaced by condensed stroma.

B. Sheep and Goats

Nutritional myopathy in **sheep** is probably more prevalent in more areas of the world than it is in cattle. The names white muscle disease, rigid lamb disease, and stiff lamb disease were coined to describe the most frequently encountered patterns in 2-to-4-week-old lambs, very often spring lambs recently turned out onto the first green pasture. Congenital nutritional myopathy does occur in lambs, but not often. The disease may occur in an outbreak among lambs from 1 day to 2 months of age or beyond. Mortality at this stage may be very low or may reach 50%. The next peak of incidence occurs at 4 to 8 months of age as weaned lambs are put onto aftermath pastures or into feedlots. Mortality is not usually very high, but the incidence of minimal clinical disease may be moderately high, and that of subclinical disease may be higher. Beyond these age groups, nutritional myopathy in more mature sheep is clinically apparent sporadically, but subclinical disease may involve from 5 to 30% cent of a group. Under special circumstances the incidence of clinical disease and mortality may rise dramatically. Thus, in various parts of the world, the disease has been precipitated by bad weather, prolonged winter feeding, subsistence on root crops, or forced activity, as well as feeding on stubble, legume pastures, dry pastures, and pastures with too much copper. Outbreaks in lambs and yearlings have also occurred on pastures on which copper had been made unavailable by topdressing with molybdenum.

Deficiency in sheep is attributed to deficiency of vitamin E or selenium, but seldom of both. In certain regions of the world, one nutrient does not seem to be able to replace the other to any extent, while in others, the addition of either vitamin E or selenium is rapidly curative.

Lesions and their corresponding clinical signs are as varied as the circumstances under which myopathy occurs. The lesions may be detectable in lamb fetuses at least 2 weeks prior to parturition but may not be present at birth. In the congenital disease, tongue and neck muscles used in suckling movements often contain the most advanced lesions (Fig. 2.42A). When the lesions occur in lambs a few days older, they are likely to be much more extensive and involve primarily the major muscles of the shoulder and thigh but also back, neck, and respiratory muscles. It has been suggested that this young age group is most susceptible because the fiber type is in transition in some muscles. The gross appearance of affected muscles is similar to that described for calves, although the likelihood of muscle mineralization is probably greater. Lambs frequently have pale muscles; consequently, the recognition of the calcified flecking in primary bundles is almost essential if diagnosis is to be made grossly with any confidence. It is necessary to monitor gross appearance by histologic examination.

In older sheep, lesions are more varied in distribution, location, and extent, but some similarity to the distribution in young animals may exist. For example, bilaterally symmetrical lesions of the thigh muscles may occur, but they may predominate in or be confined to the intermediate head of the triceps, or the tensor fascia lata. Lesions in pregnant ewes may be more or less confined to the abdominal muscles, which may rupture, allowing viscera to herniate.

Microscopically, the changes seen in affected muscle over a period of several days follow the expected sequence of calcification, phagocytosis, satellite cell proliferation, and myoblastic regenerative repair (Fig. 2.42B). All of this can be completed within about 16 to 18 days, leaving little in the way of residual lesions, but often repair lesions of several days' evolution can be seen side by side with more recent ones. Repair can take place in the face of continuing deficiency, but even normal activity during the early repair phase is likely to lead to more lesions.

Electron microscopic examination of early lesions in sheep indicates that myofilamentary changes occur prior to changes in mitochondria and other subcellular elements. There may be some genuine differences between species in this respect, since mitochondrial, not myofibrillar, changes have been observed as the first changes in pigs, rabbits, and rats under experimental conditions.

Reports of nutritional myopathy in **goats** are relatively few, but this may be due to the fact that the goat is less often part of intensive animal agriculture. Caprine nutritional myopathy has been observed in Europe, the Middle East, New Zealand, Australia, and in North America. In most instances it appears as a pasture disease with clinical and pathological changes similar to those seen in sheep.

C. Pigs

Nutritional myopathy in swine has been reported as a spontaneous disease wherever intensive pig rearing is practiced, but particularly in northern, central, and eastern continental Europe, Britain, the United States, and Canada. Classical lesions of skeletal muscles are less common than some of the other expressions of vitamin E/selenium deficiency, such as hepatic necrosis and mulberry heart disease, but systematic microscopic study of muscles has revealed a much higher incidence than was thought to exist. Since the pig is an easily managed experimental animal, much more is known about experimental deficiency in this species than in others.

Naturally occurring nutritional myopathy is not known to be congenital in pigs but piglets as young as 1 day may have extensive lesions causing paresis or paralysis. Growing pigs of all sizes up to 65 kg may be affected in outbreaks, but the most commonly affected are weaned pigs 6–20 weeks of age. A special circumstance is that in which the injection of iron-containing products precipitates an acute widespread degenerative and usually fatal myopathy in pigs a few days old (Fig. 2.43). It is believed that iron acts as a catalyst for lipid peroxidation of cell membranes. Older pigs, and particularly sows which have recently farrowed, are occasionally susceptible to widespread muscle disease of sufficient severity to cause pros-

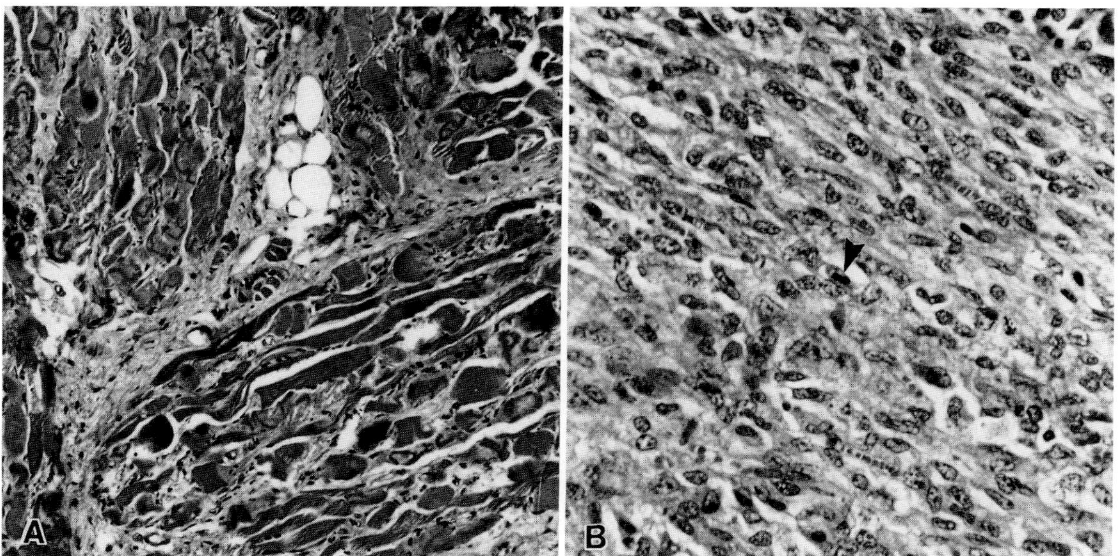

Fig. 2.42. Neonatal nutritional myopathy. Lamb. (A) Acute fragmentation, tongue muscle. (B) Repair stage with satellite cell proliferation and myoblastic regeneration. Mitotic figure (arrowhead).

tration, although more often clinical signs, if present at all, consist of lethargy and slowness of movement.

Apart from deficiencies in diet, the triggering factors in pigs seem to be relatively few. Unaccustomed exercise does not seem to play a part, but rancid or oxidized fish liver oils may do so. New grain sometimes seems to contain a myopathy-inducing factor, and pigs fed quantities of peas (*Pisum sativum*) may be particularly prone to the disease. Metals occurring as contaminants in ground mineral mixes can induce an increased requirement for vitamin E and/or selenium. Silver, copper, cadmium, cobalt, vanadium, tellurium, and zinc, and possibly other metals, in some way bind selenium or prevent its participation in

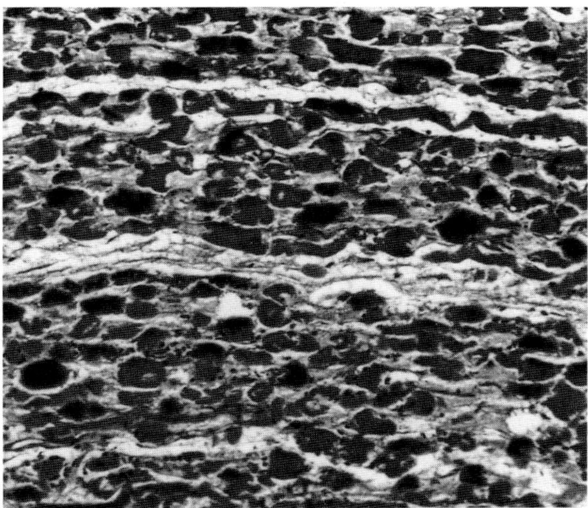

Fig. 2.43 Extensive fragmentation of iron-induced myopathy. Piglet.

protective activities. This leads to a twofold or greater demand for selenium, which can be partly alleviated by vitamin E supplementation.

Diagnosis of skeletal muscle disease in pigs, based on gross appearance only, is difficult. Mineralization of degenerate fibers is often not abundant and even when it is present and visible, the pale muscles of the pig make recognition difficult. This, and the fact that lesions in heart and liver are much more dramatic, may explain why there are relatively few reports of skeletal lesions in natural outbreaks of vitamin E/selenium deficiency disease. In the experimental disease in pigs, many skeletal lesions are microscopic, while the liver lesions can be seen grossly.

Microscopic muscle lesions are similar to those seen in lambs and calves. Type 1 fibers are principally affected, and, in view of the orderly arrangement of fibers in the pig, this leads to a distinctly central degenerative involvement of primary muscle bundles. Regeneration is often rapid and complete in surviving piglets, but the mortality rate may be as high as 80% of a litter, and usually 100% when related to iron injection. The survival rate for older pigs is higher, except for those instances in which the cardiac syndrome is prominent. Relatively few pigs survive mulberry heart disease.

Myocardial degeneration (mulberry heart disease) is the most common manifestation of vitamin E/selenium deficiency in growing weaned pigs 6–20 wk of age. It is described in Volume 3, Chapter 1, The Cardiovascular System.

The liver lesions associated with vitamin E/selenium deficiency are referred to as **toxic liver dystrophy** or **hepatosis dietetica** and are described in Volume 2, Chapter 2, The Liver and Biliary System.

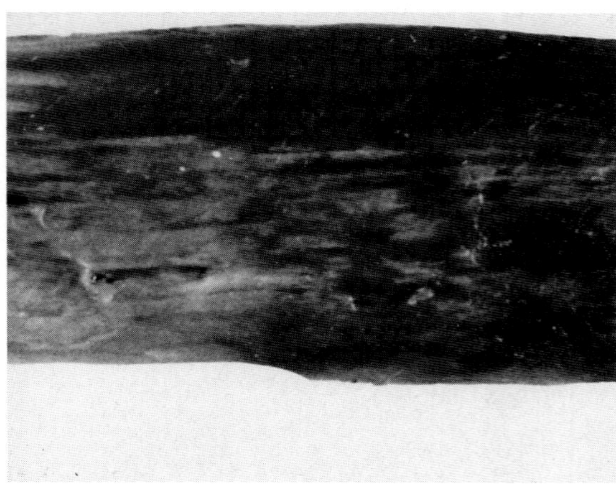

Fig. 2.44 Pallid areas of degeneration in sternocephalic muscle in myopathy. Horse.

D. Horses

A disease in foals which is, in most respects, comparable to vitamin E/selenium deficiency myopathies in other species, has been recognized for many years. The usual age range is 1 day to 12 weeks, and it may be present at birth. Cases have been reported from North America, Europe, and Australasia in most breeds of light horses and in pony, zebra, and donkey foals. The similarities of lesions and circumstances between the equine syndrome and nutritional myopathy of ruminants has led to the conclusion that it reflects a deficiency of vitamin E and/or selenium, but the disease has not been studied extensively. Selenium and glutathione peroxidase levels in affected foals are often no lower than those in healthy foals, however, and selenium and vitamin E appear to have little or no curative value, and in a few instances, little or no preventive value. Creatine phosphokinase values may be very high; over 2 million IU/liter in nonsurviving cases, and as high as 200,000 in surviving animals.

Postmortem lesions of nutritional myopathy in foals are similar in many respects to those in calves and lambs. Shoulder, neck, and thigh muscles may be bilaterally and extensively involved, and this usually accounts for inability to rise or to assume certain postures, such as the one for suckling. Involvement of digital flexors and extensors may be more frequent than in calves, but typical lesions are the chalky opaque flecking of muscle primary bundles. In a few foals and sometimes in poorly nourished stabled horses, there is involvement of the masticatory and tongue muscles. This lesion has been called **masticatory myositis** in foals and **polymyositis** in older animals, but the syndromes seem to be part of nutritional myopathy on clinical and pathological grounds (Fig. 2.44).

The distinction between true nutritional myopathy and other diseases is more difficult for older horses than it is for foals because circumstances and lesions are not so distinctive and because information is fragmentary. Nutritional myopathy occurs in animals ranging in maturity from less than 1 year to middle age. The clinical signs may be nonspecific and may be mistaken for evidence of colic, cardiac failure, wry neck, or general depression, but the most common features are stiffness and muscular weakness. Lesions are seldom widespread enough to cause severe crippling, although lethargy may be the result of skeletal muscle degeneration. A willingness to eat is retained in downer animals. The most convincing indications of nutritional myopathy are the myocardial lesions, which are similar to those seen in other species. They seem to separate the syndrome from the exertional and ischemic myopathies. Steatitis is apparently not seen in older horses with nutritional myopathy, but it is common in foals and may, in the healing stage, lead to lumpiness in subcutaneous adipose tissue.

As is the case with foals, the vitamin E and/or selenium status of horses with lesions is sometimes difficult to reconcile with information in other species. Low levels of blood selenium or glutathione peroxidase are not always present in horses with lesions. Tocopherol levels and glutathione peroxidase levels do not seem to be related in individual animals, and selenium and vitamin E may be curative or, in some cases, preventive. There is, however, indirect evidence of a need for one or both of the nutrients for optimal muscle function.

Histologic lesions in both foals and older horses are similar to those seen in calves or pigs. The repair process in foals is rapid and usually complete in 2 weeks if the foal survives, but some lasting retardation of growth may result. Older animals may have the same capability for regenerative repair, especially in subclinical disease, but it is expected that age-related limitations of the repair process may leave residual scarification if the degeneration has been severe. Residual scar tissue may be responsible for limitations of gait or deviation of the neck.

E. Other Species

Nutritional myopathy is unusual in primates and carnivores. Clinical, morphological, and therapeutic evidence suggests that it does occur.

Nutritional myopathy in ranch mink, usually accompanied by steatitis, has been a periodic but costly problem, which has largely disappeared with the introduction of commercial feeds to which vitamin E is added to counteract the effects of oxidized fish oils or other lipid peroxides. It is unlikely that the disease occurs in wild mink. Nutritional myopathy in dogs has been reported as a problem in New Zealand and circumstantially associated with selenium deficiency. Selenium appeared to be preventive.

A number of zoo animals appear to be susceptible to nutritional myopathy but details are scarce and implications, largely circumstantial. The Rottnest quokka, a small nocturnal wallaby, (*Setonix brachyurus*) and the nyala (*Tragelaphus angasi*) seem to be exquisitely susceptible. Problems of diagnosis arise in such populations in which capture myopathy is a complication (see Section X, Exertional Myopathies), but cases which circumstantially ap-

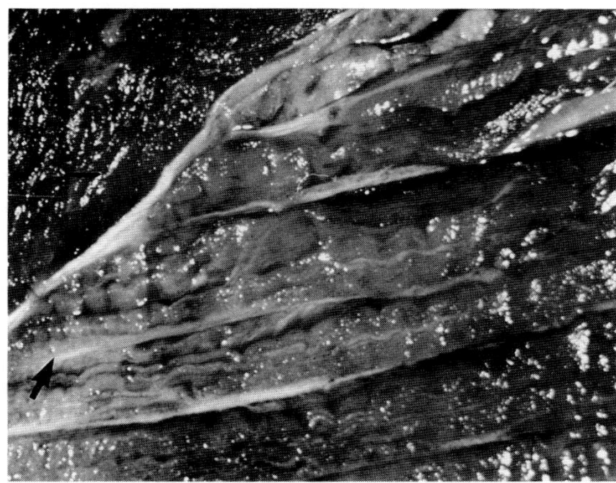

Fig. 2.45 Pallor of muscle in transport myopathy (arrow). Ox.

pear to be due to nutritional myopathy are reported in several species of gazelle and antelope in Africa, roe deer in Scotland, and white tailed deer and Rocky Mountain bighorn sheep in North America.

VIII. Transport Myopathy

This category includes a number of cases and situations involving domestic, zoo, and wild animals in which muscle disease has ostensibly been precipitated by moderate physical activity or transport stress, but in which lesions closely resemble those of nutritional myopathy. The name gives at least temporary recognition to exercise-induced or exercise-conditioned nutritional myopathy. Most of the cases in this category have already been included, by inference, in the various previous categories. Typical are the groups of feedlot cattle which show muscle stiffness after long transport (Fig. 2.45); cattle or horses which show typical lesions of nutritional myopathy, not ischemic myopathy, following anesthesia; zoo or wild animals which are tranquilized and transported in crates, yet still develop lesions of nutritional myopathy at the destination. The precipitating exercise is often extended but not severely exertional, and the lesions reflect the lack of acidotic coagulation of protein seen in azoturia. Myoglobinuria is rare, and serum muscle enzyme levels are only moderately elevated.

Bibliography

Allen, J. G. *et al.* A study of nutritional myopathy in weaner sheep. *Aust Vet J* **63:** 8–13, 1986.

Ammerman, C. B. *et al.* Contaminating elements in mineral supplements and their potential toxicity: A review. *J Anim Sci* **44:** 485–516, 1977.

Anderson, P. H. *et al.* The sequence of myodegeneration in nutritional myopathy of the older calf. *Br Vet J* **133:** 160–165, 1977.

Bergmann, V. Electron microscope findings in the skeletal musculature of sheep with enzootic muscular dystrophy. *Arch Exp Veterinaermed* **26:** 645–660, 1972.

Bradley, R. An historical account of studies on muscle. *J Comp Pathol* **99:** 290–297, 1988.

Bradley, R., and Fell, B. F. Myopathies in animals. *In* "Disorders of Voluntary Muscle," J. Walton (ed.), pp 824–872. Edinburgh and London, Churchill-Livingstone, 1981.

Bradley, R., Anderson, P. H., and Wilesmith, J. W. Changing patterns of nutritional myodegeneration (white muscle disease) in cattle and sheep in the period 1975–1985 in Great Britain. *Bov Pract* **22:** 38–45, 1987.

Butler, P., and Blackmore, D. J. Vitamin E values in the plasma of stabled thoroughbred horses in training. *Vet Rec* **112:** 60, 1983.

Clark, I. A. Tissue damage caused by free oxygen radicals. *Pathology* **18:** 181–186, 1986.

Dennis, J. M., and Alexander, R. W. Nutritional myopathy in a cat. *Vet Rec* **111:** 195–196, 1982.

Dickson, J., Hopkins, D. L., and Doncon, G. H. Muscle damage associated with injections of vitamin E in sheep. *Aust Vet J* **63:** 231–233, 1986.

Freeman, B. A., and Crapo, J. D. Biology of disease. Free radicals and tissue injury. *Lab Invest* **47:** 412–426, 1982.

Gitter, M., Bradley, R., and Pepper, R. Nutritional myodegeneration in dairy cows. *Vet Rec* **103:** 24–26, 1978.

Greig, A., and Hunter, A. R. Nutritional myopathy in feeding hogs. *Vet Rec* **107:** 62–63, 1980.

Hamliri, A. *et al.* Evaluation of biochemical evidence of congenital nutritional myopathy in two-week prepartum fetuses from selenium-deficient ewes. *Am J Vet Res* **51**: 1112–1115, 1990.

Hosie, B. D. *et al.* Acute myopathy in horses at grass in east and southeast Scotland. *Vet Rec* **119:** 444–449, 1986.

Hutchinson, L. J., Sholz, R. W., and Drake, T. R. Nutritional myodegeneration in a group of Chianina heifers. *J Am Vet Med Assoc* **181:** 581–584, 1982.

Kennedy, S., Rice, D. A., and Davidson, W. B. Experimental myopathy in vitamin E- and selenium-depleted calves with and without added dietary polyunsaturated fatty acids as a model for nutritional degenerative myopathy in ruminant cattle. *Res Vet Sci* **43:** 384–394, 1987.

Kroneman, J., and Wensvoort, P. Muscular dystrophy and yellow fat disease in Shetland pony foals. *Neth J Vet Sci* **1:** 42, 1968.

Nisbet, D. I., and Renwick, C. C. Congenital myopathy in lambs. *J Comp Pathol* **71:** 177–182, 1961.

Owen, R. *et al.* Dystrophic myodegeneration in adult horses. *J Am Vet Med Assoc* **171:** 343–349, 1977.

Patterson, D. S. P. *et al.* The toxicity of parenteral iron preparations in the rabbit and pig with a comparison of the clinical and biochemical responses to iron-dextrose in 2-day-old and 8-day-old piglets. *Zentralbl Veterinaermed* [*A*] **18:** 453, 1971.

Peet, R. L., and Dickson, J. Rhabdomyolysis in housed, fine-woolled merino sheep, associated with low plasma alpha-tocopherol concentrations. *Aust Vet J* **65:** 398–399, 1988.

Rice, D. A., and McMurray, C. H. Use of sodium-hydroxide-treated, selenium-deficient barley to induce vitamin E and selenium deficiency in yearling cattle. *Vet Rec* **118:** 173–176, 1986.

Rice, D. A. *et al.* Lack of effect of selenium supplementation on the incidence of weak calves in dairy herds. *Vet Rec* **119:** 571–573, 1986.

Roneus, B. Glutathione peroxidase and selenium in the blood of healthy horses and foals affected by muscular dystrophy. *Nord Vet Med* **34:** 350–353, 1982.

Ross, A. D. *et al.* Nutritional myopathy in goats. *Aust Vet J* **66:** 361–363, 1989.

Ruth, G. R., and Van Vleet, J. F. Experimentally induced selenium–vitamin E deficiency in growing swine: Selective destruction of type 1 skeletal muscle fibers. *Am J Vet Res* **35:** 237–244, 1974.

Schlotke, B. Muskeldystrophe bei Fohlen. *Berl Munch Tieraerztl Wochenschr* **87:** 84–86, 1974.

Smith, D. L. *et al.* A nutritional myopathy enzootic in a group of yearling beef cattle. *Can Vet J* **26:** 385–390, 1985.

Van Vleet, J. F. Comparative efficacy of five supplementation procedures to control selenium–vitamin E deficiency in swine. *Am J Vet Res* **43:** 1180–1189, 1982.

Van Vleet, J. F., Boon, D., and Ferrans, V. J. Induction of lesions of selenium–vitamin E deficiency in weanling swine fed cobalt, tellurium, zinc, cadmium, and vanadium. *Am J Vet Res* **42:** 789–799, 1981.

Wrogemann, K., and Pena, S. D. J. Mitochondrial calcium overload: A general mechanism for cell-necrosis in muscle diseases. *Lancet* **1:** 672–673, 1976.

IX. Toxic Myopathies

Several substances are capable of causing muscle fiber degeneration under experimental conditions, but few of them are available or palatable to animals. The list of phytotoxins and chemical toxins which cause naturally occurring myopathy in domestic animals is quite short, but a well-equipped veterinary hospital pharmacy is likely to contain several drugs which are myodestructive at pharmacologic doses in some species. Clinical and pathological evidence, using a twofold increase in blood creatine phosphokinase as a criterion, indicates that the following substances can produce muscle injury: amphotericin-B; diethylstilbestrol; halothane; lidocaine; D-penicillamine; prednisone; prochlorperazine; quinidine; stanozolol; suxamethonium, as well as amphetamines, barbiturates, and ethanol at higher levels.

The distinction between a toxic myopathy and a simple or conditioned nutritional myopathy is often difficult to make. For purposes of categorization, and separation of toxic from nutritional myopathy, the agents included here are those which cause segmental skeletal muscle or myocardial lesions, apparently do not act by destroying vitamin E or selenium, and are not inhibited by addition of those nutrients. The clinical signs may mimic those associated with other muscle diseases, but most of the agents also exert a persistent or generalized toxic influence. This makes them generally more lethal than nutritional deficiency and may lead to delayed effects on other tissues. The lesions associated with toxic myopathies are, in most instances grossly unremarkable, and microscopically, very difficult to differentiate from acute nutritional or exertional myopathy.

A. Gossypol Toxicity

Gossypol is a yellow, pigmented, polyphenolic substance present in cotton seeds, which is toxic to swine when cottonseed cake or meal is added to rations at a level of 10% or more. The lesion occurs after feeding of such rations for a month or more, which suggests that the toxic effects are cumulative. Gossypol is toxic to experimental lambs and calves at levels of less than 450 ppm (the level of free gossypol permitted in human and some animal foods). It produces lesions in several organs, including the heart, and death is due to cardiac failure. Affected animals are pot-bellied and poorly grown, and most die acutely. Natural outbreaks of disease in calves and lambs circumstantially link cottonseed meal with similar patterns of poor growth and sudden death. Serum enzymes are generally not significantly elevated, which seems to confirm that the heart is the only striated muscle generally affected. Other circumstantial evidence suggests, however, that in calves, skeletal muscle lesions can be locally extensive but unpredictably distributed. Both types of muscle fibers appear to be affected in a selective, segmental myopathy generally indistinguishable from other nutritional or toxic myopathies except that both types of fiber are equally involved.

B. Cassia Toxicity

In the southern United States, mature cattle and goats on pasture may ingest the beans of the senna or coffee senna (*Cassia occidentalis* or *C. obtusifolia*) late in the year after a killing frost has made the plant more palatable than normal. After eating the plant for a few days, animals develop diarrhea, show evidence of weakness, and display a swaying, stumbling gait related to the developing muscle lesions. The disease progresses rapidly, and most of the animals affected become recumbent and develop myoglobinuria and high levels of muscle enzymes in serum. Recumbent animals usually do not recover but may live for several days. The morbidity rate may reach 60%.

Postmortem lesions consist of ill-defined pallor of much of the muscle mass. Histologic changes resemble those of nutritional myopathy with minimal or no mineralization (Fig. 2.46). The destruction of muscle fibers is segmental, but sarcolemmal sheaths and muscle nuclei remain, enclosing the floccular degenerate contractile elements. Myocardium is not extensively involved, but animals dying acutely show some myocardial lesions. Apparently recovered animals have not been examined, and the specific toxin has not been characterized.

Natural outbreaks of toxicity on several pig farms occurred when grain was contaminated with *C. occidentalis* seeds. Several pigs died after a short period of reduced weight gain and a progressively wobbly, unsteady gait. Experimental reproduction of the clinical disease occurred after 40 days when as little as 1% of the diet consisted of ground seeds. Postmortem examination revealed no gross lesions, but microscopic myodegeneration of the heart and diaphragm were characterized by vacuolation and segmental hypercontraction of fibers. The lesions were unexpectedly limited considering the marked locomotor clinical signs.

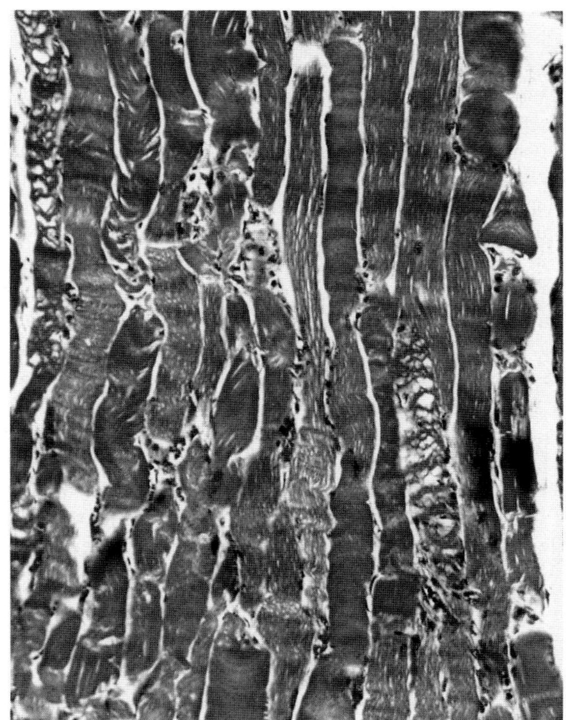

Fig. 2.46 Toxic myopathy. Myodegeneration caused by *Cassia occidentalis*. Heifer. (Courtesy of K. R. Pierce.)

C. Monensin Toxicity

Monensin is an antibiotic fermentation product of *Streptomyces cinnamonensis,* which has a growth-promoting effect in ruminants and is an efficient coccidiostat in birds and animals. Rumensin is produced commercially in very large quantities in North America and Europe, where it is added, as a concentrated premix, to pelleted or bulk feeds fed to cattle, sheep, and other ruminants, for example in zoos. Difficulties arise when monensin is fed to monogastric animals, which have a much reduced tolerance for the drug, or when human or mechanical error leads to levels of monensin that are abnormally high for the species being fed. Disease outbreaks have been recorded in horses, donkeys, mules, zebras, cattle, sheep, dogs, wallabies, camels, blesbok, Stone sheep, turkeys, and chickens. A growing number of episodes of monensin poisoning have been associated with mixing errors in packaged, pelleted, commercial animal feeds, either concentrates or final mix. Hundreds or thousands of animals have been put at risk, sometimes over wide geographic areas, each time such an event occurs. In North America alone, such mixing errors have been reported in horse feeds, cattle feeds, dog food and in connection with at least two large zoos. Some indication of susceptibility of different species is provided by the estimated 50% lethal dose (LD_{50}) for different animals. Horses and other equidae are susceptible to 2 to 3 mg per kg of body weight of monensin. Comparable figures for other species are dogs, 5–8 mg/kg; sheep and goats, 10–12 mg/kg; cattle, 20–34 mg/kg (perhaps higher); and various types of poultry, 90–200 mg/kg. Pigs, which may be given the drug for its coccidiostatic properties, or are exposed by mistake, presumably have a tolerance comparable to that of the dog. The toxic effects of monensin are potentiated by addition of tiamulin or sulphonamides in the ration for therapeutic purposes.

When a single toxic dose is fed to an animal, clinical signs of lethargy, stiffness, muscular weakness, and recumbency occur within 24 hr. Horses and other equidae are likely, in the early stages, to show marked symptoms of colic with apprehension, shifting or fidgeting, sweating, myoglobinuria, and muscle tremors. Dogs show apprehension and progressive weakness. If sublethal doses are fed, the toxic effect will be cumulative, and the clinical onset may be delayed for 2 to 3 days to weeks depending on the level, but the debility is likely to be more pronounced. Animals on low-level toxicity experiments often delay the progression of the toxic signs by greatly reducing consumption of medicated feed. Such animals frequently scour and lose weight. At dose levels capable of inducing clinical signs of toxicity in a few days, many animals show evidence of progressive cardiac failure due to a high incidence of myocardial lesions. Animals recovering from the acute disease may subsequently develop signs of progressive cardiac insufficiency, and sometimes renal failure, in addition to poor growth or poor weight gain, although the symptoms referable to skeletal muscle may disappear.

Postmortem lesions of monensin toxicity may be difficult to detect in acute cases. Skeletal muscle may lack normal rigor, and ill-defined pale streaks may be visible in both myocardium and skeletal muscle. As the clinical disease progresses, the white streaking becomes more prominent, and local atrophy of affected skeletal muscles becomes a feature. Hind-limb muscles may be the site of major degenerative changes. Lesions are only very rarely mineralized to the extent of being grossly visible.

Microscopic lesions of monensin toxicity differ very little from those of nutritional or exertional myopathy, except that mineralization of degenerate tissue usually is less marked (Fig. 2.47). One of the earliest electron-microscopically visible lesions is marked swelling and disintegration of mitochondria. Monensin is an ionophore which distorts membrane transport of sodium and potassium. This apparently leads to distortions of the electrolyte-modulated calcium gating mechanism and then mitochondrial failure, energy exhaustion, failure of calcium ion retrieval from the cytosol, and eventually myofibrillar hypercontraction and degeneration. The early mitochondrial destruction may account for the low level of calcification of the floccular muscle fragments. Both type 1 and type 2 fibers are involved.

Myofiber nuclei and satellite nuclei as well as endomysial cells apparently survive acute toxicity, and the early stages of repair are initiated during the first few days. It appears that the reparative result is not very satisfactory in spite of subjective evidence of clinical recovery in some animals. Continuing degeneration of additional muscle fi-

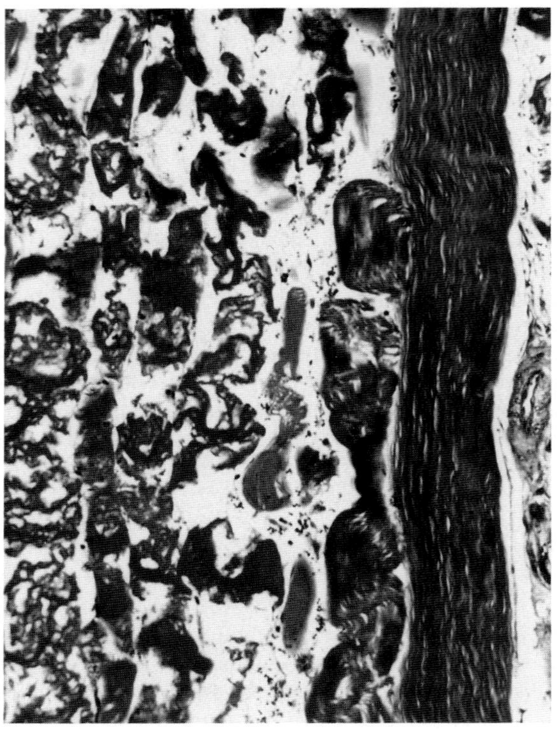

Fig. 2.47 Toxic myopathy. Monensin toxicity. Zebra.

bers, or of additional segments of fibers already damaged, seems to occur for days after monensin in tissue has been reduced to negligible levels. Muscle fiber regenerative repair ought to be uncomplicated; perhaps mitochondrial damage or altered membrane permeability have a more lasting effect in this disease than in nutritional myopathy.

Myocardial lesions in monensin toxicity are not reparable and, particularly in a growing animal, the probability of lasting cardiac malfunction is high.

D. Metallic and Nonmetallic Toxins

Various elements have been identified which, when fed to animals at appropriate levels, cause skeletal and/or myocardial degeneration. These elements fall into two groups; a metallic group including copper, cobalt, iron, silver, cadmium, zinc, and mercury, and a nonmetallic group consisting principally of selenium, tellurium, and sulfur. The intermediate stages of pathogenesis seem, in all cases, to involve the mechanisms associated with nutritional myopathy. In the case of iron, previously mentioned, the metal catalyzes lipid peroxidation of membranes and increases consumption of vitamin E and/or glutathione peroxidase. Of the elements mentioned, only selenium, iron, thallium, and perhaps sulfur and cobalt have caused natural muscle disease in animals. In some cases, metals and nonmetals form complexes to deprive cells of usable selenium or metal-containing enzymes essential for epoxide scavenging.

Bibliography

Colvin, J. M. *et al. Cassia occidentalis* toxicosis in growing pigs. *J Am Vet Med Assoc* **189:** 423–426, 1986.

Confer, A. W., Reavis, D. W., and Panciera, R. S. Light and electron microscopic changes in cardiac and skeletal muscle of sheep with experimental monensin toxicoses. *Vet Pathol* **20:** 590–602, 1983.

Dewan, M. L. *et al.* Toxic myodegeneration in goats by feeding mature fruits from the coyotillo plant (*Karwinskia humboldtiana*). *Am J Pathol* **46:** 215–226, 1965.

Henson, J. B., and Dollahite, J. W. Toxic myodegeneration in calves produced by experimental *Cassia occidentalis* intoxication. *Am J Vet Res* **27:** 947–949, 1966.

Henson, J. B. *et al.* Myodegeneration in cattle grazing cassia species. *J Am Vet Med Assoc* **147:** 142–145, 1965.

Holmberg, C. A. *et al.* Pathological and toxicological studies of calves fed a high-concentration cotton seed meal diet. *Vet Pathol* **25:** 147–153, 1988.

Hudson, L. M., Kerr, L. A., and Maslin, W. R. Gossypol toxicosis in a herd of beef calves. *J Am Vet Med Assoc* **192:** 1303–1305, 1988.

Jones, L. A. Gossypol toxicosis. *J Am Vet Med Assoc* **193:** 292–293, 1988.

Lott, J. A., and Landesman, P. W. The enzymology of skeletal muscle disorders. *CRC Crit Rev Clin Lab Sci* **20:** 153–172, 1989.

Mercer, H. D. *et al. Cassia occidentalis* toxicosis in cattle. *J Am Vet Med Assoc* **151:** 735–741, 1967.

Miller, R. E. *et al.* Acute monensin toxicosis in Stone sheep (*Ovis dalli stonei*), blesbok (*Damaliscus dorcus phillipsi*), and a Bactrian camel (*Camelus bactrianus*). *J Am Vet Med Assoc* **196:** 131–134, 1990.

Mollenhauer, H. H. *et al.* Ultrastructural observations in ponies after treatment with monensin. *Am J Vet Res* **42:** 35–40, 1981.

Morgan, S. *et al.* Clinical, clinicopathologic, pathologic, and toxicologic alterations associated with gossypol toxicosis in feeder lambs. *Am J Vet Res* **49:** 493–499, 1988.

Muylle, E. *et al.* Delayed monensin sodium toxicity in horses. *Equine Vet J* **13:** 107–108, 1981.

Nation, P. N., Crowe, S. P., and Harries, W. N. Clinical signs and pathology of accidental monensin poisoning in sheep. *Can Vet J* **23:** 323–326, 1982.

Rogers, P. A. M., Henaghan, T. P., and Wheeler, B. Gossypol poisoning in young calves. *Irish Vet J* **29:** 9–13, 1975.

Shortridge, E. H., O'Hara, P. J., and Marshall, P. M. Acute selenium poisoning in cattle. *N Z Vet J* **19:** 47–50, 1971.

Umemura, T. *et al.* Enhanced myotoxicity and involvement of both type I and II fibers in monensin-tiamulin toxicosis in pigs. *Vet Pathol* **22:** 409–414, 1985.

Van Vleet, J. F., Meyer, K. B., and Olander, H. J. Acute selenium toxicosis induced in baby pigs by parenteral administration of selenium–vitamin E preparations. *J Am Vet Med Assoc* **165:** 543–547, 1974.

Van Vleet, J. F. *et al.* Clinical, clinicopathologic and pathologic alterations in acute monensin toxicosis in cattle. *Am J Vet Res* **44:** 2133–2144, 1983.

Van Vleet, J. F., Boon, D. G., and Ferrans, V. J. Induction of lesions of selenium–vitamin E deficiency in weanling swine fed silver, cobalt, tellurium, zinc, cadmium, and vanadium. *Am J Vet Res* **42:** 789–799, 1981.

Van Vleet, J. F. *et al.* Monensin toxicosis in swine: Potentiation by tiamulin administration and ameliorative effect of treatment with selenium and/or vitamin E. *Am J Vet Res* **48:** 1520–1524, 1987.

Wilson, J. S. Toxic myopathy in a dog associated with the presence of monensin in dry food. *Can Vet J* **21:** 30–31, 1980.

X. Exertional Myopathies

The exertional myopathies include a group of diseases in which the acute lesions resemble those of acute nutritional myopathy but for which a different pathogenetic mechanism is confirmed or presumed. The initiating factor for these diseases is intensive or exhaustive activity of the major muscle masses. The initial changes take place, or are presumed to take place, in the strongly glycolytic fibers (type 2 or modified type 1 glycolytic fibers) in contrast to the primary involvement of type 1 oxidative fibers in the nutritional myopathies. It is postulated that glycogen is rapidly or aberrantly utilized and that this "burn off" generates local heat and lactate, which becomes lactic acid. The timing and sequence of events are not clear. These two products concentrated locally modify contractile proteins, reducing them to a coagulum, and local diffusion of lactate and heat leads to similar degenerative changes in adjacent muscle fibers of all types. This sequence of events greatly reduces the capacity of muscle protein to hold water; consequently, water is lost to the interstitial compartment where it raises local pressure and predisposes to ischemia, particularly if moderate muscle movement is continued (see Section V,A, Circulatory Disturbances of Muscle). For some of these diseases, particularly those in the horse and in zoo and wild animals, pronounced myoglobinemia and myoglobinuria are common. For others, particularly azoturia and capture myopathy, metabolic acidosis is likely to occur and may be fatal.

A myoglobinuric disease of pasture horses in Britain with some characteristics of azoturia/tying-up has other features which suggest that extensive myolysis may be triggered by an as yet unknown factor.

A. Azoturia

Azoturia in horses (paralytic myoglobinuria, Monday morning disease, sacral paralysis) has been known for more than a century wherever horses were used as draft animals. It is now seen less often in draft horses, but it continues to occur in working horses of other kinds and in show animals. The syndrome is seen in pleasure horses, in animals in training, and in well-trained animals participating in endurance competitions, as well as in work horses and mules in the traditional work-after-rest situation.

Paralytic myoglobinuria typically occurs very shortly after beginning work or training following some days of rest on full working rations. The carbohydrate content of the diet, once thought to be important, now seems secondary to quantity of the diet, and it is pertinent that the syndrome has been seen in horses on pasture. Weakness of the hind limbs occurs suddenly, and the animal soon becomes unable or very reluctant to move, and this is accompanied by sweating and generalized tremors. The affected muscles, which are typically those of the gluteal, femoral, and lumbar groups, are swollen and boardlike in their rigidity. Myoglobinuria appears surprisingly early in the disease, and the urine becomes dark red-brown. Severely affected horses become recumbent, a sign which is often a prelude to death from myoglobinuric nephrosis or problems associated with being down and attempting to rise. Considerable variation occurs between cases as to the nature and duration of the initiating exercise and severity of clinical signs. Recovery from mild attacks in quiet animals may occur in a few hours, but if an animal continues to struggle or go down and struggle, the recovery period is extended into days or may end in death. Atrophy of the gluteal muscles may be a feature of recovery in moderately severe cases.

Grossly visible changes in muscle are most obvious in the gluteal and lumbar regions, but lesions often are widespread. Muscles are moist, swollen, and dark, but streaks of pallor may be visible in the more extensively involved ones. If ischemic complications occur, the muscles also may show blotchy or linear hemorrhage. In animals which have survived for 2 to 3 days, muscles may become paler, and although edema may surround larger muscle divisions, the locally damaged areas appear dry compared to normal muscle.

Histologically, the lesions are characterized by hyaline change in segments of fibers, with little or no inflammatory response in the early stages. Calcification is not a feature. Where single fibers are affected, they are of type 2. In more severely affected foci, degeneration involves all fibers indiscriminately. The focal lesions are not confined by divisions between primary bundles, and large groups of degenerate fibers are interspersed with normal fibers and smaller clusters of affected fibers. Some muscle nuclei may appear pyknotic, but many remain viable through the acute stage. Within a few days, regenerative repair restores muscle fibers, unless ischemic complications allow fibrosis to develop.

Myocardium sometimes is involved in the degeneration process, but this tends to be of minor importance and rarely contributes to death. Much more significant, ultimately, is the damage done to the proximal convoluted tubules of the kidney by myoglobin or renal ischemia. Severely affected horses, particularly those which are recumbent, may develop oliguria or even anuria leading to renal failure and death (see Chapter 5, The Urinary System). Kidneys are swollen and, on cut surface, have a brown cortex with brown-red streaks in the medulla. Microscopically, the most obvious change is swelling of tubular cells and the presence of pigmented orange granules both in cells and in proteinaceous casts, which occupy the lumen. Casts are present in both proximal and distal tubules and sometimes also in Henle's loop. Interstitial edema may be present in the cortex as a result of altered tubular cell function.

B. Tying-Up

Tying-up (setfast, acute rhabdomyolysis, transient exertional rhabdomyolysis) is a common disease of riding and racing horses. It has many features in common with azoturia, but it is described separately for practical and traditional reasons and because the lesions and consequences of the disease are much milder. It is not yet clear that tying-up is only one disease since two rather different syndromes seem to occur. The first is a clinical disease which, like azoturia, causes acutely crippling signs, involves type 2 fibers primarily in a segmental degeneration, and often causes myoglobinuria. Unlike that in azoturia, the affected animal makes a rapid recovery in most instances after a brief rest and suffers very few consequences in recovery. The second syndrome appears to be similar clinically, but lesions are so transient that no regenerative repair is necessary, and again, recovery is rapid and apparently complete; pathological changes in muscle are slight and may consist only of a very local disorientation of myofilaments in some sarcomeres. It is conceivable that the two syndromes represent the upper and lower ends of one disease spectrum.

As indicated above, horses in a state of tying-up following increased, though moderate exercise are clinically difficult to distinguish from those with azoturia unless the precipitating circumstances are known. Some horses show much milder signs of distress. It is usually a recurrent disease and susceptible animals tie-up, for example, whenever they reach the midtraining stage, or are entered in competition, or foal. The disease sometimes follows a period of rest. Conventional wisdom has it that susceptible horses are those which are tense or nervous. No treatment or prophylactic therapy seems to be of much lasting effect. There is no gender or age predisposition, and there does not seem to be a way of predicting whether the acute period of tying-up will be brief or last as long as 12 to 24 hr.

Postmortem lesions are not grossly detectable, and histologic lesions are variable in quantity and distribution. Although many muscles are affected to a slight extent, the lumbar and, particularly, the gluteal muscles are usually the site of most severe lesions. In the more severe, acute tying-up cases, 1–5% of muscle fibers in a biopsy sample may be affected. Degenerate fibers are solitary, type 2 predominantly (or exclusively in some cases), and swollen (Fig. 2.48A). In longitudinal section, strong contraction bands can be seen on the shortened, contracted segments (Fig. 2.48B). Fiber contents are fractured, leaving segments of floccular protein within the basal lamina, or hyaline segments with "retraction cup or cap" ends. Myoglobinuria occurs early, and serum muscle enzyme levels start rising at 2 to 6 hr and may continue to rise for 48 hr, indicating continuing damage even as the animal is recovering clinically. The degenerative and reparative processes are essentially the same as those described for nutritional or toxic myopathy. It is possible to find acutely degenerate fibers side by side with recently repaired fibers (Fig. 2.48C).

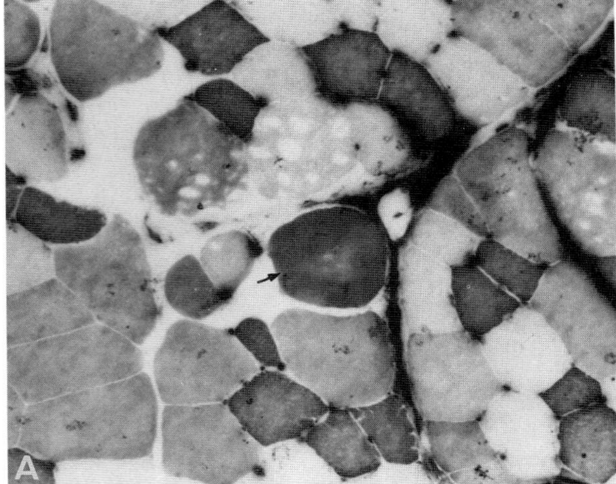

Fig. 2.48 (A) Myopathy of tying-up syndrome. Horse. Selective type 2 fiber degeneration (arrow). (B) Myopathy of tying-up syndrome. Horse. Segmental hypercontraction.

In the less severe cases, the amount of fiber damage involves less than 0.2% of fibers in irreversible change, but many more fibers may undergo some early changes, then return to normal. Serum enzymes may reach levels of only 2–10 times normal and return to normal within 12 to 24 hr.

The early irreversible fiber change as monitored by electron microscopy consists of myofibrillar waving and architectural loss, irregular mitochondrial swelling, and interfibrillar edema. After 12 hr, more marked sarcomere destruction occurs along with streaming of Z bands (Fig. 2.49). Local chemical changes in muscle consist of increased lactate, increased glucose, and lowered chloride ion levels. Metabolic acidosis apparently does not occur.

C. Azoturialike Syndromes in Other Species

Reports of azoturia in draft oxen are part of history, but more current reports of a disease in groups of cattle with

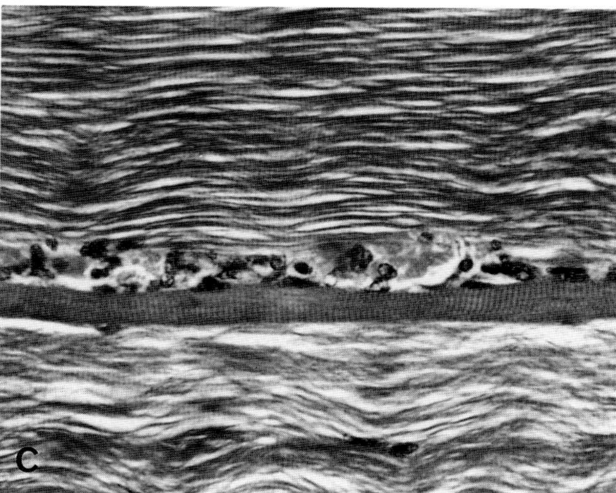

Fig. 2.48C Myopathy of tying-up syndrome. Horse. Simultaneous degeneration and repair in adjacent fibers. Tying-up episodes 12 days apart.

muscle degeneration and myoglobinuria following exercise indicates that the disease occurs even now. Vitamin E and selenium levels are apparently not low, and no myotoxin has been identified.

Reports of sheep suffering from severe myoglobinuria following chasing to exhaustion by dogs might be better categorized under capture myopathy. Further details of the disease are lacking.

Racing greyhounds sometimes display an azoturialike syndrome involving the lumbar muscles. Marked myoglobinuria is associated with clinical disease resembling tying-up.

D. Porcine Stress Syndrome

Malignant hyperthermia is an acute pharmacogenetic syndrome in which a variety of drugs used for anesthetic procedures induce metabolic acidosis, a rapid increase in body temperature, tachycardia, dyspnea, muscular rigidity, and high mortality. Humans, pigs, and dogs are affected. A naturally occurring condition, the porcine stress syndrome affects malignant hyperthermia-susceptible swine under stresses such as handling, transportation, or fighting, and may result in sudden death. The meat of affected pigs is usually pale, soft, and exudative (PSE pork). Porcine stress syndrome has been recognized in Europe and North America for a long time as "hertztod" or back muscle necrosis of pigs. These conditions are not separated in this discussion.

Susceptible pigs exhibit intense, immobilizing limb and torso muscle rigidity, respiratory difficulty, tachycardia, localized or general acidosis, and, often rapid death. Heavy-muscled pigs seem to be most susceptible to the clinical disease. The only practical way to identify susceptible live pigs is by a brief period of anesthesia with halothane. The sarcoplasmic reticulum of muscle from swine susceptible to malignant hyperpyrexia induced by halothane is deficient in inositol 1,4,5-triphosphate phosphatase activity. The deficiency leads to high intracellular concentrations of the triphosphate and of calcium ions. Halothane inhibits the enzyme and further increases the triphosphate and sarcophasmic calcium. Inositol 1,4,5-triphosphate induces release of calcium ions from vesicles of the sarcoplasmic reticulum membrane and activates calcium channels in transverse tubules and membranes. Susceptible animals develop hyperthermic rigidity, which can usually be reversed by terminating the anesthetic trial and by cooling with cold water or ice. There are degrees of susceptibility that may be demonstrated only by extensive testing. The disease occurs in all pork-producing countries of the world, and breeds affected include Landrace, Hampshire, Yorkshire, Pietrain, and crosses of these breeds.

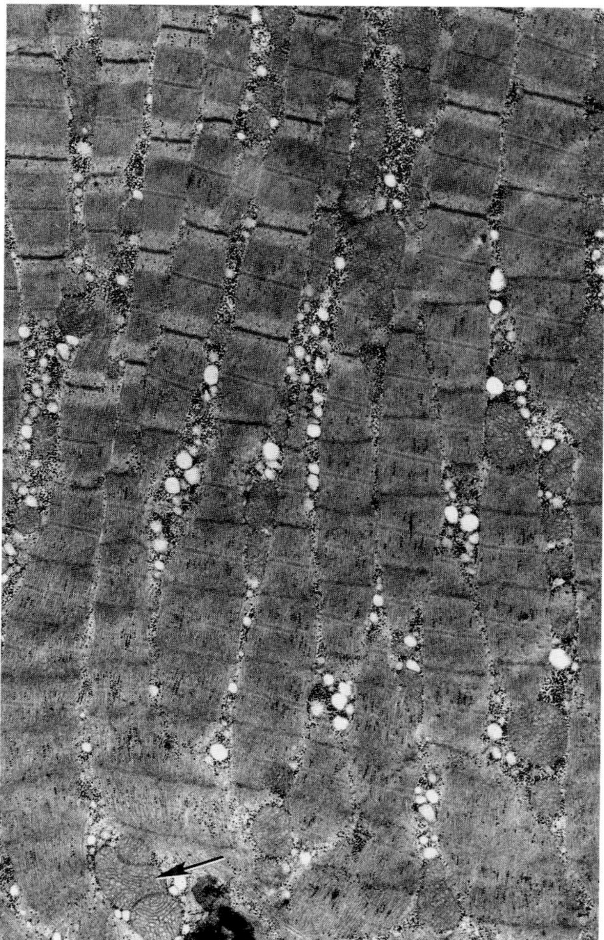

Fig. 2.49 Myopathy of tying-up syndrome. Horse. Acute myofibrillar degeneration with partial loss of Z bands and myofilaments but with preservation of mitochondria (arrow).

The number of susceptible pigs may be increasing. Current estimates indicate that 2–30% of purebred breeding pigs are susceptible to malignant hyperthermia, but no reliable survey of market pigs has been carried out. Selected carcass characteristics may be genetically linked with susceptibility to malignant hyperthermia, and this may account for the high incidence in certain herds. Deaths during transport occur commonly in pigs, but perhaps only 20–30% of them are caused by porcine stress syndrome. Diagnosis of the disease is imprecise since there may be early muscle fiber degeneration in any recently transported pig. The widespread assumption that much of the damaged or inferior pork seen in abattoirs is due to the malignant hyperthermia trait needs to be challenged.

Porcine stress syndrome appears to involve an inherited defect in the intracellular mechanism for uptake, storage, and release of calcium ions. The stimulus for precipitation of the clinical syndrome is not known, but increased physical activity, tension or stress, climatic conditions, environmental temperature, vitamin D deficiency, and energy level of the diet may play a part. In susceptible pigs, lesions are increased by exercise. The common denominator seems to be an excess of calcium ions within the muscle fibers. This in turn leads to functional uncoupling of the oxidative phosphorylation sequences, as mitochondria use the high-energy intermediates to sequester calcium. This results in a consumption of adenosine triphosphate and a rise in adenosine diphosphate and inorganic phosphates, which stimulate increased utilization of glycogen. The greatly increased glycolysis rapidly raises body heat and levels of lactate and lactic acid, which then exert secondary influences on muscle and on the acid/base homeostatic mechanisms to produce local and generalized acidosis.

It is not clear how halothane causes hyperthermia in susceptible pigs. The effect exerted by halothane may be common to all pigs, and perhaps to all other species, but malignant hyperthermia-susceptible pigs may lack a normal regulatory mechanism. The clinical manifestations of muscle rigidity appear to be induced by a failure to mobilize calcium fast enough, thereby inducing a state of hypercontraction. Acid and heat later denature myofibrillar protein, causing it to retain much less cellular water than normal. As water escapes, affected muscle becomes grossly wet and at slaughter, this fluid oozes and drips from the carcass. Eventually the carcass will contain portions of muscle which are dry and dark, and this may also be seen at slaughter if earlier episodes of malignant hyperthermia have occurred. Such lesions may be noticed when pigs are transported and then held for a few days prior to slaughter.

Postmortem examination of the muscles of susceptible pigs which have not endured an episode of hyperthermia recently, reveals normal muscles. Pigs dying of hyperthermia have pale muscles, which are wet and apparently swollen. Rigor mortis develops unusually rapidly. In addition to lesions of skeletal muscle, lesions of acute heart failure, such as pulmonary edema and congestion, hydropericardium, hydrothorax, and hepatic congestion, often are present. The muscles most likely to be affected are those of the back, loin, thigh, and shoulder. Muscles with a high proportion of type 2 fibers, such as longissimus, psoas, and semitendinosus, are most extensively and most frequently affected, and these should be examined histologically. Pigs dying as a result of transport-induced hyperthermia show similar lesions, but the changes are usually less pronounced and less extensive than those occurring with halothane induction. Hemorrhages sometimes are present in muscles, and in the warm carcass, a marked lowering of muscle pH to 5.8 or lower can be detected. On cooling, the pH rises rapidly toward neutrality. Myocardial pallor involving the ventricular muscles sometimes occurs, but the clinical signs of tachycardia are probably related to acidosis.

Microscopic examination of malignant hyperthermia susceptible animals not recently affected by hyperthermic episodes reveals normal muscle fibers, or there may be a few degenerate fibers. There is nothing distinctive about the appearance of the degenerate fibers. In pigs dying acutely of malignant hyperthermia, muscle fibers are separated by edema fluid. This is evident in rapidly fixed specimens only and may be lost in processing. Changes in muscle fibers are widespread and vary from an irregularity in sarcomere length from one region of a fiber to the next, to complete fracture and floccular disintegration. The most common changes are segmental regions of hypercontraction (Fig. 2.50) and a reticulated pattern of thin, branching bands across the fibers, interspersed with angular bands of reduced density (see Fig. 2.33B). In the former, tightly contracted sarcomeres may be visible, while in the latter, normal striations are incompletely obscured. Nuclei may appear pyknotic, but most are normal. If an animal lives somewhat longer than usual, the changes of hypercontraction progress to widespread flocculation or fragmentation. Myocardial lesions include multifocal granular degeneration of myocytes, contraction band necrosis, and myocytolysis.

There do not appear to be either morphologic or histochemical indications of susceptibility to hyperthermia in muscles of susceptible pigs. In both the natural disease and the halothane-induced disease, both type 1 and type 2 fibers are extensively and indistinguishably involved in the degenerative process. The meaning of the minor changes in capillaries, average fiber size, and motor endplates that are seen in susceptible pigs is not clear. The claim that the muscles of susceptible pigs develop continuity defects of the cell membrane immediately on exposure to halothane also needs confirmation.

E. Capture Myopathy

An acute myopathy, often associated with acute death following a chase, a struggle, or transport, occurs in many wild animals and birds. Some cases of capture myopathy represent examples of exercise-induced nutritional myopathy, but there are many in which the clinical disease is distinct from nutritional myopathy, and in which vitamin E and/or selenium do not modify the course of the disease.

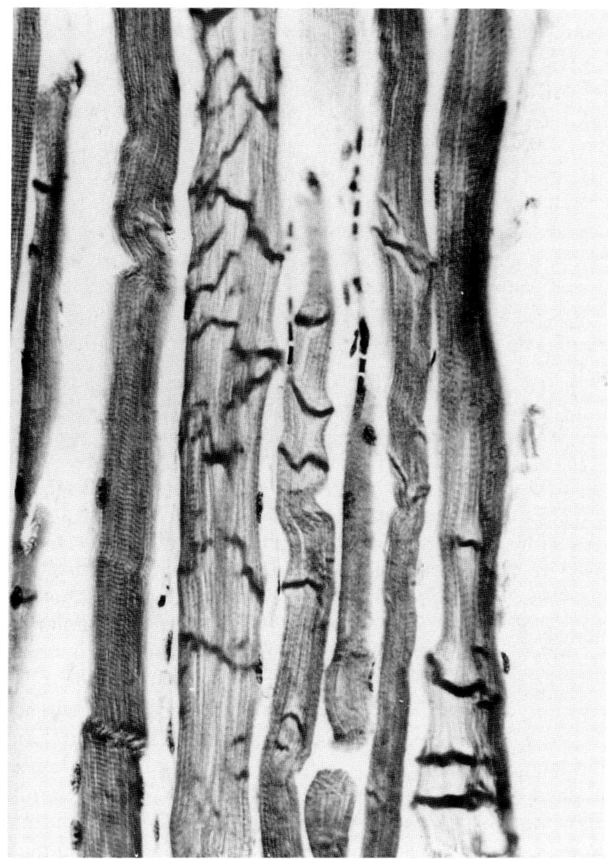

Fig. 2.50 Acute myopathy of porcine stress syndrome showing segmental hypercontraction.

Fig. 2.51 Capture myopathy. Bighorn sheep. (Courtesy of T. Spraker.)

Profound acidosis is often associated with capture myopathy, which suggests that the syndrome is akin to azoturia and porcine malignant hyperthermia, hence its inclusion here as an exertional myopathy.

Clinically affected animals show dyspnea, weakness, muscle tremors or muscle rigidity, hyperthermia, collapse, and often, acute death. Thigh muscle rupture may occur. Those which do not die acutely show myoglobinuria and elevated muscle enzymes in the blood and may subsequently die of renal failure (see Azoturia). When the animal is down, ischemic complications may be secondary (see Circulatory Disturbances of Muscle). The muscle lesions are capable of more or less complete repair over a few weeks unless infarction has intervened.

Postmortem lesions often resemble those of hyperthermia in pigs. Muscle edema is prominent in animals dying acutely. Muscles may also show indistinct pale streaks, and hemorrhagic streaking is not uncommon. Mineralization may be present in small amounts in animals which survive for at least a few hours. Many animals with capture myopathy have pale areas of necrosis in the myocardium, which become mineralized and then fibrosed if survival is prolonged.

Histologically, lesions of capture myopathy most resemble those seen in other exertional myopathies but also resemble those of nutritional or toxic myopathy (see Section X,A, Azoturia and Nutritional Myopathy) (Fig. 2.51). Recovering animals may be left with scars in skeletal muscle as well as in myocardium, and the latter may be a contributing factor to death, weeks or months after the initial acute event.

Bibliography

Aalhus, J. L. *et al.* Stunning and shackling influences on quality of porcine longissimus dorsi and semimembranosus muscles. *Meat Sci* **29:** 323–334, 1991.

Ahern, C. P. *et al.* Halothane-induced malignant hyperthermia; creatine phosphate concentration in skeletal muscle as an early indicator of the onset of the syndrome. *J Comp Pathol* **90:** 177–186, 1980.

Aldrete, J. A., and Britt, B. A. "Second International Symposium on Malignant Hyperthermia." New York, Grune and Stratton, 1978.

Bartsch, R. C. *et al.* A review of exertional rhabdomyolysis in wild and domestic animals and man. *Vet Pathol* **14:** 314–324, 1977.

Bradley, R., Wells, G. A. H., and Gray, L. J. Back muscle necrosis of pigs. *Vet Rec* **104:** 183, 1979.

Campion, D. R., and Topel, D. G. A review of the role of swine skeletal muscle in malignant hyperthermia. *J Anim Sci* **41:** 779–786, 1975.

Cardinet, G. H., Fowler, M. E., and Tyler, W. S. The effects of training, exercise, and tying-up on serum transaminase activities in the horse. *Am J Vet Res* **24:** 980–984, 1963.

Cardinet, G. H., Littrell, J. F., and Freedland, R. A. Comparative investigations of serum creatine phosphokinase and glutamic-oxaloacetic transaminase activities in equine paralytic myoglobinuria. *Res Vet Sci* **8:** 219–226, 1967.

Chalmers, G. A., and Barrett, M. W. Capture myopathy. *In* "Noninfectious Diseases of Wildlife," G. L. Hoff and J. W. Davis (eds.), pp. 84–94. Ames, Iowa, The Iowa State Press, 1982.

Duthie, G. G. *et al.* Plasma pyruvate kinase activity vs creatine kinase activity as an indicator of the porcine stress syndrome. *Am J Vet Res* **49:** 508–510, 1988.

Foster, P. S. *et al.* Inositol 1,4,5-triphosphate phosphatase deficiency and malignant hyperpyrexia in swine. *Lancet* **2**:124–127, July 5, 1989.

Freestone, J. F., and Carolson, G. P. Muscle disorders in the horse: A retrospective study. *Equine Vet J* **23:** 86–90, 1991.

Gallant, E. M. Histochemical observations on muscle from normal and malignant hyperthermia-susceptible swine. *Am J Vet Res* **41:** 1069–1071, 1980.

Gericke, M. D., and Hofmeyr, J. M. Aetiology and treatment of capture stress and myopathy in springbok, *Antidorcas marsupialis*. *S Afr J Sci* **72:** 28, 1976.

Haigh, J. C. *et al.* Capture myopathy in a moose. *J Am Vet Med Assoc* **171:** 924–926, 1977.

Harriman, D. G. F. Malignant hyperthermia myopathy—a critical review. *Br J Anaesthesiol* **60:** 309–316, 1988.

Harris, P., and Colles, C. The use of creatinine clearance ratios in the prevention of equine rhabdomyolysis: A report of four cases. *Equine Vet J* **20:** 459–463, 1988.

Harthoorn, A. M., and Van Der Walt, K. Physiological aspects of forced exercise in wild ungulates with species reference to (so-called) overstraining disease. 1. Acid-base balance and pO_2 levels in besbok, *Damaliscus dorcas phillipsi*. *J S Afr Wildl Manag Assoc* **4:** 25–28, 1974.

Harthoorn, A. M., and Young, E. A relationship between acid–base balance and capture myopathy in zebra, *Equus burchelli* and an apparent therapy. *Vet Rec* **95:** 337–342, 1974.

Heffron, J. J. A., Mitchell, G., and Dreyer, J. H. Muscle fiber type, fiber diameter, and pH values of m. longissimus dorsi of normal, malignant hyperthermia—and PSE-susceptible pigs. *Br Vet J* **138:** 45–50, 1982.

Honkavaara, M. Influence of porcine stress on blood composition and early postmortem meat quality in pigs of different halothane genotypes. *Meat Sci* **24:** 21–29, 1988.

Hoyer, von H., and Hoyer, I. Mastschweinestrass und PSE-Fleischqualitat: Eine Vitamin D—Mangelkonsequenz. *Dtsch Tierarztl Wochenschr* **89:** 482–483, 1982.

Johannsson, G., and Jonsson, L. Myocardial cell damage in the porcine stress syndrome. *J Comp Pathol* **87:** 67–74, 1977.

Johannsen, U. *et al.* Pathohistologische und elektron mikroskopische Untersuchungen der Skelettmuskulatur bei Schweinen nach derfinierter Belastung-Vergleich bewegungsramar und bewegungsreicher Haltungsformen. *Arch Exp Veterinaermed* **35:** 641–660, 1981.

Johannsen, U., Menger, S., and von Lengerken, G. Vergleichende Untersuchungen zur Ultrastruktur der Skelettmuskulatur (M. longissimus dorsi) unterschiedlich belastungsempfindlicher Schweinerasseb-Vergleigh Duroc: Pitran. *Arch Exp Veterinaermed* **36:** 357–374, 1982.

Johnson, B. D., and Perce, R. B. Unique serum isoenzyme characteristics in horses having histories of rhabdomyolysis (tying-up). *Equine Pract* **3:** 24–31, 1981.

Johnston, W. S., and Murray, I. S. Myopathy in young cattle associated with possible myoglobinuria. *Vet Rec* **97:** 176–177, 1975.

Kock, M. D. *et al.* Effects of capture on biological parameters in free-ranging bighorn sheep (*Ovis canadensis*): Evaluation of normal, stressed, and mortality outcomes and documentation of postcapture survival. *J Wildl Dis* **23**(4): 652–662, 1987.

Koterba, A., and Carlson, G. P. Acid–base and electrolyte alterations in horses with exertional rhabdomyolysis. *J Am Vet Med Assoc* **180:** 303–306, 1983.

Lewis, P. J. *et al.* Effect of exercise and preslaughter stess on pork muscle characteristics. *Meat Sci* **26:** 121–129, 1989.

Lewis, R. J. *et al.* Capture myopathy in elk in Alberta, Canada: A report of three cases. *J Am Vet Med Assoc* **171:** 927–932, 1977.

Lindholm, A., Johannson, H-E., and Kjaersgaard, P. Acute rhabdomyolysis ("tying-up") in standardbred horses. A morphological and biochemical study. *Acta Vet Scand* **15:** 325–339, 1974.

Lundström, K. *et al.* Effect of halothane genotype on muscle metabolism at slaughter and its relationship with meat quality: A within-litter comparison. *Meat Sci* **25:** 251–263, 1989.

McEwen, S. A., and Hulland, T. J. Histochemical and morphometric evaluation of skeletal muscle from horses with exertional rhabdomyolysis (tying-up). *Vet Pathol* **23:** 400–410, 1986.

Mitchell, G., and Heffron, J. J. A. Some muscle and growth characteristics of pigs susceptible to stress. *Br Vet J* **137:** 374–380, 1981.

Moore, J. N. *et al.* A review of lactic acidosis with particular reference to the horse. *J Equine Med Surg* **1:** 96–105, 1977.

Munday, B. L. Myonecrosis in free-living and recently captured macropods. *J Wildl Dis* **8:** 191–192, 1972.

O'Brien, P. J. Etiopathogenetic defect of malignant hyperthermia: Hypersensitive calcium-release channel of skeletal muscle sarcoplasmic reticulum. *Vet Res Commun* **11:** 527–559, 1987.

Peet, R. L. *et al.* Exertional rhabdomyolysis in sheep. *Aust Vet J* **56:** 155–156, 1980.

Ranatunga, K. W. Effects of acidosis on tension development in mammalian skeletal muscle. *Muscle Nerve* **10:** 439–445, 1987.

Salomon, von F.-V. *et al.* Maligne hyperthermie und morphologische parameter der skelettmuskulatur des schweines. *Mh Vet Med* **41:** 156–164, 1986.

Scholte, H. R. *et al.* Equine exertional rhabdomyolysis: Activity of the mitochondrial respiratory chain and the carnitine system in skeletal muscle. *Equine Vet J* **23:** 142–144, 1991.

Seeler, D. C., McDonell, W. N., and Basrur, P. K. Halothane- and halothane/succinylcholine-induced malignant hyperthermia (porcine stress syndrome) in a population of Ontario boars. *Can J Comp Med* **47:** 284–290, 1983.

Somers, C. J., and McLoughlan, J. V. Malignant hyperthermia in pigs: Calcium ion uptake by mitochondria from skeletal muscle of susceptible animals given neuroleptic drugs and halothane. *J Comp Pathol* **92:** 191–198, 1982.

Sosnicki, A. Histopathological observation of stress myopathy in M. longissimus in the pig and relationships with meat quality, fattening, and slaughter traits. *J Anim Sci* **65:** 584–596, 1987.

Spraker, T. R. Pathophysiology associated with capture of wild animals. *In* "The Comparative Pathology of Zoo Animals," R. J. Montali and G. Migaki (eds.), pp. 403–414. Washington, D.C., Smithsonian Institutional Press, 1980.

Swatland, H. J., and Casens, R. G. Peripheral innervation of muscle from stress-susceptible pigs. *J Comp Pathol* **82:** 229–235, 1972.

Van den Hoven, R., and Breukink, H. J. Normal resting values of plasma free carnitine and acylcarnitine in horses predisposed to exertional rhabdomyolysis. *Equine Vet J* **21:** 307–308, 1989.

Waldron-Mease, E., Raker, C. W., and Hammel, E. P. The muscular system. *In* "Equine Medicine and Surgery," 3rd Ed. R. A. Musmann, E. S. McAllister and P. W. Pratt (eds.), pp. 923–935. Santa Barbara, California, American Veterinary Publications, 1982.

Whitwell, K. E., and Harris, P. Atypical myoglobinuria: An acute myopathy in grazing horses. *Equine Vet J* **19:** 357–363, 1987.

Wobeser, G. *et al.* Myopathy and myoglobinuria in a wild white-tailed deer. *J Am Vet Med Assoc* **169:** 971–974, 1976.

XI. Myositis

Myositis means inflammation of muscle, but it is often difficult to determine whether the process is part of a classical inflammatory response, which happens to be active in muscle, or whether some components of inflammation are activated by a process which is primarily degenerative. The cause of myositis is often not evident except in the obvious cases in which the inflammatory nature of the reaction is signaled by a suppurative or granulomatous response, or where it is a component of a systemic infectious disease such as bluetongue virus infection in sheep, foot and mouth disease, or toxoplasmosis.

Infectious agents and parasites are frequently the cause of a mild myopathy (as distinct from myositis) in animals, in which case, the degenerative muscle changes are an almost incidental and often unnoticed part of systemic infection or toxemia. There are, however, some systemic and local infections which cause myositis rather than myopathy, and they will be considered here. Living muscle is an inhospitable site for almost all bacteria, and consequently myositis is rarely a complication of bacterial infections even when they are overwhelmingly septicemic or repeatedly bacteremic. Pyogenic organisms sometimes give rise to solitary muscle abscesses, particularly in pigs and goats, but bacterial polymyositis is a feature of only a few infections, such as *Haemophilus agni* in lambs, *H. somnus* in cattle, and *Actinobacillus equuli* in foals. Clostridia, on the other hand, are very well adapted to growing in muscle once damage to muscle fibers provides them with an opportunity.

A. Suppurative Myositis

Abscesses in muscle may sometimes be hematogenous in origin, but more often they result from inoculation, or from an extension of a suppurative focus in adjacent structures such as joints, tendon sheaths, or lymph nodes. The most common causes of abscesses in muscle are *Actinomyces pyogenes* in cattle and swine, *Corynebacterium pseudotuberculosis* in sheep and goats, and *Streptococcus equi* in horses. A variety of streptococci and staphylococci are retrieved from such lesions in many species.

The natural development of suppurative myositis is comparable to abscessation elsewhere. The early stages consist of a local, ill-defined, cellulitis. Healing may take place after this with a minimum of scarring, or it may proceed to the formation of a typical abscess with a liquefied center, a pyogenic membrane, and an outer fibrous sheath. The lesion may slowly organize if it is effectively sterilized, expand if it is not or, alternatively, fistulate to the surface, collapse, and heal. In the healing process, damaged muscle fibers will participate in fiber regeneration to only a very limited extent.

In cats and horses particularly, the early stage of local cellulitis may give rise to a rapidly expanding cellulitis of muscle and adjacent fibrous and fat tissue. This phlegmonous inflammation leads to extensive destruction of muscle fibers, which are subsequently mostly replaced by scar tissue. The organisms involved may vary, but *Pasteurella multocida* has been incriminated in cats, while staphylococci appear to be the most common in other domestic species.

B. Staphylococcal Granuloma

Chronic granulomas of muscle or connective tissue caused by staphylococci, and referred to as **botryomycosis** in early literature, are much less common than they once were, but they still occur, particularly in the horse and pig. They represent a persistent, low-grade infection by *Staphylococcus aureus*, but it is not clear why the same organism at other times produces a conventional abscess or a gangrenous phlegmon. In horses, lesions are most frequent on the neck and pectoral region ("breast boils"), while in the pig, castration wounds and the mammary glands are the most common sites.

The lesion begins as a microabscess around a small colony of organisms and progresses rapidly, sometimes to a very large size. The fully formed granuloma is a hard, nodular, gray-white mass of dense, fibrous tissue, irregularly cavitated by small abscesses. The abscesses may be joined by tracts, or they may fistulate to the surface. They contain a small quantity of thick, orange-yellow pus, which in turn contains minute granules. The granules consist of a central colony of the organisms, while the bulk of the granule mass is made up of "clubs" of reactive protein material; hence the typical club colony of botryomycosis. Histologically, the organisms are readily visible in tissues since they stain with hematoxylin and are relatively large. Variable numbers of neutrophils, lymphocytes, and plasma cells are present in the loose fibrous tissue outside of the club colony. Muscle is involved at the periphery of the expanding granulomatous lesion where fibrous septa surround fibers, causing atrophy and segmental degeneration.

C. Gas Gangrene and Malignant Edema

The muscles, especially if devitalized in some manner, are highly susceptible to bacteria of the genus *Clostridium*, and these organisms, when they proliferate, are highly toxigenic and cause extensive necrosis of muscle, with blood-stained edema, and the formation of gas. Once the bacteria become established, the toxins they elaborate provide a suitable and expanding environment for further bacterial growth. Death occurs as a result of systemic intoxication.

These bacteria are gram-positive bacilli, to a greater or lesser degree anaerobic, and they exist in the environment as resistant spores. Germination of the spores and vegetative growth requires fairly precise local conditions, chiefly a low oxidation–reduction potential and an alkaline pH. These conditions are best produced by deep penetrating wounds, and the lesions which result from the activity of the anaerobes are called gas gangrene, and sometimes malignant edema and anaerobic cellulitis. Since the pathogenic clostridia are frequently found in soil and feces, any contamination of an open wound is likely to introduce those potential pathogens. Although their presence in a wound always carries a threat of gas gangrene, the very great majority of wounds thus infected heal without ill effect; in these, the local conditions in the wound must be regarded as unsuitable for either germination, vegetation, or the production of toxins.

The species of the genus *Clostridium* that are of most importance as the agents of gas gangrene are *C. septicum, C. perfringens, C. novyi,* and *C. chauvoei*. These organisms not only cause gas gangrene, which is usually a mixed infection, but in animals they are, as pure infections, responsible for a number of specific diseases not associated with surface wounding and, with one exception, not associated with primary lesions in muscle. The exception is *C. chauvoei,* which causes bacterial myositis (blackleg) in ruminants. The other specific diseases include black disease caused by *C. novyi* (see Volume 2, Chapter 2, The Liver and Biliary System), braxy caused by *C. septicum,* and the clostridial enterotoxemias caused by *C. perfringens* (see Volume 2, Chapter 1, The Alimentary System).

Gas gangrene and malignant edema are essentially wound infections in which *C. septicum, C. perfringens, C. novyi, C. sordelli,* and *C. chauvoei* are the principal pathogens acting alone, or in combination with each other or with a variety of other aerobes and anaerobes, the latter being saprophytes with proteolytic and putrefactive properties. Ruminants, horses, and swine are highly susceptible to these infections, whereas carnivores are rarely affected with gas gangrene. Since deep wounds are most suitable to the development of gas gangrene, the common causes of such susceptible wounds in animals are castration, shearing, penetrating stake wounds, injuries to the female genitalia during parturition, and especially in swine, inoculation sites. The distinctive characteristics of these local infections are severe edema, the formation of bubbles of gas which give crepitation, discoloration of the overlying skin, coldness of the affected part, and in particular, the constitutional signs of profound toxemia with prostration, circulatory collapse, and sudden death.

Malignant edema is included in this title to distinguish certain cases of clostridial myositis of which gas gangrene is not a part; indeed, in animals, the nongangrenous form of clostridial myositis is much more common than the gangrenous form. Malignant edema is more typically a cellulitis than a myositis, and the muscles may escape significant injury even in fulminating, highly toxigenic infections of the sort that are fatal in 48 hr. All the factors that determine whether gangrene develops in these infections are not known. Even when the primary pathogens present are the same, the relative potencies and the amounts of the toxins produced may vary, and it is expected that accompanying nonprimary organisms will, to some extent, influence the local course of the infection. One factor which is probably of much importance in determining whether the inflammation will be confined to the connective tissues (malignant edema) or will directly involve muscle (myositis) is the adequacy of the blood supply to the muscle; if the muscle is devitalized by the initial trauma, or subsequently as a result of toxic injury to the blood vessels, the development of true gangrene is in order; in this manner, malignant edema or anaerobic cellulitis may develop into gangrene.

It will be apparent from this discussion that the pathogenesis of clostridial myositis and cellulitis is not simple and is not initiated merely by the presence of spores or vegetative forms in the wound; it may begin only when the organisms have produced enough toxin to immobilize and destroy any adjacent leukocytes and enough toxin to cause death of tissue. Once the bacteria are established and producing toxins, they are capable of creating spreading conditions suitable for their advance. The progression is longitudinal, up and down fascial planes; transverse progression is very limited. The spread is facilitated by increased capillary permeability, the edema fluid separating the muscle fibers and fascia, assisted in this by gas bubbles. This fluid also allows for further diffusion of the toxins and spread of the bacteria. Venous and capillary thrombosis result in local circulatory disturbances and these, in turn, result in further devitalization of tissue. Once the process is started, it may progress with extraordinary rapidity. If clostridial myositis is to develop in a wound, there is usually evidence of it within 24 hr.

In gas gangrene, there is extensive disintegration of muscle and saturation of the tissues with an exudate which is in part serous and in part profusely sanguineous. When lysis of exuded red cells occurs, the tissues become stained darkly with hemoglobin. The tissues have a rancid odor in the beginning and an exceedingly foul odor in the end.

Histologically, edema fluid, poor in protein, separates the muscle fibers from each other and the endomysium. The degenerating muscle fibers stain intensely with eosin, the sarcolemma and its nuclei degenerate, but the striations are unduly persistent. Such a histologic picture is always seen at the advancing margin of the lesion (Fig. 2.52). The intrinsic histiocytes of the muscle show no reaction at all, and neutrophils are never numerous; a few of the latter are loosely scattered at the advancing margin of the lesion with slightly greater numbers in the dermis,

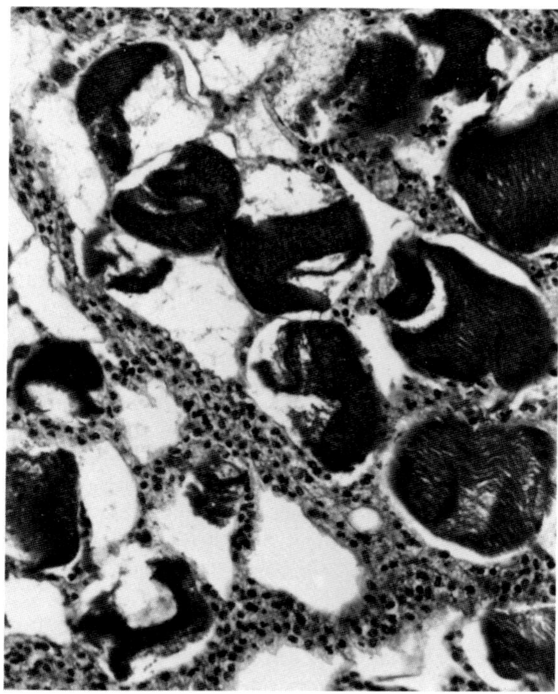

Fig. 2.52 Myositis due to *Clostridium septicum* and other organisms. Horse.

but they are rapidly and effectively immobilized and destroyed by the toxins. Deeper within the lesions, muscle fibers are fragmented, but this is probably an indication of physical forces having been applied to necrotic fiber segments at a stage in which the animal was still mobile. Fragmentation is by no means a constant feature of gas gangrene, and the absence of it at the periphery of lesions indicates only that, in the later stages of the disease, muscle activity is drastically reduced in a recumbent animal. Bacteria are seldom numerous in the lesions, but collections of them may be seen either in muscle fibers or in connective tissue. Involvement of adjacent adipose tissue by the necrotizing process liberates fat droplets, which may become embolic.

It is not uncommon for animals to die within 24 hr of the onset of local signs of gas gangrene. Invasion, sometimes massive, of the bloodstream occurs shortly before death, or shortly after, and the offending organisms can then be obtained from most tissues. As well as the local lesion, there is at postmortem severe pulmonary congestion and evidence of profound toxic degeneration of the parenchymatous organs. By very few hours postmortem, there is extensive gas formation in all organs, and they crepitate. The liver, especially, may be honeycombed with bubbles, and cut blood vessels continuously release gas bubbles.

There are some variations in the gross appearance of the lesion depending on the species of the principal pathogen. Putrefaction is the property of contaminating saprophytes, since the primary pathogens are poorly endowed in this regard; the latter are, however, saccharolytic, and that is the origin of the gas bubbles in the lesions. The exotoxin of *C. novyi* has a rather specific action on vascular endothelium and serous membranes, and relatively pure infections with this organism produce a very extensive edematous infiltration of connective tissues; there is no putrefaction, only slight or no discoloration of muscle, and the edema fluid, which is gelatinous, is quite clear or at most pink. The toxic potency of this organism is shown by the fact that in fatal infections, the organisms are so few as to be difficult to locate even in the primary focus, from which site they show little inclination to move.

A typical wound infection by *C. novyi* is that known as **swelled head** in rams, the wounds being acquired on the top of the head during fighting. This is a quickly fatal condition, death usually occurring within 48 hr. There is extensive infiltration with clear, gelatinous fluid in the tissues of the head, throat, neck, and anterior thorax, and sometimes also in the pleural and pericardial sacs. There is no discoloration of muscle and no, or scant, extravasation of erythrocytes. The bacilli are very few in number and cultivable only from the primary focus of infection. *Clostridium novyi* may also predominate in wound infections of horses and, when it does, these lesions too are a nonhemorrhagic variety of malignant edema. *Clostridium chauvoei* as a wound infection always produces heavily sanguineous edema fluid and much gas. *Clostridium septicum* produces large quantities of gelatinous exudate, which is, however, intensely stained with blood. The muscular tissue may be discolored dark red to black and, although gas is present, the bubbles are very small. Dark blood exudes freely from veins in the perimysium, and hemoglobin stains narrow collars of surrounding tissue at the margin of the lesions. *Clostridium perfringens* is the usual cause of gas gangrene in humans and produces similar lesions in animals. It shows much less tendency than *C. chauvoei* and *C. septicum* to invade the bloodstream terminally or after death. The most frequent cause of malignant edema is *C. septicum*.

D. Blackleg

Blackleg, also known as black quarter, symptomatic anthrax, and emphysematous gangrene, is a gangrenous myositis of ruminants caused by *C. chauvoei* and characterized by the activation of latent spores in muscle. This definition of blackleg separates it from gas gangrene in which, if *C. chauvoei* is involved, it is as a wound contaminant. The pathogenesis of blackleg envisages that there are spores of the organisms latent in muscle.

Blackleg can be sometimes mimicked closely by the syndrome known as stable blackleg or pseudoblackleg caused similarly by germination of latent spores of *C. septicum* in cattle. For diagnostic, prognostic, and epidemiologic reasons, it is important to make a distinction between disease of the blackleg pattern of pathogenesis and disease of the much less complex pathogenesis of gas gangrene and malignant edema. The latter diseases occur sporadically wherever cattle and sheep are raised but appear as a confined outbreak only when groups of animals have been subjected to similar traumatic procedures such as shearing wounds or intramuscular injections. Blackleg, on the other hand, in spite of a worldwide distribution, is peculiarly localized to regions, and within regions to farms. Within these locales, it is persistently but irregularly enzootic, but rather selective of its hosts, and controllable only by vaccination. The muscle lesion produced is not distinctive for the blackleg pathogenesis, nor is the presence of *C. chauvoei* in a typical gangrenous lesion, because *C. chauvoei* can also be a wound contaminant. Identification of the blackleg syndrome should be made on a freshly dead cadaver or preferably on more than one in an outbreak. In the confirmation of blackleg and of pseudoblackleg particularly, attempts at distinguishing syndromes must be tempered by the knowledge that *C. septicum* proliferates rapidly after death, whereas *C. chauvoei* does not. The probability of overgrowth is great when a few hours have elapsed between death and bacteriological examination.

Blackleg occurs most often in cattle and sheep and rarely in other domestic animals. Blackleg in **cattle** primarily affects animals 9 months to 2 years of age, with a reduced incidence at 6 to 9 months and 2 to 3 years, and an even lower incidence in animals older than 3 years. It affects animals in good condition and often selectively causes death in the best-grown or best-fattened animals in a group. Blackleg is chiefly a disease of pastured animals, with a tendency to be seasonal in summer. It is often associated with moist pastures and rapid growth of both forage and cattle, but it is also a problem on some arid ranges. Because the source of *C. chauvoei* organisms appears to be persistent on certain fields, it has been assumed that the organism is soilborne, but it is unlikely that it grows in soil. Growth does take place readily in the intestinal tract of cattle, and it is now thought that soil contamination persists by a process of constant replenishment by fecal contamination.

The detailed pathogenesis of blackleg is still somewhat uncertain, but many of the critical points in the following proposed sequence of events have been confirmed in the natural disease and in experimental infections in cattle. The infection is acquired by the ingestion of spores, and either these spores, or spores produced following one or more germinative cycles in the gut, are taken across the intestinal mucosa in some way. Macrophages may be responsible for this passage, but it may be possible for the spores to enter natural or transient apertures at tips of villi or in lymphoid crypts or be taken in by lymphoepithelial cells of the ileal domes by endocytosis. Spores are distributed to tissues where they may be stored for long periods in phagocytic cells. There may be a certain dynamic turnover, for example, in and out of Kupffer's cells in the liver, but spores can be found in many tissues of normal animals, including muscle. The latent spores in muscle are stimulated to germinate when a local event creates muscle damage or low oxygen tension. This last step is difficult to produce experimentally, but circumstantial evidence seems to confirm its inclusion in the sequence. Parallels exist in other clostridial diseases about which there is much less doubt. It may be that all that is required to establish a medium for the organisms to multiply is a small intramuscular hemorrhage or a degenerative focus initiated by traumatic damage to muscle as part of forced exercise.

The clinical manifestations of blackleg are often not observed and, because of the rapid clinical course, animals are often found dead. When animals are seen ill, symptoms consist of lameness, swelling, and crepitation of the skin over a thigh or shoulder if the lesion is superficial, and fever. It is typical for swellings to increase rapidly in size and to be hot initially and cold later. Affected animals subsequently show depression and circulatory collapse. Death rapidly ensues, and seldom does an animal survive more than 24 to 36 hr after the onset of any signs of lameness. If the muscle lesions are deep within a muscle mass or in the diaphragm, no localizing signs may be evident and no palpable changes, detectable. Similarly, lesions in the tongue, heart, or sublumbar muscles may escape clinical detection.

An animal which has died of blackleg swells and bloats rapidly, but on incision it is often not as putrid as its external appearance would suggest. Blood-stained froth flows from the nose but not usually from other orifices. Blotchy hemorrhages may be present on the conjunctivae. A poorly circumscribed swelling may be visible on superficial inspection, and crepitation may be detectable on palpation. The skin overlying crepitant swellings is taut and resonant, but normal or dark in color and of normal strength. The subcutaneous tissues and fascia around the lesion are thick with yellow gelatinous fluid, which is copiously bloodstained close to the lesion. Gas bubbles may be apparent in the fluid. The affected muscles present slightly different appearances at different distances from the center of the lesion. Toward the periphery of the lesion, the muscle is dark red, and moist with edema fluid. Toward the center, it is red-black, occasionally with putty-colored islands, and the tissue is dry, friable, and porous where gas bubbles separate the primary bundles of fibers (Fig. 2.53A,B). If this tissue is squeezed, it crepitates, and a small amount of thin red fluid oozes out. When the tissue is exposed to air, it takes on a light red color, and watery exudate drips from cut surfaces. The odor which emanates from the muscle is sweet and butyric, like rancid butter.

The initial bacterial lesion in blackleg is a cellulitis with copious edema and hemorrhage. Degeneration of the muscle fibers is caused by both diffusing toxin and injury to blood vessels, and it is probably the extent of ischemic muscle fiber death that determines how quickly and how extensively tissue gangrene extends through the muscle. Gangrenous lesions expand longitudinally with the long axis of muscles more readily than in a lateral direction, but "skip" areas may create necrotic zones which are highly irregular in contour. The expansion is enhanced by

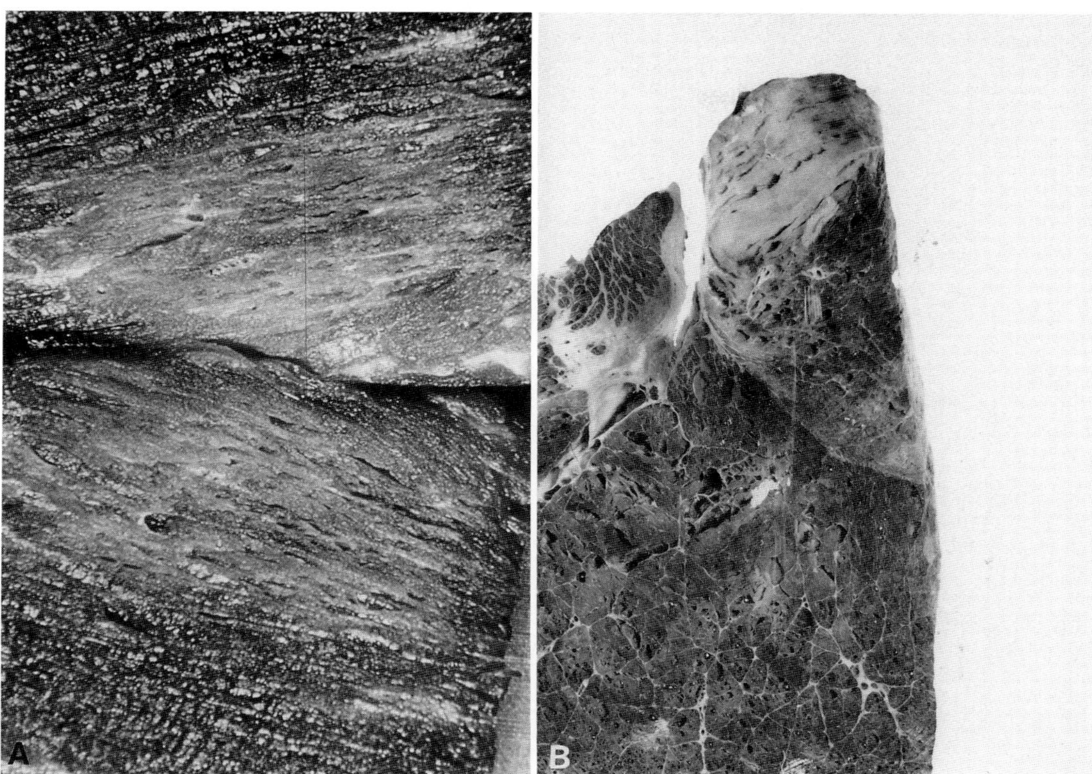

Fig. 2.53 Blackleg. Ox. (A) Dry appearance of affected muscle. (B) Muscle bundles separated by gas bubbles.

the edema fluid between fibers. At the time of death, animals with a single muscle lesion may have relatively small or very large areas of gangrene. Gas is not produced until muscle fibers die and are penetrated by bacteria and toxins. The exudate and gas bubbles separate bundles of fibers and individual fibers (Fig. 2.54), and these undergo necrosis with preservation of striations in the center of the focus, and fatty and granular degeneration toward the periphery. Leukocytes are sparse, being destroyed by diffusing toxins, and only a scattering of debris is found at the periphery of the lesion.

The lesions of blackleg are usually found in the large muscles of the pectoral and pelvic girdles, but they may be found in any striated muscle, including the myocardium. Lesions in the crura of the diaphragm and in the tongue are quite common, and if lesions are present in two or three sites simultaneously, they may be lethal before any of them is very large. This makes their clinical detection more difficult, and their detection at postmortem dependent on detailed examination of many muscles. Even with small widely separated lesions, however, the rancid butter odor may be pervasive.

In addition to the specific muscle lesions of blackleg, changes are present in the rest of the carcass. There is severe parenchymatous degeneration of liver, kidney, and endocrine glands, and although this is like conventional postmortem change, the rapidity of its development can suggest blackleg-related toxemia or bacteremia. There is often a fibrinohemorrhagic pleuritis and, as a generality, when this lesion is present without severe pneumonia, blackleg should be suspected. The parietal pleura is hemorrhagic, and large or delicate bloodstained clots of fibrin overlie the ventral mediastinum and epicardium. Pneumonia is not part of the intrathoracic lesion, but the lungs are congested, and they may be quite edematous. The myocardium may be pale and friable, or dark red; some of the latter areas contain foci of emphysematous myocarditis and necrosis. These areas give rise to fibrinohemorrhagic pericarditis; they may be primary or metastatic blackleg lesions. Endocardial lesions sometimes occur, particularly in young animals. The endocardium is hemorrhagic and may be ulcerated or contain built-up, fixed, endocardial thrombi in the atrium or on the outer wall of the right ventricle. If there is atrioventricular valve involvement, it is usually on the right side. The peritoneum is intact and normal unless an underlying myositis has extended to it. The spleen may be normal, or enlarged with congested, mushy, pulp. Pale round foci of early putrefactive necrosis may be found in the liver and kidney without reaction; these enlarge with the postmortem interval and become porous. Organisms can often be recovered from many organs and from the blood shortly after death.

Blackleg in **sheep** closely resembles the disease in cattle, and the causative organism is the same. The disease

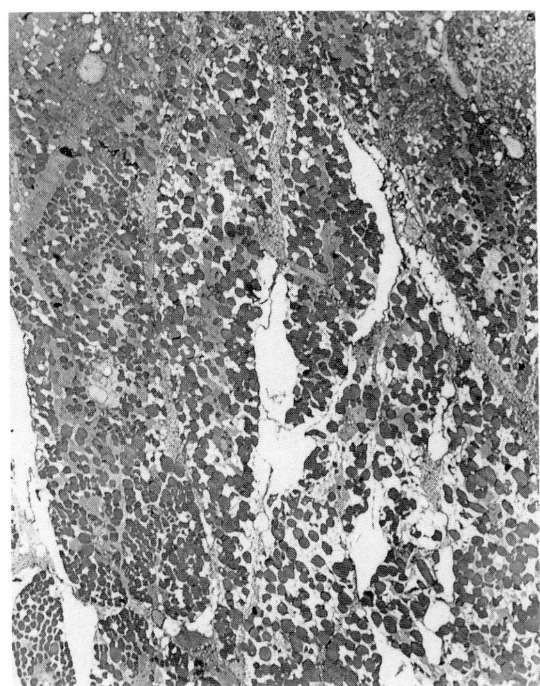

Fig. 2.54 Blackleg. Ox. Muscle bundles separated by gas bubbles.

in sheep, however, is much less common than in cattle, and, although there is some overlap in enzootic distribution, the disease in sheep usually occurs in locales quite apart from those where it occurs in cattle. The clinical signs are similar to those in cattle except that crepitation may not be palpable during life, and there is usually dark discoloration of the overlying skin. The lesions resemble those in cattle, but there are usually fewer gas bubbles, and the muscle remains more moist.

Pseudoblackleg mimics blackleg very closely. This disease of cattle produces deep lesions, like those of blackleg. A diagnosis of pseudoblackleg can be made only when there is no detectable wound, the lesion is deeply located, and *C. septicum* is demonstrated in the lesions of a carcass examined immediately after death. This latter precaution is necessary to avoid, as far as possible, misdiagnosis based on the postmortem invasion of *C. septicum* and other organs from the alimentary tract.

The lesions of pseudoblackleg differ quantitatively and somewhat qualitatively from those of blackleg. They tend to be multiple in widely separated muscles. There is very extensive bloodstained gelatinous exudate in the connective tissues, and this exudate contains only occasional small gas bubbles. The lesions may become confluent as very large patches throughout the connective tissues. By contrast, the muscle lesions, although always present, are less extensive than those in blackleg. The muscles are dark red and moist, and only after examination by multiple incisions is the lack of a distinct primary focus evident. When bubbles of gas are present in necrotic muscle, they are smaller and not as numerous as they are in blackleg.

Pseudoblackleg is reported in pigs, in which the muscle lesions are part of a generalized body invasion by *C. septicum* organisms, following entry via a primary gastric focus.

Bibliography

Buxton, A., and Fraser, G. "Animal Microbiology." Edinburgh, Blackwell Scientific Publications, 1977.

Breuhausm, B. A. *et al.* Clostridial muscle infections following intramuscular injections in the horse. *J Equine Vet Sci* **3:** 42–46, 1983.

Chou, S. M. Inclusion body myositis: A chronic persistent mumps myositis? *Hum Pathol* **17:** 765–777, 1986.

McLaughlin, S. A., Rebhun, W. C., and Van Winkle, T. J. *Clostridium septicum* infection in the horse. *Equine Pract* **1:** 17–20, 1979.

Minett, F. C. The pathogenesis of black-quarter. 1. Tissue damage and spore latency. *J Comp Pathol* **58:** 201–209, 1948.

Oakley, C. L., and Warrack, G. H. The soluble antigens of *Clostridium oedematiens* type D (*Cl. haemolyticum*). *J Pathol Bact* **78:** 543–551, 1959.

Orfeur, N. B., and Hebeler, H. F. Blackquarter in pigs. *Vet Rec* **65:** 822, 1953.

Seddon, H. R., Belschner, H. G., and Edgar, G. Blackleg in sheep in N.S.W. *Aust Vet J* **7:** 2–18, 1931.

Sterne, M., and Edwards, J. B. Blackleg in pigs caused by *Clostridium chauvoei. Vet Rec* **67:** 314–315, 1955.

E. Roeckl's Granuloma of Cattle

This nodular lesion of skeletal muscle is apparently specific for cattle, and it may be associated with similar lesions in liver, lungs, lymph nodes, and testes. The lesion is sometimes referred to as nodular necrosis, and it is well known in Europe but rare elsewhere. Included in the list of suggested causes are tuberculosis, pseudotuberculosis, sarcocystosis, blastomycosis, and larvae of *Hypoderma bovis,* but none has been regularly found in typical lesions. Acid-fast organisms are usually not present, and although *Actinomyces pyogenes* is sometimes recovered, it is not considered to be the cause of the multiple granulomas.

The lesions occur in cattle of all ages. In any single animal, nodules are all of the same size, but the size varies from one animal to the next, from 0.5 to 5.0 cm in diameter, in or under the skin. Sites of predilection are the skin around the base of the tail, the limbs, the withers, and the abdominal wall, but lesions are seldom seen deep in muscle masses.

When nodules are cut, surfaces tend to bulge and consist of three zones. The central zone is dry, dull, necrotic-looking, and often distinctly yellow. The reactive zone is gray-pink, semitranslucent, and elastic. The outer layer is thin, white fibrous tissue, which radiates out along trabecular divisions in adjacent muscle. Larger nodules may be laminated, indicating periodic growth, and small nodules may be reduced to a hard scar with or without mineralization.

Histologically, the early stages of Roeckl's granuloma are small abscesses surrounded by granulation tissue, which may contain abundant eosinophils. In other animals, the predominant inflammatory cell may be the lymphocyte. Epithelioid cells and a few giant cells may be present in the reactive zone. The capsule is made of mature collagen. These lesions seem to be the modified or exaggerated response of a sensitized animal to a persistent or repetitively introduced antigen.

F. Myositis of Dogs

Eosinophilic myositis and atrophic myositis are names applied to **myositis of the masticatory muscles of dogs,** which is characterized by progressive destruction of the muscles of mastication and leads to fixation of the jaws. The syndromes overlap in their clinical and pathologic features and may be one disease with varied expressions.

Eosinophilic myositis is the syndrome identified with the German shepherd breed, and it is expected to be associated with a prominent eosinophilia of blood and muscle tissue. A progressive course is irregularly interrupted by acute episodes. **Atrophic myositis** affects long-nosed breeds but appears otherwise to be without breed predilection; eosinophilia is present but mild, and the course is expected to be chronic and progressive but not acutely relapsing. The masseter, temporal, and pterygoid muscles are involved clinically, but lesser degrees of myositis may be detected histologically in other muscles, especially of the head and neck, in which the inflammatory change is focal and random.

The disease is apparently initiated by the abnormal formation of antibodies to the unique myosin isotypes found in the muscles of mastication of the adult dog. Thereafter a cellular response in muscle seems to predominate, rather than a humoral response. The myonecrosis is selective for the unique 2M muscle fiber type and is irregularly progressive. For historical reasons and because there seem to be some important diagnostic differences between the eosinophilic form of masticatory muscle myositis and the atrophic form, the names are retained here with knowledge that one disease process may have different faces. The progressive histological changes have much in common.

The clinical disease is expressed as recurrent attacks of pain, mandibular immobility, and sometimes swelling of affected muscles. In initial attacks, or in later ones in which muscle swelling is a feature, the eyes may protrude. Attacks, if untreated, last up to 2 to 3 weeks, and the periods between attacks may be a few weeks, months, or 2 to 3 years. Early episodes may pass undetected, but ultimately attention is drawn to the progressive inability to open or close the jaws. Muscular atrophy gradually becomes very obvious, and the head appears to have a fine foxlike contour with unusual prominence of the zygomatic arches (Fig. 2.55A). Lesions are bilaterally symmetric but not necessarily bilaterally equal. The disease, with possible rare exceptions, is progressive, and each attack presumably adds to the muscular degeneration and fibrosis in a cumulative way (Fig. 2.55B), although the long interval between early attacks may allow residual lesions to be minimized. During the acute phase, a moderate leukocytosis is often present, with 10 to 40% of the circulating white cells being eosinophils. This rises to as high as 70% in exceptional cases in German shepherds.

The progressive disease described is now less commonly seen because most cases can be managed by strategic corticosteroid treatment. Acute episodes are greatly shortened, and this seems effectively to reduce the amount of muscle fiber damage. Only a few dogs come to postmortem examination in the chronic phase.

Visible changes in the muscle vary with the stage of disease. The acutely affected muscles are swollen, dark red, doughy or hard, and the cut surface reveals hemorrhagic streaks and irregular yellow-white patches. In the late stages, there is advanced atrophy and fibrosis. Some areas consist only of pink-gray, semitranslucent, mature, connective tissue subdivided by the whiter residual tendons. Since acute episodes are recurrent, the microscopic changes seen will depend on the activity of the lesion when it is examined. The acute lesion always contains numerous eosinophils and smaller numbers of lymphocytes, neutrophils, and plasma cells. In the more chronic stages or during quiescent phases, these proportions seem to be reversed so that plasma cells predominate. The reaction in the muscle is diffuse but not uniform. Distinctly focal and spreading areas of muscle destruction and inflammatory cell aggregation seem to expand and spread in an irregular way (Fig. 2.56A). As cellular collections grow, muscle fibers in the region undergo both atrophy and degeneration. When segmental fiber degeneration occurs, the separated parts of the myofibrillar mass and the empty segment of sarcolemmal sheath are very rapidly enclosed by aggressive, invading cells, only some of which are macrophages. This vigorous invasion of fragmented fibers leads to some unusual cross-sectional structures shared with the other inflammatory myopathies, in which the center of the fiber is hollowed out and filled with invading cells (presumably $T8^{+}$ cells or killer cells), while the outer ring of the fiber consists of apparently normal myofibrils (Fig. 2.56B). There is nothing unusual about the other types of muscle fiber degeneration which occur, except that the cellular proliferation and/or infiltration is excessive and persistent. These cells often accumulate around blood vessels.

Where the reaction is acute and extensive, hyalinization of fibers occurs, and these fragments sometimes persist long after most fibers have vanished from the area. In all areas, some fibers undergo slow atrophy, a few sometimes persisting intact in a mass of connective tissue. Fibrous proliferation does occur, but at least some of the visible connective tissue in advanced atrophic cases consists of preexisting support tissue which has collapsed. At this

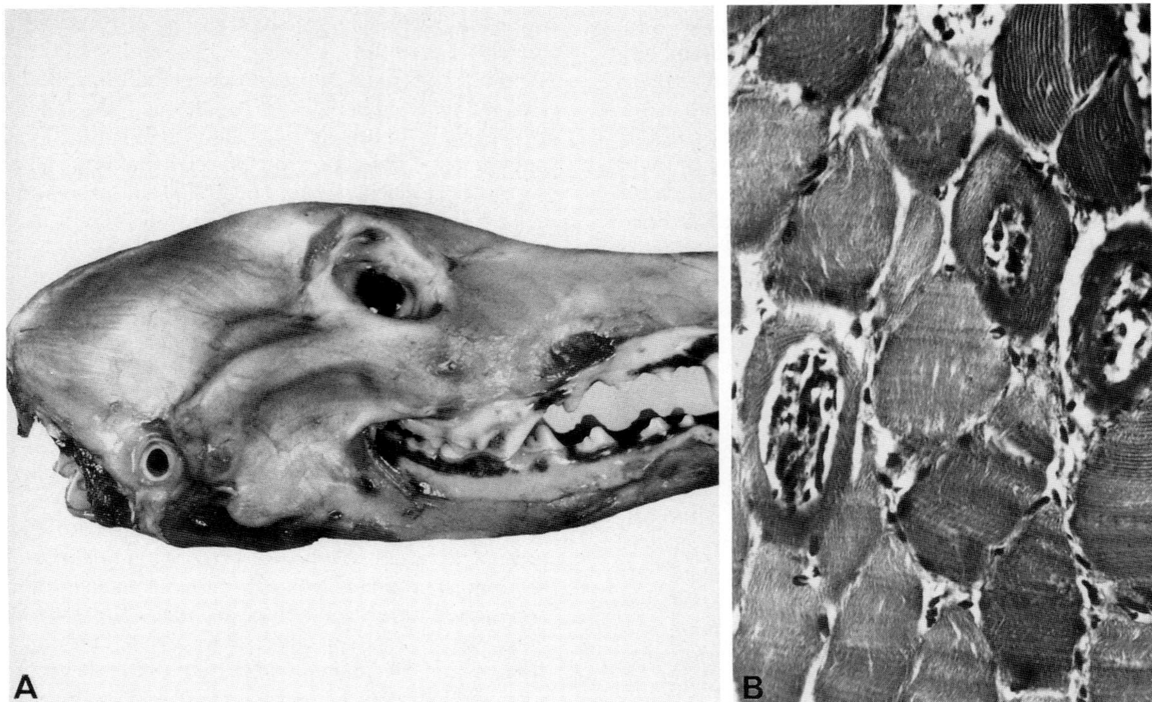

Fig. 2.55 Masticatory myopathy. Dog. (A) Severe atrophy of masticatory muscles. (B) Hollowed-out fibers.

Fig. 2.56A Masticatory myopathy. Dog. Temporal muscle. Light areas contain few residual fibers.

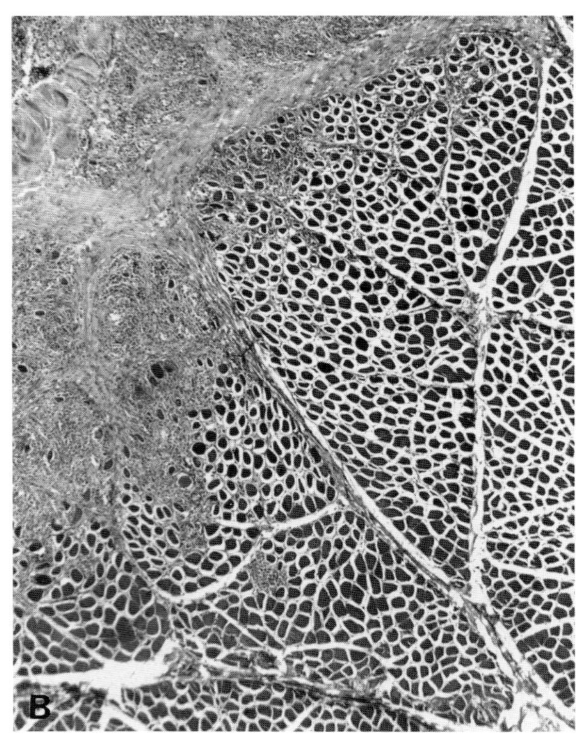

Fig. 2.56B Masticatory myopathy. Dog. Subacute focal spreading lesion.

stage in the evolution of the lesion, the cellular content is greatly reduced, but aggregates of plasma cells and lymphocytes with dark, smudgy nuclei persist. The general pattern of lesion development suggests that repair is not very effective in counteracting the bursts of degenerative activity.

Supporting structures in muscles appear to be unaffected. Nerves, end-plates, neuromuscular spindles, blood vessels, and tendons are unaltered even in chronic disease, apart from the expected modifications which accompany atrophy.

Polymyositis in dogs has been described associated with infections of *Ehrlichia canis* (see Volume 3, Chapter 2, The Hematopoietic System), and in another group of dogs with no apparent predisposing infection. Clinical signs consist of general body pain on manipulation, elevated muscle enzymes in serum, and evidence of inflammatory response in muscle biopsies. The muscles most frequently involved are the esophagus, muscles of mastication, and forelimb muscles. The infiltrative inflammatory reaction is similar to that seen in masticatory muscle myositis.

Dermatomyositis has been described as a familial disease in collies, Shetland sheepdogs, and their crosses. Lesions affect the esophagus, usually the temporalis muscle, and less often many other muscles in the body. The muscles show subacute myositis similar to other diseases in this group, but vessel wall degeneration is also common, and muscle may contain microinfarcts.

Dermatomyositis is considered an autoimmune disease. The antigens responsible for the abnormal antibody seem to be from blood vessels, rather than muscles. The disease is discussed in Chapter 5, The Skin and Appendages.

Bibliography

Buoro, I. B. J. *et al.* Polymyositis associated with *Ehrlichia canis* infection in two dogs. *J Small Anim Pract* **31:** 624–627, 1990.

Emslie-Smith, A. M., Arahata, K., and Engel, A. G. Major histocompatability complex class I antigen expression, immunolocalization of interferon subtypes, and T cell-mediated cytotoxicity in myopathies. *Hum Pathol* **20:** 224–231, 1989.

Engel, A. G., and Arahata, K. Mononuclear cells in myopathies: Quantitation of functionally distinct subsets, recognition of antigen-specific, cell-mediated cytotoxicity in some diseases, and implications for the pathogenesis of the different inflammatory myopathies. *Hum Pathol* **17:** 704–721, 1986.

Goebel, H. H., Trautmann, F., and Dippold, W. Recent advances in the morphology of myositis. *Pathol Res Pract* **180:** 1–9, 1985.

Hargis, A. M. *et al.* Postmortem findings in four litters of dogs with familial canine dermatomyositis. *Am J Pathol* **123:** 480–496, 1986.

Haupt, K. H. *et al.* Familial canine dermatomyositis: Clinicopathologic, immunologic, and serologic studies. *Am J Vet Res* **46:** 1870–1874, 1985.

Head, K. W., Maclennan, W., and Phillips, J. E. A case of eosinophilic myositis in a dog. *Br Vet J* **114:** 22–29, 1958.

Hole, N. H. Three cases of nodular necrosis (Roeckl's granuloma) in the muscles of cattle. *J Comp Pathol* **51:** 9–22, 1938.

Kornegay, J. N. *et al.* Polymyositis in dogs. *J Am Vet Med Assoc* **176:** 431–438, 1980.

Orvis, J. S., and Cardinet, G. H. Canine muscle fiber types and susceptibility of masticatory muscles to myositis. *Muscle Nerve* **4:** 354–359, 1981.

Ringel, S. P. *et al.* Quantitative histopathology of the inflammatory myopathies. *Arch Neurol* **43:** 1004–1009, 1986.

Shelton, G. D., and Cardinet, G. H., III. Pathophysiologic basis of canine muscle disorders. *J Vet Intern Med* **1:** 36–44, 1987.

Shelton, G. D. *et al.* Fiber type-specific autoantibodies in a dog with eosinophilic myositis. *Muscle Nerve* **8:** 783–790, 1985.

Shelton, G. D., Cardinet, G. H., III, and Bandman, E. Expression of fiber type-specific proteins during ontogeny of canine temporalis muscle. *Muscle Nerve* **11:** 124–132, 1988.

Walton, J. The inflammatory myopathies. *J R Soc Med* **76:** 998–1010, 1983.

Whitney, J. C. Atrophic myositis in a dog; the differentiation of this disease from eosinophilic myositis. *Vet Rec* **69:** 130–131, 1957.

G. Changes in Muscle Secondary to Systemic Infections

Lesions in muscle caused by the specific presence of an infectious agent are described elsewhere. More common than these lesions of muscle, and more difficult to explain are the degenerative changes of muscle fibers which occur in acute systemic infections. The lesions, although widespread, cannot be appreciated grossly and consist of a segmental degeneration of a few or many fibers with retraction of fiber segments. The extent of the degeneration of contractile substance may be very slight, and for this reason, floccular or mineralizing changes may be sparse. Some fibers show the distinct hyaline change of Zenker's degeneration with retraction cups visible but not frequent (Fig. 2.57). The fiber breaks lead to a minimum of reactive cellularity since there is little call for extensive cleanup by macrophages or for regenerative proliferation by satellite cells. The changes may be easily overlooked, but a clue to their presence is corrugation of only some fibers or groups of fibers in a longitudinal section of muscle. Although the corrugated fibers may appear otherwise normal, the recoil can be traced to proximal or distal transverse separations. Endomysium and blood vessels are intact, although occasionally there are petechial or larger hemorrhages, and infiltrating lymphocytes or neutrophils are few or absent. Since the degeneration is almost exclusive to sarcoplasm, survival of the animal should allow complete restoration.

XII. Parasitic Diseases of Muscle

A. Trichinellosis

Muscle is the habitat for encysted larvae of the nematode *Trichinella spiralis*, which may survive there for many years (Fig. 2.58). The muscle belongs to the animal which earlier harbored the adult worm in its duodenum, and since animal-to-animal transmission of infection is accomplished by the consumption of infected muscle, most of the species regularly involved are carnivores or scavenger species. Humans, dogs, and a variety of wild

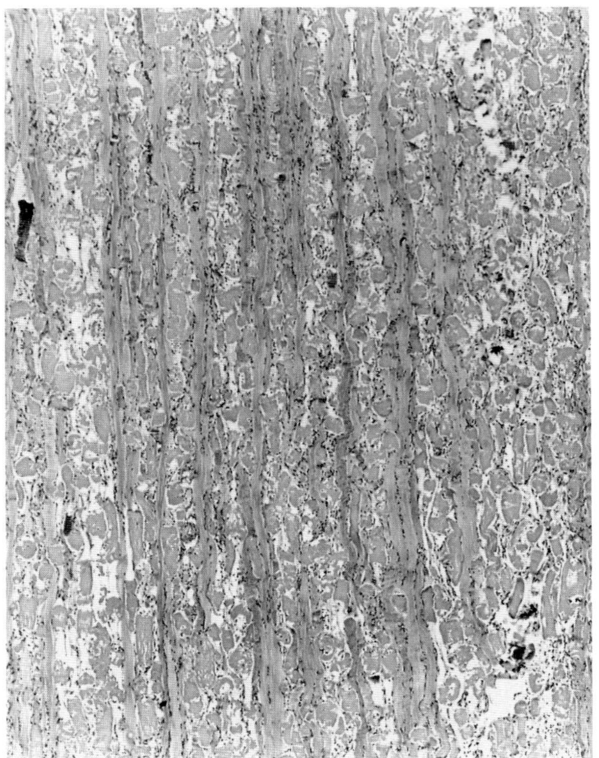

Fig. 2.57 Hyaline degeneration of muscle in salmonellosis. Pig.

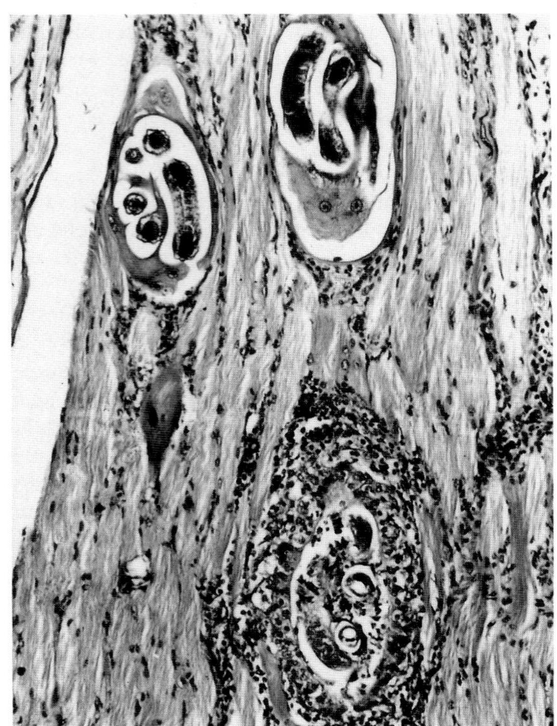

Fig. 2.58 *Trichinella spiralis*. Pig.

Canidae, cats and wild Felidae, pigs, rats, mustelids, bears, polar bears, raccoons, and mice become hosts to the adult and their persistent larvae. Other species including horses and birds may become infected when muscle tissue is included in their feed, and horsemeat has been a natural source of trichinellosis in humans. Trichinellosis is a zoonotic disease sometimes occurring in spectacular outbreaks in humans and animals. Humans become infected when they consume uncooked or incompletely cooked meat of pigs, bears, or aquatic mammals. In those regions of the world where inspection of meat in routine, the incidence of infection is very low, but even in countries such as Canada and the United States of America, the incidence in wild carnivores is 1–10%.

The parasitic cycle for *T. spiralis* begins with ingestion of infected meat fibers. Gastric juices liberate the encysted larvae, which then molt twice, grow to a length of 1 to 4 mm as threadlike, fourth-stage worms, and molt again. Maturity is reached in about 4 days following ingestion, the adults copulate, and the male dies. The ovoviviparous females penetrate via the crypts of Lieberkühn to the submucosal lymphatics where they deposit 0.1 mm long larvae into lymph vessels. The persistence of females in the duodenum is dependent on the state of surface immunity (probably IgA antibody produced locally in the gut wall) and varies from days to 5 to 6 weeks. Some larvae may be passed in feces as the female moves out of the duodenal crypts. The remaining larvae migrate with the lymph, then the blood, to reach the pulmonary and systemic circulations. Those which find their way to muscle may achieve a safe haven away from developing immunity by entering a muscle fiber; those which arrive elsewhere may survive for a brief period but are soon destroyed. In a previously sensitized host, few of the approximately 500 larvae produced by each female are able to enter a muscle fiber in time to ensure survival.

Within the muscle fiber, the larva grows, coils, and enlarges a segment of the host muscle fiber, which is induced to develop some unusual changes as the "nurse cell." Nuclei enlarge, myofibrils are greatly reduced, the basal lamina is very greatly increased in its thickness and number of folds around the affected segment of muscle fiber, and the endoplasmic reticulum, which is in intimate contact with the worm, proliferates. Mitochondria in the immediate vicinity increase in number as they are reduced in size. After a month, the larvae are up to 100 μm long and coiled in a figure eight. There is usually one per fiber. Larvae are not normally visible by naked-eye inspection of muscles unless they are old and mineralized. On routine microscopic examination of muscle, the larvae lie in a bulging glassy segment of muscle fiber, which may be loosely encircled by eosinophils, and in due course, by a scattering of lymphocytes, plasma cells, and macrophages. If the parasitized muscle segment degenerates, the larva is exposed and soon dies, to become the center of a more acute inflammatory, but still predominantly eosinophilic, reaction. Segments of muscle fiber adjacent to the encysted larva may show evidence of degeneration or subsequent regenerative repair with basophilia and centrally located nuclei. In a heavy infestation, a large proportion of the muscle fibers in predilective muscle sites may be taken up with either the parasite or adjacent reactive zones. Purely physical replacement of functional muscle accounts for most of the clinical signs of infestation when they are present, though usually they are absent.

In 2 to 3 months, the cellular reaction subsides, and the muscle fibers enclosing larvae become further modified to give the impression, on light microscopy, of a fibrous capsule. Since parasite survival can be assured only by intracellular seclusion, the "capsule" is, in reality, modified muscle cell components or, perhaps more correctly, modified satellite cells and basal lamina. It is not yet clear what role, if any, the satellite cells play outside of the regenerative process. Once the larvae are encysted in this way, further change, apart from muscle fiber degeneration, is usually confined to deposition of mineral in the encapsulating muscle structure, but this does not seem to affect parasite viability. The larvae may survive more than 20 years, although the average life span is probably a good deal less, depending on host longevity and the occurrence of fiber degeneration.

Several features of this parasitism are unexplained, including the distribution of larvae within the host. Certain muscles, such as the respiratory and masticatory muscles, are preferentially and heavily infected, while other muscles contain a reduced burden. Activity alone is not responsible because a paralyzed diaphragm is still preferentially susceptible. Concentration of larvae in preferred sites may be influenced by blood distribution, increased larval survival, the presence of some needed nutrient, or preferential shielding from normal body defense mechanisms. Heart muscle is sometimes involved, but not heavily; the muscles most involved are tongue, masseter, laryngeal muscles, diaphragm, intercostal muscles, and muscles of the eye, but no striated muscle is exempt. Since some of the selectively involved muscles are small, heavy infestation may have a significant clinical effect in the form of muscle weakness, paralysis, or reduced responsiveness. Usually parasitic infestations of muscle are asymptomatic, and this feature enhances the transfer of infection from animals to man.

Four strains of *Trichinella* are parasites of muscle. *Trichinella spiralis* is the parasite of pigs, rodents, and humans in temperate and tropical climates, and is the most prevalent strain. It is moderately resistant to short-term freezing, and its infectivity is not reduced by freezing and thawing. *Trichinella nativa* is found in colder climates and is the strain most often encountered in polar bears, bears, aquatic mammals, and Eskimos. Its cycle is similar in most respects to that of *T. spiralis,* but the larval form is much more resistant to freezing for long periods. *Trichinella nelsoni* is found in carnivores in eastern and southern Africa and also in Central and Eastern Europe. *Trichinella pseudospiralis* is found in northeastern Europe and differs from the other three strains in its failure to encyst in muscle. The strain migrates through muscle more or less continuously under experimental conditions but appears otherwise to have a cycle similar to that of *T. spiralis*. The four strains are not yet fully characterized, and the distinctions may be unnecessary; they may represent ecotypes.

B. Cysticercosis

Many of the larval forms of tapeworms of carnivores develop and are temporarily stored in the viscera or other tissues of prey species. A few have a special predilection for skeletal muscles and myocardium, and this much smaller group is dealt with here. The pathological effects of the adult tapeworms in the intestinal tract of the carnivorous host are dealt with in Volume 2, Chapter 1, The Alimentary System.

Taenia solium is a large tapeworm (up to 8.0 m long), common in many parts of the world and resident in the intestinal tract of humans and sometimes other primates. The larvae (or metacestode form) usually develop in the pig or wild pig, but for this species of tapeworm, humans can sometimes be host to both the tapeworm and the larval cysticercus. Gravid tapeworm segments are passed in feces, and because they are nonmotile, the 40,000 eggs in each segment tend to be concentrated over a small area. Susceptible pigs having access to infected human feces are easily infected. Eggs resist destruction for relatively long periods in soil, moist ground surfaces, or sewage sludge, and survive in flowing water for a while. Following ingestion, the outer shell is digested in the stomach, releasing and activating the tiny oncosphere, which penetrates the intestinal blood vessels and reaches the general circulation. Most of the larvae in the pig find their way to heart, masseter, tongue, or shoulder muscles. When they migrate in humans, they are distributed to connective tissues, brain, and viscera. The larvae become cysticerci (*Cysticercus cellulosae*), enlarging to a cyst with a single inverted scolex which, when mature, measures 1–2 cm and is easily visible between muscle fibers. Cysticerci are rapidly ensheathed at first in a loose and then a more dense, connective tissue capsule derived from endomysium. A few lymphocytes and a few or many eosinophils lie in the outer regions of the capsule. The cysticercus seems to avoid effective immune-mediated destruction for some time by converting elements of the complement system to inactive products, although in a strongly immunized animal the larvae are eventually destroyed, mineralized, and removed (Fig. 2.59A). In order to allow themselves growing room within developing inelastic collagenous capsules, the cysticerci create a crescentic zone of degenerative lysis (presumed to be induced enzymatically), and this can often be seen in histologic sections of encysted metacestodes, as can parts of the scolices and hooks in those species which have them. *Cysticercus cellulosae* has rostellar hooks. The survival time for *C. cellulosae* is not

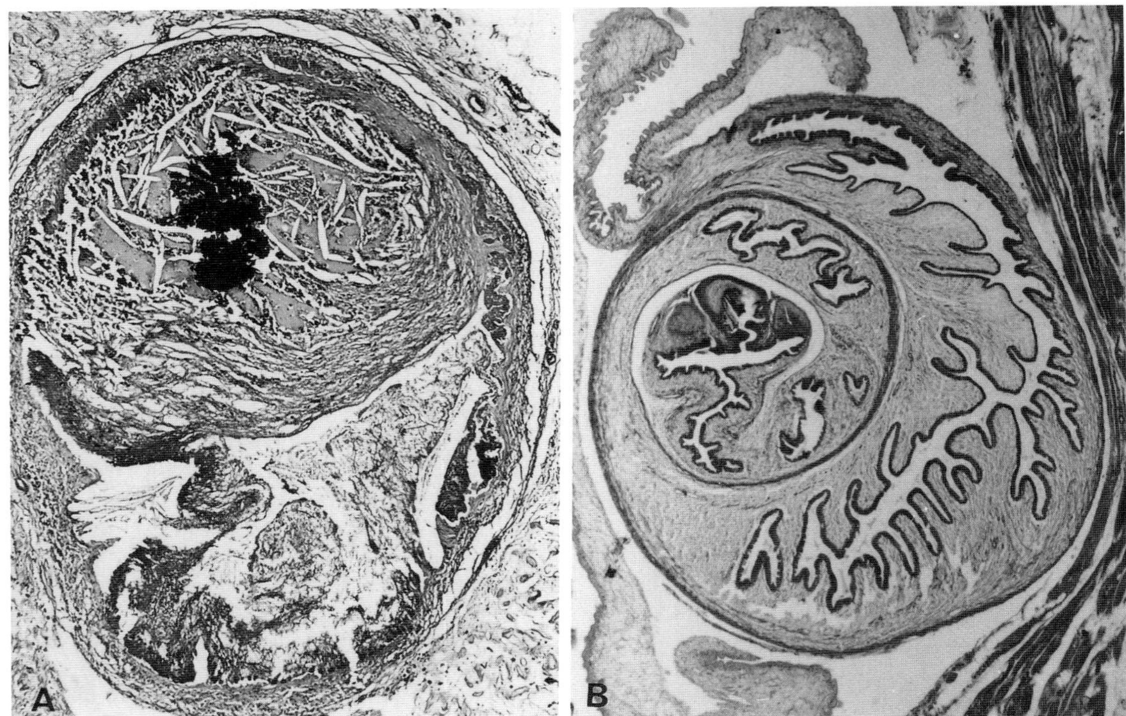

Fig. 2.59 (A) *Cysticercus cellulosae* in brain. Dog. (B) *Cysticercus ovis* in diaphragm. Sheep.

known, but the question is usually not relevant since pigs are normally slaughtered at a young age when virtually all cysticerci in muscle are viable. Humans complete the cycle and become infected when they consume raw or incompletely cooked pork.

Taenia saginata is probably the most common tapeworm in humans; its larvae are in cattle; and it is found in most regions of the world. Transfer of the very resistant eggs from the fecund proglottids to calves or cattle is often enhanced by contamination of open water by sewage, or by use of sewage as fertilizer in fields. It also occurs directly by contamination of animal feeds with human feces or from soiled human hands. The life cycle is similar to that described above for *T. solium,* and the larval form (*Cysticercus bovis*) similarly infests heart and masticatory muscles preferentially, although cysts are often widespread throughout the muscles. Histologically the reaction to the parasite involves fewer eosinophils. *Cysticercus bovis* does not have rostellar hooks. Following a period of growth and development of about 10 weeks, the larvae become infective. After about another 30 days, the cysticerci begin to die, but some larvae may be viable 9 months after infection. Death probably results from development of an immune reaction. Cattle which have acquired a resistance to the invasion of oncospheres across the gut apparently do not have an enhanced capability to cause degeneration of preexisting muscle cysticerci. Humans acquire infection when they consume inadequately cooked beef or veal.

Taenia ovis is a tapeworm commonly found in dogs and wild carnivores throughout the world, which has, as its larval form, a cysticercus (*Cysticercus ovis*) which develops in the heart and skeletal muscles of sheep and goats (Fig. 2.59B). The cycle is similar to that for *T. saginata,* but the larvae do have rostellar hooks, which may sometimes be an aid to histologic identification after the cysticercus has begun to disintegrate.

Taenia krabbei is a tapeworm in wild carnivores in temperate and arctic climates whose larval form (*Cysticercus tarandi*) is found in reindeer, gazelle, moose, and other wild ruminants. Lesions produced in the intermediate host are similar to those seen in cattle with *C. bovis,* and a similar cycle is assumed.

C. Sarcocystosis

Sarcocysts are protozoal parasites of animal muscle that in many respects resemble coccidia, the main difference being their obligatory development in two hosts. The sexual stages develop in a predator host, while the asexual phases develop in the prey animal. Some animals like the opossum and humans are vehicles for both parts of the *Sarcocystis* cycle but not for the same species, there being considerable specificity on the part of both the intermediate and definitive hosts. There is, however, some latitude; for example, several individual parasites develop in dogs, or coyotes, or foxes, or wolves as a definitive host, and a number of intermediate host species of the same general type may be "accidentally" infected at low level. Currently over 90 *Sarcocystis* species are recognized in mammals, birds, and reptiles, and 14 of these are regularly found in muscles of domestic animals as part of the intermediate host infection. Where it was once thought that the muscle phase of infection was asymptomatic and safe for the intermediate host, it is now known that particularly the schizogonous phase may cause severe clinical disease and death under both natural and experimental conditions. The production of enteric disease by the sexual phase in the predator host is considered elsewhere (see Volume 2, Chapter 1, The Alimentary System). Fetal infection is described in Volume 3, Chapter 5, The Female Genital System.

The very substantial differences between *Sarcocystis* species in respect to the intermediate host response appears to be a parasite characteristic, not a difference in host immunologic response, although clinical disease may be modified by immunity.

Consideration here is primarily given to those *Sarcocystis* infections affecting the mucles of domestic animals as intermediate hosts. Sporocysts are ingested when herbage contaminated by carnivore or human fecal material is consumed. Sporozoites are released, and these invade many tissues. One or two generations of schizogony take place within endothelial cells, the first in small arterioles, the second (if there is one) in capillaries. The second or third generation of schizonts develops within striated muscle fibers as thin-walled cysts initially containing round metrocytes, which repeatedly divide to produce numerous banana-shaped bradyzoites (Fig. 2.60). The much-enlarged mature cyst persists in muscle for long periods; the cycle is completed when muscle is consumed by the predator host, and the sexual cycle of the parasite is developed in the intestinal epithelial cells.

Clinical disease in an intermediate host may occur at either of two stages of the developmental cycle. It may take the form of fever, petechiation of mucous membranes, edema, icterus, and macrocytic hypochromic anemia 3–5 weeks after initial infection and lasting for 6 to 8 weeks during the schizogonous (parasitemic) stage. These signs seem to be related to many small episodes of intravascular coagulation, although endothelial schizonts are not the site of thrombus formation. The parasites are also present in perivascular macrophages. The second stage (in which clinical signs and death may occur) comes as the schizonts enter muscle at about 40 days, sometimes with extensive fiber degeneration and marked enzyme release, which attracts macrophages and plasma cells. Another wave of muscular disease may be associated with enlargement of the cysts in a massive infestation, and this can cause lameness. Maturation of cysts may take 60–100 days, by which time any tissue reaction has subsided; this may account for the earlier impression that the parasite was an innocuous passenger in muscle.

Three species of *Sarcocystis* have been recognized in the muscles of the **horse;** *S. bertrami, S. equicanis,* and *S. feyeri.* It is not yet clear that they are distinct. All three

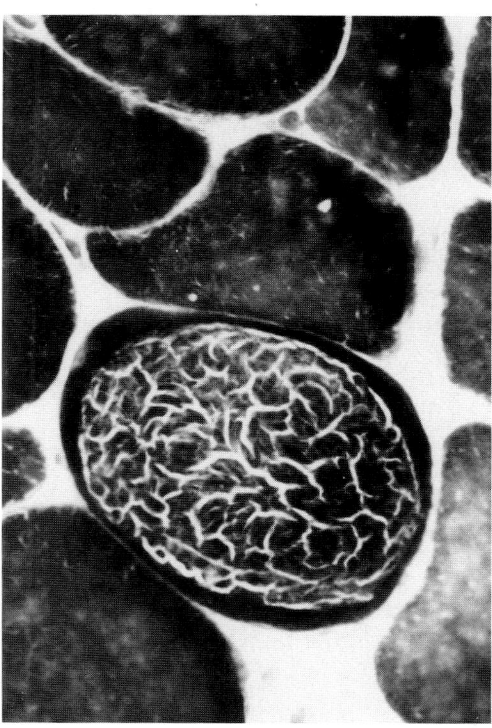

Fig. 2.60 *Sarcocystis tenella* in muscle fibers. Sheep.

complete their cycle in the dog. Of the three, *S. bertrami* seems to cause the greatest tissue response, but none seems to cause clinical disease, and all are therefore, judged "nonpathogenic." Cysts in horse muscle are microscopic in size and rarely numerous, but occasionally there is massive involvement of skeletal muscles.

Protozoal myelitis and encephalitis in the horse, previously presumed due to *Sarcocystis,* is discussed in Chapter 3, The Nervous System.

Sarcocystis species in **cattle** are presently considered to be *S. cruzi,* with a cycle completed in domestic and wild Canidae; *S. hirsuta,* with a cycle completed in the domestic cat; and *S. hominis,* with a cycle completed in humans. Two additional species are found in the water buffalo; *S. fusiformis,* with a cat–buffalo cycle, and *S. levinei* with a dog–buffalo cycle. Only *S. cruzi* seems to be capable of causing significant clinical disease in cattle. Clinical signs exhibited during the schizogonous stage may include evidence of hemolysis as previously described, as well as progressive debility, abortion, salivation, lymphadenitis, sloughing of the tip of the tail, and, in some outbreaks, death in a high proportion of affected animals. See also Eosinophilic Myositis.

Two species of *Sarcocystis* regularly affect **sheep,** namely *S. tenella* (*ovicanis*), which has a sheep–dog cycle, and *S. gigantea,* which has a sheep–cat cycle. *Sarcocystis tenella* is quite virulent in lambs if infective doses are large, and naturally occurring myelitis in mature sheep has been blamed on this organism. Chronically debilitated

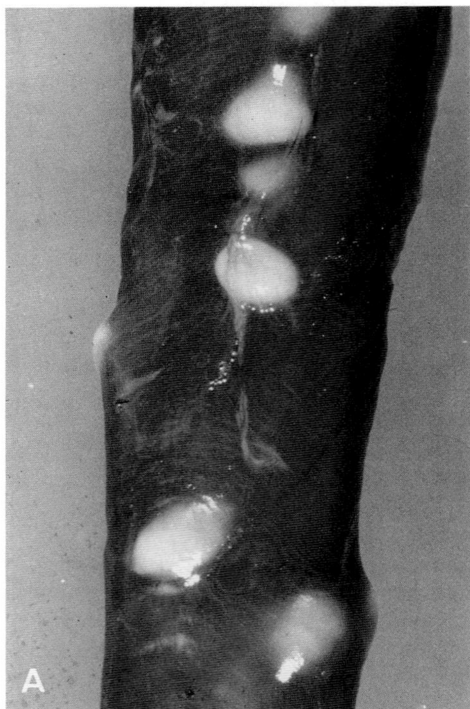

Fig. 2.61A *Sarcocystis gigantea* in esophagus. Sheep.

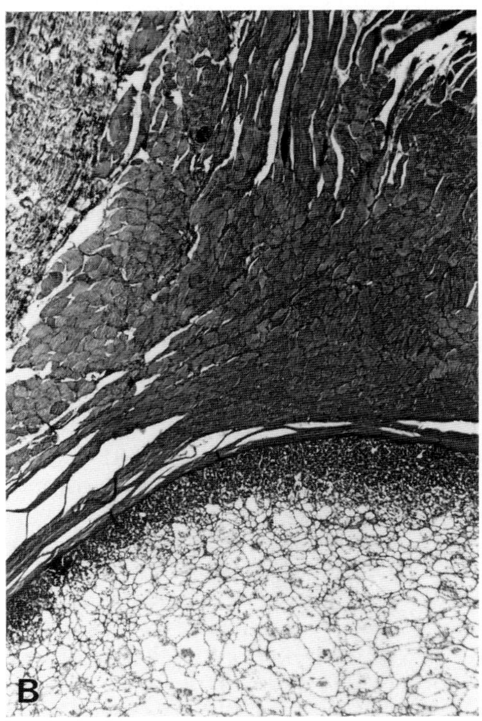

Fig. 2.61B *Sarcocystis gigantea*. Cyst margin.

sheep often have an enormous number of cysts in their muscles. This ought to have some effect on muscle performance, but lameness is not apparent. *Sarcocystis gigantea* cysts are very large (1–3 mm) and distend segments of esophageal striated muscle fibers well above the adventitial surface of the esophagus (Fig. 2.61A,B). It is the only *Sarcocystis* of domestic animals that is visible with the naked eye.

One species (*S. capracanis*), with a goat–dog cycle, is common, and another (*S. orientalis*) may also prove to have a goat–dog cycle. Neither appears to be pathogenic for **goats** as a natural infection but, experimentally, even moderate levels of infectious sporocysts led to death of some animals.

Three species of *Sarcocystis* are found in **swine;** *S. meischeriana* with a pig–dog cycle, *S. porcifelis* with a pig–cat cycle, and *S. suihominis,* which completes its cycle in humans. Of the three, only *S. meischeriana* is known to produce clinical signs of diarrhea, myositis, and lameness.

Sarcocystis organisms have been observed in the muscles of **dogs** and **cats** from time to time. It was thought that these represented erratically traveling parasites in an abnormal site, but it now appears that such cysts are morphologically unlike the species which originate in the cat or dog and complete their cycle in the sheep, horse, or cow. A new identity has not yet been given to these sarcocysts.

Sarcocystosis occurs in many wild species; for example, deer have at least five distinct species, all of which complete their cycle in the dog or in wild Canidae.

1. Histology of Infested Muscle

Histologically, *Sarcocystis* organisms are rarely accompanied by an acute inflammatory reaction, and schizonts in endothelium cause little or no evidence of endothelial cell destruction. As the organisms enter muscle, a wide range of changes may be encountered. Usually there is no muscle fiber degeneration, but there may be thin, linear collections of lymphocytes between fibers in the region. Sometimes the muscle fiber undergoes segmental hyaline change in the region of the invading parasite, and rarely, extensive floccular degeneration of muscle fibers occurs. The extent of muscle change bears little relationship to the numbers of developing cysts, but generally very low numbers of *Sarcocystis* produce no reaction.

As cysts mature, and the contained bradyzoites become more distinct, the cyst capsule within the enlarged muscle fiber becomes thicker and more clearly differentiated from the muscle sarcoplasm. Muscle sarcoplasm and muscle nuclei gather around the parasite in the early stages. In some parasitic species, the outer capsular zone develops distinct radial striations which, on electron microscopy, prove to be complex convolutions of the cyst wall. Small pores allow communication between cyst contents and muscle cell content, but apart from the obvious nutritional dependence of the parasite on muscle fiber, little is known of the biochemical interplay which must take place.

Microscopic inspection of *Sarcocystis*-infected muscle often reveals occasional degenerate parasitic cysts surrounded by variable numbers of inflammatory cells (very few of which are eosinophils), or, at a later state, macrophages and granulation tissue. It is not known whether these represent "over-age" cysts or simply random changes in an easily modified host–parasite relationship. Such reactions in muscle increase with host age.

2. *Eosinophilic Myositis of Cattle and Sheep*

This is a relatively rare condition in cattle and sheep of all ages, which has some significance for meat inspection because the lesions are usually discovered in skeletal muscle and myocardium of animals slaughtered for human consumption. Sudden deaths in cattle and sheep have been ascribed to the myocarditis. The disease is included here because, although the prevalence of sarcocysts in muscle far exceeds the prevalence of lesions, there is good evidence that eosinophilic myositis in sheep and cattle is caused by degeneration of *Sarcocystis* spp. *Sarcocystis* remnants are often found in the center of the myositis lesions, and IgE specific for the parasite is associated with degranulated eosinophils. Animals with parasites but not myositis have serum levels of IgE comparable to those in animals with lesions, suggesting that the inflammatory reaction in eosinophilic myositis is stimulated by independant parasite tube rupture or by trauma to muscle fibers. There are reports of traumatic muscle rupture in cattle initiating an eosinophilic response. A heat-stable, eosinophil-chemotactic substance has been isolated from affected bovine muscle in which eosinophilic myositis lesions were present. The substance has a molecular size less than that of bovine albumin, and its activity seems to correlate with lesion size and severity.

The gross lesions of eosinophilic myositis in cattle are characteristic, being well-demarcated, green, focal stripes or patches, which fade to off-white when exposed to air. Some lesions have a brown-green or gray-green color. Single muscles or groups of muscles may be involved, or the lesions may be widespread through all muscles, including the heart. Individual lesions may be 2–3 mm in diameter to 5 to 6 cm in both heart and skeletal muscle, and no part of the muscle mass seems to be exempt.

Histologically, both acute and more chronic, fibrous reactions may exist side by side. The reaction is characterized by large numbers of eosinophils, which impart the green color to the lesion. Dense masses of mature eosinophils separate adjacent endomysial sheaths and perimysial trabeculae. Muscle fiber degeneration is not an obvious feature of the disease, but occasional segments of fibers do undergo hypercontraction and degeneration. These may be evident later only as endomysial sheaths stuffed with eosinophils rather than muscle fibers, because muscle fiber regeneration seems to be impaired or modified.

With time, fibroplasia is evident, and all endomysial lines are exaggerated. In the chronic lesion, this becomes dense collagenous tissue as the reactive cell population changes from eosinophils to a smaller population of lymphocytes, plasma cells, and histiocytes.

In some cases, individual lesions may take on some of the characteristics of a granuloma, in which central muscle fibers along with adjacent eosinophils undergo degeneration. This central necrotic mass acquires a fringe of epithelioid cells and giant cells, and fibrous tissue becomes circumferentially oriented. The central eosinophilic mass may gather calcium salts, while the edge of the lesion consists of infiltrating eosinophils and fibroblasts which radiate into adjacent muscle. Granulomatous lesions will leave a small scar, but it is likely that the more diffuse lesions of eosinophilic myositis in cattle can disappear in time with very little residual lesion.

Eosinophilic myositis in sheep tends to occur in young animals younger than 2 years. Lesions are comparable in distribution and type to those in cattle, although the frequency of the granulomatous type of change may be higher.

D. Hepatozoonosis

Myositis caused by *Hepatozoon canis* was a consistent feature of 15 natural cases of infection in dogs in southern United States. The disease occurs in a variety of Canidae and Felidae in South Africa and the Middle East, and is becoming established in the dog and brown tick populations in North America. Animals become infected by ingesting an infected tick (*Rhipicephalus sanguineus*), allowing sporozoites to invade the gut wall and undergo schizogony in many tissues of the host, including skeletal muscle. Rupture of schizonts leads in time to an inflammatory reaction, but the subsequent stage, a large 250 μm, thick-walled, single-celled cyst does not stimulate a body response.

Affected dogs show fever, anorexia, general body pain, and gait abnormalities, and many show respiratory signs. Radiographs may show irregular periosteal proliferations. Most dogs have a marked leukocytosis and a slight increase in creatine phosphokinase levels. The organism in gametocyte form may be present in neutrophils, and muscle biopsy may demonstrate the cysts, or the typical pyogranulomatous reaction to the release of merozoites, or the schizonts themselves.

At postmortem, muscle lesions consist of multiple acute granulomas interspersed with rows of neutrophils between muscle fibers. Developing stages of the parasite are usually abundant.

E. **Toxoplasma** *and* **Neospora** *Myositis*

Two apicomplexan parasites, *Toxoplasma* and *Neospora,* in addition to *Sarcocystis,* produce myopathy in several species. Both infections are dealt with more fully elsewhere (see Volume 2, Chapter 1, The Alimentary System and Volume 1, Chapter 3, The Nervous System, respectively).

Toxoplasma gondii infections, particularly in puppies

and kittens, or in any species of farm animal naturally immunosuppressed, or in animals on immunosuppressive therapy, may produce a massive involvement of skeletal muscle fibers. In the majority of toxoplasma infections, however, muscle lesions are rare. In muscle fibers, both tachyzoites and thin-walled cysts (bradyzoites) may be present, but the former are generally transient and hard to find, and the latter are infrequent. Cysts inside muscle fibers are spherical and nonreactive.

Neospora caninum infections occur mainly in dogs, but the organisms are also found in cats, and the disease may have originated in the latter species. Many of the organisms originally identified in tissue sections as *Toxoplasma* are now recognized as *Neospora*. The two infections can be distinguished definitively by immunoperoxidase procedures, and presumptively by microscopic appearance. *Neospora* cysts are thick walled and confined to the central nervous system; *Toxoplasma* cysts are thin walled and found in several tissues, including muscle.

Neospora caninum has a much greater tendency to produce myositis because the tachyzoites form large fusiform packets in muscle after the second division stage of meronts. When these escape the muscle fiber in a partially immunized host, the inflammatory reaction, which is predominantly histiocytic and plasmacytic, is abundant and separates individual fibers. There is also segmental muscle fiber degeneration, which may be a response to direct parasite invasion, but is likely to be the result of a local release of free radicals from dying inflammatory cells. At this stage clinical signs of muscle weakness may be obvious and on postmortem, muscles over the entire body may be pale, streaky, and atrophic.

Bibliography

Barrows, P. L., Prestwood, A., and Green, C. E. Experimental *Sarcocystis suicanis* infections: Disease in growing pigs. *Am J Vet Res* **43:** 1409–1412, 1982.

Barton, C. L. *et al.* Canine hepatozoonosis: A retrospective study of 15 naturally occurring cases. *J Am Anim Hosp Assoc* **21:** 125–134, 1985.

Beech, J., and Dodd, D. C. Toxoplasmalike encephalomyelitis in the horse. *Vet Pathol* **11:** 87–96, 1974.

Bundza, A., and Feltmate, T. E. Eosinophilic myositis/lymphadenitis in slaughter cattle. *Can Vet J* **30:** 514–516, 1989

Clare, M. Three-year history of a case of eosinophilic myositis. *Vet Rec* **68:** 202–203, 1956.

Dick, T. A. *et al.* Sylvatic trichinosis in Ontario, Canada. *J Wildl Dis* **22:** 42–47, 1986.

Dies, K. Survival of *Trichinella spiralis* larvae in deep-frozen wolf tissue. *Can Vet J* **21:** 38, 1980.

Dubey, J. P., and Frenkel, J. K. Immunity to feline toxoplasmosis: Modification by administration of corticosteroids. *Vet Pathol* **11:** 350–379, 1974.

Dubey, J. P., and Lindsay, D. S. Transplacental *Neospora caninum* infection in dogs. *Am J Vet Res* **50:** 1578–1579, 1989.

Dubey, J. P. *et al.* Sarcocystosis in goats: Clinical signs and pathologic and hematologic findings. *J Am Vet Med Assoc* **178:** 683–699, 1981.

Dubey, J. P. *et al.* Newly recognized fatal protozoan disease of dogs. *J Am Vet Med Assoc* **192:** 1269–1285, 1988.

Dubey, J. P., Speer, C. A., and Fayer, R. "Sarcocystosis of Animals and Man." Boca Raton, Florida, CRC Press, 1989.

Dubey, J. P., Lindsay, D. S., and Lipscomb, T. P. Neosporosis in cats. *Vet Pathol* **27:** 335–339, 1990

Edwards, J. F. *et al.* Disseminated sarcocystosis in a cat with lymphosarcoma. *J Am Vet Med Assoc* **193:** 831–832, 1988.

Faull, W. B., Clarkson, M. J., and Winter, A. C. Toxoplasmosis in a flock of sheep: Some investigations into its source and control. *Vet Rec* **119:** 491–493, 1986.

Fayer, R., Johnson, A. J., and Lunde, M. Abortion and other signs of disease in cows experimentally infected with *Sarcocystis fusiformis* from dogs. *J Infect Dis* **134:** 624–628, 1976.

Frelier, P. F., Mayhew, I. G., and Pollock, R. Bovine sarcocystosis: Pathologic features of naturally occurring infection with *Sarcocystis cruzi*. *Am J Vet Res* **40:** 651–657, 1979.

Gajadhar, A. A., Yates, W. D. G., and Allen, J. R. Association of eosinophilic myositis with an unusual species of *Sarcocystis* in a beef cow. *Can J Vet Res* **51:** 373–378, 1987.

Gould, S. E. *et al.* Studies on *Trichinella spiralis*. *Am J Pathol* **31:** 933–963, 1955.

Granstrom, D. E. *et al.* Type I hypersensitivity as a component of eosinophilic myositis (muscular sarcocystosis) in cattle. *Am J Vet Res* **50:** 571–574, 1989.

Harcourt, R. A., and Bradley, R. Eosinophilic myositis in sheep. *Vet Rec* **92:** 233–234, 1973.

Jensen, R. *et al.* Eosinophilic myositis and muscular sarcocystosis in the carcasses of slaughtered cattle and lambs. *Am J Vet Res* **47:** 587–593, 1986.

Johnson, A. J., Hildebrandt, P. K., and Fayer, R. Experimentally induced *Sarcocystis* infection in calves: Pathology. *Am J Vet Res* **36:** 995–999, 1975.

Kirkpatrick, C. E. *et al.* *Sarcocystis* sp. in muscles of domestic cats. *Vet Pathol* **23:** 88–90, 1986.

LeCount, A. L., and Zimmermann, W. J. Trichinosis in mountain lions in Arizona. *J Wildl Dis* **22:** 432–434, 1986.

Leek, R. G., Fayer, R., and Johnson, A. J. Sheep experimentally infected with *Sarcocystis* from dogs. I. Disease in young lambs. *J Parasitol* **63:** 642–650, 1977.

Levine, N. D., and Ivens, V. "The Coccidian Parasites (Protozoa, Apicomplexa) of Carnivores." Illinois Biological Monograph No. 51, University of Illinois Press, 1979.

Levine, N. D., and Tadros, W. Named species and hosts of *Sarcocystis* (protozoa:Apicomplexa:Sarcocystidae). *Syst Parasitol* **2:** 41–59, 1981.

Lindberg, R. *et al.* Canine trichinosis with signs of neuromuscular disease. *J Small Anim Pract* **32:** 194–197, 1991.

Lindsay, D. S., and Dubey, J. P. Immunohistochemical diagnosis of *Neospora caninum* in tissue sections. *Am J Vet Res* **50:** 1981–1983, 1989.

McIntosh, A., and Miller, D. Bovine cysticercosis, with special reference to the early developmental stages of *Taenia saginata*. *Am J Vet Res* **21:** 169–177, 1960.

McManus, D. Prenatal infection of calves with *Cysticercus bovis*. *Vet Rec* **72:** 847–848, 1960.

Meads, E. B. Dalmeny disease—another outbreak—probably sarcocystosis. *Can Vet J* **17:** 271, 1976.

Oghiso, Y., and Fujiwara, K. Eosinophil chemotactic activity of muscle extracts from bovine eosinophilic myositis. *Jpn J Vet Sci* **40:** 41–49, 1978.

O'Toole, D. Experimental ovine sarcocystosis: Sequential ultrastructural pathology in skeletal muscle. *J Comp Pathol* **97:** 52–60, 1987.

O'Toole, D. *et al.* Experimental microcyst sarcocystis infection in lambs: Pathology. *Vet Rec* **119:** 525–531, 1986.

Schad, G. A. *et al. Trichinella spiralis* in the black bear (*Ursus americanus*) of Pennsylvania: Distribution, prevalence, and intensity of infection. *J Wildl Dis* **22:** 36–41, 1986.

Soule, C. *et al.* Experimental trichinellosis in horses: Biological and parasitological evaluation. *Vet Parasitol* **31:** 19–36, 1989.

Sudaric, F., and Dragicevic, R. Myositis eosinophilica kod goveda [Eosinophilic myositis in cattle]. *Veterinarski Glasnik* **28:** 135–139, 1974.

Tinling, S. P. *et al.* A light and electron microscopic study of sarcocysts in a horse. *J Parasitol* **66:** 458–465, 1980.

XIII. Neoplastic Diseases

A. *Tumors of Striated Muscle*

Primary tumors of striated muscle are rare. Malignant striated muscle tumors are twice as frequent as benign ones, and about half of the striated muscle tumors in domestic animals arise from sites other than skeletal muscles. It is reasonable to suppose that muscle cell tumors do not arise from fully differentiated muscle fibers and arise rarely from their satellite cells, but they may originate from nests of cells which retain an embryonic pluripotent capability or cells which are in a transient but committed myoblastic form. Most of the striated muscle tumors which arise from sites other than muscles originate in or adjacent to derivatives of the Wolffian or Müllerian ducts. These nonmuscle predilection sites are also sites from which leiomyomas, leiomyosarcomas, and teratomas arise in animals, and this gives further support to the notion that unusual, pluripotent cells reside there. A significant proportion of striated muscle tumors which arise in muscles grow from sites of earlier muscle fiber destruction and repair where myoblasts are likely to be present.

Rhabdomyomas and **rhabdomyosarcomas** in animals parallel the comparable tumors in humans in type, incidence, and distribution. They cover a wide range of malignancy from highly differentiated and benign at one extreme to highly malignant, metastatic, and persistent at the other. The age range tends to be lower than that for most other comparable types of neoplasms, and frequently animals affected are younger than two years. **Botryoid rhabdomyosarcomas** of the urinary bladder frequently occur in 2-year-old dogs (see Volume 2, Chapter 5, The Urinary System), and rhabdomyosarcomas have been reported in littermate kittens.

Nearly all of the benign muscle tumors seen in animals are congenital rhabdomyomas of the heart, and these compose nearly a third of all striated tumors reported. They are described in Volume 3, Chapter 1, The Cardiovascular System. Reports of four cases of benign rhabdomyomas of the larynx of dogs have identified a new type of muscle tumor. Sarcomere components were poorly organized, and they were sufficiently atypical that at least two were diagnosed originally as oncocytoma.

Malignant tumors of striated muscle arise in animals with about equal frequency from skeletal muscle and nonmuscle sites, but not from the heart. The botryoid tumors which develop in the bladder or uterus are considered elsewhere. With a few minor variations, they resemble and act like tumors which develop in skeletal muscle. Rhabdomyosarcomas of limb, neck, or head appear as hard, spherical masses deep in muscle. They are pink-gray initially, but as they increase in size, hemorrhage and necrosis modify the color. The most readily palpable or largest tumor mass may not be primary since these neoplasms metastasize back to muscle at an early stage of growth. Other tissues harbor fewer metastases. There is vigorous local invasion, and metastases may become confluent but retain nodularity.

The microscopic appearance of rhabdomyosarcomas is made extremely variable by rapid growth and a high mitotic index, a tendency to develop irregular polyploidy, and variable quantities of glycogen, endoplasmic reticulum, and contractile proteins in the sarcoplasm. Embryonal rhabdomyosarcomas show little organization and consist of lobulated sheets of neoplastic cells showing marked pleomorphism (Fig. 2.62A). Tumor cells may be small and round, oval, or racquet-shaped, or they may be very large and have multiple nuclei or a single large nucleus, which reflects a high level of ploidy. Cytoplasm may be extensively vacuolated (spider cells) or may contain well formed striated myofibrils, and in the latter instance, the tumor cells may be long and straplike. In some embryonal rhabdomyosarcomas, cells are separated by an amorphous myxoid matrix. In about half of them, striated myofibrils are easily identified by light microscopy.

The pleomorphic form of rhabdomyosarcoma is also a highly malignant variant, and cells with great variability of form grow in lobular sheets (Fig. 2.62B). The predominant cell is a small, somewhat angular and fleshy, mononuclear cell with variable nuclear size. Strap cells, giant cells, and large vacuolated spider cells may also be present, and it is usually in the larger granular cells that some fragmented myofibrils with cross striations may be seen, although only a small percentage of pleomorphic tumors can be readily confirmed by this feature. Strap cells with centrally located rows of nuclei may be found, but they rarely contain identifiable striations.

Rhabdomyosarcomas usually have a scant connective tissue stroma and are prone to outgrow their blood supply, which leads to deep necrosis. In established tumors there may also be scattered peripheral collections of lymphocytes and plasma cells, probably a response to the several abnormal antigens elaborated by these neoplasms, but there is little evidence that this response plays any part in containing sarcoma growth. Toward the outer margins of masses, small residual muscle fibers or normal muscle fibers showing segmental degeneration need to be distinguished from tumor cells.

Immunohistochemical techniques make it relatively easy to confirm that unique and/or invariable components are present in neoplastic cells without regard to their pleomorphism. Procedures for myoglobin and desmin are most useful, although desmin is also found in nonmuscle cells and smooth muscle cells, and not all rhabdomyosarcomas and rhabdomyomas have recognizable amounts of myo-

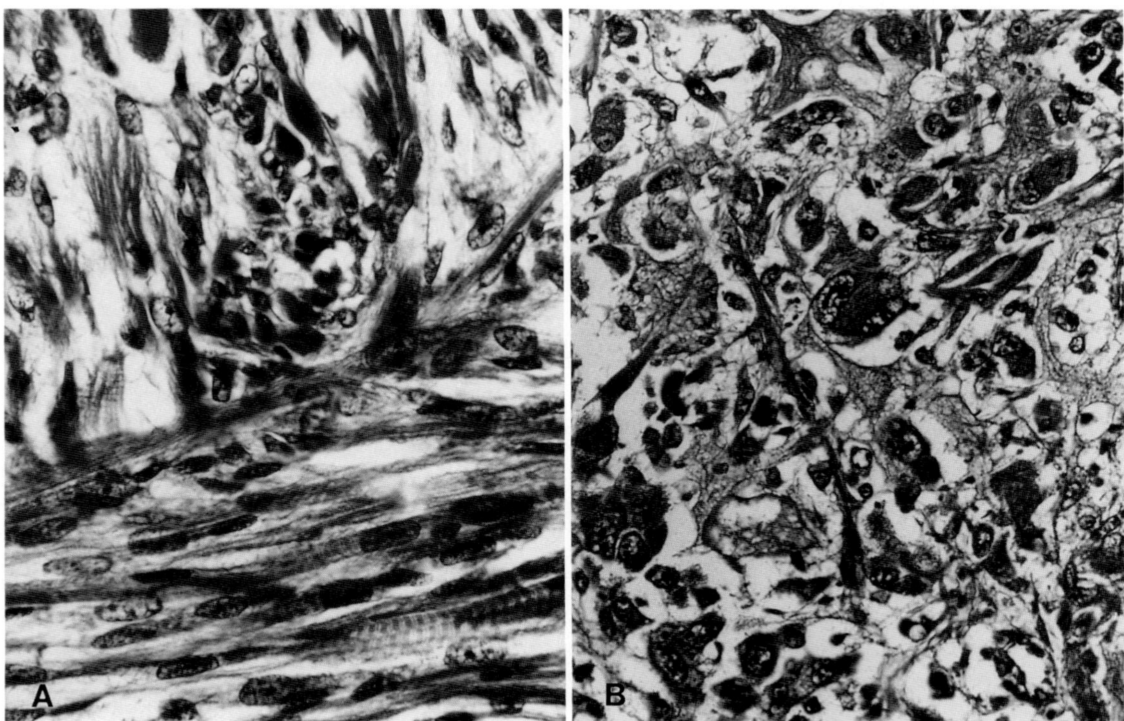

Fig. 2.62 (A) Embryonal rhabdomyosarcoma. Dog. (B) Pleomorphic rhabdomyosarcoma. Dog.

globin. Electron-microscopic identification of two or more components of normal sarcomeres is also thought to confirm the identity of suspected muscle tumors.

B. Nonmuscle Primary Tumors of Muscle

These tumors, which may be benign or malignant, arise from differentiated and undifferentiated supporting tissues of muscle and include lipomas, fibromas, myxomas, and their malignant counterparts. Tumors arising from blood vessels and nerve sheaths may also rarely appear as primary tumors within the boundaries of a muscle. All of these tumors, with the occasional exception of angiosarcomas, grow in muscle as they would in other tissues and show no special features related to their intramuscular origin. Some angiosarcomas which begin in muscle show a characteristic invasive pattern with tubular tongues of tumor cells separating and ensheathing muscle fibers but showing little tendency to destroy them or even cause significant atrophy.

C. Secondary Tumors of Muscle

Malignant melanomas, angiosarcomas, and tumors of the lymphoreticular system are observed often enough as metastases in muscles to conclude that muscle is no less vulnerable to such spread than are other tissues. Hematogenous metastases of tumors other than these are quite uncommon for reasons that are not clear, because some of the tumors that rarely establish hematogenous metastases can invade muscle from an adjacent metastatic site without difficulty. Local invasion of muscle by both epithelial and mesenchymal neoplasms is quite common but often limited. Growth is predominantly along fascial planes, which is the line of least resistance; carcinomas may also grow into degenerating muscle sheaths and extend along the sarcolemmal tubes.

Of the tumors capable of easy metastasis to muscle, lymphoid and melanotic tumors achieve widespread involvement of some muscles by processes of numerous primary metastatic emboli coupled with lateral spread. In the case of lymphoid tumors, leukemic cells collect as small linear aggregations initially, which expand to cause progressive fiber separation and fiber atrophy. In the case of malignant melanoma, the spread may be primarily lymphatic, leaving streaks of black cells, which may be seen grossly. When angiosarcomas metastasize to muscle, the result is likely to be the growth of expansive, hematomalike masses of variable size.

Bibliography

Autio-Harmainen, H. *et al.* Production of basement membrane laminin and type IV collagen by tumors of striated muscle: An immunohistochemical study of rhabdomyosarcomas of different histologic types and a benign vaginal rhabdomyoma. *Hum Pathol* **17:** 1218–1224, 1986.

Bradley, R., Wells, G. A. H., and Arbuckle, J. B. R. Ovine and porcine so-called cardiac rhabdomyoma (hamartoma). *J Comp Pathol* **90:** 551–558, 1980.

Bundtzen, J. L., and Norback, D. H. The ultrastructure of poorly differentiated rhabdomyosarcomas. *Hum Pathol* **13:** 301–313, 1982.

Clark, W. T., Shaw, S. E., and Pass, D. A. Rhabdomyosarcoma of the urethra in a dog. *J Small Anim Pract* **25:** 203–206, 1984.

Coindre, J-M. *et al.* Immunohistochemical study of rhabdomyosarcoma. Unexpected staining with S-100 protein and cytokeratin. *J Pathol* **155:** 127–132, 1988.

Cordes, D. O., and Shortridge, E. H. Neoplasms of sheep: A survey of 256 cases recorded at Ruakura Animal Health Laboratory. *N Z Vet J* **19:** 55–64, 1971.

De Jong, A. S. H., Van Kessel-Van Vark, M., and Albus-Lutter, Ch. E. Pleomorphic rhabdomyosarcoma in adults: Immunohistochemistry as a tool for its diagnosis. *Hum Pathol* **18:** 298–303, 1987.

Dodd, S., Malone, M., and McCulloch, W. Rhabdomyosarcoma in children: A histological and immunohistochemical study of 59 cases. *J Pathol* **158:** 13–18, 1989.

Hulland, T. J. Tumors of muscle. *In* "Tumors of Domestic Animals," J. Moulton (ed.). 3rd Ed., pp. 88–101. Berkeley, University of California Press, 1990.

Kelly, D. F. Rhabdomyosarcoma of the urinary bladder in dogs. *Vet Pathol* **10:** 375–384, 1973.

Liggett, A. D., Weiss, R., and Thomas, K. L. Canine laryngopharyngeal rhabdomyoma resembling an oncocytoma: Light microscopic, ultrastructural, and comparative studies. *Vet Pathol* **22:** 526–532, 1985.

Meuten, D. J. *et al.* Canine laryngeal rhabdomyoma. *Vet Pathol* **22:** 533–539, 1985.

Meyvisch, C., Thoonen, H., and Hoorens, J. The ultrastructure of rhabdomyosarcoma in a dog. *Zentralbl Veterinaermed [A]* **24:** 542–551, 1977.

TENDONS AND APONEUROSES

I. General Considerations

Tendons are derived from the same pool of embryonic mesenchymal cells as muscle fibers, and it is very likely that although commitment occurs early, the differentiation of myoblastic and tenoblastic cells occurs relatively late. In some tendons, notably the suspensory ligaments of the limbs of the foal, the first wave of differentiation is as a muscle, and only subsequently does a form of dedifferentiation (or further differentiation) produce a tendon. As might be suspected, the two end products have many structural similarities. The basic structural units of tendons are bundles of collagen, which cluster around a central elongated collection of tendon fibroblasts or tenocytes and capillaries. Multiples of these units combine to form fascicles somewhat like primary muscle bundles, and fascicles in turn combine in clusters to form the complete tendon, which is ensheathed by a looser fibrous tissue called the peritenon.

Tendons are quite completely, though sparsely, supplied with blood vessels, but in keeping with low nutritional requirements once the tendon is formed, relatively wide distances may separate adjacent parallel vascular channels in mature animals. Segments of long tendons may appear to be almost avascular at times, if flow volume is used as an indicator of vascularity.

At birth, tendons are cellular and vascular, as tenoblasts or tenocytes elaborate the orderly, synchronously kinked, parallel collagen bundles. The kinking provides a mechanism for absorbing stretch impact and provides an interlocking adhesive strength for adjacent fibers, which is enhanced by the presence of an amorphous, noncollagen ground substance, which acts as a "glue." As tendons age they change color from pearly white to yellow-tan and may acquire an even darker brown or red-gray center. Some of these color changes are related to repeated minor episodes of capillary hemorrhage, which occur even in normal, unworked horses or cattle. The distinction between normal and abnormal is often a difficult one to make, but it is normal (in the sense of usual and harmless) for tendons to undergo focal cartilaginous metaplasia. Bony metaplasia in tendons is distinctly pathologic by contrast.

An important issue in the process of tendon aging and injury relates to predisposing lesions. Cartilaginous metaplasia, ischemia, and local fibroblastic proliferations have been regarded as predisposing changes, but they are frequently found in normal animals. It now seems that preparatory events need not be postulated to explain most of the lesions seen. The predilection sites for tendon damage are predictably those with anatomic weakness or which receive disproportionate stretch forces. The subsequent changes relate to stretching of tendon collagen and vessels.

When a tendon is stretched beyond load capacity, fibrils are very likely to be pulled out of kink register, and even if no fibrils break, the tendon will be subsequently weaker at that point. Such stretch lesions are usually accompanied by ruptured collagen as well, although it may be only microscopically detectable. The sequence of events which follows involves the rupture of capillaries, release of fibrin, stimulation of tenocytes and/or peritenon cells to form myofibroblasts, and the formation of scar collagen, which in many ways resembles the original tendon. In the acute stages, this process makes the tendon swollen, warm, and painful; in the chronic stage, the tendon is larger and longer than normal. Original tendon is uniquely constructed entirely of type I collagen. When tendon repairs, the mature scar will no longer consist only of type I collagen, since 20–30% of the replacement tissue will be type III collagen, a form of fiber less able to withstand stretch forces than type I. The differences between a minor sprain and incomplete or even complete tendon rupture are differences in quantity, not form. In larger lesions, there is greater chance for tendon necrosis and sequestration, and a greater chance for the formation of a fibrin clot which is not longitudinally oriented by tension. This may lead to misalignment of the scar fibrils, adhesions to adjacent sheaths, and irregular contours of the tendon. Since myofibroblastic scar tissue is capable of contracting as it matures, the result of tendon repair is not necessarily a longer tendon but, inevitably, a weaker tendon.

II. Parasitic Diseases of Tendons and Aponeuroses

Nematodes of the family Onchocercidae make connective tissue, vessel walls, or tendons their preferred habitat. Most of these and especially those which live in tendons, belong to the genus *Onchocerca,* although not all *Onchocerca* parasites have an affinity for tendons. The adult worms, less than 1.0 mm thick and 50–80 cm long for the female, much shorter for the male, live in tendons, tendon sheaths, or connective tissues of the brisket or abdominal wall, from which site they liberate microfilariae over long periods. The larvae make their way to skin where they are picked up by a blood-sucking parasite in which the next phase of development takes place.

Three different parasites infect domesticated cattle, *Onchocerca gibsoni, O. gutturosa,* and *O. lienalis,* but full investigation may eventually increase the number. *Onchocerca gibsoni* characteristically provokes and inhabits a "worm nodule" or "worm nest" on the brisket or external surfaces of the hind limbs; it may reach 3 cm in diameter. These fibrous lesions are important in the meat industry because the lesions must be manually removed, and this is time consuming if several dozen are present. The nodules may be palpable or moveable through the skin, or they may be fixed to the dermis or ribs. Those in the flank are usually beneath the fascia lata and not externally palpable.

The adult worms inhabit fine tunnels in the discrete nodules of dense fibrous tissue where they form a tangled mass in a milky fluid. Female worms are just grossly visible, but the smaller males are not. Sometimes old hemorrhages, or mineralized remnants of worms are contained in the larger nodules, and microscopically many larvae may also be seen in the capsule as they exit, perhaps via lymphatic vessels. Microfilariae are rarely seen in lymph nodes or blood vessels, but these routes may be used to reach the skin where they collect in large numbers without stimulating a reaction. The intermediate blood-sucking host is probably a simulid or a culicoid insect.

Onchocerca gutturosa is most frequently located on the surface of the ligamentum nuchae of cattle, adjacent to the thoracic vertebral spines, and less frequently on the scapula, humerus, or femur. It is sometimes found in the horse. The adult worms, which are found in pairs, do not stimulate the formation of a nodule, as is the case for *O. gibsoni,* but lie in loose connective tissue. They apparently cause no disease or reaction, and in spite of their wide distribution around the world, they are rarely dissected out or even detected. The intermediate hosts are not all known, but simulids can transfer the infestation.

Onchocerca lienalis, which lies in delicate tunnels in the gastrosplenic ligament and in the splenic capsule of cattle, is considered to be a separate species by some, and synonymous with *O. gutturosa* by others; it is not as widespread as is *O. gutturosa,* but it is seen in Australia and North America.

Adult worms of *Onchocerca cervicalis* live between the fibers of the ligamentum nuchae over the shoulder or neck of the horse or in loose connective tissues nearby. A second very similar parasite, *O. reticulata,* resides at a much lower level of frequency in the tissue around tendon sheaths adjacent to the carpus or suspensory ligaments or at the fetlock, where a low-grade tissue reaction is initiated by the presence of adult worms. When the worms die months or years after initial infestation, they mineralize, and a poorly defined, dense, fibrous reaction engulfs them. The historical association between the presence of *O. cervicalis* and the fistulous withers or poll evil seems now not to be valid (see Chapter 1, Bones and Joints). *Onchocerca cervicalis* infestation is widely distributed. The incidence increases with age of the animal to approximately 100% of horses 10 years of age or older, and seems to have achieved highest incidence in North America. The distribution of *O. reticulata* is not well documented, but it is also widespread. The infestation usually goes undetected, and most horses are apparently asymptomatic.

The long, threadlike female worm is less than 1.0 mm thick but 50–70 cm long. Larvae are hatched from eggs prior to their release from the uterus, and they make their way through connective tissue to the skin where they presumably travel in the dermal lymph vessels. Larvae (microfilariae), which are about 4.0 μm wide and 200–250 μm long, tend to aggregate in certain skin regions, particularly the skin of the ventral abdominal midline, the inner thighs, and the eyelids. Large numbers may be found in the eyes and in selected regions of the dermis. Living microfilariae seem to stimulate no reaction whatever, but it is likely that dead larvae are capable of stimulating a response which involves lymphocytes and eosinophils. These local nodular responses may be seen in skin, accompanied by pruritus, alopecia, and scaliness, and they may be seen histologically in areas where larvae are particularly numerous. Treatment of infected horses using larvacides sometimes induces a marked edematous reaction after 24 hr in the ventral abdominal skin, and this may be interpreted as a reaction to dead parasites. The association between the presence of *O. cervicalis* larvae in the eyes of horses and periodic ophthalmia is now thought to be invalid, but some reaction in eye tissues may be related to larval death. Since larvae can live for most of a year, no tissue reaction would be expected in a very young horse. Living larvae can be quite readily detected as loosely curled structures just under the epidermis and adjacent to adnexal skin structures in histological sections.

Microfilariae of *O. cervicalis* and *O. reticulata* are transported to another horse by blood-sucking insects, in which a stage of development obligatorily occurs. Several different insects may be suitable hosts, but only *Culicoides* midges and mosquitoes of the *Anopheles* species have been confirmed in the role.

Bibliography

Baer, E., Hiltner, A., and Keith, H. D. Hierarchical structure in polymeric materials. *Science* **235:** 1015–1022, 1987.

Fackelman, G. E. The nature of tendon damage and its repair. *Equine Vet J* **5:** 141–149, 1973.

Herd, R. P., and Donham, J. C. Efficacy of ivermectin against *Onchocerca cervicalis* microfilarial dermatitis in horses. *Am J Vet Res* **44:** 1102–1105, 1983.

Lichtenfels, J. R. Helminths of domestic equids. *Proc Helminth Soc Washington* **42:** 1–92, 1975.

McCullagh, K. G., Goodship, A. E., and Silver, I. A. Tendon injuries and their treatment in the horse. *Vet Rec* **105:** 54–57, 1979.

Michna, H. Organisation of collagen fibrils in tendon: Changes induced by an anabolic steroid. *Virchows Arch (Cell Pathol)* **52:** 75–86, 1986.

Ortlepp, R. J. The biology of onchocerciasis in man and animals. *J S Afr Vet Med Assoc* **8:** 1–6, 1937.

Ottley, M. L., Dallemagne, C., and Moorhouse, D. E. Equine onchocerciasis in Queensland and Northern Territory of Australia. *Aust Vet J* **60:** 200–203, 1983.

Parker, R. B., and Cardinet, G. H., III. Myotendinous rupture of the calcanean (Achilles) mechanism associated with parasitic myositis. *J Am Anim Hosp Assoc* **20:** 115–118, 1984.

Soulsby, E. J. L. "Helminths, Arthropods and Protozoa of Domesticated Animals," 7th Ed. Philadephia, Pennsylvania, Lea & Febiger, 1982.

Stannard, A. A., and Cello, R. M. *Onchocerca cervicalis* infection in horses from the western United States. *Am J Vet Res* **36:** 1029–1031, 1975.

Webbon, P. M. A postmortem study of equine digital flexor tendons. *Equine Vet J* **9:** 61–67, 1977.

Webbon, P. M. A histological study of macroscopically normal equine digital flexor tendons. *Equine Vet J* **10:** 253–259, 1978.

Williams, I. F., Heaton, A., and McCullagh, K. G. Cell morphology and collagen types in equine tendon scar. *Res Vet Sci* **28:** 302–310, 1980.

CHAPTER 3 The Nervous System

K. V. F. JUBB
University of Melbourne, Australia

C. R. HUXTABLE
Murdoch University, Australia

I. Malformations of the Central Nervous System

Malformations of the nervous system are common in domestic animals, and their variety is perhaps greater than the variety of malformations in other tissues. There is abundant field and experimental evidence that the effects of teratogens are manifested in the nervous system with disproportionately high frequency. One explanation is that the high degree of differentiation and complexity of the nervous system gives it an increased susceptibility to developmental disturbances. The cause of malformation in individual cases is seldom determined, but there is recognition that in addition to inherited diseases there are a large number of environmental agents, both infectious and toxic, capable of causing anomalies.

Congenital abnormalities of the nervous system consist of deviations in either the nature or velocity of the developmental process. Several main patterns of abnormal development can be recognized. The largest category includes those disorders with a morphologic basis, which are a consequence of failure or disorder of structural development. Many such abnormalities are recognizable by distinctive gross changes, but others require microscopic examination. Some conditions appear to represent retardation of normal development rather than structural aberrations. Disturbance of the normal development may also manifest as premature senescence or degeneration of formed tissues such as the various forms of neuronal abiotrophy.

Some congenital abnormalities appear to represent primary disturbances of function rather than of tissue structure. The most frequently recognized functional disorders are those which arise as a consequence of inherited biochemical defects and which cause distinctive neurohistologic changes. This category is exemplified by the various lysosomal storage diseases and the leukodystrophies, which are discussed later in this chapter.

There are also a number of congenital neurologic diseases, often appearing to be inherited, for which neither an anatomical nor a biochemical basis has yet been identified. The idiopathic epileptiform conditions in various species are examples of this type.

The initial steps in the formation of the nervous system take place early in embryogenesis. As soon as the germ layers are established, a thickened band of ectoderm develops along the mid-dorsal line of the embryo. This thickened area is known as the neural plate and is the primordium for the brain and spinal cord. Differential growth at the margins and in the midplane result in folding of the plate to form a neural groove bounded on each side by an elevated neural fold. The groove continues to deepen, and the neural folds meet dorsally and fuse to form the neural tube, which simultaneously separates from the superficial ectoderm.

The establishment and further differentiation of the neural plate are clearly influenced by the precocity of the cephalic portion of the embryo, and closure of the neural tube progresses rostrally and caudally from the level of the site of development of the rhombencephalon, the most caudal division of the brain. The rostral opening or rostral neuropore closes as the brain vesicles develop. The caudal opening at the caudal extremity of the spinal cord closes later or not at all.

Cellular proliferative activity within the neural tube is

concentrated within an inner subependymal or germinal layer of actively dividing neuroectodermal cells which, with differentiation and outward migration, gives rise first to neurons and later to glia. Shortly after formation of the neural tube, the rudiments of the vertebrae and cranium appear in accord with the fundamental plan of segmental organization. The segmentation and the development of the skeletal investment of the nervous system are dependent on the developmental integrity of the neural tube. For this reason, the severe anomalies of the brain and cord tend to accompany, and are probably responsible for, many gross malformations of the axial skeleton. Anomalies of this type such as spina bifida, anencephaly, and cephalocele occur sporadically in animals but are much less common than malformations in which there is little or no abnormality of the axial skeleton.

The normal pattern of development of the nervous system is not smoothly progressive but comprises a complicated interdependent series of growth spurts of organs and tissues. Birth does not mark a single stage in the structural development of the brain, and there is considerable species variation in the stage of central nervous system development at birth. Cattle, horses, sheep, and pigs are born with nervous systems having a remarkable degree of structural and functional maturity; kittens and puppies are born with their nervous systems relatively immature. After the nervous system is fully formed, the capacity for further production of nerve cells is lost, and subsequent development comprises mainly progressive lengthening and myelination of axons and extension of neuronal dendrites and glial processes.

For each species, neural development thus proceeds along a characteristic predetermined pathway. When mature the nervous system possesses considerable inherent stability, but a time-linked process of decay is also part of the total system of development. Aging neurons tend to accumulate ceroid pigments, which almost certainly reduce their functional efficiency. There is also a continual normal loss of neurons, and in normal animals neuraxonal degeneration is an insignificant but not infrequent finding.

Abiotrophy designates the occurrence of intrinsic premature degeneration of cells and tissues. In the hereditary abiotrophies affecting the nervous systems of animals, such premature loss of vitality is manifested by accelerated and exacerbated degeneration of neurons and their processes.

Investigation of the etiology of neural effects is complicated by the fact that the same type of abnormality may be produced by both genetic and exogenous causes, and in many instances, particularly those of sporadic occurrence, the cause remains unidentified. A variety of viral transplacental infections of the fetal nervous system at critical stages of gestation may result in neurological defects. Some such viruses have the capacity to produce changes, such as symmetrical cavitating lesions which lack pathological features that suggest antecedent infection and resemble malformations thought to be of genetic or toxic cause.

It is not possible to devise a satisfactory theory of the pathogenesis of malformations of the nervous system in our present state of knowledge. Experimental studies indicate that the outcome of fetal interactions with agents potentially teratogenic for the nervous system depends on the species and age of the fetus and the nature of the agent. For the various patterns of malformations, there are corresponding critical periods during which the developing neural tissue is vulnerable. The teratogenicity of the agent is dependent on its cellular tropism, which is commonly directed toward immature rapidly dividing cells, but which often involves specific subpopulations of these cells. Thus, time sets the stage and presents the array of vulnerability, but the agent chooses from among the parts at risk those it has special affinities for damaging.

The most common mode of action for teratogens is selective destruction of cells. Such cytolytic effects have been demonstrated with neural defects induced by viruses and chemicals, as well as by physical agents, such as hyperthermia. In the case of viruses, cellular destruction can be a direct consequence of infection but may develop as part of the inflammatory reaction. The toxicities of many drugs are mediated not by the compounds themselves but by highly reactive metabolites. If they are not detoxified, these unstable metabolites interact covalently with cell macromolecules and may kill affected cells. Cytolysis seems to be the common operative mechanism in a number of important malformations which often have gross features, such as the cavitating cerebral defects and cerebellar hypoplasia. Teratogens may sometimes act by inhibition or distortion of normal cellular development and function to produce more subtle developmental deviations such as hypomyelinogenesis. Infection of calves with bovine diarrhea virus, of lambs with border disease virus, and of piglets with hog cholera virus may provide instances of non-cytolytic disturbance of fetal neural development of this type.

Some drug metabolites, rather than killing cells, result in mutations through binding to nucleoprotein. In view of the multiplicity of pathogenetic pathways which have been identified, it is not surprising that a spectrum of pathogenic effects can result from the action of the one teratogen. All of the lesions may result from an exposure, or any one effect may develop either alone or in combination. Such diversity of teratogenic expression is illustrated by the wide variety of neural and extraneural defects found in kittens born to cats exposed to griseofulvin in pregnancy. A particular abnormality may arise as an expression of many different causes, sometimes disparate in character. The cavitating cerebral defects provide examples of this.

Several attributes of fetal neural tissue are pertinent to the interpretation of malformations. Necrosis of groups of cells occurs in normal development and may be seen in association with remodeling processes. The immature nervous system does not react to damage in the same way as does adult tissue. Parts may be absorbed without trace of connective tissue or neuroglial repair. Furthermore, severe malformations are unlikely to proceed directly from destructive processes, since destruction of parts of the embryonal structure removes the inductor and inhibitor influences on neighboring cell groups. This may, in some instances, lead to arrest of normal processes and, in others, to exuberant reparative proliferation of neuroepithelium, which is indicated histologically by the formation of distinctive neuroepithelial rosettes. Thus malformations may be compounded of degeneration, necrosis, inhibition, overgrowth, and repair. There is at best only a general correlation between the nature of an anomaly and the time in development when it was initiated. In retrospective studies of malformations of unknown cause, it is possible to identify only the latest time at which the malformation may have been produced, based on the normal development of the nervous system of the species in question.

In the event of infection of the developing nervous system by potentially teratogenic viruses, the outcome is determined primarily by age-related cellular susceptibility. However, the immunological status of the fetus, particularly its capacity to produce neutralizing antibody, is likely to be important in determining the character of the neural lesions resulting from viral infection. Lesions leading to major anomalies tend to occur prior to the fetal age at which the fetus can mount a serum neutralizing-antibody response to infectious agents. The fetal brain, especially in the early stages of development, has limited inflammatory potential, but viral-induced destruction of neural tissue in the early fetus may be accompanied by an intense macrophage reaction despite lack of immunological maturity. Such necrotizing processes typically produce gross anatomical defects with little or no evidence of inflammation because of subsidence of the inflammatory changes. In older fetuses, the cytolytic processes are reduced mainly by absence of a susceptible cell population rather than by immune mechanisms, but nonsuppurative encephalitis frequently persists as a residual lesion.

Destruction of developing neural tissue may be effected by mechanisms other than those directly cytopathic for primitive neuroectodermal cells. Vascular damage with consequential edema and necrosis is one such mechanism. Although the fetus is resistant to hypoxia, except if prolonged or profound, there is evidence of the potential of circulatory disturbances in the genesis of cavitating central nervous system malformations such as hydranencephaly.

There are additional poorly understood pathogenetic influences which doubtless contribute to the outcome of fetal interactions with teratogens. Such factors include maternal and fetal genotype, and there is experimental evidence suggesting that pharmacogenetic differences among fetuses of a particular species are important determinants of the results of *in utero* drug exposure. With viral infections, teratogenic potential may vary markedly depending on the strain of the virus and on the immune status of the maternal host.

Bibliography

Brodal, A. Correlated changes in nervous tissues in malformations of the central nervous system. *J Anat Lond* **80:** 88–93, 1946.

Cho, D. Y., and Leipold, H. W. Congenital defects of the bovine central nervous system. *Vet Bull* **47:** 489–504, 1977.

Deakaban, A. S. Brain dysfunction in congenital malformations of the nervous system. *In* "Biology of Brain Dysfunction" G. E. Gaull (ed.), Vol. 3, pp. 381–423. New York, London, Plenum Press, 1973.

de Lahunta, A. Abiotrophy in domestic animals: A review. *Can J Vet Res* **54:** 65–76, 1990.

Dennis, S. M. Congenital defects of the nervous system of lambs. *Aust Vet J* **51:** 385–388, 1975.

Dickerson, J. W. T., and Dobbing, J. Prenatal and postnatal growth and development of the central nervous system of the pig. *Proc R Soc Lond (Biol)* **166:** 384–395, 1967.

Dickerson, J. W. T., Dobbing, J., and McCance, R. A. The effect of undernutrition on the postnatal development of the brain and cord in pigs. *Proc R Soc Lond (Biol)* **166:** 396–407, 1967.

Done, J. T. Congenital nervous diseases of pigs. A review. *Lab Anim* **2:** 207–217, 1968.

Done, J. T. Developmental disorders of the nervous system in animals. *Adv Vet Sci Comp Med* **20:** 69–114, 1976.

Edwards, M. J. Congenital defects in guinea pigs following induced hyperthermia during gestation. *Arch Pathol Lab Med* **84:** 42–48, 1967.

Edwards, M. J. Congenital defects due to hyperthermia. *Adv Vet Sci Comp Med* **22:** 29–52, 1978.

Friede, R. L. "Developmental Neuropathology," 2nd Ed. New York, Springer-Verlag, 1989.

Herschkowitz, N. N. Genetic disorders of brain development: Animal models. *In* "Biology of Brain Dysfunction" G. E. Gaull (ed.), Vol. 2, pp. 151–184. New York and London, Plenum Press, 1973.

James, L. F. Plant-induced congenital malformations in animals. *Wld Rev Nutr Diet* **26:** 208–224, 1977.

Kalter, H. "Teratology of the Central Nervous System." Chicago, Illinois, University of Chicago Press, 1968.

Kalter, H., and Warkany, J. Congenital malformations. Parts 1 and 2. *N Engl J Med* **308:** 424–431, 491–497, 1983.

Keeler, R. D. Livestock models of human birth defects reviewed in relation to poisonous plants. *J Anim Sci* **66:** 2414–2427, 1988.

Leipold, H. W., Dennis, S. M., and Huston, K. Congenital defects in cattle: Nature, cause and effect. *Adv Vet Sci Comp Med* **16:** 103–150, 1972.

Lemire, R. G. *et al.* "Normal and Abnormal Development of the Nervous System." Hagerstown, Maryland, Harper & Row, 1975.

McIntosh, G. H. *et al.* Foetal brain development in the sheep. *Neuropathol Appl Neurobiol* **5:** 103–114, 1979.

McIntosh, G. H. *et al.* Foetal thyroidectomy and brain development in the sheep. *Neuropathol Appl Neurobiol* **5:** 363–376, 1979.

McIntosh, G. H. *et al.* The effects of 98-day foetal thyroidectomy on brain development in the sheep. *J Comp Pathol* **92:** 599–607, 1982.

McIntosh, G. H. *et al.* The effect of maternal and fetal thyroidectomy on fetal brain development in sheep. *Neuropathol Appl Neurobiol* **9:** 215–223, 1983.

O'Hara, P. J., and Shortbridge, E. H. Congenital anomalies of the porcine central nervous system. *N Z Vet J* **14:** 13–18, 1966.

Rousseaux, C. G. Developmental anomalies in farm animals I. Theoretical considerations. *Can Vet J* **29:** 23–29, 1988.

Rousseaux, C. G., and Ribble, C. S. Developmental anomalies in farm animals II. Defining etiology. *Can Vet J* **29:** 30–40, 1988.

Scott, F. W. *et al*. Teratogenesis in cats associated with griseofulvin therapy. *Teratology* **11:** 79–86, 1974.
Shenefeldt, R. E. Animal model of human disease. Treatment of various species with a large dose of vitamin A at known stages in pregnancy. *Am J Pathol* **66:** 589–592, 1972.
Spielberg, S. P. Pharmacogenetics and the fetus. *N Engl J Med* **307:** 115–116, 1982.

A. Microencephaly

Microencephaly refers to an abnormally small brain. The diminution affects particularly the cerebrum. The hypoplastic brain is accommodated within the cranial cavity, which is often smaller than normal, so that the cerebellum appears relatively large. The gyri are of normal size but simplified pattern, and the brain stem structures are normal. The deficiency is of cerebral gray and white matter. It may be an isolated defect or associated with any of a variety of other defects. Microencephaly is manifested externally by an abnormally flattened and narrowed frontal part of the cranium. The cranial bones, particularly the frontal bones, are thicker than normal. Microencephaly occurs in fetal viral infections by Akabane virus in lambs and calves, bovine virus diarrhea virus in calves, border disease virus in lambs, and hog cholera in piglets. The condition has been experimentally produced in lambs exposed to prenatal hyperthermia.

B. Cortical Dysplasia and Malformations of Cerebral Gyri

These malformations have their teratogenetic period late in fetal life, and their occurrence is usually in association with other neural anomalies.

Cortical dysplasia encompasses defects in the architecture of the cerebral cortex. In its mildest expression, there is only histological evidence of lack of the normal orderly layered appearance of the cerebral cortex. **Neuronal heterotopia** is the presence of clusters of nerve cells at a site where they are normally absent, such as subcortical white matter. This condition represents incomplete migration of neuroblasts during fetal life and is usually associated with dysplastic development of the cortex. Scattered, rather than clustered neurons are commonly present in subcortical white matter and are not of clinical significance. The presence of aggregations of small, dark primitive neuroglial cells is a normal finding in periventricular locations and in the rhinencephalic cortex. Heterotopia may also involve glial elements, here taking the form of aberrant nests of glial cells.

In **microgyria** (polymicrogyria), the convolutions are small, unusually numerous, and there is loss of the normal gyral pattern in affected areas. The lesion may be asymmetrical or patchy, abruptly demarcated from normal cortex, but its real extent may be revealed only on cut surfaces. It may be present near the margins of cavitating lesions such as hydranencephaly. **Ulegyria** also imparts a wrinkled appearance to the cortex but arises as a consequence of scarring and atrophy in otherwise topographically normal gyri. It is a result of laminar necrosis, particularly affecting sulci, caused by prolonged ischemic/anoxic injury in the perinatal period.

In **lissencephaly** (agyria), there is persistence of the primitive pattern of the telencephalon, presumed to result from arrested migration of neuroblasts from the subependymal germinal area. The convolutions are almost entirely absent, and the brain surface may be smooth except for slight grooves in which the meningeal vessels are situated. The cerebral cortex is excessively thick. Lissencephaly has been reported in Lhasa Apso dogs.

Pachygyria (macrogyria) is characterized by excessively broad brain convolutions resulting from a reduction in the number of secondary gyri and increased depth of the gray matter underlying the smooth part of the cortex. Pachygyria is a transitional malformation of the cortex, less severe than agyria but akin to it.

C. Agenesis of the Corpus Callosum

Agenesis, partial or complete, of the corpus callosum is an uncommon anomaly but is recorded in most domestic species. It may occur alone or in association with other anomalies of the brain (Fig. 3.1). The septum pellucidum is absent as a collateral defect; the leaves are separated and displaced laterally to the roof of the lateral ventricles. There is no cingulate gyrus, the interhemispheric gyri appearing to radiate from the roof of the third ventricle.

The corpus callosum is formed by the crossing-over of fibers from one hemisphere to the other, a migration which

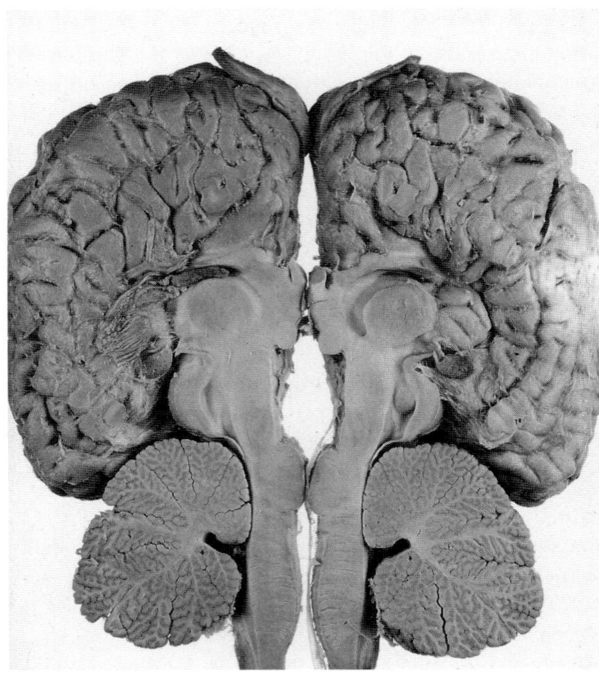

Fig. 3.1 Agenesis of corpus callosum. Horse.

begins at the lamina terminalis, which is the most anterior point of the neural tube. The migration of fibers is guided by a glial scaffold, which may be absent as the primary defect. The corpus may be entirely absent, or the commissures may be present and the corpus, defective in its central part.

D. Cystic Septum Pellucidum

The cavum septi pellucidi is a transitory space which originates within the commissural plate of the lamina terminalis by tissue resorption during embryonic development. In the normal course of development, the cavity is obliterated, resulting in a glial midline raphe. Failure of fusion of the leaflets results in a persistent cavum, which appears as a cystic fluid-filled cavity. In the most severe form of this defect, there is a rounded cavity bounded externally on its lateral borders by the ependyma of opposing lateral ventricles and lined by a network of glial fibers. In less severe forms, it presents as a cruciform or sword-shaped slit beneath the corpus callosum. This condition occasionally accompanies defects caused by fetal infections with viruses such as those of Akabane disease, bluetongue, bovine virus diarrhea, and border disease.

E. Holoprosencephaly

Holoprosencephaly refers to a spectrum of deformities involving the hemispheres and typically including aplasia of the olfactory bulbs and tracts. The dysplasia is also typically associated with facial deformities, and the extent of the nervous and facial deformities correspond closely. The most severe expression is as cyclopia, but the spectrum ranges through intermediate expressions such as cebocephaly to lesser deformities in which the orbits and nose may be normal or the nares are retruded with variably expressed clefts of lips and palate.

Cyclopia, referring to the presence of a single large median eye, merely emphasizes the most obvious and remarkable abnormality of a defect which is anatomically very complicated. The condition is not uncommon in domestic animals, especially pigs, and it can be reproduced experimentally quite simply and in many different ways.

The causes of sporadic cases are largely unknown, but many instances in humans are due to chromosomal anomalies. Varying degrees of cyclopia are present in Guernsey and Jersey fetuses which are the subjects of prolonged gestation. *Veratrum californicum* is responsible for congenital cyclopian malformations which occur endemically in lambs in the western livestock-grazing areas of the United States. The teratogenic agent in the plant is a steroidal alkaloid. It has been established that induction of the cyclopian deformity in lamb fetuses is dependent on pregnant ewes' ingesting *Veratrum californicum* on the fourteenth day of gestation. Cyclopia has also been produced experimentally in cattle and goats and, in both these species, it appears that there is a similar narrow interval of susceptibility to maternal ingestion of *Veratrum* at or about the fourteenth day of gestation. The exact mechanism of the teratogenesis is not known. The nature of the deformities suggests that the primary defect, perhaps involving a selective inhibition of mitosis, occurs at the neural plate stage of development and involves the anterior extremity of the notochord and the mesoderm immediately surrounding it. Failure of proper induction would account for the changes in the skull, in soft tissues of the face, and in the brain. The optic defect, consisting of greater-or-lesser division of a single anlage, is probably secondary to the defects of the forebrain.

The orbits may be approximated (Fig. 3.2) but, typically, there is one large orbit and a single optic foramen. Several bones, including the ethmoids, nasal septum, lacrimal bones, and premaxillae, are usually absent. The globe may be absent, rudimentary, or may form a single structure of near-normal conformation, or be partially divided or completely duplicated. The nose is a proboscis or tube which does not communicate with the pharynx and which is typically situated above the median eye. The forebrain is always severely malformed; the hemispheres are not fully cleft but are present instead as a single thin-walled vesicle with a smooth surface and common ventricle and lacking olfactory nerves and tracts, corpus callosum, septum pellucidum, and fornix.

The hindbrain is usually normal. The pituitary may be displaced or absent, and fetal gigantism associated with prolonged gestation (see Volume 3, Chapter 4, Section V

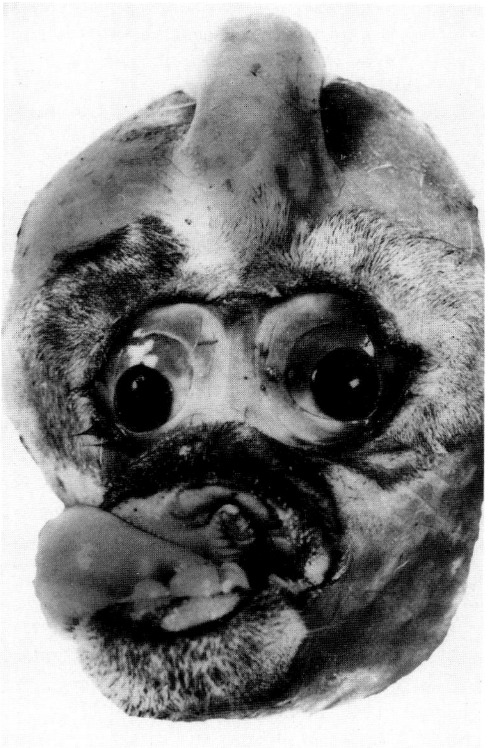

Fig. 3.2 Synophthalmos. Calf. Inherited prolonged gestation.

Diseases of the Pregnant Uterus) occurs in some deformed lambs and calves. The optic nerve is single, atrophic, or absent. The oculomotor and abducens nerves are hypoplastic or absent. There is internal hydrocephalus.

Veratrum californicum is also a cause of embryonic death. A wide range of other defects has been produced experimentally in lambs, with abbreviation of the metacarpal and metatarsal bones being the most distinctive of these.

Cebocephaly is, in many respects, anatomically comparable to cyclopia, and it probably represents a less severe expression of the same basic defects. There are two eyes, severely hypoplastic, in separate but approximated orbits. The nose is in the normal position, but deformed, and may not protrude. There is a single, small nasal cavity or a proboscis with no communication with the pharynx. The brain appears as in cyclopia but may have a slight median sulcus.

Bibliography

Binns, W., Keeler, R. F., and Balls, L. W. Congenital deformities in lambs, calves and goats resulting from maternal ingestion of *Veratrum californicum:* Hare lip, cleft palate, ataxia, and hypoplasia of metacarpal and metatarsal bones. *Clin Toxicol* **5:** 245–261, 1972.

Greene, C. E., Vandevelde, M., and Braund, K. Lissencephaly in two Lhasa Apso dogs. *J Am Vet Med Assoc* **169:** 405–410, 1976.

Hartley, W. J., and Haughey, K. G. An outbreak of micrencephaly in lambs in New South Wales. *Aust Vet J* **40:** 55–58, 1974.

Shimada, M. *et al.* The pathogenesis of abnormal cytoarchitecture in the cerebral cortex and hippocampus of the mouse treated transplacentally with cytosine arabinoside. *Acta Neuropathol* (*Berl*) **58:** 159–167, 1982.

Stewart, R. M., Richman, D. P., and Caviness, V. S. Lissencephaly and pachygyria. An architectonic and topographical analysis. *Acta Neuropathol* (*Berl*) **31:** 1–12, 1975.

F. Dysraphic States

The neural tube extends the full length of the embryo, and the defects may be cranial or spinal or concurrent in both locations.

The brain and cord are ectodermal in origin, and they are, relative to other structures, precocious in development, the precocity being expressed in time and by the influence exerted on other developing tissue, in particular the surrounding mesoderm, which is destined to be the axial skeleton. The lesions of dysraphism therefore frequently consist of combined defects of the brain and cranium and of the spinal cord and vertebral column, and there is usually a close degree of correlation between the mesodermal and ectodermal defects. The correlation is not precise, which may be explained by slight differences in timing of the primary injury or by selective injury to the mesoderm, neural crest, or both.

Human and experimental studies support a role for genetic susceptibility in the pathogenesis of the defects, but particular causative agents are not identified in the spontaneous disease. The primary defect is in orderly growth and dorsal fusion of the neural tube. The alternative view, that fusion does occur but is later reopened because of damage to midline structures or from pressure arising from obstruction to flow of cerebrospinal fluid, may be relevant in some cases.

The classification of dysraphic states traditionally emphasizes the cranial or vertebral component, but some neural lesions are not accompanied by bony defects.

1. Anencephaly

Anencephaly, which means absence of brain, is sometimes applicable literally but usually is not, because often the medulla is present and occasionally some of the mesencephalon. **Prosencephalic hypoplasia** has been used to denote cases in which a relatively normal caudal brain stem persists. In this as in other anomalies, but especially those of the nervous system, the useful nomenclature is a compromise between utility and anatomic accuracy. Anomalies of the same general or specific pattern vary considerably from case to case in the details of their expression, various combinations of typical malformations occur, and there is considerable variability in the development of morphological sequelae.

The primary defect in anencephaly is an arrest of closure of the anterior portion of the neural tube. There is, in consequence, failure of development of the cranium and, with skin lacking over the lesion, the dysplastic rudiments of neural tissue are exposed to the exterior. Anencephaly is seldom complete from the beginning, because the eyes, either well developed or present as rudimentary vesicles, are present. When the eyes are rudimentary, the ganglion cells and optic fibers may be missing. The cranial and neural defects are probably more or less proportionate. With severe degrees of anencephaly, there may be complete failure of cranial development (**acrania**) or there may be rudimentary development of the occipital and adjacent bones. In those cases of anencephaly in which the anterior portion of the brain stem is present, there may be a greater degree of cranial development but failure of fusion (**cranioschisis**). The base of the cranium is well developed, but there are a variety of anomalies of basal bones and no sella turcica. The neurohypophysis is absent, and the adenohypophysis is either absent or unidentifiable. Failure of development of the neurohypophysis can be responsible for the failure of the adenohypophysis to develop normally, and this in turn may be responsible for the prolonged gestation of some anencephalics. The caudal extent of the neural defect varies greatly. There may be involvement of the cervical vertebrae or of the entire neural tube (**craniorrhachischisis totalis**); the defect is always continuous, not segmental. Even when the spinal cord is closed, it may be hypoplastic. When the cord is well developed, it is still small because of the absence of descending tracts.

In anencephaly, the cerebral hemispheres are reduced to a tough formless mass of tissue on the exposed basal bones. The tissue is composed largely of blood vessels with some intermingled neural tissue and is termed the area cerebrovasculosa. Choroid plexuses may be recognizable, and the whole may be covered by a thin layer of squamous epithelium.

2. *Crania Bifida and Related Defects*

Encephalocele (meningoencephalocele, cephalocele) is protrusion of the brain through a defect in the cranium (crania bifida). If fluid-filled meninges only protrude, the defect is termed **meningocele.** In exencephaly, the brain may be either exposed or protruding. These anomalies can be inherited in pigs and in cats and have been associated with treatment of the pregnant queen with griseofulvin.

The morphogenesis of these defects is not simply a problem of defective ossification of the skull with secondary herniation of preformed intracranial tissue but, instead, depends on a primary defect of the neural tube, by which there is focal failure of dehiscence of the neural tube from the embryonic ectoderm and, in consequence, a focal failure of development of the skeletal encasement. The herniations are related to suture lines and are almost always median. They vary from 2 to 10 cm in diameter, and the largest diameter is always much larger than the diameter of the cranial opening. The skin forms the hernial sac, and ectopic bits of disorganized neural tissue may be attached to it; the dura mater does not form in the areas of defect. Encephalocele and meningocele occur usually in the frontal regions (Fig. 3.3A), but some are occipital (Fig. 3.3B), and these latter tend to be located below the occipital crest. Occipital encephaloceles may also be associated with spina bifida of the upper cervical region, sometimes with enlargement of the foramen magnum and absence of the arch of the atlas.

Less well known in animals are frontonasal or frontoethmoidal encephaloceles which present with facial deformities.

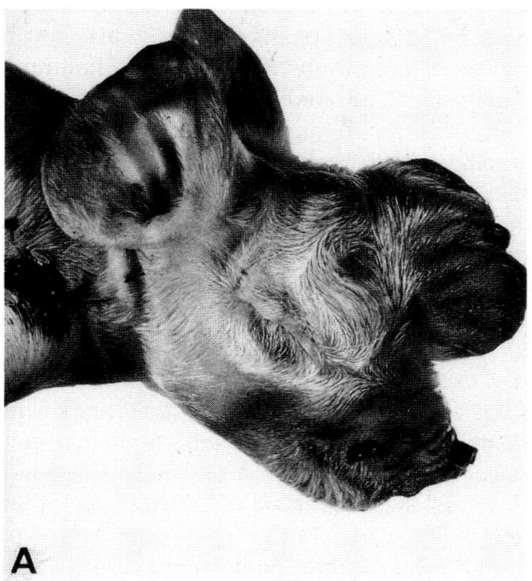

Fig. 3.3A Meningocele. Piglet.

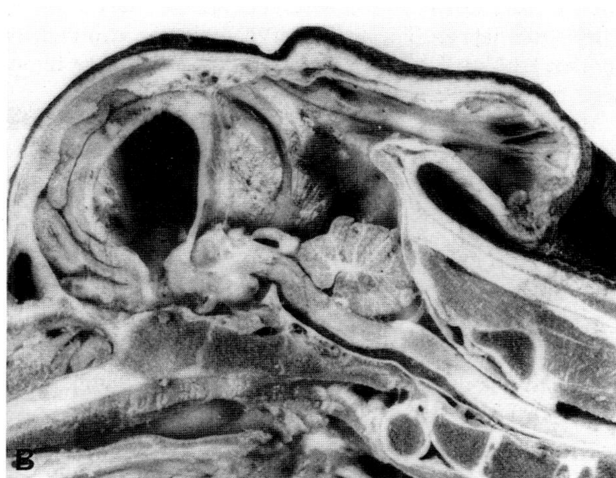

Fig. 3.3B Hydrocephalus and occipital meningocele. Calf.

3. *Spina Bifida and Related Defects*

Spina bifida in its customary usage refers to absence of the dorsal portions of the vertebrae. It is, however, a rather imperfect name, as the various forms of the defect largely represent differences in degree of defective closure of the neural tube, its separation from the ectoderm, and its induction of a skeletal investment. It is convenient to divide the defect into several classes on the basis of severity. Thus myeloschisis, spina bifida occulta, spina bifida cystica with meningocele, and spina bifida with myelomeningocele apply to the vertebral defect; amyelia, diastematomyelia, hydromyelia, and dysraphism apply to the spinal cord.

In total **myeloschisis,** tubulation does not occur, the neural plate remaining open. By total is meant a defect which involves the whole of the vertebral axis with anencephaly an expected accompaniment. There is virtually **amyelia** (absence of spinal cord), neural tissue being present only as soft red masses in the residual groove. Local myeloschisis is a localized defect due to failure of closure of the neural tube. One or more vertebral segments may be defective. The defect may occur in any portion of the vertebral axis but is expected to be lumbosacral.

The defect occurs most frequently in brachycephalic breeds of dogs and is inherited as an autosomal dominant condition in Manx cats. Affected cats are heterozygotes of variable expression, while the homozygous state is lethal.

Spina bifida occulta is perhaps the least rare form of the defect in animals, and it is occult because it is not apparent except for the presence of dimpling or deeper invagination of the skin. It may accompany defects in other remote tissues, but the defect is otherwise limited to the absence of one or more vertebral arches. The cord may be grossly normal but dysplastic microscopically, usually with diastematomyelia. The spinous processes may be bifid or absent.

In **spina bifida cystica** there are morphogenetic and anatomic parallels with the cranial defect, but because of differential growth of the vertebral and neural axes, the anteroposterior position of the skin and bone lesions may not correspond, especially in the caudal regions where the defects are expected to occur.

When **meningocele** is part of the defect, the roof of the cyst comprises skin and condensed meninges, including dura mater. The spinal cord may be normal grossly, but dysplastic segments are detected microscopically. Macroscopic lesions in the cord include partial duplication and cystic distension of the central canal which communicates with the endodural space via a cleft in the dorsal funiculi.

In **meningomyelocele,** the cyst tends to be broad based. Failure of dehiscence of neural crest from surface ectoderm provides for a central area without epithelial covering. This medullovasculosa corresponds to the cerebrovasculosa of anencephaly and consists of vascularized meninges, heterotopic cord tissue, and connective tissue. The defects in the cord include those mentioned as occurring with meningocele and, in severe cases, duplication or absence of the cord in affected segments.

4. Myelodysplasia

The most severe forms of myelodysplasia, affecting especially the potential dorsal regions of the cord, occur in association with the forms of spina bifida previously described. Myelodysplasia, not associated with the skeletal manifestations of spina bifida but probably best included in this classification, occurs quite commonly in animals, particularly in association with arthrogryposis in calves and lambs. The dysplasia, which is readily detectable by the presence of aberrant central canals (sometimes as many as six), chiefly involves limited lengths of the lumbar segment of the cord but occasionally is localized to the cervical or thoracic regions. Denervation atrophy of appendicular muscle is constantly present and is the probable basis of the arthrogrypotic changes. The degree of dysplasia is quite variable, but it is characterized by aberrations of the central canal (Fig. 3.4A), by the absence of dorsal or ventral septa, the presence of ectopic septa and clefts, and by distortion, asymmetry, and partial duplication of the ventral and dorsal horns of gray matter. In some cases the cord is duplicated completely **(diplomyelia)** (Fig. 3.4B) or partially **(diastematomyelia).**

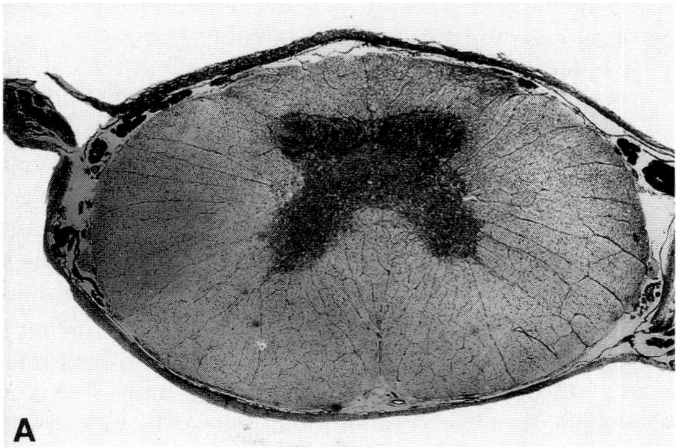

Fig. 3.4A Absence of central canal of spinal cord in dysraphism. Dog.

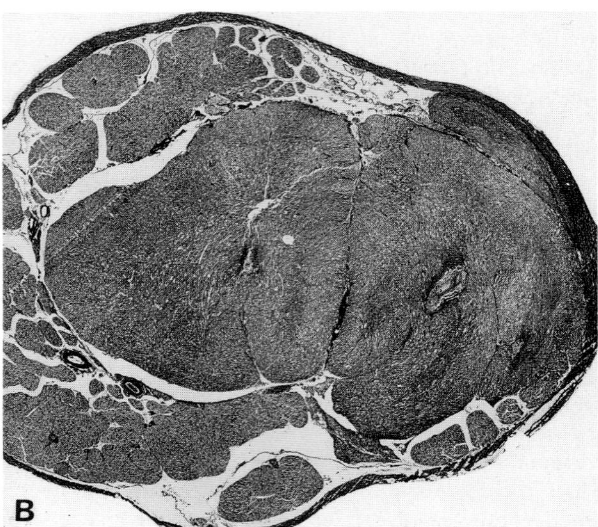

Fig. 3.4B Diplomyelia. Dog.

Syringomyelia is a tubular cavitation of the spinal cord extending over several segments; it is an uncommon anomaly in animals. Syringomyelia is rarely acquired and is best known as one of the lesions associated with the familial disorder of young Weimaraner dogs described next. When the cavitation involves the medulla, the defect is termed **syringobulbia.** Syringomyelia and syringobulbia may occur together.

Hydromyelia is the name given to a dilation of the central canal of the spinal cord. It is not commonly detected but occurs in association with spina bifida and may be a precursor lesion for syringomyelia. Both hydromyelia and syringomyelia are found in the dysraphic spinal cord of arthrogrypotic Charolais calves. Hydromyelia may be acquired as a consequence of obstruction of cerebrospinal fluid flow in the central canal of the spinal cord.

A genetically determined syndrome of pelvic limb gait disturbance in young **Weimaraner dogs** is a manifestation of a spectrum of myelodysplastic and dysraphic lesions. Affected animals exhibit a gait deficit from the time they first begin to walk, but clinical signs in some cases may be so mild as to be barely noticeable. Severely affected dogs typically are unable to completely extend the hind limbs so that their normal attitude is crouched. When walking, the hind limbs are moved together in a bunny hopping or kangaroo-gait fashion. Additionally, affected animals may show (as less constant signs) thoracolumbar scoliosis, depression of the sternum, and abnormal hair streams in the dorsal cervical region. There is no significant progression or regression of clinical signs with age, and there is not a good correlation between clinical signs and the degree of structural malformation.

Pathologically, the exposed spinal cord is normal in

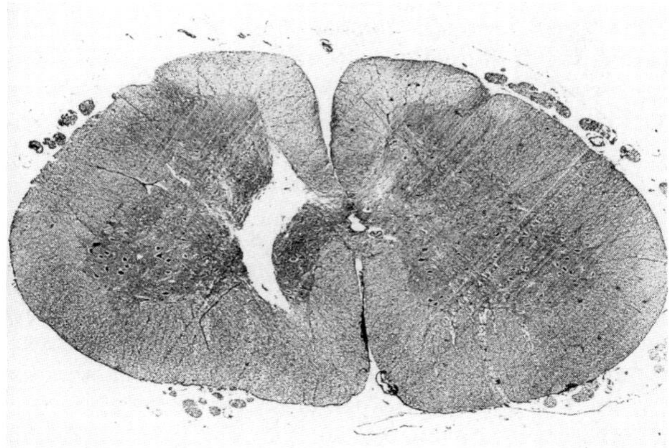

Fig. 3.5 Syringomyelia. Dog.

size and conformation to gross inspection, but a range of dysraphic defects may be present, including anomalies of the dorsal septum, which may be absent; hydromyelia and other anomalies of the central canal, which may be absent; duplication or displacement of the central canal; anomalies of extent and distribution of the central gray matter; anomalies of the ventral horns consisting of deficient delineation and development of medial cell groups, and aberrant collections of neurons and gray neuropil; and deficiency of the ventral median fissure, which may be total.

Histopathologic studies on affected embryos suggest that the primary lesion is related to aberrantly positioned mantle cells ventral to the central canal in the floor plate area.

Syringomyelia is not present until about 8 months of age and involves lumbar segments. Cavitation may be barely visible to the naked eye. Microscopically the cavity is usually found in the central gray matter dorsal and lateral to the central canal (Fig. 3.5). There is not much encroachment on the white matter except for that in the white commissures when the cavitation extends from one side to the other. Connection with the central canal is difficult to demonstrate but may be found in serial sections. Ependyma does not line the cavity, the walls of which are formed of frayed nervous tissue, with an appearance suggesting that the cord has been squashed or torn at necropsy. The tissue around the cavity is edematous and stains poorly.

5. Arnold–Chiari Malformation

The inclusion of the Arnold–Chiari malformation with the dysraphic states may require revision but it does recognize the usual concurrence of spina bifida or meningomyelocele. The defects are consistent in their pattern but variable in extent. The vermis is herniated as a tonguelike process of tissue into the foramen magnum and anterior spinal canal, where it overlies an elongated medulla, which is also displaced caudally (Fig. 3.6). The cranial nerves from the medulla alter their trajectory. The tentorium is inserted caudally toward the foramen magnum, the tentorial hiatus is shallow and wide, and the pons and occipital poles are displaced through it. The displaced occipital poles show sagittal, parallel gyri. Internal hydrocephalus may be a secondary effect.

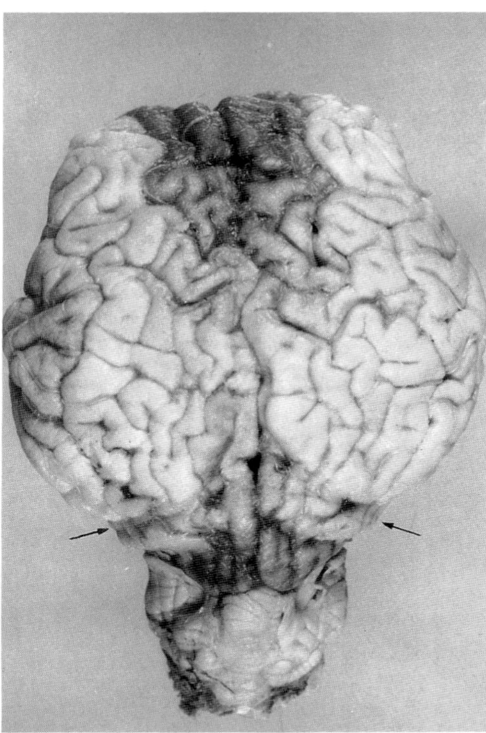

Fig. 3.6 Arnold–Chiari malformation. Calf. Cerebellar deformity and coning accompanying spina bifida. Compression and hemorrhage of the occipital lobes (arrows). Pigmentation of anterior brain is normal.

The pathogenesis of this malformation is unclear. The described changes in the nervous tissue of the posterior fossa are dominated by the features of displacement. The posterior fossa in particular, but perhaps also the anterior fossa, is too small for the normal volume of brain tissue, suggesting that there has been failure of neurogenic induction of osseous growth. The defect is observed occasionally in calves and rarely in other species.

6. Dandy–Walker Syndrome

The Dandy–Walker syndrome is included here, although it is not recognized to be associated with other dysraphic states, but may be associated with other dysplastic lesions of nervous tissue, particularly in the brain stem. There is a midline defect of the cerebellum with the vermis largely absent and the cerebellar hemispheres widely separated by a large fluid-filled cyst in an enlarged posterior fossa. The roof of the cyst, which ruptures easily when the brain is removed, consists of ependyma, a disorganized layer of glial tissue, and an outer layer of leptomeningeal tissue. The floor of the cyst is formed by an expanded fourth ventricle.

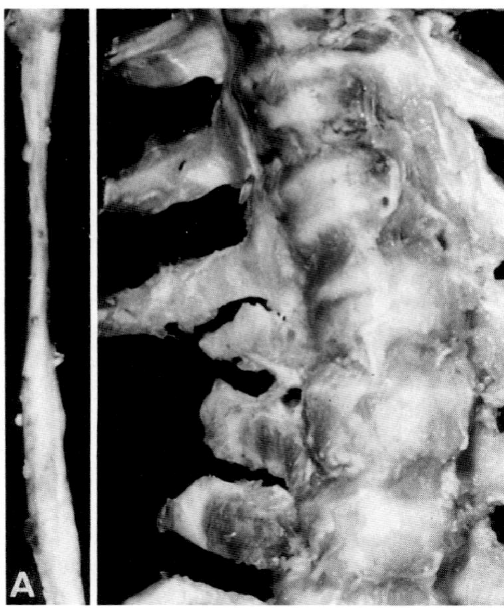

Fig. 3.7A Constriction of spinal cord in congenital scoliosis. Calf.

The pathogenesis of the defect is controversial, especially regarding the patency of the cerebellar foraminae during the early development stage and the pattern of arrested development of the rhombic lips.

G. Segmental Aplasia and Hypoplasia of the Spinal Cord

Segmental aplasia and hypoplasia may occur at any level of the spinal cord (Fig. 3.7A), but the lumbar region is most frequently affected. Some cases are associated with fetal Akabane virus infection.

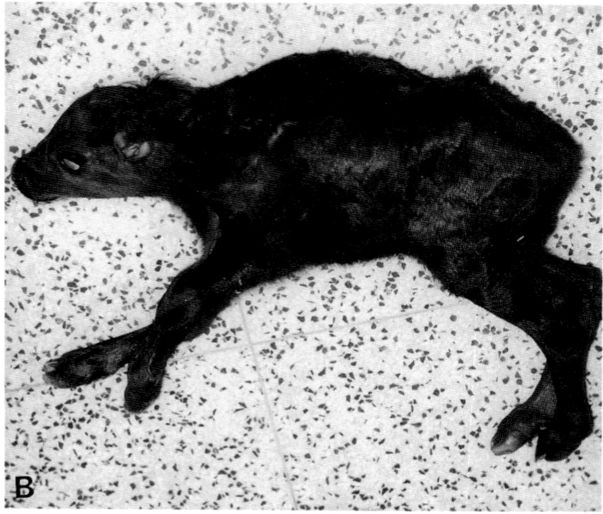

Fig. 3.7B Perosomus elumbus. Calf.

Perosomus elumbus is characterized by partial agenesis of the spinal cord. The lumbar segment is involved, and there is failure of induction of the related vertebrae. The malformation occurs in calves and lambs. The anterior part of the body is normal. Typically, the vertebral axis ends at the caudal thoracic region, and the lumbar, sacral, and coccygeal vertebrae are absent (Fig. 3.7B). The spinal cord ends in the thoracic region in a blind vertebral canal. The caudal part of the body remains attached to the anterior part by soft tissue only; the limbs are arthrogrypotic and their muscles, atrophic. Atypically, only some lumbar segments are absent, or severely hypoplastic with absence of arches and spinal muscle, and a remnant of cauda equina may then be found in the sacral region.

Sacrococcygeal agenesis occurs in association with spina bifida in Manx cats, calves, dogs, and sheep.

Bibliography

Bailey, C. L. An embryological approach to the clinical significance of congenital vertebral and spinal cord abnormalities. *J Am Anim Hosp Assoc* **11:** 426–433, 1975.

Bell, J. E., and Green, R. J. L. Studies on the area cerebrovasculosa of anencephalic fetuses. *J Pathol* **137:** 315–328, 1982.

Cho, D. G., and Leipold, H. W. Spina bifida and spinal dysraphism in calves. *Zentralbl Veterinaermed [A]* **24:** 680–695, 1977.

Cho, D. Y., and Leipold, H. W. Arnold–Chiari and associated anomalies in calves. *Acta Neuropathol (Berl)* **39:** 129–133, 1977.

Dennis, S. M., and Leipold, H. W. Anencephaly in sheep. *Cornell Vet* **62:** 273–281, 1972.

Engel, H. N., and Draper, D. D. Comparative prenatal development of the spinal cord in normal and dysraphic dogs: Embryonic stage. *Am J Vet Res* **43:** 1729–1743, 1982.

Gilbert, F. R., and Thurley, D. C. Congenital spinal cord anomaly in a piglet. *Vet Rec* **80:** 594–595, 1967.

James, C. C. M., Lassman, L. P., and Tomlinson, B. E. Congenital anomalies of the lower spine and spinal cord in Manx cats. *J Pathol* **97:** 269–276, 1969.

Kitchen, H., Murray, R. E., and Cockrell, B. Y. Animal model of human disease. Manx cat, spina bifida, sacrococcygeal agenesis. *Am J Pathol* **68:** 203–206, 1972.

Kornegay, J. N. Cerebellar vermian hypoplasia in dogs. *Vet Pathol* **23:** 374–379, 1986. (Dandy–Walker)

Leipold, H. W. *et al.* Spinal dysraphism, arthrogryposis, and cleft palate in newborn Charolais calves. *Can Vet J* **10:** 268–273, 1969.

Leipold, H. W. *et al.* Arthrogryposis and associated defects in newborn calves. *Am J Vet Res* **31:** 1367–1374, 1970.

McGrath, J. T. Spinal dysraphism in the dog. *Vet Pathol 2:* (Suppl.) 1–36, 1965.

Meyer, D. and Trautwein, G. Experimentelle Untersuchungen uber erbliche Meningocele cerebralis beim Schwein. *Vet Pathol* **3:** 529–542, 543–555, 1966.

Rokos, J. The pathogenesis of spina bifida and related malformations. *In* "Recent Advances in Neuropathology" W. T. Smith and J. B. Cavanagh (eds.), No. 1, pp. 225–245. Edinburgh, Churchill Livingstone, 1979.

Smith, M. T., and Huntington, H. W. Morphogenesis of experimental anencephaly. *J Neuropathol Exp Neurol* **40:** 20–31, 1981.

Wijeratne, W. V. S., Beaton, D., and Cuthbertson, J. C. A field occurrence of congenital meningoencephalocoele in pigs. *Vet Rec* **95:** 81–84, 1974.

Wilson, J. W. *et al.* Spina bifida in the dog. *Vet Pathol* **16:** 165–179, 1979.

Zook, B. C., Draper, D. J. and Graf-Webster, E. Encephalocele and other congenital craniofacial anomalies in Burmese cats. *Vet Med Small Anim Clin* **78:** 695–701, 1983.

H. Hydrocephalus

Hydrocephalus is characterized by an abnormal accumulation of fluid in the cranial cavity. In **internal hydrocephalus,** the fluid is within the ventricular system; in **external hydrocephalus,** the fluid is in the arachnoid space; and in **communicating hydrocephalus,** the excess fluid is present in both locations. The communicating and external types of hydrocephalus, the latter to be distinguished from cerebral atrophy, are quite rare in animals. Internal hydrocephalus, which is denoted by variably dilated ventricular cavities lined by ependyma, is quite common, and may be congenital or acquired, but both forms may be considered here for convenience.

Cerebrospinal fluid is produced by the ventricular choroid plexuses by means of the process of filtration and secretion. There are significant contributions by ependyma and extraventricular structures. Because of the permeable nature of the ependymal lining of the ventricular system, the cerebrospinal fluid is in effect an extension of the extracellular fluid of the central nervous system, and its composition is affected by metabolic and pathological changes within the brain.

The flow of fluid is from the lateral ventricle through the foramina of Monro to the third ventricle and then via the mesencephalic aqueduct to the fourth ventricle. From here, most of the fluid leaves the ventricular system and passes by way of the lateral apertures (lateral foramina of Luschka) into the subarachnoid space. The median aperture of the fourth ventricle (foramen of Magendie) is absent in animals. A small amount of fluid passes into the central canal of the spinal cord from the fourth ventricle. The greater part of the fluid flows forward into the cerebral subarachnoid spaces and basal cisterns; the balance circulates in a restricted fashion in the spinal subarachnoid space. The energy required to circulate the cerebrospinal fluid is imparted largely by the choroid plexuses through their production of fluid. The arterial pulse also contributes to cerebrospinal fluid movement through associated variations in hemispheric volume. Venous resorption of cerebrospinal fluid occurs where arachnoid villi form in the walls of the larger meningeal veins. Transfer of fluid is effected by hydrostatic pressure, with reflux being prevented by the valvular nature of the villi. The arachnoid villi are normally highly permeable, being able to permit the passage of red blood cells. Impedance of this resorption appears to contribute to the increased intracranial pressure of hypovitaminosis A in calves. Some outflow in dogs, and probably in other species as well, can occur via the cribriform plate, but this is probably a bidirectional movement of fluid, and significant outflow may possibly occur only in association with raised intracranial pressure. The spinal subarachnoid space appears to communicate freely with the lymphatic system.

Familiarity with the normal directional flow of the cerebrospinal fluid is basic to an understanding of the pathogenesis of hydrocephalus because in most cases, it is probably of obstructive origin. Certainly, an obstructive deformation can be demonstrated in most instances of acquired hydrocephalus in animals. In congenital hydrocephalus, obstruction is quite often not demonstrable, but stenotic aqueductal malformations may be found (Fig. 3.8). Obstructive hydrocephalus has been induced experimentally in many species of immature laboratory animals with a range of viruses ubiquitous in animals and humans. The selective destructive action of these viruses on ependymal cells, with subsequent reparative gliovascular proliferation, results in obstruction of cerebrospinal fluid pathways, usually in the mesencephalic aqueduct. Thus far in domestic animals, viral-induced hydrocephalus of this genesis has been demonstrated only in experimental canine parainfluenza virus infection in dogs.

Hydrocephalus is "physiological" in the early fetus when the hemispheres are largely thin-walled vesicles. Congenital hydrocephalus exaggerates this physiological degree of ventricular dilation. Even in the absence of obstruction, physiological hydrocephalus may persist or be exaggerated in instances of neural dysplasia such as, for example, in cyclopia and cebocephaly. The cavitating

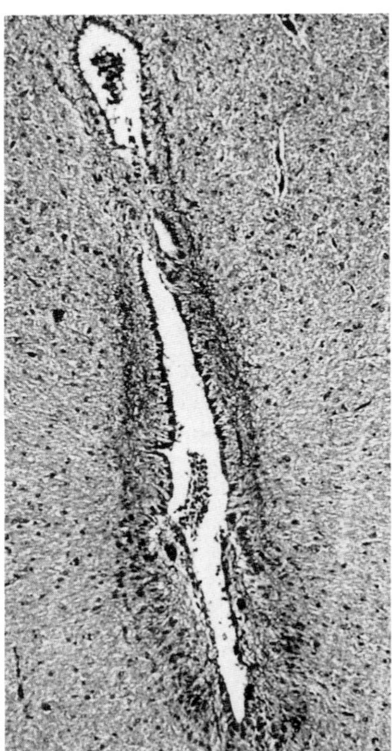

Fig. 3.8 Malformation of aqueduct. Calf.

cerebral defects, hydranencephaly and porencephaly, are associated with internal hydrocephalus. This is likely to be an ex-vacuo or compensatory hydrocephalus occurring secondary to loss of cerebral tissue, as in the inherited leukodystrophies and storage diseases, but a hydrostatic component arising from deranged cerebrospinal fluid circulatory processes is also possible. Hydrocephalus may accompany other neural anomalies, but chiefly the rather rare Arnold–Chiari and Dandy–Walker malformations, in which the hydrocephalus is obstructive in origin and associated with abnormalities of the cerebellum and medulla.

With the exception of the few examples just given, congenital hydrocephalus remains anatomically obvious but pathogenetically obscure, and such cases are common. Pups, calves, and foals are chiefly affected.

Congenital hydrocephalus is well known in pups, especially those of the brachycephalic breeds. This should not imply that the hydrocephalus is correlative to, or a product of, the brachycephaly, because within breeds there is not much variation in the degree of the skeletal defect, and no obvious relation between it and the presence and severity of hydrocephalus. Malformation of the mesencephalic aqueduct has been nominated as a significant pathogenetic factor in dogs. The anatomic expression of the hydrocephalus also varies considerably; one lateral ventricle may be involved, or both may be dilated symmetrically or asymmetrically. The third ventricle and anterior portion of the aqueduct are usually but not always involved, and the fourth ventricle is normal.

Sporadic cases of congenital hydrocephalus occur widely in cattle. Many appear to be secondary to aqueductal stenosis. An autosomal recessive gene is considered responsible for many apparently hereditary cases, but the possible roles of fetal infections and nutritional factors have not been fully evaluated.

Outbreaks of congenital hydrocephalus are recorded in calves in which slitlike deformation of the aqueduct of Sylvius was associated with lateral narrowing of the midbrain, in the peripheral areas of which there was vascular proliferation and perivascular gliosis. Hydrocephalus also occurs in association with chondrodysplasias, especially that of the "bulldog" type, but the primary neural defect in these is not known. Several hydrocephalic syndromes have been described in Hereford and shorthorn cattle; features include cerebellar hypoplasia, microphthalmia, myopathy, and ocular anomalies.

A dietary deficiency of vitamin A can, in experimental circumstances, cause congenital and neonatal hydrocephalus, but spontaneous outbreaks occur only in cattle which have fed for prolonged periods on dry pasture or in feedlots. The hydrocephalus is ascribed to functional impairment of absorption of fluid from arachnoid villi, but there is frequently severe compression of the brain and herniations in the posterior fossa, which is expected to provide mechanical obstruction to drainage.

Among the other domestic species, congenital hydrocephalus occurs sporadically in foals and piglets and with a familial incidence, possibly genetic, in pigs. The defect is quite uncommon in cats.

Acquired hydrocephalus is fairly common in animals. It does not approach in severity the congenital defect. The causes are almost always obstructive, but minor degrees of ventricular dilation occur in association with cerebral atrophy in old dogs. Meningeal lesions which destroy the arachnoid villi can lead to external hydrocephalus. This is, however, quite rare, although it has been observed in diffuse meningeal carcinomatosis. Most diffuse meningeal lesions are inflammatory, but these are typically associated with internal rather than external hydrocephalus, because meningitis tends to involve the basilar regions and the posterior fossa chiefly, and there to interfere with the patency of the foramina of Luschka. Most cases of bacterial meningitis are fatal before the changes of hydrocephalus develop unless, as frequently happens, there is concurrent inflammation of the choroid plexuses, which extends to the ependyma of the aqueduct and obstructs that channel. Even relatively chronic cases of meningitis, such as may be observed in cryptococcosis, are associated with internal rather than communicating or external hydrocephalus. Additional causes of acquired hydrocephalus are intracranial neoplasms, usually primary but sometimes metastatic, and including papillomas and carcinomas of the choroid plexus, parasitic cysts such as hydatids and coenurids, and the late effects of chronic or healed inflammations which involve the ependyma or cause inflammatory softening or atrophy of paraependymal tissue. The pyogranulomatous ependymitis and meningoencephalitis of feline infectious peritonitis occasionally results in hydrocephalus. In horses, the so-called cholesteatomas or cholesterol granulomas, which develop in the choroid plexuses of the lateral ventricles, may occlude the foramen of Monro and cause internal hydrocephalus.

Hydrocephalus does not regress. Whether it progresses or not is difficult to determine. Congenital hydrocephalus cannot be easily diagnosed in the newborn in the absence of secondary changes in the cranium, and diagnosable cases seldom live long enough for the course of the defect to be ascertained. Probably, however, congenital hydrocephalus of mild or moderate degree can remain static because, although a severe and fatal defect is common enough in puppies, hydrocephalus in brachycephalic breeds of dogs is frequently an incidental finding at postmortem, and the degree of ventricular dilation may be minor or moderate, irrespective of the age of the animal. Acquired hydrocephalus in postnatal life tends to be progressive when of the obstructive type, the course depending on the site and nature of the obstructing lesion. Compensatory hydrocephalus (occurring as a response to cerebral atrophy) is static or at the most slowly progressive; it is never severe.

Congenital hydrocephalus is frequently associated with malformation of the cranium. The degree of cranial malformation varies from a slight doming, which may be difficult to appreciate, to an enormous enlargement, which may cause dystocia. Cranial malformation is not, however, invariable, and many cases of congenital hydrocephalus of considerable severity may occur with a skull of normal contour. Whether cranial malformation occurs or not probably depends on the time of onset of the hydrocephalus relative to the degree of ossification of the cranial bones and the development and strength of the sutures, and also on the rate at which the fluid is accumulated.

It is often difficult to be certain of the presence of minor degrees of hydrocephalus. The soft brains of the newborn collapse when removed from the skull, so that dilation of ventricles may not be apparent. In older, firmer brains, asymmetry of the lateral ventricles and relative dilation of the anterior end of the aqueduct when compared with the posterior end are useful indices. The septum pellucidum, however, is the structure most sensitive to the effects of fluid accumulation; it may be fenestrated or may persist as an irregular lacework of connective tissue, but typically, it is absent. Even in mild hydrocephalus, there is usually atrophy in Ammon's horn, readily detectable by the ease with which the pyriform lobes dimple under slight pressure. With hydrocephalus of greater severity, there is ventricular dilation of corresponding degree. The lateral and third ventricles are most severely affected, and there may be no alteration in the fourth. With ventricular dilation, there is parenchymal atrophy affecting chiefly the white matter and the cerebral cortices; ventrally, the increased pressure is buttressed by the basal ganglia. The corpus callosum is elevated and thinned, and the cerebral cortices over the vertex may be reduced to thin shells of gray matter. The floor of the third ventricle is extremely thinned, the hypophysis is atrophied, and the cerebellum is compressed and displaced caudally.

The extensive cranial malformation of congenital hydrocephalus can occur only if the sutures are ununited. The temporal, frontal, and parietal bones are enlarged and thin and are separated from each other by broad membranes of connective tissue in which accessory bones may form. In these severe cases, the base of the cranium is flattened, the fossae are enlarged and smoothed out, and the orbits are separated but individually reduced in size so that the eyes may protrude.

The gray matter is remarkably resistant to the effects of the pressure exerted by the fluid, but the subcortical white matter degenerates rapidly. It is edematous, the oligodendrocytes and astrocytes are reduced in number, and compound granular corpuscles can be found sometimes in short bands lying deep and parallel to the ependyma. The ependyma, tela choroidea, and meninges are usually not significantly altered.

Bibliography

Axthelm, M. K., Leipold, H. W., and Phillips, R. M. Congenital internal hydrocephalus in polled Hereford cattle. *Vet Med Small Anim Clin* **76:** 567–570, 1981.

Baumgartner, W. K. *et al.* Acute encephalitis and hydrocephalus in dogs caused by canine parainfluenza virus. *Vet Pathol* **19:** 79–92, 1982.

Baumgartner, W. K. *et al.* Ultrastructural evaluation of acute encephalitis and hydrocephalus in dogs caused by canine parainfluenza virus. *Vet Pathol* **19:** 305–314, 1982.

Carmichael, S., Griffiths, I. R., and Harvey, M. J. A. Familial cerebellar ataxia with hydrocephalus in bull mastiffs. *Vet Rec* **112:** 354–358, 1983.

Christoferson, L. A., Leech, R. W., and Hazen, G. A. Bovine hydrocephalus in North Dakota: A survey and morphologic study. *Surg Neurol* **7:** 165–170, 1977.

Greene, H. J., Leipold, H. W., and Hibbs, C. M. Bovine congenital defects. Variations of internal hydrocephalus. *Cornell Vet* **62:** 596–615, 1974.

Higgins, R. J., Vandevelde, M., and Braund, K. B. Internal hydrocephalus and associated periventricular encephalitis in young dogs. *Vet Pathol* **14:** 236–246, 1977.

Johnson, R. T., and Johnson, K. P. Hydrocephalus following viral infection: The pathology of aqueductal stenosis developing after experimental mumps virus infection. *J Neuropathol Exp Neurol* **27:** 591–606, 1968.

Leech, R. W., Haugse, C. N., and Christoferson, L. A. Congenital hydrocephalus. Animal Model: Bovine hydrocephalus, congenital internal hydrocephalus, aqueductal stenosis. *Am J Pathol* **92:** 567–570, 1970.

Leipold, H. W., Gelatt, K. N., and Huston, K. Multiple ocular anomalies and hydrocephalus in grade beef shorthorn cattle. *Am J Vet Res* **32:** 1019–1026, 1971.

Masters, C., Alpers, M., and Kakulas, B. Pathogenesis of reovirus type 1 hydrocephalus in mice. *Arch Neurol* **34:** 18–28, 1977.

Milhorat, T. H. The third circulation revisited. *J Neurosurg* **42:** 628–645, 1975.

Sahar, A. *et al.* Spontaneous canine hydrocephalus: Cerebrospinal fluid dynamics. *J Neurol Neurosurg Psychiatry* **34:** 308–315, 1971.

Sedal, L. Cerebrospinal fluid dynamics in health and disease. *Med J Aust* **1:** 272–276, 1972.

Thompson, S. Y. Role of carotene and vitamin A in animal feeding. *World Rev Nutr Diet* **21:** 224–280, 1975.

Weller, R. O. *et al.* Experimental hydrocephalus in young dogs: Histological and ultrastructural study of the brain tissue damage. *J Neuropathol Exp Neurol* **30:** 613–627, 1971.

I. Hydranencephaly

In hydranencephaly there can be complete or almost complete absence of the cerebral hemispheres, leaving only membranous sacs filled with cerebrospinal fluid and enclosed by leptomeninges. The cranial cavity is always complete, in contrast to that in hydrocephalus, and usually of normal conformation, although occasionally there is mild doming of the skull or thickening of the cranial bones. The dorsal, and often the posterior parts of the hemispheres, are the portions most severely defective (Fig. 3.9). The leptomeninges may easily be damaged on removing the calvarium but are in their usual position and form sacs enclosing cerebrospinal fluid, the fluid occupying the space normally occupied by parenchyma. Discrete remnants of parenchyma may be present in the meninges. When the leptomeninges are incised, it is apparent that the brain stem is of near-normal conformation with well-developed hippocampus and choroid plexuses. The anterior portion of the corpus callosum and septum pellucidum

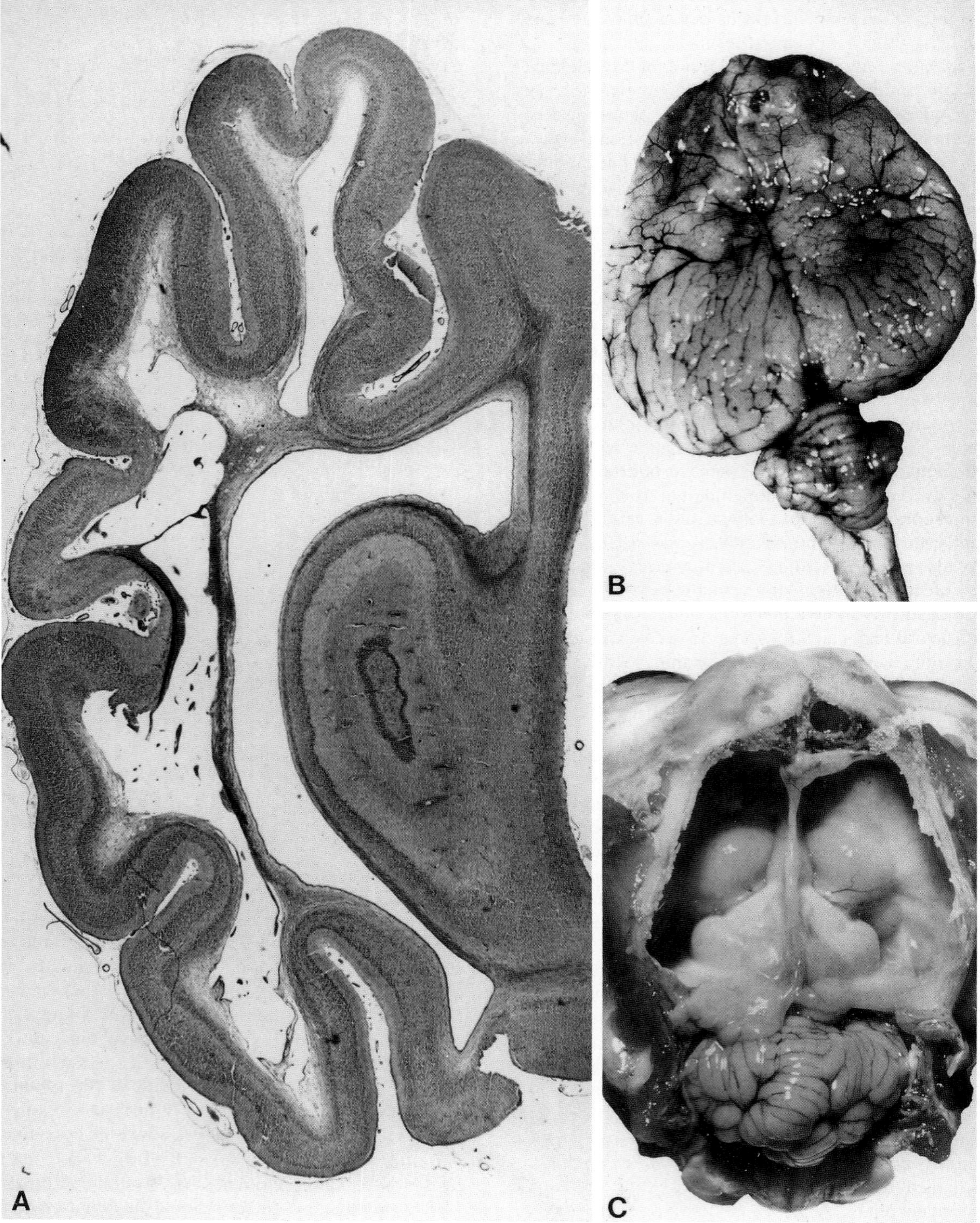

Fig. 3.9 (A) Hydranencephaly. Lamb. Swayback. (Courtesy of J. McC. Howell). (B) Hydranencephaly and cerebellar hypoplasia. Lamb. Intrauterine bluetongue virus infection. (C) Hydranencephaly. Cat. Meninges removed to expose the brainstem structures.

may be intact although attenuated. Cerebellar hypoplasia may be present as a concurrent defect, as may the histologic deficits of hypomyelinogenesis.

Hydranencephaly is the residual lesion of full-thickness necrosis of the cerebral hemisphere. In animals, the lesion develops in early fetal stages and before the mature arrangements of the cortex are present. The marginal tissue is dysplastic, flat, and microgyric, or the gyri have a radial arrangement from the defect. Although a diagnosis of hydranencephaly is readily made on macroscopic inspection, histological study may provide some insight into the nature of the disease process. The membranous coverings of the remnants of the hemispheres comprise arachnoid, pia, and a thin mantle of residual cortex in normal juxtaposition. The residual cortical tissue may be lined by attenuated ependyma displaced outward with expansion of the lateral ventricles, but this ependymal lining is frequently incomplete or absent, and the cavity abuts directly on cortical tissue, which may be unremarkable except for mild astrocytosis.

Hydranencephaly occurs in all species but is most common in calves in which it occurs either sporadically or as minor epizootics. The lesion occurs similarly although less frequently in lambs. The species occurrence of hydranencephaly reflects its etiological associations with certain viruses infecting the fetus at a critical stage of gestation. In these cases, hydranencephaly is often part of a spectrum of neural lesions, the expression of which depends mainly on the age of the fetus at infection. Those viruses which are well established as potential causes of hydranencephaly tend to be either arthropod-borne viruses such as those of Akabane disease, bluetongue, Rift Valley fever, and Wesselsbron disease, or pestiviruses such as the agents of bovine virus diarrhea and border disease.

The pathogenesis of hydranencephaly has been clarified through studies of the cavitating encephalopathies caused by fetal viral infections, particularly bluetongue and border disease in lambs. These infections at critical periods of gestation produce subventricular zones of necrosis of the developing cerebral hemispheres. These zones, which may have a vascular basis, involve the neuroblasts in their outward migration, result in cavitation, and deprive the cortex of its normal complement of neurons. These cavitations range in size from small cysts with only minor changes in the overlying cortex (porencephaly) to large confluent spaces with the hemispheres reduced to fluid-filled sacs (hydranencephaly). During the necrotizing process, there may be an intense macrophage response, but it is usual for this reaction to have subsided by the time the animal is born and becomes available for examination.

Mechanical factors also contribute to the development of hydranencephaly. A dissecting effect associated with escape of cerebrospinal fluid into the parenchyma may also be important, since segmental loss of ventricular ependyma is an early feature of the cavitating process.

Compensatory expansion of the lateral ventricles secondary to loss of brain substance occurs, and rapid expansion of the fetal calvarium during the gestation period allows stretching and rupture of residual cortical tissue. In less severely affected areas, the outer rim of cortex overlies a band of subependymal tissue, the intervening cavity being occupied by trabecular parenchymal remnants. Rosette formations of cells which resemble ependymal epithelium may be found in the subependymal area. Accumulations of mineralized debris may be present in the meninges.

J. Porencephaly

Porencephaly is cystic cavitation of the brain evolving from a destructive process in prenatal life. The defect typically involves the white matter of the cerebral hemispheres. An affected brain may contain a single cyst, or there may be multiple cystic lesions. The temporal portion of the cerebral hemispheres is an area of predilection, but porencephalic change may be found throughout the cerebral hemispheres, although typically sparing the basal nuclei. Occasionally, lesions are found in brain stem and cerebellum. The cysts are usually randomly located, but evidence of bilateral symmetry is sometimes apparent, particularly in well-developed lesions of the cerebral hemisphere. Rarely cysts may communicate with ventricular cavities or with the subarachnoid space. The cysts may be apparent from the meningeal aspect as focal fluctuant areas of attenuated cerebral cortex or as superficial, clear submeningeal cysts. The cysts range in size from microscopic dimensions to a diameter of several centimeters. The cysts are quite variable in shape, but roughly spherical or cleftlike outlines are common. On section, they are filled with clear fluid and are smooth walled, but are traversed by variable numbers of trabeculae. Less well developed lesions may appear as gelatinous softenings.

The porencephalic cavities, particularly the larger defects, are often unremarkable microscopically, being lined by a layer of flattened glia. Apart from some mild marginal astrocytosis, they show surprisingly little evidence of reactive or inflammatory change. In other cases, there is accumulation of hemosiderin-containing gitter cells about the margins of the cyst and within the cavity. The trabeculae consist of residual brain parenchyma, usually oriented about a blood vessel. Some small lesions, evident grossly as focal gelatinous areas, appear as focal leukomalacia, comprising white matter in the process of dissolution associated with accumulation of macrophages and gitter cells. Sometimes the inflammatory nature of the lesions is indicated by the presence of mild nonsuppurative meningoencephalitis with gliosis, perivascular cuffing, and focal mineralization about the margins of the cysts, and mononuclear infiltration in the meninges overlying the defects.

The pathogenetic and etiological considerations for porencephaly parallel those discussed under hydranencephaly, with porencephaly being regarded as the less severe expression of the same pathological process. Both lesions are commonly recorded in the course of outbreaks of cavitating cerebral defects and may occur together in the same brain. In the case of cavitating viral infections of the fetal brain, termination of the disease process in either porencephaly or hydranencephaly is influenced by gestational age at infection, with porencephaly tending to follow infection at a later stage than is the case with hydranencephaly.

Porencephaly is a common manifestation of prenatal infection of lambs with some strains of border disease virus and is seen in calves infected *in utero* by bovine virus diarrhea virus. The cystic or gelatinous transformations of the cerebral white matter which occur in some cases of copper deficiency in lambs (swayback) are porencephalic in nature. The lesion has also been induced experimentally in lambs by exposure of pregnant ewes to hyperthermia during the last two thirds of pregnancy.

Bibliography

Barlow, R. M. Morphogenesis of hydranencephaly and other intracranial malformations in progeny of pregnant ewes infected with pestiviruses. *J Comp Pathol* **90:** 87–98, 1980.

Halsey, J. H., Allen, N., and Chamberlin, H. R. The morphogenesis of hydranencephaly. *J Neurol Sci* **12:** 187–217, 1971.

Hartley, W. J., Alexander, G., and Edwards, M. J. Brain cavitation and micrencephaly in lambs exposed to prenatal hyperthermia. *Teratology* **9:** 299–303, 1974.

McIntosh, G. H. Foetal thyroidectomy and hydranencephaly in lambs. *Aust Vet J* **54:** 408, 1978.

Narita, M., Inui, S., and Hashiguchi, Y. The pathogenesis of congenital encephalopathies in sheep experimentally induced by Akabane virus. *J Comp Pathol* **89:** 229–240, 1979.

Osburn, B. I. *et al.* Experimental viral-induced congenital encephalopathies. I. Pathology of hydranencephaly and porencephaly caused by bluetongue vaccine virus. 2. The pathogenesis of bluetongue vaccine virus infection in fetal lambs. *Lab Invest* **25:** 197–205, 206–210, 1971.

Whittington, R. J. *et al.* Congenital hydranencephaly and arthrogryposis of Corriedale sheep. *Aust Vet J* **65:** 124–127, 1988.

K. Cerebellar Defects

Cerebellar defects are among the more important of the developmental anomalies of the nervous system because of their frequent occurrence and almost invariable accompaniment by significant and distinctive clinical manifestations. Cerebellar defects are quite common in cats and calves, relatively so in pigs, dogs, and lambs, but uncommon in foals. Most defects of morphogenesis can be divided into two broad categories, and the entities are discussed as being examples of either cerebellar hypoplasia or cerebellar abiotrophy (atrophy). This distinction into hypoplastic and atrophic types is somewhat artificial, because it is apparent that many cases are compounded of both processes.

Minor dysplastic lesions of no consequence are quite common in young animals but are fewer or less conspicuous in adults. They are most common in the flocculonodular lobules and where the cortex terminates at the peduncles. The foci are microscopic and consist of tangled islands of germinal, molecular, and granular layers with Purkinje cells haphazardly distributed. The dysplasias of the Arnold–Chiari and Dandy–Walker syndromes, of copper deficiency, and the metabolic storage disorders are discussed separately.

As is usual in teratological nomenclature, the name of the condition merely emphasizes the major component, in these cases the cerebellar defect. There may, however, be coincident malformations such as agenesis of the corpus callosum or hydranencephaly. There are always correlative changes, this term grouping together those structural changes which occur in nuclear masses which send fibers to, or receive fibers from, the cerebellum as well as in the particular tract of fibers themselves. The term implies that there is a causal connection, the cerebellar defects being primary and the subcerebellar defects being secondary, but such an implication is less likely to be valid for the cerebellar hypoplasias than for the cerebellar atrophies. Anatomic classifications of the cerebellar hypoplasias tend to take into account the distribution and severity of correlative changes, especially those in the deep cerebellar nuclei, the pontine nuclei, the inferior olives, and in the cerebellar brachia, the restiform bodies, and the spinocerebellar tracts. The pattern of correlative changes varies from case to case, and there is not a quantitative correspondence between the cerebellar defects and the correlative defects. There are also very wide variations in the severity of the cerebellar hypoplasias, which is reasonable if the severity is properly to be related to the time of onset, duration, and severity of the arrest of development. From these viewpoints, any attempt to classify cerebellar hypoplasias anatomically is artificial because the defects are quantitative.

Cerebellar hypoplasia is one of the most common congenital nervous system defects of domestic animals and is seen in all domestic species (Fig. 3.10). There is persuasive, though inconclusive, evidence for genetically determined occurrence of the disease in calves of various breeds, particularly beef shorthorns, in Arab and Arab-cross foals and in Gotland ponies, and in chow chow dogs. Cerebellar hypoplasia has been reported in piglets born to sows treated with the organophosphate trichlorfon during pregnancy and may be detected microscopically in goat kids and lambs affected by hypocuprosis. The most prevalent and best-defined cerebellar hypoplasias are those which follow infection of the developing cerebellum by certain viruses, particularly feline parvovirus, bovine virus diarrhea virus, border disease virus, and the virus of hog cholera. The occurrence of cerebellar hypoplasia as a consequence of viral infections of the developing nervous system is further discussed in the section dealing with individual viral diseases as causes of developmental anomalies of the central nervous system.

Cerebellar growth patterns determine the gestational or perinatal periods during which cerebellar hypoplasia may follow the action of a teratogen and influence the nature and scope of the structural aberrations marking the hypoplastic process. The cerebellum originates as a dorsal growth of the alar plate of the metencephalon over the fourth ventricle. Germinal cells adjacent to the fourth ven-

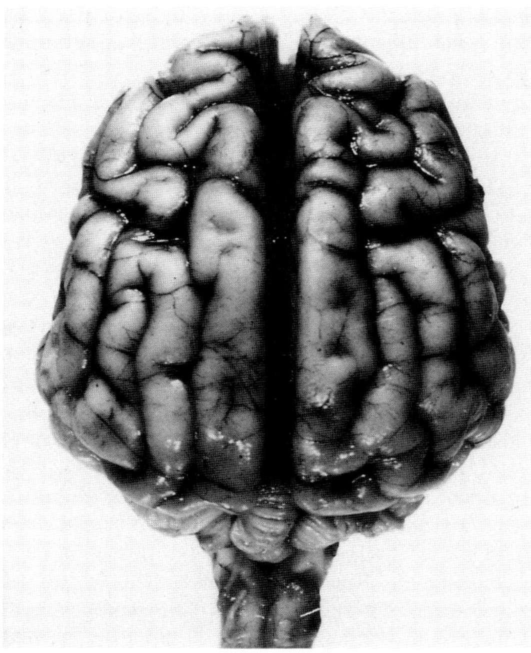

Fig. 3.10 Cerebellar hypoplasia. Dog.

tricle differentiate into neurons that migrate into the developing cerebellum to form the deep nuclei and Purkinje neurons. This differentiation occurs prior to midgestation. A population of germinal cells migrates to the surface of the developing cerebellum and forms a layer several cells thick. This is the external granular layer that covers the folia as they develop. In this layer cells proliferate rapidly and differentiate into microneurons such as basket cells, stellate cells, and granule cells, which migrate into the folium to their definitive locations. This proliferation, differentiation, and migration begins late in gestation and continues for a few weeks postnatally in most species. Normal cerebellar cytoarchitectural development, including the maturation and localization of Purkinje cells, is dependent on these microneurons' establishing orderly synaptic connections.

The actively mitotic germinative cells of the external granular layer are especially vulnerable to the effects of teratogenic agents. Feline parvovirus and the viruses of bovine virus diarrhea and border disease cause selective necrosis of external granular layer cells. That the destructive process may be augmented by vascular damage has been demonstrated for bovine virus diarrhea virus. Hog cholera virus possibly acts through inhibition of cell division and maturation. In puppies, segmental cerebellar dysplasia may result from postnatal infection with canine herpesvirus. Since the growth behavior of the subependymal cell plate has features in common with the external granular layer of the cerebellum, residual lesions attributable to this site of infection, such as cavitating defects, may be associated with cerebellar hypoplasia, particularly in cases of viral origin.

The anatomic expressions of the hypoplasia, both in the cerebellum and subcerebellar structures, are very variable, the variations probably being in degree only but not in kind. In some cases, the cerebellum may appear grossly normal, the hypoplastic defects being detectable only on microscopic examination. In such cases, the defects are irregular in distribution, although there is a more or less severe loss of Purkinje cells. The granular layer is here and there narrowed and deficient in cells, but the molecular layer is normal. Correlative lesions may be present or absent, and when present, may be asymmetrical; they occur chiefly in the deep cerebellar nuclei, olives, and cerebellar peduncles. At the other extreme, the cerebellum may be represented only by a small nubbin of tissue or two unconnected nubbins, each related to a hypoplastic peduncle (Fig. 3.11). In these severe defects, there is no folial pattern or division into lobes. Intermediate degrees of cerebellar hypoplasia are, however, the most common. It is usually possible to recognize with the naked eye that the cerebellar peduncles are diminished in size, especially the brachium pontis, that the medullary pyramids are flattened, and that the small size of the restiform bodies gives the fourth ventricle a flattened appearance. Microscopically, the cerebellar cortex is disorganized. A brief description is not possible because the pattern varies greatly from case to case and place to place in one animal. The Purkinje cells and the cells of the granular layer are the most obviously deficient and, in those which are present, regressive changes are common.

Cerebellar atrophy (or **abiotrophy**) refers to premature or accelerated degeneration of formed elements, presumably caused by some intrinsic metabolic defect. The Purkinje cells appear particularly susceptible to spontaneous

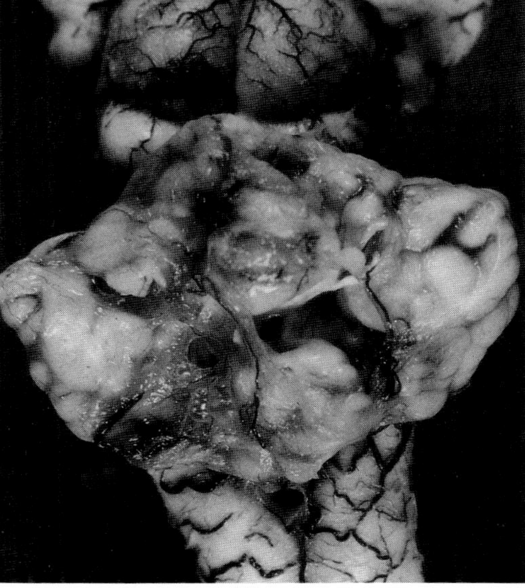

Fig. 3.11 Cerebellar dysplasia. (Cerebrum removed to expose cerebellar lesion). Ox. Asymptomatic during 5 years.

degeneration; there may be secondary depletion of granule cells, but other cortical layers are normal.

Cerebellar atrophy occurs in lambs, dogs, various breeds of calves, and piglets. The disease in lambs, which are commonly known as daft lambs, is reported only from England and Canada. It is not restricted to any breed but is presumed, the evidence being inconclusive, to be inherited. The dams are normal. Affected lambs show signs of cerebellar dysfunction at birth, which include abnormalities of muscle tone, disorders of equilibrium, and tremors. The cerebellum is, however, normal in size and gross form. The histological changes affect primarily the Purkinje and Golgi cells, and these especially in the median lobe. Degeneration and loss of these cells leave some empty baskets and a replacement astrogliosis. In the early stages of degeneration, the Purkinje and Golgi cells are shrunken and hyperchromatic or swollen and pale with cytoplasmic vacuolation. They ultimately undergo lysis and disappear, leaving a spongy zone between the granular and molecular layers. There is, simultaneously, a diminution in the population of granule cells. Regressive changes with gliosis may be apparent in the deep cerebellar nuclei and olives.

Progressive cerebellar degenerations are common in dogs, and abiotrophic defects in dogs have been recorded in a variety of breeds including Airedales, Gordon setters, rough-coated collies, Border collies, Finnish terriers, beagles, Australian kelpies, and Bernese mountain dogs. Signs in the pups become evident at about 3 months of age. The cerebellum may be normal in size and shape or slightly diminished in size and somewhat flattened. Microscopically, there is degeneration and loss of Purkinje cells and granular cells. The Purkinje cells degenerate, being either shrunken and hyperchromatic or pale, swollen, and vacuolated (Fig. 3.12A). The majority of them disappear, leaving empty baskets and a fenestrated ground layer (Fig. 3.12B). The atrophic process may show evidence of regional predilection. In Gordon setters, the dorsal portions of the median lobe and vermis are most severely affected, while in collie dogs, selective involvement of the anterior folia of the vermis is reflected in macroscopic diminution of this area. Collateral changes in the subcerebellar nuclei are minor, but in this regard it may be pertinent that affected dogs are usually killed shortly after the defect becomes apparent.

Hereditary striatonigral and **cerebello–olivary degeneration** occurs in the Kerry blue terrier. This is the only recognized naturally occurring hereditary degenerative disease involving the basal nuclei of domestic animals. Pedigrees of affected dogs indicate autosomal recessive inheritance. Clinical signs begin between 9 and 16 weeks of age and are characterized by ataxia and dysmetria. The

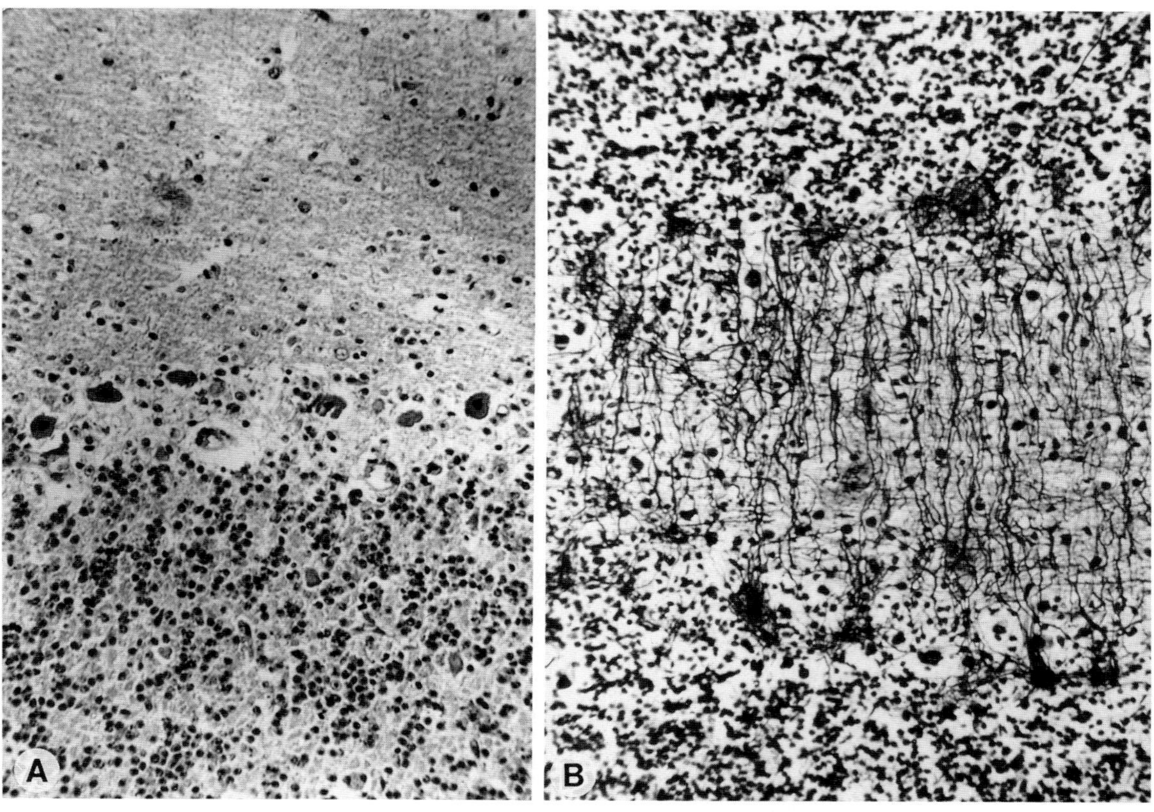

Fig. 3.12 (A) Degenerating Purkinje cells in congenital cerebellar atrophy. Dog. (B) Empty basket cells. Same case as (A).

inherent defect is neuronal degeneration, and the brain changes follow a definite anatomical and temporal pattern of development. Degeneration and loss of Purkinje cells in the cerebellar cortex is evident at the onset of clinical signs. Subsequently there is sequential involvement of the olivary nuclei, caudate nuclei, and putamen, and finally the substantia nigra. In the cerebellar cortex, chromatolytic degeneration and loss of Purkinje cells is attended by astrocytosis involving Bergmann's glia and depletion of granule cells. In the cerebellar white matter and contained nuclei, there is status spongiosus and axonal swelling. Changes in the olivary nuclei, basal nuclei, and substantia nigra tend to be symmetrical and initially (Fig. 3.13) feature neuronal degeneration, axonal swelling, spongiosus, and fibrous astrocytosis, with progression to malacia and cavitation. Olivary neurons undergo chromatolysis, while in the remaining nuclear structures, the nerve cells accumulate intracytoplasmic eosinophilic granules and undergo ischemic change.

Macroscopic evidence of the disease process in the cerebellum is limited to a modest degree of folial atrophy but, with progression of the condition, involvement of the basal and brain stem nuclei is manifest grossly by bilateral focal malacic lesions. Gross lesions are most severe in the caudate nucleus, which after 7 to 8 months of clinical illness, may be reduced to numerous microcystic cavities. It is proposed that the cerebellum and basal nuclei are the primary areas of involvement, with the lesions in the olivary nucleus and substantia nigra being attributed to transynaptic neuronal degeneration along glutaminergic neurotransmission pathways.

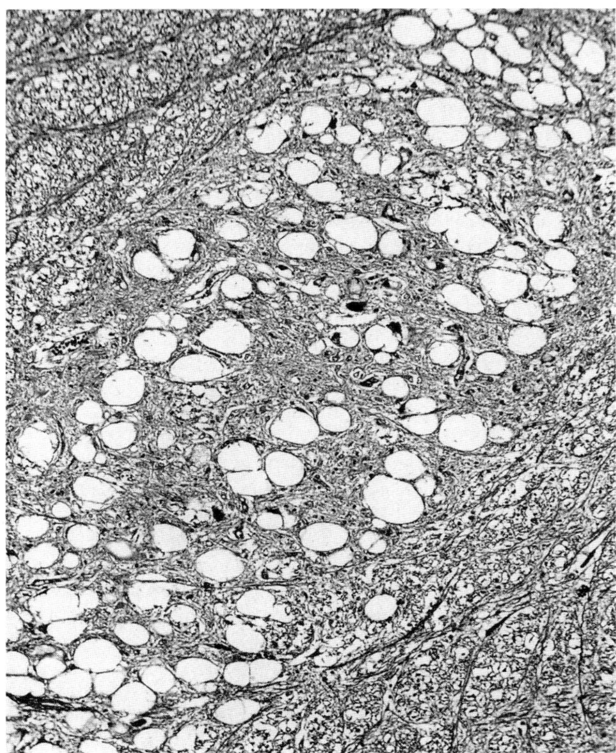

Fig. 3.13 Neuronal degeneration, spongiosus in inferior olive. Kerry blue terrier.

The Kerry blue terrier condition is not the sole cerebellar abiotrophy in which accompanying extracerebellar lesions may be found. Olivary nuclei may be affected in the Bernese running dog, the cerebral cortex in the miniature poodle, and spinal Wallerian degeneration may be found in rough-coated and Border collies and in Merino sheep. In the Swedish Lapland dog, there is a neuronal abiotrophy involving Purkinje cells; however, the effects of this are overshadowed by motor neuron degeneration, and the disease is discussed with others of that group.

Bovine familial convulsions and ataxia is a heritable disorder of purebred and crossbred Aberdeen Angus cattle in the United Kingdom. The clinical syndrome is characterized by intermittent episodic seizures in newborn and young calves and by the gradual development of ataxia with spasticity and hypermetria in calves surviving bouts of seizures extending over 2 or 3 months. The distinctive microscopic change is Purkinje cell axon swelling in the cerebellar granular layer. There is also degeneration and loss of Purkinje cells. An apparently similar disorder has been described in Charolais cattle in the United Kingdom.

Gomen disease is a cerebellar degeneration and ataxia of horses in New Caledonia. Some folial atrophy in the cerebellar vermis may be evident on gross examination. Microscopically, there is thinning of the cerebellar molecular layer and loss of Purkinje and granule cells. There is also considerable deposition of a pigment resembling lipofuscin in many of the surviving Purkinje cells as well as in the neurons of the brain and spinal cord. This material is also present in macrophages seen in areas where Purkinje cells are missing. The cause is unknown, but the disease is thought to be due to an environmental toxin.

Bibliography

Barlow, R. M. Morphogenesis of cerebellar lesions in bovine familial convulsions and ataxia. *Vet Pathol* **18:** 151–162, 1981.

Carmichael, S., Griffiths, I. R., and Harvey, M. J. A. Familial cerebellar ataxia with hydrocephalus in bull mastiffs. *Vet Rec* **112:** 354–358, 1983.

Cho, D. Y., and Leipold, H. W. Cerebellar cortical atrophy in a Charolais calf. *Vet Pathol* **15:** 264–266, 1978.

Clark, R. G. *et al.* Suspected inherited cerebellar neuroaxonal dystrophy in collie sheep dogs. *N Z Vet J* **30:** 102–103, 1982.

Cummings, J. F., and de Lahunta, A. A study of cerebellar and cerebral cortical degeneration in miniature poodle pups with emphasis on the ultrastructure of Purkinje cell changes. *Acta Neuropathol (Berl)* **75:** 261–271, 1988.

DeForest, M. E., Eger, C. E., and Basrur, P. K. Hereditary cerebellar neuronal abiotrophy in a Kerry blue terrier dog. *Can Vet J* **191:** 198–202, 1978.

de Lahunta, A. Diseases of the cerebellum. *Vet Clin North Am* **10:** 91–101, 1980.

de Lahunta, A., and Averill, D. R. Hereditary cerebellar cortical and extrapyramidal nuclear abiotrophy in Kerry blue terriers. *J Am Vet Med Assoc* **168:** 1119–1124, 1976.

de Lahunta, A. *et al.* Hereditary cerebellar cortical abiotrophy in the Gordon setter. *J Am Vet Med Assoc* **177:** 538–541, 1980.
Dungworth, D. L., and Fowler, M. E. Cerebellar hypoplasia and degeneration in a foal. *Cornell Vet* **56:** 17–24, 1966.
Finnie, E. P., and Leaver, D. D. Cerebellar hypoplasia in calves. *Aust Vet J* **41:** 287–288, 1965.
Fraser, H. Two dissimilar types of cerebellar disorders in the horse. *Vet Rec* **78:** 608–612, 1966.
Gill, J. M., and Hewland, M. Cerebellar degeneration in the Border collie. *N Z Vet J* **28:** 170, 1980.
Hartley, W. J. *et al.* Inherited cerebellar degeneration in the rough-coated collie. *Aust Vet Pract* **8:** 79–85, 1978.
Howell, J. M., and Ritchie, H. E. Cerebellar malformations in two Ayrshire calves. *Vet Pathol* **3:** 159–168, 1968.
Knecht, C. D. *et al.* Cerebellar hypoplasia in chow chows. *J Am Anim Hosp Assoc* **15:** 51–53, 1979.
Knox, B. *et al.* Congenital ataxia and tremor with cerebellar hypoplasia in piglets borne by sows treated with Neguvon vet (Metrifonate, Trichlorfon) during pregnancy. *Nord Vet Med* **30:** 538–545, 1978.
Le Gonidec, G. *et al.* A neurologic disease of horses in New Caledonia. *Aust Vet J* **57:** 194–195, 1981. (Gomen disease)
Montgomery, D. L., and Storts, R. W. Hereditary striatonigral and cerebello-olivary degeneration of the Kerry blue terrier. *Vet Pathol* **20:** 143–159, 1983.
O'Sullivan, B. M., and McPhee, C. P. Cerebellar hypoplasia of genetic origin in calves. *Aust Vet J* **51:** 469–471, 1975.
Palmer, A. C. *et al.* Cerebellar hypoplasia and degeneration in the young Arab horse: Clinical and neuropathological features. *Vet Rec* **93:** 62–66, 1973.
Swan, R. A., and Taylor, E. G. Cerebellar hypoplasia in beef shorthorn calves. *Aust Vet J* **59:** 95–96, 1982.
Thomas, J. B., and Robertson, D. Hereditary cerebellar abiotrophy in Australian kelpie dogs. *Aust Vet J* **66:** 301–302, 1989.
White, M., Whitlock, R. H., and de Lahunta, A. A cerebellar abiotrophy of calves. *Cornell Vet* **65:** 476–491, 1975.
Whittington, R. J., Morton, A. G., and Kennedy, D. J. Cerebellar abiotrophy in crossbred cattle. *Aust Vet J* **66:** 12–15, 1989.
Yamano, T. *et al.* Destruction of external granular layer and subsequent cerebellar abnormalities. *Acta Neuropathol (Berl)* **59:** 41–47, 1983.
Yasuba, M. *et al.* Cerebellar cortical degeneration in beagle dogs. *Vet Pathol* **25:** 315–317, 1988.

L. Viral Infections as Causes of Developmental Defects

1. Akabane Virus

Akabane virus, a bunyavirus, is among the most potent of the viral teratogens of domestic animals, but the infection is otherwise asymptomatic. Following maternal infection at critical stages of gestation, this agent produces a range of predominantly neural abnormalities in calves, lambs, and kids, but is best known for having produced outbreaks of arthrogryposis and hydranencephaly in calves. Epizootics of Akabane virus disease in cattle have been reported in areas of Japan, Israel, and Australia. The likely vector in Australia is the biting midge *Culicoides brevitarsus,* and the virus has also been isolated from mosquitoes in Japan. There is serological evidence that buffaloes and horses are additional vertebrate hosts. Although arthrogryposis and hydranencephaly may be the most obvious manifestations in field epizootics of bovine Akabane virus disease, a range of overlapping syndromes is observed in calves in affected herds. The pattern of fetal disease corresponds to the gestational age of the fetus at the time of infection. Infection late in gestation may cause abortion. The initial manifestation of neural abnormality in a field outbreak is the birth of incoordinate calves, and in this group, nonsuppurative encephalomyelitis is evident on histopathological examination. Microencephaly and cerebellar hypoplasia occur occasionally as manifestations of late infection. Arthrogryposis, sometimes associated with spinal deformities, appears early in the outbreak following fetal infection at 5 to 6 months of pregnancy. In arthrogrypotic calves, there is a loss of spinal ventral horn neurons, loss of myelin in the motor tracts of the spinal cord and in ventral spinal nerves, and denervation atrophy, together with fibrous and adipose replacement of skeletal musculature.

There is evidence that at least some strains of Akabane virus have the capacity to cause polymyositis, particularly in the early myotubular phase of skeletal muscle development, suggesting the possible involvement of this process in the arthrogrypotic change.

Severe hydranencephaly, manifest clinically as blindness and stupidity, is seen toward the end of the epizootic, being the result of fetal infection at 3 to 4 months of gestation. With increasing age of the fetus at the time of infection, the cavitating cerebral changes are less severe and grade toward porencephaly.

The teratogenic potential of Akabane virus in sheep and goats is qualitatively the same as that for cattle. Cavitating cerebral defects, arthrogryposis, microencephaly, and agenesis or hypoplasia of the spinal cord have been produced in lambs born of ewes inoculated between days 29 and 48 of pregnancy. Field observations in Australian flocks suggest that, in sheep, microencephaly is relatively more common as a consequence of Akabane virus infection than it is in cattle.

Bibliography

Hartley, W. J. *et al.* Pathology of congenital bovine epizootic arthrogryposis and hydranencephaly and its relationship to Akabane virus. *Aust Vet J* **53:** 319–325, 1977.
Haughey, K. E. *et al.* Akabane disease in sheep. *Aust Vet J* **65:** 136–140, 1988.
Konno, S., and Nakagawa, M. Akabane disease in cattle: Congenital abnormalities caused by viral infection. Experimental disease. *Vet Pathol* **19:** 267–279, 1982.
Konno, S., Moriwaki, M., and Nakagawa, M. Akabane disease in cattle: Congenital abnormalities caused by viral infection. Spontaneous disease. *Vet Pathol* **19:** 246–266, 1982.
Konno, S. *et al.* Myopathy and encephalopathy in chick embryos experimentally infected with Akabane virus. *Vet Pathol* **25:** 1–8, 1988.
Kurogi, H. *et al.* Congenital abnormalities in newborn calves after inoculation of pregnant cows with Akabane virus. *Infect Immun* **17:** 338–343, 1977.
McClure, S. *et al.* Maturation of immunological reactivity in the fetal lamb infected with Akabane virus. *J Comp Pathol* **99:** 133–143, 1988.

Narita, M., Inui, S., and Hashiguchi, Y. The pathogenesis of congenital encephalopathies in sheep experimentally induced by Akabane virus. *J Comp Pathol* **89:** 229–240, 1979.

Parsonson, I. M., Della-Porta, A. J., and Snowdon, W. A. Congenital abnormalities in newborn lambs after infection of pregnant sheep with Akabane virus. *Infect Immun* **15:** 254–262, 1977.

Parsonson, I. M., Della-Porta, A. J., and Snowdon, W. A. Akabane virus infection of the foetus. *Vet Microbiol* **6:** 209–224, 1981.

Parsonson, I. M. *et al.* Akabane virus infection in the pregnant ewe. 1. Growth of the virus in the foetus in the development of the foetal immune response. *Vet Microbiol* **6:** 199–207, 1981.

Parsonson, I.M. *et al.* Transmission of Akabane virus from the ewe to the early fetus (32 to 53 days). *J Comp Pathol* **98:** 2315–227, 1988.

2. Bluetongue

Hydranencephaly and porencephaly have been reported in lambs and calves whose dams received a live attenuated bluetongue vaccine or contracted bluetongue infection during pregnancy (Fig. 3.9B). Experimental studies, utilizing intrafetal inoculation at different stages of gestation, indicate that the type of congenital anomaly found depends on the fetal age at the time of inoculation. Lambs infected with bluetongue vaccine virus at 50 to 55 days of gestation develop a severe necrotizing encephalopathy and retinopathy, which at 150 days of gestation, the time of birth, is manifested as hydranencephaly and retinal dysplasia. Inoculation of lambs at 75 days of fetal age results in a multifocal encephalitis and selective vacuolation of white matter, which is manifested as porencephalic cysts in the newborn. Ocular lesions are not observed in these newborn lambs. Lesions in brains of lambs inoculated after 100 days of gestation are confined to a mild focal meningoencephalitis.

Bibliography

Anderson, C. K., and Jensen, R. Pathologic changes in placentas of ewes inoculated with bluetongue virus. *Am J Vet Res* **30:** 987–999, 1969.

Barnard, B. J. H., and Pienaar, J. G. Bluetongue virus as a cause of hydranencephaly in cattle. *Onderstepoort J Vet Res* **43:** 155–158, 1976.

Enright, F. M., and Osborn, B. I. Ontogeny of host responses in ovine fetuses infected with bluetongue virus. *Am J Vet Res* **41:** 224–229, 1980.

MacLachlan, N. J. *et al.* Bluetongue virus-induced encephalopathy in fetal cattle. *Vet Pathol* **22:** 415–417, 1985.

McKercher, D. G., Saito, J. K., and Singh, K. V. Serologic evidence of an etiologic role for bluetongue virus in hydranencephaly of calves. *J Am Vet Med Assoc* **156:** 1044–1047, 1970.

Osburn, B. I. *et al.* Experimental viral-induced congenital encephalopathies. 1. Pathology of hydranencephaly and porencephaly caused by bluetongue vaccine virus. *Lab Invest* **25:** 197–210, 1971.

Richards, W. P. C., and Cordy, D. R. Bluetongue virus infection: Pathologic responses of the nervous systems in sheep and mice. *Science* **156:** 530–531, 1967.

Richards, W. P. C., Crenshaw, G. L., and Bushnell, R. B. Hydranencephaly of calves associated with natural bluetongue virus infection. *Cornell Vet* **61:** 336–348, 1971.

Schmidt, R. E., and Panciera, R. J. Cerebral malformation in fetal lambs from a bluetongue enzootic flock. *J Am Vet Med Assoc* **162:** 567–568, 1973.

Silverstein, A. M. *et al.* An experimental virus-induced retinal dysplasia in the fetal lamb. *Am J Ophthalmol* **72:** 22–34, 1971.

3. Rift Valley Fever and Wesselsbron Viruses

These diseases are considered more fully in Volume 2, Chapter 2, Liver and Biliary System. In addition to Wesselsbron virus, there are several other flaviviruses in South Africa, including West Nile and Banzai strains, to which sheep are experimentally susceptible and which result in abortion, stillbirths, and neonatal deaths; anomalies include hydranencephaly, porencephaly, and internal hydrocephalus.

The viruses of Rift Valley fever and Wesselsbron disease are arthropod borne and tend to circulate together. They are both primarily hepatotropic, but the wild strain of Wesselsbron and attenuated strains of both are neurotropic, and the vaccine strains may be responsible for most outbreaks of congenital neurologic disease. The presenting features are similar to those of Akabane, but the high incidence of hydrops amnii and prolonged gestation is especially a feature. Wesselsbron infection and its pathogenesis are unexplained.

The destructive changes in the central nervous system produced by one or other of these viruses, or both together, can be more severe than those with other teratogenic viruses, and result in segmental aplasia of the cord, aplasia of the cerebellum, and anencephaly. It is more usual, however, that the defects include brachygnathia, hydranencephaly or porencephaly, hypoplasia of cerebellum and spinal cord, and varied musculoskeletal stigmata of arthrogryposis.

Bibliography

Barnard, B. J. H., and Voges, S. F. Flaviviruses in South Africa: pathogenicity for sheep. *Onderstepoort J Vet Res* **53:** 235–238, 1986.

Coetzer, J. A. W. Brain teratology as a result of transplacental virus infection in ruminants. *J S Afr Vet Assoc* **51:** 153–157, 1980.

Coetzer, J. A. W., and Barnard, B. J. H. Hydrops amnii in sheep associated with hydranencephaly and arthrogryphosis with Wesselsbron disease and Rift Valley fever as aetiological agents. *Onderstepoort J Vet Res* **44:** 119–126, 1977.

Coetzer, J. A. W. *et al.* Wesselsbron disease: A cause of congenital porencephaly and cerebellar hypoplasia in calves. *Onderstepoort J Vet Res* **46:** 165–169, 1979.

4. Bovine Virus Diarrhea

The diseases caused by this virus are considered in detail in Volume 2, Chapter 1, The Alimentary System.

The virus belongs to the pestivirus group, and its ecology depends on transplacental transmission and the establishment of immune tolerance and persistent infections. Horizontal transfer of infection occurs readily horizontally but is seldom clinically significant. Persistently infected fetuses which survive to become pregnant transmit the infection to the conceptus. Although viral strains differ in their pathogenicity, it is clear from both epidemiological and experimental studies that infection of susceptible cows during the early and middle stages of gestation is likely to result in either fetal death or a variety of developmental disturbances, among which neural and ocular defects predominate. Some congenitally affected calves also have erosive lesions in the upper alimentary tract and abomasum resembling those seen in adult cattle, and mandibular brachygnathism also occurs.

The outcome of infection of the fetal calf is related to gestational age, advancing fetal maturity being associated with increased resistance to the virus. Infections occurring within the first 100 days of fetal life tend to be lethal, resulting in abortion or mummification. Although the gross pathologic changes seen in these lethal infections lack unique features, the patterns of tissue response are characteristic, but are rarely seen as such fetuses die *in utero* and undergo autolysis. Experimentally it has been shown there is a necrotizing inflammatory reaction which can involve a variety of tissues. The reactive changes are dominated by mononuclear, predominantly macrophage, infiltration of hepatic portal areas, myocardium, spleen, and lymph nodes, which is reflected grossly in enlargement, nodularity, and mottling of the liver, and enlargement of spleen and lymph nodes (Fig. 3.14). The presence of growth-arrest lines in long bones suggests that the fetus undergoes one or more intrauterine crises before death. Affected fetuses may have partial alopecia, which spares the tail and the lower portion of the limbs, and the head, these being points of initial hair growth during fetal development. The microscopic skin changes which evolve from the initial necrotizing dermatitis and correlate with the alopecia are hypoplasia of hair follicles and cystic distension of adnexal glands.

It is during the 100- to 170-day period that the teratogenic effects of the virus are manifest. The period of susceptibility presumably varies with the strain of virus. Cerebellar hypoplasia is the most characteristic defect (Fig. 3.15). Gross cerebellar changes range from a more or less uniform atrophy to irregular folial atrophy and agenesis accompanied by cavitation. The hypoplastic process is compounded of the effects of necrosis of external granular layer cells and parenchymal destruction as a consequence of folial edema resulting from vasculitis. The relative contribution of these processes is quite variable, but it appears that the vasculopathy may be more prominent in older fetuses. The nature and extent of involvement of individual folia also varies considerably. The evolution of the cerebellar changes has been studied experimentally. Acute lesions are evident 2 weeks after infection. Cellular necrosis in the external granular layer is accompanied by nonsuppurative meningitis. Vasculitis, marked by endothelial proliferation and perivascular leukocytic infiltra-

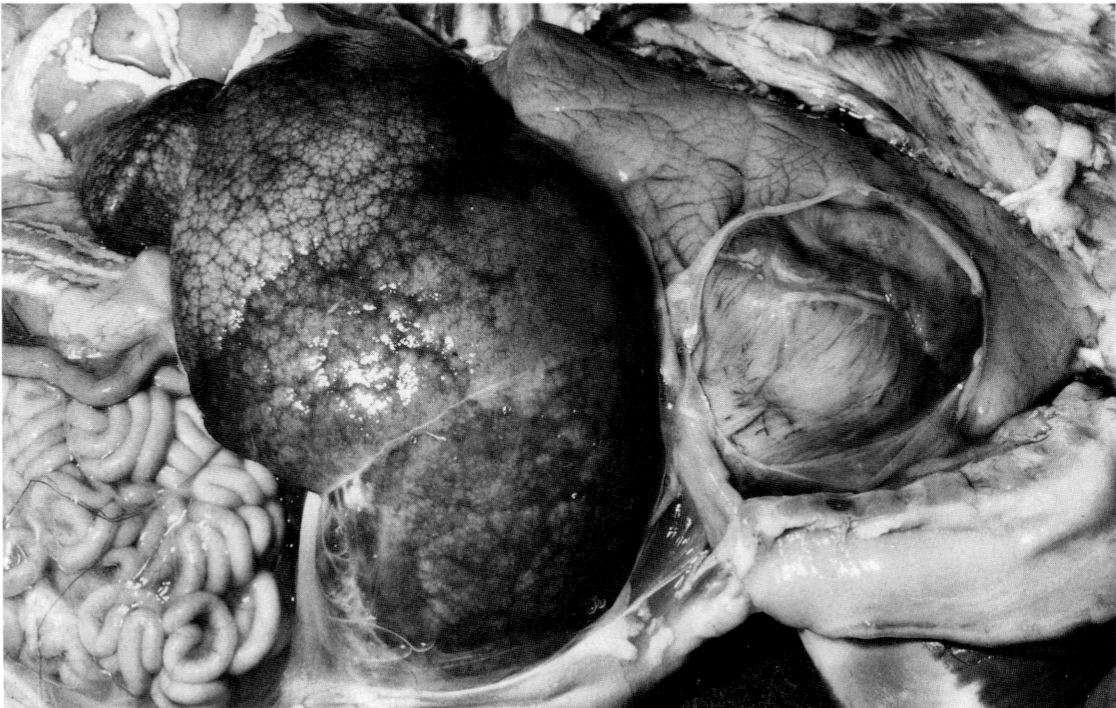

Fig. 3.14 Myocardial pallor due to myocarditis and myocardial degeneration, and chronic hepatic congestion with ascites. Bovine fetus. Bovine virus diarrhea.

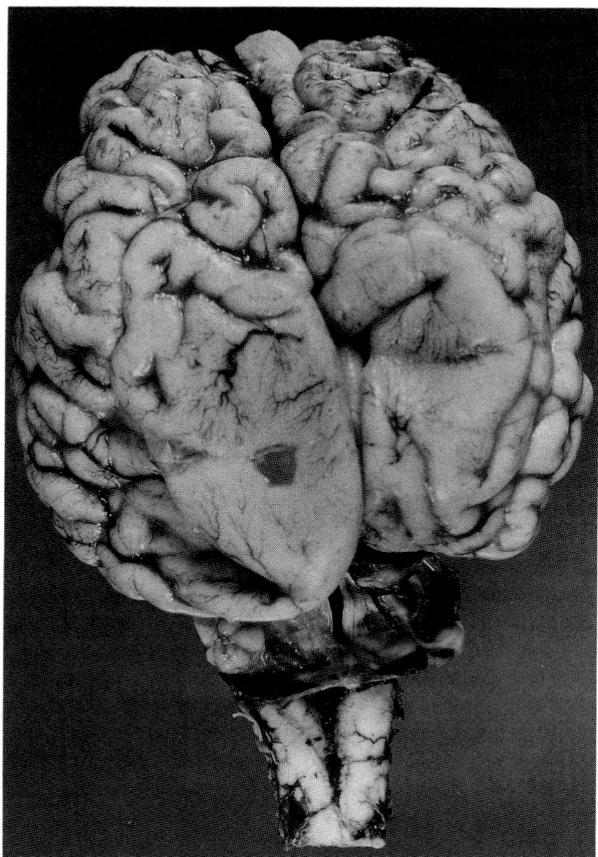

Fig. 3.15 Cerebellar hypoplasia and hydranencephaly. Ox.

tion, is associated with folial edema, and there may be focal hemorrhages in the cerebellar white matter and cortex.

Folial edema, depending on its severity, may result in total folial destruction, cavitation, or focal, often linear, areas of folial white matter deficient in myelin and axons. Where necrosis of the external granular layer predominates, the result is an irregular atrophy of affected folia. Features of the atrophic process are marked depletion of granule cells, ectopia of Purkinje cells, and the presence of swollen Purkinje cell axons in the granular layer. The evolution of the cerebellar lesion in the fetal calf extends over 6 weeks and, by 10 weeks after maternal infection, inflammatory changes are not evident in the brain.

Other central nervous system defects which may be a consequence of fetal infection are porencephaly, hydranencephaly, microencephaly, hydrocephalus, cystic septum pellucidum, and dysmyelination.

Ocular lesions commonly accompany cerebellar defects. The anomalies are represented by retinal dysplasia and atrophy, cataract, optic neuritis and atrophy, microphthalmia, and persistent pupillary membrane.

Infection of the fetus later than 170 days is unlikely to cause either intrauterine death or malformation. This increased resistance coincides with the fetus's acquiring the capacity to produce neutralizing antibody to the virus. However, the virus also has the capacity to induce more subtle developmental aberrations, such as intrauterine growth retardation and atrophy of the thymus and lymphoid tissues. The regressive changes in thymus and lymphoid tissues offer a morphological basis for the immunological suppression and tolerance phenomena associated with congenital infections with bovine virus diarrhea virus.

Bibliography

Badman, R. T. *et al.* Association of bovine viral diarrhoea virus infection to hydranencephaly and other central nervous system lesions in perinatal calves. *Aust Vet J* **57:** 306–307, 1971.

Brown, T. T. *et al.* Virus-induced congenital anomalies of the bovine fetus. 2. Histopathology of cerebellar degeneration (hypoplasia) induced by the virus of bovine viral diarrhea–mucosal disease. *Cornell Vet* **63:** 561–578, 1973.

Brown, T. T. *et al.* Pathogenetic studies of infection of the bovine fetus with bovine viral diarrhea virus. 1. Cerebellar atrophy. *Vet Pathol* **11:** 486–505, 1974.

Brown, T. T. *et al.* Pathogenetic studies of infection of the bovine fetus with bovine viral diarrhea virus. 2. Ocular lesions. *Vet Pathol* **12:** 394–404, 1975.

Brownlie, J. The pathogenesis of bovine virus diarrhoea virus infections. *Rev Sci Tech Off Int Epiz* **9:** 43–59, 1990.

Casaro, A. P. E., Kendrick, J. W., and Kennedy, P. C. Response of the bovine fetus to bovine viral diarrhea—mucosal disease virus. *Am J Vet Res* **32:** 1543–1562, 1971.

Done, J. T. *et al.* Bovine virus diarrhea–mucosal disease virus: Pathogenicity for the fetal calf following maternal infection. *Vet Rec* **106:** 473–479, 1980.

Kahrs, R. F. Effects of bovine viral diarrhea on the developing fetus. *J Am Vet Med Assoc* **163:** 877–878, 1973.

Kahrs, R. F., Scott, F. W., and de Lahunta, A. Congenital cerebellar hypoplasia and ocular defects in calves following bovine viral diarrhea–mucosal disease infection in pregnant cattle. *J Am Vet Med Assoc* **156:** 1443–1450, 1970.

Kendrick, J. W. Bovine viral diarrhea–mucosal disease virus infection in pregnant cows. *Am J Vet Res* **32:** 533–544, 1971.

Nettleton, P. F. Pestivirus infections in ruminants other than cattle. *Rev Sci Tech Off Int Epiz* **9:** 131–150, 1990.

Scott, F. W. *et al.* Cerebellar degeneration (hypoplasia), ocular lesions and fetal mummification following experimental infection with bovine viral diarrhea–mucosal disease virus. *Cornell Vet* **63:** 536–560, 1973.

Ward, G. M. *et al.* A study of experimentally induced bovine viral diarrhea–mucosal disease virus in pregnant cows and their progeny. *Cornell Vet* **59:** 525–538, 1969.

5. Border Disease

This disease of lambs is so named to reflect the first descriptions from the border counties of Britain. It is now recognized in many countries. Affected lambs showed gross tremors and long, hairy birth coats. Since the original descriptions, more protean manifestations are recognized, and goats have been added to the species naturally infected.

The disease is caused by border disease virus, a pestivirus very closely related to the virus of bovine virus diarrhea and less closely to the virus of hog cholera. These viruses possess a similar host spectrum experimentally, and interspecies transmission, especially between ruminants, does occur naturally, but the clinical expression in recipients may be modified or absent. The ecology of the border disease virus and the pathogenesis of the disease depend on the ability of the virus to cross the placenta and then to produce either disease in the fetus or a state of immunotolerance and persistent infection which allows excretion of the virus continuously in postnatal life. Strain variations of the virus, differing host responses depending on breed and genotype, and gestational age at which infection occurs contribute to the varied manifestations of the disease.

Primary infection of postnatal sheep is usually subclinical; immunity develops, and the virus is eliminated. In pregnant ewes, the virus infects the fetus within the first week of exposure and is not then influenced by the immune status of the ewe.

The border disease agent is a potential cause of a variety of developmental disorders. These include hypomyelinogenesis; cavitating cerebral defects such as porencephaly, hydranencephaly, and cystic septum pellucidum; cerebellar dysplasia; arthrogryposis and skeletal defects; mandibular brachygnathism; and thymic hypoplasia. It is in experimental infections that the pathogenicity of the virus is most freely expressed, the cavitating cerebral lesions and cerebellar dysplasia being common sequelae of experimental infections, but only occasionally encountered in the natural disease.

The virus of bovine virus diarrhea is capable of producing the border disease syndrome in sheep and goats, and the syndrome in piglets caused by congenital infection with hog cholera virus.

The ovine fetus is liable to significant damage if the dam is infected with border disease virus between days 16 and 80 of gestation. The age-related immune capacity of the fetus is the important determinant of the nature of the disease produced. In the case of infections occurring within the first half of gestation, the result may be fetal death and abortion. Alternatively, the fetus may survive, frequently carrying an immunologically tolerated infection. In this event, the virus persists in fetal tissues, and postnatally, the lamb fails to produce specific antibody. Such lambs harbor the virus for prolonged periods. Such persistently infected sheep are chronic excretors of the virus and readily transmit infection. Lateral spread is important in the field and, although the virus can be transmitted experimentally by a number of routes, the mode of lateral transfer of infection under natural conditions has not been identified.

Infections in the second half of gestation elicit both humoral and cell-mediated immune responses, and this acquisition of immunocompetence endows the fetus with substantial resistance to infection after day 80 of gestation. In the case of infections initiated between day 90 of gestation and the early days of postnatal life, the cell-mediated immune response is expressed morphologically as a nodular periarteritis. The periarteritis affects medium to small arterioles, particularly in the meninges and substance of the CNS, but also occurs mildly in a wide range of other tissues.

The developmental anomalies arising from fetal infection are also related to the gestational stage at which the fetus encounters the virus, but lesions produced also vary markedly according to virus strain, dose, route of infection, and host genetic factors. Hypomyelinogenesis, the characteristic neural lesion, is diffuse in fetuses infected early in gestation but becomes progressively milder and more restricted to higher, later-myelinating regions of the central nervous system in infections initiated later in gestation. The occurrence of porencephaly and cerebellar dysplasia has been confined to infections initiated between days 45 and 72 of gestation. The development of the cutaneous lesion requires that infection be initiated before day 80 of gestation.

The nature of the neural and cutaneous lesions is now well defined, although understanding of some pathogenetic aspects remains incomplete. The hypomyelinogenesis, and the clinical tremors, may substantially resolve during the first few months of life, notwithstanding that the animals have persistent replicating infection, suggesting that the normal processes of myelin deposition have been delayed or that cellular injury has been slowly repaired. The virus does infect myelinating oligodendroglia, astroglia, and glial progenitor cells, and it is reasonable to assume a direct effect on the differentiation or maturation of oligodendroglia. However, the virus also infects many non-neural cells, including thyroid epithelium. There are no morphological changes in the thyroid epithelium, but reduction in circulating thyroid hormone levels may contribute to the delayed maturation.

The cavitating cerebral defects and the cerebellar dysplasia arise from inflammatory destruction of developing neural elements, possibly secondary to vasculitis. These lambs may have severe locomotor and behavioral abnormalities and defects of vision, the latter probably of central origin rather than due to focal retinal dysplasias which may be present. They are serologically positive but not persistently infected.

The abnormality of birth coats occurs in fetuses infected before about 90 days of gestational age. There are no cytological changes in the papillae which can be ascribed to direct viral action. The primary follicles revert to a more primitive type, are enlarged, and produce heavily medullated fibers, which are most prominent after about 3 weeks. There are fewer secondary follicles, and their development is retarded.

Border disease infection also interferes with prenatal and postnatal development of skeleton, musculature, and viscera. Tissues most affected are those which have their main growth spurt in the fetal period. Growth-arrest lines in long bones suggest periods of interrupted intrauterine development.

Lambs which are persistently infected at birth remain so, but not all show neurological signs. Growth may be depressed as may their viability. Some sheep which are persistently infected develop oculonasal discharges and respiratory distress or severe diarrhea and die in 2 to 4 weeks with inflammatory lymphoproliferative lesions in many organs. In the brain, these reactions occur particularly in the choroid plexuses and periventricular substance. Proliferative metaplastic changes in the intestinal mucosa affect mainly the cecum and colon, with the hyperplastic glands penetrating the muscular mucosa. The delayed disease has features resembling those of mucosal disease of cattle, which is considered to be due to superinfection by a different strain of virus or by minor mutation of homologous virus in animals which are persistently infected.

Infection by border disease virus is widespread in goats, but disease attributable to this virus is not. The characteristics of the natural and experimental disease are similar to those in sheep, but spontaneous neurological disease is not a feature. The fetal goat may be much more susceptible to infection than the fetal lamb; a high incidence of fetal death, mummification, and abortion is reported.

Bibliography

Anderson, C. A. *et al.* Border disease. Virus-induced decrease in thyroid hormone levels with associated hypomyelination. *Lab Invest* **57:** 168–175, 1987.

Barlow, R. M. Morphogenesis of hydrancephaly and other intracranial malformations in progeny of pregnant ewes infected with pestiviruses. *J Comp Pathol* **90:** 87–98, 1980.

Barlow, R. M., and Paterson, D. S. P. Border disease of sheep: A virus-induced teratogenic disorder. *J Vet Med* No. 36 (Suppl.) "Advances in Veterinary Medicine." Berlin and Hamburg, Parey, 1982.

Barlow, R. M., Gardiner, A. C., and Nettleton, P. F. The pathology of a spontaneous and experimental mucosal disease-like syndrome in sheep recovered from clinical border disease. *J Comp Pathol* **93:** 451–469, 1983.

Carlsson, U. Border disease in sheep caused by transmission of virus from cattle persistently infected with bovine virus diarrhoea virus. *Vet Rec* **128:** 145–147, 1991.

Clark, G. L., and Osburn, B. I. Transmissible congenital demyelinating encephalopathy of lambs. *Vet Pathol* **15:** 68–82, 1978.

Jeffrey, M. *et al.* Immunocytochemical localization of Border disease virus in the spinal cord of fetal and newborn lambs. *Neuropathol Appl Neurobiol* **16:** 501–510, 1990.

Loken, T., Krogsrud, J., and Bjerkas, I. Outbreaks of border disease in goats induced by a pestivirus-contaminated orf vaccine, with virus transmission to sheep and cattle. *J Comp Pathol* **104:** 195–209, 1991.

Plant, J. W. *et al.* Pathology in the ovine foetus caused by an ovine pestivirus. *Aust Vet J* **60:** 137–140, 1983.

6. Hog Cholera Virus

The disease is discussed in detail in Volume 3, Chapter 1, The Cardiovascular System. Both vaccine and certain low-virulence field strains of hog cholera virus are teratogenic for the fetal piglet. Fetuses are susceptible to infection regardless of the immune status of the sow. The apparent induction of immune tolerance results in the delivery of chronically infected piglets lacking antibody. The gestational interval during which fetal piglets are susceptible to the teratogenic effects of hog cholera virus extends at least from day 10 to 97 of gestation, but the occurrence of malformation is favored by infection around day 30 of the gestation period. The most characteristic anomalies involve the nervous system. The combination of hypoplasia and dysplasia of the cerebellum and central nervous system hypomyelinogenesis, most severe in the spinal cord, is one form of the congenital tremor syndrome of piglets. Microencephaly is also a rather characteristic sequel. The mechanism by which these lesions evolve has not been identified, but their nature is compatible with a persistent neural infection resulting in selective inhibition of cell division. Additional effects noted in affected litters include fetal mummification, stillbirth, pulmonary hypoplasia, nodularity of the liver, ascites, anasarca, cutaneous purpura, arthrogryposis, and micrognathia.

Bibliography

Bradley, R. *et al.* Congenital tremor type A1: Light and electron microscopical observations on the spinal cords of affected piglets. *J Comp Pathol* **93:** 43–59, 1983.

Emerson, J. L., and Delez, A. L. Cerebellar hypoplasia, hypomyelinogenesis, and congenital tremors of pigs, associated with prenatal hog cholera vaccination of sows. *J Am Vet Med Assoc* **147:** 47–54, 1965.

Johnson, K. P. *et al.* Multiple fetal malformations due to persistent viral infection. 1. Abortion, intrauterine death, and gross abnormalities in fetal swine infected with hog cholera vaccine virus. *Lab Invest* **30:** 608–617, 1974.

Van Oirschot, J. T. Hog Cholera. *In* "Diseases of Swine" A. D. Leman, *et al.* (eds.), 6th Ed., pp. 289–300. Ames, Iowa, Iowa State University Press, 1986.

7. Feline Parvovirus

This disease is discussed in detail in Volume 2, Chapter 1, The Alimentary System. The feline parvovirus virus is pathogenic to the cerebellum of kittens before and shortly after birth, at which time the cerebellum is growing and differentiating rapidly. This virus has tropism for cells which have a high mitotic rate. The site of viral action is the external germinal layer of the cerebellum. The formation of intranuclear inclusion bodies is an early feature of the infection; they disappear by day 14 of infection. The infected cells are destroyed and with them the growth potential of the cerebellum. The Purkinje cells, which are postmitotic but immature, are also affected, although inclusion bodies do not form in them. The nuclei of the Purkinje cells show vesicular ballooning, eosinophilia, and condensation of the membrane. One or more large vacuoles form in the cytoplasm. Some Purkinje cells undergo coagulative necrosis.

In view of the tropism of parvovirus for rapidly replicating cells, a wide spectrum of abnormalities might be expected to result from infection of kittens *in utero*. The virus will cross the placenta and produce generalized infection of the fetus, as indicated by the distribution of inclusion bodies. The subependymal cell plate shares growth behavior with the external granular layer of the cerebellum, and, although heavily infected, only one instance of hydranencephaly attributable to damage at this site of infection has been recorded. In visceral organs, only slight degrees of renal hypoplasia have been observed in infected kitten fetuses.

Bibliography

Greene, C. E., Gorgacz, E. J., and Martin, C. L. Hydranencephaly associated with feline panleukopaenia. *J Am Vet Med Assoc* **180:** 767–768, 1982.

Johnson, R. H., Margolis, G., and Kilham, L. Identity of feline ataxia virus with feline panleucopenia virus. *Nature* **214:** 175–177, 1967.

Kilham, L., and Margolis, G. Viral etiology of spontaneous ataxia of cats. *Am J Pathol* **48:** 991–1011, 1966.

Kilham, L., Margolis, G., and Colby, E. D. Congenital infections of cats and ferrets by feline panleukopenia virus manifestated by cerebellar hypoplasia. *Lab Invest* **17:** 465–480, 1967.

8. Cache Valley Virus

Cache Valley virus is an arthropod-borne virus of the Bunyavirus group and, on serologic data, the most widespread of the group in North America. It is capable of infecting a variety of mammals, generally as subclinical infections, but it is occasionally teratogenic in fetal lambs. The nature of the abnormalities produced, and the pathogenesis of them, are comparable to those of Akabane virus in sheep.

Bibliography

Chung, I. *et al.* Congenital malformations in sheep resulting from *in utero* inoculation of Cache Valley virus. *Am J Vet Res* **51:** 1645–1648, 1990

Edwards, J. F. *et al.* Ovine arthrogryposis and central nervous system malformations associated with *in utero* Cache Valley virus infection: Spontaneous disease. *Vet Pathol* **26:** 33–39, 1990.

9. Chuzan Virus

Chuzan virus is a member of the Orbivirus genus, transmitted by *Culicoides* spp. It is infective for several ruminant species and in Japan has been incriminated as a teratogen of fetal cattle, with features similar to those of Akabane infection.

Bibliography

Muira, Y. *et al.* Pathogenicity of Chuzan virus, a new member of the Palyam subgroup of genus Orbivirus for cattle. *Jpn J Vet Sci* **50:** 632–637, 1988.

II. Cytopathology of Nervous Tissue

Nervous tissue is highly specialized and structurally complex, and neuropathology has always tended to be set apart as an arcane specialist area, to be entered by only a select few. The veterinary pathologist cannot, however, escape having to deal with it frequently. The rapid advancement of neuroscience over the last decade has added enormously to the understanding of pathomechanisms, and further exciting and interesting progress can be expected. What follows in this section is an attempt to provide a contemporary basis for such a general appreciation in the face of continuing rapid expansion of knowledge.

A. The Neuron

The **neuron** is the fundamental cell of the nervous system, and ultimately all neurologic disease must involve functional disturbances in neurons. In conventional histopathology, the term neuron refers to the cell body of the nerve cell, which often is only a small part of the total cell volume. For the purposes of comprehending pathogenetic mechanisms, it is important to remember always that the neuron comprises both the cell body (soma or perikaryon) and the cell processes, in particular the axon. The whole cell constitutes a structure concerned with the generation, conduction, and transmission of impulses, and in some cases a single cell performs this task over a very long distance. For example, in a lumbar dorsal root ganglion of a horse, the cell body of a sensory neuron may project a process distally to the extremity of the hind foot and centrally to the posterior brain stem, making it without doubt the largest cell in the body. If the soma were enlarged to the size of an orange, the processes would have the dimensions of a garden hose and would be over 20 km in length. Pathologic reactions within one cell may therefore be separated by considerable distance. Also, neurons function in hierarchical chains organized into anatomic systems.

The soma is the metabolic factory for the whole cell, and the great bulk of synthetic and degradative operations takes place there. The axons are serviced throughout their length by a bidirectional transport system, which moves components both away from and toward the soma (anterograde and retrograde transport respectively). The transport mechanisms are fueled by the consumption of energy along the course of the axon. Many components are moved in the anterograde direction within 25-nm vesicles by a "fast" transport system at a rate of 20 to 30 mm/hr. The motor for this system, intimately associated with the neurotubules, is an adenosine triphosphatase (ATPase) called **kinesin.** Larger vesicles of about 500 nm are transported retrogradely by this fast pathway, and the translocator has recently been identified as an isoform of **dynein,** the ATPase which powers cilia and flagellae in other types of cell. Mitochondria are also moved by fast transport, but at a somewhat slower rate. Neurofilaments, neurotubules, and some soluble proteins move by slow transport in the anterograde direction only, at a rate of about 1 to 3 mm/day, and are degraded when they reach the terminus. Should the operations of the soma or the transport systems be impaired, the health of the axon will suffer and, appropriately, in most instances the largest and longest axons are most vulnerable. Equally, when a neuron is fatally damaged the whole cell dies, and the consequent changes will spread over the whole extent of its domain. However, long lengths of the axon may degenerate without compromising the viability of the remainder of the cell and without inducing dramatic morphologic change in the soma, and there is often scope for adequate regeneration of the lost axonal extremity. These issues will be addressed more fully under **axonopathy.**

The vitality of individual neurons is sustained by their active relationship with other neurons, or other types of cells with which they interact. If a neuron dies, surviving neurons with which it has synaptic connections may regress due to lack of activation, undergo atrophy, and eventually die. This process is called **transynaptic degeneration,** and it will progress along specific anatomic pathways. It is useful also to realize that neurons are extremely diverse pharmacologically and biochemically, and this is well illustrated by the highly selective and regionalized effects of different toxins and metabolic disturbances. The selective effects of tetanus and botulinus toxins are good examples.

Material for routine diagnostic examination is in most cases fixed by immersion in formalin, and processed for embedding in paraffin: time-honored and still standard procedures. Unfortunately these procedures will usually give rise to artefacts which can be misinterpreted. This aspect is compounded by the variability in the appearance of neuronal somata between regions, between species, and according to the age of the animal.

During organogenesis, vast armies of neuroblasts proliferate, and immature neurons migrate to their final destinations. As large numbers of developing neurons are superfluous, considerable loss of cells by necrobiosis may be evident. Around the time of birth, and for varying periods afterward, the latter stages of the process may still be extant, particularly in the cerebellar cortex and paravertebral ganglia.

The mature neuron is a postmitotic cell, and throughout adult life, and especially during senescence, there is a net loss. Thus in a thorough scan of a histologic section from a senile animal, it might be expected that a few degenerate neurons will be found, even when fixation artefacts are minimal.

1. Degenerative Changes of the Nerve Cell Body

Nuclear margination. As a general rule, the neuronal nucleus is single and centrally located (Fig. 3.16), and its margination can be taken to indicate nonspecific degeneration, especially when combined with loss of staining affinity. However, an eccentric nucleus can be normal in some groups of smaller neurons, and is often present in the large cells of the periventricular gray matter, such as the mesencephalic nucleus of the trigeminal nerve and the olives. These cells also tend to have a "chromatolytic" appearance.

Chromatolysis. This term defines a change in appearance of the soma brought about by the dispersal of the rough endoplasmic reticulum (Nissl granules), and is subclassified as central or peripheral according to its locus within the cell body. Its assessment depends on an appreciation of the prominence and distribution of Nissl granules seen normally at the particular location. There are no artefactual changes that mimic chromatolysis and, provided it is accurately identified, it can always be regarded as a lesion. A number of special stains, such as cresyl violet, will demonstrate it better than routine hematoxylin and eosin.

Central chromatolysis is best appreciated in large neurons of some of the brain stem nuclei, in the spinal motor neurons, and peripheral ganglia. The chromatolytic cells are swollen and rounded out, rather than having the normal angulated appearance, and the nucleus becomes eccentric. The Nissl granules clear from the central region of the cell body, leaving this zone with a smooth ground-glass appearance (Fig. 3.16). Central chromatolysis occurs in a number of pathologic situations. It is often seen in bulbospinal motor neurons and sensory neurons whose peripherally projecting axons are injured, and especially when the injury occurs close to the cell body. The reaction follows such injury quickly, beginning within 24 hr, and becoming maximal in 1 to 3 weeks. When the dynamics of axonal flow are remembered, it can be appreciated that such an event will have a major feedback on the soma. In effect, the cell must rearrange its metabolism to adapt to its changed circumstances, and organize for a regenerative effort, involving the reconstruction of an axonal segment greater in volume perhaps than its surviving volume. In this context, central chromatolysis has been termed the **axon reaction,** and represents an anabolic adaptive response. In this reaction, the nucleus becomes extremely eccentric, and develops a prominent nucleolus and a basophilic cap of ribonucleic acid (RNA) on its cytoplasmic aspect. The rough endoplasmic reticulum (Nissl substance) disperses, and the cytoplasm becomes rich in free ribosomes, lysosomes, and mitochondria. There may also be some increase in the number of neurofilaments. All these changes reflect a shift in metabolic activity, with a switch toward increased synthesis of structural cellular proteins, and a marked decline in synthesis of transmitters. With completion of successful axonal regeneration, the chromatolytic soma returns to normal, sometimes passing through a densely basophilic phase in which it is packed with Nissl granules.

In some circumstances following axonal injury, central chromatolysis may proceed to cell death or permanent atrophy, and this is usually the case in neurons whose axons project entirely within the central nervous system. In general, the closer to the cell body the axonal lesion, the more likely is the cell to die. Those cells destined to die have a swollen achromatic cytoplasm depleted of organelles.

Central chromatolysis is induced also in more overtly neuronopathic conditions, most notably the now numerous motor neuron degenerations described in various species, and it is a feature of the pathology of perinatal copper deficiency in the sheep and goat (Fig. 3.16). Similarly, it is a striking feature in autonomic ganglia in equine and feline dysautonomias. In all such cases, the affected cells often proceed to necrosis and dissolution (Gudden's atro-

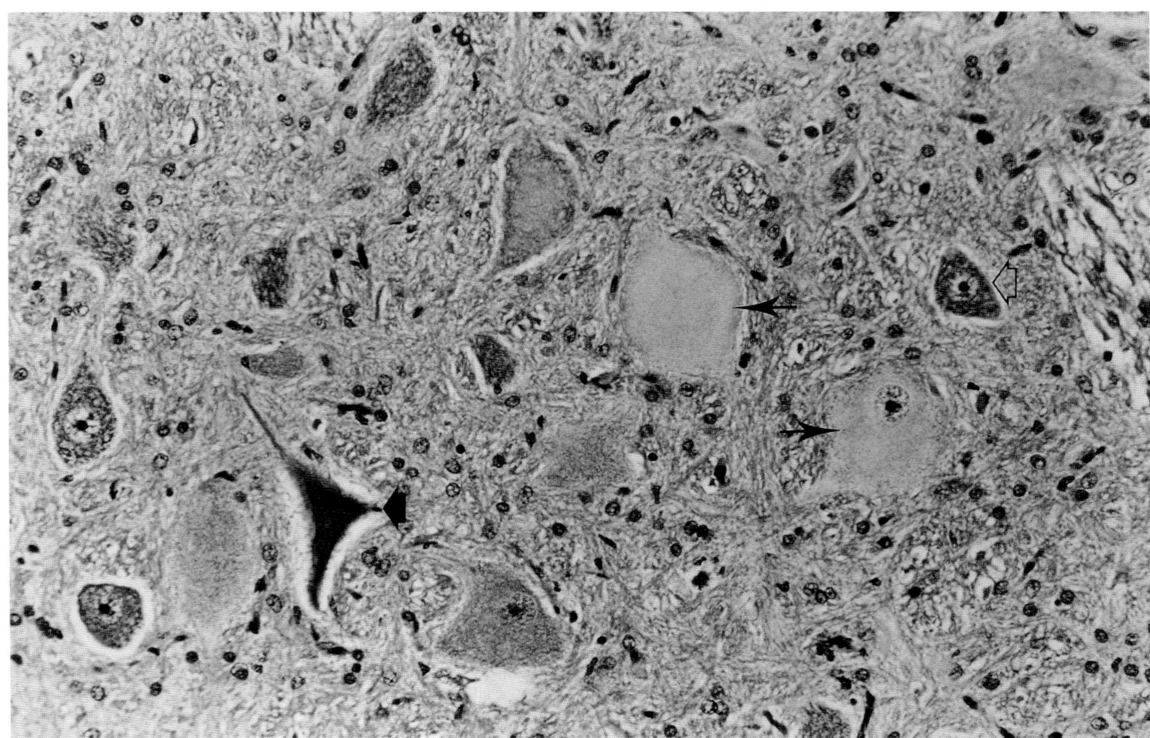

Fig. 3.16 Red nucleus. Swayback. Goat. Normal neurons (open arrow). Central chromatolysis (arrows). Shrunken, necrotic neuron (closed arrow).

phy). In many of these degenerative neuronopathies, the cytoplasmic alteration is due to massive accumulation of neurofilaments in the soma (see motor neuron disease). Nuclear margination is not as marked as in the axon reaction, and the prominent nucleolus and nuclear cap are not evident. These differences distinguish a regressive state from a regenerative one.

Peripheral chromatolysis indicates clearing of the periphery of the soma, with Nissl granules persisting around the nucleus. This change is generally associated with slight cellular shrinkage rather than swelling. It is a nonspecific lesion and can often be regarded as an early stage en route to necrosis.

In both forms of chromatolysis, microglia and astrocytes may proliferate and cover large expanses of the cell surface, thereby separating terminal boutons from the neuronal surface.

Neuronal atrophy. Loss of cytoplasmic bulk and a reduction in size (Fig. 3.17A) might be expected in situations of permanent loss of synaptic connections (see transynaptic degeneration), as when central axons undergo Wallerian degeneration but fail to regenerate.

Ischemic necrosis. This is a characteristic acute degenerative change in which the cytoplasm of the neuronal soma becomes shrunken and distinctly acidophilic (Fig. 3.17B), and the nucleus progresses through pyknosis and rhexis to lysis, leaving the coagulated cytoplasmic remnants to undergo liquefaction without undergoing phagocytosis. The remnant "ghosts" may persist for several days. Although classically described in ischemia, the reaction is not confined to that situation and may be seen in the cerebral cortex in hypoxia, hypoglycemia, the encephalopathy of thiamine deficiency, and in some chemical intoxications, such as organomercurialism and indirect salt poisoning. Following seizures, it may frequently be found in the dentate gyrus of the hippocampus and the cerebellar Purkinje cells. There are usually obvious proplastic changes in astrocytes and capillary endothelia in the vicinity of the affected neurons.

This raises the concept of **excitotoxicity,** in which neuronal degeneration and death are considered to result from excessive stimulus by an excitatory neurotransmitter. This phenomenon is thought to operate particularly in those neuronal systems utilizing glutamate as a transmitter, such as the cells in the hippocampal areas previously mentioned. Paradoxically, glutamate is potentially highly toxic to neurons, and normally is rapidly cleared by the glia following its release. Excessive release or defective clearance of glutamate from the environment of postsynaptic neurons predisposes to excitotoxicity, and such circumstances are provided by hypoxia and hypoglycemia. The pathogenesis is thought to involve ionic overloading of the cell, acute swelling, and then cell death with cytoplasmic coagulation and eosinophilia.

Neuronal necrosis may occasionally be expressed as cytoplasmic shrinkage and basophilia with nuclear disso-

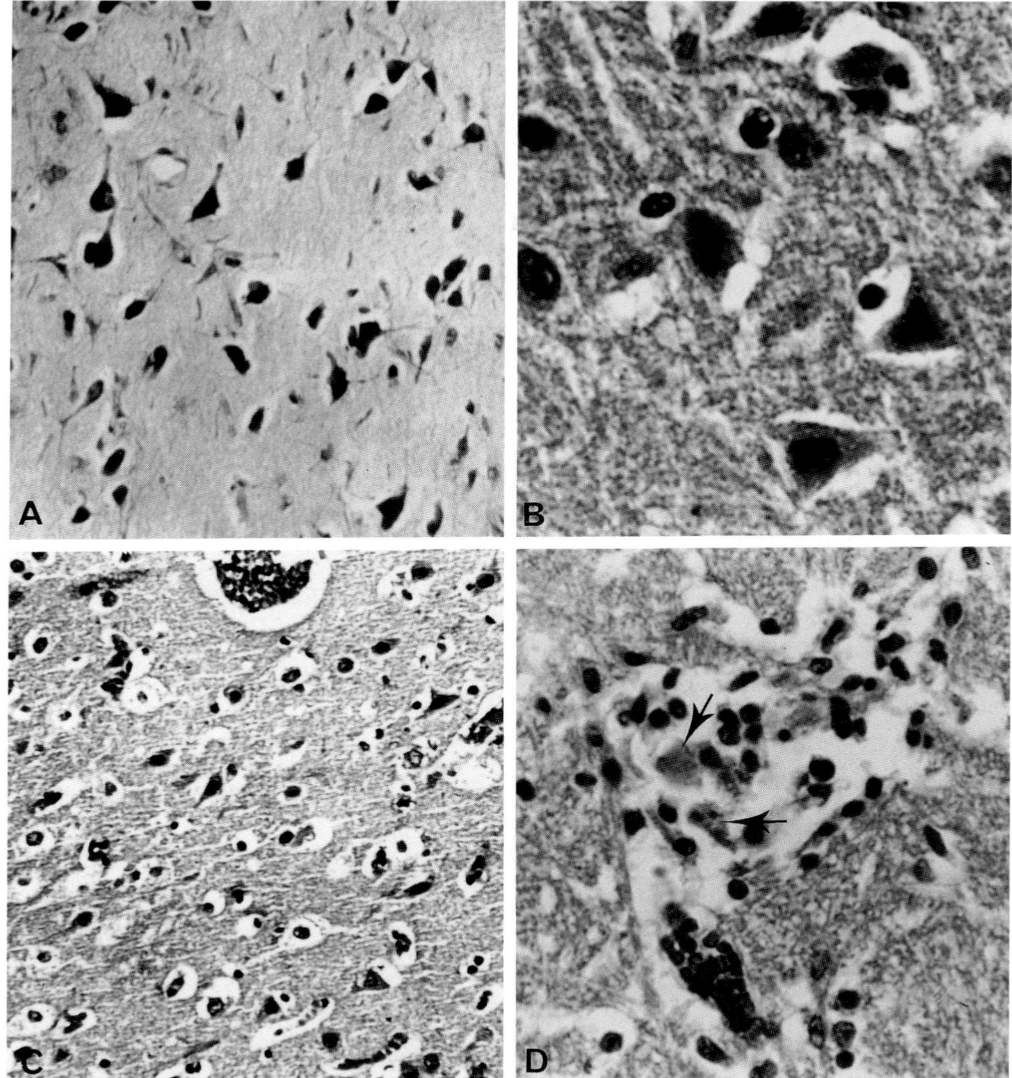

Fig. 3.17 (A) Chronic nerve cell degeneration. (B) Early ischemic nerve cell degeneration. Note cell shrinkage with condensed cytoplasm. (C) Autolyzing cerebral cortex easily misinterpreted as liquefactive necrosis. (D) Neuronophagic nodule in American encephalomyelitis. Fragmented neuron (arrows).

lution (Fig. 3.16). Shrunken **basophilic** neurons with normal nuclei are generally taken to be artefactual, and are often numerous. These "dark" neurons were once considered to be in a heightened state of physiologic activity at the time of fixation, but this idea has been abandoned since the advent of rapid perfusion fixation techniques.

Liquefactive necrosis. The characterization of this lesion is often at risk of being misinterpreted as autolytic change and fixation artefact, particularly shrinkage artefact (Fig. 3.17C). A lesion can sometimes be distinguished from an artefact by the presence or absence of significant alterations in other tissue cells, for instance, the swelling of capillary endothelium or indications of glial proliferation.

Necrosis with neuronophagia. In many viral infections, the death of neurons provokes the gathering of phagocytes around the cell body, and removal by them of the debris (Fig. 3.17D). This is usually a response of the microglia. Neuronophagia may also be seen in metabolic or toxigenic neuronal degenerations, but is generally not as extensive as that in viral infection.

Vacuolar degeneration. Neurons in the early stages of acute injury inflicted by viruses, toxins, or metabolic derangements, such as excitotoxicity, may develop numerous small cytoplasmic vacuoles, usually reflecting mitochondrial swelling. However, artefactual peripheral vacuolation is very common, giving the periphery of the cytoplasm a foamy weblike appearance. This is particularly so in the cerebellar Purkinje cells and large neurons of some of the brain stem nuclei.

Large single neuronal vacuoles, few in number, are occasionally observed in otherwise normal brains and may be seen in the red and oculomotor nuclei of cattle. When they occur at high frequency in the neurons of the medulla and midbrain, they are virtually pathognomonic of scrapie in sheep, goats, and cattle (Fig. 3.18).

Vacuolar change is strikingly evident in those lysosomal storage diseases in which the stored material is extracted during processing, or is unstained by the routine methods. In these situations, the neuron can become dramatically bloated by the accumulation of myriad secondary lysosomes which displace the normal organelles and distend the soma with a foamy mass of apparently empty vacuoles. In most cases, the storage process involves other cell types inside and outside the nervous system, and is accompanied by additional neuropathological manifestations (see neuronal storage diseases).

Storage of pigments and other materials. Neurons may accumulate large quantities of **ceroid/lipofuscin** or other pigments, either as a consequence of aging, or in storage disorders involving such substances. The pathogenesis of the ceroid/lipofuscinoses remains unclear but, as is discussed elsewhere, there is a genetic basis in some instances, while in others, unspecified environmental factors may be involved. The complex storage material accumulates as granules in a manner analogous to that of the other lysosomal storage diseases, although a clearly defined limiting membrane is not usually apparent ultrastructurally. When the process is intense, a rusty-brown discoloration of the gray matter and ganglia may be evident grossly.

Neuromelanins may also accumulate excessively in those midbrain nuclei where they are usually present in modest amounts, and extensive neuromelanosis can occur in sheep chronically poisoned by *Phalaris* spp. In extreme cases, the gray matter and ganglia may have a macroscopic greenish discoloration.

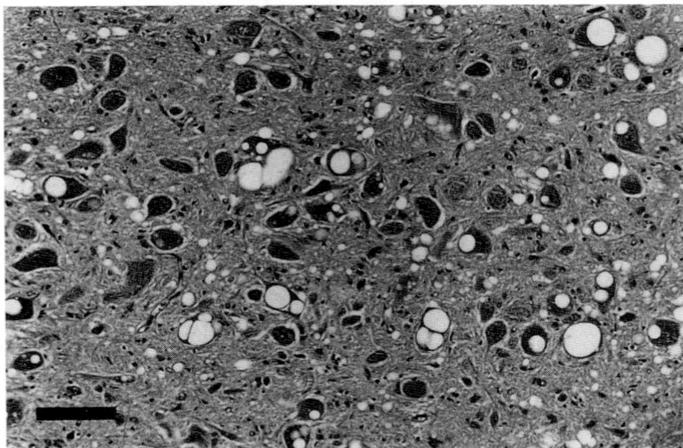

Fig. 3.18 Neuronal vacuolar degeneration. Nyala (*Tragelaphus angasi*). (Reprinted from Jeffrey, M., and Wells, G. A. H. *Vet Pathol* **25:** 398, 1988.)

Lafora bodies are basophilic to amphophilic inclusions of variable size and up to several microns in diameter; a few may be found incidentally, especially in dogs. They are composed of polyglucosans, and their pathogenesis is obscure. In the very rare **Lafora's disease** (see Storage Diseases), they occur in massive numbers throughout the brain within the neuronal soma, the dendrites, and less commonly, the axons (Fig. 3.19). In this disease, they are associated with a severe clinical state of myoclonus epilepsy.

Siderotic pigmentation of neurons, in which the cells become encrusted with basophilic complexes of iron, calcium, and phosphorus, may be found near contusions and hemorrhages. It is not known, however, whether the iron is derived from the hemoglobin or from intracellular iron-containing respiratory enzymes. The neurons may be otherwise normal, or their degeneration may produce small lakes of basophilic deposit. The latter are common in the neonatal cerebellar cortex.

Viral inclusion bodies. The best known is the Negri body of rabies, which is eosinophilic and intracytoplasmic. Herpesvirus inclusions are characteristically intranuclear, while those of the paramyxoviruses such as canine distemper may be intracytoplasmic or intranuclear. The increasing use of specific immunostains has greatly facilitated the identification of viral inclusions.

Nonviral eosinophilic cytoplasmic inclusion bodies also occur. Sometimes they are an incidental finding in otherwise normal brains at sporadic locations; in cats, they can

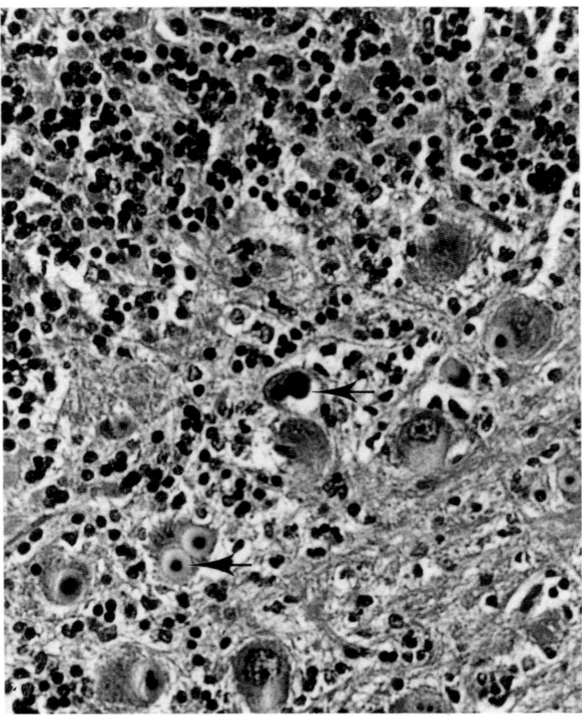

Fig. 3.19 Lafora bodies in Purkinje cells (arrows). Dog.

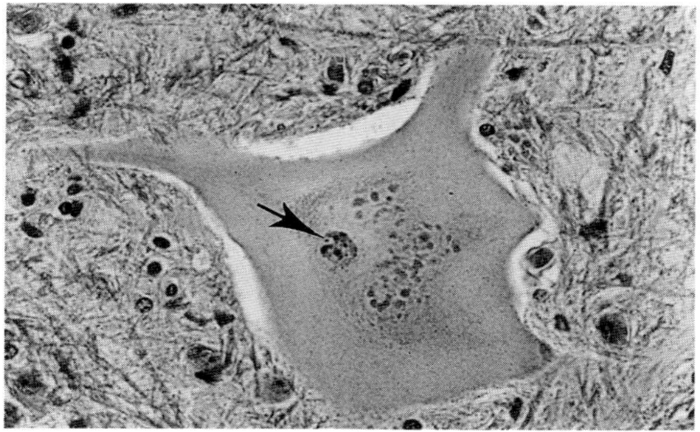

Fig. 3.20 Spinal motor neuron. Horse. Chromatolysis, nuclear dissolution (arrow), and cytoplasmic inclusions. (Motor neuron disease.)

sometimes be found in the pyramidal cells of the hippocampus and lateral geniculate nuclei.

More specific inclusions have been described in humans and animals in neuronal degenerative diseases. In humans, ultrastructural differences define several types of inclusions: **Hirano, Pick, Lewy,** and **Bunina bodies.** Hiranolike bodies, appearing histologically as elongated eosinophilic inclusions and, ultrastructurally, as masses of beaded filaments, are reported in horses, together with other structures having some of the features of Bunina bodies. Further definition of these types of inclusions in animals is required, and their pathogenetic significance is not understood (Fig. 3.20).

2. *The Axon*

The **axon** acts as the solitary efferent extension of all neurons except those sensory neurons in the spinal ganglia whose peripherally directed myelinated axons function as dendrites, in that they conduct impulses toward the cell body. Axons may branch extensively toward their terminations.

The first part of the axon is called the **initial segment,** and it has a distinctive ultrastructure, related to its being the site of membrane ion channels critical for the initiation of a propagated action potential. The axoplasm contains mitochondria, endosomes, intermediate filaments (neurofilaments), microtubules (neurotubules), and secretory vesicles or granules containing neurotransmitters appropriate for the particular cell. The axoplasm will also contain soluble macromolecules such as enzymes.

Axons are sustained by their parent cell bodies and by the cells that invest them along their course. The axoplasm is devoid of ribosomes, and axoplasmic proteins are provided by the soma. Similarly, the lysosomal apparatus is limited in the axon, in terms of digestive capacity, and many obsolete materials and organelles are returned by retrograde transport to the cell body for complete degradation. The role of neurotubules in transport mechanisms has been mentioned, and is critically important for the maintenance of the axon. Axonal diameter is distinctly reduced at the nodes of Ranvier, and these strictures are probably the reason that paranodal swellings filled with transported vesicles and organelles are a feature of many axonopathies in which transport has been disturbed.

The neurofilaments are responsible for the maintenance of axonal size and geometry, and are part of the generic cytoskeletal intermediate filament family. The proteolytic destruction of neurofilaments, triggered by the influx of calcium ions, is a common pathway for the collapse and disintegration of damaged axons. The larger axons are invested in a segmental manner by a myelin sheath, interrupted at the nodes of Ranvier, and penetrated at intervals by incisures.

All this sophistication is not resolved in routine light-microscopic examination of paraffin-embedded tissues. The course of axons can be highlighted by the use of silver staining techniques, which emphasize the shrinkage artefact and distortion produced by routine tissue preparation. Nonetheless, with experience, many lesions in paraffin-embedded tissue can be interpreted but, particularly for peripheral nerves, plastic embedding of specimens is far preferable.

Over recent decades there have been major advances in the understanding of axonal pathology, and several new concepts are firmly in place. Some axonopathic processes have been grouped according to whether they begin in the **proximal** or the **distal** portion of the axon, and whether they involve **central** or **peripheral** axons or both. Thus, for example, one may distinguish central and peripheral distal axonopathy, or a central proximal axonopathy. The principal categories of **axonopathy** can now be discussed in broad terms.

Wallerian degeneration is the term referring to the changes which follow acute focal injury to a myelinated axon, such that distal to the injury it becomes nonviable. When the soma is uninjured, there is potential for regeneration and, in peripheral nerves, this may be complete. Should the acute injury involve death of the cell body, then Wallerian degeneration of the axon will proceed as part of the dissolution of the entire neuron. The classical scenario for Wallerian degeneration is acute focal mechanical injury in a **peripheral nerve,** which effectively transects axoplasmic flow. Within 24 hr, the distal segment begins to degenerate fairly evenly along its length. Focal eosinophilic swellings occur, often containing accumulations of degenerate organelles, and then fragmentation becomes evident by 48 hr or so. There is a rapid response by the Schwann cells as the myelin sheaths are made redundant by the disintegration of the axon. Initially, myelin retracts from the nodes and then forms into ellipsoids, regarded originally as "digestion chambers" for the enzymic lysis of the axonal fragments (Fig. 3.21). The myelin itself condenses into aggregates and fragments and, together with remaining axonal debris, becomes the target of invading macrophages. Prior to this, the complex myelin lipids are progressively transformed into simpler neutral lipids over a period of 10 to 20 days, and this is reflected in the reaction to specific lipophilic stains. Macrophages

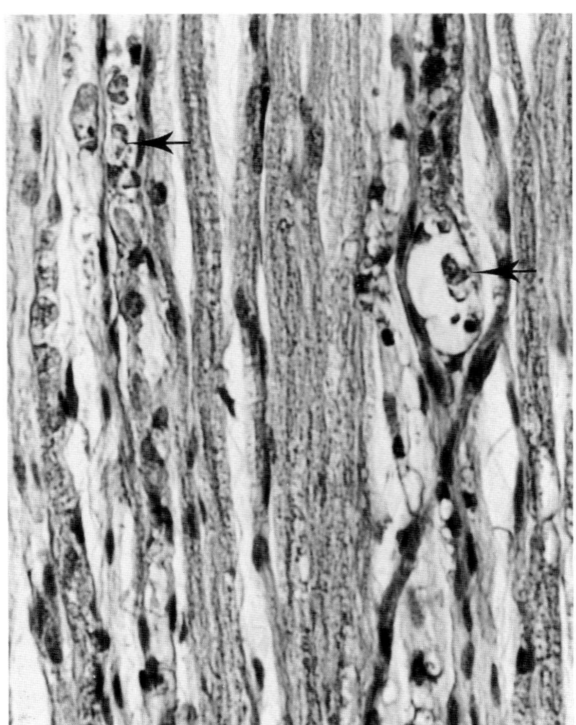

Fig. 3.21 Wallerian degeneration in peripheral nerve (arrows). Copper deficiency. Goat.

enter the sheath and soon become filled with sudanophilic droplets. These lipid-laden cells may persist in the interstitium for many weeks. Some of the myelin debris is phagocytosed by the Schwann cells themselves, and they begin to proliferate. As the debris is cleared away, proliferating Schwann cells form bands along the former course of the myelinated axons (Bungner's bands). Similar Wallerian changes occur proximal to the site of injury over several internodes. If conditions are favorable at the site of injury, sprouts from the axonal stump will find their way along the Schwann cell bands and be directed to their correct destinations. In most instances, the growing axonal sprouts advance at a rate of 2 to 4 mm/day, and the new axon will be invested by the Schwann cell cytoplasm. Sprouting from individual axons initially is multiple and, by an unknown mechanism, one sprout is selected for the completion of regeneration. Thus a new axon may arise from a different soma than did the original. The regenerated axon is remyelinated by the Schwann cells, although the new sheath is thinner than the original, and nodal length, variable and shorter. Should axonal regeneration be prevented, the Schwann cell bands persist, and endoneurial fibrosis usually develops. Abortive regeneration can lead to a tangled clump of neurites, Schwann cells, and fibrocytes at the injury site. This will happen after transection if the severed ends of the nerve fibers are separated by too great a distance.

The effectiveness of regeneration in peripheral nerves is related to the comparatively simple axon/Schwann cell relationship, the presence of a basal lamina tube around each myelinated axon, and the replicative ability and metabolic resilience of the Schwann cells.

In the **central nervous system,** these conditions do not apply. The oligodendrocyte/axon relationship is far more complex, the oligodendrocyte is relatively a poorly regenerative cell type, there is no basal lamina scaffold, and the debris from central myelin is thought to inhibit axonal sprouting. The initial regressive changes of Wallerian degeneration are similar to those described for the peripheral nerves, although they proceed over a longer time. This is because the involvement of hematogenous macrophages is slower and less intense in the central nervous system, and activated microglial cells undertake most of the work. Axonal sprouting and some remyelination can occur. However, the poverty of the regenerative response results mostly in the permanent disappearance of the axons, myelin, and oligodendrocyte cell bodies. Some of the myelin debris may be phagocytosed by reactive astrocytes, and their processes extend to fill the vacancy, creating a ramifying network of astroglial scar tissue. Wallerian degeneration in the central nervous system is most commonly seen in the spinal cord, the optic tract, and the brain stem. Probably the best known association is with the focal compressive myelopathies in the horse and dog (the wobbler syndromes).

During the active degenerative phase, recently phagocytosed and partially digested myelin debris may be distinguished in paraffin sections by the use of the luxol fast blue/periodic acid-Schiff (PAS) stain. Degenerate myelin is well visualized by the Marchi technique (Fig. 3.22A). More recently, immunostaining of myelin basic protein (MBP) has been applied. End-stage plaques of astrogliosis may be demonstrated by traditional gliophilic stains or by the use of immunostaining for glial fibrillary acidic protein (GFAP-intermediate filament protein).

The destruction of myelin in Wallerian degeneration is known as **secondary demyelination** and is to be distinguished from **primary demyelination,** in which the axon is initially undamaged (see later sections).

Distal axonopathy is a pattern of axonal disease seen in a number of chronic intoxications and genetically determined entities. It begins with degenerative changes in the distal reaches of the affected fibers and fairly characteristically involves the largest and longest, such as the proprioceptive and motor tracts of the spinal cord, the optic tract, and the recurrent laryngeal and other long peripheral nerves. Implicit in this pattern is a disturbance of anterograde axonal transport, on which the maintenance of axonal well-being depends. In general, the process begins with the formation of focal axonal swellings containing degenerate organelles. These swellings are ovoid or circular eosinophilic structures commonly referred to as **spheroids.** This may progress to axonal fragmentation and attempted regeneration, which may be abortive, and is succeeded by further degeneration. These changes may develop focally in the distal regions of the axon and may

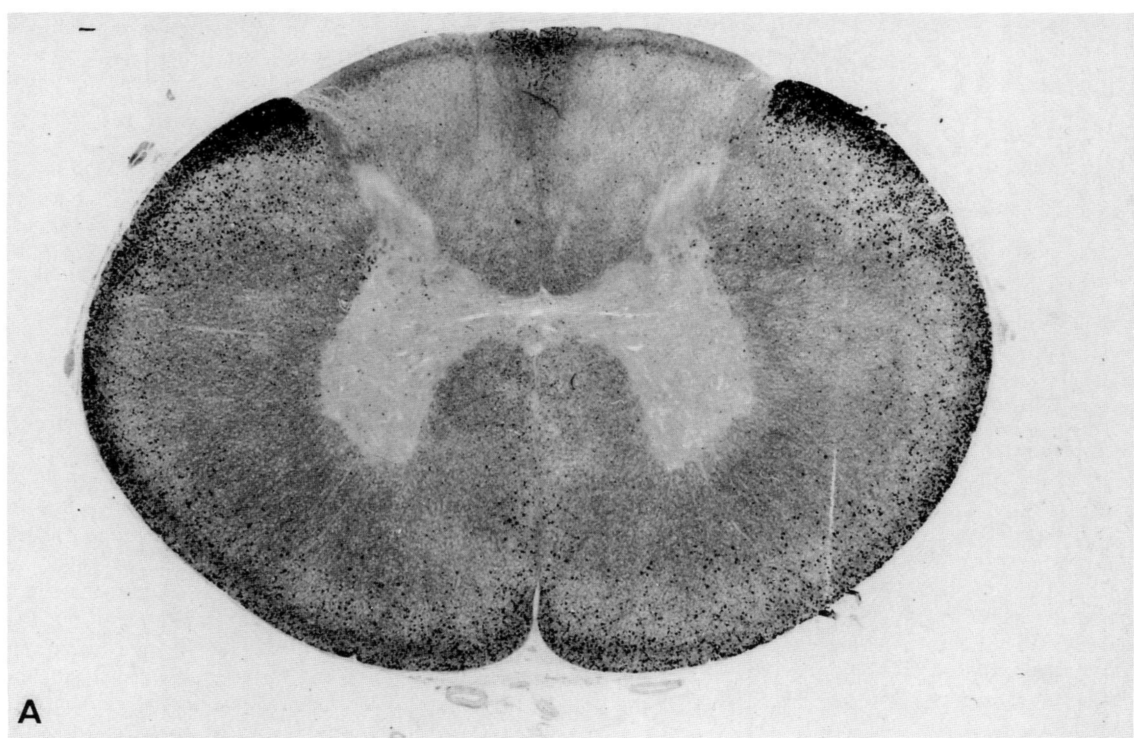

Fig. 3.22A Degeneration of myelin in spinal cord. Pig. Copper deficiency. Stained with Marchi. Note selective involvement of tracts. (Courtesy of M. D. McGavin).

extend more proximally with time. Diagnostically, the key is the recognition of the pattern of degenerative changes toward the terminations of long tracts (see Cycad Poisoning). The lesion is a feature of intoxication with certain of the organophosphates, for example.

Proximal axonopathy is the contrasting situation in which focal swellings and degeneration begin in the proximal axonal regions. It is a pattern less likely to be encountered in natural animal disease than the distal variety. It is perhaps best exemplified by the large fusiform swellings, **torpedoes,** seen on the proximal axonal segments of cerebellar Purkinje cells in the mycotoxicosis of perennial rye grass poisoning and some storage diseases (Fig. 3.22B). Such lesions may develop in central or peripheral axons and be associated with genetically determined or acquired disease. Proximal axonal swellings caused by the accumulation of neurofilaments are a feature of several neurodegenerative diseases described elsewhere in this chapter. A defect in the transport of phosphorylated neurofilaments results in the accumulation of masses of them, where they cause large, amphophilic axonal swellings. The fundamental pathogenesis remains undefined.

Axonal dystrophy is a term used to describe an axonopathic process characterized by the occurrence of large focal swellings, often concentrated in the terminals and preterminals of long axons. They are therefore frequently seen in and around relay nuclei in the brain, and in peripheral endings. The spheroids in axonal dystrophy can become extremely large, over 100 μm in diameter, and are filled with accumulations of normal organelles, degenerate organelles, and abnormal membranous and tubular structures. In hematoxylin and eosin sections, their appearance can be variable (Fig. 3.23A,B); some are densely eosinophilic, and either smooth or granular or vacuolated; others may be pale with central denser-staining cores; some may have a basophilic hue and evidence of focal mineralization. However, the swellings are not usually associated with any marked reaction on the part of surrounding elements and only rarely are seen to be undergoing fragmentation and dissolution. They are long lasting, in contrast to the spheroids of acute axonal degeneration. A thinning of the myelin sheath around spheroids will occur as lamellae slip to accommodate the focal axonal enlargement (Fig. 3.23C). The pathogenesis is unclear, but recent evidence points to a disturbance of retrograde axonal transport. The lesion is a common finding in the relay nuclei of the posterior brain stem in old age, and is a frequent accompaniment to neuronal storage diseases. It is the principal feature of diseases known as **neuroaxonal dystrophies,** of which several are recorded in the veterinary literature. In several such diseases, the topography of the axonal dystrophy seems to fit the clinical deficits, but there is evidence that, in many situations, even intense development of the lesion has no functional significance. This seems generally so in regard to axonal dystrophy in the gracilis and cuneate nuclei.

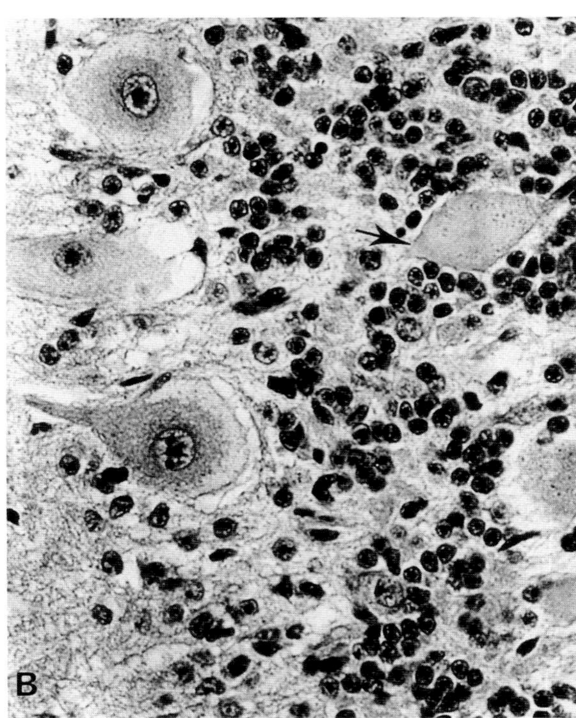

Fig. 3.22B Proximal axonal enlargement—torpedo (arrow). Purkinje cell. Mannosidosis. Aberdeen Angus calf.

B. Oligodendrocytes, Schwann Cells, and the Myelin Sheath

The oligodendrocyte is one of the close companion cells of the neuron in the central nervous system. One population of these cells occurs as satellites to nerve cell bodies and may proliferate in the event of injury to the neurons, but the role of the satellites is essentially unknown. The role of the majority of oligodendrocytes is to provide and maintain the myelin sheaths around those axons with a diameter greater than about 1 μm. They are accordingly located in the myelinated tracts among the fascicles of axons. Each mature oligodendrocyte has a compact cell body of characteristic ultrastructural appearance, and a dozen or so thin processes, each of which connects the perikaryon to a segment of myelin some distance away. Each segment of myelin covers one axonal internode, and is an extended and compacted sheet of specialized oligodendroglial plasma membrane bilayers, wound concentrically and spirally around the axon, like a rolled-up newspaper. In the formation of this compacted membrane, both the intracellular and extracellular spaces are obliterated, creating the major and minor dense lines of myelin lamellae as seen in the electron microscope. In some axons

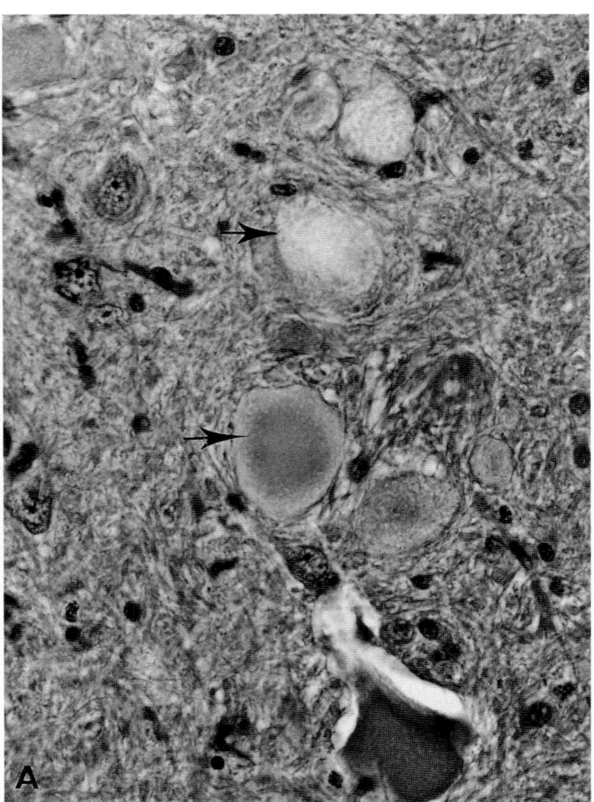

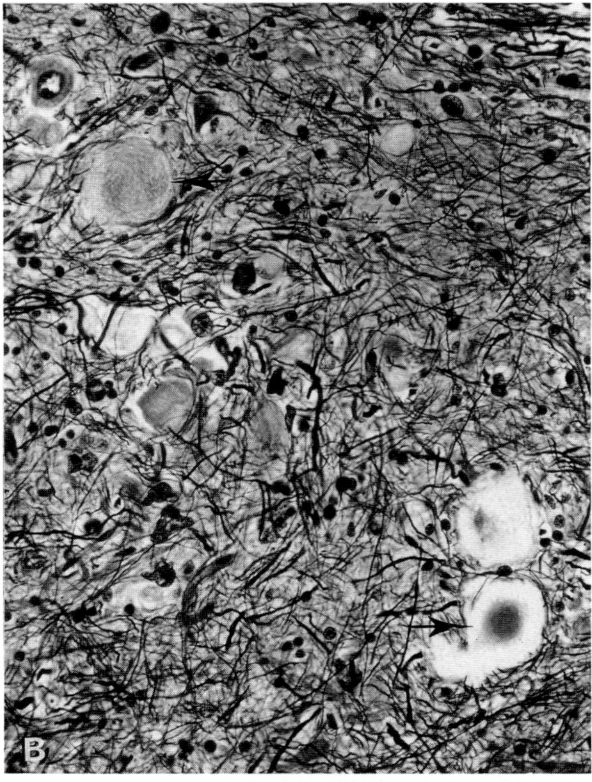

Fig. 3.23 (A and B.) Axonal dystrophy. Rottweiler. Focal axonal swellings (arrows). H&E stain (A). Silver stain reveals varying content of neurofilaments in swellings (B). (Courtesy of L. C. Cork.)

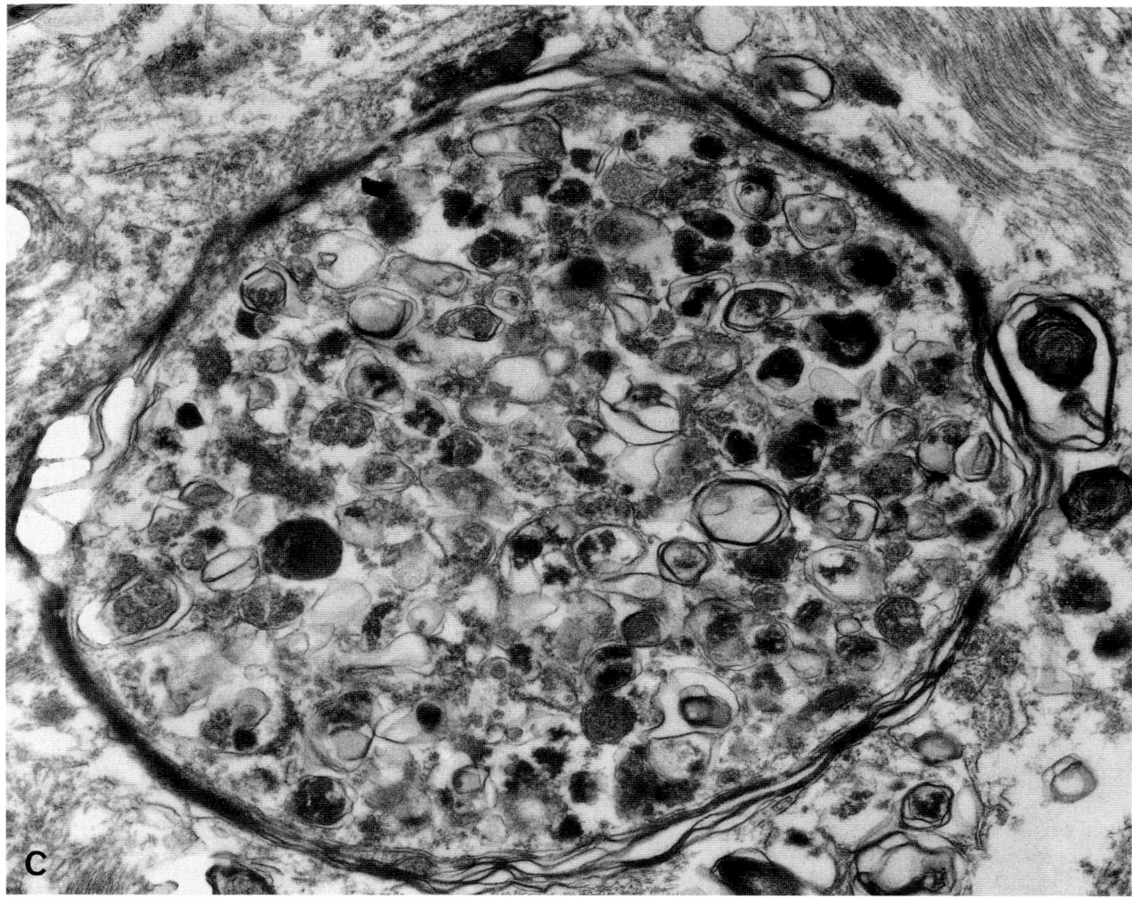

Fig. 3.23C Axonal dystrophy. Cat. Mannosidosis. Focal axonal swelling containing degenerate organelles. Note thinning of myelin sheath. (Courtesy of S. U. Walkley.)

a sheath of 100 or so bilayers may be formed. A portion of intact cytoplasm remains at the innermost and outermost lamellae, known as the "inner" and "outer" tongues respectively. Tracts of uncompacted cytoplasm course through the sheath to form the incisures, and also occur where the myelin lamellae terminate at the paranodal region, as the terminal loops. The lamellated myelin sheath is a relatively stable but plastic structure, whose lipid and protein components are supplied and turned over by the oligodendrocyte.

For light microscopy, myelin is well demonstrated by several special stains, with osmium tetroxide being particularly effective. Biochemically, central myelin is largely composed of cholesterol, galactocerebroside, and phospholipids, together with a number of distinctive protein constituents. The most abundant of these is the proteolipid protein (PLP), with lesser amounts of myelin basic protein (MBP), and myelin-associated glycoprotein (MAG). Proteolipid protein is concentrated at the intraperiod line; MBP, on the cytoplasmic face of the major dense line; and MAG, at the axoplasmic/myelin interface. They probably play an important role in the maintenance of the stability of the sheath.

It is thus apparent that one oligodendrocyte myelinates several axonal internodes, that the myelin sheath is part of the oligodendrocyte, and that death of the oligodendrocyte will result in the demise of all the myelin sheath segments supplied by that cell. However, destruction of one or more myelin sheath segments does not necessarily result in death of the parent oligodendrocyte, but may stimulate it to withdraw its remaining myelin. The dynamics of the oligodendrocyte population are still not absolutely clear, but it is becoming accepted that there is a system of undifferentiated reserve cells, able to take on to some extent regenerative and reparative tasks. These cells may originate from the perineuronal satellite oligodendroglia.

Myelination occurs relatively late in the development of the central nervous system, and the maturing oligodendrocytes invest the axons with myelin by replacing an initial ensheathment of astrocytic processes. Once this is completed, most of the cells assume the characteristics of maturity, while some do appear to remain in a less mature state. The impressive dimensions of myelin-sustaining oligodendrocytes are totally inapparent in routine histologic preparations. The cells appear in rows among the fascicles of myelinated fibers as small, round nuclei with a sparse perikaryon, and give no indication of the complex and extensive tasks they perform. There is some variability in their appearance, in that some have larger, paler nuclei than the majority, looking more like astrocytes, and these may be immature or reserve cells. Oligodendrocytes do not exhibit a range of reactions for the light-microscopist, generally undergoing rapid lysis when injured. Acute injury may be manifested by hydropic swelling of the perikaryon. On occasions, mitotic activity and an increase in numbers may be observed when a primary demyelinating process is operating, but this seems to be very rare in veterinary pathology.

The process of myelination is dependent on close interaction between the axon and the myelinating cell. The two act as a unit, and signals are exchanged between them for all aspects of the process. While the axon may survive for a long period without its myelin sheath, loss of the axon provokes immediate disintegration and removal of the myelin sheath. This situation of axonal degeneration with secondary myelin loss is termed **Wallerian degeneration,** and has been discussed under axonopathy.

Myelination in the peripheral nerves is the responsibility of the **Schwann cells,** and they have a distinctly different relationship with axons from that of the oligodendrocytes. Each peripheral internode is myelinated by a single cell, and the myelinated axon is invested by a basal lamina tube of Schwann cell origin, and by endoneurial collagen. The cell body of the Schwann cell directly apposes the axon. Peripheral myelin is also chemically distinct from central myelin, and this can be appreciated by the tinctorial difference between the two in appropriately stained sections of the spinal cord/spinal nerve interface. This chemical difference is reflected in antigenic differences; the major protein is termed P_0 and is distinct from PLP, as is the basic protein P_1 from MBP. Schwann cells are able to replicate prolifically, to phagocytose damaged myelin, and to remyelinate newly regenerated or previously demyelinated axons. This replicative ability means that the loss of a proportion of the cells may be compensated. The peripheral myelinated axon is therefore a much more resilient structure than its central counterpart. The general principles of the axon/myelin relationship, as outlined for central nervous system, still apply, however. Destruction of Schwann cells will result in the disintegration of the dependent myelin. Destruction of axons will cause myelin degradation, Schwann cell proliferation, and, in time, endoneurial fibrosis.

In paraffin sections of normal nerve, the Schwann cells appear as ovoid nuclei closely apposed to the axon. Proliferating Schwann cells in longitudinal section often appear as bands (Bungner's bands) of spindle-shaped cells resembling fibroblasts. In cross section, they form concentric whorls called onion bulbs. Nodular proliferations of Schwann cells are referred to as Reynaud bodies. Large Reynaud bodies are a common incidental finding in nerves of horses.

A number of diseases primarily involve the myelin sheath, and may frequently leave the myelinating cell body intact. These diseases usually require ultrastructural evaluation if they are to be adequately investigated.

In the **demyelinating** diseases, the sheath is removed from the axons, leaving them naked over varying lengths and providing potential for serious slowing of impulse conduction. In the peripheral nerves, the removal of myelin from randomly scattered internodes gives rise to segmental demyelination, which is best appreciated in teased fiber preparations. Demyelination is frequently carried out by macrophages, which insinuate cytoplasmic processes into the intraperiod lines and strip the sheath from the axonal internode, ingesting and digesting the myelin debris. In other situations, the myelin appears first to be disrupted by humoral factors and undergoes splitting and vesiculation prior to phagocytosis. Myelinophagy can be identified by the use of stains such as the luxol fast blue/periodic acid-Schiff technique. Degenerate myelin can be visualized by the Marchi technique (Fig. 3.22).

In the central nervous system, there is some scope for remyelination, but the complex arrangement and limited replicative capacity of the oligodendrocytes reduces the reparative potential. Regenerated myelin sheaths are thinner than the originals, appearing to the experienced eye as being too narrow for the diameter of the axon they ensheath. Remyelinated internodes are also shorter. In the peripheral nervous system, the potential for remyelination is much more favorable. Repeated bouts of demyelination may result in "onion bulb" and Reynaud body formations, and in the production of thin and irregular myelin segments.

In the **hypomyelinating** diseases, the myelinating cells fail, for various reasons, to provide adequate myelination during the development phase, and the affected individual suffers transient or permanent myelin deficiency, varying in extent and severity according to the particular disease (Fig. 3.24). The majority of these conditions involve the central nervous system only (see myelinopathies). Myelin sheaths are thin or absent, but myelinophagy is generally minimal or nonexistent. Oligodendrocytes may be few, or present in normal numbers, and may exhibit features of immaturity.

In the **dysmyelinating** diseases, there is a qualitative defect in the myelin produced, and the quantity may also be reduced. A large variety of these disorders has been produced for research purposes in inbred strains of laboratory mice.

Myelinic edema is disruption of the lamellar structure by a reopening of the extracellular space along the intraperiod line. It may occur in both central and peripheral myelin. It is caused by a number of chemical agents (hexachlorophene, for example) and leads to a spectacular state of spongy degeneration of white matter, one form of status spongiosus. With some causal agents, there are associated degenerative and reactive changes, but with other causes

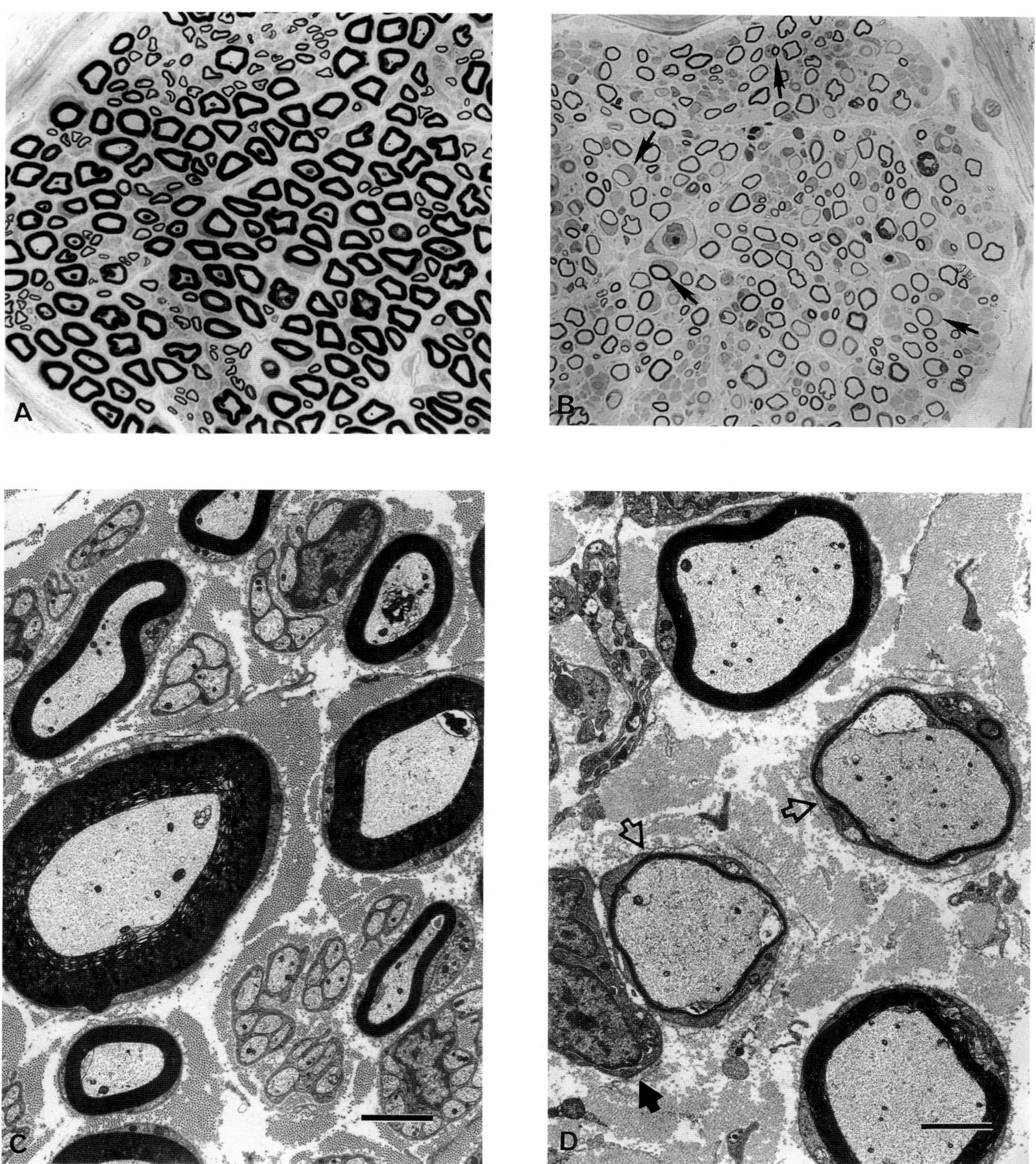

Fig. 3.24 Comparison of normal (A) and hypomyelinated (B) peripheral nerve. Voluminous Schwann cell cytoplasm (arrows). (C) and (D) corresponding ultrastructure. Normal(C); hypomyelinated (D). Thinly myelinated fibers (open arrow). Schwann cell nucleus (black arrow). Bar = 2μm. Golden retriever. Polyneuropathy. (Reprinted from Braund, K. G. *et al. Vet Pathol* **26:** 202, 1989.)

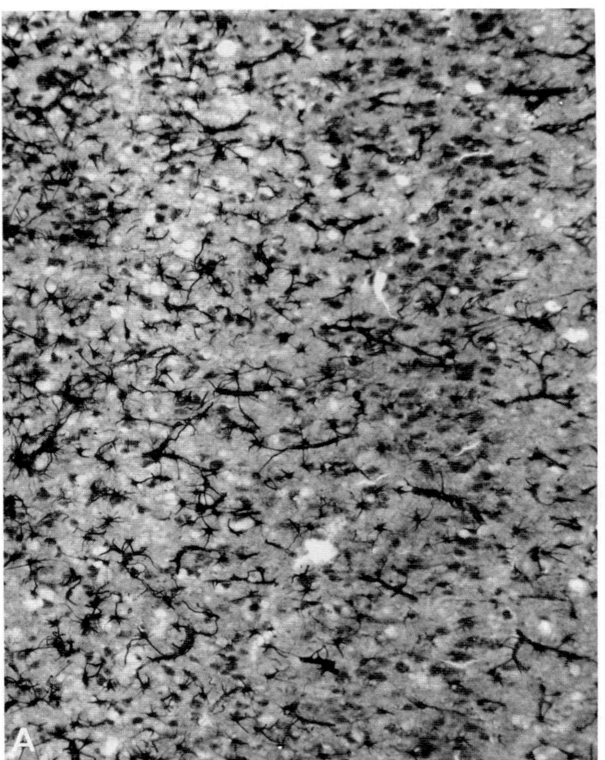

Fig. 3.25A Astrocytes. Goat. Cerebral cortex. Cajal method. (A) Normal.

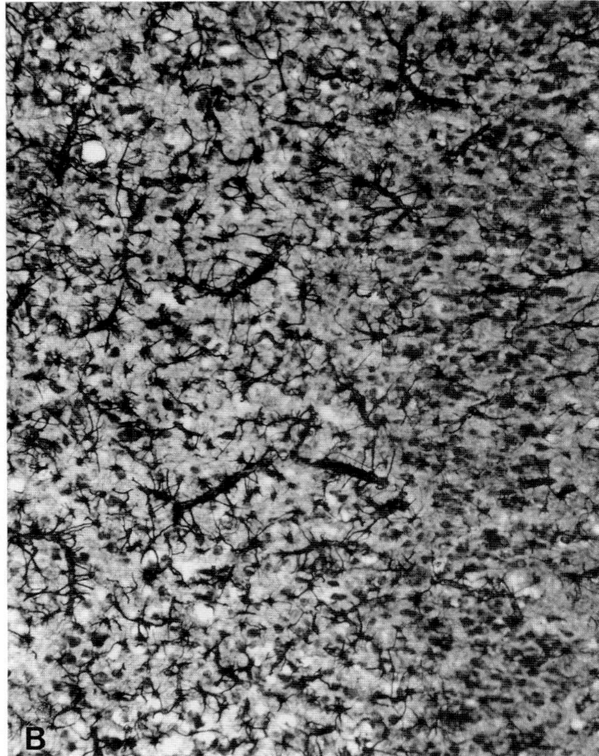

Fig. 3.25B Fibrous astrogliosis. Scrapie. (A and B reprinted from Hadlow, W. J., and Race, R. E. *Vet Pathol* **23:** 593, 1986.)

there is, remarkably, no apparent response on the part of other tissue elements, including the oligodendrocytes themselves. The lesion does not necessarily cause functional disturbances even when well developed; it seems that this may depend on the number of intact lamellae left in place. It may resolve over a period of weeks with no evidence of breakdown of the affected myelin (see Section VII,G,3, Spongiform Myelinopathies).

C. The Astrocytes

The astrocytes may be regarded as the interstitial cells of the central nervous system, as their processes occupy most of the space between and around the neuronal and oligodendroglial elements, and the perivascular and subpial zones.

Both tradition and contemporary studies define two types of astrocyte, the type 1 (protoplasmic), located mainly within the cerebral gray matter, and the type 2 (fibrous), located mainly within white matter tracts. Recent evidence suggests a common progenitor cell, the 011A cell, for the type 2 astrocyte and the oligodendrocyte. The intercellular space of the central nervous tissue is a 20-nm cleft between astrocytes and the other elements, and is interrupted by loose junctions and zonulae adherentia between the former. The astrocytic perikaryon is sparse and barely evident in routine paraffin sections, and the numerous and ramifying processes are invisible. The nuclei appear, therefore, as naked and spherical, and about the size of those of small- to medium-sized neurons; they usually lack a nucleolus, but sometimes have a chromatic dot, the centrosome.

Ultrastructurally, the astrocytic cytoplasm throughout is relatively devoid of organelles and appears largely empty and watery. The chief features are bundles of intermediate filaments, clusters of glycogen granules, and a few mitochondria and lysosomes. All capillary blood vessels in the central nervous system are closely invested by the expanded ends of astrocytic processes, the so-called **end feet.** There is also a dense network of processes at the surface beneath the pia mater, the so-called **glia limitans.** A specialized population of astrocytes occurs in the Purkinje cell layer of the cerebellum and is known as Bergmann's glia. These cells have long straight processes, which extend out through the molecular layer to the surface.

Astrocytes and their processes are well visualized by the application of special stains, such as the Cajal method (Fig. 3.25A,B); immunostaining for the intermediate filament GFAP is also becoming routine (Fig. 3.26A,B). The astrocytes are considered to play an important role in the movement of cations and water, and to be much involved in maintenance of conditions favorable for the electrical activity of neurons. Recent evidence shows the type 2

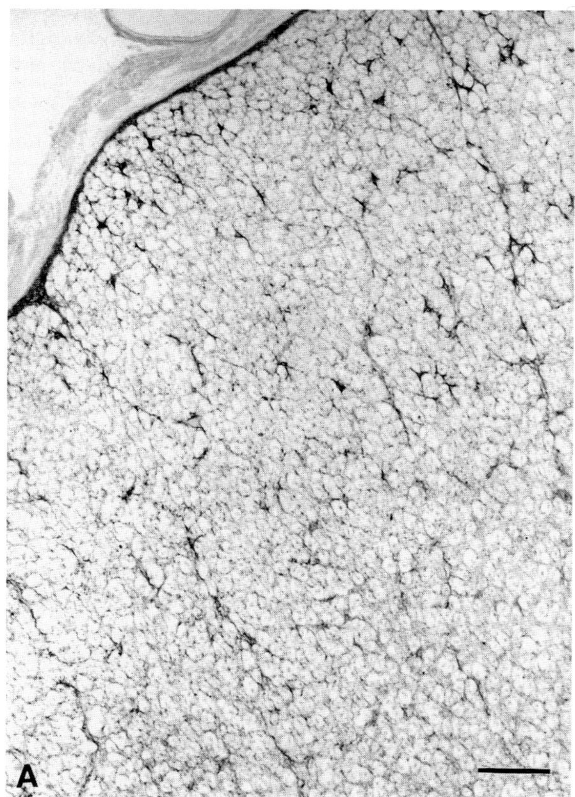

Fig. 3.26A Glial fibrillary acidic protein (GFAP) staining. Horse. Spinal cord. Normal.

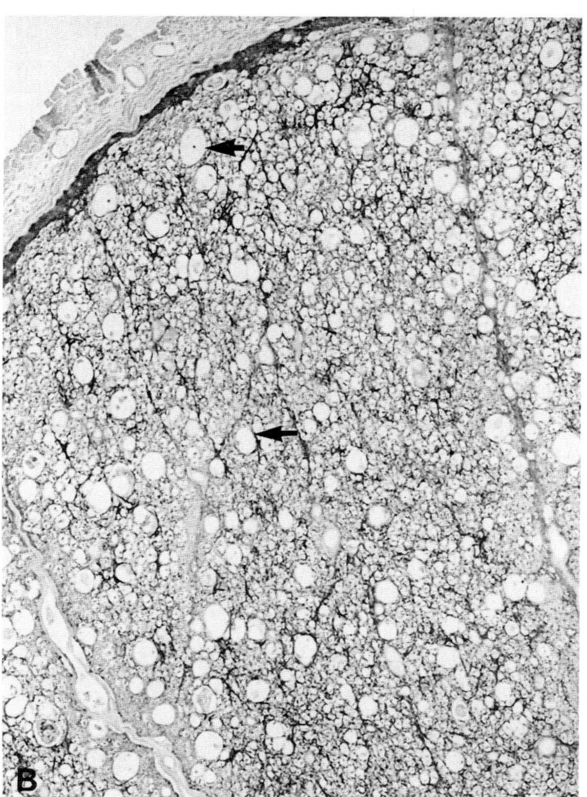

Fig. 3.26B Wobbler. Astrogliosis at site of compressive injury. Axonal degeneration (arrows). (A and B courtesy of J. V. Yovich.)

cells have a major role in this regard at the nodes of Ranvier where they form a close relationship with both the axon and the terminal myelin loops. They are also involved in the detoxification of ammonia, and perhaps other metabolites.

When lethal astrocytic injury occurs, the cytoplasm swells and becomes visible, albeit faintly, and the nucleus becomes eccentric and pyknotic. Disintegration of cytoplasm and nucleus follows rapidly. Astrocytes, however, are capable of a number of reactive responses when they, or cells around them, are damaged. A frequent response is swelling and eosinophilia of the cytoplasm, with some cells acquiring two or more nuclei. These plump reactive astrocytes are called **gemistocytes** (Fig. 3.27A). In a mild response, cytoplasmic swelling may be minimal, but some proliferation of both cells and processes generally occurs (**astrogliosis**). With the cessation of injury, cytoplasmic swelling regresses, but there may be a permanent residuum of extra cells and processes. Around the borders of severe lesions, such as malacic foci, the proliferation of processes may become extremely dense (Fig. 3.27B). Astrocytic proplasia is limited to surviving tissue, however, and postmalacic cavities cannot be filled by astrocytic processes.

Reactive astrogliosis is expected in Wallerian degeneration, following neuronal loss, and in sustained cerebral or spinal edema, and is a feature of many viral encephalitides in which viral infection of astrocytes is probably a prime stimulus. This is certainly the case in canine distemper, in which inclusion bodies are common in reactive astrocytes.

In acute cerebral or spinal vasogenic edema, there is acute astrocytic swelling in which the cells and their processes undergo a type of hydropic degeneration. This can be satisfactorily resolved only with the electron microscope, when the clear distension can be seen, particularly in the perivascular end-feet.

An acute astrocytic reaction known as the formation of **Alzheimer type 2 cells** is best exemplified by the metabolic disturbance in hepatic encephalopathy resulting from liver failure. The nucleus becomes distinctly enlarged and vesicular but remains rounded; the cytoplasm swells and may become visible. This change does not involve the generation of large masses of cytoplasmic intermediate filaments, and there is a weak reaction to GFAP immunostains. It is caused by the accumulation of ammonia and other endogenous toxins, and would therefore also be expected in cases of exogenous ammonia intoxication. In some circumstances, astrocytes are capable of phagocytosis of tissue debris, particularly myelin. This can be seen in Wallerian degeneration in the optic nerves when the bulk of myelino-

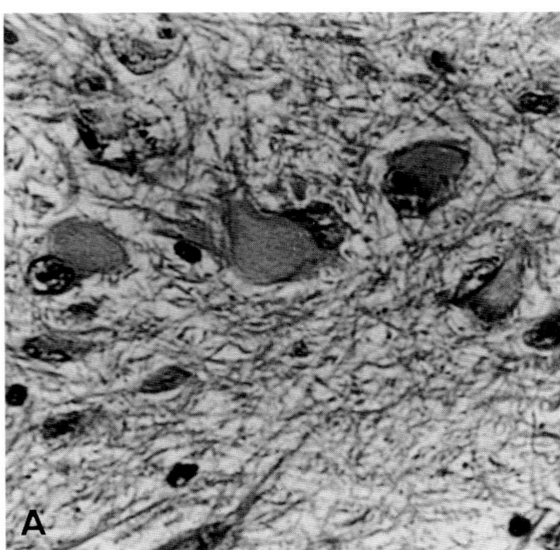

Fig. 3.27A Reactive (gemistocytic) astrocytes near malacic focus.

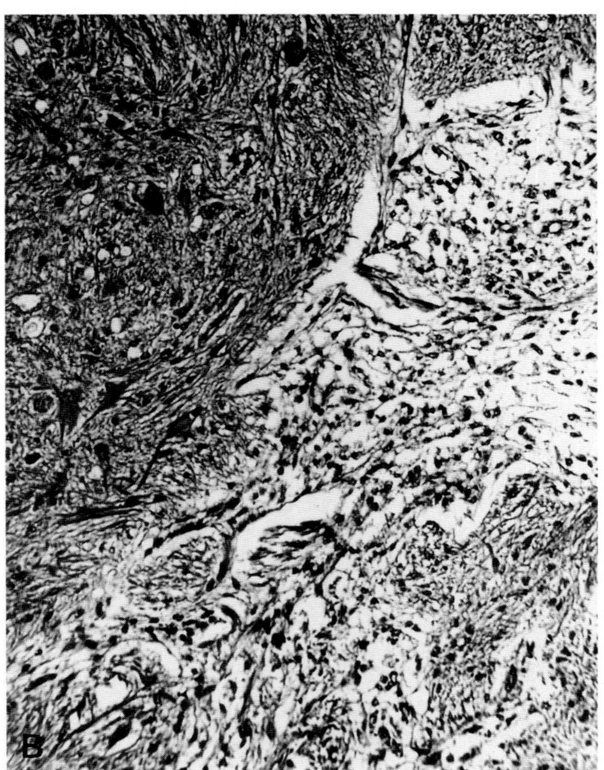

Fig. 3.27B Astrogliosis forming wall to residual cyst.

phagy may be performed by astrocytes. It has also been observed in other regions of the brain and cord.

A bizarre and rare astrocytic response is the formation of **Rosenthal fibers;** these are irregularly shaped, hyaline, eosinophilic structures of undetermined composition, formed within the cell bodies and processes. The massive production of Rosenthal fibers is a feature of **Alexander's disease,** an idiopathic entity described in humans and dogs.

D. The Microglia

The microglia are the most controversial of the cells of the nervous system, in terms of the extent of their functions and their origins. There is little doubt that they are a real entity, and their rather characteristic ultrastructure is well defined. In the normal brain and cord, they appear by routine microscopy as inconspicuous, small, hyperchromatic nuclei, often wedge shaped, and with no visible cytoplasm. However, special staining techniques reveal extensive thin, cytoplasmic processes and, in the electron microscope, dense perikaryal cytoplasm with elongated strands of rough endoplasmic reticulum and lipofuscinlike granules are characteristic. They are most frequent in the gray matter, where they may group in the vicinity of neurons, and in both gray and white matter, they are most numerous adjacent to blood vessels. They are considered to be of mesodermal origin and to form a tissue reserve of potential phagocytes. Recent evidence suggests that they are derived, during organogenesis, from blood monocytes which also give rise to the rich population of leptomeningeal and perivascular histiocytes, which can migrate into the neuropil when significant vascular damage has occurred. Other lines of evidence dispute a hematogenous origin for microglia.

The simplest microglial response to tissue injury is a hypertrophic reaction in which the nucleus becomes rounded and the cytoplasm visible as a narrow, often eccentric, eosinophilic rim. In routinely stained preparations, such cells may be difficult to distinguish from astrocytes. They may also proliferate, although their ability to do so seems limited, and many of the cells in proliferative foci are probably derived from immigrant histiocytes. Focal proliferation gives rise to nodules of 30 to 40 or more cells, whereas diffuse proliferation creates an overall impression of increased cellularity in the microscopic field (Fig. 3.28). Microglial nodules are very commonly a feature of viral encephalitides, occurring in both gray and white matter, but are not specific for viral infections. Reactive microglia may develop greatly elongated and sometimes tortuous nuclei, in which case they are called **rod cells.** These again are often seen in viral diseases.

The most vigorous response of the microglia is their transformation to macrophages, when they assume the morphology typical of cells engaged in phagocytosis. When ingesting myelin debris, their cytoplasm becomes foamy as they load themselves with lipid vacuoles. Often the nucleus becomes pyknotic, and they are referred to as **gitter cells, compound granular corpuscles,** or **fat-granule cells** (Fig. 3.29A,B) In severe lesions, many of the gitter cells will have arisen from blood monocytes as well as from microglia. Lipid-laden cells may persist for months around focal lesions, but slow migration to the perivascu-

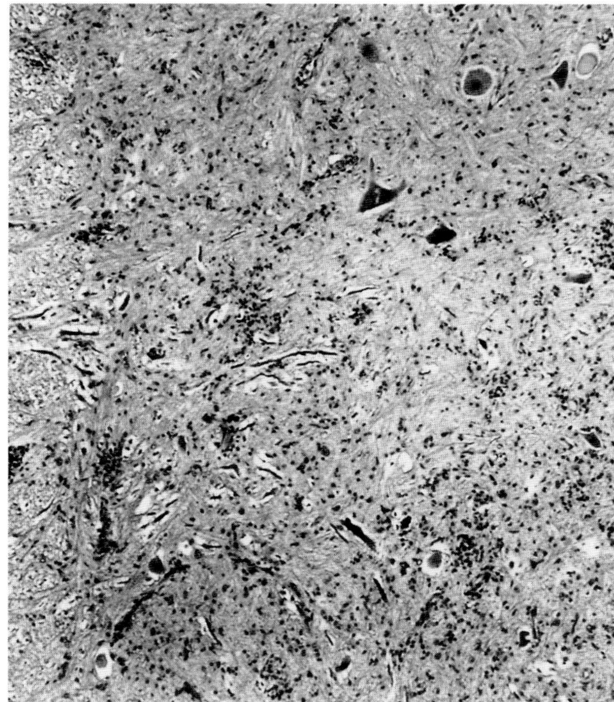

Fig. 3.28 Focal and diffuse gliosis. Poliomyelitis. Louping ill. Sheep.

lar spaces and the meninges does take place. Some of the cells may be found at these locations after even longer periods, with the ingested lipid transformed to lipofuscin.

Microglial phagocytes are usually responsible for **neuronophagia,** in which phagocytic cells gather around fragmenting degenerate neuronal cell bodies (Fig. 3.17D). This response is a feature of many viral infections in which neurons die, but is uncommon in ischemic or other forms of neuronal necrosis.

E. The Microcirculation

Some structural peculiarities of blood vessels in the central nervous system have a bearing on the development of pathologic processes. The capillaries differ from those in other tissues by being surrounded by an investment of astrocytic end-feet. Also, the endothelial cells are sealed together by tight junctions, and the basement membrane divides to incorporate pericytes into the capillary wall. This arrangement in its totality creates the well-known blood–brain barrier, which selectively limits the entry of many molecules into the neuropil in most of the brain. There are a few locales where the barrier is lacking, such as the area postrema and certain other periventricular nuclei.

The distribution of brain capillaries varies considerably, but they are more abundant in the gray matter than in the white. Their concentration is higher in some parts of the gray matter, such as the supraoptic and paraventricular nuclei and the area postrema, especially where neuroendo-

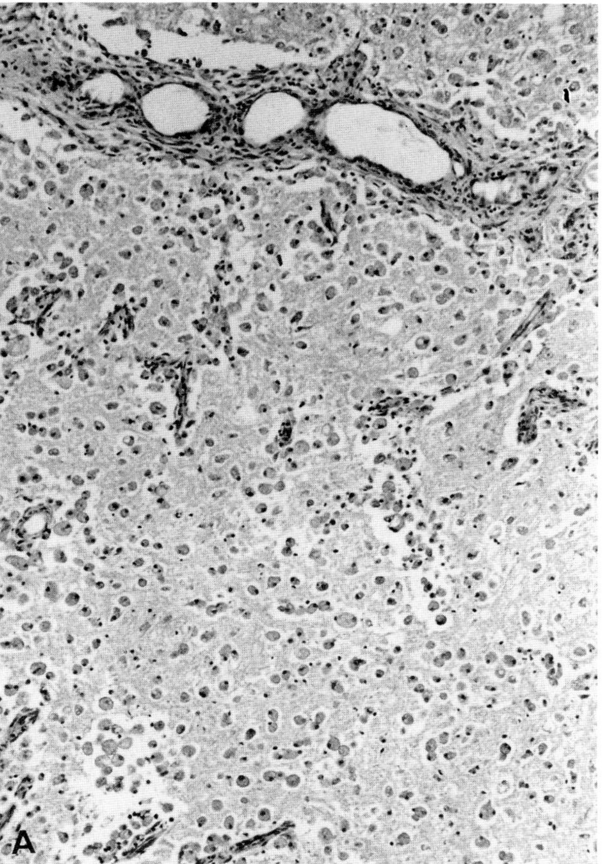

Fig. 3.29A Gitter cells and reactive capillaries in recent cerebral infarct. Dog.

crine activity is concentrated. It is in these areas that the blood–brain barrier is lacking.

So far as is known, regional variations in the concentrations of capillaries do not influence local pathologic processes. The capillary and venular endothelium is highly labile and responds to a variety of injuries by swelling and proliferating (Fig. 3.29B). However, in spite of fairly vigorous proplasia, the cerebral capillaries seem to be almost incapable of budding, so that reactive neovascularization is minimal. The formation of new capillaries is probably limited to situations in which granulation tissue is derived from the mesenchyme of the meninges.

Both arterioles and venules of the central nervous system are thin walled, especially the latter, whose walls are composed mainly of a thin layer of fibrous tissue with very little elastica and no muscle. They are thus susceptible to injury and prone to hemorrhage and, in the cerebral white matter, there is a tendency for leukocytes to sequestrate in them in bacterial infections. The veins are valveless and, as backflow of blood after death is usual, cerebral venous congestion can be difficult to assess.

Both arteries and veins have an outer adventitial layer

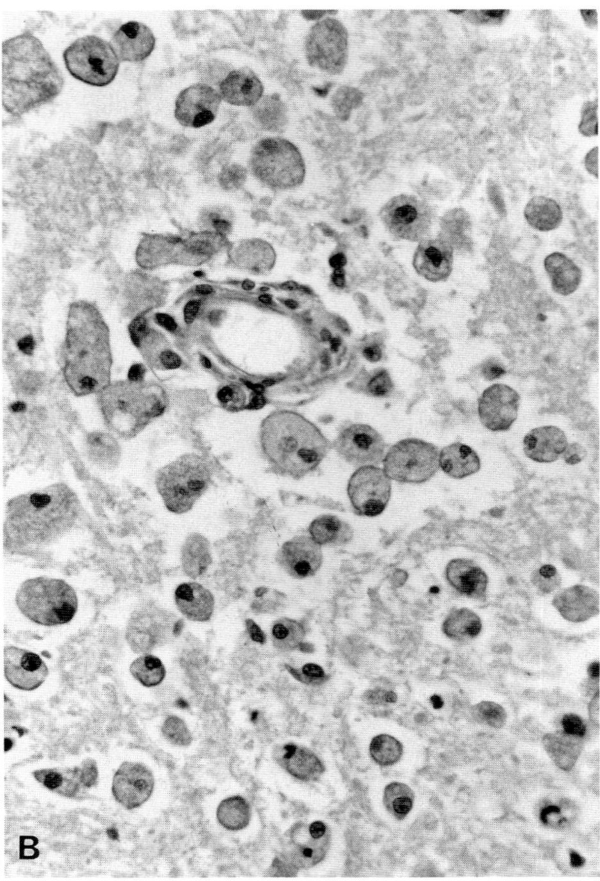

Fig. 3.29B (B) High-power view of gitter cells. Dog.

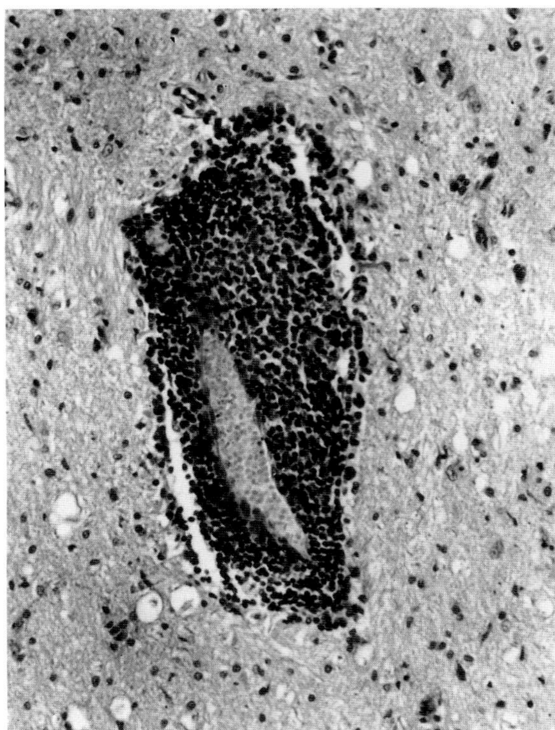

Fig. 3.30 Perivascular cuff. Dog. Encephalitis. Lymphocytes confined to Virchow–Robin space.

of variable thickness, and a perivascular **Virchow–Robin** space, which is in continuity with the subarachnoid space and is lined by an invagination of the pia-arachnoid. In diffuse inflammatory or neoplastic diseases, this space, more potential than real under normal conditions, becomes patent and accumulates reactive and invading cells. The tendency for these cells to be confined to the space gives rise to the neuropathologic term **perivascular cuffing** (Fig. 3.30). The size of cuffs usually relates to the size of the space, and they may vary from one cell thick about the smallest venules to 10 to 12 or more cells thick around the larger vessels. Perivascular cuffing is classically seen in inflammatory conditions, and all classes of reactive leukocytes may be seen, depending on the cause. While most of the cells are of hematogenous origin, there is no doubt that in some diseases, Teschen for example, they may arise largely from the proliferation of adventitial cells. Lymphoid perivascular cuffing is a feature of many viral encephalitides. Some diseases, too, are characterized by the production of cerebral and/or spinal vasculitis, but these changes are discussed in a following section.

Mention must be made of the nests of residual glia (Islands of Calleja) which are seen beneath the ependyma of the lateral ventricles and in the dentate fascia of the hippocampus. These cells are relics of the developmental period and possibly act as a supply of replacement cells in the adult. They occur in small aggregations and often as eccentric cuffs around vessels. Generally they are monomorphic and appear as small dark nuclei and may be mistaken for inflammatory cells.

Bibliography

Adams, J. H. *et al.* "Greenfield's Neuropathology," 4th Ed., New York, John Wiley, 1984.

Asbury, A. K., McKhann, G. M., and MacDonald, W. I. "Diseases of the Nervous System." Philadelphia, Pennsylvania, W.B. Saunders, 1986.

Berry, M. Regeneration in the central nervous system. *Rec Adv Neuropathol* **1:** 67–111, 1979.

Brown, A. W. Structural abnormalities in neurones. *J Clin Path* (Suppl.) (*R Coll Pathol*) **11:** 155–169, 1977.

Bunge, R. The development of myelin and myelin-related cells. *Trends Neurosci* **4:** 175–177, 1981.

de Lahunta, A. "Veterinary Neuroanatomy and Clinical Neurology," 2nd Ed. Philadelphia, Pennsylvania, Saunders, 1983.

Duffy, P. E. "Astrocytes: Normal, Reactive and Neoplastic." New York, Raven Press, 1983.

Dyck, P. J. *et al.* "Peripheral Neuropathy," 2nd Ed. Philadelphia, Pennsylvania, W.B. Saunders, 1984.

Goldman, J. F., and Yen, Shu-Hui. Cytoskeletal protein abnormalities in neurodegenerative diseases. *Ann Neurol* **19:** 209–223, 1986.

Griffin, J. W., and Watson, D. F. Axonal transport in neurological disease. *Ann Neurol* **23:** 3–13, 1988.

Hall, S. M. Regeneration in the peripheral nervous system. *Neuropathol Appl Neurobiol* **15:** 513–529, 1989.

Jordan, F. L., and Thomas, W. E. Brain macrophages: Questions of origin and interrelationship. *Brain Res Rev* **13:** 165–178, 1988.

Leiberman, A. P. The axon reaction. *Int Rev Neurobiol* **14:** 49–124, 1971.

Ludwin, S. K. Remyelination in demyelinating diseases of the central nervous system. *CRC Crit Rev Neurobiol* **3:** 1–27, 1987.

Martin, J. E., Swash, M., and Schwartz, M. S. New insights in motor neuron disease. *Neuropathol Appl Neurobiol* **16:** 97–110, 1990.

Perry, V. H., and Gordon, S. Macrophages and microglia in the nervous system. *Trends Neurosci* **11:** 273–277, 1988.

Schlaepfer, W. W. Neurofilaments: Structure, metabolism, and implications in disease. *J Neuropathol Exp Neurol* **46:** 117–129, 1987.

Waxman, S. G. The astrocyte as a component of the node of Ranvier. *Trends Neurosci* **9:** 250–253, 1986.

III. Storage Diseases

As the great majority of storage diseases involve neurons and neurologic impairment, it is appropriate for them to be discussed in this section.

Within all cells, except mature erythrocytes, normal catabolism directs a steady stream of endogenous macromolecules into vesicular compartments for degradation to simple molecules, which may be re-used or excreted. These essentially autophagic pathways, through which each cell recycles its own constituents, may also receive endogenous molecules from the extracellular milieu, taken up by endocytosis or phagocytosis. Exogenous materials may similarly be taken up.

Storage states, or disorders, are characterized by the accumulation of material(s) resistant to or exceeding the capacity of the machinery of intracellular digestion, disposal, or transport. Hemosiderosis and some types of hepatic lipidoses are common examples of the storage of physiologically normal substances due to the overloading of essentially normal, biochemically competent cells. A **storage disease,** by contrast, can be regarded as a storage process with a primary pathologic basis within the storing cells, and with potential for the perturbation of their function. The implication is that the catabolic machinery of the cells is fundamentally incompetent. This is generally demonstrable but, as is often the case in pathobiology, it is difficult to provide a neat definition which clearly distinguishes this situation in all circumstances.

Practically all cell types in the body are potentially vulnerable to storage induction, but those most vulnerable are postmitotic cells, such as neurons and cardiac myocytes. Cells in dynamic renewal systems, such as enterocytes, scarcely have the chance to become involved before their time is over.

The lysosomal apparatus provides the machinery for a great deal of intracellular degradation; most storage diseases involve intralysosomal accumulation and are hence termed **lysosomal storage diseases.** The 40 or so lysosomal acid hydrolases are capable of digesting completely the complex macromolecules synthesized for cell membranes, organelles, secretory products, etc. This enzymic destruction must be sequestered from the rest of the cell, and is carried out in vesicles provided with an ion pump in their limiting membranes, which maintains an acidic interior, or lysosol, for the optimal activity of the hydrolases.

Newly synthesized lysosomal enzymes are carried from the trans-Golgi region within primary lysosomal vesicles, and are delivered to substrate-containing vesicles (endosomes, heterophagosomes, autophagosomes, secretory vesicles) by means of fusion of vesicle membranes. The end products of digestion, which are simple lipids, amino acids, and sugars, are transported from the vesicular lysosol back into the cytosol. There may be small quantities of indigestible residues, which, in cells such as hepatocytes, may be extruded by a process of exocytosis.

The lysosomal hydrolases tend to be exoenzymes, sequentially breaking linkages at the ends of large molecules, but unable to act on linkages within them. This means that if the sequence is blocked at some point, further digestion cannot proceed. Should the digestive sequence be impaired, the cell will steadily accumulate a mass of vacuoles containing undegraded substrate. One type of molecule will be the major stored substance, but often the stored material will be somewhat heterogeneous, as the enzymes are linkage specific rather than substrate specific. Thus the morphologic hallmark of lysosomal storage disease is the presence of distended cells, crowded with vacuoles bounded by single membranes containing the stored material; these vacuoles react enzyme-histochemically for acid phosphatase or other lysosomal hydrolases. They represent the adaptive hypertrophy of the lysosomal apparatus. In some cases, the morphology and histochemical reactivity of the stored substance may fairly clearly indicate its nature, for example, glycogen. Lectin histochemistry may also be useful for characterizing stored material, exploiting the avid carbohydrate-binding properties of these agglutinating proteins in combination with a visualizing system.

There are several ways in which lysosomal digestion can be impaired. The most relevant in the context of this discussion is a deficiency in the activity of a specific lysosomal hydrolase because of a **genetic defect.** This is the basis of the inherited lysosomal storage diseases in humans and animals, most of which are transmitted as autosomal recessive and some as X-linked traits. The deficiency in activity may come about via a total absence of enzyme protein, via the production of a defective or unstable enzyme, or via the absence of a specific activator protein required by some enzymes for the initiation of activity. These activators are small-molecular-weight, heat-stable proteins, which function as detergents. They may interact with either the enzyme or its substrate. In the category of defective enzyme protein, one could also include the rare mechanism in which the enzyme lacks the specific molecular tag required to direct it, following its synthesis, to the lysosomal compartment; it is therefore immediately excreted from the cell, and the lysosomes remain deficient in that enzyme activity.

In most autosomally inherited conditions, the gene dose effect results in heterozygous individuals being generally phenotypically normal, but usually having demonstrably subnormal tissue activity of the particular enzyme in question. Tissues of homozygous affected individuals will contain swollen vacuolated cells, and chemical analysis reveals large amounts of the stored substance, and variable amounts of other metabolically related substances. Multisystem involvement is likely to occur, with storage being evident in many cell types in many organs. The cells and tissues most affected will be those most active in turning over the substrate in question, but usually the fixed and mobile macrophages are also prominently involved. This is because they avidly accumulate substrate from the tissue fluids and plasma. The process begins *in utero* and in many cases is well developed at birth, although clinical impairment may not be great at that time. The age of onset and speed of progression of disease can vary, probably on the basis of the amount of residual enzyme activity, and often involving the ratio between various isoenzymes. In several human entities, subtypes are described on this basis.

As pointed out, the *in vitro* tissue activity of the enzyme involved can usually be demonstrated to be negligible. However, if there were, for example, deficiency of an activator protein, the *in vitro* assay of tissues for the subject enzyme might reveal paradoxically high levels of activity, reflecting the hypertrophy of the lysosomal apparatus. The enzyme, however, would be inactive *in vivo* in the absence of the activator protein. In addition, and especially if the assay involves synthetic substrates, various isoenzymes may give the impression of adequate activity *in vitro,* which has no relation to the *in vivo* situation. In spite of this, assay of enzyme activity in skin fibroblasts or peripheral blood leukocytes has been diagnostically useful in many genetic storage diseases, identifying both homozygous and heterozygous individuals. More recently, with the advent of specific DNA probes, molecular genetic methods can be expected to play an increasingly important role in this area.

In an alternative mechanism of storage, an exogenous toxin may specifically inhibit a lysosomal enzyme, and temporarily induce a state analogous to a genetic enzyme deficiency. This is the established basis of at least one plant intoxication, "locoism," and is suspected in others. The morphologic and chemical characteristics are typical, but tissue activity of the subject enzyme may be quite high when it is separated from the inhibitor and assayed.

In a final general mechanism, substrate which is resistant to a normal and intact enzymic battery may enter degradative pathways. This may be an exogenous substrate, or a modified endogenous substrate. It is by this mechanism that several amphophilic drugs, such as chloroquine, have been found to induce storage diseases by complexing with endogenous molecules to produce indigestible products. Theoretically this mechanism could also have a genetic basis, by which an indigestible substrate is produced.

The lysosomal basis of some storage diseases is uncertain, and nonlysosomal entities, such as several of the glycogenoses, are clearly defined. Further, although many storage diseases have been defined in molecular terms, some have not, and in most cases, the basis of cellular dysfunction is far from clarified. There is more involved than simply the mechanical crowding out of other organelles. In most instances the storage process seems to have little primary cytotoxic effect. It seems rather that the induction of secondary and tertiary metabolic and structural effects is responsible for functional disturbances.

With regard to the **neuronal storage diseases,** in many of them the process is multisystemic, and all neurons, including those of the retina and peripheral ganglia, are involved, together with cells in most other organs. But neither of the two preceding conditions is invariable. When neurons are involved in storage disease, clinical signs of neurologic impairment eventually become evident, but, in general, this does not correlate with significant neuronal death, and the mechanisms of dysfunction are still largely unresolved and of considerable current interest. It is true that in a few storage diseases, regional neuronal death begins at an early stage and is progressive; it probably contributes significantly to functional disturbance at the end stage of many storage diseases.

As already mentioned, in the face of a progressive storage process, neurons have no recourse but to accumulate storage vacuoles until they, or the animal, die. It seems probable that there is some limited capacity to discharge some of the stored load by exocytosis but, in general, intractable constipation is inevitable as long as enzymic activity is deficient. In spite of this, the cell limits the sites of storage to the soma and some of the larger dendritic stems. As a result, the soma becomes greatly distended and the cell outline rounded and swollen, rather than angulate (Fig. 3.31). However, even within the soma, storage may be somewhat polarized, often adjacent to the axon hillock; differing planes of section may suggest that some neurons are not affected, particularly if examined early in the course of the disease. The multitude of storage vacuoles crowds and displaces other organelles, and the neuron takes on a chromatolytic and foamy appearance. Glial, endothelial, and perithelial cells are generally similarly affected.

Some recent studies in several ganglioside storage diseases have shown that certain populations of neurons, particularly in the pyramidal system of the cerebral cortex and thalamic relay nuclei, undergo a form of focal hypertrophy in order to generate more "storage space." These cells develop large swollen compartments between the axon hillock and the initial axonal segment, which have been dubbed meganeurites. In addition, cells in these regions may sprout aberrant dendritic spines, whether or not they have meganeurites. The spines have been shown to arise from the axon hillock and from meganeurites and to form synaptic contacts. The origin of the presynaptic

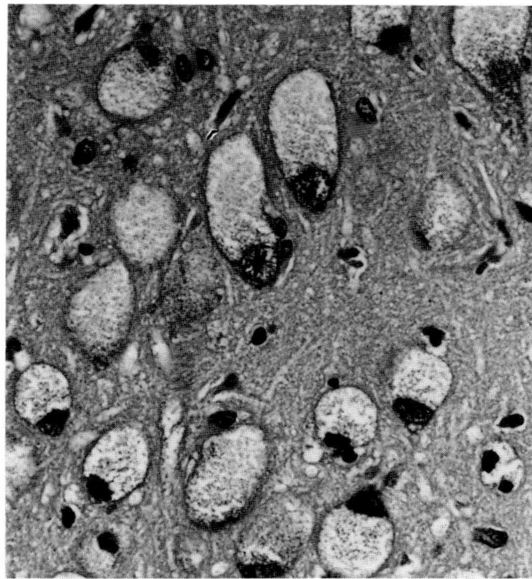

Fig. 3.31 Swollen neurons. Lysosomal storage disease. GM_1 gangliosidosis. Sheep. (Reprinted from Murnane, R. D. *et al. Vet Pathol* **28:** 332, 1991.)

elements of these contacts has not been determined, but such abnormal neuronal connections could well contribute to malfunction. These changes, dramatic as they are, cannot be appreciated without the use of special techniques, most notably the Golgi impregnation method. In addition, focal swellings may develop along the course of axons, appearing as eosinophilic spheroids, similar to those described for axonal dystrophy (Fig. 3.23). They are often very prominent in some nuclear groups, exhibiting a tendency to form in the terminal presynaptic regions of axons, but they can be seen anywhere in the white matter and also in the peripheral nerves. They do not contain specific storage material, but are crowded with degenerate organelles and/or abnormal tubules and vesicles. They probably reflect a secondary effect of the storage process on retrograde axonal transport. The functional significance of this secondary axonal dystrophy is not resolved, but it is a prominent pathologic feature in many storage diseases.

Bibliography

Dorling, P. R. Lysosomal storage diseases in animals. *In* "Lysosomes in Biology and Pathology" J. T. Dingle, R. T. Dean, W. Sly (eds.), pp. 347–379. New York, Elsevier, 1984.

Glew, R. H. *et al.* Biology of disease. Lysosomal storage diseases. *Lab Invest* **53:** 250–269, 1985.

Hannun, Y. A., and Bell, R. M. Lysosphingolipids inhibit protein kinase-C: Implications for the sphingolipidoses. *Science* **235:** 670–674, 1987.

Igisu, H., and Suzuki, K. Progressive accumulation of a toxic metabolite in a genetic leukodystrophy. *Science* **224:** 753–755, 1984.

Walkley, S. U. Pathobiology of neuronal storage disease. *Int Rev Neurobiol* **29:** 191–243, 1988.

Walkley, S. U., Baker, H. J., and Rattazzi, M. C. Initiation and growth of ectopic neurites and meganeurites during postnatal cortical development in ganglioside storage disease. *Dev Brain Res* **51:** 167–178, 1990.

Walkley, S. U. *et al.* Neuroaxonal dystrophy in neuronal storage disorders: Evidence for major GABAergic neuron involvement. *J Neurol Sci* **104:** 1–6, 1991.

A. The Inherited Storage Diseases

Virtually all the inherited storage diseases of animals are proven or assumed to be lysosomal in nature. This applies to all those to be described, with the exceptions of the canine glycogenoses analogous to types 3 and 7 glycogenoses in humans. They are classified into broad groups, according to the class of macromolecule whose degradation is defective. As many of these diseases were first described in humans, the catalog is replete with eponyms, mostly derived from the names of the eminent persons who provided those first descriptions. Within the broad groups, individual entities are defined by the nature of the dominant storage material, which reflects the specific enzymic deficit. Not all the storage diseases documented in humans have been found to have analogs in domestic animals, but the list is growing, and the prevalence of inbreeding makes it likely that this will continue. The general clinical characteristic is the onset of progressive neurologic impairment at a young age.

In the pathologic descriptions that follow, it should be appreciated that variations in the patterns of lesions are likely to surface as more cases are described. A useful practical distinction can be made between those diseases in which the vacuoles appear empty in both paraffin and resin sections (Fig. 3.32A), and those in which they contain residual material (Fig. 3.32B). The former reflect water soluble substrates leached out during processing, and point to certain diseases as described.

1. The Sphingolipidoses

This is a group of diseases whose common theme is defective degradation of a family of complex lipids, which are normal components of cell membranes. At the focal point of the degradative pathway is the compound ceramide, fatty acyl-sphingosine.

The gangliosidoses have been documented in cats (domestic and Korat), dogs (German short-haired pointer, Portuguese water dog, Japanese spaniel, and mixed breed), Friesian cattle, Suffolk sheep, and Yorkshire pigs. They are, in general, characterized clinically by the onset at an early age of discrete head and limb tremors and dysmetria. Worsening locomotor deficits and mentation terminate eventually in blindness, somnolence, seizures, and quadriplegia.

In **GM_1 gangliosidosis,** there is deficient *in vivo* activity of a β-galactosidase, and in **GM_2 gangliosidosis,** of β-hexosaminidase A. The two enzymes respectively undertake the initial two steps of ganglioside catabolism. Among the variants of the latter disease in humans are Tay–Sachs' and Sandhoff's diseases.

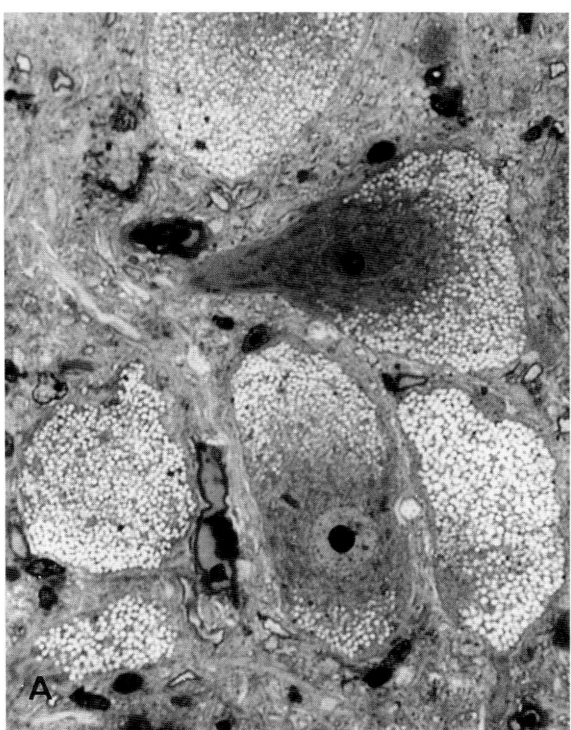

Fig. 3.32A Neuronal lysosomal storage disease (A and B). (A) Resin section showing "empty" vacuoles. Bovine mannosidosis. (Courtesy of R. D. Jolly.)

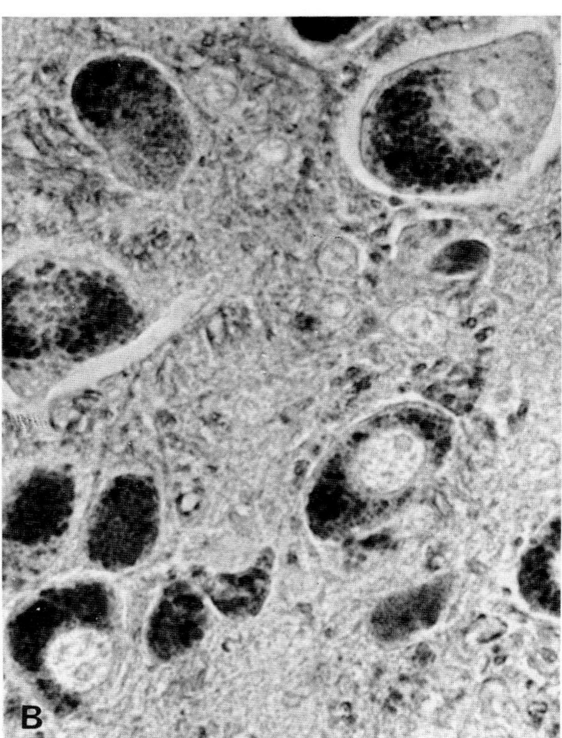

Fig. 3.32B Dense granules. Sudan black positive. Gangliosidosis. German shorthair dog. (Courtesy of E. Karbe.)

The major stored material is the ganglioside as indicated, but hepatic storage of a glycopeptide occurs in feline GM_1 and GM_2, and possibly in other instances. Tissue activities of the enzymes may be variable when assayed *in vitro*, for the reasons explained in the general introduction. All data indicate autosomal recessive inheritance.

Neuronal storage is manifested in routine paraffin sections as marked distension of the soma, with a foamy, faintly eosinophilic cytoplasm (Figs. 3.31, 3.33A). The stored material gives a strongly positive reaction to PAS in frozen sections, and is evident in plastic sections as osmiophilic granules 1–3 μm in diameter. Ultrastructurally, these are seen as characteristic membranous cytoplasmic bodies, consisting of concentric membranous whorls with a periodicity of about 5 nm (Fig. 3.33B). Vacuolated macrophages may also be found around blood vessels in the central nervous system, and storage also occurs in glial cells. Axonal spheroids may be reasonably numerous. Gliosis, demyelination, and neuronal loss are not apparent until end stage. In those instances in which hepatic storage occurs, hepatocytes and Kupffer's cells contain large, empty vacuoles, the site of storage of a water-soluble glycopeptide.

In the United States, GM_1 gangliosidosis in Suffolk sheep has recently been shown to be associated with a dual enzyme deficiency, with β-galactosidase deficiency being profound (5% residual activity), and α-neuraminidase deficiency less so (20% residual activity). Progressive ataxia in affected lambs first becomes evident at 4 to 6 months of age. Histologically, there is intense storage in most central and peripheral ganglionic neurons (Fig. 3.31), and in the kidney, liver, and lymph nodes, and cardiac Purkinje fibers. Stored material in neurons stains PAS and luxol fast blue positive, but Sudan black negative. Axonal spheroids are frequent in cerebral and cerebellar white matter.

Glucocerebrosidosis (glucosylceramidosis) has been described only in the Sydney silky terrier, and is the counterpart to Gaucher's disease of humans. It results from a deficiency in the activity of the β-glucosidase, glucosylceramidase, which catalyzes the conversion of glucosylceramide to ceramide. The former is derived from the catabolism of the gangliosides.

Storage in the dog is expressed in macrophages in the hepatic sinusoids and lymph nodes, and in neurons in some parts of the brain but, interestingly, not in Purkinje cells and not in the spinal cord. Swollen cells have weakly eosinophilic cytoplasmic vacuoles, which are PAS positive in the macrophages but PAS negative in neurons. Degenerating neurons occur in the cerebral and cerebellar cortices, and Wallerian degeneration, in related white matter.

Ultrastructurally, the macrophage storage material has a characteristic twisted, branching, tubular structure, and

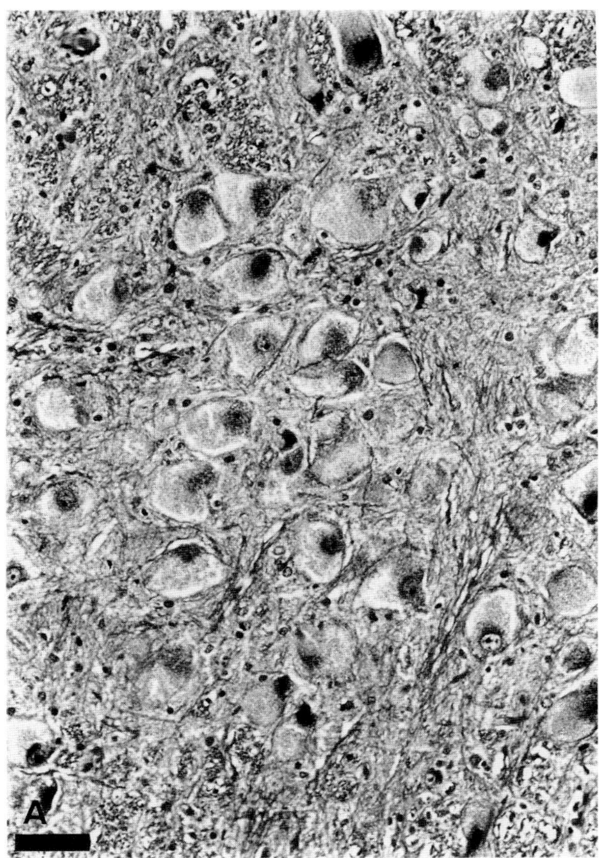

Fig. 3.33A GM_1 gangliosidosis. Portuguese water dog. Pale swollen neurons. Bar = 63 μm.

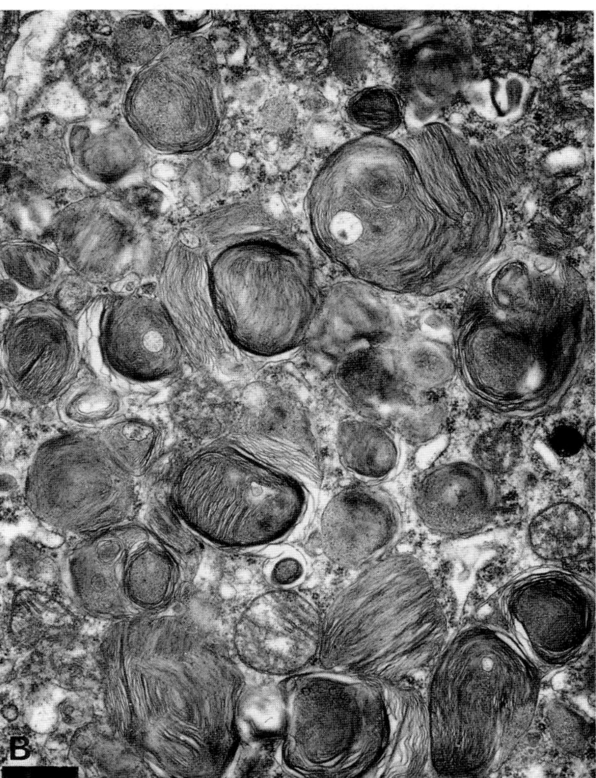

Fig. 3.33B GM_1 gangliosidosis. Portuguese water dog. Storage bodies showing concentric lamellae or parallel stacks of membranes. Bar = 0.6 μm. (Reprinted from Saunders, G. K. *et al. Vet Pathol* **25:** 265, 1988.)

the cells in the human disease are known as Gaucher cells. In neurons, the storage granules are either lamellated membranous cytoplasmic bodies, resembling the Zebra bodies of mucopolysaccharidoses, or bilamellar wisps. In Gaucher variants in humans, storage may not occur in neurons.

Sphingomyelinosis has been recognized in Siamese, domestic short-hair and Balinese cats, and the miniature poodle, and is regarded as the counterpart of Niemann–Pick disease in humans, of which there are several variants. In types A and B, deficient activity of sphingomyelinase results in the storage of sphingomyelin, cholesterol, and gangliosides in most neurons, and in macrophages in the liver, spleen, lymph nodes, adrenals, bone marrow, and lungs. However, in the type C variant, there is no apparent deficiency of sphingomyelinase activity, and the basic molecular mechanism of storage is not understood. So far in animals, all cases have been analogous to human type A, except for a domestic short-hair cat considered to have the counterpart of human type C.

At necropsy there are enlargement and pallor of the liver, enlargement of the spleen, and grayish nodules in the lungs. Microscopically the visceral organs mentioned are packed with foamy macrophages, and the neurons are distended by masses of light gray, autofluorescent granules, 1–2 μm in diameter. In frozen sections, the reactivity of vacuoles to oil-red-O and PAS stains has been found to be variable.

Type C disease is also characterized by numerous axonal spheroids in many areas of the neuraxis. Ultrastructurally, neuronal storage granules are concentric multilamellar structures or dense bodies.

Galactocerebrosidosis (globoid cell leukodystrophy; Krabbe's disease) is a member of this group, but has some unique features and does not involve neuronal storage. It is therefore discussed under myelinopathies.

Bibliography

Baker, H. J. *et al.* The gangliosidoses: Comparative features and research applications. *Vet Pathol* **16:** 635–649, 1979.

Cummings, J. F. *et al.* GM_2 gangliosidosis in a Japanese spaniel. *Acta Neuropathol (Berl)* **67:** 247–253, 1985.

Donnelly, W. J., and Sheahan, B. J. GM_1 gangliosidosis of Friesian calves: A review. *Irish Vet J* **35:** 45–55, 1981.

Hartley, W. J., and Blakemore, W. F. Neurovisceral glucocerebroside storage (Gaucher's disease) in a dog. *Vet Pathol* **10:** 191–201, 1973.

Kosanke, S. D., Pierce, K. K., and Read, W. K. Morphogenesis of light and electron microscopic lesions in porcine GM_2 gangliosidosis. *Vet Pathol* **16:** 6–17, 1979.

Lowenthal, A. C. *et al.* Feline sphingolipidosis resembling Niemann–Pick disease type C. *Acta Neuropathol (Berl)* **81:** 189–197, 1990.

Murnane R. D. *et al.* The lesions of an ovine lysosomal storage disease. *Am J Pathol* **134:** 263–270, 1989.

Murnane, R. D., Hartley, W. J., and Prieur, D. J. Similarity of lectin histochemistry of a lysosomal storage disease in a New Zealand lamb to that of ovine GM_1 gangliosidosis. *Vet Pathol* **28:** 332–335, 1991.

Neuwelt, E. A. *et al.* Characterisation of a new model of GM_2 gangliosidosis (Sandhoff's disease) in Korat cats. *J Clin Invest* **76:** 482–490, 1985.

Prieur, D. J. *et al.* Inheritance of an ovine lysosomal storage disease associated with deficiencies of β-galactosidase and α-neuraminidase. *J Hered* **81:** 245–249, 1990.

Saunders, G. K. *et al.* GM_1 gangliosidosis in Portuguese water dogs: Pathologic and biochemical findings. *Vet Pathol* **25:** 265–269, 1988.

Singer, H. S., and Cork, L. C. Canine GM_2 gangliosidosis: Morphological and biochemical analysis. *Vet Pathol* **26:** 114–120, 1989.

Snyder, S. P., Kingston, R. S., and Wenger, D. A. Animal model of human disease. Niemann–Pick disease. Sphingomyelinosis of Siamese cats. *Am J Pathol* **108:** 252–254, 1982.

Yamagami, T. *et al.* Neurovisceral sphingomyelinosis in a Siamese cat. *Acta Neuropathol (Berl)* **79:** 330–332, 1989.

2. *The Glycoproteinoses*

In this group of diseases, the unifying thread is the defective degradation of the carbohydrate component of *N*-linked glycoproteins. These carbohydrate moieties are rich in mannose, *N*-acetyl glucosamine, and fucose, and in particular diseases, this is reflected in the storage residues, which are detectable in the urine.

α-Mannosidosis has been the most economically important of the inherited storage diseases of animals, as it once occurred at high frequency in some populations of Angus cattle and the derivative Murray Grey breed. It is also recorded in the Galloway breed. The characterization and study of the disease in New Zealand has led to effective carrier detection and certification schemes for its elimination, and the incidence has declined significantly. There is an autosomal recessive inheritance mode, and affected calves have retarded growth, increasingly severe ataxia, and behavioral changes. The disease reaches end stage by about 18 months of age.

Due to the synthesis of a defective enzyme protein, such individuals have deficient lysosomal α-mannosidase activity in virtually all cells except hepatocytes. All but the final mannose molecule of the glycans destined for digestion by this enzyme are α-linked. As a result, mannose/*N*-acetylglucosamine oligosaccharides accumulate in storage vacuoles, which appear empty by light microscopy (Fig. 3.32A), as the material is extracted during tissue processing. Ultrastructurally, the vacuoles are seen to contain sparse membranous fragments and some floccular material (Fig. 3.34).

Neuronal storage is widespread and severe, but neuronal loss is not conspicuous until the terminal stages. Axonal spheroids are numerous in both gray and white matter, but especially in the cerebellar roof nuclei, the posterior brain stem proprioceptive nuclei, and on the proximal parts of Purkinje cell axons. The striking extent of storage in secretory epithelia, such as in pancreas and kidney, endothelia, fixed macrophages, and fibrocytes is best appreciated in plastic-embedded sections.

α-Mannosidosis has also been described in Persian and domestic short-hair and long-hair cats. In the first two breeds, kittens are clinically affected very early in life, having retarded growth, facial dysmorphism, ataxia, tremor, and hepatomegaly. Intense and universal neuronal storage is accompanied by hypomyelination in the cerebrum, and widespread occurrence of axonal spheroids. Extensive storage in other tissues is as described for the bovine disease.

In the domestic long-hair cat, nervous signs are reported to be milder and more slowly progressive, and there are no ocular abnormalities, hepatomegaly, myelin deficiency, or pancreatic acinar cell involvement. However, intense loss of Purkinje cells is evident, and there is great diversity in the morphology of the storage cytosomes, with many membranous cytoplasmic bodies in caudal brain stem nuclei.

β-Mannosidosis has been described in the Anglo-Nubian goat, and in Salers cattle. The final mannose residue of the glycoprotein glycans is β-linked to *N*-acetylglucosamine; its hydrolysis normally follows that of the alpha linkages described above. A deficiency of beta mannosidase leads therefore to the storage of di- or trisaccharides containing one molecule of mannose.

The disease has been most comprehensively documented in Nubian goats. There is autosomal recessive inheritance. Clinical signs are severe at birth, affected kids having small palpebral fissures, facial dysmorphisms, domed skulls, joint contractures, intention tremors, and deafness. They are generally unable to stand and, even with intensive care, will survive for only a few months.

Grossly, there are dilation of the ventricles and hypomyelination of the cerebrum and cerebellum. Recent evidence suggests that the latter is the result of congenital hypothyroidism, due to storage-mediated interference with thyroid function, and it accounts for the congenital syndrome. Microscopically, neuronal vacuolation is ubiquitous and axonal spheroids, numerous, especially in the internal capsule, cerebellar white matter, and basal ganglia. Vacuolated macrophages occur around some blood vessels in the brain, and there may be focal mineralization in the cerebellum and cerebrum. Large focal swellings filled with neurofilaments are sometimes present in the proximal axons of spinal motor neurons, and spheroids are numerous in sensory endings in mucous membranes and conjunctivae. Intense storage in most other tissues is similar to that described for α-mannosidosis, as is the ultrastructure of the storage vacuoles.

α-L-Fucosidosis is recorded in the springer spaniel. As the result of a deficiency in the activity of alpha-L-

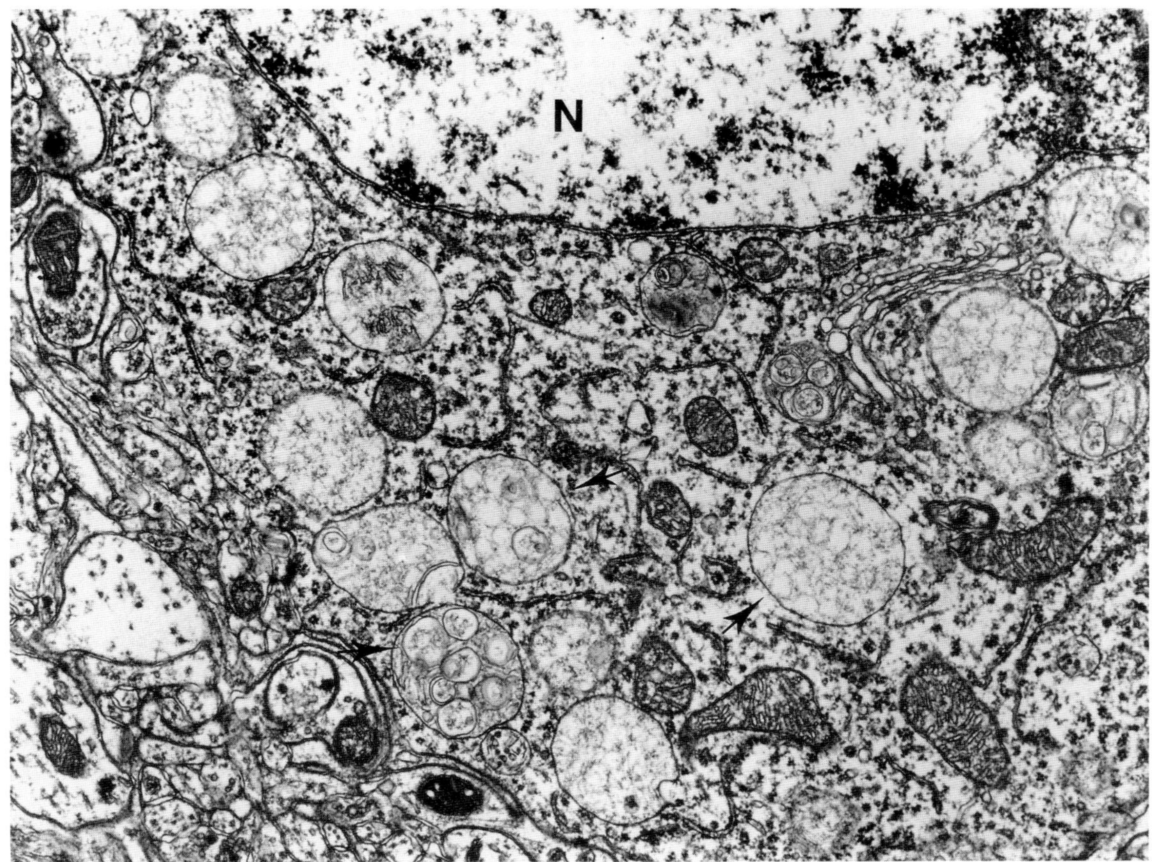

Fig. 3.34 Feline mannosidosis. Cortical neuron. Storage vacuoles (arrows) contain either sparse floccular material or membranous arrays. Nucleus (N). (Courtesy of S. U. Walkley.)

fucosidase, water-soluble, fucose-containing compounds are stored as glycosylasparagines. An autosomal recessive inheritance is established, and the disease has a somewhat delayed clinical onset, with wasting, ataxia, and proprioceptive deficits beginning at about 6 months of age, and becoming severe by about 2 years. A striking gross lesion is marked swelling, up to 10 mm diameter, of the cervical portion of the vagus nerves, the cervical nerves, and dorsal root ganglia.

As in mannosidosis, storage is intense in most tissues, cells becoming distended with apparently empty vacuoles. In the central nervous system, it is present in all neurons and in astrocytes and microglial cells; there are numerous perivascular accumulations of vacuolated macrophages. Axonal spheroids are present in the cerebellar white matter. In the thickened nerves, there is a heavy infiltration of macrophages and the accumulation of a myxoid perineurial ground substance.

Bibliography

Borland, N. A., Jerrett, I., and Embury, D. H. Mannosidosis in aborted and stillborn Galloway calves. *Vet Rec* **14:** 403–404, 1984.

Boyer, P. J. *et al.* Caprine beta-mannosidosis. Abnormal thyroid structure and function in a lysosomal storage disease. *Lab Invest* **63:** 100–106, 1990.

Cummings, J. F. *et al.* The clinical and pathological heterogeneity of feline alpha-mannosidosis. *J Vet Int Med* **2:** 163–170, 1988.

Healy, P. J., Harper, P. A., and Dennis, J. A. Phenotypic variation in bovine alpha-mannosidosis. *Res Vet Sci* **49:** 82–84, 1990.

Jolly, R. D., and Thompson, K. G. The pathology of bovine mannosidosis. *Vet Pathol* **15:** 141–152, 1978.

Jolly, R. D. *et al.* Beta-mannosidosis in a Salers calf: A new storage disease of cattle. *N Z Vet J* **38:** 102–105, 1991.

Jones, M. Z. *et al.* Caprine beta-mannosidosis: Clinical and pathological features. *J Neuropathol Exp Neurol* **42:** 268–281, 1983.

Kelly, W. R. *et al.* Canine alpha-L-fucosidosis: A storage disease of springer spaniels. *Acta Neuropathol (Berl)* **60:** 9–13, 1983.

Lovell, K. L., and Jones, M. Z. Axonal and myelin lesions in beta-mannosidosis: Ultrastructural characteristics. *Acta Neuropathol (Berl)* **65:** 293–299, 1985.

Pearce, R. D. *et al.* Caprine beta-mannosidosis: Characterisation of a model lysosomal storage disorder. *Can J Vet Res* **54:** 22–29, 1990.

Vandevelde, M. *et al.* Hereditary neurovisceral mannosidosis associated with alpha-mannosidase deficiency in a family of Persian cats. *Acta Neuropathol (Berl)* **58:** 64–68, 1982.

3. The Mucopolysaccharidoses

This group of diseases is defined by defective catabolism of the mucopolysaccharides [glycosaminoglycans (GAG)] and, as would be expected, has major expression in the skeleton and connective tissues. These diseases are therefore also discussed in relation to the bones and joints in this volume. The mucopolysaccharides are extremely large and complex molecules, and many enzymes are involved in their degradation. As a result, there are 12 entities defined in humans, but to date only four have been reported in domestic animals; in three of these, there is major involvement of neurons. Their general features include facial and skeletal dysmorphisms, degenerative joint disease, corneal clouding, thickening and distortion of heart valves, and thickening of the leptomeninges. Large quantities of heparan and dermatan sulfates are excreted in the urine.

In animals, a deficiency in the activity of **α-L-iduronidase** has been observed in the domestic short-haired cat and the Plott hound, and is regarded as the counterpart of mucopolysaccharidosis type 1 in man, variants of which include the Hurler's and Scheie syndromes. The macroscopic features described are subtended by intense storage in fibroblasts, fixed macrophages, chondrocytes, myocytes, and pericytes in most organ systems. Glandular epithelial cells are consistently less affected than mesoderm-derived cells. The involvement of chondrocytes is associated with dysplasias of endochondral ossification, and with degeneration of articular cartilage.

Neuronal storage is universal, but neuronal loss is not conspicuous. Storage vacuoles are ultrastructurally pleomorphic; they may appear to be largely empty, or to contain sparse floccular to granular amorphous material, or lamellar membranous structures termed Zebra bodies. The latter appear to be lipid in nature, and probably represent an induced secondary storage of sphingolipids. There is thus a somewhat variable reaction to tissue staining for GAG and lipid. There is some suggestion that affected cats may have a high incidence of cranial meningiomas.

A **β-glucuronidase** deficient mucopolysaccharidosis has been described in an experimental line of dogs, and is considered analogous to human mucopolysaccharidosis type VII (Sly syndrome). The general features of the clinical disease are as previously described, and there is excessive urinary excretion of chondroitin-4- and chondroitin-6-sulfates. There is widespread neurovisceral storage, with cytoplasmic inclusions appearing largely empty or containing sparse granular or lamellar material.

In the Nubian goat, a deficiency in activity of ***N*-acetylglucosamine-6-sulfatase** gives rise to a counterpart of human Sanfilippo's III-D syndrome. Clinical features encompass delayed ability of the neonate to stand and walk, persistently ataxic gait, marked bowing of the forelimbs, and corneal clouding. Gross changes include dwarfism, kyphoscoliosis, scapular hypermobility, and cartilaginous and bony abnormalities. Two main types of lysosomal storage bodies occur microscopically. The primary storage material is heparan sulfate, which appears as lucent floccular material in lysosomes in arterial smooth muscle and cardiac myocytes, fibroblasts, macrophages, hepatocytes, Kupffer's cells, and chondrocytes. Neurons, in contrast, are packed with PAS-positive multilamellar bodies representing secondary storage of gangliosides induced by interference with neuraminidase activity.

In Siamese cats, a deficiency of **arylsulphatase-B** produces a counterpart to human mucopolysaccharidosis VI (Maroteaux–Lamy syndrome). The disease has the same general features as those described, but neuronal storage does not occur. Storage in peripheral blood neutrophils is detected as metachromatic cytoplasmic granules.

Bibliography

Haskins, M. E., and McGrath, J. T. Meningiomas in young cats with mucopolysaccharidosis 1. *J Neuropathol Exp Neurol* **42:** 664–670, 1983

Haskins, M. E. *et al.* Alpha-L-iduronidase deficiency in a cat: A model of mucopolysaccharidosis 1. *Pediatr Res* **13:** 1294–1297, 1979.

Haskins, M. E. *et al.* The pathology of the feline model of mucopolysaccharidosis V1. *Am J Pathol* **101:** 657–674, 1980.

Haskins, M. E. *et al.* Mucopolysaccharidosis V1, Maroteaux–Lamy syndrome: Arylsulfatase-B deficient mucopolysaccharidosis in the Siamese cat. *Am J Pathol* **165:** 191–193, 1981.

Haskins, M. E., *et al.* Spinal cord compression and hindlimb paresis in cats with mucopolysaccharidosis V1. *J Am Vet Med Assoc* **82:** 983–985, 1983.

Jones, M. Z. *et al.* Pathogenesis of caprine *N*-acetylglucosamine-6-sulfatase deficiency (Sanfilippo's III-D syndrome): Studies of the molecular defect. *Abstr ANZ Soc Neuropathol* May, 1991.

Shull, R. M. *et al.* Animal model of human disease. Canine alpha-L-iduronidase deficiency: A model of mucopolysaccharidosis 1. *Am J Pathol* **109:** 244–248, 1982.

Shull, R. M. *et al.* Morphologic and biochemical studies of canine mucopolysaccharidosis. *Am J Pathol* **114:** 487–495, 1984.

4. The Glycogenoses

A number of enzymes are involved in the catabolism of glycogen, but only one of these, α-1,4,-glucosidase, is a lysosomal enzyme. The lysosomal pathway degrades any glycogen which finds its way into autophagic vacuoles, while the other pathways are more related to the metabolic mobilization of glycogen. In humans, eight different glycogen storage diseases have been identified, with the lysosomal type being classified as type 2, Pompe's disease. As the storage process is very widespread in this disease, it is also referred to as generalized glycogenosis. In most other types, storage is concentrated in liver and muscle.

Fully documented **α-1,4-glucosidase** deficiency has been identified in beef cattle of the shorthorn and Brahman breeds. The disease has autosomal recessive inheritance, and affected calves have severe clinical signs by about a year of age. The most damaging effects are produced in the skeletal and cardiac muscle; weakness and congestive cardiac failure are clinically dominant, and cardiomegaly and hepatomegaly, evident at necropsy. There is widespread glycogen storage, much of which occurs in typical lysosomal storage vacuoles, but some of which is intracytoplasmic. Swollen, vacuolated cells contain diastase-sensitive, PAS-positive material, which is ultrastructurally typical of glycogen. Vacuolar myopathy and cardiomyopathy are prominent (Fig. 3.35); neuronal storage is universal and severe and is accompanied by glial storage, with numerous axonal spheroids in the vestibular and cuneate nuclei, terminal fasciculus gracilis, and throughout the spinal cord gray matter (Fig. 3.22B). Mild Wallerian degeneration may be present in the lateral and ventral columns of the spinal cord and in some peripheral nerves.

Type 2 glycogenosis has also been suspected on morphologic grounds in the domestic cat, the Lapland dog, and Corriedale sheep.

A deficiency of **amylo-1,6-glucosidase** is recorded in the German shepherd dog, as a counterpart of glycogenosis type 3 (Cori's disease). Cytoplasmic glycogen storage occurs in liver, muscle, and myocardium, and in neurons and glia in the brain and spinal cord.

In the English springer spaniel dog, **phosphofructokinase** deficiency has been associated with a syndrome of compensated hemolytic anemia, but a single case report describes one affected aged dog with a myopathy in which there was storage of an amylopectinlike polysaccharide. This is considered to be analogous to human type 7 glycogenosis, involving defective glycogen brancher enzyme activity.

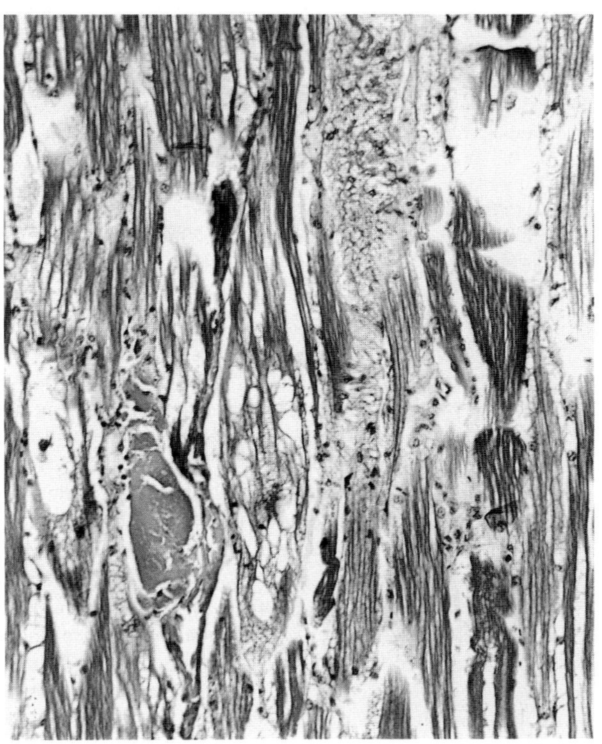

Fig. 3.35 Generalized glycogenosis. Shorthorn calf. Vacuolated and degenerate skeletal muscle fibers.

Bibliography

Ceh, L. *et al.* Glycogenosis type III in the dog. *Acta Vet Scand* **17:** 210–222. 1976.

Cook, R. D. *et al.* Changes in nervous tissue in bovine generalised glycogenosis type II. *Neuropathol Appl Neurobiol* **8:** 95–107, 1982.

Harvey, J. W. *et al.* Polysaccharide storage myopathy in canine phosphofructokinase deficiency (type VII glycogen storage disease). *Vet Pathol* **27:** 1–8, 1990.

Howell, J. McC. *et al.* Infantile and late onset form of generalised glycogenosis type II in cattle. *J Pathol* **134:** 266–277, 1981.

O'Sullivan, B. M. *et al.* Generalised glycogenosis in Brahman cattle. *Aust Vet J* **57:** 227–229, 1981.

5. The Ceroid-Lipofuscinoses

With advancing age and senility, intracytoplasmic granules of the lipopigment, lipofuscin, accumulate in neurons, fixed phagocytes, macrophages, and muscle cells. This "wear-and-tear" pigment is familiar to all pathologists, and has a characteristic histochemical profile and ultrastructure; the irregularly shaped granules have a high-density component punctuated with vacuoles of low density. For many years, in the group of diseases known as the ceroid-lipofuscinoses, it has been assumed that a similar pigment is stored in lysosomes, although the biochemical basis of storage was an enigma and remains poorly understood. However, as discussed below, more recent data strongly indicate that in many of these entities, the stored material has little to do with wear-and-tear pigment.

In animals, ceroid-lipofuscinoses of proven or presumptive inherited nature have been described in Siamese cats, South Hampshire sheep, Devon cattle, Nubian goats, and many breeds of dog including English setter, Chihuahua, dachshund, Saluki, Dalmatian, blue heeler, Border collie, Tibetan terrier, and crossbred. In general, although storage may be widespread in several organs, it is most damaging in neurons of the cerebral cortex, retina, and cerebellar Purkinje system. Thus there is frequently extensive cellular loss and atrophy in these regions, and correlating dementia, blindness, and ataxia. Macroscopically, atrophied areas may have a distinctly brownish tinge. Microscopically, the storage granules are brightly autofluorescent under ultraviolet light, pale brown-red, or colorless with hematoxylin and eosin, weakly acid-fast, magenta with PAS, and intensely positive with luxol fast blue. Ultrastructurally, they appear as membrane-bound cytosomes up to 15 nm in diameter, irregular in outline, and with a variety of forms (Fig. 3.36). Some have membranous material arranged as curvilinear bodies and fingerprint bodies, which are considered characteristic. Others may have laminated stacks of membranes akin to Zebra bodies, membranous stacks, or dense granular deposits.

The variation in features is illustrated by comparing the disease in various species and breeds. In Devon cattle, the major clinical deficit is profound blindness at about 14 months of age, with death usually due to misadventure by about 2 years. There is severe retinal atrophy but only

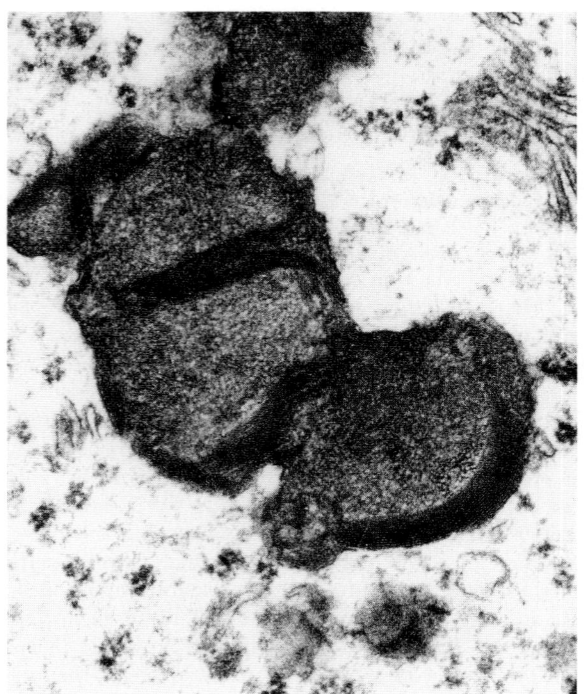

Fig. 3.36 Ceroid lipofuscinosis. Tibetan terrier. Granular and membranous storage body. Purkinje cell. (Courtesy of J. F. Cummings.)

mild loss of neurons from the cerebral and cerebellar cortices. In Tibetan terriers, blindness is again the dominant sign, but the onset is delayed to middle life and is accompanied terminally by stupor. In the Border collie and some other dog breeds, there are gait and visual deficits by 18 to 24 months of age, accompanied by increasing aggression and dementia. Retinal lesions are mild, and blindness is central in origin. Neuronal loss and gliosis are particularly severe in the cerebellar Purkinje cell layer, and significant in the limbic system.

The most thoroughly studied of the animal ceroid-lipofuscinoses is that in the South Hampshire sheep, an autosomal recessively inherited disease, analogous to Batten's disease of children. The brains of affected lambs grow normally until 4 months of age, and then undergo atrophy (Fig. 3.37). Recent data from this work indicate that 50% of the stored material is the lipid-binding protein subunit C of mitochondrial ATP synthase. This accumulates in lysosomes, complexed with a variety of lipids. The reason for the accumulation of this substrate is not yet elucidated. This work discounts previous hypotheses that the storage material is the end product of abnormal lipid peroxidation and suggests that the disease should be reclassified as a proteinosis. This idea is strengthened by the additional finding that the same storage material is present in the Devon cattle, and in some of the canine entities.

It seems likely that future research will produce further biochemical subdivisions in this group of diseases, and that the term ceroid-lipofuscinosis will be superseded by more specific names.

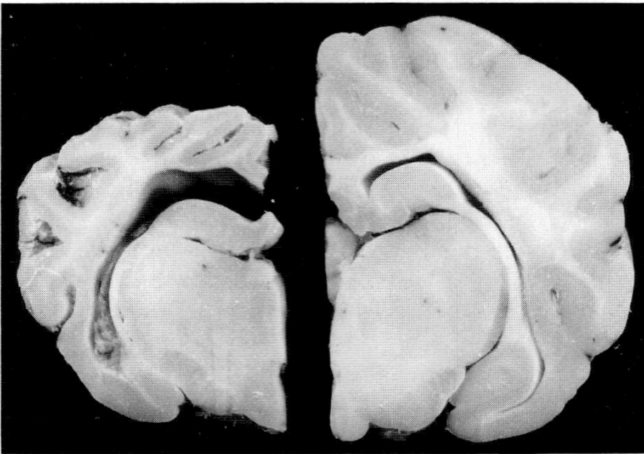

Fig. 3.37 Ceroid lipofuscinosis. South Hampshire lamb. Marked cerebral atrophy at 4 months. Control right. (Courtesy of R. D. Jolly.)

Bibliography

Appleby, E. C., Longstaffe, J. A., and Bell, F. R. Ceroid-lipofuscinosis in two Saluki dogs. *J Comp Pathol* **92:** 375–380, 1982.

Fiske, R. A., and Storts, R. W. Neuronal ceroid-lipofuscinosis in Nubian goats. *Vet Pathol* **25:** 171–173, 1988.

Harper, P. A. *et al.* Neurovisceral ceroid-lipofuscinosis in blind Devon cattle. *Acta Neuropathol (Berl)* **75:** 632–636, 1988.

Jolly, R. D. *et al.* Ceroid-lipofuscinosis (Batten's disease): Pathogenesis and sequential neuropathological changes in the ovine model. *Neuropathol Appl Neurobiol* **15:** 371–383, 1989.

Jolly, R. D. *et al.* Ovine ceroid-lipofuscinosis is a proteolipid proteinosis. *Can J Vet Res* **54:** 15–21, 1990.

Sisk, D. B. *et al.* Clinical and pathologic features of ceroid-lipofuscinosis in two Australian cattle dogs. *J Am Vet Med Assoc* **197:** 361–364, 1990.

Taylor, R. M., and Farrow, B. R. Ceroid-lipofuscinosis in Border collie dogs. *Acta Neuropathol (Berl)* **75:** 627–631, 1988.

6. *Miscellaneous Genetic Storage Diseases*

The apparent lysosomal storage of a lipid material in spinal motor neurons is described in English pointer dogs, and is discussed under motor neuron diseases.

A counterpart to human Lafora's disease has been recorded in dogs of the basset hound, poodle, and beagle breeds. Progressive myoclonus epilepsy is associated with widespread intraneuronal storage of a complex polyglucosan, which appears as characteristic Lafora bodies (Fig. 3.19). They are most numerous in Purkinje cells, and neurons of the caudate, thalamic, and periventricular nuclei. They are nonmembrane-bound spherical structures with a central basophilic core and a peripheral halo of radiating filaments. They are strongly PAS positive and are found in both perikaryon and processes; they vary greatly in size, ranging up to 15 to 20μm. Their pathogenesis is not understood, but familial association suggests an inherited metabolic defect. They are occasionally found incidentally in otherwise normal brains and spinal cords, especially in older dogs.

B. The Induced Storage Diseases

Numerous types of storage process have been successfully produced experimentally, and there is a group of naturally occurring disorders related to the ingestion of plants, in which induced storage disease is proven or suspected.

1. Swainsonine and Locoweed Toxicosis

Swainsonine, an indolizidine alkaloid, is the active principle of several species of toxic plants which have caused considerable problems for all classes of grazing livestock. These are various of the *Swainsona* spp. of Australia, and the *Astragalus* and *Oxytropis* spp., the locoweeds of North America. As a potent inhibitor of lysosomal alpha-mannosidase, swainsonine induces a form of α-mannosidosis which is a close copy of the genetic disease of cattle and cats. Continued ingestion of toxic material over a period of 4 to 6 weeks and more results in failure to thrive and, ultimately, ataxia, proprioceptive deficits, and behavioral abnormalities (''locoism'' in North America, ''pea-struck'' in Australia). Necropsy examination during, or within a short time after, exposure to the plant reveals microscopic and ultrastructural lesions identical to those of genetic α-mannosidosis. Within 2 weeks of last exposure, much of the storage disease resolves, but axonal spheroids may persist in large numbers in areas such as the cerebellar roof nuclei and posterior brain stem. Swainsonine-induced mannosidosis has been experimentally compared in the cat with genetic mannosidosis in order to determine the reversibility of changes such as meganeurite and aberrant synapse formation in higher neurons. Clinical recovery may or may not occur following cessation of exposure, and the persistence of secondary neuronal changes suggests that they, rather than the storage process, may underlie neuronal malfunction. Current evidence suggests that the exposure of young growing animals is more likely to produce irreversible disease than is exposure of adults.

The induced disease is, however, biochemically distinct from the genetic, as the alkaloid also inhibits Golgi mannosidase 2, an enzyme involved in the posttranslational trimming modifications of the glycan moiety of glycoproteins. As a result, abnormal proportions of different types of glycoproteins are produced, and the storage oligosaccharides are larger than those in the genetic disease. No modification of the storage disease appears to result from this difference.

In swainsonine intoxication of the pregnant animal, both dam and fetus are affected, and abortion and terata are well recognized in ovine locoweed toxicosis. Suppressive effects on fertility are also recognized. The relationship to these events of the storage disease on one hand, and abnormal glycoprotein processing on the other, is not elucidated.

2. Poisoning by Trachyandra *Species*

Prolonged ingestion (several weeks) of *Trachyandra divaricata* or *T. laxa* has been associated with severe neurologic disease and lipofuscinosis in sheep, horses, goats, and pigs in South Africa and Australia. The clinical syndrome is one of weakness, suggesting a neuromuscular disorder, and is often accompanied by intense **lipofuscin** storage in all central and peripheral neurons and, to a lesser extent, in macrophages of the intestinal lamina propria, hepatocytes, Kupffer's cells, and renal tubular epithelium. Pigment storage may be sufficiently intense to cause macroscopic rusty-brown discoloration of central gray nuclei and peripheral ganglia. The pigment granules have all the histochemical and ultrastructural features of lipofuscin. The clinical signs appear to be irreversible, and their relationship to the storage process is obscure; the basis of the storage process is unknown.

3. Phalaris *Poisoning*

This disease is discussed in this section because neuronal lysosomal pigment storage, although not likely to be the direct cause of clinical signs, is a prominent and diagnostically useful feature. A recent investigation suggests that nervous signs result from serotonergic effects on specific upper motor neuronal systems, and that the pigments are probably indolic metabolites which accumulate in lysosomes.

Extensive losses in sheep and rarely cattle in Australia, New Zealand, South Africa, and California have been due to grazing *Phalaris* spp., principally *P. aquatica* and *P. arundinacea,* but also *P. minor* and *P. caroliniana*. Neurotoxicity is due to a mix of methyl tryptamine and β-carboline indoleamines, chemically related to 5-hydroxytryptamine (serotonin). Intoxication is generally associated with lush growing pasture and may become evident in 3 to 10 days after initial exposure. There are two syndromes, either sudden death due to cardiac arrythmia or a subacute to chronic neurologic disease, which is the focus of our attention here. The clinical onset of this syndrome may be delayed for several months after exposure to toxic pasture has ceased. It is characterized by generalized muscle tremors progressing to stiffness, collapse on forced exercise, and tetanic seizures. In most cases, recovery does not occur, and apparent recovery is followed by relapses.

In cases examined early in the course of the disease, there may be no gross or microscopic lesions but, in general, pathologic changes are present. Most characteristically, there is storage of a granular pigment within neurons of the brain stem nuclei, spinal gray matter, and dorsal root ganglia, and in macrophages of the cerebrospinal fluid. A similar pigment is also present within renal tubular epithelial cells. When storage is intense, the affected gray matter and kidneys may have a distinct greenish discoloration on gross inspection (Fig. 3.38A). Histologically, the pigment granules have a greenish-brown color, and a typically perinuclear distribution in neurons (Fig. 3.38B), although some granules may accumulate in dendrites.

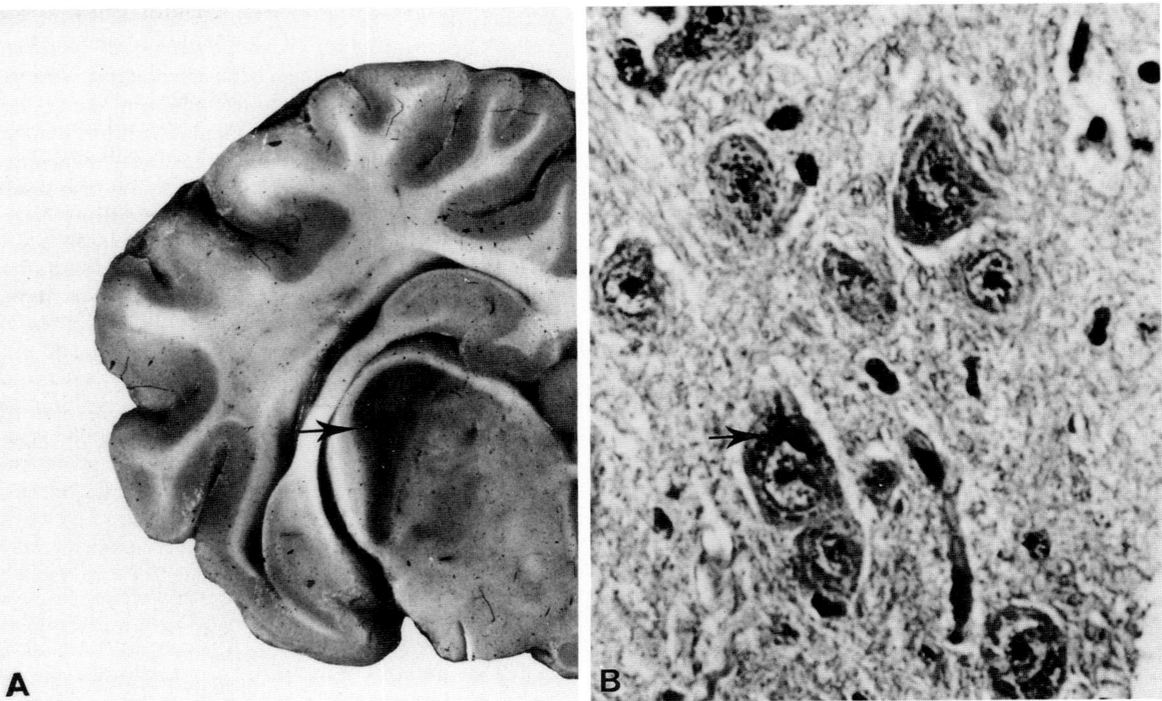

Fig. 3.38 *Phalaris* spp. poisoning. Ox. (A) Pigmentation in lateral geniculate body (arrow). (B) Pigment granules in neuronal cytoplasm (arrows).

Ultrastructurally, the storage granules are composed of concentric membranous lamellae, sometimes interspersed with fine granular material. They are membrane bound and are considered to be lysosomal in nature. In most cases, there is also Wallerian degeneration concentrated in ventral, ventromedial, and lateral funiculi throughout the spinal cord, and in the medial longitudinal fasciculus. This distribution suggests selective damage to long descending motor tracts. Severely affected areas may also have intense diffuse astrogliosis. An intense reactive astrogliosis may also be seen in ventral spinal cord gray matter, along with mild neuronal loss.

4. *Poisoning by* Solanum *Species*

Ingestion of *Solanum kwebense* or *S. fastigiatum* is associated with neuronal degeneration and loss in the cerebellum, as discussed elsewhere. There is vacuolation of Purkinje cells and some other neurons and, on morphologic grounds, the apparent lysosomal storage of a lamellated lipid material. The biochemical basis of the storage in not known.

5. *Gomen Disease*

As discussed elsewhere, this is a suspected toxicosis of horses in New Caledonia, in which there is cerebellar neuronal degeneration, and associated lipofuscin storage in neurons and phagocytic cells.

Bibliography

Bourke, C. A., Carrigan, M. J., and Dixon, R. J. The pathogenesis of the nervous syndrome of *Phalaris aquatica* toxicity in sheep. *Aust Vet J* **67:** 356–358, 1990.

Dorling, P. R., Huxtable, C. R., and Vogel, P. Lysosomal storage in *Swainsona* spp. toxicosis: An induced mannosidosis. *Neuropathol Appl Neurobiol* **4:** 285–295, 1978.

Dorling, P. R. *et al.* The pathogenesis of chronic *Swainsona* sp. toxicity. *In* "Plant Toxicology" A. A. Seawright *et al.* (eds.), pp. 255–265. Hedges Bell, Melbourne, Australia, Dominion Press, 1985.

Dorling, P. R. *et al.* Isolation of swainsonine from *Swainsona canescens:* Historical aspects. *In* "Swainsonine and Related Glycosidase Inhibitors" L. F. James, *et al.* (eds.), pp. 83–90. Ames, Iowa, Iowa State University Press, 1989.

East, N. E., and Higgins, R. J. Canary grass (*Phalaris* sp.) toxicosis in sheep in California. *J Am Vet Med Assoc* **192:** 667–669, 1988.

Huxtable, C. R. *et al.* Neurological disease and lipofuscinosis in horses and sheep grazing *Trachyandra divaricata* (branched onion weed) in south-western Australia. *Aust Vet J* **64:** 105–108, 1987.

Jian, Z. *et al.* Lafora's disease in an epileptic basset hound. *N Z Vet J* **38:** 75–79, 1990.

Molyneaux, R., and James, L. Loco intoxication: Indolizidine alkaloids of spotted locoweed (*Astragalus lentigenosis*). *Science* **216:** 190, 1982.

Newsholme, S. J., Schneider, D. J., and Reid, C. A suspected lipofuscin storage disease of sheep associated with ingestion of the plant *Trachyandra divaricata* (Jacq.) Kunth. *Onderstepoort J Vet Res* **52:** 87–92, 1985.

Walkley, S. U., and Seigel, D. A. Comparative studies of the CNS in Swainsonine-induced and inherited feline alpha mannosidosis. *In* "Swainsonine and Related Glycosidase Inhibitors" L. F. James *et al.* (eds.), pp. 57–75, Ames, Iowa, Iowa State University Press, 1989.

IV. Increased Intracranial Pressure, Cerebral Swelling, and Edema

It seems well to interpose a discussion of these phenomena here because they are common to many of the pathologic processes to be discussed later. There is normally only a narrow space separating the brain from the dura mater. Both the dura and skull are unyielding, so that only a relatively small increase in the volume of the intracranial contents is permissible without increasing intracranial pressure. When the pressure is increased, something has to yield.

The causes of increased intracranial pressure are many and varied. One component of almost all of them is edema. The edema may be more or less localized and geographically related to local lesions, or it may be diffuse. In addition to the pathogenetic mechanisms to be discussed here, acquired hydrocephalus and vitamin A deficiency in young animals can be responsible for increased intracranial pressure. In vitamin A deficiency, there is increased secretion of cerebrospinal fluid by the choroid plexus and decreased absorption by arachnoid villi, which may contribute to hydrocephalus; more important in domestic animals, however, is impaired growth and remodeling of cranium and vertebrae, which leads to disproportion between the volume of the growing nervous system and volume of the cranial cavity and spinal canal.

The local lesions which may result in **local edema** of the brain or spinal cord include neoplasms, inflammations (especially abscessation and suppurative meningitis), parasitic cysts, focal necrosis of various causes, trauma (especially when accompanied by hemorrhage and laceration), hemorrhages of parenchyma and meninges, and space-occupying lesions of the meninges, which cause pressure on the brain. In each example, the edema may be mild or extensive and may contribute more than the primary or inciting lesion to the clinical signs, to the swelling of the brain and increase of intracranial pressure.

Generalized cerebral edema and swelling of the brain also occurs in relation to a variety of systemic conditions. Some degree of swelling can be anticipated as a postmortem change, especially in the brains of young animals, and is probably to be related to the imbibition of fluid in autolysis. Cerebral edema with swelling occurs with diffuse meningitis, moderately so in the diffuse viral encephalitides, acute bacterial toxemias such as clostridial enterotoxemia, and chemical intoxications such as lead and organomercurial poisoning, and quite severely in the pathologic syndrome of sheep and cattle known as polioencephalomalacia. Generalized edema of moderate degree is expected in the many metabolic and toxic conditions which are characterized by disturbances of cellular osmoregulation, in particular, disturbances which interfere with intracellular and extracellular concentrations of sodium and potassium; the acute onset neurologic disease which occurs in salt poisoning of pigs and in water intoxication in young ruminants given water *ad libitum* after a period of deprivation would be in this category.

There are differences between gray and white matter in their susceptibility to edema, and different areas of gray substance or of white differ in their vulnerability. The considerations of edema are limited for descriptive purposes to the cytotoxic and vasogenic varieties. **Interstitial edema,** which affects the central white matter in hydrocephalus and the spinal cord in hydromyelia and syringomyelia, is considered with those diseases, as is the edema of the spongiform encephalopathies. A distinction is maintained here between **cytotoxic edema** and **vasogenic edema,** notwithstanding that the endothelial barrier between blood and brain may be functionally deranged in each type. Cytotoxic edema is intracellular and a result of deranged ionic and osmotic balance. Vasogenic edema is intercellular, reflects severe changes in vascular permeability such as may allow the passage of plasma protein, and leads to physical disruption of neuropil and fiber tracts.

Intracellular or cytotoxic edema depends on direct or indirect noxious injury to cells and interference with the mechanisms which control cell volume. For cells in non-nervous tissues, the swelling would represent the movement of water from interstitial tissues into cell cytoplasm. There is in nervous tissue, however, no significant interstitial tissue or intercellular space. The intercellular space in the brain is not more than 10–20 nm. There is a layer of material which stains as glycoprotein or mucopolysaccharide on cell membranes, but there is doubt as to whether it is the counterpart of other interstitial gels. The net increase in water in the brain in cytotoxic edema must represent a movement from plasma and the regulatory mechanisms must therefore reside in the capillary endothelium and the brain cells. The basic disturbance is in osmoregulation, which depends mainly on the efficiency of the sodium/potassium pump and ATP as an energy source.

In nonspecific cytotoxic edema, the swelling is mainly in the astrocytes. Simple swelling of astrocytes is the most obvious structural change in gray matter. Swelling of the astrocytes involves the nucleus and processes. Glycogen granules accumulate in the watery protoplasm. If mild edema persists, the astrocytes react as described in an earlier section. If the edema is severe, the nucleus is much enlarged, the chromatin is dispersed against the nuclear membrane, and the processes are voluminous. Death of acutely swollen astrocytes may not be accompanied by changes in the adjacent neuropil. Neurons may swell as a brief prelude to lysis. Satellite oligodendroglia are generally spared.

However, in nonspecific cytotoxic edema, oligodendrocytes in white matter do react. The nucleus is swollen and less dense than normal; the nucleolus is hypertrophied, and the cytoplasm is enlarged and often visible in routine material. Changes in oligodendroglial processes, which are wrapped as myelin, are difficult to evaluate histologically except in those instances of specific cytotoxic edema in which there is splitting of the intraperiod line and the accumulation of water in intermyelinic clefts.

Vasogenic edema is the term used to distinguish the extracellular accumulation of fluid and cytotoxic edema. The basis of vasogenic edema is injury to vascular endothelium of sufficient severity to allow leakage or permeation of plasma constituents including, if the injury is sufficient, plasma proteins. The fluid spreads between cells in response to the hydrostatic pressures in the cerebral circulation and those in the tissue. Vasogenic edema is a common complication of traumatic, inflammatory, and hemorrhagic lesions of the nervous system. It is not a conspicuous change in gray matter, because the dense tangle of neuropil resists the passage of fluid. There are exceptions, such as in the periventricular nuclei, where vasogenic edema occurs rather selectively in thiamine deficiency. The structure of white matter offers less resistance to the passage of edema fluid of this origin, and the comparison between the susceptibility of gray and white matter can be seen readily with local lesions near their junction in the cortex. The long fiber tracts such as the spinal cord, internal capsules, and optic tracts tend to be spared. The density and disposition of these tracts probably exert a considerable influence on the spread of edema about local lesions as, for example, the corpus callosum seems effectively to prevent spread from one hemisphere to the other.

The histologic appearances are similar to those in cytotoxic edema with the addition of dissecting changes along fiber tracts. The white matter is loosely textured, the myelinated fibers being spread apart. The spaces created may be clear or they may, depending on the degree of permeability, contain a homogeneous proteinaceous fluid (Fig. 3.39). Plasma droplets which stain brightly with PAS are frequently present in perivascular clefts (Fig. 3.40).

Diffuse cerebral edema which is mild or moderate may be difficult to recognize grossly. More severe degrees are readily recognized although easily overlooked. The brain is swollen, pale, soft, and wet and, because of its softness, the cerebral hemispheres tend to droop over the edges of the parietal bones when the cranial cap is removed. With these severe and rapid swellings, the course is short, and there may not be signs of displacement or flattening of the gyri. When the swelling is less severe, and of longer duration, the brain may be relatively firm and dry; the gyri are pale or a faint yellowish color; and the brain is displaced caudally so that it appears unusually elongate. The displacement is most obvious where it involves the cerebellum and medulla (Fig. 3.41). With moderate displacement, the posterior surfaces of the cerebellar hemispheres are depressed by contact with the occipital bones. When the displacement is of greater degree, the medulla and posterior portion of the vermis are herniated through the foramen magnum. The displaced vermis is flattened and lies like a tongue over the medulla (so-called lipping of the cerebellum). The rhomboid fossa is flattened. The anterior portion of the vermis is pressed against the anterior medullary velum and may occlude the opening of the cerebral aqueduct to cause internal hydrocephalus. The brain stem is displaced caudally, and this is especially evident in the displacement of the corpora quadrigemina

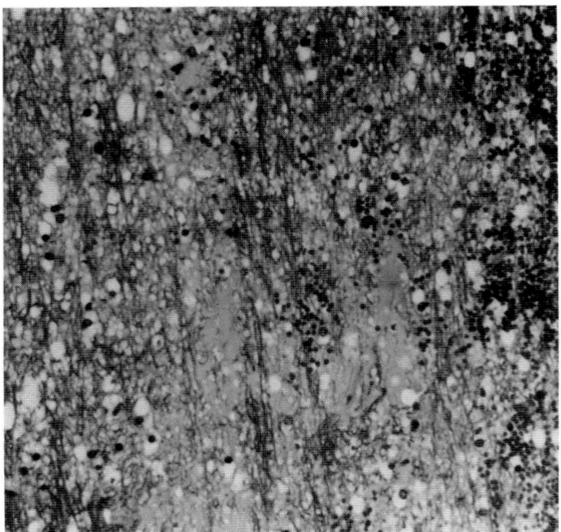

Fig. 3.39 Vasogenic edema and hemorrhage with degeneration of white matter.

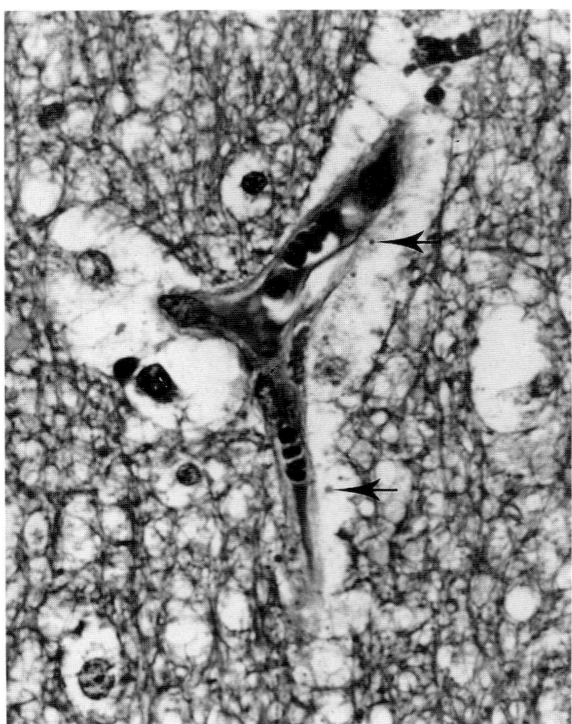

Fig. 3.40 Accumulation of clear edema with protein droplets (arrows) around vessel. Pig. Mulberry heart disease.

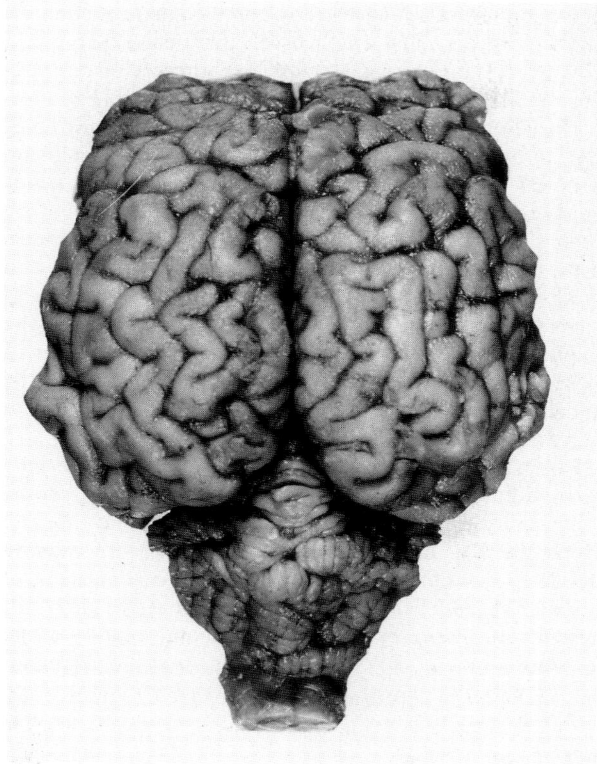

Fig. 3.41 Cerebral edema with displacement and coning of cerebellum. Calf.

well into the posterior fossa. Perhaps because the tentorium rather closely embraces the brain stem in animals, herniation of the occipital lobes through the tentorial space into the posterior fossa is seldom observed. It may, however, be observed in cerebral edema of long standing in horses and cattle, especially if the brain is fixed *in situ* so that the pressure grooves produced by the free edge of the tentorium are retained. Impaction of tissue in the tentorial space interferes with the flow of cerebrospinal fluid from the posterior to the anterior fossa and contributes to hydrocephalus.

Displacements in the anterior fossa in diffuse cerebral swelling are of lesser degree and significance. The nerves may be stretched and flattened, but pareses are seldom observed. The vasculature must be compromised in all cases. Occasionally there is thrombosis of the superior sagittal sinus with venous infarction in the dorsal cortex, ischemic necrosis in the posteromedial surface of the occipital lobe referable to compression of the posterior cerebral arteries, and pontine hemorrhage referable probably to occlusion of small veins in this area.

Localized edematous changes reach their most extensive development in the centrum semiovale of the cerebrum and in the deep white matter of the cerebellum surrounding local lesions in these areas. The edematous area may be recognized by the swelling, which may be greater in volume than that produced by the primary lesion and have a soft, depressed, and damp or watery appearance on the cut surface. The extent of the edema cannot be appreciated on gross inspection because the margins are indefinite, and the same difficulty in delineating the edematous area is experienced at microscopy. When the lesion is of prolonged duration, the extent of the edematous areas can be appreciated a little better by a yellowish discoloration which develops.

The combination of edema with a local lesion may displace the brain in one or more directions. Posterior displacement through the posterior fossa may occur as it does in diffuse swelling. The displacements may be more local, involving lateral shifts of the base of the brain or medial displacement of one hemisphere so that the falx cerebri is displaced laterally or the cingulate gyrus is herniated beneath the free margin of the falx and the lateral and third ventricles are depressed.

V. Lesions of Blood Vessels and Circulatory Disturbances

Diseases of blood vessels are considered in detail in Volume 3, Chapter 1, The Cardiovascular System. This section deals with some special features of cerebrospinal circulation and the lesions, usually ischemic or hemorrhagic, that result from vascular injury.

The blood supply to the brain is derived from the internal carotid and vertebral arteries, these sources anastomosing under the brain stem and at the circle of Willis. The major cerebral vessels, derived from the carotid and vertebral arteries, anastomose quite freely in the pia–arachnoid, but once an artery or arteriole penetrates the substance of the brain, it becomes an end artery, although there are some anastomoses at the capillary level. Although these anastomoses can be demonstrated readily, even those of arteriolar size in the meninges are probably of little value. Under normal circumstances, the cerebral arteries have rather set fields of supply. If one vessel is occluded, some collateral circulation develops, but it does not take over more than the periphery of the area of supply of the occluded vessel. There may, in the brain, be influences governing collateral circulation in addition to those operating in other tissues, but the basic considerations still apply. The development of collateral circulation will be influenced by the anatomic arrangements of vessels, the volume of ischemic tissue, the rate at which the vascular occlusion develops, the size of the occluded vessel, and, importantly, the quantity of blood flow, which is referable to the state of the systemic circulation, and the quality of blood flow, which is referable to such matters as the oxygen tension and viscosity of plasma. For complete cessation of blood flow in an area of brain, it is probably not necessary that there be complete occlusion of the corresponding artery, because flow is expected to cease, especially in peripheral twigs, while the intravascular pressure is still something above zero. For this reason, ischemic injury in the brain localized to the field of distribution of one or other major cerebral artery may result from occlusion of a carotid vessel. Extracranial anastomoses between the major arterial vessels are, however, effective in the event of vascular occlusion occurring proximal to the circle of Willis.

Arteries entering the substance of the brain are relatively small and arise at right angles from the parent vessels in the pia–arachnoid. There are abrupt changes in caliber when the meningeal vessels divide, and this provides an entrapment mechanism for large emboli. The vessels which enter the brain are progressively attenuated to capillaries in both gray and white matter, but many capillaries loop back into the cortex from near the gray–white junction. It is at the gray–white junction that many small emboli lodge, although expansion of the embolic or metastatic lesions is predominantly in the white matter (Fig. 3.42).

The arterial supply to the spinal cord is derived from the vertebral artery in the cervical region and from radicular arteries in the lumbar region anastomosing as the ventral spinal artery. There is some doubt as to the direction of flow in cervical, thoracic, and lumbar portions of the spinal artery, and it is possible that there is a border zone in the caudal cervical and cranial thoracic region that is particularly vulnerable to ischemia if flow is impaired in the vertebral artery or the posterior portions of the ventral spinal artery.

In the spinal cord, the central gray matter is supplied by branches of the ventral spinal artery which enter the ventral sulcus, and lesions of these branches affect the gray matter of the cord rather selectively. The white matter is supplied by an anastomotic complex in the meninges, which produces many small vessels that penetrate directly as end vessels and are susceptible to compression or hypotension.

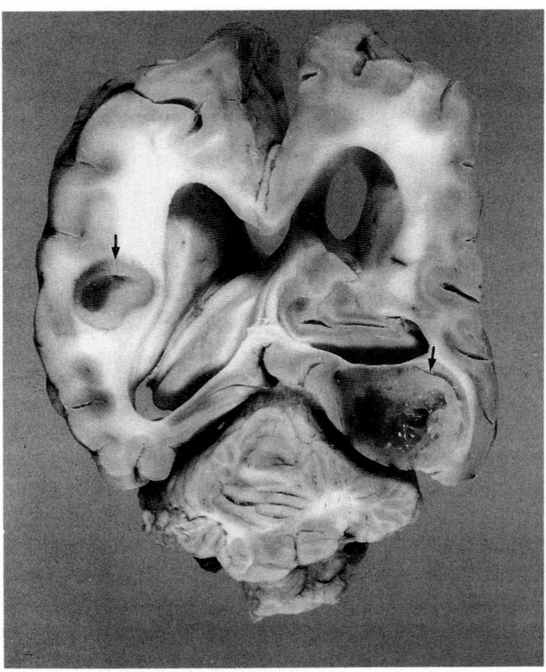

Fig. 3.42 Tumor metastases (arrows) at junction of gray and white matter. Dog.

The cerebral veins have abundant and useful anastomoses. Untoward effects are not expected to follow occlusion of single veins. The venous sinuses of the dura mater empty into the jugular veins, but they also communicate freely through the bones with extracranial veins. Because of these communications, the effects of obstruction of a dural sinus may be relatively slight, unless there is venous stagnation of the head or venous stagnation within the cranium produced by cerebral swelling or occlusion of more than one major sinus. The venous system of the spinal cord is freely anastomotic and drains via the radicular vessels to the paravertebral plexuses.

Diseases affecting the cerebrospinal vasculature may be broadly grouped into those in which there is some abnormality of the freely circulating blood, and those in which there is failure of proper circulation of healthy blood. It is with the latter group, composed largely of obstructive and hemorrhagic lesions, that we are concerned here. Diseases associated with abnormality of the circulating blood, such as the anoxias and hypoglycemia, are discussed later with neurodegenerative diseases.

A. Ischemic Lesions

The outcome of vascular obstructions depends on the type and size of the vessel obstructed, the degree and duration of ischemia, and the relative vulnerability of the tissues to anoxia. The injury may vary in severity from a temporary functional disturbance to the other extreme of infarction and necrosis. Neurons and oligodendroglia are the most sensitive of the neural structures to ischemia, astrocytes are moderately resistant, and microglia and the blood vessels are quite resistant and may survive in small areas in which all else dies. Gray matter, being possessed of a high metabolic rate and dependence on oxygen, is more sensitive than is white matter, but there are regional differences in the sensitivity of gray matter. The Purkinje cells and the neurons of the cerebral cortex are the most sensitive of all, and, within the cerebral cortex, the deeper laminae are more sensitive than the superficial laminae. The deeper laminae may be selectively destroyed in ischemia, producing a distribution of necrosis known as laminar cortical necrosis.

Obstructive lesions of cerebrospinal vessels are not commonly observed in animals, and ischemic changes may be absent even when the vessels are profoundly altered. On the other hand, lesions which are regarded as being of ischemic type are quite commonly observed in the absence of demonstrable vascular occlusion. These cases tend to be individual in their occurrence and without particular pattern.

In the cat, a syndrome called **feline ischemic encephalopathy–cerebral infarction syndrome** has been recognized. It is a disease of mature cats. The pathogenesis is unknown, and in most cases, there is no evidence of a vascular lesion. The extent and distribution of degeneration varies from case to case. The milder lesions occur in superficial cortical laminae as multiple foci. The more severe or extensive lesions tend to be in the distribution area of the middle cerebral artery and may be bilateral but asymmetrical. In some cases, there may be ischemic lesions in the brain stem.

Ischemic and hemorrhagic cerebral lesions occur in the **neonatal maladjustment syndrome** of foals, also known as the barker or convulsive foal syndrome, which is discussed with congenital atelectasis in Volume 2, Chapter 6, The Respiratory System. The nature of the circulatory derangement is not understood, but the lesions are presumed to reflect cerebral ischemia and variably delayed reflow. There is ischemic laminar necrosis of the cerebral cortex (Fig. 3.43), sometimes accompanied by necrosis in the paired gray nuclei of the midbrain and brain stem, and by multiple small hemorrhages. In some foals affected by this neonatal syndrome, there is minimal ischemic necrosis, but instead, a profuse distribution of small perivascular and petechial hemorrhages in the cerebrum, cerebellum, and brain stem.

Degenerative vascular disease of arteriosclerotic type is not important in animal disease. Atheroma occurs in some hypothyroid dogs. Siderosis of the walls of the arterioles occurs commonly as a change associated with advancing age in horses. The calcium and iron salts may be deposited in amounts capable of converting the vessels to rigid pipes (Fig. 3.44). The patency of the vessels is well maintained. There is no thrombosis, and vascular stenosis probably proceeds slowly enough for adaptive circulatory changes to occur, because small areas of softening, presumably ischemic, are unusual.

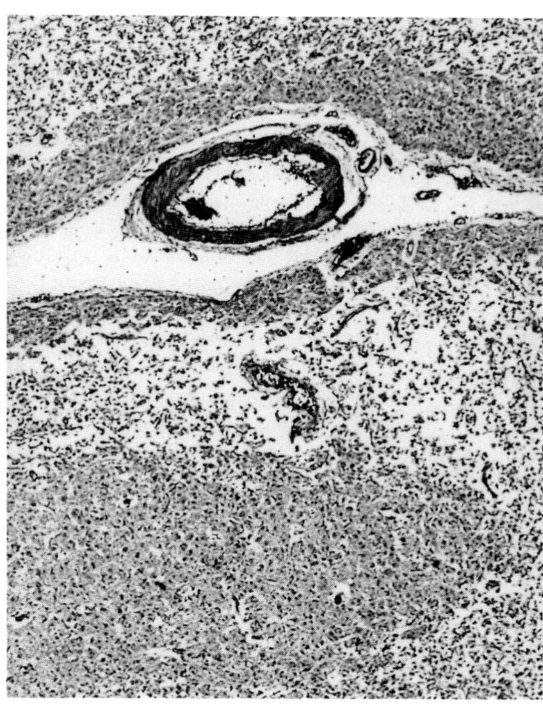

Fig. 3.43 Polioencephalomalacia of cerebral cortex. Foal. Neonatal maladjustment syndrome.

Hyaline necrosis in meningeal vessels, which is observed in swine with hepatosis dietetica, is not usually associated with secondary lesions. The hyaline necrosis in meningeal vessels in pigs with organomercurial poisoning is associated with severe cerebral injury, which may be in part due to reduced perfusion (Fig. 3.45). Amyloid degeneration of meningeal and intracerebral vessels occurs in aged dogs, and the thickenings of the meningeal arteries may be visible grossly. There may be, associated with the vascular lesions, senile argyrophilic plaques which contain amyloid deposits. The systemic distribution of amyloid is minimal. Petechial hemorrhages occur in relation to the diseased vessels in about 50% of cases and, exceptionally, there is massive hemorrhage.

Cerebrospinal angiopathy is an important cause of neurological disease in pigs. It occurs in the alimentary enterotoxemia known as edema disease of pigs with similar lesions in vessels of other tissues (Fig. 3.46), and in pigs of the particular age susceptibility to edema disease. The lesions are described in detail with Volume 2, Chapter 1, The Alimentary System. In groups of older pigs, arteritis or periarteritis can develop and may represent a chronic or persistent expression of the acute angiopathy.

Vascular permeability changes provide a basis for **annual ryegrass toxicosis** of sheep and cattle in Australia and South Africa. The distribution of the disease is governed in part by the distribution of annual ryegrasses, *Lolium rigidum* and hybrids, in winter rainfall areas, although the

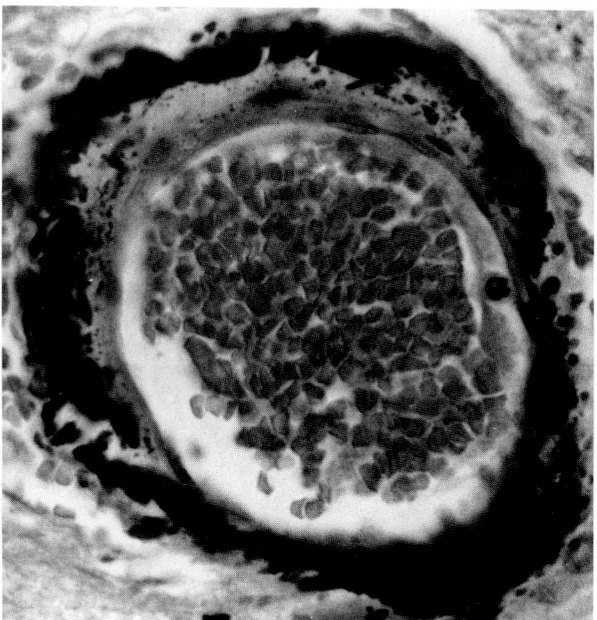

Fig. 3.44 Siderosis of cerebral vessel. Horse.

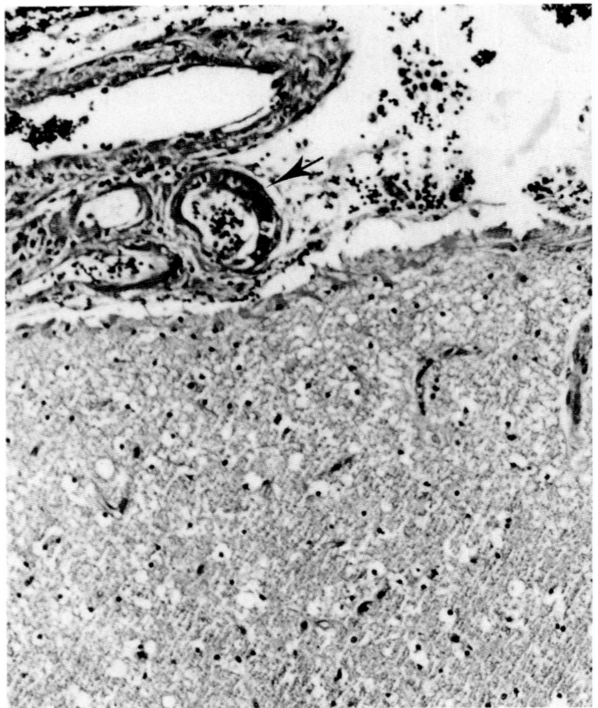

Fig. 3.45 Organomercurial poisoning. Necrosis of meningeal arterioles (arrow). Pig.

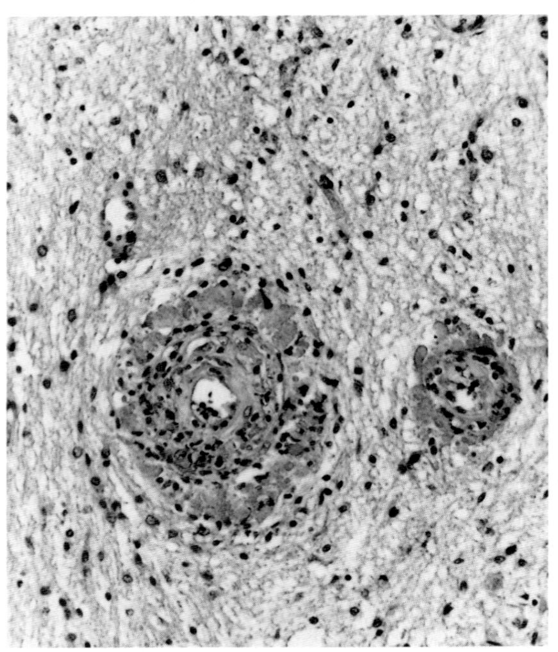

Fig. 3.46 Necrosis and inflammation of vessels in brainstem. Pig. Edema disease.

disease can occur in other areas by the use of transported fodder. The distribution is also governed by the distribution of toxin-producing *Clavibacter* (*Corynebacterium*) *rathayi* and infection by the bacterium of galls produced in the seed head by the nematode *Anguina agrostis*. In addition to *Lolium* spp., known host plants now include *Polypogon monspeliensis* (annual beard grass) and *Agrostis avenacea* (blown grass).

The nematode *Anguina agrostis* emerges from fallen galls following first autumn rains, migrates into the growing points of the ryegrass seedlings, and later penetrates the florets to produce galls. The nematodes are harmless to animals but may carry on the cuticle the bacterium which will proliferate in the gall, forming a yellow slime. The active principle, corynetoxin, is closely related to the tunicamycin antibiotics produced by some strains of *Streptomyces*.

The toxicosis is characterized clinically by neurologic signs and high mortality. Pregnant ewes may abort. The pathogenesis of abortion has not been examined, but the occurrence of hemorrhages in various organs, pulmonary edema, and swelling of endoplasmic reticulum of hepatocytes indicates a systemic intoxication, notwithstanding the prominence of neurologic signs.

The clinical signs, which are severe, include excitability, aggression in cattle, disturbances of gait, and convulsions. The clinical course may be less than 12 hr. Gross changes include pulmonary edema and a pale swollen liver, and occasionally there are hemorrhages in various organs.

Microscopic changes in nervous tissue are subtle, especially following routine fixation by immersion, which may not preserve transudates. Tracer injections show widespread alterations to cerebrovascular permeability in the brain and meninges, indicating endothelial damage and disruption of the blood–brain barrier. The perivascular transudate resembles plasma (Fig. 3.47) in staining properties and may be present as perivascular lakes or droplets which stain strongly with PAS stain. Only occasionally can fibrin be demonstrated, but extravasation of red cells from capillaries may be present in neuropil. Astrocytes are swollen with acidophilic cytoplasm, and there may be widespread necrobiosis of oligodendroglia in the cerebral gray matter.

The capillary endothelium shows ultrastructural evidence of injury. The endothelial cells are swollen and electron lucent, the cisternae of rough endoplasmic reticulum are distended, mitochondria are swollen with disorganization of cristae, and capillary lumina sometimes contain platelet aggregations. The changes are best seen in cerebellar cortex and meninges. Neuronal change is minimal, but there may be patchy loss of Purkinje cells and scattered small foci of malacia.

Cerebrospinal vasculitis, affecting both arterioles and veins, occurs in a number of specific diseases. Polyarteritis (periarteritis) nodosa appears to have some predilection for the cerebral arteries, especially in the pig, and for the spinal arteries in the dog. Vasculitis is a rather specific

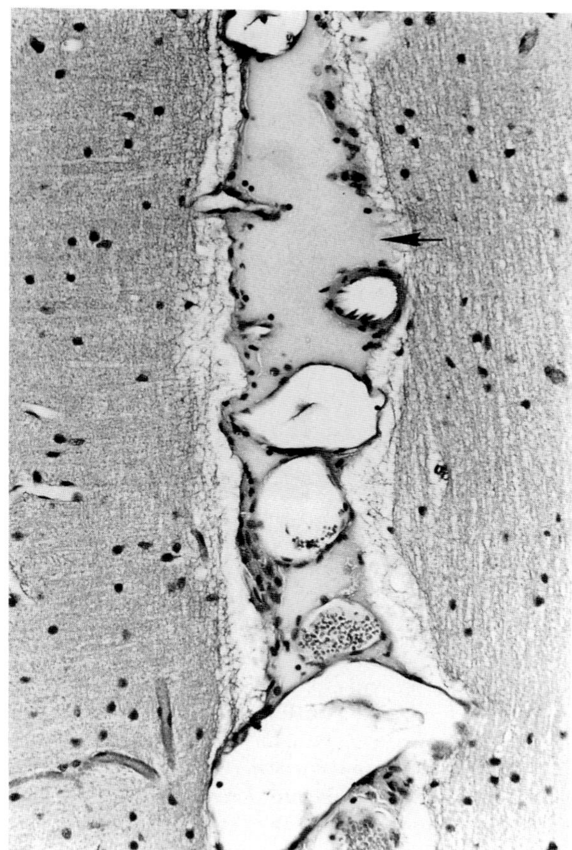

Fig. 3.47 Annual ryegrass toxicity. Vascular leakage (arrow). Leptomeninges. Cerebellum. Sheep. (Courtesy of J. McC. Howell.)

feature of hog cholera (Fig. 3.48A), sporadic bovine encephalomyelitis (Fig. 3.48B), and malignant catarrhal fever; the agents of these diseases have no predilection for neural tissue, and parenchymal degeneration, which is sometimes observed, is secondary to occlusion of vessels; this occlusion, in turn, is secondary to inflammatory lesions of the vascular adventitia. Adventitial proliferations and perivascular infiltrates are common to many of the encephalitides, both bacterial and viral, and focal softenings of the brain and cord in these inflammatory diseases can frequently be related to occlusive vasculitis. The occlusive lesions are not solely related to the acute phase of inflammation, but adventitial fibrosis and vascular stenosis may be prominent in the healed phase of encephalitis and may lead eventually to ischemic damage and softenings.

Thrombosis and **embolism** in the cerebrospinal arterioles is very seldom observed in animals. Thrombosis of an internal carotid vessel or ventral spinal artery may accompany atrial and aortic thrombosis, especially in cats (Fig. 3.49A). Bone marrow emboli form after trauma and fractures most commonly in dogs (Fig. 3.49B). Emboli composed of cartilage or nucleus pulposus occur in spinal arteries and veins in dogs and pigs and cause hemorrhagic

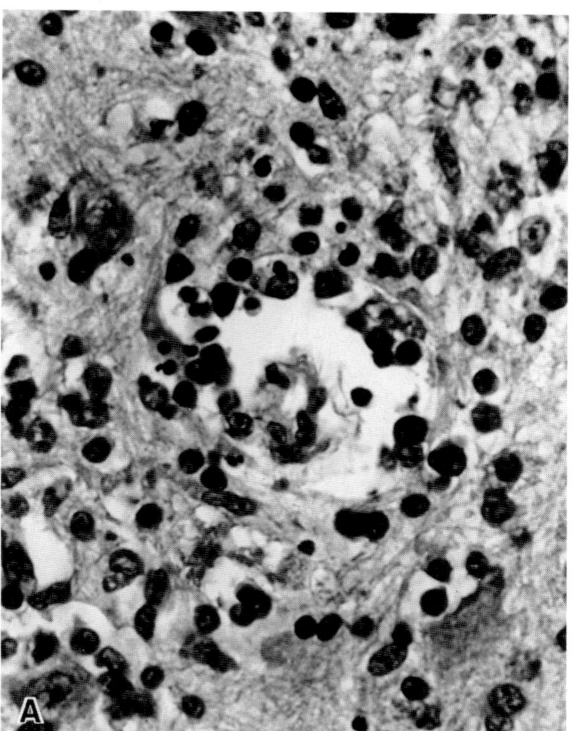

Fig. 3.48A Gliosis around degenerate vessel. Hog cholera.

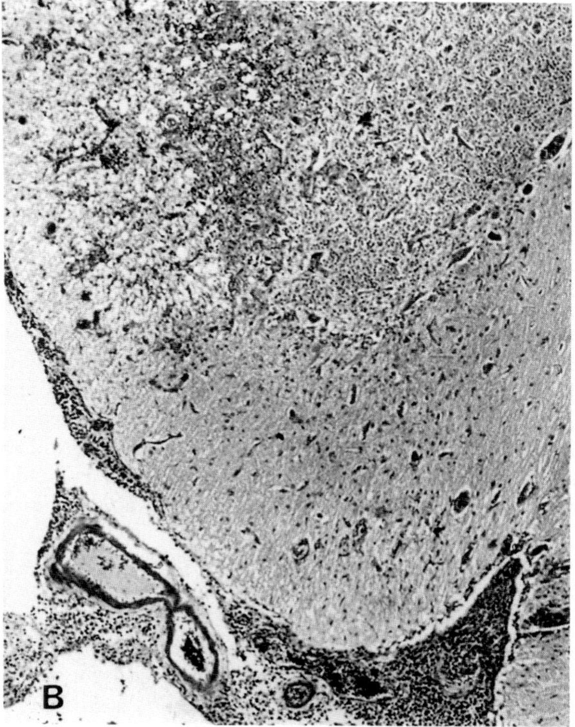

Fig. 3.48B Necrosis of cerebellar cortex and meningitis in sporadic bovine encephalomyelitis.

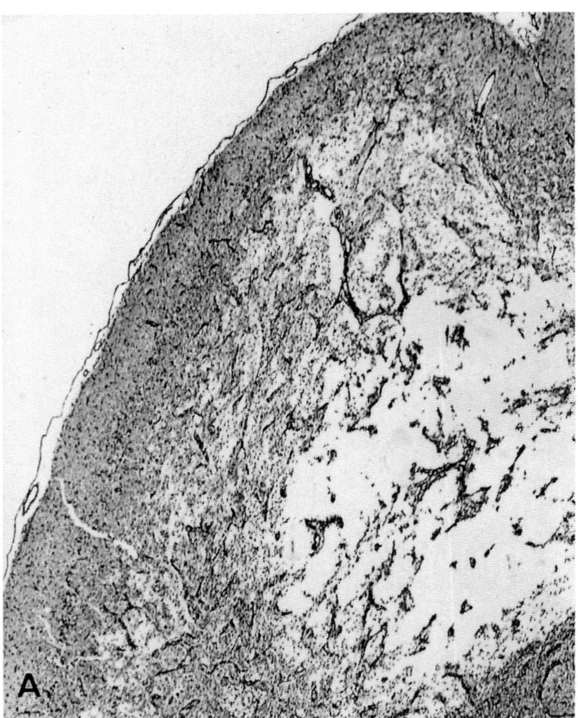

Fig. 3.49A Residual cyst in cerebral cortex, the result of carotid thrombosis and ischemic infarction. Cat.

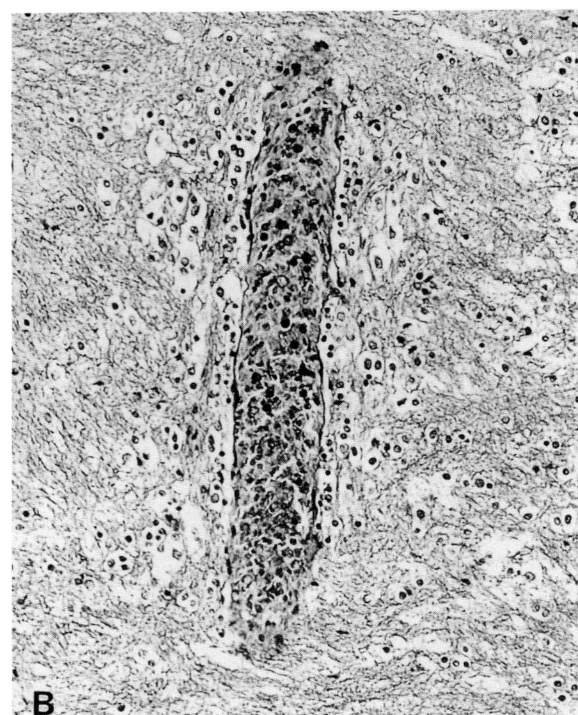

Fig. 3.49B Marrow embolus and surrounding intracellular edema following trauma and femoral fracture. Dog.

and ischemic infarcts of sudden onset (Fig. 3.50). The cervical or lumbar segments may be affected. The pathogenesis is discussed in Chapter 1, Bones and Joints. Bacterial thrombi may develop in a variety of bacteremic diseases such as erysipelas, shigellosis, pasteurellosis, and the septicemias caused by *Haemophilus, Streptococcus,* and the coliforms. Abscessation develops rapidly about these bacterial colonies, but in the early stages, perivascular zones of softening may be observed.

Perhaps the least morphological expression of arteriolar obstruction is death of neurons in a narrow zone of cortex representing the central portion of a field of distribution of a single arteriole. In the early stage, there is acute ischemic necrosis of neurons and oligodendroglia. There is no softening in these minimal lesions, but they are readily visible at low power as narrow zones of pallor. At higher magnification, the affected zone is seen to be unusually cellular, this being due in the early stages to microgliosis and later to astrogliosis.

Obstruction of an artery or arteriole results typically in infarction, and infarcts involve both gray and white matter. In the early stages, affected areas swell, and the involved gray matter may be hemorrhagic (Fig. 3.51). At the center of an infarct, there is coagulative and liquefactive necrosis, and around the periphery, there is a zone of minimal ischemic injury with the characters previously mentioned. Many capillary vessels survive, partly because of their relative insensitiveness to anoxia and partly because they benefit first from the collateral circulation which develops. Surviving vessels show endothelial and adventitial proliferation, and the astrocytes around the margin of the lesion react. The necrotic tissues are removed from the periphery by liquefaction and microglial phagocytosis, and a cyst remains (Figs. 3.46, 3.49A). The cyst contains fluid, and distended microglia, so-called compound granular corpuscles, may be found in it for very long periods. The wall of the cyst is ragged and irregular and is formed by reactive astrocytes. When the defect extends to the meninges, it may be partially filled in by proliferated leptomeningeal tissue.

Obstruction to cerebrospinal veins is the result of either pressure or inflammation with thrombosis, although rarely it may be due to neoplastic invasion and permeation of the dural sinuses. Thrombophlebitis is usually bacterial, and the associated meningitis is more conspicuous and important than the venous thromboses; isolated venous obstructions are not of much significance.

Intracranial thrombophlebitis can be due chiefly to retrograde spread of inflammation from an extracranial primary focus, but thrombophlebitis of this type is uncommon. Such primary foci may occur in the orbital or nasal cavities, the paranasal sinuses, or the middle ear. Intracranial inflammations have little tendency to cause thrombosis of the dural sinuses. In verminous infestations of the brain, especially those caused by *Strongylus* larvae in horses, there is some tendency for the larvae to migrate

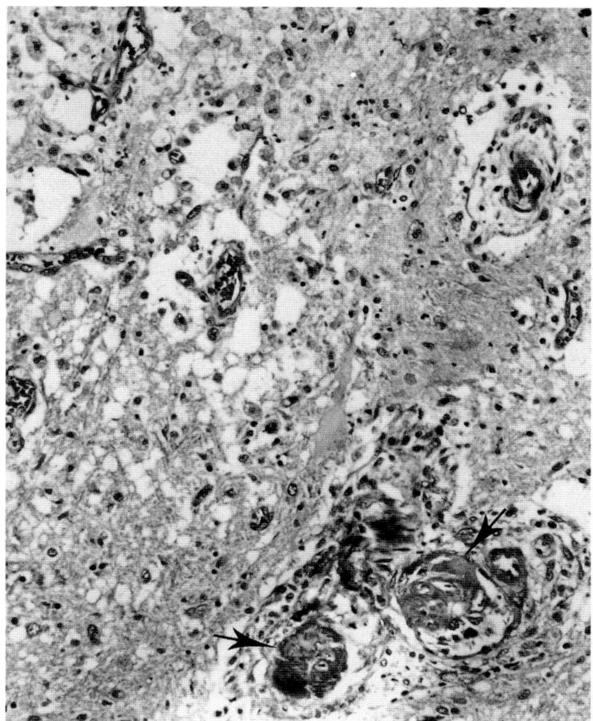

Fig. 3.50 Cartilaginous emboli (arrows) causing spinal myelomalacia. Dog.

to the superior sagittal and transverse sinuses and to cause thrombosis of these. Spinal thrombophlebitis results from ascending infections from docking wounds in lambs and from bite wounds of the tail of pigs.

Noninflammatory thrombosis of the cranial dural sinuses occurs chiefly in the superior sagittal sinus. It occurs in polioencephalomalacia of cattle, in which it is probably secondary to severe and prolonged swelling of the brain. Sinus thrombosis may also be observed following head injuries, even those not accompanied by lacerations (Fig. 3.52).

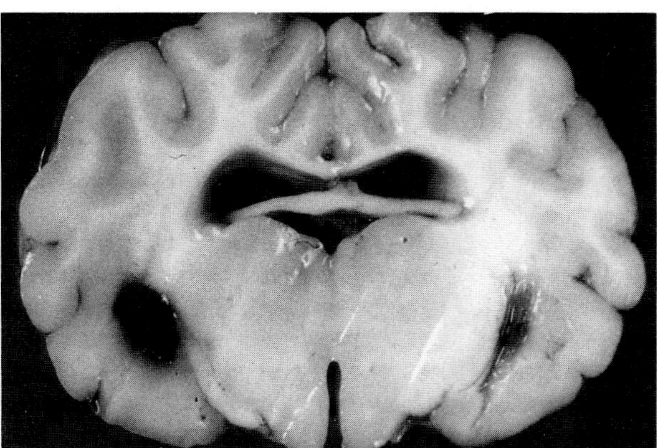

Fig. 3.51 Hemorrhage (left) and malacia (right) in cerebrum. Pig. Mulberry heart disease.

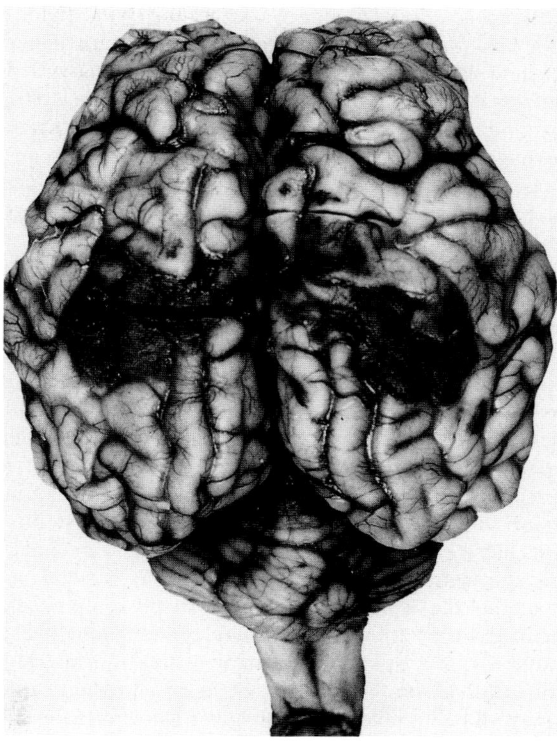

Fig. 3.52 Venous infarcts in cortex secondary to thrombosis of sagittal sinus. Calf.

Venous infarcts differ from arterial infarcts by the more extensive hemorrhage in the affected areas and by the diffusion of blood into the adjacent subarachnoid space. The superficial veins are engorged, and the congested area is edematous and swollen. The hemorrhages are perivenular and petechial or larger. They are present in the gray matter and to a lesser extent in the white matter. The subsequent course and reaction are similar to those following arterial obstruction, and the residual lesion is a depressed and shrunken or cystic area, which is dark brownish-yellow.

B. Hemorrhagic Lesions

Hemorrhages in the central nervous system may be traumatic or spontaneous and may be restricted to the meninges or the parenchyma or involve both. The causes are, in general, the same and are of much variety. They can be broadly grouped into those affecting the integrity of the vessel wall and those which reduce the coagulability of the blood. Care is necessary in deciding whether observed extravasations are significant because they can be produced readily as artefacts at postmortem. Quite extensive spread of blood into the basal meninges commonly occurs from the venous sinuses when the head is removed at the atlanto–occipital junction, and into the spinal meninges when the spinal column is transected. Hemorrhages of petechial size in the parenchyma can largely be avoided as artefacts if the brain or cord is fixed before slicing. The microscopic distinction of antemortem from postmortem hemorrhages can also be difficult because, apart from minor degenerative changes of the parenchyma, terminal hemorrhages provoke very little reaction, even on the part of the microglia.

Spontaneous hemorrhages in the brains of animals are almost invariably of petechial or slightly larger size, occurring in the hemorrhagic diatheses, especially the symptomatic purpura of septicemic infections (see Volume 3, Chapter 2, The Hematopoietic System). Such hemorrhages occur typically from capillaries as well as small venules and are found in meninges and brain. There is no specificity in the geographic distribution of purpuric hemorrhages except in infectious canine hepatitis, in which intracranial hemorrhage, although not consistently present, has a predilection for the midbrain and medulla. In focal symmetrical encephalomalacia (see clostridial enterotoxemia) the lesions are grossly visible only when hemorrhagic; they are regularly bilateral and symmetrical and have a rather specific pattern in which the internal capsules, dorsolateral thalamus, cerebral aqueduct, and pons may all be involved. Hemorrhages also characterize thiamine deficiency in dogs and cats and have a specific distribution and symmetry, being regularly present in the inferior colliculi and less consistently present in the mammillary bodies and other periventricular nuclei.

Isolated hemorrhages or hematomas are quite rare in animals. The hemorrhages may extend to the ventricles and induce hydrocephalus, or involve the central canal of the cord to produce hematomyelia.

Meningeal hemorrhages, both epidural and subdural, are common in lambs and calves which have required obstetric assistance and probably reflect hemodynamic disturbances in dystocia. Hemorrhages of similar distribution occur in calves born unassisted from cows exposed to dicumarol from moldy sweet clover.

C. Microcirculatory Lesions

A range of changes affects the microcirculation in the nervous system and may not produce ischemia or hemorrhage. They are associated with extravasation of formed or fluid elements, are not disease specific in their character although their distribution may be specific, and result from degenerative changes, especially in capillaries and venules.

Diapedesis of red cells is a common event in sudden death of many causes and is frequently mimicked in postmortem trauma. The interval between diapedesis and death may be too short for reactive changes, visible by light microscopy, to develop in the vessels. With some delay, swelling of endothelial cells and their nuclei develops. The hemorrhages may be from arterioles, with the red cells remaining in the perivascular space. Diapedesis from capillaries and small venules depends on disruption of endothelium and astrocytic processes and, depending on relative pressures, the red cells may spread in the intercellular spaces.

Diapedesis of red cells occurs in periventricular nuclei in thiamine-deficiency encephalopathy, and in the brain stem in concussion, contrecoup injury, and in hypomagnesemia in sheep and cattle (Fig. 3.53). The white matter of the cerebral and cerebellar hemispheres is susceptible to diapedesis. The hemorrhages are sometimes large enough to be visible as petechiae in diverse conditions including infectious purpuras, anoxia, fat and other microembolism, in diseases associated with erythrocyte sludging, and in disseminated intravascular coagulation.

Leakage of plasma, usually from small venules, may occur in any part of the central nervous system. It is seen most frequently surrounding areas of traumatic injury or in the previously noted conditions which lead to erythrocyte diapedesis. The plasma will contain some fibrinogen, and polymerized fibrin in its usual form may be demonstrated around the vessels. It is more usual, however, for the fibrin to be in its unpolymerized or molecular form, and the plasma both within and outside the vessels to be transformed into a homogeneous gel which stains brightly by the PAS technique. The plasma gel may appear as capillary plugs or as perivascular droplets, which appear to incite very little reaction.

Vascular injury which allows diapedesis of plasma will

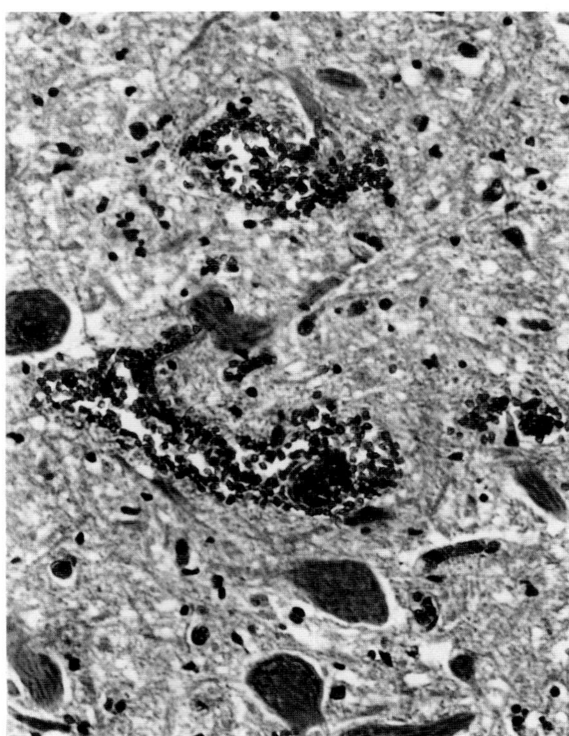

Fig. 3.53 Perivascular hemorrhage. Cow. Hypomagnesemia.

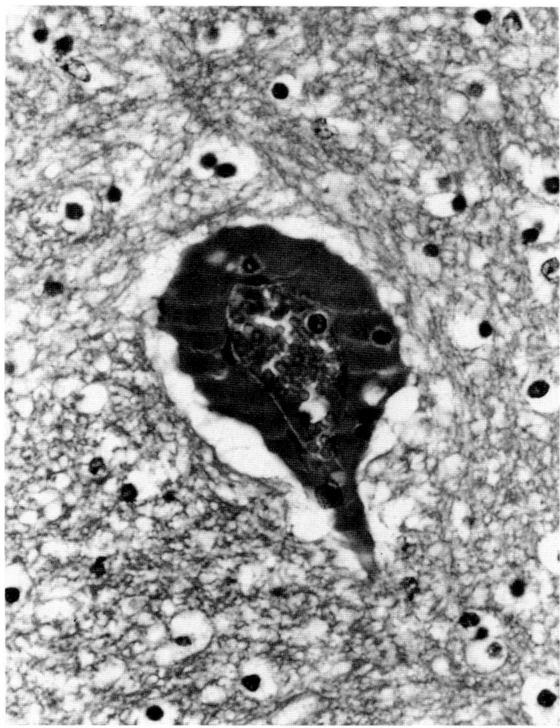

Fig. 3.54 Accumulation of proteinaceous fluid around vessel. Sheep. Focal symmetrical encephalomalacia.

also allow passage of serum or protein filtrate (Fig. 3.54). In fields of injury, the differences in levels of permeability change may be evident from the protein content of transudate. Although plasma will routinely stain because of its fibrinogen content, the amount of protein in serum may be too low for preservation and staining (Fig. 3.40). The perivascular changes occur especially about capillaries and venules in the white matter and have the characteristics described earlier for vasogenic edema. Prolonged survival or recovery may leave focal glial scars as residue (Fig. 3.27B).

Bibliography

Balentine, J. D. Pathology of experimental spinal cord trauma. I. The necrotic lesion as a function of vascular injury. II. Ultrastructure of axons and myelin. *Lab Invest* **39:** 236–253, 254-266, 1978.

Berry, P. H., Howell, J. M., and Cook, R. D. Morphological changes in the central nervous system of sheep affected with experimental annual ryegrass (*Lolium rigidum*) toxicity. *J Comp Pathol* **90:** 603–616, 1980.

Berry, P. H. *et al.* Central nervous system changes in sheep and cattle affected with natural or experimental annual ryegrass toxicity. *Aust Vet J* **56:** 402–403, 1980.

Berry, P. H. *et al.* Hepatic damage in sheep fed annual ryegrass, *Lolium rigidum*, parasitized by *Anguina agrostis* and *Corynebacterium rathayi*. *Res Vet Sci* **32:** 148–156, 1982.

Braund, K. G., and Vandevelde, M. Polioencephalomalacia in the dog. *Vet Pathol* **16:** 661–672, 1979.

Castejon, O. J. Electron microscopic study of capillary wall in human cerebral oedema. *J Neuropathol Exp Neurol* **39:** 296–328, 1980.

Cook, J. R. Fibrocartilaginous embolism. *Vet Clin North Am: Small Anim Pract* **18:** 581–591, 1988.

Dahme, E., and Schroder, B. Kongophile Angiopathie, cerebrovasculare Mikroaneurysmen and cerebrale Blutungen beim alten Hund. *Zentralbl Veterinaermed (A)* **26:** 601–603, 1979.

de Lahunta, A. Feline ischemic encephalopathy—a cerebral infarction syndrome. *In* "Current Veterinary Therapy," VI, R. W. Kirk (ed.), pp. 906–908. Philadelphia, Pennsylvania, Saunders, 1977.

Finnie, J. W., and O'Shea, J. D. Effect of tunicamycin on the blood–brain barrier and on endothelial cells in vitro. *J Comp Pathol* **102:** 363–374, 1990.

Griffiths, I. R. Some aspects of the pathology and pathogenesis of the myelopathy caused by disc protrusions in the dog. *J Neurol Neurosurg Psychiatry* **35:** 403–413, 1972.

Harding, J. D. J. A cerebrospinal angiopathy in pigs. *Pathologia Vet* **3:** 83–88, 1966.

Hartigan, P. J., and Baker, K. P. Clinical and histopathological observations on the recovery of pigs affected with cerebrospinal angiopathy. *Ir Vet J* **28:** 197–200, 1974.

Hayes, M. A. *et al.* Acute necrotizing myelopathy from nucleus pulposus embolism in dogs with intervertebral disk degeneration. *J Am Vet Med Assoc* **173:** 289–295, 1978.

Hoff, E. J., and Vandevelde, M. Case report: Necrotizing vasculitis in the central nervous systems of two dogs. *Vet Pathol* **18:** 219–223, 1981.

Schneider, D. J. First report of annual ryegrass toxicity in South Africa. *Onderstepoort J Vet Res* **48:** 251–255, 1981.

Stynes, B. A. *et al.* The production of toxin in annual ryegrass, *Lolium rigidum*, infected with a nematode, *Anguina* sp., and *Corynebacterium rathayi*. *Aust J Agric Res* **30:** 201–209, 1979.

Tengvar, C. Extensive intraneuronal spread of horseradish peroxidase from a focus of vasogenic edema into remote areas of central nervous system. *Acta Neuropathol* **71:** 177–189, 1986.

Terlecki, S., Baldwin, B. A., and Bell, F. R. Experimental cerebral ischaemia in sheep. Neuropathology and clinical effects. *Acta Neuropathol* **7:** 185–200, 1967.

Tessaro, S. V., Doige, C. E., and Rhodes, C. E. Posterior paralysis due to fibrocartilaginous embolism in two weaner pigs. *Can J Comp Med* **47:** 124–126, 1983.

Uchida, K. *et al.* Amyloid angiopathy with cerebral hemorrhage and senile plaque in aged dogs. *Jpn J Vet Sci* **52:** 605–611, 1990.

Walker, A. E., Diamond, E. L., and Mosely, J. The neuropathological findings in irreversible coma. A critique of the "respirator brain." *J Neuropathol Exp Neurol* **34:** 295–323, 1975.

Wisniewski, H. M. *et al.* Senile plaques and cerebral amyloidosis in aged dogs. *Lab Invest* **23:** 287–296, 1970.

VI. Traumatic Injuries

The importance of traumatic injuries to the head and vertebral column is in relation to the effects that such injuries have on the contained brain and spinal cord. Both the brain and spinal cord are well protected from external injurious forces by the bony encasements. The more or less rounded shape of the skull favors glancing blows and the lateral diffusion of lines of force; the diploe and sutures are capable of absorbing considerable amounts of shock;

and the internal system of bony ridges directs lines of force to the base of the skull. The spinal column is well protected by the surrounding soft tissues, by its own highly cancellous structure, system of ligaments, and intervertebral disks. The nature of acquired injuries is quite various and is determined by many factors, including the relative vulnerability of the soft tissues, the direction in which the injurious force is applied, the physical rigidity of the bones, the ability of the part to move in response to the applied force, and the mass and velocity of the force.

A. Concussion

Concussion is a transient loss of consciousness and reflex activity following a sudden injury to the head. Full recovery is expected, and it is assumed that in mild cases there is no morphological injury. In experimental animals, however, degenerative changes are found in the nuclear masses of the brain stem, and these are probably responsible for the clinical features. Following single injuries, chromatolysis develops in many of the larger neurons of the brain stem, and in a proportion of affected cells, the degeneration is progressive. With repeated episodes of concussion, affected neurons have a wider distribution, which includes the cerebral cortex, and the proportion of cells irreversibly injured is increased. These changes in the brain stem and reticular formation occur if the impact forces are substantial and the brain is subject to rapid acceleration/deceleration; in these cases direct injury in the hemispheres is to be expected. The mildest degrees of concussion produce no visible structural damage when examined with the light microscope, but there is some experimental evidence that plasma substances including proteins may be transported in vesicles through the endothelial cells, and the brain and cord may lose their capacity for autoregulation of blood flow in local areas of static injury.

Although the dynamics of the disturbances in neurons which are responsible for the concussion are unclear, some of the physical qualities of the reaction between the head and the applied force have been elucidated. The vibrations which travel back and forth in the brain from the point of impact appear not to be important. If the head is firmly immobilized, quite a considerable force may be suddenly applied to it with relatively minor effect, the force being absorbed by skull. Force of much lesser magnitude will cause concussion if the head is capable of moving in response to the blow. The principle to be obtained from these observations is that a degree of acceleration or deceleration is necessary to produce concussion, the neural injury being due to displacement of the cranium relative to its contents. The brain of an adult animal is normally a little smaller than the cranial cavity, and being suspended in the cavity, is capable of slight independent movement. If the head is freely movable and struck a heavy blow, it moves away from the point of impact and collides at the area of impact with the brain, which is momentarily static. If the skull and brain are viewed as an isolated system, it may be said that the brain moves suddenly toward the point of impact. In young animals, the brain fairly fits the cranial cavity so that relative displacement of the whole brain is expected to be minimal. The brain is, however, plastic, so that sudden acceleration or deceleration leads to internal deformation. The manner in which displacements and transient deformations of the brain result in the neuronal changes and unconsciousness is not known, but the explanation probably lies in the effect of shearing strains on neurons, axons, dendrites, synapses, and blood vessels, as well as direct pressure effects when the brain is displaced across bony ridges.

B. Contusion

In contusion, the architecture of the nervous tissue is retained, but there is hemorrhage. The hemorrhage occurs into the meninges and about the blood vessels in the parenchyma. Contusions may be diffuse or focal injuries, although often those coexist.

The pathogenetic factors in diffuse contusions are, in general, the same as those operating in causing concussion, but the applied force, the displacements, and the induced shearing and direct forces are all of greater magnitude. Typically in the diffuse injuries, some of the most severe hemorrhages occur on the surface of the brain opposite the point of impact. Hemorrhages of this distribution are known as **contrecoup,** and their development depends on the sudden movement of the brain to the point of impact with tension and tearing of pial and cortical vessels opposite the point of impact, and with direct or rotational displacement of brain over bony prominences, where the vessels become exposed to tensile and shear forces.

Focal contused injuries develop typically at the point of impact, and the mechanism of development is somewhat different from that of the diffuse injuries. In focal injuries, movement of the head is not sufficient to cause significant displacement of the brain. Instead, the applied force, usually with relatively high velocity and relatively low mass, causes fracture or deformation of the skull at the point of impact. The deformation of the skull may be only transient but sufficient to cause bruising of the tissue immediately beneath.

C. Laceration

Laceration is a traumatic injury in which there is disruption of the architecture of the tissue. The mechanics of lacerations are in general the same as those of contusion. Penetrating injuries are analogous to, but more severe than, those which result in focal contusions. Lacerations may also occur with blunt injuries of the type which cause displacement of the brain. In such cases, contrecoup lacerations occur typically on the surfaces of gyri, where these are displaced over bony prominences. Shear forces developed during deformation or molding of the brain may be adequate to cause deep hemorrhages and even the cleavage of the gray from the white matter over small areas of the cortex.

Lacerations caused by penetrating injuries are always liable to secondary infections, especially when fragments of skin and soft tissue and spicules of bone from the internal plates are displaced into the brain. In the absence of infection, repair takes place in the manner that is usual for defects of nervous tissue. The detritus of blood and nervous tissue is removed by microglia and meningothelial macrophages. If the defect is small, an astrocytic scar forms, but usually a cyst remains, lined by proliferated astrocytes. Adhesions between the glial tissue and pia mater produce meningocerebral scars, which are notably epileptogenic.

D. Fracture of the Skull

Fractures of the skull can be important in terms of the concurrent injuries to the underlying meninges and brain and because they can provide a pathway of infection to the sinuses, meninges, and brain. They are indicative of injuries produced by considerable force and, when the inner plates are fractured, of contusion or laceration of the underlying brain. Fractures of the base of the skull may involve the middle ear and allow the escape of cerebrospinal fluid and the entrance of infection. Frontal fractures involving the cribriform plate may allow cerebrospinal fluid to escape into the nasal cavity.

Fractures of the skull are usually quite easy to detect by virtue of the displacement of bone and meningeal hemorrhage. Some, however, may be difficult to detect. This applies especially to fissures which develop during transient deformation of the skull and to impacted basilar fractures, such as occur in horses which rear backward and strike the nape of the neck and occiput. With these fractures, hemorrhages may be scant, and basilar fractures are revealed only when the dura is carefully dissected away.

E. Injuries to the Spinal Cord

It is possible for direct injuries to the spinal cord to occur without obvious injury to the vertebrae. Such spinal injuries are necessarily lacerations caused by penetrating foreign bodies. Wandering parasites as causes of traumatic injury are discussed later. Much more common are indirect injuries to the cord acquired in the course of vertebral luxations or fractures with dislocation. The most common of indirect injuries to the cord are produced by extruded nucleus pulposus in dogs, and by compression of the cervical cord in the syndrome known as wobbler (see Chapter 1, Bones and Joints).

Vertebral subluxations are the result of trauma, but fracture dislocation may be pathological as well as traumatic, and frequently is in swine and ruminants when the fracture occurs through an area of vertebral osteomyelitis. The injuries to the cord occur chiefly at the time of accident and are due chiefly to the stresses to which the cord is subjected and partly to impediment to the blood supply. Cumulative injury may occur especially in dislocations, when continued pressure on the cord or stretching of the cord over bony prominences causes intermittent ischemia.

Subluxations are largely restricted to the cervical column, where there is relative mobility of the ligaments. In the thoracic and lumbar spine, comparable forces are more likely to cause fracture because of the brevity of the ligaments. The fractures may involve only a lamina, the odontoid process, or an articular process, and therefore be difficult to detect. Fracture dislocations of the vertebral column occur chiefly in the caudal cervical region and about the thoracolumbar junction. Most injurious are those fractures which involve the vertebral body because there is usually displacement of a fragment posterodorsally into the spinal canal. Both in luxation and fracture displacement, pressure may be exerted over several segments of cord by epidural and endodural blood clots.

Traumatic injuries to the spinal cord may be slight enough to be satisfactorily recoverable, or they may, at the other extreme, cause transection and an extensive length of necrosis. Early contused lesions cause swelling of the cord over one or two vertebral segments. The swelling may be easier to palpate by stroking the cord than to appreciate by inspection. Later, the cord is shrunken at the level of injury, and this also may be best detected by palpation. In the swollen zone, dark points of hemorrhage may be detected on section, and these occur usually in the central gray matter and at its junction with the white. If the demarcation of gray from white matter is obscured, it can be expected that softening of some degree is present. Cores of softening may be recognizable grossly, and the fact that they are extending up and down the cord from the point of injury is explained as a tracking of exudate, especially serum or edema fluid, along the fiber tracts and the path of least resistance since, because the meninges are usually intact, the amount of local swelling is limited.

With severe compression, the cord may be entirely necrotic at the point of injury and over several segments. Necrosis is probably due to vascular injury, and, for the same reason, isolated segments of necrosis may occur apart from the main injury. When the cord is necrotic, it is initially swollen, but after a lapse of some weeks, the dura forms a narrowed, collapsible tube containing debris.

Microscopically, in the mildest injuries, there is swelling of axons and myelin sheaths. The injured axons become beaded, but they are possibly not irreversibly injured unless fragmentation follows the beading, because functional recovery can follow minor and transient paralytic injuries. Where the axons are severed, there is bulbous swelling of the retracted ends. Degenerative neuronal changes consisting of the clumping of Nissl substance and central chromatolysis are constantly present. Ultimately, there is some loss of neurons. When softening occurs, the sequence of changes is the same as occurs in any neural lesion with loss of substance. The hemorrhage is usually slight and perivascular, but it may enter the central canal and pass along it for a considerable distance.

In long-standing spinal injuries, adhesions may be found between the meninges and between the dura mater and periosteum. In the injured zone of cord, there may be loss of fibers and myelin sheaths with astrogliosis or, in the more severe injuries with loss of substance, there may be cysts, pial–glial or collagenous scars, or the cord may be converted to a thin sclerotic band.

VII. Degeneration in the Nervous System

The foregoing sections dealing with cytopathology and the effects of circulatory disturbances and trauma are discussions of changes which are largely degenerative. The same will be true of many sections in later parts of this chapter because neural injury, irrespective of cause or severity, is characterized chiefly by degeneration. Selected for discussion in this section are those lesions and diseases of the central nervous system that are expressed as degenerations of nervous tissue.

A. Meninges

Age changes in the meninges may advance to such degree as to be pathological. Collagenous and osseous metaplasia occurs commonly in the cranial dura mater of dogs and cats. The process begins in the frontal region and extends to the temporal and occipital areas. The plaquelike thickenings of the surface meninges tend not to extend to the basal or cerebellar meninges. The ossification, which is detectable as small intradural plaques, follows hyalinization of the connective tissues and may or may not be preceded by mineralization. The dura becomes firmly adherent to the periosteum. Perhaps by an independent process, spherical calcified nodules form in the basal dura beneath the medulla in some aging dogs. Histologically, these nodules show a whorled pattern of fibroblastic cells often enclosing small concrements known as **psammoma bodies.**

Hyalinization of dural collagen occurs also in the spinal dura mater with aging and may be preliminary to the osseous metaplasia of the dura observed frequently in large breeds of dogs. The latter condition is referred to as **ossifying pachymeningitis** but, in spite of the frequency of the dural change, little is known of its pathogenesis or effects. Dural ossification is found in older age groups of some large breeds and varies in its expression from multiple separate plaques to rigid ossification over several segments of the cord. The changes are best developed in the lumbar region but may involve much of the spinal dura. The basic change appears to be degenerative and metaplastic rather than inflammatory. Locomotor dysfunction in affected animals is more likely to be attributable to concurrent spondylosis and degenerative/reactive changes in joints. The plaques consist of lamellar bone which is mature in appearance and which, from the earliest stages, contains cancellous spaces and active bone marrow.

Degenerative changes in leptomeninges are less well identified than those in the dura and are limited to collagenization and hyalinization. They are not functionally significant, but the opacities produced, especially the focal thickening in arachnoid granulations in old animals, can be misinterpreted as inflammatory change.

B. Choroid Plexuses

With advancing age, there is hyaline degeneration of the connective tissues of the choroid plexuses, especially those of the lateral ventricles (Fig. 3.55). The basement membranes of the capillaries are also involved. The hyaline tissues may become calcified.

Cholesteatosis of the choroid plexuses in the form of tumorlike nodules, usually termed cholesteatomas or cholesterol granulomas, occurs in 15 to 20% of old horses. They are more frequent in the plexuses of the fourth ventricle than in the lateral ventricles, but those in the lateral ventricles are the more important because they may attain large size and, by obstructing the foramina of Monro, cause hydrocephalus leading to dilation and pressure atrophy of the walls of the ventricles (Fig. 3.56A).

The development of cholesteatosis appears to be related in some manner to chronic or intermittent congestion and edema with congestive hemorrhages in the choroid plexuses. During an edematous episode, the plexuses are

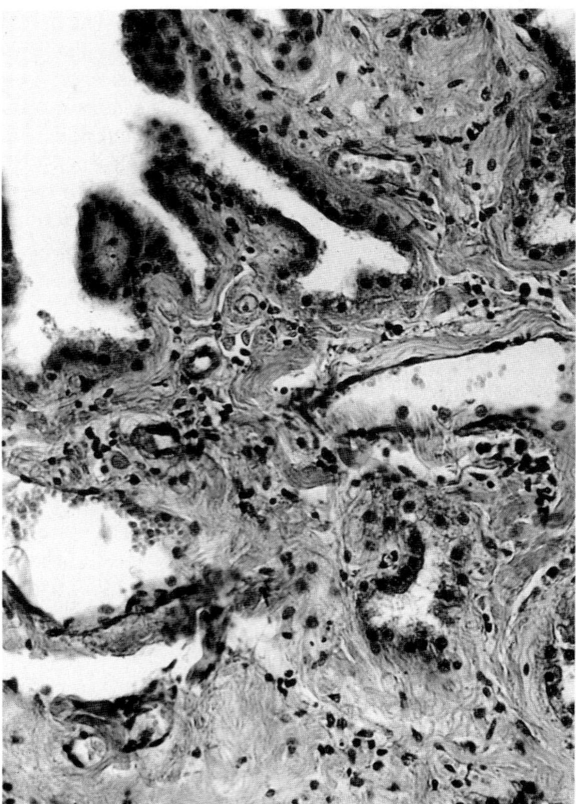

Fig. 3.55 Senile hyalinization in choroid plexus. Horse.

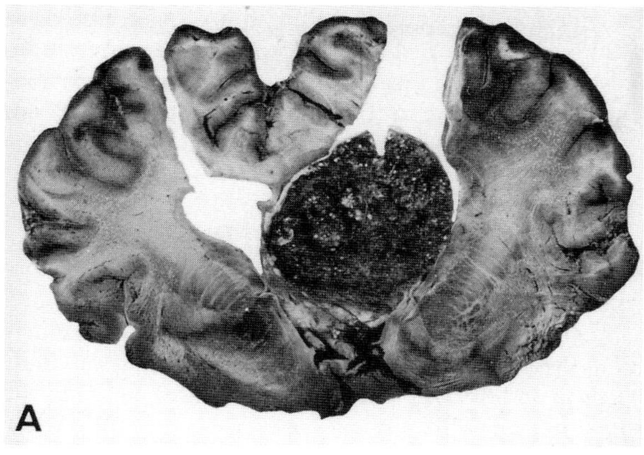

Fig. 3.56A Cholesterol granuloma in choroid plexus of right lateral ventricle. Horse. (Courtesy of P. Olafson.)

slightly swollen, yellow, and soft. The interstitial tissues are edematous and infiltrated lightly by macrophages containing lipid and hemosiderin. The crystals of cholesterol are deposited in the tissue spaces and apparently act as foreign bodies to stimulate a low-grade productive inflammation (Fig. 3.56B).

The affected plexus is swollen, and small areas or the whole of it may be occupied by the cholesterol granulomas. These are firm but crumbly, grayish nodules with a gleaming, pearly appearance on the cut surface. The crystals can be expressed from the cut surface by slight pressure.

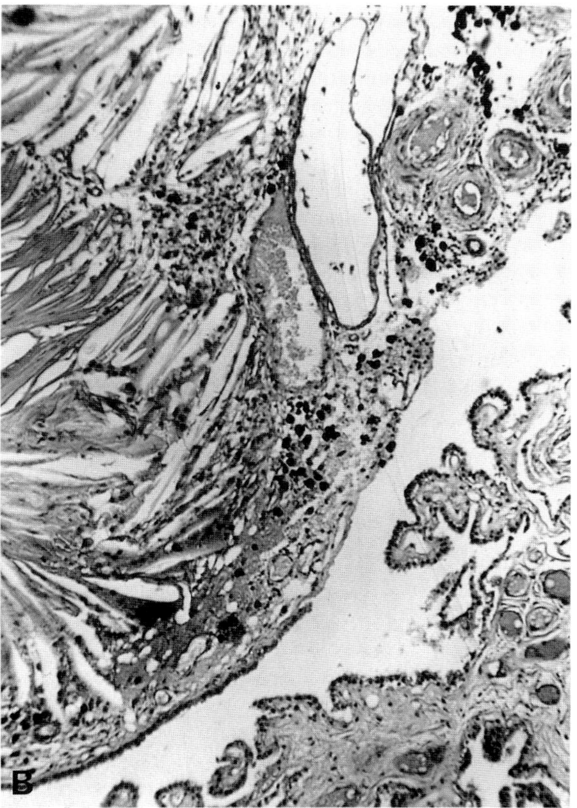

Fig. 3.56B Cholesterol granuloma in choroid plexus. Horse.

C. Atrophy in the Brain and Spinal Cord

Senile atrophy of the brain and cord is frequently obvious but seldom severe in animals. The atrophy is most marked when it is associated with the combined influences of senility and marasmus, and it occurs more frequently in sheep than in other species. The pachymeninges are thickened, and those of the cranium are partially adherent to the periosteum. The leptomeninges are thickened, tough, and granular. The brain is reduced in volume and weight and may be discolored a darker grayish-yellow. The convolutions are narrowed and wrinkled, and the sulci are shallow and widened. Both brain and cord are much firmer than is normal. The ventricles are moderately dilated. Histologically, the neurons, especially of the cortex, are shrunken, distorted and dark staining and their numbers are depleted. The amount of Nissl substance is reduced, and the neuronal nuclei stain diffusely. Lipochromes are increased in amount, and satellitosis is prominent. There is a diffuse increase in glial fibers, which is responsible for the increased firmness of the tissue. The gliosis is most prominent about Virchow–Robin spaces, and there is a general tendency for protoplasmic astrocytes to transform into fibrous ones. The Virchow–Robin spaces are dilated, and some may be visible to the naked eye. The media of the vessels contain increased amounts of collagen.

Focal atrophy of the brain and cord is the result of slight but long-continued pressure. The loss of function is often disproportionately large, as the result of acute pressure lesions, compared with the extent of pathologic change. The reverse is often true in chronic pressure, and function may be retained to a surprising degree. Hydrocephalus is common as a cause of chronic pressure, and other sources of pressure are space-occupying lesions of the meninges or bony vault.

The pathogenesis of the changes produced by chronic pressure is unclarified. Because the immediate effects of pressure are more likely to be felt by veins than by arteries or solid tissue, the atrophic changes are usually attributed to venous stasis and edema. More acute pressure effects grade into circumstances and consequences discussed under trauma. Whatever the mechanism, myelin is the most vulnerable component of the tissue and undergoes slow progressive degeneration. There is also slow progressive atrophy and loss of neurons, especially of the Purkinje cells when pressure is exerted on the cerebellum; the neurons of the deeper laminae when pressure is exerted on the cerebrum; and of the ventral horn cells in the spinal cord. The degenerating myelin is replaced by glial fibers.

Segmental cerebellar atrophy is common in pigs following a variety of nervous diseases. The anterior vermis and adjacent portions of the lateral lobes are affected, these being within the distribution of the anterior cerebellar artery. There is no acute change but, instead, a gradual loss of granular and Purkinje cells with secondary changes in the white matter and reactive gliosis. The lesion is asymptomatic.

D. Anoxia and Anoxic Poisons

The neuropathological effects of anoxia are seldom observed in animals or perhaps seldom looked for. Since most cases die acutely, there may be very little to see. The pathological effects of anoxia, not only in the central nervous system, but in general, are expressions of "vulnerability," a property which differs from organ to organ and between parts and cells of one organ. The components of the nervous system are vulnerable in a rather fixed order, which is neuron, oligodendroglia, astroglia, microglia, and blood vessels in that diminishing order of sensitivity. The neurons are not a uniform population, and there are regional differences in their susceptibility to anoxia, those of the cerebral cortex and the Purkinje cells being the most sensitive. Within the cerebral cortex, the neurons of the deeper laminae are more sensitive than those in the superficial laminae. Beneath these levels, there is a gradient of susceptibility through the geniculate bodies, hypothalamus, thalamus, paleocortex, and caudate nucleus. The various patterns of vulnerability are relative; they are established largely by experiments in animals and are subject to some variation depending in part on the experimental species and on the type of anoxia induced. They do nevertheless emphasize the problem of selective vulnerability at the cellular level. There are, in addition, broader regional differences in vulnerability to anoxia, such as those which allow parts of the cerebral cortex to be necrotic while adjacent parts survive intact, and which allow the same agent or anoxic insult to cause leukomalacia on some occasions, and poliomalacia on other occasions. The basis of these regional differences in vulnerability, and the apparent differences in the patterns of degeneration following different types of anoxia, is still to be established.

The pathological effects of anoxia on nervous tissue are of two types, which are to some extent differences in degree. The least clear morphological expression is selective neuronal necrosis. The neuronal necrosis is typically of the ischemic type, and it is followed by glial repair. Greater degrees of anoxia, sufficient to kill astroglia as well as neurons, result in softening. These are the expected changes in anoxic anoxias such as might occur with cardiac arrest or periods of apnea in grand mal epilepsy.

As a response in some episodes of anoxia, such as may be induced, for example, in the histotoxic variety of cyanide poisoning, the degeneration and necrosis may affect the white matter with only minimal changes in the gray. The reason for this is not known. The malacic changes in the white matter of the cerebral hemispheres are preceded by edema, which suggests that local circulatory disturbances contribute to its development. Clearly the attribution of selective vulnerability on the basis of local metabolic demand for oxygen is inadequate, and a fuller explanation will probably require, in addition, consideration of vascular architecture, the autoregulation of blood pressure and flow, and local metabolic factors in the maintenance of myelin and axons.

1. Cyanide Poisoning

The anoxia produced by cyanide is classified as histotoxic because the action of the poison is to inhibit intracellular respiratory enzymes. The cyanides are particularly rapid in their action, and death usually occurs from a few minutes to an hour after the onset of clinical signs. In the most acute cases, the clinical course occupies only a few minutes and is characterized by dyspnea, salivation, trembling, and recumbency; it is terminated in convulsions. The blood is bright red owing to its high degree of oxygen saturation. In cases which survive for an hour, the clinical signs are the same with, in addition, vomiting, especially in pigs, and nystagmus and cyanosis. In these cases, the blood is often dark owing to anoxemia.

The morbid changes are those associated with anoxia. Because most cases survive for half an hour to 1 hour, the blood is dark and does not clot, and the tissues are dark. There is congestion in the lungs and hemorrhages of dyspneic type in the tracheal mucosa. Ecchymoses of the epicardium are usual. In pigs, there is pulmonary edema and hydropericardium, and there may be capillary hemorrhages deep in the myocardium by the coronary grooves. Severe pulmonary edema is present in some dogs.

Descriptions of lesions in the nervous system of animals in cyanide poisoning are largely limited to experimental observations in dogs and cats. In many natural cases of the poisoning, there is no significant alteration in the brain. The degenerative lesions in the brain in experimental poisoning may involve predominantly the gray matter or the white. When in gray matter, there is patchy necrosis and laminar loss of cells in the cerebral cortex, and necrosis in the head of the caudate nucleus, paleocortex, substantia nigra, and thalamus. There is edema in the cerebral white matter, sometimes in the absence of cortical change. On the other hand, single graded exposures of rats may produce the principal lesions in white matter, especially in the corpus callosum. In this species, there are gradients of injury in the callosum reflecting gradients of susceptibility, the most severe degeneration occurring in the posterior core of the callosum. The gradients are attributed to regional differences of blood supply, which are expected to reinforce the histotoxic anoxia of cytochrome oxidase inhibition by cyanide. The earliest change described in the callosal lesion is axonal swelling producing a spongy appearance by light microscopy; alterations of myelin and oligodendrocytes are second in time.

Plants containing toxic levels of hydrocyanic acid bound as glucosides are common and widespread and are the usual source of cyanide poisoning. The compounds are in general the β-glycosides of α-hydroxynitriles and are activated by endogenous glucosidases in the plant, or in other plants, or by ruminal microorganisms. For these reasons it is chiefly in ruminants that cyanide poisoning occurs. Toxin generation of this type does not occur at the low pH of the monogastric stomach, but a lethal amount may be absorbed from the large bowel if there has been a large intake of toxigenic material.

Both sheep and cattle are capable of detoxifying cyanide in the liver to form thiocyanate. Thiocyanate itself is toxic but in a manner different from that of cyanide; if thiocyanate is present in low concentrations over prolonged periods, it is goitrogenic. The cyanide concentration in a stated species of plant varies considerably with stage and rate of growth and with the fertility of the soil. Plants which are making rapid growth contain the highest concentrations of the glucosides, especially if the growth phase follows one of retardation caused, for example, by wilting, frostbite, or close grazing. The glucosides are regarded as being metabolic by-products.

Plants of many species contain cyanogenetic glucosides, but only a few are important in this regard. The majority of cyanogenetic plants are usually harmless because of either low palatibility or low concentration of the glucosides. The important cultivated toxic plants are the sorghums, *Sorghum* spp., and the star grasses, *Cynodon* spp. Those of the cherry family are also potently cyanogenetic. Linseed concentrates can be dangerous, especially if eaten in large quantities.

Chronic cyanide intoxication is not recorded, but extended periods of grazing the cyanogenetic sorghum and sudan grass may lead to ataxia in cattle and horses, and cystitis, urinary incontinence, and abortion in mares. There is axonal degeneration and demyelination at all levels of the cord but most prominent caudally in ventral and lateral funiculi. These may be direct effects of absorbed glucoside rather than of cyanide released in the alimentary tract.

2. *Nitrate/Nitrite Poisoning*

Ruminants are also vulnerable to phytogenous **nitrates,** when intraruminal microbial conversion of nitrates to nitrites exceeds the rate of conversion of the latter to ammonia. The nitrate sources again include many crop plants as well as weeds; plant nitrate concentrations are influenced by soil nitrogen level, and physiologic stresses on the plants, such as water stress and herbicide damage. Occasionally, high nitrate concentration in drinking water is the problem. Absorbed nitrites convert hemoglobin to methemoglobin, which imparts a dark chocolate color to the blood and causes dyspnea, cyanosis, weakness, tremors, collapse, and coma. There are no gross or histologic brain lesions. The condition is discussed more fully in Volume 3, Chapter 2, The Hematopoietic System.

3. *Fluoroacetate Poisoning*

Fluoroacetate is toxic by virtue of its ability to form fluoracetyl-coenzyme A (CoA), which potently inhibits the Kreb's cycle enzymes *cis*-aconitase and succinate dehydrogenase, thereby paralyzing cellular respiration. Herbivores are usually poisoned by ingesting fluoroacetate-containing plants, whereas other animals may be affected by accident or design by being exposed to sodium fluoroacetate formulated as a pesticide. The clinical signs are of the same general character as those described previously but, in the dog, frenzied hyperactivity is reported. As with cyanide, the myocardium is highly vulnerable, and multifocal necrosis may be seen. Fluoroacetate poisoning is discussed in Volume 3, Chapter 1, The Cardiovascular System.

4. *Carbon Monoxide Poisoning*

This is rarely observed and is discussed in the interests of completeness. Carbon monoxide has an affinity for hemoglobin which is several hundred times the affinity of oxygen for hemoglobin. The carboxyhemoglobin produced prevents oxygen exchange and causes anoxic anoxia. The anoxia is reinforced by a histotoxic component due to the affinity of carbon monoxide for iron-containing respiratory enzymes.

Carboxyhemoglobin and, therefore, the blood and tissues of poisoned animals are bright pink. All organs are congested, but especially the brain, in which the veins and capillaries are much dilated. The dilated vessels are observed especially in the white matter, and hemorrhages occur from them. Acute fatalities are not expected to be associated with neural lesions. Survival is expected to be followed by complete recovery, but there may be residual lesions of anoxic type, especially neuronal loss.

Carbon monoxide poisoning is reported occasionally in animals in confined quarters which are heated by petroleum fuels. Most of the reported instances are in pigs, and poisoning is associated with a high level of stillbirth and neonatal mortality. Following experimental exposure, patchy or extensive leukomalacia may be present in the hemispheres in the newborn.

5. *Hypoglycemia*

It is useful to extend the usual classification of anoxia to include hypoglycemia, which causes a disturbance of intracellular respiration. Oxygen is available to the nervous tissue, but in the absence or reduction of the amount of substrate, the oxygen is not utilized.

Hypoglycemia may occur as a result of a functioning tumor of the pancreatic islets or as a response to insulin overdosage in the treatment of diabetes. It is also part of the metabolic disturbance of **ketosis** in cows and **pregnancy toxemia** in ewes, and arguments have been advanced for regarding the irreversibility of pregnancy toxemia as being due to hypoglycemic encephalopathy. Piglets in the first week of life readily develop hypoglycemia if there is dietary restriction from any cause. Effective gluconeogenesis does not develop in piglets until about the seventh day of life, and during this first week their glycemic levels are rather precise reflections of dietary intake. In spite of the severe convulsions and deep coma which occur in hypoglycemia, conspicuous changes do not occur in the brain. It is probable that biochemical disturbances in the neurons precede histologic signs of degeneration by a considerable period. For this reason, acute hypoglycemic death is not expected to produce cerebral lesions. When the period of coma is prolonged, it is probable that all neurons are to some extent altered, but the differentiation of hypoglycemic from autolytic and nonspecific changes can seldom be made with confidence. Although classified with the anoxias, hypoglycemia does not produce the ischemic type of neuronal degeneration characteristic of the usual sorts of anoxia. There is, instead, a tendency for the severe type of neuronal degeneration to occur, characterized by rapid and complete chromatolysis, disappearance of the cytoplasmic margins, and then fading of the cytoplasm, with pyknosis, eccentricity, and fading of the nucleus. The regional susceptibility of neurons to hypoglycemia parallels that of susceptibility to anoxia.

Bibliography

Adams, L. G. *et al.* Cystitis and ataxia associated with sorghum ingestion by horses. *J Am Vet Med Assoc* **155:** 518, 1969.

Haymaker, W., Ginzler, A. M., and Ferguson, R. L. Residual neuropathological effects of cyanide poisoning. A study of the central nervous system of 23 dogs exposed to cyanide compounds. *Milit Surg* **111:** 231–246, 1952.

Keeler, R. F., Van Kampen, K. R., and James, L. F. (eds.). "Effects of Poisonous Plants on Livestock." New York, Academic Press, 1978.

Levine, S. Experimental cyanide encephalopathy: Gradients of susceptibility in the corpus callosum. *J Neuropathol* **26:** 214–222, 1967.

McKenzie, R. A., and McMiking, L. I. Ataxia and urinary incompetence in cattle grazing sorghum. *Aust Vet J* **53:** 496–497, 1977.

Song, S-Y. *et al.* An experimental study of the pathogenesis of the selective lesion of the globus pallidus in acute carbon monoxide poisoning in cats. *Acta Neuropathol (Berl)* **61:** 232–238, 1983.

E. Malacia and Malacic Diseases

Necrosis of individual elements of nervous tissue is described earlier under cytopathology. **Encephalomalacia** and **myelomalacia** refer to necrosis in the brain and cord respectively. Malacia means softening, and is used interchangeably with that term to signify necrosis of tissue in the central nervous system. It is sometimes, as imprecise practice, used interchangeably with demyelination. There is no sharp line to be drawn between demyelination and malacia, especially in the early stages of development of the degenerative process, but the term **demyelination** should probably be restricted to those lesions in which the myelin sheath is primarily or selectively injured to leave the axons naked but intact. When the neurons and neuroglia degenerate and die as part of the primary response, the term **malacia** is applicable.

The sequence of events in softening or malacia are specific only for the tissue. The morphologic changes in necrosis, in the removal of dead tissue, and in healing are the same irrespective of the insult. The insults can be varied, and malacia, alone or as part of another change, is one of the most common of lesions in the brain and cord. It is discussed in the sections dealing with vascular accidents and trauma; it occurs in many instances of encephalomyelitis, and is discussed elsewhere in these volumes as a lesion of mulberry heart disease, of antenatal bluetongue virus infections in lambs, and of clostridial enterotoxemia in lambs which escape apoplectic death and live for some days. Malacic lesions are probably the basis of most cases of hydranencephaly.

Although the nature of the malacic process is not specific for cause, there is some specificity in the localization of lesions and in their particular pattern of distribution. These features may allow the recognition of known associations or causes, such as the nigropallidal distribution of lesions in horses poisoned by yellow star thistle and the focal symmetrical lesions of the internal capsule in lambs with clostridial enterotoxemia. Although these and some other associations have been determined, the pathogenesis of the lesions and the problems of selective injury to certain parts of the brain remain to be resolved.

The sequence of changes in malacic foci are approximately the same in all cases and are outlined below. The rapidity of change is quite variable and depends on the species and age of the animal, the location of the necrosis, the volume of tissue affected, and the inciting cause. The speed of resolution is also affected by the quality of vascular perfusion in adjacent nervous tissue, whether the necrosis is ischemic or hemorrhagic, and by the time over which the cause acts.

The malacic process, once initiated, appears to proceed very rapidly in the fetal and immature nervous system and to leave cystic structures and hydranencephaly without much evidence of continuing reaction at the time these animals are available for pathological examination (Fig. 3.9). The reasons for the rapidity of change in immature nervous tissue are not clear, but contributing factors may involve the paucity of myelin which is difficult to remove, the paucity of mature mesenchyme in meninges and about vessels, and the plasticity of vascular arrangements in the developing brain. The rapidity of change in gray matter is generally greater than that in the white. Autolytic liquefactive changes depend on continued enzymatic activity, which in turn, depends on availability of oxygen either by diffusion from surrounding tissue or by reflow in the local vessels. Small foci of necrosis are expected to resolve more quickly than are large foci. Malacic lesions in the neocortex, where diffusion from collateral vessels is available, are expected to resolve more rapidly than those in the paleocortex where the vascular supply is more strictly of the end-artery type. Some causes of malacia, such as vascular occlusion, act promptly, and the pathological changes are directly consequential. The cause may act continuously over a period, as in leukoencephalomalacia of horses, and the changes may be incremental. The process, once established, may itself initiate progressive change, as in traumatic injury to the cord in which swelling within the confines of the meninges assists the spread of edema and vascular response beyond the site of the original injury.

A malacic lesion which develops acutely may not, in the absence of hemorrhage, be demonstrable before about the twelfth hour of onset. The early change is in texture, with the affected part being soft, and in color, with the affected part being grayish. Within 2 to 3 days the malacic foci begin to disintegrate, the softness is more evident, and the surrounding tissue is swollen by edema and pale in gray matter or yellowish in white. Eventually, and subject to the above qualifications, the necrotic area liquefies, and a cyst remains. The cyst may be loculated or traversed by vascular strands which have survived the episode.

Histologically there is, in acute episodes, a reduction of staining affinity by about 12 hr. The cellular elements show the changes described earlier. The early active response involves circulating neutrophils, which may enter in large numbers, but this response is replaced in 3 to 4 days by macrophages. The first appearance of macrophages is about blood vessels, their peak activity occurs in about 2 weeks, but a few will survive for a very long time. Astrocytic gliosis replaces or surrounds the resolved necrotic area (Fig. 3.49A).

In the specific syndromes to be discussed, the degenerative changes tend to be restricted to, or to affect principally, either the gray matter or the white matter. Softening of gray matter is known as **poliomalacia,** and softening of white matter is known as **leukomalacia;** each may be qualified as to whether cerebral or spinal. The diseases to be described are specific, but malacia is not exclusive to them. Necrosis of cerebral cortex, is rare in horses except in the neonatal "barker syndrome" (Fig. 3.43). In ruminants, poliomalacia is, in the early stages, expected to respond to thiamine, but it occurs also in lead poisoning, in water intoxication, and in other, ill-defined circumstances. Lead also causes poliomalacia in dogs. Convulsive episodes in dogs, some associated with distemper, may leave malacic changes especially in parietal and temporal cortex. In pigs, polioencephalomalacia is usually due to salt poisoning, but individual cases occur in meningitis or without other association. The isolated case of malacia can be difficult to explain.

1. Focal Symmetrical Spinal Poliomalacia Syndrome

This disease, as described from Kenyan **sheep,** is characterized by focal softening of the gray matter of the spinal cord, most consistent and severe in the cervical enlargement. A similar syndrome occurs in parts of West Africa. There is no information on the cause or pathogenesis, although the distribution of necrosis in relation to the cross section of the cord and the irregular segmental involvement of the gray matter, especially in the cervical region, are consistent with a vascular component in the pathogenesis. Lesions of similar distribution and character are occasionally met with in dogs and cats with inflammatory or thrombotic occlusion of the ventral spinal artery. Affected sheep suddenly develop flaccid or spastic paresis, which always involves the forelimbs and sometimes the hind limbs as well. There are no cerebral signs. Affected animals are lambs up to 18 months of age. There are no gross lesions to be observed except in cases of long standing, in which some brownish discoloration may be noted in the malacic areas. Microscopically, there are bilateral lesions of remarkable symmetry in the ventral horns of the spinal cord. The dorsal horns are spared, as is a narrow rim of gray substance around the periphery of the ventral horns, and the commissural gray matter. The affected areas undergo dissolution with the usual reaction on the part of the microglia. At a later stage, proliferating capillaries in small numbers crisscross the microcavitations, and the astrocytes at the margins proliferate. The malacic foci are found in the cervical and lumbar enlargements as "skip" lesions involving a few segments and, when the necrosis is extensive, similar foci may be found in the medulla.

A similar syndrome has been responsible for heavy losses in native sheep in Ghana and Ivory Coast. All ages are affected but mainly adult ewes. Clinical progression is rapid from initial stumbling to ataxia and recumbency, opisthotonus and nystagmus, and ultimately flaccid paralysis. There are microscopic changes of cytotoxic edema, especially affecting oligodendrocytes widely in the nervous system and also perivascular astrocytes and capsule cells of spinal ganglia. Foci of spongy degeneration and malacia, bilateral but not always symmetrical, are of patchy distribution and most frequent in the spinal intumescence, cerebellar roof nuclei, and large nuclei of the brain stem.

Encephalomyelomalacia of similar character is reported in young individual **goats** in California. The lesions are bilaterally symmetrical and affect particularly the lumbar and cervical enlargements and the inferior colliculi. Other brain stem nuclei are inconsistently involved. The similarity of these malacic lesions to those which can be produced experimentally by nicotinamide antagonists has been noted.

Focal symmetrical spinal poliomalacia of pigs is clinically and pathologically similar to the spontaneous poliomalacias of sheep and goats. The presenting signs are spinal, with ataxia progressing to forelimb or hind-limb paresis or quadriplegia in a few days.

In field cases, malacic foci are found, symmetrically, in the ventral horns of the cervical and lumbar enlargements. The malacic foci are visible grossly as yellow-brown areas of softening or grayish depressed areas of liquefaction. The histologic changes are typical of malacia (Fig. 3.57). There is heavy loss of neurons and endothelial and glial proliferation in older lesions. Similar changes may be present in medullary nuclei.

Affected animals may show changes associated with selenium toxicoses, including scurfiness of skin, hair loss, and separation of horn at the coronet. Dietary selenium, whether as sodium selenite added to rations or the feeding of selenium-accumulator plants, reproduces the syndrome, although there is some inconsistency in the lesions

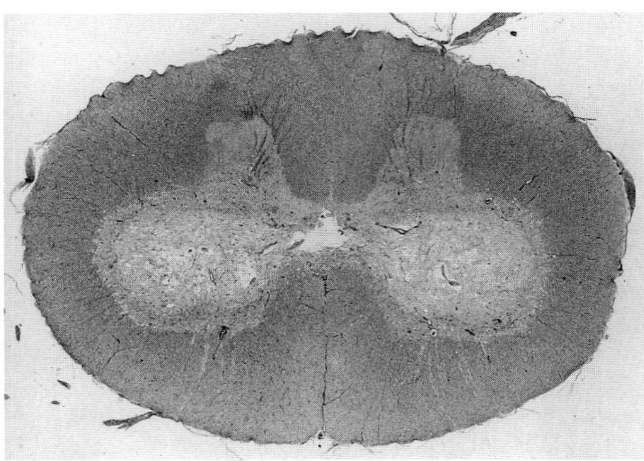

Fig. 3.57 Focal symmetrical spinal poliomalacia involving ventral horns of cervical cord. Pig. Selenium toxicity. (Courtesy of T. M. Wilson.)

produced, with degeneration being more frequent in the nuclear masses of the medulla and brain stem.

The mechanism of neurotoxicity of selenium is not known, but toxic expression is expected to be influenced by other nutritional and metabolic interactions.

Focal symmetrical poliomalacia, the lesion limited to the sacral and lumbar enlargements, is described in Ayrshire **calves.** The malacia affects the ventral horns, sparing the dorsal horns and the central gray matter. The cause is not known.

2. *Polioencephalomalacia of Ruminants*

Polioencephalomalacia, as used in practice, applies especially to softenings restricted to the cerebrocortical gray matter of laminar distribution. The lesion may alternatively be designated as cortical necrosis or laminar cortical necrosis. Necrosis of this distribution is the basis of a now well recognized syndrome of cattle, sheep, and goats known as polioencephalomalacia. It is also the lesion of salt poisoning in swine, is occasionally observed in lead poisoning of cattle, and is described among the residual neurologic lesions of cyanide poisoning. Apart from these known associations or causes, laminar cortical necrosis is observed sporadically in swine, dogs, and cats.

The disease, polioencephalomalacia of sheep, goats, cattle, and other managed ruminants, is similar in its clinical and pathologic aspects in the several species, but the course in sheep and goats is, as a rule, shorter with fewer survivors from the stage of overt brain swelling. The Merino sheep appears to be much more resistant than other breeds.

The incidence is highest in sheep in the age group from weaning to 18 months, but sporadic cases occur in older animals. This general statement of incidence applies to pastured animals as well as to those in feedlots or barns. It is a disease of young cattle, the age incidence depending on the population exposure. As originally described in Colorado, affected animals are mainly 1–2 years of age but, elsewhere, the incidence is highest in younger stock of 3 to 8 months of age. The severity of the disease is rather less in pastured than in feedlot or housed cattle and sheep. Its least clinical expression, occasionally observed in outbreaks in sheep, is blindness and dullness, a tendency to head press against obstacles, and cessation of feeding. There are no ocular abnormalities in this, or in any other, grade of the disease, and the blindness is cortical. Animals showing these mild signs usually recover completely in the course of several days. If they are killed for examination, malacic changes may be found to be minimal or absent, but there is widespread neuronal necrosis of ischemic type in the deeper laminae of the cerebral cortex. Animals affected more severely may remain on their feet and show, in addition to these signs, muscular tremors, especially of the head, and intermittent opisthotonus; if these animals survive, they become partially decorticate and remain blind and stupid.

In severe cases of polioencephalomalacia, the animals are recumbent. There is twitching of the face, ears, and eyelids, intermittent grinding of teeth, salivation, and bulbar paralysis in some. Opisthotonus is present, and some animals are convulsive. In the early stages the neurologic signs are intermittent, but later they become more constant with persistent opisthotonus and nystagmus. Flaccidity is usual although there may be transient periods of spasticity, and clonic convulsions are intermittent. Death occurs in coma after a course of one to several days. These signs are entirely referable to the nervous system; other systems are undisturbed except for the very occasional case in which massive hemoglobinuria after excess water intake precedes the onset of neurologic signs.

The lesions are qualitatively the same in all cases, but they differ in their degree and in the ease with which they may be detected, this depending on severity and duration. Young animals usually die more rapidly than old ones and may show cerebral edema and swelling only. The swelling affects the cerebral hemispheres, which are pale, slightly soft, and droop over the cut edges of the cranium. There may, in these cases of short course, be no obvious displacement of the brain. When the fresh cortex is sectioned, it usually shows a laminar paleness about 0.5 to 1 mm wide following faithfully the contour of the gray matter at the junction between gray and white. This change may be most easily visible in the gyri rather than in the depths of the sulci. Its extensiveness varies considerably from case to case but remains within the boundaries of the supply area of the middle cerebral and posterior cerebral arteries. In the anterior distribution of these vessels, the lesions are patchy but are more extensive caudally toward and over the posterior poles of the hemispheres. The dorsal surfaces of the hemispheres are involved, and the distribution is as approximately symmetrical as it can be.

In cases which survive for several days, the necrosis becomes more and more obvious. The brain is grossly swollen and, almost without exception, it is displaced caudally with herniation of the medulla and cerebellum

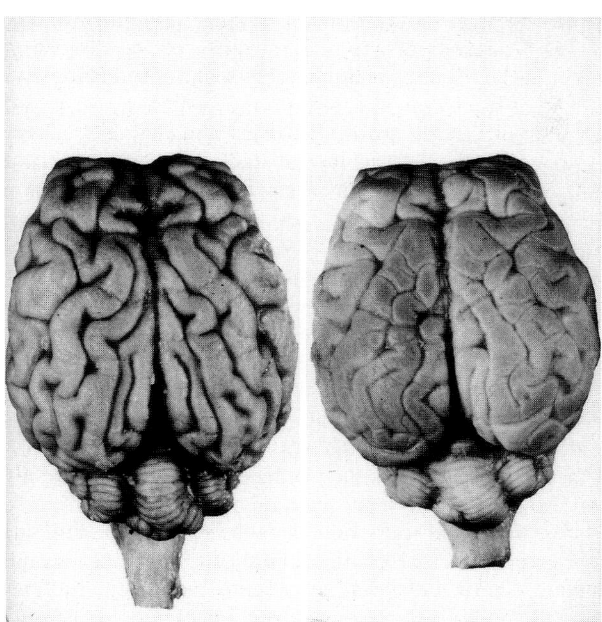

Fig. 3.58 Polioencephalomalacia. Sheep. Normal (left) and affected brains. Note flattening of gyri (right).

into the foramen magnum (Fig. 3.58). The dorsum of the hemispheres is palpably quite soft, and the normal turgidity of the brain is gone. The surfaces of the gyri are a characteristic yellowish brown, and the normal adhesion of leptomeninges to the gyral surfaces may not be present. On the cut surface of the cerebrum, there will be found areas of minimal change resembling those previously described. The necrotic zones can usually be appreciated to be narrow and pale or yellowish instead of gray, and they are friable and shrink away from the cut surface (Fig. 3.59). Lines of cleavage between the gray and the white matter can be discerned. These are present especially over and extending down the sides of the gyri but seldom extend around the depths of the sulci. In some cases, yellowish foci of softening can be found in the herniated portion of the cerebellar vermis. Hemorrhage is insignificant and usually not apparent in the cerebrum and cerebellum, but hemorrhagic foci of softening may be found in the collicular region, thalamus, and caudate nuclei. In some cases, there is thrombosis of the dorsal longitudinal sinus.

Those animals which survive for 2 weeks or longer are decorticate over more or less extensive areas. The white cores of the gyri are largely intact and project nakedly or with an irregular covering of softened friable gray matter. Over some gyri, the gray matter is intact. The subarachnoid space is widened, and the meninges droop over the enlarged sulci. Subpial cysts may be evident where the devastation is not complete and the meninges are adherent. The brain is smaller than normal and not displaced, the lateral and third ventricles are dilated, and there is an excess of cerebrospinal fluid.

The microscopic changes are of the same type in all

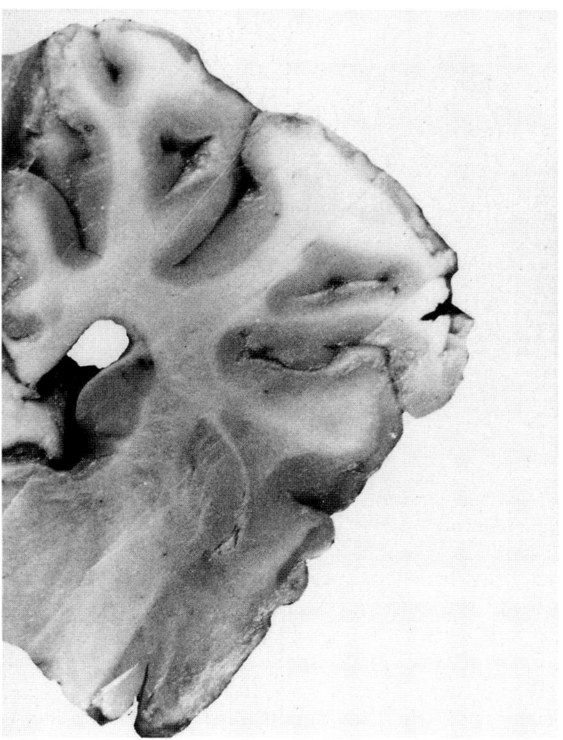

Fig. 3.59 Polioencephalomalacia. Ox. Chronic lesion. There is marked irregular narrowing of cerebral cortex.

affected parts but vary in details of distribution and severity in different areas. The least change is one of laminar necrosis of nerve cells, affecting especially the deeper laminae. The neurons are shrunken, acidophilic, and surrounded by a clear space; in 2 to 3 days, many are converted to eosinophilic globules without nuclear remnants. Healing, if it occurs, is with intense astrogliosis.

The malacic areas are laminar (Fig. 3.60). The deepest laminae are consistently involved. Frequently, the superficial lamina is involved. The softened laminae do not necessarily overlie one another, although there is usually considerable overlapping. When two distinct laminae of necrosis overlie one another, they frequently merge in the depths of the sulci to produce necrosis of the total width of the gray stratum (Fig. 3.60). Necrosis of the superficial lamina is first recognizable as a rather uniform sponginess. The layer then disintegrates, and the middle laminae, if intact, are separated from the pia by a moat of compound granular corpuscles in which vessels are very sparse (Fig. 3.61). The leptomeninges are thickened and contain many activated histiocytes.

The middle laminae may remain structurally intact for considerable periods. The neurons therein die acutely, but the glia are rather persistent. The vessels are prominent with swollen proliferating endothelial and adventitial cells and a perivascular clear space.

There is early edema and neuronal necrosis in the deepest laminae, and the edema extends to the adjacent white

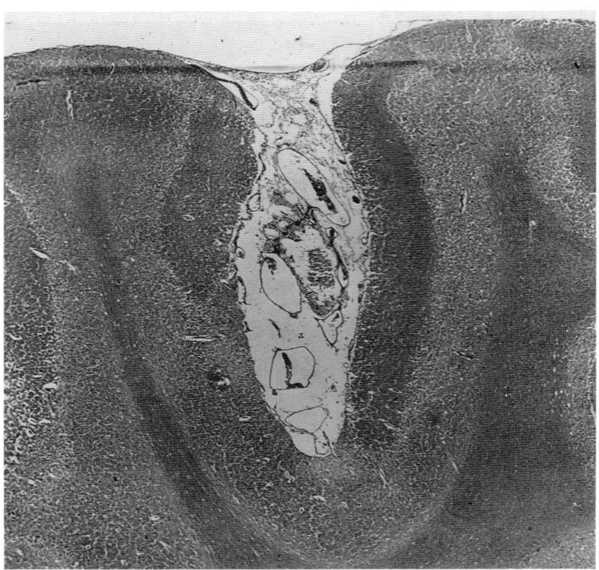

Fig. 3.60 Polioencephalomalacia. Ox. Pallid areas of degeneration involving full width of cortex in depth of sulcus.

matter, giving the zone a pale washed-out, fibrillary appearance in sections (Fig. 3.60). There is demyelination in the edematous white matter, but the glia survive fairly well, and the astrocytes react and swell. The microglia become active and concentrate in the deep laminae, which disintegrate to cleave overlying cortical remnants from the subjacent white matter (Fig. 3.62).

The macroscopic lesions observed in the cerebellum and subcortical areas are typical foci of softening. Microscopic changes of less severity are more common. They occur in the herniated portion of the vermis and consist of acute degeneration of Purkinje cells with more or less extensive cytolysis in the granular layer.

The chronic lesions of surviving animals resemble those described earlier in ischemic infarcts. The dead tissue is removed by microglia. In zones where the gray matter has been entirely necrotic, the pia may be separated from the white matter by a clear space or be in contact with the white and partially adherent as a glial–pial scar formed by connection with the proliferated astrocytes that line the defect. Where the superficial laminae have remained intact, elongate cystic spaces lined by astrocytes remain where the necrotic parenchyma has been removed.

The areas of softening quite clearly have a distribution related to the field of supply of the middle cerebral artery. When the lesion is of restricted distribution, it is related to the periphery of the field of distribution of this vessel

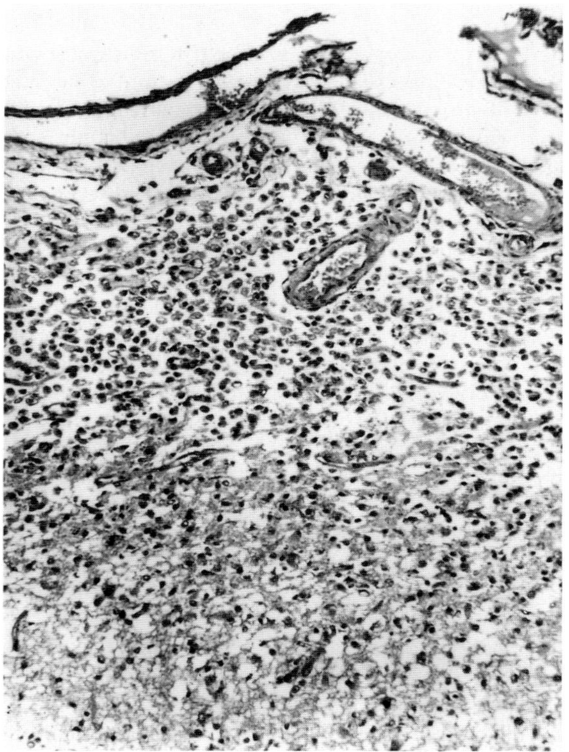

Fig. 3.61 Polioencephalomalacia. Collapse of meninges, with gitter cells replacing necrotic cortex. Ox.

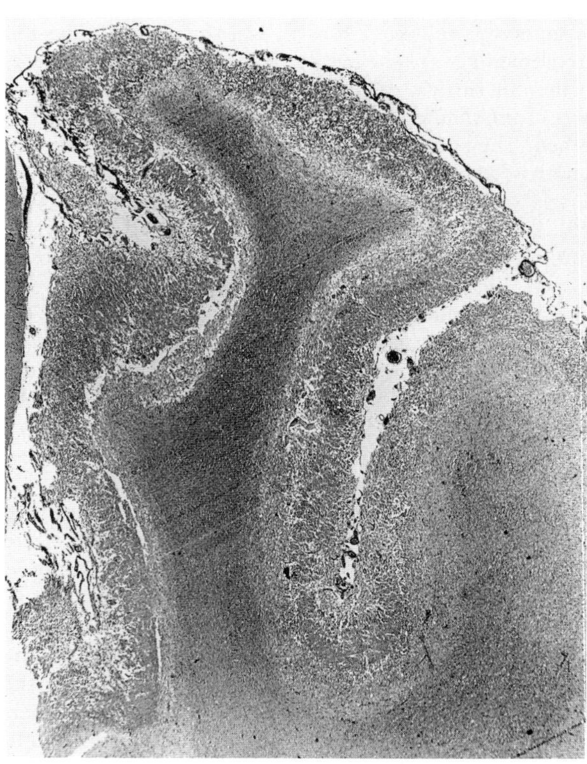

Fig. 3.62 Polioencephalomalacia. Subtotal cortical necrosis with separation. Ox.

over the dorsal cerebral cortex and, when the lesion is more extensive, it covers more of the field of supply of this vessel, but necrosis is seldom found ventral to the posterior ectosylvian fissure. The cortex anterior to the transverse sulcus and in the field of supply of the anterior cerebral artery escapes significant injury. Necrosis may be observed in the distribution area of the posterior cerebral artery, but only when the swelling and displacement is severe, a fact suggesting that necrosis in this distribution is secondary to tentorial herniation with compression of the posterior vessels against the free edge of the tentorium.

Degenerative changes in periventricular nuclei of type and distribution typical of thiamine deficiency in other species are present in some cases of polioencephalomalacia (Fig. 3.63), and their histologic appearances suggest that their development precedes the cortical necrosis. This observation applies also to the experimental disease produced by the thiamine antagonist, amprolium.

The cause of polioencephalomalacia is unknown, and it is unlikely that the condition is etiologically specific. Sporadic cases and outbreaks are observed in animals given access to water after a period of deprivation, suggesting that these cases may be expressions of water intoxication developing in the manner discussed later for salt poisoning in swine. There is a high incidence in cattle in feedlots, especially when they are fed diets based on molasses. It has been observed in sheep eating the nardoo, *Marsilea drummondi,* and associated with the thiaminase

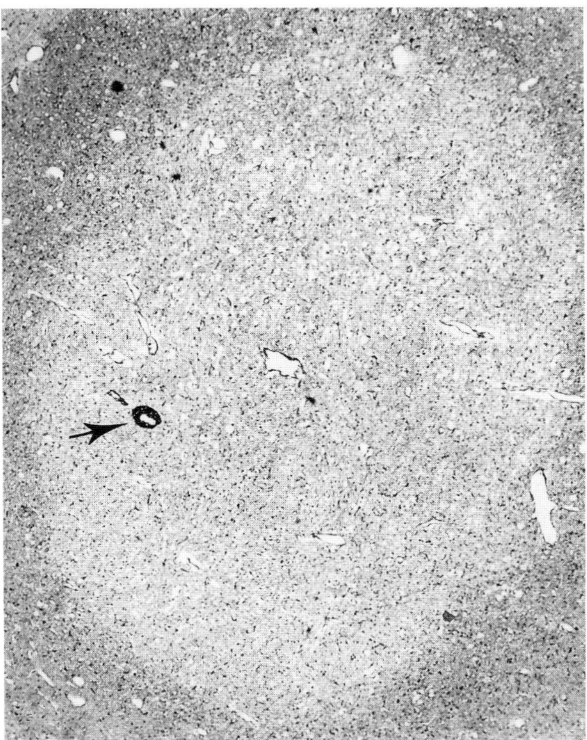

Fig. 3.63 Polioencephalomalacia. Edema in inferior colliculus with ring hemorrhage (arrow). Sheep.

present in that fern. The disease has been produced experimentally with analogs of thiamine, and by the feeding of rhizomes of bracken fern, which are known to contain thiaminase.

There are clear indications that a deficiency of thiamine or a disturbance in its metabolism is implicated in polioencephalomalacia. Thiamine, if administered early, can lead to prompt clinical recovery. Calves examined early in the course of the disease may have elevated blood pyruvate, reduced blood transketolase, and reduced levels of thiamine in liver and brain. Cerebral edema and laminar necrosis have also developed in lambs and calves given the thiamine antagonist, amprolium. Thiaminolytic enzymes can be found in rumen contents in some cases of polioencephalomalacia, and it is accepted that they are produced as exoenzymes of some rumen microbes. Degradation of thiamine may reduce the amount available and simultaneously produce analogs of the vitamin.

A variety of inorganic sulfur compounds is capable of cleaving thiamine into pyrimidine and thiazole constituents, thereby rendering it inactive. This is consistent with field evidence that diets and drinking water containing high levels of sulfates can increase the incidence of polioencephalomalacia in cattle, and in sheep fed diets high in sulfur.

3. Thiamine Deficiency

Thiamine (vitamin B_1, aneurin) is a dietary requirement of carnivores. Herbivores are capable of synthesizing their own requirements, the synthesis being microbial and taking place in the rumen, and probably in the large intestine of horses. Experimental deficiency can be produced in calves and lambs, but only when they are very young and before they have established a useful ruminal flora; neurologic signs, including ataxia, cerebellar tremors, and convulsions, occur in these experimental deficiencies (see also Polioencephalomalacia). Horses are poisoned by eating bracken fern and horsetail (*Equisetum arvense*), both of which plants contain a thiaminase. The plants are unpalatable, but poisoning may occur when they are included in pasture hays. The disease in horses is characterized by incoordination, which may be severe and lead to recumbency, and bradycardia. Affected animals respond rapidly to thiamine. The nervous system has never been suitably examined for lesions.

Thiamine deficiency has been produced in most domestic species and in others. It has been an important natural disease in foxes and cats and continues to be of some importance in mink. In these carnivorous species, the deficiency is induced by a thiamine-splitting enzyme naturally present in many species of fish. Metabolic analogs of thiamine are known and have been used experimentally to produce the deficiency, but there are no known naturally occurring competitive analogs. Processed foods for dogs and cats can easily be made deficient in thiamine, as the vitamin is susceptible to heating at 212°F or above in a medium of neutral or alkaline pH. The feeding to cats and dogs of meat which has been exposed to sulfur dioxide to preserve the fresh appearance can lead to thiamine deficiency; thiamine is destroyed by sulfates.

The syndrome, originally described as Chastek paralysis, produced by thiamine deficiency in foxes and mink is comparable to that in cats and may develop in 2 to 4 weeks of being fed the deficient diet. The onset of neurological signs is indicative of severe depletion of thiamine; the clinical and pathological consequences of marginal deficiency are not known. Initially, there is a reluctance to eat, and salivation and anorexia may be present for up to 1 week before distinct clinical signs occur. When neurologic signs first develop, they consist of ataxia, incoordination, pupillary dilation, and sluggish pupillary reflexes. Convulsions are easily induced and are characterized by strong ventroflexion of the head. At this stage, animals respond rapidly to thiamine, usually fully, but there may be residual signs of mild ataxia. After 2 to 3 days of neurological signs, animals pass into an irreversible phase of semicoma, opisthotonus, continual crying, and spasticity.

The lesions of thiamine deficiency in carnivores pass through the sequence of vacuolation of neuropil, vascular dilation, hemorrhage, and necrosis. These changes are occasionally observed in the middle laminae of the occipital and temporal cortex as poliomalacia, but the areas of remarkable vulnerability are in the periventricular gray matter. The periventricular lesions are always bilaterally symmetrical. The only consistency in the pattern of the lesions in the periventricular system is the involvement of the inferior colliculi (Fig. 3.64A). Next to the inferior colliculi, lesions are most commonly found in the medial vestibular, red and lateral geniculate nuclei, but they may be found in any of the periventricular nuclei. In contrast with Wernicke's hemorrhagic encephalitis in humans, with which the disease in carnivores can be compared, cats and foxes only irregularly develop hemorrhages in the mammillary bodies.

The initial morphological change is vacuolation in nuclei of special susceptibility. This is easiest to detect in the lateral geniculate bodies, inferior colliculi, and red nuclei (Fig. 3.64B,C,D). The vacuolation develops in cats at about the time they become susceptible to the induction of convulsions and at about the time that alterations in the permeability of the blood–brain barrier become demonstrable. This altered permeability of the vessels is limited to those nuclear masses which are known to be vulnerable to injury in this deficiency. The case may not progress beyond this stage but, instead, may pass to recovery. In the event of recovery, a quite intense astrogliosis develops and remains to indicate the past presence of thiamine deficiency. With the development of vacuolation, it is also possible to recognize vascular dilation, especially affecting venules and especially in the susceptible nuclei. Hemorrhage occurs from both capillaries and venules, consistently in fatal cases. The hemorrhages may be large enough in the colliculi and vestibular nuclei to be visible to the naked eye (Fig. 3.64A,B).

Thiamine deficiency appears to affect most severely the relay systems from the eye and ear, but the histogenesis of the changes is rather vague. Phosphorylated thiamine is the coenzyme, cocarboxylase, and it participates probably in all oxidative decarboxylations as well as in some other metabolic transformations. It is a cofactor for transketolase, α-ketoglutarate dehydrogenase, pyruvate dehydrogenase, and branched-chain α-keto acid dehydrogenase. It may also have a function in axonal conduction and synaptic transmission. In all animals in which thiamine deficiency has been produced, the deficiency is attended by an elevation of the pyruvate levels of blood. This is to be anticipated from what is known of the function of the vitamin but cannot be connected in any way to the nature of the nervous signs or the nature or distribution of the neural lesions. There is a decrease in glucose utilization by the brain, and shortly before the development of lesions, the vulnerable areas suffer a burst of metabolic activity with local production of lactate, and it is possible that the focal lesions, initially, are the result of focal accumulation of lactic acid. The bradycardia is explainable on the basis of impaired carbohydrate metabolism because cardiac muscle depends for its energy supplies on carbohydrate, chiefly pyruvate, and in this differs from skeletal muscle. Myocardial degeneration is described in the experimental disease in several species and in the spontaneous disease in cats, dogs, and foxes; focal myocardial necroses are more prominent in the right than in the left ventricle.

Although it is not possible to correlate thiamine deficiency and its lesions with the distribution and activity of pyruvate decarboxylase, there is some correlation between thiamine deficiency and the distribution of transketolase activity in the nervous system. Transketolase is a thiamine pyrophosphate-dependent enzyme active in the hexosemonophosphate shunt. The enzyme is active in white matter and apparently important in the metabolism of oligodendrocytes. During progressive thiamine depletion, brain transketolase activity declines before signs of the deficiency develop, and the decline is greater than that for pyruvate decarboxylase. Moreover, the decline of brain transketolase is greatest in those areas in which lesions develop in rats.

The relation of transketolase deficiency to the evolution of lesions is still unclear as is the sequence of morphological changes. From light microscopy, it has been inferred that the vessels are primarily injured, but clinical signs are evident before histological changes. The hemorrhages are a terminal event. Fine-structural studies suggest that changes in vessels in selective areas of injury are secondary to degenerative changes in regional glia. The glia are edematous—swollen with clear cytoplasm—and their rupture leads to increase of extracellular spaces. Additional studies have shown early damage in the neuropil with degenerative and hypertrophic changes and axonal degeneration with preservation of neurons.

4. Nigropallidal Encephalomalacia of Horses

Prolonged ingestion of yellow star thistle, *Centaurea solstitialis*, or Russian knapweed, *Centaurea repens*, pro-

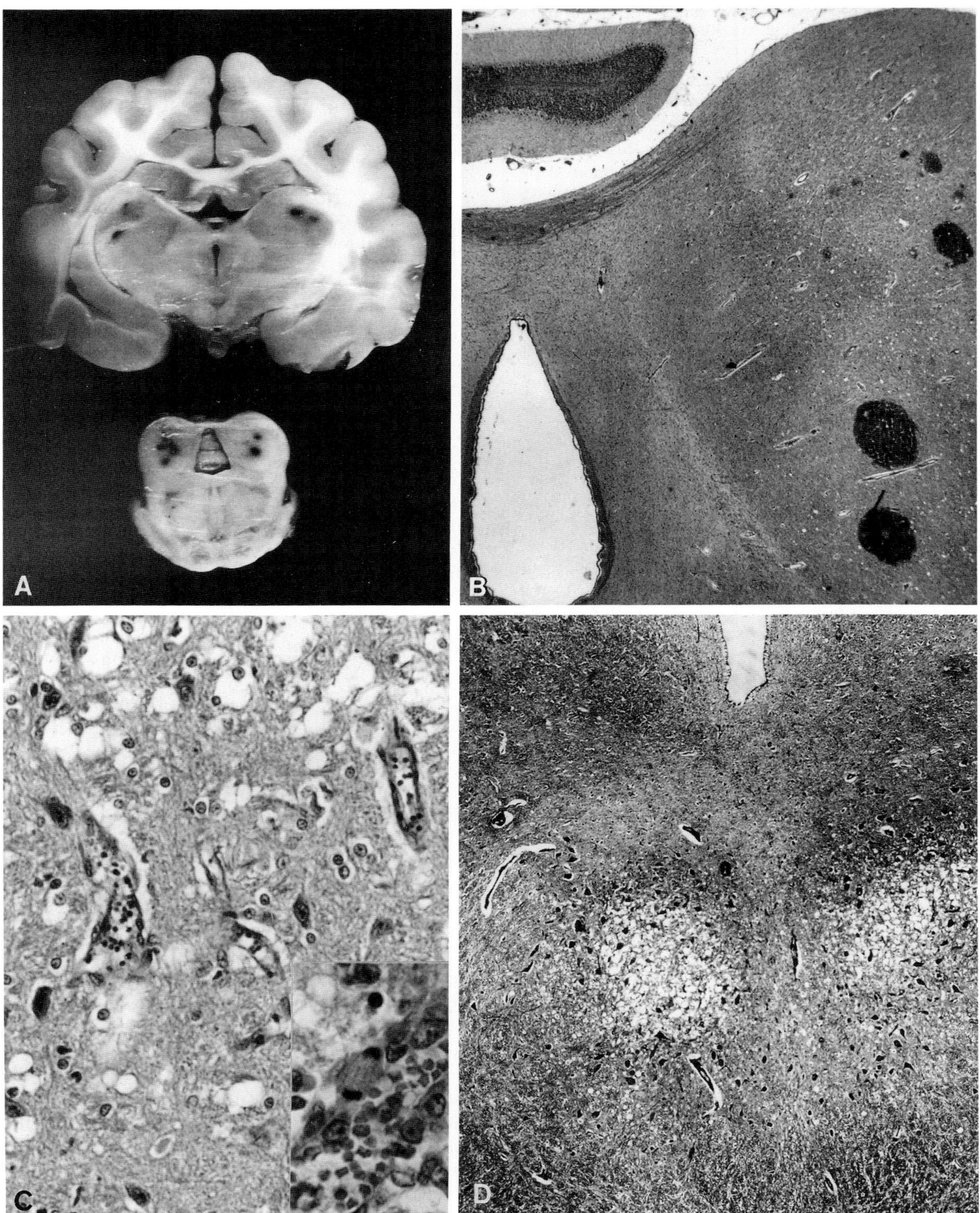

Fig. 3.64 Thiamin deficiency. Cat. (A) Hemorrhage in lateral geniculate bodies and inferior colliculi. (B) Focal hemorrhages with edema in inferior colliculus. (C) Edema, gliosis and vascular reaction in inferior colliculus. (Inset) Mitosis in endothelial cell. (D) Edema in oculomotor nuclei.

duces malacia in horses. It is a disease of dry summer pastures in which the thistle, which remains green, provides most of the forage.

The clinical onset of the disease is sudden and may occur as early as 1 month after initial exposure to star thistle. The syndrome is characterized by idle drowsiness, persistent chewing movements, and difficulty in prehension, the latter apparently related to incoordination of the lips and tongue. Sensation and reflexes are normal. Death is due to starvation, dehydration, or intercurrent disease. Malacic lesions are consistently present in the brain and affect specifically the pallidus and substantia nigra. There is some variation of the pattern depending on whether the lesions are symmetrical or not and involve both pallidus and substantia nigra or only one of the structures. Usually both structures are symmetrically involved. The pallidal foci involve the anterior portion of the globus pallidus, are lenticular in form and 1–1.5 cm in size. The core of necrotic tissue in the substantia nigra is ~0.5 cm in diameter and extends from about the level of the mammillary bodies to the point of emergence of the oculomotor nerve (Fig. 3.65). The affected areas are evident as slightly bulging, yellowish, gelatinous foci. The softening progresses rapidly, and in about 3 weeks the lesion is sharply demarcated as a pseudocystic cavity. The microscopic changes are those ordinary in malacia and do not give a clue to the immediate pathogenesis.

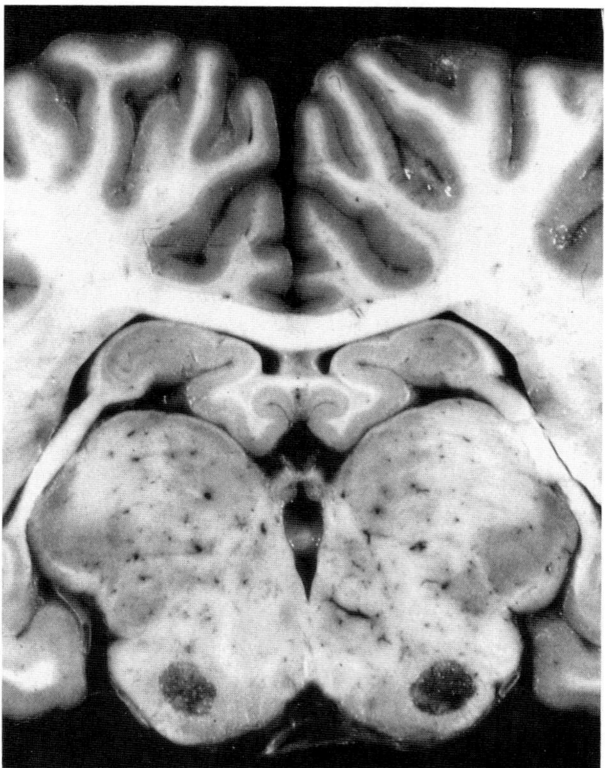

Fig. 3.65 Nigropallidal encephalomalacia. Horse. Bilaterally symmetric foci of malacia in substantia nigra.

5. Salt (Sodium Chloride) Poisoning

Salt poisoning may be a direct and immediate result of excessive ingestion of salt, or it may be indirect and delayed, developing only after several days of excessive salt intake and restricted water intake. These two types of salt poisoning are pathogenetically dissimilar, and neither will occur if animals are always provided with free access to water of low saline content.

Direct salt poisoning is largely a problem of salinity of drinking water. An excessive saline content of fodder is not a concern, except in terms of palatability, if an abundance of fresh water is available; for example, sheep may take rations containing up to 13% sodium chloride for prolonged periods without ill effects. This high tolerance for salt in fodder is utilized to restrict food intake during periods of scarcity and also to encourage water intake in the control of urolithiasis.

The salinity of drinking water is important, not only in terms of the concentration of sodium chloride or total salts, but also in terms of what salt and what acid radicals are present. It is difficult to provide figures for acceptable salinity of drinking water for livestock because of the variations in the proportions and concentrations of the various salts present, but the usual recommendation is that the concentration of sodium chloride or total salt should be less than 1.7% for sheep, 1.0% for cattle and 0.9% for horses, and preferably not more than one half of these concentrations.

Acute, direct salt poisoning occurs chiefly in cattle, especially if they are very thirsty when first given access to saline water. Poisoning may also occur in cattle if they are given free access to salt supplements after a prolonged period of salt restriction, such as occurs in cattle grazing on mountain pastures. Clinical signs in cattle are referable to the alimentary tract and nervous system. There is vomiting, polyuria, diarrhea, and abdominal pain with paresis, knuckling of the fetlocks, and blindness. Death may occur in 24 hr.

At autopsy, there is severe congestion of the mucosa of the abomasum and excessive, dark, fluid intestinal content. The alimentary changes are probably in large measure due to osmotic disturbances in the gut, but the neurologic signs are unaccounted for. Also unexplained is the development of a moderate anasarca in animals which survive for some days.

Indirect salt poisoning is a neurologic disease. There are circumstantial reasons for believing that it occurs in cattle and sheep and that it is occasionally responsible for polioencephalomalacia in these species, but the disease is proven to occur only in swine. Toxicity is related to the sodium ion; the clinical and pathologic features of this disease are not duplicated in any other.

Apparent blindness and deafness initiate the clinical syndrome in indirect salt poisoning in swine. The animal is oblivious to its environment and cannot be provoked to squeal. There is head pressing suggestive of increased intracranial pressure, arching, or pivoting, and these signs usually lead to the convulsive syndrome. The convulsions are very characteristic in their pattern and in the regularity of the time intervals in which they recur. They begin as tremors of the snout and rapidly extend as clonic spasms of the neck muscles, with jerky opisthotonus which causes the pig to walk backward and sit down. The animal passes into lateral recumbency and generalized clonic convulsions.

The lesions of this intoxication are restricted to the brain. There is moderate cerebral edema but no displacement. The basic changes in the brain are laminar loss of cortical neurons and laminar (middle laminae) malacia. These are the changes described earlier for polioencephalomalacia of cattle and sheep. The specificity of the lesion is given by the abundance of eosinophils which are infiltrated into the meninges and Virchow–Robin spaces. This particular lesion and its frequent relation to excessive intake of salt have been recognized for many decades, but until the cause was established, the syndrome was designated as eosinophilic meningoencephalitis on account of the relatively pure infiltrations of eosinophils which were found. There is a tendency for eosinophils to infiltrate in the meninges and perivascular spaces in the brains of pigs with other cerebral lesions, such as the leukomalacia of mulberry-heart disease and various encephalitides, but the combination of laminar change and cerebral eosinophilia is pathognomonic of salt poisoning.

The amount of salt in the diets of pigs varies considerably, but toxicity does not occur if plenty of water is available. The critical level of salt in the fodder is approximately 2% when water intake is restricted. Signs of poisoning may occur when, after a period of restriction, the water supply is replenished.

Although an excessive intake of salt over a period of several days sets the stage, which is reflected in elevated plasma levels of sodium and chloride, the disease is precipitated by water, so that it may equally well be regarded as a form of water intoxication. Beyond this, the pathogenesis of the lesions remains obscure. It is noteworthy that they are anoxic in character and distribution.

6. *Mycotoxic Leukoencephalomalacia of Horses*

A neurological syndrome occurring in horses forced to subsist on moldy corn for approximately 1 month or longer has been recognized in the United States since the latter part of the nineteenth century and has more recently been described from Europe, Greece, China, Argentina, and Africa. The neurologic signs are of fairly sudden onset and are characterized by drowsiness, impairment of vision, partial or complete pharyngeal paralysis, weakness, staggering, and a tendency to circle. The course is from a very few hours to about a month, but death usually occurs on the second or third day. Recovery from clinical signs apparently does not occur, chronic cases with static signs being dummies.

The importance of this disease is probably greater than the level of reporting would suggest, but much remains to be learned of the characterization of the toxic metabolites and on the spectrum of pathologic changes. It is a mycotoxicosis, the implicated fungus being *Fusarium moniliforme*. The preferred substrate for the fungus is corn (maize), and moldiness occurs in warm, moist conditions; these are the circumstances in which outbreaks of the disease occur. Sporadic cases occur in other circumstances not involving corn, but in which moldiness of fodder is involved. The neurologic disease occurs in the donkey and mule as well as in the horse.

The neurologic signs described in the spontaneous disease are related to the malacic injury. However, in the experimental disease and a small proportion of spontaneous cases, the nervous signs may include mania rather than neurologic deficit; these dramatic cases may be icteric, and the nervous signs are those usual in the horse in acute hepatic failure. In natural and experimental cases there may be hepatic lesions varying greatly in severity. The hepatic lesions when present are similar to those produced in other species by aflatoxin (see Volume 2, Chapter 2, Liver and Biliary System).

The lesion usually described is necrosis of the white matter of the cerebral hemispheres (Fig. 3.66). The surface of the brain may be unaltered on inspection, but palpable softness may be detected in the cortex overlying large areas of leukomalacia. Apparently there is not, by gross inspection, any cerebral edema or brain swelling, although the overlying gyri might be slightly flattened and discolored. The foci of softening may be of microscopic dimension, but usually they are readily visible on the cut surface. The softenings occur at random in the white matter of the cerebral hemispheres. They may be bilateral but are not necessarily so even when severe in one hemisphere; when bilateral, they are not necessarily symmetrical. The malacic foci are soft, pulpy, grayish depressions quite distinct by virtue of numerous small hemorrhages in a peripheral zone of a few millimeters' breadth. Depending on the duration, there may be a diffuse edematous swelling, yellowish in color, of the adjacent white matter.

In small foci, and probably initiating them all, there is severe edema of the white matter. The white matter is

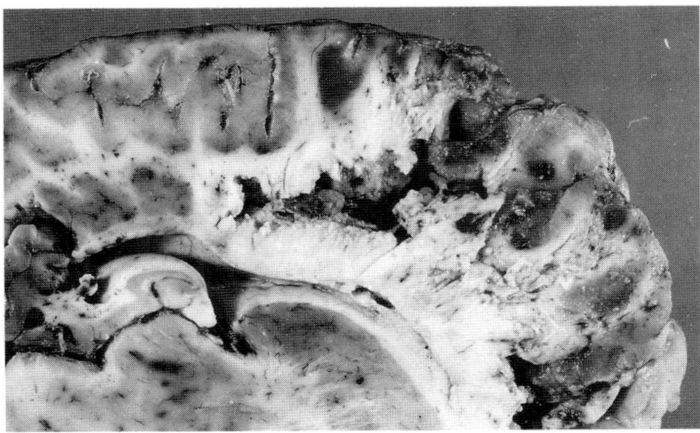

Fig. 3.66 Mycotoxic leukoencephalomalacia. Horse. Irregular malacia and hemorrhage in cerebral hemisphere. (Courtesy of T. M. Wilson.)

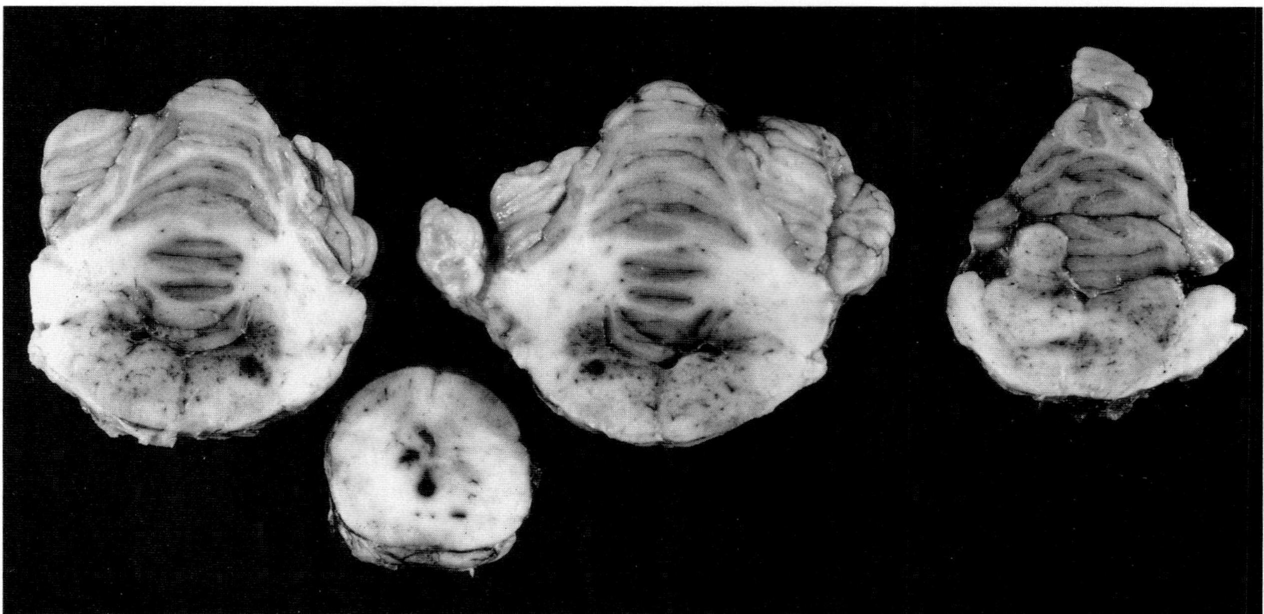

Fig. 3.67 Mycotoxic leukoencephalomalacia. Horse. Focal hemorrhage and malacia in brain stem. (Courtesy of T. M. Wilson.)

spread apart by fluid, and the myelin sheaths, axons, and glia disintegrate to form a structureless, acidophilic, semifluid mass to which the microglia react. The edematous change is not confined to the cerebrum but is reported to occur in all parts of the spinal cord. Foci of softening also occur in the brain stem and spinal cord, but here, in contrast with the brain, the necrosis affects the gray matter chiefly (Fig. 3.67).

Microscopically the areas of malacia are widely distributed and irregular in their form, mainly in the white matter, but a few small foci may be seen in the cortex. The irregular cavitations tend to surround and follow the course of the blood vessels, and there is edematous separation of tissue extending from the periphery of the cavitations (Fig. 3.68A). Cellular infiltrations, mainly of eosinophils but with some plasma cells, are present in the walls of vessels, in the perivascular spaces, and lightly in the edematous parenchyma (Fig. 3.68B). The adventitia of the vessels, especially in the brain stem, may be remarkably thickened. There is an abundance of lipofuscin pigment in macrophages.

This histological description summarizes the main lesions. Although the pathogenesis and evolution of the changes are not known in detail, it is deduced that they result from injury to blood vessels and are exaggerated expressions of the microcirculatory lesions described in an earlier section. Indeed, depending on the stage at which the brain is examined, the gamut of changes may be present, or the very acute disease may be expressed as microcirculatory lesions with edema but without macroscopic malacia.

7. Lead Poisoning

Lead is perhaps the most consistently important poison in farm animals. Poisoning is common and fatal in cattle, and less common but fatal in sheep; it is occasionally observed in horses and dogs, but must be rather rare in swine. The disease in cattle is probably always acute, whereas that in horses is usually, if not always, chronic. The signs are almost exclusively neurologic even though the amount of lead deposited in nervous tissue is relatively very small.

The usual sources of lead for cattle are paint and metallic lead in storage batteries. Adult cattle are most frequently poisoned at pasture by licking paint or putty cans from rubbish dumps. Calves are usually poisoned when from boredom or allotriophagia they lick painted pens, troughs, etc. Metallic lead does not reliably produce poisoning in ruminants under experimental conditions but frequently does so naturally, perhaps because the metal, when well weathered, contains soluble salts on its surface. Lead, chronically ingested, does accumulate in the tissues of ruminants, but it is not a cumulative poison in these species. Under conditions of chronic assimilation in ruminants, large amounts of lead may be stored in the tissues, including the brain, without causing lesions or clinical disease. Cows have been shown to tolerate 2 g of lead daily for as long as 2 years without apparent harm. Sheep may accumulate large amounts of lead in the course of grazing over abandoned lead-mining areas which have been converted to pasture, but the severe osteoporosis that may develop cannot be attributed to the accumulated lead, but may be related to copper deficiency.

The situation in horses is largely the reverse of that in ruminants because in horses lead poisoning is usually chronic. This probably represents in part a species susceptibility but is in part due to the manner of exposure in that horses are usually poisoned by the prolonged inhalation

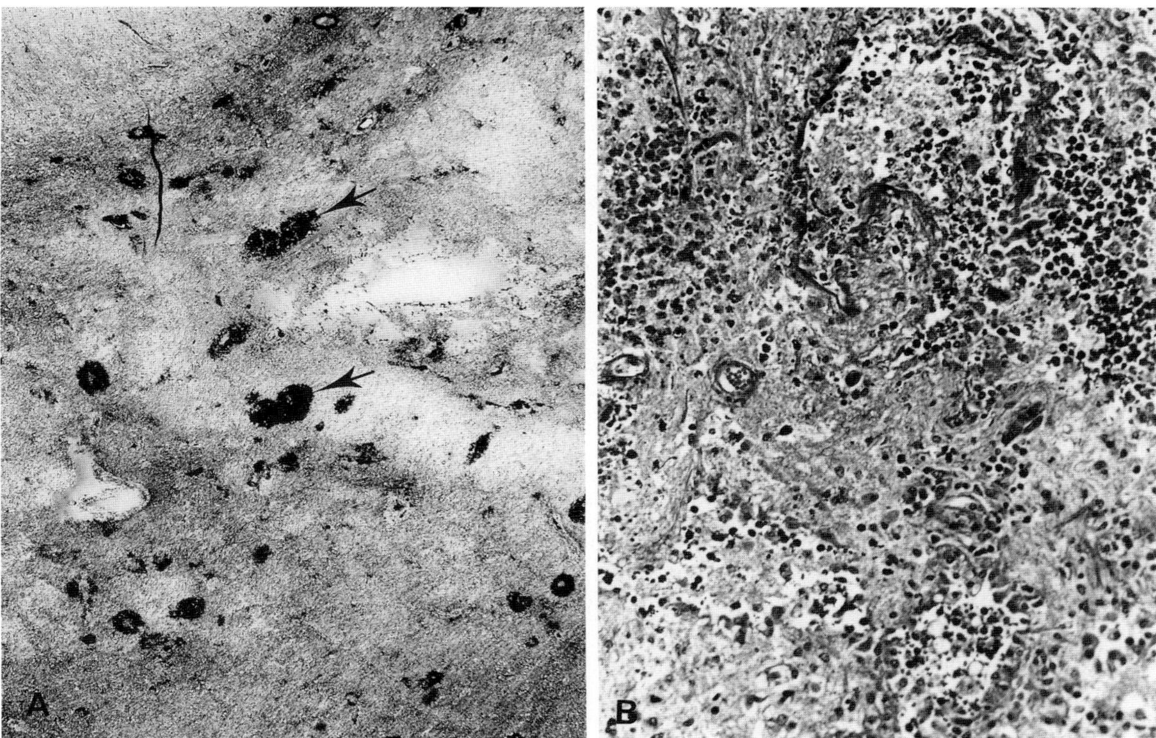

Fig. 3.68 Mycotoxic leukoencephalomalacia. Horse. (A) Malacia and hemorrhage (arrows). Myelin sheath stain. (Courtesy of E. W. Hurst). (B) Vascular degeneration in malacic foci. Infiltrating cells are eosinophils and hypertrophied microglia. (Compound granular corpuscles).

of fumes from lead smelters in which they work or by prolonged ingestion of pastures contaminated by such effluent. Contaminated pastures may contain as much as 100 to 200 parts per million (ppm) of lead in the foliage.

Lead poisoning may occur in circumstances additional to those cited. Boiled linseed oil contains lead, and it is occasionally mistakenly used as a laxative for animals. Dogs are occasionally poisoned by drinking gasoline that contains tetra-ethyl lead. Lead pipes may yield significant amounts of lead to soft water.

Lead arsenate, widely used as an orchard spray, is commonly responsible for poisoning of cattle and of sheep, but the signs and lesions refer largely to activity of the arsenate.

Lead poisoning is common and important in both humans and other animals, but in spite of this, its pathogenesis is still rather obscure. The element is usually obtained by ingestion, but only a small proportion of the ingested dose is absorbed, something on the order of 1 to 2% even of a soluble salt such as lead acetate. The limitation of absorption is due largely to formation of insoluble complexes in the alimentary tract. Absorbed lead is slowly excreted in the bile, milk, and urine, and is deposited in tissues, especially in the liver and kidney in acute poisoning and in the bones in chronic poisoning. The turnover of deposited lead is slow but continuous with gradual elimination in the bile and urine.

Lead, like other heavy metals, is usually regarded as a protoplasmic poison, which does not really say very much. Most metals exert their most deleterious effects at the sites of absorption and elimination, but lead largely spares these and affects nervous tissue with a high degree of specificity. Most metals are also enzyme inhibitors by combination with sulfhydryl groups of enzyme proteins, but this does not explain the relative neurotropism of lead.

The various species differ in their sensitivity to lead, and the clinical syndromes produced also differ somewhat, although they are always chiefly neurologic. Quotations on toxic doses for animals differ, and there seems to be remarkably little correlation between the doses of lead which will produce experimental poisoning and the very small amounts that induce toxicity naturally. Young animals are relatively more susceptible than adults.

The acute poisoning in cattle usually leads to death in 12 to 24 hr. Calves stagger, develop muscle tremors, and rapidly become recumbent. Convulsions are intermittent until death, and between convulsions there is opisthotonus, muscular tremors, champing of the jaws, and hyperesthesia to touch and sound. Adults show less tendency to early recumbency. In these there is frenzy, head pressing, and apparent blindness, with death in convulsions. When the poisoning is less acute, cattle may survive for 4 to 5 days. They are dull and apathetic, apparently blind, and without appetite. There may be salivation, intermittent grinding of teeth, and hyperesthesia, but dullness and immobility predominate. Ruminal atony is fairly constant, and dark fetid feces may be passed terminally. Death occurs quietly or in convulsions or from misadventure. The manifestations of lead poisoning in sheep are similar to those of the subacute syndrome in cattle.

The disease in horses is paralytic. When horses ingest large amounts of lead, they develop severe depression and general paralysis, sometimes with clonic convulsions and abdominal pain. In chronic poisoning, usually known as chronic plumbism, the characteristic signs are those of specific nerve pareses affecting chiefly the cranial nerves and expressed particularly as laryngeal and pharyngeal paralysis. The clinical manifestations in dogs are also neurologic, but their pattern is not characteristic. There is anorexia, emaciation, mental irritability, muscular tremors, ataxia, and intermittent convulsions.

The diagnosis of lead poisoning is necessarily chemical because lesions are either absent or nonspecific. Because of the variability of pathogenetic factors and species susceptibility previously discussed, precise figures cannot be given for the concentration of lead in tissues that can be regarded as indicating lead poisoning. As little as 4 to 7 ppm of lead has been found in the livers of horses dying of chronic intoxication. In cattle, levels of lead (on a wet-weight basis) of 40 ppm or more in the kidney and 10 ppm or more in the liver are accepted as confirmatory of lead poisoning.

Specific lesions are not observed in lead poisoning in cattle. In those cases that survive for 4 to 5 days, the ruminal contents may be foul in consequence of immotility, and the lower gut may contain a small volume of dark fetid feces, the color to be attributed to lead sulfide. There is no gastroenteritis, and any clinical signs of colic are probably to be related to nervous dysfunction. There may be mild degenerative changes in the parenchymatous organs, but these cannot be attributed to lead.

There is moderate brain swelling but probably never severe enough to cause displacement and not often severe enough to be appreciated grossly. Even microscopically, the cerebral edema is not easy to appreciate or to distinguish from early autolytic change. The capillaries and venules are congested, and there may be petechial hemorrhages. The prominence of capillaries in the gray matter is due largely to congestion, but there may be some endothelial swelling and proliferation to make them more conspicuous. Neuronal changes are equivocal. In some subacute cases in which the course is prolonged for several days, there may be laminar cortical necrosis. The best explanations of laminar necrosis are still based on ischemia-anoxia, implying that lead is not directly neurotoxic in terms of neuronal injury. Indeed, lead tends not to accumulate in nervous tissue. The capillary alterations may therefore be of considerable functional importance and probably precede the swelling of glia. The capillary changes are best seen in thick sections stained by a Nissl stain and in the cerebellum better than in the cerebrum. In the cerebellum, the vascular lesion in the molecular and Purkinje layer is often accompanied by astro- and microgliosis.

The basis of the paralytic changes in lead poisoning in horses appears not to have been described. Probably the local nerve pareses have the same histologic basis as the peripheral neuropathy of chronic plumbism in humans, in which the paralysis affects those muscles which are used most constantly and is due to segmental degeneration of axons and myelin in the distal parts of motor fibers. In humans, the peripheral neuropathy of lead poisoning is purely motor.

In dogs, there is edema of the white matter of the brain and cord. Degenerative changes in myelin sheaths occur almost universally but are somewhat less severe in heavily myelinated tracts such as the optic tract, corpus callosum, and peduncles than in lightly myelinated areas, such as the deep white matter of the cerebellum and cerebrum. There is spongy degeneration in the subthalamus, head of the caudate nucleus, and deep cortical laminae with extensive loss of neurons in such areas. Reaction on the part of astrocytes and microglia is slight.

Degenerative changes of mild nature occur in the liver and kidneys of dogs chronically poisoned with lead. The degree of change is such that it would, in routine examinations, be overlooked. Most conspicuous is enlargement and vesiculation of the nuclei in the convoluted tubules. Irregular, acid-fast, intranuclear inclusion bodies can be found in the renal tubules in some cases. These inclusions are irregularly present in cases of lead poisoning, but when present, they have diagnostic value. They may develop in any species poisoned, but in dogs, they must be distinguished from the bricklike, acidophilic inclusions that can be present in the nuclei of renal tubules and hepatic cells. These inclusions are also acid-fast, but have no specificity for lead poisoning and are present in the nuclei of apparently healthy dogs. In cattle, the degree of tubular epithelial degeneration is slight in most cases, but there is often a surprising degree of mitotic activity, and there may be severe nephrosis with extensive fibrosis in young calves.

Bibliography

Badialı, L. *et al.* Moldy corn poisoning as the major cause of an encephalomalacia syndrome in Egyptian equidae. *Am J Vet Res* **29:** 2029–2035, 1968.

Baker, D. C. *et al.* Toxicosis in pigs fed selenium-accumulating *Astragalus* plant species or sodium selenate. *Am J Vet Res* **50:** 1396–1399, 1989.

Bloom, H., Noller, B. N., and Shenman, G. A survey of blood lead levels in dogs and cats. *Aust Vet J* **52:** 312–316, 1976.

Christian, R. G., and Tryphonas, L. Lead poisoning in cattle: Brain lesions and hematologic changes. *Am J Vet Res* **32:** 203–215, 1971.

Evans, W. C. *et al.* Induction of thiamine deficiency in sheep with lesions similar to those of cerebrocortical necrosis. *J Comp Pathol* **85:** 253–267, 1975.

Gooneratne, S. R., Olkowski, A. A., and Christensen, D. A. Sulfur-induced polioencephalomalacia in sheep: Some biochemical changes. *Can J Vet Res* **53:** 462–467, 1989.

Goyer, R. A., and Rhyne, B. C. Pathological effects of lead. *Int Rev Exp Pathol* **12:** 1–77, 1973.

Hakim, A. M., and Pappius, H. M. Sequence of metabolic, clinical, and histological events in experimental thiamine deficiency. *Ann Neurol* **13:** 365–375, 1983.

Hamir, A. N. *et al.* An outbreak of lead poisoning in dogs. *Aust Vet J* **62:** 21–23, 1985.

Hammond, P. B., and Aronson, A. L. Lead poisoning in cattle and horses in the vicinity of a smelter. *Ann N Y Acad Sci* **111:** 595–611, 1964.

Harrison, L. H. *et al.* Paralysis in swine due to focal symmetrical poliomalacia: Possible selenium toxicosis. *Vet Pathol* **20:** 265–273, 1983.

Humphreys, D. J. Effects of exposure to excessive quantities of lead on animals. *Br Vet J* **147:** 18–31, 1991.

Lassen, E. D., and Buck, W. B. Experimental lead toxicosis in swine. *Am J Vet Res* **40:** 1359–1364, 1979.

Lee, J. Y. S., and Little, P. B. Studies of autofluorescence in experimentally induced cerebral necrosis in pigs. *Vet Pathol* **17:** 226–233, 1980.

Little, P. B., and Sorensen, D. K. Bovine polioencephalomalacia, infectious embolic meningoencephalitis, and acute lead poisoning in feedlot cattle. *J Am Vet Med Assoc* **155:** 1892–1903, 1969.

Marasas, W. F. O. Leucoencephalomalacia: A mycotoxicosis of equidae caused by *Fusarium moniliforme* Sheldon. *Onderstepoort J Vet Res* **43:** 113–122, 1976.

Marasas, W. F. O. *et al.* Leukoencephalomalacia in a horse induced by fumonisin B_1 isolated from *Fusarium moniliforme*. *Onderstepoort J Vet Res* **55:** 197–203, 1988.

McCulloch, E. C., and St. John, J. L. Lead-arsenate poisoning of sheep and cattle. *J Am Vet Med Assoc* **96:** 321–326, 1940.

McLeavey, B. J. Lead poisoning in dogs. *N Z Vet J* **25:** 395–396, 1977.

Okada, H. M., Chihaya, Y, and Matsukawa, K. Thiamine deficiency encephalopathy in foxes and mink. *Vet Pathol* **24:** 180–182, 1987.

Palmer, A. C., and Rossdale, P. D. Neuropathological changes associated with neonatal maladjustment syndrome in the thoroughbred foal. *Res Vet Sci* **20:** 267–275, 1976.

Palmer, A. C., Lamont, M. H., and Wallace, M. E. Focal symmetrical poliomalacia of the spinal cord in Ayrshire calves. *Vet Pathol* **23:** 506–509, 1986.

Pentschew, A. Morphology and morphogenesis of lead encephalopathy. *Acta Neuropathol* **5:** 133–160, 1965.

Pentschew, A., and Garro, F. Lead encephalomyelopathy of the suckling rat and its implications on the porphyrinopathic nervous diseases. *Acta Neuropathol* **6:** 266–278, 1966.

Read, D. H., and Harrington, D. D. Experimentally induced thiamine deficiency in beagle dogs: Clinical observations. *Am J Vet Res* **42:** 984–991, 1981.

Seidler, D., and Trautwein, G. Encephalopathie bei Thiaminmangel des Nerzes. *Acta Neuropathol (Berl)* **19:** 155–165, 1971.

Silbergeld, E. K., and Adler, H. S. Subcellular mechanisms of lead neurotoxicity. *Brain Res* **148:** 451–467, 1978.

Stevens, K. L., Riopelle, R. J., and Wong, R. Y. Repin, a sesquiterpene lactone from *Acroptilon repens* possessing exceptional biological activity. *J Nat Prod* **53:** 218–221, 1990. (Russian knapweed)

Studdert, V. P., and Labuc, R. H. Thiamine deficiency in cats and dogs associated with feeding meat preserved with sulphur dioxide. *Aust Vet J* **68:** 54–57, 1991.

Thornber, E. J. *et al.* Induced thiamin deficiency in lambs. *Aust Vet J* **57:** 21–26, 1981.

Uhlinger, C. Clinical and epidemiologic features of an epizootic of equine leukoencephalomalacia. *J Am Vet Med Assoc* **198:** 126–128, 1991.

Van Gelder, G. A. *et al.* Behavioral toxicologic assessment of the neurologic effect of lead in sheep. *Clin Toxicol* **6:** 405–418, 1973.

Wells, G. A. H., Howell, J. McC., and Gopinath, C. Experimental lead encephalopathy of calves. Histological observations on the nature and distribution of the lesions. *Neuropathol Appl Neurobiol* **2:** 175–190, 1976.

Wilson, T. M., and Drake, T. R. Porcine focal symmetrical poliomyelomalacia. *Can J Comp Med* **46:** 218–220, 1982.

Young, S., Brown, W. W., and Klinger, B. Nigropallidal encephalomalacia in horses fed Russian knapweed (*Centaurea repens L.*). *Am J Vet Res* **31:** 1393–1404, 1970.

Zook, B. C. The pathologic anatomy of lead poisoning in dogs. *Vet Pathol* **9:** 310–327, 1972.

Zook, B. C. Lead intoxication in urban dogs. *Clin Toxicol* **6:** 377–378, 1973.

F. Neurodegenerative Diseases

There are some significant neurologic diseases in which dramatic clinical disturbances are not matched by equivalent morphologic alteration in nervous tissue. For instance, toxins which interfere with synaptic function can have fatal consequences, yet leave neurons normal in appearance to routine examination. **Botulism, tetanus,** and **strychnine toxicosis** are well-known examples. In the latter two conditions, the release of spinal motor neurons from the inhibitory influence of Renshaw cells (inhibitory neurons) leads to the extensor spasms and hyperesthesia which characterize them. The activity of the inhibitory neurotransmitter glycine is blocked—in the case of tetanus by presynaptic blockade of its release, and in the case of strychnine, by post-synaptic blockade of its receptors. An inborn metabolic equivalent of these poisonings is so-called **hereditary myoclonus of polled Hereford calves,** in which there is an absence of glycine receptors on spinal motor neurons. In the Hereford disease, the calves manage to be born normally, but suffer severe tetanic spasms on stimulation, are unable to stand, and are nonviable.

A distinctive locomotor disturbance in Australian sheep (Coonabarabran staggers, cathead staggers) has been associated with grazing of the zygophyllaceous plant *Tribulus terrestris*. The clinical disease is characterized by asymmetric atrophy of pelvic limb extensor muscles, and irreversible asymmetric para- or tetraparesis without ataxia. There are no significant structural lesions in the nervous system. Evidence has been presented to suggest that the syndrome is caused by beta-carboline alkaloids acting on dopaminergic upper motor neurons in the nigrostriatum. The long-term nature of the deficits may be the result of the ability of such compounds to form adducts with DNA sequences of genes associated with the synthesis of dopamine. Locomotor disturbances have also been documented in sheep and cattle grazing other zygophyllaceous plants in North America and Africa.

A large number of plant intoxications cause acute neurologic disease without morphologic alterations. **Tremorogenic mycotoxicoses** are described in which lesions may be absent; *Claviceps paspali,* the ergot of *Paspalum,* causes intoxication in cattle, occasionally in sheep and rarely in horses.

Inherited **spasticity syndromes** are described in cattle as spastic paresis and spastic syndrome and in Scottish terriers as scotty cramp. Various inherited **myotonic syndromes** are referred to in Chapter 2, Muscles and Tendons. The common **tetanic/paretic syndromes** which accompany hypocalcemia and hypomagnesemia are not associated with significant neural lesions.

This section will deal with noninflammatory diseases in which there is selective neuronal degeneration, involving either neurons in their entirety (neuronopathies) or restricted to axons (axonopathies). In reality, this division is somewhat arbitrary, as the axon is a wholly dependent part of any neuron, but the concept is useful for the classification of various diseases.

In general, when entire neurons or just their axons are lost in these circumstances, there are reactions by adjacent tissue elements but, as there is often minimal overall tissue loss, the predominant pathologic change is microscopic. At the end stage of severe chronic diseases, macroscopic atrophy may become evident (see the cytopathology section of this chapter).

The diseases have been grouped into three categories of neuronopathies and axonopathies: central, peripheral, and central and peripheral, according to the distribution of the lesions in the central and peripheral nervous systems. The choice of category for a particular disease is based on the predominant pattern of morphologic change seen in a typical case, and the criteria are not absolute. There will inevitably be some overlap, and the naming and lesion topography of each may not necessarily be fully justified or complete. Some of the diseases are rare, and, because they are inherited, are highlighted by research workers aiming to use them as experimental model systems. New entities are progressively being identified. Sampling of the nervous system in diseases such as these should be as comprehensive as possible. The pattern and extent of lesions are important criteria in the diagnosis of known diseases, and in the documentation of new ones.

Bibliography

Bourke, C. A. A novel nigrostriatal doperminergic disorder in sheep affected by *Tribulus terrestris* staggers. *Res Vet Sci* **43:** 347–350, 1987.

Bourke, C. A., Carrigan, M. J., and Dixon, R. J. Upper motor neuron effects in sheep of some beta-carboline alkaloids identified in zygophyllaceous plants. *Aust Vet J* **67:** 248–251, 1990.

Gundlach, A. L. *et al.* Deficit of spinal cord glycine/strychnine receptors in inherited myoclonus of polled Hereford calves. *Science* **241:** 1807–1810, 1988.

Harper, P. A., Healy, P. J., and Dennis, J. A. Inherited congenital myoclonus of polled Hereford calves: A clinical, pathological, and biochemical study. *Vet Rec* **119:** 59–62, 1986.

Nicholson, S. S. Tremorgenic syndromes in livestock. *Vet Clin North Am: Food Anim Pract* **5:** 291–300, 1989.

Osweiler, G. D. *et al.* "Clinical and Diagnostic Veterinary Toxicology," 3rd Ed. Dubuque, Iowa, Kendall/Hunt Publishing Company, 1985.

Seawright, A. A. "Animal Health in Australia," 2nd Ed. vol 2. Chemical and Plant Poisons. Canberra, Australia, Aust. Gov. Printing Services, 1989.

1. Central Neuronopathies and Axonopathies

a. Compressive Optic Neuropathy Compression of optic nerves in the optic foraminae is a well-known manifestation of vitamin A deficiency in the calf and pig due to stenosis of the foraminae. In poisoning by plants of the genera *Stypandra* and *Helichrysum* and by halogenated salicylanilides, the foraminae are of normal size.

These conditions are discussed fully under myelinopathies, but one usual feature of their pathology is intense Wallerian degeneration of the optic nerves and tracts, which may be due to swelling and compression of the nerves within the optic canals. At present this pathogenesis is a presumption rather than an established fact. If such is the case, the acute phase of the lesion would be expected to show severe ischemic necrodegenerative changes involving all the nervous tissue elements localized within the optic canals, with acute axonal degeneration extending along the optic nerves, through the chiasm, and into the optic tracts of the midbrain. The retrobulbar segments of the optic nerves should be minimally affected. With the passage of time, the site of compression should exhibit malacic alterations, gitter-cell infiltration and astrogliosis, while the progression of Wallerian changes in the optic nerves and tracts should lead to extensive loss of myelinated axons and oligodendrocytes, with a dense reactive astrogliosis. In the long term, retrograde degenerative changes would be expected to lead to the loss of retinal ganglion cells and their axons.

b. Organomercurial Poisoning Mercurial compounds are cumulative poisons. The syndromes produced by the organic and inorganic salts are quite different, but the differences are probably not qualitative.

The toxicity of the organic salts depends on their solubility, and the syndrome produced depends on the size and chronicity of dosage. Poisoning by inorganic salts is expected to be by ingestion, but percutaneous absorption is possible. Poisoning by inorganic mercury is quite rare in animals. Acute toxicity following ingestion is characterized by severe abdominal pain, vomiting, and diarrhea caused by the coagulative effect of mercury in the lining of the gut. Death may occur in a few hours. If the animal survives the initial episode, death may occur several days later from nephrosis with uremia. There may at this stage be ulcerative colitis due to the concentration of mercury in the colonic mucosa, by which route it is in part eliminated. Chronic cumulative poisoning probably does not occur in animals.

Mercurialism in animals is usually caused by the organic

salts, such as ethyl mercury phosphate and mercury *p*-toluene sulfonanilide, which are applied to seed grains for the purpose of controlling fungal diseases of the germinating plants. Seed grains treated with mercury are occasionally fed accidentally, but the feeding is usually deliberate with a hope for the best. Toxicity may not be manifested if the treated grain is fed for a short period only and does not constitute more than 10% of the ration. Chronic mercurialism occurs chiefly in swine and is occasionally observed in cattle, but is most unusual in other domestic species. The manifestations are almost solely neurological.

Minamata disease is a mercurialism of cats, birds, and humans in the Minamata Bay area of Japan. It is associated with the eating of fish and shellfish from the bay. The fish contain large amounts of organomercurials, which probably spilled into the bay with industrial effluent. Similar environmental disease occurs in other countries.

The clinical syndrome of mercurialism in cattle closely resembles that of lead poisoning, but the signs, which are of sudden onset, may not occur for several weeks after the first of continuous exposures. Because degeneration of the conducting system of the heart is common, cardiac irregularities are to be expected. Signs may occur in swine as early as 15 days after being fed treated grain. There is loss of appetite, wasting, dullness, blindness, severe weakness, and incoordination, progressing rapidly until the animal can no longer stand. The dullness passes to coma, which may last for several days with intermittent episodes of clonic–tonic convulsions. Both swine and cattle may be moderately uremic.

Gross lesions are minimal. In both swine and cattle, the kidneys may be moderately swollen, pale, and wet, and in pigs there is often hydropericardium. Poisoned cattle have a mild to moderate cerebral swelling and displacement. The cerebral cortices are soft and pale, and occasionally there is juxtasagittal venous infarction. The pallor of the cortex is visible on the cut surface so that there may be no clear distinction between gray and white matter; this change is irregular in distribution. Where the gray matter is distinct from the white, a narrow pallid band separating them may be visible. In swine, the dead white color of the cortex is often striking, especially when it is emphasized by the ischemia which may be present. The appearance of the two hemispheres may be different, that uppermost being pallid and ischemic, and that of the side on which the animal is lying having congested meningeal veins. The brain is swollen, and the gyri are flattened, although this is difficult to appreciate in pigs, in which swelling is never severe.

Microscopically, there is nephrosis of mild to moderate degree, usually without clear evidence of significant tubular epithelial necrosis, although this is a matter determined by dosage. The tubular epithelium is swollen to such an extent that it may obliterate the tubular lumen. There is microscopic proteinuria. Degeneration of the Purkinje network in the heart is observed fairly consistently in cattle, but it may not be present when the course of the toxicity is short, and seems not to occur in swine. There is acidophilic, coagulative necrosis, irregularly but extensively distributed in the Purkinje network (Fig. 3.69). The degenerate substance fragments and is removed by histiocytes and replaced by connective tissue. Mineralization is common (Fig. 3.69). There are no clearly demonstrable degenerative changes in the myocardium.

Degenerative changes are consistently present throughout the nervous system in both cattle and swine. They differ quantitatively from case to case but are qualitatively the same. The lesions are rather more rapidly catastrophic in cattle than in swine. There is acute neuronal degeneration. This is of ischemic type and largely of ischemic distribution, the neurons of the middle laminae of the cerebral cortex being extensively injured. However, shrunken, acidophilic neurons can be found at all levels of the brain and cord and, with time, there is extensive cell loss from all cortical laminae. Moderate gliosis accompanies the neuronal degeneration and is expected to be most prominent in those areas of the cortical gray matter which, on gross inspection of the cut surface, are pallid. The gliosis is in part astrocytic and in part microglial. Compound granular corpuscles are not formed, the reactive microglia being prominently rod-shaped. There is edema in the subjacent white matter. In swine, the edema

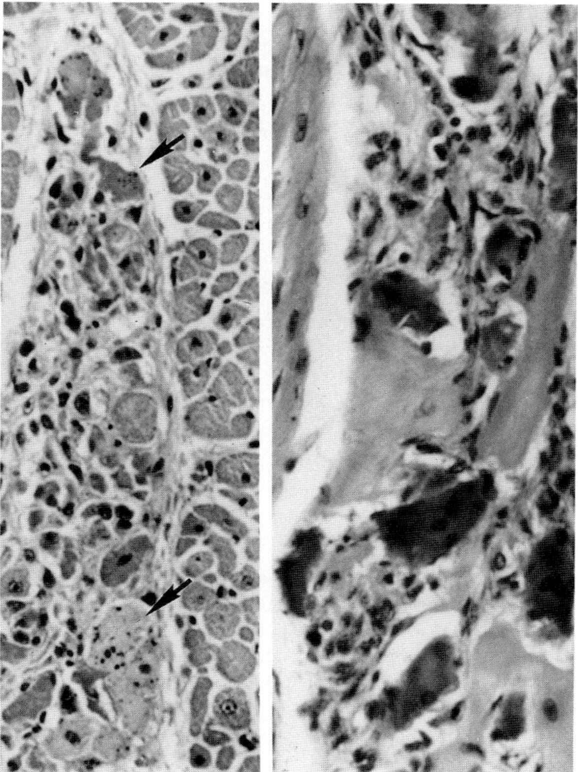

Fig. 3.69 Organomercurial poisoning. Degeneration of Purkinje network in myocardium (left, arrows). Fibrosis and mineralization of Purkinje network (right). Ox.

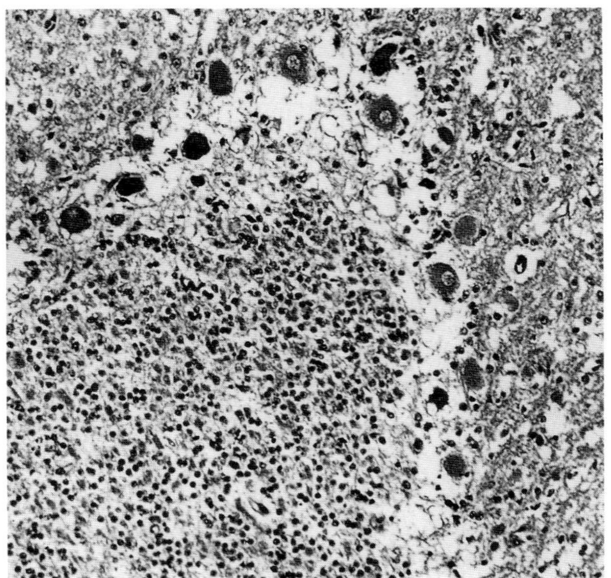

Fig. 3.70 Organomercurial poisoning. Purkinje cell necrosis and degeneration in granular layer of cerebellar cortex. Ox.

leads in time to demyelination, but in this species there is no softening. In cattle, the edema of white matter is more severe and, in the multiform layer and middle laminae, frequently produces a spongy degeneration, which may be visible grossly. This occasionally leads to laminar necrosis in cattle.

In both cattle and swine, fibrinoid necrosis of the media of leptomeningeal arteries is quite characteristic (Fig. 3.45). The change probably occurs in all cases, affecting both cerebral and spinal vessels, but it is well marked in a few cases only. Swine are more prone to the vascular lesions than are cattle. Accompanying the vascular lesions, there is an outpouring of much fibrin into the meningeal spaces.

Organomercurials appear rather selectively to damage the granule cells of the cerebellum in all species (Fig. 3.70). Cattle are more sensitive to this degeneration than are swine. In swine, the Purkinje cells remain fairly intact, but in cattle they are often lost from about the tips of the folia. The degeneration of Purkinje cells is accompanied by microgliosis in the molecular layer and may be a response to swelling and pressure rather than a direct response to mercury.

Perhaps because the course is usually short in cattle, the spinal neurons are only mildly injured. In swine, there is often extensive ischemic necrosis of individual relay neurons. This, together with peripheral neuropathy, accounts for paralytic phenomena in swine. There is axonal degeneration and demyelination in peripheral nerves, the degenerative changes being more severe when the course is prolonged.

c. *SOLANUM* SPECIES POISONING Several of the many hundreds of species of the plant genus *Solanum* are toxic to livestock by a variety of means. The disease of cattle in southern Africa known colloquially as **maldronksiekte** is associated with grazing *Solanum kwebense*. Affected animals show little abnormality at rest, but when moved, exhibit head tilt, muscle tremors, incoordination, and convulsions.

The brain at necropsy may show gross evidence of cerebellar atrophy, uniformly affecting all lobes. Histologically there is diffuse loss of Purkinje cells, gliosis in the molecular layer, and atrophy of both molecular and granular layers. Surviving Purkinje cells are swollen, with strongly acidophilic cytoplasm, which contains numerous small "empty" vacuoles. Neurons elsewhere in the brain may show similar but far less intense vacuolation. Morphologically the vacuolation is suggestive of a lysosomal storage process and, ultrastructurally, there are membranous cytoplasmic inclusions similar to those seen in the sphingolipidoses. The nature of any stored material has not been reported to date.

Similar signs and lesions in cattle have been associated with the grazing of *S. bonariensis* in Uruguay, *S. dimidiatum* in the United States, and *S. fastigiatum* in Brazil.

d. CYCAD POISONING The primitive palmlike plants of this family (genera *Cycas, Zamia, Bowenia, Macrozamia*) occur in tropical and subtropical environments, and have been associated with the intoxication of both humans and animals. The seeds and young fronds contain the toxic glycosides, cycasin and macrozamin, and, following ingestion, the hepatic metabolism of the aglycone, methylazoxymethane, may cause acute zonal hepatic necrosis. The chemical identity of neurotoxic components responsible for the syndrome remains unknown.

In tropical Australia, and perhaps elsewhere, chronic exposure of cattle is associated with a neurological syndrome known as Zamia staggers. The syndrome is characterized by pelvic limb ataxia, which may become severe. Pathologically there is a distal axonopathy in the spinal cord. As would be expected, the ascending fasciculus gracilis and dorsal spinocerebellar tracts have axonal degeneration most intense in the cervical segments, and the same is true of descending ventrolateral tracts (Fig. 3.71). During active degeneration, there will be typical Wallerian changes, leading eventually to loss of myelinated axons and a reactive fibrous astrogliosis.

e. PERENNIAL RYEGRASS AND OTHER TREMORGENS Perennial ryegrass staggers is a common disease in parts of Australia, New Zealand, and Europe, and is occasionally reported in the United States. Sheep, cattle, and horses may be affected. It occurs in the summer and autumn on dry, short pastures of *Lolium perenne,* and clinical signs appear 5–10 days following exposure. Animals develop fine head tremors and head nodding and weaving at rest and, if forced to move, have a stiff-legged, incoordinate gait and are inclined to collapse in tetanic spasms, from which they quickly recover. There is low mortality, and total recovery will take place within 3 weeks of removal

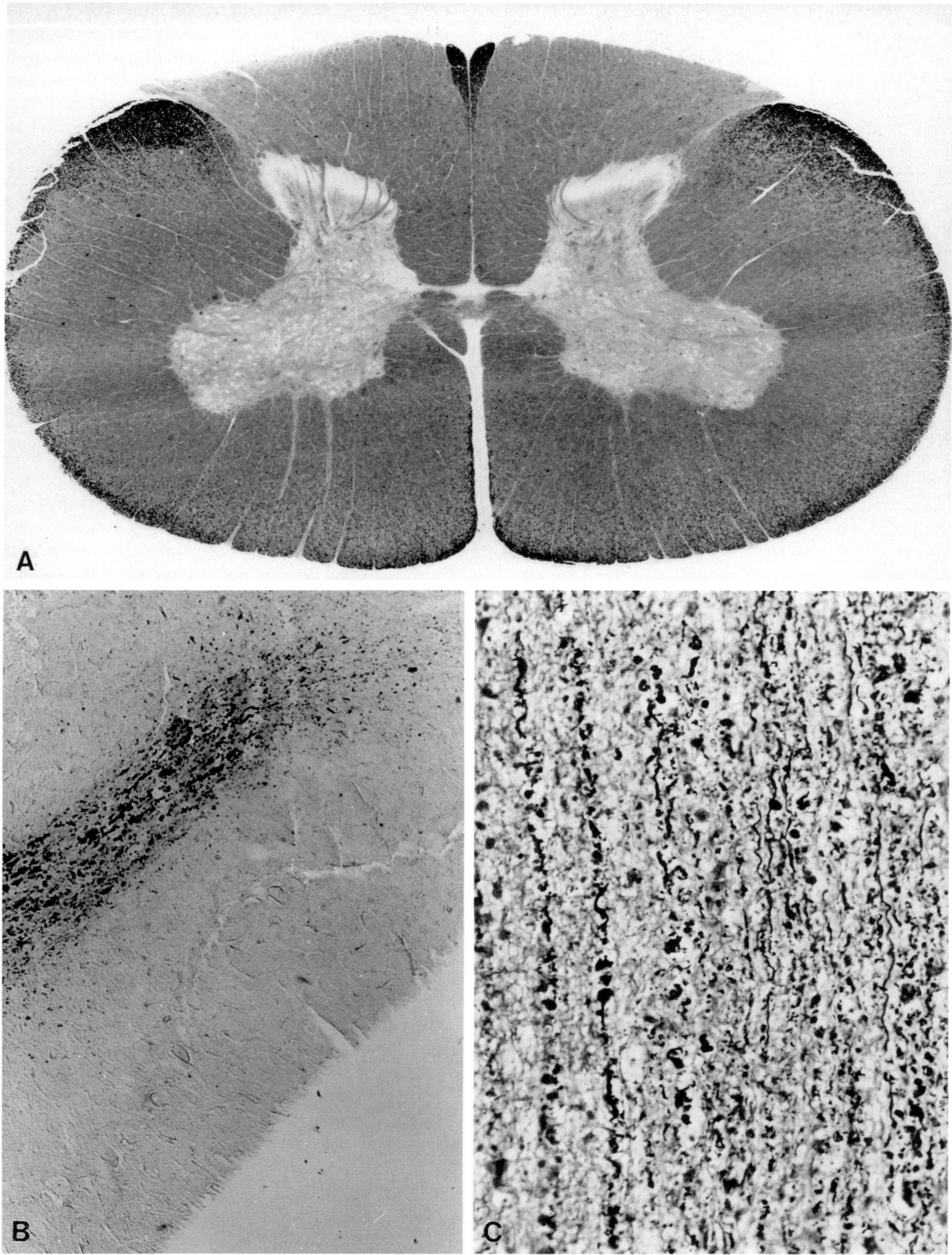

Fig. 3.71 Cycad poisoning. Ox. (Courtesy of M. D. McGavin and *Pathologia Veterinaria*). (A) C_8. Symmetric degeneration of myelin in fasciculus gracilis and dorsal spinocerebellar tracts due to distal "dying back" axonopathy. Marchi. (B) Cerebellum. The black degenerated fibers in the white matter are part of the dorsal spinocerebellar tract. Marchi. (C) Thoracic cord. Longitudinal section of fasciculus gracilis. Degenerated (black) and normal (gray) axons. Guillery, axon stain.

from toxic pasture. The disease is a mycotoxicosis due to indolic **lolitrems** produced by the endophytic fungus *Acremonium loliae*. The toxins belong to a group having similar activity, and are collectively known as **tremorgens.** The most similar disease is paspalum ergotism, in which indolic toxins are produced by the fungus *Claviceps paspali* growing on grasses of the *Paspalum* genus.

There are no gross lesions in lolitrem toxicosis, and microscopic change is limited to the occurrence of fusiform enlargement of the proximal axons of some cerebellar Purkinje cells. These axonal swellings are known as torpedoes and represent a proximal axonopathy, but their relationship to the clinical signs is unclear. The axonal lesions are best demonstrated by the use of silver staining techniques, such as the Holmes method. They have not been reported in any other tremorgenic diseases, which include several other phyto- and mycotoxicoses.

f. Equine Degenerative Myeloencephalopathy This is a chronically progressive disease of unknown cause, which has been described in several breeds and strains of equidae, including the zebra. The average age of horses at initial presentation is about 6 months, but may be as much as 24 months. The clinical presentation is dominated by disturbances of general proprioception and upper motor neuron function referable to the spinal cord and thus closely resembles a focal compressive myelopathy of the midcervical region. However, no skeletal or other gross lesions can be demonstrated.

Pathologically, there is evidence of ongoing Wallerian degeneration and post-Wallerian astrogliosis in all funiculi throughout the spinal cord, but concentrated in the ascending dorsolateral (spinocerebellar) and descending ventromedial (motor) funiculi of the anterior cervical and mid-thoracic segments. The changes are often most severe in the thoracic segments. Similar but mild changes are present in myelinated tracts of the caudal medulla and caudal cerebellar peduncles.

When well established in the thoracic cord segments, the process leads to considerable loss of myelinated axons and a dense reactive astrogliosis, which also involves the adjacent glia limitans. In some cases, the destructive process seems to reach an end point, with dense gliosis but little or no active Wallerian change. Chromatolytic and necrotic neurons, and axonal spheroids, can be found in Clarke's column (the nucleus of the dorsal spinocerebellar tract), which is located just dorsolateral and immediately adjacent to the central canal of the cord between the first thoracic and midlumbar segments. The disease involves the total destruction of many neurons in this system, and macrophages containing a lipoidal pigment can also be found in this location.

The origin of the degenerating descending axons in the ventromedial tracts is uncertain, and to date there have been no reports of neuronal degeneration in the midbrain nuclei likely to be the source of these fibers. It is possible that the descending tract lesion is purely an axonopathy. Axonal spheroids and vacuoles have also been described in the lateral cuneate and gracilis nuclei of the posterior brain stem. These are the relay nuclei receiving the long ascending proprioceptive tracts of the spinal cord. However, even though this disease occurs in young horses, the changes in these nuclei may be nonspecific, as they are commonly seen in a variety of situations, including animals with no overt clinical neurologic disease. In contrast to compressive cervical myelopathy (the wobbler syndrome), there is simultaneous involvement in multiple cord segments of ascending and descending tracts. This excludes focal compression of the cord as an etiologic factor. There is no obvious pathogenetic link between the various neuronal systems involved. The epidemiology of the disease suggests an environmental agent as the cause, and some evidence of prevention by the administration of vitamin E supplementation has been reported. The role of copper deficiency has been investigated on the basis of the similar pattern of lesions seen in delayed swayback in lambs.

g. The Axonal Dystrophies Diseases in which the pathology is characterized by axonal dystrophy, as described under cytopathology (Fig. 3.23A,B,C), have been described in sheep, horses, dogs, and cats. One of the earliest to be documented occurred in **Suffolk sheep.** In this disease, described in California, lambs normal at birth develop a progressive ataxia between about 1 and 6 months of age. The lambs eventually become recumbent after a course of 10 to 12 weeks. It is virtually certain that the disease is inherited but, to date, the mode of inheritance is unknown, other than to say that it is of recessive character.

There are no specific gross changes, and the diagnostic histologic lesion is the presence of numerous axonal spheroids, in and adjacent to several gray matter areas. The consistently affected areas are as follows: the entire spinal gray matter, with spheroids in greatest numbers at the base of the dorsal horns, in Clarke's column, and in the intermediolateral nuclei; the posterior brain stem nuclei, gracilis, cuneate, accessory cuneate, inferior olivary, lateral reticular, and lateral vestibular; the cerebellar roof nuclei with the exception of the dentate; and the anterior colliculi and lateral geniculate nuclei. In the original description of this disease, it was correctly concluded that the spheroids were the focally swollen terminations of sensory axons, and represented a genuine axonal dystrophy. The pattern of lesions indicates the involvement of two sensory systems, one visual and the other proprioceptive and possibly exteroceptive, and the clinical signs could be accounted for by perturbation of the latter.

More recently, a condition considered to be identical has been described in **Coopworth sheep** in New Zealand, and in Australia a disease in **Merino sheep** has been reported as an axonopathy, although the lesions described are consistent with axonal dystrophy. In this disease, as reported from localized areas of eastern Australia, sheep aged between 1 and 4 years develop severe posterior gait abnormalities, which are progressive and irreversible. Axonal spheroids are present in large numbers, predominantly in myelinated tracts in the midbrain and hindbrain and throughout the spinal cord, being more abundant in the dorsal funiculi than in the ventral. Wallerian changes are mild. In the brain, areas of predilection are the cerebellar peduncles, transverse pontocerebellar fibers, dorsolateral thalamic tracts, cuneate fasciculus, median longitudinal fasciculus, and corticospinal tracts. Rather than being concentrated at terminal regions, the spheroids appear to be multiple along the course of individual axons. The cause remains undefined but seems likely to be a heritable defect.

In the **Morgan horse,** a clinical syndrome of pelvic limb dysmetria and incoordination is accompanied by intense axonal dystrophy in the accessory cuneate nuclei. Although breeding experiments suggest a familial component, no definitive inheritance pattern has been demonstrated.

In **Haflinger horses,** a single report describes an ataxia syndrome first evident at about 4 months of age, with ataxia in the pelvic limbs more severe by about 2 years of age. Neuroaxonal dystrophy was evident as numerous spheroids in the nuclei gracilis, cuneate, solitary tract, and intermediomedialis, and in Clarke's column. These changes were accompanied by astrogliosis and lipofuscin pigmentation of neurons and macrophages, and significantly reduced serum tocopherol values. A familial hereditary basis is proposed.

Canine diseases classified as axonal dystrophies have been described in the Rottweiler, the collie sheep dog, and the Chihuahua.

In the **Rottweiler,** the disease is a familial progressive sensory ataxia, with a pattern of occurrence suggestive of autosomal recessive inheritance. Neurologic signs may be expressed before 12 months of age, often being first noticed as abnormal clumsiness. With time, there is steadily progressing ataxia and distinct hypermetria, particularly in the forelimbs, in which there is also toe-dragging and knuckling. Mild head tremor and incoordination may also become evident. By 4 to 6 years of age, nystagmus, crossed-extensor reflexes, and a positive Babinski sign may be present, but dogs remain alert and responsive. Strength and conscious proprioception are maintained throughout, and the signs relate mainly to a disturbance of unconscious proprioceptive input to the cerebellum.

At necropsy there may be mild patchy cerebellar atrophy, and the optic nerves appear small. The characteristic histologic feature is the presence of massive numbers of axonal spheroids (Fig. 3.23A,B,C) in the nerve root entry zone of the dorsal horn throughout the spinal cord, in the vestibular, lateral, and medial geniculate and sensory trigeminal nuclei, and in Rexed's laminae of the spinal cord. Fewer numbers of spheroids are present in the inferior olivary, trochlear, and oculomotor nuclei, and in the spinal cord ventral horn. Occasional spheroids are found in the globus pallidus, hippocampus, thalamus, hypothalamus, caudate nucleus, and reticular substance. In the cerebellum, lesions are concentrated in the vermis. Spheroids are present in the granular layer, white matter, and roof nuclei. There is some loss of Purkinje and granule cells.

The condition in the **collie sheep dog** has been reported from Australia and New Zealand, and affected animals have progressive ataxia and gait abnormalities first apparent at 2 to 4 months of age. The lesions are purely microscopic and confined to the deep cerebellar and vestibular areas. Many axonal spheroids are present in the central cerebellar white matter and adjacent roof nuclei, and in the lateral vestibular nuclei. Wallerian degeneration is minimal, and the cerebellar cortex, unaffected. The disease is strongly suspected to be inherited as an autosomal recessive trait.

In the **Chihuahua,** affected animals are normal until about 7 weeks of age, when there is sudden onset of tremor and gait disturbances. Large numbers of axonal spheroids are present in the white matter of the internal capsule, cerebellum, lateral geniculate nucleus, anterodorsal thalamus, acoustic tubercle, olivary nuclei, and the corticospinal and spinothalamic tracts. The pathogenesis remains undefined.

In the **cat,** only one condition qualifying as an axonal dystrophy has been reported. Shown to be an autosomal recessive trait, this condition was described in domestic short-haired cats in association with an unusual lilac coat color. Clinical signs first become evident at 5 to 6 weeks of age as head bobbing. In the ensuing 8 to 10 weeks, there is progressively worsening ataxia, and possibly visual impairment and vestibular deficits. The process may then stabilize, and animals, if cared for, may survive to adulthood, but are poorly grown. Gross neuropathological change is confined to slight cerebellar atrophy, most obvious in the posterior vermis. Histologically there are numerous spheroids in the inferior olivary and lateral cuneate nuclei. Lesser numbers of spheroids are in the lateral midbrain tegmentum, lateral and anterior ventral thalamic nuclei, and the cerebellar vermis. An additional feature is diffuse swelling of axons in these locations and in the medial lemniscus, medial longitudinal fasciculus, central tegmental tract, and spinal nerve dorsal roots. Neuronal loss and gliosis can be found in the inferior olivary and lateral thalamic nuclei, and in the Purkinje and granule cell layers in the cerebellar vermis. Spheroids and neuronal loss are also evident in the spiral ganglion and organ of Corti in the inner ear.

h. Congenital Axonopathy in Holstein–Friesian Calves. A series of cases is described from Australia in which calves were recumbent from birth, and exhibited spastic paresis and a variety of other neurologic signs.

The characteristic microscopic lesion is active Wallerian degeneration in all funiculi throughout the spinal cord, especially at the periphery. Similar changes are evident in the cerebellar peduncles, the median longitudinal fasciculus, spinocerebellar and rubrospinal tracts, and in the roots of cranial nerves III, V, VII, and VIII. In a few cases, mild degeneration occurs in the midbrain and some peripheral nerves. The Wallerian changes are accompanied by a very mild glial response, and it is likely that the disease begins *in utero* shortly before birth. The pathogenesis is unknown, and an autosomal recessive trait is suspected.

i. Myeloencephalopathy in Brown Swiss Cattle The first clinical sign of this disease, known colloquially as weaver syndrome, is slight ataxia, appearing at 5 to 8 months of age and worsening progressively over the ensuing 12 to 18 months. At the advanced stage, there is severe truncal ataxia and pelvic limb dysmetria, with distinct proprioceptive deficits. All this terminates in recumbency with its inevitable secondary complications. Present evidence strongly suggests an inherited autosomal recessive trait.

The major and primary lesion is in the spinal cord white matter. At all stages of the disease, there is active axonal degeneration, with both axonal lysis and spheroid formation. Axonolysis imparts an appearance of status spongiosus to affected white matter, and advanced lesions have myelin loss. The axonopathy is present in both ascending and descending tracts at all levels of the cord, but is most severe in the thoracic segment. It is suggested that the process begins in the thoracic segment and extends in antero- and retrograde directions from this site. Mild axonal lesions of a similar character occur in the brain stem, but are inconsistent in their location. However, a consistent additional lesion is degeneration and loss of Purkinje cells from the cerebellar cortex.

j. Multisystem Neuronal Degeneration of Cocker Spaniel Dogs In this condition, reported from Switzerland, red-haired cocker spaniel dogs of both sexes develop slowly progressive neurologic signs from about 6 months of age. The clinical picture is dominated by behavioral changes, disorders of gait and balance, tremors, and sometimes seizures. Ultimately, the severity of signs necessitates euthanasia.

Neuropathologic changes are bilaterally symmetrical and involve loss of neurons predominantly in the septal nuclei, globus pallidus, subthalamic nuclei, substantia nigra, tectum, medial geniculate nuclei, and cerebellar and vestibular nuclei. The neuronal loss is accompanied by gliosis and axonal spheroids. In addition, the white matter of the fimbriae of the fornix, central cerebellum, corpus callosum, thalamic striae, and subcallosal gyri have intense gliosis, axonal spheroids, loss of myelin, and perivascular macrophages. The cause remains undefined, but a genetic basis seems likely.

k. Neuronal Inclusion-Body Disease of Japanese Brown Cattle An acute neurologic disease of Japanese Brown cattle is described from the island of Kyushu, in which there is hyperexcitability, fever, profuse sweating, and usually sudden death. The cause is presently unknown, but cases recorded so far have all been in females.

There are no gross lesions of significance, but a large percentage of affected cattle have single, or sometimes multiple, eosinophilic cytoplasmic inclusion bodies in large neurons of the midbrain, pons, and medulla. The inclusions are mostly in the axon hillock region, are oval in shape, and about 18 μm in greatest diameter. Ultrastructurally, they appear as sequestrations of degenerate mitochondria, with associated aggregations of rough endoplasmic reticulum and lipofuscin bodies. They are considered to be distinct from any other type of neuronal inclusion so far described in animals or humans.

Bibliography

Baird, J. D. *et al.* Progressive degenerative myeloencephalopathy ("weaver syndrome") in Brown Swiss cattle in Canada: A literature review and case report. *Can Vet J* **29:** 370–377, 1988.

Baumgartner, W., Frese, K., and Elmada, I. Neuroaxonal dystrophy associated with vitamin E deficiency in two Haflinger horses. *J Comp Pathol* **103:** 113–119, 1990.

Chrisman, C. L. *et al.* Neuroaxonal dystrophy of Rottweiler dogs. *J Am Vet Med Assoc* **184:** 464–467, 1984.

Fukuda, T., and Kishikawa, M. Intraneuronal eosinophilic bodies of beef cattle (Japanese Brown). *Neuropathol Appl Neurobiol* **15:** 357–369, 1989.

Harper, P. A. W. Cerebellar abiotrophy and segmental axonopathy: Two syndromes of progressive ataxia of Merino sheep. *Aust Vet J* **63:** 18–21, 1986.

Harper, P. A. W., and Healy, P. J. Neurological diesease associated with degenerative axonopathy of neonatal Holstein–Friesian calves. *Aust Vet J* **66:** 143–149, 1989.

Harper, P. A. W., and Morton, A. Neuroaxonal dystrophy in Merino sheep. *Aust Vet J* **68:** 152–153, 1991.

Hooper, P. T., Best, S. M., and Campbell, A. Axonal dystrophy in the spinal cords of cattle consuming the cycad palm, *Cycas media*. *Aust Vet J* **50:** 146–149, 1974.

Jaggy, A., and Vandevelde, M. Multisystem neuronal degeneration in cocker spaniels. *J Vet Intern Med* **2:** 117–120, 1988.

Mayhew, I. G. *et al.* Equine degenerative myeloencephalopathy. *J Am Vet Med Assoc* **170:** 195–201, 1977.

Nuttall, W. D. Ovine neuroaxonal dystrophy in New Zealand. *N Z Vet J* **36:** 5–7, 1988.

Pienaar, J. G. *et al.* Maldronksiekte in cattle: A neuronopathy caused by *Solanum kwebense*. *Onderst J Vet Res* **43:** 67–74, 1976.

2. Central and Peripheral Neuronopathies and Axonopathies

a. Organophosphate Poisoning The organophosphates are a group of compounds whose toxicity is based on their ability to phosphorylate and inactivate a number of esterases for which they can act as substrates. The acute inactivation of acetylcholinesterase is a well-known effect, exploited for pesticidal and anthelmintic purposes. Our concern here is with delayed neurotoxicity, which follows 2–3 weeks after exposure and which involves the inactivation of as yet uncharacterized esterases within neurons. Inactivation of esterases tends to be irreversible, and recovery depends on synthesis of new enzyme, and this will vary with the compound and the particular enzyme(s) inactivated. There is also wide individual and species variability in sensitivity to neurointoxication. The agents themselves are a group of derivatives of phosphoric, thiophosphoric, and dithiophosphoric acids, and are used not only as poisons but also in industrial applications in hydraulic systems, as high-temperature lubricants, and as plasticizers. Delayed neurotoxicity, which may be induced by phosphates which have no acute effect on acetylcholinesterase, is particularly a property of the arylphosphates, the best known being triorthocresyl phosphate.

The clinical presentation is characterized by the onset of ataxia, weakness, and proprioceptive deficits, and ultimately paralysis. Severe dyspnea and loss of vocalization are often present in cattle and pigs, and hypomyelinogenesis is described in piglets, and laryngeal nerve degeneration, in horses given organophosphate anthelmintics.

The pathology and pathogenesis of delayed organophosphate intoxication have been studied quite thoroughly over recent decades. The compounds are not cumulatively toxic, but a single threshold dose will induce a central and peripheral distal axonopathy. Multifocal degenerative changes develop in distal regions of axons, with accumulations of organelles in the regions of proximal paranodes producing marked swellings. Axonolysis then follows, and there may be cycles of attempted regeneration and further degeneration. In the peripheral nerves, these changes are most intense in the intramuscular segments, and involvement of the recurrent laryngeal nerves makes aphonia a prominent sign in many cases. In the central nervous system, axonal degeneration is focused at the distal extremities of the long descending and ascending spinal tracts. The picture is dominated by axonal swellings, microcavitation, and secondary myelin loss.

b. Coyotillo Poisoning The coyotillo or buckthorn shrub, *Karwinskia humboldtiana,* is indigenous to the southwest United States, and its fruits are toxic to many species, including humans. Goats are most commonly affected, and intoxication is first manifested as hyperesthesia, which is followed by ataxia, gait abnormalities, and weakness, with signs developing over a couple of weeks. If the disease is mild, recovery may occur.

Pathologically there is a severe, predominantly motor, distal peripheral axonopathy, which appears to be the major pathological event, although there is significant segmental demyelination as well. Denervation atrophy becomes apparent in skeletal muscle. In the central nervous system, numerous swellings may develop in the proximal axons of cerebellar Purkinje cells, appearing in the granule cell layer and the folial white matter. Swollen axons may also be found in the lateral and ventral funiculi of the spinal cord.

c. *Aspergillus clavatus* Toxicosis An acute clinical syndrome of cattle and sheep has been reported from several countries associated with a fodder mycotoxicosis caused by *Aspergillus clavatus*. The fungus is known to produce a variety of toxic metabolites, which probably accounts for the varying reports of clinical signs and lesions. In the United Kingdom and South Africa, the clinical syndrome is dominated by the acute onset of excessive salivation and ataxia, which progresses to recumbency and death. The neuropathology is characterized by acute central chromatolysis of neurons in the red and vestibular nuclei, and in spinal ventral horn gray matter and ganglia. Wallerian degeneration and myelin edema are evident in all tracts of the spinal white matter.

d. Arsenic Poisoning Animals may be poisoned by arsenic by ingestion or percutaneous absorption, the toxic dose by the latter route being considerably less than the toxic oral dose. The commonest sources of arsenic are fluids used as insecticides and herbicides. Most arsenical dips and sprays for animals contain sodium arsenite and, after they are used, a considerable but nontoxic amount of arsenic is absorbed through the skin. Percutaneous absorption is increased if the animals are hot at dipping and kept hot afterward, and if they are not allowed to dry quickly. Absorption is rapid through shear wounds and through hyperemic skin such as of the thighs and scrotum of rams in the breeding season. Local high concentrations of arsenic on the skin also cause local acute dermatitis.

Herbicides of most importance are those containing sodium arsenite, lead arsenate, and arsenic pentoxide, and poisoning occurs usually when animals get access to recently contaminated pasture. A variety of mistakes and accidents commonly expose animals to these compounds. Paris green (cupric acetoarsenite), used as poison for grasshoppers and other parasites of plants, is occasionally responsible for poisoning. Ore deposits frequently contain large amounts of arsenic, and chronic arsenical poisoning can occur when pasture and drinking water are contaminated by the exhaust from smelters.

The toxicity of arsenicals varies considerably, depending on the solubility of the salt and, in the case of organic arsenicals, on the rate at which the arsenic is released from organic bondage. The syndromes of arsenic poisoning differ according to acuteness or chronicity and also according to whether the arsenic is organic or inorganic, but these latter differences are probably not qualitative. One, and perhaps the main, mechanism of action of arsenicals is combination with, and inactivation of, sulfhydryl groupings resulting in a general depression of metabolic activity. The organs most susceptible to the metabolic decline are the brain, lungs, liver, kidney, and alimentary mucosa. In poisoning by inorganic arsenic, the pattern of signs and lesions is the same irrespective of the route of absorption, thus indicating that, although alimentary tract signs may dominate the clinical disturbance, they are part of the systemic intoxications. In poisoning by organic arsenic, which is observed in pigs, the signs are referable entirely to the nervous system.

The ingestion of very large amounts of a soluble inorganic arsenical may result in death in less than 24 hr. There is profound depression and peripheral circulatory collapse; at postmortem in such cases, there are usually no lesions, or at most, slight edema of the abomasum.

With poisoning of somewhat lesser severity, there is a time lag of 1 to 2 days between the ingestion of arsenic and the onset of clinical signs. The onset is sudden with acute abdominal distress, nervous depression, circulatory weakness, and, after some hours, terminal diarrhea and convulsions. Some cases may survive for several days and show additional neuromuscular signs of tremor and incoordination. In chronic poisoning, the signs are nonspecific, being those of unthriftiness, capricious appetite, and loss of vigor. Pregnant cows may abort. Visible mucous membranes and the muzzle may be hyperemic and inflamed.

Whatever lesions are produced by acute poisoning by inorganic arsenicals can be explained largely on the basis of vascular injury. There is splanchnic congestion with petechial hemorrhages of serous membranes. The mucosa or submucosa of the stomach or abomasum is intensely congested, and the abomasal plicae are usually thickened by edema fluid. There are intramucosal and submucosal hemorrhages of patchy distribution, and these lead quickly to more or less extensive ulceration of the stomach and intestine. The intestinal content is very fluid and may contain shreds of mucus and detritus. There is mild, usually fatty, degeneration of the parenchymatous organs, sometimes with hepatocellular necrosis and edema of the kidney. Lesions in the brain develop in the course of 3 to 4 days and consist of moderate diffuse cerebral edema and petechiation. The hemorrhages, which are of capillary type and apparently due to necrosis of the walls of these vessels, are distributed throughout the white matter.

The anatomical changes in chronic arsenical poisoning have the same distribution as in the acute poisoning. The stomach and gut remain mildly congested, edematous and ulcerated, and there are prominent fatty changes in the heart, liver, and kidneys. Neural lesions may not be found in the central nervous system, except for those changes secondary to peripheral neuropathy. In both sensory and motor components of peripheral nerves, there is degeneration of myelin and axons.

The application of arsenic to the skin may cause acute and chronic dermatitis if cutaneous circulation is poor, but if the circulation is good, the arsenic tends to be absorbed and to cause systemic rather than local toxicity. The dermatitis, when it develops, is characterized by intense erythema, necrosis, and sloughing; the residual ulcerative lesions are indolent.

Organoarsenical phenylarsonic acid derivatives are commonly used as feed additives for swine, for growth promotion and the control of enteric disease. Poisoning is thus largely confined to this species, and is caused by accident or careless management. It must be stated, however, that the margin of safety can be quite low when arsanilic acid is used to control swine dysentery.

Two syndromes have been recognized, relating to *p*-aminophenylarsonic acid (arsanilic acid), and 3-nitro-4-hydroxyphenylarsonic acid (3-nitro).

In arsanilic acid poisoning, there is usually acute onset of cutaneous erythema, hyperesthesia, ataxia, blindness, vestibular disturbances and, terminally, muscular weakness. There are no gross neural lesions but, microscopically, mild edema of the white matter may be present in the brain and cord, and a few shrunken and degenerate neurons in the medulla. Extensive Wallerian degeneration frequently develops in the optic and peripheral nerves, but may not be present in spite of severe clinical signs.

In 3-nitro poisoning, there is a syndrome of repeated clonic convulsive seizures following exercise, with progression to paraplegia, but no blindness. Pathologically there is Wallerian degeneration consistent with a distal axonopathy in the spinal cord. The lesion is intense in the dorsal proprioceptive and spinocerebellar tracts of the cervical cord, and lateral and ventral funiculi of the posterior cord. Optic and peripheral neuropathy are mild.

e. Neonatal Copper Deficiency It is generally accepted that maternal/fetal copper deficiency is a major factor in a characteristic neurologic disease of lambs, goat kids, and piglets. The syndrome has been most studied in the lamb, and the terms swayback and enzootic ataxia, used to describe the signs shown by affected animals, have become entrenched as names for the disease. The provision of adequate copper supplementation of pregnant animals at risk can effectively eliminate the disease in their offspring, and treatment of affected animals with copper may produce some remission of signs. Supplementation of unaffected lambs at risk has also been claimed to be effective in prevention. Other manifestations of copper deficiency, such as "steely wool," osteoporosis, and hypopigmentation of black wool, could be expected in affected sheep flocks.

However, the bioavailability of and physiologic requirement for copper are influenced by many factors, and a functional deficiency state is often determined by overall availability rather than actual copper intake. Thus there may be absolute primary deficiency, or conditioned secondary deficiency, brought about by reduced absorption from the gut, reduced availability in tissues, or enhanced excretion. The interactive roles in copper metabolism of soil and dietary molybdenum, sulfate, iron, and zinc are important. Molybdenum is a prime antagonist for copper and, in the presence of adequate sulfate, limits the capacity to absorb copper from the gut and the capacity to store absorbed copper in the liver. This antagonism is unique to ruminants and is provided by the formation in the rumen of thiomolybdates, a series of anions in which sulfur progressively substitutes for oxygen in the molybdate ion. Copper complexed to thiomolybdate forms insoluble complexes which are poorly absorbed; this is primarily an effect of maximally substituted tetrathiomolybdate; lesser substituted thiomolybdates may be absorbed and be responsible for reducing copper availability at the local tissue level.

Iron is an antagonist to copper, although the mechanism is not known. Experimental exposure of ruminants to high intakes of iron induces severe hypocupremia, but a role for iron in naturally occurring copper deficiency syndromes is not of known importance.

Species and breed differences, pregnancy, plant/soil relationships, fiber content of the diet, and seasonal conditions will govern nutritional requirements. It has been shown that aspects of copper metabolism differ significantly between sheep and goats. It is also possible that undefined factors may bear significantly on metabolic availability. In spite of this, the copper content of central nervous tissue in affected neonates tends to be consistently below normal values, and represents the most reliable tissue assay. It is also true that some individuals with similarly low copper values can be clinically normal, although this holds more for goat kids and piglets than for lambs. As the ability to metabolize copper is not impaired, tissue concentrations in affected animals may return to normal fairly rapidly after dietary correction, with concentration in the liver recovering more rapidly than in the brain and spinal cord.

The effects of copper deficiency on the central nervous system occur *in utero* and during early neonatal life. Despite intensive investigation, the biological role of copper in the developing nervous system remains unclarified and a source of contention. Copper is a component of the enzymes cytochrome oxidase and superoxide dismutase, and of the protein ceruloplasmin. Interference with the functions of these has variously been proposed as the molecular basis of the disease via the suppression of mitochondrial respiration and phospholipid synthesis or via damage inflicted by superoxide radicals. It should be added that lambs with no clinical signs have been found to have neuronal degenerative lesions, and lambs with clinical signs may have minimal lesions.

Clinical swayback in lambs is described to occur in a congenital form and a delayed form, also called enzootic ataxia in which, after being normal at birth, lambs suddenly develop signs at any time between 1 week and several months of age. The clinical signs are dominated by motor disturbances, with staggering and ataxia. Congenital cases may be blind and unable to stand.

Some lambs with congenital swayback have an extensive structural lesion grossly evident in the cerebral white matter, but all have, to a degree, the neuronal degenerative changes described below for the delayed disease. The former lesion is bilateral and symmetrical gelatinous softening or cavitation (Figs. 3.9A, 3.72), which may be restricted to the occipital pole or may involve the entire corpus medullare, sparing only a thin rim of white matter adjacent to gray. Histologic and ultrastructural descriptions of the gelatinous lesion are meager, but marked edema with mild fibrillary astrogliosis and a paucity of myelin can be expected. Some myelin degradation products are usually present, but never in great quantity, and gitter cells are sparse. It seems likely that many axons are initially spared, but that rapid dissolution of all elements leads to cavitation. The pathogenesis of this lesion remains obscure; it seems that both hypomyelination and demyelination may be involved, but the basis of the tissue lysis is unexplained.

Although lambs with delayed swayback do not have lytic lesions in the cerebral white matter, they consistently have changes in both gray and white matter in other parts of the neuraxis. Most investigators consider that these changes largely develop and progress after birth, but this is not absolute, as pointed out above. Extensive Wallerian degeneration is concentrated in dorso-lateral and ventromedial tracts throughout the spinal cord (Fig. 3.22A). The pattern of tract degeneration is suggestive of a distal axonopathy. In addition, conspicuous degenerative changes are present in neurons in the red, lateral vestibular, medullary reticular, and dorsal spinocerebellar nuclei in Clarke's column, and in the spinal motor neurons, particularly in the intumescences (Fig. 3.16). Many such neurons are undergoing central chromatolysis, and some have nuclear rhexis and lysis. A few may be undergoing neuronophagia. Swollen neurons have been shown by immunocytochemistry to contain masses of phosphorylated neurofilament epitopes. Sites of neuronal loss are marked by fibrous astrogliosis.

Wallerian degeneration may be apparent in ventral spinal nerve exit zones and rootlets, and in peripheral nerves (Fig. 3.21), although this finding has not been highlighted as much as in goat kids. The pattern of axonal changes is consistent with the degenerating axons' being those arising

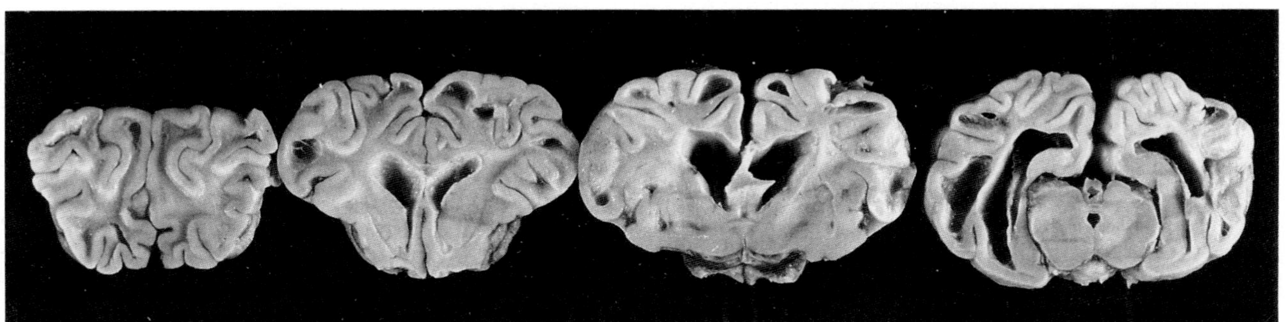

Fig. 3.72 Swayback. Lamb. Bilaterally symmetric cavitation of cerebral white matter. (From J. R. M. Innes and L. Z. Saunders, *"Comparative Neuropathology."* Academic Press, New York, 1962.)

from the nuclei where neurons are also degenerate. Current opinion seems to favor the hypothesis that the lesions represent a primary neuroaxonal degeneration with secondary myelin loss, although the myelin has been shown to be qualitatively abnormal and therefore theoretically unstable.

A very small number of lambs may have a cerebellar lesion, or cerebrocortical necrosis, as described later for goat kids. A further variant reported in lambs with delayed swayback is the occurrence of acute cerebral edema, which is sometimes unilateral and involves both gray and white matter. Small gelatinous or cystic foci may be present at the corticomedullary junction. The pathogenesis is unexplained.

There is as yet no unifying explanation for the spectrum of changes found in swayback, or for the molecular mechanisms relating to the role of copper. It has been proposed that the spectrum reflects a critical level of copper deficiency, cytotoxic at particular times when different regions of the developing brain/spinal cord are undergoing growth spurts. It has also been suggested that the lesions reflect oxidative damage concentrated in particular vascular fields.

In goat kids, the clinicopathologic spectrum is similar, but has different emphasis. The great majority of reports describe delayed swayback, with a high incidence of cerebellar degeneration/dysplasia, and of peripheral motor axon degeneration. The cerebellar changes include necrosis and dystopia of Purkinje cells, depletion of internal granule cells, and Wallerian degeneration in folial white matter. The lesions tend to be multifocal and may involve vermis and hemispheres or be restricted to the vermis. In a very few cases, congenital swayback is reported in kids, and in only two individuals were cerebral gelatinous and cavitating changes found. An additional variant reported in both lambs and kids is the occurrence of diffuse cerebrocortical necrosis.

Lesions in piglets have the same general character in regard to Wallerian changes as described for delayed swayback, but chromatolytic neurons are not evident. A swaybacklike disease has also been documented in adult captive red deer, but the role of copper is uncertain.

f. PROGRESSIVE AXONOPATHY OF BOXER DOGS A progressive disease beginning in early life, this disorder has been shown to be inherited as an autosomal recessive trait. Clinical signs first become apparent at about 2 months of age and progress fairly rapidly until 12 or 18 months, when they may either become static or advance very slowly. The dominant clinical sign is hind-limb ataxia, with proprioceptive deficits, hypotonia, and areflexia often being evident.

Degenerative lesions have been described in the spinal cord, posterior brain stem, optic nerves, spinal and cranial nerves, and major autonomic nerve trunks. Morphologic changes are most obvious in the spinal cord and posterior brain stem, and their intensity seems to parallel the clinical progression. Axonal spheroids develop in the spinal cord and are concentrated in the ventral and ventrolateral funiculi in the cervical and thoracic segments. A few axons undergoing Wallerian degeneration are also evident, but are always in a minority. Myelin sheaths are thinned around the spheroids, as would be expected, but attenuated sheaths may also occur around axonal segments not obviously swollen. Vacuolation of myelin segments is also a feature. These white matter changes have no obvious tract distribution. In the gray matter, spheroids are found in many nuclei of the posterior brain stem, and to a small extent in the spinal cord. However, from the diencephalon forward, there are virtually no gray matter lesions. Axonal swellings and myelin vacuolation are prominent in the optic nerves in the area of the chiasm.

In spinal nerve roots, there are, early in the disease, focal axonal swellings and a range of myelin abnormalities, including vacuolation, thinning, and segmental loss. With time, the involvement of ventral roots is appreciably more severe than that of dorsal roots. As the disease advances, such changes progress distally down the nerves, but axonal degeneration is accompanied by regeneration, and denervation of muscle is not significant. Changes of a similar character have been described in cranial nerves and large autonomic trunks.

The extent and nature of the lesions in this condition are not easy to explain or to match with the functional disturbances, but the clinical, electrophysiological, and pathological features have been quite thoroughly documented. Some experimental evidence is available to suggest that the myelin lesions occur in response to primary axonal changes, and represent adaptive remodeling to alterations in axonal caliber and metabolism.

g. DEGENERATIVE RADICULOMYELOPATHY OF GERMAN SHEPHERD AND OTHER DOGS This idiopathic condition, although well recognized, has remained incompletely defined. It is a slowly progressive, adult-onset disease, characterized by paraparesis and truncal ataxia.

There is some conflict in the literature on the extent and pattern of the neural lesions, with some studies describing lesions in spinal nerve roots and cord, and others restricting changes to the cord. In about 10 to 15% of cases, there is loss of patellar reflexes and degeneration in the femoral nerves. In addition, arguments have been advanced for and against a distal dying-back type of pathogenesis, and the affliction of dogs of large breeds other than the German shepherd.

Axonal degeneration and some evidence of demyelination are present in the spinal cord and probably, on occasion, in the spinal nerves. While there have been descriptions of symmetrical degenerative changes at the distal extremities of long tracts in the cord, a separate study documented myelin vacuolation, myelin deficiency, a few spheroids, and some Wallerian degeneration, distributed randomly and asymmetrically, and most intense in the thoracic segments. These differing viewpoints are not resolved, but it is generally agreed that the disease is progressive and irreversible.

h. Giant Axonal Neuropathy of the German Shepherd Dog Very few cases of this entity have been recorded, but it is well established that the disease is a distal axonopathy in which disorderly clumps of neurofilaments accumulate toward the extremities of long axons and that their most distal regions eventually degenerate completely. The neurofilamentous accumulations create the very large argentophilic axonal swellings which give the disease its name.

In the central nervous system, the giant swellings are found in the rostral regions of the ascending fasciculus gracilis and the dorsal spino-cerebellar tracts, and in the caudal regions of the descending lateral corticospinal tracts. In the peripheral nervous system, they are found in both myelinated and unmyelinated large fibers in the more distal regions of the major nerves of the limbs and the recurrent laryngeal nerves. Small focal axonal swellings are also scattered through various regions of the brain from the cerebral cortex back to the posterior brain stem, and on to the dorsal and intermediate gray columns of the cord. Some Wallerian changes accompany the small swellings in these areas.

The clinical onset is at about 15 months of age, and there is progression to paraparesis, ataxia, and megaesophagus within a few months. The disease is apparently inherited in an autosomal recessive manner.

i. Progressive Motor Neuron Diseases There is a group of neurodegenerative diseases, described in several species, which have a common clinical and pathological theme, and which can therefore be dealt with together. Diseases with this basic theme have been described in Brittany spaniels, Swedish Lapland dogs, Rottweiler dogs, domestic short-haired cats, the horse (various breeds), the Yorkshire and Hampshire pig, and the Brown Swiss calf. The clinical presentation is dominated by progressive lower motor neuron paralysis, which usually ends in tetraplegia and muscle atrophy. The pathological findings are dominated by denervation atrophy of pelvic and pectoral muscle groups, and by regressive changes in spinal motor neurons. Other neurons in the motor hierarchy may be afflicted, including the pyramidal cells of the motor cortex. In some instances, neurons in the brain stem and peripheral ganglia are also involved. In the earlier stages of neuronal degeneration, there may be chromatolytic swelling of the soma, with a reduction in basophilia, and fading of the nucleus, which tends to remain centrally located. In many of the diseases, the swollen cell bodies are often crowded with neurofilaments. Eosinophilic inclusions may be present in the cytoplasm, usually representing clustered remnants of normal organelles trapped among arrays of neurofilaments (Fig. 3.20). In some cases, these inclusions may be found to have the characteristics of Hirano, Lewy, or Bunina bodies described in human neuropathology. The neuronopathy progresses eventually to neuronal death on a cell-by-cell basis, with individual neurons in different stages of the process at any one time. Ultimately there may be considerable neuronal loss, with residual gliosis, and Wallerian change in motor nerves and spinal white matter.

These diseases are currently considered to be primary metabolic dysfunctions of the nerve cell body. While most have an apparently genetic basis and an early-age onset, a recent report describes a series of horses from the northeast United States in which the epidemiology strongly suggests an environmental cause. It should be noted also that the neuronal lesions in congenital copper deficiency have many of these features (Fig. 3.16). Variations on the basic theme are provided in terms of the age of onset and speed of progression of signs, the extent and pattern of frank neuronal degeneration, involvement of nuclei in the mid- and hind-brain, and the presence of Wallerian degeneration in the spinal white matter and peripheral nerves.

Probably the most comprehensively described condition is **hereditary spinal muscular atrophy of Brittany spaniels,** and it will serve as a generic example. This condition has received attention as a possible model for an equivalent human affliction, in which there is remorseless progression of motor paralysis (Werdnig–Hoffmann disease). In the spaniels, there is an autosomal dominant mode of inheritance and three phenotypic variants: chronic, intermediate, and accelerated. The latter leads to quadriplegia at about 3 months of age, the others over several years, if at all. At the end stage, there is pronounced denervation atrophy of pectoral and pelvic muscle groups, as well as of the tongue and masseter muscles.

The neuropathology is marked by degenerative change in spinal motor neurons, focused in the intumescences, with similar involvement of the hypoglossal and trigeminal motor nuclei. Particularly in the accelerated, homozygous variant, there are numerous pale, swollen, and chromatolytic motor neurons, with occasional cells undergoing fragmentation and necrosis. Swollen neurons are depleted of ribosomes, and a few are packed with neurofilaments (Fig. 3.73). Most characteristically, large argentophilic swellings are present in the proximal segments of many axons close to the affected cell bodies. These have been shown by electron microscopy to be tangles of disorganized neurofilaments (Fig. 3.74). All these changes are much less evident in the more chronic, heterozygous forms of the disease, and clinical weakness is only mild in many cases. Although it is established that there is a defect in the metabolism of neurofilaments as they move from the cell body into the proximal axon of individual neurons, it is not yet clear that this is the primary molecular lesion. Recent studies have shown that there is a growth arrest of spinal motor neurons and, initially, affected pups have a greater than normal number of these cells, with a shift to a smaller cell size. Ventral root axons undergo atrophy, which may be followed by loss of entire neurons.

In the **Swedish Lapland** dog, the pattern of lesions is different, there being degeneration of spinal motor neurons only in the lateral aspects of the intumescences; neuronal degeneration also occurs in spinal ganglia and cerebellar Purkinje cells, but not in brain stem nuclei. The disease has been termed **neuronal abiotrophy,** and is in-

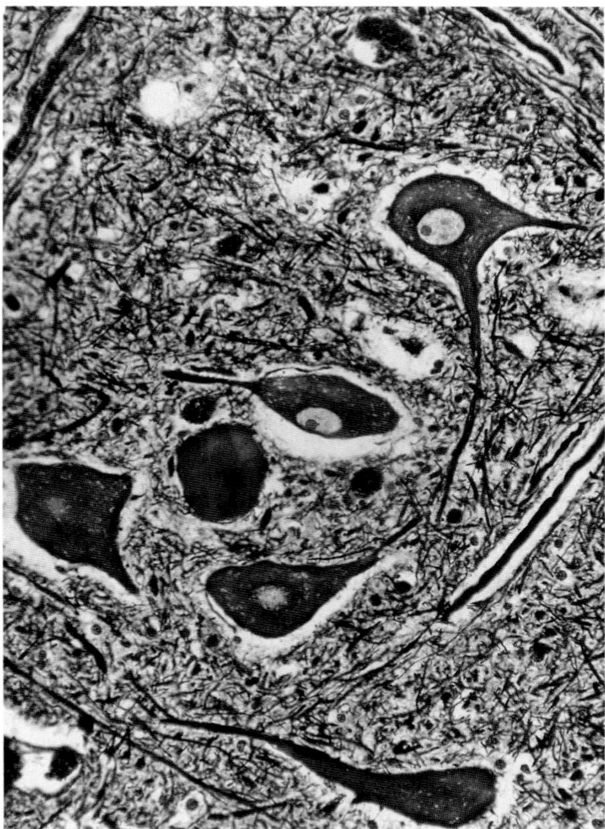

Fig. 3.73 Motor neuron disease. Brittany spaniel. Spinal motor neuron. Proximal axonal swelling packed with neurofilaments. Silver impregnation. (Courtesy of L. C. Cork.)

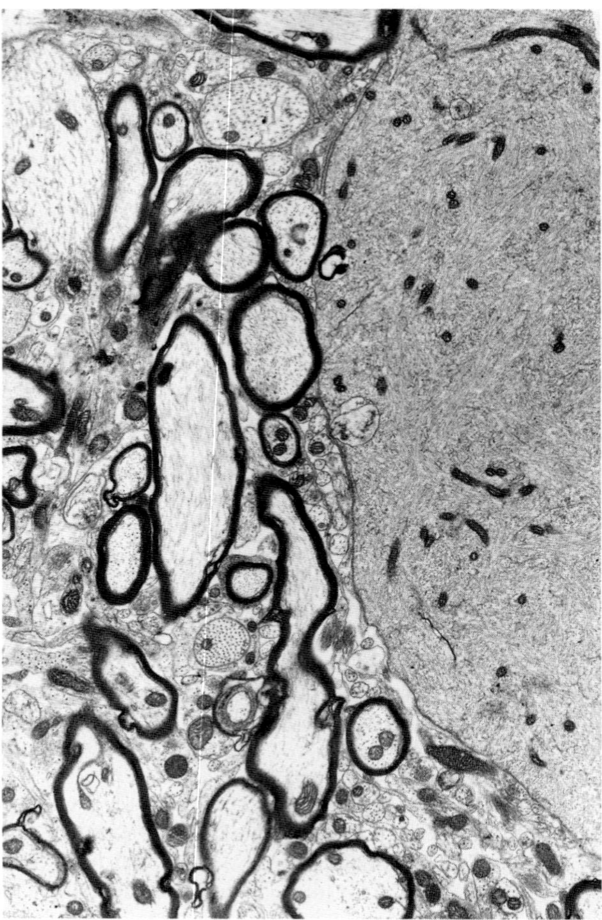

Fig. 3.74 Motor neuron disease. Brittany spaniel. Spinal motor neuron. Proximal axonal swelling packed with neurofilaments. (Courtesy of L. C. Cork.)

cluded here because of the dominance of denervation atrophy of muscle in the clinical presentation. The atrophy is concentrated in the more distal muscle groups of the limbs, in keeping with the topography of degenerate spinal motor neurons. Any cerebellar deficits are probably overridden by the lower motor neuron impairment.

In the **English pointer** dog, a strikingly different motor neuron disease has been described in which weakness and muscle atrophy become obvious at about 5 months of age and severe by about 9 months. There is correspondingly severe distal degeneration of peripheral motor nerves and denervation of muscle. However, rather than the changes described above, spinal motor neurons are filled with cytoplasmic lipid inclusions reminiscent of those seen in the gangliosidoses and mucopolysaccharidoses, and in some drug-induced conditions. No significant loss of motor neurons is apparent, and similar inclusions are present in the hypoglossal and spinal accessory nuclei, but not elsewhere.

A focal, asymmetrical spinal motor neuron degeneration with acute onset and a rapid course is described in **German shepherd** pups. The affected neurons are within the cervical spinal cord intumescence, and exhibit peripheral chromatolysis or vacuolation. There is secondary denervation and wasting of forearm muscles.

j. Neurodegeneration of Horned Hereford Calves The term shaker calf refers to a disease in which newborn horned Hereford calves are unable to stand without assistance and develop generalized fine tremors and profound muscular weakness. While they often die in the neonatal period because of secondary complications, some, after a short period of apparent remission, show relentlessly progressive spastic paraparesis, but remain alert and may survive for some months.

The disease is characterized by dramatic swelling of the cell bodies and processes of neurons throughout the spinal cord. Swelling is due to the accumulation of masses of neurofilaments, which impart a faintly fibrillar, amphophilic appearance in routine sections. This lesion involves ventral horn motor neurons, sensory neurons of the substantia gelatinosa, Clarke's column, and the intermediolateral (sympathetic) nuclei. Neuronal necrosis is minimal,

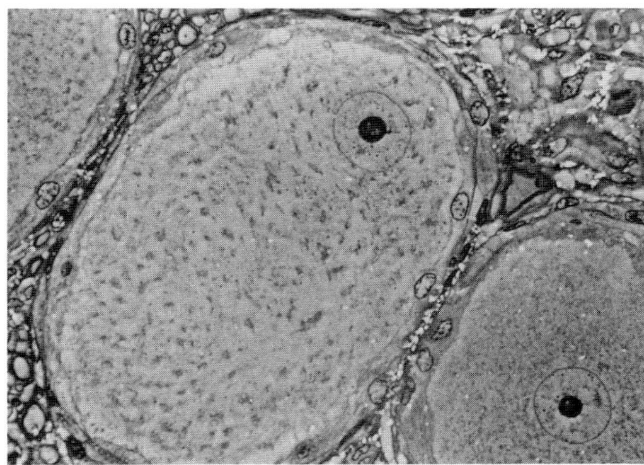

Fig. 3.75 Shaker calf. Swollen neuron. Spinal ganglion. Nissl substance dispersed by neurofilamentous masses. (Courtesy of C. G. Rousseaux.)

although the site of occasional cell loss can be found as nodules of reactive glia. Some Wallerian degeneration is evident in ventral spinal nerve roots and the ventromedial spinal white matter.

The neuronal lesion is also present in the major motor nuclei of the brain stem, the reticular formation, and the cerebellar Purkinje cells, and to a slight degree in the lateral geniculate nucleus and layer 5 of the frontal cortex. Some affected cells are evident in peripheral ganglia (Fig. 3.75), including the myenteric plexus, and in the retinal ganglion cells.

The disease is presumed to be inherited, on the basis of pedigree analysis. The fundamental metabolic defect remains to be characterized.

k. Progressive Neuronopathy of the Cairn Terrier Case reports of this multisystem chromatolytic degeneration have come from Britain, the United States, and Australia, and the clinical resemblance to globoid cell leukodystrophy emphasized. Animals of both sexes develop hind-leg weakness, quadriparesis, ataxia, head tremor, and loss of reflexes between 5 and 7 months of age.

Both central and peripheral chromatolysis of neurons are evident, and occur in Clarke's column and dorsal and ventral horn cells in the spinal cord, in sensory ganglia, and in the cuneate, lateral cuneate, glossopharyngeal, vagus, reticular, lateral vestibular, mesencephalic trigeminal, red and cerebellar roof nuclei. Wallerian degeneration is present in lateral and ventral funiculi of the cord, to varying degrees, and also in dorsal and ventral spinal nerve roots.

In a report on one case, ultrastructural studies suggested that the chromatolytic change was associated with depletion of ribosomes and increased numbers of mitochondria in perikarya, and that Wallerian degeneration was probably secondary to the metabolic disturbance in the cell body. This case was also marked by onset of signs at 11 weeks of age, bouts of cataplectic collapse, and thoracolumbar myelomalacia. The genetic basis of the disease has not yet been analyzed.

l. Primary Hyperoxaluria in the Cat In this disorder there is an inherited metabolic defect in cats causing a profound deficiency in the activity of D-glycerate dehydrogenase with an associated L-glyceric aciduria, hyperoxaluria, and the heavy deposition of oxalate crystals in the renal tubules. This leads to oxalate nephrosis and renal failure before a year of age.

An accompanying neuronal lesion occurs in the form of large swellings in the proximal axons of spinal motor neurons, ventral roots, intramuscular nerves, and the dorsal root ganglia. These swellings are caused by neurofilamentous accumulations, and are accompanied by some Wallerian degeneration in peripheral nerves. There is no obvious metabolic link between the neuronal lesions and the other biochemical disturbances, and the syndrome may represent a dual genetic defect.

m. Dysautonomias The term dysautonomia denotes a profound failure of both sympathetic and parasympathetic functions across several organ systems. In veterinary medicine there are two well-known conditions of this type: equine dysautonomia or grass sickness and feline dysautonomia or Key–Gaskell syndrome; canine cases have been reported.

Both equine and feline dysautonomia are mainly reported from Britain and western Europe; the former has been recognized for many years, the latter, only since 1981. For neither disease has the cause been identified, although a neurotoxic factor is reported in the serum of affected horses.

In the equine disease, the clinical picture is primarily the result of neurogenic obstruction of the alimentary tract with various parts of the tract involved to differing degrees, with distinct acute and chronic forms. Clinical differential diagnosis can be exceedingly difficult. The outcome is uniformly fatal.

In the cat, there is acute onset of clinical disease, which in a few cases may resolve after many months, but many animals die or require euthanasia early in the course. Onset is marked by dilated pupils, prolapsed membrana nictitans, dry mucous membranes, megaesophagus, constipation, vomiting, and dehydration.

The diseases in both species share the same basic neuropathology; in the acute phase, extensive chromatolysis and death of ganglion cells is present throughout the peripheral autonomic ganglia (Fig. 3.76), with axonal degeneration in autonomic nerve fibers. There is also neuronal degeneration in the nuclei of cranial nerves III, V, VII, and XII, the dorsal motor nucleus of the vagus, and the nucleus ambiguus. Some neuronal degeneration may be found in dorsal root ganglia, and in the ventral horn and intermediolateral areas of the spinal gray matter. In later phases of the disease, depletion of neurons at these sites

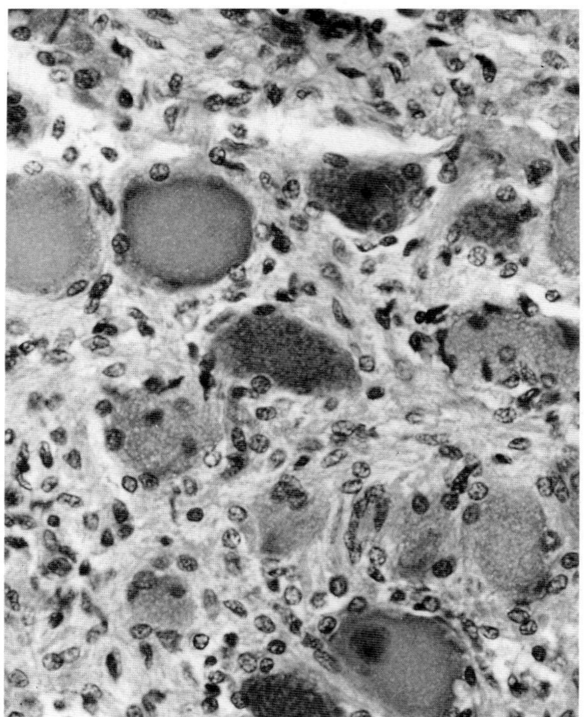

Fig. 3.76 Central chromatolysis in ganglionic neurons. Horse. Grass sickness.

is evident, with reactive and proliferative changes on the part of non-neuronal elements.

Particularly in the cat, it is considered that there is a single episode of injury to neurons, probably toxic in nature, with acute degeneration and subsequent reparative reactions if the animal survives for a number of weeks. There is a good correlation between the clinical signs and the extensive destruction in autonomic ganglia, and autonomic denervation would account for the major functional disturbances. The occasional expression of mild proprioceptive and lower motor neuron deficits may be explained by the lesions in dorsal root ganglia and spinal motor neurons, respectively.

n. Sensory Neuropathies of the Dog A small number of **English pointer** pups from a particular mating developed a syndrome of acral mutilation and analgesia. Pathologic studies revealed a reduction in size of spinal ganglia, with reduced numbers of neurons, and reduced fiber density in the dorsolateral fasciculus of the spinal cord (Lissauer's tract). These clinical and morphologic findings are consistent with a specific deficit in nociceptive pathways. It is suggested that the presumably genetically based disorder involves both hypoplasia of the system and continued degeneration postnatally.

In **long-haired dachshunds,** proprioceptive, nociceptive, and urinary deficits are associated with a distal degenerative axonopathy. Distal cutaneous nerves have marked loss of large myelinated fibers, and degenerative changes in both myelinated and unmyelinated fibers. In the spinal cord, there is axonal degeneration in the fasciculus gracilis, of greatest intensity at its distal extremity in the cervical region.

In dogs of **various breeds,** there are reports of diffuse ganglioneuritis, with destruction of primary sensory neurons and subsequent Wallerian degeneration and axonal loss in peripheral nerves in the spinal dorsal funiculi, spinal tract of the trigeminal nerve, and solitary tract. These cases are mentioned here as the extensive axonal degeneration is the more likely lesion to be routinely noted, and its pattern should immediately draw attention to the primary involvement of sensory ganglia.

Bibliography

Averill, D. R. Degenerative myelopathy in the ageing German shepherd dog: Clinical and pathologic findings. *J Am Vet Med Assoc* **162:** 1045–1051, 1973.

Barlow, R. M. Neuropathological observations in grass sickness of horses. *J Comp Pathol* **79:** 407–411, 1969.

Beck, B. E., Wood, C. D., and Whenham, G. T. Triaryl phosphate poisoning in cattle. *Vet Pathol* **14:** 128–137, 1977.

Cork, L. C. *et al.* Neurofilamentous abnormalities in motor neurons in spontaneously occurring animal disorders. *J Neuropathol Exp Neurol* **47:** 420–431, 1988.

Cork, L. C. *et al.* Hereditary canine spinal muscular atrophy: Canine motor neuron disease. *Can J Vet Res* **54:** 77–82, 1990.

Cummings, J., de Lahunta, A., and Gasteiger, E. Multisystemic chromatolytic neuronal degeneration in Cairn terriers. *J Vet Int Med* **5:** 91–94, 1991.

Cummings, J. F. *et al.* Focal spinal muscular atrophy in two German shepherd pups. *Acta Neuropathol (Berl)* **79:** 113–116, 1989.

Cummings, J. F. *et al.* Equine motor neuron disease; a preliminary report. *Cornell Vet* **80:** 357–379, 1990.

Duncan, I. D., Griffiths, I. R., and Munz, M. The pathology of a sensory neuropathy affecting long-haired dachshund dogs. *Acta Neuropath (Berl)* **58:** 141–151, 1982.

El-Hamidi, M. *et al.* Spinal muscular atrophy in Brown Swiss calves. *J Vet Med* **36:** 731–738, 1989.

Gilmour, J. S. *et al.* A fodder mycotoxicosis of ruminants caused by contamination of a distillery by-product with *Aspergillus clavatus*. *Vet Rec* **124:** 133–135, 1989.

Griffiths, I. R. *et al.* Further studies of the central nervous system in canine giant axonal neuropathy. *Neuropathol Appl Neurobiol* **6:** 421–432, 1980.

Griffiths, I. R., Sharp, N. J., and McCulloch, M. C. Feline dysautonomia (the Key–Gaskell syndrome): An ultrastructural study of autonomic ganglia and nerves. *Neuropathol Appl Neurobiol* **11:** 17–29, 1985.

Griffiths, I. R., McCulloch, M. C., and Abrahams, S. Progressive axonopathy: An inherited neuropathy of boxer dogs 4: Myelin sheath and Schwann cell changes in the nerve roots. *J Neurocytol* **16:** 145–153, 1987.

Howell, J. McC., Davison, A., and Oxberry, J. Observations on the lesions in the white matter of the spinal cord of swayback sheep. *Acta Neuropathol (Berl)* **21:** 33–41, 1969.

Hurst, E. W. The lesions produced in the central nervous system by certain organic arsenical compounds. *J Pathol* **77**:523–534, 1959.

Izumo, S. *et al.* Morphological study of the hereditary neurogenic amyotrophic dogs: Accumulation of lipid compoundlike structures in the lower motor neurons. *Acta Neuropathol (Berl)* **61:** 270–276, 1983.

Kennedy, S., Rice, D. A., and Cush, P. F. Neuropathology of experimental 3-nitro-4-hydroxyphenylarsonic acid toxicosis in pigs. *Vet Pathol* **23:** 454–461, 1986.

McKerrel, R. E. Primary hyperoxaluria (L-glyceric aciduria) in the cat: A newly recognized inherited disease. *Vet Rec* **125:** 31–34, 1989.

Montgomery, D. L., Gilmore, W. C., and Litke, L. L. Motor neuron disease with neurofilamentous accumulations in Hampshire pigs. *J Vet Diag Invest* **1:** 260–262, 1989.

Munoz-Martinez, E.J., Cueva, J., and Joseph-Nathan, P. Denervation caused by tullidora (*Karwinskia humboldtiana*). *Neuropathol Appl Neurobiol* **9:** 121–124, 1983.

Palmer, A. C., and Blakemore, W. F. A progressive neuronopathy in the young Cairn terrier. *J Small Anim Prac* **30:** 101–106, 1989.

Pletcher, J. M., and Banting, L. F. Copper deficiency in piglets characterized by spongy myelopathy and degenerative lesions in the great blood vessels. *J S Afr Vet Assoc* **54:** 45–46, 1983.

Pollin, M. M., and Griffiths, I. R. Feline dysautonomia: An ultrastructural study of neurons in the XII nucleus. *Acta Neuropathol (Berl)* **73:** 275–280, 1987.

Pollin, M. M., and Sullivan, M. Canine dysautonomia resembling the Key–Gaskell syndrome. *Vet Rec* **118:** 402–403, 1986.

Rousseaux, C. G. *et al.* "Shaker" calf syndrome: A newly recognised inherited neurodegenerative disorder of horned Hereford calves. *Vet Pathol* **22:** 104–111, 1985.

Shell, L. G., Jortner, B. S., and Leib, M. S. Familial motor neuron disease in Rottweiler dogs: Neuropathologic studies. *Vet Pathol* **24:** 135–139, 1987.

Suttle, N. F. Copper deficiency in ruminants: Recent developments. *Vet Rec* **119:** 519–522, 1986.

Suttle, N. F. The role of comparative pathology in the study of copper and cobalt deficiencies in ruminants. *J Comp Pathol* **99:** 241–258, 1988.

Wouda, W. *et al.* Sensory neuronopathy in dogs: A study of four cases. *J Comp Pathol* **93:** 437–450, 1983.

Wouda, W., Borst, G. H., and Gruys, E. Delayed swayback in goat kids, a study of 23 cases. *Vet Q* **8:** 45–56, 1986.

3. *Peripheral Axonopathies*

Some introductory comments for this section are appropriate for orientation and terminology. Lesions involving single peripheral nerves are classified as **mononeuropathies.** They are usually the result of focal compression or contusion by trauma, tumor masses, or similar lesions, and involve centrifugal Wallerian degeneration about the lesion. If several nerves are randomly involved in such a way, the term **mononeuropathy multiplex** may be applied. For the bilaterally symmetrical involvement of several nerves, the term **polyneuropathy** is employed, and carries the implication of a systemic disturbance.

From the clinical point of view, most peripheral polyneuropathies occur in association with polyneuritis or demyelination, but in this section our focus is on **noninflammatory axonopathies,** and the other types of diseases are discussed elsewhere. Intense Wallerian degeneration of peripheral motor axons will of course accompany degeneration of the ventral spinal motor nerve cell bodies, as occurs in the motor neuron diseases previously discussed. Similarly, in the very rare primary sensory polyneuropathies, peripheral sensory fibers will degenerate, but the degeneration will also extend, with the central projections of these cells, into the dorsal funiculi of the spinal cord. In some intoxications, delayed organophosphate, for example, peripheral neuropathy will be accompanied by central axonopathy, as has been illustrated. All the foregoing emphasizes the unity of the nervous system and the inconsistencies involved in deciding where neurons begin and end, but nonetheless, certain diseases have a clear central or peripheral focus.

With the exception of equine recurrent laryngeal neuropathy, non-inflammatory peripheral polyneuropathies are uncommon and are generally to be regarded as distal axonopathies of the spinal neurons. Motor axon involvement means that denervation muscle atrophy is a frequent concurrent lesion. It should be remembered too that with advancing age, subclinical degenerative changes are to be expected in peripheral nerves and spinal roots.

a. Equine Laryngeal Hemiplegia The clinical manifestations of this common and well-recognized disease are the consequence of denervation atrophy of the intrinsic muscles of the left side of the larynx. The resultant inability to adduct the arytenoid cartilage and the vocal fold leads to partial obstruction of the airway on inspiration, and inspiratory stridor on exertion, referred to as "roaring."

The underlying lesion is idiopathic degeneration of the left recurrent laryngeal nerve. Recent pathologic studies suggest a progressive degeneration of large myelinated axons, increasing in intensity toward the distal extremities of the nerve. During active degeneration, localized axonal swellings result from paranodal and internodal accumulations of granular dense bodies and degenerate mitochondria, although numerous atrophied axonal segments have also been described. Loss of axons is indicated by the presence of Bungner's bands, which may contain fragments of axonal and myelin debris, and permanent axonal loss is reflected in considerable endoneurial fibrosis. Some axonal regenerative activity may be apparent. There is also evidence of recurrent demyelination and remyelination, considered to be secondary to axonal degeneration.

The cause of this axonopathy remains unknown. Young, tall male horses with long necks seem predisposed, and there is a high incidence in thoroughbred and draft breeds. Mechanical factors operating on the left recurrent laryngeal nerve have been proposed, such as stretching of the anchored nerve, or pressure exerted where it reflects around the aorta. However, subclinical involvement of the right recurrent nerve seems to be the rule. Studies of clinically normal horses have revealed bilateral neuropathy with more severe denervation of the laryngeal adductor muscles as compared to the abductors, and this pattern also holds true for horses with laryngeal hemiplegia. The factors determining progression of the subclinical disease, the mechanism for the preferential adductor involvement, and the reason for the greater severity on the left side are unexplained.

A recent study has suggested that similar distal degenerative changes also occur in the long axons of the hind limbs, but only a few horses were sampled. Such a finding implies a systemic metabolic or toxic disorder, producing a polyneuropathy with maximal expression in the left recurrent laryngeal nerve. In New Zealand, laryngeal hemiplegia in association with stringhalt has occurred in seasonal outbreak form in horses grazing the plant *Hypochaeris radiata*. Laryngeal paralysis has occasionally been reported in cases of intoxication with lead, or organophosphates, as part of a widespread axonopathy.

No convincing evidence of lesions in the central nervous system is available at present, although they have been sought in the nucleus ambiguus, from which the recurrent laryngeal axons arise. It has been claimed that axonal spheroids in the lateral cuneate nucleus are more numerous in horses with laryngeal paralysis, but the significance of this is unclear. Such lesions are common and generally nonspecific, and it is difficult to see how they could be related pathogenetically to the rather different lesions in the motor axons supplying the larynx.

b. Canine Laryngeal Neuropathies Cases are described of neurogenic laryngeal muscle atrophy in old dogs of several breeds but, in the **Bouvier** breed, an autosomal dominant inheritance has been proposed for a condition with onset at an early age. Clinical laryngeal paresis/paralysis is evident and is accompanied by denervation atrophy of laryngeal muscles, and Wallerian degeneration of the distal recurrent laryngeal nerves.

c. Equine Suprascapular Neuropathy A mononeuropathy of the suprascapular nerve in the horse has been designated for many years by the old clinical term **sweeney.** The nerve is prone to injury at its site of reflection around the wing of the scapula, and in many cases, trauma at this site is probably the initiating cause. Evidence has been advanced to suggest, however, that entrapment of the nerve by a tendinous band may lead to degeneration in the absence of additional trauma. In cases of sufficient severity, axonal degeneration, demyelination, and endoneurial fibrosis will be associated with denervation of the spinatus muscle and chronic lameness.

d. Equine Stringhalt Stringhalt is the name given to a clinical condition of horses characterized by extreme exaggerated flexion of the hind limbs. It has been recognized for many years to occur in sporadic and epizootic forms, the latter associated with the grazing of certain plants, notably *Hypochaeris radiata,* the dandelion. The lesions are not well documented and a variety of degenerative changes of the spinal cord and peripheral nerves have been superficially reported. Most recently, some evidence for a peripheral distal axonopathy has been advanced, but more investigation is needed.

e. Endocrine Neuropathies Clinical polyneuropathy has been associated with diabetes mellitus in the dog and cat and is well recognized in humans. In the cat, hind-limb weakness, poor postural reactions, depressed patellar reflexes, and plantigrade stance are described. Axonal conduction velocity is reduced in the sciatic and ulnar nerves. Clinical remission often follows therapeutic management of the diabetic state.

The pathogenesis of diabetic neuropathy remains uncertain, but it is generally accepted that there is a distal axonopathy with cycles of degeneration/regeneration and accompanying demyelination/remyelination. The distal degeneration may be due to the impairment of axonal transport mechanisms secondary to the reduced availability of glucose for the neuron. Pathological studies in the dog and cat have been limited, but the available data are consistent with the foregoing.

A peripheral neuropathy has also been associated with canine hypothyroidism.

f. Feline Hyperlipoproteinemia A series of cats has been described with a genetically based deficiency of lipoprotein lipase activity and a resulting severe hyperchylomicronemia. Among other manifestations, there is a high incidence of mononeuropathy multiplex, with clinical palsies related to a variety of peripheral and cranial nerves and including instances of Horner's syndrome. Neurological deficits are related to multiple focal red-brown nodular masses in the perineurium of nerve trunks, which are organizing hematomas associated with a xanthomatous/granulomatous component. Xanthomatous masses extend between nerve fascicles and appear to compress and distort them, inducing Wallerian degeneration to varying degrees of severity. The xanthomas arise from the phagocytosis of cholesterol esters by macrophages. These lesions are generally located in loci where trauma is likely and also occur commonly at the emergence of spinal nerve roots.

g. Kangaroo Gait of Lactating Ewes This condition was first reported from New Zealand, and subsequently from the United Kingdom. There is a low flock incidence, and only ewes in lactation or until 1 month postweaning are affected. The clinical syndrome is consistent with bilateral radial nerve palsy, and there is a characteristic bounding gait during attempted rapid movement. In most cases, there is gradual clinical improvement and eventual recovery. The cause is not known. There have been limited pathologic studies. Extensive Wallerian degeneration with regeneration has been described in the radial nerve trunks of chronically affected ewes, but in some acutely affected animals, no radial nerve lesions could be demonstrated. In one study, there were additionally reported spongy change in the neuropil, dorsal root ganglionopathy, and neuronal degeneration in the hippocampus and cervical spinal cord, although these findings were variable.

Bibliography

Cahill, J. I., and Goulden, B. E. The pathogenesis of equine laryngeal hemiplegia—a review. *N Z Vet J* **35:** 82–90, 1987.

Cahill, J. I., Goulden, B. E., and Jolly, R. D. Stringhalt in horses: A distal axonopathy. *Neuropathol Appl Neurobiol* **12:** 459–475, 1981.

Duncan, I. D., Griffiths, I. R., and Madrid, R. E. A light and electron microscopic study of the neuropathy of equine laryngeal hemiplegia. *Neuropathol Appl Neurobiol* **4:** 483–501, 1978.

Duncan, I. D., Schneider, R. K., and Hammang, J. P. Subclinical entrapment neuropathy of the equine suprascapular nerve. *Acta Neuropathol (Berl)* **74:** 53–61, 1987.

Duncan, I. D. *et al.* Preferential denervation of the adductor muscles of the equine larynx II: Nerve pathology. *Eq Vet J* **23:** 99–103, 1991.

Haagen Venker-Van, A. J., Hartman, W., and Goedegebuure, S. A. Spontaneous laryngeal paralysis in young Bouviers. *J Am Anim Hosp Assoc* **1:** 712–720, 1978.

Jones, B. R. *et al.* Peripheral neuropathy in cats in inherited primary hyperchylomicronaemia. *Vet Rec* **119:** 268–272, 1986.

Kramek, B. A. *et al.* Neuropathy associated with diabetes mellitus in the cat. *J Am Vet Med Assoc* **184:** 42–45, 1984.

O'Toole, D. *et al.* Radial and tibial nerve pathology of two lactating ewes with kangaroo gait. *J Comp Pathol* **100:** 245–258, 1989.

Rose, R. J., Hartley, W. J., and Baker, W. Laryngeal paralysis in Arabian foals associated with oral haloxon administration. *Equine Vet J* **13:** 171–176, 1981.

G. Myelinopathies

Myelin sheaths in the central or peripheral nervous system may be the focus of various disease processes, but simultaneous central and peripheral involvement is rare. This is not surprising when one considers the fundamental differences between the two. As previously pointed out, myelin breakdown and removal in Wallerian degeneration will follow as a **secondary** consequence of axonal degeneration. On the other hand, in some inflammatory diseases, macrophages acting within the orchestration of the immune system will strip myelin from axons, leaving the latter intact for a time at least. In this context, antibodies or inflammatory mediators may bind to and destabilize the structure of the myelin lamellae, causing disruption of the sheath and provoking its phagocytosis. Such **primary demyelination** occurs, for example, in the central nervous system in canine distemper, and in the peripheral nerves in "coonhound paralysis." The demyelinating aspects of these entities will be discussed elsewhere.

Attention in this section is on noninflammatory diseases in which some disorder of myelin formation, maintenance, or stability is the primary event. It should always be remembered that the sheath is a specialized extended process of oligodendrocyte or Schwann cell plasma membrane, and myelinopathies are therefore part of the cytopathology of these cells.

1. The Hypomyelinations/Dysmyelinations

The great majority of these diseases are restricted to the central nervous system, and they have been described in most domestic species with the notable exception of the horse. Although in many cases a genetic basis has been demonstrated or strongly implied in dysmyelinogenesis, several viruses have been implicated, and the possibility of toxic or nutritional factors should be kept in mind. A common theme is sex-linked inheritance and affected males.

Myelinogenesis begins some time after the middle of gestation and continues in the postnatal period for varying times depending on the species. It is more advanced at birth in those species in which the young are able to stand and walk soon after, for it correlates with the overall maturity of the nervous system.

The process requires a complex unfolding of events in order to be successful. In the first place, there must be differentiation of competent myelinating cells in sufficient number, and they must migrate to, recognize, and contact the target axons appropriately. Second, it has been clearly shown that the axon itself must send a specific signal to the myelinating cell to initiate its investment. The diameter of the axon dictates whether or not it is myelinated and how thick the sheath will be. The threshold size is about 1 μm in the central nervous system and 2 μm in the peripheral nerves. Finally, the molecular components of the myelin must be produced and delivered to their correct sites in the membrane. One or several of these processes may be perturbed to give rise to diseases characterized by hypomyelinogenesis.

Myelination does not occur synchronously throughout the nervous system, but in a distinctly regional sequence. Thus lesions of this type may involve some tracts more than others, or completely spare some tracts. There may be a complete absence of myelin, or an inadequate amount of myelin which may be either chemically normal or abnormal. For the latter case, the term **dysmyelination** has been coined but, as these conditions are characterized by a reduced quantity of myelin, **hypomyelination** is a suitable generic term. As would be expected, such diseases are manifested early in life, and a very common and dominant clinical feature is the onset of a severe generalized tremor syndrome. This has led to the use of names such as congenital tremor, shaker, or trembler to describe the clinical state. The severity of the clinical signs can vary from life threatening to mild.

The deficiency of myelin may in some instances be permanent, while in others it seems that myelination may be delayed, but eventually proceeds to the extent that clinical deficits resolve. In general, the pathologic picture is dominated by a paucity of myelin with little or no evidence of the breakdown and removal of previously formed sheaths. There may be other accompanying structural defects, notably cerebellar hypoplasia, particularly when a teratogenic virus is the cause.

When routine morphologic evaluations are being made on very young animals, reference should be made to age-matched controls. In the dog, for instance, very little myelin is normally present prior to 2 weeks of post-natal age, whereas in the sheep, myelination begins at about day 50 of gestation. The investigation of such conditions makes use of traditional stains for myelin and axons, plastic-embedded sections for light and electron microscopy, and lately, some of the immunostaining techniques for marker antigens, such as myelin basic protein. There needs to be some familiarity with the normal morphologic features of early myelination. Some of these are mentioned in the disease descriptions that follow.

a. CANINE HYPOMYELINOGENESIS A number of canine hypomyelinogeneses have been documented, and several provisional conclusions can be made. All but one of the diseases involve the central myelin only, and it appears that they are not pathogenetically identical and can be divided into two broad groups according to severity. The most severe are those described in the Samoyed, springer spaniel and Dalmatian breeds.

In the **Samoyed,** severe generalized tremors become apparent at about 3 weeks of age, with a predominance of males being affected. Inability to stand leads to severe incapacitation and a high mortality rate. Profound hypomyelination is suggested on gross inspection of the brain by a lack of contrast between gray and white matter throughout the central nervous system. The sparing of the peripheral nervous system is well appreciated macroscopically by comparing the myelin-deficient optic nerves to the other cranial nerves, in particular the adjacent oculomotor nerve.

By routine light microscopy, there is normal peripheral myelin but an almost total absence of central myelin, and a diffuse microgliosis. Silver stains suggest axons to be present in normal number. Electron microscopy reveals a few axons to have thin and poorly compacted myelin sheaths, the presence of numerous astroglial processes, and a large number of microglia. Oligodendrocytes are greatly reduced in number and appear immature. Axons, by contrast, are normal in number and morphology, but the great majority are devoid of a myelin sheath. It has been suggested that a central pathogenetic event is retardation of gliogenesis, with oligodendrocytes failing to differentiate fully. A genetic basis is suspected.

In the **springer spaniel,** severe generalized tremor in male pups is evident at 10 to 12 days of age (shaking pups). Central hypomyelination is profound but less severe in the spinal cord than in the brain and optic nerves. In contrast to that in the Samoyed, microgliosis is not prominent, but in other respects, the gross and histologic features are similar. The electron microscope, however, resolves some further differences. For example, some thin myelin sheaths are present and have lamellae of normal configuration. Some axons have a few internodes sheathed, many of which are shorter than would be expected for the axonal caliber, and many of which are single units terminating as "heminodes." Although some immature oligodendrocytes are evident, so too are normally mature ones. In animals 2 months old or older, there are also hallmarks of the early stages of normal myelination which persist well beyond their usual period of the first month of life. These include such features as large amounts of oligodendrocyte cytoplasm in lateral loops, and outwardly terminating lateral loops. In addition, many oligodendrocytes have lysosomal digestive vacuoles thought to relate to the degradation of abnormal myelin components. Impairment of stem cell division and oligodendrocyte differentiation and metabolism have been suggested to underlie the hypomyelination. An X-linked genetic defect has been established.

The disease in the **Dalmatian** is documented only as a single case report, involving a male pup which developed severe tremors in the neonatal period and was destroyed at 8 weeks of age. Profound deficiency of central myelin was accompanied by reduced numbers of oligodendroglia, with a few axons having extremely thin and poorly compacted sheaths.

Disorders of lesser severity are described in the chow chow, Weimaraner, Bernese mountain dog and in two crossbred (lurcher) dogs.

Affected **chow chow** dogs have marked impairment from 2 weeks of age, but progressive improvement leads to a virtual absence of clinical deficits by the end of the first year of life. At all stages, the animals remain ambulatory, bright, and responsive, in spite of a pronounced hypermetric gait with a distinctive rocking horse motion when trying to initiate movement. Tremors and head-bobbing are exacerbated by excitement and disappear at rest.

In pups examined at about 3 months of age, no gross lesions are reported, but profound hypomyelination is evident histologically in the central nervous system. This is particularly so in the cerebral subcortical white matter, cerebellar folia, ventral half of the cerebral peduncles, optic tracts, and the peripheral zones of the ventral and lateral funiculi of the spinal cord. By contrast, the cerebellar peduncles and fasciculus proprius of the spinal cord are relatively well myelinated. Electron microscopy reveals mostly thin myelin sheaths in least-affected areas, and mostly naked axons in those worst affected. Oligodendrocytes are present in normal numbers and generally appear morphologically normal, although stellate cells containing intermediate filaments may be found.

In older dogs in clinical remission, the brain is well myelinated apart from some mildly deficient foci in subcortical white matter and corpus callosum. However, myelin deficiency is still marked in the lateral and ventral funiculi of the cord, which also has a few degenerate axons. Ultrastructurally, these areas contain a few thinly or well-myelinated axons separated by masses of astrocytic processes. Indications of immaturity are provided by poorly compacted sheaths and some massively oversized sheaths, which fold away from the axon in redundant loops.

To date it has not been possible to establish a definite heritable basis for this disease, but it is suggested to be the result of a genetically determined delay in oligodendrocyte maturation.

The **Weimaraner** syndrome is closely similar but has been studied pathologically only in the early phase. The tremor syndrome, in evidence at 3 weeks of age, seems to resolve by a year. Pups necropsied at 4 to 5 weeks of age have no gross abnormalities, but histologic staining reactions suggest myelin deficiency, and this is particularly obvious at the periphery of the lateral and ventral funiculi of the spinal cord. This contrasts with the relatively well myelinated dorsal columns. Oligodendrocytes are reduced in number. Ultrastructurally, there is considerable evidence of myelin immaturity as previously described. Astrocytic fibers are prominent, and oligodendrocyte morphology seems normal. A heritable basis has been proposed but not confirmed, and delayed oligodendrocyte differentiation suggested as the underlying defect.

The **Bernese mountain dog trembler pup** has a fine head and limb tremor which subsides with sleep and improves substantially by 9 to 12 weeks of age. There are no gross lesions, and hypomyelination is concentrated in the spinal cord, where axons are thinly sheathed by morphologically normal myelin. Oligodendrocytes in this case appear to be increased in numbers and morphologically normal. An autosomal recessive mode of inheritance has been proposed. **Lurcher** pups were reported to have unmyelinated and thinly myelinated axons in the peripheral sections of the lateral funiculi of the spinal cord. The changes are subtle and not readily apparent unless plastic-embedded sections are employed. These animals are of interest as they are crossbred, making a genetic cause a more remote possibility.

The sole reported occurrence of **peripheral hypomyelination** in domestic animals involves the **golden retriever** breed. Two littermates, one male and one female, had ataxia, weakness, a crouched stance, and pelvic-limb muscle atrophy at about 2 months of age. Peripheral nerve samples revealed axons of all calibers to be thinly myelinated and Schwann cells to be increased in number and hypertrophied (Fig. 3.24). There were no indications of demyelination.

b. PORCINE HYPOMYELINOGENESIS Congenital tremor syndrome is well recognized in pigs, and in some cases, there are no morphologic lesions. In terms of hypomyelination, five distinct entities have been described, two genetically determined, two caused by viral infection, and one by intoxication.

In the **Landrace** breed, hypomyelinogenesis has been attributed to an X-linked recessive inheritance, thus afflicting male neonates and having the potential to affect crossbreeds. There is myelin deficiency throughout the central nervous system but most obvious in the spinal cord, cerebellum, and cerebral gyri. There is a reduction in oligodendrocyte numbers; many small and medium-sized axons are unmyelinated or only thinly so. Large-diameter axons appear normally myelinated.

In the **British saddleback** breed, there is an autosomal recessive inheritance mode. Hypomyelination is present throughout the neuraxis with axons of all sizes being affected. Many sheaths are poorly compacted and vacuolated. Oligodendrocytes are numerous, and many appear immature ultrastructurally, and sometimes contain cytoplasmic intermediate filaments. Still others contain autophagosomes and dense bodies. The neuropil also contains excessive astrocytic fibers and lipid-laden macrophages. These findings, together with biochemical evidence of myelin degradation, suggest that newly formed myelin is unstable and is rapidly broken down.

Transplacental infection in the middle trimester with the **hog-cholera virus** is capable of causing significant hypomyelinogenesis in addition to other neural lesions in the fetus, most notably cerebellar hypoplasia. The cerebellar changes may be focal and are most likely to be detected in sagittal sections of the vermis and hemispheres. The myelin lesion is generally similar to that seen in lambs with Border disease.

An as yet uncharacterized virus, separate from the hog-cholera virus, has been cited as the cause of congenital tremor and hypomyelination in piglets.

A clinicopathologic syndrome similar to that produced by the hog-cholera virus occurs in piglets of sows exposed to the organophosphate **trichlorfon** during critical stages of pregnancy.

c. OVINE AND CAPRINE HYPOMYELINOGENESIS The lamb has been quite extensively studied in regard to myelination problems, one determining factor being the recognition of **Border disease** in the 1950s and the subsequent interest it generated. It has become the model for the several pestivirus-induced myelin disorders of animals. The Border disease virus can cause a transplacental infection, which, if on or after day 50, has the potential to produce a range of terata. In the classical syndrome, newborn lambs are often referred to as hairy shakers, due to the combination of generalized tremor and a hairy birth coat. There is diffuse hypomyelination in the central nervous system, which is especially prominent in the spinal cord, particularly the more caudal regions. Some myelination is present, but has ultrastructural features of immaturity, in that lamellae remain uncompacted, and axons retain the angulated profiles they normally acquire just before myelination and lose soon after. There are also compact sheaths abnormally thin for axonal diameter, and intramyelinic and periaxonal vacuoles.

While glial cell numbers may appear normal by light microscopy, there is in fact a deficiency of mature oligodendrocytes and an increase in the number of microglia. Many of these latter are packed with lipid droplets, the origin of which does not appear to be degraded myelin. It is suggested that there is a delay in the maturation of oligodendrocytes, and that although the time of onset of myelination is close to normal, its rate is greatly reduced. Some recent evidence has raised the interesting possibility that the blockade of glial maturation may be the result of virus-induced fetal hypothyroidism. Abundant viral antigen can be found in the thyroid, and circulating thyroxin is significantly depressed.

Severe hypomyelinogenesis is a feature of neonatal goat kids with **β-mannosidosis.** Hypothyroidism has recently been proposed as the underlying mechanism, as a consequence of the lysosomal storage disease.

d. BOVINE HYPOMYELINOGENESIS Myelination in calves proceeds between the twentieth gestational and the eighth postnatal weeks. A situation analogous to that in piglets and lambs may develop in calves exposed to bovine virus diarrhea infection *in utero*. A similar spectrum of terata and spinal hypomyelinogenesis may follow infection after day 100 of gestation.

2. *Leukodystrophic and Myelinolytic Diseases*

This section will address those diseases characterized by the degeneration and loss of myelin, not as a result of, or in association with, inflammation, but due to some failure of the myelinating cells to sustain and maintain their sheaths in an intact and ordered condition. The implication is that sheaths are initially formed, but then deteriorate, leaving their axons intact for a time at least. The use of terminology in this area reflects the stubborn refusal diseases often display to the human determination to fit them into neat categories. For our purposes, **primary demyelination** is taken to imply the removal from around intact axons of structurally and chemically normal myelin, usually by macrophages and often in an inflammatory setting. The term **leukodystrophy** is designed for diseases with a heritable basis, early onset in life, lack of inflammation, symmetry of lesions, and in which some inherent qualitative defect in myelin (dysmyelinogenesis) leads to its dissolution and removal. **Myelinolysis** refers to the initial disruption of myelin structure by extensive decompaction of lamellae as a prelude to its breakdown and removal, and as such could be involved in many processes. There is scope for considerable overlap and liberal interpretation of definitions. The diseases discussed in this section will not include any which could easily be classified as demyelinating, in the sense previously indicated, or as malacic in the sense indicated elsewhere in this chapter, in spite of the propensity for focal softening and even cavitation in some of the leukodystrophies.

a. GLOBOID CELL LEUKODYSTROPHY This disease, also known as **galactocerebrosidosis** and **Krabbe's disease,** has been documented in the Cairn and West Highland white terrier, miniature poodle, bluetick hound, basset hound, domestic cat, and polled Dorset sheep. It is due to a genetically determined deficiency (presumed autosomal recessive in most cases) in the activity of lysosomal galactocerebroside β-galactosidase (galactosylceramidase), and as such falls within the sphingolipidosis group of the lysosomal storage diseases. However, it has special characteristics which make it more appropriate to be discussed as a leukodystrophy, and it was so named before the lysosomal defect was recognized. Neurologic impairment begins at an early age and progresses rapidly to a fatal conclusion. The enzymic deficit blocks the catabolism of the galactocerebrosides (galactosylceramides). These compounds are major components of myelin, and the disorder thus principally affects the metabolic well-being of the oligodendrocytes and Schwann cells.

Early in the disease, during active myelination, the myelinating cells store galactocerebroside within lysosomes. However, the enzyme is also involved in the breakdown of other metabolites, most notably galactosylsphingosine (psychosine), which is also normally synthesized by oligodendrocytes. This substance is highly cytotoxic to oligodendrocytes when it accumulates, and causes their extensive degeneration and death. Myelination ceases, and formed myelin degenerates. During this process, macrophages accumulate to ingest the degenerate myelin but are also unable to degrade galactocerebroside. They give rise to the distinctive, swollen, PAS-positive globoid cells, which form large cuffs around blood vessels in the central white matter, and are also found in the leptomeninges and in the endoneurium of peripheral nerves (Fig. 3.77). Their presence at the latter site is useful for premortem diagnosis by nerve biopsy. At end stage, the pathology of the disease is marked by the presence of these cells, together with diffuse demyelination, axonal loss, and dense astrogliosis in the brain and spinal cord.

b. CAVITATING LEUKODYSTROPHY OF THE DALMATIAN DOG Reported only from Norway, this leukodystrophy

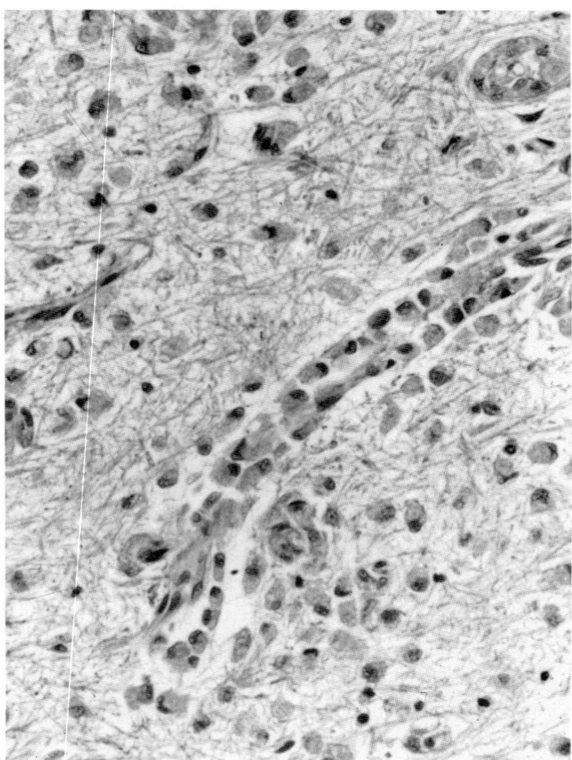

Fig. 3.77 Globoid cell leukodystrophy. Cairn terrier. Macrophages, filled with degenerate myelin products, aggregating around vessels.

has a probable autosomal recessive inheritance and a clinical onset at around 3 to 6 months of age. The signs are progressive and variable, and include either gait deficit or visual impairment or both. The former begins as pelvic and then thoracic limb ataxia and dysmetria, with eventual paraparesis. Mental status remains unimpaired throughout.

The neuropathology is characterized by bilateral, symmetrical, focal areas of intense loss of myelin in the centrum semiovale, corpus callosum, internal capsule, caudate nucleus, optic nerves, and thoracic cord white matter adjacent to the ventral horns. Initially axons remain intact. The myelin destruction appears to begin with lamellar vesiculation followed by phagocytosis, and gitter cells are abundant. Fibrous astrogliosis becomes prominent around advanced plaques. The intensity of the process is variable, but lesions are usually visible macroscopically in the central cerebral white matter, appearing as grayish depressed areas or, in advanced cases, as cavitations. In some animals, the brain is small, and the lateral ventricles, enlarged.

c. HEREDITARY MYELOPATHY OF THE AFGHAN HOUND The initial reports of this disease described an acute necrotizing myelomalacia, but subsequent studies indicated a mechanism involving florid myelinolysis and cavitation, but essentially sparing axons. It has an apparent recessive inheritance, preadult onset, symmetry of lesions, and is noninflammatory, all of which allow for classification as a leukodystrophy. The age of clinical onset is between 3 and 12 months, and posterior ataxia and weakness progress rapidly to paraplegia in a few days. Within 2 weeks there is thoracic limb weakness and phrenic paralysis.

The lesions are focused in the thoracic spinal cord where cribriform and spongiotic changes are present in all funiculi (Fig. 3.78B) and are grossly discernible as discoloration and softening or cavitation. Caudally the lesions extend into the lumbar segments, but only in the ventral funiculi; they extend cranially to the midcervical level in the dorsal and/or ventral funiculi (Fig. 3.78A). In some cases there may be lesions in the superior olivary nuclei. The sparing of axons is reflected in the paucity of distal Wallerian degeneration in tracts ascending or descending through the lesions.

Light and electron microscopic examinations have revealed initial vacuolation of myelin by splitting of lamellae at the intraperiod line and expansion of the extracellular space. Fragmented and degenerate myelin is phagocytosed by gitter cells, leaving surviving axons and reactive astrocytes within microcavities. The cavities are traversed by blood vessels associated with delicate glial strands. Some focal axonal swelling and disintegration are evident, but myelinolysis following vacuolation is the dominant change. The fundamental pathomechanisms remain undefined.

d. LEUKOENCEPHALOMYELOPATHY OF THE ROTTWEILER DOG Some difficulties arise in classifying this disease as a true leukodystrophy, as the onset is delayed until after 12 months of age. However, there is the probability of autosomal recessive inheritance, and the lesions involve progressive noninflammatory symmetrical myelin degeneration and removal. Clinically there are slow, but relentlessly progressive, ataxia, hypermetria, and paresis of all limbs, especially the forelimbs, and ultimately severe proprioceptive loss.

The principal locus of the disorder is the cervical spinal cord, which on gross inspection may have extensive dull white discoloration of the dorsal and lateral funiculi. Histologic examination reveals lesions to be present also in the deep cerebellar white matter, and in other regions of the cord. In these areas there is significant loss of myelin, with preservation of most axons, either naked or thinly myelinated, and fibrous astrogliosis. There is some evidence that vesicular degeneration of myelin is an early

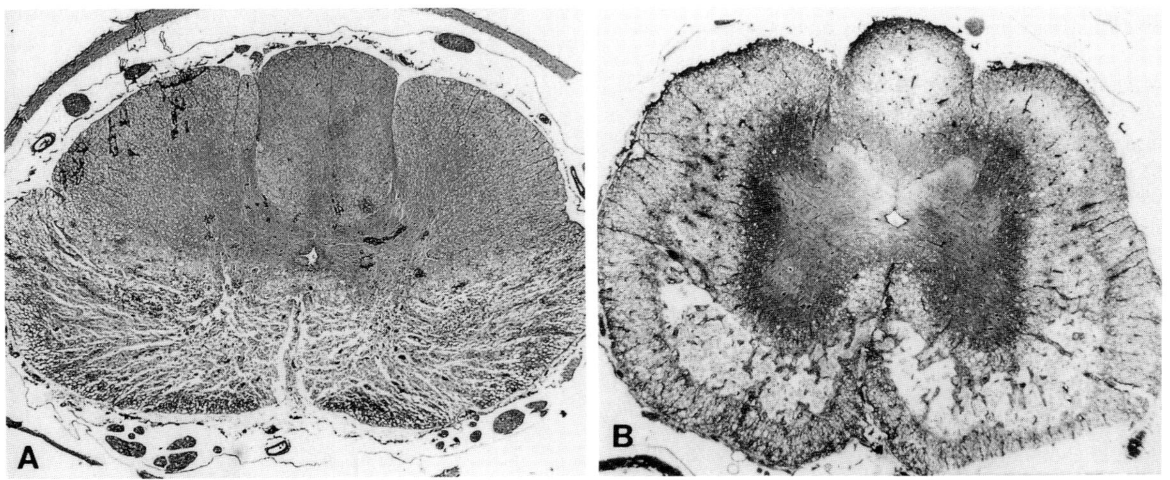

Fig. 3.78 Afghan myelopathy. (A) Caudal cervical cord. (B) Thoracic cord.

change. In the cervical cord, there may additionally be focal microcavitation deep within the affected areas, edema, numerous gitter cells, and vascular prominence. A narrow rim of intact white matter usually survives immediately deep to the glia limitans. These severe changes may on occasion extend into the thoracic cord and pyramidal tracts.

e. Progressive Spinal Myelinopathy of Murray Grey Cattle Reported in calves of the Murray Grey breed in Australia, this disease may afflict animals at birth, or not become expressed until about 12 months of age. The clinical syndrome is one of spinal ataxia, manifested as incoordination of the pelvic limbs and lateral swaying of the hindquarters at rest. A consistent sign is collapse of one hind limb, with a tendency to fall to one side. These pelvic limb deficits are progressive and lead to increasing impairment.

The characteristic lesions are microscopic only, symmetrical, and consistently occur in the lateral and ventral funiculi throughout the spinal cord. There is a distinct deficit of myelin in the lateral funiculi beneath the dorsal root entry zone, and in ventral funiculi adjacent to the ventral median fissure. In some cases, the fasciculi gracilis and cuneatus are also involved. The process appears to involve ballooning degeneration of myelin sheaths, which may impart a distinctly spongy appearance to the lesion. The progressive loss of myelin is not accompanied by large numbers of myelinophages, and Wallerian degeneration is minimal. There is substantial astrogliosis and thickening of the glia limitans.

In the brain, similar white matter lesions may be found in the spinocerebellar tracts, tectospinal tracts, and medial longitudinal fasciculus. Some neurons undergoing central chromatolysis are present in ventral spinal gray matter, Clarke's column, the red nuclei, lateral vestibular nuclei, reticular formation, and cerebellar roof nuclei.

The general similarity to the pathology of ovine swayback has been noted by investigators, but an autosomal recessive inheritance pattern has been demonstrated, and copper status has been found to be normal.

f. Progressive Ataxia of Charolais Cattle This disease is pathologically unique among the domestic species, and appears to represent a progressive inability on the part of some oligodendrocytes to maintain the paranodal extremities of their myelin domains. The disease is presumably genetically determined, but has been reported in three-quarter crossbred animals. Both sexes may be affected, and clinical signs become evident between 8 and 24 months of age.

The clinical spectrum involves increasing ataxia and dysmetria, head tremor, and some tendency to aggressive behavior. The mental status remains bright and alert. Affected females often urinate in a series of short spurts. The clinical signs progress slowly but steadily, and the animals can be expected to become recumbent by 3 years of age at the latest.

There are no specific gross lesions, but the microscopic findings are distinctive and unprecedented, and are located mainly in the cerebellar white matter and peduncles, the internal capsule, corpus callosum, optic tract, lateral lemniscus, median longitudinal fasciculus, pontine decussation, and ventral and lateral funiculi of the spinal cord. In these white matter tracts, there are multifocal, granular, eosinophilic plaques about 30 μm in diameter (Fig. 3.79). The plaques are traversed by axons and stain with many, but not all, of the features of normal myelin. There is no sign of myelin degradation or phagocytosis. Immediately around the plaques there appear to be an increased number of astrocytes and oligodendrocytes. Some very minor axonal degeneration may be found in some plaques and in normal white matter. It has been concluded that early plaques can be distinguished from old ones. Electron microscopy reveals them to be complex. The ultrastructure suggests great disorder at the myelin paranodes, with relatively normal internodal regions. New plaques reveal axons encompassed, near abnormally long nodes of Ranvier, by hypertrophied oligodendrocyte tongues and processes within a thin myelin sheath. Old plaques contain demyelinated axons (some of which are swollen), surrounded by disorganized myelin lamellae and masses of oligodendrocyte processes. It is suggested that each plaque represents the territory of one oligodendrocyte, and reflects its failure to maintain normal paranodal myelin loops, in the course of which a massive hypertrophy of its processes occurs.

g. Multifocal Symmetrical Myelinolytic Encephalopathy of Simmental and Limousin Calves Reports from Australia, New Zealand, and Britain describe similar diseases in calves of these breeds, in which spongy vacuolation of white matter progresses focally to lysis and cavitation, but with the initial sparing of many axons and

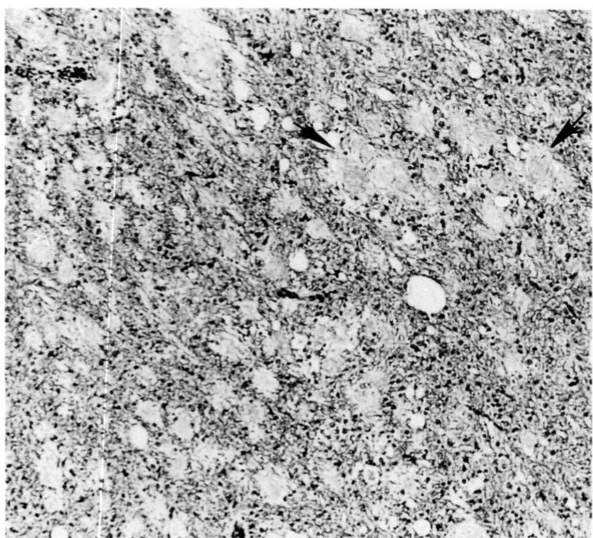

Fig. 3.79 Plaques in white matter (arrows). Progressive ataxia of Charolais cattle.

nerve cell bodies. In Limousin calves, such foci are present in the cerebellar peduncles and optic chiasm; in Simmental calves, in the internal capsule, caudate nucleus, putamen, periaqueductal white matter, lateral cuneate nucleus, and inferior olive. In Limousins, blindness and dysmetria appear at about 1 month of age, and some animals develop seizures and opisthotonus, whereas others remain stable. In Simmentals, the clinical onset is between 5 and 8 months of age. There is progressive ataxia, weight loss, and eventual dullness and emaciation, with death usual by 12 months of age. Investigation of these diseases is still at a preliminary stage. The pathogenesis is uncertain, and the classification as myelinolytic disease, preliminary.

h. FELINE SPINAL MYELINOPATHY An adult-onset condition involving progressive noninflammatory myelin loss in the spinal cord has been reported in cats in California. Myelin deficiency develops in the absence of significant axonal degeneration, and is accompanied by astrogliosis. The lesions occur in all funiculi, and are most intense in the thoracic and lumbar segments. The pathomechanism is undefined.

i. HYPERTROPHIC NEUROPATHY OF THE TIBETAN MASTIFF This recessively inherited condition affects the peripheral nerves and is considered to reflect a primary metabolic defect of Schwann cells, with failure to maintain myelin during axonal elongation in postnatal growth. The central nervous system is not involved. The clinical onset is consistently between 7 and 10 weeks of age. There is rapid progression of generalized weakness with hyporeflexia, slowed nerve-conduction velocity, and ultimately tetraplegia in many cases.

In peripheral nerves and their spinal roots, there are changes consistent with demyelination, remyelination, and endoneurial fibrosis. Thus many axons are denuded, although some have a thin sheath; proliferating Schwann cells form onion-bulbs, and the endoneurial collagen is prominent. Axonal degeneration is not a feature. The most characteristic features are revealed ultrastructurally. Schwann cell cytoplasm contains dense accumulations of 6- to 7-nm filaments, occurring at sites where the myelin sheath has separated along the major dense lines, and in the adaxonal cytoplasm. There are also anomalous incisure patterns within the sheaths, with many incisural openings staggered through the entire thickness of the sheath.

Bibliography

Anderson, C. A. *et al.* Border disease virus-induced decrease in thyroid hormone levels with associated hypomyelination. *Lab Invest* **57:** 168–175, 1987.

Binkhorst, G. J. *et al.* Neurological disorders, virus persistence, and hypomyelination in calves due to intrauterine infections with bovine virus diarrhoea virus. *Vet Q* **5:** 145–155, 1983.

Bjerkas, I. Hereditary "cavitating" leucodystrophy in Dalmatian dogs. *Acta Neuropathol (Berl)* **40:** 163–169, 1977.

Braund, K. G. *et al.* Congenital hypomyelinating polyneuropathy in two golden retriever littermates. *Vet Pathol* **26:** 202–208, 1989.

Cooper, B. J. *et al.* Defective Schwann cell function in canine inherited hypertrophic neuropathy. *Acta Neuropathol (Berl)* **63:** 51–56, 1984.

Cummings, J. F., and deLahunta, A. Hereditary myelopathy of Afghan hounds, a myelinolytic disease. *Acta Neuropathol (Berl)* **42:** 173–181, 1978.

Cummings, J. F. *et al.* Tremors in Samoyed pups with oligodendrocyte deficiencies and hypomyelination. *Acta Neuropathol (Berl)* **71:** 267–277, 1986.

Daniel, R. C., and Kelly, W. R. Progressive ataxia in Charolais cattle. *Aust Vet J* **58:** 32, 1982.

Done, J. T. The congenital tremor syndrome in pigs. *Adv Vet Sci* **16:** 98–102, 1976.

Duncan, I. D. Abnormalities of myelination of the central nervous system associated with congenital tremor. *J Vet Int Med* **1:** 10–23, 1987.

Griffiths, I. R., Duncan, I. D., and McCulloch, M. Shaking pups: A disorder of central myelination in the spaniel dog II: Ultrastructural observations on the white matter of the cervical spinal cord. *J Neurocytol* **10:** 847–858, 1981.

Harper, P. A. *et al.* Multifocal symmetrical encephalopathy in Simmental calves. *Vet Rec* **124:** 121–122, 1989.

Harper, P. A. *et al.* Multifocal encephalopathy in Limousin calves. *Aust Vet J* **67:** 111–112, 1990.

Jortner, B. S., and Jonas, A. M. The neuropathology of globoid-cell leukodystrophy in the dog. *Acta Neuropathol (Berl)* **10:** 171–182, 1968.

Kornegay, J. N. Hypomyelination in Weimaraner dogs. *Acta Neuropathol (Berl)* **72:** 394–401, 1987.

Luttgen, P. J., Braund, K. G., and Storts, R. W. Globoid cell leukodystrophy in a basset hound. *J Small Anim Pract* **24:** 153–160, 1983.

Mayhew, I. G. *et al.* Tremor syndrome and hypomyelination in lurcher pups. *J Small Anim Pract* **25:** 551–559, 1984.

Pritchard, D. H., Napthine, D. V., and Sinclair, A. J. Globoid cell leukodystrophy in polled Dorset sheep. *Vet Pathol* **17:** 399–405, 1980.

Richards, R. B., and Edwards, J. R. A progressive spinal myelinopathy in beef cattle. *Vet Pathol* **21:** 35–41, 1986.

Slocombe, R. F., Mitten, R., and Mason, T. A. Leucoencephalomyelopathy in Australian Rottweiler dogs. *Aust Vet J* **66:** 147–150, 1989.

Vandevelde, M. *et al.* Dysmyelination in chow chow dogs: Further studies in older dogs. *Acta Neuropathol (Berl)* **55:** 81–87, 1981.

3. Spongiform Myelinopathies

There is a group of diseases in which the dominant pathological feature is dramatic vacuolation of myelin, which often occurs without any overt indication of large-scale myelin breakdown or phagocytosis. There is still the familiar problem, however, of deciding how to categorize some diseases which have overlapping features of myelin vacuolation and myelin degeneration, and it is acknowledged that some of the conditions described could be placed elsewhere.

Myelin vacuolation is one of several different morphological changes encompassed by the term status spongiosus of nervous tissue (Fig. 3.80). Although producing a striking light microscopic picture when well developed,

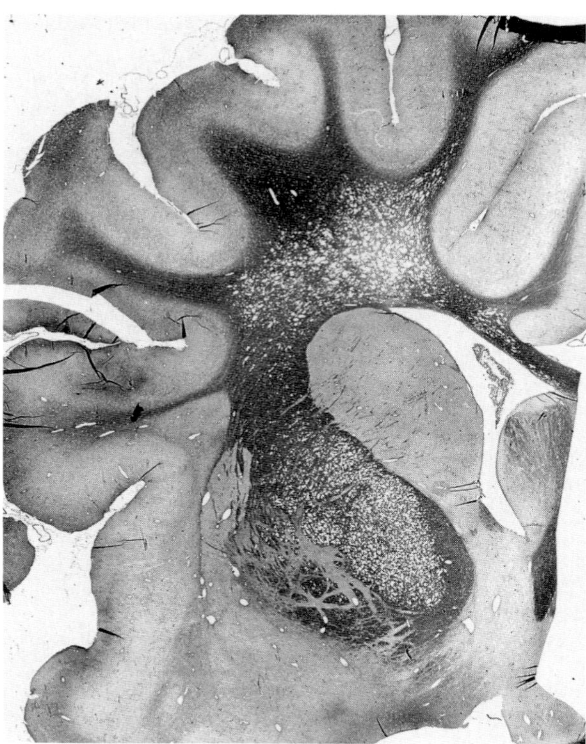

Fig. 3.80 Spongiform myelinopathy. Silver fox. Status spongiosus. Central cerebral white matter. (Reprinted from Hargen, G. and Bjerkas, I. *Vet Pathol* **27:** 187, 1990.)

electron microscopy is required for its fine definition (Fig. 3.81). Histologic changes suggestive of this lesion are a fairly common postmortem artefact.

Vacuolation of myelin may come about in several ways, the most frequent of which involves separation of lamellae along the intraperiod line, thereby reopening the extracellular space originally obliterated within the spiraling processes of the myelinating cell. This mechanism can produce vacuoles within the sheath at multiple levels and multiple loci along an internode. In experimental situations, this type of vacuolation has been associated with an increase in tissue water and electrolytes, consistent with simple edema, and accounts for the brain swelling which may occur in severe cases. The vacuoles therefore contain no stainable material. A variety of associated functional disturbances have been described but, remarkably, the lesion may have little functional impact, even when quite intense and prolonged. It seems likely that the myelin lesion is one facet of a complex metabolic disturbance, whose ramifications vary according to basic causes. In humans, the classic disease of this type is **Canavan's disease,** with congenital and infantile forms due to an autosomal recessive trait, and a rare juvenile form with no demonstrable familial association.

Vacuolation may also arise by separation of lamellae at the major dense line, reopening the intracellular compartment of the myelinating cell, or by ballooning of the periaxonal space. Depending on the particular disease, the vacuolation of the myelin may or may not be accompanied by structural changes in other elements of the tissue, including the myelinating cell bodies, and, in general, these are more likely to occur if vacuolation has been prolonged. In some instances, too, prolongation of the process is associated with a reduced quantity of myelin. In those cases in which there is no indication of myelin degradation, its steady withdrawal by the normal catabolic pathways with concurrent suppression of synthesis is implied.

The spongiform myelinopathies of animals fall into two broad groups: those which are **idiopathic,** and those which have a defined **metabolic or toxic** basis.

a. Idiopathic Spongiform Myelinopathies This group is documented in a small number of case reports involving calves, pups, and kittens, and most are suspected to have a heredofamilial basis. Several conditions involving newborn **Hereford** calves have been described over the years, with some conflicting and confusing aspects. Earlier reports proposed a condition, dubbed hereditary neuraxial edema, to be a single entity with a variable clinical and pathological expression. Some recent studies have gone a long way toward resolving the confusion, and several distinct diseases have now been dissected from this syndrome. Notable among these are **inherited congenital myoclonus** and **branched-chain ketoacid dehydrogenase deficiency(BCKAD).** The former is recognized in several countries in newborn polled Herefords which have violent myoclonic muscle spasms which prevent them from standing. There are no structural lesions, but biochemical studies have revealed an absence of glycinergic receptors on spinal inhibitory neurons. The functional disturbance is thus analogous to strychnine poisoning and is equally fatal.

From New Zealand a disease of horned **Hereford** calves has been named **congenital brain edema.** Newborn calves are unable to stand after birth and have coarse tonic muscle contractions. Vacuolation of myelin is diffuse, extends into the gray matter, and is accompanied by elevated brain water content. Hydropic degeneration of astrocytes and expansion of the extracellular space are detectable ultrastructurally. There is also considered to be a deficiency of myelin.

In Britain, a very similar picture is recorded in **polled Hereford** calves. There may remain an entity in Hereford calves, severely neurologically impaired at birth, in which myelin vacuolation is extensive in the central nervous system, is confined to the white matter, and is not associated with hypomyelinogenesis. For the time being, such cases could be classified as hereditary neuraxial edema in the absence of any demonstrable metabolic disturbance suggestive of BCKAD or other aminoacidopathy.

In **Samoyed** pups, there is the onset of a generalized tremor syndrome in the first few weeks of life, and a severe and ultimately fatal outcome. Myelin vacuolation is diffuse

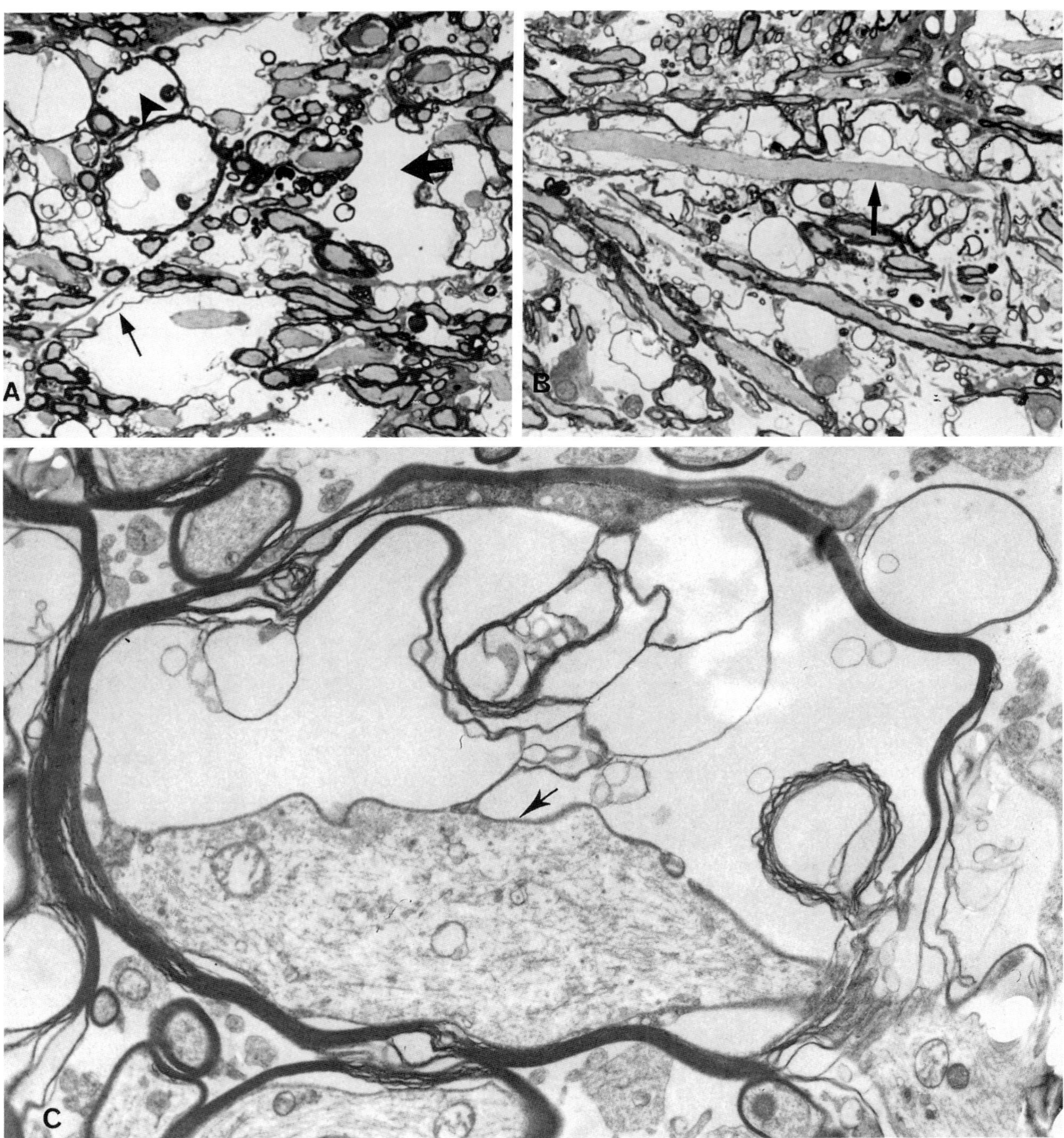

Fig. 3.81 Spongiform myelinopathy. Silver fox. (A, B, C.) Cross and longitudinal sections reveal intact axons amid disrupted myelin. Myelin vacuolation originating at intraperiod line and periaxonal space. (Reprinted from Hargen, G., and Bjerkas, I. *Vet Pathol* **27:**187, 1990).

throughout the central nervous system, and there is no change in any other tissue element.

Labrador pups are described to have initial episodes of extensor rigidity, tremor, and dorsal flexion of the neck at 4 to 6 weeks of age, and then progressive dysmetria. Intense vacuolation of myelin is confined to the white matter in the central nervous system, but occurs to a mild degree in peripheral nerves as well. Throughout the central nervous system, the myelin lesion is accompanied by fibrous astrogliosis and prominence of capillary blood vessels.

In a **silky terrier** pup, a myoclonic syndrome is described, with vacuolation of myelin occurring in the brain but not the spinal cord, together with the presence of Alzheimer type 2 astrocytes.

In **Egyptian Mau** kittens, there is progressive ataxia and hypermetria. Vacuolar change extends from white into gray matter throughout the central nervous system.

A recent account is given of a hereditary nervous disease in farmed silver foxes in Norway. The presenting sign is caudal limb ataxia appearing between 2 and 4 months of age, and progressing over the next 4 to 8 weeks, after which time clinical improvement seems to occur.

No gross lesions are present, and histologically there is symmetrical myelin vacuolation affecting white matter of the cerebrum, cerebellum, brain stem, and spinal cord (Fig. 3.80). The extent and severity of vacuolation vary from case to case. In long-standing cases, vacuolation seems to resolve to a large extent, with a residual intense astrogliosis.

Ultrastructurally, early features of the disease include intramyelinic vacuolation due to lamellar separation at the intraperiod line, large cytoplasmic vacuoles in oligodendrocyte cytoplasm, demyelination, expansion of extracellular space, and astrocytic hypertrophy (Fig. 3.81). Late in the disease, there is evidence of remyelination in gliotic areas.

b. Toxic/Metabolic Spongiform Myelinopathies

i. *Hepatic and Renal Encephalopathy* The endogenous intoxications associated with hepatic and, to a lesser extent, renal failure, are recognized to induce brain lesions of this type, the former quite frequently, the latter less so.

The familiar clinicopathologic term, **hepatic encephalopathy,** designates a complex autointoxication, in which accumulations of ammonia and other metabolites reflect the inability of the liver to carry out its normal detoxifying role. This may in turn reflect reduced liver mass, portosystemic shunting, or both. It must be emphasized that the neural lesions of hepatic encephalopathy are several and variable, as will be discussed.

Clinical evidence of nervous dysfunction may be accompanied by extensive and well-developed spongy vacuolation of myelin (Fig. 3.82), which tends to be most intense at the junction of the cerebral cortex and adjacent white matter, and often around the deep cerebellar nuclei. The lesion is to be expected in ruminants with subacute or chronic phyto- or mycotoxic liver injury, and in small animals with developmental portosystemic shunts or acquired liver disease. There may also be Alzheimer type 2 astroglial cells, although, with the notable exception of the horse, this is generally less a feature in animals than it seems to be in humans. In the horse, hepatic encephalopathy is characterized by the presence of Alzheimer type 2 cells, with no significant myelin vacuolation.

Although the myelin vacuolation may be induced experimentally with ammonia alone, the syndrome is not a simple ammonia intoxication, and is generally held to be a multifactorial metabolic disturbance.

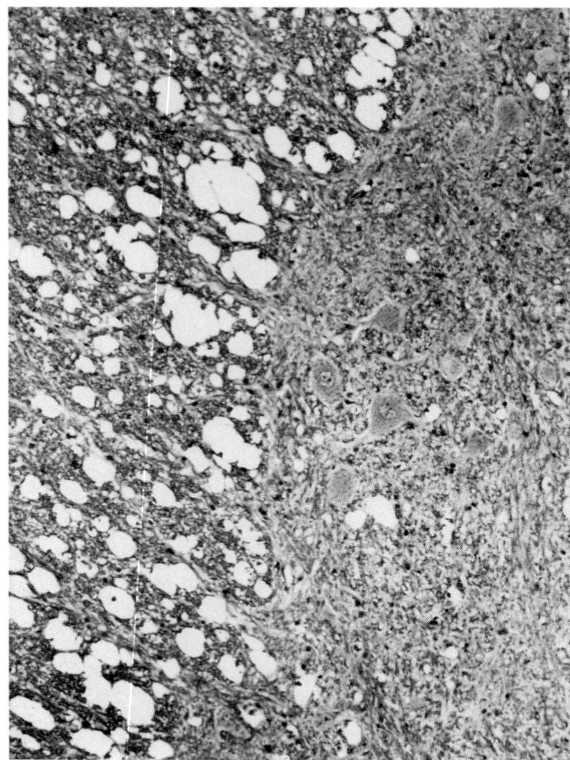

Fig. 3.82 Hepatic encephalopathy. Intramyelinic edema and spongiform change surrounding spinal gray matter. Sheep.

ii. *Branched-Chain α-Ketoacid Decarboxylase Deficiency* This disease, also known as **maple syrup urine disease,** has been described in Hereford, polled Hereford, and polled shorthorn calves, and is suspected of being inherited as an autosomal recessive character. The molecular nature of the gene mutation has recently been elucidated for polled Hereford calves. The metabolic defect leads to the accumulation of the branched-chain aminoacids leucine, isoleucine, and valine and their respective ketoacids, ketoisocaproic, keto-β-methylvaleric, and ketoisovaleric acids. The disease may be expressed *in utero,* and calves become dull and recumbent by 2 to 4 days of age, and finally develop opisthotonus. The urine in many cases has a smell of maple syrup, the characteristic which gave the disease its common name in affected children. Branched-chain aminoacids are found to be in abnormally high concentrations in plasma, cerebrospinal fluid, and tissues.

Pathologically, there may be gross evidence of brain swelling expressed as flattened cerebral gyri and, histologically, severe spongy vacuolation of myelin in the central nervous system. The spongiotic change is pronounced in the large myelinated tracts of the cerebral hemispheres and cerebellum (Fig. 3.83), and in myelinated tracts abutting brain stem nuclei and spinal gray matter. Most of the spinal white matter is not affected. In cases which survive for a week or more, a modest number of axonal spheroids

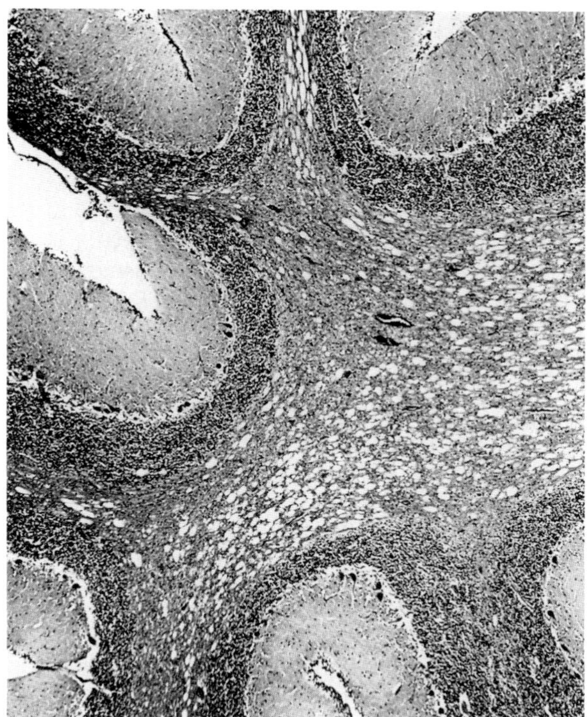

Fig. 3.83 Maple syrup urine disease. Calf. Status spongiosus. Central cerebellar white matter.

may be found in affected white matter. The brain water content is elevated, and the electron microscope reveals splitting of myelin lamellae at the intraperiod line. The molecular mechanism of the intoxication remains to be precisely defined, but probably involves toxic metabolites.

iii. *Hexachlorophene Toxicosis* Hexachlorophene is a polychlorinated phenolic compound with useful topical antiseptic properties, and is sometimes given orally against *Fasciola hepatica*. Intoxication has usually been associated with repeated application to the skin of very young animals, whence absorption is rapid, and systemic accumulation may cause effects within several days. The clinical features of severe intoxication are shaking, shivering, excitement, tonic–clonic convulsions, and terminal coma.

The major lesion is spongy vacuolation of central and peripheral myelin (Fig. 3.84), with lamellar splitting at the intraperiod line, and no associated degenerative or reactive changes. This lesion may persist for many weeks, even after clinical signs have abated. Under experimental conditions, vacuolation of the retinal photoreceptor outer segments is also evident, but this has not been a reported feature of accidental intoxication.

iv. *Halogenated Salicylanilide Toxicosis* The halogenated salicylanilides are a group of long-acting anthelmintic agents for use in sheep and goats, and include rafoxanide, clioxanide, and closantel. Accidental overdosing has led to reports of intoxication expressed as ataxia, depression, and blindness, which may be permanent. The response appears to be inconsistent, and unknown predisposing factors are possibly involved.

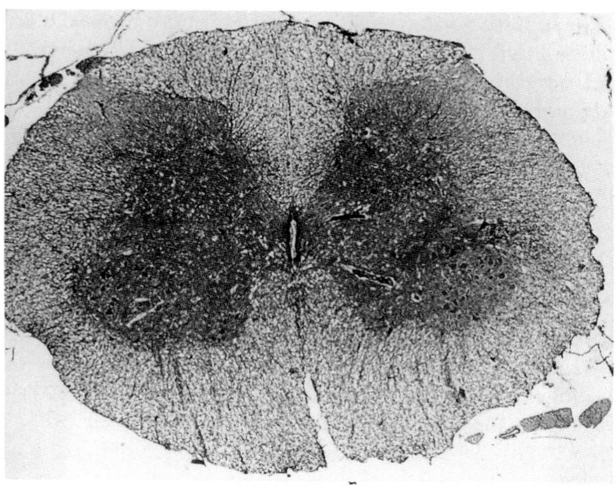

Fig. 3.84 Intramyelin edema. Hexachlorophene poisoning. Dog. Diffuse spongiform change. Spinal cord white matter.

Microscopically, there is spongy vacuolation of myelin in both the central nervous system and peripheral nerves, without any accompanying reaction or degeneration. The lesion in the brain is most intense in the brain stem. This change may persist for several weeks. Acute degenerative changes may be found in the optic nerves and in the retinal ganglion cell layer, where hemorrhage has also been described. The syndrome has some similarities to stypandrol poisoning, described in the following section.

v. *Stypandrol Toxicosis* The binaphthalene tetrol, stypandrol, is the toxic principle of the Australian lileacious *Stypandra* spp. (blindgrass) and also of the Asiatic *Hemerocallis* spp., hence the alternative name hemerocallin. Ingestion of toxic strains of such plants by grazing animals or birds has an acute effect, usually manifested as moderate depression and blindness, although a high dose may produce prostration and death.

In acutely affected animals, myelin vacuolation caused by lamellar splitting at the intraperiod line is extensive throughout the central nervous system, and may persist for many weeks or even months, without apparent functional effect. Blindness, however, tends to be permanent and is associated with Wallerian degeneration of the optic nerves and tracts, and degeneration of retinal photoreceptors. Although not established with certainty, it seems that the degeneration of the optic nerves may be the result of compression within the optic canals secondary to swelling caused by the myelin edema, and that there is a separate direct toxic effect on the photoreceptors.

vi. Helichrysum *Species Toxicosis* The many plant species of this genus are widely distributed, and distinct syndromes of toxicity in ruminants are reported from Australia and South Africa: *H. blandowskianum* in Australia produces acute hepatic necrosis, presumptive hepatic encephalopathy, and spongiform change in the white matter of the brain stem, with a high mortality rate; *H. argyrosphaerum* in South Africa is associated with paresis, paralysis, and permanent blindness.

Intense spongy vacuolation of myelin in the brain is sufficient to cause grossly visible swelling of the optic nerves and chiasm, and gelatinous swelling of the periventricular white matter and corpus callosum. Focal hemorrhages may occur in the optic nerves and chiasm. Microscopically, the spongiform myelin lesion in the optic nerves is accompanied by Wallerian degeneration, suggesting compression within the optic canals. Myelin vacuolation is less severe in the spinal cord and peripheral nerves. The syndrome has similarities to stypandrol toxicosis, but the active principle is as yet undefined.

Bibliography

Basson, P. A. *et al.* Blindness and encephalopathy caused by *Helichrysum argyrosphaerum* D.C. (Compositae) in sheep and cattle. *Onderst J Vet Res* **42:** 135–148, 1975.

Bei Zhang *et al.* Premature translation termination of the pre-E1a subunit of the branched chain α-ketoacid dehydrogenase as a cause of maple syrup urine disease in polled Hereford calves. *J Biol Chem* **265:** 2425–2427, 1990.

Hagen, G., and Bjerkas, I. Spongy degeneration of white matter in the central nervous system of silver foxes (*Vulpes vulpes*). *Vet Pathol* **27:** 187–193, 1990.

Hagen, G., Blakemore, W. F., and Bjerkas, I. Ultrastructural findings in spongy degeneration of white matter in silver foxes (*Vulpes vulpes*). *Acta Neuropathol (Berl)* **80:** 590–596, 1990.

Hall, G. A., and Reid, I. M. The effects of hexachlorophene on the nervous system of sheep. *J Pathol* **114**: 241–246, 1974.

Harper, P. A., Healy, P. J., and Dennis, J. A. Maple syrup urine disease (branched chain ketoaciduria). *Am J Pathol* **136:** 1445–1447, 1990.

Hooper, P. T. Spongy degeneration in the central nervous system of domestic animals. *Acta Neuropathol (Berl)* **31:** 325–334, 1975.

Jolly, R. D. Congenital brain oedema of Hereford calves. *J Pathol* **113:** 199–204, 1974.

Kelly, D. F., and Gaskell, C. J. Spongy degeneration of the central nervous system in kittens. *Acta Neuropathol (Berl)* **35:** 151–158, 1976.

Sheahan, B. J. *et al.* Structural and biochemical changes in a spinal myelinopathy in twelve English foxhounds and two harriers. *Vet Pathol* **28:** 117–124, 1991.

Zachary, J. F., and O'Brien, D. P. Spongy degeneration of the central nervous system in two canine littermates. *Vet Pathol* **22:** 561–571, 1985.

4. Nonmyelinic Spongiform Encephalomyelopathies

The term status spongiosus applies to a variety of lesions in which the microscopic appearance of the neural tissue is transformed by numerous vacuoles or microcavities. As discussed, one form of this state results from the vacuolation of myelin, and is the most common. This section will discuss diseases in which the change results from either the swelling of astrocytes or the vacuolation of neurons and neurites. Astrocytic swelling may occur nonspecifically in, or adjacent to, any area of neural tissue acutely injured by trauma, or ischemia, or in which vascular permeability is disturbed. In the absence of such circumstances, its occurrence suggests some direct toxic or metabolic influence on the astrocytes themselves. It can thus be a feature in hepatic encephalopathy. Spongiform vacuolation of neurons and their processes may occur to a limited degree as an incidental change in the red nucleus and nuclei of the posterior brain stem, and also as an artefact of fixation. It needs to be interpreted against this background, and is usually considered in the context of the diagnosis of scrapie and related conditions. It is distinctly different from the foamy vacuolation seen in the lysosomal storage diseases.

a. Bovine Citrullinemia This disease is a fulminating neurologic affliction of newborn Holstein–Friesian calves, resulting from a blockade of urea cycle metabolism caused by a deficiency of arginosuccinate synthetase. The enzyme deficiency leads to hyperammonemia and citrullinemia, and is due to an autosomal recessive genetic defect.

Calves are affected within the first week of life, the clinical features being depression, head-pressing, stupor, convulsions, terminal coma, and death. There are no gross neural lesions, but microscopically there is mild to moderate spongy vacuolation of the deep laminae of the cerebral cortex. The spongy change is due to astrocytic swelling. Hydropic astrocytes have enlarged vesicular nuclei, but their cell bodies and processes are vacuolated and give rise to the spongiform appearance of the neuropil, the creation of perineuronal spaces, and the enlargement of pericapillary spaces. The astrocytic reaction is believed to represent a stress on these cells in their attempts to detoxify ammonia. Astrocytes play a major role in ammonia detoxification via utilization of glutamine synthetase and glutamate dehydrogenase activities. In spite of the marked degenerative changes in astrocytes, other elements of the tissue remain morphologically normal, but neuronal function is obviously grossly disturbed.

Affected calves also have pale yellow-ochre discoloration of the liver, which is correlated microscopically and ultrastructurally with hydropic degeneration of hepatocytes.

b. Scrapie and Related Diseases This group of diseases, of which ovine scrapie is the archetype, composes a distinct and unique class. Scrapielike diseases have been documented in North America in ranch mink (transmissible mink encephalopathy), and in captive mule deer, black-tailed and white-tailed deer, and Rocky Mountain elk (chronic wasting disease), and in captive zoo ruminants in the United Kingdom (Fig. 3.18). Most recently, bovine spongiform encephalopathy has been described as an addition to this group in Britain, along with a few cases in domestic cats. It appears probable at this stage that all these scrapielike conditions in animals are in fact caused by the ovine scrapie agent rather than by similar species-specific agents.

In humans, Creutzfeldt–Jakob disease, Kuru and

Gerstmann-Straussler disease have many similarities to scrapie, but there is no convincing evidence that there is human–animal cross infection, or that the agents are identical. In all these human, and most of the animal diseases listed, scrapie immunoreactive amyloid can be demonstrated in affected brains, and scrapie-associated fibrils (SAF) extracted.

Scrapie in sheep, and to a small extent in goats, is endemic in parts of Europe and the United States. There is no doubt that it is an infectious disease, but the nature of the causal agent remains a matter of intense controversy. To date, it has defied precise definition. It provokes no detectable immune response but is titratable in terms of infectivity. Its status as a virus remains uncertain; it is not known whether the agent contains nucleic acids. This has resulted in the prion hypothesis, which proposes a proteinaceous infectious particle derived from a normal host protein. Although there are serious conceptual difficulties for the idea of a protein able to direct its own replication, the prion hypothesis is far from defunct at the time of this writing. In addition, it is not discounted that there may be a small nucleic acid genome associated with a host membrane protein, and this forms the basis of the virino hypothesis.

It is well established that the infectious particle, whatever its nature, is associated with the protease-resistant core of a protein, which is an isoform of a normal cell surface sialoglycoprotein. The normal isoform is designated PrP^{c} and the disease-associated isoform, PrP^{sc}. Evidence suggests that both are derived from a single messenger RNA (mRNA), and that the PrP gene product has an important biological function. The expression of PrP^{sc} probably represents aberration of posttranslational processing.

Electron-microscopic examination of detergent extracts of infected brain reveals characteristic SAF, which have PrP^{sc} as their major component. In many cases, plaques of this material may be present in the neuropil, as histochemically detectable amyloid deposits, and it can be detected more sensitively by specific immuno-staining as scrapie immunoreactive amyloid. Cerebrovascular amyloidosis is also usually demonstrable in ovine scrapie.

Scrapie can be experimentally transmitted from sheep to other species by a variety of routes, and a great body of investigative work has been carried out in both mice and sheep. This has established the importance of genetic factors; host genes influencing incubation time and susceptibility have been identified in both mouse and sheep, and work on mice has also identified different strains of scrapie and some evidence for a genome for the agent.

The morbidity of the natural disease is low, except in some closed stud flocks, and clinically affected animals are usually in the 2- to 5-year-old group. There is a long incubation period, and the disease, once clinically manifest, is progressive and ultimately fatal in from 10 days to several months. Infection occurs mostly around the time of birth, and probably via ingestion, but the question of vertical modes of transmission remains unclarified. The likelihood of ewe-to-lamb transmission is highly significant, in that a ewe could give birth to and infect several lambs before showing clinical signs. The agent proliferates initially in the lymphoid tissues and lower intestine, but may take up to 2 years to reach the nervous system. A further 2 years may lapse before clinical signs appear. The methods of spread of the agent between both cells and organ systems are still enigmatic, although there is good evidence that the nervous system is first infected via the splanchnic nerves, which carry the agent to the thoracic spinal cord. Once within the central nervous system, the agent seems to be of no consequence until it reaches clinical target areas, where the expression of PrP^{sc} in select neurons could presumably lead to progressive dysfunction.

Both clinical signs and pathologic changes are somewhat variable but, in general, affected sheep are initially alert but excitable, tremble when excited, and may have seizures. Later, paresthesias may appear, manifested as agitated rubbing against posts and trees, and nibbling at the feet and legs when lying down, behavior which gave rise to the colloquial name scrapie. Self-traumatization can cause extensive loss of wool and abrasions of the skin. There is also progressive dysmetria and emaciation, and finally paralysis and death.

Apart from emaciation and self-trauma, there are no significant gross lesions, and no inflammatory changes. The most characteristic finding is the presence of large intraneuronal vacuoles in the medullary reticular, medial vestibular, lateral cuneate, and papilliform nuclei. The neuronal vacuolation may however be present throughout the brain stem and spinal cord (Fig. 3.85A). The vacuoles contain no stainable material and may considerably distort and displace the normal organelles. Although the vacuolated neurons do not generally appear degenerate in any other way, shrunken and apparently degenerate neurons may be found in mesencephalic, medullary, and deep cerebellar nuclei, and in the intermediolateral nucleus of the spinal cord. In late stages, neuronal loss may be evident, and a diffuse fibrillary astrogliosis, present mainly in the mesencephalon (Fig. 3.85B) and the outer layers of the cerebellar cortex. All these lesions are bilaterally symmetrical, and the cerebral cortex is rarely involved (Fig. 3.25A,B).

Spongy vacuolation of the neuropil in gray matter is due to vacuolation of neuronal processes but, although prominent in mouse scrapie, it is seldom a feature of the natural disease in sheep. Recent ultrastructural studies of mouse scrapie have suggested that vacuoles may arise as a result of incorporation of abnormal membrane into organelles in neuronal perikaryons and processes, and probably also in oligodendrocyte cytoplasm and myelin.

Bovine spongiform encephalopathy (BSE, mad cow disease) exploded into prominence as a virgin ground epizootic in Britain in the mid-1980s. Previously healthy dairy cattle aged between 3 and 6 years become apprehensive, hyperesthetic, and dysmetric. The animals display fear and aggressive behavior, with progressive gait distur-

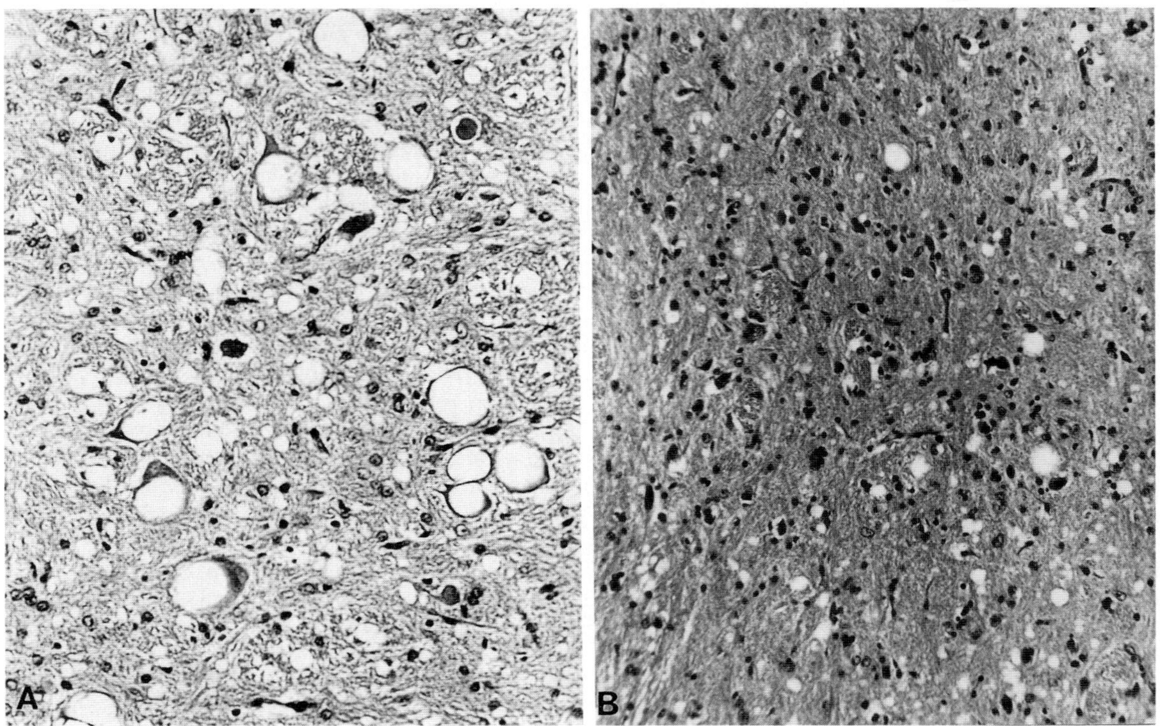

Fig. 3.85 Scrapie. Sheep. (A) Vacuolated neurons in brainstem. (B) Vacuolated neurons, loss of neurons, and gliosis in medulla.

bances leading to frequent falling. Eventually, frenzied episodes or recumbency necessitate euthanasia. The epidemiologic data point to the infection of cattle with the ovine scrapie agent via meat-meal made from sheep offal. The infectivity of offal is linked to the cessation of the practice of hydrocarbon solvent extraction of fat from meat and bone meal in rendering plants. It is still uncertain whether transmission from dam to calf will occur, or whether the disease will disappear in a dead-end host, as a result of stringent regulation of feed supplements. If the cow is a dead-end host, the current prediction is a declining incidence in the mid 1990s, given an incubation period of 2.5 to 8 years.

The pathologic features of BSE resemble those of scrapie. Scrapie-associated fibrils and immunoreactive amyloid are demonstrable. Vacuolation of neuronal cell bodies and processes is particularly prominent in the dorsal vagal, medullary reticular, vestibular, solitary, spinal trigeminal, and red nuclei. The severity of neuropil and neuronal somatic vacuolation varies independently between anatomic locations, and the latter is sometimes accompanied by ceroid-lipofuscin granules. Occasional necrotic neurons and axonal spheroids may be present, as may a mild astrogliosis.

In the United Kingdom, it has recently been confirmed that spongiform encephalopathy has been naturally transmitted to domestic cats, increasing the concerns about possible spread to other species, including humans. In this context, it is reassuring that transmissible mink encephalopathy is considered to be a dead-end scrapie infection, acquired by the recycling of sheep meat products.

c. Neuronal Vacuolar Degeneration of Angora Goats A scrapielike neuronopathy is described in young Angora goats in Australia (where scrapie is not present), and is presumed to be genetically determined. Animals become ataxic at about 3 months of age, and progress to severe paresis. No significant gross lesions are present, but there is spectacular vacuolation of large neurons in the red and other brain stem nuclei, and in the spinal motor neurons. Some Wallerian changes are present in the brain stem, spinal cord, and peripheral nerves.

d. Focal Spongiform Encephalopathy of Dogs A disease was originally described in **bull mastiffs** as familial cerebellar ataxia with hydrocephalus, but the main pathologic feature is spongy vacuolation of several gray matter nuclei. Clinical signs emerge between 6 and 9 weeks of age, and include ataxia and visual impairment, and a number of other variable deficits.

Macroscopically, there is moderate to severe dilation of the ventricular system. Microscopically, bilaterally symmetrical spongy vacuolation and gliosis are evident in all three deep cerebellar nuclei, the posterior colliculi and, to a lesser extent, the lateral vestibular nuclei. The vacuolation is accompanied by axonal spheroids, but nerve cell bodies appear normal. Some of the vacuoles appear to involve myelin sheaths, but the lesions are essentially confined to the gray matter, and the cytologic basis of the vacuolation is not precisely defined. An autosomal recessive genetic defect has been proposed. Similar vacuolating lesions of these gray matter areas have been described in **Saluki** dogs in Canada.

Bibliography

Aldhous, P. Spongiform encephalopathy found in a cat. *Nature* **345:** 194, 1990.

Dennis, J. A. *et al.* Molecular definition of bovine argininosuccinate synthetase deficiency. *Proc Natl Acad Sci USA* **86:** 7947–7951, 1989

Fraser, H. The pathology of natural and experimental scrapie. *In* "Slow Virus Diseases of Animals and Man" R. H. Kimberlin (ed.), pp. 267–305. Amsterdam, North Holland, 1976.

Guiroy, D. C. *et al.* Topographic distribution of scrapie amyloid-immunoreactive plaques in chronic wasting disease in captive mule deer (*Odocoileus hemionus*). *Acta Neuropathol (Berl)* **81:** 475–478, 1991.

Healy, P. J., Harper, P. A., and Dennis, J. A. Bovine citrullinaemia: A clinical, pathological, biochemical, and genetic study. *Aust Vet J* **67:** 255–258, 1990.

Jeffrey, M., Scott, J. R., and Fraser, H. Scrapie inoculation of mice: Light and electron microscopy of the superior colliculi. *Acta Neuropathol (Berl)* **81:** 562–571, 1991.

Kimberlin, R. H. Transmissible encephalopathies in animals. *Can J Vet Res* **54:** 30–37, 1990.

Lancaster, M. J., Gill, I. J., and Hooper, P. T. Progressive paresis in Angora goats. *Aust Vet J* **64:** 123–124, 1987.

Marsh, F. R. The subacute spongiform encephalopathies. *In* "Slow Viral Diseases of Animals and Man" R. H. Kimberlin (ed.), pp. 360–380. Amsterdam, North Holland, 1976.

Petrie, L., Heath, B., and Harold, D. Scrapie: Report of an outbreak and a brief review. *Can Vet J* **30:** 321–327, 1989.

Wells, G. A. *et al.* Bovine spongiform encephalopathy: Diagnostic significance of vacuolar changes in selected nuclei of the medulla oblongata. *Vet Rec* **125:** 521–524. 1989.

Wilesmith, J. W., Ryan, J. B., and Atkinson, M. J. Bovine spongiform encephalopathy: Epidemiological studies on the origin. *Vet Rec* **128:** 199–203, 1991.

VIII. Inflammation in the Central Nervous System

Inflammation of the brain is **encephalitis;** of the spinal cord, **myelitis;** of the ependyma, **ependymitis;** of the choroid plexus, **choroiditis;** and of the meninges, **meningitis,** qualified as **leptomeningitis** when it involves the pia–arachnoid and as **pachymeningitis** when it involves the dura. This is the area of neuropathology which is of most veterinary importance because it embraces many of the transmissible highly fatal infections of animals, because even the sporadic infections are common, and because there is always a pressing need for the pathologist to separate the inflammations into general or specific etiologic categories. Most of the specific infectious diseases to be described in this chapter are caused by agents which demonstrate a remarkable or specific neurotropism. These are, however, only a segment of the list of agents that could be drawn up to include those which commonly, occasionally, or rarely involve the central nervous system. Throughout these volumes, frequent reference is made to, or descriptions given of, the neurologic lesions which are part of, or complications of, many specific infectious diseases caused by pathogenic microorganisms of all classifications.

Inflammatory processes in the central nervous system are basically the same as those in other tissues, but they derive some specific features from the special responsiveness and anatomic arrangements of the central nervous system. It is important to recognize the criteria of inflammation in the central nervous system and then to presume from that, as far as is possible, the nature of the infecting organisms, because opportunities to intervene clinically are brief, and the time frame for clinical intervention is less than that required for secure microbiological diagnosis.

Some of the causes of encephalomyelitis, or meningitis, are quite explicit in their expression, either because the organism itself is visible, as is expected of *Cryptococcus* as an example, or inclusion bodies are manifest, as in canine distemper and pseudorabies. In the absence of such specific indicators, less specific suggestions as to cause recognize that different agents have predilections for different parts of the central nervous system, and it is frequently necessary to act on presumptions drawn from the nature of the lesion and from its distribution in the brain or cord. The regional priorities are identified in discussion of each disease.

Considerably less is known of factors which influence the localization and course of infections in the nervous system than the little that is known of these factors in other tissues. It is frequently stated that the nervous system and its leptomeninges offer less resistance to infection than do other tissues of the body, the statement applying especially to infections by bacteria and fungi. There is some basis for the statement. The cerebral parenchyma and meningeal spaces are excellent places for propagating a variety of microorganisms and even tissue grafts, and a variety of organisms which are usually and truly regarded as feeble pathogens can produce severe and progressive lesions in the central nervous system. But the situation is much more complicated than is so far implied. As indicated previously, the variety of bacterial and mycotic infections which may invade and injure the brain and meninges is very wide, and so also is the variety of viruses. With some of these infections, the course is to a large extent predictable. Thus in Glasser's disease it is expected that *Haemophilus parasuis* will cause meningitis in about 80% of fatal cases, and in fatal streptococcal infections of newborn calves, meningitis is expected. But unpredictability is much more in evidence; thus, a small necrobacillary lesion of the larynx or lung of a calf may produce one or several cerebral abscesses, while systemic necrobacillosis seldom involves the brain, and bacterial valvular endocarditis may produce hundreds of embolic abscesses in the brain but, in fact, usually does not produce any. The lack of predictability applies not only to bacterial infections, but also to all classes of infectious agents.

The problem is not only one of why and how the CNS, of known high vulnerability to infection, is so frequently spared in systemic diseases; it is also one of why and how it is infected hematogenously. There is presently no better knowledge of why four of five cases of Glasser's disease will have meningitis than why one of four will not, even though in the one case the pathologic syndrome is otherwise fully developed.

Infections which, in other organs or tissues, may be inconsequential and even asymptomatic frequently cause death or permanent disability when they involve the nervous system. There are several contributing factors, the most important of which is the indispensability of most portions of the central nervous system. The nervous tissue cannot reconstitute itself, but it may, if the lesion develops slowly, manage a considerable degree of functional compensation. Vascular responses may be more an impediment than a help in the reaction to inflammations, because they lead consistently to edema and brain swelling, which spread the consequences of the inflammation far from the active focus. Vascular proliferation and fibrous-tissue encapsulation may develop only when the inflammatory reaction involves the meninges and the larger blood vessels, because these are the only source of reticulin and collagenous tissue. There is still doubt as to how viruses spread in the nervous tissue, but bacteria and fungi can spread rapidly and extensively in the fluid of the ventricular system and meninges. Spread of infection from the ventricular fluid to the periventricular veins occurs readily across the ependyma, and spread from the meninges to the brain, or vice versa, can occur via the Virchow–Robin spaces. The special drainage of cerebrospinal fluid is such that the drainage of exudate rich in fibrin and cells is very poor.

The structure of nervous tissue and meninges limits the anatomical types of inflammation that may occur. Fibrinous inflammations are confined to the meninges and larger perivascular spaces. Fibrin is usually indicative of a bacterial infection, but there are exceptions. Fibrinopurulent, largely fibrinous, exudate is caused by the mycoplasmas; fibrinous exudation is typical of the meningeal reaction in malignant catarrhal fever of cattle, and it is occasionally observed in organomercurial poisoning of swine and cattle. Hemorrhagic inflammations are not common except as examples of symptomatic purpura in infections such as porcine erysipelas and infectious canine hepatitis and in the infarcts of embolic infections. Hemorrhage is characteristic of helminthic infections, but these lesions are perhaps best regarded as traumatic malacias. Suppurative and granulomatous inflammations are the usual response to bacterial and mycotic infections. Viral infections are characterized by an inflammatory response designated as nonsuppurative; this reaction is described in more detail later, and here it is necessary to note only that it is typically composed of neuronal degeneration, perivascular cuffing by mononuclear cells, and focal or diffuse glial proliferations.

Applying these broad criteria, there is seldom much difficulty in deciding on the class of the infecting agent. There is occasionally some difficulty in distinguishing between the reactions to degeneration and to viral infections. While suppuration does not occur in either of these processes, an early infiltration of neutrophils occurs in acute degenerations, such as the malacias, and in the first stages of many cases of viral encephalomyelitis, a few neutrophils can be found migrating in affected gray matter. Acute demyelinating processes are associated with, and probably stimulate, perivascular cuffing. The cuffs tend to be quite distinct from those in viral infections by being relatively very thick (up to 10 or more cells thick) and to be composed of pure populations of lymphocytes. If there are proliferated adventitial cells with plasma cells or other leukocytes mixed in the cuffs of lymphocytes, then the cuffs are likely to be a response to infection. Only three diseases in animals are caused by viruses and characterized by demyelination, and these, namely canine distemper, leukoencephalomyelitis of goats, and visna of sheep, have their own distinguishing characteristics. A possible fourth is the so-called old dog encephalitis. Glial responsiveness occurs in a variety of degenerative and viral lesions, but glial nodules are characteristic only of an infectious process, usually viral but occasionally rickettsial or bacterial.

A. Pyogenic Infections

The brain and spinal cord are protected against direct penetration of infection by the skeletal encasement and by the dura mater. Of these two, the dura mater provides the more reliable barrier. The barriers are highly efficient, but the nervous system is at risk when pyogenic processes are active in its neighborhood. Suppurative osteomyelitis of the skull is as aggressive and persistent as is this process in other bones, and the emissary veins in the diploë are potential routes of retrograde infection. The dura is almost impermeable to inflammatory processes, but it is vulnerable wherever it is penetrated by nerve trunks; the vulnerability is highest in such locations as at the cribriform plate and temporal bone, where the dura is fused with the periosteum, and there is no potential epidural space. If the dura is invaded, the external layer of the arachnoid provides a barrier and, if this too is invaded, the pia mater offers a barrier to spread of infection to brain and cord. In spite of the anatomic delicacy of the leptomeninges, the barriers they offer are quite substantial and, as an example, purulent leptomeningitis seldom spreads directly to the brain or cord. The disposition of barriers influences the progression of hematogenous infections as well as those of direct extension. For these reasons, it is possible to discuss pyogenic infections according to whether they are epidural, subdural, leptomeningeal, or intramedullary, the latter being inflammations in the brain and cord. Before discussing localizations, however, it is necessary to discuss briefly the various routes of infection of the nervous system.

Infectious agents and inflammatory processes may take one of four routes to the nervous system. They may invade via the peripheral nerves. This is the route that some viruses of high neurotropism are expected to take, invasion probably occurring along axons (see later). Of the bacterial infections, there is some evidence that *Listeria monocytogenes* can invade along the cranial nerves. Infection may be implanted directly in penetrating injuries or during surgical procedures. Direct extension of infection may occur from adjacent structures either by erosion of the bone or extension along the nerves. The most common primary sites providing for spread of this sort are the middle ear and tympanic bulla, the paranasal sinuses, and the ethmoid cells. The most common, and indeed a very common, route is hematogenous. The great majority of hematogenous infections are arterial, but a few are venous, involving both cranial and paravertebral veins. The potential importance of veins as pathways of infection is due to the large number of emissaries from the dural sinuses through the skull and the extensive anastomotic system of valveless veins, which permits reflex flow in many directions.

1. Epidural Abscess

Epidural abscess may result from hematogenous dissemination of pyogenic organisms to the epidural space or the bony encasement, by direct implantation in traumatic injuries, or by direct invasion from an adjacent structure or natural cavity. Cranial epidural abscesses are usually small and flattened between the dura and bone. They may be, however, quite large and cause bulging of the cranium and local compression of the brain. The spread of the process within the epidural space may be limited by encapsulating fibrous tissue or by the natural adhesions of the dura to the sutures. When the basal epidural space is invaded, whether from the middle ear, sella turcica, or one of the basal foramina, encapsulation is usually not observed, the pus dissecting its way beneath the dura mater until the animal dies.

Most cranial epidural abscesses arise by direct extension from one of the paranasal sinuses. Spread from the middle ear and through the cribriform plate is usually directly to the leptomeninges and brain. Occasionally, epidural suppuration is observed to have tracked from a retropharyngeal or nodal abscess through the cranial foramina.

Spinal epidural abscesses are usually secondary to osteomyelitis of the vertebral bodies and usually caused by *Actinomyces pyogenes*. They are observed most frequently in ruminants and swine. The abscess is usually small, and the dura is rarely penetrated. Most of these processes are remarkably indolent, but occasionally there is no confinement of the pus, and it tracks up and down the vertebral canal in the epidural space. Local venous invasion of the epidural space, leptomeninges, or cord is commonly observed in lambs as a complication of docking and in pigs as a complication of tail biting, but these are usually spreading suppurations. Migrating grass awns cause osteomyelitis and epidural abscess in the thoracolumbar regions in dogs. Spinal epidural abscesses usually cause compression malacia of the cord, if the osteomyelitis from which they originate does not first lead to pathologic fracture with displacement.

2. Subdural Abscess

Subdural suppurations are seldom observed. They are the result of local penetration, perhaps most frequently from the paranasal sinuses. They may be extensions from epidural abscesses or, because subdural suppurations are prone to cause local phlebitis, they may give rise to epidural suppuration. Subdural infections are likely to spread widely via the veins or after penetrating the outer layer of the arachnoid. The spread is also permitted by the slowness with which leptomeningeal fibrous tissue develops to encapsulate the reaction.

3. Leptomeningitis

Leptomeningitis is usually but not always purulent. Serocellular meningitis is described later with viral inflammations; hemorrhagic meningitis occurs in septicemic anthrax and very seldom in other septicemias; a purely fibrinous variety of inflammation occurs in malignant catarrhal fever and seldom in anything else; and granulomatous inflammations occur in infections of this type, such as tuberculosis and cryptococcosis. These nonsuppurative inflammations of the leptomeninges are described with the diseases of which they may be part. They are mentioned here to draw attention to the wide variety of infectious agents which may localize in and produce inflammation of the leptomeninges. We have also mentioned or discussed purulent leptomeningitis with the many specific diseases of which it can be part but dwell on the purulent process here because it is common, lacks specificity, and is suitable for a "type" description.

Purulent leptomeningitis may arise by direct extension from an adjacent structure. Extension from an epidural abscess or inflammation may result in diffuse leptomeningitis but, in most of the few cases of this origin, the leptomeningitis is local and overshadowed by the brain abscess that usually forms. Leptomeningitis may arise by local extension from a brain abscess, either by direct permeation or by spread in the Virchow–Robin spaces. Such an origin is observed frequently in listeriosis and in association with very large cerebral abscesses, but most cerebral abscesses, which are usually small, track inward to the white matter rather than out to the meninges. The reverse sequence in which meningitis spreads to the brain and cord is not unusual in granulomatous infections but is distinctly unusual in suppurative infections, and when it occurs, it consists of invasion of the surface of the brain by neutrophils (Fig. 3.86A) rather than of abscessation.

Both purulent meningitis and cerebral abscesses are usually of hematogenous origin, but they are seldom concurrent, and one usually finds meningitis alone or

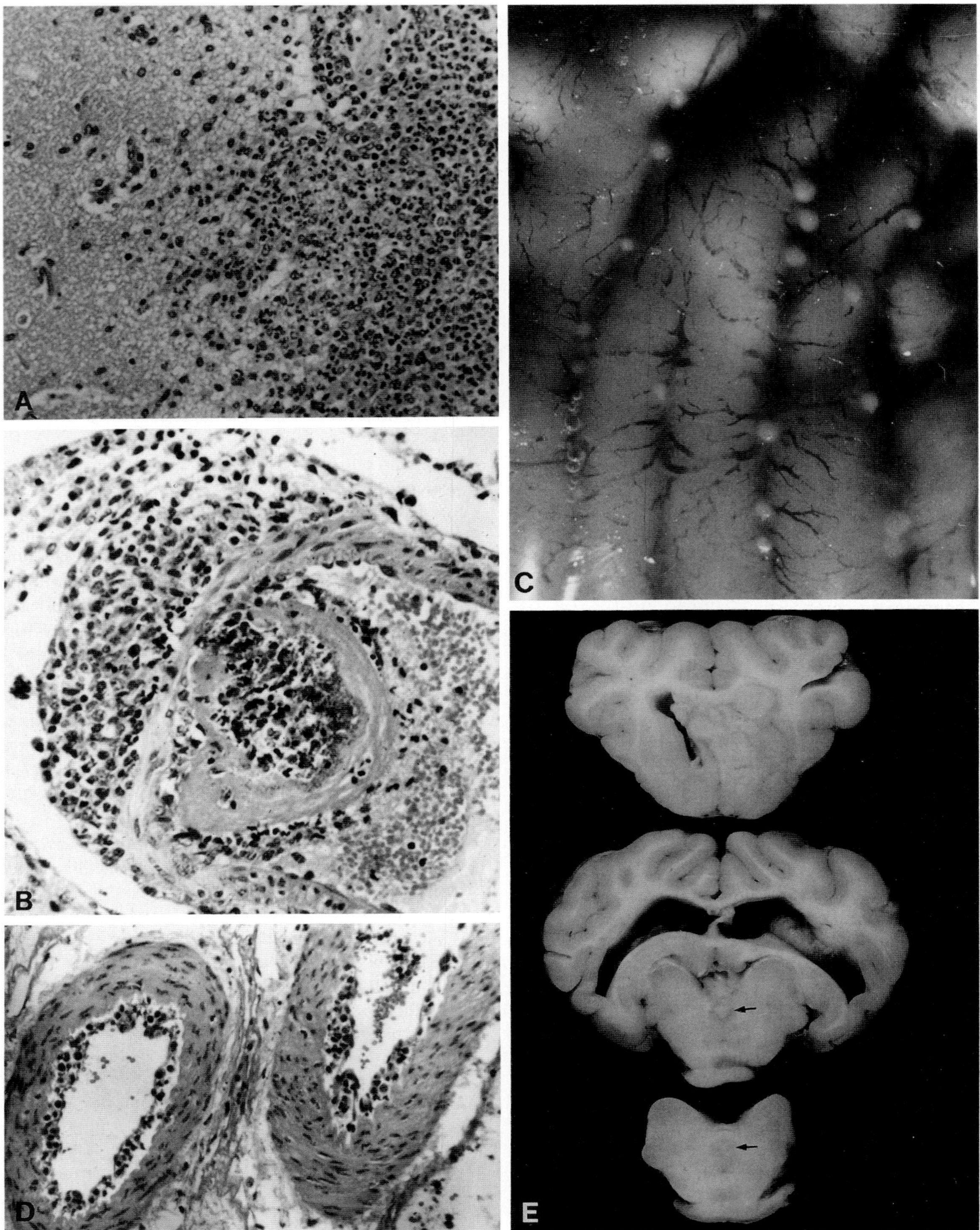

Fig. 3.86 (A) Suppurative meningitis with extension into cerebellar cortex. (B) Embolic meningeal arteritis from streptococcal endocarditis. Pig. (C) Miliary meningeal abscesses. Ox. *Actinomyces pyogenes*. (D) Meningeal arteritis with leukocytes beneath endothelium in coliform meningitis. Calf. (E) Frontal sections to show periventricular abscess, early hydrocephalus, and occlusive ependymitis of aqueduct (arrows) secondary to streptococcal choroiditis. Pig.

abscessation alone. There are occasional exceptions including thromboembolic lesions and leptomeningitis complicated by choroiditis. Septic emboli are prone to localize in the brain but may localize in meningeal vessels (Fig. 3.86B). Although those in meningeal vessels may lead to diffuse suppurative meningitis, they are usually quickly walled off and prevented from spreading (Fig. 3.86C,D). In the other exception, that of choroiditis concurrent with leptomeningitis, encephalitis or cerebral abscesses may develop because choroiditis leads to exudation into the cerebrospinal fluid, and the ependyma is virtually no barrier to infection. Even this combination of pathologic processes is unusual and, when choroiditis is complicated by ependymitis and suppurative encephalitis (Fig. 3.86E), there is usually little or no meningitis. It is frequently argued that whether encephalitis develops by spread of meningitis is largely a question of time, but it is not unusual for animals with purulent meningitis to survive for a week, and sometimes they survive much longer. There is ample time for the inflammation to spread throughout the cranial and spinal meninges—and to the brain if it were going to do so—but it seldom does, and there is seldom anatomic justification for the common diagnosis of suppurative meningoencephalitis.

These considerations mention only a few of the discrepant phenomena in cerebrospinal suppurative meningitis. Most cases of cerebrospinal suppurative meningitis are of hematogenous origin, but virtually nothing is known of factors that govern their localization. There are some remarkable consistencies and inconsistencies, and their explanation must await a clearer knowledge of both host and microbial factors in infectious disease. In the section on inflammatory diseases of joints and synovial structures, attention is drawn to frequent concurrence of polyarthritis with choroiditis and leptomeningitis in fibrinopurulent infections. Erysipelas, in which arthritis is a typical lesion, is an exception in that, although it may produce septicemic and embolic lesions in the brain, it does not cause suppurative meningitis. Neonatal streptococcal infections in calves, lambs, and piglets (but not in foals) frequently produce a combination of polyarthritis, purulent leptomeningitis, choroiditis, and, in calves only, endophthalmitis (Fig. 3.87A,B). *Streptococcus pneumoniae* usually produces a fulminating septicemia in calves, characterized only by very acute splenitis, but if the course of this infection is less fulminating than usual, polyarthritis and meningitis can be found. Colibacillosis of protracted course commonly causes meningitis and polyarthritis in neonatal calves and lambs, which is well developed, but even in fulminating infections, the slight changes of early inflammation can be found in these structures. On the other hand, both coliform and streptococcal infections in calves and piglets may avoid the joints and meninges and localize instead in the choroid plexuses and spread from there to the ventricles and brain, or localize in the brain alone. The coliforms and streptococci behave differently in calves with respect to the eyes; a combination of synovial, meningeal, and intraocular localizations is almost invariably of streptococcal origin (and the lesions can be well developed within 12 hr of birth, or even at birth according to farmers); the coliforms can, but only seldom do, cause endophthalmitis.

Pasteurella haemolytica and *P. multocida* are usually regarded only as causes of fibrinous pneumonia and hemorrhagic septicemia, but they are responsible for localized infections in other locations, including the meninges, in ruminants. When polyarthritis is present in septicemic or pulmonary pasteurellosis, fibrinosuppurative leptomeningitis can also be anticipated, but meningitis may also occur without synovial localizations. Isolated cases and limited outbreaks of entirely meningeal pasteurellosis are observed in cattle and sheep, usually young ones. The course is asymptomatic until the meningitis develops. The infection must be blood-borne, but that there is no clinical evidence of the bacteremic phase is a curious phenomenon of the infection; it has its parallels, however, such as in the outbreaks of abortion in cows and the sporadic abortions in cows, ewes, and sows caused by *Pasteurella* in which, except for abortion, the dam shows no evidence of the bacteremic phase.

Once infectious agents gain access to the leptomeninges there is little resistance to spread in the meningeal spaces, and the inflammatory process becomes more or less diffuse in most cases. When the inflammatory process, excepting some which are granulomatous, remains localized, it is probable that only the inflammatory reaction, and not the infection, reaches the meningeal spaces. The cerebrospinal fluid is an excellent culture medium for many sorts of bacteria, and these spread rapidly in the fluid, assisted to some slight extent by normal flow, so that, although meningitis may appear grossly to have a limited distribution, its true distribution can be determined only microscopically.

The apparent gross distributions of meningeal inflammations vary somewhat with the cause. Thus in listeriosis, the process is confined largely or entirely to the meninges covering the medulla oblongata and upper cord, and in Glasser's disease and in malignant catarrhal fever, the exudate is concentrated over the cerebellum and occipital poles of the cerebrum. In pyogenic meningitis, the basal meninges show the most obvious changes.

In the first day or so of suppurative meningitis before exudation is clearly recognizable, the meninges may be faintly opaque and hyperemic (Fig. 3.87C). After a few days, the appearance of the brain and cord is typical. The basal cisterns, which accumulate the most exudate, are filled with creamy pus or with a grayish-yellow fibrinopurulent exudate. The extreme exudation in these cisterns is due in part to their large size, but in part also to sedimentation of particulate exudate. The exudate is in the arachnoid spaces, and there is little if any on the outer surface of this membrane (Fig. 3.87D). The arachnoid appears stretched. It is easy to overlook even copious exudates because their color is not very different from that of the brain. A useful clue is that even the largest basilar vessels and the trunk of the oculomotor nerve are partially or

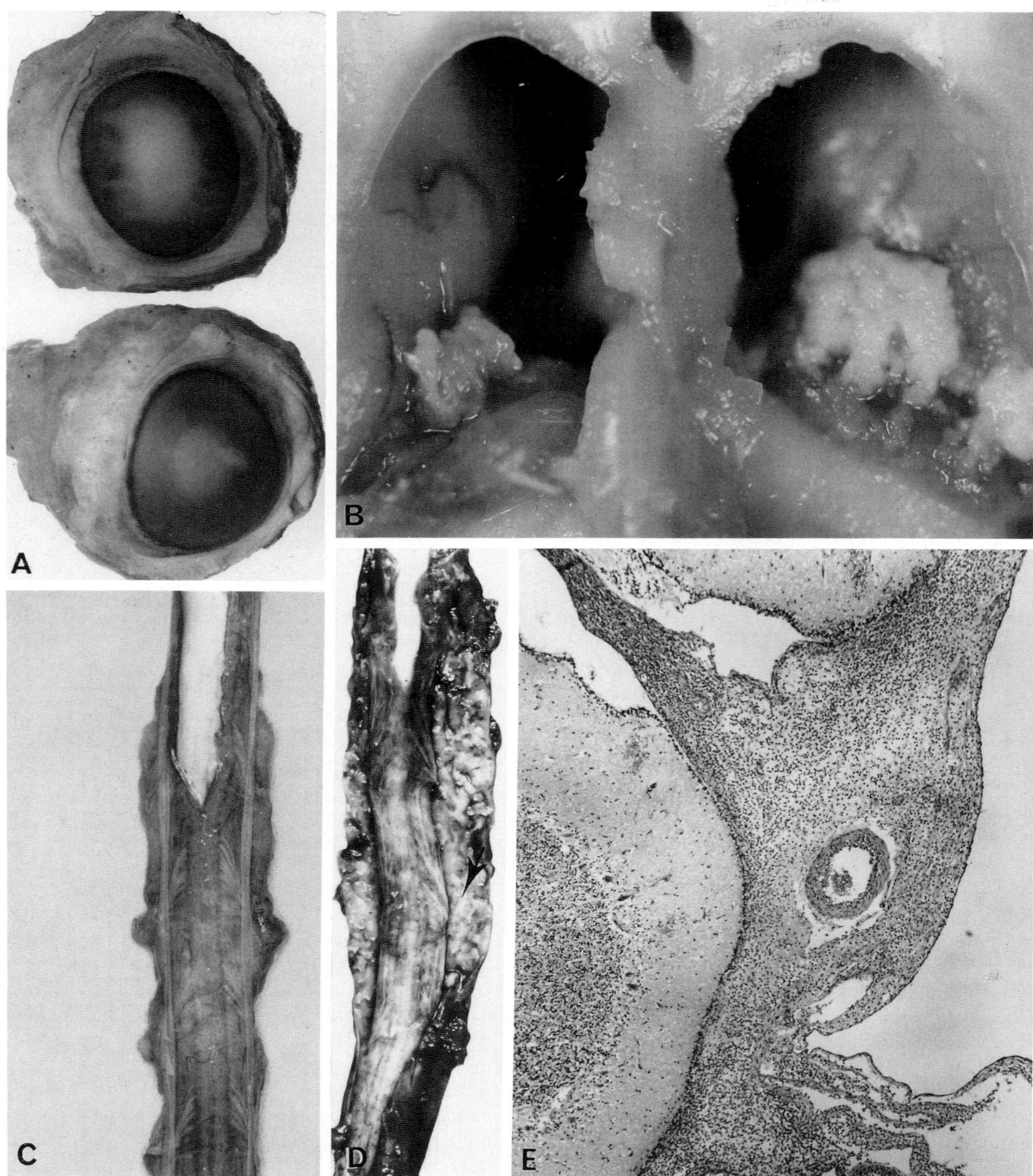

Fig. 3.87 (A) Hypopyon in streptococcal endophthalmitis. Calf. (B) Horizontal slice through lateral ventricles to show streptococcal choroiditis and acquired hydrocephalus. (C) Spinal leptomeningitis. Calf. *Escherichia coli*. (The dura is incised and reflected full length, the congested leptomeninges are incised at top.) (D) Spinal meningitis. Dog. Dura reflected to expose exudate (arrow). (E) Fibrinopurulent exudate in leptomeninges. Pig. *Haemophilus parasuis*.

completely buried and obscured by exudate, and the filling in of basal sinuses and grooves obliterates the normal topography. Over the hemispheres the exudate is usually confined to the fissures, where the arachnoid space is wide, and spares the surfaces of the gyri, where the arachnoid space is narrow. There is seldom frank pus over the hemispheres except in cases of unusually long survival.

The severe degree of exudation described is what is usually seen in animals. In very acute or early cases, the exudation may be considerably less, and detectable only as a congestion and cloudiness of the basal meninges extending toward the convexity of the hemispheres as fine grayish sleeves about the arteries and veins. On careful inspection by naked eye, almost every case of purulent meningitis can be detected, but the microscope may be necessary to confirm some cases.

The brain is swollen in every acute case of pyogenic meningitis, and the swelling is frequently severe enough to cause displacement with coning of the cerebellum. The pathogenesis of the edema and swelling is not known. The edema affects the white matter. It is possible that obstruction of the meningeal orifices of Virchow–Robin spaces by exudate and stasis of flow of meningeal fluid may contribute to the edema. The brain itself is normal except for softness and swelling and the rare cortical infarcts in the cerebrum or cerebellum.

Choroiditis commonly complicates leptomeningitis. It is usually quite obvious when it affects the plexuses of the fourth ventricles, but that affecting the plexuses of the lateral ventricle is not apparent until the ventricles are opened, although it may be expected if cloudy fluid escapes from the third ventricle when the infundibular process is opened. The ventricular fluid is cloudy, flakes of exudate overlie the plexuses and float in the fluid, and smeary sediments of pus lie on the walls of the ventricles (Fig. 3.87B). The exudate may be impacted in the aqueduct, occluding it and leading to ependymitis. Internal hydrocephalus then develops rapidly.

Microscopically, pyogenic meningitis does not differ in its character from pyogenic inflammations in other loose tissues, such as the lung. A few mononuclear cells are mixed with a very large number of neutrophils in the arachnoid spaces (Fig. 3.87E). The amount of fibrin in the exudate varies. There may be some infiltration for a short distance along the Virchow–Robin spaces about veins (Fig. 3.88). The pia mater as a rule remains intact, and it is only in exceptional cases that some microbial activity is observed in the adjacent parenchyma, or the pia is eroded to allow neutrophils to invade the surface of the brain. When the choroid plexuses are involved, they are swollen and infiltrated with leukocytes, many of them mononuclear. The plexus epithelium is eroded and covered by a deposit of fibrin and cells. The ependyma is also eroded, there is edema of subependymal tissues, and the surrounding veins are infiltrated.

Internal hydrocephalus is a sequel to ependymitis and occlusion of the aqueduct, as a result of which the lateral and third ventricles are dilated (Fig. 3.89). This condition is a complication of choroiditis, but hydrocephalus may also be a sequel without choroiditis being present. The medullary foramina are frequently occluded as a result of inflammation in the tela choroidea, so that cerebrospinal fluid cannot escape from the ventricular system to the arachnoid spaces; the hydrocephalus is noncommunicating. Obstruction to the flow of fluid in the arachnoid spaces may occur as a result of brain swelling with impaction in the tentorial incisure, heavy deposits of exudate in the arachnoid space, or as a result of meningeal adhesions and

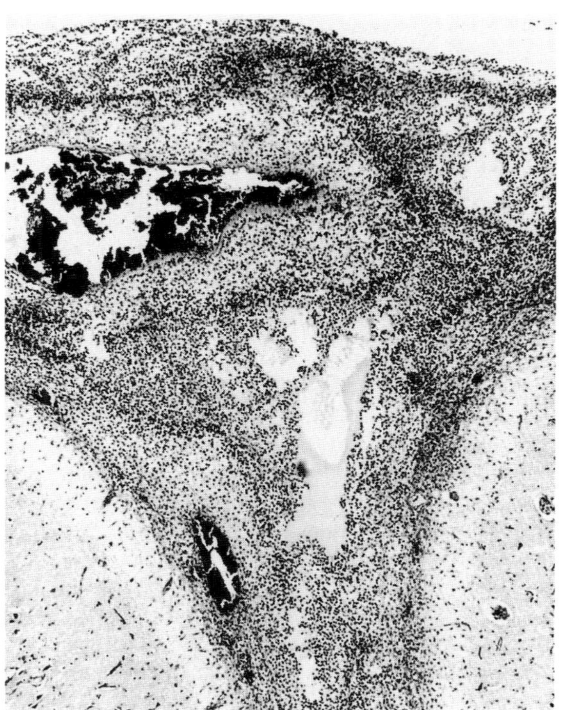

Fig. 3.88 Purulent meningitis caused by *Haemophilus* spp. Pig.

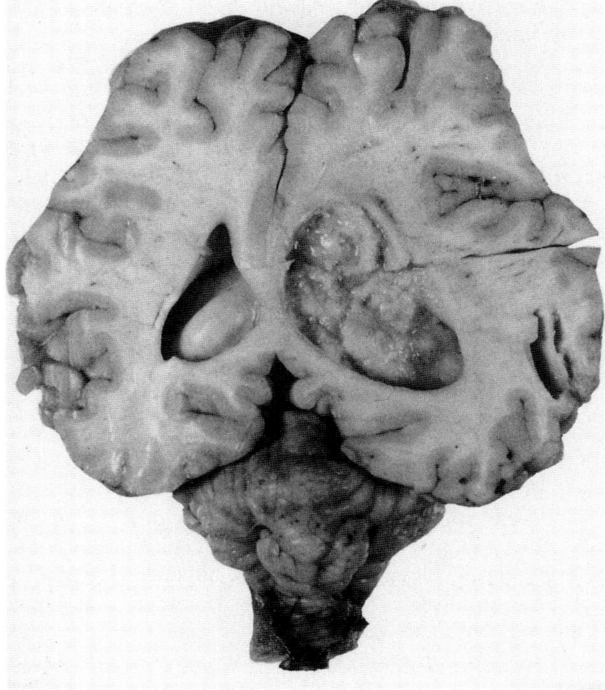

Fig. 3.89 Necrobacillary abscess arising in choroiditis. Ox. Swelling of hemisphere and posterior displacement with coning of cerebellum.

thickenings in chronic inflammations. The hydrocephalus of this pathogenesis is communicating.

Chronic pyogenic leptomeningitis is rarely observed in animals. The process may sterilize itself or be sterilized by antibiotics, but much of the injury is established in the early stages of the process and, once the diagnosis is evident clinically, death is the expected outcome. The early injury is exaggerated by the persistence of exudate even after the infection is controlled because there is no free drainage from the meningeal spaces. Healing occurs only after there has been considerable destruction of the meningeal framework with repair by fibrous tissue. The meningeal adhesions may produce cystic loculations in the arachnoid space, and obliterate the medullary foramen or basal arachnoid space and cause lingering death from hydrocephalus.

In a purulent process, leptomeningeal vasculitis commonly occurs. The expected sequelae of venous or ischemic infarcts are seldom observed even in those few cases with vasculitis in which thrombi can be found. This apparent discrepancy is probably due to the rate at which the vascular obstructions develop and the type and size of vessel. Cases in which suppurative thrombophlebitis of the sagittal or transverse sinuses, or both, can be readily observed may not have venous infarcts in the brain. Thrombosis in the afferent circulation usually involves the smallest of meningeal vessels and usually develops at a rate which allows collateral circulation to develop. Inflammation of larger arteries (Fig. 3.86D), in which the endothelium is dissected from the intima by leukocytes, has not been observed to lead to thrombosis.

4. Septicemic Lesions, Septic Embolism, and Cerebral Abscess

It is approximately correct to state that the central nervous system is injured to some extent in every episode of septicemia or sustained bacteremia. The simplest and most common type of injury is inflicted on the venules, especially those of the cerebral white matter and to a lesser extent those of the cerebellar white matter. Injury of this type is not of much significance. There is sludging of leukocytes and probably of erythrocytes in these vessels, associated usually with degenerative and reactive changes in the endothelium. In the symptomatic purpuras, it is not usually possible to find the exact site from which the injured vessels bleed in most organs; in the brain, it is frequently possible to find the site of diapedesis. Associated with the endothelial injury there is often leakage of plasma into the perivascular space and, given time, adventitial proliferation and infiltration of a few lymphocytes or other leukocytes into the Virchow–Robin spaces. Commonly, the injury is more severe, the vessel wall is totally necrotic in its cross section, and in these lesions, a few bacteria may be demonstrable (Fig. 3.90).

Septic embolism in the central nervous system may be a complication of active endocarditis. In such cases the

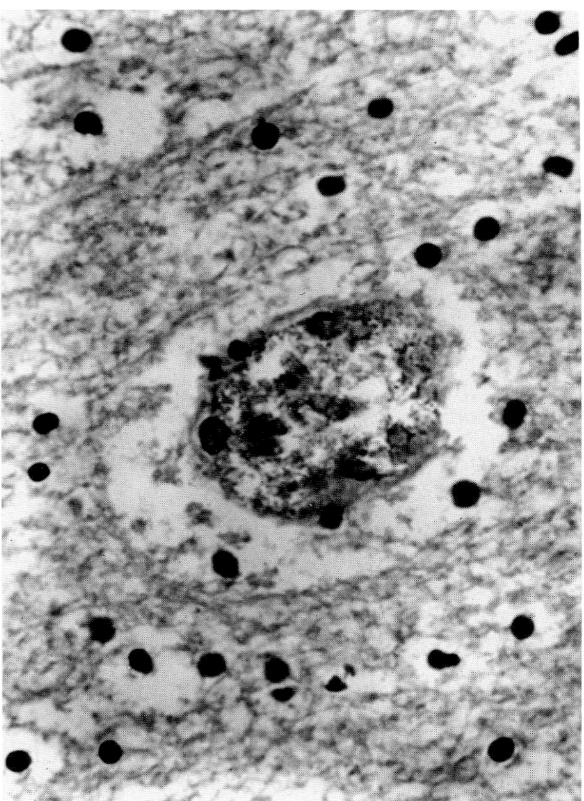

Fig. 3.90 Necrosis of cerebral venule in septicemia (*Escherichia coli*). Calf.

bacteria are usually gram-positive. *Erysipelothrix rhusiopathiae* does this fairly commonly in pigs, streptococci do it in all species, and *Actinomyces* (*Corynebacterium*) *pyogenes* may do it in cattle. Other bacteria and other sites of origin are less commonly implicated.

The toxigenicity and pathogenicity of many infections by gram-negative bacteria in septicemic phase seem to depend largely on the number of organisms present, as has been demonstrated in sheep for septicemic pasteurellosis caused by *Pasteurella haemolytica.* In septicemic pasteurellosis of sheep, in infections of foals by *Actinobacillus equuli,* in infections of sheep by *Haemophilus agni* and closely related organisms, in the septicemic diseases of cattle caused by *Haemophilus somnus,* and in some coliform infections of calves, the blood literally swarms with bacteria, because common to these infections is massive and widespread bacterial embolism. These are the principal infections in which bacterial embolism occurs in the central nervous system without there being a demonstrable primary focus such as endocarditis, pneumonia, or an abscess somewhere. The problem of cerebral embolism in these infections is not solely one of quantitative factors determined by the height of the bacteremia; qualitative factors of unknown nature are also involved.

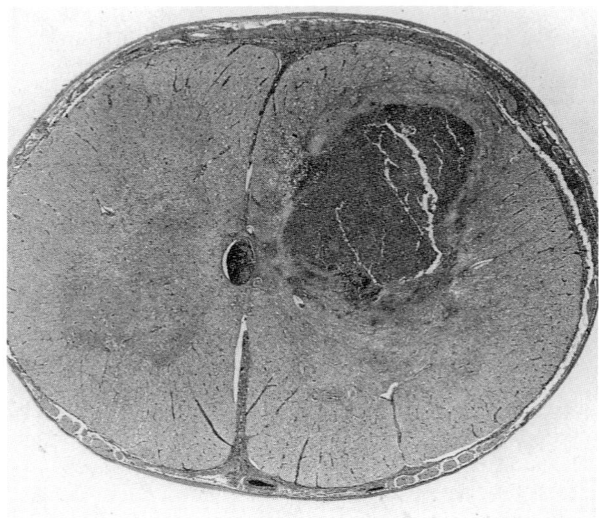

Fig. 3.91 Meningitis, tracking abscess in dorsal horn, and ependymitis, complicating docking wound. Lamb.

Septic thromboemboli and **bacterial emboli** lodge in the small cerebral vessels but, whereas thromboemboli tend to lodge particularly in small arterioles, bacterial emboli lodge in capillaries and venules. There are different consequences depending on the vessels involved, although the consequences are not particularly significant because, in either location, abscessation occurs. Arterial emboli frequently cause ischemic infarcts, and venous and capillary emboli cause hemorrhagic infarcts. The venular lesions also extend rapidly to involve large veins, whereas local spread in arteries does not occur.

Cerebral abscesses may arise in embolism, by direct implantation in wounds, or by direct invasion of the brain from an adjacent structure. Leptomeningitis rather rarely leads to abscessation, whereas choroiditis commonly leads to periventricular abscess. Abscesses in the spinal cord are seldom sought or observed; they may be hematogenous via arteries or veins (Fig. 3.91), and rarely do they enter through the dura.

Abscesses of hematogenous origin may occur anywhere in the brain, but there are two areas of remarkable predilection, these being the hypothalamus and cerebral cortex at the junction of gray and white matter (Fig. 3.92A,B). *Listeria monocytogenes* is an exception because it always demonstrates an affinity for the reticular formation in the brain stem. There may be one abscess only in the brain, or they may be multiple, especially when due to bacterial embolism. Multiple abscesses of septic thromboembolic origin tend often to be localized in one area of the brain. Some care is necessary in specifying the origin of multiple adjacent foci of suppuration because the production of satellites, each of which may be larger than the primary, is a natural attribute of brain abscesses.

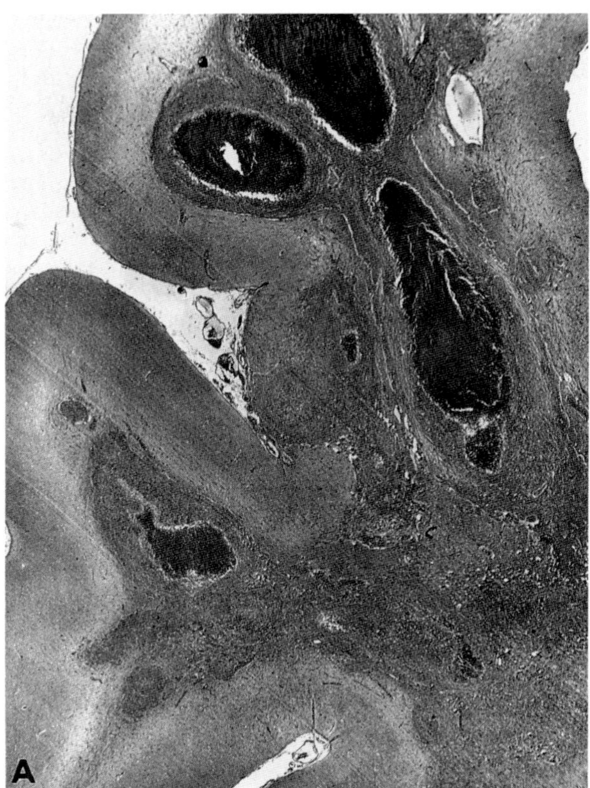

Fig. 3.92A. Hematogenous abscesses in cerebrum. Pig.

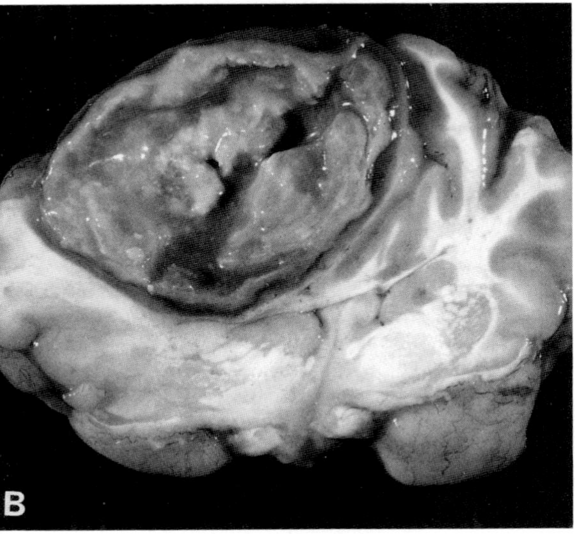

Fig. 3.92B Abscess distorting left cerebral hemisphere. Goat. *Pasteurella multocida* and *Fusobacterium necrophorum*.

Abscesses arising by direct invasion of the brain may develop in any location. There are two common sites of invasion, namely the cribriform plate and inner ear, and two somewhat less common, namely the hypophyseal fossa and the paranasal sinuses. Hypothalamic abscesses are observed occasionally in cattle and dogs. At least some of those in dogs are caused by minute foreign bodies migrating with phlebitis through the orbital fissure or foramen rotundum and carrying actinomycetes. The importance of the cribriform plate and internal ear in the development of cerebral abscess is due to the frequency with which infections occur in those locations, that in neither site is there an actual or potential epidural space to protect the brain, and because nerves and vessels enter and leave through both.

Frontal abscesses of sinal origin occur occasionally in cattle as a complication of dehorning wounds, but abscesses in this site occur most commonly in sheep in which sinusitis, especially in the ethmoid cells, develops as a suppurative complication of myiasis (*Oestrus ovis*). *Actinomyces* (*Corynebacterium*) *pyogenes* is the organism usually present. The olfactory bulb is destroyed, and the first ventricle is opened, the infection then spreading to the substance of the hemisphere (Fig. 3.93). There is little tendency to spread into the meninges from the point of entry or to invade the cortical gray matter, although both layers are usually secondarily involved by expansion of the abscess later.

Abscesses commonly develop at about the cerebellopontine angle as complications of suppurative otitis media. These are usually problems of pharyngitis, infection spreading via the Eustachian tube to the middle ear and from there to the brain, by either erosion of the bulla or extension along natural foramina.

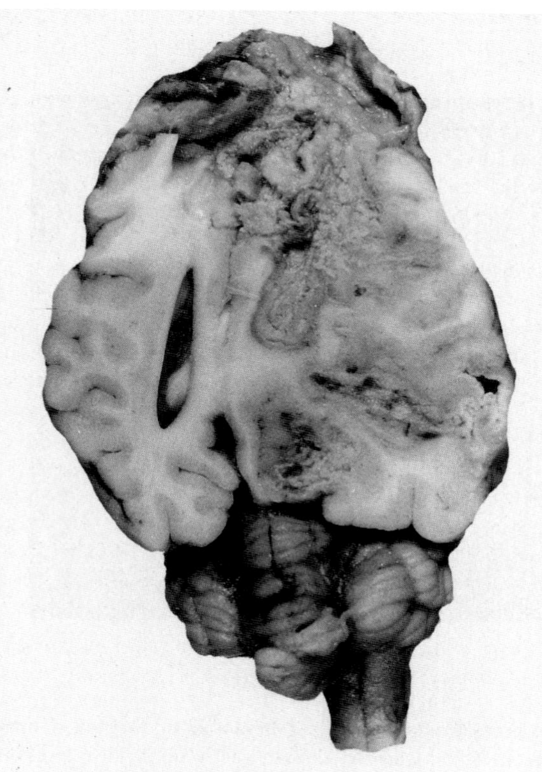

Fig. 3.93 Corynebacterial abscess arising from right ethmoid. Sheep.

Cerebellopontine abscesses are rarely, if ever, observed in horses in spite of the frequency of pharyngitis in this species. This exemption of horses may be due to the diversion of exudates from the Eustachian tube to the guttural pouches. Otogenic cerebral infections are very seldom observed in dogs. Those observed have come via the auditory meatus rather than the Eustachian tube, and the cerebral reaction is nonspecifically granulomatous, usually without pus. Otitis media caused by *Pasteurella multocida* is fairly common as a complication of chronic cases of upper respiratory infection in cats and, when the infection extends to the cranial cavity, it produces a diffuse purulent leptomeningitis and not, as expected, a cerebellopontine abscess. Most otogenic infections occur in sheep and swine; a few are observed in cattle, and in each species, the outcome is an abscess of the brain.

Otic infections are commonly bilateral so that otogenic abscesses may be also. The usual organism in these abscesses is *Actinomyces* (*Cornyebacterium*) *pyogenes* alone or mixed with *Pseudomonas aeruginosa, Pasteurella multocida,* and mixed cocci. Affected pigs usually have several other chronic debilitating infections at the same time so that several animals in the herd may have otogenic abscesses at the same time. This pattern of infection in calves is uncommon and sporadic. In sheep, on the other hand, the disease may be observed in limited outbreaks, and there is an association with grazing on rough, dry, mature summer pastures. The reasons for this association are unknown.

Established abscesses are found much more commonly in the white matter than in the gray, in spite of the fact that they usually begin in the gray. There is no suitable explanation for this tendency to track into white matter, but once within white matter, they do permeate along fiber tracts (Fig. 3.94). As a result of this activity, satellite abscesses are to be expected, sometimes chains of them. Connecting purulent tracts may be very thin and difficult to find.

An abscess may begin as an intense accumulation of neutrophils in and around a thrombosed vessel (Fig. 3.95A), or it may begin in a focus of septic encephalitis in which neutrophils lightly infiltrate a zone of early softening. The principal differences between a cerebral abscess and an abscess in some other location is the vulnerability of the surrounding tissue to edema, which, in nervous tissue, can destroy it, and the very slowness with which encapsulation occurs (Fig. 3.95B). The meninges and larger blood vessels are the only sources of fibroblastic tissue for encapsulation.

In the early stages, the abscess cavity contains a liquefying center, and the margins are irregular and poorly defined even microscopically. The surrounding brain tissue is edematous and infiltrated by neutrophils. The microglia and vessels are fairly resistant and reactive in a narrow peripheral zone, but the neurons and neuroglia degener-

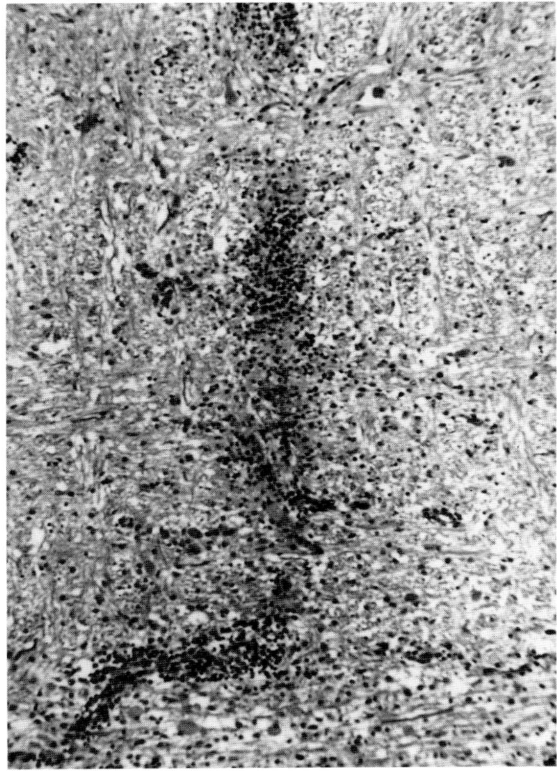

Fig. 3.94 Spreading tract in suppurative (streptococcal) encephalitis. Pig.

ate. The majority of abscesses develop rather slowly and later become encapsulated. The capsule seems often to be formed more by condensation of vessels around an expanding focus than by proliferating fibroblastic tissue, but both contribute; the result is an irregular capsule thicker on the meningeal than on the ventricular aspect. The majority of capsules are very distinct and 1–3 mm wide. Their distinctness is due to the paucity of reticulin or collagen spreading out into the surrounding parenchyma or into the abscess cavity. In old abscesses with shrinkage, the core may separate from the capsule, and the capsule may separate extensively from the surrounding cerebral substance. There is yellowish discoloration of the surrounding tissue due to edema, the edematous zone being often much larger than the abscess itself. The microscopic structure of the wall of a chronic abscess does not differ much from case to case except in the thickness of the capsule. A narrow zone of histiocytes or compound granular corpuscles faces the neutrophilic debris. The next zone is a laminated layer of collagenous fibers between which are rows of compound granular corpuscles laden with debris. The outer zone is vascular, the vessels being large and usually nonreactive. The surrounding nervous tissue is severely edematous with degeneration of myelin and fibers. The astrocytes are swollen and reactive about encapsulated abscesses, but it takes a long time for their fibrils to begin to intertwine with capsular reticulin. The veins are heavily cuffed by lymphocytes.

When abscesses are multiple, death is the outcome after a short course. When they are isolated, they may permit prolonged survival. The course is usually short with medullary abscess, even when small, because it, or more usually the edema it provokes, interferes with vital centers. Abscesses in the hypothalamus or cerebrum may track through the white matter to the ventricles to produce pyencephaly, which is quickly fatal (Fig. 3.96). Large abscesses ultimately expand to the meninges to produce adhesive meningitis. Many abscesses act as space-occupying lesions either by virtue of their size or the edema they provoke or both. The consequences of space occupation depend on the site and size of the lesion.

a. Listeriosis Listeriosis is caused by *Listeria monocytogenes*. The specific name, *monocytogenes*, was inspired by the observation of a monocytosis provoked by sublethal doses of the organism in rabbits and guinea pigs. Monocytosis has been observed in dogs also, but it does not occur in domestic ruminants or swine.

Listeriosis is of worldwide distribution, possibly excepting the tropics. The organism has been isolated from diseased mammals and birds of many species and produces septicemia, meningitis, and abortion in humans. In spite of its importance, very little is known of its epidemiology and the pathogenesis of infection. The organism is commonly isolated from tissues of normal animals, including tonsils and other gut-associated lymphoid tissue, and in large numbers from the feces of ruminants. It is remarkably viable in the external environment, being able to survive in dried media for several months and in suitably moist soil for ~1 year. The disease is most common in winter and early spring, but there is no information on factors responsible for this seasonal occurrence. Stress-related factors may alter the balance between the host and the bacterium, a facultative intracellular organism, and by suppressing cell-mediated resistance, allow the development of bacteremia. Organisms invading along peripheral nerves may be separated from the host's immune system.

Listeriosis as encephalitis may be sporadic or occur in outbreaks in which the morbidity may be 10% or higher, and outbreaks usually are associated with heavy feeding of silage. This association of outbreaks of cerebral listeriosis with silage feeding is a circumstantial observation, but the association is so common that silage is fed as a calculated risk and removed from the ration when the first case occurs. The association with silage may indicate an acquired susceptibility of the animals or the provision of a growth medium which leads to heavy infection pressure. The organism will multiply in spoiled, incompletely fermented silage with a pH of 5.5 or above.

Listeriosis behaves as three separate diseases or syndromes. They seldom overlap, so that each syndrome probably has a separate pathogenesis. The three recognized syndromes are infection of the pregnant uterus with abortion, septicemia with miliary visceral abscesses, and

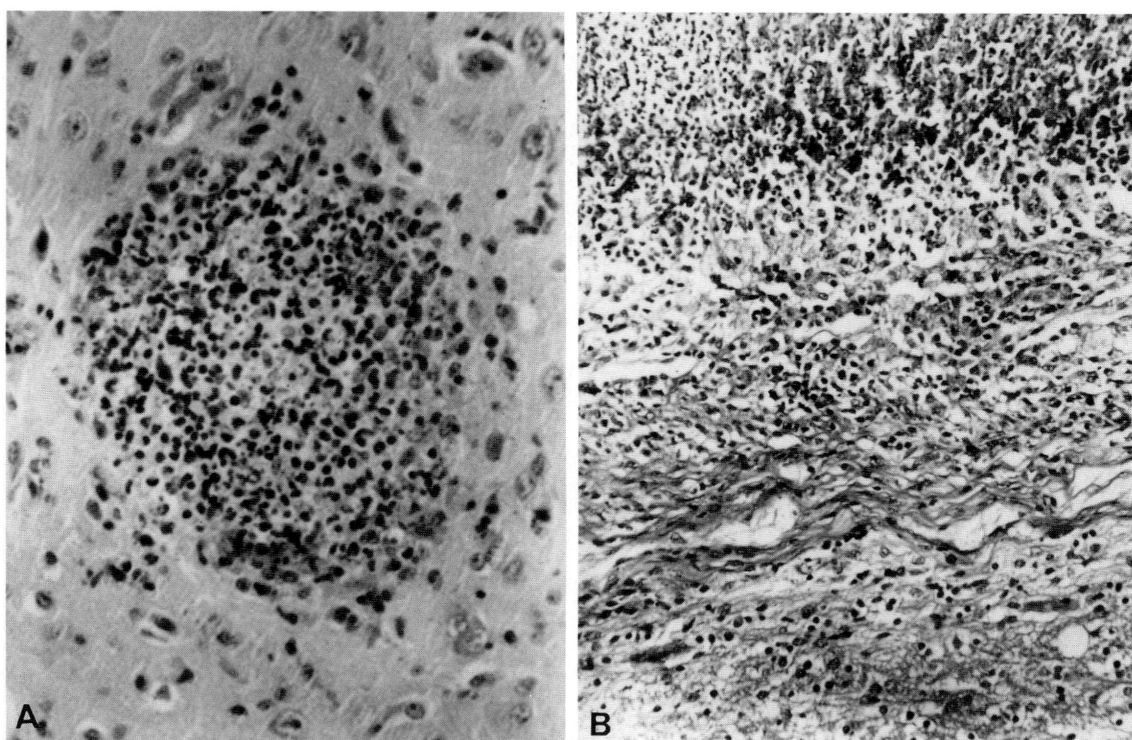

Fig. 3.95 (A) Early cerebral abscess from septic embolus. Pig. (B) Structure of wall of cerebral abscess. Abscess (above) is separated from normal brain by a thin capsule.

encephalitis. The uterine infection is discussed with Diseases of the Pregnant Uterus (Volume 3, Chapter 5, Female Genital System). Additional syndromes of clinical significance in ruminants include conjunctivitis, possibly from contaminated silage dust, and mastitis.

Knowledge of the pathogenesis of listeriosis is no better than epidemiology, and we discuss the unsettled problems here with reference to ruminants, the species most commonly affected. The septicemic disease with visceral, principally hepatic, abscessation does occur in ruminants but probably only in neonates. The visceral infection, although seldom with abscessation, is also the pattern in fetuses, so that neonatal infections are continuations of intrauterine infections in most cases.

Aborting ruminants are not usually ill, and abortion is usually late in gestation. The uterine infection is probably hematogenous, the bacteremic phase is asymptomatic, and localization occurs only in the uterus. Infection of the uterine contents can be established quite readily by oral exposure of pregnant animals and by intravenous inoculation.

The pathogenesis of encephalitis in listeriosis is not the rather simple matter that the pathogenesis of the intrauterine infection is. In animals, *Listeria* has a remarkable affinity for the brain stem, the lesions being most severe in the medulla and pons and less severe anteriorly in the thalamus and posteriorly in the cervical parts of the spinal cord. Whether or not this localization, which is consistent and specific in natural infections, can be hematogenous, or whether there is some other way, is uncertain. The bulk of evidence is against the hematogenous route. Intravenous and oral dosing of pregnant ruminants will regularly produce intrauterine infection but seldom produce intracranial infection and probably never produce encephalitis of the specific distribution. When a pregnant animal dies of encephalitic listeriosis, the uterine contents are usually sterile. Listeriosis, being so often a disease of the winter season, occurs when ewes are pregnant, but it is expressed in an individual as encephalitis or abortion and only very rarely as overlapping syndromes. The specific distribution of listerial encephalitis in the natural disease is inconsistent with a hematogenous infection. Bacteremias with localization in the central nervous system regularly cause meningitis, choroiditis, and cerebral abscesses, and this is what *L. monocytogenes* does, but only as an experimental hematogenous infection.

Local invasion is the only alternative to hematogenous invasion. Encephalitis of typical distribution has been produced, albeit irregularly, only by conjunctival or intranasal instillation or by injection into facial tissues. Usually, this form of exposure produces local inflammation and a bacteremia, but there is evidence that infection can pass along cranial nerves to the medullary centers. This implicates the trigeminal nerve particularly. There is a correlation between listeric encephalitis and trigeminal neuritis and

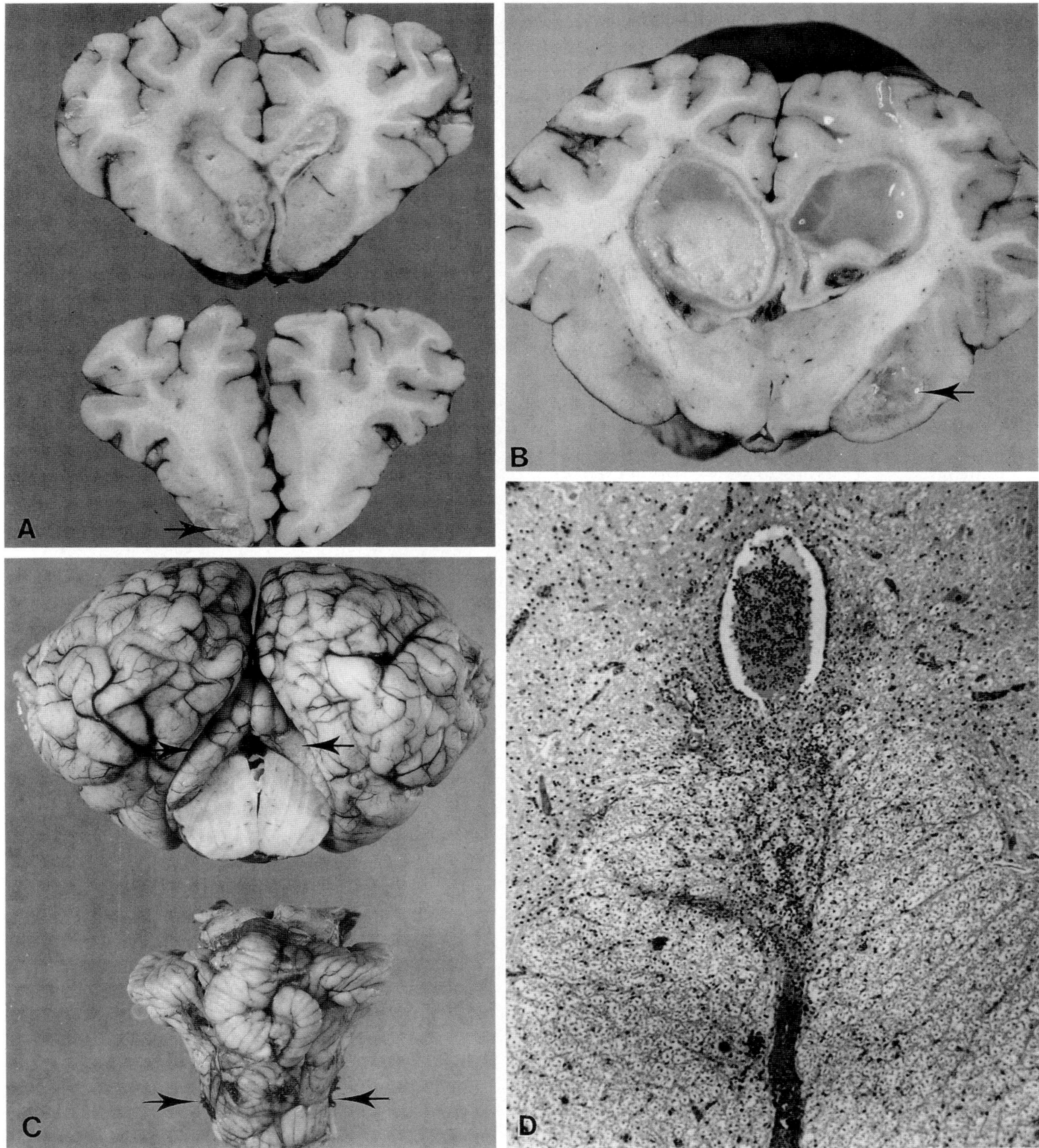

Fig. 3.96 Pyencephaly. Calf. (A) Pus in lateral ventricles (above) and left olfactory ventricle (below, arrow). (B) Dilation of lateral ventricles by fibrinopurulent exudate (arrow). (C) Brain swelling with cerebellar coning and hemorrhage (arrows) and subtentorial herniation of hemispheres (arrows). (D) Inflammatory exudate in aqueduct.

between the severity of the trigeminal neuritis and homolateral lesions in the brain stem. The organism has been demonstrated in myelinated axons of the trigeminal nerve and fiber tracts in the brain stem and in the cytoplasm of medullary neurons. The initial point of entry may be provided by physical injury to skin or mucosa by, for example, plant fiber or erupting teeth.

Conjunctivitis and keratitis follow experimental conjunctival exposure, but this should not imply that the endophthalmitis of the natural disease is produced by local invasion. Rhinitis is clinically apparent in many cases of encephalitic listeriosis, but histologic evidence does not support the idea that the olfactory nerves are a route of invasion of the brain. The position with respect to other cranial nerves and the internal ear has not been examined.

The neurological signs and lesions of listeriosis in the various ruminants are qualitatively the same, differing only in severity. The signs are combinations of mental confusion and depression, head pressing, and paralysis of one or more medullary centers. Characteristically, there is deviation of the head to one or other side without rotation of the head; when such an animal moves it does so in circles, hence the name circling disease. There is frequently unilateral paralysis of the seventh nerve causing drooping of an ear, eyelid, and lips. There may also be paralysis of the masticatory muscles and of the pharynx. Purulent endophthalmitis, which is usually unilateral, is often present and has caused listeriosis to be confused with malignant catarrhal fever. The course of the disease in sheep and goats is a few hours to 2 days; survival occurs but usually with neurologic handicaps.

Listerial encephalitis occurs almost solely in adult ruminants. Systemic listeriosis is occasionally observed in lambs and calves up to 1 week of age and in others which are several months of age. It is the bacteremic disease which occurs in the early neonatal period, and this is characterized by bacterial colonization and abscess formation. The abscesses are miliary in distribution, very numerous in the liver, but much less numerous in the heart and other viscera.

Listeriosis in swine is comparable to the disease in ruminants, but is relatively rare. Outbreaks of encephalitis may be observed with lesions of usual distribution in the brain stem. Alternatively, there may be abortion and neonatal death. The usual expression of the disease is visceral with miliary abscesses in the liver and heart.

The patterns of listeriosis in other domestic species appear to follow the general scheme but are rarely observed. The encephalitic form in adult horses, the visceral form in foals, and abortions in mares are reported. Several cases of encephalitis caused by *L. monocytogenes* are reported in dogs.

Gross lesions are usually not observed in the brain in listerial encephalitis. Occasionally, the medullary meninges are thickened by a greenish gelatinous edema, and grayish foci of softening may be found in the cross section of the medulla. The initial lesions are parenchymal; involvement of the meninges, which is almost constant, is secondary to the parenchymal lesions. Mild meningitis commonly affects the cerebellum and anterior cervical cord and less commonly is found in patches over the cerebrum and down the spinal cord. The characteristic parenchymal lesion is a microabscess. It may begin in a tiny collection of neutrophils (Fig. 3.97A) but more usually

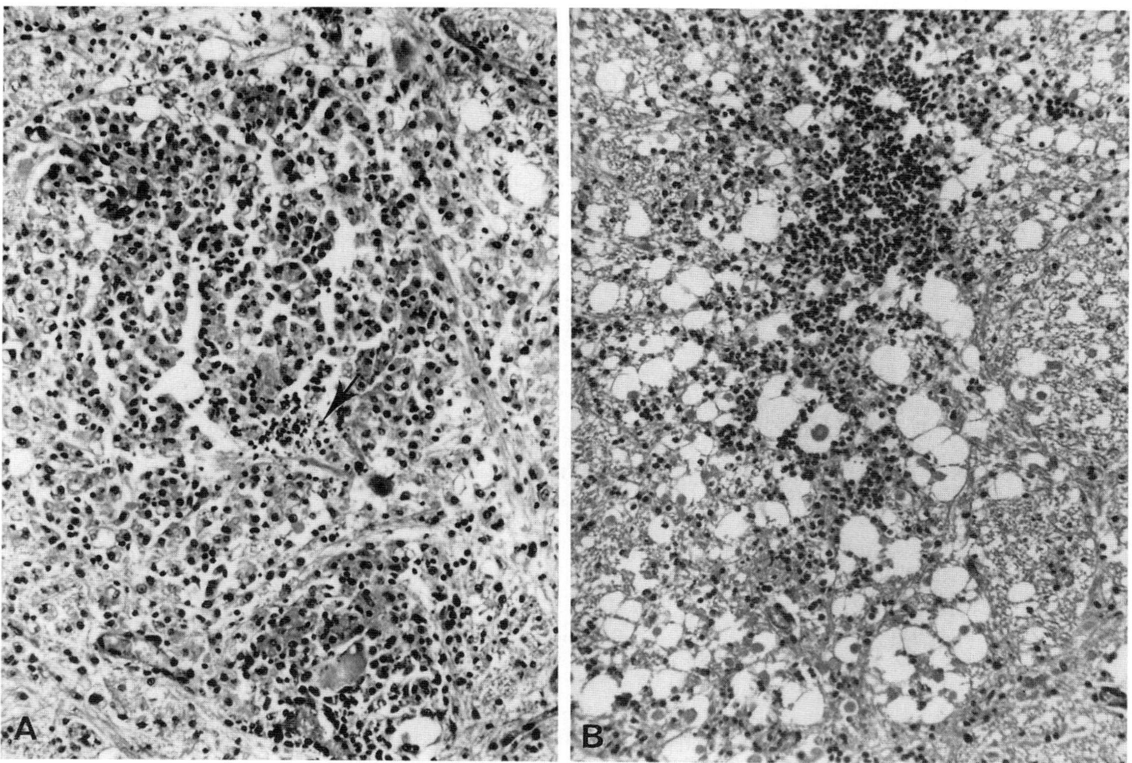

Fig. 3.97 (A) Microglial reaction with small focus of suppuration (arrow) in listeriosis. Sheep. (B) Suppurative encephalitis in listeriosis. Ox.

does so in a minute focus of microglial reaction. The glial nodules may persist as such, the cells taking on the characters of histiocytes, but the tendency is always for the nodules to be infiltrated by neutrophils and for their centers to liquefy (Fig. 3.97B). The focal lesions do not expand much, but suppurative foci may streak through the white matter. Apparently, the organism is not highly toxigenic because the parenchyma surrounding the glial nodules and focal abscesses may be little changed. Commonly, however, the white matter is edematous and rarefied. Such areas may be large and lightly, but diffusely, infiltrated by neutrophils and hypertrophied microglia. Focal softenings occur and may coalesce. They are related to vessels that are occluded by inflammatory and thrombotic changes.

Acute vasculitis with exudation of fibrin occurs in the white matter in relation to suppurative foci. The vasculitis is secondary to drainage in the Virchow–Robin spaces from the primary parenchymal foci. It is in this manner that the meningeal infiltrates develop. Perivascular cuffing is heavy. The cuffs are composed mainly of lymphocytes and histiocytes with a few admixed neutrophils and eosinophils. In some cases, the granulocytes predominate.

b. *HAEMOPHILUS SOMNUS* INFECTIONS IN CATTLE Fastidious gram-negative organisms that produce disease in ruminants include the unrecognized taxa "*Haemophilus somnus*" in cattle, and "*H. agni*" and "*Histophilus ovis*" in sheep. They are regarded as species-specific strains of the same organism for which a proper name is not established, although *H. ovis* has priority if the genus were to be recognized. *Haemophilus somnus* will be used here for the bovine strains, which show considerable pathogenic variability. The organism resembles members of the genus *Haemophilus* in most respects and has special nutrient requirements that can be satisfied by blood and yeast derivatives, but these requirements are chemically undefined. They are not the chemically defined X and Y requirements of the genus *Haemophilus* so it is likely that the name will remain invalid.

Thrombotic meningoencephalitis (TME), the best-known manifestation of disease caused by this organism, was first reported in 1956, and over the next 20 years became a common and important disease of cattle in feedlots in North America. The disease has also been identified in Germany, Italy, Scotland, Switzerland, and New Zealand, and remains an important disease of young cattle, but acute and chronic infections involving other organs are also economically significant.

Haemophilus somnus is regularly isolated from the respiratory tract of healthy animals, and commonly resides in the urogenital tracts of male and female cattle. The mechanism, site, and circumstance by which it invades the bloodstream are not known, but it is possible that organisms in genital discharges or aerosolized urine invade via the respiratory tract. There are virulent and nonviru-

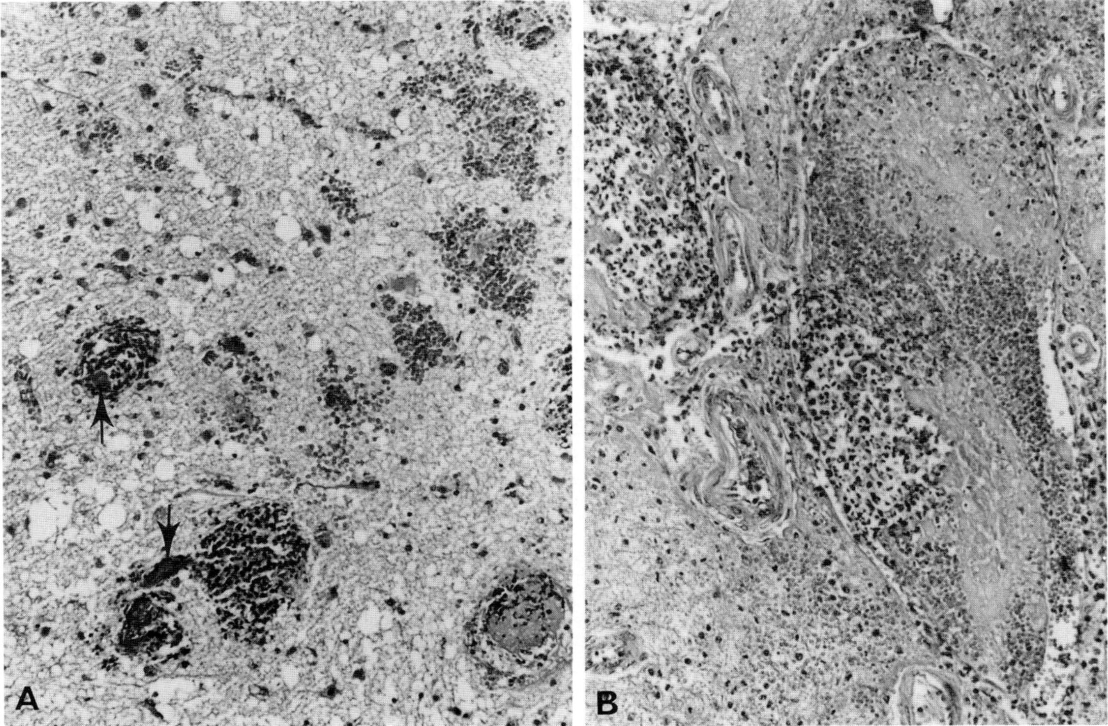

Fig. 3.98 *Haemophilus somnus* infection. Ox. (A) Thrombosis, hemorrhage, and bacterial colonies in cerebral white matter. (B) Thrombosis of meningeal vessels.

lent biotypes; the ability to survive and proliferate in bovine monocytes may be a virulence mechanism, and may also contribute to the development of chronic disease. Calves are infected by carrier cows in the first months of life, and they in turn disseminate the infection in feedlots.

The syndrome caused by *Haemophilus somnus* is a septicemia that may cause acute death or may localize in one or several organs causing subacute or chronic, fatal or nonfatal disease. The septicemic phase of the disease is brief, and in some cases the infection is controlled at this stage without localization, but often it leads to cerebral vasculitis with acute thrombosis, hemorrhage, and necrosis (Fig. 3.98A,B). The cerebral vessels are particularly vulnerable to damage by the organism, but vasculitis can develop in most organs. Vasculitis and thrombosis probably result from bacterial adherence to endothelial cells, causing them to contract and expose the subendothelial collagen, thus initiating thrombosis. Bacterial embolism does not occur but may be simulated microscopically by intravascular proliferation of bacteria at sites of thrombosis. Thus the former name thromboembolic meningoencephalitis is inaccurate.

Septicemia caused by *Haemophilus somnus* develops in cattle of various ages maintained under various management systems, but in North America the disease is predominantly one of young cattle in feedlots. It is most common in early winter, shortly after susceptible animals are moved there from pastures. Infection occurs in a large proportion of animals, but disease occurs in a minority. In these, the septicemic phase is accompanied by fever and stiffness with few other signs.

Cerebral localization produces a variety of neurologic signs, and without treatment, affected animals become comatose and die within 1 to 2 days. The cerebral lesions are distinctive, and are visible in a large majority of untreated cases. The cerebrospinal fluid is cloudy and may contain pus and fibrin. Scattered throughout the brain and spinal cord are multiple foci of hemorrhage and necrosis (Fig. 3.99A,B). These foci may be 1–30 mm in diameter, and range from the bright red of recent hemorrhage to dark red-brown in older lesions. They have near-random distribution throughout the brain, but there may be some predilection for the thalamus and junction of the gray and white matter of the cerebral cortex. Meningitis is usually visible grossly and is most easily identified over the hemorrhagic foci. In animals that survive for a day or more, diffuse purulent leptomeningitis involves the basilar portions of the brain, and in these animals, the older parenchymal lesions may have begun to soften.

The histologic lesions are similar in all organs but are usually most severe in the brain and consist of an intense vasculitis with thrombosis and extension of the inflammation into the surrounding parenchyma, with or without infarction (Fig. 3.98A,B). Small venules are primarily affected, with thrombi often containing colonies of bacteria. The inflammatory response is neutro-

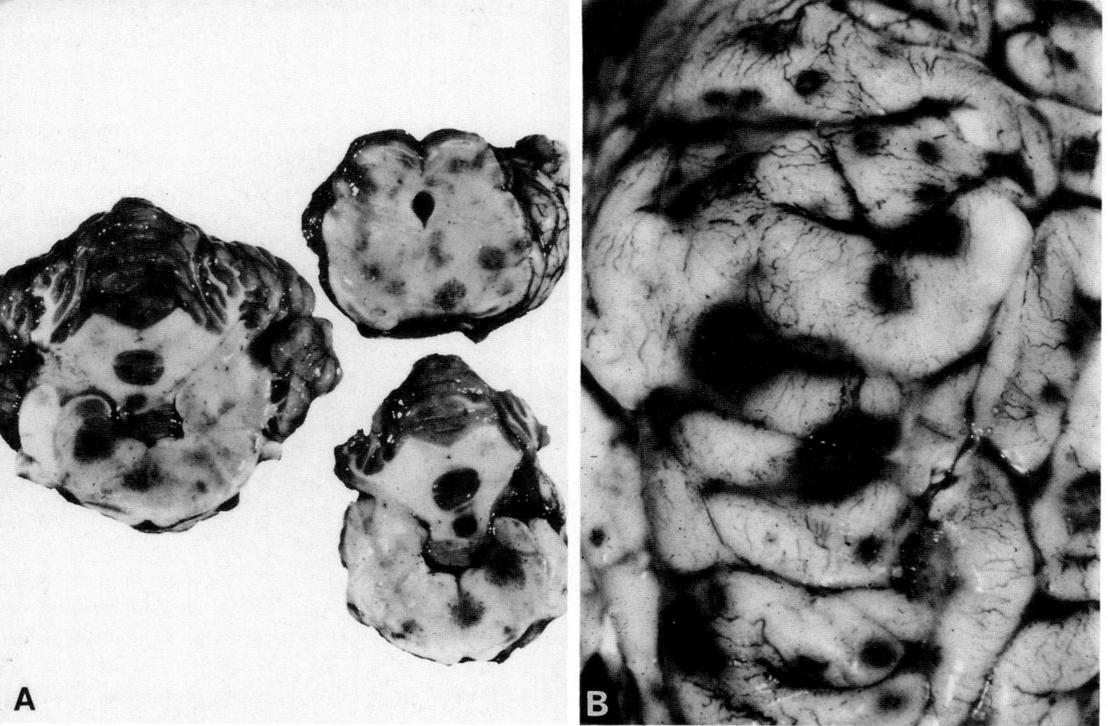

Fig. 3.99 *Haemophilus somnus* infection. Ox. (A) Multiple areas of hemorrhage and necrosis in brain stem and (B) cerebral cortex.

philic, and the cerebral lesions are quickly converted to abscesses.

Although cerebral vascular localization is a common and most dramatic result of the septicemia, petechiation and evidence of inflammation are often visible throughout the body, even in animals that die from fulminating disease. Foci of inflammation are easiest to identify in the renal medulla, skeletal muscle, lung, and laryngeal mucosa, but may also be visible in myocardium, intestine, and urinary bladder. Subacute to chronic disease develops commonly in some organs, especially joints, lungs, and heart, and is becoming an important manifestation of *Haemophilus somnus* infections.

In the acute disease, there is excess fluid in many joints, particularly the atlanto–occipital, where the capsule is distended by fluid that contains fibrin and sometimes blood. The synovial membranes and connective tissues around the affected joints are edematous and petechiated. The organism is sensitive to antibiotics, and the course of the disease is often modified by therapy. The joints of animals that have survived several days contain thick mats of fibrin and pus, and the periarticular tissues are browned by old hemorrhage. Erosion of articular cartilages is rare.

Haemophilus somnus has been isolated from cattle with a variety of types of pneumonias, and although its role in these pneumonias is not always clear, it is an acknowledged pulmonary pathogen (Fig 3.100). Pneumonia is an uncommon feature of the acute encephalitic form of the disease, but the lung can be involved as part of widespread vasculitis. It is not clear whether all or some of the respiratory tract lesions result from septicemia or inhalation. Experimentally some strains of *H. somnus* that have reduced encephalitogenic potential are capable of producing fibrinous pneumonia and suppurative bronchopneumonia, of the types usually associated with *Pasteurella* spp. The organism is commonly associated with these types of pneumonias in feedlot cattle, the acute fibrinous pneumonia being morphologically indistinguishable from that caused by *Pasteurella haemolytica*. *Haemophilus somnus* occasionally causes fibrinous pleuritis unaccompanied by pneumonia. Ulcers of the larynx occur in other diseases of feedlot animals, but, like atlanto–occipital arthritis, they are a regular enough feature of *Haemophilus somnus* infection to serve as a valuable clue to the diagnosis and prompt a search for lesions in the brain. Similarly, retinal hemorrhages are grossly visible in a percentage of animals, and are of diagnostic assistance clinically and when gross brain lesions are absent.

A major manifestation of *H. somnus* infection in some parts of North America is myocardial localization following asymptomatic septicemia. Infarction, myocarditis, or abscess formation may result, and can lead to cardiac failure with or without mural and valvular endocarditis. In many animals, myocardial abscesses are found only when the myocardium of animals with chronic pneumonia is incised. Abscesses are most common in the left ventricular free wall, particularly in the papillary muscles.

Haemophilus somnus causes sporadic disease involving many other organs including the ear, mammary gland, and those of the male and female genital tracts. It can also colonize the pregnant uterus and produce fetal disease and abortion (see Volume 3, Chapter 4, The Female Genital System). The reported cases have not involved animals with cerebral localization, and may result from asymptomatic septicemia or transcervical invasion.

Fig. 3.100 *Haemophilus somnus*. Ox. Bronchopneumonia. (Courtesy of S. C. Groom.)

Bibliography

Andrews, J. J. *et al.* Microscopic lesions associated with the isolation of *Haemophilus somnus* from pneumonic bovine lungs. *Vet Pathol* **22:** 131–136, 1985.

Blenden, D. C., and Szatalowicz, F. T. Ecologic aspects of listeriosis. *J Am Vet Med Assoc* **151:** 1761–1766, 1967.

Bowmer, E. J., Conklin, R. H., and Steele, J. H. Listeriosis. *In* "CRC Handbook Series in Zoonoses" J. H. Steel (ed.), Section A, Vol. 1, pp.423–443. Boca Raton, Florida, CRC Press, 1982.

Charles, G. *Escherichia coli* infection in lambs. *Aust Vet J* **33:** 329–330, 1957.

Charlton, K. M. Spontaneous listeric encephalitis in sheep. *Vet Pathol* **14:** 429–434, 1977.

Charlton, K. M., and Garcia, M. M. Spontaneous listeric encephalitis and neuritis in sheep. *Vet Pathol* **14:** 297–313, 1977.

Cordy, D. R., and Osebold, J. W. The neuropathogenesis of

listeria encephalomyelitis in sheep and mice. *J Infect Dis* **104:** 164–173, 1959.
Emerson, F. G., and Jarvis, A. A. Listeriosis in ponies. *J Am Vet Med Assoc* **152:** 1645–1646, 1968.
Gates, G. A., Blenden, D. C., and Kintner, L. D. Listeric myelitis in sheep. *J Am Vet Med Assoc* **150:** 200–204, 1967.
Gray, M. L., and Killinger, A. H. *Listeria monocytogenes* and listeric infections. *Bacteriol Rev* **30:** 309–382, 1966.
Harris, F. W., and Janzen, E. D. The *Haemophilus somnus* disease complex (Hemophilosis): A review. *Can Vet J* **30:** 816–822, 1989.
Kidd, A. R. M., and Terlecki, S. Visceral and cerebral listeriosis in a lamb. *Vet Rec* **78:** 453–454, 1966.
Lederer, J. A., Brown, J. F., and Czuprynski, C. J. "*Haemophilus somnus*", a facultative intracellular pathogen of bovine mononuclear phagocytes. *Infect Immun* **55:** 381–387, 1987.
Otter, A., and Blakemore, W. F. Observations on the neural transport of *Listeria monocytogenes* in a mouse model. *Neuropathol Appl Neurobiol* **15:** 590, 1989.
Piechulla, K. *et al.* Deoxyribonucleic acid relationships of "*Histophilus ovis/Haemophilus somnus*", "*Haemophilus haemoglobinophilus*" and "*Actinobacillus seminis*". *Int J System Bacteriol* **36:** 1–7, 1986.
Schuh, J. *Haemophilus somnus* myocarditis versus thrombotic meningoencephalitis in western Canada. *Can Vet J* **32:** 439, 1991.
Shand, A., and Markson, L. M. Bacterial meningoencephalitis in calves (*Pasteurella* infection). *Br Vet J* **109:** 491–495, 1953.
Stephens, L. R. *et al.* Infectious thromboembolic meningoencephalitis in cattle: A review. *J Am Vet Med Assoc* **178:** 378–384, 1981.
Stuart, F. A. *et al.* Experimental *Haemophilus somnus* infection in pregnant cattle. *Br Vet J* **146:** 57–66, 1990.
Woodbine, M. (ed.). "Problems of Listeriosis." Leicester, England, Leicester University Press, 1976.

B. Viral Infections

Viral infections of the central nervous system are common, usually being part of systemic infections rather than examples of a specialized affinity for nervous tissue. The virus may not have an affinity for neural tissue but injures it by virtue of its proximity to infected mesenchymal tissues, as is the case in diseases such as hog cholera and infectious canine hepatitis. There are other viruses (which are discussed below) that are infective for neural cells, and their route of invasion of the nervous system is of importance.

The **routes of invasion of the nervous system** available to viruses are via nerves, including the olfactory tracts, and from the blood. Centripetal spread along nerve trunks occurs for both fixed and street rabies virus. Centripetal spread in axons does not depend on viral replication in axons, but passage is provided by the mechanisms of fast axoplasmic transport. Following muscular inoculation, the virus may replicate in myocytes, and centripetal spread can occur in sensory fibers from muscle and tendon spindles and in motor fibers from muscle end-plates. Herpes simplex can ascend sensory fibers apparently from specific receptor sites at synapses, and this is one of the routes used by pseudorabies and some other herpes infections in animals. Invasion by the olfactory route is theoretically more simple, since olfactory receptors extend into and beyond the olfactory epithelium, and a cuff of arachnoid extends through the cribriform plate to the olfactory submucosa. This arrangement would allow viruses to spread along nerves without first penetrating to the submucosa or, having penetrated to the submucosa, to reach the cerebrospinal fluid directly. This may be the route taken by infectious bovine rhinotracheitis virus, Teschen virus, and pseudorabies virus in young pigs. The olfactory route could be used following intranasal exposure or following hematogenous seeding of the olfactory epithelium or mucosa.

For most viruses, spread to the central nervous system is hematogenous after multiplication in some other tissue and the development of a viremia of sufficient magnitude and duration. Viruses are susceptible to removal from blood by histiocytes, but viremia can be sustained if the viruses are small, associated with cells of blood, or replicated in endothelium or lymphatic tissues. Some viruses, such as canine distemper, infect the vascular endothelial cells, whereas others infect surrounding glia, suggesting transport across endothelium in pinocytotic vesicles. Invasion across the choroid plexus is not impeded by the fenestrated endothelial cells there but would require active infection of the epithelium and ependyma. There are areas of selective permeability to indicator dyes in the brain, such as the tuber cinereum and area postrema, but there is no evidence that these are selectively used by viruses.

The means by which viruses **spread in nervous tissue** are not known. Rapid dissemination can occur in the cerebrospinal fluid, but some viruses, such as rabies, extend rapidly through the parenchyma. The rate of spread of rabies virus, assuming cell-to-cell growth and transmission, is probably more rapid than its generation time would permit, and permeation of the limited extracellular spaces of the brain by large viruses seems unlikely.

The separation of the specialized cells of the nervous system into neurons, oligodendrocytes, astrocytes, and ependymocytes understates the diversity of cell populations in the brain and cord. Different cell populations may show different susceptibilities to infection by different viruses. Thus the pseudorabies virus is nonselective in the destruction of cells in the anterior cortex of piglets, feline parvovirus selects rapidly proliferating cells of ependyma and cerebellar cortex, and the enteroviruses of pigs appear to affect particularly the spinal motor neurons.

The degeneration of neurons, the reactivity of the glia, and the perivascular reactions are the general hallmarks of viral infection of the central nervous system and imply a sequence of viral cytopathogenicity and reaction to cellular degeneration. However, viruses which produce manifestations deviating from this rather simple system are many. The viruses of hog cholera and bluetongue at a suitably early fetal stage of development may produce **malformations** characterized by degeneration of tissue without reaction; later when the animal is immunologically competent, these viruses may produce ordinary encephalomyelitis. Infectious feline enteritis virus causes cerebellar hypoplasia in the neonate but is without effect at a later stage of development. Many of these virus-induced malformations have in the past been attributed to toxic or genetic effects.

Some viruses may produce **persistent tolerated infection** without clinical signs or lesions. At the other end of the spectrum is rabies virus, which can cause death with minimal cytopathic effect and reaction, or no morphological change at all. Some virus infections are characterized by long latency and slow attrition, scrapie and visna of sheep being the standard examples expressed as either degeneration or chronic inflammation. **Transformation** of nervous system cells by viruses can be demonstrated *in vitro,* and a variety of tumors of the nervous system can be produced by viruses in experimental animals, but they have, as yet, no natural counterpart. The contribution of an immunological response to both the progress and morphology of encephalomyelitis is very difficult to determine. The development of inflammation and neurological disease in murine lymphocytic choriomeningitis infection depends on immunological competence, and the encephalitis of malignant catarrhal fever may be approximately of this type. The parainfectious encephalitides of humans following infection with measles, vaccinia, and varicella have no established counterparts in domestic species, although **postvaccinal encephalitis** after rabies immunization is occasionally observed in dogs.

1. General Pathology

Viral infections of the central nervous system produce what is designated in practice as a nonsuppurative inflammation, a term which includes quite a variety of quantitative and qualitative changes, but which implies something more specific than a literal translation would suggest. The changes are not specific for viral infections, being produced in some bacterial infections such as salmonellosis in swine, in the rickettsial infection, "salmon poisoning" of dogs, and probably in other rickettsioses. Nor are the lesions qualitatively the same in all viral infections, a fact that is of considerable usefulness in differential diagnosis. The differences are due in part to inherent characters of the agents, routes of invasion, patterns of localization, success of viral replication and release, duration and degree of cellular injury, and host defense reactions. As well as the differences, there are also the very considerable similarities which allow them all to be included with the **nonsuppurative encephalomyelitides.**

The nervous system is not homogeneous with respect to its susceptibility to viral infections. Viruses differ in their affinity for different types of cells and for different areas of the brain, an affinity which can be modified experimentally by adaptation and possibly also by the development of variant virus strains in natural infections. The distribution of lesions in the different infections is a reflection of the varied affinities and is useful in differential diagnosis but only generally applicable. The distribution of lesions or even of inclusion bodies is only a very rough guide to the distribution and activity of the virus.

Changes in the vessel walls and the **Virchow–Robin space** are almost constant in encephalitis, and when present, are usually the most striking microscopic changes. This depends on the accumulation of cells in the adventitia of the vessel and in the perivascular space of Virchow–Robin. Generally, the accumulated cells are leukocytes, but in some diseases there may be very few hematogenous elements recognizable, the cuffs being composed instead of cells of histiocytic type, appearing to have proliferated *in situ* from adventitial elements. These latter frequently fragment to resemble degenerate neutrophils if the postmortem interval is prolonged. Injury to the wall proper of the vessel is not constant, but when it occurs, it may affect the arteries as a hyalinizing or fibrinoid change. This is characteristic of malignant catarrhal fever, for example, and occurs also in equine encephalomyelitis. The endothelium may be selectively injured in hog cholera and infectious canine hepatitis; the local responses are different and are described elsewhere with the diseases. The lesions associated with equine herpesvirus infection may be limited to quite subtle endothelial swelling and proliferation in small blood vessels. When, in inflammation, the perivascular cuffs are large enough to disorganize the wall of the vessel and compress it, the endothelium frequently shows signs of swelling and proliferation, and there may also be altered adhesiveness so that leukocytes and red cells tend to stick to it. Thrombosis is incidental and very seldom observed, but compression of the vessels probably accounts for the ischemic parenchymal softenings that occur.

Infiltrating cells are predominantly lymphocytes, and they accumulate in the perivascular spaces (Fig. 3.30). The earliest cells in the perivascular spaces may be neutrophils; in acute lesions, some of these can be found wandering in parenchyma or grouped in dense clusters. They soon disappear but may be found later for a short time in areas which soften. Lymphocytes remain in the cuffs and are admixed after a week or so with an increasing number of plasma cells and macrophages. A few eosinophils may also be found in pigs.

Perivascular cuffing is not specific for viral infections. It is also a reaction to degeneration of neural tissue.

Glial reactions occur if the parenchyma is injured, even though the injury may not be appreciable, but the reaction may be absent in those infections which selectively involve the vessel walls. In inflamed areas, the oligodendroglia degenerate. The astrocytes degenerate or react depending on how strictly and severely they are injured. The stimuli for astrocytic reaction are discussed early in this chapter. The gliosis, which is so frequently a feature of nonsuppurative encephalomyelitis, is almost solely microglial. The gliosis may be diffuse or focal but is commonly both (Fig. 3.28). The diffuse gliosis may be more apparent than real if due more to hypertrophy than to proliferation of these glia. Focal gliosis may occur anywhere in the parenchyma and may be related to small vessels and injury to microvasculature. When there are more than a dozen or so cells in such foci, some of the cells are likely to be lymphocytes, and occasionally there are a few plasma cells. The microglia in the center of a focus are frequently degenerate. There are two specifically named forms of microgliosis: neuronophagic nodules are foci of microglia about degenerating neurons (Fig. 3.17D), and glial shrubbery is an accumulation of these cells in the molecular layer of the cerebellum in relation to degenerating Purkinje cells (Fig. 3.101).

Neuronal changes must be the principal determinants of the outcome of the infection. The distribution of neuronal degeneration is to some slight extent specific, but the extensiveness is nonspecific and varies considerably. The morphologic features of the degenerate cells are nonspecific. As a rule, the more severe degrees of neuronal degeneration are caused by viruses that are highly neurotropic, such as rabies, but the severity of neuronal degeneration is not a dependable lead in differential diagnosis. The fact that rabies virus can be lethal without morphologic change in the nervous system raises the possibility that other viruses may cause severe encephalopathy without inflammatory change. The usual form of neuronal degeneration is central chromatolysis, which may extend to completion with the cell, then swollen, pale, and devoid of a nucleus. This is typical of the axonal reaction to injury, but the axons are intact. Many neurons appear as if coagulated, being shrunken, rounded, and isolated, and staining darkly with eosin. The nucleus is pyknotic or has disappeared. The coagulated neuron stimulates the formation of the neuronophagic nodule, but not all necrotic neurons elicit the reaction. Usually, intact neurons can be found adjacent to degenerate ones.

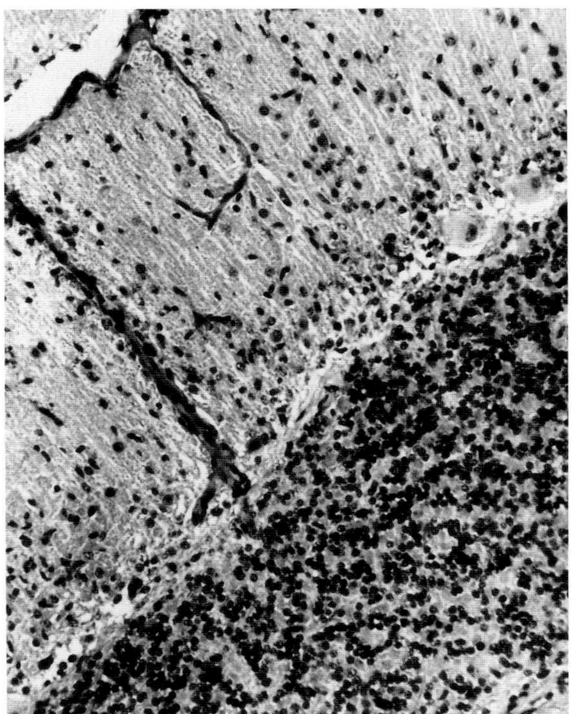

Fig. 3.101 Encephalitis in louping ill. Sheep. Destruction of Purkinje cells and gliosis in molecular layer of cerebellum (glial shrubbery).

Lesions in the white matter occur consistently even though many of the viruses are supposed to be specifically tropic for the gray matter. Cuffing reactions are to be expected even if degeneration is limited to adjacent gray matter. Microgliosis is focal rather than diffuse, and the nodules are small. Some degree of disintegration of myelin is inevitable, but conspicuous demyelination is a feature only of canine distemper, leukoencephalitis of goats, and visna. Occasionally, there is focal softening with cyst formation in the white matter, apparently independent of vascular lesions. Both acute and chronic demyelinating diseases can be caused by viral infections. It is possible that demyelination in some diseases represents an immunopathologic reaction to viral infection, but this has not been established. Myelin disruption appears to be directly associated with viral infection of oligodendrocytes.

Meningitis is seldom severe except in local distributions, and these tend to be in relation to underlying parenchymal lesions. Some agents, of which typical examples are those of sporadic bovine encephalomyelitis (Fig. 3.48B) and malignant catarrhal fever, have a selective affinity for meningeal structures in the central nervous system. Others, such as the virus of canine distemper, share the affinities so that leptomeningitis is part of the primary response. In the more purely neurotropic infections, such as rabies, the meningitis, or more precisely the meningeal infiltration, is due probably to drainage of products of reaction via the Virchow–Robin spaces from the brain or cord. The reacting cells in the meninges are of the same type as those in the brain, and they float freely in the arachnoid spaces.

No virus is known to be tropic for peripheral nerves in the sense of selectively producing direct lesions in them. Degenerative changes occur fairly early in the end-plates and terminal parts of axons when the central cell body is destroyed. Foci of microglia and lymphocytes occur in the nerve roots quite commonly. Inflammation of the paravertebral ganglia, and especially the trigeminal (Gasserian) ganglia, is characteristic of rabies and of Teschen disease and related infections in pigs, but the frequency and distribution of this lesion in other viral encephalomyelitides and in other ganglia is unknown.

Inclusion bodies may form in neurons, neuroglia, or microglia, and other mesenchymal cells. Inclusion bodies in the nervous system are usually acidophilic. They may be found only in neurons, as in rabies, in glia, or in both. They may be cytoplasmic, nuclear, or both. Intranuclear inclusions, which must be distinguished from altered nucleoli, have considerable specificity, and some, but rather less, reliance can be placed on intracytoplasmic inclusions. Cytoplasmic viral inclusions must be distinguished from normal inclusions. The problems of determination of inclusion bodies are discussed further in relation to the specific diseases.

2. Rabies

Rabies is an ancient disease which is still endemic in many parts of the world. It has been eradicated from Great Britain and Scandinavia and has not occurred in Australia, New Zealand, and some islands. Eradication has not been successful where the disease is established in multiple host systems.

The virus belongs to the genus Lyssavirus, which also includes Mokola, Lagos bat, Kotonkan, Obodhiang, and Duvenhage virus. These latter named viruses are of African origin, but the Duvenhage strain has become widespread in insectivorous bats in continental Europe. These types should be regarded as able to produce rabieslike disease in mammals.

"Fixed" rabies virus, which is the basis of vaccine strains, is a laboratory biotype stabilized in its properties by serial intracerebral passage. It is highly neurotropic, is not secreted in saliva, and does not produce Negri bodies. "Street" virus is the feral biotype which circulates in enzootics and epizootics. In addition to neurotropism, it is tropic for salivary glands, in which it reaches high concentration, and possibly in other mucus-secreting epithelia, and produces Negri bodies. Antigenic analysis of street virus by monoclonal antibodies has demonstrated heterogeneity, with 12 or more types in bats and at least 5 in terrestrial mammals. The significance of the variations for epidemiology and pathogenesis of infection is unclear, but they may well represent ecotypes compartmentalized to particular host/vector species in particular geographic areas, and they may remain stable over long periods, that is for decades. The existence of ecotypes may represent the natural equivalent of the laboratory-induced "fixed" virus. The properties of "fixation" have wider implications than solely concern for the safety of vaccines, and some degree of fixation to host probably occurs in nature as a result of continuous intraspecies passage and may influence the pathologic expression of the disease.

Notwithstanding the variability of rabies virus, the disease produced is one disease to which all mammals are susceptible, although there are species differences in susceptibility. The disease can be regarded as one of carnivores because it is almost always transmitted naturally only by bites, and humans, herbivores, etc., are dead-end hosts. There are exceptions to the rule that rabies is transmitted by bites: oral infection can occur in diverse species, and the application of modified virus in "baits" utilizes this potential; aerosol infection can occur, as in dense congregations of colonial bats in bat caves, probably as droplet infection from salivary secretions; and a variety of aberrant circumstances may provide transfer opportunities for infectious virus, as has been reported for corneal transplants.

The reservoir hosts vary from time to time and from region to region. The principal reservoir vectors are foxes and skunks in the United States, with the raccoon becoming more important on the Atlantic seaboard; foxes and dogs in northern Canada; foxes moving from east to west in Europe; wolves in eastern Europe and Iran; jackals in India and northern Africa; the mongoose and genet cat in South Africa; and the mongoose in the Caribbean. Sylvatic vectors are responsible for most transmissions to humans and domestic animals in countries where dog populations are controlled. In tropical areas where domestic and feral dogs are not controlled, these animals are the principal hazards for humans and livestock.

Bats present a special epidemiologic problem. Fructivorous and insectivorous bats as well as vampires are capable of transmitting rabies. Vampire bats are inhabitants of South and Central America extending into northern Mexico and are, historically, responsible for a high incidence of rabies in mammals, especially cattle, but including humans. Much remains to be learned of the bat–virus relationship using modern knowledge and techniques to reassess, in particular, questions of dormancy, recovery from neurologic disease, and salivary secretion of virus. When clinically affected, vampires manifest the disease as the furious form and show unnatural daylight activity. Transmission of rabies by insectivorous bats to mammals, including humans, is documented for a wide range of animals, but the cases are individual, and there is nothing to suggest that bats act as incendiaries to establish enzootics in terrestrial mammals.

The establishment of infection ordinarily depends on inoculation of the virus into a wound, such usually being inflicted by the bite of a rabid animal. Contamination of a fresh wound by infected saliva or tissues is much less dangerous. The virus replicates first in myocytes near the site of inoculation and enters local neuromuscular and neurotendinous spindles. Much is unknown as to the local events at the site of inoculation or of the role of the complete virion or defective particles at the site of inoculation in determining the extraordinary variability of incubation periods. Spread to the sensory paravertebral ganglia can be very rapid, or it may be delayed for many months. The virus travels along peripheral nerves to the central nervous system, and it may then travel centrifugally along all peripheral nerves, so that in fatal cases, the virus may be found in the central and peripheral nervous systems and in other tissues and milk. This pattern of migration along peripheral nerves is a property of the strictest of neurotropic viruses. Pseudorabies probably does the same thing, and the agent of equine encephalomyelitis, when rendered highly neurotropic by repeated intracerebral passage, is also capable of using nerves as a route of invasion of the nervous system. Viral tropisms are relative, however, and mutable. In mammals, the rabies virus has an affinity also for salivary glands in which replication occurs in the acini, the virus being released directly into the ducts. In bats the affinity for salivary glands may be greater than the affinity for neural tissue.

It is seldom that a clinical diagnosis of rabies can be made with confidence. The terms furious rabies and dumb rabies place emphasis on particular features within a spectrum of behavioral changes and are inappropriate for noncarnivorous animals. Aberrant behavioral patterns can be recognized in affected animals during epizootics. The period of salivary excretion of virus before the onset of neurological signs is expected to be not more than a few days, vampire bats possibly excepted, and the duration of the clinical disease is expected to be a few days only. Once expressed clinically as neurologic disease, rabies is almost invariably fatal; recovery with or without neurologic deficit is quite rare but has been observed in several species following experimental exposure. Progressive infection and clinical disease do not inevitably follow exposure; up to 25% of feral populations may have specific antibodies as evidence that proliferation of the infecting inoculum was sufficient to provoke an immune response without progressing to neurologic disease.

The lesions of rabies, when present, are typical of the nonsuppurative encephalomyelitides with ganglioneuritis and parotid adenitis. Inflammatory changes are usually present but, notably, they may be very mild or absent. To some extent at least, the severity of lesions reflects the duration of the clinical disease. In the nervous system, inflammatory and degenerative changes are most severe from the pons to the hypothalamus and in the cervical spinal cord, with relative sparing of the medulla. This relative sparing of the medulla appears to apply to all the domestic species. The most severe lesions of the disease are generally found in dogs, whereas other species, especially ruminants, which are highly susceptible, may show little more than an occasional vessel with a few cuffing lymphocytes and a few very small glial nodules (Babes' nodules in this disease), and this in spite of having numerous Negri bodies. These reactive phenomena probably reflect largely the degree of neuronal degeneration, and this may be remarkably slight in herbivores and remarkably severe in dogs.

The reaction is typically one of perivascular cuffing and focal gliosis. The cuffs are one to several cells thick and composed solely of lymphocytes (Fig. 3.102A); ring hemorrhages confined largely to the perivascular space are common about cuffed vessels. Hemorrhages are occasionally severe enough to be visible grossly in the spinal cords of horses and cattle (Fig. 3.102B). The Babes' nodules are composed of microglia, and they occur in both white and gray matter. The nodules vary greatly in size, some containing only six or seven cells and some containing a hundred or more. Diffuse as well as focal gliosis occurs in areas of gray matter such as the pons and in the spinal cord, both horns of the latter being involved.

Neuronal degeneration in carnivores may be very ex-

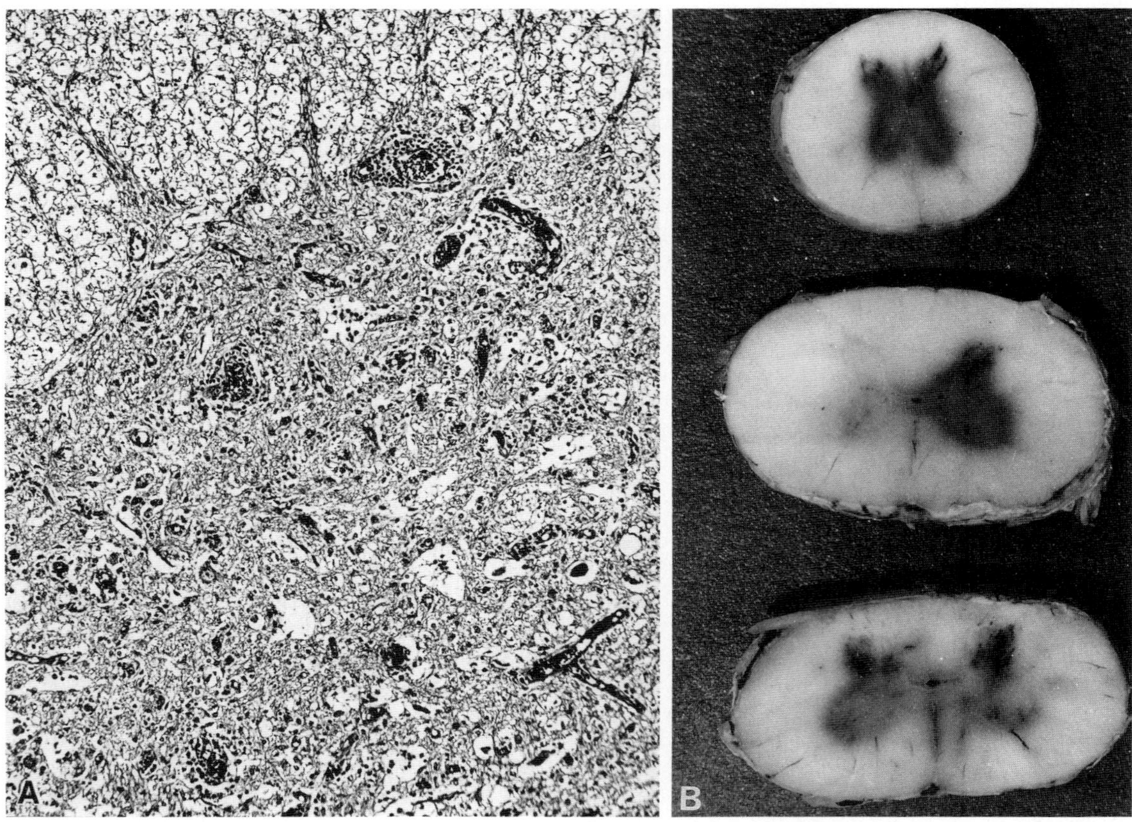

Fig. 3.102 Rabies. (A) Perivascular cuffs and focal gliosis. Horse. (B) Hemorrhage in gray matter of spinal cord.

tensive and quite out of proportion to the observed reactive changes but may be very slight in pigs and herbivores. The specificity of the neuronal changes and of the whole pathologic picture depends on the inclusion bodies of Negri. These are always intracytoplasmic and are present most commonly in the hippocampus of carnivores and in the Purkinje cells of herbivores. Fixed virus does not produce Negri bodies, and street virus fails to do so in as many as 30% of cases. Neurons of any distribution may contain inclusion bodies, but they tend to be scarce where the inflammatory reaction is severe. Indeed Negri bodies may be found only in neurons which are otherwise histologically normal; they are not present in degenerate neurons. They have also been found, but rarely, in ganglion cells of the adrenal medulla, salivary glands, and retina. While the number of Negri bodies has little relation to the length of the incubation period, there is a relation to the duration of the clinical disease. They may not be found if the animal is killed instead of being allowed to die. They are produced consistently in white mice, the usual test animal.

Negri bodies are round or oval structures usually ~2–8 μm in diameter (Fig. 3.103A). They are plastic, their shape being molded to their environment. Those in the dendrites, seldom observed except in Purkinje cells, are oval, and those in the cell body are usually rounded. There may be one or more per cell, and affected cells are otherwise only little changed. The inclusions are surrounded by a clear thin halo. Nonspecific homogeneous inclusions may be found in the pyramidal cells of the hippocampus in cats, skunks, and dogs. There may be several such inclusions per cell, but each is minute, not measuring more than 1.5 μm. In old sheep and cattle the larger neurons, especially of the medulla and cord, may contain nonspecific inclusions. These have a dustlike distribution, are usually numerous, and are brightly acidophilic, angulated, and approximately 1.0 μm in size. Nonspecific inclusions, which are indistinguishable from Negri bodies by light microscopy, occur in the lateral geniculate neurons in cats. Fluorescent antibody techniques are required for positive identification and are essential in the rare chronic cases which may not yield virus on mouse inoculation.

If there is no ganglioneuritis in the paravertebral ganglia, then the possibility of the animal's having rabies is very remote. If there is ganglioneuritis, it may be part of rabies or something else. Pigs, for example, get ganglioneuritis in the Teschen group of infections. Inflammatory changes in the trigeminal (Gasserian) ganglion in rabies may be present without inflammatory or neuronal changes being clearly evident in the brain. The ganglionic changes are of the same character as those in the brain, being characterized by acute degeneration of the ganglion cells, proliferation of the capsule cells, and microglial nodules (Fig. 3.103B).

The natural transmission of rabies depends on virus being present in the saliva and, therefore, in the salivary glands. Fixed virus has no affinity for the salivary glands, and none is present in some cases of infection with the

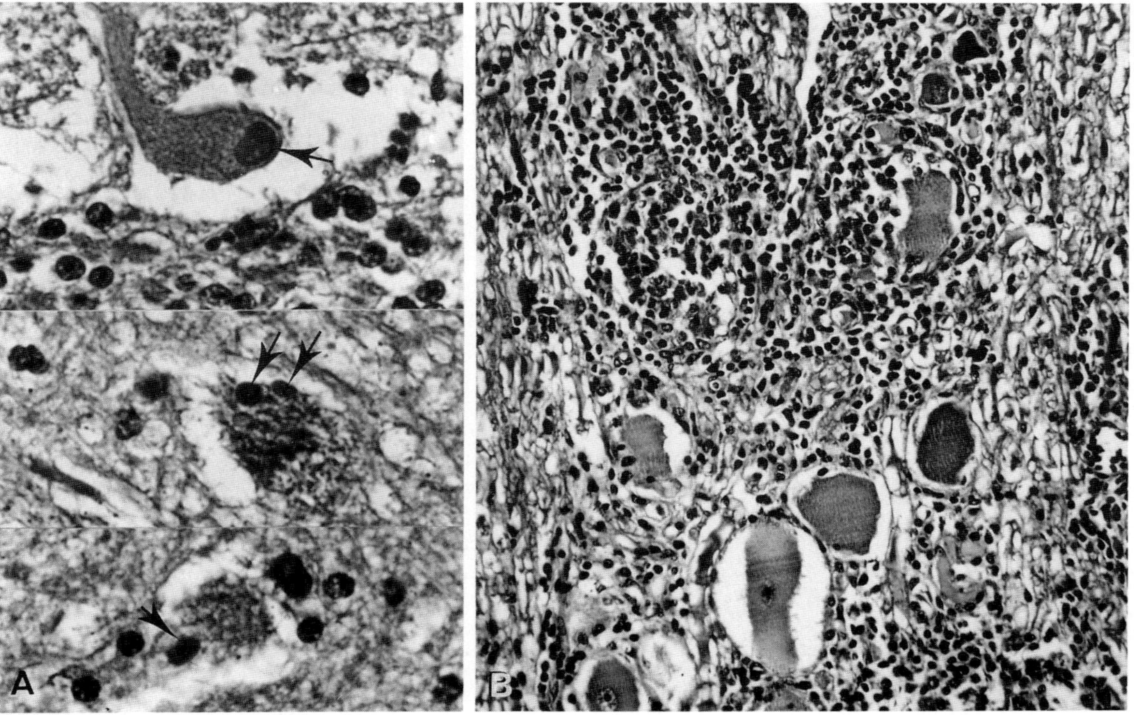

Fig. 3.103 Rabies. (A) Composite picture showing Negri bodies (arrows) in neuronal cytoplasm. (B) Severe nonsuppurative inflammation of trigeminal ganglion.

street virus. Degenerative changes are reported in the epithelium of the mandibular salivary gland, but not in the parotid, in dogs.

The diagnosis of rabies is made microscopically, utilizing fluorescent antibody labeling, or by animal inoculation, and is likely to be missed unless there is clinical suspicion or unless brains are examined routinely. There is seldom anything even suggestive in the general postmortem, but self-inflicted wounds and foreign bodies in the stomach of a carnivore should always be regarded with suspicion.

Bibliography

Acha, P. N. Epidemiology of paralytic bovine rabies and bat rabies. *Bull Off Int Epiz* **67:** 343–382, 1967.

Baer, G. M., Shanthaveerappa, T. A., and Bourne, G. H. Studies on the pathogenesis of fixed rabies virus in rats. *Bull WHO* **33:** 783–794, 1965.

Barnard, B. J. H., and Voges, S. F. A simple technique for the rapid diagnosis of rabies in formalin-preserved brain. *Onderstepoort J Vet Res* **49:** 193–194, 1982.

Barnard, B. J. H. *et al.* Non-bite transmission of rabies in Kudu (*Tragelaphus strepsiceros*). *Onderstepoort J Vet Res* **49:** 191–192, 1982.

Beran, G. W. Rabies and infections by rabies-related viruses. *In* "CRC Handbook Series in Zoonoses." Section B, Vol 1, pp. 57–135. Boca Raton, Florida, CRC Press, 1981.

Cameron, A. M., and Conroy, J. D. Rabieslike neuronal inclusions associated with a neoplastic reticulosis in a dog. *Vet Pathol* **11:** 29–37, 1974.

Charlton, K. M., Casey, G. A., and Campbell, J. B. Experimental rabies in skunks: Mechanisms of infection of the salivary gland. *Can J Comp Med* **47:** 363–369, 1983.

Charlton, K. M. *et al.* Experimental rabies in skunks and foxes. Pathogenesis of the spongiform lesions. *Lab Invest* **57:** 634–645, 1987.

Clark, H. F. and Prabhakar, B. S. Rabies. *In* "Comparative pathology of Viral Diseases," Vol II, pp. 165–214. R. G. Olsen *et al.* (eds.). Boca Raton, Florida, CRC Press, 1985.

Fekadu, M., Chandler, F. W., and Harrison, A. K. Pathogenesis of rabies in dogs inoculated with Ethiopian rabies virus strain. Immunofluorescence histologic, and ultrastructural studies of the central nervous system. *Arch Virol* **71:** 109–126, 1982.

Johnson, R. T. Experimental rabies. Studies of cellular vulnerability and pathogenesis using fluorescent antibody staining. *J Neuropathol* **24:** 662–674, 1965.

Khan, M. A., Diesch, S. L., and Goyal, S. M. Current status of rabies. *Int J Zoon* **13:** 215–229, 1986.

Murphy, F. A. *et al.* Comparative pathogenesis of rabies and rabieslike viruses. *Lab Invest* **28:** 361–376, 1973.

Murphy, F. A. *et al.* Experimental chronic rabies in the cat. *Lab Invest* **43:** 231–241, 1980.

Rosatte, R. C. Bat rabies in Canada: History, epidemiology, and prevention. *Can Vet J* **28:** 754–756, 1987.

Rupprecht, C. E., and Dietzschold B. Perspectives on rabies virus pathogenesis. *Lab Invest* **57:** 603–606, 1987.

Soave, O. A. Transmission of rabies to mice by ingestion of infected tissue. *Am J Vet Res* **27:** 44–46, 1966.

Vaughan, J. B., Gerhardt, P., and Newell, K. W. Excretion of street rabies virus in the saliva of dogs. *J Am Med Assoc* **193:** 363–368, 1965.

3. Aujeszky's Disease

Aujeszky's disease is also known as pseudorabies, mad itch, infectious bulbar paralysis, and porcine herpesvirus infection. The common domestic species are naturally susceptible, although there are very few reports in horses and goats. Progressive infections do not occur in humans. The disease is reported from Europe, in the eastern countries of which it is a serious problem, and from the British Isles, North Africa, China, New Zealand, South America, and the United States of America, where outbreaks are sporadic in nature, usually involving few animals, but sometimes very serious in individual herds of pigs. Natural infections have also been observed in rats and mice and various species of wildlife and on fur farms. Of the laboratory animals, the rabbit is the most susceptible and is preferred for identification of the virus because of the fairly consistent development of intense local pruritus following subcutaneous inoculation. Guinea pigs are less susceptible and may resist subcutaneous inoculation but succumb to intracerebral and, occasionally, to intraperitoneal inoculation.

The virus belongs to the herpes group but is unusual for a member of that group in its relative lack of host specificity and by being spread laterally as well as vertically in swine. A number of strains have been identified, which exhibit a wide range of virulence.

The virus is capable of surviving for several months in dried tissues and for approximately 2 months in infected premises. It is moderately resistant to phenol and has contaminated batches of hog-cholera vaccine. Probably the virus is maintained in enzootic areas in swine, for which it is highly contagious but usually asymptomatic, and in brown rats. Transmission can occur by ingestion, but the usual method of spread between pigs is thought to be by contact of infective secretions with nasal mucosa or abraded skin. Animals are susceptible to intranasal inoculation and, irrespective of the route of infection, the virus can be found in nasal secretions. The virus may also be present in saliva and urine, but on this the reports are at variance. It is also present in blood, but this is of no significance for epidemiology or transmission. The infection will occur in pigs by contact very readily and probably by direct nose-to-nose transmission, but it does not appear to be contagious between individuals of other species, and they probably acquire their infection by contact with swine or, possibly, rats. Pigs may harbor virus for many months in tonsils and nasopharyngeal secretions after exposure, but in other domestic species, the virus is fairly strictly neurotropic, and therefore is not excreted unless given experimentally in large doses. Ingestion of infected pig meat is the usual source of infection for dogs and cats. Cattle and sheep may become infected by direct contact with carrier swine or by aerosol exposure, but there is strong circumstantial evidence implicating contaminated feed.

Latency, such as characterizes many herpesvirus infections, has been demontrated in pigs, with the virus being persistent in trigeminal ganglia and tonsils for more than a year after experimental infection. Such animals may intermittently shed virus in nasopharyngeal secretions.

The pathogenesis of the infection following local inoculation is well established for the rabbit and is probably comparable in other species. The virus causes a local reaction at the site of inoculation, if percutaneous, and then spreads centripetally along the related nerve to the spinal cord; it then spreads outward again along other peripheral nerves as other segments of cord are progressively invaded by spread within the central nervous system. Because of the progressive advance of infection along the cord, death may occur before demonstrable amounts of virus reach the brain and before lesions have time to develop there. Intracerebral inoculation produces encephalitis, and virus spreads to the cord and centrifugally along peripheral nerves to an extent which depends on survival time. Because the virus also circulates in the blood, there is some possibility but no evidence that it invades the brain directly, the evidence instead suggesting that it localizes in viscera and invades the nervous system along autonomic nerves. Following nasal or intraocular exposure, the virus spreads along the related nerves. The route of invasion following ingestion is not known. Transplacental infections occur in pigs, causing abortion in about 50% of sows pregnant in the first month, and the delivery of macerated, mummified, and normal fetuses when infection occurs at later stages of gestation.

The characteristic clinical sign of Aujeszky's disease in animals other than pigs is an intense cutaneous irritation developing at the point of inoculation or at the terminal distribution of a nerve trunk which passes the point of inoculation. This does not occur until the virus reaches the related segment of cord.

The signs and course of Aujeszky's disease in pigs are very variable, the majority of cases being a mild febrile illness without pruritus or nervous signs. In these cases, recovery is expected in a week or so. Sows may subsequently produce mummified litters. Age is, however, a very important factor governing the severity of the disease in swine; the mortality rate in nursing pigs and young weaners may be very high. Very young sucklings do not show specific nervous signs but rapidly become prostrate and die in 12 to 24 hr. In slightly older piglets, incoordination progresses rapidly to paralysis with muscular twitchings, tremors, and convulsions. Some pigs showing severe signs of encephalitis recover. Experimental peripheral inoculation will produce asymptomatic meningitis and encephalitis in swine, the inflammatory lesions being of quite considerable severity.

The disease in older pigs is often characterized by fever, rhinitis, and coughing. There may be generalized pruritus in natural cases, but it is not severe, being expressed usually by rubbing of nose or head. Fetal resorption, mummification, stillbirths, and abortions are frequently reported.

In other domestic species, Aujeszky's disease is usually sporadic, although significant outbreaks are recorded in sheep and cattle. The mortality rate is very high. Death may occur without signs of illness or within 1 to 2 days of the onset of clinical signs. There is fever, and the itching may be on any part of the body but is most frequently about the head or hind limbs. Other neurological signs are variable but constantly present. The clinical signs in dogs and cats are not well documented. Dogs may show extreme dullness and tachypnea, salivation, and diffuse pruritus. Vomiting and diarrhea may be severe. The variability of signs in dogs and cats may be due to lesions in the autonomic system. Death occurs within 24 to 48 hr of the onset of clinical signs.

There are no specific gross lesions of Aujeszky's disease. At the site of cutaneous infection, there is acute serofibrinous inflammation in the dermis and subcutis. It is not clear how much of this is attributable to the virus and how much, to trauma. The virus is present in the exudate. Gross changes may be seen in young pigs. There may be necrosis of tonsils and posterior fauces and sometimes of trachea and esophagus. Rhinitis with patchy epithelial necrosis is common. The lungs may be edematous with foci of hemorrhagic necrosis, and focal necrosis may be seen in liver, spleen, and adrenals.

The neuropathology of Aujeszky's disease in domestic animals is incompletely studied. The gray matter especially is affected, but death may occur before there are clear indications of neuronal degeneration or inflammatory reaction in the brain. Even when exudative changes are evident, neuronal degeneration may be mild, but it is usually severe (Fig. 3.104A,B). With naturally acquired infections, the inflammatory changes are of nonsuppurative type as described earlier, but following intracerebral inoculation, the reaction is much more severe and is characterized by meningitis and encephalitis, in which neutrophils and eosinophils may be abundant. There is severe ganglioneuritis in paravertebral ganglia. The specificity of the reaction in the brain depends on the development of acidophilic intranuclear inclusion bodies in neurons and astroglia (Fig. 3.104C). These inclusion bodies occur in all species including pigs, in which originally they were not found; fixation in a mercurial fixative is helpful for their demonstration. Inclusions in swine are solid and amphophilic, resembling those of herpes, but in other species the inclusions are granular and often small and multiple in an affected nucleus. By any route of infection, piglets tend to develop a panencephalitis with most severe lesions in the cerebral cortex (Fig. 3.104B); in other domestic species, the distribution of lesions in the central nervous system is local to, and determined by, the route of exposure.

The virus of Aujeszky's disease does not exhibit the high degree of neutrotropism that the virus of rabies does, although this probably varies with the host species and strain of virus. In the rabbit, for example, it produces inclusion bodies in mesenchymal cells of the meninges and in Schwann cells, focal necrosis in the liver, spleen, and adrenals, and hemorrhage and edema in the lungs. It is reported in dogs and cats to produce acute myocarditis after peripheral inoculation. In piglets dying of the natural infection, there may be focal necrosis in the liver, lung, spleen, and adrenals. Little attention has been given to extraneural lesions or localizations in this infection, and

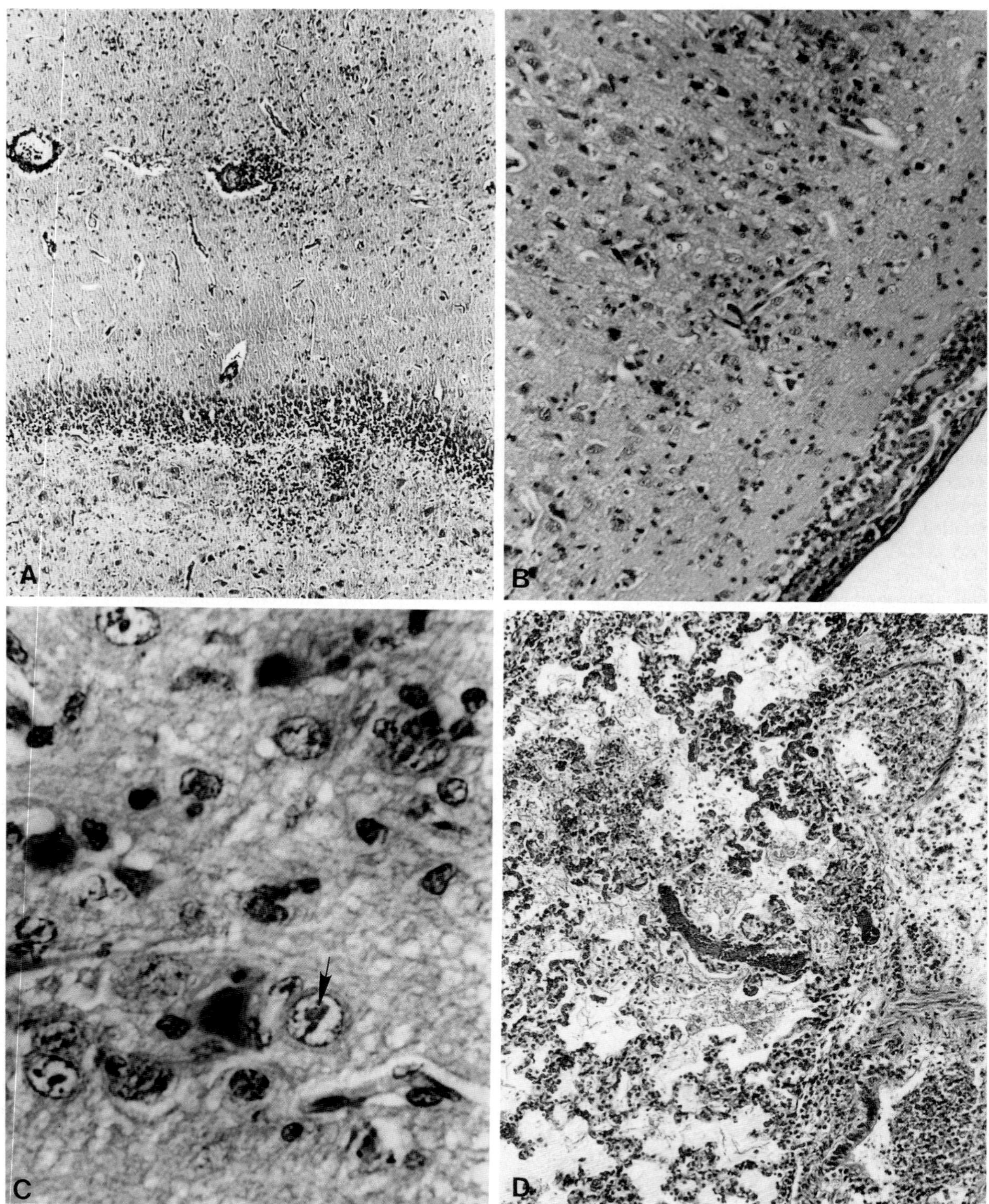

Fig. 3.104 Aujeszky's disease. Pig. (A) Perivascular cuffing, focal and diffuse gliosis in dentate gyrus. (B) Meningitis and necrosis of cells in cerebrum. (C) Neuronal necrosis and irregular inclusion bodies in nuclei (arrow). (D) Necrotizing bronchopneumonia.

such studies might provide an explanation for the fatal outcome of infections in which virus is not present in the brain; there is, as yet, no satisfactory explanation for these occurrences, but it might reside in involvement of the peripheral nervous system. Pulmonary lesions may be mild or severe. Edema and mild cellular infiltration may be diffuse, and there may be focal or confluent necrotizing, hemorrhagic pneumonia (Fig. 3.104D). Hemorrhage and necrosis are present in lymph nodes, and foci of necrosis may be found in tonsils, liver, spleen, and adrenal. In aborted or stillborn piglets which are suitable for examination, there is usually no evidence of encephalitis, but foci of necrosis may be found in liver and other parenchymatous tissues together with focal bronchiolar necrosis and interstitial pneumonia.

Bibliography

Beran, G. W. *et al.* Persistence of pseudorabies virus in infected swine. *J Am Vet Med Assoc* **176:** 998–1000, 1980.

Bergmann, V., and Becker, C.-H. Untersuchungen zur Pathomorphologie und Pathogenese der Aujeszkyschen Krankheit. *Pathol Vet* **4:** 97–119,493–512, 1967.

Corner, A. H. Pathology of experimental Aujeszky's disease in piglets. *Res Vet Sci* **6:** 337–343, 1965.

Crandall, R. A. Pseudorabies (Aujeszky's disease). *Vet Clin North Am* **4:** 321–331, 1982.

Ducatelle, R., Coussement, W., and Hoorens, J. Immunoperoxidase study of Aujeszky's disease in pigs. *Res Vet Sci* **32:** 294–302, 1982.

Masic, M., Ercegan, M., and Petrovic, M. Die Bedeutung der Tonsillen fur die Pathogenese und Diagnose der Aujeszkyschen Krankheit bei Schweinen. *Zentralbl Veterinaermed* **12B:** 398–405, 1965.

McCracken, R. M., and Dow, C. An electron microscopic study of Aujeszky's disease. *Acta Neuropathol* **25:** 207–219, 1973.

Nara, P. L. Porcine herpesvirus. *In* "Comparative Pathobiology of Viral Diseases," Vol 1, pp. 89–113. R. D. Olsen *et al.* (eds.). Boca Raton, Florida, CRC Press, 1985.

Narita, M. *et al.* Necrotizing enteritis in piglets associated with Aujeszky's disease virus infection. *Vet Pathol* **21:** 450–452, 1984.

Olander, H. J. *et al.* Pathologic findings in swine affected with a virulent strain of Aujeszky's virus. *Vet Pathol* **3:** 64–82, 1966.

Schmidt, S. P. *et al.* A necrotizing pneumonia in lambs caused by pseudorabies virus (Aujeszky's disease virus) *Can J Vet Res* **51:** 145–149, 1985.

Van Oirschot, J. T. (ed.). "Vaccination and Control of Aujeszky's Disease." Dordrecht, Netherlands, Kluwer Academic Publishers, 1988.

Wohlegemuth, K. *et al.* Pseudorabies virus associated with abortion in swine. *J Am Vet Med Assoc* **172:** 478–479, 1979.

4. *Hemagglutinating Encephalomyelitis Virus of Pigs*

The virus is classified with the corona group and can be isolated from the respiratory tract of normal pigs. There is serological evidence of wide distribution in many swine-raising areas. The disease occurs in piglets 1–3 weeks of age and follows a clinical course of about 3 days to 3 weeks. The mortality rate is very high, and survivors are usually unthrifty. The clinical signs are dominated by anorexia and vomiting after an incubation period of about 4 days, and this syndrome is called vomiting and wasting disease. Neurologic signs are present in some outbreaks and consist of stilted gait, hyperesthesia, progressive paresis, and convulsions in some cases.

Lesions of the central nervous system may be found in some affected piglets which do not show clinical signs of nervous disease. The frequency with which inflammatory change is found is quite variable for unknown reasons. The lesions when present are those of nonsuppurative encephalitis affecting particularly the gray matter of medulla and brain stem. In such cases there is inflammation of the trigeminal, paravertebral, and autonomic ganglia.

Following exposure, replication of virus occurs in the epithelium of nose, tonsil, lung, and small intestine. Spread to the central nervous system appears to be along nerves rather than hematogenously. Viral antigen is detectable in the ganglia during the incubation period of the disease. In the brain it is first demonstrable in trigeminal and vagal sensory nuclei, with later rostral spread in brain stem. Viral replication occurs in the myenteric plexus of the stomach, and involvement of the autonomic system probably can explain the predominant clinical signs of vomition and constipation.

5. *Enterovirus Encephalitis of Pigs*

Encephalomyelitis of pigs caused by enteroviruses occurs in many countries and is distinguishable only by a study of the agents. Outbreaks of the disease in various countries, and perhaps with deviant clinical patterns, have led to the development of local names Teschen disease, Talfan disease, poliomyelitis suum, benign enzootic paresis, Ontario encephalomyelitis, and polioencephalomyelitis. Many porcine enteroviruses are capable of causing neurologic disease by intracerebral inoculation, but ability to produce encephalitis does not appear to be a natural attribute of most strains.

Porcine enteroviruses are ubiquitous but limited in their pathogenicity to the one host. At least 11 serological subtypes are distinguished by virus-neutralization tests, and infection by one serotype does not confer protection against infection by another. Infection follows the fecal–oral route, and initial replication occurs in the tonsils and the intestinal epithelium, especially of ileum and colon; the enteric phase is not clinically significant or accompanied by tissue change. Viremia occurs regularly in infection by some serotypes and may lead to localization in the pregnant uterus and death of fetuses. Five serogroups are known to cause neurologic disease, but serological subtypes are not, however, correlated with virulence.

The highly virulent Teschen virus belongs to serogroup 1 and is probably limited to Europe. Less virulent strains including those responsible for Talfan disease, benign enzootic paresis, and Ontario encephalomyelitis are widely distributed, possibly excepting Asia. Teschen disease is of high morbidity and high mortality, affecting all age groups, and expressed clinically with convulsions, opisthotonus, nystagmus, and coma. Death commonly occurs in 3 to 4 days. Survivors may have residual paralysis. Less virulent strains produce lower morbidity and mortality, the clinical syndrome expressed as paresis, and ataxia which seldom progresses to complete paralysis.

Infection with virulent virus may produce nervous signs as soon as 6 days after exposure. The virus is present in large amounts in tonsils and cervical lymph nodes by 24 hr, and in the mesenteric nodes and feces by 48 hr. Multiplication in the alimentary tract is followed by a brief period of viremia, during which nervous infection occurs. The infection is asymptomatic in the absence of neurological signs, and up to 95% of exposed pigs develop latent or inapparent infections.

The pathological changes in these syndromes are the same, although minor differences in severity and distribution of the lesions are reported. There are no gross changes. The histologic changes are those of nonsuppurative polioencephalomyelitis extending throughout the cerebrospinal axis from the olfactory bulbs to the lumbar cord. Any series of cases of each of the syndromes provides a continuous spectrum of severity and distribution of the lesions. Lymphocytic meningitis is mild in the cerebral meninges and usually overlies areas of parenchymal injury in the cerebrum. In weaners and older animals in which the course is more prolonged than in very young pigs, an intense lymphocytic meningitis develops over the cerebellum, usually in conjunction with inflammatory lesions in the underlying molecular layer (Fig. 105A). Cerebellar meningitis is very slight if the course of the disease is 4–5 days or less, so that, although emphasized in reports of Teschen disease, it is not a feature in very young animals in which the course is short. The most severe lesions occur in the brain stem from the hypothalamus through the medulla, and decrease in intensity and diffuseness down the spinal cord. There is relative sparing of the cerebral and cerebellar cortices, but the deep substance of the cerebellum is consistently and often severely involved. The lesions in the spinal cord in each syndrome are largely confined to the gray matter, particularly the ventral horns, but may selectively involve the dorsal horns in very young pigs. Lesions are consistently present in the dorsal root ganglia, and especially the trigeminal ganglia (Fig. 3.105B).

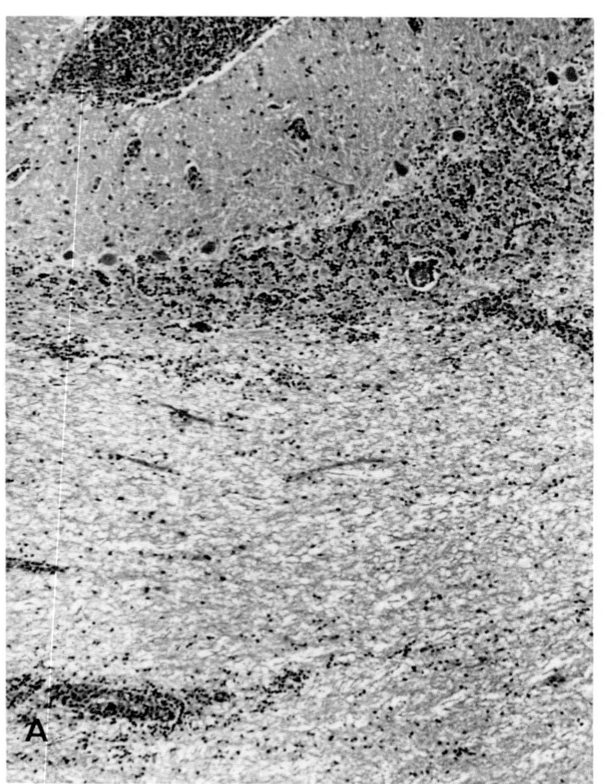

Fig. 3.105A Nonsuppurative meningitis and encephalitis (cerebellum) in enterovirus encephalomyelitis. Pig.

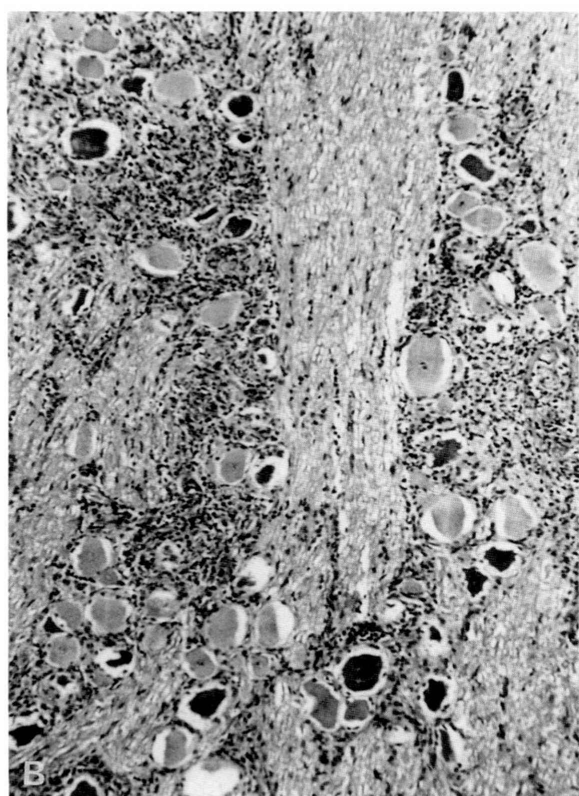

Fig. 3.105B Ganglioneuritis in enterovirus encephalomyelitis. Pig.

Bibliography

Andries, K., and Pensaert, M. B. Immunofluorescence studies on the pathogenesis of hemagglutinating encephalomyelitis virus infection in pigs after oronasal inoculation. *Am J Vet Res* **41:** 1372–1378, 1980.

Appel, M., Greig, A. S., and Corner, A. H. Encephalomyelitis of swine caused by a hemagglutinating virus. IV. Transmission studies. *Res Vet Sci* **6:** 482–489, 1965.

Dardiri, A. H., Seibold, H. R., and DeLay, P. D. The response of colostrum-deprived, specific pathogen-free pigs to experimental infection with Teschen disease virus. *Can J Comp Med* **30:** 71–81, 1966.

Forman, A. J., Pass, D. A., and Connaughton, I. D. The characterization and pathogenicity of porcine enteroviruses isolated in Victoria. *Aust Vet J* **58:** 136–142, 1982.

Koestner, A., Kasza, L., and Holman, J. E. Electron microscopic evaluation of the pathogenesis of porcine polioencephalomyelitis. *Am J Pathol* **49:** 325–337, 1966.

Long, J. F. Pathogenesis of porcine polioencephalomyelitis. *In* "Comparative Pathobiology of Viral Diseases," Vol. 1, pp. 179–197. R. G. Olsen *et al.* (eds.). Boca Raton, Florida, CRC Press, 1985.

Meyvisch, C., and Hoorens, J. An electron microscopic study of experimentally induced HEV encephalitis. *Vet Pathol* **15:** 102–113, 1978.

Mills, J. H. L. and Nielsen, S. W. Porcine polioencephalomyelitides. *Adv Vet Sci* **12:** 33–104, 1968.

Narita, M. *et al.* Demonstration of viral antigen and immunoglobulin (IgG and IgM)in brain tissue of pigs experimentally infected with haemagglutinating encephalomyelitis virus. *J Comp Pathol* **100:** 119–128, 1989.

Shadduck, J. A, Koestner, A., and Kasza, L. Host range-studies of two porcine polioencephalomyelitis viruses. *Am J Vet Res* **27:** 473–476, 1966.

Werdin, R. E., Sorensen, D. K., and Stewart, W. C. Porcine encephalomyelitis caused by hemagglutinating encephalomyelitis virus. *J Am Vet Med Assoc* **168:** 240–246, 1976.

6. Louping Ill

Louping ill is a tick-borne viral encephalomyelitis of sheep, which has been enzootic in England, Scotland, and Northern Ireland for more than a century and may be present in Spain. Cattle, horses, goats, and deer pastured with affected sheep sometimes contract the disease, and nonfatal human infections are known. Outbreaks in piglets have followed the feeding of raw meat from lambs, and a case has been described in a dog.

The virus of louping ill belongs to a family of tick-borne flaviviruses which are closely interrelated antigenically but distinguishable by an appropriate set of serological tests. The group includes louping ill, central European tick-borne fever, Omsk hemorrhagic fever, Kyasanur forest disease of India, far eastern Russian tick-borne fever, Powassan virus, and Langat virus of Malaysia. Excepting louping ill, these are diseases primarily of humans. The Russian viruses do not appear to be pathogenic for sheep or goats. Their natural resistance is perhaps related to heavy and persistent endemic infections; experimental inoculation of some of these viruses will produce encephalitis in sheep.

The tick responsible for transmission in Great Britain is *Ixodes ricinus,* the castor-bean tick. Other species of *Ixodes,* and perhaps other arthropods, are potential vectors. *Ixodes ricinus* is parasitic on a variety of mammals in addition to sheep and on birds, but whether any of these act as reservoirs is not known. The larval and nymphal ticks acquire the virus when feeding on infected sheep and transmit the infection to new hosts in the succeeding nymphal and adult phases. Because of its natural mode of transmission, louping ill is most prevalent in early summer and early autumn when the ticks are active.

Although tick transmission is the usual mode of infection, the disease can be contracted in humans, monkeys, and mice by inhalation of infective droplets, but this route is not thought to be important naturally. Rabbits and guinea pigs are not susceptible even by intracerebral inoculation.

Louping ill is a systemic infection and, while it remains so, the disease is mildly febrile but otherwise of no consequence. When it invades the nervous system and produces signs of encephalitis, the mortality rate is very high. The morbidity in endemic areas is quite low, and the disease is largely confined to older lambs and yearlings whose colostral immunity is not reinforced by natural exposure. The incidence of encephalitis in naturally acquired systemic infection is not known, but experimental subcutaneous inoculation of sheep produces encephalitis in less than 50% of those exposed. There is evidence that the virus may take one of two routes into the brain, the neural infection being either directly hematogenous with growth of the virus across the vessels, or indirectly from the blood, the virus localizing in nasal structures and entering the brain via the olfactory nerves. Under experimental conditions, a number of factors may facilitate entry of virus into the brain, and facilitation is apparently given in natural cases by concurrent tick-borne fever, a rickettsiosis also transmitted by *I. ricinus,* and known to impair humoral and cellular defense mechanisms.

The viremic phase of louping ill is clinically silent or a febrile phase with dullness. Recovery may occur in a couple of days and leave solid immunity. If neurologic signs are to develop, they do so at about the fifth day and are characterized by incoordination of gait, tremors, cerebellar ataxia, and terminal paralysis.

There are no gross lesions. The disease is an acute polioencephalomyelitis. There is mild leptomeningitis corresponding to areas of inflammation of the parenchyma. Inflammation of the cerebellar leptomeninges may be quite severe when the cerebellar cortex is acutely affected (Fig. 3.101). The inflammatory lesions are rather more obvious than occur in most viral encephalitides but are of the usual type, although unusually large numbers of neutrophils may be present in very severe cases, and largely restricted to gray matter, although cuffing and focal gliosis occur in the white matter. Neuronal degeneration may be severe, and neuronophagia, prominent. Some degree of selective vulnerability of the Purkinje and Golgi cells of the cerebellum is generally accepted and, although it can be demonstrated in many cases, its detection probably depends to a large extent on the duration of active infection. The spinal lesion is a poliomyelitis affecting particularly the ventral horns (Fig. 3.28). Inclusion bodies have not been observed in sheep, but acidophilic, intracytoplasmic inclusions in neurons of the brain stem and cord are reported in experimentally affected monkeys and mice.

Bibliography

Bannatyne, C. C. *et al.* Louping-ill virus infection of pigs. *Vet Rec* **106:** 13, 1980.

Reid, H. W. The epidemiology of louping ill. *In* "Tick-Borne Diseases and Their Vectors" pp. 501–504. J. K. H. Wilde, (ed.). Edinburgh, Centre for Tropical Veterinary Medicine, 1978.
Timoney, P. J. Susceptibility of the horse to experimental inoculation with louping-ill virus. *J Comp Pathol* **90:** 73–86, 1980.
Wilson, D. R. Studies in louping-ill. 1. Cultivation of the louping-ill virus *in vitro*. *J Comp Pathol* **55:** 250–267, 1945.

7. Japanese Encephalitis

Japanese encephalitis virus is an arthropod-borne flavivirus related to St. Louis, Murray Valley, and West Nile viruses. All are important human pathogens, but only Japanese encephalitis causes disease in domestic animals. The virus is found through much of eastern Asia from the southeast of the Soviet Union to Indonesia and from Japan to India, where the disease is endemic, with dramatic annual epidemics. In humans there is a high ratio of subclinical to overt infections with a case fatality rate of 10 to 15% and a high incidence of residual neurologic defects in survivors. Transplacental infection followed by abortion occurs in humans, and this is also the most serious expression of the disease in pigs, the domestic species most importantly infected.

The virus is transmitted mainly by the mosquito, *Culex tritaeniohynchus*, but other species of this genus and of the genera *Aedes* and *Armigeres* may be important, as the virus is known to be vertically transmitted through some of them. Many species of animals and birds are susceptible to mosquito-borne infection and develop antibody responses in timing with suitable climatic and habitat cycles. The reservoir hosts are not known but may well be birds, although the domestic fowl is probably not a host. The pig is a very important domestic animal in many of the endemic areas and is an important amplifier host for the virus, developing sustained viremia of sufficient titer to infect feeding mosquitoes, and indeed is probably the preferred host for *C. tritaeniohynchus*.

Most horses, pigs, and cattle in endemic areas possess neutralizing antibodies against the virus. Intranasal and intracerebral inoculation can produce fatal encephalitis in calves, but natural cases of encephalitis in this species are quite rare. Among animals infected naturally with the virus, only horses and donkeys develop clinical encephalitis. There are no clinical signs of encephalitis in pigs, but pregnant susceptible sows may produce stillborn piglets. Infected stillborn and neonatal pigs may show hydrocephalus, cerebellar hypoplasia and hypomyelinogenesis, and anasarca; histological changes are restricted to the nervous system and may include non-suppurative encephalitis. The lesions in these neonatal and stillborn piglets probably reflect the timing of infection in relation to the development of immune competence.

Severe epidemics of this encephalitis have occurred in horses in Japan. The lesions are confined to the central nervous system, and in quality and distribution are the same as those produced by the American viruses. Inclusion bodies are not reported in the Japanese disease. Encephalitis occurs in piglets up to 6 months of age as a nonsuppurative inflammation which is diffuse in the brain and cord, but in the cerebellum, it affects rather selectively the molecular and Purkinje layers. The histologic pattern of Japanese encephalitis in pigs is apparently similar to that of Teschen and related diseases, but it has not yet been reported in comparative detail.

Bibliography

Huang, C. H. Studies of Japanese encephalitis in China. *Adv Vir Res* **27:** 71–101, 1982.
Rosen, L. The natural history of Japanese encephalitis virus. *Annu Rev Microbiol* **40:** 395–414, 1986.
Rosen, L. *et al.* Experimental vertical transmission of Japanese encephalitis virus by *Culex tritaeniorhynchus* and other mosquitoes. *Am J Trop Med Hyg* **40:** 548–556, 1989.
Umenai, T. *et al.* Japanese encephalitis: Current worldwide status. *Bull WHO* **63:** 625–631, 1985.

8. American and Venezuelan Encephalitis of Horses

Epidemics of disease in horses known to be an encephalomyelitis have been recognized in the United States of America since about 1930, but what was probably the same disease, unrecognized as to its true nature, has been known for most of this century. The causative viruses were first isolated from affected horses, and the disease became known as equine encephalomyelitis, infectious equine encephalomyelitis, and American equine encephalomyelitis. As the name implies, horses were originally regarded as the primary hosts, but it is now known that horses are accidental and unfortunate hosts, that birds are the most common vertebrate reservoir hosts, and that mosquitoes are the principal vectors. Horse and human, in both of which species the disease is of very considerable importance, are now known to be, in terms of transmission and often literally as well, dead-end hosts in which the titer of virus in blood is ordinarily too low to be a source of infection for mosquitoes.

Not all birds are capable of reacting as reservoir hosts. Red-winged blackbirds, cardinals, sparrows, cedar waxwings, and the captive Chinese pheasant are highly susceptible to infection and nearly always die. Many other species, including adult domestic fowl and turkeys, are not sickened by the infection, although fatalities can be produced in the young of these species. In most eastern areas of the United States, the virus cycles between water birds and the mosquito, *Culiseta melanura,* which does not feed on large mammals. Western encephalitis covers most of the United States where the cycle is between wild birds and the mosquito, *Culex tarsalis*. This mosquito does feed readily on animals, and human and animal cases occur regularly. The Venezuelan virus cycles between the mosquito and small rodents in the Caribbean areas, but the Venezuelan virus does attain titers in horses sufficient to infect the vector mosquitoes.

Mosquitoes once infected are known to remain so for life, and there is evidence that the virus is capable of multiplying in the insects. There are, however, ecological problems of considerable magnitude still to be solved regarding the reservoir–vector relationship. Some mosquitoes cannot be infected, some can be infected but cannot transmit the infection, and those species which can be infected differ in the amount of virus necessary to infect them and in their capability for transmitting the infection. Many species of mosquitoes are selective in their range of hosts. It is now known that some species of mosquitoes in each of the genera *Aedes, Culex, Anopheles,* and *Culiseta* are potential vectors and that, of these, *Culex* is the most important. Arthropods other than mosquitoes may also be of some, but lesser, importance. The virus has been found in chicken mites (*Dermanyssus gallinae*), chicken lice (*Menopon pallidum, Eomenocanthus stramineus*), and assassin bugs (*Triatoma sanguisuga*). The spotted-fever tick, *Dermacentor andersoni,* is capable of transmitting the infection stage to stage and hereditarily.

The ecological and epidemiological problems are not restricted to the American scene. One immunogenic type of the American virus, the Western type, is a natural infection of horses and rodents in Czechoslovakia and the second immunogenic type, the Eastern type, has been found in a monkey in the Philippines, and there is serological evidence of equine infection in those islands. The Venezuelan and Eastern viruses are antigenically related, and there is an overlapping distribution of these viruses in South America. In North America, the distribution of Eastern and Western viruses is fairly well delineated. The Western type occurs primarily west of the Mississippi with isolated occurrences in the eastern states up to the Atlantic seaboard; the Eastern type is largely confined to the Atlantic and Caribbean coasts and occurs sporadically as far west as the Mississippi Valley.

Although humans and horses are the principal mammalian victims, other species are susceptible. Pigs readily develop asymptomatic infections, but because they do not have a viremia, they are not of significance for natural propagation of the virus. Calves are susceptible to intracerebral inoculation but recover in 2 weeks. Guinea pigs and white mice are highly susceptible, rabbits are less so, and sheep, dogs, and cats are refractory.

There is no information on why the infection, which must cycle naturally in birds and mosquitoes, intermittently spills over into human and equine populations. The problem is rather more difficult even than this because the question remains why the infection is systemic and uncomplicated in some, perhaps in most, cases and in others, and often in outbreaks, its behavior is altered, and it invades the nervous system. Individual outbreaks may involve humans only or horses only or both species more or less at the same time. Being arthropod borne, the diseases occur in seasonal patterns and are most prevalent in late summer.

Young horses are more susceptible than the old. Initially, there is viremia with fever and depression, usually unnoticed. The animal may then recover, or the virus may invade the nervous system, by which time the fever has subsided. The neurologic signs are characterized by derangements of consciousness and terminal paralysis. There may be early restlessness with compulsive walking, often in circles. There is central blindness. The animal becomes somnolent and assumes unnatural postures. At this stage, the course may remain static and the animal lives as a "dummy," or paralysis may develop, often first affecting cranial nerves, but later general and flaccid. The signs are largely cortical, and the cortex is the principal site of the lesions. The course, if fatal, is usually 2–4 days.

There are no gross changes. The microscopic changes are limited almost exclusively to gray matter (Fig. 3.17D). When the course is short, 1 day or less, the reaction is largely on the part of neutrophil leukocytes. These diffusely infiltrate the gray matter and may be found in foci suggestive of malacia (Fig. 3.106). There is early microglial reaction to produce rod cells. The endothelial cells, especially of veins, are swollen, and hyaline or granular thrombi are common in these vessels. There are narrow cuffs composed of lymphocytes and neutrophils with perivenous hemorrhage and edema. After a couple of days, the neutrophils disappear, the cuffs are composed of lymphocytes, and there are both focal and diffuse microglial proliferations as in the standard nonsuppurative reactions. Hurst described intranuclear inclusions similar to those in Borna disease; although there can be no doubt that they occur, many observers have failed to find them.

The most severe lesions are in the cerebral cortex, especially the frontal, rhinencephalic, and occipital areas, with lesions of lesser intensity in the pyriform lobes. Severe lesions are also present in thalamus and hypothala-

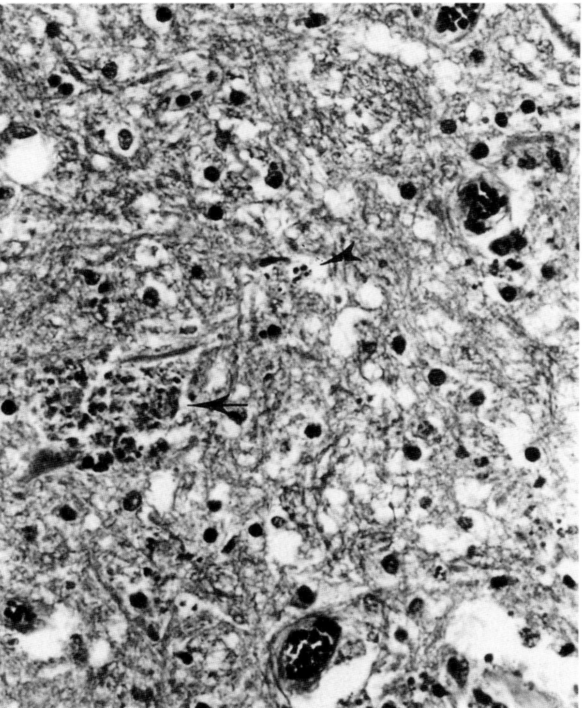

Fig. 3.106 American encephalomyelitis. Horse. Neuronal necrosis (arrow), glial necrosis (arrowhead) in cerebrum.

mus. From the thalamus caudally, the intensity of inflammation diminishes but reveals no selectivity for particular nuclear masses. The cerebellum is less severely injured than other portions, although inflammatory changes may be found in the deep nuclei and spottily in the cortex. Mild changes occur in both dorsal and ventral horns of the cord, but their distribution is irregular. The trigeminal ganglia are not affected.

Bibliography

Johnson, K. M., and Martin, D. H. Venezuelan equine encephalitis. *Adv Virus Res* **19:** 76–116, 1974.

Kissling, R. E., and Chamberlain, R. W. Venezuelan equine encephalitis. *Adv Vet Sci* **11:** 65–84, 1967.

Sudia, W. D., and Newhouse, V. F. Epidemic Venezuelan equine encephalitis in North America. A summary of virus–vector–host relationships. *Am J Epidemiol* **101:** 1–13, 1975.

Walton, T. E. *et al.* Experimental and epizootic strains of Venezuelan equine encephalomyelitis virus. *J Infect Dis* **128:** 271–282, 1973.

9. Borna Disease

Borna is a village of Saxony, Germany, and it gives its name to this viral encephalomyelitis of horses, which was originally enzootic there. Curiously, the disease is still confined more or less to the same localities in Central Europe. Encephalomyelitis has been observed in sheep in contact with horses dying of Borna disease, and the virus has been isolated from sheep with encephalomyelitis. Indeed Borna disease is listed as a common cause of death in sheep in endemic localities. The rabbit is also infected naturally. Experimentally many species are susceptible, and propagation can be obtained in cell cultures. The virus in culture remains strictly cell associated and is still unclassified.

Few encephalitides had been so long known and so little known as Borna disease, which situation changed with the description of Near Eastern encephalomyelitis and evidence that it is akin to, and probably identical with, Borna disease. In this view, the virus of Borna disease is an arborvirus, which in the Near East is transmitted primarily by the tick *Hyalomma anatolicum anatolicum*. The virus can pass transovarially to the larvae of this tick, and regular recovery can be made from wild birds. The problem of alternative reservoirs and vectors, especially in eastern Europe and Russia, is still to be clarified. The disease in the Near East has involved horses, donkeys, cattle, and sheep in natural outbreaks and is readily transmissible intracerebrally to rabbits, sheep, and chickens, with Joest–Degen bodies developing in rabbits. Kittens are susceptible to oral infection. The Near Eastern disease in horses is seasonal, corresponding with tick activity, whereas in sheep, outbreaks are influenced by seasonal migration of flocks. It appears that most tick-transmitted infections in horses, and possibly in ruminants, produce an inapparent reaction and a long period of viremia.

The nature of the Borna virus is unclear. In a number of biological and physicochemical characteristics it is conventional, but virus particles cannot be seen in cells, *in vivo* or *in vitro,* which exhibit specific antigenicity, and there is no demonstrated serological relationship to any other virus. The rabbit is very susceptible to intracerebral inoculation, and the virus can be passaged then to rats, mice, and other species. Much interest has been generated in aspects of the pathogenesis of the infection in these laboratory species. How much of this experimental information is relevant to the feral virus and to disease in domestic species is unexamined. On the assumption that parallels exist, the salient features of the laboratory disease are summarized. They take no account of Near Eastern encephalomyelitis because no recent account is available.

Borna disease virus replicates in various cell cultures which produce a virus-specific protein which is detectable immunohistochemically and is indicative of infectivity. It has exclusive affinity for the nervous system, in which it induces persistent infection of neurons, in which the infection advances centripetally via neural pathways to the brain and then centrifugally. The outcome of the infection appears to be determined largely by immunological events. Infection provokes both humoral and cellular immune responses. The humoral responses appear to have little effect on the progress of infection, which depending on the system used, may lead to progressive inflammatory disease or to permanent persistent infection and degenerative neural phenomena. The latter are typical of experimental infection in mice and in rats exposed when immature and in which inflammatory infiltrates are absent. The inflammatory process appears to be mediated by cellular immune responses directed against and leading to deletion of infected neurons. Persistent infection may depend on transfer of virus from degenerating neurons to adjacent astrocytes or Schwann cells.

Borna disease occurs chiefly in young horses, and the mortality rate is about 90%. The incubation period is not less than 4 weeks and introduces a clinical syndrome which is purely neurologic but of varied course, death occurring in 1 to 3 weeks. There is pharyngeal paralysis, hyperesthesia, muscular tremors, and spasms; blindness is common. Drowsiness and flaccid paresis develop terminally.

There are no gross lesions. The histologic changes are those of nonsuppurative encephalitis dominated by perivascular cuffs, often very thick and spreading to the adjacent parenchyma. They occur predominantly in gray matter but also involve fiber tracts related to affected areas of gray matter. Parenchymal cellular reaction involves mainly microglia. Neuronal degeneration and neuronophagia are frequent, and inclusion bodies (Joest–Degen bodies) form. The inclusion bodies are mainly in nuclei, especially in the hippocampus, and are very occasionally cytoplasmic. They stain well and reddish with Giemsa and have a clear halo.

The distribution of lesions in Borna disease differs from that in other equine encephalomyelitides and parallels closely the distribution of viral antigen, as displayed by immunohistochemistry, and the distribution of infectivity, as determined by titration in cell cultures. The distribution is axial, with the most severe alterations in the mesencephalon, diencephalon, and hypothalamus. In cortical structures, the most severe damage is in the laterobasal convolutions, including the pyriform lobe, and in the hippocampus. The dorsal cerebrum and the cerebellum are relatively spared. Lesions may be present in optic nerves and retina.

The distribution and character of lesions produced in laboratory animals are similar to those in the horse; by extrapolation, the lesions in sheep would also be similar, although a detailed description is not available. There is paravertebral ganglioneuritis in experimental laboratory animals and inflammatory change in nerve roots; probably this occurs in the natural disease also.

Bibliography

Carbone, K. M., Moench, T. R., and Lipkin, W. I. Borna disease virus replicates in astrocytes, Schwann cells, and ependymal cells in persistently infected rats: Location of viral genomic and messenger RNAs by *in situ* hybridization. *J Neuropathol Exp Neurol* **50:** 205–214, 1991.

Daubney, R., and Mahlau, E. A. Viral encephalomyelitis of equines and domestic ruminants in the Near East: Part II. *Res Vet Sci* **8:** 419–439, 1967.

Duchala, C. S. *et al.* Preliminary studies on the biology of Borna disease virus. *J Gen Virol* **70:** 3507–3511, 1989.

Gosztonyi, G., and Ludwig, H. Borna disease of horses. An immunohistological and virological study of naturally infected animals. *Acta Neuropathol* **64:** 213–221, 1984.

Ihlenburg, H. Experimentelle Prufung der Empfanglicheket der Katze fur das Virus der Bornaschen Krankheit. *Arch Exp Veterinaermed* **20:** 859–864, 1966.

Ludwig, H., Bode, L., and Gosztonyi, G. Borna disease: A persistent virus infection of the central nervous system. *Prog Med Virol* **35:** 107–151, 1988.

Magrassi, F., Leonardi, G., and Scanu, A. Osservazioni si un encefalopatia virale des gatto. *Boll Soc Ital Biol Sper* **27:** 1233, 1951.

Martin, L. A., and Hinterman, J. Une maladie non decrite du chat; la myelite infectieuse. *Arch Inst Pasteur, Maroc* **5:** 64–73, 1955.

Matthias, D. Der Nachweis von latent infizierten Pferden, Schafen, und deren Bedeutung als Virusreservoir bei der Bornaschen Krankheit. *Arch Exp Veterinaermed* **8:**506–511, 1954.

Mayr, A., and Danner, K. Borna—a slow virus disease. *Comp Immun Microbiol Infect Dis* **1:** 3–14, 1978.

McGaughey, C. A. Infectious myelitis of felines. *Ceylon Vet J* **1:** 34–36, 1953.

Rott, R. *et al.* Borna disease, a possible hazard for man? *Arch Virol* **118:** 143–149, 1991.

10. Lentivirus Encephalitis

The diseases with which we are concerned here are visna of sheep and caprine arthritis–encephalitis of goats. Typically for this type of virus infection, the virus is highly cell associated and replicates only slowly; infection persists for the life of the animal; the incubation period before seroconversion may be several months and before clinical disease may be months or years; the clinical disease is progressive; and the lesions are dominated by active mononuclear inflammatory cells.

In both natural hosts, sheep and goats, four clinical and pathological syndromes are recognized, namely **mastitis, arthritis, interstitial pneumonia (maedi and ovine progressive pneumonia),** and **encephalomyelitis (visna of sheep).** Cross-infections are possible experimentally, which, in the systems so far examined, produce persistent infection without disease, suggesting that such infections are only to a very limited extent permissive.

Within particular endemic situations, any one or combination of the four syndromes may be present, and when in combination, usually one syndrome is predominant. The reasons for this are unknown, but the viral pathogenesis is quite complex, mutational changes in envelope proteins are frequent, and there is epidemiological evidence for a strong influence of host genotypes.

Although viral antigen can be demonstrated in a variety of cells, productive infections are probably limited to the widely distributed cells of the monocyte–macrophage system. Virus production occurs only in the mature cells of this series, which suggests that viral persistence may depend on infection of a stem cell population. How macrophage infection translates into disease and why disease is limited to certain tissues remain unknown but may be related to the different functional characteristics macrophages adopt in different tissues, including presentation of antigen to lymphocytes. Natural transmission of infection is also by monocytes, mainly in colostrum and milk, but other routes which present infected macrophages are possible.

As a natural disease, **visna** occurs in sheep of both sexes, but clinical signs are seldom, if ever, observed in animals younger than 2 years. The onset is very insidious. The earliest sign may be a barely perceptible posterior ataxia and a fine trembling of the lips. The first sign to be noticed may be extensor paralysis of the hind limbs. Once paralytic signs are evident, a fatal outcome appears certain. There is no fever and no sign of cerebral dysfunction, and death is due to starvation or secondary infection.

The incidence of visna is relatively low. The disease appears to be readily transmissible. The experimental disease parallels the natural one, and pathologic changes are progressive throughout the long incubation period. The agent can be recovered from the brain as early as 16 days after inoculation and throughout the long incubation period and the period of clinical disease. While present, it is active even though there may be no clinical sign of neurologic deficit. The course of the infection can be followed fairly well by routine examinations of the cerebrospinal fluid.

Normal sheep are expected to have not more than 50 cells/mm^3 of spinal fluid; in visna, the number of cells, and these are chiefly lymphocytes, is elevated sometimes to 1000 or more. After intracerebral inoculation, there is a latent period of up to 8 weeks, after which the number of cells in spinal fluid begins to increase. The cell count may remain at a high level for several months without other signs of disease. Thereafter, the animal may recover, as indicated by a decrease in the cell count of the spinal fluid, or the cell count may remain high; paralysis may develop; and death may follow.

The disease in the brain is chronic and demyelinating. The active inflammatory process is acute, and in cell cultures, the virus produces acute cytopathic effects. It is difficult therefore to explain the long active course of the disease in the central nervous system. Productive infections do occur in cultured cells, but it is difficult to demonstrate extracellular virus *in vivo*. Viral antigen cannot readily be demonstrated, and viral particles are difficult to demonstrate by electron microscopy. The difficulties may derive from the demonstration that most virus is present as provirus DNA, and that productive infection in the sheep is probably very limited. Infection is accompanied by the production of neutralizing antibody, but immune mechanisms may be ineffective because of the selection of mutants which resist neutralization. Mutants have been demonstrated to be fully virulent, and antigenic change could create the necessary conditions for the relapsing course of the disease.

There are no gross neural changes in this disease, and the histologic change is one of a patchy demyelinating encephalomyelitis. The distribution of lesions, involving principally the white matter, is unlike the distribution produced by other neurotropic viruses. There is a mild to severe mononuclear type of cerebrospinal meningitis. The parenchymal lesion may be well established by 1 to 2 months, and these early lesions are intensely inflammatory with perivascular cuffing and gliosis. They reveal clearly that the process begins in, and immediately beneath, the ependyma diffusely throughout the cerebrospinal axis. In this early stage, the myelinated fibers in the inflammatory foci remain remarkably intact; the gray matter of the cord is irregularly but often intensely affected by a nonsuppurative reaction even 2–3 months after inoculation. In the paralytic and terminal stages of the disease, the periventricular destruction of white matter in the cerebrum and cerebellum is extensive, and in some sections of the brain, especially in the cerebellum, almost every bit of white matter is destroyed, leaving the gray matter free.

Destruction of myelinated fibers in the spinal cord is patchy and not due to progressive spread of the pericentral inflammation. The demyelinated plaques are characteristically peripheral and triangular in shape with a base on the pia mater. Although dorsal and lateral tracts are most frequently involved, there is no selectivity for particular fiber tracts and no symmetry. The degenerating foci are almost malacic in their severity, and the plaques contain numerous reactive microglia and astroglia. The spinal nerve roots share in the degenerative process. Germinal centers may form in the choroid plexus.

There is no explanation for the periventricular distribution of the brain lesions in this disease, and their pathogenesis is obscure. The severity of inflammatory lesions corresponds, in experimental disease, with the titer of virus inoculated. In areas of intense inflammation, liquefactive foci of necrosis occur in the white matter, and the loss of myelin is expected to be of Wallerian type. In the spinal cord, evidence of remyelination can be found, indicating that oligodendrocytes are not target cells and that demyelination may be primary.

Caprine arthritis-encephalitis appears to be widely distributed, but the expression of the infection is highly variable, and many infected goats show little or no clinical disease. Clinical disease of the nervous system affects kids 2–4 months of age and is frequently fatal. Animals which develop the early nervous disease or have early inapparent infections tend to develop synovitis and periarthritis in adulthood (see Chapter 1, Bones and Joints).

The clinical signs are referable to motor spinal dysfunction without signs of cerebral disease. Onset is indicated by hind-limb lameness and ataxia with paresis, which progresses over several weeks to paralysis. The inflammatory lesions in the central nervous system may remain active for several years in goats which survive (Fig. 3.107A,B). In the early clinical phase of the disease, changes are widely distributed in the white matter of the brain and cord, particularly in the subependyma and beneath the pia in the cord. The distribution and character of the lesions in the nervous tissue in the goat are, in general, similar to those in visna of sheep. There is, however, less tendency for the periventricular lesions to progress to gross cavitation of cerebral white matter. Instead, there is a tendency for the inflammatory and myelinoclastic areas to increase in number and severity caudally from the mesencephalon. As in visna, the spinal cord changes are discontinuous and, where present, involve the myelin in subpial plaques or in one or more quadrants of the cord. The extent of perivascular infiltration by mononuclear cells (Fig. 3.107B) is also greater in kids than in sheep.

In addition to the encephalomyelitis, mastitis, and arthritis of this disease, interstitial pneumonia occurs in some natural and experimental cases (Fig. 3.107C,D). The pulmonary lesions are similar to those of maedi of sheep and are fully developed only in adults. The frequency with which early pulmonary changes resolve or progress is not known.

Bibliography

Cheevers, W. P., and McGuire, T. C. The Lentiviruses: Maedi/visna, caprine arthritis-encephalitis, and equine infectious anemia. *Adv Vir Res* **34:** 189–211, 1989.

Cheevers, W. P. *et al.* Chronic disease in goats infected with two isolates of caprine arthritis–encephalitis lentivirus. *Lab Invest* **58:** 510–517, 1988.

Dickson, J., and Ellis, T. Experimental caprine retrovirus infection in sheep. *Vet Rec* **125:** 649, 1989.

Georgsson, G. *et al.* Expression of viral antigens in the nervous system of visna-infected sheep: An immunohistochemical study on experimental visna induced by virus strains of increased neurovirulence. *Acta Neuropathol* **77:** 299–306, 1989.

Lavimore, M. D. *et al.* Lentivirus-induced lymphoproliferative disease. Comparative pathogenicity of phenotypically distinct ovine lentivirus strains. *Am J Pathol* **130:** 80–90, 1988.

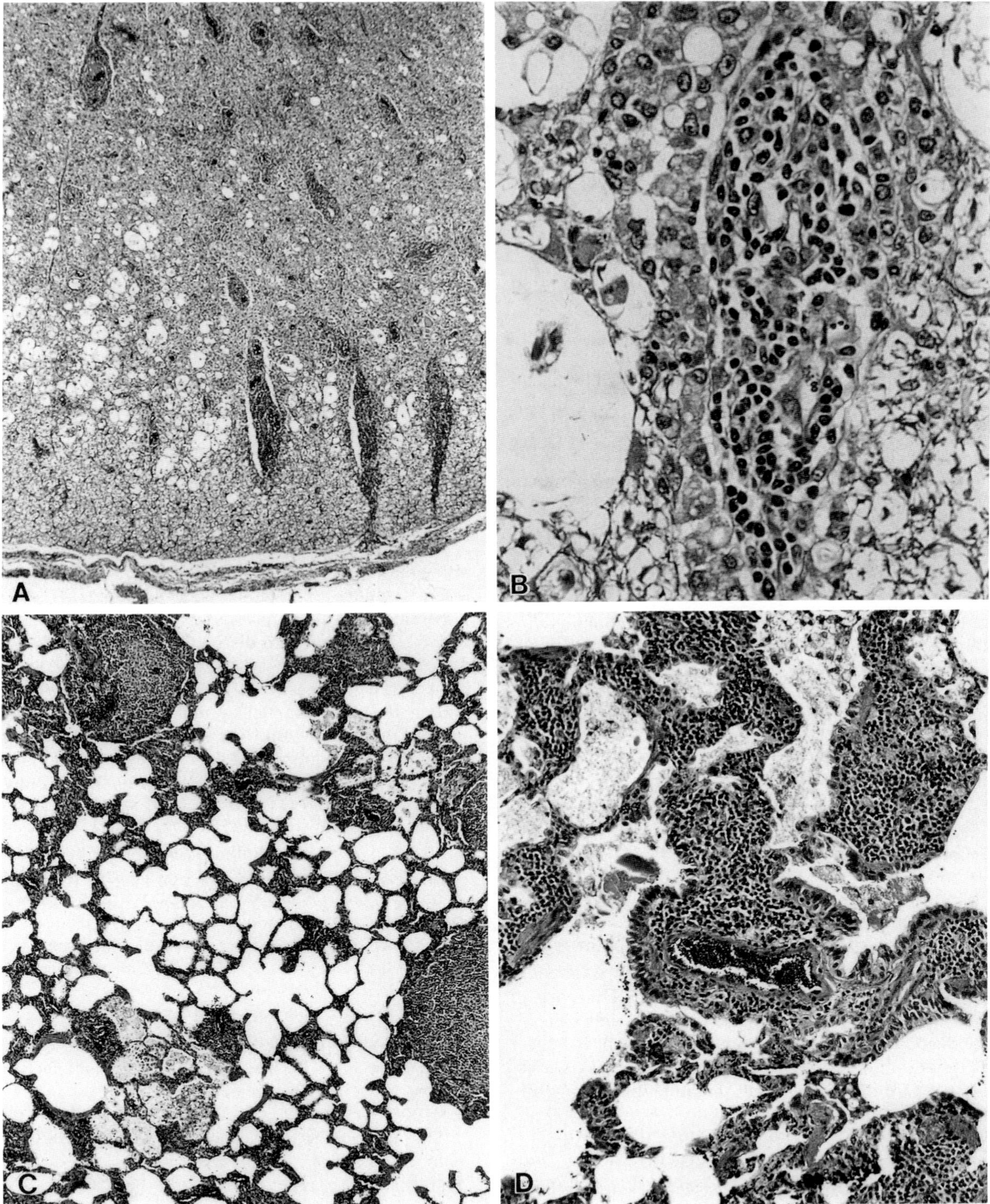

Fig. 3.107 Caprine arthritis–encephalitis. (A) Leukomyelitis. (B) Detail of perivascular reaction. Note macrophages bordering vessel. (C) Prominent perivascular and peribronchiolar lymphoid cuffs and focal interstitial pneumonia. (D) Detail of (C) showing lymphocytes around vessels and bronchioles, and in alveolar walls.

Narayan, O. Lentiviruses are etiological agents of chronic diseases in animals and acquired immunodeficiency syndrome in humans. *Can J Vet Res* **54:** 42–48, 1990.

Peterhans, E. *et al.* Lentiviren bei Schaf und Ziege: Eine Literaturubersicht. *Schweiz Arch Tierheilk* **130:** 681–700, 1988.

Zink, M. C., Yager, J. A., and Myers, J. D. Pathogenesis of caprine arthritis–encephalitis virus. Cellular localization of viral transcripts in tissues of infected goats. *Am J Pathol* **136:** 843–854, 1990.

11. Paramyxovirus Encephalomyelitis of Pigs

Outbreaks of this disease have been recorded in Mexico since 1980. The virus has the characters of the parainfluenza group, but its relationship to other members, in particular to Sendai and the hemagglutinating virus of Japan which is reported to cause pneumonia and encephalitis in pigs, is not reported.

Outbreaks of the disease appear to be self-limiting in commercial pigs. In pregnant sows, the infection may be subclinical or responsible for fetal death, mummification, and stillbirths and for the occasional appearance of corneal opacity in the sow. Piglets up to 2 weeks of age are most susceptible with up to 50% morbidity and very high fatality. The clinical signs are of encephalomyelitis leading to death within 2 to 4 days, although subclinical infections are also frequent and may be manifested only by corneal opacity.

The lesion is a typical nonsuppurative encephalomyelitis affecting mainly gray matter of thalamus, midbrain, and cortex. Inclusion bodies have not been demonstrated. Anterior uveitis is mild, the inflammatory cells congregating in the iridocorneal angle and the corneoscleral junction. The corneal opacity is due to edema.

Bibliography

Stephan, H. A., Gay, G. M., and Ramirez, T. C. Encephalomyelitis, reproductive failure and corneal opacity (blue eye) in pigs, associated with a paramyxovirus infection. *Vet Rec* **122:** 6–10, 1988.

12. Iriki Virus Encephalitis

A number of viruses belonging to the Simbu subgroup of *Bunyavirus* can infect cattle but, apart from Akabane, they are not known as significant pathogens. Iriki virus, which is serologically related to Akabane, has caused epizootic encephalitis in calves in Japan. The lesions have all the attributes of nonsuppurative encephalomyelitis and are most severe in the midbrain, medulla oblongata, and ventral horns of the spinal cord.

Bibliography

Miyazato S. *et al.* Encephalitis of cattle caused by Iriki isolate, a new strain belonging to Akabane Virus. *Jpn J Vet Sci* **51:** 128–136, 1989.

13. Bovine Herpesvirus Encephalitis

A bovine herpesvirus encephalitis was first recognized in Australia in 1962. It was subsequently identified in North America, Hungary, Italy, Scotland, Argentina, and Brazil. On the basis of serum neutralization and immunofluorescence assays, it was presumed that the herpesvirus involved was bovine herpesvirus-1 (BHV-1), the infectious bovine rhinotracheitis virus, but it has been found that the herpesviruses from bovine encephalitis in Australia and several other outbreaks have restriction endonuclease DNA fingerprints substantially different from those of BHV-1. It is believed that the encephalitis-producing herpesvirus (bovine herpesvirus-5) is a distinct virus and that bovine herpesvirus encephalitis is unrelated to BHV-1 infection. This distinction has not been proven in all cases, but no genomically identified isolate of BHV-1 has been made from cases of bovine encephalitis.

Bovine herpesvirus encephalitis occurs as a sporadic disease or as outbreaks in young animals—calves and yearlings. The morbidity in herds may be as high as 50%, but is usually much lower; few recognizably sick animals survive. The affected animals are febrile, develop labored breathing and nasal and ocular discharges. Neurologic signs consist of incoordination, blindness, and muscular tremors. Death usually occurs 4–5 days after signs become obvious. Affected animals waste rapidly during the course of the disease, but no specific gross lesions are present.

Bovine herpesvirus encephalitis involves both gray and white matter with widespread neuronal necrosis and malacia. A lymphocytic, histiocytic leptomeningitis and thick perivascular cuffs are present as well as focal and diffuse gliosis. The lesions are distinctive because of the severe cytonecrotic changes, which are particularly prominent in the cerebral hemispheres. The encephalitis gets its specificity from the intranuclear inclusion bodies which are present in nuclei of neurons and astrocytes. Electron microscopy shows these inclusion bodies to contain herpesvirus particles. The bovine herpesvirus-5 virus has also been isolated from aborted bovine fetuses.

Bibliography

Carrillo, B. J., Pospischil, A., and Dahme, E. Pathology of a bovine viral necrotizing encephalitis in Argentina. *Zbl Vet Med B* **30:** 161–168, 1983.

Schudel, A. A. *et al.* Infections of calves with antigenic variants of bovine herpesvirus-1 (BHV-1) and neurological disease. *J Vet Med B* **33:** 303–310, 1986.

Studdert, M. J. Bovine encephalitis herpesvirus. *Vet Rec* **125:** 584, 1989.

C. Protozoal Infections

The cerebral complications of infections by protozoa such as *Babesia, Theileria, Trypanosoma,* and *Toxoplasma* are discussed elsewhere in these volumes. *Acanthamoeba* can produce embolic meningoencephalitis of dogs as part of a generalized infection.

From time to time and in individual cases, pathologists observe sporozoan parasites in neural tissues of fetuses, neonates, and adults, and lesions presumed to be in consequence of their presence. There are, however, difficulties in specific identification of the parasites and in attribution of pathogenicity. The syndromes considered here are reasonably defined but are subject to revision as the parasites are identified and their life cycles, clarified.

1. Encephalitozoonosis

Encephalitozoon (*Nosema*) *cuniculi* is a microsporidian parasite capable of establishing infection in a wide variety of mammalian species and birds. It is rarely a zoonotic infection. Endemic infection is common in colonies of laboratory rodents in which the clinical consequences are mild, but the pathologic changes may confuse other studies. Among domestic species, the disease is of interest in carnivores, especially farmed foxes, in which serious mortalities occur, and occasionally in dogs and mink. The incidence of subclinical infection is not known, but a serological survey of an unselected population of stray dogs, among which clinical disease was not observed, identified 10–15% of seropositive animals.

The organism is an obligate intracellular parasite with a direct life cycle. It develops in parasitophorous vacuoles in cells of many tissues, especially endothelial cells, but is most easily found in brain and kidney in acute active infections. In chronic infections, the organisms can be sparse or impossible to find in microscopic sections, although it seems that animals once infected remain permanently so and excrete the organism mainly in urine.

Clinical disease occurs in dog and fox pups. In both hosts, transplacental infections appear to be important, but oral transmission, as by ingestion of infected rabbit carcasses, may occur. Experimental infection of mature dogs does not lead to clinical disease.

Tissue changes in encephalitozoonosis are most prominent in brain and kidney, but the organism selectively parasitizes vascular endothelium, and the segmental vasculitis which results is responsible for lesions in many tissues. Gross lesions may be limited to the kidneys as severe, nonsuppurative interstitial nephritis (Fig. 3.108A). Organisms are abundant in sections of kidney early in the disease, but at later stages they are difficult to find. They are especially numerous in the epithelium and lumen of tubules, in glomerular capillaries, and in the interstitium, and are present in small vessels and in the media and adventitia of intrarenal arteries. Fibrinoid necrosis affects some glomeruli, and the arterial lesions resemble those of periarteritis nodosa. Focal hepatic necrosis and nonsuppurative portal infiltrations are associated with organisms in hepatocytes and Kupffer's cells and with nodular vasculitis in the triads. Focal myocardial necrosis and inflammation is frequently associated with vasculitis.

The lesions in the nervous system are those of a widespread nonsuppurative meningoencephalomyelitis. The severity of lesions varies unpredictably in different parts of the nervous system, reflecting the random localization of the organism and the irregular distribution of inflamma-

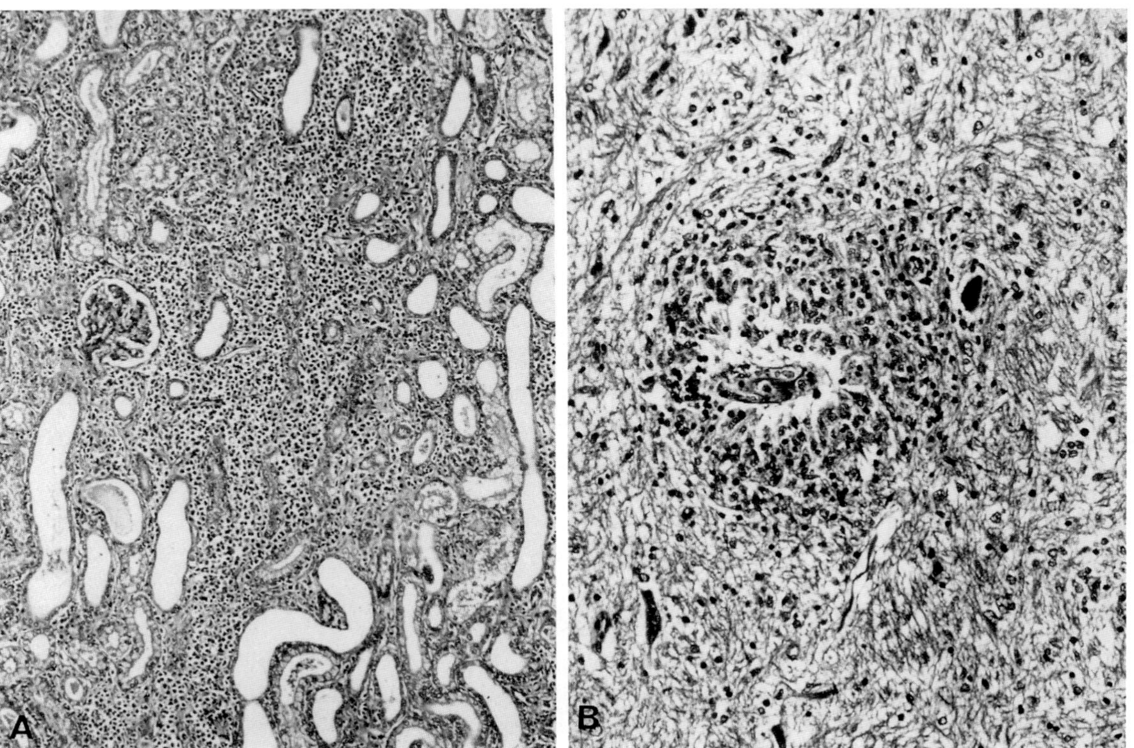

Fig. 3.108 (A) Diffuse interstitial nephritis in encephalitozoonosis. Dog. (B) Focal granulomatous encephalitis. *Encephalitozoon* spp.

tory vascular change. Focal gliosis and microscopic granulomas surround small vessels (Fig. 3.108B). About larger vessels showing segmental fibrinoid change, mononuclear cells form cuffs involving the adventitia and perivascular space and eventually assume an epithelioid cell appearance. There is astrocytosis in the surrounding parenchyma. The vascular lesions in the meninges in the acute disease resemble those of polyarteritis nodosa and become dominated by sclerotic changes in the chronic disease in which perivascular cuffing and granulomatous reactions persist.

Puppies which survive the early clinical disease may remain stunted and develop progressive renal disease. It is possible that encephalitozoonosis, as for any sporadic disease, is underdiagnosed, especially in chronic infections in which the organism is difficult to demonstrate. Immunohistochemical methods help to identify sparse organisms and to distinguish them from similar parasites, particularly *Toxoplasma* and *Neospora*.

2. Equine Protozoal Encephalomyelitis

The disease is known only from natural infections diagnosed postmortem in the Americas. Most frequently, young adult horses are affected by progressive asymmetric neurologic signs mainly referable to the spinal cord but in some cases including cranial nerve pareses or atrophy of muscle groups. Lesions are confined to the central nervous system and are more extensive in the cord. Their specificity depends on histologic demonstration of the organism, which is classified with *Sarcocystis*.

There are multiple foci of necrotizing, nonsuppurative inflammation affecting both gray and white matter, the large foci visible grossly. Cuffs form of lymphocytes and plasma cells and the tissue infiltrates include neutrophils, eosinophils, and occasional giant cells . The organism is found in the cytoplasm of neurons, mononuclear cells, and vascular endothelium. The individual organism is banana-shaped, resembling the motile stages of coccidia, and the multinucleate stages resemble schizonts.

3. Neosporosis

Neospora caninum is a coccidian-type protozoan of demonstrated pathogenicity in dogs, cats, rats, mice, sheep, horses, and cattle and of very wide distribution. Neither the life cycle nor the source of infection is known. Transplacental infection does occur and can occur repetitively in dogs and would account for the neonatal and juvenile disease, but it is not known whether disease in adults represents new or activated latent infection.

Neospora caninum cannot be reliably distinguished from *Toxoplasma gondii* in histologic sections of pathological material: the manifestations of infection, whether in fetuses, neonates, or adults, are basically the same; and many pathologic diagnoses of toxoplasmosis are, retrospectively, infections by *Neospora*.

The disease is as protean as toxoplasmosis in its manifestations and the species-by-species occurrence seems also to be approximately the same, but species susceptibility will need to be further examined, especially of those such as pigs and sheep in which fetal and neonatal death have been seen as important expressions of toxoplasmosis. The distribution of *Neospora* infection on serological evidence is much less than that of *Toxoplasma* in dogs.

The necrotizing inflammatory lesions can be widely disseminated in organs and tissues. In acute generalized infections, widespread injury can occur in all parenchymatous organs and in lymph nodes. In neonates and juveniles, the clinical signs may refer in particular to asymmetric lesions of the spinal cord and, in dogs, to myositis and polyradiculitis. The necrotizing lesions are usually fatal; in less severe disease, the lesions become more granulomatous in patterns appropriate to the tissue and may then be confined to muscle, the peripheral and central nervous systems, and the eyes.

The specificity of the lesions of *Neospora* infection depend on demonstration of the organism in tissue, on immunochemical reactions, and on ultrastructure of the organism. It is intracellular in neurons and other cells, multiplies by endodyogeny, has no parasitophorous vacuole, and has numerous rhoptries. Apparently, cysts form only in neural tissue. The cyst wall at 1 to 4 μm is thicker than that of *Toxoplasma* and encloses slender PAS-positive bradyzoites. The cysts may be up to 40 μm in diameter.

Bibliography

Bjerkas, I. Brain and spinal cord lesions in encephalitozoonosis in blue foxes. *Acta Pathol Microbiol Immunol Scand* **95:** 269–279, 1987.

Bjerkas, I., and Nesland, J. M. Brain and spinal cord lesions in encephalitozoonosis in the blue fox. *Acta Vet Scand* **28:** 15–22, 1987.

Botha, W. S., Ven Dellen, A. F., and Stewart, C. G. Canine encephalitozoonosis in South Africa. *J S Afr Vet Assoc* **50:** 135–144, 1979.

Boy, M. G., Galligan, D. T., and Divers, T. J. Protozoal encephalomyelitis in horses: 82 cases (1972–1986). *J Am Vet Med Assoc* **196:** 632–634, 1990.

Cummings, J. F. *et al.* Canine protozoan polyradiculoneuritis. *Acta Neuropathol* **76:** 46–54, 1988. (*Neospora*)

Davis, S. W., Daft, B. N., and Dubey, J. P. *Sarcocystis neurona* cultured *in vitro* from a horse with equine protozoal myelitis. *Equine Vet J* **23:** 315–317, 1991.

Dubey, J. P., and Lindsay, D. S. Neosporosis in dogs. *Vet Parasitol* **36:** 147–151, 1990.

Dubey, J. P., and Porterfield, M. L. *Neospora caninum* (Apicomplexa) in an aborted equine fetus. *J Parasitol* **76:** 732–734, 1990.

Dubey, J. P. *et al.* Newly recognised fatal protozoan disease of dogs. *J Am Vet Med Assoc* **192:** 1269–1285, 1988. (*Neospora*)

Dubey, J. P. *et al.* Repeated transplacental transmission of *Neospora caninum* in dogs. *J Am Vet Med Assoc* **197:** 857–860, 1990.

Dubey, J. P. *et al. Neospora caninum* encephalomyelitis in a British dog. *Vet Rec* **126:** 193–194, 1990.

Dubey, J. P., Lindsay, D. S., and Lipscomb, T. P. Neosporosis in cats. *Vet Pathol* **27:** 335–339, 1990.

Fayer, R. *et al.* Epidemiology of equine protozoal myeloencephalitis in North America based on histologically confirmed cases. *J Vet Int Med* **4:** 54–57, 1990.

Hollister, W. S. *et al.* Prevalence of antibodies to *Encephalitozoon cuniculi* in stray dogs as determined by ELISA. *Vet Rec* **124:** 332–336, 1989.

McCully, R. M. *et al.* Observations on the pathology of canine microsporidiosis. *Onderstepoort J Vet Res* **45:** 75–92, 1978.

O'Toole, D., and Jeffrey, M. Congenital sporozoan encephalomyelitis in a calf. *Vet Rec* **121:** 563–566, 1987.

Parish, S. M. *et al.* Myelitis associated with protozoal infection in newborn calves. *J Am Vet Med Assoc* **191:** 1599–1600, 1987.

Pearce, J. R. *et al.* Amebic meningoencephalitis caused by *Acanthamoeba castellani* in a dog. *J Am Vet Med Assoc* **87:** 951–952, 1985.

Szabo, J. R., and Shadduck, J. A. Experimental encephalitozoonosis in neonatal dogs. *Vet Pathol* **24:** 99–108, 1987.

Szabo, J. R., and Shadduck, J. A. Immunologic and clinicopathologic evaluation of adult dogs inoculated with *Encephalitozoon cuniculi*. *J Clin Microbiol* **26:** 557–563, 1988.

Tiner, J. D. Birefringent spores differentiate *Encephalitozoon* and other microsporidia from coccidia. *Vet Pathol* **25:** 227–230, 1988.

Van Dellen, A. F. Light and electron microscopical studies on canine encephalitozoonosis: Cerebral vasculitis. *Onderstepoort J Vet Res* **45:** 165–186, 1978.

D. Parasitic Infestations

Nothing is known of what motivates and directs the migration of larval parasites. Those which migrate somatically are apt to go astray, and this appears especially likely when they wander in an alien host. Aberrant pathways include the nervous system with such frequency as to suggest that nematodes have a special propensity for wandering in the central nervous system. Whether this is indeed the case still remains to be proven, but parasitic migrations in nervous tissue are more likely to be symptomatic than aberrant migrations in other tissues, and there is an impressive list of parasites which, often or occasionally, have been found in brain or cord. Many of these infestations and the parasites in question are discussed elsewhere in these volumes.

Coenurus cerebralis is fairly common in the brains of sheep in Europe, less common in other herbivores, and rather rare in horses and humans. About 40% of pigs harboring *Cysticercus cellulosae* have cysts in the meninges and brain as well as in the muscle, and the same species has been identified in the brains of dogs; possibly *Cysticercus bovis* and other cysticerci will invade nervous tissue with comparable frequency. Apparently, hydatids are seldom found in brain.

Angiostrongylus cantonensis is a metastrongylid lungworm whose only known definitive host is the rat. It is widely distributed in the warm Pacific regions, but its distribution is much more limited than that of the gastropod intermediate hosts and the rat.

The parasite resides in the pulmonary arteries of the rats, the eggs lodge as emboli in alveolar capillaries, and the larvae, which hatch in about 6 days, follow the tracheal–intestinal route to the exterior. The first-stage larvae actively penetrate terrestrial and aquatic slugs and snails, which act as intermediate hosts. Transport hosts for third-stage larvae include frogs, crabs, and prawns. In addition to the rat, dogs, humans, and occasionally other species are infected by eating intermediate or transport hosts and, possibly, directly by ingestion of infective larvae which have emerged from intermediate hosts.

Ingested larvae enter and are dispersed by the circulation to many tissues but predominantly to brain, kidney, and muscle. Molting larvae in the brain produce a mild to severe inflammatory reaction before reentering the venous circulation for return to the pulmonary arteries.

Aberrant infections are important in humans and dogs and are reported in horses and macropods. The human disease, eosinophilic meningoencephalitis, is usually mild and without sequelae, but infection in dogs can be accompanied by ascending paralysis. Those larvae which enter the brain in dogs are probably inhibited in their development and destroyed there (Fig. 3.109A).

The lesions are granulomatous, randomly distributed in the cord and brain, and are most frequent and severe in the cord (Fig. 3.109B). Rarely, degenerate parasites are present in the granulomas, but apparently viable worms in the tissue are not accompanied by an inflammatory reaction. Eosinophils infiltrate the granulomas but are more numerous in affected meninges.

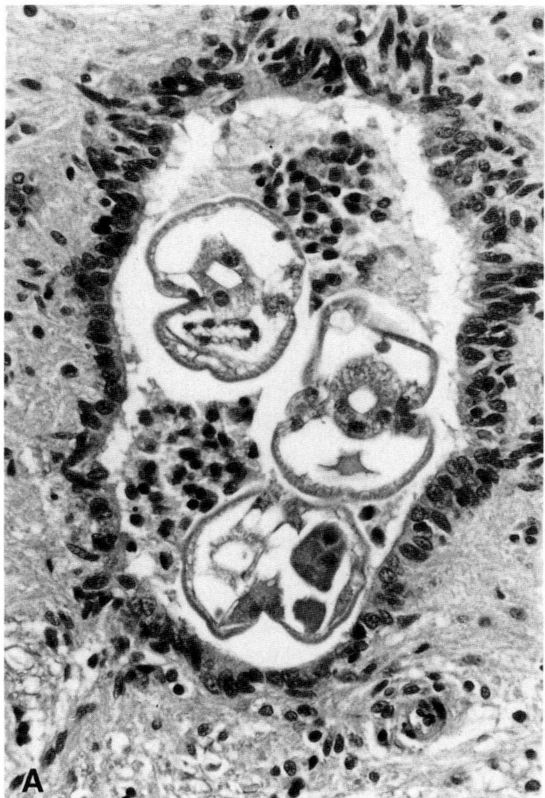

Fig. 3.109A Larvae of *Angiostrongylus* sp. in central canal of spinal cord. Dog.

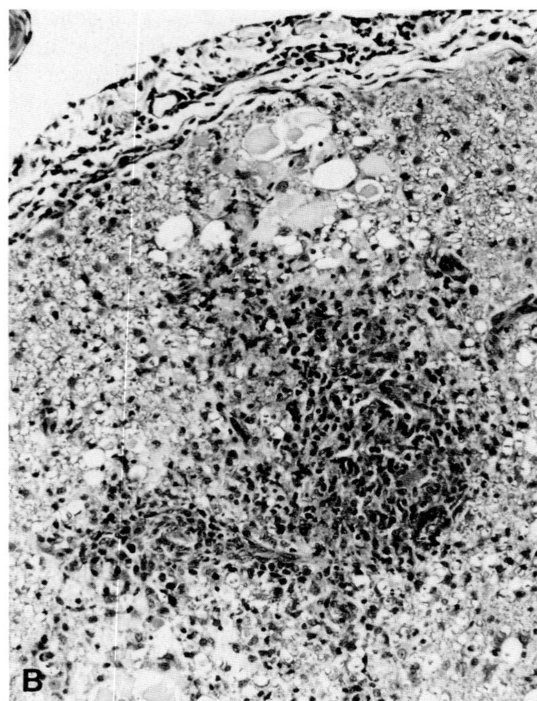

Fig. 3.109B Granuloma formation, Wallerian degeneration and spinal meningitis. *Angiostrongylus*. Dog.

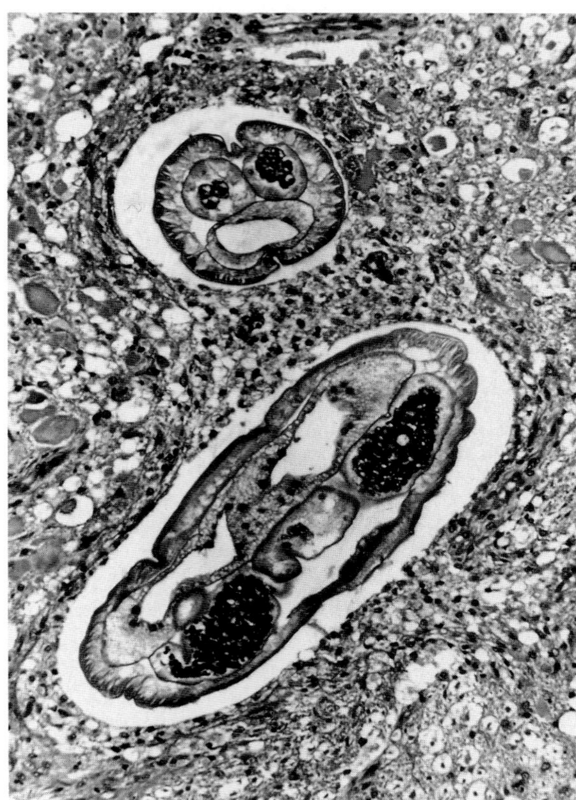

Fig. 3.110 *Pneumostrongylus* in spinal cord of sheep.

Pneumostrongylus (*Parelaphostrongylus*) *tenuis* is a metastrongylid parasite of the white-tailed deer, *Odocoileus virginianus,* in North America. The intermediate hosts are terrestrial slugs and snails. Ingested larvae reach the spinal cord of the deer in ~10 days, but their route of migration is not known. They develop for up to 1 month in the dorsal horns of the cord at all levels and then migrate into the meningeal spaces. Some penetrate the dural veins and sinuses and mature. Eggs or larvae are carried in venous blood to the lungs. The larvae do very little to the cord in white-tailed deer, but the reaction is more severe in other species, including the sheep (Fig. 3.110). *Elaphostrongylus panticola* and *E. rangifera* of deer in Northern Europe and Russia have a life cycle similar to that of *P. tenuis,* but aberrant infections have not been reported.

Setaria digitata (see Peritoneum) is a parasite of cattle. Microfilariae can be carried to horses, sheep, and goats by mosquitoes, and immature worms wandering in the brain and spinal cord are responsible for the disease of horses known as kumri. *Setaria marshalli* is reported to affect sheep in Russia. *Halicephalobus* (*Micronema*) *deletrix* is a rhabditiform nematode which is accidentally, but rarely, a parasite. Massive intracranial invasion is reported in horses. The syndrome is acute and of short duration. There are focal arachnoid hemorrhages and patchy meningeal thickenings. All stages of the parasitic life cycle may be found among the specimens in the brain, most easily in perivascular spaces (Fig. 3.111). Parasitic granulomas in the kidney may accompany the cerebral invasion.

Gurltia paralysans, found in the spinal veins of cats, is reported as being responsible for a high incidence of paralysis in this host, and *Angiostrongylus vasorum* has caused hemorrhagic malacia in the brains of dogs, but there is some doubt whether these worms are in their proper hosts.

Larval worms may also migrate aberrantly in the central nervous systems of their natural hosts. *Stephanurus dentatus* quite frequently invades the spinal canal and may even encyst in the meninges in pigs. *Strongylus* spp. occasionally invade the brains of horses; Fig. 3.112 shows the types of lesions that occur, and, although not identified in this case, the larvae were probably of *S. vulgaris,* because these were identified in thrombi in the aortic bulb and carotid artery. Ascarids have a propensity for wandering in the brains of alien hosts and occasionally do so in their natural hosts.

Trematodes apparently have little tendency to invade nervous tissue. *Troglotrema acutum* may invade the brain from its normal habitat in the paranasal sinuses. The eggs of the lung flukes, *Paragonimus* spp., have been observed in the brains of dogs, possibly arriving there as emboli. The only larval arthropods of interest are *Hypoderma bovis,* which normally migrates through the spinal canal; *Oestrus ovis,* which may invade the brain from the nasal sinuses; and *Cuterebra* in the dog and cat.

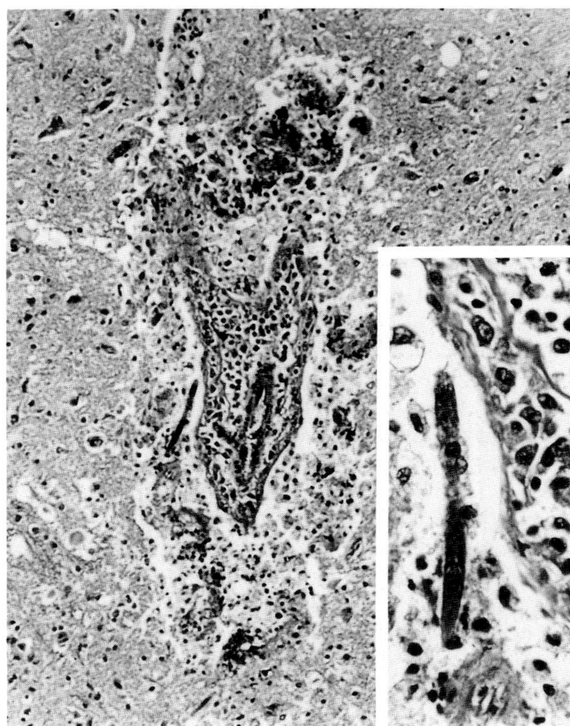

Fig. 3.111 Perivascular inflammation in brain. Horse. *Halicephalobus (Micronema) deletrix.* (Inset) Parasite with cellular reaction.

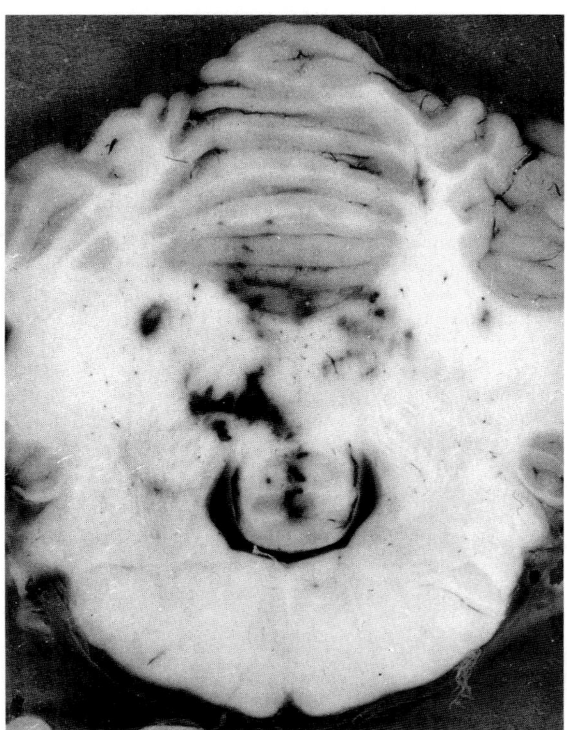

Fig. 3.112 Hemorrhagic tracks in cerebellar white matter probably produced by *Strongylus vulgaris.* Horse.

The few parasites specifically mentioned are the most important in terms of neuropathology. Occasionally helminth larvae are discovered accidentally but rarely identified in sections of brain or cord, and it is somewhat more common to find lesions typical of those produced by migratory parasites without being able to locate the parasite. Some parasites, such as *Elaphostrongylus,* usually remain in the central nervous system, whereas others, such as ascarids and strongyles, can be expected to keep moving. Finding the parasite is, therefore, largely a matter of luck, even when it is sought very early after the onset of clinical signs.

The lesions produced in nervous tissue by migratory larvae are mainly malacic and fairly distinctive in their pattern. *Coenurus cerebralis* produces, in the invasion phase, a purulent meningoencephalitis and later acts as a space-occupying lesion, but other invading parasites produce mainly traumatic lesions with very little reactionary inflammation except for a few eosinophils. The lesions produced by nematodes are sometimes grossly visible as hemorrhagic foci or narrow, slightly tortuous tracks. Brownish, hemorrhagic discoloration depends on the parasite hitting a vein or arteriole, and it appears that some worms have a tendency to migrate along veins. There may be only one or several such tracks in the central nervous system, and they occur quite at random. Microscopically, the lesion is an irregular focus or pathway of traumatic malacia into which some hemorrhage may have occurred. There may be slight cellular infiltration in the adjacent meninges or nerve roots. The track is liquefied, and its margins not sharp, and apart from lymphocytes, gitter cells, and a few eosinophils, there is no significant reaction in the damaged tissue or in the adjacent vessels. The disruption, which is not selective in the tissues destroyed, leads to microcavitation. The disrupted axons, swollen, tortuous, and as globose fragments, persist for some time in the microcavitations (Fig. 3.109B).

Bibliography

Adcock, J. L., and Hibler, C. P. Vascular and neurophthalmic pathology of elaeophorosis in elk. *Vet Pathol* **6:** 185–213, 1969.

Alicata, J. E. *Angiostrongylus cantonensis* (Nematoda: Metastrongylidae) as a causative agent of eosinophilic meningoencephalitis of man in Hawaii and Tahiti. *Can J Zool* **40:** 5–8, 1962.

Anderson, R. C. The development of *Pneumostrongylus tenuis* in the central nervous system of the white-tailed deer. *Vet Pathol* **2:** 360–379, 1965.

Basson, P. A. *et al.* Meningeal setariosis: Report on two cases in antelopes. *J S Afr Vet Med Assoc* **37:** 249–253, 1966.

Baumann, R., and Bohm, L. K. Cysticercose des Lendenmarkes beim Hunde. *Wien Tierarztl Mschr* **28:** 7–13, 1941.

Beaver, P. C. Wandering nematodes as a cause of debility and disease. *Am J Trop Med* **6:** 433–440, 1957.

Bohm, L. K., and Supperer, R. Untersuchungen uber *Setariem* (Nematoda) bei Heimischen Wiederkauern und deren Beziehung zur "epizootischen cerebrospinalen Nematodiasis" (Setariosis). *Z Parasitenk* **17:** 165–174, 1955.

Cooley, A. J., Clemmons, R. M., and Gross, T. L. Heartworm disease manifested by encephalomyelitis and myositis in a dog. *J Am Vet Med Assoc* **190:** 431–432, 1987.

Darien, B. J., Belknap, J., and Nietfeld, J. Cerebrospinal fluid changes in two horses with central nervous system nematodiasis (*Micronema deletrix*). *J Vet Int Med* **2:** 201–205, 1988.

Innes, J. R. M., and Pillai, C. P. Kumri, so-called lumbar paralysis of horses in Ceylon (India and Burma) and its identification with cerebrospinal nematodiases. *Br Vet J* **111:** 223–235, 1955.

Jortner, B. S. *et al.* Lesions of spinal cord parelaphostrongylosis in sheep. Sequential changes following intramedullary larval migration. *Vet Pathol* **22:** 137–140, 1985.

Kadenatsii, A. N. *Setaria marshalli* infection in sheep. *Trud Omsk Vet Inst* **15:** 137–141, 1957.

Kennedy, P. C., Whitlock, J. H., and Roberts, S. J. Neurofilariosis, a paralytic disease of sheep. *Cornell Vet* **42:** 118–124, 1952.

Krishnamurty, D. *Cysticercus cellulosae,* their incidence in canines. *Indian Vet J* **25:** 367–370, 1949.

Kurtz, H. J., Loken, K., and Schlotthauer, J. C. Histopathologic studies on cerebrospinal nematodiasis of moose in Minnesota naturally infected with *Pneumostrongylus tenuis*. *Am J Vet Res* **27:** 548–557, 1966.

Lehmensick, R. Ueber die Veranderungen am Iltisschadel durch den Befall mit *Troglotrema acutum*. *Z Parasitenk* **12:** 659, 664, 1942.

Little, P. B. Cerebrospinal nematodiasis in Equidae. *J Am Vet Med Assoc* **160:** 1407–1413, 1972.

Mackerras, M. J., and Sanders, D. F. The life history of the rat lung-worm *Angiostrongylus cantonensis*. *Aust J Zool* **3:** 1–21, 1955.

Mason, K. V. Canine neural angiostrongylosis: The clinical and therapeutic features of 55 natural cases. *Aust Vet J* **64:** 201–203, 1987.

Mason, K. V. *et al.* Granulomatous encephalomyelitis of puppies due to *Angiostrongylus cantonensis*. *Aust Vet J* **52:** 295, 1976.

Mohiyudden, S. Enzootic bovine paraplegia in some Malnad tracts (hilly and heavy rainfall region) of Mysore State with particular reference to cerebrospinal nematodiasis as its probable cause. *Indian J Vet Sci* **26:** 1–20, 1956.

Nettles, V. F., and Prestwood, A. K. Experimental *Parelaphostrongylus andersoni* infections in white-tailed deer. *Vet Pathol* **13:** 381–393, 1976.

Sartin, E. A. *et al.* Cerebral cuterebrosis in a dog. *J Am Vet Med Assoc* **189:** 1338–1339, 1986.

Shoho, C. Helminthic diseases of the central nervous system and intrauterine fetal infections by helminths. *Rev Iber Parasit,* Special volume, 927–951, 1955.

Sprent, J. F. A. On the invasion of the central nervous system by nematodes. I. The incidence and pathological significance of nematodes in the central nervous system. II. Invasion of the central nervous system in ascariasis. *Parasitology* **45:** 31–40, 41–55, 1955.

Thomas, J. S. Encephalomyelitis in a dog caused by *Baylisascaris* infection. *Vet Pathol* **25:** 94–95, 1988.

Wright, J. D. *et al.* Equine neural angiostrongylosis. *Aust Vet J* **68:** 58–60, 1991.

E. Miscellaneous Inflammatory Diseases

1. Sporadic Bovine Encephalomyelitis

Sporadic bovine encephalomyelitis is caused by organisms of the genus *Chlamydia*. The disease occurs in the United States, Japan, Czechoslovakia, and Australia and, although nothing is known of the epidemiology, the disease probably has a worldwide distribution. Cattle only are affected, and calves younger than 6 months are most susceptible. Other domestic species are resistant to the infection, and of the laboratory species, only the guinea pig and hamster have been shown to be usefully susceptible. As a rule, sporadic bovine encephalomyelitis is indeed sporadic, affecting only a few animals in herd. Occasionally as many as 50% are affected, and ~50% of these die. In recovered cases there are no neurologic sequelae. The clinical syndrome is not particularly characteristic. It is composed of moderate fever and signs of catarrhal inflammation of the respiratory tract. There is some stiffness, weakness of the hind limbs with staggering and knuckling of the fetlocks, and muscle tremors. There is some dullness; signs of excitement are not present. Death occurs in a few days to a few weeks.

The gross morbid change which suggests a diagnosis of sporadic bovine encephalomyelitis is serofibrinous inflammation of the serous membranes and synoviae. This is most consistently a peritonitis, and in ~50% of fatal cases, there is also pleuritis and pericarditis. The meninges appear congested and edematous. Microscopically, there is a rather severe and diffuse meningoencephalomyelitis (Fig. 3.48B). The meningitis is most severe about the base of the brain. The reactive cells are almost solely mononuclears of histiocytic and plasma types, with only a few neutrophils. These cells infiltrate the meninges and perivascular spaces and mix with reactive adventitial cells of the vessel walls. The vascular endothelium proliferates secondary to lesions in the vascular walls and, as a result of these, ischemic changes may occur in the parenchyma. Reactive microglial nodules are widespread in the brain, and many of these probably develop about obliterated terminal arterioles.

The diagnosis of sporadic bovine encephalomyelitis depends on demonstration of the causative agent. Elementary bodies produced by this organism can be found by appropriate staining techniques in the cytoplasm of mononuclear cells in the exudates in the meninges and from serosal membranes and in microglia of nodules. These elementary bodies are not numerous and may be easier to find in infected eggs and guinea pigs.

Bibliography

Menges, R. W., Harshfield, G. S., and Wenner, H. A. Sporadic bovine encephalomyelitis. Studies on pathogenesis and etiology of the disease. *J Am Vet Med Assoc* **122:** 249–294, 1953.

Omori, T., Ishii, S., and Matumoto, M. Miyagawenellosis of cattle in Japan. *Am J Vet Res* **21:** 564–573, 1960.

Wenner, H. A. *et al.* Sporadic bovine encephalomyelitis. 2. Studies on etiology. Isolation of nine strains of an infectious agent from naturally infected cattle. *Am J Hyg* **57:** 15–29, 1953.

2. "Old Dog" Encephalitis

"Old dog" encephalitis is rather rare. Most cases occur in dogs past middle age, but it has, however, been observed in dogs as young as 1 year. The disease is of insidious onset and is characterized by circling, swaying, and weaving. Compulsive walking with pushing against fixed objects is typical, but there is neither paralysis nor convulsions. The disease progresses over 3 or 4 months to coma or termination.

The cause of the disease is unknown but is reasonably suspected of being associated with infection by the virus of canine distemper or related agent. Old dog encephalitis is not simply a progression of the encephalomyelitis of canine distemper. Virus has not been isolated from affected animals except by explantation of affected brain and then only with difficulty, and the disease is not transmissible by direct inoculation. Inclusion bodies are readily found in some cases, and their structure is identical with paramyxovirus nucleocapsids of canine distemper virus in nervous tissue. Antigen which responds to fluorescent antibody prepared against canine distemper virus is abundant in cells of the gray matter, and serum antibody titers can be very high.

The lesions are confined to the brain, which appears slightly reduced in size. The ventricles are moderately dilated. The reaction is nonsuppurative, qualitatively always the same, but varying in degree. The most obvious change is cuffing, and the cuffs are remarkable for their large size and the purity of the lymphocytic populations in them. Plasma cells are present in small numbers. The infiltrating cells are confined to the Virchow–Robin space and seldom spread into the parenchyma. The large cuffs occur in both gray and white matter but are most common at the junction of these two zones. Focal gliosis does not occur in this disease, but there is some proliferation of astrocytes about vessels and neurons. There is uniform and rather diffuse atrophic sclerosis of the cerebral white matter, which gives an impression of gliosis, but astrogliosis is not prominent. There is some demyelination producing typical punched-out areas in the white matter, and distorted myelin sheaths in the heavily myelinated tracts are quite extensive when specially stained. Lymphocytes may be found in the choroid plexuses where they are inserted into the brain, and about vessels where they enter the parenchyma.

Nerve cells, especially in Ammon's horn and the pons, reveal chromatolysis with only a few remnants of Nissl substance in the periphery of the cytoplasm. The chromatolytic cytoplasm is slightly acidophilic. The neuronal nuclei in the forebrain are remarkably swollen in most of the altered nerve cells. In occasional nuclei, there is pinkish inclusionlike material. Neuronophagia does not occur. The astrocytic nuclei are remarkably swollen, have an irregular outline, and may contain traces of pinkish deposit in the nucleoplasm. In a proportion of cases, possibly those of longest duration, prominent intranuclear and cytoplasmic eosinophilic inclusion bodies may easily be found.

Virus which has been isolated by the explantation technique has been identified as canine distemper virus, albeit with some minor differences in polypeptide composition. These findings raise the question as to whether this persistent neuronal infection is caused by a mutant feral strain or is the result of mutation of an established infection which did not produce any of the stigmata of classical canine distemper. The role of immune processes in mediating the pathologic changes is also unclear.

3. Meningoencephalitis of Pug Dogs

This disease is reported only from the United States and in dogs of this one breed, but it is also known to occur in Canada, Australia, and New Zealand. Although the lesion is basically inflammatory with perhaps an overlay of changes which may reflect repetitive convulsions, the geographic and breed distribution is not consistent with an infectious cause, although this is probable. There is a wide age range. Generalized convulsions, and their aftermath, dominate the clinical picture, which may include lethargy, ataxia, and progression to coma. The clinical signs refer essentially to cortical disease, which progresses rapidly, that is over a few weeks, but which may extend to several months.

The lesions are particularly in the cerebral cortex, and are bilateral but asymmetric, often confluent over large areas, and extend to the adjacent white matter with relative sparing of the deeper periventricular tissue. The geography of the lesions therefore helps to distinguish this disease from other encephalitides of the dog. Grossly, localized swellings in the cerebrum contribute to asymmetry, and malacic foci may be seen as typical yellowish areas of softening (Fig. 3.113) or, in cases of longer duration, as tiny cystic cavities.

The histological changes are necrotizing and with an affinity for the hemispheres. Numerous foci of meningitis, characterized by infiltrations of lymphocytes, plasma cells, and monocytes (Fig. 3.114), diminish caudally and may be absent in the posterior fossa and spinal cord. These infiltrates breach the pial barrier and are destructive to

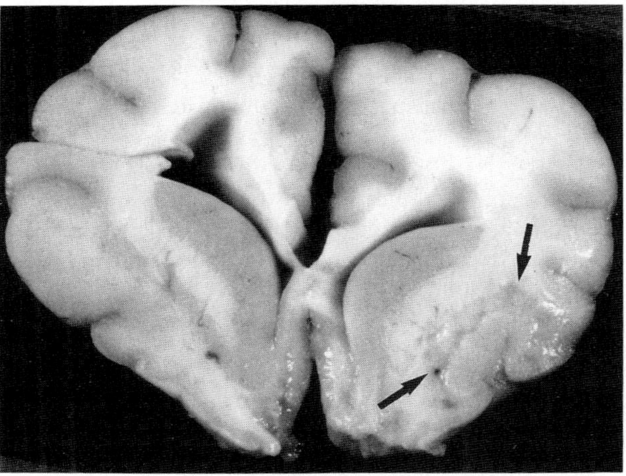

Fig. 3.113 Pug dog meningoencephalitis. Malacic focus in ventrolateral cerebral hemisphere (arrows). (Reprinted from Bradley, G. A. *Vet Pathol* **28:** 91, 1991.)

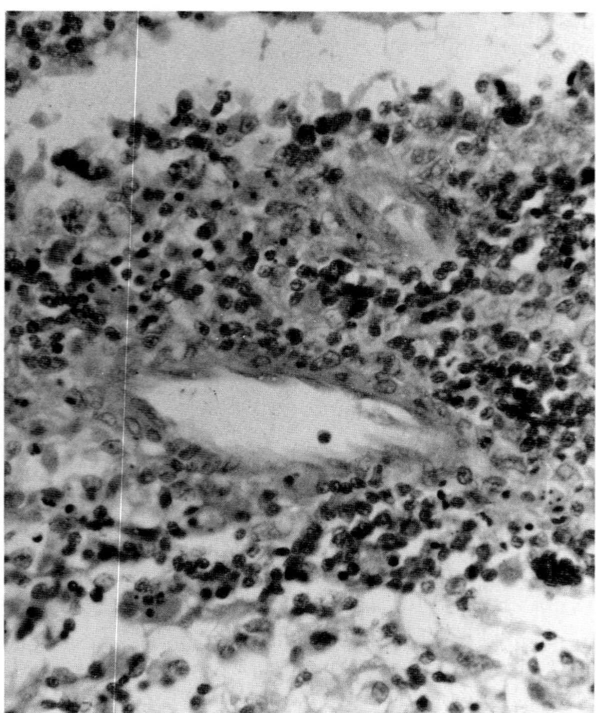

Fig. 3.114 Pug dog meningoencephalitis. Meningeal arteriole surrounded by mixed mononuclear infiltrate.

the superficial cortex to an extent and severity which is unusual. The evidence of cerebral necrosis extends from selective neuronal necrosis to areas of malacia, the latter especially in chronic cases. Vascular endothelium in the cortex is reactive and associated with edema, occasional petechiae, and diffuse accumulation of mononuclear cells in parenchyma and vascular cuffs.

The differential features of this meningoencephalitis, in addition to its nonsuppurative nature, are the malacic degenerations and predilection for the cerebral cortex. The cause is unknown, the disease is sporadic, and the evolution of the lesion is not described. The limitation of the disease to pug dogs may not be sustained.

4. Granulomatous Encephalitis

Granulomatous encephalitis of dogs is distinguishable from the usual range of granulomatous inflammations which may occur in disseminated infections and is separable from those conditions designated as reticuloses, which have histologic features of neoplasia. Diseases within this spectrum are not rare in dogs. They share certain features, including patchy distribution in the brain and cord, preferential involvement of white matter, expanding perivascular accumulations of cells without the other usual evidence of neuroparenchymal inflammation, and confinement to the central nervous system.

Variations in the distribution and extent of the lesions result in a variety of clinical signs. Spinal lesions may be associated with ataxia, paresis, or paralysis (Fig. 3.115A). Lesions in the brain stem frequently produce signs of vestibular dysfunction. Changes of behavior, forced movement and circling, depression, and convulsions occur with supratentorial lesions. Macroscopic lesions may not be evident, but in those cases in which the cellular aggregations become confluent, there may be irregular areas of malacia (Fig. 3.115B), or the volume of tissue clearly increased and asymmetrical.

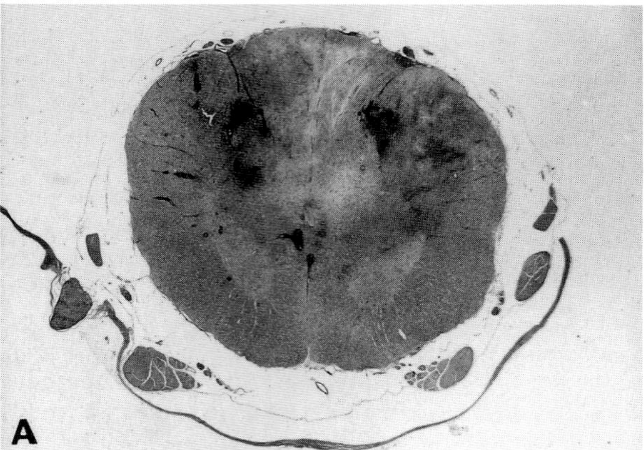

Fig. 3.115A Granulomatous encephalitis. Dog. Focal lesions in spinal cord. (Courtesy of J. B. Thomas.)

The minimal lesion is cuffing of vessels by lymphocytes and plasma cells with small eccentric clumps of macrophages. The macrophages increase in number and may come to compose the cuff, appearing as discrete granulomas which, depending on plane of section, may appear to be in the parenchyma. Occasional mitoses are present in these cells. Further transformation to epithelioid cells occurs later. Where the cuffing response is severe, edema and necrosis may occur in the adjacent white matter, leading to a spillover of mononuclear cells into the parenchyma and to the usual reactive changes. Large malacic foci are unusual. In cases of prolonged duration, confluence of lesions occurs, and reparative responses include the deposition of abundant reticulin and collagen oriented in perivascular arrangements.

The histologic changes are patchy in distribution. There may be very few foci, or they may be disseminated, or they may be localized to one area and confluent, or there may, in the same animal, be both confluent and disseminated distributions. The essential histologic feature is the perivascular aggregation of cells rather selectively in the white matter. Involvement of meninges is also patchy and often related to lesions of white matter directly underlying. The perivascular aggregations expand in concentric arrangements and displace surrounding parenchyma (Fig. 3.116A). The surrounding white matter becomes edematous, and gliosis develops. Macrophages remove disintegrating white substance. The cellular processes of aggregation and reaction merge.

It is in the cytological characterizations of the cell aggre-

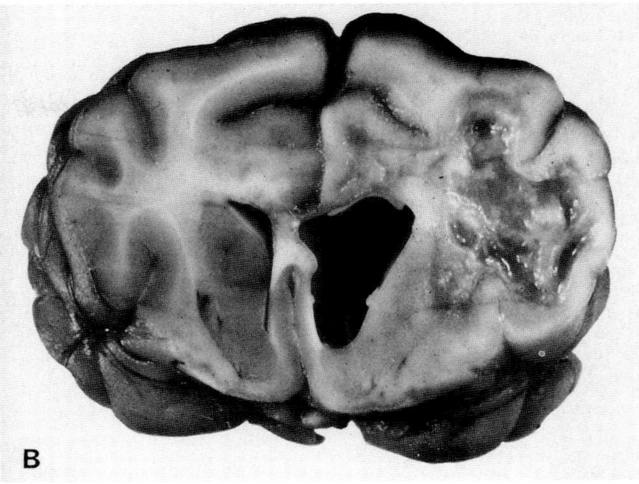

Fig. 3.115B Granulomatous encephalitis. Dog. Malacia of cerebrum and hydrocephalus secondary to infarction.

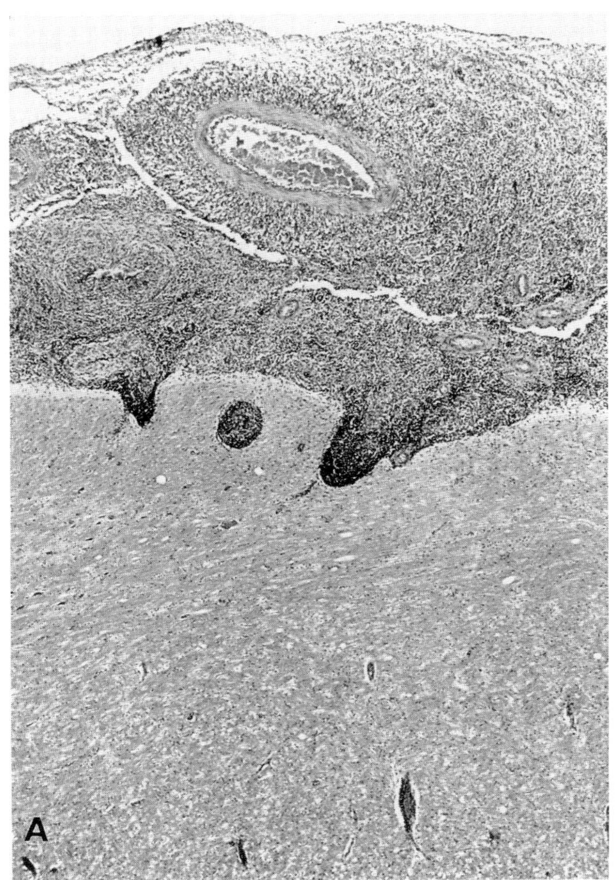

Fig. 3.116A Granulomatous encephalitis. Dog. Dense perivascular meningeal infiltrate.

gates that the difficulties arise. In many cases, and perhaps in the majority of cases, the cells are easy to classify, being well differentiated as lymphocytes, monocytes, plasma cells, and histiocytes (Fig. 3.116B). Granulocytes may be present but are not numerous, and histiocytes often show epithelioid transformation and may form small syncytial masses. In other cases there are present, in addition to the cells which can be readily identified, large immature cells of reticulohistiocytic type in which mitoses may be few or many, but the suggestion that neoplastic reticulosis is at one end of a spectrum of changes must await the application of new cytologic techniques.

5. *Postinfectious Encephalomyelitis*

Neurological disease which follows, after a variable period, common viral infections or vaccination exposure is well known in children as postinfectious or postvaccinal encephalitis. The common pathologic basis is a demyelinating inflammatory process dominated by mononuclear inflammatory cells with a distinctive perivenous distribution. Examples which meet the criteria have occurred in relation to rabies vaccination in dogs, when such vaccines were prepared in neural tissue, and the pathologic process is the same as that in experimental allergic encephalomyelitis. The histologic changes are widely distributed in the brain and cord and affect mainly the white matter. Perivascular infiltrates of lymphocytes, plasma cells, and monocytes/macrophages widely distend the space and spread into the surrounding parenchyma. Proliferation of adventitial cells may be prominent. Although much descriptive emphasis has been given to demyelination, this is restricted to the perivascular areas of infiltration and to surrounding areas showing the usual degenerative and reactive changes. The lesions are not readily distinguished in dogs from granulomatous encephalitis.

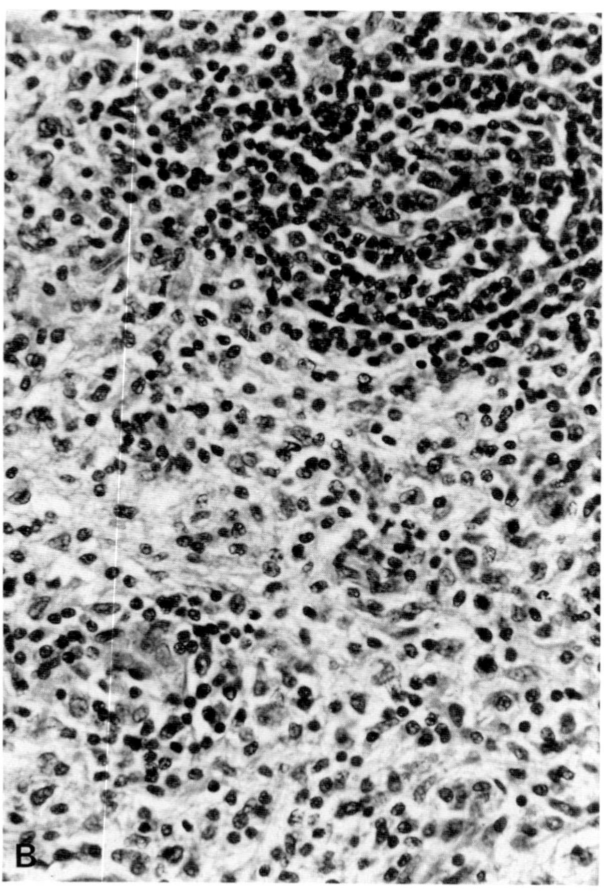

Fig. 3.116B Granulomatous encephalitis. Dog. Detail of cellular infiltrate in brain.

6. *Coonhound Paralysis*

Coonhound paralysis affects some dogs which are exposed to raccoons. Signs develop within 7 to 10 days as an ascending flaccid paralysis, which leads to quadriplegia and rapid atrophy of muscle. There are no cerebral signs. Some dogs die of respiratory paralysis, but the majority will recover slowly if nursing is adequate. Dogs which have recovered appear to be of increased sensitivity to subsequent exposure but may survive several bouts of paralysis. The disease has been transmitted using pooled saliva of raccoons, but the responsible substance or agent has not been identified. Pathologic changes are found in the ventral roots of spinal nerves and in peripheral nerves. There is leukocytic infiltration of mononuclear and plasma cells around venules, but the extent of infiltration is variable and not correlated with the course of the illness or the severity of nerve degeneration. There is primary and Wallerian degeneration afflicting ventral roots in particular, with axonal reaction in motor neurons and atrophy of denervation in muscle. This idiopathic polyradiculoneuritis is the most common inflammatory condition of peripheral nerves in dogs.

7. *Neuritis of the Cauda Equina*

There is a form of polyneuritis in horses in which the presenting signs are referable to the sacrococcygeal nerves and include perineal anesthesia, tail paralysis, urinary incontinence, fecal retention, weakness, atrophy of coccygeal muscles, and in long-standing cases, atrophy of the muscles of the hind limbs. The neuritis is progressive. The cause is unknown, but the nature of the reaction suggests a role for immune mediation.

Although the pathologic changes emphasize the sacral and caudal spinal nerve roots, there may simultaneously be asymmetric pareses of other nerves producing isolated limb pareses, and paresis or paralysis referable to cranial nerves. The lesions in these other nerves are similar in character to, but much milder than, those in caudal nerve roots. Inflammatory changes are present in sensory and some autonomic ganglia, but changes in the spinal cord are limited to those which reflect peripheral nerve injury.

The gross changes affect in particular the extradural parts of the sacral and coccygeal nerves and may extend through the intervertebral foraminae into the adjacent muscle. The roots are thickened and fusiform and usually discolored by recent or old hemorrhage. The intradural segments of affected nerves are discolored but not usually enlarged.

Microscopically there is granulomatous inflammation with extensive fibrosis. The thickening and discoloration are attributable to hemorrhage, proliferation of epineural tissue, and inflammatory cell infiltrates. The infiltrating cells are lymphocytes, plasma cells, and macrophages frequently disposed as to form granulomas, often with central epithelioid and giant cells; granulocytes do not feature in the infiltrates. Degenerative and regenerative changes are present in the myelin and axons of affected roots and appear to be more closely associated with endoneurial and perineurial fibroplasia than with leukocytic infiltrates.

Bibliography

Braund, K. G. *et al.* Granulomatous meningoencephalomyelitis in six dogs. *J Am Vet Med Assoc* **172:** 1195–1200, 1978.

Cordy, D. R. Canine granulomatous meningoencephalitis. *Vet Pathol* **16:** 325–333, 1979.

Cordy, D. R., and Holliday, T. A. A necrotising meningoencephalitis of pug dogs. *Vet Pathol* **26:** 191–194, 1989.

Cummings, J. F., de Lahunta, A., and Timoney, J. F. Neuritis of the cauda equina, a chronic idiopathic polyradiculoneuritis in the horse. *Acta Neuropathol (Berl)* **46:** 17–24, 1979.

Cummings, J. F. *et al.* Coonhound paralysis. Further clinical studies and electron microscopic observations. *Acta Neuropathol (Berl)* **56:** 167–178, 1982.

Cummings, J. F., de Lahunta, A., and Mitchell, W. J. Ganglioradiculitis in the dog. *Acta Neuropathol (Berl)* **60:** 29–39, 1983.

Fankhauser, R., Fatzer, R., and Luginbuhl, H. Reticulosis of the central nervous system (CNS) in dogs. *Adv Vet Sci Comp Med* **16:** 35–71, 1972.

Hall, W. W., Imagawa, D. T., and Choppin, P. W. Immunological evidence for the synthesis of all canine distemper virus polypeptides in chronic neurological disease in dogs. Chronic distemper and old dog encephalitis differ from SSPE in man. *Virology* **98:** 283–287, 1979.

Imagawa, D. T. *et al.* Isolation of canine distemper virus from dogs with chronic neurological diseases. *Proc Soc Exp Biol Med* **164:** 355–362, 1980.

Jervis, G. A., Burkhart, R. L., and Koprowski, H. Demyelinating encephalomyelitis in the dog associated with antirabies vaccination. *Am J Hyg* **50:** 14–26, 1949.

Lincoln, S. D. *et al.* Etiologic studies of old dog encephalitis. I. Demonstration of canine distemper viral antigen in the brain in two cases. *Vet Pathol* **8:** 1–8, 1971.

Lincoln, S. D. *et al.* Studies on old dog encephalitis. II. Electron microscopic and immunohistologic findings. *Vet Pathol* **10:** 124–129, 1973.

Thomas, J. B., and Eger, C. Granulomatous meningoencephalomyelitis in 21 dogs. *J Small Anim Pract* **30:** 287–293, 1989.

Vandevelde, M. *et al.* Chronic canine distemper virus encephalitis in mature dogs. *Vet Pathol* **17:** 17–29, 1980.

Wright, J.A., Fordyce, P., and Edington, N. Neuritis of the cauda equina in the horse. *J Comp Pathol* **97:** 667–675, 1987.

IX. Neoplastic Diseases of the Nervous System

Primary neoplastic disease of the central nervous system of animals is rare in some species but not in the dog and cat. Neoplasms derived from virtually every cell type in the nervous system are recorded. Meningeal sarcomas, perivascular sarcomas (distinct from the reticuloses), malignant melanotic tumors, and hemangiosarcomas illustrate the range of nonspecific primary neoplasms. In dogs and cats, the tumors lend themselves to the general schemes of classification, but in other species, the few tumors which have been observed, and which are thought to be neuroectodermal in origin, do not fit readily into accepted classifications.

The tumors may be congenital, and these do not differ in their characteristics from similar tumors in adult animals, with the possible exception of medulloblastoma, which is thought to be derived from the residual cells of the external granular layer of the cerebellum; the craniopharyngiomas, which are thought to arise from remnants of Rathke's pouch, which forms the adenohypophysis; the chordoma, which is thought to arise from remnants of notochord; the intracranial teratoma, which is probably derived from germ cells; and cystic epidermoid tumors, which are thought to result from the inclusion of surface ectodermal cells at the time of closure of the neural groove. The relatively high incidence of tumors in the boxer dog and the Boston terrier indicates a hereditary predisposition, which these breeds also have for endocrine neoplasia. The conventional tumors of the nervous system are selected for discussion here. Tumors of the pituitary and nonchromaffin paraganglia are discussed in Volume 3, Chapter 3, The Endocrine System.

A. Chordoma

Chordomas are rare tumors in animals, with the exception of mink and European ferrets. In humans, they are slow-growing persistent tumors mainly of the sacrococcygeal region, with some occurring in paraspinal and cranial regions. On the basis of histological characters, the tumor is presumed to arise from notochordal remnants. Grossly, they are gelatinous, gray, friable, and lobulated. Histologically, the lobules are not encapsulated but are well defined by bands of connective tissue. The principal cell types are large, clear, and vacuolated with a botanical appearance and some resemblance to cartilage. The cytoplasm contains large clear vacuoles with distinct boundaries. At the periphery of the lobules there are smaller, stellate cells with eosinophilic cytoplasm; mitotic figures are rare in these, but they may be the germinative cells. There may be islands of bone or cartilage. Mucinous substance can be demonstrated in the cytoplasmic vacuoles of the larger cells and between the cells; the smaller peripheral cells may contain PAS-positive granules. Differentiation from mucinous chondrosarcoma can be difficult without assistance of immunochemical methods.

Too few chordomas have been described in animals to allow a statement on their natural history. They represent a significant proportion of primary malignant bone tumors in humans and occur in regions where small rests of notochordal tissue are common. Such rests have not been described in domestic animals.

B. Epidermoid Tumors

Epidermoid cysts in the brain are confined to the fourth ventricle and environs. They are rare, probably represent surface ectoderm misplaced at the period of closure of the neural groove, and are described only in humans and dogs. The cystic structure is lined by squamous epithelium, and the cavities contain keratinaceous debris. They occur in young dogs and may develop to several centimeters in diameter. The clinical signs are determined by the location of the tumor.

C. Craniopharyngioma

Aberrant differentiation of Rathke's epithelium is common in dogs, expressed in cystic structures lined by respiratory-type epithelium. The craniopharyngioma, in contrast, consists of clumps of epithelial cells palisaded on collagenous stroma. Keratinization may be present to form pearls, as in well-differentiated squamous carcinoma. Degenerative foci contain cholesterol crystals and blood pigments.

D. Pineal Tumors

These tumors are rare. They are described in horses, cattle, and dogs. The diagnosis is, quite reasonably, based on the site of the tumor and its replacement of the pineal body. Other positive identifying characteristics are absent. Microscopically, there is a resemblance to medulloblastoma. The tumor may extend into midbrain and thalamus. Teratomatous tumors with characteristics of the gonadal teratomas are not described in the pineal gland of animals.

E. Germ Cell Tumors

Germ cell tumors are presumed to arise from embryonic germ cells which, intended for the developing gonad, may become widely distributed. They are responsible for a spectrum of tumors in humans, usually in the midline and according to histological features designated as seminoma, choriocarcinoma, entodermal sinus tumor, and teratoma. They are revealed early in life, and the preferred locations in the cranial cavity are in the pineal region or in the hypothalamus above the sella. These are rare tumors in dogs. Their location and growth pattern are similar to those of pituitary adenomas, and some may be misdiagnosed as craniopharyngioma.

The reported canine cases were suprasellar and classified on the basis of location, admixture of distinct cell types and patterns, and positive immunochemical staining for alpha fetoprotein. Sheets and nests of cells resembling seminoma were admixed with areas of vacuolated hepatoid cells, glandular formations similar to those in gonadal teratomas, and occasional foci of squamous epithelium.

F. Spinal Nephroblastoma

These are single intradural masses occurring in young dogs in the region between the tenth thoracic and second lumbar segments, and affected animals present with signs of cord compression. The histologic appearance is of glandular areas intermingling with cellular areas. The glandular areas consist of rosettes and tubules, the latter tortuous, branching, and sometimes papilliferous with infoldings reminiscent of embryonic glomerular capsules (Fig. 3.117). The cellular areas contain densely packed cells of blastema appearance with ovoid clear nuclei and indistinct cytoplasm. Some streaming is evident, but muscle fibers are not present. In some of these tumors, glomerular structures can be recognized with confidence as can tubular cross sections strongly suggestive of distal renal tubules. Presumably these tumors arise from ectopic embryonic remnants, but notably, the kidneys do not contain neoplasm, although nephroblastomas are well recognized in dogs. From time to time, undifferentiated tumors of embryonic type and possibly derived from remnants of pronephros are seen in the cervical region of dogs and sheep.

G. Granular Cell Tumor

The granular cell tumors are, histologically, quite distinctive, and there is little cytologically to distinguish the benign and malignant varieties. An earlier designation as granular cell myoblastoma recognized their preponderance in limbs of humans and a resemblance to myoblasts, but the latter origin, notwithstanding that they are well known in the tongue, is untenable in the light of their wide distribution as solitary tumors of soft tissue. The designation alveolar soft-part sarcoma is probably the best for the present. The histologic characters of the tumor are repetitive from case to case, and they may be primary in a multiplicity of sites. They are included here because they do occur in the meninges and spinal nerve roots of dogs, and the notion is attractive that some, at least, of the tumors are derived in common with cells of nerve sheaths and other neural crest mesenchyme.

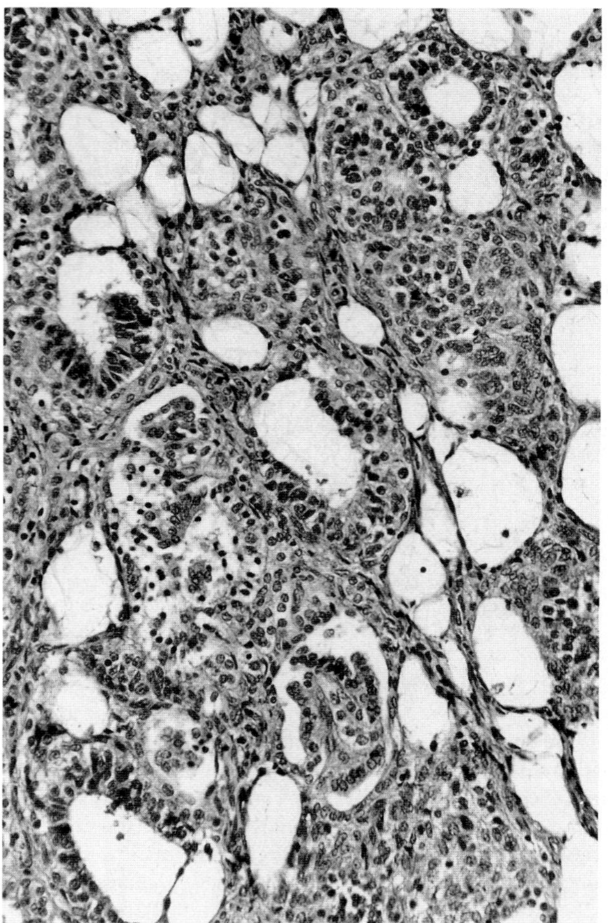

Fig. 3.117 Spinal nephroblastoma. Dog. Neoplastic cells showing tubular structures with infoldings.

The tumors are usually small and well circumscribed but not well encapsulated. The cells are large and rounded; the cytoplasm is clear, apart from fine granules which may be few or numerous; and the nuclei are small and rounded. The cytoplasmic granules are acidophilic and PAS-positive. Delicate stroma in which amyloid may be deposited separates the cells individually or in small groups. Stigmata of anaplasia are absent.

H. Mesodermal Tumors

The principal mesodermal tumors are the sarcomas and meningiomas.

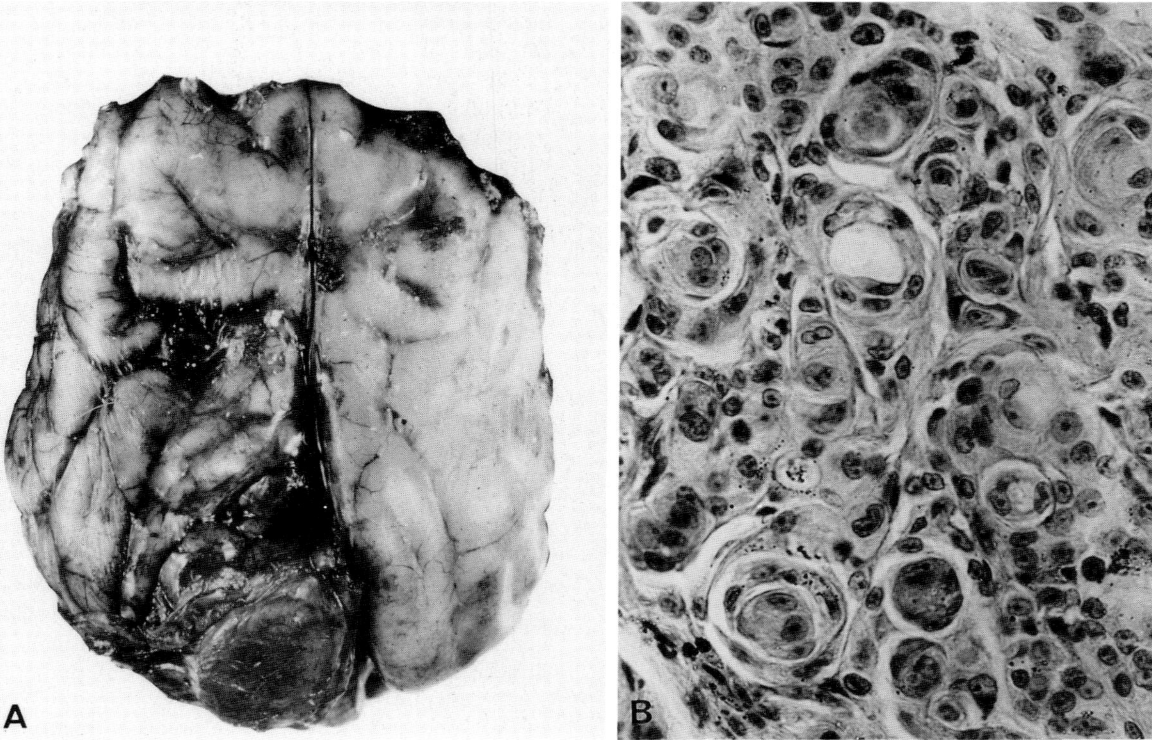

Fig. 3.118 (A) Meningioma over left occipital cortex. Dog. (B) Histologic appearance of psammomatous meningioma.

Sarcomas may provide circumscribed or diffuse involvement of the meninges, or they may form local lesions of the brain or cord. They are not rare. Those of the brain and cord retain a perivascular orientation, even although they are infiltrative. Histologically, some are acceptably fibrosarcoma, but some are more pleomorphic and anaplastic spindle cell tumors. The histogenesis of these tumors is uncertain. They may arise from fibroblasts, pericytes, or resident reticuloendothelial cells of the monocyte–macrophage series. Some of the more anaplastic astroglial tumors provoke a remarkable degree of capillary endothelial hyperplasia and fibrosis of the vascular adventitia in brain and meninges. The proliferating fibroblastic cells are atypical, their appearance consistent with fibrosarcoma.

Meningiomas are alternatively known as arachnoid fibroblastomas, a term that is acceptable to those pathologists who regard the meninges as being of mesenchymal rather than neuroectodermal origin. This is one of the commonest of intracranial and intraspinal tumors in humans and is relatively common in cats and dogs, rare in cattle, and not recorded in horses. It rises within the meninges, usually in close association with the dura, and grows expansively, compressing but seldom invading the brain (Fig. 3.118A). Those in humans are presumed, by virtue of their site and structure, to arise mainly from the arachnoid granulations and also from islands of arachnoid cells which lie in the inner surface of the dura mater. The meningiomas observed in animals conform in histologic type to those observed in humans. They may be multiple in cats.

Meningiomas are globular, ovoid, or tuberous, sometimes plaquelike, well circumscribed, and have a smooth surface. They are gray, sometimes yellowish, on cut surface, firm, and may be gritty. There are several histologic varieties. The meningotheliomatous or epithelioid meningioma is diffusely cellular with the cells in sheets or pseudoalveoli. The stroma varies in amount, is richly vascular, and may contain much collagen. The cells are large with abundant finely granular pale cytoplasm without a distinct margin, and the nuclei are spherical or ovoid and rather vesicular. The psammomatous meningioma has the same general features as the foregoing, but the cells arrange themselves in whorls (Fig. 3.118B). In the center of a whorl, lamellar hyaline tissue forms, derived possibly from cells, stroma, or a blood vessel. As the hyaline focus expands, it tends to be impregnated with salts of calcium and iron to form psammoma bodies. The fibroblastic meningioma is similar to fibroblastic tumors elsewhere.

Angioblastic meningiomas are highly vascular with prominent endothelial cells in formed vessels and lining vascular clefts. The vessels are surrounded by spindle cells, giving a distinct resemblance to hemangiopericytoma.

Most intracranial meningiomas are benign, as indicated by low frequency of metastases and failure to invade the brain. By these criteria, ~2.5% are malignant. In contrast, extracranial meningiomas, which occur mainly in the paranasal region and orbit, are anaplastic and locally aggressive.

I. Vascular Tumors

Secondary hemangiosarcomas of brain and cord, in multiple foci, are not uncommon, especially in dogs. Primary vascular tumors and vascular malformations are rare. There is difficulty in separating hemangioma from highly vascular meningiomas; both contain endothelial cells, pericytes, and stromal cells, the latter of uncertain parentage.

Meningioangiomatosis is a rare benign lesion, best regarded as a malformation or hamartoma producing circumscribed plaques on the surface of the brain. Blood vessels are in excess in the lesions and are cuffed by proliferating cells which are considered to be meningothelial. The lesion does extend into the underlying neural substance, which shows mixed degenerative and reactive changes.

J. Neuroglial Tumors

Primary tumors within the brain and cord are gliomas in which groups are astrocytoma, oligodendroglioma, ependymoma, and medulloblastoma. Gliomatous tumors as a group are not by any means rare. A number of them are discovered accidentally at postmortem during routine examination of brains. The signs of their presence are quite variable and depend on their location, with signs referable to dysfunction of a special part of the brain and the general effects of increased intracranial pressure. The tumors themselves are space occupying and may cause displacements of intracranial structures, and to this is added the effect of edema, which is more severe immediately about the tumor.

Of the domestic species, glial tumors are observed most frequently in dogs, and this probably reflects their true relative incidence in domestic animals. Glial tumors are observed occasionally in cattle and cats and rather rarely in other species. Certain breeds of dogs are more likely to develop brain tumors than are others. Not enough data have been accumulated to be able to state this predisposition precisely, but one is much less surprised to find a brain tumor in a brachycephalic breed than in any other.

In humans, the different glial tumors have a predilection for certain areas of the brain. Medulloblastomas are expected to be midline tumors of the cerebellum; the more malignant astrocytomas, such as would be subclassified as glioblastoma multiforme, most often develop in the cerebrum; another variety of astrocytoma subclassified as polar spongioblastoma has a predilection for the optic nerve and pons; and the oligodendroglioma usually arises in the hemispheral white matter. These distributions are not absolute, but they do suggest that the morphology, and probably also the histologic malignancy, of a glial tumor is conditioned to some extent by the environment in which it grows, and such must have some bearing on interpretation of histogenesis and, therefore, on classification.

The conventional histomorphologic classification is continued here. It has established therapeutic and prognostic value, notwithstanding its imperfections. The application of immunochemical techniques designed to identify antigenic markers specific to cell types is finding widespread application and is particularly useful in studies of embryogenesis and glial cell lineages. However, differentiating and mature glia are quite different from anaplastic types, and immunochemical procedures are, at present, quite unreliable for neoplastic cells. They do, however, support the more conventional evidence of the interrelatedness of the neuroglia, ependyma, astrocytes, and oligodendrocytes and give comfort to the pathologist who recognizes that different cell types predominate in different parts of one tumor, that the cellular profile can be quite different in subsequent biopsies, and that some tumors are of mixed cellular composition.

The proliferative capacities of astrocytes in pathologic conditions have long been recognized. That reparative proliferation of oligodendrocytes can occur is now accepted. The notion of stem or reserve glial cells seems unnecessary. It remains, however, that differentiated neuronal cells are postmitotic, with the probable exception of only olfactory neurons. The origin of microglial cells present in parenchyma, vascular adventitia, and meninges remains vexing. For the present, the evidence of their derivation from bone marrow as representatives of the monocyte–macrophage or reticuloendothelial system is accepted.

The glial tumors are a very pleomorphic group, and special stains have virtually no advantage over routine stains in sorting them out. To take care of their histologic variety and the older view that they represented diverse differentiations of embryonic cells, a complicated histologic classification was erected to include all varieties. Even with these complicated classifications, tumors occurred which could not be fitted. The tendency nowadays is to regard most neoplasms as arising by dedifferentiation of differentiated elements and to classify them accordingly. The medulloblastoma is accepted as being derived from embryonic cells of some sort, but the other glial tumors are regarded as arising from differentiated glia; the present tendency is to classify them accordingly as being either astrocytomas, oligodendrogliomas, or ependymomas. Each class of tumor can then be further subdivided or graded according to the degree of histologic malignancy. There is little doubt that this type of classification has histogenetic validity, and it appears to represent a considerably simplified view because recognizable oligodendrogliomas and ependymomas are readily recognizable, and just about all of the remainder are astrocytomas or "glioma, undifferentiated." There are, however, still considerable difficulties in making rigid classifications because of the facts that many, and perhaps most, of these tumors are mixed, and individual tumors change their character. Gliomas which are acceptably pure oligodendrogliomas or ependymomas do occur, but it has been known that glial tumors may be composed of cells of more than one origin. The admixing of, for example, oligodendrocytes and astrocytes may be quite intimate, but there is a tendency for the different lineages to occupy different parts of the tumor mass. The classification depends on which cell type predominates. Human glial tumors, especially astrocytic tumors, may have areas of varied malignancy in a single tumor, and the histologic appearance of the tumor varies from time to time. Animal tumors reveal the same diversity.

The assessment of malignancy of the glial tumors is also a problem of some magnitude, and one to which the usual criteria of malignancy are not entirely applicable. Metastases do occur outside the cranial cavity, but, for some unknown reasons, they are so rare (unless surgery has given them a way out) that any report of them is necessarily first regarded with considerable scepticism. Metastasis within the cranial cavity, chiefly as local meningeal implants, is also uncommon. Malignancy therefore must be assessed on cytological characters and, in an organ such as the brain, also in the relation of the tumor to some part of the brain with indispensable function. The determination of malignancy on cytological characters necessarily depends on sampling the tumor widely enough to find its most malignant part. Another factor that must be taken into consideration is the diffuseness of the neoplastic change; the concept of fields of neoplasia applies to neural as well as to other tissue and, in the astrocytomas particularly, it appears that malignant dedifferentiation may occur simultaneously over rather broad fields in appearances which microscopically suggest quite distant infiltration.

1. Astrocytoma

Astrocytoma is the most common of the primary intracranial tumors. The type has been found in dogs, cats, and cattle and may occur in the brain or cord. The various glioblastomas, spongioblastomas, astroblastomas, and qualified astrocytomas belong in this category, representing different predominating histologic patterns. The gross appearance varies, depending largely on the degree of malignancy. These tumors can be very difficult to detect grossly, especially when they involve white matter or grow slowly. They are then white and, because of their firmness, may be more readily palpable than visible. Their presence may be suspected only by deviation of some architectural feature (Fig. 3.119B,D). Larger and more malignant tumors are prone to vascular accidents and necrosis, and they are then easy to see (Fig. 3.119E), but the margins are never discrete, especially not when they are surrounded by edematous tissue. The extent of the tumor is always much greater than can be appreciated grossly.

Histologically, these tumors are very diverse. Even the more benign of them are subtly invasive. These may appear, microscopically, as an increased population of fibrous astrocytes which individually are not clearly malignant (Fig. 3.119A). In those which are slightly more malignant, the population is more dense, the nuclei are a little larger and darker and show slight but definite variations in size and shape but no mitoses. The cells are recognizable as astrocytes. The walls of the vessels may be slightly thickened. With greater degrees of malignancy, hemorrhage and necrosis are expected, and the adventitial and endothelial cells of the vessels proliferate. In these, only a few cells are recognizable as astrocytes. Pleomorphism, giant nuclei, and multinucleated giant cells are common. Mitotic figures are common and atypical. The cell processes are disorganized but may show a tendency to palisade around blood vessels. In some tumors of moderate undifferentiation, the cells, often in fairly discrete areas, have abundant acidophilic cytoplasm but few or no processes; these are the gemistocytic astrocytomas (Fig. 3.119C), and these cells are probably to be regarded as degenerative types.

2. Oligodendroglioma

This is the easiest of the glial tumors to recognize even when growing rapidly. Grossly, it usually appears well demarcated, being grayish, soft, and almost fluctuating (Fig. 3.120A). Hemorrhage and necrosis occur but are unusual. The tumor is densely cellular with almost no stroma. The nuclei are remarkably uniform and like those of normal oligodendroglia in size and shape. The cytoplasm does not stain, but its membrane does, so that the nucleus seems to lie in a clear polyhedral or rounded halo (Fig. 3.120B). These tumors occur in white matter, and those near the third ventricle may contain areas distinguishable as astrocytoma or ependymoma. Mucinous degeneration and cyst formation may occur in these tumors, and calcification may occur in some of them. There are no clear indices of malignancy, although all must be regarded as malignant.

3. Ependymoma

This tumor necessarily arises in relation to ependyma about the ventricles and the central canal of the cord. Most arise about the third ventricle. They are gray and fleshy but may be dark from blood and hemorrhage if they project into a ventricle.

The tumors are usually densely cellular, and those arising about the third ventricle may be difficult to distinguish from undifferentiated pituitary tumors. The nuclei are small, dark, and regular, and the cytoplasm has no distinct boundaries. Pseudorosettes form around blood vessels, resembling some types of astrocytoma, and tubular cavities may form lined by cells of epithelial appearance. Branching papillary stroma may be covered by recognizable ependymal cells (Fig. 3.121). The cells are bound together by desmosomes, blepharoplasts are present, and cilia may be seen on cells lining cavities or covering papillae. Ependymomas of the spinal cord may be more papilliferous than those in the brain, with the tumor cells attached to fronds, which are supported by a delicate stroma and

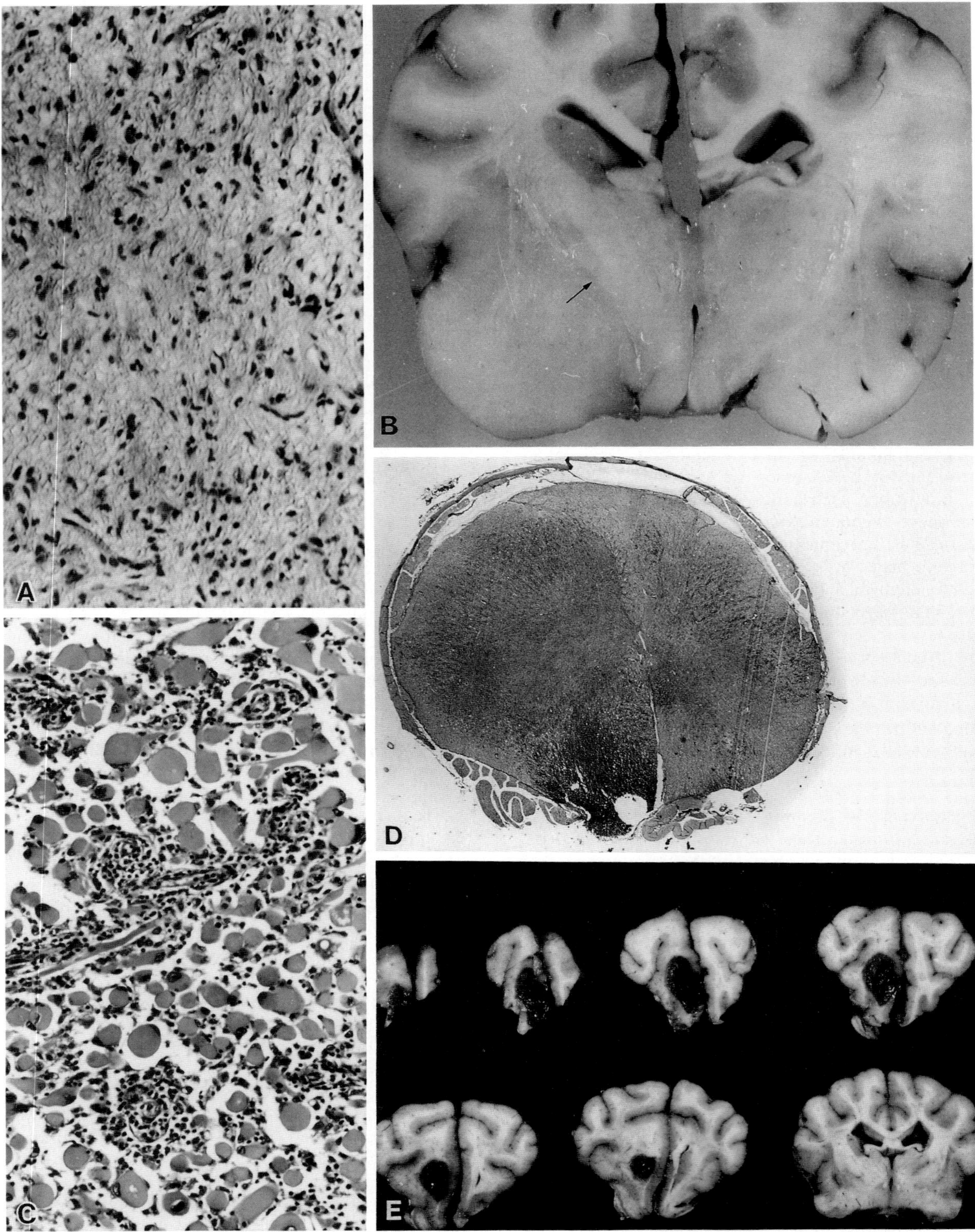

Fig. 3.119 (A) Fibrous astrocytoma. Dog. (B) Astrocytoma in piriform lobe. Dog. Note lack of definition but displacement of internal capsule (arrow) by the homogeneous tumor. (C) Gemistocytic astrocytoma. Dog. (D) Astrocytoma of spinal cord. Dog. (E) Hemorrhagic astrocytoma of left frontal lobe. Dog.

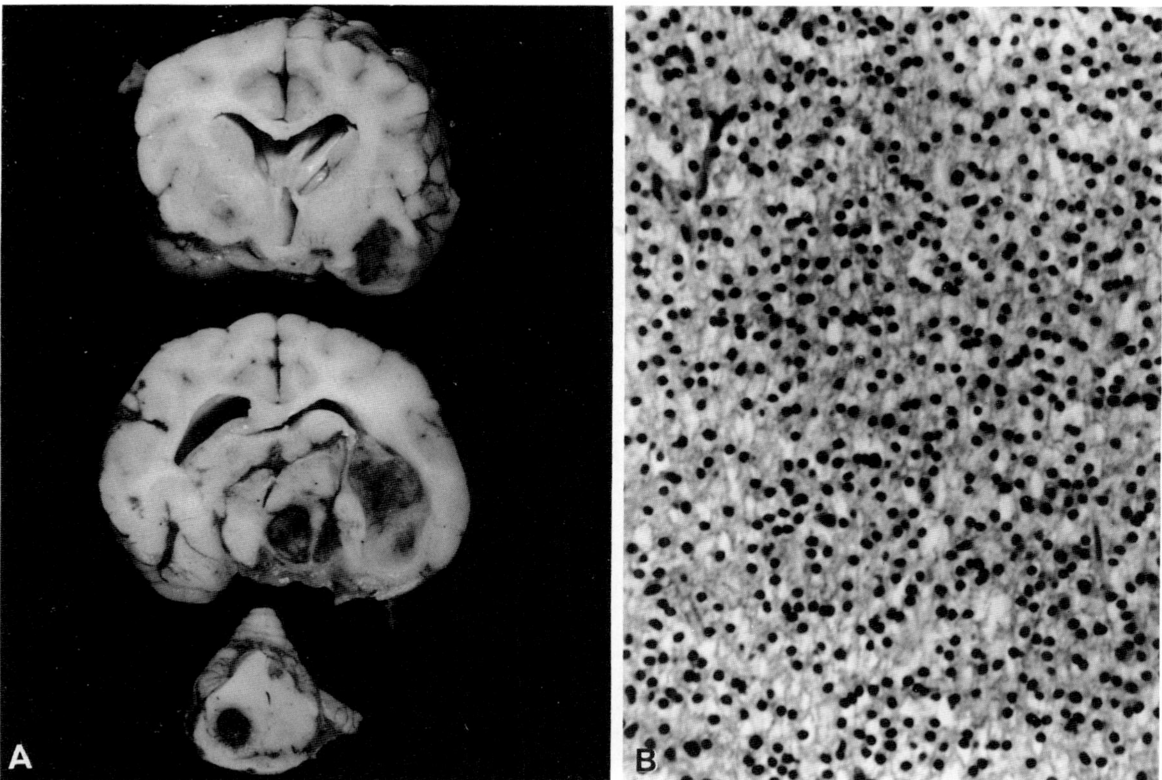

Fig. 3.120 (A) Oligodendroglioma involving right side of brain stem and piriform lobe. Dog. (Bottom section reversed to show caudal extension in brain stem). (B) Histologic pattern of oligodendroglioma.

embedded in mucinous intercellular stroma. These may be difficult to distinguish from neuroblastoma.

4. Neuronal Tumors

Medulloblastoma, well known in children, is rare in animals, but occurs in the young of several species (Fig. 3.122), mainly calves and dogs. There is no such cell as a "medulloblast," the name having been dreamed up for this tumor of unknown parentage. It is currently thought to arise from the undifferentiated cells found in neonatal life beneath the cerebellar pia mater, which are thought to be the precursors of the cerebellar cortex. These tumors grow rapidly. Histologically, they are densely cellular with scant stroma and few vessels (Fig. 3.123A). The cells are small and classically carrot-shaped with oval or elongate nuclei and the cytoplasm tapering at one pole. They are also supposed to produce small perivascular palisades and pseudorosettes (Fig. 3.123B). These classical features are not always present, and then there is nothing by which this tumor can be identified except its location in the cerebellum. Tumors of this histological appearance do occur in other sites in the brain of calves.

Neuroblastomas are rare. Histologic criteria to distinguish them from other cellular neurogenic tumors are not satisfactory, and identification must ultimately depend on other means. They are thought to arise from primitive neuroepithelial cells with differentiation toward postmitotic neuroblasts. The histologic appearances are similar to those of medulloblastoma, consisting of masses of small rounded cells which resemble lymphocytes with hyperchromatic nuclei and scant cytoplasm. The presence of rosettes and pseudorosettes is helpful.

Neuroblastomas, because of their derivation, are limited to the young, with the exception of olfactory neuroblastoma, which may occur at later ages. The olfactory tumors are uncommon, but have been identified in various animal species, mainly in dogs and cats. The olfactory epithelium is unique among neuronal structures in that the basal cells retain the ability to divide and differentiate to become sustentacular cells or bipolar neurosensory cells. The tumors are locally aggressive and may penetrate the cribriform plate. The dense cellular population is homogeneous, arranged in sheets or clusters but also forming true and pseudorosettes. They may be palisaded on trabeculae, oval in shape with scant cytoplasm. Cytoplasmic processes may form an abundant, delicate fibrillary matrix.

The detection of type C retroviral particles identified with feline leukemia in spontaneous olfactory neuroblastomas of cats is of interest.

Ganglioneuroma and its undifferentiated counterpart, the neuroblastoma (sympathoblastoma), are embryonic

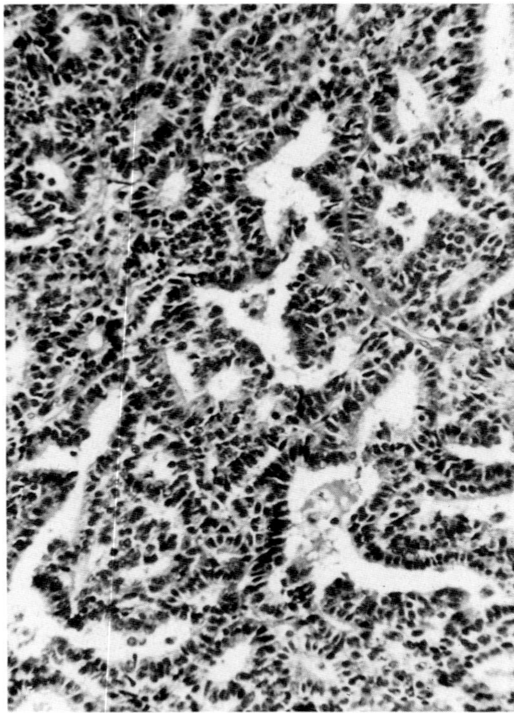

Fig. 3.121 Ependymoma. Dog. Typical branching papillary structure.

tumors related to the sympathetic nervous system and adrenal medulla. The ganglioneuroma is quite rare in animals.

The few ganglioneuromas observed in cattle have developed in relation to the abdominal sympathetic plexuses. The degree of differentiation of these tumors varies considerably, and they are a mixture of ganglion cells, Schwann cells, and nerve fibers. The ganglion cells show

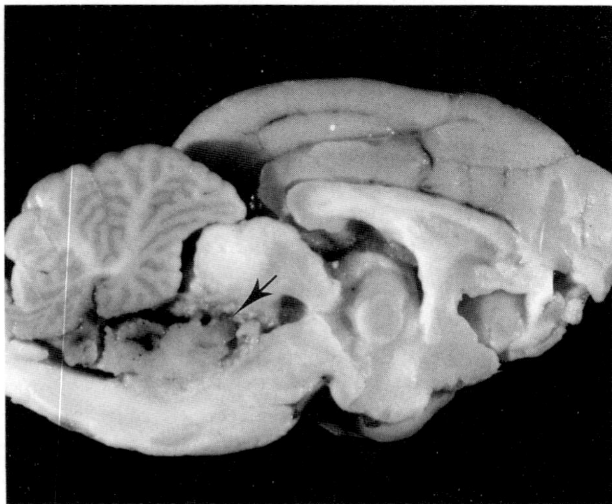

Fig. 3.122 Medulloblastoma between cerebellum and fourth ventricle with extension into colliculus (arrow). Cat.

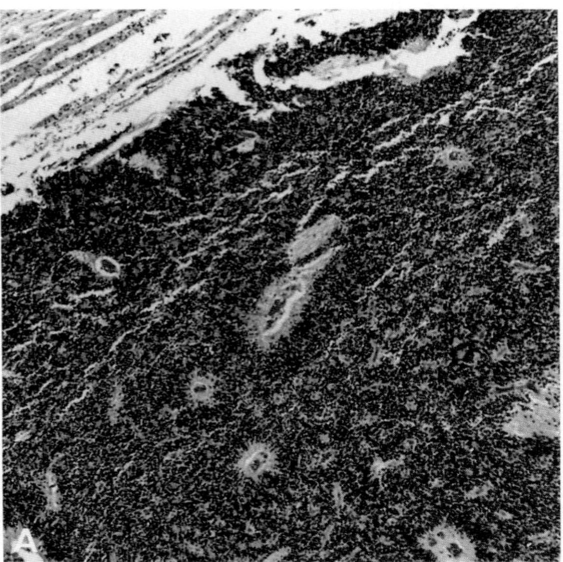

Fig. 3.123A Medulloblastoma. Histologic patterns of tumor in a calf. (Courtesy of M. D. McGavin). Pseudorosettes are visible.

different degrees of differentiation, from primitive neuroblasts to some that are remarkably mature. The mature cells are likely to be impregnated with calcium and the stromal cells pigmented with melanin. The adrenal tumors are probably better regarded as hamartomatous malformations rather than as neoplasms. A similar interpretation may be appropriate for solitary or diffuse lesions of the intestinal muscularis.

K. Choroid Plexus Tumors

These are rare. They may be papillomas or carcinomas. They are vascular papillary growths which implant widely

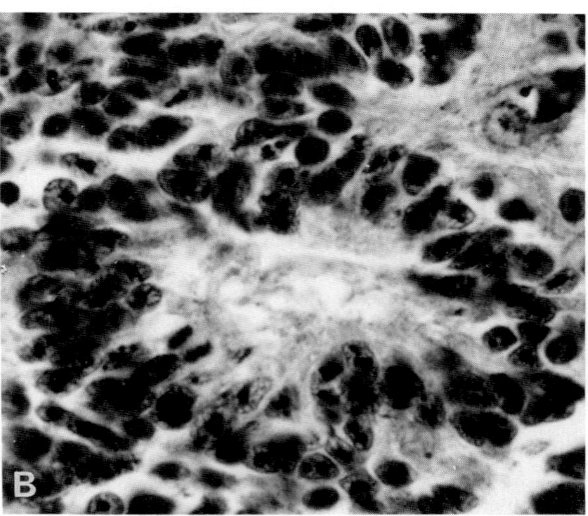

Fig. 3.123B Medulloblastoma. Pseudorosettes are detailed.

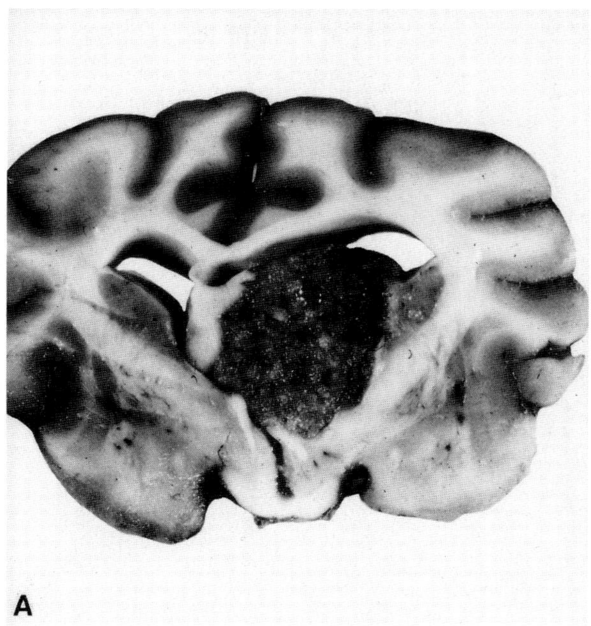

Fig. 3.124A Choroid plexus papilloma. Dog.

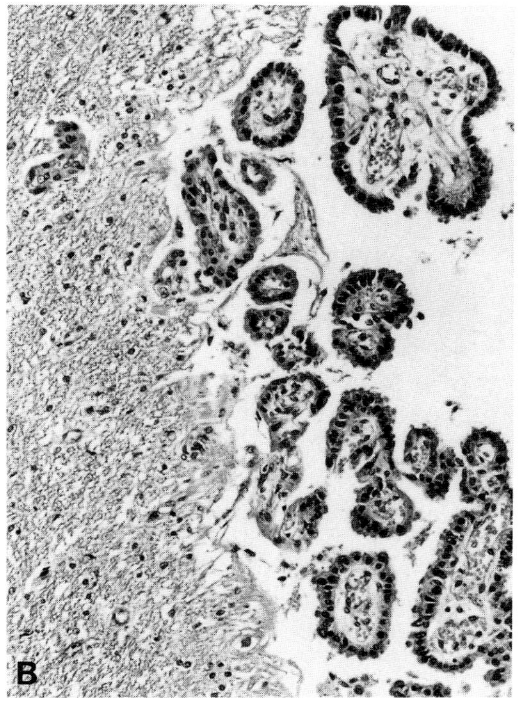

Fig. 3.124B Carcinoma of choroid plexus. Tumor is invading periventricular white matter. Dog.

on the meninges. The cells retain recognizable choroidal character (Fig. 3.124A,B). Internal or communicating hydrocephalus is a complication.

L. Schwann Cell Tumors

The Schwann cell of the peripheral nervous system is accepted as the origin of schwannomas, neurofibromas, neurofibrosarcomas, and neurilemmomas. Most of these tumors are seen in the skin. The dermal tumors do pose problems of differentiation from fibrous and perivascular lesions (see Chapter 5, The Skin and Appendages). In skin, they do show some significant differences from the tumors of nerve trunks and roots.

Schwannomas may be largely solitary infiltrating lesions at any site on a nerve trunk. Those distant from the central nervous system are not encapsulated or well defined, are difficult to dissect cleanly, and have an ordinary fibrous appearance and texture. Schwannomas of nerve roots tend to be well-defined fusiform tumors. They probably arise from a single nerve and extend proximally and distally in conjunction with the nerve but external to it. In this way they may extend through the intervertebral foraminae and extend also to involve other nerves which have a plexus arrangement, such as the brachial plexus. They may arise within and remain within the dura mater, and such tumors may be globose rather than fusiform and soft and discolored from hemorrhage. Expansive intradural growth may be slow and compress the brain or cord, but some of these tumors are malignant and invasive (Fig. 3.125D) and metastasize particularly to the lung.

Any cranial or spinal nerve root may be the site of growth, but in the dog, in which they are not rare, the brachial plexus is most frequently involved. In the thoracic region and on the acoustic nerve (Fig. 3.125A), the tumors tend to originate within the dura. The histologic appearance varies within and between tumors depending on the presence or absence of anaplastic change and the extent of hemorrhage, degeneration, fibrosis, and mineralization. Blood vessels are often prominent but ill formed. The anaplastic tumors consist of closely packed cells with oval or elongate nuclei, which give an impression of interlacing streams not supported or guided by reticulin or collagen (Fig. 3.125B). In areas of tumors which are not anaplastic, two patterns are dominant and referred to as Antoni A and Antoni B. Antoni A arrangements are repetitive and give the tumor its character. Uniform, fusiform cells are arranged as bands, herringbones, whorls, or palisades (Verocay bodies) (Fig. 3.125C). Antoni B tissue is degenerative and may predominate in some sections. It is loose and myxoid, sometimes hyalinized and may be sparsely cellular (Fig. 3.125E). Microcysts may be present in areas of myxoid change. Cartilaginous and osseous metaplasia occur infrequently.

Neurofibromatosis of cattle is a well-recognized entity. It is common in abattoir material from old animals but has been observed in very young calves. The skin may be affected, but the lesions are usually restricted to deeper nerves of the thoracic wall and viscera. The brachial plexus, intercostal nerves, hepatic autonomic plexus, the epicardial plexus and autonomic nerves of the mediastinum are those most frequently affected in various collec-

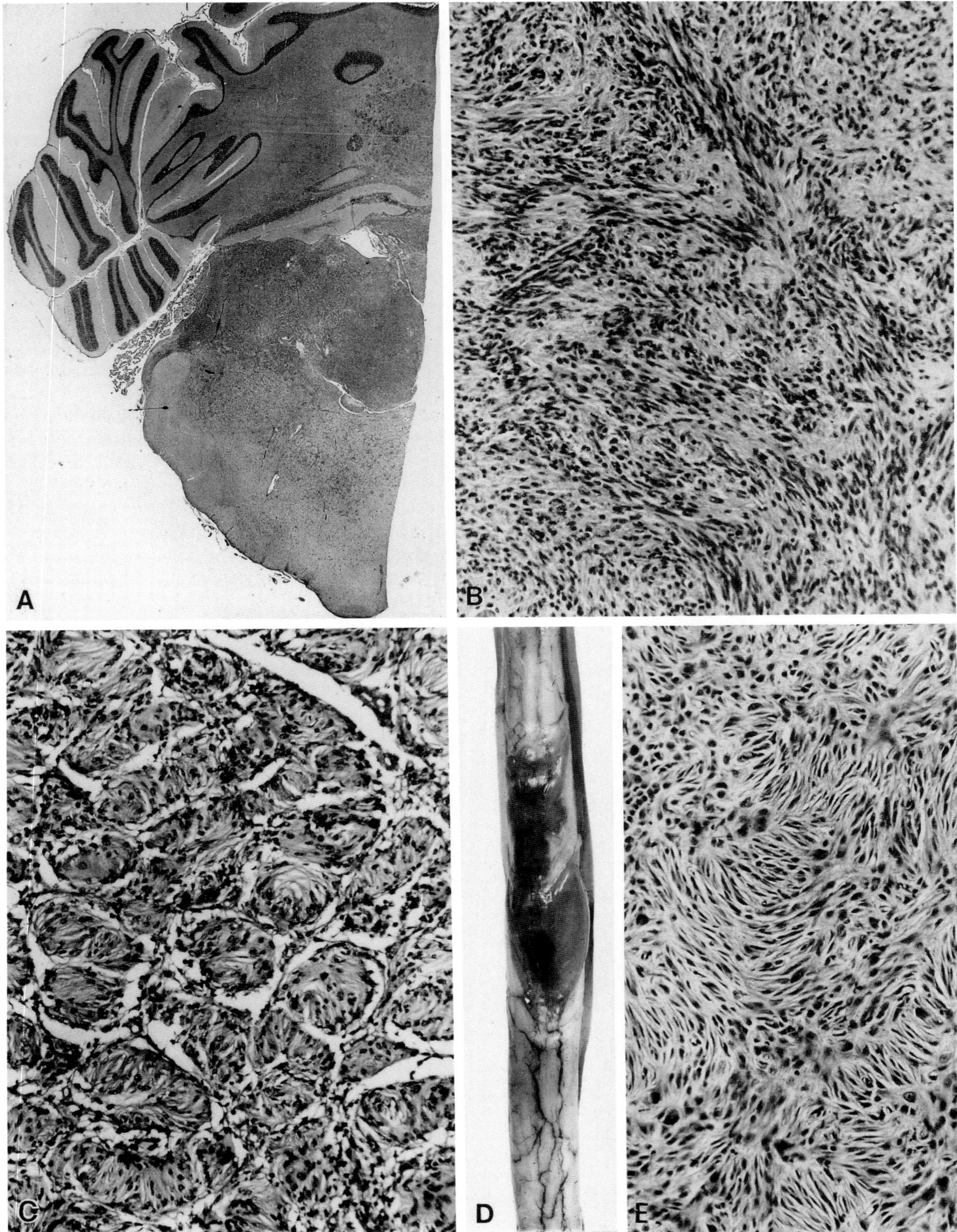

Fig. 3.125 (A) Acoustic schwannoma at cerebellopontine angle and filling fourth ventricle. Dog. (B) Section of (A). (C) Patterns resembling Verocay bodies in neurofibromatosis. Ox. (D) Malignant schwannoma. Spinal cord. Dog. (E) Schwannoma. Cat.

tive patterns. Sympathetic ganglia, especially the stellate and others of the thorax, are also frequently involved. Affected nerves are thickened, firm, and gray and may bear yellow-gray nodules. Affected ganglia may be enlarged to several centimeters and appear lobulated on section. The histologic appearances are as described above for schwannomas, but anaplastic change is rare (Fig. 3.46C).

Bibliography

Baumgartner, W., and Peixoto, P. V. Immunohistochemical demonstration of keratin in canine neuroepithelioma. *Vet Pathol* **24:** 500–503, 1987.

Bistner, S. R. *et al.* Neuroepithelial tumor of the optic nerve in a horse. *Cornell Vet* **73:** 30–40, 1983.

Bradley, R. L., Withrow, S. J., and Snyder, S. P. Nerve sheath tumors in the dog. *J Am Anim Hosp Assoc* **18:** 915–921, 1982.

Canfield, P. A light microscopic study of bovine peripheral nerve-sheath tumours. *Vet Pathol* **15:** 283–291, 1978.

Canfield, P. The ultrastructure of bovine peripheral nerve-sheath tumours. *Vet Pathol* **15:** 292–300, 1978.

Cordy, D. R. Vascular malformations and hemangiomas of the canine spinal cord. *Vet Pathol* **16:** 275–282, 1979.

Cox, N. R., and Powers, R. D. Olfactory neuroblastomas in two cats. *Vet Pathol* **26:** 341–343, 1989.

Dahme, E., and Schiefer, B. Intracranielle geschwulste bei Tieren. *Zentralbl Veterinaermed* **7:** 341–363, 1960.

Fagerland, J. A., and Greve, J. H. Unusual psammoma bodies in an extra-cranial syncytial meningioma from a dog. *Vet Pathol* **17:** 45–52, 1980.

Fingeroth, J. M. *et al.* Spinal meningiomas in dogs: 13 cases (1972–1987). *J Am Vet Med Assoc* **191:** 720–726, 1987.

Fischer, K. Subependymale Zellproliferationen und Tumor-disposition brachycephaler Hunderassen. *Acta Neuropathol* **8:** 242–254, 1967.

Fankhauser, R., and Frauchiger, E. Mikrogliomatose beim Hund. *Dtsch Tieraerztl Wochenschr* **74:** 142–146, 1967.

Fankhauser, R., Luginbuhl, H. and McGrath, J. T. Tumors of the nervous system. *Bull WHO* **50:** 53–69, 1974.

Geib, L. W. Ossifying meningioma with extracranial metastasis in a dog. *Vet Pathol* **3:** 247, 254, 1966.

Haskins, M. E., and McGrath, J. T. Meningiomas in young cats with mucopolysaccharidosis I. *J Neuropathol Exp Neurol* **42:** 664–670, 1983.

Jacob, K. L. Cerebrale Gliome bei Hunden. *Zentralbl Veterinaermed* **10:** 945–962, 1959.

Jolly, R. D., and Alley, M. R. Medulloblastoma in calves. *Vet Pathol* **6:** 463–468, 1969.

Josephson, G. K. A., and Little, P. B. Four bovine meningeal tumours. *Can Vet J* **31:** 700–703, 1990.

Koestner, A. Primary lymphoreticuloses of the nervous system in animals. *Acta Neuropathol (Berl) (Suppl.)* **VI:** 85–89, 1975.

Kornegay, J. N., and Gorgacz, E. J. Intracranial epidermoid cysts in three dogs. *Vet Pathol* **19:** 646–650, 1982.

Leestma, J. E. Brain tumors. *Am J Pathol* **100:** 243–315, 1980.

Luginbuhl, H. Geschwulste des Zentralnervensystems bei Tieren. *Acta Neuropathol (Suppl)* **1:** 9–18, 1962.

Palmer, A. C. Clinical and pathological features of some tumours of the central nervous systems in dogs. *Res Vet Sci* **1:** 36–46, 1960.

Patnaik, A. K. *et al.* Paranasal meningioma in the dog: A clinicopathologic study of ten cases. *Vet Pathol* **23:** 362–368, 1986.

Patnaik, A. K., Kay, W. J., and Hurvitz, A. I. Intracranial meningioma: A comparative pathologic study of 28 dogs. *Vet Pathol* **23:** 369–373, 1986.

Polak, M. On the true nature of the so-called medulloblastomas. *Acta Neuropathol* **8:** 84–95, 1967.

Rabotti, G. F. *et al.* Induction of multiple brain tumours (gliomata and leptomeningeal sarcomata) in dogs by Rous sarcoma virus. *Nature* **209:** 884–886, 1966.

Raimondi, A. J., and Beckman, F. Perineurial fibroblastomas: Their fine structure and biology. *Acta Neuropathol* **8:** 1–23, 1967.

Ribas, J. L. *et al.* A histochemical and immunocytochemical study of choroid plexus tumors of the dog. *Vet Pathol* **26:** 55–64, 1989.

Ribas, J. L. *et al.* Comparison of meningio-angiomatosis in a man and a dog. *Vet Pathol* **27:**369–371, 1990.

Scherer, H. J. Structural development in gliomas. *Am J Cancer* **34:** 333–351, 1938.

Schrenzel, M. D. *et al.* Type C retroviral expression in spontaneous feline olfactory neuroblastomas. *Acta Neuropathol* **80:** 547–553, 1990.

Simon, J., and Albert, L. T. Neuroblastomas in dogs. *J Am Vet Med Assoc* **136:** 210–214, 1960.

Stebbins, K. E., and McGrath, J. T. Meningio-angiomatosis in a dog. *Vet Pathol* **25:** 167–168, 1988.

Sullivan, D. J., and Anderson, W. A. Tumors of the bovine acoustic nerve. *Am J Vet Res* **19:** 842–852, 1958.

Summers, B. A. *et al.* A novel intradural extramedullary spinal cord tumor in young dogs. *Acta Neuropathol (Berl)* **75:** 402–410, 1988. (Nephroblastoma)

Valentine, B. A. *et al.* Suprasellar germ cell tumors in the dog: A report of five cases and review of the literature. *Acta Neuropathol* **76:** 94–100, 1988.

Vandevelde, M., Higgins, R. J., and Greene, C. E. Neoplasms of mesenchymal origin in the spinal cord and nerve roots of three dogs. *Vet Pathol* **13:** 47–58, 1976.

Vandevelde, M., Braund, K. G., and Hoff, E. J. Central neurofibromas in two dogs. *Vet Pathol* **14:** 470–478, 1977.

Vuletin, J. C., Friedman, H., and Gordon, W. Extraneuraxial canine meningioma. *Vet Pathol* **15:** 481–487, 1978.

Willard, M. D., and de Lahunta, A. Microgliomatosis in a schnauzer dog. *Cornell Vet* **72:** 211–219, 1982.

Zimmerman, H. M. The nature of gliomas as revealed by animal experimentation. *Am J Pathol* **31:**1–29, 1955.

Zaki, F. A., and Hurvitz, A. I. Spontaneous neoplasms of the central nervous system of the cat. *J Small Anim Pract* **17:** 773–782, 1976.

CHAPTER 4 The Eye and Ear

BRIAN P. WILCOCK
Ontario Veterinary College, Canada

THE EYE

I. General Considerations

The role of the veterinary pathologist in the diagnosis of ocular disease is greatly influenced by the accuracy with which the eye can be examined in the living animal. Some of its tissues, such as the lens, do not lend themselves well to pathologic techniques. Some focal lesions, especially of the posterior segment, may be readily visible and magnified with the ophthalmoscope but difficult to locate histologically. The reluctance of many pathologists to embrace ophthalmic pathology stems from the disappointing quality of sections made from formalin-fixed globes processed by routine methods, and from unfamiliarity with the complex terminology shared by clinical ophthalmologists and ophthalmic pathologists. At least equally daunting is the need to be familiar with the ever-growing list of inherited disorders that occur in purebred dogs, the species which indisputably dominates veterinary ophthalmology. In many instances, the correct diagnosis requires the correlation of the structural lesion with the age, breed, and specific clinical features of the disease to a much greater degree than is the case with most other body systems.

The eye undergoes very rapid postmortem change that not only obscures subtle degenerative lesions but also mimics genuine developmental or degenerative diseases. Even with a globe obtained within minutes of death or surgical removal, improper handling of the specimen frequently results in a section of poor quality. The globe can be speedily and gently removed by grasping the third eyelid with forceps and applying traction to the globe while making a circumferential incision at the fornix. Blunt curved scissors inserted through this incision may be used to sever extraocular muscles and optic nerve, and allow the globe to be removed from the orbit. All orbital fat and extraocular muscles should be gently removed from the sclera to permit rapid penetration of fixative to the retina. The choice of fixative depends on the disease suspected and on the type of examination to which the eye will be subjected. Formalin has the advantage of ready availability, little danger of overfixation, and adequate preservation of color and macroscopic detail for photography. Also, it permits localization in the bisected globe of lesions identified ophthalmoscopically, and the use of electron microscopy should such examination be warranted by the findings of light microscopy. However, formalin penetrates the sclera slowly, and there are postmortem changes, including retinal detachment, even in globes fixed immediately after death or surgery. Rapid-acting fixatives such as Zenker's, Helly's or Bouin's are preferred for globes in which preservation of histologic detail is paramount. All render the globe and its refractive media opaque and less suitable for macroscopic photography than does formalin. All require strict attention to the duration of fixation and thorough postfixation washing in water (Zenker's, Helly's) or 70% ethanol (Bouin's). Regardless of the method of fixation, all eyes benefit from hardening in 70% ethanol over about 24 hr to prevent retinal detachment when trimming the globe for embedding. A mixture of equal parts cold 4% buffered glutaraldehyde and 10% neutral formalin has been recommended as an ocular fixative for both light and electron microscopy.

In all domestic animals, the preferred section for histology is made from a midsagittal slab which includes optic nerve, thereby allowing examination of both tapetal and nontapetal fundus in the same section.

Bibliography

Dodds, W. J. *et al.* The frequencies of inherited blood and eye diseases as determined by genetic screening programs. *J Am Anim Hosp Assoc* **17:** 697–704, 1981.

Duke-Elder, S. "System of Ophthalmology," Vol. I–XV. St. Louis, Missouri, Mosby, 1964.

Fine, B. S., and Yanoff, M. "Ocular Histology," 2nd Ed. Hagerstown, Maryland, Harper & Row, 1979.

Gelatt, K. N. Feline ophthalmology. *Compend Cont Ed* **1:** 576–583, 1979.
Gelatt, K. N. (ed.). "Veterinary Ophthalmology." Philadelphia, Pennsylvania, Lea & Febiger, 1981.
Jensen, H. E. Histological changes in the canine eye related to aging. *Proc Am Coll Vet Ophthalmol,* 3–15, 1974.
Martin, C. L., and Anderson, B. G. Ocular anatomy. *In* "Veterinary Ophthalmology" K. N. Gelatt (ed.), pp. 58–64. Philadelphia, Pennsylvania, Lea & Febiger, 1981.
Peiffer, R. L. (ed.). "Comparative Ophthalmic Pathology." Springfield, Illinois, Charles C Thomas, 1983.
Prince, J. H. *et al.* "Anatomy and Histology of the Eye and Orbit in Domestic Animals." Springfield, Illinois, Charles C Thomas, 1960.
Rubin, L. F. "Atlas of Veterinary Ophthalmoscopy." Philadelphia, Pennsylvania, Lea & Febiger, 1974.
Rubin, L. F. "Inherited Eye Disease in Purebred Dogs." Philadelphia, Pennsylvania, Williams & Wilkins, 1989.
Saunders, L. Z., and Rubin, L. F. "Ophthalmic Pathology of Animals." Basel, Switzerland, Karger, 1975.
Smolin, G., and O'Connor, G. R. "Ocular Immunology," Philadelphia, Pennsylvania, Lea & Febiger, 1981.
Spencer, W. H. (ed.). "Opththalmic Pathology." Philadelphia, Pennsylvania, W. B. Saunders, 1985.
Stryer, L. The molecules of visual excitation. *Sci Am,* July 1987, 42–50.
Walls, G. L. (ed.). "The Vertebrate Eye and its Adaptive Radiation." New York, Hafner Publishing, 1967.
Yanoff, M., and Fine, B. S. "Ocular Pathology." 3rd Ed., Philadelphia, Pennsylvania, J. B. Lippincott, 1989.

II. Developmental Anomalies

Ocular developmental defects are common in domestic animals, particularly in purebred dog breeds in which extensive linebreeding has been used to increase the predictability of the phenotype. Many of the defects involve the eyelids and result from accentuation of anatomic peculiarities of the breed, such as entropion from deliberate enophthalmos or misdirected hairs from overly prominent facial folds. Such anomalies are clinically obvious and amenable to surgery, and rarely require the attention of a pathologist.

Anomalies of the globe are usually multiple, which reflects the interdependence of the various parts of the developing eye. Without proper consideration of ocular embryology, the lesions found in anomalous eyes can be a catolog of observations rather than predictable results of single errors in organogenesis. It is also important to recognize the differences in normal ocular structure among the various species, and the different rates at which mature form is attained. For example, the retina of carnivore eyes remains immature until about 6 weeks postnatally, whereas that of ruminants and horses is mature at birth.

The primary optic vesicle is an evagination of the forebrain that, with differential growth of brain and surface ectoderm, becomes separated from the presumptive diencephalon by the optic stalk. The apposition of primary optic vesicle to overlying surface ectoderm induces a focal ectodermal thickening, the lens placode. The placode grows to form a primitive lens vesicle. It is the developing lens which orchestrates the invagination of the optic vesicle to form the bilayered optic cup and bring the lining neuroectoderm into the apposition that provides the future photoreceptor and pigment epithelial layers. Surrounding the optic cup is a mass of mesenchyme, derived from neural crest, that will form the vascular and fibrous tunics of the eye under the induction of the differentiating neuroectoderm (Fig. 4.1). Ocular adnexa and muscles form independently and seem not to require normal development of the globe, as evidenced by the presence of normal lacrimal gland, lids, and extraocular muscles in most cases of severe microphthalmos.

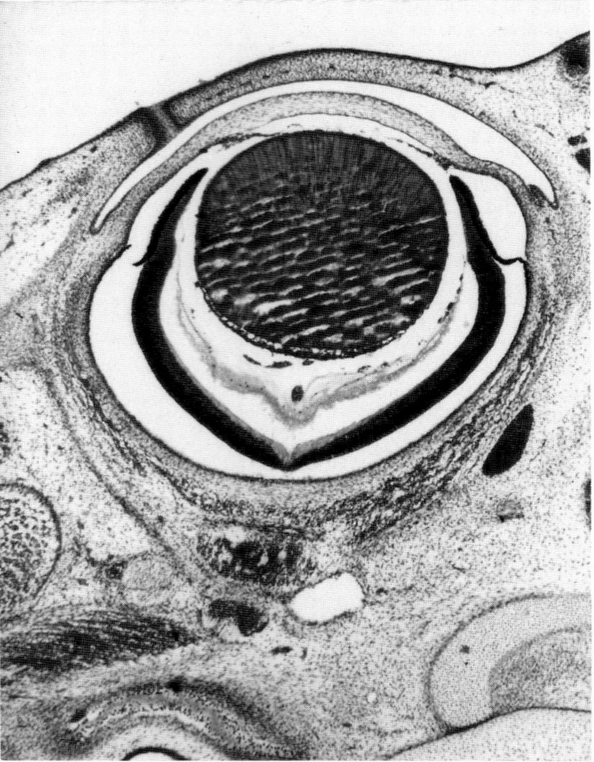

Fig. 4.1 Canine embryo, 34 days gestation. Lids fused, cornea fully formed. Large lens surrounded by complete vascular tunic derived posteriorly from hyaloid artery and anteriorly from the future pupillary membrane. Iris not yet formed. Retina almost fills cavity of optic vesicle.

Bibliography

Aguirre, G. D., Rubin, L. F., and Bistner, S. I. Development of the canine eye. *Am J Vet Res* **33:** 2399–2414, 1972.
Bellhorn, R. W. A survey of ocular findings in 16- to-24-week-old beagles. *J Am Vet Med Assoc* **162:** 139–141, 1973.
Bellhorn, R. W. A survey of ocular findings in eight- to-ten-month-old beagles. *J Am Vet Med Assoc* **164:** 1114–1116, 1974.
Bistner, S. I. Embryology of the canine and bovine eyes. *In* "Veterinary Ophthalmology," K. N. Gelatt (ed.), pp. 3–11. Philadelphia, Pennsylvania, Lea & Febiger, 1981.

Bistner, S. I., Rubin, L. F., and Aguirre, G. Development of the bovine eye. *Am J Vet Res* **34:** 7–12, 1973.

Donovan, A. The postnatal development of the cat retina. *Exp Eye Res* **5:** 249–254, 1966.

Duke-Elder, S. "System of Ophthalmology," Vol. III, parts 1 and 2. St. Louis, Missouri, Mosby, 1964.

Garner, A., and Griffiths, P. Bilateral congenital ocular defects in a foal. *Br J Ophthalmol* **53:** 513–517, 1969.

Gelatt, K. N., Leipold, H. W., and Huston, K. Congenital ophthalmic anomalies in cattle. *Mod Vet Pract* **57:** 105–109, 1976.

Greene, H. J. *et al.* Congenital defects in cattle. *Ir Vet J* **27:** 37–44, 1973.

Hu, F., and Montagna, W. The development of pigment cells in the eyes of rhesus monkeys. *Am J Anat* **132:** 119–132, 1971.

Huston, R., Saperstein, G., and Leipold, H. W. Congenital defects in foals. *Equine Med Surg* **1:** 146–161, 1977.

Mann, I. "Developmental Abnormalities of the Eye," 2nd Ed. Philadelphia, Pennsylvania, Lippincott, 1957.

Martin, C. L., and Anderson, B. G. Ocular anatomy. *In* "Veterinary Ophthalmology" K. N. Gelatt (ed.), pp. 58-64. Philadelphia, Pennsylvania Lea & Febiger, 1981.

Pei, Y. F., and Rhodin, J. A. G. The prenatal development of the mouse eye. *Anat Rec* **168:** 105–125, 1970.

Priester, W. A. Congenital ocular defects in cattle, horses, cats, and dogs. *J Am Vet Med Assoc* **160:** 1504–1511, 1972.

Riis, R. C. Equine ophthalmology. *In* "Veterinary Ophthalmology" K. N. Gelatt (ed.), pp. 509–570. Philadelphia, Pennsylvania, Lea & Febiger, 1981.

Saunders, L. Z., and Rubin, L. F. "Ophthalmic Pathology of Animals. An Atlas and Reference Book." Basel, Switzerland, Karger, 1975.

Selby, L. A., Hopps, H. C., and Edmonds, L. D. Comparative aspects of congenital malformations in man and swine. *J Am Vet Med Assoc* **159:** 1485–1490, 1971.

Weidman, T. A., and Kuwabara, T. Development of the rat retina. *Invest Ophthalmol Vis Sci* **8:** 60–69, 1969.

A. *Defective Organogenesis*

Failure of the eye to attain even the stage of optic cup is not a rare occurrence and is usually of unknown cause. The defect is usually bilateral but asymmetrical, and the severity of the defect relates to the stage of organogenesis at which the insult occurred. Failure of formation of the primary vesicle, or its early and complete regression, is true anophthalmos and is very rare. Failure of optic vesicle invagination gives rise to the very rare congenital cystic eye. Incomplete invagination results in congenital retinal nonattachment. Failure of division (or subsequent fusion) of the optic primordium as it grows from the telencephalon results in cyclopia, or synophthalmos, a single dysplastic midline globe.

1. Anophthalmos and Microphthalmos

Anophthalmos, total absence of ocular tissue, is a very rare lesion and almost all cases described are more correctly termed severe microphthalmos in that some vestige of eye is found in serially sectioned orbital content. The usefulness of distinguishing between the two is questionable, and many authors have adopted the term clinical anophthalmos for all such cases. Concurrent anomalies of skeletal and central nervous systems are common.

Macroscopic examination of orbital content usually reveals a normal lacrimal gland and vestigial extraocular muscles. The globe is usually recognized as an irregular mass of black pigment, with structures such as cornea or optic nerve variably recognizable. Histologically, there is almost always a mass of pigmented neurectoderm, reminiscent of ciliary processes, and some effort at retinal differentiation. There is frequently some remnant of lens, a finding which suggests regression of an embryonic globe that had reached at least the stage of optic cup. One or more plates of cartilage, presumably derived from third eyelid analog, are common.

2. Cyclopia and Synophthalmos

Damage to the prosencephalon prior to the outgrowth of the optic vesicles may result in improper separation of paired cranial midline structures, including eyes. Cyclopia defines a fetal malformation characterized by a single median orbit containing a single globe. Most specimens have some duplication of intraocular structures such as lens, iris, or hyaloid vessels and are thus more properly considered incomplete separation or early fusion (synophthalmos) (Fig. 4.2). Some specimens have two dysplastic globes within a single orbit. Severe cranial anomalies accompany cyclopia and synophthalmos, including absent or deformed ears, a median proboscis, cranioschisis, cleft palate, and brain anomalies ranging from microcephaly to hydranencephaly and hydrocephalus.

Cyclopianlike malformations have been reported in sheep, chickens, and dogs, and as inherited defects in cattle, with the most thoroughly documented cases being

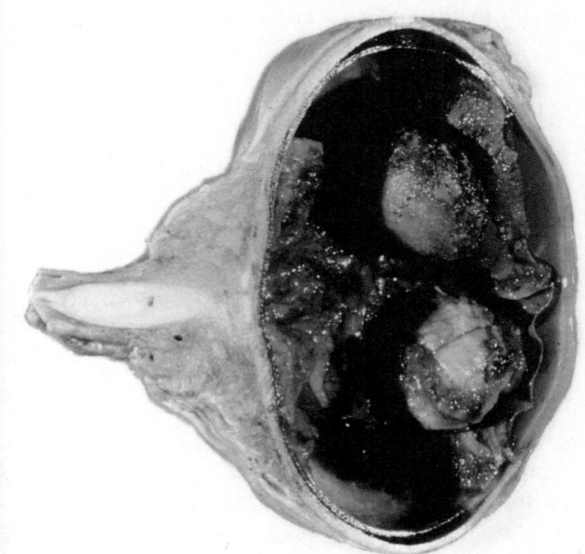

Fig. 4.2 Globe from typical cyclopian calf. Duplication of lens and pupil indicates synophthalmos rather than true cyclopia.

in sheep grazing alpine pastures rich in the legume *Veratrum californicum*. Fresh and dried plants contain three steroidal alkaloids—jervine, cyclopamine, and cyloposine—capable of damaging the developing neural groove of the fetal lamb. Ewes eating the plant on gestational day 15 have lambs with the cyclopian malformation, for it is at that time that the neural groove has formed and the first cranial somites are forming. A similar syndrome has been produced in kids and calves by maternal feeding of the plant on day 14 of gestation. Ingestion of the alkaloids prior to day 15 in sheep may cause fetal death but no anomalies, and exposure soon after day 15 may cause various skeletal abnormalities but not cyclopia.

In naturally occurring outbreaks, affected lambs have deformities ranging from cyclopia with microcephaly to relatively normal lambs with harelip and cleft palate. Prolonged gestation is common in the case of severely malformed fetuses.

3. Cystic Eye and Retinal Nonattachment

Failure of apposition of the optic vesicle to the cranial ectoderm results in failure of lens induction, which in turn removes the major stimulus for the invagination of the optic vesicle to form the optic cup. Persistence of the primary optic vesicle is seen as a cystic eye (Fig. 4.3), consisting of a scleral sheet lined by neurectoderm of variable neurosensory and pigmentary differentiation. The absence of lens and of bilayered iridociliary epithelium distinguish this rare lesion from the more common dysplastic eye of secondary microphthalmos.

Incomplete invagination of the optic vesicle allows persistence of the cavity of the primary optic vesicle and prevents attachment of the presumptive neurosensory retina to the developing retinal pigment epithelium. In the postnatal globe, retinal nonattachment cannot easily be distinguished from acquired retinal separation (Fig. 4.4). In each instance, the retina is extensively folded and may have improper differentiation of neuronal layers. The diagnosis of retinal nonattachment is assisted if there is also lack of apposition between the two layers of neurectoderm covering the anterior uvea (destined to be iridal and ciliary epithelium) and if retinal rosettes are evident. In addition, since nonattachment is an early and fundamental error in organogenesis, such eyes usually lack a lens and probably will be microphthalmic with multiple anomalies.

4. Coloboma

The mildest and latest defect in organogenesis results from failure of complete fusion of the lips of the embryonic fissure, a slitlike but normal channel in the floor of the optic cup and stalk through which the vasoformative mesoderm and stromal mesenchyme enter the globe. Failure of closure of the fissure may occur anywhere along its length, but the channel persists most frequently as a notchlike defect of the caudal pole at, or just ventral to, the optic disk and is lined by an outpouching of dysplastic neurectoderm. If the defect is sufficiently large, the outpouching

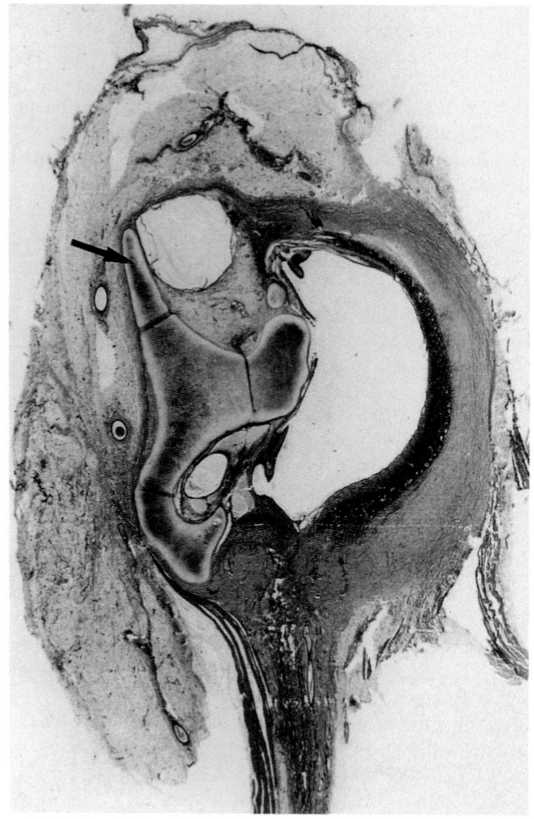

Fig. 4.3 Severe microphthalmos. Foal. There is no lens, no apparent attempt at invagination, and no neurosensory retinal differentiation. Persistence of the cavity of the optic vesicle qualifies this as a cystic eye. Cartilage plate (arrow) is probably analog of third eyelid.

of neurectoderm induces a similar bulge in the sclera, termed **scleral ectasia** (Fig. 4.5). Occasionally such ectasias are so large as to form a retrobulbar cyst as large as the globe itself (Fig. 4.6). Regardless of size, the lining of the scleral coloboma is formed by neurectoderm that bulged through the defect in the optic cup. Abortive neurosensory differentiation within the cyst wall is common (Fig. 4.7).

Colobomas occur in all domestic species but are especially frequent in collie dogs as one manifestation of the collie eye anomaly. They are rare in horses and cattle, and in both species, most cases have been reported in blue-eyed or incompletely albino animals. In Charolais cattle, colobomas of or near the optic disk are inherited as an autosomal dominant trait with incomplete penetrance. The lesion is bilateral but not necessarily equal in severity. Colobomas of iris or eyelid also occur but are rarely significant to ocular function and even more rarely receive histologic examination. Colobomas not aligned with the embryonic fissure also occur, but their pathogenesis is unknown.

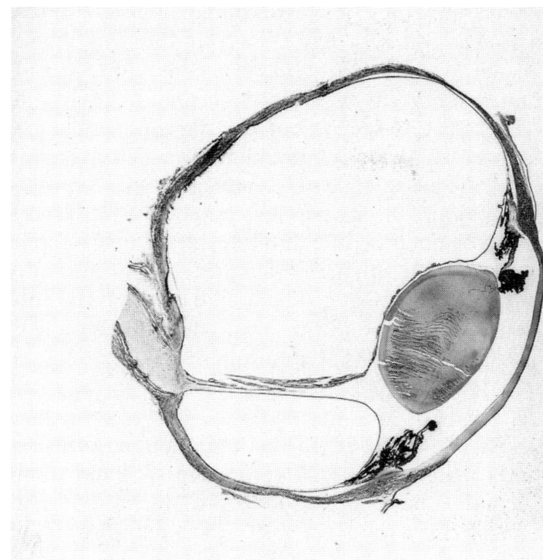

Fig. 4.4 Microphthalmos. Foal. There is congenital retinal detachment.

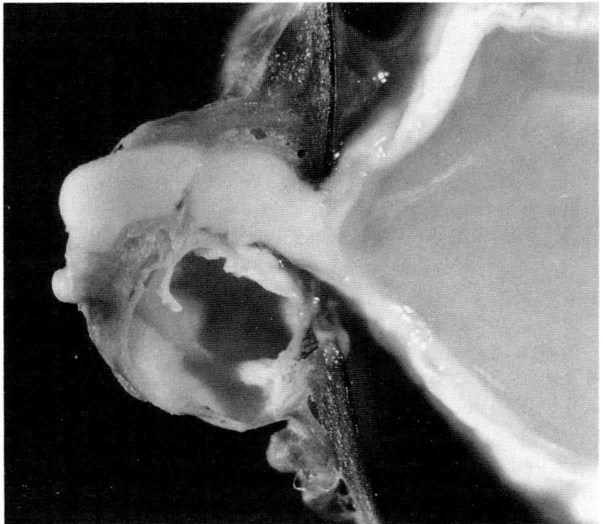

Fig. 4.5 Scleral ectasia and retinal separation. Collie. Cavity within the ectasia is analogous to the cavity of the primary optic vesicle.

B. Defective Differentiation

Subsequent to formation of the optic cup, ocular differentiation involves continued differentiation of neurectoderm into retinal and uveal neuroepithelium, and induction of primitive periocular mesenchyme to form the fibrous and vascular tunics of the globe. The normal development of retinal pigment epithelium from the neurectoderm of the posterior half of the optic vesicle seems prerequisite for these differentiations to occur. Aberrant differentiation of the surface ectoderm destined to form corneal epithelium and lens is infrequent except for those *in utero* degenerative diseases of the lens which lead to congenital cataract.

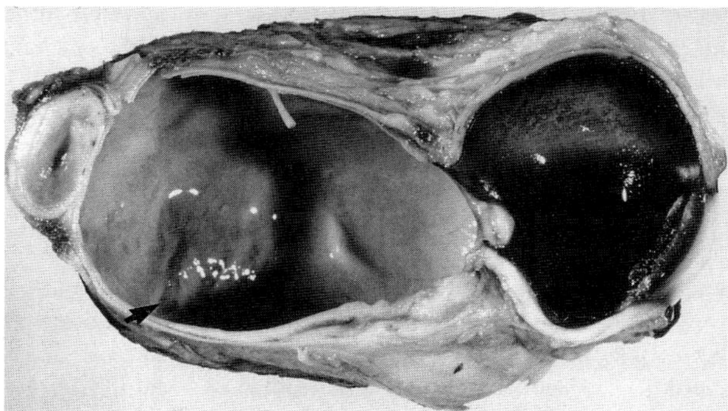

Fig. 4.6 Retrobulbar cyst (arrow) formed by coloboma and massive scleral ectasia. Calf. Globe is small. Retina is completely separated.

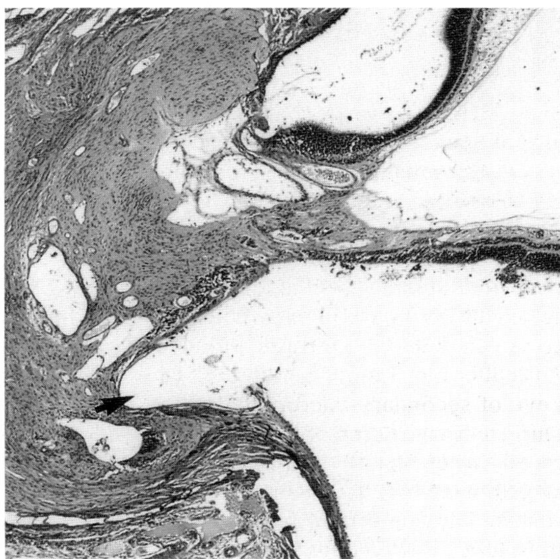

Fig. 4.7 Coloboma (arrow) at the optic disk. Collie pup with collie eye anomaly. Dysplastic neuroectoderm lines the defect and attempts to form sensory retina.

It is traditional to present specific ocular anomalies as they relate to structures of the adult eye, and thus as anomalies of cornea, iris, lens, retina, and so on. This approach correlates well with the clinical examination of the eye, but an understanding of the evolution of anomalies may best be gained if seen in terms of defective differentiation of the germ layers. For this reason, anomalies of ocular differentiation are presented here as anomalies of mesenchyme, ectoderm, and neurectoderm. Ocular neurectoderm is the primary organizer of ocular differentiation.

Bibliography

Bendixen, H. C. Littery occurrence of anophthalmia or microphthalmia together with other malformations in swine—presumably due to vitamin A deficiency of the maternal diet. *Acta Pathol Microbiol Scand (Suppl.)* **54:** 161–179, 1944.

Binns, W. *et al.* A congenital cyclopian-like malformation in lambs. *J Am Vet Med Assoc* **134:** 180–183, 1959.

Binns, W. *et al.* Chronologic evaluation of teratogenesis in sheep fed *Veratrum californicum*. *J Am Vet Med Assoc* **147:** 839–842, 1963.

Hale, F. The relation of vitamin A to anophthalmos in pigs. *Am J Ophthalmol* **18:** 1087–1093, 1935.

Leipold, H. W., and Huston, K. Congenital syndrome of anophthalmia—microphthalmia with associated defects in cattle. *Vet Pathol* **5:** 407–418, 1968.

Leipold, H. W., Gelatt, K. N., and Huston, K. Multiple ocular anomalies and hydrocephalus in grade beef shorthorn cattle. *Am J Vet Res* **32:** 1019–1026, 1971.

McCormack, J. Typical colobomas in Charolais bulls. *Vet Med Small Anim Clin* **72:** 1626–1628, 1977.

Rosenfeld, I., and Beath, O. A. Congenital malformations in the eyes of sheep. *J Agric Res* **75:** 93–103, 1947.

Rubin, L. F. Hereditary retinal detachment in Bedlington terriers: A preliminary report. *Small Anim Clin* **3:** 387–389, 1963.

Saunders, L. Z., and Fincher, M. G. Hereditary multiple eye defects in grade Jersey calves. *Cornell Vet* **41:** 351–366, 1957.

Wheeler, C. A., and Collier, L. L. Bilateral colobomas involving the optic discs in a quarter horse. *Eq Vet J (Suppl.)* **10:** 39–41, 1990.

1. *Anomalies of Mesenchyme*

After formation of the optic cup and separation of the lens vesicle, the periocular mesenchyme undergoes a complex series of migrations, differentiations, and atrophies that determines the final structure of the vascular and fibrous tunics of the globe. At the anterior edge of the optic cup, successive waves of mesenchymal invasion form corneal endothelium, corneal stroma, and the anterior half of the transient perilenticular vascular network. The posterior half is formed by invasion of vasoformative mesoderm and mesenchyme through the embryonic fissure to form the extensive hyaloid artery system (Fig. 4.1). Another mesenchymal wave accompanies the ingrowth of the neurectoderm at the anterior lip of the optic cup to form the iris stroma. Its peripheral portion later atrophies to form the porous filtration angle of the anterior chamber, a process that may not be completed in carnivores until several weeks after birth. The choroid and sclera are induced by the developing retinal pigment epithelium to form from the mesenchyme surrounding the caudal half of the optic cup.

Anomalies of mesenchyme may result from defective ingrowth or differentiation, as with choroidal and iris hypoplasia, but more frequent are the defects associated with incomplete atrophy of the normally transient embryonal mesenchyme of the intraocular vasculature or filtration angle, such as occurs with persistent pupillary membrane and some primary glaucomas.

a. CHOROIDAL HYPOPLASIA. This is a relatively common lesion in the eye of dogs by virtue of its prevalence in the collie breed as the hallmark of collie eye anomaly, and a very similar syndrome occurs in Australian shepherd dogs and Shetland sheepdogs. It is also seen in a variety of dog breeds in association with genes for color dilution (merle, dapple, and harlequin). The hypoplasia is thought to result from induction failure by a defective retinal pigment epithelium. The basic defect is not clearly established but may be related to defective pigmentation, a suggestion supported by the prevalence of iris and choroidal hypoplasia in white animals of all species, especially those with blue irises. Some degree of retinal dysplasia is also common, an observation which bears on the role of normally developing retinal pigment epithelium in ocular differentiation. Even in otherwise normal (nonwhite) animals with a blue iris, there is usually hypoplasia of the tapetum and choroid.

b. COLLIE EYE ANOMALY. This is a common disease of smooth and rough collies, first reported in 1953 and at one time estimated to have affected 90% of North American collies. During the period 1975–1979, the defect was still present in over 70% of 20,000 collies examined in a voluntary screening program. Prevalence in Europe and the United Kingdom is lower (30–60%). The basic defect, patchy to diffuse choroidal hypoplasia, is inherited as an autosomal recessive trait, but the numerous associated defects are more unpredictable in their familial pattern. Similar syndromes are reported in Border collies, Shetland sheepdogs, and Australian shepherd dogs, and are probably of similar pathogenesis. The prevalence in these breeds, as in rough collies, has marked geographic variation.

The ophthalmoscopic findings include one or more of retinal vessel tortuosity, focal to diffuse choroidal and tapetal hypoplasia, typical coloboma, and retinal separation with intraocular hemorrhage. Other observations that are occasionally made in eyes of affected dogs are enophthalmos, microphthalmos, and corneal stromal mineralization. The disease is always bilateral but not necessarily equal. Even the mild, visually insignificant lesion of focal choroidal hypoplasia is genetically significant.

Macroscopic examination of the bisected globe reveals abnormal pallor of the posterior segment of the globe. If the globe is transilluminated, the sclera and choroid are focally or diffusely more translucent than normal. The pallor and translucency imply choroidal hypoplasia. Within or adjacent to the optic disk there may be a colobomatous pit of variable size, the lining of which is continuous with the retina. Accompanying the larger type of pit is a bulge in the overlying sclera, called **scleral ectasia** or **posterior staphyloma.** If there is retinal separation, it is usually complete, with the only sites of attachment being at the abnormal optic disk and at the ora ciliaris (Fig. 4.8). In such cases, there may be extensive intravitreal hemorrhage and retinal tears. Almost all collie eyes with retinal separation have large optic disk colobomas. Detachment from the ora ciliaris may also occur, leaving the folded retina on the floor of the globe (Fig. 4.9).

The histologic lesion found in all affected eyes is choroidal hypoplasia, which virtually always is diffuse, despite

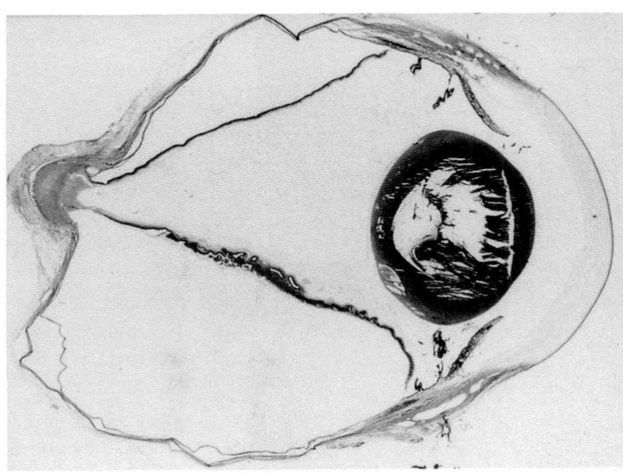

Fig. 4.8 Retinal separation. Collie eye anomaly. There is coloboma at the optic disk. Choroid and sclera are thin. Retinal folding does not constitute dysplasia.

ophthalmoscopic observation of a lesion that appears to be only focal. The choroid is thin and poorly pigmented, and the tapetum is thinner than normal or even absent (Fig. 4.10A,B). Retinal pigment epithelium is poorly pigmented even in nontapetal fundus and may be vacuolated. Because the choroid and tapetum in the normal dog do not reach adult thickness until about 4 months postpartum, aged-matched control eyes are essential if overinterpretation of normal choroidal immaturity is to be avoided.

Histologic examination of eyes with optic disk colobomas reveals the bulging of dysplastic neurectoderm, continuous with retina, into the pit in the nerve head. The neurectoderm may show jumbled differentiation into ganglion cells, photoreceptor rosettes, glial cells, or pigment epithelium. Rosettes are common in the neurosensory ret-

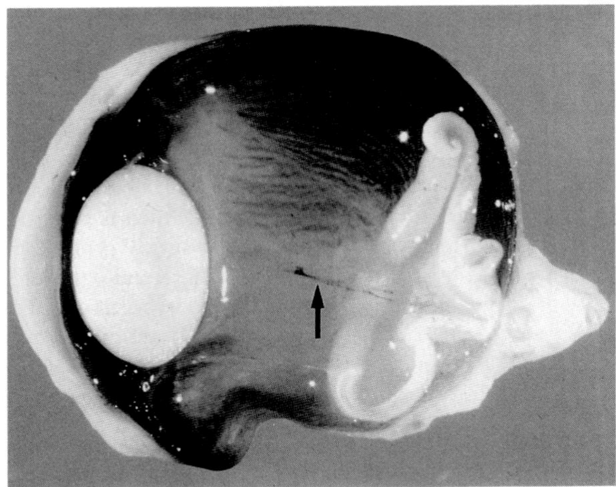

Fig. 4.9 Retinal separation from ora ciliaris. Collie eye anomaly. Note hypopigmented choroid and prominent hyaloid artery (arrow).

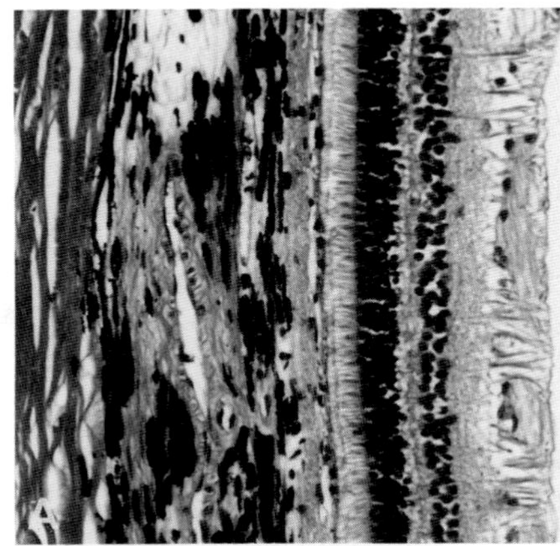

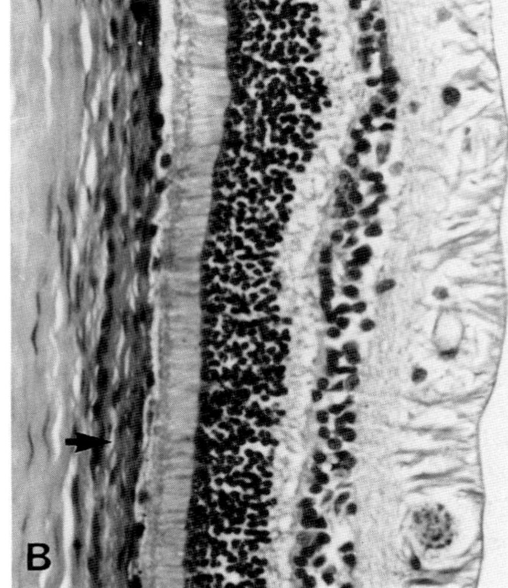

Fig. 4.10 (A) Normal posterior pole of the globe of 13-week-old puppy. Tapetum present but thin (normal for puppy). Choroidal thickness approximates that of retina. (B) Posterior pole just dorsal to the optic disk of 13-week-old collie with collie eye anomaly. Tapetum is absent. Choroid (arrow) severely hypoplastic.

ina adjacent to affected disks or embedded in the optic disk itself. In some specimens, there are degenerative retinal lesions overlying severely hypoplastic choroid. Edematous clefts are seen in the nerve fiber layer, and ganglion cells may be severely vacuolated.

Other retinal lesions include retinal folds and detachment. The folds are seen on histologic section as tubes of fully differentiated retina cut in cross section or tangentially, and are thought to represent folds in a neurosensory retina that at least temporarily has grown in excess of the space available for it within the optic cup. These folds correspond to the clinically detectable vermiform streaks on the fundus, and gradually disappear as the dog (and eye) matures, allowing the growth of scleral shell to catch up with that of retina. Presumably it is a more severe imbalance in growth of inner and outer layers of the optic cup which predisposes to retinal separation in about 10%

of eyes with this syndrome, as a retina which is too small attempts to stretch from optic disk to ora ciliaris by the shortest route.

Focal fibroblastic metaplasia and mineralization is occasionally seen in the subepithelial corneal stroma of dogs with collie eye anomaly, but a similar defect is seen in anomalous eyes of other breeds; a genetic link to the collie eye defect is not established. Tortuosity of retinal veins, a controversial clinical lesion sometimes considered part of collie eye anomaly, has no described histologic counterpart.

The earliest lesion of this anomaly is defective differentiation of primitive retinal pigment epithelium to form rosettelike structures near the optic disk. Proper differentiation of both pigment epithelium and neurosensory retina requires obliteration of the lumen of the primary optic vesicle, which allows the two neurectodermal layers to come into apposition. Whether the earliest lesion of collie eye anomaly results from inherently defective differentiation of pigment epithelium or from imperfect apposition of the two neurectodermal layers has not been resolved, but a central role of the pigment epithelium in determining ocular morphology suggests that the primary defect is in maturation of the presumptive retinal pigment epithelium. Anomalous development of choroid and sclera, including coloboma, is not seen in fetuses up to 45 gestational days but is seen in neonates. This suggests that the defect is in choroidal maturation rather than in formation. Another manifestation of mesenchymal maldevelopment is delayed atrophy and remodeling in the anterior chamber. The filtration angle may be closed, iris stroma may be attached to the corneal endothelium by a mesenchymal bridge, and remnants of anterior perilenticular mesenchyme are unusually prominent. Pigmentation of iridal neurectoderm is sparse. As these neonatal anterior segment lesions are not seen at the age (8–20 weeks) when puppies are examined ophthalmoscopically, it is presumed that they reflect only delayed mesenchymal remodeling.

2. *Defects of Anterior Segment Mesenchyme*

Hypoplasia of the iris is a rare defect that may occur alone or in conjunction with multiple ocular defects. It is relatively most frequent in horses, where it may be inherited and associated with cataract and conjunctival dermoids. The defect presumably results from incomplete inward migration of the anterior lip of the optic cup, with resultant lack of a neurectodermal scaffold to guide the migration of presumptive iris stroma. The hypoplasia is usually severe, and most cases are clinically described as aniridia. Histologic examination of such eyes usually reveals the vestigial iris as a triangular mesenchymal stump covered posteriorly by normal-appearing pigmented epithelium (Fig. 4.11A). The trabecular meshwork within the filtration angles may be malformed, but the ciliary apparatus is usually normal. The lens often is cataractous (Fig. 4.11B) and sometimes ectopic or hypoplastic. Glaucoma has been described as a sequel in horses (but not in other species), and would be an expected sequel in severely affected eyes in any species.

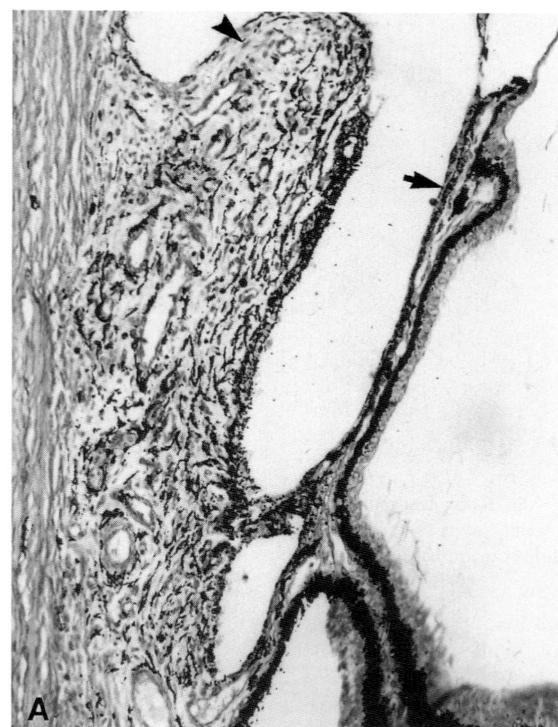

Fig. 4.11A Iris hypoplasia (arrow) and goniodysgenesis (arrowhead). Dog. Stroma and epithelium are inadequate.

Hypopigmentation of the iris may be unilateral (heterochromia iridis) or bilateral (ocular albinism). The iris is normal except for incomplete development of pigment granules in the cytoplasm of otherwise normal stromal and epithelial cells. Tapetum and, less reliably, choroid of affected eyes usually are hypoplastic.

Incomplete atrophy of the anterior chamber mesenchyme is relatively common in dogs and occurs occasionally in other domestic species. During organogenesis, three waves of mesenchyme migrate between the surface ectoderm and the anterior rim of the optic cup. The first two waves form corneal endothelium and stroma. The third wave forms a sheet stretching across the face of the lens and future iris. It differentiates centrally into the anterior half of the perilenticular vascular tunic and atrophies late in gestation or in the early postnatal period. Failure to atrophy results in **persistent pupillary membrane.** The more peripheral portions of the third mesenchymal wave differentiate into iris stroma and trabecular meshwork, the latter by a combination of atrophy and remodeling that continues until the third postnatal week in dogs.

Much less common are those defects grouped under the general category of anterior segment dysplasia or **anterior segment cleavage syndromes,** which include multiple

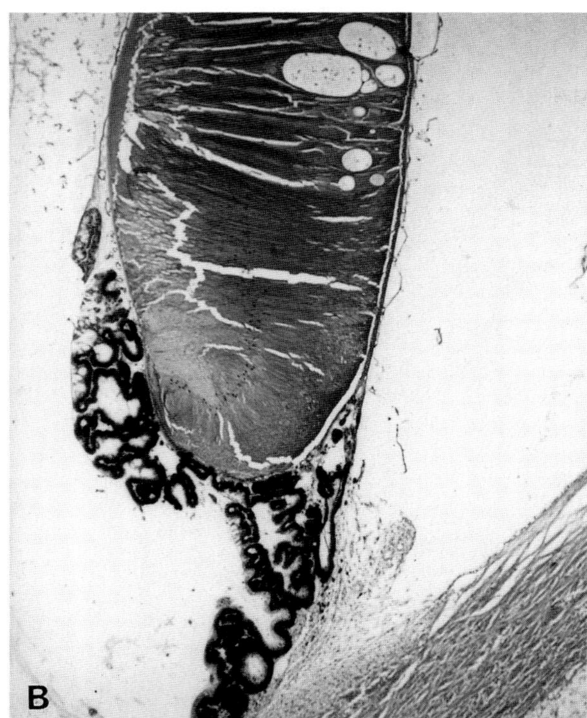

Fig. 4.11B Iris hypoplasia, congenital cataract and dysplasia of ciliary processes. Piglet, one of three affected.

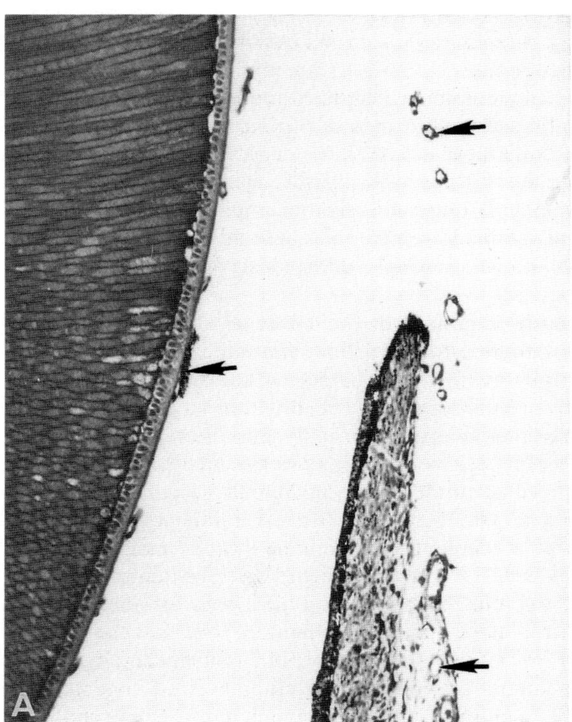

Fig. 4.12A Persistence of anterior tunica vasculosa lentis. Portions are seen as fine vascular channels on anterior lens capsule, in the pupil, and on anterior surface of iris (arrows).

anomalies of cornea, lens, and anterior uvea that stem from disordered development of anterior segment mesenchyme. Such eyes are commonly microphthalmic and usually have microphakia, cataract, and congenital corneal opacities at sites of congenital synechiae.

Persistent pupillary membrane refers to the delayed or incomplete atrophy of the anterior perilenticular vascular network that, in the fetus, originates from the minor arterial circle of the iris and invests the developing lens. Atrophy is frequently incomplete at birth and, in dogs, persistent remnants are common up to about 6 months of age. These insignificant and usually bloodless strands are seen as short, threadlike protrusions from the area of the minor arterial circle (iris collarette), and they may insert elsewhere on the iris, cross the pupil, or extend blindly into the anterior chamber (Fig. 4.12A). Persistent pupillary membranes achieve clinical significance in two ways. First, the size and number of strands crossing the pupil may be such that vision is obstructed. Second, strands that contact lens or cornea are associated with focal dysplasia of lens or corneal endothelium, clinically seen as opacity (Fig. 4.12B).

Histologic descriptions are mainly from studies in basenji dogs, in which persistent pupillary membrane occurs as an autosomal recessive trait of variable penetrance. In this breed, atrophy of the pupillary membrane is abnormally slow even in dogs free of the defect in adult life, and remnants in puppies up to 8 months old are common. The membranes are seen as thin endothelial tubes, invested with a thin adventitial stroma, extending from vessels in the iris stroma near the collarette. The tubes are usually empty, but in severely affected eyes may contain erythrocytes, and the adventitia may contain melanin. The tubes

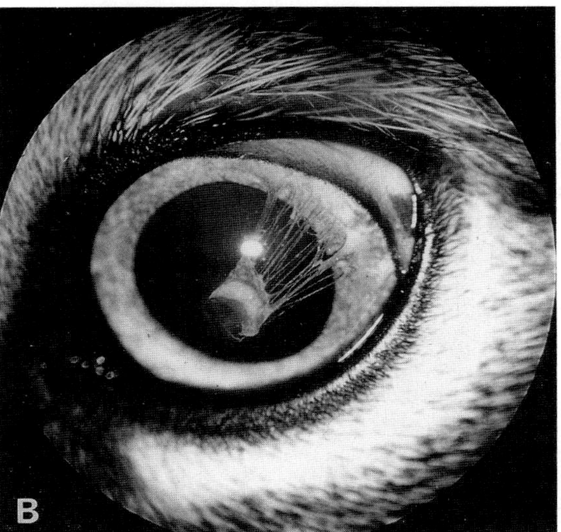

Fig. 4.12B Persistent pupillary membrane. Dog. Central crescent insertion is on the anterior pole of the lens.

weave in and out of the plane of section en route to corneal, iridic, or lenticular insertions. At sites of corneal insertion, corneal endothelium is either absent or dysplastic, with the latter manifested as fibrous metaplasia. Descemet's membrane is malformed or absent in the areas of attachment, and there is associated deep stromal corneal edema to account for the clinically observed, minute gray stromal opacities. Contact with the lens is accompanied by similar epithelial and basement membrane dysplasia, resulting in one or more epithelial, subcapsular, or polar cortical cataracts.

Maldevelopment of the **filtration angle (goniodysgenesis)** occurs as a prevalent, inherited condition in dogs and in severely anomalous eyes of animals of any species. The defect may result from incomplete atrophy of mesenchyme that normally fills the fetal iridocorneal angle (Fig. 4.13A,B), or may represent a true dysplasia of the fibrillar condensation destined to form the pectinate ligament. In either event, the result is seen as an imperforate or inadequately perforated mesodermal sheet separating anterior chamber from the trabecular meshwork. The only lesion may be a pectinate ligament that is thicker, more heavily pigmented, and less fenestrated than normal. Alternatively, the trabecular meshwork may appear as a solid mesenchymal mass that may not be distinguishable from postglaucoma compression and fibrosis of a developmentally normal filtration angle (see Glaucoma).

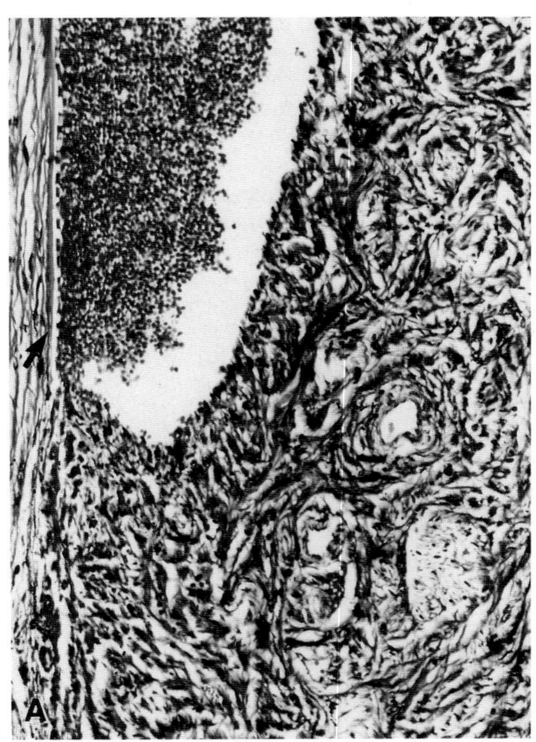

Fig. 4.13A Goniodysgenesis. Calf. Note dense mesoderm in the area that should be the open lattice of the trabecular meshwork. Termination of Descemet's membrane, where the pectinate ligament should insert (arrow).

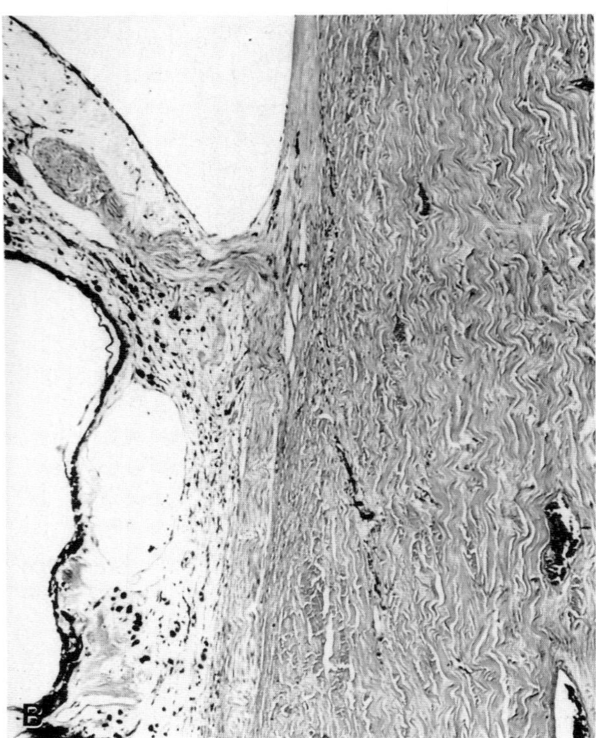

Fig. 4.13B Goniodysgenesis. Ciliary cleft is filled with primitive mesenchyme and has no development of trabecular meshwork.

3. Incomplete Atrophy of Posterior Segment Mesenchyme

Incomplete atrophy of posterior segment mesenchyme may result in the mild and common lesion of **persistent hyaloid artery,** or in the much rarer but clinically more significant lesions of **persistent posterior perilenticular vascular tunic** with or without concurrent persistence of the primary vitreous. The hyaloid artery and its branches are formed from mesenchyme that enters the optic cup through the embryonic fissure prior to its closure. The vessel traverses the optic cup from optic disk to lens, there it ramifies over the posterior lens surface and joins with the branches of the anterior chamber pupillary membrane to form a complete perilenticular vascular tunic. As with its anterior chamber counterpart, the hyaloid system undergoes almost complete atrophy before birth. Persistence of some vestige into adult life is common and clinically insignificant. In ruminants the most common remnant is Bergmeister's papilla, a cone of glial tissue with a vascular core which extends from optic disk for a few millimeters into the vitreous. In calves until about 2 months of age, the vestigial hyaloid system may still contain blood. In carnivores, it is the anterior perilenticular portion that normally persists for

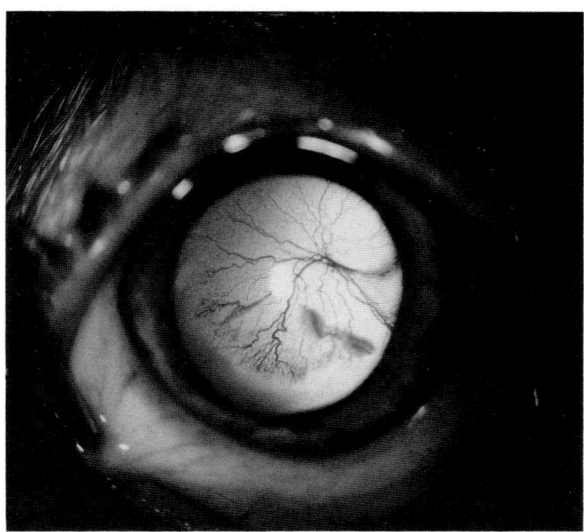

Fig. 4.14 Persistence of hyaloid artery and posterior tunica vasculosa lentis. Dog. Posterior polar cataract is the almost inevitable consequence as seen here.

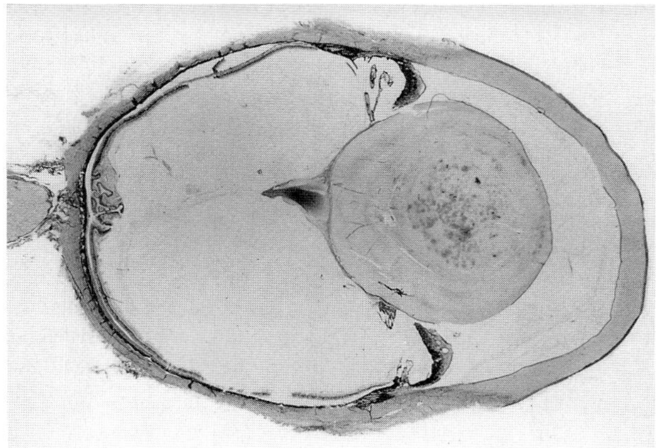

Fig. 4.15 Persistent hyperplastic primary vitreous. A fibrocartilaginous plaque adheres to an elongated lens (lenticonus) and extends to the posterior pole. Retina is dysplastic near the optic disk.

several weeks postnatally. Bloodless remnants of the main hyaloid artery are common in dogs and ruminants up to several years of age.

Much less common is undue persistence and even hyperplasia of the anterior end of the hyaloid system, the **posterior tunica vasculosa lentis** (Fig. 4.14). In humans the retained tissue may be predominantly fibrous and is thought to be a metaplastic derivative of the neurectodermal primary vitreous which accompanies the hyaloid vasculature. This rare anomaly, also called **persistent hyperplastic primary vitreous,** is typically unilateral in humans and is accompanied by microphthalmos, microphakia, retinal detachment, shallow anterior chamber, and embryonic filtration angles. The several reports of this anomaly in dogs have described a unilateral or bilateral retrolental vascular or fibrovascular network, usually without any reported anomalies other than the expected posterior polar cataract. Such lesions are better described as persistent posterior tunica vasculosa lentis. In Doberman pinschers and Staffordshire bull terriers, the defect is inherited and forms a spectrum that includes persistent pupillary membrane, cataract, lenticonus, and microphthalmia, as well as persistence of variable amounts of primary vitreous and posterior tunica vasculosa lentis (Fig. 4.15). The defects are detected as early as gestational day 30, at which time hyperplasia of posterior tunica vasculosa lentis is already obvious. Posterior polar cataracts and preretinal membranes are observed by day 37.

Bibliography

Arnbjerg, J., and Jensen, O. A. Spontaneous microphthalmia in two Doberman puppies with anterior chamber cleavage syndrome. *J Am Anim Hosp Assoc* **18:** 481–484, 1982.

Barnett, K. C., and Grimes, T. D. Unilateral persistence and hyperplasia of the primary vitreous in the dog. *J Small Anim Pract* **14:** 561–565, 1973.

Barnett, K. C., and Knight, G. C. Persistent pupillary membrane and associated defects in the basenji. *Vet Rec* **85:** 242–249, 1969.

Bertram, T., Coignoul, F., and Cheville, N. Ocular dysgenesis in Australian shepherd dogs. *J Am Anim Hosp Assoc* **20:** 177–182, 1984.

Bistner, S., Shaw, D., and Riis, R. C. Diseases of the uveal tract (Part I). *Compend Cont Ed* **1:** 868–876, 1979.

Boevé, M. H., van der Linde-Sipman, T., and Stades, F. C. Early morphogenesis of persistent hyperplastic tunica vasculosa lentis and primary vitreous. *Invest Ophthalmol Vis Sci* **29:** 1076–1086, 1988.

Creel, D. Inappropriate use of albino animals in research. *Pharmacol Biochem Behav* **12:** 969–977, 1980.

Eriksson, K. Hereditary aniridia with secondary cataract in horses. *Nord Vet Med* **7:** 773–793, 1955.

Gelatt, K. N. The canine glaucomas. *In* "Veterinary Ophthalmology" K. N. Gelatt (ed.), pp. 390–434. Philadelphia, Pennsylvania, Lea & Febiger, 1981.

Gelatt, K. N., and McGill, L. D. Clinical characteristics of microphthalmia with colobomas of the Australian shepherd dog. *J Am Vet Med Assoc* **162:** 393–396, 1973.

Gelatt, K. N., Huston, K., and Leipold, H. W. Ocular anomalies in incomplete albino cattle. *Am J Vet Res* **30:** 1313–1316, 1969.

Grimes, T. D., and Mullaney, J. Persistent hyperplastic primary vitreous in a greyhound. *Vet Rec* **85:** 607–611, 1969.

Joyce, J. R. *et al.* Iridal hypoplasia (aniridia) accompanied by limbic dermoids and cataracts in a group of related quarterhorses. *Eq Vet J (Suppl)* **10:** 26–28, 1990.

Latshaw, W. K., Wyman, M., and Benzke, W. G. Embryologic development of an anomaly of ocular fundus in the collie dog. *Am J Vet Res* **30:** 211–217, 1969.

Lucus, D. R. Ocular associations of dappling in the coat colours of dogs. II. Histology. *J Comp Pathol* **53:** 260–266, 1954.

Martin, C. L. Development of pectinate ligament structure of the dog: Study by scanning electron microscopy. *Am J Vet Res* **35:** 1433–1439, 1974.

Peiffer, R. L., Gelatt, K. N., and Gwin, R. M. Persistent primary vitreous and pigmented cataract in a dog. *J Am Anim Hosp Assoc* **13:** 478–480, 1977.

Pruett, R. C., and Schepens, C. L. Posterior hyperplastic primary vitreous. *Am J Ophthalmol* **69:** 535–543, 1970.

Rebhun, W. C. Persistent hyperplastic primary vitreous in a dog. *J Am Vet Med Assoc* **169:** 620–622, 1976.

Reese, A. B. Persistent hyperplastic primary vitreous. *Am J Ophthalmol* **40:** 317–331, 1955.

Roberts, S. R. Color dilution and hereditary defects in collie dogs. *Am J Ophthalmol* **63:** 1762–1775, 1967.

Roberts, S. R., and Bistner, S. I. Persistent pupillary membrane in basenji dogs. *J Am Vet Med Assoc* **153:** 533–542, 1968.

Roberts, S. R., Dellaporta, A., and Winter, F. C. The collie ectasia syndrome. Pathologic alterations in the eyes of puppies one to fourteen days of age. *Am J Ophthalmol* **61:** 1458–1466, 1966.

Rubin, L. F., Nelson, E. J., and Sharp, C. A. Collie eye anomaly in Australian shepherd dogs. *Prog Vet Comp Ophthalmol* **1:** 105–108, 1991.

Smelser, G. K., and Azanics, V. The development of the trabecular meshwork in primate eyes. *Am J Ophthalmol* **71:** 366–385, 1971.

Stades, F. C. Persistent hyperplastic tunica vasculosa lentis and persistent hyperplastic primary vitreous in 90 closely-related Doberman pinschers: Clinical aspects. *J Am Anim Hosp Assoc* **16:** 739–751, 1980.

Van der Linde-Sipman, J. S., Stades, F. C., and DeWolff-Rouendaal, D. Persistent hyperplastic tunica vasculosa lentes and persistent hyperplastic primary vitreous in the Doberman pinscher. Pathological aspects. *J Am Anim Hosp Assoc* **19:** 791–802, 1983.

Yakely, W. L., *et al.* Genetic transmission of an ocular fundus anomaly in collies. *J Am Vet Med Assoc* **152:** 457–461, 1968.

4. Anomalies of Neurectoderm

Included under this heading are anomalies of retina, optic nerve, and of neuroepithelium of iris and ciliary body. Of these, retinal anomalies are by far the most frequent and most significant.

a. Retinal Dysplasia Retinal dysplasia is a general term denoting abnormal retinal differentiation, characterized by jumbling of retinal layers and by glial proliferation. Retinal dysplasia is common in dogs and in cattle, and results from failure of proper apposition of the two layers of the optic cup, from failure of proper induction by an inherently defective retinal pigment epithelium, or from necrosis of the developing retina. Regardless of pathogenesis, the dysplastic retina is characterized by retinal folds, retinal rosettes, patchy to diffuse blending of nuclear layers, loss of retinal cells, and glial scars. The folds and rosettes are the histologic counterparts of the vermiform streaks seen on the fundus with the ophthalmoscope. The hallmark of retinal dysplasia is the rosette, composed of a central lumen surrounded by one to three layers of neuroblasts. The three-layered rosette is the most common in naturally occurring cases in animals, and shows more or less complete retinal differentiation. Most such rosettes are probably retinal folds cut transversely. The lumen contains pink fibrils resembling photoreceptors and is bounded by a thin membrane resembling the normal outer limiting membrane. One- and two-layered rosettes are encountered infrequently and consist of a lumen surrounded by undifferentiated neuroblasts.

There are good examples of each class of retinal dysplasia among naturally occurring diseases of domestic animals in which the retinal lesion is the major or sole ocular change. In addition, anomalous retinal differentiation is to be expected in any severely anomalous globe, such as those with cyclopia, microphthalmos, or large colobomas. In dogs, the location, ophthalmoscopic appearance, and effect on vision vary from breed to breed but tend to be uniform within each breed, a fact used by clinical ophthalmologists when attempting to distinguish inherited dysplasias from those occurring as isolated anomalies or as sequelae to *in utero* infections.

The first type, that resultant from retinal nonattachment, should probably be considered extensive retinal folding rather than true dysplasia, unless actual disorganization and blending of neural layers can be demonstrated. There may be associated cataract. Cases with retinal nonattachment have extensively folded retinas, since the distance to be traversed from disk to ora ciliaris in a straight line is shorter than the convex route taken by attached retina, and redundant retina is obliged to fold upon itself.

The second type, which is probably the one seen with greatest frequency in dogs and in the greatest variety of breeds, is retinal folding and subtle disorganization of the outer nuclear layer. The defect may be nonprogressive and clinically insignificant or progress to complete retinal separation and blindness, with the outcome greatly influenced by the specific breed in which the defect occurs. In English springer spaniels, in which this type of dysplasia has been most extensively studied, changes are seen as early as gestational day 45 and always by day 55. Focal infolding of the neuroblastic layer away from the retinal pigment epithelium and focal loss of the junctions between the neuroblasts (the outer limiting membrane) are the early changes, followed by overt focal retinal separation and extensive retinal folding. In all breeds in which this type of dysplasia has been adequately studied, it is inherited as a simple autosomal trait. The primary defect is probably related to faulty induction by the pigment epithelium, resulting in a focal dysregulation of retinal growth that leads to the retinal folding.

Retinal dysplasia occurs in combination with chondrodysplasia in several dog breeds, but particularly in Labrador retrievers. Cataract and persistent hyaloid remnants may accompany the retinal lesion. In this breed, all the defects are the result of a single gene, with recessive effects on the skeleton and incompletely dominant effects on the eye.

Another common example of retinal dysplasia, thought to be secondary to defects of retinal pigment epithelium, is seen in collie eye anomaly, in which foci of dysplastic retina are common in the peripapillary retina, in the optic disk, and in the wall of the colobomas. Some of these foci, which may include rosettes, are thought to represent dysplastic differentiation of primitive retinal pigment epithelium. Others, however, routinely disappear as the eye grows, suggesting that the folding results from a temporary imbalance between retinal and scleral growth. Since the other defects of choroidal and scleral formation and maturation in this syndrome are attributed to faulty induction by a defective retinal pigment epithelium (RPE), it is reasonable to attribute the retinal folding to the same mechanism.

In the third category, retinal necrosis, are found the most complex examples of retinal dysplasia in domestic animals. A wide variety of viral and physico–chemical insults to the embryonic eye may cause retinal maldevelopment, but naturally occurring examples are almost exclusively viral. Since the carnivore retina continues to develop for about 6 weeks after birth, the opportunity is great for injury postnatally to produce retinal maldevelopment in puppies and kittens. Retinal maturation is most rapid in central (peripapillary) retina and progressively less toward the periphery, so that occasionally dysplastic lesions may be encountered only in peripheral retina, suggesting a viral (or other) injury quite close to the 6-week-old limit for dysplasia of this pathogenesis. Mature retina will scar but will not develop lesions of dysplasia.

The specific viruses implicated in domestic animals are bovine virus diarrhea virus in cattle, bluetongue virus in sheep, herpesvirus in dogs, and both parvovirus and leukemia virus in cats. The histologic lesion is similar for all diseases, with variation in lesions caused by the same virus in one species as great as the variation caused by different viruses in different species. The most significant clue suggesting viral rather than genetic cause is the presence of residual inflammation and postnecrotic scarring in retina, optic nerve, and perhaps subtly, in choroid. Injured retinal pigment epithelium undergoes one or more of reactive hyperplasia, migration into injured retina as discrete pigmented cells in areas of scarred retina, or metaplastic formation of multilayered fibroglial plaques in place of normal simple cuboidal epithelium (Fig. 4.16A,B). Disorganization of nuclear layers and rosette formation are seen, as in other types of dysplasia.

Infection of calves with bovine virus diarrhea virus between 79 and 150 gestation days is the most frequently encountered and thoroughly studied retinal dysplasia of known viral etiology. Work with other viruses has been too limited to allow definition of the susceptible period in fetal development or of the full range of resultant lesions. The limited descriptions of the other viral-induced retinal lesions suggest that the sequence of events is probably quite similar for all such agents.

The initial ocular lesion is nonsuppurative panuveitis and retinitis with multifocal retinal and choroidal necrosis. The acute inflammatory disease gradually subsides over several weeks, and most cases of spontaneous abortion or neonatal death retain scant vestige of previous inflammation. Those ocular structures already well differentiated at the time of the endophthalmitis (cornea, uvea, optic nerve) may undergo atrophy and scarring or be left virtually untouched. Other tissues such as retina are actively

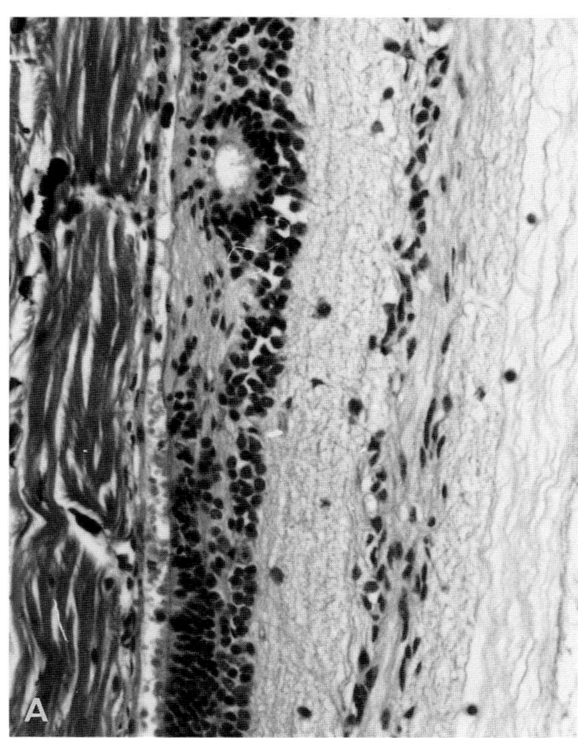

Fig. 4.16A Retinal dysplasia. BVD virus. Calf. Focal retinal scar with loss of outer nuclear layer and photoreceptors. Abortive regeneration with small rosette. Blending of inner and outer nuclear layers. Note lack of ganglion cells.

differentiating and exhibit a combination of this atrophy and scarring as well as abortive regeneration and arrested differentiation. Retinal pigment epithelium in most examples (bluetongue being an apparent exception) is infected and subsequently injured. The result is a patchy alternation of abortive retinal regeneration, hyperplastic pigment epithelium, and postnecrotic glial scarring (Fig. 4.17). The lesions are usually more severe in nontapetal retina and are bilateral but not necessarily symmetrical. It seems reasonable to speculate that those naturally occurring cases in which the dysplasia is confined to peripheral retina represent late viral infection when only peripheral retina is still differentiating.

Because the virus has affinity for other neural tissues, all calves with retinal dysplasia induced by bovine viral diarrhea virus also have cerebellar atrophy, and some have hydrocephalus or hydranencephaly. A similar association with hydrocephalus and other brain anomalies has been described for feline parvovirus infection in cats, bluetongue virus infection in sheep, and in a possibly hereditary syndrome in white shorthorn and Hereford cattle. In the latter two instances, the involvement of virus could not be excluded based on published information.

b. Optic Nerve Hypoplasia Hypoplasia is the most common anomaly of the optic nerve. The defect may be

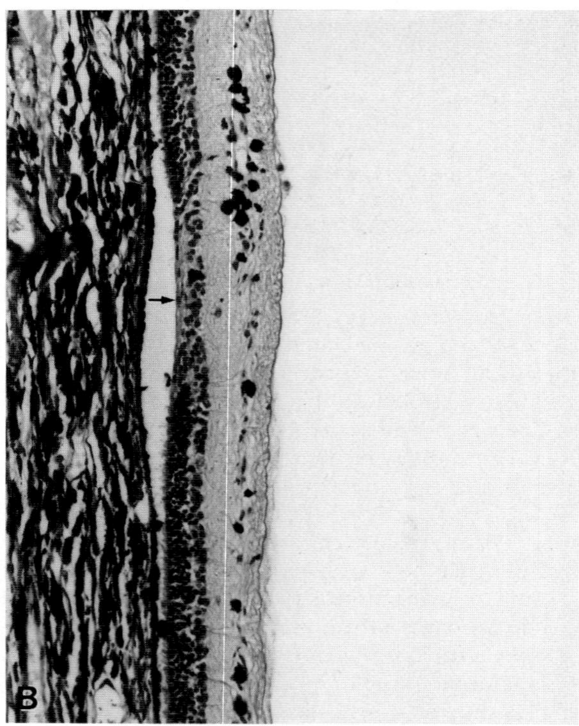

Fig. 4.16B Calf, BVD. Pigment-laden cells from retinal pigment epithelium have migrated into ganglion cell layer. Postnecrotic scar (arrow) with loss of photoreceptors and outer nuclear layer.

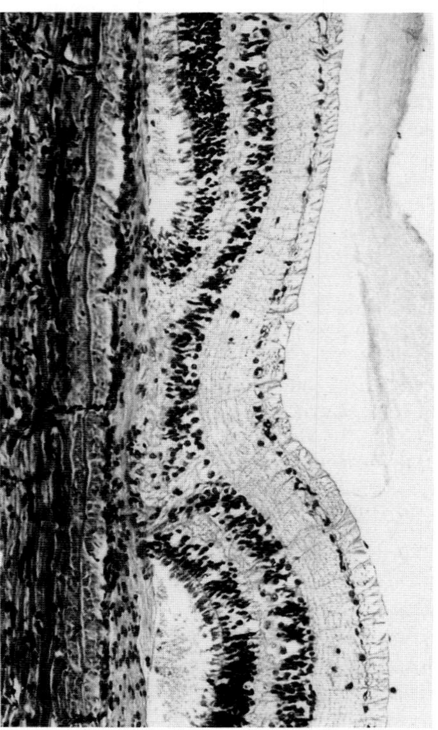

Fig. 4.17 Focal chorioretinal scar with loss of outer nuclear layer and fibrous metaplasia of adjacent retinal pigment epithelium. Calf, prenatal BVD infection.

unilateral or bilateral, and usually occurs in eyes with other anomalies and particularly in eyes with retinal dysplasia. In most instances, the so-called hypoplasia is more likely to be atrophy as the inevitable result of the destruction of ganglion cells in viral, toxic, genetic, or idiopathic retinal disease (Fig. 4.18). The only clear example of an alternative pathogenesis is that associated with maternal deficiency of vitamin A in cattle, in which atrophy of the developing optic nerve results from failure of remodeling of the optic nerve foramen and subsequent stenosis. A similar lesion occurs in pigs, but in that species, hypovitaminosis A seems more indiscriminately teratogenic, and optic nerve hypoplasia is accompanied by diffuse ocular dysplasia and multiple systemic anomalies. Hypoplasia is a relatively frequent clinical diagnosis in toy breeds of dogs, without apparent visual defects (and thus rarely receives histologic examination). Most examples are probably hypomyelination of the optic disk, which results from premature halt of myelinated nerve fibers at, or posterior to, the lamina cribrosa. The opposite, with myelin extending too far into the nerve fiber layer of the peripapillary retina, is also seen in dogs and is a frequent but insignificant occurrence in horses.

Inherited optic nerve hypoplasia is documented in one strain of laboratory mice, although it may accompany inherited retinal dysplasia or multiple inherited anomalies

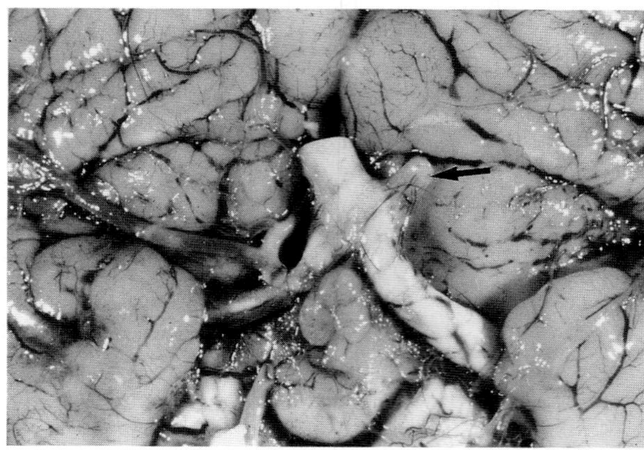

Fig. 4.18 Optic chiasm. Foal with unilateral secondary (degenerative) microphthalmos. Small left optic nerve (arrow) due to prenatal atrophy following ganglion cell destruction.

in any species. Histologic examination of affected eyes reveals few if any ganglion cells and a thin and moth-eaten nerve fiber layer.

Bibliography

Albert, D. M. Retinal neoplasia and dysplasia. I. Induction by feline leukemia virus. *Invest Ophthalmol* **16:** 325–337, 1977.

Albert, D. M. *et al.* Canine herpes-induced retinal dysplasia and associated ocular anomalies. *Invest Ophthalmol* **15:** 267–278, 1976.

Ashton, N., Barnett, K. C., and Sachs, D. D. Retinal dysplasia in Sealyham terriers. *J Pathol Bact* **96:** 269–272, 1968.

Barnett, K. C. Comparative aspects of canine hereditary eye disease. *Adv Vet Sci* **20:** 39–67, 1976.

Barnett, K. C., Bjorck, G. R., and Kock, E. Hereditary retinal dysplasia in the Labrador retriever in England and Sweden. *J Small Anim Pract* **10:** 755–759, 1969.

Barnett, K. C., and Grimes, T. D. Bilateral aplasia of the optic nerve in a cat. *Br J Ophthalmol* **58:** 663–667, 1974.

Barrie, K. N., and Gelatt, K. N. Diseases of the canine posterior segment: The ocular fundus. *In* "Veterinary Ophthalmology" K. N. Gelatt (ed.), pp. 480–484, Philadelphia, Pennsylvania, Lea & Febiger, 1981.

Bistner, S. I., Rubin, L. F., and Saunders, L. Z. The ocular lesions of bovine viral diarrhea-mucosal disease. *Vet Pathol* **7:** 275–286, 1970.

Brown, T. T. *et al.* Pathogenetic studies of infection of the bovine fetus with bovine viral diarrhea virus. II. Ocular lesions. *Vet Pathol* **12:** 393–404, 1975.

Carrig, C. B. *et al.* Retinal dysplasia associated with skeletal abnormalities in Labrador retrievers. *J Am Vet Med Assoc* **170:** 49–57, 1977.

Gelatt, K. N. Inherited retinopathies in the dog. *Compend Cont Ed* **1:** 307–313, 1979.

Gelatt, K. N., Leipold, H. W., and Coffman, J. R. Bilateral optic nerve hypoplasia in a colt. *J Am Vet Med Assoc* **155:** 627–631, 1969.

Greene, H. J., and Leipold, H. W. Hereditary internal hydrocephalus and retinal dysplasia in shorthorn calves. *Cornell Vet* **64:** 367–375, 1974.

Hayes, K. C., Nielsen, G. W., and Eaton, H. D. Pathogenesis of the optic nerve lesion in vitamin A-deficient calves. *Arch Ophthalmol* **80:** 777–787, 1968.

Heywood, R., and Wells, G. A. H. A retinal dysplasia in the beagle dog. *Vet Rec* **87:** 178–180, 1970.

Keller, W. F. *et al.* Retinal dysplasia in English springer spaniels. *Proc Am Coll Vet Ophthalmol* 63–66, 1977.

Kennedy, P. C., Kendrick, J. W., and Stormont, C. Adenohypophyseal aplasia, an inherited defect associated with abnormal gestation in Guernsey cattle. *Cornell Vet* **47:** 160–178, 1957.

Kern, T. J., and Riis, R. C. Optic nerve hypoplasia in three miniature poodles. *J Am Vet Med Assoc* **178:** 49–54, 1981.

Lahav, M., and Albert, D. M. Clinical and histopathologic classification of retinal dysplasia. *Am J Ophthalmol* **75:** 648–667, 1974.

Lavach, J. D., Murphy, J. M., and Severin, G. A. Retinal dysplasia in the English springer spaniel. *J Am Anim Hosp Assoc* **14:** 192–199, 1978.

MacMillan, A. D. Retinal dysplasia in the dog and cat. *Vet Clin North Am* **10:** 411–415, 1980.

Nelson, D. L., and MacMillan, A. D. Multifocal retinal dysplasia in field trial Labrador retrievers. *Proc Am Coll Vet Ophthalmol* 11–23, 1978.

O'Toole, D. *et al.* Retinal dysplasia of English springer spaniel dogs: Light microscopy of the postnatal lesions. *Vet Pathol* **20:** 298–211, 1983.

Percy, D. H. *et al.* Lesions in puppies surviving infection with canine herpes virus. *Vet Pathol* **8:** 37–53, 1971.

Percy, D. H., Scott, F. W., and Albert, D. M. Retinal dysplasia due to feline panleukopenia virus infection. *J Am Vet Med Assoc* **167:** 935–937, 1975.

Sheffield, B., and Fishman, D. A. Intercellular junctions in the developing neural retina of the chick embryo. *Ztschr Zellforsch* **104:** 405–418, 1970.

Silverstein, A. M., Osburn, B. I., and Prendergast, R. A. The pathogenesis of retinal dysplasia. *Am J Ophthalmol* **72:** 13–21, 1971.

Silverstein, A. M. *et al.* An experimental virus-induced retinal dysplasia in the fetal lamb. *Am J Ophthalmol* **72:** 22–34, 1971.

Whitely, H. E. Dysplastic canine retinal morphogenesis. *Invest Ophthalmol Vis Sci* **32:** 1492–1498, 1991.

5. Anomalies of Surface Ectoderm

From fetal surface ectoderm are derived corneal epithelium, lens, lacrimal apparatus, and the epithelial portions of the eyelids and associated adnexa. Seldom are anomalies of these structures the subject of histopathologic study, inasmuch as they are clinically obvious and of significance only if they result in corneal irritation, impaired vision, or unacceptable appearance.

Excessively large or small palpebral fissures are part of current fashion in some dog breeds. Micropalpebral fissure frequently leads to entropion as the lid margin curls inward, and resultant corneal abrasion necessitates surgical correction. Congenital entropion also occurs as sporadic flock epizootics in lambs, but whether this is a structural deformity or the result of eyelid spasm is unclear. Entropion associated with microphthalmos occurs in all species. Other eyelid defects include colobomas, which are focal to diffuse examples of eyelid agenesis, and delayed separation of the eyelid fusion, which is the normal state during organogenesis.

Disorders of cilia are very common in dogs but uncommon in other species. Congenital defects include one or more of ectopic cilia, misdirected but otherwise normal cilia (trichiasis), the occurrence of a second row of cilia from the orifice of normal or atrophic Meibomian glands (distichiasis), and excessively large cilia (trichomegaly). In each instance, the significance of the anomaly depends on the presence or absence of corneal irritation.

The lacrimal gland and its ducts develop from an isolated bud of surface ectoderm and, although anomalies must surely exist, they have not been investigated. Failure of patency of the lacrimal puncta occurs in dogs and horses and manifests as excessive tearing. Ectopic or supernumerary openings have been reported in dogs and in cattle.

a. Corneal Anomalies Primary corneal maldevelopment is rare in all species. The category may be expanded to include corneal dystrophies, defined as bilateral, inherited, and usually central corneal opacities that, despite their typically adult onset, presumably have a congenital basis. These rare lesions will be discussed with degenerative diseases of the adult cornea.

Corneal anomalies may be ectodermal or mesenchymal, and may affect one or more of corneal size, shape, or transparency. **Microcornea** refers to a small but histologically normal cornea in an otherwise normal globe. A small cornea occurring in a microphthalmic globe is expected and does not merit a separate description. Mild microcornea of no clinical significance is reportedly common in certain dog breeds. **Megalocornea** has not been reported in domestic animals except in predictable association with congenital buphthalmos.

Dermoid is a congenital lesion of cornea or conjunctiva characterized by focal skinlike differentiation, and as such is properly termed a **choristoma.** They occur in all species. There is one report of a geographically high prevalence of multiple and sometimes bilateral dermoids as an inherited phenomenon in polled Hereford cattle in the American midwest, but ordinarily they seem to occur as single, random anomalies of unknown pathogenesis. Defective induction (skin instead of corneal epithelium) by the invading corneal stromal mesenchyme is the most popular speculation.

The degree of differentiation varies, but most consist of stratified squamous keratinized and variably pigmented epithelium overlying an irregular dermis containing hair, sweat glands, and sebaceous glands. Very rarely, cartilage or bone is seen. The degree of adnexal differentiation varies widely but may approach that of normal skin (Fig. 4.19A,B,C). At the edge of the dermoid, the dermal collagen reorients to blend with the regular stroma of cornea, and the epidermis transforms itself to corneal epithelium. Surgical removal may be for cosmetic reasons, or may be required if dermoid hairs irritate the cornea or if the position of the dermoid interferes with vision. In most instances of corneal dermoid, the choristoma is attached to the surface of a corneal stroma of normal thickness, so excision of the dermoid should not risk perforation of the globe.

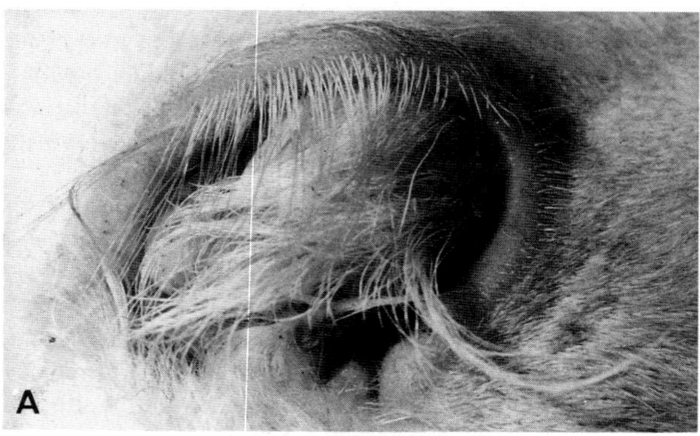

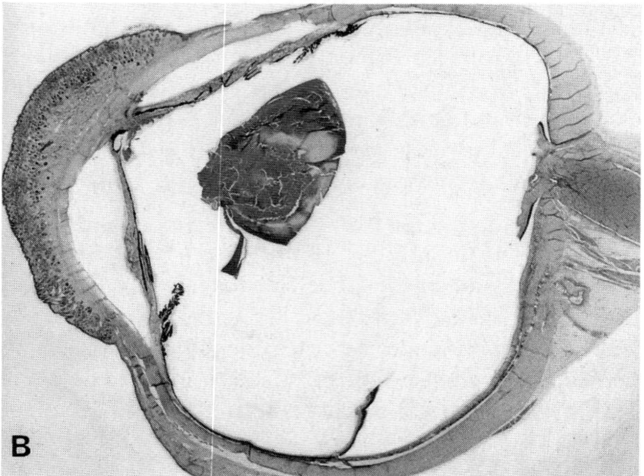

Fig. 4.19 Corneal dermoid, calf. (A) Notchlike defect in lower lid is a coloboma. (B) Globe from (A). Anterior rupture of lens capsule with well-organized anterior synechia, probably from foreign-body perforation.

Congenital corneal opacities are usually caused by anomalous formation of the anterior chamber, particularly congenital anterior synechiae and persistent pupillary membranes. Adherence of anterior chamber structures to the corneal endothelium, or perhaps their interposition during ingrowth of the corneal endothelium, results in focal absence of the corneal endothelium and disorganization of adjacent corneal stroma. Grossly, the affected cornea has deep stromal opacity caused by stromal edema or fibrosis in the area of the defective endothelium. Pigment, originating from adherent uveal strands, may be found in the corneal stroma. The opacity may be diffuse or focal, depending on the extent of uveal–corneal adhesion. Diffuse, congenital corneal opacity occurs in Holstein–Friesian cattle in England and Germany. The histologic lesion is diffuse corneal edema, but its pathogenesis is unknown. The cornea remains permanently opaque.

Corneal opacity caused by noncellular depositions occurs in dogs and is usually of adult onset despite an apparently genetic basis. The exception is multifocal, subepithelial deposition of basophilic, periodic acid-Shiff positive material in the corneas of puppies with collie eye anomaly or other mesodermal dysgeneses. The material is of unknown origin and may be the histologic counterpart of the transient, multifocal, subepithelial opacities seen quite commonly in 2 to 3-week-old puppies whose eyes are otherwise normal and thus unavailable for histologic examination.

b. ANOMALIES OF LENS The lens may be abnormally small, abnormally shaped, ectopic, or cataractous. Of these, only ectopia and cataract are common.

Aphakia is the congenital absence of lens, and it may be primary or secondary. It is claimed that primary aphakia is possible only in a rudimentary globe because of the central role of lens in the induction of invagination of the primary optic vesicle. Any globe with the structure of optic cup, regardless of how dysplastic, must have had a lens early in organogenesis, and its absence later must be the result of degeneration. This assumption is an extrapolation from work done many years ago in chicken embryos; while no work has been published to refute this contention, there is no work in mammals to confirm it. In the one report of aphakia in modern literature that includes histologic examination, several other puppies had small lenses, and all had invaginated optic cups with iris and retina. There was no conclusion about the nature of the injury to the developing eyes or the timing of such injury.

Microphakia, or congenitally small lens, is reported in

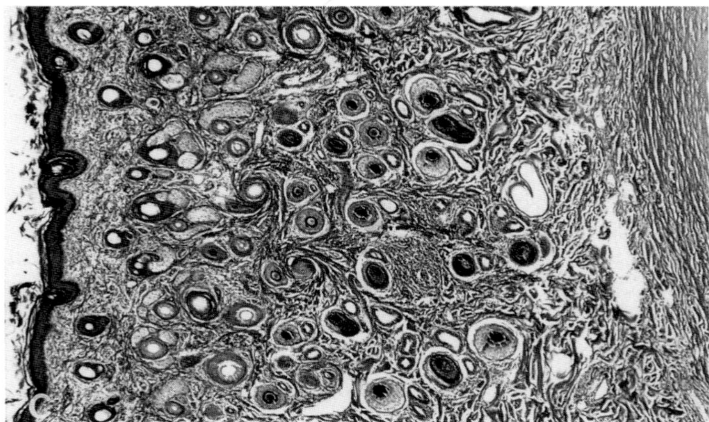

Fig. 4.19C Same eye as in (A) and (B). Development approaches that of normal skin. Note abrupt transition to dense regular corneal stroma at deep margin of dermoid.

dogs, calves, and cats, but is nonetheless rare. Most such reports describe the defect in association with ectopia lentis, microphthalmos, and anterior chamber mesenchymal anomalies. Such lenses are spherical and almost always are cataractous.

Lenticonus and **lentiglobus** are rare defects of lens shape characterized by an abrupt change in capsular configuration so that the lens acquires a globular or conical protrusion. The defect is usually polar and, in animals, usually posterior. From scattered and very old descriptions, it is difficult to define the "typical" histology of such lesions or their pathogenesis. The defect usually appears as a focal overgrowth of cortical lens fibers covered by thin posterior lens capsule and retained posterior epithelium. Of four relatively recent descriptions, all of dog eyes, all had congenital cataract but only in one did the cataract involve the protruding lens fibers themselves. Other ocular lesions reported include hyperplasia of tunica vasculosa lentis, rupture of the lens protrusion, and dysplasia of ciliary epithelium. At least in Doberman pinscher dogs, the posterior lentiglobus or lenticonus accompanying hyperplastic tunica vasculosa lentis appears to be an acquired defect caused by the abnormal fibrovascular elements adherent to lens.

Congenitally ectopic lenses occur in all species but are relatively common only in dogs and horses. Much more common than congenital luxations are spontaneous luxations in adult dogs, which may be associated with acquired lesions of the zonule. The reason for the particular susceptibility of small terriers and poodles to spontaneous lens luxation is unknown.

Congenital cataract occurs in most severely anomalous eyes but may occur as an isolated ocular lesion (Fig. 4.20). When cataract is present in eyes with multiple anomalies, it usually results from persistence of some part of perilenticular vasoformative mesoderm, but may also result from intraocular inflammation or toxic degeneration. Persistence of pupillary membrane or hyaloid system frequently results in multiple epithelial defects and subcapsular opacities at the sites of mesodermal contact with the lens.

In dogs, congenital primary cataracts are frequently hereditary but, as with corneal and retinal diseases, most hereditary cataracts are not congenital. Subtle, nonprogressive nuclear or cortical opacities are common but clinically insignificant in dogs and are of unknown pathogenesis. Primary and usually diffuse cataract is the most

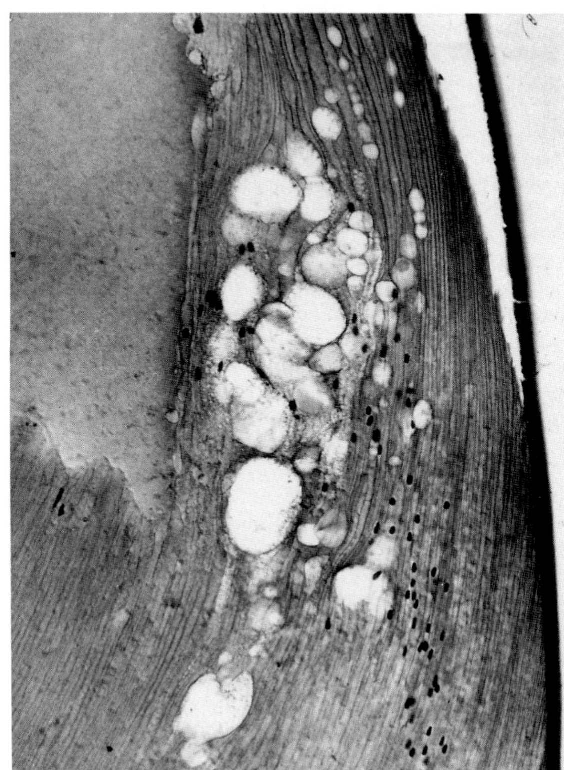

Fig. 4.20 "Bladder cells" in congenital cataract. Dog.

common ocular anomaly of horses. The pathogenesis is unknown, but there is usually no other ocular lesion. Congenital nuclear cataracts have been described as an inherited lesion in Morgan horses in the United States of America.

Congenital cataract is rare in cattle, swine, sheep, and goats. In cattle, hereditary congenital cataract occurs in Holstein–Friesians and in Jerseys and is thought to be an autosomal recessive trait. It is also seen as an infrequent result of fetal infection with the virus of bovine virus diarrhea.

There is a single report of bilateral, complete cataracts in a litter of Persian kittens, but there are no examples in swine or small ruminants except in association with multiple ocular defects.

The pathology of congenital cataract is the same as that of acquired cataract and is discussed later. It may be nuclear, cortical or capsular, focal or diffuse, stationary or progressive.

Bibliography

Aguirre, G. D., and Bistner, S. I. Microphakia with lenticular luxation and subluxation in cats. *Vet Med Small Anim Clin* **66:** 498–500, 1973.

Aguirre, G. D., and Bistner, S. I. Posterior lenticonus in the dog. *Cornell Vet* **63:** 455–461, 1973.

Barrie, K. P. *et al.* Posterior lenticonus, microphthalmia, congenital cataracts and retinal folds in an old English sheepdog. *J Am Anim Hosp Assoc* **15:** 715–717, 1979.

Barkyoumb, S. D., and Leipold, H. W. Nature and cause of bilateral ocular dermoids in Hereford cattle. *Vet Pathol* **21:** 316–324, 1984.

Beech, J., Aguirre, G., and Gross, S. Congenital nuclear cataracts in the Morgan horse. *J Am Vet Med Assoc* **184:** 1363–1365, 1984.

Brightman, A. H., Everitt, J., and Bevier, G. Epibulbar solid dermoid choristoma in a pig. *Vet Pathol* **22:** 292–294, 1985.

Gelatt, K. N. Cataracts in cattle. *J Am Vet Med Assoc* **159:** 195–200, 1971.

Gwin, R. M., and Gelatt, K. N. The canine lens. *In* "Veterinary Ophthalmology" K. N. Gelatt (ed.), pp. 435–473. Philadelphia, Pennsylvania, Lea & Febiger, 1981.

Lavach, J. D., and Severin, G. A. Posterior lenticonus and lenticonus internum in a dog. *J Am Anim Hosp Assoc* **13:** 685–687, 1977.

Martin, C. L. Zonular defects in the dog: A clinical and scanning electron microscopic study. *J Am Anim Hosp Assoc* **14:** 571–579, 1978.

Martin, C. L., and Leipold, H. W. Aphakia and multiple ocular defects in Saint Bernard puppies. *Vet Med Small Anim Clin* **69:** 448–453, 1974.

Olesen, H. P., Jensen, O. A., and Norn, M. S. Congenital hereditary cataract in cocker spaniels. *J Small Anim Pract* **15:** 741–749, 1974.

Peiffer, R. L. Bilateral congenital aphakia and retinal detachment in a cat. *J Am Anim Hosp Assoc* **18:** 128–130, 1982.

Peiffer, R. L., and Gelatt, K. N. Congenital cataracts in a Persian kitten. *Vet Med Small Anim Clin* **70:** 1334–1335, 1975.

Rubin, L. F. Cataracts in golden retrievers. *J Am Vet Med Assoc* **165:** 457–458, 1974.

Rubin, L. F., Koch, S. A., and Huber, R. J. Hereditary cataracts in miniature schnauzers. *J Am Vet Med Assoc* **154:** 1456–1458, 1969.

Takei, Y., and Mizuno, K. Electron microscopic studies on zonules. *Graefes Arch Klin Exp Ophthalmol* **202:** 237–244, 1977.

Trelstad, R. L., and Coulombre, A. J. Morphogenesis of the collagenous stroma in the chick cornea. *J Cell Biol* **50:** 840–858, 1971.

III. Ocular Adnexa

The adnexa include eyelids, nictitating membrane, and lacrimal and accessory lacrimal glands. Developmental, degenerative, inflammatory, and neoplastic diseases of these structures are commonly encountered in clinical practice, but only the neoplasms and proliferative inflammatory lesions are regularly submitted for histologic examination.

A. Eyelids

Disorders of size and configuration of the palpebral fissure are common in purebred dogs, as are anomalies of number or placement of cilia. None requires histologic evaluation.

The outer lid surface is skin and thus may suffer any of the afflictions of that tissue so that the evaluation of blepharitis uses the nomenclature and logic of dermatopathology (see Chapter 5). The inner (bulbar) surface is conjunctiva, and its diseases are discussed later.

External hordeolum or **stye** is a suppurative adenitis of the adnexal glands of Zeis or Moll. **Internal hordeolum** is suppurative inflammation of the meibomian gland. With persistent inflammation, the leakage of sebaceous secretion into adjacent soft tissue stimulates a granulomatous inflammation not unlike its epidermal counterpart, furunculosis. In this location, the nodular firm swelling is called a **chalazion.** Grossly, chalazion may be confused with meibomian adenoma and, in fact, frequently accompanies such adenomas.

B. Lacrimal System

Acquired disease of the lacrimal system is probably quite common in dogs if one includes keratoconjunctivitis sicca (see Keratitis) and eversion of the gland of the third eyelid.

Dacryoadenitis is inflammation of the lacrimal gland, and may result from involvement in orbital cellulitis or orbital trauma, spread from severe intraocular inflammation, incidental involvement in systemic diseases such as malignant catarrhal fever, feline infectious peritonitis, and canine distemper; or apparently specific immunologic assault. Specific dacryoadenitis caused by a coronavirus is extremely common in laboratory rats, in which acute necrotizing inflammation of lacrimal, Harderian, and salivary glands results in eventual fibrosis and squamous metaplasia of affected glands. Residual lesions in mildly affected rats are multiple lymphoid aggregates in the glandular interstitium. Similar changes are often seen in dogs with keratoconjunctivitis sicca, and in the absence of demonstrated viral cause, are assumed to represent autoimmune lacrimal adenitis. The analogous lesion in humans with Sjögren's syndrome is associated with influx of numerous T-helper cells into the gland, but no studies have yet been published to prove this immune pathogenesis for canine lacrimal adenitis and atrophy. However, the efficacy of cyclosporine, which acts primarily by suppression of T-helper cells, in reversing canine lacrimal adenitis provides evidence for such a pathogenesis.

Protrusion of the **nictitans gland** is quite common in dogs, and is thought to reflect a congenital laxity in the connective tissue anchoring the gland to the cartilage of the third eyelid. Because the resultant eversion is unsightly and resembles a neoplasm, these lesions frequently are excised, even although the membrana nictitans may be normal except for overlying conjunctival inflammation from exposure and abrasion. Since this gland sometimes supplies a significant proportion of total lacrimal secretion, its surgical removal may be followed by keratoconjunctivitis sicca in dogs that have less than optimal function of the primary lacrimal gland. In dogs with keratoconjunctivitis sicca, the gland may suffer the same lymphocytic interstitial adenitis, fibrosis, and atrophy as affects the lacrimal gland itself.

C. Conjunctiva

At the orifice of the meibomian glands, the epidermis of the lid undergoes abrupt transition to the pseudostratified columnar mucous membrane typical of the palpebral and bulbar conjunctivae. Goblet cells increase in number from the lid margin to the fornix, but ordinarily are absent in bulbar conjunctiva. Lymphoid aggregates are common in the subepithelial connective tissue, particularly below the bulbar conjunctiva and the inner aspect of the nictitating membrane. These aggregates are more prominent in the conjunctiva of horses than in other domestic species. Whether this is normal or a reflection of increased antigenic stimulation of the conjunctiva in the dusty environment of many horse stables is unknown. The transition from conjunctival to corneal epithelium occurs at, or slightly central to, the corneoscleral junction, and is marked by gradual loss of pigment, rete ridges, subepithelial blood vessels, and lymphoid tissue.

The general pathology of the conjunctiva is similar to that of other mucous membranes. Acute conjunctival injury, whether physical, chemical, or microbial, results in hyperemia and unusually severe edema. Evacuation of goblet cells and cellular exudation from the very labile conjunctival vessels add to the excessive lacrimation caused by any ocular irritation. The ocular discharge progresses from serous to mucoid and perhaps purulent with increasing severity of insult. Chronic irritation results in epithelial hyperplasia, hyperplasia of goblet cells and lymphoid aggregates, or even squamous metaplasia progressing to keratinization. The goblet cell hyperplasia is a very uncommon lesion when compared to squamous metaplasia and lymphoid hyperplasia. Lymphoid hyperplasia may be so marked as to result in grossly visible white nodules that may require surgical or chemical removal to reduce irritation of the adjacent cornea. While lymphoid hyperplasia is characteristic of a number of etiologically specific conjunctival diseases, it is best considered a nonspecific response to any chronic antigenic stimulation. Conjunctivitis frequently accompanies other ocular disease, notably keratitis, uveitis, and glaucoma. Conversely, conjunctival inflammation may spread to cornea, uvea, and orbit, although only secondary corneal involvement is common.

The causes of conjunctivitis include every class of noxious stimulus, including allergy and desiccation. Alleged bacterial causes, based on isolations from conjunctival swabs, are usually not distinguishable from the normal varied conjunctival flora. At least in dogs, the isolation of gram-negative organisms, especially coliforms, *Pseudomonas,* and *Proteus,* should be considered significant in light of the almost exclusively gram-positive flora of normal conjuctiva. Conjunctivitis occurs in a wide variety of multisystem diseases such as canine distemper and ehrlichiosis, equine viral arteritis and babesiosis, bovine virus diarrhea, malignant catarrhal fever, hog cholera, rinderpest, African swine fever, and others. Conjunctivitis accompanies most viral and allergic diseases of the upper respiratory tract. Only those diseases in which conjunctivitis is prominent or the only sign are discussed here.

Infectious bovine rhinotracheitis is usually accompanied by serous to purulent conjunctivitis that can be confused clinically with infectious bovine keratoconjunctivitis caused by *Moraxella bovis*. However, corneal involvement with rhinotracheitis is uncommon and is never the central suppurating ulcer typical of infectious keratoconjunctivitis. In an unpredictable number of animals, multifocal white glistening nodules, 1–2 mm in diameter, may be seen on the palpebral or bulbar conjunctiva. They appear as early as 3 days after instillation of virus into conjunctival sac, and represent hyperplastic lymphoid aggregates. Overlying conjunctiva may be ulcerated, and the defect filled with fibrin. Infectious bovine rhinotracheitis is discussed with the Respiratory (Volume 2, Chapter 6), Alimentary (Volume 2, Chapter 1), and Female Genital Systems (Volume 3, Chapter 4).

Feline infectious conjunctivitis is common and is caused by mycoplasma, chlamydia, or one of the upper respiratory tract viruses (herpesvirus, calicivirus, reovirus). Combined infections occur and are perhaps the rule. Usually the condition occurs in association with upper respiratory or oral lesions that may suggest an etiologic diagnosis. *Mycoplasma felis* or *M. gateae* may cause conjunctivitis unassociated with other signs in immunosuppressed cats, but instillation of organisms into the conjunctival sac of cats without prior corticosteroid administration does not cause disease. The conjunctivitis is pseudodiphtheritic and initially is unilateral. Histologically there is nonspecific erosive and suppurative conjunctivitis. Diagnosis requires the demonstration of coccoid bodies in the periphery of conjunctival epithelial cells.

1. Parasitic Conjunctivitis

Parasitic conjunctivitis is relatively common worldwide and may be caused by members of the genera *Thelazia, Habronema, Draschia, Onchocerca,* and several members of the family Oestridae. Of these, only *Thelazia* is truly an ocular parasite; the others cause eyelid, conjunctival, or orbital disease incidentally in the course of larval migration.

Members of the genus *Thelazia* are thin, rapidly motile nematodes 7–20 mm in length that inhabit the conjunctival sac and lacrimal duct of a variety of wild and domestic mammals. Their prevalence is much greater than the prevalence of conjunctivitis, suggesting that their number must be greater than usual before signs of conjunctival irritation are observed. The genus is found worldwide, and a listing of every species in every host is not justified here. The commonest species associated with conjunctivitis in domestic animals are *T. lacrymalis* in horses in Europe and North America, *T. rhodesis* in ruminants worldwide and *T. californiensis* in many species including dog, cat, bear, coyote, deer, and humans. Female worms are viviparous, and larvae free in lacrimal secretions are consumed by flies of the genus *Musca,* in which they develop for 15 to 30 days. The third-stage infective larvae migrate to the fly's proboscis and are returned to the conjunctival sac as the fly feeds.

Ocular habronemiasis results from deposition of larvae by the fly intermediate host, usually *Musca domestica* or *Stomoxys calcitrans,* in the moisture of the medial canthus of horses. Larvae of *Habronema muscae, H. microstoma,* or *Draschia* (*Habronema*) *megastomum* are the culprits. The burrowing larvae cause an ulcerative, oozing lesion about 0.5–1.0 cm in diameter at the medial canthus, which becomes progressively more nodular as granulomatous reaction to the larvae mounts. Mineralized granules may be found within the lesion along with caseous debris, liquefaction, and viable larvae. The histologic lesion is similar to that of cutaneous habronemiasis, namely chronic granulomatous inflammation surrounding live or dead larvae and eosinophils.

2. Ophthalmomyiasis

A syndrome of periocular and even intraocular invasion by fly larvae occurs in various species, including humans. Its various manifestations are known collectively as ophthalmomyiasis. Specific oculovascular myiasis, uitpeuloog or gedoelstial myiasis, is a disease of domestic ruminants and horses caused by invasion and migration of larvae of *Gedoelstia* spp. of Oestridae. The *Gedoelstia* are parasites of the blue wildebeest and hartebeest, the larvae being deposited in the eye, rather than in the nares, as is the habit of *Oestrus ovis*. The most important member of the genus in terms of frequent aberrant parasitism in domestic species is *G. hassleri,* which, in its natural antelope host, migrates to the nasal cavity via the vascular system and cerebral meninges and subdural space. The parasitism is not clinically significant in the antelope, but in domestic species which are aberrant hosts, severe ocular and neural disease occurs, sometimes on a large scale. The disease is seasonal and occurs particularly in domestic ruminants in contact with wildebeest.

The ocular lesions vary from a transient mild conjunctivitis to a destructive ophthalmitis with orbital or periorbital edema or abscessation affecting one or both eyes. Neurological signs of varied pattern are common in sheep, partly due to the larvae directly and partly to thrombophlebitis marking their route of invasion. Thrombosis may be very extensive, may involve the jugular vessels and endocardium, and may cause sudden death when coronary vessels are affected.

Larval migration may be into conjunctival sac, orbital tissues, or into the eye itself. In the last instance, ophthalmomyiasis interna, the globe is often destroyed by the larval penetration. However, a syndrome of relatively harmless larval migration in the subretinal space or within vitreous has been reported in humans. The characteristic subretinal linear tracks may be accompanied by focal retinal separation, preretinal and subretinal hemorrhage, and focal proliferations of retinal pigment epithelium. Two reported cases in cats had similar subretinal tracks, hyperplasia of pigment epithelium, and retinal hemorrhages. In one, the live motile larva was detected either on the face of, or just within, the retina. Subsequent examination failed to detect the larva, and the eye lesions resolved except for the subretinal tracks and pigment clumps.

The penetration is usually by a single larva despite numerous eggs or larvae within conjunctiva. The larva may die within the globe or continue its migration by uneventful exit from the globe via sclera, optic nerve, or vessel adventitia.

3. Allergic Conjunctivitis

Presumed allergic conjunctivitis occurs in all species, but is most likely to be investigated in dogs. Rarely is a specific allergen identified and, like its counterparts in allergic skin diseases, the diagnosis is based on the failure to demonstrate infectious or mechanical causes, response to corticosteroid therapy, and sometimes a convincing association with environmental changes. Biopsy is rarely warranted but, when taken during the acute disease, may show epithelial changes ranging from erosion to hyperplasia to squamous metaplasia, with eosinophils around dilated subepithelial blood vessels and percolating throughout the epithelium. More chronic lesions, which are the more usual to be biopsied, have squamous metaplasia and lymphocytic–plasmacytic linear, perivascular, or nodular infiltrates. The linear infiltrates predominate in a poorly characterized interface plasmacytic conjunctivitis more or less specific for German shepherd dogs. The bulbar surface of third eyelid is the favorite location, and many believe this lesion (sometimes referred to as plasmoma) to be the conjunctival variant of pannus keratitis.

Bibliography

Barrie, K. P., and Gelatt, K. N. Diseases of the eyelids (part I). *Compend Cont Ed* **1**: 405–410, 1979.

Barrie, K. P., and Parshall, C. J. Eyelid pyogranulomas in four dogs. *J Am Anim Hosp Assoc* **14**: 433–438, 1979.
Basson, P. A. Studies on specific oculo-vascular myiasis of domestic animals (uitpeuloog): I. Historical review. II. Experimental transmission. III. Symptomatology, pathology, aetiology and epizootiology. *Onderstepoort J Vet Res* **29**: 81–87, 203–209, 211–240, 1962.
Basson, P. A. Gedoelstial myiasis in antelopes in southern Africa. *Onderstepoort J Vet Res* **33**: 77–91, 1966.
Hughes, J. P., Olander, H. J., and Wada, M. Keratoconjunctivitis associated with infectious bovine rhinotracheitis. *J Am Vet Med Assoc* **145**: 32–39, 1964.
Lavach, J. D., and Gelatt, K. N. Diseases of the eyelids (part II). *Compend Cont Ed* **1**: 485–492, 1979.
Mason, G. I. Bilateral ophthalmomyiasis interna. *Am J Ophthalmol* **91**: 65–70, 1981.
Patton, S., and McCracken, M. D. The occurrence and effect of thelazia in horses. *Equine Pract* **3**: 53–57, 1981.
Raphel, C. F. Diseases of the equine eyelid. *Compend Cont Ed* **4:** S14–S21, 1982.

IV. The Cornea

The cornea of domestic mammals is a horizontal ellipse varying from 0.6 to 2.0 mm in thickness among the various species. In general, the larger and older the animal, the thicker the cornea. It appears as a structural and physiologic modification of sclera, and when chronically injured may lose the specialized features of cornea and resemble limbic sclera both ophthalmoscopically and histologically. Embryologically, however, the epithelium is derived from surface ectoderm, and the stroma is from neural crest mesenchyme, in contrast to the vascular–mesenchymal (non-neural) origin for sclera.

The major attribute of cornea is its clarity, and the loss of clarity is the most obvious indicator of corneal disease. The clarity results from several highly specialized anatomic and physiologic features: an unusually regular, nonkeratinized, and nonpigmented surface epithelium; an avascular, cell-poor stroma composed of very thin collagen (mostly type I) fibrils arranged in orderly lamellae (Fig. 4.21A); and a high degree of stromal dehydration maintained primarily by an Na-K-dependent adenosine triphosphatase (ATPase) pump in the cell membrane of the corneal endothelium. This dehydration is passively protected by the hydrophobic corneal epithelium and by the lack of stromal vascularity.

The reaction of cornea to injury is strongly influenced by these anatomic and physiologic features. The acutely injured cornea cannot respond with acute inflammation because it lacks blood vessels. Instead, edema is the hallmark of early corneal injury. The edema may result from injury to the corneal epithelium or endothelium (Fig. 4.21B,C), and is described below. With long-standing corneal disease, the cornea may undergo metaplasia to resemble limbic sclera and thus acquire the full range of inflammatory responses available to vascularized tissue. The chronically irritated epithelium undergoes epidermal metaplasia with the appearance of rete ridges, basilar pigmentation, and surface keratinization. The stroma acquires a capillary network and dermislike irregular fibroplasia. These changes, while they enable the cornea to survive in a hostile environment and to combat the inflammatory stimulus, also deprive it of its transparency.

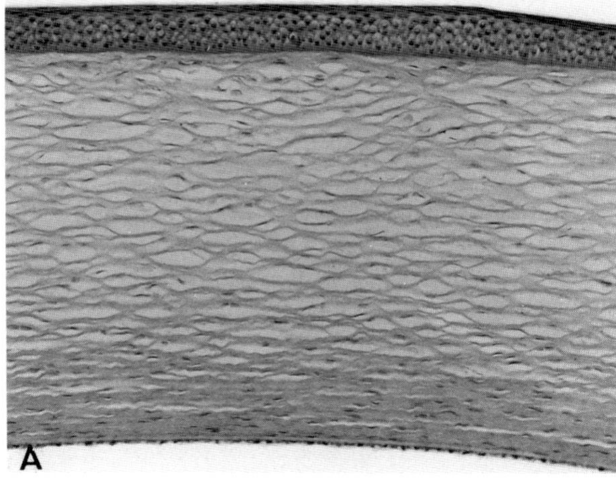

Fig. 4.21A Normal canine cornea. Uniform, nonkeratinized epithelium. Stroma is poorly cellular and compact. The corneal endothelium is frequently torn or missing as the result of sectioning artefact.

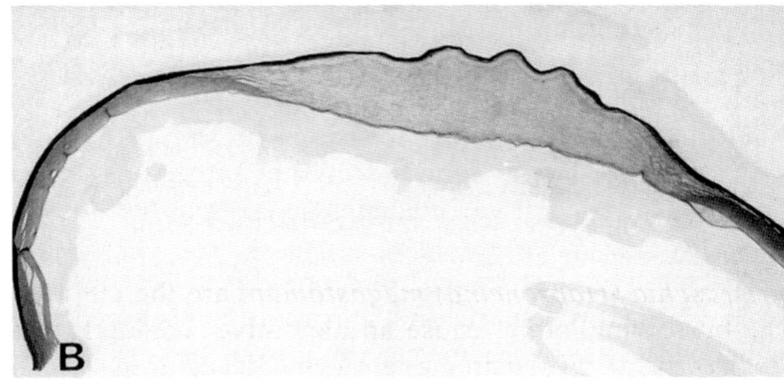

Fig. 4.21B Central corneal edema from abrasion of the corneal endothelium by lens. Foal, congenital anterior lens luxation.

Corneal injury may result from physical or chemical trauma, microbial agents, increased intraocular pressure and, rarely, from inborn errors of metabolism. Specific features of some of these injuries will be discussed later, but those features common to most corneal injuries are presented here.

A. Corneal Edema

Corneal edema occurs rapidly following injury and results from imbibition of lacrimal water through damaged corneal epithelium or failure of electrolyte (and thus water) extrusion by the corneal endothelium. If the epithelial or endothelial defect is focal, the resultant edema is limited to the stroma adjacent to the defect. The edematous cornea is clinically opaque, and may be up to five times its normal

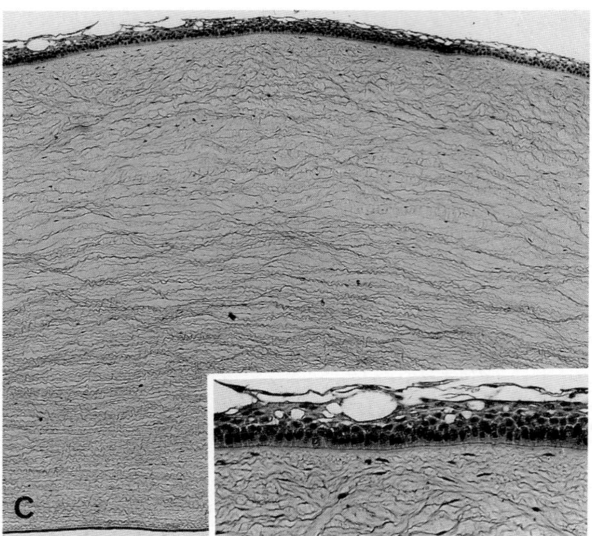

Fig. 4.21C Edema of cornea and epithelial keratitis. Ox. Phenothiazine photosensitivity. Inset. Detail of epithelium.

thickness (Fig. 4.21B). Edematous stroma stains less intensely than normal, and collagen lamellae are separated into a fine feltwork of pale-staining fibrils by excessive hydration of the proteoglycan ground substance. Percolation of stromal fluid into the epithelium results in the intercellular and intracellular edema known as **bullous keratopathy.**

Edema may also be part of more chronic corneal disease. Corneal vascularization in response to severe injury is accompanied by edema, as the porous new capillaries leak fluid into the interstitial spaces. A small amount of peripheral corneal edema frequently accompanies the peripheral stromal vascularization seen in chronic anterior uveitis of any cause. Sometimes the edema is unexpectedly diffuse, severe, and may persist even after the uveitis itself has subsided. Such eyes have a neutrophilic or lymphocytic destructive endothelialitis, with leukocytes interspersed among the vacuolated, pyknotic endothelial cells (see later under anterior uveitis). Other examples of corneal edema are seen in glaucoma and anterior segment anomalies. In the former, it is assumed that the high aqueous pressure drives fluid into the hydrophilic corneal stroma to a degree that overcomes the endothelial ion pump that dehydrates the stroma under normal conditions. In anterior segment anomalies, persistent pupillary membranes or congenital anterior synechiae cause focal defects in endothelial continuity and thus focal opacities due to deep stromal edema.

Persistent corneal edema seems to predispose to stromal vascularization and fibrosis, but numerous experimental models show that edema *per se* stimulates neither. A natural example of virtually permanent corneal edema occurs in Boston terrier and Chihuahua dogs with endothelial dystrophy, where neither fibrosis nor vascularization occurs despite years of severe diffuse stromal edema. When vascularization and fibrosis occur, they are in response to cytokines released by damaged epithelium, stromal keratocytes, or immigrant leukocytes rather than the edema itself.

B. Corneal Wounds

The healing of corneal wounds varies with the depth of penetration. Those defects involving epithelium alone, or epithelium and superficial stroma, heal by epithelial sliding followed by mitosis. The sliding begins within a few hours and is greatly enhanced by the secretion of fibronectin from adjacent injured epithelium. Mitotic activity, in contrast, is delayed for about 24 hr, is stimulated by epidermal growth factor derived from the injured epithelium and the normal tear film, and is most marked in the corneal basal cells near the limbus. Small defects are covered by flattened epithelial cells from adjacent normal cornea, and such shallow lesions heal completely by subsequent mitosis of basilar epithelium to rebuild epithelial thickness. Even if such abrasions affect the entire corneal surface, sliding and subsequent mitosis from the bulbar conjunctiva eventually lead to corneal restitution. Healing of shallow, uninfected corneal ulcers is rapid. For example, 7-mm ulcers heal within a mean of 11 days in horses.

Initially the epithelium has the characteristics of conjunctiva, including pigment, but within a few weeks it adopts a corneal epithelial configuration. Shallow defects in superficial stroma are filled by epithelial cells, creating an epithelial facet that is permanent but clinically insignificant. Epithelial adhesion to the underlying stroma remains fragile for 6 to 8 weeks until the hemidesmosomal attachments of epithelium to basal lamina reform, and until the new epithelium secretes type VII collagen fibrils that anchor the basal lamina to the stroma. In the interim, the cells adhere to a mixture of fibrin and fibronectin derived from the inflamed conjunctival vessels via the tear film or from the injured cornea itself. In many cases the only evidence of previous shallow ulceration is a thickened basal lamina resulting from secretion by the regenerating epithelium, and gentle undulation of the normally flat epithelial–stromal interface.

Deeper defects that include more than the outer third of stroma must heal by epithelial sliding and replication combined with stromal fibroplasia (Fig. 4.22A). Within a few hours of the insult, neutrophils reach the wound via the tear film, attracted by proteases released by injured epithelium. They migrate into the stroma and control bacterial contamination, degrade damaged collagen, and stimulate both fibroplasia and vascularization via production of various cytokines, especially basic fibroblast growth factor. Repair of the stroma is invariably by fibroplasia and never restores the stroma to complete normalcy. Viable stromal cells (keratocytes) adjacent to the wound undergo fibroblastic metaplasia and secrete large amounts of sulfated ground substance, particularly chondroitin sulfate. Histiocytes that slowly accumulate in the injured stroma may also assume the morphologic characteristics of fibro-

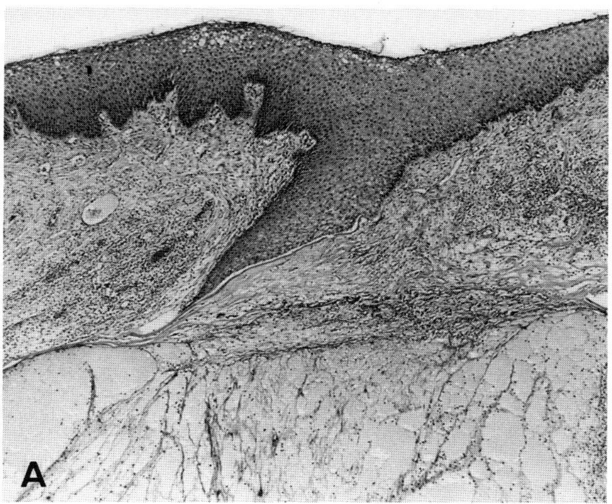

Fig. 4.22A Corneal perforation. Steer. Defect filled by downgrowth of hyperplastic corneal epithelium. Adjacent stroma is vascularized and chronically inflamed. There is anterior synechia.

Fig. 4.22B Corneal epidermalization and chronic superficial stromal inflammation with vascularization. Anterior synechia adherent by fibrous plaque that is partly formed by metaplastic corneal endothelium.

blasts. The stimulus for the fibroblasts to form, enlarge, and begin the production of new collagen and ground substance apparently also comes from proliferating reparative epithelium, which produces fibroblast/angioblast stimulatory cytokines. If the defect is not covered by epithelium, or if the animal is neutropenic, stromal fibroplasia is markedly retarded. These events initially occur without stromal vascularization, and nonseptic corneal wounds, even if deep, may heal without vascularization if they do so rapidly. The fibroblastic repair tissue gradually becomes less cellular, the collagen fibrils reorient to resemble more closely the parallel arrays of normal stroma, and the ground substance gradually reverts from an embryonic configuration dominated by chondroitin sulfate to the normal predominance of keratan sulfate. Complete restitution of normal stroma, however, never occurs (Fig. 4.22B), although the residual scar may be subtle and better detected by clinical examination than by histology.

A corneal perforation heals, as does a deep but incompletely penetrating wound, except for the involvement of corneal endothelium and Descemet's membrane. The cut edges of Descemet's elastic membrane retract from the wound, and the transcorneal gap is initially plugged with fibrin. Surface epithelium grows inward along the cut surface of the stroma and is inhibited only by contact with viable corneal endothelium. As with the surface epithelium, the corneal endothelium attempts to bridge the defect by sliding over the fibrin scaffold to restore endothelial continuity. The cells may enlarge severalfold to compensate for endothelial cell loss. Replacement by mitosis begins within about 24 hr in some experimental models, but the regenerative capability of the corneal endothelium in adult animals of most domestic species is very limited, and repair occurs by endothelial sliding and hypertrophy. So potent is this capability that normal stromal dehydration can be maintained even in the face of a 50% reduction in endothelial cell density. The cut ends of Descemet's membrane make no apparent effort at regrowth, but rather the endothelium gradually secretes a new membrane, which may eventually fuse with the old or remain separated from it by a layer of fibrous tissue.

The sequence of epithelial sliding and regeneration, remodeling stromal fibrosis, and endothelial repair is not uniformly successful. Large gaping wounds fill with proliferating epithelium and stromal fibrous tissue, which may protrude through the defect in Descemet's membrane and endothelium and into the anterior chamber. The fibroblasts, most of which are probably derived from keratocytes but which may also evolve via endothelial metaplasia, tend to grow along the posterior surface of Descemet's membrane. Regenerating or sliding endothelium is then separated from the coiled remnants of the original Descemet's membrane by a dense fibrous layer, called a **retrocorneal membrane.** Eventually, the corneal endothelium may resume continuity on the posterior surface of this membrane, secrete a new Descemet's membrane, and result in a cornea with two separate Descemet's membranes. Rarely the downgrowth of surface epithelium may gain access to the anterior chamber, in which it grows uninhibited as if in an organ-culture chamber. Glaucoma from overgrowth of the filtration angle or pupil is the usual consequence.

Bibliography

Bahn, C. F. *et al*. Postnatal development of corneal endothelium. *Invest Ophthalmol Vis Sci* **27**: 44–51, 1986.

Bellhorn, R. W., and Henkind, P. Superficial pigmentary keratitis in the dog. *J Am Vet Med Assoc* **149**: 173–175, 1966.

Brogdon, J. D. *et al*. Effect of epidermal growth factor on healing of corneal endothelial cells in cats. *Am J Vet Res* **50**: 1237–1243, 1989.

Cameron, J. D., Flaxman, B. A., and Yanoff, M. *In vitro* studies of corneal wound healing: Epithelial–endothelial interactions. *Invest Ophthalmol* **13**: 575–579, 1974.

Capella, J. A. Regenerating of endothelium in diseased and injured corneas. *Am J Ophthalmol* **74**: 810–817, 1972.

Catcott, E. J., and Griesemer, R. A. A study of corneal healing in the dog. *Am J Vet Res* **15**: 261–265, 1954.

Cintron, C., Covington, H. I., and Kublin, C. L. Morphologic analyses of proteoglycans in rabbit corneal scars. *Invest Ophthalmol Vis Sci* **31**: 1789–1798, 1990.

Fujikawa, L. S. *et al*. Fibronectin in healing rabbit corneal wounds. *Lab Invest* **45**: 120–129, 1981.

Gipson, I. K., Spurr-Michaud, S. J., and Tisdale, A. S. Anchoring fibrils form a complex network in human and rabbit cornea. *Invest Ophthalmol Vis Sci* **28**: 212–220, 1987.

Hanna, C. Proliferation and migration of epithelial cells during corneal wound repair in the rabbit and rat. *Am J Ophthalmol* **61**: 55, 1966.

Kay, E. P,. Nimni, M. E., and Smith, R. E. Modulation of endothelial cell morphology and collagen synthesis by polymorphonuclear leukocytes. *Invest Ophthalmol Vis Sci* **25**: 502–512, 1984.

Khodadoust, A. A. *et al*. Adhesion of regenerating corneal epithelium. The role of the basement membrane. *Am J Ophthalmol* **65**: 339–348, 1968.

Kitazawa, T. *et al*. The mechanism of accelerated corneal epithelial healing by human epidermal growth factor. *Invest Ophthalmol Vis Sci* **31**: 1773–1778, 1990.

Kuwabara, T., Perkins, D. G., and Cogan, D. G. Sliding of the epithelium in experimental corneal wounds. *Invest Ophthalmol* **15**: 4–14, 1976.

Landshman, N. *et al*. Relationship between morphology and functional ability of regenerated corneal endothelium. *Invest Ophthalmol Vis Sci* **29**: 1100–1109, 1988.

Mathers, W. D. *et al*. Dose-dependent effects of epidermal growth factor on corneal wound healing. *Invest Ophthalmol Vis Sci* **30**: 2403–2406, 1989.

Matsuda, H., and Smelser, G. K. Electron microscopy of corneal wound healing. *Exp Eye Res* **16:** 427–442, 1973.

Maurice, D. M. The transparency of the corneal stroma. *Vision Res* **10**: 107–108, 1970.

Neaderland, M. H. *et al*. Healing of experimentally induced corneal ulcers in horses. *Am J Vet Res* **48**: 427–430, 1987.

Soubrane, G. *et al*. Binding of basic fibroblast growth factor to normal and neovascularized rabbit cornea. *Invest Ophthalmol Vis Sci* **31**: 323–333, 1990.

Van Horn, D. L. *et al*. Regenerative capacity of the corneal endothelium in rabbit and cat. *Invest Ophthalmol* **16**: 597–613, 1977.

Watanabe, K., Nakagawa, S., and Nishida, T. Stimulatory effects of fibronectin and EGF on migration of corneal epithelial cells. *Invest Ophthalmol Vis Sci* **28**: 205–211, 1987.

C. Corneal Dystrophy

Corneal dystrophy should refer to bilateral, inherited, but not necessarily congenital, defects in structure or function of one or more corneal components. These are recognized most frequently as corneal opacities, deposits, or erosions. All are uncommon, but the least uncommon is bilateral recurrent central corneal erosion in middle-aged or old dogs, and occasionally in horses or cats. Stromal dystrophies are discussed with corneal deposits next.

Epithelial-stromal dystrophy (recurrent erosion syndrome) in dogs was first described in boxer dogs (hence the name boxer ulcer) and, while boxers and related breeds may be predisposed, similar recurrent erosions are encountered in a wide variety of breeds. The clinical syndrome is distinctive, characterized by a shallow central corneal erosion with scant edema and no vascularization. The lesion refuses to heal, or repeatedly reulcerates, because of poor adhesion of the epithelium to the underlying stroma or basal lamina. The defect appears not to be in epithelial healing *per se*, since sliding and mitotic activity are normal in affected dogs. Keratectomy specimens reveal poorly adherent hyperplastic epithelium at the ulcer margins, usually with multiple clefts separating epithelium from stroma even in areas distant from the obvious ulcer. The basal lamina is usually not visible with light microscopy, and the epithelium appears to be attempting to adhere to a thin zone of hypocellular, pale-staining stroma. The observation of pyknotic and lytic keratocyte nuclei within this superficial zone suggests that the basic defect is degeneration of the superficial stroma, so that epithelial hemidesmosomes and anchoring collagen fibrils have no firm anchor. Very chronic cases usually acquire superficial stromal granulation tissue appropriate to any chronic ulceration, but its onset is much delayed in comparison to infectious or traumatic ulcers.

Corneal endothelial dystrophy occurs in Boston terriers, Chihuahuas, and several other dog breeds, and causes slowly progressive bilateral corneal edema in mature dogs. The edema usually begins adjacent to the lateral limbus and may initially be unilateral and unaccompanied by other clinical signs. Later, epithelial fluid bullae may rupture to cause painful corneal ulcers and associated inflammation. Despite the persistent stromal edema, fibrosis and vascularization do not occur unless rupture of epithelial bullae initiates keratitis. The primary lesion is spontaneous necrosis of corneal endothelium followed by hypertrophy and sliding of viable endothelium. A marked progressive decrease in overall endothelial cell density results, eventually, in what usually is severe bilateral corneal edema. The reason for the endothelial cell death is unknown. Focal irregularities in Descemet's membrane occur in areas of endothelial loss, presumably a result of new basement membrane production by adjacent reactive endothelium.

A rare, juvenile-onset, genetically transmitted endothelial dystrophy in Manx and domestic short-hair cats is manifest as bilateral, progressive central epithelial and stromal edema. Fluid accumulates within superficial stroma and within the epithelium. Primary morphologic abnormalities are not described in the Manx, but in short-hairs there is irregularity and vacuolation of corneal endothelium.

D. Corneal Stromal Depositions

Deposition of mineral, lipid, or pigment within the cornea may be primary or occur secondary to chronic corneal injury in any species.

Corneal pigmentation often accompanies chronic corneal irritation in dogs and less frequently in other species, particularly horses. The pigment is melanin and is found in the basal layer of the corneal epithelium and in the superficial stroma. It is the result of a progressive ingrowth of new germinal cells that have retained pigment from the bulbar conjunctiva. The clinical name, pigmentary keratitis, is purely descriptive (Fig. 4.23). The corneal epithelium is invariably hyperplastic and often has the other features of chronically irritated cornea, such as rete ridge formation and keratinization, that characterize corneal epidermalization. There is usually evidence of chronic stromal inflammation, including vascularization. Infrequently, nonepithelial corneal pigmentation is the residual lesion of uveal (iris) adherence to cornea, with uveal melanin left behind as the adhesion resolves.

Corneal lipidosis occurs as apparently spontaneous crystalline corneal stromal opacities in dogs, as a result of persistent hypercholesterolemia, and as part of chronic stromal inflammation. The deposits are usually mixtures of cholesterol, phospholipids, and neutral fats. When the deposition is bilateral and unassociated with previous keratitis or serum lipid abnormality, the lesion qualifies as **corneal stromal dystrophy.** Crystalline lipid-rich corneal dystrophy is common in young Siberian huskies and occurs sporadically in other dog breeds, notably collies, Airedale terriers, beagles, and cavalier King Charles spaniels.

Diets high in cholesterol produce diffuse corneal stromal lipidosis in rabbits, as well as focal lipid deposits in uveal epithelium and stroma. While hyperlipemia is not a feature of most cases of corneal lipidosis in dogs (most of which are spontaneous dystrophies), circumferential peripheral stromal lipidosis is reported in German shepherd dogs with hyperlipoproteinemia resulting from hypothyroidism.

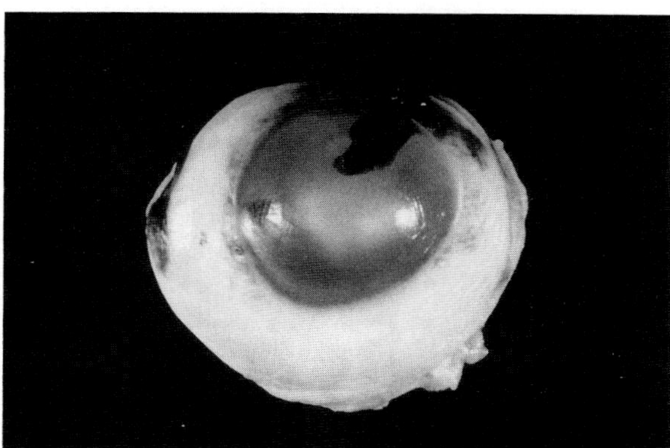

Fig. 4.23 Corneal pigmentation. Pug dog, a breed with normally bulging eyes.

Regardless of pathogenesis, the histologic lesion is similar. Cholesterol crystals and lipid vacuoles are found principally in anterior stroma, and are sometimes surrounded by lipid-laden macrophages and variable numbers of other leukocytes. Vascularization is often present, but its pathogenesis is unknown.

Mineral deposition occurs primarily in the anterior stroma and the epithelial basement membrane. Predisposing corneal changes include desiccation, anesthesia, edema, or inflammation. There are many methods for inducing deposition of calcium salts, but stromal edema seems to be the common denominator in almost all cases. The edema may result from corneal epithelial desiccation (exposure keratitis), uveitis, deliberate corneal trauma, or chemical injury. Hypercalcemia from vitamin D toxicity or hyperparathyroidism exacerbates the mineralization and is essential to lesion development in some experimental models.

An unidentified corneal deposition is often seen in canine eyes suffering from multiple anomalies, particularly those involving uvea. Similar deposits are seen, with less regularity, in the horizonal midportion of the cornea of many normal puppies. Fine basophilic periodic acid-Schiff positive linear deposits are associated with the epithelial basement membrane or superficial stroma. There is some disarray of superficial stromal fibers but no inflammation. The nature and pathogenesis of the depòsits are unknown, but most disappear after a few months.

E. Corneal Degeneration

Corneal degeneration is a vague term sometimes used to describe those corneal lesions characterized by noninflammatory loss of epithelial or stromal viability. Diseases such as keratoconjunctivitis sicca and pannus keratopathy are sometimes considered primary degenerative lesions, but their principal manifestation is inflammatory, and they are discussed under keratitis.

The only degenerative, noninflammatory acquired corneal lesion presented here is the **corneal sequestrum** of **cats.** This lesion, also called corneal mummification or corneal nigrum, is initially seen as a central, nonulcerated brown focus in one or both corneas of cats of any age or breed. Persian and Siamese cats are more frequently affected than are other breeds. Histologically, the lesion is bland corneal epithelial desiccation that may be mistaken as artefact. The stroma is hyalinized, featureless, and orange-brown. In older lesions the deep margin of the focus may be marked by a zone of reactive mononuclear leukocytes and, perhaps, a few giant cells. The epithelium is usually absent over the central portion of the biopsy specimen, because lesions have usually ulcerated by the time keratectomy is done. The nature of the pigment and the pathogenesis of this unique disease are unknown. The sequestrum may eventually slough, and the defect heals by granulation, although most lesions are treated by excision before that stage is reached.

Bibliography

Aguirre, G. D., Rubin, L. F., and Harvey, C. E. Keratoconjunctivitis sicca in dogs. *J Am Vet Med Assoc* **158:** 1566–1579, 1971.

Cooley, P. L., and Dice, P. F. Corneal dystrophy in the dog and cat. *Vet Clin North Am: Small Anim Pract* **20:** 681–692, 1990.

Cooley, P. L., and Wyman, M. Indolent-like corneal ulcers in three horses. *J Am Vet Med Assoc* **188:** 295–297, 1986.

Crispen, S. M. Corneal dystrophies in small animals. *Vet Annual* **22:** 298–310, 1982.

Crispen, S. M., and Barnett, K. C. Arcus lipoides corneae secondary to hypothyroidisim in the Alsatian. *J Small Anim Pract* **19:** 127–142, 1978.

Ekins, M. B., Waring, G. O., and Harris, R. R. Oval lipid corneal opacities in beagles, Part II: Natural history over four years and study of tear function. *J Am Anim Hosp Assoc* **16:** 601–605, 1980.

Formston, C. *et al.* Corneal necrosis in the cat. *J Small Anim Pract* **15:** 19–25, 1974.

Gelatt, K. N., and Samuelson, D. A. Recurrent corneal erosions and epithelial dystrophy in the boxer dog. *J Am Anim Hosp Assoc* **18:** 453–460, 1982.

Gwin, R. M. Primary canine corneal endothelial cell dystrophy: Specular microscopic evaluation, diagnosis, and therapy. *J Am Anim Hosp Assoc* **18:** 471–479, 1982.

Gwin, R. M., and Gelatt, K. N. Bilateral ocular lipidosis in a cottontail rabbit fed an all-milk diet. *J Am Vet Med Assoc* **171:** 887–889, 1977.

Harrington, G. A., and Kelly, D. F. Corneal lipidosis in a dog with bilateral thyroid carcinoma. *Vet Pathol* **17:** 490–493, 1980.

Kirschner, S. E., Niyo, Y., and Betts, D. M. Idiopathic persistent corneal erosions: Clinical and pathological findings in 18 dogs. *J Am Anim Hosp Assoc* **25:** 84–90, 1989.

Martin, C. L., and Dice, P. F. Corneal endothelial dystrophy in the dog. *J Am Anim Hosp Assoc* **18:** 327–336, 1982.

McMillan, A. D. Crystalline corneal opacities in the Siberian husky. *J Am Vet Med Assoc* **175:** 829–832, 1979.

Souri, E. The feline corneal nigrum. *Vet Med Small Anim Clin* **70:** 531–534, 1975.

Startup, F. G. Corneal necrosis and sequestration in the cat: A review and record of 100 cases. *J Small Anim Pract* **29:** 476–486, 1988.

F. Corneal Inflammation

Corneal inflammation is called keratitis and is traditionally divided into epithelial, stromal (interstitial), and ulcerative keratitis. Most lesions reaching a pathologist are ulcerated or show extensive stromal scarring below a healed ulcer. Regardless of cause, corneal inflammation initially follows the stereotyped sequence of edema and leukocyte immigration from tears and distant limbic venules. With severe lesions, corneal stromal vascularization, fibrosis, and epithelial metaplasia with pigmentation may occur.

Keratitis usually results from physical, chemical, or microbial injury to the cornea, but the cornea may also be affected by extension of disease from elsewhere in the eye or adnexa or conjunctiva. The stroma and endothelium may become involved in diseases of the uvea by extension via the aqueous or by direct extension from iris root or ciliary apparatus across the limbus. Purely stromal keratitis is uncommon except as an extension from a severe anterior uveitis.

Purely **epithelial keratitis** is rarely encountered in histologic preparations, because the clinical lesion either is transient and progresses to ulceration (as in acute keratoconjunctivitis sicca) or is so mild that eyes are unavailable for histologic examination. **Superficial punctate keratitis** is, in fact, noninflammatory and consists of multiple fine epithelial opacities that are probably foci of epithelial hydropic change. Intercellular fluid accumulation (bullous keratopathy) is seen as a sequel to corneal edema.

Stromal keratitis is subdivided into superficial and deep. Superficial stromal keratitis is common only in dogs, particularly in German shepherds, as a chronic, nonulcerative proliferative inflammation termed **pannus keratitis,** chronic superficial keratitis, or Uberreiter's syndrome. The clinical disease is distinctive. The early lesion is seen in dogs of either sex, usually in early middle age, as a vascularized opacity growing into the corneal stroma from the limbus. The ingrowth is bilateral, although not always of simultaneous onset, and most frequently originates from the ventrolateral limbus. There is no ulceration, but pigmentation is often marked. The untreated lesion eventually infiltrates the entire cornea, converting the superficial stroma to an opaque membrane resembling granulation tissue.

The histologic appearance varies with the duration of the lesion. The initial lesion is a superficial stromal infiltration of mononuclear cells, especially plasma cells. Subsequently, there is progressive vascularization and fibroplasia in the superficial third of the stroma, accompanied by

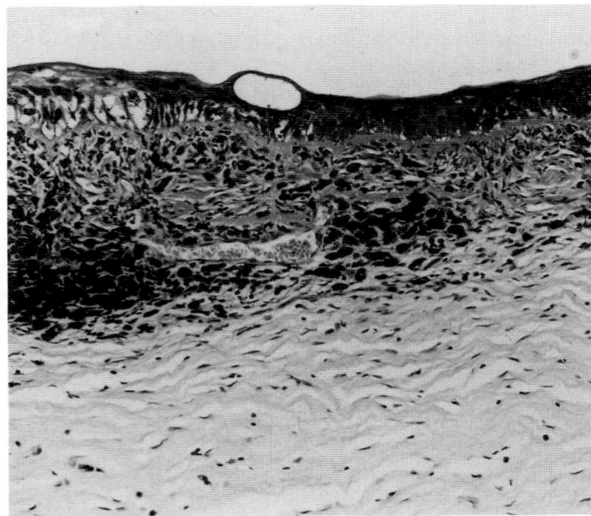

Fig. 4.24 Superficial stromal keratitis (of pannus). Dog. Epithelial hyperplasia, chronic superficial stromal inflammation and pigmentation.

epithelial hyperplasia and pigmentation that may include the stroma (Fig. 4.24). The deep stroma is never affected.

The pathogenesis of the condition is unknown, but an immune reaction to altered corneal epithelial antigens is hypothesized. Its response to continuous corticosteroid administration supports this hypothesis, although direct immunofluorescence tests for intraepithelial or basement membrane immunoglobulin are negative. Infectious agents are not consistently isolated. A histologically similar lesion of the bulbar conjuctiva of third eyelid occurs in the same breed (so-called plasmoma) and may reflect the same (unknown) pathogenesis.

Nonulcerative **deep stromal keratitis** may result from extension from anterior uveitis or from endothelial damage by uveal inflammation, trauma, or glaucoma. Inflammations extending from the anterior uvea are distinctly laminar and initially consist of perilimbic edema and leukocytosis. Later, corneal stromal vascularization occurs as a laminar, perilimbic "brush border" of vessels extending from the vascular plexus at the base of iris and ciliary body. As the uveitis subsides, the corneal lesion regresses until only the empty ghosts of these vessels remain. This lesion is particularly common in the eyes of horses which have suffered one or more bouts of equine recurrent ophthalmitis.

Ulcerative keratitis includes a large group of lesions caused by physical and chemical trauma, desiccation, bacterial or viral infection, and rarely from primary degeneration of the corneal epithelium itself. Regardless of cause, the loss of epithelium initiates a predictable series of corneal reactions caused by tear imbibition, local production of cytokines, and opportunistic microbial contamination of the wound. Imbibition causes superficial stromal edema below the ulcer and is followed by immigration of neutrophils from the tear film and, later, from the limbus. The leukocytes, although somewhat protective against opportunistic pathogens, also add their collagenases, proteases, and stimulatory cytokines to the wound and thereby may contribute to its progression. Epithelial and stromal repair proceeds as already described for corneal wound healing, but the repair fails in those cases in which microbial contamination is well established or in which the cause of the initial ulceration has not been corrected. Common examples of the latter are found in dogs in which corneal trauma by misdirected cilia or facial hair, or desiccation due to lacrimal gland dysfunction, persists.

The usual role of bacteria and fungi in the pathogenesis of corneal ulceration is opportunistic. However, these opportunists contribute significantly to the perpetuation and worsening of the lesion. Proteases and collagenases of microbial, leukocytic, or corneal origin progressively liquefy corneal stroma, a process termed **keratomalacia** (Fig. 4.25). Ulcers contaminated by *Pseudomonas* and *Streptococcus* spp. are particularly prone to rapid liquefaction because of the potent collagenases and proteases produced by these organisms. *Pseudomonas* ulcers have been extensively investigated because of the devastating liquefaction of cornea that commonly accompanies this infection. The bacteria themselves produce numerous proteases and other toxins, which may be important in the establishment of the early infection, but most of the characteristic stromal malacia results from the action of proteases originating from leukocytes, reactive corneal epithelium, or injured stroma. The stroma contains a variety of proenzymes (for collagenases, elastases, gelatinases, and other stromal lysins) that are cleaved by the *Pseudomonas* toxins to produce the active enzymes. Which toxins are produced, and in what quantities, is very strain dependent. The stepwise degradation of stroma is seen histologically as a featureless eosinophilic coagulum, which occurs with progressive septic ulcers regardless of the species of bacterium. The neutrophils may encircle the liquefying focus as a thick wall of live and fragmented cells. The resulting lesion is then called a **ring abscess** (Fig. 4.26) and is seen

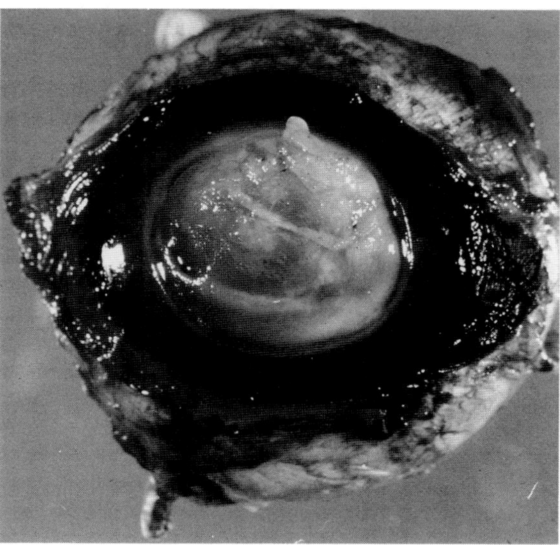

Fig. 4.25 Keratomalacia. Horse with *Pseudomonas* keratitis.

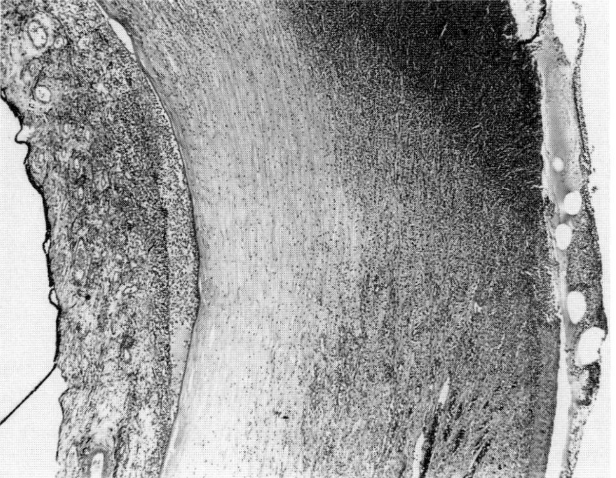

Fig. 4.26 Ring abscess. Calf. Edematous cornea is ulcerated and infiltrated by neutrophils. Leukocytes are within and on the surface of the iris which is adherent to the cornea.

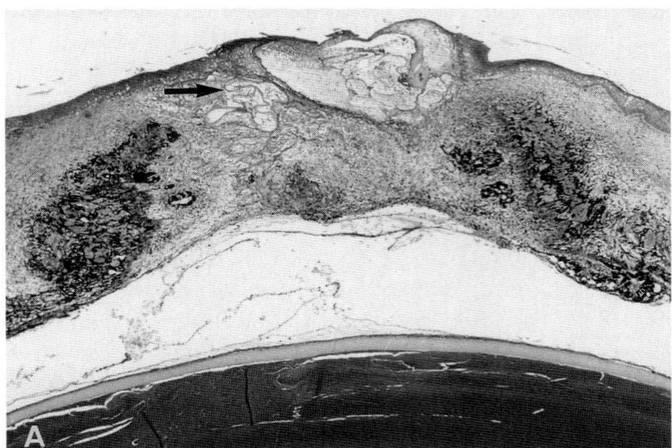

Fig. 4.27A Corneal epithelium is attempting to heal across a fibrin mass plugging the defect. Iris is incorporated into lesion and will form anterior staphyloma. Note coiled remnant of Descemet's membrane (arrow).

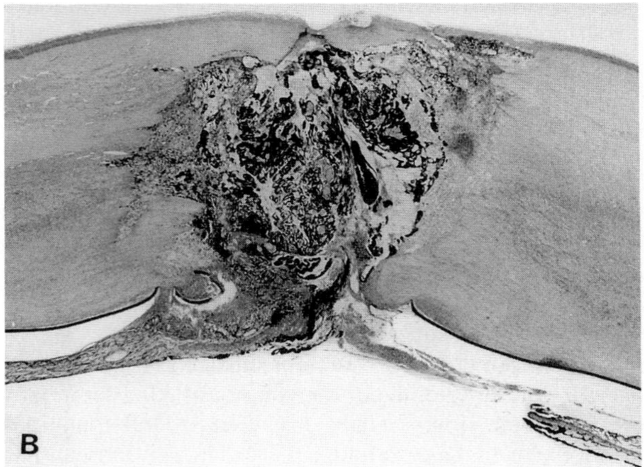

Fig. 4.27B Iris entrapped within cornea following perforation of ulcer (iris prolapse).

more commonly in cattle than any other species, perhaps because of the prevalence of untreated, contaminated corneal ulcers in that species and the prevalence of septic corneal perforation.

The sequelae of ulcerative keratitis involve cornea, conjunctiva, and uvea. The ulcer itself may heal with vascularization and scarring proportional to the severity of the initial lesion. It may persist as a stubborn but nonprogressive lesion, or it may progress to involve more of the stroma and epithelium. Stromal liquefaction that reaches Descemet's membrane results in its forward bulging as a **descemetocele.** This membrane, although resistant to penetration of the microbial agents themselves, is apparently permeable to inflammatory mediators and microbial toxins which diffuse into the anterior chamber. These chemicals, combined with a vasoactive sensory neural reflex from irritated cornea, are responsible for the vasodilation and exudation in anterior uvea which are seen histologically in virtually all globes with deep ulcerative keratitis. Even in nonperforating keratitis, the anterior uveal inflammation may result in sufficient fibrin exudation so as to predispose to focal adhesions, called synechiae, of iris to the injured cornea. In the case of corneal perforation, the iris flows forward to plug the defect and subsequently becomes incorporated into the corneal scarring. This defect is a permanent anterior synechia but is usually called **anterior staphyloma,** meaning a fibrous tunic (i.e., corneal) defect lined by uvea (i.e., iris) (Fig. 4.27A,B).

The conjunctiva is involved in almost all instances of keratitis, either as a victim of the same injury or as the nearest vascularized tissue to the diseased cornea. Hyperemia, cellular exudation, and lymphofollicular hyperplasia are common as the conjunctiva responds to the diffusion of inflammatory mediators of microbial, leukocytic, and tissue origin from the injured cornea.

1. Keratoconjunctivitis Sicca

Necrosis of corneal epithelium caused by desiccation is seen as a common and usually spontaneous condition in dogs, but desiccation secondary to exophthalmos or failure of the blinking reflex is seen in all species. Acute desiccation keratitis is particularly common in calves moribund as a result of neonatal diarrhea or meningoencephalitis, and is seen as bilateral, large, shallow, central corneal ulcers. Depending on the length of time between development of the ulcer and the animal's death, the epithelial loss may be accompanied by stromal edema and neutrophil infiltration. A similar lesion occurs in animals subjected to general anesthetics when failure to blink prevents adequate distribution of tears.

Desiccation keratitis may follow destruction or denervation of lacrimal or accessory lacrimal gland in any species by orbital inflammation, drugs, neoplasia, or trauma. Squamous metaplasia with resultant inadequacy of secretion may be seen with chronic deficiency of vitamin A. Specific lacrimal adenitis with subsequent atrophy is well recognized with coronavirus infection in rats and may be seen in the acute or chronic phases of canine distemper. Similar adenitis probably occurs with other viruses and in other species, but such lesions are poorly documented. Transient keratoconjunctivitis sicca may accompany acute herpetic keratoconjunctivitis in cats.

Keratoconjunctivitis sicca is encountered more commonly in dogs than in any other species, with an overall prevalence in North America of about 1%. Most cases are chronic, progressive, and idiopathic. The reason for greater than expected prevalence in certain breeds (English bulldog, Lhasa Apso, Shih Tzu, West Highland white terrier, and others) is unknown. Because the disease is amenable to medical or surgical management, few specimens are available for histologic examination until the very chronic stages. At this time the lacrimal gland is atrophic with interstitial lymphoid infiltration and fibrosis, but provides no clue as to the initial lesion. The ability of certain immune modulators, notably cyclosporine, to reverse the disease points to some kind of immune-mediated phenomenon, perhaps autoimmunity.

The corneal changes vary with the severity and rapidity of onset of lacrimal deficiency. In acute disease with marked lacrimal deficiency, clinical signs of ulcerative keratitis may occur. The corneal epithelium is thinned, has numerous hydropically degenerate cells, and may suffer full-thickness ulceration. The accompanying stromal changes, including eventual vascularization and fibrosis, are those of ulcerative keratitis. More commonly in dogs, however, the desiccation is not absolute (at least initially), and the epithelial response is protective epidermalization without prior ulceration. Keratinization, marked hyperplasia with rete ridge formation, and pigmentation are commonly seen. Stromal inflammation and vascularization are usually superficial, resulting in a lesion very similar to pannus keratitis. Squamous metaplasia may also occur in the bulbar conjunctiva. The conjunctivitis that clinically is the earliest lesion of keratoconjunctivitis sicca is rarely available for histologic examination.

2. Herpetic Keratitis of Cats

Feline herpetic keratitis caused by feline herpesvirus-1 is seen either as the sole ocular lesion or in concert with conjunctivitis. Clinical signs associated with herpesvirus infections in cats include conjunctivitis, keratitis, rhinotracheitis, and, in neonates, systemic disease with encephalitis and necrosis in visceral organs. Acquired immunity alters the manifestations of the disease and results in different lesions predominating in different age groups. Keratitis is commonest in adult cats and seems to result from activation of latent infection during concurrent immunosuppressive disease or corticosteroid therapy. Concurrent mild respiratory disease may be present. In contrast, the infection in adolescent cats causes a nonspecific bilateral erosive conjunctivitis without keratitis. Intranuclear inclusions are numerous within cells prior to sloughing, and leukocytes are sparse until ulceration permits opportunistic contamination. Upper respiratory disease is almost always present and is typically more severe than that in adults. In adult cats the disease is often unilateral and primarily corneal. It probably reflects recrudescence of latent infection. The typical corneal lesions are multifocal minute corneal erosions and ulcers which have a tendency to coalesce into branching dendritic ulcers. Severe or recurrent lesions in immunosuppressed cats may result in underlying stromal keratitis with lymphocytic infiltration, persistent edema, and vascularization. Inclusion bodies, intranuclear and typical of herpesviruses, are sometimes found in degenerating epithelium at the ulcer's margin. Lesion development is preceded by viral replication within otherwise normal corneal epithelium, and only in immunosuppressed cats is viral antigen abundant within the stroma.

3. Feline Eosinophilic Keratitis

Another uniquely feline ocular lesion is seen clinically as unilateral or bilateral proliferative, superficial stromal keratitis. There is no breed, age, or sex predilection, and no known association with other ocular or systemic disease. Since diagnosis is made by cytologic or histologic examination, this disease is more likely to be seen by pathologists than most other corneal disorders. Scrapings of the surface of the lesion reveal numerous eosinophils and fewer mast cells and other mononuclear leukocytes. Eosinophils may be less conspicuous on histologic examination of keratectomy specimens, perhaps because most seem determined to emigrate through the epithelium and into the tear film. Instead, the stromal lesion is a mixture of macrophages, plasma cells, fibroblasts and, unpredictably, mast cells and eosinophils. The latter are least frequent in older lesions. No bacterial or fungal agents have been seen. While there are histologic similarities to cutaneous eosinophilic ulcer and linear granuloma, no statistical association has been proven, and the lack of understanding of even the cutaneous eosinophilic lesions makes such attempted comparisons of very limited value.

Bibliography

Austad, R., and Oen, E. O. Chronic superficial keratitis (keratitis superficialis chronica) in the dog. I. A review of the literature. *J Small Anim Pract* **19:** 197–210, 1978.

Bedford, P. G. C., and Longstaffe, J. A. Corneal pannus (chronic superficial keratitis) in the German shepherd dog. *J Small Anim Pract* **20:** 41–56, 1979.

Brown, S. I., Bloomfield, S. E., and Tam, W. I. The cornea-destroying enzyme of *Pseudomonas aeruginosa*. *Invest Ophthalmol* **13:** 174–180, 1974.

Gelatt, K. N. Corneal diseases in the dog. *Compend Cont Ed* **1:** 78–84, 1979.

Iglewski, B. H., Burns, R. P., and Gipson, I. K. Pathogenesis of corneal damage from pseudomonas exotoxin A. *Invest Ophthalmol* **16:** 73–76, 1977.

Jones, G.E . *et al.* Mycoplasmas and ovine keratoconjunctivitis. *Vet Rec* **99:** 137–141, 1976.

Kessler, E., Mondino, B. J., and Brown, S. I. The corneal response to *Pseudomonas aeruginosa:* Histopathological and enzymatic characterization. *Invest Ophthalmol* **16:** 116–125, 1977.

Nasisse, M. P. *et al.* Experimental ocular herpesvirus infection in the cat. *Invest Ophthalmol Vis Sci* **30:** 1758–1768, 1989.

Paulsen, M. E. *et al.* Feline eosinophilic keratitis: A review of 15 clinical cases. *J Am Anim Hosp Assoc* **23:** 63–68, 1987.

Slatter, D. H. *et al.* Ubereitter's syndrome (chronic superficial keratitis) in dogs in the Rocky Mountain area—a study of 463 cases. *J Small Anim Pract* **18:** 757–772, 1977.

Steuhl, K-P. *et al.* Relevance of host-derived and bacterial factors in *Pseudomonas aeruginosa* corneal infections. *Invest Ophthalmol Vis Sci* **28:** 1559–1568, 1987.

4. Mycotic Keratitis

Mycotic keratitis is not a specific disease but is often viewed as such because of its consistently poor response to therapy and tendency to progress to corneal perforation. The offending fungus is usually a member of the normal conjunctival flora, and its role in the disease is that of opportunistic contaminant. *Aspergillus* is the most frequent isolate. The condition most often occurs in eyes with traumatic corneal injury, particularly if they have been receiving long-term antibiotic-corticosteroid therapy. Horses seem particularly prone to mycotic keratitis, perhaps related to the mold-laden, dusty environment in which many horses are housed; only rarely does the lesion occur in dogs or cats. Fungi of the genera *Aspergillus* and *Penicillium* are most commonly isolated. Since virtually all stabled horses have fungi as part of their conjunctival flora, seeing the hyphae within the corneal stroma is required for the diagnosis. Isolation from a corneal swab or shallow scraping is not adequate.

The typical early lesion is deep ulcerative keratitis, specific only in the fungi observed within the lesion. Some chronic lesions are exclusively stromal, probably the result of epithelial healing of the initial penetration or because therapy eliminated the infection in the superficial stroma. For whatever reason, the typical equine eye enucleated for mycotic keratitis has an intense neutrophil-rich deep stromal keratitis with several characteristic features: the neutrophils are karyorrhectic, the inflammation is most intense immediately adjacent to Descemet's membrane, and frequently there is lysis of the normally resistant Descemet's membrane, with spillage of the corneal inflammation into the anterior chamber. Fungi are numerous within the malacia of the deep stroma and within Descemet's membrane itself, but rarely if ever are seen within the anterior chamber. They are sparse or absent within the superficial half of the stroma, which explains why corneal scrapings or even keratectomy specimens may fail to reveal the agent. The reason for the apparent targeting of Descemet's membrane is not known, but the presence of the apparent tropism even in untreated eyes suggests that it is a genuine tropism and not just persistence of a previously generalized stromal infection in the site least likely to be reached by topical fungicides.

Bibliography

Bistner, S. I., and Riis, R. C. Clinical aspects of mycotic keratitis in horses. *Cornell Vet* **69:** 364–374, 1979.

Chandler, F. W., Kaplan, W., and Ajello, L. Mycotic keratitis. *In* "Histopathology of Mycotic Diseases," pp. 83–84. Chicago, Illinois, Year Book Medical Publishers, 1980.

Hodgson, D. R., and Jacobs, K. A. Two cases of *Fusarium* keratomycoses in the horse. *Vet Rec* **110:** 520–522, 1982.

Moore, C. P. *et al.* Bacterial and fungal isolates from Equidae with ulcerative keratitis. *J Am Vet Med Assoc* **182:** 600–603, 1983.

Moore, C. P. *et al.* Prevalence of ocular microorganisms in hospitalized and stabled horses. *Am J Vet Res* **49:** 773–777, 1988.

5. *Infectious Bovine Keratoconjunctivitis*

This disease vies with squamous cell carcinoma as the most important disease of the bovine eye. It occurs worldwide, is most prevalent in summer due to the increase in fly vectors, and has a clinical expression that ranges from initial conjunctivitis and ulcerative keratitis to iris prolapse, glaucoma, and phthisis bulbi. The prevalence of severe sequelae reflects inadequate management of the disease rather than any special virulence of this agent as compared to other infectious causes of keratitis in other species.

The disease behaves as an infectious epizootic within a susceptible population, frequently affecting over 50% of the cattle at risk within 2 weeks of the initial clinical case. Shedding of virulent organisms by a carrier animal is thought to be the usual route of introduction into a previously unexposed group, although a role for various mechanical or biological vectors is also assumed.

Moraxella bovis has been confirmed as the most important causative agent, although agents including *Mycoplasma bovoculi, Mycoplasma conjunctivae, Acholeplasma laidlawii,* and bovine herpesvirus may contribute to lesion severity. Earlier skepticism about the virulence of *M. bovis,* based on the unreliability of reproduction of the disease, isolation of the organism from apparently healthy cattle, and failure of isolation from some overtly affected cattle, has been overcome by detailed information on the pathogenesis of the disease. It is now clear that virulence of *M. bovis* is associated with hemolytic, leucocytolytic, piliated strains, which predominate in the eyes of only affected cattle. Nonpiliated, nonhemolytic strains predominate in healthy cattle and are probably part of the normal conjunctival flora. The use of immunofluorescence has demonstrated *M. bovis* in many of the naturally occurring cases for which the results of culture were negative. In naturally occurring outbreaks, the number of isolations of hemolytic *M. bovis* falls to almost zero as the outbreak wanes, but a few chronically affected carriers remain as the most important source of virulent bacteria for outbreaks of disease in the next summer.

In addition to variation in the virulence of different strains of *M. bovis,* sunlight, dust, and perhaps, concurrent infection with infectious bovine rhinotracheitis virus increase the severity of the disease. Calves are usually affected more severely than cattle older than 2 years, although absolute resistance to infection seems fragile. The protective effect of serum antibody against the disease is controversial. Specific immunoglobulin A (IgA) is found in tears of infected calves, and there is substantial evidence that locally produced IgA is strongly protective.

Following experimental inoculation of virulent *M. bovis* onto the cornea, pilus-mediated adhesion and production of bacterial cytotoxin result in microscopic ulceration in as little as 12 hr. Initial adhesion is to older surface epithelium ("dark cells") and results in the development of microscopic pits in the cell surface. *Moraxella* is found within degenerate epithelial cells, but it is not known whether invasion is necessary for subsequent cellular destruction. In field epizootics, the earliest lesion is bulbar conjunctival edema and hyperemia, followed in 24 to 48 hr by the appearance of a shallow, central corneal ulcer. The ulcer is a small (less than 0.5 cm) focus of epithelial necrosis that may appear as erosion, vesicle or full-thickness epithelial loss. In untreated animals destined to develop the full clinical expression, the ulcer enlarges, deepens, and frequently attracts enough neutrophils to qualify as a corneal abscess. Stromal liquefaction ensues, probably as a result of neutrophil lysis, which is itself initiated by *Moraxella*-derived leukotoxins. By the end of the first week, there is extensive stromal edema and vascularization extending from the limbus. As with any severe ulcerative keratitis, the subsequent progression or regression of the lesion varies with each case as modifications by therapy, opportunistic bacterial and fungal contamination, trauma, inflammation, and immunity interact. Keratomalacia frequently leads to forward coning of the weakened cornea (keratoconus). In most instances, whether treated or not, the cornea heals by sloughing of necrotic tissue and filling of the defect by granulation tissue. Re-epithelialization may take up to a month, leaving a cornea that is slightly coned and variably scarred. The scarring often is scant and interferes little with vision in spite of the severity of the primary lesion.

Less satisfactory sequelae, while not common in relation to the overall disease prevalence, are still relatively common. Sterile anterior uveal inflammation may result in focal or generalized adherence of iris to cornea (anterior synechia) or lens (posterior synechia). Descemetocele may progress to corneal rupture, which in turn may lead to phthisis bulbi or resolve by sealing with prolapse of the iris. Synechia and staphyloma may lead to impairment of aqueous drainage and thus to the lesions of glaucoma.

6. *Infectious Keratoconjunctivitis of Sheep and Goats*

Epizootics of conjunctivitis and keratitis in sheep and goats share many of the features of the bovine disease: summer prevalence, rapid spread, and exacerbation by dust, sunlight, and flies. Feedlot lambs seem particularly susceptible. Unlike that of bovine keratoconjunctivitis, the range of clinical signs and proposed causes suggests that there may in fact be several different diseases. Many agents including bacteria, mycoplasmas, chlamydiae, and rickettsiae have been suggested as causes, but various mycoplasmas and *Chlamydia psittaci* may be the important agents. The lesions caused by *M. mycoides* var. *capri* in goats and *M. conjunctivae* var. *ovis* in sheep are similar but usually milder than those caused by *Moraxella bovis* in cattle. This is particularly true of goats, in which deep corneal ulceration is uncommon.

Keratoconjunctivitis associated with *Chlamydia psittaci* is usually predominantly conjunctivitis. Initial chemosis and reddening are followed by massive lymphofollicular hyperplasia in bulbar conjunctiva and nictitating membrane. Keratitis may occur, but ulceration is seldom prominent. Animals with conjunctivitis may have concurrent polyarthritis from which chlamydiae can be isolated.

Bibliography

Brown, J. F., and Adkins, T. R. Relationship of feeding activity of face fly to production of keratoconjunctivitis in calves. *Am J Vet Res* **33:** 2551–2555, 1972.

Jones, G. E. *et al.* Mycoplasmas and ovine keratoconjunctivitis. *Vet Rec* **99:** 137–141, 1976.

Kagnoyera G., George, L. W., and Munn, R. Light and electron microscopic changes in cornea of healthy and immunomodulated calves infected with *Moraxella bovis*. *Am J Vet Res* **49:** 386–395, 1988.

Kagonyera, G. M., George, L. W., and Munn, R. Cytopathic effects of *Moraxella bovis* on cultured bovine neutrophils and corneal epithelial cells. *Am J Vet Res* **50:** 10–17, 1989.

Nayar, P. S. G., and Sanders, J. R. Antibodies in lacrimal secretions of cattle naturally or experimentally infected with *Moraxella bovis*. *Can J Comp Med* **39:** 32–40, 1975.

Pedersen, K. B., Froholun, L. O., and Bovre, K. Fimbriation and colony type of *Moraxella bovis* in relation to conjunctival colonization and development of keratoconjunctivitis in cattle. *Acta Pathol Microbiol Scand* **80:** 911–918, 1972.

Pugh, G. W., and Hughes, D. E. Comparison of the virulence of various strains of *Moraxella bovis*. *Can J Comp Med* **34:** 333–340, 1970.

Pugh, G. W. *et al.* Infectious bovine keratoconjunctivitis: Comparison of a fluorescent antibody technique and cultural isolation for the detection of *Moraxella bovis* in eye secretions. *Am J Vet Res* **38:** 1349–1352, 1977.

Pugh, G. W. *et al.* Experimentally induced infectious bovine keratoconjunctivitis: Resistance of vaccinated cattle to homologous and heterologous strains of *Moraxella bovis*. *Am J Vet Res* **37:** 57–60, 1976.

Rogers, D. G., Cheville, N. F., and Pugh, G. W., Jr. Pathogenesis of corneal lesions caused by *Moraxella bovis* in gnotobiotic calves. *Vet Pathol* **24:** 287–295, 1987.

Trotter, S. L. *et al.* Epidemic caprine keratoconjunctivitis: Experimentally induced disease with a pure culture of *Mycoplasma conjunctivae*. *Infect Immunol* **18:** 816–822, 1977.

Vandergaast, N., and Rosenbusch, R. F. Infectious bovine keratoconjunctivitis epizootic associated with area-wide emergence of a new *Moraxella bovis* pilus type. *Am J Vet Res* **50:** 1438–1441, 1989.

Whitley, R. D., and Albert, R. A. Clinical uveitis and polyarthritis associated with *Mycoplasma* species in young goats. *Vet Rec* **115:** 217–218, 1984.

Wilcox, G. E. Infectious bovine keratoconjunctivitis: A review. *Vet Bull* **38:** 349–360, 1968.

Williams, L. W., and Gelatt, K. N. Food animal ophthalmology. *In* "Veterinary Ophthalmology" K.N. Gelatt (ed.), pp. 614–621. Philadelphia, Pennsylvania Lea & Febiger, 1981.

Wilt, G. R., Wu, G., and Bird, C. Characterization of the plasmids of *Moraxella bovis*. *Am J Vet Res* **50:** 1678–1683, 1989.

V. The Lens

The lens is a flattened sphere of epithelial cells suspended in the pupillary aperture by an equatorial row of suspensory zonules radiating from the basement membrane of the nonpigmented ciliary epithelium in the valleys between the ciliary processes and from pars plana. The morphologic reaction of lens to injury is very limited due to the simplicity of its structure and physiology, and its lack of vascularity.

The lens is entirely epithelial. Outermost is a thick, elastic capsule, which is the basement membrane produced by the underlying germinal epithelial cells. The capsule is thickest at the anterior pole and becomes progressively thinner over the posterior half of the lens. The capsule in the neonate is thin, but it thickens progressively throughout life.

Below the capsule is a layer of simple cuboidal lens epithelium, which, in all but fetal globes, is found below

Fig. 4.28 Normal canine lens with characteristic regularity of surface epithelium and lens fibers.

the capsule of only the anterior half of the lens. The apex of these cells faces inward toward the lens nucleus. At the equator, these germinal cells extend into the lens cortex as the nuclear bow, an arc of cells being progressively transformed from cuboidal germinal epithelium to the elongated spindle shape of the mature lens fibers (Fig. 4.28). The bulk of the lens is composed of onionlike layers of elongated epithelial cells anchored to each other by interlocking surface ridges, grooves, and protuberances. These elongated fibers contain no nucleus and few cytoplasmic organelles, relying almost entirely on anaerobic glycolysis for energy. Since the lens cannot shed aging fibers as does skin or intestine, these cells are compacted into the oldest central part of the lens, the nucleus. The continuous accumulation of these old desiccated fibers with altered crystalline protein results in the common but visually insignificant aging change of nuclear sclerosis.

Although in many ways similar to cornea in structure and function, the optical clarity of lens rests not with the regularity of its fibers but in its high percentage of cytoplasmic soluble crystalline protein and paucity of light-scattering nuclei or mitochondria. The lens is about 35% protein, the highest of any tissue, and over 90% of it is the soluble crystalline variety. Insoluble high-molecular-weight protein (albuminoid) is found in the nucleus and cell membranes. Opacity of lens is associated, at least in some cases, with decreasing concentrations of crystalline and increasing albuminoid protein, the latter insoluble in water and optically opaque. Many of the insults that result in degeneration of lens ultimately interfere with its nutrition. Since it is avascular in the postnatal animal, the lens relies entirely on the aqueous for the delivery of nutrients and removal of metabolic wastes. Glaucoma, ocular inflammation, metabolic disorders, and various toxins share the common feature of altering the amount or quality of lenticular nutrition by altering the flow or composition of the aqueous humor.

A. Ectopia Lentis

The only lenticular defects of importance in domestic animals are those affecting location, configuration, and clarity. Those affecting configuration are usually developmental defects and are discussed in earlier sections. Dislocations of the lens may be congenital or acquired, and the latter include spontaneous dislocations and those secondary to trauma and glaucoma. Apparently spontaneous dislocations are encountered most frequently in middle-aged (3–8 years) terrier dogs in which an inherited predisposition to bilateral zonular rupture exists. The pathogenesis of the defect has been best studied in the Tibetan terrier, in which the zonules develop in a dysplastic, reticulate fashion that precedes luxation by several years. Traumatic dislocation is usually via blunt trauma, notably automobile accidents. The dislocation may be partial (subluxation) or complete (luxation), and in the latter instance, the free lens may damage corneal endothelium or vitreous causing edema and liquefaction, respectively. Such lenses may be surgically removed but seldom receive histopathologic examination. Anterior luxation frequently results in glaucoma, perhaps caused by anterior prolapse of the vitreous into the pupil. Lens luxation is inexplicably uncommon in cats. It is seen in middle-aged cats as unilateral and usually anterior luxation. One third of cases occur in eyes with no other observed lesion, while the remainder occur in eyes with preexistent uveitis or glaucoma, or a history of trauma.

Bibliography

Curtis, R. Lens luxation in the dog and cat. *Vet Clin North Am: Small Anim Pract* **20:** 755–773, 1990.

Curtis, R., and Barnett, K. C. Primary lens luxation in the dog. *J Small Anim Pract* **21:** 657–668. 1980.

Martin, C. L. Zonular defects in the dog: A clinical and scanning electron microscopic study. *J Am Anim Hosp Assoc* **14:** 571–579, 1978.

B. Cataract

Cataract is the most common and most important disorder of lens. Cataract means lenticular opacity, and is usually prefixed by adjectives relating to location, maturity, extent, suspected cause and ophthalmoscopic appearance (Fig. 4.29). Of these adjectives, only those of location and, to some extent, maturity are useful in histologic description. The simple structure of the lens results in stereotyped reaction to injury that provides few clues as to pathogenesis. The histology of cataract is the histology of the general pathology of the lens and includes germinal epithelial hyperplasia and metaplasia, hydropic change, fiber necrosis, and occasionally, deposition of calcium salts or cholesterol. Unless permitted by invasion through a capsular

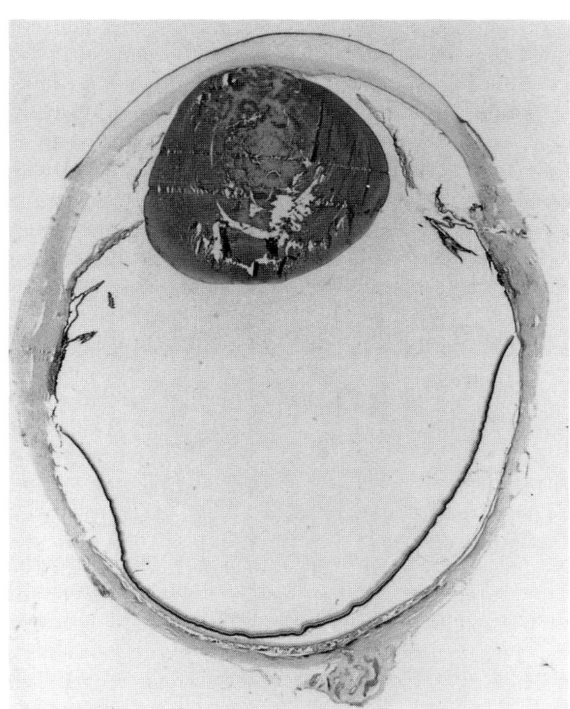

Fig. 4.29 Anterior polar pyramidal cataract. Puppy. Anterior bulging of liquefied lens has destroyed corneal endothelium, resulting in diffuse corneal edema.

tear, inflammation cannot occur within the avascular and totally epithelial lens.

Epithelial hyperplasia or **metaplasia** is the usual histologic counterpart of anterior subcapsular cataract, and is usually seen following focal trauma, or adherence of the iris or of persistent pupillary membranes to the anterior surface of the lens. Initial epithelial degeneration or necrosis is followed by hyperplasia and sometimes by fibrous metaplasia. The resultant epithelial plaque lies just under the anterior lens capsule. The innermost epithelial layer may remain basilar in type rather than fibroblastic. Each epithelial layer, even if metaplastic, secretes a new basement membrane that separates each layer from the adjacent layers. The end result is a focal plaque formed by multiple, sandwiched layers of flattened epithelium and basement membrane at the anterior pole of the lens. Remnants of adherent iris or pupillary membrane, including pigment, may complicate the histologic appearance.

A more common manifestation of epithelial hyperplasia is **migration** from the equator to line the posterior capsule. This reestablishment of the fetal morphology is seen most commonly with any chronic cataract in young animals, whose epithelial cells perhaps retain greater migratory ability. Adjacent cortical fibers are usually degenerate.

Degeneration and subsequent **fragmentation** or **liquefaction** of lens fibers is the most common lesion of cataract. Cataractous fragmentation must be distinguished from the almost unavoidable artifactual fragmentation of fibers that occurs in histologic sections. Degenerate fibers break into pieces that rapidly contract to acquire rounded ends, in contrast to the sharp, jagged ends of artifactually shattered fibers. As fiber fragmentation progresses, the fragments liquefy and assume a spherical shape. These proteinaceous spheres are referred to as **morgagnian globules.** Clefts and protein lakes appear between fibers, presumably the result of complete liquefaction of fibers. Some of the clefts are probably the result of osmotic fluid imbibition by the cataractous lens. The osmosis results from protein denaturation into more numerous, smaller peptides and from degeneration of the capsular epithelium in which resides the Na-K-dependent ATPase osmotic pump critical to normal lens hydration. Abortive efforts at new fiber formation by lens epithelium results in the formation of large, foamy nucleated cells called **bladder cells,** pathognomonic of cataract (Fig. 4.20). In advanced cataracts the degenerate fibers may liquefy to the extent that their low-molecular-weight end products diffuse through the semipermeable capsule, resulting in the spontaneous clearing of the opaque lens typical of the hypermature cataract. Histologically, such lenses consist of a dense, eccentric residual nucleus in a lake of proteinaceous fluid, surrounded by a wrinkled capsule. Deposition of calcium salts is seen rarely. Such hypermature cataracts are often accompanied by a lymphocytic–plasmacytic iridocyclitis, presumably in response to the leaking lens protein (see phacolytic uveitis).

A different picture is seen in lenses in which degeneration is associated with rupture of the capsule, as occurs with ocular trauma. First, massive release of more or less native lens protein at the time of rupture may cause a severe perilenticular nonsuppurative endophthalmitis about 10 days after the initial trauma (see phacoclastic uveitis). Second, the capsular rent permits leukocytes to enter the lens to speed the dissolution of lens fibers. Fibroblastic metaplasia of lens epithelium may result in cartilage or even bone within the lens. Even after total destruction of the lens fibers, the durable capsule will be found somewhere in the anterior or posterior chamber as a curled eosinophilic mass, often encapsulated in fibrous tissue probably derived from surviving lens epithelium or from injured ciliary epithelium. Such remnants distinguish lenticular rupture with subsequent dissolution from true developmental aphakia.

The sequence of histologic change in cataract is the same regardless of cause, and thus diagnosis of cause can be made only in light of patient data or concurrent ocular disease. In dogs, for example, familial cataracts may be congenital or of later onset. Specific examples may typically occur alone or with other ocular lesions, and occur at an age, in a location and with a progression sufficiently characteristic to allow presumptive diagnosis of a breed-specific syndrome. Cataract also occurs secondary to glaucoma, endophthalmitis, ocular trauma, and anterior segment anomalies, and observation of these latter defects permits presumptive diagnosis of the pathogenesis of the accompanying cataract.

Cataract may result from exposure of the lens to a wide variety of physical and chemical insults, such as solar or other irradiation, cold, increased intraocular pressure, toxins, nutritional excesses and deficiencies, nearby inflammation, and direct trauma. The list of potential cataractogenic chemical toxins grows daily and includes food additives, chemotherapeutic agents, and by-products of ocular inflammation. The pathogenesis of the cataract is not determined for more than a few such insults, but a common denominator seems to be the ability to upset the precarious balance between substrate supply and enzymic activity within the almost exclusively anaerobic lens. This imbalance results in degeneration of fibers, accumulation of nonmetabolized substrate, or production of abnormal metabolites. The latter two classes of products may be cytotoxic or osmotically active, thus drawing water into the critically dehydrated lens and causing opacity.

Most cataracts in humans and animals are not identified as being caused by a single insult, but are assumed to represent the result of years of accumulated and perhaps synergistic cataractogenic activity of environmental, dietary, and inborn insults. The majority of cataracts seen in veterinary practice fall into one of three categories: inherited, postinflammatory, and idiopathic. In reality, the large group of inherited cataracts in dogs is of unknown pathogenesis, although extrapolation from knowledge of similar cataracts in rodents and humans suggests inborn errors of lenticular metabolism are at fault. Postinflammatory cataracts result from injury to lenticular epithelium by adjacent inflammation, interference with aqueous production, composition, and flow, and accumulation of toxic bacterial, leukocytic, and plasma by-products in the lenticular environment. Adherence of iris to lens (posterior synechia) inevitably causes a focal subcapsular cataract.

Other than these broad categories, there are a few naturally occurring examples of cataract about which there is some understanding.

Diabetic cataract develops in about 70% of spontaneously diabetic dogs. The opacity is bilateral and begins in the cortex at the equator. Progression to complete cortical opacity usually occurs within a few weeks. The pathogenesis of the cataract has traditionally been ascribed to the excessively high level of glucose within the aqueous. Glucose is normally the major energy source for lens fibers, with most of it used to fuel the Embden–Meyerhof pathway of anaerobic glycolysis. When the rate-limiting enzyme of this pathway, hexokinase, is saturated with glucose, the back-up of glucose is shunted to alternative metabolic pathways. Chief among these is the sorbitol pathway, activated in the rabbit lens by glucose concentrations of greater than 90 mg/dl. In this pathway, the excess glucose is converted by an aldose reductase to the polyalcohol, sorbitol, which is then slowly reduced to a ketose. Because this second reaction is much slower than the first, sorbitol may accumulate to very high concentrations within the lens, and osmotically attracts water even to the point of hydropic cell rupture. Under experimental conditions at least, the early cataract may be reversed if aqueous sugar levels are reduced to normal, but the later cataract is irreversible.

However, osmotic events alone are not enough to explain all of the structural and metabolic changes in sugar-induced cataracts. The efficacy of antioxidants in ameliorating such cataracts, the nature of intralenticular biochemical alterations, and detection of increased intralenticular oxidants all point to some kind of oxidative damage as an additional promoter of cataract.

Galactose-induced cataracts probably have the same complex and incompletely understood pathogenesis as the diabetic cataract and are seen in orphaned kangaroos and wallabies raised on cows' milk, as well as in a host of experimental models. Since marsupial milk is much lower in lactose than is bovine milk, the enzymically ill-equipped neonate develops osmotic diarrhea from undigested lactose and galactose in the intestine, and some excess galactose enters the aqueous humor. The lens, deficient in the enzymes to utilize the galactose by converting it to glucose-6-phosphate for anaerobic glycolysis, shunts the galactose via aldose reductase to its polyalcohol, dulcitol, which acts osmotically as does sorbitol to disrupt lens fibers. Cataract reported in puppies and wolf cubs fed commercial milk replacer, or in kittens on feline milk replacer, has been attributed to deficiency of arginine, although in several case reports the specific dietary error was not identified. Cataract due to **dietary deficiency** of any of several sulfur-containing amino acids, zinc, or vitamin C occurs in farmed fish, and many models of nutritional cataract exist in various laboratory animals.

Various forms of irradiation cause cataract. The lens absorbs most of the ultraviolet and short-wavelength visible blue light that would otherwise damage the retina. At least in humans, the chronic exposure to such irradiation is thought to be important in the pathogenesis of senile cataract. **Sunlight-induced cataract** has been described several times in farmed fish but not yet for other domestic animals as a naturally occurring phenomenon. Absorption of ultraviolet or near-ultraviolet wavelengths by lens epithelium nucleic acids or lenticular aromatic amino acids results in photochemical generation of free radicals and peroxidative damage to numerous structural components of the lens.

A similar pathogenesis probably explains the development of cataract in animals irradiated as part of cancer therapy. In one study, 28% of dogs receiving **megavoltage x-radiation** for nasal carcinoma developed diffuse cortical cataract within 12 months of irradiation. In humans, the risk of cataract is dose related, and reaches virtual certainty with dosages of 800 to 1500 centigray or rads, whereas rodents require at least twice that dosage. The dogs in the study cited received between 3680 and 5000 centigrays. Antioxidants such as vitamin E or C, or hypoxia, are significantly protective against several models of light- or other irradiation-induced cataract, providing further support for the common denominator of oxidative stress in the pathogenesis of such cataracts.

The aminoglycoside antibiotic and anthelmintic **hygro-**

mycin B has been shown to induce posterior cortical and subcapsular cataracts in sows, but not boars, fed the drug continuously for 10 to 14 months. The effect is dose-dependent and perhaps even cumulative. Pigs fed the same therapeutic daily dose, but consuming the drug on an 8-week-on, 8-week-off basis in accordance with the manufacturer's recommendations, do not develop cataracts. The pathogenesis of the cataract is unknown, but a partial inhibition of hygromycin-induced cataracts *in vitro* by addition of vitamin E suggests that peroxidative damage to lens fiber membranes may be important. Deafness in pigs, and also dogs, caused by hygromycin B is discussed with the Ear.

Bibliography

Bhuyan, K. C., and Bhuyan, D. K. Molecular mechanism of cataractogenesis: III. Toxic metabolites of oxygen as initiators of lipid peroxidation and cataract. *Curr Eye Res* **3:** 67–81, 1984.

Glaze, M. B., and Blanchard, G. L. Nutritional cataracts in a Samoyed litter. *J Am Anim Hosp Assoc* **19:** 951–954, 1983.

Martin, C. L., and Chambreau, T. Cataract production in experimentally orphaned puppies fed a commercial replacement for bitch's milk. *J Am Anim Hosp Assoc* **18:** 115–118, 1982.

Poston, H. A. *et al.* The effect of supplemental dietary amino acids, minerals, and vitamins on salmonids fed cataractogenic diets. *Cornell Vet* **67:** 472–509, 1977.

Rathbun, W. B. Biochemistry of the lens and cataractogenesis: Current concepts. *Vet Clin North Am: Small Anim Pract* **10:** 377–398, 1980.

Roberts, S. M. *et al.* Ophthalmic complications following megavoltage irradiation of the nasal and paranasal cavities in dogs. *J Am Vet Med Assoc* **190:** 43–47, 1987.

Varma, S. D. *et al.* Oxidative stress on lens and cataract formation: Role of light and oxygen. *Curr Eye Res* **3:** 35–58, 1984.

Woollard, A. C. S. *et al.* Abnormal redox status without increased lipid peroxidation in sugar cataract. *Diabetes* **39:** 1347–1352, 1990.

VI. The Uvea

The uvea is the vascular tunic of the eye. It is derived from the primitive neural crest mesenchyme surrounding the primary optic cup (only the vascular endothelium is mesodermal). Its differentiation is guided by the retinal pigment epithelium. Anteriorly, the mesenchyme accompanies the infolding of the neurectoderm at the anterior lip of the optic cup to form the stroma of the iris and ciliary processes. That portion of anterior periorbital mesenchyme not accompanying these neurectodermal ingrowths remains to form the ciliary muscle and trabecular meshwork. Posteriorly it forms the choroid and sclera. In all domestic mammals except the pig, the choroid undergoes further differentiation to produce the tapetum lucidum dorsal to the optic disk. Defects in the development of the retinal pigment epithelium (including its cranial specialization as iridic and ciliary epithelium) inevitably result in defective induction or differentiation of the adjacent uvea.

The mature uvea includes iris, ciliary body, and choroid, the last divided into vascular portion and tapetum lucidum. The filtration angle is shared by iris, ciliary body, and sclera. Its diseases are discussed under the heading of glaucoma.

The **iris** is the most anterior portion of the uveal tract. It is a muscular diaphragm separating anterior from posterior chamber, forming the pupil and resting against the anterior face of the lens. The bulk of the iris is stroma of mesenchymal origin, with melanocytes, fibroblasts, and endothelial cells its major constituents. There is neither epithelium nor basement membrane along its anterior face, but rather a single layer of tightly compacted fibrocytes and melanocytes.

The posterior surface of the iris is formed by the double layer of neurectoderm from the anterior infolding of the optic cup. The two layers are heavily pigmented and are apposed apex to apex, with the basal aspect of the posterior epithelium facing the posterior chamber, and separated from it by a basement membrane. The basilar portion of the anterior epithelium, in contrast, is differentiated to form the smooth muscle fibers of the dilator muscle of the iris. These fibers lie along the posterior aspect of the iris stroma immediately adjacent to the epithelium. The constrictor muscle is found deeper within iris stroma but only in the pupillary third to quarter of the iris. The iris epithelium is rather loosely adherent between layers and between adjacent cells of the same layer, so that cystic separation occurs quite commonly. Numerous spaces reminiscent of bile canaliculi lie between adjacent cells and communicate freely with the aqueous humor of the posterior chamber.

The **ciliary body** extends from the posterior iris root to the origin of neurosensory retina. Like iris, it consists of an inner double layer of neuroepithelium and an outer mesenchymal stroma. The epithelial cells are oriented apex to apex and separated from the posterior chamber and vitreous by a basal lamina. Only the outer epithelial layer is pigmented. The ciliary body is divided into an anterior pars plicata and a posterior pars plana, the latter blending with retina at the ora ciliaris retinae. The pars plicata consists of a circumferential ring of villuslike epithelial ingrowths supported by a fibrovascular core, called ciliary processes. External to the ciliary processes the mesenchyme forms a ring of smooth muscle, the ciliary muscle, responsible for putting traction on the lens zonules and effecting the changes in lens shape necessary for visual accommodation. The muscle in domestic animals, particularly ungulates, is poorly developed, and accommodation is thought to be minimal in these species. The lens zonules anchor in the basal lamina of the nonpigmented ciliary epithelium, particularly of the pars plana and within the crypts between ciliary processes.

The **choroid** is the posterior continuation of the stroma of the ciliary body. The posterior continuations of the inner and outer layers of ciliary epithelium are retina and retinal pigment epithelium, respectively, with the transition made rather abruptly at the ora ciliaris retinae. The choroid consists almost entirely of blood vessels and melanocytes, except for the postnatal metaplasia to tapetum dorsal to the optic disk. The choroid is thinnest peripherally, thickest at the posterior pole, blends indistinctly with sclera externally, and is separated from the retinal pigment epithelium internally by a complex basal lamina called Bruch's membrane.

The **general pathology** of **uvea** includes anomalous or incomplete differentiation, degeneration, inflammation, and neoplasia. Anomalies have been previously discussed, and uveal neoplasms are considered in the section on ocular neoplasia. Uveal degenerations, except as a sequel to uveitis, are poorly documented. **Idiopathic atrophy** of the **iris** is described in Shropshire sheep as a bilateral defect obvious by 1 to 2 years of age. About 25% of the iris is converted to full- or partial-thickness holes. Those of partial thickness are spanned by a posterior bridge of iris epithelium. The eye is otherwise normal except for rudimentary corpora nigra. The pathogenesis of the apparently spontaneous atrophy is unknown. Similar atrophy is seen in middle-aged Siamese cats and in several breeds of small dogs (poodles, Chihuahuas, miniature schnauzers). The pathogenesis is unknown, and there are no published descriptions of the microscopic lesions in dogs or cats.

Multifocal cystic separation of the posterior iris epithelium is common in old dogs, and occasionally may be seen clinically as one or more translucent black cysts attached to the posterior iris or freely floating in the aqueous. Whether the cysts are truly degenerative, or represent residual lesions of fluid exudation from an undetected iritis, is unknown.

A. Uveitis

Uveal inflammation is common and may result from ocular trauma, noxious chemicals, infectious agents, neoplasia, or immunologic events. In addition, corneal injury may cause hyperemia and increased permeability of anterior uveal vessels either by percolation of bacterial toxins or inflammatory mediators into the aqueous, or by stimulation of a vasoactive sensory reflex via the trigeminal nerve. The uvea may be the initial site of inflammation, as in localization of infectious agents, or may become involved as the nearest vascular tissue capable of responding to injury of the lens, cornea, or ocular chambers. Conversely, the uvea seldom undergoes inflammation without affecting adjacent ocular structures.

The vocabulary of uveitis and its sequelae is complex. **Anterior uveitis** describes inflammation of iris and ciliary body. **Posterior uveitis** involves ciliary body and choroid, with panuveitis occasionally used to designate diffuse uveitis. **Chorioretinitis** describes inflammation of choroid and, usually less severely, overlying retina. **Endophthalmitis** is inflammation of uvea, retina, and ocular cavities, with **panophthalmitis** reserved for inflammation that has spread to involve all ocular structures including sclera. The usefulness of such terminology is doubtful when one considers the vascular unity of the uvea and its intimate association with other ocular tissues. Uveal exudation inevitably leads to protein and cellular exudation into the aqueous and vitreous and thus technically is endophthalmitis. Uveal vessels permeate the sclera as a normal anatomic feature and thus provide an easy route for uveal leukocytes to enter the sclera. By convention, the choice of diagnostic classification is strongly influenced by clinical severity, with anterior uveitis the mildest and panophthalmitis the most severe lesion.

Ocular inflammation is further classified as suppurative, granulomatous or nonsuppurative, nongranulomatous. The last is a peculiar historical term in human ophthalmology. It is equivalent to lymphocytic–plasmacytic inflammation and will be so described here. The usefulness of such classification in predicting causes decreases as the lesion ages and as the events of host immune response blend with the initial inflammation. Furthermore, the reaction may differ between anterior and posterior segments, with choroiditis much more commonly lymphocytic than suppurative, despite concurrent anterior uveal suppuration. Specific examples of uveitis are presented later. Discussed here are the features common to uveal inflammation and its sequelae, regardless of cause.

Acute uveitis involves the usual sequence of protein-rich fluid exudation followed by emigration of leukocytes, typically neutrophils. In the iris, the fluid and cells readily percolate through the loose stroma to enter the anterior chamber as the clinically observed aqueous flare and hypopyon, so that large numbers of neutrophils are rarely seen within the iris itself. In ciliary processes there usually is severe stromal edema, perhaps a consequence of the initial inability of the serous exudate to pass through the tight intercellular junctions of the ciliary epithelium. Choroid exhibits the most convincing vascular engorgement as well as edema, with the latter frequently seeping through the retinal pigment epithelium to cause serous retinal separation.

Leukocytes are initially **neutrophils** in uveitis of bacterial origin, such as in the neonatal septicemias of calves, foals, and pigs. Neutrophils also predominate in acute mild neurogenic uveitis associated with corneal epithelial injury and in the acute phase of phacoclastic uveitis. In very mild uveitis they are found marginated along the endothelium of iris and ciliary venules, in perivascular adventitia, adherent to ciliary processes, and in the filtration angle (Fig. 4.30). Neutrophils rapidly degenerate within the aqueous to assume an unsegmented globular morphology. Clumps may adhere to the corneal endothelium as keratic precipitates, settle ventrally within the anterior chamber as hypopyon, or plug the filtration spaces, possibly (but rarely) to cause glaucoma if the plugging is extensive. Fibrin exudation may accompany the acute inflammation, but fibrinolysis is very efficient within the anterior chamber so that glaucoma rarely results.

Nonsuppurative uveitis usually is dominated by lymphocytes and plasma cells. It may occur simply as a chronic form of what was initially a suppurative uveitis, but is more frequently seen as the typical manifestation of immune-mediated uveitis, ocular trauma, viral and mycotic uveitis, phacolytic uveitis, and uveitis accompanying

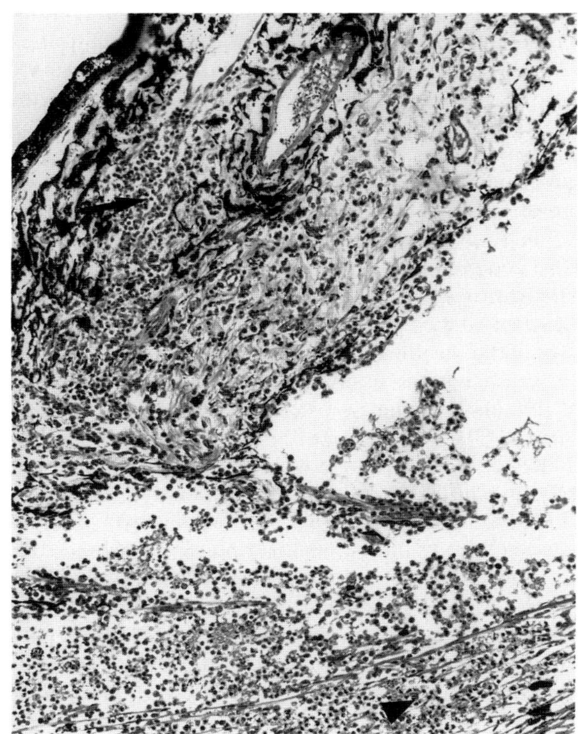

Fig. 4.30 Anterior uveitis. Dog. Leukocytes within stroma of iris (arrow) and peripheral cornea (arrowhead), and within filtration angle.

intraocular neoplasia. Inasmuch as the inflammation in most of these examples is probably of immunologic pathogenesis, it is probably more accurate to consider all nonsuppurative uveitis as immune mediated, with the prefix idiopathic or the name of the inciting antigen. Dogs, cats, and horses are the species most frequently affected, and in these species, the inciting agent is usually unknown.

The eye is an immunologically privileged site, with no resident lymphocytes, no antigen-processing macrophages or dendritic cells, and no lymphatic drainage. These peculiarities, plus the blood–eye barrier created by the tight intercellular junctions of iris endothelium, ciliary nonpigmented epithelium, and retinal pigmented epithelium, fostered the mistaken belief that many intraocular antigens were sufficiently sequestered as to be seen as "nonself" in the event of their release into systemic circulation. The numerous diseases characterized by lymphocytic–plasmacytic uveitis in the absence of an identified infectious agent have thus been broadly grouped as examples of autoimmune reaction to sequestered uveal, lenticular, or retinal antigens.

More recent studies have demonstrated that these supposedly unique and sequestered antigens are neither unique nor completely sequestered. Antigens identical to some of the lenticular or uveal antigens, for example, are found in nonsequestered tissues elsewhere in the body. Antigens inoculated into anterior chamber induce a systemic humoral and T-cell response, clearly pointing to at least some leakiness in the blood–eye barrier. A variety of experiments have lead to the suggestion of a carefully regulated system of ocular immunity, termed **anterior chamber-associated immune deviation.** In this system, intraocular antigens are somehow processed within the eye before draining from the trabecular meshwork into systemic circulation. These antigens, on reaching the spleen, initiate a typical humoral immune response but an atypical cell-mediated immune response. Proliferation of cytotoxic and suppressor T cells is enhanced, but those T cells committed to the production of cytokines as part of delayed hypersensitivity are suppressed. The theoretical result is that when these activated lymphocytes return to the eye, they are only of the types destined to produce the most localized and specific effects on the offending antigen, with the least nonspecific "bystander" injury. As a further safety measure to prevent unnecessarily damaging immune-mediated injury, both uveal tissue and aqueous humor contain cytokines [transforming growth factor β (TGF-β) is one] that inhibit activation of T lymphocytes.

Whatever their type, the splenic lymphocytes reach the eye about 1 week after experimental introduction of antigen into anterior chamber. Typically the lymphocytes are seen as perivascular aggregates in iris stroma, in ciliary body and, less obviously, in choroid and even retina. In long-standing cases (which are most likely to receive histologic examination), the aggregates may be very large and resemble lymphoid follicles (Fig. 4.31). As in other tissues, amplification of the immune response results in recruitment of lymphocytes that are not necessarily specific for the inciting antigen. The polyclonal nature of these lymphocytes is probably important in the typically recurrent nature of uveitis in all species. Once established in the eye, these cells respond to a diverse range of circulating antigens that enter the eye through a blood–eye barrier disrupted by the previous bout of inflammation. It is thus possible, or even probable, that chronic, recurrent uveitis results not from persistence of a single antigen, or repeated exposure to the same antigen, but is a stereotyped ocular response to activation of any one of its many acquired lymphoid populations by a variety of circulating antigens or native ocular antigens.

Granulomatous uveitis is distinguished from simple lymphocytic–plasmacytic uveitis by the conspicuous presence of epithelioid macrophages and, occasionally, giant cells. Ocular localization of some species of dimorphic fungi or of algae, helminths, or mycobacteria may cause granulomatous ophthalmitis, as may lens rupture and the Vogt–Koyanagi–Harada-like syndrome in dogs.

The major significance of uveitis is its effect on adjacent nonuveal tissues. Some effects result from the accumulation of acute exudates or chemical by-products of inflammation, but most result from the later organization of exudates and proliferative events of wound healing within ocular cavities.

Corneal changes include edema and peripheral stromal hyperemia (ciliary flush). The edema results from corneal

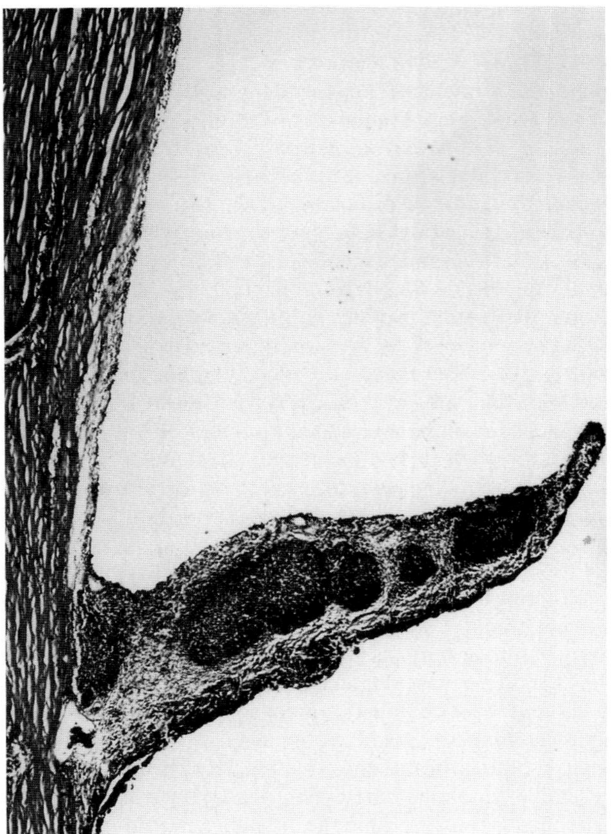

Fig. 4.31 Lymphonodular iritis and secondary glaucoma. Cat. The iridocorneal angle has been dislocated far posterior to the termination of Descemet's membrane (angle recession).

endothelial damage or as part of the reaction of limbic blood vessels to the inflammatory mediators released from the adjacent uvea. In the former instance, damage may be the direct result of the agent causing uveitis, as occurs in infectious canine hepatitis or feline infectious peritonitis. It may also occur as a result of an immune response to endothelial cells containing antigens of these infectious agents, to cross-reaction between microbial and corneal endothelial antigens, or as a nonspecific response to the presence of the chemical by-products of inflammation within the anterior chamber. Similar by-products mediate the acute inflammatory response in the nearby limbic and conjunctival vasculature, leading to edema in the peripheral corneal stroma. Hyperemia of this limbic network also results in the circumferential peripheral corneal stromal hyperemia, resembling a brush border, that is a clinical hallmark of anterior uveitis. In eyes with chronic uveitis, corneal edema may also result from glaucoma or from anterior synechia. Persistent edema may lead to stromal fibrosis, vascularization, bullous keratopathy, and the risk of ulceration. Limbic hyperemia may give way to peripheral corneal stromal vascularization, again presumed to be merely a response to the spillover of angiogenic cytokines from the chronic intraocular inflammation.

The accumulation of fibrin, leukocytes, and erythrocytes in the aqueous may result in plugging of the filtration angle and subsequent glaucoma. The infrequent observation of this sequel suggests either unusual potency of the fibrinolytic system within the aqueous or the inability of exudates to plug more than the most ventral portion of the circumferential angle. Much more common is the organization of inflammatory exudates on the surface of iris or ciliary body. Adherence of iris to lens (**posterior synechia**) is more common than adherence to cornea (**anterior synechia**) because of the normally intimate association of the lens and iris. If the posterior synechia involves the circumference of the iris, the pupillary flow of aqueous is blocked, posterior chamber pressure rises, and the iris bows forward (**iris bombé**) and may actually adhere anteriorly to the cornea. Glaucoma results from pupillary block, peripheral anterior synechia, or both. In severe and prolonged anterior uveitis, there may be development of a fibrovascular membrane on the iris face, which may span the pupil to cause pupillary block (occlusio pupillae) or cover the face of the pectinate ligament to cause neovascular glaucoma (Fig. 4.32). Alternatively, the membrane may contract on the face of the iris resulting in infolding of the pupillary border to adhere to the anterior (**ectropion uveae**) or posterior (**entropion uveae**) iris surface. Atrophy of iris may follow severe and necrotizing inflammation, and some examples can be distinguished from idiopathic and senile atrophy by the observation of residual lesions of the previous uveitis, such as lymphoid aggregates, focal synechiae, and uveal hyalinization.

The ciliary apparatus suffers the same range of chronic lesions as does iris. Deposition of PAS-positive hyaline material along the luminal surface of the ciliary epithelium is particularly frequent in horses. It appears to be depos-

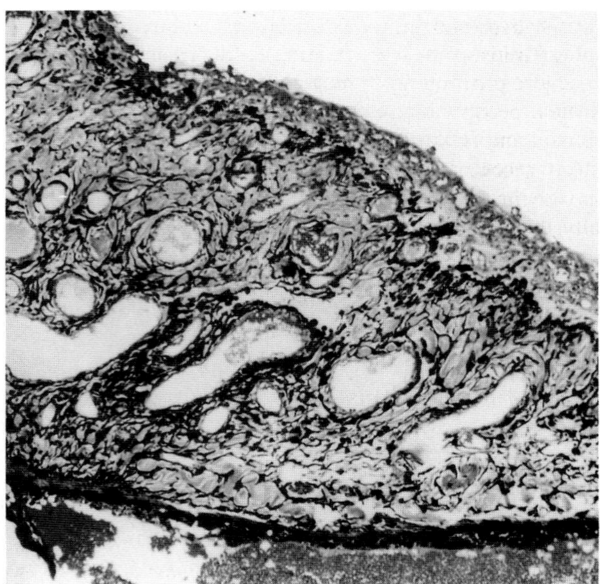

Fig. 4.32 Preiridal fibrovascular membrane.

ited in the cytoplasm of the nonpigmented epithelium, and may represent aberrant basement membrane. It is not fibrin. Organization of exudate within the posterior chamber or vitreous results in a retrolental fibrovascular membrane, called a **cyclitic membrane,** which stretches around the ciliary body and across the back of the lens. Vitreous is almost always liquefied as a result of the severe uveitis, and continued contraction of fibrin in the posterior chamber and vitreous causes a separation of retina. Histologic examination of most cyclitic membranes reveals a fibrovascular retrolental membrane incorporating lens into its anterior face and a folded, degenerate retina in its posterior surface.

The residual lesions of chronic choroiditis include focal lymphoid aggregates and scarring. Tapetum usually remains unaffected. As choroiditis severe enough to evoke these lesions will almost invariably have involved retina and retinal pigment epithelium, the residual scar will involve these structures. Chorioretinal scars are seen as focal fibrous chorioretinal adhesions in place of normal retinal pigment epithelium. Because these scars prevent the involved retina from separating as part of processing artifact, they frequently appear as "spot welds" along an otherwise artifactually detached retina. Retinal pigment epithelium may be hypertrophic or hyperplastic, particularly if retina has been chronically separated by choroidal effusion. The fibroblastlike cells forming the scar may be derived from retinal Müller cells, choroidal fibroblasts, or metaplasia of retinal pigment epithelium, the last being the major source.

Cataract is a common sequel to uveitis, either as a result of uveal adhesions to lens surface, altered aqueous flow with lenticular malnutrition, exposure to injurious inflammatory by-products, or increased aqueous pressure in postinflammatory glaucoma.

Phthisis bulbi describes a hypotonic, shrunken, structurally disorganized eye that is the end stage of severe ophthalmitis. Phthisis is seen most commonly as a sequel to severe prolonged suppurative septic ophthalmitis from corneal perforation. Cornea and sclera are thickened by fibrosis and leukocytic infiltration, and ocular content is barely recognizable. Mineralization and even ossification may occur, but cartilage is absent (unlike that in congenitally dysplastic globes). A shrunken, end-stage eye that contains ocular structures with at least recognizable orientation is properly termed **atrophia bulbi.** The term is seldom used, but atrophia is much more common than true phthisis bulbi.

Bibliography

Bistner, S., Shaw, D., and Riis, R. C. Diseases of the uveal tract (part I). *Compend Cont Ed* **1:** 868–875, 899–906, 1979.

Davidson, M. G. *et al.* Feline anterior uveitis: A study of 53 cases. *J Am Anim Hosp Assoc* **27:** 77–83, 1991.

Peiffer, R. L., Wilcock, B. P., and Yin, H. The pathogenesis and significance of pre-iridal fibrovascular membrane in domestic animals. *Vet Pathol* **27:** 41–45, 1990.

Swanson, J. F. Ocular manifestations of systemic disease in the dog and cat. *Vet Clin North Am: Small Anim Pract* **20:** 849–867, 1990.

1. Immune-Mediated Uveitis

The humoral and cellular events of immunologic reaction may occur within the eye in response to endogenous or exogenous antigen. The normal eye contains no lymphoid tissue and, following initial antigenic challenge, must rely on diffusion of antigen to the spleen before effector lymphocytes enter the eye. The lymphocytes do so as perivascular aggregates throughout the uvea. Subsequent exposure to the sensitizing antigen results in one or several of hypersensitivity reactions I through IV. For reasons previously stated (see anterior chamber-associated immune deviation), it is likely that the prolonged disruption of the blood–eye barrier following any uveitis allows a variety of circulating antigens to contact the polyclonal lymphoid population newly established within the uvea, and perpetuate the uveitis.

There is no clear distinction between immune-mediated uveitis and uveitis traditionally ascribed to a specific causative agent. Except for rapidly progressing bacterial uveitis following hematogenous localization or penetrating injury, virtually all uveitis probably has an immune component superimposed on initial nonspecific inflammation. Even traumatic uveitis probably permits unusually large amounts of endogenous ocular antigens to enter venous drainage and to reach the spleen and other lymphoid tissue. Types III and IV hypersensitivity have been induced in various laboratory animals using tissue-specific antigens of photoreceptor, uveal, lens, and corneal origin. Lens-induced uveitis and an idiopathic uveitis associated with dermal depigmentation in dogs (the so-called Vogt–Koyanagi–Harada syndrome) are naturally occurring examples of uveitis induced by endogenous ocular antigens. Recent demonstration of strong cross-reactivity between leptospiral antigens and equine corneal endothelium serves to further obscure the distinction between infectious and immune-mediated ocular disease.

Included in this section are those diseases that are exclusively or predominantly immune mediated. In general, all are chronic, nonsuppurative, and diffuse affections of the uvea in which the infiltrating leukocytes are predominantly lymphocytes and plasma cells in perivascular collars. Clinically, they are either continuously progressive syndromes or subject to periodic irregular clinical exacerbations and remissions. The histologic lesion is continuously present, and may eventually lead to the formation of uveal lymphoid nodules.

Examples of immune-mediated uveitis should be divided into those in which the antigen is known by culture, morphology, or history, and those of unknown cause. In the former category are many examples of viral, mycotic, protozoan, and helminthic uveitis, postvaccinal uveitis in dogs, and at least some cases of recurrent ophthalmitis in horses. The idiopathic group includes the majority of canine and feline cases. Lens-induced uveitis is a frequent and special syndrome that is, at least in part, a manifestation of endogenous uveitis.

a. CANINE ADENOVIRUS Infectious canine hepatitis virus (canine adenovirus, type 1) is the best-documented cause of immune-mediated uveitis in domestic animals. The systemic disease is discussed in Volume 2, Chapter 2, Liver and Biliary System. During the acute viral stage of the disease, viral replication within endothelium and stromal phagocytes of the uvea results in a primary mild nonsuppurative uveitis that usually is clinically undetected. Inoculation of field virus into the anterior chamber of dogs and foxes may result in viral inclusion bodies within corneal endothelium and subsequent edema, but edema is not a feature of the active stage of naturally occurring disease. During the convalescent phase of the disease, or 6–7 days after vaccination with a modified live virus, a small percentage of dogs develop anterior uveitis, endothelial damage, and corneal edema that is a manifestation of type III hypersensitivity to persistent viral antigen in which complement fixation attracts neutrophils. The proteases of neutrophils are responsible for the cell injury.

The histologic lesion is bilateral but not usually of equal intensity so that clinically apparent disease may be unilateral. Corneal edema results from diffuse hydropic degeneration of corneal endothelium and secondary stromal edema. In a small percentage of affected dogs, the damage is so persistent as to cause interstitial keratitis and permanent fibrosis. Whether this sequel results from unusually persistent antigen, unusually severe endothelial damage or age-dependent variation in endothelial regenerative ability is unknown. Intranuclear inclusion bodies of adenovirus type may be seen in a few degenerate endothelial cells. There is an accompanying anterior uveitis, with lymphocytes and plasma cells around vessels in iris and ciliary body, in the filtration angle, and adherent to cornea as keratic precipitates. Choroidal involvement is mild or absent. Sequelae such as synechia or angle obstruction with debris are infrequent, occurring in fewer than 5% of affected eyes. In most dogs, whether recovering from natural or vaccine-induced infection, the ocular reaction subsides within 3 to 4 weeks.

b. EQUINE RECURRENT OPHTHALMITIS (PERIODIC OPHTHALMIA) This is a worldwide and important cause of blindness in horses and mules. The blindness results from repeated attacks of anterior uveitis occurring at unpredictable intervals and with increasing severity. With each attack there is increasing involvement of posterior uvea, retina, and optic nerve, and increasingly frequent sequelae of cataract, lens luxation, synechiae, retinal separation, and interstitial keratitis. Despite the frequent observation of posterior synechiae, glaucoma is rarely reported. It is speculated that aqueous drainage in horses relies less on the trabecular meshwork and more on uveal resorption than is true of dogs or cats. The disease may initially be unilateral but eventually affects both eyes. Blindness is usually a late sequel, but may occur early in the disease if exudative choroiditis causes retinal separation.

Gross lesions of the acute disease are typical of anterior uveitis in any species: serous conjunctivitis, chemosis, circumcorneal ciliary hyperemia, corneal edema, and plasmoid aqueous and vitreous, with fibrin and leukocytes in the aqueous. Clinically, such animals are often systemically ill as detected by fever, decreased appetite, and depression. Lacrimation and photophobia are usually marked. Subsequent attacks tend to become increasingly severe, and resolution of the gross lesions between attacks is less complete. Such horses, during the quiescent period, may have one or more of peripheral corneal vascularization with fibrosis and persistent edema; irregular thickening and pigmentation of iris; multiple posterior synechiae; patchy residual uveal pigment on lens capsule; and peripapillary retinal hyperreflectivity suggesting retinal scarring.

The microscopic lesions depend on the stage of the disease and represent a continuum from anterior uveitis to endophthalmitis with retinal scarring, or even phthisis bulbi. The earliest lesion is anterior uveal inflammation that is transiently neutrophilic but rapidly becomes predominantly lymphocytic. Ciliary processes are most obviously affected. Edema, fibrin, and leukocytes distend the stroma, and PAS-positive hyaline material obscures the nonpigmented epithelium (Fig. 4.33). Leukocytes and fibrin lie in the anterior chamber and in the filtration angle. In the eyes of horses with a history of several attacks of uveitis, the exudate in active phases of the disease is almost purely lymphocytic–plasmacytic and is found about vessels of choroid, retina, and optic nerve as well as anterior uvea (Fig. 4.34A). Peripheral corneal vascularization, both from conjunctival and limbic ciliary vessels, becomes increasingly prominent and extends farther toward the center of the cornea (Fig. 4.34B). Edema accompanies the newly formed vessels. The chorioretinitis may be sufficiently exudative to cause multifocal retinal separation. As these severe uveal lesions regress during clinically quiescent periods, they leave behind characteristic residual changes. Relatively early in the disease there is the development of perivascular lymphoid aggregates in iris and ciliary body, which persist and may even form true lymphoid nodules (Fig. 4.34C,D). The ciliary processes may remain thickened by fibrous organization of stromal edema, and a hyaline membrane often seems to cover the ciliary epithelium. This material, in fact, lies within the apical cytoplasm of the nonpigmented ciliary epithelium and may be the histologic counterpart of crystalline protein inclusions seen in this epithelium ultrastructurally. Small blood vessels persist along corneal stromal lamellae, and there is subtle fibrous disorganization of the stroma, the result of previous edema. Peripapillary chorioretinal scarring is seen as focal retinal photoreceptor loss, jumbling of layers, and gliosis. Adjacent retinal pigment epithelium may be hypertrophic or hyperplastic, and a focal cluster of lymphocytes in the nearby choroid is common.

Choroidal vessels are unusually thick-walled due to

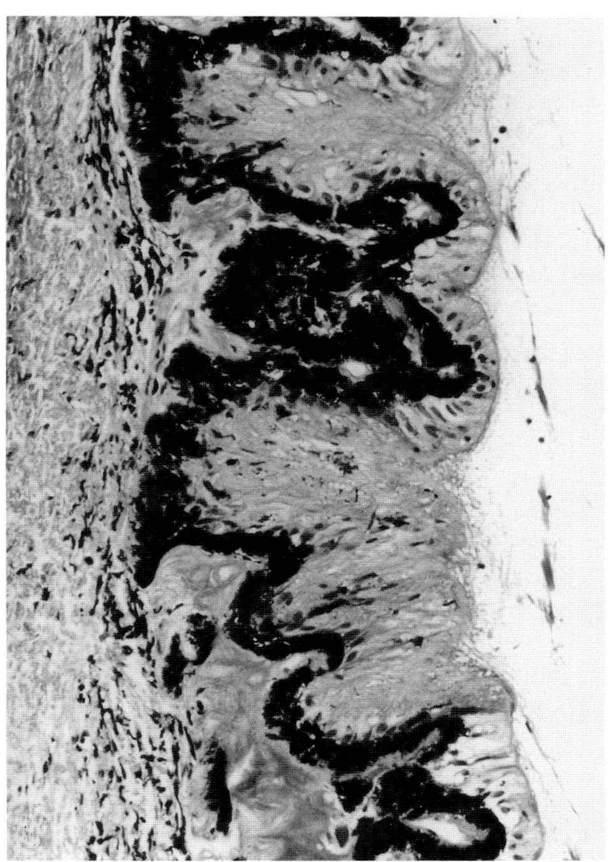

Fig. 4.33 Hyalinization of inner nonpigmented ciliary epithelium. Equine recurrent ophthalmitis.

Fig. 4.34A Choroiditis and inflammatory infiltrate of pars plana. Equine recurrent ophthalmitis.

edema or fibrin, the latter probably analogous to the hyalinization described in several reports. Increased vascular permeability persists even in quiescent periods, with loss of the blood–aqueous barrier demonstrated by fluorescein angiography. Whether or not this vascular alteration participates in the perpetuation of the uveitis is unknown, but it is known that such alterations predispose to localization of circulating immune complexes and subsequent type III hypersensitivity-induced inflammation (Auer reaction).

Focal retinal detachments may reattach by fibrous organization of subretinal exudate or may progress to total separation with a barely recognizable retina adherent to posterior lens capsule. Gliosis and lymphocytic aggregates may be found within proximal optic nerve. Scarring in optic disk and adjacent retina often is clinically obvious and may occur in horses with no other lesions of uveitis, leading to speculation that it is not really linked to, or at least not specific for, equine recurrent uveitis. Lesions of such sequelae as cataract, chronic conjunctivitis, and glaucoma are described elsewhere.

The causes and pathogenesis of recurrent equine ophthalmitis have not been intensively studied, in contrast to the abundance of opinion expressed in reviews or texts. The almost universal opinion is that the disease is the result of hypersensitivity to exogenous antigen. The most frequently cited antigens are *Leptospira* and dead microfilariae of *Onchocerca cervicalis*. The recent demonstration of antigenic cross-reaction between equine corneal endothelium and several common leptospiral serovars suggests that accidental autoimmunity may participate in some of the lesions. Cross-reaction with other ocular antigens has not been studied, but the repeated observation that lesion development follows the development of serum or aqueous antibody titers to *Leptospira* makes an immune pathogenesis very likely for this disease syndrome.

There seems little doubt that *Leptospira*, particularly *L. pomona*, can initiate uveitis in horses and in humans. In both species the uveitis develops as a sequel to infection, delayed by weeks or years. The initial suspicion of this association was the observation, as early as 1948, that horses with uveitis frequently had very high serum and aqueous agglutination titers against *L. pomona*. This observation has since been repeatedly confirmed, and has been supported by the observation of uveitis in 22 of 36 eyes of Shetland ponies inoculated subcutaneously with small numbers of *L. pomona*. All ponies underwent subsequent leptospiremia, but none developed ocular lesions until 50 weeks after inoculation. The lesions were typical of anterior uveitis, and six eyes progressed to phthisis bulbi after repeated bouts of uveitis. In 18 eyes there were central retinal scars typical of the naturally occurring disease. Although there seems

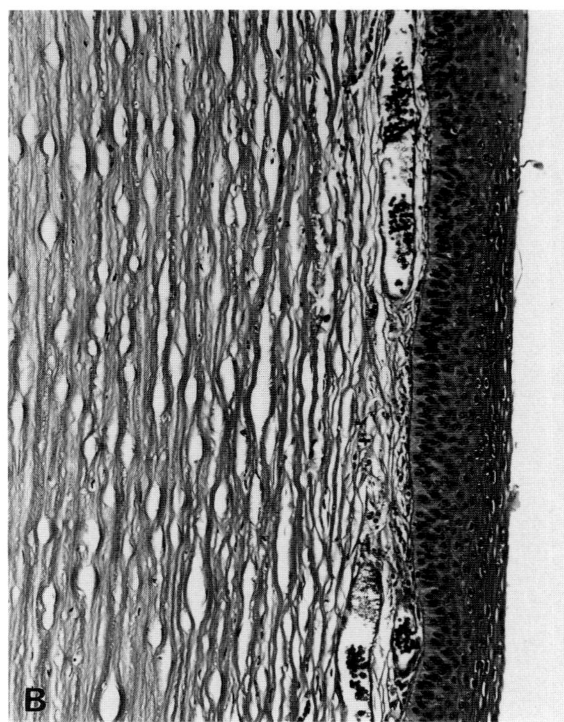

Fig. 4.34B Peripheral corneal stromal vascularization and subtle fibrosis in a horse with equine recurrent ophthalmitis.

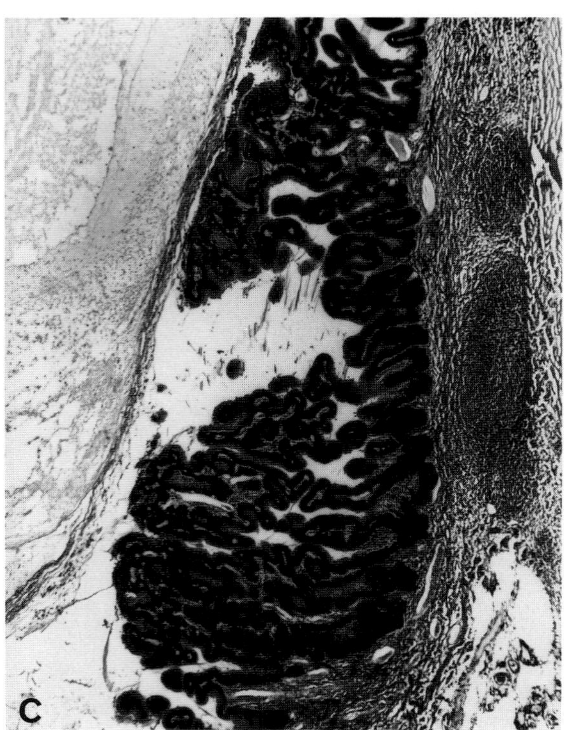

Fig. 4.34C Equine recurrent ophthalmitis. Lymphoid nodules in iris stroma.

little doubt that hypersensitivity to *L. pomona* can cause the syndrome of equine recurrent ophthalmitis, it is unlikely to be the only cause. In a recent survey in Florida, only 1 of 10 horses with uveitis had a positive microagglutination titer against *L. pomona*. Eight had cutaneous onchocerciasis, but that was not significantly different from the overall 60% prevalence of that parasite in the sample population. Typical of the long-standing controversy about this disease is a 1990 survey of horses with uveitis in Virginia and Maryland, in which 52 of 80 affected horses had positive serum titers to one or more leptospiral serovars, and a statistically significant association between uveitis and positive titer was found for the serovars *pomona* and *autumnalis*.

Onchocerca cervicalis infection of the eye is considered briefly with helminthic uveitis. Reaction to dead microfilariae within uveal tissues is considered by some an important cause of equine recurrent uveitis, although the high prevalence of this parasite in the horse population makes such claims difficult to support statistically. The enthusiasm for this pathogenesis may be generated, in part, by the importance of onchocerciasis as a leading cause of blinding keratitis in people.

c. Bovine Malignant Catarrhal Fever-Associated Uveitis The presence of severe uveitis is an important clue in the clinical differentiation of malignant catarrhal fever from other bovine systemic disorders, particularly from mucosal disease. The histologic lesions within the eye resemble those elsewhere in the body: arterial necrosis and perivascular and intramural lymphocytic accumulations. The presence of mitotic figures among the lymphoid cells is distinctive. The arteritis usually is most obvious in the iris, but may be seen affecting arterioles or venules in retina, choroid, meninges of optic nerve, or even peripheral cornea. There is marked corneal edema with a ring of peripheral corneal stromal vascularization, clinically seen as a dark red circumferential brush border of straight vessels in the perilimbal cornea. Blood vessels in the conjunctiva and even in the newly vascularized cornea may be targets for the disease, so that the edema and hemorrhage of vessel injury are added to the nonspecific lesions of conjunctivitis and peripheral keratitis that accompany uveitis of any cause in all species.

Infiltration of lymphocytes among corneal endothelial cells is associated with patchy necrosis of that layer, which may also contribute to the corneal edema (Fig. 4.35). A layer of mononuclear leukocytes enmeshed in fibrin often is adherent to the aqueous face of the corneal endothelium.

Even the very early lesions are lymphocytic. In vessels the first changes involve subendothelial and adventitial lymphocytic and lymphoblastic accumulation, with little necrosis. Despite long-standing speculation for an immune-complex pathogenesis for the vasculitis, proof is lacking. Deposition of immunoglobulin or complement is

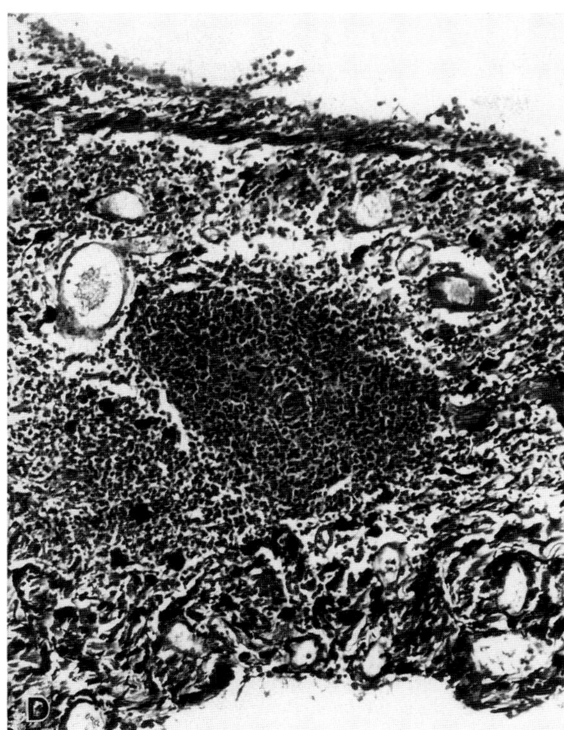

Fig. 4.34D Equine recurrent ophthalmitis. Lymphoid nodules in iris stroma.

not a significant feature of the vascular lesion within the eye, and a T cell-dependent, type IV immune pathogenesis has been suggested.

d. Feline Infectious Peritonitis-Associated Uveitis The coronavirus of feline infectious peritonitis causes diffuse uveitis that is probably immune mediated (see Volume 2, Chapter 4, The Peritoneum). The frequency of ocular lesions is unknown because the eyes are not regularly examined in cats with the disease. Estimates based on clinical examination range from about 10% in an outbreak to 50% of unselected clinical cases. Most cats which die of the disease have ocular involvement as detected by coagulation of aqueous with acidic fixatives (indicating increased aqueous protein). The histologic evidence for inflammation in some eyes may be subtle indeed, and such eyes usually reach postmortem without clinically detected uveitis. Conversely, some cats develop severe uveitis attributed to this virus by clinical, serologic, or histologic evaluation, without concurrent evidence of the disease elsewhere.

The typical histologic lesion, as is the case elsewhere in the body, varies with time and location. The leukocytic infiltration is most extensive in ciliary body and adjacent limbic sclera, and is usually a rather even mixture of neutrophils, lymphocytes, plasma cells, and macrophages. In some eyes the infiltrate is purely histiocytic. The inflammatory cell population often becomes more purely

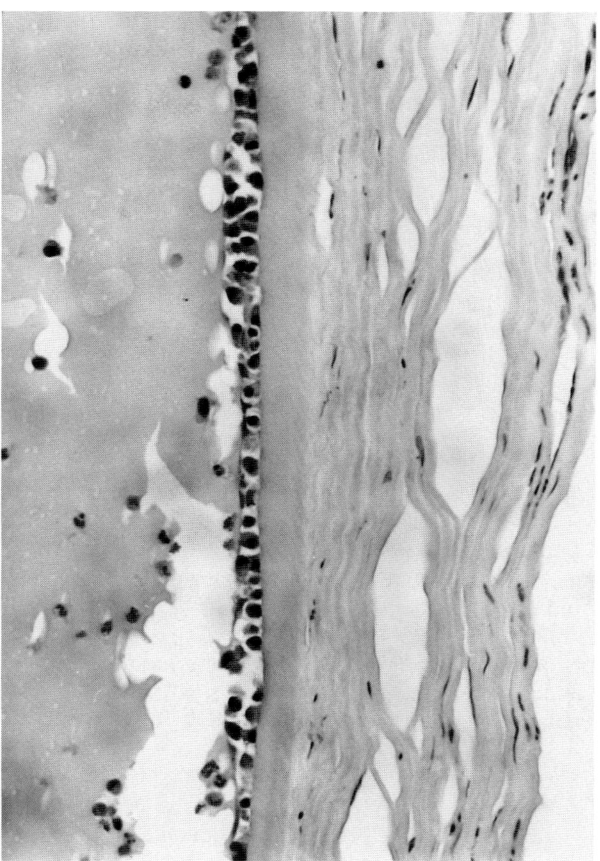

Fig. 4.35 Lymphocytes as part of a presumed immune-mediated endothelialitis obscure the few remaining corneal endothelial cells. Bovine malignant catarrhal fever.

lymphoid in the choroid and more neutrophilic in the anterior chamber. Perivascular lymphocytic–plasmacytic aggregates are common in retrobulbar connective tissue and in the optic nerve sheath, and in the retina. In the retina the accumulations are larger and are more likely to involve a true phlebitis than is the case with the subtle perivascular retinitis that is a frequent and nonspecific accompaniment to most forms of anterior uveitis. Sequelae to the uveitis are rarely seen, either because the cats are in the late stages of the disease when ocular lesions are examined or because euthanasia halts its progression. Retinal separation with serous subretinal exudate is occasionally observed. The presence of large globular accumulations of macrophages and neutrophils adherent to the corneal endothelium (**keratic precipitates**) is an important clinical hallmark of the disease, and is useful histologically as well. Neutrophilic endothelialitis with severe corneal edema may also occur.

e. Idiopathic Lymphocytic Uveitis of Dogs and Cats Chronic lymphocytic–plasmacytic uveitis is frequent in dogs and cats. Most cases are of unknown cause, in that affected animals are otherwise healthy, the eye itself contains no visible causal agent, and serologic tests are inconclusive. The last point is the most hotly debated, particularly for cats, in which there is serologic evidence for involvement of toxoplasmosis or feline immunodeficiency virus in the pathogenesis of the uveitis. Other studies still conclude that about 80% of all uveitis in cats is of unknown cause, but many cats had not been tested for antibody to the immunodeficiency virus.

The lesions are similar in dogs and cats. Perivascular accumulations of lymphocytes and plasma cells are seen throughout the uvea and, with less regularity, around small vessels in the retina. Ciliary body tends to have the greatest accumulation. Formation of lymphoid nodules may occur in chronic and severe cases, and for unknown reasons, these are seen much more often in cats than in dogs. The presence of this lymphonodular anterior uveitis is highly correlated with the development of glaucoma in cats but not in dogs (see succeeding sections).

Idiopathic lymphocytic uveitis is presumed to be immune mediated, but the identity of the antigen or antigens is unknown. In dogs the lesion may be confused with the mild uveitis that accompanies maturing cataracts (see Phacolytic Uveitis), while in cats the alternative diagnoses include feline infectious peritonitis, toxoplasmosis, and lymphoma.

f. GRANULOMATOUS UVEITIS IN DOGS (VOGT–KOYANAGI–HARADA SYNDROME) Despite the exotic-sounding name, this disease is relatively frequent in those areas in which the most susceptible breeds (Akitas, Siberian huskies, Samoyeds) are popular. The clinical syndrome of facial dermal depigmentation and severe bilateral uveitis is distinctive, although many dogs examined for the uveitis are not noted to have skin lesions. The canine syndrome closely parallels the human disease, except for the encephalitis that is the least frequent part of the human syndrome and has not been confirmed in dogs. The human disease is most prevalent in Asians; the predilection in dogs for the Japanese Akita is a fascinating but unexplained parallel.

The histologic lesion is a destructive granulomatous endophthalmitis with abundant dispersal of melanin. The melanin-laden retinal pigmented epithelium seems to be especially susceptible. Retinal detachment and destructive granulomatous inflammation are seen in advanced cases. The lesion is distinguished from the more prevalent systemic mycotic diseases by the predilection for pigmented tissues within the eye, the lack of visceral involvement, and the distinctive skin lesions, if they are present (see Chapter 5, The Skin and Appendages).

The pathogenesis of the human disease is thought to involve cell-mediated immune reaction to uveal (or epidermal) melanin.

Bibliography

Bistner, S., and Shaw, D. Uveitis in the horse. *Compend Cont Ed* **2**: S35–S43, 1980.

Bussanich, M. N., Rootman, J., and Dolman, C. L. Granulomatous panuveitis and dermal depigmentation in dogs. *J Am Anim Hosp Assoc* **18**: 131–138, 1982.

Cousins, S. W., and Streilein, J. W. Flow cytometric detection of lymphocyte proliferation in eyes with immunogenic inflammation. *Invest Ophthalmol Vis Sci* **31**: 2111–2122, 1990.

Gwin, R. M. *et al.* Idiopathic uveitis and exudative retinal detachment. *J Am Anim Hosp Assoc* **16:** 163–170, 1980.

Lindley, D. M., Boosinger, T. R., and Cox, N. R. Ocular histopathology of Vogt–Koyanagi–Harada-like syndrome in an Akita dog. *Vet Pathol* **27**: 294–296, 1990.

Liu, S. H., Prendergast, R. A., and Silverstein, A. M. The role of lymphokines in immunogenic uveitis. *Invest Ophthalmol Vis Sci* **24**: 361–367, 1983.

Morter, R. L. *et al.* Experimental equine leptospirosis (*Leptospira pomona*). *Proc 68th Ann Mtg of U.S. Livestock Sanitary Assoc* 147–152, 1964.

Parma, A. E. *et al.* Experimental demonstration of an antigenic relationship between *Leptospira* and equine cornea. *Vet Immunol Immunopathol* **10**: 215–224, 1985.

Peiffer, R. L., Jr., and Wilcock, B. P. Histopathologic study of uveitis in cats: 139 cases (1978–1988). *J Am Vet Med Assoc* **198**: 135–138, 1991.

Roberts, S. R. Etiology of equine periodic ophthalmia. *Am J Ophthalmol* **55**: 1049–1955, 1963.

Schmidt, G. M. *et al.* Equine ocular onchocerciasis: Histopathologic study. *Am J Vet Res* **43**: 1371–1375, 1982.

Streilein, J. W. Anterior chamber associated immune deviation: The privilege of immunity in the eye. *Surv Ophthalmol* **35**: 67–73, 1990.

Whitely, H. E. *et al.* Ocular lesions of bovine malignant catarrhal fever. *Vet Pathol* **22**: 219–225, 1985.

Williams, R. D. *et al.* Experimental chronic uveitis—ophthalmic signs following equine leptospirosis. *Invest Ophthalmol* **10**: 948–954, 1971.

g. LENS-INDUCED UVEITIS Uveitis in response to leakage of lens material is seen in all species, but is most frequent by far in dogs. The term **lens-induced uveitis** encompasses two very different syndromes—phacolytic and phacoclastic uveitis—that differ markedly in clinical severity, in histopathology, and in pathogenesis.

Phacolytic uveitis is a mild lymphocytic–plasmacytic anterior uveitis that occurs in response to the leakage of denatured lens protein through an intact lens capsule, which occurs regularly in the course of maturation of cataracts toward total fiber liquefaction. The inflammation is readily controlled by routine therapy, so the pathologist is likely to encounter this lesion only as an incidental finding in eyes with hypermature cataracts that were enucleated for reasons unrelated to the uveitis. The lesion is identical to that described for idiopathic (immune-mediated) uveitis, except it is always mild. Its pathogenesis is unknown. The lens leaks small denatured lens proteins that are not immunogenic but are, perhaps, direct inflammatory stimulants with lymphocytic chemotactic properties.

Phacoclastic uveitis is, at least histologically, a more complicated disease that follows rupture of a normal lens in an unknown percentage of cases. The rupture is usually from corneal penetration by a thorn, quill, bullet, or cat claw, and thus usually is of the anterior capsule. The clinical syndrome is distinctive: corneal perforation and mild traumatic uveitis that are successfully managed by conventional therapy, followed by the sudden reappearance of a severe, intractable uveitis 10–14 days after the initial injury. Poor response to medical therapy and the eventual development of glaucoma or phthisis bulbi prompt enucleation, so that phacoclastic uveitis is one of the most prevalent ocular diseases to be submitted for histologic examination.

The macroscopic changes in the bisected globe are diagnostic: the lens is flattened in its anteroposterior dimension, and there frequently is a wedge of opacification extending from the anterior capsule toward the nucleus (Fig. 4.36A). Usually there is posterior synechia, iris bombé, and the various other lesions of any severe uveitis (Fig. 4.36B).

The histologic lesions vary considerably depending on duration and, probably, on the amount of lens protein that escaped through the rupture site. There are often complex lesions that result from a combination of the direct effects of trauma, immunologic reaction to massive release of lens protein, reparative proliferation of metaplastic lens and/or iridociliary epithelium, and possible contributions by corneal wound healing, sepsis, and glaucoma.

The simplest and presumably earliest lesion of phacoclastic uveitis occurs at the site of capsular perforation. The edges of the capsule are retracted and coiled outward, and a wedge of neutrophils and liquified lens material extends from the perforation toward the nucleus. The inflammation outside of the lens is usually distinctly perilenticular and involves a mixture of neutrophils and macrophages in the anterior and posterior chambers, and a lymphocyte-dominated reaction within the uveal stroma.

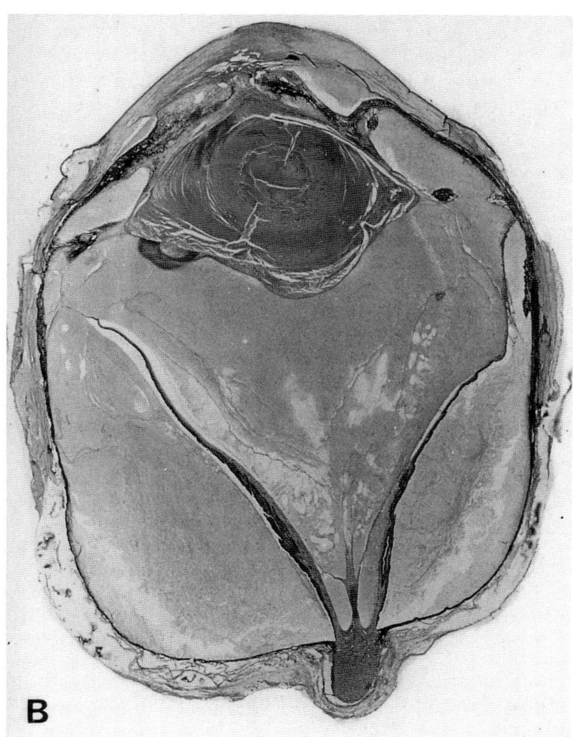

Fig. 4.36B Phacoclastic uveitis with posterior synechiae, iris bombe, and a serous endophthalmitis with complete retinal detachment. Note continuity between transcorneal scar and perilenticular fibroplasia.

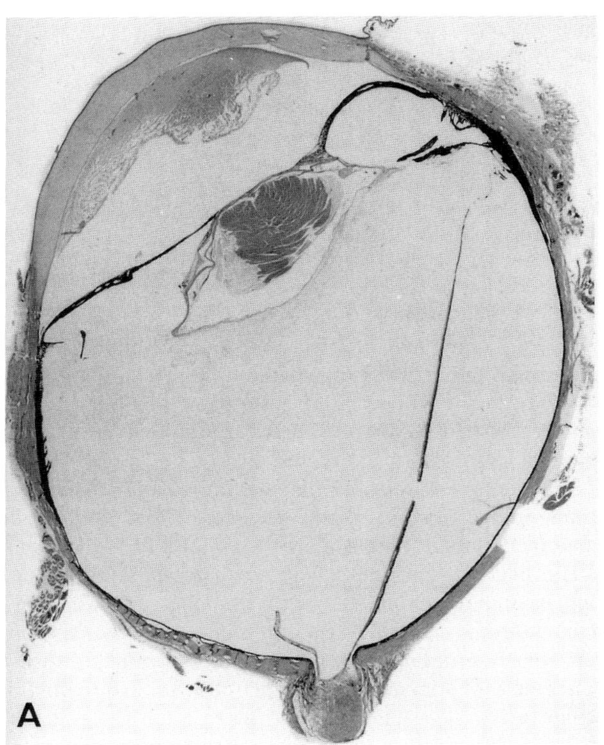

Fig. 4.36A Posterior synechia, iris bombe, and glaucomatous cupping of optic disk following traumatic corneal and lenticular perforation. The anterior–posterior flattening of the liquified lens is typical.

Older lesions (which predominate in most enucleated globes) are dominated by the perilenticular proliferative changes of wound repair. There is proliferation and fibroblastic metaplasia of lens epithelium adjacent to the perforation, which escapes from the lens to ramify over the lens surface and frequently incorporates lens, ciliary processes, and iris leaves into a large fibrous mass that obstructs aqueous outflow. Metaplasia of ciliary epithelium or recruitment of fibroblasts from uveal stroma may contribute to the proliferation (Fig. 4.37). Many such specimens contain little evidence of inflammation other than the fibroplasia, probably because of very extensive antiinflammatory therapy that is, in hindsight, useless against the proliferative events that doom the eye to glaucoma or phthisis.

Phacoclastic uveitis is an important complication of cataract surgery in which fragments of lens cortex or epithelium may be left in the eye. These initiate the same inflammatory and proliferative reaction as described, and

Fig. 4.37 Phacoclastic uveitis. Severed ends of anterior lens capsule (arrow) are fimbriated. Typical fibrous proliferation fuses cataractous lens to the (pigmented) iris.

the complications are as refractory to conventional antiinflammatory therapy as is the naturally occurring disease.

The immune pathogenesis of phacoclastic uveitis has been extensively studied, but with no universally accepted conclusion. The current theory is that the release of massive amounts of lens protein antigen overwhelms the splenic T-cell tolerance to small amounts of lens antigen. Recruitment of lens-sensitized lymphocytes into the perilenticular uvea then initiates both the pyogranulomatous perilenticular inflammation and the proliferative events of healing that, unfortunately, doom the eye. This pathogenesis, if true, explains the typical delay between injury and reaction, and why rapid surgical removal of lens is preventive. It may also explain the unpredictability of phacoclastic uveitis, especially following small perforations in puppies, which seem often to heal uneventfully. Even in adult dogs the disease is unpredictable, so owners' questions about the risk of phacoclastic uveitis as the justification for surgical removal of a perforated lens cannot be answered with certainty. One recent study found lens removal shortly after perforation prevented serious complications in 6 of 7 dogs thus treated, whereas 5 of 6 dogs treated with aggressive medical therapy lost the eye to complications of the uveitis.

Two interesting variations on what is basically a canine scheme are seen in rabbits and cats. Rabbits suffer what appears to be spontaneous lens capsule rupture of a previously normal lens. The response is a well-contained perilenticular granulomatous inflammation very similar to human phacoanaphylactic uveitis. Cats occasionally develop lesions similar to those in dogs, but also develop a unique feline primary intraocular pleomorphic sarcoma that may arise from metaplastic lens epithelium or from other transformed epithelial elements in the reparative reaction (see Feline Primary Ocular Sarcoma).

Bibliography

Davidson, M. G. *et al.* Traumatic anterior lens capsule disruption. *J Am Anim Hosp Assoc* **27**: 410–414, 1991.

Dietz, H. H., Jensen, O. A., and Wissler, J. Lens-induced uveitis in a domestic cat. *Nord Vet Med* **37**: 10–15, 1985.

Fischer, C. A. Lens-induced uveitis. *In* "Comparative Ophthalmic Pathology" R. L. Peiffer, Jr. (ed.), pp. 254–263. Springfield, Illinois, Charles C Thomas, 1983.

Misra, R. N., Rahi, A. H. S., and Morgan, G. Immunopathology of the lens: II. Humoral and cellular immune responses to homologous lens antigens and their roles in ocular inflammation. *Br J Ophthalmol* **61**: 285–296, 1977.

Murphy, J. M. *et al.* Sequelae of extracapsular lens extraction in the normal dog. *J Am Anim Hosp Assoc* **16**: 47–51, 1980.

Paulsen, M. E. *et al.* The effect of lens-induced uveitis on the success of extracapsular cataract extraction: A retrospective study of 65 lens removals in the dog. *J Am Anim Hosp Assoc* **22**: 49–55, 1986.

Rahi, A. H. S., Misra, R. N., and Morgan, G. Immunopathology of the lens: I. Humoral and cellular immune responses to heterologous lens antigens and their roles in ocular inflammation. *Br J Ophthalmol* **61**: 164–176, 1977.

Wilcock, P. B., and Peiffer, R. L., Jr. The pathology of lens-induced uveitis in dogs. *Vet Pathol* **24**: 549–553, 1987.

2. *Endophthalmitis*

a. Bacterial Endophthalmitis Bacteria may enter the eye hematogenously or via penetrating wounds. Those arriving hematogenously cause their initial lesion in ciliary body or, less frequently, in choroid. Those arising via penetration usually incite the initial reaction in anterior chamber, particularly if the penetration is via perforation of an ulcerative keratitis. Most are suppurative, and their extent and severity vary with the size of inoculum, virulence of the agent, and host response and its duration.

The list of organisms capable of causing endophthalmitis is long. It is probably true that any bacterium capable of bacteremia or septicemia can cause endophthalmitis. Particularly prominent are the streptococci and coliforms in neonatal septicemia (Fig. 4.38). The failure to detect ocular lesions in such animals is more often the result of the brief, fatal course of the disease than of specific ocular resistance. The ocular lesion may be very mild, better detected by opacification of the plasmoid aqueous in Bouin's or Zenker's fixative than by histologic examination. Histology may reveal only edema of ciliary processes

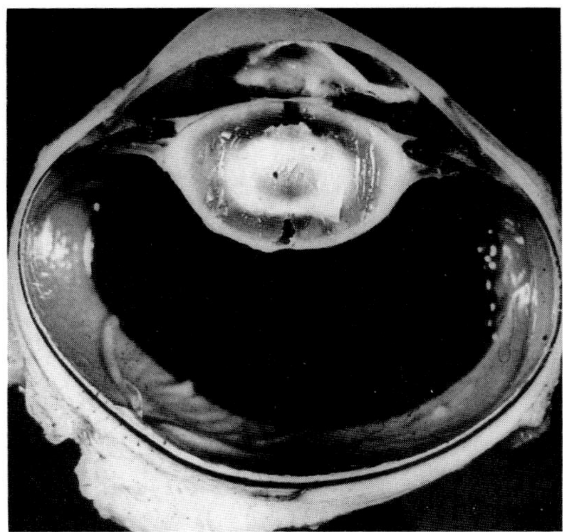

Fig. 4.38 Acute streptococcal ophthalmitis with corneal edema and hypopyon. Calf.

with a few neutrophils along the capillary endothelium or enmeshed in filaments among ciliary processes. The best-known bovine septicemic disease, infectious thrombotic meningoencephalitis, is seen as focal rather than diffuse chorioretinitis.

Exceptions to the generalization that bacterial endophthalmitis is suppurative occur if infection is caused by bacteria which, in other tissues, incite lymphocytic or even granulomatous inflammation. **Ocular tuberculosis** is largely of historical interest. It occurred as part of generalized systemic disease, and the typical tubercles were most numerous in the choroid. *Mycobacterium tuberculosis* var. *bovis* was the usual isolate except in cats, in which the human strain was common (Fig. 4.39). In cats, ocular tuberculosis may also occur as keratoconjunctivitis without uveal involvement.

Brucella canis may cause chronic lymphocytic endophthalmitis that is probably immunologically mediated. Agglutinating titers for *B. canis* antigen in aqueous exceed those of serum, and the ocular lesions are similar to those of equine recurrent ophthalmitis.

Listeria monocytogenes often causes endophthalmitis in association with meningoencephalitis in ruminants (see Chapter 3, The Nervous System). The condition is unilateral, and the pathogenesis is obscure.

Uveitis caused by the rickettsias of **Rocky Mountain spotted fever** and **ehrlichiosis** are discussed under Retinitis.

b. Mycotic Endophthalmitis Fungi may affect the eye as causes of keratitis, orbital cellulitis, or endophthalmitis. Only rarely do the fungi causing keratitis or orbital infection penetrate the fibrous tunic to cause intraocular disease. However, hematogenous uveal localization is rather common in the course of systemic mycoses caused by *Cryptococcus neoformans* and *Blastomyces dermatitidis* and, less regularly, with *Coccidioides immitis* and *Histoplasma capsulatum*. In immunodeficient animals, one might expect occasionally to detect endophthalmitis as part of generalized disease caused by saprophytic fungi such as *Aspergillus* or *Candida*.

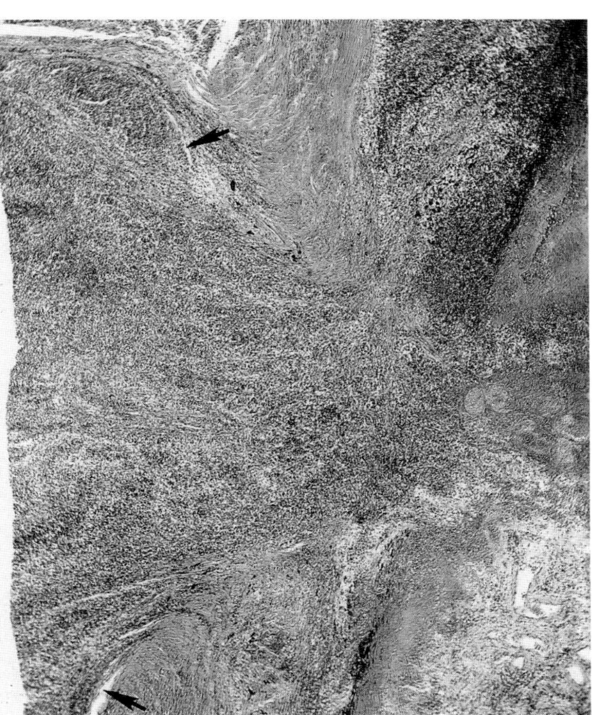

Fig. 4.39 Tuberculous ophthalmitis. Cat. Exudate in vitreous and detaching retina (arrows).

The frequency with which endophthalmitis accompanies systemic mycosis is unknown and probably varies with the specific agent, the species affected, and whether the disease is in an endemic area or is a sporadic occurrence. Hematogenous ocular mycosis is found almost exclusively in dogs, except for cryptococcosis, which is more common in cats. *Blastomyces* and *Cryptococcus* are more likely to invade the eye in the course of generalized infection than are *Coccidioides* or *Histoplasma*, and occurrence in nonendemic areas is strongly linked to prolonged systemic corticosteroid therapy. Involvement is bilateral but not necessarily equal. Blastomycosis, cryptococcosis, and coccidioidomycosis are discussed with the Respiratory System (Volume 2, Chapter 6), and histoplasmosis with the Hematopoietic System (Volume 3, Chapter 2).

Blastomycosis is the most frequently reported cause of intraocular mycosis in dogs. It is rare in cats. Between 20 and 26% of dogs with the systemic disease are blind or have grossly observed ocular lesions, suggesting that intraocular involvement would be recognized more often if histologic examinations were routinely done. The clinical ocular disease is severe diffuse uveitis, frequently with retinal separation.

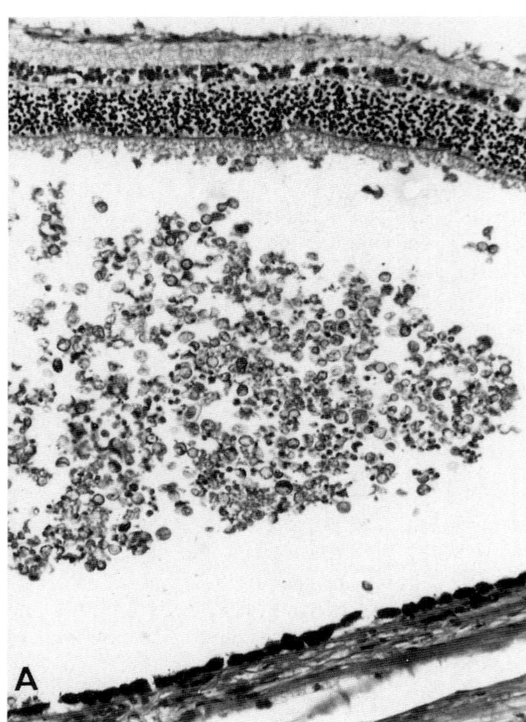

Fig. 4.40A Subretinal exudate containing *Blastomyces dermatitidis*. Dog receiving long-term corticosteroid therapy.

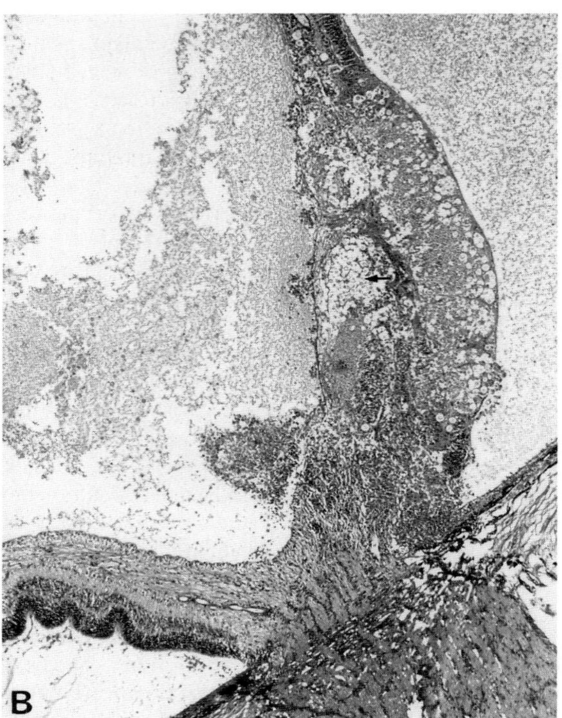

Fig. 4.40B Retinal separation and focal necrotic retinitis. Cat, *Cryptococcus*. Note soap-bubble appearance (arrow) and minimal host reaction.

The histologic appearance is of diffuse pyogranulomatous or granulomatous endophthalmitis with retinitis, exudative retinal separation, and commonly, granulomatous optic neuritis. Choroiditis is often more pronounced than is the anterior uveitis. The greatest accumulation of leukocytes often is in the subretinal space enlarged by exudative retinal detachment. The causative diagnosis depends on the demonstration of the spherical-to-oval, thick-walled yeasts in vitreous aspirates or in the histologic section. They are usually most numerous in the subretinal exudate, but are rare in anterior chamber or in retina itself (Fig. 4.40A). The organisms are free or within macrophages, are 5–20 μm in diameter and show occasional broad-based budding. Extremes in sizes may result in yeasts from 2 to 30 μm in diameter. The eye may, in addition, have the full spectrum of corneal, lenticular, and glaucomatous sequelae, as expected of any severe uveitis. Panophthalmitis with orbital cellulitis is seen in about one third of enucleated globes.

Cryptococcosis is similar to blastomycosis in that the lesions are predominantly within retina, choroid, and optic nerve. However, infection of the eye may arise either hematogenously or by extension from the brain via optic nerves, and lesions are often conspicuously lacking in cellular host reaction. Large collections of poorly stained pleomorphic yeasts, surrounded by wide capsular halos, impart a typical soap-bubble appearance to the histologic lesions (Fig. 4.40B). The yeasts vary in size, but most are 4–40 μm in diameter. Round, oval, and crescentic forms are seen. In some animals, however, a granulomatous reaction mimicking that of blastomycosis can be found. In such lesions the organism typically is scarce.

The frequency with which *Coccidioides immitis* infects the eye appears to be low, ~2%, despite the prevalence of generalized infection in endemic areas. The ocular lesion resembles blastomycosis in that pyogranulomatous reaction occurs around fungal spherules. The reaction is predominantly purulent around newly ruptured spherules, gradually becoming granulomatous as the released endospores mature. The lesion tends to be more destructive than other mycoses, usually spreading to involve sclera and even episclera in a suppurative panophthalmitis.

Histoplasma capsulatum is a common cause of generalized mycosis in dogs but is rare in other domestic species. It has a predilection for lymphoid tissue and other tissues rich in phagocytes such as lung and liver, and perhaps this preference accounts for the paucity of ocular involvement in spontaneous disease. In dogs and cats infiltrative choroiditis or panuveitis occurs and is dominated by plasma cells and by macrophages filled with the organisms. Retinal separation, plasmoid vitreous, and optic neuritis also develop. The reaction tends to target the choroid, and to be less destructive than either blastomycosis or coccidioidomycosis.

Prototheca are colorless saprophytic algae capable of

causing enteric, cutaneous, mammary, or generalized granulomatous disease in a variety of mammalian species. Ocular lesions have been described only in dogs with the disseminated form of the disease. The lesions are bilateral and may vary from lymphocytic–plasmacytic to granulomatous panuveitis with optic neuritis and exudative retinal separation. The host response is usually quite mild. The lesions resemble ocular mycosis, particularly cryptococcosis, and are distinguished only by the observation of the pleomorphic algae. In histologic section, the algae are free or within phagocytes. The organisms are spherical to oval, from 2 to 20 μm in diameter with a refractile, PAS-positive and argyrophilic cellulosic cell wall. Each cell consists of granular, weakly basophilic cytoplasm surrounding a central nucleus. *Prototheca* produces by asexual multiple fission, so that multiple daughter cells form within a single cell wall. One or two cycles of nuclear division without cytoplasmic cleavage may produce transient multinucleated cells before eventual cytoplasmic division results in up to eight daughter cells. Each daughter cell acquires a capsule, resulting in a parent cell crisscrossed by septations that represent the cell walls of maturing daughter cells. Rupture of the parent cell wall releases the unicellular autospores. Collapsed, crumpled, and seemingly empty cell walls are visible in histologic section. There is no budding as with *Blastomyces* and *Cryptococcus*.

Those canine isolates that were definitively identified were *P. zopfii*. An enteric route of entry is probable inasmuch as necrotic enteritis is a feature of the disease (see Volume 2, Chapter 1, The Alimentary System). Immunodeficiency may be prerequisite for dissemination of the organism. Lesions are found in many visceral organs, skin, and lymph nodes in most cases. The reaction is granulomatous but is usually minimal in comparison to the large number of organisms.

c. Protozoan Endophthalmitis While ocular lesions have been reported in infections caused by protozoa of the genera *Toxoplasma, Leishmania, Encephalitozoon, Besnoitia* and *Trypanosoma,* only *Toxoplasma* specifically causes intraocular lesions, although some cases so diagnosed may have been caused by *Neospora*. Most of the others cause keratoconjunctivitis that occasionally extends to anterior or generalized uveitis in which the causal agent may be found. The exception is *Encephalitozoon* that may induce periarteritis within the uvea and retina as it does elsewhere.

Toxoplasmosis affecting the eye is, as elsewhere, much more frequently suspected than proven, and clinical diagnoses greatly outnumber those confirmed by histopathology. The histologic lesion is usually in retina, uvea, or extraocular muscles, and varies from focal, acute coagulative necrosis to granulomatous or lymphocyte-rich inflammation. The organisms are seen most easily as intracellular pseudocysts during acute disease or as true cysts during remission. The more noxious merozoites are found only with difficulty as 7- to 9-μm crescentic, basophilic bodies within phagocytes, or free amid necrotic debris.

The histologic changes of ocular toxoplasmosis have received various interpretations, and valid differences probably exist between species and between individuals of differing immune status. In humans, the disseminated disease is usually congenital, and the ocular lesion is multifocal necrotic retinitis, in which free or encysted *Toxoplasma gondii* are found. There may be lymphocytes and plasma cells in adjacent choroid. In human adults, the ocular lesion is predominantly a lymphocytic–plasmacytic choroiditis, suggesting that the pathogenesis is related more to host immune response to the previously encountered, ubiquitous antigen than it is to local infection.

Lesions analogous to human congenital toxoplasmosis occur in young cats as multiple foci of retinal necrosis. Choroiditis may be present and is lymphocytic-plasmacytic, and anterior uveitis is seen in only 20–30% of such cases. Much more common than this classical retinochoroiditis, however, is lymphocytic–plasmacytic anterior uveitis with serologic evidence for active *Toxoplasma* infection. The role of toxoplasmosis in feline anterior uveitis is controversial. Retrospective histologic studies list the majority of such cases as idiopathic and are presumed to be immune mediated, with antigen or antigens unknown. In the several large published studies, there has not been a single case confirmed by observation of merozoites or cysts in the eye; even serologic evidence in these studies pointed to toxoplasmosis in only 1–2% of cases. In contrast, one study reported evidence of anterior uveal production of toxoplasma-specific antibody in 32 of 69 cats with anterior uveitis, and another report confirmed that anterior uveitis is indeed the most frequent manifestation of toxoplasmosis in cats, seen in 60% of cats with confirmed toxoplasmosis. Nonetheless, evidence remains less than conclusive about the role of *T. gondii* in the prevalent and enigmatic syndrome of anterior uveitis in this species. It may be that the local production of *Toxoplasma* antibody in cats with uveitis is merely the result of nonspecific recruitment of *Toxoplasma*-sensitized lymphocytes into the chronically inflamed uvea, and that lymphoid aggregates in such eyes are producing a whole range of antibody quite irrelevant to the original cause of the uveitis.

The situation in other species is not clear, but the prevalence of ocular lesions seems quite low. Lymphocytic cyclitis and multifocal necrotic retinitis are most frequently described and are usually seen together. Multifocal choroiditis, not necessarily adjacent to retinal lesions, is lymphocytic-plasmacytic in most species but granulomatous in sheep. A more common lesion is severe myonecrosis in extraocular muscles associated with free and encysted *Toxoplasma*. Toxoplasmosis is discussed in more detail in the Alimentary System (Volume 2, Chapter 1).

d. Parasitic Endophthalmitis Many parasites are found incidentally in the eye, including echinococcosis in primates and cysticercosis in swine, multifocal ischemic chorioretinitis and optic neuritis in elk caused by occlusive vasculitis due to microfilariae of *Elaeophora schneideri*, and uveitis associated with fortuitous localization of larvae of *Toxocara canis* or other ascarids, *Angiostrongylus vasorum*, *Dirofilaria immitis*, and *Onchocerca cervicalis*. In addition, adults of *Setaria* spp. are occasionally found within the eye of horses. The long threadlike worms are seen floating within the aqueous, and the uveitis that results seems to be the result of mechanical irritation. The only specific intraocular parasitism is seen with the lens fluke of fish (*Diplostomum spathaceum*), which, after penetrating the skin, seeks the lens with remarkable speed and specificity. The principal lesion is cataract induced by the intralenticular presence of hundreds of larvae awaiting ingestion by fish-eating birds for completion of their life cycle, but infected fish may have larvae arrested in many other ocular or extraocular locations.

Chronic mild anterior uveitis is reported to accompany ectopic localization of immature *Dirofilaria immitis* within aqueous and vitreous cavities of dog eyes. Pathological studies are sparse, but endophthalmitis is reported in which anterior synechiae, subretinal exudate, and early cyclitic membrane may accompany the numerous vitreal and subretinal larval nematodes.

Ocular onchocerciasis affects humans and horses. The human disease is endemic in Africa and Central America and is one of the most frequent causes of blindness in the world. The microfilariae of the causal agent, *Onchocerca volvulus*, are transmitted by *Simulium* spp. flies to affect the skin, eyelids, and corneas of children and young adults. The microfilariae are found throughout the eye, but the lesion of greatest visual significance is diffuse sclerosing superficial stromal keratitis complicated by anterior uveitis with synechiae and eventual glaucoma.

Equine onchocerciasis has some similarities. The parasite, *O. cervicalis*, is of worldwide distribution, and surveys from the United States of America document the prevalence of dermal infection in horses as varying from 48 to 96%. About one half of the infected horses have microfilariae in conjunctiva or sclera. The microfilariae enter the eye only incidentally in the migration from the ligamentum nuchae to the subcutis. The ocular sites of greatest concentration are the peripheral cornea and the lamina propria of the bulbar conjunctiva near the limbus. The microfilariae in the cornea are associated with a superficial stromal keratitis resembling the disease of man, albeit much milder. Some of the horses also have anterior uveitis typical of equine recurrent ophthalmitis, prompting theories that *Onchocerca* is one cause of this disease. The microfilariae can be recovered from the conjunctiva and eyelids of horses with uveitis, keratoconjunctivitis, and eyelid depigmentation, but are recovered with equal frequency from horses with no ocular disease and no microscopic reaction to the worms.

Ocular manifestations of **visceral larva migrans** in humans are associated with larvae of *Toxocara canis* or, perhaps more frequently, of the raccoon roundworm *Baylisascaris procyonis*. The unilateral granulomatous fundic lesions are caused by a single wandering larva, and are relatively common in children but have rarely been described in nonhuman subjects, despite the rather common occurrence of ascarid-induced granulomas in canine kidneys, lungs, or livers. The paucity of reports may not reflect the actual prevalence of disease in specific canine populations. One large survey of working sheepdogs in New Zealand recorded a 39% prevalence of lesions attributed to visceral larva migrans, contrasted to a 6% prevalence in similar dogs living in urban environments. The active lesions were lymphocytic and granulomatous uveitis, nonsuppurative retinitis, and peripapillary nontapetal retinal necrosis. Inactive lesions involved choreoretinal scars and multifocal chronic retinal separations in dogs older than 3 years. Larvae most compatible with *Toxocara canis* were seen in sections of some acutely affected eyes. The high prevalence in these dogs was tentatively ascribed to the feeding of uncooked frozen mutton that may have contained *T. canis* larvae as part of a dog–sheep–dog life cycle. A report of similar lesions in Border collies in the United States was associated with the feeding of raw pork.

Ocular disease may also result from the intraocular migration of fly larvae. This syndrome, termed *internal ophthalmomyiasis*, is discussed with diseases of conjunctiva.

Bibliography

Bistner, S., Shaw, D., and Riis, R. C. Diseases of the uveal tract (part III). *Compend Cont Ed* **2:** 46–53, 1980.

Buyukmihci, N. C., and Moore, P. F. Microscopic lesions of spontaneous ocular blastomycosis in dogs. *J Comp Pathol* **97:** 321–328, 1987.

Buyukmihci, N., Rubin, L. F., and DePaoli, A. Protothecosis with ocular involvement in a dog. *J Am Vet Med Assoc* **167:** 158–161, 1975.

Carlton, W. W., and Austin, L. Ocular protothecosis in a dog. *Vet Pathol* **10:** 274–280, 1973.

Chandler, F. W., Kaplan, W., and Ajello, L. "Histopathology of Mycotic Diseases." Chicago, Illinois, Year Book Medical Publishers, 1980.

English, R. V. *et al.* Intraocular disease associated with feline immunodeficiency virus infection in cats. *J Am Vet Med Assoc* **196:** 1116–1119, 1990.

Flamm, H., and Zehetbauer, G. Die Listeriose des Auges im Tierversuch. *Graefe's Arch Ophthalmol* **158:** 122–135, 1956.

Gwin, R. M. *et al.* Ocular lesions associated with *Brucella canis* infection in a dog. *J Am Anim Hosp Assoc* **16:** 607–610, 1980.

Hughes, P. L., Dubielzig, R. R., and Kazacos, K. R. Multifocal retinitis in New Zealand sheep dogs. *Vet Pathol* **24:** 22–27, 1987.

Johnson, B. W. *et al.* Retinitis and intraocular larval migration in a group of border collies. *J Am Anim Hosp Assoc* **25:** 623–629, 1989.

Migaki, G. *et al.* Canine protothecosis: Review of the literature and report of an additional case. *J Am Vet Med Assoc* **181:** 794–797, 1982.

Piper, R. C., Cole, C. R., and Shadduck, J. A. Natural and experimental ocular toxoplasmosis in animals. *Am J Ophthalmol* **69:** 662–668, 1970.

Trevino, G. S. Canine blastomycosis with ocular involvement. *Pathol Vet* **3:** 651–658, 1966.

Vainisi, S. J., and Campbell, L. H. Ocular toxoplasmosis in cats. *J Am Vet Med Assoc* **154:** 141–152, 1969.
Van Kruiningen, H. J., Garner, F. M., and Schiefer, B. Protothecosis in a dog. *Pathol Vet* **6:** 348–354, 1969.
Wilder, H. C. Nematode endophthalmitis. *Trans Amer Acad Ophthalmol Oto-laryngol* **55:** 99–109, 1950.

B. Glaucoma

Glaucoma is a pathophysiologic state characterized by prolonged increase in intraocular pressure. Although such increases of pressure may theoretically result from increased production or decreased removal of aqueous, only the latter is known to occur. The lesions in glaucomatous eyes include those related to the pathogenesis of the glaucoma and those resulting from the glaucoma itself. Glaucoma occurs commonly in dogs, less commonly in cats, occasionally in horses, and rarely in other species. Because most affected eyes eventually require enucleation, glaucoma is one of the most frequent ocular conditions examined histologically.

The lesion predisposing to glaucoma may be the result of antecedent ocular disease, particularly anterior uveal inflammation with posterior or anterior synechiae. Such cases are termed secondary glaucoma. Primary glaucoma describes those cases without evidence of prior ocular disease and, in practical terms, is synonymous with malformation of the filtration angle. Primary glaucoma is seen almost exclusively in dogs and vies with neoplasia as the most frequent cause of glaucoma in dogs.

Because the pathogenesis of glaucoma so frequently involves developmental or acquired distortion of the filtration angle, a description of that structure is appropriate here.

The **filtration apparatus** is a series of mesenchymal sieves that occupies the iridocorneal angle, and extends circumferentially around the globe. These sieves appear to form by rarefaction of the same mesenchyme that forms iris stroma, and its rarefaction continues (at least in carnivores) for several weeks after birth. This area of perforated mesenchyme is the ciliary cleft, bordered externally by sclera, posteriorly by the muscles of the ciliary body, and internally by the iris stroma. Its anterior border is the **pectinate ligament,** which is visible clinically as a series of cobweblike branching cords (carnivores) or a fenestrated sheet (ungulates) stretching from the termination of Descemet's membrane to the anterior portion of the iris root. They consist of collagenous cords covered by a very thin endothelium, with a thin intervening layer of basement membranelike material. The endothelium is continuous with the corneal endothelium, and the collagenous core is continuous with corneal stroma.

Aqueous humor percolating through the pectinate ligament into the ciliary cleft must then pass through mesenchymal sieves consisting of collagenous cords covered by phagocytic and pinocytotic endothelium, called the **trabecular meshwork.** The large, open network of cords occupying most of the ciliary cleft is the uveal trabecular meshwork, and external to it is a more compressed network called the corneoscleral trabecular meshwork. Ordinarily, aqueous humor produced by the ciliary processes passes through the pupil, through the pectinate ligament, and then through the uveal and corneoscleral trabecular meshworks en route to the scleral venous plexus that will return the aqueous to the systemic circulation. Improper development or acquired obstruction of any part of this drainage pathway may result in glaucoma, but one must remember that the ciliary cleft extends 360° around the iridocorneal angle. Blockage of most of it is required for the development of glaucoma, and this assessment is virtually impossible with two-dimensional histologic examination. It is quite common to encounter dog eyes with maldeveloped filtration angles in both portions contained in a histologic section, yet with no evidence of glaucoma. Examination of the circumference of the angle with a dissecting microscope or scanning electron microscope in such cases often reveals the maldevelopment to be segmental and thus not a cause for glaucoma.

Differences exist among species in the finer details of angle structure and in the degree to which alternative routes of aqueous outflow are utilized (Fig. 4.41A,B). The horse, for example, has very thick pectinate fibers, an inconspicuous corneoscleral trabecular meshwork and

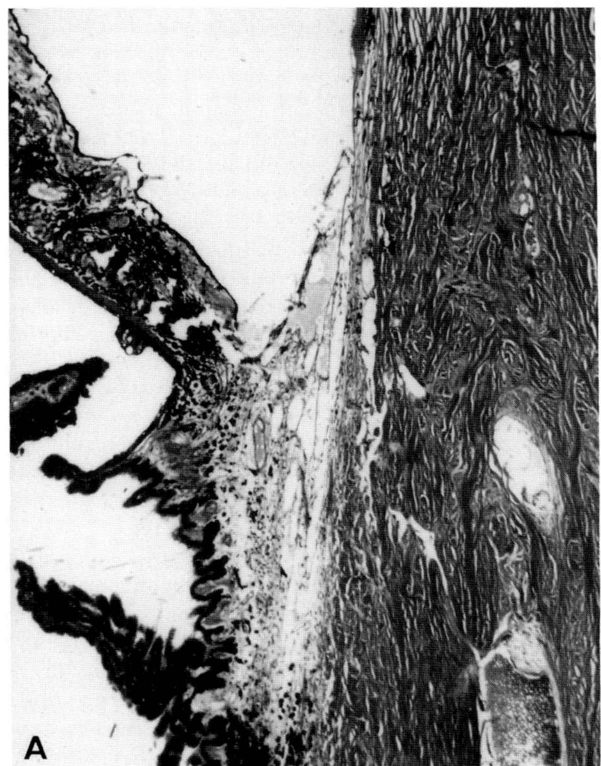

Fig. 4.41A Normal canine filtration angle showing fenestrated pectinate ligament inserting at the termination of Descemet's membrane. Large vessels are part of scleral venous plexus.

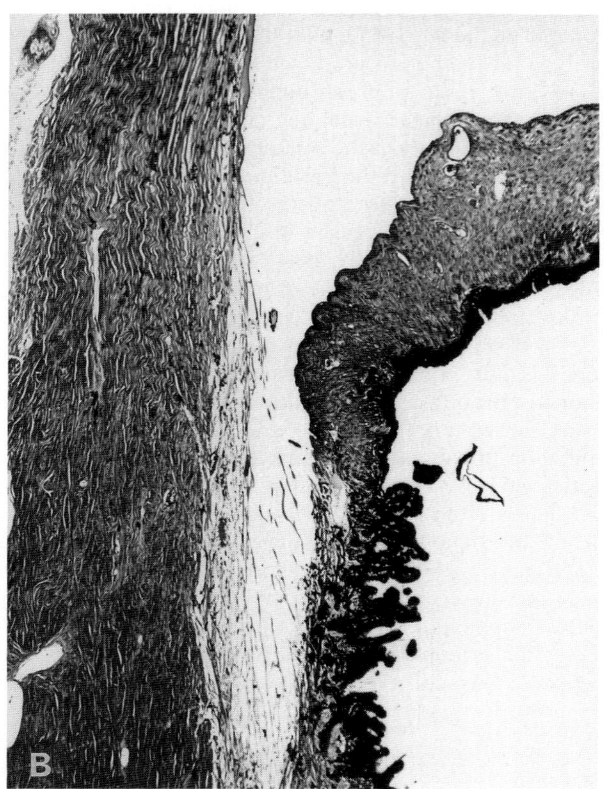

Fig. 4.41B Normal feline filtration angle.

scleral venous plexus, and alternative routes of aqueous outflow (into iris stroma or caudally through ciliary muscle into choroid) that are probably much more important than similar routes in dogs or cats. In contrast, the cat has extremely delicate pectinate fibers, a very large, open ciliary cleft, a conspicuous scleral venous plexus, and minimal (about 3% of aqueous outflow) reliance on alternative outflow pathways (Fig. 4.41B). In dogs alternative drainage routes account for 15 to 25% of all outflow. The existence of these alternative routes may explain the absence of glaucoma in some eyes (especially in horses) in which the angle changes would ordinarily have resulted in glaucoma, and may even explain the presence of glaucoma in eyes with apparently normal angles but lesions affecting portions of these other potential drainage routes.

Primary glaucoma is most frequently encountered in dogs. Although theoretically primary glaucoma may have no visible angle lesion (which is frequently the case in humans), in dogs there is almost always a readily detected maldevelopment. The one exception is primary open-angle glaucoma in beagle dogs, in which there is no visible antecedent lesion. The broad term **goniodysgenesis** encompasses all developmental defects of the filtration angle, of which two types account for most canine cases. The most prevalent is continuation of mature iris stroma across the trabecular meshwork to insert into the termination of Descemet's membrane. Some consider this an example of dysplasia of pectinate ligament, and the term **imperforate pectinate ligament** is widely used. The band usually is much broader than pectinate ligament, no matter how poorly perforated, but the term seems well entrenched. It is seen as a breed-related and thus presumably inherited defect in Bouvier des Flandres, Basset hounds, American cocker spaniels, Dandie Dinmont terriers, Siberian huskies, Samoyeds, and numerous other breeds. The defect, and the resultant glaucoma, are occasionally seen in mixed-breed dogs. The dysplasia is usually bilateral but not necessarily of equal extent, so that the glaucoma often is initially present only in one eye. The prevalence of the iridopectinate dysplasia is much higher than the prevalence of glaucoma, and even in dogs with very extensive dysplasia that should seemingly eliminate almost all aqueous drainage, the onset of glaucoma is not until several years of age. Age-related changes in outflow resistance in alternative routes of aqueous outflow have been postulated as the explanation, but no proof exists.

The second major type of goniodysgenesis is seen as an apparent arrest in the maturation of the trabecular meshwork so that the ciliary cleft is filled with dense tissue resembling primitive anterior uveal mesenchyme (Fig. 4.13A,B). This may occur in conjunction with iris hypoplasia or anterior chamber cleavage syndromes, but it may exist as an isolated defect.

Secondary glaucoma occurs most commonly as the result of peripheral anterior synechiae, with the root of the iris effectively sealing the filtration angle. The synechiae may result from iris bombé, from expanding neoplasia behind or within the iris, or from primary adhesion of an inflamed iris to the cornea. Because the iris does not normally contact the cornea, anterior synechia by the last mechanism is frequent only as a consequence of corneal perforation, in which case the iris flows forward to seal the defect and may then adhere diffusely to the corneal endothelium (Fig. 4.42A). Other causes of secondary glaucoma include occlusion of the trabecular meshwork by preiridal fibrovascular membrane, inflammatory debris, or tumor cells. Lens luxation may precipitate glaucoma by allowing vitreous to occlude the pupil, by stimulating anterior uveitis, or by trapping iris against cornea. Rarely, lens swelling with cataract (intumescent cataract) seems to occlude the pupil. In small terrier dogs with an inherited tendency to luxation, there is an unusually high prevalence of glaucoma. Removal of lens prior to the onset of glaucoma prevents the expected glaucoma, seemingly establishing the causal role of the luxation. Posterior synechiae may also occlude the pupil (particularly as part of phacoclastic uveitis), but posterior synechiae usually cause glaucoma via iris bombé and thus a circumferential peripheral anterior synechia.

There are substantial differences among species in what mechanisms of glaucoma predominate. In **dogs,** goniodysgenesis, posterior synechiae with iris bombé, anterior uveal melanoma, and anterior lens luxation are the leading causes. In **cats,** diffuse iris melanoma and chronic idiopathic anterior uveitis (by an unknown mechanism) are

Fig. 4.42A Glaucoma caused by swelling of cataractous lens (intumescent cataract) and resultant functional anterior synechia. Dog. Note thin sclera typical of buphthalmos secondary to glaucoma.

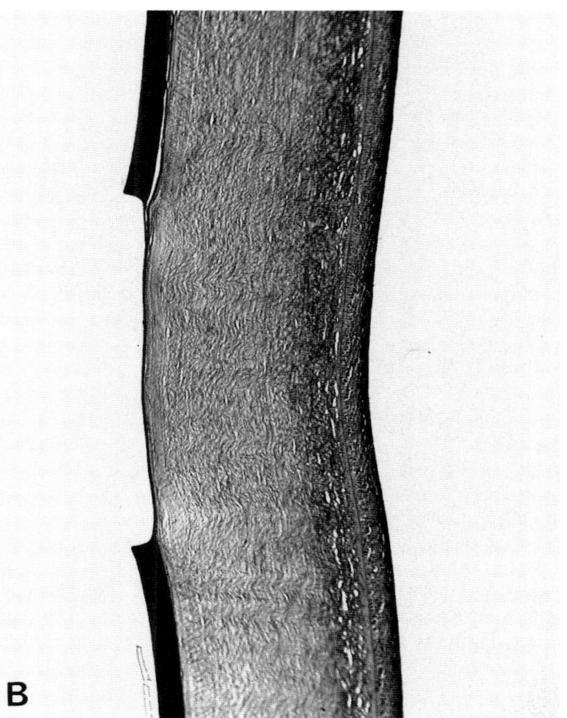

Fig. 4.42B Focal break in Descemet's membrane (corneal stria) commonly seen in horses and dogs with glaucoma.

the only prevalent causes of glaucoma, with posterior synechiae very uncommon. In **horses,** the presence of preiridal fibrovascular membranes across the pectinate face (neovascular glaucoma) is the most frequent cause, although the stimulus for the membrane development is unknown in most reported cases. In those horses with glaucoma in which the membrane had not apparently crossed the pectinate ligament, the glaucoma may have been caused by obstruction of alternative (i.e., iridal) routes of aqueous outflow, which are assumed to be more important in horses than in dogs or cats.

Lesions that develop as a result of glaucoma vary with the duration and severity of the glaucoma and the distensibility of the globe, and affect virtually all parts of the globe. Enlargement of the globe **buphthalmos** or **megaloglobus**) occurs most readily in young animals or in those species with thin scleras such as cats and laboratory animals. In cornea, increased aqueous pressure forces fluid into the corneal stroma resulting in diffuse edema and eventual fibrosis and vascularization. If buphthalmos occurs, corneal stretching results in rents in Descemet's membrane, visible clinically as corneal striae (Fig. 4.42B). These are relatively most frequent in horses and least prevalent in cats with glaucoma. Failure of lids to cover the enlarged globe permits corneal desiccation and eventual ulceration with all its sequelae. Cataract is usual, presumably the result of stagnation of aqueous and subsequent lens malnutrition. Iris and ciliary body undergo bland atrophy, most obvious as thinning and flattening of ciliary processes. Collapse of the ciliary cleft and trabecular meshwork itself is frequent and makes evaluation of these structures for possible goniodysgenesis very difficult. Increased prominence of PAS-positive material in linear arrays resembling posterior migration of Descemet's membrane is probably

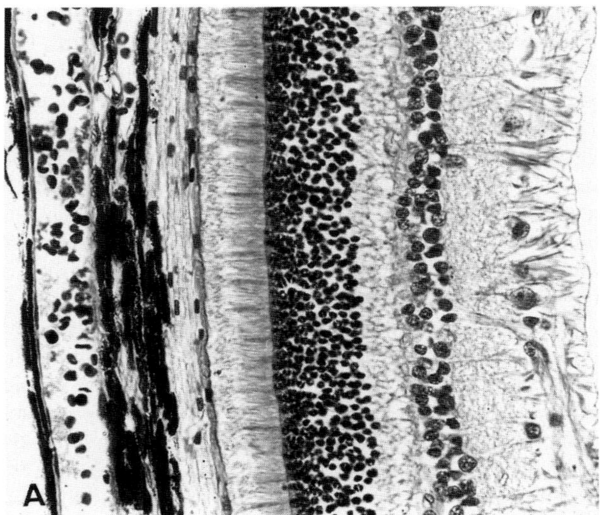

Fig. 4.43A Normal canine retina overlying tapetum lucidum.

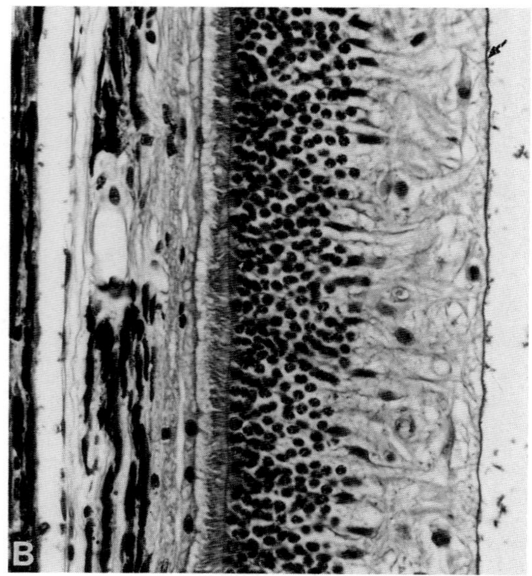

Fig. 4.43B Early glaucomatous retinopathy, dog. Ganglion cells absent. Remnant of inner nuclear layer remains. Inner limiting membrane is abnormally prominent (arrow) over atrophic nerve fiber layer.

a response of the endothelium covering the trabecular beams to the glaucoma, but its possible primary role in the pathogenesis of glaucoma has not been adequately investigated.

The retinal lesion is characteristic and is the most reliable method of diagnosis of glaucoma based solely on histologic criteria. Atrophy begins in nerve fiber and ganglion cell layers, making glaucoma the only naturally occurring cause of inner retinal atrophy other than the rare instances of traumatic or neoplastic disruption of optic nerve (Fig. 4.43A,B). Loss of nerve fibers unmasks the normally inconspicuous Müller fibers, a lesion that may be more easily seen and more confidently interpreted than the loss of the nerve fibers or ganglion cells themselves (Fig. 4.44). This is particularly true in cats, in which the ganglion cells persist with considerable tenacity under circumstances that would, in dogs, have progressed to a very obvious atrophy. With increasing duration or severity, the inner nuclear layer and its axons and dendrites also atrophy, resulting in thinning of the inner nuclear layer and the blending of this layer with the outer nuclear layer as the plexiform layers (the axons and dendrites of the nuclear layer cells) rarefy.

Eventually the retina exists only as a thin glial scar with scattered remnants of outer nuclear layer and melanin-laden phagocytes derived from retinal pigmented epithelium. The retina overlying tapetum is less severely affected than nontapetal retina. Excavation ("cupping") of the optic disk is a pathognomonic lesion when present, but its absence does not rule out the diagnosis. It occurs by two mechanisms, either (or both) of which may explain the cupping in an individual eye. Particularly in animals with a thin sclera and lamina cribrosa, the elevation of pressure may cause rapid posterior bowing of the lamina, resulting in visible cupping without apparent nerve atrophy. This is frequently seen in cats and rabbits, but not in ungulates with their thick, rigid lamina. In all species, cupping also occurs by axonal loss from the optic nerve. It has been suggested that the posterior bowing of the lamina cribrosa contributes to the later atrophy of nerve by mechanical pinching of axons or blood vessels as they pass through the distorted lamina (Fig. 4.45A,B). This cupping is distinguished from coloboma by the absence of dysplastic neurectoderm lining the defect and the presence of inner retinal atrophy. The pathogenesis of the retinal and optic disk atrophy is controversial. Either pressure-induced retinal ischemia or interference with axoplasmic flow in the axons of the ganglion cells en route to the optic nerve may be responsible. Why the dorsal (tapetal) retina is more resistant is unknown.

Fig. 4.44 Glaucomatous retinal atrophy. Ganglion cells are absent, inner nuclear layer is sparse, and Müller's fibers (arrows) unduly obvious due to loss of nerve fiber layer.

Bibliography

Bedford, P. G. C. The clinical and pathological features of canine glaucoma. *Vet Rec* **108:** 53–58, 1980.

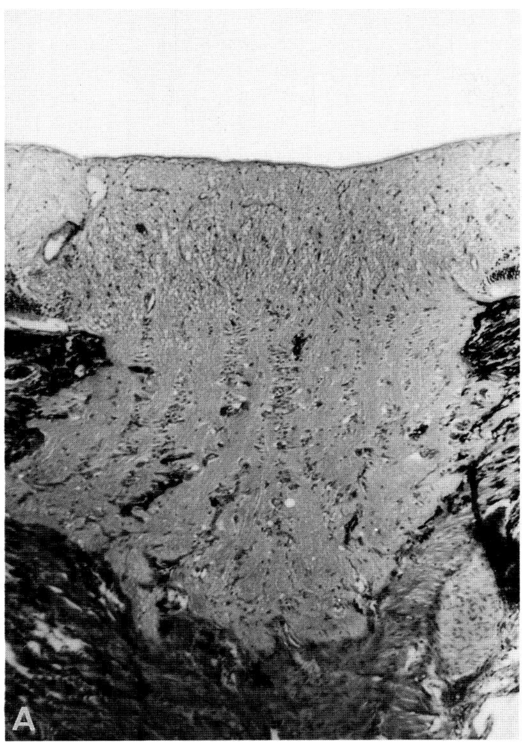

Fig. 4.45A Normal canine optic disk.

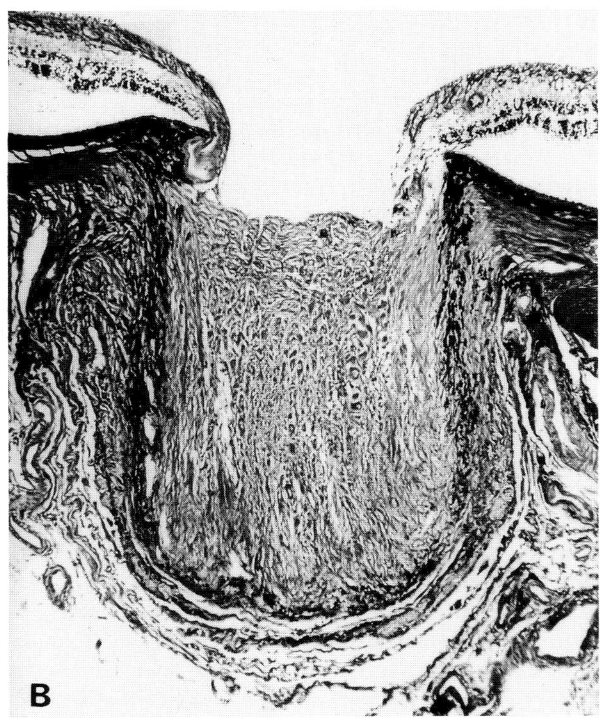

Fig. 4.45B Cupping of optic disk and rarefaction of the nerve. Chronic glaucoma.

Brooks, D. E. Glaucoma in the dog and cat. *Vet Clin North Am: Small Anim Pract* **20:** 775–797, 1990.

De Geest, J. P. *et al.* The morphology of the equine iridocorneal angle: A light and scanning electron microscopic study. *Eq Vet J* **10:** 30–37, 1990.

Martin, C. L. The pathology of glaucoma. *In* "Comparative Ophthalmic Pathology" R. L. Peiffer (ed.), pp. 137–169. Springfield, Illinois, Charles C Thomas, 1983.

Quigley, H. A., and Addicks, E. M. Chronic experimental glaucoma in primates. II. Effect of extended intraocular pressure elevation on optic nerve head and axonal transport. *Invest Ophthalmol* **19:** 137–152, 1980.

Smelser, G. K., and Azanics, V. The development of the trabecular meshwork in primate eyes. *Am J Ophthalmol* **71:** 366–385, 1971.

Smith, P. J., Samuelson, D. A., and Brooks, D. E. Aqueous drainage patterns in the equine eye: Scanning electron microscopy of corrosion cast. *J Morphol* **198:** 33–42, 1988.

van der Sinde-Sipman, J. S. Dysplasia of the pectinate ligament and primary glaucoma in the Bouvier des Flandres dog. *Vet Pathol* **24:** 201–206, 1987.

Wilcock, B. P., Brooks, D. E., and Latimer, C. A. Glaucoma in horses. *Vet Pathol* **28:** 74–78, 1991.

Wilcock, B. P., Peiffer, R. L., Jr., and Davidson, M. G. The causes of glaucoma in cats. *Vet Pathol* **27:** 35–40, 1990

VII. The Retina

When involution of the primary optic vesicle brings into apposition the anterior and posterior poles, the anterior (innermost) neurectodermal layer undergoes mitotic replication and subsequent specialization to form the nine layers of the neurosensory retina. The outermost neurectoderm remains as a relatively unspecialized simple cuboidal layer, the retinal pigment epithelium. Although it is traditionally considered the tenth retinal layer, its structure, function, and reaction to injury are unlike those of neurosensory retina, and it is best discussed separately. In this discussion, retina refers only to the neurosensory retina.

In the fixed, bisected globe, the retina is seen as a thin, opaque membrane lining the posterior half of the globe between vitreous and choroid. It joins the darkly pigmented pars plana of the ciliary body at an abrupt transitional point called the **ora ciliaris retinae.** In all but the best-preserved specimens, the retina is separated artifactually from the retinal pigment epithelium and adjacent choroid, remaining adherent only at the ora ciliaris and at the optic disk.

Histologically, the neurosensory retina begins abruptly at the ora ciliaris as a multilayered continuation of the inner layer of the ciliary epithelium. In dogs and sometimes in horses, the layers here are poorly defined and photoreceptors, sparse. The peripheral retina is only about half the thickness (100 μm) and has half the photoreceptor density (250,000/mm^3) of the central retina, with fewer nuclear and plexiform elements and a thin innermost nerve fiber layer.

The retina consists of three structural components: neurons, glia, and vasculature. The neurons are the functional elements and transmit the photoactivated electrical impulse from photoreceptor process to occipital cortex. The photoreceptor is the *raison d'être* of the entire globe. It is a sensory, apical cytoplasmic process of the neurons forming the outer nuclear layer. These processes, called rods or cones, based on their shape and ultrastructural composition, extend from the outer nuclear layer toward the choroid. They are enveloped by a glycosaminoglycan interphotoreceptor matrix, and interdigitate with apical processes of the retinal pigment epithelium, but no actual adhesions exist between the two layers. Within the outer segment of the photoreceptors are stacks of collapsed, disklike spheres which contain the photoactive chemicals. The disks within rod outer segments are constantly produced basally and shed apically at the rate of 80 to 100 disks per day, with an outer segment turnover time of 6 days in dogs. Effete disk debris is engulfed and degraded by the retinal pigment epithelium. Such turnover has not been demonstrated in the outer segment of cones, which have stacks of lamellae formed by infoldings of the plasma membrane. In addition, cones appear to be of many different types within a single retina. It is probably the ratio of different cones sensitive to different wavelengths of light that permits the visual cortex to discriminate color. In general, fish, amphibia, reptiles, and birds have excellent color discrimination. Ungulates can distinguish yellows, blues, and, variably, green and red. Carnivores have very limited color perception as far as can be determined.

Other retinal layers are best described in terms of function. The photoelectric stimulus originating in the photoreceptor outer segment is transmitted through the outer nuclear layer and along the axons of the photoreceptor nuclei to the bipolar and horizontal neurons of the inner nuclear layer. The accumulation of outer nuclear layer axons and inner nuclear layer dendrites forms the outer synaptic or plexiform layer. The inner nuclear layer contains the nuclei of the bipolar, horizontal, amacrine, and glial (Müller) cells. The bipolar cells receive impulses from the photoreceptors and relay them to ganglion cells. The bipolar cells also stimulate the horizontal cells, which transmit the impulse horizontally to excite adjacent bipolar cells. Amacrine cells counterbalance the bipolar cells in that their stimulation releases an inhibitor of ganglion cell excitation. The glial cells are primarily structural support cells, whose processes traverse the retina to form the retinal scaffold, and their anterior and posterior terminations fuse to form the inner and outer limiting membranes. The axons of the bipolar and amacrine cells, dendrites of ganglion and horizontal cells, and glial processes form the thick, inner synaptic or plexiform layer. The ganglion cell layer is the thinnest and innermost of the neuronal layers. Large, granular neurons form a single and often sparse layer, supplemented by a few astrocytes, that become bilayered in the area centralis to accommodate the marked increase in photoreceptor density. The density of ganglion cells predicts, in a general way, visual acuity. They are most closely packed in animals requiring fine visual discrimination (most birds, predatory fish, many reptiles). In contrast, they are sparse in ungulates, who do not feed by sight, but who flee at anything that moves. Their axons form the nerve fiber layer, which gradually increases in thickness toward the optic disk. In most animals, the fibers are not myelinated until they reach the optic disk. The nerve fiber layer is separated from the vitreous by an internal limiting membrane formed by the terminations of the Müller fibers and a true basal lamina.

The organization of the retinal vasculature is an important variable in ophthalmoscopic examination, both because vascular abnormalities are frequent signposts of disease and because normal species variation can erroneously be diagnosed as disease. Carnivores, ruminants, and swine have large venules and smaller arterioles radiating from the optic disk to peripheral retina. The horse has about 60 thin, short vessels extending from the disk for about 5.0 mm into surrounding retina. In dogs and cats, the major vessels lie within the deep half of the nerve fiber layer and the ganglion cell layer. In ruminants and pigs, the vessels are very superficial and bulge into the vitreous, covered only by a thin layer of nerve fibers and basal lamina. The retinal vessels form an end-artery circulation which supplies the inner layers of the retina. The photoreceptors and outer nuclear layers are avascular and receive nutrients primarily by diffusion from choroid. Such dependence cannot be absolute (except in horses) because degeneration of these outer layers is surprisingly slow (weeks to months) following retinal separation. In contrast, occlusion of a retinal vessel results in focal infarction of the inner retina within less than 1 hour.

The blood vessels also participate in the blood–eye barrier similar to that already described for uvea. The tight endothelial junctions and junctions between adjacent retinal pigmented epithelial cells conspire to create a retina that is immunologically isolated from nonocular tissues in a manner similar to that described for uvea. Like that in uvea, such a barrier is probably not absolute, and the various retinal antigens are not likely to be totally sequestered or absolutely unique to retina. Saline extracts of retina yield the retinal S antigen, and the interphotoreceptor retinoid-binding protein is another antigen that may be important in the initiation or perpetuation of degenerative (see sudden acquired retinal degeneration in Section VII,B,1) or inflammatory retinopathies.

The retinal pigmented epithelium extends from the ora ciliaris to optic disk as the posterior continuation of the outer layer of ciliary epithelium. It forms a simple cuboidal epithelial layer that is separated from the choroid by a complex basal lamina, Bruch's membrane. The apical border interdigitates with the photoreceptors, with an average of about 30 photoreceptors contacting a single pigment epithelial cell, but forms no junctional complexes. The inclusion of the adjective "pigmented" is somewhat a misnomer in domestic species except the pig, inasmuch as the epithelium overlying the tapetum contains no cytoplasmic pigment granules. This seemingly insignificant layer plays a major role in embryologic induction of the eye as previously described, and also plays a crucial role in the nurturing of the photoreceptors throughout life. The pigment epithelium engulfs and degrades obsolete rod and cone outer segments, absorbs light to protect photoreceptors, synthesizes and degrades part of the glycosaminoglycan matrix enveloping photoreceptor outer segments, and participates in the vitamin A–rhodopsin cycle. Which of these functions are most essential for photoreceptor health is still unclear.

The ocular fundus is a clinical term describing those ophthalmoscopically visible portions of the posterior globe, excluding vitreous. The fundus is commonly divided into dorsal tapetal and ventral nontapetal fundus, with the optic disk usually at the junction of the two. Retina, although almost transparent, does absorb some incident and reflected light to somewhat dull the fundus reflection ophthalmoscopically. Areas of retinal atrophy absorb less light and are seen as areas of increased tapetal reflectivity. Pre- or subretinal exudates, conversely, increase light absorption and are seen as focal fundic opacities. Developmental or acquired absence of tapetum allows black choroidal pigment to be seen. More severe choroidal lesions may be seen as red choroidal vasculature or even pink sclera obscured by variable amounts of residual pigment. Particularly in dogs, cats, and horses, selective breeding has made hypoplastic variations in amount and pigmentation of choroid and tapetum normal for particular breed or color varieties.

The **general pathology** of **retina** is often said to resemble that of brain. While this is undoubtedly true, inasmuch as retina is merely an extension of the brain, the prevalence of such lesions as malacia, nonsuppurative perivascular cuffing, and proliferative microgliosis within the retina is very low compared to the brain. Although no actual data are published, most animals with encephalitis do not, in fact, have concurrent retinitis. Retinal inflammation is most often the result of spread from choroid or across the vitreous from anterior uvea. Degenerations are much more common than inflammations and are not usually accompanied by inflammatory reaction. The mature mammalian retina has no capacity for regeneration of entire neural cells, although photoreceptor outer segments and glia may be replaced if destroyed in the course of degenerative or inflammatory disease. Even the fetal retina has poor regenerative capacity, as evidenced by the prevalence of retinal dysplasia following prenatal or neonatal retinal injury. Retinal repair is by proliferation of inner layer astrocytes, which eventually form a dense glial scar. Occasionally the astrocytes proliferate along the vitreal face of the retina, forming a preretinal fibroglial membrane. Similar subretinal membranes are seen with chronic detachments and originate from retinal pigment epithelium or Müller cells. The retinal pigment epithelium retains mitotic ability. When injured, these cells respond with hypertrophy, hyperplasia, and fibrous metaplasia. The presence of pigment in the neuroretina is frequent in instances of retinal atrophy, most probably derived from migration of retinal pigmented epithelial cells into the adjacent retina.

Autolytic changes are visible within retina within 30 min of death and within a few hours are of sufficient magnitude to interfere with the diagnosis of retinal degenerations. The earliest histologic change is pyknosis of a few nuclei in outer and inner nuclear layers, and loss of uniform density of the photoreceptor layer. Progressive dissolution of the photoreceptor outer segments results in retinal separation. Nuclear layer pyknosis and ganglion cell chromatolysis are widespread within 4 to 6 hr. By 12 hr, the retinal separation is complete, and the extensively folded retina, with autolytic photoreceptors, may mimic genuine retinal separation. The extensive pyknosis within both nuclear layers distinguishes the two, being absent in antemortem separation (see the following for other criteria). By 18 hr after death, the retina is represented by a barely separable bilayer of pyknotic nuclei suspended in a pale, eosinophilic foamy matrix representing fragmented nerve fiber and plexiform layers.

Bibliography

Barrie, K. N., and Gelatt, K. N. Diseases of the canine posterior segment: The ocular fundus. *In* "Veterinary Ophthalmology" K. N. Gelatt (ed.), pp. 480–484. Philadelphia, Pennsylvania Lea & Febiger, 1981.

Bellhorn, R. W., Murphy, C. J., and Thirkill, C. E. Anti-retinal immunoglobulins in canine ocular diseases. *Sem Vet Med Surg (Small Anim)* **3:** 28–32, 1988.

Buyukmihci, N., and Aguirre, G. Rod disc turnover in the dog. *Invest Ophthalmol* **15:** 579–584, 1976.

Donovan, A. The postnatal development of the cat retina. *Exp Eye Res* **5:** 249–254, 1966.

Young, R. W., and Bok, D. Participation of the retinal pigment epithelium in the rod outer segment renewal process. *J Cell Biol* **42:** 392–403, 1969.

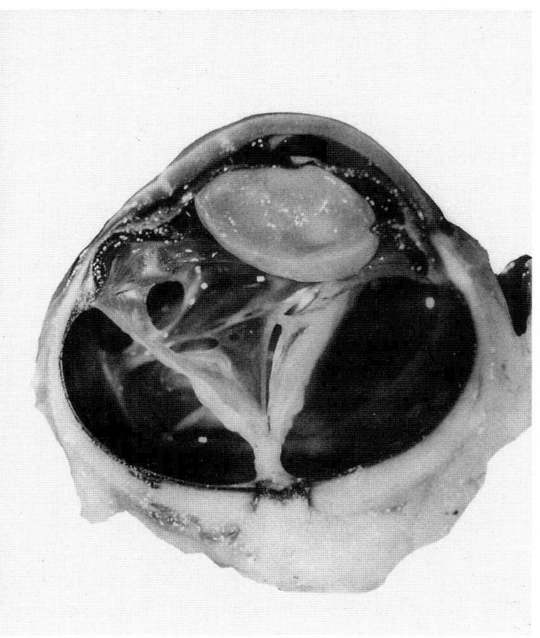

Fig. 4.46 Healed ophthalmitis with retinal detachment, synechia, and displacement of lens.

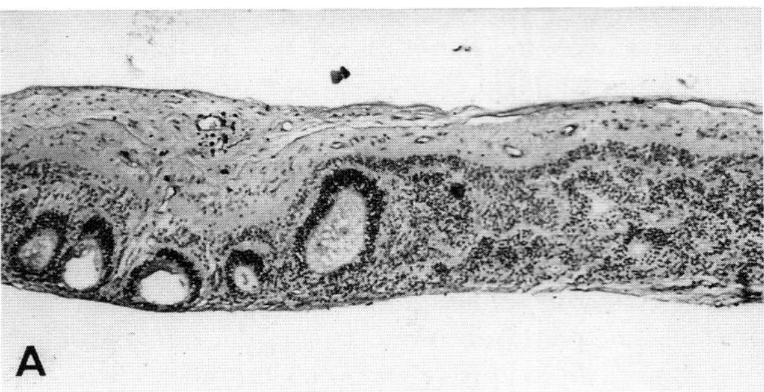

Fig. 4.47A Cystic degeneration and gliosis in spontaneously detached retina. Dog.

A. Retinal Separation

The retina is firmly attached in the globe only at the ora ciliaris and at optic disk. When the retina separates, it does so by cleaving photoreceptors from their interdigitations with the retinal pigment epithelium. Separation may occur as the result of accumulation of inflammatory exudates (Fig. 4.46), transudates, tumor cells, or helminths between pigment epithelium and photoreceptors, by contraction of a cyclitic membrane, or by leakage of liquefied vitreous through retinal tears. Such tears may result from orbital trauma or from progression of peripheral cystic retinal degeneration (Fig. 4.47A). The latter is relatively common in humans but not in domestic animals, despite the frequent occurrence of microcystoid retinal degeneration in the peripheral retina of dogs (Fig. 4.47B) and, less often, of horses.

The diagnosis of retinal separation in fixed specimens is complicated by the ease with which retinal separation can be induced by delayed fixation or improper handling of globes. The credibility of the diagnosis is greatly enhanced by the presence of subretinal exudates or cyclitic membranes (Fig. 4.48), but in their absence, the diagnosis rests on the observation of photoreceptor outer segment

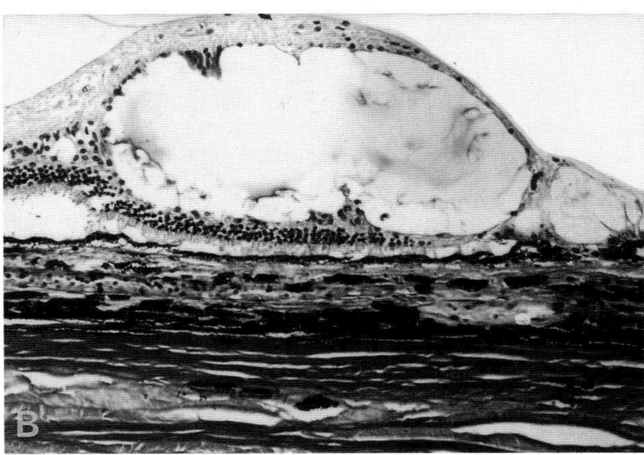

Fig. 4.47B Peripheral microcystoid retinal degeneration. Dog. Such change is very common and of no apparent functional significance.

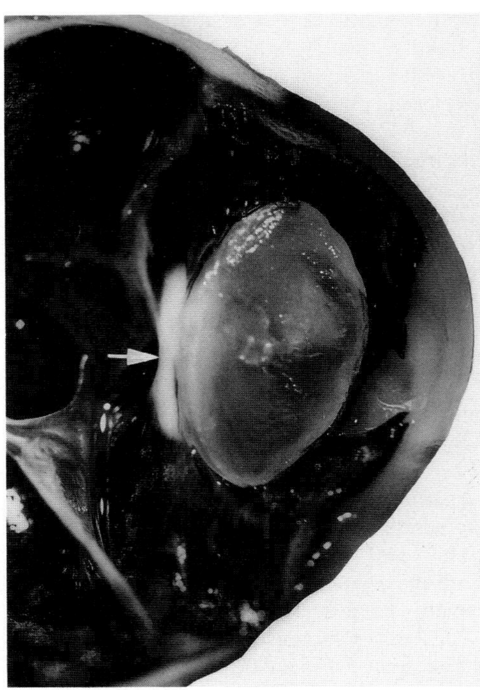

Fig. 4.48 Healed ophthalmitis. Ox. Cyclitic membrane (arrow) behind lens.

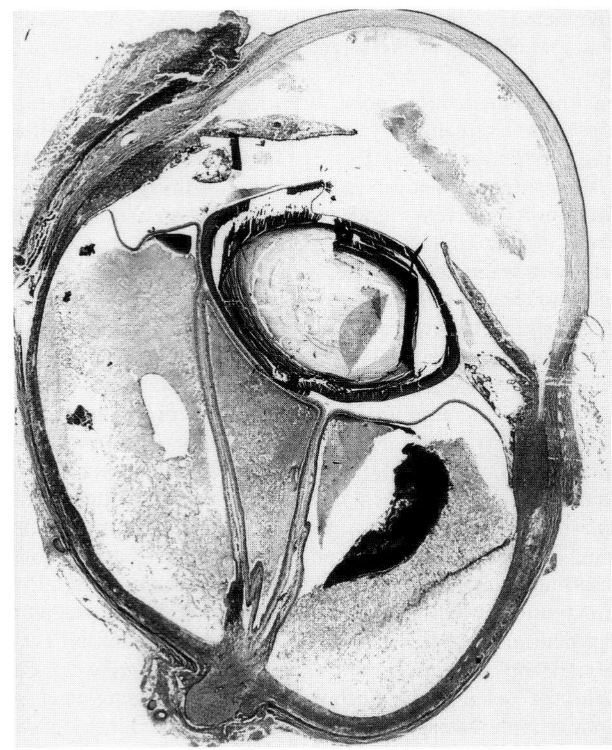

Fig. 4.49 Complete retinal exudative separation. Dog with metastatic choroidal melanoma.

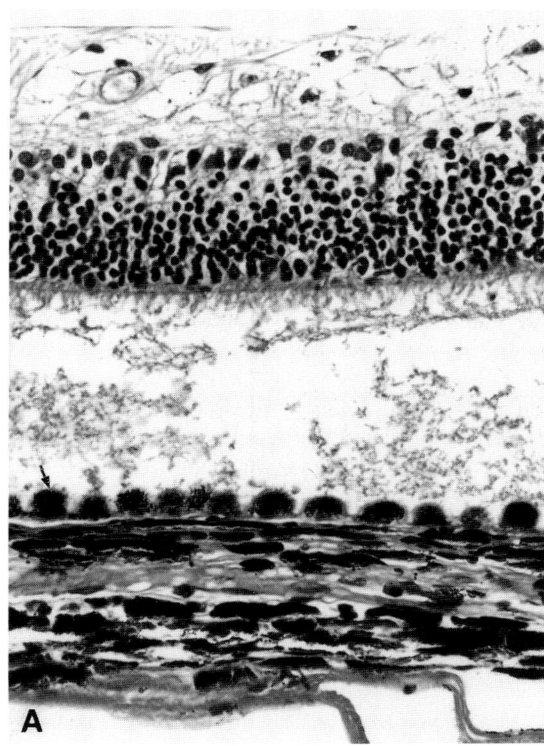

Fig. 4.50A Serous retinal separation. Dog. Hypertrophic retinal pigment epithelium; photoreceptors atrophic; loss of nuclei from inner and outer nuclear layers. Blending of nuclear layers due to atrophy of outer plexiform layer.

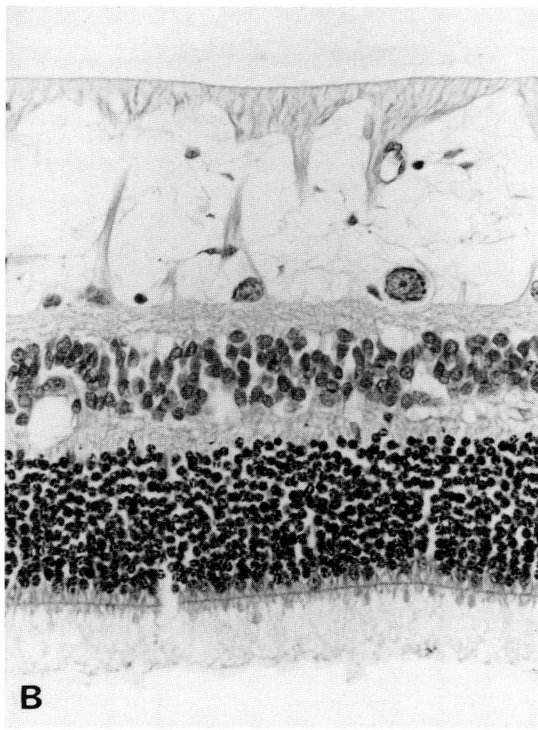

Fig. 4.50B Prolonged retinal separation with retinal edema and photoreceptor degeneration. Radial Müller's cell fibers anchor nerve fiber layer to ganglion cell layer. Retinal outer limiting membrane is prominent because of photoreceptor loss.

degeneration, hypertrophy and hyperplasia of pigmented epithelium, and the development of marked edema in inner nuclear, ganglion cell and inner plexiform layers (Figs. 4.49, 4.50A,B). **Hypertrophy** of the **retinal pigmented epithelium** is the most rapid change, occurring within a few hours after separation. The edematous changes are visible with the light microscope as early as 3 days following experimentally induced separation in owl monkeys. Coalescence of the edema creates a virtual cleavage of inner from outer retina, called **retinoschisis.** The cleavage is spanned by the radial Müller fibers, which seem the only anchor holding retina together. **Photoreceptor degeneration** is slower to appear under the light microscope, with loss of outer segments (probably the most subtle change that can be unequivocally diagnosed with routine light microscopy) visible by about 14 days. Inner segments and the cell bodies of the outer nuclear layer are almost unaffected and may remain so for months, suggesting that their maintenance is not so intimately linked to the pigment epithelium as is the case with the outer segments. This temporal hierarchy of change permits reasonably accurate aging of retinal separations, sometimes a necessary or at least interesting assessment in eyes enucleated after numerous clinical examinations or manipulation. The outer retinal lesions are apparently not ischemic, inasmuch as there is very little necrosis and no similarity to the lesion induced by retinal artery occlusion. Perhaps the outer layers can survive by diffusion of oxygen and nutrients from the subretinal fluid or from vascularized inner layers, and indeed the speed of photoreceptor atrophy varies with the height of the separation. An exception is seen in horses, inasmuch as the horse retina depends almost entirely on choroidal diffusion for oxygenation. Separation in this species results in rapid, full-thickness retinal infarction. A very frequent lesion in horses is focal, linear, or multifocal chorioretinal glial scarring with pigment migration and fibrous metaplasia of pigment epithelium, a lesion which is probably a healed infarct following traumatic separation or thromboembolic disease.

Bibliography

Aaberg, T. M., and Machemer, R. Correlation of naturally occurring detachments with long-term retinal detachment in the owl monkey. *Am J Ophthalmol* **69:** 640–650, 1970.

Anderson, D. H. *et al.* The onset of pigment epithelial proliferation after retinal detachment. *Invest Ophthalmol Vis Sci* **21:** 10–19, 1981.

Anderson, D. H. *et al.* Retinal detachment in the cat: The pigment epithelial–photoreceptor interface. *Invest Ophthalmol Vis Sci* **24:** 906–926, 1983.

Anderson, D. H. *et al.* Morphological recovery in the reattached retinal. *Invest Ophthalmol Vis Sci* **27:** 168–183, 1986.

Erickson, P. A. *et al.* Retinal detachment in the cat: The outer nuclear and outer plexiform layers. *Invest Ophthalmol Vis Sci* **24:** 927–942, 1983.

Machemer, R. Experimental retinal detachment in the owl monkey. II. Histology of retina and pigment epithelium. *Am J Ophthalmol* **66:** 396–410, 1968.

Machemer, R., and Lagua, H. Pigment epithelium proliferation in retinal detachment (massive periretinal proliferation). *Am J Ophthalmol* **80:** 1–23, 1975.

B. Retinal Degeneration

Retinal degeneration, more commonly called retinal atrophy, may result from senile change, nutritional deficiency, metabolic disorder, or injury caused by infectious, chemical, or physical agents. With the exception of the previously described glaucomatous retinal atrophy, virtually all are initially degenerations of photoreceptor outer segments or of retinal pigmented epithelium, and many retinal atrophies of different pathogenesis have similar histologic appearance. The similarities become even stronger as the lesions progress to the severity usually encountered in enucleated eyes from clinically affected animals. Nonetheless, it is useful to review the subject and to discuss the differences between some of the best-studied examples of naturally occurring retinal atrophies. Most frequent are the inherited retinopathies of dogs, grouped under the name **progressive retinal atrophy.** Less common are retinal degenerations caused by deficiencies of taurine, vitamin E, or vitamin A, by excessive visible light, or by several toxic or metabolic diseases.

1. Inherited Retinal Atrophies in Dogs

Progressive retinal atrophy describes, admittedly with some inaccuracies, a large group of bilateral retinal diseases in dogs. They share the clinical features of being bilateral, progressing to blindness and being unassociated with inflammatory or other ocular disease. More than 100 breeds have been identified as affected, although there is little published information as to the prevalence within various breeds. All thus far studied are inherited as an autosomal recessive trait. Some are juvenile onset degenerations that result from a congenital biochemical defect and are thus properly termed **photoreceptor dysplasias.** Photoreceptors never reach proper ultrastructural or physiologic maturity, and affected dogs may be blind by a year or two of age. Irish setters, collies, Norwegian elkhounds, and miniature schnauzers are the best-studied breeds that are affected, each with slightly different clinical expression and biochemical abnormality. Some initially affect only rods, but most affect both rods and cones. Alaskan malamute dogs have what seems to be a purely cone dysplasia, leaving dogs visually impaired in daylight but with good night (i.e., rod-dependent) vision.

A quite separate group of diseases are currently considered to be true degenerations, in that photoreceptor development seems normal. It may be, however, that even these inherited atrophies will be shown to have a developmental biochemical defect. The trend to date has been to reclassify more and more degenerations as dysplasias, in parallel with the use of more sensitive investigative techniques.

All of these different diseases may have significant differences in pathogenesis, but by the time the eyes are removed from impaired or totally blind dogs, the histologic and ultrastructural lesions are similar. At this stage, the old name of **progressive retinal atrophy** continues to be used as a catch-all. The light-microscopic lesion is degeneration of photoreceptors beginning dorsolateral to the optic disk. Over months or years the photoreceptor loss extends, and there is secondary atrophy of nuclear and plexiform layers of retina (Fig. 4.51). Eventually, in dogs permitted to live long enough, the retina remains as a poorly cellular glial scar. Despite the many similarities in clinical and histologic features, the importance of these retinopathies in the study of retinal disease in general warrants some more-detailed explanation of the best-described variants.

Retinal atrophy in Irish setters is described as rod–cone dysplasia, inherited as a simple autosomal recessive trait. Dogs homozygous for the defect have arrested differentiation of the rod external segments. Cones are less affected. The defect is detected ultrastructurally as early as 16 days after birth, at which time in the normal retina, the outer rod segments should be developing adjacent to the pigment epithelium. Arrested development is followed by degeneration of inner rod segments, so that there is diffuse loss of all rod photoreceptors by 12 weeks of age. This is followed by loss of cones and of outer nuclear layer. By the time

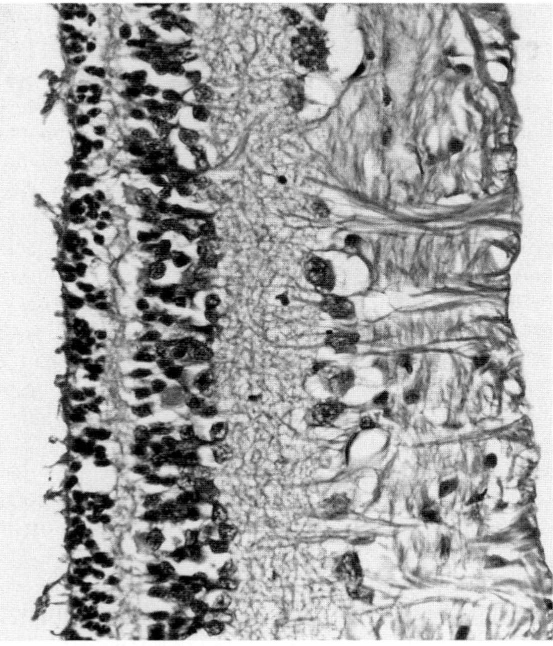

Fig. 4.51 Progressive retinal atrophy. English cocker spaniel. Diffuse atrophy of photoreceptors and outer nuclear layer.

the dog is about 1 year of age, there is diffuse atrophy of the outer nuclear layer, and the inner nuclear layer is in direct contact with the pigment epithelium. The inner retinal layers are unaffected. Dogs show visual deficits in dim light as early as 6 weeks of age and are usually blind by 1 year. The biochemical defect is a marked deficiency of the phosphodiesterase responsible for the continuous hydrolysis of cyclic guanine monophosphate within outer segments. While the function of the enzyme in this site is not fully understood, the resultant excess of cyclic guanine monophosphate (cGMP) is toxic to photoreceptors *in vitro*, and is known to cause arrested development or degeneration of rod outer segments *in vivo*. In the Irish setter, the substrate levels are about 10 times higher in affected than in control dogs, and the elevation precedes morphologic change in the photoreceptors. There are other biochemical retinal abnormalities (in rhodopsin and in membrane lipids), but it is not known whether they are primary abnormalities or merely effects of the cGMP phosphodiesterase deficiency. The basic defect is in the gene encoding the outer segment-specific phosphodiesterase.

Rough collies have a very similar rod–cone dysplasia, also inherited as an autosomal recessive tract. The progression of the lesion is slightly slower than that in Irish setters, but the biochemical defects are virtually identical. In neither instance is it clear how photoreceptor death leads to death of all the outer nuclear layer neurons.

Retinal atrophy in Norwegian elkhounds resembles that of the Irish setter in most respects. Onset of visual deficits in dim light is early, progression to blindness is only slightly less rapid, and the histologic retinal lesion in blind dogs is almost identical to that in setters. Ultrastructurally and biochemically, however, the two diseases are distinct. In elkhounds the primary lesion is restricted to rods and is not associated with elevated cGMP. The rods are not halted in their development but develop imperfectly. There is disorientation of lamellar disks and eventual disintegration. Light microscopic lesions appear at about 6 months of age and progress to complete photoreceptor loss by 1 to 2 years. Atrophy progresses to affect outer nuclear, outer plexiform, and inner nuclear layers. Eventually the retina, as in any of the canine familial retinopathies, may remain only as a thin glial scar with disorganized neuronal clumps. A second type of rod–cone dysplasia exists in this breed, with rapid progression to blindness by 12 to 18 months of age. Photoreceptor growth is erratic and apparently uncoordinated, but there is also dysplasia of rod and cone axonal synaptic junctions in the outer plexiform layer, which results in greatly reduced transmission of impulses from the photoreceptors.

Retinal atrophy in miniature poodles is classified as a true degeneration, in that photoreceptor differentiation seems to be normal until 6 to 9 weeks of age. After this time, and progressing at an unpredictable rate, there is disk disorganization and plasma membrane fragmentation in rod outer segments, visible as early as 15 weeks of age in the retina adjacent to the optic nerve. Peripheral retina is unaffected at this early stage, but by several years of age, the entire retina is affected. Cones are affected in a similar but milder fashion later in the disease course. The reason for the fragmentation is unknown, but it is known that affected dogs have abnormally slow outer segment turnover (about 40% of normal) prior to any observed structural change. Affected poodles have evidence for decreased rate of incorporation of docosahexanoic acid, the major long-chain structural membrane fatty acid, into rod outer segments. It may be that the decreased disk turnover permits older membranes to unduly persist and to be peroxidatively damaged *in situ*. The histologic lesion is identical to that of Irish setters or Norwegian elkhounds. Cataracts are present in many poodles with retinal atrophy. Because of wide variation in disease progression, dogs may not be noticed to have dim-light deficiencies until middle, or even old, age.

Retinal atrophy in Alaskan malamute dogs is a progressive cone dysplasia, inherited as a simple recessive trait. Although now rare because of elimination by selective breeding, it remains an interesting example of specific cone dysfunction. Affected dogs are noticed to have poor vision in bright light as early as 8 weeks of age. Night vision is, and remains, clinically normal, as does the ophthalmoscopic appearance of affected eyes. The ultrastructural lesion is disorganization and loss of cones, with rods normal. Adult dogs have no cone outer segments, and atrophic inner segments, but no change in rods or outer nuclear layer.

Central progressive retinal atrophy of dogs (pigmented epithelial dystrophy) denotes a peculiar lesion in retinal pigment epithelium of dogs that apparently results from defective intracellular phagocytosis of shed photoreceptor outer segments. Normal pigment epithelium engulfs and enzymatically degrades this material, resulting in a gradual buildup of intracellular lipopigments throughout life. In dogs with this pigment epithelial dystrophy, membrane peroxidation is excessive, and lipopigments accumulate to excess. Associated with the pigmentary accumulation, the epithelial cells hypertrophy. Photoreceptor outer segments adjacent to hypertrophic pigment epithelium degenerate. As the lesion progresses, hypertrophy and hyperplasia of epithelium give rise to dysplastic pigmented cell clumps. Within such clusters there may be an eosinophilic, hyaline, periodic acid-Schiff positive concretion resembling drusen. This material, rather frequent in ophthalmic specimens from humans with a variety of degenerative retinal or choroidal diseases, is a concretion of excess basal lamina produced by the pigment epithelium. The eventual histologic lesion in affected dogs is a monolayer of hypertrophic, lipochrome-rich pigment epithelial cells with multifocal hyperplastic clumps. Retina has atrophy of photoreceptors and outer nuclear layer, and some irregular gliosis. Pigment-laden cells may invade the retina.

The disease is sporadic and of unpredictable clinical

progression. The prevalence is, for example, much higher in Great Britain than in North America, where it is rare. Retrieving and herding dogs are most frequently affected. The ophthalmoscopic lesion of irregular black mottling begins near the optic disk and may progress to generalized pigment mottling interspersed with the increased reflectivity of atrophic retina. The mode of transmission is unknown. An interesting speculation, based on morphologic similarities, is that the disease represents a defect in vitamin E metabolism within pigment epithelium.

Sudden acquired retinal degeneration is an enigmatic, rapidly progressing photoreceptor degeneration that is histologically identical to the inherited progressive retinal atrophies. Blindness occurs very rapidly (over a period of a few days to a few weeks). Affected dogs are adult or even elderly, and the disease can affect any breed or crossbreed. The fundoscopic lesion is bilaterally symmetrical and diffuse across the retina, but histologic studies of the early lesions are very few. The cause is unknown, but the presence of the retinal disease is linked to systemic signs of polyuria, polydipsia, and elevated serum cholesterol and alkaline phosphatase. Some, but not even the majority, of the affected dogs have adrenal cortical hyperfunction. How this malfunction causes the irreversible retinopathy, if indeed it does, is unknown. One small study demonstrated circulating, complement-fixing antibody to retinal S-antigen and interphotoreceptor retinoid-binding protein, raising the possibility that the disease is a cytotoxic autoimmune phenomenon.

2. Inherited Retinopathies in Cats

Inherited retinal dysplasias and degenerations have been reported in a variety of cat breeds, but only in the Abyssinian breed has the syndrome been adequately studied. In this breed there are two different diseases: early-onset rod–cone dysplasia and late-onset retinal degeneration affecting rods much sooner than cones. The early-onset dysplasia is inherited as an autosomal dominant trait. It is histologically and ultrastructurally similar to the disease in Irish setter dogs, and a similar defect in the activity of cGMP phosphodiesterase has been reported. Affected cats are blind by a few months of age.

The late-onset retinal degeneration is inherited as an autosomal recessive, and affected cats progress slowly to blindness by 5 to 10 years of age. The earliest structural changes are in rod outer segments in peripheral retina, with jumbling of the rod disks and patchy blunting of the photoreceptors themselves. Only after many years is there histologically detected diffuse photoreceptor loss. A much more prevalent feline retinopathy, caused by a deficiency of dietary taurine, is discussed later.

3. Inherited Night Blindness in Horses

This poorly documented disease affects the Appaloosa breed and is probably inherited, and is seen as night blindness with daylight vision that is usually, but not always, normal. No structural lesion is seen in retinas of affected horses, and functional studies point to a defect in intraretinal synaptic transmission in outer plexiform or inner nuclear layers rather than a defect in photoreceptors.

Bibliography

Aguirre, G. Inherited retinal degeneration in the dog. *Trans Am Acad Ophthalmol Otolaryngol* **81:** 667–676, 1976.

Aguirre, G. D., and Rubin, L. F. Pathology of hemeralopia in the Alaskan malamute dog. *Invest Ophthalmol* **13:** 231–235, 1974.

Aguirre, G. D., and Rubin, L. F. Rod-cone dysplasia (progressive retinal atrophy) in Irish setters. *J Am Vet Med Assoc* **166:** 157–164, 1975.

Aguirre, G. *et al.* Rod-cone dysplasia in Irish setters: A defect in cyclic GMP metabolism in visual cells. *Science* **201:** 1133–1134, 1978.

Barnett, K. C. Canine retinopathies—I. History and review of the literature. *J Small Anim Pract* **6:** 41–55, 1965.

Buyukmihci, N., Aguirre, G., and Marshall, M. Retinal degenerations in the dog. II. Development of the retina in rod–cone dysplasia. *Exp Eye Res* **30:** 575–591, 1980.

Chader, G. J. Animal mutants of hereditary retinal degeneration: General considerations and studies on defects in cyclic nucleotide metabolism. *Prog Vet Comp Ophthalmol* **1:** 109–126, 1991.

Cogan, D. G., and Kuwabara, T. Photoreceptive abiotrophy of the retina in the elkhound. *Pathol Vet* **2:** 101–128, 1965.

Goldman, A. I., and O'Brien, P. J. Phagocytosis in the retinal pigment epithelium of the RCS rat. *Science* **201:** 1023–1025, 1978.

Liu, Y. P. *et al.* Involvement of cyclic GMP phosphodiesterase activator in an hereditary retinal degeneration. *Nature* **280:** 62–64, 1979.

Lolley, R. N., and Farber, D. B. A proposed link between debris accumulation, guanosine 3′-5′ cyclic monophosphate changes and photoreceptor cell degeneration in retinas of RCS rats. *Exp Eye Res* **22:** 477–486, 1976.

Millichamp, N. J. Retinal degeneration in the dog and cat. *Vet Clin North Am: Small Anim Pract* **20:** 799–836, 1990.

Parry, H. B. Degeneration of the dog retina. VI. Central progressive atrophy with pigment epithelial dystrophy. *Br J Ophthalmol* **38:** 653–668, 1954.

West-Hyde, L., and Buyukmihci, N. Photoreceptor degeneration in a family of cats. *J Am Vet Med Assoc* **181:** 243–247, 1982.

Witzel, D. A. *et al.* Night blindness in the Appaloosa: Sibling occurrence. *J Am Anim Hosp Assoc* **13:** 383–386, 1977.

4. Light-Induced Retinal Degeneration

Light of various wavelengths has a variety of injurious effects on cornea, lens, or retina that vary with the wavelength, duration, and intensity of the light. The effects also vary with a large but poorly understood group of animal variables that include ocular pigmentation, habitat, previous experience with photoperiod, nutrition, body temperature, age and, most obviously, species. The wavelength of light has the greatest effect; short wavelengths in the ultraviolet and blue range (up to about 475 nm) have the greatest energy per photon and are the most damaging. Fortunately, most of these wavelengths are absorbed by cornea and lens, so that their lethal effects on retina are seldom seen. They may cause corneal epithelial injury or cataract, although these effects are apparently rare in domestic animals (see Section V,B, Cataract).

In humans, accidental exposure to light from arc welding, solar eclipses, or ophthalmic examination or operating equipment (including lasers) creates the potential for rapid injury from mechanical disruption or heat. Although such damage is certainly possible in other animals, most naturally occurring lesions result from the additive effects of much less intense ultraviolet and short-wavelength visible light because of unnatural photoperiods. Animals with poorly pigmented eyes, and those adapted for nocturnal vision, are most susceptible. Susceptibility also increases with age and with temperature.

The initial lesion is disruption of rod outer-segment disks, with eventual destruction of all photoreceptors and their nuclei. Because the lesion is identical to most inherited, nutritional, and toxic retinopathies, the diagnosis is made on the circumstantial evidence of abnormally bright light, abnormally long light photoperiod, or a rapid change in photoperiod. Most instances occur with rodents or fish kept in continuous fluorescent light. Albino rodents or deepwater fish are, predictably, the most susceptible.

The mechanism by which visible light of moderate intensity damages the retina is still incompletely understood, and different experimental models give rise to different theories. Most studies use blue light in the 400- to 475-nm range which, unlike shorter ultraviolet wavelengths, is not filtered out by cornea or lens. The most popular theory is that of light-induced oxidation of the very abundant polyunsaturated long-chain fatty acids of the rod disks, with the generation of free radicals to then cause cell membrane damage. This theory gains support from studies showing a protective effect by vitamin C or E, and enhanced injury under conditions of retinal hyperoxia.

5. *Nutritional Retinopathy*

Nutritional causes of retinal degeneration include deficiencies of vitamins C, A, or E, and the amino acid, taurine. Retinal atrophy and cataracts have been seen in fish with a dietary deficiency of vitamin C. The lesions were thought to be light induced, with the fish unusually susceptible because of the deficiency in the antioxidant effects of the vitamin. The ocular lesions of vitamin E deficiency resemble those of retinal pigment epithelial dystrophy and were referred to briefly under that heading. Pups fed severely deficient diets develop night blindness within about 6 weeks, and an extinguished electroretinogram suggestive of diffuse photoreceptor damage. These last two effects are not seen in naturally occurring central retinal atrophy. Retinopathy has been described only in primates and dogs fed rations deliberately and severely deficient in vitamin E. Lipofuscin, seen as eosinophilic cytoplasmic inclusions, accumulated to excess in the pigment epithelium, and was followed by hypertrophy of the pigment epithelium and degeneration of photoreceptor outer segments. Eventually there were full-thickness central retinal atrophy and some small foci of retinal separation.

a. Hypovitaminosis A Retinopathy caused by hypovitaminosis A is seldom encountered except in growing cattle or swine kept in confinement and fed a ration deficient in the vitamin over months or years. Grains other than corn (maize) are very poor sources of vitamin A, and the level in corn falls markedly with prolonged storage. Green pasture is very rich in carotene, which is converted to vitamin A by intestinal epithelium. Hay that is excessively dry, leached by rain, cut late in the year, or stored for prolonged periods is a much less adequate source, but in most pastured animals the liver reserves are sufficient to prevent clinical signs of deficiency for at least 6 months and often up to 2 years. Young, rapidly growing animals have greater requirements and smaller stores of the vitamin and are thus more susceptible than adults.

Hypovitaminosis A affects bone remodeling and causes epithelial cell atrophy and defects in synthesis of rhodopsin. Ocular lesions can result from each of these three defects.

As previously discussed, maternal deficiency of vitamin A causes blindness in offspring due to defective remodeling of optic nerve foraminae and subsequent ischemic or pressure atrophy of the nerve. In piglets, there may be massive ocular dysplasia and such anomalies as cleft palate, skeletal deformities, hydrocephalus, epidermal cysts, genital hypoplasia, and anomalous hearts. Optic nerve atrophy is preceded by optic disk swelling (papilledema) and followed by atrophy of nerve fiber and ganglion cell layers. This sequence of events may occur if very young animals are on deficient diets, with the optic nerve changes being caused in part by stenosis of optic foramen and in part by increased cerebrospinal fluid pressure that itself results from atrophy and metaplasia of arachnoid villi. The papilledema precedes optic nerve necrosis and is reversible. The corneal lesions of hypovitaminosis A have received scant attention and are seldom seen in natural outbreaks.

The acquired ocular effect of hypovitaminosis A involves photoreceptor outer segments. The ophthalmoscopic lesion is multifocal retinal atrophy and scarring in animals with slow or absent pupillary light reflex and apparent blindness. The histologic lesion is patchy-to-diffuse photoreceptor atrophy, which first affects the rod outer segments. Night blindness is thus the initial complaint and is often the chief complaint about a deficient herd. Eventually the atrophy affects all photoreceptors and their nuclei, and may progress to full-thickness atrophy with scarring. The lesions have been produced in all domestic species on specially formulated diets, but naturally occurring retinal lesions are almost restricted to cattle with chronic deficiencies.

The pathogenesis of the photoreceptor atrophy demonstrates the structure:function interdependence of retinal cells. Vitamin A is converted to retinene and then to the glycoprotein rhodopsin. Rhodopsin is stored as a component of the lamellar disks of the outer segment. Light initiates a physicochemical change in rhodopsin, resulting in a cascade of events culminating in the hyperpolarization of the outer segment membrane. The resultant electrical impulse is transmitted to bipolar cells, ganglion cells, and then to brain. The deficiency of vitamin A necessarily results in a deficiency of rhodopsin. The corresponding ultrastructural lesion is swelling, then fragmentation of lamellar disks that can be reversed by therapy with vitamin A unless inner segments have also been affected. Regeneration simulates normal development and requires about 2 weeks to rebuild outer segments completely. Vitamin A is discussed further with Bones and Joints in Chapter 1.

b. TAURINE-DEFICIENCY RETINOPATHY Retinal degeneration caused by taurine deficiency is seen only in cats, although taurine is the predominant free amino acid in the retina of other species. Among domestic mammals, only the cat seems unable to synthesize taurine from cysteine in amounts adequate for retinal function. Taurine is considered a dietary essential for cats, and its deficiency results in a characteristic central retinal atrophy and in cardiomyopathy (see Volume 3, Chapter 1, The Cardiovascular System).

The ocular lesion of taurine deficiency was first detected in cats fed semipurified diets in which casein was the only protein. After several months, such cats developed focal retinal atrophy adjacent to the optic disk, which progressed to generalized retinal atrophy. Supplementation with taurine halted but did not reverse the lesion, presumably because photoreceptor nuclei or inner segments already had been damaged. The naturally occurring disease of cats, called **feline central retinal degeneration,** is described in a variety of breeds. The lesion is bilateral, dorsolateral to the optic disk, and is usually unassociated with visual impairment. The histologic lesion is photoreceptor degeneration, initially targeting cone outer segments but eventually affecting rods as well. The rods of the peripheral retina are the last to degenerate. Taurine also seems essential for membrane integrity of the tapetal reflective rodlets, so that dissolution of the membrane surrounding these crystalline intracytoplasmic inclusions is another characteristic lesion. Some of the cases of idiopathic central atrophy are associated with the feeding of dry dog food, which is low in taurine. Most, if not all, examples of feline central retinal atrophy are probably due to taurine deficiency.

Less clear is the association of taurine deficiency with diffuse retinal atrophy in cats (Fig. 4.52). Familial atrophy occurs in Abyssinian and Persian cats, but most cases are of unknown cause. Continued deficiency of taurine leads to diffuse retinal atrophy and thus might be responsible. Many cases of central atrophy remain static for years, perhaps as the permanent scar of a temporary dietary deficiency.

6. *Toxic Retinopathies*

Experimental toxic retinopathies have been caused by many chemicals and toxic plants, but only a few toxic plants cause important diseases of domestic animals.

Bracken fern (*Pteridium aquilinum*) causes a progressive retinal degeneration in sheep in several areas of Great Britain. The common name bright blindness refers to pupillary dilation and tapetal hyperreflectivity of the severely affected sheep. The disease has been seen only in flocks grazing hills rich in bracken fern, and has been reproduced by prolonged feeding of the fern to sheep. A similar syndrome has been noted in cattle during long-term exposure to the fern. The lesion is usually seen in middle-aged or older sheep as bilateral and initially central tapetal hyperreflectivity. Diffuse involvement follows. The histologic lesion is nonspecific, consisting of photoreceptor outer segment degeneration progressing to depletion of all retinal layers.

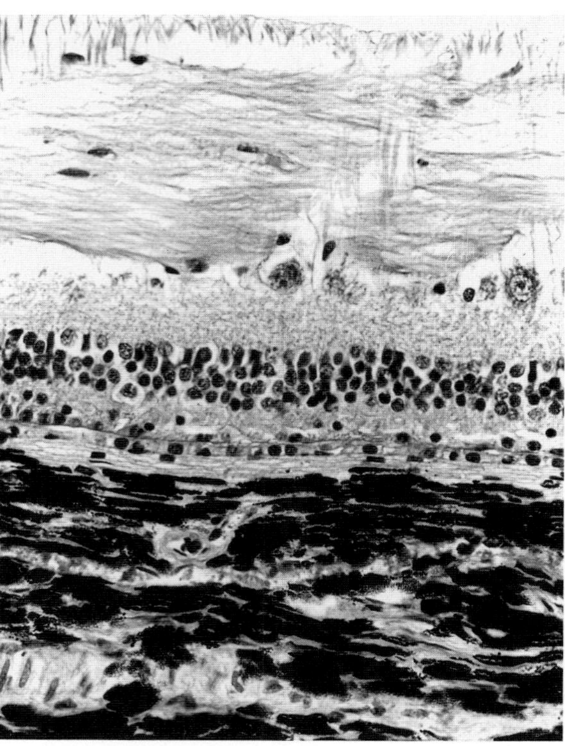

Fig. 4.52 Idiopathic diffuse retinal atrophy. Cat. Note complete atrophy of photoreceptors and depletion of outer nuclear layer.

Blindness is one of the features of intoxication with **locoweed,** *Astragalus* and *Oxytropis* spp., in the United States of America, darling pea, *Swainsona* spp., and blind grass, *Stypandra* spp. in Australia, and selenium indicator plants worldwide.

Astragalus and *Swainsona* cause a neurovisceral lysosomal storage disease analogous to genetically transmitted mannosidosis (see Chapter 3, The Nervous System). All members of the genus *Swainsona* contain an indolizidine alkaloid, swainsonine, that is a potent inhibitor of lysosomal mannosidase. At least some *Astragalus* spp. contain a similar alkaloid. Chronic ingestion of the plant occurs in cattle, sheep, and horses forced to eat the plants on dry pastures where nothing more palatable is available. Affected animals develop behavioral abnormalities and defects of gait and vision. The histologic lesion consists of widespread cytoplasmic vacuolation in most organs due to the intralysosomal accumulation of mannose-rich oligosaccharides. Onset of clinical signs may require several months of heavy *Swainsona* ingestion, but ultrastructural vacuolation is seen within a few days. The ocular lesion is, as elsewhere in the central nervous system, vacuolation of neuronal cytoplasm and, later, axonal degeneration. The vacuolation is readily reversible on cessation of ingestion of the plant and seems not to be the lesion responsible for clinical signs. The axonal degeneration is not reversible and is probably the more important lesion. Whether blindness is retinal or central in origin is unknown.

Poisoning with *Stypandra* spp. occurs in sheep and goats on dry pastures in southwestern Australia. The plant is among the first to reappear after autumn rains end the drought, and is eaten if nothing better is available. Acute intoxication is frequently fatal. Animals surviving the acute stage become blind and ataxic. In retina there is diffuse photoreceptor atrophy and patchy hyperplasia of retinal pigment epithelium. Axonal degeneration is found within the optic nerve and elsewhere within the central nervous system.

The colloquial term blind staggers refers to chronic intoxication of sheep and cattle with plants known to accumulate organic selenium selectively. Affected animals wander aimlessly, become weak and ataxic and are finally paralyzed prior to death. There is some question as to whether blindness is genuine or merely the result of stupor. Ocular lesions are not described. The syndrome of blind staggers does not occur in experimental selenium toxicity, and it is possible that the syndrome is of much more complex pathogenesis than simple selenium toxicity. Plants of the genera *Astragalus* and *Oxytropis* are selenium accumulators as well as sources of swainsoninelike alkaloids.

7. *Miscellaneous Retinopathies*

Retinal lesions are found in a number of metabolic disorders and systemic states. Best known among these is diabetes mellitus, but retinal lesions are found also in any of the neuronal storage diseases, coagulation disorders, anemia, disseminated intravascular coagulation, hyperviscosity syndrome, and hypertension, and following excessive exposure to oxygen or light.

Diabetes mellitus is the major cause of blindness in humans in North America. The cause of the blindness is chorioretinal vascular disease with subsequent retinal degeneration. The characteristic lesions are seen only in patients with diabetes of 10 to 15 years' duration. Even though virtually all chronic diabetics develop some retinal lesions, less than 10% become blind. Blindness is strongly predictive of the development of fatal diabetic nephropathy. Lesion development is not prevented by insulin replacement. Other ocular lesions include cataract, rubeosis iridis, and glycogen-induced vacuolation of iris epithelium and massive thickening of the ciliary basal lamina. The corneal epithelium may be unduly fragile, and tear production may be reduced.

The retinal lesion in humans is mostly the result of microvascular disease. Loss of retinal pericytes, development of microaneurysms, thickening of capillary basal lamina, and retinal hemorrhages constitute the early, degenerative phase of the retinopathy. This is followed by a proliferative phase in which more capillary aneurysms, arteriolar–venular shunts, and neovascularization occur as the presumed responses to retinal ischemia. The neovascularization is initially bland and confined to retina, but later there is extension into preretinal vitreous with accompanying fibroplasia (retinitis proliferans). Hemorrhages and hyalinized collections of leaked plasma are common in the retina.

In nonprimates, the naturally occurring retinopathy is seen only in dogs and, even then, infrequently. This low frequency may be due to the fact that affected dogs do not live long enough for the retinal disease to develop. In dogs deliberately made diabetic and kept for up to 6 years, microvascular lesions typical of human diabetes occur. Pericyte loss is accompanied by capillary aneurysms, reactive endothelial proliferation, and perivascular plasmoid exudates or hemorrhages.

Retinal hemorrhages are seen in a variety of primary clotting disorders, in thrombocytopenia of any cause, and in degenerative or inflammatory vascular disorders. Massive hemorrhage may occur from completely separated retinas. The best-known examples in veterinary medicine are multifocal hemorrhage from vessels damaged in the course of thrombotic meningoencephalitis of cattle, and with Rocky Mountain spotted fever or ehrlichiosis of dogs. Apparently unique to cats is multifocal retinal hemorrhage observed in severe anemia. The lesions heal with scarring if the cat survives the anemia, suggesting that the hemorrhage is only the most visible manifestation of multifocal and probably ischemic retinopathy. Similar retinal atrophy, but without hemorrhage, has been seen in horses following massive but sublethal blood loss as in surgery or from nasal hemorrhage subsequent to severe cranial trauma. Affected eyes have multifocal retinal atrophy and hyperpigmentation. Similar lesions may also result from focal retinal separation (which causes infarction in horses), or from thromboembolic consequences of bacteremia. Retinal hemorrhage, and sometimes retinal infarcts, occur in an unknown percentage of animals with disseminated intravascular coagulation of any pathogenesis. Horses seem particularly susceptible, perhaps because of their poorly vascularized retina (Fig. 4.53). Rarely, retinal infarcts are caused by neoplastic emboli.

Senile retinopathy is characterized by microcystoid degeneration, which is very common in dogs from middle age onward (Fig. 4.47B). A similar lesion is found occasionally in horses. The lesion affects peripheral retina adjacent to ora ciliaris and for a variable distance medially. There is formation of small cystic spaces within inner nuclear and plexiform layers, fusion of inner and outer nuclear layers, pigment cell accumulation and haphazard atrophy and mingling of nuclei in a manner simulating peripheral retinal dysplasia. If the cysts rupture to the vitreal face, the retina external to the cyst is seen as an

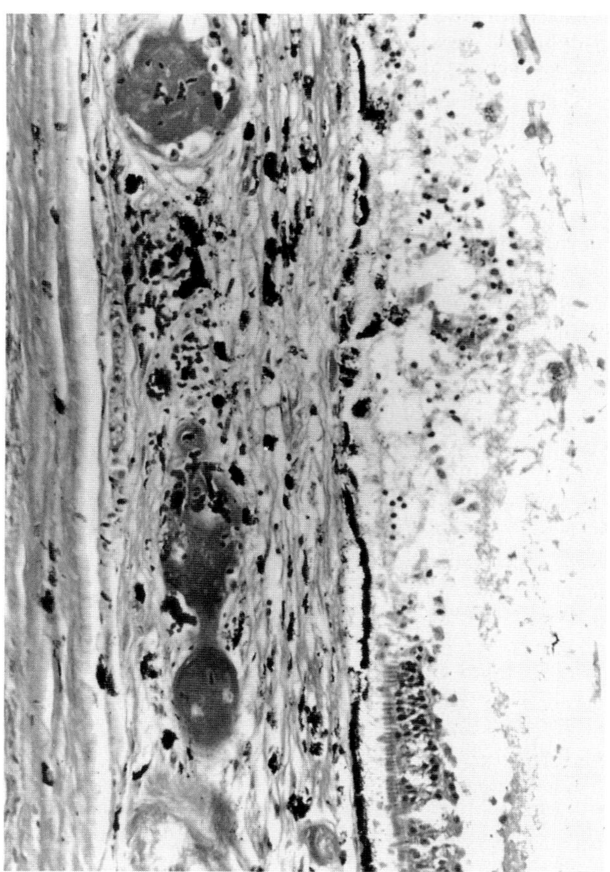

Fig. 4.53 Choroidal thrombosis and vasculitis. Horse. Idiopathic purpura hemorrhagica. Note infarction of adjacent retina.

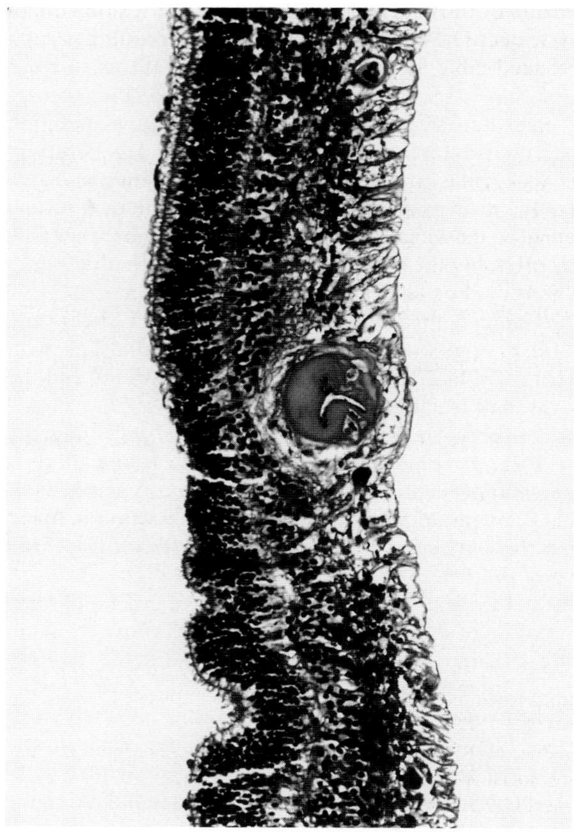

Fig. 4.54 Hypertensive retinopathy, dog. Hyalinized, thick-walled retinal arteriole and focal retinal degeneration.

atrophic hole. Such holes are foci of extreme retinal thinning, so that only the external limiting membrane separates vitreous from pigment epithelium. The pigment epithelium and choroid are usually normal, but they too may show atrophy and fibrosis. Intermingling of cysts and holes in peripheral retina is common. Complete breaks are uncommon and do not usually lead to retinal separation as occurs in humans.

Multifocal coalescing peripheral retinal atrophy is very frequent in very old dogs and horses, and is of no apparent visual importance.

Hypertensive retinopathy is in most cases associated with chronic renal failure. At least 60% of dogs with chronic renal failure are hypertensive. Dogs and cats are most frequently affected.

The macroscopic ocular lesions include retinal or preretinal hemorrhage, retinal edema, and retinal detachment. The histologic lesions are primarily in retinal and choroidal vessels, which have lesions varying from fibrinoid necrosis of tunica media to medial hypertrophy with adventitial fibrosis. Changes that are probably secondary to vessel damage include localized retinal necrosis, exudative retinal separation with resultant atrophy of photoreceptors and hypertrophy of retinal pigmented epithelium, and intraretinal hemosiderin deposition (Fig. 4.54). Vascular lesions and associated necrosis may also occur in anterior uvea. Eyes that are eventually enucleated or obtained at necropsy may have a variety of other lesions that probably occur secondary to chronic retinal detachment and chronic intraocular hemorrhage. Most notable among these is preiridal fibrovascular membrane and its resultant hyphema or neovascular glaucoma.

The early lesions, likely to be seen only under experimental conditions, are the result of exaggerated autoregulatory vasoconstriction in response to the systemic hypertension. Sustained vasoconstriction leads to ischemic necrosis of the deprived retina or choroid, as well as necrosis of vascular endothelium distal to the constricted precapillary sphincters. The histologic consequences are focal retinal necrosis, and leakage of plasma or even erythrocytes through damaged endothelium. This leakage causes intramural fibrinoid change in the vessels and edema or hemorrhage in adjacent retina.

Many of the neuronal storage diseases cause retinal lesions identical to those in the brain. The list of those with described ocular lesions probably reflects those in which the eyes have been examined rather than a true reflection of those diseases in which ocular lesions do, or do not, occur. Those interested should consult a useful, referenced table in the text by Slatter.

Bibliography

Aguirre, G. D. Retinal degeneration associated with the feeding of dog food to cats. *J Am Vet Med Assoc* **172:** 791–796, 1978.

Andersen, A. C., and Hart, G. H. Histological changes in the retina of the vitamin A-deficient horse. *Am J Vet Res* **4:** 307–317, 1943.

Bellhorn, R. W., Aguirre, G. D., and Bellhorn, M. B. Feline central retinal degeneration. *Invest Ophthalmol* **13:** 608–616, 1974.

Bradley, R., Terlecki, S., and Clegg, F. G. The pathology of a retinal degeneration in Friesian cows. *J Comp Pathol* **92:** 69–83, 1982.

Dorling, P. R., Huxtable, C. R., and Vogel, P. Lysosomal storage in *Swainsona* spp. toxicosis: An induced mannosidosis. *Neuropathol Appl Neurobiol* **4:** 285–295, 1978.

Gwin, R. M. *et al.* Hypertensive retinopathy associated with hypothyroidism, hypercholesterolemia, and renal failure in a dog. *J Am Anim Hosp Assoc* **14:** 200–209, 1978.

Ham, W. T. *et al.* The nature of retinal radiation damage: Dependence on wavelength, power level, and exposure time. *Vision Res* **20:** 1105–1111, 1980.

Hayes, K. C., Carey, R. E., and Schmidt, S. Y. Retinal degeneration associated with taurine deficiency in the cat. *Science* **188:** 949–951, 1975.

Huxtable, C. R., and Dorling, P. R. Swainsonine-induced mannosidosis. *Am J Pathol* **107:** 124–126, 1982.

Kremer, I. *et al.* Oxygen-induced retinopathy in newborn kittens. *Invest Ophthalmol Vis Sci* **28:** 126–130, 1987.

Lanum, J. The damaging effects of light on the retina. Empirical findings, theoretical and practical implications. *Surv Ophthalmol* **22:** 221–249, 1978.

Lerman, S. Light-induced changes in ocular tissues. *In* "Clinical Light Damage to the Eye" D. Miller (ed.), Chapter 10, pp. 183–215. New York, Springer-Verlag, 1987.

Michels, M., and Sternberg, P., Jr. Operating microscope-induced retinal phototoxicity: Pathophysiology, clinical manifestations, and prevention. *Surv Ophthalmol* **34:** 237–252, 1990.

Noell, W. K. Possible mechanisms of photoreceptor damage by light in mammalian eyes. *Vision Res* **20:** 1163–1171, 1980.

Riis, R. C. *et al.* Vitamin E deficiency retinopathy in dogs. *Am J Vet Res* **42:** 74–86, 1981.

Schaller, J. P. *et al.* Induction of retinal degeneration in cats by methylnitrosourea and ketamine hydrochloride. *Vet Pathol* **18:** 239–247, 1981.

Schmidt, S. Y., Berson, E. L., and Hayes, K. C. Retinal degeneration in cats fed casein. I. Taurine deficiency. *Invest Ophthalmol* **15:** 47–52, 1976.

Slatter, D. "Fundamentals of Veterinary Ophthalmology," 2nd Ed., Philadelphia, Pennsylvania, W. B. Saunders, 1990.

Toole, D. O., Miller, G. K., and Hazel, S. Bilateral retinal microangiopathy in a dog with diabetes mellitus and hypoadrenocorticism. *Vet Pathol* **21:** 120–121, 1984.

Tulsiani, D. R. P., Harris, T. M., and Touster, O. Swainsonine inhibits the biosynthesis of complex glycoproteins by inhibition of Golgi mannosidase II. *J Biol Chem* **257:** 7936–7939, 1982.

Van Donkersgoed, J., and Clark, E. G. Blindness caused by hypovitaminosis A in feedlot cattle. *Can Vet J* **29:** 925–927, 1988.

Van Kampen, K. R., and James, L. R. Ophthalmic lesions in locoweed poisoning of cattle, sheep, and horses. *Am J Vet Res* **32:** 1293–1295, 1971.

Zuclich, J. A. Ultraviolet induced damage in the primate cornea and retina. *Curr Eye Res* **3:** 27–34, 1984.

C. Retinitis

Retinitis as the sole ocular lesion is rare but may occur in animals with neurotropic virus infections, with toxoplasmosis, and with thrombotic meningoencephalitis of cattle (Fig. 4.55). In the latter disease, however, it is more usual to find the typical thrombotic, inflammatory lesions in choroid as well as retina. Their character is identical to the lesions in the brain. The multifocal chorioretinal scars expected as sequelae are seldom seen, perhaps because cattle with neurologic and ocular lesions almost inevitably die. The prevalence of the ocular lesion, useful as an aid in the clinical diagnosis, is estimated at 30 to 50% in animals with the septicemic form of the disease, and as high as 65% in experimentally infected calves.

Multifocal viral retinitis with the same histologic features as the respective brain lesions occurs in animals with scrapie, hog cholera, rabies, Teschen disease, Borna disease, pseudorabies in pigs, and canine distemper. Undoubtedly the list is incomplete. The ocular lesions associated with canine distemper will be described here because

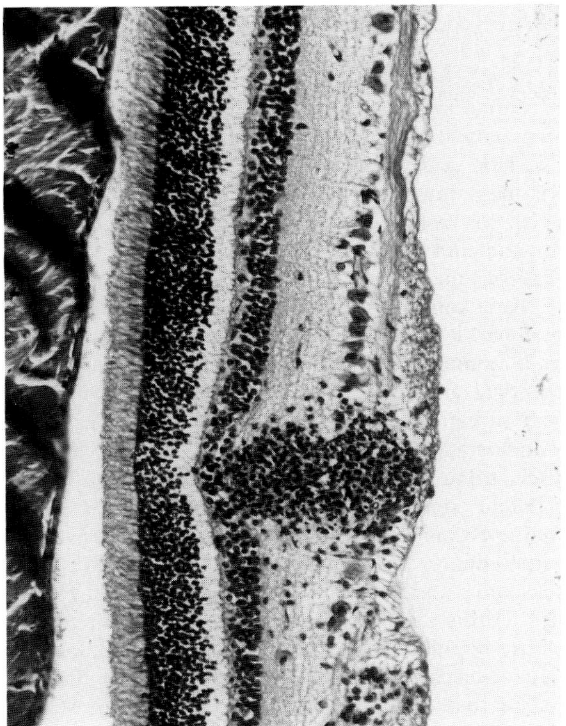

Fig. 4.55 Focal destructive retinitis. *Haemophilus somnus.* Steer.

it is a classic example of inflammatory and postinflammatory retinopathy of viral etiology; canine distemper is discussed with the Respiratory System (Volume 2, Chapter 6).

1. Canine Distemper

Retinal and optic nerve lesions occur in most dogs with naturally occurring distemper. The lesions most often are degenerative rather than inflammatory, although some of the degenerative changes may have been sites of inflammation earlier in the disease course.

Acute lymphocytic–plasmacytic chorioretinitis and optic neuritis are found in about 25% of dogs submitted for laboratory confirmation of the disease. Random perivascular cuffing, edema, focal exudative retinal separation, and hypertrophy of retinal pigment epithelium are present. Eosinophilic intranuclear inclusion bodies occur in ganglion cells or astrocytes in 30 to 40% of the cases, which is the only etiologically specific change in what is an otherwise nonspecific picture shared by many systemic infections.

The more prevalent lesions are multiple random foci of retinal degeneration and scarring. These usually affect the full thickness of retina and are most likely sequelae to the previous undetected retinitis. Such foci often contain numerous melanin-laden cells, probably derived from migration of adjacent, injured retinal pigment epithelium. Occasionally only the outer nuclear layer and photoreceptors are missing, probably a sequel to focal exudative retinal detachment.

Optic nerve lesions of one type or another are present in all dogs with ocular lesions. Nonsuppurative neuritis, astrocytic scarring, and demyelination similar to that in brain are the three most frequent changes. In those dogs suffering only the demyelinating disease, the ocular lesions may be inapparent, or there may be demyelination of optic nerve and ganglion cell degeneration.

2. Tick-Transmitted Infections

Infection with the tick-transmitted rickettsial agents of Rocky Mountain spotted fever (*Rickettsia rickettsii*) or canine ehrlichiosis (*Ehrlichia canis*) cause ocular lesions in dogs. The clinical and histologic ocular lesions are virtually identical and occur in a high percentage (80% for Rocky Mountain spotted fever) of dogs with active infection. Most of the lesions result from injury to vascular endothelium parasitized by the rickettsiae, and multifocal hemorrhage, edema, and vascular necrosis occur in all parts of the eye. Multifocal retinal hemorrhage, perivascular retinal edema, and necrosis of endothelium in retinal venules and arterioles are the characteristic retinal changes. Although often listed along with other agents as a cause of anterior uveitis or endophthalmitis, most naturally occurring infections have clinical signs attributable only to the vascular injury rather than a genuine uveal inflammation. There is one report of unusually severe uveitis occurring 14–28 days after experimental infection with *Rickettsia rickettsii*, following the disappearance of all other signs of the acute systemic disease. Dogs thus affected had a neutrophilic and lymphocytic destructive vasculitis, assumed to represent a type III immune reaction to parasitized endothelium.

Bibliography

Barnett, K. C., and Palmer, A. C. Retinopathy in sheep affected with natural scrapie. *Res Vet Sci* **12:** 383–385, 1971.

Davidson, M. G. *et al.* Ocular manifestations of Rocky Mountain spotted fever in dogs. *J Am Vet Med Assoc* **194:** 777–781, 1989.

Dukes, T. W. The ocular lesions in thromboembolic meningoencephalitis (ITEME) of cattle. *Can Vet J* **12:** 180–182, 1971.

Jubb, K. V., Saunders, L. Z., and Coates, H. V. The intraocular lesions of canine distemper. *J Comp Pathol* **67:** 21–29, 1957.

VIII. Optic Nerve

The optic nerve is a white fiber tract of brain formed by the outgrowth of ganglion cell axons from the eye through sievelike perforations in posterior polar sclera, called the **lamina cribrosa.** The axons travel within a preformed neurectodermal tube formed by the primary optic stalk to reach the optic chiasm and then the lateral geniculate body. The neurectoderm lining the optic stalk induces the surrounding mesenchyme to form the three meningeal layers, similar to and continuous with those of brain itself. Later differentiation of neurectoderm produces the astrocytes and oligodendroglia that, together with the ganglion cell, axons, and fibrovascular septa from pia mater, form the substance of the optic nerve. The optic disk is the intraocular portion of the nerve and is the only portion available to ophthalmoscopic examination. It is formed by the convergence of ganglion cell axons prior to their exit via the lamina cribrosa. The axons of the nerve fiber layer are unmyelinated, and at what point (relative to lamina cribrosa) the axons become myelinated determines the ophthalmoscopic appearance of the optic disk. Histologically, the disk is unmyelinated in most domestic species except the dog, contains abundant glia, and may have a small paracentral excavation—the physiologic cup—from which Bergmeister's papilla originates. A few pigmented cells are commonly seen, as are small neuroblastic clusters, both probably minor anomalies of retinal differentiation but of no significance.

There is considerable variation in the normal histology of the optic nerve among animals of different species and ages. Optic disk myelination has already been mentioned. The lamina cribrosa is formed by heavy fibrous trabeculae in horses, dogs, and cattle and is therefore more obvious than in cats and laboratory animals. Fibrous septa within the nerve are prominent in cattle and horses, and their similarity to the axons in hematoxylin and eosin sections may mask a pathologic paucity of nerve fibers. The fibrous tissue reportedly increases with age.

The general pathology of optic nerve shares features of both retinal and neural disease. Because it is in direct continuity with both structures via its axons, and with brain via the perineural cerebrospinal fluid, it is common that optic nerve be affected by diseases of either retina or brain. Thus optic neuritis is expected in at least a proportion of animals suffering with inflammation of retina or neural white matter, and optic nerve atrophy inevitably follows loss of ganglion cells. Fortuitous hematogenous localization of infectious agents or tumor cells may occur in optic nerve as anywhere else.

Papilledema is hydropic swelling of the optic disk. It may result from extraocular events that cause an increase in cerebrospinal fluid pressure within the optic nerve or from local vascular leakage. The former is usually associated with retrobulbar tissue masses, but is also seen with intracranial neoplasms and with hypovitaminosis A. Ocular hypotony may cause optic disk edema as a result of decreased tissue hydrostatic pressure. Serous inflammation within the nerve also results in papilledema. Papilledema is a common clinical diagnosis that rarely is available for histologic examination.

Optic neuritis is a term sometimes used rather broadly to describe both inflammatory and degenerative diseases of the nerve. Optic neuritis is seen clinically as swelling, hyperemia, and focal hemorrhage within the optic disk. Affected animals, usually dogs or horses, are blind when the lesion is bilateral. Although described as a clinical entity not associated with other ocular lesions, histopathologic confirmation is lacking. Optic neuritis may, of course, accompany any case of retinitis or endophthalmitis.

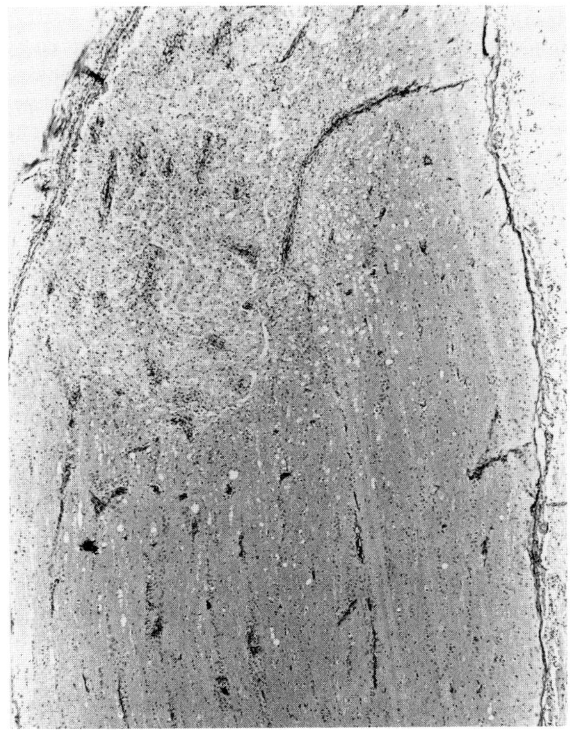

Fig. 4.56 Axonal degeneration in the retrobulbar, intraorbital optic nerve. Dog, 11 days after being struck by a car.

The pattern of inflammation within the nerve may provide clues as to the pathogenesis of the neuritis. Perineuritis, or optic nerve leptomeningitis, is typical of meningeal spread of bacterial meningitis from the brain. Toxoplasmosis and cryptococcosis frequently cause multifocal and nonselective lesions within the extraocular nerve, as does canine distemper. Optic neuritis originating as endophthalmitis is usually restricted to the optic disk. Feline infectious peritonitis is frequently associated with perineuritis and optic neuritis in which the mononuclear aggregates are around blood vessels in the meninges and in the extensions of the meninges into the nerve.

Chronic optic neuritis, like its counterpart in the brain, is characterized by focal gliosis, astrocytic scarring, and secondary axonal degeneration. The loss of axons may be partially masked by the increased prominence of glia and pial septa.

Degeneration of the optic nerve is part of optic neuritis, glaucoma, and chronic, severe retinal atrophy of any cause. Initiation of gliosis and fibrosis may eventually make the chronic degenerative lesion indistinguishable from that of chronic inflammation. The most frequently diagnosed example is that following trauma to one or both nerves in dogs or cats struck by cars (Fig. 4.56). The gross lesion may be avulsion or contusion. Injury to the nerve may be instantaneous, as caused by tearing or complete severance, or may result from vascular injury with slightly delayed ischemic necrosis. In severed nerves, there is disintegration of the distal axons back to the lateral geniculate body. The proximal portion of each affected axon dies back to the ganglion cell, which eventually also dies. The inner nuclear layer remains unaffected, a useful criterion to distinguish traumatic, "die back," ganglion cell atrophy from that of glaucoma.

Degeneration of optic nerve also occurs in calves deficient in vitamin A, and in ruminants ingesting male fern or hexachlorophene. Ingestion of male fern, *Dryopteris*, on pasture or as a taenicidal extract causes papilledema and subsequent optic nerve demyelination when ingested in large amounts. Retina may be unaffected early, but ganglion cell atrophy occurs eventually. Hexachlorophene administered to calves or sheep as an anthelmintic causes edema and then atrophy and gliosis of optic nerve.

Proliferative optic neuropathy is an unusual lesion of horses. Anecdotal descriptions are numerous, but histologic descriptions are few. The lesion is a raised, gray mass on the surface of the optic disk, unassociated with visual deficit. The mass is composed of spherical mononuclear cells with hyperchromatic, eccentric nuclei and foamy eosinophilic cytoplasm (Fig. 4.57). Some of these cells are also found within extraocular optic nerve. The cytoplasmic content may be stored lipid, but its origin is not known. The described lesion bears much resemblance to the proliferation of myelin-laden macrophages that occurs in and on optic nerves injured by trauma or ischemia. Also, the distinction between the proliferative optic neuropathy and gliomas or granular cell tumors described in various reports is unclear.

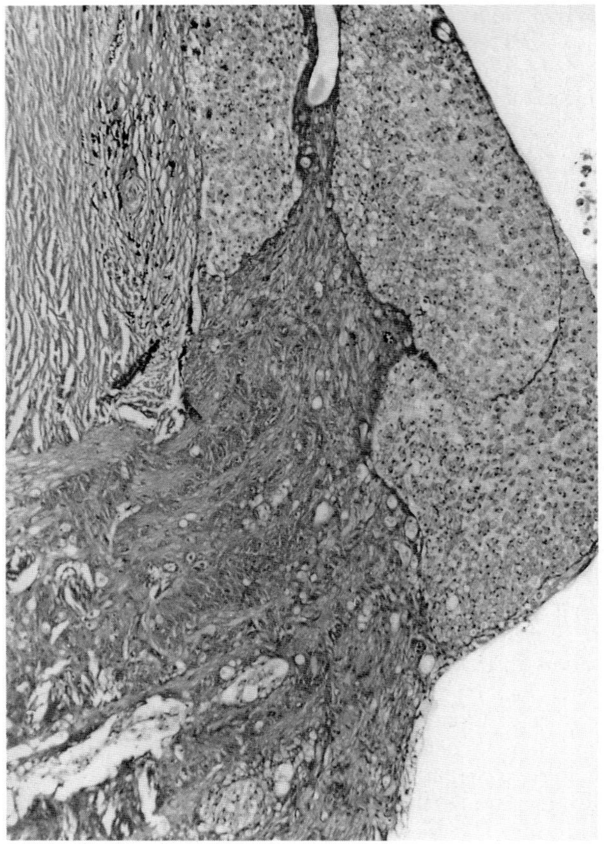

Fig. 4.57 Equine proliferative optic neuropathy. The identity of the foamy cells is much debated.

Bibliography

Bistner, S. *et al.* Neuroepithelial tumor of the optic nerve in a horse. *Cornell Vet* **73:** 30–40, 1983.

Gelatt, K. N. *et al.* Optic disc astrocytoma in a horse. *Can Vet J* **12:** 53–55, 1971.

Nafe, L. A., and Carter, J. D. Canine optic neuritis. *Compend Cont Ed* **3:** 978–984, 1981.

Saunders, L. Z., and Rubin. L.F. "Ophthalmic Pathology of Animals." Basel, Switzerland, Karger, 1975.

Saunders, L. Z., Bistner, S. I., and Rubin, L. F. Proliferative optic neuropathy in horses. *Vet Pathol* **9:** 368–378, 1972.

Vestre, W. A., Turner, T. A., and Carlton, W. W. Proliferative optic neuropathy in a horse. *J Am Vet Med Assoc* **181:** 490–491, 1982.

IX. The Sclera

The limbus marks the transition from the avascular, nonpigmented, and very orderly cornea to the vascularized, pigmented, and interwoven fibrous tissue that identifies sclera. The sclera forms the posterior two thirds of the fibrous tunic of the eye, blending with choroid on its inner aspect and orbital fascia exteriorly. Its thickness increases with age and varies considerably among domestic species. In cattle and horses, it is thickest at the posterior pole (2.2 mm in cattle and 1.3 mm in horses) and thinnest at the orbital equator (1.0 mm in cattle, about 0.5 mm in horses). In dogs and cats, it is much thinner, about 0.3 mm at the posterior pole and 0.1 mm at the equator, varying somewhat with age and globe size. In carnivores, however, there is a circumferential ring of thickened (1 mm) sclera at the limbus, in which is buried the venous plexus receiving aqueous drainage. The sclera is perforated by numerous vessels and nerves, the most notable of which are the optic nerve and limbic scleral venous plexus.

The optic nerve fibers exit the globe through extensive scleral fenestrations called the lamina cribrosa. Diseases of the sclera are few in comparison to diseases of other ocular structures. Most are inflammatory and arise by extension from within the globe or from orbital cellulitis. The efficiency with which the sclera resists inflammatory spread is evidenced by the infrequency of panophthalmitis as opposed to endophthalmitis, and the even greater infrequency of intraocular involvement resulting from orbital inflammation. When the sclera is involved in inflammatory disease originating within the eye, its initial involvement is seen histologically as leukocytes in perivascular adventitia, which is in direct communication with the choroid. A similar phenomenon is seen in scleral extension of choroidal neoplasms, in which collars of tumor cells surround scleral vessels but show little inclination to infiltrate directly into scleral connective tissue.

Nodular fasciitis (nodular granulomatous episcleritis) is the most prevalent primarily scleral disease of dogs. It occurs rarely in cats. It is the proliferative, nodular lesion of the limbus that has been variously termed nodular fasciitis, nodular scleritis or episcleritis, fibrous histiocytoma, proliferative keratoconjunctivitis, conjunctival granuloma, and collie granuloma. The various names reflect the spectrum of clinical presentations of this lesion. The variations are treated as a single entity, nodular fasciitis, in this discussion.

The usual macroscopic lesion is a firm, painless, moveable, nodular swelling, 0.5–1.0 cm in diameter, below the bulbar conjunctiva at, or just posterior to, the limbus. Infiltrative extension of the mass into the peripheral corneal stroma is accompanied by edema and vascularization (Fig. 4.58A). Although the temporal limbus unilaterally is the most frequent site for initial occurrence, other common locations include third eyelid and elsewhere along the limbus. Third eyelid involvement is often bilateral and occurs almost exclusively in rough collies. Limbic and nictitans involvement may occur in the same dog and even in the same eye.

Ocular nodular fasciitis behaves as would a locally infiltrative neoplasm. Extension is usually into peripheral corneal stroma and posteriorly into sclera, episclera, and Tenon's capsule. The tissue of origin of this lesion is unresolved. Fibrous tissue of sclera, episclera, and Tenon's capsule have all been suggested. Lesions involving the third eyelid, or the rare case of palpebral subconjunctival origin, probably originate from the fascia native to

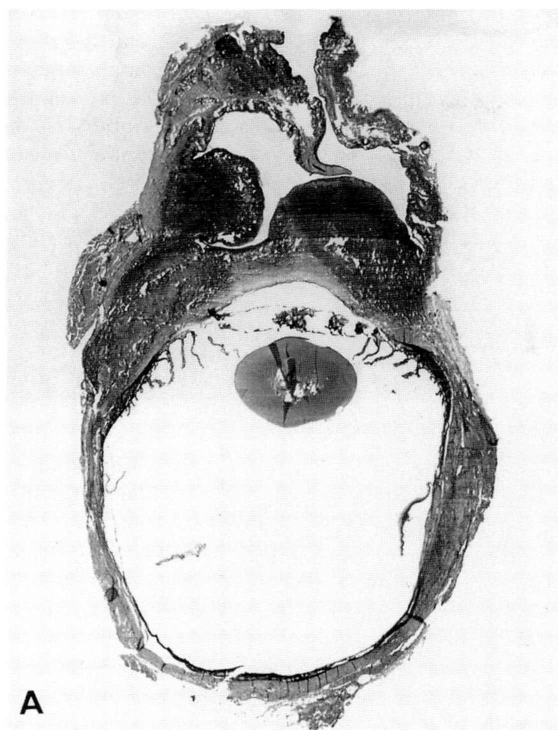

Fig. 4.58A Nodular fasciitis. Dog. Infiltrative scleral mass of 2 years' duration.

those structures. Histologic examination distinguishes this lesion from extension of intraocular tumors or the rare scleral sarcomas. Occasionally, infiltration of therapeutically refractory fasciitis is so extensive, both into cornea and sclera, as to require enucleation.

Histologically, the lesion is a proliferative, nonencapsulated mixture of neocapillaries, spindle cells, and mononuclear leukocytes (Fig. 4.58B). The spindle cells may be fibroblasts, histiocytes, or a mixture of both. The spindle cells are haphazardly arranged and, despite a fibrous appearance to the section, surprisingly little collagen is demonstrated by special stains except in coarse septa that may dissect the mass into irregular lobules. Reticulin, however, is abundant. The mononuclear leukocytes are found loosely throughout the mass but are usually most numerous near its periphery. When present in cornea, the above cell mixture affects stroma but spares the epithelium and an adjacent zone of subepithelial stroma.

Necrotizing scleritis is a rare lesion seen in dogs as a poorly delineated inflammatory and proliferative lesion of anterior sclera. The disease incites much more inflammatory reaction, as measured by clinical criteria, than does nodular fasciitis. The lesion consists of coalescing scleral granulomas centered around remnants of denatured, refractile collagen. Eosinophils sometimes are seen in the centers as well. Some cases have diffuse granulomatous inflammation rather than discrete granulomas. The lesion tends to slowly spread circumferentially and posteriorly

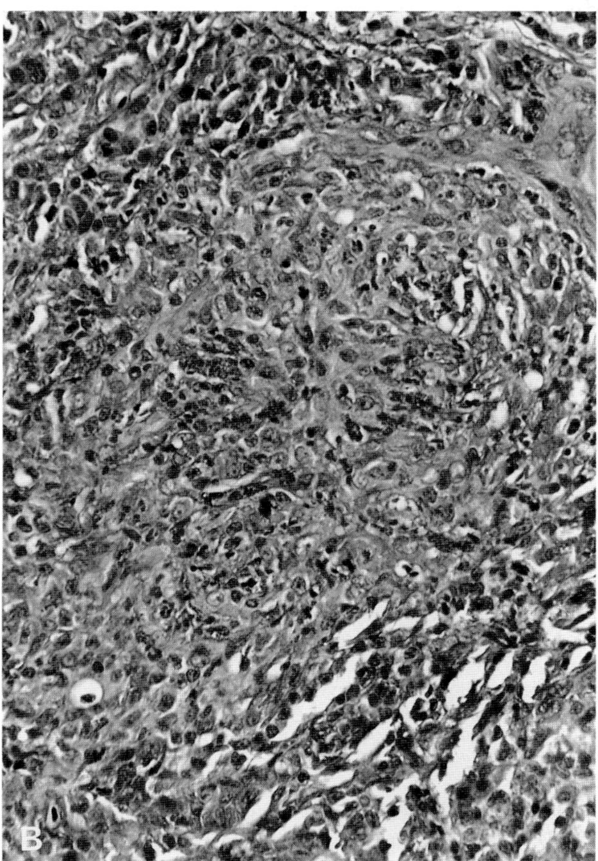

Fig. 4.58B Higher magnification, nodular fasciitis.

to involve the entire sclera, and involvement of uvea and even retina with granulomas eventually occurs. Bilateral involvement is usual, but not necessarily at the same time. Response to antiinflammatory therapy is poor. No etiologic agent has been seen.

X. The Orbit

Diseases of the orbit are few and relatively uncommon in domestic animals except for those resulting from trauma. Systemic diseases of bone, muscle, blood vessels, and nerves may incidentally affect orbital components. Orbital fat fluctuates with nutritional status, contributing to the enophthalmos of malnourished animals. Ordinarily, however, orbital disease arises by extension of inflammatory lesions from the mouth, paranasal sinuses, or from penetrating wounds through periorbital soft tissue. Extension from intraocular inflammation is surprisingly rare, a tribute to the barrier offered by the sclera. Conversely, orbital disease rarely invades the globe. Metastatic orbital neoplasia is rare except for lymphoma of cattle and of cats. While theoretically the orbit may suffer from primary neoplasia of any of the bony or soft tissues within it, such occurrences are rare. Of these, ill-defined spindle-cell sarcomas and lacrimal gland tumors in dogs are the most common (see Ocular Neoplasia).

Orbital cellulitis is the commonly used term to describe pyogenic orbital inflammation. The cause is usually bacterial, and the pathogenesis involves extension from nearby inflammation of paranasal sinuses, molar tooth socket, or periorbital soft tissue. Only rarely does uncontrolled endophthalmitis spread through the sclera into the orbit. Bacteremic localization within the orbit, while presumably occurring as do such localizations elsewhere, is seldom detected except perhaps for *Streptococcus equi* infection in young horses.

Orbital inflammation most frequently results from penetrating foreign bodies, whether by direct penetration or particle migration from conjunctival sac or pharynx. Horses seem particularly prone. Aberrant localization by nematode parasites (*Dirofilaria immitis, Ancylostoma caninum*) or *Diptera* larvae is reported.

Bibliography

Gwin, R. M., Gelatt, K. N., and Peiffer, R. L. Ophthalmic nodular fasciitis in the dog. *J Am Vet Med Assoc* **170:** 611–614, 1977.

Smith, J. S., Bistner, S., and Riis, R. Infiltrative corneal lesions resembling fibrous histiocytoma: Clinical and pathologic findings in six dogs and one cat. *J Am Vet Med Assoc* **169:** 722–726, 1976.

XI. Ocular Neoplasia

Although the eye is the site of a wide range of primary and metastatic neoplasms, only a few are of sufficient prevalence or importance to justify discussion here. Metastatic ocular neoplasia is reported rather infrequently but it is common when sought. Multicentric lymphoma in cats, dogs, and cattle regularly involves the eye, although in cattle the retrobulbar tissue is preferred over the eye itself. Carcinomas are reported more frequently than sarcomas, and this probably reflects the greater prevalence and metastatic potential of carcinomas. Uveal vessels are the usual sites of lodgement, and ocular disease may result from vessel occlusion, or from inflammation in response to tumor antigen, or to necrosis of either tumor or damaged host tissue. Hyphema is more common in eyes with tumor-induced uveitis than with uveitis of other causes and is therefore a diagnostically useful sign.

Primary ocular tumors may arise from the eyelids and adnexa, from optic nerve, or from the globe. Those of the globe may originate from any of the tissues, but only those from uveal melanoblasts and neurectoderm are anything other than rare. Most primary intraocular tumors are benign in terms of histologic appearance or potential for metastasis. Dogs and cats are frequently affected, but primary intraocular neoplasms are inexplicably rare in other domestic species.

The most important primary ocular neoplasms are squamous cell carcinoma, meibomian adenoma, melanoma, and ciliary adenoma. Other tumors are uncommon.

Bibliography

Blodi, F. C., and Ramsey, F. K. Ocular tumors in domestic animals. *Am J Ophthalmol* **64:** 627–633, 1967.

Dubielzig, R. R. Ocular neoplasia in small animals. *Vet Clin North Am: Small Anim Pract* **20:** 837–848, 1990.

Gwin, R. M., Gelatt, K. N., and Williams, L. W. Ophthalmic neoplasms in the dog. *J Am Anim Hosp Assoc* **18:** 853–866, 1982.

Lavach, J. D., and Severin, G. A. Neoplasia of the equine eye, adnexa and orbit: A review of 68 cases. *J Am Vet Med Assoc* **170:** 202–203, 1977.

Williams, L. W., Gelatt, K. N., and Gwin, R. M. Ophthalmic neoplasms in the cat. *J Am Anim Hosp Assoc* **17:** 999–1008, 1981.

A. *Squamous Cell Carcinoma*

Squamous cell carcinoma arises from the conjunctival epithelium of the limbus, third eyelid, or eyelid in cattle, horses, cats, and dogs, in that order of frequency. Bovine ocular squamous cell carcinoma is the most common and most economically significant neoplasm of domestic animals. Its relative rarity in dogs is peculiar and unexplained.

Bovine ocular squamous cell carcinoma, which is largely restricted to the Hereford breed, also occurs in other breeds of cattle, as well as Indian water buffalo, sheep, and cattalo (Fig. 4.59). It has a prevalence shown to be related directly to exposure to ultraviolet radiation, and less directly to lack of pigment in lids and conjunctiva. Nevertheless, variation in prevalence in different lines of Herefords in the same district has led to speculation that other genetic factors within the breed may influence susceptibility. The question has been further widened by demonstration of papillomaviruses in some of the papillomatous precursor lesions which eventually transform into squamous cell carcinoma. Similar papillomaviruses, as well as being the causative agents of cutaneous warts, have been demonstrated in bovine alimentary papillomas in Scotland, and viral DNA persists in the squamous cell

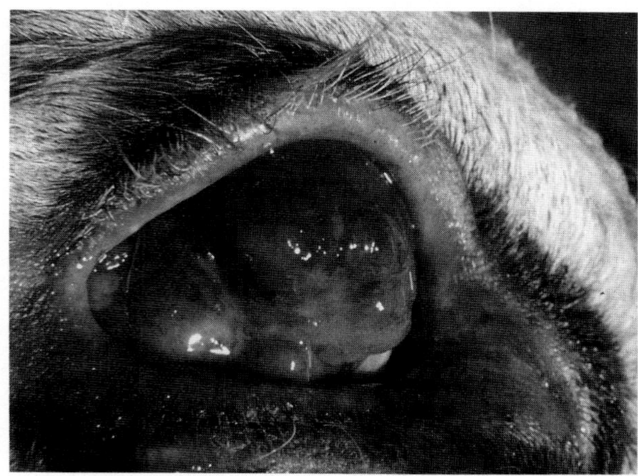

Fig. 4.59 Squamous cell carcinoma. Ox.

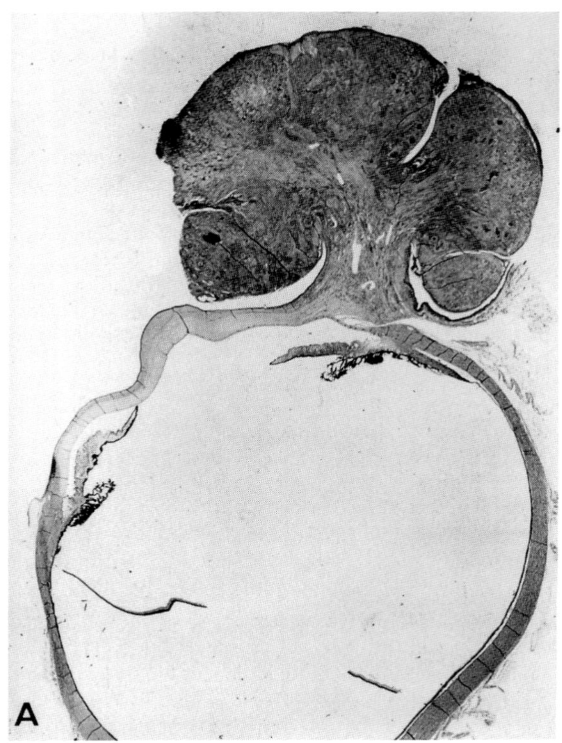

Fig. 4.60A Squamous papilloma of corneoscleral junction. Ox.

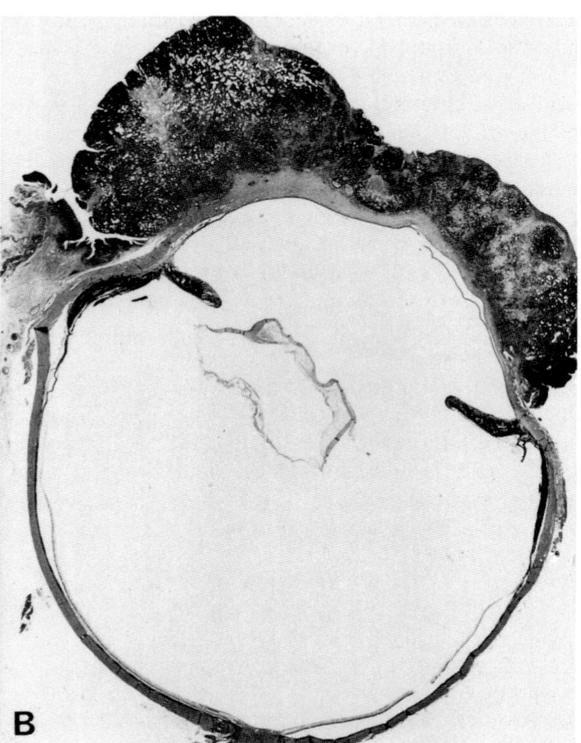

Fig. 4.60B Squamous cell carcinoma arising from bulbar conjunctiva.

carcinomas which arise from these papillomas in cattle grazing pastures that contain bracken fern. It remains to be determined whether or not there is any relationship between a viral component of the ocular carcinoma and the fact that in many cases the tumor regresses after immunotherapy. At least at the moment, no viral particle or viral genome has been consistently demonstrated in ocular squamous cell carcinomas in any species. Environmental co-carcinogens such as those in bracken have not yet been implicated in the induction of ocular tumors.

The tumor in all species develops through a series of premalignant stages, called epidermal plaques and papillomas, before proceeding over months or years to carcinoma *in situ* and to invasive carcinoma (Figs. 4.59, 4.60A,B). Spontaneous regression of the precancerous lesions may occur with an estimated frequency of 25 to 50%. At least in cattle, plaques are much more common (about 6 : 1) than papillomas or outright carcinomas. The epidermal plaque is characterized by marked acanthosis, with variable presence of keratinization, dyskeratosis, and epidermal downgrowth into the subconjuctival connective tissue. Invasion through basal layer or basement membrane is not seen. Papilloma also involves acanthosis but, in addition, there is marked para- and hyperkeratosis with papillary projections supported by a vascularized connective tissue core. Papillomas may be up to 3.0 cm in diameter, pedunculated or sessile, and are often ulcerated. Carcinoma *in situ* arises by focal or multifocal transformation of increasingly dysplastic cell nests in the deep layers of plaques or papillomas. Fully developed carcinoma has squamous cell invasion across the basement membrane. Tumor invasion is almost always accompanied by an intense lymphocytic–plasmacytic infiltration, presumably the host response to tumor antigen. It is assumed that this response is responsible for regression of some of the precursor lesions, although spontaneous regression of fully developed carcinoma is rare. Stimulation of immune-mediated rejection by intra lesional inoculation of antigenic tumor extracts or nonspecific lymphocyte stimulants induces partial or total regression of small tumors.

Histologically, ocular squamous cell carcinoma resembles similar tumors in other sites and ranges from well-differentiated carcinomas with keratin pearl formation to anaplastic carcinomas with marked nuclear size variation and mononuclear tumor giant cells. Metastatic or invasive potential has not been correlated with histologic criteria, but there is a correlation between site of origin and subsequent behavior. Most surveys in cattle identify the bulbar conjunctiva of the limbus as the most frequent site of origin, estimated at about 70% of all occurrences. Some surveys consider nictitating membrane as next most frequent, with palpebral conjunctiva of the true eyelid as third. Other reports claim nictitans origin to be uncommon and eyelid tumors to be as common as those of limbic origin. Tumors probably do not arise from cornea unless

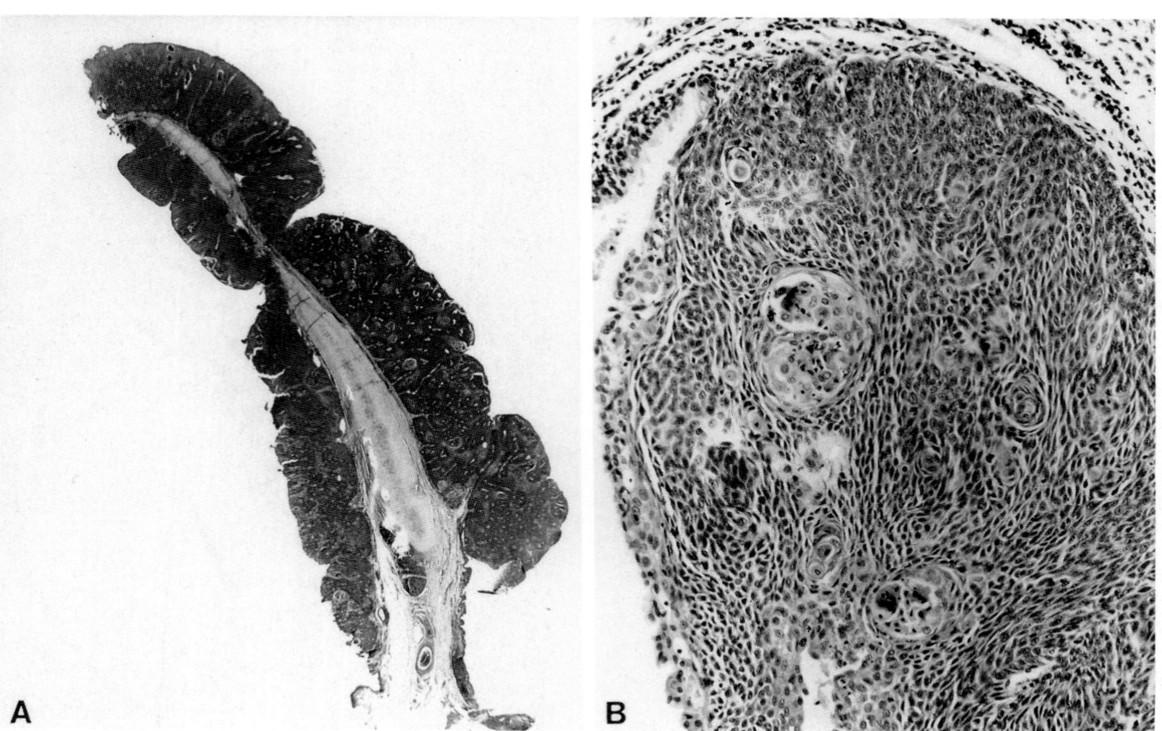

Fig. 4.61 (A) Early squamous cell carcinoma involving conjunctiva of nictitating membrane. Horse. (B) Histology of (A). Disorderly and defective maturation, premature keratinization, and invasion across the basement membrane.

it has previously been vascularized. Tumors arising at the limbus are confronted by the dense and poorly vascularized connective tissue of sclera and peripheral cornea, which retards metastasis to extraocular sites. Invasion of corneal stroma, sclera, and anterior chamber occurs slowly. Tumors arising from the nictitans extend to the root of the membrane and then to the cartilage and bone of the orbit and internal nares. Metastasis probably will eventually occur in all instances, with parotid lymph node the initial site. Wide dissemination to thoracic and abdominal organs has been reported and is probably limited only by the limited longevity of the animal.

Squamous cell carcinoma of the **equine** eye is much less thoroughly documented but is quite common. In contrast to cattle, the preferred site is the edge of the third eyelid, followed by limbic bulbar conjunctiva (Fig. 4.61). Bilateral involvement is seen in 15 to 20%. All breeds may be affected, and the mean age of affected horses is about 9 years. The same range of precancerous lesions occurs in horses as in cattle. Prognosis is strongly influenced by therapy, but the untreated neoplasm is slow to metastasize, and even then, it is usually only to local lymph nodes. Retrospective studies document 10–15% of equine ocular squamous cell carcinomas to have regional or distant spread, but the data do not consider duration of the disease prior to therapy (Fig. 4.62).

In **cats,** ocular squamous cell carcinoma most frequently affects the skin or palpebral conjunctiva of the eyelids. White cats are particularly susceptible, and squamous cell carcinomas in these animals may occur simultaneously or sequentially on eyelids, ear pinnae, nose, and lips. The early lesion is one of sunlight-induced epithelial necrosis, and even the early neoplasm may be ulcerated and inflamed to a degree that may mask its neoplastic character and delay appropriate therapy. Growth tends to be circumferential around the lid margins, resulting in a palpebral fissure bordered by a thickened, red, and ulcerated tumor.

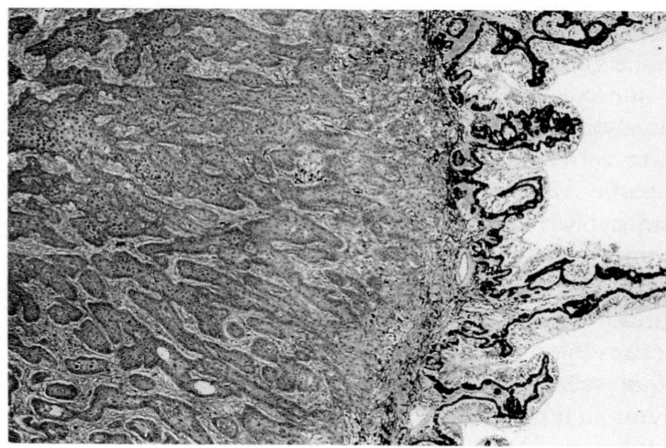

Fig. 4.62 Scleral squamous cell carcinoma has grown inward to approach ciliary processes, an unusual behavior for these normally exophytic tumors. Horse.

Metastasis to local lymph nodes occurs late in the course of the disease.

Squamous cell carcinoma in **dogs** infrequently involves the eye. Proliferative eyelid or conjunctival growths in dogs are much more likely to be meibomian adenomas, viral papillomas, or nodular fasciitis. In one study of 202 canine eyelid neoplasms, squamous cell carcinoma accounted for only 2% of lesions. Precancerous changes probably occur but, in contrast to cattle and horses, eyelid papillomas in dogs are usually benign and nonprogressive lesions.

Bibliography

Anderson, D. E., and Badzioch, M. Association between solar radiation and ocular squamous cell carcinoma in cattle. *Am J Vet Res* **52:** 784–787, 1991.

Anson, M. A., Benfield, D. A., and McAdaragh, J. P. Bovine herpes virus-5 (DN-599) antigens in cells derived from bovine ocular squamous cell carcinoma. *Can J Comp Med* **46:** 334–337, 1982.

Atluru, D., Johnson, D. W., and Muscoplat, C. C. Tumor-associated antigens of bovine cancer eye. *Vet Immunol Immunopathol* **3:** 279–286, 1982.

Dugan, S. J. *et al.* Prognostic factors and survival of horses with ocular/adnexal squamous cell carcinoma: 147 cases (1978–1988). *J Am Vet Med Assoc* **198:** 298–303, 1991.

Russell, W. O., Wynne, E. S., and Loquvam, G. S. Studies on bovine ocular squamous cell carcinoma (cancer eye). I. Pathological anatomy and historical review. *Cancer* **9:** 1–52, 1956.

Spradbrow, P. B., and Hoffman, D. Bovine ocular squamous cell carcinoma. *Vet Bull* **50:** 449–459, 1980.

Williams, L. W., and Gelatt, K. N. Ocular squamous cell carcinoma. *In* "Veterinary Ophthalmology" K. N. Gelatt (ed.), pp. 622–632. Philadelphia, Pennsylvania, Lea & Febiger, 1981.

B. Meibomian Adenoma

Meibomian adenoma is the most common ocular neoplasm of dogs, accounting for at least 50% of eyelid tumors. It is comparable in most respects to sebaceous adenomas found elsewhere in the skin, and many authors have abandoned the name, meibomian adenoma, in favor of the latter term. However, these tumors are specifically of meibomian gland and not of other eyelid sebaceous glands and regularly have histologic features that are infrequently seen in their cutaneous counterparts. In the eyelid tumors, the lobules of foamy, eosinophilic sebaceous cells are surrounded by basal (reserve) cells that are regularly a prominent part of the tumor (Fig. 4.63). In many instances, the basal cell component is so prominent as to cause diagnosis to be made of basal cell tumor with sebaceous differentiation or even sebaceous adenocarcinoma. None metastasizes, and even the so-called carcinomas show little inclination for invasive growth. Other distinctive features are the regular occurrence of squamous metaplasia appropriate to sebaceous gland ducts, and a marked lymphocytic–plasmacytic infiltrate within the stroma between tumor lobules and around the tumor margins, and localized granulomatous response (chalazion) to leaking secretion. The basal cells in many of these tumors are heavily pigmented, with mature melanocytes interspersed with the basal cells and melanophages prominent in the stroma.

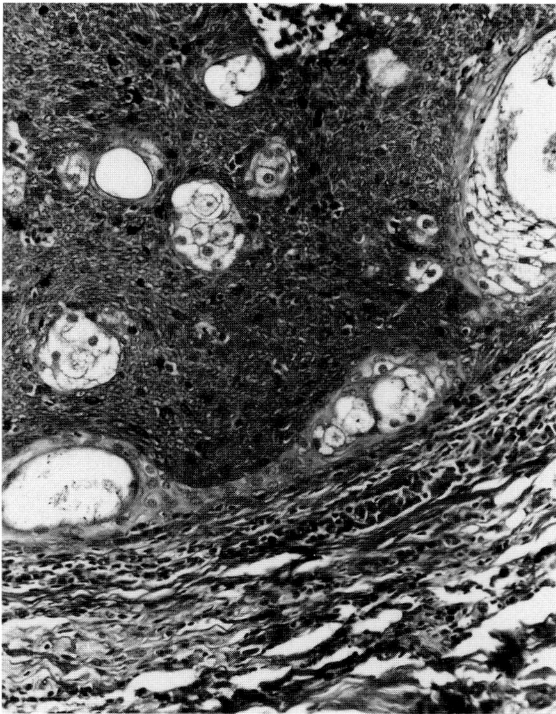

Fig. 4.63 Meibomian adenoma. Dog. Note sebaceous cells, basal (reserve) cells, melanin, abundant mononuclear leukocytic infiltration.

C. Other Adnexal and Conjunctival Tumors

A wide range of neoplasms has been reported to occasionally affect the conjunctiva or adnexa of domestic animals. Other than meibomian adenoma, squamous cell carcinoma, or papilloma (previously described) or melanoma (to be described) in their respective species, all are uncommon to rare.

Hemangiomas and hemangiosarcomas occur anywhere in the conjunctiva, most frequently on the perilimbal bulbar conjunctiva of dogs and horses. They resemble vascular tumors elsewhere, although some of the canine tumors occur in the very superficial plexus adjacent to a hyperplastic conjunctival epithelium, and have been called angiokeratomas. Their behavior is the same as that of hemangioma. Malignant variants seem more prevalent in horses, where more than half of the reported cases are solid hemangiosarcomas with a very aggressive local infiltration and, usually, distant metastasis. There is much speculation, but no proof, that sunlight is important in their causation.

Adenocarcinoma of the gland of the third eyelid occurs as a nodular swelling in very old dogs (mean age 11.5 years). They are locally infiltrative, recur after attempted resection, but are cured by complete removal of the third eyelid. Only chronically neglected cases metastasize to lung after a very protracted local expansion. Histologically these are tubular carcinomas with abundant squamous metaplasia. They should not be confused with the prominence of the gland that occurs with prolapse of the gland (cherry eye) or with lymphocytic interstitial adenitis.

Lymphoma may have several ocular manifestations, the most frequent of which are diffuse uveal metastases as part of generalized lymphoma in dogs or cats, or as retrobulbar tumor in cattle. It may occasionally occur as a conjunctival disease as part of generalized lymphoma or as a mucocutaneous manifestation of epitheliotrophic lymphoma. There are several reports of conjunctival or third eyelid lymphoma occurring in horses as an apparently isolated lesion cured by local excision.

Bibliography

George, C., and Summers, B. A. Angiokeratoma: A benign vascular tumor of the dog. *J Small Anim Pract* **31:** 390–392, 1990.

Glaze, M. B. *et al.* A case of equine adnexal lymphosarcoma. *Eq Vet J* **10:** 83–84, 1990.

Håkanson, N., Shively, J. N., and Merideth, R. E. Granuloma formation following subconjunctival injection of triamcinolone in two dogs. *J Am Anim Hosp Assoc* **27:** 89–92, 1991.

Hargis, A. M., Lee, A. C., and Thomassen, R. W. Tumor and tumorlike lesions of perilimbal conjunctiva in laboratory dogs. *J Am Vet Med Assoc* **173:** 1185–1190, 1978.

Krehbiel, J. D., and Langham, R. F. Eyelid neoplasms of dogs. *Am J Vet Res* **36:** 115–119, 1975.

Moore, P. F., Hacker, D. V., and Buyukmihci, N. C. Ocular angiosarcoma in the horse: Morphological and immunohistochemical studies. *Vet Pathol* **23:** 240–244, 1986.

Roberts, S. M., Severin, G. A., and Lavach, J. D. Prevalence and treatment of palpebral neoplasms in the dog: 200 cases (1975–1983). *J Am Vet Med Assoc* **189:** 1355–1359, 1986.

Wilcock, B. P., and Peiffer, R. L. Adenocarcinoma of the gland of the third eyelid in seven dogs. *J Am Vet Med Assoc* **193:** 1549–1550, 1988.

D. Melanotic Tumors

Our understanding of the biology of primary ocular melanomas in animals has suffered greatly from the premature assumption that they are similar to the much-studied human ocular melanomas, and for many years the descriptions and prognoses reflected the human experience. Perhaps because of this error, a great many enucleated globes were available for histologic evaluation, inadvertently allowing the flurry of retrospective studies that eventually established ocular melanomas in dogs and cats as distinct entities quite different in structure and behavior from the human neoplasms.

Primary melanomas of the eye or adnexa are common in dogs and cats, rare in horses, and almost nonexistent in other domestic species. Even in dogs and cats, the prevalence and behavior of the various types of ocular melanomas differ markedly, so that generalizations must be studiously avoided.

1. Canine Ocular Melanomas

In **dogs,** melanomas arise in the skin of the lid margin, in conjunctiva, in anterior uvea, in the pigmented line demarcating the limbus, and in the choroid (Fig. 4.64A,B). The tumors arising in each location represent a distinct biologic group with a natural behavior and therapeutic protocol quite different from melanomas arising at one of the other sites.

Melanomas of the haired skin of the lid margin are the second most prevalent eyelid tumor in dogs. They are rare in this site in other species. They are almost invariably benign and closely resemble benign melanomas elsewhere in the skin.

Melanomas arising in the conjunctiva are very infrequent compared to all other ocular melanomas. The few case reports do not allow for generalizations about behavior, but they often are histologically and behaviorally malignant. Their relationship to eyelid melanoma seems analogous to those of the lip, where melanomas of the haired exterior lip are benign like most skin melanomas, and those of the mucous membrane of the inside of the lip are malignant like those elsewhere in the mouth. Conjunctival melanomas may appear as well-pigmented tumors of bland, plump melanocytes with little anisokaryosis or mitotic activity, but often are poorly melanotic and have invasive clusters of angular epithelioid cells with marked anisocytosis and numerous mitotic figures. Local recurrence and spread after excision is frequent, and metastasis to lung has been reported.

Limbal (epibulbar) melanoma is a histologically and behaviorally benign tumor of the melanocytes normally found in an oblique line that demarcates the junction of corneal stroma with sclera at the limbus. The tumor is composed of large plump melanocytes with a central nucleus and abundant cytoplasmic pigment. Mitotic figures are absent, and nuclear variation is minimal. The tumor grows outward as a protruding spherical nodule, hence the alternative name of epibulbar melanoma. There may be nodular expansion into peripheral cornea, but virtually never into the uvea or anterior chamber. Tumors with similar clinical and histologic appearance that do invade the globe should be assumed to have originated as anterior uveal melanomas, and the scleral extension of such tumors along the adventitia of the scleral venous plexus merely mimics the outflow of many other materials from within the eye.

Anterior uveal melanomas (melanocytomas) of dogs is the most frequent intraocular tumor in that species. It is topographically, histologically, and behaviorally unrelated to human epithelioid ocular melanomas, to which it was long compared. The typical tumor arises from melanocytes of the iris root or adjacent ciliary body, and is composed of variable proportions of lightly pigmented spindle cells and heavily pigmented plump melanocytes identical

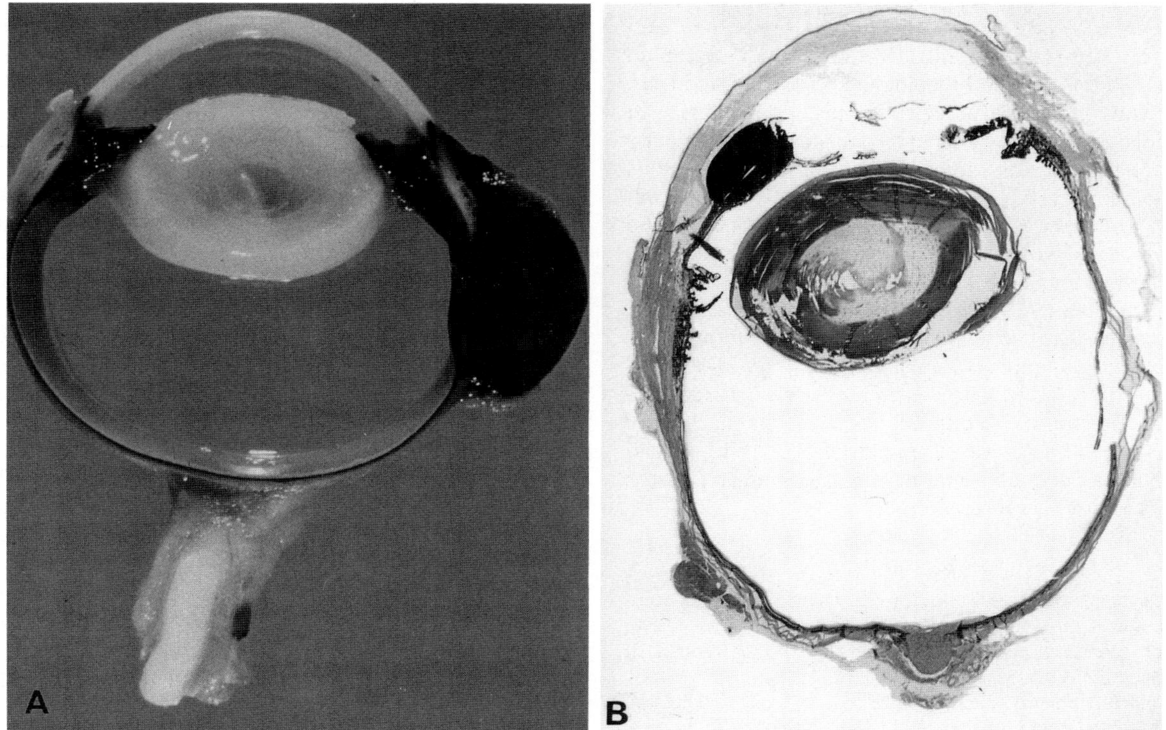

Fig. 4.64 (A) Ciliary melanoma at the limbus, which resembles an epibulbar melanoma grossly. (B) Benign iris melanoma (nevus), dog iris.

to those of limbal melanomas. The spindle cells are assumed to be the proliferative population, and the plump cells probably represent the mature, end-stage melanocyte with a storehouse of cytoplasmic pigment.

The diagnosis itself presents no problem, but offering an accurate prognosis is more complex. About 15% of all canine anterior uveal melanomas are histologically malignant, and one third of these have been confirmed to be behaviorally malignant by virtue of extraocular metastases. The overall prevalence of behavioral malignancy is thus about 5%, and this small group can be predicted by mitotic index. Histologically malignant tumors are dominated by the spindle cells rather than the plump cells, are more lightly pigmented, have much more anisokaryosis and more mitotic figures than the benign tumors. Of these, mitotic index is the most reliable predictor of behavior. Benign tumors have virtually no mitotic figures. Those confirmed as behaviorally malignant have a mitotic index of 3 or more (usually much more!).

Even benign melanomas are significant to the eye, spreading transclerally and circumferentially within the globe. Glaucoma from occlusion of ciliary cleft is seen in about half the dogs presented for enucleation, and is probably the eventual fate of all eyes with this neoplasm. Uveitis from tumor necrosis or hyphema from tumor-induced uveal neovascularization are other frequent accompaniments.

Clinically insignificant **iris nevi or freckles** occur in dogs as nonprogressive pigmented spots. Their only significance is to cause premature and unnecessary enucleation. Histologically, the lesions are well circumscribed clusters of bland melanocytes adjacent to the anterior border layer of the iris.

Choroidal melanomas account for about 80% of all human ocular melanomas, but are rare in dogs. The few reported were discovered as incidental findings on fundoscopic examination, and grew very slowly. They seem very similar to the benign melanomas of limbus or anterior uvea: well-pigmented, cytologically bland, and cause clinical signs only by their slow expansion into adjacent sensitive tissues. In the case of choroidal melanomas, such expansion results in localized retinal detachment and tumor infiltration of overlying retina or adjacent optic nerve. One of five dogs in one report became blind because of complete retinal separation from the choroid.

2. *Feline Ocular Melanomas*

Ocular melanomas in cats are very different from their canine counterparts. Cats very rarely have eyelid or conjunctival melanomas, paralleling the paucity of melanomas reported in other cutaneous or mucous membrane sites in this species. The few reported conjunctival melanomas have been histologically and behaviorally very malignant. Limbal melanomas, histologically and behaviorally identical to those in dogs, are occasionally seen. The major presentation, however, is of diffuse iris thickening and

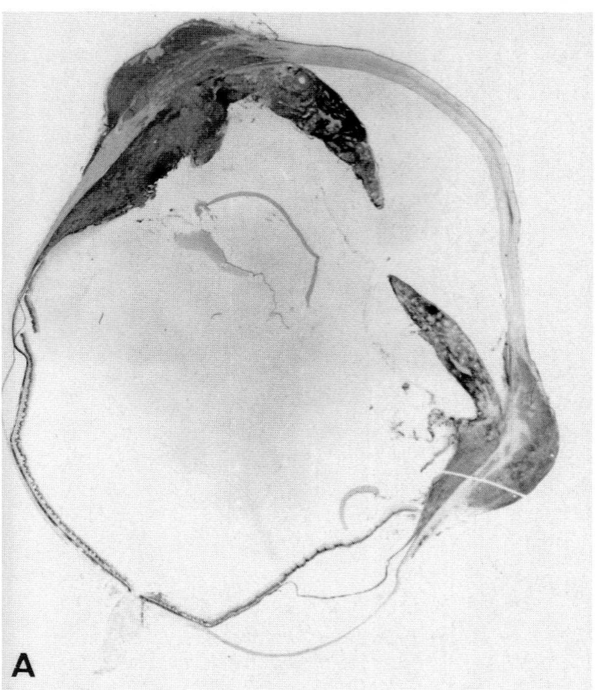

Fig. 4.65A Diffuse iris melanoma. Cat. The scleral infiltration is a grim prognostic sign in this species.

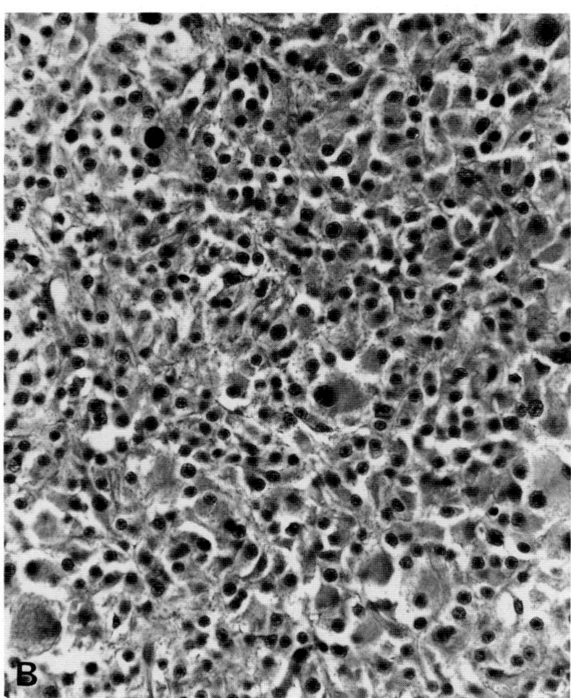

Fig. 4.65B Pleomorphic round cells. Iris melanoma. Cat. Pigmentation is light and mononuclear giant cells are conspicuous.

glaucoma—clinical findings typical of the diffuse iris melanoma unique to cats (Fig. 4.65A). Only rarely does one encounter a well-pigmented, plump-cell melanoma analogous to the common canine uveal melanoma.

The **diffuse iris melanoma** was at first considered an interesting but benign lesion causing diffuse thickening and hyperpigmentation of the iris. Subsequent retrospective studies have been unanimous in documenting a 50 to 60% metastatic rate. Because cats seem even more elusive than dogs in terms of follow-up studies, the actual number of cases with good postoperative data is small, and considerable debate persists among ophthalmologists about the real rate of metastasis of these tumors.

Histologically, these tumors diffusely infiltrate the stroma of the iris, the ciliary cleft, and then the overlying sclera, peripheral cornea, and ciliary body. They are notoriously pleomorphic, and are apt to be misdiagnosed by those pathologists not aware of this disease. Tumor cells vary from spindle-shaped cells to multinucleated epithelioid cells (Fig. 4.65B). Pigmentation often is light, and the cytoplasm may be foamy and eosinophilic. Balloon cells with foamy cytoplasm and very distinct cell boundaries are frequent in some tumors. The accurate prediction of tumor behavior is compromised, in all published studies, by the low percentage of affected cats available for follow-up. Metastasis has been correlated with large tumor size, intrascleral spread, and mitotic index.

Like dogs, cats have focal-to-coalescing hyperpigmentation of the anterior iris stroma. Some cases, as in dogs, are nonprogressive and harmless. Others seem slowly to coalesce and thicken, at which time they are indistinguishable from the most benign variants of diffuse iris melanoma.

3. Equine Ocular Melanomas

If one excludes the chance occurrence of gray-horse skin melanomas on the eyelid, primary ocular melanomas in horses are rare. Almost all reported cases have been in gray horses, many quite young (younger than 8 years). Most involve anterior uvea and are histologically similar to the benign uveal melanomas of dogs. None has metastasized.

There are two reports of benign limbal melanomas in horses, histologically similar to those in dogs.

Bibliography

Acland, G. M. *et al.* Diffuse iris melanoma in cats. *J Am Vet Med Assoc* **176:** 52–56, 1980.

Barnett, K. C., and Platt, H. Intraocular melanomata in the horse. *Eq Vet J* **10:** 76–82, 1990.

Belkin, P. V. Malignant melanoma of the bulbar conjunctiva in a dog. *Vet Med Small Anim Clin* **70:** 957–958, 1975.

Cook, C. S. *et al.* Malignant melanoma of the conjunctiva in a cat. *J Am Vet Med Assoc* **5:** 505–506, 1985.

Duncan, D. E., and Peiffer, R. L. Morphology and prognostic indicators of anterior uveal melanomas in cats. *Prog Vet Comp Ophthalmol* **1:** 25–32, 1991.

Gelatt, K. N., Johnson, K. A., and Peiffer, R. L. Primary iridal

pigmented masses in three dogs. *J Am Anim Hosp Assoc* **15:** 339–344, 1979.
Harling, D. E. *et al.* Feline limbal melanoma: Four cases. *J Am Anim Hosp Assoc* **22:** 795–802, 1986.
Hirst, L. W., and Jabs, D. A. Benign epibulbar melanocytoma in a horse. *J Am Vet Med Assoc* **183:** 33–334, 1983.
Hogan, R. N., and Albert, D. M. Does retinoblastoma occur in animals? *Prog Vet Comp Opthalmol* **1:** 73–82, 1991.
Patnaik, A. K., and Mooney, S. Feline melanoma: A comparative study of ocular, oral, and dermal neoplasms. *Vet Pathol* **25:** 105–112, 1988.
Peiffer, R. L. The differential diagnosis of pigmented ocular lesions in the dog and cat. *Calif Vet* **5:** 14–18, 1981.
Saunders, L. Z., and Barron, C. N. Primary pigmented intraocular tumors in animals. *Cancer Res* **18:** 234–245, 1958.
Schäffer, E. H., and Funke, K. Das primär-intraokulare maligne melanom bei hund und katze. (Primary malignant ocular melanoma in dogs and cats). *Tierärztl Praxis* **13:** 343–359, 1985.
Wilcock, B. P., and Peiffer, R. L., Jr. Morphology and behavior of primary ocular melanomas in 91 dogs. *Vet Pathol* **23:** 418–424, 1986.

E. Tumors of Ocular Neurectoderm

These tumors include adenoma and carcinoma of mature ciliary epithelium and medulloepithelioma and retinoblastoma from embryonic neurectoderm. The prevalence of these neoplasms is second to that of anterior uveal melanomas, although it is perhaps underestimated because the most common examples are small, slowly expansive ciliary adenomas that are unlikely to cause clinical signs. They are relatively common in dogs, rare in cats, and virtually unknown in other species except for medulloepitheliomas in horses.

Ciliary adenoma is the most common of this group (Fig. 4.66A,B). It is a well-differentiated papillary or tubular adenoma arising from the nonpigmented inner layer of ciliary epithelium. Most originate from the pars plicata, but occasionally the histologic evidence points to origin from posterior iris epithelium. The tumor cells resemble mature ciliary epithelium and usually have very little associated stroma. Nuclei are basilar, regular, and are surrounded by eosinophilic cytoplasm (Fig. 4.67A). The tumor cells are not pigmented, although melanophages are occasionally seen within tumor stroma. They make an abundance of basal lamina oriented, as in normal ciliary epithelium, toward the inside of the eye. Its abundance, easily seen with periodic acid-Schiff reagent, is useful in distinguishing ciliary tumors from carcinomas metastatic to the eye. Examples which have little tubular or papillary organization, or have locally invasive growth, have traditionally been called **ciliary carcinomas.** Metastasis is exceedingly rare, so there seems little point in debating the histological criteria that justify the appellation of malignancy.

Even small ciliary body tumors may cause hyphema or glaucoma. This paradox, unexplained by tumor mass or by particularly abundant tumor vasculature, is attributed to this tumor's strong propensity to induce preiridal fibrovascular membranes (Fig. 4.67B). Ciliary body tumors are

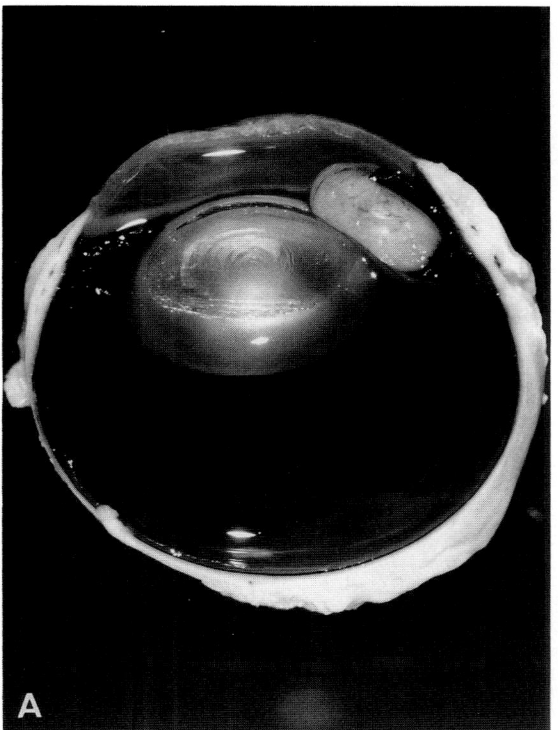

Fig. 4.66A Ciliary adenoma. Dog.

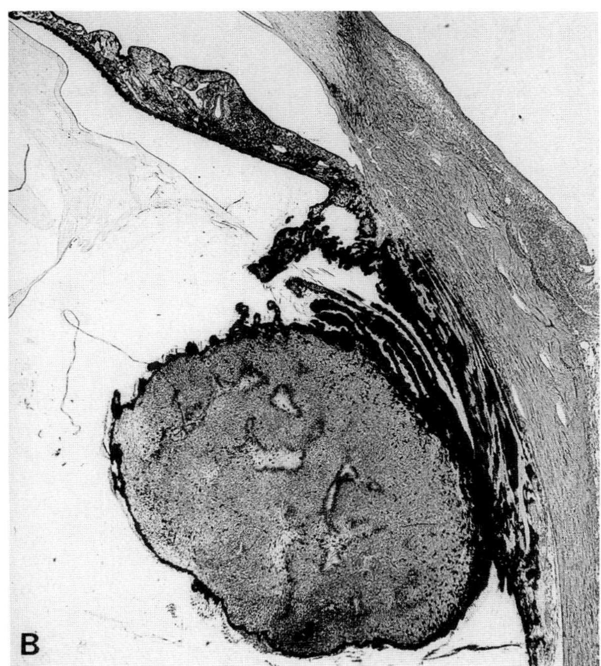

Fig. 4.66B Ciliary adenoma. Dog.

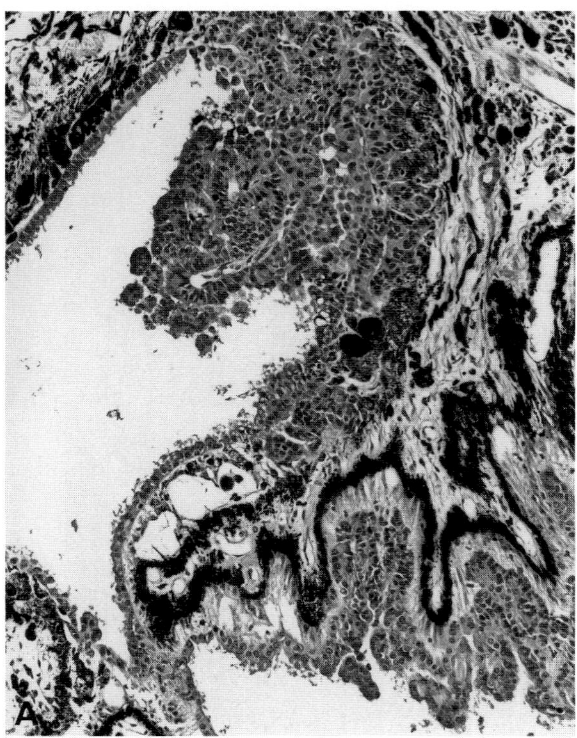

Fig. 4.67A Ciliary adenoma. Dog. Note well-differentiated tubular proliferation.

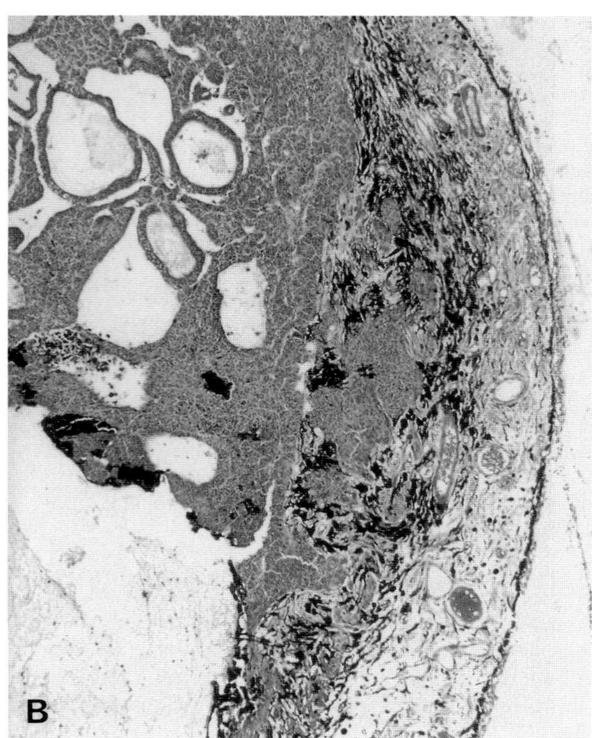

Fig. 4.67B Ciliary adenoma. Dog. Note accompanying preiridal vascular membrane.

more likely to induce such neovascularization than is any other ocular disease. Large tumors may, of course, induce glaucoma by such mechanisms as pupillary block or infiltration of ciliary cleft.

Medulloepitheliomas and retinoblastomas arise from the primitive neurectoderm of the optic cup. Retinoblastoma is the second most frequent neoplasm of children, yet critical review of the veterinary literature fails to reveal a single acceptable diagnosis of this tumor. Conversely, medulloepitheliomas are rare in children, but examples have been observed in animals, mainly in the horse, in which these rare neoplasms are probably the most common primary intraocular tumor. The neoplasm may originate from any portion of embryonic neurectoderm and may show differentiation into any neurectodermal derivative; retina, ciliary epithelium, vitreous, or neuroglia (Fig. 4.68A,B). The typical neoplasm is a loose network of branching cords of small basophilic neuroblasts resembling those of embryonic retina. Mitotic figures are numerous. The cords have definite polarity: they rest upon a basement membrane analogous to the inner limiting membrane of retina, and some have adjoining apical terminal bars analogous to outer limiting membrane. A typical feature is the clustering of neuroblasts around an empty central lumen defined by the terminal bars, creating a true rosette (Fig. 4.69). The basilar portion of the rosette or cord faces a hyaluronic acid-rich myxoid matrix analogous to the vitreous. This histologic feature may be found in only a few foci within a huge mass that is otherwise composed of poorly differentiated, neuroblastlike cells, and has abundant necrosis. Many tumors also contain foci of cartilage, skeletal muscle, or brain tissue and are classified as teratoid medulloepitheliomas. Metastases are not recorded.

Bibliography

Bellhorn, R. W. Ciliary body adenocarcinoma in the dog. *J Am Vet Med Assoc* **159:** 1124–1128, 1971.

Broughton, W. L., and Zimmerman, L. E. A clinicopathologic study of 56 cases of intraocular medulloepitheliomas. *Am J Ophthalmol* **85:** 407–418, 1978.

Eagle, R. C., Font, R. L., and Swerczek, T. W. Malignant medulloepithelioma of the optic nerve in a horse. *Vet Pathol* **15:** 488–494, 1978.

Lahav, M. *et al.* Malignant teratoid medulloepithelioma in a dog. *Vet Pathol* **13:** 11–16, 1976.

Langloss, J. M., Zimmerman, L. E., and Krehbiel, J. D. Malignant intraocular teratoid medulloepithelioma in three dogs. *Vet Pathol* **13:** 343–352, 1976.

Wilcock, B., and Williams, M. M. Malignant intraocular medulloepithelioma in a dog. *J Am Anim Hosp Assoc* **16:** 617–619, 1980.

F. Feline Posttraumatic Sarcoma

This syndrome seems unique to cats. As the name implies, eyes subjected to trauma, especially penetrating

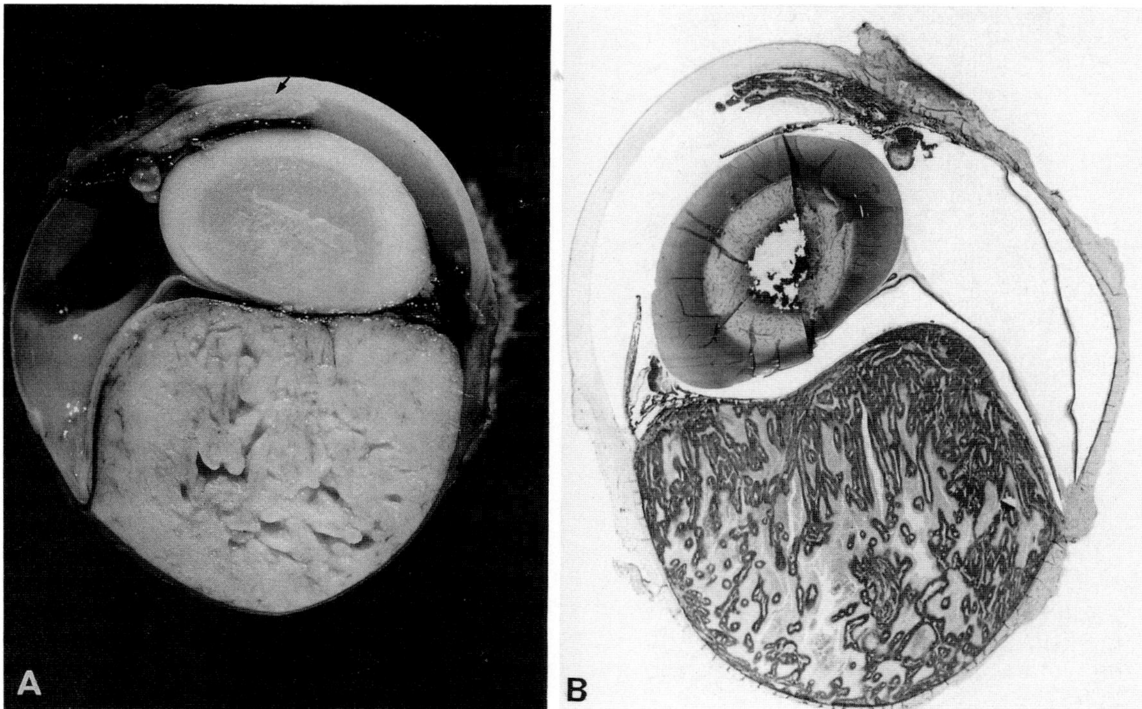

Fig. 4.68 Medulloepithelioma. Dog. (A) The main tumor lies between choroid and retinal pigment epithelium. There is also tumor in anterior chamber (arrow). (B) Histologic section of (A). Note tubular latticework of the tumor.

injury, are prone to develop pleomorphic spindle cell sarcomas that destroy the globe and have substantial risk of metastasis. The interval between injury and observed tumor varies from 5 months to 11 years. Those skeptical about claiming such neoplasia to be the result of an injury 10 years previously prefer to call these tumors primary ocular sarcomas, although such lag times are common in experimental models of carcinogenesis. The risk for injured eyes to develop sarcoma is unknown. Almost all recorded cases have perforated lenses, leading to speculation that these tumors represent malignant transformation of the perilenticular fibroplasia and epithelial metaplasia that characterizes phacoclastic uveitis in all species. Of relevance to ocular surgeons is the development of sarcomas in cat eyes receiving prosthetic implants, presumably viewed by the eye as yet another form of trauma.

The tumor itself varies from fibrosarcoma to osteosarcoma to giant-cell tumor, varying even within the same eye. The tumor tends to line the inside of the eye, and then extend via scleral venous plexus or optic nerve to involve the orbit. Most cases are presented with advanced disease, and follow-up data to document the prevalence of metastasis are scant. Available evidence uniformly documents a metastatic prevalence of at least 60%.

Bibliography

Dubielzig, R. R. *et al.* Clinical and morphologic features of posttraumatic ocular sarcomas in cats. *Vet Pathol* **27:** 62–65, 1990.

Peiffer, R. L., Monticello, T., and Bouldin, T. W. Primary ocular sarcomas in the cat. *J Small Anim Pract* **29:** 105–116, 1988.

G. Optic Nerve Tumors

Although the optic nerve and adjacent retina can presumably develop all of the neoplasms of the central nervous system (excepting those from tissues like ependyma that are not present in the eye), documented examples are few indeed. Most are reported as individual case reports prior to the era of immunohistochemical markers that would have permitted more precise classification.

Primitive neuroepithelial tumors of optic nerve are occasionally seen in young dogs and in horses. They are composed of nests, cords, and rosettelike structures formed by small hyperchromatic neuroblastic cells with a very high mitotic index. Very rapid spread throughout the orbit and into brain occurs in affected dogs and in the one published equine case.

Spindle cell tumors, considered by some to be optic nerve meningiomas or neurofibrosarcomas, will be considered.

H. Orbital Neoplasms

Tumors may be primary within the orbit or arise by extension from adjacent structures or by hematogenous localization. They usually produce deviation or protru-

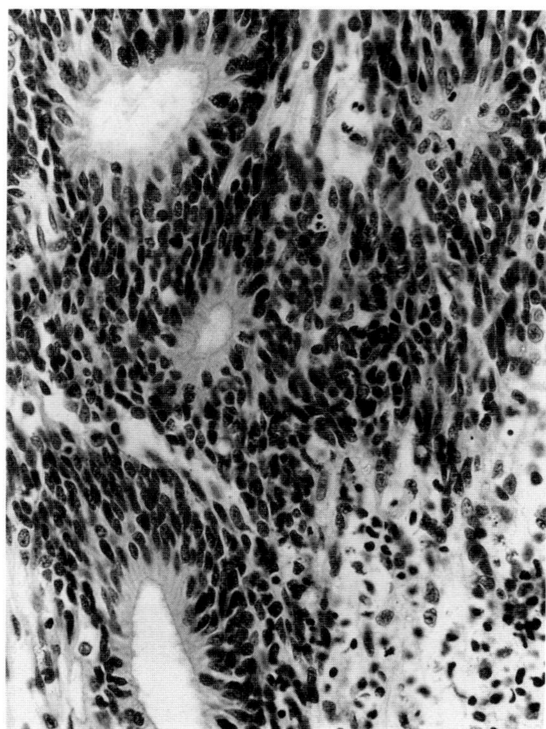

Fig. 4.69 Medulloepithelioma. Dog. Note rosette with rudimentary formation of retinal outer limiting membrane and photoreceptors.

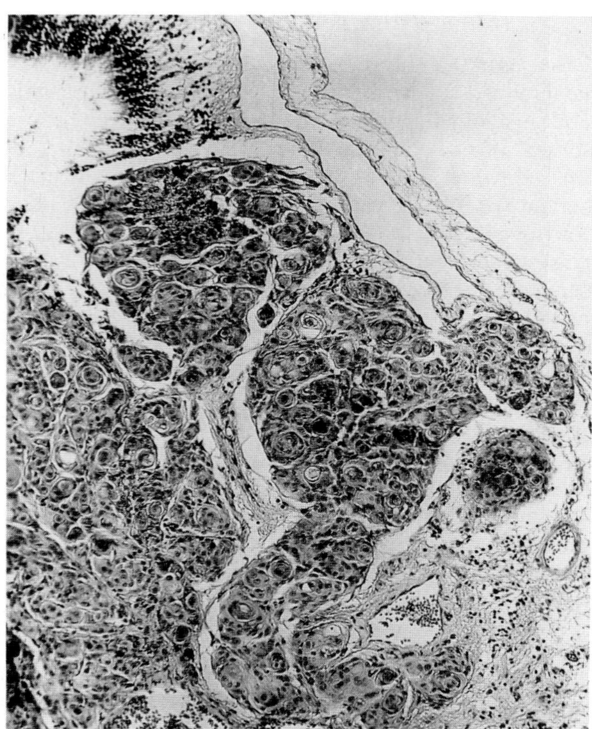

Fig. 4.70 Meningioma of optic nerve invading eye. Dog.

sion of the globe with secondary desiccation keratoconjunctivitis. In dogs, sarcomas and carcinomas are of almost equal prevalence, and primary orbital tumors are approximately equal in frequency to those arising by extension from nose, mouth, or distant sites. Of the primary orbital tumors, sarcomas are much more prevalent than epithelial tumors. The sarcomas are a bewildering array of locally infiltrative spindle cell tumors of unknown origin, with the abundance of diagnoses probably reflecting the diversity of pathologists' opinions rather than actual proof of histologic identity (Fig. 4.70). Metastasis is rare, but their infiltrative growth habit in this difficult site makes eventual elective euthanasia a frequent outcome. One tumor that does have a distinctive appearance is multilobular osteoma arising from the bones of the orbit, identical in appearance and behavior to this tumor elsewhere in the canine skull.

Most primary epithelial tumors of the canine orbit are lacrimal adenocarcinomas, which are locally invasive, recur after attempted resection, but are not noted to metastasize in what usually is a brief postoperative follow-up period. Tumors infiltrating the orbit from the nearby zygomatic salivary gland are similar histologically and behaviorally.

In other species, metastatic lymphoma is the only significant orbital tumor.

THE EAR

I. General Considerations

Diseases of the ear of domestic animals are of small interest to most veterinary pathologists. Of the three broad categories—defects of hearing, otitis media, and otitis externa—only otitis media in farm animals regularly receives some attention, but even that is usually restricted to macroscopic examination at necropsy. Otitis externa is almost exclusively the realm of the clinical practitioner, although hyperplasia of chronically inflamed epithelium within the external ear canal may simulate neoplasia and be submitted for histologic interpretation. Disorders of hearing certainly occur in domestic animals, but their prominence in the medical literature is the result more of their usefulness as models for human deafness than of their intrinsic importance in animals. The growing availability of equipment for evaluation of hearing may, however, see a resurgence of interest in deafness in specific breeds or colors of dogs and cats. The tedious task of preparing adequate sections of the cochlea and labyrinth when related to the relative importance or prevalence of the diseases makes examination of the inner ear a rare event in most veterinary institutions. The gross and microscopic anatomy of the ear in the various species is reviewed in specialized texts, and the appropriate references are included in the bibliography.

II. External Ear: Otitis Externa

Diseases of the external ear may be a local manifestation of generalized skin disease or may specifically affect the ear because of some anatomic or physiologic peculiarity. Disorders of skin that only incidentally affect the ear are discussed with diseases of the Skin and Appendages (Chapter 5).

Ear-tip necrosis may result from frostbite, cold-agglutinin disease, ergot poisoning, thrombosis during the course of septicemia, or from trauma inflicted by the animal itself or cannibalistic herdmates. Particularly in pigs, the prevalence of atrophic and misshapen ears may be quite high in individual herds. This usually results from septicemia, sarcoptic mange, or cannibalism. In some instances, however, no such history can be elicited, and a syndrome of infectious ear necrosis has been proposed. Culture of ears showing marginal lesions results in isolation of a variety of bacterial agents, and claims for a primary bacterial etiology are not convincing. The one that comes closest is the spirochete *Borrelia suilla*. It usually acts as an opportunistic contaminant of abraded skin in pigs, giving rise to localized (usually ear margins) or generalized ulcerative dermatitis.

Auricular hematoma occurs as a consequence of trauma, usually from excessive head shaking by dogs with otitis externa. Dogs and pigs with pendulous ears are particularly prone to hematoma formation, but cats are occasionally affected as well. Hematomas usually develop on the concave side of the pinna and are initially fluctuant but become firm as the hematoma (or blood-contaminated seroma) organizes. As it is converted to granulation tissue by fibroblastic and capillary ingrowth, the lesions become hard. Subsequent fibroblastic contraction may result in disfigurement of the pinna.

The location of the initial damage is unclear. Subcutaneous and subperichondrial sites are suggested but a predominantly intrachondral location seems most likely. The cartilage plate of the pinna is cleaved by a longitudinal fracture and then by the hemorrhage. Granulation tissue forms at the interface of blood and cartilage, and the lesion is eventually converted to a fibrous scar. Late in the reparative process, regeneration of cartilage occurs adjacent to perichondrium or the ruptured edge of the cartilage. The intrachondral location of the hematomas is thought to be the result of rupture of vessels as they pass through minute foraminae in the cartilage plate. The conventional view that the initial lesion results from shear forces or outright trauma caused by head shaking has recently been challenged by one study of 40 dogs and 20 cats with auricular hematoma. No association with ear configuration or otitis externa was found, but most affected animals had serum and local tissue changes interpreted as evidence for a lupuslike autoimmune disease. The link between this proposed pathogenesis and the observation of the early cartilaginous fractures and hemorrhages was not made.

Dermatologic disease of the ear pinna may be part of a generalized skin disorder, but a few diseases seem unique to the ear. Pigs with chronic sarcoptic mange may have gross lesions only on the ears, consisting of reddening, crusting, and thickening of the pinna. Short-haired dogs with pendulous ears (dachshunds, pointers, bloodhounds) are predisposed to a chronic dermatitis of the ear margin, variously termed marginal auricular dermatosis or ear margin seborrhea. The gross lesion is a multifocal, greasy, gray, nodular encrustation that may coalesce to cause thickening of the entire ear margin. The histologic lesion resembles that of seborrhea elsewhere with hyperkeratosis, acanthosis, and a mild superficial perivascular dermatitis in which mononuclear leukocytes predominate.

Chronic ulcerative dermatitis of the ear margin of white cats (feline solar dermatitis) is discussed below as a premalignant lesion under squamous cell carcinoma, as is a similar disease of sheep.

Alopecia of the pinna occurs in dogs and in Siamese cats. In dogs, the alopecia appears gradually and progresses slowly. In cats the lesion may wax and wane at irregular intervals throughout life. Microscopically there is bland pilosebaceous atrophy resembling endocrine alopecia, but the precise cause is unknown.

Pinnal vasculitis is surprisingly frequent and resembles the lesion of cutaneous vasculitis elsewhere. In its mildest form, it may cause only pilosebaceous atrophy and thus may mimic the lesions of idiopathic pinnal alopecia when examined clinically.

Otitis externa is a very common disorder in dogs and cats, and is probably equally common in goats. In dogs, the pathogenesis is complex, while in cats and goats, the ear mite is by far the major initiating factor. Otitis externa in dogs is most prevalent in breeds with pendulous ears or with abundant hair within the ear canal, which implies that inadequate circulation of air and entrapment of moisture are important predispositions to the disease. Foreign bodies, usually foxtail or grass awns, may be important mechanical irritants, creating a breach for microbial infection, but since otitis is common in city dogs, other factors apparently are involved. The role of microorganisms in the pathogenesis of otitis externa is, as elsewhere, intimately linked to environmental circumstances that permit their uncontrolled proliferation. The bacteria and fungi cultured from diseased ears almost invariably are members of the normal aural flora. *Staphylococcus* spp., *Pseudomonas* spp., *Proteus* spp., and the yeast *Malassezia pachydermatis* (syn. *Pityrosporum canis*) are cultured much more frequently or in greater numbers from ears with otitis than from normal ears, but this may relate as much to their resistance to antibiotic treatment as to their causative role. The lipid-rich environment of the ear canal favors the lipophilic *Malassezia*, which is the best candidate for a primary pathogen in the causation of canine otitis externa, inasmuch as it frequently is the only pathogen isolated, and affected ears improve dramatically with antifungal therapy.

Otitis externa in species other than dogs is closely correlated to infestation with **ear mites,** and even in dogs, acariasis is an important cause of otitis.

Otodectes cyanotis is the ear mite of carnivores. The importance of the mite in initiating otitis externa is not entirely clear because the ear canal, once inflamed, becomes an environment unsuitable for the mites and thus gives rise to misleadingly low estimates of the prevalence of ear mites in dogs with otitis. As few as five mites can initiate otitis externa in dogs, with bacterial and mycotic opportunists rapidly masking the role of mites in the development of the lesion. Estimates vary from 2 to 50% prevalence of mites in dogs with otitis externa.

In cats, the role of mites is firmly established. In contrast to dogs, all cats have erect and sparsely-haired pinnae, and the role of ear carriage and hair is negligible. There is a poor correlation between the number of mites, the severity of inflammation in the external meatus, and clinical signs.

The mites are obligate parasites, spending their entire 3- to 4-week life cycle on the host. Transmission between animals of the same or different species readily occurs, and spread of the mites from their preferred aural niche to paws as the animal scratches, or tail as it sleeps with tail curled to touch the ear, is occasionally seen.

The mechanism by which the mites initiate otitis is controversial. Some maintain that the mites feed only on epithelial debris, while others claim that the mites pierce the epithelium and ingest blood and lymph. The latter claim is supported by the demonstration of host-specific serum components within the mites, but such components could exist in exudates, and should not be considered proof of penetrating feeding behavior.

The variation in the number of mites necessary to cause clinically obvious otitis, the species variation, and some tenuous age resistance to disease (if not infection) suggest an allergic basis for the otitis. Most cats, and perhaps dogs, probably are infected with mites at some time. They may react to subsequent minor infection with immune-mediated inflammation of a severity not predicted by the small number of mites within the ear canal.

Psoroptes cuniculi is the ear mite of rabbits that also affects goats, horses, and deer. Its prevalence among domestic animals seems to be greatest in goats, ranging from about 20 to 80%. Careful examination may be required to find the mites. Clinical signs in goats, if present at all, are usually mild, consisting of ear twitching and head shaking. The ear canal of goats is the best site for isolation of several pathogenic *Mycoplasma* species in clinically normal animals. The same mycoplasmas were isolated from ear mites (*P. cuniculi* or *Raillietia caprae*) of infected goats, fueling speculation that the mites are important vectors of the mycoplasmas in endemic herds. Similar coexistence of ear mites and mycoplasmas has been observed in cattle, but identification of the bacteria within the mites was not reported.

Psoroptes cuniculi may be quite common in horses (20% of horses in one Australian report), but it is unclear how frequently it causes clinical signs. Head shaking and resentment of handling of the ears are the usual signs.

Raillietia auris, the ear mite of cattle, occurs in most areas of the world. Infection may be common but is rarely of clinical significance. Manifestations of disease may follow extension of the otitis externa through the tympanum, and affected animals show head tilt and circling due to otitis interna.

Otobius megnini, the spinose ear tick, is parasitic on domestic ruminants, pigs, horses, and dogs, and cats in restricted geographic areas. Infestation may be so heavy that the ear canals are full of ticks, making diagnosis easy. In light infestations, they occur deep in the folds of the ear, at the bottom of the external meatus. Only the larvae and nymphs are parasitic. The larvae attach themselves to the skin below the hairline and, biting through the skin, suck lymph until they are engorged. The parasites irritate the external auditory meatus, and the ensuing exudate may completely fill it. In addition, secondary bacterial infection may occur in the inflamed areas around the bite wounds, and this infection may extend downward and cause otitis media. The microscopic appearance of the ear in this infestation has not been described.

Otitis externa is a nonspecific lesion. The relationship between type of exudate and causative agent has not been critically evaluated. The initial macroscopic lesion is hyperemia of the external auditory meatus followed by accumulation of serum, cerumen, leukocytes, and epithelial debris. A predominance of leukocytes tends to produce a suppurative exudate, whereas cerumen yields a dry, dark brown, crumbly accumulation.

The histopathology of otitis externa varies with the duration of the inflammation and is rather typical inflammation of any epithelial surfaces. Only chronic lesions are likely to be seen by the pathologist. Such specimens show epidermal hyperplasia with acanthosis, atrophy of hair follicles, hyperkeratosis, and parakeratosis. Crusts of inflammatory debris adhere to the epithelial surface. Ulceration may be present, particularly if *Pseudomonas* is predominant and *Malassezia* yeasts may be seen in the surface keratin or debris. The dermis contains numerous lymphocytes, plasma cells, mast cells, and neutrophils, the latter usually within dilated venules or adjacent to ulcers. Neutrophils are very sparse in mite-induced otitis. In very chronic and severe examples, dermal fibrosis is marked, and ossification may occur. A feature of chronic otitis externa is increased production of cerumen. Histologically there is hyperplasia of the normally very large sebaceous glands and cystic hyperplasia of the coiled, tubular, apocrine ceruminous glands, which are distended by eosinophilic cerumen. The combination of epithelial hyperplasia, glandular hyperplasia, and stromal fibroplasia may create proliferative lesions that occlude the ear canal and simulate neoplasia.

Bibliography

Cook, R. W. Ear mites (*Raillietia manfredi* and *Psoroptes cuniculi*) in goats in New South Wales. *Aust Vet J* **57:** 72–75, 1981.

Cottew, G. S. Mycoplasmas in ears. *Aust Vet J* **62:** 420, 1985.

Cottew, G. S., and Yeats, F. R. Mycoplasmas and mites in the ears of clinically normal goats. *Aust Vet J* **59:** 77–81, 1982.

Dubielzig, R. R., Wilson, J. W., and Seireg, A. A. Pathogenesis of canine aural hematomas. *J Am Vet Med Assoc* **8:** 873–876, 1984.

Fernando, S. D. A. Certain histopathologic features of the external auditory meatus of the cat and dog with otitis externa. *Am J Vet Res* **28:** 278–282, 1967.

Gedek, B. *et al.* The role of *Pityrosporum pachydermatis* in otitis externa of dogs: Evaluation of a treatment with miconazole. *Vet Rec* **104:** 138–140, 1979.

Kowalski, J. J. The microbial environment of the ear canal in health and disease. *Vet Clin North Am: Small Anim Pract* **18:** 743–754, 1988.

Kuwahara, J. Canine and feline aural hematoma: Clinical, experimental, and clinicopathologic observations. *Am J Vet Res* **47:** 2300–2308, 1986.

Larsen, S. Intrachondral rupture and hematoma formation in the external ear of dogs. *Pathol Vet* **5:** 442–450, 1968.

Littlejohn, A. I. Psoroptic mange in the goat. *Vet Rec* **82:** 148–155, 1968.

Menzies, G. C. The cattle ear mite, *Raillietia auris* (Leidy, 1872) in Texas. *J Parasitol* **43:** 200, 1957.

Powell, M. B. *et al.* Reaginic hypersensitivity in *Otodectes cynotis* infection of cats and mode of mite feeding. *Am J Vet Res* **41:** 877–882, 1980.

van der Gaag, I. The pathology of the external ear canal in dogs and cats. *Vet Q* **8:** 307–327, 1986.

Weisbroth, S. H. *et al.* Immunopathology of naturally occurring otodectic otoacariasis in the domestic cat. *J Am Vet Med Assoc* **165:** 1088–1093, 1974.

Williams, J. F., and Williams, C. S. F. Psoroptic ear mites in dairy goats. *J Am Vet Med Assoc* **173:** 1582–1583, 1978.

III. Middle Ear: Otitis Media

Otitis media is inflammation of the tympanic cavity within the temporal bone. Its cause is almost always bacterial, the organisms reaching the poorly drained cavity via the Eustachian tube or following perforation of the tympanum. In swine and lambs, the infection usually ascends from the pharynx via the Eustachian tube. In dogs and cats, chronic otitis externa is the major predisposition, although the tympanum is quite resistant to inflammatory lysis. There are insufficient cases recorded in other species to permit generalization. In all species there is circumstantial evidence for hematogenous localization of infection in the middle and inner ear, perhaps occurring most often in pigs.

Otitis media as a clinically obvious entity is most frequent in feeder pigs. The infection is usually unilateral and associated with hemolytic streptococci. The clinical signs of head tilt, circling, and ataxia suggest involvement of the inner ear in most instances. Otitis media in swine may occur as small epizootics involving a dozen or more pigs. The basis for the clustering of the disease is unproven, but its sporadic association with atrophic rhinitis suggests spread from the upper respiratory tract. In lambs, the lesion is usually clinically undetected and unilateral, and often occurs in association with pneumonia. *Pasteurella hemolytica* is the usual isolate from both ear and lung. Bottle feeding also increases the prevalence of otitis media in lambs.

Regardless of the route of entry, the lesion and its progression are similar. The epithelium lining the tympanic cavity is hyperemic, edematous, and may be ulcerated. Neutrophils exuding from the reactive vessels under the epithelium enter the tympanic cavity, joining the initially serous or serofibrinous exudate to make it progressively more purulent. Exudate may temporarily drain into the pharynx via the Eustachian tube, which is soon sealed by inflammatory swelling of its epithelium. In severe infections, the exudate escapes via inflammatory lysis of the tympanum or, rarely, the bone on the ventral floor of the tympanic bulla. Chronic inflammations are characterized by inspissation of exudate, lysis of the ossicles, and occasionally, the tympanum, and spread to inner ear and brain stem.

Bibliography

Jensen, R. *et al.* Middle ear infection in feedlot lambs. *J Am Vet Med Assoc* **181:** 805–807, 1982.

Macleod, N. S. M., Wiener, G., and Barlow, R. M. Factors involved in middle ear infection (otitis media) in lambs. *Vet Rec* **91:** 360–362, 1972.

Olson, L. D. Gross and microscopic lesions of middle and inner ear infections in swine. *Am J Vet Res* **42:** 1433–1440, 1982.

IV. Inner Ear

A. Otitis Interna

Otitis interna is almost always the result of infection spreading from the middle ear. The inflammation is usually of bacterial origin and thus suppurative. Ascension via the eighth nerve is relatively frequent and results in focal suppurative meningitis or encephalitis in the region of the pons. The usual clinical syndrome is vestibular dysfunction, described in a subsequent section.

B. Deafness

Deafness is difficult to assess in animals, and its diagnosis by simply observing behavioral abnormalities is almost impossible unless the animal is totally and bilaterally deaf. The recent introduction of electrodiagnostic tests has identified a higher prevalence of deafness, especially in dogs, than was suspected. Conductive deafness results from interference with the conduction of sound to the sensory end organ (of Corti) by diseases of external or middle ear. Alternatively, **sensorineural deafness** results from maldevelopment or degeneration of the sensory organ, eighth nerve, or auditory pathways within the brain. The last is very unusual and, because of the multitude of possible pathways, is seen only in massive destructive lesions with their neurologic signs overshadowing the hearing loss. Nerve deafness usually involves the organ of Corti, and is by far the most prevalent type of deafness encountered in animals by virtue of hereditary deafness in several breeds of dogs, and in dogs or cats with color-dilution anomalies. Conductive deafness may result from obliteration of the external auditory meatus by chronic proliferative inflammation or tumor, from inflammatory or traumatic rupture of the tympanum, or from entrapment of the ossicles in exudates or granulation tissue. Rarely there is destruction of the middle or inner ears by osteomyelitis or neoplasia.

The anatomy of the auditory apparatus is complex. The sensory fibers of the eighth cranial nerve terminate at the base of sensory hair cells within the organ of Corti, the latter a sensory specialization of the epithelium lining the cochlear duct. Excitation of the hair cells results from pulsations within the endolymph fluid that fills the entire membranous labyrinth, including cochlear duct. The precise mechanism of such excitation remains unknown. Sound waves in the environment reach the endolymph via the tympanum and ossicles. Vibration of the tympanum is transmitted to the ossicles, and vibration of the ossicles causes vibration of the oval window which separates the foot plate of the stapes from the endolymph. Any lesion that interferes with vibration of the tympanum or ossicle interferes with the establishment of fluid waves within the endolymph. Traumatic fracture of ossicles, rupture of tympanum, or dampening of vibration by exudates within middle ear can do this.

1. Hereditary Cochleosaccular Degeneration

Hereditary deafness in association with incomplete pigmentation of hair coat and uvea is seen in cats, dogs, mink, and mice. Incompletely documented examples exist in other domestic species. The pigmentary defect is not true albinism but is white spotting or merling, the distinction being the absence of melanocytes in white areas in the latter instance and functionally defective melanocytes in the former. In some instances the hearing defect is associated with inheritance of the merling gene. All homozygotes and many of the heterozygotes are deaf and have some degree of iris heterochromia. In other instances there is heterochromia iridis but no apparent coat color dilution (Dalmatian dogs), while in others, both eyes and coat are phenotypically normal. Even when the coat color is white, as in deaf bull terriers, the genetic basis for the white coat and the associated deafness need not be the merling gene. Out of this confusing picture one should rescue the concept that color-dilute animals have ocular heterochromia and deafness much more frequently than normal animals, but that other genetic bases for deafness also exist. Ocular anomalies and deafness commonly are encountered in the same animal, probably the result of the inductive influence of pigmentation on both organs. For example, all white cats with blue irises are deaf, as are many Dalmatian dogs with iris heterochromia. Several dog breeds, particularly Dalmatians and English setters, have a prevalence of early-onset deafness in excess of that predictable from their prevalence in the overall canine population. Prominent also are merle dogs such as Australian shepherds, Australian blue heelers, old English sheepdogs, and Great Danes.

The fully developed lesion of cochleosaccular degeneration is atrophy of the sensory and supporting cells of the organ of Corti and the saccular macula, collapse of the dorsal or lateral walls of the cochlear and saccular membranous labyrinth, and secondary degeneration of the neurons within the spiral ganglion. The osseous portion of the labyrinth is unaffected, as is the vestibular portion of the inner ear. With minor variations, all cases in dogs and cats are similar. Initial structure and function are normal, but there is failure to achieve normal maturation and, subsequently, degeneration occurs.

The carnivore ear is completely developed at birth and continues to mature for 2 (organ of Corti) to 4 (stria vascularis) weeks. Thereafter, the epithelial lining of the cochlear duct is incapable of mitotic regeneration. In white kittens, the ear is morphologically and physiologically normal at birth but shows arrested development as early as 1 week of age. At this time there is inward sagging of the free dorsolateral wall of the cochlear duct (Reissner's membrane) and saccule, and hydropic change in the stria vascularis. The degeneration within the organ of Corti and adjacent nutritive stria vascularis is so rapid that there is no consensus as to the initial or causal lesion. It may be that degeneration of the stria vascularis results in ischemic degeneration of the avascular organ of Corti, or there may be a primary defect within the sensory cells themselves resulting from an inborn metabolic error.

2. Senile Deafness

Many animals become hard of hearing as they reach old age. This phenomenon, called presbyacusis, is more frequently observed in old dogs because dogs tend to be kept well into their advanced years and because hearing loss is more readily noticed in them than in other species. In humans the loss of hearing is progressive from about the fortieth year of life and particularly affects hearing of high tones. The cause seems to be inherent age-related degeneration of the epithelial tissues within the cochlea and of the spiral ganglion, a process that may be accelerated by excessive noise, arteriosclerosis, and nutritional factors. The essential lesions are atrophy of all epithelial structures within the cochlear duct and the associated auditory nerves, as well as neuronal atrophy within the spiral ganglion. Occasionally, the major lesion is atrophy within the stria vascularis or the basal membrane supporting the organ of Corti.

Presbyacusis in animals has been studied in guinea pigs, rats, mice, dogs, and cats. Degeneration of the hair cells in the organ of Corti and loss of neurons from the spiral ganglion are the common denominators in all such instances, and presumably is the aging change common to all mammalian species. Stria vascularis atrophy has not been seen in these animals.

3. Acoustic and Chemical Ototoxicity

Noise, either as a sudden loud noise or as moderate but prolonged environmental background, causes degeneration of the sensory hair cells within the organ of Corti. Loud noise causes hair cell necrosis and even outright disruption of the organ of Corti or Reissner's membrane by mechanical trauma, mediated via fluid waves within the endolymph that must be the otic equivalent of tidal waves. Environmental noise is frequently an occupational hazard. The eventual lesion is hair cell necrosis, but whether this results from repeated sublethal microtrauma by noise peaks or by interference with hair cell metabolism is not known. That animals are susceptible to such trauma is well established in experimental models, but investigation of naturally occurring examples is not reported.

The list of chemicals that are ototoxic is very long, and is even longer if idiosyncratic drug injury is included. Only a few major examples likely to be encountered in veterinary practice are included here.

Aminoglycoside antibiotics (gentamicin, streptomycin, kanamycin, neomycin, and others) are all nephrotoxic and ototoxic in proportion to their blood levels and duration of administration. Overdosing, or administration to animals with decreased renal function, markedly increases the risk of toxic injury to the inner ear. Clinical signs of vestibular dysfunction precede evidence of hearing impairment. The initial lesion affecting hearing is degeneration of the apical portion of cochlear hair cells. The earliest visible lesion is mitochondrial swelling and increased number of myelin figures. The early lesion is reversible but later becomes permanent, presumably as mitochondrial swelling leads to structural disintegration of cellular respiratory enzymes and cell death. Cats are particularly susceptible, and vestibular toxicity (defects of posture, balance, and gait) precedes clinical evidence of hearing loss. Ordinarily, these signs appear only after several weeks of aminoglycoside therapy.

The **diuretics** furosemide, bumetanide, and ethacrynic acid are chemically related, and all are ototoxic to dogs and cats. Other domestic species have not been tested. Electrophysiologic evidence of hearing impairment occurs within a few hours of even a single high dose but is unaccompanied by structural changes within the cochlea. The exception is ethacrynic acid intoxication, in which edema of the stria vascularis is seen ultrastructurally. There is a corresponding alteration in the composition of the endolymph produced by the stria. The hearing impairment, which occurs frequently in humans receiving therapeutic doses of the drugs, is assumed to result from structural or physiologic lesions in the hair cells nourished by this abnormal endolymph.

Acetylsalicylic acid (aspirin) and its derivatives are ototoxic for humans and several laboratory animal models, but no structural lesion has yet been detected.

The antibacterial–anthelmintic agent, **hygromycin B,** as well as causing cataracts in swine (see the Eye) also causes permanent deafness in dogs if therapeutic dosages are given. The drug is not approved for use in dogs. Deafness is reported occasionally in swine receiving the medication in excess of recommended levels. Histologic descriptions of cochlear lesions are not available.

The antiseptic combination of chlorhexidine and cetrimide (Savlon) is a widely used antiseptic that occasionally is used to cleanse the external ear canal. If used in dogs or cats with a ruptured typanic membrane, this solution is toxic to both vestibular and cochlear cells, although the clinical signs in affected animals are primarily vestibular.

4. Other Causes of Deafness

Deafness is reported in humans with various storage diseases, but few animal counterparts have been studied for this specific defect. Goats with β-mannosidosis have deformed ear pinnae, bony exostoses within the tympanic cavity, and deforming accumulation of oligosaccharides within lysosomes of many tissues of middle and inner ear. Epithelial and mesothelial cells of Reissner's membrane, most structural cells of the organ of Corti, and cochlear neurons of the spiral ganglion are all affected, even at a few days of age.

C. Vestibular Dysfunction

Vestibular dysfunction is characterized by head tilt and falling toward the affected side, ataxia without weakness, and nystagmus. Clinical signs are most obvious with unilateral disease. The lesion may be in brain or in the vestibular apparatus, or both. Animals with vestibular dysfunction caused by brain lesions, as in listeriosis or canine distemper, usually show other signs of neurologic dysfunction. Since vestibular signs are more readily detected than is partial hearing loss, mild lesions of the inner ear are more frequently associated with vestibular abnormalities than with hearing deficits.

The causes of peripheral vestibular disease are the same as those causing deafness: uncontrolled otitis media, trauma, invasive neoplasia, and a number of drugs. Congenital vestibular disease has been reported in several dog breeds, but no morphologic observations were given. Nonspecific destruction of part or all of the vestibular apparatus as a sequel to otitis media is by far the most common cause of labyrinthitis in all species.

Idiopathic vestibular disease occurs in old dogs and in cats of any age. In old dogs it is often mistakenly diagnosed as an acute cerebrovascular accident (stroke), but there is no demonstrable brain lesion in such dogs, and most have rapid, spontaneous remission of clinical signs unless prevented by premature euthanasia. No histologic examination of the labyrinth has been reported. An equally mysterious acute vestibular dysfunction affects cats of any age. Recovery usually occurs over a few days to weeks. No lesions are reported.

Bibliography

Bedford, P. G. C. Congenital vestibular disease in the English cocker spaniel. *Vet Rec* **105:** 530–531, 1979.

Blauch, B., and Martin, C. L. A vestibular syndrome in aged dogs. *J Am Anim Hosp Assoc* **10:** 37–40, 1974.

Braniš, M., and Burda, H. Inner ear structure in the deaf and normally hearing Dalmatian dog. *J Comp Pathol* **95:** 295–299, 1985.

Brown, R. D. *et al.* Comparative acute ototoxicity of intravenous

bumetanide and furosemide in the pure-bred beagle. *Toxicol Appl Pharmacol* **48:** 157–169, 1979.
Coleman, J. W. Hair cell loss as a function of age in the normal cochlea of the guinea pig. *Acta Otolaryngol* **82:** 33–40, 1976.
Deol, M. S. The relationship between abnormalities of pigmentation and of the inner ear. *Proc R Soc Lond* **175:** 201–217, 1970.
Gallé, H. G., and Vanker-van Haagen, A. J. Ototoxicity of the antiseptic combination chlorhexidine/cetrimide (Savlon): Effects on equilibrium and hearing. *Vet Q* **8:** 56–60, 1986.
Hayes, H. M. *et al.* Canine congenital deafness: Epidemiologic study of 272 cases. *J Am Anim Hosp Assoc* **17:** 473–476, 1981.
Igarashi, M. *et al.* Inner ear anomalies in dogs. *Arch Otorhinolaryngol* **81:** 249–255, 1972.
Knowles, K. *et al.* Reduction of spiral ganglion neurons in the aging canine with hearing loss. *J Vet Med A* **36:** 188–199, 1989.
Liberman, M. C., and Kiang, N. Y. S. Acoustic trauma in cats. Cochlear pathology and auditory-nerve activity. *Acta Otolaryngol (Suppl)* **358:** 1–63, 1978.
Mair, I. W. S. Hereditary cochleosaccular degeneration. *In* "Spontaneous Animal Models of Human Disease" E. H. Andrews, B. C. Ward, and N. H. Altman (eds.), Vol. 1, pp. 86–88. New York, Academic Press, 1979.
Mair, I. W. S. Hereditary deafness in the Dalmatian dog. *Arch Otorhinolaryngol* **212:** 1–14, 1976.
Quick, C. A. Chemical and drug effects on inner ear. *In* "Otolaryngology" M. M. Paparella and D. A. Shumrick (eds.), Vol. 2, pp. 391–406. Philadelphia, Pennsylvania, Saunders, 1973.
Render, J. A., Lovell, K. L., and Jones, M. Z. Otic pathology of caprine β-mannosidosis. *Vet Pathol* **25:** 437–442, 1988.
Watson, A. D. J., and Burnett, D. C. Hygromycin B and deaf dogs. *Aust Vet J* **66:** 302–303, 1989.
Wilkes, M. K., and Palmer, A. C. Congenital deafness in dobermans. *Vet Rec* **118:** 218–219, 1986.
Wright, Ch. G. Neural damage in the guinea pig cochlea after noise exposure. A light microscopic study. *Acta Otolaryngol* **82:** 82–94, 1976.

V. Neoplasms and Similar Lesions

Neoplasms of the ear include those capable of affecting skin elsewhere, as well as primary tumors of ceruminous glands and, very rarely, tumors of Eustachian or auditory epithelium and of cranial nerves. Inflammatory polyps are included in this section because of their gross resemblance to neoplasia.

Squamous cell carcinoma is by far the most important skin tumor affecting the ear, although the predilection of canine histiocytoma for the ear pinna probably makes it the most frequent tumor affecting the pinna. Squamous cell carcinoma of the ear occurs commonly in white cats, with a frequency more than 10 times that of nonwhite cats, and in sheep in sunny climates. In cats, the tumor has a long precancerous phase consisting of erythematous, ulcerative dermatitis of the ear margin, known as feline solar or actinic dermatitis because of its association with exposure to sunlight. Other sparsely haired areas, such as nose, lips, and eyelids, may be affected. Histologically, the lesion consists of multifocal coalescing epidermal necrosis overlying a diffuse superficial dermal infiltrate of lymphocytes and plasma cells. The lesion waxes and wanes with the intensity of sunlight. Epidermal hyperplasia proceeds to dysplasia, carcinoma *in situ*, and invasive squamous cell carcinoma in a manner analogous to bovine ocular squamous cell carcimona. The lesion is bilateral but not necessarily of synchronous progression. The progression to invasive neoplasia usually occurs over 3 or 4 years. Metastasis is a late occurrence and prevented (at least in the case of ears) by amputation. Rarely, squamous cell carcinoma occurs within the external ear canal or even within the tympanic cavity of the middle ear. Local infiltration results in damage to the seventh cranial nerve with resultant signs of vestibular dysfunction and facial paralysis.

Neoplasms of the ceruminous glands occur in dogs and cats. The tumors are very similar to those of apocrine sweat glands. The ceruminous glands are modified sweat glands within the deep portion of the external auditory meatus. Tumors are relatively more prevalent in cats than in dogs, and in both species they tend to occur in very old animals. In dogs most are benign, but in cats about half are histologically malignant. The adenomas are smooth, nodular, or pedunculated masses seldom exceeding 1.0 cm in diameter. The epithelium overlying the tumor is intact unless there is concurrent otitis externa. Ceruminous gland tumors cannot be distinguished in the live animal from cystic dilatation with epithelial hyperplasia typical of chronic otitis externa. Carcinomas may occasionally invade from the auditory meatus into the region of the parotid salivary gland or into bone, and when very anaplastic, may be confused with salivary carcinoma. The diagnosis of carcinoma usually is based on histologic criteria of anaplasia and local invasiveness rather than on behavioral evidence of metastasis.

Histologically, adenomas are well-differentiated tubular and cystic growths. The epithelial cells are cuboidal and eosinophilic. They may be flattened when the tubular or acinar lumen is dilated. The most typical feature is the presence within lumina of deeply eosinophilic or orange, colloidlike secretion typical of cerumen. Mixed ceruminous tumors analogous to mixed apocrine sweat glands or mammary tumors occur infrequently. Carcinomas do not differ markedly from adenomas but have less secretion and more cellular anaplasia, and show invasion by tumor cells into an abundant fibrous stroma rich in mononuclear leukocytes. Mixed tumors with cartilage and bone are described.

As expected, a variety of other neoplasms have been described in the external ear canal or tympanic cavity of dogs and cats. All would seem to be rare, and no prognostic statements are justified by the small numbers. In one small series in dogs, eight of eleven middle ear neoplasms originated in the external ear and perforated the tympanic membrane. Two were papillary adenomas formed by ciliated columnar epithelium and goblet cells, thought to have arisen from the epithelium of the dorsal portion of the tympanic cavity. The final case was an anaplastic carcinoma of unknown origin involving oropharynx and ear.

Paraganglioma and fibroma in dogs, and fibrosarcoma in cats, are represented by single case reports.

Inflammatory polyps are observed in dogs and more frequently in cats with clinical signs of head shaking, head tilt, ataxia, or nystagmus. These may be the most common "tumor" of the feline ear canal. In cats, the lesion is a loose mass of connective tissue containing numerous small blood vessels and mononuclear leukocytes, covered by epithelium that may be either stratified, nonkeratinized squamous, or simple to bilayered ciliated columnar. Often the ciliated epithelium is found only focally, but nonetheless it is the characteristic feature distinguishing this lesion from nonspecific proliferations of glands and connective tissue seen in many cases of chronic otitis externa. The presence of this ciliated epithelium is used to support theories that such polyps originate from the Eustachian tube, but only rarely has that opinion been confirmed by careful dissection. An alternative origin could be from ciliated epithelium of the tympanic cavity itself. Some of these polyps are grotesque, protruding from outer ear, hanging into the oropharynx, or even protruding through the nose. A familial occurrence has been seen in Abyssinian and Himalayan kittens, further confusing the debate about the cause of these distinctive lesions.

Under the umbrella of inflammatory polyp fall many of the tumorlike proliferations excised from the external ear canal of dogs (rarely cats) with chronic otitis externa. The lesion consists of variable proportions of hyperplastic surface epithelium, hyperplastic or dysplastic sebaceous and ceruminous glands, fibroplasia, and leukocytes. The rupture of the glands often adds the lesions of sterile foreign body periadenitis to the chronic inflammation associated with the bacteria, yeast, or foreign bodies causing the initial otitis itself.

Another nodular lesion, clinically resembling neoplasia, is the dentigerous cyst of heterotopic polyodontia in foals, seen as a draining nodule on the anterior aspect of the base of the pinna.

Bibliography

Bradley, R. L. *et al.* Nasopharyngeal and middle ear polypoid masses in five cats. *Vet Surg* **14:** 141–145, 1985.

Harvey, C. E., and Goldschmidt, M. H. Inflammatory polypoid growths in the ear canal of cats. *J Small Anim Pract* **19:** 669–677, 1978.

Indrieri, R. J., and Taylor, R. F. Vestibular dysfunction caused by squamous cell carcinoma involving the middle ear and inner ear in two cats. *J Am Vet Med Assoc* **184:** 471–473, 1984.

Legendre, A. M., and Krahwinkel, D. J., Jr. Feline ear tumors. *J Am Anim Hosp Assoc* **17:** 1035–1037, 1981.

Little, C. J. L., Pearson, G. R., and Lane, J. G. Neoplasia involving the middle ear cavity of dogs. *Vet Rec* **124:** 54–57, 1989.

Patnaik, A. K., and Birchard, S. J. Canine paraganglioma: A case report. *J Small Anim Pract* **26:** 681–687, 1985.

Pulley, L. T., and Stannard, A. A. Tumors of skin and soft tissue. *In* "Tumors in Domestic Animals" 2nd Ed., J. E. Moulton (ed.), pp. 58–59. Berkeley, California, University of California Press, 1978.

Rogers, K. S. Tumors of the ear canal. *Vet Clin North Am: Small Anim Pract* **18:** 859–868, 1988.

Stanton, M. E. *et al.* Pharyngeal polyps in two feline siblings. *J Am Vet Med Assoc* **186:** 1311–1313, 1985.

Stone, E. A., Goldschmidt, M. H., and Littman, M. P. Squamous cell carcinoma of the middle ear in a cat. *J Small Anim Pract* **24:** 647–651, 1983.

CHAPTER 5 The Skin and Appendages

JULIE A. YAGER
University of Guelph, Canada

DANNY W. SCOTT
Cornell University, U.S.A.

With a contribution by
Brian P. Wilcock
Ontario Veterinary College, Canada

I. General Considerations

The integument forms the anatomic boundary between the animal and its external environment and is the largest of the body systems. Its crucial functions include protection against external physical and chemical damage, thermoregulation, conservation of critical substances, particularly water, synthesis of vitamin D, secretory activity, immunosurveillance, and a vital sensory function, which makes the skin one of the principal organs of communication between the animal and its surroundings.

The intact skin is remarkably resistant to the wide range of physicochemical and microbiological insults to which it is exposed continuously. The skin maintains a macroscopically normal appearance over a wide range of environmental conditions and heals quickly after mild physical damage. Any breach in the integrity of the skin or any alteration in its barrier function may tip the balance in favor of potential pathogens. Simple wetting of the skin and fleece, for example, predisposes a sheep to bacterial fleece rot, which in turn may provide an opportunity for fatal fly strike. The skin may also reflect systemic diseases from generalized infections to nutritional deficiencies, and the development of cutaneous lesions often depends on both local and systemic factors. For example, abnormalities in keratinization due to systemic metabolic changes may alter the surface microenvironment allowing sufficient bacterial proliferation to cause disease.

Formerly, the epidermis was thought of as the passive player in inflammatory reactions of the integument. That concept is no longer tenable. The keratinocyte is capable of producing, with appropriate stimuli, a great variety of the inflammatory mediators. The keratinocyte can also express receptors for these molecules and thus function in an autocrine fashion. Interleukin 1 (IL-1), an important up-regulator of inflammation, is produced constitutively by keratinocytes but, unless there is epidermal damage, IL-1 is lost, along with the exfoliating surface cells, and homeostasis is maintained. Other cytokines that have been demonstrated as products of the activated keratinocyte include IL-3, IL-6, IL-8, tumor necrosis factor, and colony stimulating factors. Not only can keratinocytes influence the movement of inflammatory cells into the epidermis by cytokines such as IL-8, but, by the expression of adhesion molecules such as intercellular adhesion molecule (ICAM)-1 under the stimulus of gamma interferon, cause retention of lymphocytes through interaction with lymphocyte function-associated antigen (LFA)-1. Many of the diseases we fail to understand may prove to be the result of abnormal production of these cytokines or an abnormal regulation of receptor expression.

Many skin diseases of food-producing animals are of considerable economic significance. Examples include sarcoptic mange in pigs, fly strike and foot rot in sheep, and dermatophilosis in tropical cattle. Dermatological diseases also form a large proportion of the typical case load in small animal veterinary practice. Reflecting the regrettable inbreeding that is so widely practiced, many of these diseases are breed associated, and some are clearly inherited. In the 1980s there was a tremendous expansion in the number of dermatological diseases recognized. This was not always accompanied by a concomitant increase in the understanding of their cause and pathogenesis. Skin biopsy is now a routine procedure in veterinary medicine, and pathologists in diagnostic laboratories must be knowledgeable in this area.

It is very important that the pathologist be familiar with the normal skin structure, not only in the different species but also in different anatomical locations. Animal skin differs morphologically from that of humans, making extrapolation from human histology texts misleading. Species-to-species variation is also marked in terms of epidermal thickness, dermal thickness, types and arrangements of hair follicles, and adnexal structures. Similarly, there is substantial variation in morphology between regions of the body in the individual animal. Tissue sections from species-, age-, sex-, and site-matched controls may be required before subtle lesions, such as alterations in dermal collagen, can be appreciated.

The integument is composed of epidermis, dermis, hair follicles, adnexal glands, and the subcutis. The appendages include nails, hooves, and claws.

Bibliography

Ackerman, A. B. "Histologic Diagnosis of Inflammatory Skin Diseases," Philadelphia, Pennsylvania, Lea & Febiger, 1978.

Fitzpatrick, T. B. *et al.* "Dermatology in General Medicine," 3rd Ed. New York, McGraw-Hill, 1987.

Gross, T. L., Irhke P. J., and Walder, E. J. "Veterinary Dermatopathology: A Macroscopic and Microscopic Evaluation of Canine and Feline Skin Disease," St. Louis Missouri, Mosby–Year Book, 1992.

Kupper, T. S. Production of cytokines by epithelial tissues. A new model for cutaneous inflammation. *Am J Dermatopathol* **11:** 69–73, 1989.

Lever, W. F., and Schaumburg-Lever, G. "Histopathology of the Skin," 6th Ed. Philadelphia, Pennsylvania, Lippincott, 1983.

Luger, T. A., and Schwarz, T. Evidence for an epidermal cytokine network. *J Invest Dermatol* **95:** 100S–104S, 1990.

Muller, G. H., Kirk, R. W., and Scott, D. W. "Small Animal Dermatology," 3rd Ed. Philadelphia, Pennsylvania, W.B. Saunders, 1989.

Nickoloff, B. J., Griffiths, C. E. M., and Barker, J. N. W. N. The role of adhesion molecules, chemotactic factors, and cytokines in inflammatory and neoplastic skin disease—1990 update. *J Invest Dermatol* **94:** 151S–157S, 1990.

Pinkus, H., and Mehregan, A. H. "A Guide to Dermatohistopathology," 3rd Ed. New York, Appleton-Century-Crofts, 1976.

Scott, D. W. Feline dermatology 1900–1978: A monograph. *J Am Anim Hosp Assoc* **16:** 331–459, 1980.

Scott, D. W. "Large Animal Dermatology," Philadelphia, Pennsylvania, W.B. Saunders, 1988.

Scott, D. W. Feline dermatology 1986–1988: Looking to the 1990s through the eyes of many counselors. *J Am Anim Hosp Assoc* **26:** 515–537, 1990.

Spearman, R. I. C. The biochemistry of skin disease. *Mol Aspects Med* **5:** 63–126, 1982.

A. *Epidermis*

The epidermis varies in thickness according to the species and to the anatomic site. In general, the thickness of the epidermis varies inversely with hairiness. For example, in the cat, the glabrous skin of footpad and planum nasale may be 900 μm in thickness compared to 25 μm in haired skin. Of the domestic animal species, pigs have the thickest epidermis in haired skin, at 70 to 140 μm.

The epidermis may be viewed as a vast holocrine gland in continual production. The basal cells are the only mitotically active cells in the stratified squamous epithelium. Keratinocytes leaving the stratum basale undergo terminal differentiation during their outward migration. The end product, the dead squames of the stratum corneum, sloughs at the skin surface. The process of keratinization is a genetically programmed set of complex biochemical reactions, which include (1) the formation and subsequent stabilization of fibrillar and matrix prekeratin proteins, which are coded for by separate genes; (2) differential expression of cell-surface proteins, including adhesion molecules; (3) formation of a chemically resistant protein envelope, thus giving the stratum corneum many of its mechanical properties; and (4) concomitant breakdown of the nucleus and cellular organelles of the keratinocytes by hydrolytic enzymes released into the cytoplasm of the fully differentiated cells.

The **stratum basale** is the deepest and germinative level of the epidermis. It consists of a single layer of cells, which are usually more cuboidal than columnar in animal skin. While basal cells are mitotically active, mitotic figures are rarely seen in normal animal skin. This is partly because of the rate of turnover and also because of the effect of circadian rhythm, which tends to synchronize mitoses with sleeping periods in some animal species, notably rodents. Basal cells express a specific set of keratin filaments, K5 and K1. They also express surface receptors belonging to the integrin and cadherin families. These molecules mediate cell-to-cell and cell-to-substrate adherence and have profound influence on keratinocyte growth and differentiation. Not surprisingly, as the cells differentiate, the types of keratin formed and the cell-surface molecules expressed alter.

The **stratum spinosum** is composed of one or more layers of polygonal cells. The artefactual shrinkage of the cells in fixed preparations, except at the frequent desmosomal junctions, gives rise to the intercellular bridges or spinous projections for which the layer is named. These are not very prominent in the epidermis of normal haired animal skin. The stratum spinosum is usually between 2 and 5 cells thick in hairy skin, thinner in dogs and cats than in cattle, horses, and pigs, but may be up to 19 cell layers thick in areas such as the canine or feline footpad or the nasal planum.

Ultrastructurally, the cytoplasm of the stratum spinosum cells is filled with keratin filaments, anchored to the cytoplasmic membrane at the desmosomal junctions. These intermediate filaments are approximately 8 nm in diameter and comprise acidic and basic subunits, which assemble in pairs. The subunits of keratin vary not only with the state of differentiation in normal epidermis but with the anatomical site and the physiological state of the epidermis. While normal suprabasal cells express the keratin pair K1 and K10, skin undergoing hyperplasia expresses K6 and K16. The precursor proteins for the cornified envelope are also produced in the stratum spinosum, but are inactive.

The **stratum granulosum** is often discontinuous and only one cell thick in the haired skin of animals but may be 2–4 cells thick around hair follicle ostia and up to 8 cells thick in footpads. It is composed of nucleated, flattened keratinocytes, which are distinguished by the blue-black granules of keratohyalin. These contain profilaggrin, the precursor of the matrix protein which eventually glues together the keratin filaments.

The **stratum corneum** is composed of many layers of flattened, terminally differentiated, dead cells. The conventional fixation and embedding techniques used to prepare microscopic sections result in the loss of up to 50% of this layer. In unfixed canine skin, frozen sections show the stratum corneum to be up to 47 cell layers thick, compared with the 7–10 layers in fixed tissue. The thickness of the stratum corneum also varies with breed within species.

In the stratum corneum, profilaggrin forms filaggrin, which heavily cross-links macrofilaments of keratin. The cornified cells are bound together by an insoluble lipid-rich extracellular matrix as bricks are bound by mortar. Transglutaminase enzymes are critical to the cross-linking of the cornified envelope precursor proteins, such as involucrin, keratolinin, and loricrin. These form, along with the intercellular lipids, the critical permeability barrier. The types and distribution of these proteins and lipids will vary somewhat between the species. In sheep, the outer layers of stratum corneum are covered by a lipid layer 9–14 μm thick, representing emulsified suint and sebum.

B. Immigrant Cells

Keratinocytes make up 85% of the epidermis; the remaining cells resident in the epidermis are melanocytes (5–8%), Langerhans' cells (5%), Merkel cells, and indeterminate dendritic cells. The dendritic melanocytes and Langerhans' cells are not connected by intercellular junctions, and there are no desmosomal connections to adjacent keratinocytes. Merkel cells, which occur in tylotrich pads, have desmosomal attachments to adjacent keratinocytes. They contain typical dense-core neurosecretory cytoplasmic granules and function as mechanoreceptors.

The **melanocytes** are numerous in pigmented skin but sparse or absent in nonpigmented skin. Cats, for example, have very few epidermal melanocytes in haired skin, but melanocytes are found in prepuce, footpads, and planum nasale. Melanocytes are numerous in the hair follicles, except in white animals. Melanin may be present in dermal macrophages in normal animals, but the presence of melanophages often indicates a pathologic process in the overlying epithelium, involving destruction of melanin-containing basal cells. The melanocytes lie in the basal layer, sometimes appearing in hematoxylin and eosin stained sections as "clear" cells. Their dendritic nature is more evident on silver staining or by ultrastructural examination. The melanocytes extend their dendritic processes around adjacent keratinocytes, forming an "epidermal-melanin unit." Melanin is transferred to the keratinocytes, and the pigment migrates to the surface with the differentiating cells. In unfixed tissue, melanocytes may be identified by the Dopa reaction, which imitates *in vitro* the physiologic melanin formation *in vivo*. Melanocytes also stain positively with antibodies to S-100, but this reaction is not limited to melanocytes.

Langerhans cells have been identified in animal skin, although they do not always possess the typical "tennis racket" organelle (Birbeck granule) which is so characteristic of the ultrastructure of the human equivalent. Birbeck granules have been identified in bovine Langerhans cells but not in those of the dog or cat. Langerhans cells are members of the monocyte–macrophage system and are the major antigen-presenting cells of the epidermis. They reside not only in the epidermis but also migrate, after antigenic stimulation, to the local lymph nodes. They are pivotal cells in contact sensitization. Langerhans cells are not distinguished on routine sections but are stained specifically by gold chloride impregnation. They are also argyrophilic but are not stained with the argentaffin reaction. Various cytochemical reactions may be used to identify these cells in different animal species. Bovine Langerhans cells, for example, stain for adenosine triphosphatase and alkaline phosphatase. Using the latter reaction, a density of ~1600 Langerhans cells/mm^2 was identified in bovine epidermis. Langerhans cells typically express class II major histocompatability complex (MHC) and CD1a molecules. Class II antigens have been detected on canine and bovine Langerhans cells, and CD1a molecules have been demonstrated on canine Langerhans cells.

Intraepidermal lymphocytes, members of the skin immune system, are present in domestic animals, but have not been studied extensively. In humans and rodents, intraepidermal lymphocytes represent ~2% of all T cells in the skin. Being CD8 positive, they function as suppressor/cytotoxic cells. T cells bearing the γ/δ receptor are highly epidermotropic in the mouse and reside in the outer layers of the epidermis. They are phylogenetically more primitive than the α/β T cells and probably function in non-MHC-restricted antigen recognition and subsequent cytotoxicity. Cutaneous T cells probably monitor the skin surface in the same manner as long-lived recirculating T cells monitor the mucosal surfaces of the gastrointestinal and respiratory tracts.

C. Dermoepidermal Junction

This is typically straight in hairy animal skin in contrast to the undulating border of human skin, largely due to the anchoring function of the hair follicles, which probably renders epidermal rete ridges unnecessary. The pig, with its sparse hair coat, has better developed rete ridges than the other domestic animal species. Rete ridges are also present in nonhaired areas such as footpads and nasal planum. The dermoepidermal junction is marked by the 40- to 60-nm thick basement membrane zone. It is composed of the following layers: lamina lucida externa, lamina densa, lamina lucida interna, and anchoring filaments. The basement membrane zone is usually indistinct in routinely stained sections of hairy skin, except in the horse, but may be demonstrated by periodic-acid Schiff or by silver stains. These stain, respectively, the glycosaminoglycan layer immediately beneath the submicroscopic basal lamina (collagen type IV), and the fine reticulin fiber network beneath the glycosaminoglycan layer. Ultrastructurally, the epidermis is seen to be joined to the dermis by a series of hemidesmosomes and criss-crossing anchoring fibrils (collagen type VII), which allow for the shearing movement which occurs at the skin surface. The basement membrane zone also contains laminin, nidogen, fibronectin, heparan sulfate, and osteonectin. Of these, the interaction of fibronectin with its receptor on cells of the stratum basale is particularly important in attachment of the epidermis to the dermis. Both the keratinocytes and the dermal fibroblasts contribute to the formation of the many components of the basement membrane zone.

D. Dermis

This is divided into superficial and deep layers. The superficial dermis is equivalent to the papillary dermis of humans in its structure, having finer fibers of collagen than the deeper layer. The dermis is composed of collagen and elastin fibers embedded in a glycosaminoglycan-rich ground substance. Collagen, elastin, and most of the glycosaminoglycans are produced predominantly by the dermal fibroblasts.

Collagens are a superfamily of closely related but genetically distinct proteins. Dermal collagen fibers are predominantly of type I and type III collagen. Collagen types IV and VII are constituents of the basement membrane zone, the latter comprising the anchoring fibrils. Collagen types IV and V are also present around blood vessels and hair follicles. Type V has been described forming a network around fiber bundles of type I and III collagen. Type VI is distributed throughout the dermis. The second fibrillar component of the dermis, elastic fibers, compose less than 1% of the dry weight.

The fine collagen bundles in the superficial dermis tend to be parallel to the surface. In the deep dermis, the collagen bundles are approximately three times thicker than those of the superficial dermis and form a closely packed and interwoven layer. The elastin fibers do not stain routinely but may be demonstrated by Verhoeff's stain or orcein-Giemsa. The orcein-Giemsa stain is particularly useful, not only because it demonstrates elastic fibers, but also because it stains the metachromatic granules of mast cells, differentiates smooth muscle from collagen, and demonstrates keratinized cells, bacteria, and fungi better than hematoxylin and eosin.

The principal **glycosaminoglycans** of skin are hyaluronic acid and dermatan sulfate, with contributions from chondroitin-4 and chondroitin-6 sulfate and heparan, the last released from mast cells. Most of these are associated with proteins and are thus termed proteoglycans. The large proteoglycan polymers are interspersed within the fibrous matrix and fill the interstitial spaces. The extracellular matrix is no longer considered merely an inert structural support. Differentiated cell behavior is dependent on complex interactions of cells with the surrounding matrix.

The proteoglycans are important in maintaining the normal hydration of the dermis. Hygroscopic properties of molecules such as hyaluronic acid are responsible. When water is lost on routine tissue processing, the glycosaminoglycan residues shrink to very fine filaments. These are extremely sparse in normal skin, where they are concentrated around blood vessels and adnexal structures. The filaments usually require alcian blue, colloidal iron, or toluidine blue for their demonstration.

The dermis contains cellular elements in addition to the fibroblasts. **Mast cells** reside around blood vessels and are extremely variable in number at different anatomical locations and in different species. The cat skin typically has numerous dermal mast cells, up to 20 per high-power field. In dogs, the normal accepted range is 4–12 cells per high-power field. Dermal mast cells are of the connective tissue type. The granules, in formalin-fixed tissue, stain metachromatically with toluidine blue. Some staining heterogeneity has been demonstrated, however, in mast cell populations in normal and atopic dogs. Dermal mast cells are thought to play an important role in the initiation of inflammatory and immune responses in the skin, because they secrete, on appropriate stimulation, tumor necrosis factor-α. This induces the adhesion molecule ELAM-1 on vascular endothelial cells, promoting adhesion of immune cells to dermal vessels. T lymphocytes, dermal dendritic cells, and macrophages are present around postcapillary venules; proper differentiation of these mononuclear cells may require cell-surface marker reagents not yet widely available for many of the domestic animal species. Although not prominent in hematoxylin and eosin-stained sections of normal skin, these cells are functionally very important in the skin immune system. Eosinophils and neutrophils are very rare in normal skin.

The **cutaneous vasculature** is divided into intercommunicating superficial, middle, and deep plexuses. The deep plexus occurs at the subcutaneous–dermal junction, the middle plexus lies at the level of the sebaceous glands, and the superficial plexus sends capillary loops to the dermoepidermal junction. Lymphatics are present in the superficial dermis and around adnexae but usually are not visible in paraffin sections of normal skin. Nerve fibers, in general, follow the blood vessels.

E. Hair Follicle

In all species of domestic animals, follicles or groups of follicles tend to occur in triads. There are two types of follicles, simple and compound. The first refers to a single primary follicle. Compound follicles comprise a grouping of primary follicles surrounded by smaller secondary follicles. Simple hair follicles occur in cattle and horses, while sheep, goats, dogs, and cats have compound hair follicles. In sheep, compound follicles comprise a large central primary hair with two or more lateral primaries, each surrounded by their secondary hairs. In dogs, the primary (P) and secondary (S) hairs usually emerge from a common follicular infundibulum. The number and arrangement of the secondary hair follicles in the compound group vary between species, between breeds within a species, and at different anatomic locations in an individual animal. For example, in fine-wooled Merino sheep the S : P ratio may be 20 : 1 compared to 4 : 1 in mountain breeds of sheep, and the compound follicles on the ventrum of the cat show a higher S : P ratio than those of the dorsum. Hair follicles undergo cyclic growth, divided into three phases: anagen (actively growing), catagen (transitional), and telogen (resting). The base of the anagen follicle rests at the junction of the dermis and subcutis in the canine and feline integument, whereas it usually lies in the mid-dermis of horses and cattle. The telogen follicle bases are found closer to the epidermis than the anagen follicle bases in all species.

The **anagen hair follicle** comprises the dermal papilla, the hair matrix within the hair bulb, the hair shaft, and the inner and the outer root sheaths. The hair matrix surrounds the inductive dermal papilla with its primitive mesenchymal cells. In pigmented hairs, large numbers of melanocytes lie between the hair matrix cells. The hair matrix produces both the hair and the internal root sheath. The internal root sheath, which comprises three layers (a cuticle, Huxley's, and Henle's layers), hardens first so as to form a rigid tube of internal root-sheath protein, which protects the zone of keratinization in the newly forming hair shaft. Huxley's layer contains large, brightly eosinophilic granules known as trichohyalin. These granules contain a high concentration of arginine, a precursor of citrulline, which is probably involved in cross-linking of the keratins of the internal root sheath. The cuticle of the internal root sheath, which does not stain in routine sections, interdigitates with the cuticle of the hair shaft. The hair shaft has, in addition to the outer cuticle, a cortex and a medulla. Fetal (lanugo) hairs and smaller secondary hairs in the goat lack the medulla. The cortex consists of fully keratinized spindle-shaped cells oriented parallel to the hair shaft. These do not stain with eosin and contain most of the pigment granules from which the hair derives its color. Hair keratin is hard or α keratin. It is composed of 7.5-nm diameter filamentous proteins rich in cysteine and tyrosine. The filaments are embedded in globular matrix proteins. There are marked species differences in the composition of the filamentous keratin proteins, while the composition of the globular proteins is more constant. The medulla is also composed of nonstaining keratinized cells which, in some species, contain vacuoles and glycogen and, sometimes, melanin. The internal root sheath breaks up at the level of attachment of the arrector pili muscle to the external root sheath. The pilar canal above is lined by the external root sheath, which is an invagination of the surface epithelium. In the proximal portion or infundibulum, it undergoes keratinization in the manner of the surface epithelium, usually with a well-developed granular layer. In the mid-portion, known as the isthmus, the keratinization is designated tricholemmal, and lacks a granular layer. In the deep or inferior portion of the follicle, and adjacent to the internal root sheath, the external root sheath is not keratinized, and the cells have a clear cytoplasm because of high glycogen content. The external root sheath is continuous with the base of the hair bulb, which is continuous with the hair matrix. The basement membrane zone of the external root sheath is thin in anagen hairs. It is invested by a sheath of dense fibrous connective tissue.

Catagen, the intermediate involuting phase, is relatively rapid so that few hairs in this stage are seen in normal skin (less than 7%). Following cessation of mitosis in the hair matrix, the last keratinized portion of hair shaft forms a brushlike fringe of eosinophilic keratin, which interdigitates with tricholemmal keratin to form the hair club. The epithelial structures distal to this are resorbed by autophagy, and the basement membrane zone of the degenerating follicle becomes folded and wrinkled to form the thick, hyaline membrane known as the glassy membrane. Eventually, the dermal papilla moves up to the hair germ.

Telogen or resting hairs are attached to the hair germ by the club, which holds the hair in place until it is dislodged by the next-generation hair. The dermal papilla remains as a small ball of dermal mesenchyme beneath the inconspicuous hair germ. These germ cells have been considered to represent the resting progenitor cell population for the next cycle of hair growth. It is now hypothesized that the cells of the "bulge" region of the external root sheath contain the progenitor cells for the matrix of the next growth cycle.

Specialized hairs include sinus hairs and tylotrich hairs. They occur in all species. The best-known examples of sinus hairs are vibrissae or whiskers on the face, but sinus hairs occur at other sites, for example, on the palmar aspect of the carpus in cats. The distinctive anatomical feature of the very large sinus hair follicle is an endothelial-lined blood sinus, which lies between the external sheath of the follicle and the outer part of a connective tissue capsule. Projecting into the upper part of the sinus is the sinus pad. The lower part of the sinus is traversed by trabeculae and supplied by sensory nerve fibers. The sinus hairs function as slow-adapting mechanoreceptors. Tylotrich hairs are scattered among normal hairs, and function as rapid-adapting mechanoreceptors. They are large primary follicles that have a ring of neurovascular tissue at the level of the sebaceous gland. The tylotrich follicle is associated with a tylotrich pad. These are focal areas of epidermal thickening subtended by a layer of well-vascularized and innervated connective tissue. The myelinated nerves arborize in the connective tissue but become unmyelinated just before ending on a plaque enclosed within specialized neurosecretory cells (Merkel cells) located in the basal layer of the epithelium. The tylotrich pads serve as rapid-adapting mechanoreceptors.

F. Sebaceous Glands

The sebaceous glands are holocrine glands which develop as part of the hair follicle complex. Each primary hair has its own sebaceous gland, which usually comprises 2–3 simple alveoli opening via squamous epithelium-lined ducts into the upper part of the hair follicle. The secondary hair follicles usually share sebaceous glands. Free sebaceous glands may occur on glabrous skin, such as the lip and anus. The sebaceous glands vary in size and structure between different species and at various anatomical sites within a species. In general, the larger sebaceous glands are found where hair follicle density is lowest. Particularly large accumulations of sebaceous glands may be afforded a special name, as in the submental organ of the cat and the tail gland in the dog. Some of these glands produce semiochemicals which mediate important aspects of animal behavior. Other functions attributed to sebum include inhibition of surface microbial growth, prevention of microbial invasion, protection against high temperatures by promoting light reflection from a glossy coat, and protection against surface desiccation.

The alveoli of the sebaceous glands are lined by cuboidal basal epithelial cells. These reserve cells differentiate into vacuolated sebocytes, which then degenerate in the necrotic zone to form the sebum. However, the ultrastructural observation that the cells on the lateral surfaces of the alveoli undergo keratinization has cast some doubt on the accepted mode of sebaceous secretion. Keratinized cells are unlikely to transform back into lipid-laden sebocytes. It is hypothesized, instead, that sebum is produced in a column from cells at the base of the alveolus, in a manner somewhat analogous to the production of hair.

G. Sweat Glands

There are two types of coiled tubular sweat glands in animals. The most numerous type develops embryologically as part of the hair follicle complex. They are found in all haired skin areas, although associated only with primary hair follicles. These glands have been known as **apocrine glands** because, on light microscopy, the mechanism of secretion appears to be a pinching off of apical blebs of cytoplasm. More recent studies have shown that this is largely an artefact. Sweat production from these glands results from the combined processes of cell death (holocrine secretion), vesicle exocytosis, active ion and water transport, and a minor contribution from microapocrine blebbing.

The second type of sweat gland is of more recent phylogenetic occurrence. These glands develop independent of the hair follicle and are thus found only on glabrous skin, such as footpad and hoof in animals. They have been designated the **eccrine gland,** on the basis of assumptions about the mode of secretion. However, as with the apocrine gland, the eccrine gland adopts a variety of secretory modes, including microapocrine blebbing. It appears that the terminology based on function is no longer tenable. We have adopted the proposal that the sweat glands be renamed, with the terms **paratrichial** to replace apocrine and **atrichial** for eccrine.

There are more similarities than dissimilarities between the structures of the two types of sweat glands. Both are tubular structures, consisting of a fundus and a duct. The shape of the fundus and the degree of coiling vary between species. The fundus of both peritrichial and atrichial sweat glands is lined by an inner secretory epithelium and an outer myoepithelial layer. The myoepithelium forms a complete layer in the paratrichial glands of ruminants but is discontinuous in the horse. The height of the epithelium has been assumed to change with the secretory status of the cell but, in the cow, the flat secretory cells of the paratrichial glands do not alter in height when active. The ducts of both types of glands are lined by a bilayer of cuboidal epithelial cells. The duct of the paratrichial gland traverses the external root sheath and discharges into the infundibulum, whereas the duct of the atrichial gland traverses the epidermis and opens directly to the skin surface. Specialized nasolabial glands occur in the dermis of the ruminant muzzle. These mucus-secreting acinar glands are analogous to salivary glands.

H. Arrector Pili Muscle

The arrector pili muscles are smooth muscles which insert on a bulge in the external root sheath of the hair follicles. They are largest in the areas where the hairs are often erect, such as along the dorsum in cats and dogs. The muscle fibers are often vacuolated, particularly in older animals.

I. Subcutis

The subcutis is composed of lobules of adipose tissue supported by a fibrous tissue stroma containing inconspicuous blood vessels and nerves. In the dog and cat, the junction between the dermis and subcutis undulates as the adipose tissue extends superficially to invest the base of anagen hair follicles.

Bibliography

Al-Bagdadi, F. K., Titkemeyer, C. W., and Lovell, J. E. Histology of the hair cycle in male beagle dogs. *Am J Vet Res* **40:** 1734–1741, 1979.

Amakiri, S. F. Melanin and Dopa-positive cells in the skin of tropical cattle. *Acta Anat* **103:** 434–444, 1979.

Amakiri, S. F., Ozoya, S. E., and Ogunnaike, P. O. Nerves and nerve endings in the skin of tropical cattle. *Acta Anat* **100:** 391–399, 1978.

Blazquez, N. B. *et al.* A pheromonal function for the perineal skin glands in the cow. *Vet Rec* **123:** 49–50, 1988.

Britt, A. G. *et al.* Structure of the epidermis of Australian Merino sheep over a 12-month period. *Aust J Biol Sci* **38:** 165–174, 1985.

Bryan, L.A. *et al.* Immunocytochemical identification of bovine Langerhans cells by use of a monoclonal antibody directed against class II MHC antigens. *J Histochem Cytochem* **36:** 991–995, 1988.

Cotsarelis, G., Sun, T.-T., and Lavker, R. M. Label-retaining cells reside in the bulge area of pilosebaceous unit: Implications for follicular stem cells, hair cycle, and skin carcinogenesis. *Cell* **61:** 1329–1337, 1990.

Emerson, J. L., and Cross, R. F. The distribution of mast cells in normal canine skin. *Am J Vet Res* **26:** 1379–1382, 1965.

Forest, J. W., Fleet, M. R., and Rogers, G. E. Characterization of melanocytes in wool-bearing skin of Merino sheep. *Aust J Biol Sci* **38:** 245–257, 1985.

Goldsberry, S., and Calhoun, M. L. The comparative histology of the skin of Hereford and Aberdeen Angus cattle. *Am J Vet Res* **20:** 61–68, 1959.

Hauser, C. *et al.* Interleukin-1 is present in normal human epidermis. *J Immunol* **136:** 3317–3323, 1986.

Hay, E. D. Extracellular matrix. *J Cell Biol* **91:** 205s–233s, 1981.

Jenkinson, D. M. The topography, climate, and chemical nature of the mammalian skin surface. *Proc R Soc Edin* **79B:** 3–22, 1980.

Jenkinson, D. M. Sweat and sebaceous glands and their function in domestic animals. *In* "Advances in Veterinary Dermatology," Vol. 1, C. von Tscharner and R.E.W. Halliwell (eds.), pp. 229–251. London, Baillière Tindall, 1990.

Jenkinson, D. M., and Lloyd, D. H. The topography of the skin surface of cattle and sheep. *Br Vet J* **135:** 376–379, 1979.

Jenkinson, D. M., Montgomery, I., and Elder, H. Y. The ultrastructure of the sweat glands of the ox, sheep, and goat during sweating and recovery. *J Anat* **129:** 117–140, 1979.

Jenkinson, D. M. *et al.* Comparative studies of the ultrastructure of the sebaceous gland. *Tissue Cell* **17:** 683–698, 1985.

Kozlowski, G. P., and Calhoun, M. L. Microscopic anatomy of the integument of sheep. *Am J Vet Res* **30:** 1267–1279, 1969.

Lindholm, J. S. *et al.* Variation of skin surface lipid composition among mammals. *Comp Biochem Physiol* **69B:** 75–78, 1981.

Lloyd, D. H., Amakiri, S. F., and Jenkinson, D. M. Structure of the sheep epidermis. *Res Vet Sci* **26:** 180–182, 1979.

Lloyd, D. H., and Garthwaite, G. Epidermal structure and surface topography of canine skin. *Res Vet Sci* **33:** 99–104, 1982.

Lloyd, D. H., Dick, W. D. B., and Jenkinson, D. M. Structure of the epidermis in Ayrshire bullocks. *Res Vet Sci* **26:** 172–179, 1979.

Lloyd, D. H., Dick, W. D. B., and Jenkinson, D. M. The effects of some surface-sampling procedures on the stratum corneum of bovine skin. *Res Vet Sci* **26:** 250–252, 1979.

Lovell, J. E., and Getty, R. The hair follicle, epidermis, dermis, and skin glands of the dog. *Am J Vet Res* **18:** 873–885, 1957.

Marcarian, H. Q., and Calhoun, M. L. Microscopic anatomy of the integument of adult swine. *Am J Vet Res* **27:** 765–772, 1966.

Meyer, W., Schwarz, R., and Neurand, K. The skin of domestic mammals as a model for the human skin, with special reference to the domestic pig. *Curr Prob Dermatol* **7:** 39–52, 1978.

Montagna, W., and Parakkal, P. F. "The Structure and Function of Skin," 3rd Ed. New York, Academic Press, 1974.

Montgomery, I., Jenkinson, D. M., and Elder, H. Y. The effects of thermal stimulation on the ultrastructure of the fundus and duct of the equine sweat gland. *J Anat* **135:** 13–28, 1982.

Munger, B. L., and Ide, C. The structure and function of cutaneous sensory receptors. *Arch Histol Cytol* **51:** 1–34, 1988.

Oldberg, A. *et al.* Structure and function of extracellular matrix proteoglycans. *Biochem Soc Trans* **18:** 789–792, 1990.

Reynolds, A. J., and Jahoda, C. A. B. Hair follicle stem cells? A distinct germinative epidermal cell population is activated *in vitro* by the presence of hair dermal papilla cells. *J Cell Sci* **99:** 373–385, 1991.

Sar, M., and Calhoun, M. L. Microscopic anatomy of the integument of the common American goat. *Am J Vet Res* **27:** 444–456, 1966.

Sato, K. *et al.* Biology of sweat glands and their disorders. I. Normal sweat gland function. *J Am Acad Dermatol* **20:** 537–565, 1989.

Scott, D. W. The biology of hair growth and its disturbances. *In* "Advances in Veterinary Dermatology," Vol. 1, C. von Tscharner and R.E.W. Halliwell (eds.), pp. 3–33. London, Baillière Tindall, 1990.

Strickland, J. H., and Calhoun, M. L. The integumentary system of the cat. *Am J Vet Res* **24:** 1018–1029, 1963.

Suter, M. M. *et al.* Keratinocyte differentiation in the dog. *In* "Advances in Veterinary Dermatology," Vol. 1, C. von Tscharner and R.E.W. Halliwell (eds.). pp. 252–264. London, Baillière Tindall, 1990.

Talukdar, A. H. A histological study of the dermoepidermal junction in the skin of horse. *Res Vet Sci* **15:** 328–332, 1973.

Talukdar, A. H., Calhoun, M. L., and Stinson, A. W. Sweat glands of the horse: A histologic study. *Am J Vet Res* **31:** 2179–2190, 1970.

Talukdar, A. H., Calhoun, M. L., and Stinson, A. W. Microscopic anatomy of the skin of the horse. *Am J Vet Res* **33:** 2365–2390, 1972.

Uitto, J., Olsen, D. R., and Fazio, M. J. Extracellular matrix of the skin: 50 years of progress. *J Invest Derm* **92:** 61S–77S, 1989.

Webb, A. J., and Calhoun, M. L. The microscopic anatomy of the skin of mongrel dogs. *Am J Vet Res* **15:** 274–280, 1954.

Wollina, U., Berger, U., and Mahrle, G. Immunohistochemistry of porcine skin. *Acta Histochem* **90:** 87–91, 1991.

Wynn, P. C. *et al.* Characterization and localization of receptors for epidermal growth factor in ovine skin. *J Endocrinol* **121:** 81–90, 1989.

II. Dermatohistopathology

Dermatohistopathology has a specialized vocabulary. The definitions employed in this chapter are elaborated on in succeeding sections.

A. Epidermal Changes

Hyperkeratosis is an increased thickness of the stratum corneum, and it may be absolute if the epithelium is of normal or increased thickness, or it may be relative to a thinned underlying epidermis. The type of hyperkeratosis is further specified as **orthokeratotic (anuclear)** and **parakeratotic (nucleated)** (Fig. 5.1A,B). Ortho- and parakeratotic hyperkeratosis are commonly referred to as hyperkeratosis and parakeratosis, respectively. Other adjectives describing hyperkeratosis include basket-weave (e.g., dermatophytosis, endocrinopathies), compact (e.g., lichenoid dermatoses, cutaneous horns) and laminated (e.g., ichthyosis).

Ortho- and parakeratotic hyperkeratosis may be seen as alternating layers in the stratum corneum. This implies episodic changes in epidermopoiesis. If the changes are generalized, the lesions appear as horizontal layers. If the changes are focal, the resultant lesion is a vertical defect in the stratum corneum. Ortho- and parakeratotic hyperkeratosis are common findings in any chronic dermatosis. They simply imply altered epidermopoiesis, whether inflammatory, hormonal, neoplastic, or developmental in origin.

The presence of diffuse parakeratotic hyperkeratosis is most consistent with ectoparasitism, zinc-responsive dermatoses, some vitamin A-responsive dermatoses, thallotoxicosis, dermatophilosis, and dermatophytosis.

Focal parakeratotic hyperkeratosis overlying epidermal papillae (parakeratotic caps) wherein the subjacent dermal papillae are edematous (papillary squirting) is seen in primary idiopathic seborrheic dermatitis of dogs and horses.

Diffuse orthokeratotic hyperkeratosis suggests endocrinopathies, nutritional deficiencies, secondary seborrheas, and developmental abnormalities (ichthyosis, hypotrichosis, color mutant alopecia). Orthokeratotic hyperkeratosis which is disproportionately severe in the hair follicles suggests vitamin A-responsive dermatosis, vitamin A deficiency, acne, schnauzer comedo syndrome, and follicular dysplasia syndromes.

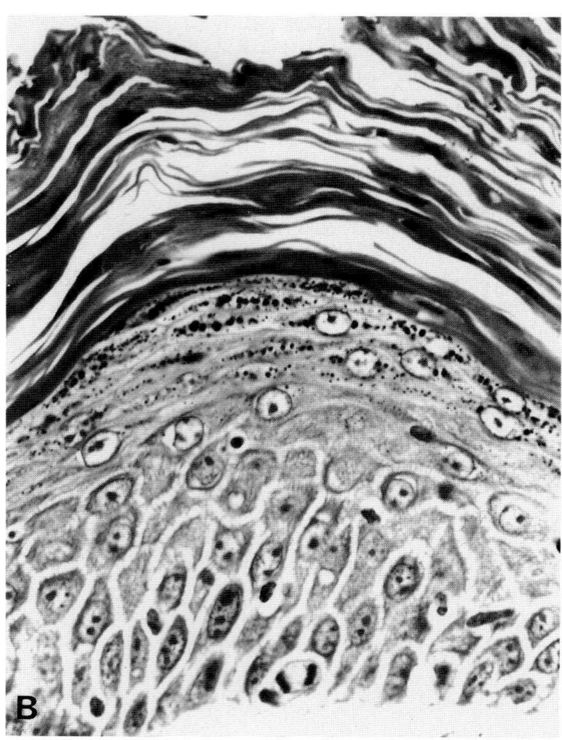

Fig. 5.1B Epidermal changes. Parakeratotic hyperkeratosis, hypergranulosis, and spongiosis.

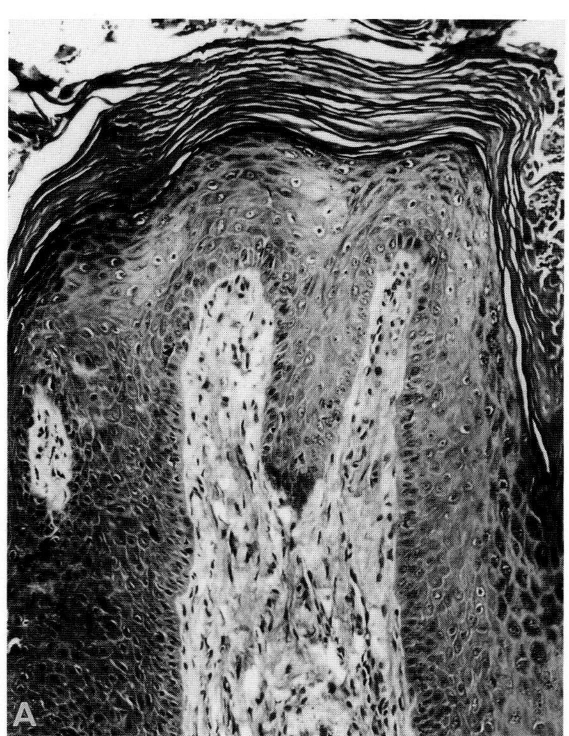

Fig. 5.1A Epidermal changes. Orthokeratotic hyperkeratosis, rete ridge formation, and acanthosis.

Hypokeratosis is a decreased thickness of the stratum corneum. It is much less common than hyperkeratosis, and reflects an exceptionally rapid epidermal turnover time and/or decreased cohesion between cells of the stratum corneum. Hypokeratosis may be found in seborrheic and other exfoliative skin disorders. It may also be produced by excessive surgical preparation of the biopsy site or by friction and maceration in intertriginous areas.

Dyskeratosis is premature and faulty keratinization of individual cells. Dyskeratosis is used less commonly to indicate a general fault in the keratinization process and, thus, the state of the epidermis as a whole. Dyskeratotic cells are characterized by eosinophilic, swollen cytoplasm and condensed, dark-staining nuclei (Fig. 5.2). Such cells are often difficult to distinguish from necrotic keratinocytes on light-microscopic examination. The judgment usually rests on whether the rest of the epithelium is keratinizing or dying. Dyskeratosis may be seen in a number of inflammatory dermatoses, including the pemphigus complex, the lichenoid dermatoses, zinc-responsive dermatoses, and some vitamin A-responsive dermatoses.

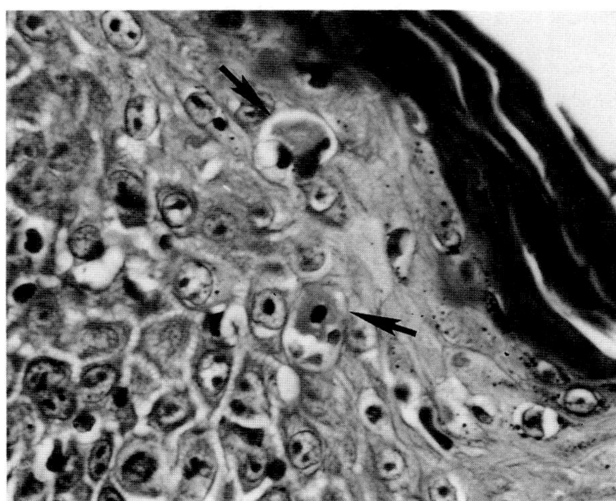

Fig. 5.2 Epidermal changes. Dyskeratosis. Premature keratinization. Note dark cytoplasm and pyknotic nucleus (arrow).

Dyskeratosis also occurs in neoplastic dermatoses, especially papilloma, squamous cell carcinoma, keratoacanthoma, and warty dyskeratoma.

Hypergranulosis and **hypogranulosis** indicate an increased or decreased thickness of the stratum granulosum. Hypergranulosis may be seen in any dermatosis in which there is epidermal hyperplasia and orthokeratotic hyperkeratosis. Hypogranulosis is often seen in dermatoses in which there is parakeratotic hyperkeratosis.

Hyperplasia is an increased thickness of the noncornified epidermis due to an increased number of epidermal cells. The term acanthosis is often used interchangeably with hyperplasia. However, **acanthosis** specifically indicates an increased thickness of the stratum spinosum, and is usually due to hyperplasia and occasionally to hypertrophy of cells of the stratum spinosum. Epidermal hyperplasia is often accompanied by rete ridge formation in which "pegs" of epidermis appear to project downward into the underlying dermis. Rete ridges are not found in normal haired skin of domestic animals, except for the pig.

Epidermal hyperplasia may be further specified as irregular (uneven, elongated, pointed rete ridges with an obliterated or preserved rete-papillae configuration); regular (more or less evenly thickened epidermis); psoriasiform (more or less evenly elongated rete ridges, which are clubbed and/or fused at their bases); papillated (digitate projections of the epidermis above the skin surface); and pseudocarcinomatous (Fig. 5.3) or pseudoepitheliomatous (extreme, irregular hyperplasia, which may include increased mitoses, squamous eddies, and horn pearls, thus resembling squamous cell carcinoma; however, there is no cellular atypia, and the basement membrane zone is not breached). These five forms of epidermal hyperplasia may be seen in various combinations in the same specimen.

Epidermal hyperplasia is a common feature of virtually any chronic inflammatory process. Pseudocarcinomatous

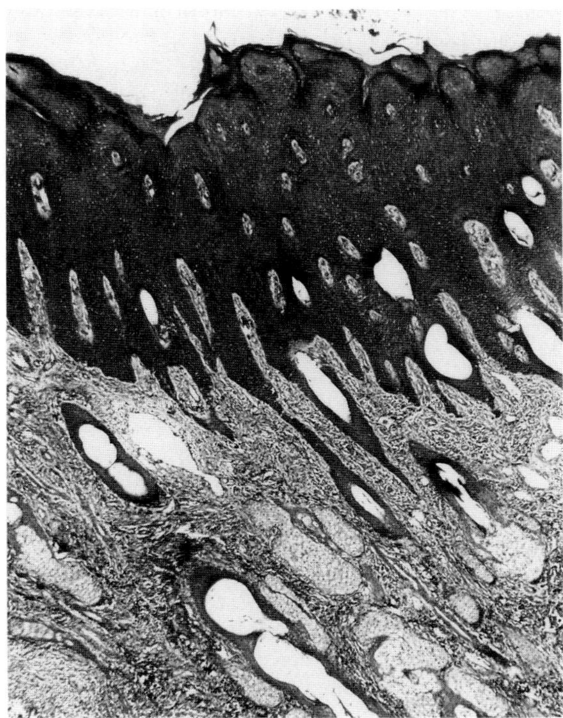

Fig. 5.3 Epidermal changes. Pseudocarcinomatous hyperplasia.

hyperplasia is most commonly associated with underlying chronic dermal suppurative, granulomatous, or neoplastic processes and ulcers. Papillated hyperplasia is most commonly seen with neoplasia, callosities, epidermal nevi, seborrheic dermatitis, and zinc-responsive dermatoses. Psoriasiform hyperplasia is most commonly seen with porcine pityriasis rosea, psoriasiform lichenoid dermatosis of springer spaniels, parapsoriasis, chronically traumatized lesions, and epitheliotropic lymphoma.

Hypoplasia is decreased thickness of the noncornified epidermis due to a decreased number of cells. **Atrophy** is decreased thickness of the noncornified epidermis due to a decreased size of cells. An early sign of epidermal hypoplasia or atrophy is the loss of the rete ridges in areas of skin where they are normally present. Epidermal hypoplasia and atrophy are uncommon. The extremely thin normal epidermis of domestic animals makes atrophy or hypoplasia very difficult to detect without micromorphometry. They are seen with endocrine and some immune-mediated dermatoses.

Necrosis is the death of cells or tissues within living tissue (Fig. 5.4) and is judged, on light microscopy, primarily on the basis of nuclear morphology. Necrolysis is often used synonymously with necrosis, but implies a separation and sequestration of necrotic tissue. Together with nuclear changes of necrosis, individual necrotic keratinocytes are identified by loss of intercellular bridges with resultant rounding-up of the cell, and a normal-sized or swollen eosinophilic cytoplasm.

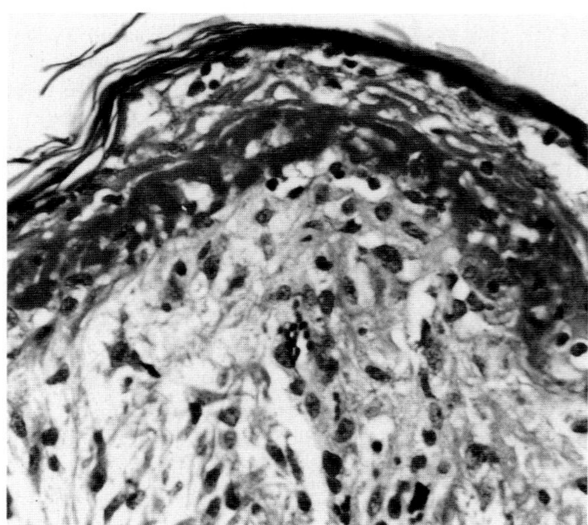

Fig. 5.4 Epidermal changes. Full thickness necrosis.

Epidermal necrosis may be focal, as in drug eruptions, microbial infections, erythema multiforme, graft-versus-host disease, and lichenoid dermatoses. Or the necrosis may be more extensive due to physical and chemical injury, to interference with vascular supply, or to immunologic mechanisms as in erythema multiforme and toxic epidermal necrolysis. Satellite cell necrosis is characterized by individual keratinocyte necrosis in association with contiguous (satellite) lymphoid cells, and is seen in many lichenoid dermatitides.

Intercellular edema (spongiosis) of the epidermis is characterized by a widening of the intercellular spaces with accentuation of the intercellular bridges, giving the involved epidermis a spongy appearance (Fig. 5.5A). Severe intercellular edema may lead to rupture of the intercellular bridges and the formation of spongiotic vesicles within the epidermis (Fig. 5.5B). Severe spongiotic vesicle formation may disrupt the basement membrane zone in some areas, giving the appearance of subepidermal vesicles. Intercellular edema is a common feature of acute or subacute inflammatory dermatoses. Diffuse spongiosis, which also involves the hair follicle outer root sheath, may be seen with the feline eosinophilic plaque or granuloma, zinc-responsive dermatoses, seborrheic dermatitis, and epidermal dysplasia with *Malassezia pachydermatis* infection in West Highland white terriers. Spongiosis of the upper one half of the epidermis, which is overlaid by marked diffuse parakeratotic hyperkeratosis, is seen in the canine hepatocutaneous syndrome.

Intracellular edema (hydropic degeneration, vacuolar degeneration, ballooning degeneration) of the epidermis is characterized by increased size, cytoplasmic pallor, and displacement of the nucleus to the periphery of affected cells (Fig. 5.6). Severe intracellular edema may result in reticular degeneration and intraepidermal vesicles.

Intracellular edema is a common feature of any acute or subacute inflammatory dermatosis. Caution must be

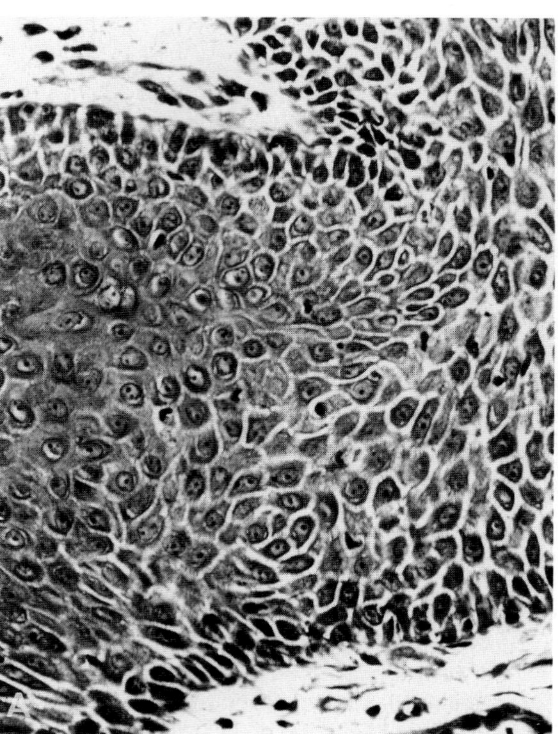

Fig. 5.5A Epidermal changes. Spongiosis.

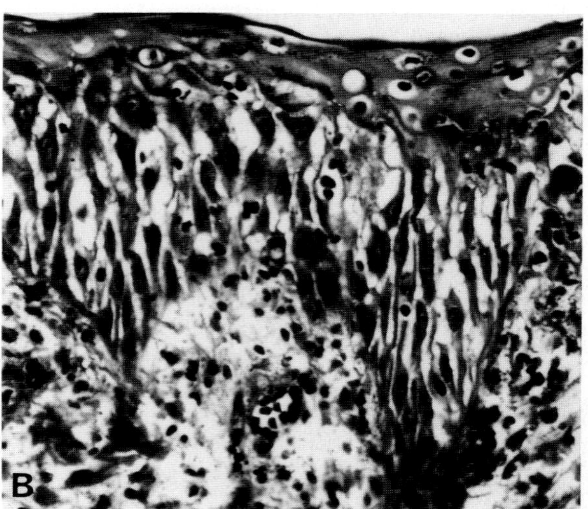

Fig. 5.5B Epidermal changes. Intercellular edema, almost to point of vesicle formation.

exercised not to confuse freezing artefact, delayed fixation artefact, and the intracellular accumulation of glycogen in the outer root sheath of normal hair follicles or in injured epidermis with intracellular edema.

Ballooning degeneration is a specific type of degenerative change in epidermal cells, and is characterized by swollen eosinophilic cytoplasm without vacuolation, en-

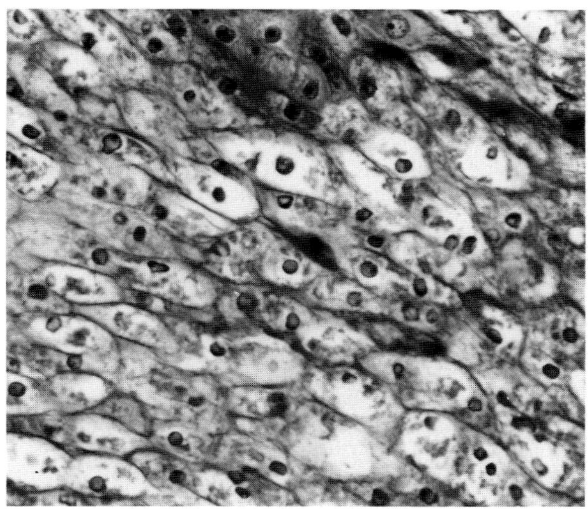

Fig. 5.6 Epidermal changes. Intracellular edema.

larged, condensed, or occasionally multiple nuclei, and a loss of cohesion resulting in acantholysis. Ballooning degeneration is a specific feature of viral infections, particularly of herpesviruses. The term has been used to describe, in addition, keratinocytes with pale swollen cytoplasm, such as occur in poxvirus infections.

Reticular degeneration is caused by severe intracellular edema of epidermal cells, wherein the cells burst, resulting in multilocular intraepidermal vesicles whose septa are formed by resistant cell walls. It may be seen with any acute or subacute inflammatory dermatosis, but is especially common in dermatophilosis and acute contact dermatitis.

Hydropic degeneration of the basal epidermal cells describes intracellular edema restricted to cells of the stratum basale (Fig. 5.7). This process may also affect the basal cells of the outer root sheath of hair follicles. Hydropic degeneration of basal cells is usually focal, but if severe and extensive, may result in intrabasal clefts or vesicles, or subepidermal clefts or vesicles due to dermo-epidermal separation. Hydropic degeneration of basal cells is an uncommon finding and is usually associated with idiopathic lichenoid dermatoses, drug eruptions, lupus erythematosus, erythema multiforme, toxic epidermal necrolysis, dermatomyositis, and epidermolysis bullosa simplex.

Acantholysis is a loss of cohesion between epidermal cells resulting in intraepidermal clefts, vesicles, and bullae (Fig. 5.8). This process may also involve the outer root sheath of hair follicles and glandular ductal epithelium. Acantholysis is further specified by reference to the level at which it occurs, i.e., subcorneal, intragranular, intraepidermal, or suprabasilar. Acantholysis may be caused by severe spongiosis, ballooning degeneration, proteolytic enzymes released by neutrophils or eosinophils in inflammatory processes, developmental defects as in bovine familial acantholysis, and neoplastic transformation as in squamous cell carcinoma, actinic keratosis, and warty dyskeratoma. Marked acantholysis is seen with the autoimmune pemphigus complex and occasionally with subcorneal pustular dermatosis.

Exocytosis is the migration of inflammatory cells and/or erythrocytes through the intercellular spaces of the epidermis. Exocytosis of inflammatory cells is a common feature of any inflammatory dermatosis. Exocytosis of erythrocytes implies purpura, severe vasodilatation, or trauma. When spongiosis is associated with the exocytosis of predominantly eosinophils, it is often referred to as eosinophilic spongiosis, and may be seen in ectoparasitism, pemphigus, pemphigoid, sterile eosinophilic pustulosis, feline eosinophilic plaque, eosinophilic granuloma, hypereosinophilic syndrome, equine multisystemic

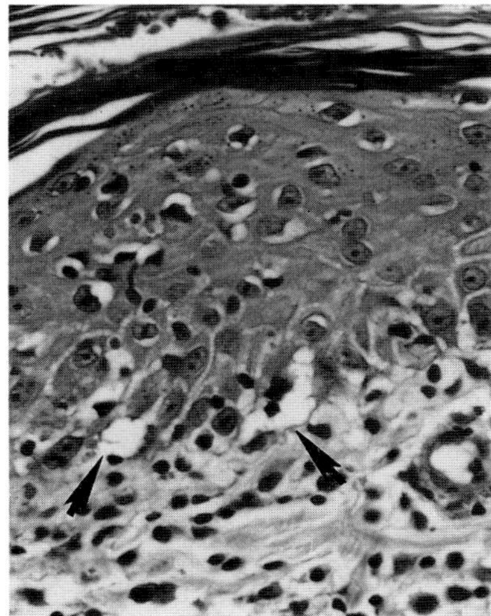

Fig. 5.7 Epidermal changes. Hydropic degeneration of basal keratinocytes (arrows).

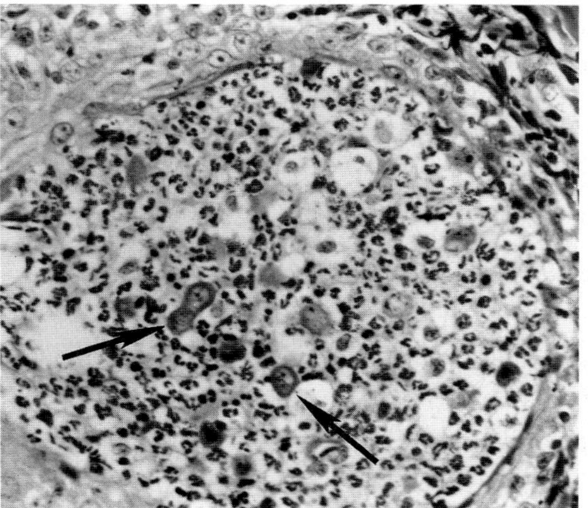

Fig. 5.8 Epidermal changes. Acantholysis. Free-floating keratinocytes with normal nuclear morphology (arrows).

eosinophilic epitheliotropic disease, and porcine pityriasis rosea.

Clefts or lacunae are slitlike spaces, which do not contain fluid, within the epidermis or at the dermoepidermal junction. Clefts may be caused by acantholysis or hydropic degeneration of basal cells. However, clefts may also be handling artefacts.

Microvesicles, vesicles, and **bullae** refer to microscopic and macroscopic fluid-filled, relatively acellular spaces within or below the epidermis. Such lesions are often called blisters. They may be caused by severe intercellular or intracellular edema, ballooning degeneration, acantholysis, hydropic degeneration of basal cells, subepidermal edema, or other factors resulting in dermoepidermal separation, such as the autoantibodies in bullous pemphigoid. Microvesicles, vesicles, and bullae may be subcorneal, intragranular, intraepidermal, suprabasilar, intrabasal, or subepidermal in location. When these lesions contain larger numbers of inflammatory cells, they may be referred to as vesicopustules.

Microabscesses and **pustules** are microscopic or macroscopic intraepidermal and subepidermal cavities filled with inflammatory cells. Microabscesses and pustules are further described on the basis of location and cell type: spongioform pustule of Kogoj—a multilocular accumulation of neutrophils within and between keratinocytes, especially those of the stratum granulosum and stratum spinosum, in which the cell boundaries form a sponge-like network; Munro's microabscess—a small, desiccated accumulation of neutrophils within or below the stratum corneum; Pautrier's microabscess—a small, focal accumulation of abnormal lymphoid cells typical of epitheliotropic lymphoma; and eosinophilic microabscess.

Hyperpigmentation refers to excess melanin deposited within the epidermis and, often, concurrently in dermal melanophages. Hyperpigmentation may be focal or diffuse, and confined to the stratum basale or present throughout all epidermal layers. It is a common nondiagnostic finding in chronic inflammatory and hormonal dermatoses, as well as in some developmental and neoplastic disorders. Hyperpigmentation must always be cautiously assessed with regard to the animal's normal pigmentation.

Hypopigmentation refers to decreased melanin in the epidermis. It may be associated with congenital or acquired idiopathic defects in melanization (leukoderma, vitiligo), toxic effects of certain chemicals on melanocytes (e.g., monobenzyl ether of dihydroquinone in rubbers and plastics), inflammatory disorders which affect melanization or destroy melanocytes, hormonal disorders, and dermatoses featuring hydropic degeneration of basal cells (e.g., lupus erythematosus). In those hypopigmented dermatoses associated with hydropic degeneration of basal cells, the underlying superficial dermis usually reveals pigmentary incontinence.

A **crust** is a consolidated, desiccated surface mass composed of varying combinations of keratin, serum, cellular debris, and often microorganisms. Crusts are further described on the basis of their composition: serous (mostly serum), hemorrhagic (mostly blood), cellular (mostly inflammatory cells), serocellular (exudative; a mixture of serum and inflammatory cells), and palisading (alternating horizontal rows of parakeratotic hyperkeratosis and pus, as seen in dermatophilosis and dermatophytosis).

Crusts merely indicate a prior exudative process and are rarely of diagnostic significance. However, crusts should be closely examined for dermatophyte spores and hyphae, filaments and coccoid elements of *Dermatophilus congolensis,* and large numbers of acantholytic keratinocytes, which are indicators of pemphigus complex or subcorneal pustular dermatosis. Bacteria and bacterial microcolonies and yeasts are common inhabitants of surface debris and are of no diagnostic significance.

Epidermal collarette refers to the formation of elongated, hyperplastic rete ridges at the lateral margins of a pathologic process, which appear to curve inward toward the center of the lesion. Epidermal collarettes may be seen with neoplastic, granulomatous, and suppurative dermatoses.

Horn cysts (keratin cysts) are circular cystic structures surrounded by flattened epidermal cells. They contain concentrically arranged lamellar keratin. Horn cysts are features of trichoepitheliomas and basal cell tumors. **Pseudohorn cysts** are cystlike structures formed by the irregular invagination of a hyperplastic, hyperkeratotic epidermis. They are seen in numerous hyperplastic or neoplastic epidermal dermatoses. **Horn pearls** (squamous pearls) are focal, circular, concentric layers of squamous cells showing gradual keratinization toward the center, often accompanied by cellular atypia and dyskeratosis. Horn pearls are features of squamous cell carcinoma, keratoacanthoma, and pseudocarcinomatous hyperplasia. **Squamous eddies** are whorl-like patterns of squamoid cells with no atypia, dyskeratosis, or central keratinization. Squamous eddies are features of numerous neoplastic and hyperplastic epidermal disorders.

Dells are small depressions or hollows in the surface of the epidermis, which are usually associated with focal epidermal atrophy and orthokeratotic hyperkeratosis. They may be seen in lichenoid dermatoses, especially in lupus erythematosus.

Epidermolytic hyperkeratosis (granular degeneration) is characterized by (1) perinuclear clear spaces in the upper epidermis; (2) indistinct cell boundaries formed either by lightly staining material or by keratohyaline granules peripheral to the perinuclear clear spaces; (3) a markedly thickened stratum granulosum, and (4) orthokeratotic hyperkeratosis. It may be seen in certain types of ichthyosis, linear epidermal nevi, actinic keratoses, seborrheic keratoses, papillomas, keratinous cysts, and squamous cell carcinoma.

Epidermal mast cells are frequently seen in biopsies from cats with inflammatory dermatoses. They are found within the epidermis as well as the hair follicle outer root sheath, and are most commonly found in diseases of immune-mediated origin, especially those with tissue eosinophilia.

B. Dermal Changes

Dermal collagen is subject to a number of pathologic changes. Collagen may undergo hyalinization (confluence and increased eosinophilic, glassy, refractile appearance, as in chronic inflammation and connective tissue diseases); fibrinoid degeneration (deposition of, or replacement with, a brightly eosinophilic, fibrillar, or granular substance resembling fibrin, as in connective tissue diseases); lysis (a homogeneous, eosinophilic change with complete loss of structural detail, seen with microbial and parasitic infections, ischemia, and equine axillary nodular necrosis); degeneration (a structural and tinctorial change characterized by slight basophilia, a granular appearance and frayed edges of collagen fibrils, seen with feline, canine, and equine eosinophilic granuloma); dystrophic mineralization (deposition of calcium salts as basophilic, amorphous, granular material along collagen fibrils as in hyperglucocorticoidism and, rarely, idiopathy); atrophy (thin collagen fibrils and decreased fibroblasts, with resultant decreased dermal thickness, as seen with hormonal dermatoses); disorganization and fragmentation (cutaneous asthenia), and alignment in vertical streaks (elongated, parallel strands of collagen in the superficial dermis, perpendicular to the epidermal surface, as seen in chronically rubbed, licked, or scratched skin, such as in acral lick dermatitis and chronic hypersensitivity reactions).

Fibroplasia is the formation and development of fibrous tissue in increased amounts. The term is often used synonymously with granulation tissue. Fibroplasia is characterized by a fibrovascular proliferation. The blood vessels with prominent endothelial cells are oriented roughly perpendicular to the surface of the skin. The new collagen fibrils, with prominent fibroblasts, are oriented roughly parallel to the surface of the skin. Edema and inflammatory cells are constant features of fibroplasia. **Desmoplasia** usually refers to fibroplasia induced by neoplastic processes.

Fibrosis is a later stage of fibroplasia. Increased numbers of fibroblasts and collagen fibrils are the characteristic findings. There is little or no inflammation. **Sclerosis** (scar) may be the end point of fibrosis. Increased numbers of collagen fibrils have a thick, eosinophilic, hyalinized appearance, and the number of fibroblasts is greatly reduced.

Papillomatosis refers to the projection of dermal papillae above the surface of the skin, resulting in an irregular, undulating configuration of the epidermis. Papillomatosis is often associated with epidermal hyperplasia, and is seen with chronic inflammatory and neoplastic dermatoses. **Villi** are dermal papillae covered by one to two layers of epidermal cells, which project into the base of a vesicle or bulla. Villi are seen in pemphigus vulgaris and warty dyskeratoma, and occasionally in actinic keratosis and squamous cell carcinoma. **Festoons** are dermal papillae, devoid of attached epidermal cells, which project into a vesicle or bulla. Festoons are seen in porphyria, bullous pemphigoid, epidermolysis bullosa, and drug eruption.

Pigmentary incontinence refers to the presence of melanin granules free within the subepidermal dermis and within dermal macrophages (melanophages). Pigmentary incontinence may be seen with any process that damages the stratum basale and the basement membrane zone, especially hydropic degeneration of basal cells (lichenoid dermatoses, lupus erythematosus, dermatomyositis, erythema multiforme, epidermolysis bullosa simplex). Melanophages may be seen in chronic inflammatory conditions in which melanin production is greatly increased.

Dermal edema is recognized by dilated lymphatics (not visible in normal skin), widened spaces between blood vessels and perivascular collagen (perivascular edema), or widened spaces between large areas of dermal collagen (interstitial edema). The dilated lymphatics and widened perivascular and interstitial spaces may or may not contain a lightly eosinophilic, homogeneous, frothy-appearing proteinaceous fluid. Dermal edema is a common feature of any inflammatory dermatosis. Severe edema of the superficial dermis may result in subepidermal vesicles and bullae, necrosis of the overlying epidermis, and predisposition to artefactual dermoepidermal separation during handling and processing of biopsy specimens. Severe edema of the superficial dermis may result in vertical orientation and stretching of collagen fibers, producing the gossamer (weblike) collagen effect seen with erythema multiforme and severe urticaria.

Mucinous degeneration (myxedema, myxoid degeneration, mucoid degeneration, mucinosis) is characterized by increased amounts of an amorphous, stringy, granular, basophilic material, which separates, thins, or replaces dermal collagen fibrils and surrounds blood vessels and appendages in hematoxylin-eosin-stained sections. Only small amounts of mucin are visible in normal skin, mostly around appendages and blood vessels. Mucin is more easily demonstrated with stains for acid mucopolysaccharides, such as alcian blue and colloidal iron. Mucinous degeneration may be seen as a focal process in numerous inflammatory, neoplastic, and developmental dermatoses. Diffuse mucinous degeneration is a feature of hypothyroidism, acromegaly, lupus erythematosus, idiopathic mucinoses, and the normal skin of the Chinese Shar-Pei dog.

A ***Grenz* zone** is a marginal zone of relatively normal collagen that separates the epidermis from an underlying dermal alteration. A *Grenz* zone may be seen in neoplastic and granulomatous disorders.

Follicular epithelium is affected by most of the histopathologic changes described for the epidermis. Follicular keratosis, plugging, and dilatation are common features of such diverse conditions as inflammatory, hormonal, and developmental dermatoses. **Perifolliculitis, folliculitis,** and **furunculosis** (penetrating or perforating folliculitis) refer to varying degrees of follicular inflammation. Follicular atrophy refers to the gradual involution and disappearance characteristic of hormonal and nutritional dermatoses, follicular dystrophies, and ischemia. Hair follicles should be closely examined, and the phase of growth, determined. A predominance of telogen hair follicles is characteristic of hormonal and nutritional dermatoses and states of telogen defluxion (stress, disease, drugs). **Follicular dystrophy** refers to the presence of incompletely and/or abnormally formed hair follicles such as seen in hypotrichoses and color-dilution (mutant) alopecia. Perifollicular fibrosis is seen in chronic folliculitides, canine dermatomyositis, and granulomatous sebaceous adenitis. Dystrophic mineralization of the hair follicle basement membrane zone and subsequent transepithelial elimination of the mineral is seen in the calcinosis of canine hyperglucocorticoidism and as a senile change in dogs, especially poodles. The finding of large numbers of bacteria or yeasts in noninflamed hair follicles in dogs almost always indicates the presence of clinically relevant infection.

Sebaceous and **sweat glands** may be involved in various dermatoses. Sebaceous glands may be involved in many suppurative and granulomatous inflammations (sebaceous adenitis). In dogs and cats, an idiopathic granulomatous to pyogranulomatous sebaceous adenitis is seen, which, in the late stages, is characterized by complete absence of sebaceous glands. They may become atrophic and cystic in hormonal dermatoses, occasional chronic inflammatory processes, and as a senile change. Sebaceous glands may also become hyperplastic in chronic inflammatory dermatoses and in senile nodular sebaceous hyperplasia. Sebaceous gland atrophy and hyperplasia must be carefully assessed with regard to the site of collection of the specimen.

Paritrichial (apocrine) sweat glands are commonly involved in suppurative and granulomatous dermatoses (**hidradenitis**). Periglandular accumulation of plasma cells is commonly seen in chronic infections, lichenoid dermatoses, and acral lick dermatitis. They may become dilated or cystic in many inflammatory, developmental, and hormonal dermatoses, and as a senile change. The light-microscopic recognition of peritrichial gland atrophy is a moot point, as dilated tubules lined by flattened epithelial cells may be a feature of the normal postsecretory state. Furthermore, the secretory cells of bovine peritrichial sweat glands are normally low cuboidal.

Cutaneous blood vessels show several changes, including dilatation (ectasia), endothelial swelling, hyalinization, fibrinoid degeneration, vasculitis, thromboembolism, and extravasation (diapedesis) of erythrocytes (purpura).

The **subcutaneous fat** (panniculus adiposus, hypodermis) is subject to the connective tissue and vascular changes previously described. It is frequently involved in suppurative and granulomatous dermatoses. In addition, subcutaneous fat may exhibit inflammatory changes (panniculitis, steatitis) without significant involvement of the overlying dermis and epidermis (sterile nodular panniculitis, feline nutritional steatitis, bacterial or fungal panniculitis, lupus erythematosus panniculitis, erythema nodosum, subcutaneous fat sclerosis), and may atrophy in various hormonal, inflammatory, and idiopathic dermatoses. Fat micropseudocyst formation and lipocytes containing radially arranged needle-shaped clefts are seen with subcutaneous fat sclerosis and feline idiopathic sterile panniculitis.

C. Miscellaneous Changes

Apoptotic bodies (colloid bodies, hyaline bodies, filamentous bodies, Civatte bodies) are degenerate basal epidermal cells which appear as round, homogeneous, eosinophilic bodies in the stratum basale or just below. Apoptotic bodies are features of all lichenoid dermatoses.

Thickening of the basement membrane zone appears as focal, linear, homogeneous, eosinophilic bands below the stratum basale. The basement membrane zone is well demonstrated with periodic acid-Schiff stain. Thickening of the basement membrane zone is a feature of lichenoid dermatoses, especially lupus erythematosus.

Dysplasia is a faulty or abnormal development of individual cells. The term also describes abnormal development of the epidermis as a whole. Dysplasia may be a feature of neoplastic, hyperplastic, and developmental dermatoses.

Nests (theques) are well-circumscribed clusters or groups of cells within the epidermis and/or the dermis. Nests are seen in some neoplastic and hamartomatous dermatoses such as melanocytic nevi and melanomas.

Multinucleated epidermal giant cells may be found in viral and neoplastic skin disorders and in non-neoplastic dermatoses characterized by epidermal hyperplasia, dyskeratosis, and chronic pruritus.

Lymphoid nodules are well-circumscribed, rounded, dense, usually perivascular accumulations of predominantly mature lymphocytes in the deep dermis and/or subcutis. They are uncommon and seen most frequently in conjunction with immune-mediated dermatoses, dermatoses associated with tissue eosinophilia, and in panniculitis. They are also prominent in insect-bite granuloma (pseudolymphoma).

Transepidermal elimination is a mechanism by which foreign or altered constituents can be removed from the dermis. The process involves unique morphologic alterations of the surface epidermis or hair follicle outer root sheath, which form a channel and, therefore, facilitate extrusion.

Papillary squirting is present when superficial dermal papillae are edematous and contain dilated vessels, and when the overlying epidermis is also edematous, parakeratotic, and contains exocytosing leukocytes. "Squirting" papillae are a feature of seborrheic dermatitis and zinc-responsive dermatoses.

Flame figures are areas of altered collagen surrounded by eosinophils and eosinophil cytoplasmic granules. In chronic lesions, the eosinophil content decreases, histiocytes increase in number, and palisading granulomas may be formed. Flame figures may be seen in eosinophilic granuloma, sterile eosinophilic pustulosis, and insect/arthropod bite reactions.

Nevus literally means spot or birthmark. The term is often used clinically to describe any congenital skin lesion, and histologically for cells (nevus cells, nevocytes) that compose the common pigmented mole (nevus pigmentosus) of humans. A nevus is a circumscribed stable malformation of the skin, congenital or tardive in onset, consisting of local excess of one or several of the normal mature constituents of the skin. The term nevus should always be used with a modifier such as melanocytic, epidermal, vascular, connective tissue, organoid, etc.

A **hamartoma** is a tumorlike malformation composed of an abnormal mixture of tissue elements or an abnormal proportion of a single element. Unlike a choristoma, the components of a hamartoma are normal to the location.

D. Cellular Infiltrates

Dermal cellular infiltrates are described in terms of the type(s) of cell(s) present and the pattern(s) of cellular infiltration. In general, cellular infiltrates are either monomorphous (one cell type) or polymorphous (more than one cell type). Further clarification as to the predominant cells is accomplished with terms like lymphocytic, histiocytic, neutrophilic, eosinophilic, and plasmacytic.

Cellular infiltration is usually perivascular, periappendageal (perifollicular and periglandular), lichenoid (assuming a bandlike configuration which parallels the epidermis), nodular, or diffuse. The cell type(s) and pattern(s) of infiltration are important diagnostic clues to many dermatoses.

E. Pattern Analysis

Pattern analysis is based on the recognition of 10 different non-neoplastic reaction patterns in the skin as seen at scanning magnification. Pattern analysis supplants vague, archaic pathologic diagnoses such as chronic nonspecific dermatitis and subacute nonsuppurative dermatitis with morphologic diagnoses which immediately generate a list of differential diagnoses.

1. Perivascular Dermatitis

In perivascular dermatitis, the predominant inflammatory reaction is centered around the dermal blood vessels. Most perivascular dermatitides involve predominantly the superficial dermal blood vessels. Concurrent involvement of the deep dermal blood vessels suggests a systemic disease (infection, immune mediated). In the horse, most perivascular dermatitides are both superficial and deep.

Most perivascular dermatitides are caused by ectoparasitism, hypersensitivity reactions, viral infections, dermatophilosis, dermatophytosis, nutritional deficiencies, and diseases of altered keratinization (seborrhea). Any perivascular dermatitis containing numerous eosinophils should first be suspected of representing ectoparasitism, endoparasitism, or a hypersensitivity reaction. Focal areas of epidermal edema, eosinophilic exocytosis, and necrosis (''epidermal nibbles'') are suggestive of ectoparasitism. Other perivascular dermatitides that may contain numerous eosinophils include porcine pityriasis rosea, zinc-responsive dermatoses, equine multisystemic eosinophilic, epitheliotropic disease, and feline hypereosinophilic syndrome.

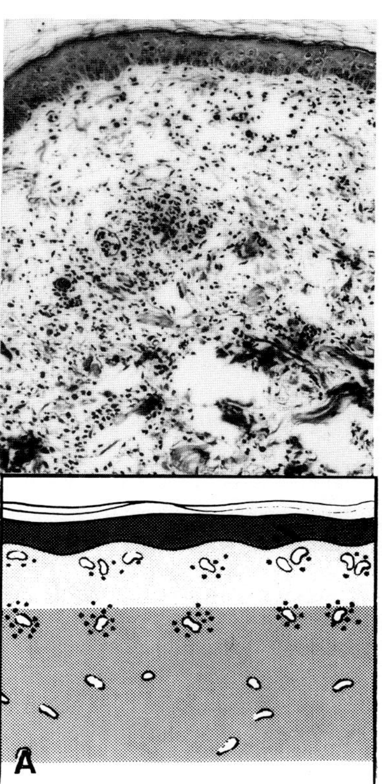

Fig. 5.9 Schematic and histologic appearances of dermatitis, (A) through (F). (A) Superficial perivascular dermatitis with minimal epidermal change. Note marked dermal edema.

Perivascular dermatitis is subdivided on the basis of accompanying epidermal changes into three types.

In **pure perivascular dermatitis,** there are few or no epidermal changes. The most common dermatoses in this category include acute hypersensitivity reactions and urticaria (Fig. 5.9A).

Spongiotic perivascular dermatitis is characterized by varying degrees of spongiosis and spongiotic vesicle formation (Fig. 5.9B). Severe spongiotic vesiculation may disrupt the basement membrane zone, resulting in subepidermal vesicles. The epidermis usually shows varying degrees of hyperkeratosis and hyperplasia. The most common dermatoses in this category include hypersensitivity reactions, contact dermatitis, ectoparasitism, dermatophytosis, dermatophilosis, and viral infections. Diffuse spongiosis, wherein the hair follicle outer root sheath is also involved, suggests feline eosinophilic plaque, feline eosinophilic granuloma, zinc-responsive dermatoses, seborrheic dermatitis, and epidermal dysplasia, and *Malassezia pachydermatis* infection of West Highland white terriers.

Hyperplastic perivascular dermatitis is characterized by varying degrees of epidermal hyperplasia and hyperkeratosis with little or no spongiosis (Fig. 5.9C). This is a common, nondiagnostic, chronic reaction pattern. The most common dermatoses in this category are chronic

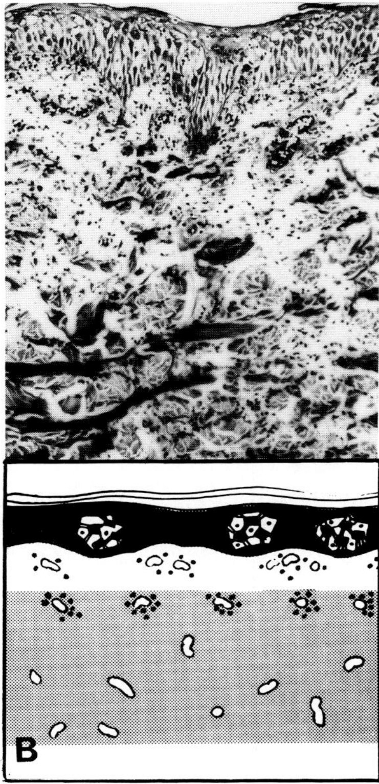

Fig. 5.9B Spongiotic dermatitis. Dermal edema and epidermal spongiosis.

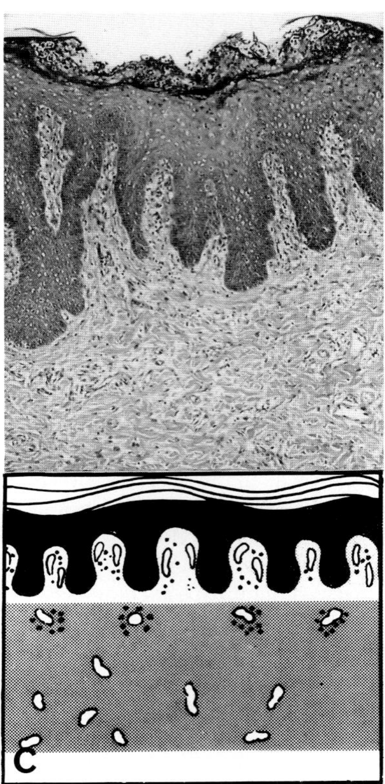

Fig. 5.9C Hyperplastic dermatitis. Orthokeratotic hyperkeratosis, acanthosis, rete ridge formation, dermal fibrosis and mononuclear cell infiltration.

hypersensitivity reactions, diseases of altered keratinization, dermatophytosis, dermatophilosis, viral infections, psychogenic dermatoses (e.g., acral lick dermatitis), and feline indolent ulcer.

2. *Interface Dermatitis*

Although interface dermatitis was originally described as a subdivision of perivascular dermatitis, the authors believe it is a special category unto itself. First, this reaction pattern is rarely if ever precisely perivascular. Second, it is virtually always associated with a unique group of dermatoses.

In interface dermatitis, the dermoepidermal junction is obscured by hydropic degeneration of epidermal basal cells, a lichenoid (bandlike) cellular infiltrate (usually lymphoplasmacytic), or both (Fig. 5.9D). Apoptotic bodies, satellite cell necrosis, and pigmentary incontinence are commonly seen. The hydropic type of interface dermatitis is seen with drug eruptions, lupus erythematosus, dermatomyositis, erythema multiforme, toxic epidermal necrolysis, epidermolysis bullosa simplex, bovine viral diarrhea, rinderpest, bovine pseudolumpy skin disease, graft-versus-host reactions, and occasionally vasculitis. The lichenoid type may be seen with drug eruptions, lupus erythematosus, idiopathic lichenoid dermatoses, lichenoid keratoses, the Vogt–Koyanagi–Harada syndrome, pemphigus, pemphigoid, lichenoid psoriasiform dermatosis of springer spaniels, malignant catarrhal fever, and epitheliotropic lymphoma. Occasionally, lichenoid reactions are seen in association with staphylococcal infections and ectoparasitism (cheyletiellosis, scabies) in dogs. In these cases, the lichenoid cellular infiltrate also contains numerous neutrophils and/or eosinophils, and other histopathologic findings suggestive of these disorders may be present. In the Vogt–Koyanagi–Harada syndrome, the lichenoid cellular infiltrate contains a preponderance of large histiocytes, which often contain lightly sprinkled melanin. Focal thickening and smudging of the basement membrane zone is suggestive of lupus erythematosus or dermatomyositis.

3. *Vasculitis*

Vasculitis (Fig. 5.9E) may be classified on the basis of the dominant inflammatory cell within vessel walls. There are neutrophilic, eosinophilic, lymphocytic, and mixed types.

Neutrophilic vasculitis may be leukocytoclastic (associated with karyorrhexis of neutrophils resulting in "nuclear dust") or nonleukocytoclastic, and is seen with connective tissue disorders (lupus erythematosus, rheumatoid arthritis, dermatomyositis), hypersensitivity reactions (drug eruptions, infections, toxins), polyarteritis nodosa, canine staphylococcal hypersensitivity, Rocky Mountain spotted

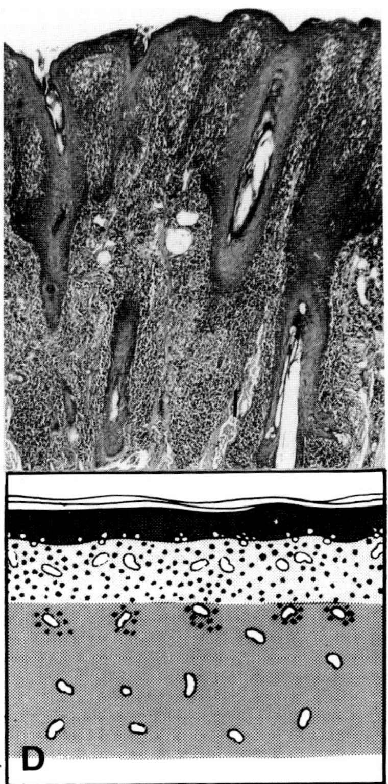

Fig. 5.9D Interface dermatitis. Lichenoid infiltrate obscures the dermo-epidermal junction.

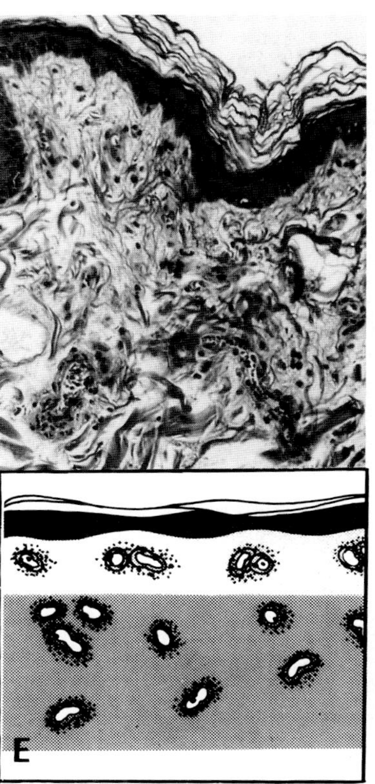

Fig. 5.9E Vasculitis. Note inflammatory cells in vessel wall. (Courtesy of R. Seiler).

fever, equine purpura hemorrhagica, hog cholera, septicemia, thrombophlebitis, and as an idiopathic disorder.

Lymphocytic vasculitis may be seen with drug eruptions, ectoparasitism, malignant catarrhal fever, equine viral arteritis, and as an idiopathic disorder.

Eosinophilic vasculitis has been reported as an idiopathic, presumably immune-mediated disorder in the horse.

Other vasculopathies wherein actual vasculitis is not usually present include ergotism, fescue toxicosis, and equine photoactivated vasculitis.

4. Nodular and Diffuse Dermatitis

Nodular dermatitis denotes discrete clusters of inflammatory cells (Fig. 5.9F). Such dermal nodules are usually multiple but may occasionally be large and solitary. Diffuse dermatitis denotes a cellular infiltrate so dense that discrete cellular aggregates are no longer easily recognized. Nodular and diffuse dermatitis may be characterized by predominantly neutrophilic, histiocytic, eosinophilic, or mixed cellular infiltrates. **Neutrophils** predominate in dermal abscesses often associated with infectious agents including bacteria, fungi, *Prototheca*, protozoa, and *Mycoplasma*. Abscesses may be sterile, as in foreign body reactions, equine axillary nodular necrosis, and the sterile pyogranuloma syndrome.

Histiocytes predominate in granulomatous inflammation, which is typically chronic. Granulomatous infiltrates containing large numbers of neutrophils are frequently called pyogranulomatous. While all granulomatous dermatitis is nodular or diffuse in pattern, not all nodular and diffuse dermatitides are granulomatous. **Granulomas** are nodular or tumorlike masses of granulomatous inflammation. They may be subclassified as tuberculoid (a central zone of neutrophils and necrosis surrounded by histiocytes and epithelioid cells, which are, in turn, surrounded by giant cells, then a layer of lymphocytes, then an outer layer of fibroblasts) or sarcoidal (naked epithelioid cells). Tuberculoid granulomas may be seen in tuberculosis, feline leprosy, atypical mycobacterial infection, and *Corynebacterium pseudotuberculosis* infections. Sarcoidal granulomas may be seen in equine sarcoidosis, canine sterile sarcoidal granulomas, and foreign-body reactions. "Palisading" granulomas are characterized by the alignment of histiocytes like staves around a central focus of collagen degeneration (feline, canine, and equine eosinophilic granuloma, equine mastocytoma), parasite or fungus (habronemiasis, pythiosis, conidiobolomycosis, basidiobolomycosis, demodicosis), fibrin (rheumatoid nodule), lipids (xanthoma), or other foreign material (e.g., calcium, as in dystrophic calcinosis cutis and calcinosis circumscripta). Granulomas and pyogranulomas which track hair follicles resulting in large, vertically oriented (sausage-shaped) lesions are typical of the sterile granuloma/pyogranuloma syndrome of dogs and cats.

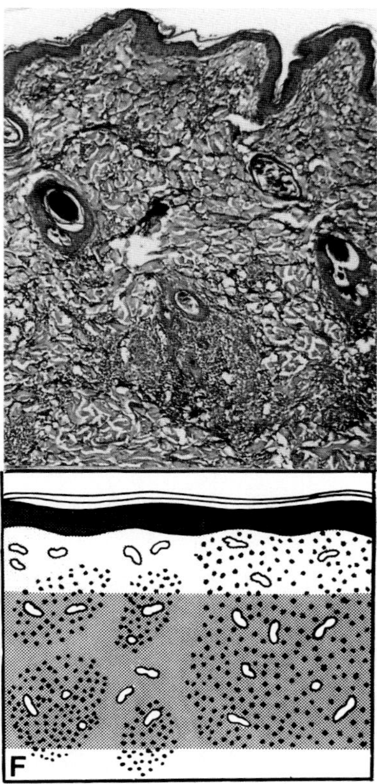

Fig. 5.9F Nodular to diffuse dermatitis. Dermis infiltrated with foamy macrophages. Panniculus is spared.

Nodular and diffuse dermatitis are often associated with certain unusual inflammatory cell types (Fig. 5.9F). **Foam cells** are histiocytes with vacuolated cytoplasm due to their contents (lipid, debris, microorganisms). **Epithelioid cells** are histiocytes with elongated or oval vesicular nuclei and abundant finely granular, eosinophilic cytoplasm with ill-defined cell borders. **Multinucleated giant cells** are histiocytic variants which assume three morphologic forms: Langhans' (nuclei form a circle or semicircle at the periphery of the cell), foreign-body (nuclei are scattered throughout the cytoplasm), and Touton (nuclei form a wreath which surrounds a central, homogeneous, amphophilic core of cytoplasm which is, in turn, surrounded by abundant foamy cytoplasm). In general, these three forms of giant cells have little diagnostic specificity, although the Touton variety is strongly indicative of xanthomas, and the Langhans' type suggests the need for an acid-fast stain.

Eosinophils may predominate in feline, canine, and equine eosinophilic granuloma, in certain parasitic dermatoses (habronemiasis, elaephoriasis, parafilariasis, dirofilariasis, dracunculiasis), where hair follicles have ruptured, and in hairy vetch toxicosis. Mixed cellular infiltrates are most commonly neutrophils and histiocytes (pyogranuloma), and eosinophils and histiocytes (eosinophilic granuloma), or a combination of the three cell types.

Plasma cells are common components of nodular and diffuse dermatitis in domestic animals, and are of no particular diagnostic significance. They may contain eosinophilic, intracytoplasmic inclusions, which are called Russell bodies. These accumulations of glycoprotein are largely globulin and may be large enough to displace the cell nucleus.

Reactions to ruptured hair follicles are the most common cause of nodular and diffuse pyogranulomatous dermatitis in domestic animals, and any such lesion should be examined for keratinous and epithelial debris and serially sectioned to rule out this possibility. All other nodular and diffuse dermatitides should be cultured, examined in polarized light for foreign material, and stained for bacteria and fungi. In general, microorganisms are most likely to be found near areas of suppuration and necrosis.

5. Intraepidermal Vesicular and Pustular Dermatitis

Vesicular and pustular dermatitides show considerable microscopic and macroscopic overlap. This is because vesicles in domestic animals tend to accumulate leukocytes very early and rapidly. Thus, vesicular dermatitides in these species frequently appear pustular or vesiculopustular both macroscopically and microscopically.

Intraepidermal vesicles and pustules (Fig. 5.10A) may be produced by intercellular and/or intracellular edema (any acute to subacute dermatitis reaction), ballooning degeneration (viral infections), acantholysis (pemphigus due to autoantibodies, microbial infections due to the proteolytic en-

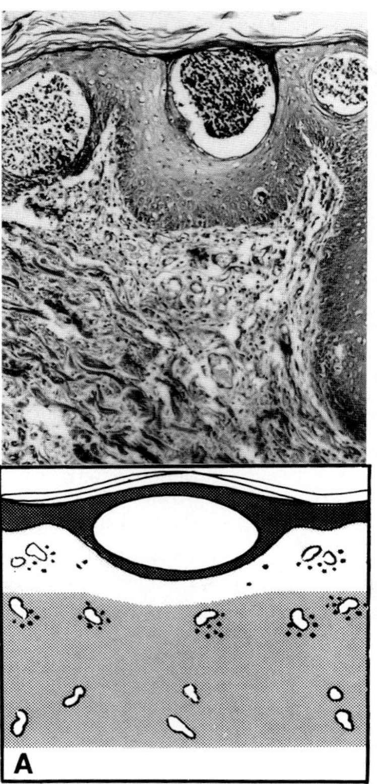

Fig. 5.10 Schematic and histologic appearances of dermatitis, (A) through (F). (A) Intraepidermal vesiculopustular dermatitis.

zymes from neutrophils and eosinophils, subcorneal pustular dermatosis, bovine exfoliative dermatitis, sterile eosinophilic pustulosis, developmental defect in bovine familial acantholysis) and hydropic degeneration of basal cells (lupus erythematosus, epidermolysis bullosa simplex, dermatomyositis, drug eruptions). It is useful to classify intraepidermal vesicular and pustular dermatitides on the basis of their location within the epidermis (Table 5.1).

6. Subepidermal Vesicular and Pustular Dermatitis

Subepidermal vesicles and pustules may be formed through hydropic degeneration of basal cells (lupus erythe-

TABLE 5.1

Histopathologic Classification of Intraepidermal and Subepidermal Pustular and Vesicular Diseases

Anatomic location[a]	Other helpful findings
INTRAEPIDERMAL	
Subcorneal	
Microbial infection	Neutrophils, microorganisms (bacteria, fungi) ± mild acantholysis, focal necrosis of epidermis
Canine subcorneal pustular dermatosis	Neutrophils ± moderate acantholysis
Canine sterile eosinophilic pustulosis	Eosinophils ± mild follicular involvement
Pemphigus foliaceus	Marked acantholysis, neutrophils and/or eosinophils, ± follicular involvement
Pemphigus erythematosus	Marked acantholysis, neutrophils and/or eosinophils, ± follicular involvement, ± lichenoid infiltrate
Canine linear IgA dermatitis	Neutrophils, ± mild acantholysis
Bovine exfoliative dermatitis	Neutrophils
Systemic lupus erythematosus	Neutrophils, ± mild acantholysis, interface dermatitis
Intragranular	
Pemphigus foliaceus	Marked acantholysis, granular "cling-ons," neutrophils and/or eosinophils, ± follicular involvement
Pemphigus erythematosus	Marked acantholysis, granular "cling-ons," neutrophils and/or eosinophils, ± lichenoid infiltrate
Pemphigus vegetans	Marked acantholysis, eosinophils, papillomatosis
Epitheliotropic lymphoma	Atypical lymphoid cells, Pautrier's microabscesses
Bovine familial acantholysis	Marked acantholysis
Porcine dermatosis vegetans	Eosinophils and neutrophils, papillomatosis
Viral dermatoses	Ballooning degeneration, ± inclusion bodies, ± acantholysis
Spongiotic dermatitis	Eosinophilic spongiosis suggests ectoparasitism, pemphigus, pemphigoid, canine sterile eosinophilic pustulosis
Canine hepatocutaneous syndrome	Diffuse parakeratosis, marked edema of upper epidermis
Suprabasilar	
Pemphigus vulgaris	Marked acantholysis, ± follicular involvement
Intrabasal	
Lupus erythematosus	Interface dermatitis, ± thickened basement membrane zone, ± dermal mucinosis
Dermatomyositis	Interface dermatitis, ± thickened basement membrane zone, ± dermal mucinosis, ± perifollicular fibrosis
Epidermolysis bullosa simplex	Little or no inflammation
Erythema multiforme	Interface dermatitis, marked single-cell necrosis of keratinocytes
Toxic epidermal necrolysis	Full-thickness coagulation necrosis of epidermis, little or no inflammation
Graft-versus-host disease	Interface dermatitis
SUBEPIDERMAL	
Bullous pemphigoid	Subepidermal vacuolar alteration, variable inflammation
Porphyria	Little or no inflammation, festooning, hyalinization of blood vessel walls
Epidermolysis bullosa	Little or no inflammation, ± hydropic degeneration
Lupus erythematosus	Interface dermatitis, ± thickened basement membrane zone, ± dermal mucinosis
Dermatomyositis	Interface dermatitis, ± thickened basement membrane zone, ± dermal mucinosis, ± perifollicular fibrosis
Erythema multiforme	Interface dermatitis, marked single-cell necrosis of keratinocytes
Toxic epidermal necrolysis	Full-thickness coagulation necrosis of epidermis, little or no inflammation
Severe subepidermal edema or cellular infiltration	
Severe spongiosis	Spongiotic vesicles

[a] Drug eruptions can mimic virtually all of these reaction patterns.

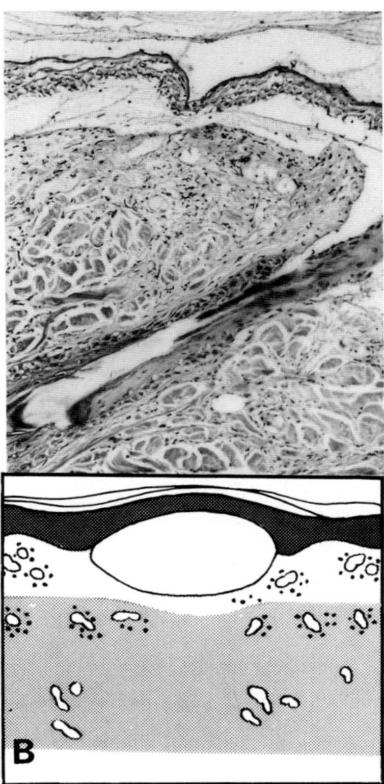

Fig. 5.10B Subepidermal vesicular dermatitis.

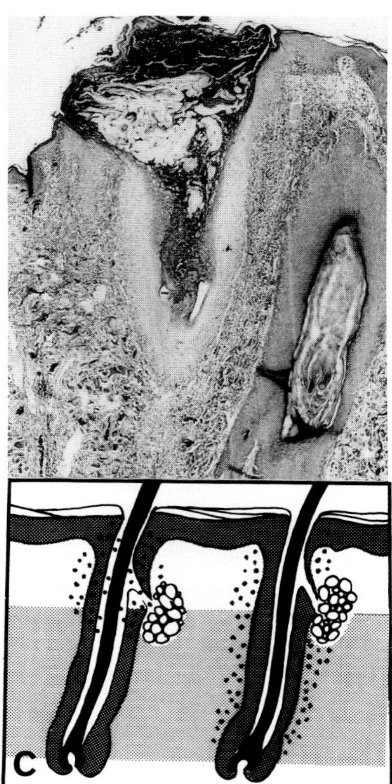

Fig. 5.10C Folliculitis.

matosus, epidermolysis bullosa simplex, dermatomyositis, drug eruption, toxic epidermal necrolysis), dermoepidermal separation (Fig. 5.10B) (bullous pemphigoid, erythema multiforme, epidermolysis bullosa, drug eruption, porphyria), severe subepidermal edema and/or cellular infiltration (especially urticaria, cellulitis, vasculitis, ectoparasitism), and severe intercellular edema with disruption of the basement membrane zone (spongiotic perivascular dermatitis) (Table 5.1). Caution is warranted when examining older lesions, as reepithelialization may result in subepidermal vesicles and pustules assuming an intraepidermal location. Such reepithelialization is usually recognized as a single layer of flattened, elongated basal epidermal cells at the base of the vesicle or pustule.

7. Perifolliculitis, Folliculitis, and Furunculosis

Perifolliculitis means accumulation of inflammatory cells around a hair follicle. Strictly defined, perifolliculitis requires exocytosis of these cells through the follicular epithelium. The term is often used more loosely to describe those commonly occurring superficial perivascular dermatitides in which the perifollicular vascular plexus is preferentially involved. We suggest the term **mural folliculitis** to describe those lesions in which the wall of the follicle is targeted. **Folliculitis** implies the accumulation of inflammatory cells within follicular lumina (Fig. 5.10C). **Furunculosis** (penetrating or perforating folliculitis) signifies hair follicle rupture. Perifolliculitis, folliculitis, and furunculosis usually represent a pathologic continuum, and all may be present in the same specimen. Follicular inflammation is a common gross and microscopic finding in dogs, often occurring as a secondary complication in pruritic dermatoses (e.g., hypersensitivity reactions and ectoparasitism), diseases of altered keratinization, and hormonal dermatoses.

Follicular inflammation may be caused by bacteria, fungi (dermatophytes, yeasts), parasites (*Demodex* spp., *Pelodera strongyloides, Stephanofilaria* spp.) and, rarely, canine atopy, food allergy, and seborrheic dermatitis. The folliculitides associated with bacteria, fungi, and parasites are usually suppurative initially, whereas those occasionally associated with atopy, food allergy, and seborrheic dermatitis are usually predominantly spongiotic (small numbers of exocytosing mononuclear cells and/or neutrophils). Any chronic folliculitis, particularly where there is furunculosis, can become pyogranulomatous or granulomatous.

Furunculosis, regardless of the initiating cause, is frequently associated with moderate to marked tissue eosinophilia, presumably as a reaction to liberated follicular contents. Idiopathic sterile eosinophilic folliculitides may be seen in cattle and dogs (sterile eosinophilic pustulosis). Insect stings are postulated as the cause of eosinophilic folliculidites affecting the noses of dogs. In cats and horses, sterile eosinophilic folliculitis may be seen in conjunction with hypersensitivity reactions (atopy, food allergy, onchocerciasis, equine eosinophilic granuloma, *Culicoides* hypersensitivity, flea-bite hypersensitivity). Equine unilateral papular dermatosis is also characterized by eosinophilic folliculitis and furunculosis.

Mural folliculitis is a feature of pemphigus, and these disorders should be considered if there is severe acantholysis. The hair follicle outer root sheath may be involved in the hydropic degeneration and lichenoid cellular infiltrates of lupus erythematosus, drug eruptions, erythema multiforme, and idiopathic lichenoid dermatoses.

A lymphocytic perifolliculitis directed at the bulb of anagen hair follicles is characteristic of alopecia areata. A mild perifolliculitis associated with marked mucinosis of the outer root sheath is characteristic of alopecia mucinosa. Perifollicular fibrosis is seen with chronic folliculitides, dermatomyositis, and chronic granulomatous sebaceous adenitis.

8. Fibrosing Dermatitis

Fibrosis marks the resolving stage of an intense, destructive inflammatory reaction or signifies an ongoing, more insidious, inflammatory process. Fibrosis which is recognizable histologically does not necessarily produce a visible clinical scar. Ulcers limited to the upper portion of the superficial dermis do not result usually in scarring, whereas virtually all ulcers that extend into the deep dermis proceed to fibrosis and clinical signs of scarring. Fibrosing dermatitis follows many severe insults to the dermis and is often of minimal diagnostic value. The most common causes of fibrosing dermatitis include furunculosis, vascular disease, lymphedema, equine exuberant granulation tissue, dermatomyositis, lupus erythematosus, photodermatitis, acral lick dermatitis, and morphea (localized scleroderma). Caprine parelaphostrongylosis is characterized by fibrosing dermatitis and focal hydropic degeneration of epidermal basal cells. In cats, a common chronic ulcerative dermatitis is characterized by linear, bandlike, subepidermal fibrosis which extends peripheral to the ulcer (Fig. 5.10D).

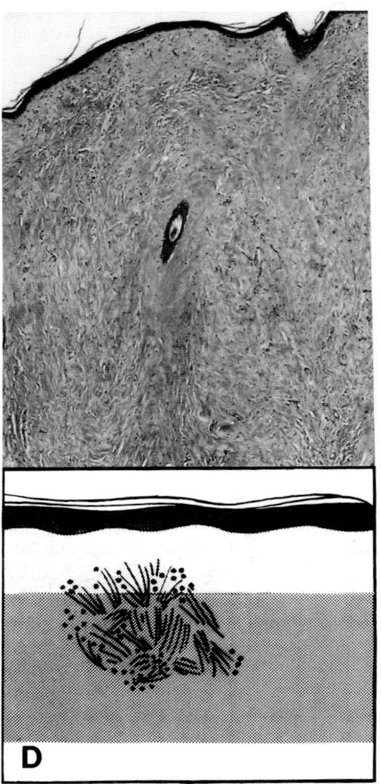

Fig. 5.10D Fibrosing dermatitis. Superficial dermis is diffusely fibrotic.

9. Panniculitis

The panniculus is commonly involved as an extension of dermal inflammatory processes, especially suppurative and granulomatous dermatoses, and there is usually some deep dermal involvement with primary panniculitides. Panniculitis may be caused by infectious agents, foreign bodies, vitamin E deficiency, trauma, pancreatic disease, vasculitis, drug eruption, and lupus erythematosus. However, the majority of panniculitides are sterile and idiopathic.

Panniculitis has been divided into **lobular** (fat lobules are primarily involved), **septal** (interlobular connective tissue septa are primarily involved), and **diffuse** (both anatomic areas involved) types. However, these anatomical divisions are rarely of diagnostic significance and, in fact, all three patterns may be seen in a single lesion from the same patient. In addition, although panniculitides may exhibit remarkable cytologic variability (granulomatous, pyogranulomatous, suppurative, necrotizing, lymphoplasmacytic, eosinophilic, fibrosing, vaso-occlusive), the various cytologic types appear to offer little clinical, diagnostic, therapeutic, or prognostic information. Thus, most panniculitides, regardless of cause, look identical histologically (Fig. 5.10E).

In dogs, one form of panniculitis mimics a rare form of lupus erythematosus (lupus erythematosus panniculitis or lupus profundus). It is characterized by early septal inflammation (lymphohistioplasmacytic), leukocytoclastic vasculitis, mucinous degeneration, lymphoid nodules, and later lobular extension with minimal fat necrosis. A similar microscopic lesion occurs at sites of previous rabies vaccination.

In cats, a rare subcutaneous sclerosis is characterized by severe septal fibrosis, minimal granulomatous inflammation and fat necrosis, and numerous fat micropseudocysts and lipocytes containing radially arranged needle-shaped clefts.

10. Atrophic Dermatosis

Atrophic dermatosis is characterized by varying degrees of epithelial and connective tissue atrophy (Fig. 5.10F). The most common disorders in this category are endocrine, nutritional, and developmental dermatoses and telogen defluxion. Atrophic dermatoses show variable combinations of the following histopathologic findings: diffuse orthokeratotic hyperkeratosis, epidermal atrophy,

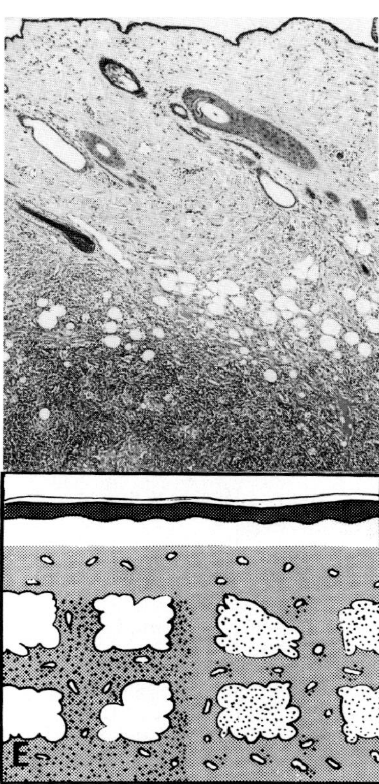

Fig. 5.10E Panniculitis, lobular type. Cells infiltrate panniculus. Dermis relatively spared.

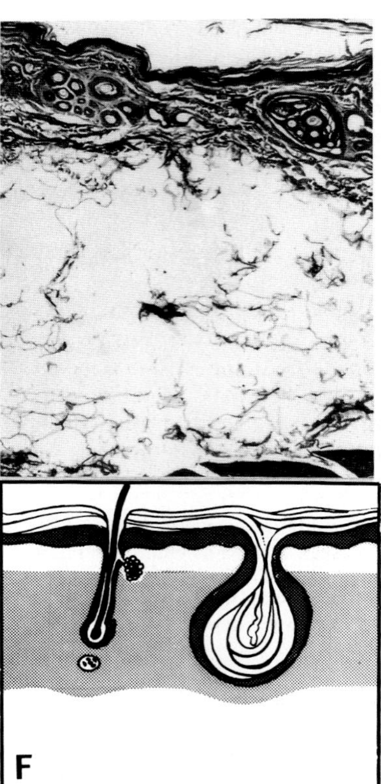

Fig. 5.10F Endocrine skin lesion. Thin dermis, hyperkeratosis, pilosebaceous atrophy, follicular keratosis, and plugging.

follicular keratosis, follicular atrophy, inactive or telogen hair follicles, pilar canals devoid of hairs, excessive trichilemmal keratinization (flame follicles), epithelial melanosis, and sebaceous gland atrophy. Inflammatory changes are frequent and potentially misleading in the atrophic dermatoses, and reflect the common occurrence of secondary bacterial infection and/or seborrhea with these disorders. Findings suggestive of specific endocrinopathies include diffuse mucinous degeneration (hypothyroidism, acromegaly), dystrophic mineralization (hyperadrenocorticism), decreased amount of dermal elastin (hyposomatotropism, hyperadrenocorticism), hypertrophied and/or vacuolated arrector pili muscles (hypothyroidism), dermal atrophy (hyperadrenocorticism, hyposomatotropism), and absence of arrector pili muscles (hyperadrenocorticism).

Findings suggestive of nutritional disorders include misshapen, corkscrew hairs (protein deficiency, vitamin C-responsive dermatosis). Developmental disorders (hypotrichosis, follicular dysplasia, color mutant alopecia) are often characterized by follicular dysplasia and anomalous deposition of melanin.

11. Mixed Reaction Patterns

Because diseases are a pathologic continuum, reflecting various combinations of acute, subacute, and chronic changes, and because animals can have more than one dermatosis at the same time, it is common to see one or two biopsies from the same animal which show two or more reaction patterns. For instance, it is common to see an overall pattern of perivascular dermatitis (due to hypersensitivity reactions or ectoparasitism) or atrophic dermatosis (due to endocrinopathy) with a subordinate, focal pattern of folliculitis, furunculosis, or intraepidermal pustular dermatitis (due to secondary bacterial infection). Likewise, one can find multiple patterns from one or two biopsies from an animal with a single disease (e.g., vasculitis, diffuse necrotizing dermatitis, and fibrosing dermatitis in a patient with necrotizing vasculitis). In two retrospective clinicopathologic studies on cattle and horses, respectively, mixed reaction patterns were encountered in 13 and 16.3% of the cases. It is important always to consider the relative diagnostic importance of the patterns present. Interface dermatitis, for example, is a pattern of great diagnostic significance, pointing to a group of specific diseases and normally would take precedence over a pattern such as perivascular dermatitis.

Bibliography

Goldschmidt, M. H. Small animal dermatopathology: "What's old, what's new, what's borrowed, what's useful." *Sem Vet Med Surg* **2:** 162–165, 1987.

Grunwald, M. H., Lee, J. Y. Y., and Ackerman, A. B. Pseudocarcinomatous hyperplasia. *Am J Dermatopathol* **10:** 95–103, 1988.

Kanerva, L. Electron microscopic observations of dyskeratosis, apoptosis, colloid bodies, and fibrillar degeneration after skin irritation with dithranol. *J Cut Pathol* **17:** 373–44, 1990.

Scott, D. W. Lymphoid nodules in skin biopsies from dogs, cats, and horses with nonneoplastic dermatoses. *Cornell Vet* **79:** 267–272, 1989.

Scott, D. W. Excessive trichilemmal keratinization (flame follicles) in endocrine skin disorders of the dog. *Vet Dermatol* **1:** 37–40, 1989.

Scott, D. W. Epidermal mast cells in the cat. *Vet Dermatol* **1:** 65–69, 1990.

Seaman, W. J., and Chang, S. H. Dermal perifollicular mineralization of toy poodle bitches. *Vet Pathol* **21:** 122–123, 1984.

Steffen, C. Dyskeratosis and the dyskeratoses. *Am J Dermatopathol* **10:** 356–363, 1988.

Torres, S. M., and Sanchez, J. L. Cutaneous plasmacytic infiltrates. *Am J Dermatopathol* **10:** 319–329, 1988.

Woo, T. Y., and Rasmussen, J. E. Disorders of transepidermal elimination. Part 1. *Int J Dermatol* **24:** 267–279, 1985.

Wood, C. *et al.* Eosinophilic infiltration with flame figures. A distinctive tissue reaction seen in Well's syndrome and other diseases. *Am J Dermatopathol* **8:** 186–193, 1986.

Yager, J. A., and Wilcock, B. P. Skin biopsy: Revelations and limitations. *Can Vet J* **29:** 969–972, 1988.

III. Congenital and Hereditary Diseases of Skin

It is convenient to consider congenital and hereditary disorders of the skin together, but the congenital dermatoses are not necessarily hereditary and vice versa. Congenital hypotrichosis may reflect congenital hypothyroidism of environmental iodine deficiency or inherited thyroidal enzymatic defects. Congenital hypertrichosis occurs in cases of prolonged gestation, both sporadic and genetic. Fetal dermatitis is common in epizootic abortion of cattle and in some cases of mycotic abortion in cattle. Examples of inherited defects of skin that are not congenital include symmetrical alopecia of Holstein cattle and color-mutant alopecia of dogs, conditions which are not expressed until the animals are young adults. Many common skin disorders, although not considered to be hereditary, probably reflect altered constitutional states of genetic origin. Examples include canine atopy and seborrheic diseases. These, and hereditary tumors, congenital porphyria, tyrosinemia, lymphedema, congenital and heritable endocrinopathies, and some heritable disorders of pigmentation are discussed elsewhere in these volumes.

A. *Epitheliogenesis Imperfecta*

Congenital inherited discontinuities of the squamous epithelium of skin and oral mucous membranes occur occasionally in calves and piglets, rarely in foals and lambs, and extremely rarely in puppies and kittens. In all species, the lesions are sharply demarcated, variable-sized defects in the epithelium. The exposed dermis or submucosa is easily traumatized and secondarily infected, predisposing the animal to septicemia. Fetuses with extensive lesions may be aborted. In these cases, there is scarcely any normal epidermis; the hooves are incompletely cloven and without horn, the external ears are deformed and fused with the occipital skin, the lips are fused to the gums, and the eyelids are fused to the sclera. Small lesions are not inconsistent with life and may heal with scar tissue.

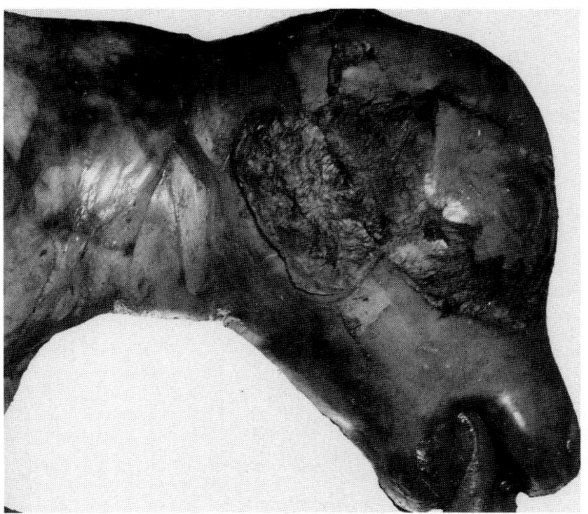

Fig. 5.11 Epitheliogenesis imperfecta. Calf.

In **cattle,** the condition occurs in Holstein–Friesians, Ayrshires, Jerseys, European breeds, shorthorns, Aberdeen Angus, and the *Bos indicus* Sahiwal (Fig. 5.11). The mode of inheritance, where established, is autosomal recessive. The cutaneous defects have a predilection for the distal extremities, particularly the claws and the skin over fetlocks and knees, but may involve muzzle, nostril, cheeks, ear margins, tongue, and hard palate. Lesions in Jerseys are usually extensive and are often accompanied by brachygnathia inferior and atresia ani. Histologically, there is an absence of epithelium and adnexal structures, with an abrupt transition between affected and normal skin. Defects in the lipid and collagen biosynthetic capabilities of dermal fibroblasts from a calf with epitheliogenesis imperfecta have been reported. These findings, along with histologic evidence of clefting at the dermoepidermal junction in marginal epidermis, suggest that at least some of these conditions may represent more than an epithelial defect and may come to be reclassified as one of the epidermolysis bullosa disorders.

The condition in **foals** has a similar predilection for the hooves, limbs, and tongue and may involve the anterior esophagus. It is perhaps more usual for epitheliogenesis imperfecta in **swine** to be expressed as one large or several small defects on the trunk, usually with an almost symmetrical distribution (Fig. 5.12). This pattern occurs in individual piglets or with a familial incidence. An alternative pattern affects the dorsal surface or the anterior ventral surface of the tongue. There may be additional epithelial defects in the oral cavity and at the coronets. Some, perhaps all, of these piglets have concurrent congenital hydroureter and hydronephrosis. Epitheliogenesis imperfecta occurs widely but with low incidence in several

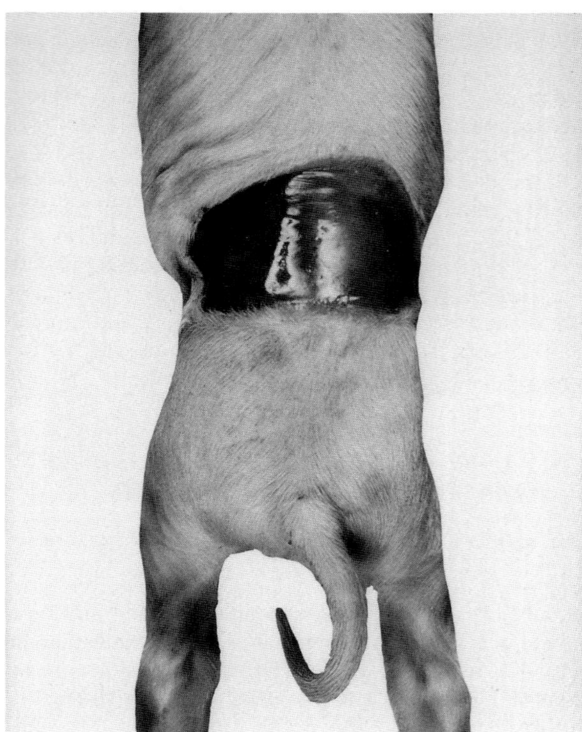

Fig. 5.12 Epitheliogenesis imperfecta. Pig.

breeds of **sheep** without a clear indication as to whether it is genetically determined. There are epithelial defects in the buccal cavity, especially on the dorsal free portion of the tongue. The hooves, which are well formed, are detached and shed very shortly after birth. There are no epithelial defects about the coronet, as there are in calves, and the defect is probably in the laminae of the hoof and periople.

Bibliography

Bentinck-Smith, J. A congenital epithelial defect in a herd of Berkshire swine. *Cornell Vet* **41:** 47–51, 1951.

Crowell, W. A., Stephenson, C., and Gosser, H. S. Epitheliogenesis imperfecta in a foal. *J Am Vet Med Assoc* **168:** 56–58, 1976.

Fordyce, G. *et al.* The prevalence of epitheliogenesis imperfecta in Sahiwal cattle and their crosses in a North Queensland beef herd. *Aust J Agric Res* **38:** 427–435, 1987.

Frey, J. *et al.* Collagen and lipid biosynthesis in a case of epitheliogenesis imperfecta in cattle. *J Invest Dermatol* **93:** 83–86, 1989.

Gupta, B. N. Epitheliogenesis imperfecta in a dog. *Am J Vet Res* **34:** 443–444, 1973.

Hutt, F. B., and Frost, J. N. Hereditary epithelial defects in Ayrshire cattle. *J Hered* **39:** 131–137, 1948.

Jayasekara, M. U., and Leipold, H. W. Epitheliogenesis imperfecta in shorthorn and Angus cattle. *Zentralbl Veterinaermed* [A] **26:** 497–501, 1979.

Munday, B. L. Epitheliogenesis imperfecta in lambs and kittens. *Br Vet J* **126:** 47, 1970.

Wipprecht, C., and Horlacher, W. R. A lethal gene in Jersey cattle (epitheliogenesis imperfecta). *J Hered* **26:** 363–368, 1935.

B. Wattles

These structures, which are not unlike those normally seen in goats, occur occasionally in **pigs** of either sex. They are up to 6 cm long and hang from the underside of the mandible. The appendages have a cartilaginous core. The defect is supposedly due to a single dominant gene. Similar, although smaller, structures have been reported rarely in Merino, Dorset Down, and Karakul **sheep**.

Bibliography

Lancaster, M. J., and Medwell, W. D. Neck wattles in lambs. *Aust Vet J* **68:**75–76, 1991.

Roberts, E., and Morrill, C. C. Inheritance and histology of wattles in swine. *J Hered* **35:**149–151, 1944.

C. Ichthyosis

Ichthyosis, so named because the cutaneous lesions resemble fish scales, encompasses a heterogeneous group of rare genodermatoses affecting cattle, dogs, pigs, mice, and llamas. Attempts to correlate the animal disorders with the several forms of human ichthyosis have lead to inconsistencies in classification and nomenclature, particularly in the bovine conditions. However, the constant feature of these disorders, despite varying anatomic patterns and severity, is marked lamellar hyperkeratosis. The pathogenesis probably involves accentuated corneocyte cohesion, which interferes with the normal desquamative processes of the stratum corneum. The underlying biochemical disorders are unknown; the defects are likely to vary between the species and in the various expressions of ichthyosis within a species.

Two main forms of ichthyosis have been reported in **cattle. Ichthyosis fetalis** is incompatible with life, and affected animals are aborted or survive no more than a few days after birth. It is most akin to the rare human disorder, the harlequin fetus. The skin is covered by large horny

Fig. 5.13A Ichthyosis congenita. Sheets of adherent keratin in axilla. Calf.

plates separated by deep fissures which correspond to normal cleavage lines of the skin. There is an absence of hair, but short stubble may line the bottom of the fissures or be trapped in the thick, hyperkeratotic plaques. The tight, inelastic skin causes eversions at mucocutaneous junctions. Small ear pinnae are often seen. Histologically, there is marked orthokeratotic hyperkeratosis of the laminated type, affecting both surface and follicular epithelia. Ichthyosis fetalis has been described in Norwegian red poll, Friesian, and Brown Swiss calves. The mode of inheritance in the former is autosomal recessive.

Ichthyosis congenita is a less severe expression of the defect, and affected calves live longer. It has been reported in Jerseys (Fig. 5.13A), Pinzgauer, Chianina, and Holstein–Friesians. In one report, a preponderance of males among affected Holstein calves suggested an incomplete sex linkage, but in 22 German Pinzgauer calves, inheritance was considered to be autosomal recessive. The lesions are most severe on the limbs, abdomen, and muzzle. Hairlessness is not an initial feature, but alopecia may develop. The thick, curling squames are readily visible where the hair is short, as behind the muzzle and in the axilla. Pinzgauer calves had associated microtia, cataracts, and thyroid abnormalities.

Ichthyosis in **dogs** is rare. It has been reported in the terrier, Doberman pinscher, golden retriever, cocker spaniel, and mixed breeds. It is presumed to be a simple autosomal recessive trait, since the parents of affected pups are normal. Lesions are present at birth and are generalized, although flexural surfaces and footpads are more severely affected. Dry gray scales, sometimes forming verrucous or featherlike projections, are tightly adherent to the underlying, sometimes hyperpigmented skin. There is accompanying alopecia and lichenification. Erythroderma may be seen. Histologically, the lesions include marked orthokeratotic hyperkeratosis, sometimes showing digitate projections, hypergranulosis, and follicular keratosis. Vacuolation of keratinocytes and reticular degeneration may occur, prompting the comparison with human epidermolytic hyperkeratosis. (Fig. 5.13B).

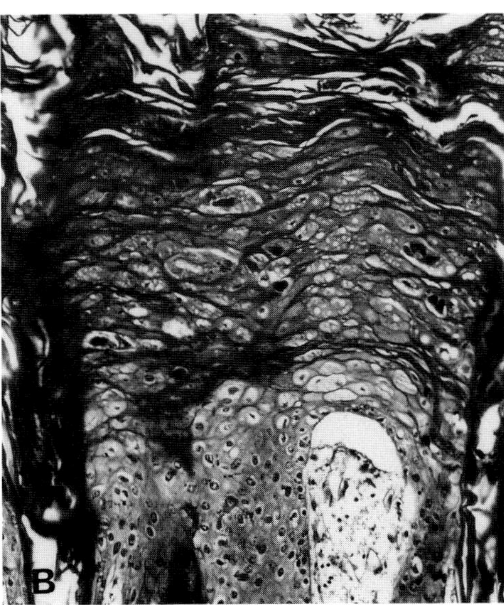

Fig. 5.13B Hyperkeratosis and intracellular edema. Ichthyosis congenita.

Ultrastructural studies of ichthyosiform **llama** skin showed a retention of lipid vacuoles within and between cells of the laminated stratum corneum. An increase in membrane-coating granules was noted also. The former lesion has been described in human lamellar ichthyosis; the latter has not. Abnormal lipid metabolism may contribute to the excessive cohesiveness characteristic of ichthyosis.

Bibliography

Baker, J. R., and Ward, W. R. Ichthyosis in domestic animals: A review of the literature and a case report. *Br Vet J* **141:** 1–8, 1985.

Belknap, E. B., and Dunstan, R. W. Congenital ichthyosis in a llama. *J Am Vet Med Assoc* **197:** 764–767, 1990.

Julian, R. J. Ichthyosis congenita in cattle. *Vet Med* **55:** 35–41, 1960.

Scott, D. W. Congenital ichthyosis in a dog. *Comp Anim Pract* **19:** 7–11, 1989.

Williams, M. L., and Elias, P. M. Genetically transmitted, generalized disorders of cornification. The ichthyoses. *Dermatol Clin* **5:** 155–178, 1987.

D. Hereditary Zinc Deficiency

Congenital and hereditary zinc-deficiency disorders affecting the skin have been reported in cattle and bull terrier dogs.

Lethal trait A-46, also known as hereditary parakeratosis and hereditary thymic hypoplasia, was originally described in black pied Danish cattle of Friesian descent. It occurs in Friesian cattle in Great Britain and Europe and has also been described in shorthorn beef calves in the United States. The trait is inherited as an autosomal recessive. Lethal trait A-46 is a zinc-deficiency disease secondary to intestinal malabsorption of zinc. Affected animals have reduced plasma zinc concentration, and oral zinc supplementation reverses the parakeratotic lesions. Withdrawal of zinc treatment results in a decrease in plasma zinc and a recrudescence of clinical signs. This condition is analogous to the human condition, acrodermatitis enteropathica. Both human and bovine diseases are probably associated with a deficiency of zinc ligands important in the intestinal transport of zinc.

Gross lesions in calves commence at 4 to 8 weeks of age as crusting and alopecia around the muzzle, eyes, base of ears, and between the mandibular rami. The lesions spread to involve the neck and the skin over the joints. Affected calves also salivate excessively and may develop conjunctivitis, rhinitis, bronchopneumonia, or diarrhea. The animals are unthrifty and, without treatment, usually die before they reach 6 months of age. The striking internal lesion observed at necropsy is thymic hypoplasia.

The characteristic microscopic lesion in the skin of lethal A-46 calves is marked diffuse parakeratotic hyperkeratosis, typical of zinc-deficiency states. The thymic lesion is a severe depletion of cortical lymphocytes. There is a resultant hypoplasia of the secondary lymphoid organs, such as spleen, lymph nodes, and mucosal lymphoid aggregates. Lymphocyte numbers and function are normal at birth, but activity decreases with the development of the zinc deficiency.

Lethal acrodermatitis of bull terrier dogs is also characterized by parakeratotic skin lesions, retarded growth, and immunologic deficiencies. Reduction in plasma zinc levels has been demonstrated, but the disorder is not responsive to parenteral or oral zinc supplementation and so differs from lethal trait A-46 of cattle. The trait is autosomal recessive.

Skin lesions develop at 6 to 10 weeks, although one report is of an affected pup at 3 days of age. The scale-crusted lesions have a predilection for areas of abrasion, including the digits and footpads, elbows, hocks (Fig. 5.14), periocular and perioral skin, and muzzle. Nail dystrophy and paronychia occur, and pustular lesions of secondary pyoderma are common. Respiratory tract infections, diarrhea, ocular lesions, and behavioral changes accompany the cutaneous manifestations. The median survival time is approximately 7 months. At necropsy, in addition to the cutaneous lesions, there is thymic and lymph node hypoplasia. Bronchopneumonia is the most frequent immediate cause of death. Microscopic lesions in the skin are severe, diffuse parakeratotic hyperkeratosis with pustular dermatitis due to secondary pyoderma.

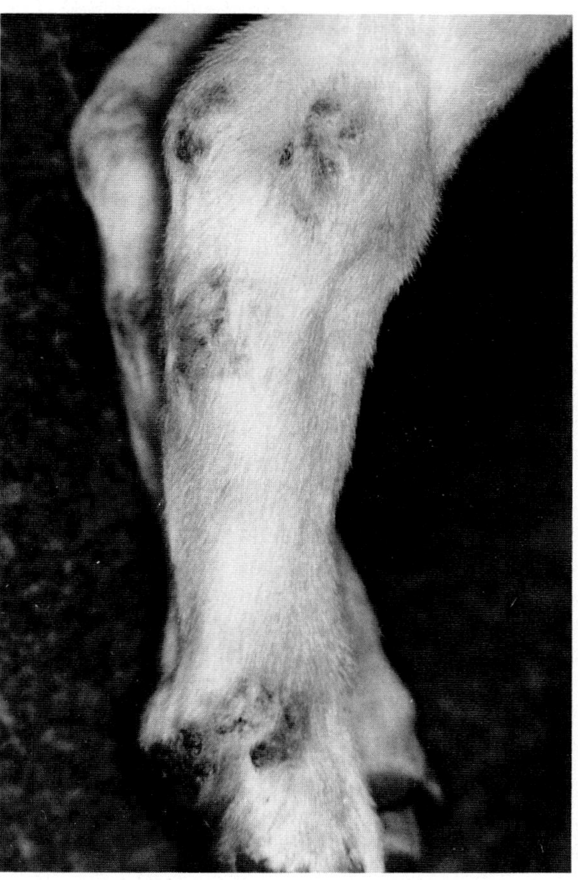

Fig. 5.14 Lethal acrodermatitis of bull terriers. (Courtesy of B. Smits and *Veterinary Dermatology*).

The pathogenesis of the disease is unknown. Plasma zinc levels are variable. Serum alkaline phosphatase is reduced in those dogs with lowered serum zinc, as would be expected. Immunological abnormalities, including reduced lymphocyte blastogenesis and serum immunoglobulin levels, are consistent with experimental zinc deficiency. However, administration of zinc by various routes has not been able to halt the course of the disease. A defect in the cellular metabolism of zinc has been proposed as the underlying defect.

Bibliography

Adrichem, P. W. M., Van Leeuwen, J. M., and Van Kluijve, J. J. Parakeratosis of the skin of calves. *Neth J Vet Sci* **4:** 57–63, 1971.

Andresen, E. *et al.* Evidence of a lethal trait A-46, in black pied Danish cattle of Friesian descent. *Nord Vet Med* **22:** 473–485, 1970.

Brummerstedt, E. *et al.* The effect of zinc on calves with hereditary thymus hypoplasia (lethal trait A46). *Acta Pathol Microbiol Scand* **79:** 686–687, 1971.

Flagstad, T. Lethal trait A46 in cattle: Intestinal zinc absorption. *Nord Vet Med* **28:** 160–169, 1976.

Jezyk, P. F. *et al.* Lethal acrodermatitis in bull terriers. *J Am Vet Med Assoc* **188:** 833–839, 1986.

Perryman, L. E. *et al.* Lymphocyte alterations in zinc-deficient calves with lethal trait A46. *Vet Immunol Immunopathol* **21:** 329–248, 1989.

Price, J., and Wood, D. A. Zinc-responsive parakeratosis and ill-thrift in a Friesian calf. *Vet Rec* **110:** 478, 1982.

Vogt, D. W., Carlton, C. G., and Miller, R. B. Hereditary parakeratosis in shorthorn beef calves. *Am J Vet Res* **49:** 120–121, 1988.

E. Congenital and Tardive-Onset Hypotrichosis

Partial to complete absence of hair has been reported as a congenital defect in all domestic species, but is most frequent in cattle. Congenital hypotrichosis is often hereditary and, in some instances, the affected animals have been given breed status, as in the Chinese crested dog and Large White Ulster pig. Sometimes, selecting for desirable traits such as diluted coat color has resulted in unwanted side effects, as in color-mutant alopecia in dogs. The degree of hairlessness extends from total absence (atrichia) to a mild, progressive patchy alopecia.

Partial hypotrichosis is often symmetrical in distribution. The remaining hair may be brittle, easily epilated, fine, or curly. In some syndromes, such as congenital black hair follicular dysplasia and color-mutant alopecia in dogs, it is related to the pigmentation of the skin. Hair follicles may be totally lacking or reduced in number; often they are present in normal or even increased numbers but fail to develop normally. The latter may be better termed follicular dystrophies. Furthermore, some of the delayed-onset alopecias, although developmental, are not congenital; they would more properly be classified as follicular abiotrophies.

Various types of heritable congenital hypotrichosis occur in **cattle**. Affected breeds include Ayrshire, Holstein–Friesian, Jersey, Guernsey, Hereford, Normandy–Maine–Anjou cross Charolais, and Pinzgauer. Hereditary hypotrichosis is usually due to a single autosomal recessive gene, as in Jerseys, Guernseys, Herefords, and Holstein–Friesians, but a single sex-linked partially dominant gene is held responsible for streaked hairlessness of Holstein–Friesians. An X-linked recessive gene is responsible for hypotrichosis in Normandy–Maine–Anjou–Charolais cattle.

Hypotrichosis may be accompanied by other developmental defects, such as incisor anodontia in Friesian and Normandy–Maine–Anjou–Charolais breeds or oligodontia seen in a Simmental × red Holstein with a chromosomal anomaly (Xq deletion). Hypotrichosis may be associated with lethal characteristics, as in streaked hairlessness of Holstein–Friesians and a very severe hypotrichosis of Jersey cattle.

Congenital hypotrichosis of **nongenetic** origin is associated with iodine deficiency and goiter, with adenohypophyseal hypoplasia in Guernseys and Jerseys, and with the maternal ingestion of *Veratrum album* in Japanese cattle. These calves also have small bodies, short limbs, hypophyseal aplasia, and teeth covered by gingiva. Intrauterine infection with the virus of bovine virus diarrhea may cause patchy hypotrichosis in calves (Fig. 5.15).

The gross appearance in bovine hypotrichosis is variable, probably reflecting the various underlying genetic deficits. A lethal condition in Holstein–Friesians is accompanied by generalized severe alopecia. Hair is present only in sharply defined regions including the muzzle, eyelids, pastern, tip of the tail, ears, and umbilicus. Normal numbers of follicles are present but are hypoplastic. In Guernseys, the calves are born haired, but the coat is progressively lost, again sparing the eyelids, muzzle, and distal extremities. Symmetrical alopecia in Holstein–Friesians commences approximately 6 weeks to 6 months after birth, affecting first the head, neck, and back, and progressing distally along limbs and caudally to the tail.

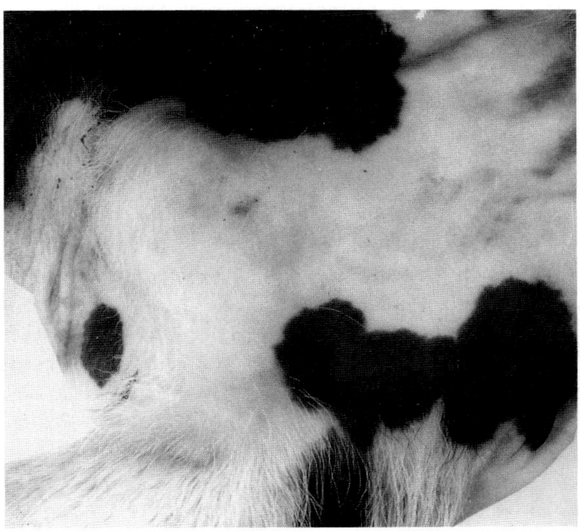

Fig. 5.15 Congenital alopecia. Calf. Note absence of hair over scapula and humerus.

Several, possibly related, hypotrichotic conditions, described in polled and horned Herefords, are associated with changes in the character of the coat. One such condition reported as semihairlessness is inherited as a simple autosomal recessive trait. More recently, a disorder reported as congenital hypotrichosis describes a range of clinical appearances. Some affected animals are born with a sparse, soft, and curly coat, which becomes progressively alopecic; some animals are born almost completely alopecic; and some have a thin woolly coat, which is retained. The least-affected areas of the alopecic animals are the tail switch, ears, axillae, and prepuce, but even here the hair is thin and easily epilated. Histologic and ultrastructural studies indicate a follicular dystrophy with markedly abnormal trichohyaline granules and degenerative changes in the Huxley's layer of the internal root sheath. Failure to form the inner root sheath leaves the newly keratinized hair shaft with no protective tube as it grows up the pilar canal. Hair shafts are dysplastic and fragmented, showing vacuolar changes in the medulla and cortex. Affected hairs are more soluble than normal. Ultrastructurally, the abnormally large trichohyaline granules lack the normal micro- and macrofilaments.

A condition in American and Canadian polled Hereford calves, termed **progressive alopecia, congenital anemia, and dyskeratosis** has some similarities but some important differences. It too is characterized by a sparse, short, kinked or curly coat at birth, which progressively develops into generalized alopecia. Both conditions are also associated with abnormally prominent foreheads. However, in the latter condition, affected calves are also anemic due to bone marrow dyserythropoiesis. The histologic lesions in skin also differ; while there is dyskeratosis of Huxley's layer of the inner root sheath, abnormal trichohyaline granules are not described, and additional dyskeratotic changes are present in surface epidermis, in the outer root sheath of hair follicles, and in the ruminal epithelium. The dyskeratotic epidermal lesions and the giant trichohyaline granules may coexist.

The so-called **woolly hair coat syndrome**, reported from Australia, predominantly affects polled Herefords but has been described also in the horned type. The coat is abnormally short and curly, but this phenotypic change does not precede the progressive alopecia usually seen in affected North American Herefords, and thus the condition is not truly hypotrichotic. It should be noted, however, that some of the calves described as hypotrichotic in the United States also retained their curly coats until death. In the Australian cattle, the coat changes are a marker for lethal cardiomyopathy (see Volume 3, Chapter 1, The Cardiovascular System).

The **baldy calf syndrome** was described initially in Canadian Holstein–Friesian cattle. The inheritance is autosomal recessive, and both sexes are affected. The gross lesions are postnatal in onset and comprise patches of scaly, thickened, partially alopecic skin on the neck, shoulders, periocular region, and over bony prominences. The cutaneous lesions have a gross appearance and distribution similar to those of lethal trait A-46, and the lesions are characterized histologically by marked parakeratosis. Excessive salivation and progressive emaciation are other common clinical features. However, in baldy calves, there is a failure to develop horn buds, medial curling of the pinnae, and a marked overgrowth of the hoof not seen with the lethal A-46 trait. The thymus is normal. While hepatic zinc levels may be depressed in baldy calves, plasma levels are usually normal or elevated. The baldy calf, unlike lethal A-46 affected animals, does not respond to oral zinc supplementation, and animals usually die before maturity. Animals showing partial affliction may survive beyond maturity. The histomorphology of the lesions in skin and their progressive development await description. **Inherited epidermal dysplasia** has been suggested as an alternative name for the condition.

Hypotrichosis is rare in **dogs** except for the Chinese crested, Turkish naked, Peruvian hairless, African sand dog, Xoloitzcuintl, and Abyssinian breeds, which have been selected for the trait. These animals are bald, apart from sparse hair tufts on the tail, head, and feet. Histologically, there is complete aplasia of hair follicles. The gene responsible is an incomplete autosomal dominant, and the defect is an arrested development of the hair follicles at about day 38 of gestation.

Sporadic examples of congenital hypotrichosis have been reported in a variety of dog breeds and probably represent a heterogeneous collection of genetic defects. The alopecia may be generalized or regional. Diffuse congenital alopecia has been described in four male basset hound littermates and in a family of beagles. The beagle disorder, termed **congenital hypotrichosis universalis**, has been likened to human male-pattern baldness in that primary hair follicles remain, but produce very fine hair shafts.

Regional hypotrichosis, termed variously **congenital symmetrical alopecia**, **congenital hypotrichosis,** and **congenital ectodermal defect,** has been reported in the whippet, cocker spaniel, Belgian shepherd, black Labrador, and several miniature poodles. The alopecia affects predominantly the ventrum, dorsum of the head, dorsal pelvic region, and posterior–medial aspects of the thighs. The skin becomes hyperpigmented and scaly. Histologically, there is a marked reduction or complete absence of both primary and secondary hair follicles. The absence of adnexal glands is reported in some cases but not others. With the exception of one female Labrador, only males are affected, suggesting a sex-linked mode of inheritance. Anomalies of dentition have been associated with symmetrical alopecias in poodles and in a Labrador.

A **follicular dysplasia syndrome** has been described in Siberian husky dogs. The onset of the alopecia is tardive. Hair loss is symmetrical and predominantly affects the guard hairs. The undercoat develops an abnormal rusty color and becomes woolly. Regrowth of hair, such as at sites of previous biopsy, is retarded. Histologically, there is an increase in the number of primary hairs in catagen, normally a transient stage of the hair follicle cycle. These catagen hairs are morphologically abnormal, having an excessive degree of tricholemmal cornification. Such hairs are indistinguishable from the catagen arrest follicles described as flame follicles in hyposomatotropism and castration-responsive dermatosis. Other lesions include mild orthokeratotic hyperkeratosis, sebaceous gland atrophy, and fragmentation of the hair shafts within the follicular infundibula.

The most common form of inherited alopecia in the dog is the late-onset, color-dilution alopecia, to be discussed.

In **cats,** a rare but severe form of hypotrichosis has been named alopecia universalis (Sphinx cat, Canadian hairless cat). This trait has been, regrettably, sought after by some breeders. The animals have virtually no hair except for a few short facial hairs and vibrissae. The skin is thickened and greasy. Histologically there are secondary hyperplastic changes in the epidermis and an absence of primary hair follicles. Sebaceous and sweat glands are present but open directly onto the surface of the skin.

Hereditary hypotrichosis is an autosomal recessive trait in Siamese cats. The animals are born with an abnormally fine hair coat, which is shed by about 2 weeks of age. Some hair growth occurs by 8 to 10 weeks, but total alopecia develops by 6 months. Histologically, there are small, poorly developed primary hair follicles, most of which are in telogen phase and devoid of hair. Hereditary hypotrichosis has also been reported in the Devon rex breed.

Congenital hypotrichosis in **swine** may be associated with iodine deficiency, with exposure to modified hog cholera virus *in utero,* or may be hereditary. Inherited hypotrichosis has been attributed to an autosomal dominant trait, which is probably lethal when homozygous. Histologically, hair follicles are reduced 50–70%. Recessive hairlessness also occurs in pigs.

Congenital hypotrichosis occurs in polled Dorset **lambs,** but although suspected as hereditary, the genetic basis has not been determined. Skin biopsy reveals hypoplastic follicles. In **goats,** congenital hypotrichosis has been reported very rarely.

Bibliography

Becker, R. B., Simpson, C. F., and Wilcox, C. J. Hairless Guernsey cattle. Hypotrichosis—a nonlethal character. *J Hered* **54:** 3–7, 1963.

Bracho, G. A. *et al.* Further studies of congenital hypotrichosis in Hereford cattle. *Zbl Vet Med* **31:** 72–80, 1984.

Braun, U. *et al.* Hypotrichose und Oligodontie, verbunden mit einer Xq-deletion, bei einem Kalb der Schweizerischen Fleckviehrasse. *Tierarztl Prax* **16:** 39–44, 1988.

Carbrey, E. *et al.* Transmission of hog cholera by pregnant sows. *J Am Vet Med Assoc* **149:** 23–30, 1966.

Chastain, C. B., and Swayne, D. E. Congenital hypotrichosis in male basset hound littermates. *J Am Vet Med Assoc* **187:** 845–846, 1985.

David, L. T. Histology of the skin of the Mexican hairless swine (*Sus scrofa*). *Am J Anat* **50:** 283–292, 1932.

Dolling, C. H. S., and Brooker, M. G. A viable hypotrichosis in poll Dorset sheep. *J Hered* **57:** 87–90, 1966.

Eldridge, F. E., and Atkeson, F. W. Streaked hairlessness in Holstein–Friesian cattle. A sex-linked lethal character. *J Hered* **44:** 265–271, 1953.

Hanna, P. E., and Ogilvie, T. H. Congenital hypotrichosis in an Ayrshire calf. *Can Vet J* **30:** 249–250, 1989.

Holmes, J. R., and Young, G. B. Symmetrical alopecia in cattle. *Vet Rec* **66:** 704–706, 1954.

Hutt, F. B. A note on six kinds of genetic hypotrichosis in cattle. *J Hered* **54:** 186–187, 1963.

Itakura, C., Nakano, D., and Goto, M. Histopathologic features of the skin of hairless newborn pigs with goiter. *Am J Vet Res* **40:** 111–114, 1979.

Jayasekara, M. U., Leipold, H. W., and Cook, J. E. Pathological changes in congenital hypotrichosis in Hereford cattle. *Zentralbl Veterinaermed* [A] **26:** 744–753, 1979.

Jubb, T. F. *et al.* Inherited epidermal dysplasia in Holstein–Friesian calves. *Aust Vet J* **67:** 16–18, 1990.

Kislovsky, D. Inherited hairlessness in the goat with some observations on parallel mutation. *J Hered* **28:** 265–267, 1937.

Kunkle, G. A. Congenital hypotrichosis in two dogs. *J Am Vet Med Assoc* **185:** 84–85, 1984.

Letard, E. Hairless Siamese cats. *J Hered* **29:** 173–175, 1931.

Post, K., Dignean, M. A., and Clark, E. G. Hair follicle dysplasia in a Siberian husky. *J Am Anim Hosp Assoc* **24:** 659–662, 1988.

Roberts, E., and Carroll, E. The inheritance of hairlessness in swine. *J Hered* **22:** 125–132, 1931.

Robinson, R. The Canadian hairless or Sphinx cat. *J Hered* **64:** 47–49, 1973.

Rose, R., Smith, J. E., and Leipold, H. W. Increased solubility of hair from hypotrichotic Herefords. *Zentralbl Veterinaermed* **30:** 363–368, 1983.

Rose, R., Smith, J. E., and Leipold, H. W. Role of arginine-converting enzyme in hypotrichosis of Hereford cattle. *Zentralbl Veterinaermed* **30:** 369–372, 1983.

Shand, A., and Young, G. B. A note on congenital alopecia in a Friesian herd. *Vet Rec* **76:** 907–909, 1964.

Steffen, D. J. *et al.* Congenital anemia, dyskeratosis, and progressive alopecia in polled Hereford calves. *Vet Pathol* **28:** 234–240, 1991.

Stogdale, L., Botha, W. S., and Saunders, G. N. Congenital hypotrichosis in a dog. *J Am Anim Hosp Assoc* **18:** 184–187, 1982.

Thomsett, L. F. Congenital hypotrichia in the dog. *Vet Rec* **73:** 915–917, 1961.

Wijeratne, W. V. S. *et al.* A genetic, pathological, and virological study of congenital hypotrichosis and incisor anodontia in cattle. *Vet Rec* **122:** 149–152, 1988.

F. Hypotrichosis Associated with Pigmentary Alterations

Canine color-dilution alopecia occurs in blue and red Doberman pinschers, dachshunds, whippets, standard poodles, and fawn Irish setters. Puppies are born with normal hair, but over the first 6 months, a slowly progressive alopecia develops, usually with secondary seborrhea and folliculitis. Examination of plucked hairs reveals irregular hair shafts with distortions at sites of abnormal pigment clumping. Histologically, the hair follicles are mostly in telogen phase. The melanin pigment within the hair shaft is coarsely clumped and irregularly dispersed (Fig. 5.16). Remnants of dystrophic, poorly keratinized hair shafts may be seen in follicular infundibula, which are cystically distended with keratin. Melanin-containing macrophages are frequently present around the base of hair follicles. Melanin-bearing cells are also prominent in the surface epithelium. Ultrastructurally, while melanocytes contain the full range of developing melanosomes (stages I–IV), there are excessive numbers of stage IV melanosomes and many macromelanosomes, suggesting fusion. In contrast, the surrounding keratinocytes contain few melanosomes, suggesting that the genetic defect may also encompass a failure of melanin transfer within the epidermal-melanin units.

A similar disorder has been described in **cattle** as **cross-**

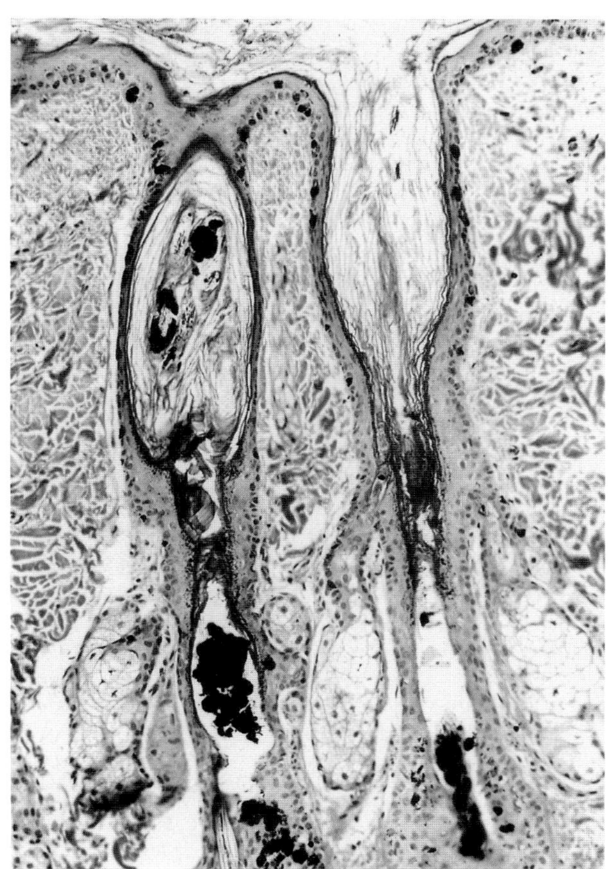

Fig. 5.16 Color-dilution alopecia. Dog. Note the clumped melanin in dystrophic hair shafts, follicular keratosis, and prominent melanocytes in surface epidermis.

related congenital hypotrichosis. Occurring in crosses involving several European breeds such as the Simmental, Gelbvieh, and Charolais, most affected calves are, however, of Simmental–Angus and Simmental–Holstein lineage. The resulting color-dilute animals have short, sparse, and curly hair coats, with a diminished or absent tail switch. Ear, muzzle, and ventrum are the most severely affected areas. In a second form of the disorder, the color-dilute hair is short, curly, but not sparse, and the white areas are unaffected. The histologic lesions are virtually identical to those of canine color-mutant alopecia. Scanning electron microscopy confirms the marked abnormalities in profile and surface morphology of affected hairs. Presumably, it is these defects which lead to breakage and contribute to the alopecia.

A condition known as **congenital black hair follicular dysplasia** has been described in the **dog,** including mixed breeds, **cow, and horse.** The puppies are born with normal coats, and the hair over the nonpigmented skin remains long and lustrous. By a month of age, however, the black skin is dry, scaling, and covered with sparse, short, bristly dull hairs. Histologically the hair follicles are in telogen phase, and there is marked follicular keratosis. The hair shafts and keratin plugs contain aggregates of melanin, and there are numerous melanophages around the inactive hair bulbs. The lesions resemble those of color-mutant alopecia. The mode of inheritance is not known, but an autosomal recessive trait is suggested. The gross and histologic lesions in the cow are very similar to those described in dogs. White areas are normal.

Bibliography

Austin, V. Blue dog syndrome. *Mod Vet Pract* **56:** 31–34, 1975.

Ayers, J. R. *et al*. Pathological studies of cross-related congenital hypotrichosis in cattle. *J Vet Med* **36:** 447–456, 1989.

Carlotti, D. N. Canine hereditary black hair follicular dysplasia and colour-mutant alopecia: Clinical and histopathological aspects. *In* "Advances in Veterinary Dermatology," Vol. 1, C. von Tscharner and R.E.W. Halliwell (eds.), pp. 43–46. London, Baillière Tindall, 1990.

Conroy, J. D., Rasmusen, B. A., and Small, E. Hypotrichosis in miniature poodle siblings. *J Am Vet Med Assoc* **166:** 697–699, 1975.

Gosselin, Y., Papageorges, M., and Teuscher, E. Black hair follicular dysplasia in a dog. *Canine Pract* **9:** 8–15, 1982.

Guaguere, E., and Janin, A. Histopathological and ultrastructural study on the blue Doberman syndrome. *In* "Advances in Veterinary Dermatology," Vol. 1, C. von Tscharner and R.E.W. Halliwell (eds.), pp. 395–396. London, Baillière Tindall, 1990.

Miller, W. H. Alopecia associated with coat-color dilution in two Yorkshire terriers, one saluki, and one mix-breed dog. *J Am Anim Hosp Assoc* **27:** 39–43, 1991.

Miller, W. H., and Scott, D. W. Black-hair follicular dysplasia in a Holstein cow. *Cornell Vet* **80:** 273–277, 1990.

Ostrowski, S., and Evans, A. Coat-color-linked hair follicle dysplasia in "buckskin" Holstein cows in central California. *Agri Pract* **10:** 12–13, 1989.

Selmanowitz, V. J., Markofsky, J., and Orentreich, N. Black-hair follicular dysplasia in dogs. *J Am Vet Med Assoc* **171:** 1079–1081, 1977.

G. Hypertrichosis

Hypertrichosis occurs in neonatal calves and lambs following episodes of maternal hyperthermia. Idiopathic congenital hypertrichosis is reported in the pig and horse. Hereditary hypertrichosis is an autosomal dominant trait in European Friesian cattle. The most common veterinary example of hypertrichosis is that associated with pituitary tumors in old horses; it has also been reported with chronic foot-and-mouth disease virus infection in cattle.

Border disease is caused by a teratogenic pestivirus (Togaviridae), which is closely related antigenically to the agents of hog cholera and bovine virus diarrhea. Border disease virus primarily affects **sheep.** Naturally occurring infections are reported in goats, but seldom cause skin lesions in that species. Details are given in Chapter 3, The Nervous System.

H. Epidermolysis Bullosa

Epidermolysis bullosa refers to a **group of inherited mechanobullous diseases** whose common feature is the formation of cutaneous blisters following minor trauma. Epidermolysis bullosa in humans embraces a complex set of diseases which are heterogeneous, both clinically and genetically. Seventeen subtypes are described in humans, classification being based on the anatomical location of the defect at the dermoepidermal junction, clinical features, and mode of inheritance. In domestic animals, various forms of epidermolysis bullosa have been identified in certain breeds of sheep, dogs, horses, and cattle. Attempts have been made to compare the animal conditions with those of humans, but it is likely that the underlying genetic defects will be heterogeneous in animals.

Sheep. A disorder resembling human **epidermolysis bullosa dystrophica** has been reported in Suffolk and South Dorset Down breeds of sheep in New Zealand, in Swiss Weisses Alpenschaf lambs, and Scottish blackface and their crosses in Great Britain. The ovine conditions much resemble the recessive form of dystrophic epidermolysis bullosa of humans, in that the subepidermal cleft forms beneath the lamina densa of the basement membrane zone. Recently, immunofluorescence and biochemical studies of the Swiss lambs have confirmed an absence of collagen VII, the main structural component of the anchoring fibrils of the basement membrane zone.

The lesions in lambs develop as early as 6 hr postparturition. There is a predilection for areas exposed to mechanical damage, such as skin over bony prominences, lips, and nasolabial plane. Unlike dogs, sheep develop lesions in the oral cavity with erosions occurring on the mucosa of tongue, gingiva, and hard palate. Lesions also develop at the coronary band, causing loosening and shedding of the hooves and leading to the colloquial name of red foot. The disease is fatal. Histologically, the subepidermal bullae are filled with proteinaceous fluid and erythrocytes and are roofed by the full thickness of the epidermis, including the periodic-acid Schiff positive basement membrane.

This disease awaits distinction from epitheliogenesis imperfecta of lambs, as described earlier.

Cattle. An autosomal dominant trait, termed **congenital bovine epidermolysis,** and resembling human epidermolytic epidermolysis bullosa, also known as epidermolysis bullosa simplex, has been described in calves sired by one Simmental bull in Ireland. Ulcers were present on tongue, lips, gingiva, muzzle, distal extremities, and at sites of minor abrasion. Progressive alopecia developed in similar locations, but bullae were not seen, probably because of the thinner nature of bovine haired skin. Histologically, the subepidermal clefts occurred in both surface and follicular epidermis and were associated with vacuolar change in basal keratinocytes. The periodic-acid Schiff positive basement membrane zone was attached to the floor of the clefts.

A **mechanobullous disease with sub-basilar separation**, reported in Texan Brangus cattle, is similar to the recessive form of dystrophic epidermolysis in people. In contrast to the Simmental disease, the inheritance was autosomal recessive, the clinical lesions were more severe, and the disease was rapidly fatal. Lesions were most severe over the distal limbs and pressure points, and nasolabial plane. The hooves separated at the coronary band and exposed the corium. Ulcers were present on the tongue and hard palate. Histologically, the cleft was subepidermal, and the resulting bullae contained proteinaceous fluid and a few inflammatory cells. The basement membrane remained attached to the keratinocytes and thus formed the roof of the bullae. Ultrastructural examination did not detect any abnormalities in tonofilaments or desmosomes.

Familial acantholysis is a disease of Aberdeen Angus calves in New Zealand, and is probably an autosomal recessive trait. It is clinically similar to that described in the Brangus but histologically and ultrastructurally different. The oral mucosa and the skin over the carpus, metacarpophalangeal joints, phalanges, and coronary band are ulcerated and inflamed. Involvement of the coronary band leads to partial separation of the hoof. Large sheets of oral mucosa can be peeled away from the underlying tissue. Ulceration of the proximal esophagus may occur. The hair of the entire body is readily epilated, and epidermal tissue adheres to the base of the tufts. Histologically, the lesions include suprabasilar clefts with loss of adhesion between cells of the basal layer and stratum spinosum. Electron-microscopic examination reveals an absence of desmosomal junctions and an anomalous development and clumping of their associated tonofilaments. These may be seen as inclusion bodies on light microscopy.

Horses. A dermatosis resembling **junctional epidermolysis bullosa** is recognized in Belgian foals. Lesions are present at birth or develop soon after. Multiple asymmetrical cutaneous erosions may be present all over the body, but are most severe on pressure points and at mucocutaneous junctions (Fig. 5.17). Flaccid bullae may be present in the oral mucosa. Coronary band lesions often lead to separation of the hoof. Corneal erosions and dystrophic teeth are further indications of the general ectodermal defect. The disease is incompatible with life. Histologically, multiple clefts separate the dermis and epidermis with minimal lytic change of the basal keratinocytes and without inflammation. Once ulceration has occurred, secondary bacterial infection stimulates marked inflammatory change. Ultrastructural studies have shown that the cleft occurs along the lamina lucida with the disruption of basilar hemidesmosomes. The mode of inheritance is not established, but is probably autosomal recessive.

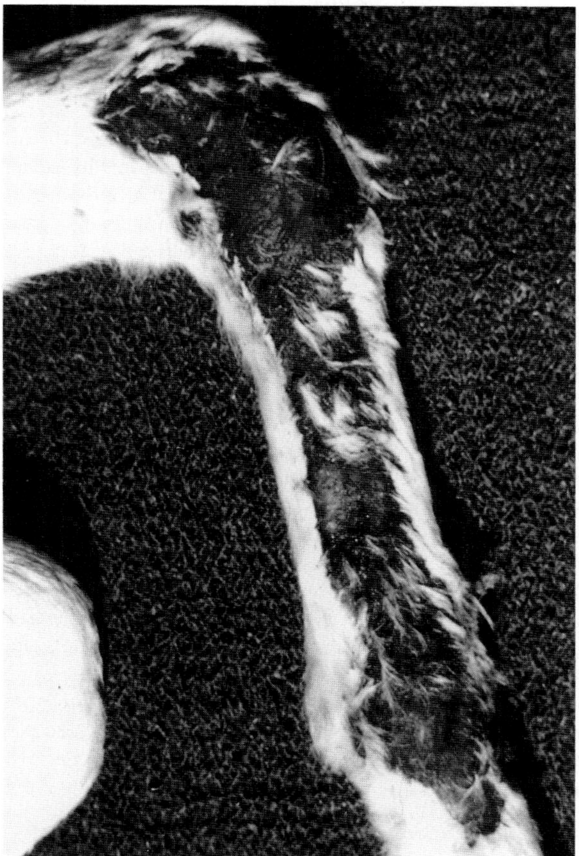

Fig. 5.17 Lacerations on the leg. Belgian foal. Junctional epidermolysis bullosa.

Dogs. A disease of collie dogs, originally named **epidermolysis bullosa simplex,** has more recently been reclassified as a mild form of dermatomyositis (see the following). In the original report of epidermolysis bullosa simplex, bullae were created by application of mild friction to the skin. This is not readily accomplished in dogs with dermatomyositis. The possibility remains that the two diseases are different, although expressing several common features.

A disease resembling **junctional epidermolysis bullosa** has been described in the toy poodle. The lesions chiefly affect footpads and oral mucosa. The histologic lesions are subepidermal, noninflammatory clefts and bullae; the ultrastructural findings indicate cleavage through the lamina lucida, leaving the lamina densa attached to the dermis.

Bibliography

Alley, M. R., O'Hara, P. J., and Middleberg, A. An epidermolysis bullosa of sheep. *N Z Vet J* **22:** 55–59, 1974.

Bassett, H. A congenital bovine epidermolysis resembling epidermolysis bullosa simplex of man. *Vet Rec* **121:** 8–11, 1987.

Bruckner-Tuderman, L., Guscetti, F., and Ehrensperger, F. Animal model for dermolytic mechanobullous disease: Sheep with recessive dystrophic epidermolysis bullosa lack collagen VII. *J Invest Dermatol* **96:** 452–458, 1991.

Cooper, T. W., Bauer, E. A., and Briggaman, R. A. The mechanobullous diseases (epidermolysis bullosa). *In* "Dermatology in General Medicine. Textbook and Atlas," 3rd Ed. T.B. Fitzpatrick *et al.*(eds.), pp. 610–626. New York, McGraw-Hill, 1987.

Dunstan, R. W. *et al.* A disease resembling junctional epidermolysis bullosa in a toy poodle. *Am J Dermatopathol* **10:** 442–447, 1988.

Johnson, G. C. *et al.* Ultrastructure of junctional epidermolysis bullosa in Belgian foals. *J Comp Pathol* **98:** 329–336, 1988.

Jolly, R. D., Alley, M. R., and O'Hara, P. J. Familial acantholysis of Angus calves. *Vet Pathol* **10:** 473–483, 1973.

Kohn, C. W. *et al.* Mechanobullous disease in two Belgian foals. *Eq Vet J* **21:** 297–301, 1989.

McTaggart, H. S. Red-foot disease of lambs. *Vet Rec* **94:** 153–159, 1974.

Scott, D. W., and Schultz, R. D. Epidermolysis bullosa simplex in the collie dog. *J Am Vet Med Assoc* **171:** 721–727, 1977.

Thompson, K. G. *et al.* A mechanobullous disease with subbasilar separation in Brangus calves. *Vet Pathol* **22:** 283–285, 1985.

Fig. 5.18A Familial canine dermatomyositis. Alopecia and crusting of face. Puppy.

I. Familial Canine Dermatomyositis

This is a common dermatosis of collie and Shetland sheep dogs in the United States and Canada and probably occurs wherever the breeds exist. It has been recognized in cross-bred dogs with collie or sheltie progenitors. In the collie, and probably in the sheltie, the disease is an autosomal dominant with variable expression. The discovery of muscle lesions in dogs with skin lesions similar to but reportedly more inflammatory than those of epidermolysis bullosa simplex led to the renaming of this disorder as dermatomyositis. Analogies between human juvenile dermatomyositis and the canine disease have been drawn, but there are significant differences between the human and canine diseases, in the distribution of the muscle groups affected, in the presence or absence of elevated muscle enzymes, and in the nature of skin and muscle lesions.

Skin lesions have a delayed onset, usually at a few months of age. Lesions develop on the dorsum of the nose, inner aspects of the pinna, perioral and periocular regions, tail tip, and over bony prominences of the limbs (Fig. 5.18A). Initially, lesions are characterized by erythema, crusting, vesiculation, and erosions. Chronic residual lesions are characterized by alopecia, hypo- or hyperpigmentation, and scaling. Mucosal lesions have been noted in one incross experimental breeding.

Many dogs outgrow the disease. Those that do not show reduced growth rate and may develop diseases suggesting diminished immunocompetence, such as septicemia, generalized demodecosis, and pyoderma. Secondary amyloidosis has been reported.

Muscle lesions are detected clinically several weeks after the onset of the skin lesions. Necropsy studies indicated that the gross muscle lesions of pallor and softening are most severe at 5 months of age. Muscles of mastication and the superficial portions of muscles below the elbow and stifle are most affected, reflecting a similar regional distribution to the skin lesions. In severely affected animals, involvement of the esophageal muscle may lead to megaesophagus and death by aspiration pneumonia. In less severely affected animals, there may be complete recovery, or muscle atrophy may be residual. Most dogs with both skin and muscle manifestations of the disease have hyperplastic peripheral lymph nodes. Muscle lesions are not apparent in many dogs; either they do not occur or are too subtle for clinical or electromyographic detection.

Histologically, the cutaneous lesions vary according to the severity and stage of the disease. Unless secondary pyoderma or demodecosis intervenes, the skin lesions of dermatomyositis are more degenerative than inflammatory. Early lesions, in mildly affected dogs, are often subtle. They may be restricted to single-cell necrosis or vacuolar change of scattered basal keratinocytes, with occasional noninflammatory clefts forming at the dermoepidermal junction (Fig. 5.18B). Clefts may develop into bullae, filled with serum and containing moderate numbers

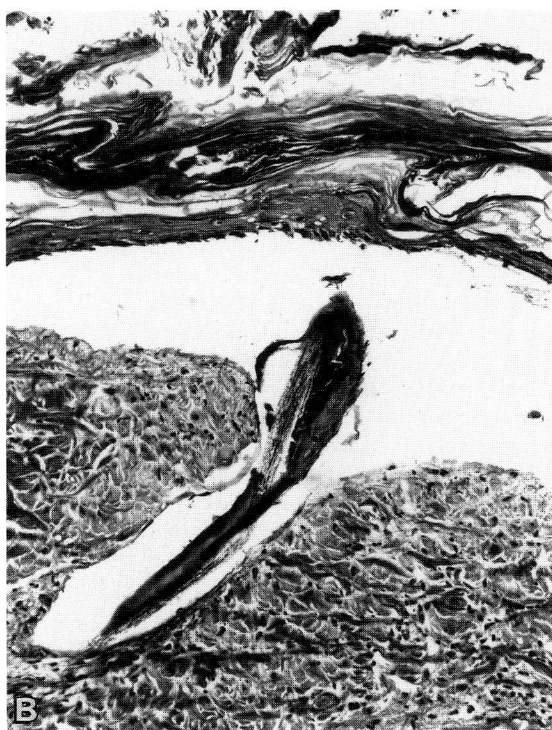

Fig. 5.18B Familial canine dermatomyositis. Lesion induced by applying friction to the skin.

of inflammatory cells, and roofed by an epithelium in which individual basal cell keratinocytes are necrotic. Clefts and bullae may be found in surface and follicular epithelia, and are commonly induced at the edges of the section by the shearing action of the biopsy punch. Intraepithelial pustules are an uncommon finding. Chronic lesions are a fibrosing dermatitis, with only mild pigmentary incontinence to indicate the previous basal keratinocyte damage. Hair follicles are often atrophic, appearing to fade from the section (Fig. 5.19A, B).

Perifolliculitis and perifollicular fibrosis, comprising a condensation of perifollicular fibroblasts and indeterminate mononuclear cells, have been described. Mild chronic vasculitis, characterized by the presence of scattered intramural lymphocytes, pyknosis of nuclei, and eosinophilic degeneration of the media of small arterioles, may be found in many lesions, although it may require diligent searching of serial sections. The deep dermis and subcutis have a generalized but mild increase in cellularity. Arteritis has been described also in vessels of muscle, bladder and spermatic cord. In deep skin biopsies, lesions may be seen in the underlying cutaneous musculature (see Chapter 2, Muscles and Tendons).

The pathogenesis of canine familial dermatomyositis is not understood. The human disease is considered one of the idiopathic inflammatory myopathies and an autoimmune connective tissue disorder. In dogs, the onset and severity of the skin lesions have been correlated with

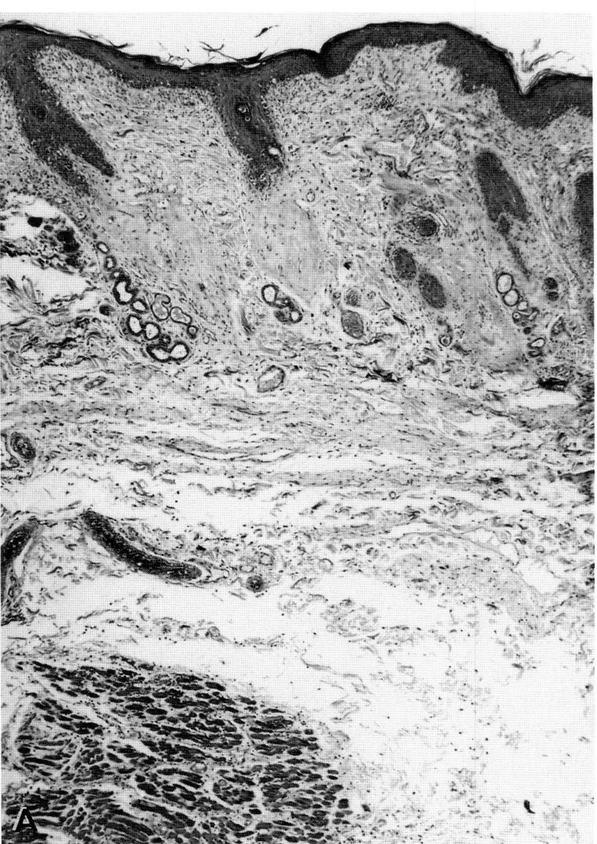

Fig. 5.19A Dermatomyositis. Dog. Low power showing follicular atrophy, hydropic interface dermatitis, and myositis.

elevations in serum immunoglobulin and antigen–antibody complex levels, suggesting an immune pathogenesis, but the possibility of epiphenomena cannot be ruled out. Genetic predisposition and environmental triggers, including infectious agents, are both important in the human disease. Viral infections, particularly picornaviruses, have been implicated in some human cases, either as a direct cause of myositis or by triggering autoimmunity. Crystalline viruslike arrays may be found in the endoplasmic reticulum of endothelial cells in affected muscle of collie dogs, similar to those seen in myocytes in some human cases. Finally, arteritis is seen in human juvenile dermatomyositis. The vasculitis seen in the canine lesions, albeit mild, could lead to low-grade anoxia, which might account for the atrophic lesions in epidermis, hair follicles and, possibly, muscle. The reason for the vascular lesions remains unknown.

Bibliography

Gross, T. L., and Kunkle, G. A. The cutaneous histology of dermatomyositis in collie dogs. *Vet Pathol* **24:** 11–15, 1987.

Hargis, A. M., and Haupt, K. H. Review of familial canine dermatomyositis. *Vet Ann* **30:** 277–282, 1990.

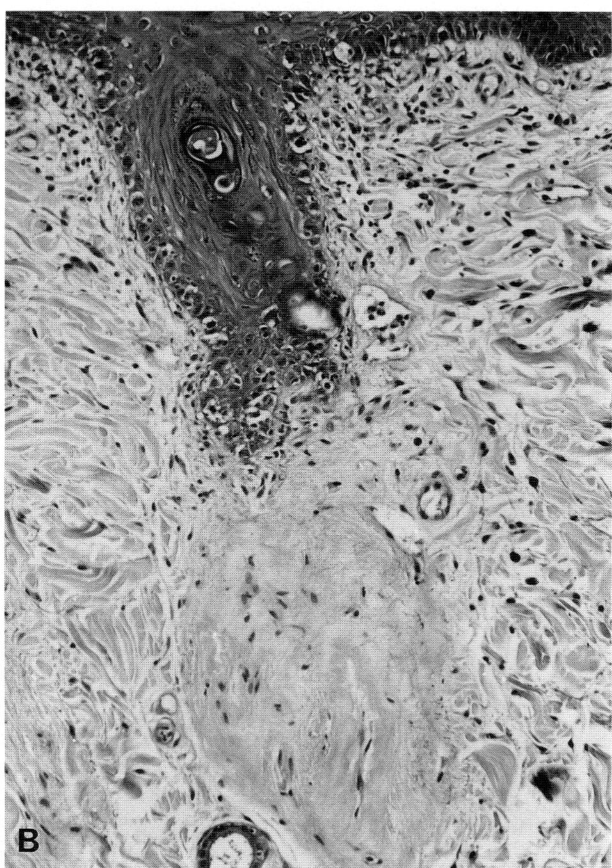

Fig. 5.19B Dermatomyositis. Dog. Higher magnification showing follicular atrophy, fibrosis, and hydropic interface dermatitis.

Hargis, A. M. *et al.* Severe secondary amyloidosis in a dog with dermatomyositis. *J Comp Pathol* **100:** 427–433, 1989.

Haupt, K. H. *et al.* Familial canine dermatomyositis: Clinical, electrodiagnostic, and genetic studies. *Am J Vet Res* **46:** 1861–1875, 1985.

Plotz, P. H. *et al.* Current concepts in the idiopathic inflammatory myopathies: Polymyositis, dermatomyositis, and related disorders. *Ann Intern Med* **111:** 143–157, 1989.

J. Hereditary Collagen Dysplasia

Hereditary collagen dysplasia (dermatosparaxis, cutis asthenia, cutis hyperelastica, Ehlers–Danlos syndromes) in humans and animals comprises a complex group of hereditary disorders of connective tissue, which result in a decrease in the tensile strength of affected tissues. The predominant clinical signs in animals are cutaneous fragility and hyperextensibility, but subclinical lesions have been demonstrated in bone and deeper connective tissues. A variety of defects in collagen biosynthesis, fibrillogenesis, and in fiber maintenance are responsible for the different forms of the disease in animals and humans. A defect in one connective tissue molecule also has consequences for the structural properties of other connective tissue components. Some human and animal forms of the disease appear analogous, but it is becoming apparent that even some of the defined types of human Ehlers–Danlos are not homogeneous.

The first and best-understood form of collagen dysplasia in animals, dermatosparaxis, has been likened to human Ehlers–Danlos type VII, but it is likely that there is more than one type of defect in the human disease. One is in the enzyme procollagen *N*-peptidase; the other is a structural defect in the alpha chain of type 1 procollagen, which may interfere with the action of the enzyme. It has been emphasized that there is only a limited repertoire of morphologic changes possible in collagen fibers and that similar structural abnormalities may result from different molecular defects. Thus, clinically similar syndromes with similar light-microscopic, and even ultrastructural, lesions may result from different mutations. Conversely, similar genetic defects may be expressed differently phenotypically. For example, bovine dermatosparaxis and one form of Ehlers–Danlos VII in humans share the same deficiency in procollagen *N*-peptidase, but the human disease is characterized by joint laxity rather than skin fragility. Because these diseases are heterogeneous in their clinical, genetic, and biochemical features, and until these are better defined, it is proposed to use the collective term collagen dysplasia.

Collagen dysplasias occur in cattle, sheep, dogs, cats, horses, rabbits, mink, and possibly, pigs. The clinical signs common to all are cutaneous fragility, hyperextensibility, and laxity. These signs vary in severity between the different animal syndromes, no doubt reflecting the type and degree of defective collagen metabolism in each. Sheep tend to be the most severely affected, followed by cattle, dogs, cats, and horses. However, within species there is a variation in clinical severity, depending on the particular defect. The fragility of affected skin renders it extremely susceptible to laceration. Dermatosparactic lambs develop oral lesions simply from the trauma associated with nursing, and affected cats induce skin lacerations in the process of normal grooming. The tensile strength of dysplastic skin in affected dogs is 5–10% normal. Poor wound healing is not a feature of the animal diseases, but scars in affected animals are fragile ("papyraceous" or "cigarette-paperlike"). Hyperextensibility is readily demonstrated by pulling up a skin fold; a skin extensibility index has been developed to evaluate clinically affected dogs. Several types of Ehlers–Danlos syndrome of humans exhibit hyperextensible and hypermobile joints. In general, joint involvement is not a feature of the animal diseases, although hypermobile joints occur in one collagen dysplasia of lambs.

Diagnosis of collagen dysplasia syndromes is made on the basis of appropriate clinical signs and the demonstration of a morphologic or biochemical abnormality in collagen formation. On light microscopy, some syndromes will show an obvious depletion of dermal collagen and a morphologic abnormality in the collagen fibers present, but many will not. In some, the dermis will be obviously thinner; in some, it may be thicker because of accumulated proteoglycan matrix. In some forms of collagen dysplasia, the collagen bundles stain red rather than blue with Masson's trichrome, but this is not pathognomonic for collagen dystrophies. The light-microscopic lesions may be subtle, requiring a section of skin from an age- and breed-matched control animal for their appreciation.

Ultrastructural evaluation is usually essential to diagnosis. Fibrils may be uniformly reduced in diameter or may show a variation in diameter; the cross-sectional profiles may be abnormal (hieroglyphic fibrils in dermatosparaxis in sheep, calves, and dogs); fibers may be fused or branched, resulting in abnormally large cross-sectional diameters; fibers may be unraveled or loosely packed, fragmented, or twisted (as in the dominant forms of collagen dysplasia in dogs).

Acquired collagen disorders should not be confused with the heritable collagen dysplasias. Most notably, cats with hyperadrenocorticism may develop fragile skin. Histologic examination reveals a marked thinning of the dermis, and a reduction in the number of collagen bundles and in their diameter. An unusual localized collagen disorder in an old English setter presented as marked laxity of the skin of the head, neck, and shoulders. The collagen bundles were fragmented, thin, and separated by increased ground substance. Ultrastructurally, the collagen bundles were unraveled, and the endoplasmic reticulum of the fibroblasts had markedly dilated cisternae. An ischemic pathogenesis was suggested.

Cattle. The condition in cattle is usually termed **dermatosparaxis,** which literally means torn skin. Dermatosparaxis is a recessive trait. It has been described in the White-Blue cattle in Belgium, and in the Hereford, Charolais, and Simmental breeds. Affected cattle have abnormally high concentrations of procollagen in the dermis because of a deficiency in an enzyme, amino-terminal procollagen peptidase, which excises the amino-terminal propeptide from the procollagen molecule. The amino-terminal propeptide is an extension peptide which inhibits the formation of collagen fibrils until the soluble procollagen molecule has been secreted into the extracellular space. Deficiency of this enzyme and the resultant accumulation of procollagen interferes with the packing of collagen into the fibrils and ultimately the fibers which confer tensile strength on the dermis.

The light microscopic lesions include a reduction in the number and diameter of collagen fibers, an unraveling of collagen fibers, and a disorganization of fiber arrangement in the dermis. The normal regular bundles of parallel, cylindrical fibers, arranged in register, are replaced by loose, twisted ribbons of irregular fibrils. Cross-sectional profiles of these abnormal fibers have been likened to hieroglyphics.

Sheep. A recessive form of collagen dysplasia resembling bovine **dermatosparaxis** is seen in Dala sheep in Norway and in Border Leicester–Southdown crossbred sheep in Australia. The underlying defect in the Norwegian sheep is a deficiency of amino-terminal procollagen peptidase. The pathogenesis may be more than a simple deficiency in the enzyme; fibroblasts from dermatosparactic sheep fail to contract collagen gels and show a reduced attachment to collagenous substrates. Dermatosparactic ovine fibroblasts lack a 34-kDa protein which is important in binding the cells to the collagen. The accumulation of procollagen in dermatosparactic skin could result both from a deficiency of the procollagen *N*-peptidase and from a deficiency of the collagen-binding proteins which act as a presenter of the procollagen to the enzyme. The clinical disease is more severe in sheep than cattle. The predominant lesions are skin fragility and resultant lacerations. Affected lambs tear their skin while suckling and develop fatal bacterial septicemias in the first few weeks of life. The histologic and ultrastructural lesions are very similar to those in cattle.

The syndrome in the Australian sheep is probably of different pathogenesis. There is a one-third reduction in collagen content in comparison to controls, but there is no excess of dermal procollagen, indicating that a deficiency of procollagen peptidase is unlikely. The Australian sheep also have hypermobile joints. Histologically, there is a depletion of collagen bundles in the dermis, which is particularly marked in the Australian lambs. The collagen bundles are short and loosely fibrillar. The spaces left by the collagen depletion are filled with glycosaminoglycan ground substance. Ultrastructural lesions are similar to those in the cattle.

Another fragile-skin disease has been reported in New Zealand Romney lambs. The disease is clinically less severe than dermatosparaxis and does not show the typical ultrastructural changes.

Dogs. A recessive form of collagen dysplasia, resembling bovine dermatosparaxis, has been recognized in a dog. The more common form of collagen dysplasia in dogs is inherited as an autosomal dominant trait. This condition, often colloquially termed "rubber puppy" syndrome, occurs in several breeds, including springer spaniels and in mongrels (Fig. 5.20). The underlying defect is one of fiber packing. The biochemical basis for the collagen packing defect has not been elucidated. It is probable that there

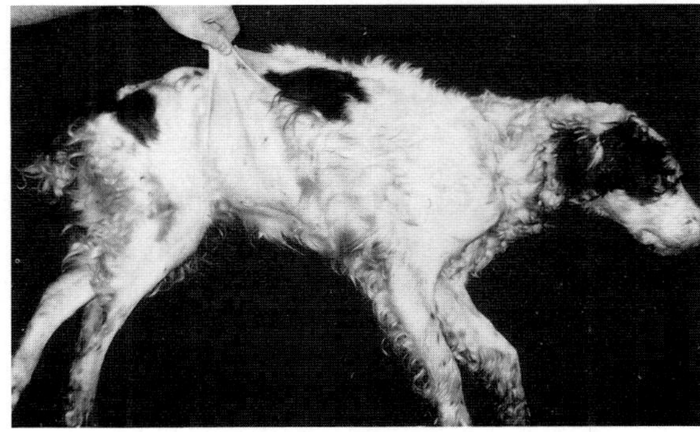

Fig. 5.20 Cutaneous asthenia. Hyperextensibility of skin. Dog.

are minor variations in the metabolic defects between each family of affected dogs, but all result in altered fibrillogenesis. No abnormalities in collagen biosynthesis are recognized, making it unlikely that mutations in the collagen genes are responsible. Some significant changes in the glycosaminoglycan, hyaluronic acid, and in the proteoglycan, proteodermatan sulfate, are detectable in these dogs. The latter is normally closely associated with the surface of collagen fibers. It is possible that these matrix molecules are involved in regulation of collagen fibrillogenesis, and therefore an abnormality in their production could play a role in the collagen-packing defect.

The predominant clinical characteristic is cutaneous fragility, although laxity and hyperextensibility of the skin are also present. The tensile strength of affected skin may be only 5–10% of that of normal littermates. Radiographic and microradiographic studies reveal subclinical involvement of bone.

The microscopic lesions in the canine syndrome may be difficult to appreciate without a site-matched specimen from a littermate. The dermis is usually thinner in affected dogs, collagen bundles are variably depleted, and there may be fragmentation, irregularity, and a loose interweaving of the bundles. These changes are best seen in skin which has been previously traumatized. Abnormal and normal fibers stain similarly with hematoxylin-eosin and cannot be distinguished with polarized light microscopy. However, with Masson's and Gomori's trichrome stains, abnormal bundles stain red, instead of the normal blue or green. This change is not specific to collagen dysplasia. Ultrastructural examination is required to identify the collagen defects. Even then, 70% of the collagen fibers may be morphologically normal. The abnormal fibers are disorganized and composed of loosely packed fibrils. Branching and fusion results in some abnormally large diameter fibers. Segmental change within normal fibers probably acts as a weak link. This would account for the very marked decrease in tensile strength despite there being a majority of normal collagen fibers.

Cats. Collagen dysplasia, similar to the canine disease in that it is transmitted as an autosomal dominant trait and results from a collagen-packing defect, has been recognized in cats (Fig. 5.21). There are no histologic lesions, but ultrastructurally there is a bimodal distribution of normal and abnormal collagen fibers in the deep dermis with extreme disorganization in the packing of fibrils within the abnormal fibers.

A recessive syndrome in the Himalayan cat appears analogous to dermatosparaxis in calves and lambs in that the biochemical defect is a procollagen peptidase deficiency. Only a small proportion of collagen molecules are affected, and the clinical expression is mild. Histologically, the dermis is thinner than normal and contains small, irregularly oriented collagen bundles and tight knots of randomly tangled fibers. Scanning and transmission electron-microscopic studies reveal tangled, nonparallel arrays of fibrils in the abnormal fibers. Hieroglyphic fibers, seen in the ruminant diseases, are not a feature in this cat.

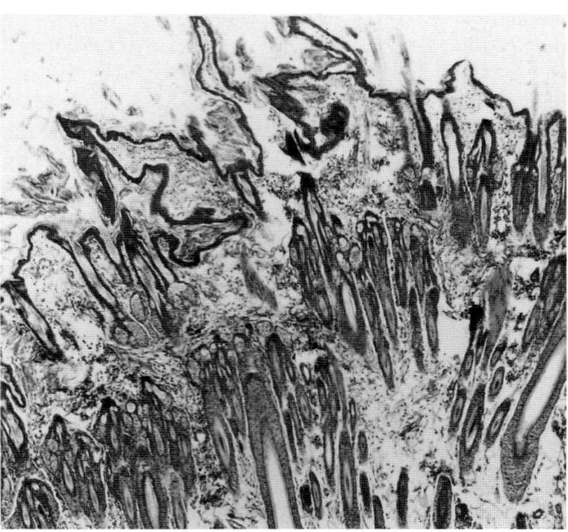

Fig. 5.21 Collagen dysplasia. Cat. Deficit of dermal collagen indicated by approximation of hair follicles.

Lesions, characterized clinically by fragile skin and histologically by a marked depletion of collagen fibers, have been seen in cats with hyperadrenocorticism.

Horses. Collagen dysplasia occurs in Quarter horses. An autosomal recessive inheritance has been suggested. The clinical signs are cutaneous fragility (Fig. 5.22A) and hyperextensibility. Unlike the collagen dysplasias in other species, the equine disorder is characterized by localized areas of abnormal skin. There are no significant microscopic lesions, although the dermis may be thinner than normal. The ultrastructural lesions show an increase in

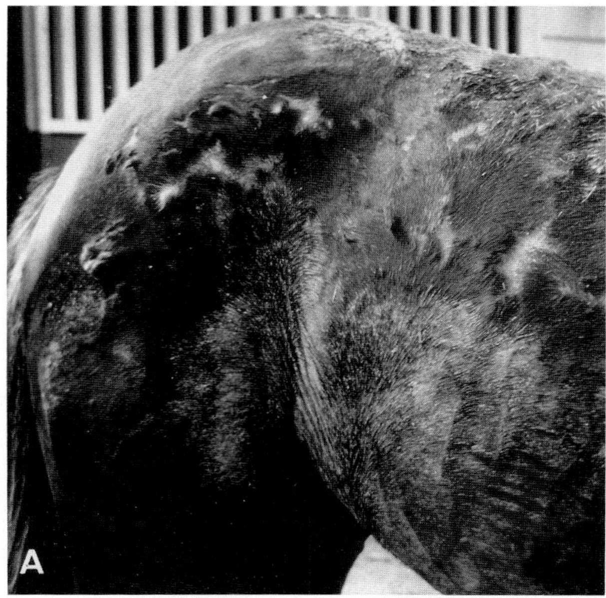

Fig. 5.22A Cutaneous fragility in a horse with collagen dysplasia. (Courtesy of K. Thompson).

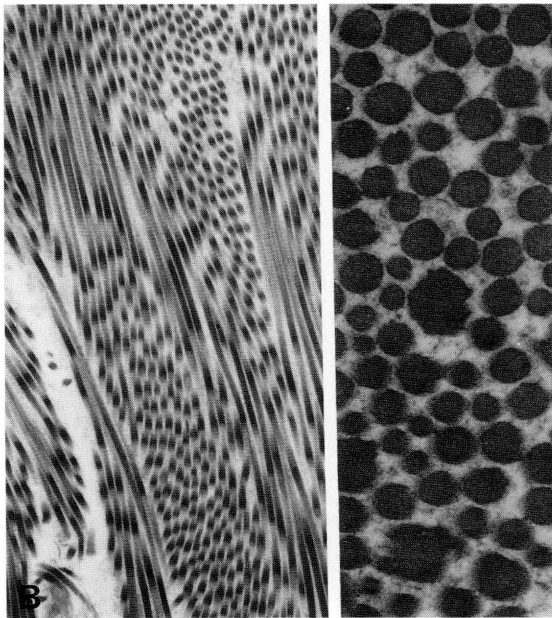

Fig. 5.22B Collagen dysplasia. Horse. Electron micrographs showing (left) twisted collagen fibrils within fiber bundles and (right) variation in fibril diameter and irregular shape of some large fibrils. (Courtesy of M. H. Hardy and O. Vrablic.)

the range of fibril diameters and a collagen-packing defect (Fig. 5.22B). Increased acid solubility of collagen suggests that the defect arises at the level of assembly or stabilization of fibrils and/or fibers.

The lesions in an Arabian-crossbred filly with cutaneous fragility and hyperextensibility differ from those in Quarter horses. The dermal collagen fibers are thinner than normal and have abnormal red cores when stained with Masson's trichrome. Ultrastructurally, collagen fibers are fragmented, and there is a marked increase in granular matrix. The cisternae of the fibroblast endoplasmic reticulum are distended, and the lysosomes contain collagen fragments. The relationship of this condition to heritable collagen dysplasia remains unknown.

Pigs. A condition termed cutis hyperelastica is described in a litter of Large White-Essex pigs. Affected piglets showed hyperextensible skin along the back, thorax, and flanks, and multiple shallow depressions and areas of alopecia and scarring 3–7 cm wide. The significant microscopic lesion, reported without illustration, is a marked increase in dermal elastic fibers.

Bibliography

Bavinton, J. H., Peters, D. E., and Ramshaw, J. A. M. A morphologic study of a mild form of ovine dermatosparaxis. *J Invest Dermatol* **84:** 391–395, 1985.

Byers, P. H. Inherited disorders of collagen gene structure and expression. *Am J Med Genet* **34:** 72–80, 1989.

Cahill, K. I. *et al.* A collagen dysplasia in a greyhound bitch. *N Z Vet J* **28:** 203–204, 213, 1980.

Clark, R. G., Bissett, A. B., and Rao, S. N. A fragile skin condition in Romney lambs. *N Z Vet J* **25:** 213, 1977.

Collier, L. L., Leathers, C. W., and Counts, D. F. A clinical description of dermatosparaxis in a Himalayan cat. *Fel Pract* **10:** 25–36, 1980.

Counts, D. F., Knighten, P., and Hegreberg, G. Biochemical changes in the skin of mink with Ehlers–Danlos syndrome: Increased collagen biosynthesis in the dermis of affected mink. *J Invest Dermatol* **69:** 521–526, 1977.

Counts, D. F. *et al.* Dermatosparaxis in a Himalayan cat: I. Biochemical studies of dermal collagen. *J Invest Dermatol* **74:** 96–99, 1980.

Fjolstad, M., and Helle, O. A hereditary dysplasia of collagen tissues in sheep. *J Pathol* **112:** 183–187, 1974.

Freeman, L. J., Hegreberg, G. A., and Robinette, J. D. Ehlers–Danlos syndrome in dogs and cats. *Sem Vet Med Surg* **2:** 221–227, 1987.

Gunson, D. E., Halliwell, E. W., and Minor, R. R. Dermal collagen degradation and phagocytosis. Occurrence in a horse with hyperextensible fragile skin. *Arch Dermatol* **120:** 599–604, 1984.

Hanset, R., and Lapiere, C. M. Inheritance of dermatosparaxis in the calf. *J Hered* **65:** 356–358, 1974.

Hardy, M. H. *et al.* An inherited connective tissue disease in the horse. *Lab Invest* **59:** 253–262, 1988.

Hegreberg, G. A., Padgett, G. A., and Henson, J. B. Connective tissue disease of dogs and mink resembling Ehlers–Danlos syndrome of man. III. Histopathologic changes of the skin. *Arch Pathol* **90:** 159–166, 1970.

Helle, O., and Ness, N. N. A hereditary skin defect in sheep. *Acta Vet Scand* **13:** 443–445, 1972.

Holbrook, K. A., and Byers, P. H. Skin is a window on heritable disorders of connective tissue. *Am J Med Genet* **34:** 105–121, 1989.

Holbrook, K. A. *et al.* Dermatosparaxis in a Himalayan cat: II. Ultrastructural studies of dermal collagen. *J Invest Dermatol* **74:** 100–104, 1980.

Jayasekara, M. U., Leipold, H. W., and Phillips, R. Ehlers–Danlos syndrome in cattle. *Zeitsch Tierzucht Zuchtungsbiol* **96:** 100–107, 1979.

Lapiere, C. M., Lenaers, A., and Kohn, L. D. Procollagen peptidase: An enzyme excising the coordination peptides of procollagen. *Proc Natl Acad Sci USA* **68:** 3054–3058, 1971.

Lenaers, A. *et al.* Collagen made of extended-chains, procollagen, in genetically defective dermatosparaxic calves. *Eur J Biochem* **23:** 533–543, 1971.

Lerner, D. J., and McCracken, M. D. Hyperelastosis cutis in 2 horses. *J Eq Med Surg* **2:** 350–352, 1978.

Mauch, C. *et al.* A defective surface collagen-binding protein in dermatosparactic sheep fibroblasts. *J Cell Biol* **106:** 205–211, 1988.

McOrist, S. *et al.* Ovine skin collagen dysplasia. *Aust Vet J* **59:** 189–190, 1982.

Minor, R. R. Collagen metabolism: A comparison of diseases of collagen and diseases affecting collagen. *Am J Pathol* **98:** 225–280, 1980.

Minor, R. R. *et al.* Genetic diseases of connective tissues in animals. *Curr Probl Dermatol* **17:** 199–215, 1987.

O'Hara, P. J. *et al.* A collagenous tissue dysplasia of calves. *Lab Invest* **23:** 307–314, 1970.

Patterson, D. F., and Minor, R. R. Hereditary fragility and hyperextensibility of the skin of cats. *Lab Invest* **37:** 170–179, 1977.

Scott, D. V. Cutaneous asthenia in a cat, resembling Ehl-

ers–Danlos syndrome in man. *Vet Med Small Anim Clin* **69:** 1256–1258, 1974.
Van Halderen, A., and Green, J. R. Dermatosparaxis in White Dorper sheep. *J S Afr Vet Assoc* **59:** 45, 1988.
Wick, G., Olsen, B. R., and Timpl, R. Immunohistologic analysis of fetal and dermatosparactic calf and sheep skin with antisera to procollagen and collagen type I. *Lab Invest* **39:** 151–156, 1978.

K. Dermatosis Vegetans

This hereditary disease of Landrace pigs originated in Scandinavia in the 1940s and spread to England, Canada, and Australia via importation. Sporadic cases still occur. The trait is inherited as an autosomal recessive, and it is expressed by singular cutaneous lesions, deformities of the feet, and a giant-cell pneumonitis. The anatomical changes vary in their degree of expression; they may be present at birth or develop when the animal is 2–3 months old, but usually they are evident by the third week of life.

The cutaneous changes begin as erythematous papules over which there is slight scaling. The papules expand at the periphery leaving the center depressed, thus resembling pityriasis rosea. The lesions expand and coalesce to form variable-sized plaques, which may involve much of the body (Fig. 5.23A). These have a rough brown-black central area of scale-crust and a raised erythematous border. The lesions become increasingly lichenified and crusted, and in several weeks they are papillomatous. If the animal survives, the cutaneous lesions regress over 4 to 6 months.

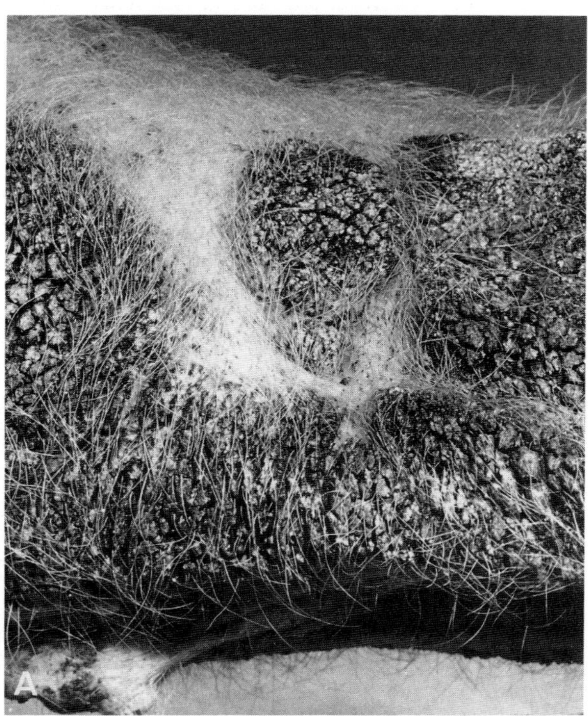

Fig. 5.23A Dermatosis vegetans. Pig. Coalesced, crusted lesions on abdominal wall. (Courtesy of D. H. Percy.)

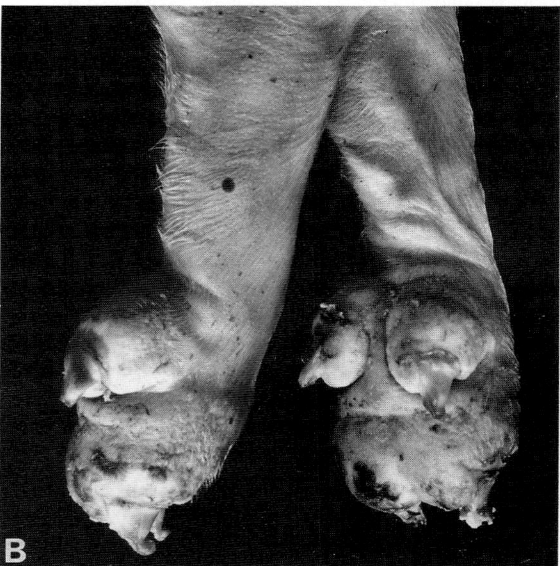

Fig. 5.23B Dermatosis vegetans. Pig. Congenital lesions of feet. (Courtesy of J. A. Flatla.)

The foot lesions are always present at birth (Fig. 5.23B). There is edematous swelling of the coronary region, where the skin is also erythematous and covered with yellow greasy exudate. The horn of the wall is thickened and bears rings parallel to the coronary groove. The edematous swelling eventually disappears, but the changes in the hoof persist.

Pigs with dermatosis vegetans are unthrifty, and most die in 5 to 6 weeks or even later after the cutaneous lesions have resolved. Almost all that survive for a week or more after birth develop progressive dyspnea, which is due to a diffuse interstitial pneumonia of proliferative giant-cell type.

The microscopic lesions in the skin constitute an intraepidermal pustular dermatitis. The pustules occur at various levels and are of the eosinophilic spongiform type. Other epidermal lesions include orthokeratotic and parakeratotic hyperkeratosis, acanthosis, rete-ridge formation, spongiosis, and neutrophilic and eosinophilic exocytosis. Dermal lesions include edema, hyperemia, and perivascular accumulations of both mononuclear and granulocytic inflammatory cells, particularly eosinophils.

Bibliography

Flatla, J. L., Hansen, M. A., and Slagsvold, P. Dermatosis vegetans in pigs. Symptomatology and genetics. *Zentralblat Veterinaermed* **8:** 25–42, 1961.
Percy, D. H., and Hulland, T. J. The histopathological changes in the skin of pigs with dermatosis vegetans. *Can J Comp Med* **33:** 48–54, 1969.

Webb, R. F., and Bourke, C. A. Dermatosis vegetans in pigs. *Aust Vet J* **64:** 287–288, 1987.

L. Dermoid Cyst of Rhodesian Ridgeback Dogs

The dermoid cyst of Rhodesian ridgeback dogs is a congenital defect caused by incomplete separation of the neuroectoderm and epithelial ectoderm. The mode of inheritance has not been established, but it may be a simple recessive characteristic.

Dermoid sinuses occur on the dorsal midline at cervical, anterior thoracic, or sacrococcygeal locations. There are single, or occasionally multiple, sinus tracts, which may extend into the spinal canal and attach to the dura mater. Less complete sinuses extend only to the supraspinous ligament or may be attached to it by a fibrous ligament. Some forms have no connection to underlying tissue. A tuft of hair characteristically emerges from the funnel-like opening, but if bacterial infection occurs, purulent exudate may drain from the sinus. The sinus tract is lined by stratified squamous epithelium with adnexal structures and is filled with keratin, sebum, and hair.

Bibliography

Hathcock, J. T., Clampett, E. G., and Broadstone, K. V. Dermoid sinus in a Rhodesian ridgeback. *Vet Med Small Anim Clin* **74:** 53–56, 1979.

M. Cutaneous Mucinosis of Shar-Pei Dogs

In the Chinese Shar-Pei dog, a severe dermatosis, known as cutaneous mucinosis or congenital myxedema, has been elevated to breed status. The mucinosis is responsible for the turgid, wrinkled skin characteristic of the breed. The trait is variably expressed, but in its most severe form is manifested as cutaneous "blisters," which represent pools of subepidermal mucin. The mucinosis may partially resolve with age and may be reduced with corticosteroid drugs.

Histologically, the primary lesion is dermal thickening due to increased accumulation of glycosaminoglycan, causing separation of dermal collagen bundles in superficial and deep dermis. Because of the hygroscopic nature of the glycosaminoglycan, much of the space represents water removed during processing, and the mucin appears only as a lacy network of fine basophilic strands. Selective use of special stains, such as alcian blue at pH 2.5 (positive) and periodic-acid Schiff (negative), reveals the acid nature of the increased dermal mucin. Ultrastructural studies reveal amorphous granular material dilating the cisternae of dermal fibroblasts and aggregating in the interstitium. The abnormal microenvironment of the wrinkled skin predisposes these dogs to severe pyoderma and chronic demodecosis.

Focal cutaneous mucinosis occurs rarely in breeds other than the Shar-pei. More generalized acquired cutaneous mucinosis is usually associated with hypothyroidism in the dog.

Bibliography

Dillberger, J. E., and Altman, N. H. Focal mucinosis in dogs: Seven cases and a review of cutaneous mucinoses of man and animals. *Vet Pathol* **26:** 132–139, 1986.
Truhan, A. P., and Roenigk, H. H. The cutaneous mucinoses. *J Am Acad Dermatol* **14:** 1–18, 1986.

N. German Shepherd Dermatofibrosis

A dominant trait affecting German shepherd dogs, dermatofibrosis occurs mostly in Europe but is also reported in Canada and the United States of America. The onset of the skin lesions is delayed until middle age. Lesions are multiple cutaneous and subcutaneous firm nodules of mature fibrous connective tissue, usually measuring from a few mm to 4 cm. The extremities are initially affected, but lesions are progressive, and in severely affected dogs are generalized. The nodules are covered with a normal hair coat unless the dogs traumatize the lesions, causing erosions, alopecia, and hyperpigmentation. In milder cases, the lesions are detected only by palpation. On cut section, the nodules are composed of dense, white fibrous tissue. Lenticular masses may be found in fascial planes between muscle bundles.

Histologically, the nodules are composed of mature fibrous connective tissue, which smooths the normally convoluted contours of the epidermis and displaces resident hair follicles and adnexal glands. The deep dermis is markedly thickened as the fibrous tissue extends into the subcutis. The collagen bundles stain normally with hematoxylin and eosin and trichrome stains. Fibroblasts, fibrocytes, and blood vessels are not increased in number, and there is usually no significant inflammation. The collagen bundles are ultrastructurally normal.

The skin lesions are a marker for an internal malignancy. In a large study in Norway, 43 of 45 German shepherd dogs with renal adenocarcinoma showed the cutaneous lesions previously described. The tumors are typically bilateral, cystic, and very large, measuring up to 27 cm diameter and weighing up to 3 kg. The cysts contain brown-red viscid fluid. Approximately 50% of the tumors are metastatic; however, a benign adenoma and a unilateral non-neoplastic cyst have been reported in two dogs. Histologically, the tumors are well differentiated cystic adenocarcinomas. Multiple leiomyomas of uterine and vaginal smooth muscle are frequently associated with the cutaneous masses and kidney tumors.

The relationship of the skin lesions and the development of the renal neoplasms is not understood. The proliferation of fibrous connective tissue may represent a paraneoplastic process in which growth factors, such as transforming growth factor-beta (TGFβ), are produced by the renal tumor.

The disease has been seen in crossbreed dogs with German shepherd parentage. A disorder with similar histologic lesions in the skin is reported in the German shorthaired pointer at 8 weeks of age. Solitary cutaneous nodules with similar histology are common in many breeds of dog; these are of varied pathogenesis and do not predict internal malignancy.

Bibliography

Gilbert, P. A., Griffin, C. E., and Walder, E. J. Nodular dermatofibrosis and renal cystadenoma in a German shepherd dog. *J Am Anim Hosp Assoc* **26:** 253–256, 1990.

Jones, B. R., Alley, M. R., and Craig, A. S. Cutaneous collagen nodules in a dog. *J Small Anim Pract* **26:** 445–451, 1985.

Lium, B., and Moe, L. Hereditary multifocal renal cystadenocarcinomas and nodular dermatofibrosis in the German shepherd dog: Macroscopic and histopathologic changes. *Vet Pathol* **22:** 447–455, 1985.

Suter, M., Lott-Stolz, G., and Wild, P. Generalized nodular dermatofibrosis in six Alsatians. *Vet Pathol* **20:** 632–634, 1983.

O. Collagen Disorder of the Footpads of German Shepherd Dogs

This disorder has been observed in German shepherd pups from a few weeks to a few months of age. The disease appears to be inherited as an autosomal recessive trait with several dogs affected in a litter. Lesions mainly affect the footpads, which become tender and may ulcerate. Nodules may be palpable in the footpads and on the hocks. Histologically, there is a nodular-to-diffuse dermatitis in which inflammation is centered on small fragments of degenerate collagen. The inflammatory infiltrate is composed of neutrophils and numerous mononuclear cells. Marked edema and fibrin deposition are present in the dermis, indicating severe vascular damage. Degenerative vascular lesions are described. The lesions may regress spontaneously, but some dogs subsequently develop fatal renal amyloidosis by 2 to 3 years of age.

IV. Disorders of Epidermal Differentiation

The epidermis is of ectodermal origin and normally undergoes an orderly pattern of proliferation, differentiation, and keratinization. Basal cells are the only mitotically active cells, and postmitotic keratinocytes leave the basal cell layer to migrate in a highly regulated fashion upward through the different layers of the stratum spinosum to finally shave off from the cornified layer as dead keratinized cells. In normal canine skin, the migration from the basal to the cornified layer requires 22 days, as measured by tritiated thymidine incorporation studies. However, this epidermal turnover is markedly shortened in various skin diseases; for example, in seborrheic skin disease, it is 7–8 days. This defined process of proliferation and differentiation is associated not only with changes in keratinocyte shape and orientation but also with profound changes in intracellular proteins, cell-surface molecules, and intercellular contacts. These changes include the differential production of keratin filaments in proliferating cells, the assembly of adhesion molecules and intercellular junctions, and the expression of differentiation-specific molecules including involucrin, loricrin, and transglutaminase necessary for the formation of the cornified envelope. The different keratins are often used as markers of differentiation. Whereas basal cells express keratin K5 and K14, differentiated suprabasal cells produce keratins K1 and K10. Hyperproliferative skin as seen in skin disease in turn expresses a new set of keratin filaments not seen in normal skin, K6 and 16.

Basal cells produce basement membrane components (as do dermal fibroblasts) and are firmly anchored on the basement membrane via hemidesmosomes and adhesion junctions. Integrin adhesion molecules are important structural components of these junctions, mediating tight contacts to extracellular molecules such as fibronectin and collagen. To enable upward migration of cells into the stratum spinosum, rearrangement and formation of new adhesion complexes are necessary. Desmosomes and adhesion junctions enable the cells to form the tightly sealing stratified squamous epithelium. In the stratum spinosum, several new proteins (involucrin, loricrin, keratolinin) are formed that are necessary for the formation of the cornified envelope. In the stratum granulosum, these proteins are cross-linked for the formation of the cornified envelope. Many of these proteins are eventually secreted into the intercellular space together with cholesterol and other lipids to form the insoluble extracellular matrix of the cornified layer. In the stratum granularum, keratohyaline granules are formed that contain, as a major component, filaggrin, a molecule that cross-links keratin filaments. This is necessary for the formation of the sturdy, mechanically resistant cornified layer.

Proliferation and differentiation of epidermal cells are mutually exclusive. After several rounds of mitosis, basal cells undergo an irreversible growth arrest before they begin to differentiate. Various growth factors and inhibitors have been found to influence keratinocyte proliferation and differentiation and are thought to work together in a highly regulated, synergistic way. Numerous nutritional factors are also important for normal epidermal turnover, including fatty acids, vitamins A and B, zinc, and copper. In addition, many hormones (such as corticosteroids and thyroid hormone) have an effect on keratinocyte proliferation and differentiation.

In various disease processes, this highly regulated epidermal turnover is altered. Little is known about altered expression of the different molecules in diseases of keratinization. In humans, increased levels of TGFα and IL-8 as well as increased expression of epidermal growth factor (EGF) receptors, have been associated with psoriasis. Increased proliferation and/or decreased desquamation of cells lead to thickening of the epidermis.

Bibliography

Boyer, B., and Thiery, J. P. Epithelial cell adhesion mechanisms. *J Membr Biol* **112:** 97–108, 1989.

Fuchs, E. Keratins as biochemical markers of epithelial differentiation. *Trends in Genetics* **4:** 277–281, 1988.

Fuchs, E. Epidermal differentiation: The bare essentials. *J Cell Biol* **111:** 2807–2814, 1990.

Hohl, D. Cornified cell envelope. *Dermatologica* **180:** 201–211, 1990.

Huang, N.-N., Wang, D., and Heppel, L. A. Extracellular ATP is a mitogen for 3T3, 3T6, and A431 cells and acts synergistically with other growth factors. *Proc Natl Acad Sci USA* **86:** 7904–7908, 1989.

Mehrel, T. *et al.* Identification of a major keratinocyte cell envelope protein, loricrin. *Cell* **61:** 1103–1112, 1990.

Norris, D. A., Huff, J. C., and Weston, W. L. The state of research in cutaneous biology: A perspective in the 50th anniversary year of the Society of Investigative Dermatology and the Journal of Investigative Dermatology. *J Invest Dermatol* **92:** 179S–197S, 1989.

Richter, K. H. *et al.* Epidermal G_1-chalone and transforming growth factor beta are two different endogenous inhibitors of epidermal cell proliferation. *J Cell Physiol* **142:** 496–504, 1990.

Rothe, M. Growth factors. *Arch Dermatol* **125:** 1390–1397, 1989.

Suter, M. M. *et al.* Differential expression of cell-surface antigens on canine keratinocytes defined by monoclonal antibodies. *J Histochem Cytochem* **38:** 541–549, 1990.

Thacher, S. M. Purification of keratinocyte transglutaminase and its expression during squamous differentiation. *J Invest Dermatol* **92:** 578–584, 1989.

Yaar, M., Palleroni, A. V., and Gilchrest, B. A. Normal human keratinocytes contain an interferonlike protein that may modulate their growth and differentiation. *Ann N Y Acad Sci* **548:** 299–311, 1988.

A. Seborrheic Skin Diseases: Primary Idiopathic Seborrhea

The term seborrhea is entrenched in veterinary literature as a broad classification for a variety of clinical syndromes ranging from simple dandruff to severe inflammation with scaling and crusting. Seborrhea literally means abnormal flow of sebum. However, the major clinical abnormality in seborrheic skin diseases is altered keratinization.

Most seborrheic skin diseases are secondary to such disorders as hormonal imbalances (especially hypothyroidism, hyperadrenocorticism, and sex hormone imbalances), ectoparasitism (especially cheyletiellosis, pediculosis, and demodicosis), endoparasitism, dermatophytosis, hypersensitivities (inhalant, dietary, drug), abnormal lipid metabolism (malabsorption, liver or pancreatic disease, diabetes mellitus), dietary deficiencies (fatty acids, protein, vitamin A, zinc), chronic catabolic states, and environmental factors (especially hot, dry conditions). Severe seborrheic skin disease may also be a manifestation of systemic lupus erythematosus, pemphigus foliaceus, epitheliotropic lymphoma, and internal malignancy.

Whatever the underlying cause, seborrheic skin is characterized by certain abnormalities. The first is altered keratinization with or without altered glandular function. This is reflected grossly by varying degrees of scaling and crusting, with or without greasiness. Second, the surface lipids, whether the seborrheic skin is dry or greasy, have an increased percentage of free fatty acids and cholesterol, and a decreased percentage of diester waxes. Third, the altered keratinization and lipid film are accompanied by a marked increase in the number of surface bacteria per unit of skin. In addition, the bacterial flora usually switches from the normally nonpathogenic resident micrococci, corynebacteria, and coagulase-negative and coagulase-positive staphylococci to a pure, heavy, potentially pathogenic population of coagulase-positive staphylococci. Thus a seborrheic skin, regardless of the underlying cause, is often complicated by secondary bacterial disease.

Clinically, seborrheic skin disease is often separated into three morphologic types. **Seborrhea sicca** is characterized by dry skin with focal or diffuse flaking and accumulations of white to gray nonadherent scales. **Seborrhea oleosa** is characterized by focal or diffuse scaling associated with excessive lipid production that produces yellowish to brownish material that adheres to the skin and hair. **Seborrheic dermatitis** is characterized by scaling and greasiness with gross evidence of local or diffuse inflammation. Pruritus and/or secondary bacterial infection is often present with all three forms. This clinical categorization has little significance in terms of differential diagnosis, and animals may be dry in an area of the body, and greasy in another. Although seborrheic skin disease has been described in all domestic species, it is most common in dogs, cats, and horses.

Seborrheic skin disease is common in **dogs**. Primary idiopathic seborrheic skin disease occurs in many breeds. The association with particular breeds and also an early age of onset suggest that primary canine seborrhea may have an inherited basis. Cocker spaniels and springer spaniels tend to have a greasy, inflammatory form of seborrhea with numerous hyperkeratotic plaques, comedones, and follicular casts. Inflammatory ceruminous otitis externa is also a constant finding. In cocker spaniels, cell-proliferation kinetics indicate that seborrheic individuals have increased epithelial cell proliferation in the epidermis, hair follicle infundibulum, and sebaceous gland. In addition, recombinant grafting studies have shown that the hyperproliferative epidermis from seborrheic cocker spaniels remains hyperproliferative, suggesting that the abnormality may be in the keratinocyte itself. Other breeds with primary, greasy seborrhea include the basset hound, West Highland white terrier, German shepherd, dachshund, and Chinese Shar-pei. Breeds having a dry form of primary seborrhea include the Irish setter, Doberman pinscher, dachshund, and West Highland white terrier. Primary seborrhea in German shepherds, West Highland white terriers, and Labrador retrievers is often very inflammatory, lichenified, and pruritic. Seborrheic skin disease is typically more pronounced on the face, pinnae, trunk, pressure points, intertriginous areas, mucocutaneous areas, and paws. In Labrador retrievers, the distribution is often strikingly ventral (water-line disease).

Primary seborrhea is uncommon in cats and horses. Primary seborrhea oleosa, presumably of autosomal recessive inheritance, has been described in Persian **cats**. Severely affected kittens show lesions at 3 to 4 days of age, and develop progressively severe, generalized greasiness, matting of the haircoat, rancid odor, comedones, alopecia, and ceruminous otitis externa. Pruritus is absent. A milder form of the disease is recognized in 6- to 8-week-old kittens, with mild to moderate greasiness of the skin and haircoat.

Primary seborrhea in the **horse** occurs in both dry and greasy forms, and tends to be restricted to the mane and tail. Pruritus is absent.

Primary idiopathic seborrhea is usually characterized histopathologically by superficial perivascular dermatitis. Epidermal hyperplasia is usually mild to moderate and papillated in configuration. A pronounced keratinization defect is present, typified by alternating vertical tiers of orthokeratotic and parakeratotic hyperkeratosis. The parakeratosis is typically found overlying the epidermal papillae (parakeratotic "caps"). The underlying dermal papillae are often edematous, leading to spongiosis and leukocytic exocytosis of the overlying epidermis ("papillary squirting"). Spongioform or Munro's microabscesses may be seen in conjunction with the parakeratosis. The perivascular inflammatory cells include variable combinations of lymphocytes, neutrophils, plasma cells, histiocytes, and mast cells. Since secondary bacterial infection is common, subordinate patterns of suppurative folliculitis, furunculosis, perifolliculitis, and intraepidermal pustular dermatitis are frequently seen.

Secondary seborrheic skin disease is also characterized by a pronounced keratinization defect, orthokeratotic and/or parakeratotic in nature. The overlying histopathologic reaction pattern, however, reflects the underlying disease process (hypersensitivity, ectoparasitism, endocrinopathy, etc.).

Bibliography

Baker, B. B., and Maibach, H. I. Epidermal cell renewal in seborrheic skin of dogs. *Am J Vet Res* **48:** 726–727, 1987.

Ihrke, P. J. *et al.* Microbiology of normal and seborrheic canine skin. *Am J Vet Res* **39:** 1487–1489, 1978.

Kwochka, K. W. Pathophysiology, differential diagnosis, and medical management of keratinization defects in the dog. *Proc Can Acad Vet Dermatol* **3:** 2–14, 1989.

Kwochka, K. W., and Rademakers, A. M. Cell proliferation kinetics of epidermis, hair follicles, and sebaceous glands of cocker spaniels with idiopathic seborrhea. *Am J Vet Res* **50:** 1918–1922, 1989.

Paradis, M., and Scott, D. W. Hereditary primary seborrhea oleosa in Persian cats. *Feline Pract* **18:** 17–20, 1990.

B. Acne

Acne is seen in short-coated breeds of **dogs**, especially English bulldogs, boxers, Great Danes, and Doberman pinschers. It usually occurs between 3 and 12 months of age with no sex predilection and occasionally persists into adult life. The etiology and pathogenesis are unknown, but may be similar to that in humans, involving increased circulating levels of androgens and acne-prone skin. The lesions of canine acne are asymptomatic papules and pustules which arise from comedones. Typically, the skin and lips are involved. Secondary bacterial folliculitis and furunculosis may occur. Histologically, there is marked orthokeratotic hyperkeratosis, dilatation and plugging of hair follicles, and variable inflammation.

Acne is common in **cats** and has no sex or breed predilections. It usually occurs in mature cats and persists for life. The cause is unknown. The lesions of feline acne are asymptomatic comedones on the skin and lips. Occasionally, secondary bacterial folliculitis and furunculosis develop. Histologically, marked orthokeratotic hyperkeratosis, dilatation and plugging of hair follicles, with variable inflammation are seen.

C. Schnauzer Comedo Syndrome

This condition occurs only in the miniature schnauzer breed. Either sex may be affected, and the condition usually develops early in life. The exclusive occurrence in schnauzers and the early onset suggest that this syndrome may be a developmental dysplasia of hair follicles with an inherited basis. Clinically, the condition is characterized by multiple asymptomatic comedones over the dorsal midline. Occasionally, secondary bacterial folliculitis and furunculosis may develop. Histologically, there is marked orthokeratotic hyperkeratosis, dilatation and plugging of hair follicles (comedones), and variable inflammation.

D. Tail-Gland Hyperplasia

Many **dogs** have an oval area of skin on the dorsal surface of the tail above the fifth to seventh coccygeal vertebrae referred to as the tail (supracaudal, preen) gland. Microscopically, this area is characterized by large, densely packed perianal ("hepatoid") and sebaceous glands. In some dogs, especially adult to aged males, this area enlarges. The enlargement is usually firm to slightly spongy, and associated with partial alopecia, scaling and greasiness. At this stage, the lesion is asymptomatic. Occasionally, the lesions become cystic and/or secondarily infected, or neoplastic.

In most instances, canine tail-gland hyperplasia is associated with hormonal imbalances, especially elevated levels of blood testerosterone or hypothyroidism. Histologically, the lesions are characterized by marked hyperplasia of the perianal gland component, with a variable inflammatory response.

The entire dorsal surface of the tail in **cats** is replete with large, densely packed sebaceous glands. In some cats, especially sexually active males of the Persian, Siamese, and rex breeds, this area becomes clinically seborrheic, whereupon a brownish to blackish, greasy keratosebaceous material accumulates on the hairs of the skin surface. Unless secondarily infected, the condition is asymptomatic. The cause of feline tail-gland hyperplasia is unknown, and the colloquialism "stud tail" is misleading, as the condition is also seen in intact females and neutered males and females. Histologically, the condition is characterized by marked hyperplasia of sebaceous glands, with variable orthokeratotic hyperkeratosis and inflammation.

Bibliography

Scott, D. W., and Reimers, T. J. Tail gland and perianal gland hyperplasia associated with testicular neoplasia and hypertestosteronemia in a dog. *Canine Pract* **13:** 15–17, 1986.

E. Canine Nasodigital Hyperkeratosis

Canine nasodigital hyperkeratosis is characterized by increased amounts of horny tissue originating from and tightly adherent to the epidermis of the footpads ("hard-pad disease") and/or nasal planum. This disorder may be seen in association with canine distemper, pemphigus foliaceus, systemic or discoid lupus erythematosus, zinc-responsive dermatoses and as an idiopathic senile change. Severe digital hyperkeratosis occurs as an hereditary disorder in Irish terriers. Histologically, nasodigital hyperkeratosis is characterized by marked orthokeratotic and/or parakeratotic hyperkeratosis, and irregular to papillated epidermal hyperplasia. Age, breed, and sex predilections and other histopathologic findings reflect those of the underlying diseases.

F. Keratoses

Keratoses are firm, elevated, circumscribed areas of excessive keratin production. In humans, keratoses are common and of numerous types. Keratoses are uncommonly reported in domestic animals. Actinic keratoses are discussed elsewhere.

Equine linear keratosis is a readily recognizable dermatosis. No sex predilection is apparent, but the Quarter horse breed is most commonly affected, and most animals develop the condition at 1 to 5 years of age. The cause of linear keratosis is unknown, but the common occurrence in Quarter horses, especially closely related animals, suggests a hereditary component. The skin lesions consist of one or more vertically oriented linear areas of alopecia, scaling, and crusting, which are commonly seen on the neck, shoulder, and lateral thorax. The lesions are usually unilateral. Pruritus and pain are absent. The lesions persist for life. Histopathologic findings consist of orthokeratotic hyperkeratosis and irregular to papillated epidermal hyperplasia, with or without a mild superficial perivascular dermatitis, featuring lymphocytes and histiocytes.

Equine cannon keratosis is a readily recognizable disease of the horse, also known by the colloquial name of stud crud. Despite this inference, cannon keratosis also occurs in mares. The lesions consist of vertically oriented, moderately well demarcated areas of alopecia, scaling, and crusting on the cranial surface of the rear cannon bones. The lesions are usually bilateral. Pruritus and pain are absent. Cannon keratosis persists for life. Histopathologic findings include orthokeratotic and/or parakeratotic hyperkeratosis, irregular to papillated epidermal hyperplasia, and mild superficial perivascular dermatitis, featuring lymphocytes and histiocytes.

Seborrheic keratoses have been recognized in dogs. They are of unknown cause, and have nothing to do with seborrhea. They may be single or multiple and have no apparent age, breed, sex, or site predilections. The lesions are elevated plaques or nodules with a hyperkeratotic, often greasy surface. They are frequently hyperpigmented. Pruritus and pain are usually absent. Histologically, seborrheic keratoses are characterized by orthokeratotic hyperkeratosis, epidermal hyperplasia (basaloid and squamoid), and papillomatosis.

Lichenoid keratoses have been reported as single to multiple wartlike papules or hyperkeratotic plaques on the lateral surface of the pinna in dogs. Histologically, lichenoid keratosis is characterized by irregular to papillated epidermal hyperplasia, moderate to marked orthokeratotic and/or parakeratotic hyperkeratosis, and a lichenoid inflammatory infiltrate predominantly consisting of lymphocytes and plasma cells.

Cutaneous horns are recognized occasionally in all domestic species. Some are of unknown cause, but others originate from papillomas, basal cell tumors, squamous cell carcinomas, or other keratoses. In cattle, sheep, and goats, cutaneous horns may arise in lesions of dermatophilosis. In the cat, multiple cutaneous horns on the footpads have been reported in association with feline leukemia virus infection; feline leukemia virus was cultured from the horns, and type C viral particles were seen in the lesions with the electron microscope.

Cutaneous horns may be single or multiple, and have no apparent age, breed, sex, or site predilections. The lesions are firm, well-circumscribed hornlike projections from the skin. They may be small (1 mm diameter × 5 mm length) or quite large (3 cm diameter × 12 cm length). Histologically, cutaneous horns are characterized by extensive, compact, laminated, orthokeratotic and/or parakeratotic hyperkeratosis. The base of the horn must be inspected for the possible underlying cause.

Bibliography

Anderson, W. I., Scott, D. W., and Luther, P. B. Idiopathic benign lichenoid keratosis on the pinna of the ear in four dogs. *Cornell Vet* **79:** 179–184, 1989.

Center, S. A., Scott, D. W., and Scott, F. W. Multiple cutaneous horns on the footpads of a cat. *Feline Pract* **12:** 26–30, 1982.

Scott, D. W. Equine linear keratosis. *Eq Pract* **7:** 39–42, 1985.

G. Granulomatous Sebaceous Adenitis

Granulomatous to pyogranulomatous sebaceous adenitis is an uncommon, sterile, idiopathic dermatosis of dogs and cats. There is no sex predilection, and the disorder tends to affect young adult to middle-aged animals. Although many breeds and mongrels may be affected, there are breed predilections for standard poodles, vizslas, Samoyeds, akitas, and chow chows. Genetic factors may be involved.

Short-coated animals usually develop more or less symmetrical annular areas of alopecia and scaling, especially over the head and trunk. Longer-coated animals develop more or less symmetrical multifocal to generalized areas of patchy alopecia and brittle to broken hairs encircled by yellowish to brownish keratosebaceous casts (follicular casts). Inflammation and pruritus are usually minimal unless the condition is complicated by secondary bacterial infection.

Early histopathologic findings include focal areas of granulomatous to pyogranulomatous perifollicular dermatitis centered on sebaceous glands. Variable degrees of superficial perivascular dermatitis and keratinization defect (orthokeratotic and/or parakeratotic) accompany these findings. Follicular hyperkeratosis may be prominent. Chronic lesions are characterized by variable superficial perivascular dermatitis and a moderate to marked keratinization defect, which involves the skin surface as well as the hair follicle. Often sebaceous glands are completely absent. Perifollicular fibrosis and a sparse infiltrate of mononuclear cells may mark the site of sebaceous gland destruction.

Granulomatous sebaceous adenitis may not be a single entity, with a single pathogenesis. The granulomatous lesions tend to be pronounced in the vizsla, often involving the deep dermis, whereas, in the standard poodle, hyperkeratosis is the most obvious histologic abnormality. The hyperkeratosis may be secondary to immune-mediated destruction of the sebaceous glands. Alternatively sebaceous adenitis could be primarily a keratinization defect.

Bibliography

Carothers, M. A. *et al.* Cyclosporin-responsive granulomatous sebaceous adenitis in a dog. *J Am Vet Med Assoc* **198:** 1645–1648, 1991.

Guaguère, E., Alhaidari, Z., and Magnol, J. P. Adénite sébacée granulomateuse: A propos de trois cas. *Point Vét* **21:** 107–111, 1989.

McAllister, M. M. Adenohypophysitis associated with sebaceous gland atrophy in a dog. *Vet Pathol* **28:** 340–341, 1991.

Rosser, E. J. *et al.* Sebaceous adenitis with hyperkeratosis in the standard poodle: A discussion of 10 cases. *J Am Anim Hosp Assoc* **23:** 341–345, 1987.

Scott, D. W. Granulomatous sebaceous adenitis in dogs. *J Am Anim Hosp Assoc* **22:** 631–634, 1986.

Scott, D. W. Adénite sébacée pyogranulomateuse stérile chez un chat. *Point Vét* **21:** 107–111, 1989.

H. Vitamin A-Responsive Dermatosis

Vitamin A-responsive seborrheic dermatoses are reported most commonly in cocker spaniels, but occur in other breeds (see Section VIII, Nutritional Diseases of the Skin). There is, however, no clinicopathologic evidence to suggest that these animals have a vitamin A deficiency.

I. Lichenoid–Psoriasiform Dermatosis of Springer Spaniels

This idiopathic, probably genetically programmed dermatosis is seen in 4- to 18-month-old springer spaniels of either sex. Asymptomatic, generally symmetric erythematous, hyperkeratotic, lichenoid papules and plaques usually begin on the pinnae and groin and progressively involve large areas of the body. Secondary bacterial infection is common.

Histopathologic findings include a perivascular to lichenoid dermatitis with areas of psoriasiform epidermal hyperplasia, intraepidermal microabscesses (containing eosinophils and neutrophils), and Munro's microabscesses. Chronic lesions frequently show papillated epidermal hyperplasia, papillomatosis, and moderate to marked orthokeratotic and/or parakeratotic hyperkeratosis.

Bibliography

Gross, T. L. *et al.* Psoriasiform–lichenoid dermatitis in the springer spaniel. *Vet Pathol* **23:** 76–78, 1986.

Mason, K. V., Halliwell, R. E. W., and McDougal, B. J. Characterization of lichenoid–psoriasiform dermatosis of springer spaniels. *J Am Vet Med Assoc* **189:** 897–901, 1986.

J. Ear-Margin Dermatosis

This is a common, idiopathic, seborrheic disorder that affects the edges of the pinnae of several breeds of dogs with pendulous pinnae, especially the dachshund. There is no sex predilection, and the disorder usually begins in young adults. Initial scaling of the ear margins is followed by waxy keratosebaceous accumulations and alopecia. The disease is symmetrical and asymptomatic. The dermatosis may be complicated by secondary bacterial infection and fissures, at which point ulceration, oozing, crusting, pain, and pruritus may be seen.

Histopathologic findings include marked orthokeratotic and/or parakeratotic hyperkeratosis overlying a mild superficial perivascular dermatitis.

K. Psoriasiform Exfoliative Dermatoses

A rare, idiopathic exfoliative dermatitis resembling large plaque parapsoriasis in humans has been described in dogs and cats. Large erythematous, scaly, irregular plaques are found in a more or less symmetric distribution over the trunk and proximal limbs. The lesions are asymptomatic unless complicated by secondary bacterial infection. Histopathologic findings include a superficial perivascular to lichenoid dermatitis associated with regular to psoriasiform epidermal hyperplasia, prominent multifocal parakeratosis, and a diffuse exocytosis of lymphocytes. The lymphocytes tend to be arranged in a row within the stratum basale of the epidermis.

An idiopathic exfoliative dermatitis has also been described in goats. Scaling, crusting, and alopecia are most severe over the ventrum, perineum, face, and pinnae. Pruritus is absent. Histopathologic findings include a superficial perivascular dermatitis with psoriasiform epidermal hyperplasia.

Bibliography

Jeffries, A. R., Jackson, P. G. G., and Casas, F. C. Alopecic exfoliative dermatitis in goats. *Vet Rec* **121:** 576, 1987.

Scott, D. W. Exfoliative dermatoses in a dog and a cat resembling large plaque parapsoriasis in humans. *Compan Anim Pract* **2:** 22–29, 1988.

L. *Epidermal Dysplasia with* **Malassezia pachydermatis** *Infection in West Highland White Terriers*

This is an idiopathic, presumably genetic disorder of West Highland white terriers of both sexes. Fifty percent of affected dogs first manifest clinical signs at younger than 1 year of age. Diffuse erythema, scaling, and intense pruritus progressively involve the entire body surface. Chronic cases become alopecic, lichenified, hyperpigmented, very greasy, and rancid smelling. Secondary bacterial infections are common.

Histopathologic findings are characterized by a hyperplastic superficial perivascular dermatitis with marked orthokeratotic and/or parakeratotic hyperkeratosis, which also involves hair follicles. The epidermis is diffusely spongiotic and exhibits lymphocytic exocytosis. The epidermis and follicular outer root sheath epithelium show numerous fingerlike projections into the surrounding connective tissue (epidermal "buds") and focal areas of epidermal dysplasia. Numerous budding *Malassezia* yeasts are present in surface and follicular keratin. Mast cells may be aligned in a row in the immediate subepithelial connective tissue.

Bibliography

Scott, D. W., and Miller, W. H. Epidermal dysplasia and *Malassezia pachydermatis* infection in West Highland white terriers. *Vet Dermatol* **1:** 25–36, 1989.

V. Disorders of Pigmentation

Melanin pigments are chiefly responsible for the coloration of the hair, skin, and eyes. Melanins also play important roles in photoprotection and in free-radical scavenging. Melanins embrace a wide range of pigments including the brown-black eumelanins, yellow or red-brown pheomelanins, and other pigments whose physicochemical natures are intermediate. Pigment types in horses, sheep, goats, and llamas have been analyzed. Despite the different properties of the various melanins, they all arise from a common metabolic pathway in which dopaquinone is the key intermediate. Tyrosine is converted to Dopa, which is then oxidized to dopaquinone. Both reactions are catalyzed by the same copper-containing enzyme, tyrosinase. The subsequent steps in the process of melanogenesis apparently occur spontaneously, although three factors, described in rodents, may regulate melanin biosynthesis after the tyrosinase steps.

Melanin is synthesized by melanocytes, dendritic cells of neuroectodermal origin which reside in the stratum basale of the surface epidermis, in the external root sheath, and in the hair matrix of hair follicles. Melanocytes are also found in the dermis, especially perivascularly. Melanogenesis takes place in membrane-bound organelles called melanosomes. These are designated types I through IV according to maturation. Melanosomes originate from the Golgi apparatus where the tyrosinase enzyme is packaged. Type I melanosomes contain no melanin and are electron lucent. As melanin is progressively laid down on protein matrices, melanosomes become increasingly electron dense. At the same time, they migrate to the periphery of the dendritic extensions, where transfer of melanin to adjacent epidermal cells takes place. Transfer involves the endocytosis of the tips of the dendrites and incorporation of type IV melanosomes to adjacent keratinocytes.

Melanocytes express receptors for alpha-melanocytic stimulating hormone (MSH), probably the most important hormone regulating melanogenesis. Alpha-MSH stimulates tyrosinase activity and the peripheral migration of melanosomes along dendrites. Adrenocorticotropic hormone, estrogens, and testosterone also exert local effects on pigmentation.

The phenotypic coloration of skin and hair is reliant on the interaction of melanocytes and keratinocytes, known as the functional epidermal melanin unit. Disorders of pigmentation involve alteration of either or both of these components.

Bibliography

Guaguère, E., Alhaidari, Z., and Ortonne, J. P. Troubles de la pigmentation melanique en dermatologie des carnivores. 1re partie: Eléments de physiopathologie. *Point Vét* **17:** 549–557, 1985.

Sponenberg, D. P. *et al.* Pigment types of various color genotypes of horses. *Pigm Cell Res* **1:** 410–413, 1988.

Sponenberg, D. P. *et al.* Pigment types in sheep, goats, and llamas. *Pigm Cell Res* **1:** 414–418, 1988.

Pawelek, J. *et al.* New regulators of melanin biosynthesis and the autodestruction of melanoma cells. *Nature* **286:** 617–619, 1980.

A. *Hyperpigmentation*

Acquired hyperpigmentation of the skin (**melanoderma**) is encountered frequently, is usually a result of minor or chronic irritation and may be accompanied by mild hyperkeratosis. Both melanosis and hyperkeratosis are common responses to mild injuries by agents as diverse as mites and irradiation. Hypermelanosis results from an increased rate of melanosome production, an increase in melanosome size, or an increase in the degree of melanization of the melanosome. It may be associated with an increase in the number of melanocytes, as occurs following trauma and ultraviolet light exposure. It is likely that inflammatory mediators play a role in stimulating melanocyte proliferation and up-regulating receptors for MSH; prostaglandin D_2, for example, stimulates pigment cell proliferation, and IL-1 affects MSH receptor expression *in vitro*. Basic fibroblast growth factor (bFGF) has been shown to be a mitogen for human melanocytes. Proliferating human epidermal cells in culture produce bFGF, perhaps illustrating one possible mechanism behind the hyperplastic and hyperpigmented lesions that typify many chronic dermatoses.

Acquired hyperpigmentation may also involve hair (**melanotrichia**). This is usually seen as a result of inflammatory skin disorders, especially those caused by biting insects in the horse, and has been described in white Merino sheep exposed to ultraviolet light.

Bibliography

Bolognia, J., Murray, M., and Pawelek, J. UVB-induced melanogenesis may be mediated through the MSH-receptor system. *J Invest Dermatol* **92:** 651–656, 1989.

Forrest, J. W., and Fleet, M. R. Pigmented spots in the wool-bearing skin of white Merino sheep induced by ultraviolet light. *Aust J Biol Sci* **39:** 125–136, 1986.

Guaguère, E. *et al.* Troubles de la pigmentation mélanique en dermatologie des carnivores. 3. Hypermélanoses. *Point Vét* **18:** 699–709, 1987.

Halaban, R. *et al.* Basic fibroblast growth factor from human keratinocytes is a natural mitogen for melanocytes. *J Cell Biol* **107:** 1611–1619, 1988.

Nordlund, J. J., Collins, C. E., and Rheins, L. A. Prostaglandin E_2 and D_2 but not MSH stimulate the proliferation of pigment cells in the pinnal epidermis of the DBA/2 mouse. *J Invest Dermatol* **86:** 433–437, 1986.

1. Focal Macular Melanosis

Lesions resembling **lentigo** in humans occur in domestic animals. The major significance lies in confusion with melanoma. **Cats** with orange, cream, and tricolored coloration appear predisposed to lentigo simplex. In **dogs**, a generalized, juvenile-onset disease, named lentiginosis profusa, is apparently inherited as a dominant trait in pugs, but the lesions occur in many breeds sporadically. Merino **sheep** may acquire pigmented macules, particularly after shearing. Lesions are concentrated on the back, suggesting a role for sunlight exposure. Experimental exposure to ultraviolet light induced lesions as early as 10 days postirradiation. Grossly, lentigines are flat or slightly raised, black macules. In cats, the mucocutaneous junctions of the mouth, eye, and nose, and the footpads are involved. Generalized lesions have been described in a silver cat. In dogs, the ventrum is often involved. The lesions may increase in size and number with age. Histologically, the lesions are restricted to the epidermis, which shows focal hyperplasia, hypergranulosis, rete ridge formation, and marked melanosis. In lesions from the Merino sheep, the number of epidermal melanocytes increased at the dermoepidermal junction and also appeared in the normally nonpigmented outer root-sheath epithelium.

Bibliography

Nash, S., and Paulsen, D. Generalized lentigines in a silver cat. *J Am Vet Med Assoc* **196:** 1500–1501, 1990.

Scott, D. W. Lentigo simplex in orange cats. *Compan Anim Pract* **1:** 23–25, 1987.

Van Rensburg, I. B. J., and Briggs, O. M. Pathology of canine lentiginosis profusa. *J S Afr Vet Assoc* **57:** 159–161, 1986.

2. Acromelanism

Acromelanism is seen in Siamese and Himalayan cats whose coat color appears to be influenced by external temperature (high temperatures producing light hairs, low temperatures producing dark hairs) and factors affecting heat production and loss (alopecia, inflammation). These phenomena appear to be associated with temperature-dependent enzymes involved in melanin synthesis.

3. Canine Acanthosis Nigricans

Canine acanthosis nigricans is an idiopathic dermatitis, characterized by progressive hyperpigmentation, alopecia, and lichenification. The lesions are roughly bilaterally symmetrical and typically commence in the axillae, spreading to involve proximal limbs, ventral abdomen, neck, and inguinal area. Seborrhea and bacterial pyoderma are frequent complications. Histological examination reveals a hyperplastic dermatitis with orthokeratotic and parakeratotic hyperkeratosis, acanthosis, and rete ridge formation. All layers of the epidermis are heavily melanized. Spongiosis, neutrophilic exocytosis, and serous crust may also be present. The dermal inflammatory reaction is pleomorphic in cell type and superficial in location.

The primary or idiopathic form of acanthosis nigricans occurs predominantly in dachshunds. The pathogenesis of idiopathic acanthosis nigricans is unknown. The microscopic lesions are inflammatory, which is at odds with the most widely held theory of endocrine dysfunction. Deficiency of thyroid hormones, thyroid-stimulating hormone, thyrotropin-releasing factor, and melatonin have been postulated, but there is little supporting evidence for these. In view of the early age of onset (usually <1 year old) and the predilection for dachshunds, canine idiopathic acanthosis nigricans may be a heritable disorder.

Secondary acanthosis nigricans is associated with atopic disease, friction in obese dogs, and hormonal disorders such as hypothyroidism. In humans, some forms of acanthosis nigricans are associated with internal malignancies; a similar correlation has not been proven in dogs.

Bibliography

Anderson, R. K. Canine acanthosis nigricans. *Compend Cont Ed* **1:** 466–471, 1979.

B. Leukoderma and Leukotrichia

Reduction in pigmentation of the skin is **leukoderma,** and of the hair is **leukotrichia.** Leukoderma and leukotrichia may occur independently. They may result from a decrease in melanin (**hypomelanosis**), a complete absence of melanin (**amelanosis**), or from a loss of existing melanin (**depigmentation**). These events result either from an absence of the pigment-synthesizing melanocytes or from a failure of melanocytes to produce normal amounts of melanin or to transfer it to adjacent keratinocytes.

1. Hereditary Hypopigmentation

a. PIEBALDISM In this example of the first kind of hypomelanosis, the white patches in piebald animals and humans result from a congenital failure of melanocytes to migrate from the neural crest to the integument. Piebaldism is further subdivided into variegated and nonvariegated types. The Dalmatian dog is an example of nonvariegated piebaldism. The black spots develop as a result of melanocytes migrating postnatally from the walls of blood vessels or from nerve endings. Variegated piebaldism occurs in animals with the "merle" gene, for example, merle collies and harlequin Great Danes. In these animals, the white areas contain no melanocytes. The color-dilute gray areas contain some melanocytes, but these have sparse melanosomes with an abnormal morphology. Hereditary deafness may accompany the pigmentary disorders. In cats, a syndrome which has been likened to the human Waardenburg–Klein syndrome is characterized by a white coat, hypo- or heterochromia of the iris, and deafness. The trait is autosomal dominant with incomplete penetrance. Deafness is also seen in piebald dogs, like the Dalmatian, and in merled breeds, like the collie.

Bibliography

Schaible, R. H., and Brumbaugh, J. A. Electron microscopy of pigment cells in variegated and nonvariegated, piebald-spotted dogs. *Pigm Cell* **3:** 191–200, 1976.

b. ALBINISM Albinism is reported in all domestic animals. Two forms of albinism in horses are lethal traits. The melanocytes in albino animals and humans are normally distributed but are defective in function. Albinism may result from an inability to synthesize tyrosinase or, in the presence of the enzyme, from a failure of melanosome melanization or autodegradation of melanin. The extent of the biochemical defect varies, so that albinism covers a spectrum from amelanosis through graded pigmentary dilution.

Chediak–Higashi syndrome in humans, Hereford cattle, Persian cats, mink, and various other animal species is an example of partial albinism. Although melanin is produced, the basic membrane defect in this syndrome leads to the formation of giant melanosomes, which are passed with difficulty to the keratinocytes. The clumping of these giant melanosomes produces the color-dilution effect. Chediak–Higashi syndrome is discussed in Volume 3, Chapter 2, The Hematopoietic System.

Maltese dilution in cats is an autosomal recessive trait (dd) responsible for color dilute-blue and cream cats and blue-point Siamese. It is a generalized albinism but without ocular involvement. The pale coloration is due to clumping of melanin granules. No lesions have yet been associated with this phenotype.

Cyclic hematopoiesis (cyclic neutropenia), a lethal hereditary disease of collie dogs, is caused by an autosomal recessive gene with a pleiotropic effect on coat color dilution. Affected dogs are silver-gray. The hematological aspects of this disease are considered in Volume 3, Chapter 2, The Hematopoietic System. The abnormal hair pigmentation results from the diminished formation of melanin from its precursor tyrosine rather than from pigment clumping. The normal collie coat color is restored in animals receiving bone marrow transplants to correct the cyclic hematopoiesis.

Tyrosinase deficiency, a rare disease in chow chows, is characterized by depigmentation of the buccal mucosa, tongue, and hair. The defect is due to a transient tyrosinase deficiency. The integument and appendages regain their normal pigmentation within a few months.

Color-dilution alopecia, a tardive-onset hypotrichosis associated with color-dilution traits in the dog, is discussed under Congenital and Hereditary Diseases of the Skin.

Bibliography

Adalsteinsson, S. Albinism in Icelandic sheep. *J Hered* **68:** 347–349, 1977.

Lund, J. E., and Barkman, D. Color dilution in the gray collie. *Am J Vet Res* **35:** 265–268, 1974.

Prieur, D. J., and Collier, L. L. Chediak–Higashi syndrome of animals. *Am J Pathol* **90:** 533–536, 1978.

Prieur, D. J., and Collier, L. L. Maltese dilution of domestic cats. A generalized cutaneous albinism lacking ocular involvement. *J Hered* **75:** 41–44, 1984.

Roberts, S. R. Color dilution and hereditary defects in collie dogs. *Am J Ophthalmol* **63:** 1762–1775, 1967.

Schneider, J. E., and Leipold, H. W. Recessive lethal white in two foals. *J Eq Med Surg* **2:** 479–482, 1978.

Scott, D. W. Vitiligo in the Rottweiler. *Canine Pract* **15:** 22–25, 1990.

Slominski, A., Paus, R., and Bomirski, A. Hypothesis: Possible role for the melatonin receptor in vitiligo: Discussion paper. *J R Soc Med* **82:** 539–541, 1989.

Yang, T. J. Recovery of hair coat color in gray collie (cyclic neutropenia)—normal bone marrow transplant chimeras. *Am J Pathol* **91:** 149–152, 1978.

Zelickson, A. S. *et al.* The Chediak–Higashi syndrome: Formation of giant melanosomes and the basis of hypopigmentation. *J Invest Dermatol* **49:** 575–581, 1967.

2. Acquired Hypopigmentation

This follows damage to the epidermal melanin unit by various insults, including trauma, inflammation, radiation, contactants, autonomic nervous system disorders, endocrinopathies, infections, and nutritional deficiencies. In general, **horses** tend to depigment following cutaneous injury, whereas the skin of other animals tends to become hypermelanotic.

Examples of depigmenting **equine** diseases include onchocerciasis, *Culicoides* hypersensitivity, ventral midline dermatitis, and coital vesicular exanthema. Aural plaques in horses are hyperkeratotic, depigmented plaques on the inner aspect of the pinnae. Although once considered to result from blackfly bites, these are verruca plana (flat warts). Depigmenting lesions in horses may result from contact with equipment such as rubber bit guards or crupper straps or with feed buckets. Monobenzene ether of hydroquinone, a common ingredient in rubber, inhibits melanogenesis.

In **dogs,** hypopigmentation occurs in immune-mediated diseases in which there is destruction of the basal keratinocytes. Examples include lupus erythematosus, drug eruptions, bullous pemphigoid, and the various forms of pemphigus. Acquired depigmentation of the lips and/or nose also occurs in dogs as a result of contact with plastic or rubber dishes or toys containing dihydroquinone monobenzene ethers. Microbial lesions, such as deep pyoderma, may heal with depigmentation. Depigmenting lesions have been noted also in canine leishmaniasis and in dermatophytosis caused by *Microsporum persicolor*. Epitheliotropic lymphoma often presents with depigmenting, ulcerative lesions of skin and mucocutaneous junctions. Cutaneous depigmentation has been described in association with internal neoplasms, including Sertoli cell tumor of the testis, but the relationship between the cutaneous lesion and the neoplasia is poorly understood. Subcutaneous injection of corticosteroid or progesterone hormones may lead to focal hypopigmentation in the dog.

Unilateral periocular depigmentation is described in a **cat** with Horner's syndrome. Focal depigmentation of the nose and lips may follow upper respiratory tract viral infections.

a. VITILIGO Literally meaning blemish, vitiligo is an acquired hypomelanosis of humans and animals, characterized by gradually expanding pale macules that are often symmetrical or segmental in distribution. It is rarely a hereditary disease and, when it is, the onset is tardive. Vitiligo, unlike albinism or piebaldism, is the result of depigmentation. The reason for the degeneration of existing melanocytes in affected skin is not known. An autoimmune pathogenesis is possible, based on the clinical association of human vitiligo with other autoimmune disorders, the presence of lymphocytes at the advancing edge of the lesion, and the demonstration of antibodies to melanin-producing cells in human, canine, and equine forms of the disorder. Another possibility is a failure, possibly through environmental insult, of the mechanisms which normally protect melanocytes from the toxic effects of the melanin precursors, including free radicals. In this model, the activation of an autoimmune process is secondary. There is little evidence to support a third, neurogenic hypothesis, which postulates the release of toxic factors from peripheral nerves.

Canine vitiligolike diseases have been described in Belgian tervuren, Doberman pinscher, Newfoundland, Rottweiler, German shepherd, and old English sheepdogs. Vitiligo in a dachshund developed concurrently with juvenile-onset diabetes mellitus. The condition best characterized is seen in Belgian tervurens. The depigmentation in this breed occurs chiefly on the pigmented skin and mucous membranes of the face and mouth in young adult dogs. Histologic examination of affected skin shows an epithelium devoid of both pigment granules and Dopa-positive cells. Electron microscopy confirms the lack of melanocytes in the lesions; their place is taken by Langerhans' or indeterminate dendritic cells. Antimelanocytic antibodies have been demonstrated in affected dogs and not in normal animals.

Arabian fading syndrome is an idiopathic noninflammatory leukoderma in Arabian horses, and occasionally other breeds, which is not associated, despite the connotations of its name, with constitutional signs. The depigmentation, which has been likened to vitiligo, commences in young animals as round, depigmented macules or patches on the muzzle, lips, around the eyes, and occasionally the anus, vulva, preputial sheath, and hooves. There may be repigmentation, but the lesions are usually permanent. Increased incidence in certain family lines suggests genetic predilection. The etiology is unknown, but an autoimmune pathogenesis is supported by the detection of circulating antimelanocytic antibodies.

In **cattle,** vitiligolike lesions have been described in Holstein–Friesians and in black Japanese cattle. In the latter, depigmentation developed in conjunction with hyperthyroidism.

Siamese cats may develop vitiligo. Antibodies to an 85-kDa surface antigen of melanocytes were demonstrated in four cats with vitiligo. No antibodies were detected in three normal Siamese cats tested.

Bibliography

Guaguère, E., Alhaidari, Z., and Ortonne, J.-P. Troubles de la pigmentation mélanique en dermatologie des carnivores. 2. Hypomélanoses et amélanoses. *Point Vét* **18:** 5–13, 1986.

Jimbow, K. *et al.* Some animal models of human hypomelanotic disorders. *Pigm Cell* **3:** 367–377, 1976.

Lamoreux, M. L., Gerrity, L., and Boissy, R. E. Animal models of human disease: Vitiligo in several animal species. *Comp Pathol Bull* **19:** 3–5, 1987.

Mahaffey, M. B., Yarbrough, K. M., and Munnell, J. F. Focal loss of pigment in the Belgian tervuren dog. *J Am Vet Med Assoc* **173:** 390–396, 1978.

Naughton, G. K. *et al.* Antibodies to surface antigens of pigmented cells in animals with vitiligo. *Proc Soc Exp Biol Med* **181:** 423–426, 1986.

Scott, D. W. "Large Animal Dermatology." Philadelphia, Pennsylvania, W.B. Saunders, 1988.

b. VOGT-KOYANAGI-HARADA SYNDROME A depigmenting condition in dogs partially resembles an extremely rare condition in humans, the Vogt–Koyanagi–Harada syndrome. The cause is unknown, although an autoimmune pathogenesis has been postulated. The canine lesions comprise bilateral panuveitis (see Chapter 4, The Eye and Ear) and cutaneous depigmentation, chiefly of the lips, nose, and eyelids. The scrotum and footpads are less often affected. Leukotrichia is a common finding around the areas of leukoderma. Occasionally, depigmented lesions become ulcerated, erythematous, or crusted. The histologic pattern of the cutaneous lesion is an interface dermatitis with minimal degeneration of basal keratinocytes, and a lichenoid inflammatory cell infiltrate in which large histiocytic cells usually predominate. Pigmentary incontinence is a common finding, but vacuolar change of the basal keratinocytes is not. This, and the histiocytic nature of the inflammatory infiltrate, are major features of differentiation from discoid and systemic lupus.

Bibliography

Bussanich, M. N., Rootman, J., and Dolman, C. L. Granulomatous panuveitis and dermal depigmentation in dogs. *J Am Anim Hosp Assoc* **18:** 131–138, 1982.

Morgan, R. V. Vogt–Koyanagi–Harada syndrome in humans and dogs. *Compend Cont Ed* **11:** 1211–1218, 1989.

c. Leukotrichia **Reticulated leukotrichia,** colloquially known as tiger stripe, is recognized in the Standardbred, Thoroughbred, and Quarter horse breeds. The lesions occur predominantly in yearlings and comprise linear crusts arranged in a cross-hatch pattern on the dorsal midline from the withers to the tail. Crusting is followed by alopecia and a regrowth of permanently white hair. The underlying skin is normally pigmented. The etiology and pathogenesis are unknown. **Spotted leukotrichia** occurs in the horse as multiple, often somewhat symmetrical, small circular areas of white hair. The spots occur most commonly on the rump and thorax, and Arabians have a predilection. The etiology and pathogenesis are unknown. **Hyperesthetic leukotrichia,** so called because the lesions are extremely painful, has been reported only in Californian horses. Single- or multiple-crusted lesions occur on the dorsal midline and heal leaving permanently white hairs. Leukotrichia, also termed poliosis, has been reported in **dogs** in association with Vogt–Koyanagi–Harada syndrome, tyrosinase deficiency in chow chows, and as an idiopathic, possibly heritable condition in a litter of Labrador retrievers. In the last example, the condition reversed.

Bibliography

Stannard, A. A. Hyperesthetic leukotrichia. *In* "Current Therapy in Equine Medicine 2" N.E. Robinson (ed.), pp. 647–648. Philadelphia, Pennsylvania, W.B. Saunders, 1987.

White, S. D., and Batch, S. Leukotrichia in a litter of Labrador retrievers. *J Am Anim Hosp Assoc* **26:** 319–321, 1990.

d. Alopecia Areata and Universalis These refer respectively to localized and generalized forms of an idiopathic condition of humans. A disease resembling alopecia universalis has been seen in a dog and horse; alopecia areata has been described in dog, cow, and horse. The predominant lesion is alopecia, but leukotrichia may affect the regrowing hair. These disorders are discussed in more detail under Immune-Mediated Diseases of the Skin (Section X).

e. Copper Deficiency This is the only significant pigmentary disorder of the integument in food-producing animals. Copper deficiency may be simple or conditioned by other dietary substances, particularly sulfate and molybdenum. Since copper is an essential constituent of tyrosinase, deficient animals show depigmentation of hair or wool. Affected cattle with normally black coats become rusty-brown and develop "spectacle" lesions round the eyes. Black sheep develop intermittent bands of light-colored wool corresponding to periods of restricted availability of copper (Fig. 5.24). The deficiency of copper also affects the physical nature of the wool or hair. In sheep, the wool has less crimp, prompting the colloquial name of string or steely wool. The straightness of the wool is due to inadequate keratinization, probably caused by imperfect oxidation of sulfhydryl groups in prekeratin, a process that involves copper (Fig. 5.25).

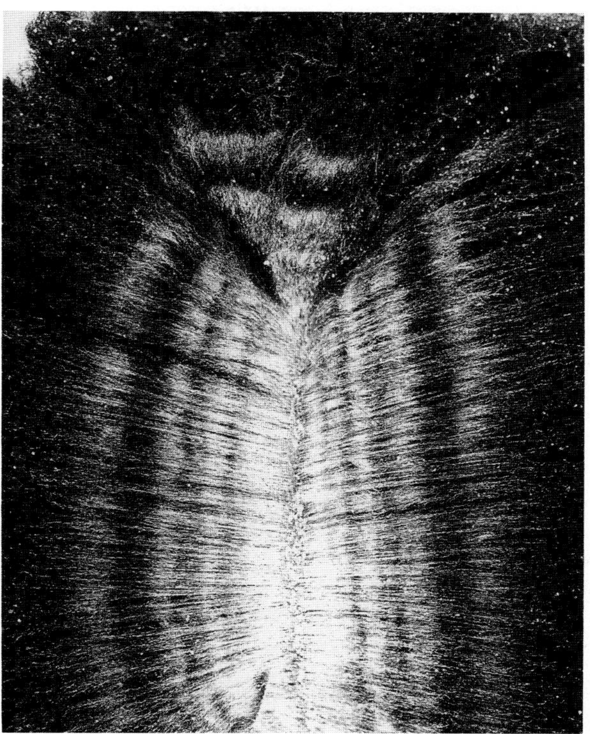

Fig. 5.24 Fleece of black-wooled sheep showing bands of achromotrichia corresponding to periods of molybdenum administration. (Courtesy of W. J. Hartley.)

VI. Physicochemical Diseases of Skin

The integument, with its large surface area in direct contact with the environment, is extremely vulnerable to chemical and physical injuries. The physical stresses include friction, pressure, vibration, electricity, high and low ambient temperatures, humidity, visible light, and ultraviolet, infrared, and ionizing radiation. Cutaneous reactions to visible light are discussed separately. Chemical toxins may exert their effect directly or indirectly, as in thallium poisoning.

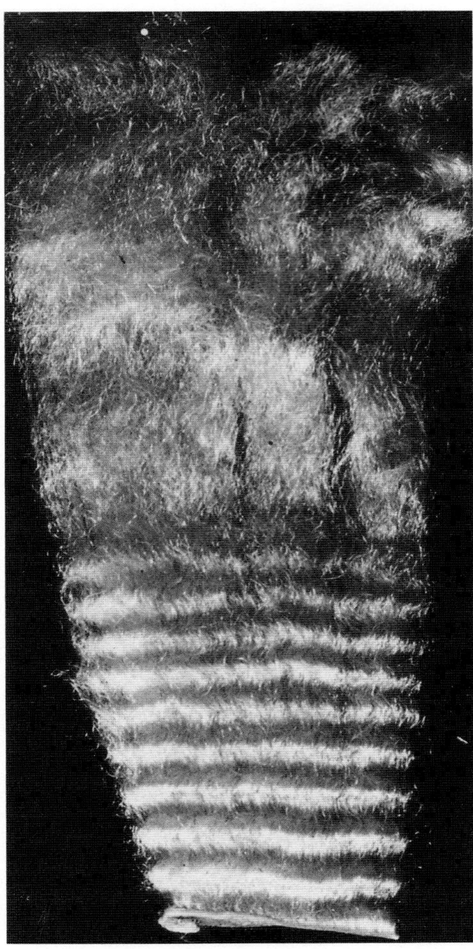

Fig. 5.25 Staple from sheep that received 1.0 mg copper per day, then 10 mg per day. Note stringiness and lack of crimp on deficient regime. (Courtesy of W. J. Hartley.)

A. Physical Injury to Skin

1. Mechanical, Frictional, Traumatic, and Psychogenic Injury

The hair coat of most domestic animals protects them from the blisters that so commonly develop in human skin subjected to prolonged pressure or frictional contact with a hard surface. The following examples tend to occur in heavy animals, in animals immobilized for various reasons such as paralysis, or in those exposed to a physically inimical environment.

Callosities occur when pressure or friction is applied to a localized area of skin. They are characterized by epidermal proliferation with prominent hyperkeratosis and represent a protective response of the integument to the physical injury. Grossly, callosities are well-circumscribed, raised, alopecic, gray, keratinous plaques. They tend to occur over bony prominences, particularly the hocks, elbows, lateral surfaces of the digits, and on the sternum. Callosities can develop in all domestic animal species, but are most common in dogs, particularly the giant breeds, and in pigs housed on concrete floors with inadequate bedding. Callosities are subject to secondary pyoderma and, in the pig, bursitis may also develop.

A **hygroma** is a false or acquired bursa which develops subcutaneously over bony prominences. They are most common in the giant breeds of dogs at pressure points such as the lateral aspect of the elbow, the greater trochanter of the femur, and the tuber coxae. Usually pressure induces a protective callus, but in some animals, persistent decubitus ulcers or recurrent hematoma formation eventually lead to the induction of a hygroma. The gross lesion is a variable-sized cystic cavity separated from the skin by loose connective tissue. The wall of the hygroma is dense connective tissue, which may have a smooth or a villous inner lining. The contents are mucinous and yellow to red, depending on the degree of hemorrhage. Histologically, the wall is composed of granulation tissue of varying maturity. A flattened layer of fibroblasts may give the appearance of an epithelial lining. The cavity may contain clumps of fibrin. Organization of fibrin deposits at the margin of the cavity gives rise to the grossly apparent villous projections. Occasionally these undergo cartilaginous metaplasia.

Decubitus ulcers are the result of ischemic necrosis which follows application of constant pressure to a localized area of skin. Consequently, decubitus ulcers are seen most often in animals that are recumbent for prolonged periods, for example, neurological or surgical patients. The "downer" cow is particularly prone to decubitus ulcers, as are emaciated or debilitated animals.

Intertrigo refers to localized dermatitis affecting folded areas of skin. The combined effect of friction, heat, maceration, bacterial proliferation, and irritation by retained secretions leads to superficial inflammation. Examples in dogs include facial, lip, vulvar, and tail-fold dermatitis. Body-fold dermatitis is particularly common in Chinese Shar-pei puppies. Udder–thigh dermatitis occurs in dairy cows, one German study reporting a prevalence of ~1% in first-calf heifers. Udder edema is a predisposing factor. Lesions are erythematous and swollen, sometimes ulcerated, and often emit an unpleasant odor. Healing may take up to 12 weeks.

Intensive rearing systems for **swine** production have potentiated the occurrence of a variety of traumatic lesions. Those in piglets probably result from contact with concrete floors in farrowing crates. The knees are the most common site affected, followed by the fetlocks and hocks. The nipples, particularly the anterior pair, are often involved, as is the tail. The lesions occur within a few hours of birth as circumscribed red macules followed by necrosis, ulceration, and crusting. The ulcers heal over 3 to 4 weeks, leaving no permanent defect. The supernumerary digits of the hind legs are subject to trauma in sows housed on concrete slats. Trauma resulting from vices such as tail biting, ear biting, and flank biting are increasingly common in growing pigs under intensive rearing systems of management. Wounds from a variety of causes, including fight wounds, often develop into subcutaneous abscesses in pigs. Cutaneous necrosis over the shoulder blades of sows may be caused by the drip-watering system used for cooling. An **ulcerative dermatitis** in **Belgian Landrace sows** occurs as a sporadic problem. Lesions occur on the ear margins, anterior aspects of the limbs, and around the teats. Focal dermoepidermal separation and lesions described as dyskeratotic epidermal cells prompted the comparison with erythema multiforme and drug reaction, but negative immunofluorescence tests rendered unlikely any of the common immune-mediated diseases. Ultrastructural examination ruled out the collagen dystrophies and heritable lesions of the dermoepidermal junction. The pathogenesis remains unknown. Another disorder, ear-tip necrosis, is discussed under The Ear (Chapter 4).

Cutaneous wounds are common in **horses** and can be largely attributed to the flighty temperament of the species. Exuberant granulation tissue, "proud flesh," is a relatively frequent and serious sequel to wounds of the distal limbs. A poor circulation, minimal soft tissue, lack of adequate drainage, and a tendency for excessive movement predisposes the distal limbs to the development of excess granulation tissue. The gross lesion, a tumorlike mass of red-brown tissue, must be distinguished from equine sarcoid, cutaneous habronemiasis, mycoses, and squamous cell carcinoma. Histologically, the presence of immature capillaries and capillary loops arranged perpendicular to the elongated fibroblasts and newly synthesized collagen are distinctive features. A superficial layer of granulation tissue may form in association with any of these conditions, and the entire lesion should be examined before a diagnosis is made.

Decubitus ulcers may develop on pressure points when horses are given insufficient bedding; saddle sores and girth galls result from poorly fitting tack.

Cats are particularly prone to subcutaneous abscesses or cellulitis as a result of fight wounds. The cat bite produces a puncture-type wound which seals over and enables the introduced bacteria (chiefly oral flora) to multiply in the damaged tissue.

Some cats, particularly Siamese, Burmese, Himalayan, and Abyssinian breeds, develop a **psychogenic** dermatitis known also as neurodermatitis and feline hyperesthesia syndrome. Psychogenic alopecia and dermatitis has two clinical forms. In one, affected cats lick and chew at a single site creating a well-demarcated erythematous, ulcerated lesion of variable size, which usually is located on an extremity, the abdomen, or flank. The lesion grossly resembles those of eosinophilic ulcer or plaque. The second, or alopecic form, is characterized by partial alopecia, broken hairs, and normal skin. The lesions often occur as a "stripe" along the dorsal midline but may involve perineal, genital, posteriomedial thigh, or abdominal areas. This distribution pattern is usually symmetrical, resembling feline endocrine alopecia, but the hairs in psychogenic alopecia are broken off rather than easily epilated as occurs in endocrine alopecia. Diagnosis is largely dependent on history and clinical signs. Biopsied skin in the alopecic form is usually normal, although wrinkling of the outer root sheath and intra- and perifollicular hemorrhage may reflect the trauma applied to the hairs. The inflammatory form has no distinctive histologic features, being an ulcerative, hyperplastic, superficial perivascular dermatitis.

Injection-site reactions are not uncommon, with both subcutaneously administered vaccines and therapeutic drugs responsible. The nodules or small tumors, which may ulcerate or fistulate, are often suspected to be neoplasms. Histologically, the lesion is composed predominantly of lymphocytes. These may be arranged around a central core of caseous necrosis. Irregular refractile or faintly basophilic crystals are often embedded in the eosinophilic debris. Plasma cells, macrophages, and multinucleate giant cells are also present, but in lesser numbers than the lymphocytes. The strong antigenic stimulus provided by the exogenous antigen sometimes results in the formation of germinal centers. These lesions, sometimes termed pseudolymphoma, may be differentiated from genuine lymphoma by the heterogeneity of the cell population and the lack of anaplastic characteristics in the lymphoid cells.

In **dogs,** traumatic injury to the skin is quite common and is often associated with compound fractures sustained in motor vehicle accidents. Dog-fight wounds tend to be tears rather than punctures, as occur in cats; consequently abscesses are a less common sequel. A plethora of **foreign bodies** may penetrate the canine integument, two of the more dramatic examples being foxtails and porcupine quills. The external ear canal and interdigital webs are favored sites.

Injection-site reactions also occur sporadically in the dog. Lesions are similar to those described in the cat. A particular reaction to rabies antigen is discussed under Immune-Mediated Diseases (Section X).

Myospherulosis is a rare form of foreign body reaction in which endogenous erythrocytes interact with an exogenous substance, such as antibiotics or ointments, or with endogenous fat. The lesions are subcutaneous nodules composed of sheets of large macrophages in which the cytoplasm is filled with homogeneous eosinophilic spherules. Negative for special stains such as periodic acid-Schiff, the spherules stain for endogenous peroxidase, thus establishing their identity as erythrocytes.

Dogs also develop **psychogenic dermatitis,** including foot chewing and licking, tail biting and flank sucking. The gross lesions may be slight, but the superficial excoriations often develop into pyoderma with secondary bacterial infection.

Acral lick dermatitis, otherwise known as lick granuloma or neurodermatitis, is a relatively common psychogenic dermatitis of dogs, particularly in large, active breeds that are prone to suffer boredom. Males are affected twice as frequently as females. The areas traumatized by persistent licking and chewing are most commonly the anterior carpus and metacarpus, followed by the radius, tibia, and metatarsus. Erythema and epidermal excoriations give rise to a single, well-circumscribed, ulcerated, oval plaque. Occasionally lesions are multiple.

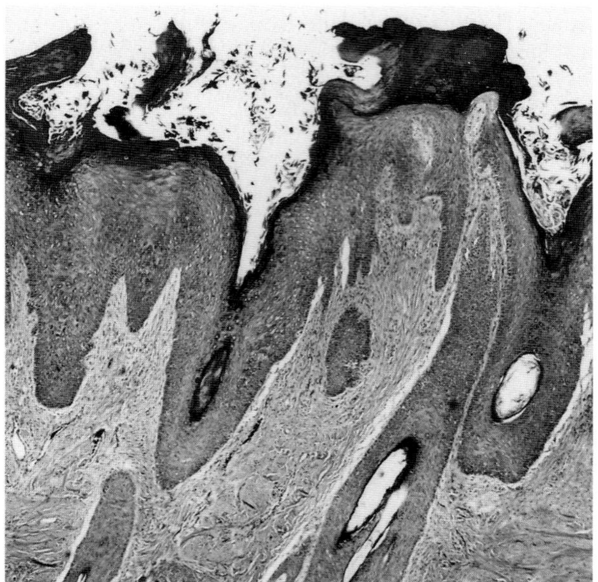

Fig. 5.26 Acral lick dermatitis. Dog. Hyperplastic dermatitis.

Secondary bacterial infection and pyoderma may result, and regional lymph nodes are often swollen. The lesions heal, leaving a well-circumscribed alopecic plaque, often with peripheral hyperpigmentation. Histologically, acral lick dermatitis is hyperplastic superficial perivascular dermatitis with marked orthokeratotic and parakeratotic hyperkeratosis, acanthosis, and rete ridge formation (Fig. 5.26). There is a sharp transition between the hyperplastic and ulcerated epidermis, which is covered by a thick scale-crust. Superficial dermal fibrosis is usually marked, and collagen fibers in dermal papillae are often arranged perpendicular to the surface epithelium. This vertical streaking of collagen (Fig. 5.27) is thought to result from chronic irritation. Perifolliculitis, folliculitis, and sometimes furunculosis may accompany the more superficial lesions. Plasmacytic infiltrates often surround the paratrichial sweat glands. Sebaceous glands and hair follicles appear hyperplastic.

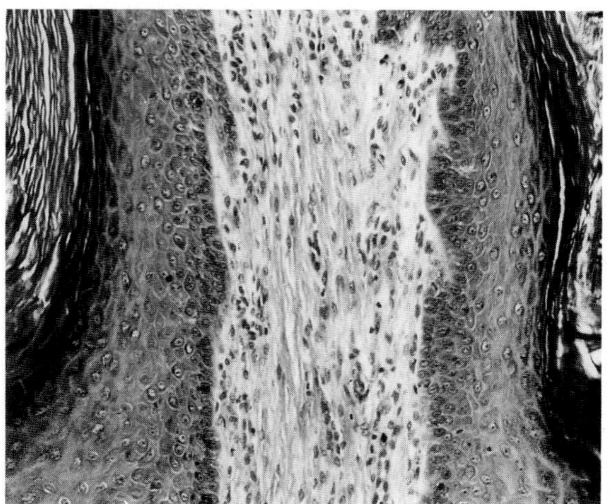

Fig. 5.27 Vertical alignment of collagen in acral lick dermatitis.

Traction alopecia has been reported in dogs. It results from low-grade local ischemia induced by traction. The traction force is applied by elastic ties, which some owners use to pull the forelock hair into a topknot. The lesions are focal patches of cutaneous atrophy and alopecia, which may become eroded and crusted. Histologically, the lesion is an atrophic dermatosis. The epidermis is thin and may occasionally show single-cell necrosis, erosion, or ulceration. The hair follicles are inactive and atrophic, appearing faded. Inflammation is minimal and is restricted to areas of surface ulceration.

Pyotraumatic dermatitis is a common complication of allergic dermatitis or any pruritic dermatosis. Breeds with a thick undercoat, such as German shepherds, Labradors, and St. Bernards, are particularly susceptible. Lesions, which are extremely painful, occur at sites of self-trauma, such as the dorsal rump in a flea-infested animal. Grossly, the initial lesions are erythematous, exudative, sharply demarcated patches, spreading extensively if not treated. Alopecia and hyperpigmentation are the typical sequelae. Pyotraumatic dermatitis is characterized histologically by epidermal ulceration. The denuded dermis is covered by a thick serocellular inflammatory crust. The predominantly neutrophilic reaction tends to be restricted to the area beneath the ulcer. Folliculitis and furunculosis are not features of classical pyotraumatic dermatitis.

Tail-tip necrosis is a disease of feedlot beef **cattle** in which slatted floor housing has been shown to be an important causal factor. The pathogenesis is presumed to be ischemia, secondary to compression of, or blunt trauma to, the more proximal parts of the tail. Clinically, there is alopecia, scaling, and crusting. Ulceration and suppuration are frequent sequelae. In early lesions, histologic examination reveals only perivascular edema and hemorrhage. In fully developed lesions, there is dermal scarring, follicular atrophy, vascular wall hypertrophy, and fragmentation of extravasated erythrocytes.

Bibliography

Brennan, K. E., and Ihrke, P. J. Grass awn migration in dogs and cats. A retrospective study of 182 cases. *J Am Vet Med Assoc* **182:** 1201–1204, 1983.

Coignoul, F. L., Bertram, T. A., and Martineau, G. P. Pathology of an ulcerative dermatitis in Belgian Landrace sows. *Vet Pathol* **22:** 306–310, 1985.

Hargis, A. M. *et al.* Myospherulosis in the subcutis of a dog. *Vet Pathol* **21:** 248–251, 1984.

Hendrick, M. J., and Dunagan, C. A. Focal necrotizing granulomatous panniculitis associated with subcutaneous injection of rabies vaccine in cats and dogs: 10 cases (1988–1989). *J Am Vet Med Assoc* **198:** 304–305, 1991.

Penny, R. H. C. *et al.* Clinical observations of necrosis of the skin of suckling piglets. *Aust Vet J* **47:** 529–537, 1971.

Reinke, S. I. *et al.* Histopathologic features of pyotraumatic dermatitis. *J Am Vet Med Assoc* **190:** 57–60, 1987.
Rosenkrantze, W. S., Griffin, C. E., and Walder, E. J. Traction alopecia in the canine: Four case reports. *Cal Vet* **43:** 7–8, 12, 1989.
Scott, D. W., and Walton, D. K. Clinical evaluation of a topical treatment for canine acral lick dermatitis. *J Am Anim Hosp Assoc* **20:** 565–570, 1984.
Sigmund, von H. M., Klee, M., and Schels, H. Udder–thigh dermatitis of cattle: Epidemiological, clinical, and bacteriological investigations. *Bov Pract* **18:** 18–23, 1983.

2. *Cold Injury*

Most cold-induced cutaneous lesions (frostbite) result not only from direct freezing and disruption of the cells, but more important, from vascular injury and resultant tissue anoxia. In experimental frostbite lesions in Hanford miniature swine, vacuolation of keratinocytes was the earliest change in the epidermis, followed by spongiosis, epidermal necrosis, and separation of the necrotic epithelium from the dermis. Hyperemia and hemorrhage were also early lesions. Inflammatory changes, comprising neutrophilic infiltration and necrotizing vasculitis, occurred between 6 and 46 hr post injury. Thrombosis of small arterioles increased in severity up to 1 week post injury. By 2 weeks, considerable epithelial regeneration had taken place, either as a complete replacement or as crescents beneath the necrotic epidermis.

Cutaneous injury resulting from cold is uncommon in well-nourished, healthy domestic animals. Well-acclimatized long-haired animals can tolerate temperatures of −50°C for indefinite periods. Cold injury occurs most commonly on the tips of the ears and tail of cats, the scrotum of male dogs and bulls, and the tips of the ears, tail, and teats in cattle. The teats are particularly vulnerable if cows are turned out into the cold with wet udders. The gross lesions include alopecia, scaling, and pigmentary alterations in the skin or hair, or both. In severe cases, the ischemic necrosis results in dry gangrene and sloughing of the affected part.

Bibliography

Schoning, P., and Hamlet, M. P. Experimental frost-bite in Hanford miniature swine. I. Epithelial changes. *Br J Exp Pathol* **70:** 41–49, 1989.
Schoning, P., and Hamlet, M. P. Experimental frost-bite in Hanford miniature swine. II. Vascular changes. *Br J Exp Pathol* **70:** 51–57, 1989.

3. *Thermal Injury*

Heat may be applied to the skin in a variety of forms, and depending on duration and intensity, will produce mild to severe necrotizing lesions. Dry heat causes desiccation and carbonization, while moist heat causes "boiling" or coagulation. Thermal injury in domestic animals may be caused by scalds, for example, domestic pets accidentally scalded by hot water or oil; flame, as in barn fires; friction, as in rope scalds; electrical burns, as in cats or dogs chewing electrical wires; heat, as in injuries from heating pads; and lightning. Occasionally animals struck by lightning show a jagged line of singed hair running down a shoulder or flank (Fig. 5.28). This finding is valuable in establishing an otherwise difficult diagnosis.

Burns are classified into four degrees according to depth of injury. If the epidermis alone is affected, the burn is first degree. The heated areas are erythematous and edematous as a result of vascular reaction in the dermis, but vesicles do not form. The epithelial cells show no morphological sign of injury, although there may be surface desquamation after a few days. In the second-degree burn, the coagulative effects of heat on the epidermal cells are evident. The cytoplasm of the epithelial cells is hypereosinophilic, and the nuclei are shrunken or karyorrhexic. The vascular changes are more prominent than in lesser burns, with marked dermal edema and spongiosis. Vesicles and bullae form in the epidermis, often at the dermoepidermal junction (Fig. 5.29). The bullae contain serum, granular debris, and leukocytes. Healing can be complete if secondary infection does not lead to deeper injury. In third-degree burns, the destructive effect of the heat extends full thickness, causing coagulation necrosis of connective tissues, blood vessels, and adnexa. Heat of sufficient in-

Fig. 5.28 Path of lightning strike. Cow. (Courtesy of M. A. Hayes.)

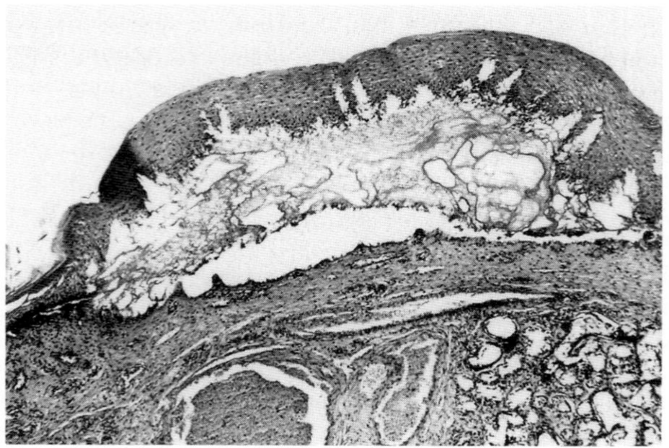

Fig. 5.29 Subepidermal bulla caused by burn.

tensity or duration to penetrate this deeply usually desiccates and chars the outer epidermis. The epidermis is coagulated, and the same change in the dermis produces a swollen amorphous agglomeration of the connective tissues, about which there is an acute inflammatory reaction. The necrotic tissue sloughs, and the defect is filled in by granulation tissue. Permanent scarring with loss of adnexa results. Fourth-degree burns are similar in character to those of third-degree burns, but penetrate below the dermis to and beyond the subcutaneous fascia; their local consequences depend on what lies underneath.

Bibliography

Gieser, D. R., and Walker, R. D. Management of large animal thermal injuries. *Compend Cont Ed* **7:** S69–S78, 1985.

Scarratt, W. K. *et al.* Cutaneous thermal injury in a horse. *Eq Pract* **6:** 13–17, 1984.

Swaim, S. F., Lee, A. H., and Hughes, K. S. Heating pads and thermal burns in small animals. *J Am Anim Hosp Assoc* **25:** 156–162, 1989.

B. Chemical Injury to Skin

1. Primary Contact Irritant Dermatitis

One of the important functions of the skin is to provide protection against external noxious agents. The stratum corneum is the major protective barrier. Its integrity is highly dependent on its water content, which in turn is protected by the lipids of the stratum corneum.

Penetration of the skin is enhanced by physical damage to the stratum corneum, chemical alteration of the barrier components, overhydration of the stratum corneum by excess moisture, increased temperature, or alterations of the skin pH. Irritant substances may induce damage by altering the water-holding capacity of the keratin layer or by penetrating the epidermis and directly damaging the cells. Irritant substances vary markedly in their potency. Some, such as strong acids or alkalis, induce immediate and severe tissue damage; others, such as mild detergents or soaps may require repetitive applications and covering of the area to assist penetration before lesions develop.

Primary irritant contact dermatitis must be distinguished from allergic contact dermatitis (discussed in Section X, Immune-Mediated Dermatoses). Irritant contact dermatitis is the more common condition in animals. It may occur in any species but is most frequent in the horse, cow, and dog. The types of agents capable of causing direct cutaneous damage include acids, such as carbolic or sulfuric, alkalis, cresol tars, paints, kerosene, turpentine, antiseptics, and insecticides. An example of the last mentioned is "flea collar" dermatitis of dogs and cats. Feces and urine are also potential irritants.

The distribution of gross lesions of irritant contact dermatitis in the dog and cat typically involves the glabrous skin of the ventral abdomen, axilla, medial thigh, perianal and perineal areas, footpads (Fig. 5.30), ventral tail, chin, and inner aspect of the pinnae. The hairy skin is involved only if the irritant substance is in an aerosol or liquid form. The "flea collar" dermatitis lesions occur as a band around the neck corresponding to the position of the offending collar. In horses, lesions occur most commonly on the muzzle, lower limbs, and in areas of contact with the riding tack. Horses with diarrhea may develop a severe irritant dermatitis on the soiled perineum.

The gross lesions, which typically have a sudden onset, are marked erythema, swelling, a transient papular–vesicular stage which leads to ulceration and, in severe cases, sloughing of the affected skin (Fig. 5.30). The sequelae include alopecia, scarring, and alteration in skin and hair pigmentation. In most species of domestic animals, hyperpigmentation occurs, but in horses, leukoderma or leukotrichia may be a permanent result of irritant contact dermatitis.

The histologic lesions of irritant contact dermatitis are not pathognomonic, nor can they be distinguished from those of allergic contact dermatitis (Fig. 5.31). Both are characterized by superficial perivascular dermatitis, either spongiotic or hyperplastic. The inflammatory infiltrate is variable in nature, probably reflecting such factors as chronicity, self-trauma and secondary infection. The diagnosis of irritant contact dermatitis depends largely on the history

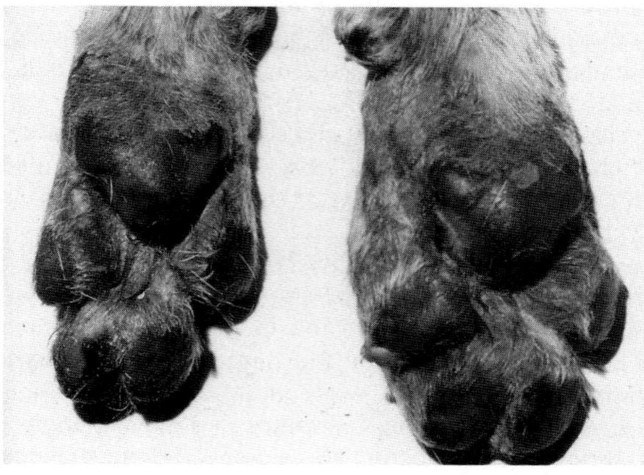

Fig. 5.30 Irritant (alkali) contact dermatitis. Dog. Footpad swelling and erythema. (Courtesy of R. P. Johnson.)

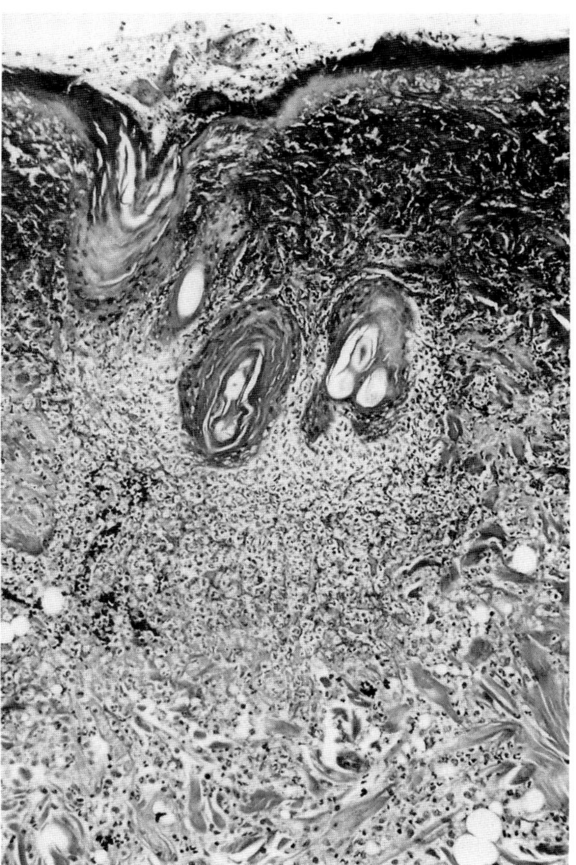

Fig. 5.31 Irritant contact dermatitis. Varsol burn. Dog. Coagulation necrosis.

and the clinical signs, particularly the distribution of the lesions.

An unusual irritant contact dermatitis caused by percutaneous absorption of a landscaping product containing calcium chloride has occurred naturally and has been reproduced experimentally in the dog. The multifocal, flat-topped, and centrally ulcerated papules and small nodules predominantly affect glabrous skin, such as the lips, axilla, inguinal, and interdigital skin. Histologically, the lesions comprise multifocal granulomata centered on degenerate, mineralized collagen fibers. Ultrastructural examination of experimental lesions indicated that mineral was deposited within the collagen bundles within 24 hr of initial skin contact. The main differential diagnosis is calcinosis cutis of hyperadrenocorticism, to which the lesions are histologically identical.

In **swine,** cutaneous erythema and pruritus has been observed within 48 hr of tiamulin administration. The most severely affected pigs were recumbent and developed a fatal necrolytic dermatitis. The areas most severely affected were those in contact with feces and urine. Histologic lesions included full-thickness epidermal necrosis, intraepidermal pustules, and serocellular crusting. It was hypothesized that the lesions represent a severe form of irritant contact dermatitis to tiamulin or one of its metabolites in the excreta. Skin lesions regressed when the drug was withdrawn. An eosinophilic dermatitis was observed in pigs following heavy salting of the pen floors.

Bibliography

Andersson, P., and Petaja, E. Profound eosinophilic dermatitis in swine caused by sodium chloride. *Nord Vet Med* **20:** 706–707, 1968.

Hunziker, N. Histology of contact dermatitis in humans and experimental animals. *In* "Advances in Modern Toxicology," Vol. 4, Dermatotoxicity and Pharmacology. F.N. Marzulli and H.I. Maibach (eds.), pp. 373–412. Washington, D.C., Hemisphere Publishing, 1977.

Kunkle, G. A. Contact dermatitis. *Vet Clin North Am* **18:** 1061–1068, 1988.

Lapèrle, A. Dermatite aigue chez des porcs traités à la tiamuline. *Med Vét Quebec* **20:** 20–22, 1990.

Lindberg, M. Studies on the cellular and subcellular reactions in epidermis of irritant and allergic dermatitis. *Acta Derm-Venereol (Suppl.)* **105:** 7–37, 1982.

Muller, G. H. Contact dermatitis in animals. *Arch Dermatol* **96:** 423–426, 1967.

Nesbitt, G. H., and Schmitz, J. A. Contact dermatitis in the dog: A review of 35 cases. *J Am Anim Hosp Assoc* **13:** 155–163, 1977.

Norsworthy, G. D. Localized disinfectant toxicity. *Feline Pract* **7:** 49–50, 1977.

Paradis, M., and Scott, D. W. Calcinosis cutis secondary to percutaneous penetration of calcium carbonate in a Dalmatian. *Can Vet J* **30:** 57–59, 1989.

Pilarczyk, J. P. Chemical burn and toxicity in a dog from misuse of a flea dip. *Canine Pract* **6:** 51–52, 1979.

Schick, M. P., Schick, R. O., and Richardson, J. A. Calcinosis cutis secondary to percutaneous penetration of calcium chloride in dogs. *J Am Vet Med Assoc* **191:** 207–211, 1987.

2. *Thallotoxicosis*

The heavy metal thallium is a potent toxin, particularly the thallous salt. Thallium has been used extensively as a rodenticide. Its sale for this purpose is banned in some countries, including the United States of America, but elsewhere it is still employed, and its use may increase as pests develop resistance to more selective poisons.

In domestic animals, accidental or malicious thallium poisoning is rare. It occurs chiefly in dogs, less often in cats, and is reported in sheep, cattle, and pigs. The minimum toxic dose is 8.0 mg per kg of body weight, but the poison is cumulative. Absorption occurs rapidly from the gastrointestinal tract. Percutaneous and respiratory tract absorption is also possible. The toxin is disseminated widely in the body and is persistent, being very slowly excreted in feces and urine.

The mechanism of toxicity is not fully understood. There are two main hypotheses. The first holds that thallium exerts its toxic effect by combining with sulfhydryl groups, a mechanism common to many heavy metals. The second, which is based on the similarity of ionic radii between thallium and potassium, suggests that thallium may replace potassium in many critical biochemical functions, thus acting as a general cellular poison. The toxic effect may be in part due to thallium interacting adversely with derivatives of riboflavin.

The clinical effects depend on the dose and rapidity of administration. Acute thallotoxicosis is characterized by severe gastrointestinal irritation and neurological signs, including motor paralysis, which may lead to death by respiratory failure. Animals may survive the acute episode to develop the chronic syndrome or may bypass the acute disease altogether. The chronic syndrome is characterized by cutaneous, renal, and neurological abnormalities, progressive debilitation, and death.

The cutaneous lesions develop between 7 and 10 days after ingestion of thallium and principally affect frictional areas. The pattern of skin involvement in cats and dogs is characteristic, beginning at the commissures of the lips or nasal cleft, occasionally on the ear margins, and expanding over the face and head. The mucous membranes are characteristically brick-red in color and may be ulcerated. Lesions also develop on the interdigital skin, footpads, axillae, inguinal areas, perineum, and lateral extensor surfaces. The lesions are marked erythema, scaling, alopecia, exudation, and crusting. The layers of scale-crust exfoliate with attached hairs to leave a raw, oozing surface. The paws often become very swollen. In more chronic cases, thick scales on the footpads resemble "hard pad" disease, conventionally associated with canine distemper. In less severely affected animals, ease of depilation may be the only clinical indication of thallium poisoning.

The pathogenesis of the alopecia is not fully understood. Whereas thallium enters the hair, as do other heavy metals, by binding to sulfhydryl groups in the keratin, this is unlikely to be destructive to the hair follicle. Probably thallium interferes with the energy metabolism in the rapidly dividing matrix cells of anagen follicles. In experimental infections of rats, a rapid decline in the mitotic rate is followed by necrosis of the matrix cells within 48 hr of intoxication. The follicle passes into an abnormal catagen phase followed by complete involution (telogen). However, no club attachment is formed, and the hairs are rapidly shed. If the animal survives, hair growth is resumed. Thallium also severely alters the keratinization process, and both the surface and external root-sheath epithelium demonstrate marked parakeratotic hyperkeratosis.

The microscopic lesions in the skin are dominated by massive, diffuse parakeratotic hyperkeratosis, which affects both surface and external root-sheath epithelia. There is accompanying follicular plugging, hypogranulosis, and epidermal hyperplasia. Neutrophil exocytosis and the formation of spongiform pustules occur in both surface and follicular epithelia (Fig. 5.32). Partial- or full-thickness necrosis of the surface epithelium may also occur. The dermal lesions include marked hyperemia, edema, erythrocytic exocytosis, and infiltration of neutrophils and mononuclear cells. There may be focal necrosis of sweat and sebaceous glands. Hair follicles are mostly

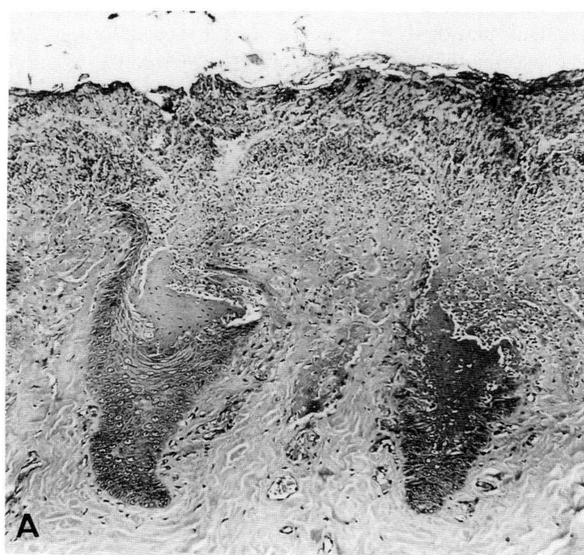

Fig. 5.32A Thallium toxicity. Dog. Hyperplastic dermatitis with necrosis.

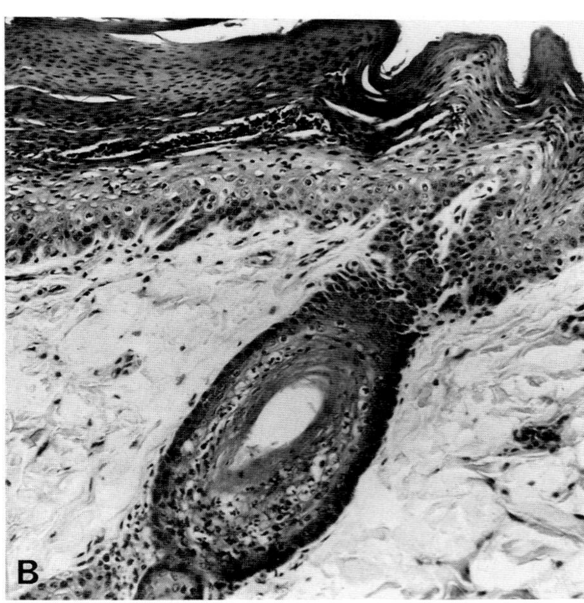

Fig. 5.32B Thallium toxicity. Dog. Marked parakeratosis.

in catagen or telogen. Degenerative changes are noted in anagen follicles.

Histologic lesions in other tissues include multifocal necrosis of both myocardial and skeletal muscle fibers, nephrosis, pulmonary edema, reticuloendothelial hyperplasia, and lymphoid depletion of spleen and lymph nodes. Secondary bacterial bronchopneumonia may occur as a result of the damage to ciliated epithelia and resultant disturbance of the mucociliary apparatus. Hemorrhagic gastroenteritis occurs in acute thallotoxicosis. Focal suppurative pancreatitis is described in several animals, but its causal relationship to thallium is not established. Ulcerative esophagitis follows dilation secondary to neuronal damage. Lesions in the central nervous system include neuronal chromatolysis, neuronophagia, and severe edema with little glial reaction. Myelinated peripheral nerves have degenerative lesions, including focal distension of myelin sheaths and swelling and occasional fragmentation of axons.

The differential diagnosis of the cutaneous lesions includes autoimmune diseases, such as pemphigus vulgaris and bullous pemphigoid, and immune-mediated diseases, such as toxic epidermal necrolysis. Candidiasis and lymphosarcoma must also be ruled out. The microscopic lesions are characteristic, but confirmation requires the demonstration of thallium in the urine.

Bibliography

Cavanagh, J. B., and Gregson, M. Some effects of a thallium salt on the proliferation of hair follicle cells. *J Pathol* **125:** 179–191, 1978.

Chandler, H. A., and Scott, M. A review of thallium toxicology. *J R Nav Med Serv* **72:** 75–79, 1986.

Coyle, V. Diagnosis and treatment of thallium toxicosis in a dog. *J Small Anim Pract* **21:** 391–397, 1980.

Douglas, K. T., Bunn, M. A., and Baindur, S. R. Thallium in biochemistry. *Inter J Biochem* **22:** 429–438, 1990.

Schwartzman, R. M., and Kirschbaum, J. O. The cutaneous histopathology of thallium poisoning. *J Invest Dermatol* **39:** 169–173, 1962.

Zook, B. C., and Gilmore, C. E. Thallium poisoning in dogs. *J Am Vet Med Assoc* **151:** 206–217, 1967.

Zook, B. C., Holzworth, J., and Thornton, G. W. Thallium poisoning in cats. *J Am Vet Med Assoc* **153:** 285–299, 1968.

3. Arsenic Toxicosis

Arsenical poisoning is an important toxicosis in domestic animals, particularly cattle and dogs. In humans, inorganic arsenic is associated with a variety of dermatoses, usually developing as a result of local irritation rather than from systemic poisoning. Similar contact lesions occur in animals sprayed or dipped in a concentrated arsenic solution. Contact lesions occur also in dogs lying on heavily contaminated ground. Lesions include erythema and epidermal necrosis, leading to the formation of indolent ulcers. Lesions may affect oral cavity, lips, other mucocutaneous junctions, and the feet. Chronic arsenic poisoning in farm animals is associated with ill thrift and a dry, seborrheic, alopecic coat.

Bibliography

Bruere, S. N. Arsenical poisoning in farm dogs. *N Z Vet J* **28:** 220, 1980.

Evinger, J. V., and Blakemore, J. C. Dermatitis in a dog associated with exposure to an arsenic compound. *J Am Vet Med Assoc* **184:** 1281–1282, 1984.

Selby, L. A. *et al.* Epidemiology and toxicology of arsenic poisoning in domestic animals. *Environ Health Perspect* **19:** 183–189, 1977.

4. Mercury Toxicosis

Organomercurial toxicosis in domestic animals is associated chiefly with neurologic and renal disorders and is discussed with the appropriate systems. In chronic poisoning in **cattle**, skin manifestations including pustules, ulcers, hyperkeratosis, and depilation of hair at the tail head are described, but their pathogenesis is poorly understood. **Horses** ingesting mercury-treated seed grain develop total body depilation, followed by loss of the long hairs of mane, tail, and forelock. The hooves are not affected, and the cutaneous lesion is mild scaling. Experimental chronic methylmercury intoxication in horses produces an exudative dermatitis, but histologic lesions are not described. Local toxic contact dermatitis follows application of mercurial-containing counterirritants to the legs.

Bibliography

Irving, F., and Butler, D. G. Ammoniated mercury toxicity in cattle. *Can Vet J* **16:** 260–264, 1975.

Seawright, A. A., Roberts, M. C., and Costigan, P. Chronic methylmercurialism in a horse. *Vet Hum Toxicol* **20:** 6–9, 1978.

5. Cutaneous Iodism

Generalized seborrhea sicca is reported in horses and cattle accidentally overdosed with iodine-containing drugs or medicated feed. In experimental toxicosis of calves, the cutaneous lesions were limited to scaly patches without alopecia. Conversely, suspected iodism in a horse produced generalized alopecia, sparing only the face, mane, and tail.

Bibliography

Fadok, V. A., and Wild, S. Suspected cutaneous iodism in a horse. *J Am Vet Med Assoc* **10:** 1104–1106, 1983.

Mangkoewidjojo, S., Sleight, S. D., and Convey, E. M. Pathologic features of iodide toxicosis in calves. *Am J Vet Res* **41:** 1057–1061, 1980.

6. Selenium Toxicosis

Selenium toxicosis occurs in horses, cattle, sheep, and pigs, chiefly as a result of the ingestion of plants which have accumulated toxic levels of selenium, but occasionally as a result of accidental overdosage of a selenium supplement. Selenium is widely distributed in soils at concentrations which range from 0.01 parts per million (ppm) or less to 500 ppm or more. The areas of high soil concentration are known as poison strips. They are particularly extensive in Wyoming and South Dakota in the United States of America and in western Canada, but occur in parts of Australia, Ireland, Mexico, and Israel, among others.

Plants are divisible into seleniferous and nonseleniferous species. **Seleniferous plants** can selectively concentrate selenium in their foliage and seeds as compared with nonseleniferous species grown under the same conditions. The seleniferous species are subdivided into obligate accumulators, which require high levels of selenium for survival, and facultative accumulators. The former, which include members of the genus *Astragalus*, *Machaeranthera*, and *Stanleya*, may accumulate over 1000 ppm selenium. Because of their high requirement for selenium they are known as indicator species. Facultative or secondary selenium accumulators, such as *Aster*, *Atriplex*, *Castilleja*, and *Gutierrezia*, take up lesser amounts of selenium. Many nonseleniferous weeds, crop plants, and grasses are capable of passively accumulating selenium if growing on soils with a high selenium content. Indicator plants also increase the availability of selenium to non-seleniferous plants by converting insoluble selenites to soluble selenates and returning these to the soil. Selenium poisoning can occur whenever seleniferous plants are eaten, irrespective of levels of selenium in the soil, and it can occur whenever the levels of water-soluble selenium in the soil are high, irrespective of the botanical composition.

Seleniferous plants are not palatable, and indigenous stock learn to avoid them. Selenium poisoning occurs chiefly in animals which are traveling or which have been recently introduced, and in indigenous animals forced in times of scarcity to eat the seleniferous plants. Clinically, there are acute and chronic syndromes associated with the ingestion of seleniferous plants such as *Astragalus* and *Oxytropis*, but the plants contain toxic factors other than selenium. The acute toxicity causes severe gastrointestinal and cardiovascular signs, with mortality in some instances approaching 100%.Two different syndromes have been described as chronic selenium poisoning. **Blind staggers** is characterized by neurological signs which probably are not due to selenium alone but to other toxic principles in the seleniferous plants, such as miserotoxin. The second syndrome is called **alkali disease** because it was originally believed that the pH of the selenium-rich soils was a factor in its pathogenesis. Unlike blind staggers, alkali disease is reproducible in ungulates fed sublethal concentrations of selenium. Chronic experimental poisoning fails to induce cutaneous lesions in swine.

Horses and cattle chronically intoxicated with selenium become emaciated and develop partial alopecia and a general roughness of coat. There is loss of the long hairs in the mane, forelock, and tail of horses (leading to the name bob-tail disease), and loss of the long tail hairs in cattle. Sheep do not show cutaneous lesions, although in Australia fleece shedding has been attributed to selenium toxicity on some properties. Selenium toxicity is also suspected in alopecias of the beard and flanks of goats in the western United States. In all species, lesions commencing at the coronary band may lead to separation and shedding of the hoof or to the formation of dystrophic grooves which parallel the coronary band (Fig. 5.33). Each groove corresponds to a period of poisoning. The presence of internal lesions, such as nephrosis, myocardial degeneration, and hepatic fibrosis in chronically poisoned animals, is subject to conflicting reports.

The mechanism by which selenium might exert its effects on the integument and appendages is not known, but conceivably, being competitive with sulfur, it modifies the

Fig. 5.33 Selenium poisoning producing characteristic rings and grooves in hoof wall. Horse. (Courtesy of Queensland Department of Agriculture.)

structure of keratin. Relatively large amounts of selenium are deposited in hair after poisoning.

Diagnosis requires the demonstration of high levels of selenium in the blood (>2 ppm) or hair (>10 ppm) but there is marked individual variation, and some animals with high levels of selenium may not show signs of toxicosis. Dietary levels of selenium should also be determined.

Bibliography

Casteel, S. W. *et al*. Selenium toxicosis in swine. *J Am Vet Med Assoc* **186:** 1084–1085, 1985.

Crinion, R. A. P., and O'Connor, J. P. Selenium intoxication in horses. *Irish Vet J* **32:** 81–86,1978.

Gardiner, M. R. Chronic selenium toxicity studies in sheep. *Aust Vet J* **42:** 442–448, 1966.

Harr, J. R., and Muth, O. H. Selenium poisoning in domestic animals and its relationship to man. *Clin Toxicol* **5:** 175–186, 1972.

Herigstad, R. R., Whitehair, C. K., and Olson, O. E. Inorganic and organic selenium toxicosis in young swine: Comparison of pathologic changes with those of swine with vitamin E–selenium deficiency. *Am J Vet Res* **34:** 1227–1238, 1973.

James, L. F., and Shupe, J. L. Selenium accumulators. *In* "Current Veterinary Therapy. Food Animal Practice," 2nd Ed. J.L. Howard (ed.), pp. 394–396. Philadelphia, Pennsylvania, W.B. Saunders, 1986.

Knott, S. G., and McCray, C. W. Two naturally occurring outbreaks of selenosis in Queensland. *Aust Vet J* **35:** 161–165, 1959.

Maag, D. D., Osborn, J. S., and Clopton, J. R. The effect of sodium selenite on cattle. *Am J Vet Res* **21:** 1049–1053, 1960.

Miller, W. T., and Schoening, H. W. Toxicity of selenium fed to swine in the form of sodium selenite. *J Agric Res* **56:** 831–842, 1938.

Rosenfeld, I., and Beath, O. A. Pathology of selenium poisoning. *Wyoming Agric Exp Stn Bull* No. 275, p. 27, 1946.

Traub-Dargatz, J. L., and Hamar, D. W. Selenium toxicity in horses. *Compend Cont Ed* **8:** 771–776, 1986.

Van Kampen, K. R., and James, L. F. Manifestations of intoxication by selenium-accumulating plants. *In* "Effects of Poisonous Plants on Livestock" R.F. Keeler, K.R. Van Kampen, and L.F. James (eds.), pp. 135–138. New York, Academic Press, 1978.

7. Organochlorine and Organobromine Toxicoses

Organochlorine and organobromine compounds implicated in toxicities causing, among others, cutaneous lesions, include chlorinated naphthalenes, polybrominated biphenyls (PBBs), and dibenzofurans. Polychlorinated biphenyls (PCBs) are the cause of an important industrial dermatitis of humans known as chloracne.

Chlorinated naphthalene toxicosis (X-disease, bovine hyperkeratosis) is largely of historical interest. During the 1940s and early 1950s, chlorinated naphthalene was a common additive in many petroleum products. As such, it was a frequent feed contaminant, and poisonings of cattle were common. The most serious losses occurred when lubricants containing pentachloronaphthalene were used in feed-pelleting machines. Chlorinated naphthalene can also produce cutaneous lesions by local administration, but systemic signs are absent. Chlorinated naphthalene was used in wood preservatives, and animals housed in barns made of wood treated with these preservatives developed the characteristic skin changes. These chemicals are no longer in use, and there has not been an important outbreak in recent years. However, between 1948 and 1952, the losses from mortality alone in the United States of America were estimated at $2 to 4 million per annum. The disease also occurred in Germany, Australia, and New Zealand.

The poison is cumulative and, whether it is ingested as a single dose or over a considerable period, the disease produced is chronic. The first sign of poisoning is a decrease in vitamin A levels in the plasma. The lesions of chlorinated naphthalene toxicity result from a poorly understood interference with vitamin A metabolism and resemble the lesions of vitamin A deficiency.

Cattle are the most susceptible species. The first obvious clinical sign is increased lacrimation and salivation. The appetite is lost, and affected animals become depressed and emaciated. The skin of the neck, shoulders, and perineum becomes thickened, with marked scaling. Lesions gradually generalize, sparing only the legs. In the most chronic cases, the skin is lichenified, alopecic, and covered in deeply fissured plaques of hyperkeratotic scale. Horn growth is interfered with, and an indentation is present at the base. The skin lesions do not become well developed until 2 to 3 months after ingestion and, if the dose is large enough, the animal may die before the skin lesions are severe. Concurrent severe infections with bovine papular stomatitis virus, originally considered part of the primary disease process, probably reflect altered mucosal resistance or decreased immunocompetence.

The histologic lesions are marked hyperkeratosis of surface epithelia and follicular keratosis. Squamous metaplasia of the columnar epithelial lining of ducts and glands occurs throughout the body.

Cutaneous lesions in **cats** exposed to wood preservatives have been ascribed to chlorinated naphthalene toxicity. The lesions include bilateral alopecia and encrustations on the eyelid and around the nostrils.

Polybrominated biphenyls and **tetrachlorodibenzo-*p*-dioxin (TCDD)** have been blamed for episodes of accidental poisoning of cattle and horses, but responsibility for the cutaneous lesions was unconfirmed.

Bibliography

Case, A. C., and Coffman, J. R. Waste oil: Toxic for horses. *Vet Clin North Am* **3:** 273–277, 1973.

Fries, G. F. The PBB episode in Michigan: An overall appraisal. *CRC Crit Rev Toxicol* **16:** 105–156, 1985.

Hansel, W., Olafson, P., and McEntee, K. The isolation and identification of the causative agent of bovine hyperkeratosis (X-disease) from a processed wheat concentrate. *Cornell Vet* **44:** 94–101, 1955.

Olson, C. Bovine hyperkeratosis (X-disease, highly chlorinated naphthalene poisoning). Historical review. *Adv Vet Sci* **13:** 101–157, 1969.

Wagener, K. Possible effects of German wood preservative on cats and dogs, with special reference to hyperkeratosis. *J Am Vet Med Assoc* **20:** 145–147, 1952.

8. Mimosine Toxicosis

Mimosine is a toxic amino acid occurring in *Mimosa pudica* and *Leucaena leucocephala* (formerly known as *L. glauca*). The latter is a pantropical shrub legume which has potential as a pasture plant in subtropical and tropical regions. Poisoning, characterized by alopecia, occurs in a number of countries and goes by a number of local names: *jumbey* in the West Indies, *lamtoro* in Indonesia, and *koa haole* in Hawaii. The mechanism of toxicity is complex and probably affects various cells differently. Mimosine may act as an amino acid antagonist, by complexing with pyridoxal phosphate, or by chelating heavy metals, thus inhibiting metalloenzymes.

Mimosine toxicity occurs in horses, cattle, pigs, and sheep and has been experimentally reproduced in laboratory animals. There is considerable variation in the toxicity of the plant in different parts of the world. The disease tends to occur in countries of the South Pacific region and parts of Africa.

Horses appear to be most susceptible and lose their hair, especially the long hair of the mane and tail. In severe cases, there is patchy loss of hair above and below the hocks and knees and on the flanks and neck. Disturbed growth at the coronet and periople may produce dystrophic rings on the hoofs. There is loss of condition and weakness in horses, which perhaps is attributable to malnutrition rather than to mimosine.

Prolonged ingestion of *L. leucocephala* by cattle is associated with poor weight gain, alopecia, and with goiter, which is not prevented by iodine supplementation. The goiter is caused by a rumen transformation product of mimosine, 3-hydroxy-4(1H)-pyridone (DHP), which prevents iodine binding in the thyroid gland. Susceptibility to mimosine toxicity correlates with an absence of rumen bacteria capable of detoxifying DHP; toxicosis may be prevented by inoculating DHP-degrading bacteria into the rumen.

Mimosine has a marked depilatory action on the fleece of sheep. A single oral dose of 400 to 650 mg/kg body weight allowed the fleece to be removed by hand 14 days later. Low concentrations of mimosine are potent inhibitors of DNA synthesis in sheep skin slices, including the wool follicles.

Mimosine toxicity causing depilation in pigs is reported from Indonesia and the Bahamas. A Bahamian belief that pigs having a lot of long hair are ill-thrifty led to the practice of intentional poisoning. Experimental toxicity in swine produces fetal resorption in sows without integumentary lesions. Regional differences in the plants or the pigs may account for the conflicting findings.

Bibliography

Hammond, A. C. *et al.* Prevention of leucaena toxicosis of cattle in Florida by ruminal inoculation with 3-hydroxy-4-(1H)-pyridone-degrading bacteria. *Am J Vet Res* **50:** 2176–2180, 1989.

Hegarty, M. P. Toxic amino acids of plant origin. *In* "Effects of Poisonous Plants on Livestock" R.F. Keeler, K.R. Van Kampen, L.F. James (eds.), pp. 575–585. New York, Academic Press, 1978.

Letts, G. A. *Leucaena glauca* and ruminants. *Aust Vet J* **39:** 287–288, 1963.

Montagna, W., and Yun, J. S. The effect of seeds of *Leucaena glauca* on the hair follicles of the mouse. *J Invest Dermatol* **40:** 325–332, 1963.

Mullenax, C. H. Observations on *Leucaena glauca. Aust Vet J* **39:** 88–91, 1963.

Owen, L. N. Hair loss and other toxic effects of *Leucaena glauca* ("jumbey"). *Vet Rec* **70:** 454–456, 1958.

Reis, P. J., Tunks, D. A., and Hegarty, M. P. Fate of mimosine administered orally to sheep and its effectiveness as a defleecing agent. *Aust J Biol Sci* **28:** 495–501, 1975.

Seawright, A. A. *Leucaena glauca* in Queensland. *Aust Vet J* **39:** 211, 1963.

Wayman, O., Iwanaga, I. I., and Hugh, W. I. Fetal resorption in swine caused by *Leucaena leucocephala* (Lam.) de Wit. in the diet. *J Anim Sci* **30:** 583–588, 1970.

9. Gangrenous Ergotism and Fescue Toxicosis

These two conditions can be considered together because the lesions of chronic ergotism caused by *Claviceps purpurea* and those of poisoning by tall fescue, *Festuca arundinacea,* are identical.

Ergotism is the oldest known mycotoxicosis. The ergot of *Claviceps* spp. fungi is a compacted mass of hyphae, known as the sclerotium, which develops in the seed heads of many species of grasses and cereal grains and completely replaces the ovary. Ergotism is the disease which results from ingestion of toxic alkaloids produced by the fungi. The alkaloids are derivatives of lysergic acid and include ergotamine, ergometrine, and ergotoxine, which is itself a composite of three alkaloids. The quantity and spectrum of alkaloids in the ergots vary considerably with the strain of fungus, type of plant, season of the year, climatic conditions, and other regional factors. The ergots also produce a variety of amines, such as histamine, acetylcholine, and other nitrogenous compounds with physiological activity.

Of the various pharmacologic effects exerted by the ergot alkaloids, the most important in the pathogenesis of gangrenous ergotism is direct stimulation of adrenergic nerves supplying arteriolar smooth muscle. This produces marked peripheral vasoconstriction. Arteriolar spasm and damage to capillary endothelium leads to thrombosis and ischemic necrosis of tissues.

Gangrenous ergotism caused by *C. purpurea* is a disease mainly of cattle. It may occur in animals at pasture but is more common in animals stall-fed infected grain. Gangrenous ergotism represents the chronic form of intoxication by ergot-producing fungi. Chronic ergotism develops after a week of feeding contaminated grain and begins with acute lameness with redness and swelling of the extremities. The hind legs are more frequently affected than the forelegs. Lesions seldom extend above the fetlock (Fig. 5.34), but ischemic necrosis may extend to the midmetatarsus. The feet become cold and insensitive with dry necrosis and a prominent line of separation between viable and dead tissue. The necrotic tissue may slough eventually. Ergotism also causes dry gangrene of the tips of the ears and tail. Gangrenous ergotism has been described in goats feeding on ergot-infested pasture. The toxicosis can be produced experimentally in sheep, but the syndrome is quite different from that in cattle, being characterized by ulceration of the tongue and of the alimentary mucosae. Sows are relatively resistant but may develop agalactia as a result of central inhibition of prolactin secretion.

Fescue toxicosis, also known as fescue foot, is a disease of cattle characterized in the acute form by dry gangrene

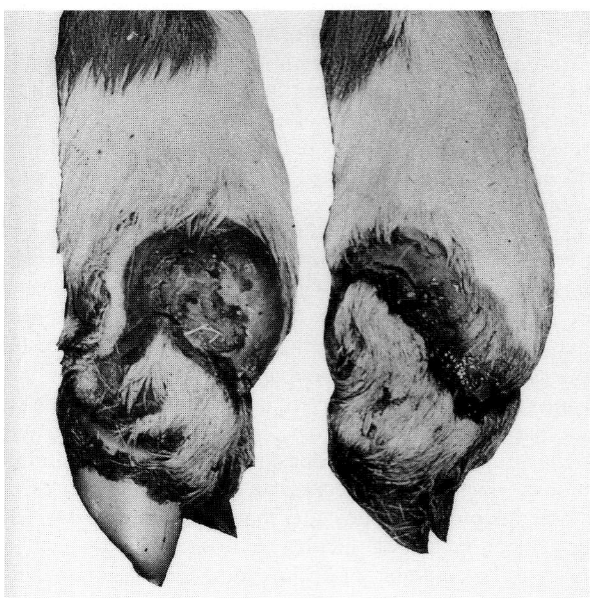

Fig. 5.34 Ischemic necrosis. Ergotism. Calf.

of the extremities commencing 2 weeks after ingestion of the tall fescue grass, *Festuca arundinacea*. This grass is a common pasture plant in the United States and is usually harmless. Under certain poorly understood conditions, the plant is toxic. A correlation exists between the length of time of stockpiling and the frequency of outbreaks. Fescue foot tends to occur with the onset of colder weather, indicating that low ambient temperatures may contribute to its development. The nature of the toxin in fescue poisoning is not known. There is a correlation between infection of tall fescue with the endophytic fungus *Acremonium coenophalium* and disease outbreaks, suggesting that the toxicant may be a mycotoxin. The acute syndrome in cattle is virtually identical to ergotism. A chronic disease, known as summer slump, is associated with marked necrosis of abdominal fat and severe weight loss. Chronic disease in horses is also associated with weight loss, and mares may show agalactia.

Bibliography

Burfening, P. J. Ergotism. *J Am Vet Med Assoc* **163:** 1288–1290, 1973.

Goodman, A. A. Fescue lameness. *North Am Vet* **33:** 701–702, 1952.

Hibbs, C. M., and Wolf, N. Ergot toxicosis in young goats. *Mod Vet Pract* **63:** 126–128, 1982.

Maag, D. D., and Tobiska, J. W. Fescue lameness in cattle. Ergot alkaloids in tall fescue grass. *Am J Vet Res* **17:** 202–204, 1956.

Nordskog, A. W., and Clark, R. T. Ergotism in pregnant sows, female rats, and guinea pigs. *Am J Vet Res* **6:** 107–116, 1946.

Siegel, M. R. *et al.* A fungal endophyte in tall fescue; incidence and dissemination. *Phytopathology* **74:** 932–937, 1984.

Williams, M. *et al.* Induction of fescue foot syndrome in cattle by fractionated extracts of toxic fescue hay. *Am J Vet Res* **36:** 1353–1357, 1975.

Woods, A. J., Jones, J. B., and Mantle, P. G. An outbreak of gangrenous ergotism in cattle. *Vet Rec* **78:** 742–749, 1966.

10. Tricothecene Toxicoses

Stachybotryotoxicosis is caused by macrocyclic tricothecene toxins produced by the fungus *Stachybotrys* spp. Ulcerative and necrotizing lesions of the skin and mucous membranes have been reported in horses, cattle, sheep, and pigs, chiefly from Russia and eastern Europe. Initial lesions affect the lips, buccal commissures, and nostrils. Marked edema of the face may follow. Death follows development of hemorrhagic diathesis, enteritis, and septicemia. At necropsy, lesions (in addition to the hemorrhagic diathesis) include ulceration of mucosal surfaces of the alimentary system, pneumonia, renal infarcts, multifocal hepatic necrosis, and lymphadenitis. In many instances, these lesions may represent secondary mycotic or bacterial involvement.

T-2 toxin is a highly irritant **trichothecene mycotoxin** from *Fusarium* molds on grain, and causes cutaneous ulceration when applied locally to the skin of pigs. Experimental feeding of T-2-contaminated feed, in combination with aflatoxin, induces crusting and ulceration of the lips, snout, buccal commissures, and prepuce. The pathogenesis of the lesion is thought to be contact irritant dermatitis due directly to the T-2 toxin or to a urinary metabolite, HT-2 toxin. Various other trichothecene mycotoxins are also cutaneous irritants and may cause vomition or feed refusal.

Bibliography

Harvey, R. B. Cutaneous ulceration and necrosis in pigs fed aflatoxin- and T-2 toxin-containing diets. *J Vet Diagn Invest* **2:** 227–229, 1990.

Hintikka, E. L. Stachybotryotoxicosis in cattle and captive animals: Stachybotryotoxicosis in horses: Stachybotryotoxicosis in sheep: Stachybotryotoxicosis in swine. *In* "Mycotoxic Fungi, Mycotoxins, and Mycotoxicoses" T.D. Wyllie and L.G. Morehouse (eds.), pp.152–161, 181–185, 203–207, 268–273. New York, Dekker, 1978.

11. Miscellaneous Toxicities

Cutaneous lesions were associated with a toxicosis of beef cattle developing 2 weeks after the cattle were turned onto a pasture of **hairy vetch** (*Vicia villosa* Roth). More recent reports in South African cattle have implicated other species of *Vicia*. Initial lesions were papulonodular eruptions affecting the perineum, udder, tail head, and neck. Since the lesions were **pruritic,** secondary excoriation, alopecia, and lichenification developed. Histologically, the skin lesions were perivascular dermatitis in which eosinophils and mononuclear cells predominated. At necropsy, nodular and diffuse inflammatory lesions involved a wide range of tissues but were most severe in myocardium, kidney, lymph nodes, thyroid, and adrenal glands. The inflammatory reaction was granulomatous and eosinophilic. Attempts to reproduce the condition were not successful, not perhaps surprising in light of the widespread use of this legume in pasture, hay, and silage. Lesions, similar to those described in cattle, have been reported in a horse pastured on *Vicia villosa*.

Pyrexia with dermatitis in dairy cows is a syndrome with some similarities to that described as hairy vetch toxicity. It has been reported in the United States, England, Wales, France, and the Netherlands. Friesian dairy cows developed pruritic papular eruptions affecting the head and neck, tail head, and udder. Secondary lesions were due to self-trauma. In another outbreak, hemorrhages were a prominent clinical sign. On necropsy, multifocal inflammatory cell infiltrates were present in the kidney, myocardium, and liver. These were predominantly mononuclear with variable numbers of eosinophils and multinucleate giant cells.

The cause of these diseases remains unclear, nor is it certain whether they represent one entity or a common tissue reaction to a variety of insults. The episode in Wales was associated with the introduction of a new silage additive into several farms. The outbreak in the Netherlands was associated with the feeding of **di-ureido-isobutane** (DUIB) in the seed cake. This condition was reproduced in 2 cows by feeding a DUIB-containing diet for 1 month. Histologically, the lesions of the Dutch outbreak also resembled those of the putative hairy vetch toxicity. These bovine conditions are pathologically similar to idiopathic eosinophilic dermatitis of the horse (see Section XVIII, Miscellaneous Skin Disorders).

Cats consuming decomposed scallops from a seafood-processing plant developed necrosis, with subsequent sloughing, of pinnae and digits.

Bibliography

Anderson, C. A., and Divers, T. J. Systemic granulomatous inflammation in a horse grazing hairy vetch. *Am J Vet Med Assoc* **183:** 569–570, 1983.

Breukink, D. H. *et al.* Pyrexia with dermatitis in dairy cows. *Vet Rec* **103:** 221–222, 1978.

Dyson, D. A., and Reed, J. B. H. Haemorrhagic syndrome of cattle of suspected mycotoxic origin. *Vet Rec* **100:** 400–402, 1977.

Gourreau, J. M. *et al.* Le syndrome dermatite pruringineuse de la vache latière. *Point Vét* **18:** 135–140, 1986.

Green, J. R., and Kleynhans, R. Suspected vetch (*Vicia benghalensis L*) poisoning in a Friesland cow in the Republic of South Africa. *J S Afr Vet Assoc* **60:** 109–110, 1989.

Panciera, R. J., Johnson, L., and Osburn, B. I. A disease of cattle grazing hairy vetch pasture. *J Am Vet Med Assoc* **148:** 804–808, 1966.

Thomas, G. W. Pyrexia with dermatitis in dairy cows. *In Pract* **1:** 16–18, 1979.

VII. Actinic Diseases of Skin

The radiant energy of the sun includes components that are potentially harmful to the mammalian skin. This radiation is known as **actinic radiation,** and its acute effect is the well-known **sunburn** reaction. **Phototoxicity** or, as it is more commonly termed in veterinary medicine, **photosensitization,** is essentially an exacerbated form of sunburn, caused by the activation of photodynamic chemicals in the skin by radiation of an appropriate wavelength. The chemical may be exogenous or endogenous and may reach the skin via various routes, as will be described. **Photoallergy** is distinct from phototoxicity; it occurs when the photoproduct of an exogenous chemical acts as an antigen. Photoallergic reactions require sensitizing exposure to the drug or chemical and are more clinically diverse. Photoallergies have not been conclusively documented in animals. Derivatives of quinadoxin used as additives in swine rations have been investigated for photoallergic properties because of lesions in stockmen handling the products. The **skin cancers** induced or exacerbated by actinic radiation are considered in Section XIX,A with Tumors of the Skin.

A. Direct Effect of Solar Radiation

Most of the direct photobiologic reactions in the skin are induced by shortwave ultraviolet radiation in the 290- to 320-nm band (UV-B). The ozone layer strongly absorbs the very damaging wavelengths below 290 (UV-C). Longer wavelengths of 320 to 400 nm constitute UV-A, the least harmful form of ultraviolet radiation, unless in the presence of photodynamic agents whose action spectra fall in this range. The integument is normally protected against the deleterious effects of ultraviolet radiation by the hair coat, the stratum corneum, and melanin pigmentation. In the stratum corneum, urocanic acid, a catabolite of filaggrin, strongly absorbs ultraviolet light. Melanin both absorbs and scatters UV radiation and, being able to trap free radicals, is also important in minimizing the deleterious effects of incident photons.

The mechanisms of the reaction commonly termed **sunburn** are poorly understood. The initial transient erythema is probably due to a direct heating effect and possibly photobiologic effects of UV-B acting directly on dermal capillaries. Two theories have been proposed for the pathogenesis of the delayed erythema reaction. The first postulates direct damage to endothelial cells; the second evokes the release of cytokines from the radiation-damaged keratinocytes. Ultraviolet radiation has been shown to increase the production of keratinocyte-derived cytokines. Ultraviolet light also induces adaptive responses in the epidermis, in particular epidermal hyperplasia and alterations in melanin pigmentation. An immediate pigment darkening is thought to be due to changes in existing melanin and the delayed or "tanning" reaction to stimulation of melanogenesis and proliferation of melanocytes. Ultraviolet light depresses local and systemic immune responses. Depression of contact hypersensitivity responses is associated with a reduction in the number of Langerhans' cells and inactivation of their antigen-presenting functions. Long-term effects of ultraviolet irradiation include degenerative changes in the dermis (solar elastosis) and epidermis (solar keratosis). Nucleoprotein is susceptible to ultraviolet radiation damage resulting in mitotic inhibition and, if extensive enough, cell death. Sublethal damage may promote mutagenesis or carcinogenesis. Ultraviolet radiation is not only a tumor initiator and promoter but also may alter immunologic reactivity in favor of the growth of the tumor, through the induction of suppressor T cells.

Potentially, all animals are susceptible to the acute and chronic effects of actinic radiation, but the protection afforded by the hair coat, stratum corneum, and skin pigmentation is normally sufficient. The following conditions typically affect animals whose anatomical defences are poor, either by lacking skin pigmentation or hair cover. It is also possible that an increasing prevalence of sun-induced dermatoses and tumors may be seen in animals, particularly those living at high altitudes or low latitudes. This trend has been noted in humans, coincident with the depletion of the ozone layer and a consequent increase in the intensity of ultraviolet radiation reaching the earth's surface.

Solar dermatitis, or sunburn, occurs most frequently in cats, pigs, and goats. The lesions in **cats** typically affect the tips of the ears of white, blue-eyed animals. Less frequently the eyelids, nose, and lips are involved. The initial lesion is erythema followed by alopecia, scaling, and crusting. The ear tip may curl over. Lesions are exacerbated each summer, often eventuating in malignant transformation into squamous cell carcinoma. Primary phototoxicity in swine occurs in white or light-colored **pigs.** While any age group may be affected, the condition is most severe in suckling and weaner pigs. Occasionally, badly affected ears may slough. Light-colored **goats** are also prone to solar dermatitis. The udders are particularly susceptible when does are turned out into strong sunlight after a winter indoors.

The characteristic microscopic feature of sun-induced epithelial damage is the presence of degenerate keratinocytes, known as sunburn cells. Representing necrotic keratinocytes, these may be induced within 30 min of sun exposure. They have hypereosinophilic cytoplasm, pyknotic nuclei, and lie singly or in clusters in the outer stratum spinosum. Other lesions include spongiosis, vacuolization of keratinocytes, loss of the granular layer, and, in severe burns, vesiculation. Dermal hyperemia and edema are prominent features. In mild lesions, there is a slight increase in mononuclear cells; in severe burns, there is marked vascular damage, erythrocyte extravasation, and neutrophilic exocytosis.

Solar elastosis, a hallmark of chronic sun exposure in humans, has been described only rarely in dogs, sheep, and horses. The lesions usually occur in conjunction with solar radiation-associated neoplasms, particularly squamous cell carcinomas. Solar elastosis is seen, in hematoxylin and eosin-stained sections, as scattered or agglomerated thick, irregular, basophilic degenerate elastic fibers.

Solar keratoses, common precancerous skin lesions in humans, occur in cats and dogs. A chronic dermatosis, resembling solar keratosis in humans, developed in beagle **dogs** housed in outdoor kennels with reflective, light-colored flooring. In northern parts of Australia, white dogs, particularly bull terriers, develop solar keratoses which often progress to squamous cell carcinoma. These lesions frequently involve the lightly haired areas, particularly the ventral abdomen and flank. Such lesions are probably related to the basking behavior exhibited by the animals. Early lesions of erythema and scaling evolve into thick, rough, erythematous, crusted patches and plaques. Hemorrhagic bullae may develop. Histologically, the early lesions have many of the features of sunburn, including epidermal hyperplasia, spongiosis, acute dermal inflammation and focal necrotic keratinocytes. More chronic lesions show pronounced epidermal hyperplasia with **dysplasia,** ortho- and parakeratotic hyperkeratosis, perivascular mononuclear cell infiltrates, and dermal scarring, but seldom develop significant solar elastosis, as typifies human solar keratoses. Lesions frequently progress to invasive squamous cell carcinoma. Solar keratoses may also develop cutaneous horns.

Bibliography

Campbell, G. A., Gross, T. L., and Adams, R. Solar elastosis with squamous cell carcinoma in two horses. *Vet Pathol* **24:** 463–464, 1987.

Hargis, A. M. A review of solar-induced lesions in domestic animals. *Compend Cont Ed* **3:** 287–296, 1981.

Irving, R. A., Day, R. S., and Eales, L. Porphyrin values and treatment of feline solar dermatitis. *Am J Vet Res* **43:** 2067–2069, 1982.

Knowles, D. P., and Hargis, A. M. Solar elastosis associated with neoplasia in two Dalmatians. *Vet Pathol* **23:** 512–514, 1986.

Mason, K. V. The pathogenesis of solar-induced skin lesions in bull terriers. *Proc American Academy of Veterinary Dermatology and American College of Veterinary Dermatology*, Phoenix, Arizona, p. 12, 1987.

Montagna, W., Kirchner, S., and Carlisle, K. Histology of sun-damaged human skin. *J Am Acad Dermatol* **21:** 907–918, 1989.

Roberts, L. K. *et al.* Ultraviolet light and modulation of the immune response. *In* "Immune Mechanisms in Cutaneous Disease" D.A. Norris (ed.), pp. 167–218. New York, Marcel Dekker, 1989.

B. Photosensitization

Photosensitivity refers to an enhanced susceptibility of the skin to actinic radiation induced by the local presence of a **photodynamic agent.** A photodynamic agent, or chromophore, has a chemical configuration which enables it to absorb specific wavelengths of ultraviolet or visible light, known as the action spectrum. The action spectrum of many photodynamic agents lies beyond the UV-B range, and thus renders normally harmless incident radiation damaging. Activation of the photodynamic agent raises it to a metastable triplet state. The activated photodynamic substance may react directly with the biologic substrate or with molecular oxygen, producing reactive oxygen intermediates such as superoxide anion, singlet oxygen, and hydroxyl radical. In addition, oxygen free radicals may be formed indirectly, as the result of a calcium-dependent, protease-mediated activation of xanthine oxidase in the skin. Release of reactive species initiates chain reactions which damage macromolecules including nucleic acids, proteins, and lipoproteins. The major targets of phototoxic reactions include the nucleus, the cell membrane, and subcellular organelles, particularly lysosomes and mitochondria. The mechanisms of tissue injury vary with the particular photodynamic agent involved, and these are generally poorly characterized. Most work has been done on the psoralens, because of their extensive use in photochemotherapy in humans. Mechanisms of psoralen-UVA-induced photosensitization are complex, involving oxygen-independent photoconjugation of psoralen to DNA, which results in the inhibition of DNA synthesis and cell proliferation, and oxygen-dependent mechanisms responsible for the erythematous and edematous reactions in skin. In experimental porphyrin-induced phototoxicity, primary mitochondrial damage is thought to lead to a cascade of deleterious reactions, resulting in the formation of xanthine oxidase-derived oxygen free radicals.

The photodynamic agent usually reaches the skin via the systemic circulation, although there are some instances in which exposure is by percutaneous absorption.

Such photodynamic agents cause local reactions. The agent may originate externally, or it may be an endogenous substance which has accumulated to an abnormal degree as a result of metabolic dysfunction. The three categories of photosensitization are classified according to the source of the agents. In primary or type I photosensitization, the photodynamic agents are exogenous. Aberrant endogenous pigment synthesis is responsible for type II photosensitization. Phylloerythrin, a degradation product of chlorophyll, induces type III photosensitization. This is known also as hepatogenous photosensitization because it depends on the failure of the liver to eliminate phylloerythrin. A fourth group contains those examples of photosensitization for which the pathogenesis is presently undetermined.

The gross lesions are similar for all forms of photosensitization. They occur on those areas of the body most exposed to sunlight and lacking protective fleece, hair coat, or skin pigmentation. In **cattle,** any area of light-colored skin is susceptible. This is best demonstrated in broken-colored animals such as Holsteins, in which the white skin is affected, but the black is spared. The relatively hairless skin of the teats, udder, perineum, and muzzle is also affected. The ventral surface of the tongue is frequently affected in cattle if the muzzle is being constantly exposed during licking. In **sheep,** the susceptible sites are the ears, eyelids, face, muzzle, and coronets, although the back may be affected in animals with an open fleece or those that have been shorn closely. The very marked edema of the ears in sheep causes them to droop, and swelling of the muzzle may cause dyspnea (Fig. 5.35A); the disease in sheep is appropriately known as big or swelled head. The udders and teats of dairy **goats** are predisposed. In **horses,** lesions are most common on the face and distal extremities but may affect any white skin. Lesions in **pigs** are uncommon and have a predilection for the ears, eyelids, udder, back, and white skin. Photosensitization is extremely rare in **dogs** and **cats**. Two white-nosed working Border collie sheep dogs in New Zealand developed lesions resembling photosensitization. Lesions in Siamese cats develop secondary to congenital porphyria.

The initial reaction in photosensitization is erythema, followed by edema, which is more prominent in sheep than in cattle. The lesions are intensely pruritic, causing rubbing, scratching, and kicking at affected parts. There is marked exudation and extensive necrosis. Affected skin becomes dry and sloughs in desiccated sheets (Fig. 5.35B). Necrosis is frequently seen on the upper surfaces of the ears of sheep; the tips typically curl upward as a result of mummification. Although there is swelling of the eyelids and some lacrimation, the eyes are not involved except when phenothiazine is the photosensitizing agent. Among the more obscure manifestations of photosensitization is the convulsive reaction of some sheep and cattle, photosensitized by ingestion of St. John's wort, on contact with cold water. Icterus may or may not be present depending on the type of photosensitization. Icterus typically is associated with hepatogenous photosensitization, but hepato-

Fig. 5.35A Photosensitization. Swelling of ears, eyelids, and lips. Acute change. Lamb. (Courtesy of W. J. Hartley.)

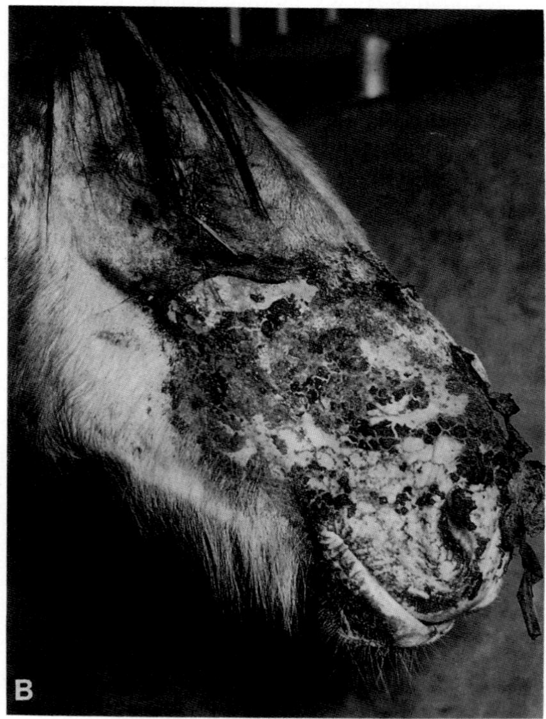

Fig. 5.35B Photosensitization. Necrosis of skin. Horse.

genous photosensitization may occur in its absence. In severe episodes of photosensitization, animals may die. This is more often the result of concomitant damage to other organs, particularly the liver, than to cutaneous damage alone. Injury to erythrocytes in cutaneous circulation may produce severe hemolysis.

1. Primary Photosensitization

Primary photosensitization results from ingestion and direct absorption of photodynamic agents. These are chiefly obtained from plants, in particular St. John's wort (*Hypericum perforatum*), buckwheat (*Fagopyrum* spp.), spring parsley (*Cymopterus watsoni*) and *Ammi* spp. Drug-induced photosensitization may follow the administration of the anthelmintic phenothiazine.

Photosensitization induced by St. John's wort affects horses, cattle, sheep, and goats. The disease is known as hypericism, and the active principle is **hypericin,** a helianthrone pigment. The agent is present as a red fluorescent pigment in the leaves of the plant. Hypericin is present at all stages of plant growth, but significant amounts are consumed by livestock only when the plant is prolific or other feed is scarce.

Photosensitization induced by buckwheat affects sheep, pigs, cattle, goats, and horses. The photodynamic agent **(fagopyrin)** is also a red fluorescent pigment of the helianthrone family.

Photosensitization occurs in cattle, sheep, and poultry as a result of **furocoumarin** ingestion. Plants such as spring parsley (*Cymopterus watsoni*), *Ammi majus* (bishop's weed) and *Thamnosma texana* (Dutchman's breeches) contain furocoumarins, of which the **psoralens** are important. The furocoumarins differ from the helianthrone photosensitizing pigments by inducing, additionally, corneal edema and keratoconjunctivitis. Furocoumarins are also formed in parsnips and celery as phytoallexins when the plants become infected with certain types of fungi. These have been associated with **phytophotocontact dermatitis** in pigs in New Zealand. Lesions were vesicular, affecting only the dorsal aspect of the snout. Onset of the lesions was associated with contact with green vegetable material containing parsnips (*Pastinaca sativa*) and celery (*Apium graveolens*), followed by exposure to sunlight. Meteorological data showed that each incident was preceded by increased solar-radiation levels. The lesions were reproduced by rubbing the snouts and feet of white pigs with the leaves of the fungus-infected celery and parsnips before exposing the areas to UV light. *Cymopterus watsoni* causes phytophotodermatitis in sheep in Utah and Nevada. Lesions principally affected the nonwooled areas, such as the muzzle, lips, and udder. High lamb mortality may be incurred from mismothering. Plants of the Umbelliferaceae evoke phytophotodermatitis in humans, especially those in horticultural occupations such as celery workers. The pathogenesis of phytophotocontact dermatitis involves the adsorption onto the skin of psoralens, which become activated by UV light of the 334- to 440-nm range.

Texan cattle and deer develop a primary photosensitization after consuming moldy leaves of *Cooperia pedunculata*. In addition to skin lesions, keratitis occurs frequently and may lead to blindness.

Phenothiazine photosensitization is also characterized by corneal edema and keratoconjunctivitis. The unusual location of the lesions has been explained by the secretion of the ruminal metabolite, phenothiazine sulfoxide, in tears and aqueous humor. Lesions occur most commonly in calves but also in sheep, swine, and birds. Pigs develop cutaneous lesions more frequently than do sheep or cattle, probably because the activating radiation is more able to penetrate the integument. The greater susceptibility of calves has been ascribed to a relatively inefficient conversion of the photodynamic sulfoxide metabolite back to phenothiazine in the liver. This conversion depends on mixed-function oxidase enzymes and is performed effectively in ovine liver.

Several other chemicals are capable of causing photosensitization, including sulfonamides and tetracyclines. These drugs cause photosensitization in humans and induce experimental disease but have not been associated with problems in livestock under natural conditions.

2. Photosensitization Due to Defective Pigment Synthesis

Photosensitization due to endogenous pigment accumulation is the result of abnormal porphyrin metabolism. The photodynamic agents include uroporphyrin I, coproporphyrin I, and protoporphyrin III. They accumulate in the blood and tissues when there is a malfunction in heme biosynthesis because of an enzyme deficiency. The type I porphyrins accumulate when there is a deficiency of uroporphyrinogen III cosynthetase and a resultant failure of heme feedback suppression, as occurs in congenital erythropoietic porphyria in cattle, humans, and probably Siamese cats. Protoporphyrin IX accumulates when there is a deficiency of ferrochelatase (heme synthetase), as occurs in erythropoietic protoporphyria in cattle and humans.

Bovine congenital erythropoietic porphyria is the result of a deficiency in uroporphyrinogen III cosynthetase, a key enzyme in heme biosynthesis. The condition is inherited as a simple recessive trait affecting many breeds including shorthorn, Ayrshire, Holstein, and Jamaican cattle. It has also been reported in crossbred cattle. The disease is known as osteohemochromatosis and pink tooth, both suggested by the reddish-brown coloration of porphyrin pigments in dentin and bone. The pigment is also deposited in other tissues, but the discoloration may be obvious only in lungs, spleen, and kidney, in which it is deposited in the interstitial tissue and tubular epithelium. The pigments are excreted in the urine, hence the alternative names porphyrinuria and hematoporphyrinuria. Affected urine is an amber to brown color, which darkens on exposure to light and fluoresces bright red on exposure to ultraviolet radiation. Affected teeth and bones also fluoresce.

The cutaneous lesions result from the photodynamic

properties of the accumulated porphyrins, in particular the **uroporphyrins,** which absorb UV-A radiation. Reactive oxygen species directly induced by the porphyrins or, possibly, via activation of xanthine oxidase in the skin are responsible for the cell-membrane damage. The gross cutaneous lesions are typical of photosensitization. The microscopic lesions closely resemble those of erythropoietic porphyrias in humans. The chief lesions are subepidermal clefts, hyalinization of dermal capillary walls, and a minimal infiltrate of inflammatory cells. The basement membrane zone lines the base of the subepidermal cleft, in some instances covering small projections of dermal papillae (festoons). Festoons are a more prominent feature of the human lesion since dermal papillae are better developed in human skin.

The anemia of bovine congenital erythropoietic porphyria is discussed in Volume 3, Chapter 2, The Hematopoietic System.

Bovine erythropoietic protoporphyria is inherited as a recessive trait in Limousin cattle in the United States. It differs from bovine congenital porphyria in that photodermatitis is the sole clinical manifestation of the disease. Animals do not have discolored teeth, anemia, or urine porphyrin excretion. The enzyme defect is a deficiency of ferrochelatase, which allows protoporphyrin IX to accumulate in blood and tissues. Heterozygotes have a reduced (50%) ferrochelatase activity.

Porphyria of **swine** is inherited as a dominant characteristic. Although it mimics certain aspects of bovine erythropoietic porphyria, photosensitization does not occur, even in white-skinned animals. The defect in porcine porphyria is not known.

Photosensitization occurs in Siamese cats with excessive accumulation of uroporphyrinogen I, coproporphyrinogen I, and protoporphyrins in blood, urine, feces, and tissues. The defect is presumably a deficiency of uroporphyrinogen III cosynthetase, as has been established in humans and cattle.

3. Hepatogenous Photosensitization

The most common form of photosensitization in domestic animals is secondary to hepatic injury, which alters the liver's capacity to excrete the potent photodynamic agent, **phylloerythrin.** Phylloerythrin is formed from chlorophyll in the intestinal tract by microbial action and is transported to the liver in the portal circulation. Hepatocytes take up the phylloerythrin and excrete it into the bile. One of the earliest signs of liver cell damage is a reduced ability to transport and excrete phylloerythrin. The circulating phylloerythrin is rather poorly excreted by the kidney, and excretion is further decreased by mild renal tubular damage in some toxicoses. The phylloerythrin accumulates in tissues, including the skin, where its photodynamic properties induce typical lesions, provided the animal is exposed to sufficient solar radiation of the appropriate wavelength.

Toxic plants and mycotoxins account for most cases of hepatogenous photosensitization. A few of the many plants implicated include *Panicum* spp. grasses, lantana (*Lantana camara*) and bog asphodel (*Narthecium ossifragum*). Some plants work in combination; black sagebrush appears to precondition sheep to photosensitization caused ultimately by *Tetradymia* spp. The mycotoxin sporidesmin from *Pithomyces chartarum* causes **facial eczema,** an economically important photosensitization of sheep and cattle in Australia, New Zealand, and South Africa. **Geeldikkop,** a disease characterized by hepatogenous photosensitization, is associated with extensive losses among sheep and goats in South Africa (see Volume 2, Chapter 2, Liver and Biliary System). Photosensitization has been associated with chemically induced hepatopathies, such as that caused by carbon tetrachloride. Hepatitis of infectious origin, for example leptospirosis, has been suspected to cause photosensitization. Photosensitization tends, however, to occur when the hepatic damage is generalized. Severe focal necrotizing lesions do not generally cause photosensitization because there is enough hepatic reserve to remove the phylloerythrin from the circulation. Mild diffuse hepatic damage can induce photosensitization, particularly if ambient solar radiation levels are high and shade is unavailable. A hepatogeneous photosensitization, secondary to a presumed genetic defect in phylloerythrin transport, has been reported in Corriedale lambs. Hepatogenous photosensitization is discussed in Volume 2, Chapter 2, Liver and Biliary System.

Bibliography

Allison, A. C., Magnus, I. A., and Young, M. R. Role of lysosomes and of cell membranes in photosensitization. *Nature* **209:** 874–878, 1966.

Athar, M. *et al.* A novel mechanism for the generation of superoxide anions in hematoporphyrin derivative-mediated cutaneous photosensitization. *J Clin Invest* **83:** 1137–1143, 1989.

Betty, R. C., and Trikojus, V. M. Hypericin and a nonfluorescent photosensitive pigment from St. John's wort *(Hypericum perforatum). Aust J Exp Biol Med Sci* **21:** 175–182, 1943.

Binns, W., James, L. F., and Brooksby, W. *Cymopterus watsoni:* A photosensitizing plant for sheep. *Vet Med Small Anim Clin* **59:** 375–379, 1964.

Brenner, D. A., and Bloomer, J. R. Comparison of human and bovine protoporphyria. *Yale J Biol Med* **52:** 449–454, 1979.

Carmichael, W. W. Chemical and toxicological studies of the toxic freshwater cyanobacteria *Microcystis aeruginosa, Anabaena flos-aquae,* and *Aphanizomenon flos-aquae. S Afr J Sci* **78:** 367–372, 1982.

Clare, N. T. Photosensitization in diseases of domestic animals. *Commonw Agric Bur Rev Ser* No.3, Commonwealth Bureau of Animal Health, 1952.

Cornelius, C. E., Irwin, M. A., and Osburn, B. I. Hepatic pigmentation with photosensitivity. A syndrome in Corriedale sheep resembling Dubin–Johnson syndrome in man. *J Am Vet Med Assoc* **146:** 709–713, 1965.

De Vries, H. *et al.* Photochemical reactions of quindoxin, olaquindox, carbadox, and cyadox with protein, indicating photoallergic properties. *Toxicology* **63:** 85–95, 1990.

Dickie, C. W., and Berryman, J. R. Polioencephalomalacia and photosensitization associated with *Kochia scoparia* consumption in range cattle. *J Am Vet Med Assoc* **175:** 463–465, 1979.

Dollahite, J. W., *et al.* Photosensitization in lambs grazing kleingrass. *J Am Vet Med Assoc* **171:** 1264–1265, 1977.

Dollahite, J. W., Younger, R. L., and Hoffman, G. O. Photosensitization in cattle and sheep caused by feeding *Ammi majus* (greater Ammi; bishop's weed). *Am J Vet Res* **39:** 193–197, 1978.

Enzie, F. D., and Whitmore, G. E. Photosensitization keratitis in young goats following treatment with phenothiazine. *J Am Vet Med Assoc* **123:** 237–238, 1953.

Fourie, P. J. J. The occurrence of congenital porphyrinuria (pink tooth) in cattle in South Africa (Swaziland). *Onderstepoort J Vet Sci* **7:** 535–566, 1936.

Galitzer, S. J., and Oehme, F. W. Photosensitization: A literature review. *Vet Sci Commun* **2:** 217–230, 1978.

Glastonbury, J. R. W., and Boal, G. K. Geeldikkop in goats. *Aust Vet J* **62:** 62, 1985.

Gordon, H. McL., and Green, R. J. Phenothiazine photosensitization in sheep. *Aust Vet J* **27:** 51–52, 1951.

Ivie, G. W. Toxicological significance of plant furocoumarins. *In* "Effects of Poisonous Plants on Livestock" R.F. Keeler, K.R. Van Kampen, and L.F. James (eds.), pp. 475–486. New York, Academic Press, 1977.

Jorgensen, S. K. Congenital porphyria in pigs. *Br Vet J* **115:** 160–175, 1959.

Laksevela, B., and Dishington, I. W. Bog asphodel (*Northecium ossifragum*) as a cause of photosensitization of lambs in Norway. *Vet Rec* **112:** 375–378, 1983.

Montgomery, J. F., Oliver, R. E., and Poole, W. S. H. A vesiculobullous disease in pigs resembling foot and mouth disease. I. Field cases. *N Z Vet J* **35:** 21–26, 1987.

Montgomery, J. F., Oliver, R. E., and Poole, W. S. H. A vesiculobullous disease in pigs resembling foot and mouth disease. II. Experimental reproduction of the lesion. *N Z Vet J* **35:** 27–30, 1987.

Morison, W. L., Parish, J. A., and Epstein, J. H. Photoimmunology. *Arch Dermatol* **115:** 350–355, 1979.

Oertli, E. H. *et al.* Phototoxic effect of *Thamnosma texana* (Dutchman's breeches) in sheep. *Am J Vet Res* **44:** 1126–1129, 1983.

Pathak, M. A. Molecular aspects of drug photosensitivity with special emphasis on psoralen photosensitization reaction. *J Nat Cancer Inst* **69:** 163–170, 1982.

Rowe, L. D. Photosensitization problems in livestock. *Vet Clin North Am* **5:** 301–323, 1989.

Rowe, L. D. *et al.* Photosensitization of cattle in southeast Texas: Identification of phototoxic activity against *Cooperia pedunculata*. *Am J Vet Res* **48:** 1658–1661, 1987.

Sassa, S. *et al.* Accumulation of protoporphyria IX from D-aminolevulinic acid in bovine skin fibroblasts with hereditary erythropoietic protoporphyria. *J Exp Med* **153:** 1094–1101, 1981.

Scott, D. W., Mort, J. D., and Tennant, B. C. Dermatohistopathologic changes in bovine congenital porphyria. *Cornell Vet* **69:** 145–158, 1979.

Spikes, J. D. Porphyrins and related compounds as photodynamic sensitizers. *Ann N Y Acad Sci* **244:** 496–508, 1975.

Wender, S. H. Action of photosensitizing agents isolated from buckwheat. *Am J Vet Res* **7:** 486–489, 1946.

With, T. K. A short history of porphyrins and the porphyrias. *Int J Biochem* **11:** 189–200, 1980.

With, T. K. Porphyria in animals. *Clin Hematol* **9:** 345–370, 1980.

Witzel, D. A., Dollahite, J. W., and Jones, L. P. Photosensitization in sheep fed *Ammi majus* (bishop's weed) seed. *Am J Vet Res* **39:** 319–320, 1978.

4. Photoaggravated Dermatoses

In humans, several autoimmune dermatoses are exacerbated by exposure to UV light. These include pemphigus, lupus erythematosus, and bullous pemphigoid. A similar relationship has been proposed in the analogous canine diseases.

A poorly understood disease in the horse, **photoactivated vasculitis** affects only the white-haired extremities. The suggested pathogenesis is an immune-mediated vasculitis, in which the immune complexes may be acting as photodynamic agents. Photoactivated vasculitis is not merely a photosensitization reaction, because other white areas on the horse are unaffected, and the lesions do not always regress with cessation of sunlight exposure. The lesions often affect the heels and have to be differentiated from the other manifestations of the "greasy heel" complex. The acute lesions are well demarcated, erythematous, oozing, and crusted; the chronic lesions are hyperkeratotic plaques. Histologically, there is intense dermal edema, vascular dilation, and a subtle small vessel leukocytoclastic vasculitis, affecting only the superficial plexus. Thrombi may be seen occasionally.

Bibliography

Stannard, A. A. Photoactivated vasculitis. *In* "Current Therapy in Equine Medicine II" N.E. Robinson, (ed.), p647. Philadelphia, Pennsylvania, W.B. Saunders, 1987.

VIII. Nutritional Diseases of Skin

The elasticity of the skin, the orderly maturation of the epidermis, and the quality and luster of the horny appendages reflect in a sensitive but general manner the well-being of an animal. This applies equally to nutritional diseases and diseases of other causes. The general metabolic transformations which take place in the skin are not qualitatively different from those in other tissues, but there are some quantitative differences, such as the high requirements and turnover of sulfur-containing amino acids in the elaboration of keratin. In most metabolic disturbances and deficiencies of essential nutrients, whether from dietary lack, malabsorption, or the action of antimetabolites, changes will be reflected in the skin. The molecular basis for these skin lesions is, however, poorly understood. There are only a few syndromes that are of natural occurrence and sufficiently clearly defined to warrant discussion here. A larger number can be produced experimentally.

A. Protein–Calorie Deficiency

Starvation or protein–calorie malnutrition results in changes in the skin, the first being the disappearance of subcutaneous fat. Even though water intake may not be restricted, there is reduced hydration of the connective tissues of the subcutis and dermis, and the skin wrinkles and loses its elasticity. As hair growth and keratinization draw heavily on the daily protein intake of an animal, an early sign of starvation is the development of a dull, dry, and often brittle hair coat. Alopecia develops as a thinning of the hair rather than baldness, and seasonal shedding may cease or be prolonged. In pigs, the hair may become long and shaggy. Undernutrition of the ewe, between 115 and 135 days of gestation, decreases the number of secondary wool follicles in the developing fetus. Most experimental work on the effects of nutrition on hair growth have been performed in sheep, the purpose being either to improve wool production or to investigate chemical methods of shearing. Changing sheep from a low-protein to a high-protein diet led to a 33% increase in fleece production, due mostly to an increase in the rate of mitosis in the hair bulb. The major nutritional limitation to wool production is the amount and type of amino acid intake, specifically cysteine. Recent experiments in gene manipulation have investigated the possibility of providing the sheep with the enzymes necessary for cysteine biosynthesis. Conversely, specific amino acid deficiencies have been investigated as potential replacements for mechanical shearing. Deficiencies of methionine and lysine weaken the fleece and inhibit wool growth. Analogs of essential amino acids, such as ethionine and methoxine, cause a cessation of wool growth.

The skin may atrophy as other organs do or it may, as is often the case in swine, become mildly hyperkeratotic and assume a dirty, dry, scaly appearance. The skin of a malnourished animal is of increased susceptibility to bacterial infection and to parasitic infestations and their effects. The effects of starvation on other organs such as liver, pancreas, bone, and bone marrow are discussed elsewhere.

B. Fatty Acid Deficiency

Fatty acid deficiency may occur in all domestic species associated with general dietary deficiency, malabsorption, or liver disease. It is manifested chiefly by diffuse scaling and alopecia. The scaliness is initially dry but becomes oily. Otitis externa may be an accompanying lesion, and the skin is susceptible to secondary bacterial infections. Histologic lesions include epidermal hyperplasia, orthokeratotic, or parakeratotic hyperkeratosis and hypergranulosis. The pathologic mechanisms underlying the epidermal hyperproliferation are not well understood. Experimental studies demonstrated an increase in epidermal DNA synthesis and a decrease in prostaglandin E and F levels in the skin of essential fatty acid-deficient mice. The lower prostaglandin levels probably reflect a lack of precursor arachidonic acid. Deficiency of prostaglandin E_2 could influence epidermal cell kinetics through its effect on ratios of cyclic adenosine monophosphate (AMP) to guanosine monophosphate (GMP). Topical application of prostaglandin E_2 or arachidonic acid to experimentally fatty acid-deficient rats corrects scaling within 2 weeks.

Experimental essential fatty acid deficiency in **pigs** is produced by feeding less than 0.06% fat. Cutaneous lesions include brown encrustation of the ears, axillae, and flanks, diffuse scaling, and alopecia. A seborrheic dermatosis in **cats** characterized by dry scaly skin, and alopecia is responsive to fatty acid supplementation but is not likely to be the result of a true deficiency. Experimental essential fatty acid deficiency in cats produces dry seborrheic coats. Diets containing linoleic acid and linolenic acids as the sole source of essential fatty acids also induce deficiency. Cats are obligate carnivores, as they lack delta-6-desaturase, the enzyme responsible for converting these 18 carbon fatty acids to longer chain fatty acids. Arachidonic acid is an essential fatty acid for the cat.

C. Hypovitaminoses and Vitamin-Responsive Dermatoses

Cutaneous lesions may occur as manifestations of **hypovitaminoses** associated with deficiencies of vitamins A, C, and E, riboflavin, pantothenic acid, biotin, and niacin. Most of these are described in experimentally induced deficiencies. Many of the naturally occurring hypovitaminoses are probably not the result of a single vitamin deficiency but represent the cumulative effect of several inadequacies of the diet.

Vitamin A is involved in cellular growth and differentiation, as well as having specialized functions in visual processes and reproduction. Vitamin A has a controlling effect on epithelial differentiation. Deficiency of vitamin A causes squamous epithelia to become hyperkeratotic and secretory epithelia to undergo squamous metaplasia.

Hypovitaminosis A has been reported in all species of domestic animals, although many accounts are anecdotal. It may be secondary to dietary deficiency, decreased intestinal absorption, liver disease, or toxicities such as chlorinated naphthalene toxicity of cattle. The cutaneous lesions of hypovitaminosis A in **cattle** are a marked scaling and crusting dermatitis; in the **pig,** follicular hyperkeratosis; and in **cats,** scaling, follicular plugging, and alopecia.

Vitamin A-responsive dermatoses occur in **dogs.** In one syndrome, cocker spaniels are predisposed, probably because of a congenital abnormality of epidermopoiesis and keratinization (see Diseases of Keratinization). The lesions include hyperkeratotic plaques, follicular plugging, and the formation of keratin fronds. Ventral and lateral chest and abdomen are sites of predilection. There may be an accompanying otitis externa and pyoderma. Histologically, the predominant lesion is marked follicular keratosis. Vertically oriented keratin casts protrude from the follicular ostia. There is mild to moderate epidermal hyperplasia and surface hyperkeratosis of the basket-weave type. Dermal inflammation is mild, mononuclear, and perivascular, unless secondary bacterial infection intervenes to produce suppurative folliculitis and/or furunculosis. A more severe syndrome, in a Labrador pup, presented with generalized seborrhea oleosa, crusting, and patchy alopecia (Fig. 5.36A,B). The histologic lesions reflected a severe disturbance in keratinization, with marked epidermal hyperplasia, alternating orthokeratotic and parakeratotic hyperkeratosis, follicular keratosis, and dyskeratosis (Fig. 5.36C,D). The sebaceous glands were hyperplastic. Very

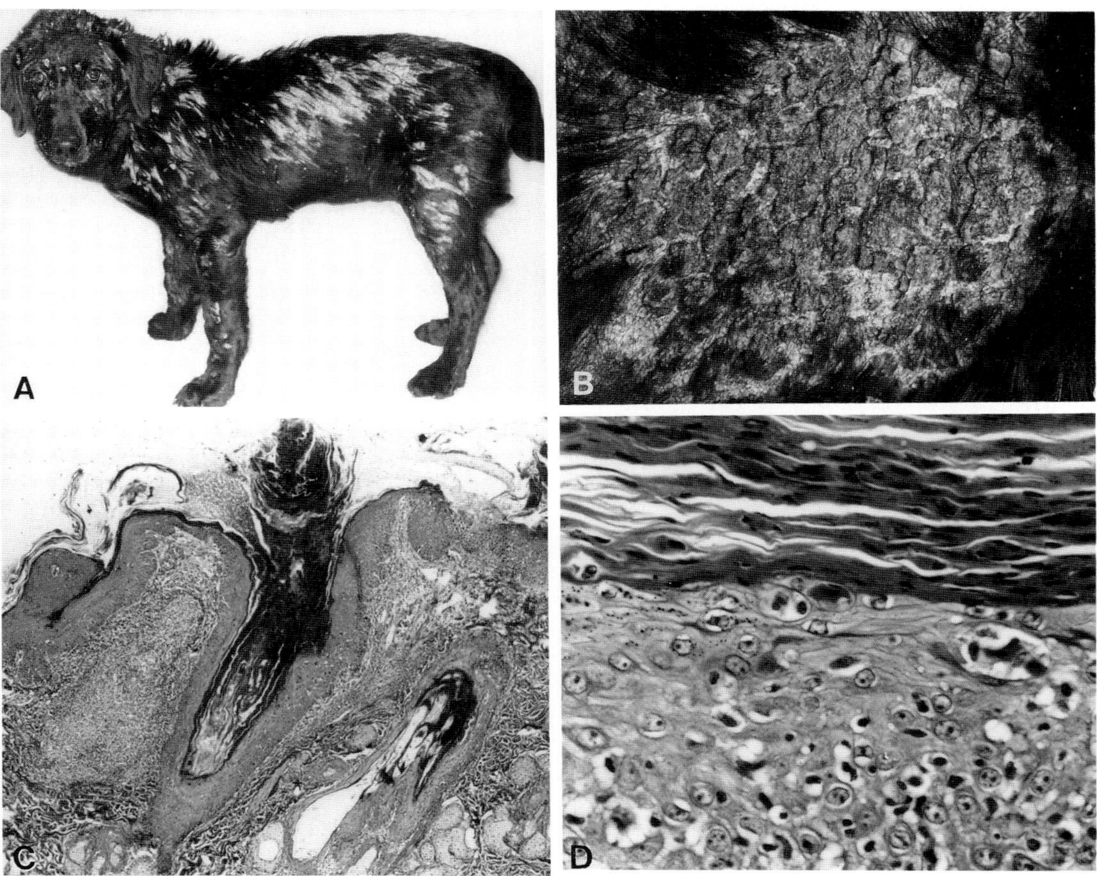

Fig. 5.36 Vitamin A-responsive seborrhea. Dog. (A) and (B) Alopecia, scaling, and crusting. (C) Follicular keratosis and acanthosis. (Courtesy of W. M. Parker, M. H. Hardy, and the *J Am Anim Hosp Assoc*) (D) Parakeratotic hyperkeratosis, dyskeratosis, acanthosis, and mononuclear cell exocytosis.

large doses of oral vitamin A were necessary to maintain a normal coat throughout the dog's life. These diseases are not vitamin A deficiencies *per se,* in that plasma levels of vitamin A are within the normal range. The fact that therapy is effective can be attributed more to the "normalizing" effect of vitamin A (and retinoids) on cellular differentiation in the epidermis.

The **B-vitamin complex** is essential to the maintenance and proper functioning of many important metabolic pathways. These vitamins interact with each other, with vitamin C, and with fat-soluble vitamins. Single deficiencies of these vitamins are rare.

Riboflavin deficiency (vitamin B_2) is mostly a problem in swine and chickens fed grain rations with borderline concentrations of the vitamin. Ruminants do not become deficient, because of rumen synthesis of the B-complex vitamins. Young calves, however, may develop deficiency if deprived of milk or an appropriate replacer. Animals develop hyperemia around the lips, nose, and buccal mucosa, diarrhea, weight loss, and generalized alopecia. Cutaneous lesions described in hyporiboflavinosis include a scaling and ulcerative dermatitis in the pig, erythema, scaling, and dry hair coat on the ventral abdomen and hind legs in the dog, and alopecia in the cat.

Pantothenic acid deficiency is best documented in pigs but may occur in dogs and calves. Pantothenic acid deficiency in pigs produces progressive alopecia with dermatitis and ulceration, in addition to the general effects of weight loss, diarrhea, and neurologic signs. Young preruminant calves may show dermatologic signs, which include alopecia, a roughened coat, and dermatitis. Leukotrichia has been described in dogs.

Biotin deficiency rarely occurs spontaneously, except in intensively reared swine, as the vitamin is widely distributed in feeds. Deficiencies usually develop in the course of general dietary insufficiency but may be induced by feeding raw egg whites, which contain avidin, a substance which renders biotin unavailable. Prolonged oral antibiotic therapy has been associated with biotin deficiencies. Cutaneous lesions reported in biotin-deficient **pigs** include alopecia, pustular dermatitis, and cracked hooves. Cracks affect toes, side walls, and heels, causing lameness and significant economic loss, as claw lesions can account for up to one third of all cullings in sows and fattening pigs. Microscopic lesions include epidermal hyperplasia, orthokeratotic and parakeratotic hyperkeratosis, epidermal necrosis and pustules, and folliculitis. Experimental deficiencies in the **cat** cause seborrhea and leukotrichia and, in severe cases, generalized papulocrustous dermatitis. A deficiency attributed, in part, to a lack of vitamin B, probably biotin, was reported in **lambs** reared artificially on reconstituted cow's milk. Alopecia was largely due to a fleece in which individual fibers were thin, weak, and straight. Erythroderma correlated with superficial capillary dilation, a lesion seen in a biotin-responsive disease of human infants. Histologically and ultrastructurally, the wool fibers had reduced numbers of cortical cells. Histochemical stains showed a delay in the incorporation and oxidation of sulfydryl groups. Most follicles remained in anagen, but hair bulbs were smaller than those in controls and developed autophagic vacuoles. Supplementation with B-group vitamins partially restored the fleece. Biotin deficiency reportedly causes a dry, brittle hair coat, scaling, and leukotrichia in **dogs.** An uncontrolled clinical study on the effect of biotin supplementation on a range of skin diseases in dogs claimed a 60% cure rate.

Niacin deficiency occurs spontaneously in pigs fed a corn ration, which renders the vitamin unavailable to the animal. Cutaneous lesions include alopecia and a crusting dermatitis. Niacin deficiency in dogs induces reddening and ulceration of the oral mucous membranes, resembling human hyponiacinosis, known as pellagra. Reference has been made to pruritic dermatitis of the hind legs and abdomen in niacin-deficient dogs.

Pyridoxine deficiency (vitamin B_6), produced experimentally in cats, was reported to cause a dull, unkempt coat with scaliness and alopecia of the head and extremities. Histologically, the alopecic areas had few hair follicles, and all were in telogen. Haired areas showed follicular hyperkeratosis and epidermal hyperplasia. Bacterial colonization of follicular infundibula occasionally led to folliculitis and furunculosis. Since pyridoxine may be indirectly involved in zinc transport, through its effect on tryptophan metabolism, these effects might be due to alteration in the levels of zinc.

Vitamin C-responsive disease occurs in calves fed inadequate milk replacer. The animals have low levels of ascorbic acid in the plasma, and lesions respond to parenteral administration of the vitamin. The lesions commence on the head and neck, extending rapidly to the trunk and limbs. Heavy scaling appears first, followed by alopecia. The denuded skin is erythematous, thin, and highly susceptible to even minor injury. Purpura is prominent, particularly on the extremities. Histologically, there is orthokeratotic hyperkeratosis, follicular keratosis, marked vascular congestion, and perifollicular hemorrhage. On necropsy, hemorrhages were found in other organs. In piglets, low vitamin C levels in a feeder ration were associated with a "scurfy" skin condition.

Vitamin E deficiency is discussed with the Cardiovascular System (Volume 3), with Muscles and Tendons (this volume), and with the Liver and Biliary System (Volume 2). Here we concentrate on "yellow-fat disease" (steatitis), which involves subcutaneous fat and may present clinically as a skin disease. The yellowness of the fat is due to the deposition of ceroid.

Nutritional panniculitis occurs in **cats, mink, foals,** and **swine.** The disease is associated with the feeding of fish meal, fish offal, or other products with a high concentration of unsaturated fatty acids. Destruction of vitamin E also occurs in food which, through improper processing or storage, becomes rancid. The disease is not, however, the result of a simple deficiency, since cats fed diets deficient in vitamin E but also lacking in unsaturated fatty acids do not develop panniculitis. In mink and foals, it may be associated with degeneration of muscles, and in swine it may occur alone or be associated with any one or a combination of ulceration of the squamous mucosa of the stomach, muscle degeneration, and hepatosis dietetica. The disease is frequently fatal in **cats** after a short clinical course in which there is progressive depression; the only characteristic feature clinically is a palpable thickening and increased firmness of the subcutaneous fat, easiest to detect in the inguinal region. The changes are not confined to the subcutaneous tissues but affect all fat depots. The color of the fat varies from gray to lemon-yellow to orange, and the fat is indurated and sometimes edematous.

The initial histologic change is the deposition of globules of ceroid in the interstitial tissue. This, together with a fishy odor, may be all that is observed in swine, in which the fat is soft and gray rather than yellow. In most affected cats, fat necrosis stimulates panniculitis, which is initially neutrophilic but becomes granulomatous. Numerous macrophages, and occasionally giant cells, ingest the ceroid pigment. Ceroid may be found also in macrophages of the liver, spleen, and lymph nodes. Ceroid, a variant of lipofuscin, is acid fast and autofluorescent.

Vitamin E-responsive dermatosis has been described in the **goat.** Kids and adults, on a selenium-deficient diet, developed periorbital alopecia and generalized seborrhea. The hair coat was dull and brittle. Histologically, the lesions were orthokeratotic hyperkeratosis and mild superficial perivascular dermatitis, in which mononuclear cells predominated.

Experimental vitamin E deficiency in dogs induced seborrhea, which was initially dry but became greasy, generalized erythroderma, and alopecia. Secondary pyoderma developed. Dogs also showed defective T-lymphocyte function. Histologically, the epidermis was hyperplastic, spongiotic, and focally parakeratotic. Dermal blood vessels were dilated, and the superficial dermis was edematous. Restoration of Vitamin E to the diet reversed the cutaneous lesions within 10 weeks.

Alopecia in **calves** associated with the feeding of milk substitute was attributed, in part, to vitamin E deficiency. The calves had low levels of serum vitamin E, and the hair regrew after vitamin E therapy was instigated for concurrent muscular dystrophy.

Bibliography

Brooks, P. H., Smith, D. A., and Irwin, V. C. R. Biotin-supplementation of diets; the incidence of foot lesions, and the reproductive performance of sows. *Vet Rec* **101:** 46–50, 1977.

Carey, C. J., and Morris, J. G. Biotin deficiency in the cat and the effect on hepatic propionyl CoA carboxylase. *J Nutr* **107:** 330–334, 1977.

Chapman, R. E., and Black, J. L. Abnormal wool growth and alopecia of artificially reared lambs. *Aust J Biol Sci* **34:** 11–26, 1981.

Coffin, D., and Holzworth, J. "Yellow fat" in two laboratory cats: Acid-fast pigmentation associated with a fish-base ration. *Cornell Vet* **44:** 63–71, 1954.

Davis, C. L., and Gorham, J. R. The pathology of experimental and natural cases of "yellow fat" disease in swine. *Am J Vet Res* **15:** 55–59, 1954.

De Jong, M. F., and Sytsema, J. R. Field experience with *d*-biotin supplementation to gilt and sowfeeds. *Vet Q* **5:** 58–67, 1983.

Dodd, D. C. *et al.* Muscle degeneration and yellow fat disease in foals. *N Z Vet J* **8:** 45–50, 1960.

Elias, P. M. *et al.* Retinoid effects on epidermal structure, differentiation, and permeability. *Lab Invest* **44:** 531–540, 1981.

Frigg, M., Schulze, J., and Volker, L. Clinical study on the effect of biotin on skin conditions in dogs. *Schweizer Arch Tierheilkunde* **131:** 621–625, 1989.

Gershoff, S. N. Vitamin A deficiency in cats. *Lab Invest* **6:** 277–240, 1957.

Goodwin, R. F. W. Some clinical and experimental observations on naturally occurring pantothenic acid deficiency in pigs. *J Comp Pathol* **72:** 214–232, 1962.

Gorham, J. R., Boe, N., and Baker, G. A. Experimental "yellow fat" disease in baby pigs. *Cornell Vet* **41:** 323–338, 1951.

Hansen, A. E., Sinclair, J. G., and Wiese, H. F. Sequence of histologic changes in skin of dogs in relation to dietary fat. *J Nutr* **52:** 451–554, 1954.

Hutchison, G., and Mellor, D. J. Effects of maternal nutrition on the initiation of secondary wool follicles in foetal sheep. *J Comp Pathol* **93:** 577–583, 1983.

Hynd, P. I. Effects of nutrition on wool follicle cell kinetics in sheep differing in efficiency of wool production. *Aust J Agric Res* **40:** 409–417, 1989.

Ihrke, P. J., and Goldschmidt, M. H. Vitamin A-responsive dermatosis in the dog. *J Am Vet Med Assoc* **182:** 687–690, 1983.

Johnson, W. S. Vitamins A and C as factors affecting skin condition in experimental piglets. *Vet Rec* **79:** 363–365, 1966.

Klein-Szanto, A. J. P., Martin, D. H., and Pine, A. H. Cutaneous manifestations in rats with advanced vitamin A deficiency. *J Cutan Pathol* **7:** 260–270, 1980.

Loewenthal, L. A., and Montagna, W. Effects of caloric restriction on skin and hair growth in mice. *J Invest Dermatol* **24:** 429–433, 1955.

McLaren, D. S. Cutaneous changes in nutritional disorders. *In* "Dermatology in General Medicine," 3rd Ed. T.B. Fitzpatrick *et al* (eds.), pp. 1601–1613. New York, McGraw-Hill, 1987.

Pritchard, G. C. Alopecia in calves associated with milk substitute feeding. *Vet Rec* **112:** 435–436, 1983.

Reis, P. J. *et al.* Effects of methoxinine, an analogue of methionine, on the growth and morphology of wool fibers. *Aust J Biol Sci* **39:** 209–223, 1986.

Scott, D. W. Vitamin C-responsive dermatosis in calves. *Bov Pract* **2:** 22–27, 1981.

Scott. D. W. Vitamin A-responsive dermatosis in the cocker spaniel. *J Am Anim Hosp Assoc* **22:** 125–129, 1986.

Scott, D. W., and Sheffy, B. E. Dermatosis in dogs caused by vitamin E deficiency. *Comp Anim Pract* **1:** 42–46, 1987.

Spratling, F. R. *et al.* Experimental hypovitaminosis-A in calves. Clinical and gross postmortem findings. *Vet Rec* **77:** 1532–1542, 1965.

Ward, K. A. *et al.* The regulation of wool growth in transgenic animals. *In* "Advances in Veterinary Dermatology," Vol. 1, C. von Tscharner and R.E.W. Halliwell (eds.), pp. 70–76. London, Baillière Tindall, 1990.

Wilkinson, G. T. Nutritional deficiencies in the cat. *Vet Ann* **21:** 183–189, 1981.

D. Mineral Deficiency

Iodine, cobalt, copper, and zinc deficiencies may lead to integumentary lesions. Iodine deficiency is discussed with the Endocrine Glands (Volume 3). The effects of copper deficiency on wool and hair are referred to under Disorders of Pigmentation. Cobalt deficiency causes a progressive, wasting disease in ruminants. There are nonspecific changes in the wool or hair coat, including lack of growth and increased fragility.

Zinc is an essential trace element. It occurs in all body tissues, particularly bones, teeth, muscle, and integument. In the integument, a substantial proportion of zinc is in the wool or the hair coat. Zinc is a component of many important metalloenzymes and is a cofactor for many others. It exerts its primary effect through zinc-dependent enzymes which regulate RNA and DNA metabolism. Zinc thus plays a role in all metabolic processes involved with tissue growth, maturation, and repair. Zinc also modulates many aspects of the immune and inflammatory responses. The relationship between the changes in particular tissue enzyme activities brought about by zinc deficiency and clinical manifestations of the deficiency syndrome are not, however, well understood.

Zinc deficiency causes anorexia, alterations in food utilization, growth retardation, reproductive disorders, depression of the immune response, hematologic abnormalities, depression of central nervous system development, decreased wound healing, and keratinization defects in epidermis, hair, wool, and horny appendages. The disease occurs in all domestic species but is of most significance in the pig and dog.

1. Parakeratosis in Swine

Zinc-responsive dermatosis in swine (parakeratosis) became an important clinical entity in the 1950s, coincident with and related to the widespread introduction of dry meal feeding. The cause is not a simple deficiency. The availability of dietary zinc is adversely affected by the presence of phytic acid, as can occur in soybean protein rations, a high concentration of calcium, a low concentration of free fatty acids, alterations in intestinal flora, and the presence of bacterial and viral enteric pathogens such as transmissible gastroenteritis virus. Zinc deficiency may induce a secondary vitamin A deficiency as a result of its effect on appetite and food utilization. Economic losses are due to depression of growth rate, but, with improved

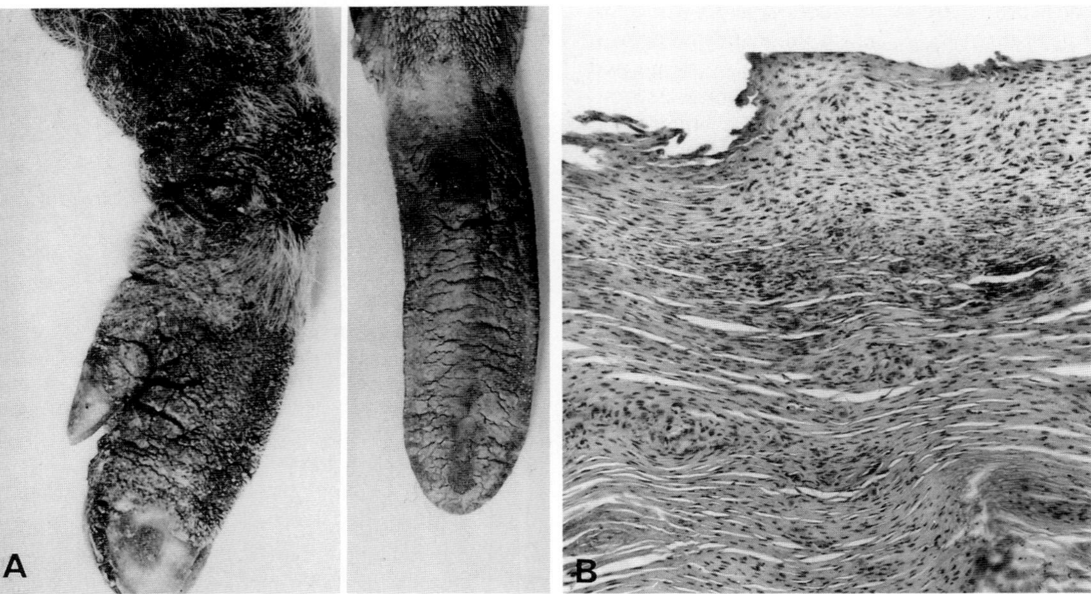

Fig. 5.37 (A) Zinc-responsive disease involving foreleg and tongue. Pig. (B) Parakeratotic hyperkeratosis in zinc responsive dermatosis.

management techniques, parakeratosis is no longer a major problem.

Parakeratosis occurs in young, growing pigs, 2–4 months of age. The initial gross lesions are erythematous macules on the ventral abdomen and medial surface of the thigh. The lesions develop into papules, which become covered with a gray-brown, dry, roughened scale-crust that may reach 5–7 mm in thickness. Deep fissures penetrate the crust and are filled with a brown-black detritus, which is composed of sebum, sweat, soil, and other debris (Fig. 5.37A). These areas are susceptible to secondary bacterial infection, often leading to pyoderma or subcutaneous abscessation. Lesions are roughly symmetrical and have a predilection for the lower limbs, particularly over the joints, around the eyes, ears, snout, scrotum, and tail. In severely affected animals, lesions may become generalized. Pruritus is not a feature of parakeratosis. The dorsal surface of the tongue is "furred" (Fig. 5.37A), and the esophageal mucosa loses its normal smooth sheen and becomes dull, white, and lusterless.

The microscopic **lesion** in the skin is a very marked hyperplastic dermatitis with diffuse parakeratotic hyperkeratosis (Fig. 5.37B). The oral mucous membranes also demonstrate a hyperplastic epithelium. Acanthosis, elongation of rete ridges, and mitotic figures in the basal keratinocytes are regular features of the hyperplastic response of the epidermis. The dermal lesions in uncomplicated parakeratosis include variable vasodilation and a mild to moderate, predominantly mononuclear cell, perivascular infiltrate. With bacterial infection, there may be nodular or diffuse suppurative dermatitis, suppurative folliculitis, perifolliculitis, or furunculosis.

Parakeratosis in swine must be differentiated grossly from sarcoptic mange and exudative epidermitis. The former is usually intensely pruritic, and the latter usually occurs in a younger age group, and the scale-crust is greasy rather than dry. Parakeratosis is rarely fatal unless from toxemia or septicemia secondary to cutaneous bacterial infections or by exacerbation of intercurrent infections such as pneumonia. Affected pigs recover rapidly on zinc supplementation.

Bibliography

Kernkamp, H. C. H., and Ferrin, E. F. Parakeratosis in swine. *J Am Vet Med Assoc* **123:** 217–220, 1953.

Luecke, R. W. *et al.* Calcium and zinc in parakeratosis of swine. *J Anim Sci* **16:** 3–11, 1957.

Whitenack, D. L., Whitehair, C. K., and Miller, E. R. Influence of enteric infection on zinc utilization and clinical signs and lesions of zinc deficiency in young swine. *Am J Vet Res* **39:** 1447–1454, 1978.

2. *Canine Zinc-Responsive Dermatoses*

Experimental zinc deficiency in the dog produces emaciation, general debility, conjunctivitis, keratitis, and skin lesions. The skin lesions comprise focal areas of erosion and crusting with a tendency to occur bilaterally on areas of contact or friction, particularly on the extremities. The experimental canine deficiencies have less extensive gross lesions and lesser degrees of parakeratosis than occur in zinc-deficient pigs.

Naturally occurring zinc-responsive dermatoses in the dog fall under two syndromes. **Syndrome 1** primarily affects **Siberian huskies,** but has been seen in Alaskan malamutes, and other breeds not of Arctic origin, such as Great Danes. The cutaneous lesions usually become manifest before the dogs are 1 year of age. Older dogs may develop lesions during times of stress such as pregnancy, lactation, or intercurrent disease. Lesions comprise a scaling and crusting dermatitis with a predilection for the face, particularly around the eyes, lips, nose, pressure points, and footpads. Secondary pyoderma is not uncommon. The pathogenesis of the syndrome is not well established. Alaskan malamutes with chondrodysplasia have zinc-responsive spermatozoal defects and reduced zinc absorption from the intestinal tract; possibly malabsorption is responsible for the dermatological disease.

Syndrome 2 occurs in puppies of any breed and is associated with a relative deficiency of zinc, probably secondary to excessively high levels of calcium and/or phytates in the diet in rapidly growing animals. In the United States, this disease has been associated with the feeding of generic dog food, and in Britain and Sweden with the feeding of soy- and/or cereal-based diets. Recovery quickly follows restoration of a diet meeting the National Research Council standards for canine nutrition or zinc supplementation of cereal diets. Now that dog food manufacturers are aware of the problem, the disease is less common.

The gross lesions are multiple, roughly bilaterally symmetrical, scaling and crusted plaques, particularly affecting the muzzle, pressure points, distal extremities, and trunk. There is extreme thickening and fissuring of the footpads and, sometimes, the planum nasale. Secondary pyoderma may develop, and puppies often show a local lymphadenopathy.

The histologic lesions are usually typical of zinc deficiency, with epidermal hyperplasia, marked parakeratotic hyperkeratosis affecting surface epithelium and that of the follicular ostia, and a mild to moderate superficial perivascular mononuclear or eosinophil-rich dermatitis. In a series of 13 affected dogs, the keratinization defect was sufficiently severe to induce dyskeratotic changes at all levels of the epidermis. However, in a report from Scotland, orthokeratotic hyperkeratosis alone was found. Initial reports of experimental zinc deficiency in the dog also failed to detect parakeratotic lesions, and the lesions reported in a more recent publication on experimental canine zinc deficiency are much less severe than those in the clinical cases. It is possible that dietary factors other than zinc are involved in some of these diseases, particularly those responding to dietary changes and not just to zinc supplementation. A vitamin A-responsive dermatosis in a young Labrador retriever had histologic lesions very similar to those described in American dogs with generic dog food disease.

A study on the zinc concentration in serum, leukocytes, and hair of dogs with zinc-responsive dermatoses and normal controls found a significant difference between the levels in hair and serum, but not in leukocytes. However, serum and plasma zinc levels are not an accurate method of assessing zinc status. Values, even in a normal animal, are subject to marked variation according to sex, age, stress, and concurrent disease. Contamination during sample collection and processing is another major problem, rendering serum or plasma zinc determination an unreliable test for deficiency disease.

Bibliography

Brown, R. G. *et al.* Alaskan malamute chondrodysplasia. V. Decreased gut zinc absorption. *Growth* **42:** 1–6, 1978.

Degryse, A.-D. *et al.* Recurrent zinc-responsive dermatosis in a Siberian husky. *J Small Anim Pract* **28:** 721–726, 1987.

Fadok, V. A. Zinc-responsive dermatosis in a Great Dane: A case report. *J Am Anim Hosp Assoc* **18:** 409–414, 1982.

Ohlen, B., and Scott, D. W. Zinc-responsive dermatitis in puppies. *Canine Pract* **13:** 6–10, 1986.

Sanecki, R. K., Corbin, J. E., and Forbes, R. M. Tissue changes in dogs fed a zinc-deficient ration. *Am J Vet Res* **43:** 1642–1646, 1982.

Smith, J. C. The vitamin A-zinc connection: A review. *Ann N Y Acad Sci* **355:** 62–75, 1980.

Sousa, C. A. *et al.* Dermatosis associated with feeding generic dog food: 13 cases (1981–1982). *J Am Vet Med Assoc* **192:** 676–680, 1988.

Thoday, K. L. Diet-related zinc-responsive skin disease in dogs: A dying dermatosis? *J Small Anim Pract* **30:** 213–215, 1989.

Van den Broek, A. H. M. Diagnostic value of zinc concentrations in serum, leucocytes, and hair of dogs with zinc-responsive dermatosis. *Res Vet Sci* **44:** 41–44, 1988.

Van den Broek, A. H. M., and Thoday, K. L. Skin disease in dogs associated with zinc deficiency: A report of five cases. *J Small Anim Pract* **27:** 313–323, 1986.

3. Zinc Deficiency in Cats

Experimental diets deficient in zinc resulted in a thin hair coat, scaliness, and ulceration of the lips. Histologically, typical parakeratotic hyperkeratosis was apparent.

Bibliography

Kane E. *et al.* Zinc deficiency in the cat. *J Nutr* **111:** 488–495, 1981.

4. Zinc Deficiency in Ruminants

The clinical signs and gross lesions of zinc deficiency are similar in ruminants.

Experimentally induced zinc deficiency in **sheep** causes lesions in the integument and horny appendages. The wool fibers are thin, lose their crimp, and are easily epilated. The entire fleece may be shed. Wool eating and excess salivation are prominent clinical signs. The fleece may be stained with a red-brown water-soluble pigment. Crusting and scaling lesions develop around the eyes, nose, hooves, scrotum, and on pressure points. The normal ringed structure of the horns is lost, and the horns are shed. Abnormal growth of the hooves may lead to soreness and the adoption of a kyphotic stance. Parakeratosis is prominent histologically. Similar gross lesions have been reported in natural episodes of zinc deficiency in sheep, but histologically the hyperkeratosis was predominantly of the orthokeratotic rather than parakeratotic type. It has been suggested that the parakeratotic lesions are most pronounced at sites of friction, pressure, or trauma.

Zinc-responsive dermatoses in **goats** are reported from the United States, Europe, and Australia, sometimes associated with depressed reproductive efficiency. Experimental zinc deficiency of goats produces the typical weight loss, alopecia, and parakeratotic scaling dermatosis described in other ruminant species.

Naturally occurring zinc-deficiency dermatoses are uncommon in **cattle,** but have been reported, for example, in grazing cattle from Guyana and housed dairy cows in Finland. Experimental deficiencies produce cutaneous lesions similar to those in other species. The skin generally is scaly, partially alopecic, hard and dry, with fissuring and exudation most prominent on the limbs and the perineum and about the mouth and eyes. The hereditary zinc deficiencies of cattle are discussed with Congenital and Hereditary Diseases of Skin.

Bibliography

Blackmon, D. M., Miller, W. J., and Morton, J. D. Zinc deficiency in ruminants. *Vet Med* **62:** 265–270, 1967.

Hanson, L. J., Sorensen, D. K., and Kernkamp, H. C. H. Essential fatty acid deficiency—its role in parakeratosis. *Am J Vet Res* **19:** 921–930, 1958.

Miller, J. K., and Miller, W. J. Experimental zinc deficiency and recovery of calves. *J Nutr* **76:** 467–474, 1962.

Miller, W. J. *et al.* Experimentally produced zinc deficiency in the goat. *J Dairy Sci* **47:** 556–559, 1964.

Mills, C. F., and Dalgarno, A. C. The influence of dietary calcium concentration on epidermal lesions of zinc deficiency in lambs. *Proc Nutr Soc* **26:** 19, 1967.

Mills, C. F. *et al.* Zinc deficiency and the zinc requirements of calves and lambs. *Br J Nutr* **21:** 751–768, 1967.

Nelson, D. R. *et al.* Zinc deficiency in sheep and goats: Three field cases. *J Am Vet Med Assoc* **184:** 1480–1485, 1984.

Pierson, R. E. Zinc deficiency in young lambs. *J Am Vet Med Assoc* **149:** 1279–1282, 1966.

Reuter, R. *et al.* Zinc-responsive alopecia and hyperkeratosis in Angora goats. *Aust Vet J* **64:** 351–352, 1987.

Suliman, H. B. *et al.* Zinc deficiency in sheep: Field cases. *Trop Anim Health Prod* **20:** 47–51, 1988.

Suttle, N. F. Problems in the diagnosis and anticipation of trace element deficiencies in grazing livestock. *Vet Rec* **119:** 148–152, 1986.

Underwood, E. J. "Trace Elements in Human and Animal Nutrition." New York, Academic Press, 1977.

IX. Endocrine Diseases of Skin

Endocrine dermatoses share many similar gross features, including hypotrichosis or alopecia that is frequently bilaterally symmetric, variable degrees of pigmentary disturbance (hypermelanosis or hypomelanosis), a hair coat that is often coarse, dull, dry, brittle, easily epilated, and fails to regrow normally after being clipped, and variable degrees of seborrheic skin disease and/or bacterial pyoderma.

Endocrine dermatoses also have many histologic findings in common. These include orthokeratotic hyperkeratosis, follicular keratosis, follicular dilatation, follicular atrophy, absence of hairs in hair follicles, a predominance of telogen hair follicles, excessive trichilemmal keratinization of hair follicles, epidermal atrophy, epidermal hypermelanosis, and sebaceous gland atrophy (Fig. 5.38). These lesions suggest endocrine skin disease, but are not pathognomonic. Confirmation of an endocrine dermatosis requires demonstration of a hormone deficiency or excess, lasting response to specific therapy, and demonstration of appropriate lesions in endocrine glands. Secondary seborrheic skin disease and/or bacterial pyoderma often complicate endocrine dermatoses. They result in varying degrees of inflammation, from superficial perivascular dermatitis (hyperplastic, spongiotic) to perifolliculitis, folliculitis, and furunculosis.

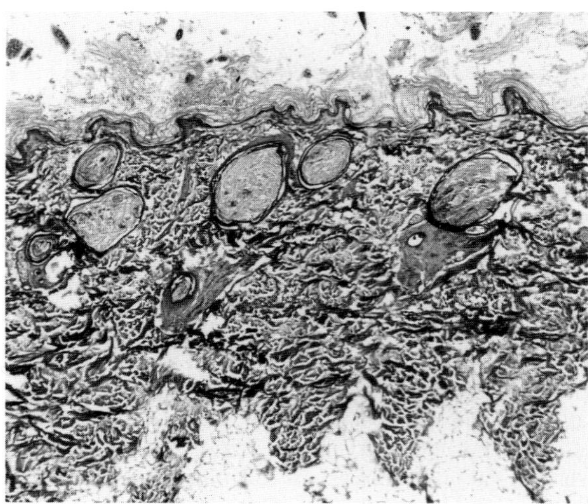

Fig. 5.38 Endocrine skin disease. Orthokeratotic hyperkeratosis, follicular keratosis, pilosebaceous atrophy.

Hypothyroidism is the most common endocrine skin disorder of the **dog.** It is associated with inadequate serum concentrations of thyroxine and triiodothyronine. Canine hypothyroidism may be caused by lymphocytic thyroiditis, idiopathic thyroid atrophy, pituitary neoplasia, hypopituitarism, bilateral thyroid neoplasia, developmental defects of the thyroid gland, iodine deficiency, hypothalamic defects, and iatrogenic mechanisms (surgery, drugs). It is most common in middle-aged dogs and has no sex predilection. Chow chow, Great Dane, Irish wolfhound, boxer, English bulldog, dachshund, Afghan, Newfoundland, Alaskan malamute, Doberman pinscher, Brittany spaniel, poodle, golden retriever, Irish setter, and miniature schnauzers are predisposed.

Gross cutaneous changes in canine hypothyroidism may include hypotrichosis or alopecia (focal, multifocal, or generalized; symmetric or asymmetric); rare hypertrichosis; a coarse, dull, dry, brittle, easily epilated hair coat, which often fails to regrow after clipping; pigmentary disturbances; normal or thick skin, which may be swollen by myxedema and cool to the touch; easy bruising; poor wound healing; gynecomastia, and frequent secondary seborrheic skin disease (dry or oily) and/or bacterial pyoderma.

Histologically, canine hypothyroidism is characterized by varying degrees of cutaneous atrophy with or without

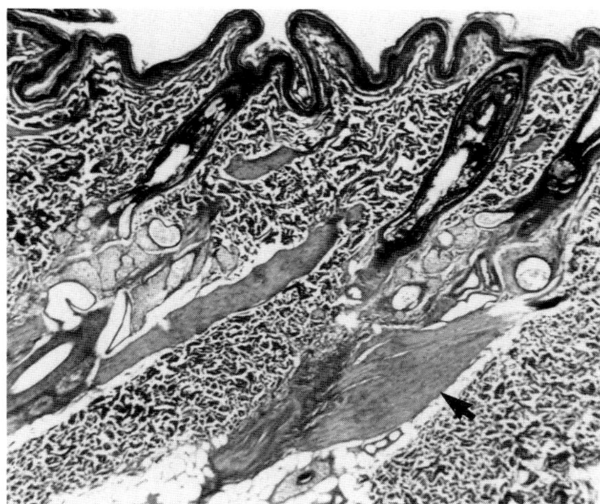

Fig. 5.39 Endocrine skin disease. Hypothyroidism. Dog. Follicular keratosis, follicular atrophy, prominent arrector pili muscles (arrow).

inflammation. Skin changes found to be indicative of hypothyroidism include vacuolated arrector pili muscles (present in ~74% of the cases studied), hypertrophied arrector pili muscles (~43%), thickened dermis (~41%), and increased dermal mucin (~30%) (Fig. 5.39).

Hypothyroidism occurs less frequently in other domestic animals, and usually in association with iodine deficiency and goiter. In **Merino sheep** and **Afrikander cattle,** a hereditary defect in the biosynthesis of thyroid hormone produces symmetric hypotrichosis and thick, myxedematous, wrinkled skin. In **goats,** hypothyroidism occurs in a mixed strain of Saanen and dwarf goats in association with a hereditary congenital thyroglobulin deficiency. Gross cutaneous changes include bilaterally symmetric hypotrichosis and thick, myxedematous, scaly skin. Histologic findings in caprine hypothyroidism include orthokeratotic hyperkeratosis, follicular keratosis, diffuse dermal mucinous degeneration (myxedema), and dermal thickening.

Hyperadrenocorticism is the second most common endocrine skin disorder of the **dog**. This disease is associated with excessive exposure to endogenous or exogenous glucocorticoids. Canine hyperadrenocorticism may be caused by idiopathic bilateral adrenocortical hyperplasia, functional pituitary neoplasms, functional adrenocortical neoplasms, "ectopic" production of an adrenocorticotropiclike substance by nonpituitary, nonadrenal neoplasms, and iatrogenic mechanisms. It is most common in middle-aged dogs and has no sex predilection. The boxer, Boston terrier, dachshund, and poodle are predisposed.

Gross cutaneous changes in canine hyperadrenocorticism may include symmetric hypotrichosis or alopecia; a coarse, dull, dry, brittle, easily epilated hair coat, which often fails to regrow after clipping; pigmentary disturbances; thin, hypotonic skin; easy bruising; poor wound healing; calcinosis cutis; multiple phlebectasias, and secondary seborrheic skin disease including comedone formation and/or bacterial pyoderma (Fig. 5.40A,B).

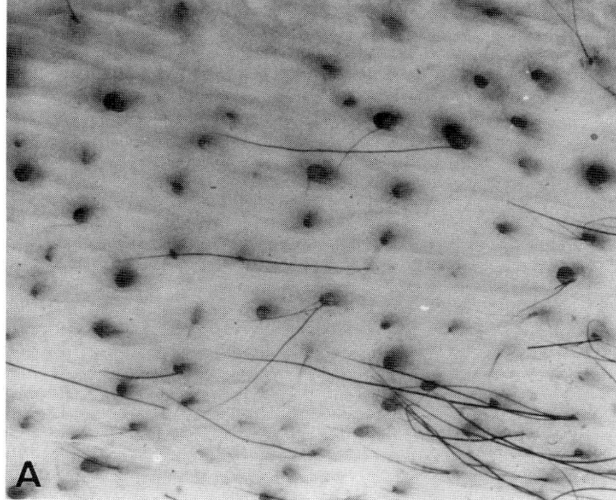

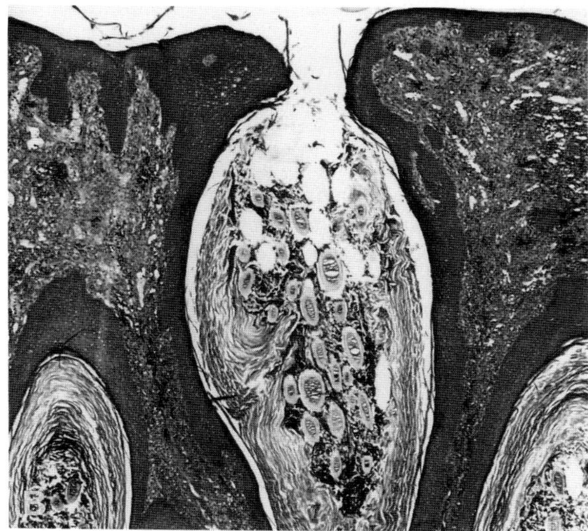

Fig. 5.40 Endocrine skin disease. Comedones. (A) Characteristic of hyperadrenocorticism. (B) Histology.

Histologically, canine hyperadrenocorticism is characterized by varying degrees of cutaneous atrophy with or without inflammation. Lesions indicative of hyperadrenocorticism include dystrophic mineralization involving dermal collagen and epidermal and follicular basement membrane zone (up to 67% of the cases) (Fig. 5.41), dermal thinning (~51%) and absence of arrector pili muscles (~33%). Cutaneous phlebectasias are characterized by marked dilatation and congestion of superficial dermal capillaries lined with a single layer of flattened endothelial cells (macular phase), to a lobular proliferation of dilated and congested but normal-appearing superficial dermal capillaries (papular phase). The intercapillary connective tissue stroma is often edematous, and the overlying epidermis often surrounds the vascular lesion like a collarette. Dermal perifollicular mineralization is a senile change in toy poodles.

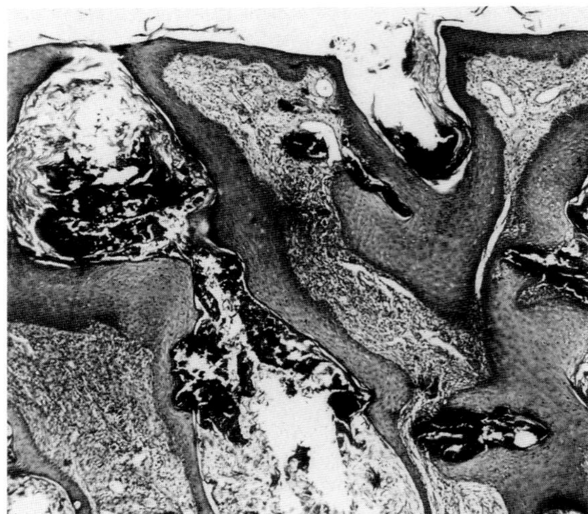

Fig. 5.41 Endocrine skin disease. Calcinosis cutis. Dog. Hyperadrenocorticism. Mineralization in dermis and hair follicles.

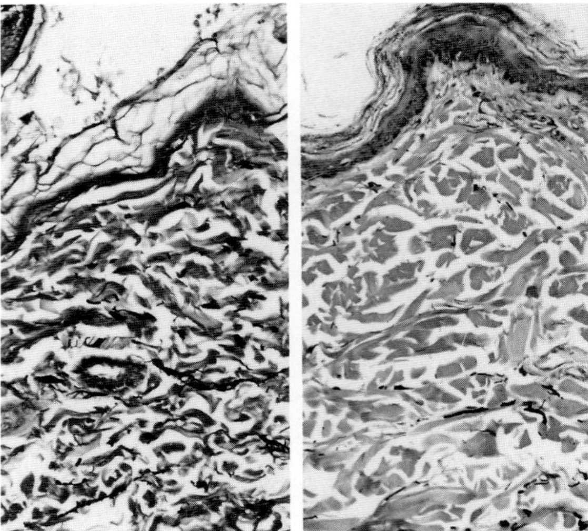

Fig. 5.42 Endocrine skin disease. Normal dermis (left) and dermis showing decreased elastin (right) specific for hyposomatotropism. Acid–orcein–Giemsa stain.

Hyperadrenocorticism occurs rarely in **cats,** in association with bilateral adrenocortical hyperplasia, pituitary or adrenocortical neoplasia, and iatrogenic mechanisms. Gross and histopathologic findings in affected skin are similar to those reported in dogs, but calcinosis cutis and phlebectasias are not reported in cats. Cutaneous fragility, manifesting as skin lacerations, is occasionally a feature of feline hyperadrenocorticism. Histologically, there is a marked reduction of dermal collagen.

Hyperadrenocorticism occurs in **horses** in association with functional pituitary neoplasms. The disease affects primarily aged horses and has no breed or sex predilection. Gross cutaneous changes may include a coarse, brittle, long, shaggy hair coat (hirsutism); dry or oily skin; episodic hyperhidrosis; poor wound healing; secondary skin infections (especially dermatophilosis), and rare xanthomas.

Hyposomatotropism occurs in **dogs** with congenital pituitary dwarfism and as an adult-onset disorder in otherwise normal dogs. In the German shepherd and Carelian bear dog, pituitary dwarfism is thought to be inherited as a simple autosomal recessive condition and is usually associated with a cystic Rathke's cleft. Adult-onset hyposomatotropism is usually seen in male dogs of the chow chow, keeshond, Pomeranian, and poodle breeds.

Gross cutaneous changes in canine hyposomatotropism may include symmetric hypotrichosis or alopecia; a coarse, dull dry, brittle, easily epilated hair coat; hyperpigmentation; normal or thin, hypotonic skin; and secondary seborrhea sicca. Histologic changes found to be indicative of hyposomatotropism include decreased amount and size of dermal elastin fibers (Fig. 5.42). However, dermal elastin abnormalities in dogs with adult-onset hyposomatotropism are not commonly seen until alopecia has been present for 2 years or longer. Flame follicles, which are inactive follicles with excessive trichilemmal keratinization, are often present but are not specific for hyposomatotropism.

Hypersomatotropism (acromegaly) is occasionally recognized in **dogs**. Excessive somatotropin (growth hormone) production is associated with progestagen administration or with the metestrus (luteal) phase of the estrous cycle in intact female dogs. Cutaneous changes include thick, folded, myxedematous skin, especially on the head, neck, and distal extremities. The hair coat may be long and thick, and the nails may exhibit rapid overgrowth.

Histologic findings in canine hypersomatotropism include variable degrees of orthokeratotic hyperkeratosis, hypergranulosis, and regular to irregular epidermal hyperplasia. Dermal collagen and fibroblasts are increased in amount and number, respectively. The dermis is often thickened, dense, and cellular in appearance. Myxedema is present in about a third of cases.

Hyperestrogenism is a rare cause of skin disease in **dogs**. It is seen in middle-aged to older intact male dogs with functional testicular neoplasms (especially Sertoli cell neoplasms). Boxer, Shetland sheepdog, Cairn terrier, Pekingese, collie, and Weimaraners are predisposed. Hyperestrogenism is also seen in middle-aged to older intact female dogs with polycystic ovaries or functional ovarian neoplasms.

Cutaneous changes in canine hyperestrogenism may include symmetric hypotrichosis or alopecia; a coarse, dull, dry, brittle, easily epilated hair coat, which often fails to regrow following clipping; pigmentary disturbances (especially macular melanosis of the genital skin); gynecomastia; vulvar enlargement; linear preputial dermatosis, and secondary seborrheic skin disease. Histologic findings include varying degrees of cutaneous atrophy with or without inflammation.

Diethylstilbestrol administration has been reported to

cause alopecia in the dog. Hair regrew when the drug was withdrawn. Histologic findings were typical of endocrine dermatoses.

The most common dermatological manifestations of **diabetes mellitus** in dogs and cats are seborrheic skin disease with thin and hypotonic skin (usually dry and scaly), varying degrees of hypotrichosis, and secondary bacterial pyoderma. Rarely, pruritus vulvae and xanthomatosis occur in association with canine diabetes mellitus. The histologic findings in canine and feline diabetes mellitus have not been reported.

Estrogen-responsive dermatosis is a rare skin disorder seen predominantly in prematurely ovariectomized bitches. It also occurs in bitches prior to their first estrus, during pseudopregnancy, and in association with estrous cycle abnormalities. Hypoestrogenism has not been documented in the vast majority of cases. The dermatosis is characterized by symmetric hypotrichosis or alopecia, easily epilated hair coat in affected areas, and infantile nipples and vulva. Histologic findings include orthokeratotic hyperkeratosis, follicular keratosis, follicular atrophy and dilatation, and hair follicles in telogen.

Testosterone-responsive dermatosis is a rare skin disorder seen in mature male dogs either with normal, atrophied, or cryptorchid testicles, with testicular neoplasms, or following castration. Hypoandrogenism has not been documented. The dermatosis is characterized by symmetric hypotrichosis or alopecia; a coarse, brittle, dull, dry, easily epilated hair coat; thin and hypotonic skin, and seborrhea sicca. Histologic findings include orthokeratotic hyperkeratosis, follicular keratosis, follicular dilatation and atrophy, epidermal atrophy, hair follicles predominantly in telogen, dermal thinning, and sebaceous gland atrophy.

Dermatoses associated with testicular neoplasia are not uncommon in the dog. Such dermatoses include the feminization syndrome classically ascribed to a functional Sertoli cell neoplasm (see the preceding section), and various dermatoses associated with "functional" interstitial cell neoplasms, seminomas, or Sertoli cell neoplasms. Such dogs manifest varying degrees of symmetric hypotrichosis or alopecia, coat color changes, pigmentary disturbances (especially macular melanosis of the genital skin), linear preputial dermatosis, perianal gland and tail gland hyperplasia, and secondary seborrhea (dry or greasy). Histopathologic findings include varying degrees of cutaneous atrophy.

Castration-responsive dermatosis is an uncommon, idiopathic dermatosis of dogs. Alaskan malamutes, Siberian huskies, keeshonds, and chow chows appear to be predisposed. Affected dogs have symmetric loss of primary hairs and retention of secondary hairs, which give the coat a woolly appearance; eventually alopecia, especially prominent on the caudomedial thighs, rump, tail, and neck; coat color changes, and pigmentary disturbances of the skin. The testicles are grossly and histologically normal. Histopathologic findings in affected skin show varying degrees of cutaneous atrophy. Flame follicles are often present.

Seasonal flank alopecia has been described in ovariohysterectomized dogs, especially boxers and English bulldogs. Affected dogs develop recurrent symmetric alopecia and intense hyperpigmentation localized more or less to the flank region. The alopecia begins in March or April (spring) and spontaneously resolves in July or August. The cause is unknown. Histopathologic findings included varying degrees of cutaneous atrophy and severe melanosis of epidermis, hair follicle outer root sheath, sebaceous glands, and sebaceous ducts.

An endocrine dermatosis has been reported in **Pomeranians,** which may be associated with a partial **deficiency of the 21-hydroxylase enzyme.** Affected dogs have symmetric alopecia of the trunk, caudal thighs and ventral neck; high baseline plasma adrenocorticotropic hormone (ACTH) concentrations, and increased production of progesterone, 17-hydroxyprogesterone, and dehydroepiandrosterone sulfate or androstenedione after ACTH administration. Histopathologic findings include varying degrees of cutaneous atrophy.

Feline idiopathic symmetric alopecia is a rare skin disorder of castrated males and ovariohysterectomized females. The disease has been known as feline endocrine alopecia, although there is little convincing evidence for endocrine dysfunction. Neither hypoandrogenism nor hypoestrogenism has been documented. Thyroid function tests are normal.

The dermatosis is characterized by symmetric hypotrichosis, which begins in the genital and perineal regions and may progress to involve the tail, ventral abdomen, caudomedial thighs, lateral thorax, and caudomedial front limbs. Hairs in affected areas are easily epilated. Histologic findings include predominantly telogen-phase hair follicles and an absence of hairs in many follicles.

The lesions of symmetric alopecia are not restricted to this idiopathic disorder. A careful clinical evaluation of 51 cats with symmetric alopecia revealed an underlying flea infestation in almost half of the animals. Other diseases which may present with symmetric alopecia include atopy, food allergy, psychogenic alopecia, telogen defluxation, and some ectoparasitic diseases.

Bibliography

Barsanti, J. A. *et al.* Diethylstilbestrol-induced alopecia in a dog. *J Am Vet Med Assoc* **182:** 63–64, 1983.

Gross, T. L., and Ihrke, P. J. The histologic analysis of endocrine-related alopecia in the dog. *In* "Advances in Veterinary Dermatology" Vol 1 C. von Tscharner and R.E.W. Halliwell, (eds.), pp. 75–88, London, Baillière Tindall, 1990.

Jones, B. R. *et al.* Cutaneous xanthomata associated with diabetes mellitus in a cat. *J Small Anim Pract* **26:** 33–41, 1985.

Kwochka, K. W., and Short, B. G. Cutaneous xanthomatosis and diabetes mellitus following long-term therapy with megestrol acetate in a cat. *Compend Cont Ed* **6:** 185–192, 1984.

Miller, W. H. and Buerger, R. G. Cutaneous mucinous vesiculation in a dog with hypothyroidism. *J Am Vet Med Assoc* **196:** 757–759, 1990.

Nelson, R. W., Feldman, E. C., and Smith, M. C. Hyperadrenocorticism in cats: Seven cases (1978–1987). *J Am Vet Med Assoc* **193:** 245–250, 1988.

Rosser, E. J. Castration-responsive dermatosis in the dog. *In* "Advances in Veterinary Dermatology" Vol 1 C. von Tscharner and R.E.W. Halliwell, (eds.), pp. 34–42. London, Baillière Tindall, 1990.

Schmeitzel, L. P., and Lothrop, C. D. Hormonal abnormalities in Pomeranians with normal coat and in Pomeranians with growth hormone-responsive dermatosis. *J Am Vet Med Assoc* **197:** 1333–1341, 1990.

Scott, D. W. Histopathologic findings in endocrine skin disorders of the dog. *J Am Anim Hosp Assoc* **18:** 173–183, 1982.

Scott, D. W. Cutaneous phlebectasias in cushingoid dogs. *J Am Anim Hosp Assoc* **21:** 351–354, 1985.

Scott, D. W. Excessive trichilemmal keratinization (flame follicles) in endocrine skin disorders of the dog. *Vet Dermatol* **1:** 37–40, 1989.

Scott, D. W. Seasonal flank alopecia in ovariohysterectomized dogs. *Cornell Vet* **80:** 187–195, 1990.

Scott, D. W., and Concannon, P. W. Gross and microscopic changes in the skin of dogs with progestogen-induced acromegaly and elevated growth hormone levels. *J Am Anim Hosp Assoc* **19:** 523–527, 1983.

Scott, D. W., and Walton, D. K. Hyposomatotropism in the mature dog: A discussion of 22 cases. *J Am Anim Hosp Assoc* **22:** 467–473, 1986.

Scott, D. W., Manning, T. O., and Reimers, T. J. Iatrogenic Cushing's syndrome in the cat. *Feline Pract* **12:** 30–36, 1982.

Seaman, W. J., and Chang, S. H. Dermal perifollicular mineralization of toy poodle bitches. *Vet Pathol* **21:** 122–123, 1984.

Thoday, K. L. Aspects of feline symmetric alopecia. *In* "Advances in Veterinary Dermatology" Vol 1 C. von Tscharner and R.E.W. Halliwell, (eds.), pp. 47–69, London, Baillière Tindall, 1990.

White, S. D. *et al.* Cutaneous markers of canine hyperadrenocorticism. *Compend Cont Ed Pract Vet* **11:** 446–463, 1989.

X. Immune-Mediated Dermatoses

A. *Hypersensitivity Dermatoses*

The hypersensitivity or allergic dermatoses account for many important skin diseases in animals. Flea-bite hypersensitivity, atopy (allergic inhalant dermatitis), and food hypersensitivity are important skin diseases of dogs and cats. Insect (especially *Culicoides* spp.) hypersensitivity and atopy are important skin disorders in the horse. These hypersensitivity disorders are rarely reported in ruminants and swine.

There are four basic types of hypersensitivity. These are all involved singly or in combination in the pathogenesis of allergic dermatoses. In **type 1** (anaphylactic) hypersensitivity, the reaction of specific antigens with immunoglobulin E (IgE) (occasionally IgG) bound to tissue mast cells or circulating basophils induces degranulation and release of pharmacologically active molecules. The central mechanism in **type 2** (cytotoxic) hypersensitivity is cell lysis by the membrane attack complex of the activated complement system. **Type 3** hypersensitivity (immune complex) depends on the formation with antigen excess of soluble immune complexes, which activate complement and aggregate platelets with resulting release of various chemical mediators of inflammation. **Type 4** hypersensitivity (cell mediated) results when T lymphocytes, stimulated by specific antigen, release a variety of lymphokines, particularly macrophage-activating factors. Traditionally certain histologic patterns have been associated with each of these categories. Capillary dilation, edema, mast cell degranulation, and eosinophil infiltration are associated with type 1; vasculitis and neutrophilic infiltration, with type 3; and perivascular mononuclear infiltrates, with type 4. That this is an oversimplification has become clear with continuing investigation into the mechanisms of hypersensitivity. Categories overlap, mechanisms are interdependent, and species differences are common. Lesions in naturally occurring allergic dermatoses seldom reflect a specific pathogenesis, and biopsy of suspected cases is used mainly to rule out other diseases.

Bibliography

Halliwell, R. E. W. Clinical and immunological aspects of allergic skin diseases in domestic animals. *In* "Advances in Veterinary Dermatology," Vol. 1, C. von Tscharner and R.E.W. Halliwell (eds.), pp. 91–116. London, Baillière Tindall, 1990.

Halliwell, R. E. W., and Gorman, N. T. "Veterinary Clinical Immunology." Philadelphia, Pennsylvania, W.B. Saunders, 1989.

Muller, G. H., Kirk, R. W., and Scott, D. W. "Small Animal Dermatology IV. Philadelphia, W.B. Saunders, 1989.

Norris, D. A. "Immune Mechanisms in Cutaneous Disease." New York, Marcel Dekker, 1989.

1. Urticaria and Angioedema

Urticaria (hives, heat bumps, food bumps, protein bumps) refers to transient, well-circumscribed, edematous, often pruritic swellings of the dermis. Angioedema (angioneurotic edema) is a related condition but of greater severity since the subcutis is also affected. If the perilaryngeal tissues are involved, death may ensue by asphyxiation. Urticaria may be induced immunologically, usually by type 1 hypersensitivity but occasionally by type 3 mechanisms. Initiators include drugs and vaccines, food, inhalant and contact allergens, infectious agents, and insect bites or stings. Non-immunological urticaria may involve physical factors such as cold, solar radiation, heat and pressure, mast cell-releasing agents such as radiocontrast media, agents affecting arachidonic acid metabolism, such as aspirin, and psychogenic factors. Urticaria occasionally is associated with internal malignancy. Dermatographism is a unique form of urticaria occurring soon after exposure to minor skin pressure or irritation. The final change, regardless of inciting agent and pathogenesis, is a localized alteration of vascular permeability with consequent edema.

Urticaria and angioedema occur most frequently in horses (Fig. 5.43) and dogs, occasionally in pigs, and rarely in cats and ruminants. Recognized causes of hypersensitivity urticaria in the dog include vaccines, food,

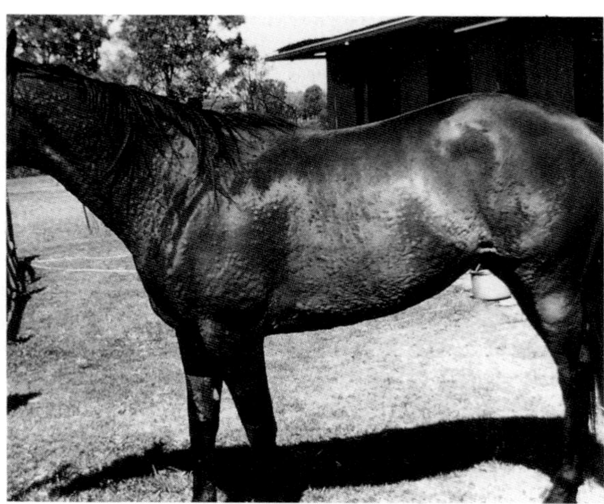

Fig. 5.43 Urticaria. Horse. Transient multiple irregular plaques. (Courtesy of J. D. Baird.)

drugs, infections, stinging and biting insects, intestinal parasites, and plants. Nonimmunologically mediated urticaria in the dog has been ascribed to cold, dermatographism, sunlight, psychogenic factors, and to administration of a radiographic contrast medium, which presumably had a histamine-releasing action. Equine urticaria may be provoked by foods, drugs (particularly penicillin), infections, insect or arthropod envenomation, intestinal parasitism, dermatographism, vaccines, plants, atopy, and psychogenic factors. Urticaria has been reported in cattle associated with drugs (especially penicillins), insect and arthropod stings and bites, infections, vaccines, foods, plants, and dermatographism. It also occurs as part of a generalized hypersensitivity disease in high-producing dairy cows (especially Jerseys and Guernseys) which become sensitized to autologous milk protein. Urticarial reactions in pigs are seen with insect bites or stings, foods, infections (especially erysipelas), drugs, vaccines, and plants. Urticaria in cats has been associated with vaccines, drugs, foods, stinging and biting insects, and intestinal parasites.

The gross lesions of urticaria are multiple, transient, well-circumscribed, flat-topped edematous papules or plaques causing focal elevation of the hair coat. Local heat is absent, and the lesions pit with digital pressure. The lesions are often seen on the face, lateral neck, shoulders, and thorax. The wheals are of variable size, usually round, but annular, arciform, and serpiginous forms develop, especially in horses. Some animals also show signs of edema around the prepuce, vulva, or anus.

The microscopic lesions of urticaria are marked edema of the superficial and middle dermis, indicated by separation of collagen bundles and dilation of lymphatics, variable hyperemia, and a mild to moderate perivascular infiltrate, chiefly lymphocytic. Eosinophils are often prominent in equine and ruminant lesions. Mast cells usually are not increased. The reaction is dermal in urticaria, and involves the dermis and subcutis in angioedema. Vasculitis is not a feature of urticaria in animals.

The diagnosis relies on the history and typical signs, particularly the transient nature of the wheals. In horses and ruminants, fly bites may induce lesions that resemble urticarial wheals, but these are usually dome-shaped and have a central crust, whereas urticarial plaques are flat and not crusted.

Bibliography

Cornick, J. L., and Brumbaugh, G. W. Dermatographism in a horse. *Cornell Vet* **79:** 109–116, 1989.

Rosenberg, M. *et al.* Acute urticaria and bronchospasm following radiocontrast media in a dog. *J Allergy Clin Immunol* **59:** 339–340, 1977.

Willemse, A. Acquired cold urticaria in a dog. *J Am Anim Hosp Assoc* **18:** 961–964, 1982.

2. *Atopic Dermatitis*

Atopy is defined as "the inherited predisposition to develop reaginic antibodies to environmental antigens that results in allergic disease" (Halliwell and Gorman, p. 608). In humans, asthma and allergic rhinitis are the major clinical manifestations of the atopic state. The integument is the chief target organ in atopic dogs, cats, and horses. Allergic rhinitis occurs in cattle.

The pathogenesis of atopy is complex and remains poorly understood. The production of homocytotropic IgE antibodies to ordinarily innocuous stimuli results, in the atopic individual, in immediate and late-phase type 1 hypersensitivity reactions in the target organ. Following allergen binding to specific IgE on the surface of tissue mast cells, degranulation releases a plethora of preformed and newly synthesized inflammatory mediators, including histamine and the leukotrienes. Histamine levels are demonstrably higher in the skin of atopic humans and dogs. While the mast cell is the cell classically associated with type 1 hypersensitivity reactions, many other cell types, including macrophages, eosinophils, some lymphocytes, and platelets have receptors for IgE. Basophils may also be important effector cells. Allergen-specific basophil degranulation has been demonstrated in atopic dogs, correlating well with positive intradermal allergen tests. Antigen-independent histamine hyperreleasability from basophils and mast cells has been demonstrated in humans and in dogs. The effect has been attributed to increased phosphodiesterase activity with a resultant reduction of cyclic AMP. Cyclic AMP hyporesponsiveness is now thought to be a secondary phenomenon, resulting from chronic mediator release. More recently, histamine-releasing factors, produced by a variety of cell types, have been demonstrated in humans. These react with IgE only from atopic individuals (designated IgE +), suggesting that IgE may be heterogeneous. Eosinophils are thought to play an important role in the perpetuation of the lesion and the induction of tissue injury in atopic diseases in humans, including atopic dermatitis. The role of the eosinophil has not been investigated in atopic animals.

The evidence for type 1 hypersensitivity in canine atopy

includes positive immediate hypersensitivity reactions to intradermal skin tests, the passive transfer of reactivity by serum (Prausnitz–Küstner reaction), the reduction in Prausnitz–Küstner reaction when serum is absorbed with anti-IgE, demonstration of allergen-specific IgE in serum, and successful hyposensitization of atopic dogs with appropriate antigens. Unlike those in atopic humans, serum IgE levels in canine atopics are not significantly elevated above those of the normal population. Compared to that in humans, the normal level of IgE in dogs is high, probably reflecting response to endoparasites. An increase in serum IgE does not constitute atopy, for, as indicated above, IgE may be heterogeneous. Furthermore, serum IgE levels do not necessarily reflect the severity of the dermatitis; IgE is not the only homocytotropic antibody isotype involved in atopy: IgG_1 (IgG_d) might act as an allergen-specific, short-term, skin-sensitizing antibody in atopic dogs. The skin-sensitizing antibodies in cats, sheep, and horses have not been characterized.

Cell-mediated immune functions are also defective in atopic humans and dogs. The defects include a diminished response to contact sensitization and a decrease in histamine-induced suppression of mitogen-stimulated lymphocyte blastogenesis reactions *in vitro*. Because these changes are also mimicked in dogs rendered experimentally hypersensitive to *Ascaris* antigen or ovalbumin, they are thought to result from the type 1 hypersensitivity reaction rather than from a defect inherent in the atopic state. Only two tests of cell-mediated function differ between the atopic and experimentally hypersensitive dogs: a reduction in concanavalin A-induced suppression at supraoptimal levels of mitogen and an enhancement of the cutaneous reaction to the injection of cutaneous mitogens. The former abnormality precedes the onset of dermatitis in atopic children, and the latter abnormality suggests that an important dysregulation of immune function in the target organ may be of diagnostic use.

The combination of exaggerated IgE-mediated hypersensitivity responses and functional abnormalities in T-lymphocyte function suggests that the basis of atopy may be a defect in T-cell regulation of IgE production. Abnormal suppressor T-lymphocyte function, as has been demonstrated in atopic humans and dogs, is one potential mechanism. Attention has focused on lymphocyte-generated cytokines in the regulation of immunoglobulin isotype production in B lymphocytes, particularly interleukin (IL)-4. Interleukin-5, a growth and activation factor for eosinophils, is also increased in the skin of atopic humans. It is likely that production of other cytokines, including those which inhibit IL-4-mediated IgE synthesis, such as γ-interferon and prostaglandin E_2, may be altered in atopic individuals.

The route of allergen exposure in canine atopy is considered to be respiratory. Evidence in favor of this hypothesis is the induction of typical pruritic skin lesions in sensitized Basenji–greyhounds following inhalation provocation tests. The mechanism whereby the allergen localizes on the respiratory mucosa but elicits a response in the skin has not been resolved. However, studies in human atopic dermatitis have implicated the Langerhans cell, which expresses low-affinity receptors for IgE (CD23), as the major presenter of allergen, suggesting that contact is the major route of exposure.

Canine atopy is a common condition accounting for ~10% of all dermatological disease. It has a familial pattern, although all breeds, including mongrels, may be affected and there is no good correlation with dog leukocyte antigen (DL-A) types. The following breeds are predisposed: Cairn terrier, West Highland white terrier, Scottish terrier, wire-haired fox terrier, Lhasa apso, Dalmatian, pug, Boston terrier, golden retriever, Irish setter, boxer, English setter, Labrador retriever, miniature schnauzer, and Chinese Shar-pei. There is a higher incidence of atopy in bitches than in male dogs. The disease usually appears between 1 and 3 years of age, but the range is 6 months to 6 years. Depending on the source of allergen, canine atopy may be seasonal or nonseasonal. Important allergens include house dust, human dander, molds, weeds, feathers, grasses, and pollens, and possibly marijuana. The major clinical sign is pruritus, frequently manifest as foot chewing and face rubbing. Saliva causes yellow-brown discoloration of the coat in white-haired breeds.

Pruritus initiates an itch–scratch cycle which is central to the pathogenesis of the lesions. Gross lesions primarily affect the face (Fig. 5.44), the ears, the feet, and the axillae. In chronically affected dogs, lesions may become generalized. They develop as a result of self-trauma and vary from erythema, excoriation, crusting, and traumatic alopecia to scaling, hyperpigmentation, and lichenification. Secondary bacterial pyoderma occurs in about a third of cases. Approximately half of atopic dogs show conjunctivitis and otitis externa, but asthma and rhinitis are rare. Seborrhea and hyperhidrosis affect ~10% of atopic dogs.

Cutaneous syndromes associated with **feline atopy** are extremely variable and include facial pruritus, pinnal pruritus, miliary dermatitis, eosinophilic granuloma complex lesions (indolent ulcer, eosinophilic plaque, eosinophilic collagenolytic granuloma), symmetric hypotrichosis or al-

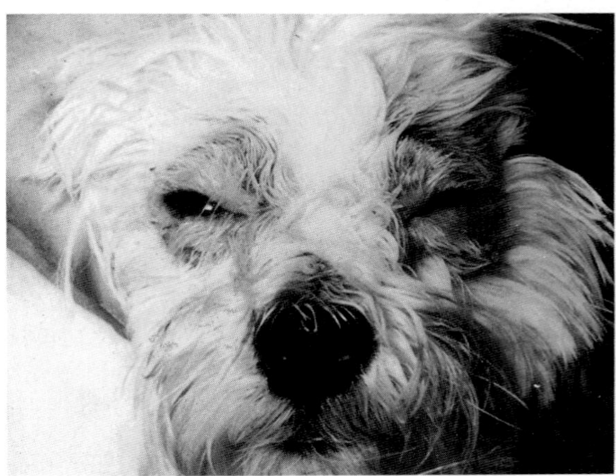

Fig. 5.44 Atopy. Dog.

opecia, urticaria, and generalized pruritus. In addition, these cutaneous reaction patterns can occur in various combinations in the same individual.

Atopic horses may have symmetric pruritus of the face, ears, ventrum, legs, or tail head, or may be presented for recurrent seasonal or nonseasonal urticaria. Arabians and Thoroughbreds may be predisposed.

Presumptive atopy was reported in **Suffolk sheep,** which had a recurrent, seasonal, pruritic skin disorder affecting the face, ears, ventrum, and hindquarters.

The histologic pattern of atopy in dogs, cats, and sheep is perivascular dermatitis, of little diagnostic specificity. The most consistent lesions are vasodilation and diffuse edema of the superficial dermis, with a mild to moderate perivascular accumulation of mononuclear cells. Granulocytes are usually in the minority unless secondary lesions are present, when neutrophils may be numerous. One author reports neutrophils in 80 of 100 biopsies from atopic dogs. In lesional and nonlesional skin of atopic **dogs,** the numbers of mast cells are significantly higher than those in normal controls. Eosinophils are reportedly scarce in the canine lesions, as in human atopic dermatitis. However, in human lesions, major basic protein is abundant, indicating that eosinophil degranulation is widespread. Similar studies need to be performed in canine lesions. Quantitative studies showed atopic dogs to have significantly more eosinophils in nonlesional than in site-matched biopsies from normal dogs, but the absolute numbers were low, making this of doubtful assistance in the assessment of routine biopsies. Sebaceous glands are measurably larger than those of normal dogs. These morphometric studies also failed to reveal significant epidermal hyperplasia and no increase in dermal–epidermal folding, supporting a previous contention that elongation of rete ridges is not a feature of atopy. The lesions become more heterogeneous with secondary infection, excoriation, seborrhea, or superimposed flea bite hypersensitivity or ectoparasitism.

In **cats** and **sheep,** eosinophils are often the predominant inflammatory cell type. In addition, cats may have typical eosinophilic granuloma complex lesions or focal areas of eosinophilic folliculitis. The epidermis is usually hyperplastic. Spongiosis, leukocytic exocytosis, crusting, orthokeratotic and parakeratotic hyperkeratosis, and follicular keratosis are variable findings. In **horses,** the perivascular reaction is usually superficial and deep, and eosinophils dominate. In addition, focal areas of eosinophilic folliculitis may be seen.

Major and minor criteria have been suggested for the establishment of a **diagnosis** of canine atopy in a manner similar to that employed in humans. Histologic lesions are not among these criteria, as they are not a specific indicator of atopy.

Bibliography

Brown, M. A., and Hanifin, J. M. Atopic dermatitis. *Curr Opin Immunol* **2:** 531–534, 1990.

Butler, J. M. *et al.* Pruritic dermatitis in asthmatic Basenji–greyhound dogs: A model for human atopic dermatitis. *J Am Acad Dermatol* **8:** 33–38, 1983.

Carlotti, D., and Prost, C. L'atopie féline. *Point Vét* **20:** 777–784, 1988.

Evans, A. G. Allergic inhalant dermatitis attributable to marijuana exposure in a dog. *J Am Vet Med Assoc* **195:** 1588–1590, 1989.

Kapsenberg, M. L. *et al.* Aberrant T-cell regulation of IgE production in atopy. *Eur Respir J* **4**(Suppl. 13): 27S–30S, 1991.

Leiferman, K. M. Eosinophils in atopic dermatitis. *Allergy* **44:** 20–26, 1989.

Lichenstein, L. M. Histamine-releasing factors and IgE heterogeneity. *J Allergy Clin Immunol* **83:** 814–820, 1988.

Nimmo-Wilkie, J. S. *et al.* Morphometric analyses of the skin of dogs with atopic dermatitis and correlations with cutaneous and plasma histamine and total serum IgE. *Vet Pathol* **27:** 179–186, 1990.

Nimmo-Wilkie, J. S. *et al.* Abnormal cutaneous response to mitogens and a contact allergen in dogs with atopic dermatitis. *Vet Immunol Immunopathol* **28:** 97–106, 1991.

Schwartzman, R. M. Immunologic studies of progeny of atopic dogs. *Am J Vet Res* **45:** 375–378, 1984.

Scott, D. W., and Campbell, S. G. A seasonal pruritic dermatitis resembling atopy in sheep. *Agri-Pract* **8:** 46–49, 1987.

Scott, D. W., Walton, D. K., and Slater, M. R. Miliary dermatitis. A feline cutaneous reaction pattern. *Proc Ann Kal Kan Sem* **2:** 11–18, 1986.

Scott, D. W., Miller, W. H., Jr., and Shanley, K. J. Sterile eosinophilic folliculitis in the cat: An unusual manifestation of feline allergic skin disease? *Compan Anim Pract* **19:** 6–11, 1989.

Turner, C. R. *et al.* Dermal mast cell releasability and end organ responsiveness in atopic and nonatopic dogs. *J Allergy Clin Immunol* **83:** 643–648, 1989.

Willemse, A. *et al.* Allergen specific IgG_d antibodies in dogs with atopic dermatitis as determined by the enzyme-linked immunosorbent assay (ELISA). *Clin Exp Immunol* **59:** 359–363, 1985.

Willemse, T. Atopic skin disease: A review and a reconsideration of diagnostic criteria. *J Small Anim Pract* **27:** 771–778, 1986.

3. Food Hypersensitivity

Food hypersensitivity is uncommon in dogs and cats, and rare in horses and cattle. The mechanisms of food hypersensitivity may involve any of the classic types, although type I hypersensitivity is thought to be the major mechanism in humans. The immunopathogenesis of food hypersensitivity in animals has not been elucidated. Immediate (within minutes to hours) and delayed (within several hours to days) reactions to foods have been reported. Many putative food allergies are probably nonimmunologic. Termed adverse reactions to food, these include idiosyncratic reactions, toxicities, and pharmacological effects, such as in foods containing vasoactive amines. Provocation tests to establish a definitive diagnosis are rarely undertaken. In dogs, foods implicated in food hypersensitivity include beef, dairy products, soy, and wheat; in cats, they include fish, liver, chicken, and dairy products. The clinical signs of food hypersensitivity are nonseasonal and often poorly responsive to glucocorticoid therapy.

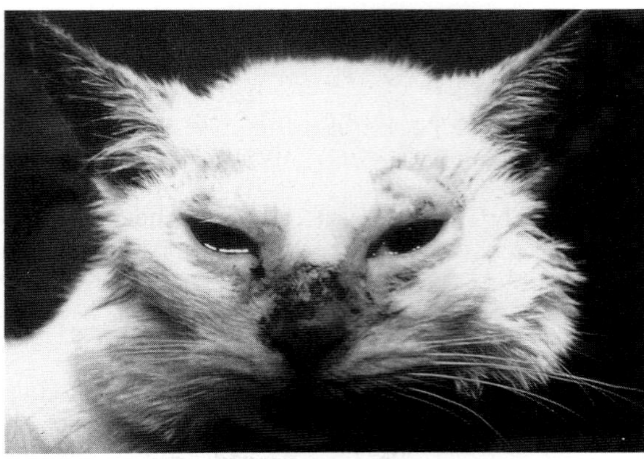

Fig. 5.45 Food hypersensitivity dermatitis. Periocular and nasal alopecia, erythema, and crusting. Kitten.

Cutaneous syndromes associated with food hypersensitivity in **cats** are identical to those described previously for feline atopy: facial pruritus, pinnal pruritus, generalized pruritus, miliary dermatitis, eosinophilic granuloma complex lesions, symmetric hypotrichosis or alopecia, and urticaria (Fig. 5.45). Concurrent diarrhea is present in less than 10% of the cases.

In **dogs,** food hypersensitivity can mimic numerous cutaneous syndromes: atopy, scabies, flea-bite hypersensitivity, pruritic generalized folliculitis or seborrhea, recurrent pyotraumatic dermatitis (hot spots), urticaria, pruritic otitis externa, and generalized pruritus. Concurrent gastrointestinal disturbances are present in about 10% of the cases.

In **horses**, food hypersensitivity may be characterized by generalized pruritus, urticaria, or tail rubbing.

The histologic findings of food hypersensitivity in dogs and cats are those of perivascular dermatitis, as described for atopy. Lymphocytes and neutrophils are commonly seen in both species, but numerous eosinophils are seen frequently only in cats. In addition, cats may have typical eosinophilic granuloma complex lesions or eosinophilic folliculitis. As in atopy, the basic pattern of perivascular dermatitis may be complicated by secondary infection, seborrhea or ectoparasitism.

Bibliography

Carlotti, D. N., Remy, I., and Prost, C. Food allergy in dogs and cats. A review and report of 43 cases. *Vet Dermatol* **1:** 55–62, 1990.

Gross, T. L., Kwochka, K. W., and Kunkle, G. A. Correlation of histologic and immunologic findings in cats with miliary dermatitis. *J Am Vet Med Assoc* **189:** 1322–1325, 1986.

Jeffers, J. G., Shanely, K. J., and Meyer, E. K. Diagnostic testing of dogs for food hypersensitivity. *J Am Vet Med Assoc* **198:** 245–250, 1991.

White, S. D. Food hypersensitivity in 30 dogs. *J Am Vet Med Assoc* **188:** 695–698, 1986.

White, S. D., and Sequoia, D. Food hypersensitivity in cats: 14 cases (1982–1987). *J Am Vet Med Assoc* **194:** 692–695, 1989.

4. Allergic Contact Dermatitis

Allergic contact dermatitis represents the classical delayed hypersensitivity reaction, mediated by T lymphocytes. The initiating substances are haptens, which are not immunogenic until conjugated with a carrier protein. The carrier is usually an epidermal protein. The Langerhans' cell is responsible for antigen presentation both locally in the skin and by migration to regional lymph nodes. Upon reexposure to the allergen, antigen-specific memory T lymphocytes are recruited into the skin. Mast cells and possibly basophils have been implicated in early events in the immunopathogenesis of allergic contact dermatitis. Irritant contact dermatitis has classically been considered distinct from contact hypersensitivity reactions. However, some contact allergens have significant irritant properties.

Contact hypersensitivity is rarely documented in dogs, cats, and horses. This may be due, in part, to the protective nature of the hair coat. It may also demonstrate species differences in the potential to mount delayed hypersensitivity reactions in the skin. Dogs, for example, have proved difficult to sensitize to dinitrochlorobenzene. Some of the putative cases of contact hypersensitivity have not been confirmed by positive patch tests. One well-documented case in the dog describes contact hypersensitivity to *Tradescantia fluminensis* (wandering Jew plant). Other substances associated with contact hypersensitivity in dogs and cats include various topical medicaments (especially shampoos, neomycin), plants, home furnishings, and cleaning products. A unique situation is the depigmenting dermatitis of the nose and lips seen with contact reactions to plastics and rubbers (food bowls, chewy toys). The depigmentation may be a direct toxic effect, rather than the result of allergy. German shepherds and Labrador retrievers may be predisposed to develop contact hypersensitivity reactions. In horses, contact hypersensitivity has been reported in association with various topical medicaments (especially insect repellents), pasture plants, and tack items.

The predominant sign in contact hypersensitivity is a pruritic, erythematous, maculopapular dermatitis. Lesions have a distribution which reflects the area of contact and tends to involve the glabrous skin, unless the allergen is in a liquid or aerosol form. The lesions are complicated by self-trauma with up to 40% of dogs showing secondary bacterial pyoderma. Chronic lesions are crusting, scaling, lichenification, partial to complete alopecia, and hyperpigmentation. Leukoderma rather than hyperpigmentation often occurs in the reaction to plastic food containers.

Contact hypersensitivity has a superficial perivascular pattern, spongiotic in the acute phase and hyperplastic in the chronic phase. Lymphocytes and histiocytes are the predominant inflammatory cells. It is not possible to distinguish primary irritant contact dermatitis from contact hypersensitivity microscopically. In humans, the two may be distinguished by biopsy of the site of a positive patch test, but this is not the case in dogs. A recent examination of the lesions in positive patch tests from spontaneously hypersensitive dogs revealed significant differences from the human reactions. Epidermal necrosis, sometimes full thickness, occurred frequently and the predominant inflammatory cells were neutrophils, rather than lymphocytes.

Bibliography

Kunkle, G. A., and Gross, T. L. Allergic contact dermatitis to *Tradescantia fluminensis* (Wandering Jew) in a dog. *Compend Cont Ed Pract Vet* **5:** 925–930, 1983.

Thomsen, M. K., and Kristensen, F. Contact dermatitis in the dog. A review and a clinical study. *Nord Vet Med* **38:** 129–147, 1986.

Thomsen, M. K., and Thomsen, H. K. Histopathological changes in canine allergic contact dermatitis patch test reactions: A study on spontaneously hypersensitive dogs. *Acta Vet Scand* **30:** 379–384, 1989.

5. Flea-Bite Hypersensitivity

Flea-bite hypersensitivity is the most common form of allergic dermatitis in the dog and cat. The fleas chiefly responsible are *Ctenocephalides felis* and *C. canis*. *Pulex irritans* is sometimes involved. Although the flea bite itself is an irritant, due to the release of histamine-like agents, proteolytic enzymes, and anticoagulants, even a massive flea infestation produces few lesions in a normal animal. In a sensitized animal, however, very severe lesions result from the presence of a very few fleas. Most dogs and cats which are flea saliva-hypersensitive have immediate skin-test reactions to the intradermal injection of flea antigen, indicating a type 1 hypersensitivity reaction. Allergen-specific IgE is detectable in serum. A number of protein antigens, between 18,000 and 45,000 daltons, have been identified as allergenic in dogs, with a major allergen of 30,000 to 32,000 (nonhaptenic). Flea-hypersensitive dogs also usually have delayed skin-test reactions, indicating a type 4 hypersensitivity reaction. In 10 to 20% of patients, only delayed reactions occur. Other experimental aspects of skin testing and dermatopathology in dogs with flea-bite hypersensitivity suggest that late-phase immediate hypersensitivity (IgE-mediated) reaction and cutaneous basophil hypersensitivity may play a role in the pathogenesis of the lesions. Experimental induction of flea-bite hypersensitivity in dogs did not produce the orderly sequence of events seen in the classic studies performed in guinea pigs. Immediate and delayed reactions commenced in random sequence. Dogs continually exposed to fleas either fail to develop lesions of flea-bite hypersensitivity or develop them later and to a lesser degree, suggesting that these dogs become partially or completely immunologically tolerant. The atopic state appears to predispose dogs to the development of flea-bite hypersensitivity.

There is no breed predilection for flea-bite hypersensitivity in dogs or cats, and the disease is not correlated consistently with coat type. Although dogs and cats may develop the hypersensitivity at any age, it is rare for clinical signs to appear before 6 months. There is a tendency for clinical signs to be more severe in summer and autumn, when fleas are most active, but this depends on geographic location and the degree of infestation of the environment. The initial clinical sign is pruritus. Flea bite hypersensitivity, unlike atopy, has as a primary lesion an erythematous papule or wheal, but secondary lesions are rapidly induced by self-mutilation. In the acute stage, secondary lesions are pustules, crusts, alopecia, and irregular patches of moist dermatitis, which follow constant licking. Chronic lesions show varying degrees of alopecia, scaling, crusting, lichenification, and hyperpigmentation. Acute and chronic lesions often coexist. The distribution of the lesions strongly suggests the diagnosis, reflecting the most favored feeding sites of the flea. These sites are the dorsal lumbosacrum, caudomedial thigh and caudoventral abdomen. The head and neck are affected quite frequently in the cat. Lesions may become generalized in severely affected animals. In cats, symmetric hypotrichosis or alopecia and eosinophilic granuloma complex lesions are other manifestations of flea-bite hypersensitivity. The European rabbit flea (*Spilopsyllus cuniculi*) produces pruritic papulocrustous lesions along the margins of the ear pinnae in cats.

The histology of flea-bite hypersensitivity is characterized by varying degrees of hyperplastic superficial perivascular dermatitis. One of the most consistent features is the predominance of eosinophils in the dermal reaction. In biopsies taken from dogs with experimentally induced flea-bite hypersensitivity 4–18 hr after biting, eosinophils accounted for up to 20% of the cellular infiltrate. Lymphocytes and histiocytes may dominate in more chronic (type 4 hypersensitivity) reactions, but eosinophils are still prominent. The epidermis shows moderate to marked hyperplasia with elongation of the rete ridges, orthokeratotic and/or parakeratotic hyperkeratosis, and variable spongiosis, erosion, or ulceration. Intraepidermal eosinophilic microabscesses, with or without surrounding epidermal necrosis, are a common finding, as in other ectoparasitisms. The dermis is edematous in active lesions, and superficial dermal fibrosis is a common feature of the chronic lesions. Histologic evidence of secondary bacterial infection and/or seborrhea often complicates the picture. In cats, histologic findings typical of the eosinophilic granuloma complex or eosinophilic folliculitis may be present.

Bibliography

Gross, T. L., and Halliwell, R. E. W. Lesions of experimental flea-bite hypersensitivity in the dog. *Vet Pathol* **22:** 78–81, 1985.

Halliwell, R. E. W., and Schemmer, K. R. The role of basophils in the immunopathogenesis of hypersensitivity to fleas (*Ctenocephalides felis*) in dogs. *Vet Immunol Immunopathol* **15:** 20–212, 1987.

Halliwell, R. E., Preston, J. F., and Nesbitt, J. G. Aspects of the immunopathogenesis of flea allergy dermatitis in dogs. *Vet Immunol Immunopathol* **17:** 483–494, 1987.

Studdert, V. P., and Arundel, J. H. Dermatitis of the pinnae of cats in Australia associated with the European rabbit flea (*Spilopsyllus cuniculi*). *Vet Rec* **123:** 624–625, 1988.

6. Culicoides *Hypersensitivity*

Some **horses** develop a hypersensitivity to biting midges of the genus *Culicoides,* leading to a chronic, recurrent, pruritic, and alopecic dermatitis. The disease is known as Queensland itch in Australia, sweet itch in the United Kingdom, kasen in Japan, dhobie itch in the Philippines, and summer eczema, summer sores, or allergic urticaria in Europe. No doubt many of the more than 2000 species of *Culicoides* are capable of inducing hypersensitivity, but some species have been definitely implicated including *C. brevitarsus* in Australia, *C. obsoletus* in Canada, *C. varipennis, C. spinosus, C. stellifer,* and *C. insignis* in the United States, *C. pulicaris* in Britain, and *C. lupicaris, C. nubeculosis, C. imicola,* and *C. punctatus* in Israel. *Culicoides* hypersensitivity represents type 1 and type 4 hypersensitivity reactions to components of insect saliva. Both immediate and late-phase IgE-mediated hypersensitivity are demonstrated on intradermal testing using extracts of *Culicoides* adults. The immediate reaction is transferrable to the skin of a normal horse with serum (Prausnitz–Küstner reaction). Fluctuations in both numbers of peripheral blood eosinophils and levels of serum histamine have been correlated with the biting time of the midges (usually dawn and dusk).

Culicoides hypersensitivity occurs sporadically, often with a familial history. In Europe, Icelandic horses are strongly disposed to develop *Culicoides* hypersensitivity, and an immunogenetic study established a correlation between certain histocompatibility antigens and the incidence of the disease in this breed. Horses do not usually develop signs until 2 years of age, although hypersensitivity is reported in a 5-month-old foal. Depending on geographic location, the disease is seasonal or shows seasonal exacerbation.

The typical gross lesions occur on the dorsum of the neck and at the tail head (Fig. 5.46), reflecting the biting habits of the midges. Pruritus is intense. Lesions may affect the poll and ears, and spread from the withers onto the shoulders and from the tail to the perineum and rump. The long hairs of the mane and tail are often broken off, resulting in a "buzzed-off mane" or "rat-tail" appearance. Certain species of *Culicoides* feed along the ventral body wall, and in a susceptible horse, lesions develop at this site. While ventral midline dermatitis has been associated more commonly with bites from the horn fly, a survey of 101 horses in Florida found ventral midline lesions in 30% and both dorsal and ventral lesions in 50%. The primary lesion induced by the *Culicoides* is a discrete papule, made visible by the overlying erect hairs. Most of the lesions are self-inflicted as a result of extreme pruritus. Erosion, ulceration, crusting, and scaling are also induced by self-mutilation. Lichenification and alopecia may be very marked in chronically affected animals.

The histologic appearance of the acute natural primary lesion, and of biopsies taken from the sites of a positive intradermal test, is indicative of a type 1 hypersensitivity reaction. The pattern is usually a superficial and deep perivascular dermatitis with a predominance of eosinophils, and edema and hyperemia in the superficial dermis. Erosion or ulceration may be present. In chronic lesions, hyperplastic changes predominate, including rete ridge formation, often to the point of pseudocarcinomatous hyperplasia, and orthokeratotic and parakeratotic hyperkeratosis. The dermis is usually edematous, and eosinophils and lymphocytes remain prominent components of the inflammatory cell population. Fibrosis may be mild to marked depending on the degree and chronicity of self-inflicted trauma. In one study, the following focal, secondary histologic patterns were found: eosinophilic folliculitis (24% of the cases), intraepidermal pustular (neutrophilic) dermatitis (19%), and palisaded eosinophilic granuloma (9%).

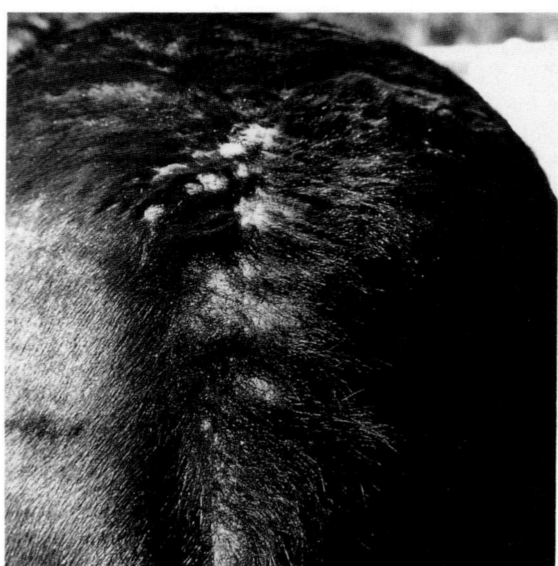

Fig. 5.46 *Culicoides* hypersensitivity. Horse. Alopecia, scaling, crusting at tail head. Note "rat tail" appearance.

A seasonal pruritic dermatitis of **sheep** was attributed to *Culicoides* hypersensitivity. Affected sheep had varying degrees of erythema, crusting, hyperkeratosis, lichenification, and alopecia on the teats, ventral abdomen, medial thighs, medial forelegs, pinnae, lips, nose, and periocular area. *Culicoides obsoletus* was trapped from the environment, and a whole-body antigen of this insect produced immediate and delayed reactions when injected intradermally into affected sheep. Skin biopsy revealed a hyperplastic superficial perivascular dermatitis in which eosinophils and lymphocytes dominated.

Bibliography

Baker, K. P., and Quinn, P. J. A report on clinical aspects and histopathology of sweet itch. *Eq Vet J* **10:** 243–248, 1978.

Braverman, Y. Preferred landing sites of *Culicoides* species (Diptera: Ceratopogonidae) on a horse in Israel and its relevance

to summer seasonal recurrent dermatitis (sweet itch). *Eq Vet J* **20:** 426–429, 1988.
Connan, R. M., and Lloyd, S. Seasonal allergic dermatitis in sheep. *Vet Rec* **123:** 335–337, 1988.
Fadok, V. A., and Greiner, E. C. Equine insect hypersensitivity: Skin test and biopsy results correlated with clinical data. *Eq Vet J* **22:** 236–240, 1990.
Morrow, A. N., Baker, K. P., and Quinn, P. J. Skin lesions of sweet itch and the distribution of dermal mast cells in the horse. *J Vet Med B* **34:** 347–355, 1987.
Riek, R. F. Studies on allergic dermatitis ("Queensland itch") of the horse. 1. Descriptions, distribution, symptoms, and pathology. *Aust Vet J* **29:** 177–184, 1953.
Riek, R. F. Studies on allergic dermatitis (Queensland itch) of the horse: The aetiology of the disease. *Aust J Agric Res* **5:** 109–129, 1954.
Scott, D. W. Histopathologie cutanée de l'hypersensibilité aux piqûres de *Culicoides* chez le cheval. *Point Vét* **22:** 583–588, 1990.

7. *Staphylococcal Hypersensitivity*

Staphylococcal hypersensitivity in the **dog** is a rare dermatologic syndrome believed to result from an allergic reaction to antigens of *Staphylococcus intermedius*. The pathogenesis is unknown, although type 3 and type 4 reactions have been postulated on the basis of biopsy findings and intradermal skin tests. Dogs which develop bacterial hypersensitivity usually have a preexisting or concurrent seborrheic or bacterial skin disease. Bacterial hypersensitivity may also accompany other allergies (atopy, food) and hypothyroidism. The lesions of bacterial hypersensitivity appear suddenly on the ventral abdomen and spread rapidly to involve the lateral thorax and dorsal thoracolumbar area. Lesions may become generalized. Three categories of gross lesions are described, probably reflecting the degree of hypersensitivity and duration of the reaction. The **hemorrhagic bulla** is the least common lesion and forms a 1- to 3-cm diameter red-purple vesicle or bulla. An identical lesion develops at the site of a positive intradermal test. The **erythematous pustule** is the most common

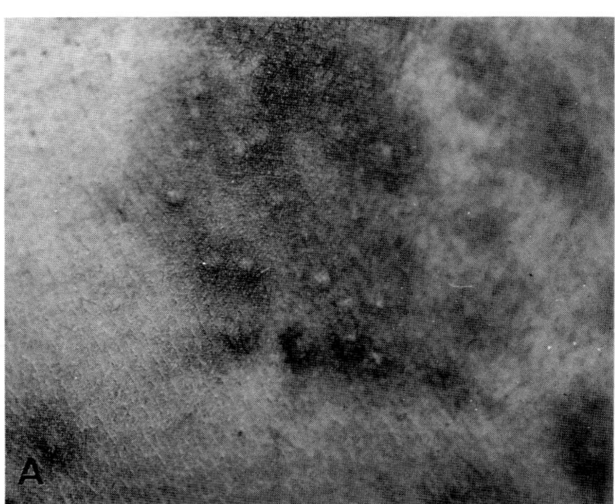

Fig. 5.47A Bacterial hypersensitivity. Erythematous pustules.

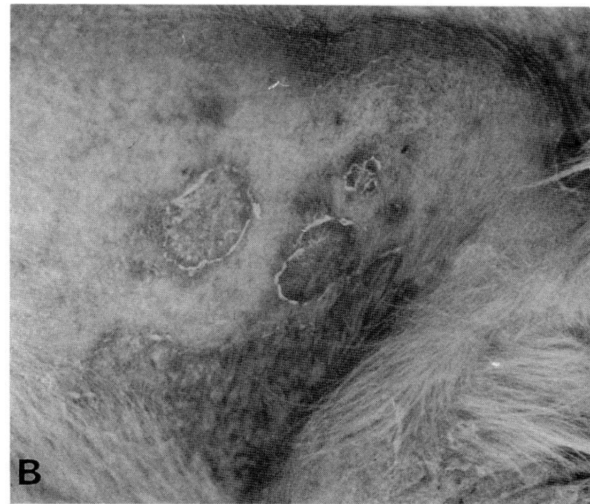

Fig. 5.47B Bacterial hypersensitivity. Annular erythematous lesions with epithelial collarette.

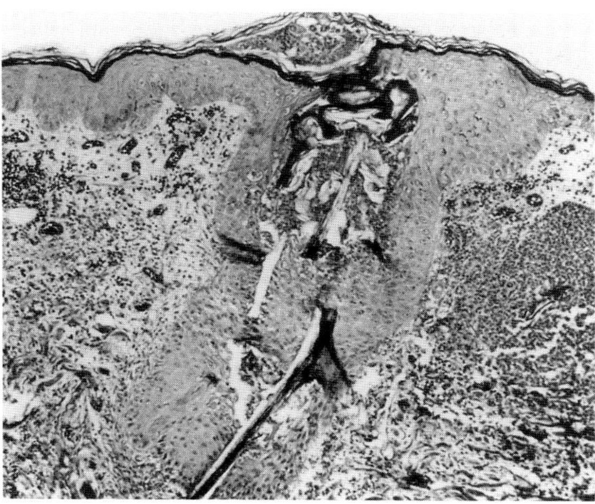

Fig. 5.48 Histology of Fig. 5.47A. Subcorneal pustule at follicle orifice, hemorrhage, edema, and vascular dilation in dermis.

manifestation of bacterial hypersensitivity and forms a small pustule on a halo of intensely erythematous skin (Fig. 5.47A). The **seborrheic plaque** is a chronic erythematous, scaling plaque up to 10 cm in diameter, which has a tendency to heal from the center, often with hyperpigmentation (Fig. 5.47B).

Microscopically, bacterial hypersensitivity is characterized by varying degrees of vasculitis and suppurative dermatitis (intraepidermal pustular dermatitis, folliculitis, furunculosis). There is marked endothelial swelling, vascular engorgement, and edematous separation of the vessel wall with accompanying dermal edema, extravasation of erythrocytes, lymphatic dilation, neutrophil margination, and exocytosis. The vasculitis is usually mixed (neutrophils and lymphocytes), and fibrinoid degeneration is uncommon. The distinguishing feature of the hemorrhagic bulla is extensive extravasation of erythrocytes. Occasionally, bullous separation of the dermoepidermal junction is seen. The erythematous pustule, usually subcorneal, is located at the infundibulum of hair follicles (Fig. 5.48). The pustules contain neutrophils and a few acantholytic keratinocytes. Erythrocytic extravasation and exocytosis occur in the erythematous pustule but to a lesser extent than in the hemorrhagic bulla. The seborrheic plaque is the least distinctive of the three types of lesions. Hyperplastic epidermal changes are prominent in the seborrheic plaque, including rete ridge formation, hypergranulosis, and parakeratotic hyperkeratosis.

Bibliography

Scott, D. W., MacDonald, J. M., and Schultz, R. D. Staphylococcal hypersensitivity in the dog. *J Am Anim Hosp Assoc* **14:** 766–779, 1978.

8. Hormonal Hypersensitivity

Skin lesions resulting from hypersensitivity to endogenous gonadal hormones occur rarely in the dog. Results of intradermal skin tests suggest that type 1 and type 4 hypersensitivity reactions are involved. Hormonal hypersensitivity primarily affects bitches which usually have a history of abnormal estrous cycles or pseudopregnancy. Skin lesions develop during estrus or pseudopregnancy and are exacerbated with each episode. Lesions are bilaterally symmetric, initially involving the perineal, genital, caudomedial thigh, and rump areas. With time, lesions become generalized. The lesion is a pruritic, papulocrustous eruption. Histologic changes include varying degrees of superficial perivascular dermatitis in which either neutrophils or mononuclear cells predominate.

9. Drug Eruption

Drug eruptions are occasionally reported in dogs and cats, and rarely in other domestic animals. Drugs responsible for skin eruptions may be administered orally, topically, or by injection. Drug hypersensitivities are believed to involve all 4 types of hypersensitivity reactions.

Any drug may cause an eruption, and no specific type of reaction is consistently produced by any one drug. Thus, drug eruption can mimic virtually any dermatosis. The most common drugs recognized to produce hypersensitivity reactions in domestic animals are the sulfonamides (especially trimethoprim potentiated) and penicillins. Erythema multiforme and toxic epidermal necrolysis have been most commonly seen with trimethoprim-potentiated sulfonamides, cephalosporins, and levamisole. Diethylcarbamazine and 5-fluorocytosine have been associated with fixed drug eruption on the scrotum of dogs. The ulcerative lesions healed with hyperpigmentation when the drug was withdrawn, but recrudescence at the same site, with vesiculation, occurred when the dog was rechallenged. The mechanism underlying fixed drug eruption is unknown. In humans the epidermal invasion of T cells in fixed drug eruptions is associated with the expression of the intracellular adhesion molecule ICAM-1 on the surface of lesional keratinocytes. Cyclosporin has been reported to cause a lymphoplasmacytoid dermatitis with malignant features, usually manifest as a solitary plaque or nodule. Urticaria and angioedema have been associated with levamisole, barbiturates, and some antibiotics. Drug eruptions manifesting as exfoliative erythroderma have been seen in dogs treated with levamisole and lincomycin. Cats with thymoma may present with an exfoliative dermatitis which histologically, resembles erythema multiforme or graft-versus-host reactions. Drug eruptions resembling bullous pemphigoid clinically, histologically, and immunohistochemically have been associated with triamcinolone. It is quite possible that some cases described as bullous pemphigoid actually represent drug eruptions. Pemphigus foliaceus-like drug eruptions have been described in cats treated with ampicillin or cimetidine and dogs receiving trimethoprim-sulfonamide. Lesions were classical subcorneal, acantholytic pustules, with additional vasculitis in some cases. Systemic lupuslike drug reactions have been reported in dogs and in cats. Cutaneous vasculitis caused by immune-complex deposition may be initiated by drug administration. Sulphadiazine administration in Dobermans causes a poorly defined skin rash as well as ocular, joint, kidney, and hematological abnormalities suggestive of systemic vasculitis.

Just as the clinical morphology of drug eruptions varies greatly, so do the histologic findings: perivascular dermatitis (allergylike), hydropic and/or lichenoid interface dermatitis (erythema multiforme, toxic epidermal necrolysis, lupus erythematosuslike), vasculitis, intraepidermal vesiculopustular dermatitis (pemphiguslike), and subepidermal vesicular dermatitis (pemphigoidlike).

Bibliography

Evans, A. G., and Stannard, A. A. Idiopathic multiple cutaneous seromas in a horse: A possible manifestation of drug eruption. *Eq Pract* **12:** 27–34, 1990.

Mason, K. V. Subepidermal bullous drug eruption resembling bullous pemphigoid in a dog. *J Am Vet Med Assoc* **190:** 881–886, 1987.

Mason, K. V. Fixed drug eruption in two dogs caused by diethylcarbamazine. *J Am Anim Hosp Assoc* **24:** 301–303, 1988.

Mason, K. V. Cutaneous drug eruptions. *Vet Clin North Am* **20:** 1633–1653, 1990.

Mason, K. V., and Day, M. J. A pemphigus foliaceuslike eruption associated with the use of ampicillin in a cat. *Aust Vet J* **64:** 223–224, 1987.

McEwan, N.A. *et al.* Drug eruption in a cat resembling pemphigus foliaceus. *J Small Anim Pract* **28:** 713–720, 1987.

Medleau, L. *et al.* Trimethoprim-sulfonamide-associated drug eruptions in dogs. *J Am Anim Hosp Assoc* **26:** 305–311, 1990.

Peterson, M. E., Kintzer, P. P., and Hurvitz, A. I. Methimazole treatment of 262 cats with hyperthyroidism. *J Vet Intern Med* **2:** 150–157, 1988.

Shiohara, T. *et al.* Fixed drug eruption. Expression of epidermal keratinocyte adhesion molecule-1 (ICAM-1). *Arch Dermatol* **125:** 1371–1376, 1989.

VanHees, J. *et al.* Levamisole-induced drug eruptions in the dog. *J Am Anim Hosp Assoc* **21:** 255–260, 1985.

10. Endoparasite Hypersensitivity

Dermatitis resulting from allergy to endoparasites is a poorly documented condition in dogs and cats. It is postulated that IgE-mediated immune reactions to gastrointestinal nematode antigens may induce hypersensitivity reactions in the skin. Pruritus is the predominant clinical sign and is associated either with a generalized papulocrustous eruption or with apparently normal skin. Occasionally urticarial lesions are present. Histologic findings include varying degrees of superficial perivascular dermatitis with eosinophils prominent.

Dirofilaria immitis microfilariae have been associated with a pruritic papulonodular dermatitis in dogs. The lesions are most commonly found on the head and face, over the trunk, and on proximal limbs. The lesions, which vary from 0.5 to 3.0 cm in diameter, are well circumscribed, firm, alopecic, and often ulcerated. Histologic examination reveals a superficial and deep perivascular dermatitis to nodular pyogranulomatous dermatitis. Numerous eosinophils are usually present. Microfilariae are found intra- and/or extravascularly within the lesions.

Bibliography

Scott, D. W., and Vaughn, T. C. Papulonodular dermatitis in a dog with occult filariasis. *Compan Anim Pract* **1:** 31–35, 1987.

B. Autoimmune Dermatoses

The autoimmune skin diseases are uncommon to rare, but merit detailed consideration because their characteristic microscopic lesions enable specific diagnosis to be made on biopsy. The selection of a fresh lesion is crucial. Demonstration of a tissue-bound or circulating autoantibody using appropriate immunological tests may be helpful in confirming the diagnosis of an autoimmune skin disease. However, such tests (e.g., direct and indirect immunofluorescence testing, immunohistochemistry) are fraught with interpretational pitfalls (false-positive or false-negative test results) and should never be interpreted in the absence of histologic examination. Demonstration of autoantibodies does not necessarily confirm causative roles for these antibodies. In many of the putative cases of autoimmune skin disease in animals, the criteria for autoimmunity have not been fully met.

Bibliography

Haines, D. M., Cooke, E. M., and Clark, E. G. Avidin–biotin–peroxidase complex immunohistochemistry to detect immunoglobulin in formalin-fixed skin biopsies in canine autoimmune skin disease. *Can J Vet Res* **51:** 104–109, 1987.

Kalaher, K. M., Scott, D. W., and Smith, C. A. Direct immunofluorescence testing of normal feline nasal planum and footpad. *Cornell Vet* **80:** 105–109, 1990.

Moore, F. M. *et al.* Localization of immunoglobulin and complement by the peroxidase antiperoxidase method in autoimmune and nonautoimmune canine dermatopathies. *Vet Immunol Immunopathol* **14:** 1–9, 1987.

Scott, D. W. Autoimmune skin disease in the horse. *Eq Pract* **11:** 20–32, 1989.

Scott, D. W. *et al.* Pitfalls in immunofluorescence testing in dermatology. III. Pemphiguslike antibodies in the horse and direct immunofluorescence testing in equine dermatophilosis. *Cornell Vet* **74:** 305–311, 1984.

Scott, D. W. *et al.* Pemphiguslike antibodies in the goat. *Agri-Pract* **8:** 9–10, 1987.

Scott, D. W. *et al.* Immune-mediated dermatoses in domestic animals: Ten years after—part I and part II. *Compend Cont Ed Pract Vet* **9:** 424–435 and 539–553, 1987.

Werner, L. L., Brown, K. A., and Halliwell, R. E. W. Diagnosis of autoimmune skin disease in the dog: Correlation between histopathologic, direct immunofluorescent, and clinical findings. *Vet Immunol Immunopathol* **5:** 47–64, 1984.

1. Pemphigus

Pemphigus refers to a group of autoimmune skin diseases characterized clinically by pustules, vesicles, bullae, erosions, and ulcers, histologically by loss of adhesion between cells (acantholysis), and immunologically by the development of autoantibodies (''pemphigus antibodies'') directed against surface antigens of keratinocytes of various stratified squamous epithelia including skin, mucocutaneous junctions, oral mucosa, esophagus, and vagina (Table 1). The antigens against which the antibodies are directed are integral membrane glycoproteins and vary depending on the form of pemphigus. In pemphigus foliaceus of humans, the antigen has a molecular mass of 160 kDa and has been shown to be a cadherin adhesion molecule, desmoglein I, located in desmosomes. The human antigen of pemphigus vulgaris has a molecular mass of 130 kDa and seems to have some homology with desmoglein, but it is located in the interdesmosomal plasma membrane. In the vast majority of cases, the stimulus triggering the immune response is unknown, although drugs such as penicillamine may precipitate clinically, histologically, and immunologically classical cases of pemphigus, which regress when the drug is withdrawn. Exposure to ultraviolet light is known to exacerbate the disease. The binding of pemphigus autoantibodies to keratinocyte antigens is associated with the synthesis and secretion of urokinase-type plasminogen activator (uPA), a serine protease that in turn activates plasminogen. It also is associated with loss of cohesion between keratinocytes, which leads to the primary lesion, a vesicle. There are two hypotheses to explain the pathogenesis of vesicle formation. The first postulates a direct interruption of cell adhesion via binding of the autoantibody to the antigen, which is an adhesion molecule. The second proposes that uPA, by activating plasminogen, indirectly induces the cleaving of intercellular contacts. The second hypothesis is based on the fact that anti-uPA antibodies or the specific inhibitor of uPA, plasminogen activator inhibitor type-2, inhibit lesion formation *in vitro*.

Pemphigus is characterized by the deposition of immunoglobulin, with or without complement, on the surface of keratinocytes. The typical chicken wire pattern may be demonstrated by direct immunofluorescence testing (Fig. 5.49) or immunohistochemistry. Unfortunately, false-negative (poor lesion selection, prior glucocorticoid therapy) and false-positive (any dermatosis in which spongiosis is present) reactions are frequent. Hence, these

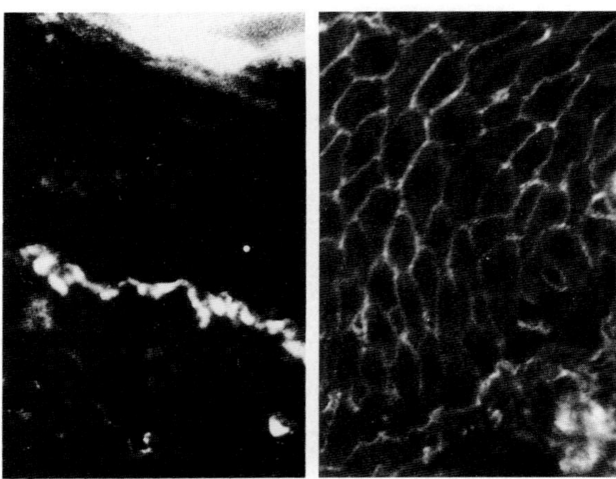

Fig. 5.49 Bullous pemphigoid. Direct immunofluorescent tests. (Left) Linear band of immunoglobulin at dermoepidermal junction in bullous pemphigoid. Note intercellular deposition of immunoglobulins characteristic of all forms of pemphigus (right).

immunopathologic tests can be interpreted only in the light of clinical and histopathologic findings. The autoimmune character of pemphigus in humans depends on demonstrating pemphigus autoantibodies in the serum by indirect immunofluorescence testing. Indirect immunofluorescence testing is usually negative, however, in domestic animals with pemphigus. In addition, "pemphiguslike" antibodies can be occasionally found in nonpemphigus diseases.

Different types of pemphigus are recognized. These are divided according to the level within the epidermis at which acantholysis occurs, the clinical findings, and other immunologic findings. The development of acantholysis at different epidermal levels appears to be related to recognition of different antigens of the keratinocyte plasma membrane by the autoantibodies.

a. PEMPHIGUS VULGARIS This is the most severe, but, despite the appellation "vulgaris," is not the most common variant in the dog and cat. There are no apparent age, breed, or sex predilections. The fragile vesicles or bullae in the epidermis are transient and readily rupture to leave roughly circular, shallow, flat-based erosions or ulcers. The margins often develop epidermal collarettes. Firm sliding pressure to adjacent unaffected skin may induce fresh vesicle formation or dislodge the skin (the Nikolsky sign). Lesions involve mucous membranes (Fig. 5.50), mucocutaneous junctions, and skin, especially in the inguinal and axillary areas. Oral involvement is present in ~90% of the cases, and in 50% of the cases, the lesions commence in the mouth. Involvement of the nailbeds is also seen. Microscopically there is suprabasilar acantholysis with intraepidermal clefts, vesicles, or bullae between the stratum basale and the stratum spinosum (Fig. 5.51). The basal keratinocytes, although separated from each other following disruption of intercellular contacts, are anchored to the basal lamina (like "a row of tombstones"). The outer layers of the epidermis form the roof of the vesicle, which contains a few acantholytic keratinocytes, singly or in clumps, but few or no inflammatory cells. The process may extend into the hair follicle epithelium (Fig. 5.52). The dermal reaction varies from a mild perivascular accumulation of mononuclear cells to a moderately intense lymphocytic–plasmacytic lichenoid interface dermatitis.

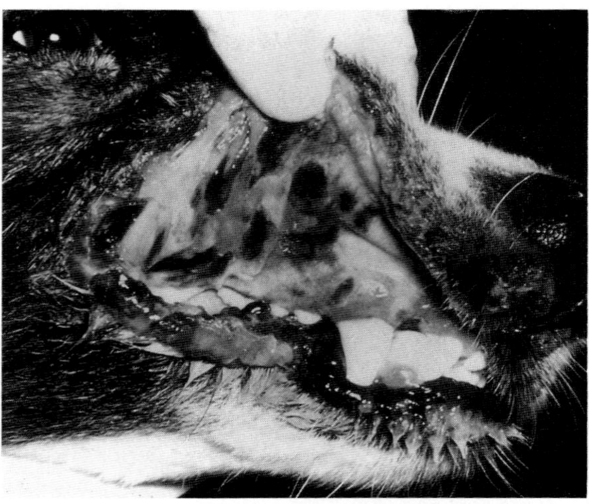

Fig. 5.50 Pemphigus vulgaris. Flaccid bullae in oral mucosa. (Courtesy of W. M. Parker and the *Canadian Veterinary Journal*.)

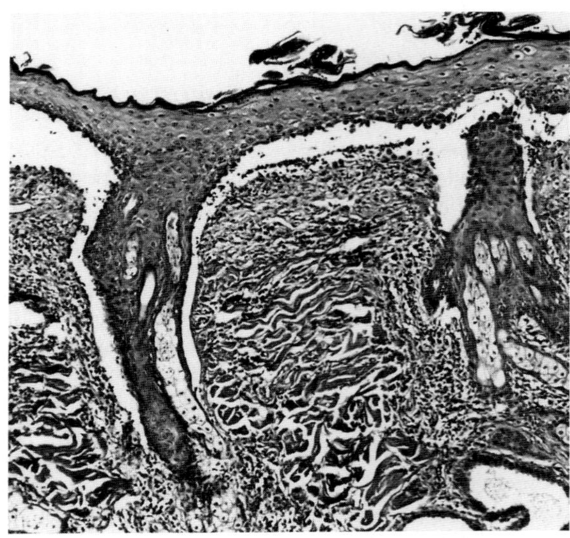

Fig. 5.51 Pemphigus vulgaris. Suprabasilar acantholysis producing intraepithelial cleft. (Courtesy of R. Seiler.)

b. PEMPHIGUS VEGETANS This is an extremely rare, benign variant of pemphigus vulgaris, reported only in dogs. The skin lesions are chronic, generalized, proliferative, scaly, and crusted. Verrucous vegetations studded with small pustules are characteristic. The predominant histo-

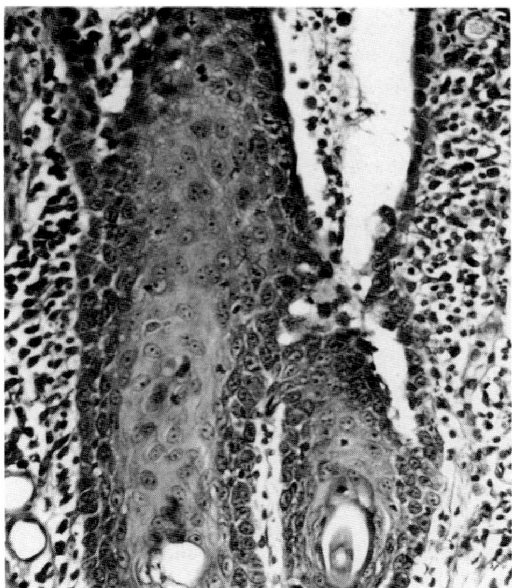

Fig. 5.52 Pemphigus vulgaris. Hair follicle involvement with lichenoid infiltrate.

logic lesions are numerous large, intraepithelial pustules composed almost entirely of eosinophils and a few acantholytic epithelial cells. These pustules occur at several levels through the surface epithelium and also involve the follicular epithelium. These lesions are present on a background of papillated epidermal hyperplasia.

c. PEMPHIGUS FOLIACEUS In pemphigus foliaceus, the acantholytic process occurs at a superficial level in the epidermis, the lesions being typically exfoliative rather than erosive or ulcerative. Pemphigus foliaceus is the most common form of pemphigus in the dog, cat, and horse, and also occurs in goats. There are no sex or age predilections, but Akitas, chow chows, dachshunds, bearded collies, Newfoundlands, Doberman pinschers, schipperkes, and Appaloosa horses appear predisposed. The gross lesions are similar in all species and are usually generalized but do not involve mucosae. Transient vesicopustules rupture to form very shallow erosions, which become covered by thick crust and scale (Fig. 5.53A,B). Alopecia is often marked. Lesions in the dog and cat tend to appear first on the nose and spread to periorbital areas, the ears, neck, and ventral abdomen. Involvement of the feet may produce marked papillary hyperkeratosis and crusting of the footpads, and there is a predilection for the nail beds, occasionally leading to sloughing of the nail. In horses, lesions often begin on the face or distal extremities, and may be localized to coronets. Generalized involvement is also seen. The histologic pattern of pemphigus foliaceus is an acantholytic subcorneal (Fig. 5.54) or intragranular pustular dermatitis. Either neutrophils or eosinophils may predominate. At the base of the pustule, acantholytic keratinocytes continue to detach and enter the pustule. Ruptured pustules form a thick inflammatory crust in which

Fig. 5.53 Pemphigus foliaceus. Horse. (A) Generalized exfoliative dermatitis and alopecia. (B) Detail of skin in (A) showing massive scale crust.

acantholytic cells are prominent. The external root sheath of the hair follicle undergoes a similar acantholytic process. Rarely, a lichenoid inflammatory infiltrate is seen.

Superficial bacterial pyoderma (impetigo or superficial folliculitis), subcorneal pustular dermatosis, sterile eosinophilic pustulosis, and linear IgA dermatosis show a histologic pattern similar to that of pemphigus foliaceus. Differentiation is on the basis of numbers of acantholytic keratinocytes in the pustule or crust (numerous in pemphigus foliaceus, absent to numerous in subcorneal pustular dermatosis, few or absent in bacterial pyoderma, sterile eosinophilic pustulosis, and linear IgA dermatosis), involvement of the hair follicle epithelium (not present in impetigo or subcorneal pustular dermatosis), and the presence of "cling-on" stratum granulosum cells (present only in pemphigus foliaceus).

d. PEMPHIGUS ERYTHEMATOSUS This may represent a benign form of pemphigus foliaceus, and an intermediate form between pemphigus and lupus erythematosus. It has been reported in dogs and cats. Collies and German shepherds are predisposed. The gross lesions closely resemble those described for pemphigus foliaceus except that they are usually restricted to the face and ears. Lesions usually commence on the dorsum of the nose and may extend to the periorbital skin and the pinna. Leukoderma of the

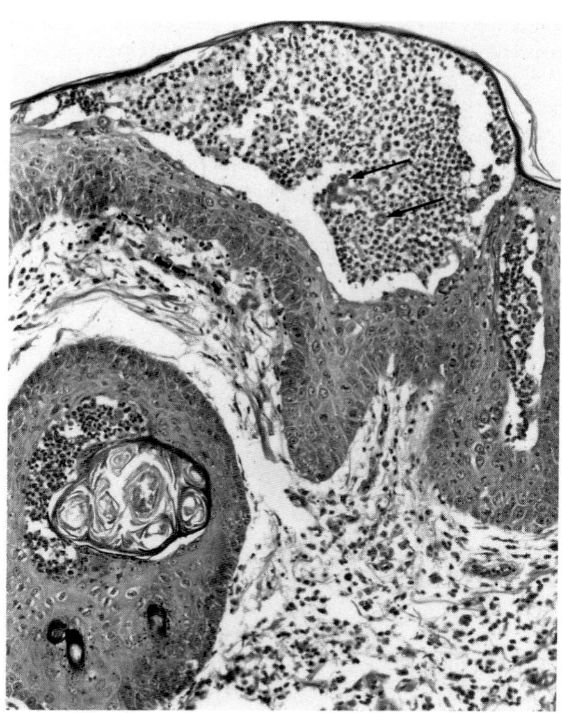

Fig. 5.54 Pemphigus foliaceus. Subcorneal pustular dermatitis with acantholysis. Note hair follicle involvement and dark-staining acantholytic keratinocytes (arrows) within pustule.

nasal planum is seen in chronic cases. The histologic lesions are similar to those of pemphigus foliaceus, but there are usually very prominent mural follicular acantholytic pustules, the epidermis may show papillated hyperplasia, and a lichenoid interface dermatitis is frequently found. The diagnosis rests on the immunologic findings. The direct immunofluorescence test or immunohistochemistry reveals, in addition to the diffuse cell surface fluorescence pattern typical of the pemphigus group, a linear band of immunoglobulin, with or without complement, deposited at the basement membrane zone. The latter pattern is typical of lupus erythematosus. Dogs and cats with pemphigus erythematosus may, in addition, have positive antinuclear antibody tests.

Bibliography

August, J. R., and Chickering, W. R. Pemphigus foliaceus causing lameness in four dogs. *Compend Cont Ed* **7:** 894–902, 1985.

Bennett, D. *et al.* Two cases of pemphigus erythematosus (the Senear–Usher syndrome) in the dog. *J Small Anim Pract* **26:** 219–227, 1985.

Manning, T. O. *et al.* Pemphigus diseases in the feline: Seven case reports and discussion. *J Am Anim Hosp Assoc* **18:** 433–443, 1982.

Pfeiffer, C. J., Spurlock, S., and Ball, M. Ultrastructural aspects of equine pemphigus foliaceuslike disease. Report of cases. *J Submicrosc Cytol Pathol* **20:** 453–461, 1988.

Rubinstein, N., and Stanley, J. R. Pemphigus foliaceus antibodies and a monoclonal antibody to desmoglein 1 demonstrate stratified squamous epithelia-specific epitopes of desmosomes. *Am J Dermatopathol* **9:** 510–514, 1987.

Scott, D. W. *et al.* Pitfalls in immunofluorescence testing in dermatology. II. Pemphiguslike antibodies in the cat, and direct immunofluorescence testing of normal dog nose and lip. *Cornell Vet* **73:** 275–279, 1983.

Scott, D. W. *et al.* Pemphigus foliaceus in a goat. *Agri-Pract* **5:** 38–45, 1984.

Stanley, J. R., Koulu, L., and Thivolet, C. Pemphigus vulgaris and pemphigus foliaceus autoantibodies bind different molecules. *Clin Res* **32:** 616A, 1984.

Stannard, A. A., Gribble, D. H., and Baker, B. B. A mucocutaneous disease in the dog, resembling pemphigus vulgaris in man. *J Am Vet Med Assoc* **166:** 575– 582, 1975.

Suter, M. M. *et al.* Pemphigus in the dog: Comparison of immunofluorescence and immunoperoxidase method to demonstrate intercellular immunoglobulins in the epidermis. *Am J Vet Res* **45:** 367–369, 1984.

Suter, M. M. *et al.* Ultrastructural localization of pemphigus vulgaris antigen on canine keratinocytes *in vivo* and *in vitro*. *Am J Vet Res* **51:** 507–511, 1990.

Wilkinson, J. E. *et al.* Role of plasminogen activator in pemphigus vulgaris. *Am J Pathol* **134:** 561–569, 1989.

2. *Bullous Pemphigoid*

Bullous pemphigoid is a chronic, autoimmune skin disease characterized clinically by vesicles, bullae, and ulcers, histologically by the subepidermal location of these vesicles and bullae, and immunologically by the presence of an antibody to the bullous pemphigoid antigen in the basement membrane zone (Fig. 5.49). The antigen recognized by human bullous pemphigoid autoantibodies is a 230-kDa protein; the C terminus of this protein shares considerable homology with desmoplakin I, a desmosomal plaque molecule. The mechanism of dermoepidermal separation is thought to be indirect, resulting from complement activation and resultant leukocyte chemotaxis and activation. Immunofluorescence testing localizes C3, as well as autoantibody, to the basement membrane zone. The possibility that separation of basal cells from the underlying dermis may be the direct result of antibody binding to the hemidesmosomes has not been ruled out. In almost all cases, the stimulus triggering the immune response is unknown, although drugs may precipitate clinically, histologically, and immunopathologically classical cases of pemphigoid, which regress when the drug is withdrawn. Exposure to ultraviolet light is known to exacerbate bullous pemphigoid.

Bullous pemphigoid occurs in dogs and horses, and has no age or sex predilection. Collies appear predisposed. The gross lesions are vesiculobullous, ulcerative, and crusted in nature. The lesions affect mucous membranes, particularly of the oral cavity, mucocutaneous junctions, and skin of the face (Fig. 5.55A), ears, axilla, groin, and paws. Grossly, bullous pemphigoid is usually indistinguishable from pemphigus vulgaris. Histologically, there is a subepidermal bulla filled with a fibrin net and variable numbers of neutrophils, eosinophils, and mononuclear cells (Fig. 5.55B). Acantholytic keratinocytes are not present. The basal cells which line the roof of the bulla are not

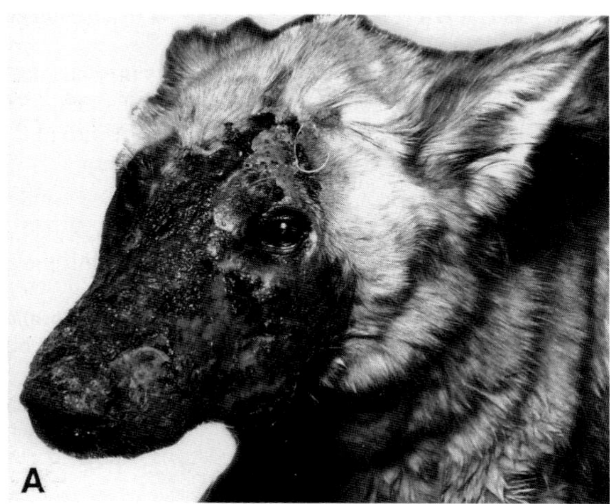

Fig. 5.55A Bullous pemphigoid. Severe facial lesion in affected dog.

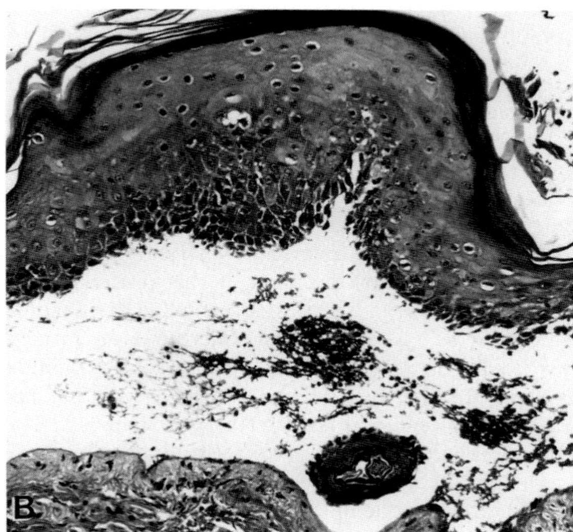

Fig. 5.55B Bullous pemphigoid. Subepidermal vesiculation. Floor of vesicle is formed by the basement membrane zone. Note lack of acantholysis.

initially degenerate. The floor of the bulla is lined by the basement membrane zone. There is often marked subepidermal vacuolar alteration. The dermis is usually markedly edematous, capillaries are dilated with swollen endothelial cells, and there is a perivascular accumulation of neutrophils, eosinophils, and mononuclear cells. A mild to moderate lichenoid interface dermatitis may be present. Eosinophil infiltration is not a marked feature in canine cases, although it is one of the hallmarks of the human disease for which it is named. Diseases to consider in differential diagnosis are systemic and discoid lupus erythematosus, drug eruptions, epidermolysis bullosa, dermatomyositis, cutaneous porphyria, erythema multiforme, toxic epidermal necrolysis, and severe urticaria.

Classically, bullous pemphigoid is characterized by the presence of immunoglobulin and/or complement deposited at the basement membrane zone within skin lesions. Unfortunately, false-negative (poor lesion selection, prior glucocorticoid therapy) and false-positive (lupus erythematosus, linear IgA dermatosis, many other inflammatory dermatoses) reactions are common. In addition, IgM may be found at the basement membrane zone of nasal planum in normal dogs and cats, and footpads from normal dogs. Hence, interpretation is impossible without accurate clinical information and appropriate histopathologic material.

Bibliography

Scott, D. W. Pemphigoid in domestic animals. *Clin Dermatol* **5:** 155–162, 1987.

Stanley, J. R. Pemphigus and pemphigoid as paradigms of organ-specific, autoantibody-mediated diseases. *J Clin Invest* **83:** 1443–1448, 1989.

Tanaka, T. *et al.* cDNA cloning of bullous pemphigoid (BP) antigen reveals structural and sequence homology with desmoplakin (DP) I. *J Invest Dermatol* **94:** 583, 1990.

3. *Lupus Erythematosus*

Lupus erythematosus occurs in two forms: systemic lupus erythematous, affecting multiple tissues including the skin, and discoid lupus erythematosus, which is localized to the skin. While the two conditions share cutaneous lesions and certain immunological abnormalities, it is not certain that the one is a localized form of the other. Progression of discoid lupus to systemic lupus rarely, if ever, occurs.

Systemic lupus erythematosus (SLE) is a multisystem disease characterized by the presence of B-cell hyperactivity and formation of a variety of autoantibodies. It is considered to result from a general failure of immune autoregulation, including defects of suppressor T-cell function and dysregulation of endogenous cytokines such as IL-6. There is a genetic predisposition to the disease, which is potentiated by environmental factors, including exposure to ultraviolet radiation and drugs. Infectious agents, in particular C-type RNA viruses, have been suggested as initiating agents. Sex hormones and nutrition are known to modulate disease activity. Loss of immune autoregulation results in the production of antibodies against a range of membrane and soluble antigens. The most characteristic of these are the antinuclear antibodies directed against double-stranded DNA, RNA, nucleoprotein, and histone-related antigens. Antibodies are also detected against erythrocytes, leukocytes, platelets, organ-specific antigens, clotting factors, and IgG (rheumatoid factor). While some of these antibodies may be responsible for tissue injury, the main tissue damage is effected by the antigen–antibody complexes which deposit at various sites throughout the body and incite a type 3 hypersensitivity reaction. In the skin, the antigen–antibody complexes are located under the basement membrane zone of the epidermis and to a lesser extent in the walls of the dermal blood vessels. The current hypothesis for the pathogenesis of the ultraviolet-exacerbated cutaneous lesions is as follows: (1) ultraviolet light induces the translocation of antigens normally expressed only intracellularly, such as extractable nuclear antigens, to the keratinocyte cell membrane; (2) specific autoantibodies to these antigens attach to keratinocytes, inducing keratinocyte death by antibody-dependent cellular cytotoxicity mediated by lymphocytes or monocytes; and (3) injured keratinocytes release a plethora of cytokines, which account for the resultant localization of the lymphohistiocytic infiltrate.

Canine and feline systemic lupus occurs without clear age or sex predilections. In humans, there is a clear preponderance of SLE in females. Only one series of cases in dogs reports a predilection for intact females. Collies, Shetland sheepdogs, German shepherds, and Siamese cats appear predisposed to SLE. Attempts to establish colonies of affected dogs by breeding affected parents has met with mixed success. Polyarthritis, anemia, thrombocytopenia, stomatitis, glomerulonephritis, and dermatitis are the most common disease manifestations. Dermatologic signs occur in ~33% of affected dogs and cats. Exposure to ultraviolet radiation exacerbates the skin lesions. Gross skin lesions are extremely variable, from a mucocutaneous, ulcerative dermatitis resembling pemphigus vulgaris and bullous pemphigoid to generalized seborrheic conditions of little specificity. There is a predilection for the face, ears, and distal limbs (Fig. 5.56). Footpads and nail beds may be involved. Secondary bacterial pyodermas are common. Occasionally dogs develop a symmetric scaling, erythematous and alopecic maculopapular lesion extending over the bridge of the nose onto the cheeks, resembling in distribution the classical malar rash of human systemic lupus.

Horses with systemic lupus erythematosus have manifested polyarthritis, anemia, thrombocytopenia, proteinuria, fever, and weight loss. Skin lesions have included lymphedema of the extremities, panniculitis, alopecia, leukoderma, and scaling of the face, neck, and trunk, and generalized exfoliative dermatitis.

The microscopic lesions of systemic lupus are variable. The critical diagnostic pattern is an interface dermatitis with hydropic degeneration of basal cells (hydropic) and a lymphohistioplasmacytic infiltrate at the dermoepidermal junction (lichenoid) (Fig. 5.57A,B). The same changes often affect follicular epithelium. Pigmentary incontinence, a consequence of the basal cell degeneration, is a diagnostically helpful histologic lesion, as is marked subepidermal vacuolar change. Occasionally this change is sufficiently severe to cause cleft or bulla formation between the dermis and epidermis. In the acute stage, there is moderate to marked edema, capillary dilation, extravasation of leukocytes and erythrocytes, and variable lymphohistiocytic infiltrates around vessels of superficial and deep capillary plexuses. Mucinous degeneration may be prominent. As the lesion becomes chronic, there is focal thickening of the basement membrane zone reflecting ac-

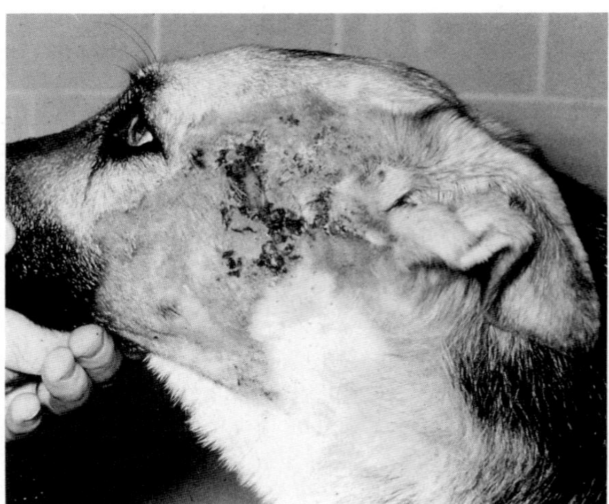

Fig. 5.56 Lupus erythematosus. Gross lesion.

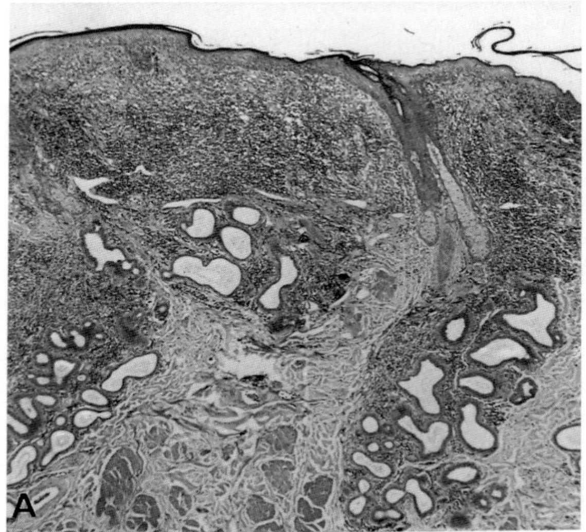

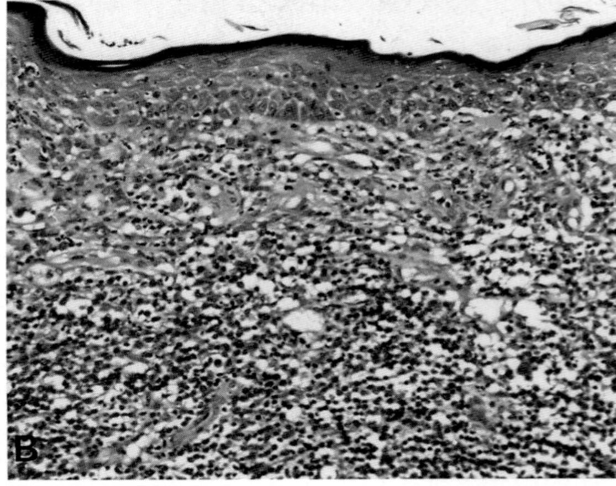

Fig. 5.57 Lupus erythematosus. (A) and (B) Histology showing interface dermatitis with lichenoid band at dermoepidermal junction and following adnexa.

cumulation of antigen–antibody complexes. Fibrinoid deposits may also occur around superficial blood vessels, and occasionally leukocytoclastic vasculitis develops. Other lesions include ortho- or parakeratotic hyperkeratosis, epidermal hyperplasia, dyskeratosis, adnexal atrophy, and dermal fibrosis. A rare histologic variant, which occurs in the dog and cat, is a subcorneal pustular dermatitis. The pustules, which contain numerous acantholytic epithelial cells and neutrophil leukocytes, closely resemble those seen in pemphigus foliaceus. This lesion may coexist with the more typical interface dermatitis or occur alone.

The localized cutaneous or **discoid** form of **lupus erythematosus** is described most commonly in the dog, and rarely in horses and cats. The disease is more prevalent in bitches. There is a marked breed predilection in collies, Shetland sheepdogs, Siberian huskies, and German shepherds. Sunlight exacerbates the lesions, which typically affect the nasal planum in dogs and cats, and the face and neck in horses. The lesions include erythema, depigmentation, scaling, crusting, alopecia, and occasionally ulceration. Rarely, dogs are presented with only nasodigital hyperkeratosis. Histologic findings are as previously described for systemic lupus.

Lupus panniculitis (lupus profundus) is characterized by well-circumscribed subcutaneous nodules over the trunk and proximal extremities. It is rarely reported in dogs. Histologically, it is initially a septal panniculitis with dense nodular infiltrates of lymphocytes and plasma cells and fewer macrophages. The fat lobules may undergo necrosis, often represented by hyalinization without mineralization. Usually the dermis and epidermis show the typical microscopic features of lupus, including hydropic degeneration of basal cells, pigmentary incontinence, thickened basement membrane zone, and sclerosis of the dermis. Abundant mucinous degeneration is usually prominent. Leukocytoclastic vasculitis may be present in the interlobular septa. Histologically, lupus panniculitis is distinctive, but a rabies vaccine-induced lymphocytic panniculitis that is virtually indistinguishable histologically oocurs in dogs.

Bibliography

Geor, R. J. *et al.* Systemic lupus erythematosus in a filly. *J Am Vet Med Assoc* **197:** 1489–1492, 1990.

Goldman, M. Cytokines in the pathophysiology of systemic lupus erythematosus. *Autoimmunity* **8:** 175–179, 1990.

Lee, L. A., and Norris, D. A. Mechanisms of cutaneous tissue damage in lupus erythematosus. *In* "Immune Mechanisms in Cutaneous Disease" D.A. Norris (ed.), pp. 359–386. New York, Marcel Dekker, 1989.

Linker-Israeli, M. *et al.* Elevated levels of endogenous IL-6 in systemic lupus erythematosus. A putative role in pathogenesis. *J Immunol* **147:** 117–123, 1991.

Rosencrantz, W. S. *et al.* Histopathological evaluation of acid mucopolysaccharide (mucin) in canine discoid lupus erythematosus. *J Am Anim Hosp Assoc* **22:** 577–584, 1986.

Scott, D. W. *et al.* Canine lupus erythematosus. I. Systemic lupus erythematosus. II. Discoid lupus erythematosus. *J Am Anim Hosp Assoc* **19:** 461–479 and 481–488, 1983.

C. Other Immune-Mediated Dermatoses

1. Cold Agglutinin Disease

Cutaneous signs associated with cryoglobulins and cryofibrinogens result from vascular insufficiency (obstruction, stasis, spasm, thrombosis). Cold agglutinin disease has been reported rarely to cause skin disease in dogs and cats. Lesions include erythema, purpura, cyanosis, necrosis, ulceration, and occasionally sloughing, and are precipitated or exacerbated by exposure to cold. The paws, pinnae, nose, and tip of tail are typically involved. Skin biopsy usually reveals necrosis, ulceration, and often secondary suppurative changes. Fortuitous sections may show thrombosed or necrotic blood vessels, or blood vessels containing an amorphous eosinophilic substance.

2. Graft-versus-Host Disease

Graft-versus-host disease occurs as a complication of bone marrow transplantation. It has been recorded in dogs and horses. The disease results from donor T-lymphocyte responses to recipient transplantation antigens. Principal target organs are the skin, liver, and intestinal tract. Skin lesions include generalized, multifocal alopecia, exfoliative erythroderma, and ulcerative dermatitis. Histologic findings include varying degrees of hydropic and/or lichenoid interface dermatitis with marked "satellite cell necrosis" of keratinocytes. Follicular lesions were prominent in one bone-marrow-transplanted beagle with clinical alopecia. The lichenoid dermatitis affected the follicular wall and the base of the hair follicle, resembling the histologic lesions of alopecia areata.

3. Erythema Multiforme

Erythema multiforme is an uncommon disorder in the dog, cat, horse, and cow. Despite recognition of multiple etiologic and triggering factors, the pathogenesis is not understood. The most common predisposing factors in humans and animals are drugs and infections. An immunologic pathogenesis has been inferred from the histologic appearance of the lesions, in which lymphocytes are often adjacent to dying epidermal cells, suggesting direct cellular cytotoxicity. Occurrence of similar lesions in experimental or naturally occurring graft-versus-host disease further enhances the hypothesis that the lesion is one of cell-mediated immunity. Immunoglobulin and complement deposits in skin lesions and elevated circulating immune-complex levels are probably epiphenomena. The possibility also exists that the drug or infectious agent may be able to induce keratinocytes to undergo apoptosis without the assistance of the immune system.

The gross lesions are characterized by an acute onset of symmetric erythematous macules and papules (Fig. 5.58A). These spread peripherally and clear centrally to produce annular, arciform, and polycyclic patterns (Fig. 5.58B). Urticarial plaques with similar configurations and vesiculobullous lesions also occur. Lesions are not pruritic, and there are no systemic signs, other than those attributable to the underlying cause. Lesions tend to in-

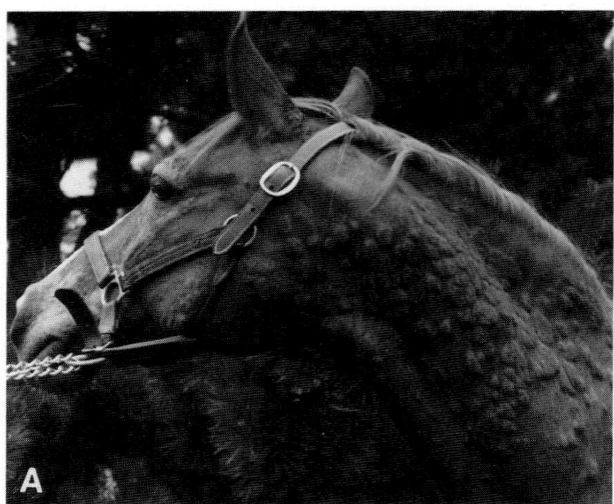

Fig. 5.58A Erythema multiforme. Horse. Multiple irregular plaques that are not transient as they are in horse with urticaria (Fig. 5.43).

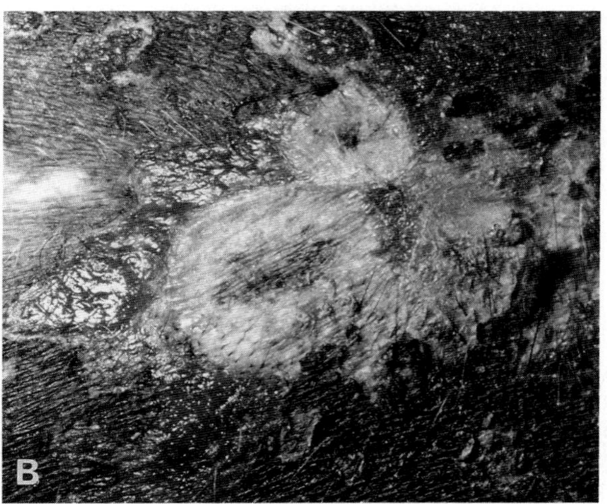

Fig. 5.58B Gross lesions of erythema multiforme.

volve mucocutaneous junctions, the ventral trunk, proximal limbs, and occasionally the nose and footpads. Histologically, the lesions show the characteristic interface dermatitis of hydropic subtype. Necrotic keratinocytes are scattered throughout the epidermis and through the adnexal epithelia. A sparse lymphohistiocytic infiltrate occurs at the dermoepidermal junction (Fig. 5.59A) and around superficial blood vessels. As the lesion ages, the infiltrate may become more dense, subepidermal bullae may develop, and the epidermal necrosis may become confluent (Fig. 5.59B). Vasculitis is not a feature. The histologic pattern is shared by acute lupus erythematosus, less severe forms of toxic epidermal necrolysis, and drug eruptions. A less common finding in erythema multiforme is severe dermal edema with vertical orientation of dermal collagen and subepidermal vesiculation.

Bibliography

Huff, J. C. Immunologic mechanisms of erythema multiforme. *In* "Immune Mechanisms in Cutaneous Disease" D.A. Norris (ed.), pp. 387–402. New York, Marcel Dekker, 1989.

Medleau, L. *et al.* Erythema multiforme and disseminated intravascular coagulation in a dog. *J Am Anim Hosp Assoc* **26:** 643–646, 1990.

Scott, D. W., Miller, W. H., and Goldschmidt, M. H. Erythema multiforme in the dog. *J Am Anim Hosp Assoc* **19:** 453–459, 1983.

4. *Toxic Epidermal Necrolysis*

Toxic epidermal necrolysis is characterized by disseminated erythema, widespread bullous necrosis of epidermis and mucous membranes, and severe toxicity. Like erythema multiforme, it is most often associated with drug intake or concurrent infections, but many cases are idiopathic. In humans the relationship between toxic epidermal necrolysis and severe forms of erythema multiforme, known as erythema multiforme major or Stevens–Johnson syndrome, has been a matter of controversy.

The lesions of erythema multiforme major show areas of full-thickness epidermal necrosis as in toxic epidermal necrolysis, but differences in the clinical course, response to therapy, and fatality rate suggest that they are separate entities.

The pathogenesis of toxic epidermal necrosis is not known. Cell-mediated immune mechanisms have been postulated as the basis for necrosis. Lymphopenia and alterations in the T-helper and T-suppressor lymphocyte ratios occur in human patients. Drugs associated with toxic epidermal necrolysis, however, do not induce lymphocyte transformation *in vitro*. Furthermore, the widespread confluent epidermal necrosis, in the virtual absence of infiltrating inflammatory cells, argues against an effector role for cytotoxic lymphocytes. As in erythema multiforme, lesional deposition of antibody and complement most likely represent epiphenomena.

Toxic epidermal necrolysis is a serious, acute-onset disease in dogs and cats. It has been associated with concurrent infectious disease, neoplasia, and drug eruption. Drugs implicated include ampicillin, cephaloridine, levamisole, 5-fluorocytosine, and hetacillin. There is severe systemic illness and significant mortality. Lesions are usually generalized on mucous membranes, mucocutaneous junctions, and hairy skin. There is ulceration, often with an epidermal collarette, crusting, scaling, and secondary pyoderma. The skin is extremely painful and fragile. A positive Nikolsky sign is often present. Vesiculation may occur but is often transient. Grossly, toxic epidermal necrolysis may be difficult to distinguish from pemphigus vulgaris, bullous pemphigoid, systemic lupus erythematosus, thallium poisoning, cutaneous lymphosarcoma, and candidiasis.

Histologically, it is characterized by full thickness coagulative epidermal necrosis, which may extend into the external root sheath of the hair follicle. Separation of the necrotic epidermis occurs at the dermoepidermal junction

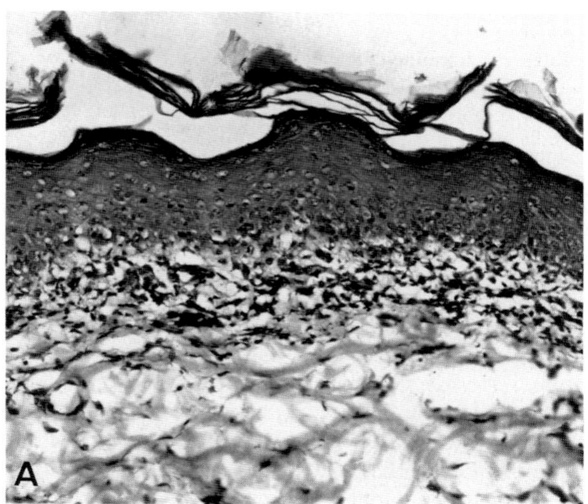

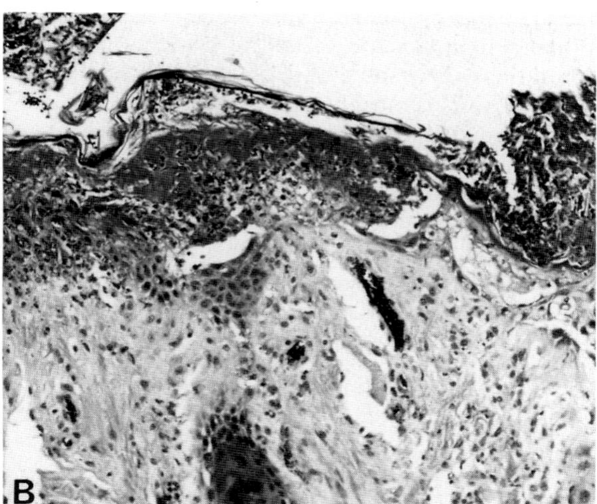

Fig. 5.59 (A) Histology of erythema multiforme showing interface dermatitis. (B) Histology of erythema multiforme showing partial- to full-thickness epidermal necrosis.

and leads to subepidermal vesiculation. The dermal inflammation is minimal. The reaction, when present, is perivascular or rarely lichenoid and chiefly lymphohistiocytic. Lesions with less than full-thickness epidermal necrosis and more intense dermal inflammation probably cannot be distinguished from erythema multiforme.

Bibliography

Rachofsky, M. A., Chester, D. K., and Read, W. K. Toxic epidermal necrolysis. *Compend Cont Ed* **11:** 840–845, 1989.

Scott, D. W. *et al.* Toxic epidermal necrolysis in two dogs and a cat. *J Am Anim Hosp Assoc* **15:** 271–279, 1979.

5. Vasculitis

Vasculitis is considered an uncommon cutaneous syndrome in dogs and horses, and is rarely seen in cats, pigs, and cattle. It is, however, probably more common than is diagnosed. The pathomechanism of most cutaneous vasculitides is assumed to involve type 3 hypersensitivity reactions. In **dogs,** cutaneous vasculitis has been associated with drugs, infections (Rocky Mountain spotted fever, staphylococcosis), malignancies, and connective tissue diseases (lupus erythematosus, rheumatoid arthritis). Focal cutaneous vasculitis has been reported at the site of rabies vaccination in dogs. A proliferative thrombovascular necrosis has been recognized in dogs, with wedge-shaped areas of necrosis and ulceration on the apical margins of the pinnae (Fig. 5.60). In **horses,** cutaneous vasculitis has been seen with numerous infections, especially strangles, equine influenza, equine viral arteritis, and *Corynebacterium pseudotuberculosis.* Equine purpura hemorrhagica is an acute, usually streptococcal (strangles)-related, leukocytoclastic vasculitis characterized clinically by urticaria and extensive edema of the distal limbs, ventrum, and head. These swellings may progress to exudation and sloughing. A photoactivated vasculitis has been reported in mature horses that develop severe edema in the nonpigmented areas of the distal limbs during the summer. Approximately 50% of cases of vaculitis in dogs and horses are idiopathic. Cutaneous vasculitis in **pigs** is most commonly associated with *Erysipelothrix rhusiopathiae* infection. An outbreak of leukocytoclastic cutaneous vasculitis in newborn **calves** in Belgium could not be ascribed to any drug or infectious agent.

Skin lesions include palpable purpura, hemorrhagic bullae, edema, necrosis, and well-demarcated ulcers, especially involving the paws, pinnae, lips, tail, and oral mucosa.

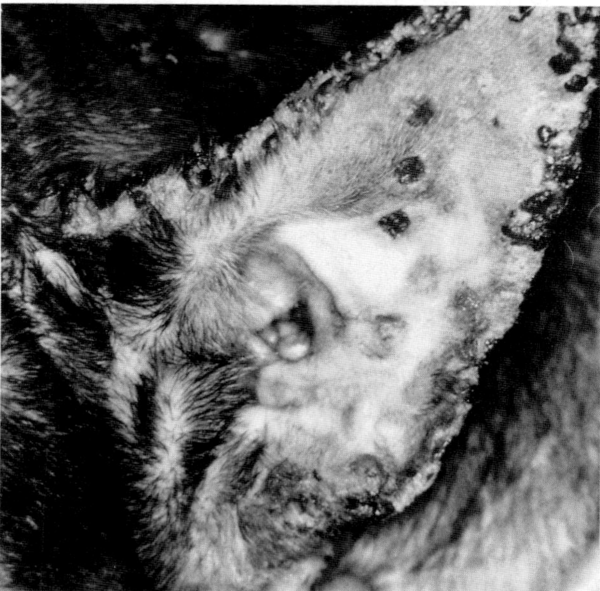

Fig. 5.60 Cutaneous vasculitis. Dog. Typical punched out ulcers on the ear tips.

Histologic findings include varying degrees of neutrophilic and/or lymphocytic vasculitis, possibly reflecting the age of the lesions and the types of immune reactants. Fibrinoid necrosis is most commonly seen in horses. Eosinophils may be the predominant cells in some cases of equine vasculitis. Involvement of deep dermal blood vessels may suggest systemic disease. Canine proliferative thrombovascular necrosis is characterized by arteriolar proliferation, sclerosis, hyaline degeneration, and thrombosis. Vasculitis is not seen.

Direct immunofluorescence testing or immunohistochemistry may demonstrate immunoglobulin and/or complement in vessel walls and occasionally at the basement membrane zone. Positive tests are most likely in lesions less than 24 hr old.

Bibliography

Crawford, M. A., and Foil, C. S. Vasculitis: Clinical syndromes in small animals. *Compend Cont Ed Pract Vet* **11:** 400–415, 1989.

De Geest *et al.* An outbreak of cutaneous vasculitis in newborn calves. *Vlaams Diergeneeskd Tijdschr* **59:** 7–11, 1990.

Morris, D. D. Cutaneous vasculitis in horses: 19 cases (1978–1985). *J Am Vet Med Assoc* **191:** 460–464, 1987.

6. Rabies Vaccine-Induced Vasculitis in Dogs

This lesion is being recognized with increasing frequency in those countries where routine rabies vaccination by the subcutaneous route is practiced. All breeds are affected, but poodles are predisposed. Lesions, which develop at sites of previous vaccination, are alopecic, hyperpigmented patches of atrophic skin (Fig 5.61). Lesions may appear months after the vaccine was administered. The histologic appearance is highly characteristic, although the vascular lesions are subtle (Fig. 5.62A). Venules, arterioles, and small arteries develop a very mild chronic lymphocytic vasculitis, characterized by thickening of the vessel wall, a few intramural mononuclear inflammatory cells, scattered nuclear debris, and variable perivascular mononuclear infiltrates. Occasionally a more florid leukocytoclastic vasculitis is seen (Fig. 5.62B). The epidermis may show vacuolar change in the basal layer and pigmentary incontinence. The dermis is atrophic, sometimes mucinous, and hair follicles are in telogen and pale staining (faded). A diffuse increase in mononuclear

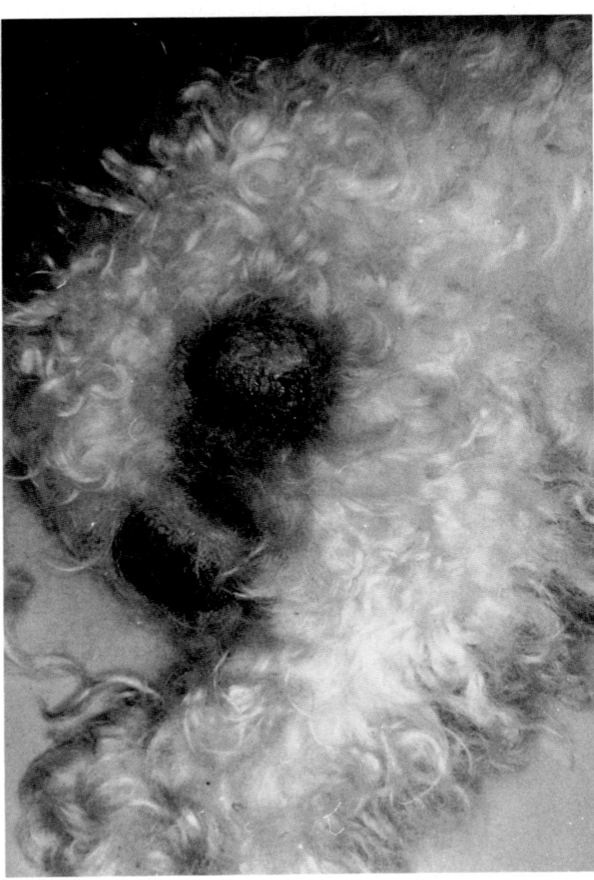

Fig. 5.61 Hyperpigmented, alopecic patch on posterior thigh. Poodle. Rabies vaccine-induced lesion. (Courtesy of L. Schmeitzel.)

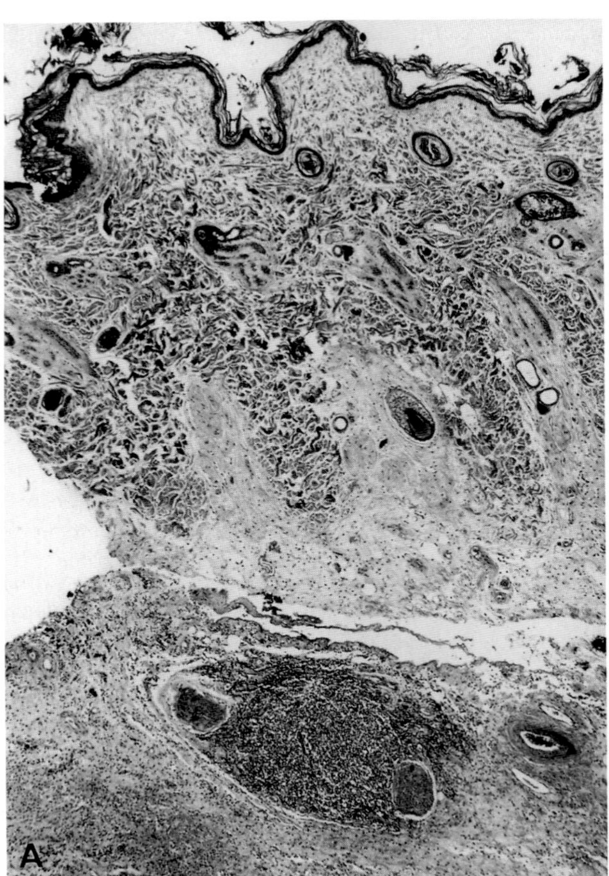

Fig. 5.62A Rabies vaccine-induced vasculitis. Dog. Degenerate and inflamed vessel.

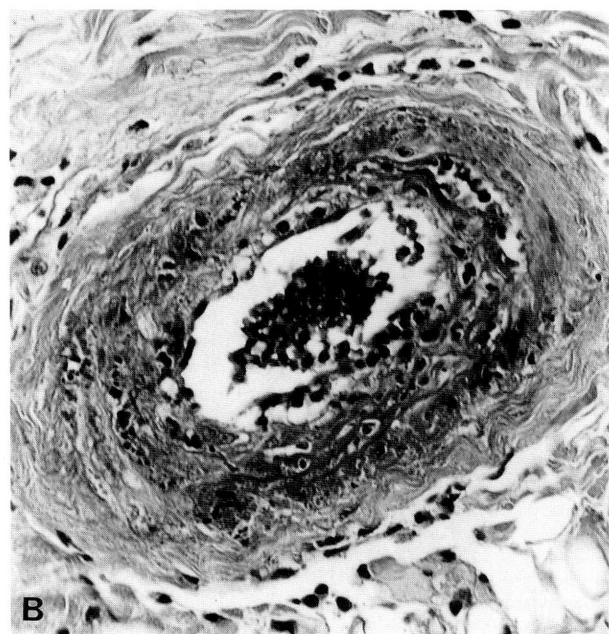

Fig. 5.62B Rabies vaccine-induced vasculitis. Dog. Note the lymphocytic panniculitis and follicular atrophy.

cells throughout the dermis may be accompanied by lymphocytic panniculitis. Immunofluorescence staining has identified rabies antigen in the vessels and epithelium of the hair follicles. A low-grade immune-mediated vasculitis with resultant tissue anoxia leading to the atrophic changes in the overlying skin has been suggested as the pathogenesis. Differential diagnosis includes dermatomyositis, traction alopecia, and lupus profundus.

Bibliography

Wilcock, B. P., and Yager, J. A. Focal cutaneous vasculitis and alopecia at sites of rabies vaccination in dogs. *J Am Vet Med Assoc* **188:** 1174-1177, 1986.

7. Linear IgA Dermatosis

Linear IgA dermatosis is a very rare sterile superficial pustular dermatitis of dachshunds. There is no apparent age or sex predilection. Clinically, the dermatosis is characterized by multifocal-to-generalized pustules that evolve into annular areas of erosions, epidermal collarettes, crusts, alopecia, and hyperpigmentation. The trunk is typically involved. Pruritus is absent or minimal.

Histologically there is a subcorneal pustular dermatitis and superficial folliculitis. Neutrophils are the dominant cell type, and acanthocytes are in small numbers. Direct immunofluorescence testing reveals the deposition of IgA, with or without complement, at the basement membrane zone.

Bibliography

Scott, D. W, Manning, T. O., and Lewis, R. M. Linear IgA dermatoses in the dog: Bullous pemphigoid, discoid lupus erythematosus, and a subcorneal pustular dermatitis. *Cornell Vet* **72:** 394–402, 1982.

8. Vogt–Koyanagi–Harada Syndrome

This syndrome, seen only in dogs and humans, is characterized by the concurrent acute onset of bilateral uveitis and depigmentation of the nose, lips, eyelids, and occasionally the footpads and anus. Although the cause of the disorder is unknown, a cell-mediated hypersensitivity to melanin has been hypothesized. It is discussed further under Disorders of Pigmentation.

9. Plasma Cell Pododermatitis

This is a rare disorder of cats of all breeds, ages, and sexes. Although the pathogenesis is unknown, the tissue plasmacytosis, the hypergammaglobulinemia, and the beneficial response to immunomodulating drugs suggest an immune-mediated basis. In addition, occasional cats have other immunologic abnormalities, such as positive antinuclear antibody tests, Coombs-positive anemia, glomerulonephritis, and positive direct immunofluorescence testing (immunoglobulin deposited at the basement membrane zone).

Clinically, plasma cell pododermatitis begins as a soft, nonpainful swelling of multiple footpads on multiple paws. The central metacarpal or metatarsal pads are usually most severely affected. Affected pads feel soft and mushy, and the surface is crosshatched with white scaly striae. In some cases, one or more pads may become ulcerated or develop fleshy granulomatous proliferations, which may hemorrhage.

Histologically, early lesions are characterized by a superficial and deep perivascular dermatitis with plasma cells predominating. Later, there is a diffuse plasmacytic dermatitis. Russell bodies are numerous. Leukocytoclastic vasculitis is occasionally seen. Ulcerated or proliferative lesions show varying degrees of superimposed suppurative-to-pyogranulomatous inflammation.

Bibliography

Gruffydd-Jones, T. J., Orr, C. M., and Lucke, V. M. Footpad swelling and ulceration in cats: A report of five cases. *J Small Anim Pract* **21**: 381–389, 1980.

Taylor, J. E., and Schmeitzel, L. P. Plasma cell pododermatitis with chronic footpad hemorrhage in two cats. *J Am Vet Med Assoc* **197**: 375–377, 1990.

10. Lichenoid Dermatoses

Lichenoid dermatoses are rare skin disorders of dogs and cats. There is no apparent age, breed, or sex predilection. The dermatoses are characterized by the usually asymptomatic, symmetric, grouped, flat-topped papules that develop a scaly to markedly hyperkeratotic surface. Lesions may coalesce to form hyperkeratotic, alopecic plaques. The distribution of lesions is quite variable.

Histologically these dermatoses are characterized by lichenoid interface dermatitis. Hydropic degeneration is often minimal. Epidermal hyperplasia and orthokeratotic-to-parakeratotic hyperkeratosis are usually prominent. If focal areas of suppurative epidermitis and/or suppurative folliculitis are present, a lichenoid tissue reaction in response to staphylococcal infection should be suspected.

Bibliography

Buerger, R. G., and Scott, D. W. Lichenoid dermatitis in a cat: A case report. *J Am Anim Hosp Assoc* **24**: 55–59, 1988.

Scott, D. W. Lichenoid reactions in the skin of dogs: Clinicopathologic correlations. *J Am Anim Hosp Assoc* **20**: 305–317, 1984.

11. Cutaneous Amyloidosis

Cutaneous amyloidosis occurs, rarely, in horses and dogs. The cutaneous lesions are not usually associated with known triggering factors in horses, but in dogs, cutaneous amyloidosis has been associated with monoclonal gammopathy, dermatomyositis, and is seen, not infrequently, in the stroma of plasmacytomas of skin and oral cavity (see Tumors of the Skin).

In **horses,** the lesions are multiple, asymptomatic papules, nodules, and plaques, which are most commonly seen on the head, neck, shoulders, and pectoral region. The lesions are firm, well circumscribed and 0.5–10 cm in diameter. The overlying skin and haircoat are normal. The initial lesions may be urticarial in type. The cutaneous lesions may be accompanied by similar nodules in the respiratory mucosa and associated lymph nodes but are seldom associated with systemic amyloidosis. Histologic findings include a nodular-to-diffuse granulomatous dermatitis and panniculitis. Large extracellular deposits of amyloid appear as variable-sized areas of homogeneous, amorphous, hyaline, eosinophilic material, which may contain clefts or fractures. Multinucleated histiocytic giant cells are usually numerous. The physical structure of amyloid gives it special properties, such as apple green birefringence when Congo red-stained sections are polarized. The amyloid fibrils (AL amyloid) are derived from monoclonal immunoglobulin light chains.

In **dogs** with monoclonal gammopathy, purpuric lesions are seen, and cutaneous hemorrhage is easily induced by minor external trauma. The superficial dermis contains an amorphous, homogeneous, eosinophilic substance, and the walls of blood vessels in the involved area are thickened by deposition of the same substance.

Bibliography

Schwartzman, R. M. Cutaneous amyloidosis associated with a monoclonal gammopathy in a dog. *J Am Vet Med Assoc* **185:** 102–104, 1984.

Stunzi, H. *et al.* Systemische Haut-und Unterhautamyloidose beim Pferd. *Vet Pathol* **12:** 405–414, 1975.

12. Alopecia Areata

Alopecia areata is a rare dermatosis of dogs, cats, horses, and cattle. It is characterized by focal or multifocal, asymptomatic, well-circumscribed, annular areas of noninflammatory alopecia. The affected skin appears clinically normal. Lesions are most commonly seen on the face, neck, and trunk. There is no apparent age, breed, or sex predilection.

Early, clinically active lesions are characterized by an accumulation of lymphocytes ("swarm of bees") around the inferior segment of anagen hair follicles—a peribulbar lymphocytic perifolliculitis. This change may be difficult to demonstrate, requiring multiple biopsies from the periphery of early, expanding lesions. Histologic findings in chronic, clinically static lesions are nondiagnostic, revealing a predominance of catagen and telogen hair follicles and follicular atrophy, but may be diagnosed as an endocrine skin disorder.

Bibliography

Scott, D. W., and Gourreau, J. M. Alopecia areata (pelade) chez les bovins. *Point Vét* **22**: 671–674, 1990.

13. Localized Scleroderma

Localized scleroderma (morphea)is a rare disease of dogs. It is characterized by asymptomatic, well-demarcated, sclerotic plaques that are alopecic, smooth, and shiny. Lesions tend to be linear and occur on the trunk and limbs. Affected dogs are otherwise healthy. A diffuse fibrosing dermatitis is seen histologically. The entire dermis and subcutis are replaced by dense collagen bundles. Pilosebaceous units are essentially absent. A mild superficial and deep perivascular accumulation of lymphohistiocytic cells is present.

Bibliography

Scott, D. W. Localized scleroderma (morphea) in two dogs. *J Am Anim Hosp Assoc* **22:** 207–211, 1986.

14. Colostrum-Transferred Exfoliative Dermatitis in Calves

Two calves from one cow, but sired by different bulls, developed erythematous and vesicular lesions on the muzzle from days 1 to 4, followed by generalized exfoliative lesions from days 4 to 25, and alopecia from days 25 to 50. Regrowth of hair began at about day 45. When colostrum from this cow was fed to an unrelated newborn calf, the same sequence of events occurred. Histologically the lesions were spongiotic, neutrophilic, subcorneal pustular dermatitis. Circumstantial evidence implicates colostral transfer in the pathogenesis of the lesions. While the recovery correlates with the half-life of maternally derived IgG, an immune pathogenesis remains speculative.

Bibliography

Bassett, H. Bovine exfoliative dermatitis: A new bovine skin disease transferred by colostrum. *Irish Vet J* **39:** 106–110, 1985.

XI. Viral Diseases of Skin

Cutaneous lesions occur in the course of a number of viral diseases in domestic animals. Viruses may induce skin lesions after local infection, but the intact integument is resistant to viral penetration, and injection via an arthro-

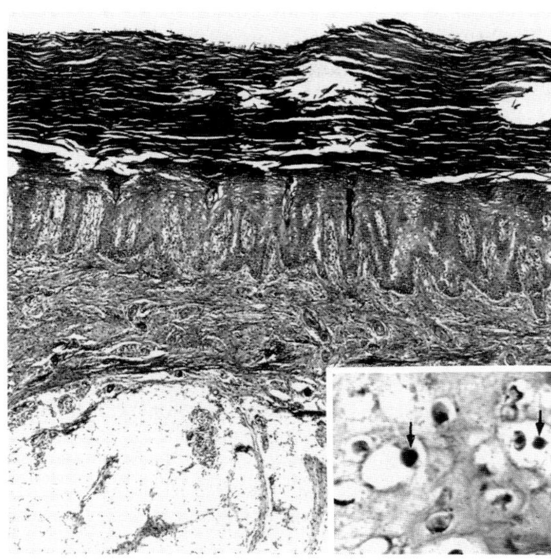

Fig. 5.63 Digital hyperkeratosis. "Hard-pad disease." Canine distemper. Orthokeratotic hyperkeratosis. (Inset) Intracellular edema of spinosum cells with intracytoplasmic inclusion bodies (arrows).

pod bite or introduction through a cutaneous wound is a prerequisite for infection. Examples of local viral infection include papillomas induced by the Papovaviridae, bovine mammillitis induced by a herpesvirus, and the so-called milker's nodule in humans caused by a parapoxvirus. Much more often, viruses localize in the skin during the viremic phase of a systemic infection. Examples include some poxvirus infections, malignant catarrhal fever, and the vesicular stomatitides such as foot and mouth disease. Pantropic viruses, such as those of canine distemper and hog cholera, may cause cutaneous lesions, but most viruses causing cutaneous lesions are epitheliotropic. Some epitheliotropic viruses, in particular the Poxviridae, have a predilection for the epithelium of the skin. Others, including the viruses associated with the mucosal diseases, cause primary lesions in the alimentary tract with lesser involvement of the skin.

A "rash," comprising erythematous macules due to long-lasting dilation of dermal blood vessels, is often associated with systemic viral disease in humans. Such lesions are rarely seen in animals, but may be hidden by the hair coat. An exception is the cutaneous erythema occurring in hog cholera and African swine fever.

Cutaneous viral diseases are more common in food-producing animals than in pets. Some of these diseases, notably sheeppox, cause significant mortality. Others have an economic impact because of their deleterious effect on production. Herpes mammillitis and pseudocowpox, for example, reduce milk yield in dairy cattle, and contagious pustular dermatitis affects the growth rate of lambs by interfering with their ability to suck. A few of the large animal viral dermatoses are extremely mild, for example molluscum contagiosum in the horse. Viral diseases with cutaneous manifestations are rare in dogs and cats. Canine distemper is associated with nasodigital hyperkeratosis, so-called hard pad disease (Fig. 5.63) and pustular dermatitis. In cats, rare manifestations of cutaneous disease occur with feline herpesvirus infection and feline calicivirus infection.

Bibliography

Cheville, N. F. Cytopathology in viral diseases. *Monogr Virol* **10:** 1–236, 1975.

Fenner, F. *et al.* "Veterinary Virology." New York, Academic Press, 1987.

Gibbs, E. P. J., Smale, C. J., and Voyle, C. A. Electron microscopy as an aid to the rapid diagnosis of virus diseases of veterinary importance. *Vet Rec* **106:** 451–458, 1980.

Odend'hal, S. "The Geographic Distribution of Animal Viral Diseases." New York, Academic Press, 1983.

Timoney, J. F. *et al.* "Hagan and Bruner's Microbiology and Infectious Diseases of Domestic Animals." 8th Ed. Ithaca, New York, Comstock Publishing, 1988.

A. Poxvirus Infections

The Poxviridae are a large family of complex DNA viruses. Highly epitheliotropic, they cause cutaneous and systemic disease in birds, wild and domestic mammals, and humans. Of the domestic species, dogs appear least susceptible to pox infections. Some members of the Poxviridae, including sheeppox, ectromelia, monkeypox, and the now eradicated human smallpox, cause severe systemic disease. Others cause mild, localized disease, for example, pseudocowpox, which chiefly affects the teats of milking cows. A few poxviruses are associated with hyperplastic or neoplastic conditions, such as molluscum contagiosum in horses and the Shope fibroma virus of rabbits.

The Poxviridae share group-specific nucleoprotein antigens. Animal poxviruses are in the subfamily Chordopoxvirinae. The genera include *Orthopoxvirus* (cowpox, vaccinia, rabbit pox, monkeypox, buffalopox, camelpox), *Avipoxvirus* (fowlpox, pigeonpox, and many others), *Capripoxvirus* (sheeppox, goatpox, and lumpy skin disease virus), *Leporipoxvirus* (myxoma, rabbit, hare, and squirrel fibromas), *Suipoxvirus* (swinepox), and *Parapoxvirus* (contagious pustular dermatitis, papular stomatitis, pseudocowpox). The genera *Yabapoxvirus* and *Molluscipoxvirus* have been proposed but not yet approved.

Many of the poxviruses are host specific, but the orthopoxviruses, such as cowpox and vaccinia, affect a wide range of species. Some poxviruses, for example pseudocowpox, are zoonotic. Infection is usually achieved by cutaneous or respiratory routes. Poxviruses, whether acquired by the subcutaneous or respiratory routes, usually gain access to the systemic circulation via the lymphatic system, although multiplication at the site of injection in the skin may lead to direct entry into the blood and a primary viremia. A secondary viremia disseminates the virus back to the skin and to other target organs.

Poxviruses induce lesions by a variety of mechanisms.

Degenerative changes in the epithelium are caused by virus replication and induce vesicular lesions typical of many poxvirus infections. Degenerative lesions in the dermal or submucosal tissues sometimes result from ischemia secondary to vascular damage, due to virus multiplication in endothelial cells. Poxvirus infections also induce proliferative lesions. Poxviruses, such as contagious pustular dermatitis, replicating in the epidermis typically induce hyperplasia along with degenerative changes. The host cell DNA synthesis is stimulated before the onset of cytoplasmic virus-related DNA replication. Proliferative changes may be explained by a gene, present in several poxviruses including molluscum contagiosum, whose product has significant homology with epidermal growth factor. Poxviruses also encode for functions which may counteract host defences. These include genes related to those encoding the SERPINs (a superfamily of related proteins important in regulating serine protease enzymes which mediate kinin, complement, fibrinolytic, and coagulation pathways) and genes encoding anti-interferon activities.

Pox lesions have a typical developmental sequence. They commence as erythematous macules, become papular, and then vesicular. The vesicular stage is well developed in some pox infections, such as sheeppox, and is transient or nonexistent in others, such as contagious pustular dermatitis. Vesicles develop into umbilicated pustules with a depressed center and a raised, often erythematous border. This lesion is the so-called pock. The pustules rupture and a crust forms on the surface. This crust may become very thick, as in lesions of contagious pustular dermatitis. Lesions heal and often leave a residual scar. The mucosal lesions are briefly vesicular and develop into ulcers rather than pustules.

Histologically, pox lesions begin as epidermal cytoplasmic swelling and vacuolation, usually first affecting the cells of the outer stratum spinosum. There is evidence, from experimental studies with the virus of contagious pustular dermatitis, that postinjury proliferating keratinocytes are the target for viral replication. Rupture of the damaged keratinocytes produces multiloculated vesicles, so-called reticular degeneration. The early dermal lesions include edema, vascular dilation, a perivascular mononuclear cell infiltrate, and a variable neutrophilic infiltrate. Neutrophils migrate into the epidermis and aggregate in vesicles to form microabscesses. Large intraepidermal pustules may form and sometimes extend into the superficial dermis. There is usually marked epithelial hyperplasia and sometimes pseudocarcinomatous hyperplasia of the adjacent epithelium. This contributes to the raised border of the umbilicated pustule. Rupture of the pustule produces inflammatory crust, often colonized on its surface by bacteria.

Poxvirus lesions often contain characteristic intracytoplasmic inclusion bodies. These are single or multiple and of varying size and duration. The more prominent inclusions are designated type A. They are eosinophilic, reflecting their high protein content, and weakly Feulgen positive. Smaller, basophilic Feulgen positive type B bodies also occur and represent the site of virus replication.

Diagnosis of poxvirus infections is usually based on typical clinical appearance and may be supported by characteristic histologic lesions. Demonstration of the virus by electron microscopy confirms a poxvirus etiology but may not differentiate morphologically similar viruses such as the closely related orthopoxviruses. Definitive identification of specific viruses requires their isolation and then identification by serologic procedures.

Bibliography

Baxby, D. Poxvirus hosts and reservoirs. *Arch Virol* **55:** 169–179, 1977.

Buller, R. M., and Palumbo, G. J. Poxvirus pathogenesis. *Microbiol Rev* **55:** 80–122, 1991.

Fenner, F., Wittek, R., and Dumbell, K. R. "The Orthopoxviruses." New York, Academic Press, 1989.

Mims, C. A. Pathogenesis of rashes in virus diseases. *Bacteriol Rev* **30:** 739–760, 1966.

1. Parapoxvirus *Diseases*

a. CONTAGIOUS PUSTULAR DERMATITIS Contagious pustular dermatitis is a poxvirus disease of sheep and goats, with incidental infections occurring in humans, cows, wild ruminants and, very rarely, dogs. The disease is caused by a parapoxvirus closely related to pseudocowpox and bovine papular stomatitis. Synonyms for contagious pustular dermatitis include contagious ecthyma, infectious labial dermatitis, soremouth, scabby mouth, and orf. Orf is used to describe human infections.

The disease occurs wherever sheep or goats are raised. While live-virus vaccines control the disease, they also ensure its continuance by perpetuating infection in the environment. The economic significance of contagious pustular dermatitis results chiefly from loss of condition, since affected animals neither suck nor graze. Morbidity in a susceptible population may reach 90%, but mortality

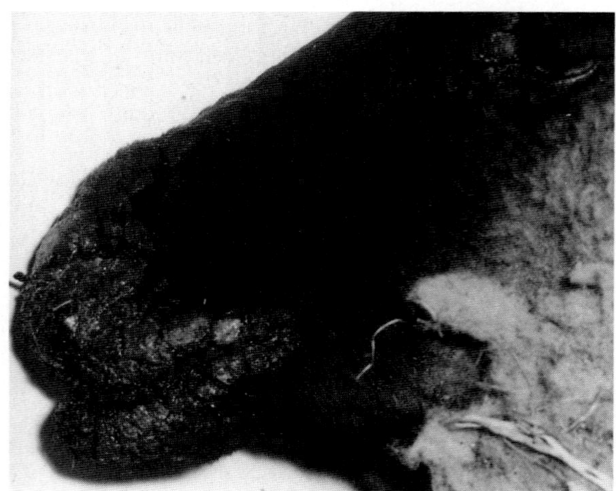

Fig. 5.64 Contagious pustular dermatitis. Scabby lesions at margins of lips.

rarely exceeds 1% unless secondary infection intervenes. In an outbreak in Brazil, 93% mortality in kids was reported. Mortality often results from the invasion of primary lesions by the larvae of the screwworm fly (*Cochliomyia hominivorax*) or by bacteria such as *Fusobacterium necrophorum* and occasionally *Dermatophilus congolensis*. Cellulitis may complicate pedal lesions, mastitis may complicate mammary lesions, and necrotizing stomatitis and aspiration pneumonia may complicate oral lesions.

Contagious pustular dermatitis affects sheep and goats of all breeds. It is predominantly a disease of lambs and kids. Infection is established through cutaneous abrasions, particularly those associated with dry and prickly pasture or forage. Clinically affected lambs may transmit the virus to the udder of the ewe. The virus is hardy and probably persists in the environment indefinitely in crust material shed from affected animals. Chronically infected, reinfected or, possibly, latently infected carrier animals may allow the virus to persist in a flock for several years.

The gross lesions usually commence at the commissures of the lips and spread around the lip margins to the muzzle (Fig. 5.64). Primary lesions sometimes occur on the face about the eyes. In severe cases, lesions may develop on the gingiva, dental pad, palate, and tongue (Fig. 5.65A,B). Lesions mainly confined to the tongue must be differentiated from those of foot-and-mouth disease. The buccal lesions are raised, red or gray foci with a surrounding zone of hyperemia. Very rarely, lesions extend to the esophagus, rumen, and omasum and are recorded in lungs and heart, and in the lower alimentary canal, causing ulcerative gastroenteritis. Lesions on the limbs are less common than on the lips and tend to involve the coronet, interdigital cleft, and bulb of the heels. They may extend,

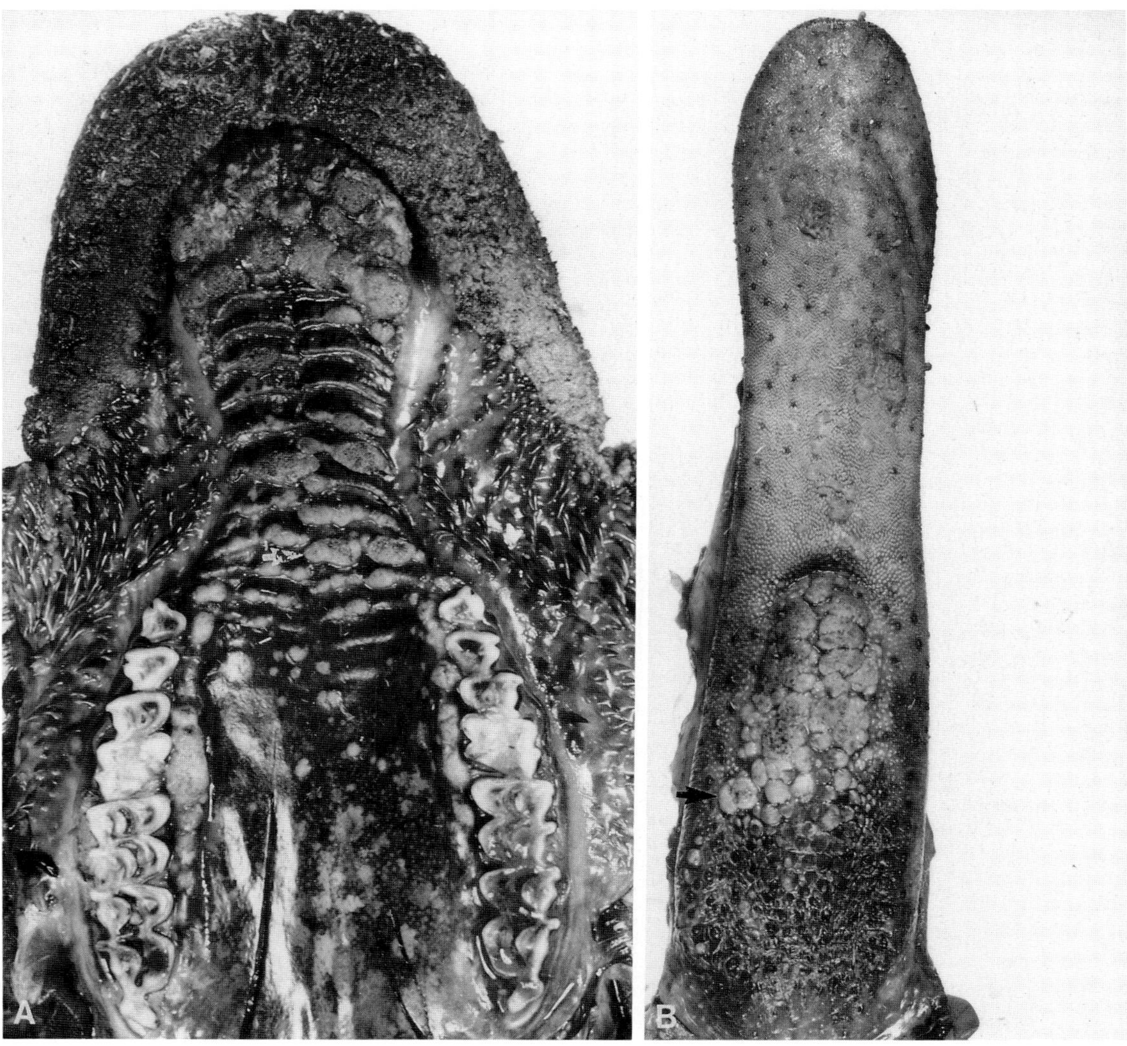

Fig. 5.65 Contagious pustular dermatitis. (A) Lesions on palate (arrow), dental pad, and lips. (B) Proliferative lesions (arrow). Tongue.

in severe cases, to the knee or hock on the posterior aspect of the leg. Lesions of the mammary gland affect the teats and adjacent skin of the udder. Lesions may develop in other areas of sparsely wooled skin, such as the inner thigh, axilla, the edge of wounds in recently earmarked lambs, tail-dock sites or at the margins of the Mules' operation incision. Proliferative lesions affecting predominantly the head, neck, and body have been described in Nubian goats.

The lesions develop through the typical pox phases but are much more proliferative. The vesicular stage is transient, and pustules are flat rather than umbilicated. The most significant feature of the gross lesion is the layer of thick brown-gray crust, which may be elevated 2–4 mm above the skin surface. Depending on the degree of secondary infection, regression is complete by 4 weeks. Papillomatous growths, resulting from continued epidermal proliferation, sometimes occur.

The microscopic lesions of contagious pustular dermatitis are characterized by vacuolation and swelling of the keratinocytes in the stratum spinosum, reticular degeneration, marked epidermal proliferation, intraepidermal microabscesses, and accumulation of scale-crust. In experimentally abraded sheep skin, the active site of viral replication was found to be the newly proliferative keratinocyte population, growing up under the superficial necrotic layer. By ~30 hr postinfection, keratinocyte swelling and vacuolation commences in the outer stratum spinosum. It is accompanied by cytoplasmic basophilia, which corresponds ultrastructurally to an increased complement of polyribosomes, presumably active in viral protein synthesis. Basophilic intracytoplasmic inclusion bodies are reported as early as 31 hr postinfection. By 72 hr postinfection, the keratinocytes show nuclear pyknosis and marked hydropic change, leading to reticular degeneration. The term ballooning degeneration is often used, but the keratinocytes do not have the homogeneous eosinophilic cytoplasm typical of this condition. At this time intracytoplasmic eosinophilic inclusion bodies appear. The inclusion bodies persist for 3 to 4 days, as long as the hydropic cells are found. The proliferative reaction in the epidermis is underway by 55 hr postinfection with mitotic figures numerous in the stratum basale. By 3 days postinfection, the epithelium is 3–4 times normal thickness, and the rete ridges are markedly elongated (Fig. 5.66). Pseudocarcinomatous hyperplasia is common.

The dermal lesions include superficial edema, marked capillary dilation, and a perivascular mononuclear infiltration. Neutrophils migrate into the areas of reticular degeneration and form microabscesses, which are initially roofed by the stratum corneum but which subsequently rupture. A thick layer of scale-crust is built up, composed of ortho- and parakeratotic hyperkeratosis, proteinaceous fluid, degenerating neutrophils, cellular debris, and bacterial colonies. The subsequent microscopic appearance of the lesions depends on the degree of secondary bacterial infection.

Contagious pustular dermatitis occurs in a variety of

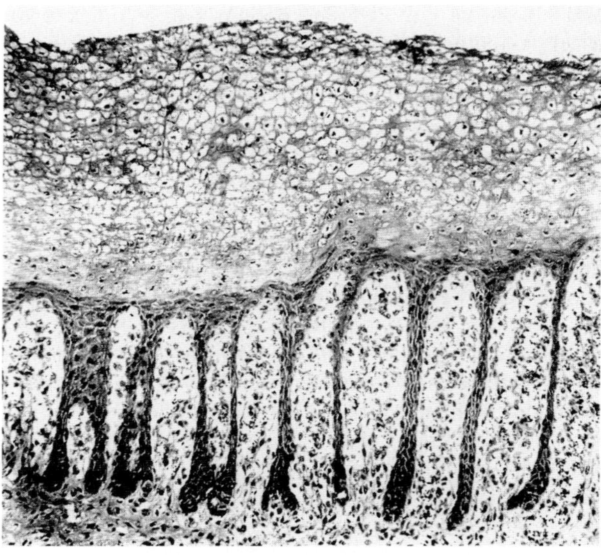

Fig. 5.66 Contagious pustular dermatitis. Histology of lesion showing epithelial proliferation, increased vascularity, and intracellular edema.

wild and captive ungulates. Dogs may acquire infection by eating infected lamb carcasses. Lesions resemble those in sheep.

Bibliography

Chubb, R., and Couch, A. Detection of antigen by immunofluorescence in orf virus lesions in sheep. *Vet Rec* **116:** 546–547, 1985.

Coats, J. W., and Hoff, S. Contagious ecthyma: An unusual distribution of lesions in goats. *Can Vet J* **31:** 209–210, 1990.

Darbyshire, J. H. A fatal ulcerative mucosal condition of sheep associated with the virus of contagious pustular dermatitis. *Br Vet J* **117:** 97–105, 1961.

Kluge, J. P., Cheville, N. F., and Peery, T. M. Ultrastructural studies of contagious ecthyma in sheep. *Am J Vet Res* **33:** 1191–1200, 1972.

Leavell, U. W. *et al.* Orf. Report of 19 human cases with clinical and pathological observations. *J Am Med Assoc* **204:** 657–664, 1966.

McKeever, D. J. *et al.* Studies of the pathogenesis of orf virus infection in sheep. *J Comp Pathol* **99:** 317–328, 1988.

Robinson, A. J., and Balassu, T. C. Contagious pustular dermatitis (orf). *Vet Bull* **51:** 771–782, 1981.

Wilkinson, G. T., Prydie, J., and Scarnell, J. Possible "orf" (contagious pustular dermatitis, contagious ecthyma of sheep) infection in the dog. *Vet Rec* **87:** 766–767, 1970.

b. Ulcerative Dermatosis of Sheep This disease of the epidermis of sheep is caused by an unclassified poxvirus, which is similar to the agent of contagious pustular dermatitis, but the viruses do not cross-protect. The disease occurs in South Africa, where it is known as pisgoed or pisgras, in the United Kingdom as a contagious venereal infection, and in the United States, where it is known as lip and leg ulceration, anovulvitis, infectious balanoposthitis and ulcerative vulvitis. The various names indicate the essential features of the disease, which are ulcerative papules on the lips, face, legs, feet, vulva, prepuce, and occasionally, the glans penis. The genital lesions are transmissible by coitus.

Presumably, infection results from viral contact with damaged skin. The pathologic process is primarily ulcerative, the ulcers varying in size up to 4 to 5 cm in diameter and 3 to 5 mm in depth. Pus covers the granulation tissue at the base of the ulcer and underlies a scab which is thin, brown, and bloody, and unlike the thick parakeratotic crusts which develop in contagious pustular dermatitis. The underlying dermis is diffusely swollen, especially in distensible parts such as the vulva and prepuce. The lesions on the hairy parts of the face tend to be fairly well circumscribed, but those of the feet tend to spread, especially on the interdigital skin. The vulval lesions usually begin on the tip and spread around the margins of the lips. An ulcerative ring tends to form around the preputial orifice, but the preputial mucosa is spared. Lesions on the glans penis remain moist. The urethral process may become necrotic. The labial and pedal lesions have to be distinguished from those of contagious pustular dermatitis and foot and mouth disease, and the preputial lesions have to be distinguished from noncontagious forms of balanoposthitis. Detailed descriptions of the histopathology of ulcerative dermatitis are lacking; the lesions are supposedly distinguishable from those of contagious pustular dermatitis by the lack of epithelial hyperplasia.

Bibliography

Flook, W. H. An outbreak of venereal disease among sheep (ulcerative dermatosis). *J Comp Pathol* **16:** 374–375, 1903.

Roberts, R. S., and Bolton, J. F. A venereal disease of sheep. *Vet Rec* **57:** 686–687, 1945.

Steyn, D. G. Pisgoed or pisgras (ulcerative dermatosis). *16th Rep Direct Vet Services Anim Indust,* Union S Afr, pp. 417–420, 1930.

Trueblood, M. S., and Chow, T. L. Characterization of the agents of ulcerative dermatosis and contagious ecthyma. *Am J Vet Res* **24:** 42–46, 1963.

Trueblood, M. S. Relationship of ovine contagious ecthyma and ulcerative dermatosis. *Cornell Vet* **56:** 521–526, 1966.

Tunnicliff, E. A. Ulcerative dermatosis of sheep. *Am J Vet Res* **10:** 240–249, 1949.

c. PSEUDOCOWPOX. Pseudocowpox is caused by a parapoxvirus. The virus is closely related to bovine papular stomatitis virus and contagious pustular dermatitis virus. The infection is common and often inapparent. The disease affects chiefly milking cows. Morbidity in a herd approaches 100% but only 10–15% of cows are affected at any one time. The economic significance lies in the effect on milk production, either as a result of sore teats or because of secondary bacterial mastitis. Lesions, which affect the teats (Fig. 5.67), udder, and perineum, commence as erythematous macules but do not form the umbilicated pustules seen in cowpox and vaccinia infections. Instead, an ulcer develops and is covered by a characteristic ring or horseshoe-shaped crust. Transmission to humans induces the "milker's nodule."

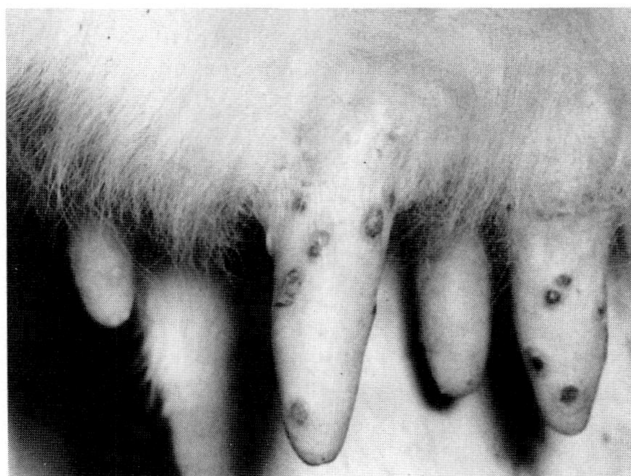

Fig. 5.67 Pseudocowpox, teats. Cow.

d. BOVINE PAPULAR STOMATITIS Bovine papular stomatitis virus occasionally causes lesions on the teats of cows with suckling calves and on the lips and muzzle of the calves. Lesions more typically affect the oral cavity of young cattle (see Volume 2, Chapter 1, The Alimentary System). Transmission to humans induces lesions identical to "milker's nodule" caused by pseudocowpox infection.

Bibliography

Cheville, N. F., and Shey, D. J. Pseudocowpox in dairy cattle. *J Am Vet Med Assoc* **150:** 855–861, 1967.

Friedman-Kien, A. E., Rowe, W. P., and Banfield, W. G. Milker's nodules: Isolation of a poxvirus from a human case. *Science* **140:** 1335–1336, 1963.

Gibbs, E. P. J. Viral diseases of the skin of the bovine teat and udder. *Vet Clin North Am* **6:** 187–202, 1984.

Huck, R. A. A paravaccinia virus isolated from cows' teats. *Vet Rec* **78:** 503–504, 1966.

Moscovici, C. *et al.* Isolation of a viral agent from pseudocowpox disease. *Science* **141:** 915–916, 1963.

Nagington, J., Lauder, I. M., and Smith, J. S. Bovine papular stomatitis, pseudocowpox, and milker's nodules. *Vet Rec* **81:** 306–313, 1967.

2. Orthopoxvirus *Diseases*

a. COWPOX Cowpox and cowpoxlike viruses belong to the genus *Orthopoxvirus*. They are closely related to, but antigenically different from, vaccinia virus. Cowpox affects a range of species including wild and domestic Felidae, cattle, rodents, humans, and several zoo and circus animals, including elephants and rhinoceros. This broad host range raises a problem in nomenclature. To avoid the confusion of terms like carnivorepox and ratpox, Fenner suggests modifying cowpox with the appropriate adjective, for example feline cowpox. The epidemiology of cowpox is thought to involve a virus reservoir in small wild mammals, from which other indicator animals accidentally become infected. However, cat-to-human transmission has been reported.

Since the first report in 1978, **feline cowpox** infection has been documented with increasing frequency within its geographical range, namely Great Britain and Europe. That a wild mammal is the reservoir of infection is supported by serologic evidence. Furthermore, cowpoxlike virus has been isolated from wild rodents in Russia. Infection is thought to result from injuries incurred during hunting. Supportive evidence comes from epidemiological studies, which show a higher prevalence in free-ranging, country-dwelling cats and a seasonal incidence in the autumn, a time when the small-mammal population is at its highest. Furthermore, the primary lesions have a predilection for the face and forepaws.

Domestic cats develop cutaneous and, occasionally, respiratory lesions. Approximately 50% of cats show a single primary cutaneous lesion, usually on the head, neck, forelimbs, or paws. The initial lesion is ulcerative and has been described as bitelike. Lesions heal by granulation but may occasionally develop into abscesses or cellulitis. The secondary cutaneous lesions develop between 4 and 16 days after the onset of the primary lesion. Generalization from primary inoculation sites in the skin has also been demonstrated in experimentally infected cats. Mucocutaneous junctions, oral mucosa, and the tongue may also be involved. The secondary lesions are multiple (usually more than 10 lesions). They commence as firm 2- to 3-mm nodules, enlarging over 2 to 3 days to form 0.5- to 2-cm circular, ulcerated papules or plaques. Lesions are rarely vesicular except on the oral mucosa and inner aspect of the pinna. The ulcers heal with thick gray crusts over 2 to 3 weeks. Cellulitis may develop, but generally, secondary bacterial infection is not a significant complication. Corticosteroid or progesterone therapy may be responsible for converting a local infection into a generalized one or exacerbating the secondary lesions in some animals. Although fatal cowpox infection has been seen in cats with concurrent feline immunodeficiency virus (FIV) infection, there is no convincing evidence to show a direct relationship. Approximately 30% of cowpox-infected cats concurrently infected with FIV have uneventful recoveries.

The disease is rarely fatal. Approximately 20% of cats show some signs of malaise in the acute stage of secondary lesion development, presumably associated with a viremic phase, and a few exhibit clinical signs of upper respiratory tract disease. Lower respiratory tract lesions, which are rare in domestic cats, include pleural effusion and localized areas of cream-colored consolidation in the ventral lung lobes. They are thought to develop from systemic spread rather than from primary respiratory infection.

The microscopic lesions in naturally infected cats are focal, sharply demarcated ulcers covered by a fibrinonecrotic exudate. The ulcers may extend to the deep dermis, subcutis, or even muscle. An intense dermal inflammatory cell infiltrate of neutrophils and mononuclear cells may be associated with the base of the ulcers. Eosinophilic, homogeneous, intracytoplasmic inclusion bodies, 3–7 μm in diameter, occur in keratinocytes in the hyperplastic epithelium at the margin of the ulcers and in the epithelium of the external root sheath and sebaceous gland. Epidermal lesions typical of poxvirus infections develop in cats experimentally infected intravenously or by skin scarification. These lesions include focal hyperplasia, with reticular degeneration and multilocular vesiculation. Epidermal cells bordering the vesicles contain eosinophilic intracytoplasmic inclusion bodies. Viral inclusions also occur in macrophages, fibroblasts, and endothelial cells. The extensive necrosis is probably ischemic in origin, following viral damage to endothelial cells. The pulmonary lesion is a necrotizing alveolitis in which eosinophilic inclusion bodies are present in the degenerating cells.

The histologic lesion is diagnostic. Electron-microscopic examination of scabs reveals typical orthopox virions. Cowpox infection may be confirmed by immunoperoxidase staining of lesions or virus isolation. Serologic tests may be helpful in establishing retrospective diagnoses, as virus-neutralizing antibodies are persistent for several years.

There have been outbreaks of severe orthopox infections in several species of **nondomestic Felidae** in English and Russian zoos. Lions, cheetahs, pumas, jaguars, and ocelots have been affected. Rats have been implicated as the source of infection in the Russian outbreaks. Two forms of the disease are recognized. The cutaneous form is rarely fatal. The lesions are ulcerative and crusted, as described for domestic cats. The respiratory form is uniformly fatal as a severe fibrinous and necrotizing bronchopneumonia and pleuritis. A virus closely resembling cowpox is responsible.

Cowpox, despite the name, is not endemic in **cattle**. Infections in cattle occur only rarely and only in Europe. Outbreaks reported from other geographic locations can be ascribed to vaccinia infection; the lesions induced by the two orthopoxviruses are clinically indistinguishable. Typical poxvirus lesions develop on the teats and udder, and occasionally at other sites such as the muzzle of suckling calves. The infection is transmissible to humans, causing lesions resembling those of primary vaccination.

Bibliography

Baxby, D. Is coxpox misnamed? A review of ten human cases. *Br Med J* **1:** 1379–1381, 1977.

Baxby, D. *et al.* An outbreak of cowpox in captive cheetahs: Virological and epidemiological studies. *J Hyg [Camb]* **89:** 365–372, 1982.

Boerner, F., Jr. An outbreak of cowpox, introduced by vaccination, involving a herd of cattle and a family. *J Am Vet Med Assoc* **64:** 93–97, 1923.

Downie, A. W. A study of the lesions produced experimentally by cowpox virus. *J Pathol Bacteriol* **48:** 361–378, 1939.

Gaskell, R. M. *et al.* Natural and experimental poxvirus infection in the domestic cat. *Vet Rec* **112:** 164–170, 1983.

Marennikova, S. S., and Shelukhina, E. M. White rats as a source of pox infection in Carnivora of the family Felidae. *Acta Virol* **20:** 442, 1976.

Marennikova, S. S. *et al.* Outbreak of pox disease among Carnivora (Felidae) and Edentata. *J Infect Dis* **135:** 358–366, 1977.
Thomsett, L. R., Baxby, D., and Denham, E. M. H. Cowpox in the domestic cat. *Vet Rec* **103:** 567, 1978.

b. VACCINIA Vaccinia, the type species for the *Orthopoxvirus* genus, does not cause natural infection in domestic animals. The virus, whose exact origins remain a mystery, was used widely to protect human beings against variola or smallpox. Incidental infections in animals, cattle, horses, and pigs were transferred from vaccinated humans. The lesions in cattle are indistinguishable from those of cowpox, and those in swine are indistinguishable from those of swinepox. When vaccinia is inoculated onto scarified skin of horses, papular lesions result, resembling a naturally occurring poxvirus infection described in horses in the United States. Furthermore, inoculation of the skin of the flexor surface of the pastern produces lesions resembling the classical grease heel form of horsepox. Vaccinia infections of animals are rare now that smallpox has been eradicated and routine vaccination has been discontinued.

Bibliography

Studdert, M. J. Experimental vaccinia virus infection of horses. *Aust Vet J* **66:** 157–159, 1989.

c. BUFFALOPOX Buffalopox is caused by an orthopoxvirus virus so closely related to vaccinia it may be regarded as a subspecies. Contact with vaccinated humans was considered the typical mode of transmission; however, outbreaks of buffalopox continue despite the discontinuation of smallpox vaccination. This strengthens the hypothesis that buffalopox is a distinct virus, being maintained in the buffalo population. The disease occurs in epidemic form in India, where it is of economic importance. It has been reported in Indonesia, Italy, Pakistan, Russia, and Egypt. Zebu cattle are apparently refractory to infection. The lesions predominantly affect the teats, udder, medial aspects of the thighs, lips, and muzzle, but may be generalized.

Bibliography

Baxby, D., and Hill, B. J. Characteristics of a new poxvirus isolated from Indian buffaloes. *Arch Gesamte Virusforsch* **35:** 70–79, 1971.
Fenner, F., Wittek, R., and Dumbell, K. R. "The Orthopoxviruses." New York, Academic Press, 1989.
Lal, S. M., and Singh, I. P. Buffalopox—a review. *Trop Anim Health Prod* **9:** 107–112, 1977.
Magsood, M. Generalized buffalo-pox. *Vet Rec* **70:** 321–322, 1958.

d. CAMELPOX Camelpox, a distinct species of *Orthopoxvirus,* causes severe disease in dromedary camels. The disease is widespread in northern and eastern Africa and mideastern Asia but has not been reported in Australia. Mortality may be high in calves, but the chief effect is a decrease in milk production and loss of condition. The lesions affect both skin and mucous membranes and follow the usual pattern of pox lesions. Lesions tend to be most concentrated around the mucocutaneous junctions of the face but may be found on the neck and forelegs. Generalized skin lesions occur more typically in calves. Fatalities are usually associated with secondary bacterial infection leading to septicemia, a phenomenon more prevalent in the rainy season.

Bibliography

Baxby, D. Smallpoxlike virus from camels in Iran. *Lancet* **2:** 1063–1065, 1972.
Davies, F. G., Mungai, J. N., and Shaw, T. Characteristics of a Kenyan camelpox virus. *J Hyg* **75:** 381–385, 1975.
Kriz, B. A study of camelpox in Somalia. *J Comp Pathol* **92:** 1–8, 1982.

e. HORSEPOX, UASIN GISHU DISEASE, AND EQUINE MOLLUSCUM CONTAGIOSUM Horsepox, a common disease of the horse in the eighteenth and nineteenth centuries, is rare today. It appears that the horse is an alternative host for the orthopoxviruses that infect humans and cattle. Experimental infection of horses with vaccinia reproduces the grease heel lesions of Jenner's horsepox and the more generalized form known as equine papular dermatitis. Orthopoxviruses have been isolated from equine papular dermatitis and from Uasin Gishu disease in Kenya. Uasin Gishu virus is closely related to vaccinia and cowpox; the virus of equine papular dermatitis has not been fully characterized.

Poxviral lesions in the horse take several clinical forms. Jenner originally described an exudative dermatitis of the flexor aspects of the hind pasterns. The condition is colloquially named **grease heel** because thick, yellow, greaselike exudate mats the hair. Unfortunately, poxvirus infection is only one manifestation of this clinical entity, sparking considerable controversy as to the true nature of equine pox. A second form has a predilection for the **muzzle and buccal cavity.** The lesions, which develop in the typical sequence of a pock, affect the inner surface of the lips and cheeks, the gums, and the ventral surface of the tongue. Following the development of the buccal eruptions, crops of pocks may appear in the anterior nares, on the face, and on other parts of the body. It is a benign infection, sometimes seriously complicated by bacterial contamination.

The third manifestation takes the form of generalized papular eruptions. In the United States and Australia, the disease is known as **equine papular dermatitis.** This is a highly contagious disorder, spreading by direct contact and through infected harness, bedding, and grooming tools. The lesions are firm papules, up to 0.5 cm diameter, which tend to develop on the lateral neck and shoulder and thorax but become generalized. The papules become crusted, and the crusts slough to leave circular alopecic patches. Resolution may take 6 weeks. **Uasin Gishu disease** from Kenya is also characterized by generalized skin lesions, but the disease differs clinically from equine papular dermatitis. Lesions begin similarly, as small papules, but develop into crusted papillomatous proliferations, up to 2 cm diameter. Lesions eventually resolve, but the disease may continue for 2 years.

Yet another form of proliferative poxvirus lesion in the horse has been likened to human **molluscum contagiosum**. This mildly contagious, self-limiting cutaneous infection is caused by a poxvirus, tentatively classified into the genus *Molluscipoxvirus*. This virus has not yet been cultivated *in vitro*, but has many characteristics similar to vaccinia. The proliferative nature of the lesions may be explained by the presence of a DNA sequence encoding a conserved domain of epidermal growth factor. A similar sequence has been described in some orthopoxviruses, including vaccinia. Cutaneous lesions resembling the human condition clinically, microscopically, and ultrastructurally occur in horses, macropods (kangaroos and quokkas), and chimpanzees. The small, self-limiting lesions are easily overlooked and usually are found incidentally at autopsy or surgery. The equine lesions may be localized to the penis, prepuce, axillary and inguinal areas, and muzzle. Concurrent systemic disease, such as granulomatous enteritis, may predispose to more widespread distribution. Commencing as multiple, circular, smooth-surfaced, gray-white, 1- to 2-mm papules, the lesions become umbilicated and develop a central pore from which a tiny caseous plug is extruded.

The microscopic lesions of molluscum contagiosum are highly characteristic. Well-demarcated foci of epidermal hyperplasia and hypertrophy form pear-shaped lobules in the superficial dermis. The individual keratinocytes are markedly swollen and contain large intracytoplasmic inclusions known as molluscum bodies (Fig. 5.68). These occur initially as eosinophilic, floccular aggregates in the cells of the inner stratum spinosum. As the keratinocytes move toward the surface, the inclusions grow in size and density, compressing the nucleus against the cytoplasmic membrane until it is a thin crescent. The inclusion becomes increasingly basophilic so that cells of the stratum corneum contain deep purple molluscum bodies. These exfoliate through a pore which forms in the stratum corneum and enlarges into a central crater. There is usually no dermal reaction. The molluscum bodies are easily identified in cytologic preparations. The inclusions contain myriad poxvirus particles at various developmental stages, but to date no attempts have been made to culture these viruses, presumably because the diagnosis has been postmortem. The histology of Uasin Gishu disease is identical to that of molluscum contagiosum.

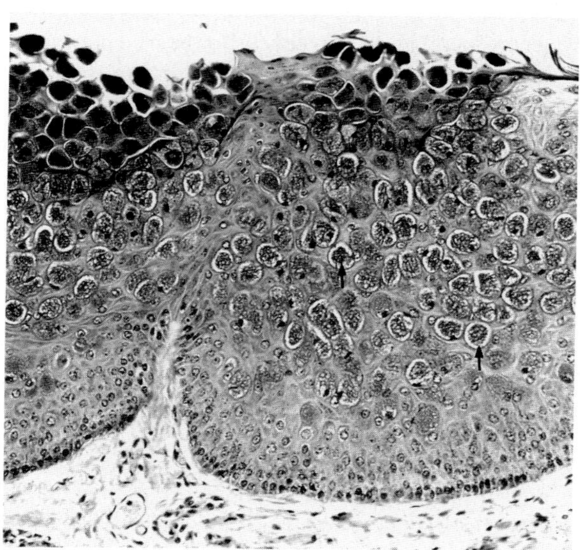

Fig. 5.68 Molluscum contagiosum. Horse. Focal epidermal hyperplasia. Note intracytoplasmic inclusion bodies—"molluscum bodies" (arrows).

Bibliography

Bagnell, B. G., and Wilson, G. R. Molluscum contagiosum in a red kangaroo. *Australas J Dermatol* **15:** 115–120, 1974.

Cooley, A. J. *et al.* Molluscum contagiosum in a horse with granulomatous enteritis. *J Comp Pathol* **97:** 29–34, 1987.

Eby, C. H. A note on the history of horsepox. *J Am Vet Med Assoc* **132:** 420–422, 1958.

Kaminjolo, J. S., and Winqvist, G. Histopathology of skin lesions in Uasin Gishu skin disease of horses. *J Comp Pathol* **85:** 391–395, 1975.

Kaminjolo, J. S. *et al.* Uasin Gishu skin disease of horses in Kenya. *Bull Anim Health Prod Afr* **23:** 225–233, 1975.

McIntyre, R. W. Virus papular dermatitis of the horse. *Am J Vet Res* **10:** 229–232, 1949.

Moens, Y., and Kombe, A. H. Molluscum contagiosum in a horse. *Eq Vet J* **20:** 143–145, 1988.

Porter, C. D., and Archard, L. C. Characterization and physical mapping of *molluscum contagiosum* virus DNA and location of a sequence capable of encoding a conserved domain of epidermal growth factor. *J Gen Virol* **68:** 673–682, 1987.

Rahaley, R. S., and Mueller, R. E. Molluscum contagiosum in a horse. *Vet Pathol* **20:** 247–250, 1983.

3. Capripoxvirus *Diseases*

Sheeppox, goatpox, and lumpy skin disease of cattle are caused by viruses of the genus *Capripoxvirus*. The exact relationship between these viruses has been controversial. It is now believed that they represent strains of a single virus. The evidence includes antigenic and biochemical similarity, a high degree of nucleotide sequence homology, lack of absolute host specificity in most strains, evidence of recombination in the field, and demonstration of cross-infection and cross-protection. Nevertheless, most strains of *Capripoxvirus* show definite host preferences. Recently, restriction endonuclease digest pattern studies have enabled the placement of strains into groups which correlate with the host species from which they were originally isolated.

Bibliography

Gerson, P. D., and Black, D. N. A comparison of the genomes of capripoxvirus isolates from sheep, goats, and cattle. *Virology* **164:** 341–349, 1988.

Kitching, R. P., and Taylor, W. P. Clinical and antigenic relationship between isolates of sheep and goat pox viruses. *Trop Anim Health Prod* **17:** 64–74, 1985.

Kitching, R. P., Bhat, P. P., and Black, D. N. The characterisation of African strains of capripoxvirus. *Epidemiol Infect* **102:** 335–343, 1989.

a. SHEEPPOX. Sheeppox is the most serious of the pox diseases of domestic animals. It exists currently in Africa, Asia, and the Middle East where, despite attempts at vaccination, it is responsible for cycles of epidemic disease followed by periods of endemic maintenance with low morbidity. The disease is exotic to the Americas, Australia, and New Zealand. Eradication measures eliminated the disease from Britain in the midnineteenth century but have only recently been successful in Eastern European countries. Sheeppox causes extensive economic loss through high mortality, reduced meat, milk, or wool yields, commercial inhibitions from quarantine requirements, and the cost of disease-prevention programs.

Transmission of infection is by direct contact with diseased sheep or indirect contact via a contaminated environment. Insect transmission has been demonstrated experimentally. The virus is resistant to desiccation and remains viable for up to 2 months on wool or 6 months in dried crust. There are breed differences in disease susceptibility. Fine-wooled Merino sheep are particularly sensitive, while breeds native to endemic areas, such as Algerian sheep, are comparatively resistant. Sheeppox occurs in all ages of sheep, but the disease is most severe in lambs, with the mortality reaching 80–100%. A high level of background immunity, such as occurs in endemic areas of Kenya, is associated with a low mortality, even in the young.

Sheeppox is a systemic disease. Infection is usually by the respiratory route but may occur through skin abrasions. The incubation period is 4–7 days and is followed by a leukocyte-associated viremia. The virus localizes in many organs, including the skin, where the virus concentration is highest 10–14 days postinfection. The initial clinical signs are fever, lacrimation, salivation, serous nasal discharge, and hyperesthesia. The skin lesions develop 1–2 days later. They have a predilection for the sparsely wooled areas and typically involve eyelids, cheeks, nostrils, vulva, udder, scrotum, prepuce, ventral surface of the tail, and the medial thigh.

The macroscopic lesions follow the typical pattern for pox infections. Sheeppox lesions have a prominent vesicular stage (Fig. 5.69A,B). The vesicles are umbilicated and, being multilocular, yield only a small amount of fluid if punctured. Occasionally a large vesicle forms as a result of cleavage of necrotic epidermis from underlying dermis. The pustule stage is characterized by the formation of a thin crust. In severely affected animals, the lesions coalesce. There may be marked, gelatinous dermal edema. Highly susceptible animals often develop hemorrhagic papules early in the course of the disease and, later, ulcerative lesions in the gastrointestinal and respiratory tracts. Approximately one third of animals develop multiple pulmonary lesions, which are foci of pulmonary consolidation (Fig. 5.70A). The kidneys have multifocal, circular, fleshy nodules throughout the cortices (Fig. 5.70B).

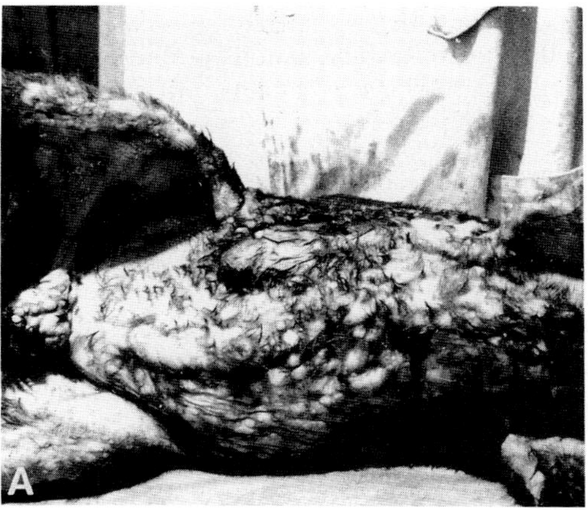

Fig. 5.69A Sheeppox. Disseminated irregular swellings involving ventral abdomen and scrotum.

Fig. 5.69B Sheeppox. Close-up of papules. (A and B courtesy of X. Ivanov.)

Healing of the skin lesions is slow, taking up to 6 weeks. A scar may remain. In the milder form of the disease, seen in endemic areas, the full range of pox lesions does not develop. Instead, epidermal proliferation produces papules covered by scale-crust, which heal with desquamation in a few days. Such lesions often occur on the ventral surface of the tail.

Sheeppox lesions have the typical epithelial changes for the group, including marked hydropic degeneration of stratum spinosum keratinocytes, microvesiculation,

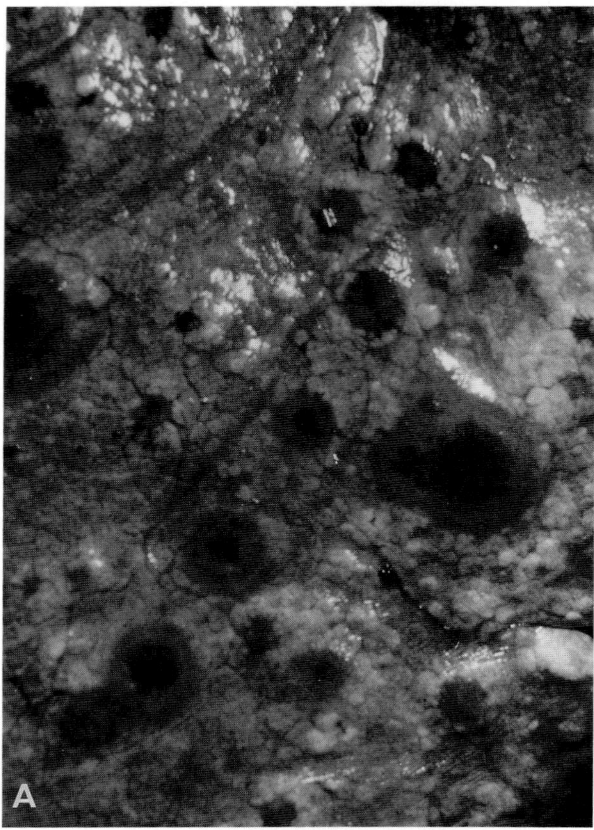

Fig. 5.70A Sheeppox. Multiple foci of pulmonary consolidation. (Courtesy of C. C. Brown).

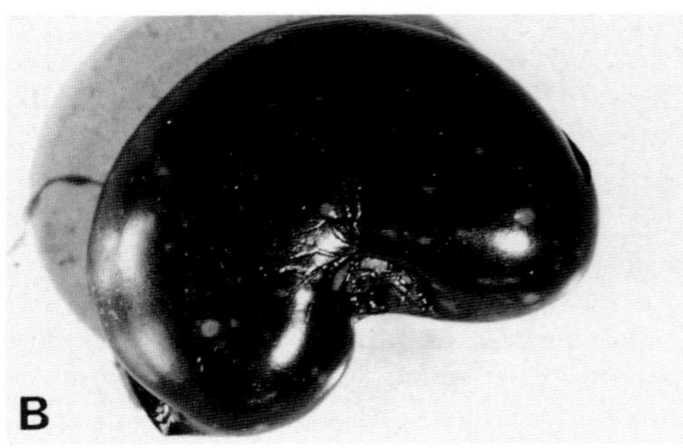

Fig. 5.70B Multifocal fleshy nodules in kidney. (Courtesy of X. Ivanov.)

eosinophilic intracytoplasmic inclusion bodies, and epidermal hyperplasia (Fig. 5.71A,B). The lesions affect both surface epithelium and that of the hair follicles. There are, in addition, marked dermal lesions reflecting the systemic route of cutaneous involvement and possibly implicating immune-mediated lesions in addition to those caused by direct viral damage. The initial dermal lesions, corresponding to the macroscopic erythematous macule, are marked edema, hyperemia, and neutrophilic exocytosis.

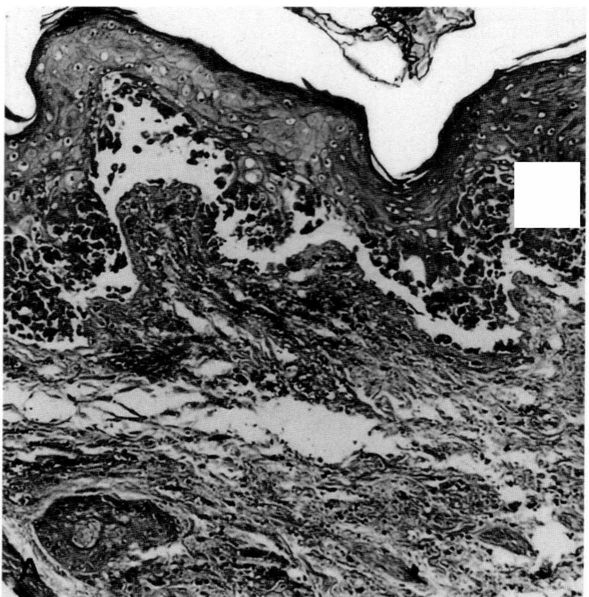

Fig. 5.71A Sheeppox. Necrosis of deep layers of epidermis with early separation.

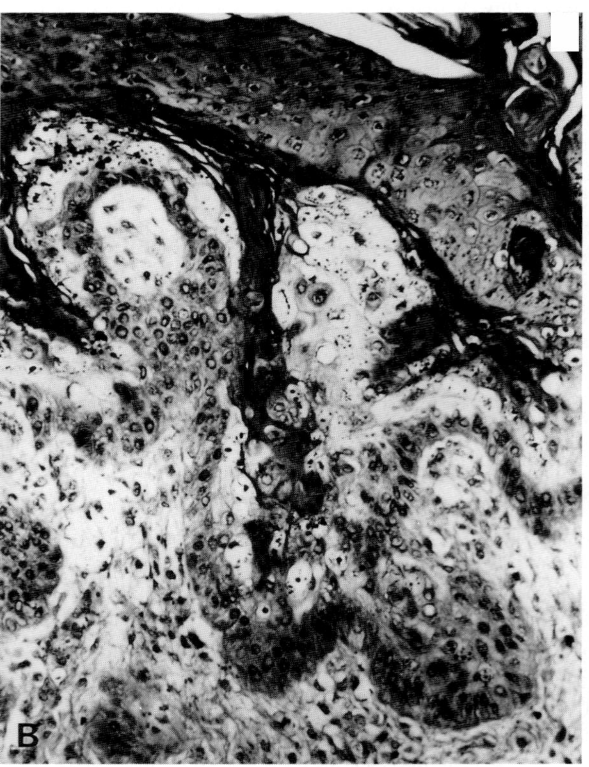

Fig. 5.71B Sheeppox. Proliferation of basal epithelium and intraepidermal vesicle formation.

During the papular stage, large numbers of mononuclear cells accumulate in the increasingly edematous dermis. These cells, first described by Borrel, are called cellules claveleuses or sheeppox cells and are characteristic of the disease. The nuclei of sheeppox cells are vacuolated and have marginated chromatin. The vacuolated cytoplasm contains single, occasionally multiple, eosinophilic intracytoplasmic inclusion bodies. Sheeppox cells are virus-infected monocytes, macrophages, and fibroblasts, but not endothelial cells. Approximately 10 days postinfection and corresponding with the most prominent epithelial lesions and peak of skin infectivity, a severe necrotizing vasculitis develops in arterioles and postcapillary venules. Virus particles have not been identified in endothelial cells, and the vasculitis may be due to immune-complex deposition. Ischemic necrosis of the dermis and overlying epidermis follows.

The pulmonary lesions are proliferative alveolitis and bronchiolitis with focal areas of caseous necrosis. Alveolar septal cells contain intracytoplasmic inclusion bodies. Histologic lesions, characterized by the accumulation of sheeppox cells, may involve heart, kidney, liver, adrenals, thyroid, and pancreas (Fig. 5.72).

The course and outcome of sheeppox depends not only on the usual host–virus relationship but also on the nature and location of secondary infections. The virus itself may cause death during the febrile, eruptive phase of the disease. Of great importance, however, are the secondary bacterial infections, which rapidly develop in the necrotic tissue of the pocks. Death is often due to bacterial septicemia or pneumonia.

b. GOATPOX Goatpox occurs in North Africa and the Middle East. A benign form of goatpox occurs in California and Sweden. The clinical signs of goatpox vary in different geographic areas. The disease has many parallels with sheeppox, but is generally milder with a low mortality rate (5%), although generalized eruption with mortality rates approaching 100% cent may occur. The cutaneous lesions have a predilection for the same areas as for sheeppox. In nursing kids, lesions may appear on the buccal mucosa or anterior nares. In animals with higher levels of resistance, the lesions may be confined to the udder, teats, inner aspects of thighs, or ventral surface of the tail.

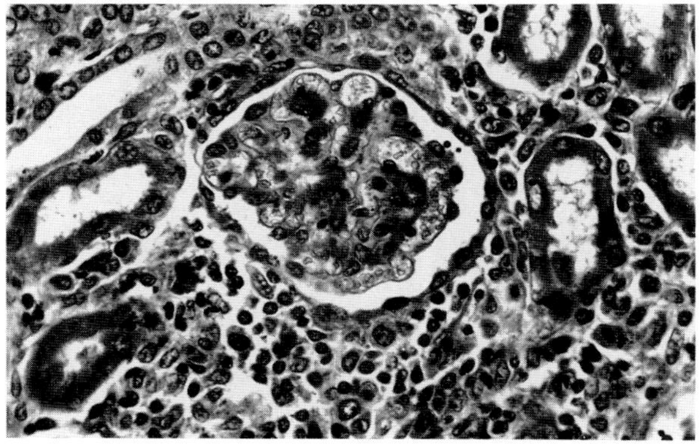

Fig. 5.72 Sheeppox. Focal cellular infiltration in kidney. (Courtesy of X. Ivanov.)

Bibliography

Bennett, S. C. J., Horgan, E. S., and Mensur, A. H. The pox diseases of sheep and goats. *J Comp Pathol* **54:** 131–159, 1944.

Davies, F. G. Characteristics of a virus causing a pox disease in sheep and goats in Kenya with observations on the epidemiology and control. *J Hyg [Camb]* **76:** 163–171, 1976.

Davies, F. G. Sheep and goat pox. *In* "Virus Diseases of Food Animals. A World Geography of Epidemiology and Control" E.P.J. Gibbs (ed.), pp. 733–750. London, Academic Press, 1981.

Kitching, R. P., and Mellor, P. S. Insect transmission of capripoxvirus. *Res Vet Sci* **40:** 255–258, 1986.

Murray, M., Martin, W. B., and Koylu, A. Experimental sheeppox. A histologic and ultrastructural study. *Res Vet Sci* **15:** 201–208, 1973.

Plowright, W., MacLeod, W. G., and Ferris, R. D. The pathogenesis of sheeppox in the skin of sheep. *J Comp Pathol* **69:** 400–413, 1959.

Ramachandran, S. Observations on the histopathology of lung lesions in experimental pox infection of sheep and goats. *Ceylon Vet J* **15:** 78–82, 1967.

Singh, I. P., Pandey, R., and Srivastava, R. N. Sheeppox: A review. *Vet Bull* **49:** 145–154, 1979.

c. LUMPY-SKIN DISEASE Lumpy-skin diease, the *Capripoxvirus* infection of cattle and buffalo, is characterized by the eruption of multiple, well-circumscribed skin nodules, accompanied by fever, ventral edema, and generalized lymphadenopathy. Lumpy-skin disease is confined to the African continent and Madagascar.

Cattle of all ages, sex and breeds are affected, although the disease is more severe in Channel Island breeds. *Bos indicus* and *Bos taurus* cattle are equally susceptible. The disease occurs in epidemics—a notable one in 1944 affected 8 million cattle. Infection is transmitted mechanically by a variety of biting insects. Epidemics tend to follow periods of prolonged rainfall, which favor population increases in vector species. A forest maintenance cycle, probably involving Cape buffalo, is thought to be the reservoir of infection in the interepidemic periods. No reservoir host apart from cattle has been identified. A synergistic relationship between the virus and *Dermatophilus congolensis* infection has been postulated.

The morbidity is extremely variable, and inapparent infection is not uncommon. The mortality is usually low, around 1%, but may be greater than 50%. Economic losses accrue from debilitation, loss of milk and meat production, damage to hides, and reproductive wastage due to fever-associated abortions and temporary sterility in bulls.

The natural incubation period of lumpy-skin disease is 2–4 weeks, but this may be halved in experimental infections. In severely affected animals, the development of large numbers of cutaneous lesions over most of the body is preceded by fever, marked weight loss, profuse salivation, oculonasal discharge, ventral edema, and generalized lymphadenopathy. In the mild disease, there may be few isolated nodules and no prodromal fever. The cutaneous lesions are firm, circumscribed, flat-topped nodules 0.5–5.0 cm in diameter (Fig. 5.73). They may coalesce. The nodules have a creamy-gray color on cut section and involve the full width of the cutis, extending into the subcutis and occasionally adjacent muscles. Nodules affecting the scrotum, perineum, udder, vulva, glans penis, eyelids, and conjunctiva are usually flatter and in nonpigmented tissue are surrounded by a zone of intense hyperemia. The fate of the nodules varies. Typically, they undergo central necrosis and sequestration, but some may resolve rapidly and completely, and others may fail to separate but, instead, become indurated and persist as hard intradermal lumps for many months. Sequestration is preceded by central necrosis in the nodule and occurs rapidly. Separation of the epidermis around the margin of the nodule exposes a rim of dermal granulation tissue. As the process of separation extends into the dermis, the nodule comes to contain a core or sequestrum of necrotic material ("sit-fast"), which is cone shaped and flat topped. When the sequestrum is removed, a deep ulcer remains, which is slowly filled with granulation tissue. Secondary bacterial infections develop in the necrotic cores of the nodules and contribute very significantly to the seriousness of the disease. Large craterous ulcers develop, and lead to lymphangitis and lymphadenitis. Local extension of lesions causes blindness, tenosynovitis, arthritis, or mastitis.

The mucous membranes of the upper respiratory and upper alimentary tracts often develop multiple, discrete ulcerative lesions, irrespective of the number of cutaneous nodules. Those in the respiratory tract may cause swelling sufficient to result in severe dyspnea and asphyxia. Aspiration may lead to pneumonia or, if the animal recovers, scarring may cause stenosis of the anterior portion of the trachea. Nodules occasionally occur in parenchymal organs including kidneys, lungs, and testes.

Although the virus is introduced percutaneously, the infection is systemic. A leukocyte-associated viremia disseminates the virus to the various tissues, including the skin, where greatest virus concentration occurs 9–12 days postinfection. The virus infects a wide range of cells including keratinocytes, mucous and serous glandular epithelium, fibrocytes, skeletal muscle, macrophages, pericytes, and endothelial cells. Damage to the endothelial cells causes vasculitis, which is central to the pathogenesis of the lumpy-skin disease lesions.

The acute lesions consist of vasculitis, lymphangitis, thrombosis, marked dermal edema sometimes inducing dermoepidermal separation, and infarction. The epidermis shows the typical hydropic changes associated with poxvirus infection (Fig. 5.74). Intracytoplasmic, eosinophilic, homogeneous, occasionally granular inclusion bodies occur in endothelial cells, pericytes, keratinocytes, macrophages, and fibroblasts. Virions in various stages of development are present in these inclusion-containing cells and in peripheral nerves. Neutrophils, macrophages, and occasionally eosinophils migrate into the dermis in the acute lesions to be replaced as the lesion ages by a predominantly mononuclear cell population. The infarcted tissue

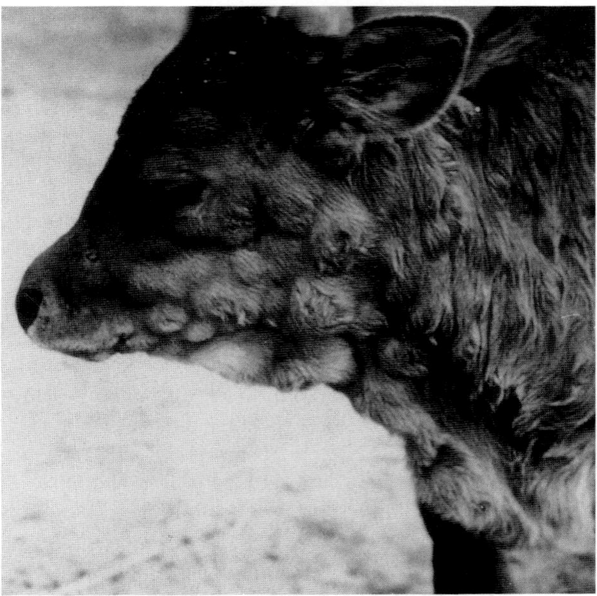

Fig. 5.73 Lumpy-skin disease. Circumscribed, nodular lesions of skin. (Courtesy of C. C. Brown.)

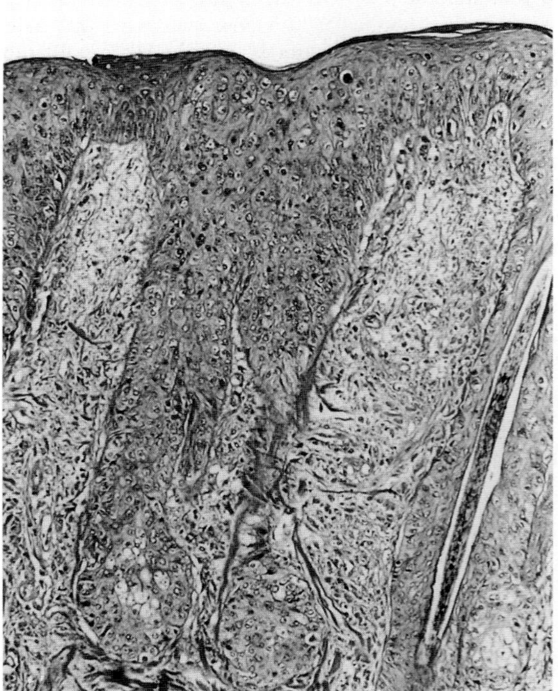

Fig. 5.74 Lumpy-skin disease. Proliferation of epidermis with rete ridge formation and hydropic change in acute disease.

is sequestered and surrounded by granulation tissue. Inclusion bodies are absent from the resolving lesions but may be present in adjacent skin or sebaceous glands. The lymph nodes are edematous and hyperplastic.

The chief differential diagnosis is pseudolumpy skin disease caused by a herpesvirus identical to the bovine herpes mamillitis virus but originally known as the Allerton virus. Pseudolumpy skin disease is a milder condition clinically, and the nodules are superficial, resembling only the early stage of lumpy-skin disease. Confirmation of the latter is best achieved by demonstration of poxvirus particles in fresh or formalin fixed tissue.

Bibliography

Burdin, M. L. The use of histopathological examinations of skin material for the diagnosis of lumpy-skin disease in Kenya. *Bull Epizoot Dis Afr* **7:** 27–36, 1959.

Capstick, P. B. Lumpy-skin disease—experimental infection. *Bull Epizoot Dis Afr* **7:** 51–62, 1959.

Davies, F. G. Lumpy-skin disease. *In* "Virus Diseases of Food Animals. A World Geography of Epidemiology and Control" E.P.J. Gibbs (ed.), pp. 751–766. London, Academic Press, 1981.

Prozesky, L., and Barnard, B. J. H. A study of the pathology of lumpy-skin disease in cattle. *Onderstepoort J Vet Res* **49:** 167–175, 1982.

Weiss, K. E. Lumpy-skin disease virus. *Virol Monogr* **3:** 111–131, 1968.

Woods, J. A. Lumpy skin disease—a review. *Trop Anim Health Prod* **28:**11–17, 1988.

4. *Suipoxvirus Disease*

The host-specific poxvirus, **swinepox,** is the chief cause of pox lesions in swine. In the past, vaccinia was also responsible. The disease occurs worldwide and is endemic to areas of intensive swine production. The disease has received relatively little attention as it is usually mild, and mortality is negligible. It affects chiefly young, growing piglets but occurs in neonates, indicating that transplacental infection is possible. Normally, swinepox is transmitted by contact. The virus is resistant and persists in dried crust from infected animals. The sucking louse *Haematopinus suis* often acts as a mechanical vector and also assists infection by causing skin trauma. The gross lesions typically affect the ventral and lateral abdomen, lateral thorax, and medial forearm and thigh (Fig. 5.75). Occasionally, lesions on the dorsum predominate. In severe infection, lesions may be generalized and rarely involve the oral cavity, pharynx, esophagus, stomach, trachea, and bronchi. The morphology of the gross lesions follows the typical pattern of pox infection. The erythematous papules usually transform into umbilicated pustules without a significant vesicular stage. The inflammatory crust eventually sheds to leave a white macule. Grossly, swinepox must be differentiated from the vesicular diseases, hog cholera, pityriasis rosea, dermatosis vegetans, and sunburn.

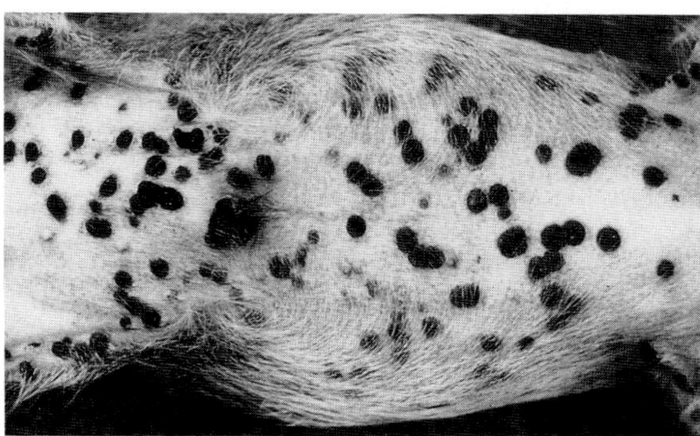

Fig. 5.75 Swine pox. Gross lesions on ventral body.

The histologic lesions also follow the pattern for pox infections (Fig. 5.76A,B). The eosinophilic, intracytoplas-

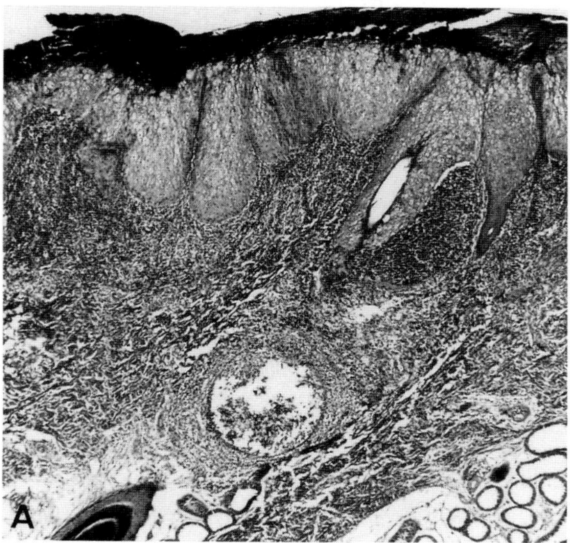

Fig. 5.76A Swine pox. Histology showing acanthosis due to intracellular edema, serous crust, and diffuse dermatitis.

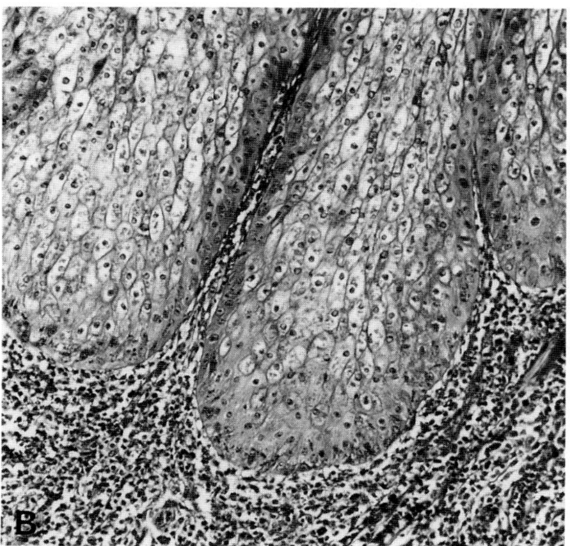

Fig. 5.76B Swine pox. Histology as in (A).

mic inclusion bodies are quite transient and are not found in older lesions. Vacuoles develop in the nuclei of infected stratum spinosum keratinocytes early in the course of swinepox infection.

Vaccinia causes swinepox lesions which are clinically difficult to differentiate from the suipox infection. Vaccinia infections have a shorter incubation period and smaller, more transient lesions. Histologically the lesions are very similar, except that nuclear vacuoles are not observed. The diseases are definitively differentiated only by virological examination.

Bibliography

Cheville, N. F. The cytopathology of swinepox in the skin of swine. *Am J Pathol* **49:** 339–352, 1966.

DeBoer, G. F. Swinepox. Virus isolation, experimental infections and the differentiation from vaccinia virus infections. *Arch Virol* **49:** 141–150, 1975.

Garg, S. K., Chandra, R., and Rao, V. D. P. Swinepox. *Vet Bull* **59:** 441–448, 1989.

Kasza, L. Swinepox. *In* "Diseases of Swine" A. D. Leman, *et al.* (eds.), 6th Ed. pp. 315–321. Ames, Iowa, Iowa State Univ. Press, 1986.

Kasza, L., and Griesemer, R. A. Experimental swinepox. *Am J Vet Res* **23:** 443–451, 1962.

Miller, R. B., and Olson, L. D. Epizootic of concurrent cutaneous streptococcal abscesses and swinepox in a herd of swine. *J Am Vet Med Assoc* **172:** 676–680, 1978.

Neufeld, J. L. Spontaneous pustular dermatitis in a newborn piglet associated with a poxvirus. *Can Vet J* **22:** 156–158, 1981.

B. Herpesvirus Infections

The Herpesviridae is a large family of enveloped DNA viruses responsible for important animal diseases, of which relatively few are primary skin diseases. Bovine mammillitis of cattle is the major cutaneous infection. Skin lesions occur in systemic herpesvirus infections such as malignant catarrhal fever in cattle and occasionally in infectious bovine rhinotracheitis. Rarely are skin lesions associated with feline herpesvirus infection, usually in the absence of respiratory lesions. Vesicles, pustules, ulcers, and depigmentation are seen on the genital organs and occasionally the lips and nostrils of horses with equine herpesvirus-3 infection (equine coital exanthema) and in bovine herpesvirus-1 infectious vulvovaginitis and balanoposthitis.

1. Bovine Herpesvirus-2 Diseases

a. Pseudolumpy-Skin Disease Bovine herpesvirus-2 (BHV-2) is a member of the Alphaherpesvirinae subfamily and is antigenically related to human herpes simplex virus. First isolated in Africa in 1957 from lesions resembling lumpy-skin disease, it was named the Allerton virus. The associated disease was subsequently named pseudolumpy-skin disease to differentiate it from the *Capripoxvirus* disease. Pseudolumpy-skin disease is characterized by a generalized eruption of superficial cutaneous nodules which develop a central depression but which heal without scar formation and do not produce the deep necrotic sequestra of true lumpy-skin disease.

Bibliography

Alexander, R. A., Plowright, W., and Haig, D. A. Cytopathogenic agents associated with lumpy-skin disease of cattle. *Bull Epizoot Dis Afr* **5:** 489–492, 1957.

Huygelen, C. *et al.* Allerton virus, a cytopathic agent associated with lumpy-skin disease. 2. Inoculation of animals with tissue culture passaged virus. *Zentralbl Veterinaermed* **7:** 755–760, 1960.

b. Bovine Herpes Mammillitis Bovine herpes, or ulcerative mammillitis, is a localized form of BHV-2 infection seen sporadically in the United States, Canada, Great Britain, Europe, Africa, and Australia. Serologic surveys indicate that infection is much more common than the disease. In Africa, several species of wild animals have antibody titers to the virus, although clinical disease is not seen.

Herpes mammillitis is chiefly a disease of lactating dairy cows but occurs in heifers about to calve and in beef animals. The incidence is usually sporadic with occasional local epidemics in fully susceptible herds. In previously exposed herds, the disease affects only the recently introduced nonimmune, first-calf heifers. There is no mortality. The economic significance lies in the effect on milk production of teat lesions or secondary bacterial mastitis, which complicate ~20% of cases.

Intact teat skin is refractory to virus penetration, indicating that some form of teat trauma precedes infection. Transmission of the virus is presumed to involve mechanical vectors, particularly the milking machine, but biting flies, such as *Stomoxys calcitrans,* have also been implicated. In the latter instance, it is difficult to explain the localization of lesions to the teat and mammary gland. Local tissue temperature has, however, been shown to be critical in the pathogenesis. Experimental cutaneous inoculation of BHV-2 results in higher virus titers and larger and more persistent lesions when the site is kept cold. The increased prevalence of disease in the autumn months may also relate to the temperature sensitivity of the virus. The source of infection within a herd is not known, but latency, a characteristic of the Herpesviridae, is likely to be important. Latency has been demonstrated in experimental BHV-2 infections.

The macroscopic lesions affect the teats, less frequently the udder, and occasionally the perineum of lactating cows. Transmission of infection to nursing calves may result in ulcerative lesions of the muzzle, chin, lips, and occasionally the oral cavity. Teat lesions develop after a 3–7 day incubation period. The teat becomes very swollen and painful and develops 1- to 2-cm diameter plaques. Vesicles are rare. The epidermis in the center of the plaque becomes necrotic and sloughs to expose an erythematous, irregular-shape ulcer. Exuded serum mixed with blood forms a thin brown crust, which is easily displaced by the teat cups. The lesions heal beneath the crust, and there is no residual scar. Lesions on the udder are often diffuse, no residual scar. Lesions on the udder are often diffuse, giving rise to the term "gangrene of the udder." Regional lymph nodes are swollen in the early stages.

The microscopic lesions are characterized by the formation of epithelial syncytia containing prominent intranuclear eosinophilic inclusion bodies. The inclusion bodies are typical Cowdry type A. The nuclear chromatin is marginated, and the eosinophilic inclusions are surrounded by a clear halo. The inclusions are numerous from the time of the first macroscopic lesion until the fifth day. Thereafter, they are very difficult to find. Syncytial cell formation commences early in groups of cells in the stratum basale and inner stratum spinosum. The process extends to involve the full thickness of the epidermis, the outer root sheath of the hair follicle infundibulum, and the sebaceous gland. By the fifth day after macroscopic lesions develop, the epidermis is necrotic, although the outlines of syncytial cells are apparent. The inclusion bodies lose their sharp outline and fill the entire nucleus. Some are free in the necrotic epidermis, released from fragmenting nuclei. The necrotic epidermis and adnexa are infiltrated with large numbers of neutrophils. Loss of the necrotic epithelium leaves an ulcer covered by a hemorrhagic and fibrinous exudate and a band of degenerating neutrophils. Deeper, there is dermal edema and a predominantly mononuclear cell infiltrate. In the healed lesion, the epithelium is reestablished; there is superficial dermal fibrosis, and a perivascular infiltrate of predominantly mononuclear cells.

The diagnosis of herpes mammillitis is confirmed by isolation of the virus. A rapid provisional diagnosis may be made by examining clinical material by electron microscopy. Biopsy is diagnostic if lesions are collected before the fifth day. Cytology smears prepared from an early lesion are diagnostic if syncytial cells containing eosinophilic intranuclear inclusion bodies are found. Serologic tests allow a retrospective diagnosis provided paired samples are collected and a rising titer identified.

Experimental intravenous inoculation of **sheep** with BHV-2 results in lesions resembling those of similarly inoculated cattle. Naturally occurring outbreaks of dermatitis of the pasterns caused by BHV-2 have been described in captive Dahl sheep. Goats will develop skin lesions only following intravenous inoculation.

Bibliography

Castrucci, G. *et al.* Studies on the pathogenesis of bovid herpesvirus-2 infection. *In* "Proceed. XIIth World Congress on Diseases of Cattle, the Netherlands." Vol. II, pp. 950–954. Utrecht, The Netherlands.

Castrucci, G. *et al.* Attempts to reactivate bovid herpesvirus-2 in experimentally infected calves. *Am J Vet Res* **41:** 1890–1893, 1980.

Cillu, V., and Castrucci, G. Infection of cattle with bovid herpesvirus 2. *Folia Vet Latina* **6:** 1–44, 1976.

Gibbs, E. P. J., Johnson, R. H., and Osborne, A. D. The differential diagnosis of viral infections of the bovine teat. *Vet Rec* **87:** 602–609, 1970.

Gibbs, E. P. J., Johnson, R. H., and Osborne, A. D. Experimental study of the epidemiology of bovine herpes mammillitis. *Res Vet Sci* **14:** 139–144, 1973.

Haig, D. A. Production of generalized skin lesions in calves inoculated with bovine herpes mammillitis virus. *Vet Rec* **80:** 311–312, 1967.

Letchworth, G. J., and Carmichael, L. E. Local tissue temperature: A critical factor in the pathogenesis of bovine herpesvirus 2. *Infect Immun* **43:** 1072–1079, 1984.

Martin, W. B., and Scott, F. M. M. Latent infection of cattle with bovid herpesvirus 2. *Arch Virol* **60:** 51–58, 1979.

Westbury, H. A. Infection of sheep and goats with bovid herpesvirus 2. *Res Vet Sci* **31:** 353–357, 1981.

2. *Bovine Herpesvirus-4 Diseases*

A herpesvirus serologically indistinguishable from DN599 strain (bovine herpesvirus-4) has been isolated from udder lesions on dairy cows (**mammary pustular dermatitis**) in Iowa and South Dakota. The teats are not involved. The macroscopic lesions are vesicles, pustules, ulcers, and crusts, originally 2–4 mm in diameter, becoming larger with coalescence. The microscopic lesion is an intraepidermal pustular dermatitis.

Bibliography

Osorio, F. A., and Reed, D. E. Experimental inoculation of cattle with bovine herpesvirus-4: Evidence for a lymphoid-associated persistent infection. *Am J Vet Res* **44:** 980, 1983.

Reed, D. E., Langpap, T. J., and Anson, M. A. Characterization of herpesviruses isolated from lactating dairy cows with mammary pustular dermatitis. *Am J Vet Res* **38:** 1631–1636, 1977.

Thiry, E. *et al.* Bovine herpesvirus-4 (BHV-4) infections of cattle. *In* "Herpesvirus Diseases of Cattle, Horses, and Pigs" G. Wittman (ed.), pp. 96–115. Boston, Kluwer Academic Publishers, 1989.

3. *Feline Herpesvirus*

Feline herpesvirus 1, a well-recognized pathogen of the upper respiratory tract is commonly associated with oral ulceration and, on rare occasions, with lesions of the skin, including the footpads. In one instance, three cats developed myriad shallow, irregular-shape ulcers up to 0.5 cm in size following ovariohysterectomy. Lesions were generalized, sparing only the surgical preparation area. In these rare cases, virus is typically cultured from the skin and throat, but signs of upper respiratory disease are usually absent. The pathogenesis of the skin lesions is unknown but probably relates to local infection by heavily infected oral secretions in a stressed animal. Viremic spread cannot be ruled out, but persistently infected cats do not develop cutaneous lesions during episodes of virus recrudescence.

Bibliography

Flecknell, P. A. *et al.* Skin ulceration associated with herpesvirus infection in cats. *Vet Rec* **104:** 313–315, 1979.

Johnson, R. P., and Sabine, M. The isolation of herpesviruses from skin ulcers in domestic cats. *Vet Rec* **89:** 360–363, 1971.

C. Retroviral Infections

Several skin disorders have been associated with retrovirus infections caused by feline leukemia virus (FeLV) and feline immunodeficiency virus (FIV) in the cat. Since both these viruses are immunosuppressive, the dermatologic conditions are considered secondary. However, cutaneous keratin horns seen in FeLV-infected cats have been represented as a primary effect of viral infection of keratinocytes. Recurrent pyoderma and paronychia have been attributed to the immunosuppressive effects of FeLV. Bacterial skin disease affecting the face and cellulitis associated with ear tags are described in the acute stage of experimental FIV infection in specific pathogen free kittens. Dermatologic diseases associated with naturally occurring chronic FIV infection include generalized demodectic mange, notoedric mange, cowpoxvirus infections, pustular pyoderma, atypical mycobacteriosis, miliary dermatitis, and abscesses. Cats that are FIV positive are more frequently diagnosed as having subcutaneous abscesses and cellulitis than are noninfected cats. Since the virus is shed in high titer in the saliva, and bite wounds are thought to be an important mode of transmission, this finding may simply reflect a greater tendency for cats that fight to become FIV-positive cats. Many of the associations made between retrovirus infection and specific diseases are anecdotal.

Bibliography

Fleming, E. J. *et al.* Clinical, hematologic, and survival data from cats infected with feline immunodeficiency virus (1983–1988). *J Am Vet Med Assoc* **199:** 913–916, 1991.

Medleau, L. Recently described feline dermatoses. *Vet Clin North Am* **20:** 1615–1632, 1991.

D. Parvoviral Infections

More typically associated with reproductive problems, **porcine parvovirus** has been implicated in outbreaks of vesicular and ulcerative dermatitis and glossitis in 1- to 4-week-old piglets in the midwestern United States. A variety of clinical signs including diarrhea and sneezing accompanied the skin lesions. Small slitlike erosions, ruptured vesicles, and extensive ulceration were seen on the tongue, lips, snout, coronary band, and interdigital spaces. Severe lesions sometimes led to separation and sloughing of the hoof wall. Intact vesicles were seen rarely. Sporadic cases have been reported in which the lesions were exudative rather than ulcerative. They were grossly indistinguishable from those of exudative epidermitis. Parvovirus has been cultured from skin and internal organs of affected piglets, and parvoviral antigen has been detected in hair follicles of lesional skin. Skin lesions were reproduced with tissue culture parvovirus but were not so severe as those induced with crude suspensions prepared from the skin lesions. In the sporadic cases, *Staphylococcus hyicus* and swinepox were also recovered. It is likely that the severe clinical disease results from dual bacterial and viral infection.

Bibliography

Kresse, J. I. *et al.* Parvovirus infection in pigs with necrotic and vesicle-like lesions. *Vet Microbiol* **10:** 525–531, 1985.

Whitaker, H. K., Neu, S. M., and Pace, L. W. Parvovirus infection in pigs with exudative skin disease. *J Vet Diag Invest* **2:** 244–246, 1990.

XII. Bacterial Diseases of Skin

The healthy intact integument is remarkably resistant to bacterial invasion. Colonization of the skin by pathogenic bacteria is usually limited by a lack of moisture, constant desquamation of surface corneocytes, and ecological pressures exerted by the normal flora. The normal flora is divided into a permanent resident population, which forms stable colonies on the outer stratum corneum, hair follicle infundibula, and hair shafts, and a more diverse transient population, which multiplies temporarily but is not maintained. The ability to produce antibiotics and lipases which split sebum into fatty acids toxic to nonresidents helps the resident flora keep its competitive advantage. Once the skin barrier is damaged, bacteria from the transient flora and occasionally the resident flora may assume pathogenicity.

The resident flora in domestic animals generally includes coagulase-negative staphylococci, *Micrococcus* spp., *Corynebacterium* spp., *Streptococcus* spp., and *Acinetobacter* spp. In dogs (*Staphylococcus intermedius*) and swine (*S. hyicus*), coagulase-positive staphylococci are part of the resident flora. In large animals, qualitative studies have shown that the skin and hair coat are a veritable "cesspool" of bacteria.

The resistance of the skin to bacterial infection obtains only for as long as the skin is structurally and physiologically intact. Wherever the microclimate of the skin surface is altered, the resident flora is disturbed. Abnormal, but not visibly infected, skin supports a denser population of bacteria, and the population alters to favor more pathogenic species, such as coagulase-positive staphylococci and Gram-negative transients. Factors which assist bacterial colonization and invasion include continued moisture, leading to maceration of the stratum corneum and leaching of the protective lipid coat, alterations in keratinization (such as in endocrine and seborrheic skin disorders), frictional damage, physical irritation due to external parasites and their secretions, self-induced damage (pruritic dermatoses such as those associated with hypersensitivity reactions and ectoparasites), accumulated dirt, excess sweating, and direct trauma such as bite wounds or penetrating foreign bodies (Fig. 5.77). Follicular hyperkeratosis, com-

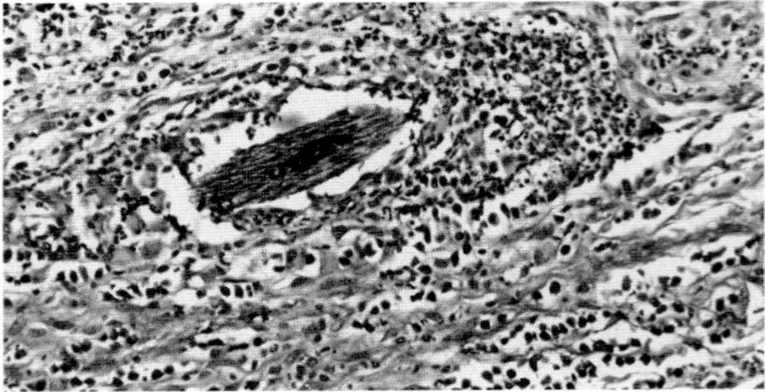

Fig. 5.77 Hair fragment in dermis surrounded by granulocytes and macrophages.

mon to many chronic skin conditions, predisposes to bacterial colonization of hair follicles and folliculitis. Once the infection is established, its outcome depends on the nature of the organism and on the host's defenses. A wide variety of immunologic defects, hereditary or acquired, may predispose an animal to the development of bacterial skin disease. Examples of hereditary defects include IgA deficiency in beagles and the Chinese Shar-pei, complement (C3) deficiency in Brittany spaniels, granulocytopathy syndromes in Irish setters and Holstein–Friesian cattle, combined immunodeficiency (T and B lymphocytes) in basset hounds and Arabian horses, cyclic hematopoiesis in gray collies, and the Chediak–Higashi syndrome in cattle and Persian cats. Acquired immunosuppressive states include those associated with various diseases (systemic viral, protozoal, or fungal infections, generalized demodicosis, hyperadrenocorticism, diabetes mellitus, hypothyroidism), nutritional deficiencies (especially protein, zinc, vitamin E), or drugs (especially glucocorticoids and anticancer agents).

Bacterial infection of the skin is often called **pyoderma,** classified as primary or secondary, superficial or deep. However, it is likely that all bacterial pyodermas are secondary to some exogenous and/or endogenous triggering factor(s). Bacterial pyodermas are more common in dogs than in other domestic species.

Coagulase-positive staphylococci are the bacteria most often isolated from bacterial pyoderma in dogs (*S. intermedius*), horses (*S. aureus, S. intermedius*) and ruminants (*S. aureus*). *Staphylococcus hyicus* is responsible for exudative epidermitis in piglets and may cause superficial pyoderma in horses and cattle. *Dermatophilus congolensis* causes a superficial pyoderma in many animal species. *Pseudomonas aeruginosa,* an opportunistic invader of damaged skin along with *Proteus* spp. and other Gram-negative transients, also causes a specific superficial pyoderma in sheep, known as fleece-rot. *Corynebacterium pseudotuberculosis* causes folliculitis and ulcerative lymphangitis in horses. An unidentified *Corynebacterium* recovered from the hair of beagle dogs with alopecia is the only bacterial infection specifically of hair in animals. Most bacteria are potential secondary invaders: cat-fight abscesses yield *Bacteroides* spp., *Fusobacterium* spp., *Pasteurella multocida*, β-hemolytic streptococci and *Peptostreptococcus anaerobius,* which are part of the feline oral flora; *Fusobacterium necrophorum* in young piglets and the spirochete *Borrelia suilla* are often wound invaders; and many bacteria, including *Nocardia* spp. (Fig. 5.78), *Actinomyces* spp., *Actinobacillus* spp., *Mycobacterium* spp., and coagulase-positive staphylococci, have the potential to cause pyogranulomatous deep dermal or subcutaneous lesions.

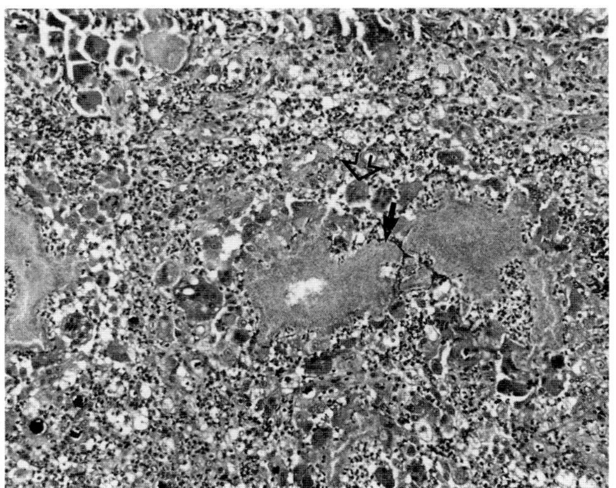

Fig. 5.78 Granulomatous dermatitis in nocardiosis. Cat. Note bacterial colonies (arrow) surrounded by giant cells (open arrow).

Bibliography

Cox, H. U. *et al.* Temporal study of staphylococcal species on healthy dogs. *Am J Vet Res* **49:** 747–751, 1988.

Devriese, L. A., Nzuambe, D., and Godard, C. Identification and characteristics of staphylococci isolated from lesions and normal skin of horses. *Vet Microbiol* **10:** 269–277, 1985.

Jansen, B. C., and Hayes, M. The relationship between the skin and some bacterial species occurring on it in the Merino. *Onderstepoort J Vet Res* **54:** 107–111, 1987.

Jenkinson, D. M. The topography, climate, and chemical nature of the mammalian skin surface. *Proc R Soc Edin* **79B:** 3–22, 1980.

Lloyd, D. H. The inhabitants of the mammalian skin surface. *Proc R Soc Edin* **79B:** 25–42, 1980.

Love, D. N. *et al.* Isolation and characterisation of bacteria from abscesses in the subcutis of cats. *J Med Microbiol* **12:** 207–212, 1979.

Love, D. N., Johnson, J. L., and Moore, L. V. H. *Bacteroides* species from the oral cavity and oral-associated diseases of cats. *Vet Microbiol* **19:** 275–281, 1989.

Medleau, L., and Blue, J. L. Frequency and microbial susceptibility of *Staphylococcus* spp. isolated from feline skin lesions. *J Am Vet Med Assoc* **193:** 1080–1081, 1988.

Noble, W. C. Bacterial skin infections in domestic animals and man. *In* "Advances in Veterinary Dermatology," Vol. 1, C. von Tscharner and R. E. W. Halliwell, (eds.), pp. 311–326. London, Baillière Tindall, 1990.

Specht, T. E. *et al.* Skin pustules and nodules caused by *Actinomyces viscosus* in a horse. *J Am Vet Med Assoc* **198:** 457–459, 1991.

A. *Superficial Bacterial Pyoderma*

Superficial bacterial pyodermas involve the epidermis and/or superficial portion of the hair follicle, usually heal without scarring, are of short duration, do not involve the regional lymph nodes, and are not usually associated with systemic illness. Superficial bacterial pyodermas are characterized by the formation of papules, transient pustules, and annular areas of crusting, scaling, and alopecia, sometimes located at the ostia of hair follicles. Histologically they take the form either of an intraepidermal pustular dermatitis (Fig. 5.79) and/or a superficial folliculitis. The predominant inflammatory cell present is the neutrophil.

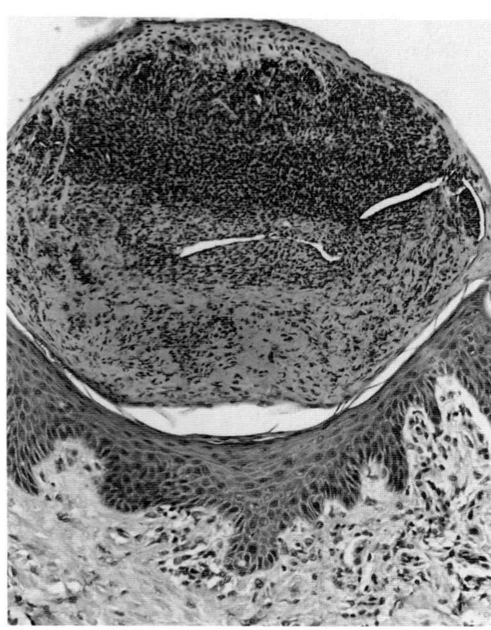

Fig. 5.79 Histology of superficial pyoderma. Pig. Intraepidermal pustular dermatitis.

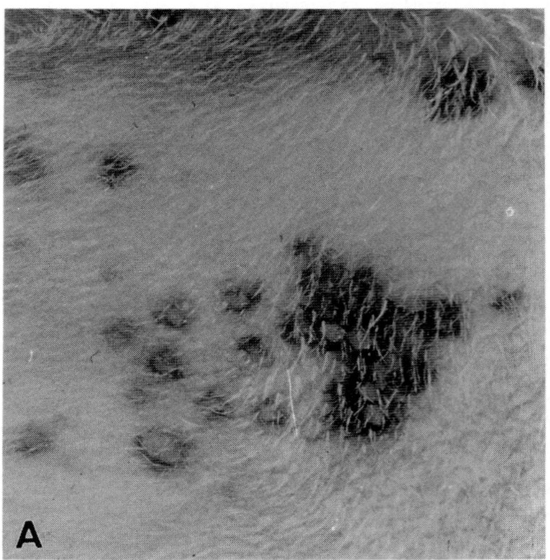

Fig. 5.80A Superficial pustules in impetigo. Puppy.

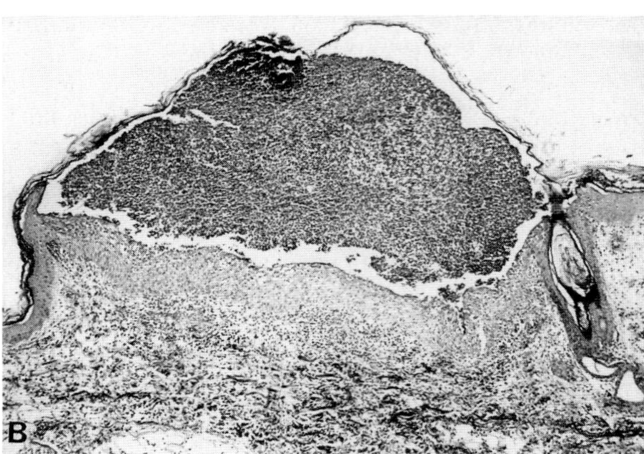

Fig. 5.80B Pustular bacterial dermatitis impetigo. Puppy.

Despite the bacterial cause, often the bacterial colonies are not readily apparent in histologic sections. *Dermatophilus congolensis* is an exception.

1. Impetigo

Impetigo is a nonfollicular subcorneal pustular dermatitis caused by coagulase-positive staphylococci and rarely by *Streptococcus* spp. It occurs commonly in puppies, occasionally in piglets and kittens, and affects the udders of lactating dairy cows and goats. Puppies between 2 weeks and 1 year of age are affected. Crowded or unhygienic conditions, poor nutrition, or debilitating diseases may be predisposing factors. The lesions occur on glabrous skin of the axilla, groin, and ventral abdomen. Erythematous macules or papules, 5–10 mm in diameter, develop into superficial pustules containing honey-colored exudate (Fig. 5.80A). These rupture to form shallow erosions with an erythematous border and a light yellow crust. The final stage is a hyperpigmented macule. Impetigo in kittens commences on the back of the neck and may spread to the withers, head, ventral chest, and abdomen. *Pasteurella multocida* and β-hemolytic streptococci are cultured from the superficial pustules. The lesions may develop from excessive wetting of the kitten's neck as it is transported by the queen. The lesions are similar morphologically to those of puppies.

Impetigo in dairy cows is contagious, often associated with staphylococcal mastitis, and may spread to the hands of milkers. Pustules, 2–4 mm in diameter, form around the base of the teats and may extend over the surface of the udder. Impetigo in goats similarly affects the udder, may spread to the perineum and ventral tail, may predispose to staphylococcal mastitis, and may spread to the hands of humans.

Histologically, impetigo in all species is characterized by intraepidermal pustular dermatitis (Fig. 5.80B). Follicles are spared. The pustules are usually subcorneal and contain mostly neutrophils, which may or may not contain intracellular bacteria. Acantholysis is minimal to absent. Focal epidermal necrosis surrounding the pustules is not uncommon.

2. Exudative Epidermitis of Pigs

Exudative epidermitis is an acute, rapidly progressive, often fatal, superficial pyoderma of the suckling or early-weaned piglet. It occurs in all countries undertaking large-scale swine production. It has been reported under several names, including seborrhea oleosa, impetigo contagiosa suis, and greasy pig disease. The causative organism is *Staphylococcus hyicus*. The disease chiefly affects piglets from 5 to 35 days of age, although mild cases occur in older pigs. The morbidity varies from 10 to 100%, and the mortality, from 5 to 90%, with an average of 25%.

Staphylococcus hyicus is readily isolated from the lesion of exudative epidermitis and is capable of reproducing the disease experimentally. The latter may be accomplished by scarification, intradermal, subcutaneous, and intramuscular inoculation or by instillation into the conjunctiva. Atraumatic inoculation will induce successful colonization of the skin and, depending on bacterial strain, clinical disease. Intravenous inoculation of *S. hyicus* generates a polyarthritis as well as dermatitis. Typical signs are also produced by the subcutaneous inoculation of bacteria-free culture supernatant of *S. hyicus,* suggesting that the syndrome is caused by a staphylococcal exotoxin (exfoliatin) analogous to that causing the staphylococcal scalded-skin syndrome seen in children. These toxins, which exhibit a distinct sequence homology to mammalian serine proteases, bind to filaggrin in the keratohyaline granules of the stratum granulosum. The exfoliative toxin of *S. hyicus* has been partially purified; the 27-kDa toxin is heat labile and targets cells of the stratum granulosum. The organism may be cultured from the integument of normal pigs; thus, skin colonization alone is not sufficient to precipitate disease. Factors which could tip the host–parasite balance toward the bacterium include immaturity of the protective mechanisms in the juvenile skin or immune system, breaks in the skin barrier, intercurrent infections, a lack of competing inhibitory flora, and nutritional deficiencies.

The gross lesions in the acute disease have a sudden onset and appear first around eyes, ears, snout, and lips and spread to legs, ventral thorax, and abdomen. The initial lesion is a focal erosion of the stratum corneum followed by an accumulation of yellow-brown exudate at the base of hairs. Small papules with a surrounding hyperemic zone develop adjacent to the follicular orifices. Lesions usually become generalized over 24 to 48 hr, covering the body in a distinctive, thick, yellow-brown, greasy, and malodorous exudate (Fig. 5.81A). The underlying skin is very erythematous. If the piglet survives for 4 to 5 days, the exudate dries, and deep cracks and fissures form in the blackish-brown crust. Removal of plaques of crust exposes a raw surface with the hairs remaining in place. Occasionally, subcutaneous abscesses and/or necrosis of the ears and tail are seen. Regional lymph nodes may develop a suppurative lymphadenitis. Some animals show ulcerative glossitis and stomatitis. Death is due to a combination of dehydration, protein and electrolyte loss, and cachexia. In the subacute form of exudative epidermitis, the onset is more gradual, and the lesions remain localized to the snout, ears, feet, and carpi. Affected skin is markedly thickened with lichenification and prominent scaling. Mortality in these animals is low, but recovery is slow, and growth is severely retarded.

Many pigs dying of the generalized disease have distinctive alterations in the urinary tract, consisting of extensive hydropic change in the epithelial cells of the collecting tubules and renal pelvis. The degenerate desquamated cells may be seen grossly as linear intratubular deposits in the renal papillae and as a sediment in the pelvis and lower urinary tract. The reason for degeneration of urinary epithelium is not known but may be related to electrolyte disturbances. The sediment is sometimes responsible for occluding the ureters and, with concurrent bacteremia, pyelonephritis develops.

Histologically, the earliest lesion of exudative epidermitis is a subcorneal pustular dermatitis involving the interfollicular epidermis. Extension of infection into the follicular infundibulum induces an accompanying superficial suppurative folliculitis. In the well-developed lesions of exudative epidermitis, there is thick surface crust composed of orthokeratotic and parakeratotic hyperkeratosis, neutrophilic microabscesses, serous lakes, and numerous colonies of gram-positive cocci (Fig. 5.81B). Groups of

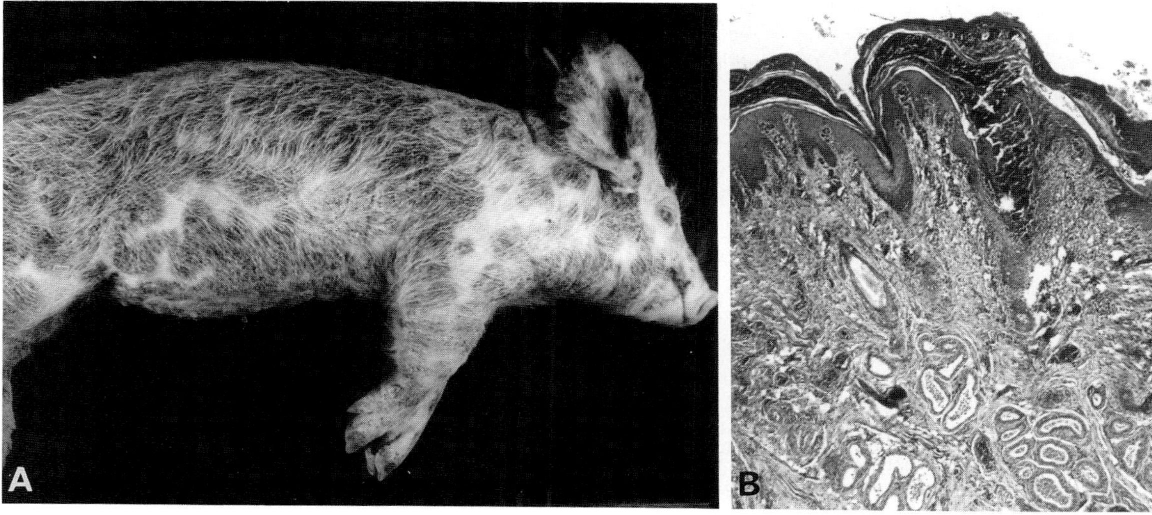

Fig. 5.81 (A) Greasy-pig disease. (B) Histology of greasy-pig disease.

cells of the outer stratum spinosum show intracellular edema, and there is generalized spongiosis. Neutrophilic exocytosis is prominent, both through the surface epithelium and into the infundibula of hair follicles. Spongiform pustules form at all levels of the epidermis and in the outer root sheath of hair follicles. Bacterial colonies are demonstrable in the follicular keratin and on the surface of hair shafts. The deep follicle is usually spared. The epidermis is hyperplastic with elongation of the rete ridges. Mitoses are numerous in the stratum basale. The dermis is edematous, with marked capillary dilation and a neutrophilic and occasionally eosinophilic perivascular infiltrate. As the lesion becomes more severe, epidermal erosion and ulceration may occur with involvement of the dermis in a diffuse suppurative dermatitis. More chronic lesions show less exudation and prominent hyperplastic changes in the epidermis, particularly marked irregular hyperplasia, pseudocarcinomatous hyperplasia, and parakeratotic hyperkeratosis. There is a predominantly mononuclear, perivascular inflammatory cell infiltrate in the dermis.

The gross lesions of exudative epidermitis are very distinctive, and in young piglets the disease is not readily confused with others. In older pigs with subacute disease, exudative epidermitis must be distinguished from sarcoptic mange (usually pruritic) and zinc-eficiency dermatitis.

In the **horse,** *S. hyicus* has been recovered occasionally from alopecic, exudative, and crusted lesions on the limbs and caudal area of the pastern ("grease heel"). While the bacterium was isolated in company with other pathogens, such as *S. aureus* or *D. congolensis,* experimental inoculation of scarified skin induced similar, localized lesions in a normal horse.

Bibliography

Andrews, J. J. Ulcerative glossitis and stomatitis associated with exudative epidermitis in suckling swine. *Vet Pathol* **16:** 432–437, 1979.

Dancer, S. J. *et al.* The epidermolytic toxins are serine proteases. *FEBS Lett* **268:** 129–132, 1990.

Devrises, L. A. *et al. Staphylococcus hyicus* in skin lesions of horses. *Eq Vet J* **15:** 263–265, 1983.

Lloyd, D. H. *et al.* Colonisation of gnotobiotic piglets by *Staphylococcus hyicus* and the development of exudative epidermitis. *Micro Ecol Health Dis* **3:** 15–18, 1990.

Mebus, C. A., Underdahl, N. R., and Twiehaus, M. J. Exudative epidermitis. Pathogenesis and pathology. *Pathol Vet* **5:** 146–163, 1968.

Obel, A.-L., and Nicander, L. Epithelial changes in porcine exudative epidermitis. An ultrastructural study. *Pathol Vet* **7:** 329–345, 1970.

Sato, H. *et al.* Isolation of exfoliative toxin from *Staphylococcus hyicus* subsp. *hyicus* and its exfoliative activity in the piglet. *Vet Microbiol* **27:** 263–275, 1991.

3. Dermatophilosis

Dermatophilosis (cutaneous streptothricosis, rain scald, rain rot, lumpy wool, mycotic dermatitis) is an acute or chronic, superficial pyoderma caused by the actinomycete *Dermatophilus congolensis*. The disease affects a wide range of species in all age groups. It occurs most often in cattle, horses, sheep, and goats and rarely in dogs, cats, pigs, and humans.

The disease is most prevalent in tropical or subtropical climates with high ambient temperatures and a monsoonal rain pattern, such as occur in parts of Africa, South America, Australia, New Guinea, New Zealand, and India. The disease occurs sporadically in the United Kingdom, Europe, United States, and Canada. The economic losses associated with dermatophilosis are substantial in the tropical countries where the disease is endemic. The losses in cattle are chiefly due to direct damage to the hides, deterioration of general condition, loss of meat production, and reduced milk production. In sheep, the losses relate mainly to the downgrading of the wool. Dermatophilosis also predisposes to cutaneous myiasis in sheep. In both species, the disease may increase susceptibility to secondary pathogens. Mortality is usually very low in sheep, except for occasional outbreaks in lambs. It may be as high as 10% in cattle, particularly if they are debilitated initially.

Dermatophilus congolensis is a pleomorphic Gram-positive bacterium which grows as a filamentous, branching septate mycelium. Septation occurs both transversely and longitudinally, giving rise to small coccoid spores known as zoospores (Fig. 5.82). The dormant zoospores are able to survive in the dried exudate on infected animals, resisting desiccation and extremes of temperature for several months. Under favorable conditions of temperature and moisture, the zoospores become motile

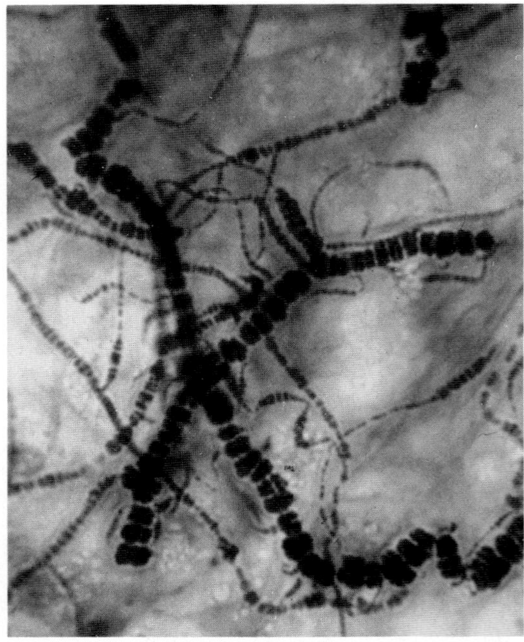

Fig. 5.82 Branching filaments with transverse and longitudinal septation in smear of *Dermatophilus congolensis*.

and migrate to a new location, germinate, and commence filamentous growth. The organism does not appear to survive for long periods in the soil or on inert objects. The principal source of infection in dermatophilosis is the infected animal, including the healthy carrier and the apparently recovered animal. In endemic areas, up to 50% of apparently healthy cattle may be carriers of the bacterium, which persists in the ostia of hair follicles.

Dermatophilus congolensis cannot penetrate healthy skin. The hair or wool coat, the superficial sebaceous film, and the stratum corneum form a barrier to its progress, which must be damaged for disease to occur. The two most important predisposing factors in the pathogenesis of dermatophilosis are prolonged wetting and mechanical damage to the skin. The incidence of the disease in endemic areas is well correlated with the wet season. The intensity of the rainfall is more significant than the total precipitation. Humidity alone is insufficient to precipitate lesions in an infected animal. Outbreaks of dermatophilosis may occur in sheep and cattle following "jetting" with a high-pressure stream of water. Deep saturation of the skin affects both stratum corneum and the protective lipid layer. Moisture causes the individual corneocytes to swell and detach from each other, and more intense wetting leaches out the protective sebaceous film from the intercellular spaces of the stratum corneum. The moisture also activates latent zoospores and aids in their transportation to distant sites on the body.

Rainfall may also act indirectly to increase the range and activity of potential arthropod vectors. These arthropod vectors are more important in the endemic tropical and subtropical areas than in temperate zones. Ticks, particularly *Amblyomma* and *Hyalomma* spp., act as vectors, and the lesions in tropical cattle often occur at the preferred sites of attachment by the tick. The organism has been isolated from the tick mouthparts and from other biting insects such as mosquitoes and flies. In Zambia, the stable fly, *Stomoxys* spp., is more important in the epizootiology of the disease than are tick vectors. The organism may also be transported mechanically by a wide range of nonhematophagous arthropods including *Musca domestica, Lucilia cuprina,* and *Calliphorinae* spp. The sheep ked, *Melophagus ovinus,* is probably a vector. Arthropod vectors act not only as agents of transmission, but more important, also assist in breaching the epidermal barrier. Many other forms of minor trauma predispose to dermatophilosis, including shear wounds in sheep, particularly if the sheep are put through a dip after shearing; cattle grazing on thorny pastures may develop the disease, and lesions in the axilla and groin of sheep appear to be associated with the awns of tall summer grasses. Concurrent poxvirus infection in 40 of 50 naturally occurring cases of dermatophilosis in Nigerian cattle has prompted the suggestion of synergistic infection.

Once epidermal defenses are breached and the zoospores germinate, branching filaments extend through the epidermis but do not usually penetrate the basement membrane zone. Keratinolytic activity may aid in colonization.

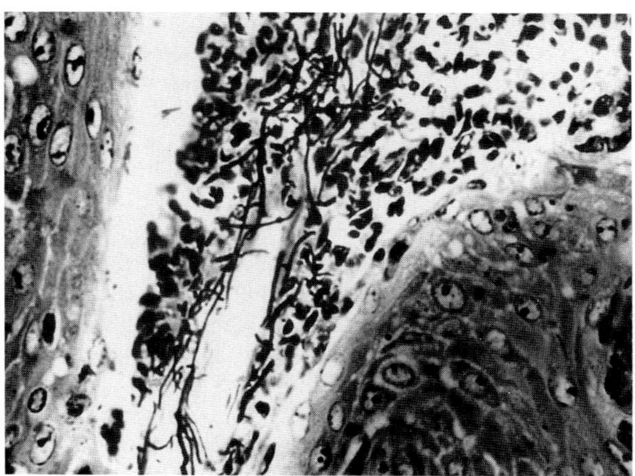

Fig. 5.83 Mycelium of *Dermatophilus congolensis* at orifice of hair follicle.

Extensive proliferation occurs in the outer root sheath of the hair follicle, which probably represents the initial site of infection in the naturally occurring disease (Fig. 5.83). The bacterial invasion stimulates an acute inflammatory reaction, characterized by superficial dermal edema and the accumulation of large numbers of neutrophils. Neutrophilic exocytosis also leads to microabscess formation in the surface epidermis and the external root sheath. The inflammatory response, particularly the neutrophil component, inhibits deeper penetration by *D. congolensis*. By 24 hr postinfection, epidermal regeneration from adjacent external root sheath epithelia is established, and by 36 hr, the epidermis is completely restored. Since the bacterial presence has stimulated premature keratinization in the original epidermis, the bacteria now invade the newly formed epithelium from a reservoir in the follicle sheath, and another burst of dermal inflammation is stimulated. Such cyclic events build up a thick scale-crust, composed of alternating layers of parakeratotic and orthokeratotic hyperkeratosis with bands of degenerating neutrophils (Fig. 5.84). Bacterial filaments within the crust are largely composed of multiple rows of coccoid bodies, which eventually develop into motile zoospores (Fig. 5.82). Pathologically, dermatophilosis typically appears as a hyperplastic superficial perivascular dermatitis with a palisading crust and prominent reticular degeneration of the superficial epidermis. Other reaction patterns less commonly observed include suppurative epidermitis and suppurative folliculitis. Rarely, deep granulomatous or pyogranulomatous dermatitis and panniculitis may be seen, especially in cats and ruminants.

The gross lesions of dermatophilosis in **cattle** usually commence along the dorsal midline between the withers and rump, extending laterally onto the flanks, thoracic wall, shoulders, and neck (Fig. 5.85A,B). In cattle from tick-infested regions, lesions often involve the axilla, groin, or scrotal skin. Lesions in calves may commence on the muzzle and spread over the head and neck. Isolated

Fig. 5.84 Proliferation of epidermis, hyperkeratosis with abscess formation, and inflammatory cell infiltration of dermis and epidermis. *Dermatophilus congolensis.*

lesions on the poll and ears are also recognized in calves. Generalized lesions may occur. The initial lesion in dermatophilosis is a small papule, which may be evident only by palpation. A slight serous exudate then forms about a group of hair shafts, causing the hairs to mat together and stand erect. The number of primary lesions is highly variable and, when the atmosphere and skin are dry, the lesions may not progress. In hot, humid conditions, new lesions develop and coalesce to produce an irregular mosaic which may involve most of the skin. The characteristic lesions are raised, roughly circular, thick, lamellated, gray-brown scale-crusts, which are penetrated by tufts of hair. In very severe lesions, the hairs may be buried in thick plaques of scale-crust (Fig. 5.85B). When individual crusts are forcibly detached, the hair is also epilated. The undersurface of the scale-crust is typically concave and moist, with little or no pus. The underlying epidermis is moist and erythematous, and in older lesions, small hillocks of granulation tissue may be present. Ultimately the scale-crusts detach to leave a scaling, alopecic epidermis, but as hair follicles are not destroyed, regrowth of the hair coat is possible. Oral lesions of dermatophilosis occur rarely in cattle.

The gross lesions in **sheep** fit into three categories, each reflecting the body region involved and not implying a differing pathogenesis. Lesions of the woolly areas are

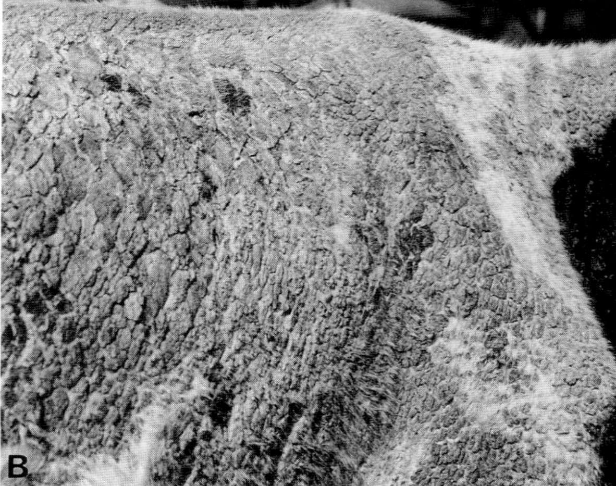

Fig. 5.85 *Dermatophilus congolensis* infection. Cow. (A) and (B) Appearance of skin lesions.

known as mycotic dermatitis or lumpy wool disease. They typically affect the dorsum, neck, and lateral thorax. The lesions evolve in the same manner as those in cattle but often escape detection for many months because the exudate does not reach the tip of the staple. The scale-crusts in sheep are often pyramidal because of the spread of the lesion as the crust is formed, and may be up to 3 cm thick. In other respects, the scale-crust resembles that of bovine dermatophilosis. Not all cases of dermatophilosis are progressive. Some become inactive in the early exudative stage, and the thin shields of dry, gray or brown scale-crust are carried away from the skin by the growth of wool.

Dermatophilosis affecting the hairy skin of the distal extremities of lambs and weaners in the United Kingdom and Australia has been named strawberry footrot, although the disease has no relationship to footrot. Small, raised, dome-shaped scale-crusts form on the leg, particularly on the anterior aspect of the pastern, but may expand and coalesce to cover the leg to the level of the hock or knee. Removal of the dense scale-crust reveals shallow ulcers with hemorrhagic pits, supposedly resembling a strawberry.

Dermatophilosis affecting the hairy skin of the face, ears, or scrotum has not been given a separate title. Hard, discrete, amber-colored scale-crusts form on the haired areas and tend to be flatter than those on wooled areas. Severe generalized disease occurs occasionally in young sucklings. Over a few days, most of the body is covered with a sticky, brownish exudate, which mats the wool to the tips. The exudate dries and fissures. Secondary bacterial invasion is common, and the disease is often fatal.

The gross lesions in **horses** also have a predilection for the dorsum, particularly the saddle area, loin, and rump, as well as areas of coat drainage posterior to the shoulder and down the flanks. Horses with lesions on the distal limbs, particularly around the caudal pastern area (so-called grease heel), often have a history of grazing or working in wet, marshy, or muddy environments. The ventral abdominal skin may be affected where continual splashing occurs. Local lesions also occur on the muzzle, nostril, and ears. In severe cases, particularly in foals, lesions may extend over the whole body. The typical lesions are multiple, randomly scattered, oval-shaped plaques composed of matted hair and exudate, which give the impression that large drops of rain have scalded the skin. A more diffuse form produces a wet paintbrush effect. Lesions are not pruritic but may be painful in the acute stage. The scale-crusts develop as described in the bovine disease and have the same clinical appearance when removed, showing the typical concave base through which the tuft of hair emerges. There is yellow-green pus under the surface of the crust in early lesions in horses, but the underside of the crusts is dry in older lesions. Large scale-crusts slough to leave an alopecic patch, which becomes haired as recovery occurs.

Lesions in **goats** occur around the mouth, nose, base of beard, ears, or on the dorsal midline, rump, and caudal aspect of the thighs. Lesions resembling strawberry footrot in sheep also develop.

In **dogs and cats,** dermatophilosis is rare. Lesions may affect the dorsum and resemble those in cattle and horses. In cats, the organism may cause deep pyogranulomatous fistulous lesions, especially in the region of the popliteal lymph node. Tongue lesions have also been described. The organism is cultured occasionally from tonsillar abscesses in **pigs** and from subcutaneous and lymph node granulomas in ruminants. Concurrent *D. congolensis* and *S. hyicus* infections have been reported in piglets.

Bibliography

Baker, G. J., Breeze, R. G., and Dawson, C. O. Oral dermatophilosis in a cat: A case report. *J Small Anim Pract* **13:** 649–653, 1972.

Bida, S. A., and Dennis, S. M. Sequential pathological changes in natural and experimental dermatophilosis in Bunaji cattle. *Res Vet Sci* **22:** 18–22, 1977.

Carakostas, M. C., Miller, R. I., and Woodward, M. G. Subcutaneous dermatophilosis in a cat. *J Am Vet Med Assoc* **185:** 675–676, 1984.

Ellis, T. M., Sutherland, S. S., and Gregory, A. R. Inflammatory cell and immune function in Merino sheep with chronic dermatophilosis. *Vet Microbiol* **21:** 79–93, 1989.

Gibson, J. A., Thomas, R. J., and Domjahn, R. L. Subcutaneous and lymph node granulomas due to *Dermatophilus congolensis* in a steer. *Vet Pathol* **20:** 120–122, 1983.

Hänel, H. *et al.* Quantification of keratinolytic activity from *Dermatophilus congolensis*. *Med Microbiol Immunol* **180:** 45–51, 1991.

Harriss, S.T. Proliferative dermatitis of the legs ("strawberry footrot") in sheep. *J Comp Pathol* **58:** 314–328, 1948.

Hyslop, N. St. G. Dermatophilosis (streptothricosis) in animals and man. *Comp Immunol Microbiol Infect Dis* **2:** 389–404, 1980.

Isitor, G. N. *et al.* Frequency of involvement of pox virions in lesions of bovine dermatophilosis. *Trop Anim Health Prod* **20:** 2–10, 1988.

Kaplan, W., and Johnston, W. J. Equine dermatophilosis (cutaneous streptotrichosis) in Georgia. *J Am Vet Med Assoc* **149:** 1162–1171, 1966.

Lloyd, D. H., and Jenkinson, D. M. The effect of climate on experimental infection of bovine skin with *Dermatophilus congolensis*. *Br Vet J* **136:** 122–134, 1980.

Lomax, L. G., and Cole, J. R. Porcine epidermitis and dermatitis associated with *Staphylococcus hyicus* and *Dermatophilus congolensis* infections. *J Am Vet Med Assoc* **183:** 1091–1092, 1983.

Philpott, M., and Ezeh, A. O. The experimental transmission by *Musca* and *Stomoxys* species of *D. congolensis* infection between cattle. *Br Vet J* **134:** 515–520, 1978.

Roberts, D. S. The histopathology of epidermal infection with the actinomycete *Dermatophilus congolensis*. *J Pathol Bacteriol* **90:** 213–216, 1965.

Samui, K. L., and Hugh-Jones, M. E. The financial and production impacts of bovine dermatophilosis in Zambia. *Vet Res Commun* **14:** 357–365, 1990.

Sutherland, S. S., Ellis, T. M., and Edwards, J. R. Evaluation of vaccines against ovine dermatophilosis. *Vet Microbiol* **27:** 91–99, 1991.

4. Miscellaneous Superficial Pyodermas

Equine pastern dermatitis (so-called grease heel, scratches, mud fever) is a secondary bacterial pyoderma of the caudal pastern and heel of horses. There are many potential causes: primary irritant contact dermatitis, contact hypersensitivity, staphylococcal folliculitis, dermatophilosis, dermatophytosis, chorioptic mange, trombiculidiasis, horsepox, photosensitization, vasculitis, and pemphigus foliaceus. Horses kept under wet, muddy, or unsanitary conditions may be more susceptible. The lesions include erythema, swelling, serous exudation which mats the hair, crusting, erosions, and eventually ulceration. Chronic lesions include lichenification, fissuring, and exuberant granulation tissue.

Ovine fleece-rot, sometimes known as water-rot or weather stain, occurs on adult sheep with a staple long enough to maintain moisture. Moisture predisposes to fleece rot by inducing an acute inflammatory reaction in its own right and by encouraging the proliferation of bacteria on the skin surface. Chief of these is *Pseudomonas aeruginosa,* which retards multiplication of other species by producing growth inhibitors. *Pseudomonas* does not invade the epidermis, but instead it produces soluble enzymes such as lecithinase, proteases, elastase, and a hemolysin, which probably exacerbate the inflammation induced by wetness. The combined effects produce an acute exudative dermatitis. Plasma proteins and neutrophils leak through the epidermal surface and mat the wool staple to form a gray-brown pultaceous mass. The fleece may be stained green by the pigment pyocyanin, produced by *P. aeruginosa.* The so-called pink rot is apparently caused by a spore-bearing bacillus which produces a red pigment. Fleece-rot is of economic importance, partly because the wool loses some of its value, but more important, because it predisposes to cutaneous myiasis. Variations in fleece attributes are thought to confer differential susceptibility to the disease but high- and low-responder genetic strains of sheep also react differently to intradermal bacterial antigen, suggesting that differences in the immune response may also contribute. Skin biopsy reveals suppurative epidermitis and suppurative superficial folliculitis.

Trichomycosis axillaris is a bacterial infection of the hair shafts of humans. A similar infection with *Corynebacterium* spp. has been reported in the dog. Cutaneous changes included a diffuse, irregular, and patchy alopecia. Hairs were broken off, and hair shafts contained small, firm nodules. Microscopically, the nodules were masses of bacteria that surrounded the hair shafts.

Pyotraumatic dermatitis is discussed under Physicochemical Diseases of Skin.

Bibliography

Buck, G. E., Stewart, D. D., and Diamond, S. S. Isolation of a *Corynebacterium* from beagle dogs affected with alopecia. *Am J Vet Res* **35:** 297–299, 1974.

Chin, J. C., and Watts, J. E. Dermal and serological response against *Pseudomonas aeruginosa* in sheep bred for resistance and susceptibility to fleece rot. *Aust Vet J* **68:** 28–31, 1991.

Hollis, D. E., Chapman, R. E., and Hemsley, J. A. Effects of experimentally induced fleece-rot on the structure of the skin of Merino sheep. *Aust J Biol Sci* **35:** 545–556, 1982.

Raadsma, H. W. Fleece rot and body strike in Merino sheep. III. Significance of fleece moisture following experimental induction of fleece rot. *Aust J Agric Res* **40:** 897–912, 1989.

B. Deep Bacterial Pyoderma

Deep pyodermas involve the hair follicle, dermis, and the panniculus. They frequently involve the regional lymph nodes and may be associated with systemic illness. Deep pyodermas are characterized by the formation of papules, nodules, fistulae, and ulcers (Figs. 5.86, 5.87A,B). Pruritus and pain are variable. Histologically, they take the form of deep folliculitis (Fig. 5.88A), furunculosis (Fig. 5.88B), nodular to diffuse dermatitis, or panniculitis. Chronic lesions may be predominantly a fibrosing dermatitis with massive destruction of epidermal appendages. The causative organism may or may not be demonstrable histologically. In dogs, horses, and goats, furunculosis is usually accompanied by tissue eosinophilia, presumably directed at liberated follicular contents (keratin, etc.). Such reactions are often misinterpreted as hypersensitivity reactions or probable parasite migration.

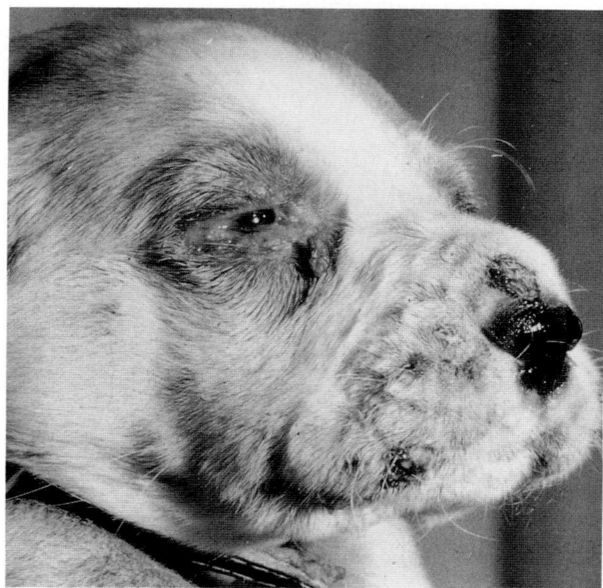

Fig. 5.86 Juvenile cellulitis. Dog.

1. Staphylococcal Folliculitis and Furunculosis

Staphylococcal folliculitis and furunculosis are common in the dog, horse, goat, and sheep, and uncommon in the cow, cat, and pig. In **dogs,** lesions may be localized to areas such as the bridge of the nose, muzzle, chin, paws, or pressure points, or be widespread. **German shep-**

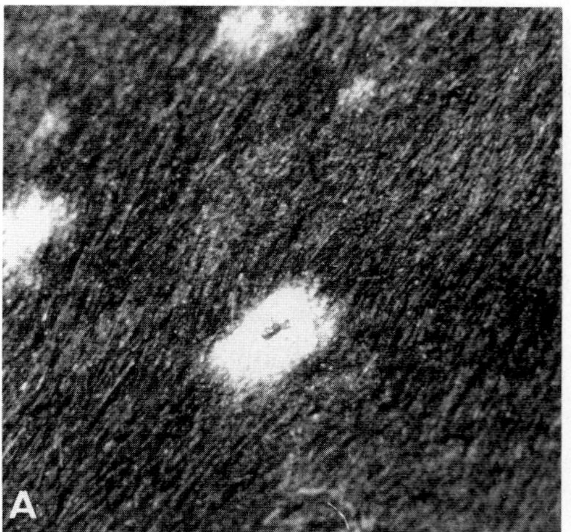

Fig. 5.87A Bacterial folliculitis. Horse. Alopecic foci with central crusts.

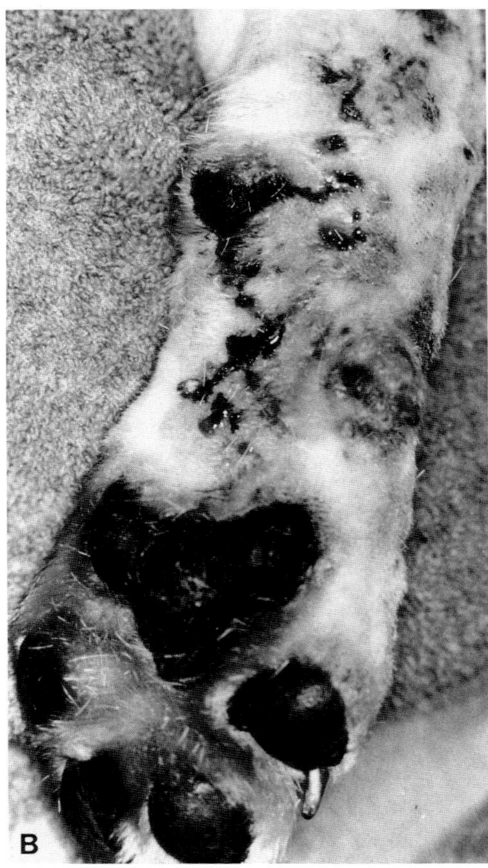

Fig. 5.87B Deep pyoderma. Dog.

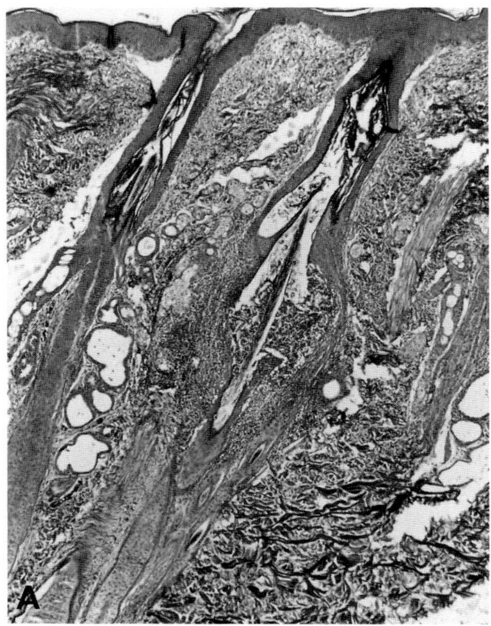

Fig. 5.88A Histology of folliculitis.

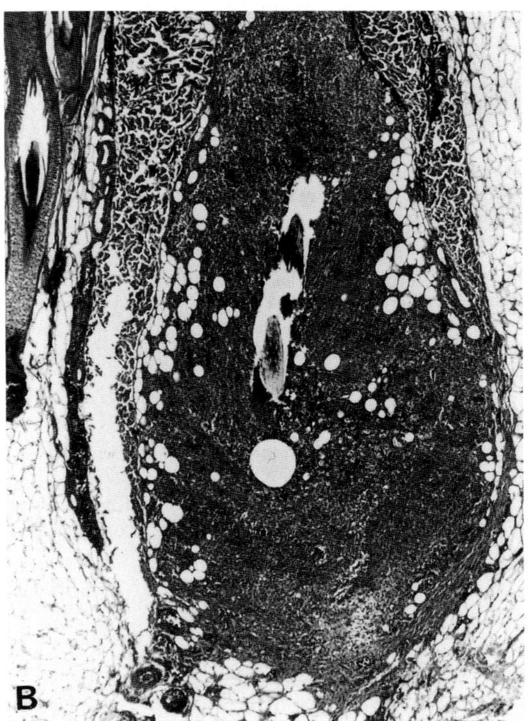

Fig. 5.88B Histology of furunculosis.

herd dog pyoderma is a unique, deep, staphylococcal pyoderma which typically involves the rump, caudolateral thighs, flanks, and ventral abdomen. Susceptibility to this infection appears to be inherited, probably as an autosomal recessive trait, and the condition is usually incurable. Neither bacterial nor flea allergies appear to be predisposing factors, and general immunological functions in affected dogs are normal. Neutrophils exhibit normal chemotactic and bactericidal functions *in vitro*.

Equine staphylococcal pyoderma is most commonly seen in areas under the saddle (saddle boils, saddle scab) and tack. Less commonly, lesions may be localized to the caudal aspect of the pasterns (grease heel) or the tail (tail worm, tail pyoderma).

In **goats,** staphylococcal pyoderma is most commonly seen on the udder, ventral abdomen, medial thighs, perineum, face, pinnae, and distal limbs.

Ovine staphylococcal pyoderma may be seen as a self-limited infection of the lips and perineum of 3- to 4-week-old lambs (plooks). It also occurs as a severe facial dermatitis in sheep of all ages, but especially in adult ewes prior to lambing. This facial form appears to be contagious, and spread through the flock has been attributed to infection of face and head abrasions while animals are feeding at troughs or fighting. Staphylococcal pyoderma may also affect the distal limbs or teats.

Staphylococcal folliculitis and furunculosis are occasionally seen in **cattle** (especially tail, perineum, scrotum, face), cats (chin, dorsal head, neck, back, rump) and pigs (especially hindquarters, abdomen, and chest).

Bibliography

Barta, O. *et al.* Lymphocyte transformation suppression caused by pyoderma—failure to demonstrate it in uncomplicated demodectic mange. *Comp Immunol Microbiol Infect Dis* **6:** 9–17, 1983.

Krick, S. A., and Scott, D. W. Bacterial folliculitis, furunculosis, and cellulitis in the German shepherd dog: A retrospective analysis of 17 cases. *J Am Anim Hosp Assoc* **25:** 23–30, 1989.

Parker, B. N. J., Bonson, M. D., and Carroll, P. N. Staphylococcal dermatitis in unweaned lambs. *Vet Rec* **113:** 570–571, 1983.

Scott, D. W., and Manning, T. O. Equine folliculitis and furunculosis. *Equine Pract* **2(6):** 11–32, 1980.

Synge, B. A., Scott, F. M. M., and MacDougall, D. C. Dermatitis of the legs of sheep associated with *Staphylococcus aureus*. *Vet Rec* **116:** 459–460, 1985.

White, S. D. Pyoderma in five cats. *J Am Anim Hosp Assoc* **27:** 141–146, 1991.

Wisselink, M. A. *et al.* German shepherd dog pyoderma: A genetic disorder. *Vet Q* **11:** 161–164, 1989.

Wisselink, M. A. *et al.* Investigations on the role of flea antigen in the pathogenesis of German shepherd dog pyoderma (GSP). *Vet Q* **12:** 21–28, 1990.

Wisselink, M. A. *et al.* Investigations on the role of staphylococci in the pathogenesis of German shepherd dog pyoderma (GSP). *Vet Q* **12:** 29–34, 1990.

2. Miscellaneous Deep Pyodermas

Ulcerative lymphangitis is a bacterial infection of the cutaneous lymphatics in horses and cattle and, rarely, sheep and goats. The condition is most commonly associated with poor hygiene and management and insect transmission. The most commonly isolated organism is *Corynebacterium pseudotuberculosis;* many other numerous bacteria are occasionally isolated. Lymphangitis is also seen in the course of equine glanders (*Pseudomonas mallei*) and bovine nocardiosis or "farcy" (*Nocardia farcinica*). Lesions consist of firm to fluctuant nodules, which may abscess and ulcerate, and corded lymphatics, most commonly on the distal extremities. Skin biopsy reveals nodular-to-diffuse pyogranulomatous dermatitis and lymphangitis.

Porcine spirochetosis is an ulcerative or granulomatous dermatitis of swine caused by the interplay of the spirochete *Borrelia suilla,* poor hygiene, and cutaneous wounds. Lesions characterized by erythema, edema, necrosis, ulceration, and grayish-brown gelatinous exudate are most commonly seen on the head, ears, gums, shoulders, flanks, and scrotum of young piglets. Skin biopsy reveals diffuse dermatitis (suppurative, necrotizing, pyogranulomatous, or fibrosing) and vasculitis. Spirochetes may be demonstrated with silver stains, such as Warthin–Starry.

Bibliography

Harcourt, R. A. Porcine ulcerative spirochaetosis. *Vet Rec* **92:** 647–648, 1972.

3. Cellulitis

Cellulitis is a severe, deep, suppurative infection in which the area of infection is poorly confined and tends to dissect through tissue planes. There may be extensive edema, and the skin is often friable, darkly discolored, devitalized, and sloughs easily. Numerous aerobic and anaerobic bacteria may produce cellulitis. Histologically, cellulitis is characterized by diffuse suppurative-to-pyogranulomatous-to-necrotic dermatitis.

Some of the more commonly seen cellulitides in domestic animals include (1) staphylococcal (dogs, Thoroughbred horses); (2) corynebacterial (*Corynebacterium pseudotuberculosis*); (3) clostridial (malignant edema, blackleg, big head); and (4) *Rhodococcus equi,* typically a pathogen of the equine lung which has been recovered from cutaneous abscesses and cellulitis in young horses. The bacterial infection may, in some instances, be facilitated through cutaneous penetration by the larvae of *Strongyloides westeri*. *Rhodococcus equi* is also a rare cause of subcutaneous abscesses in cats. An uncharacterized mycoplasma has been recovered from cellulitis in cats.

Juvenile cellulitis (Fig. 5.86) is discussed in Section XVIII, Miscellaneous Skin Disorders.

Bibliography

Elliott, G., Lawson, G. H. K., and Mackenzie, C. P. *Rhodococcus equi* infection in cats. *Vet Rec* **118:** 693–694, 1986.

Keane, D. P. Chronic abscesses in cats associated with an organism resembling *Mycoplasma*. *Can Vet J* **24:** 289–291, 1983.

Markel, M. D., Wheat, J. D., and Jang, S. S. Cellulitis associated with coagulase-positive staphylococci in racehorses: Nine cases (1975–1984). *J Am Vet Med Assoc* **189:** 1600–1603, 1986.

Perdrizet, J. A., and Scott, D. W. Cellulitis and subcutaneous abscesses caused by *Rhodococcus equi* infection in a foal. *J Am Vet Med Assoc* **190:** 1559–1561, 1987.

Valberg, S. J., and McKinnon, A. O. Clostridial cellulitis in the horse: A report of five cases. *Can Vet J* **25:** 67–71, 1984.

C. Cutaneous Bacterial Granulomas

Many bacteria are capable of inducing granulomatous reactions in the dermis and subcutaneous tissues in most animals. Trauma is usually involved in the pathogenesis of the lesions. Many of the bacteria involved are of low virulence, sometimes saprophytes, introduced accidentally from the environment. The lesions are cutaneous or subcutaneous nodules (Fig. 5.89), which may ulcerate or develop fistulous tracts to the skin surface. Histologically, there is a diffuse or nodular granulomatous, often pyogranulomatous dermatitis and panniculitis. Diagnosis requires cytologic examination or biopsy, and diagnosis is aided by special stains including the Brown and Brenn tissue Gram stain, Ziehl–Neelsen's acid-fast stain, Fite's modified acid-fast stain, and auramine-rhodamine fluorescence for mycobacteria. In some atypical mycobacterial infections, frozen sections are necessary to demonstrate the agent. Bacteriologic examination, which may require special procedures, and occasionally animal transmission studies are necessary to confirm individual diagnoses. Bacterial cutaneous granulomas caused by *Actinobacillus lignieresii, Actinomyces* spp., *Nocardia* spp. (Fig. 5.78),

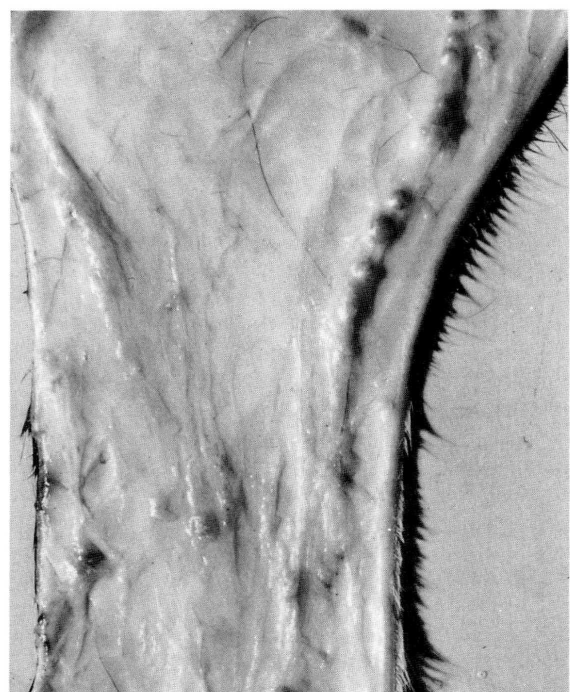

Fig. 5.89 Skin lesion tuberculosis. Ox. Underside of hide showing row of granulomatous nodules in subcutis.

and *Mycobacterium bovis* are considered elsewhere. *Mycobacterium tuberculosis*, on rare occasions, causes cutaneous lesions in dogs.

Bibliography

Davenport, D. J., and Johnson, G. C. Cutaneous nocardiosis in a cat. *J Am Vet Med Assoc* **188:** 728–729, 1986.

Foster, E. S. *et al.* Cutaneous lesion caused by *Mycobacterium tuberculosis* in a dog. *J Am Vet Med Assoc* **188:** 1188–1190, 1986.

1. Feline Leprosy

The cat appears to be more susceptible to the development of mycobacterial granulomas in the skin than are other animals. This may be because cat bites produce puncture wounds which deposit organisms deep in the dermis or subcutis. The geographic distribution of feline leprosy is strongly regional. Feline leprosy occurs in New Zealand, on the Pacific coast of the United States and Canada, and in eastern Australia, and has been reported in Great Britain, the Netherlands, France, and Germany. Attempts to define the species of *Mycobacterium* responsible have been hampered by the inability to culture the organism. *M. lepraemurium,* the agent of murine leprosy, may be the cause of feline leprosy.

Most cats affected are younger than 3 years, and it is possible that feline leukemia virus-associated immunosuppression is involved in the pathogenesis. Lesions occur anywhere on the body but are most common on the head and limbs. They also develop on the buccal mucosa, tongue, and nares. The lesions are single or multiple, indurated, subcutaneous, or mucosal nodules, sometimes with an ulcerated surface. On cut section the nodules are roughly hemispherical, 1–3 cm in diameter, and tan in color. Regional lymph nodes are usually enlarged. Histologically, the lesions take one of two forms, with intermediate stages (Fig. 5.90A,B). These patterns probably reflect differences in the cell-mediated immune status and have been likened to the tuberculoid and lepromatous forms of human leprosy, which reflect the patient's ability to mount an immune response to *M. leprae*. The tuberculoid response infers a high degree of cell-mediated immunity. The lesions are epithelioid granulomas with central areas of caseous necrosis, which seldom mineralize. Large numbers of lymphocytes and some plasma cells surround the granulomas and blood vessels. Acid-fast bacteria are present in substantial numbers only in the areas of caseous necrosis. The lepromatous form is a diffuse granulomatous dermatitis and panniculitis, in which sheets of large epithelioid macrophages predominate, with occasional multinucleated histiocytic giant cells but relatively few lymphocytes and plasma cells. The macrophages contain massive numbers of acid-fast bacilli, often arranged in parallel bundles. Lesions of intermediate type have fewer bacteria and a heavier population of lymphocytes and plasma cells. Inflammatory cell infiltration of cutaneous nerves and neutrophilic infiltration occur in both forms of feline leprosy, but the granulomatous neuritis of human leprosy in not

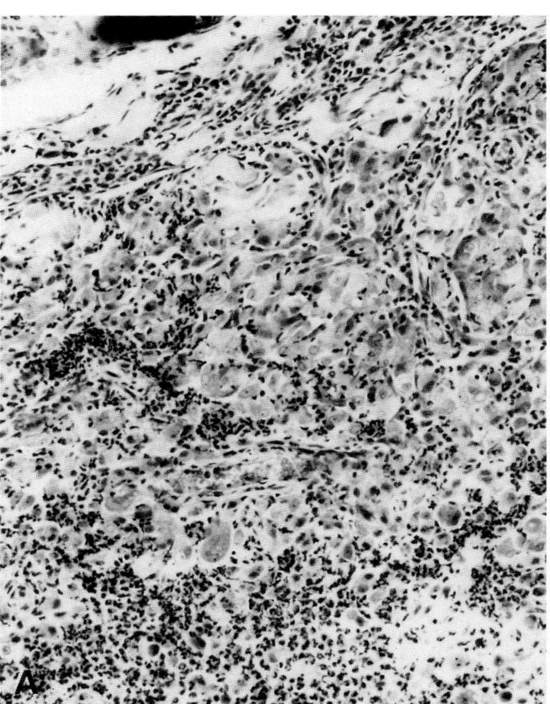

Fig. 5.90A Feline leprosy. Mingled accumulations of neutrophils and macrophages in skin lesion.

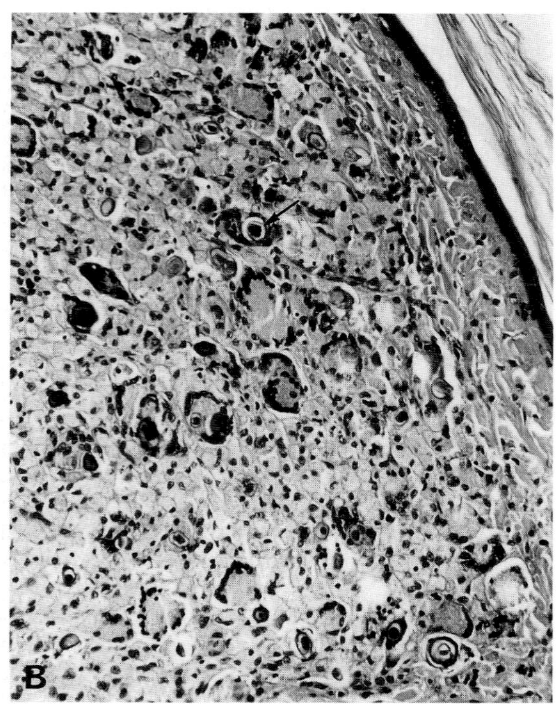

Fig. 5.90B Feline leprosy. Macrophages, concretions (arrow), and multinuclear giant cells.

seen. Regional lymph nodes often show a granulomatous lymphadenitis with large numbers of intracellular acid-fast bacilli.

Diagnosis of feline leprosy is based on the demonstration of a granulomatous reaction associated with the presence of acid-fast bacilli, and the failure to culture *Mycobacterium bovis* and the type IV fast-growing atypical mycobacteria (see the following sections). The bacteria in most feline leprosy lesions are acid-alcohol fast and stain well with Ziehl–Neelsen. If the staining is faint, the Fite modified acid-fast stain gives better results.

Bibliography

McIntosh, D. W. Feline leprosy: A review of 44 cases from western Canada. *Can Vet J* **23:** 291–295, 1982.

Schiefer, H. B., and Middleton, D. M. Experimental transmission of a feline mycobacterial skin disease (feline leprosy). *Vet Pathol* **20:** 460–471, 1983.

Thompson, E. J., Little, P. B., and Cordes, D. O. Observations of cat leprosy. *N Z Vet J* **27:** 233–235, 1979.

2. Atypical Mycobacterial Granulomas

These lesions occur in many species, including cattle, pigs, dogs, cats, and humans. The causative agents are facultative pathogens from the family Mycobacteriaceae. They have been called atypical or anonymous mycobacteria, but a more accurate term is opportunist, which reflects the predisposing factors necessary for infection and disease to occur. Since these organisms are mainly soil and water contaminants, infection is most often achieved by wound contamination. Runyon classified mycobacteria into four groups. Cutaneous lesions have been associated with *M. marinum*, a Group I photochromogenic organism, *M. xenopi* from Group III, and a variety of Group IV organisms, which are characterized by rapid growth *in vitro*. Examples are *M. fortuitum, M. smegmatis, M. chelonei*, and *M. thermoresistible*. The granulomas induced by atypical mycobacteria must be distinguished from true tuberculous lesions and feline leprosy by bacteriological examination.

The so-called **skin tuberculosis** lesions in cattle came to attention in western countries during tuberculosis-eradication campaigns because of their association with false-positive tuberculin reactions. The lesions are caused by saprophytic atypical mycobacteria and usually occur on the lower legs. They are single or often multiple, ulcerated cutaneous and subcutaneous nodules between 1 and 8 cm in diameter. The cut surface shows caseous, purulent, and occasionally mineralized foci. Lesions often involve the lymphatic chain (Fig. 5.89), but unlike *M. bovis* infections, the regional lymph nodes are not affected. The microscopic lesions are indistinguishable from tuberculous granulomas. Acid-fast bacilli that cannot be differentiated morphologically from *M. bovis* are usually present in small numbers but have not been cultivated in most cases. *Mycobacterium kansasii* has been isolated from some lesions.

Feline atypical mycobacterial granulomas are caused by *M. fortuitum, M. xenopi, M. smegmatis, M. chelonei, M. thermoresistible*, or *M. phlei*. The lesions are usually located in areas of previous trauma, such as bite wounds in the dorsal lumbosacral region or ventral abdomen. Lesions of *M. smegmatis* appear to have a particular predilection for the inguinal fat pads of obese cats. Macroscopically, the lesions comprise single or multiple cutaneous plaques and subcutaneous nodules 0.5–3.0 cm in diameter, which ulcerate and ooze small amounts of purulent exudate. The cut surface is often perforated by fine fistulae. In *M. smegmatis* infection, free fat globules and foci of fat necrosis occur in the subcutaneous fat layer. The microscopic lesions are nodular to diffuse pyogranulomatous dermatitis and panniculitis (Fig. 5.91). A smooth-bordered extracellular lipid vacuole, surrounded by a narrow rim of degenerating neutrophils, occupies the center of many of the nodules of epithelioid macrophages. Small bundles of Gram-variable, weakly acid-fast, filamentous, beaded bacteria are rarely demonstrated within the lipid vacuoles. *Mycobacterium smegmatis* has not been identified within epithelioid macrophages or neutrophils, prompting the suggestion that the organism is protected from phagocytosis as long as it persists in the lipid globules. Cytologic examination of smears prepared from exudates shows a pyogranulomatous inflammatory reaction and rarely the acid-fast bacilli. While the histologic appearance is suggestive of the diagnosis, the bacteria are often difficult to find, and confirmation requires culture and identification of the organism.

Atypical mycobacteria occasionally cause subcutane-

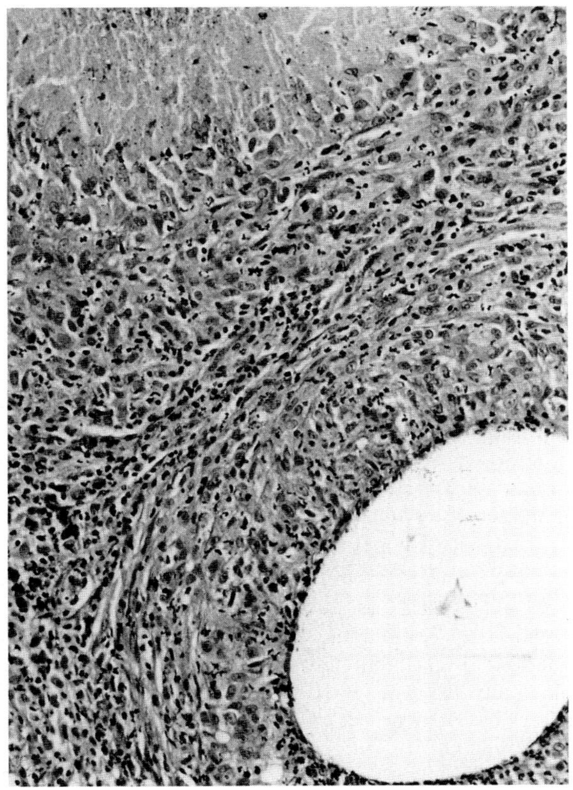

Fig. 5.91 *Mycobacterium smegmatis* panniculitis. Cat. Caseation, above, with accumulation of mixed inflammatory cells. Organisms were demonstrated in the fat globule (below right).

ous granulomas in **dogs.** *Mycobacterium chelonei* and *M. fortuitum* are isolated from some lesions. There is often a history of previous trauma to the area. The lesions chiefly affect the head, particularly the ears and the trunk. The microscopic lesions are nodular to diffuse pyogranulomatous dermatitis and panniculitis. The granulomas are predominantly epithelioid with no multinucleated histiocytic giant cells and little caseous necrosis. Sometimes large numbers of intracellular acid-fast bacilli are present, as in cat leprosy. In cases associated with *M. chelonei* and *M. fortuitum*, acid-fast organisms may be impossible to identify in paraffin-processed tissue. Ziehl–Neelsen or Kinyoun stains applied to frozen sections may demonstrate small numbers of organisms within lipid vacuoles.

Bibliography

Gross, T. L., and Connelly, M. R. Nontuberculous mycobacterial skin infections in two dogs. *Vet Pathol* **20:** 117–119, 1983.

Kunkle, G. A. *et al.* Rapidly growing mycobacteria as a cause of cutaneous granulomas: Report of five cases. *J Am Anim Hosp Assoc* **19:** 513–521, 1983.

Wilkinson, G. T., Kelly, W. R., and O'Boyle, D. Pyogranulomatous panniculitis due to *Mycobacterium smegmatis. Aust Vet J* **58:** 77–78, 1982.

3. Bacterial Pseudomycetoma

Bacterial pseudomycetoma (botryomycosis) is a chronic granulomatous disease caused by bacteria, usually coagulase-positive staphylococci. The condition occurs in all domestic animals and humans. Lesions usually follow some form of skin trauma and involve the deep dermis and subcutis, occasionally extending to the muscle and adjacent bone. The macroscopic lesions are single or multiple subcutaneous nodules, which often ulcerate and exude a purulent exudate containing tiny, white granules, the so-called grains, and are most commonly seen over the distal extremities. The microscopic lesion is a pyogranulomatous dermatitis and panniculitis. The predominant inflammatory cells are epithelioid macrophages and neutrophils, but multinucleated histiocytic giant cells, lymphocytes, and plasma cells will be present, according to the stage of the lesion. There is often extensive fibrosis. The macroscopic grains correspond microscopically to round or oval bodies in which a central core of bacterial colony is embedded in a homogeneous matrix. The bacterial colonies are usually gram-positive cocci. A homogeneous eosinophilic substance, known as Splendore–Hoeppli material, may coat the grains and take the form of a radiating corona of club-shaped bodies. The presence of Splendore–Hoeppli material is not unique to bacterial pseudomycetoma and may be seen in other microbial infections and foreign body reactions. It stains positively with periodic-acid Schiff and probably represents glycoprotein antigen–antibody complexes. Macroscopic tissue grains and microscopic club colonies are not restricted to bacterial pseudomycetoma. Differentiation from eumycotic and actinomycotic mycetomas in hematoxylin and eosin-stained tissue sections may be difficult and requires special stains.

Bibliography

Donovan, G. A., and Gross, T. L. Cutaneous botryomycosis (bacterial granulomas) in dairy cows caused by *Pseudomonas aeruginosa. J Am Vet Med Assoc* **184:** 197–199, 1984.

Scott, D. W. Bacterial pseudomycosis (botryomycosis) in the horse. *Equine Pract* **10:** 150–19, 1988.

Walton, D.K ., Scott, D. W., and Manning, T. O. Cutaneous bacterial granuloma (botryomycosis) in a dog and cat. *J Am Anim Hosp Assoc* **19:** 537–541, 1983.

D. Bacterial Pododermatitis in Ruminants

1. Footrot

Footrot in sheep, goats, and cattle is a contagious, mixed bacterial infection of the digits, in which the obligate parasite *Bacteroides nodosus* is both the transmitting agent and the principal pathogen. Footrot should not be used to designate those noncontagious mixed bacterial infections of the digit in which *Fusobacterium necrophorum* is the principal pathogen. These conditions, which include interdigital dermatitis, toe abscess, and infectious bulbar necrosis of sheep and foul-in-the-foot of cattle, are more correctly termed necrobacillosis of the foot.

The pathogenesis of footrot has been extensively studied in sheep. An initial insult to the interdigital skin is usually provided by water maceration brought about by grazing sheep on lush, continually damp pasture. Footrot is thus particularly prevalent in areas of high rainfall and temperate climate. Penetration of the skin by *Strongyloides* larvae may also predispose to foot rot. The primary colonization of the damaged epidermis is by *F. necrophorum*, supported by aerobic diphtheroids, coliforms, and other bacteria originating from the skin or feces. Footrot lesions then develop if the animal harbors *B. nodosus*. The anaerobic bacterium is an obligate parasite of the ruminant foot, surviving in small lesions in carrier animals or in clinically normal interdigital skin. It has the ability to remain viable for prolonged periods with very restricted nutrients but has very limited survival in the environment. Footrot is a synergistic infection, chiefly between *B. nodosus* and *F. necrophorum*, but also involving *Actinomyces pyogenes* and other aerobic and anaerobic bacteria of fecal origin. The chief contribution of *B. nodosus* is the production of serine and nonserine proteases which facilitate bacterial penetration of the epidermal matrix. It also contributes a heat-stable soluble factor, which enhances the growth, and hence virulence, of *F. necrophorum*. *Fusobacterium necrophorum* is chiefly responsible for the tissue destruction and resultant inflammatory response. This organism also produces an exotoxin destructive to leukocytes. Various other species of bacteria, growing opportunistically in the necrotic debris, help generate an anaerobic environment essential for the survival of the two chief pathogens.

Footrot in **sheep** may be virulent or benign. **Benign footrot** is also known as nonprogressive footrot or scald. The conditions differ in the extent of involvement of the epidermal matrix of the hoof and hence the degree of horn separation. This in turn is related to bacterial virulence. *Bacteroides nodosus* recovered from lesions of benign footrot has significantly less proteolytic activity than do bacteria recovered from virulent lesions. Isolates from benign lesions specifically lack elastase enzymes. The proteolytic capabilities of the organism determine the extent of invasion of the epidermal matrix and hence, indirectly, the degree of horn separation. The proteolytic enzymes do not directly mediate the horn separation by keratolytic action, but rather they aid bacterial penetration of the epidermis. The associated destruction of the stratum spinosum and granulosum and the resultant inflammatory reaction cause the separation of the horny part of the hoof from the underlying soft tissue.

Virulent footrot in sheep commences as an interdigital dermatitis in the axial bulbar notch, extending forward and caudally around the bulb of the heel. In the early stages, the interdigital skin is pale, swollen, and moist, and the surface is macerated. There may be a margin of hyperemia adjacent to the periople. Separation of the horn is usually evident ~7 days later and is accompanied by severe lameness. The separation commences at the skin–horn junction, spreading to the bulb and then to the sole. From the sole, the process spreads to the wall of the hoof involving the axial and abaxial surfaces of both digits. A small amount of gray, greasy malodorous exudate accumulates in the cleft beneath the underrun horn. There is no significant suppuration. The germinative layers of the epithelium are not destroyed by the process, so that attempted regeneration of the horn is continuous, but the new horn is rapidly destroyed. If the infection subsides, there is frequently an overgrowth of horn from the injured epidermis. Chronically affected feet become long and misshapen.

Virulent footrot is an economically important disease. Although it is not in itself fatal, footrot lesions are subject to secondary myiasis. Lameness or recumbency result in generalized debility with a considerable loss of body weight. A 10% depression in wool growth has been documented in experimentally infected sheep.

Benign footrot of sheep is indistinguishable from the early stages of virulent footrot, but erosive lesions are confined usually to the skin in the posterior portion of the interdigital cleft. Occasionally there is separation of the soft horn on the heel and the posterior portion of the sole, but separation of the hard horn does not occur in benign footrot. In subacute and chronic forms of benign footrot, there is marked hyperkeratosis. The bacterium persists in the thickened stratum corneum, causing little or no inflammatory reaction. Lameness, if it occurs, is mild and temporary, and the effect on production is minimal.

Benign footrot is distinguished from other types of ovine interdigital dermatitis by the demonstration of *B. nodosus* in the lesion by smear or cultural techniques. Affected animals are lame but not as severely as in virulent footrot. Benign footrot may persist for months if environmental conditions permit but usually rapidly regresses after topical therapy or if sheep are placed in a dry environment.

Footrot in **cattle** is of less significance than that in sheep and resembles the benign form of the ovine condition. Infection is much more common than the disease. Between 34 and 75% of apparently normal cattle harbor *B. nodosus* in the hoof skin. With one exception, these isolates of *B. nodosus* have a low proteolytic index and resemble the bacteria isolated from benign footrot in sheep. Experimental transmission of these isolates to sheep and cattle induces a mild interdigital dermatitis. Naturally occurring lesions are also usually restricted to the interdigital skin, which is eroded, ulcerated, and sometimes deeply fissured. The fissures contain serous exudate and a small amount of foul-smelling gray pus. The lesions may extend to the heels with separation of the soft horn, but extensive underrunning of the horn does not occur. Chronic lesions are more proliferative. Thick layers of deeply fissured hyperkeratotic scale may project 1–2 cm above the skin surface. The hyperkeratotic scale forms an ideal habitat facilitating persistence of the organism. Chronic cases are associated with no clinical abnormality or with mild lameness. In the acute disease, lameness may be pronounced.

Bibliography

Egerton, J. R., and Parsonson, I. M. Benign foot-rot—a specific interdigital dermatitis of sheep associated with infection by less proteolytic strains of *Fusiformis nodosus*. *Aust Vet J* **45:** 345–349, 1969.

Egerton, J. R., Roberts, D. S., and Parsonson, I. M. The aetiology and pathogenesis of ovine foot-rot. I. A histologic study of the bacterial invasion. *J Comp Pathol* **79:** 207–216, 1969.

Gordon, L. M., Yong, W. K., and Woodward, C. A. M. Temporal relationships and characterization of extracellular proteases from benign and virulent strains of *Bacteroides nodosus* as detected in zymogram gels. *Res Vet Sci* **39:** 165–172, 1985.

Laing, E. A., and Egerton, J. R. The occurrence, prevalence and transmission of *Bacteroides nodosus* infection in cattle. *Res Vet Sci* **24:** 300–304, 1978.

Roberts, D. S., and Egerton, J. R. The aetiology and pathogenesis of ovine foot-rot. II. The pathogenic association of *Fusiformis nodosus* and *F. necrophorus*. *J Comp Pathol* **79:** 217–227, 1969.

Stewart, D. J. The role of elastase in the differentiation of *Bacteroides nodosus* infections in sheep and cattle. *Res Vet Sci* **27:** 99–105, 1979.

Stewart, D. J., Kortt, A. A., and Lilley, G. G. New approaches to footrot vaccination and diagnosis utilizing the proteases of *Bacteroides nodosus*. *In* "Advances in Veterinary Dermatology," Vol 1. C von Tscharner and R. E. W. Halliwell (eds.), pp. 359–369. London, Baillière Tindall, 1990.

Wilkinson, F. C., Egerton, J. R., and Dickson, J. Transmission of *Fusiformis nodosus* infection from cattle to sheep. *Aust Vet J* **46:** 382–384, 1970.

2. *Necrobacillosis of the Foot*

There are several mixed-bacterial infections affecting the feet of ruminants in which *Fusobacterium necrophorum* is the principal pathogen, although it is not necessarily the initiating agent. These infections, in contrast to footrot, do not involve *B. nodosus* and are not contagious.

Ovine interdigital dermatitis is seen in footrot-free flocks running on wet or swampy grounds. The disease is an acute necrotizing dermatitis of the interdigital skin. The condition clinically resembles benign footrot, although it is more transient. It is distinguished from benign footrot on bacteriological findings.

Foot abscess (infective bulbar necrosis or heel abscess) is a mixed bacterial infection of the heel of heavy adult sheep. Usually only one digit of one foot is involved. Ewes lambing on wet pastures are typically affected. The lesions develop as an extension of ovine interdigital dermatitis and result from a synergistic infection involving *F. necrophorum* and *Actinomyces pyogenes*. Lesions usually occur on the hind feet and particularly on the medial claw. The process is necrotizing, and the lesions contain grayish blood-stained necrotic debris rather than pus. A dark red line separates living from dead tissue. The infection often spreads to involve the second interphalangeal joint and associated ligaments. The pastern swells and may develop discharging sinuses. There is severe lameness. Healing is slow and may result in permanent deformity of the foot.

Toe or lamellar abscess of sheep is more prevalent in high-rainfall areas and tends to affect adult sheep whose hooves are overgrown or of bad conformation. The lesions are often caused by *F. necrophorum*. The organisms probably gain entrance through minute fissures in the horn, or at the toe, where the wall may become separated from the sole at the "white line" by impacted mud. Usually only one claw is affected. Pus forming beneath the hard horn is released when the hoof is pared. If unattended, the suppurative process may extend along the sensitive laminae of the sole or break out at the coronet. There is extensive suppurative and necrotic laminitis and, often, necrosis of the tip of the distal phalanx.

Necrobacillosis of the bovine foot, also known as foul-in-the-foot, panaritium or fouls, is a mixed bacterial infection in which *F. necrophorum* and *Bacteroides melaninogenicus* are usually the chief pathogens. Necrobacillosis is often initiated by trauma. A severe cellulitis develops in the interdigital tissues, producing marked swelling in the interdigital cleft. The swelling often spreads to involve the heel, fetlock, and occasionally higher. Sinuses discharging necrotic material may open up in the interdigital space and above the coronet.

Bibliography

Egerton, J. R., Yong, W. K., and Riffkin, G. G. "Footrot and Foot Abscess of Ruminants." Boca Raton, Florida, CRC Press, 1989.

Parsonson, I. M., Egerton, J. R., and Roberts, D. S. Ovine interdigital dermatitis. *J Comp Pathol* **77:** 309–313, 1967.

Roberts, D. S., Graham, N. P. H., and Egerton, J. R. Infective bulbar necrosis (heel-abscess) of sheep, a mixed infection with *Fusiformis necrophorus* and *Corynebacterium pyogenes*. *J Comp Pathol* **78:** 1–8, 1968.

E. Skin Lesions in Systemic Bacterial Disease

Cutaneous lesions may also develop in the course of systemic bacterial infection. Such lesions are often recognized in swine, probably because there are many bacterial septicemias in pigs, and because the lesions are easy to see. Skin lesions are often associated with Gram-negative septicemias, particularly salmonellosis. Erythema, cyanosis, and ischemic necrosis chiefly affect the extremities and result from tissue hypoxia due to endotoxin-induced venous thrombosis. Cutaneous lesions associated with Gram-positive septicemias include the pathognomonic rhomboidal, palpable, purpuric plaques of swine erysipelas, known as diamond skin lesions (Fig. 5.92). Skin biopsy reveals neutrophilic vasculitis, cutaneous ischemia, and necrotizing hidradenitis.

Septicemias, which occur occasionally in companion animals, may result in septic thrombosis of cutaneous vessels, with resultant deep, necrotizing skin lesions. Skin lesions may result from the vasculitis of Rocky Mountain spotted fever, caused by *Rickettsia rickettsii*. Edema and dermatitis of the scrotum and deep dermal necrosis of ear tips, nasal planum, nipples, and digits have been described. *Brucella canis* was isolated from chronic, ulcerative, nodule-to-plaquelike lesions over the hocks and carpi of a dog. Skin biopsy revealed nodular-to-diffuse pyogranulomatous dermatitis with numerous lymphoid nodules.

In horses, the purpura developing occasionally as a

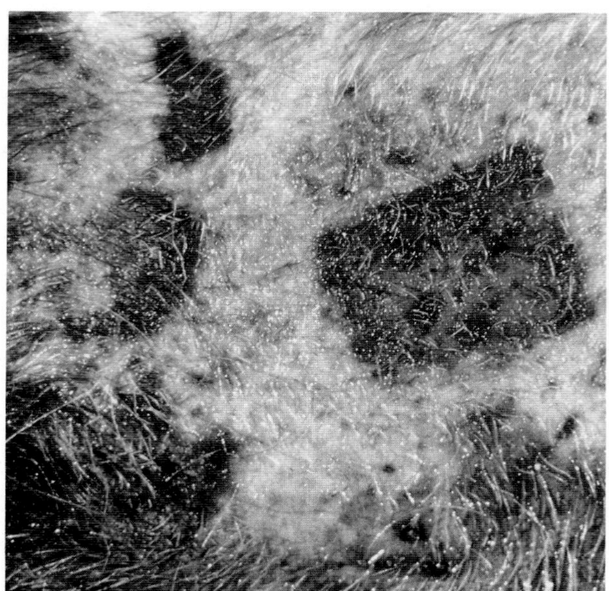

Fig. 5.92 Classical diamond skin lesions in *Erysipelothrix rhusiopathiae* infection. Pig.

Fig. 5.93A Crytococcosis. Cat. Ulcerative lesions on face and nose.

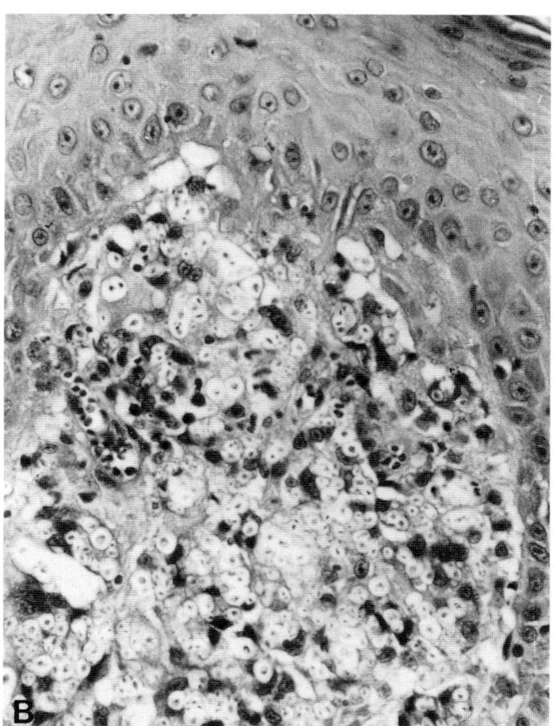

Fig. 5.93B Cryptococcosis. Cat. Organisms beneath epidermis with minimal inflammatory reaction.

sequel to *Streptococcus equi* infection is thought to develop as a result of immune-complex vasculitis.

Bibliography

Dawkins, B. G. *et al.* Pyogranulomatous dermatitis associated with *Brucella canis* infection in a dog. *J Am Vet Med Assoc* **181:** 1432–1433, 1982.

Weiser, I. B., and Greene, C. E. Dermal necrosis associated with Rocky Mountain spotted fever in four dogs. *J Am Vet Med Assoc* **195:** 1756–1758, 1989.

XIII. Mycotic Diseases of Skin

Fungal diseases are divided into three categories: the cutaneous mycoses, the subcutaneous or intermediate mycoses, and the deep or systemic mycoses. The **cutaneous mycoses** are caused by fungi which invade the skin and its appendages to a very limited extent. The most important of the superficial mycoses, dermatophytosis, is caused by fungi which parasitize the nonviable keratinized tissue. The **intermediate mycoses** are caused by a wide range of saprophytic fungi, which induce disease only when introduced into the subcutaneous tissues by penetrating cutaneous wounds. These organisms, although capable of forming successful local infections, are not able to spread systemically. In the **deep** or **systemic mycoses,** cutaneous lesions, characterized grossly by nodules, ulcers, fistulas, and abscesses and histologically by granulomatous or pyogranulomatous dermatitis and panniculitis, occur as part of the generalized disease (Fig. 5.93A,B). **Systemic mycoses** include opportunistic infections of immunosuppressed hosts by organisms such as *Cryptococcus* and *Paecilomyces* spp. and the true pathogenic fungal infections such as **blastomycosis, coccidioidomycosis,** and **histoplasmosis.** These conditions are discussed elsewhere according to their principal systemic lesions.

Severity of mycotic disease is the outcome of the host–fungus interplay. True pathogenic fungi, capable of causing systemic mycoses in apparently normal animals, show thermal dimorphism, which enables growth at higher temperatures, and have evolved a variety of mechanisms, such as inhibition of the respiratory burst of phagocytic cells, to evade the host defenses. Less-virulent organisms may cause disease only in weakened hosts. *Candida,* for example, causes severe disease only in immunosuppressed or otherwise debilitated animals.

While the gross lesions of cutaneous mycoses are seldom diagnostic, the size, morphology, and tinctorial qualities of fungi render histologic examination of lesions a useful diagnostic aid. Special stains such as periodic acid-Schiff and Gomori's methenamine-silver are usually helpful. The latter may be combined with hematoxylin and eosin to demonstrate both fungal agent and tissue response. Microscopic examination may establish not only that a lesion is of fungal origin, but may enable the classification of the agent into a general group (e.g., the dematiaceous or pigmented fungi), or into a genus (e.g., *Aspergillus* and *Candida*), or to the species level (e.g., *Blastomyces dermatitidis*). The diagnosis should be confirmed by fungal culture and identification.

Bibliography

Ainsworth, G. C., and Austwick, P. K. C. "Fungal Diseases of Animals." Farnham Royal, Slough, England, Commonw. Agric. Bur., 1973.

Chandler F. W., and Watts, J. C. "Pathologic Diagnosis of Fungal Infections." Chicago, Illinois, American Society of Clinical Pathologists Press, 1987.

Chandler, F. W., Kaplan, W., and Ajello, L. "A Color Atlas and Textbook of the Histopathology of Mycotic Diseases." London, Wolfe Medical Publishers, 1980.

Connole, M. D. Current status of the ecology, epidemiology of animal mycoses: with special reference to Queensland, Australia. *In* "Recent Advances in Medical and Veterinary Mycology" K. Iwata (ed.), pp. 123–128. Baltimore, Maryland, University Park Press, 1977.

Euzeby, J. The laboratory diagnosis of cutaneous and subcutaneous mycoses in animals. *Folia Vet Lat* **7:** 111–129, 1977.

Jungerman, P. F., and Schwartzman, R. M. "Veterinary Medical Mycology." Philadelphia, Pennsylvania, Lea & Febiger, 1972.

Rippon, J. W. "Medical Mycology. The Pathogenic Fungi and the Pathogenic Actinomycetes," 3rd Ed. Philadelphia, Pennsylvania, W.B. Saunders, 1988.

Smith, J. B. "Opportunistic Mycoses of Man and Other Animals." Wallingford, C.A.B. International Mycological Institute, 1989.

A. Cutaneous Mycoses

These mycoses are caused by fungi which attack the outer layers of the skin, the hair, and the horny appendages, such as nail, hoof, and claw. They include dermatophytosis, caused by pathogenic keratinolytic fungi of the genera *Microsporum, Trichophyton,* and *Epidermophyton,* and dermatomycosis, infections caused by nondermatophytic fungi.

1. Dermatomycosis

a. CANDIDIASIS This is an opportunistic infection, chiefly of mucosal surfaces, caused by *Candida* spp., which are normal inhabitants of the skin and the gastrointestinal tract. *Candida* infections are predisposed to by debilitation, immunosuppression, and long-term use of antibiotics. Skin lesions of candidiasis occur occasionally in dogs, cats, pigs, and horses. Skin lesions in dogs include gray-white mucoid plaques and ulcers on mucocutaneous junctions, multifocal patches of exudative dermatitis, onychomycosis and otitis externa. Other lesions of candidiasis are considered in Volume 2, Chapter 1, The Alimentary System.

Bibliography

Pichler, M. E., Gross, T. L., and Krol, W. R. Cutaneous and mucocutaneous candidiasis in a dog. *Compend Cont Ed* **7:** 225–230, 1985.

Reynolds, I. M., Miner, P. W., and Smith, R. E. Cutaneous candidiasis of swine. *J Am Vet Med Assoc* **152:** 182–186, 1968.

b. MALASSEZIASIS *Malassezia pachydermatis* (*Pityrosporum canis*), a commensal organism on the canine skin, is associated with otitis externa (see The Ear, Chapter 4). The role of this fungus in the causation of chronic skin disease in the dog is more controversial. The fungus probably acts as a significant opportunistic pathogen in skin damaged from a variety of other causes, including chronic atopy. It is an important contributor to the lesions described in West Highland white terriers as epidermal dysplasia syndrome. In humans, the causal relationship between pityrosporal yeasts and seborrheic dermatitis is now generally accepted; it is likely that *Malassezia* similarly will gain acceptance as a cause of seborrheic dermatitis in dogs. In a recent report of 11 canine cases, lesions attributed to *Malassezia* were described as regional or generalized, erythematous, hyperpigmented, scaly, alopecic, and lichenified. Only antifungal therapy resolved the dermatitis. Histologically, there was severe hyperplastic superficial perivascular dermatitis with marked orthokeratotic and parakeratotic hyperkeratosis, irregular epidermal hyperplasia and spongiform pustules. Oval budding yeasts were found in the scale-crust, often in high numbers and in combination with bacterial colonies. Large numbers of yeasts were also demonstrable in cytological preparations. *Malassezia furfur,* the agent of tinea versicolor in humans, has been proposed as the cause of circular lesions of hypopigmentation on the teats and udders of goats.

Bibliography

Bliss, E. L. Tinea versicolor dermatomycosis in the goat. *J Am Vet Med Assoc* **184:** 1512–1514, 1984.

Dufait, R. *Pityrosporum canis* as a cause of canine chronic dermatitis. *Vet Med Small Anim Clin* **78:** 1055–1057, 1983.

Mason, K. V., and Evans, A. G. Dermatitis associated with *Malassezia pachydermatis* in 11 dogs. *J Am Anim Hosp Assoc* **27:** 13–20, 1991.

2. Dermatophytosis

Dermatophytosis results from a superficial infection of the keratinized layers of the skin and its appendages by a group of taxonomically related mycelial fungi which have in common the ability to parasitize keratinized tissues of the living host. The disease occurs worldwide and is common in humans and animals. The fungal species involved may show some regional variation, but in general, the clinically important species are cosmopolitan.

The dermatophytes belong to three genera, *Microsporum, Trichophyton,* and *Epidermophyton*. They are, however, commonly categorized on an ecological basis. The **anthropophilic** species, chiefly *Epidermophyton,* are adapted to humans, only rarely infect animals, and have lost their saprophytic properties. **Zoophilic** species principally affect animals. Some species, such as *Microsporum canis,* have become so well adapted to their animal host that inapparent carriers are the rule. Such animals are important sources of infection for humans, and dermatophytosis is an important zoonosis. **Geophilic** dermatophytes, such as *Microsporum gypseum,* normally reside in the soil but are capable of infecting animals and humans. Geophilic fungi belonging to these genera, but which lack the pathogenic ability to invade the keratinized structures of the living host, are not considered dermatophytes.

The dermatophytes do not invade living tissue but remain confined to the keratinized layers, which they attack with proteolytic enzymes having keratinolytic activity. In well-adapted hosts, the fungi induce little reaction. With a low level of host–parasite adaptation, an inflammatory and immunologic reaction is induced by fungal products permeating the underlying dermis. These products are chiefly extracellular proteinases such as keratinase, collagenase, and elastase. Part of the adaptation process may involve the secretion, by the host, of substances which inhibit the production of these fungal enzymes.

Dermatophytosis may be transmitted experimentally by rubbing infected hairs, cutaneous scales, or cultivated organisms onto abraded skin. Natural infection is by contact, which may be direct or indirect by exposure to contaminated objects such as grooming equipment or tack. The organisms are resistant in the environment for extended periods, but the reservoir of infection for the zoophilic species is the inapparent carrier animal. The initiation of infection requires alteration of the stratum corneum, either by slight trauma or by continued moisture and maceration. Some dermatophytes produce lipases, which assist penetration of the lipid coat of the epidermis. Young animals are more susceptible to dermatophyte infection than are adults. This may be due to age-related differences in the skin physiology, such as pH, or to lack of specific immunity, or to both. Severe dermatophyte infection in older animals usually signals an underlying systemic disease or reduced immunologic competence.

If infection is established, branching septate hyphae colonize the surface stratum corneum, the follicular infundibula, and the hair shafts. The boring hyphae penetrate the hair cuticle and tunnel extensively through the cortex. They behave as saprophytes in the telogen hairs but penetrate only to the upper limit of the zone of active keratinization in anagen hairs. The hyphae break up into round or oval arthrospores, either within the hair (endothrix) or on its external surface (ectothrix). In general, *Microsporum* infections are ectothrix and *Trichophyton* are both endo- and ectothrix.

The initial reaction to dermatophyte invasion has been likened to an irritant contact dermatitis, but one caused by substances like trichophytin, which are released by the fungus. This results in increased epidermopoiesis, which is reflected histologically by moderate to marked hyperplasia, often with rete ridge formation, and by ortho- and parakeratotic hyperkeratosis (Fig. 5.94). The surface epidermis and proximal external root sheath are both affected. It has been suggested that the increased epidermal turnover constitutes an important mechanism for the elimination of the fungus. The early invasion of keratinized structures is accompanied by infiltration of both dermis and epidermis with moderate numbers of mononuclear cells and neutrophils. As the lesions develop, more pronounced exocytosis of neutrophils into the fungus-laden cornified layer forms subcorneal and intracorneal microabscesses. Dermatophytes such as *T. mentagrophytes* may induce leukocyte attraction indirectly by induction of complement activation or through the elaboration of low-molecular-weight chemotactic factors. It is also likely that the damaged keratinocytes contribute directly by the production of chemotactic cytokines. While the neutrophil is usually the dominant granulocyte, in cattle and in dogs significant numbers of eosinophils may be seen, particularly in the later stages. **Folliculitis** and **pyogranulomatous furunculosis** often develop in severe dermatophyte infections in which host and parasite are poorly adapted (e.g., *T. terrestre* infection in dogs and cats). The hyperkeratosis of the pilar canal and follicular plugging common to most dermatophyte infections predisposes to secondary bacterial infection, which may be manifested as a suppurative folliculitis or pyogranulomatous furunculosis, or both. Not infrequently in dogs, **trichogranulomas** are caused by the implantation of dermatophyte-infected hairs into the dermis. A rare and unique form of dermatophytic infection,

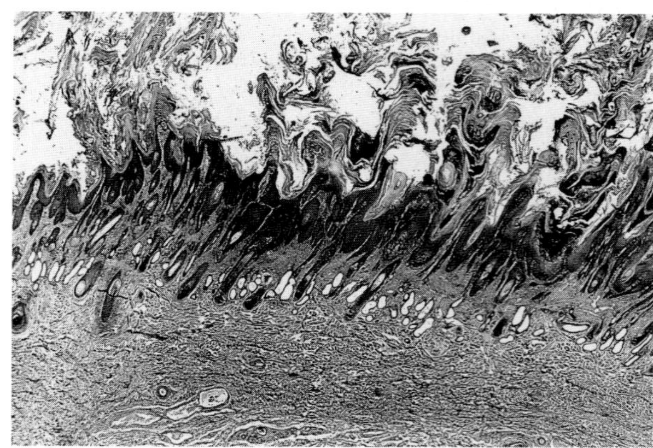

Fig. 5.94 Ringworm. Acanthosis and hyperkeratosis in an area of ringworm. Calf.

the **dermatophytic pseudomycetoma,** has been recognized, chiefly in Persian cats. Here, hyphae form tissue "grains" in the deep dermis and subcutaneous fat (Fig. 5.95A,B).

The gross appearance of the individual lesions of ringworm are variable, but their character is similar, consisting of different degrees of erythema, follicular papules, scaling, crusting, and alopecia. The alopecia results from breakage of the brittle, parasitized hair shafts. The classical ring lesion caused by central clearing and peripheral expansion is the exception rather than the rule in many infections. Erythematous, ulcerated nodules due to secondary furunculosis or trichogranulomas may be mistaken for neoplasms, such as the canine histiocytoma. Infection of the nail, onychomycosis, is manifested by brittle, cracked, or malformed nails.

Dermatophytosis is a self-limiting infection. Lesions take from a few weeks to several months to regress, depending on the species of fungus and degree of host adaptation. Reinfection is not common, but when it occurs, the lesions are smaller and have a shorter course. Resistance is better stimulated by zoophilic species, which tend to cause a more intense inflammatory reaction. The mechanisms of resistance and immunity are only poorly understood. Cell-mediated immunity is generally considered more important than antibody-mediated reactions. Nonspecific serum factors, such as transferrin, which limits availability of iron, and α_2-macroglobulin, which inhibits the proteolytic enzymes of the dermatophytes, may also contribute to resistance.

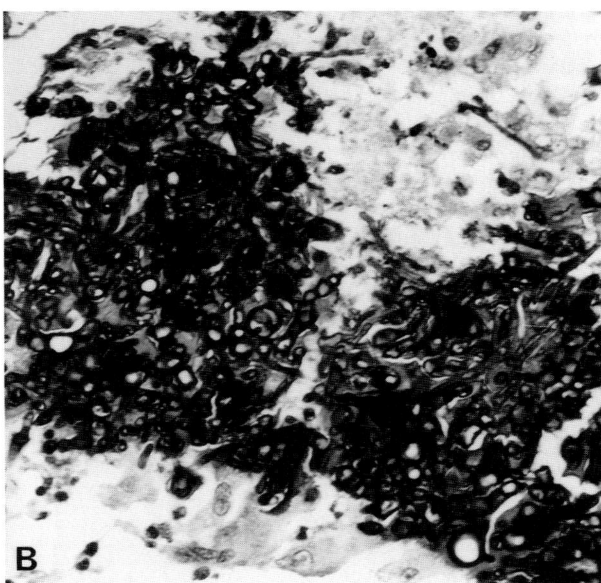

Fig. 5.95B *Microsporum canis* pseudomycetoma. Persian cat. High-power appearance of hyphae.

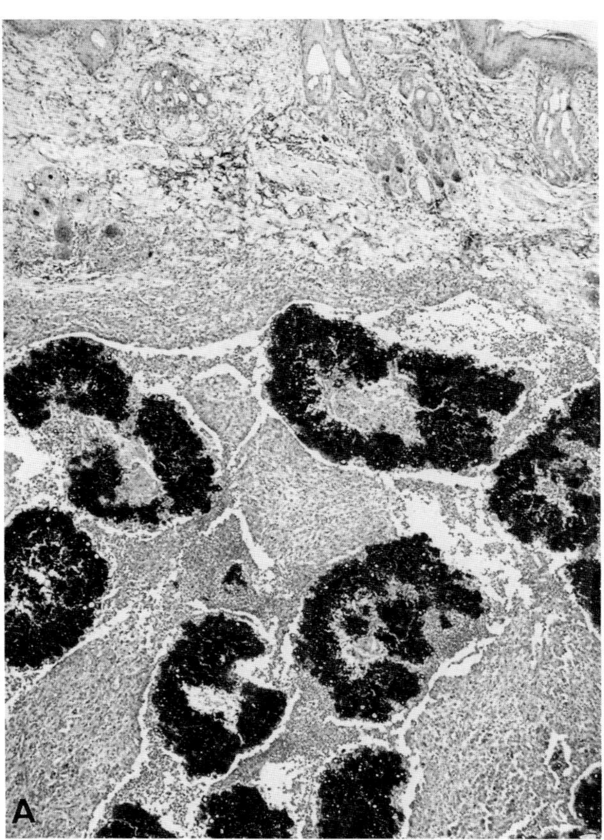

Fig. 5.95A *Microsporum canis* pseudomycetoma. Persian Cat. PAS stains the fungal hyphae in a granulomatous panniculitis.

Histologically, there is no single pattern of inflammation associated with dermatophyte infection. However, a high index of suspicion should be attached to lesions of folliculitis and furunculosis, and to lesions with marked ortho- and parakeratotic hyperkeratosis, particularly if associated with neutrophilic microabscesses. Generally, *Microsporum* infections are characterized by a mosaic of polygonal arthrospores in an ectothrix pattern, whereas *Trichophyton* infections are characterized by chains of round arthrospores in both endo- and ectothrix patterns. Sections showing trichogranulomas or pyogranulomatous dermatitis should be checked carefully for the presence of free-lying parasitized hair shafts.

Diagnosis of dermatophytosis is based on the history, clinical lesions, microscopic examination of skin scrapings and plucked hairs, ultraviolet light examination of hair, skin biopsy, and fungal culture. In some species, such as cattle, a presumptive diagnosis is possible on the basis of gross lesions. In other species, particularly the dog and cat, the protean manifestations of dermatophytosis necessitate the use of laboratory tests to establish the diagnosis. The Wood's lamp delivers ultraviolet light, which, when directed onto infected hairs, causes the metabolites of some dermatophytes to fluoresce. Although the test is simple and rapid, it is applicable only to *M. canis, M. audouinii, M. distortum,* and *T. schoenleinii.* Less than half of *M. canis* infections show positive fluorescence. False positives may also be induced by lint, white hairs, scale-crust, and various medicaments. The microscopic examination of plucked hairs and epithelial debris, after clearing with potassium hydroxide, reveals broken hairs, fungal arthrospores, and hyphae. Artefacts such as fat globules and the normal medullary pigment granules must be recognized. Examination of plucked hairs should be used in conjunction with culture techniques, as it is only 60–70% accurate alone. Histopathology is accurate in about 80% of cases. Both arthrospores and hyphae are readily demonstrated in tissue sections (Fig. 5.96A,B). While the fungal elements may be identified in hematoxylin and eosin-stained sections (pale blue arthrospores, clear hyphae in relief against the eosinophilic keratin background), they are more easily identified using special fungal stains such as periodic-acid Schiff, Gomori's methenamine silver, and acid orcein-Giemsa.

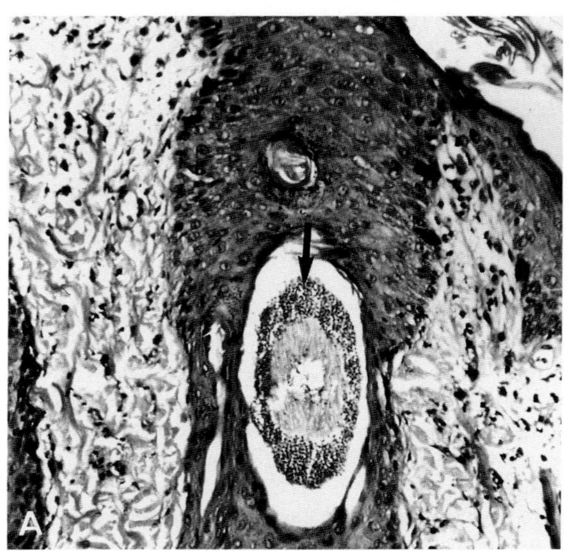

Fig. 5.96A Ringworm. Organisms (arrow) around hair shaft. Dog.

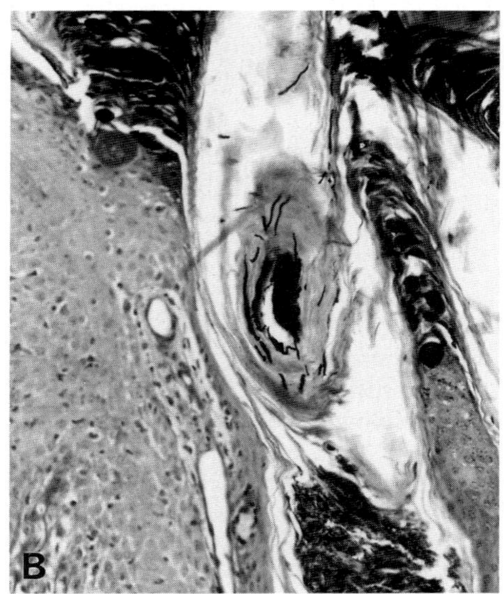

Fig. 5.96B Ringworm. Filamentous organisms in follicle. Calf. Hyperkeratosis and acanthosis. Silver impregnation.

Bibliography

Ackerman, A. B. Subtle clues to diagnosis by conventional microscopy. Neutrophils within the cornified layers as clues to infection by superficial fungi. *Am J Dermatopathol* **1:** 69–75, 1979.

Hutton, R. D., Kerbs, S., and Yee, K. Scanning electron microscopy of experimental *Trichophyton mentagrophytes* infections in guinea pig skin. *Infect Immun* **21:** 247–253, 1978.

Jones H. E. Cell-mediated immunity in the immunopathogenesis of dermatophytosis. *Acta Derm Venereol* (*Stockh*) (Suppl.) **121:** 73–83, 1986.

Kanbe, T., Suzuki, S., and Tanaka, K. Structural differentiation in the frond and boring hypha of the dermatophyte *Microsporum canis* invading human hair *in vitro*. *J Electron Microsc* **35:** 38–46, 1986.

Matsumoto, T., and Ajello, L. Current taxonomic concepts pertaining to the dermatophytes and related fungi. *Int J Dermatol* **26:** 491–499, 1987.

Otcenasek, M. Ecology of the dermatophytes. *Mycopathologia* **65:** 67–72, 1978.

Pepin, G. A., and Oxenham, M. Zoonotic dermatophytosis (ringworm). *Vet Rec* **118:** 110–111, 1986.

Philpot, C. M. Geographical distribution of the dermatophytes: A review. *J Hyg* (*Camb*) **80:** 301–313, 1978.

Sohnle, P. G. Dermatophytosis. *In* "Immunology of the Fungal Diseases" R. A. Cox, (ed.), pp. 1–27. Boca Raton, Florida, CRC Press, 1989.

Tagami, H. *et al.* Analysis of transepidermal leukocyte chemotaxis in experimental dermatophytosis in guinea pigs. *Arch Dermatol Res* **273:** 205–217, 1982.

Takiuchi, I. *et al.* Isolation of an extracellular proteinase (keratinase) from *Microsporum canis*. *Sabouraudia* **20:** 281–288, 1978.

Thomsett, L. R. The diagnosis of ringworm infection in small animals. *J Small Anim Pract* **18:** 803–814, 1977.

Wright, A. I., and Allingham, R. Diagnosing ringworm. *Vet Rec* **98:** 411–412, 1976.

a. Equine Dermatophytosis Dermatophytosis or girth itch is a common disease in most areas of the world where horses are raised. It is particularly prevalent in large establishments where tack, blankets, and grooming equipment are communal. The infection predominantly occurs in young horses (75% are younger than 4 years). It is more common during hot and humid weather in subtropical and tropical regions, and in winter in colder climates (closed housing, crowding).

Trichophyton equinum is the most common pathogen in the horse. Others, in decreasing order of prevalence, include *M. equinum, M. canis, T. mentagrophytes, T. verrucosum*, and *M. gypseum*. Most horses, particularly those in training, first develop lesions on the girth or saddle area. This is related to the abrasive action of the tack. Multiple lesions spread over the shoulder, neck, back, and loins. *Microsporum gypseum,* a geophilic fungus, is associated with lesions on the thorax, abdomen, and legs.

The most typical lesions commence as small papules, which develop into plaques covered with erect hairs matted together with a small amount of exudate. The infected hairs break or are easily epilated to leave a small patch of alopecia 1–4 cm in diameter, covered by gray, powdery scale-crust. The lesions expand peripherally and may become confluent, producing a moth-eaten appearance. Lesions caused by *M. gypseum* and *M. canis* are smaller, usually less than 1 cm and are less scaly than those of *T. equinum*. Other manifestations include generalized scaling without alopecia, folliculitis and furunculosis, urticarial swellings, and lesions restricted to the caudal aspect of the pastern; the latter are erroneously diagnosed as grease heel (see Section XII, Bacterial Diseases of the Skin).

b. Bovine Dermatophytosis Ringworm is very common in cattle and *Trichophyton verrucosum* is the most frequent isolate. Others include *T. mentagrophytes, T. equinum, M. gypseum, M. nanum,* and *M. canis*. The disease is most prevalent in young, housed animals. A higher incidence is associated with larger groups of cattle and the winter season. The sites of predilection are the head, particularly the periocular region in calves; neck, particularly the intermaxillary space and dewlap in bulls; and thorax and limbs in cows and heifers. In young, poorly nourished animals, the lesions may spare only the limbs. Variations in lesion distribution in animals of different age and sex have been correlated with behavioral patterns. Individual lesions are well circumscribed, alopecic, roughly circular, 2–6 cm in diameter, and covered with a very thick layer of gray-white scale-crust (Fig. 5.97A). The lesions caused by *T. mentagrophytes* are smaller and less crusted than those caused by *T. verrucosum*. Vaccination for dermatophytosis is used effectively in Norway, where the disease is notifiable.

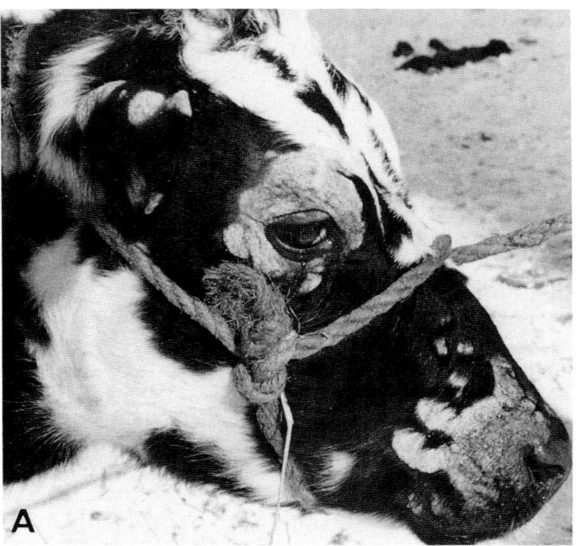

Fig. 5.97A Ringworm. Calf. Gray-white, rough, hairless lesion on ear, muzzle, and periorbital skin.

c. Caprine and Ovine Dermatophytosis In both sheep and goats, dermatophytosis is generally rare, despite the finding of a wide range of keratophilic fungi, including some pathogenic species, on the hooves, horns, and fleece of normal animals. Those reports of sporadic outbreaks tend to come from Africa, Asia, and eastern Europe. The agent involved is usually *T. verrucosum*. Lesions tend to affect the hairy rather than wooled areas, particularly the face, and are typically alopecic and crusted. However, in an Australian flock which developed an endemic infection with *M. canis,* the lesions affected the wooled areas, particularly the rump and flanks. The staples were matted with a brown exudate, and the underlying skin was erythematous, crusted, and scaly. Histologically, the wool fibers were heavily infested with the fungus. In goats, lesions affect mainly the ears, head, and neck, but all areas of the body may be involved. The lesions resemble those of cattle.

d. Porcine Dermatophytosis The disease is uncommon in pigs. The lesions may be confused with those of juvenile pustular psoriasiform dermatitis (pityriasis rosea; see Miscellaneous Skin Diseases). Ringworm has a predilection for the back of the ears, the dorsolateral neck and back, but may occur anywhere. The lesions are roughly circular, erythematous patches or plaques up to 10 cm diameter, with a slightly raised border. There is moderate scaling but no obvious loss of hair. Healing is slow, taking up to 6 months. The most common pathogen is *M. nanum*.

e. Canine Dermatophytosis The widespread belief that the disease is common may have arisen through incorrect interpretation of ring-shaped lesions of various types, improper interpretation of ultraviolet fluorescence of hair, and false-positive cultures using dermatophyte test medium. The condition chiefly affects young dogs; in older animals, it is often associated with debilitating disease, lack of immune competence, or infection with dermatophytes from wild mammals, so-called sylvatic ringworm. The most frequently isolated species is *M. canis,* followed by *M. gypseum* and *T. mentagrophytes*. The exact prevalence of individual species varies with geographic area. There are occasional reports of infection with species such as *T. terrestre* and *T. rubrum;* multiple infections occur rarely. *Trichophyton mentagrophytes* transmitted from mice and voles and *T. erinacei* transmitted from hedgehogs are the most frequent causes of sylvatic ringworm in Britain. Sylvatic ringworm is more prevalent in hunting breeds, such as the Jack Russell terrier.

The lesions are extremely variable with few conforming to the classic type. They are usually multiple and may become generalized, as is typically the case in sylvatic

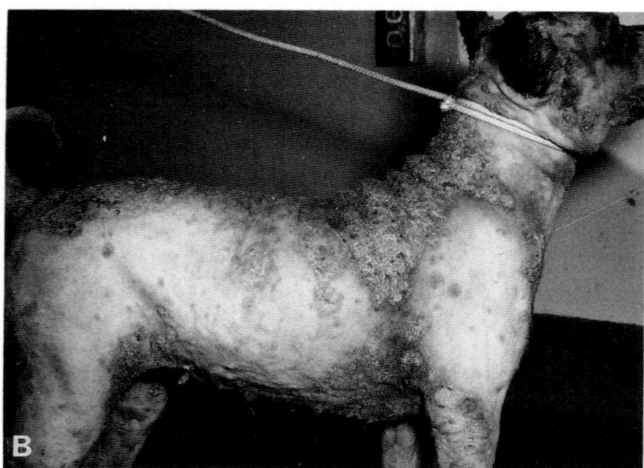

Fig. 5.97B Ringworm. Thickening and corrugation of skin. Dog.

ringworm (Fig. 5.97B). Lesions comprise variously sized and shaped areas of alopecia with scaling, crusting, and variable erythema. Dermatophytosis may also take the form of folliculitis and furunculosis, including lesions resembling nasal pyoderma and pododermatitis. Occasionally the traumatic implantation of fragments of dermatophyte-infected hair evokes a pyogranulomatous dermal reaction characterized by pruritus, alopecic nodules, or plaques, which must be differentiated from neoplasia by microscopic evaluation. Dermatophytosis caused by *M. canis* is usually self-limiting; sylvatic ringworm, by contrast, may persist for many years and is difficult to treat.

f. Feline Dermatophytosis Dermatophytosis is a common infection in the cat, caused almost exclusively by *Microsporum canis*. The cat is the natural host for *M. canis* and acts as a source of infection for other animals and humans. The level of parasitic adaptation is reflected by the high percentage of asymptomatic carrier cats. *Microsporum gypseum* and *T. mentagrophytes* are the next most frequent causes of feline dermatophytosis; other fungi such as the geophilic *M. terrestre* are occasional pathogens. Seasonal and geographic variability in the prevalence of *M. canis* infection of cats has been reported in the United States. The disease tends to occur more frequently in hot and humid climates and in the summer and autumn.

The lesions of feline dermatophytosis are extremely variable. They are often localized, but may be multifocal or generalized. They commonly commence on the head and limbs. The lesions vary from small areas of alopecia or simply broken hairs to nodular, ulcerated, and fistulous lesions resembling human "kerion." The range of manifestations includes the "classic" ringworm lesion, localized or generalized papular crusts, and/or vesicular dermatitis with or without alopecia, folliculitis, paronychia, and multifocal or generalized seborrhea sicca.

Dermatophytic pseudomycetoma is a rare manifestation of *M. canis* infection. The fact that it occurs almost exclusively in the Persian cat raises the possibility of a genetic deficit in the specific or nonspecific cutaneous defenses of these animals. Gross lesions are nodular or tumorlike with draining fistulous tracts. The pathognomonic histologic lesion is the presence, deep in the dermis and subcutis, of aggregates of compact mycelia within a granulomatous tissue reaction (Fig. 5.95). The fungi may be surrounded by amorphous eosinophilic material representing the Splendore–Hoeppli reaction. The overlying skin may show parasitized hair shafts in the absence of inflammation or the more typical folliculitis. Fungi in the superficial lesions have normal morphology. Invasion of hyphae through the external root sheath has been described. The apparent growth of dermatophytic fungi in viable tissue has led to some controversy as to the pathogenesis of this lesion.

Bibliography

Abdel-Hafez, A. I. I., Moharram, A. M., and Abdel-Gawad, K. M. Survey of keratinophilic and saprobic fungi in the cloven hooves and horns of goats and sheep from Egypt. *J Basic Microbiol* **30:** 13–20, 1990.

Ajello, L., Kaplan, W., and Chandler, F. W. Dermatophyte mycetomas: Fact or fiction? *In* "Proceedings of the Fifth International Conference on the Mycoses," pp.135–140, Publication No. 396, Washington, D.C., Pan American Health Organization, 1980.

Ali-Shtayeh, M. S. *et al.* Keratinophilic fungi on sheep hairs from the west bank of Jordan. *Mycopathologia* **106:** 95–101, 1989.

Bourdin, M. *et al.* Premiére observation d'un mycétome à *Microsporum canis* chez un chat. *Rec Méd Vét* **151:** 475–480, 1975.

Bubash, G. R., Ginther, O. J., and Ajello, L. *Microsporum nanum:* First recorded isolation from animals in the United States. *Science* **143:** 366–367, 1964.

Chineme, C. N., Adekeye, J. O., and Bide, S. A. Ringworm caused by *Trichophyton verrucosum* in young goats. A case report. *Bull Anim Health Prod Afr* **29:** 75–78, 1981.

Connole, M. D., and Baynes, I. D. Ringworm caused by *Microsporum nanum* in pigs in Queensland. *Aust Vet J* **42:** 19–24, 1966.

Edwardson, J. An outbreak of ringworm in a group of young cattle. *Vet Rec* **104:** 474–477, 1979.

Gudding, R., and Naess, B. Vaccination of cattle against ringworm caused by *Trichophyton verrucosum*. *Am J Vet Res* **47:** 2415–2417, 1986.

Holfeld, N. W., Schiefer, B., and Boycott, B. R. Granulomatous dermatitis in carnivores associated with dermatophyte infection. *Can Vet J* **21:** 103–105, 1980.

Jackson, R. B., Peel, B. F., and Donaldson-Wood, C. Endemic *Microsporum canis* infection in a sheep flock. *Aust Vet J* **68:** 122, 1991.

Kane, J., Padhye, A. A., and Ajello, L. *Microsporum equinum* in North America. *J Clin Microbiol* **16:** 943–947, 1982.

Lepper, A. W. D. Experimental bovine *Trichophyton verrucosum* infection. The cellular responses in primary lesions of the skin resulting from surface or intradermal inoculation. *Res Vet Sci* **16:** 287–298, 1974.

Medleau, L., and White-Weithers, N. E. Dermatophytosis in cats. *Compend Cont Ed* **13:** 557–560, 1991.

Miller, W. H., and Goldschmidt, M. H. Mycetomas in the cat caused by a dermatophyte: A case report. *J Am Anim Hosp Assoc* **22:** 255–260, 1986.

Pandey, V. S., and Mahin, L. Observations on ringworm in goats caused by *Trichophyton verrucosum*. *Br Vet J* **136:** 198–199, 1980.

Pascoe, R. R. The epidemiology of ringworm in racehorses caused by *Trichophyton equinum* var. *autotrophicum*. *Aust Vet J* **55:** 403–407, 1979.

Power, S. B., and Malone, A. An outbreak of ringworm in sheep in Ireland caused by *Trichophyton verrucosum*. *Vet Rec* **121:** 218–220, 1987.

Scott, D. W., Kirk, R. W., and Bentinck-Smith, J. Dermatophytosis due to *Trichophyton terrestre* infection in a dog and cat. *J Am Anim Hosp Assoc* **16:** 53–59, 1980.

Stankushev, K., and Duparinova, M. Carriers of dermatophytes and *Candida* species amongst pigs. *Vet Med Nauki* **13:** 75–80, 1976.

Wilkinson, G. T. Multiple dermatophyte infections in a dog. *J Small Anim Pract* **20:** 111–115, 1979.

Wright, A. I. Ringworm in dogs and cats. *J Small Anim Pract* **30:** 242–249, 1989.

Yager, J. A. *et al.* Mycetoma-like granuloma in a cat caused by *Microsporum canis*. *J Comp Pathol* **96:** 171–176, 1986.

B. Subcutaneous Mycoses

The subcutaneous mycoses constitute a heterogeneous group of diseases in which the traumatic implantation of the fungal agent induces slowly progressive dermal or subcutaneous lesions. The involvement of contiguous bone may lead to osteomyelitis. In some diseases, for example, sporotrichosis, there is extension along the lymphatic chain, but only very rarely does a subcutaneous mycosis become systemic.

The subcutaneous mycoses include eucomycotic mycetoma, sporotrichosis, subcutaneous phaeohyphomycosis, subcutaneous zygomycosis, and pythiosis. Rhinosporidiosis, an intermediate mycosis of the nasal mucosa, is discussed in Volume 2, Chapter 6, The Respiratory System. With the exception of the mycetomas, these groups are defined on the basis of fungal classification or characteristics.

1. Eumycotic Mycetoma

Mycetoma is a clinical syndrome characterized by the triad of tumefaction, draining sinuses, and the presence of macroscopic granules in the sinus exudate. These granules, also known as grains, vary in size, shape, texture, and color according to the species of microorganism causing the mycetoma. Grains are composed of compact mycelial filaments, sometimes in a matrix of amorphous "cement." The causative organisms are a diverse group including filamentous bacteria (actinomycotic mycetoma) and fungi (eumycotic mycetoma). The actinomycotic mycetomas have granules containing mycelial filaments less than 1 μm in diameter, whereas the eumycotic mycetoma have granules containing septate, often bizarre mycelia greater than 2 μm in diameter and sometimes large numbers of chlamydospores.

Authenticated eumycotic mycetomas are rare in animals. Most reports are from the United States, with a few from Germany, South Africa, and Australia. The causative fungi are saprophytic organisms; disease is opportunistic, requiring cutaneous injury. *Curvularia geniculata* is the most commonly implicated fungus, followed by *Pseudoallescheria boydii*. Horses and dogs are the species chiefly affected. Some reported cases, particularly those attributed to *Exserohilum rostratum* (*Dreschlera rostrata*) and *Bipolaris* (*Dreschlera*) *spicifera*, should be reclassified as phaeohyphomycosis, because lesions lack the typical granules.

The macroscopic lesions are nodules which slowly enlarge into tumorlike growths. Fistulous tracts develop and discharge serosanguinous or thin purulent fluid containing the characteristic granules. The cut surface shows dense gray-white connective tissue perforated by necrotic sinuses containing the fungal granules. These granules may also be embedded in the solid tissue. The granules are white, red, yellow, brown, or black, soft or hard in texture, and range in size from 2.5 μm to 2.0 mm. The black or brown granulated mycetomas, known as black-grain mycetomas, are caused by pigmented fungi such as *Curvularia* spp. The lesions are usually single but may be multiple. Equine lesions involve the external nares, commissure of the mouth, neck, shoulder, girth, and proximal extremities. Canine lesions occur mostly on the feet, but abdominal lesions have been reported following surgery or intestinal perforation by plant material. Disseminated disease may develop from subcutaneous or visceral lesions.

In microscopic section, the mycetoma granules are scattered through a densely fibrotic, chronically inflamed connective tissue stroma. The granules lie in the center of small abscesses, representing cross sections of purulent sinus tracts. They are occasionally surrounded by epithelioid macrophages. The granules are composed of an organized mass of fungal hyphae, with or without chlamydospores. The mycelium, depending on the fungal species, may be embedded in lightly eosinophilic material known as cement. The granules may be surrounded by amorphous eosinophilic material with club-shaped projections. This substance probably represents deposition of insoluble antigen–antibody complexes and is known as the Splendore–Hoeppli reaction. In *Curvularia* mycetoma, the black-walled fungus forms a partial rim around the edge of the granule, taking up a scroll- or C-shaped configuration. The abscesses are enclosed by a zone of epithelioid macrophages and mature collagen or vascular granulation tissue, depending on the age of the lesion. The stroma is heavily infiltrated with macrophages, multinucleate giant cells, and smaller numbers of plasma cells and lymphocytes.

Diagnosis of mycetoma depends on demonstration of the typical granules within the appropriate clinical lesion. The physical characteristics of the granules, which are specific to a fungus or group of fungi, may suggest a tentative diagnosis. However, attribution of the causative species usually requires culture or, in tissue sections, identification by immunofluorescence techniques.

Bibliography

Allison, N. *et al.* Eumycotic mycetoma caused by *Pseudallescheria boydii* in a dog. *J Am Vet Med Assoc* **194:** 797–799, 1989.

Baszler, T. *et al.* Disseminated pseudallescheriasis in a dog. *Vet Pathol* **25:** 95–97, 1988.

Boomker, J., Coetzer, J. A. W., and Scott, D. B. Black grain mycetoma (maduromycosis) in horses. *Onderstepoort J Vet Res* **44:** 249–252, 1977.

Bridges, C. H. Maduromycotic mycetomas in animals. *Curvularia geniculata* as an etiologic agent. *Am J Pathol* **33:** 411–427, 1957.

Brodey, R. S. *et al.* Mycetoma in a dog. *J Am Vet Med Assoc* **151:** 442–451, 1967.

2. Subcutaneous Phaeohyphomycosis

The distinguishing feature of this group of subcutaneous mycoses is that they are caused by various species of **dematiaceous,** or pigmented, **fungi.** The fungal elements in tissue are septate, dark-colored hyphae, which are dispersed or occur in small groups. If the fungal mycelium forms a macroscopic granule, the lesion is classified as a mycetoma. Some species of dematiaceous fungi occasionally induce mycetoma.

Subcutaneous phaeohyphomycosis occurs sporadically in **cats, horses,** and **cattle.** Rare cases are described in **dogs.** The taxonomy of these pigmented fungi and the terminology of the diseases they cause appear to be in a continual state of flux. Fungal species implicated in animal disease include *Exophiala jeanselmei* (*Phialophora jeanselmei, P. gougerotii*), *Exophiala spinifera, Phialophora verrucosa, Bipolaris* (*Drechslera, Helminthosporum*) *spicifera, Moniella suaveolens, Cladosporium* sp., *Wangiella* (*Hormodendrum*) sp., *Pseudomicrodochium suttonii, Curvularia geniculata, C. lunata,* and *Exserohilum rostratum* (*Drechslera rostrata*). The cutaneous lesions in cattle may be in conjunction with mycotic nasal granulomas. *Peyronella glomerata* was isolated from superficial hyperkeratotic lesions on the ear pinnae of wild goats in the United Kingdom.

The gross lesions are single or multiple subcutaneous nodules which sometimes ulcerate or develop sinus tracts. There are no macroscopic granules. Lesions tend to occur on the face or the distal extremities in the cat, but in the horse, lesions are more often on the body. Systemic infections occasionally occur, but they do not usually spread from cutaneous lesions. The route of exposure is probably respiratory, and the animal is usually immunologically compromised or debilitated.

The histologic lesions of phaeohyphomycosis are characterized by the presence of yellow-brown fungal hyphae in the tissues. These are best seen in unstained sections. Fungal stains, while delineating the mycelia, mask the pigment. The septate hyphae are irregular in width, length, and branching pattern. Constrictions of thick septa produce dilated segments up to 25 μm in diameter, which resemble chlamydospores. True chlamydospores are occasionally present. The hyphae are scattered in cystic cavities or in microabscesses and are occasionally surrounded by immune precipitates as in club colonies. The microabscesses are surrounded by a zone of granulation tissue infiltrated with epithelioid macrophages and multinucleate, histiocytic giant cells which often contain fungal elements. In well-established lesions, dense bands of mature collagen separate the pyogranulomas.

The diagnosis of phaeohyphomycosis is made by demonstrating the pigmented fungi. Not all species, however, have well-pigmented hyphae and detection in hemotoxylin and eosin-stained sections may be difficult. Fontana–Masson silver stain for melanin may aid in detecting lightly pigmented strains. Specific fungal identification requires culture.

Bibliography

Beale, K. M., and Pinson, D. Phaeohyphomycosis caused by two different species of *Curvularia* in two animals from the same household. *J Am Anim Hosp Assoc* **26:** 67–70, 1990.

Kaplan, W. *et al.* Equine phaeohyphomycosis caused by *Drechslera spicifera. Can Vet J* **16:** 205–208, 1975.

Kettlewell, P., McGinnis, M. R., and Wilkinson, G. T. Phaeohyphomycosis caused by *Exophiala spinifera* in two cats. *J Med Vet Mycol* **27:** 257–264, 1989.

Kwochka, K. W. *et al.* Canine phaeohyphomycosis caused by *Drechslera spicifera:* A case report and literature review. *J Am Anim Hosp Assoc* **20:** 625–633, 1984.

McGinnis, M. R. Chromoblastomycosis and phaeohyphomycosis: New concepts, diagnosis medical therapy. *J Am Acad Dermatol* **8:** 1–16, 1983.

McGinnis M. R., Rinaldi, M. G., and Winn, R. E. Emerging agents of phaeohyphomycosis: Pathogenic species of *Bipolaris* and *Exserohilum. J Clin Microbiol* **24:** 250–259, 1986.

Patton, C. S. *Helminthosporium speciferum* as the cause of dermal and nasal maduromycosis in a cow. *Cornell Vet* **67:** 236–244, 1977.

3. Sporotrichosis

Sporotrichosis is a chronic disease, primarily of skin and subcutaneous tissues, caused by the dimorphic fungus, *Sporothrix schenckii.* The organism is saprophytic, occurring in soil and in decaying plant material. Infection usually follows contamination of cutaneous wounds but may follow inhalation. Transmission from cats to humans is an important exception, in which direct contact may confer contagion. Systemic disease occurs rarely. In species other than the cat, usually only debilitated or immunosuppressed subjects are susceptible.

Sporotrichosis occurs in horses, mules, cats, dogs, cattle, and buffalo. The disease in animals is uncommon but is important because of potential transmission to humans. In experimental infections of cats, the organism may become sequestered in apparently recovered animals but is reactivated on administration of corticosteroids. The fungus is present in internal organs and colonic feces as well as in cutaneous lesions.

The gross lesions in horses tend to occur on the distal extremities. In dogs and cats, primary lesions often occur

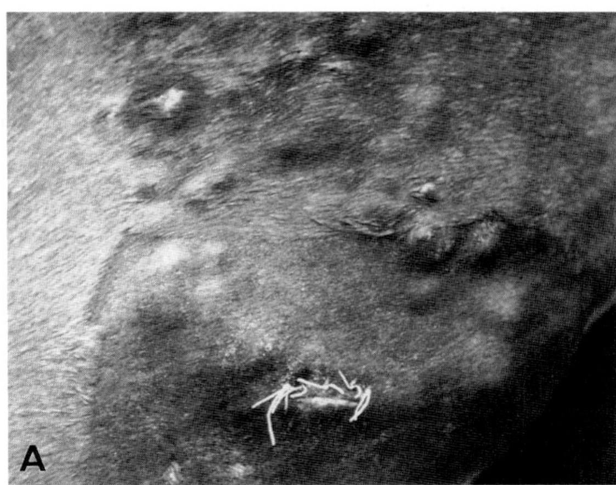

Fig. 5.98A Sporotrichosis. Nodular dermal lesions. Horse. (Courtesy of P. A. Taylor.)

on the head or the distal extremities and spread to other body sites. There are three clinical syndromes associated with sporotrichosis. The mildest form is cutaneous. It is characterized by the formation of single or multiple subcutaneous nodules 1–5 cm in diameter, which tend to ulcerate, discharging a small amount of thick, red-brown purulent exudate. Cavitation may expose underlying tendons and bones. The lesions have a chronic course. An unusual manifestation described only in the dog is otitis externa. The cutaneous–lymphatic form, the most common in equine infections, is characterized by the formation of multiple nodules, arranged in lines along the course of lymphatics, which become thick and corded (Fig. 5.98A). Ulceration of the nodules leaves deep, crateriform lesions which heal slowly. The systemic disease has disseminated lesions occurring in a wide range of tissues. Cats are most prone to develop systemic disease; horses appear resistant. Immunosuppression does not appear to be a prerequisite in the cat. Experimental cutaneous infections in normal cats showed a 50% rate of systemic spread.

The microscopic lesion of sporotrichosis is a diffuse and/or nodular pyogranulomatous dermatitis and panniculitis (Fig. 5.98B). Neutrophilic microabscesses lie in sheets of epithelioid macrophages, with scattered histiocytic, multinucleate giant cells, lymphocytes, and plasma cells. There is usually marked fibrosis around the granulomatous infiltrate. The epidermis may be ulcerated or, if intact, hyperplastic. The fungi grow in tissues in yeast form. In the lesions of dogs and horses, organisms may be extremely sparse and sometimes are demonstrated only by fluorescent-antibody techniques or by culture. Typically, cutaneous lesions in cats contain large numbers of organisms, a feature more usually associated with the disseminated form of sporotrichosis.

The yeasts are round, oval, or cigar-shaped, single or budding cells, 2–6 μm in diameter. The cigar-shaped forms, considered classical for sporotrichosis, may not be found regularly in tissues. The organism may be intracellular following engulfment by phagocytes, or extracellular, often in the center of microabscesses. Eosinophilic "asteroid" bodies may be present. Once thought to be specific

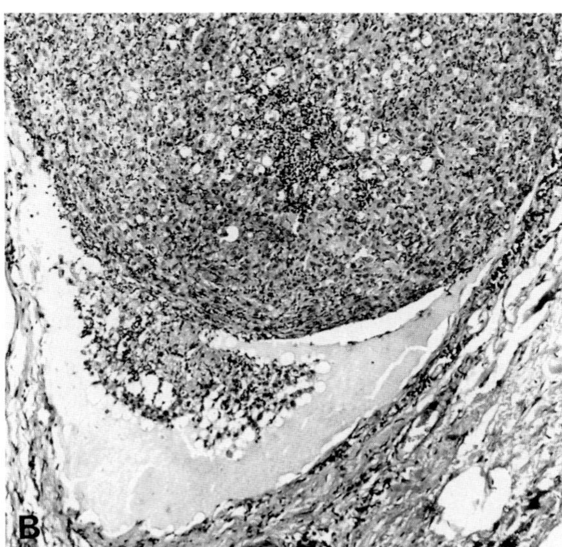

Fig. 5.98B Sporotrichosis. Granuloma with central neutrophil focus. The reaction is extending into lymphatic. (Courtesy of P. A. Taylor.)

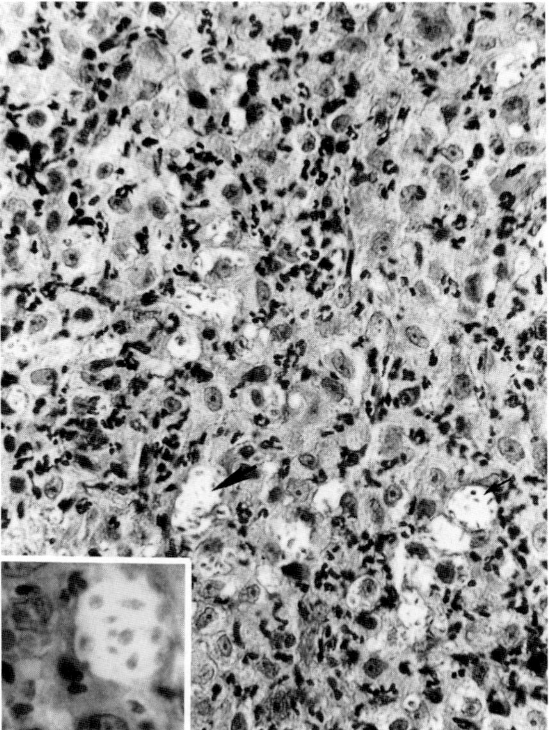

Fig. 5.99 Sporotrichosis. Detail of inflammatory reaction. Mingled neutrophils and histiocytes with organisms in vacuoles (arrow and inset). (Courtesy of P. A. Taylor.)

for sporotrichosis, they simply represent another manifestation of the Splendore–Hoeppli antigen–antibody reaction.

The diagnosis of sporotrichosis in the dog and horse may be made difficult by the paucity of organisms in the lesions. Since the yeast cells stain poorly with hematoxylin and eosin (Fig. 5.99), Gomori's methenamine silver may be required, and serial sections are often necessary. Indirect immunoperoxidase or direct fluorescent-antibody techniques using antiserum against *Sporothrix schenckii* may facilitate diagnosis in sections. Cytologic assessment of impression smears from exudate may be helpful in making a presumptive diagnosis, particularly in feline cases, which typically contain numerous organisms. Intraperitoneal or intrascrotal inoculation of mice with infective material induces a granulomatous inflammatory reaction containing abundant organisms. The most reliable method of diagnosis, however, is fungal isolation and identification.

Bibliography

Dion, W. M., and Speckmann, G. Canine otitis externa caused by the fungus *Sporothrix schenkii*. *Can Vet J* **19:** 44–45, 1978.

Dunstan, R. W. *et al.* Feline sporotrichosis: A report of five cases with transmission to humans. *J Am Acad Dermatol* **15:** 37–45, 1986.

Fishburn, F., and Kelly, D. C. Sporotrichosis in a horse. *J Am Vet Med Assoc* **151:** 45–46, 1967.

Humphreys, F. A., and Helmer, D. E. Pulmonary sporotrichosis in a cattle beast. *Can J Comp Med* **7:** 199–204, 1943.

Lurie, H. I. Histopathology of sporotrichosis: Notes on the nature of the asteroid body. *Arch Pathol* **75:** 421–437, 1963.

Macdonald, E., Ewert, A., and Reitmeyer, J. C. Reappearance of *Sporothrix schenckii* lesions after administration of Solu-Medrol to infected cats. *Sabouraudia* **18:** 295–300, 1980.

Scott, D. W., Bentinck-Smith, J., and Hagerty, G. F. Sporotrichosis in three dogs. *Cornell Vet* **64:** 416–426, 1974.

Travassos, L. R., and Lloyd, K. O. *Sporothrix schenckii* and related species of *Ceratocystis*. *Microbiol Rev* **44:** 683–721, 1980.

4. Zygomycosis and Pythiosis

Zygomycosis and pythiosis (oomycosis) constitute a diverse group of mycotic diseases, formerly known as **phycomycosis,** a term with no taxonomic equivalent. Zygomycosis is caused by members of the class Zygomycetes, chiefly by species in the orders Entomophthorales and Mucorales. Zygomycosis encompasses mucormycosis (see Volume 2, Chapter 2, The Alimentary System), and entomophthoromycosis. In the latter group, *Conidiobolus coronatus* and *Conidiobolus lamprauges* cause granulomatous lesions in the nasal mucosa and surrounding skin of humans and horses. *Basidiobolus haptosporus* is associated with subcutaneous disease in horses.

Pythiosis encompasses those diseases caused by organisms originally designated *Hyphomyces destruens* but now known as *Pythium* spp. Demonstration of a sporangial structure from which biflagellate zoospores are formed has caused the organisms to be reclassified from the Fungi to the Protista, where they are placed in the phylum Oomycetes. These diseases have in common the presence in tissue of broad, irregular, and rarely septate hyphae.

Cutaneous lesions caused by zygomycetes and oomycetes occur chiefly in the **horse** and rarely in the **dog** and **cow.** Two organisms have been associated most frequently with a disease known as equine phycomycosis, Florida horse leeches, bursattee, or swamp cancer, namely *Basidiobolus haptosporus,* a member of the Entomophthales, and *a Pythium* sp. identified as *Pythium insidiosum.* Since the lesions caused by the zygomycetes and by *Pythium insidiosum* in the horse are very similar, they will be discussed together.

a. Equine Pythiosis and Subcutaneous Zygomycosis Occurrence is chiefly in tropical areas, particularly Australia, India, Indonesia, South America, Florida, and Texas. In northern Australia, *Pythium insidiosum* is the most common cause, accounting for about four times as many cases as *Basidiobolus haptosporus.*

The organisms gain entry through cutaneous wounds. The habitat of the causative organism is thought to influence lesion distribution. *Pythium insidiosum,* which has an aquatic habitat, causes lesions chiefly on the lower limbs but also on the ventral abdomen and chest. Horses are thought to acquire the infection when standing in swampy areas for extended periods. *Basidiobolus haptosporus* is a soil and vegetation saprophyte. Lesions occur on the lateral head, chest, and neck. Infection is probably acquired when horses roll or lie on contaminated ground. The equine disease is associated with intense pruritus.

The macroscopic lesions are chronic, rapidly progressive, subcutaneous nodules or tumorous masses, which may reach 30 cm in diameter. The surface is alopecic, scarred, and ulcerated, and receives the openings of multiple draining sinuses (Fig. 5.100A,B). The cut surface is dense, cream-colored, fibrous tissue coursed by irregular

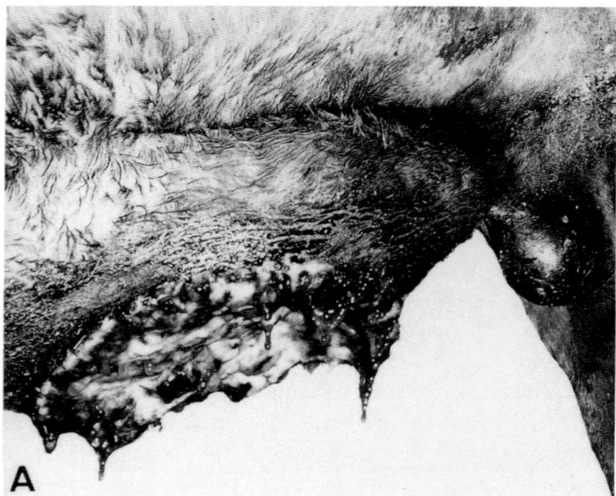

Fig. 5.100A Equine phycomycosis. Ulcerating, draining lesion on ventral abdominal skin.

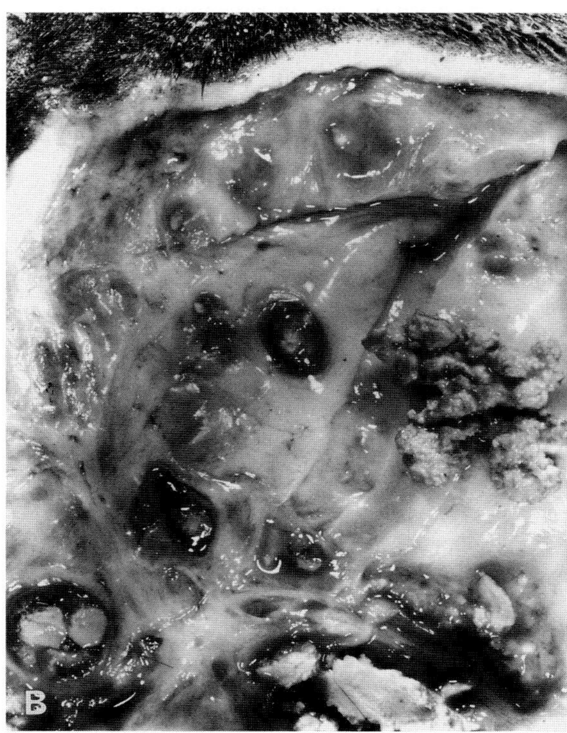

Fig. 5.100B Equine phycomycosis. Margin of ulcerating lesion showing granulation tissue and multiple light-colored necrotic cores (kunkers).

channels filled with a core of gray, white, or cream-colored necrotic coagula known as leeches in the United States and kunkers in Australia. In pythiosis, the coagula may be sequestered and, if removed, resemble a piece of branching coral. The lesions of basidiobolomycosis expand less rapidly and are generally smaller. Coagula are also smaller and branch rarely. A discontinuous band of coagulum, at the interface of superficial edematous connective tissue and deeper, fibrous tissue, marks the level of fungal invasion in basidiobolomycosis. Spread to local lymph nodes and to internal organs, such as the lung, has been reported in pythiosis on rare occasions.

Microscopically, the coagula are composed of necrotic tissue and cell debris, derived partially from degenerating eosinophils. In early lesions, outlines of collagen bundles and small blood vessels may remain. The presence of hyphae in the walls of these necrotic blood vessels has prompted the suggestion that ischemia may contribute to the necrosis. The necrotic core is surrounded by a zone of eosinophils and neutrophils blending into broad bands of dense well-vascularized connective tissue (more prominent in pythiosis), infiltrated with macrophages (Fig. 5.101A), multinucleate, histiocytic giant cells (numerous in basidiobolomycosis and rare in pythiosis), neutrophils, and eosinophils.

The causative fungi grow in tissue as very broad, irregular hyphae, which stain poorly with routine stains. The hyphae occur within the coagula; in pythiosis, hyphae are typically peripheral (Fig. 5.101B). The hyphae of *Basidiobolus*, when stained with Gomori's methenamine silver, vary from 5–20 μm, whereas those of *Pythium* are narrower, at 2 to 6 μm. The hyphal walls of *Pythium* are much thicker than those of *Basidiobolus*. Furthermore, individual hyphae of *Basidiobolus haptosporus* may be enveloped in eosinophilic material, which represents the Splendore–Hoeppli reaction.

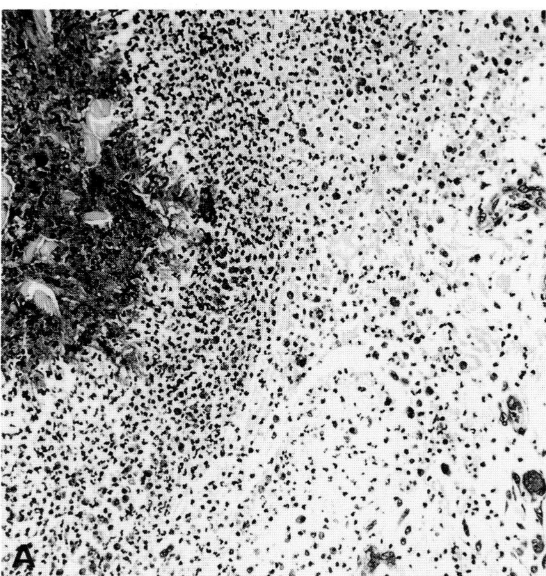

Fig. 5.101A Equine phycomycosis. Section through edge of necrotic core (above, left).

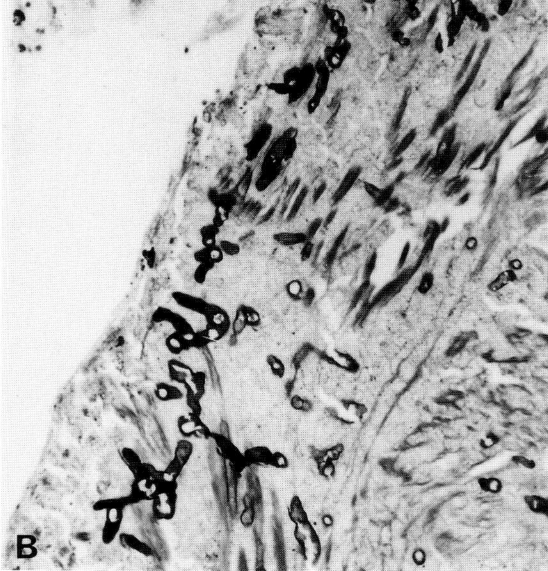

Fig. 5.101B Equine phycomycosis. Fungal hyphae in edge of lesion.

Diagnosis of equine pythiosis is suggested by the typical gross lesions, particularly if large kunkers occur, but must be differentiated from cutaneous habronemiasis. In cutaneous habronemiasis, the foci of necrosis are yellow, small, and do not form branched structures. Histologically the lesions of cutaneous habronemiasis have small foci of eosinophilic necrosis, which are frequently calcified; the fibrosis is extensive, and lymphoid follicles may be prominent. While definitive diagnosis of pythiosis is by culture, an immunodiffusion test may assist in those cases in which culture has failed.

b. CANINE PYTHIOSIS *Pythium insidiosum* is responsible for subcutaneous granulomatous disease in the dog. The condition is seen only in warm climates where affected dogs have access to standing bodies of water. In at least two of the reported cases, pythiosis developed at sites of previous trauma. The gross lesions in the canine disease differ in one substantial respect from those of the horse by lacking kunkers. The histologic pattern is nodular to diffuse pyogranulomatous dermatitis. Fungal invasion of vessels, thrombosis, and subsequent necrosis are common findings. The fungi are readily demonstrated by Gomori's methenamine silver stain but may be difficult to detect in routinely stained sections. Primary intestinal and disseminated disease have been described and are discussed elsewhere.

c. BOVINE PYTHIOSIS Lesions in the few bovine cases confirmed by culture showed a predilection for the lower limbs. Kunkers were not a feature of the gross lesions. Histologically the fungi were embedded in eosinophilic material resembling that of the Splendore–Hoeppli reaction.

Bibliography

Austwick, P. K. C., and Copland, J. W. Swamp cancer. *Nature* **250:** 84, 1974.

Bridges, C. H., and Emmons, C. W. A phycomycosis of horses caused by *Hyphomyces destruens*. *J Am Vet Med Assoc* **138:** 579–589, 1961.

Campbell, C., K. Pythiosis. *Eq Vet J* **22:** 227–278, 1990.

Mendoza, L., and Alfaro, A. A. Equine pythiosis in Costa Rica: Report of 39 cases. *Mycopathologia* **94:** 123–129, 1986.

Mendoza, L., Kaufman, L., and Standard, P. G. Immunodiffusion test for diagnosing and monitoring pythiosis in horses. *J Clin Microbiol* **23:** 813–816, 1986.

Mendoza, L., Kaufman, L., and Standard, P. G. Antigenic relationship between the animal and human pathogen *Pythium insidiosum* and nonpathogenic *Pythium* species. *J Clin Microbiol* **25:** 2159–2162, 1987.

Miller, R. I., and Campbell, R. S. F. The comparative pathology of equine cutaneous phycomycosis. *Vet Pathol* **21:** 325–332, 1984.

Miller, R. I., Olcott, B. M., and Archer, M. Cutaneous pythiosis in beef calves. *J Am Vet Med Assoc* **186:** 984–985, 1985.

O'Neill Foil, C. S. *et al.* A report of subcutaneous pythiosis in five dogs and a review of the etiologic agent *Pythium* spp. *J Am Anim Hosp Assoc* **20:** 959–966, 1984.

XIV. Protozoal Diseases of Skin

Cutaneous lesions occur in several **systemic protozoal** infections including **leishmaniasis** in the dog and cat and infection in the horse by several trypanosomes. Cutaneous lesions have also been associated with *Theileria* and *Babesia* infections. *Trypanosoma equiperdum* and dourine are discussed in Volume 3, Chapter 4, The Female Genital System, and the rest of these diseases are discussed in Volume 3, Chapter 2, The Hematopoietic System.

A. Besnoitiosis

Members of the genus *Besnoitia* are responsible for a serious disease of cattle and horses, and more rarely, goats and sheep. The genus is currently classified under the family Sarcocystidae, in the subfamily Toxoplasmatinae.

Besnoitia have a two-host life cycle. The definitive hosts are felids, and the intermediate hosts vary with the parasitic species. For *B. besnoiti,* the intermediate host is the ox, and for *B. bennetti,* it is the horse. These two organisms are morphologically identical and are distinguished only by their host range. Sexual reproduction occurs in the intestinal tract of the definitive host, and the sporulated oocysts are passed in the feces. The exact mode of infection of the intermediate hosts is not known, but contamination of watering places and mechanical transmission by *Glossina palpalis* may be involved. The life cycle in the intermediate host is characterized by cyst formation. These occur mostly in the dermis, subcutis, fascia, muscle, mucosa of the upper respiratory tract, pharynx, and conjunctiva. Infection, however, may be generalized and is typically so in some infections in rodents and wild animals. The tissue cysts represent parasitized host cells—the fibroblast in *B. wallacei* infection of rodents and the histiocyte in *B. besnoiti* experimental infection of rabbits. The bradyzoites multiply in cellular vacuoles and induce hyperplastic and hypertrophic changes in the host cells. These often divide to form multinucleate cells. The enlarging mass of crescent-shaped bradyzoites compresses the cell cytoplasm and nuclei into a thin rim forming an inner coat to the cyst. A hyalinized collagenous cyst wall is laid down around the parasitized cell. The cysts, which measure up to 500 μm with a 10–30 μm thick wall, are visible to the naked eye. The definitive host is infected by ingesting parasitized tissue from the intermediate host.

Bovine besnoitiosis is a serious disease in South Africa, being associated with severe loss of condition in affected cattle, mortality up to 10%, and marked damage to the hide. The disease also occurs in central and northern parts of Africa, southern Europe, Asia, South America, and the Russia. After an incubation period of ~1 week, a pyrexic phase develops in which animals become anorexic, depressed, and reluctant to move. Approximately 1–4 weeks after the onset of the pyrexic phase and corresponding with cutaneous cyst formation, there is generalized lymphadenopathy, edematous swellings of the extremities, and severe systemic signs. Pregnant cows may abort at this time. The second stage is known as the depilatory stage since it is characterized by marked alopecia, thickening of the skin, exudation, and fissuring. The cysts may be visible macroscopically in the scleral conjunctiva or nasal mucosa as small, round, white foci. Animals lose condition, and up to 10% mortality may occur at this point. The third phase is characterized by dry seborrhea. Animals remain unthrifty for an extended period and rarely regrow a normal hair coat. The skin remains alopecic, lichenified, and scaly.

Histologically, the acute febrile stage is accompanied by epidermal hyperplasia, marked hyperemia, dermal edema, and perivascular accumulations of lymphocytes, plasma cells, and large histiocytes, which will become hosts to the parasites. The crescent-shaped trophozoites occur in the arterioles and lymphatics and are free in the tissue spaces. Occasionally they may be detected in macrophages. As the parasites become encysted, the inflammation and edema diminish. The mature cysts incite little or no cellular reaction unless they rupture; a necrotizing or granulomatous response ensues around the collapsed hyaline capsule. Numerous eosinophils are present in these reactions. The mature cyst wall has four distinct layers. Outermost is a condensed, hyalinized, laminated, birefringent layer of collagen fibers. Next is a very thin, homogeneous intermediate zone. The third layer is the cytoplasm of the host cell, and in this layer lie the several giant, vesicular but compressed host cell nuclei. A thin inner membrane, probably condensed cell cytoplasm, encloses the dense mass of 5- to 7-μm crescentic bradyzoites. These may be separated from the wall by an artefactual shrinkage space. Lesions in other tissues include focal disseminated myositis, keratitis, periostitis, endostitis, lymphadenitis, pneumonia, periorchitis, orchitis, epididymitis, arteritis, and perineuritis.

Equine besnoitiosis is a similar disease but has not been so well characterized as the bovine condition. **Caprine besnoitiosis** occurs in Iran in both wild and domestic species. The *Besnoitia* cysts are observed in the skin, blood vessels, epididymis, and testes. The pathological changes are comparable to those of *B. besnoiti* infection in cattle. In an outbreak affecting more than 500 domestic goats in Kenya, ocular cysts were the most common finding, but cysts were found in many body systems. Dorper **sheep** were also affected in that outbreak. A disease resembling besnoitiosis was reported in New Zealand lambs.

Bibliography

Bigalke, R. D. The artificial transmission of *Besnoitia besnoiti* (Marotel 1912) from chronically infected to susceptible cattle and rabbits. *Onderstepoort J Vet Res* **34:** 303–316, 1967.

Bigalke, R. D., and Naude, T. W. The diagnostic value of cysts in the scleral conjunctiva in bovine besnoitiosis. *J S Afr Vet Med Assoc* **33:** 21–27, 1962.

Bigalke, R. D. *et al.* Studies on the relationship between *Besnoitia* of blue wildebeest and impala, and *Besnoitia besnoiti* of cattle. *Onderstepoort J Vet Res* **34:** 7–28, 1967.

Binninger, C. E., and McGuire, T. C. Atypical globidiosis in a lamb. *J Am Vet Med Assoc* **151:** 606–608, 1967.

Bwangamoi, O. Besnoitiosis and other skin diseases of cattle (*Bos indicus*) in Uganda. *Am J Vet Res* **29:** 737–743, 1968.

Bwangamoi, O., Carles, A. B., and Wandera, J. G. An epidemic of besnoitiosis in goats in Kenya. *Vet Rec* **125:** 461, 1989.

Cheema, A. H., and Toofanian, F. Besnoitiosis in wild and domestic goats in Iran. *Cornell Vet* **69:** 159–168, 1979.

Frenkel, J. K. *Besnoitia wallacei* of cats and rodents: With a reclassification of other cyst-forming Isosporoid Coccidia. *J Parasitol* **63:** 611–628, 1977.

Hicks, B. R. *Besnoitia*-like infection in lambs. *N Z Zool* **9:** 47, 1982.

McCully, R. M. *et al.* Observations on *Besnoitia* cysts in the cardiovascular system of some wild antelopes and domestic cattle. *Onderstepoort J Vet Res* **33:** 245–276, 1966.

Schulz, K. C. A. A report on naturally acquired besnoitiosis in bovines with special reference to its pathology. *J S Afr Vet Med Assoc* **31:** 21–35, 1960.

Terrell, T. G., and Stookey, J. L. *Besnoitia bennetti* in two Mexican burros. *Vet Pathol* **10:** 177–184, 1973.

Wallace, G. D., and Frenkel, J. K. *Besnoitia* species (Protozoa, Sporozoa, Toxoplasmatidae): Recognition of cyclic transmission by cats. *Science* **188:** 369–371, 1975.

B. Miscellaneous Coccidian Parasites

Caryospora spp., apicomplexan parasites whose primary hosts are reptiles and raptors, rarely cause pyogranulomatous dermatitis in puppies. Immunosuppression and concurrent disease, such as canine distemper, probably play a facilitatory role. Lesions involve skin and draining lymph nodes and comprise diffuse pyogranulomatous dermatitis. Macrophages contain large numbers of intracellular organisms, including schizonts, gamonts, oocysts, and caryocysts. Caryocysts have a thin cyst wall enclosing the host cell nucleus and contain up to three sporozoites. Not all stages of the life cycle may be present in the tissue sections, precluding a microscopic diagnosis in some cases. Immunohistochemical studies identified the agent in one case as *C. bigenetica*. Experimental oral infection of immunosuppressed puppies with *C. bigenetica* induced typical skin lesions affecting muzzle, periocular skin, footpads, ears, and abdomen within 10 days of inoculation.

Neospora caninum, a coccidial parasite of dogs, typically causes systemic and neurologic disease. In one dog, multifocal ulcerated cutaneous lesions were the major presenting sign. Histologic lesions were described as pyogranulomatous, eosinophilic, necrotizing, and hemorrhagic. Numerous tachyzoites, 4–7 μm × 1.5–5 μm, were present in macrophages and neutrophils and rarely in endothelial cells and fibroblasts. Tissue cysts were not found in the cutaneous lesions.

An unidentified *Sarcocystis*-like protozoan has been associated with multiple cutaneous abscesses and disseminated visceral lesions in a dog. The skin lesions were diffuse, necrotizing, hemorrhagic, and suppurative. Large numbers of protozoa were present, mostly in macrophages and neutrophils, and occasionally in fibroblasts and endothelial cells. Some vessels contained thrombi, and there was associated dermal and epidermal infarction. The organism did not stain with antisera to the other apicomplexan parasites so far identified as causing dermatitis in dogs, namely *Neospora caninum* and *Caryospora* sp.

Bibliography

Dubey, J. P. *et al.* Newly recognized fatal protozoan disease of dogs. *J Am Vet Med Assoc* **192:** 1269–1285, 1988.

Dubey, J. P. *et al. Caryospora*-associated dermatitis in dogs. *J Parasitol* **76:** 552–556, 1991.

Dubey, J.P. *et al.* Fatal cutaneous and visceral infection in a Rottweiler dog associated with a *Sarcocystis*-like protozoon. *J Vet Diagn Invest* **3:** 72–75, 1991.

XV. Algal Diseases of Skin: Protothecosis

Protothecosis is caused by achloric algae of the genus *Prototheca,* which are thought to be mutant forms of the green algae, *Chlorella*. The organisms are saprophytic and ubiquitous, occupying habitats as diverse as acidic lake water, tree sap, feces of various animals, and potato skins. *Prototheca* spp. are opportunistic and very rarely infect animals. Naturally occurring protothecosis was first described in cattle with mastitis. In dogs and cats, it is usually systemic, with gastrointestinal and ocular involvement, but primary cutaneous disease has been reported from the United States, Australia, and Britain. Species implicated in animal disease include *Prototheca wickerhamii* and *Prototheca zopfii*.

Cutaneous protothecosis occurs in the cat and dog as multiple small gray-white or tan-colored, ill-defined nodules. Histologically, the reaction is diffuse granulomatous dermatitis and panniculitis in which the causative organisms are numerous. The organisms vary from 2 to 20 μm in diameter and resemble yeasts, for which they were originally mistaken. Their presence in large numbers gives the tissue section a vacuolated appearance, as the organisms do not stain with hematoxylin and eosin. The thick cell wall and the internal septations stain well with Gomori's methenamine silver and Gridley's fungal stains. Some large round bodies, up to 30 μm in diameter, contain up to 8 endospores. Although many organisms are free, some are contained in epithelioid macrophages and multinucleate, histiocytic giant cells. Lymphocytes, plasma cells, and neutrophils are present in focal aggregates throughout the granulomatous reaction. Diagnosis is confirmed by culture or by immunofluorescence techniques.

Bibliography

Chandler, F. W., Kaplan, W., and Callaway, C. S. Differentiation between *Prototheca* and morphologically similar green algae in tissue. *Arch Pathol Lab Med* **102:** 353–356, 1978.

Dillberger, J. E. *et al.* Protothecosis in two cats. *J Am Vet Med Assoc* **192:** 1557–1559, 1988.

Kaplan, W. Protothecosis and infections caused by morphologically similar green algae. *In* "The Black and White Yeasts." Pan Am Health Org. Publ. 256 Proc. IV Int. Conf. Mycoses, Washington, D.C., WHO, 1978.

Macartney, L., Rycroft, A. N., and Hammil, J. Cutaneous protothecosis in the dog: First confirmed case in Britain. *Vet Rec* **123:** 494–496, 1988.

Migaki, G. *et al.* Canine protothecosis: Review of the literature and report of an additional case. *J Am Vet Med Assoc* **181:** 794–797, 1982.

XVI. Arthropod Ectoparasites

Of the parasitic arthropods, only a small fraction are parasites of domestic animals, but the harmfulness of these is quite out of proportion to their number. Some, such as the mange mites, are pathogens in their own right, but the majority of them owe their immense importance to their ability to act as mechanical or biological transmitters for many pathogenic viruses, rickettsiae, bacteria, spirochetes, protozoa, and helminths. The problems of vectors are discussed in relation to the specific diseases throughout these volumes.

The parasites of concern to us here belong to the two large classes, Insecta and Arachnida. The class Insecta contains four important orders, Diptera (flies), Siphonaptera (fleas), Mallophaga (biting lice), and Siphunculata (sucking lice). The class Arachnida contains the order, Acarina, in which ticks and mites are classified. For data on biological characteristics and classification, reference should be made to texts on entomology.

Bibliography

Arundel, J. H., and Sutherland, A. K. "Animal Health in Australia. Volume 10. Ectoparasitic Diseases of Sheep, Cattle, Goats and Horses."Australian Government Publishing Service, 1988.

Baron, R. W., and Weintraub, J. Immunological responses to parasitic arthropods. *Parasitol Today* **3:** 77–81, 1987.

Dobson, K. J. External parasites. *In* "Diseases of Swine," 6th Ed. A. D. Leman, *et al* (eds.), pp. 664–675. Ames, Iowa, Iowa State Univ. Press, 1986.

Foil, L., and Foil, C. Parasitic skin diseases. *Vet Clin N Am: Eq Pract* **2:** 403–437, 1986.

Soulsby, E. J. L. "Helminths, Arthropods, and Protozoa of Domesticated Animals," 7th Ed. London, Baillière Tindall, 1982.

Urquhart, G. M. *et al.* "Veterinary Parasitology." Essex, England, Longman Scientific and Technical, 1987.

A. *Flies*

The parasitic insects of the order Diptera represent varying degrees of adaptation to a parasitic existence. Some, such as *Musca,* are facultative feeders. Some, such as the Simuliidae and parasitic species of Culicidae and Ceratopogonidae, are obligate bloodsuckers, although usually only the females draw blood. At the other end of the spectrum are the Oestridae, whose larvae are obligate parasites, and some members of the Hippoboscidae, which are obligate parasites in the adult stage.

Because of the variety of parasitic modes, it is not possible to generalize on the effects of flies on domestic animals, nor, with the exception of a few obligate parasites, is it possible to be specific, because little information is available on primary pathogenicity. Flies adversely affect domestic animals by causing annoyance which occasionally leads to fatality, but more often to loss of production or injury; by direct toxicity, which may be fatal following massive insect attack; by indirect toxicity, due to the deposition of larva into damaged skin (myiasis); by local irritant effects causing dermatitis, which may predispose to secondary bacterial infection or to myiasis; by injection of antigens, which induce hypersensitivity reactions; by blood-feeding activities, which cause anemia; and by the mechanical transmission of other pathogens. For example, it has been estimated that tabanid flies could transmit 35 pathogens including the virus of equine infectious anemia and trypanosomes.

Animal annoyance, so-called fly-worry, is an important source of economic loss to the cattle, sheep, and to a lesser extent, swine industries. Fly worry is caused by the biting flies such as the horn fly [*Haematobia* (*Lyperosia*) *irritans*], the stable fly (*Stomoxys calcitrans*), and the horse flies (several genera in the family Tabanidae), and by nonbiting species such as the house flies (*Musca* spp.), *Hypoderma* spp. and the sheep-head fly (*Hydrotaea irritans*). The nonbiting flies cause annoyance by clustering around the eyes and nostrils, where they feed on lacrimal or nasal secretions (*Musca autumnalis* and *Hydrotaea irritans*) or by other means such as simulating the sound of a bumble bee (*Hypoderma* spp.). Fly-worry occasionally induces such apprehension that the animals run aimlessly (gadding), and severe injury or death may result from misadventure. Deaths are usually sporadic, but impressive levels of mortality have been reported. Of much greater economic importance is the loss of production associated with fly worry. When the insects are numerous, they cause very considerable annoyance to livestock and interfere with feeding and resting and cause reduced milk production and weight gain.

Mortality may arise from direct toxic effect as well as from misadventure. Death may be the result of urticarial swelling of the head and neck, of shock, or of suffocation when massive numbers of flies swarm in the nostrils. Mosquitoes, especially vicious species such *Aedes vigilex,* may cause significant mortality among piglets and puppies. The Simuliidae (black flies) are responsible for massive animal mortalities, particularly in river valleys following extensive flooding, when the insect population expands. Simuliid flies exert systemic effects through inoculation of a heat-stable toxin, which causes increased vascular permeability and abnormalities in cardiorespiratory function, and which may cause death.

The hematophagous flies seldom cause serious loss of blood. Anemia may result from heavy infestations by *Haematobia irritans,* mosquitoes, the sheep ked *Melophagus ovinus,* and *Stomoxys calcitrans,* which may ingest as much as 16 mg blood per feed.

Local irritant effects result from the injection of salivary fluids into the host. Very little is known of the nature of the cutaneous lesions produced in animals by these insects. The character and severity of the local lesions vary. Pruritus is often intense, resulting in secondary traumatic lesions. The primary lesions are usually erythematous papules or wheals, often surrounding a central bleeding point (mosquitoes) or small puncture wounds (biting flies). The wheals are usually transient but may persist for several weeks. The puncture wounds often develop an exudative crust. Histologically there may be intraepithelial eosinophilic spongiform pustules or focal areas of epidermal necrosis indicating the penetration point. The dermal reaction is superficial perivascular in pattern and contains predominantly eosinophils, lymphocytes, and plasma cells. Occasionally there is acute necrosis of the surface of the papules, including both dermis and epidermis.

The injected salivary substances are irritant and many are allergenic, and hypersensitivity reactions probably contribute to the severity of the local lesions caused by a variety of biting flies. Hypersensitivity reactions to *S. calcitrans* are recognized in cattle; affected animals develop coalescing blisters on the forelimbs. An important allergic dermatitis of horses is caused by hypersensitivity to *Culicoides* spp. (see Section X, Immune-Mediated Diseases of Skin).

Rather more important than the bloodsucking or biting flies are those species whose larvae are highly destructive facultative or obligate parasites. Infestations with such larvae cause myiasis, which is discussed next.

Bibliography

Campbell, J. B. *et al.* Effects of stable flies on weight gains and feed efficiency of calves on growing or finishing rations. *J Econ Entomol* **70:** 592–594, 1977.

Cheng, T. H. The effect of biting fly control on weight gain in beef cattle. *J Econ Entomol* **51:** 275–278, 1958.

Foil, L., and Foil, C. Dipteran parasites of horses. *Eq Pract* **10:** 21–38, 1988.

Foil, L. D. *et al.* The role of horn fly feeding and the management of seasonal equine ventral midline dermatitis. *Eq Pract* **12:** 6–14, 1990.

Hunter, A. R. Sheep headfly disease in Britain. *Vet Rec* **97:** 95–96, 1975.

Moorhouse, D. E. Cutaneous lesions on cattle caused by stable fly. *Aust Vet J* **48:** 644, 1972.

Ruhm, W. Black-flies (Simuliidae, Diptera), a cause of annoyance and injury to livestock. *Vet Med Rev* **1:** 38–50, 1983.

Steelman, C. D. Effects of external and internal arthropod parasites on domestic livestock. *Ann Rev Entomol* **21:** 155–178, 1976.

Stork, M. G. The epidemiological and economic importance of fly infestation of meat- and milk-producing animals in Europe. *Vet Rec* **105:** 341–343, 1979.

Titchener, R. N., Newbold, J. W., and Wright, C. L. Flies associated with cattle in southwest Scotland during the summer months. *Res Vet Sci* **30:** 109–113, 1981.

1. Myiasis

Myiasis is the infestation of the tissue of living animals with the larvae of dipterous flies. The larvae may be facultative or obligate parasites. The important families are Cuterebridae, Gasterophilidae (stomach bots of horses), Oestridae (nasal bots, warbles), and Calliphoridae (blow-

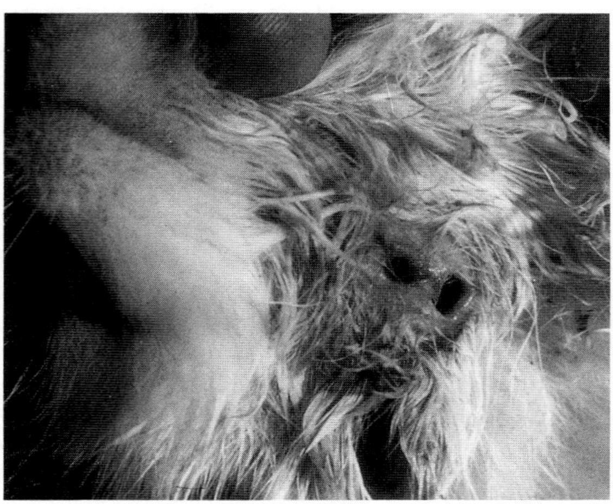

Fig. 5.102 *Cuterebra* infestation with draining lesions. Cat.

flies). Only those flies whose larvae cause cutaneous or subcutaneous lesions are discussed here. Nasal and stomach bots are described elsewhere.

a. CUTEREBRA The larvae of *Cuterebra* spp. are obligate parasites of rodents and rabbits, but occasionally aberrant infestations occur in cats and rarely in dogs, pigs, and humans. The larvae penetrate lacerated skin, producing initially firm, then fluctuant, cystlike subcutaneous abscesses in which the larvae mature. The larvae breathe through a pore in the skin, through which they are visible. In cats, the larvae have a predilection for the neck area (Fig. 5.102), often over the submandibular gland, but swellings also occur in the scrotal region. Larvae may also enter via natural orifices and may give rise to lesions in aberrant locations such as the pharynx, nasal cavity, and brain.

Bibliography

Catts, E. P. Biology of New World bot flies: Cuterebridae. *Annu Rev Entomol* **27:** 313–338, 1982.

Kazacos, K. R. *et al. Cuterebra* sp. as a cause of pharyngeal myiasis in cats. *J Am Anim Hosp Assoc* **16:** 773–776, 1980.

b. WARBLES Warbles caused by *Hypoderma bovis* and *Hypoderma lineatum* (Diptera: Oestridae) occur chiefly in cattle, although horses, sheep, and humans are occasionally affected. *Przhevalskiana silensus,* reported under a variety of synonyms, affects sheep and goats in Asia and eastern Europe. In South and Central America, *Dermatobia hominis* affects cattle, sheep, goats, pigs, and humans. There are no warble flies in Australia.

The warble flies are also known as heel flies because the eggs are deposited predominantly on the hair of the legs. The larvae emerge 4–6 days later and burrow into the skin, causing minimal irritation. The larvae migrate along fascial planes, leaving tracks of greenish gelatinous material known as butcher's jelly. The first instar larvae of *H. bovis* overwinter in the epidural fat, while those of *H. lineatum* develop in the esophageal submucosa. In the esophageal lesions, the collagen bundles around the first instar larvae of *H. lineatum* appear fragmented, as if undergoing enzymatic digestion. A collagenase has been isolated from *H. lineatum*. In the epidural lesions of *H. bovis,* it is the fat tissue which appears necrotic. In the spring, the larvae migrate dorsally to the subcutaneous tissue of the back to form subcutaneous nodules ~3 cm diameter with a central pore for respiration. The lesions, which are known as warbles, last for 4 to 6 weeks, during which the larvae undergo two molts. The mature third-instar larvae emerge from the breathing hole and pupate in the soil. In horses, the lesions occur in the saddle region and are often blind, in that the larvae do not complete their development. Fatalities resulting from aberrant migration into the central nervous system are reported in horses.

Histologically, the cellular reaction is predominantly eosinophilic and lymphocytic. The eosinophilic infiltrate gives butcher's jelly its greenish coloration. However, the most intense inflammatory reactions occur at sites of previous migration rather than around the viable larvae, suggesting that in naive hosts, the parasites depress any effective host responses. Proteinases with the capacity to cleave the third component of bovine complement have been isolated from the first instar larvae of *H. lineatum*. Such enzymes could well ablate the host's inflammatory responses. The actual warble is lined by a wall of granulation tissue which matures to form a connective tissue capsule in which lie islands of eosinophils. The cystic cavity between the cuticle of the parasite and the granulation tissue fills with fibrin and a few inflammatory cells, chiefly eosinophils. Cuticle sloughed during ecdysis, or remnants of dead larvae, incite a marked foreign-body giant-cell reaction. Once the larvae emerge, the cavity is repaired by fibrosis, but small foreign-body granulomas may persist for months.

Warbles are economically important. The buzzing of the adult flies disturbs cattle, causing considerable loss in milk and meat production. Larval tracks in the tissues decrease carcass value, and the larval-induced holes markedly depreciate the value of the hide. Larval rupture, either accidental or deliberate, may induce a fatal anaphylactic reaction. This may result from systemic effects of the warble toxin, from type I hypersensitivity reactions, or a combination of both. Serological tests are available for the detection of *H. bovis* and *H. lineatum* infestations. These are being used to monitor the progress of eradication schemes such as that being carried out in Great Britain. A joint Canadian–U.S. pilot study successfully eliminated *H. lineatum* and *H. bovis* in the test areas by a combination of chemical control and sterile insect release.

Bibliography

Anderson, P. H., and Kirkwood, A. C. A reaction in cattle to toxins of *Hypoderma bovis* (warble fly) larvae. *Br Vet J* **124:** 569–575, 1968.

Andrews, A. H. Warble fly: The life cycle, distribution, economic losses, and control. *Vet Rec* **103:** 348–353, 1978.

Baron, R. W. Cleavage of purified bovine complement component C^3 in larval *Hypoderma lineatum* (Diptera: Oestridae) hypodermins. *J Med Entomol* **27:** 899–904, 1990.

Eyre, P., Boulard, C., and Deline, T. Local and systemic reactions in cattle to *Hypoderma lineatum* larval toxin: Protection by phenylbutazone. *Am J Vet Res* **42:** 25–28, 1981.

Guo, R. M., and Fu, G. Z. Report on the infestation of *Hypoderma lineatum* infestation in sheep. *Chin J Vet Med* **12:** 16–17, 1986.

Hadlow, W. J., Ward, J. K., and Krinsky, W. L. Intracranial myiasis by *Hypoderma bovis* (Linnaeus) in a horse. *Cornell Vet* **67:** 272–281, 1977.

Hadwen, S., and Bruce, E. A. Anaphylaxis in cattle and sheep produced by the larvae of *Hypoderma bovis, H. lineatum*, and *Oestrus ovis*. *J Am Vet Med Assoc* **51:** 15–44, 1917.

Klein, S. A., Bowe, M., and Klein, K. K. Adverse selection and warble flies in Alberta. *Can J Ag Econ* **37:** 111–123, 1989.

Kunz, S. E. *et al.* Use of sterile insect releases in an IPM program for control of *Hypoderma lineatum* and *H. bovis* (Diptera: Oestridae): A pilot test. *J Med Entomol* **27:** 523–529, 1990.

Lecroisey, A. *et al.* Complete amino acid sequence of the collagenase from the insect *Hypoderma lineatum*. *J Biol Chem* **262:** 7546–7551, 1987.

Sinclair, I. J. *et al.* The serological incidence of *Hypoderma bovis* in cattle in England and Wales in spring 1989. *Vet Rec* **126:** 327–329, 1990.

Wolfe, L. S. Observations on the histopathological changes caused by the larvae of *Hypoderma bovis* (L.) and *Hypoderma lineatum* (devill.) (Diptera: Oestridae) in tissues of cattle. *Can J Anim Sci* **39:** 145–157, 1959.

c. CALLIPHORINE MYIASIS This occurs in all animal species but is most common in sheep, particularly in Australia, where it is of major economic importance. The flies involved are members of the subfamily Calliphorina (blowflies). A facultative parasitic mode is adopted by the larvae (maggots), which is an adaptation of their beneficial and important role in the breakdown of carrion. The transition from the wool of dead sheep to the soiled, wet wool on a live animal is not a great one. Important genera are *Lucilia, Calliphora, Microcalliphora, Phormia,* and *Chrysomia*. The species of fly involved in fly strike differs with geographic location: *Lucilia cuprina* is the most important primary fly in Australia; *Phormia regina,* in United States and Canada; and *Lucilia sericata,* in Great Britain. Several different species of flies are involved in the development of the lesion of cutaneous myiasis. The primary flies, such as *L. cuprina,* are capable of initiating a strike on living sheep. The secondary flies, such as *Chrysomyia rufifaces,* are not able to initiate a strike, but greatly exacerbate the lesions initiated by the primary fly. They may also displace the maggots developing from the eggs laid by the primary fly. The tertiary flies, such as the house-fly *Musca domestica,* attack at a later stage and do not contribute significantly to the skin damage.

The development of fly strike in sheep depends on abundance of primary flies, susceptible sheep, and moisture. The prevalence of the disease tends to follow the increase and decrease in the population of primary flies, which in turn depends on the climatic zone. In general, the flies require warm and moist but not hot conditions. Thus there is usually a double wave of primary flies, peaking in the spring and autumn. Certain breeds of sheep, in particular the fine-wooled Merino, have an inherent predisposition to attract fly strike. The character of the fleece and conformational features, such as skin wrinkling, allow retention of moisture or predispose to fecal or urinary soiling, which are all initiating factors in fly strike. Moisture is the third important factor, whether provided by rain, dew, urine, sweat, or inflammatory exudate. Moisture predisposes to bacterial proliferation. These bacteria are often of fecal or urinary origin, but *Pseudomonas aeruginosa* (of fleece rot) is of major importance. *Dermatophilus congolensis* may be involved occasionally.

The odor induced by the bacterial proliferation and resultant inflammatory exudate attracts the primary flies, which deposit batches of 50 to 200 ova in the matted wool. The larvae emerge within 12 to 24 hr and grow rapidly, feeding on inflammatory exudates. The primary larvae secrete proteolytic enzymes, including collagenases, which liquefy the host tissues and provide predigested nutrients. Antibodies to these enzymes are being tested for inhibitory effects against larval growth, the aim being development of a fly-strike vaccine. The cutaneous necrosis which results attracts the secondary flies to oviposit. The resulting larvae tunnel into the adjacent viable tissue and markedly expand the lesion. The putrefactive odor attracts more flies, and the process is further exacerbated. The lesions of fly strike may result in death from shock, debilitation, toxemia, or bacterial septicemia.

The lesions are most common in the perineum (breech strike), particularly in sheep with a narrow conformation and/or marked skin wrinkling, which favor urine or fecal soiling. Lesions may affect the preputial orifice (pizzle strike) particularly in animals with narrow urethral orifices, which predispose to urine soiling. Rams with deep head folds may develop poll strike, possibly predisposed to by fight wounds. Wound strike occasionally follows castration or tail docking and body strike follows prolonged wetting, which in turn predisposes to fleece rotor dermatophilosis.

The initial gross lesion is a patch of dark-brown, moist wool which has a foul smell. The wool is often very hot as a result of putrefaction and inflammation caused by the larvae. As the process advances, the maggots burrow under the skin, causing irregular ulcers with scalloped edges. The lesions are irritating and pruritic. Occasionally the maggots migrate deeper into the muscle.

The economic losses of fly strike in sheep result from death, disfigurement, depreciation of the fleece, and costs associated with prevention and treatment. In other species, cutaneous myiasis affects only debilitated animals in dirty conditions.

Bibliography

Gherardi, S. G. *et al.* Field observations on body strike in sheep affected with dermatophilosis and fleece-rot. *Aust Vet J* **60:** 27–28, 1983.

Hall, C. A., Martin, I. C. A., and McDonell, P. A. Patterns of wool moisture in Merino sheep and its association with blowfly strike. *Res Vet Sci* **29:** 181–185, 1980.

Harwood, R. F. Myiasis. *In* "Entomology in Human and Animal Health," 7th Ed. pp. 296–318. New York, Macmillan, 1979.

Sandeman, R. M. Prospects for the control of sheep blowfly strike by vaccination. *Int J Parasitol* **20:** 537–541, 1990.

Watts, J. E., Murray, M. D., and Graham, N. P. H. The blowfly strike problem of sheep in New South Wales. *Aust Vet J* **55:** 325–334, 1979.

Watts, J. E. *et al.* The significance of certain skin characters of sheep in resistance and susceptibility to fleece-rot and body strike. *Aust Vet J* **56:** 57–63, 1980.

d. SCREWWORM MYIASIS This is also caused by members of the Calliphorinae, but they differ from the blowflies in that the larvae are obligate parasites, invading wounds on live animals. The American species of screwworm flies are *Callitroga* (*Cochliomyia*) *hominivorax* and *Callitroga macellaria*. The African and Asian screwworm fly is *Chrysomyia bezziana*. The disease occurs in Africa, Asia, Central and South America, and Mexico, but has been virtually eradicated in the United States and Mexico, following a program in which release of massive numbers of irradiated male flies rendered the annual breeding a sterile one. An outbreak of screwworm in Libya in 1988, seen as a major threat to the livestock of southern Europe and Africa, was dealt with in a similar fashion. International air travel proffers a source of spread to susceptible countries.

Screwworm myiasis affects all domestic animals and humans and is an important cause of mortality in wildlife. The flies oviposit in cutaneous wounds, such as those caused by castration, dehorning, branding, or accidental injuries. The navel of neonatal calves, the perineum of recently calved cows, and tick bites are also favorable sites for oviposition. The larvae penetrate and liquefy the tissue with the aid of proteolytic enzymes. A blood-stained fluid, often containing incompletely digested shreds of tissue, oozes from the wound, which contains clusters of voraciously feeding larvae. A distinctive and particularly evil odor emanates from the lesion. The lesions are extremely painful and may expand rapidly, leading to death in untreated animals.

Bibliography

Humphrey, J. D., Spradbery, J. P., and Tozer, R. S. *Chrysomyia bezziana:* Pathology of Old World screw-worm fly infestations in cattle. *Exp Parasitol* **49:** 381–397, 1980.

Krafsur, E. S., Whitten, C. J., and Novy, J. E. Screwworm eradication in North and Central America. *Parasitol Today* **3:** 131–137, 1987.

Rajapaksa, N., and Spradbery, J. P. Occurrence of the Old World screw-worm fly *Chrysomyia bezziana* on livestock vessels and commercial aircraft. *Aust Vet J* **66:** 21–23, 1989.

Readshaw, J. L. Screwworm eradication a grand delusion? *Nature* **320:** 407–410, 1986.

Sutherst, R. W., Spradbery, J. P., and Maywald, G. F. The potential geographical distribution of the Old World screw-worm fly, *Chrysomyia bezziana*. *Med Vet Entomol* **3:** 273–280, 1989.

2. Sheep Ked Infestation

Melophagus ovinus (Diptera: Hippoboscidae) is a wingless fly which causes a chronic, pruritic dermatitis of sheep. Goats are also affected. Of worldwide distribution, the disease's chief economic importance is the associated loss of wool production.

Melophagus ovinus is an obligate ectoparasite. The eggs develop within the female to the larval stage. After parturition, the female attaches its larva to wool fibers with the aid of a sticky substance. The immotile larva transforms into a chestnut-brown pupa ~3–4 mm long. The pupal stage lasts 3–5 weeks, and the adult keds live 4–5 months. They prefer the sides of the neck and body and are difficult to detect in fully fleeced animals. The adults feed actively on blood. While anemia may develop in severe ked infestations, the more significant lesions are the result of the severe pruritus, which causes the sheep to rub and bite and thus damage the fleece. The adult fly excreta stains the wool, further reducing its value. The irritation induced by the bites also affects weight gain.

Histologic lesions reported are superficial and deep perivascular dermatitis with eosinophils and lymphocytes predominating. Fibrinoid necrosis of small arterioles is also described.

Bibliography

Arundel, J. H., and Sutherland, A. K. "Animal Health in Australia. Volume 10. Ectoparasitic Diseases of Sheep, Cattle, Goats and Horses." Australian Government Publishing Service, 1988.

Nelson, W. A., and Bainborough, A. R. Development in sheep of resistance to the ked *Melophagus ovinus* (L.). III. Histopathology of sheep skin as a clue to the nature of resistance. *Exp Parasitol* **13:** 118–127, 1963.

3. Hornfly Dermatitis

Although mainly a pest of cattle, the horn fly (*Haematobia irritans*) is one cause of ventral midline dermatitis in the horse. Horn flies, clustering on the ventral abdomen (and occasionally on the neck), produce bites marked by tiny drops of dried blood. A few days later pruritic, scaling, alopecic patches develop. These become lichenified and heal with either leukoderma or with melanosis. The lesions are often single, usually well circumscribed, and occur near the umbilicus. Lesions of *Culicoides* hypersensitivity (see Section X, Immune-Mediated Dermatoses) and onchocerciasis may also occur on the ventrum, but these are diffuse, often extending from the axillae to the groin. Both *Culicoides* hypersensitivity and onchocerciasis are sporadic diseases, whereas up to 80% of horses in a group may be affected with horn-fly-bite dermatitis.

4. Mosquito-Bite Dermatitis

Although a common occurrence in all species, mosquito bites in **cats** may induce a reaction which is histologically indistinguishable from some forms of the eosinophilic granuloma complex. The disease is seasonal. Clinically, cats develop crusted, ulcerated, and sometimes hypopigmented lesions on the bridge of the nose, the ears, and footpads. Histologic lesions include intraepidermal eosinophilic microabscessation, diffuse eosinophil and mast cell-rich dermatitis, and eosinophilic granulomas around focal areas of collagen denaturation. Some cats show immediate hypersensitivity reactions to intradermal injection of mosquito extract, indicating an underlying allergic process.

Bibliography

Mason, K. V., and Evans, A. G. Mosquito bite-induced eosinophilic dermatitis in cats. *J Am Vet Med Assoc* **12:** 2086–2088, 1991.

B. Lice

Lice are host-specific, obligate parasites of the class Insecta. Two orders of lice are recognized. The Mallophaga are biting lice which have mouth parts specially adapted for chewing the epithelial debris of the skin which constitutes their food. They include species parasitic for birds and mammals. The Anoplura are bloodsucking lice and have mouth parts adapted to this purpose. They are parasitic only for mammals. Lice cannot live away from their hosts for more than a few days. Consequently, spread of infestation occurs mainly by direct contact among hosts. Pigs and humans are parasitized only by sucking lice, birds and cats are parasitized only by biting lice, and the other domestic animals are parasitized by both types. Because various species of lice have adapted to different microenvironments within the host pelage, it is possible for an animal to carry several species at once.

Infestation with lice is **pediculosis.** It tends to be a seasonal problem, being worse in winter. The signs associated with pediculosis are extremely variable. Low infestations may be unaccompanied by clinical signs in carrier animals. Most lesions result from skin irritation and resultant pruritus. They include alopecia alone, papulocrustous dermatitis, and damage to wool or hide caused by rubbing or biting. Sucking lice may induce anemia, which is occasionally fatal in heavily infested animals. Loss of weight and decreased milk production are associated with the constant irritation seen in some lice infestations.

Lice are almost always host specific, but *Heterodoxus longitarsus,* normally parasitic on kangaroos, has become an important ectoparasite of Australian dogs. Because of host specificity, it is advantageous to consider pediculosis of the different hosts in turn.

Two species of lice occur on **horses,** namely *Haematopinus asini,* a sucking louse, and *Damalinia equi,* a biting louse. The populations of lice fluctuate considerably, being highest when the hair is long, as in winter, or in debilitated animals which have not shed their hair. In warm weather, the populations decline, but some lice persist in the long hair of the mane and tail. Both species of louse induce skin irritation. The lesions of pediculosis, rough coat, variable alopecia, and self-excoriation result from the animal rubbing or biting at the irritated areas.

Haematopinus eurysternus, the short-nosed louse, *H. quadripertusus,* the tail-switch louse, *H. tuberculatis,* the buffalo louse, *Linognathus vituli,* the long-nosed louse, and *Solenoptes capillatus* are sucking lice of **cattle.** *Damalinia bovis* is the one biting species. The various species have preferred habitats. *Damalinia bovis* tend to cluster about the poll, forehead, neck, back, and rump; *Linognathus* and *Solenoptes* prefer the head, neck, and dewlap. There may be antagonism between the species; in dual infestation with *D. bovis* and *L. vituli,* the former tends to occupy the dorsal half and the latter the ventral areas. Some lice are widely distributed; others tend to cluster in groups. Infestations are quite common, particularly in colder weather or seasons, but, unless heavy, are not particularly deleterious. Poorly fed, overcrowded, and unthrifty animals are more susceptible to heavy infestations. Conversely, heavy infestations may indicate underlying disease, as debilitated or ill animals cease normal grooming activities. Lesions reflect pruritus. Bovine pediculosis has little deleterious effect on weight gain and other production parameters. Economic consequences are rather due to deterioration of hide quality, damage to fences, and the costs of treatment. The exception is *Hematopinus eurysternus* infestation, which may cause anemia and death.

The species of sucking lice affecting **sheep** include *Linognathus ovillus* (the face or blue louse), *L. africanus* (also called the blue louse), and *L. pedalis* (the foot louse). *Linognathus ovillus* is not very pathogenic. *Linognathus pedalis* characteristically infests the hairy skin on the legs. Infestations are frequently light and may be confined to one limb, as the lice form localized clusters. Rarely, heavy infestations spread to adjacent scrotal or abdominal skin, causing irritation and dermatitis secondary to self-trauma. While goat and sheep lice are considered host specific, there are reports of naturally occurring transmission of the goat louse *Damalinia caprae* to sheep and experimental transmission of *D. ovis* to goats.

Damalinia ovis, a biting louse, is a common and serious ectoparasite in sheep. Populations of *D. ovis,* on an individual or in a flock, can build up very quickly. The numbers of lice fluctuate with the season, increasing usually in late winter and early summer. The population declines in the summer, probably because the lice, which are sensitive to heat and low humidity, cannot survive in the body fleece where temperatures at the tip of the staple may reach 48°C. There are also unexplained individual differences among sheep in their susceptibility, but in general, unthrifty sheep carry the heaviest infestation. The lice feed on loose skin scales and sebaceous secretion, and there is a correlation between the degree of scurf and the size of the louse population. The highest concentration of lice occurs along the middorsal line, chiefly over the withers, but the parasites range over the entire body. Eggs are attached to the wool fibers close to the body surface. The infestation has serious consequences because of the marked pruritus it induces. The cause of the pruritus is unknown but is thought to be more than simple mechanical irritation. Affected sheep rub, scratch, and bite at the skin, resulting in severe damage to the fleece. Focal crusting may be associated with some louse infestations. Economic loss is due to reduced fleece quality, cost of prevention, and mortality from secondary myiasis. There is, however, no evidence to support the claim that sheep louse infestation leads to unthriftiness.

In **goats,** the sucking louse, *Linognathus stenopsis,* is more pathogenic than the biting lice, *Damalinia caprae, D. limbata,* and *D. crassipes.* The hair coat of Angora goats may be seriously damaged by the irritation-induced pediculosis.

One species, *Haematopinus suis,* a sucking louse, is parasitic on **pigs.** Preferential sites include the ears and skin folds of the neck, axillae, and inguinal areas. Infestations are often severe, but with the exception of nursing piglets, anemia is not severe. In white-skinned animals, numerous small puncta may be seen, especially in scalded carcasses. Constant irritation from lice also interferes with growth rate and efficiency of food conversion. The lice are vectors for swine pox, African swine fever, and *Eperythrozoon suis.*

Linognathus setosus is a sucking louse, and *Trichodectes canis* and *Heterodoxus spiniger* are biting lice of **dogs.** Pediculosis is a rare disease in pet dogs. Breeds with moderately long, fine hair may provide a more favorable environment for lice, and the disease is more prevalent in the cooler winter months. The biting lice cause pruritus, which may be associated with mild to moderately severe papulocrustous dermatitis or with patchy alopecia. Infestation in the absence of pruritus may be an incidental finding.

Only one species, the biting louse, *Felicola subrostratus,* occurs on **cats.** Infestation may be an incidental finding, or it may be associated with mild pruritus in the absence of lesions. Occasionally there is generalized scaling (seborrhea sicca) or multifocal or generalized papulocrustous dermatitis.

Bibliography

Arundel, J. H., and Sutherland, A. K. "Animal Health in Australia. Volume 10. Ectoparasitic Diseases of Sheep, Cattle, Goats and Horses." Australian Government Publishing Service, 1988.

Britt, A. G. *et al.* Effects of the sheep-chewing louse (*Damalinia ovis*) in the epidermis of the Australian Merino. *Aust J Biol Sci* **39:** 137–143, 1986.

Chalmers, K., and Charleston, W. A. G. Cattle lice in New Zealand: Observations on the prevalence, distribution and seasonal patterns of infestation: Observations on the biology and ecology of *Damalinia bovis* and *Linognathus vituli*: Effects on host live weight gain and hematocrit levels. *N Z Vet J* **28:** 198–200, 214–216, 235–237, 1980.

Collins, R. C., and Dewhirst, L. W. Some effects of the sucking louse *Haematopinus eurysternus* in cattle on unsupplemented range. *J Am Vet Med Assoc* **146:** 129–132, 1965.

Cummins, L. J., and Graham, J. F. The effect of lice infestation on the growth of Hereford calves. *Aust Vet J* **58:** 194–196, 1982.

Murray, M. D. Arthropods—The pelage of mammals as an environment. *Int J Parasitol* **17:** 191–195, 1987.

Nelson, W. A., Shemarchuk, J. A., and Haufe, W. O. *Haematopinus eurysternus:* Blood of cattle infested with the short-nosed cattle louse. *Exp Parasitol* **28:** 263–272, 1970.

O'Callaghan, M. G., Moore, E., and Langman, M. *Damalinia caprae* infestations on sheep. *Aust Vet J* **65:** 66, 1988.

Oormazdi, H., and Baker, K. P. Studies on the effects of lice on cattle. *Br Vet J* **136:** 146–153, 1980.

Sinclair, A. N. Crusts on the epidermis of some louse-infested Merino sheep. *Aust Vet J* **66:** 151–154, 1989.

Sinclair, A. N., Butler, R. W., and Picton, J. Feeding of the chewing louse *Damalinia ovis* (Schrank) (Phthiraptera: Trichodectidae) on sheep. *Vet Parasitol* **30:** 233–251, 1989.

Wilkinson, F. C., Chaneet, G. C., and Beetson, B. R. Growth of populations of lice, *Damalinia ovis,* on sheep and their effects on production and processing performance of wool. *Vet Parasitol* **9:** 243–252, 1982.

C. Fleas

Fleas are ubiquitous and obligate parasites. They are not permanent parasites and often leave one host for another. They are not host specific, but there is probably some host preference. Fleas are chiefly a problem in cats, dogs, pigs, and humans.

Fleas are the most common ectoparasites of **cats** and **dogs.** The most important species are *Ctenocephalides felis,* the cat flea, and *Ctenocephalides canis,* the dog flea. However, infestations also occur with *Pulex irritans* (human flea), *Leptopsylla segnis* (rat flea), *Echidnophaga gallinacea* (chicken stick-tight flea), *Spilopsyllus cuniculi* (European rabbit flea), and *Ceratophyllus* spp. (bird and hedgehog fleas).

The clinical manifestations of flea infestation are highly variable. Some animals, despite heavy infestations, remain asymptomatic carriers. Some animals may develop **flea-bite dermatitis,** which is a reaction to the many irritant substances in the flea's saliva, but the vast majority of animals which develop lesions do so because of **hypersensitivity reactions** to allergenic components of the flea saliva. **Flea-allergy dermatitis** is extremely common and very important disease of the dog and cat; it is discussed in detail under Immune-Mediated Dermatoses (Section X). Finally, the bloodsucking activities of fleas may induce blood-loss anemia in heavily infested animals, particularly in kittens, puppies, or debilitated adults.

Lesions affecting the pinna of cats may be caused by *Spilopsyllus cuniculi.* Typically, hunting cats acquire the infestation from their prey, namely rabbits and hares. Macroscopic lesions are crusted, alopecic patches on both aspects of the pinna. Histologically, basophils and eosinophils are prominent in the dermal inflammatory cell infiltrate.

The two fleas most commonly associated with **swine** are the human flea (*P. irritans*) and the chicken stick-tight flea (*E. gallinacea*). The cat flea (*C. felis*) occasionally infests piglets. In Africa, *Tunga penetrans,* the chigoe flea, has been associated with swine infestations, although it is chiefly a human parasite. The female flea burrows into the skin, causing ulcerative lesions. The skin around the coronary band, on the scrotum, and on the snout is favored. Infestations of the teat canal have been associated with agalactia in sows.

Heavy infestations with *Ctenocephalides* spp. in Africa lead to anemia, reduced weight gain, and even death in **sheep** and **goats,** particularly in the young. Fleas occasionally infest horses (*T. penetrans* and *E. gallinacea*) and cattle (*C. felis*). Heavy infestation with *C. felis* was reported to cause mortality in calves, lambs, and kids in Israel.

Bibliography

Blackmon, D. M., and Nolan, M. P. *Ctenocephalides felis* infestation. *Agri-Pract* **5:** 6–8, 1984.

Cooper, J. E. An outbreak of *Tunga penetrans* in a pig herd. *Vet Rec* **80:** 365–366, 1967.

Fagbemi, B. O. Effect of *Ctenocephalides felis strongylus* infestation on the performance of West African dwarf sheep and goats. *Vet Q* **4:** 92–95, 1982.

Kalkofen, U. P., and Greenberg, J. *Echidnophaga gallinacea* infestation in dogs. *J Am Vet Med Assoc* **165:** 447–448, 1974.

Kristensen, S., Haarlov, N., and Mourier, H. A study of skin diseases in dogs and cats. IV. Patterns of flea infestation in dogs and cats in Denmark. *Nord Vet Med* **30:** 401–413, 1978.

Kwochka, K. W. Fleas and related diseases. *Vet Clin North Am* **17:** 1235–1262, 1987.

Obasaju, M. F., and Otesile, E. B. *Ctenocephalides canis* infestation of sheep and goats. *Trop Anim Health Prod* **12:** 116–118, 1980.

Studdert, V. P, and Arundel, J. H. Dermatitis of the pinnae of cats in Australia associated with the European rabbit flea (*Spilopsyllus cuniculi*). *Vet Rec* **123:** 624–625, 1988.

Yeruham, L., Rosen, S., and Hadani, A. Mortality in calves, lambs, and kids caused by severe infestation with the cat flea *Ctenocephalides felis* (Bouché, 1835) in Israel. *Vet Parasitol* **30:** 351–356, 1989.

D. Mites

1. Sarcoptic Mange

Sarcoptes scabiei (Acarina: Sarcoptidae) is responsible for scabies in humans and sarcoptic mange in domestic animals worldwide. It is a common ectoparasite in swine. The disease occurs in cattle and goats but is not as important as psoroptic mange. The disease is rare in horses and sheep. Ovine sarcoptic mange has not been reported in North America. Sarcoptic mange is common in the dog and may be underdiagnosed. It occurs rarely in the cat, with three documented cases to date. So-called feline scabies is caused by *Notoedres*. The economic importance of sarcoptic mange in food-producing animals, chiefly pigs, is due to depressed growth rate and decreased rates of food conversion. Production studies in experimentally infected pigs report conflicting results, but field studies, particularly those demonstrating the effect of therapy, support the contention that the disease is of economic significance. Sarcoptic mange is a notifiable disease in many countries.

Through host adaptation, *S. scabiei* has become divided into morphologically indistinguishable varieties which rarely cross infect. Each variety is named for its host; thus *S. scabiei* var. *equi* is usually confined to the horse but may live temporarily on cattle or humans. Humans are quite readily parasitized by most of the animal-adapted varieties. However, in most cross-infections, the parasites remain on the skin surface and do not complete their life cycle.

In the normal host, the parasite completes it life cycle in tunnels burrowed into and under the stratum corneum. After mating in a molting pocket close to the surface, the female burrows through the stratum corneum to feed on cells of the stratum granulosum and stratum spinosum. The excavation is achieved by cutting mouthparts and cutting hooks on the legs. In swine, this phase of infestation takes ~3 weeks. Epidermal cell damage induces proliferative changes in the surrounding keratinocytes so that the surface openings of the tunnels become sealed with thick parakeratotic scale-crust. In swine this process takes a further 3–4 weeks. After 7 weeks of infestation, the crust falls off, and the mites vacate the tunnels. Approximately 40–50 ova are laid in the burrows at a rate of 1 to 3 per day. They develop through the larval and nymphal stages in the same tunnel or in new ones, to reach maturity in 10 to 15 days, depending on the species. Both parasites and ova have poor viability in the external environment. The disease, which is highly contagious, is transmitted largely by direct contact but may occur following indirect contact with contaminated objects such as bedding.

The pathogenesis of lesions in *S. scabiei* infestation is due to direct damage inflicted by the parasite mechanically and by the irritant effects of its secretions and excreta, and by an allergic reaction developed against components of the mite or one or more of the extracellular products of the parasite. Evidence for an allergic pathogenesis in animals comes chiefly from experimental infestations of swine. Initial lesions in pigs are due to parasite invasion and are localized and nonpruritic. After 7 to 11 weeks, there is a generalized urticarial eruption associated with extreme pruritus. This eruption coincides with the development of immediate and delayed hypersensitivity reactions and peripheral eosinophilia. The delayed hypersensitivity response is dependent on antigen dose, whereas the immediate hypersensitivity response is independent of the degree of antigen exposure. Lesions regress between 12 and 18 weeks after the initial infestation.

The variability of the clinical manifestations of sarcoptic mange probably reflects individual variations in the duration and intensity of the hypersensitivity reaction and in the related capacity of the host to limit parasitic multiplication. Most *Sarcoptes scabiei*-related diseases are caused chiefly by the allergic reaction, and the lesions are the result of self-trauma induced by severe pruritus. However, animals with a weak hypersensitivity reaction may exhibit a severe crusting dermatitis characterized by the presence of large numbers of mites. This type of disease is typically seen in poorly nourished animals or animals debilitated by coexisting disease. An example is the chronic form of scabies in dogs associated with long-term corticosteroid therapy. This has been referred to as Norwegian-type scabies, following the human nomenclature for severe scabies in immunosuppressed people. Diminished levels of hypersensitivity are correlated also with the development of chronic infections in swine, known as hyperkeratotic mange. This manifestation of mange is sporadic and affects chiefly breeding adults.

The primary parasite-related lesions of sarcoptic mange are erythematous macules or papules, which develop a local scale-crust in reaction to the burrowing mites. The thickness of the overlying scale-crust is proportional to the number of tunnels beneath. In poorly nourished animals or immunosuppressed animals, which develop massive mite infestations, the lesions are characterized by alopecia, marked lichenification, accumulation of thick gray scale-crust, and fissuring. The lesions of sarcoptic mange hypersensitivity largely result from self-trauma induced by pruritus. In the early stages, these include erythematous papules, excoriations, hemorrhagic crusts, and patchy alopecia. Chronic hypersensitivity lesions include marked alopecia, scaling, and lichenification.

The distribution of the lesions is characteristic in the various species. In pigs (Fig. 5.103), the mites have a predilection for the inner surface of the pinna where they cause primary lesions. The papular lesions associated with the allergic reactions are located chiefly on the rump, flank, and abdomen, and the secondary changes due to chronic self-trauma follow a similar pattern. In the chronic hyperkeratotic form of the disease, heavily crusted lesions develop over the whole body but are most severe on the head, neck, and legs. In **dogs,** the preferred sites are the lateral elbows, hocks, ventral thorax, and lateral margin of the pinna. There may be an associated peripheral lymphadenopathy. The lesions may become generalized in untreated dogs or dogs inappropriately treated with corticosteroids. In **cattle,** the lesions chiefly affect the neck, head, and sacral areas but may become generalized. The disease in **goats** also has a predilection for the head but may involve the whole body. In the **sheep,** only the haired areas develop lesions, particularly the lips, nostrils, external surface of the pinna, and occasionally the legs. Generalized lesions occur in the more hairy desert sheep of the Sudan. Lesions in the **horse** begin on the head and neck and may extend to involve most of the body but seldom the legs or mane.

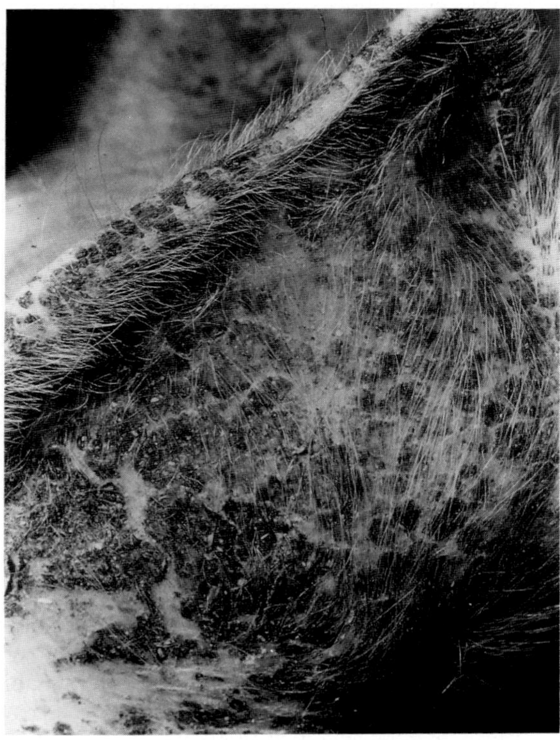

Fig. 5.103 Sarcoptic mange. Pig.

Histologically the lesions vary with the balance between allergic reaction and parasitic infestation. In immunosuppressed animals, large numbers of adult mites lie in burrows lined by parakeratotic stratum corneum. There is a thick scale-crust composed of hyperkeratotic and parakeratotic hyperkeratosis, serum lakes, and neutrophilic debris. The epidermis is markedly hyperplastic, with prominent rete ridge formation and/or pseudocarcinomatous hyperplasia. Dermal lesions include variable vasodilation, endothelial swelling, edema, fibrosis, perivascular mononuclear cell and eosinophilic infiltration, and neutrophilic exocytosis, depending on the development of secondary bacterial infection or epidermal excoriation. In acute allergic reactions, the lesions are marked vasodilation, dermal edema, perivascular lymphocytic and eosinophilic infiltration, epidermal spongiosis, and hyperplasia. Mites are few. The chronic allergic lesions reflect continued trauma with dermal fibrosis, epidermal hyperplasia, and a predominantly mononuclear cell perivascular infiltrate.

Diagnosis of the typical allergic form of the disease depends chiefly on the clinical signs of extreme pruritus and the nature and distribution of the cutaneous lesions. Mites are characteristically difficult to demonstrate, either in skin scrapings or in microscopic section. Approximately two thirds of affected dogs fail to yield parasites even when multiple scrapings are performed. The microscopic lesions are not diagnostic, being indistinguishable from other allergic dermatoses. The most useful diagnostic procedure is often response to therapy. In the chronic form of sarcoptic mange associated with poorly developed hypersensitivity reactions, mites are plentiful in scrapings and in tissue section.

Bibliography

Abu-Samra, M. T., Hago, B. E. D., and Aziz, M. A. Sarcoptic mange in sheep in the Sudan. *Ann Trop Med Parasitol* **75:** 639–645, 1981.

Anderson, R. K. Norwegian scabies in a dog: A case report. *J Am Anim Hosp Assoc* **17:** 101–104, 1981.

Arends, J. J., Stanislaw, C. M., and Gerdon, D. Effects of sarcoptic mange on lactating swine and growing pigs. *J Anim Sci* **68:** 1495–1499, 1990.

Arlian, L. G. Biology, host relations, and epidemiology of *Sarcoptes scabiei*. *Annu Rev Entomol* **34:** 139–161, 1989.

Cargill, C. F., and Dobson, K. J. Experimental *Sarcoptes scabiei* infestation in pigs: (1) Pathogenesis. (2) Effects on production. *Vet Rec* **104:** 11–14, 33–36, 1979.

Davis, D. P., and Moon, R. D. Density of itch mite, *Sarcoptes scabiei* (Acari: Sarcoptidae) and temporal development of cutaneous hypersensitivity in swine mange. *Vet Parasitol* **36:** 285–293, 1990.

Davis, D. P., and Moon, R. D. Dynamics of swine mange: A critical review of the literature. *J Med Entomol* **27:** 727–737, 1990.

Folz, S. D. Canine scabies (*Sarcoptes scabiei* infestation). *Compend Cont Ed* **6:** 176–180, 1984.

Hollanders, W., and Vercruysse, J. Sarcoptic mite hypersensitivity: A cause of dermatitis in fattening pigs at slaughter. *Vet Rec* **126:** 308–310, 1990.

Jackson, P. G. G., Richards, H. W., and Lloyd, S. Sarcoptic mange in goats. *Vet Rec* **112:** 330, 1983.

Kershaw, A. *Sarcoptes scabiei* infestation in a cat. *Vet Rec* **124:** 537–538, 1989.

Martineau, G., Van Neste, D., and Charette, R. Pathophysiology of sarcoptic mange in swine—Part I and II. *Compend Cont Ed* **9:** F51–F57, F93–F97, 1987.

Morsy, G. H., and Gaafar, S. M. Responses of immunoglobulin-secreting cells in the skin of pigs during *Sarcoptes scabiei* infestation. *Vet Parasitol* **33:** 165–175, 1989.

Morsy, G. H., Turek, J. J., and Gaafar, S. M. Scanning electron microscopy of sarcoptic mange lesions in swine. *Vet Parasitol* **31:** 281–288, 1989.

Nusbaum, S. R. *et al. Sarcoptes scabiei bovis*—a potential danger. *J Am Vet Med Assoc* **166:** 252–256, 1975.

Roncalli, R. A. The history of scabies in veterinary and human medicine from biblical to modern times. *Vet Parasitol* **25:** 193–198, 1987.

Sheahan, B. J. Pathology of *Sarcoptes scabiei* infection in pigs. I. Naturally occurring and experimentally induced lesions. II. Histologic, histochemical and ultrastructural changes at skin test sites. *J Comp Pathol* **85:** 87–95, 97–110, 1975.

Smith, H. J. Transmission of *Sarcoptes scabiei* in swine by fomites. *Can Vet J* **27:** 252–254, 1986.

Wooten-Saadi, E. *et al.* Growth performance and behavioral patterns of pigs infested with sarcoptic mites (Acari: Sarcoptidae). *J Econ Entomol* **80:** 625–628, 1987.

2. *Notoedric Mange*

Notoedres cati (Acarina: Sarcoptidae) is predominantly a parasite of the cat and rabbit, although infestations of dogs and humans may occur. The disease in cats is uncommon to rare, although there are some endemic areas of higher prevalence. The mite has a life cycle similar to that of *S. scabiei*. The infestation is highly contagious, with transmission chiefly by direct contact. The major clinical sign is pruritus. The lesions in cats commence on the head and ears, particularly on the margin of the pinna, but may extend to the neck, paws, or become generalized. Lesions include partial alopecia, thickening and wrinkling of the skin, and in chronic cases, the formation of tightly adherent yellowish-gray crusts (Fig. 5.104A) There may be an accompanying regional lymphadenopathy. Lesions in dogs are indistinguishable from sarcoptic mange. Histologically, the lesion is a hyperplastic, eosinophil-rich, superficial perivascular dermatitis with focal parakeratosis. The diagnosis is based on history, clinical signs, and demonstration of typical mites in section (Fig. 5.104B) or skin scrapings. These are reported to be plentiful, but we suggest that feline notoedric mange may follow the pattern of sarcoptic mange in other species, and allergic disease associated with the presence of very few mites may be more frequent than is realized.

Bibliography

Scott, D. W. Feline dermatology. Parasitic disorders. *J Am Anim Hosp Assoc* **16:** 365–374, 1980.

3. *Psoroptic Mange*

The psoroptic mites (Acarina: Psoroptidae) infest sheep, cattle, horses, rabbits, and goats, as well as other nondomestic species. Humans are not susceptible. Psoroptic mites are reputed to be more host specific than sarcoptic mites, but several reports of interspecies transfer suggest that this is not the case. *Psoroptes ovis* infests both sheep and cattle, and mites transferred from one host species readily cause disease in the other. *Psoroptes cuniculi* affects the ears of several species, including rabbit, horse, donkey, goat, and sheep. *Psoroptes natalensis* affects cattle in South Africa and South America. *Psoroptes equi* is the body mite of horses. Some authors recognize *P. hippotis, P. caprae,* and *P. bovis*.

Psoroptic mange is a serious disease in cattle and sheep. Bovine psoroptic mange showed a recrudescence in North America during the 1970s and early 1980s but has been brought under control by the effective use of ivermectin. Ovine psoroptic mange has been eradicated from many countries, including Australia and New Zealand. Psoroptic mange remains a reportable disease in several countries. The economic importance in sheep and cattle results from a marked decrease in weight gain, reduced milk production, reduced fleece weight and quality, occasional mortality, and costs related to prevention and eradication campaigns.

The psoroptic mites do not burrow into the outer epidermis, as do the sarcoptic mites, but instead complete their life cycle on the skin surface. The conventional wisdom has it that the mite receives its nutrition by piercing the epidermis and sucking tissue fluids and lymph. Actually lipids from the stratum corneum provide a major source of nutrients in the early stages of infestation, probably supplemented by serous and hemorrhagic inflammatory exudates in the later stages. There is little evidence for penetration of the viable epidermis.

Psoroptic mange is characterized by an intensely pruritic dermatitis. The pathogenicity of the mite has been attributed to its local irritant effect on the epidermis, but this does not readily explain the marked loss of condition induced by *Psoroptes* infestation in some species, particularly cattle. The detrimental systemic effects may derive directly or indirectly from a chronic hypersensitivity reac-

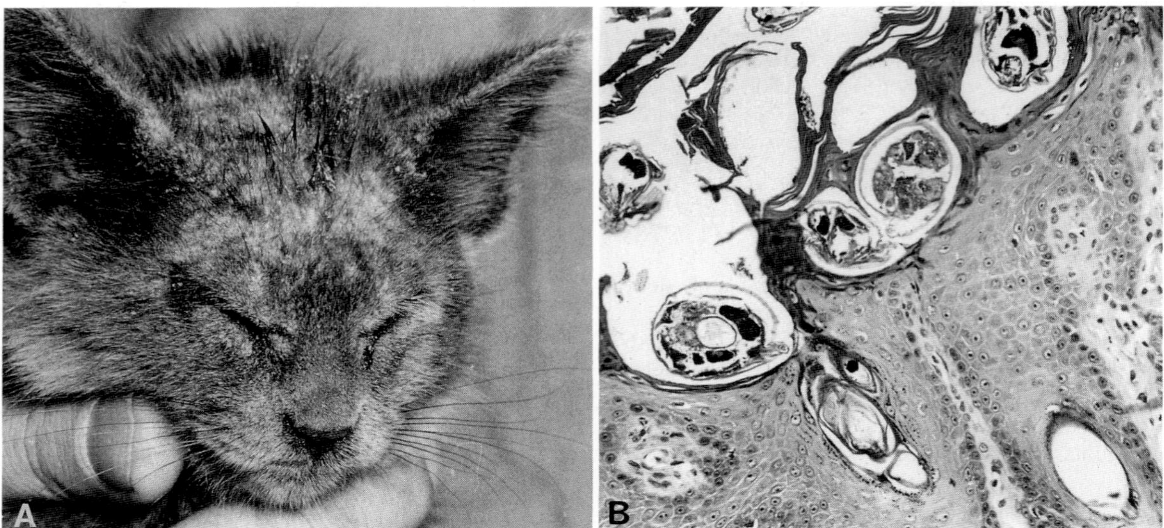

Fig. 5.104 Notoedric mange. Cat. (A) Gross appearance. (B) Mites in keratin on surface of thickened epidermis.

tion rather than from local dermatitis. The histologic lesions, in which the predominant inflammatory cells in the superficial dermis are eosinophils, mast cells, and lymphocytes, are consistent with an allergic pathogenesis. Constant pruritus resulting from allergy markedly reduces feed intake, and secondary bacterial infection or myiasis may further contribute to loss of condition. There is little evidence to support the hypothesis that the mites inoculate a toxic compound along with their saliva.

Histologically, the lesions are similar in all species. The pattern of inflammation is superficial perivascular dermatitis with predominantly spongiotic, exudative, or hyperplastic reactions depending on chronicity. The eosinophil is the most numerous of the infiltrating leukocytes, followed by lymphocytes, other mononuclear cells, and mast cells. Dermal edema is usually marked. Mites are present both on top of and under the surface scale-crust. Sebaceous gland hyperplasia has been described in lesions in sheep and cattle.

Psoroptic mange in **sheep,** also known as sheep scab, may occur as a latent infection in which mites persist in the ears, infraorbital fossae, inguinal and perineal folds, and at the base of the horns. In rams, mites may be found on the scrotum or prepuce in small, dry lesions. Latency occurs in the summer months when the fleece microclimate is less favorable to parasite proliferation. In autumn and winter or with debilitation of the host, the parasitic population expands, and lesions are induced. The withers and sides are particularly affected. The initial lesions are papules ~0.5 cm in diameter covered with a yellow serous crust, which may matt the fleece. The individual lesions expand at the periphery and may coalesce to become diffuse over most of the body surface. The main damage to the fleece is caused by self-trauma induced by the severe pruritus. Affected sheep scratch, kick, rub, and tear out the fleece with their teeth.

The disease in **cattle** also diminishes in the summer months as a result of decreasing mite populations, in turn at least partly attributable to increased self-grooming. Previous exposure to mites, although not conferring a solid resistance to reinfection, also limits the mite population, chiefly by reducing the rate of oviposition in the females. Lesions in reinfested cattle, occur earlier and progress more slowly, reflecting the decreased mite population. The lesions in naturally affected cattle usually commence about the poll, withers, or at the base of the tail (Fig. 5.105). The lesions result chiefly from persistent licking, rubbing, and scratching induced by the pruritus. The infested areas are fairly well defined by a sharp margin between alopecic and normal skin. The alopecic areas become lichenified and covered by dry gray crusts and scales. Severely affected calves may develop mild anemia, lymphopenia, and a marked neutropenia. Experimentally infected calves develop eosinophilia which parallels the clinical course of infection.

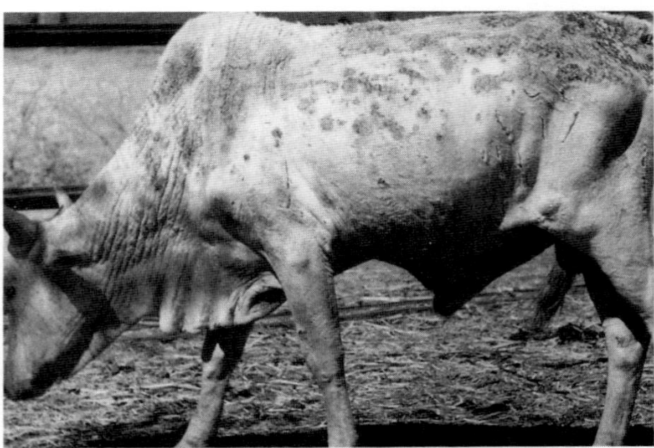

Fig. 5.105 Psoroptic mange. Cow. (Courtesy of F. I. Awad.)

In **goats,** *P. cuniculi* is known as the ear canker mite because of its predilection for the external auditory meatus. Mite infestation induces head shaking and occasionally crusted lesions on the inner surface, and alopecia on the outer surface of the pinna. In debilitated or stressed animals, thick brown-yellow, dry scale-crust accumulates on the inner aspect of the pinna and, rarely, spreads to involve the poll, body, and the legs. Concurrent mycoplasma and *P. cuniculi* infections have been described in goats' ears; however, mycoplasma may be cultured from the ears of clinically normal animals, placing some doubt on the significance of the finding.

In **horses,** *P. cuniculi* is found quite frequently in the ear. Infestations have been associated with head shakers in the United Kingdom and Australia, but are often subclinical. *Psoroptes equi* infestations are rare; lesions are crusted papules with alopecia, and the preferred sites are at the base of the mane, forelock, and tail.

Bibliography

Abu-Samra, M. T., Abbas, B., and Ibrahim, K. E. E. Five cases of psoroptic mange in the domestic donkey (*Equus asinus asinus*) and treatment with ivermectin. *Eq Vet J* **19:** 143–144, 1987.

Cook, R. W. Ear mites (*Raillieta manfredi* and *Psoroptes cuniculi*) in goats in New South Wales. *Aust Vet J* **57:** 72–75, 1981.

Downing, W. The life history of *Psoroptes communis* var. *ovis* with particular reference to latent or suppressed scab. *J Comp Pathol Ther* **49:** 183–209, 1936.

Fisher, W. F., and Wright, F. C. Effects of the sheep scab mite on cumulative weight gains in cattle. *J Econ Entomol* **74:** 234–237, 1981.

Guillot, F. S. Population increase of *Psoroptes ovis* (Acari: Psoroptidae) on stanchioned cattle during summer. *J Med Entomol* **18:** 44–47, 1981.

Hourrigan, J. L. Spread and detection of psoroptic scabies of cattle in the United States. *J Am Vet Med Assoc* **175:** 1278–1280, 1979.

Kirkwood, A. C. Effect of *Psoroptes ovis* on the weight of sheep. *Vet Rec* **107:** 469–470, 1980.

Losson, B., Detry-Pouplard, M., and Pouplard, L. Haematological and immunological response of unrestrained cattle to *Psoroptes ovis*, the sheep scab mite. *Res Vet Sci* **44:** 197–201, 1988.

Lucas, K. M., and Roberts, F. H. S. Psoroptic otacariasis of the horse. *Aust Vet J* **22:** 186, 1946.

Meleney, W. P. Control of psoroptic scabies on calves with ivermectin. *Am J Vet Res* **43:** 329–331, 1982.

Pascoe, R. R. Mites in "head shaker" horses. *Vet Rec* **107:** 234, 1980.

Roberts, I. H., and Meleney, W. P. Variations among strains of *Psoroptes ovis* (Acarina: Psoroptidae) on sheep and cattle. *Ann Entomol Soc Am* **64:** 109–116, 1971.

Sinclair, A. N., and Filan, S. J. Lipid ingestion from sheep epidermis by *Psoroptes ovis* (Acari: Psoroptidae). *Vet Parasitol* **31:** 149–164, 1989.

Stromberg, P. C. *et al.* Systemic pathologic responses in experimental *Psoroptes ovis* infestation of Hereford calves. *Am J Vet Res* **47:** 1326–1331, 1986.

Stromberg, P. C., and Guillot, F. S. Pathogenesis of psoroptic scabies in Hereford heifer calves. *Am J Vet Res* **50:** 594–601, 1989.

Williams, J. F., and Williams, C. S. F. Psoroptic ear mites in dairy goats. *J Am Vet Med Assoc* **173:** 1582–1583, 1978.

Wilson, G. I., Blachut, K., and Roberts, I. H. Infectivity of scabies mites, *Psoroptes ovis* (Acarina: Psoroptidae), to sheep in naturally contaminated enclosures. *Res Vet Sci* **22:** 292–297, 1977.

Wright, F. C., and DeLoach, J. R. Feeding of *Psoroptes ovis* (Acari: Psoroptidae) on cattle. *J Med Entomol* **18:** 349–350, 1981.

4. Chorioptic Mange

Chorioptic mange, caused by *C. bovis* (Acarina: Psoroptidae), affects horses, cattle, sheep, and goats. Not host specific, *C. bovis* is an obligate parasite which lives on the surface of the skin. The mite populations tend to fluctuate considerably as a result of host and environmental factors. Inapparent infections allow persistence in the population. Clinically affected animals are pruritic and have papular, crusted, scaly, alopecic, and/or lichenified lesions, depending on the duration of the disease and the degree of self-trauma inflicted.

The disease in **cattle** predominantly affects housed dairy cows and is most prevalent in winter. Subclinical infections are probably quite common. The major clinical sign is pruritus, but this is not so severe as that in sarcoptic or psoroptic mange. Chorioptic mange is in general a less serious condition than psoroptic mange in cattle, although a syndrome of highly irritant coronitis was associated with falling milk production. The typical distribution of lesions is perineum, udder, caudal areas of thigh, and rump. Lesions are predominantly alopecia, lichenification, and wrinkling of the skin.

The disease in **horses** is uncommon. As indicated by the colloquial name leg mange, lesions occur preferentially on the lower limb around the fetlock but may extend proximally to the thigh and ventral abdomen. The lesions are most severe in winter, as in cattle.

In **goats,** lesions originally were described as commencing on the neck and spreading to the back, root of tail, and lateral body. Another pattern, in which lesions affect the coronet, pasterns, and lower limbs, appears to be more typical. The face, udder, and scrotum also may be affected.

In **sheep,** *C. bovis* mites prefer the distal extremities, particularly the pastern and interdigital skin of the hind limbs. The scrotum may be affected and the resultant scrotal dermatitis may lead to temporary infertility. Chorioptic mange has been eradicated from the sheep population in the United States. It is a reportable disease in some countries.

Bibliography

Heath, A. C. G. The scrotal mange mite, *Chorioptes bovis* (Hering 1845) on sheep: Seasonality, pathogenicity, and intraflock transfer. *N Z Vet J* **26:** 299–300, 309–310, 1978.

McKenna, C. T., and Pulsford, M. F. A note on the occurrence of *Chorioptes communis* var. *ovis* on sheep in south Australia. *Aust Vet J* **23:** 146–147, 1947.

Sweatman, G. K. Life history, non-specificity, and revision of the genus *Chorioptes*, a parasitic mite of herbivores. *Can J Zool* **35:** 641–689, 1957.

5. *Otodectic Mange*

Otodectes cynotis (Acarina: Psoroptidae) is an obligate parasite of the external skin surface of dogs and cats. While the mite may be found at several body sites, its preferred habitat is the external ear canal. The major lesion is thus otitis externa (see Chapter 4, The Eye and Ear). Focal, erythematous, alopecic, or excoriated lesions occur occasionally on the face, feet, neck, or tail head.

6. *Cheyletiellosis*

Members of the genus *Cheyletiella* (Acarina: Cheyletidae) affect dogs, cats, rabbits, wild animal species, and incidentally, humans. There are three species involved: *C. parasitivorax* is chiefly a parasite of rabbits, although formerly considered a canine pathogen; *C. yasguri* is now regarded as the major canine cheyletiellid, and the species most commonly associated with feline infestation is *C. blakei*. Host specificity is weak, and cross-infestations are not uncommon. The mites are obligate parasites, completing their life cycle on the skin surface in ~35 days. The infestation is transmitted by direct contact, frequently from a carrier female to her litter. The mites usually survive for only 1–2 days away from the host (slightly longer at 4°C), making indirect transmission possible but less likely.

The pathogenicity of cheyletiellid infestation is controversial. The presence of the mite on naive hosts usually, but not always, induces hyperkeratosis, which may or may not be associated with pruritus. Adults, possibly because of acquired immunity, usually have asymptomatic infestations.

The gross lesions in both the dog and cat reflect the mite's predilection for the dorsal midline. Lesions often commence over the caudal back and progress anteriorly but may become generalized. The typical lesion is a moderate to marked exfoliation of small, dry, white scales (seborrhea sicca). The mites crawl in pseudotunnels in the loose keratin debris, and their movement has produced the colloquial name walking dandruff for cheyletiellosis. Cats may develop, in addition, focal, multifocal, or generalized erythematous papules or crusted lesions. Occasionally animals show pruritus in the absence of scaling. Diagnosis depends on demonstration of the mites. Skin scraping or, better, brush techniques, yield both mites and ova. The latter are usually attached to hairs. Biopsy, which shows a spongiotic, hyperplastic, superficial perivascular dermatitis with variable numbers of eosinophils, is nondiagnostic, since mites are usually not demonstrable. Humans in contact with affected animals often develop a pruritic maculopapular rash on the arms and trunk. As the mites do not complete their life cycle on human skin, these lesions regress once the animal is treated.

Bibliography

Ayalew, L., and Vaillancourt, M. Observations on an outbreak of infestation of dogs with *Cheyletiella yasguri* and its public health implications. *Can Vet J* **17:** 184–190, 1976.

Fox, J, G., and Reed, C. *Cheyletiella* infestation of cats and their owners. *Arch Dermatol* **114:** 1233–1234, 1978.

McKeever, P. J., and Allen, S. K. Dermatitis associated with *Cheyletiella* infestation in cats. *J Am Vet Med Assoc* **174:** 718–720, 1979.

Schmeitzel, L. P. Cheyletiellosis and scabies. *Vet Clin North Am* **18:** 1069–1076, 1988.

7. *Psorergatic Mange*

Psorergates ovis (Acarina: Cheyletidae) is a parasite of the integument of the sheep. The disease occurs in Australia, New Zealand, the United States, South Africa, and Argentina. The mite, which is much smaller than sarcoptid mites, is an obligate parasite and goes through its life cycle on the skin surface in the loose keratin debris. The mite does not penetrate deeper than the stratum corneum. Infestations occur predominantly on the dorsum. Seasonal influences greatly affect mite populations, with lowest numbers occurring in the summer and highest in the spring.

Mite infestations induce pruritus; the pathogenesis is not known, although hypersensitivity reactions have been postulated. Lesions result from the sheep biting and pulling at the fleece. Fleece damage is more severe in the fine-wooled Merino than in the coarse-wooled breeds. The lesions occur on the flanks, thighs, and lateral body wall and comprise bleached, twisted tufts, which give the fleece a ragged, tasseled appearance.

A mite named *P. bos* has been isolated from nonpruritic, scaling, and alopecic lesions in cattle in New Mexico and Texas.

Bibliography

Johnson, P.W. *et al.* The prevalence of itchmite, *Psorergates ovis*, among sheep flocks with a history of fleece derangement. *Aust Vet J* **67:** 117–120, 1990.

Malan, F. S., and Roper, N. A. The seasonal incidence of the sheep itch mite, *Psorergates ovis* Womersly under subtropical conditions. *J S Afr Vet Assoc* **53:** 171–174, 1982.

Roberts, I. H., and Meleney, W. P. Psorergatic acariasis of cattle. *J Am Vet Med Assoc* **146:** 17–23, 1965.

Sinclair, A. The epidermal location and possible feeding site of *Psorergates ovis*, the sheep itch mite. *Aust Vet J* **67:** 59–62, 1990.

8. *Demodectic Mange*

Demodex mites (Acarina: Demodicidae) are normal inhabitants of the hair follicles or sebaceous glands in all species of domestic animals and in humans. The mites found in the different hosts are regarded as separate species, although they are similar morphologically. The mites are generally named for the host species, as in *D. canis* of dogs and *D. bovis* of cattle. *Demodex phylloides* of swine, *D. ghanensis* of cattle, *D. caballi* of horses, and *D. aries* of sheep are exceptions in the nomenclature. It was considered unusual for an animal to host more than one species of *Demodex*, but synhospitality may in fact be the rule rather than the exception.

Demodex spp. are obligate parasites, completing their

life cycle in the hair follicle or its adnexae. They are rapidly killed by desiccation on the surface of the skin, but mites move from follicle to follicle, and it is probably at this time that transmission to another host takes place. Transmission usually occurs by direct contact from the dam to her offspring during nursing in the neonatal period.

Demodex mites are part of the normal fauna of the skin in most, if not all, animal species. This implies that small numbers of mites exist in harmony with the host, and it is only when the equilibrium between the host and parasite is altered in favor of the mite that excessive proliferation occurs, and lesions of demodectic mange are produced. The most severe expression of demodectic mange is seen in the dog as a generalized dermatitis (Fig. 5.106A), which is on occasion fatal. Multiple lesions occur commonly in cattle, less commonly in pigs and goats, but with no systemic consequences. The mites of sheep and cats rarely assume pathogenicity.

Demodectic infestation in **dogs** begins when puppies are infected from the bitch via direct skin contact during nursing in the neonatal period. Mites are first observed on the muzzles of newly infected pups. Occasionally this normally benign parasitic infestation is converted into a pathogenic one because of two interrelated factors, genetic predisposition and selective or partial states of immunodeficiency. The disease is more prevalent in particular breeds of purebred dogs and often affects litters of certain bitches which, though clinically normal themselves, produce successive litters of affected puppies. It appears that puppies genetically predisposed to develop demodectic mange have a selective deficit in cell-mediated immunity, which permits massive proliferation of the mite population. Genetic susceptibility may be linked to certain histocompatibility phenotypes in dog. The disease itself, in its complicated form, appears to induce a secondary immunodeficiency state. Dogs with chronic generalized pustular demodicosis show reduced cell-mediated immune responsiveness as measured by depressed lymphocyte blastogenesis reactivity. This change has been attributed to the secondary bacterial infection rather than to the mite infestation *per se*.

The disease in dogs takes two clinical forms. Localized or squamous demodicosis occurs in young dogs 3–10 months of age and is usually self-limiting. These dogs do not have depressed cell-mediated immune responsiveness and can respond to intradermal challenge with a crude *Demodex* antigen. The lesions are single or multiple, well-circumscribed, erythematous, scaly, and alopecic patches, usually affecting the head around the lips and eyes or on the extremities.

The generalized form usually affects young dogs. In older dogs, severe demodicosis may be an indication of serious internal disease and concomitant immunosuppression. The lesions of generalized demodicosis are particularly severe on the face, forelimbs, and feet. In some dogs, lesions may be confined to the feet (Fig. 5.106B). Lesions include patchy to diffuse alopecia, erythema, scaling, and crusting, with or without lesions of secondary bacterial pyoderma. Peripheral lymphadenopathy occurs in ~50% of affected dogs. Often the dogs are depressed, febrile, and debilitated, and may die.

Histologically, the localized form of the disease is characterized by a predominantly lymphocytic–plasmacytic perifolliculitis. Marked follicular hyperkeratosis is associated with the presence of variable numbers of mites in the upper third of the follicle. Pigmentary incontinence around the affected follicle, at approximately the level of the sebaceous gland duct, may provide a helpful clue in those cases in which few mites are present in the section. The external root sheath may be hyperplastic, spongiotic, or show hydropic degeneration. Lymphocytes are present in early lesions, eosinophils are present in small numbers, and there may be mast cell hyperplasia. Small granulomas, sometimes containing remnants of mites or merely eosinophilic and mineralized debris, may be the only lesions in resolving cases.

The microscopic lesions of the generalized form vary

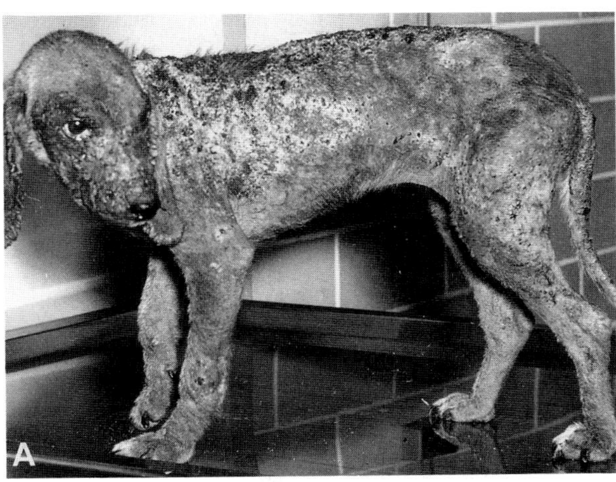

Fig. 5.106A Generalized demodectic mange. Dog.

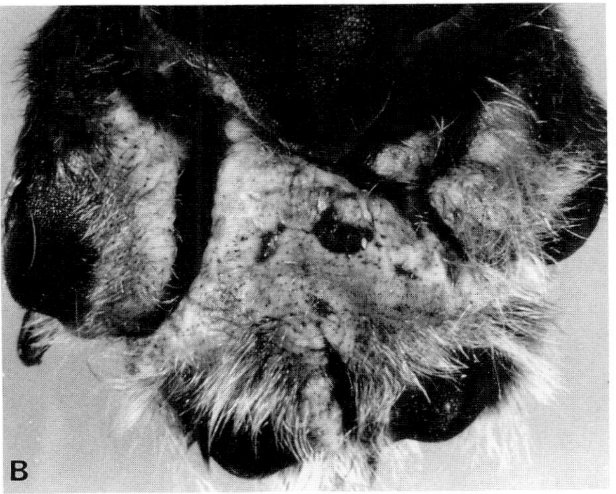

Fig. 5.106B Pododermatitis in demodecosis. Dog.

depending on the presence and extent of secondary bacterial infection and the generation of deep pyoderma. Typically, large numbers of mites occupy the hair follicles at all levels and also occlude the opening of the sebaceous gland into the pilar canal (Fig. 5.107). Marked follicular hyperkeratosis causes follicular plugging. Bacterial proliferation within the plugged follicle often induces a neutrophilic folliculitis. The combined effects of follicular keratosis, mite proliferation, and folliculitis lead to follicular rupture and release of mites, bacteria, keratin, sebum, and other irritant products into the dermis. The bacteria, chiefly *Staphylococcus* spp., induce a suppurative dermatitis often with abscessation. The keratin and other irritant substances stimulate a granulomatous reaction, chiefly of epithelioid macrophages, but a few multinucleate giant cells may be present. This pyogranulomatous furunculosis of demodectic mange differs from most other types of furunculosis in that eosinophils are rare or absent in the reaction. It has been postulated that the absence of eosinophils is a morphologic expression of the acquired immunodeficiency state. Epidermal lesions include hyperplasia, orthokeratotic and parakeratotic hyperkeratosis, and variable spongiosis, neutrophilic exocytosis, ulceration, and inflammatory crusting. Chronic lesions also have marked dermal fibrosis, often with obliteration of adnexa. Mites or fragments of mites are found in the subcapsular zone of regional lymph nodes associated with a local granulomatous inflammatory response. These do not indicate active invasion but rather passive transport to the node, via lymphatic channels.

Demodectic mange in **cattle** occurs worldwide. Three species are responsible: *D. bovis, D. ghanensis,* and one unnamed parasite. Economic significance lies largely in the damage that mite infestation produces in hides. In some parts of Africa and Madagascar, demodectic mange in cattle may become generalized and fatal, this outcome being predisposed for by other debilitating conditions such as malnutrition, tick-worry, and tropical heat.

The typical gross lesions are multiple cutaneous papules or nodules, usually between 2 and 4 mm in diameter but occasionally reaching 1.0 cm or more. Nodules vary in numbers from a few to several hundred (Fig. 5.108A,B). The nodules are visible in smooth-coated cattle, often indicated by overlying tufts of erect hairs. In rough-coated cattle such as Herefords, detection usually requires palpation. The preferred sites are the shoulders, neck, dewlap, and muzzle, but in heavy infestations, nodules may be present over most of the body. The content of the nodules is thick, waxy, or caseous material, sometimes stained with blood. The contents may liquefy and discharge to the surface, forming a thick crust, or rupture of the nodule into

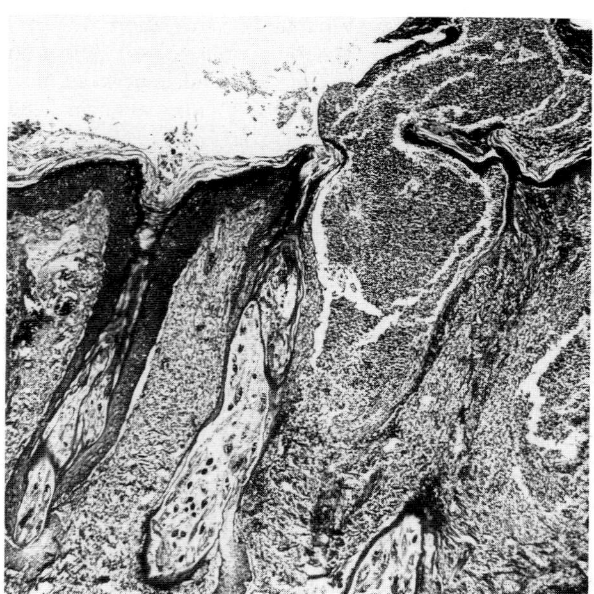

Fig. 5.107 Mites packed within hair follicles. Dog.

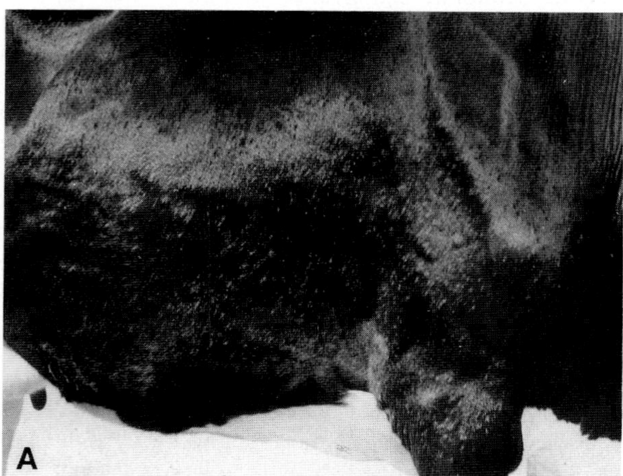

Fig. 5.108A Demodectic mange. Cow.

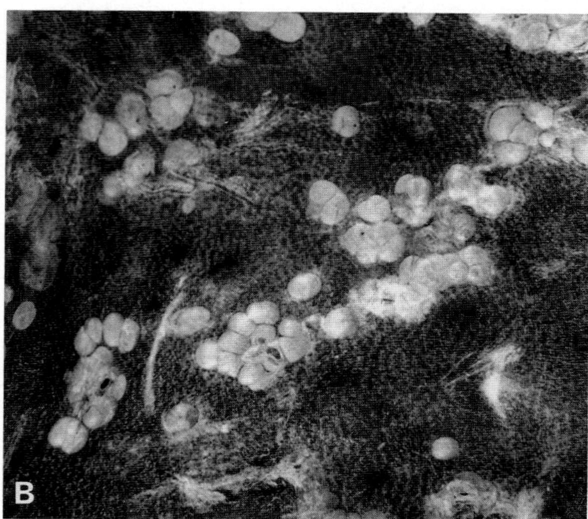

Fig. 5.108B *Demodex bovis* nodules beneath hide.

the dermis may generate an abscess or a granulomatous reaction.

Histologically, the nodules are follicular cysts lined by a flattened squamous epithelium and filled with keratin squames and large numbers of demodicid mites (Fig. 5.109). Adult parasites occur occasionally in sebaceous glands and rarely in apocrine sweat glands. A mild mononuclear cell infiltrate may occur around the epithelial lining. Rupture of a follicular cyst induces a marked nodular granulomatous reaction in which degenerating and occasionally mineralized segments of parasites and keratin debris are surrounded by epithelioid macrophages, multinucleate giant cells, lymphocytes, plasma cells, and eosinophils.

The lesions in **goats,** caused by *D. caprae,* are similar to those described for cattle in both distribution and morphology. Some affected goats are mildly depressed, inappetent, and have decreased milk production. Generally, the chief economic significance of caprine demodicosis lies in damage to the hides.

Demodectic mange is rare in **sheep.** *Demodex ovis* infestation is probably not uncommon in the medium- to coarse-wooled sheep, affecting the meibomian glands of the eyelid and the sebaceous glands of the primary follicles on the body, particularly the neck, flank, and shoulders. *Demodex aries* infests the large sebaceous glands of the vulva, prepuce, and nostrils. Grossly, lesions may be papular, nodular, and rarely pustular. *Demodex ovis* infestation has been associated with matted fleece ("stringy wool").

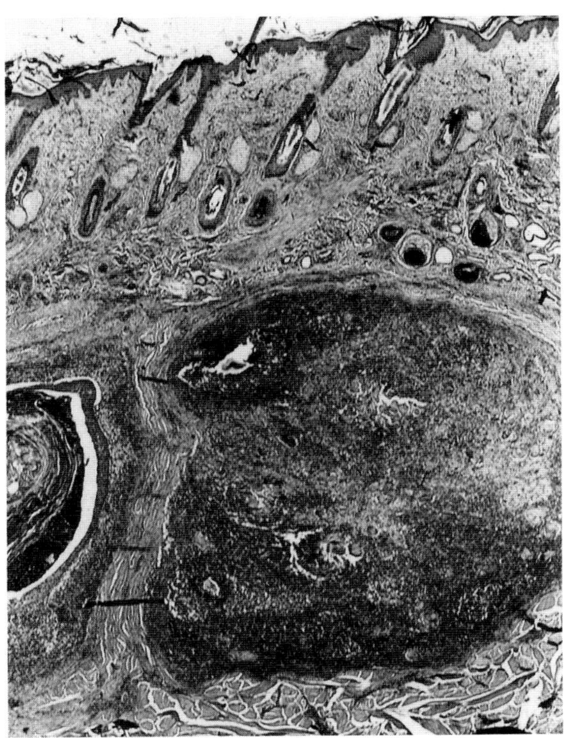

Fig. 5.109 Histologic appearance of nodules.

Histologically, the mites are present in sebaceous glands or the pilar canal, occasionally inciting folliculitis or furunculosis.

Whereas *Demodex* spp. are found commonly in the eyelid glands of the **horse,** demodectic mange is a very rare disease. *Demodex caballi* parasitizes the pilosebaceous units of the eyelids and muzzle, whereas *D. equi* infestation occurs over the body. Lesions associated with the latter occur on the head, neck, and shoulder but may become generalized. They include papules, nodules, and occasionally pustules. A patchy to diffuse alopecia with marked scaling occurs in the "squamous" form of the disease.

Demodex phylloides of **pigs** is not uncommon, the mites residing in the pilosebaceous units on the eyelids with no clinical effect. Demodectic or follicular mange in pigs is also not uncommon but is relatively unimportant in comparison to sarcoptic mange. The principal economic loss results from extensive trimming of affected carcasses. The lesions typically involve the ventral abdomen, ventral neck, eyelids, and snout. They commence as small red macules, developing into cutaneous nodules covered by surface scale. Incision of the nodules releases thick white caseous debris, which is full of mites. The histologic lesions are as described in cattle.

Cats rarely develop demodectic mange. Two species of *Demodex* mites have been associated with the 32 cases described in the literature up to 1990. One is *D. cati,* which resides in the hair follicles and sebaceous glands; the other is as yet unnamed and is found on the skin surface in pits in the stratum corneum. Unlike that in dogs, there is no age predisposition. The gross lesions may be single or multiple areas of alopecia, erythema, and scaling, which resemble the localized form of canine demodicosis. Lesions have a predilection for the eyelid, head, and neck. Generalized lesions may occur in association with systemic disease, such as feline leukemia virus or feline immunodeficiency virus infection, hyperadrenocorticism, or diabetes mellitus, which presumably suppress normal cell-mediated immune responses. The lesions are multifocal to generalized areas of alopecia and erythema, with variable degrees of scaling, and papulocrustous dermatitis. Secondary pyoderma is not a common accompaniment. A third syndrome, otitis externa, has also been associated with *D. cati* infestation. Histologic lesions tend to be limited to follicular and surface hyperkeratosis, sometimes with follicular atrophy. Inflammatory cell infiltration is minimal in *D. cati* lesions; some eosinophil, mast cell, and neutrophil infiltrates have been described in infestations with the unnamed species, but the reaction was never severe, and concurrent allergic disease could not be ruled out.

Bibliography

Baker, K. P. The histopathology and pathogenesis of demodecosis of the dog. *J Comp Pathol* **79:** 321–327, 1969.

Besch, E. D., and Griffiths, H. J. Demonstration of *Demodex equi* (Railliet 1895) from a horse in Minnesota. *J Am Vet Med Assoc* **128:** 82–83, 1956.

Carter, H. B. A note on the occurrence of the follicle mite (*Demodex* sp.) in Australian sheep. *Aust Vet J* **18:** 120–124, 1942.

Chapman, R. E. A clinical manifestation in wool of demodectic infestation of sheep. *Aust Vet J* **49:** 595–596, 1973.

Chesney, C. J. Demodicosis in the cat. *J Small Anim Pract* **30:** 689–695, 1989.

Conroy, J. D., Healey, M. C., and Bane, A. G. New *Demodex* spp. infesting a cat: A case report. *J Am Anim Hosp Assoc* **18:** 405–407, 1982.

Desch, C.E . *Demodex aries* sp. nov., a sebaceous gland inhabitant of the sheep, *Ovis aries*, and a redescription of *Demodex ovis* Hirst, 1919. *NZ J Zool* **13:** 367–375, 1986.

Fisher, W. F. Natural transmission of *Demodex bovis*. *J Parasitol* **69:** 223–224, 1973.

Harland, E. C., Simpson, C. F., and Neal, F. C. Demodectic mange of swine. *J Am Vet Med Assoc* **159:** 1752, 1971.

Healey, M. C., and Gaafar, S. M. Immunodeficiency in canine demodectic mange. I. Experimental production of lesions using antilymphocyte serum. II. Skin reactions to phytohemagglutinin and concanavalin A. *Vet Parasitol* **3:** 121–131, 133–140, 1977.

Koutz, F. R., Groves, H. F., and Gee, C. M. A survey of *Demodex canis* in the skin of clinically normal dogs. *Vet Med* **55:** 52–53, 1960.

Murray, M. D. A clinical case of demodectic mange in a sheep. *Aust Vet J* **35:** 93, 1959.

Murray, M. D., Nutting, W. B., and Hewetson, R. W. Demodectic mange of cattle. *Aust Vet J* **52:** 49, 1976.

Nutting, W. B. Biology and pathology of hair follicle mites (demodicidae). *In* "Biology and Pathology of Hair Follicle Mites" L. C. Parish, W. B. Nutting, and R. M. Schwartzman (eds.), pp. 181–199. New York, Praeger, 1983.

Oduye, O. O. The pathology of bovine demodecosis in Nigeria. *Bull Anim Health Prod Afr* **23:** 45–50, 1975.

Rapp, J., and Koch, F. Demodicosis in the pig. *Vet Med Rev* **1:** 67–69, 1979.

Scott, D. W. Canine demodicosis. *Vet Clin North Am* **9:** 79–92, 1979.

Sheahan, B. J., and Gaafar, S. M. Histologic and histochemical changes in cutaneous lesions of experimentally induced and naturally occurring canine demodicidosis. *Am J Vet Res* **31:** 1245–1254, 1970.

Thomson, J. R., and Mackenzie, C. P. Demodectic mange in goats. *Vet Rec* **111:** 185, 1982.

White, S. D. *et al.* Generalized demodicosis associated with diabetes mellitus in two cats. *J Am Vet Med Assoc* **191:** 448–450, 1987.

Wilkie, B. N., Markham, R. J. F., and Hazlett, C. Deficient cutaneous response to PHA-P in healthy puppies from a kennel with a high prevalence of demodicosis. *Can J Comp Med* **43:** 415–419, 1979.

Williams, J. F., and Williams, C. F. Demodicosis in dairy goats. *J Am Vet Med Assoc* **180:** 168–169, 1982.

9. *Trombiculidiasis*

The nymphs and adults of the trombiculid mites are free living or parasitize plants or other arthropods; the larvae are parasitic and are known as harvest mites or chiggers. The parasitic larvae are six-legged and resemble minute red or yellow spiders just visible to the naked eye. The mites attach themselves to the skin and make a channel into the epidermis, called a stylostome, through which salivary enzymes are injected and digested tissue fluids are withdrawn. The mites engorge to twice their original size over a period of 3 to 5 days, after which they drop off and complete their life cycle in the soil. An intensely pruritic dermatitis develops at the sites of attachment, probably as a result of an allergic reaction to the salivary secretions delivered through the stylostome.

Wild vertebrates are the usual host for the trombiculid mite larvae but food-producing domestic animals, pets, and humans may be accidentally infested. The disease tends to have a seasonal incidence, occurring in the late summer and autumn when climatic conditions favor an expansion of the mite population. Factors such as soil type also influence the prevalence of trombiculidiasis in different geographical regions.

Neotrombicula autumnalis, the European harvest mite, attacks most domestic species. *Trombicula sarcina,* an Australian species known as the leg-itch mite, is an important parasite of sheep. *Eutrombicula alfreddugesi* (North American chigger), *Euschoengastia latchmani,* and *Walchia americana* are some of the species implicated in trombiculidiasis in cats, dogs, and horses.

The lesions tend to occur in areas close to ground contact. In sheep, the larvae attach preferentially on the skin of the caudal pastern. The interdigital web is the predilection site in the dog. In a massive infestation reported in two dogs, temporary hind-limb paresis developed. The mite in horses is known as the "heel-bug" because of its tendency to parasitize the feathered area of the pastern. Lesions may also occur along the mane and at the tail head. Face, particularly lip, involvement is not infrequent. The lesions are intensely pruritic. In cats, lesions affect the paws, head, and ears, but an atypical generalized form of the disease may occur in association with *W. americana* infestation.

The gross lesions in all species are small, erythematous papules on which are clustered tiny (0.2–0.4 mm) bright red, orange, or yellow mites. The mites leave a small, shallow ulceration which oozes a watery discharge and becomes crusted. Pruritus induces marked self-trauma, which may incite secondary bacterial infection and pyoderma. Histologically, the mite is present in tunnels within the stratum spinosum or in the stratum corneum. The presence of the mites induces both hyperplastic and degenerative changes in the epidermis. The inflammatory reaction is superficial perivascular in pattern, with eosinophils and mast cells predominating.

Bibliography

Crogan, W. E. Trombidiasis of sheep. *Aust Vet J* **25:** 103–104, 1949.

Flemming, E. J., and Chastain, C. B. Miliary dermatitis associated with *Eutrombicula* infestation in a cat. *J Am Vet Med Assoc* **27:** 529–531, 1991.

Fraser, A. C. Heel-bug in the thoroughbred horse. *Vet Rec* **50:** 1455–1458, 1938.

Prosl, H., Rabitsch, A., and Brabenetz, J. *Neotrombicula autum-*

nalis (harvest mite) in veterinary medicine. Nervous symptoms in dogs following massive infestation. *Tierarztliche Praxis* **13:** 57–64, 1985.

Ridgway, J. R. An unusual skin condition in thoroughbred horses. *Vet Rec* **92:** 382, 1973.

10. Other Mite-Induced Dermatoses

The cat fur mite, *Lynxacarus radovsky*, infests cats in the United States, Australia, Fiji, South America, and the Caribbean. The mite attaches to the hair shafts rather than to the skin surface. The mite infestation mimics seborrhea sicca, giving the coat a characteristic "salt and pepper" appearance. The infestation is not usually associated with lesions, although crusted, exudative, and pruritic lesions have been described.

The straw-itch mite (*Pyemotes tritici*) is normally found in straw or grain, where it parasitizes the larvae of soft-bodied grain insects. The parasite may occasionally infest humans and lower animals, causing pruritic dermatitis. Mildly pruritic lesions may develop in horses fed infested hay. Multiple papules and wheals may occur on the neck, withers, and thorax. *Acarus* (*Tyroglyphus*) *farinae* and *Acarus* (*Tyroglyphus*) *longior* may cause a pruritic, exudative, crusting, and alopecic dermatosis in horses exposed to contaminated grain or hay.

Cats and dogs may be infested on rare occasions with the poultry mite, *Dermanyssus gallinae*. The infestation resembles that of cheyletiellosis, but pruritic eruptions have been described.

Sheep may, on rare occasions, become infested with the stored product mite, *Sancassania berlesei*. These mites cannot infest dry skin; infestations are secondary to other conditions such as myiasis.

Bibliography

Arundel, J. H., and Sutherland, A. K. "Animal Health in Australia. Volume 10. Ectoparasitic Diseases of Sheep, Cattle, Goats and Horses." Australian Government Publishing Service, 1988.

Barton, N. J., Stephens, L. R., and Dombrow, R. Infestation of sheep with the stored product mite *Sancassania berlesei* (Acaridae). *Aust Vet J* **65:** 140–143,1988.

Bowman, W. L., and Domrow, R. The cat fur-mite (*Lynxacarus radovsky*) in Australia. *Aust Vet J* **54:** 403, 1978.

Greve, H., and Gerrish, R. R. Fur mites (*Lynxacarus*) from cats in Florida. *Fel Pract* **11:** 28–30, 1981.

Kunkle, G. A., and Greiner, E. C. Dermatitis in horses and man caused by the straw itch mite. *J Am Vet Med Assoc* **181:** 467–469, 1982.

Norvall, J., and McPherson, E. A. Dermatitis in the horse caused by *Acarus farinae*. *Vet Rec* **112:** 385–386, 1983.

Ramsay, G. W., Mason, P. C., and Hunter, A. C. Chicken mite (*Dermanyssus gallinae*) infesting a dog. *N Z Vet J* **23:** 155–156, 1975.

E. Ticks

Ticks belong to the suborder Ixodoidea in the class Arachnida. They are divided into two suborders, the Argasidae and the Ixodidae. The Argasidae are the so-called soft ticks, lacking the scutum which characterizes the Ixodidae. Included in this group is *Argas persicus*, a complex of tick species which are important parasites of birds. *Otobius megnini*, known as the spinose ear-tick, is parasitic to all domestic animals, causing severe parasitic otitis externa and predisposing to bacterial infection or myiasis. Most of the pathogenic species of tick are found in the Ixodidae.

Ticks are most important as vectors for a large number of serious viral, rickettsial, bacterial, and protozoal diseases of domestic animals. Ticks also harm their hosts more directly by causing local injury at the site of attachment. If infestation is heavy, fatalities may result. The local injury may predispose to secondary bacterial infection and to screwworm myiasis. Heavy infestations are also capable of causing anemia as a result of the bloodsucking activities of the ticks, but the Ixodidae, which engorge only once at each instar, are much less important in this respect than are the argasid ticks, which as adults engorge repeatedly. Tick bites may also induce hypersensitivity reactions. Several species of ixodid ticks are capable of causing paralysis of the host, including *Ixodes rubicundus* of South Africa, *I. holocyclus* of Australia and Asia, and *Dermacentor andersoni* of North America.

The local reaction to ticks is variable, depending on properties of the tick, for example, its ability to secrete prostaglandins, and on host factors, in particular the level of tick resistance. Primary tick-bite lesions are papules and wheals, which may develop crusts, erosions, and ulcers and lead to focal alopecia. In contrast, tick-sensitized hosts produce extremely erythematous reactions. Histologically, the primary lesions are less severe and develop more slowly than those of sensitized animals. For example, in experimental *Ixodes holocyclus* infestations of cattle, the inflammatory cell reaction to the tick mouthparts embedded deeply into the dermis is not marked, even at 40 hr postattachment. The lesions are restricted to the immediate feeding site and are predominantly neutrophilic. By contrast, previously sensitized cattle show epidermal spongiosis and vesiculation at some distance from the actual point of penetration as early as 1 hr postattachment. By 12 hr, intraepidermal vesicles, bullae, and microabscesses containing basophils with eosinophils and neutrophils are prominent. Basophils, many degranulated, are numerous in the edematous dermis from 1 hr postattachment, reflecting the important role cutaneous basophil hypersensitivity plays in tick immunity.

Severe hypersensitivity reactions to *Boophilus microplus* have been described in the horse in Australia. Within 30 min of reinfestation, sensitized horses develop intensely pruritic papules and wheals, chiefly on the lower legs and muzzle.

Bibliography

Allen, J. R., Doube, B. M., and Kemp, D. H. Histology of bovine skin reactions to *Ixodes holocyclus* Neumann. *Can J Comp Med* **41:** 26–35, 1977.

Allen, J. R., Khalil, H. M., and Graham, J. E. The location of

tick salivary antigens, complement and immunoglobulin in the skin of guinea-pigs infested with *Dermacentor andersoni* larvae. *Immunology* **38:** 467–472, 1979.
Brown, S. J. Highlights of contemporary research on host immune responses to ticks. *Vet Parasitol* **28:** 321–334, 1988.
Brown, S. J., and Askenase, P. W. Immune rejection of ectoparasites (ticks) by T-cell and IgG antibody recruitment of basophils and eosinophils. *Fed Proc* **42:** 1744–1749, 1983.
McLaren, D. J., Worms, M. J., and Askenase, P. W. Cutaneous basophil associated resistance to ectoparasites (ticks). Electron microscopy of *Rhipicephalus appendiculatus* larval feeding sites in actively sensitized guinea pigs and recipients of immune serum. *J Pathol* **139:** 291–308, 1983.
Purnell, R. E. Tick-borne diseases. *Br Vet J* **137:** 221–240, 1981.
Riek, R. F. Allergic reactions in the horse to infestation with larvae of *Boophilus microplus*. *Aust Vet J* **30:** 142–144, 1954.
Schleger, A. V. *et al. Boophilus microplus:* Cellular responses to larval attachment and their relationship to host resistance. *Aust J Biol Sci* **29:** 499–512, 1976.
Tatchell, R. J., and Moorhouse, D. E. The feeding processes of the cattle tick *Boophilus microplus* (Canestrini) II. The sequence of host-tissue changes. *Parasitology* **58:** 441–459, 1968.

XVII. Helminth Diseases of Skin

The skin is the natural portal of entry of a number of metazoan parasites which have their final habitat in the gastrointestinal tract or elsewhere. As a rule, those infective larvae which can invade percutaneously are not selective in their choice of skins, so that infestation of alien hosts occurs. Such parasites are quite varied in their nature and include the infective larvae of the trematodes, *Schistosoma,* and of the nematodes of various genera including *Strongyloides, Gnathostoma, Ancylostoma, Bunostomum, Uncinaria,* and others. Infective larvae of the filariids such as *Dirofilaria* and *Setaria* are deposited in the skin by the biting insects that are their intermediate hosts. The first percutaneous invasion of one of these parasites in its natural hosts takes place very quickly (for example, the larvae of *Strongyloides* and *Bunostomum* reach the dermis in 15 min) and without provoking a significant reaction. Repeated invasions in a natural host or single invasion in an unnatural host are met with some resistance, which is manifested as an acute dermatitis limited to the invaded area. The cutaneous lesions are, except under experimental circumstances, seldom observed in animals. It is on the glabrous, nonpigmented skin of humans that they are easily observed and well recognized as the so-called creeping eruption. There the larvae produce acutely inflamed, serpiginous, vesicopapular tracts which may advance several centimeters a day. Usually, aberrant larvae die in the skin, but some enter the vessels and become lodged in the lungs or other tissues.

The cutaneous lesions produced by the blood flukes and those nematodes which pass through the skin on the way to the gut are discussed elsewhere with the mature parasites. To be discussed in more detail here are those helminthic infestations which remain more or less localized to the dermis. The dermatitis produced by the larvae of *Elaeophora* is discussed in Volume 3, Chapter 1, The Cardiovascular System.

A. Cutaneous Habronemiasis

This common disease of horses is caused by the aberrant deposition of the larvae of the spirurid nematodes *Habronema majus* (*microstoma*), *H. muscae,* and *Draschia megastoma* at cutaneous or mucocutaneous sites by transmitting flies. The adult worms normally develop in the stomach of horses following the ingestion of infective larva (see Volume 2, Chapter 1, The Alimentary System). The cutaneous disease is more prevalent in summer when the transmitting flies, *Musca* spp. and *Stomoxys calcitrans,* are active.

The location of the lesions reflects the areas which attract these flies. The most usual sites are, therefore, the medial canthus of the eye (Fig. 5.110), which is constantly moistened by tears, the glans penis and prepuce, and any cutaneous wound. Since lacerations are more common on the distal extremities, these too are predilection sites for habronemiasis. Although lacerations facilitate habronemiasis, the fly bite itself is sufficient to initiate an infestation.

The gross lesions are rapidly progressive and proliferative in nature, comprising ulcerated, tumorous masses of red-brown granulation tissue. The surface is friable and hemorrhages readily. Lesions may be single or multiple and range in size from 5 to 15 cm in diameter and 0.5 to 1.5 cm in depth. They are often irregular in shape in the early stages but become circular as they enlarge. On cut section, multiple, small (1–5 mm) yellow-white, caseous, and occasionally gritty foci are scattered through the granulation tissue. These are often confused with the kunkers of pythiosis, but lack the characteristic branching pattern of the true kunker. In the deeper parts of the lesion, the more mature connective tissue has a dense, white appearance. The lesions on the conjunctiva and eyelids do not usually exceed 2 cm. Commencing with a

Fig. 5.110 Cutaneous habronemiasis. Horse.

serous conjunctivitis, small ulcerated, proliferative nodular lesions develop on the mucous membrane of the third eyelid and at the medial canthus. Lacrimal duct involvement characteristically produces a lesion 2–3 cm below the medial canthus. The entire conjunctiva may be affected, resulting in profuse lacrimation, photophobia, chemosis, and inflammation of the eyelid. On cut section, the nodular lesions of brown-red granulation tissue contain the typical caseous or mineralized foci. Involvement of the penis and prepuce may cause prolapse of the urethral process and dysuria. On rare occasions, *Habronema* granulomas are found in the lung.

Histologically, multiple aggregates of degenerate eosinophils are scattered randomly throughout a collagenous connective tissue stroma of variable maturity. The larvae or their remnants, which may be mineralized, occur in the center of these foci. Epithelioid macrophages and multinucleate giant cells often surround degenerating larvae. However, larvae may be rare or absent in many caseous foci. The fibrous connective tissue is heavily and diffusely infiltrated with eosinophils and with lesser numbers of lymphocytes and plasma cells. The surface of the lesion is usually covered with a fibrinonecrotic exudate overlying a highly vascular granulation tissue infiltrated with neutrophils.

Since the gross lesions of cutaneous habronemiasis may resemble those of exuberant granulation tissue, botryomycosis, pythiosis, equine sarcoid, and squamous cell carcinoma, the diagnosis requires biopsy and histologic evaluation. Squamous cell carcinoma and equine sarcoid are readily distinguished histologically. Exuberant granulation tissue lacks the caseous foci of degenerating eosinophils and larvae. Special stains will reveal the causative agents in botryomycosis (see Section XII, Bacterial Diseases of Skin) and pythiosis (see Section XIII, Mycotic Diseases of Skin). Nodular collagenolytic granuloma (nodular necrobiosis) presents a different clinical appearance but may be confused histologically with habronemiasis (see Section XVIII, Miscellaneous Skin Disorders). Habronemiasis may complicate a preexisting lesion; secondary *Habronema* infestations occur with pythiosis, *Corynebacterium pseudotuberculosis* infection and in skin tumors, particularly squamous cell carcinoma of the penis.

Cutaneous habronemiasis has been reported in a **dog.** Lesions developed on the face but, unlike those of the equine disease, were not characterized by rapid proliferation of granulation tissue. The dog was housed under unsanitary conditions in the company of several heavily parasitized ponies.

Bibliography

Murray, D. R., Ladds, P. W., and Campbell, R. S F. Granulomatous and neoplastic diseases of the skin of horses. *Aust Vet J* **54:** 338–341, 1978.

Reddy, A. B., Gaur, S. N. S., and Sharma, U. K. Pathological changes due to *Habronema muscae* and *Draschia megastoma* infection in equines. *Indian J Anim Sci* **46:** 207–210, 1976.

Reid, C. H. Habronemiasis and Corynebacterium in chest abscesses in Californian horses. *Vet Med Small Anim Clin* **60:** 233–242, 1965.

Sanderson, T. P., and Niyo, Y. Cutaneous habronemiasis in a dog. *Vet Pathol* **27:** 208–209, 1990.

Vasey, J. R. Equine cutaneous habronemiasis. *Compend Cont Ed* **3:** S290–2956, 1981.

B. Stephanofilariasis

Members of the genus *Stephanofilaria* (Spirurida: Setariidae) are parasites of **cattle.** All stephanofilarid parasites cause similar cutaneous lesions, but the species are geographically separated, and the lesions occur on different parts of the host's body. *Stephanofilaria stilesi* occurs in North America, affecting the abdominal skin near the midline; in Australian cattle, initial lesions developed at the medial canthus of the eye; *S. dedoesi* in Indonesia causes dermatitis, known locally as "cascado," on the sides of the neck, dewlap, withers, and around the eyes; *S. assamensis* causes "hump-sore" of Zebu cattle in India and Russia; *S. kaeli* causes dermatitis of the legs of cattle from the Malay Peninsula; *S. dinniki* affects the shoulder of black rhinoceros; *S. zaheeri* causes dermatitis of the ears of buffaloes, and *S. okinawaensis* causes lesions on the teats and muzzle of cattle in Japan. *Stephanofilaria kaeli, S. assamensis,* and *S. dedoesi* have been reported as causes of crusting dermatitis in **goats.** There is, however, some doubt as to the authenticity of *S. kaeli, S. assamensis, S. okinawaensis,* and *S. dedoesi* as separate species. They cannot be separated morphologically, and it has been suggested that they should be synonymized.

Stephanofilariasis is transmitted by flies. The hornfly, *Haematobia irritans,* is the vector of *S. stilesi,* and the buffalo fly, *Haematobia irritans exigua,* is the major vector for stephanofilariasis in Australia. *Musca conducens* transmits *S. assamensis* and *S. kaeli.* The flies ingest microfilariae when feeding on cutaneous lesions. After a period of development in the fly, the infective stage is deposited into the skin by the biting species of fly or onto cutaneous wounds in the case of the nonbiting vectors such as *M. conducens.* The adults of *S. stilesi* live in cystic diverticula off the base of the hair follicles. The parasites are very small, with males reaching 3 mm and females 8 mm in length.

The initial macroscopic lesions in *S. stilesi* infestations are circular patches 1 cm or less in diameter in the skin of the ventral midline, in which the hairs are moist and erect, and the underlying epidermis is spotted with small hemorrhages and droplets of serum. These initial foci enlarge and coalesce, sometimes to produce a lesion 25 cm or more in diameter. As the foci enlarge, new spots of hemorrhage and exudation develop at the periphery, while in the central areas, the hair is shed and the exudate builds up into scabs or rough dry crusts through which the few remaining hairs penetrate (Fig. 5.111). The lesions are mildly pruritic, and they may be aggravated by rubbing. In the healing stage, the affected areas remain as alopecic, lichenified plaques.

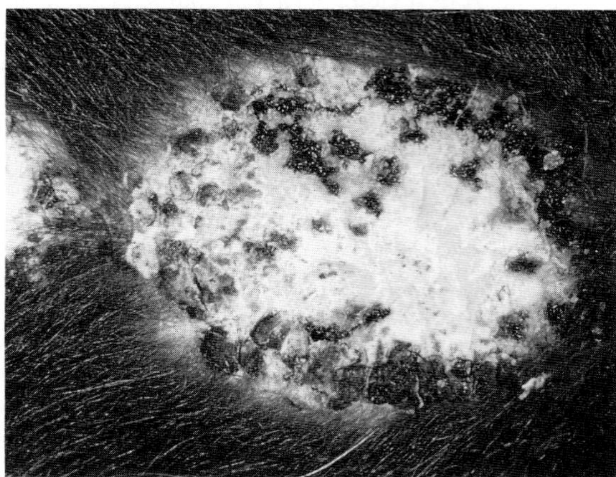

Fig. 5.111 Stephanofilariasis. Cow. Gross lesion on ventral midline.

Histologically, sections of adult parasites may be seen in the cystic diverticula from hair follicles or lying free in the adjacent dermis (Fig. 5.112A). They may be identified as *Stephanofilaria* spp. with some confidence if larvae are seen in the uteri of females, since this parasite is viviparous; if ova and not larvae are found, it is more likely that the parasite is of the genus *Rhabditis*. Microfilariae occur free in the dermis or in dermal lymphatics, enclosed within their own vitelline membranes (*S. stilesi*), or may be found free or unhatched in surface exudate (*S. kaeli*). There is little dermal reaction to the microfilariae or to the adults enclosed in cystic hair follicles, but the presence of adults in the dermis stimulates a mononuclear inflammatory reaction. There is an accompanying superficial and deep perivascular dermatitis characterized by accumulations of eosinophils and mononuclear cells, chiefly lymphocytes (Fig. 5.112B). The epidermis is hyperplastic, spongiotic, and often covered by orthokeratotic and parakeratotic hyperkeratosis and inflammatory crust. Spongiform microabscesses containing eosinophils and mononuclear cells are also described; such lesions are more typically associated with the bites of arthropod ectoparasites. It is difficult to assess the relative contributions made to the lesion by the stephanofilarial parasite and by the bites of the fly which acts as the vector.

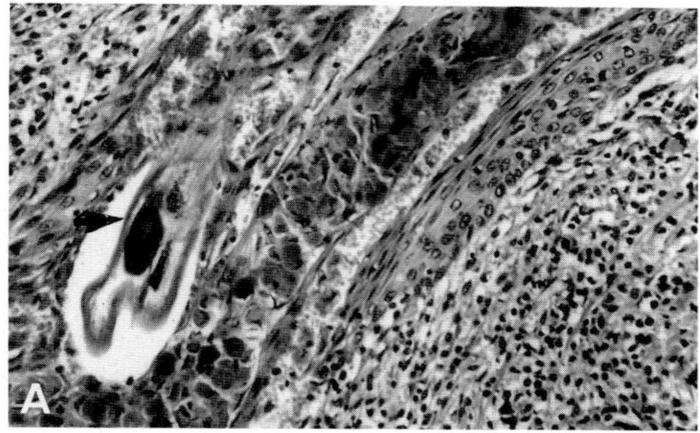

Fig. 5.112A Stephanofilariasis. Cow. Section of worm in follicle (arrow).

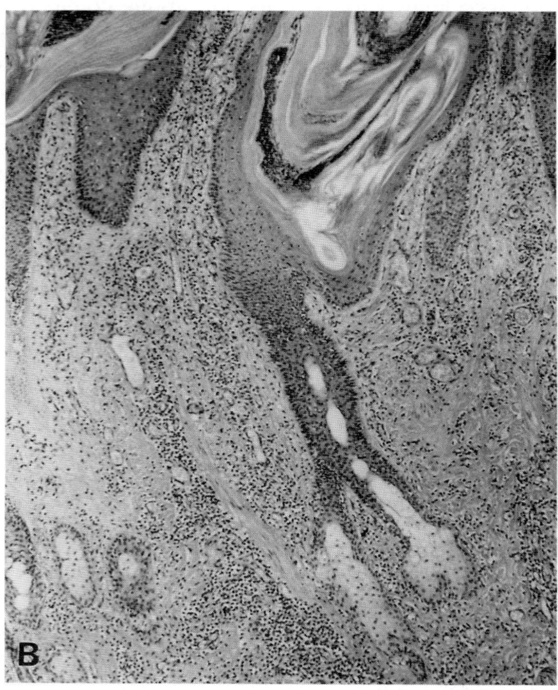

Fig. 5.112B Stephanofilariasis. Cow. Appearance of dermatitis reaction.

Diagnosis is usually made on the basis of the typical gross lesions, by histologic examination of biopsy specimens, or by demonstration of microfilariae by deep scrapings.

Bibliography

Agrawal, M. C., Vegad, J. L., and Dutt, S. C. Pathology of naturally occurring *Stephanofilaria zaheeri* Singh 1958 infection in buffaloes. *Indian J Anim Sci* **48:** 261–265, 1978.

Dies, K. H., and Pritchard, J. Bovine stephanofilarial dermatitis in Alberta. *Can Vet J* **26:** 361–362, 1985.

Johnson, S. J. Stephanofilariasis—a review. *Helminthol Abst* **56:** 287–289, 1987.

Levine, N. D., and Morrill, C. C. J. Bovine stephanofilarial dermatitis in Illinois. *J Am Vet Med Assoc* **127:** 528–530, 1955.

Loke, Y. W., and Ramachandran, C. P. The pathology of lesions in cattle caused by *Stephanofilaria kaeli* Buckley, 1937. *J Helminthol* **41:** 161–166, 1967.

Singh, S. N. On a new species of *Stephanofilaria* causing dermatitis of buffaloes' ears in Hyderabad (Andhra Predesh) India. *J Helminthol* **32:** 239–250, 1958.

Ueno, H., Chibana, T., and Yamashiro, E. Occurrence of chronic dermatitis caused by *Stephanofilaria okinawaensis* on the teats of cows in Japan. *Vet Parasitol* **3:** 41–48, 1977.

C. *Onchocerciasis*

Onchocerca spp. are filarial nematodes (Spirurida: Onchocercidae) which affect cattle, horses, sheep, goats, and humans. The adult worms reside in various connective tissue locations, chiefly ligaments, tendons, and subcutaneous nodules, but also perimysial sheaths, cartilage, and aorta. The microfilariae migrate to the dermis where they lie free in tissue spaces or lymphatics. They are aspirated from the dermis by the intermediate hosts, which are insects of the families Simuliidae (black flies, gnats) and Ceratopogonidae (midges). The prevalence of infection is very high in cattle and horses, ranging up to 100% in some studies.

Onchocerca cervicalis occurs worldwide and is the major species affecting the horse in North America. The adults reside in the ligamentum nuchae. *Onchocerca gutturosa* more commonly affects cattle but also is recovered from horses. It has been suggested that *O. lienalis*, which predominantly resides in the gastrosplenic ligament, is synonymous with *O. gutturosa*. *Onchocerca reticulata* occurs in horses in Europe and Asia. The adults live in the flexor tendons and suspensory ligaments, particularly of the forelimbs. *Onchocerca gibsoni* infests cattle in Australia, Asia, and Africa. *Onchocerca ochengi* infests cattle in Africa.

Most animals show no reaction to the presence of living microfilariae in the dermis. The reason for the lack of host reaction to the living microfilariae is unknown. Dead microfilariae are, however, highly phlogistic. An adverse reaction to microfilaricide therapy, called the Mazotti reaction, is due to the intense inflammatory reaction engendered by the dead parasites. *Onchocerca* microfilariae cause cutaneous lesions in some horses; in cattle, the cutaneous lesions are caused by the adults.

1. Equine Cutaneous Onchocerciasis

The pathogenicity of *Onchocerca* microfilariae, originally questioned because the parasites could be readily demonstrated in the skin of clinically normal horses, became accepted when it was shown that ivermectin treatment induces rapid and concurrent disappearance of cutaneous lesions and microfilariae. However, adults in the ligamentum nuchae are not affected by a single dose of ivermectin and, despite the widespread use of this drug, the prevalence of adult parasites in horses has not diminished. It has been postulated that individual hypersensitivity to microfilarial antigens is the reason for the sporadic occurrence of equine skin lesions.

Areas affected are the face, neck, medial aspects of the forelimbs, ventral thorax, and abdomen. Lesions on the face may be diffuse, periorbital, or central. The central lesions, which often involve the facial markings, are very characteristic of onchocerciasis. The lesions on the neck are focal and tend to occur at the base of the mane. The lesions on the ventral thorax and abdomen are often diffuse, in comparison with the localized lesions of horn fly-induced ventral midline dermatitis. The ventral lesions are indistinguishable grossly from those of *Culicoides* hypersensitivity, but onchocerciasis does not cause lesions at the base of the tail. It is likely that onchocerciasis and *Culicoides* hypersensitivity can exist simultaneously.

Cutaneous lesions include partial or complete alopecia, scaling, crusting, and leukoderma. Secondary excoriations or ulcerative dermatitis are induced by self-trauma in pruritic animals. Ocular lesions are discussed with diseases of the eye (Chapter 4). The microscopic lesions associated with microfilarial infection vary from none to a superficial and deep perivascular dermatitis that is predominantly lymphocytic and eosinophilic. Microfilariae may be present in large numbers or may be very sparse. They are best recovered from unfixed biopsies which are minced and incubated in saline at 37°C.

Viable adult parasites in the nuchal ligament are not associated with significant lesions; in older horses, there is an increased frequency of caseated, mineralized, and granulomatous lesions, associated with death of the parasite. *Onchocerca gutturosa* is also found in the nuchal ligament of horses in Australia. Histologic examination revealed that this parasite does not penetrate the elastic tissue of the ligament, and the inflammatory reaction is localized around the worm. Calcification, not caseation, is the fate of degenerate worms in contrast to *O. cervicalis*.

2. Bovine Onchocerciasis

Gross lesions in cattle infested with *O. gibsoni* are multiple subcutaneous papules and nodules of ~2 cm diameter, although lesions up to 9 cm are recorded. Lesions may be hard or soft depending on the degree of calcification and fibrosis or the degree of caseation and suppuration. Lesions predominantly affect the brisket but also stifle and hip. Histologic assessment of *O. gibsoni* nodules revealed dead worms and associated degenerative changes, such as mineralization in ~30% of nodules. Eosinophilic infiltration was more marked around viable worms, with eosinophils apparently adherent to the cuticle. In *O. gutturosa* infestations, the worm is found in the nuchal ligament and connective tissue around tendons and ligaments of the shoulder, hip, and stifle. It does not form nodules and is thus easily overlooked. In *O. ochengi* infestations, scrotum, udder, flanks, side, and head are affected.

Bibliography

Camp, C. J., and Leid, R. W. Equine complement activation as a mechanism for equine neutrophil migration in *Onchocerca cervicalis* infections. *Clin Immunol Immunopathol* **26:** 277–286, 1983.

Cheema, A. H., and Ivoghli, B. Bovine onchocerciasis caused by *Onchocerca armillata* and *O. gutturosa*. *Vet Pathol* **15:** 495–505, 1978.

Ferenc, S. A. *et al. Onchocerca gutturosa* and *Onchocerca lienalis* in cattle: Effect of age, sex, and origin on prevalence of onchocerciasis in subtropical and temperate regions of Florida and Georgia. *Am J Vet Res* **47:** 2266–2268, 1986.

Henson, P. M., MacKenzie, C. D., and Spector, W. G. Inflammatory reactions in onchocerciasis: A report on current knowledge and recommendations for further study. *Bull WHO* **57:** 667–671, 1979.

Ladds, P. W., Nitisuwirjo, S., and Goddard, M. E. Epidemiological and gross pathological studies of *Onchocerca gibsoni* infection in cattle. *Aust Vet J* **55:** 455–462, 1979.

Lees, M. J., Kleider, N., and Tuddenham, T. J. Cutaneous onchocerciasis in the horse: Five cases in southwestern British Columbia. *Can Vet J* **24:** 3–5, 1983.

Lyons, E. T., Drudge, J. H., and Tolliver, S. C. Verification of ineffectual activity of ivermectin against adult *Onchocerca* spp. in the ligamentum nuchae of horses. *Am J Vet Res* **49:** 983–985, 1988.

Mellor, P. S. Studies on *Onchocerca cervicalis* Railliet and Henry 1910: I. *Onchocerca cervicalis* in British horses. II. Pathology in the horse. *J Helminthol* **47:** 97–110, 111–118, 1973.

Nitisuwirjo, S., and Ladds, P. W. A quantitative histopathological study of *Onchocerca gibsoni* nodules in cattle. *Tropenmed Parasitol* **31:** 467–474, 1980.

Ottley, M. L., Dallemagne, C., and Moorhouse, D. E. Equine onchocerciasis in Queensland and the Northern Territory of Australia. *Aust Vet J* **60:** 200–203, 1983.

Pollitt, C. C. *et al.* Treatment of equine onchocerciasis with ivermectin paste. *Aust Vet J* **63:** 152–156, 1986.

Schiller, E. L., D'Antonio, R., and Figueroa Marroquin, H. Intradermal reactivity of excretory and secretory products of onchocercal microfilariae. *Am J Trop Med Hyg* **29:** 1215–1219, 1980.

Schmidt, G. M. *et al.* Equine onchocerciasis: Lesions in the nuchal ligament of midwestern U.S. horses. *Vet Pathol* **19:** 16–22, 1982.

Thomas, A. D. Skin lesions in cases of onchocerciasis in horses in Northern Transvaal, South Africa. *Trans R Soc Trop Med Hyg* **52:** 298, 1958.

Webster, W. A., and Dukes, T. W. Bovine and equine onchocerciasis in eastern North America with a discussion on cuticular morphology of *Onchocerca* spp. in cattle. *Can J Comp Med* **43:** 330–332, 1979.

D. Parafilariasis

There are two species of interest in the genus *Parafilaria. Parafilaria multipapillosa* occurs in horses in Eastern Europe. *Parafilaria bovicola* is endemic in cattle in Africa, India, and parts of Europe. The parasite was introduced into Sweden in 1978, probably following importation of infected cattle. Both species are thin, threadlike worms, 2–7 cm in length. The adults inhabit the subcutaneous and intermuscular connective tissues, producing nodules 1–2 cm in diameter. In the spring and summer, the nodules rapidly enlarge, burst open, hemorrhage, and heal. This coincides with the migration of the fully gravid female into the more superficial dermis to ovideposit and to release the infective microfilariae. The intermediate hosts, flies such as *Haematobia atripalpis* in Russia and *Musca* spp., are infected when they feed from these bleeding points, known as "blood nodules."

Bibliography

Bech-Nielsen, S. *et al. Parafilaria bovicola* (Tubangui 1934) in cattle: Epizootiology-vector studies and experimental transmission of *Parafilaria bovicola* to cattle. *Am J Vet Res* **43:** 948–954, 1982.

Bech-Nielsen, S., Sjogren, U., and Lundquist, H. *Parafilaria bovicola* (Tubangui 1934) in cattle: Epizootiology-disease occurrence. *Am J Vet Res* **43:** 945–947, 1982.

De Jesus, Z. Haemorrhagic filariasis in cattle caused by a new species of *Parafilaria. Philipp J Sci* **55:** 125–130, 1934.

Nevill, E. M. Experimental transmission of *Parafilaria bovicola* to cattle using *Musca* species (Subgenus Eumusca) as intermediate hosts. *Onderstepoort J Vet Res* **46:** 51–57, 1979.

Niilo, L. Bovine hemorrhagic filariasis in cattle imported into Canada. *Can Vet J* **9:** 132–137, 1968.

Sundquist, B. *et al.* Preparation and evaluation of the specificity of *Parafilaria bovicola* antigen for detection of specific antibodies by ELISA. *Vet Parasitol* **28:** 223–235, 1988.

E. Pelodera *Dermatitis*

Some of the biological characteristics of the Rhabditidae have been discussed with the principal parasitic genus, *Strongyloides,* under parasitic diseases of the intestine. Here it is necessary to describe only the cutaneous lesions produced occasionally by the small free-living worms of this family. Those worms which are found in the lesions are usually classified as *Pelodera* (*Rhabditis*) *strongyloides.*

Pelodera dermatitis occurs most commonly in dogs, occasionally in cattle, and rarely in horses and sheep. The development of dermatitis to a clinical degree probably requires exposure to many larvae. These worms live as saprophytes in warm, moist soil which is rich in organic matter, and significant infestations probably require that the host's skin should be continually moist and filthy. The lesions develop on contact areas, particularly at the margins of areas caked with dirt. Affected dogs often have a history of being bedded on straw. An acute dermatitis develops, commonly with suppurative folliculitis, and is due to the combined stimuli of the parasites, bacterial folliculitis, and of the scratching, rubbing, and licking prompted by severe itchiness, probably induced by an allergic reaction to the parasite. The gross lesions include erythema, papules, excoriations, scaling, exudation, and crusting, with partial to complete alopecia. Pustules may occur, particularly in dogs. Histologically, the worms are in the lumina of the hair follicles or in the dermis, surrounded by an intense eosinophilic inflammatory reaction.

Bibliography

Farrington, D. O., Lundvall, R. L., and Greve, J. H. *Pelodera strongyloides* dermatitis in a horse in Iowa. *Vet Med Small Anim Clin* **71:** 1199–1202, 1976.

Horton, M. L. Rhabditic dermatitis in dogs. *Mod Vet Pract* **61:** 158–159, 1980.

Rhode, E. A. *et al.* The occurrence of *Rhabditis dermatitis* in cattle. *North Am Vet* **34:** 634–637, 1953.

F. Other Nematodes

Dracunculus medinensis (Spirurida: Dracunculidae) is the "guinea worm" of humans in Asia and Africa. It has been introduced into America, the West Indies, and Fiji. The parasite has been reported from various animals, including horses, cattle, and cats. *Dracunculus insignis* is the species which occurs in dogs and wild carnivores in

North America. Infection is particularly prevalent in raccoons and mink, which appear to be the natural definitive species. The parasite occurs typically in the subcutaneous tissues of the limbs. The mature female may measure up to 70 cm in length, resulting in the formation of a 2- to 4-cm nodule. The gravid female produces an intraepidermal bulla with her anterior end. Rupture of the bulla forms a shallow ulcer from which a milky exudate drains. When these lesions contact water, the worm is stimulated to release very large numbers of larvae. The intermediate host is *Cyclops*, but frogs may act as paratenic hosts. The final host is infested by ingesting frogs or water containing infected copepods. The adult worms mature in the connective tissue of the host in approximately 1 year. Fine needle aspiration of the cutaneous nodules reveals *Dracunculus* larvae, approximately 500 μm long and covered by a striated cuticle. The adults lie in a pseudocyst lined by fibrous connective tissue and infiltrated with eosinophils, lymphocytes, and multinucleate giant cells.

Dirofilaria immitis microfilariae are associated rarely with cutaneous lesions (see Immune-Mediated Diseases of Skin). Aberrent adult *D. immitis* have been recovered from cutaneous abscesses and interdigital cysts in parasitized dogs. *Dirofilaria repens* occurs in the subcutaneous connective tissues of dogs in Mediterranean countries and tropical parts of the Orient. The intermediate hosts are probably mosquitoes. The microfilariae occur in the dermal lymphatics. This parasite is not known to be pathogenic. *Dipetalonema reconditum* occurs in the subcutis of dogs. Its only significance is in the differential diagnosis of *Dirofilaria immitis* infestations by identification of microfilariae in peripheral blood.

Suifilaria suis occurs in pigs in South Africa. The worms are 2–4 cm long and live in the subcutaneous and intermuscular connective tissues, sometimes producing small whitish nodules. The female is oviparous, and the eggs are released to the surface via small vesicular eruptions in the epidermis. The remainder of the life cycle is unknown.

Three species of *Brugia* (formerly *Wuchereria*) are known in animals, but their pathogenicity is not recorded. They inhabit the lymphatics. *Brugia patei* occurs in carnivores in Kenya; *B. pahangi* has a wide host range but develops chiefly in carnivores in Malaysia; and the nonperiodic biotype of *B. malayi* exists widely as a zoonosis in Malaysia.

Parelaphostrongylus tenuis, a common parasite of white-tailed deer, is an occasional cause of neurological disease in goats. Some affected goats also develop an unusual dermatitis, often restricted to one side of the body. Lesions are vertically oriented, alopecic, ulcerated, crusted, or scarred linear tracks on the shoulder, thorax, or flanks. The lesions have been explained tentatively on the basis of ganglioneuritis leading to irritation of dermatomes. Histologically, the lesion is a fibrosing dermatitis with focal areas of basal keratinocyte hydropic degeneration.

Anatrichosoma cuteum was the cause of ulcerative pododermatitis in a South African cat, and an unidentified *Anatrichosoma* sp. was detected in a small cutaneous nodule on the dorsum of a young dog. The nematodes, in both cases, were within epidermal microabscesses.

Bibliography

Coles, L. D., North, C., and Schillhorn van Vreen, T. W. Adult *Dirofilaria immitis* in hind-leg abscesses of a dog. *J Am Anim Hosp Assoc* **24:** 363–365, 1988.

Crichton, V. F., and Beverley-Burton, M. Distribution and prevalence of *Dracunculus* spp. (Nematoda: Dracunculoidea) in mammals in Ontario. *Can J Zool* **52:** 163–167, 1974.

Elkins, A. D., and Berkenblit, M. Interdigital cyst in the dog caused by an adult *Dirofilaria immitis*. *J Am Anim Hosp Assoc* **26:** 71–72, 1990.

Hendrix, C. M. *et al. Anatrichosoma* sp. infection in a dog. *J Am Vet Med Assoc* **191:** 984–985, 1987.

Lange, A. L. *et al. Anatrichosoma* sp. infestation in the footpads of a cat. *J S Afr Vet Assoc* **51:** 227–229, 1980.

Panciera, D. L., and Stockham, S. L. *Dracunculus insignis* infection in a dog. *J Am Vet Med Assoc* **192:** 76–78, 1988.

XVIII. Miscellaneous Skin Disorders

A. Laminitis

Laminitis literally refers to inflammatory changes in the laminae of the hoof. However, the important forms of laminitis in horses and cattle are probably primarily ischemic diseases. Apart from laminitis of traumatic origin, laminitis usually represents a local manifestation of a generalized metabolic disturbance. It occurs in all hoofed species, but particularly affects horses and cattle. Laminitis is described as one of the most devastating of equine diseases. Equine laminitis is sporadic, although fat, underexercised ponies are especially predisposed. The disease occurs sporadically in dairy cows, heifers, fattening cattle, and young bulls. A heritable form has been reported in Jersey cattle in South Africa, the United States of America, and the United Kingdom.

The etiologic and predisposing factors in laminitis vary, but it is likely that they initiate a single common pathway which results in ischemic damage to the digital laminae. In horses, the most frequent precipitating event, and one which has been utilized for experimental purposes, is alimentary carbohydrate overload. Other associations of laminitis include various forms of acute colic, toxemias, excessive water intake immediately after exercise, excessive consumption of lush pasture, and drug therapy. Black walnut (*Juglans nigra*) toxicity causes natural disease in the horse. It too has been exploited as an experimental model, without the side effects of colic, diarrhea, and shock that accompany carbohydrate overload. Repeated trauma to the foot, as in training, may cause a form of laminitis known as "road founder." In cattle, carbohydrate overload is also an important predisposing cause of laminitis. Others include metritis, mastitis, and ketosis.

The pathogenesis of metabolic laminitis remains incompletely understood, although more is known of the hemodynamic events. It is generally agreed that hemodynamic alterations in the digital microvasculature result in ischemic necrosis of the soft lamina in acute disease and proliferative responses in chronic disease. In the acute stages, local ischemia due to decreased digital capillary perfusion occurs. The dorsal laminae have been shown, by angiographic and morphologic studies, to be the last structures in the hoof to receive blood and to have minimal collateral circulation. They are thus more likely to be susceptible to ischemia. The hemodynamic event thought to initiate the deleterious sequence is digital vasoconstriction, leading to an increase in postcapillary resistance and, probably, edema. Edema would further impede digital microcirculation and predispose to microthrombosis and arteriovenous shunting. Catecholamines are likely to be responsible in part for the potent vasoconstriction in acute laminitis, but the nature of the initiating stimuli and the roles of other vasoactive mediators are as yet obscure.

Disorders of intravascular coagulation may have a role in the pathogenesis, but it is not certain whether the coagulopathy precedes the vascular lesion or is a sequel to it. The reported alterations in coagulation parameters in laminitis vary. However, heparin significantly reduces the incidence of lameness in ponies with experimentally induced carbohydrate-overload laminitis.

The role of endotoxemia and lactic acidosis in the genesis of laminitis is also controversial. The hypothesis states that in grain overload, the rising levels of lactic acid decrease the intracecal pH. This, and the endotoxin released from destruction of the gram-negative resident flora, combine to damage the cecal mucosa and permit the absorption of endotoxin. Endotoxin, being a potent stimulus for the release of vasoactive mediators, could theoretically induce digital vasoconstriction and set in motion the sequence of events leading to laminitis. Laminitic lesions have been induced following ruminal infusion of lactic acid in sheep but not in cattle. Although lactic acid levels are increased in the cecum of horses fed a laminitis-inducing diet, blood lactate levels rise only slightly in experimental carbohydrate overload in horses. Furthermore, classical endotoxemia rarely develops in naturally occurring laminitis, and experimental administration of endotoxin has not produced laminitis. Horses with black walnut-induced laminitis do not display systemic signs.

The demonstration of impaired response to insulin in obese ponies and in ponies which have suffered from laminitis has given rise to the hypothesis that insulin intolerance leads to laminitis through a sequence of increasing thromboxane A_2 levels, platelet aggregation, and subsequent peripheral vasoconstriction.

Acute laminitis presents as a sudden lameness affecting all feet, or just the fore or hind feet. An increase in the hoof wall temperature and a bounding digital pulse occur, indicating marked vascular engorgement in the hoof tissues. In horses, pain is often severe enough to provoke systemic disturbances. A section through an acutely affected hoof reveals little gross alteration beyond congestion of the laminar dermis and occasionally hemorrhage. There is no hoof deformity, although the skin above the coronary band may be swollen. Histologically, the acute changes in the dermis are hyperemia, hemorrhage, marked edema, and occasionally thrombosis. Mitotic figures are present in the epidermal laminae of horses with Grade II laminitis induced by black walnut at 12 hr; no mitoses occur in normal controls. By 84 hr, and with the development of Grade III laminitis, the architecture of the dorsal laminae is disorganized, and the tips of the primary laminae are necrotic. Islands of proliferating epithelial cells lying in a mildly reactive dermis show high mitotic activity. In cattle, swelling of endothelial and medial cells in small arteries and arterioles and edema of the perineurium of small nerves also occur. Inflammatory cell infiltration either is absent or is limited to moderate perivascular accumulations of mononuclear cells and few focal accumulations of neutrophils. Epidermal changes vary from hydropic degeneration of the basal keratinocytes to extensive coagulation necrosis of the secondary epidermal laminae.

Chronic laminitis is defined clinically as the phase commencing after several days of lameness or when rotation of the third phalanx is first radiographically evident. The rotation has been attributed to loss of the interlocking force normally supplied by the epidermal laminae. In chronic cases, the ventral deviation is caused also by irregular hyperplasia of epidermal laminae placing a wedge of epidermis between the phalanx and the immoveable hoof wall. The weight of the animal, the leverage force placed on the toe, and the pulling force of the deep digital flexor tendon contribute mechanically to the rotation. In severely affected animals, the third phalanx may penetrate the sole, which becomes convex and droops. The laminar degeneration occurs mainly on the anterior aspect of the hoof so that, although the hoof may spread, it is usual for the toe to turn up and the anterior aspect of the hoof to become concave and wrinkled by encircling horizontal ridges.

The chief microscopic lesion in equine laminitis is marked irregular hyperplasia of the epidermal laminae. The regenerating secondary laminae may not regain their orderly arrangement, but instead form irregular and anastomosing epidermal cords. The epidermal laminae, both primary and secondary, become markedly hyperkeratotic. The reason for the hyperkeratosis is not known. Both physical and physiologic influences on keratogenesis are likely to be altered in chronic laminitis. Similar epidermal lesions occur in chronic bovine laminitis, although parakeratotic hyperkeratosis develops in addition to the orthokeratotic hyperkeratosis. Alterations in the dermal vasculature are most prominent in cattle. Moderate to marked arteriolosclerosis and arteriosclerosis occur in chronic bovine laminitis, especially in the solar dermis. Other changes in laminitis in cattle include chronic dermal granulation tissue, organized and recanalized vascular thrombi, perineural fibrosis, and perivascular accumulations of macrophages, often containing hemosiderin.

Laminitis, not related to traumatic and metabolic episodes, occurs in all species but is important only in ungulates. Erysipelas in lambs, the different causative types of footrot in pigs, sheep, and cattle, and bluetongue in sheep are examples of diseases in which degenerative and inflammatory changes occur in the laminae of the hoof.

Bibliography

Budras, K. D., Hullinger, R. L., and Sack, W. O. Light and electron microscopy of keratinization on the laminar epidermis of the equine hoof with reference to laminitis. *Am J Vet Res* **50:** 50–60, 1989.

Field, J. R., and Jeffcott, L. B. Equine laminitis—another hypothesis for pathogenesis. *Med Hypoth* **30:** 203–210, 1989.

Galey, F. D. *et al.* Black walnut (*Juglans nigra*) toxicosis: A model for equine laminitis. *J Comp Pathol* **104:** 313–326, 1991.

Hunt, R. J. The pathophysiology of acute laminitis. *Compend Cont Ed* **13:** 1003–1009, 1991.

Leach, D. H., and Oliphant, L. W. Ultrastructure of the equine hoof wall secondary epidermal lamellae. *Am J Vet Res* **44:** 1561–1570, 1983.

Moore, J. N., Allen, D., and Clark, E. S. Pathophysiology of acute laminitis. *Vet Clin North Am* **5:** 67–72, 1989.

Pollitt, C. C., and Molyneux, G. S. A scanning electron microscopical study of the dermal microcirculation of the equine foot. *Eq Vet J* **22:** 79–87, 1990.

B. Eosinophilic and Collagenolytic Dermatitides

This section includes some very common but poorly understood diseases of the cat and horse. It is possible that we are dealing with a common reaction pattern evoked by a variety of different etiologic stimuli. The horse and the cat may well respond with eosinophils, where other species would utilize the neutrophil. In evidence, it has been shown that equine neutrophils and eosinophils respond equally well *in vivo* to chemotactic stimuli which would typically attract only neutrophils in other species.

Collagenolysis is the second feature of some, but not all, of these diseases. Bundles of dermal collagen may be hypereosinophilic, fragmented, lysed, and/or mineralized. Degenerate collagen forms the nidus of inflammatory reactions, which may be eosinophilic, granulomatous, or neutrophilic (see Section III,O, Collagen Disorder of the Footpads of German Shepherd Dogs). In the collagenolytic diseases dominated by the eosinophil, flame figures are found. These are eosinophilic accumulations, somewhat resembling flickering flames, which surround the collagen fiber. In humans, this material has been shown to be major basic protein and other constituents of eosinophil granules. It has been hypothesized that these eosinophil products may be responsible for the collagen degradation. A role for adsorbed foreign antigens, particularly of ectoparasite origin, has been proposed in attracting eosinophils to the site and in stimulating their degranulation, much as occurs when eosinophils react against the cuticle of invading parasites like *Onchocerca*.

Bibliography

Brehmer-Andersson, E. *et al.* The histopathogenesis of the flame figures in Well's syndrome based on five cases. *Acta Derm Venereol (Stockh)* **66:** 213–219, 1986.

McEwen, B. J. *et al.* The response of the eosinophil in acute inflammation in the horse. *In* "Advances in Veterinary Dermatology" C. von Tscharner and R. E. W. Halliwell (eds.), pp.176–194. London, Baillière Tindall, 1990.

1. Feline Eosinophilic Granuloma with Collagenolysis

These lesions are common in the cat. They were incorporated traditionally into the eosinophilic granuloma complex—a heterogeneous group of cutaneous, mucocutaneous, and oral lesions subdivided into the indolent ulcer (rodent ulcer), eosinophilic plaque, and linear granuloma. There is little justification, other than precedent, for considering that these entities are related. They are dissimilar morphologically and histologically, and no common pathogenetic mechanism or causative agent has been demonstrated. Apparent outbreaks of eosinophilic granuloma suggest an infectious agent, but despite diligent searching, only an unidentified filamentous bacterium and calicivirus have been detected in the lesions. Their significance is doubtful. The association between eosinophilic granuloma and feline leukemia virus infection is also not proven. The lesions also often occur in cats with demonstrated allergy. Antiepithelial antibodies have been detected, but few believe the lesion to be autoimmune. In some cases, the lesion may result from arthropod bites, either due to direct injury or, more likely, to hypersensitivity reactions. Eosinophils are major effector cells in reaction to parasites, mediating direct damage through release of major basic protein and eosinophilic cationic protein. In one convincing demonstration, a mosquito was allowed to bite the nose of a sensitive cat. Subsequent biopsy of the site revealed typical microscopic lesions of eosinophilic granuloma with collagenolysis. It is, however, most likely that eosinophilic granuloma with collagenolysis represents a tissue reaction pattern to a variety of injuries.

The macroscopic appearance varies. Lesions often affect the posterior aspect of the hind leg but also the medial aspect of the foreleg. The gross appearance is a well-circumscribed, sometimes linear, yellow-pink firm plaque. Lesions are 2–4 mm wide but may be up to 10 cm in length. The surface of the linear granuloma is rarely ulcerated (Fig. 5.113A) but may be partially alopecic. Oral cavity lesions are common, with tongue, gingiva, or palate being involved. Lesions affecting the upper lip (Fig. 5.113B) are well-circumscribed, sometimes bilateral, oval ulcers with a brown-red, shiny surface and raised border. These last-mentioned lesions correspond to the indolent ulcer.

Histologically, a nodular dermatitis is associated with multiple, somewhat linear foci of collagen degeneration. The collagen fibers may be frayed, granular, hyalinized, mineralized, or lysed and replaced by eosinophilic or mucinous amorphous material. Surrounding the foci of collagen degeneration are dense accumulations of histiocytes, eosinophils, and plump fibroblasts. Sometimes the collagen fibers are effaced by an eosinophilic coagulum (flame figures). Palisading epithelioid macrophages and multinucleate giant cells may ring some foci of degenerate collagen and the eosinophilic debris. In some lesions, showing the

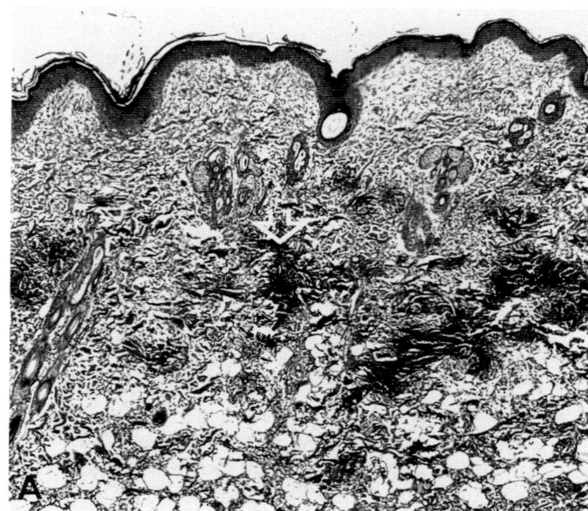

Fig. 5.113A Feline eosinophilic granuloma. Linear granuloma. Note intact epidermis and multifocal degeneration of dermal collagen (arrow).

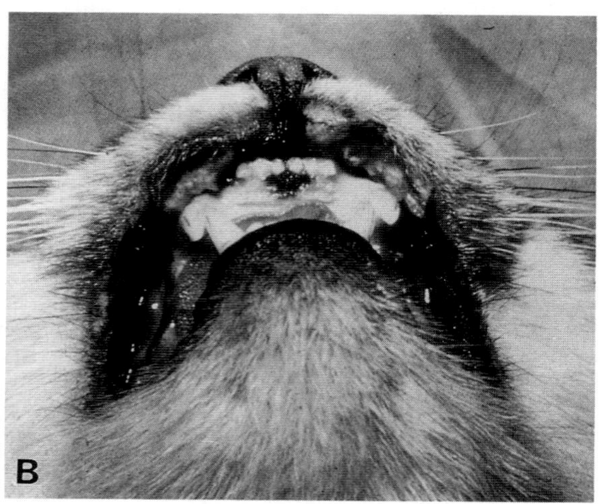

Fig. 5.113B Feline eosinophilic granuloma. Indolent ulcer. (From *Small Animal Dermatology*, 3rd Ed., G. H. Muller, R. W. Kirk, and D. W. Scott, 1983. Courtesy W. B. Saunders.)

the gross morphology of indolent ulcer, the microscopic lesions are lymphoplasmacytic and neutrophilic, rather than eosinophilic and granulomatous.

Bibliography

Rosencrantz, W. Eosinophilic granuloma complex (confusion). *Vet Focus* **1:** 29–32, 1989.

Russell, R. G., Slattum, M. M., and Abkowitz, J. Filamentous bacteria in oral eosinophilic granulomas of a cat. *Vet Pathol* **25:** 249–250, 1988.

Scott, D. W. Observations on the eosinophilic granuloma complex in cats. *J Am Anim Hosp Assoc* **11:** 261–270, 1975.

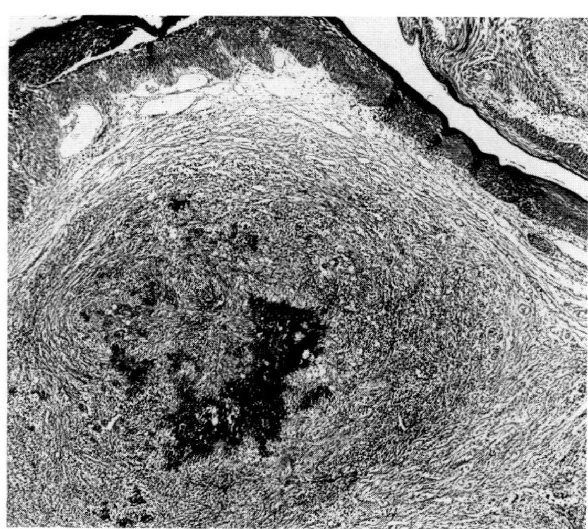

Fig. 5.114 Eosinophilic granuloma in oral cavity. Siberian husky dog.

2. Canine Eosinophilic Granuloma with Collagenolysis

These lesions are rare in the dog, although there is an apparent predilection for the Siberian husky breed. Lesions are grossly and histologically almost identical to those in the cat (Fig. 5.114). Oral cavity lesions are most common, with irregular, often ulcerated plaques and nodules present on the ventral and lateral surface of the tongue or on the palatine mucosa. The less frequent cutaneous form is characterized by multiple papules, nodules, and plaques chiefly affecting the ventral abdomen, flanks, and prepuce. Rarely, solitary lesions are seen in the external ear canal. Both oral and cutaneous forms of eosinophilic granuloma affect young, predominantly male dogs. Rarely, mucosal and cutaneous lesions coexist.

Bibliography

Potter, K. A., Tucker, R. D., and Carpenter, J. L. Oral eosinophilic granuloma of Siberian huskies. *J Am Anim Hosp Assoc* **16:** 595–600, 1980.

Scott, D. W. Cutaneous eosinophilic granulomas with collagen degeneration in the dog. *J Am Anim Hosp Assoc* **19:** 529–532, 1983.

3. Equine Nodular Collagenolytic Granuloma

Frequently encountered in skin biopsies from horses, this lesion has been known variously as nodular necrobiosis and eosinophilic granuloma. Grossly, it is characterized by multiple cutaneous nodules. The lesions tend to occur over the withers, back, and lateral neck. They are firm, nonpruritic nodules from 0.5 to 5 cm in diameter. The overlying skin and hair coat are usually normal. Microscopically there is collagen degeneration and an eosinophilic and granulomatous response, as described. Mineralization of the degenerate collagen is often prominent in equine lesions. The cause is not known, but arthropod injury is suspected. Histologically, the lesions have to be

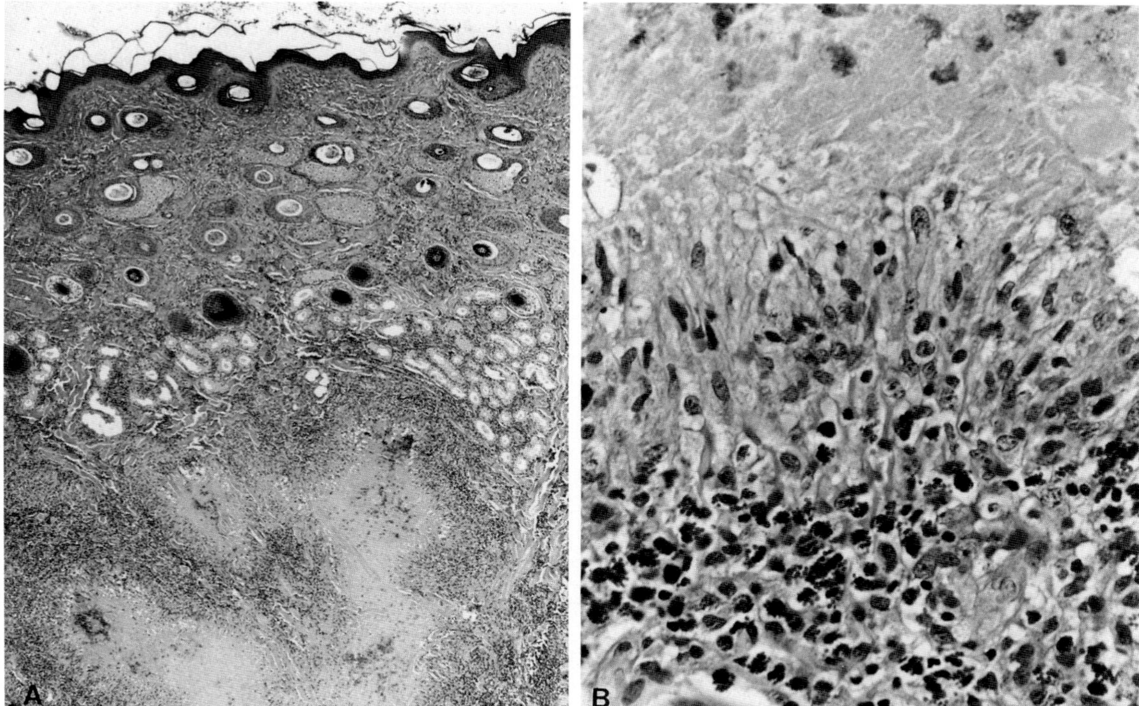

Fig. 5.115 Equine nodular collagenolytic granuloma. (A) Foci of collagen degeneration deep in dermis. (B) Detail showing degenerate collagen, and remnants of eosinophils surrounded by epithelioid macrophages and eosinophils.

differentiated from habronemiasis and from hypoderma nodules (Fig. 5.115A,B).

4. *Feline Eosinophilic Plaque*

Eosinophilic plaque is a common cutaneous and occasional oral cavity lesion of cats. These lesions are often associated with underlying hypersensitivity disorders: atopy, food hypersensitivity, or flea-bite hypersensitivity. There is a growing consensus that these lesions represent a severe manifestation of the allergic response to flea, food, or environmental antigens. Infectious agents are not thought to play a role in pathogenesis, and attempts to transfer the lesions to another site by autologous transfer have not been successful. Most lesions occur on the abdomen and medial thighs. Lesions may be single or multiple and are characterized by well-circumscribed, 0.5–7 cm diameter, raised, round-to-oval, erythematous, oozing-to-ulcerated plaques, which are severely pruritic. Blood eosinophilia is a constant feature.

Histologic findings vary from a superficial and deep perivascular dermatitis to a diffuse dermatitis, in which eosinophils and mast cells predominate (Fig. 5.116A,B). Diffuse spongiosis, which involves epidermis and hair follicle outer root sheath, is common, as are eosinophilic microabscesses and lymphoid nodules. Some authors report collagenolytic foci in lesions which otherwise comply with the diagnosis of eosinophilic plaque.

Bibliography

Gross, T. L., Kwochka, K. W., and Kunkle, G. A. Correlation of histologic and immunologic findings in cats with miliary dermatitis. *J Am Vet Med Assoc* **189:** 1322–1325, 1986.

Moriello, K. A. *et al.* Lack of autologous tissue transmission of eosinophilic plaques in cats. *Am J Vet Res* **51:** 995–998, 1990

5. *Multisystemic Eosinophilic Epitheliotropic Disease of the Horse*

An uncommon, possibly seasonally occurring condition of predominantly young horses, multisystemic eosinophilic epitheliotropic disease presents clinically with either cutaneous or gastrointestinal signs, usually accompanied by severe weight loss. The form in which the alimentary tract is principally affected has been reported as eosinophilic enteritis. The gross cutaneous lesions compose a generalized, exfoliative and alopecic dermatitis which tends to commence on the head and limbs but becomes rapidly generalized. Erosive and ulcerative lesions affecting the coronary band, the lips, and oral cavity also develop early in the course of the disease. Pitting edema of the ventral chest, abdomen, and distal limbs is common. The condition is fatal.

Histologically, the pattern of lesion is a superficial and deep perivascular dermatitis in which eosinophils, plasma cells, and lymphocytes often form dense cuffs around the vessels. Within the hyperplastic epidermis, foci of eosino-

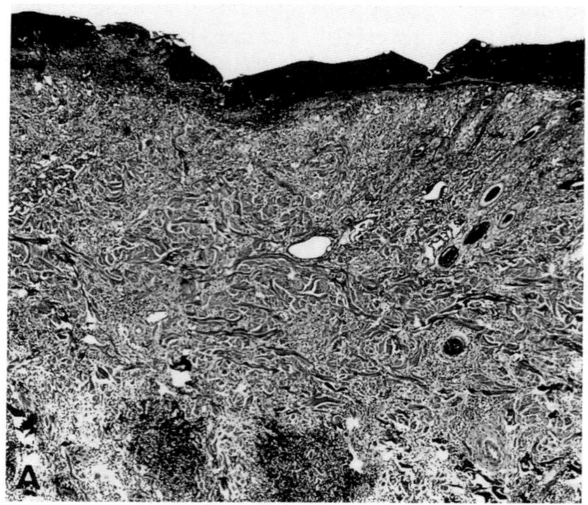

Fig. 5.116A Feline eosinophilic plaque. Diffuse dermatitis with surface ulceration.

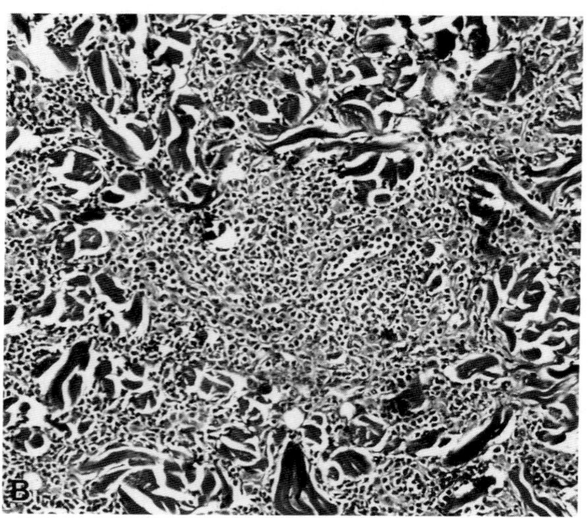

Fig. 5.116B Detail of (A) showing eosinophils and mast cells.

philic spongiosis and eosinophilic subcorneal pustules containing acantholytic cells may occur and need to be differentiated from those of pemphigus foliaceus, another of the generalized, exfoliative dermatitides in the horse. Lymphoplasmacytic and eosinophilic perivasculitis, sometimes accompanied by eosinophilic granulomas and marked fibrosis, occur in epithelial tissues throughout the body. Most notably affected is the pancreas, in which acini may be virtually destroyed, replaced by fibrosis, duct proliferation, and eosinophil infiltration. Other tissues involved are the liver, salivary glands, gastrointestinal tract, including oral cavity and esophagus, uterus, bile duct, and bronchial epithelium.

The pathogenesis of the disease is unknown. One hypothesis postulates infection with an epitheliotropic virus. The lesions have some histologic similarities to those of malignant catarrhal fever, but no virus has yet been identified in the equine lesions. Other possibilities include toxicity (the lesions resemble hairy vetch toxicosis) or hypersensitivity reactions to endogenous or exogenous antigen.

Bibliography

Guaguère, E. *et al.* Cas clinique: Dermatite éosinophilique généralisée et entérocolite éosinophilique chez un cheval. *Point Vét* **20:** 863–868, 1988.

Nimmo-Wilkie, J. S. *et al.* Chronic eosinophilic dermatitis: A manifestation of a multisystemic, eosinophilic, epitheliotropic disease in five horses. *Vet Pathol* **22:** 297–305, 1985.

6. Hypereosinophilic Syndrome

The hypereosinophilic syndrome is a rare disorder of **cats** characterized by persistent idiopathic eosinophilia associated with a diffuse infiltration of various organs by mature eosinophils. Rarely, affected cats may have a generalized, severely pruritic, maculopapular erythema. Histologically, a superficial and deep perivascular dermatitis containing numerous eosinophils is seen.

Bibliography

Harvey, R. G. Feline hypereosinophilia with cutaneous lesions. *J Small Anim Pract* **31:** 453–456, 1990

7. Sterile Eosinophilic Folliculitis

Sterile eosinophilic folliculitides are rare dermatoses that have been reported in the dog, cat, cow, and horse. The pathogenesis of these disorders is unknown. There is no apparent age, breed, or sex predilection. Except in cats, in which an underlying hypersensitivity disorder is always present, affected animals are otherwise normal. We speculate that some of these disorders may be the result of insect bites.

In **dogs,** two syndromes have been described. **Sterile eosinophilic pustulosis** is characterized by the acute onset of a multifocal-to-generalized pruritic, erythematous, papulopustular eruption. Secondary lesions include annular erosions and epidermal collarettes, crusts, alopecia, and hyperpigmentation. The trunk and head are commonly affected. Affected dogs usually have blood eosinophilia. **Sterile eosinophilic pinnal folliculitis** is characterized by a nonseasonal, nonpruritic, bilaterally symmetric papulopustular eruption on the lateral surface of the pinnae. Both conditions are characterized histologically by eosinophilic folliculitis and furunculosis. In addition, in sterile eosinophilic pustulosis, subcorneal eosinophilic pustules (nonfollicular) and perifollicular flame figures may be seen.

In **cats,** sterile eosinophilic folliculitis appears to be an unusual manifestation of underlying atopy, flea-bite hypersensitivity, or food hypersensitivity. Pruritic, erythematous papules and pustules and a disproportionately severe symmetric traumatic hypotrichosis are present over the abdomen, caudomedial thighs, and rump. Affected cats have marked blood eosinophilia.

In **cattle,** sterile eosinophilic folliculitis is characterized by multiple annular areas of crusts, hyperkeratosis, and alopecia. The lesions are most commonly present on the head, neck, and trunk. Affected cattle do not have blood eosinophilia. The condition is usually misdiagnosed as dermatophytosis.

In **horses,** eosinophilic folliculitis and furunculosis may be seen as focal histologic changes in association with numerous specific diseases: eosinophilic granuloma, onchocerciasis, *Culicoides* hypersensitivity, and atopy. In **unilateral papular dermatosis,** eosinophilic folliculitis and furunculosis are the major histologic lesions. Clinically, the small, (2–4 mm) papular, sometimes crusted lesions affect the lateral thorax and neck of only one side of the body.

Bibliography

Carlotti, D., Prost, C., and Magnol, J. P. La maladie d'Ofugi (pustulose éosinophilique stérile). A propos d'une observation chez un Pinscher. *Prat Med Chirurg Anim Cie* **24:** 131–138, 1989

Scott, D. W. Sterile eosinophilic pustulosis in dog and man: Comparative aspects. *J Am Acad Dermatol* **16:** 1022–1026, 1987.

Scott, D. W. Canine sterile eosinophilic pinnal folliculitis. *Compan Anim Pract* **2:** 19–22, 1988.

Scott, D. W., Miller, W. H., and Shanley, K. J. Sterile eosinophilic folliculitis in the cat: An unusual manifestation of feline allergic skin disease? *Compan Anim Pract* **19:** 6–11, 1989.

Scott, D. W., Walton, D. K., and Guard, C. L. Sterile eosinophilic folliculitis in cattle. *Agri-Pract* **7:** 8–14, 1986.

Walton, D. K., and Scott, D. W. Unilateral papular dermatosis in the horse. *Eq Pract* **4:** 15–21, 1982.

C. Sterile Granulomas and Pyogranulomas

Sterile granulomatous or pyogranulomatous dermatoses are rarely reported in dogs, cats, and horses. Absence of microbial agents and foreign material and good response to glucocorticoid therapy have suggested an immune-mediated pathogenesis.

1. Sterile Pyogranuloma Syndromes

The sterile granuloma/pyogranuloma syndrome occurs in **dogs** of all ages and both sexes but collies, boxers, Great Danes, Weimaraners, and golden retrievers appear to be predisposed. Most dogs have multiple asymptomatic papules, nodules, or plaques on the face and/or feet. Pedal lesions are frequently secondarily infected, ulcerative, and fistulous. Histologic findings include large perifollicular granulomas or pyogranulomas, which are elongated and vertically oriented. They track hair follicles but do not invade them.

Sterile granulomatous or pyogranulomatous dermatitis has also been reported in **cats.** Older male cats had multiple, pruritic, erythematous-to-violaceous papules, nodules, and plaques, especially on the head and pinnae. Histologically these lesions were characterized by perifollicular pyogranulomatous dermatitis. Middle-aged female cats had pruritic, bilaterally symmetric, erythematous-to-purpuric plaques in the temporal regions. Histologically these lesions were characterized by diffuse granulomatous inflammation wherein multinucleated histiocytic giant cells, unexplained purpura and erythrophagocytosis, and a superficial dermal Grenz zone were prominent features.

Bibliography

Carpenter, J. L. *et al.* Idiopathic periadnexal multinodular granulomatous dermatitis in 22 dogs. *Vet Pathol* **24:** 5–10,1987.

2. Sarcoidal Granulomatous Disease

Sarcoidosis in humans is a systemic granulomatous disease of undetermined etiology. The granulomas are characteristic, being composed predominantly of epithelioid macrophages with few lymphocytes (naked granulomas). A sterile sarcoidal granulomatous dermatitis has been described in **dogs.** Affected animals had multiple erythematous papules, nodules, and plaques, which were neither pruritic nor painful. The lesions most commonly affected the neck and trunk. Nodular-to-diffuse sarcoidal granulomatous inflammation was present histologically.

Equine generalized granulomatous disease is characterized by an exfoliative dermatitis, wasting, and sarcoidal granulomatous inflammation in multiple organ systems. The dermatosis usually begins as scaling, crusting and alopecia on the face or limbs and progresses to generalized exfoliative dermatitis. Histologic findings include multifocal, often perifollicular and mid-to-deep dermal nodules of sarcoidal granulomatous inflammation. Multinucleated histiocytic giant cells are constantly seen, often in large numbers.

Bibliography

Scott, D. W. Autoimmune skin diseases in the horse. *Equine Pract* **11:** 20–32, 1989.

Scott, D. W., and Noxon, J. O. Sterile sarcoidal granulomatous skin disease in three dogs. *Canine Pract* **15:** 11–18, 1990.

3. Nodular Panniculitis

Sterile or nodular panniculitis is a rare condition affecting **dogs, cats,** and **horses.** The term describes a specific lesion but does not indicate a single disease entity. In animals, most lesions are idiopathic, although sterile panniculitis has been associated with immunologic disorders (lupus erythematosus and erythema nodosum in the dog), nutritional deficiencies (vitamin E deficiency in cats), trauma, foreign bodies, and pancreatic disease. There is no age or breed predisposition in cats or horses but, in dogs, affected animals are younger than one year, and the dachshund breed is overrepresented.

The gross lesions are subcutaneous nodules which may become cystic, ulcerate, or develop fistulous tracts. Lesions are often multiple and may be grouped or distributed widely. There is a tendency for the canine lesions to affect the trunk. The microscopic lesion is a lobular panniculitis, which varies from pyogranulomatous to granulomatous to fibrotic, depending on the stage of the lesion. Sterile panniculitis is indistinguishable histologically from the panniculitides of infectious cause.

Bibliography

Baker, B. B., and Stannard, A. A. Nodular panniculitis in the dog. *J Am Vet Med Assoc* **167:** 752–755, 1975.
Karcher, L. F. *et al.* Sterile nodular panniculitis in five horses. *J Am Vet Med Assoc* **196:** 1823–1826, 1990.
Shanley, K. J., and Miller, W. H. Panniculitis in the dog: A report of five cases. *J Am Anim Hosp Assoc* **21:** 545–550, 1985.

D. Necrotizing Panniculitis

Necrotizing cutaneous lesions are rare complications of pancreatic disease in the dog. Cases have been reported in a dog with pancreatic carcinoma and one with pancreatic nodular hyperplasia. The pathogenesis probably pertains to the release of pancreatic lipases.

The lesions are multifocal, erythematous, nonpruritic cutaneous nodules, which may ulcerate centrally and discharge seropurulent exudate. Histologically, the lesion is a necrotizing panniculitis in which mineralization of the lipocytes is a predominant feature. The initial inflammatory reaction is neutrophilic but may become granulomatous with time. Panniculitis secondary to pancreatitis and pancreatic neoplasia is well recognized in humans.

Bibliography

Mason, K. V. Disseminated necrotizing panniculitis associated with pancreatic carcinoma in a dog. *Proc Am Acad Vet Derm*, St Louis, Missouri, p. 61, 1989.
Moreau, P. M. *et al.* Disseminated necrotizing panniculitis and pancreatic nodular hyperplasia in a dog. *J Am Vet Med Assoc* **180:** 422–425, 1982.

E. Canine Juvenile Cellulitis

This disease, also known as puppy strangles or puppy pyoderma, is, despite the name, an idiopathic disease. A bacterial etiology was discounted once it was determined that lesions are sterile and that the disease responds to corticosteroids better than to antibiotics alone. Infectious agents have been suspected because several members of a litter may be affected. However, viruses have not been cultured from lesions, and inoculation of infected tissue into normal puppies failed to transfer the disease. Depressed *in vitro* lymphocyte blastogenesis responses have been reported but are likely to represent the result, not the cause of the disease. Typically, puppies younger than 4 months are affected. Lymphadenopathy of the mandibular nodes is common and may precede the onset of the skin lesions. Lymphadenitis is described in non-palpable nodes distant to the skin lesions, such as the popliteal, and in a clinically normal pup from an affected litter. The disease is likely to be systemic; constitutional signs such as malaise are typical, and arthritis affecting multiple joints has been reported in three dogs with juvenile cellulitis. Cutaneous lesions comprise papules, pustules, crusts, alope-

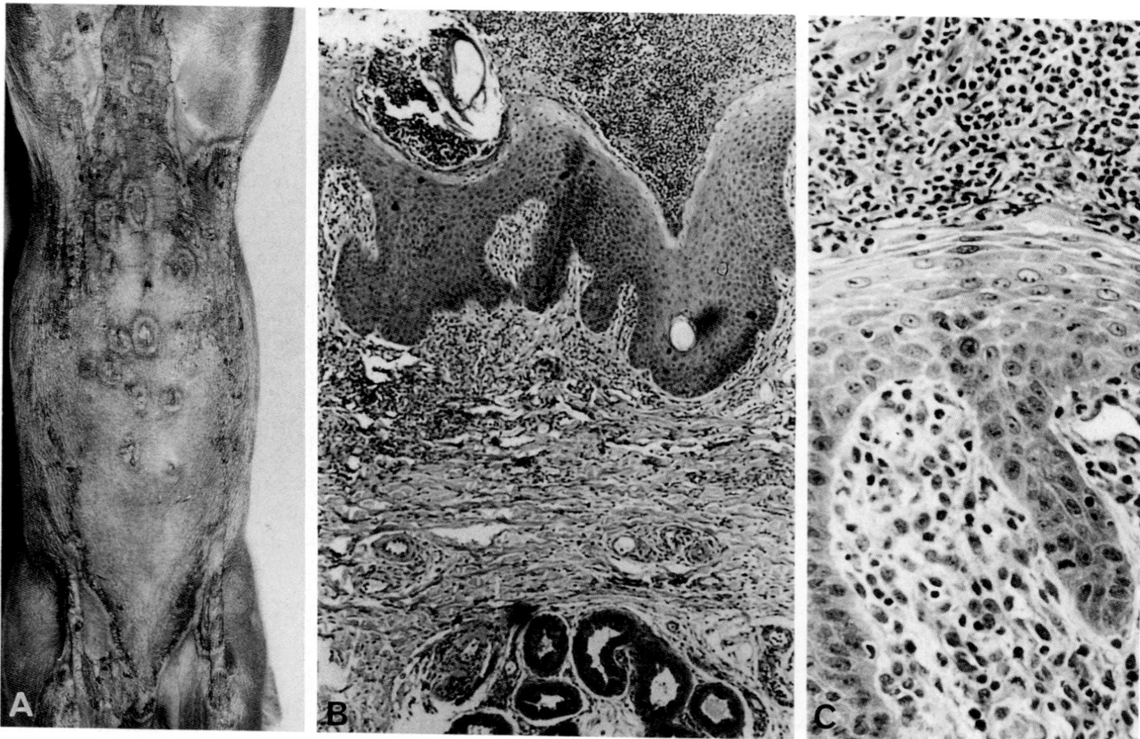

Fig. 5.117 Pityriasis rosea. Pig. (A) Circular, ring, and coalescing lesions involving ventral abdomen and proximal limbs. (Courtesy of J. A. Flatla.) (B) and (C) Hyperplastic dermatitis with heavy cellular infiltrations of inflammatory cells.

cia, and very marked edema. The muzzle, face, and ears are chiefly affected (Fig. 5.86). Histologically, the lesions are pyogranulomatous nodular to diffuse dermatitis with furunculosis and pyogranulomatous lymphadenitis.

Bibliography

Reimann, K. A. *et al.* Clinicopathologic characterization of canine juvenile cellulitis. *Vet Pathol* **26:** 499–504, 1989.
White, S. D. *et al.* Juvenile cellulitis in dogs: 15 cases (1979-1988). *J Am Vet Med Assoc* **195:** 1609–1611, 1989.

F. Follicular Mucinosis (Alopecia Mucinosa)

A disease resembling follicular mucinosis in humans has been described in the **cat.** The lesions in two cats were well-demarcated patches of alopecia and scaling, chiefly affecting the head, neck, and shoulders. Histologically, there was mucinous degeneration of the outer root sheath of the follicular infundibulum. Both cats developed epitheliotropic lymphoma within several months of the initial biopsy. In humans, approximately one third of patients develop mycosis fungoides.

Bibliography

Mehregan, D. A. *et al.* Follicular mucinosis: Histopathologic review of 33 cases. *Mayo Clin* **66:** 387–390, 1991.
Scott, D. W. Feline dermatology 1983–1985: "The secret sits." *J Am Anim Hosp Assoc* **23:** 255–274, 1987.

G. Pityriasis Rosea (Porcine Juvenile Pustular Psoriasiform Dermatitis)

This disease of weanling pigs was originally named pityriasis rosea. Because the clinical signs and gross lesions bear little relationship to those of the human disease for which pityriasis rosea was originally named, the new designation, porcine juvenile pustular psoriasiform dermatitis, has been suggested. The disease is of no significance, except esthetic. The cause is not known. A hereditary predisposition has been suggested but not proven.

The disease begins with small, scaly, erythematous papules on the skin of the abdomen and inner thighs. The papules expand centrifugally to produce at first scaly plaques, and later, when the central areas return to normal, ring-shaped lesions which are erythematous and scaly (Fig. 5.117A). As the rings expand, they coalesce to produce mosaic patterns and may extend to the sides and perineum.

The acute histologic lesion is superficial and deep perivascular dermatitis with eosinophils, neutrophils, and mononuclear cells present. Spongiosis and leukocytic exocytosis lead to the formation of spongiform pustules. Superficial epidermal necrosis may extend into the ostia of the hair follicles. As the lesions heal, marked regular psoriasiform hyperplasia and parakeratotic scale crusts are predominant (Fig. 5.117B,C).

Bibliography

Dunstan, R. W., and Rosser, E. J. Does pityriasis rosea occur in pigs? *Am J Dermatopathol* **8:** 86–89, 1986.

H. Superficial Necrolytic Dermatitis (Hepatocutaneous Syndrome)

This is a rare, but histologically distinctive disease of the dog, with the clinical lesions and biopsy findings closely resembling human necrolytic migratory erythema. This disease is most frequently associated with glucagon-secreting tumors of the pancreas but, more recently, has been recognized in patients with diabetes mellitus and hepatic failure. The first cases of superficial necrolytic dermatitis in the dog were attributed to diabetes mellitus. The disease has been described since in dogs with glucagon-producing alpha-cell pancreatic tumors. Superficial necrolytic dermatitis in the dog is, however, most frequently associated with an underlying chronic hepatitis, hence the alternative appellation, hepatocutaneous syndrome. Diabetes and hyperglucagonemia are thought to be secondary to the hepatitis. The cause of the hepatitis is generally unknown, although mycotoxins were suspected in one dog.

The pathogenesis of superficial necrolytic dermatitis is unknown, but hepatic dysfunction and derangements of glucose and amino acid metabolism are clearly involved. Elevated glucagon levels alone are unlikely to be directly responsible for the skin lesions, as both dogs and humans may develop the disease in their absence, and dermatitis is not an inevitable result of the hyperglucagonemic state. Hypoaminoacidemia, due to sustained gluconeogenesis, is documented in both canine and human patients, and it

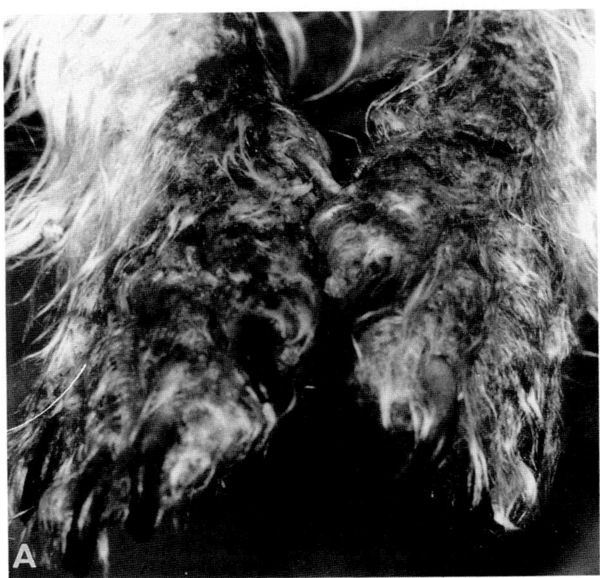

Fig. 5.118A Superficial necrolytic dermatitis (hepatocutaneous syndrome). Dog. Scaling, crusting lesions on feet.

has been postulated that it may deplete epidermal proteins and induce epidermal necrosis. Other aspects of hepatic function may be involved. The lesions have many similarities to zinc-responsive dermatoses.

Lesions have a roughly bilateral distribution. The muzzle, lips, periocular skin, edges of the pinnae, distal extremities, ventrum, and points of pressure or friction, such as the hocks, and the external genitalia are typically affected. Oral and mucocutaneous lesions are occasionally reported. Lesions are erythematous, erosive, ulcerative, and crusted. The footpads are markedly hyperkeratotic (Fig. 5.118A).

The histologic lesions are virtually pathognomonic (Fig. 5.118B). The distinctive feature is a band of hydropic, pale-staining keratinocytes in the upper half of a usually acanthotic stratum spinosum. Both intra- and intercellular edema contribute to the epidermal pallor. As these cells degenerate, clefts and vesicles may form in the outer stratum spinosum. Neutrophils may accumulate to form subcorneal pustules. The stratum corneum is diffusely and markedly parakeratotic. The epithelium of the follicular infundibulum is also affected. The basal cell layer is basophilic and hyperplastic, forming small rete ridges. Dermal inflammation is minimal, predominantly mononuclear, and perivascular. In eroded lesions, neutrophilic exocytosis is prominent, and inflammatory crust covers the surface.

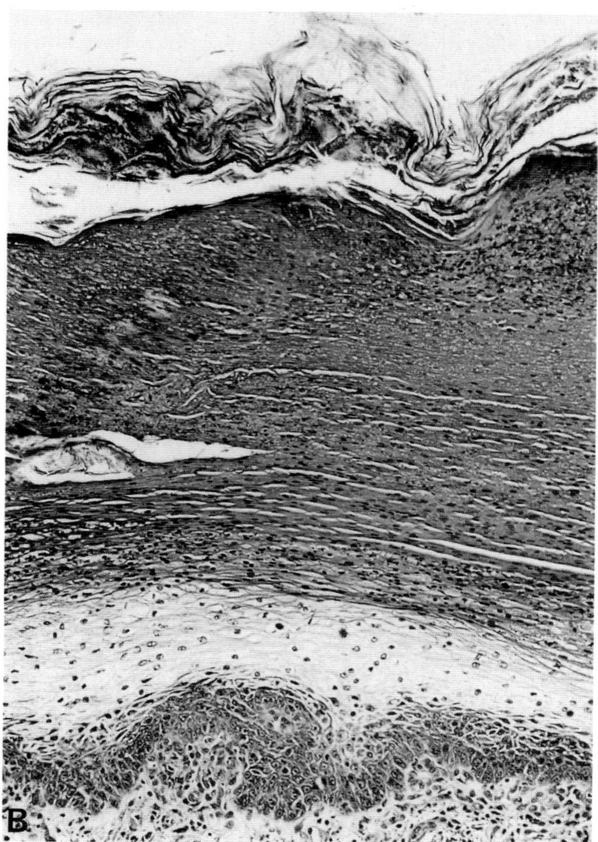

Fig. 5.118B Superficial necrolytic dermatitis (hepatocutaneous syndrome). Dog. Note the typical intracellular edema in the stratum spinosum and the thick layer of parakeratotic and necrotic stratum corneum.

The diagnosis is based on typical histologic findings. However, the pathognomonic epidermal edema may not be present in every biopsy. Differential diagnoses include other parakeratotic, hyperplastic dermatitides, such as the zinc-responsive dermatoses, lethal acrodermatitis of bull terriers, and thallium toxicity.

Bibliography

Gross, T. L. *et al.* Glucagon-producing pancreatic endocrine tumors in two dogs with superficial necrolytic dermatitis. *J Am Vet Med Assoc* **197:** 1619–1622, 1990.

Kasper, C. S., and McMurry, K. Necrolytic migratory erythema without glucagonoma versus canine superficial necrolytic dermatitis: Is hepatic impairment a clue to pathogenesis? *J Am Acad Dermatol* **25:** 534–541, 1991.

Little, C. J. L., McNeil, P. E., and Robb J. Hepatopathy and dermatitis in a dog associated with the ingestion of mycotoxins. *J Small Anim Pract* **32:** 23–26, 1991.

Miller, W. H. *et al.* Necrolytic migratory erythema in dogs: A hepatocutaneous syndrome. *J Am Anim Hosp Assoc* **26:** 573–581, 1990.

Walton, D. K. *et al.* Ulcerative dermatosis associated with diabetes mellitus in the dog: A report of four cases. *J Am Anim Hosp Assoc* **22:** 79–88, 1986.

XIX. Neoplastic Diseases of Skin and Mammary Gland Contributed by Brian P. Wilcock

The skin is the most common site for neoplasia in dogs, horses, cattle, and, arguably, in cats. There is considerable variation in the reported prevalence of skin tumors, the ratio of malignant to benign tumors, and the relative prevalence of different histologic types. Part of this variation may be genuine and reflect differences in genetic susceptibility and environmental carcinogen exposure of different animal populations, but much of the variation probably stems from the lack of uniform criteria for assignment of a given tumor to a given diagnostic category.

Primary skin tumors can be divided into those of ectodermal and those of mesodermal origin. Those of ectodermal origin are subdivided into tumors of the epidermis and of the adnexa. Those of mesoderm are divided according to the structural elements of dermis (fibrous tissue, muscle, fat, blood vessels) and those of the leukocyte-related cells of the dermis (histiocytes, mast cells, lymphocytes). Melanocytic tumors occupy a special category of their own. In general, ectodermal neoplasms are behaviorally benign, and those of adnexa are almost exclusively so. Many types of mesodermal tumors, in contrast, are histologically malignant and regularly exhibit locally infiltrative growth and occasionally will metastasize.

Clinical texts put considerable emphasis on age, site, breed, or species predilections for some tumors. Although such predilections may be biologically interesting and provide insight into fundamentals of carcinogenesis, they are of little use to the diagnostic pathologist. In a few instances, the behavior of a given neoplasm differs among species or different locations, and only for these tumors will such data be given here. Comments that a specific type of tumor is prevalent or rare in a given species are occasionally included to help those not familiar with skin tumors focus their diagnostic decision making first on the most probable diagnoses.

Bibliography

Baker, J. R., and Leyland, A. Histological survey of tumours of the horse, with particular reference to those of the skin. *Vet Rec* **96:** 419–422, 1975.

Carpenter, J. L., Andrews, L. K., and Holzworth, J. Tumors and tumor-like lesions. *In* "Diseases of the Cat," Vol. 1, J. Holzworth (ed.) pp. 406–596. Philadelphia, Pennsylvania, W. B. Saunders, 1987.

Conroy, J. D. Canine skin tumors. *J Am Anim Hosp Assoc* **19:** 91–114, 1983.

Cotchin, E. A general survey of tumours in the horse. *Equine Vet J* **9:** 16–21, 1977.

Koestner, A. Prognostic role of cell morphology of animal tumors. *Toxicol Pathol* **13:** 90–117, 1985.

Miller, M. A. *et al.* Cutaneous neoplasia in 340 cats. *Vet Pathol* **28:** 389–395, 1991.

Moriello, K. A., and Rosenthal, R. C. Clinical approach to tumors of the skin and subcutaneous tissues. *Vet Clin North Am: Small Anim Pract* **20:** 1163–1190, 1990.

Nielsen, S. W., and Cole, C. R. Cutaneous epithelial neoplasms of the dog—a report of 153 cases. *Am J Vet Res* **21:** 931–948, 1960.

Priester, W. A., and Mantel, N. Occurrence of tumors in domestic animals. Data from 12 United States and Canadian colleges of veterinary medicine. *J Nat Cancer Inst* **47:** 1333–1344, 1971.

Pulley, L. T., and Stannard, A. A. Tumors of the skin and soft tissues. *In* "Tumors in Domestic Animals," 3rd Ed. rev., J. E. Moulton (ed.), pp. 23–87. Berkeley, University of California Press, 1990.

Scott, D. W. Feline dermatology 1900–1978: A monograph. Neoplastic disorders. *J Am Anim Hosp Assoc* **3:** 419–425, 1980.

Scott, D. W. "Large Animal Dermatology." Philadelphia, Pennsylvania, W. B. Saunders, 1988.

Sundberg, J. P. *et al.* Neoplasms of Equidae. *J Am Vet Med Assoc* **170:** 150–152, 1977.

Withrow, S. J., and MacEwen, E. G. (eds.). "Clinical Veterinary Oncology." Philadelphia, Pennsylvania, J. B. Lippincott, 1989.

A. Tumors of the Epidermis

Tumors of the epidermis include squamous papilloma, squamous cell carcinoma, basal cell tumors, and intracutaneous cornifying epithelioma (keratoacanthoma). We include here as well a variety of keratin-filled cystic lesions that can be confused with true neoplasms. The decision to classify a given tumor as one or another of the epidermal or adnexal tumors is often quite arbitrary. A neoplasm of the multipotent germinal cells of the epidermis may show a variety of differentiations to resemble a variety of mature tissues. Sometimes the differentiation is only in one direction and results in a distinctive group of tumor cells, making precise identification unequivocal. At other times, the tumor cells may differentiate toward several skin structures, forming squamous cells, sebaceous glands, or hair. We then have the choice of calling such a tumor a basal cell tumor with squamous and adnexal differentiation, or to choose the most aggressive or dominant example from among the various observed differentiations and to so name the tumor.

1. Epidermal Cyst

Epidermal cysts are seen as single or multiple, smooth, spherical dermal nodules seldom larger than 1 cm in diameter. Each nodule consists of an orderly wall of stratified squamous epithelium identical to epidermis, with gradual keratinization at its luminal surface (Fig. 5.119A,B). The

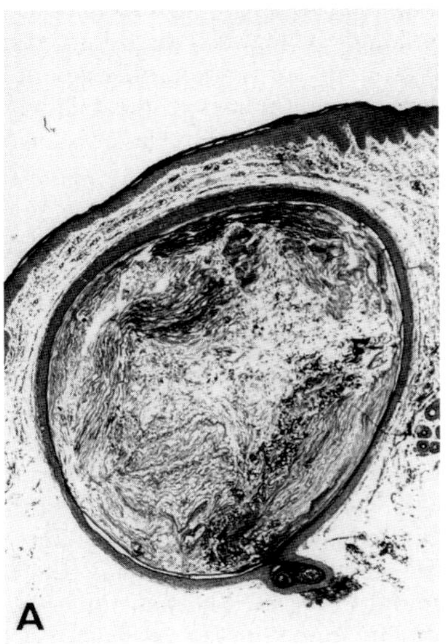

Fig. 5.119A Epidermal cyst.

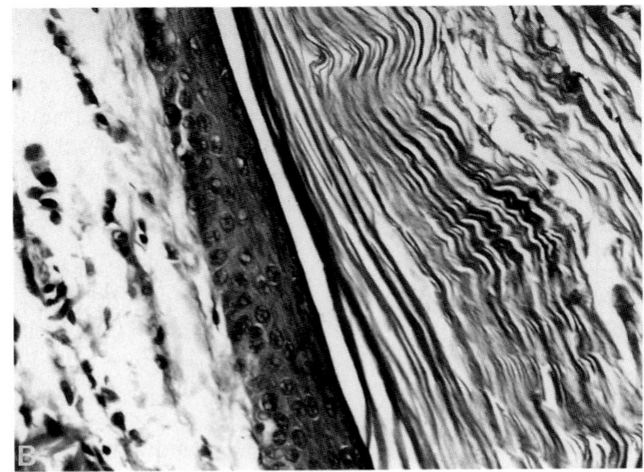

Fig. 5.119B Epidermal cyst with simple wall, gradual keratinization.

cyst lumen is filled with concentric laminations of keratin. There is usually neither calcification nor inflammation within the cyst. The basal layer of the cyst wall abuts dermal collagen in a smooth line, analogous to the normal dermal–epidermal junction. In long-standing lesions, the epithelial wall may become very thin, and rupture may release entrapped keratin to stimulate pyogranulomatous dermal inflammation. The origin of epidermal cysts can seldom be determined from the biopsy sample. Some apparently arise from dilatation of the infundibulum of occluded hair follicles, and indeed occasionally we can detect superficial dermal scarring which may support this hypothesis. As well, such cysts may contain fragments of mature hair shafts. Other cysts arise from traumatic, developmental, or surgical implantation of epidermal fragments into dermis or subcutis. Epidermal fragments which are traumatically implanted ordinarily atrophy and disappear; their persistence and development may be governed by host factors, which would account for their frequency in dogs and some strains of sheep. Implantation in sheep is caused by penetrating grass seeds. Persistent cysts may develop into squamous carcinomas.

Dilated pore of Winer is a flask-shaped epidermal cyst on the head or neck of middle-aged or old cats. It is connected to the skin surface by a narrow pore. It differs from epidermal cyst (including follicular cysts) by virtue of a characteristic thickening of the cyst wall near its base, where the stratified keratinizing epithelium develops very regular rete ridges in parallel columns. For practical purposes, it is yet another variant of epidermal/follicular cyst.

Dermoid cysts arise by developmental failure of epidermal closure along embryonic fissures that maroons an island of multipotential ectoderm within the subcutis. Some, such as those occurring along the dorsal midline of Rhodesian ridgeback dogs, actually retain a sinus pore to the skin surface. Dermoid cysts are easily distinguished from the epidermal cysts by their complex basal cell wall, which has adnexal differentiation. Primitive or mature hair follicles and sebaceous glands are the rule, with sweat gland differentiation theoretically possible but rarely seen. There is usually less keratinization (of epidermal type) than in other cysts, and the basal cell–dermal junction is very irregular. Specimens with only hair follicle differentiation may be confused with unilocular trichoepitheliomas, but the dermoid cyst has gradual keratinization rather than abrupt tricholemmal keratinization.

2. *Papilloma and Fibropapilloma*

Cutaneous papillomas are benign proliferative epithelial neoplasms that have a complex etiology and pathogenesis. The differences in site preference, clinical course, and histology of such lesions have been made more understandable by the discovery that most papillomas are caused by infection with a host- and site-specific papillomavirus of the family Papovaviridae. Papillomaviruses are associated with skin papillomas in all domestic species except the cat. Cattle, horses, and dogs are most frequently affected. In each species there are several viruses distinguishable by gene sequencing, and in most instances, each identified virus type has a preferred site for replication and, except for bovine papillomaviruses, seems highly species specific. In general, papillomaviruses induce two types of cutaneous lesions—squamous papilloma and fibropapilloma—but lesions of intermediate type are common. Not all papillomas are caused by viruses, and the proof of viral cause is a complex matter since the papillomaviruses do not grow in tissue culture. Advances in immunohistochemical localization of the viral antigens or detection via gene probes has greatly increased the sensitivity of viral detection, permitting retrospective studies on fixed tissues for the presence of the viral genome.

At least six different papillomaviruses occur in **cattle.** Three (bovine papillomavirus types 1, 2, and 5) cause cutaneous fibropapillomas, while the other three (types 3, 4, and 6) cause papillomas of skin (3, 6) or esophagus (4). For both lesion types, affected cattle are younger than 2 years and have single or multiple lesions that usually spontaneous regress within a year of appearance (Fig. 5.120A).

The typical papilloma is a 1- to 2-cm wartlike, filoform, exophytic mass composed of hyperplastic epidermis supported by thin, inconspicuous dermal stalks. Most of the hyperplasia is due to marked expansion of stratum spinosum. Many of the cells have hydropic ballooning of their cytoplasm, large eosinophilic keratohyaline granules, and vesicular nuclei that contain virus particles on electron microscopy and viral antigen detectable by immunofluorescence. True viral inclusion bodies are seldom seen in paraffin-embedded sections of natural disease.

Microscopic lesions typical of so-called fibropapillomas include the features of acanthosis, hyperkeratosis, and down-growth of rete ridges (Fig. 5.120B), but the dermal proliferation predominates. The proliferating cell is a large, plump fibroblast. The cells are arranged in haphazard whorls and fascicles rather than in perpendicular sheets, as in granulation tissue. In some the epidermal proliferation is minimal and is seen only as slight acanthosis and accentuation of rete pegs, while in others, the hyperplasia resembles full-fledged papillomas. The rela-

Fig. 5.120A Multiple viral papillomas (warts). Steer.

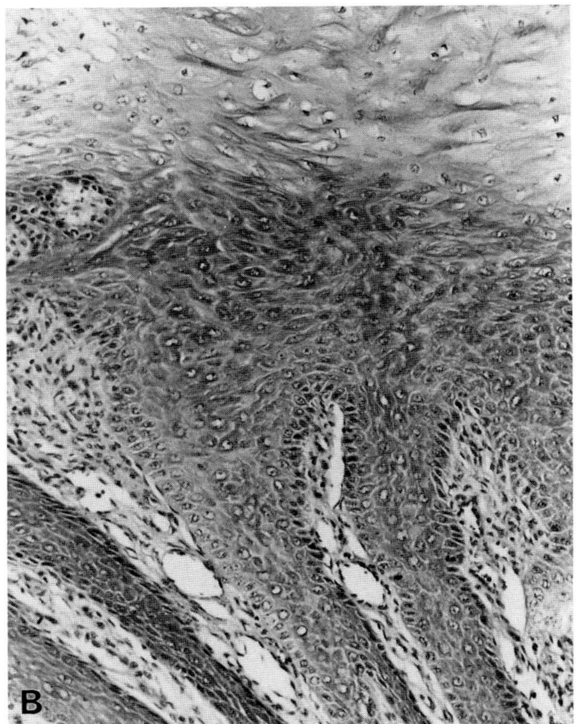

Fig. 5.120B Hyperplasia and vacuolation of the stratum spinosum, bovine squamous papilloma.

tionship between these two extremes of morphologic expression is clarified somewhat by sequential observations on experimentally transmitted lesions in cattle and horses. In calves inoculated by epidermal scarification, there is an initial mild fibroblastic proliferation in the papillary dermis ~25 days after inoculation. By 45 days, epidermal hyperplasia, predominantly acanthosis, accompanies the fibrosis and thereafter increases in prominence relative to the connective tissue. The connective tissue in these squamous papillomas is mature, hyalinized, and cell poor. In contrast, intradermal inoculation of the same virus isolate initially causes a mononuclear leukocytic inflammatory reaction, but is followed by a papillary fibroplasia by day 25. The proliferations remain mostly fibroblastic, and small lesions (80%) never induce epidermal hyperplasia. Only in a few larger (more than 1 cm) nodules does focal acanthosis develop and progress to the severe hyperplasia characteristic of a papilloma in most species. The lesions produced represent a continuum in which all of the morphologic variants of flat or raised, pedunculated or sessile, typical or atypical bovine papillomas can be found. Resolving lesions are characterized by thinning of the stratum spinosum despite persistence of the thick stratum corneum, and hyalinization of the dermal collagen. Fibrocyte nuclei become scarce and pyknotic. Tumor regression is believed to be mediated via cellular immunity, inasmuch as humoral immunity is not effective in initiating tumor regression; vaccines that induce humoral immunity are of little efficacy; and animals with defective cell-mediated immunity do not reject cutaneous papillomas. Mononuclear leukocytes are numerous at the margins, and to a lesser degree within the stroma, of regressing tumors. While some of the variation in anatomic localization or histologic pattern appears to result from management practices and routes of inoculation, the different viral strains themselves seem to have site preferences and typically induce one or the other histologic reaction (papilloma or fibropapilloma).

Lesions indistinguishable from proven virally induced fibropapillomas are common in **horses** and are termed sarcoids. Bovine papillomavirus genome is detected in at least some typical sarcoids, and the lesion has been reproduced by intradermal inoculation of bovine papilloma virus. Viral genome consistent with either one of the closely related bovine papillomaviruses types 1 and 2 has been found in equine sarcoids. No evidence of equine cutaneous papillomavirus was found. The gross and histologic features are those of a fibropapilloma. The tumors are raised, pedunculated, or sessile (Fig. 5.121), and may be either verrucous (papillary) or fibroblastic. Multiple sites are common, and the tumors may be found in horses of any age or breed and in several animals in a group. Histologically, sarcoids are distinctive; the epidermis is at least focally acanthotic and usually sends very long, thin cords of pseudoepitheliomatous hyperplasia into the proliferating dermal connective tissue. Ulceration is common. The dermal proliferation usually is of tightly whorling plump spindle cells with abundant ground substance (Fig. 5.122). The superficial layer of cells is usually oriented at right angles to the basilar epidermal layer. Collagen production is less than that in a fibroma, and nuclei are much more frequent. Anaplasia is absent, and mitoses are typically few. Recurrence following surgery is common, but metastasis has not been reported. The histologic diagnosis may be difficult in older or traumatized lesions in which the peculiar association of thin, sinuous rete down-growths and superficial dermal fibroplasia is obscured by ulceration and granulation tissue. Sarcoids are discussed further in the section on spindle cell tumors of skin.

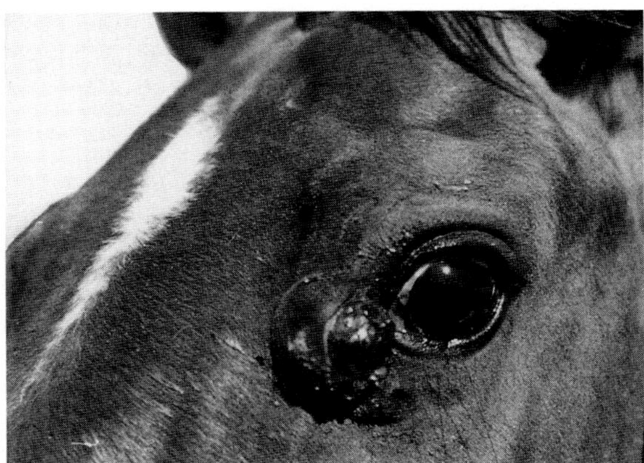

Fig. 5.121 Equine sarcoid.

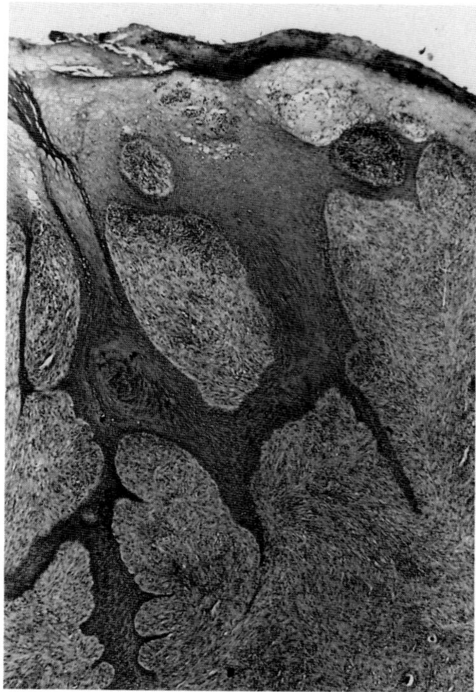

Fig. 5.122 Epidermal hyperplasia and associated fibroblastic proliferation. Equine sarcoid.

True squamous papillomas occasionally occur in horses, representing only about 5% of all equine tumors submitted for histologic examination. That estimate is probably low because of the innocuous nature of the tumor and the reliability of clinical diagnosis. As with the same tumor in cattle, they occur as single or multiple masses in young horses, and usually regress within a year (Fig. 5.123). Muzzle is the most prevalent site, but tumors may occur anywhere, including genital mucous membranes. They are histologically identical to squamous papillomas in calves. Two different viral genome sequences have been described in horses. Unlike the bovine viral strains, however, the equine virus appears to be species specific.

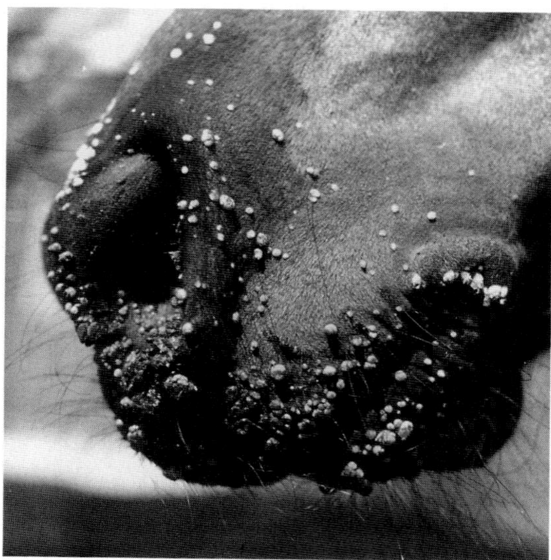

Fig. 5.123 Multiple viral papillomas. Equine muzzle.

Papillomas and fibropapillomas have rarely been recorded in **sheep.** There is one report of multiple cutaneous papillomas in a sheep in England in which papillomalike virus was observed and subsequently transmitted to other sheep.

Multiple cutaneous papillomas have been described in **goats,** but circumstantial evidence points toward an actinic rather than viral cause. No virus has been demonstrated by immunofluorescence, electron microscopy, or *in situ* hybridization.

Dogs are affected by at least five different clinicopathologic syndromes associated with papillomalike lesions: multiple oral papillomas in young dogs, solitary cutaneous papillomas in dogs of any age, venereal papillomas, eyelid or conjunctival papillomas, and sarcoidlike fibropapillomas. Eyelid papillomas may appear as part of juvenile papillomatosis or as solitary lesions in older dogs. Only the oral papillomatosis has been well studied. The typical transient squamous papillomas contain viruslike basophilic intranuclear inclusions that are typical of papillomavirus with electron microscopy, immunofluorescence, and peroxidase–antiperoxidase labeling. The virus is similar to bovine papillomavirus type 1. Evidence of viral pathogenesis for the other papillomalike lesions is not so strong. Some solitary cutaneous papillomas and venereal papillomas contain papillomaviral genome, but their relationship to each other or to the virus of oral papillomatosis is unknown. The presence of virus in the sarcoidlike lesions has not been investigated.

Bibliography

Amtmann, E., Muller, H., and Sauer, G. Equine connective tissue tumors contain unintegrated bovine papilloma virus DNA. *J Virol* **35:** 962–964, 1980.

Barthold, S. W., and Olson, C. Common membrane neoantigens on bovine papilloma virus-induced fibroma cells from cattle and horses. *Am J Vet Res* **39:** 1643–1645, 1978.

Barthold, S. W. *et al.* Atypical warts in cattle. *J Am Vet Med Assoc* **165:** 276–280, 1974.

Cheville, N. F., and Olson, C. Epithelial and fibroblastic proliferation in bovine cutaneous papillomatosis. *Pathol Vet* **1:** 248–257, 1964.

Duncan, J. R. *et al.* Persistent papillomatosis associated with immunodeficiency. *Cornell Vet* **65:** 205–211, 1975.

Ford, J. N. *et al.* Evidence for papilloma viruses in ocular lesions in cattle. *Res Vet Sci* **32:** 257–259, 1982.

Fujimoto, Y., and Olson, C. The fine structure of the bovine wart. *Pathol Vet* **3:** 659–684, 1966.

Fulton, R. E., Doane, F. W., and Macpherson, L. W. The fine structure of equine papillomas and the equine papilloma virus. *J Ultrastruct Res* **30:** 328–343, 1970.

Garma-Avina, A. Equine congenital cutaneous papillomatosis: A report of 5 cases. *Equine Vet J* **13:** 59–61, 1981.

Gibbs, E. P. J., Smale, C. J., and Lawman, M. J. P. Warts in

sheep. Identificaton of a papilloma virus and transmission of infection to sheep. *J Comp Pathol* **85:** 327–334, 1975.
Jablonska, S., Orth, G., and Lutzner, M. A. Immunopathology of papillomavirus-induced tumors in different tissues. *Springer Semin Immunopathol* **5:** 33–62, 1982.
Lancaster, W. D., Olson, C., and Meinke, W. Bovine papilloma virus: Presence of virus-specific DNA sequences in naturally occurring equine tumors. *Proc Natl Acad Sci USA* **74:** 524–528, 1977.
Lancaster, W. D., and Olson, C. Animal papillomaviruses. *Microbiol Rev* **46:** 191–207, 1982.
Olson, R. O., Olson, C., and Easterday, B. C. Papillomatosis of the bovine teat (mammary papilla). *Am J Vet Res* **43:** 2250–2252, 1982.
Orth, G. *et al.* Identification of papilloma viruses in butchers' warts. *J Invest Dermatol* **76:** 97–102, 1981.
Pulley, L. T., Shively, J. N., and Pawlicki, J. J. An outbreak of bovine cutaneous fibropapillomas following dehorning. *Cornell Vet* **64:** 427–434, 1974.
Rebhun, W. C. *et al.* Interdigital papillomatosis in dairy cattle. *J Am Vet Med Assoc* **177:** 437–440, 1980.
Spradbrow, P. B. Papillomaviruses, papillomas and carcinomas. *In* "Proceedings No. 60, Advances in Veterinary Virology," T. G. Hungerford (director), pp. 60:15–60:20, University of Sydney, Sydney, Australia, 1982.
Spradbrow, P. B., and Ford, J. Bovine papillomavirus type 2 from bovine cutaneous papillomas in Australia. *Aust Vet J* **60:** 78–79, 1983.
Sundberg, J. P. Papillomaviruses in animals. *In* "Papillomaviruses and Human Disease," Syrjänen, K., Gissman, L., and Koss, L. G. (eds.), pp. 40–103. Berlin, Springer-Verlag, 1987.
Vanselow, B. A., and Spradbrow, P. B. Papillomaviruses, papillomas, and squamous cell carcinomas in sheep. *Vet Rec* **110:** 561–562, 1982.
Watrach, A. M. The ultrastructure of canine cutaneous papilloma. *Cancer Res* **29:** 2079–2084, 1969.

3. *Intracutaneous Cornifying Epithelioma (Keratoacanthoma)*

Intracutaneous cornifying epithelioma is a benign cystic tumor of the skin of dogs. The classical tumor consists of an invagination of epidermis to line a single 1- to 3-cm keratin filled cyst, which communicates with the skin surface via a pore. The common sites are dorsum of the neck and back. Usually there is only a single tumor, but in Norwegian elkhounds, a breed predisposed to these tumors, they may be multiple or occur in succession.

The gross lesion is a slightly raised firm nodule that may have a central craterous pore from which keratin can be expressed. The histologic lesion is a simple or multiloculated cavity filled with keratin and lined by a complex wall of basal cells (Fig. 5.124A,B). The basal cells have gradual and complete maturation to stratum corneum, but may become somewhat disoriented so as to form keratin pearls within the wall as well as shedding keratin into the lumen. The basilar layer is orderly, and while it may show blunt cordlike expansion into surrounding compressed dermis, there should not be invasion of tumor cells across a basal lamina to be free within the dermis. If seen, that feature justifies a diagnosis of squamous cell carcinoma. There is no sebaceous or hair differentiation by the basal cells.

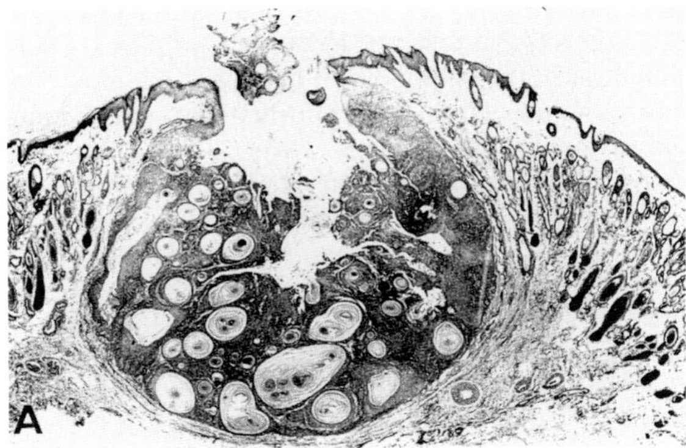

Fig. 5.124A Intracutaneous cornifying epithelioma. Note pore to skin surface.

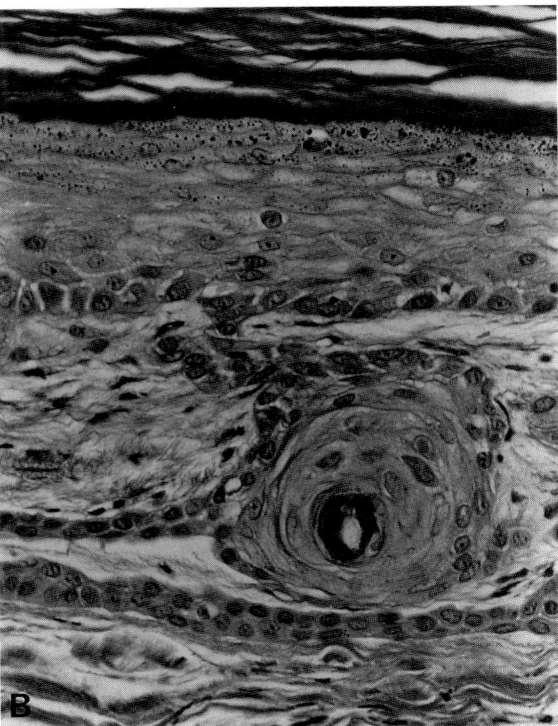

Fig. 5.124B Detail of wall of intracutaneous cornifying epithelioma. Keratinization is gradual, and microcysts are common within the tumor wall.

Focal rupture is common, inciting a granulomatous or pyogranulomatous reaction in surrounding dermis.

4. *Squamous Cell Carcinoma*

Squamous cell carcinoma is a relatively common, locally invasive, and occasionally metastatic neoplasm of most domestic species. Sunlight is probably the most important carcinogenic stimulus for these tumors and accounts for the preference of squamous cell carcinoma for the eyelid and conjunctiva of cattle and horses, the ear pinna of cats and sheep, and the vulva of cattle, goats, and recently sheared sheep (Fig. 5.125A,B). Chronic exposure to sunlight has also been proven to cause squamous cell carcinoma of the lightly haired, poorly pigmented abdominal and juxtanasal skin of dogs.

The progression from solar keratosis to carcinoma occurs over several years and in many instances never attains the full status of carcinoma. In humans, the prevalence of squamous cell carcinoma closely parallels the yearly hours and intensity of ultraviolet irradiation and the sunbathing habits of the population. The same is proven of some animal examples, and is probably true of all. In carcinoma of the ear pinna in Australian sheep, for example, the incidence increases with age to a maximum of 12% in 12-year-old sheep, and the prevalence increases with changes in latitude and altitude that favor maximal exposure to sunlight. The same is true of squamous carcinoma of the eyelid. Hair coat and skin pigmentation are highly protective against tumor development.

In addition to sunlight, carcinogens contained in tobacco, coal tar and soot, arsenic, and smegma have been shown experimentally or by epidemiologic inference to cause squamous cell carcinoma of skin or other tissues. The influence of epidermal injury *per se* in the initiation of cancer is still unsettled, but there is support for this theory in the greater risk of squamous carcinoma at sites of ear notching, branding, and perhaps, chronic inflammation. The last is relevant to the prevalence of squamous cell carcinoma of the nail bed of dogs and, less often, cats. Dogs seem less susceptible to actinic squamous cell carcinoma than are other domestic species, but develop digital carcinoma at sites of chronic paronychia. Whether the lesion progresses from inflammation to neoplasia, or is neoplastic from its inception and merely mimics inflammation, is unresolved.

The appearance of squamous cell carcinoma depends on whether it develops *de novo* or has been preceded by a chronic, precancerous skin lesion. In the former instance, the macroscopic lesion is usually a firm, white, poorly demarcated dermal mass that is ulcerated and streaked with red. In some locations (eye, penis), it is raised and papillary, even though its surface is still ulcerated. In examples following chronic inflammation, as with carcinoma of the ear pinna or nail bed, the tumor may be hidden by the reddening, crusting, and ulceration of the inflammatory lesion. Microscopic examination of such lesions is often dominated by hyperkeratosis, parakeratosis, ulceration, acanthosis, and superficial dermal scarring, with only a few foci of unequivocal neoplasia in which atypical squamous cells have invaded across the basal lamina of the hyperplastic rete pegs. The precancerous plaque and papilloma lesions of ocular squamous cell carcinoma have been described under Ocular Neoplasia (Chapter 4, Section XI).

The histologic diagnosis of squamous cell carcinoma usually is simple. The tumor cells most resemble those of normal stratum spinosum but have vesicular nuclei with one or more very prominent nucleoli. Keratinization and intercellular bridges are seen in all but the most anaplastic specimens, clearly indicating their epidermal origin. Cyto-

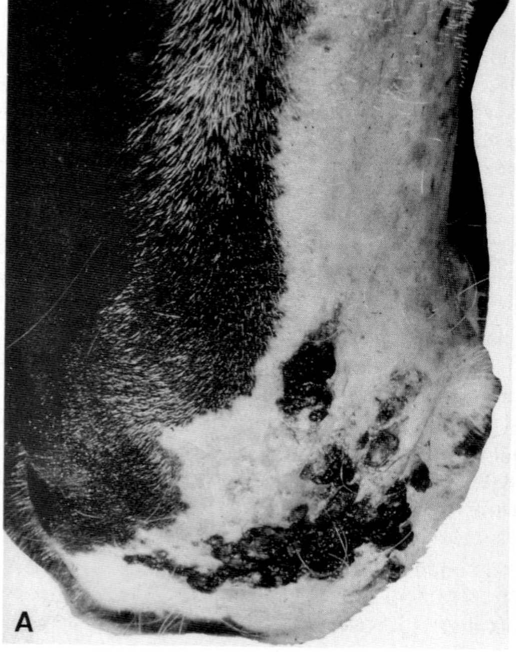

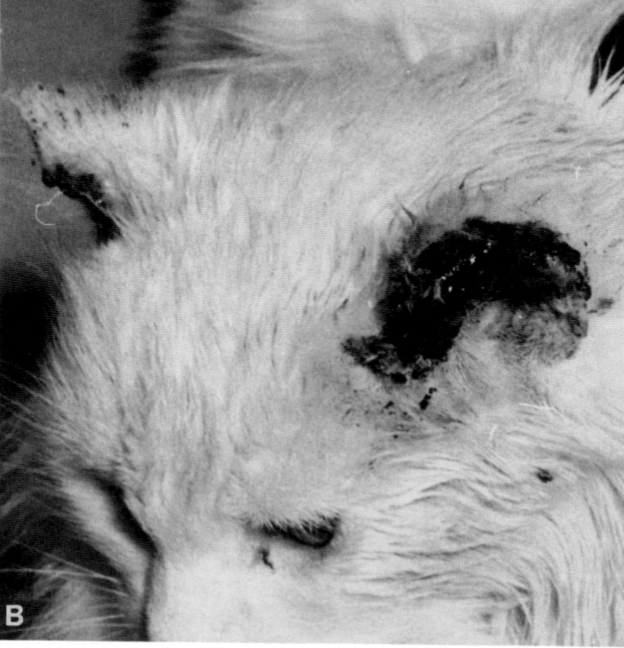

Fig. 5.125 (A) Squamous cell carcinoma involving unpigmented skin of horse's nose. (B) Squamous cell carcinoma involving unpigmented skin of cat's ears.

plasm is usually abundant and eosinophilic. With increasing anaplasia, tumors exhibit wide variation in nuclear size, decreasing cytoplasmic mass and increased basophilia, disappearance of intercellular bridges and obvious keratin formation, and the presence of mononuclear or even multinucleated giant tumor cells. Variation in the prominence of these cytologic features has not, as yet, been correlated with biologic behavior.

The growth habit of squamous cell carcinoma is characteristic and distinguishes this tumor from basal cell tumors, intracutaneous cornifying epithelioma, and reactive pseudoepitheliomatous hyperplasia. The essential diagnostic criterion is the presence of polyhedral cells, resembling those of stratum spinosum, on the dermal side of the basal cell layer and basal lamina. Other benign and reactive lesions may show apparently invasive growth and even unusual cell morphology within the epidermis, but only squamous carcinoma lacks the cushion of basal cells and basal lamina between the tumor cells and the dermis. In the typical squamous cell carcinoma, the tumor cells spread through the dermis as slender anastomosing cords, with some cells falling off the cords to remain as apparently isolated islands in the dermal stroma. Keratinization within such cords or islands results in laminated keratin "pearls" surrounded by tumor cells. Mitotic figures are usually numerous and are in proportion to the degree of anaplasia. Although some smaller, basal-like cells may be present, there is never the orderly maturation from a continuous basal layer to a stratum corneum as seen in normal skin or in benign lesions. Some tumors, especially those of actinic origin with a long precancerous phase, are accompanied by abundant mononuclear leukocytic inflammation. Others, such as those of the digit, are both inflamed and very scirrhous. In such tumors, the neoplastic cells may sometimes be obscured by inflammatory cell influx, difficult to distinguish from the reactive fibroblasts, or largely destroyed by the ulceration of the inflamed mass.

The behavior of squamous cell carcinoma of the skin is usually that of locally destructive spread. Its metastatic potential is low, with certain qualifications depending on location. Squamous carcinomas arising from internal sites such as tonsil, gastric epithelium, and urinary bladder do not share this relatively innocuous behavior. Those initiated by sunlight are slow to metastasize, then usually only to local lymph nodes. In contrast, those originating on the canine digit may be more prone to metastasize, but even these are cured by amputation in virtually all but the most neglected cases. Occurrence on multiple digits simultaneously or consecutively is seen in both dogs and cats in a low percentage of cases.

Carcinoma of the horn of cattle is almost exclusive to castrated male adult cattle in India and neighboring countries. The tumor gradually infiltrates and destroys the horn core and may invade adjacent sinuses and cranial bone. The histology is of well-differentiated squamous cell carcinoma, but neither the site of origin nor the reason for its peculiar site and sex preference is understood. Theories about chronic trauma to the huge horns are not now in vogue.

Bibliography

Hargis, A. M., Thomassen, R. W., and Phemister, R. G. Chronic dermatosis and cutaneous squamous cell carcinoma in the beagle dog. *Vet Pathol* **14:** 218–228, 1977.

Hargis, A. M., and Thomassen, R. W. Solar keratosis (solar dermatosis, senile keratosis) and solar keratosis with squamous cell carcinoma. *Am J Pathol* **94:** 193–196, 1979.

Hawkins, C. D., Swan, R. A., and Chapman, H. M. The epidemiology of squamous cell carcinoma of the perineal region of sheep. *Aust Vet J* **57:** 455–457, 1981.

Kern, W. H., and McCray, M. K. The histopathologic differentiation of keratoacanthoma and squamous cell carcinoma of the skin. *J Cutan Pathol* **7:** 318–325, 1980.

Ladds, P. W., and Entwistle, K. W. Observations on squamous cell carcinomas of sheep in Queensland, Australia. *Br J Cancer* **35:** 110–114, 1977.

Liu, S., and Hohn, R. B. Squamous cell carcinoma of the digit of the dog. *J Am Vet Med Assoc* **153:** 411–424, 1968.

Madewell, B. R., Conroy, J. D., and Hodgkins, E. M. Sunlight–skin cancer association in the dog: A report of three cases. *J Am Vet Med Assoc* **182:** 171, 1983.

Madewell, B. R. *et al.* Multiple subungual squamous cell carcinomas in five dogs. *J Am Vet Med Assoc* **180:** 731–734, 1982.

Strafuss, A. C., Cook, J. E., and Smith, J. E. Squamous cell carcinoma in dogs. *J Am Vet Med Assoc* **168:** 425–427, 1976.

5. Basal Cell Tumors

Tumors of the multipotential germinal epidermal cells are common in dogs and cats but are rare in other domestic species. They are the most common skin tumor of cats. Those proliferations that remain undifferentiated are termed basal cell tumors. Those that exhibit differentiation toward basal cell-derived structures such as sebaceous glands, hair follicles, or even keratinized epidermis are given a wide variety of names depending on the predominant direction of such differentiation. There is, in fact, no objective point at which a basal cell tumor with a hint of adnexal differentiation becomes an adnexal tumor with a prominent basal cell component. Human basal cell tumors most frequently occur in lightly pigmented skin exposed to sunlight, the latter known to favor the development of the tumor. No such predilection is proven for dogs and cats, although actinic squamous carcinomas of eyelid (particularly in horses) often are basal cell rich.

Histologically, basal cell tumors are composed of cords or nests of cells resembling those of normal stratum germinativum. Each has an oval, deeply basophilic nucleus, a single nucleolus, and scant, eosinophilic cytoplasm with indistinct cell boundaries. In contrast to normal stratum germinativum, there are no intercellular bridges, and it is presumed that the cell of origin is a primordial cell common to epidermis and the skin appendages. It is for this reason that many tumors show some differentiation toward hair follicles or sebaceous glands, or have focal keratinization, while remaining predominantly undifferentiated. The cells extend into the dermis as ribbons or islands, or grow expansively as solid sheets or multiple small cysts lined by basal cells. Multiple growth patterns are frequently seen within the same tumor. There is no prognostic difference between the different forms, all being locally expansive but not exhibiting metastasis despite the frequency of mitotic figures.

The most common architectural patterns are the ribbon and cystic forms, the latter particularly common in cats. In the ribbon form, cords of basal cells, 2–4 cells thick, form sinuous tracts into the dermis. They are supported by a definite fibrous stroma, and it is common that the outermost layer of basal cells is oriented perpendicular to the stroma in a manner resembling a picket fence. The amount and density of stroma varies widely, with some extreme examples (medusa and morphea types) having abundant, mucinous stroma simulating the myoepithelial component of mixed tumors. The basal cells form solid cords and not tubules, thereby distinguishing them from the less common, well-differentiated sweat gland tumors. Although the basal cell ribbons are technically invasive, they do so as blunt cohesive cords, and the overall tumor outline is well circumscribed. This is in contrast to the single-cell invasion and stromal disruption by squamous cell carcinoma.

The cystic form may occur as a single large cyst, but multiple small cysts are more common (Fig. 5.126). Each is formed by several layers of undifferentiated basal cells with no inner limiting membrane and usually no adnexal or squamous differentiation. The cysts are usually filled with eosinophilic debris, and most probably form by central necrosis of an initially solid mass of basal cells. Only a very few have sebaceous differentiation within the viable cyst wall. Those with abundant keratinization, sometimes termed keratotic basal cell tumors, are primitive hair tumors (trichoepithelioma), inasmuch as the abrupt keratinization is without a granular cell layer and resembles more the abortive formation of hair than it does of epidermis. There is no reliable way to distinguish the two tumors, and since the prognosis is the same, no urgent need to do so.

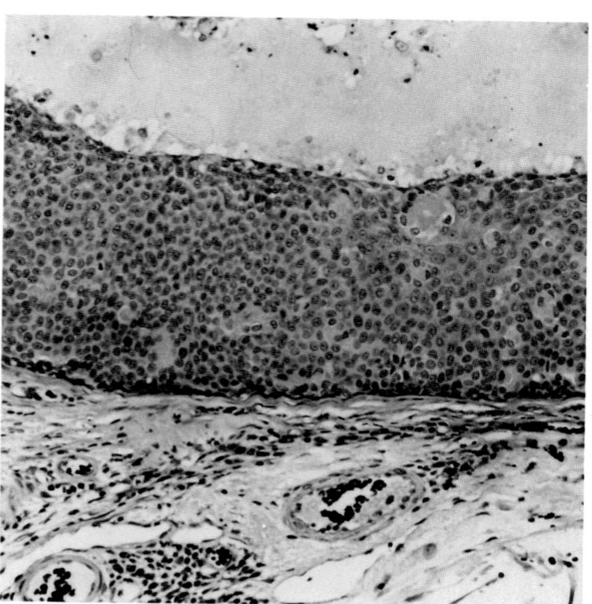

Fig. 5.126 Cystic basal cell tumor.

Basal cell tumors do not metastasize and rarely recur if initial excision is complete. Multiple tumors (rarely more than two) may occur in cats. There appears no danger of transformation of a bona fide basal cell tumor into squamous cell carcinoma, but we must be cautious in the interpretation of early actinic squamous cell carcinomas, which may resemble solid basal cell tumors except for a few foci of unequivocal atypia and single cell invasion. The old category of basosquamous carcinoma is no longer in vogue for such tumors, which behaviorally are squamous cell carcinomas.

6. *Adnexal Tumors*

Tumors of the cutaneous adnexa are very prevalent in dogs, are occasionally encountered in cats, but are rare in other domestic species. The vast majority are behaviorally benign, and there frequently is overlap among supposedly different tumor types within the same lump, so that the current fascination with ever-increasing refinements of tumor classification is hard to justify. Here we shall accept the premise that adnexal tumors derive from a primitive germinal cell capable of multiple different maturational paths within the same tumor. The classification used here is simple and traditional, lumping tumors of similar appearance and behavior as a single entity unless there is prognostic or therapeutic justification for splitting them. It is particularly clear in the case of sebaceous and hair follicle tumors that attempting to impose finite and arbitrary divisions on a biological continuum is unsound.

Sebaceous gland tumors include sebaceous hyperplasia, adenoma, and the rare carcinoma. A lesion intermediate between adenoma and carcinoma has been called sebaceous epithelioma, which is a basal cell tumor with extensive sebaceous differentiation.

Sebaceous gland hyperplasia is seen as multiple, raised, multilobulated masses within the superficial dermis below a focally hyperplastic, hyperkeratotic epidermis. Many are ulcerated by continued tumor growth or by trauma. Most are less than 1 cm in diameter and are most frequently submitted with clinical diagnoses of multiple papillomas. Histologically, each tumor is regular and typically consists of rather symmetrical hyperplasia of a single sebaceous gland clustered about a keratinized sebaceous duct (Fig. 5.127). In some specimens more than one gland seems involved. The sebaceous cells are fully mature, and the basal (reserve) cell population is inconspicuous. Surrounding dermis and adnexa are compressed. The hyperplasias are multiple in about 40% of affected dogs, and may occur in multiple crops.

Sebaceous adenoma is divided by some into adenoma and epithelioma. The former resembles sebaceous hyperplasia in that most of the cells are mature sebaceous cells, but differs in that there are many more basal (reserve)

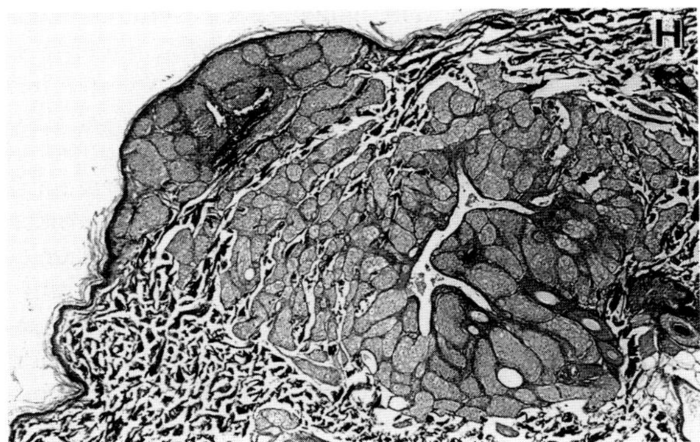

Fig. 5.127 Sebaceous hyperplasia about a single excretory duct.

cells than in hyperplasia, and the lobular proliferation is greater, less symmetrical, and not necessarily related to a single sebaceous duct. In many instances, the distinction from hyperplasia is arbitrary, and a study of the photographs in published surveys makes it clear that the distinctions are not uniformly applied. Many nodules change character from area to area.

Sebaceous epithelioma is a term used by some to describe basal cell tumors with prominent sebaceous differentiation (Fig. 5.128). Many of these tumors also have foci of squamous differentiation and some lobules of very mature sebaceous cells identical to lobules of sebaceous hyperplasia. Tumors of this type are particularly frequent on the eyelid margin, where they originate from the meibomian glands. In this location they frequently are pigmented and may be mistaken clinically for melanomas. Sebaceous epitheliomas (or basal cell tumors) are benign and virtually always are solitary.

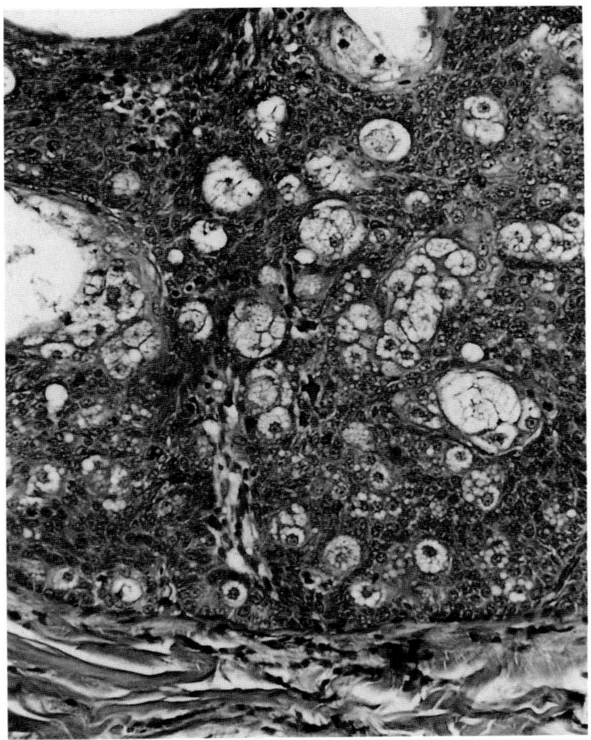

Fig. 5.128 Basal cell tumor with sebaceous differentiation (sebaceous epithelioma). Dog.

Sebaceous carcinomas are occasionally encountered in dogs and rarely in cats. In contrast to the arbitrary and biologically questionable divisions applied to the benign sebaceous tumors, true sebaceous carcinomas are histologically distinctive. These are locally infiltrative, solitary, poorly circumscribed tumors made up of pleomorphic cells that show some evidence of sebaceous differentiation. Lobule formation is often present but is only a feeble effort. Mitotic figures, anisocytosis, anisokaryosis, and hyperchromasia are present as the cytologic counterparts of malignancy. The tumor is distinguished from basal cell tumors with sebaceous differentiation in that the basal cell tumor has a preponderance of basal cells with variable amounts of orderly and complete sebaceous maturation. The sebaceous carcinoma lacks these two distinct and mature populations, with all of the cells seemingly disordered attempts at sebaceous cells. Mitotic figures may be numerous in the basal cell tumors, but there are no other cytologic suggestions of malignancy.

The prognosis following wide surgical excision seems to be good, although no large series has been published.

7. *Perianal Gland Tumors*

Perianal gland tumors are common in intact, aged male dogs. They occur occasionally in females and rarely in castrated males. They have not been reported in other domestic species.

Most perianal gland tumors are called adenomas, even though the lobular development is so uniform that the tumor is more likely to be hormone-dependent hyperplasia. The strong predilection for intact male dogs and the regression of about 95% of these tumors following castration also points to hyperplasia rather than true neoplasia. Nonetheless, perianal adenoma seems firmly entrenched, and the difference in terminology is of no practical importance. The tumor growths most commonly occur in the perianal area, but about 10% occur in sites of ectopic perianal gland tissue such as tail, flank, back, prepuce, and even chin. Of these, the ventral skin of the tail is the most frequent location.

The gross appearance of **perianal gland adenomas** is of one or more raised, rubbery masses that may grow to 10 cm or more in diameter. Ulceration and secondary infection are common, and can serve to confuse the diagnosis. The typical tumor consists of multiple tan lobules separated by a delicate but definite collagenous stroma.

The histologic lesion is similar to normal perianal gland. Well-differentiated tumors consist of lobular proliferation of polygonal cells with abundant, finely vacuolated eosinophilic cytoplasm and a central, small, round, vesicular nucleus. The cells resemble hepatocytes, and these tumors are sometimes called hepatoid gland tumors. At the periphery of each lobule is a rim of basal reserve cells, and in some tumors, these cells may predominate over the fully differentiated hepatoid cells. Squamous metaplasia and squamous pearl formation are common and are no cause for concern (Fig. 5.129). Mitotic figures are very infrequent. Local excision, combined with castration, is curative.

Perianal gland adenocarcinomas account for no more than 5% of all perianal tumors if rigorous diagnostic criteria are used. They are relatively more common in castrated males and in females. They exhibit the usual features of malignancy: variation in nuclear and cytoplasmic mass, poor cytoplasmic differentiation, frequent mitotic figures, and local or distant spread. Lymphatic invasion must be diagnosed with caution since compression of reserve cells by lobular expansion may simulate a tumor embolus surrounded by lymphatic endothelium. Direct extension into the pelvic canal or metastasis to regional lymph nodes occurs as a late event in chronically neglected tumors. There are no reports correlating histologic grade with prognosis. One retrospective study correlated behavior with clinical staging. Three quarters of tumors less than 5 cm in diameter, and without extensive local infiltration, were cured by excision. Larger tumors, or invasive ones, were five to eight times more likely to cause problems by recurrence or continued invasive growth. Only 14% had clinical evidence of metastasis, all interpreted as being in the late stages of this slow-moving disease. Castration has no demonstrated effect on tumor behavior.

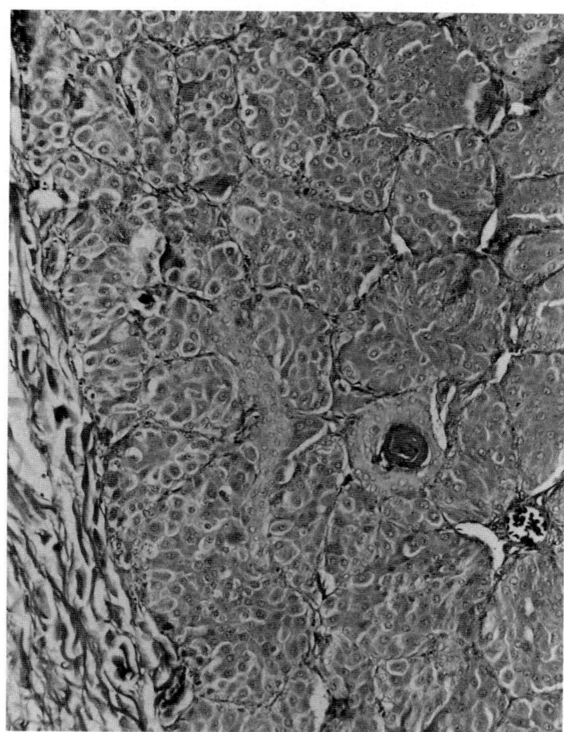

Fig. 5.129 Lobular arrangement of cuboidal hepatoid cells and squamous metaplasia. Canine perianal gland adenoma.

8. Hair-Follicle Tumors

Hair-follicle (pilar) neoplasms are common in dogs but apparently are very rare in other domestic species. Their classification in humans is very complex, and seems unwarranted in animals since all hair-follicle tumors share the same benign clinical course. As with other differentiated basal cell tumors, there are many examples of intermediate or multiple differentiation so that published prevalence data are strongly biased by where the authors choose to draw the lines of taxonomic separation.

Trichoepithelioma is a tumor of primitive hair germ that exhibits incomplete differentiation toward the formation of hair shafts. They occur as single (rarely multiple) skin tumors anywhere on the body, with a slight preference for the back. The typical macroscopic appearance is of a very firm, white multilobulated and encapsulated tumor that may be mineralized. Histologically, the usual pattern consists of multiple nodules of basal cells; the center of each nodule undergoes abrupt keratinization without interposition of prickle or granular cells (Fig. 5.130). The outer layer of basal cells may abut the fibrous stroma or send out basal cell ribbons into the stroma in a manner similar to that of basal cell tumors. Rupture of some of the keratinaceous cysts stimulates a pyogranulomatous inflammation within the stroma and even within the cyst, sometimes accompanied by mineralization and foreign-body giant cells. Mineralization and inflammation are more typical of the closely related pilomatrixoma.

Trichoepitheliomas are distinguished from all other epidermal or appendage tumors by the abrupt central keratinization and the folliclelike basal cell nests which surround it. Other tumors or cysts simulating trichoepithelioma lack one or both of these features.

Tricholemmoma is a rare pilar tumor of dogs, which architecturally resembles trichoepithelioma. The major difference lies in the cytoplasmic character of the basal cells attempting to form the hair follicles. In tricholemmomas, the outermost layer of cells has marked cytoplasmic

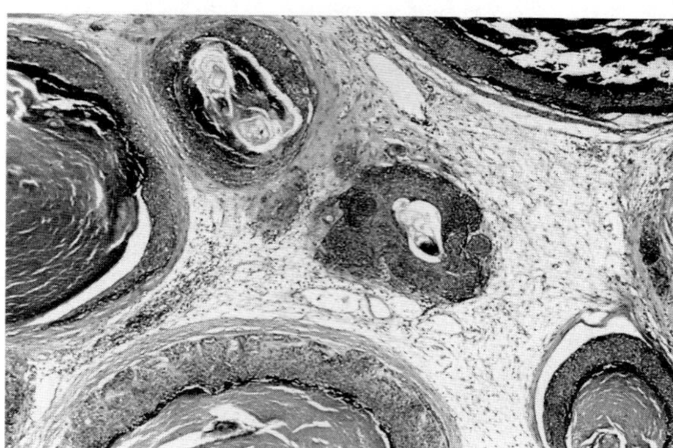

Fig. 5.130 Trichoepithelioma. Dog. Multiple cysts with abrupt keratinization but no shadow cells.

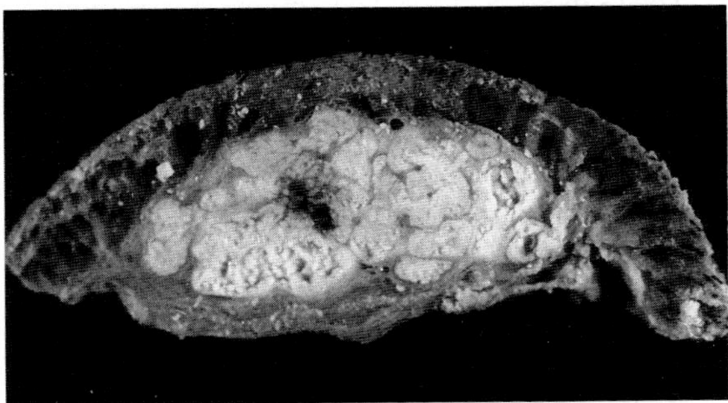

Fig. 5.131 Pilomatrixoma. Dog. Multilobulated, mineralized, and well-demarcated keratinized dermal tumor.

glycogen storage, seen in histologic section as cytoplasmic clearing. These are thought to represent differentiation toward outer root-sheath cells. Each tumor follicle is surrounded by a thick, homogeneous basal lamina resembling the vitreous sheath of the normal follicle. The behavior of these rare tumors is benign and nonrecurring.

Pilomatrixoma macroscopically resembles trichoepithelioma (Fig. 5.131) but is often more heavily mineralized and consists of fewer, larger cysts than does trichoepithelioma and shows less tendency to mimic basal cell tumor. Pilomatrixomas, also known as epitheliomas of Malherbe or necrotizing and calcifying epithelioma, are believed to be derived from primitive hair matrix and thus show incomplete differentiation toward hair cortex. The typical microscopic morphology is of one or more large, thick-walled cysts partially filled with so-called shadow or ghost cells (Fig. 5.132A,B). These cells are flattened, eosinophilic epithelial cells with a central empty halo in place of the lysed nucleus. They are analogous to immature hair cortex cells and contain keratin of hairlike configuration rather than that of epidermis. The cyst wall is composed of multiple layers of basal cells, showing a sudden zonal degeneration to form the central laminations of shadow cells. Mineralization of shadow cells is very common, and cyst rupture results in the pyogranulomatous inflammation

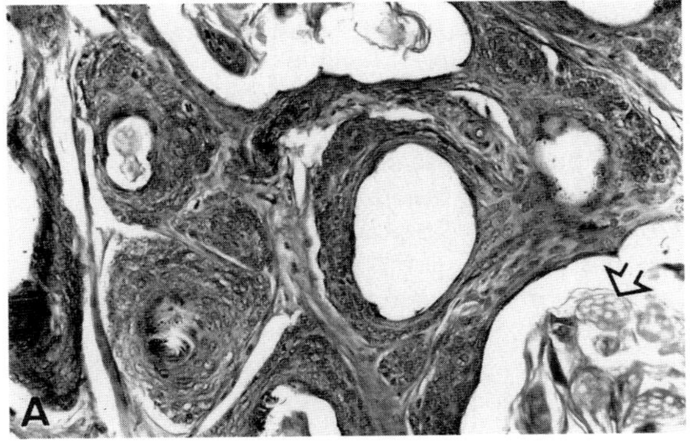

Fig. 5.132A Pilomatrixoma. Dog. Multiple lobules of basal cells with abrupt keratinization and a few luminal shadow cells (arrow).

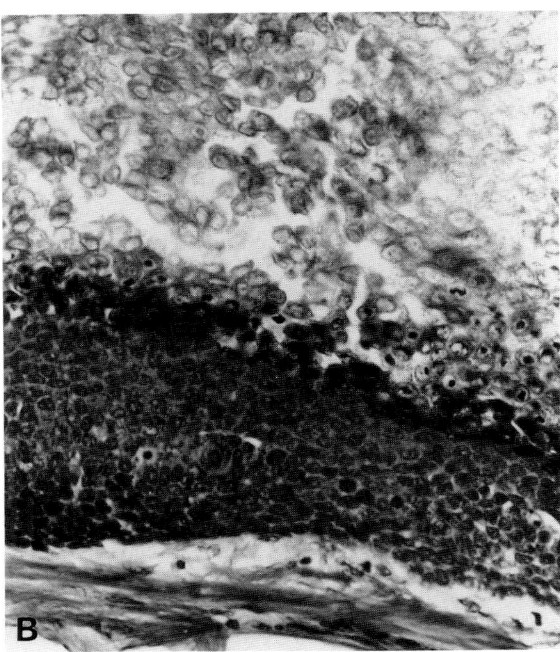

Fig. 5.132B Pilomatrixoma. Dog. Detail of lobule wall showing shadow cells (left).

that is very typical of these tumors. Foci of necrosis within the tumor are frequently seen, and calcification or even ossification within the stroma occurs occasionally.

Pilomatrixomas are almost always single and benign. The very few reports of behaviorally malignant tumors describe tumors not readily distinguished at initial presentation from benign tumors.

Calcinosis circumscripta is included here because of its macroscopic resemblance to hair tumors. It occurs quite commonly in dogs as soft white deposits in the subcutis. There is one report in a cat. The lesions are almost invariably related to the skin but may occur in the tongue or in paravertebral soft tissue. They are found chiefly in large dogs, and have a predilection for the limbs, especially the subcutis of the footpads and over bony prominences. They form firm bulging masses which, when incised or ulcerated, discharge a white pastelike material, which consists largely of calcium salts. Although the small lesions form in the subcutis, they may later encroach on the dermis. The cut surface reveals lakes of chalk-white material embedded in a variable amount of fibrous tissue. Histologically, the deposits consist of fine or coarse basophilic granules. The masses are surrounded by a cellular zone of variable width in which giant cells and large macrophages are prominent but among which there are also lymphocytes and plasma cells. Some of the marginal giant cells and macrophages are degenerate and undergoing calcification. The lakes and cellular borders are surrounded by a zone of condensed fibrous tissue.

The pathogenesis of calcinosis circumscripta is not known. The name apocrine cystic calcinosis has been applied to the lesion with the suggestion that the sequence involves trauma and reactive hyperplasia of apocrine glands and dystrophic mineralization of excessive, and probably abnormal, apocrine secretion. A relationship to apocrine glands is not easy to trace and does not account for the frequent occurrence of the lesion on the tongue, gums, and occasionally elsewhere in the mouth. The calcareous deposits are very similar to those of "putty brisket" of cattle, in which there is no glandular contribution. Any explanation of pathogenesis must take account of the facts that the lesion occurs predominantly in dogs younger than 2 years, and that more than 50% of affected animals are German shepherds. Calcinosis circumscripta in horses is discussed with Bones and Joints (Chapter 1).

9. *Sweat Gland Tumors*

Tumors of the paratrichial (apocrine) sweat glands are the least frequent of the adnexal skin tumors in dogs. They are relatively more common in cats. There is a single report of a mixed sweat gland tumor in a bull. Tumors of the eccrine sweat glands are rare but histologically distinctive tumors occurring in the foot pads.

Paratrichial cystic dilatation occurs in two forms: cystic dilatation without apparent hyperplasia, and dilatation with epithelial hyperplasia. The former is relatively common in dog skin, particularly in areas of dermal scarring. The implication is that such dilatation is the result of duct obstruction, a reasonable hypothesis. The normal columnar epithelial lining is attenuated in proportion to the degree of dilatation. The hyperplastic cyst is lined by 1 to 2 layers of tall columnar secretory epithelium resembling that of normal sweat gland. It is probable that many examples of cystic hyperplasia have been classified as cystadenomas. The observation of very orderly proliferation, especially if in multiple sites, should lead to a diagnosis of hyperplasia. In some dogs the hyperplasia occurs very extensively, sometimes almost universally.

Paratrichial sweat gland adenomas include cystic and papillary types, the latter merely a variant of the former. The cystic lesion consists of one or more cavities lined by well-differentiated columnar epithelium, often with the apical secretory blebs typical of normal glands. Many tumors have at least some tubules lined by an orderly bilayer of cuboidal epithelium, perhaps an effort by the tumor cells to recreate the structure of the duct of the normal gland (Fig. 5.133). Proliferation of epithelium may result in intraluminal papillary growths supported by a delicate fibrous stroma. The proliferation may fill the cyst lumen.

Paratrichial sweat gland carcinomas include papillary and tubular types, the latter clearly predominating. The tubular carcinomas are locally aggressive growths, spreading through the dermis, subcutis, and muscle from the primary focus. Distant metastasis seems a late feature of the disease progression, and may be more likely in cats than in dogs.

The histology of papillary carcinomas differs from that of their benign counterparts in the usual criteria of nuclear variation, cellular anaplasia, loss of polarity, and local stromal invasion. The tubular carcinomas have focal cystic or solid areas, and are typically very scirrhous. Identification is based on those tubules retaining differentiation as to resemble normal sweat glands, particularly the presence of secretory blebs at the luminal surface. The sweat gland carcinoma is the only primary tubular adenocarcinoma of skin.

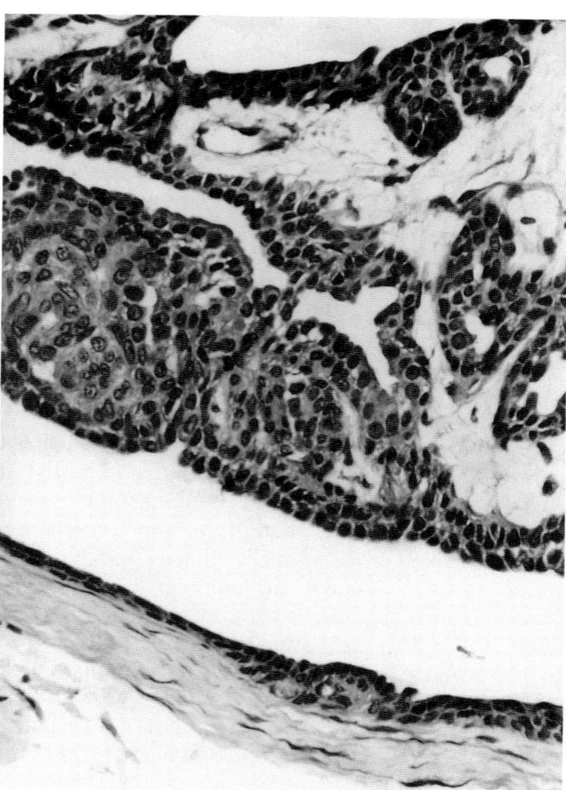

Fig. 5.133 Papillary paratrichial (apocrine) adenoma. Note tendency toward regimented, bilayered epithelium.

Mixed sweat gland tumors are very rare. The proliferation involves both epithelial cells and the periglandular myoepithelial sheath, and may be analogous to the mixed mammary tumor of dogs. Mucinous, chondroid, and osseous metaplasia within the tumor stroma is allegedly of myoepithelial origin. Too few have been reported to allow any useful discussion of their biologic behavior. Tumors of **atrichial (eccrine) glands** are rare but do occur as adenomas or scirrhous tubular carcinomas within the loose connective tissue of the pad of dogs and cats.

Carcinoma of the apocrine glands of the anal sac is an uncommon but very malignant tumor of old female dogs. Males are rarely affected. The tumor histologically varies from tubular to solid adenocarcinoma, which almost invariably spreads to regional lymph nodes and then various viscera. Over half have already metastasized prior to diagnosis. Tumor growth is often directed inward so that a grossly visible perianal mass is seen in only about 50%

of cases. The unusual feature of the tumor is its ability regularly to induce hypercalcemia in affected dogs (see Volume 3, Chapter 3, The Endocrine Glands).

Bibliography

Bevier, D. E., and Goldschmidt, M. H. Skin tumors in the dog. Part 1. Epithelial tumors and tumorlike lesions. *Compend Cont Ed* **3:** 389–398, 1981.

Case, M. T. *et al.* Metastasis of a sebaceous gland carcinoma in the dog. *J Am Vet Med Assoc* **154:** 661–664, 1969.

Diters, R. W., and Goldschmidt, M. H. Hair follicle tumors resembling tricholemmomas in six dogs. *Vet Pathol* **20:** 123–125, 1983.

Garma-Avina, A. and Valli, V. E. Mixed sweat gland tumor in a bull (a case report). *Vet Med Small Anim Clin* **76:** 557–559, 1982.

Gordon, L. R. The cytology and histology of epidermal inclusion cysts in the horse. *J Equine Med Surg* **2:** 371–374, 1978.

Headington, J. T. Tumors of the hair follicle. A review. *Am J Pathol* **85:** 480–505, 1976.

Jabara, A. G., and Finnie, J. W. Four cases of clear-cell hidradeno-carcinomas in the dog. *J Comp Pathol* **88:** 525–532, 1978.

Luderschmidt, C., and Plewig, G. Circumscribed sebaceous gland hyperplasia: Autoradiographic and histoplanimetric studies. *J Invest Dermatol* **70:** 207–209, 1978.

Luther, P. B., Scott, D. W., and Buerger, R. G. The dilated pore of Winer—An overlooked cutaneous lesion of cats. *J Comp Pathol* **101:** 375–379, 1989.

Meuten, D. J. Hypercalcemia associated with an adenocarcinoma derived from the apocrine glands of the anal sac. *Vet Pathol* **18:** 454–471, 1981.

Neilsen, S. W., and Aftosmis, J. Canine perianal gland tumors. *J Am Vet Med Assoc* **144:** 127–135, 1964.

Ross, J. T. *et al.* Adenocarcinoma of the apocrine glands of the anal sac in dogs: A review of 32 cases. *J Am Anim Hosp Assoc* **27:** 349–355, 1991.

Rudolph, R., Gray, A. P., and Leipold, H. W. Intracutaneous cornifying epithelioma ("keratoacanthoma") of dogs and keratoacanthoma of man. *Cornell Vet* **67:** 254–264, 1977.

Rudolph, R. L. *et al.* Morphology of experimentally induced so-called keratoacanthomas and squamous cell carcinomas in 2 inbred lines of *Mastomys natalensis*. *J Comp Pathol* **91:** 123–134, 1981.

Schuh, J. C. L., and Valentine, B. A. Equine basal cell tumors. *Vet Pathol* **24:** 44–49, 1987.

Strafuss, A. C. Basal cell tumors in dogs. *J Am Vet Med Assoc* **169:** 322–324, 1976.

Strafuss, A. C. Sebaceous gland adenomas in dogs. *J Am Vet Med Assoc* **169:** 640–642, 1976.

Vail, D. M. *et al.* Perianal adenocarcinoma in the canine male: A retrospective study of 41 cases. *J Am Anim Hosp Assoc* **26:** 329–334, 1990.

Wilson, G. P., and Hayes, H. M. Castration for treatment of perianal gland neoplasms in the dog. *J Am Vet Med Assoc* **174:** 1301–1303, 1979.

Zenoble, R. D., Crowell, W. A. and Rowland, G. N. Adenocarcinoma and hypercalcemia in a dog. *Vet Pathol* **16:** 122–123, 1979.

B. Melanomas

Melanomas are common in dogs, gray horses, and in some lines of miniature swine. They are seen occasionally as congenital lesions in Duroc (and occasionally other) pigs, and in goats and calves. They are uncommon in cats and sheep. Melanomas usually are cutaneous neoplasms but can occur wherever melanocyte clusters are found. Eye and mouth are common sites, but meninges, bone, aorta, and so on are all reported sites. Those of the oral cavity and inner surface of lips of dogs are almost invariably malignant both histologically and behaviorally, with a median survival of only ~14 weeks after tumor removal. Conversely, those of the canine eye and eyelid are almost invariably benign. The behavior of melanomas of the canine digit has caused considerable debate, and conflicting data supporting or refuting claims of greater metastatic potential for melanomas in this site have been published. In our opinion, melanomas arising in the area of the toe are not inherently more aggressive than histologically similar skin melanomas elsewhere, but they more often have histologic features of malignancy than do melanomas arising elsewhere in skin.

Melanomas in **dogs** represent from 6 to 20% of all skin tumors, with most reports in the 6–10% range. The macroscopic appearance is not useful in establishing either a diagnosis or a prognosis except in the most general of ways. Most tumors smaller than 1.0 cm in diameter are benign, and most greater than 2.5 cm are malignant. Most are raised, firm, black nodules, but amelanotic, soft, and flat tumors also occur. Basal cell and adnexal tumors may be heavily pigmented and are to be distinguished from melanomas, as must the many tumors which induce hyperpigmentation of overlying epidermis.

The histologic identification of melanomas can be difficult because the patterns they form can vary widely, but almost all melanomas have some junctional activity, or some cytoplasmic pigmentation, and tend to assume a cellular morphology and growth pattern intermediate between carcinoma and sarcoma.

Benign melanoma with junctional activity is analogous to the human compound junctional nevus. It is a commonly observed pattern in dogs, particularly in melanomas of the eyelid. The characteristic lesion consists of nests of 3 to 20 oval, heavily pigmented melanocytes within the epidermis near the dermal–epidermal junction (Fig.

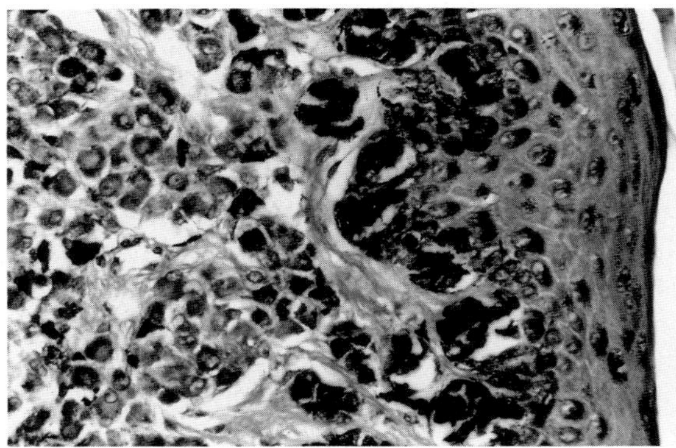

Fig. 5.134 Junctional activity. Benign melanoma.

5.134). From these junctional nests, clusters of melanocytes extend into the dermis in a manner suggesting they have "dropped off" the junctional clusters. The cells in these dermal nests are usually oval (epithelioid) near the epidermis but become more spindlelike and less pigmented as they extend deeper into the dermis. Although they are not encapsulated, there is normally clear demarcation between the expansive tumor and the normal dermis. The overlying epidermis is usually thin and hairless because of follicular obliteration by the tumor. Proper histologic classification often requires bleaching of the heavy pigment until nuclear morphology can be seen. In important contrast to malignant epithelioid melanomas with junctional activity, the cells of the benign junctional melanoma are very uniformly oval, with fine nuclear chromatin, essentially no nuclear size variation, and very few mitotic figures.

Benign dermal melanomas are thought to arise from nests of melanocytes left behind in the migration of neural crest precursor cells to the skin. The tumor cells are either fibroblastlike spindle cells arranged in interlacing fascicles or plump epithelioid cells arranged in small clusters surrounded by a reticulin net. Artefactual shrinkage of the nests of cells away from the fine stroma during histologic processing often creates the impression of tumor emboli within dermal lymphatics. Clues to the benign nature of this group come from the regularity of nuclear morphology, paucity of mitotic figures, and absence of endothelial lining in the alleged lymphatics. In some specimens there is a zone of normal dermis separating tumor from the epidermis. The degree of pigmentation is variable among tumors and within the same tumor. Those of spindle cell type are often only lightly pigmented, but demonstration of the pigment is essential if this variant is to be readily distinguished from other dermal spindle cell tumors. Dermal melanomas regularly are infiltrated by macrophages, which become heavily laden with pigment, usually more abundant and in larger granules than in the tumor cells themselves. Bleaching to determine the typical, cleaved nuclear morphology of histiocytes may be necessary if positive identification of these cells is important.

Malignant melanomas are descriptively divided into epithelioid or spindle cell or mixtures of the two. Tumors with a predominantly epithelioid morphology are by far the most frequent. Regardless of type, about three quarters of the histologically malignant melanomas result in death of the dog within 2 years of excision, with a median survival of ~1 year. The range for survival times is very wide.

The main decision about suspected malignant melanomas is not what specific classification should be applied but whether the tumor is, in fact, a melanoma and a malignant one. The criteria for malignancy do not differ from those applicable to other tumors. Weight is put on variation in nuclear size and shape, the presence of coarse and very basophilic chromatin, and observation of mitotic figures. Invasion of upper layers of epidermis, as opposed to basilar junctional activity, or cordlike invasion of dermis are useful but often difficult to distinguish amid the ulceration and tumor-induced inflammation seen in some melanomas. The presence of bizarre, angular giant tumor cells with one or several nuclei is clear evidence of malignancy. Observation of even a few mitotic figures, rare in benign melanomas, is an important clue.

The identification of the tumor as a melanoma may be more difficult. Junctional activity, if present, is almost unique to melanomas (epitheliotropic lymphoma is the alternative). A tendency for the tumor cells, whether spindle or epithelioid, to form tight clusters surrounded by a reticulin wall is also quite typical. Cellular pleomorphism and a tendency for angular cytoplasmic outline are other useful features but are not always present. Cytoplasmic pigmentation is by far the best criterion if present. Almost all melanomas have some pigmentation, especially near the epidermis, and all have the enzymic ability to form pigment. The former may be seen on hematoxylin and eosin sections as subtle cytoplasmic bronzing or muddiness. Alternatively, the cytoplasm of most cells may contain only a few small brown granules. Detection of such subtle pigmentation may be enhanced by ammoniacal silver stains, by fluorescence microscopy of formalin-fixed, deparaffinized unstained sections or by the Dopa reaction occurring in fresh tumor. The last test measures the enzymic capability of melanocytes to convert tyrosine to melanin by simulating that reaction with dihydroxyphenylalanine (Dopa) instead of tyrosine as the substrate. Unfortunately, fresh unfixed tissue is essential. A useful method is cytologic examination performed on tumor aspirates or impression smears, in which cytoplasmic melanin is usually more obvious than in histologic material. The presence of S-100 antigen is useful, but not infallible, as it occurs in many other tumor types.

Melanomas in **horses** occur in two forms: rare idiopathic dermal melanomas analogous to benign dermal melanoma of dogs, and dermal melanocytosis, which is questionably neoplastic. The latter condition is common and biologically unique but poorly understood. The condition affects horses, and rarely mules, that are born dark gray or black but become progressively less pigmented as they age. Tumorlike accumulations of melanocytes and melanophages develop as the horse ages and as the hair and skin color fades. Melanotic masses are common in gray-white horses older than 6 years and reach ~80% prevalence in aged populations. It has been proposed that all gray horses will develop melanocytosis if they live long enough.

The macroscopic lesions are seen as one or more firm, raised, smooth, gray-black nodules in the skin of perineum, ventral tail, external genitalia, or parotid region. Other areas may occasionally be the primary site. Initial tumors may be single or multiple, and may enlarge to several kilograms. Tumors may spread locally and may apparently metastasize to various organs. Whether the metastases are truly the result of tumor spread or are focal accumulations of melanophages representing a pigmentary storage disease remains uncertain.

The tumors develop in the dermis as heavily pigmented

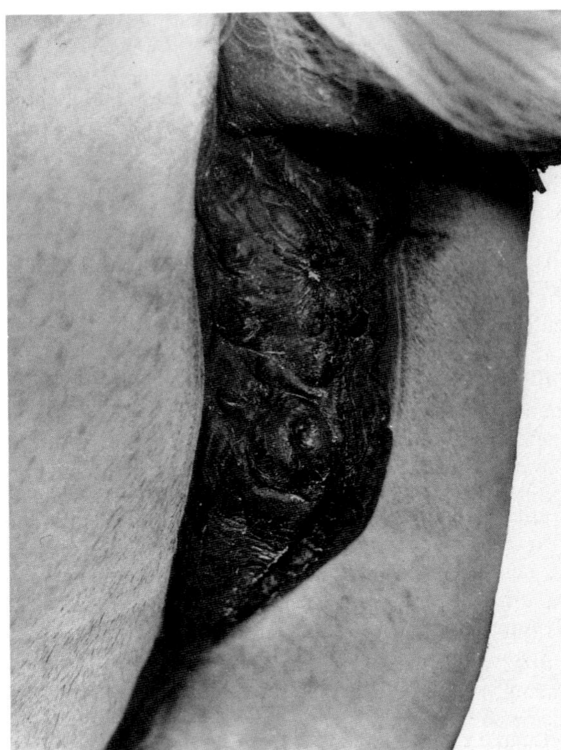

Fig. 5.135 Malignant melanoma of perineum. Mare.

masses, the cellular details totally obscured by pigment. The cells represent a mixture of benign-appearing melanocytes and melanophages. There is no junctional activity, and the tumors are histologically comparable to benign dermal melanomas in dogs except for the unusually heavy pigmentation. When visceral tumors are examined ultrastructurally, most of the cells are melanophages and benign-appearing melanocytes. This has led to speculation that the apparent metastases are more correctly ectopic melanocytic nests stimulated to produce excess melanin just as are the more familiar skin masses. Because the bulk of each tumor consists of melanophages, this disease has been compared to various visceral storage diseases rather than true neoplasia. In this theory, the rare development of unequivocal malignancy from skin or visceral foci represents nothing more than malignant transformation *in situ* of a hyperplastic population (Fig. 5.135).

Benign melanomas that are either congenital or of juvenile onset occasionally occur in horses of any breed or coat color. They are solitary, do not recur after complete excision, and histologically resemble canine dermal melanomas.

Melanomas in **swine** occur as congenital lesions in Duroc, Hormel, and Sinclair breeds of swine, the last two genetically related strains of miniature swine widely used in biomedical research. Melanomas are occasionally seen in other swine breeds and appear to represent the same disease as in the miniature swine. Tumors may be single or multiple, affect skin only or involve multiple internal organs, and be obvious or occult at birth. The initial skin lesion is a flat black spot which spreads within the skin before becoming a nodule. Histologically each tumor begins as a focus of melanocytic hyperplasia within the basal layer. Continued growth leads to a benign junctional melanocytoma with some atypical nuclear morphology within the junctional clusters suggesting early malignancy. The tumor then invades superficial dermis to resemble benign junctional melanoma. It is at this stage that most tumors are examined. Later, continued dermal invasion leads to lymphatic permeation and possibly, systemic metastases in some affected pigs, although there is no proof that the visceral tumors do not develop *in situ* as spontaneous examples of tumor genesis. Tumor cells are usually well pigmented, and both epithelioid and spindle cells are usually seen. A remarkable feature of the swine melanomas is the high prevalence of spontaneous regression that occurs presumably as the result of the cytotoxic effect of the numerous tumor-specific T lymphocytes which infiltrate these tumors. Tumor necrosis, depigmentation, and dermal fibrosis follow the infiltration, resulting in a residual dermal scar laden with melanophages. Regression does not always occur, and young adult swine may have developed lesions of sufficient size in brain, heart, spinal cord, or other sites as to cause organ dysfunction.

In **cattle** melanomas occur as large, benign growths in young animals, although occurrences both in neonates and aged animals are reported. Benign junctional and benign cellular dermal types have been described, but most descriptions make no mention of histologic pattern.

Melanomas in the skin of **cats** are rare and resemble those of dogs. Most are behaviorally benign, and there are no published data about histologic predictors of malignancy for the feline tumors.

Bibliography

Aronsohn, M. G., and Carpenter, J. L. Distal extremity melanocytic nevi and malignant melanomas in dogs. *J Am Anim Hosp Assoc* **26:** 605–612, 1990.

Balch, C. M. *et al.* A multifactorial analysis of melanoma. II. Prognostic factors in patients with stage I (localized) melanoma. *Surgery* **86:** 343–351, 1979.

Bostock, D. E. Prognosis after surgical excision of canine melanomas. *Vet Pathol* **16:** 32–40, 1979.

Case, M. T. Malignant melanoma in a pig. *J Am Vet Med Assoc* **144:** 254–256, 1964.

Conroy, J. D. Melanocytic tumors of domestic animals with special reference to dogs. *Arch Dermatol* **96:** 372–380, 1967.

Flatt, R. E., Nelson, L. R., and Middleton, C. C. Melanotic lesions in the internal organs of miniature swine. *Arch Pathol* **93:** 71–75, 1972.

Foley, G. L., Valentine, B. A., and Kincaid, A. L. Congenital and acquired melanocytomas (benign melanomas) in eighteen young horses. *Vet Pathol* **28:** 363–369, 1991.

Garma-Aviana, A., Valli, V. E., and Lumsden, J. H. Cutaneous melanomas in domestic animals. *J Cutan Pathol* **8:** 3–24, 1981.

Gebhart, W., and Niebauer, G. W. Beziehungen zwischen pigmentschwund und melanomatose an beispiel des lipizzanerschimmels. *Arch Dermatol Res* **259:** 29–42, 1977.

Hjerpe, C. A., and Theilen, G. H. Malignant melanomas in porcine littermates. *J Am Vet Med Assoc* **144:** 1129–1130, 1964.

Inoshita, T., and Youngberg, G. A. Fluorescence of melanoma cells. A useful diagnostic tool. *Am J Clin Pathol* **78:** 311–315, 1982.

Jones, D. H., and Amoss, M. S. Cell-mediated immune response in miniature Sinclair swine bearing cutaneous melanomas. *Can J Comp Med* **46:** 209–211, 1982.

Kraft, I., and Frese, K. Histological studies on canine pigmented moles. The comparative pathology of the naevus problem. *J Comp Pathol* **86:** 143–155, 1976.

Levene, A. Disseminated dermal melanocytosis terminating in melanoma. A human condition resembling equine melanotic disease. *Br J Dermatol* **101:** 197–205, 1979.

Montes, L. F., Vaughan, J. T., and Ramer, G. Equine melanoma. *J Cutan Pathol* **6:** 234–235, 1979.

Oxenhandler, R. W. *et al.* Malignant melanoma in the Sinclair miniature swine. An autopsy study of 60 cases. *Am J Pathol* **96:** 707–720, 1979.

Patnaik, A. K., and Mooney, S. Feline melanoma: A comparative study of ocular, oral, and dermal neoplasms. *Vet Pathol* **25:** 105–112, 1988.

Poppema, S. *et al. In situ* analysis of the mononuclear cell infiltrate in primary malignant melanoma of the skin. *Clin Exp Immunol* **51:** 77–82, 1983.

Traver, D. S. *et al.* Epidural melanoma causing posterior paresis in a horse. *J Am Vet Med Assoc* **170:** 1400–1403, 1977.

C. Spindle Cell Tumors

Tumors that arise from spindle-shaped cells of the dermis and subcutis are common in dogs, cats, and horses. They are uncommon to rare in the other domestic species. Any attempt to classify such tumors must be prefaced by the admission that a morphologic classification based on histologic appearance has only a limited chance of reflecting the true cellular origin of the tumor.

Most collagen-producing spindle cell tumors have historically been classified as one variant or another of fibroma, but most spindle-shaped cells—whether they are histiocytes, Schwann cells, mesothelium, lipocytes, or others—can form collagen and result in histologic replicas of true fibromas. Conversely, a classification based on tumor histogenesis is simply not possible using standard histologic techniques. Even electron microscopy cannot always distinguish among all such "facultative" fibroblasts, nor are histochemical techniques rewarding in more than a few instances. The confusion exists for malignant as well as benign growths. The former tend to remain anonymous as primitive, nondescript mesenchymal cells, while the latter converge toward fibroblastlike cells.

This group includes a variety of neoplasms and proliferative growths that grow as nonencapsulated but well-demarcated proliferations of interwoven spindle-shaped cells with a collagenous matrix. The standard growth pattern is of interlacing bundles of cells displacing normal dermis or subcutis. Epidermal invasion is exceedingly rare even with the malignant variants.

Cutaneous spindle cell tumors include fibroma, fibrous histiocytoma, leiomyoma, schwannoma, hemangiopericytoma, and their malignant counterparts. Also included because of microscopic resemblance to one or more of the above are equine sarcoids, collagen nevi and keloidlike lesions, and polypoid acrochordons (skin tags). Papillomavirus-induced fibropapillomas and spindle cell dermal melanomas, which should also be included, have been described in earlier paragraphs.

1. Benign Noninfiltrative Spindle Cell Tumors

The diagnosis of **fibroma** has been used in all species to describe nodular benign dermal masses of relatively mature fibroblasts producing abundant mature collagen. The cellular proliferation is typically arranged in broad, streaming interlacing bundles rather than tight whorls as in schwannomas. Blood vessels are inconspicuous, which helps distinguish fibroma from hemangiopericytoma and granulation tissue. The cells show almost no nuclear variation, single inconspicuous nucleoli, and no mitotic figures. Fibromas have no malignant potential. The histologic separation of fibromas from other benign fibrous proliferations, such as collagen nevus or hypertrophic scar, may in some instances be difficult and arbitrary.

Myxoma is a fibroma in which the tumor cells have the stellate morphology of primitive mesenchymal cells. The stroma, instead of being collagenous and hard, is a slimy, mucinous mixture of reticulin and delicate collagen fibers embedded in a hyaluronic acid-rich glycosaminoglycan matrix. The stroma of a true myxoma is pale blue in hematoxylin and eosin-stained sections, and stains with alcian blue and metachromatically with toluidine blue. Portions of many spindle cell tumors are loosely arranged as the result of degeneration and edema, and the adjective myxomatous is sometimes used descriptively for such areas. The differential diagnoses for myxoma include myxedematous variants of any cutaneous spindle cell tumor, and the extreme nodular accumulations of mucin seen in cutaneous mucinosis of Chinese Shar-pei and other excessively wrinkled, loose-skinned breeds of dogs.

Collagen nevi are relatively frequent in dogs as small, firm, usually flat dermal nodules that are most common on the legs and may be multiple. Each nevus represents a localized overgrowth of resident dermal fibrous tissue. The usual lesion consists of a poorly cellular mass of very mature but normally aligned collagen (Fig. 5.136). The overlying epidermis is often atrophic. Other elements of the skin—hair follicles and their adnexa—may also be hypertrophic, suggesting that the pathogenesis of these small hamartomas is excessive local action of growth promoters or, alternatively, inadequate growth inhibitor. Multiple collagen nevi occur as an inherited disorder in German shepherd dogs, apparently linked to concurrent renal adenocarcinoma (see dermatofibrosis).

Hypertrophic scar resembles fibromas and collagen nevi, but we can usually find distorted remnants of hair follicles or residual islands of inflammation to point to a previous inflammatory or traumatic event. Depending on the cause and age of the lesion, other features like pseudoepitheliomatous hyperplasia (of lick granuloma) or em-

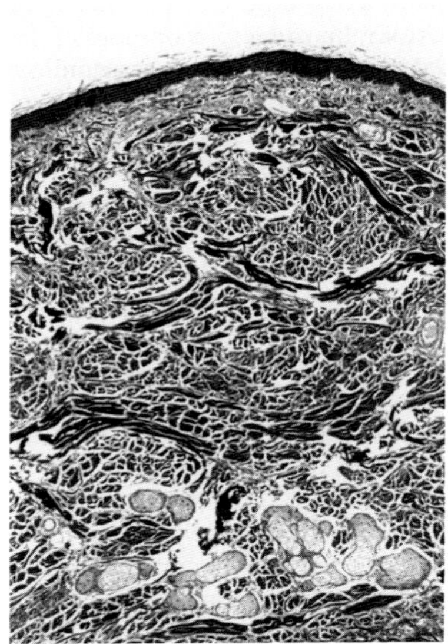

Fig. 5.136 Collagen nevus with only a few adnexal remnants. Dog.

bedded fragments of hair (of old trichogranuloma) may be present.

Acrochordon, or **skin tag,** is seen in dogs as a small, pendulous flap of skin. The skin is usually wrinkled and hairless, and may be hyperpigmented. Histologically, such tags consist of normal or atrophic epidermis overlying a core of vascularized collagenous connective tissue and are sometimes referred to as pendulous soft fibromas. Adnexa are absent.

Fibrous polyp resembles skin tag, but the fibrovascular core is the dominant component. The name is purely descriptive, and nothing is known of its pathogenesis.

2. Locally Infiltrative or Malignant Spindle Cell Tumors

Locally aggressive spindle cell tumors are common in the dermis or subcutis of the dog. These tumors have been variously diagnosed, on the basis of real or perceived differences, as hemangiopericytomas, schwannomas, neurofibromas, and neurofibrosarcomas. A few have been called malignant fibrous histiocytomas. As a group they are more highly cellular, more regimented, and less collagenous than simple fibromas or fibrosarcomas. The typical or diagnostic histology of each has been described and prevalence data thus derived, but the histologic criteria have not survived the scrutiny of the electron microscope. Thus, tumors regarded histologically as hemangiopericytomas may be composed of cells ultrastructurally typical of Schwann cells, or vice versa. The neurofibroma and neurofibrosarcoma are of uncertain status, but the ultrastructural evidence favors an origin from Schwann cells. With reservations about the true identity of each, they are described below.

Hemangiopericytomas are seen mainly in dogs and occasionally in horses; tumors of similar pattern in cats have a high mitotic rate and usually are classed as fibrosarcomas. They are relatively common, occur in older dogs of any breed, and have a marked predilection for the skin of the legs, particularly the thigh. Although the histologic identity may be controversial, its existence as a clinical entity is unquestionable. The gross lesion is a multinodular firm swelling within deep dermis or subcutis, nonencapsulated but usually appearing well circumscribed. This appearance at surgery is deceptive, and seldom does histologic examination confirm the surgeon's impression of complete excision. They exhibit locally invasive growth and often recur after surgical excision. Metastasis is very rare but can occur from unusually anaplastic specimens. The expected histologic appearance is of a dense spindle cell tumor in which the cells form tight whorls about capillary lumina (Fig. 5.137). Some tumors have alternating dense and loose areas. The dense areas are composed of plump spindle cells with only slight collagenous stroma, arranged in short, curved bundles resembling thumbprints. Whorling about vessels is considered the diagnostic feature and is best seen in loose, edematous areas of the tumor. In such areas, the tumor cells that are not in whorls may appear as plump epithelioid cells with abundant eosinophilic cytoplasm. Multinucleated giant cells may be regionally numerous. Biopsies from tumors recurring after initial excision often contain cells exhibiting greater variation in nuclear size and chromatin pattern, and more mitotic activity, than the initial biopsy. Lymphoid aggregates are found in some specimens, but there is no information about their prognostic significance.

Schwannomas are reported in all species but are commonly diagnosed only in dogs and cattle. Those that are clearly related to nerves are described in Chapter 3, The Nervous System. In cattle the lesions are multiple and most commonly affect intrathoracic autonomic nerves as gray-white nodular thickenings. They are seen most fre-

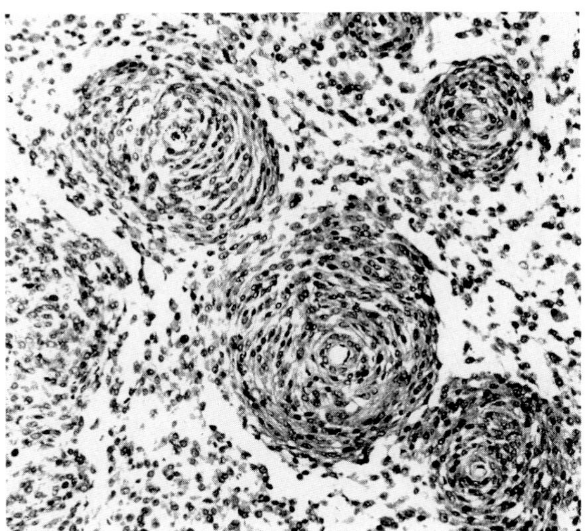

Fig. 5.137 Hemangiopericytoma. Dog. Concentric whorls of plump spindle cells around capillary lumina.

quently as unexpected lesions in healthy animals at slaughter. In dogs, the diagnosis of schwannoma has been used most confidently for those spindle cell tumors clearly associated with nerve trunks; almost all such reports are of nodular tumors affecting cervical spinal nerves.

The distinction between hemangiopericytoma and schwannoma, and between either of these and other connective tissue tumors, can be made ultrastructurally and with histochemical or immunohistochemical techniques. Schwann cells contain cholinesterase, and each is surrounded by a very prominent basal lamina. Pericytes have extensive cytoplasmic processes, prominent intercellular junctional complexes, and contain well-defined cytoplasmic filaments. In both tumors, collagen production is usually scant, but this is not entirely reliable.

The need to distinguish between these tumors, and therefore to employ special investigative techniques, is not warranted by prognostic data. Postsurgical behavioral data on these "spindle cell tumors," specifically diagnosed as hemangiopericytoma, fibrosarcoma, schwannoma, or neurofibrosarcoma on histologic criteria (Fig. 5.138), indicated no difference between local growth patterns, tendency to recur, or prevalence of metastasis for tumors of similar mitotic index. The degree of anaplasia reflected by high mitotic index, rather than tumor identity, determines the likelihood of recurrence or metastasis. In one large study, tumors with more than one mitotic figure per high-power field had an incidence of recurrence of 60% within 3 years in contrast to only 25% for tumors of lower mitotic index. Most recurrences were within less than one year of initial excision. Another study had a recurrence rate of only 21%, and no association between histologic features and subsequent behavior as proven. The mean interval from initial excision to recurrence was 21 months.

Amputation neuroma is a seemingly painful, firm swelling at the site of previous tail amputation in dogs. The histologic lesion is nodular aberrant proliferation of small myelinated nerve bundles randomly distributed amid abundant connective tissue. It is not neoplastic, but may have some histologic or clinical similarities to schwannomas, and is thus presented here. A similar lesion is seen at sites of digital neurectomy in horses.

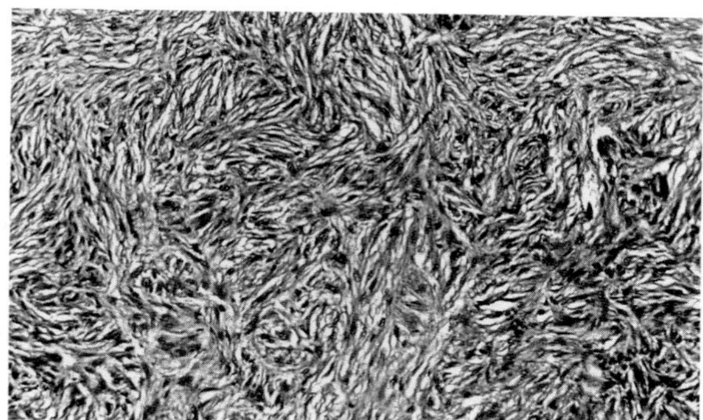

Fig. 5.138 Tight interlacing thumbprint pattern shared by several types of locally infiltrative cutaneous spindle cell tumors.

Sarcoids in horses and mules are very prevalent, transplantable dermal spindle cell tumors. They affect horses of all ages, breeds, and colors without preferences, occur on legs or head (Fig. 5.121) much more frequently than elsewhere, and are multiple in 15 to 33% of affected horses. Similar lesions are very common in deer in northeastern North America and perhaps elsewhere, caused by a papillomavirus similar to bovine papillomavirus types 1 and 2. There is no similarity between these lesions and human sarcoids. The lesions in horses and mules may regress over a course of many years. Recurrence at the site of excision occurs in ~50% of cases, usually within a few months. The rate of recurrence can be greatly reduced by the use of immunostimulants to cell-mediated immunity following surgery.

There is mounting evidence to support a causative role for the bovine papillomavirus in the production of these tumors. Inoculation of bovine papillomavirus type 1 into the dermis of horses results in a lesion qualitatively similar to naturally occurring sarcoids. Using different inocula and different routes, a variety of dermal fibroblastic or acanthotic epidermal lesions has been produced, although none has been an exact copy of the natural lesion. Considering the wide range of lesions caused by the virus even in its natural bovine host, this may not be a genuine difference but one induced by dose or route of inoculation. Papillomavirus genome, or at least fragments of it, are demonstrable in sarcoids, although entire virions are not. Because horses with sarcoids do not have serum-neutralizing antibody against bovine papillomavirus nor are they immune to tumor development following experimental inoculation of the virus, the causative role of the virus is not proven.

The macroscopic lesion resembles warts or exuberant granulation tissue. Histologically, the lesion combines epithelial and fibroblastic proliferation in a characteristically intimate association (Fig. 5.122). The epidermis is hyperplastic and sends long tentacles into the tangled, whorling spindle cells of the dermis. The plump fibroblasts are always closely applied to the epidermis, often perpendicular to the basal layer. Variants in which the epidermis predominates in the papillomalike fashion are occasionally encountered. Ulceration is common and may obscure the epidermal hyperplasia. Mitotic figures may be numerous in young, fast-growing tumors, but metastasis does not occur.

Fibrous histiocytoma is a controversial diagnosis sometimes made on cutaneous spindle cell tumors with cytologic features that resemble fibroblasts in some areas and histiocytes in others. The diagnosis has, at least in some circles, become popular in human dermatopathology because many fibrous skin nodules in humans contain immunohistochemical indications of histiocytic origin. This may simply reflect the multipotentiality of the primitive dermal mesenchymal cell that is the presumed origin for many of these tumors.

Most reports have been of so-called malignant fibrous histiocytoma, and most have been in cats. Depending on the specific report, features like storiform whorls of fibroblastic cells, foci of histiocytelike cells, lymphoid aggregates, and multinucleated cells have all been described. Those tumors in cats in which gigantic multinucleated cells are the predominant feature have been considered by some to be identical to the giant-cell tumor of soft parts or of tendon sheath, depending on initial location. Giant cells with dozens or even hundreds of nuclei are the unmistakable hallmark of these locally infiltrative sarcomas. Since these controversial and rare tumors appear to represent a histologic variant in the continuum of locally invasive "mesenchymomas" with no unique prognostic implications, there seems little point in embroiling ourselves in what, in human pathology, has been a long and fruitless controversy. The tumor described in dogs and cats bears no clinical or histologic resemblance to the much more prevalent benign cutaneous histiocytoma or to the rare malignant histiocytosis. The conjunctival lesion that has been called fibrous histiocytoma does not resemble any of the skin tumors, and is discussed in Chapter 4, The Eye and Ear.

Fibrosarcoma, like its benign counterpart, has tended to be a diagnostic catchall. In it are deposited most anaplastic spindle cell sarcomas that cannot be identified as tumors of specific histogenesis. Nonetheless, many have enough differentiation histochemically and ultrastructurally to qualify as fibroblastic, even when the cell of origin is not truly a fibroblast. Fibrosarcomas are common in dogs and cats but are uncommon in other species.

The macroscopic appearance is of a white, poorly demarcated multinodular tumor anywhere in the body but with a predilection for the subepithelial connective tissue of the nose, mouth, and skin. Most are firm due to heavy collagen content, but myxomatous variants occur. Histologically, fibrosarcomas consist of haphazard bundles of plump spindle cells invading normal dermis. The degree of differentiation is a continuous gradation, with some tumors resembling fibromas and others so anaplastic as to be unrecognizable as fibroblastic. Collagen production occurs in all, but some collagen is found in most sarcomas and is not diagnostically useful when seen only in scant amounts. Fibrosarcomas regularly have numerous mitotic figures, marked variation in nuclear size and shape, and may have giant cells with one or several nuclei. The behavior is reasonably well correlated to the degree of anaplasia. Locally invasive growth is the rule, with metastasis occurring in only ~10% of cases. Recurrence after initial excision, or metastasis, are almost exclusively from those tumors with high mitotic index and marked cellular pleomorphism. Fibrosarcoma may be difficult to distinguish from fibrous dysplasias, reactive granulation tissue (especially if irradiated), malignant variants of hemangiopericytoma/schwannoma, the rare amelanotic spindle cell melanoma, and the fibroblastic stromal reaction incited by many invasive carcinomas. The haphazard cell arrangement, purity of the cell population (albeit pleomorphic), and destructive growth habit are useful but not infallible criteria.

Fibrosarcoma in **cats** exists in two forms. In cats older than 5 years the fibrosarcoma may be multiple, rather anaplastic, and caused by one of several strains of feline sarcoma virus. The causal virus is a recombinant hybrid of host genome and feline leukemia virus that is found only in cats serologically positive for feline leukemia virus. Kittens infected at more than a few months of age are likely to be immune and resist tumor development. Even in cats with developing tumors, development of immune reaction to tumor-specific antigen often leads to spontaneous tumor regression. Viral antigen and oncornavirus particles are routinely demonstrable within such tumors.

Solitary fibrosarcomas in older cats occur much more frequently than the multicentric form. The tumor is similar in all aspects to its canine counterpart and has no relationship to the feline sarcoma, or its parent leukemia virus. It is locally infiltrative, and prognosis depends on the ease with which complete excision can be achieved. Mitotic index influences the rapidity of local recurrence but not the overall prevalence of recurrence. Only ~10% have proven metastases, most often to lung.

3. Other Mesenchymal Tumors

Lipoma and **liposarcoma** occur in all species. Lipomas are localized nodules of otherwise normal-appearing fat. It is only their macroscopic appearance as a lump which has justified their traditional inclusion as neoplasms. Necrosis, hemorrhage, and fibrosis may occur in larger, traumatized lipomas. They are common and often multiple in old dogs, but are seen in such diverse species as budgerigars, rats, mice, cattle, horses, and primates.

Liposarcomas are distinctly rare even in dogs. The tumor typically is pleomorphic, with undifferentiated spindle cells mixed with recognizable lipocytes and cells of intermediate differentiation. Stains for neutral fats are often helpful in establishing the cytoplasmic vacuolations to be lipid, but a little lipid does not prove the cell to be lipocytic in origin, as fat may accumulate as a degenerative change or by phagocytosis.

The tumor may occur anywhere, but the subcutis is the usual location. Stellate, spindle-shaped, or round cells have finely vacuolated eosinophilic cytoplasm which stains positively for fat. Coalescence of the vacuoles gives rise to single large cytoplasmic vacuoles, which displace the nuclei in a manner similar to that of normal lipocytes. Primitive mesenchymal cells and fibroblastlike cells are often seen in areas of poor, or perhaps different, differentiation. From the limited information available, metastasis is very uncommon.

Well-differentiated but **infiltrative lipomas** of dogs do occur. The tissue resembles normal fat but infiltrates and apparently causes the destruction of muscle and connective tissue. Because of its behavior, the tumor almost always recurs and may eventually become inoperable as it infiltrates vital areas such as intercostal muscles, jugular groove, or tendons.

Bibliography

Bevier, D. E. and Goldschmidt, M. H. Skin tumors in the dog. Part II. Tumors of the soft (mesenchymal) tissues. *Compend Cont Ed* **3:** 506–514, 1981.

Bostock, D. E., and Dye, M. T. Prognosis after surgical excision of fibrosarcomas in cats. *J Am Vet Med Assoc* **175:** 727–728, 1979.

Bostock, D. E., and Dye, M. T. Prognosis after surgical excision of canine fibrous connective tissue sarcomas. *Vet Pathol* **17:** 581–588, 1980.

Bradley, R. L., Withrow, S. J., and Snyder, S. P. Nerve sheath tumors in the dog. *J Am Anim Hosp Assoc* **18:** 915–921, 1982.

Brown, N. O. *et al.* Soft-tissue sarcomas in the cat. *J Am Vet Med Assoc* **173:** 744–749, 1978.

Burns, B. F., and Evans, W. K. Tumours of the mononuclear phagocyte system: A review of clinical and pathological features. *Am J Hematol* **13:** 171–184, 1982.

Canfield, P. A light microscopic study of bovine peripheral nerve sheath tumors. *Vet Pathol* **15:** 283–291, 1978.

Gleiser, C. A. *et al.* Malignant fibrous histiocytoma in dogs and cats. *Vet Pathol* **16:** 199–208, 1979.

Gross, T. L., and Carr, S. H. Amputation neuroma of docked tails in dogs. *Vet Pathol* **27:** 61–62, 1990.

Hardy, W. D., Jr. The feline sarcoma viruses. *J Am Anim Hosp Assoc* **17:** 981–997, 1981.

Kim, K., and Goldblatt, P. J. Malignant fibrous histiocytoma. Cytologic, light-microscopic and ultrastructural studies. *Acta Cytol* **26:** 507–511, 1982.

Kramek, B. A., Spackman, C. J. A., and Hayden, D. W. Infiltrative lipoma in three dogs. *J Am Vet Med Assoc* **186:** 81–82, 1985.

Lagace, R., Schurch, W., and Seemayer, T. A. Myofibroblasts in soft tissue sarcomas. *Virchows Arch* **389:** 1–11, 1980.

Lattes, R. Malignant fibrous histiocytoma. A review article. *Am J Surg Pathol* **6:** 761–771, 1982.

McChesney, A. E. *et al.* Infiltrative lipoma in dogs. *Vet Pathol* **17:** 316–322, 1980.

Murphy, J. M. *et al.* Immunotherapy in ocular equine sarcoid. *J Am Vet Med Assoc* **174:** 269–272, 1979.

Postorino, N. C. *et al.* Prognostic variables for canine hemangiopericytoma: 50 cases (1979–1984). *J Am Anim Hosp Assoc* **24:** 501–509, 1988.

Ragland, W. L., McLaughlin, C. A., and Spencer, G. R. Attempts to relate bovine papilloma virus to the cause of equine sarcoid: Horses, donkeys, and calves inoculated with equine sarcoid extracts. *Equine Vet J* **2:** 168–172, 1970.

Ragland, W. L., Keown, G. H., and Spencer, G. R. Equine sarcoid. *Equine Vet J* **2:** 2–11, 1970.

Ross, J., Hendrickson, M. R., and Kempson, R. L. The problem of the poorly differentiated sarcoma. *Semin Oncol* **9:** 467–483, 1982.

Snover, D. C., Sumner, H. W., and Dehner, L. P. Variability of histologic pattern in recurrent soft tissue sarcomas originally diagnosed as liposarcoma. *Cancer* **49:** 1005–1015, 1982.

Strafuss, A. C., and Bozarth, A. J. Liposarcoma in dogs. *J Am Anim Hosp Assoc* **9:** 183–187, 1973.

Vilanova, J. R. *et al.* Benign schwannomas: A histopathological and morphometric study. *J Pathol* **137:** 281–286, 1982.

4. Vascular Tumors

Vascular tumors originating in skin are very common in dogs, less frequent in cats, and only occasionally seen in the other species. There are three described benign tumors (hamartoma, hemangioma, and angiokeratoma) that have arbitrary histologic distinctions of dubious credibility. All are benign. Hemangiosarcoma is behaviorally malignant and can be a challenging histologic diagnosis if lumina are not well formed.

Hemangioma is a solitary nonencapsulated but well-circumscribed nodule that may occur at any level of the skin but is most frequent in the deep dermis. It consists of closely approximated endothelial channels formed by relatively normal-appearing endothelium. The vessel walls vary greatly in thickness but are composed only of connective tissue, never muscle, as in normal vessels of this size (Fig. 5.139). The collagen often is abundant and hyalinized, and numerous mast cells are sometimes seen within the collagen. Lumina vary from capillary to cavernous, and always contain blood. Thrombosis may be seen. The critical feature for diagnosis is a proliferation of endothelial cells that form channels unaccompanied by the muscle or pericytes that support normal vessels of similar caliber.

Angiokeratoma is described in dogs as a hemangioma of the immediately subepithelial dermis (or subconjunctival lamina propria). The adjacent epithelium is markedly hyperplastic and may intrude among the channels, interpreted by some as vascular invasion into the epithelium.

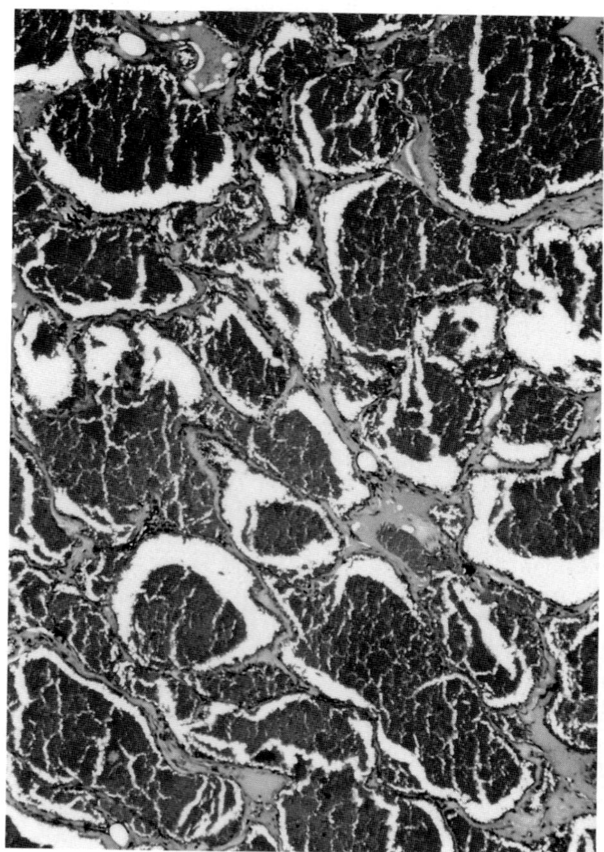

Fig. 5.139 Cutaneous hemangioma. Dog.

The supporting photos have not been convincing. Conjunctival epithelium (the site of most reported cases) regularly responds with hyperplasia to underlying tumors, and there is no compelling reason to introduce this human term into the veterinary literature. The same two general patterns occur in horses, with those in the superficial dermis inducing marked epithelial hyperplasia (verrucous hemangioma). Occurrence of congenital hemangiomas in horses has been reported.

Vascular hamartomas occur in some intriguing, repeatable sites in several species, but are also seen as isolated lesions with no apparent site predilection. In general, they consist of localized dermal excesses of primitive to well-formed blood vessels. Some instances seem to be in response to arteriovenous shunts, some to the angiogenesis of chronic irritation, and others are apparently spontaneous excesses of no known cause. The supposed difference from hemangioma is that the vessels in the hamartoma are structurally normal, with all supporting elements, so that the proliferation is truly of blood vessels rather than just of endothelium. The distinction sounds better on paper than it often appears in the tissue.

Bovine cutaneous angiomatosis is a mixed capillary, arteriolar, and venular hamartoma in the dermis of cattle. Single or multiple lesions are fragile, exophytic soft red masses up to several centimeters in diameter in young adult cattle. Fibroblasts and leukocytes often are present, but the regular orientation as in granulation tissue is not seen.

Similar lesions, usually without the inflammation, occur as poorly circumscribed tumors in the dermis of the canine and equine extremities. Some are confirmed as local angiogenic responses to acquired arteriovenous shunts, but most are not investigated as to cause and may be genuine, atypical hemangiomas. They do not form discrete masses that displace normal dermis as do most hemangiomas, but occur as numerous scattered clusters of variable-size and sometimes thick-walled vessels.

Vascular hamartomas or hemangiomas occur in the scrotal skin of dogs and boars as solitary or multiple plaques consisting of hemangiomalike capillary clusters below hyperplastic epithelium. The described lesions bear striking similarity to the verrucous hemangioma of horses and the angiokeratoma of dogs, underlining the confusion about the true nature of this tumor. All are benign.

Hemangiosarcoma is a frequent and very malignant neoplasm in dogs and cats, is occasionally seen in horses, but is rare in other species. Skin is the third most common origin (next to spleen and heart) in dogs, and is probably the most common origin in cats. The skin may also receive metastases from hemangiosarcomas primary in spleen or heart.

Histologically it consists of hyperchromatic pleomorphic spindle cells that form cleftlike channels reminiscent of blood vessels. The easiest diagnoses are those in which vessel formation is obvious, whereas some cannot easily be distinguished from other spindle cell tumors such as fibrosarcoma. Mitotic figures may be few or numerous, and are unreliable for making the diagnosis. Positive histochemical staining for the highly conserved von Willebrand's (factor VIII) antigen is close to foolproof in cases in which channel formation is equivocal. The tumors are very infiltrative locally, and readily recur following local excision. Data for the behavior of cutaneous hemangiosarcoma are surprisingly sparse. Local recurrence and infiltrative growth occurs in almost all affected cats, with a mean interval of 4 months between initial excision and recurrence. Visceral metastasis, however, is infrequent. Recurrence rates in dogs have not been published, and retrospective studies have concentrated on the more prevalent and rapidly fatal splenic or cardiac tumors. Too few equine cases have been adequately followed to allow any behavioral conclusions, other than to confirm that local recurrence and metastasis do occur.

Lymphangioma and **lymphangiosarcoma** are rare but are seen most frequently in dogs. Histologically they resemble hemangiomas and hemangiosarcoma except for the absence of blood (and of factor VIII antigen). The lymphangiosarcomas behave very aggressively with widespread and apparently rapid metastasis.

Bibliography

Arp, L. H., and Grier, R. L. Disseminated cutaneous hemangiosarcoma in a young dog. *J Am Vet Med Assoc* **185:** 671–673, 1984.

Brown, N. O., Patnaik, A. K., and MacEwen, E. G. Canine hemangiosarcoma: Retrospective analysis of 104 cases. *J Am Vet Med Assoc* **186:** 56–58, 1985.

George, C., and Summers, B. A. Angiokeratoma: A benign vascular tumour of the dog. *J Small Anim Pract* **31:** 390–392, 1990.

Hargis, A. M., and McElwain, T. F. Vascular neoplasia in the skin of horses. *J Am Vet Med Assoc* **184:** 1121–1124, 1984.

Rudd, R. G. *et al.* Lymphangiosarcoma in dogs. *J Am Anim Hosp Assoc* **25:** 695–698, 1989.

Scavelli, T. D. *et al.* Hemangiosarcoma in the cat: Retrospective evaluation of 31 cases. *J Am Vet Med Assoc* **187:** 817–819, 1985.

D. Tumors of Leukocyte-Related Cells of the Dermis

1. Mast Cell Tumors

Single or multiple nodular dermal proliferations of mast cells occur in all domestic species but are most common in dogs (Fig. 5.140). When the mast cells are present in large numbers and as an essentially pure population, a diagnosis of mast cell tumor (mastocytoma) is usually made. Some, more cautiously, consider such growths as cutaneous mastocytosis until their neoplastic nature is confirmed by invasive growth or morphologic evidence of anaplasia. As heavy accumulations of mast cells may occur in a variety of parasitic, mycotic, allergic, and idiopathic inflammatory syndromes, such caution is warranted.

Mast cell tumors account for 15 to 20% of canine skin tumors and are the most frequent malignant or potentially malignant tumor in the skin of dogs. Their macroscopic appearance varies widely with their stage of progression

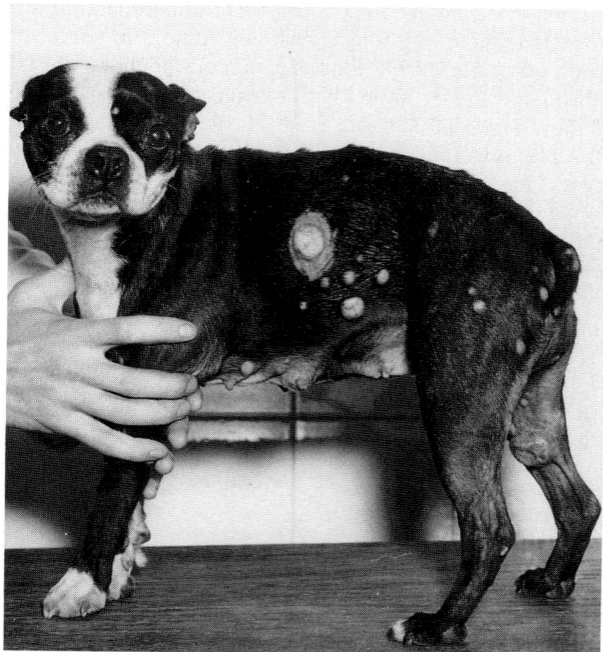

Fig. 5.140 Multiple mast cell tumors. Dog.

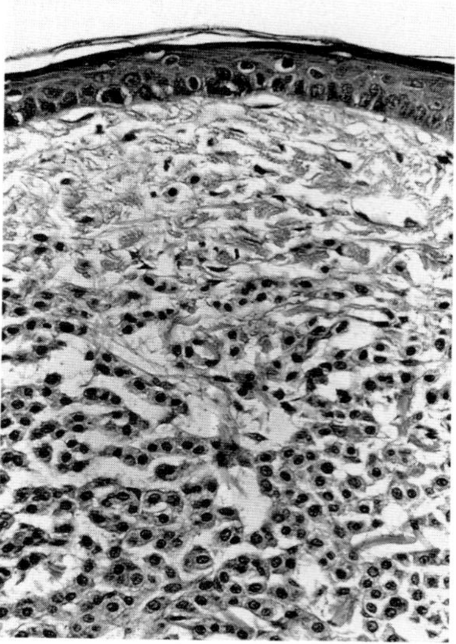

Fig. 5.141 Short cords of uniform round cells with sparing of superficial dermis. Canine mast cell tumor.

and degree of histologic differentiation, but the most common lesion is a rubbery, nodular nonencapsulated dermal mass ~3 cm in diameter. There is no consistently reported site of predilection, but boxer dogs are greatly predisposed. The mean age of affected dogs is ~8 years, a clinically useful contrast to the 2.5 years mean age of dogs affected with a grossly similar and even more common tumor, canine cutaneous histiocytoma.

The histologic lesion is seldom a diagnostic challenge except in the very poorly differentiated tumors. The mast cells form diffuse sheets or densely packed cords of round cells with central round nucleus, abundant granular basophilic cytoplasm, and indistinct cell membrane (Fig. 5.141). Scattered diffusely among the tumor cells are mature eosinophils. Even solid tumors tend to have some areas composed of cords of tumor cells alternating with collagen bundles, particularly at the infiltrative border of the tumor. They are never encapsulated or even well demarcated except at their superficial edge. Here they are typically separated from an intact epidermis by a thin zone of compressed dermis. Ulceration in large or traumatized tumors may bring the skin surface down to contact the tumor. Other histologic features, varying from common to rare, include fibrinoid necrosis of vessel walls, eosinophilic or mononuclear leukocytic vasculitis, and denaturation of collagen fibers.

These tumors vary in both cytologic type and overall growth pattern. The cytologic type is prognostically important. While the overall prevalence of euthanasia due to the neoplasm is about 50% within 4 years, there is a huge difference in behavior depending on the histologic grade of the tumor. Grading is based on a rather subjective evaluation of three criteria: mitotic index, degree of granularity, and degree of anisocytosis, and ranges from grade 1 (well-differentiated) to grade 3 (poorly differentiated). The grade 1 tumor consists of mast cells with a central round nucleus, single small nucleolus, and abundant, well-granulated cytoplasm. There is very little cellular variation, and mitotic figures are almost never seen. Tumors of this type account for ~50% of the mast cell tumors of dogs, and their behavior is usually benign. Excision is usually curative, with over 90% of affected dogs surviving more than 4 years.

The cells of undifferentiated (grade 3) tumors exhibit poor cytoplasmic granulation, large nucleus in relationship to cytoplasmic volume, irregular nuclear shape, moderate variation in nuclear size, several nucleoli, and the occasional binucleate cell or mitotic figure. Numerous mitoses (>4 per high-power field) are not a common feature of even the most anaplastic of mast cell tumors. The cytoplasmic granulation may be inapparent or visible only as fine cytoplasmic dusting, and may require staining with toluidine blue to demonstrate the red, metachromatic granules. About 20% of the canine tumors fall into this prognostic category. Almost all succumb to extensive local recurrence or metastatic disease within a few months of initial resection. The ability of radiation therapy or chemotherapy to significantly alter this grim prognosis has not been adequately documented.

The architectural variations seem unrelated to cytologic grade. Eosinophils are few to numerous. Edema may be very severe, giving the tumor an appearance that macroscopically and microscopically resembles acute inflammation, and clinically is confused with lipoma. Foci of tumor necrosis, of vascular necrosis, of mineralization, and of collagen-induced granuloma are all occasionally encountered.

Recurrence is common because of the infiltrative, poorly delineated growth habit which predisposes to recurrence unless initial excision is unusually wide. Metastases occur first to local lymph nodes and then to various organs, apparently favoring those with an abundant phagocytic component. Spleen is most frequently affected, with liver and bone marrow next.

Other clinical signs may occasionally result from the release of histamine or other vasoactive products from the mast cells. Gastroduodenal ulceration is relatively frequent in dogs with disseminated disease, occurring in perhaps 25% of cases. Histamine release stimulates the specific H_2 gastric parietal cell receptors, resulting in increased acid secretion and perhaps local mucosal ischemia. Ulceration follows and may lead to fatal exsanguination. Hypotensive shock from massive synchronous degranulation, as may occur with cryosurgery, is a rarely reported complication.

Mast cell tumors in **cats** exist as primary cutaneous mast cell tumors and as visceral mastocytosis. They are separate diseases. The metastatic potential of cutaneous mast cell tumors is very low (~5%), and those destined for behavioral malignancy are easily detected by anisocytosis, hyperchromasia, and a mitotic index of >2. Apparent recurrence at the surgical site or elsewhere in the skin is seen in 25 to 50% of the cases, but most of these probably represent multicentric origin.

The lesion presents as one or several firm, raised nodules seldom more than 2 cm in diameter. The head and neck are preferred sites. Histologically the cells are very characteristic. They are extraordinarily uniform, polygonal to round, and grow in a diffuse sheet interrupted only by small clusters of lymphocytes. The cytoplasm is clear or only faintly basophilic. Obvious granularity is infrequent. The nucleus is round, central, and relatively hyperchromatic in comparison to that of canine mast cells. Even with metachromatic stains such as Giemsa or toluidine blue, cytoplasmic granules may stain poorly, yet ultrastructurally they are abundant. Eosinophils are usually present but scattered sparsely, and initially may not be obvious. The tumor is never encapsulated but is less infiltrative histologically than is its canine counterpart. Much less frequent is a second histologic type seen most often in young cats as multiple, simultaneous or sequential tumors. Siamese cats are predisposed. The tumor mast cells resemble histiocytes, and even toluidine blue staining may be equivocal. The cells are confirmed as mast cells on electron-microscopic examination. Other features—eosinophils, lymphoid aggregates, well-circumscribed growth habit, and benign behavior—are similar to the more usual mast cell tumor of cats previously described.

Mast cell tumors in **horses** have been cautiously termed cutaneous mastocytosis, but the growths are comparable to the cutaneous tumors of other species. Both grossly and microscopically, the equine tumors may resemble the well-differentiated canine tumor, although collagen denaturation, vascular necrosis, and focal mineralization are more prominent. Although mast cell tumors in some respects resemble such lesions as cutaneous onchocerciasis, habronemiasis, or idiopathic collagen necrobiosis, none of these three is characterized by sheets of mast cells. Multiple congenital mast cell tumors which spontaneously regress have been reported.

In **pigs**, mast cell nodules have been described as tumors and as multifocal inflammatory aggregates perhaps in response to *Eperythrozoon*. Their morphology resembles that of well-differentiated feline tumors. In the pigs with multiple skin lesions, visceral aggregates were also found.

In **cattle**, scant data suggest the cutaneous tumors are usually multiple and are associated with visceral mast cell aggregates, although purely cutaneous tumors have been reported. There are more reports of cutaneous metastases of multicentric visceral tumors than there are of primary skin disease. Cattle of any age, including calves, may be affected.

Bibliography

Bostock, D. E. The prognosis following surgical removal of mastocytomas in dogs. *J Small Anim Pract* **14:** 27–40, 1973.

Buerger, R. G., and Scott, D. W. Cutaneous mast cell neoplasia in cats: 14 cases (1975–1985). *J Am Vet Med Assoc* **190:** 1440–1443, 1987.

Bundza, A., and Dukes, T. W. Cutaneous and systemic porcine mastocytosis. *Vet Pathol* **19:** 453–455, 1982.

Cheville, N. F. *et al*. Generalized equine cutaneous mastocytosis. *Vet Pathol* **9:** 394–407, 1972.

Hill, J. E., Langheinrich, K. A., and Kelley, L. C. Prevalence and location of mast cell tumors in slaughter cattle. *Vet Pathol* **28:** 449–450, 1991.

Holzinger, E. A. Feline cutaneous mastocytomas. *Cornell Vet* **63:** 87–93, 1973.

McGavin, M. D., and Leis, T. J. Multiple cutaneous mastocytomas in a bull. *Aust Vet J* **44:** 20–22, 1968.

Patnaik, A. K., Ehler, W. J., and MacEwen, E. G. Canine cutaneous mast cell tumor: Morphologic grading and survival time in 83 dogs. *Vet Pathol* **21:** 469–474, 1984.

Shaw, D. P., Buoen, L. C., and Weiss, D. J. Multicentric mast cell tumor in a cow. *Vet Pathol* **28:** 450–452, 1991.

Tams, T. R. Canine mast cell tumors. *Compend Cont Ed* **3:** 869–878, 1981.

Ward, J. M., and Hurvitz, A. I. Ultrastructure of normal and neoplastic mast cells of the cat. *Vet Pathol* **9:** 202–211, 1972.

Wilcock, B. P., Yager, J. A., and Zink, M. C. The morphology and behavior of feline cutaneous mastocytomas. *Vet Pathol* **23:** 320–324, 1986.

2. Benign Cutaneous Histiocytoma

Benign cutaneous histiocytoma is a very common skin tumor of young dogs. About 70% occur in dogs younger than 4 years. The occasional specimen does originate from a 9- to 10-year-old animal, in which instance great caution must be used to distinguish histiocytoma from cutaneous lymphoma or plasmacytoma. The tumors have a predilection for the head, and especially the ear pinna, but may occur anywhere. They are not found in extracutaneous sites.

The macroscopic lesion is a single, firm, nonulcerated, dome-shaped mass, usually 1–2 cm in diameter. On cut surface it is white, firm, well delineated, but nonencapsulated. The tumor is almost never multiple, and recurrence after excision is rare. We know surprisingly little of their undisturbed biological behavior, since published reports are of excised tumors. All are supposed to undergo spontaneous immune-mediated regression. Metastasis does not occur.

The histologic appearance is unique among tumors of any species. Most specimens exist as a solid sheet of plump, slightly elongated histiocytes within superficial dermis. The tumor surface is in intimate contact with the basal epidermal layer, a useful difference from mast cell tumors. The epidermis often is thin as if stretched by the growing tumor, yet thin cords of pseudoepitheliomatous hyperplasia infiltrate the superficial portion of the tumor. Within the main body of the tumor the cell boundaries are indistinct. The identity of the cells is best appreciated at the tumor borders where cell density is less. Some tumors are less dense throughout, in which case the histiocytes are arranged in coalescing clusters of a dozen or more cells surrounded by a reticulin or thin collagenous stroma. Nuclear pleomorphism and high mitotic index are typical of this tumor, causing the unwary to diagnose malignancy. Despite the apparent anaplasia, the mitoses are normal and the nuclear chromatin is fine, nucleoli are absent or inconspicuous, and the nuclear membrane, very regular. These are cytologically typical of benign cells. Cleaved, bean-shaped nuclei are frequently seen and are a useful clue as to the histiocytic identity of the cell.

The histologic appearance of regressing tumors can be confusing and, when advanced, is not clearly distinguishable from inflammation. At the deep border of the tumor, there is a bandlike infiltration of lymphocytes and plasma cells, the former predominating. The leukocytes gradually infiltrate the tumor, and are scattered throughout as loosely nodular aggregates. Occasionally, tumor cell necrosis can be seen adjacent to some of these nodules. In tumors in advanced regression, there are a few clumps of histiocytes scattered in a hypocellular, mildly sclerotic dermis, but lymphocytes and plasma cells predominate. It is typical that the last vestiges of tumor, and thus the greatest aggregation of leukocytes, are seen around the deep portion of dermal appendages and simulate chronic perifolliculitis.

Histiocytoma is distinguished from anaplastic, poorly granulated mast cell tumors by the presence in the former of numerous mitotic figures, cleaved nuclei, and intimate epidermal association. Eosinophils, vascular necrosis, and collagen denaturation are absent, although these are not invariably present in mast cell tumors either.

3. Other Canine Histiocytic Tumors

At least four skin tumors in addition to the benign histiocytoma of dogs have been called histiocytoma. The fibrous histiocytoma has been discussed with spindle cell tumors. Remaining are the syndromes of cutaneous histiocytosis, malignant histiocytosis, and systemic histiocytosis.

Benign cutaneous histiocytosis is seen as multiple dermal (rarely subcutaneous) histiocytic nodules in young dogs of any breed. The lesions may wax and wane for several years, but eventually disappear. Each histologic lesion resembles the much more prevalent solitary benign cutaneous histiocytoma, and a history of multiple nodules is essential for a correct diagnosis. Epitheliotropic lymphoma is the main alternative, with a much worse prognosis.

Systemic histiocytosis is a familial disease of Bernese mountain dogs in which nodular perivascular infiltrates of large but apparently not neoplastic histiocytes occur in various tissues. Skin and peripheral lymph nodes are most frequently affected. The disease typically occurs in young (2–5 years) male dogs, and usually has unpredictable remissions and exacerbations not apparently influenced by therapy. Most affected dogs eventually are euthanized after years of disease.

The histologic lesion consists of angiocentric (both intramural and perivascular) nodular accumulations of large but bland histiocytes accompanied by a smaller population of lymphocytes and granulocytes. The lesions are most severe in the subcutaneous tissue, accompanied by fibrosis as the lesion ages. Similar nodules in lymph nodes result in effacement of normal architecture; unlike reactive histiocytic lymphadenopathies, the subcapsular sinuses are not appreciably involved. Diagnosis on skin biopsy alone would be difficult because of the multitude of other histiocyte-rich nodular, deep dermatoses. The presence of lesions of similar type in lymph nodes, viscera, conjunctiva, and bone marrow greatly strengthens the diagnosis.

Malignant histiocytosis is also seen as a familial disease in male Bernese mountain dogs. The obvious assumption that the two syndromes represent the extremes of a single disease has not yet been proven. In malignant histiocytosis, affected dogs are older (mean age of onset is 7 years), and the skin is an infrequent and incidental target. Most lesions are in viscera, especially lung, and consist of noncohesive aggregations of pleomorphic, anaplastic histiocytes supported by a delicate fibrovascular stroma. Numerous multinucleated cells, cells with bizarre nuclei, and mitotic figures are seen, which is in clear contrast to systemic histiocytosis and to other canine lymphoproliferative diseases. Syndromes similar to both systemic and malignant histiocytosis are occasionally seen in other breeds of dogs, even crossbreds.

Bibliography

Cockerell, G. L., and Slauson, D. O. Patterns of lymphoid infiltrate in the canine cutaneous histiocytoma. *J Comp Pathol* **89:** 193–203, 1979.

Duncan, J. R., and Prasse, K. W. Cytology of canine cutaneous round cell tumors. Mast cell tumor, histiocytoma, lymphosarcoma and transmissible venereal tumor. *Vet Pathol* **16:** 673–679, 1979.

Glick, A. D., Holscher, M., and Campbell, G. R. Canine cutane-

ous histiocytoma: Ultrastructural and cytochemical observations. *Vet Pathol* **13:** 374–380, 1976.
Howard, E. B., and Nielsen, S. W. Cutaneous histiocytomas of dogs. *Natl Cancer Inst Monogr* **32:** 321–327, 1969.
Kelly, D. F. Canine cutaneous histiocytoma. A light- and electron-microscopic study. *Pathol Vet* **7:** 12–27, 1970.
Mays, M. B. C., and Bergeron, J. A. Cutaneous histiocytosis in dogs. *J Am Vet Med Assoc* **188:** 377–381, 1986.
Moore, P. F. Systemic histiocytosis of Bernese mountain dogs. *Vet Pathol* **21:** 554–563, 1984.
Moore, P. F., and Rosin, A. Malignant histiocytosis of Bernese mountain dogs. *Vet Pathol* **23:** 1–10, 1986.
Taylor, D. O. N., Dorn, C. R., and Luis, O. H. Morphologic and biologic characteristics of the canine cutaneous histiocytoma. *Cancer Res* **29:** 83–92, 1969.

Cutaneous plasmacytoma is a benign tumor that is common in dogs and rare in cats. It has not been described in other species. It occurs in middle-aged or old dogs with a marked predilection for feet, ear canal, and mouth. The typical tumor is a small spherical mass grossly similar to benign cutaneous histiocytoma, but its histology is distinctive. Sheets of pleomorphic round cells are divided into solid lobules of 10 to 20 cells by a fine fibrous stroma. The cells often have marked variation in nuclear size and degree of basophilia, with binucleation or multinucleation, and mitotic figures are quite prevalent. At least some of the cells retain a perinuclear halo (suggesting a Golgi zone), "clockface" nucleus, and basophilic cytoplasm considered typical of plasma cells. Russell bodies are sometimes seen even in very atypical cells, and the cytoplasm of most cells is pyroninophilic. Cells recognizable as plasma cells are most easily found near the periphery. Electron microscopy or immunohistochemistry for canine immunoglobulin confirms the diagnosis, but seldom are these tests necessary. Occasionally a particularly anaplastic example must be distinguished from amelanotic epithelioid melanoma. The distinction is critical, since even bizarre plasmacytomas are cured by excision. Amyloid is found in a small percentage (perhaps up to 10% of plasmacytomas. They bear no apparent relationship to multiple myeloma. Multiple tumors, rarely more than two to three, have been reported, as has a very low prevalence (~5%) of local recurrence.

Bibliography

Baer, K. E. *et al.* Cutaneous plasmacytomas in dogs: A morphologic and immunohistochemical study. *Vet Pathol* **26:** 216–221, 1989.
Lucke, V. M. Primary cutaneous plasmacytomas in the dog and cat. *J Small Anim Pract* **28:** 49–55, 1987.
Rakich, P. M. *et al.* Mucocutaneous plasmacytomas in dogs: 75 cases (1980–1987). *J Am Vet Med Assoc* **194:** 803–810, 1989.
Sandusky, G. E., Carlton, W. W., and Wightman, K. A. Diagnostic immunohistochemistry of canine round cell tumors. *Vet Pathol* **24:** 495–499, 1987.

4. Cutaneous Lymphomas

Cutaneous lymphoma includes at least two diseases: examples of systemic lymphoma in which the skin is involved as the sole, initial, or most obvious tumor site, and primary epitheliotropic lymphoma that originates in skin and only reluctantly spreads to other tissues. The former resembles multicentric lymphoma in its cytologic and histologic details, and will be described only briefly. The assumption that the nonepitheliotropic cutaneous lymphomas are all B-cell tumors like their histologically similar visceral or nodular counterparts is attractive but unproven, because most reports of these rare lesions predate the use of such testing.

Nonepitheliotropic cutaneous lymphoma in **cattle** is a rare disease of young animals unassociated with the bovine leukemia virus. Numerous small flat dermal or subcutaneous nodules wax and wane for months, with most animals reportedly developing nodal and visceral lymphoma late in the disease course. The histology is typical of bovine lymphoma elsewhere: obliterative proliferation of lymphocytes in the deep dermis. Detailed histologic or cytologic characterization has not been published, and most descriptions predate the availability of immunohistochemical techniques for identification of lymphocytic subsets.

The situation in **horses** appears to be similar to that in cattle, in that skin nodules are rarely a prominent manifestation of lymphoma, and most cases seem part of generalized, multisystem disease. Less than 5% of equine lymphomas are classified as cutaneous. The two best descriptions are of pleomorphic histiocytic lymphoma limited to subcutaneous nodules. The nodules contain a mixture of small lymphoid cells with numerous mitotic figures, large mononuclear histiocytes, and a reticular framework of stellate stromal cells. Both reports describe unidentified coryneform bacteria within the nodules, but homogenates of infected tissue induced no lesions in other horses. The pleomorphic character of cutaneous lymphoma in horses is similar to that of other forms of lymphoma in this species. Distinguishing the lesion from granulomatous inflammation may as be difficult as it is for lymphoma in other sites in horses.

Dogs and **cats** have lesions similar to those in cattle and horses. The range of histologic subtypes parallels that found in lymphoma in lymph nodes or viscera. Perhaps because of a more extensive literature, there are, however, reports of nonepitheliotropic lymphoma affecting only the skin. Permanent cure was accomplished by excision of the nodule or nodules. The histologic lesion is usually typical of the diffuse pattern of poorly differentiated lymphocytic lymphoma (Rappaport's classification), with sheets of lymphocytes growing as an obliterative, poorly circumscribed mass in the deep dermis or subcutis. Epidermal involvement by tumor is not seen.

Epitheliotropic lymphoma is relatively frequent in dogs, but has been rarely reported in cats, calves, and various laboratory animals. It probably occurs in all species. The neoplastic cell has been identified as a T lymphocyte, and the disease in these species is a very close histologic parallel to the human disease.

In humans, the historic subdivisions into mycosis fungoides, pagetoid reticulosis, Woringer–Kolopp disease, and others have been replaced by the belief that all represent different temporal or clinical presentations of the same T-cell epitheliotropic malignancy. In animals the same range of clinical and histologic appearance occurs, greatly influenced by the time, during the long disease course, at which the observations are made. The following description is for dogs. Too few cases have been described in other domestic species to allow for meaningful generalizations.

The disease typically is diagnosed in very old (older than 10 years) dogs, although its insidious nature may have delayed diagnosis so that the onset of clinical signs (interpreted as inflammation) was really at a much younger age. The clinical presentation varies greatly in the same dog at different times and among different dogs. Generalized cutaneous hyperemia, exfoliative dermatitis, dermal plaques or nodules, and mucocutaneous ulcers may occur simultaneously or in any sequence. Clinical diagnosis is impossible, but the histopathology is distinctive when fully developed. A linear band of pleomorphic mononuclear leukocytes occurs within the superficial dermis and within the surface, follicular, and sweat gland epithelium. At least initially, many of these cells are non-neoplastic lymphocytes, but even in the early lesions, a distinctive neoplastic population can be seen. The neoplastic cells vary from small (4–5 μm) to larger (12 μm) lymphocytes with irregular nuclear outline. The larger cells, in particular, may have a hyperchromatic, convoluted nucleus, although in dogs these "mycosis" cells are less numerous than those in the human disease. They tend to become more numerous as the disease progresses. The key to the diagnosis is the epidermotropism. At all stages of the disease, lymphocytes are numerous within the epithelium as individual cells with a small clear halo of spongiosis, or as clusters of atypical lymphocytes within a small vesicle. The latter structures, often termed Pautrier's microabscesses, are pathognomonic (Fig. 5.142). Extension of tumor more deeply into the skin occurs in those dogs developing grossly palpable nodules or plaques. The prominence of the irregular, large, pale lymphocytes and the epitheliotropism distinguish it from lymphoma metastatic to skin, and from other lymphoproliferative lesions.

The prognosis is grave. Solitary lesions may be cured by excision, and there are preliminary reports that radiation may be of benefit. Both are inappropriate for the usual, generalized disease. Chemotherapeutic regimens useful for other lymphomas are of no value. The usual course of disease is probably several years, but survival after diagnosis (which admittedly is usually late in the disease progression) is only a few months because of the pain associated with the mucocutaneous lesions. Spread to skin-associated lymph nodes occurs only late in the disease. Visceral involvement is rare, perhaps because euthanasia halts the natural disease progression. Leukemia analogous to human Sézary syndrome, in which the large convoluted T lymphocytes are in the blood, is a rare occurrence.

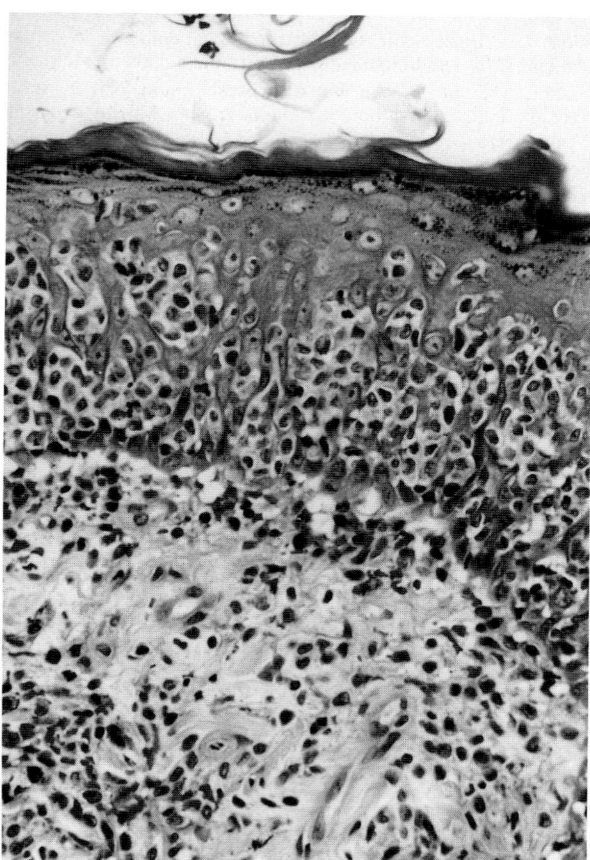

Fig. 5.142 Pleomorphic round-cell tumor infiltrating dermis and epidermis. Epitheliotropic lymphoma.

The histologic lesion in cattle and cats is identical to that in dogs. The range of clinical presentation and the prognosis (other than poor) is yet to be established in these species.

Pseudolymphoma is a term borrowed from human medicine. While not particularly informative, it is an efficient description for those lesions that, on first glance, can be mistaken for lymphoma. Most lesions that might be included under the heading are nodular lymphocytic–plasmacytic lesions of panniculitis, vasculitis, or perifolliculitis within the deep dermis or subcutis, each of which has been described in earlier sections of this chapter. Nodules induced by vaccine reactions or arthropod stings are encountered rather frequently in dogs, alarming to owners because of what seems to be sudden appearance and rapid growth. In all instances, the distinction from lymphoma rests with the observation of nonlymphoid inflammatory cells, the presence of plasmacytic maturation, and absence of cytologic atypia. Based solely on published descriptions, the differentiation from so-called histiocytic cutaneous lymphoma in horses, in which pleomorphism seems to be the rule, could be difficult.

Bibliography

Baker, J. L., and Scott, D. W. Mycosis fungoides in two cats. *J Am Anim Hosp Assoc* **25:** 97–101, 1989.

Beal, K. M. *et al.* An unusual presentation of cutaneous lymphoma in two dogs. *J Am Anim Hosp Assoc* **26:** 429–432, 1990.

Brown, N. O. *et al.* Cutaneous lymphosarcoma in the dog: A disease with variable clinical and histologic manifestations. *J Am Anim Hosp Assoc* **16:** 565–572, 1980.

Caciolo, P. L. *et al.* Cutaneous lymphosarcoma in the cat: A report of nine cases. *J Am Anim Hosp Assoc* **20:** 491–496, 1984.

Chu, A. C., and MacDonald, D. M. Pagetoid reticulosis: A disease of histiocytic origin. *Br J Dermatol* **103:** 147–157, 1980.

Conroy, J. D. Canine cutaneous lymphosarcoma. *Vet Clin North Am: Small Anim Pract* **9:** 141–143, 1979.

DeBoer, D. J., Turrel, J. M., and Moore, P. F. Mycosis fungoides in a dog: Demonstration of T-cell specificity and response to radiotherapy. *J Am Anim Hosp Assoc* **26:** 566–572, 1990.

Detilleux, P. B., Cheville, N. F., and Sheahan, B. J. Ultrastructure and lectin histochemistry of equine cutaneous histiolymphocytic lymphosarcomas. *Vet Pathol* **26:** 409–419, 1989.

Doe, R., Zackheim, H. S., and Hill, J. R. Canine epidermotropic cutaneous lymphoma. *Am J Dermatopathol* **10:** 80–86, 1988.

Goldschmidt, M. H., and Bevier, D. E. Skin tumors in the dog. Part III. Lymphohistiocytic and melanocytic tumors. *Compend Cont Ed* **3:** 588–594, 1981.

Grekin, D. A., and Zackheim, H. S. Mycosis fungoides. Symposium on cutaneous signs of systemic disease. *Med Clin North Am* **64:** 1005–1016, 1980.

Legendre, A. M., and Becker, P. U. Feline skin lymphoma: Characterization of tumor and identification of tumor-stimulating serum factor(s). *Am J Vet Res* **40:** 1805–1807, 1979.

Thrall, M. A. *et al.* Cutaneous lymphosarcoma and leukemia in a dog resembling Sézary syndrome in man. *Vet Pathol* **21:** 182–186, 1984.

Wilcock, B. P., and Yager, J. A. The behavior of epidermotropic lymphoma in twenty-five dogs. *Can Vet J* **30:** 754–756, 1989.

Zenoble, R. D., and George, J. W. Mycosis fungoides-like disease in a dog. *J Am Anim Hosp Assoc* **16:** 203–208, 1980.

Zwahlen, R. D., Tontis, A., and Schneider, A. Cutaneous lymphosarcoma of helper/inducer T-cell origin in a calf. *Vet Pathol* **24:** 504–508, 1987.

E. Tumors of the Mammary Gland

Mammary masses are very prevalent in dogs and cats, but are rare in other domestic species. Because of their prevalence in these species, and because they offer a relevant model for human breast cancer, the accumulated literature is vast. Happily, this is one of the few instances in which there has been focus on prognosis rather than description, so that previous complex classification schemes have been replaced by some very simple and behaviorally predictive classifications. Mammary tumors in cats have traditionally been considered quite different from those in dogs because in cats over 90% are behaviorally malignant. This separation need not be maintained for pathologists, for the predictors of behavior that exist in excised canine tumors are equally valid for cats. It is merely that the predictors of malignancy are found much more often in feline tumors than in their canine counterparts. Since some of the non-neoplastic proliferations require therapy other than (and less radical than) mastectomy, it is important that they be correctly interpreted in biopsy specimens. There is no prognostic or therapeutic difference among the benign neoplasms, and their descriptions simply convey the range of appearances of these tumors. The behavior of mammary carcinomas depends almost entirely on a single variable: their degree of invasiveness within the host tissue surrounding the tumor. The degree of cytologic or architectural maturity, and overall tumor size, also exert statistically significant effects on prognosis.

1. Mammary Hyperplasia and Dysplasia

Mammary ductal ectasia or **hyperplasia** occurs frequently in bitches and occasionally in queens. Single or multiple ducts are distended with proteinaceous fluid with some cellular debris. Their leakage may induce sterile granulomatous inflammation rich in ceroid-laden macrophages. The epithelium usually is columnar and may even have the beginnings of papillary proliferation, more compatible with some type of endocrine-responsive hyperplasia than with an obstructive ectasia. They may spontaneously regress, and ovariectomy reliably causes their regression. Progestagen administration induces them. It is not known whether they predispose to, or even progress to, neoplasia, although they would seem fertile soil for neoplastic transformation.

Mammary fibroadenomatous hyperplasia (mammary hypertrophy, fibroadenoma) occurs very commonly in cats. It is occasionally seen in bitches, but much less frequently than mammary alveolar (lobular) hyperplasia. The lesion usually occurs in young intact females, affects one or more glands, and may regress spontaneously or require ovariectomy. It can be induced in spayed females or in male cats by administration of progesterone-containing compounds. It is histologically distinctive: minor proliferation of mammary ducts lined by simple columnar epithelium, surrounded by massive and concentric proliferation of edematous stroma compatible with myoepithelium (Fig. 5.143). The prominence of the stroma is in sharp contrast to mammary neoplasia in cats which almost always is solid or alveolar carcinoma with little stroma.

Mammary lobular hyperplasia, in contrast, consists of patchy hyperplasia of secretory acini indistinguishable from the normal lactating gland (Fig. 5.144). It occurs in intact, nonpregnant and nonlactating bitches, and occasionally in queens. Serial sectioning of the mammary glands reveals numerous small (1–4 mm diameter) hyperplastic foci in 50% of bitches, even at 3 years of age. They are most numerous in the caudal two glands, paralleling the greater prevalence of neoplasia in these glands. Circumstantial evidence, and much theory of carcinogenesis, supports the view that such lesions should be considered preneoplastic changes. Indeed, many hyperplastic lobules contain dysplastic foci with some degree of epithelial jum-

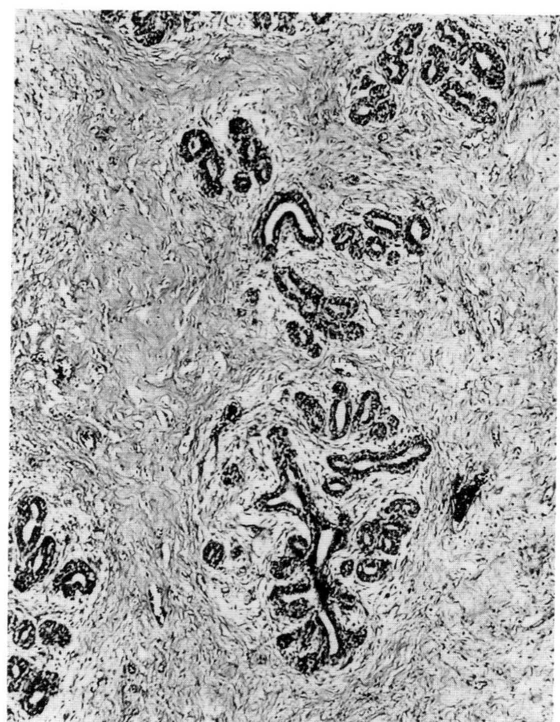

Fig. 5.143 Fibroadenomatous hyperplasia. Mammary gland. Cat.

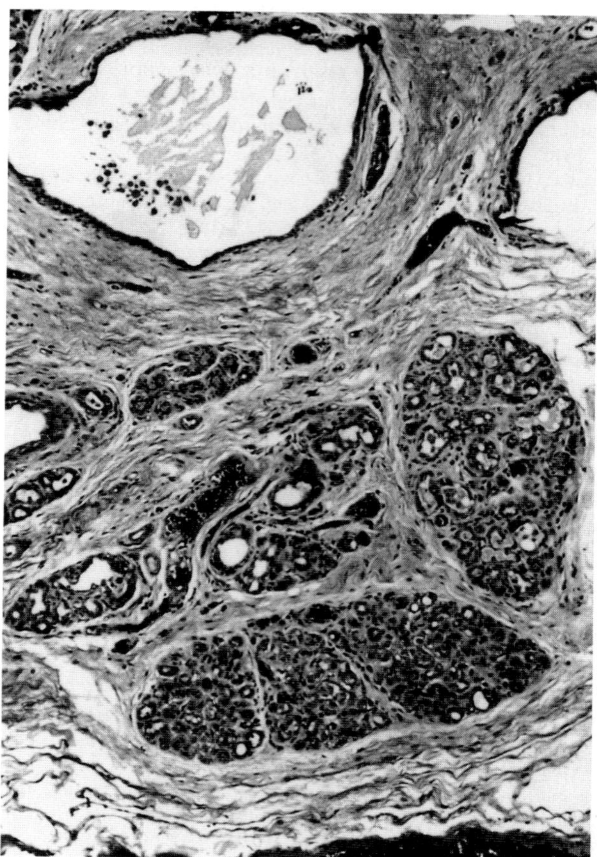

Fig. 5.144 Canine mammary lobular hyperplasia.

bling and hyperchromasia, blurring the distinction between hyperplasia and *in situ* neoplasia.

2. *Mammary Neoplasia in Dogs and Cats*

The mammary gland of the bitch and cat is a composite of several glands, each with its own duct system and secretory lobules. The lobules undergo cyclic changes during reproductive cycles; ductular proliferation is active in pregnancy; and alveolar differentiation is completed in late pregnancy and lactation. The cycles of proliferation and involution do not occur uniformly or synchronously in all lobules of a gland, producing a histologic picture which can be quite variable in a single section, and even more so if there are interspersed hyperplastic and dysplastic changes. It is probable that cyclic changes in stromal tissues also occur during growth and involution of secretory tissue. The intralobular and interlobular connective tissue is light and fibrillary in functional lobules, and becomes remodeled and condensed to poorly cellular collagen during involution and senile atrophy. Desmoplasia is a feature of some malignant lesions as a response of the connective tissue. Inflammatory fibrosis is commonly present in response to superimposed infections, especially in those cases in which papillary growths obstruct duct lumina. Osseous metaplasia may occur in the stroma of glands with neoplastic lesions.

The myoepithelial cells form a continuous layer around the teat sinus and ducts and a discontinuous layer about the acini. They can be distinguished from fibrocytes by staining for alkaline phosphatase, for which fibrocytes are negative. Abnormal proliferation of myoepithelial cells probably does not occur independent of dysplastic or neoplastic change in epithelial cells. The proliferating myoepithelial cells form expansile masses in the stroma and, most abundantly, in the connective cores of papillary growths in the ducts, but without breaching the basement membrane. They secrete a mucinous ground substance, giving a myxoid appearance, and the masses become progressively, in benign tumors, converted to hyaline cartilage. The cartilage may undergo endochondral ossification to contribute a second bony component to the tumor mass. The contribution of myoepithelial cells to the mass of the tumor may be insignificant in early lobular or papillary growths and in anaplastic carcinomas but may predominate in many tumors of benign nature.

The histologic architecture of mammary epithelial tumors in the bitch will be compounded by variable admixtures of the alterations previously described and sometimes further complicated by hemorrhage and necrosis and by squamous metaplasia in injured cisterns. The more benign epithelial patterns may be clearly lobular, reproducing adenomatous acini with infolding of epithelium or with the acinar lumen obliterated by solid masses of cells, or complicated papillary growths may be present in dilated ducts. The malignant tumors also reflect a lobular or ductual origin, although this may be obscured in larger masses. Those of lobular origin simulate adenocarcinomas of other glandular tissue, but typically they are solid masses of cells with scant stroma. Necrosis of central cells in the solid cords produces a suggestion of glandular structures. Irregular branching tubular patterns are usually associated with scirrhous stroma. The papillary carcinomas retain that structure except when compressed within the confines of the duct. The duct-like structures are often cystic and contain detached tumor cells singly or in small clumps.

Numerous retrospective studies from around the world, using traditional histologic and cytologic criteria for malignancy, are surprisingly uniform in reporting a benign : malignant ratio of almost exactly 1 : 1 in dogs, and about 1 : 9 in cats. However, several credible studies have shown that, in dogs, these traditional classification criteria have overestimated true behavioral malignancy by at least 50%. Some studies claim behavioral malignancy in as few as 10% of all canine mammary neoplasms.

Mammary epithelial neoplasms represent a histologic continuum from the most benign to the most malignant (Fig. 5.145A,B). The variations are numerous, and large neoplasms often have several regional variations in histologic pattern. Rather than describe in detail all patterns and their statistical correlation to subsequent behavior, it is more useful to emphasize those gross or histologic changes that should be added to the benign or malignant sides of the decision-making scales. All data are for post-excisional behavior of epithelial neoplasms, as there are no data for untreated tumors. The behavior of rare mammary sarcomas is described later.

Presence of invasion by the tumor into surrounding tissue is the single most important predictor of behavioral malignancy. It is a very reliable observation that can be made prior to any surgical procedure, when the clinician is attempting to choose the best surgical approach. Lymphedema, satellite nodules, and firm cords radiating from the tumor across the midline are bad prognostic signs. In dogs, 80% of animals with invasive tumors will be dead because of tumor (including by elective euthanasia) within 2 years, and most of these will be within the first postoperative year. Metastases to lung (80%) and lymph nodes (65%) are frequently found at necropsy. Conversely, 80% of dogs with noninvasive tumors—regardless of any other histologic

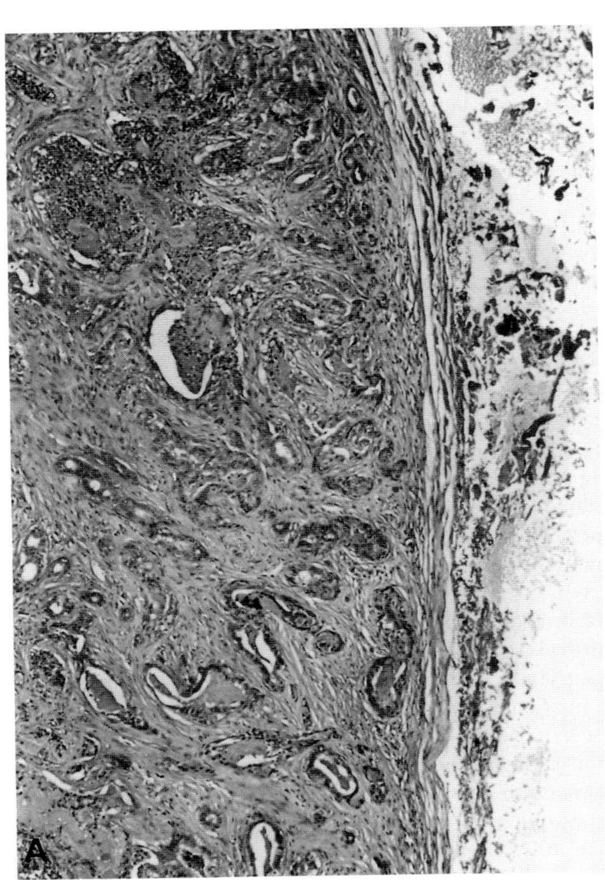

Fig. 5.145A Canine mammary tubular adenoma, nonencapsulated but very well circumscribed.

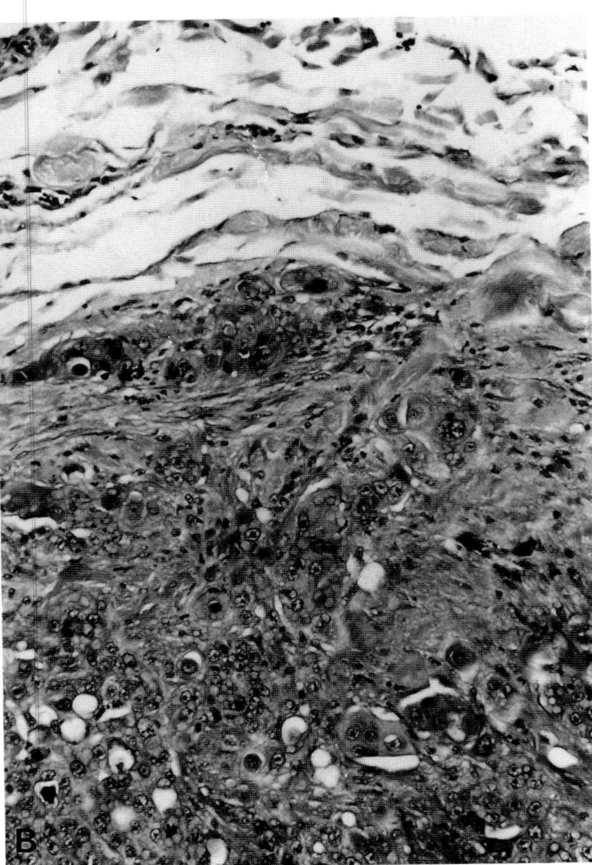

Fig. 5.145B Anaplastic mammary carcinoma, bitch.

features—will be alive. The presence of tumor emboli in lymphatics or veins is absolute evidence for local invasion (Fig. 5.146A). In its absence, care must be taken to ensure that the invasion is really into surrounding normal tissue, and not just into the abundant stroma accompanying the tumor itself. The presence of invasion influences other prognostic variables: its presence reduces by more than 50% the median survival times predicted by histologic pattern, and its absence virtually eliminates the influence of tumor size on the prognosis in dogs (but perhaps not in cats in which noninvasive tumors are rare).

Size exerts a profound influence in both dogs and cats. In cats with tumors < 2 cm in diameter at the time of surgery, the median postoperative survival is 3 years. With tumors greater than 3 cm in diameter, the median survival is <6 months. In dogs, size must be first correlated with histologic pattern. Invasive tumors over 5 cm in diameter are uniformly fatal within a year, while tumors over 5 cm that are not invasive are almost all behaviorally benign. Tumors that have been present long enough to be that large, and yet still have exhibited no clinically obvious spread, are almost certain to be benign mixed mammary tumors (Fig. 5.146B).

Histologic pattern, independent of invasiveness, is important in both species. Using the architectural patterns of papillary, tubular, solid, and anaplastic as indicators of cellular maturity in **dogs,** there are significant reductions in 1-year and median survival times as the cellular pattern

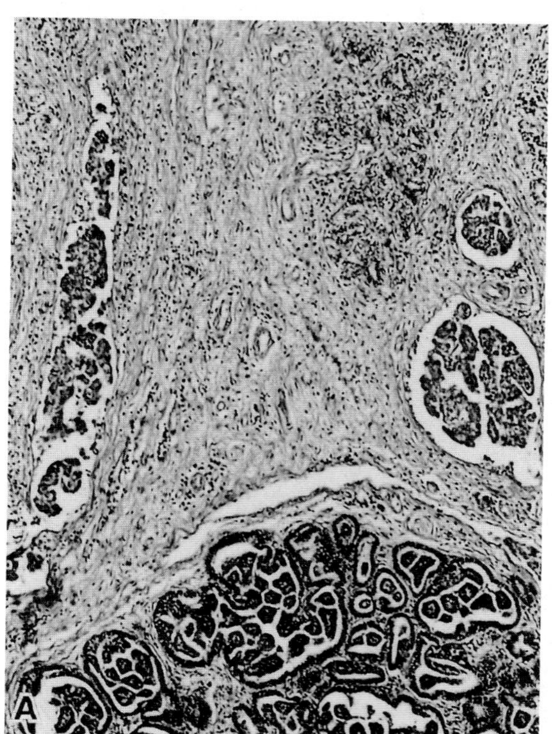

Fig. 5.146A Papillary adenocarcinoma with lymphatic spread. Mammary gland. Bitch.

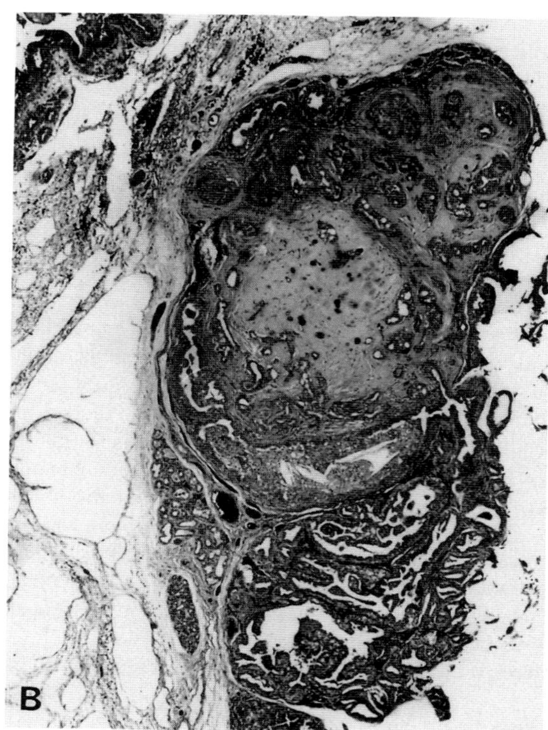

Fig. 5.146B Canine benign mixed mammary tumor with typical mixture of tubular, papillary, and cartilaginous elements.

deteriorates toward anaplastic (where there is virtually 100% 1-year mortality). It is with the most prevalent patterns (papillary and tubular) that pathologists have greatest difficulty in distinguishing benign from malignant tumors. The lack of invasion and the presence of abundant myoepithelial stroma—often with cartilaginous or osseous metaplasia—are almost infallible indicators of a good prognosis. Papillary and tubular patterns with abundant stroma are analogous to the benign mixed mammary tumor of traditional classifications, accounting for at least 80% of all mammary tumors in bitches. It is almost never seen in cats. In **cats** initial conclusions that the histologic pattern influenced survival were subsequently modified, and the latest data point to mitotic index rather than histologic maturity *per se* as a significant prognostic indicator.

Other histologic features like squamous metaplasia, presence of necrosis, or magnitude of inflammation have not had a statistically significant influence on behavior in most studies.

In summary, mammary epithelial neoplasms in bitches are behaviorally benign in at least 80% of cases. Behavioral malignancy is predicted best by invasive growth habit, with median survival also influenced by degree of histologic maturity and overall tumor size. In cats, more than 90% are behaviorally malignant, with the expected histologic findings of local intravascular emboli and invasive growth (Fig. 5.147A,B). Histologic maturity influences the median survival time but not the eventual out-

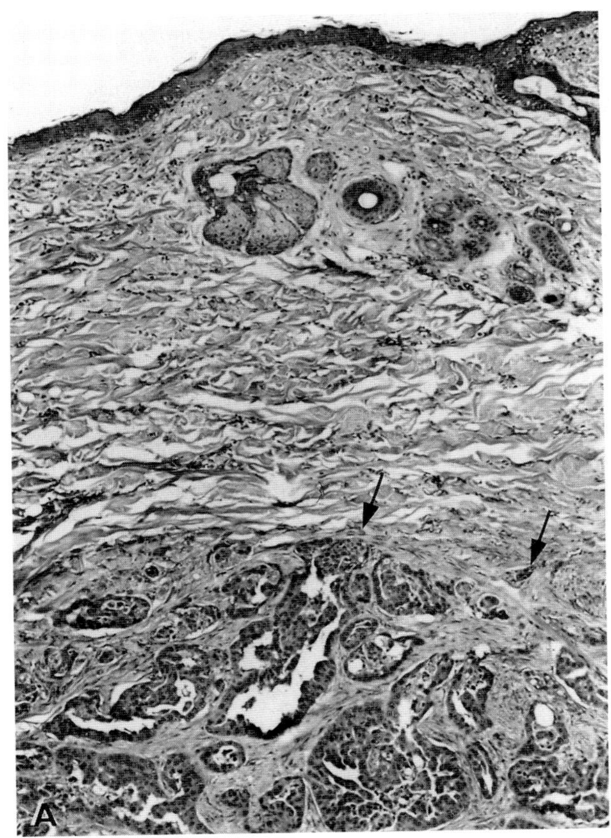

Fig. 5.147A Scirrhous feline mammary carcinoma with incipient and established peripheral invasion (arrows).

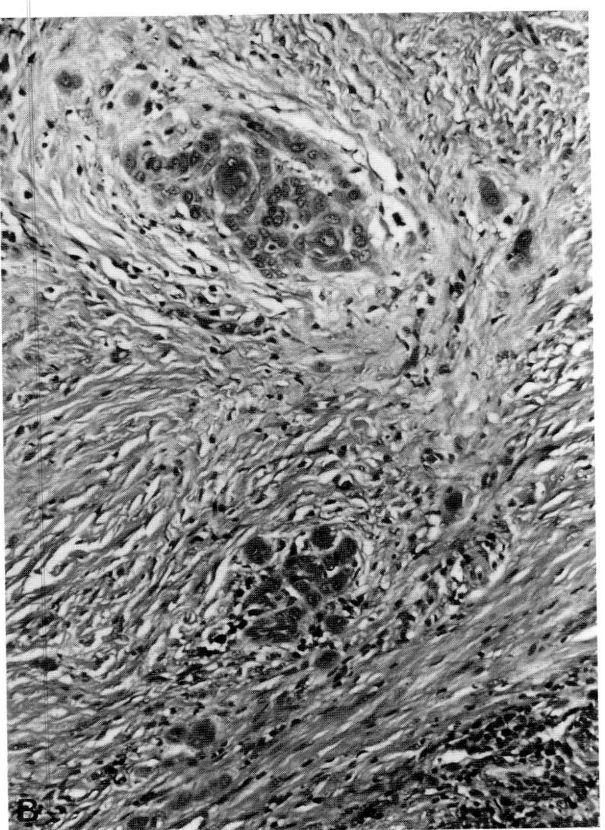

Fig. 5.147B Higher magnification of (A), with poorly cohesive islands of tumor cells invading the dermis.

come. Early excision of small (less than 8 cm^3 volume) tumors is associated with a 50% 1-year survival and a 3-year median survival, regardless of other variables.

Primary mammary sarcomas are infrequent in dogs and very rare in cats. They presumably arise from the fibrous stroma of the gland or of the adjacent subcutaneous tissue. It is impossible (and irrelevant) to distinguish sarcomas of mammary origin from those that happen to arise in soft tissue adjacent to the gland. Histologic change suggestive of malignancy is occasionally seen in the myoepithelial stroma, or in the metaplastic bone or cartilage of mixed mammary tumors (so-called malignant mixed tumors). If such tumors metastasize, it is virtually always the epithelial component that spreads.

The sarcomas most often are osteosarcomas or fibrosarcomas. Very rare tumors that resemble myoepithelium, with no observed tubular epithelial component, occur in both dogs and cats. They resemble smooth muscle tumors. Metastasis seems to be rare, but credible data on this tumor are scarce.

3. Mammary Tumors in Other Species

The rarity of mammary tumors in ruminants, horses, and pigs is difficult to explain, since at least cows and mares frequently are kept well into their theoretical tumor years."

In **mares,** the few reported tumors have been scirrhous solid carcinomas which have been extremely invasive locally, and all have had widespread metastasis.

In **cows,** the literature is old and of variable quality in that many diagnoses had no histologic confirmation. Most descriptions are of fibropapillomalike growths in the teat canal or large ducts, but everything from fibroma, squamous cell carcinoma, and osteoma to fibrosarcoma has been diagnosed. Many of the polypoid teat or cisternal growths were probably hyperplastic inflammatory polyps. Goats and water buffalo are cited in review articles as having a very low prevalence of mammary neoplasia as well.

Bibliography

Allen, H. L. Feline mammary hypertrophy. *Vet Pathol* **10:** 501–508, 1973.

Allen, S. W., and Mahaffey, E. A. Canine mammary neoplasia: Prognostic indicators and response to surgical therapy. *J Am Anim Hosp Assoc* **25:** 540–546, 1989.

Bostock, D. E. Canine and feline mammary neoplasma. *Br Vet J* **142:** 506–515, 1986.

Brodey, R. S., Goldschmidt, M. H., and Roszel, J. R. Canine mammary gland neoplasms. *J Am Anim Hosp Assoc* **19:** 62–90, 1983.

Cameron, A. M., and Faulkin, L. J. Hyperplastic and inflammatory nodules in the canine mammary gland. *J Natl Cancer Inst* **47:** 1277–1287, 1971.

Hayden, D. W. *et al.* Feline mammary hypertrophy/fibroadenoma complex: Clinical and hormonal aspects. *Am J Vet Res* **42:** 1699–1703, 1981.

Hellmen, E., and Lindgren, A. The expression of intermediate filaments in canine mammary glands and their tumors. *Vet Pathol* **26:** 420–428, 1989.

McEwen, E. G. *et al.* Prognostic factors for feline mammary tumors. *J Am Vet Med Assoc* **185:** 201–204, 1984.

Moulton, J. E. Tumors of the mammary gland. *In* "Tumors of Domestic Animals" J. E. Moulton (ed.), 3rd Ed., pp. 518–552. Berkeley, University of California Press, 1990.

Moulton, J. E., Rosenblatt, L. S., and Goldman, M. Mammary tumors in a colony of beagle dogs. *Vet Pathol* **23:** 741–749, 1986.

Povey, R. C., and Osborne, A. D. Mammary gland neoplasia in the cow. A review of the literature and report of a fibrosarcoma. *Pathol Vet* **6:** 502–12, 1969.

Warner, M. R. Age incidence and site distribution of mammary dysplasias in young beagle bitches. *J Natl Cancer Inst* **57:** 57–61, 1976.

Weijer, K., and Hart, A. A. M. Prognostic factors in feline mammary carcinoma. *J Natl Cancer Inst* **70:** 709–716, 1983.

Weijer, K. *et al.* Feline malignant mammary tumors. I. Morphology and biology: Some comparisons with human and canine mammary carcinomas. *J Natl Cancer Inst* **49:** 1697–1704, 1972.

Index

Boldface, major discussion; f, figure; t, table.

C

D

E

G

H

I

J

K

L

M

O

P

Q

R

S

T

U

V